Family name	Functional group structure[a]	Simple exa[obscured]	[obscured]
Sulfone	$—C—S^{2+}(:\ddot{O}:^-)_2—C—$	$H_3C—S^{2+}(O^-)_2—CH_3$	[obscured] Dimethyl sulfone
Thiol	$—C—\ddot{S}—H$	$H_3C—SH$	*-thiol* Methanethiol
Carbonyl, $—C(=\ddot{O})—$			
Aldehyde	$—C—C(=\ddot{O})—H$	$H_3C—C(=O)—H$	*-al* Ethanal (Acetaldehyde)
Ketone	$—C—C(=\ddot{O})—C—$	$H_3C—C(=O)—CH_3$	*-one* Propanone (Acetone)
Carboxylic acid	$—C—C(=\ddot{O})—\ddot{O}H$	$H_3C—C(=O)—OH$	*-oic acid* Ethanoic acid (Acetic acid)
Ester	$—C—C(=\ddot{O})—\ddot{O}—C—$	$H_3C—C(=O)—O—CH_3$	*-oate* Methyl ethanoate (Methyl acetate)
Amide	$—C—C(=\ddot{O})—\ddot{N}H_2$, $—C—C(=\ddot{O})—\ddot{N}(—C)—H$, $—C—C(=\ddot{O})—\ddot{N}(—C)—$	$H_3C—C(=O)—NH_2$	*-amide* Ethanamide (Acetamide)
Carboxylic acid chloride	$—C—C(=\ddot{O})—Cl$	$H_3C—C(=O)—Cl$	*-oyl chloride* Ethanoyl chloride (Acetyl chloride)
Carboxylic acid anhydride	$—C—C(=\ddot{O})—\ddot{O}—C(=\ddot{O})—C—$	$H_3C—C(=O)—O—C(=O)—CH_3$	*-oic anhydride* Ethanoic anhydride (Acetic anhydride)

Organic Chemistry

Fourth Edition

Organic Chemistry

Fourth Edition

John McMurry
Cornell University

Brooks/Cole Publishing Company

I(T)P™ An International Thomson Publishing Company

Pacific Grove • Albany • Bonn • Boston • Cincinnati • Detroit • London • Madrid • Melbourne
Mexico City • New York • Paris • San Francisco • Singapore • Tokyo • Toronto • Washington

For more information contact:

BROOKS/COLE PUBLISHING COMPANY
511 Forest Lodge Road
Pacific Grove, CA 93950
USA

International Thomson Publishing Europe
Berkshire House 168-173
High Holborn
London WC1V 7AA
England

Thomas Nelson Australia
102 Dodds Street
South Melbourne, 3205
Victoria, Australia

Nelson Canada
1120 Birchmount Road
Scarborough, Ontario
Canada M1K 5G4

International Thomson Editores
Campos Eliseos 385, Piso 7
Col. Polanco
11560 México D. F. México

International Thomson Publishing GmbH
Königswinterer Strasse 418
53227 Bonn
Germany

International Thomson Publishing Asia
221 Henderson Road
#05-10 Henderson Building
Singapore 0315

International Thomson Publishing Japan
Hirakawacho Kyowa Building, 3F
2-2-1 Hirakawacho
Chiyoda-ku, Tokyo 102
Japan

Printed in the United States of America

10 9 8 7 6 5

Library of Congress Cataloging-in-Publication Data

McMurry, John.
Organic chemistry / John McMurry. — 4th ed.
p. cm.
Includes index.
ISBN 0-534-23832-7
1. Chemistry, Organic. I. Title.
QD251.2.M43 1996 95-44992
547—dc20 CIP

THIS BOOK IS PRINTED ON ACID-FREE RECYCLED PAPER

Preface

I've been asked hundreds of times over the past ten years why I wrote this book. It wasn't because there aren't other perfectly acceptable organic chemistry textbooks out there, and it wasn't because I thought I'd get rich. I wrote this book because I love writing. I get great pleasure and satisfaction from taking a complicated subject, turning it around until I see it clearly from a new angle, and then explaining it in simple words. I write to explain chemistry to students today the way I wish it had been explained to me years ago.

The tremendous response to three previous editions has been very gratifying and suggests that this book has served students well. I hope you will find that this fourth edition of *Organic Chemistry* builds on the proven strengths of the first three and serves students even better. I have made every effort to make this fourth edition as effective, clear, and readable as possible, to show the beauty and logic of organic chemistry, and to make it enjoyable to learn.

Organization and Teaching Strategies

As in the previous editions, I use a dual organization in this fourth edition, blending the traditional functional-group approach with a mechanistic approach. The primary organization of this book is by functional group, beginning with the simple (alkenes) and progressing to the more complex. Students new to the subject and not yet versed in the subtleties of mechanisms prefer this organization because it is straightforward. In other words, the *what* of chemistry is easier for most new students to grasp than the *why*. Within this primary organization, however, I place heavy emphasis on explaining the fundamental mechanistic similarities of reactions. This emphasis is particularly evident in the chapters on carbonyl-group chemistry (Chapters 19–23) where mechanistically related reactions like the aldol and Claisen condensations are covered together. By the time students reach this material, they have seen all the common mechanisms, and the value of mechanisms as an organizing principle has become more evident.

The Lead-Off Reaction: Addition of HBr to Alkenes Students naturally attach great importance to a text's lead-off reaction because it is the first reaction they see and is discussed in such detail. I use the addition of HBr to an alkene as the lead-off to illustrate general principles of organic chemistry because it has many advantages: it is relatively straightforward; it involves a common but important functional group; no prior knowledge of chirality or kinetics is needed to understand it; and, most importantly, it is a *polar* reaction. As such, I believe that electrophilic addition reactions represent a much more useful and realistic introduction to functional-group chemistry than a lead-off such as radical alkane chlorination.

Reaction Mechanisms In the first edition, I introduced an innovative format for explaining reaction mechanisms that met with an enthusiastic response. Now set off by a color panel, mechanisms shown in this format have the reaction steps printed vertically while the changes taking place in each step are explained next to the reaction arrow. This format allows the reader to see easily what is occurring at each step in a reaction without having to jump back and forth between structures and text. Pages 157 and 199 show examples.

Organic Synthesis Organic synthesis is treated as a teaching device that helps students organize and deal with a large body of factual information (the same kind of skill so critical in medicine). Two sections, the first in Chapter 8 (Alkynes) and the second in Chapter 16 (Chemistry of Benzene), explain the thought processes involved in working synthesis problems. The value of starting from what is known and logically working backwards is given particular emphasis.

Modular Presentation Topics are arranged in a modular way wherever possible. Thus, the chapters on simple hydrocarbons are grouped together (Chapters 3–8), the chapters on spectroscopy are grouped together (Chapters 12–14), and the chapters on carbonyl-group chemistry are grouped together (Chapters 19–23). I believe that this organization brings to these subjects a cohesiveness not found in other texts and allows the instructor the flexibility to teach in an order different from that presented in this book.

Basic Learning Aids Clarity of explanation and smoothness of information flow are crucial requirements for any textbook. In writing and revising this text, I consistently aim for summary sentences at the beginning of paragraphs, lucid explanations, and smooth transitions between paragraphs and between topics. New concepts are introduced only when they are needed, not before, and are immediately illustrated with concrete examples. Frequent cross-references to earlier (but not later) material are given, and numerous summaries are provided to draw information together, both within and at the ends of chapters. In addition, the back of this book contains a wealth of material helpful for learning organic chemistry, including a large glossary, an explanation of how to name polyfunctional organic compounds, and answers to most in-text problems. For still further aid, an accompanying *Study Guide and Solutions Manual* provides a summary of name reactions, a summary of methods for preparing functional groups, a summary of functional-group reactions, and a summary of the uses of important reagents.

Changes and Additions in the Fourth Edition

The primary reason for preparing a new edition is to keep the book up-to-date, both in its scientific coverage and its pedagogy. My overall aim has been to retain and refine the features that made earlier editions so successful, while adding new ones.

- **The writing** has again been revised at the sentence level, paying particular attention to such traditionally difficult subjects as stereochemistry and nucleophilic substitution reactions.
- **The artwork** has been redone, and many new computer-generated molecular models have been added to aid in three-dimensional perception. Figures frequently present structures in several different formats, side by side, so that students learn structures thoroughly and become used to the various ways in which chemists graphically represent their work. Look at pages 27 and 121 to see some examples.
- **Stereo views** of computer-generated ball-and-stick molecular models have been added as an aid for three-dimensional perception, using the stereo viewer bound into the back of the book. There are examples on pages 29 and 111.
- **The organic chemistry of metabolic pathways** is presented in an entirely new chapter (Chapter 30). Several of the most important pathways—glycolysis, the citric acid cycle, gluconeogenesis, and others—are dissected and analyzed according to the organic reaction mechanisms by which the various steps occur. This new chapter will be of particular interest to the large number of premedical and biology students who take the organic chemistry course.
- **Interlude boxes** at the end of each chapter present interesting applications of organic chemistry relevant to the main chapter subject. Including topics from science, industry, and day-to-day life, these applications enliven and reinforce the material presented in each chapter. Some Interlude topics address environmental concerns and examine popular assumptions about chlorinated organic compounds or chemical toxins in food and the water supply. Other Interludes discuss such topics as the 1995 announcement of a polyyne form of carbon, the development of insect antifeedants for use as pesticides, a look at the facts about vitamin C, and the chemistry of chiral drugs.

- **Biologically important organic reaction mechanisms** are specially identified by the use of a margin icon. Students often wonder about what topics are "important," and this icon helps biologically inclined students answer that question. See page 412, for example.
- **Spectra** are all new, have been redrawn for clarity and accuracy, and are presented with a light color background, coded by type of spectra. Some examples are on pages 428 and 436.
- **NMR spectroscopy** (Chapter 13) has been completely revised and updated. Proton NMR spectroscopy is now presented first, and the use of the DEPT technique for carbon NMR is emphasized.
- **New problems** have been added at the end of each chapter, including a new kind of problem called "A Look Ahead," in which students are challenged to extend their thinking by applying concepts just learned in one chapter to the subject of a future chapter.
- **A review of reaction mechanisms** is given in a new Appendix C in which mechanisms are referenced to their presentation in the main text.

To facilitate the changes outlined above, some material from the previous edition has been compressed, several infrequently used reactions (such as the Hunsdiecker reaction) have been deleted, and other material has been moved. The material on polymer chemistry, formerly in a late, separate chapter, is now integrated into Chapters 7, 14, and 21 to ensure its coverage.

A Complete Ancillary Package

For Instructors

Test Items
by Tammy Tiner, Texas A & M University
- Contains over 1000 short-answer and multiple-choice test items
- Computerized versions are also available

Transparencies
- A complete new set of 125 full-color transparencies covering most topics in the course.

Computerized Test Items
- Contains over 1000 short-answer and multiple-choice test items
- Available for DOS, Windows, and Macintosh platforms
- Print version also available

CD-ROM
- Includes animations of reaction mechanisms and selected illustrations from the text.

For Students

Study Guide and Solutions Manual
by Susan McMurry
- Includes answers to all in-text and end-of-chapter problems and explains in detail how answers are obtained.
- Also contains a summary of name reactions, a summary of methods for preparing functional groups, a summary of functional-group reactions, a summary of the uses of important reagents, and tables of spectroscopic information.

Beaker 2.1
- Software that is an expert system for the organic chemistry student. Available for DOS/Windows and Macintosh platforms.
- Sophisticated, yet easy to use program for exploring organic chemistry principles, for studying and solving, sketching and analyzing molecular structures, for constructing NMR spectra, for performing reactions, and more.

Organic Chemistry Toolbox
by Norbert J. Pienta, University of North Carolina, Chapel Hill
- A text-specific program for constructing molecular models, drawing Lewis dot structures, creating animations of reactions, solving chemistry problems, and studying structures.
- Available for DOS/Windows and Macintosh platforms.

Acknowledgments

It is a great pleasure to thank the many people whose help and suggestions were so valuable in preparing this fourth edition. Foremost is my wife, Susan, who read, criticized, and improved the manuscript, and who was my constant companion throughout all stages of this book's development. Among the reviewers providing valuable comments were Clair Cheer, San Jose State University; Dale Drueckhammer, Stanford University; Douglas Dyckes, University of Colorado, Denver; Kent Gates, University of Missouri, Columbia; Dan Harvey, University of California, San Diego; John Hogg, Texas A & M University; Jack Kampmeier, University of Rochester; John A. Landgrebe, University of Kansas; James Long, University of Oregon; Guy Mattson, University of Central Florida; David McKinnon, University of Manitoba; Kirk McMichael, Washington State University; Roger Murray, University of Delaware; Carmelo Rizzo, Vanderbilt University; Peter W. Slade, University College of Fraser Valley; Ronald Starkey, University of Wisconsin, Green Bay; J. William Suggs, Brown University; Michelle Sulikowski, Texas A & M University; Joyce Takahashi, University of California, Davis; Bruce Toder, State University of New York, Brockport; and Barbara J. Whitlock, University of Wisconsin, Madison.

Reaction Mechanisms

Interludes

Brief Contents

Contents

1 Structure and Bonding 1

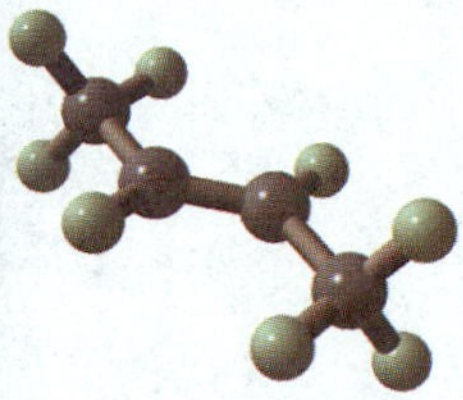

2 Bonding and Molecular Properties 38

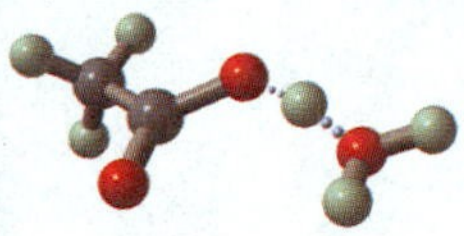

3 The Nature of Organic Compounds: Alkanes and Cycloalkanes 70

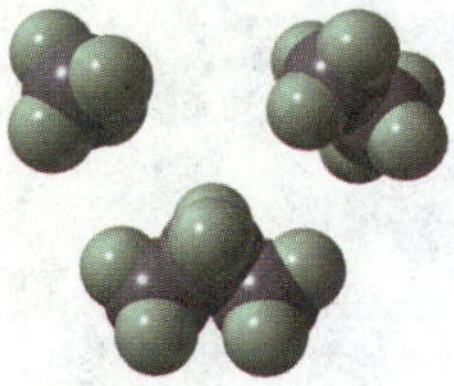

4 Stereochemistry of Alkanes and Cycloalkanes 106

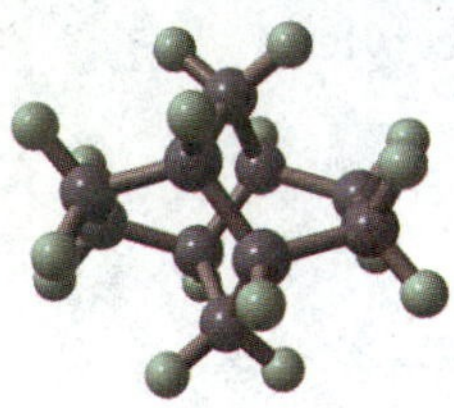

5 An Overview of Organic Reactions 146

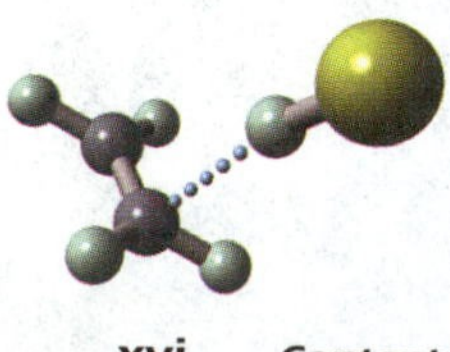

6 Alkenes: Structure and Reactivity 178

7 Alkenes: Reactions and Synthesis 220

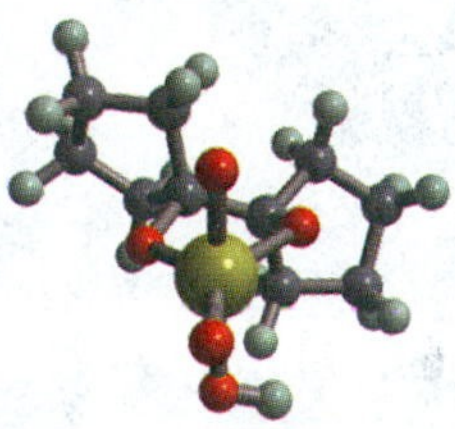

8 Alkynes 264

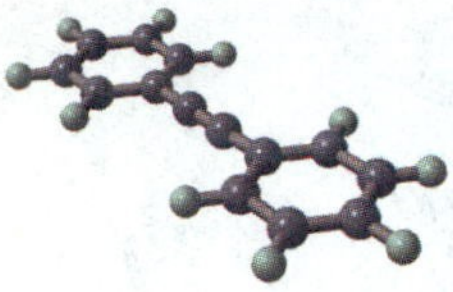

9 Stereochemistry 294

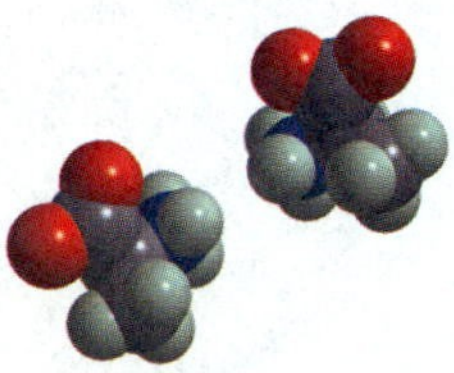

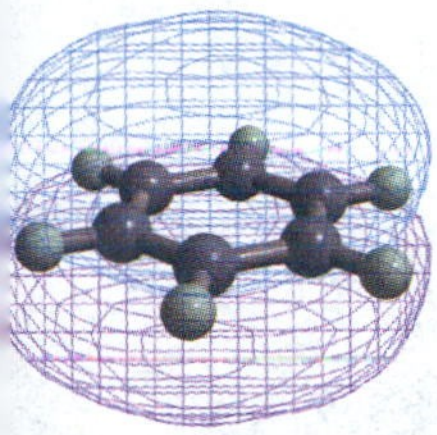

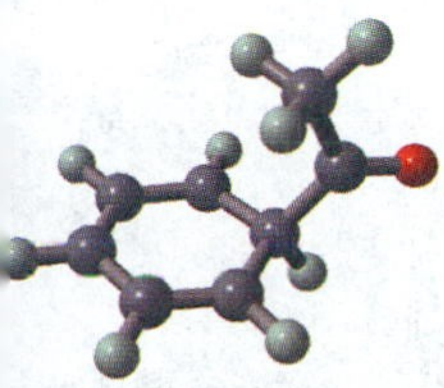

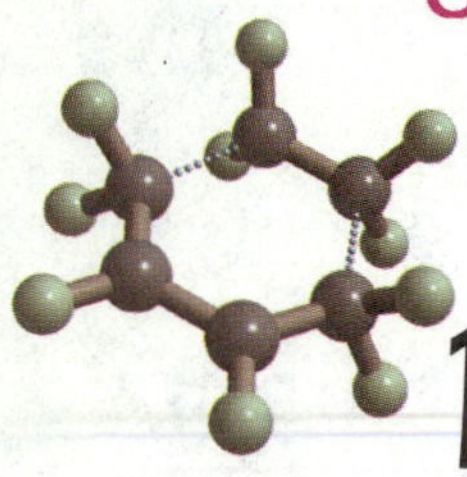

17 Alcohols and Thiols 632

18 Ethers, Epoxides, and Sulfides 676

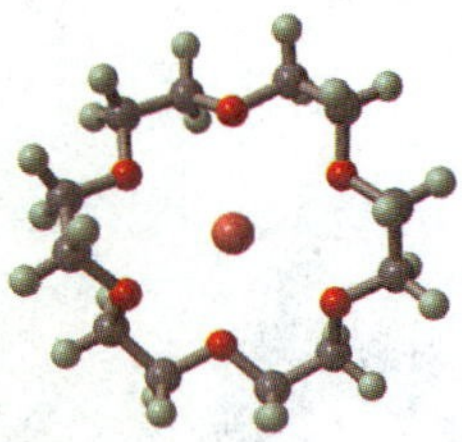

Chemistry of Carbonyl Compounds: An Overview 705

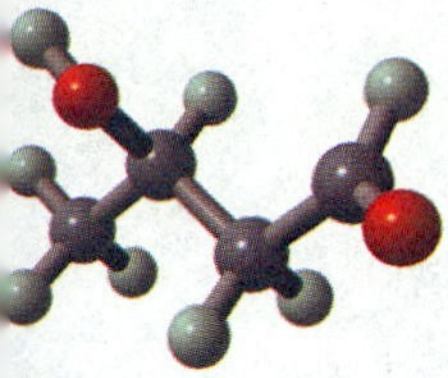

19 Aldehydes and Ketones: Nucleophilic Addition Reactions 716

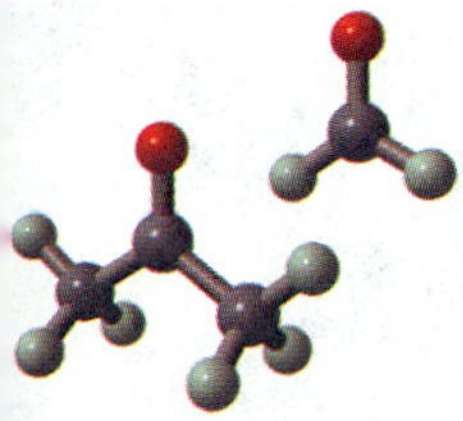

20 Carboxylic Acids 774

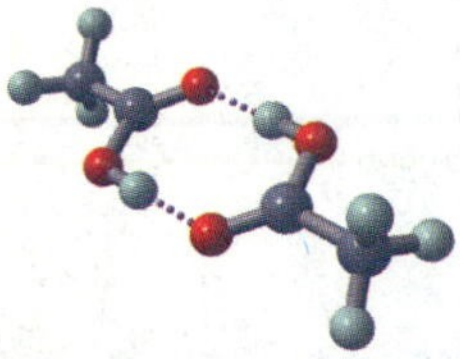

21 Carboxylic Acid Derivatives and Nucleophilic Acyl Substitution Reactions 802

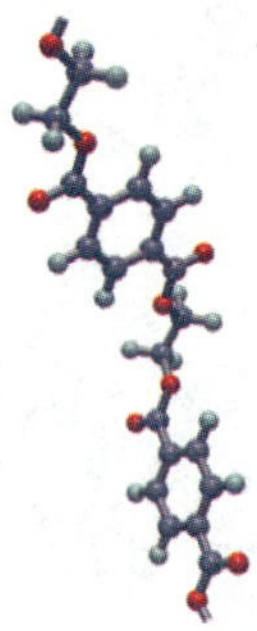

22 Carbonyl Alpha-Substitution Reactions 860

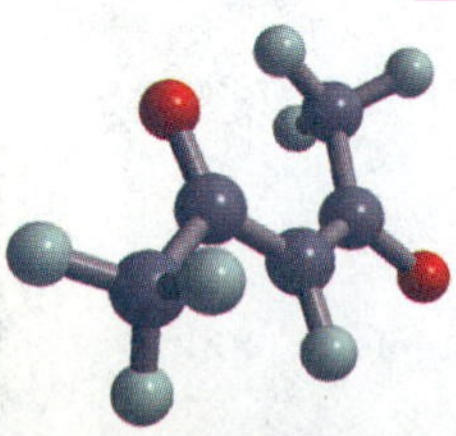

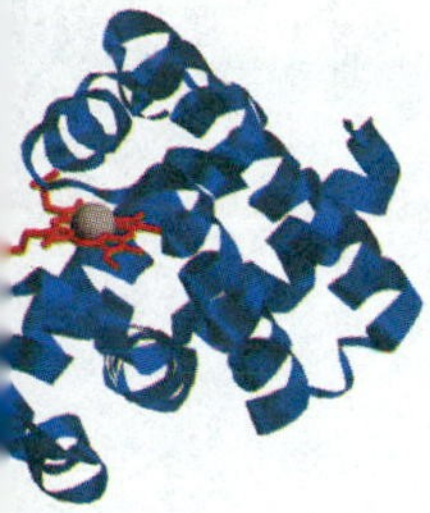

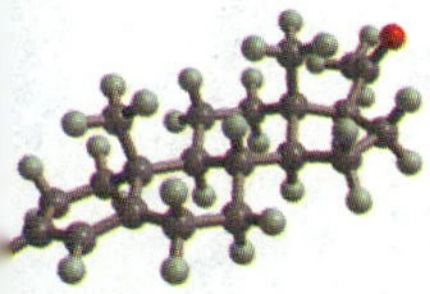

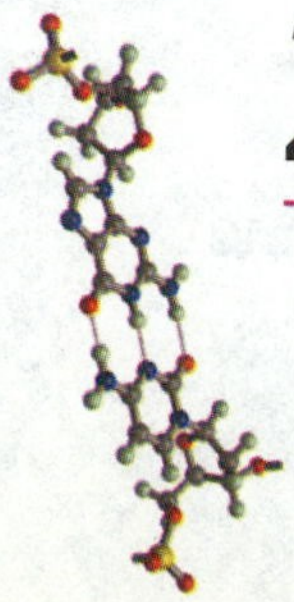

A Note for Students

We have the same goals. Yours is to learn organic chemistry; mine is to help you learn. I've already done the best I can with my part, and now it's going to take some work from you. The following suggestions should prove helpful.

- **Don't Read the Text Immediately.** As you begin each new chapter, look it over first. Read the introductory paragraphs, find out what topics will be covered, and then turn to the end of the chapter and read the summary. You'll be in a much better position to learn the material if you know where you're going.
- **Work the Problems.** There are no shortcuts; working problems is the only way to learn organic chemistry. The practice problems show you how to approach the material, the in-text problems at the ends of most sections provide immediate practice, and the end-of-chapter problems provide both additional drill as well as some real challenges. Short answers to in-text problems are given at the back of the book; full answers and explanations for all problems are given in the accompanying *Study Guide and Solutions Manual*.
- **Use the Study Guide.** The *Study Guide and Solutions Manual* that accompanies this text gives complete solutions to all problems as well as a wealth of supplementary material. Included are a summary of how to prepare functional groups, a summary of the reactions that functional groups undergo, a summary of important reagents, a summary of name reactions, and much more. This material can be extremely useful, both as a source of information and as a self-test, particularly when you're studying for an exam. Find out now what's there so you'll know where to go when you need help.
- **Ask Questions.** Faculty members and teaching assistants are there to help you. Most of them will turn out to be genuinely nice people with a sincere interest in helping you learn.
- **Use Molecular Models.** Organic chemistry is a three-dimensional science. Although this book uses stereo views and many careful drawings to help you visualize molecules, there's no substitute for building a molecular model and turning it around in your own hands.

Good luck! I sincerely hope you enjoy learning organic chemistry and come to see the beauty and logic of its structure. I've heard from many students who used the first three editions of this book and would be glad to receive more comments and suggestions from those who use this new edition.

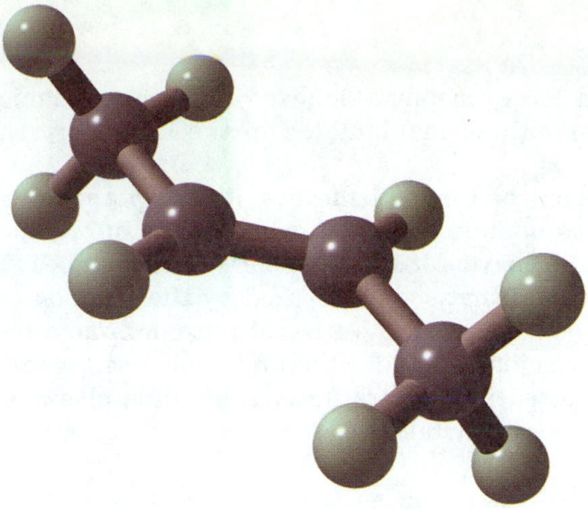

2-Butene, $CH_3CH{=}CHCH_3$, has both planar and tetrahedral carbon atoms.

1

Structure and Bonding

What is organic chemistry, and why should you study it? The answers are all around you. Every living organism is made of organic chemicals. The proteins that make up your hair, skin, and muscles; the DNA that controls your genetic heritage; the foods you eat; the clothes you wear; and the medicines you take are all organic chemicals. Anyone with a curiosity about life and living things must have a fundamental understanding of organic chemistry.

The foundations of organic chemistry date from the mid-eighteenth century, when chemistry was evolving from an alchemist's art into a modern science. At that time, unexplainable differences were noted between substances obtained from living sources and those obtained from minerals. Compounds from plants and animals were often difficult to isolate and purify. Even when pure, they were often difficult to work with and tended to decompose more easily than compounds from minerals. The Swedish chemist Torbern Bergman was the first to express this difference between "organic" and "inorganic" substances in 1770, and the term *organic chemistry* soon came to mean the chemistry of compounds from living organisms.

To many chemists at the time, the only explanation for the different behavior of organic and inorganic compounds was that organic compounds contained a peculiar "vital force" as a result of their origin in living sources.

One consequence of this vital force, chemists believed, was that organic compounds could not be prepared and manipulated in the laboratory as could inorganic compounds.

Although the vitalistic theory had many believers, its acceptance was by no means universal, and it's doubtful that the development of organic chemistry was much delayed. As early as 1816, the theory received a heavy blow when Michel Chevreul[1] found that soap, prepared by the reaction of alkali with animal fat, could be separated into several pure organic compounds, which he termed "fatty acids." Thus, for the first time, one organic substance (fat) had been converted into others (fatty acids plus glycerin) without the intervention of an outside vital force.[2]

$$\text{Animal fat} \xrightarrow[H_2O]{NaOH} \text{Soap} + \text{Glycerine}$$

$$\text{Soap} \xrightarrow{H_3O^+} \text{“Fatty acids”}$$

A little more than a decade later, the vitalistic theory suffered still further when Friedrich Wöhler[3] discovered in 1828 that it was possible to convert the "inorganic" salt, ammonium cyanate, into the previously known "organic" substance, urea.

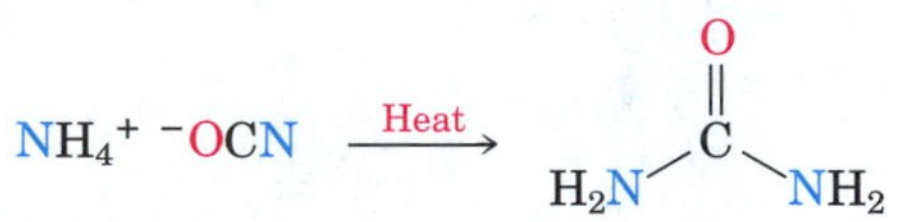

Ammonium cyanate **Urea**

By the mid-nineteenth century, the weight of evidence was clearly against the vitalistic theory, and William Brande[4] wrote in 1848 that "no definite line can be drawn between organic and inorganic chemistry. . . . Any distinctions . . . must for the present be merely considered as matters of practical convenience calculated to further the progress of students."

Chemistry today is unified. The same principles that explain the simplest inorganic compounds also explain the most complex organic ones. The only distinguishing characteristic of organic chemicals is that all contain the element carbon. Nevertheless, the division between organic and inorganic chemistry, which began for historical reasons, maintains its "practical convenience . . . to further the progress of students."

Organic chemistry, then, is the study of carbon compounds. Carbon, atomic number 6, is a second-row element whose position in the periodic table is shown in Figure 1.1. Although carbon is the principal element in

---

[1]Michel Eugène Chevreul (1786–1889); b. Angers, France; educated at Paris, Muséum d'Histoire Naturelle: professor of physics, Lycée Charlemagne (1813); professor of chemistry (1830).

[2]In the equations that follow, a single arrow, $\longrightarrow$, is used to indicate an actual reaction. Later in this book you'll also see forward and backward arrows, $\rightleftharpoons$, indicating equilibrium, and an offset double arrow, $\Longrightarrow$, indicating a multistep transformation whose individual steps aren't shown.

[3]Friedrich Wöhler (1800–1882); b. Eschersheim; studied at Heidelberg (Gmelin); professor, Göttingen (1836–1882).

[4]William Thomas Brande (1788–1866); b. London; lecturer in chemistry, London (1808); Royal Institution (1813–1854).

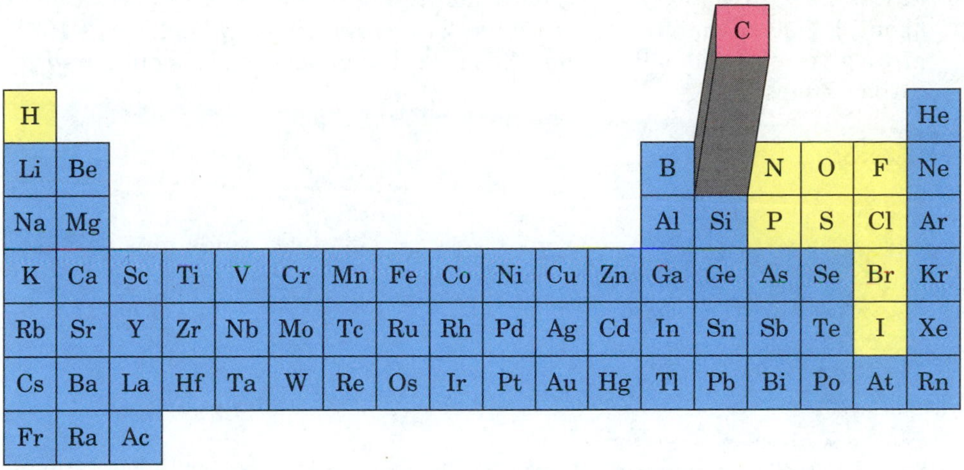

Figure 1.1 The position of carbon in the periodic table. Other elements commonly found in organic compounds are shown in yellow.

organic compounds, most also contain hydrogen, and many contain nitrogen, oxygen, phosphorus, sulfur, chlorine, or other elements.

Why is carbon special? What is it that sets carbon apart from all other elements in the periodic table? The answers to these questions derive from the unique ability of carbon atoms to bond together, forming long chains and rings. Carbon, alone of all elements, is able to form an immense diversity of compounds, from the simple to the staggeringly complex: from methane with one carbon, to DNA with up to tens of *billions*.

Nor are all carbon compounds derived from living organisms. Modern chemists are highly sophisticated in their ability to synthesize new organic compounds in the laboratory. Medicines, dyes, polymers, plastics, food additives, pesticides, and a host of other substances are all prepared in the laboratory. Organic chemistry is a science that touches the lives of everyone. Its study can be a fascinating undertaking.

1.1 Atomic Structure

Before beginning a study of organic chemistry, let's review some general ideas about atoms and bonds. Atoms consist of a dense, positively charged *nucleus* surrounded at a relatively large distance by negatively charged *electrons* (Figure 1.2). The nucleus consists of subatomic particles called *neutrons*, which are electrically neutral, and *protons*, which are positively charged. Though extremely small—about 10^{-14} to 10^{-15} meter (m) in diameter—the nucleus nevertheless contains essentially all the mass of the atom.

Electrons have negligible mass and circulate around the nucleus at a distance of approximately 10^{-10} m. Thus, the diameter of a typical atom is about 2×10^{-10} m, often expressed as 2 *angstroms*[5] (Å), where 1 Å = 10^{-10} m. To give you an idea how small this is, a thin pencil line is about 3 *million* carbon atoms wide.

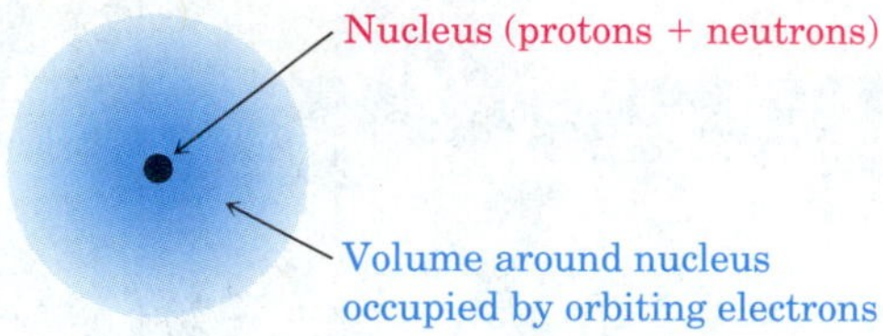

Figure 1.2 A schematic view of an atom. The dense, positively charged nucleus contains most of the atom's mass and is surrounded by negatively charged electrons.

An atom is described by its **atomic number (*Z*)**, which gives the number of protons in the atom's nucleus, and its **mass number (*A*)**, which gives the total of protons plus neutrons. All the atoms of a given element have the same atomic number—1 for hydrogen, 6 for carbon, 17 for chlorine, and so on—but they can have different mass numbers depending on how many neutrons they contain. The average mass number of a great many atoms of an element is called the element's **atomic weight**. Because it's an average, atomic weight is usually not an integer—1.008 for hydrogen, 35.453 for chlorine, and so on.

1.2 Atomic Structure: Orbitals

How are the electrons distributed in an atom? A major breakthrough in our understanding of atomic structure occurred in 1926 when the theory of **quantum mechanics** was proposed independently by Paul Dirac, Werner Heisenberg, and Erwin Schrödinger.[6] All three formulations are mathematical expressions for describing the electronic structure of atoms, but Schrödinger's is the one most commonly used by chemists.

According to Schrödinger, the motion of an electron around a nucleus can be described mathematically by what is known as a **wave equation**—the same kind of expression used to describe the motion of waves in

[5]The angstrom is still used by most organic chemists, even though it has been replaced in SI (Système International) units by the picometer (pm), where 1 pm = 10^{-12} m and 1 Å = 100 pm. Because of the easy decimal conversion, only angstrom measurements will be given in this book.

[6]Erwin Schrödinger (1887–1961); b. Vienna, Austria; University of Vienna (1910); assistant, University of Vienna (1910); assistant to Max Wein, University of Stuttgart, Germany (1920); professor of physics, Universities of Zürich, Berlin, Graz, and Dublin; Nobel Prize in physics (1933).

a fluid. The solution to a wave equation is called a **wave function**, or **orbital**, and is denoted by the Greek letter psi, ψ. A good way of viewing an orbital is to think of it as a mathematical expression whose square (ψ^2) predicts the volume of space around a nucleus where an electron can be found. Although we don't know the exact position of an electron at a given moment, the orbital tells us where we would be most likely to find it.

We can think of an orbital as a time-lapse photograph of an electron's movement around the nucleus. Such a photograph would show the orbital as a blurry cloud indicating the region of space around the nucleus where the electron has been. This electron cloud doesn't have a sharp boundary, but for practical purposes we can set the limits by saying that an orbital represents the space where an electron spends most (90–95%) of its time.

What do orbitals look like? There are four different shapes for orbitals, denoted *s*, *p*, *d*, and *f*. Of the four, we'll be concerned primarily with *s* and *p* orbitals because these play the most important role in organic chemistry. The *s* orbitals are spherical, with the nucleus at their center; *p* orbitals have dumbbell shapes; and four of the five *d* orbitals have cloverleaf shapes, as shown in Figure 1.3. (The fifth *d* orbital has the shape of a donut wrapped around the middle of an elongated dumbbell.)

An *s* orbital

A *p* orbital

A *d* orbital

Figure 1.3 Computer-generated representations of *s*, *p*, and *d* orbitals. The *s* orbitals are spherical, the *p* orbitals are dumbbell-shaped, and four of five *d* orbitals are cloverleaf-shaped.

An atom's electrons can be thought of as being grouped in different layers, or **shells**, around the nucleus. The farther a shell is from the nucleus, the more electrons it can hold and the greater the energies of those electrons. Thus, an atom's lowest-energy electrons occupy the first shell, which is nearest the nucleus and has a capacity of only two electrons. The second shell is farther from the nucleus and can hold eight electrons; the third shell is still farther from the nucleus and can hold eighteen electrons; and so on, as shown in Figure 1.4 (p. 6).

Different shells have different numbers and kinds of orbitals, each of which can hold a pair of electrons. As indicated in Figure 1.4, the two lowest-energy electrons in an atom are in the first shell and occupy a single *s* orbital, denoted 1*s*. Next in energy after the 1*s* electrons are the two 2*s* electrons. Because they're higher in energy, 2*s* electrons are farther from the positively charged nucleus on average, and their spherical orbital is somewhat larger than that of 1*s* electrons.

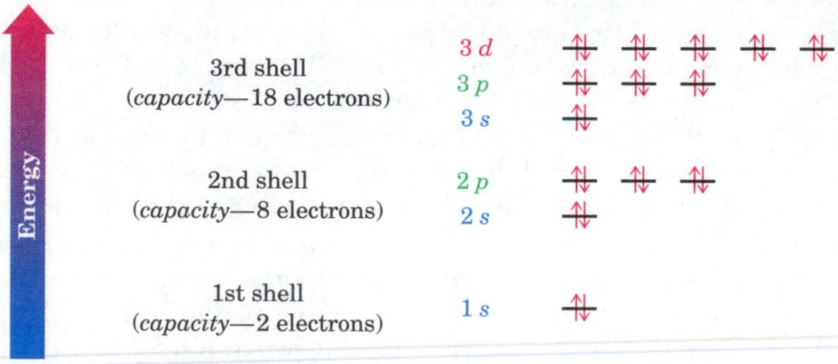

Figure 1.4 The distribution of electrons in an atom. The first shell holds a maximum of two electrons in one 1*s* orbital; the second shell holds a maximum of eight electrons in one 2*s* and three 2*p* orbitals; the third shell holds a maximum of eighteen electrons in one 3*s*, three 3*p*, and five 3*d* orbitals; and so on. The two electrons in each orbital are represented by up and down arrows ⇅.

The six 2*p* electrons are next higher in energy. As Figure 1.5 indicates, there are three 2*p* orbitals, which are equal in energy and are oriented in space so that each is perpendicular to the other two. They are denoted $2p_x$, $2p_y$, and $2p_z$ to indicate on which axis they lie. Note that the plane passing between the two lobes of a *p* orbital lies in a region of zero electron density. Such a region of zero electron density is called a **node**, and the plane is called a **nodal plane**. Nodes have important consequences with respect to chemical reactivity that we'll take up in Chapter 31.

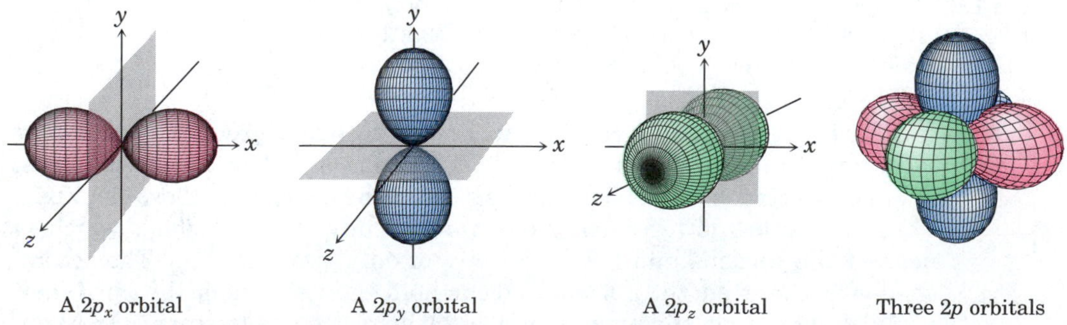

Figure 1.5 Shapes of the 2*p* orbitals. Each of the three mutually perpendicular dumbbell-shaped orbitals has a nodal plane passing between its two lobes.

Still higher in energy are the 3*s* orbital (spherical), three 3*p* orbitals (dumbbell-shaped), 4*s* orbital (spherical), and five 3*d* orbitals. As previously mentioned, we won't be too concerned with *d* orbitals, but you might note that the 3*d* orbital shown in Figure 1.3 has four lobes and two nodal planes.

1.3 Atomic Structure: Electron Configurations

The lowest-energy arrangement, or **ground-state electron configuration**, of an atom is a description of the orbitals that the atom's electrons occupy. We can predict this arrangement by following three rules:

1. The lowest-energy orbitals fill up first according to the order 1*s* → 2*s* → 2*p* → 3*s* → 3*p* → 4*s* → 3*d*, as shown in Figure 1.4.
2. Only two electrons can occupy an orbital, and they must be of opposite spin[7] (a statement called the **Pauli exclusion principle**).
3. If two or more empty orbitals of equal energy are available, one electron occupies each with their spins parallel until all are half-full (a statement called **Hund's rule**).

Some examples of how these rules apply are shown in Table 1.1. Hydrogen, for instance, has only one electron, which must occupy the lowest-energy orbital. Thus, hydrogen has a 1*s* ground-state configuration. Carbon has six electrons and the ground-state configuration $1s^2\ 2s^2\ 2p_x\ 2p_y$.[8]

Table 1.1 Ground-State Electron Configurations of Some Elements

Element	*Atomic number*	*Configuration*	*Element*	*Atomic number*	*Configuration*
Hydrogen	1	1*s* ↑	Lithium	3	2*s* ↑ 1*s* ↑↓
Carbon	6	2*p* ↑ ↑ — 2*s* ↑↓ 1*s* ↑↓	Neon	10	2*p* ↑↓ ↑↓ ↑↓ 2*s* ↑↓ 1*s* ↑↓
Sodium	11	3*s* ↑ 2*p* ↑↓ ↑↓ ↑↓ 2*s* ↑↓ 1*s* ↑↓	Argon	18	3*p* ↑↓ ↑↓ ↑↓ 3*s* ↑↓ 2*p* ↑↓ ↑↓ ↑↓ 2*s* ↑↓ 1*s* ↑↓

[7]For the purposes of quantum mechanics, electrons are considered to spin around an axis in much the same way that the earth spins. This spin can have two orientations, denoted as up ↑ and down ↓.

[8]A superscript is used to represent the number of electrons in a particular orbital. For example, $1s^2$ indicates that there are two electrons in the 1*s* orbital. No superscript is used when there is only one electron in an orbital.

PROBLEM

1.1 How many electrons does each of these elements have in its outermost electron shell?
(a) Potassium (b) Aluminum (c) Krypton

PROBLEM

1.2 Give the ground-state electron configuration for each of these elements:
(a) Boron (b) Phosphorus (c) Oxygen (d) Chlorine

1.4 Development of Chemical Bonding Theory

By the mid-nineteenth century, the new science of chemistry was developing rapidly, and chemists had begun to probe the forces holding molecules together. In 1858, August Kekulé[9] and Archibald Couper[10] independently proposed that, in all organic compounds, carbon has four "affinity units." That is, carbon is *tetravalent*; it always forms four bonds when it joins other elements to form stable compounds. Furthermore, said Kekulé, carbon atoms can bond to each other to form extended chains of carbon atoms linked together.

Shortly after the tetravalent nature of carbon was proposed, extensions to the Kekulé–Couper theory were made when the possibility of *multiple* bonding between atoms was suggested. Emil Erlenmeyer[11] proposed a carbon-to-carbon triple bond for acetylene, and Alexander Crum Brown[12] proposed a carbon-to-carbon double bond for ethylene. In 1865, Kekulé provided another major advance when he suggested that carbon chains can double back on themselves to form rings of atoms.

Although Kekulé was correct in describing the tetravalent nature of carbon, chemistry was still viewed in a two-dimensional way until 1874. In that year, Jacobus van't Hoff[13] and Joseph Le Bel[14] added a third dimension to our ideas about molecules by proposing that the four bonds of carbon are not oriented randomly but have a specific spatial direction. Van't Hoff went

---

[9]Friedrich August Kekulé (1829–1896); b. Darmstadt; University of Giessen (1847); studied under Liebig, Dumas, Gerhardt, and Williamson; assistant to Stenhouse, London; professor, Heidelberg (1855), Ghent (1858), and Bonn (1867).

[10]Archibald Scott Couper (1831–1892); b. Kirkintilloch, Scotland; studied at the universities of Glasgow and Edinburgh (1852) and with Würtz in Paris; assistant in Edinburgh (1858).

[11]Richard A. C. E. Erlenmeyer (1825–1909); b. Wehen, Germany; studied in Giessen and in Heidelberg; professor, Munich Polytechnicum (1868–1883).

[12]Alexander Crum Brown (1838–1922); b. Edinburgh; studied at Edinburgh, Heidelberg, and Marburg; professor, Edinburgh (1869–1908).

[13]Jacobus Hendricus van't Hoff (1852–1911); b. Rotterdam; studied at Polytechnic at Delft, Leyden, Bonn, Paris, and received doctorate at Utrecht (1874); professor, Utrecht, Amsterdam (1878–1896), Berlin; Nobel Prize (1901).

[14]Joseph Achille Le Bel (1847–1930); b. Péchelbronn, Alsace; studied in the École Polytechnique and at the Sorbonne; industrial consultant.

even further and suggested that the four atoms to which carbon is bonded sit at the corners of a regular tetrahedron, with carbon in the center.

A representation of a tetrahedral carbon atom is shown in Figure 1.6. Note the conventions used to show three-dimensionality: Solid lines represent bonds in the plane of the page, heavy wedged lines represent bonds coming out of the page toward the viewer, and dashed lines represent bonds receding back behind the page, away from the viewer. These representations will be used throughout the text.

A tetrahedron

Stereo View

Figure 1.6 Van't Hoff's tetrahedral carbon atom. The heavy wedged line comes out of the plane of the paper; the normal lines are in the plane; and the dashed line goes back behind the plane of the page.

PROBLEM

1.3 Draw a molecule of chloroform, $CHCl_3$, using wedged, normal, and dashed lines to show its tetrahedral geometry.

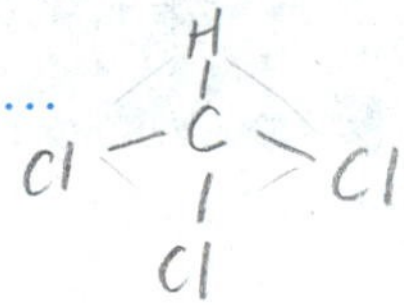

1.5 The Nature of Chemical Bonding: Ionic Bonds

Why do atoms bond together, and how does quantum mechanics describe bonding? The *why* question is relatively easy to answer: Atoms bond together because the compound that results is more stable (has less energy) than the separate atoms. Just as water always flows downhill, energy is always released and flows *from* the system when a chemical bond is formed. Conversely, energy is absorbed and must be added *to* the system when a chemical bond is broken. The *how* question is more difficult. To answer it, we need to know more about the properties of atoms.

We know through observation that eight electrons (an octet) in the outermost shell (the **valence shell**) impart special stability to the noble-gas elements in group 8A of the periodic table: Ne (2 + 8); Ar (2 + 8 + 8);

Kr (2 + 8 + 18 + 8). We also know that the chemistry of many main-group elements is governed by a tendency for them to take on the electron configuration of a noble gas. The alkali metals in group 1A, for example, have a single electron in their valence shells. By losing this electron, they can achieve a noble-gas configuration.

The amount of energy needed to remove an electron from an isolated atom in the gas phase is called the atom's **ionization energy**, abbreviated E_i. Alkali metals, at the far left of the periodic table, give up a valence electron easily and have low ionization energies. Elements at the middle and far right of the periodic table hold their valence electrons more tightly, give them up less readily, and therefore have higher values of E_i. Figure 1.7 displays trends in E_i data for the entire periodic table.

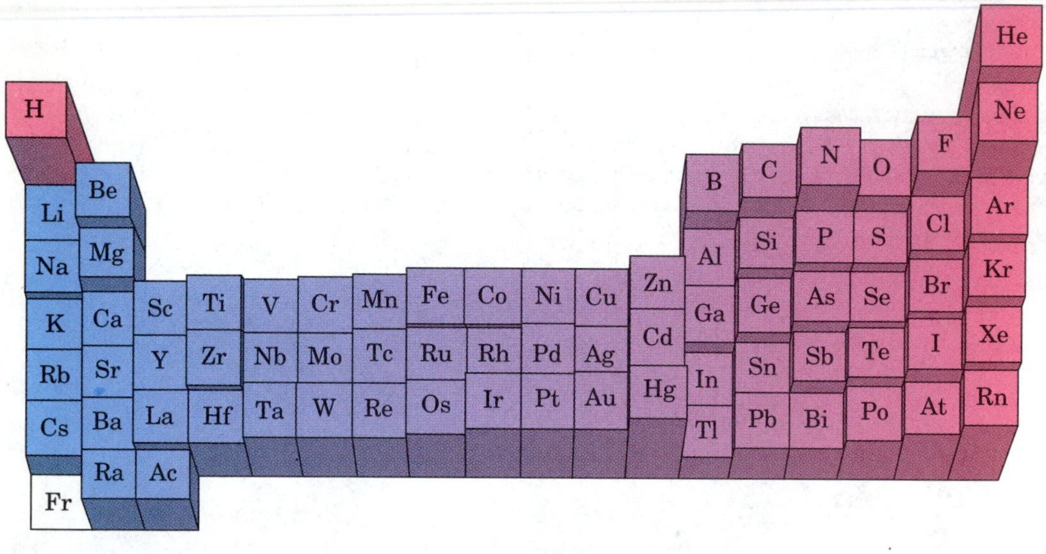

Figure 1.7 Relative ionization energies of elements. Metallic elements on the left of the periodic table have lower ionization energies and lose an electron more easily than nonmetallic elements on the right.

Just as the metals on the left of the periodic table tend to form positive ions by losing an electron, the halogens and other reactive nonmetals on the right of the table tend to form *negative* ions by *gaining* one or more electrons. The measure of this tendency of an isolated atom in the gas phase to gain an electron is called the atom's **electron affinity**, abbreviated E_{ea}. Energy is released when an electron is added to most elements, and most E_{ea}'s are therefore negative. Figure 1.8 displays relative electron affinities.

The simplest kind of chemical bonding is that between a metal with a low E_i and a nonmetal with a large negative E_{ea}. For example, when sodium

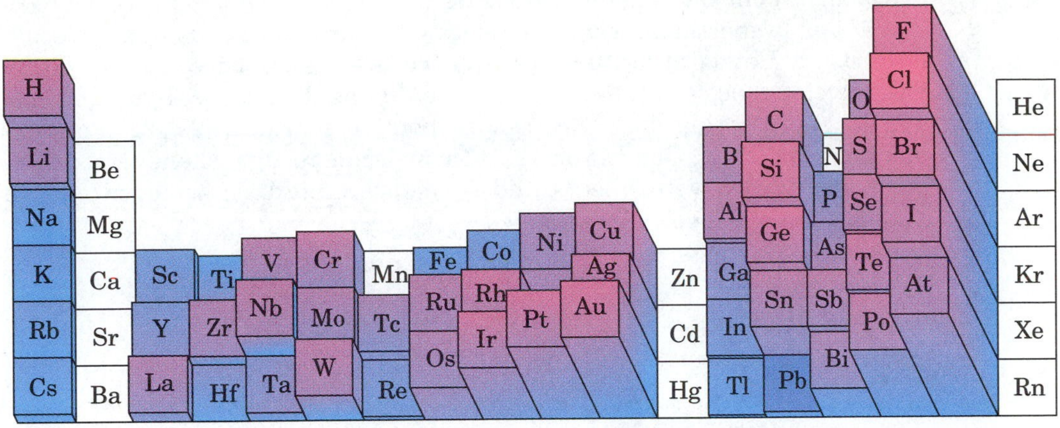

Figure 1.8 Relative electron affinities—the larger the peak, the more negative the electron affinity. Metallic elements on the left of the periodic table have less negative electron affinities than nonmetallic elements on the right and therefore gain an electron less easily. Elements shown in white have positive electron affinities.

metal [E_i = 496 kJ/mol (118.5 kcal/mol)][15] reacts with chlorine gas [E_{ea} = −349 kJ/mol (−83.3 kcal/mol)], sodium gives an electron to chlorine forming positively charged Na^+ ions and negatively charged Cl^- ions. The NaCl product is said to be an **ionic solid** and to have **ionic bonding**. That is, the ions are held together by electrostatic attraction between unlike charges. A similar situation exists for many other metal salts such as potassium fluoride (K^+F^-), lithium bromide (Li^+Br^-), and so on.

1.6 The Nature of Chemical Bonding: Covalent Bonds

We've just seen that metals on the far left and nonmetals on the far right of the periodic table form ionic bonds by gaining or losing electrons to achieve a noble-gas configuration. How, though, do elements in the middle of the periodic table form bonds? Look at the carbon atom in methane, CH_4, for example. Certainly the bonding in methane is not ionic, because it would be very difficult for carbon ($1s^2\,2s^2\,2p^2$) either to gain or to lose *four* electrons to achieve a noble-gas configuration.[16]

In fact, carbon bonds to other atoms, not by donating electrons, but by *sharing* them. Such a shared-electron bond, first proposed in 1916 by G. N.

[15]Organic chemists have been slow to adopt SI units, preferring to use kilocalories (kcal) rather than kilojoules (kJ) as a measure of energy. This book will use dual units, with values shown in both kJ/mol and kcal/mol: 1 kJ/mol = 0.239 kcal/mol; 1 kcal/mol = 4.184 kJ/mol.

[16]The electron configuration of carbon can be written either as $1s^2\,2s^2\,2p^2$ or as $1s^2\,2s^2\,2p_x\,2p_y$. Both notations are correct, but the latter is more informative because it indicates that two of the three equivalent p orbitals are half filled.

Lewis,[17] is called a **covalent bond**. The collection of atoms held together by covalent bonds is called a **molecule**.

A simple shorthand way of indicating the covalent bonds in a molecule is to use a **Lewis structure**, or **electron-dot structure**, in which the valence electrons of an atom are represented as dots. Thus, hydrogen has one dot representing its 1*s* electron, carbon has four dots ($2s^2\ 2p^2$), oxygen has six dots ($2s^2\ 2p^4$), and so on. A stable molecule results whenever a noble-gas configuration with filled *s* and *p* valence orbitals is achieved for the atoms, as in the following examples:

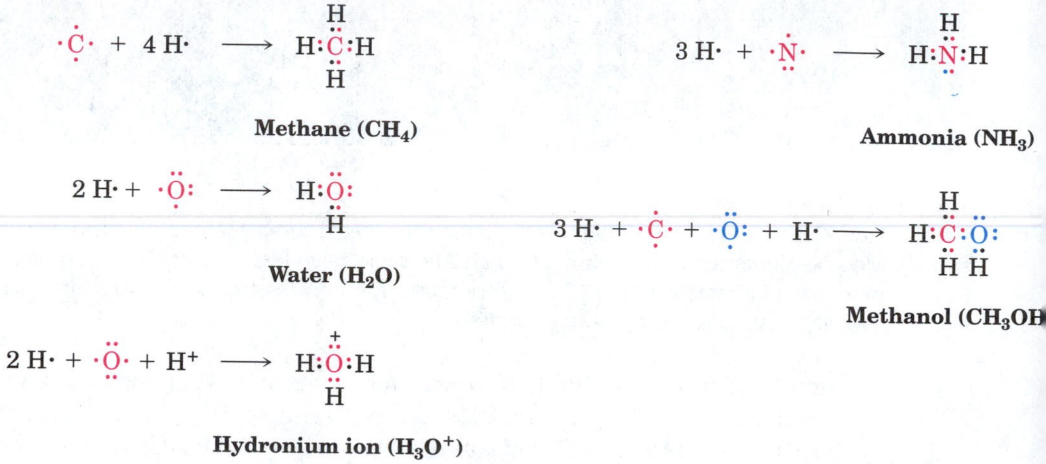

The number of covalent bonds an atom forms depends on how many valence electrons it has. Atoms with one, two, or three valence electrons form one, two, or three bonds, but atoms with four or more valence-shell electrons form as many bonds as they need electrons to fill the *s* and *p* levels of their valence shells and reach an octet. Thus, boron has three valence electrons ($2s^2\ 2p^1$) and forms three bonds, as in BF_3; carbon ($2s^2\ 2p^2$) fills its valence shell by forming four bonds, as in CH_4; nitrogen ($2s^2\ 2p^3$) forms three bonds, as in NH_3; and oxygen ($2s^2\ 2p^4$) forms two bonds, as in H_2O.

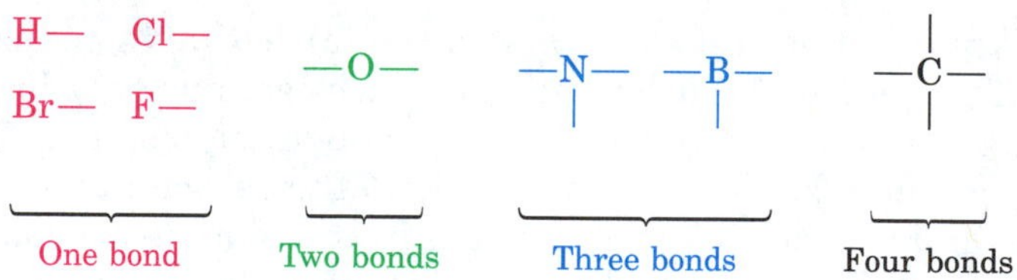

Valence electrons that are not used for bonding are called **nonbonding electrons**, or **lone-pair electrons**. The nitrogen atom in ammonia, for instance, shares six of its eight valence electrons in three covalent bonds with hydrogens, and has its remaining two valence electrons in a nonbonding lone pair.

---

[17]Gilbert Newton Lewis (1875–1946); b. Weymouth, Massachusetts; Ph.D. Harvard (1899); professor, Massachusetts Institute of Technology (1905–1912), University of California, Berkeley (1912–1946).

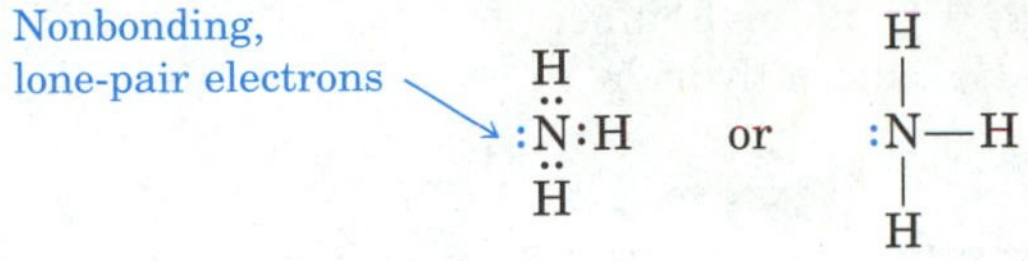

Ammonia

Lewis structures are valuable because they make electron bookkeeping possible and act as reminders of the number of valence electrons present. Simpler still is the use of **"Kekulé" structures**, also called **line-bond structures**, in which a two-electron covalent bond is indicated as a line drawn between atoms. Lone pairs of nonbonding valence electrons are often ignored when drawing line-bond structures, but it's still necessary to keep them in mind. Some examples are shown in Table 1.2.

Table 1.2 Lewis and Kekulé Structures of Some Simple Molecules

Name	*Lewis structure*	*Kekulé structure*	*Name*	*Lewis structure*	*Kekulé structure*
Water (H_2O)	H:Ö:H (with lone pair below O)	H—O—H	Methane (CH_4)	H:C:H with H above and below C	H—C—H with H above and below C
Ammonia (NH_3)	H:N:H with H above N, lone pair below N	H—N—H with H above N	Methanol (CH_3OH)	H:C:Ö:H with H above and below C, lone pairs on O	H—C—O—H with H above and below C

PRACTICE PROBLEM

How many hydrogen atoms does phosphorus bond to in forming phosphine, $PH_?$?

Solution Phosphorus is in group 5A of the periodic table and has five valence electrons. It needs to share three more electrons to make an octet, and therefore bonds to three hydrogen atoms, giving PH_3.

PROBLEM

1.4 What are likely formulas for the following substances?
(a) $CCl_?$ 4 (b) $AlH_?$ 3 (c) $CH_?Cl_2$ 2 (d) $SiF_?$ 4 (e) $CH_3NH_?$ 2

PROBLEM 4 4 7 5 7 1 3 2

1.5 Write both Lewis and line-bond structures for these substances, showing all nonbonded electrons:
(a) $CHCl_3$, chloroform (b) H_2S, hydrogen sulfide

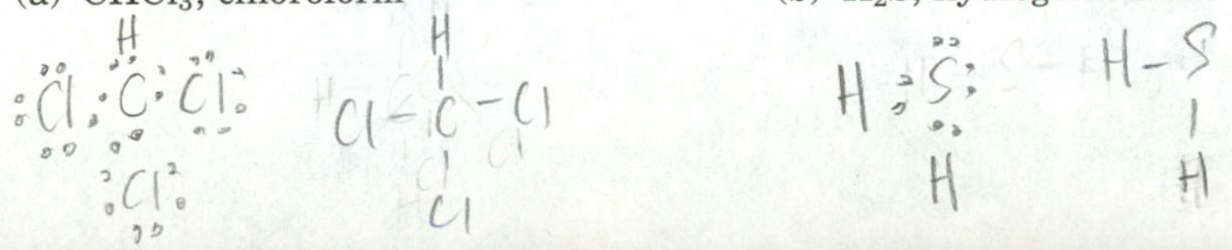

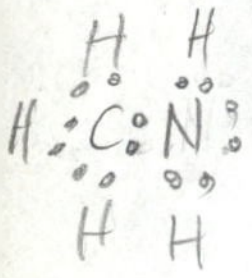

(c) CH_3NH_2, methylamine
(d) BH_3, borane
(e) NaH, sodium hydride
(f) CH_3Li, methyllithium

PROBLEM

1.6 Which of the following substances would you expect to have covalent bonds and which ionic bonds? Explain.
(a) CH_4 (b) CH_2Cl_2 (c) LiI (d) KBr (e) $MgCl_2$ (f) Cl_2

PROBLEM

1.7 Write both a Lewis structure and a line-bond structure for ethane, C_2H_6.

1.7 Describing Covalent Bonds: Valence Bond Theory and Molecular Orbital Theory

How do covalent bonds form? Two models have been developed to describe covalent bond formation, *valence bond theory* and *molecular orbital theory*. Each model has its strengths and weaknesses, and chemists use them interchangeably depending on the circumstances. Valence bond theory is the more easily visualized of the two, so most of the descriptions we'll be using in this book derive from that approach. We'll take a brief look at both theories and then return for a second look at molecular orbital theory in Section 1.10.

Valence Bond Theory

According to **valence bond theory**, a covalent bond results when two atoms approach each other closely so that a singly occupied orbital on one atom *overlaps* a singly occupied orbital on the other atom. The now-paired electrons in the overlapping orbitals are attracted to the nuclei of both atoms and thus bond the atoms together. In the H_2 molecule, for example, the H–H bond results from the overlap of two singly occupied hydrogen 1*s* orbitals:

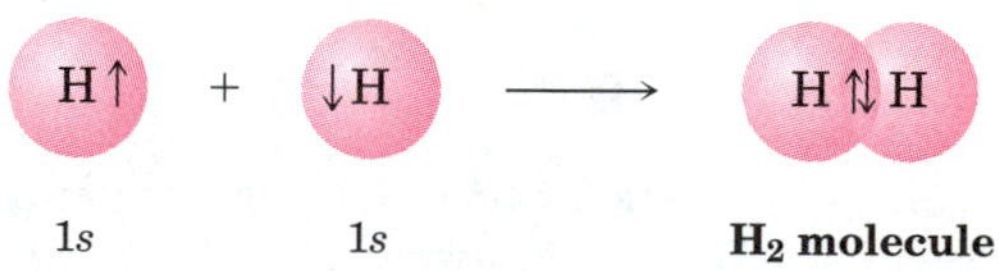

The strength of a covalent bond depends on the amount of orbital overlap—the greater the overlap, the stronger the bond. This means that bonds formed by overlap of other than *s* orbitals have a directionality to them. In HF, for example, the covalent bond involves overlap of a hydrogen 1*s* orbital (nondirectional) with a fluorine 2*p* orbital (directional) and forms along the *p* orbital axis.

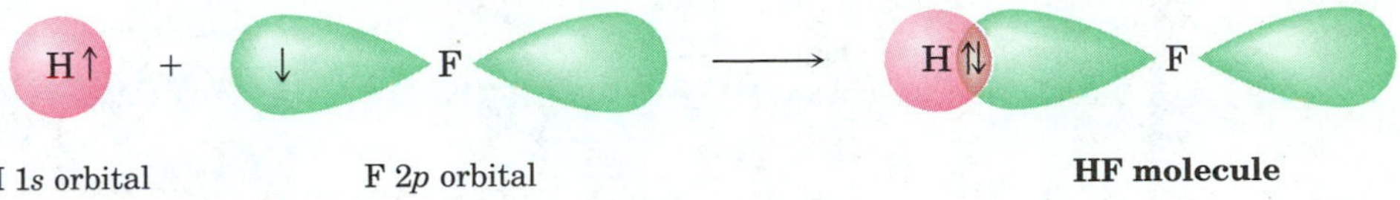

The key ideas of valence bond theory can be summarized as follows:

1. Covalent bonds are formed by overlap of atomic orbitals, each of which contains one electron of opposite spin.
2. Each of the bonded atoms maintains its own atomic orbitals, but the electron pair in the overlapping orbitals is shared by both atoms.
3. The greater the amount of orbital overlap, the stronger the bond.

The bond in the H_2 molecule has the elongated egg shape we might get by pressing two spheres together. If a plane were to pass through the middle of the bond, the intersection of the plane and the overlapping orbitals would be a circle. In other words, the H–H bond is *cylindrically symmetrical*, as shown in Figure 1.9. Such bonds, which are formed by the head-on overlap of two atomic orbitals along a line drawn between the nuclei, are called **sigma (σ) bonds**. Although σ bonds are the most common kind, we'll see shortly that there is another type as well.

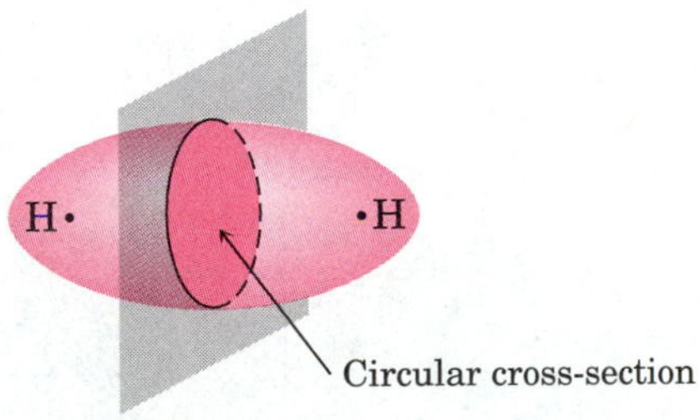

Figure 1.9 The cylindrical symmetry of the H–H bond. The intersection of a plane cutting through the orbital is a circle.

During the reaction 2 H· → H_2, 436 kJ/mol (104 kcal/mol) of energy is released. Because the product H_2 molecule has 436 kJ/mol less energy than the starting 2 H·, we say that the product is more stable than the starting material and that the new H–H bond has a **bond strength** of 436 kJ/mol. In other words, we would have to put 436 kJ/mol of energy *into* the H–H bond to break the H_2 molecule apart into two H atoms. Figure 1.10 shows the relative energy levels of the different species.

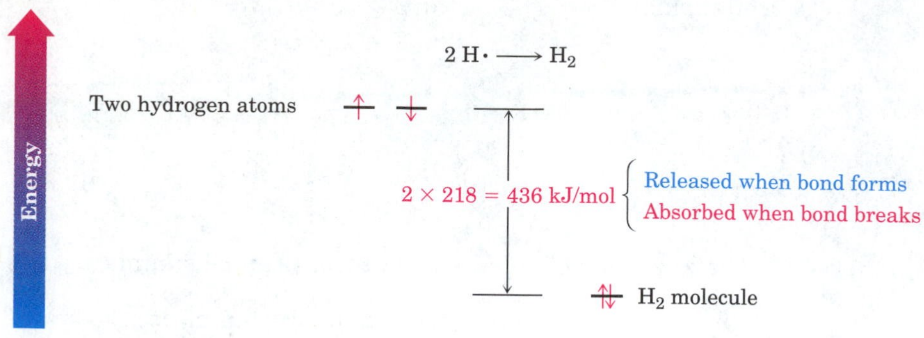

Figure 1.10 Energy levels of H atoms and the H_2 molecule. Because the H_2 molecule is lower in energy by 436 kJ/mol (104 kcal/mol) than the two H atoms, 436 kJ/mol of energy is released when the H–H bond forms.

How close are the two nuclei in the H_2 molecule? If they are too close, they will repel each other because both are positively charged, yet if they are too far apart, they won't be able to share the bonding electrons. Thus, there is an optimum distance between nuclei that leads to maximum stability (Figure 1.11). Called the **bond length**, this distance is 0.74 Å in the H_2 molecule. Every covalent bond has both a characteristic bond strength and bond length.

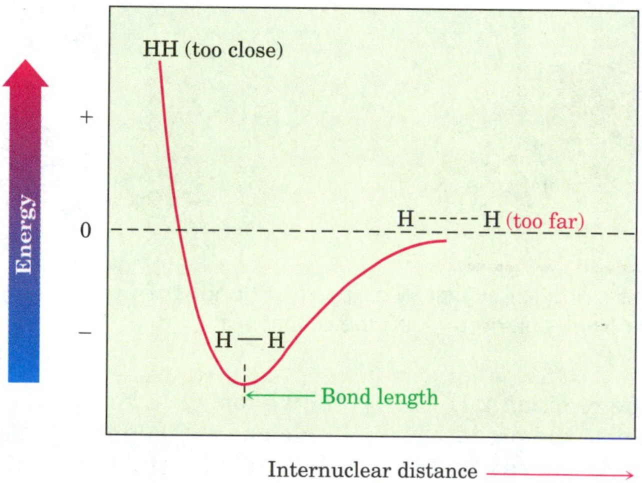

Figure 1.11 A plot of energy versus internuclear distance for two hydrogen atoms. The distance at the lowest-energy point is called the bond length.

Molecular Orbital Theory

Molecular orbital (MO) theory describes covalent bond formation as a mathematical combination of atomic orbitals (wave functions) to form *molecular orbitals*, so called because they belong to an entire molecule rather than to an individual atom. Just as an *atomic* orbital describes a region of space around an *atom* where an electron is likely to be found, so a *molecular* orbital describes a region of space in a *molecule* where electrons are most likely to be found. Like an atomic orbital, a molecular orbital has a specific size, shape, and energy. In the H_2 molecule, for example, two singly occupied 1*s* atomic orbitals combine. There are two ways for the orbital combination to occur—an additive way and a subtractive way. The additive combination leads to formation of a molecular orbital that is roughly egg-shaped, while the subtractive combination leads to formation of a molecular orbital that has a node between nuclei (Figure 1.12).

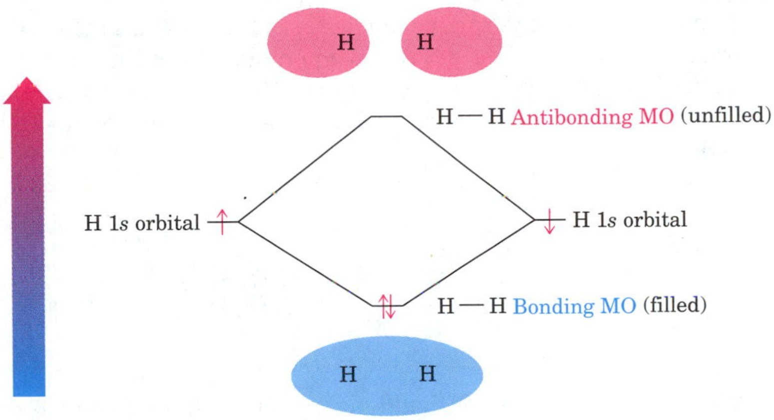

Figure 1.12 Molecular orbitals of H_2. Combination of two hydrogen 1*s* atomic orbitals leads to two H_2 molecular orbitals. The lower-energy, bonding MO is filled, and the higher-energy, antibonding MO is unfilled.

The additive combination is lower in energy than the two hydrogen 1*s* atomic orbitals and is called a **bonding MO**. Any electrons it contains spend most of their time in the region between the two nuclei, helping to bond the atoms together. The subtractive combination is higher in energy than the two hydrogen 1*s* orbitals and is called an **antibonding MO**. Any electrons it contains can't occupy the central region between the nuclei where there is a node and can't contribute to bonding.

The key ideas of molecular orbital theory can be summarized as follows:

1. Molecular orbitals are to molecules what atomic orbitals are to atoms. Molecular orbitals describe regions of space in a molecule

where electrons are most likely to be found, and they have a specific size, shape, and energy level.

2. Molecular orbitals are formed by combining atomic orbitals. The number of MO's formed is the same as the number of atomic orbitals combined.
3. Molecular orbitals that are lower in energy than the starting atomic orbitals are bonding; MO's that are higher in energy than the starting atomic orbitals are antibonding; and MO's with the same energy as the starting atomic orbitals are nonbonding.

1.8 Hybridization: *sp*³ Orbitals and the Structure of Methane

The bonding in the hydrogen molecule is fairly straightforward, but the situation is more complicated in organic molecules with tetravalent carbon atoms. Let's start with a simple case and consider methane, CH_4. Carbon has four electrons in its valence shell and can form four bonds to hydrogens. In Lewis structures:

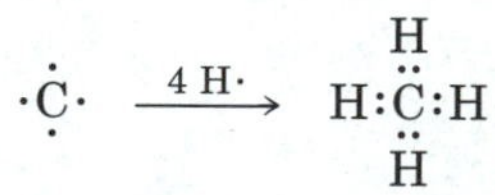

What are the four C–H bonds in methane like? Because carbon uses two kinds of orbitals (2*s* and 2*p*) to form bonds, we might expect methane to have two kinds of C–H bonds. In fact, though, a large amount of evidence shows that all four C–H bonds in methane are identical and are spatially oriented toward the corners of a regular tetrahedron (see Figure 1.6). How can we explain this?

The answer was provided in 1931 by Linus Pauling,[18] who showed mathematically how an *s* orbital and three *p* orbitals can combine, or **hybridize**, to form four equivalent atomic orbitals with tetrahedral orientation. Shown in Figure 1.13, these tetrahedrally oriented orbitals are called ***sp*³ hybrids**.[19]

---

[18]Linus Pauling (1901–1994); b. Portland, Oregon; Ph.D. California Institute of Technology (1925); professor, California Institute of Technology (1925–1967); University of California, San Diego; Professor Emeritus, Stanford University (1974–1994); Nobel Prize (1954, 1963).

[19]The superscript "3" in the name *sp*³ hybrid indicates that 3 *p* atomic orbitals combine to form the hybrid, not that 3 electrons occupy it.

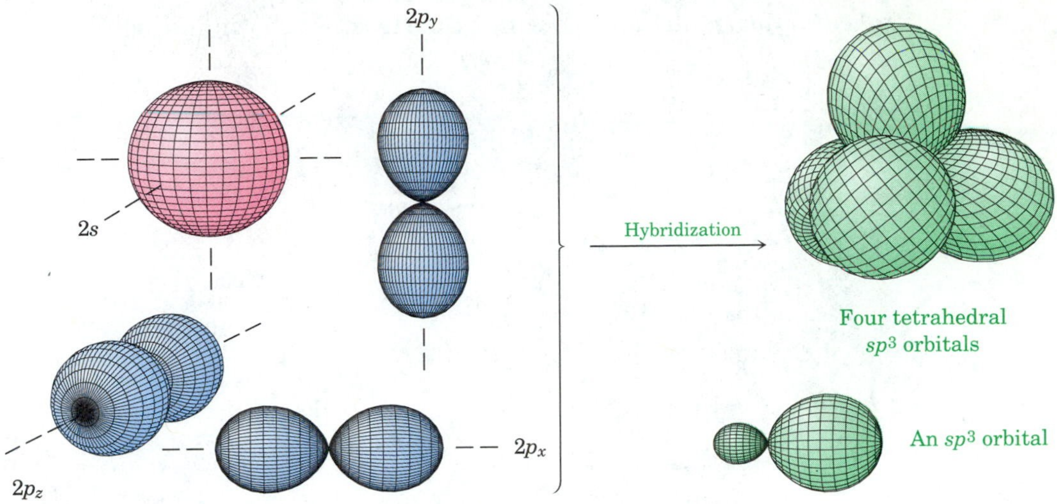

Figure 1.13 Four sp^3 hybrid orbitals are formed by combination of an atomic *s* orbital and three atomic *p* orbitals. The sp^3 hybrids are oriented to the corners of a regular tetrahedron. Because the two lobes of the *p* atomic orbitals have different algebraic signs, they can combine with the *s* orbital in either an additive or a subtractive way, thus making the resultant sp^3 hybrid unsymmetrical and giving it a directionality.

The concept of hybridization explains *how* carbon forms four equivalent tetrahedral bonds but doesn't explain *why* it does so. Looking at an sp^3 hybrid orbital from the side suggests the answer. When an *s* orbital hybridizes with three *p* orbitals, the resultant hybrid orbitals are unsymmetrical about the nucleus. One of the two lobes of an sp^3 orbital is much larger than the other and can therefore overlap better with the orbital from another atom when it forms a bond. As a result, sp^3 hybrid orbitals form stronger bonds than do unhybridized *s* or *p* orbitals.

The asymmetry of sp^3 orbitals arises because of a property of orbitals that we have not yet considered. When the wave equation corresponding to a *p* orbital is solved, the two lobes have opposite algebraic signs, + and −. Thus, when a *p* orbital hybridizes with an *s* orbital, one *p* lobe is *additive* with the *s* orbital, but the other *p* lobe is *subtractive*. The resultant hybrid orbital is therefore strongly oriented in one direction, as shown in Figure 1.13.

When the four identical orbitals of an sp^3-hybridized carbon atom overlap with the 1*s* orbitals of four hydrogen atoms, four identical C–H bonds are formed and methane results. Each C–H bond in methane has a strength of 438 kJ/mol (105 kcal/mol) and a length of 1.10 Å. Because the four bonds have a specific geometry, we also can define a property called the **bond angle**. The angle formed by each H–C–H is exactly 109.5°, the so-called *tetrahedral angle*. Methane thus has the structure shown in Figure 1.14.

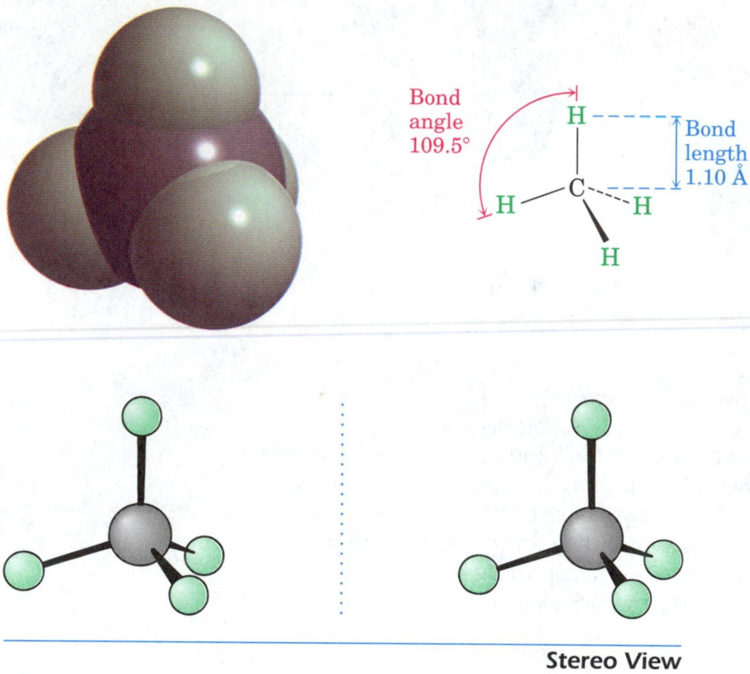

Figure 1.14 The structure of methane. The drawings are computer-generated.

1.9 The Structure of Ethane

The same kind of hybridization that explains the methane structure also explains how carbon atoms can bond together to make possible the many millions of known organic compounds. Ethane, C_2H_6, is the simplest molecule containing a carbon–carbon bond:

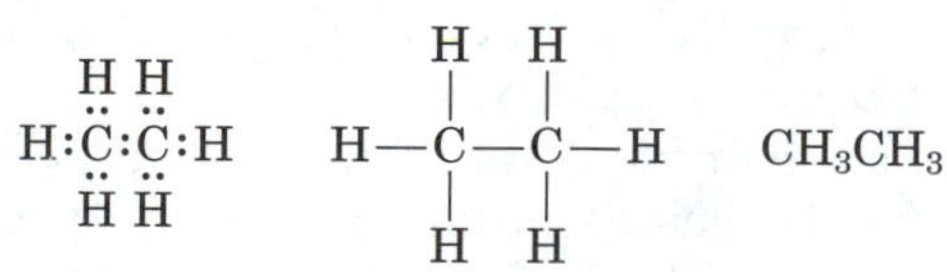

Some representations of ethane

We can picture the ethane molecule by imagining that the two carbon atoms bond to each other by σ overlap of an sp^3 hybrid orbital from each. The remaining three sp^3 hybrid orbitals on each carbon overlap with hydrogen 1*s* orbitals to form the six C–H bonds, as shown in Figure 1.15. The C–H bonds in ethane are similar to those in methane, though a bit weaker—420 kJ/mol (100 kcal/mol) for ethane versus 438 kJ/mol for methane. The C–C bond is 1.54 Å long and has a strength of 376 kJ/mol (90 kcal/mol). All the bond angles of ethane are near the tetrahedral value of 109.5°.

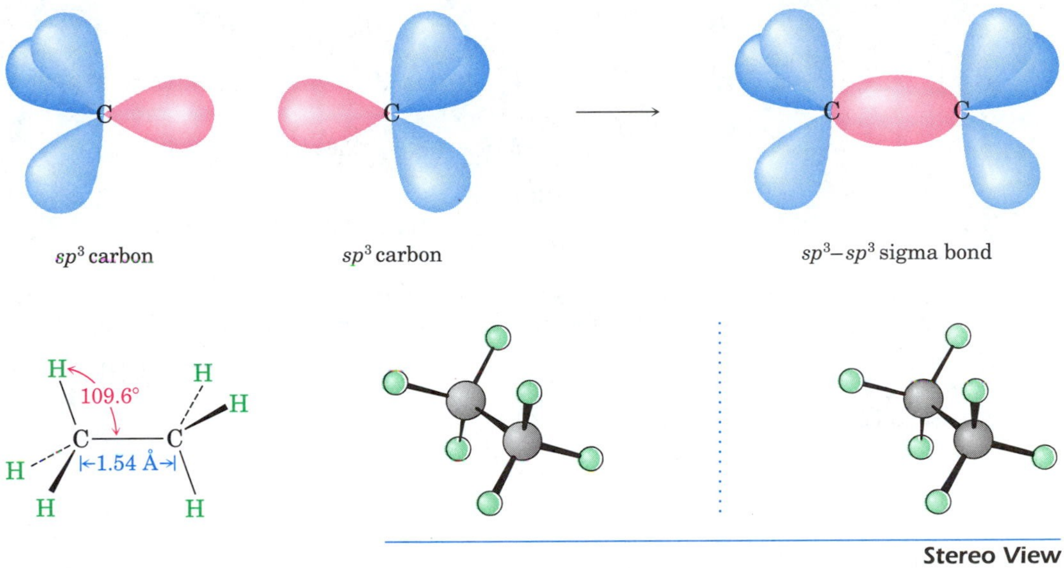

Figure 1.15 The structure of ethane. The carbon–carbon bond is formed by σ overlap of two carbon sp^3 hybrid orbitals. (For clarity, the smaller lobes of the sp^3 hybrid orbitals are not shown.)

PROBLEM

1.8 Draw a line-bond structure for propane, $CH_3CH_2CH_3$. Predict the value of each bond angle, and indicate the overall shape of the molecule.

PROBLEM

1.9 Why can't an organic molecule have the formula C_2H_7?

each carbon can only have 4 bonds

1.10 Hybridization: sp^2 Orbitals and the Structure of Ethylene

Although sp^3 hybridization is the most common electronic state of carbon, it's not the only possibility. Look at ethylene, C_2H_4, for example. It was recognized over 100 years ago that ethylene carbons can be tetravalent only if the two carbon atoms share *four* electrons and are linked by a *double* bond. Furthermore, ethylene is known to be planar (flat) and to have bond angles of approximately 120°.

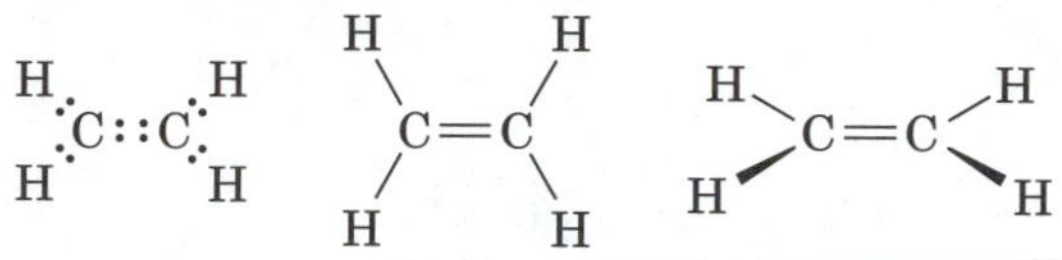

Top view   Side view

Ethylene

When we discussed sp^3 hybrid orbitals in explaining the bonding in methane, we said that all four of carbon's outer-shell atomic orbitals combine

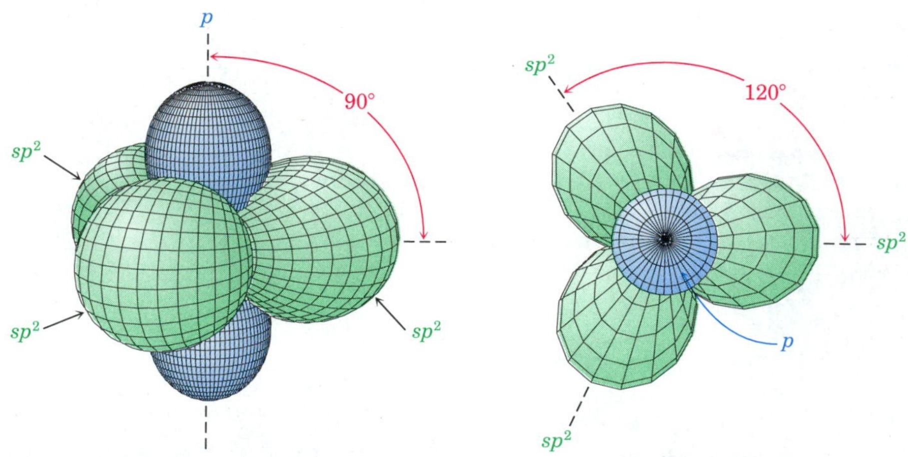

Side view   Top view

Figure 1.16 An sp^2-hybridized carbon. There are three equivalent sp^2 hybrid orbitals lying in a plane at angles of 120° to each other and a single unhybridized p orbital perpendicular to the sp^2 plane.

to form four equivalent sp^3 hybrids. Imagine instead that the 2*s* orbital combines with only *two* of the three available 2*p* orbitals. Three hybrid orbitals called **sp^2 hybrids** result, and one 2*p* orbital remains. The three sp^2 orbitals lie in a plane at angles of 120° to one another, with the remaining *p* orbital perpendicular to the sp^2 plane, as shown in Figure 1.16.

When two sp^2-hybridized carbons approach one another, they form a σ bond by sp^2–sp^2 overlap. At the same time, the unhybridized *p* orbitals on each carbon approach with the correct geometry for *sideways* overlap, leading to the formation of what is called a **pi (π) bond**. Note that the π bond has regions of electron density on either side of a line drawn between nuclei but has no electron density directly between nuclei. The combination of an sp^2–sp^2 σ bond and a 2*p*–2*p* π bond results in the sharing of four electrons and the formation of a C=C double bond (Figure 1.17).

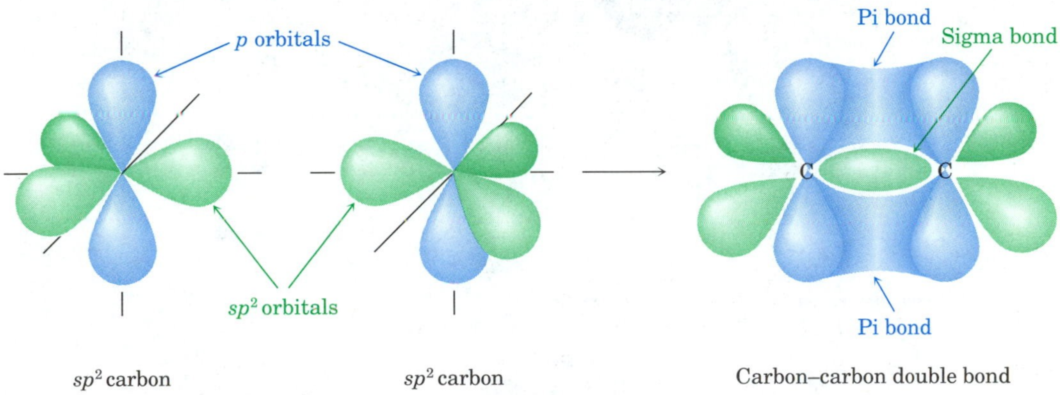

Figure 1.17 Orbital overlap of two sp^2-hybridized carbons to form a C=C double bond. One part of the double bond results from σ (head-on) overlap of sp^2 orbitals, and the other part results from π (sideways) overlap of unhybridized *p* orbitals. The π bond has regions of electron density on either side of a line drawn between nuclei.

To complete the structure of ethylene, four hydrogen atoms form σ bonds to the remaining four sp^2 orbitals. Ethylene has a planar structure with H-C-H and H-C-C bond angles of approximately 120° (the H-C-H bond angles are 116.6°, and the H-C-C bond angles are 121.7°). Each C–H bond has a length of 1.076 Å and a strength of 444 kJ/mol (106 kcal/mol).

As you might expect, the carbon–carbon double bond in ethylene is both shorter and stronger than the single bond in ethane because it results from the sharing of four electrons rather than two. Ethylene has a C=C bond length of 1.33 Å and a strength of 611 kJ/mol (146 kcal/mol) versus a C–C bond length of 1.54 Å and a strength of 376 kJ/mol for ethane. Note, though, that the strength of a carbon–carbon double bond is less than twice that of a single bond because the overlap in the π part of the double bond is not as effective as the overlap in the σ part. The structure of ethylene is shown in Figure 1.18.

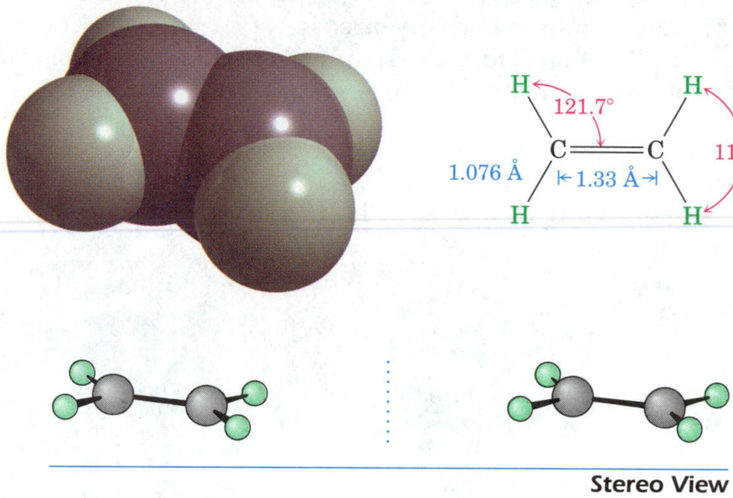

Figure 1.18 The structure of ethylene. The computer-generated structures show only the connections between atoms and do not explicitly show the C=C double bond.

We said in Section 1.7 that chemists use two models for describing covalent bonds, *valence bond theory* and *molecular orbital theory.* Having seen a valence bond description of the double bond in ethylene, let's also look at a molecular orbital description.

Just as bonding and antibonding σ molecular orbitals result from the combination of two *s* atomic orbitals in H_2, so bonding and antibonding π molecular orbitals result from the combination of two *p* atomic orbitals in ethylene. As shown in Figure 1.19, the π bonding MO has no node between nuclei and results from an additive combination of *p* orbital lobes with the same algebraic sign. The π antibonding MO has a node between nuclei and results from a subtractive combination of lobes with different algebraic signs. Only the bonding MO is filled; the higher-energy, antibonding MO is vacant.

We'll come back to this molecular orbital description of multiple bonding in future chapters, particularly when we discuss compounds with more than one double bond.

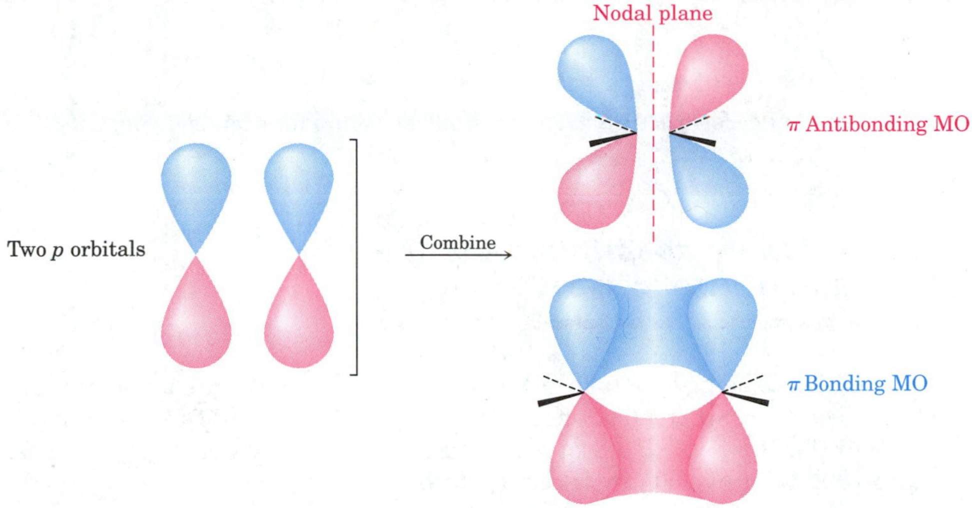

Figure 1.19 A molecular orbital description of the C–C π bond. The π bonding MO results from an additive combination of atomic orbitals and is filled. The π antibonding MO results from a subtractive combination of atomic orbitals and is unfilled.

PRACTICE PROBLEM

Formaldehyde, CH_2O, contains a carbon–*oxygen* double bond. Draw Lewis and line-bond structures of formaldehyde, and indicate the hybridization of the carbon atom.

Solution There is only one way that two hydrogens, one carbon, and one oxygen can combine:

Lewis structure Line-bond structure

Like the carbon atoms in ethylene, the doubly bonded carbon atom in formaldehyde is sp^2-hybridized.

PROBLEM

1.10 Draw all the bonds in propene, $CH_3CH{=}CH_2$. Indicate the hybridization of each carbon, and predict the value of each bond angle.

PROBLEM

1.11 Answer Problem 1.10 for 1,3-butadiene, $H_2C{=}CH{-}CH{=}CH_2$.

PROBLEM

1.12 Draw both a Lewis structure and a line-bond structure for acetaldehyde, CH_3CHO.

1.11 Hybridization: *sp* Orbitals and the Structure of Acetylene

In addition to being able to form single and double bonds by sharing two and four electrons, carbon also can form a *triple* bond by sharing six electrons. To account for the triple bond in a molecule such as acetylene, C_2H_2, we need a third kind of hybrid orbital, an ***sp* hybrid**.

$$H{:}C{:::}C{:}H \qquad H{-}C{\equiv}C{-}H$$

Acetylene

Imagine that, instead of combining with two or three *p* orbitals, a carbon 2*s* orbital hybridizes with only a single *p* orbital. Two *sp* hybrid orbitals result, and two *p* orbitals remain unchanged. The two *sp* orbitals are linear, or 180° apart on the *x*-axis, while the remaining two *p* orbitals are perpendicular on the *y*-axis and the *z*-axis, as shown in Figure 1.20.

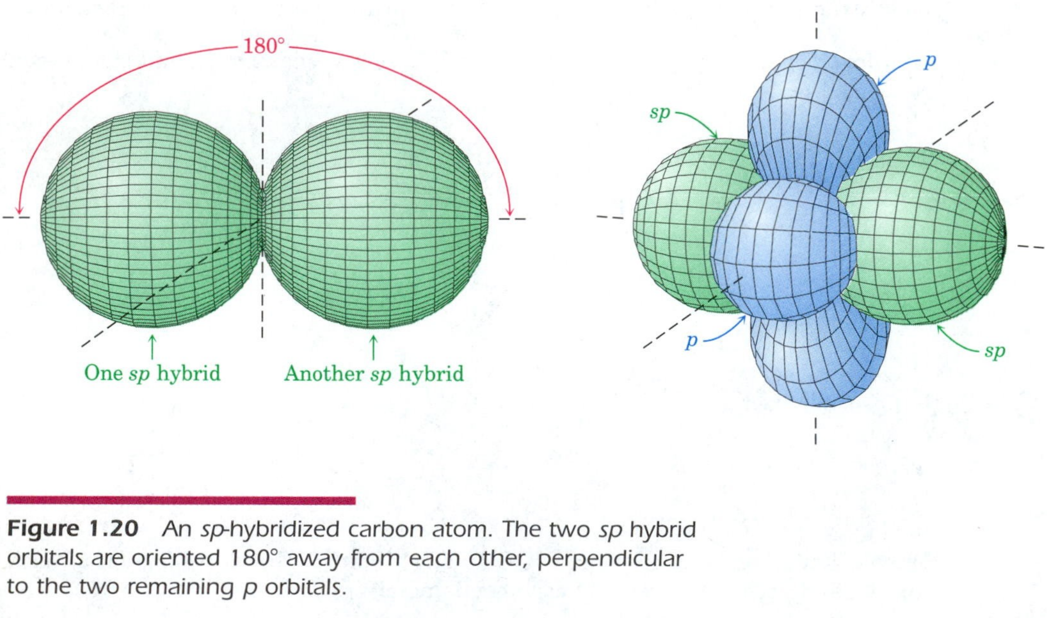

Figure 1.20 An *sp*-hybridized carbon atom. The two *sp* hybrid orbitals are oriented 180° away from each other, perpendicular to the two remaining *p* orbitals.

When two *sp*-hybridized carbon atoms approach each other, *sp* hybrid orbitals from each carbon overlap head-on to form a strong *sp*–*sp* σ bond. In addition, the p_z orbitals from each carbon form a p_z–p_z π bond by sideways overlap, and the p_y orbitals overlap similarly to form a p_y–p_y π bond. The net effect is the sharing of six electrons and formation of a net carbon–carbon triple bond. The remaining *sp* hybrid orbitals each form a σ bond with hydrogen to complete the acetylene molecule (Figure 1.21).

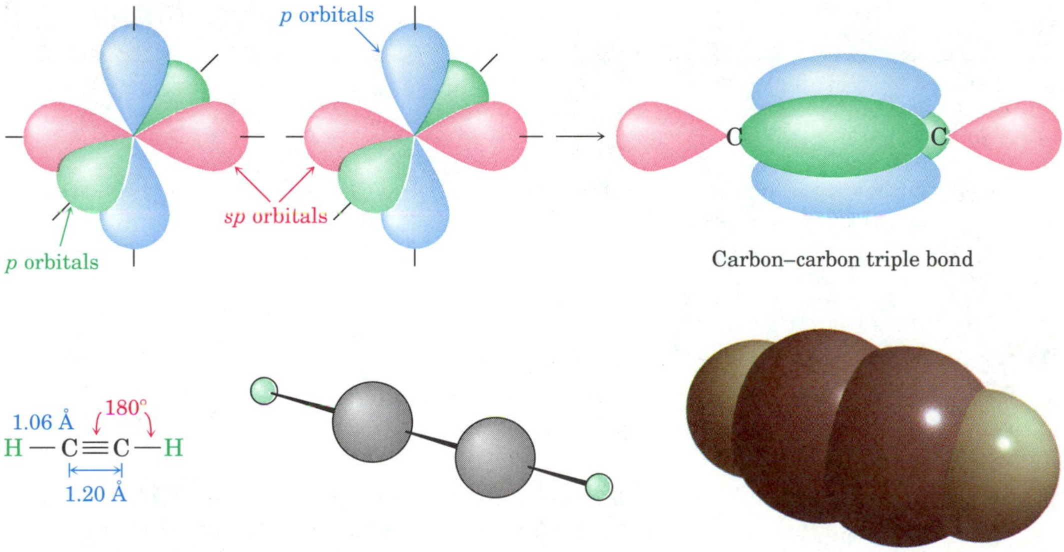

Figure 1.21 The structure of acetylene. The two *sp*-hybridized carbon atoms are joined by one *sp*–*sp* σ bond and two *p*–*p* π bonds.

Because of *sp* hybridization, acetylene is a linear molecule with H-C-C bond angles of 180°. The carbon–hydrogen bond in acetylene has a length of 1.06 Å and a strength of 552 kJ/mol (132 kcal/mol). The carbon–carbon bond length is 1.20 Å and its strength is about 835 kJ/mol (200 kcal/mol), making the triple bond in acetylene the shortest and strongest of any carbon–carbon bond. A comparison of *sp*, sp^2, and sp^3 hybridization is given in Table 1.3.

Table 1.3 Comparison of C–C and C–H Bonds in Methane, Ethane, Ethylene, and Acetylene

		Bond strength		
Molecule	*Bond*	*(kJ/mol)*	*(kcal/mol)*	*Bond length (Å)*
Methane, CH_4	$C_{sp^3}—H_{1s}$	438	105	1.10
Ethane, CH_3CH_3	$C_{sp^3}—C_{sp^3}$	376	90	1.54
	$C_{sp^3}—H_{1s}$	420	100	1.10
Ethylene, $H_2C{=}CH_2$	$C_{sp^2}{=}C_{sp^2}$	611	146	1.33
	$C_{sp^2}—H_{1s}$	444	106	1.076
Acetylene, $HC{\equiv}CH$	$C_{sp}{\equiv}C_{sp}$	835	200	1.20
	$C_{sp}—H_{1s}$	552	132	1.06

PROBLEM

1.13 Draw a line-bond structure for propyne, $CH_3C{\equiv}CH$. Indicate the hybridization of each carbon, and predict a value for each bond angle.

1.12 Hybridization of Other Atoms: Nitrogen, Oxygen, and Boron

The concept of hybridization described in the previous four sections is not restricted to carbon compounds. Covalent bonds formed by other elements in the periodic table also can be described using hybrid orbitals. Look at the nitrogen atom in ammonia, NH_3, for example. A nitrogen atom has five outer-shell electrons and therefore forms three covalent bonds to complete its valence electron octet.

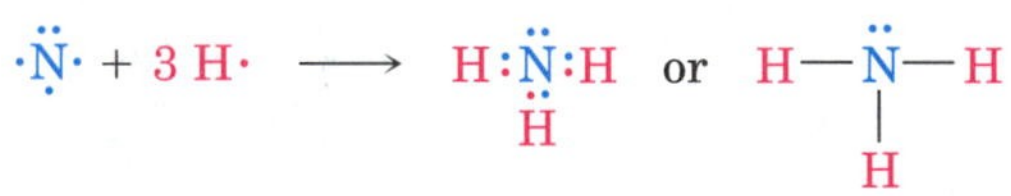

The experimentally measured H-N-H bond angle in ammonia is 107.3°, close to the tetrahedral value of 109.5° found in methane. We therefore assume that nitrogen hybridizes to form four sp^3 orbitals, exactly as carbon does. One of the four sp^3 orbitals is occupied by two nonbonding electrons, and the other three each have one electron. Sigma overlap of these three half-filled nitrogen sp^3 hybrid orbitals with hydrogen 1*s* orbitals completes the ammonia molecule (Figure 1.22). The N–H bond length is 1.008 Å, and the bond strength is 449 kJ/mol (107 kcal/mol). Note that the unshared lone pair of electrons in the fourth sp^3 hybrid orbital occupies just as much space as an N–H bond does and is very important to the chemistry of ammonia.

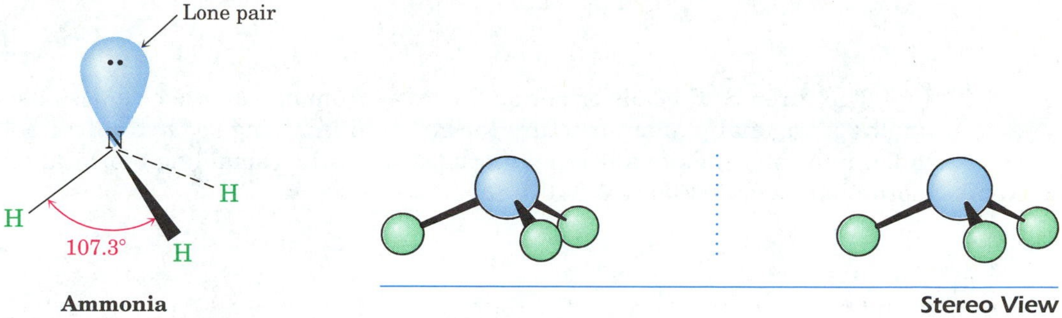

Figure 1.22 Hybridization of nitrogen in ammonia. The nitrogen atom is sp^3-hybridized, resulting in H-N-H bond angles of 107.3°.

Like the carbon atom in methane and the nitrogen atom in ammonia, the oxygen atom in water is sp^3-hybridized. Because an oxygen atom has six outer-shell electrons, however, it forms only two covalent bonds and has two lone pairs (Figure 1.23). The H-O-H bond angle in water is 104.5°, somewhat less than the 109.5° tetrahedral angle expected for sp^3-hybridization. This somewhat diminished bond angle is probably due to a repulsive interaction between the two lone pairs that forces them apart, thereby compressing the H-O-H angle. The O–H bond length is 0.958 Å, and the bond strength is 498 kJ/mol (119 kcal/mol).

Figure 1.23 The structure of water. The oxygen atom is sp^3-hybridized and has two lone pairs of electrons. The H-O-H bond angle is 104.5°.

The last example of orbital hybridization we'll consider is that of the boron atom in boron trifluoride, BF_3. Because boron has only three valence electrons ($1s^2\ 2s^2\ 2p_x$), it can form only three bonds and can't reach a stable valence electron octet.

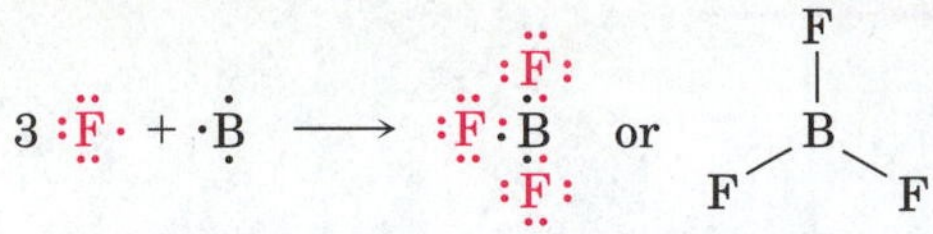

The three B–F bonds orient as far away from one another as possible, giving a trigonal planar structure for BF_3 and implying sp^2 hybridization. Each fluorine bonds to a boron sp^2 orbital, with the remaining *p* orbital on boron left vacant (Figure 1.24).

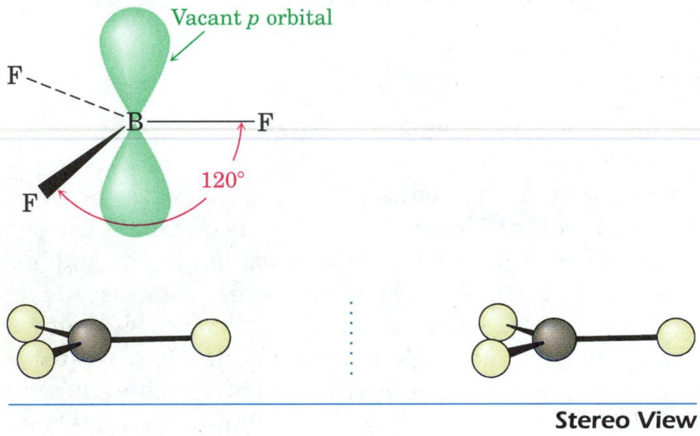

Figure 1.24 The structure of boron trifluoride. The boron atom is sp^2-hybridized and has a vacant *p* orbital perpendicular to the BF_3 plane.

PROBLEM

1.14 Draw Lewis and line-bond structures for formaldimine, CH_2NH. How many electrons are shared in the C–N bond? What is the hybridization of the nitrogen atom?

PROBLEM

1.15 What geometry do you expect for each of the following?

(a) The oxygen atom in methanol, $H_3C—\ddot{\underset{..}{O}}—H$

(b) The nitrogen atom in trimethylamine, $H_3C—\ddot{N}(CH_3)—CH_3$

(c) The phosphorus atom in $:PH_3$

INTERLUDE

Chemical Toxicity and Risk

Charred meat contains cancer-causing substances called *polycyclic aromatic hydrocarbons.* Are these ribs safe to eat?

We hear and read a lot these days about the dangers of "chemicals"—about pesticide residues on food, unsafe medicines, and so forth. What's a person to believe?

Life is not risk-free—we all take many risks each day. We decide to ride a bike rather than drive, even though there is a ten times greater likelihood per mile of dying in a bicycling accident than in a car. We may decide to smoke cigarettes, even though it increases our chance of getting cancer by 50%. Making judgments that affect our health is something we do every day without thinking about it.

But what about risks from chemicals? Risk evaluation is carried out by exposing test animals (usually rats) to a chemical and then monitoring for signs of harm. To limit the expense and time needed, the amounts administered are hundreds or thousands of times greater than those a human might normally encounter. Once the animal data are available, the interpretation of those data involves many assumptions. If a substance is harmful to animals, is it necessarily harmful to humans? How can a large dose for a small animal be translated into a small dose for a large human? As pointed out by the sixteenth-century Swiss physician Paracelsus, "The dose makes the poison." All substances, including water and table salt, are toxic to some organisms to some extent, and the difference between help and harm is a matter of degree.

The standard method for evaluating acute chemical toxicity, as opposed to long-term toxicity, is to report an *LD_{50} value*, the amount of

Table 1.4 Some LD_{50} Values

Substance	*LD_{50} (g/kg)*	*Substance*	*LD_{50} (g/kg)*
Aflatoxin B_1	4×10^{-4}	Formaldehyde	2.4
Aspirin	1.7	Sodium cyanide	1.5×10^{-2}
Chloroform	3.2	Sodium cyclamate	17
Ethyl alcohol	10.6		

(continued)➤

a substance per kilogram body weight that is lethal to 50% of the test animals. The LD_{50} values of various substances are shown in Table 1.4.

How we respond to risk is strongly influenced by familiarity. The presence of chloroform in municipal water supplies—at a barely detectable level of 0.000 000 01%—has caused an outcry in many cities, yet chloroform has a lower acute toxicity than aspirin. Many foods contain natural ingredients far more toxic than synthetic food additives or pesticide residues, but the ingredients are ignored because the foods are familiar. Peanut butter, for example, contains tiny amounts of aflatoxin, a far more potent cancer threat than sodium cyclamate, an artificial sweetener that has been banned because of its "risk."

All decisions involve tradeoffs. Does the benefit of a pesticide that will increase the availability of food outweigh the health risk to 1 person in 1 million who are exposed? Do the beneficial effects of a new drug outweigh a potentially dangerous side effect to a small number of users? The answers aren't always obvious, but it's the responsibility of legislators and well-informed citizens to keep their responses on a factual level rather than an emotional one.

Summary and Key Words

Organic chemistry is the study of carbon compounds. Although a division into organic and inorganic chemistry occurred historically, there is no scientific reason for the division.

An atom is composed of a positively charged nucleus surrounded by one or more negatively charged electrons. The electronic structure of an atom can be described by the Schrödinger **wave equation**, in which electrons are considered to occupy **orbitals** centered around the nucleus. Different orbitals have different energy levels and different shapes. For example, *s* orbitals are spherical and *p* orbitals are dumbbell-shaped. The **electron configuration** of an atom can be found by assigning electrons to the proper orbitals, beginning with the lowest-energy orbitals.

There are two kinds of chemical bonds—**ionic bonds** and **covalent bonds**. Ionic bonds are based on the electrostatic attraction of unlike charges and are commonly found in inorganic salts. Covalent bonds are formed when an electron pair is shared between atoms. According to **valence bond theory**, electron sharing occurs by overlap of two atomic orbitals. According to **molecular orbital theory**, bonds result from the combination of atomic orbitals to give molecular orbitals, which belong to the entire molecule. Valence bond theory is particularly suitable for describing single bonds, and molecular orbital theory is particularly suitable for describing multiple bonds. Bonds that have a circular cross-section and are formed by head-on overlap are called **sigma (σ) bonds**; bonds formed by sideways overlap of *p* orbitals are called **pi (π) bonds**.

Carbon uses hybrid orbitals to form bonds in organic molecules. When forming only single bonds (tetrahedral geometry), carbon is sp^3-hybridized and has four equivalent ***sp*³ hybrid orbitals**. When forming a double bond (planar geometry), carbon is *sp*²-hybridized, has three equivalent ***sp*² hybrid orbitals**, and has one unhybridized *p* orbital. A carbon–carbon double bond results when two *sp*²-hybridized carbon atoms bond together. When forming a triple bond (linear geometry), carbon is *sp*-hybridized, has two equivalent ***sp* hybrid orbitals**, and has two unhybridized *p* orbitals. A carbon–carbon triple bond results when two *sp*-hybridized carbon atoms bond together.

Other atoms such as nitrogen, oxygen, and boron also hybridize to form stronger bonds. The nitrogen atom in ammonia and the oxygen atom in water are *sp*³-hybridized; the boron atom in boron trifluoride is *sp*²-hybridized.

Working Problems

There is no surer way to learn organic chemistry than by working problems. Although careful reading and rereading of this text is important, reading alone isn't enough. You must also be able to work with the information you've read and be able to use your knowledge in new situations. Working problems gives you practice at doing this.

Each chapter in this book provides many problems of different sorts. The in-chapter problems are placed for immediate reinforcement of ideas just learned. The end-of-chapter problems provide additional practice and are of two types: drill and thought. Early problems are primarily of the drill type, providing an opportunity for you to practice your command of the fundamentals. Later problems tend to be more thought-provoking, and many are challenges to your depth of understanding.

As you study organic chemistry, take the time to work the problems. Do the ones you can, and ask for help on the ones you can't. If you're stumped by a particular exercise, check the accompanying *Study Guide and Solutions Manual* for an explanation that will help clarify the difficulty. Working problems takes effort, but the payoff in knowledge and understanding is immense.

ADDITIONAL PROBLEMS

1.16 How many valence electrons does each of the following atoms have?
(a) Magnesium (b) Sulfur (c) Bromine

1.17 Give the ground-state electron configurations for the following elements. For example, carbon is $1s^2\,2s^2\,2p^2$.
(a) Sodium (b) Aluminum (c) Silicon (d) Calcium

1.18 What are likely formulas for the following molecules?
(a) $AlCl_?$ (b) $CF_2Cl_?$ (c) $NI_?$

1.19 Write Lewis (electron-dot) structures for the following molecules:
(a) H—C≡C—H (b) AlH_3 (c) $CH_3—S—CH_3$
(d) Cl—C(=O)—Cl (e) $H_2C=CH—CH=CH_2$ (f) $CH_3—C(=O)—O—H$

1.20 Write a Lewis structure for acetonitrile, $H_3C{-}C{\equiv}N$. How many electrons does the nitrogen atom have in its outer shell? How many are bonding, and how many are nonbonding?

1.21 What is the hybridization of each carbon atom in acetonitrile (Problem 1.20)?

1.22 Draw both a Lewis structure and a line-bond structure for vinyl chloride, C_2H_3Cl, the starting material from which PVC [poly(vinyl chloride)] plastic is made.

1.23 Fill in any unshared electrons that are missing from the following line-bond structures:

(a) $CH_3{-}O{-}CH_3$

(b) 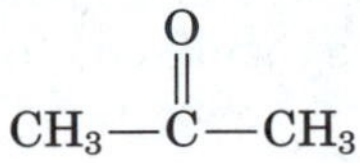

(c) 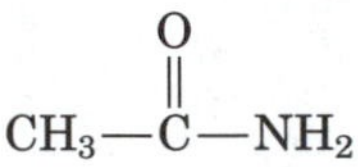

(d) CH_2ClF

1.24 Convert the following line-bond structures into molecular formulas:

(a) 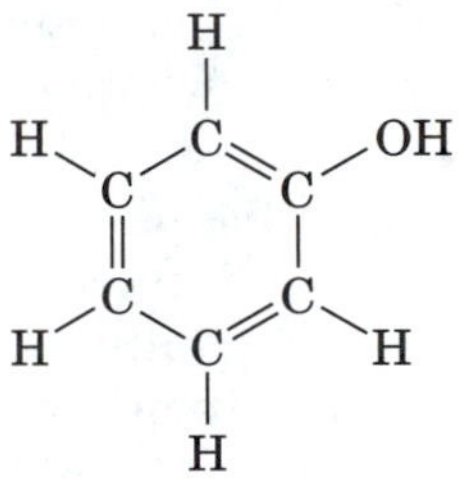

Phenol

(b) 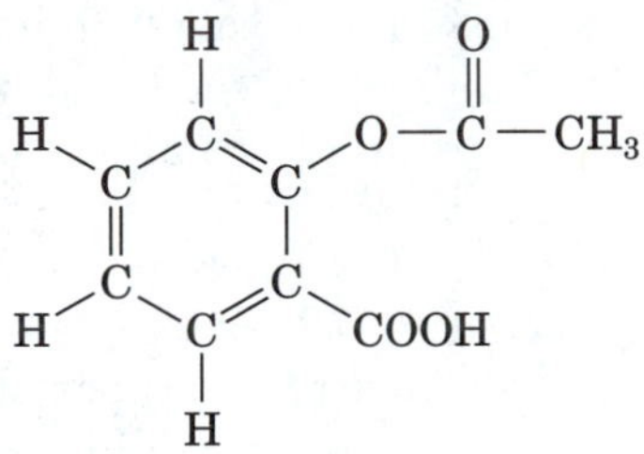

Aspirin

(c) 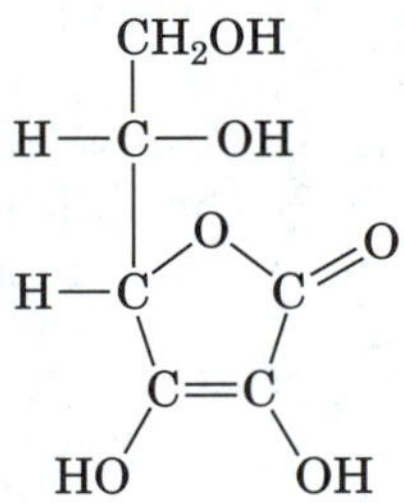

Vitamin C

(d) 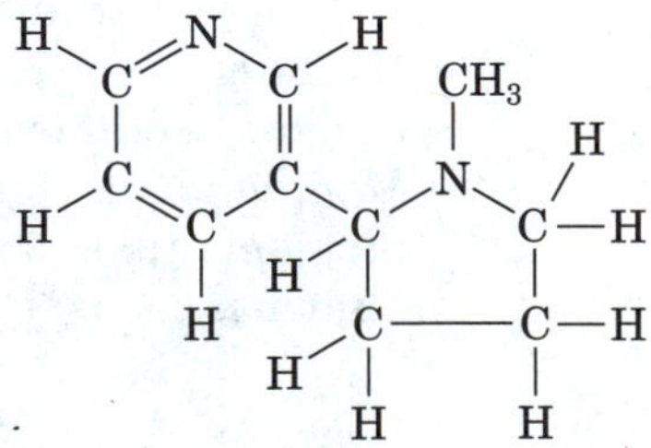

Nicotine

(e) 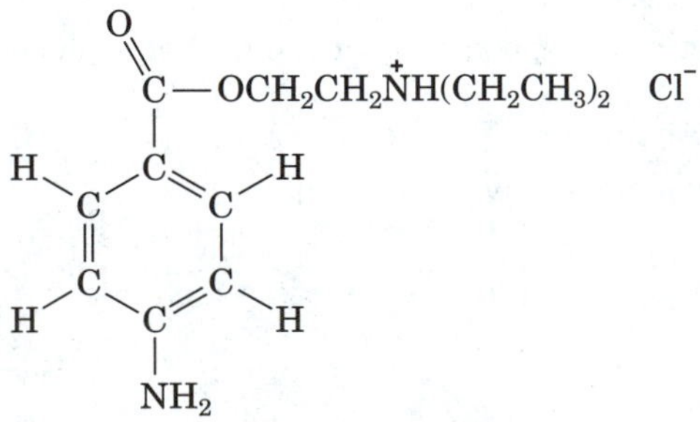

Novocain

(f) 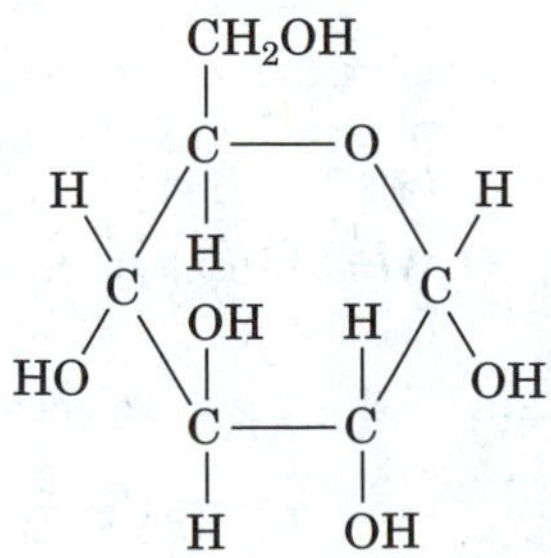

Glucose

1.25 Convert the following molecular formulas into line-bond structures that are consistent with valence rules.

(a) C_3H_8
(b) CH_5N
(c) C_2H_6O (2 possibilities)
(d) C_3H_7Br (2 possibilities)
(e) C_2H_4O (3 possibilities)
(f) C_3H_9N (4 possibilities)

1.26 What kind of hybridization do you expect for each carbon atom in the following molecules?

(a) Propane, $CH_3CH_2CH_3$
(b) 2-Methylpropene, $(CH_3)_2C{=}CH_2$
(c) 1-Buten-3-yne, $H_2C{=}CH{-}C{\equiv}CH$
(d) Cyclobutene,

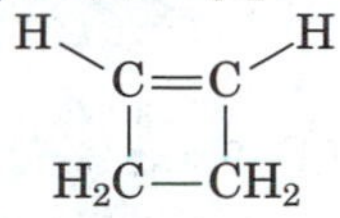

(e) Dimethyl ether, CH_3OCH_3

1.27 What is the overall shape of benzene, and what hybridization do you expect for each carbon?

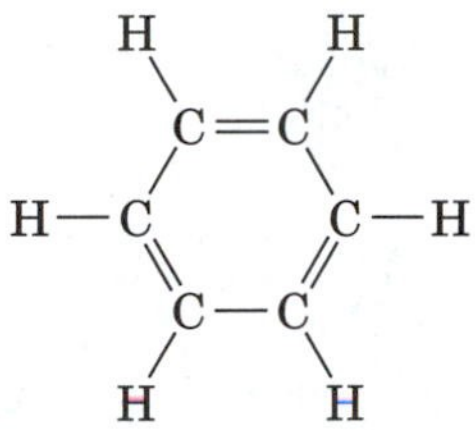

Benzene

1.28 What bond angles do you expect for the following?

(a) The C-O-C angle in CH_3OCH_3
(b) The C-N-C angle in CH_3NHCH_3
(c) The C-N-H angle in CH_3NHCH_3
(d) The C-B-C angle in $(CH_3)_3B$

1.29 What kind of hybridization do you expect for the following? (See Problem 1.28.)

(a) The oxygen atom CH_3OCH_3
(b) The nitrogen in CH_3NHCH_3
(c) The boron in $(CH_3)_3B$

1.30 What shape do you expect the following species to have?

(a) The ammonium ion, NH_4^+
(b) Trimethylborane, $(CH_3)_3B$
(c) Trimethylphosphine $(CH_3)_3P$
(d) Formaldehyde, $H_2C{=}O$

1.31 Draw a three-dimensional representation of the oxygen-bearing carbon atom in ethanol, CH_3CH_2OH, using the standard convention of solid, wedged, and dashed lines.

1.32 Consider the molecules SO_2 and SO_3 and the ion SO_4^{2-}. Write Lewis structures and predict the shape of each.

1.33 Draw line-bond structures for these molecules:

(a) Acrylonitrile, $H_2C{=}CHCN$
(b) Ethanol, CH_3CH_2OH
(c) Butane, $CH_3CH_2CH_2CH_3$

1.34 Sodium methoxide, $NaOCH_3$, contains both covalent and ionic bonds. Which do you think is which?

1.35 What kind of hybridization do you expect for each carbon atom in the following molecules?

(a) Acetic acid,

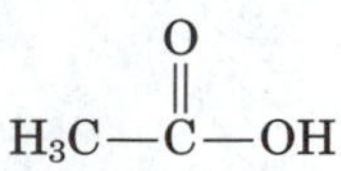

(b) 3-Buten-2-one,

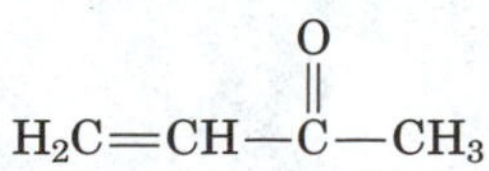

(c) Acrylonitrile, $H_2C{=}CH{-}C{\equiv}N$

(d) Benzoic acid,

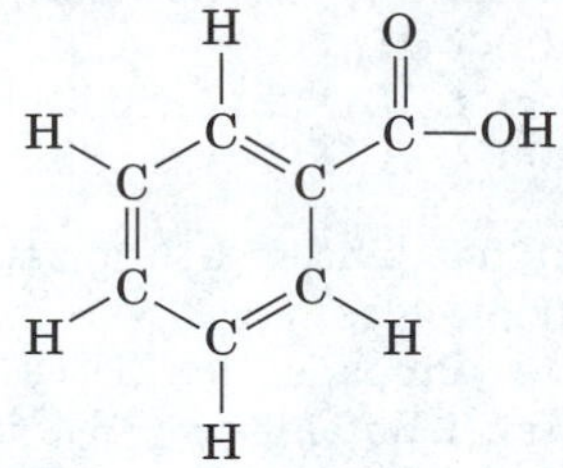

1.36 What bond angles do you expect for the following?

(a) The C–N–H angle in aniline,

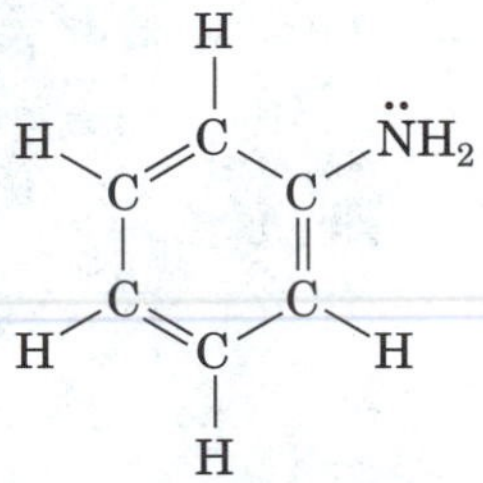

(b) The C–N–C angle in pyridine,

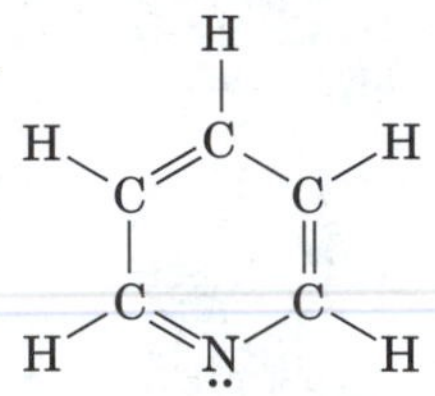

(c) The C-Be-C angle in dimethylberyllium, CH_3BeCH_3

(d) The C-P-C angle in trimethylphosphine, $(CH_3)_3P$

1.37 What kind of hybridization do you expect for each of the following? (See Problem 1.36.)

(a) The nitrogen atom in aniline

(b) The nitrogen atom in pyridine

(c) The beryllium atom in CH_3BeCH_3

(d) The phosphorus atom in $(CH_3)_3P$

1.38 Identify the bonds in the following compounds as either ionic or covalent.

(a) NaCl (b) CH_3Cl (c) Cl_2 (d) HOCl

1.39 Allene, $H_2C{=}C{=}CH_2$, is somewhat unusual in that it has two adjacent double bonds. Draw an orbital picture of allene. Is the central carbon atom sp^2- or sp-hybridized? What about the hybridization of the terminal carbons? What shape do you predict for allene?

1.40 Allene (see Problem 1.39) is related structurally to carbon dioxide, CO_2. Draw an orbital picture of CO_2, and identify the hybridization of carbon.

1.41 Complete the Lewis electron-dot structure of caffeine, showing all lone-pair electrons, and identify the hybridization of the indicated atoms.

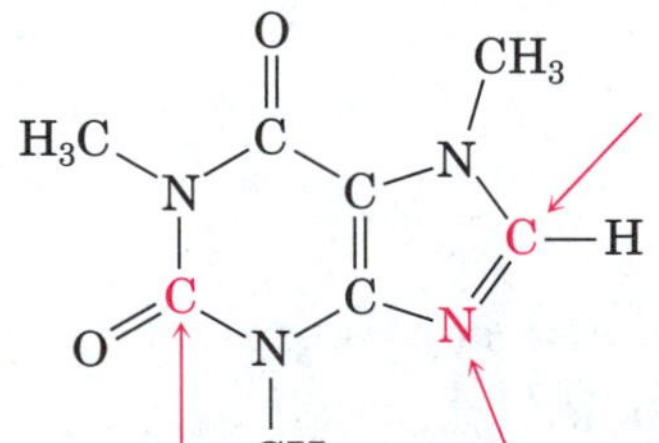

Caffeine

1.42 Although almost all stable organic species have tetravalent carbon atoms, species with trivalent carbon atoms are known to exist. *Carbocations* are one such class of compounds.

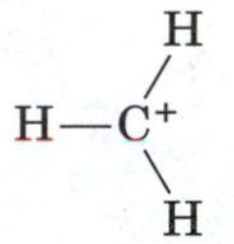

A carbocation

(a) What is the relationship between a carbocation and a trivalent boron compound such as BF_3?
(b) How many valence electrons does the positively charged carbon atom have?
(c) What hybridization do you expect this carbon atom to have?
(d) What geometry does the carbocation have?

1.43 A *carbanion* is a species that contains a negatively charged, trivalent carbon.
(a) What is the relationship between a carbanion and a trivalent nitrogen compound such as NH_3?

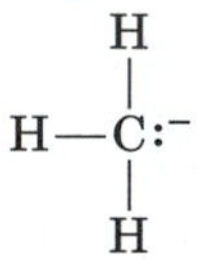

A carbanion

(b) How many valence electrons does the negatively charged carbon atom have?
(c) What hybridization do you expect this carbon atom to have?
(d) What geometry does the carbanion have?

1.44 Divalent species called *carbenes* are capable of fleeting existence. For example, methylene, $:CH_2$, is the simplest carbene. The two unshared electrons in methylene can be either spin-paired in a single orbital or unpaired in different orbitals. Predict the type of hybridization you expect carbon to adopt in singlet (spin-paired) methylene and triplet (spin-unpaired) methylene. Draw pictures of each, and identify the types of carbon orbitals present.

1.45 Propose structures for molecules that meet the following descriptions:
(a) Contains two sp^2-hybridized carbons and two sp^3-hybridized carbons
(b) Contains only four carbons, all of which are sp^2-hybridized
(c) Contains two sp-hybridized carbons and two sp^2-hybridized carbons

A Look Ahead

1.46 There are two different substances with the formula C_4H_{10}. Draw them both, and tell how they differ.

1.47 There are two different substances with the formula C_3H_6. Draw them both, and tell how they differ.

1.48 There are two different substances with the formula C_2H_6O. Draw them both, and tell how they differ.

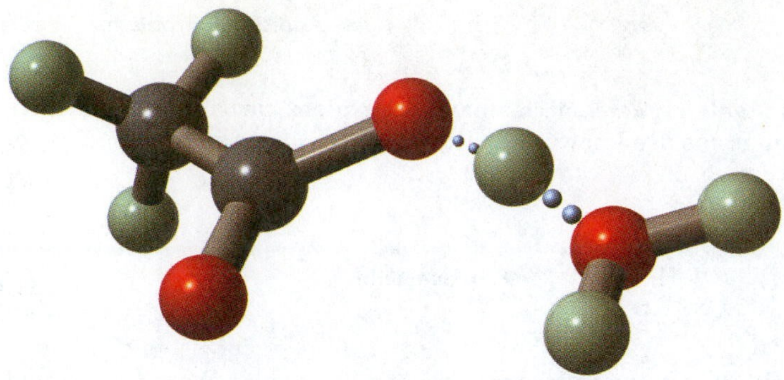

Acetic acid transfers a proton to water in aqueous solution.

2

Bonding and Molecular Properties

We saw in the last chapter how covalent bonds between atoms are described, and we looked at the hybrid orbital model used to portray most organic molecules. Before going on to a systematic study of complex organic substances, however, we still need to review a few fundamental topics. In particular, we need to look more closely at how electrons are distributed in covalent bonds and at some of the consequences that arise when the bonding electrons are not shared equally between atoms.

2.1 Polar Covalent Bonds: Electronegativity

Up to this point, we've viewed chemical bonding in an either/or manner: A given bond is either covalent or ionic. It's more accurate, though, to look at bonding as a continuum of possibilities, from a perfectly covalent bond with a symmetrical electron distribution on the one hand, to a perfectly ionic bond between positive and negative ions on the other (Figure 2.1).

The carbon–carbon bond in ethane is electronically symmetrical and therefore fully covalent; the two bonding electrons are equally shared by the two equivalent carbon atoms. The bond in sodium chloride, by contrast, is largely ionic.[1] Sodium has transferred an electron to chlorine to give Na^+ and Cl^- ions, which are held together in the solid by electrostatic attraction.

[1]Even the bond in $Na^+ Cl^-$ is only about 80% ionic rather than 100%.

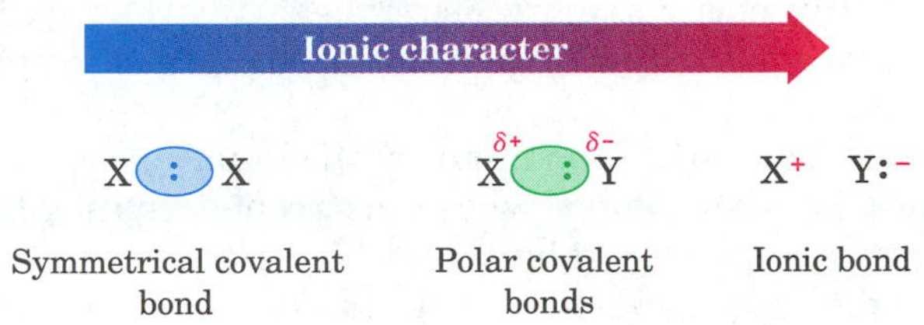

Figure 2.1 The continuum in bonding from covalent to ionic as a result of unsymmetrical electron distribution. The symbol δ (lowercase Greek delta) means *partial* charge, either partial positive (δ+) or partial negative (δ−).

Between these two extremes lie the great majority of chemical bonds, in which the electrons are attracted *somewhat* more strongly by one atom than by the other. We call such bonds in which the electron distribution is unsymmetrical **polar covalent bonds**.

Bond polarity is due to differences in **electronegativity**, the intrinsic ability of an atom to attract electrons in a covalent bond. Metals on the left side of the periodic table attract electrons only weakly, whereas the halogens and other reactive nonmetals on the right side of the periodic table attract electrons strongly. As shown in Figure 2.2, electronegativity values are given

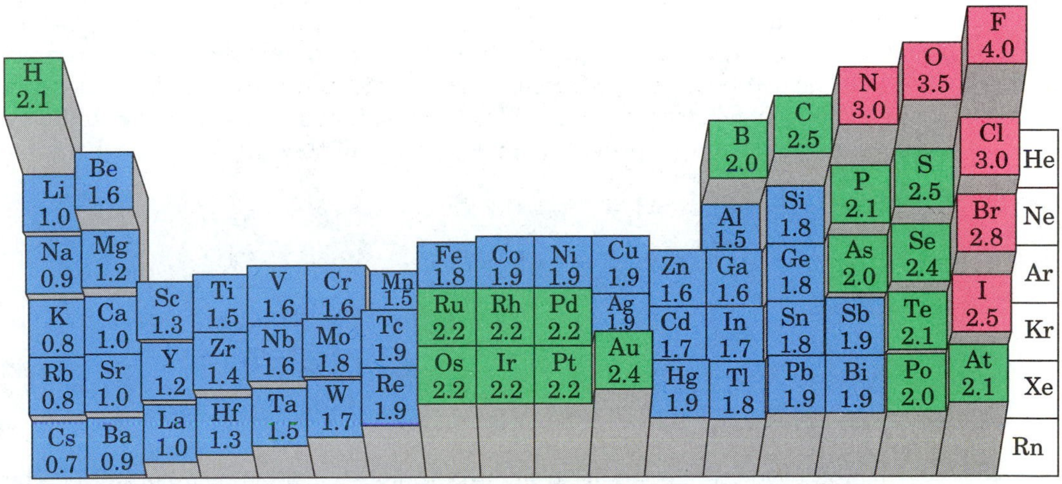

Figure 2.2 Electronegativity trends in the periodic table. Elements on the right side of the table are more electronegative than elements on the left side. The values are on an arbitrary scale, with H = 2.1 and F = 4.0. Carbon has an electronegativity value of 2.5. Any element more electronegative than carbon has a value greater than 2.5, and any element less electronegative than carbon has a value less than 2.5.

on an arbitrary scale, with H = 2.1 and F = 4.0. Carbon has an electronegativity value of 2.5. Any element more electronegative than carbon has a value greater than 2.5, and any element less electronegative than carbon has a value less than 2.5.

As a general rule, bonds between atoms with similar electronegativities are covalent, bonds between atoms whose electronegativities (EN) differ by less than 2 units are polar covalent, and bonds between atoms whose electronegativities differ by more than 2 units are largely ionic. Carbon–hydrogen bonds, for example, are relatively nonpolar because carbon and hydrogen have similar electronegativities. Bonds between carbon and more electronegative elements such as oxygen, fluorine, and chlorine, by contrast, are polarized so that the bonding electrons are drawn away from carbon toward the electronegative atom. This leaves carbon with a partial positive charge, denoted by $\delta+$, and the electronegative atom with a partial negative charge, $\delta-$ (δ is the lowercase Greek letter delta). For example, the C–Cl bond in chloromethane is polar covalent:

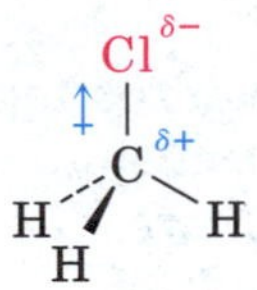

Chloromethane

Chlorine: EN = 3.0
Carbon: EN = 2.5
ΔEN = 0.5

The arrow ↦ is used to indicate the direction of bond polarity. By convention, *electrons move in the direction of the arrow.* The tail of the arrow (which looks like a plus sign) is electron-poor ($\delta+$), and the head of the arrow is electron-rich ($\delta-$).

Bonds between carbon and less electronegative elements are polarized so that carbon bears a partial negative charge and the other atom bears a partial positive charge. Organometallic compounds, such as tetraethyllead, the "lead" in gasoline, provide good examples of this kind of polar bond.

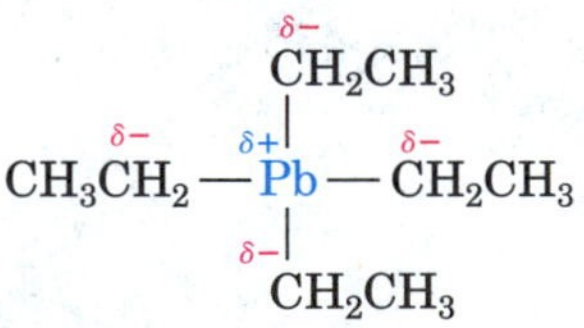

Tetraethyllead

Carbon: EN = 2.5
Lead: EN = 1.9
ΔEN = 0.6

When speaking of an atom's ability to polarize a bond, we often use the term **inductive effect**. An inductive effect is simply the shifting of electrons in a bond in response to the electronegativity of nearby atoms. Metals such as lithium and magnesium inductively donate electrons, whereas electronegative nonmetals such as oxygen and chlorine inductively withdraw electrons. Inductive effects play a major role in understanding chemical reactivity, and we'll use them many times throughout this text to explain a variety of chemical phenomena.

PROBLEM.

2.1 Which element in each of the following pairs is more electronegative?
(a) Li or H (b) Be or Br (c) Cl or I (d) C or H

PROBLEM.

2.2 Use the $\delta+/\delta-$ convention to indicate the direction of expected polarity for each of the bonds indicated.
(a) $H_3C{-}Br$ (b) $H_3C{-}NH_2$ (c) $H_3C{-}Li$ (d) $H_2N{-}H$
(e) $H_3C{-}OH$ (f) $H_3C{-}MgBr$ (g) $H_3C{-}F$

PROBLEM.

2.3 Use the electronegativity values shown in Figure 2.2 to rank the following bonds from least polar to most polar:
(a) $H_3C{-}Li$ (b) $H_3C{-}K$ (c) $H_3C{-}F$ (d) $H_3C{-}MgBr$ (e) $H_3C{-}OH$

2.2 Polar Covalent Bonds: Dipole Moment

Since individual bonds are often polar, molecules as a whole are often polar also. Overall molecular polarity results from the summation of all individual bond polarities and lone-pair contributions in the molecule. The measure of this net molecular polarity is a quantity called the *dipole moment*. As a practical matter, strongly polar substances are often soluble in polar solvents such as water, whereas nonpolar substances are insoluble in water.

Dipole moments can be thought of in the following way: Assume that there is a center of mass of all positive charges (nuclei) in a molecule and a center of mass of all negative charges (electrons) in the molecule. If these two centers don't coincide, then the molecule has a net polarity. The **dipole moment**, μ (Greek mu), is defined as the magnitude of the charge Q at either end of the molecular dipole times the distance r between the charges, $\mu = Q \times r$. Dipole moments are expressed in *debyes*[2] (D), where $1\ \text{D} = 3.336 \times 10^{-30}$ coulomb meters (C·m) in SI units. For example, the charge on an electron is 1.60×10^{-19} C. Thus, if one proton and one electron were separated by 100 pm (a bit less than the length of an average covalent bond), the dipole moment would be 1.60×10^{-29} C·m, or 4.80 D.

$$\mu = Q \times r$$

$$\mu = (1.60 \times 10^{-19}\ \text{C})(100 \times 10^{-12}\ \text{m})\left(\frac{1\ \text{D}}{3.336 \times 10^{-30}\ \text{C·m}}\right) = 4.80\ \text{D}$$

It's relatively easy to measure dipole moments, and values for some

[2]Peter Joseph Wilhelm Debye (1884–1966); b. Maastricht, Netherlands; Ph.D. Munich (1910); professor of physics, Zürich, Utrecht, Göttingen, Leipzig, Berlin; professor of chemistry, Cornell University (1936–1966); Nobel Prize (1936).

common substances are given in Table 2.1. Once the dipole moment is known, it's then possible to calculate the amount of charge separation in a molecule. In chloromethane, for example, the measured dipole moment is $\mu = 1.87$ D. If we assume that the contributions of the nonpolar C–H bonds are small, then most of the chloromethane dipole moment is due to the C–Cl bond. Since the C–Cl bond length is 178 pm (1.78 Å), the dipole moment of chloromethane would be $1.78 \times 4.8 \text{ D} = 8.5 \text{ D}$ if the C–Cl bond were ionic (that is, if a full negative charge on chlorine were separated from a full positive charge on carbon by a distance of 178 pm). But because the actual dipole moment of chloromethane is only 1.87 D, the C–Cl bond is only about $(1.87/8.54)(100\%) = 22\%$ ionic. Thus, the chlorine atom in chloromethane has an excess of about 0.2 electron, and the carbon atom has a deficiency of about 0.2 electron.

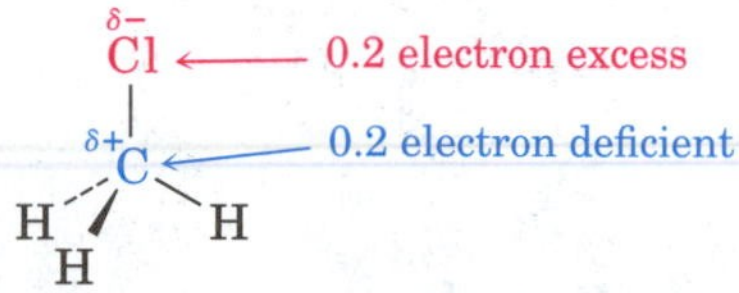

Chloromethane (μ = 1.87 D)

Water and ammonia have relatively large dipole moments (Table 2.1) because oxygen and nitrogen are both more electronegative than hydrogen. In addition, the lone-pair electrons on the oxygen atom of water and the nitrogen atom of ammonia make large contributions to overall dipole moments because they have no atom attached to them to "neutralize" their negative charge.

Table 2.1 Dipole Moments of Some Compounds

Compound	*Dipole moment (D)*	*Compound*	*Dipole moment (D)*
NaCl	9.0	NH_3	1.47
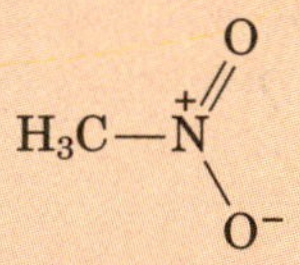 **Nitromethane**	3.46	CH_4	0
		CCl_4	0
		CH_3CH_3	0
CH_3Cl	1.87	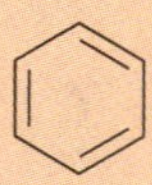 **Benzene**	0
H_2O	1.85		
CH_3OH	1.70	BF_3	0
$H_2C{=}\overset{+}{N}{=}N^-$ **Diazomethane**	1.50		

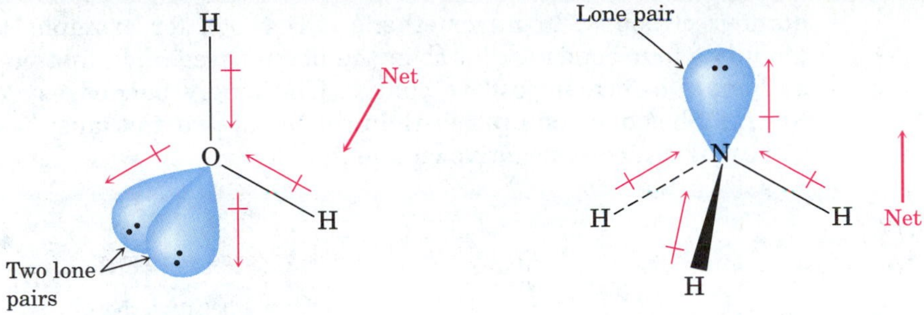

Water, H_2O (μ = 1.85 D)　　**Ammonia, NH_3 (μ = 1.47 D)**

By contrast with water and ammonia, methane, tetrachloromethane, and ethane have zero dipole moments. Because of the symmetrical structures of these molecules, the individual bond polarities exactly cancel.

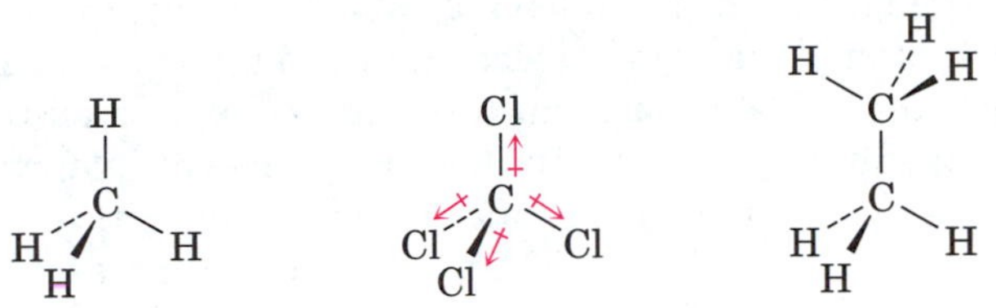

Methane (μ = 0 D)　　**Tetrachloromethane (μ = 0 D)**　　**Ethane (μ = 0 D)**

PROBLEM...

2.4 Account for the observed dipole moments of methanol (CH_3OH, 1.70 D) by using an arrow to indicate the direction in which electron density is displaced.

PROBLEM...

2.5 Carbon dioxide, CO_2, has zero dipole moment even though carbon–oxygen bonds are strongly polarized. Explain.

PROBLEM...

2.6 Make three-dimensional drawings of the following compounds, and predict whether each has a dipole moment. If you expect a dipole moment, show its direction.
(a) $H_2C{=}CH_2$　(b) $CHCl_3$　(c) CH_2Cl_2　(d) $H_2C{=}CCl_2$

2.3 Formal Charges

Closely related to the ideas of bond polarity and dipole moment is the occasional need to assign *formal charges* to specific atoms within a molecule. This is particularly true for atoms that have an apparently "abnormal" number of bonds. In nitromethane (CH_3NO_2), for example, the nitrogen atom has four bonds rather than the usual three and must be represented as having a formal positive charge. The singly bonded oxygen atom, by contrast, has one bond rather than the usual two and must be represented as having a formal negative charge.

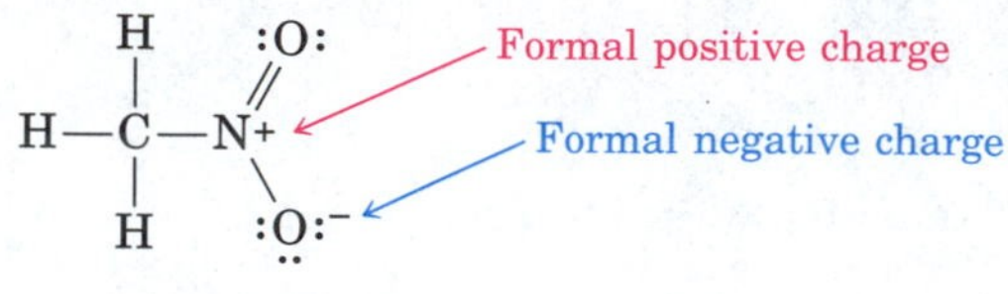

Nitromethane

Formal charges result from a kind of electron "bookkeeping" and can be thought of in the following way. A typical covalent bond is formed when each atom donates one electron. Although the bonding electrons are shared by both atoms, each atom can still be considered to "own" one electron for bookkeeping purposes. In methane, for example, carbon owns one electron in each of the four bonds, for a total of four. Since a neutral, isolated carbon atom has four valence electrons, and since the carbon atom in methane still owns four, the methane carbon is neutral and has no formal charge.

The same is true for the ammonia nitrogen atom, which has three covalent N–H bonds. Atomic nitrogen has five valence electrons, and the ammonia nitrogen also has five—one from each of three shared N–H bonds plus two in the lone pair. Thus, the nitrogen atom in ammonia is neutral and has no formal charge.

The situation is different in nitromethane. Atomic nitrogen has five valence electrons, but the nitromethane nitrogen owns only *four*—one from the C–N bond, one from the N–O single bond, and two from the N=O double bond. Thus, the nitrogen has formally lost an electron and must therefore have a positive charge. A similar calculation for the singly bonded oxygen atom shows that it has formally gained an electron and must have a negative charge. (Atomic oxygen has six valence electrons, but the singly bonded oxygen in nitromethane has seven—one from the O–N bond and two from each of three lone pairs.)

To express the calculations in a general way, we can say that the **formal charge** on an atom is equal to the number of valence electrons in a neutral, isolated atom minus the number of electrons owned by that atom in the molecule:

$$\textbf{Formal charge} = \begin{pmatrix}\text{Number of}\\ \text{valence electrons}\\ \text{in free atom}\end{pmatrix} - \begin{pmatrix}\text{Number of}\\ \text{valence electrons}\\ \text{in bound atom}\end{pmatrix}$$

$$= \begin{pmatrix}\text{Number of}\\ \text{valence}\\ \text{electrons}\end{pmatrix} - \begin{pmatrix}\text{Half of}\\ \text{bonding}\\ \text{electrons}\end{pmatrix} - \begin{pmatrix}\text{Number of}\\ \text{nonbonding}\\ \text{electrons}\end{pmatrix}$$

For the nitromethane nitrogen,

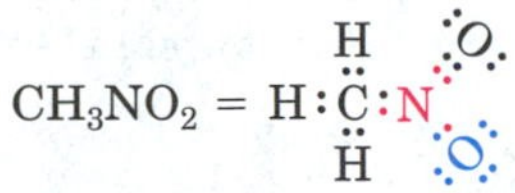

Nitrogen valence electrons	=	5
Nitrogen bonding electrons	=	8
Nitrogen nonbonding electrons	=	0

$$\text{Formal charge} = 5 - \tfrac{8}{2} - 0 = +1$$

For the singly bonded nitromethane oxygen,

Oxygen valence electrons	=	6
Oxygen bonding electrons	=	2
Oxygen nonbonding electrons	=	6

$$\text{Formal charge} = 6 - \tfrac{2}{2} - 6 = -1$$

A summary of commonly encountered formal charges and the bonding situations in which they occur is given in Table 2.2.

Table 2.2 A Summary of Formal Charges on Atoms

Atom	C			N			O		
Structure	—C+— (3 bonds)	—C— (4 bonds)	—C:−— (3 bonds, 1 lone pair)	—N+— (4 bonds)	—N:— (3 bonds, 1 lone pair)	—N−— (2 bonds, 2 lone pairs)	—O+— (3 bonds, 1 lone pair)	—O— (2 bonds, 2 lone pairs)	—O:− (1 bond, 3 lone pairs)
Number of bonds	3	4	3	4	3	2	3	2	1
Lone pairs	0	0	1	0	1	2	1	2	3
Formal charge	+1	0	−1	+1	0	−1	+1	0	−1

Molecules such as nitromethane, which are neutral overall but have plus and minus charges on individual atoms, are said to be **dipolar**. Dipolar character in molecules often has important chemical consequences, and it's helpful to be able to identify and calculate the charges correctly.

PROBLEM. . . . .

2.7 Dimethyl sulfoxide, a common solvent, has the structure indicated. Show why dimethyl sulfoxide must have formal charges on S and O.

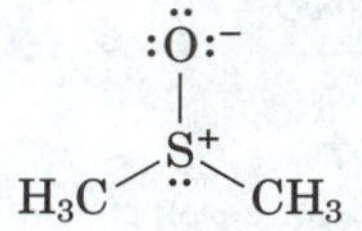

Dimethyl sulfoxide

PROBLEM. . . . .

2.8 Calculate formal charges for the atoms in these molecules:

(a) Diazomethane, $H_2C{=}N{=}\ddot{N}$:
(b) Acetonitrile oxide, $H_3C{-}C{\equiv}N{-}\ddot{O}$:
(c) Methyl isocyanide, $H_3C{-}N{\equiv}C$:

. . . .

2.4 Chemical Structures and Resonance

Most substances can be represented without difficulty by the Lewis structures or Kekulé line-bond structures we've been using up to this point, but an interesting problem arises in some cases. Take the acetate ion, CH_3COO^-, for example. When we draw a Lewis structure for the acetate ion, we need to show a double bond to one oxygen and a single bond to the other. But which oxygen is which? Should we draw a double bond to the "top" oxygen and a single bond to the "bottom" oxygen, or vice versa?

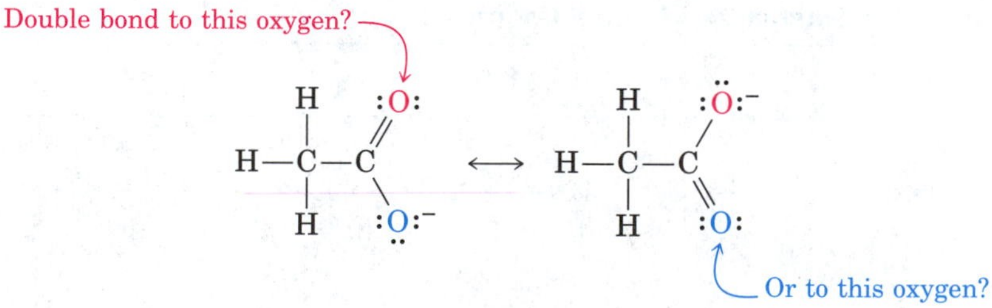

Acetate ion

Although the two oxygen atoms in the acetate ion appear different in Lewis structures, experiments show that they are actually equivalent. Both carbon–oxygen bonds, for example, are 1.27 Å in length, midway between the length of a typical C–O single bond (1.35 Å) and a typical C=O double bond (1.20 Å). In other words, *neither* of the two Lewis structures for the acetate ion is correct by itself; the true structure is intermediate between the two. The two individual Lewis structures are called **resonance forms**, and their relationship is indicated by the double-headed arrow between them. *The only difference between resonance forms is in the distribution of valence electrons. The atoms themselves occupy exactly the same place in both resonance forms.*

The best way to think about resonance forms is to realize that a species like the acetate ion is no different from any other substance. An acetate ion doesn't jump back and forth between two resonance forms, spending part of its time looking like one and the rest of its time looking like the other. Rather, the acetate ion has a single unchanging structure that is a *hybrid* of the two forms and has characteristics of both.

The difficulty in visualizing the resonance concept is that we can't draw an accurate picture of a species like the acetate ion because the familiar Kekulé structures we typically use to represent organic molecules don't work well for resonance hybrids. The difficulty, however, lies with the *representation* of the substance, not with the substance itself.

2.5 Drawing and Interpreting Resonance Forms

Resonance is an extremely useful concept, which we'll return to on numerous occasions throughout the rest of this book. We'll see in Chapter 15, for example, that the six carbon–carbon bonds in so-called *aromatic* compounds such as benzene are equivalent because benzene is a hybrid of two resonance forms. Though each individual resonance form seems to imply that benzene has alternating single and double bonds, neither form is correct by itself. The true benzene structure is a hybrid of the two individual forms, and all six carbon–carbon bonds are equivalent.

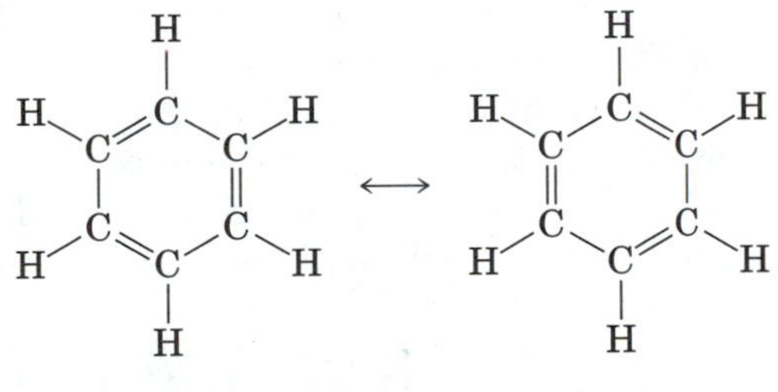

Benzene (two resonance forms)

When first dealing with resonance theory, it's useful to have a set of guidelines that describe how to draw and interpret resonance forms. The following rules should prove helpful:

1. *Individual resonance forms are imaginary, not real.* The real structure is a composite, or resonance hybrid, of the different forms. Substances such as the acetate ion and benzene are no different from any other substance. They have single, unchanging structures, and they do not switch back and forth between resonance forms. The only difference between these and other substances is in the way they must be represented on paper.
2. *Resonance forms differ only in the distribution of their π or nonbonding electrons.* Neither the position nor the hybridization of atoms changes from one resonance form to another. In the acetate

ion, for example, the carbon atom is sp^2-hybridized and the oxygen atoms remain in exactly the same place in both resonance forms. Only the positions of the π electrons in the double bond and the nonbonding electrons in a lone pair differ from one form to another:

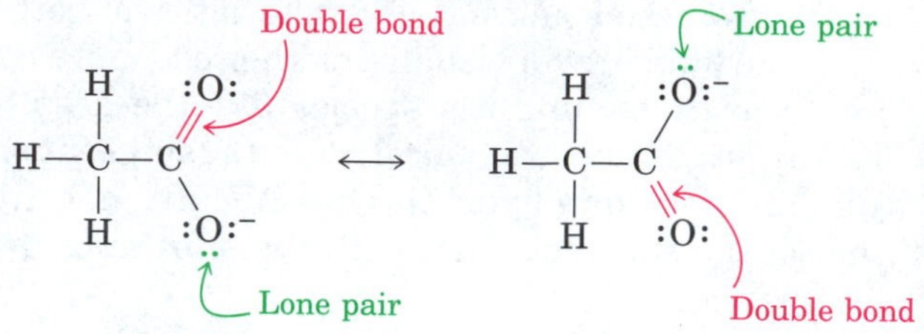

Similarly in benzene: The π electrons in the double bonds move, but the carbon and hydrogen atoms remain in place.

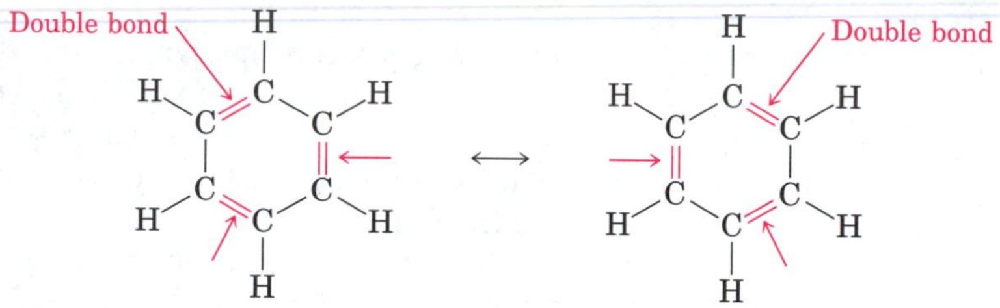

3. *Different resonance forms of a substance don't have to be equivalent.* For example, we'll see later that compounds like acetone (a common industrial solvent) are converted into anions by reaction with a strong base. The resultant acetone anion has two resonance forms: one contains a carbon–*oxygen* double bond and the other contains a carbon–*carbon* double bond. Even though the two resonance forms aren't equivalent, they both contribute to the overall resonance hybrid.

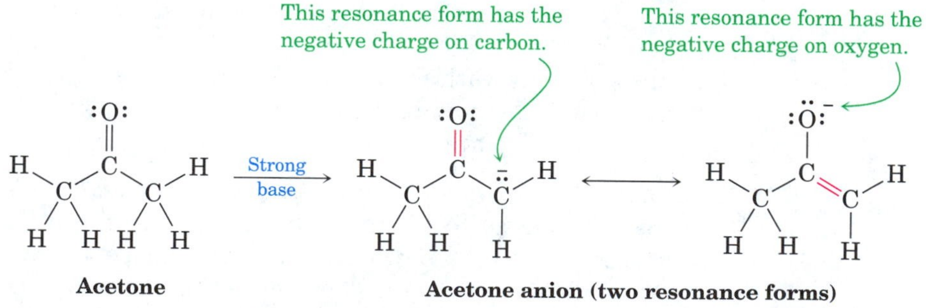

When two resonance forms are nonequivalent, the actual structure of the resonance hybrid is closer to the more stable form than to the less stable form. Thus, we might expect the acetone anion to look more like the resonance form that places the negative charge on the electronegative oxygen atom rather than on a carbon atom.

4. *Resonance forms must be valid Lewis structures and obey normal rules of valency.* A resonance form is like any other structure: The octet rule still applies. For example, one of the following structures for the acetate ion is not a valid resonance form because the carbon atom has five bonds and ten electrons:

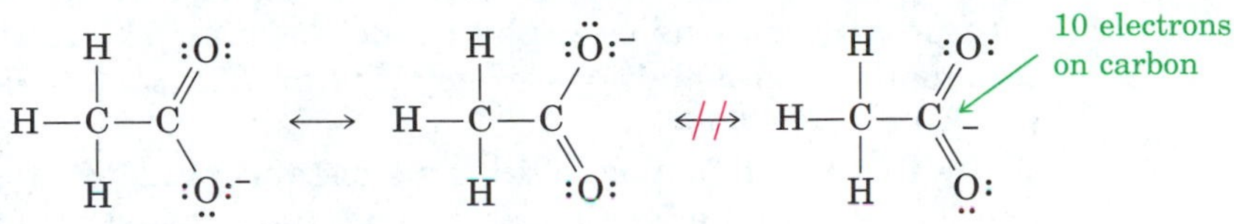

Acetate ion

NOT a valid resonance form

5. *The resonance hybrid is more stable than any individual resonance form.* In other words, resonance leads to stability. The greater the number of resonance forms, the more stable the substance. We'll see in Chapter 15, for instance, that a benzene ring is much more stable than might be expected.

PRACTICE PROBLEM

Draw as many resonance forms as you can for the carbonate ion, CO_3^{2-}.

Solution Just as the acetate ion has its negative charge distributed over two oxygen atoms in two resonance forms, the carbonate ion has its negative charges distributed over three oxygen atoms in three resonance forms. The difference between the forms is in the position of the double bond and the lone-pair electrons.

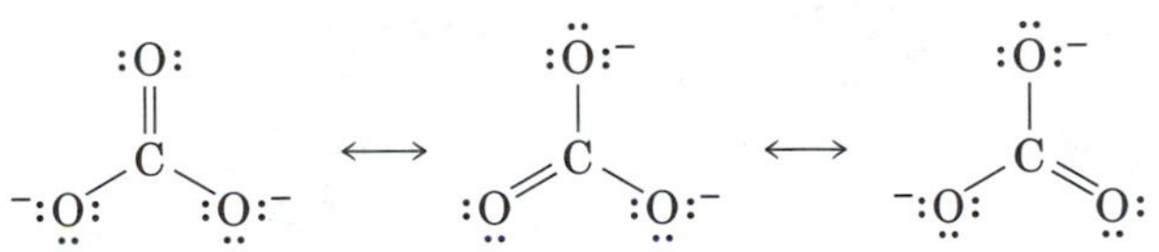

Carbonate ion (three resonance forms)

PROBLEM

2.9 Draw as many resonance structures as you can for these species:

(a) The nitrate ion, NO_3^-

(b)

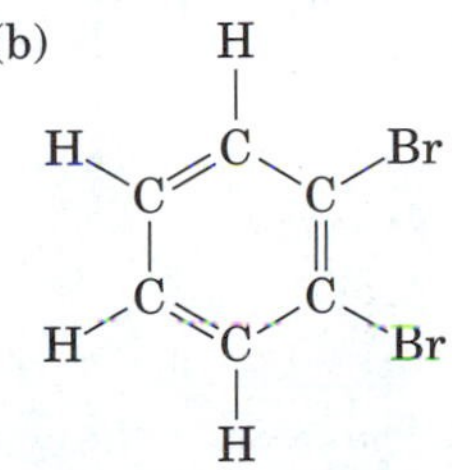

***ortho*-Dibromobenzene**

(c) The allyl cation, $H_2C{=}CH{-}CH_2^+$

(d) Hydrazoic acid, $:N{\equiv}\overset{+}{N}{-}\ddot{\underset{\cdot\cdot}{N}}{-}H$

2.6 Acids and Bases: The Brønsted–Lowry Definition

Yet another important concept related to electronegativity and polarity is that of *acidity* and *basicity*. We'll soon see that the acid–base behavior of organic molecules helps explain much of their chemistry. There are two frequently used definitions of acidity, the *Brønsted–Lowry definition* and the *Lewis definition*.

A **Brønsted–Lowry acid** is a substance that donates a proton (hydrogen ion, H^+), and a **Brønsted–Lowry base** is a substance that accepts a proton.[3] When HCl gas dissolves in water, for example, an acid–base reaction occurs. A polar hydrogen chloride molecule donates a proton, and a water molecule accepts the proton, yielding Cl^- and H_3O^+. Chloride ion (Cl^-), the product that results when the acid HCl loses a proton, is called the **conjugate base** of the acid; H_3O^+, the product that results when the base H_2O gains a proton, is called the **conjugate acid** of the base. Other common mineral acids such as H_2SO_4 and HNO_3 behave similarly, as do organic acids such as acetic acid, CH_3COOH.

In a general sense,

$$\underset{\textbf{Acid}}{\mathrm{H{-}A}} + \underset{\textbf{Base}}{:\mathrm{B}} \rightleftharpoons \underset{\textbf{Conjugate base}}{\mathrm{A}:^-} + \underset{\textbf{Conjugate acid}}{\mathrm{H{-}B^+}}$$

For example,

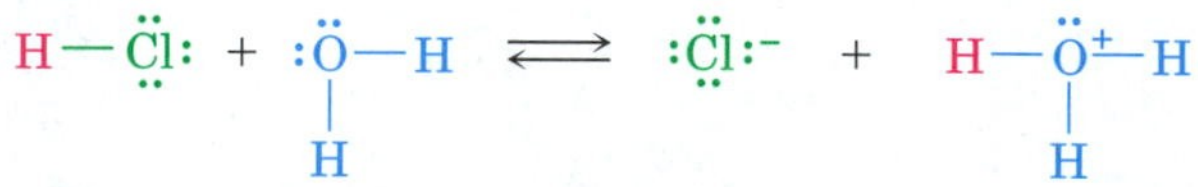

Acid **Base** **Conjugate base** **Conjugate acid**

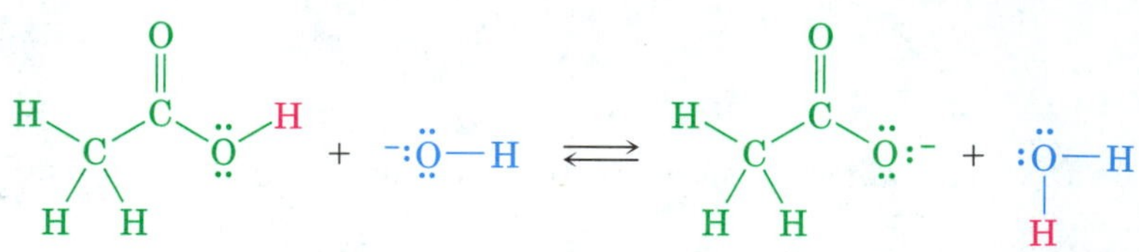

Acid **Base** **Conjugate base** **Conjugate acid**

---

[3]The name *proton* is often used as a synonym for the hydrogen ion, H^+, because loss of the valence electron from a neutral hydrogen atom leaves only the hydrogen nucleus—a proton. Bare protons are far too reactive to exist in solution by themselves, however, and are always bonded to some other species. In the presence of water, for example, a proton bonds to the oxygen atom, forming H_3O^+.

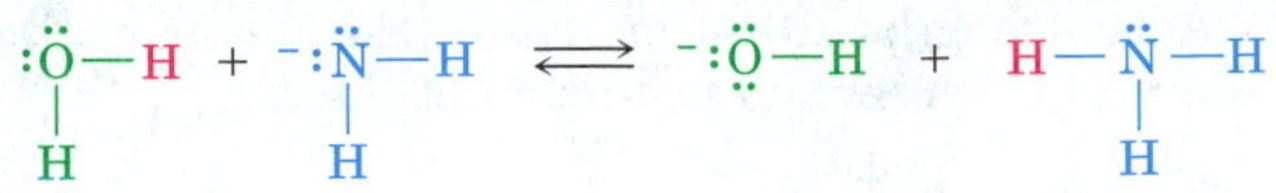

Note in the above examples that water can act *either* as an acid or as a base. In its reaction with HCl, water is a base that accepts a proton to give the hydronium ion, H_3O^+. In its reaction with amide ion, $^-NH_2$, however, water is an acid that donates a proton to give ammonia, NH_3, and hydroxide ion, HO^-.

Acids differ in their proton-donating ability. Stronger acids such as HCl react almost completely with water, whereas weaker acids such as acetic acid (CH_3COOH) react only slightly. The exact strength of a given acid in water solution is described using the standard expression for the equilibrium constant, K_{eq}:

$$HA + H_2O \rightleftharpoons A^- + H_3O^+$$

$$K_{eq} = \frac{[H_3O^+][A^-]}{[HA][H_2O]}$$

where HA represents any acid.[4]

In the dilute aqueous solution normally used for measuring acidity, the concentration of water, $[H_2O]$, remains nearly constant at approximately 55.6 M. We can therefore rewrite the equilibrium expression using a new quantity called the **acidity constant, K_a**. The acidity constant for any generalized acid HA is simply the equilibrium constant multiplied by the molar concentration of pure water, 55.6 M:

$$HA + H_2O \rightleftharpoons A^- + H_3O^+$$

$$K_a = K_{eq}[H_2O] = \frac{[H_3O^+][A^-]}{[HA]}$$

Stronger acids have their equilibria toward the right and thus have larger acidity constants, whereas weaker acids have their equilibria toward the left and have smaller acidity constants. The range of K_a values for different acids is enormous, running from about 10^{15} for the strongest acids to about 10^{-60} for the weakest. The common inorganic acids such as H_2SO_4, HNO_3, and HCl have K_a's in the range 10^2–10^9, while organic carboxylic acids have K_a's in the range 10^{-5}–10^{-15}. As you gain more experience in later chapters, you'll develop a rough feeling for which acids are "strong" and which are "weak" (always remembering that the terms are relative).

[4]Remember that brackets [ ] around a substance mean that the concentration of the enclosed species is given in moles per liter (mol/L), M.

Acid strengths are normally expressed using pK_a values, where the pK_a is the negative common logarithm of the K_a:

$$pK_a = -\log K_a$$

A stronger acid (larger K_a) has a *smaller* pK_a, and a weaker acid (smaller K_a) has a *larger* pK_a. Table 2.3 lists the pK_a's of some common acids in order of their strength. A more comprehensive table is given in Appendix B.

Table 2.3 Relative Strengths of Some Common Acids and Their Conjugate Bases

	Acid	*Name*	pK_a	*Conjugate base*	*Name*	
Weaker acid	CH_3CH_2OH	Ethanol	16.00	$CH_3CH_2O^-$	Ethoxide ion	Stronger base
	H_2O	Water	15.74	HO^-	Hydroxide ion	
↓	HCN	Hydrocyanic acid	9.31	CN^-	Cyanide ion	↑
	CH_3COOH	Acetic acid	4.76	CH_3COO^-	Acetate ion	
	HF	Hydrofluoric acid	3.45	F^-	Fluoride ion	
	HNO_3	Nitric acid	−1.3	NO_3^-	Nitrate ion	
Stronger acid	HCl	Hydrochloric acid	−7.0	Cl^-	Chloride ion	Weaker base

Notice in Table 2.3 that there is an inverse relationship between the acid strength of an acid and the base strength of its conjugate base. To understand this relationship, think about what is happening to the acidic proton: A *strong acid* is one that loses its proton easily, meaning that its conjugate base has little affinity for the proton and is therefore a *weak base*. A *weak acid* is one that loses its proton with difficulty, meaning that its conjugate base has a high affinity for the proton and is therefore a *strong base*. The fact that HCl is a strong acid, for example, means that Cl^- does not hold the proton tightly and is thus a weak base. Water, however, is a weak acid, meaning that OH^- holds the proton tightly and is a strong base.

PROBLEM. . . . .

2.10 Formic acid, HCOOH, has pK_a = 3.75, and picric acid, $C_6H_3N_3O_7$, has pK_a = 0.38. Which is the stronger acid?

PROBLEM. . . . .

2.11 Amide ion, H_2N^-, is a much stronger base than hydroxide ion, HO^-. Which would you expect to be a stronger acid, NH_3 or H_2O? Explain.

. . . .

2.7 Predicting Acid–Base Reactions from pK_a Values

Compilations of pK_a values like those in Table 2.3 are useful for predicting whether a given acid–base reaction will take place. In general, an acid will donate a proton to the conjugate base of any acid with a higher pK_a. Conversely, the conjugate base of an acid will abstract a proton from any acid with a lower pK_a. For example, the data in Table 2.3 indicate that hydroxide ion will react with acetic acid, CH_3COOH, to yield acetate ion, CH_3COO^-, and water. Since water (pK_a = 15.74) is a weaker acid than acetic acid (pK_a = 4.76), hydroxide ion has a greater affinity for a proton than acetate ion has.

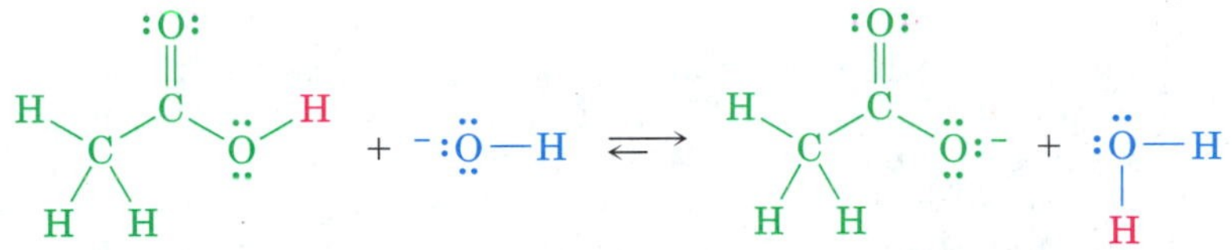

Acetic acid (pK_a= 4.76) **Hydroxide ion** **Acetate ion** **Water (pK_a= 15.74)**

An easier way to predict acid–base reactivity is to remember that the products of an acid–base reaction must be more stable than the reactants for reaction to occur. In other words, the product acid and base must be weaker and less reactive than the starting acid and base. In the reaction of acetic acid with hydroxide ion, for example, the product conjugate base (acetate ion) is weaker than the starting base (hydroxide ion), and the product conjugate acid (water) is weaker than the starting acid (acetic acid).

Stronger acid **Stronger base** **Weaker acid** **Weaker base**

PRACTICE PROBLEM. . . . .

Water has pK_a = 15.74, and acetylene has pK_a = 25. Which of the two is more acidic? Does hydroxide ion react with acetylene?

$$H{-}C{\equiv}C{-}H + H{-}O^- \xrightarrow{?} H{-}C{\equiv}C{:}^- + H{-}O{-}H$$

Solution In comparing two acids, the one with the lower pK_a is stronger. Thus, water is a stronger acid than acetylene. Since water gives up a proton more easily than acetylene does, the HO^- ion must have less affinity for a proton than the $HC{\equiv}C{:}^-$ ion has. In other words, the anion of acetylene is a stronger base than hydroxide ion, and the reaction will not proceed as written.

PRACTICE PROBLEM...

According to the data in Table 2.3, acetic acid has $pK_a = 4.76$. What is its K_a?

Solution Since pK_a is the negative logarithm of K_a, it's necessary to use a calculator with an ANTILOG or INV LOG function. Enter the value of the pK_a (4.76), change the sign (−4.76), and then find the antilog (1.74×10^{-5}). Thus, $K_a = 1.74 \times 10^{-5}$.

PRACTICE PROBLEM...

Use the K_a value for acetic acid calculated in the preceding Practice Problem to find the concentration of H_3O^+ in a 0.100 M solution of acetic acid.

Solution Let the concentration of H_3O^+ be represented by $[X]$. Then the concentration of acetate ion is also $[X]$, because there is a 1:1 ratio of H_3O^+ and acetate, and the concentration of acetic acid is $[0.100 - X]$. But since acetic acid is a weak acid, we know that the value of X is small, and we can assume that $[0.100 - X]$ is approximately 0.100. The equilibrium equation then becomes

$$K_a = \frac{[H_3O^+][\text{Acetate}]}{[\text{Acetic acid}]}$$

$$1.74 \times 10^{-5} = \frac{[X][X]}{[0.100 - X]} = \frac{[X]^2}{[0.100 - X]} \approx \frac{[X]^2}{0.100}$$

so

$$[X]^2 = 1.74 \times 10^{-6} \quad \text{and} \quad X = 1.32 \times 10^{-3}\ \text{M}$$

The H_3O^+ concentration is therefore 1.32×10^{-3} M, corresponding to pH = 2.9.

PROBLEM...

2.12 Will either of the following reactions take place as written, according to the pK_a data in Table 2.3?

(a) $HCN + CH_3COO^-\ Na^+ \xrightarrow{?} Na^+\ ^-CN + CH_3COOH$

(b) $CH_3CH_2OH + Na^+\ ^-CN \xrightarrow{?} CH_3CH_2O^-\ Na^+ + HCN$

PROBLEM.

2.13 Ammonia, NH_3, has $pK_a \approx 36$ and acetone has $pK_a \approx 19$. Will the following reaction take place?

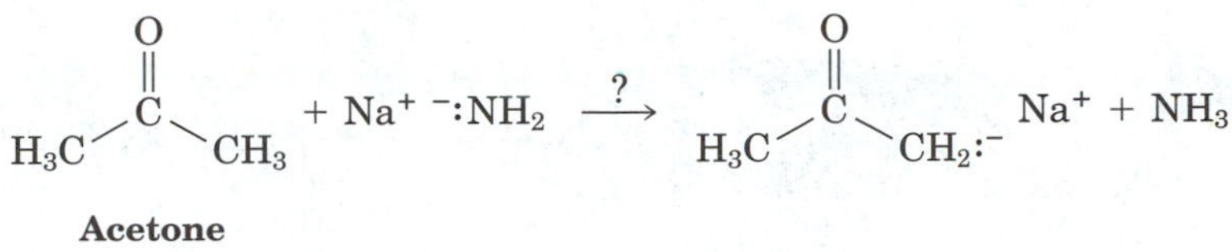

PROBLEM.

2.14 What is the K_a of HCN if its pK_a is 9.31?

PROBLEM.

2.15 What is the pH of a 0.500 M aqueous solution of benzoic acid, $pK_a = 4.19$?

2.8 Acids and Bases: The Lewis Definition

The Brønsted–Lowry definition of acidity discussed in the previous two sections encompasses all compounds containing hydrogen. Of even more use, however, is the *Lewis definition* of acids and bases, which is not limited to compounds that gain or lose protons. A **Lewis acid** is a substance that accepts an electron pair, and a **Lewis base** is a substance that donates an electron pair. As a result of this electron donation from a base to an acid, a covalent bond is formed.

Lewis Acids

The fact that a Lewis acid must be able to accept an electron pair means that it must have a vacant, low-energy orbital (or a polarized bond to hydrogen so that H^+ can be lost). Thus, the Lewis definition of acidity is much broader than the Brønsted–Lowry definition and includes many other species in addition to H^+. For example, various metal cations such as Mg^{2+} are Lewis acids because they accept a pair of electrons when they form a bond to a base. In the same way, compounds of group 3A elements such as BF_3 and $AlCl_3$ are Lewis acids because they have unfilled valence orbitals and can accept electron pairs from Lewis bases, as shown in Figure 2.3. Similarly, many transition-metal compounds, such as $TiCl_4$, $FeCl_3$, $ZnCl_2$, and $SnCl_4$, are Lewis acids.

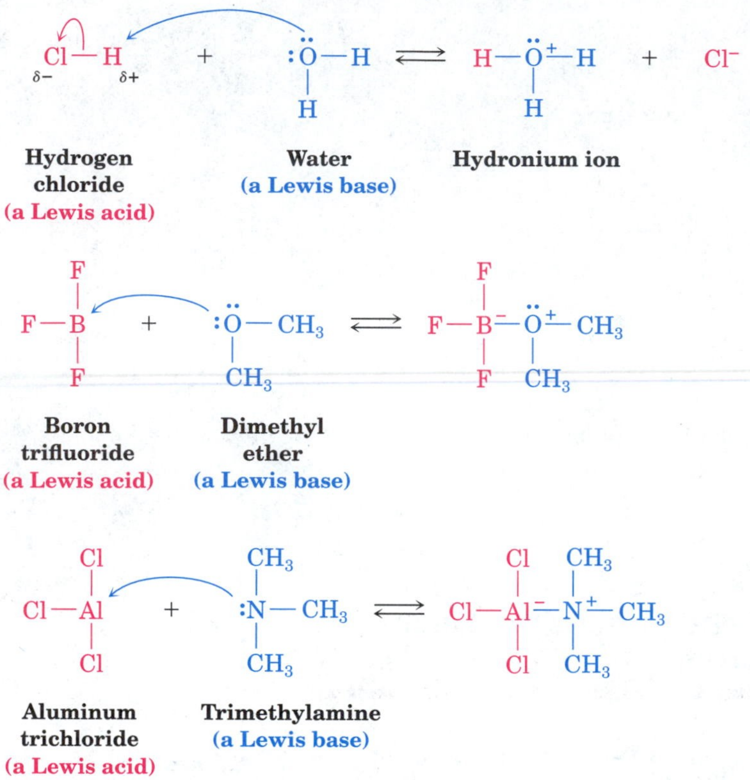

Figure 2.3 The reactions of some Lewis acids with some Lewis bases. The Lewis acids accept an electron pair; the Lewis bases donate a pair of nonbonding electrons. Note how the flow of electrons from the Lewis base to the Lewis acid is indicated by the curved arrows.

Look closely at the acid–base reactions in Figure 2.3 and note how they are shown. In the first reaction, the Lewis base water abstracts H^+ from the strongly polarized hydrogen chloride molecule. In each of the remaining two reactions, a Lewis base donates an electron pair to a vacant valence orbital of a Lewis acidic boron or aluminum atom. In all three reactions, a curved arrow indicates the direction of electron-pair flow from the electron-rich Lewis base to the electron-poor Lewis acid. This kind of arrow is used extensively in organic chemistry and always has the same meaning: A pair of electrons moves *from* the atom at the tail of the arrow *to* the atom at the head of the arrow. *A curved arrow indicates the movement of electrons, not the movement of atoms.*

Some further examples of Lewis acids are shown on the following page:

Some Lewis acids

Some neutral proton donors:

H_2O HCl HBr HNO_3 H_2SO_4

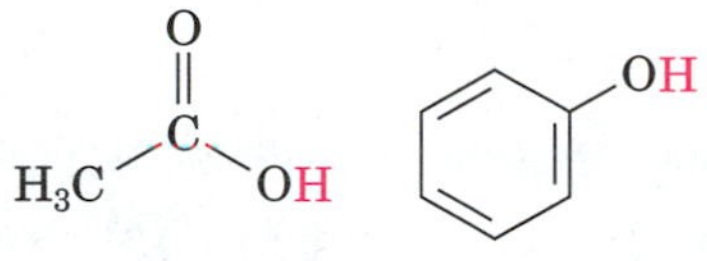

CH_3CH_2OH

A carboxylic acid **A phenol** **An alcohol**

Some cations:

Li^+ Mg^{2+} Br^+

Some metal compounds:

$AlCl_3$ BF_3 $TiCl_4$ $FeCl_3$ $ZnCl_2$

Lewis Bases

The Lewis definition of a base as a compound with a pair of nonbonding electrons that it can use to bond to a Lewis acid is similar to the Brønsted–Lowry definition. Thus, H_2O, with its two pairs of nonbonding electrons on oxygen, acts as a Lewis base by donating an electron pair to a proton in forming the hydronium ion, H_3O^+:

Acid **Lewis base** **Hydronium ion**

In a more general sense, most oxygen- and nitrogen-containing organic compounds are Lewis bases because they have lone pairs of available electrons. Divalent oxygen compounds each have two lone pairs of electrons, and trivalent nitrogen compounds have one lone pair. Note in the following examples that some compounds can act as both acids and bases, just as water can. Alcohols and carboxylic acids, for instance, act as acids when they donate a proton but as bases when their oxygen atom accepts a proton.

Some Lewis bases

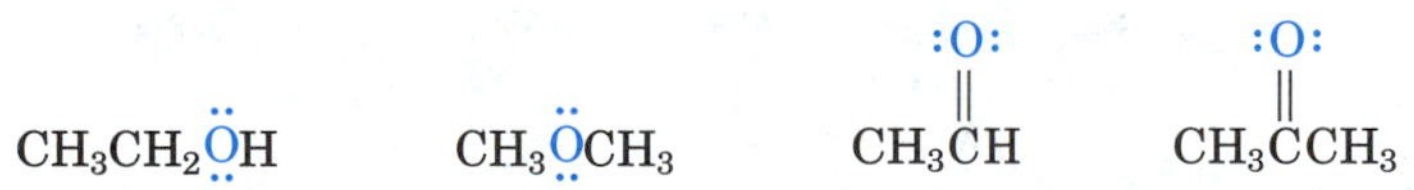

An alcohol **An ether** **An aldehyde** **A ketone**

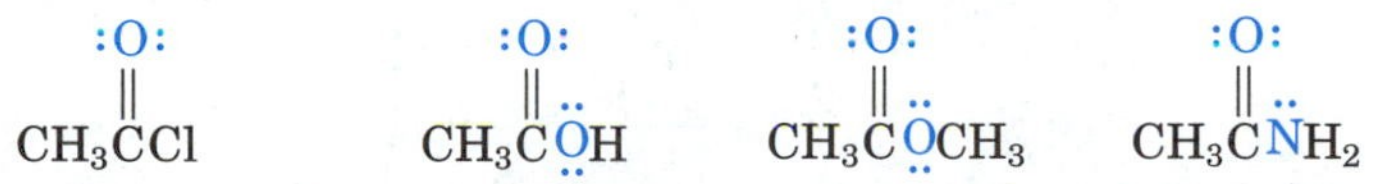

An acid chloride **A carboxylic acid** **An ester** **An amide**

An amine **A sulfide**

For example:

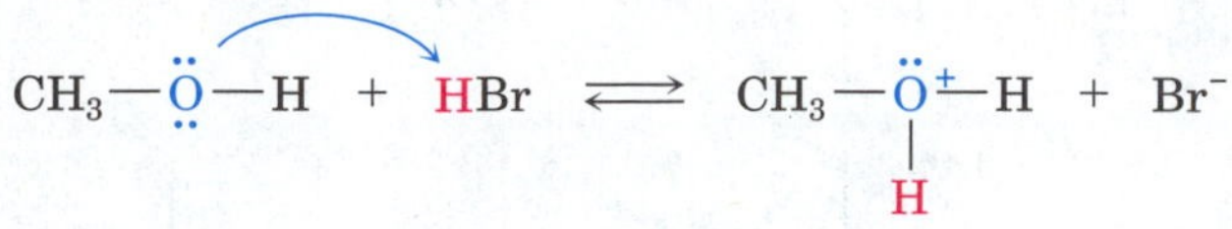

Methyl alcohol (base) **Hydrogen bromide (acid)** **Methyloxonium bromide**

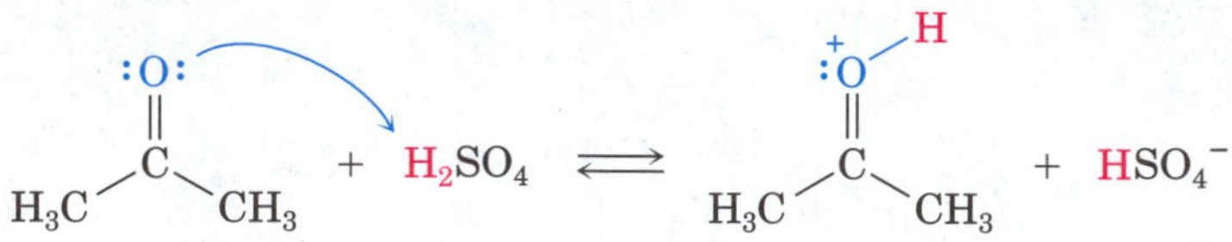

Acetone (base) **Sulfuric acid**

Notice in the list of Lewis bases given above that some compounds, such as carboxylic acids, esters, and amides, have more than one atom with a lone pair of electrons and can therefore react at more than one site. Acetic acid, for example, can be protonated either on the doubly bonded oxygen atom or on the singly bonded oxygen atom:

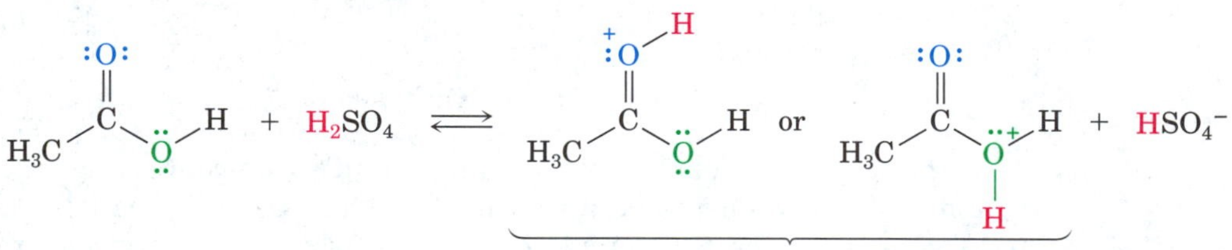

Acetic acid (base) **Sulfuric acid** **Protonated acetic acid**

Reaction normally occurs only once in such instances, and the more stable of the two possible protonation products is formed. (For acetic acid, protonation actually occurs on the doubly bonded oxygen.)

PRACTICE PROBLEM

Show how acetaldehyde can act as a Lewis base.

Solution The oxygen atom of acetaldehyde has two lone pairs of electrons that it can donate to a Lewis acid such as H^+.

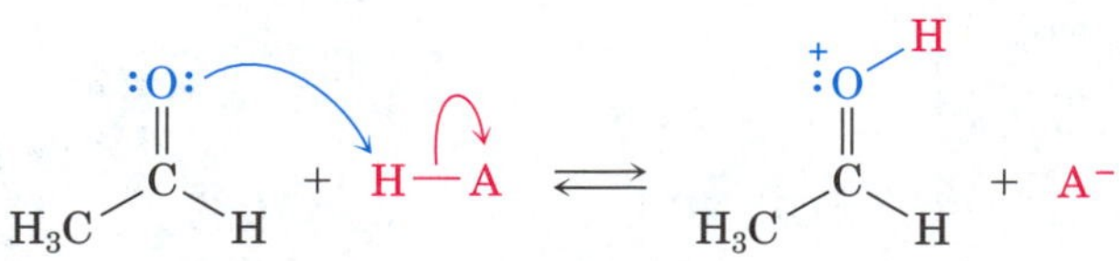

Acetaldehyde

PROBLEM

2.16 Show how the species in part (a) can act as Lewis bases in their reactions with HCl, and show how the species in part (b) can act as Lewis acids in their reaction with OH^-.

(a) CH_3CH_2OH, $HN(CH_3)_2$, $P(CH_3)_3$ (b) H_3C^+, $B(CH_3)_3$, $MgBr_2$

PROBLEM

2.17 Explain by calculating formal charges why the following acid–base complexes have the charges indicated.

(a) 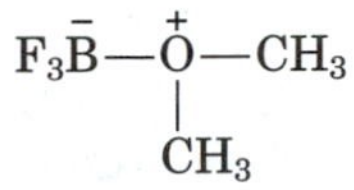

(b) 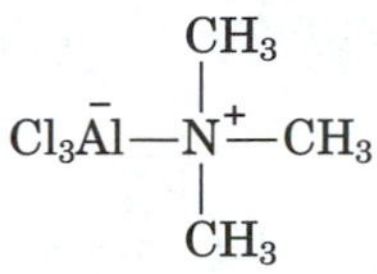

PROBLEM

2.18 Boron trifluoride reacts with formaldehyde to give an acid–base complex. Which partner is the acid and which is the base?

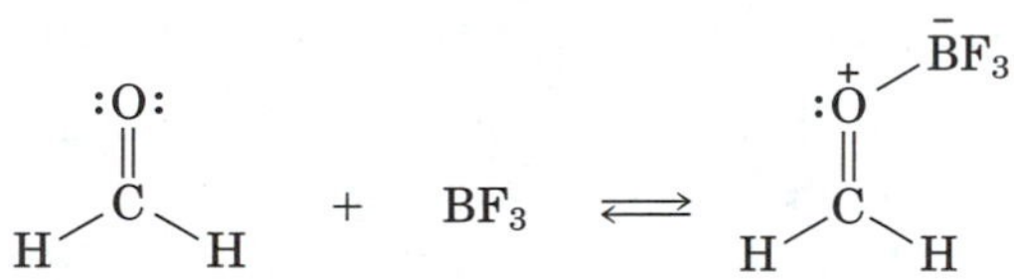

Formaldehyde **Boron trifluoride**

2.9 Drawing Chemical Structures

In the Kekulé structures we've been drawing up to this point, a line between atoms represents the two electrons in a covalent bond. Such structures have been used for many years and comprise a universal chemical language. Two chemists from different countries may not understand each other's spoken words, but a chemical structure means the same to both.

Most organic chemists find themselves drawing many structures each day, and it would soon become awkward if every bond and atom had to be indicated. For example, vitamin A, $C_{20}H_{30}O$, has 51 different chemical bonds uniting the 51 atoms. Vitamin A can be drawn showing each bond and atom, but doing so is a time-consuming process, and the resultant drawing is difficult to read (see Table 2.4). Chemists have therefore devised several shorthand ways for writing structures. In **condensed structures**, carbon–hydrogen and carbon–carbon bonds aren't shown; instead, they're understood. If a carbon has three hydrogens bonded to it, we write CH_3; if a carbon has two hydrogens bonded to it, we write CH_2; and so on. The compound called 2-methylbutane, for example, is written as follows:

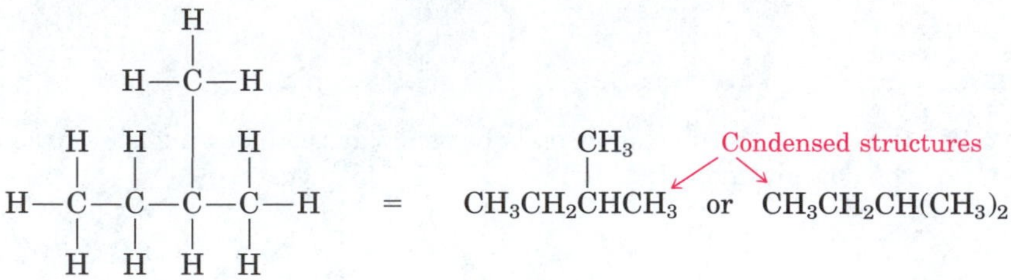

2-Methylbutane

Notice that the horizontal bonds between carbons aren't shown in condensed structures—the CH_3, CH_2, and CH units are simply placed next to each other—but the vertical carbon–carbon bond in the first condensed structure above is shown for clarity. Notice also that in the second condensed structure, the two equivalent CH_3 units attached to the CH carbon are grouped together as $(CH_3)_2$.

Simpler still is the use of **skeletal structures** such as those shown in Table 2.4. The rules for drawing skeletal structures are straightforward:

1. Carbon atoms aren't usually shown. Instead, a carbon atom is assumed to be at each intersection of two lines (bonds) and at the end of each line. Occasionally, a carbon atom might be indicated for emphasis or clarity.
2. Hydrogen atoms bonded to carbon aren't shown. Since carbon always has a valence of 4, we mentally supply the correct number of hydrogen atoms for each carbon.
3. Atoms other than carbon and hydrogen *are* shown.

Table 2.4 gives some examples of how these rules are applied.

PROBLEM...

2.19 Convert these skeletal structures into molecular formulas:

(a) 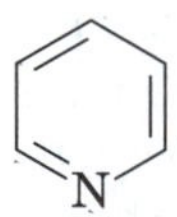

Pyridine

(b) 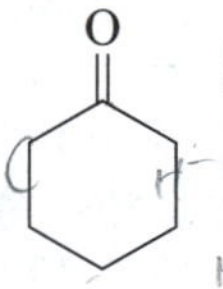

Cyclohexanone

(c) 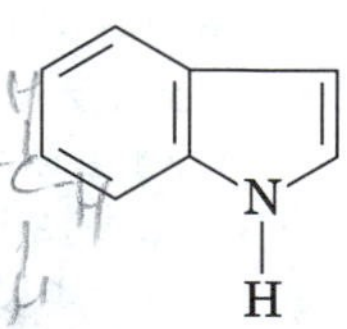

Indole

PROBLEM...

2.20 Propose skeletal structures for compounds that satisfy the following molecular formulas (there is more than one possibility in each case):

(a) C_5H_{12} (b) C_2H_7N (c) C_3H_6O (d) C_4H_9Cl

....

Table 2.4 Kekulé and Skeletal Structures for Several Compounds

Compound	*Kekulé structure*	*Skeletal structure*
Butane, C_4H_{10}	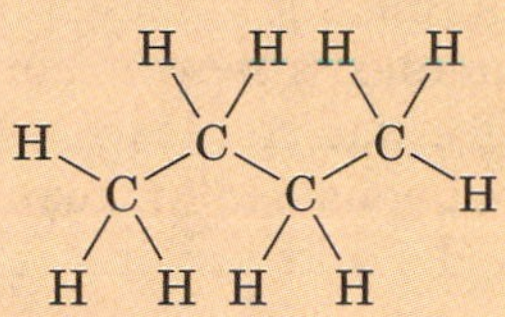	
Isoprene, C_5H_8	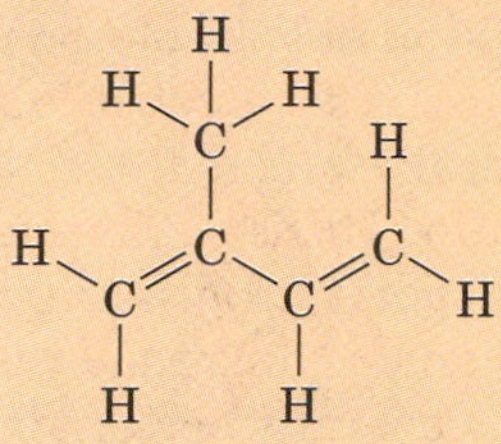	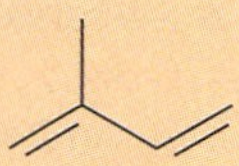
Methylcyclohexane, C_7H_{14}	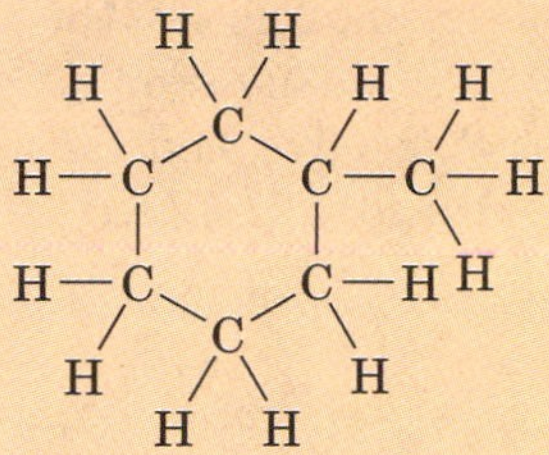	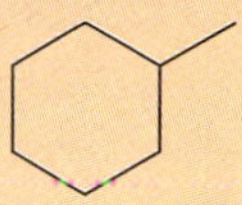
Phenol, C_6H_6O	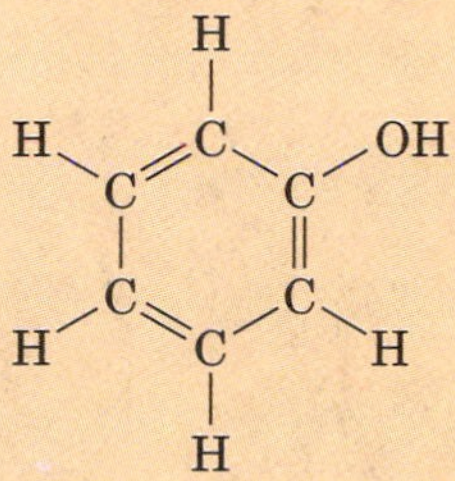	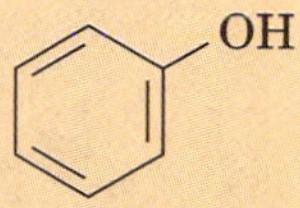
Vitamin A, $C_{20}H_{30}O$	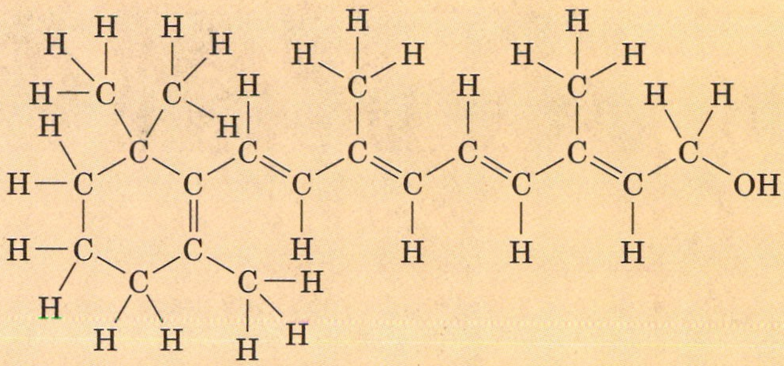	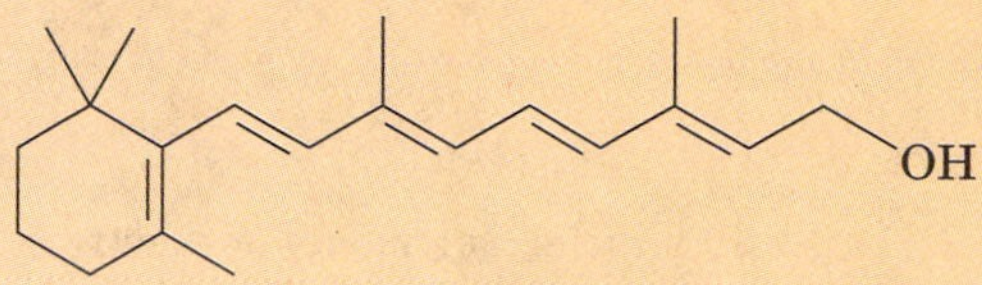

2.10 Molecular Models

Organic chemistry is a three-dimensional science, and molecular shape is often critical in determining the chemistry a compound undergoes. One particularly helpful technique for learning organic chemistry is to use molecular models. With practice, you can learn to see many spatial relationships even when viewing two-dimensional drawings, but there's no substitute for building a molecular model and turning it in your hands to get different perspectives.

Many kinds of models are available, some at relatively modest cost, and everyone should have access to a set of models while studying this book. Research chemists generally prefer to use either space-filling models such as Corey–Pauling–Koltun (CPK™) Molecular Models, or skeletal models such as Dreiding Stereomodels™. Both are expensive but are precisely made to reflect accurate bond angles, intramolecular distances, and atomic radii. Space-filling models are generally better for examining the crowding within a molecule, while skeletal models let the user measure bond angles and interatomic distances more easily. For student use, ball-and-stick models are generally the least expensive and most durable. Figure 2.4 shows two kinds of models of acetic acid, CH_3COOH.

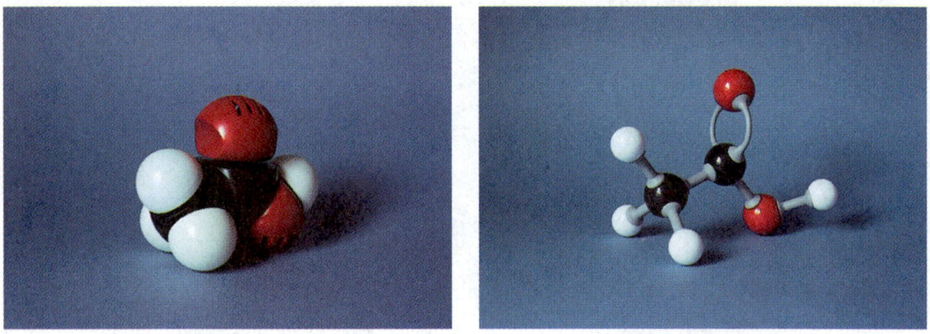

(a) (b)

Figure 2.4 Molecular models of acetic acid, CH_3COOH: (a) space-filling; (b) ball-and-stick.

PROBLEM

2.21 Build a molecular model of ethane, $H_3C{-}CH_3$, and sight along the C–C bond to see the relationships between hydrogens on the different carbons.

INTERLUDE

Alkaloids: Naturally Occurring Bases

The characteristic odor of ripe fish is due to methylamine, CH_3NH_2.

Just as ammonia, NH_3, is a weak base, there is a large group of nitrogen-containing organic compounds called *amines* that behave as weak bases. In the early days of organic chemistry, basic amines derived from natural sources were known as "vegetable alkali," but they are now referred to as **alkaloids**. The study of alkaloids provided much of the impetus for the growth of organic chemistry in the nineteenth century, and it remains today a fascinating area of research.

Alkaloids vary widely in structure, from the simple to the enormously complex. The odor of rotting fish, for example, is caused by methylamine, a simple relative of ammonia in which one of the NH_3 hydrogens has been replaced by an organic CH_3 group. (The use of acidic lemon juice to mask fish odors is, in fact, an acid–base reaction.)

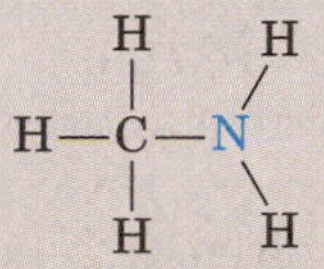

Methylamine
(found in rotting fish)

Many alkaloids have pronounced biological properties, and many of the pharmaceutical agents used today are derived from naturally occurring amines. Morphine and related alkaloids from the opium poppy, for instance, are used for pain relief; ephedrine from the Chinese plant *Ephedra sinica* is used as a bronchodilator and decongestant; and atropine from the flowering plant *Atropa belladonna*, commonly called the deadly nightshade, is used as an antispasmodic agent for the treatment of colitis.

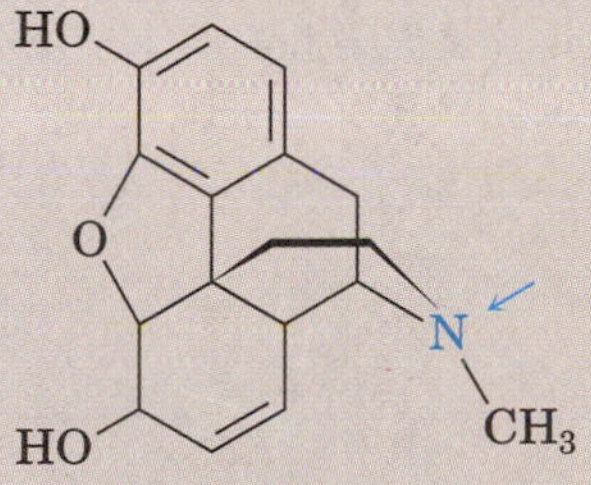

Morphine

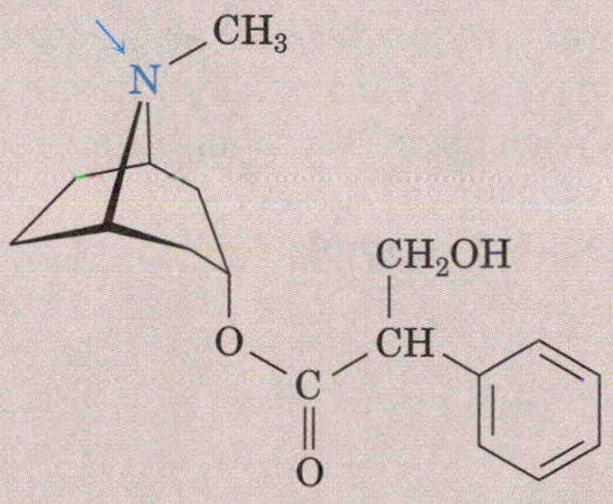

Atropine

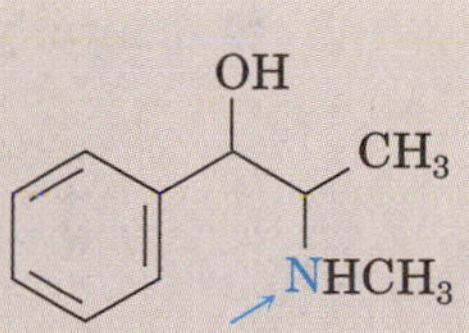

Ephedrine

Summary and Key Words

Organic molecules often have **polar covalent bonds** as a result of unsymmetrical electron sharing caused by the intrinsic **electronegativity** of atoms. For example, a carbon–chlorine bond is polar because chlorine attracts the shared electrons more strongly than carbon does. Carbon-metal bonds, by contrast, are usually polarized in the opposite sense because carbon attracts electrons more strongly than most metals. Carbon–hydrogen bonds are relatively nonpolar. Many molecules as a whole are also polar owing to the cumulative effects of individual polar bonds and electron lone pairs. The polarity of a molecule is measured by its **dipole moment**, μ.

Plus (+) and minus (−) signs are used to indicate the presence of **formal charges** on atoms in molecules. Assigning formal charges to specific atoms is a bookkeeping technique that makes it possible to keep track of the valence electrons around an atom.

$$\textbf{Formal charge} = \begin{pmatrix}\text{Number of valence electrons} \\ \text{owned by free atom}\end{pmatrix} - \begin{pmatrix}\text{Number of valence electrons} \\ \text{owned by bonded atom}\end{pmatrix}$$

Some substances, such as the acetate ion and benzene, can't be represented by a single Lewis or line-bond structure and must be considered as a composite, or **resonance hybrid**, of two or more structures, neither of which is correct by itself. The only difference between two resonance forms is in the location of their π or nonbonding electrons. The nuclei remain in the same places in both structures.

Acidity and basicity are closely related to polarity and electronegativity. A **Brønsted–Lowry acid** is a compound that can donate a proton (hydrogen ion, H^+), and a **Brønsted–Lowry base** is a compound that can accept a proton. The exact strength of a Brønsted–Lowry acid or base is expressed by its **acidity constant**, K_a, or by the negative logarithm of the acidity constant, pK_a. The higher the pK_a, the weaker the acid. More useful is the Lewis definition of acids and bases. A **Lewis acid** is a compound that has a low-energy empty orbital that can accept an electron pair; BF_3, $AlCl_3$, and H^+ are examples. A **Lewis base** is a compound that can donate an unshared electron pair; NH_3 and H_2O are examples. Most organic molecules that contain oxygen and nitrogen are Lewis bases.

Organic molecules are usually drawn using either condensed structures or skeletal structures. In **condensed structures**, carbon–carbon and carbon–hydrogen bonds aren't shown. In **skeletal structures**, only the bonds and not the atoms are shown. A carbon atom is assumed to be at the ends and at the junctions of lines (bonds), and the correct number of hydrogens is mentally supplied. For example:

Cyclohexene

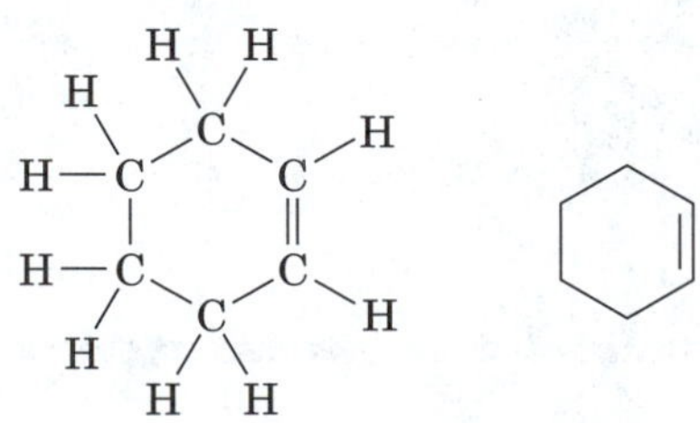

ADDITIONAL PROBLEMS

2.22 Convert the following structures into skeletal drawings.

(a)

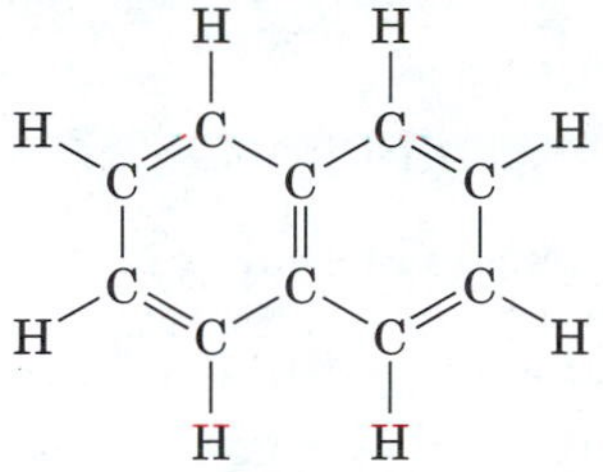

Naphthalene

(b)

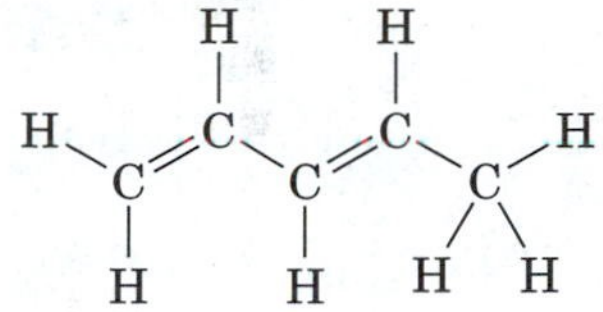

1,3-Pentadiene

(c)

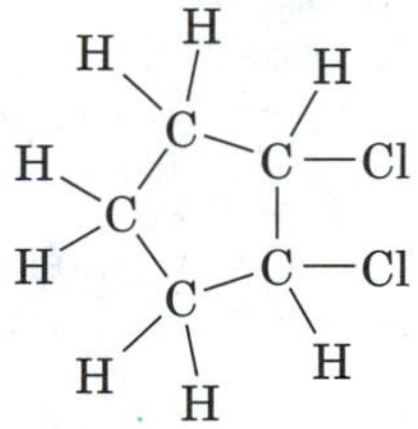

1,2-Dichlorocyclopentane

(d)

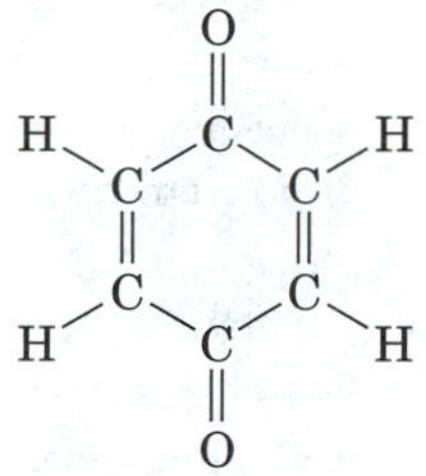

Quinone

2.23 Convert the following skeletal drawings into Kekulé structures that show all carbons and hydrogens:

(a)

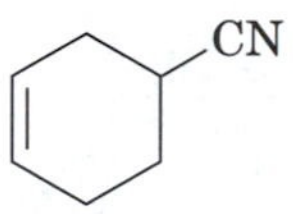

(b)

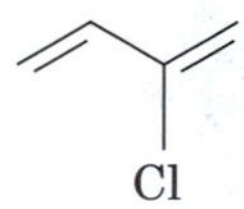

(c)

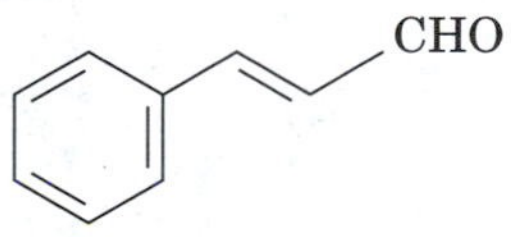

(d)

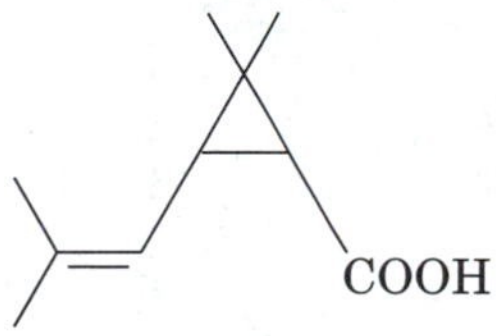

2.24 Calculate the formal charges on the atoms indicated in red.

(a) $(CH_3)_2\ddot{O}BF_3$

(b) $H_2\ddot{C}{-}N{\equiv}N{:}$

(c) $H_2C{=}N{=}\ddot{N}{:}$

(d) $:\ddot{O}{=}\ddot{O}{-}\ddot{O}{:}$

(e)

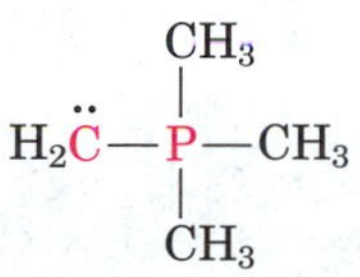

(f)

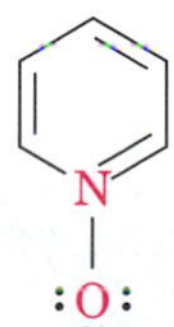

2.25 Use the electronegativity table (Figure 2.2) to predict which bond in each of the following sets is more polar.

(a) $H_3C{-}Cl$ and $Cl{-}Cl$
(b) $H_3C{-}H$ and $H{-}Cl$
(c) $HO{-}CH_3$ and $(CH_3)_3Si{-}CH_3$
(d) $H_3C{-}Li$ and $Li{-}OH$

2.26 Indicate the direction of bond polarity for each compound in Problem 2.25.

2.27 Which of the following molecules have dipole moments? Indicate the expected direction of each.

(a) 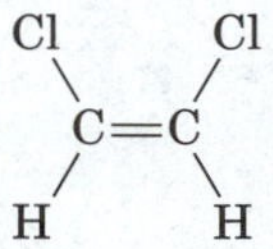

cis-1,2-Dichloroethylene

(b) 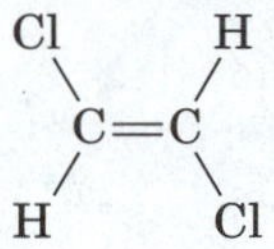

trans-1,2-Dichloroethylene

(c) LiH

(d) $F_3B—N(CH_3)_3$

(e) 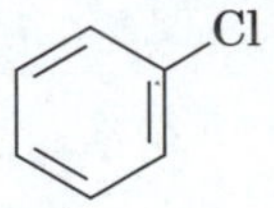

(f) 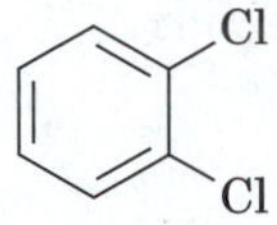

2.28 Explain the observation that phosgene, $Cl_2C{=}O$, has a smaller dipole moment than formaldehyde, $H_2C{=}O$.

2.29 The dipole moment of HCl is 1.08 D, and the H–Cl bond length is 1.36 Å. What percent ionic character does the H–Cl bond have?

2.30 How can you explain the fact that fluoromethane (CH_3F, μ = 1.81 D) has a smaller dipole moment than chloromethane (CH_3Cl, μ = 1.87 D), even though fluorine is more electronegative than chlorine? (*Remember*: $\mu = Q \times r$.)

2.31 Identify the most electronegative element in each of these molecules:
(a) CH_2FCl (b) $FCH_2CH_2CH_2Br$ (c) $HOCH_2CH_2NH_2$ (d) CH_3OCH_2Li

2.32 Which of the following pairs of structures represent resonance forms?

(a) 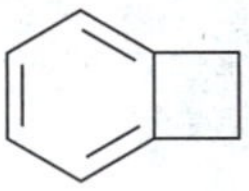 and 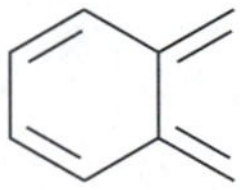

(b) 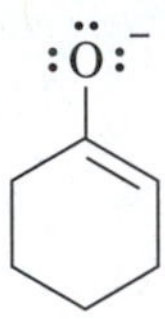 and 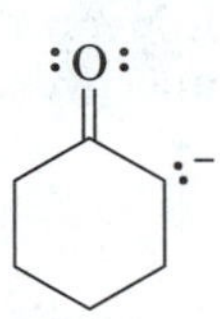

(c) 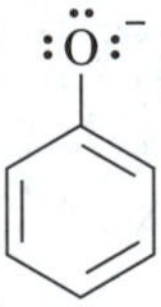 and 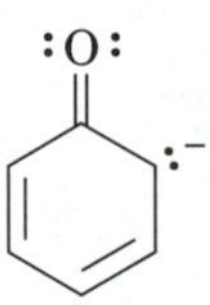

(d) 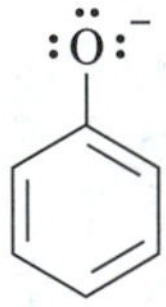 and 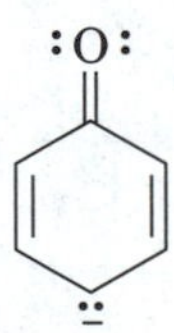

2.33 Draw as many resonance structures as you can for the following species.

(a) 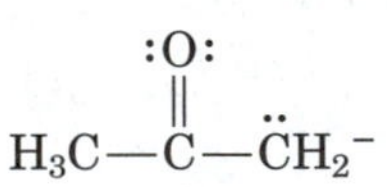

(b) 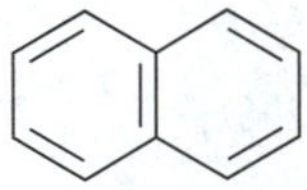

(c) 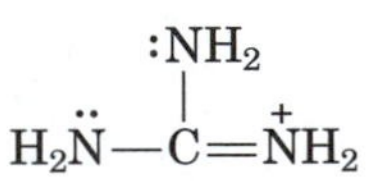

(d) $H_3C—\ddot{\underset{..}{S}}—\overset{+}{C}H_2$

(e) $H_2C{=}CH—CH_2^+$

(f) $H_2C{=}CH—CH{=}CH—\overset{+}{C}H—CH_3$

2.34 Cyclobutadiene is a rectangular molecule with two shorter double bonds and two longer single bonds. Why do the following structures *not* represent resonance forms?

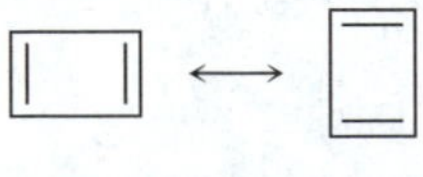

2.35 Alcohols can act either as weak acids or as weak bases, just as water can. Formulate the reactions of methyl alcohol, CH_3OH, with a strong acid such as HCl and with a strong base such as $Na^+ \ ^-NH_2$.

2.36 How can you explain the fact that the O–H hydrogen in acetic acid is more acidic than any of the C–H hydrogens? (*Hint*: Consider bond polarity.)

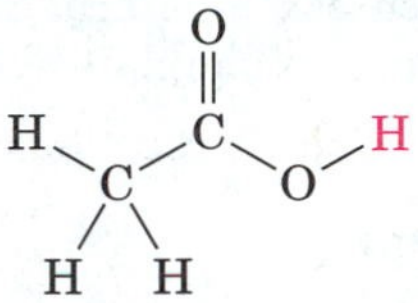

Acetic acid

2.37 Which of the following reagents are likely to act as Lewis acids and which as Lewis bases?
(a) $AlBr_3$ (b) $CH_3CH_2NH_2$ (c) BH_3 (d) HF (e) CH_3–S–CH_3 (f) $TiCl_4$

2.38 Draw Lewis electron-dot structures for each of the molecules in Problem 2.37. Make sure that you indicate the unshared electron pairs where present.

2.39 Write the products of these acid–base reactions:
(a) $CH_3OH + H_2SO_4 \rightleftharpoons$ (b) $CH_3OH + NaNH_2 \rightleftharpoons$
(c) $CH_3NH_3^+ \ Cl^- + NaOH \rightleftharpoons$

2.40 Assign formal charges to the atoms in each of the following molecules:

(a) 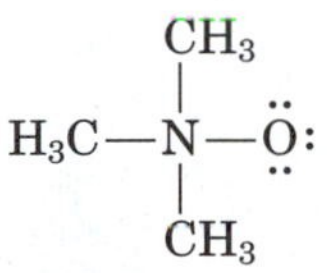 (b)  (c) 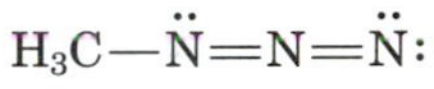

2.41 Rank the following substances in order of increasing acidity:

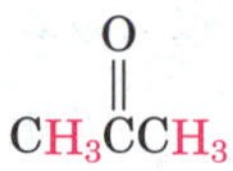 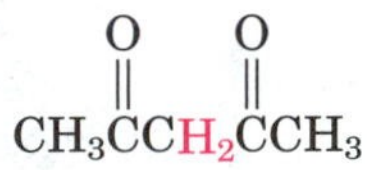 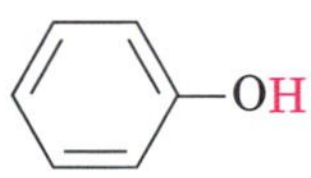 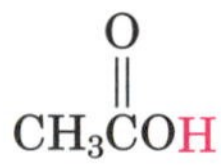

Acetone	**2,4-Pentanedione**	**Phenol**	**Acetic acid**
(pK_a = 19)	(pK_a = 9)	(pK_a = 9.9)	(pK_a = 4.76)

2.42 Which, if any, of the four substances in Problem 2.41 are strong enough acids to react almost completely with NaOH? (The pK_a of H_2O is 15.7.)

2.43 The ammonium ion (NH_4^+, pK_a = 9.25) has a lower pK_a than the methylammonium ion ($CH_3NH_3^+$, pK_a = 10.66). Which is the stronger base, ammonia (NH_3) or methylamine (CH_3NH_2)? Explain.

2.44 Is *tert*-butoxide anion a strong enough base to react with water? In other words, can a solution of potassium *tert*-butoxide in water be prepared? (The pK_a of *tert*-butyl alcohol is approximately 18.)

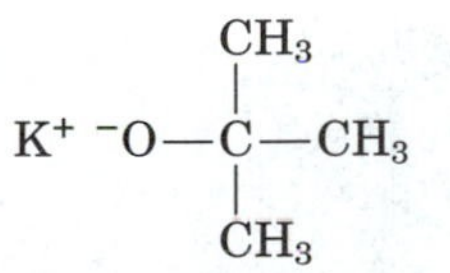

Potassium *tert*-butoxide

2.45 Calculate K_a values from the following pK_a's:
(a) Acetone, pK_a = 19.3 (b) Formic acid, pK_a = 3.75

2.46 Calculate pK_a values from the following K_a's:
(a) Nitromethane, $K_a = 5.0 \times 10^{-11}$ (b) Acrylic acid, $K_a = 5.6 \times 10^{-5}$

2.47 What is the pH of a 0.050 M solution of formic acid (see Problem 2.45)?

2.48 Sodium bicarbonate, $NaHCO_3$, is the sodium salt of carbonic acid (H_2CO_3), $pK_a = 6.37$. Which of the substances shown in Problem 2.41 will react with sodium bicarbonate?

2.49 Assume that you have two unlabeled bottles, one of which contains phenol ($pK_a = 9.9$) and one of which contains acetic acid ($pK_a = 4.76$). In light of your answer to Problem 2.48, propose a simple way to determine what is in each bottle.

2.50 Identify the acids and bases in these reactions:

(a) $CH_3OH + H^+ \longrightarrow CH_3\overset{+}{O}H_2$

(b) $CH_3OH + {}^-NH_2 \longrightarrow CH_3O^- + NH_3$

(c) $CH_3\overset{O}{\overset{\|}{C}}CH_3 + TiCl_4 \longrightarrow CH_3{-}\overset{\overset{+}{O}{-}\overset{-}{Ti}Cl_4}{\overset{\|}{C}}{-}CH_3$

(d) 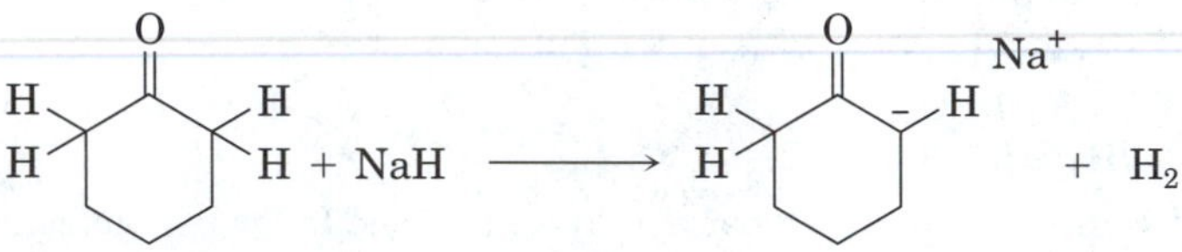

(e) 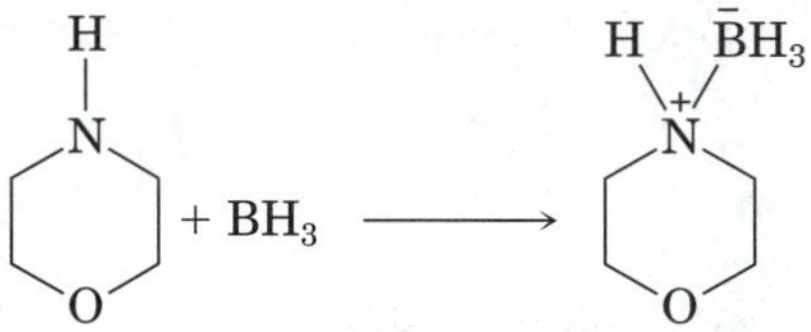

(f) 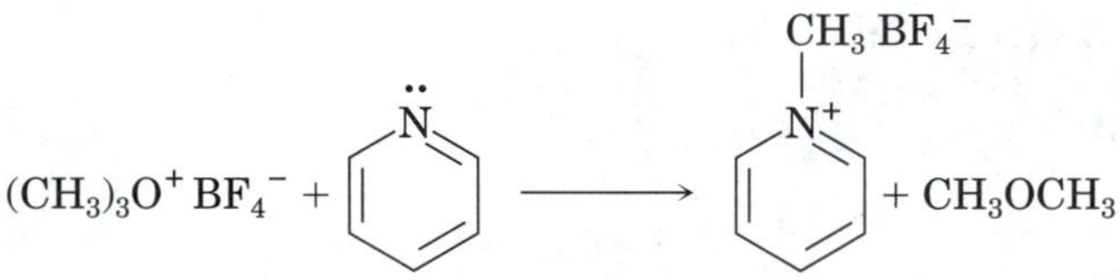

2.51 Which of the following pairs represent resonance structures?

(a) $CH_3C{\equiv}\overset{+}{N}{-}\ddot{\underset{..}{O}}{:}^-$ and $CH_3\overset{+}{C}{=}\ddot{N}{-}\ddot{\underset{..}{O}}{:}^-$

(b) $CH_3\overset{:O:}{\overset{\|}{C}}{-}\ddot{\underset{..}{O}}{:}^-$ and $:\overset{-}{C}H_2\overset{:O:}{\overset{\|}{C}}{-}\ddot{\underset{..}{O}}{-}H$

(c) 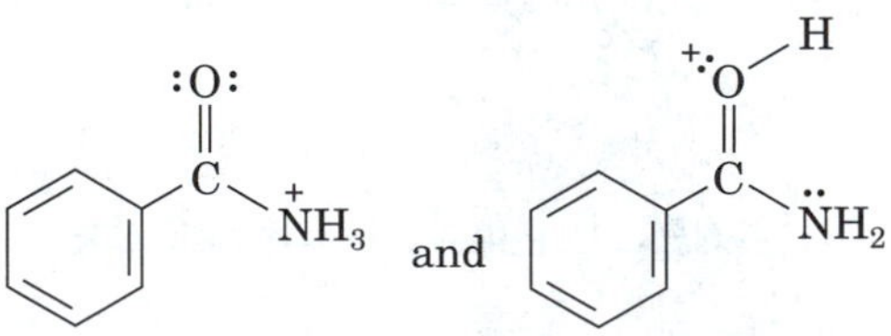

(d) 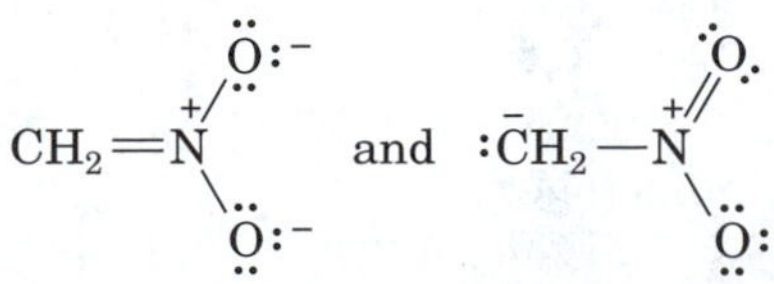

2.52 Draw as many resonance structures as you can for these species:

(a) Nitromethane, 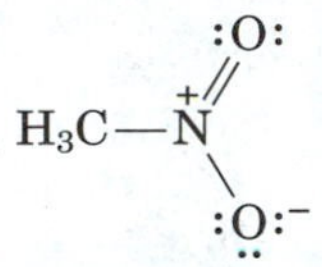

(b) Ozone, $:\ddot{O}=\overset{+}{\ddot{O}}-\ddot{O}:^-$

(c) Diazomethane, $H_2C=\overset{+}{N}=\overset{-}{\ddot{N}}:$

2.53 Dimethyl sulfone has dipole moment $\mu = 4.4$ D. Calculate the formal charges present on oxygen and sulfur, and suggest a geometry for the molecule that is consistent with the observed dipole moment.

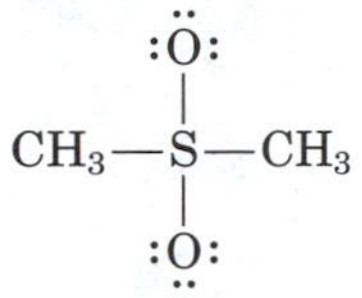

Dimethyl sulfone

A Look Ahead

2.54 We'll see in the next chapter that organic molecules can be classified according to the *functional groups* they contain, where a functional group is a collection of atoms with a characteristic chemical reactivity. Use the electronegativity values given in Figure 2.2 to predict the polarity of the following functional groups.

(a) 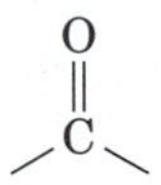

Ketone

(b) 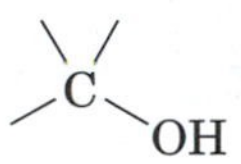

Alcohol

(c) 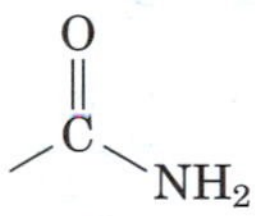

Amide

(d) $-C\equiv N$

Nitrile

2.55 Carboxylic acids, which contain the —COOH functional group, are stronger acids than alcohols, which contain the —OH functional group. Acetic acid (CH_3COOH), for example, is approximately 10^{10} more acidic than methanol (CH_3OH). Draw the structures of the anions resulting from loss of H^+ from acetic acid and methanol, and explain the large difference in acidity.

Methane, ethane, and propane are the three simplest alkanes.

3

The Nature of Organic Compounds: Alkanes and Cycloalkanes

According to *Chemical Abstracts*, a publication that abstracts and indexes the chemical literature, there are more than 11 million known organic compounds. Each of these compounds has its own physical properties, such as melting point and boiling point, and each has its own chemical reactivity.

Chemists have learned through many years of experience that organic compounds can be classified into families according to their structural features and that the members of a given family often have similar chemical reactivities. Instead of 11 million compounds with random reactivity, there are a few dozen families of organic compounds whose chemistry is reasonably predictable. We'll study the chemistry of specific families throughout the rest of this book, beginning in the present chapter with a look at the simplest family, the *alkanes*.

3.1 Functional Groups

The structural features that make it possible to classify compounds by reactivity are called functional groups. A **functional group** is a part of a larger molecule; it is composed of an atom or a group of atoms that have a characteristic chemical behavior. Chemically, a given functional group behaves in nearly the same way in every molecule it's a part of. For example, one of the simplest functional groups is the carbon–carbon double bond. Because the electronic structure of the carbon–carbon double bond remains essentially the same in all molecules where it occurs, its chemical reactivity also remains the same. Ethylene, the simplest compound with a double bond, undergoes reactions that are remarkably similar to those of cholesterol, a much more complicated molecule that also contains a double bond. Both, for example, react with bromine to give products in which a bromine atom has added to each of the double-bond carbons (Figure 3.1).

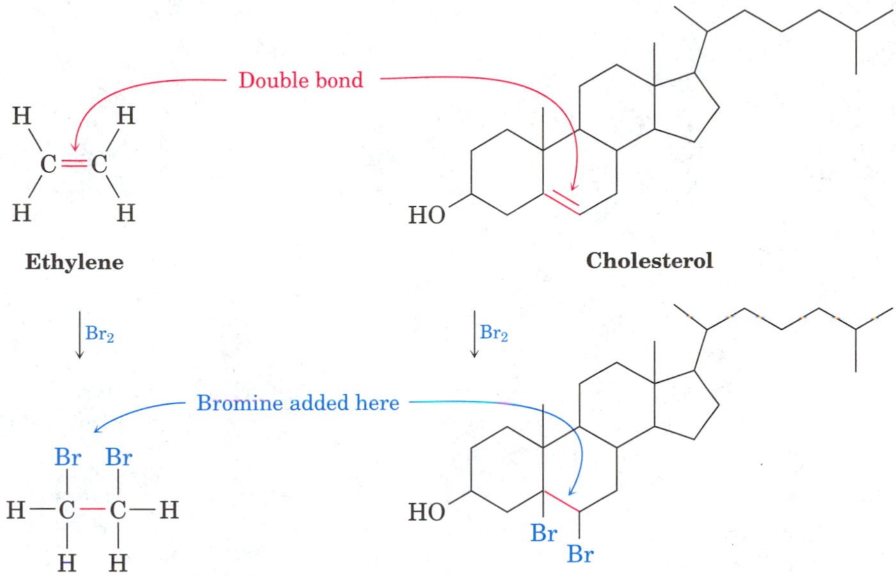

Figure 3.1 The reactions of ethylene and cholesterol with bromine. In both cases, bromine reacts with the C=C double-bond functional group in exactly the same way. The size and nature of the remainder of the molecule are unimportant.

The example shown in Figure 3.1 is typical: *The chemistry of every organic molecule, regardless of size and complexity, is determined by the functional groups it contains*. Table 3.1 lists many of the common functional groups and gives simple examples of their occurrence. Look carefully at this table to see the many types of functional groups found in organic compounds. Some functional groups, such as those in alkenes, alkynes, and aromatic rings, have only carbon–carbon double or triple bonds; others have halogen; and still others have oxygen, nitrogen, or sulfur. Much of the chemistry you'll be studying in the remainder of this book is the chemistry of these functional groups.

Table 3.1 Structures of Some Common Functional Groups

Family name	*Functional group structure*[a]	*Simple example*	*Name ending*
Alkane	(Contains only C—H and C—C single bonds)	CH_3CH_3	*-ane* Ethane
Alkene	$>C{=}C<$	$H_2C{=}CH_2$	*-ene* Ethene (Ethylene)
Alkyne	—C≡C—	H—C≡C—H	*-yne* Ethyne (Acetylene)
Arene	Six-membered ring of alternating C=C and C—C bonds, each C bearing one substituent bond	C_6H_6 (six-membered ring of alternating C=C and C—C bonds, one H on each C)	None Benzene
Halide	—C—X: (X = F, Cl, Br, I)	H_3C—Cl	None Chloromethane
Alcohol	—C—O—H	H_3C—O—H	*-ol* Methanol
Ether	—C—O—C—	H_3C—O—CH_3	*ether* Dimethyl ether
Amine	—C—N—H (with H on N), —C—N—H, —C—N—	H_3C—NH_2	*-amine* Methylamine
Nitrile	—C—C≡N:	H_3C—C≡N	*-nitrile* Ethanenitrile (Acetonitrile)
Nitro	—C—N^+(=O:)—O^-	H_3C—N^+(=O)—O^-	None Nitromethane
Sulfide	—C—S—C—	H_3C—S—CH_3	*sulfide* Dimethyl sulfide
Sulfoxide	—C—S^+(O^-)—C—	H_3C—S^+(O^-)—CH_3	*sulfoxide* Dimethyl sulfoxide

Family name	*Functional group structure*[a]	*Simple example*	*Name ending*
Sulfone	$-C-S^{2+}(:\ddot{O}:^-)_2-C-$	$H_3C-S^{2+}(O^-)_2-CH_3$	*sulfone* Dimethyl sulfone
Thiol	$-C-\ddot{S}-H$	H_3C-SH	*-thiol* Methanethiol
Carbonyl, $-C(=O)-$			
Aldehyde	$-C-C(=O)-H$	$H_3C-C(=O)-H$	*-al* Ethanal (Acetaldehyde)
Ketone	$-C-C(=O)-C-$	$H_3C-C(=O)-CH_3$	*-one* Propanone (Acetone)
Carboxylic acid	$-C-C(=O)-\ddot{O}H$	$H_3C-C(=O)-OH$	*-oic acid* Ethanoic acid (Acetic acid)
Ester	$-C-C(=O)-\ddot{O}-C-$	$H_3C-C(=O)-O-CH_3$	*-oate* Methyl ethanoate (Methyl acetate)
Amide	$-C-C(=O)-\ddot{N}H_2$ $-C-C(=O)-\ddot{N}(-)-H$ $-C-C(=O)-\ddot{N}(-)-$	$H_3C-C(=O)-NH_2$	*-amide* Ethanamide (Acetamide)
Carboxylic acid chloride	$-C-C(=O)-Cl$	$H_3C-C(=O)-Cl$	*-oyl chloride* Ethanoyl chloride (Acetyl chloride)
Carboxylic acid anhydride	$-C-C(=O)-\ddot{O}-C(=O)-C-$	$H_3C-C(=O)-O-C(=O)-CH_3$	*-oic anhydride* Ethanoic anhydride (Acetic anhydride)

[a]The bonds whose connections aren't specified are assumed to be attached to carbon or hydrogen atoms in the rest of the molecule.

It's a good idea at this point to memorize the structures of the functional groups shown in Table 3.1 so that they'll be familiar when you see them again. They can be grouped into several categories, as described below.

Functional Groups with Carbon–Carbon Multiple Bonds

Alkenes, alkynes, and arenes (aromatic compounds) all contain carbon–carbon multiple bonds. Alkenes have a double bond, alkynes have a triple bond, and aromatic rings have three alternating double and single bonds in a six-membered ring of carbon atoms. Because of their structural similarities, these compounds also have chemical similarities.

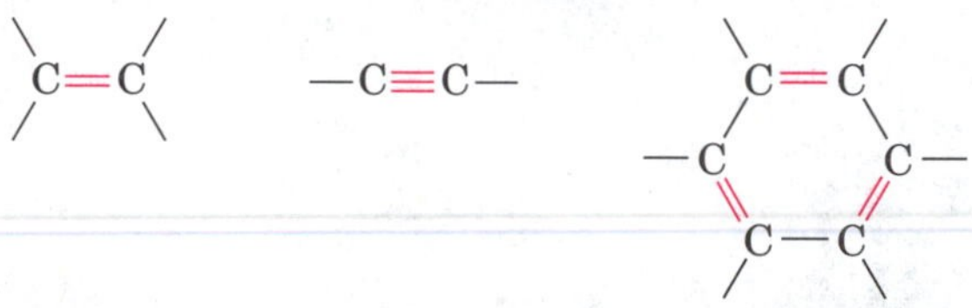

Alkene **Alkyne** **Arene (aromatic ring)**

Functional Groups with Carbon Singly Bonded to an Electronegative Atom

Alkyl halides, alcohols, ethers, amines, thiols, and sulfides all have a carbon atom singly bonded to an electronegative atom. Alkyl halides have a carbon atom bonded to halogen, alcohols have a carbon atom bonded to the oxygen of a hydroxyl group (–OH), ethers have two carbon atoms bonded to the same oxygen, amines have a carbon atom bonded to a nitrogen, thiols have a carbon atom bonded to an –SH group, and sulfides have two carbon atoms bonded to the same sulfur. In all cases, the bonds are polar, with the carbon atom bearing a partial positive charge ($\delta+$) and the electronegative atom bearing a partial negative charge ($\delta-$).

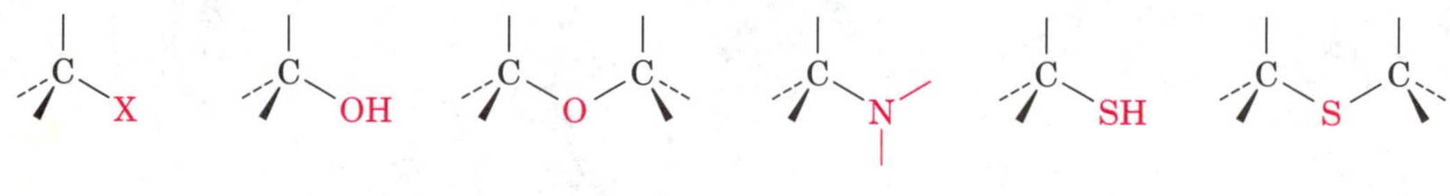

Alkyl halide **Alcohol** **Ether** **Amine** **Thiol** **Sulfide**

Functional Groups with a Carbon–Oxygen Double Bond (Carbonyl Groups)

Note particularly in Table 3.1 the different families of compounds that contain the **carbonyl group, C=O** (pronounced car-bo-**neel**). Carbon–oxygen double bonds are present in some of the most important compounds in organic chemistry. These compounds are similar in many respects but differ depending on the identity of the atoms bonded to the carbonyl-group carbon. Aldehydes have a hydrogen bonded to the C=O, ketones have two carbons bonded to the C=O, carboxylic acids have an –OH group bonded to the C=O, esters have an ether-like oxygen bonded to the C=O, amides have an amine-like nitrogen bonded to the C=O, acid chlorides have a chlorine bonded to the C=O, and so on.

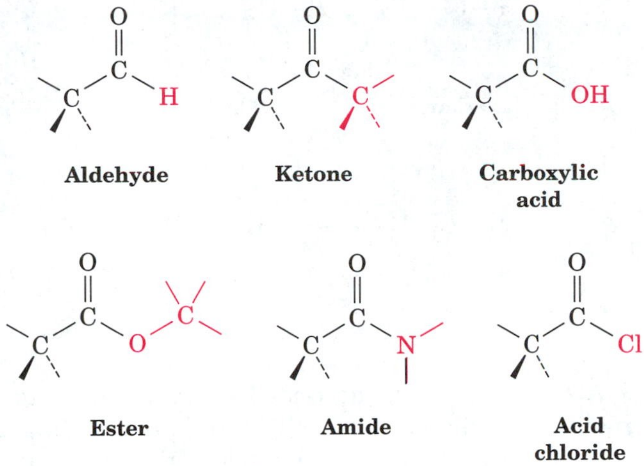

PROBLEM

3.1 Identify the functional groups in each of the following molecules:

(a) 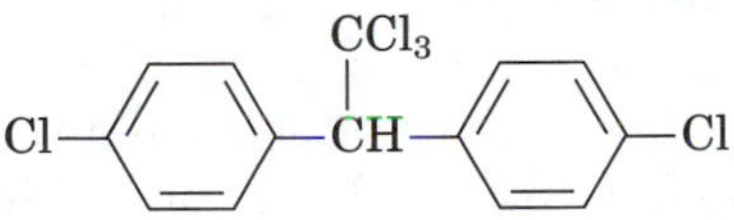

DDT

(b) 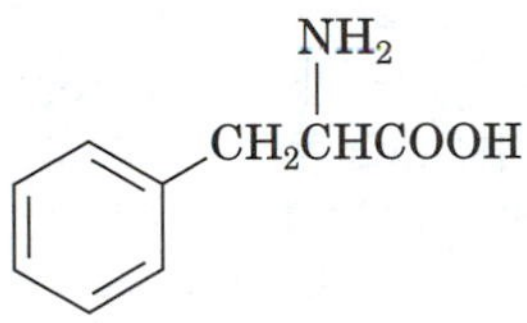

Phenylalanine

(c) 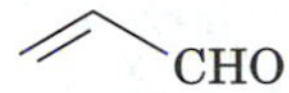

Acrolein

(d) 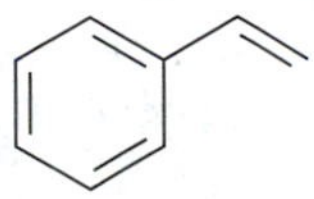

Styrene

PROBLEM

3.2 Propose structures for simple molecules that contain these functional groups:
(a) Alcohol (b) Aromatic ring (c) Carboxylic acid
(d) Amine (e) Both ketone and amine (f) Two double bonds

3.2 Alkanes and Alkane Isomers

We saw in Section 1.9 that the carbon–carbon single bond in ethane results from σ (head-on) overlap of carbon sp^3 orbitals. If we imagine joining three, four, five, or even more carbon atoms by carbon–carbon single bonds, we can generate the large family of molecules called **alkanes**.

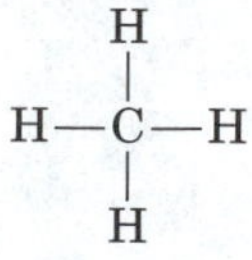

Methane

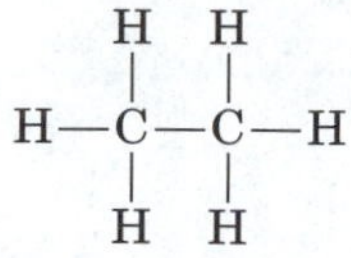

Ethane

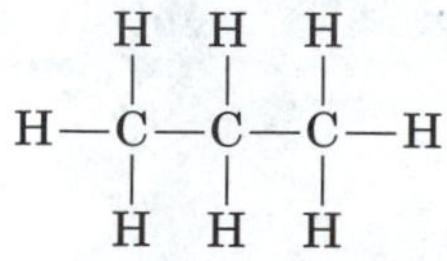

Propane

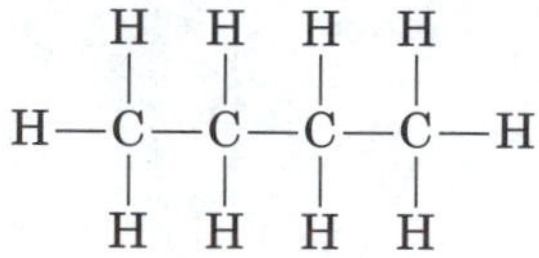

Butane

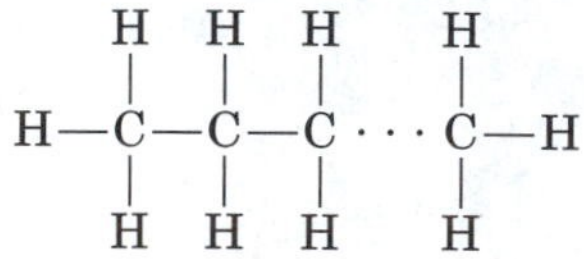

C_nH_{2n+2}

Alkanes are often described as *saturated hydrocarbons*—**hydrocarbons** because they contain only carbon and hydrogen; **saturated** because they have only C–C and C–H single bonds and thus contain the maximum possible number of hydrogens per carbon. They have the general formula C_nH_{2n+2}, where n is any integer. Alkanes are also occasionally referred to as **aliphatic** compounds, a name derived from the Greek *aleiphas*, meaning "fat." We'll see later that animal fats contain long carbon chains similar to alkanes.

Think about the ways that carbon and hydrogen might combine to make alkanes. With one carbon and four hydrogens, only one structure is possible: methane, CH_4. Similarly, there is only one possible combination of two carbons with six hydrogens (ethane, CH_3CH_3) and only one possible combination of three carbons with eight hydrogens (propane, $CH_3CH_2CH_3$). If larger numbers of carbons and hydrogens combine, however, more than one kind of molecule can result. For example, there are *two* molecules with the formula C_4H_{10}: The four carbons can be in a row (butane), or they can branch (isobutane). Similarly, there are three C_5H_{12} molecules, and so on for larger alkanes.

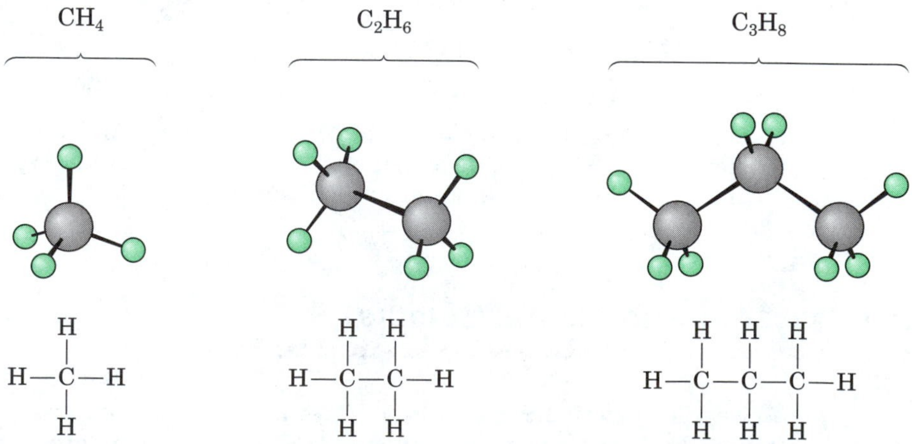

Methane **Ethane** **Propane**

C_4H_{10}

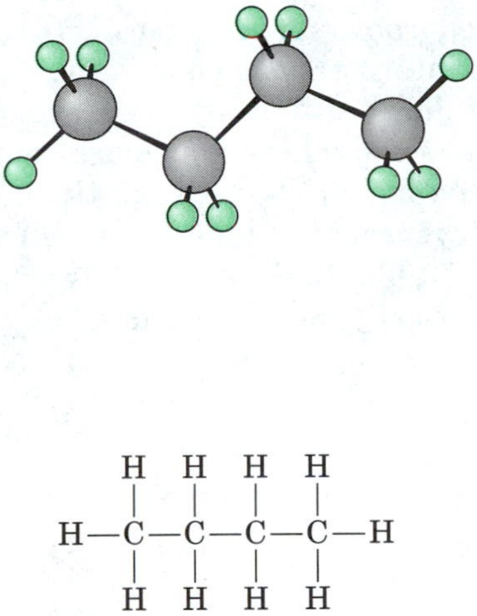

Butane

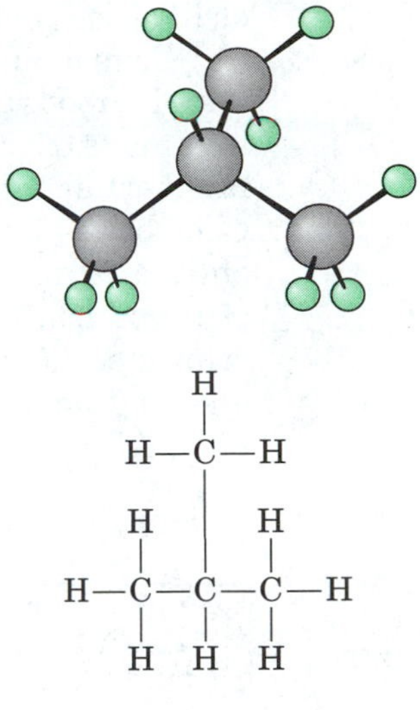

Isobutane
(2-Methylpropane)

C_5H_{12}

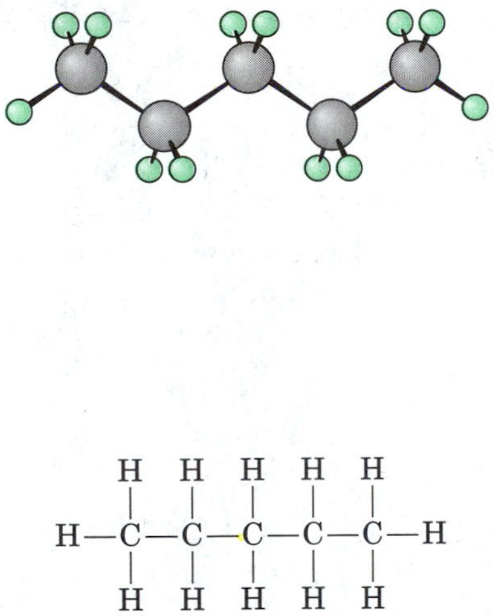

Pentane

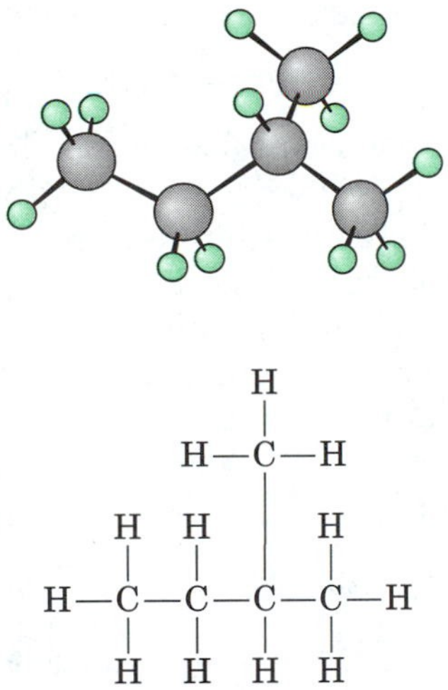

2-Methylbutane

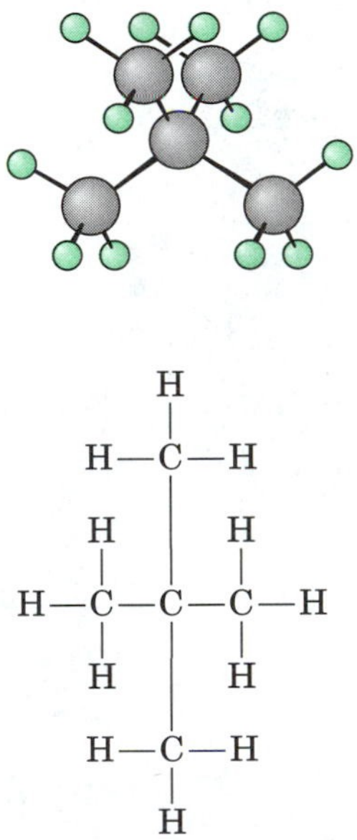

2,2-Dimethylpropane

Compounds like butane, whose carbons are connected in a row, are called **straight-chain alkanes**, or **normal alkanes**. On the other hand, compounds like 2-methylpropane (isobutane), whose carbon chains branch, are called **branched-chain alkanes**. The difference between the two is that you can draw a line connecting all the carbons of a straight-chain alkane without retracing your path or lifting your pencil from the paper. For a branched-chain alkane, however, you either have to retrace your path or lift your pencil from the paper to draw a line connecting all the carbons.

Compounds like the two C_4H_{10} molecules, which have the same formula but different structures, are called *isomers*, from Greek *isos* + *meros* meaning "made of the same parts." **Isomers** are compounds that have the same numbers and kinds of atoms but differ in the way the atoms are arranged. Compounds like butane and isobutane, whose atoms are connected differently, are called **constitutional isomers**. We'll see shortly that other kinds of isomerism are also possible, even among compounds whose atoms are connected in the same order. As Table 3.2 shows, the number of possible alkane isomers increases dramatically as the number of carbon atoms increases.

Table 3.2 Number of Alkane Isomers

Formula	*Number of Isomers*	*Formula*	*Number of Isomers*
C_6H_{14}	5	$C_{10}H_{22}$	75
C_7H_{16}	9	$C_{15}H_{32}$	4,347
C_8H_{18}	18	$C_{20}H_{42}$	366,319
C_9H_{20}	35	$C_{30}H_{62}$	4,111,846,763

Constitutional isomerism is not limited to alkanes—it occurs widely throughout organic chemistry. Constitutional isomers may have different carbon skeletons (as in isobutane and butane), different functional groups (as in ethyl alcohol and dimethyl ether), or different locations of a functional group along the chain (as in isopropylamine and propylamine). Regardless of the reason for the isomerism, constitutional isomers always have the same formula but a different connection of atoms.

Different carbon skeletons: $CH_3CH(CH_3)CH_3$ **2-Methylpropane (Isobutane)** and $CH_3CH_2CH_2CH_3$ **Butane**

Different functional groups: CH_3CH_2OH **Ethyl alcohol** and CH_3OCH_3 **Dimethyl ether**

Different position of functional groups: $CH_3CH(NH_2)CH_3$ **Isopropylamine** and $CH_3CH_2CH_2NH_2$ **Propylamine**

A given alkane can be arbitrarily drawn in many ways. For example, the straight-chain, four-carbon alkane called butane can be represented by any of the structures shown in Figure 3.2. These structures don't imply any particular three-dimensional geometry for butane; they only indicate the connections among its atoms. In practice, we soon tire of drawing all the bonds in a molecule and usually refer to butane by the condensed structure, $CH_3CH_2CH_2CH_3$, or even more simply as n-C_4H_{10}, where n signifies *normal*, straight-chain butane.

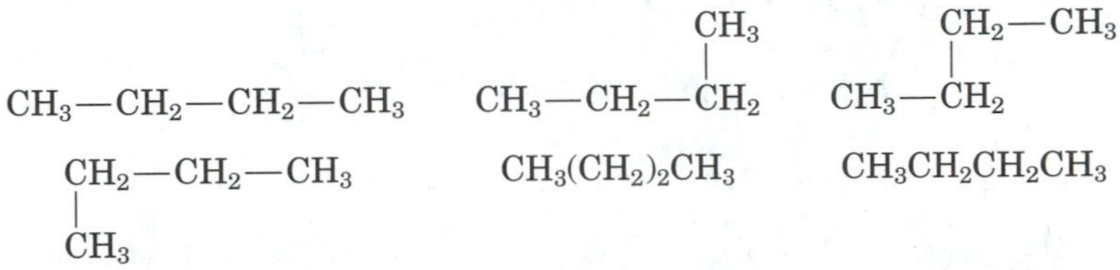

Figure 3.2 Some representations of butane, C_4H_{10}. The molecule is the same regardless of how it's drawn. These structures imply only that butane has a continuous chain of four carbon atoms.

Straight-chain alkanes are named according to the number of carbon atoms in the chain, as shown in Table 3.3. With the exception of the first four compounds—methane, ethane, propane, and butane—whose names have historical roots, the alkanes are named based on Greek numbers according to how many carbons the molecule has. The suffix *-ane* is added to the end of each name to indicate that the molecule identified is an alkane. Thus, pent*ane* is the five-carbon alkane, hex*ane* is the six-carbon alkane, and so on. The names of at least the first ten should be memorized.

Table 3.3 Names of Straight-Chain Alkanes

Number of Carbons (n)	*Name*	*Formula* (C_nH_{2n+2})	*Number of Carbons (n)*	*Name*	*Formula* (C_nH_{2n+2})
1	Methane	CH_4	9	Nonane	C_9H_{20}
2	Ethane	C_2H_6	10	Decane	$C_{10}H_{22}$
3	Propane	C_3H_8	11	Undecane	$C_{11}H_{24}$
4	Butane	C_4H_{10}	12	Dodecane	$C_{12}H_{26}$
5	Pentane	C_5H_{12}	13	Tridecane	$C_{13}H_{28}$
6	Hexane	C_6H_{14}	20	Icosane	$C_{20}H_{42}$
7	Heptane	C_7H_{16}	21	Henicosane	$C_{21}H_{44}$
8	Octane	C_8H_{18}	30	Triacontane	$C_{30}H_{62}$

PRACTICE PROBLEM

Propose structures for two isomers with the formula C_2H_6O.

Solution We know that carbon forms four bonds, oxygen forms two, and hydrogen forms one. Putting the pieces together yields two isomeric structures:

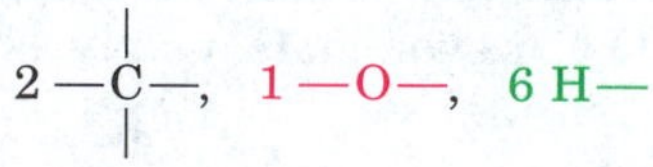

give

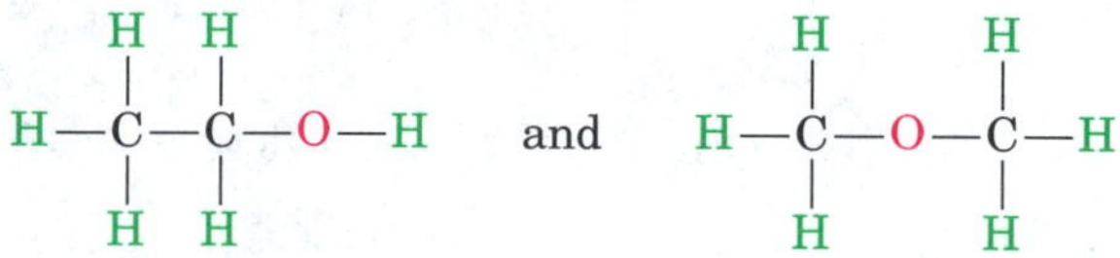

PROBLEM

3.3 Draw structures of the five isomers of C_6H_{14}.

PROBLEM

3.4 There are seven constitutional isomers with the formula $C_4H_{10}O$. Draw as many as you can.

PROBLEM

3.5 Propose structures that meet the following descriptions:
(a) Two isomeric esters with formula $C_5H_{10}O_2$
(b) Two isomeric nitriles with formula C_4H_7N

PROBLEM

3.6 How many isomers are there with the following structures?
(a) Alcohols with formula C_3H_8O
(b) Bromoalkanes with formula C_4H_9Br

3.3 Alkyl Groups

If a hydrogen atom is removed from an alkane, the partial structure that remains is called an **alkyl group**. Alkyl groups are named by replacing the *-ane* ending of the parent alkane with an *-yl* ending. For example, removal of a hydrogen from methane, CH_4, generates a *methyl* group, $-CH_3$, and removal of a hydrogen from ethane, CH_3CH_3, generates an *ethyl* group, $-CH_2CH_3$. Similarly, removal of a hydrogen atom from the end carbon of any *n*-alkane gives the series of straight-chain *n*-alkyl groups shown in Table 3.4. Combining an alkyl group with any of the functional groups listed earlier makes it possible to generate and name many thousands of compounds. For example:

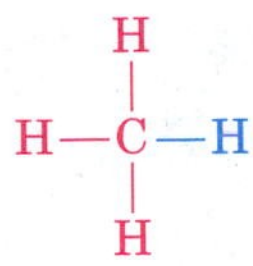

Methane

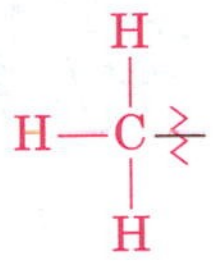

A methyl group[1]

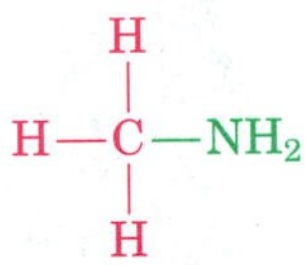

Methylamine

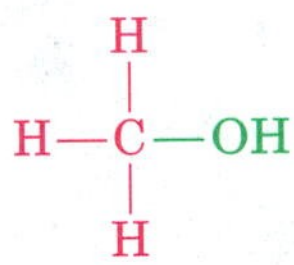

Methyl alcohol

Table 3.4 Some Straight-Chain Alkyl Groups

Alkane	*Name*	*Alkyl group*	*Name (abbreviation)*
CH_4	Methane	$-CH_3$	Methyl (Me)
CH_3CH_3	Ethane	$-CH_2CH_3$	Ethyl (Et)
$CH_3CH_2CH_3$	Propane	$-CH_2CH_2CH_3$	Propyl (Pr)
$CH_3CH_2CH_2CH_3$	Butane	$-CH_2CH_2CH_2CH_3$	Butyl (Bu)
$CH_3CH_2CH_2CH_2CH_3$	Pentane	$-CH_2CH_2CH_2CH_2CH_3$	Pentyl

Just as *n*-alkyl groups are generated by removing a hydrogen from an *end* carbon, branched alkyl groups are generated by removing a hydrogen atom from an *internal* carbon. Two 3-carbon alkyl groups and four 4-carbon alkyl groups are possible (Figure 3.3).

[1]The symbol $\wr$ is used throughout this book to indicate that the partial structure shown is bonded to another, unspecified group.

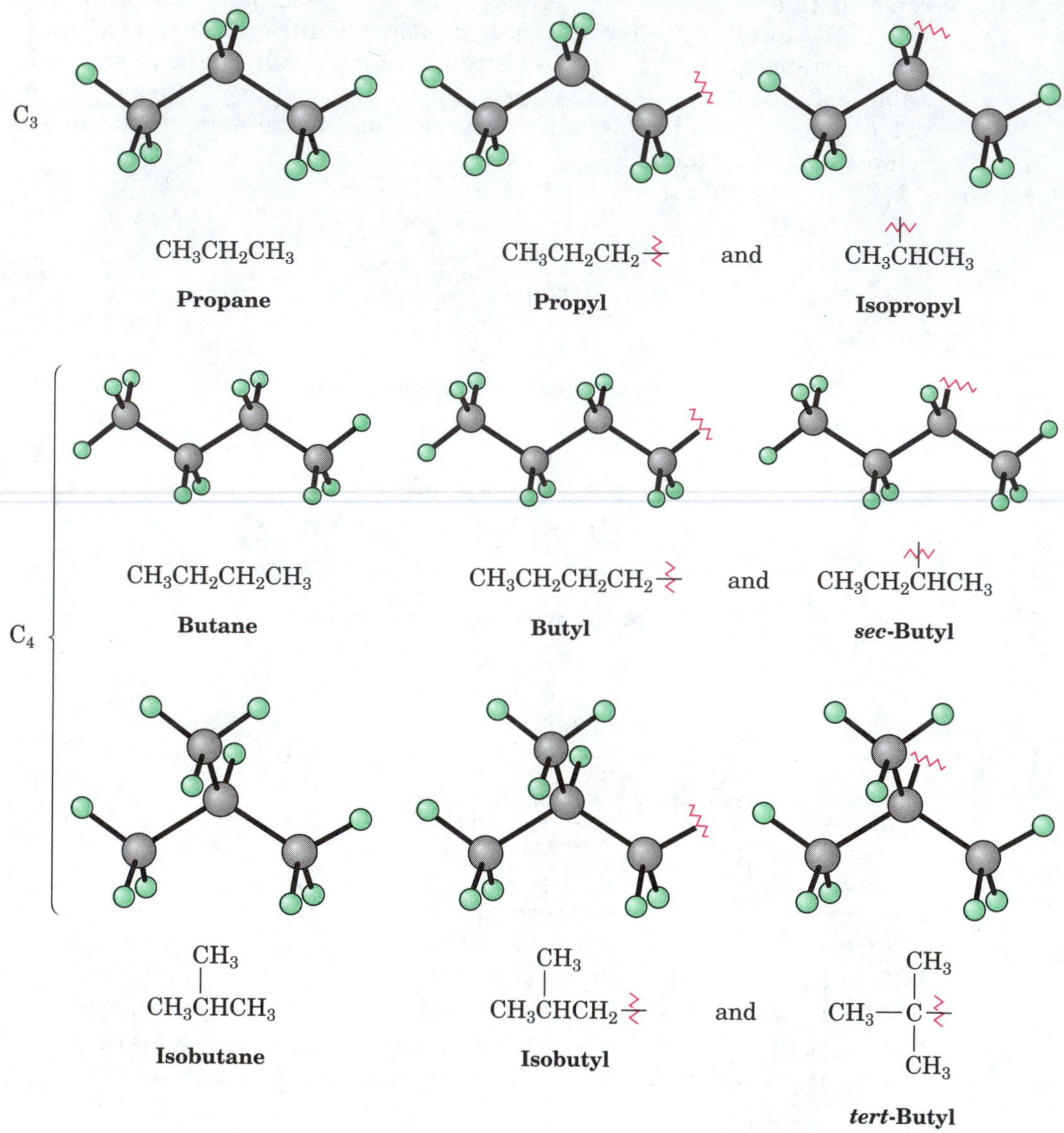

Figure 3.3 Generation of straight-chain and branched-chain alkyl groups from *n*-alkanes.

One further word of explanation about naming alkyl groups: The prefixes *sec* (for secondary) and *tert* (for tertiary) used for the C_4 alkyl groups in Figure 3.3 refer to the degree of alkyl substitution at the carbon atom in question. There are four possible degrees of alkyl substitution for carbon, denoted 1° (primary), 2° (secondary), 3° (tertiary), and 4° (quaternary):

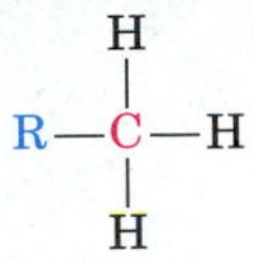

Primary carbon (1°) is bonded to one other carbon

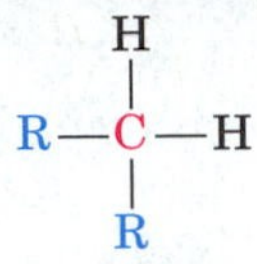

Secondary carbon (2°) is bonded to two other carbons

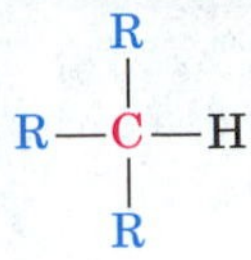

Tertiary carbon (3°) is bonded to three other carbons

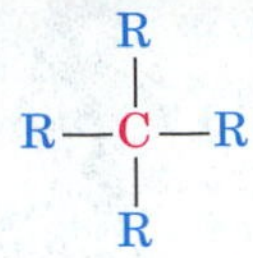

Quaternary carbon (4°) is bonded to four other carbons

The symbol R is used to represent a generalized alkyl group. The R group can be methyl, ethyl, propyl, or any of a multitude of others. You might think of R as representing the Rest of the molecule, which we aren't bothering to specify because it's not important. The terms *primary, secondary, tertiary,* and *quaternary* are routinely used in organic chemistry, and their meanings should become second nature. For example, if we were to say "The product of the reaction is a primary alcohol," we would be talking about the *general class of compounds* that has an alcohol functional group (–OH) bonded to a primary carbon atom, RCH_2OH:

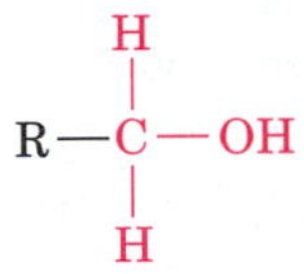

General class of primary alcohols, RCH_2OH

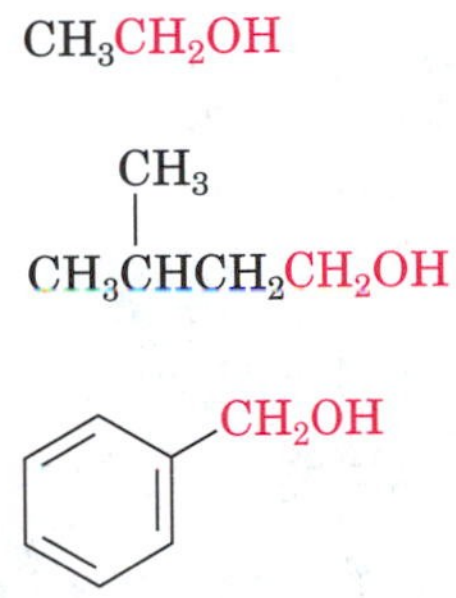

Some specific examples of primary alcohols, RCH_2OH

In addition, we also speak about hydrogen atoms as being primary, secondary, or tertiary. Primary hydrogen atoms are attached to primary carbons (RCH_3), secondary hydrogens are attached to secondary carbons (R_2CH_2), and tertiary hydrogens are attached to tertiary carbons (R_3CH).

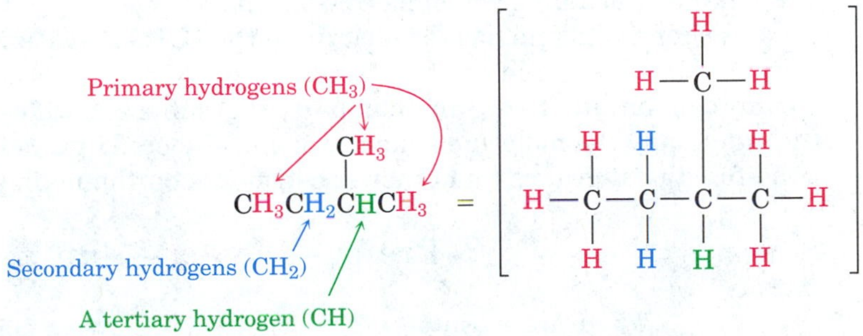

PROBLEM.

3.7 Draw the eight five-carbon alkyl groups (pentyl isomers).

PROBLEM.

3.8 Draw structures of alkanes that meet these descriptions:
(a) An alkane with two tertiary carbons
(b) An alkane that contains an isopropyl group
(c) An alkane that has one quaternary and one secondary carbon

PROBLEM.

3.9 Identify the carbon atoms in the following molecules as primary, secondary, tertiary, or quaternary:

(a)
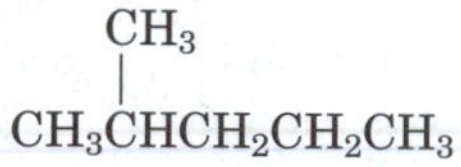
(b)
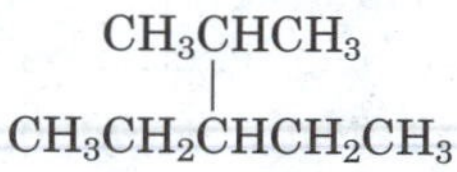
(c)
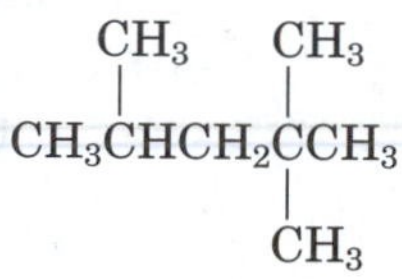

PROBLEM.

3.10 Identify the hydrogen atoms on the compounds shown in Problem 3.9 as primary, secondary, or tertiary.

3.4 Naming Alkanes

In earlier times, when relatively few pure organic chemicals were known, new compounds were named at the whim of their discoverer. Thus, urea (CH_4N_2O) is a crystalline substance isolated from urine; morphine ($C_{17}H_{19}NO_3$) is an analgesic (painkiller) named after Morpheus, the Greek god of dreams; and barbituric acid is a tranquilizing agent named by its discoverer in honor of his friend Barbara.

As the science of organic chemistry slowly grew in the nineteenth century, so too did the number of known compounds and the need for a systematic method of naming them. The system of nomenclature we'll use in this book is that devised by the International Union of Pure and Applied Chemistry (IUPAC, usually spoken as **eye**-you-pac).

A chemical name has three parts in the **IUPAC system**: prefix, parent, and suffix. The parent selects a main part of the molecule and tells how many carbon atoms are in that part; the suffix identifies the functional-group family the molecule belongs to; and the prefix gives the location(s) of the functional groups and other substituents on the main part.

Prefix—Parent—Suffix

Prefix	Parent	Suffix
↗ Where are substituents?	↑ How many carbons?	↖ What family?

As we cover new functional groups in later chapters, the applicable IUPAC rules of nomenclature will be given. In addition, Appendix A gives an overall view of organic nomenclature and shows how compounds that contain more than one functional group are named. For the present, let's see how to name branched-chain alkanes.

All but the most complex branched-chain alkanes can be named by following four steps. For a very few compounds, a fifth step is needed.

1. Find the parent hydrocarbon:
   (a) Find the *longest continuous chain of carbon atoms* present in the molecule, and use the name of that chain as the parent name. The longest chain may not always be apparent from the manner of writing; you may have to "turn corners."

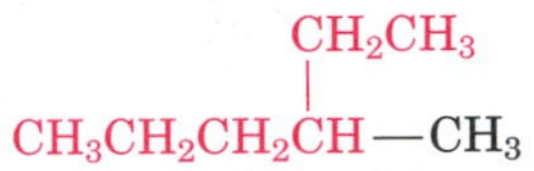

Named as a substituted hexane

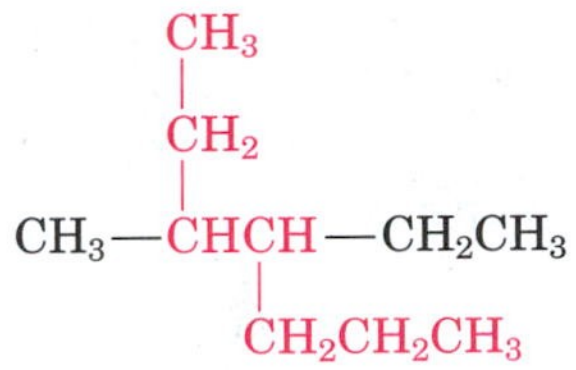

Named as a substituted heptane

   (b) If two different chains of equal length are present, choose the one with the larger number of branch points as the parent:

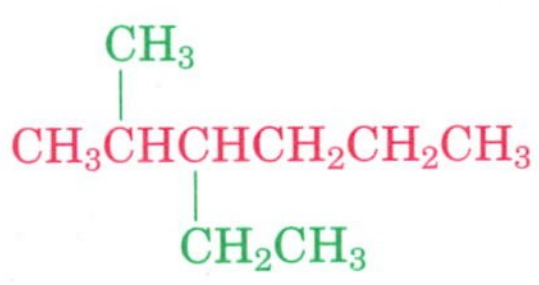

Named as a hexane with *two* substituents

NOT

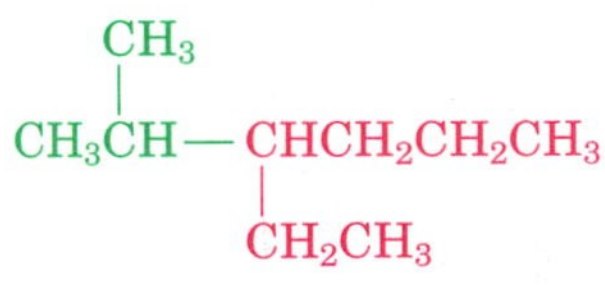

as a hexane with *one* substituent

2. Number the atoms in the main chain:
   (a) Beginning at the end *nearer the first branch point*, number each carbon atom in the parent chain:

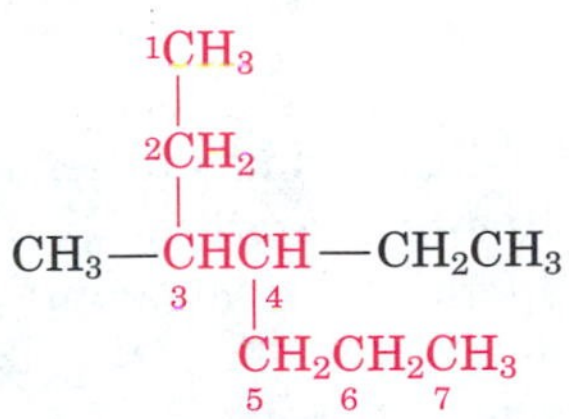

NOT

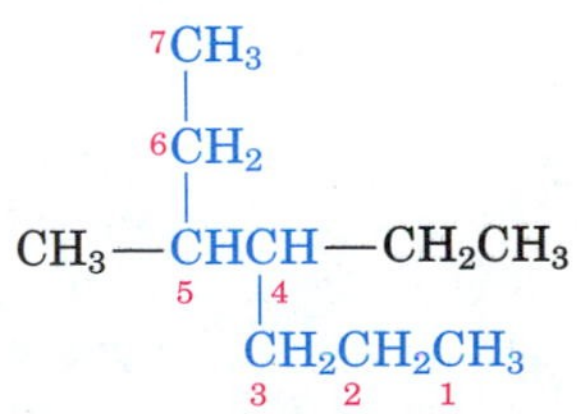

The first branch occurs at C3 in the proper system of numbering, not at C4.

(b) If there is branching an equal distance away from both ends of the parent chain, begin numbering at the end nearer the *second* branch point:

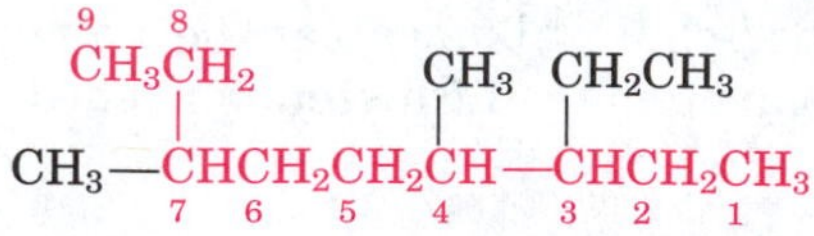

NOT

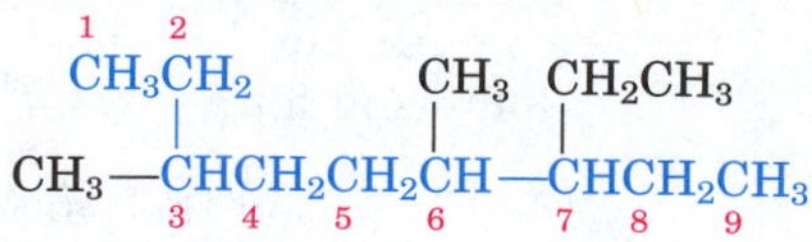

3. Identify and number the substituents:
   (a) Assign a number to each substituent according to its point of attachment to the main chain:

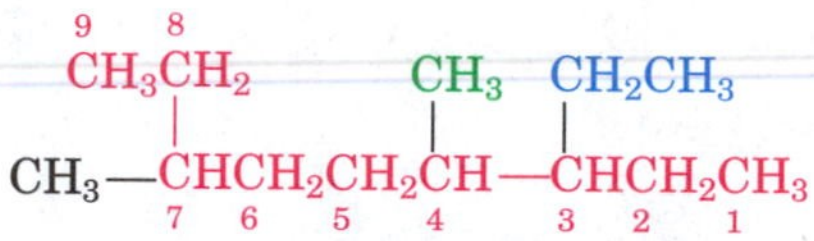

Named as a nonane

Substituents: On C3, CH_2CH_3 (3-ethyl)
On C4, CH_3 (4-methyl)
On C7, CH_3 (7-methyl)

   (b) If there are two substituents on the same carbon, assign them both the same number. There must be as many numbers in the name as there are substituents.

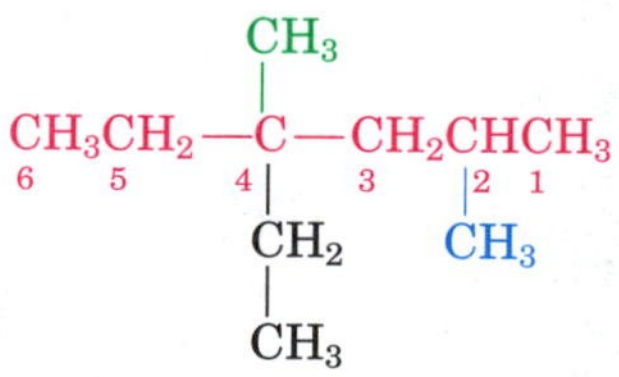

Named as a hexane

Substituents: On C2, CH_3 (2-methyl)
On C4, CH_3 (4-methyl)
On C4, CH_2CH_3 (4-ethyl)

4. Write the name as a single word, using hyphens to separate the different prefixes and using commas to separate numbers. If two or more different substituents are present, cite them in alphabetical order. If two or more identical substituents are present, use one of the prefixes *di-*, *tri-*, *tetra-*, and so forth. Don't use these prefixes for alphabetizing purposes, however.

Full names for some of the examples we have been using follow.

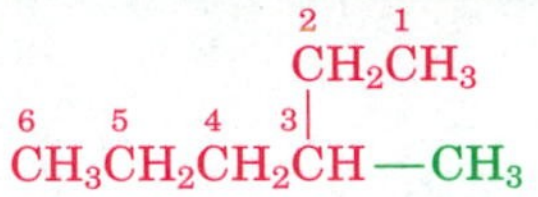

3-Methylhexane

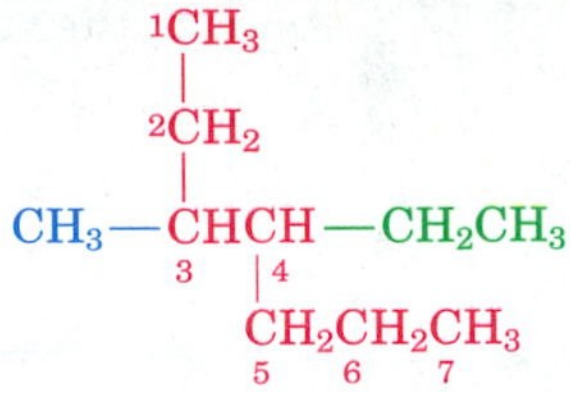

4-Ethyl-3-methylheptane

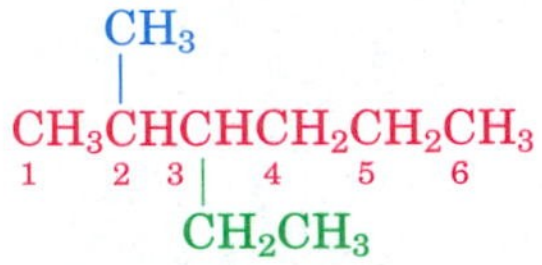

3-Ethyl-2-methylhexane

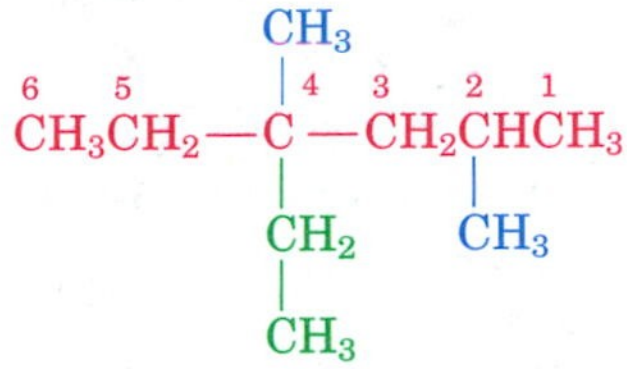

4-Ethyl-2,4-dimethylhexane

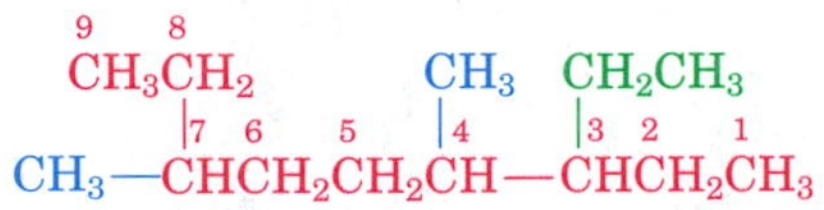

3-Ethyl-4,7-dimethylnonane

In some particularly complex cases, a fifth step is necessary. It occasionally happens that a substituent of the main chain has sub-branching:

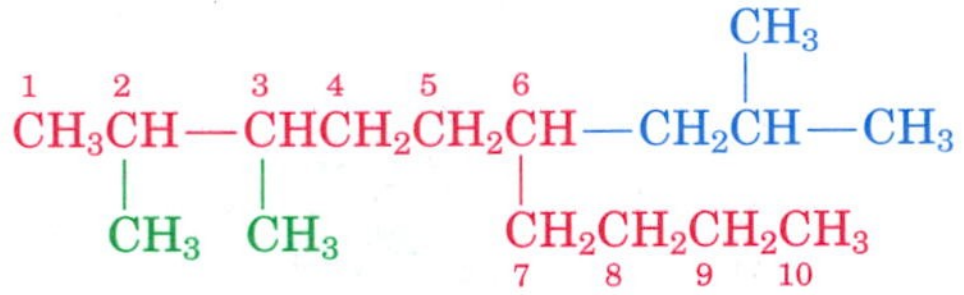

Named as a 2,3,6-trisubstituted decane

In this case, the substituent at C6 is a four-carbon unit with a sub-branch. To name the compound fully, the sub-branched substituent must first be named.

5. A complex substituent is named by applying the first four steps above, just as though the substituent were itself a compound. For the compound shown above, the complex substituent is a substituted propyl group:

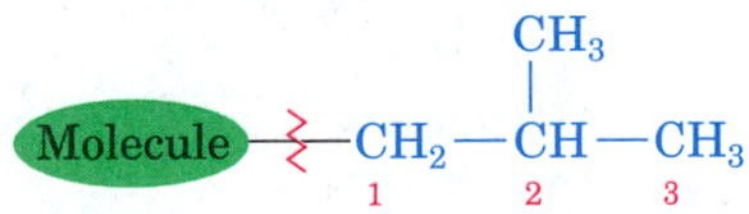

We begin numbering *at the point of attachment* to the main chain and find that the complex substituent is a 2-methylpropyl group. To avoid confusion, this substituent name is set off in parentheses when naming the complete molecule:

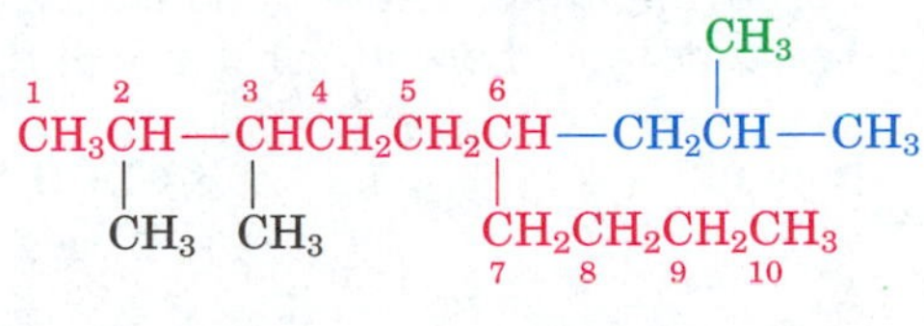

2,3-Dimethyl-6-(2-methylpropyl)decane

As a further example:

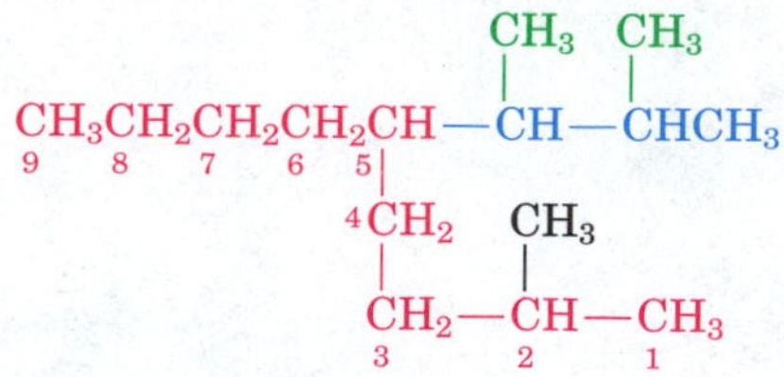

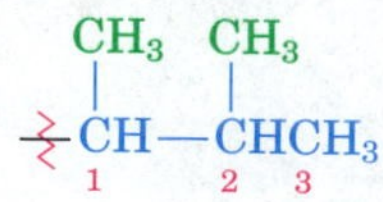

5-(1,2-Dimethyl propyl)-

2-Methyl-5-(1,2-dimethyl propyl)nonane

For historical reasons, some of the simpler branched-chain alkyl groups also have nonsystematic, or *common*, names, as noted earlier in Section 3.3.

1. Three-carbon alkyl group:

Isopropyl (*i*-Pr)

2. Four-carbon alkyl groups:

***sec*-Butyl (*sec*-Bu)**

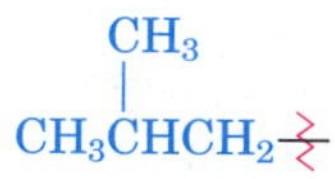

Isobutyl

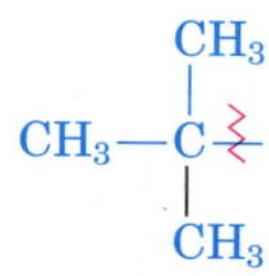

***tert*-Butyl (*t*-Butyl or *t*-Bu)**

3. Five-carbon alkyl groups:

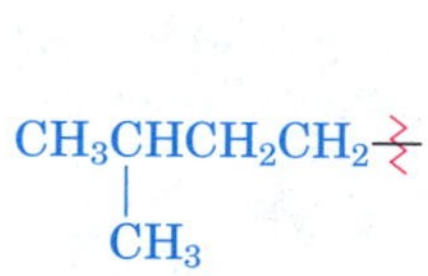

Isopentyl, also called isoamyl (*i*-amyl)

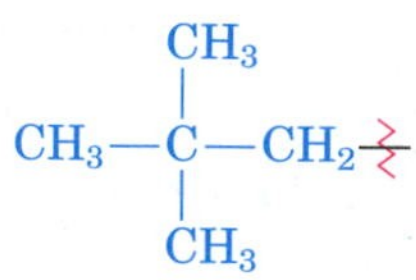

Neopentyl

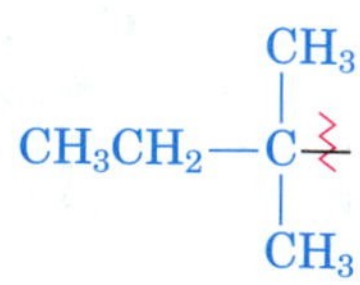

***tert*-Pentyl, also called *tert*-amyl (*t*-amyl)**

The common names of these simple alkyl groups are so well entrenched in the chemical literature that IUPAC rules make allowance for them. Thus, the following compound is properly named *either* 4-(1-methylethyl)heptane or 4-isopropylheptane. There is no choice but to memorize these common names; fortunately, there aren't many of them.

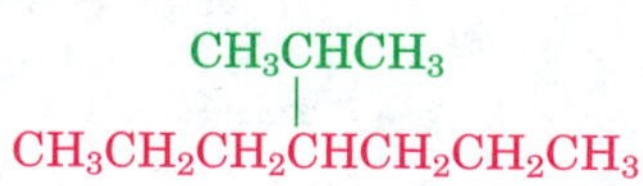

4-(1-Methylethyl) heptane or 4-Isopropyl heptane

When writing an alkane name, the prefix *iso-* is considered part of the alkyl group name for alphabetizing purposes, but the hyphenated prefixes *sec-* and *tert-* are not. Thus, isopropyl and isobutyl are listed alphabetically under *i*, but *sec*-butyl and *tert*-butyl are listed under *b*.

PRACTICE PROBLEM

What is the IUPAC name of this alkane?

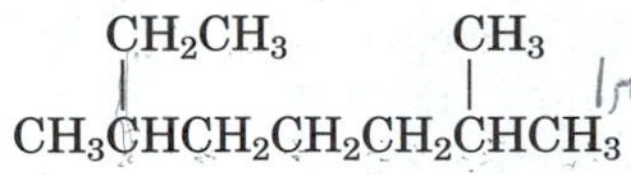

Solution The molecule has a chain of eight carbons (octane) with two methyl substituents. Numbering from the end nearer the first methyl substituent indicates that the methyls are at C2 and C6, giving the name 2,6-dimethyloctane.

PRACTICE PROBLEM

Draw the structure of 3-isopropyl-2-methylhexane.

Solution First, look at the parent name (hexane) and draw its carbon structure:

C—C—C—C—C—C **Hexane**

Next, find the substituents (3-isopropyl and 2-methyl), and place them on the proper carbons:

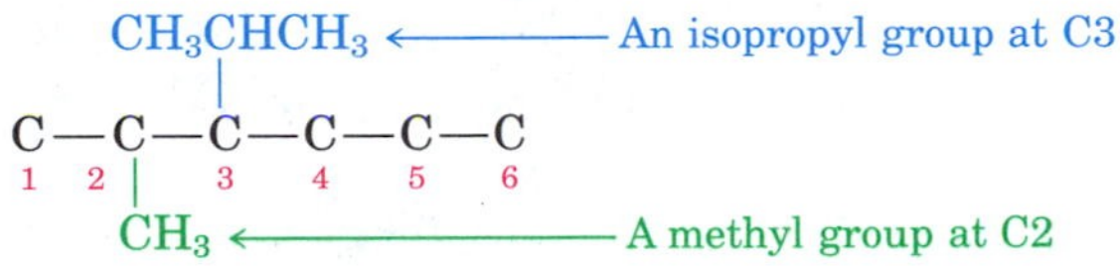

Finally, add hydrogens to complete the structure:

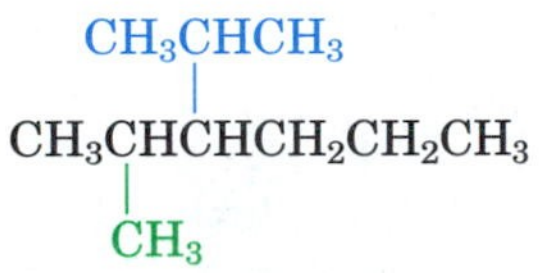

3-Isopropyl-2-methylhexane

PROBLEM

3.11 Give IUPAC names for these compounds:

(a) The three isomers of C_5H_{12}

(b) 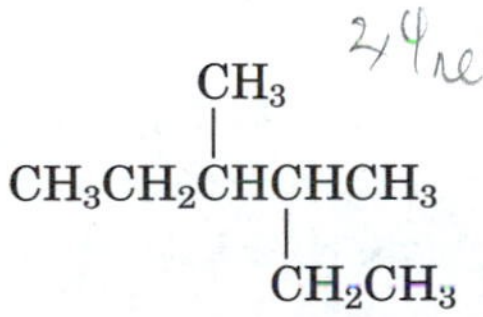

(c) 

(d) 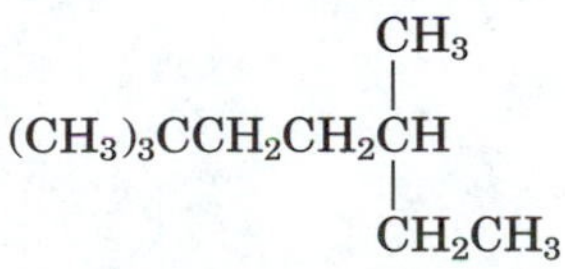

PROBLEM

3.12 Draw structures corresponding to the following IUPAC names:
(a) 3,4-Dimethylnonane
(b) 3-Ethyl-4,4-dimethylheptane
(c) 2,2-Dimethyl-4-propyloctane
(d) 2,2,4-Trimethylpentane

PROBLEM

3.13 The following names are incorrect. Draw the structures they represent, explain why the names are incorrect, and give correct names.
(a) 1,1-Dimethylpentane
(b) 3-Methyl-2-propylhexane
(c) 4,4-Dimethyl-3-ethylpentane
(d) 5-Ethyl-4-methylhexane
(e) 2,3-Methylhexane
(f) 3-Dimethylpentane

PROBLEM

3.14 Give systematic names for the eight five-carbon alkyl groups you drew in Problem 3.7.

3.5 Properties of Alkanes

Alkanes are sometimes referred to as **paraffins**, a word derived from the Latin *parum affinis*, meaning "slight affinity." This term aptly describes their behavior, for alkanes show little chemical affinity for other substances and are chemically inert to most laboratory reagents. Alkanes do, however, react with oxygen, chlorine, and a few other substances under appropriate conditions.

Reaction with oxygen occurs during combustion in an engine or furnace when the alkane is used as a fuel. Carbon dioxide and water are formed as products, and a large amount of heat is released. For example, methane (natural gas) reacts with oxygen according to the equation

$$CH_4 + 2\ O_2 \longrightarrow CO_2 + 2\ H_2O + 890\ \text{kJ/mol}\ (213\ \text{kcal/mol})$$

The reaction of an alkane with chlorine occurs when a mixture of the two is irradiated with ultraviolet light (denoted $h\nu$, where ν is the Greek letter nu). Depending on the relative amounts of the two reactants and on the time allowed, a sequential substitution of the alkane hydrogen atoms by chlorine occurs, leading to a mixture of chlorinated products. Methane, for example, reacts with chlorine to yield a mixture of CH_3Cl, CH_2Cl_2, $CHCl_3$, and CCl_4. We'll explore the details of this reaction in Section 5.3.

$$CH_4 + Cl_2 \longrightarrow CH_3Cl + HCl$$
$$\xrightarrow{Cl_2} CH_2Cl_2 + HCl$$
$$\xrightarrow{Cl_2} CHCl_3 + HCl$$
$$\xrightarrow{Cl_2} CCl_4 + HCl$$

Alkanes show regular increases in both boiling point and melting point as molecular weight increases (Figure 3.4), an effect that can be explained by the presence of weak **van der Waals forces** between molecules. These intermolecular forces, which operate only over very small distances, result from induced polarization of the electron clouds in molecules.

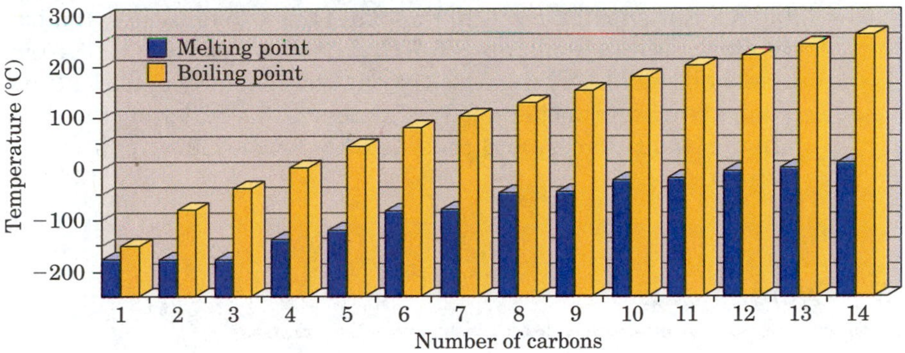

Figure 3.4 A plot of melting and boiling points versus number of carbons for the C_1–C_{14} alkanes. There is a regular increase with molecular size.

Although the average electron distribution in a nonpolar molecule like an alkane is uniform over time, the distribution at any given instant is nonuniform. One side of a molecule may, by chance, have a slight excess of electrons relative to the opposite side, giving the molecule a temporary dipole moment. This temporary dipole in one molecule causes a nearby molecule to adopt a temporarily opposite dipole, with the result that a tiny electrical attraction is induced between the two (Figure 3.5).

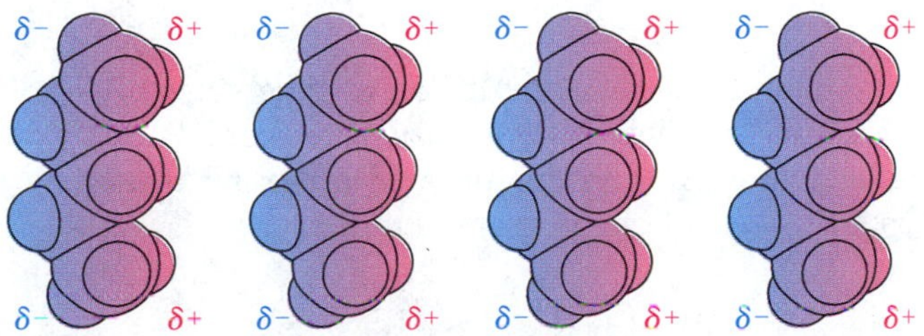

Figure 3.5 Attractive van der Waals forces are caused by temporary dipoles in molecules, as shown in these space-filling models of pentane.

Temporary molecular dipoles have a fleeting existence and are constantly changing, but the cumulative effect of an enormous number of them produces attractive forces sufficient to cause a substance to be in the liquid state. Only when sufficient energy is applied to overcome these forces does the liquid boil. As you might expect, van der Waals forces increase as molecule size increases, accounting for the higher boiling points of larger alkanes.

Another interesting effect seen in alkanes is that increased branching lowers the alkane's boiling point. Thus, pentane has no branches and boils at 36.1°C, isopentane (2-methylbutane) has one branch and boils at 27.85°C, and neopentane (2,2-dimethylpropane) has two branches and boils at 9.5°C. Similarly, octane boils at 125.7°C, whereas isooctane (2,2,4-trimethylpentane) boils at 99.3°C. Because they are more nearly spherical than straight-chain alkanes, branched-chain alkanes have smaller surface areas, smaller van der Waals forces, and consequently lower boiling points than straight-chain alkanes.

3.6 Cycloalkanes

We've discussed only open-chain alkanes up to this point, but chemists have known for over a century that compounds with *rings* of carbon atoms also exist. Such compounds are called **cycloalkanes** or **alicyclic compounds** (**ali**phatic **cyclic**). Since cycloalkanes consist of rings of $-CH_2-$ units, they have the general formula $(CH_2)_n$, or C_nH_{2n}, and are represented by polygons in skeletal drawings:

Cyclopropane **Cyclobutane** **Cyclopentane**

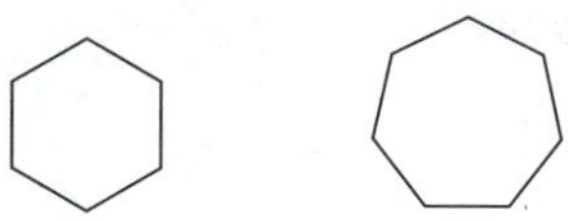

Cyclohexane **Cycloheptane**

Alicyclic compounds with many different ring sizes abound in nature. For example, chrysanthemic acid contains a three-membered (cyclopropane) ring. Various esters of chrysanthemic acid occur naturally as the active insecticidal constituents of chrysanthemum flowers.

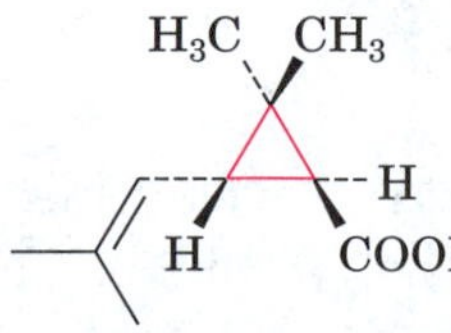

Chrysanthemic acid

Prostaglandins, such as PGE_1, contain a five-membered (cyclopentane) ring. Prostaglandins are potent hormones that control a wide variety of physiological functions in humans, including blood platelet aggregation, bronchial dilation, and inhibition of gastric secretions.

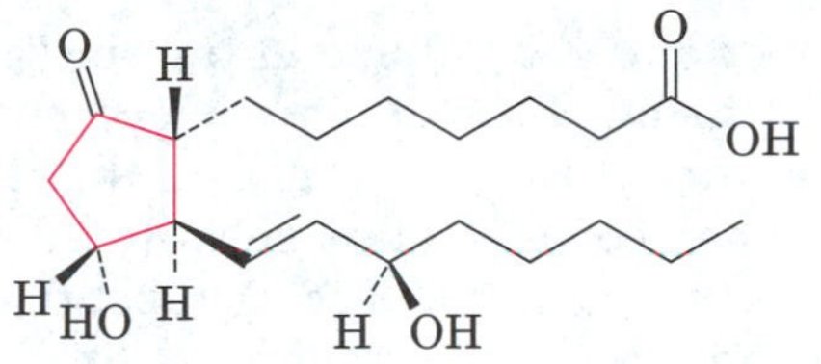

Prostaglandin E_1 (PGE_1)

Steroids, such as cortisone, contain four rings joined together—three six-membered (cyclohexane) and one five-membered (cyclopentane) ring:

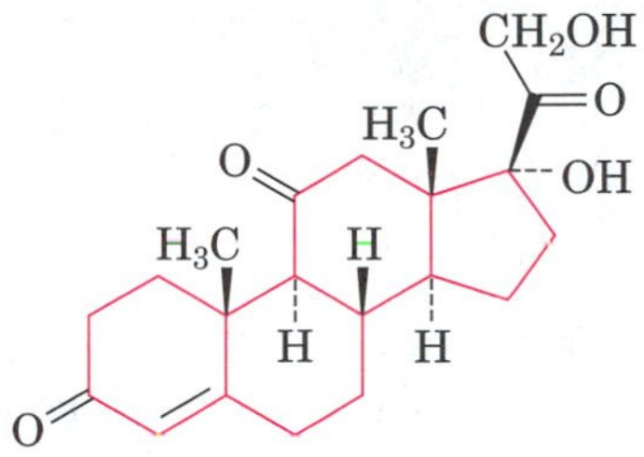

Cortisone

The melting points and boiling points of some simple unsubstituted cycloalkanes are shown in Figure 3.6. Melting points are affected irregularly by increasing molecular weight, probably because the different shapes of the various cycloalkanes cause differences in the efficiency with which molecules pack together in crystals. Boiling points, however, show a regular increase with molecular weight.

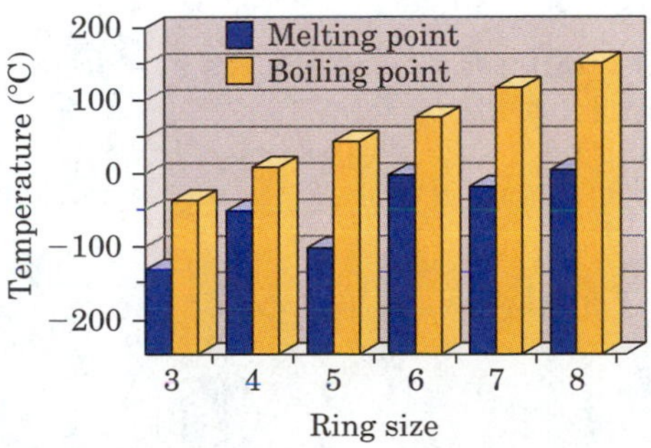

Figure 3.6 Melting points and boiling points for cycloalkanes, cyclo-$(CH_2)_n$.

3.7 Naming Cycloalkanes

Substituted cycloalkanes are named by rules similar to those given previously for open-chain alkanes. For most compounds, there are only two rules:

1. Count the number of carbon atoms in the ring and the number in the largest substituent. If the number of carbon atoms in the ring is equal to or greater than the number in the largest substituent, the compound is named as an alkyl-substituted cycloalkane. If the number of carbons atoms in the largest substituent is greater than the number in the ring, the compound is named as a cycloalkyl-substituted alkane. For example:

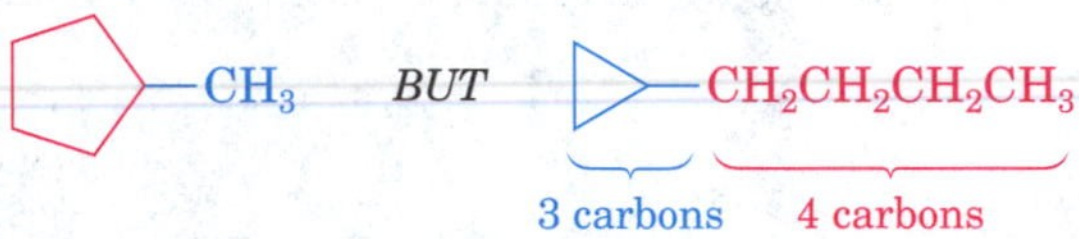

Methylcyclopropane **1-Cyclopropylbutane**

2. For alkyl-substituted cycloalkanes, start at a point of attachment and number the substituents on the ring so as to arrive at the lowest sum:

1,3-Dimethylcyclohexane (Sum: 1 + 3 = 4)

NOT

1,5-Dimethylcyclohexane (Sum: 1 + 5 = 6)

(a) When two or more different alkyl groups are present, number them by alphabetical priority:

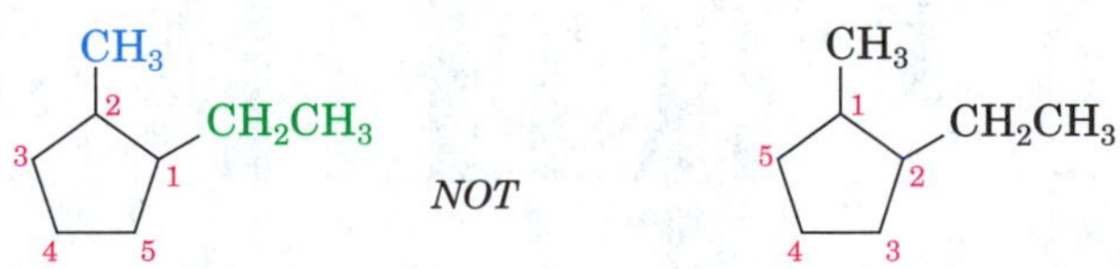

1-Ethyl-2-methylcyclopentane NOT **2-Ethyl-1-methylcyclopentane**

(b) Halogens, if present, are treated exactly like alkyl groups:

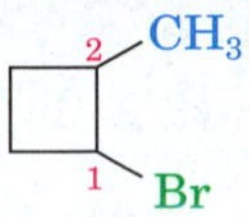

NOT

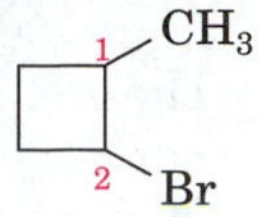

1-Bromo-2-methylcyclobutane — **2-Bromo-1-methylcyclobutane**

Some additional examples follow:

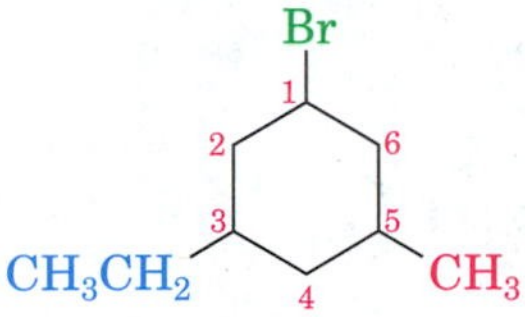

1-Bromo-3-ethyl-5-methylcyclohexane

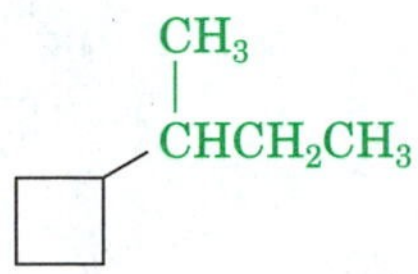

(1-Methylpropyl)cyclobutane (or *sec*-Butylcyclobutane)

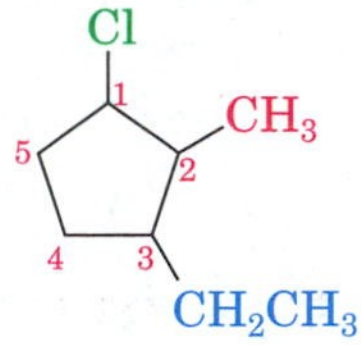

1-Chloro-3-ethyl-2-methylcyclopentane

PROBLEM

3.15 Give IUPAC names for the following cycloalkanes:

(a)

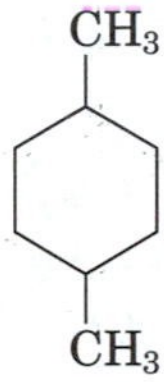

(b)

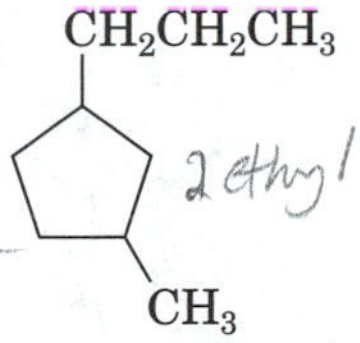

(c)

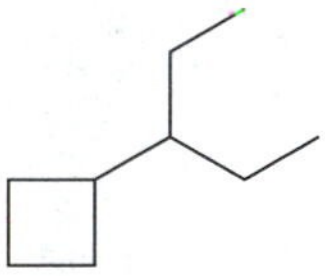

(d)

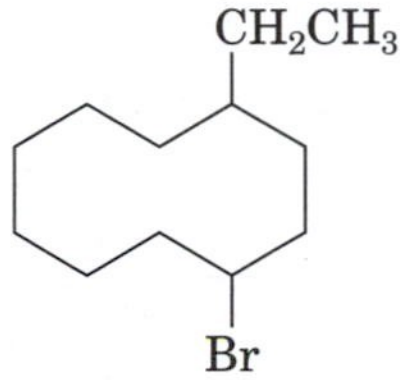

(e)

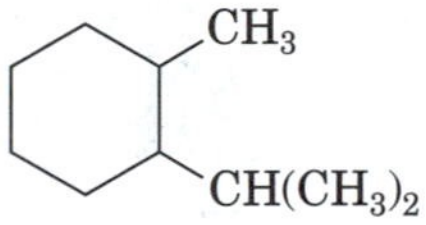

(f)

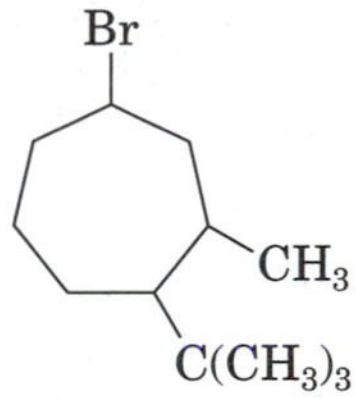

PROBLEM

3.16 Draw structures corresponding to the following IUPAC names:

(a) 1,1-Dimethylcyclooctane
(b) 3-Cyclobutylhexane
(c) 1,2-Dichlorocyclopentane
(d) 1,3-Dibromo-5-methylcyclohexane

3.8 Cis–Trans Isomerism in Cycloalkanes

In many respects, the chemistry of cycloalkanes is similar to that of open-chain, acyclic alkanes. Both classes of compounds are nonpolar and are chemically inert to most reagents. There are, however, some important differences.

One difference is that cycloalkanes are less flexible than their open-chain counterparts. To see what this means, think about the nature of carbon–carbon single bonds. We know from Section 1.7 that σ bonds are cylindrically symmetrical. In other words, the intersection of a plane cutting through a carbon–carbon single-bond orbital looks like a circle. Because of this cylindrical symmetry, *rotation* is possible around carbon–carbon bonds in open-chain molecules. In ethane, for example, rotation around the C–C bond occurs freely, interconverting different geometric arrangements of the hydrogens (Figure 3.7).

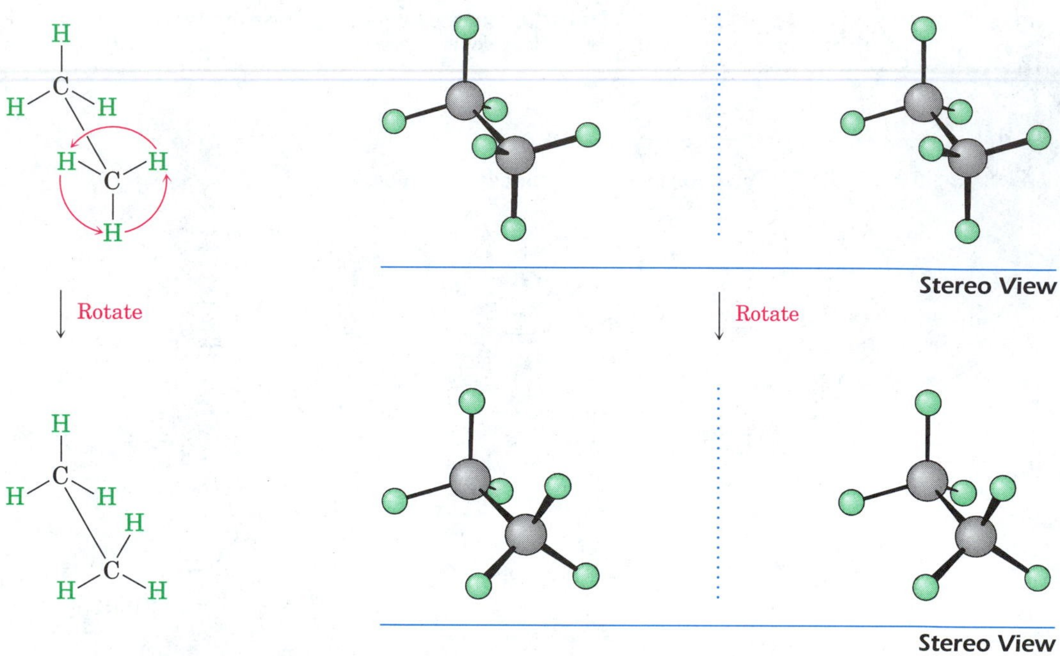

Figure 3.7 Free rotation occurs around the carbon–carbon single bond in ethane because of σ bond cylindrical symmetry.

In contrast to the free rotation around single bonds in open-chain alkanes, there is much less freedom in cycloalkanes because of their geometric constraints. Cyclopropane, for example, must be a rigid, planar molecule (three points define a plane). No bond rotation can take place around a cyclopropane carbon–carbon bond without breaking open the ring (Figure 3.8).

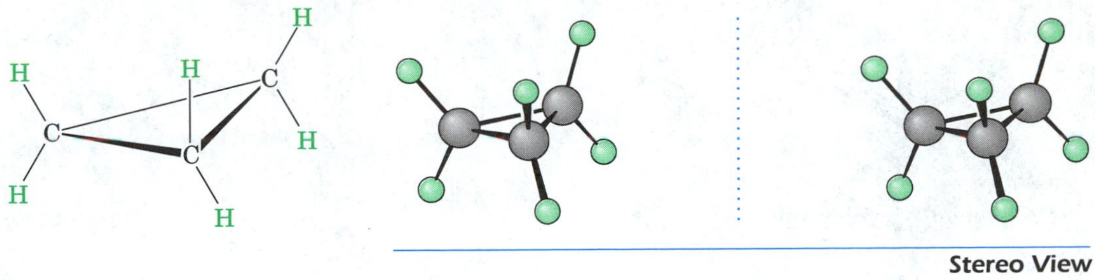

Figure 3.8 The structure of cyclopropane. No rotation is possible around the carbon–carbon bonds without breaking open the ring.

Larger cycloalkanes have increasingly more freedom, and the very large rings (C_{25} and up) are so floppy that they are nearly indistinguishable from open-chain alkanes. The common ring sizes (C_3, C_4, C_5, C_6, C_7), however, are severely restricted in their molecular motions.

Because of their cyclic structure, cycloalkanes have two sides, a “top” side and a “bottom” side, leading to the possibility of isomerism in substituted cycloalkanes. For example, there are two different 1,2-dimethylcyclopropane isomers, one with the two methyls on the same side of the ring, and one with the methyls on opposite sides (Figure 3.9, p. 98). Both isomers are stable compounds; neither can be converted into the other without breaking and reforming chemical bonds. Make molecular models to prove this to yourself.

Unlike the constitutional isomers butane and isobutane (see Figure 3.3), which have their atoms connected in a different order, the two 1,2-dimethylcyclopropanes have the *same* order of connection but differ in the spatial orientation of their atoms. Compounds that have their atoms connected in the same order but differ in three-dimensional orientation, are called **stereoisomers**.

Constitutional isomers (different connections between atoms)

$$CH_3{-}\underset{}{\overset{CH_3}{\overset{|}{C}H}}{-}CH_3 \quad \text{and} \quad CH_3{-}CH_2{-}CH_2{-}CH_3$$

Stereoisomers (same connections but different three-dimensional geometry)

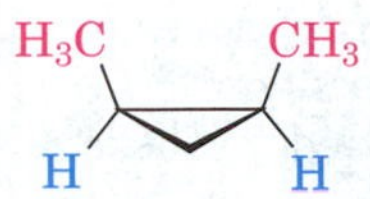

and

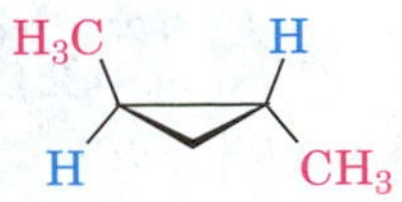

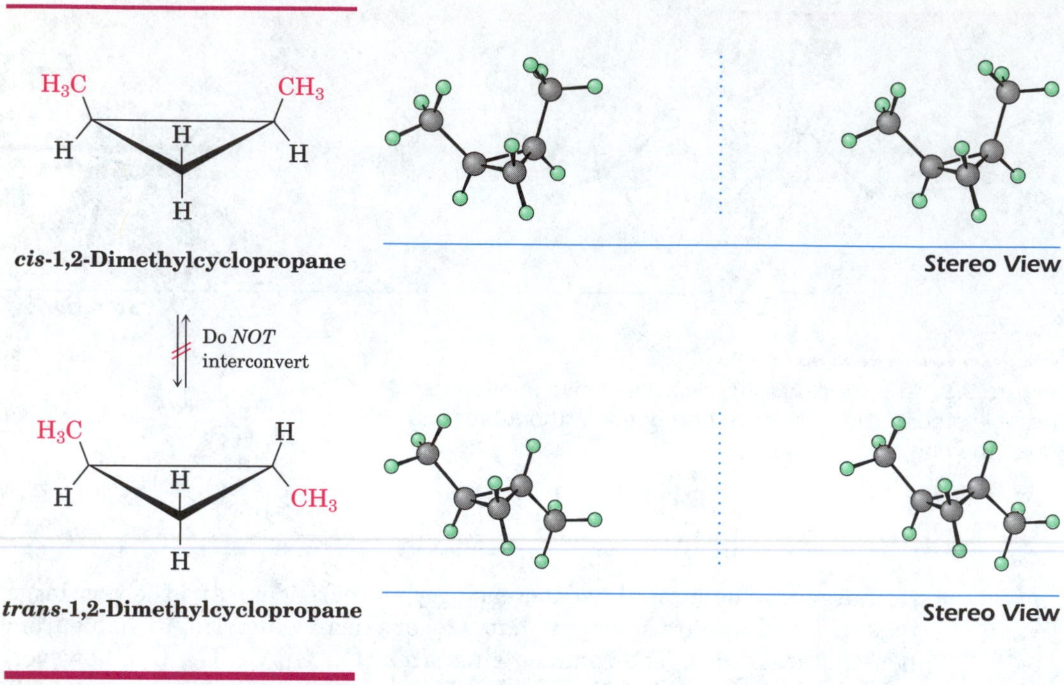

Figure 3.9 There are two different 1,2-dimethylcyclopropane isomers, one with the methyl groups on the same side of the ring and the other with the methyl groups on opposite sides of the ring.

The 1,2-dimethylcyclopropanes are special kinds of stereoisomers called **cis–trans isomers**. The prefixes *cis-* (Latin, "on the same side") and *trans-* (Latin, "across") are used to distinguish between them. Cis–trans isomerism is a common occurrence in substituted cycloalkanes.

cis-1,3-Dimethylcyclobutane

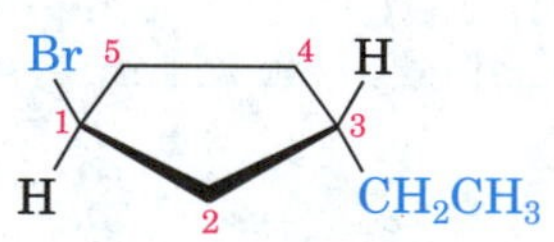

trans-1-Bromo-3-ethylcyclopentane

PROBLEM

3.17 Draw the structures of the following molecules:
(a) *trans*-1-Bromo-3-methylcyclohexane
(b) *cis*-1,2-Dimethylcyclopentane
(c) *trans*-1-*tert*-Butyl-2-ethylcyclohexane

INTERLUDE

Petroleum

The world's natural gas and petroleum deposits represent by far the largest source of alkanes. Laid down eons ago, these deposits are derived from the decomposition of plant and animal matter, primarily of marine origin. *Natural gas* consists chiefly of methane, but also contains ethane, propane, butane, and isobutane. *Petroleum* is a complex mixture of hydrocarbons that must be refined into fractions before it can be used.

Refining begins by distillation of crude oil into three principal cuts: straight-run gasoline (bp 30–200°C), kerosene (bp 175–300°C), and gas oil (bp 275–400°C). Finally, distillation under reduced pressure gives lubricating oils and waxes, and leaves an undistillable tarry residue of asphalt (Figure 3.10).

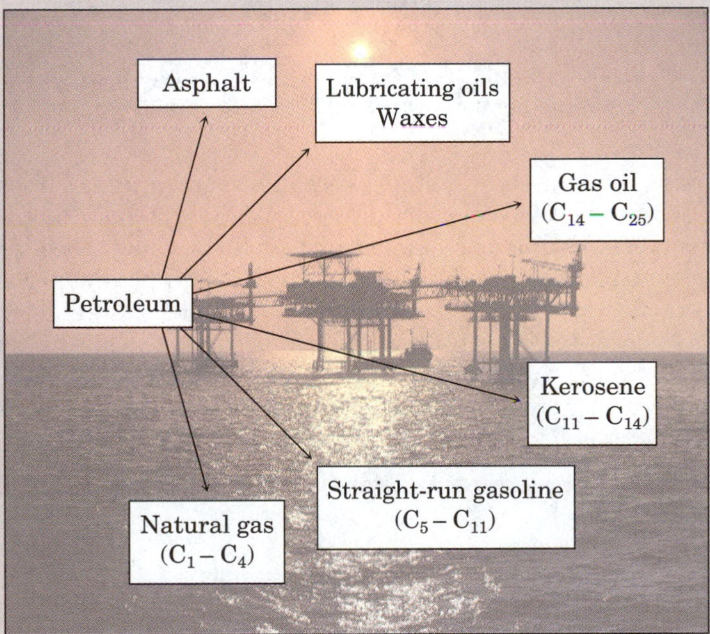

Figure 3.10 The products of petroleum refining.

The distillation of crude oil is only the first step in gasoline production. Straight-run gasoline turns out to be a poor fuel because of *engine knock*. In the typical four-stroke automobile engine, a piston draws a

(continued)➤

mixture of fuel and air into a cylinder on its downward stroke and compresses the mixture on its upward stroke. Just before the end of the compression, a spark plug ignites the mixture and combustion occurs, driving the piston downward and turning the crankshaft.

Not all fuels burn equally well, though. When poor fuels are used, combustion can be initiated in an uncontrolled manner by a hot surface in the cylinder before the spark plug fires. This preignition, detected as an engine knock, can destroy the engine by putting irregular forces on the crankshaft and raising engine temperature.

The *octane number* of a fuel is the measure by which its antiknock properties are judged. It was recognized long ago that straight-chain hydrocarbons are far more prone to induce engine knock than are highly branched compounds. Heptane, a particularly bad fuel, is assigned a base value of 0 octane number; 2,2,4-trimethylpentane (commonly known as isooctane) has a rating of 100.

$CH_3CH_2CH_2CH_2CH_2CH_2CH_3$

Heptane
(octane number = 0)

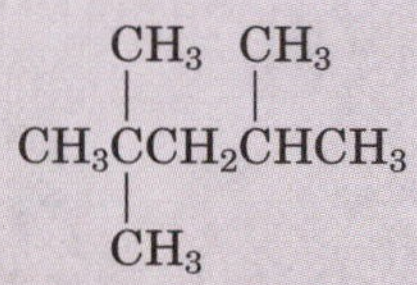

2,2,4-Trimethylpentane
(octane number = 100)

Because straight-run gasoline has a high percentage of unbranched alkanes and is therefore a poor fuel, petroleum chemists have devised several methods for producing higher-quality fuels. One of these methods, called **catalytic cracking**, involves taking the high-boiling kerosene cut (C_{11}–C_{14}) and "cracking" it into smaller molecules suitable for use in gasoline. The process takes place on a silica–alumina catalyst at temperatures of 400–500°C, and the major products are light hydrocarbons in the C_3–C_5 range. These small hydrocarbons are then catalytically recombined to yield useful C_7–C_{10} alkanes.

Summary and Key Words

A **functional group** is an atom or group of atoms within a larger molecule that has a characteristic chemical reactivity. Because functional groups behave approximately the same way in all molecules where they occur, the chemical reactions of an organic molecule are largely determined by its functional groups.

Alkanes are a class of **hydrocarbons** with the general formula C_nH_{2n+2}. They contain no functional groups, are relatively inert, and can be either **straight-chain** (**normal alkanes**) or **branched**. Alkanes are named by a series of **IUPAC rules** of nomenclature. Compounds that have the same chemical formula but different structures are called **isomers**. More specifically, compounds such as butane and isobutane, which differ in their connections between atoms, are called **constitutional isomers**.

Cycloalkanes contain rings of carbon atoms and have the general formula C_nH_{2n}. Although free rotation is possible around C–C single bonds in open-chain alkanes, rotation is greatly reduced in cycloalkanes. Disubstituted cycloalkanes can therefore exist as **cis–trans isomers**. The cis isomer has both substituents on the same side of the ring; the trans isomer has substituents on opposite sides of the ring. Cis–trans isomers are just one kind of **stereoisomers**—isomers that have the same connections between atoms but differ in their three-dimensional arrangements.

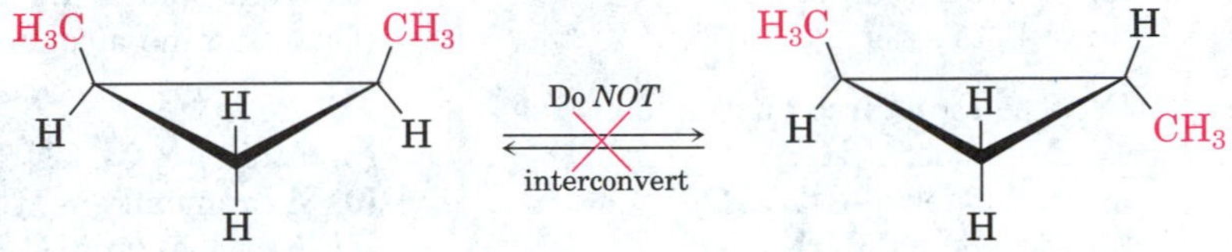

***cis*-1,2-Dimethylcyclopropane**
(methyl groups on same side of ring)

***trans*-1,2-Dimethylcyclopropane**
(methyl groups on opposite sides of ring)

ADDITIONAL PROBLEMS

3.18 Locate and identify the functional groups in the following molecules:

(a)

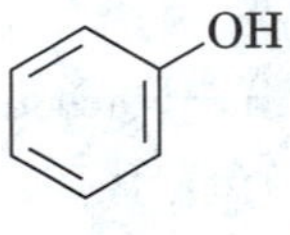

Phenol

(b)

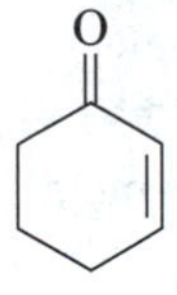

2-Cyclohexenone

(c)

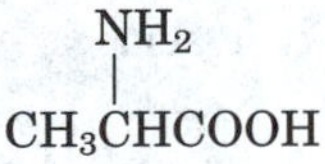

Alanine

(d)

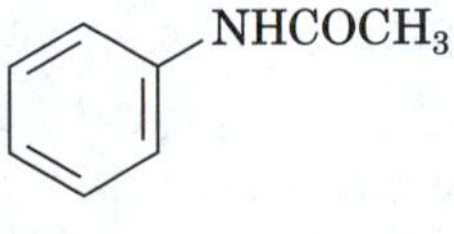

Acetanilide

(e)

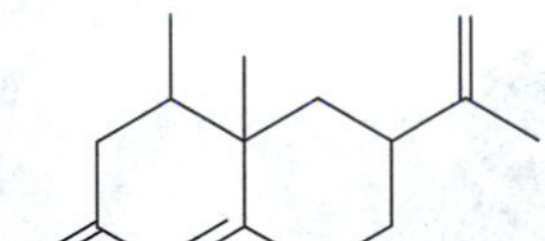

Nootkatone (from grapefruit)

(f)

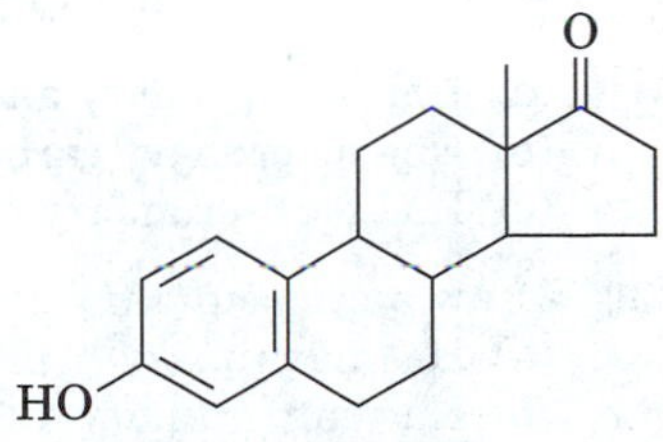

Estrone

(g)

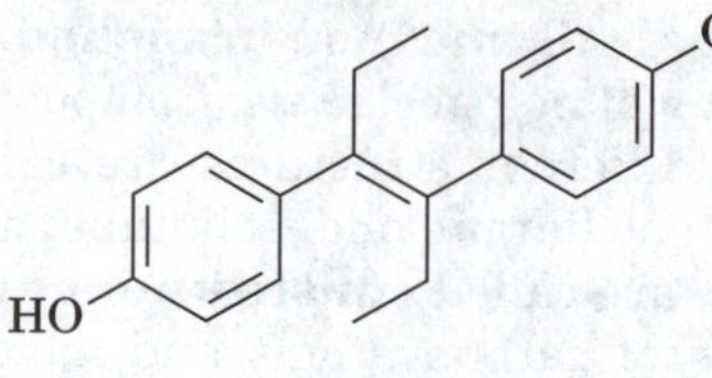

Diethylstilbestrol

(h)

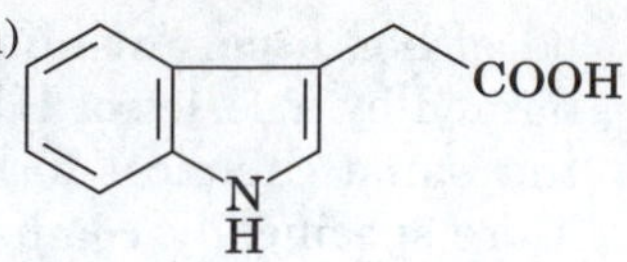

3-Indoleacetic acid

3.19 Draw structures that meet these descriptions (there are many possibilities):
(a) Three isomers with the formula C_8H_{18}
(b) Two isomers with the formula $C_4H_8O_2$

3.20 Draw structures of the nine isomers of C_7H_{16}.

3.21 Propose structures that meet these descriptions:
(a) A ketone with five carbons
(b) A four-carbon amide
(c) A five-carbon ester
(d) An aromatic aldehyde
(e) A keto ester
(f) An amino alcohol

3.22 Propose structures for the following:
(a) A ketone, C_4H_8O
(b) A nitrile, C_5H_9N
(c) A dialdehyde, $C_4H_6O_2$
(d) A bromoalkene, $C_6H_{11}Br$
(e) An alkane, C_6H_{14}
(f) A cycloalkane, C_6H_{12}
(g) A diene (dialkene), C_5H_8
(h) A keto alkene, C_5H_8O

3.23 Draw as many compounds as you can that fit these descriptions:
(a) Alcohols with formula $C_4H_{10}O$
(b) Amines with formula $C_5H_{13}N$
(c) Ketones with formula $C_5H_{10}O$
(d) Aldehydes with formula $C_5H_{10}O$
(e) Esters with formula $C_4H_8O_2$
(f) Ethers with formula $C_4H_{10}O$

3.24 Draw compounds that contain the following:
(a) A primary alcohol
(b) A tertiary nitrile
(c) A secondary bromide
(d) Both primary and secondary alcohols
(e) An isopropyl group
(f) A quaternary carbon

3.25 Draw and name all monobromo derivatives of pentane, $C_5H_{11}Br$.

3.26 Draw and name all monochloro derivatives of 2,5-dimethylhexane, $C_8H_{17}Cl$.

3.27 Predict the hybridization of the carbon atom in each of the following functional groups:
(a) Ketone
(b) Nitrile
(c) Carboxylic acid
(d) Ether

3.28 Draw structural formulas for the following:
(a) 2-Methylheptane
(b) 4-Ethyl-2,2-dimethylhexane
(c) 4-Ethyl-3,4-dimethyloctane
(d) 2,4,4-Trimethylheptane
(e) 3,3-Diethyl-2,5-dimethylnonane
(f) 4-Isopropyl-3-methylheptane

3.29 Draw a compound that:
(a) Has only primary and tertiary carbons
(b) Has no primary carbons
(c) Has four secondary carbons

3.30 Draw a compound that:
(a) Has no primary hydrogens
(b) Has only primary and tertiary hydrogens

3.31 For each of the following compounds, draw an isomer with the same functional groups.

(a) 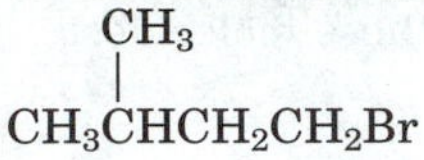

(b) 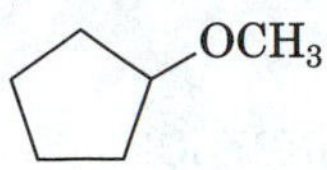

(c) $CH_3CH_2CH_2C{\equiv}N$

(d) 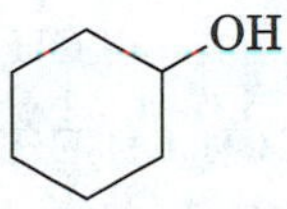

(e) CH_3CH_2CHO

(f) 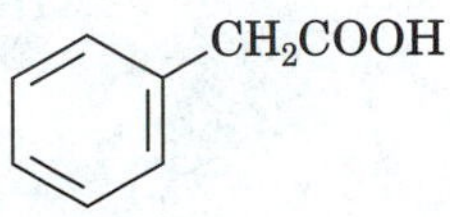

3.32 In each of the following sets, which Kekulé structures represent the same compound, and which represent different compounds?

(a) 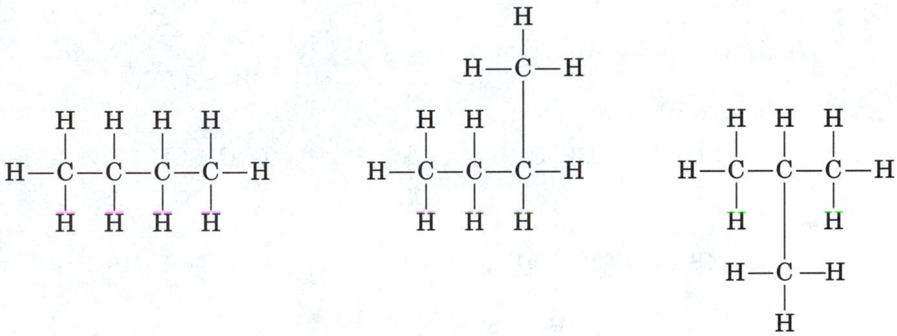

(b) 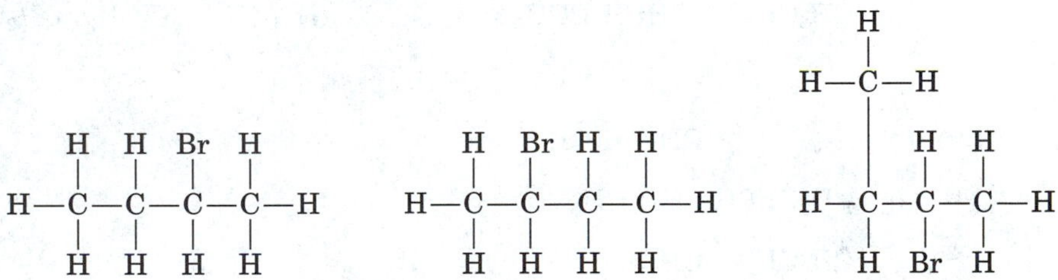

(c) 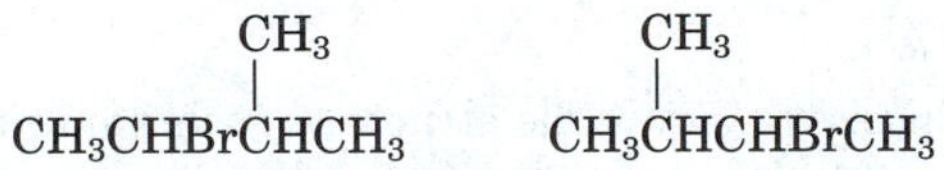 $(CH_3)_2CHCHBrCH_2CH_3$

(d) 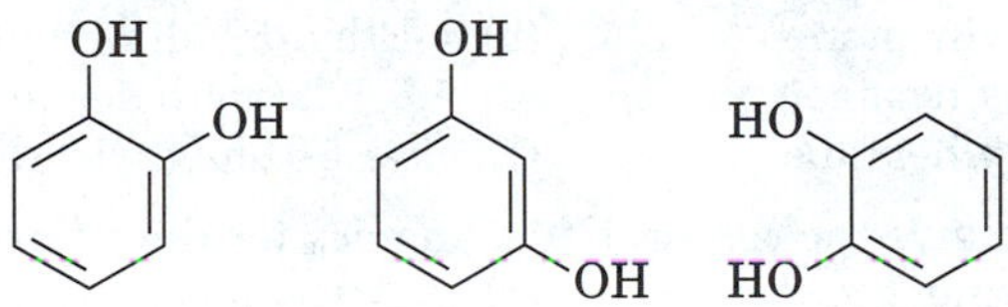

(e) 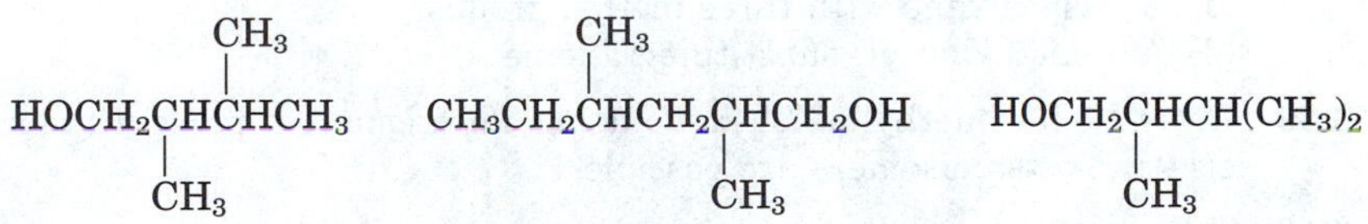

3.33 Draw structures for these compounds:

(a) *trans*-1,3-Dibromocyclopentane
(b) *cis*-1,4-Diethylcyclohexane
(c) *trans*-1-Isopropyl-3-methylcycloheptane
(d) Dicyclohexylmethane

3.34 Identify the kinds of carbons (1°, 2°, 3°, or 4°) in these molecules:

(a) 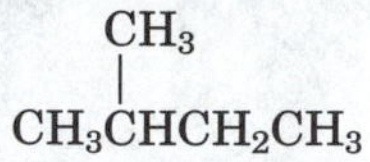

(b) $(CH_3)_2CHCH(CH_2CH_3)_2$

(c) 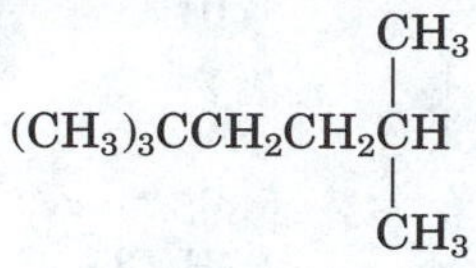

(d) 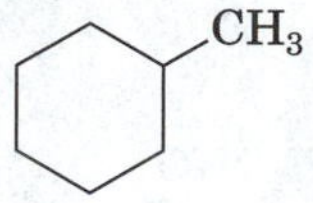

(e) 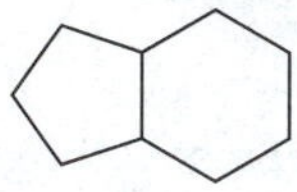

(f) 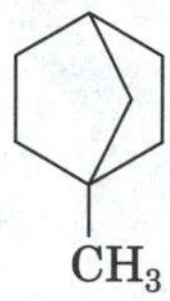

3.35 Name the following compounds:

(a) 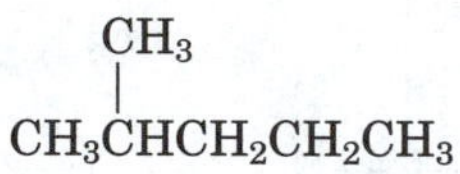

(b) $CH_3CH_2C(CH_3)_2CH_3$

(c) $(CH_3)_2CHC(CH_3)_2CH_2CH_2CH_3$

(d) 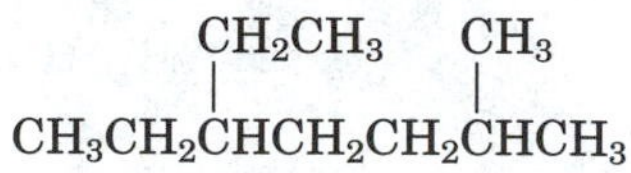

(e) 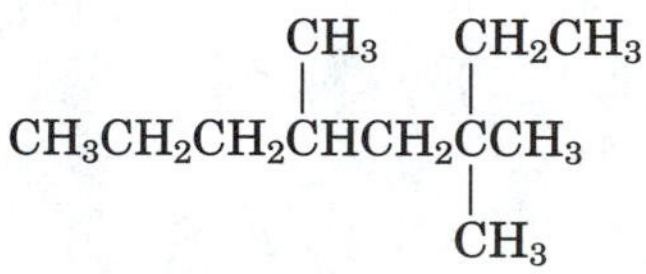

(f) $(CH_3)_3CC(CH_3)_2CH_2CH_2CH_3$

(g) 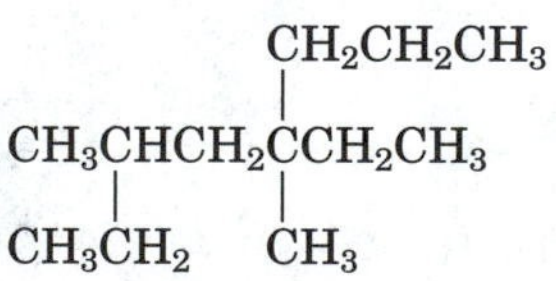

3.36 Name the five isomers of C_6H_{14}.

3.37 The following names are incorrect. Draw the structure each name represents, tell what is wrong with each name, and give the correct IUPAC name.

(a) 2,2-Dimethyl-6-ethylheptane
(b) 4-Ethyl-5,5-dimethylpentane
(c) 3-Ethyl-4,4-dimethylhexane
(d) 5,5,6-Trimethyloctane
(e) 2-Isopropyl-4-methylheptane
(f) *cis*-1,5-Dimethylcyclohexane

3.38 Propose structures and give the correct IUPAC names for the following:

(a) A dimethylcyclooctane
(b) A diethyldimethylhexane
(c) A cyclic alkane with three methyl groups
(d) A (3-methylbutyl)-substituted alkane

3.39 Draw 1,3,5-trimethylcyclohexane using a hexagon to represent the ring. How many cis–trans stereoisomers are possible?

3.40 Tell whether the following pairs of compounds are identical, constitutional isomers, or stereoisomers.

(a) *cis*-1,3-Dibromocyclohexane and *trans*-1,4-dibromocyclohexane
(b) 2,3-Dimethylhexane and 2,5,5-trimethylpentane

(c)

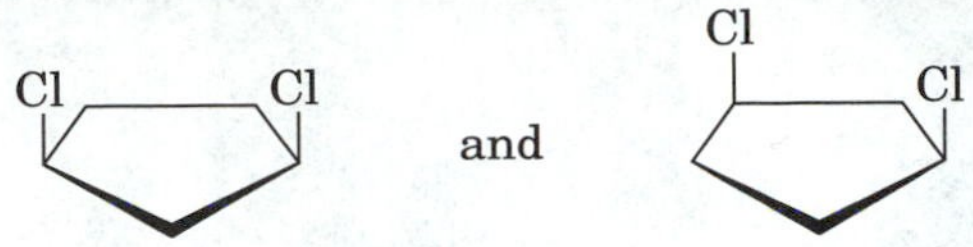

3.41 Draw two constitutional isomers of *cis*-1,2-dibromocyclopentane.

3.42 Draw a stereoisomer of *trans*-1,3-dimethylcyclobutane.

3.43 Malic acid, $C_4H_6O_5$, has been isolated from apples. Since this compound reacts with 2 equivalents (equiv) of base, it is a dicarboxylic acid.
(a) Draw at least five possible structures.
(b) If malic acid is a secondary alcohol, what is its structure?

3.44 Cyclopropane was first prepared by reaction of 1,3-dibromopropane with sodium metal. Formulate the cyclopropane-forming reaction and then predict the product of the following reaction. What geometry do you expect for the product? (Try building a molecular model.)

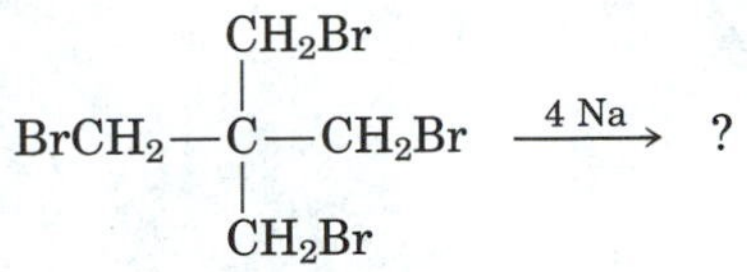

3.45 Formaldehyde, $H_2C{=}O$, is known to all biologists because of its usefulness as a tissue preservative. When pure, formaldehyde *trimerizes* to give trioxane, $C_3H_6O_3$, which, surprisingly enough, has no carbonyl groups. Only one monobromo derivative ($C_3H_5BrO_3$) of trioxane is possible. Propose a structure for trioxane.

3.46 Draw the three cis–trans isomers of menthol.

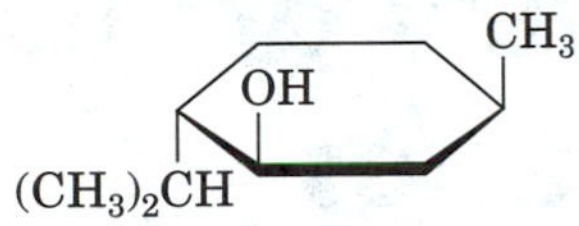

Menthol

3.47 Draw the five cycloalkanes with the formula C_5H_{10}.

A Look Ahead

3.48 There are two different substances named *trans*-1,2-dimethylcyclopentane. Make molecular models and see if you can find the relationship between them. We'll explore this kind of isomerism in Chapter 9.

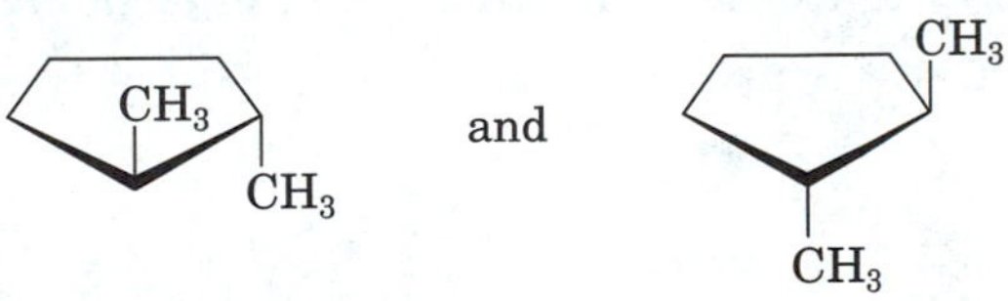

3.49 We'll see in the next chapter that cyclohexane has a puckered, chair-like shape rather than a flat shape. Why?

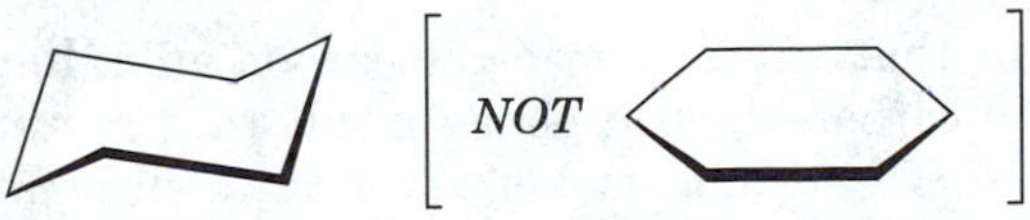

Cyclohexane

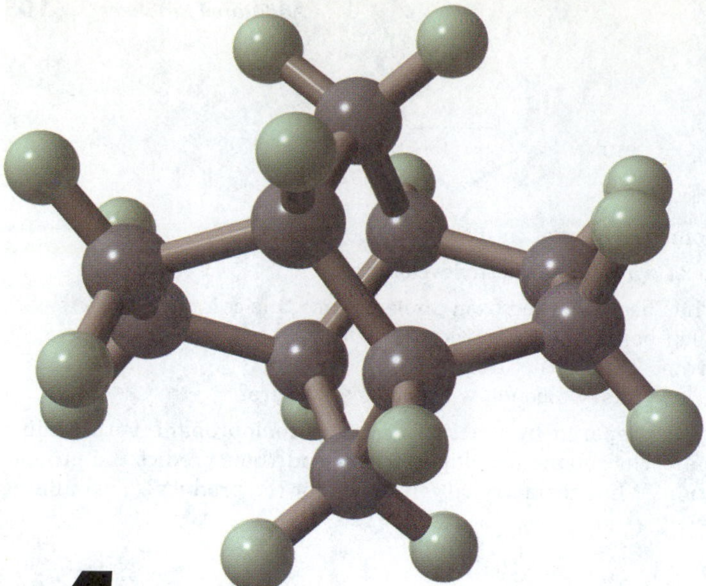

This polycyclic alkane is known by the common name *twistane.*

4

Stereochemistry of Alkanes and Cycloalkanes

Up to this point, we have mostly viewed molecules in a two-dimensional way and have given little thought to any chemical consequences that might arise from the spatial arrangement of atoms in molecules. Now it's time to add a third dimension to our study. **Stereochemistry** is the branch of chemistry concerned with the three-dimensional aspects of molecules.

4.1 Conformations of Ethane

We know that an sp^3-hybridized carbon atom has tetrahedral geometry and that the carbon–carbon bonds in alkanes result from σ overlap of carbon sp^3 orbitals. Let's now look into the three-dimensional consequences of such bonding. What are the spatial relationships between the hydrogens on one carbon and the hydrogens on a neighboring carbon?

We saw in Section 3.8 that free rotation can occur around carbon–carbon single bonds in open-chain molecules such as ethane because of σ bond cylindrical symmetry. Orbital overlap in the C–C single bond is exactly the same for all geometric arrangements of the atoms (Figure 4.1). The different arrangements of atoms that result from rotation about a single bond are called **conformations**, and a specific conformation is called a **conformer** (**confor**mational iso**mer**). Unlike constitutional isomers, different conformers can't usually be isolated, because they interconvert too rapidly.

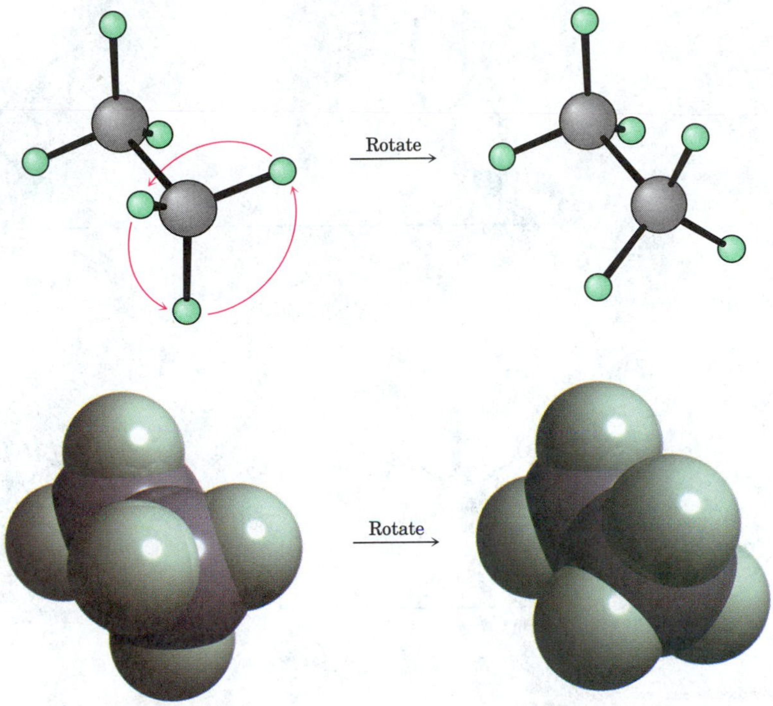

Figure 4.1 Some conformations of ethane. Rapid rotation around the carbon–carbon single bond interconverts the different conformers.

Chemists represent conformational isomers in two ways, as shown in Figure 4.2. **Sawhorse representations** view the carbon–carbon bond from an oblique angle and indicate spatial orientation by showing all the C–H bonds. **Newman[1] projections** view the carbon–carbon bond directly end-on and represent the two carbon atoms by a circle. Bonds attached to the

[1]Melvin S. Newman (1908–1993); b. New York; Ph.D. (1932), Yale University; professor, Ohio State University (1936–1973).

front carbon are represented by lines going to the center of the circle, and bonds attached to the rear carbon are represented by lines going to the edge of the circle. The advantages of Newman projections are that they're easy to draw and the relationships among substituents on the different carbon atoms are easy to see.

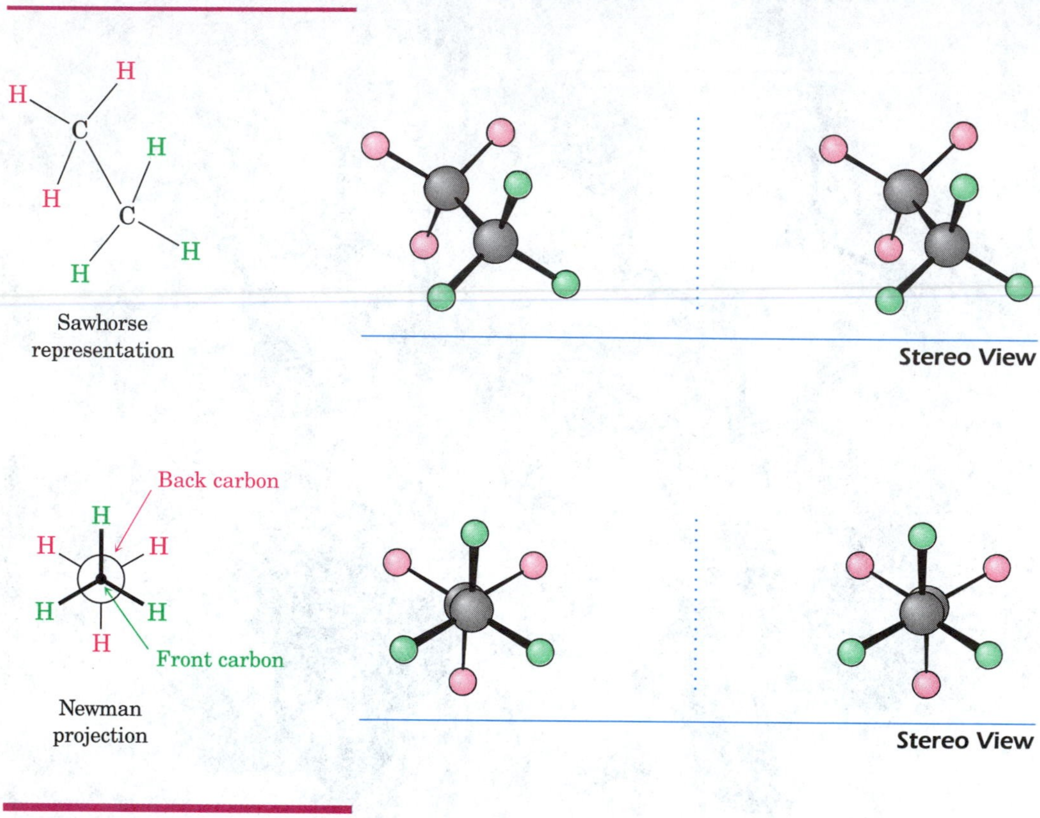

Figure 4.2 A sawhorse representation and a Newman projection of ethane. The sawhorse projection views the molecule from an oblique angle, while the Newman projection views the molecule end-on.

In spite of what we've just said about σ bond symmetry, we don't actually observe *perfectly* free rotation in ethane. Experiments show that there is a slight (12 kJ/mol; 2.9 kcal/mol) barrier to rotation and that some conformations are more stable than others. The lowest-energy, most stable conformation is the one in which all six carbon–hydrogen bonds are as far away from each other as possible (**staggered** when viewed end-on in a Newman projection). The highest-energy, least stable conformation is the one in which the six carbon–hydrogen bonds are as close as possible (**eclipsed** in a Newman projection). Between these two limiting conformations are an infinite number of other possibilities. Since the barrier to rotation is 12 kJ/mol, and since the barrier is caused by three equal hydrogen–hydrogen eclipsing interactions, we can assign a value of approximately 4.0 kJ/mol (1.0 kcal/mol) to each single interaction.

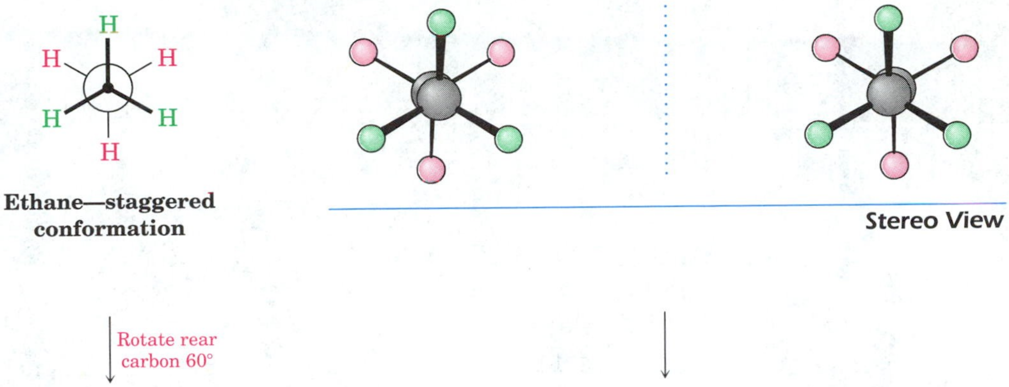

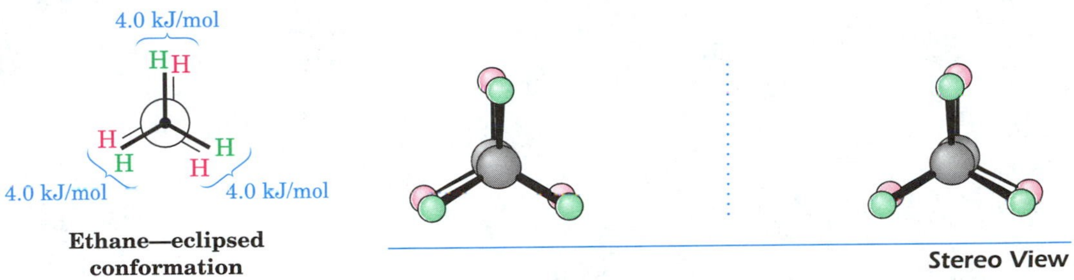

The 12 kJ/mol of extra energy present in the eclipsed conformation of ethane is called **torsional strain**. The barrier to rotation that results from torsional strain can be represented on a graph of potential energy versus degree of rotation in which the angle between C–H bonds on front and back carbons (the *dihedral angle*) goes full circle from 0° to 360° when seen end-on. Energy minima occur at staggered conformations, and energy maxima occur at eclipsed conformations, as shown in Figure 4.3 (p. 110).

To what is torsional strain due? The cause has been the subject of some controversy, but most chemists now believe that torsional strain is due to the slight repulsion between electron clouds in the carbon–hydrogen bonds as they pass by each other at close quarters in the eclipsed conformer. Calculations indicate that the internuclear hydrogen–hydrogen distance in the staggered conformer is 2.55 Å but that this distance decreases to about 2.29 Å in the eclipsed conformer.

PROBLEM

4.1 Build a molecular model of ethane, and look at the interconversion of staggered and eclipsed forms.

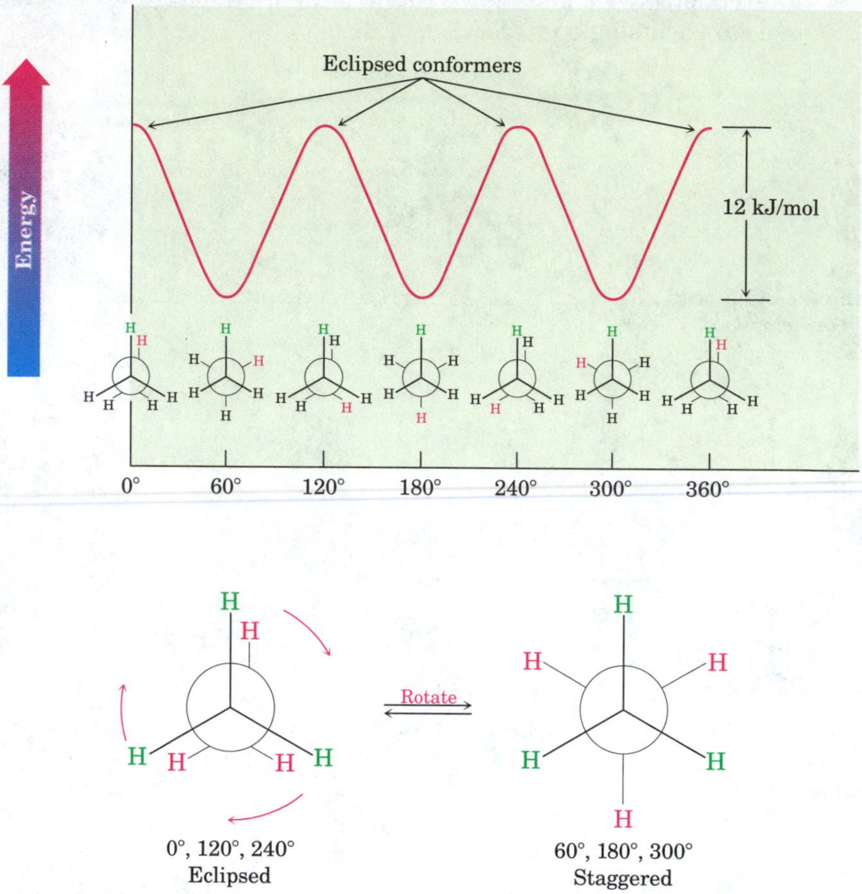

Figure 4.3 A graph of potential energy versus bond rotation in ethane. The staggered conformers are 12 kJ/mol lower in energy than the eclipsed conformers.

4.2 Conformations of Propane

Propane is the next higher member in the alkane series, and there is again a torsional barrier that results in slightly hindered rotation around the carbon–carbon bonds. The barrier is slightly higher in propane than in ethane—14 kJ/mol (3.4 kcal/mol) versus 12 kJ/mol. In the eclipsed conformer of propane, there are two ethane-type hydrogen–hydrogen interactions and one additional interaction between a carbon–hydrogen bond and a carbon–carbon bond. Since each eclipsing hydrogen–hydrogen interaction has an energy "cost" of 4.0 kJ/mol, we can assign a value of 14 − (2 × 4.0) = 6.0 kJ/mol (1.4 kcal/mol) to the eclipsing interaction between the carbon–carbon bond and the carbon–hydrogen bond (Figure 4.4).

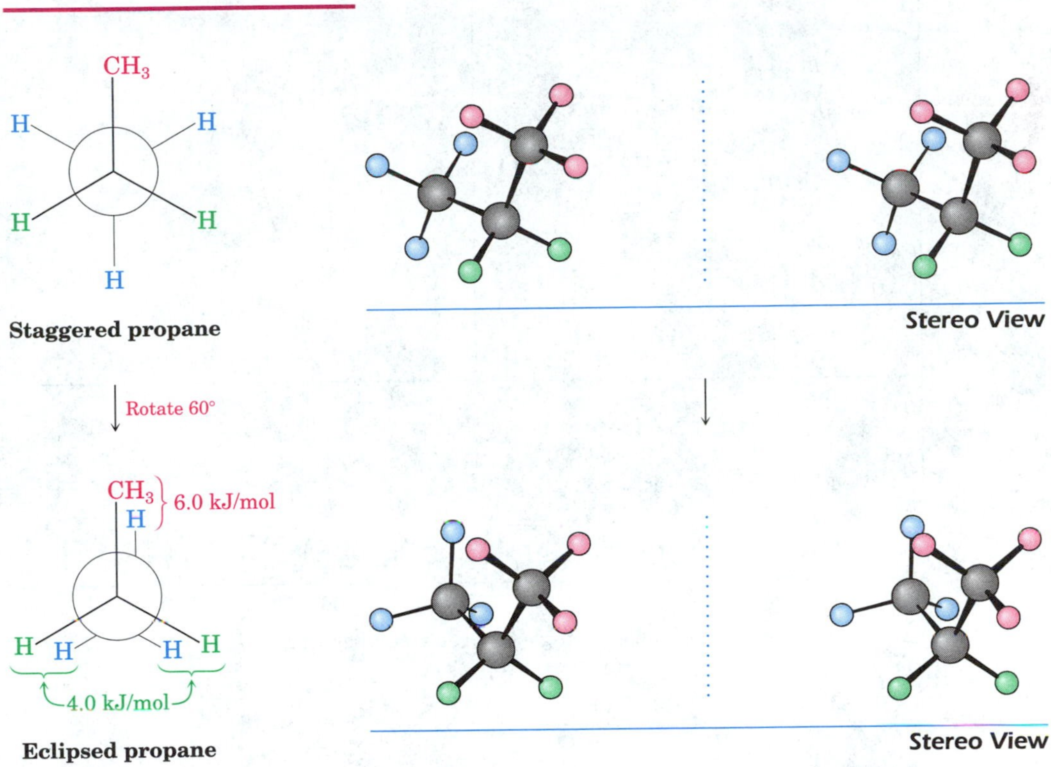

Figure 4.4 Newman projections of propane showing staggered and eclipsed conformations. The staggered conformer is lower in energy by 14 kJ/mol.

PROBLEM.

4.2 Make a graph of potential energy versus angle of bond rotation for propane, and assign values to the energy maxima.

4.3 Conformations of Butane

The conformational situation becomes more complex as the alkane becomes larger. In butane, for instance, a plot of potential energy versus rotation about the C2–C3 bond is shown in Figure 4.5.

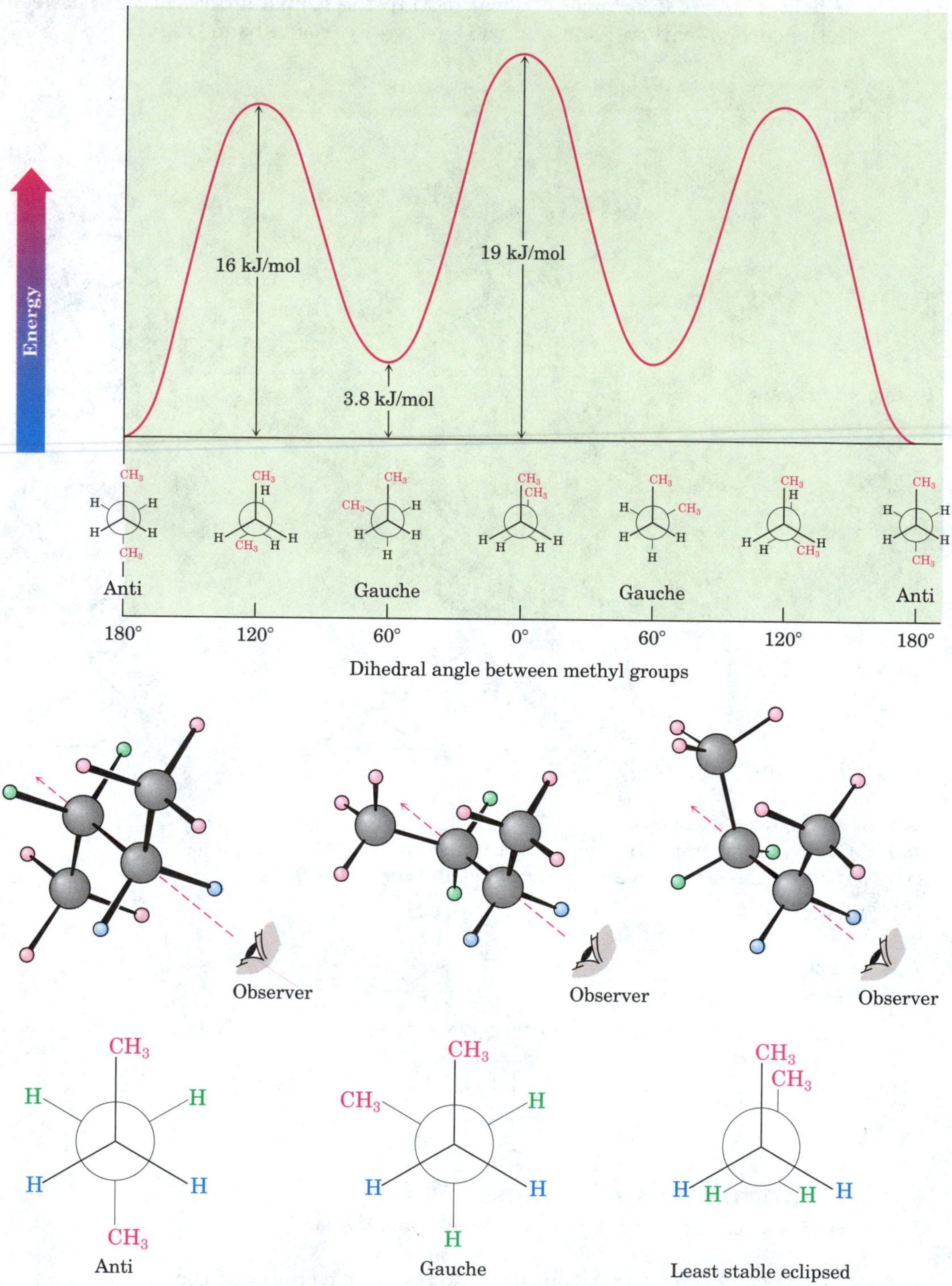

Figure 4.5 A plot of potential energy versus rotation for the C2–C3 bond in butane. The energy maximum occurs when the two methyl groups eclipse each other, and the energy minimum occurs when the two methyl groups are far apart (anti).

Not all staggered conformations of butane have the same energy, and not all eclipsed conformations are the same. The lowest-energy arrangement, called the **anti conformation**, is the one in which the two large methyl groups are as far apart as possible: 180°. As rotation around the C2–C3 bond occurs, an eclipsed conformation is reached in which there are two methyl–hydrogen interactions and one hydrogen–hydrogen interaction. If we assign the energy values for eclipsing interactions that were previously derived from ethane and propane, we might predict that this eclipsed conformation should be more strained than the anti conformation by 2 × 6.0 kJ/mol (two methyl–hydrogen interactions) plus 4.0 kJ/mol (one hydrogen–hydrogen interaction), or a total of 16 kJ/mol (3.8 kcal/mol). This is exactly what is found.

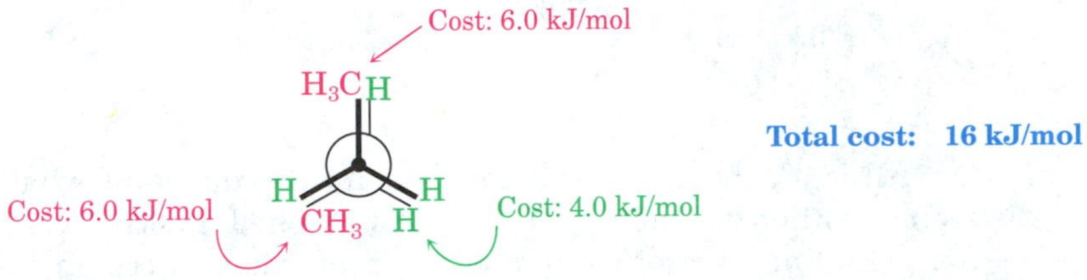

As bond rotation continues, an energy minimum is reached at the staggered conformation where the methyl groups are 60° apart. Called the **gauche conformation**, it lies 3.8 kJ/mol (0.9 kcal/mol) higher in energy than the anti conformation *even though it has no eclipsing interactions*. This energy difference is due to the fact that the hydrogen atoms of the methyl groups are near each other in the gauche conformation, resulting in *steric strain*. **Steric strain** is the repulsive interaction that occurs when atoms are forced closer together than their atomic radii allow. It's the result of trying to force two atoms to occupy the same space (Figure 4.6).

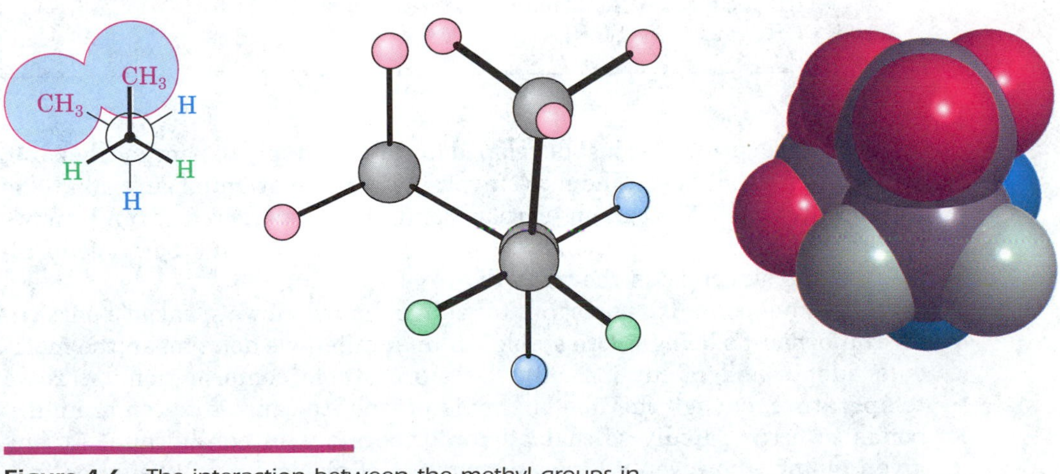

Figure 4.6 The interaction between the methyl groups in gauche butane. Steric strain results because the two methyl groups are too close together.

As the dihedral angle between the methyl groups approaches 0°, an energy maximum is reached. Because the methyl groups are forced even closer together than in the gauche conformation, substantial amounts of both torsional strain and steric strain are present. A total strain energy of 19 kJ/mol (4.5 kcal/mol) has been estimated for this conformation, allowing us to calculate a value of 11 kJ/mol (2.5 kcal/mol) for the methyl–methyl eclipsing interaction: total strain (19 kJ/mol), less strain of two hydrogen–hydrogen eclipsing interactions (2 × 4.0 kcal/mol), equals 11 kJ/mol.

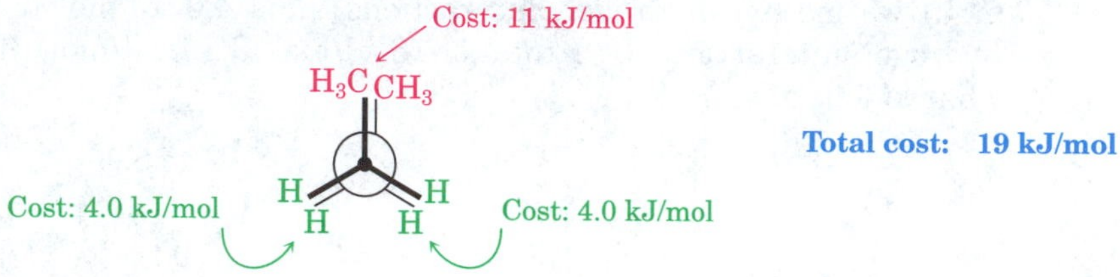

After 0°, the rotation becomes a mirror image of what we've already seen. Another gauche conformation is reached, another eclipsed conformation, and finally a return to the anti conformation occurs.

The notion of assigning definite energy values to specific interactions within a molecule is a very useful one that we'll return to later in this chapter. A summary of what we've seen thus far is given in Table 4.1.

Table 4.1 Energy Costs for Interactions in Alkane Conformers

		Energy cost	
Interaction	*Cause*	*(kJ/mol)*	*(kcal/mol)*
H ↔ H eclipsed	Torsional strain	4.0	1.0
H ↔ CH_3 eclipsed	Mostly torsional strain	6.0	1.4
CH_3 ↔ CH_3 eclipsed	Torsional plus steric strain	11	2.6
CH_3 ↔ CH_3 gauche	Steric strain	3.8	0.9

The same principles just developed for butane apply to pentane, hexane, and all higher alkanes. The most favorable conformation for any alkane is the one in which the carbon–carbon bonds have staggered arrangements and in which large substituents are arranged anti to each other. A generalized alkane structure is shown in Figure 4.7.

One final point: It's important to realize that when we speak of a particular conformer as being "more stable" than another, we don't mean the molecule adopts and maintains only the more stable conformation. At room temperature, enough thermal energy is present to cause rotation around σ bonds to occur rapidly so that all conformers are in equilibrium. At any given instant, however, a larger percentage of molecules will be found in a more stable conformation than in a less stable one.

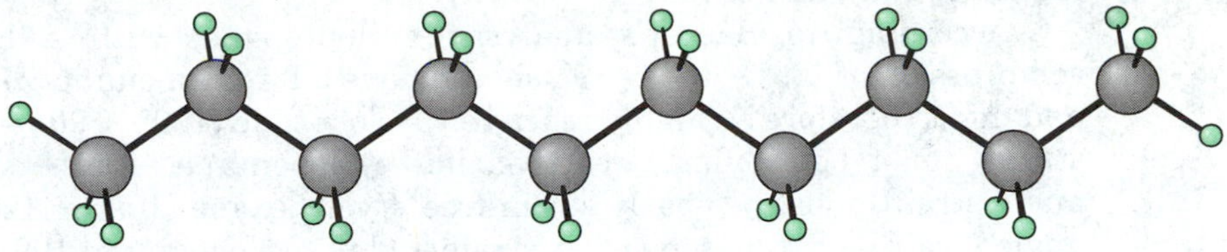

Figure 4.7 The most stable alkane conformation is the one in which all substituents are staggered and the carbon–carbon bonds are arranged anti, as shown in this structure of decane.

PROBLEM.

4.3 Sight along the C2–C3 bond of 2,3-dimethylbutane, and draw a Newman projection of the most stable conformation.

PROBLEM.

4.4 Consider 2-methylpropane (isobutane). Sighting along the C2–C1 bond:
(a) Draw a Newman projection of the most stable conformation.
(b) Draw a Newman projection of the least stable conformation.
(c) Make a graph of energy versus angle of rotation around the C2–C1 bond.
(d) Since a hydrogen–hydrogen eclipsing interaction "costs" 4.0 kJ/mol and a hydrogen–methyl eclipsing interaction costs 6.0 kJ/mol, assign relative values to the maxima and minima in your graph.

4.4 Conformation and Stability of Cycloalkanes: The Baeyer Strain Theory

Chemists in the late 1800s knew that cyclic molecules existed, but the limitations on ring sizes were unclear. Numerous compounds containing five-membered and six-membered rings were known, but smaller and larger ring sizes had not been prepared. For example, no cyclopropanes or cyclobutanes were known, despite many efforts to prepare them.

A theoretical interpretation of this observation was proposed in 1885 by Adolf von Baeyer.[2] Baeyer suggested that, since carbon prefers to have tetrahedral geometry with bond angles of approximately 109°, ring sizes other than five and six may be too *strained* to exist. Baeyer based his hypothesis on the simple geometric notion that a three-membered ring (cyclopropane) should be an equilateral triangle with bond angles of 60°, a four-membered ring (cyclobutane) should be a square with bond angles of 90°, a

[2]Adolf von Baeyer (1835–1917); b. Berlin; Ph.D. Berlin (1858); professor, Berlin, Strasbourg (1872–1875), Munich (1875–1917); Nobel Prize (1905).

five-membered ring (cyclopentane) should be a regular pentagon with bond angles of 108°, and so on.

According to Baeyer's analysis, cyclopropane, with a bond-angle compression of 109° − 60° = 49°, should have a large amount of *angle strain* and must therefore be highly reactive. Cyclobutane (109° − 90° = 19° angle strain) must be similarly reactive, but cyclopentane (109° − 108° = 1° angle strain) must be nearly strain-free. Cyclohexane (109° − 120° = −11° angle strain) must be somewhat strained, but cycloheptane (109° − 128° = −19° angle strain) and higher cycloalkanes must have bond angles that are forced to be too large. Carrying this line of reasoning further, Baeyer suggested that very large rings should be impossibly strained and incapable of existence.

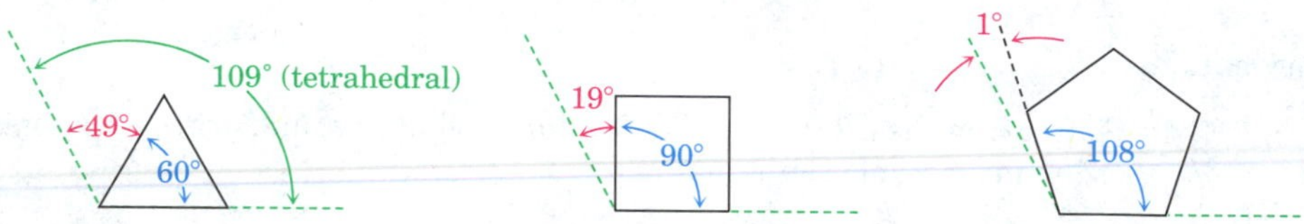

Cyclopropane **Cyclobutane** **Cyclopentane**

Although there is some truth to Baeyer's suggestion about angle strain in small rings, he was wrong in believing that small and large rings are too strained to exist. Rings of all sizes from 3 through 30 and beyond can now be prepared. Nevertheless, the concept of **angle strain**—the strain induced in a molecule when a bond angle deviates from the ideal tetrahedral value—is a very useful one. Let's look at the facts.

4.5 Heats of Combustion of Cycloalkanes

To measure the amount of strain in a compound, we must first measure the total energy of the compound and then subtract the energy of a strain-free reference compound. The difference between the two values should represent the amount of extra energy in the molecule due to strain.

The simplest way to determine cycloalkane strain energies is to measure their **heats of combustion**, the amount of heat (energy) released when a compound burns completely with oxygen:

$$—(CH_2)_n— + n\left(\tfrac{3}{2}\,O_2\right) \longrightarrow n\,CO_2 + n\,H_2O + \text{Heat}$$

The more energy (strain) a compound contains, the more energy (heat) is released on combustion. If we compare the heats of combustion of two isomeric substances, more energy is released during combustion of the more strained substance because that substance has more energy to begin with.

Because the heat of combustion of a hydrocarbon depends on its molecular formula, it's more useful to look at heats of combustion per CH_2 unit (Figure 4.8). In this way, the size of the hydrocarbon is not a factor, and we

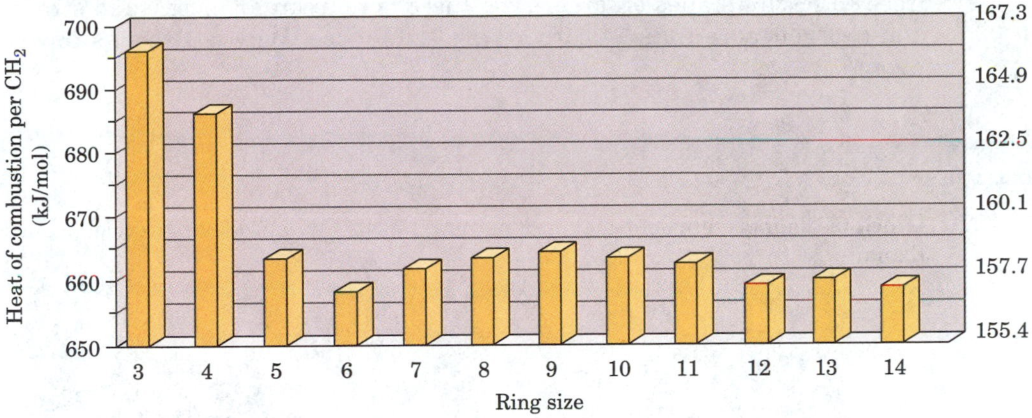

Figure 4.8 Heats of combustion per CH for cycloalkanes as a function of ring size. By convention, the amount of energy released in an exothermic reaction has a negative sign. Thus, these heats of combustion are negative values.

can compare cycloalkane rings of different sizes to a standard, strain-free, acyclic alkane. Figure 4.9 shows the results of this comparison. Total strain energies are calculated by taking the difference between sample heat of combustion per CH_2 and reference heat of combustion per CH_2 and multiplying by the number of carbons, n, in the ring.

The data in Figure 4.9 show clearly that Baeyer's theory is incorrect. Cyclopropane and cyclobutane are indeed quite strained, just as predicted,

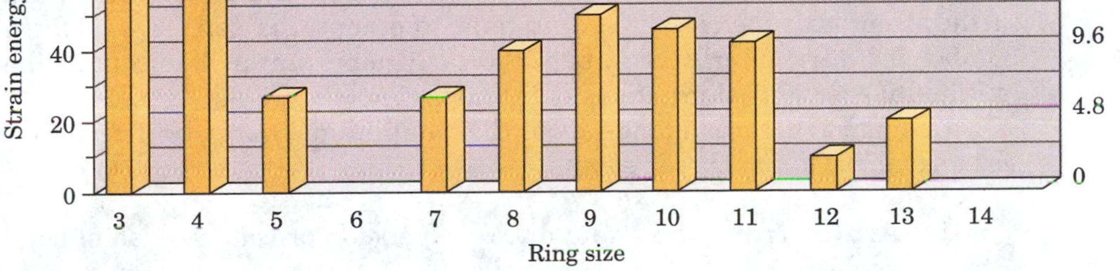

Figure 4.9 Cycloalkane strain energy as a function of ring size. Note that cyclohexane rings are strain-free.

but cyclopentane is more strained than predicted, and cyclohexane is strain-free. For cycloalkanes of larger size, there is no regular increase in strain, and rings of more than 14 carbons are strain-free. Why is Baeyer's theory wrong?

PROBLEM

4.5 Which is the more efficient fuel on a per-gram basis, cyclopropane or cyclohexane? Explain.

4.6 The Nature of Ring Strain

Baeyer was wrong for a very simple reason: He assumed that rings are flat. In fact, though, most cycloalkanes *aren't* flat; they adopt puckered three-dimensional conformations that allow bond angles to be nearly tetrahedral. Only for three- and four-membered rings is the concept of angle strain important.

Several factors in addition to angle strain are involved in determining the shape and total strain energy of cycloalkanes. One such factor is the barrier to bond rotation (torsional strain) encountered earlier in the discussion of alkane conformations (Section 4.1). We said at that time that open-chain alkanes are most stable in a staggered conformation and least stable in an eclipsed conformation. A similar conclusion holds for cycloalkanes: Torsional strain is present in cycloalkanes if any neighboring C–H bonds eclipse each other. For example, cyclopropane must have considerable torsional strain (in addition to angle strain), because C–H bonds on neighboring carbon atoms are eclipsed (Figure 4.10). Larger cycloalkanes minimize torsional strain by adopting puckered, nonplanar conformations.

In addition to angle strain and torsional strain, *steric strain* is yet a third factor that contributes to the overall strain energy of cycloalkanes. As in gauche butane (Section 4.3), two nonbonded groups repel each other if they approach too closely and attempt to occupy the same space. Such nonbonded steric interactions are particularly important in determining the minimum-energy conformations of medium-ring (C_7–C_{11}) cycloalkanes.

In summary, cycloalkanes adopt their minimum-energy conformations for a combination of three reasons:

1. **Angle strain**, the strain due to expansion or compression of bond angles
2. **Torsional strain**, the strain due to eclipsing of bonds on neighboring atoms
3. **Steric strain**, the strain due to repulsive interactions when atoms approach each other too closely

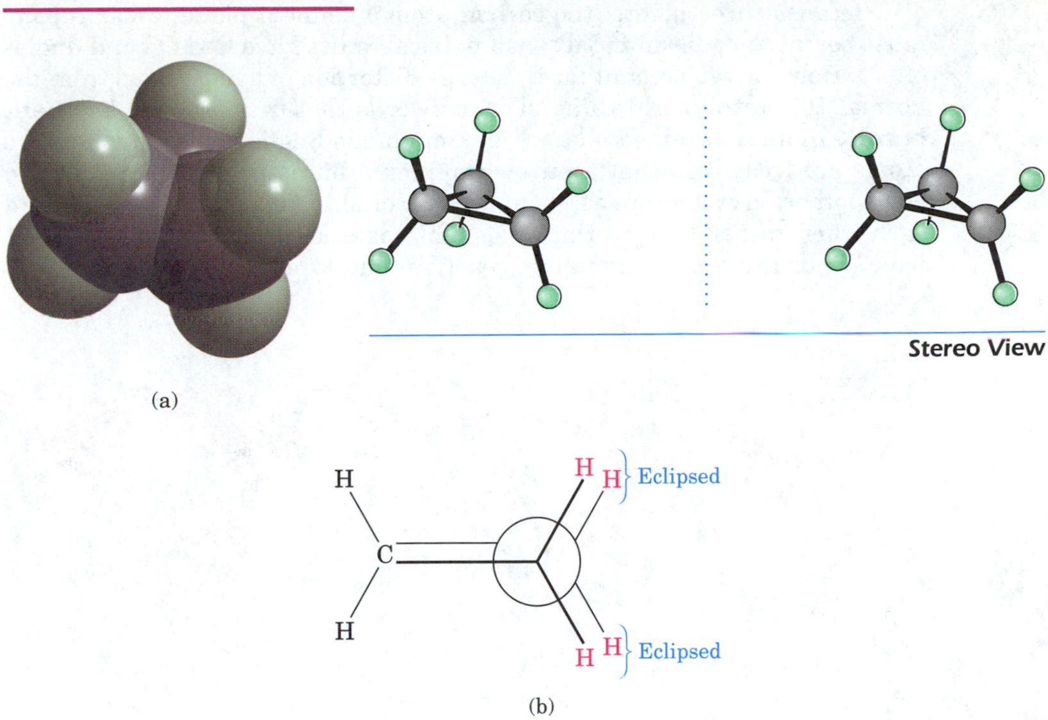

Figure 4.10 The conformation of cyclopropane. Part (b) is a Newman projection along a C–C bond, showing the eclipsing of neighboring C–H bonds.

PROBLEM

4.6 Each hydrogen–hydrogen eclipsing interaction in ethane costs about 4.0 kJ/mol. How many such interactions are present in cyclopropane? What fraction of the overall 115 kJ/mol (27.6 kcal/mol) strain energy of cyclopropane is due to torsional strain?

PROBLEM

4.7 *cis*-1,2-Dimethylcyclopropane has a larger negative heat of combustion than *trans*-1,2-dimethylcyclopropane. How can you account for this difference? Which of the two compounds is more stable?

4.7 Cyclopropane: An Orbital View

Cyclopropane, a colorless gas (bp = −33°C), was first prepared by reaction of sodium with 1,3-dibromopropane:

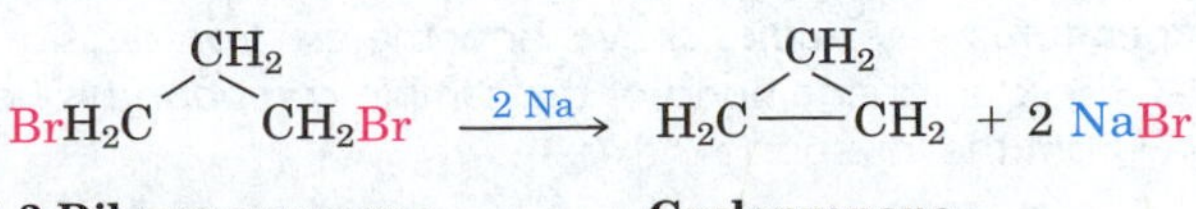

1,3-Dibromopropane **Cyclopropane**

Because three points (the carbon atoms) define a plane, cyclopropane must be flat and, assuming it's symmetrical, must have C-C-C bond angles of 60°. How can we account for this large distortion of bond angles from the normal 109° tetrahedral value? The answer is that cyclopropane has **bent bonds**. In an unstrained alkane, maximum bonding is achieved when two atoms are located so that their overlapping orbitals point directly toward one another. In cyclopropane, though, the orbitals can't point directly toward each other; rather, they overlap at a slight angle. The result is that cyclopropane bonds are weaker and more reactive than typical alkane bonds.

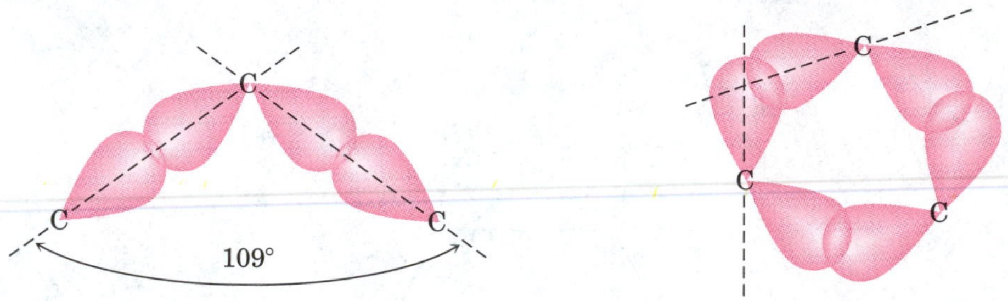

(a) A typical alkane C–C bond (b) A bent cyclopropane C–C bond

Physical evidence for the presence of bent bonds in cyclopropanes has been provided by careful, low-temperature X-ray studies, which are able to map the electron density of bonds. As shown in Figure 4.11, the electron

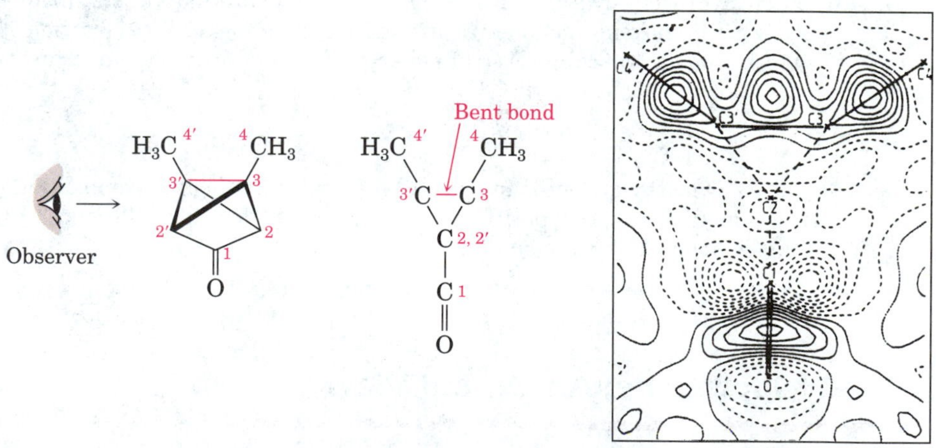

Figure 4.11 An electron-density map provided by low-temperature X-ray studies shows how the electron density in the C3–C3′ cyclopropane bond of the sample compound is bent away from the internuclear axis.

density in the C3–C3′ cyclopropane bond of the substance examined is strongly displaced outward from the internuclear axis. In contrast, the electron densities in the "normal" C–C bonds attaching the methyl groups to the ring are perfectly centered on their internuclear axes.

4.8 Conformations of Cyclobutane and Cyclopentane

Cyclobutane has less angle strain than cyclopropane but has more torsional strain because of its larger number of ring hydrogens. As a result, the total strain for the two compounds is nearly the same—110.4 kJ/mol (26.4 kcal/mol) for cyclobutane versus 114.9 kJ/mol (27.5 kcal/mol) for cyclopropane. Studies show that cyclobutane is not quite flat but is slightly bent so that one carbon atom lies about 25° above the plane of the other three (Figure 4.12). The effect of this slight bend is to *increase* angle strain but to *decrease* torsional strain, until a minimum-energy balance between the two opposing effects is achieved.

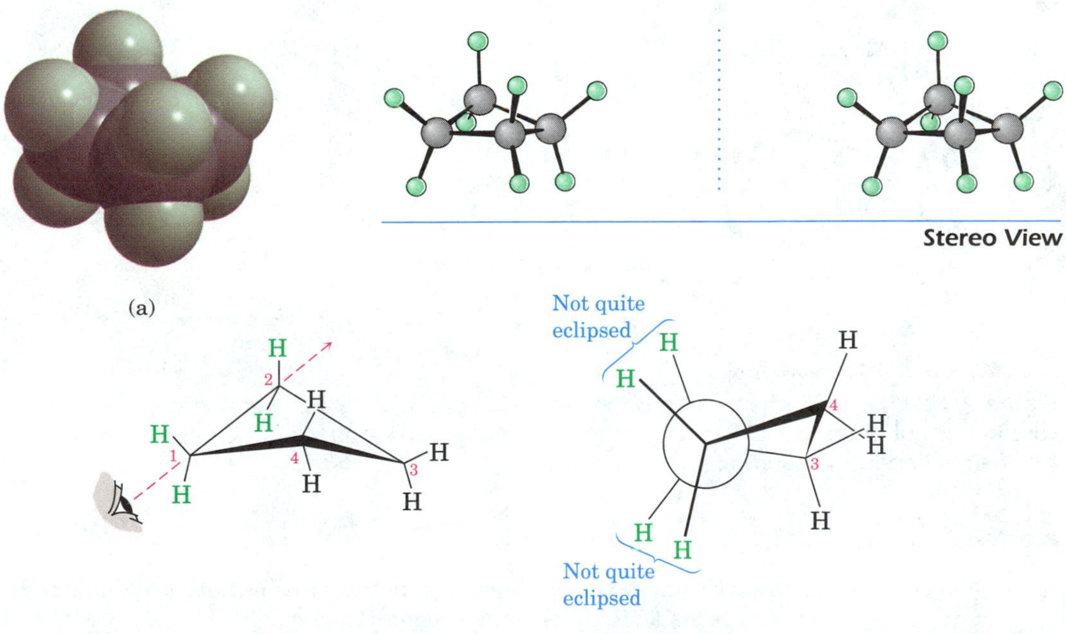

Figure 4.12 The conformation of cyclobutane. Part (a) shows computer-generated molecular models. Part (c) is a Newman projection along the C1–C2 bond, showing that neighboring C–H bonds are not quite eclipsed.

Cyclopentane was predicted by Baeyer to be nearly strain-free, but combustion data indicate a total strain energy of 26.0 kJ/mol (6.2 kcal/mol). Although cyclopentane has practically no angle strain, the 10 pairs of neighboring hydrogens introduce considerable torsional strain into a planar conformation. As a result, cyclopentane adopts a puckered, out-of-plane conformation that strikes a balance between increased angle strain and decreased torsional strain. Four of the cyclopentane carbon atoms are in approximately the same plane, with the fifth carbon atom bent out-of-plane. Most of the hydrogens are nearly staggered with respect to their neighbors (Figure 4.13).

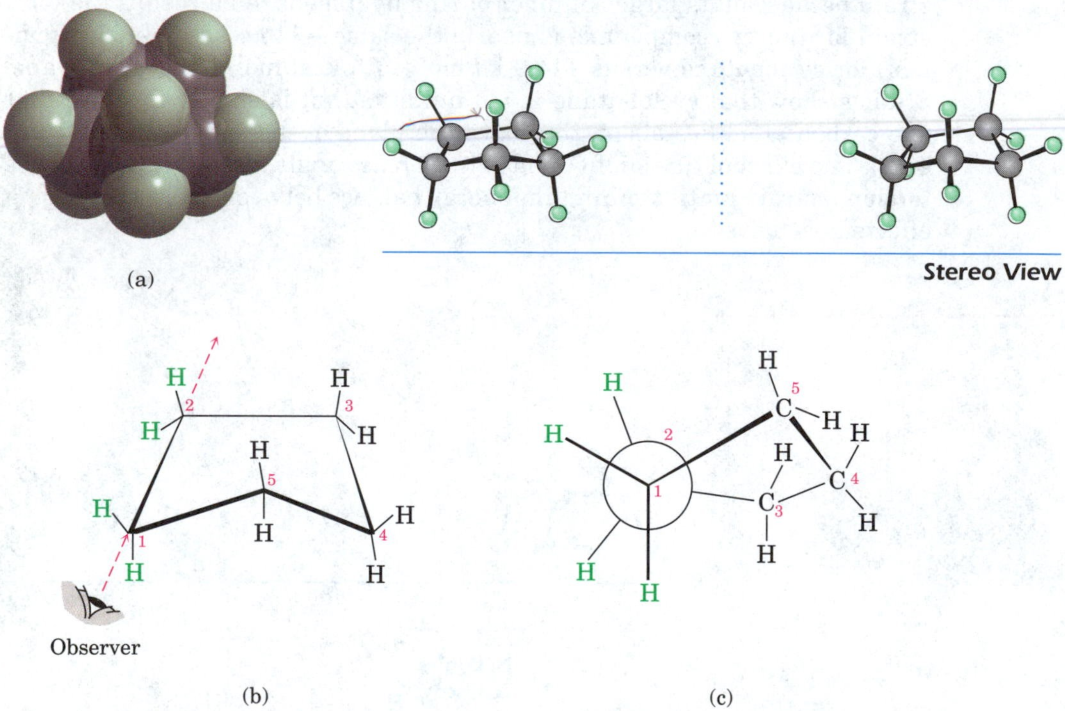

Figure 4.13 The conformation of cyclopentane. Carbons 1, 2, 3, and 4 are nearly planar, but carbon 5 is out of the plane. Part (c) is a Newman projection along the C1–C2 bond showing that neighboring C–H bonds are nearly staggered.

PROBLEM. . . . .

4.8 *cis*-1,2-Dimethylcyclobutane is less stable than its trans isomer, but *cis*-1,3-dimethylcyclobutane is more stable than its trans isomer. Explain.

PROBLEM. . . . .

4.9 Draw the most favorable conformation of *cis*-1,3-dimethylcyclobutane (see Problem 4.8).

PROBLEM. . . . .

4.10 How many hydrogen–hydrogen eclipsing interactions would be present if cyclopentane were planar? Assuming an energy cost of 4.0 kJ/mol for each eclipsing interaction, how much torsional strain would planar cyclopentane have? How much of this strain is relieved by puckering if the measured total strain of cyclopentane is 26.0 kJ/mol?

4.9 Conformations of Cyclohexane

Substituted cyclohexanes are the most common cycloalkanes because of their wide occurrence in nature. A large number of compounds, including many important pharmaceutical agents, contain cyclohexane rings.

Combustion data show that cyclohexane is strain-free, with neither angle strain nor torsional strain. How can this be? The answer was first suggested in 1890 by Hermann Sachse[3] and later expanded on by Ernst Mohr.[4] Cyclohexane is not flat as Baeyer assumed; instead, it is puckered into a three-dimensional conformation that relieves all strain. The C-C-C angles of cyclohexane can reach the strain-free tetrahedral value if the ring adopts a **chair conformation**, so-called because of its similarity to a lounge chair—a back, a seat, and a foot-rest (see Figure 4.14). Furthermore,

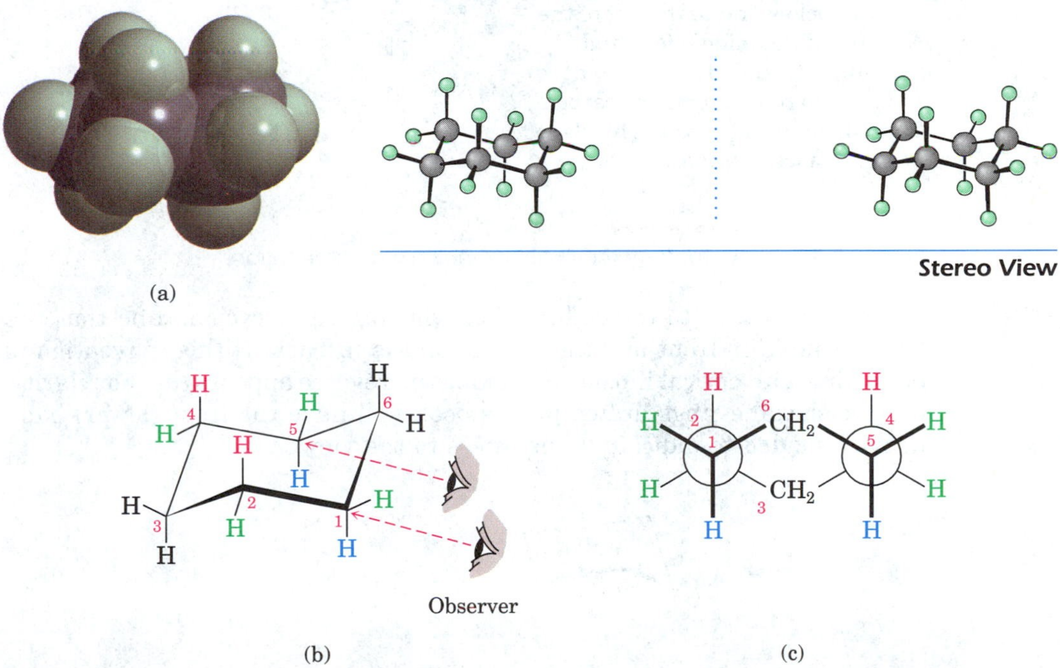

Figure 4.14 The strain-free chair conformation of cyclohexane. All C-C-C bond angles are 111.5° (close to the ideal 109.5° tetrahedral angle), and all neighboring C–H bonds are staggered.

[3]Hermann Sachse (1862–1893); b. Berlin; Ph.D. Berlin (1889); assistant, Technische Hochschule Charlottenburg-Berlin.

[4]Ernst Mohr (1873–1926); b. Dresden; Ph.D. Kiel (1897); professor, University of Heidelberg.

sighting along any one of the carbon–carbon bonds in a Newman projection shows that chair cyclohexane has no torsional strain; all neighboring C–H bonds are staggered.

The easiest way to visualize chair cyclohexane is to build your own molecular model. Two-dimensional drawings such as Figure 4.14 are useful, but there is no substitute for holding, twisting, and turning a three-dimensional model in your hands. The chair conformation of cyclohexane can be drawn by following the three steps shown in Figure 4.15.

1. Draw two parallel lines, slanted downward and slightly offset from each other. This means that four of the cyclohexane carbon atoms lie in a plane.
2. Locate the topmost carbon atom above and to the right of the plane of the other four and connect the bonds.
3. Locate the bottommost carbon atom below and to the left of the plane of the middle four and connect the bonds. Note that the bonds to the bottommost carbon atom are parallel to the bonds to the topmost carbon.

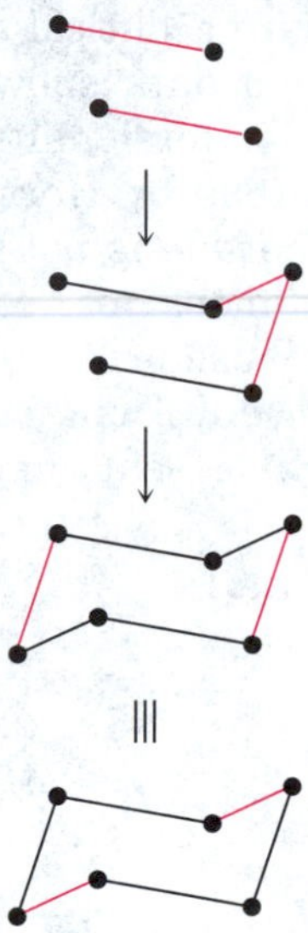

Figure 4.15 How to draw the cyclohexane chair conformation.

It's important to remember when viewing chair cyclohexane that the lower bond is in front and the upper bond is in back. If this convention is not defined, an optical illusion can make the reverse appear true. For clarity, the cyclohexane rings drawn in this book will have the front (lower) bond heavily shaded to indicate its nearness to the viewer.

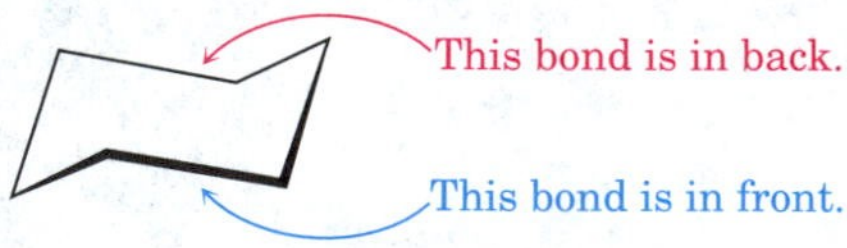

4.10 Axial and Equatorial Bonds in Cyclohexane

The chair conformation of cyclohexane leads to many consequences. For example, we'll see in Section 11.12 that the chemical behavior of many substituted cyclohexanes is directly controlled by their conformation.

Another consequence of the chair conformation is that there are two kinds of positions for hydrogen atoms on the ring: **axial positions** and **equatorial positions** (Figure 4.16). Chair cyclohexane has six axial hydrogens that are perpendicular to the ring (parallel to the ring *axis*) and six equatorial hydrogens that are in the rough plane of the ring (around the ring *equator*).

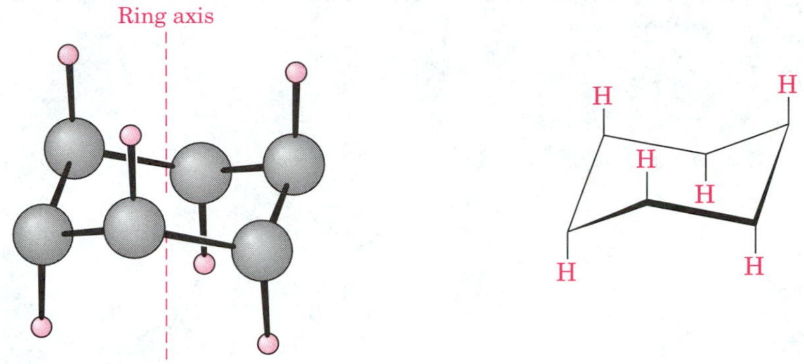

(a) Six *axial* hydrogens (parallel to the ring axis)

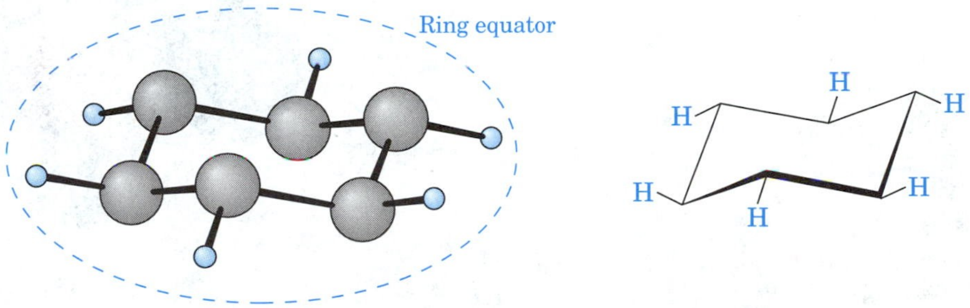

(b) Six *equatorial* hydrogens (around ring equator)

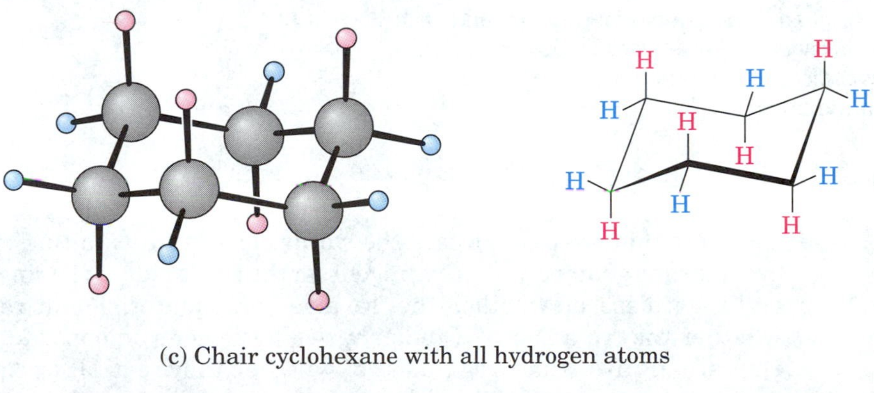

(c) Chair cyclohexane with all hydrogen atoms

Figure 4.16 Axial and equatorial hydrogen atoms in chair cyclohexane. The six axial hydrogens are parallel to the ring axis, and the six equatorial hydrogens are in a band around the ring equator.

Look carefully at the disposition of the axial and equatorial hydrogens in Figure 4.16. Each carbon atom in cyclohexane has one axial and one equatorial hydrogen. Furthermore, each side of the ring has both axial and equatorial hydrogens in an alternating arrangement. For example, if the top side of the cyclohexane ring shown in Figure 4.17 has axial hydrogens on carbons 1, 3, and 5, then it has equatorial hydrogens on carbons 2, 4, and 6. Exactly the reverse is true for the bottom side: Carbons 1, 3, and 5 have equatorial hydrogens, but carbons 2, 4, and 6 have axial hydrogens.

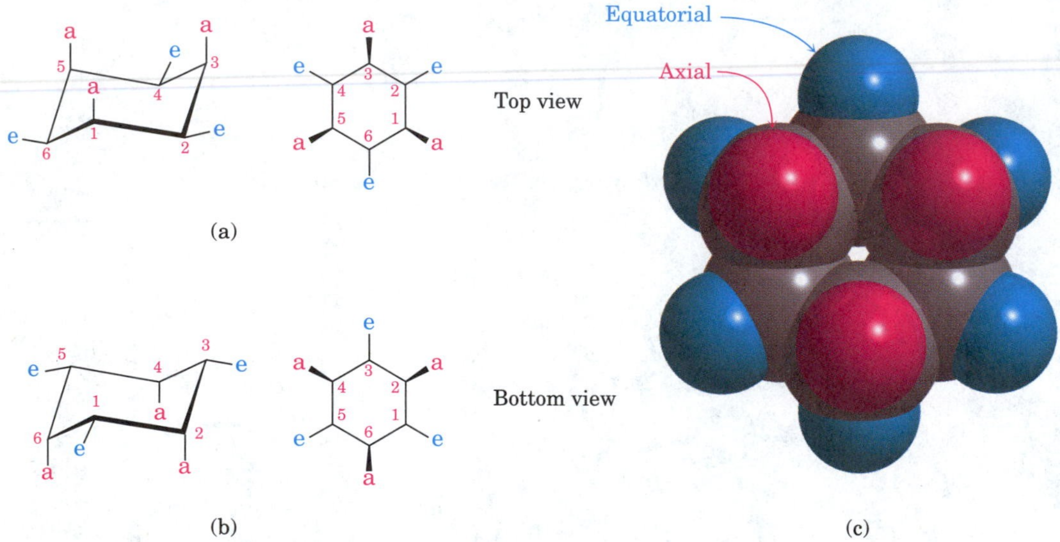

Figure 4.17 Alternating axial and equatorial positions in chair cyclohexane. Each carbon atom has one axial and one equatorial position, and each side has alternating axial/equatorial positions. Part (c) is a picture of a cyclohexane ring viewed from the bottom.

Note that we haven't used the words *cis* and *trans* in this discussion of cyclohexane geometry. Two hydrogens on the same side of the ring are always cis, regardless of whether they're axial or equatorial and regardless of whether they're adjacent. Similarly, two hydrogens on opposite sides of the ring are always trans, regardless of whether they're axial or equatorial.

Axial and equatorial bonds can be drawn following the procedure outlined in Figure 4.18. (Look at a molecular model as you practice.)

Axial bonds: The six axial bonds, one on each carbon, are parallel and alternate up–down.

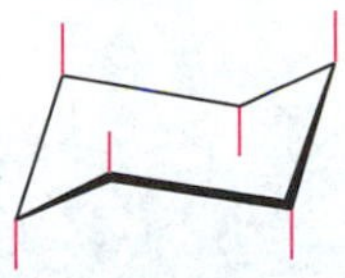

Axial bonds

Equatorial bonds: The six equatorial bonds, one on each carbon, come in three sets of two parallel lines. Each set is also parallel to two ring bonds. Equatorial bonds alternate between sides around the ring.

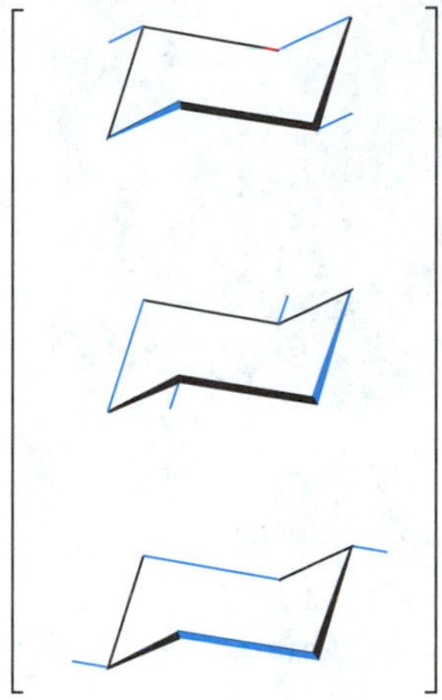

Equatorial bonds

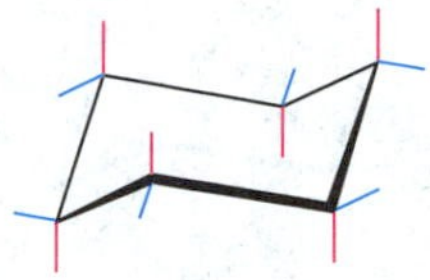

Completed cyclohexane

Figure 4.18 A procedure for drawing axial and equatorial bonds in chair cyclohexane.

4.11 Conformational Mobility of Cyclohexane

Because chair cyclohexane has two kinds of positions, axial and equatorial, we might expect to find two isomeric forms of a monosubstituted cyclohexane. In fact, though, there is only *one* methylcyclohexane, *one* bromocyclohexane, *one* cyclohexanol, and so on, because cyclohexane rings are *conformationally mobile* at room temperature. The two chair conformations readily interconvert, resulting in the exchange of axial and equatorial positions. This interconversion of chair conformations, usually referred to as a **ring-flip**, is shown in Figure 4.19. Molecular models show the process more clearly, and you should practice ring-flipping with models.

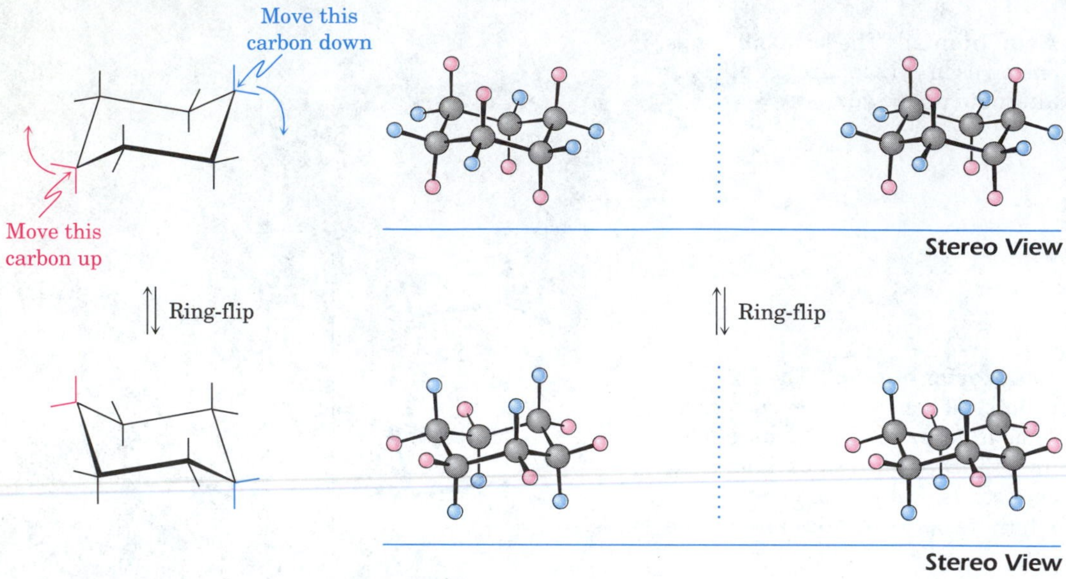

Figure 4.19 A ring-flip in chair cyclohexane interconverts axial and equatorial positions.

A chair cyclohexane can be ring-flipped by keeping the middle four carbon atoms in place while folding the two ends in opposite directions. The net result of a ring-flip is that axial and equatorial positions interconvert. An axial substituent in one chair form becomes an equatorial substituent in the ring-flipped chair form, and vice versa. For example, axial methylcyclohexane becomes equatorial methylcyclohexane after ring-flip. Since the energy barrier to chair–chair interconversion is only about 45 kJ/mol (10.8 kcal/mol), the process is extremely rapid at room temperature. We therefore see only what appears to be a single structure, rather than distinct axial and equatorial isomers.

PRACTICE PROBLEM

Draw 1,1-dimethylcyclohexane, indicating whether each methyl group is axial or equatorial.

Solution First draw a chair cyclohexane ring, and then put two methyl groups on the same carbon. The methyl group in the rough plane of the ring is equatorial, and the other (above or below the ring) is axial.

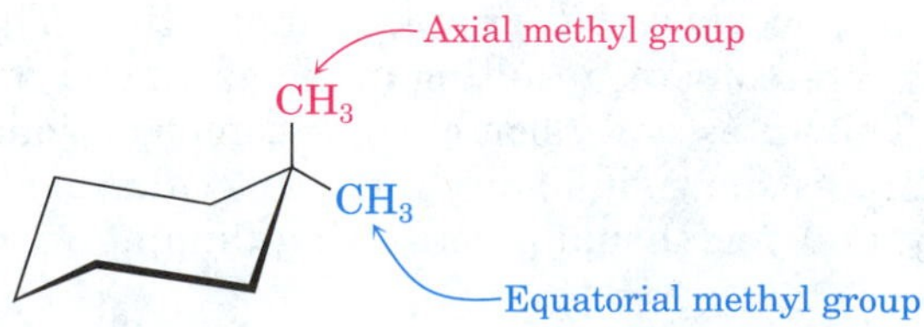

PROBLEM...

4.11 Draw two different chair conformations of bromocyclohexane, showing all hydrogen atoms. Identify each substituent as axial or equatorial.

PROBLEM...

4.12 Why must a *cis*-1,2-disubstituted cyclohexane, such as *cis*-1,2-dichlorocyclohexane, have one group axial and one group equatorial?

PROBLEM...

4.13 Why must a *trans*-1,2-disubstituted cyclohexane have either both groups axial or both groups equatorial?

PROBLEM...

4.14 Draw two different chair conformations of *trans*-1,4-dimethylcyclohexane, and label all positions as axial or equatorial.

....

4.12 Conformations of Monosubstituted Cyclohexanes

Although cyclohexane rings rapidly flip between conformations at room temperature, the two conformers of a monosubstituted cyclohexane aren't equally stable. In methylcyclohexane, for example, the equatorial conformer is more stable than the axial conformer by 7.6 kJ/mol (1.8 kcal/mol). Similarly for other monosubstituted cyclohexanes: A substituent is almost always more stable in an equatorial position than in an axial position.

It's possible to calculate the percentages of two isomers at equilibrium using the equation $K = e^{-(\Delta E/RT)}$, where K is the equilibrium constant between isomers, $e = 2.718$ (the base of natural logarithms), ΔE is the energy difference between isomers, T is the kelvin temperature, and $R = 8.314$ J/(K·mol), as shown in Figure 4.20 (p. 130). For example, an energy difference of 7.3 kJ/mol (1.7 kcal/mol) means that about 95% of methylcyclohexane molecules have the methyl group equatorial at any given instant, and only 5% have the methyl group axial.

The energy difference between axial and equatorial conformers is due to steric strain caused by so-called **1,3-diaxial interactions**. That is, the axial methyl group on C1 is too close to the axial hydrogens three carbons away on C3 and C5, resulting in 7.6 kJ (1.8 kcal/mol) of steric strain (Figure 4.21).

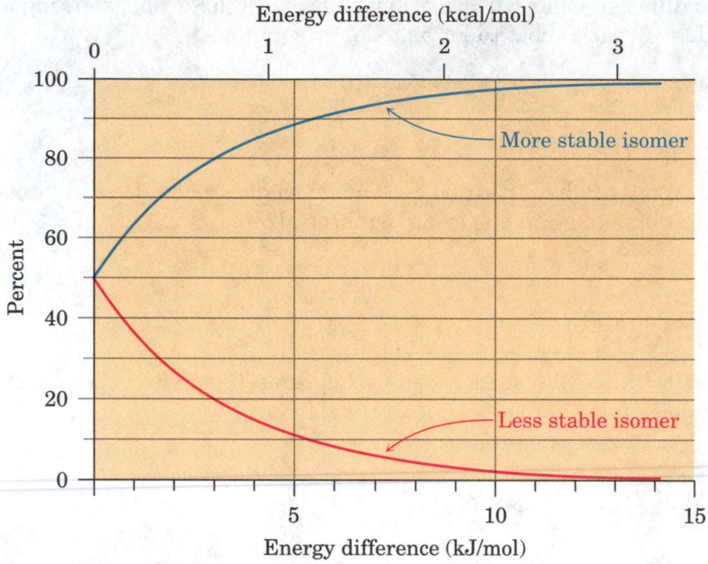

Figure 4.20 A plot of the percentages of two isomers at equilibrium versus the energy difference between them. The curves are calculated from the equation $K = e^{-(\Delta E/RT)}$, where K is the equilibrium constant between isomers, $e = 2.718$ (the base of natural logarithms), ΔE is the energy difference between isomers, T is the kelvin temperature, and $R = 8.314$ J/(K·mol).

1,3-Diaxial steric strain is already familiar—we've seen it before as the steric strain between methyl groups in gauche butane (Section 4.3). Recall that gauche butane is less stable than anti butane by 3.8 kJ/mol (0.9 kcal/mol) because of steric interference between hydrogen atoms on the two methyl groups. Comparing a four-carbon fragment of axial methylcyclohexane with gauche butane shows that the steric interaction is the same in both cases (Figure 4.22, p. 132). Because methylcyclohexane has two such interactions, though, it has $2 \times 3.8 = 7.6$ kJ/mol of steric strain.

Sighting along the C1–C2 bond of axial methylcyclohexane shows that the axial hydrogen at C3 has a gauche butane interaction with the axial methyl group at C1. Sighting similarly along the C1–C6 bond shows that the axial hydrogen at C5 also has a gauche butane interaction with the axial methyl group at C1. Both of these interactions are absent in equatorial methylcyclohexane, and we therefore find an energy difference of 7.6 kJ/mol (1.8 kcal/mol) between the two forms.

What is true for methylcyclohexane is also true for other monosubstituted cyclohexanes: A substituent is almost always more stable in an equatorial position than in an axial position. The exact amount of 1,3-diaxial steric strain in a specific compound depends, of course, on the nature and size of the axial group; Table 4.2 (p. 132) lists strain values for some common

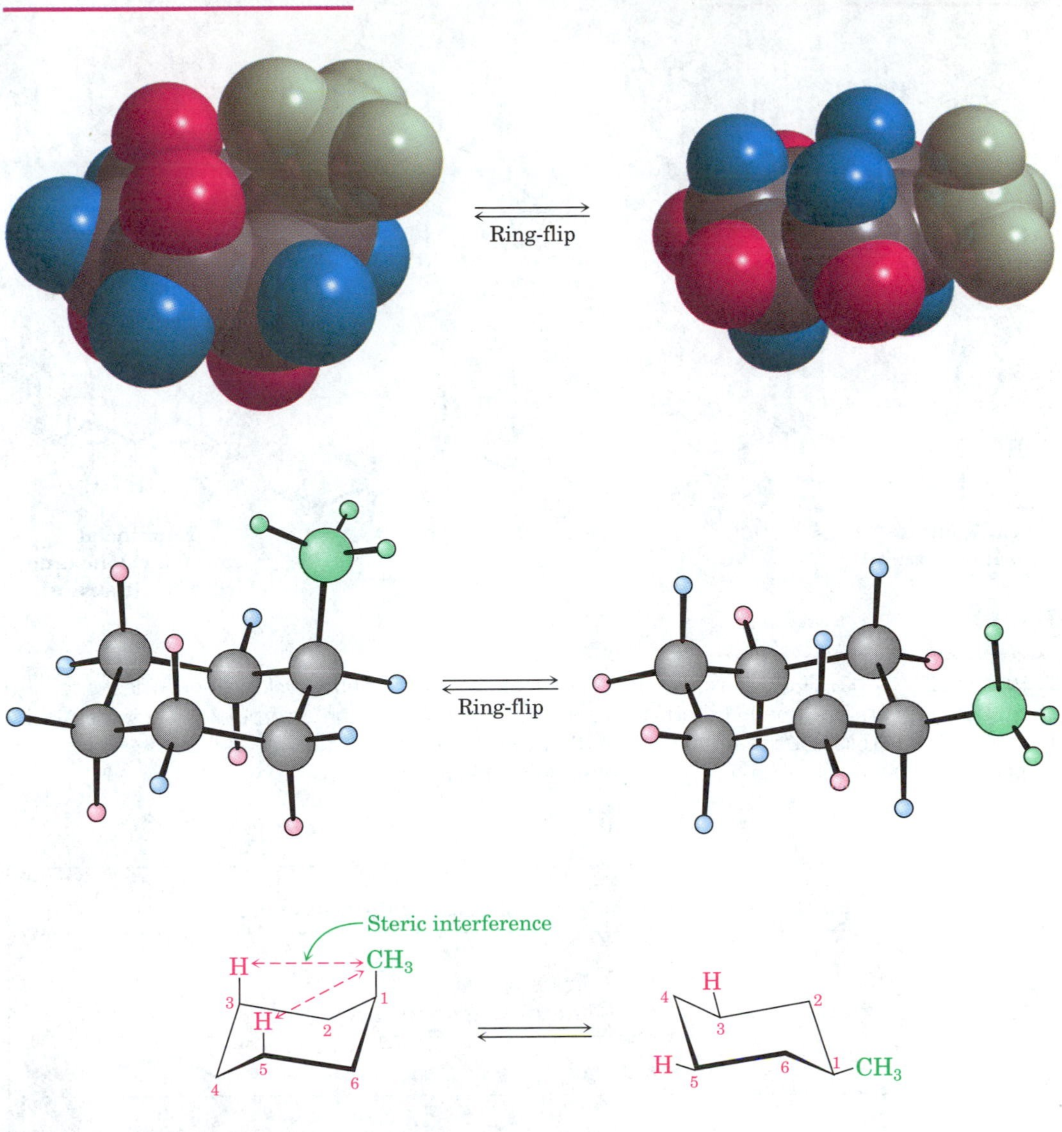

Figure 4.21 Interconversion of axial and equatorial methylcyclohexane, as represented in several formats. The equatorial conformer is more stable than the axial conformer by 7.6 kJ/mol.

substituents. As you might expect, the amount of steric strain increases through the series $H_3C- < CH_3CH_2- < (CH_3)_2CH- \ll (CH_3)_3C-$ in parallel with the increasing bulk of the successively larger alkyl groups. Note that the values in Table 4.2 refer to 1,3-diaxial interactions of the indicated group with a *single* hydrogen atom. These values must be doubled to arrive at the amount of strain in a monosubstituted cyclohexane.

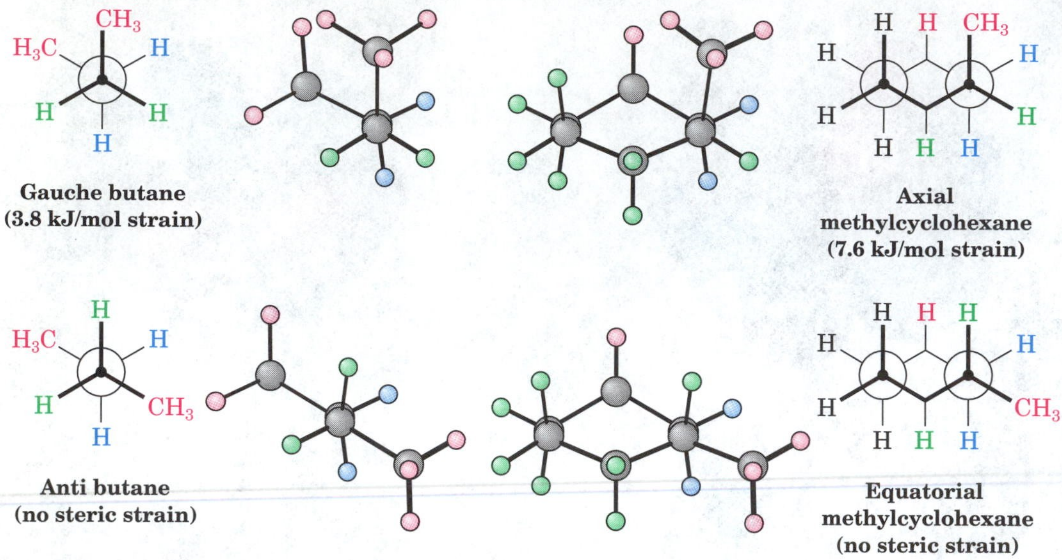

Figure 4.22 Origin of 1,3-diaxial cyclohexane interactions in methylcyclohexane. The steric strain between an axial methyl group and axial hydrogen atoms three carbons away is identical to the steric strain in gauche butane. (Note that the CH_3 group in methylcyclohexane is displaced slightly away from a true axial position to minimize strain.)

Table 4.2 Steric Strain in Monosubstituted Cyclohexanes

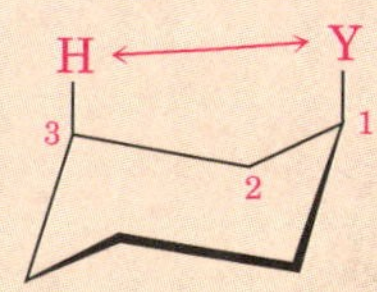

	Strain of one H–Y *1,3-diaxial interaction*	
Y	*(kJ/mol)*	*(kcal/mol)*
—F	0.5	0.12
—Cl	1.0	0.25
—Br	1.0	0.25
—OH	2.1	0.5
—CH_3	3.8	0.9
—CH_2CH_3	4.0	0.95
—$CH(CH_3)_2$	4.6	1.1
—$C(CH_3)_3$	11.4	2.7
—C_6H_5	6.3	1.5
—COOH	2.9	0.7
—CN	0.4	0.1

PROBLEM.

4.15 How can you account for the fact (see Table 4.2) that an axial *tert*-butyl substituent has much larger 1,3-diaxial interactions than isopropyl, but isopropyl is fairly similar to ethyl and methyl? Use molecular models to help with your answer.

PROBLEM.

4.16 Why do you suppose an axial cyano substituent causes practically no 1,3-diaxial steric strain (0.4 kJ/mol)?

PROBLEM.

4.17 Look at Figure 4.20 and estimate the percentages of axial and equatorial conformers present at equilibrium in bromocyclohexane.

4.13 Conformational Analysis of Disubstituted Cyclohexanes

Monosubstituted cyclohexanes usually have the substituent in an equatorial position. In disubstituted cyclohexanes, however, the situation is more complex because the steric effects of both substituents must be taken into account. All steric interactions in both possible chair conformations must be analyzed before deciding which conformation is favored.

Let's look at 1,2-dimethylcyclohexane. There are two isomers, *cis*-1,2-dimethylcyclohexane and *trans*-1,2-dimethylcyclohexane, which must be considered separately. In the cis isomer, both methyl groups are on the same side of the ring, and the compound can exist in either of the two chair conformations shown in Figure 4.23 (p. 134). (It's often easier to see whether a compound is cis- or trans-disubstituted by first drawing the ring as a flat representation and then converting to a chair conformation.)

Both chair conformations of *cis*-1,2-dimethylcyclohexane shown in Figure 4.23 have one methyl group axial and one methyl group equatorial. The conformation on the left has an axial methyl group at C2, which has 1,3-diaxial interactions with hydrogens on C4 and C6. The ring-flipped conformation on the right has an axial methyl group at C1, which has 1,3-diaxial interactions with hydrogens on C3 and C5. In addition, both conformations have gauche butane interactions between the two methyl groups. *The two conformations are exactly equal in energy,* with a total steric strain of 3×3.8 kJ/mol = 11.4 kJ/mol (2.7 kcal/mol).

In *trans*-1,2-dimethylcyclohexane, the two methyl groups are on opposite sides of the ring, and the compound can exist in either of the two chair conformations shown in Figure 4.23. The situation here is quite different from that of the cis isomer. The trans conformation on the left in Figure 4.23 has both methyl groups equatorial and therefore has only a gauche

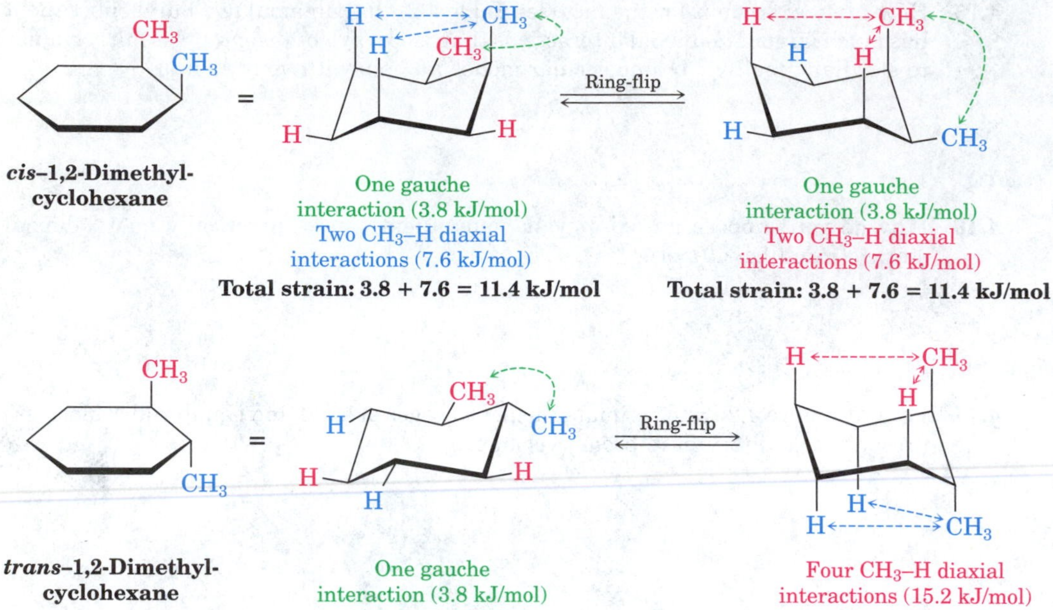

Figure 4.23 Conformations of *cis-* and *trans*-1,2-dimethylcyclohexane. In the cis isomer, the two chair conformations are equal in energy because each has one axial methyl group and one equatorial methyl group. In the trans isomer, the conformation with both methyl groups equatorial is favored by 11.4 kJ/mol (2.7 kcal/mol) over the conformation with both methyl groups axial.

butane interaction between methyls (3.8 kJ/mol) but no 1,3-diaxial interactions. The conformation on the right, however, has both methyl groups axial. The axial methyl group at C1 interacts with axial hydrogens at C3 and C5, and the axial methyl group at C2 interacts with axial hydrogens at C4 and C6. These four 1,3-diaxial interactions produce a steric strain of 4×3.8 kJ/mol $= 15.2$ kJ/mol and make the diaxial conformation $15.2 - 3.8 = 11.4$ kJ/mol (2.7 kcal/mol) less favorable than the diequatorial conformation. We therefore predict that *trans*-1,2-dimethylcyclohexane will exist almost exclusively (>99%) in the diequatorial conformation.

The same kind of **conformational analysis** just carried out for *cis-* and *trans*-1,2-dimethylcyclohexane can be done for any substituted cyclohexane, such as *cis*-1-*tert*-butyl-4-chlorocyclohexane in the following practice problem. It turns out that the large amount of steric strain caused by an axial *tert*-butyl group effectively holds the cyclohexane ring in a single conformation. Chemists sometimes take advantage of this steric locking to study the chemical reactivity of an immobile cyclohexane ring.

PRACTICE PROBLEM

What is the most stable conformation of *cis*-1-*tert*-butyl-4-chlorocyclohexane, and by how much is it favored?

Solution First draw the two chair conformations of the molecule. *cis*-1-*tert*-Butyl-4-chlorocyclohexane can exist in either of the two chair conformations indicated:

$2 \times 1.0 = 2.0$ kJ/mol steric strain $\qquad$ $2 \times 11.4 = 22.8$ kJ/mol steric strain

In the left-hand conformation, the *tert*-butyl group is equatorial and the chlorine is axial. In the right-hand conformation, the *tert*-butyl group is axial and the chlorine is equatorial. These conformations aren't of equal energy because an axial *tert*-butyl substituent and an axial chloro substituent produce different amounts of steric strain. Table 4.2 shows that the 1,3-diaxial interaction between a hydrogen and a *tert*-butyl group costs 11.4 kJ/mol (2.7 kcal/mol), whereas the interaction between a hydrogen and a chlorine costs only 1.0 kJ/mol (0.25 kcal/mol). An axial *tert*-butyl group therefore produces $(2 \times 11.4 \text{ kJ/mol}) - (2 \times 1.0 \text{ kJ/mol}) = 20.8$ kJ/mol (4.9 kcal/mol) more steric strain than does an axial chlorine, and the compound preferentially adopts the conformation with the chlorine axial and the *tert*-butyl equatorial.

PROBLEM

4.18 Draw the most stable chair conformation of the following molecules, and estimate the amount of 1,3-diaxial strain in each.
(a) *trans*-1-Chloro-3-methylcyclohexane
(b) *cis*-1-Ethyl-2-methylcyclohexane
(c) *cis*-1-Bromo-4-ethylcyclohexane
(d) *cis*-1-*tert*-Butyl-4-ethylcyclohexane

4.14 Boat Cyclohexane

In addition to chair cyclohexane, a second conformation called **boat cyclohexane** is also free of angle strain. We haven't paid it any attention thus far, however, because boat cyclohexane is less stable than chair cyclohexane (Figure 4.24).

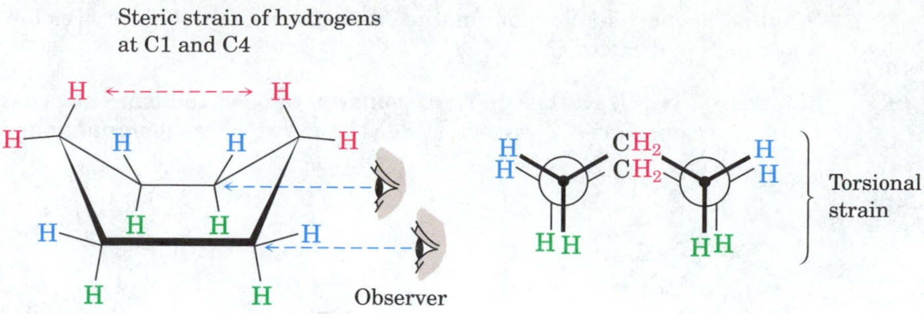

Figure 4.24 The boat conformation of cyclohexane. There is steric strain and torsional strain in this conformation but no angle strain.

There are two "kinds" of carbon atoms in boat cyclohexane. Carbons 2, 3, 5, and 6 lie in a plane, with carbons 1 and 4 above the plane. The inside hydrogen atoms on carbons 1 and 4 approach each other closely enough to produce considerable steric strain, and the four pairs of hydrogens on carbons 2, 3, 5, and 6 are eclipsed. Thus, boat cyclohexane has both steric strain

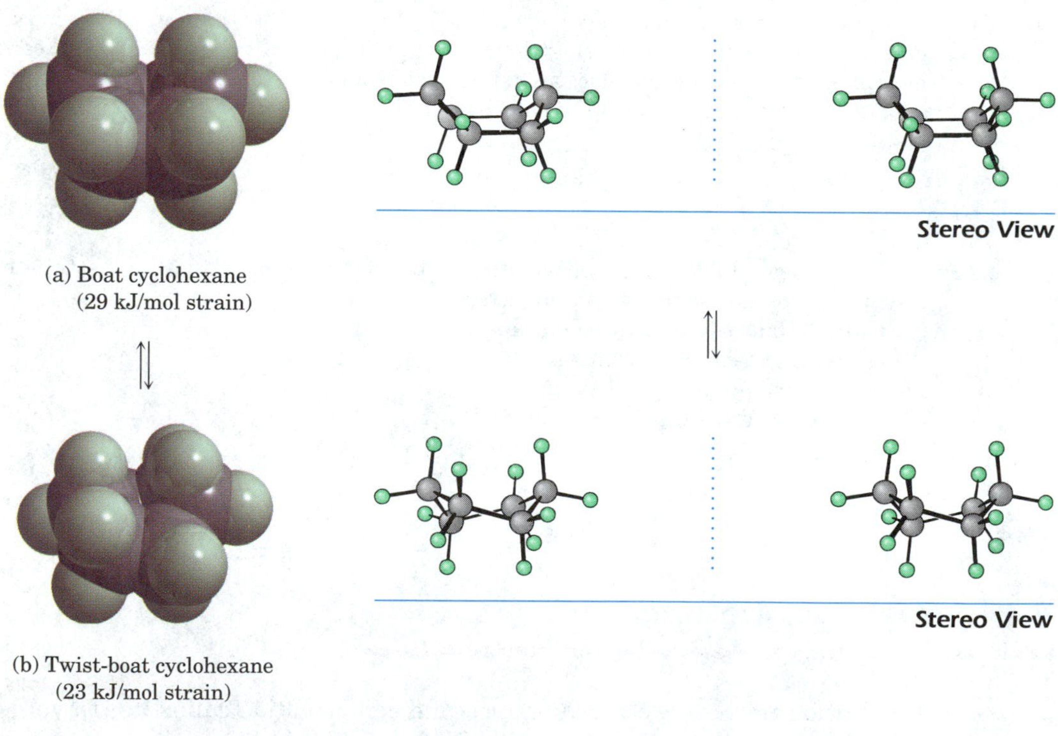

Figure 4.25 Boat and twist-boat conformations of cyclohexane. The twist-boat conformation is lower in energy than the boat conformation by 6 kJ/mol. Both conformations are much more strained than chair cyclohexane.

and torsional strain. The Newman projection in Figure 4.24, obtained by sighting along the C2–C3 and C5–C6 bonds, shows this eclipsing clearly.

Boat cyclohexane is approximately 29 kJ/mol (7.0 kcal/mol) less stable than chair cyclohexane, although this value is reduced to about 23 kJ/mol (5.5 kcal/mol) by twisting slightly, thereby relieving some torsional strain (Figure 4.25). Even the **twist-boat conformation** is still much more strained than the chair conformation, though, and molecules adopt this geometry only under special circumstances.

PROBLEM.

4.19 *trans*-1,3-Di-*tert*-butylcyclohexane is one of the few molecules that exists largely in a twist-boat conformation. Why? Draw the likely twist-boat conformation.

4.15 Conformations of Polycyclic Molecules

The last point we'll consider about cycloalkane stereochemistry is to see what happens when two or more cycloalkane rings are fused together to construct a **polycyclic** molecule—for example, decalin:

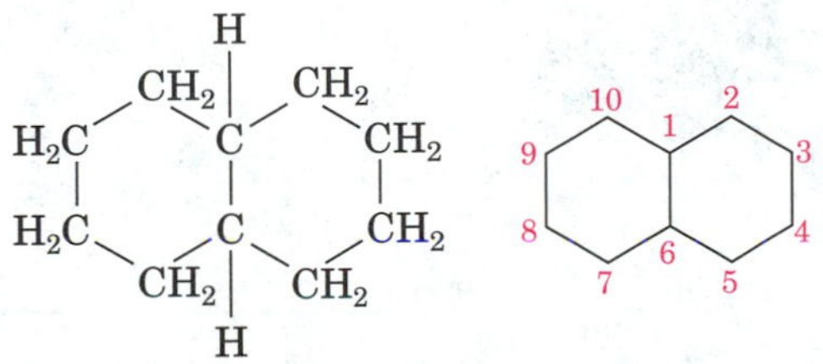

Decalin (two fused cyclohexane rings)

Decalin consists of two cyclohexane rings joined to share two carbon atoms (the *bridgehead* carbons, C1 and C6) and a common bond. Decalin can exist in either of two isomeric forms, depending on whether the rings are trans fused or cis fused. In *trans*-decalin, the hydrogen atoms at the bridgehead carbons are on opposite sides of the rings; in *cis*-decalin, the bridgehead hydrogens are on the same side. Figure 4.26 shows how both compounds can be represented using chair cyclohexane conformations. Note that *trans*- and *cis*-decalin are not interconvertible by ring-flips or other rotations. They are cis–trans stereoisomers (Section 3.8) and have the same relationship to each other that *cis*- and *trans*-1,2-dimethylcyclohexane have.

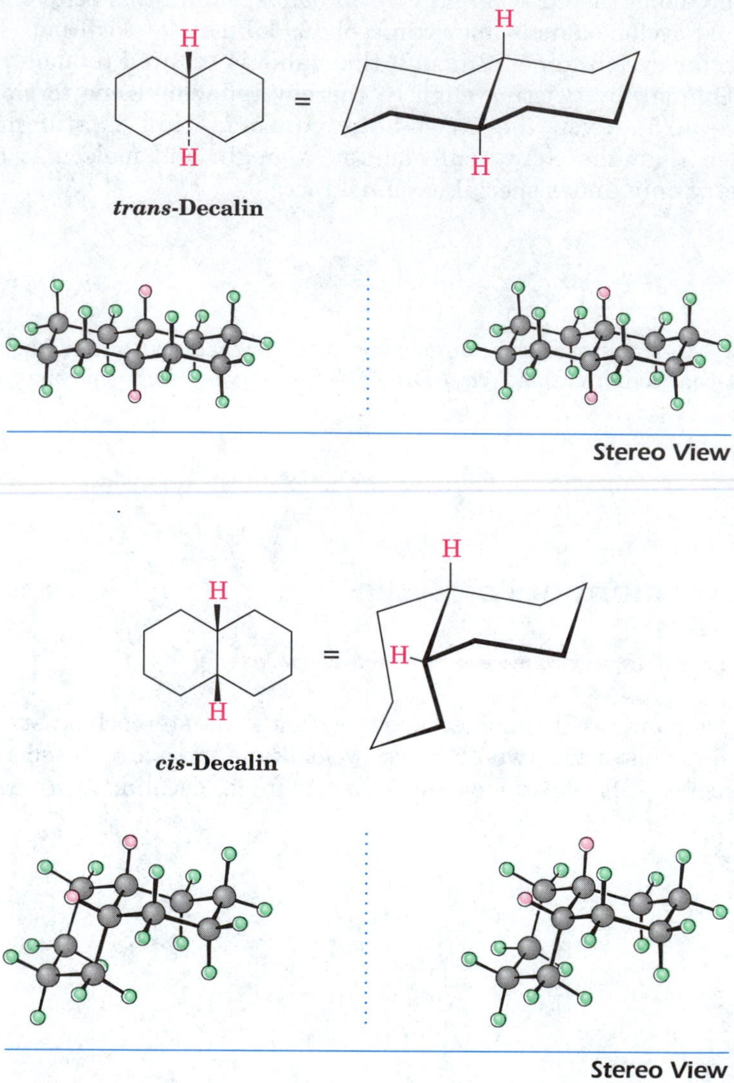

Figure 4.26 Representations of *trans-* and *cis-*decalin. The hydrogen atoms at the bridgehead carbons are on the same side in the rings in the cis isomer but on opposite sides in the trans isomer.

Polycyclic compounds are common, and many valuable substances have fused-ring structures. For example, the steroids—such as cholesterol—have four rings fused together—three six-membered and one five-membered. Though steroids look complicated compared with cyclohexane or decalin, they consist simply of chair cyclohexane rings locked together. The same principles that apply to the conformational analysis of simple cyclohexane rings apply equally well (and often better) to steroids.

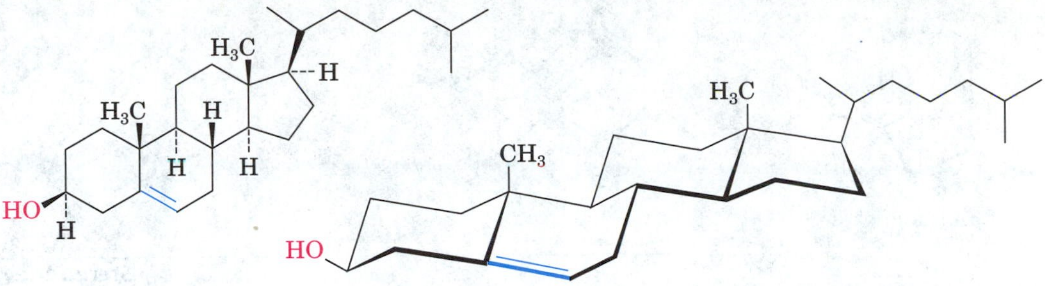

Cholesterol (a steroid)

Another common fused-ring system is the norbornane, or bicyclo[2.2.1]-heptane, structure. Like decalin, norbornane is a *bicycloalkane*, so-called because *two* rings would have to be broken open to generate an acyclic structure. Its systematic name, bicyclo[2.2.1]heptane, reflects the fact that the molecule has seven carbons, is bicyclic, and has three bridges of 2, 2, and 1 carbon atoms connecting the two bridgehead carbons.

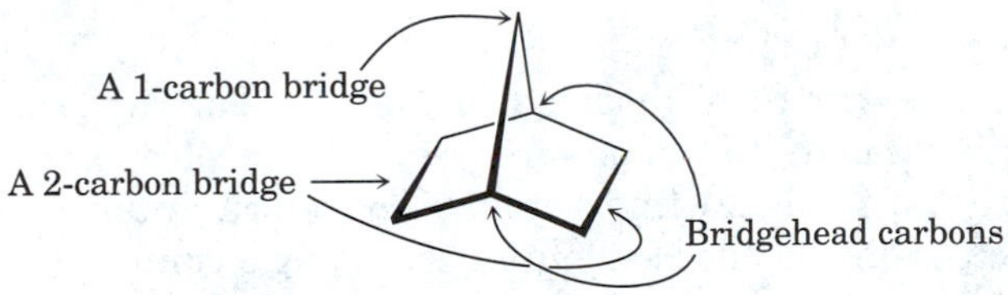

Norbornane
(Bicyclo[2.2.1]heptane)

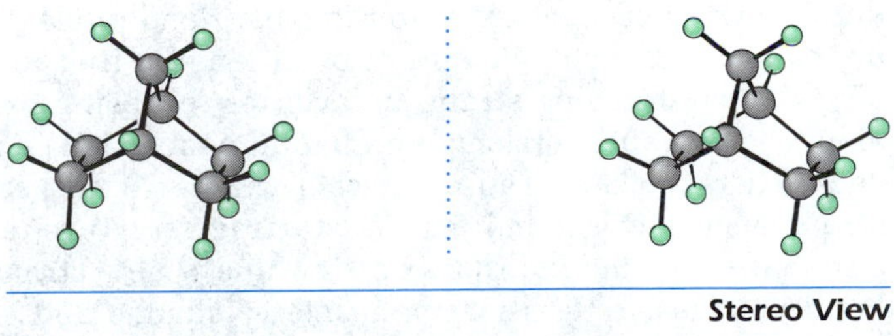

Norbornane has a conformationally locked boat cyclohexane ring in which carbons 1 and 4 are joined by an extra $-CH_2-$ group. Note how, in drawing this structure, the bond crossing in front of another is indicated by a break in the rear bond. Making a molecular model is particularly helpful when trying to see the three-dimensionality of norbornane.

Substituted norbornanes, such as camphor (shown at the top of the next page), are found widely in nature, and many have been important historically in developing organic structural theories.

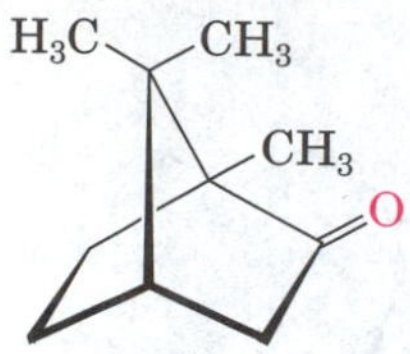

Camphor

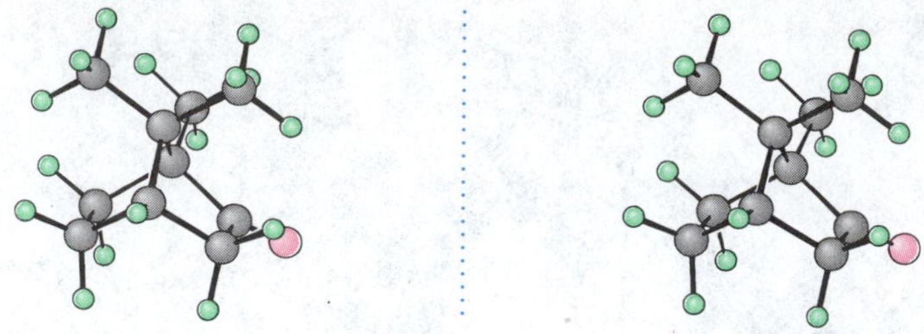

Stereo View

PROBLEM.

4.20 Which isomer is more stable, *trans*-decalin or *cis*-decalin? Explain.

INTERLUDE

Molecular Mechanics

All the structural models of molecules portrayed in this book are computer-drawn. To make sure they accurately portray bond angles, bond lengths, torsional interactions, and steric interactions, the geometry of each molecule has been optimized using a commercially available *molecular mechanics* program developed by N. L. Allinger of the University of Georgia.

The idea behind molecular mechanics is to input a rough geometry for a molecule and then calculate a total strain energy for that starting geometry, using mathematical equations that assign energy values to specific kinds of molecular interactions. Bond angles that are too large or too small cause angle strain; bond lengths that are too short or too long cause stretching strain; unfavorable eclipsing interactions around single bonds cause torsional strain; and nonbonded atoms that approach each other too closely cause steric, or van der Waals, strain. With a total strain energy calculated for the starting structure, a small change is automatically made to the geometry, and a new strain energy is calculated. Successive iterations of geometry change and energy calculation ultimately converge on a minimum-energy geometry corresponding to the most favorable conformation of the molecule.

Molecular mechanics calculations have proven to be enormously useful in organic chemistry, particularly in pharmaceutical research where the complementary fit between a drug molecule and a receptor molecule in the body is often a key to designing new pharmaceutical agents. Morphine and other opium alkaloids, for instance, have a certain three-dimensional shape (Figure 4.27) that allows them to nestle into complementary shaped cavities on opiate receptor proteins in the brain. With

(continued)➤

this shape known, other molecules calculated to have similar shapes can be designed, leading to the possibility of enhanced biological activity.

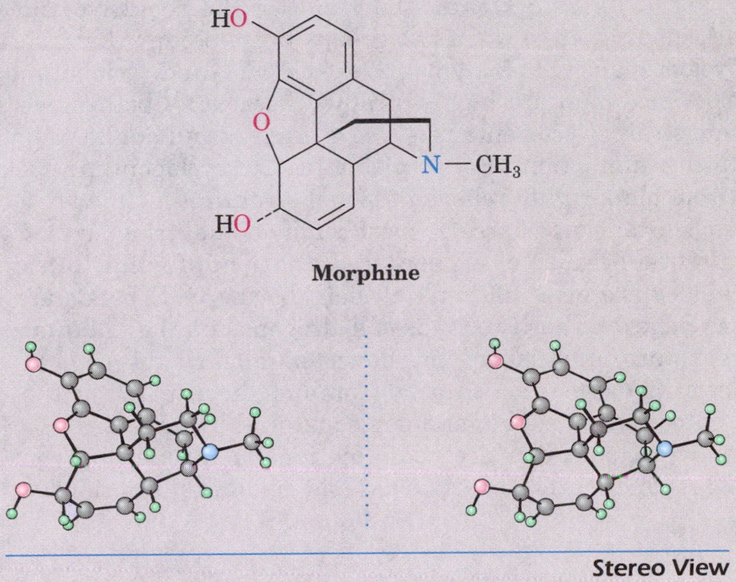

Figure 4.27 The structure of morphine and a stereoview of its minimum-energy conformation, as calculated by molecular mechanics.

Summary and Key Words

Carbon–carbon single bonds in alkanes are formed by σ overlap of carbon sp^3 hybrid orbitals. Rotation is possible around σ bonds because of their cylindrical symmetry, and alkanes therefore have a large number of rapidly interconverting **conformations**. **Newman projections** make it possible to visualize the spatial consequences of bond rotation by sighting directly along a carbon–carbon bond axis. The **staggered** conformation of ethane is 12 kJ/mol (2.9 kcal/mol) more stable than the **eclipsed** conformation because of **torsional strain**. In general, any alkane is most stable when all its bonds are staggered.

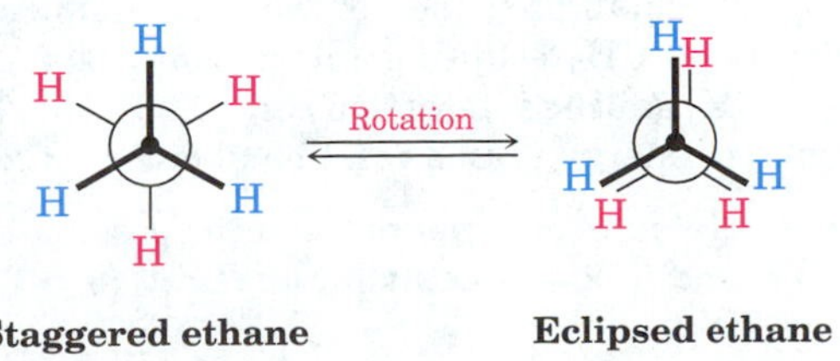

Cycloalkanes have special characteristics that affect their properties. Combustion studies indicate, for example, that not all cycloalkanes are equally stable. Three kinds of strain contribute to the overall energy of a cycloalkane: (1) **angle strain**, the resistance of a bond angle to compression or expansion from the normal 109° tetrahedral value; (2) **torsional strain**, the energy cost of having neighboring C–H bonds eclipsed rather than staggered; and (3) **steric strain**, the result of the repulsive van der Waals interactions that arise when two groups try to occupy the same space.

Cyclopropane (114.9 kJ/mol; 27.5 kcal/mol) and cyclobutane (110.4 kJ/mol; 26.4 kcal/mol) are highly strained because of both angle strain and torsional strain. Cyclopentane is free of angle strain but has a large number of eclipsing interactions. Both cyclobutane and cyclopentane pucker slightly away from planarity to relieve torsional strain.

Cyclohexane rings are the most common of all ring sizes. Cyclohexane is strain-free because of its puckered **chair conformation**, in which all bond angles are near 109° and all neighboring C–H bonds are staggered. Chair cyclohexane has two kinds of hydrogens: **axial** and **equatorial**. Axial hydrogens are oriented up and down, parallel to the ring axis, whereas equatorial hydrogens lie in a belt around the equator of the ring. Each carbon atom has one axial and one equatorial hydrogen.

Chair cyclohexanes are conformationally mobile and can undergo a **ring-flip**, which interconverts axial and equatorial positions:

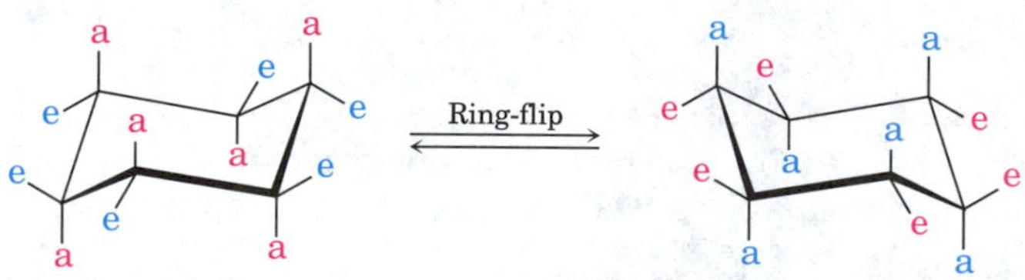

Substituents on the ring are more stable in the equatorial position, because axial substituents cause **1,3-diaxial steric strain**. The amount of 1,3-diaxial strain caused by an axial substituent depends on its bulk.

ADDITIONAL PROBLEMS

4.21 Define the following terms:
(a) Angle strain (b) Steric strain (c) Torsional strain
(d) Heat of combustion (e) Conformation (f) Staggered
(g) Eclipsed (h) Gauche butane

4.22 Consider 2-methylbutane (isopentane). Sighting along the C2–C3 bond:
(a) Draw a Newman projection of the most stable conformation.
(b) Draw a Newman projection of the least stable conformation.
(c) If a $CH_3 \leftrightarrow CH_3$ eclipsing interaction costs 11 kJ/mol (2.5 kcal/mol) and a $CH_3 \leftrightarrow CH_3$ gauche interaction costs 3.8 kJ/mol (0.9 kcal/mol), make a quantitative plot of energy versus rotation about the C2–C3 bond.

4.23 What are the relative energies of the three possible staggered conformations around the C2–C3 bond in 2,3-dimethylbutane?

4.24 Construct a qualitative potential-energy diagram for rotation about the C–C bond of 1,2-dibromoethane. Which conformation would you expect to be more stable? Label the anti and gauche conformations of 1,2-dibromoethane.

4.25 Which conformation of 1,2-dibromoethane (Problem 4.24) would you expect to have the larger dipole moment? The observed dipole moment is $\mu = 1.0$ D. What does this tell you about the actual structure of the molecule?

4.26 The barrier to rotation about the C–C bond in bromoethane is 15 kJ/mol (3.6 kcal/mol).
(a) What energy value can you assign to an H–Br eclipsing interaction?
(b) Construct a quantitative diagram of potential energy versus amount of bond rotation for bromoethane.

4.27 Define the following terms:
(a) Axial bond
(b) Equatorial bond
(c) Chair conformation
(d) 1,3-Diaxial interaction

4.28 Draw a chair cyclohexane ring and label all positions as axial or equatorial.

4.29 Why is a 1,3-cis disubstituted cyclohexane more stable than its trans isomer?

4.30 Why is a 1,2-trans disubstituted cyclohexane more stable than its cis isomer?

4.31 Which is more stable, a 1,4-trans disubstituted cyclohexane or its cis isomer?

4.32 *N*-Methylpiperidine has the conformation shown. What does this tell you about the relative steric requirements of a methyl group versus an electron lone pair?

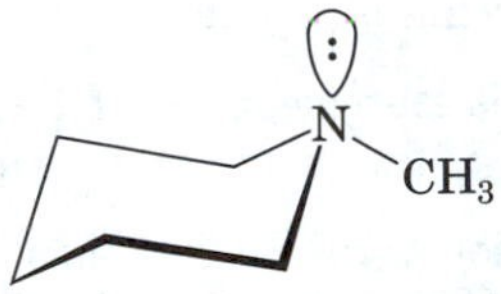

***N*-Methylpiperidine**

4.33 Draw the two chair conformations of *cis*-1-chloro-2-methylcyclohexane. Which is more stable, and by how much?

4.34 Draw the two chair conformations of *trans*-1-chloro-2-methylcyclohexane. Which is more stable, and by how much?

4.35 β-Glucose contains a six-membered ring in which all the substituents are equatorial. Draw β-glucose in its more stable chair conformation.

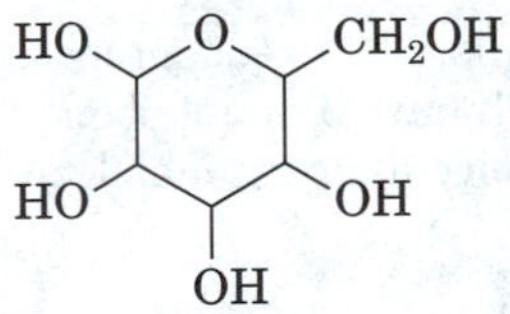

Glucose

4.36 From the data in Figure 4.20 and Table 4.2, estimate the percentages of molecules that have their substituents in an axial orientation for these compounds:
(a) Isopropylcyclohexane
(b) Fluorocyclohexane
(c) Cyclohexanecarbonitrile, $C_6H_{11}CN$
(d) Cyclohexanol, $C_6H_{11}OH$

4.37 Assume that you have a variety of cyclohexanes substituted in the positions indicated. Identify the substituents as either axial or equatorial. For example, a 1,2-cis relationship means that one substituent must be axial and one equatorial, whereas a 1,2-trans relationship means that both substituents are axial or both are equatorial.
(a) 1,3-Trans disubstituted
(b) 1,4-Cis disubstituted

(c) 1,3-Cis disubstituted
(d) 1,5-Trans disubstituted
(e) 1,5-Cis disubstituted
(f) 1,6-Trans disubstituted

4.38 The diaxial conformation of *cis*-1,3-dimethylcyclohexane is approximately 23 kJ/mol (5.4 kcal/mol) less stable than the diequatorial conformation. Draw the two possible chair conformations, and suggest a reason for the large energy difference.

4.39 Approximately how much steric strain does the 1,3-diaxial interaction between the two methyl groups introduce into the diaxial conformation (see Problem 4.38)?

4.40 In light of your answer to Problem 4.39, draw the two chair conformations of 1,1,3-trimethylcyclohexane and estimate the amount of strain energy in each. Which conformation is favored?

4.41 Draw 1,3,5-trimethylcyclohexane using a hexagon to represent the ring. How many cis–trans stereoisomers are there?

4.42 Which of the 1,3,5-trimethylcyclohexane stereoisomers you drew in Problem 4.41 is the most stable?

4.43 We saw in Problem 4.20 that *cis*-decalin is less stable than *trans*-decalin. Assume that the 1,3-diaxial interactions in *trans*-decalin are similar to those in axial methylcyclohexane (one $CH_3 \leftrightarrow H$ interaction costs 3.8 kJ/mol; 0.9 kcal/mol), and calculate the magnitude of the energy difference between *cis*- and *trans*-decalin.

4.44 Using molecular models as well as structural drawings, explain why *trans*-decalin is rigid and cannot ring-flip, whereas *cis*-decalin can easily ring-flip.

4.45 How many cis–trans stereoisomers of 1,2,3,4,5,6-hexachlorocyclohexane are there? Draw the structure of the most stable isomer.

4.46 Increased substitution around a bond leads to increased strain. Take the four substituted butanes listed below, for example. For each compound, sight along the C2–C3 bond and draw Newman projections of the most stable and least stable conformations. Use the data in Table 4.1 to assign strain energy values to each conformation. Which of the eight conformations is most stable? Which is least stable?
(a) 2-Methylbutane
(b) 2,2-Dimethylbutane
(c) 2,3-Dimethylbutane
(d) 2,2,3-Trimethylbutane

4.47 One of the two chair structures of *cis*-1-chloro-3-methylcyclohexane is more stable than the other by 15.5 kJ/mol (3.7 kcal/mol). Which is it? What is the energy cost of a 1,3-diaxial interaction between a chlorine and a methyl group?

4.48 The German chemist J. Bredt proposed in 1935 that bicycloalkenes such as 1-norbornene, which have a double bond to the bridgehead carbon, are too strained to exist. Make a molecular model of 1-norbornene and explain Bredt's proposal.

1-Norbornene

4.49 Tell whether each of the following substituents on a steroid is axial or equatorial. (A substituent that is "up" is on the top side of the molecule as drawn, and a substituent that is "down" is on the bottom side.)
(a) Substituent up at C3
(b) Substituent down at C7
(c) Substituent down at C11

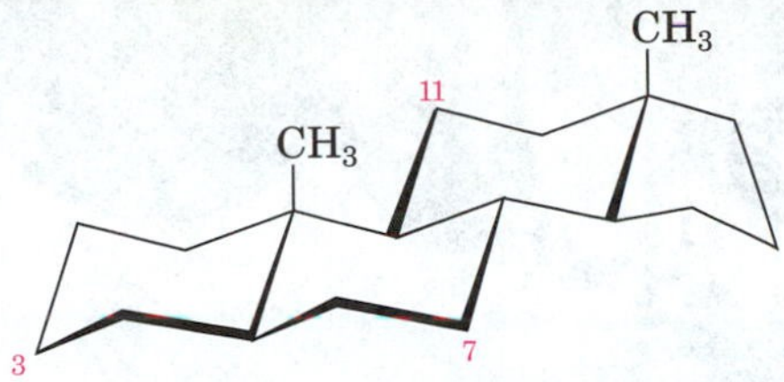

4.50 Amantadine is an antiviral agent that is active against influenza A infection. Draw a three-dimensional representation of amantadine showing the chair cyclohexane rings.

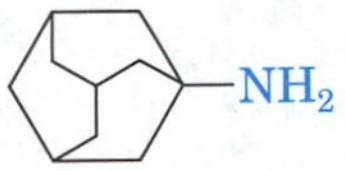

Amantadine

A Look Ahead

4.51 We'll see in Chapter 11 that alkyl halides undergo an *elimination* reaction to yield alkenes on treatment with strong base. For example, chlorocyclohexane gives cyclohexene on reaction with $NaNH_2$:

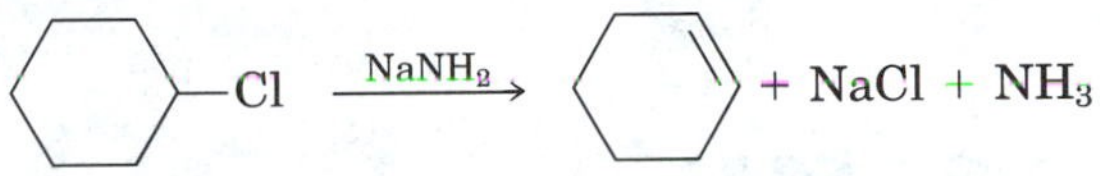

If axial chlorocyclohexanes are generally more reactive than their equatorial isomers, which do you think would react faster, *cis*-1-*tert*-butyl-2-chlorocyclohexane or *trans*-1-*tert*-butyl-2-chlorocyclohexane? Explain.

4.52 We'll see in Chapter 19 that ketones react with alcohols to yield products called *acetals*. Why does the all-cis isomer of 4-*tert*-butylcyclohexane-1,3-diol react readily with acetone and an acid catalyst to form an acetal, but other stereoisomers do not react?

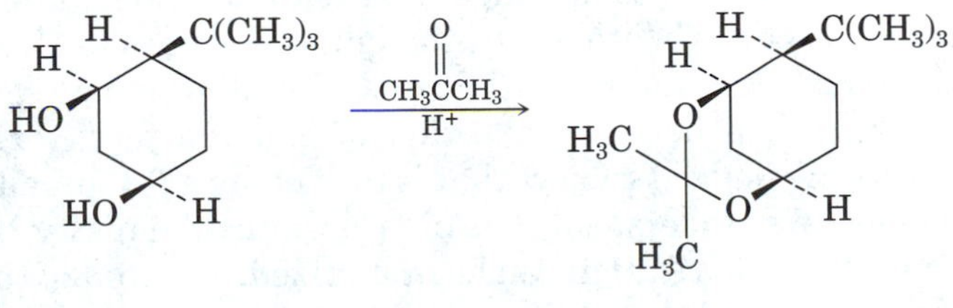

An acetal

In formulating your answer, draw the more stable chair conformations of all four stereoisomers and the product acetal. Use molecular models for help.

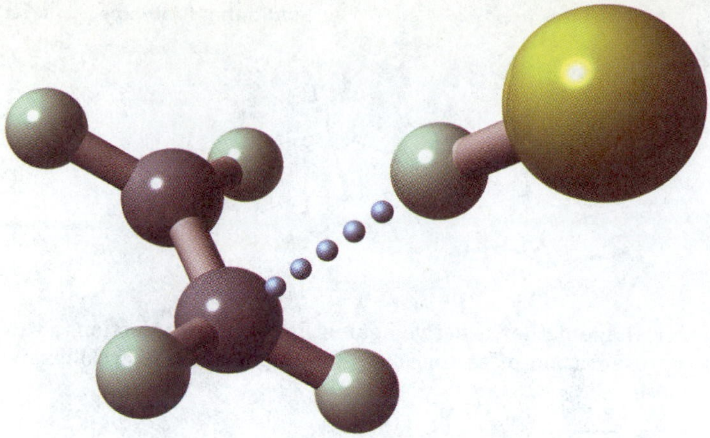

Ethylene is protonated by reaction with HBr, yielding a carbocation intermediate.

5

An Overview of Organic Reactions

When first approached, organic chemistry can seem like a bewildering collection of millions of compounds, dozens of functional groups, and a huge number of reactions. With study, though, it becomes evident that there are only a few fundamental ideas that underlie all organic reactions.

Far from being a collection of isolated facts, organic chemistry is a beautifully logical subject that is unified by a few broad themes. When these themes are understood, learning organic chemistry becomes much easier and rote memorization can be minimized. The aim of this book is to describe the themes and clarify the patterns that unify organic chemistry. We'll begin by taking an overview of the fundamental kinds of organic reactions that take place and seeing how reactions can be described.

5.1 Kinds of Organic Reactions

Organic chemical reactions can be organized in two ways: by *what kinds* of reactions occur and by *how* reactions occur. Let's look first at the kinds of reactions that take place. There are four general types of organic reactions: additions, eliminations, substitutions, and rearrangements.

Addition reactions occur when two reactants add together to form a single new product with no atoms "left over." We can generalize the process as

These reactants add together . . . $A + B \longrightarrow C$ . . . to give this single product.

An example of an addition reaction that we'll be studying soon is the reaction of an alkene, such as ethylene, with HBr to yield an alkyl bromide:

These two reactants . . .

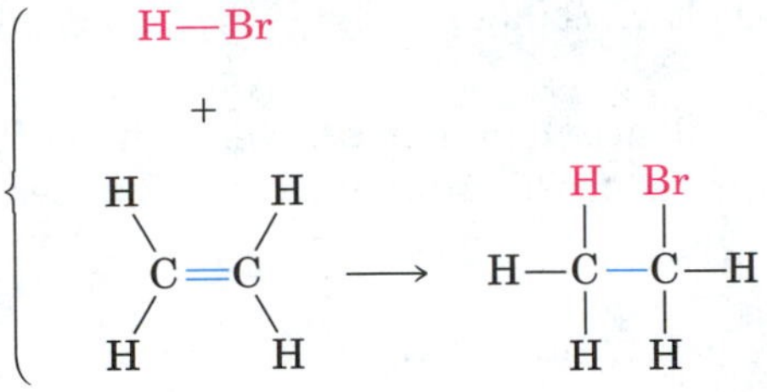

. . . add to give this product.

Ethylene (an alkene) **Bromoethane (an alkyl halide)**

Elimination reactions are, in a sense, the opposite of addition reactions. Eliminations occur when a single reactant splits into two products:

This one reactant . . . $A \longrightarrow B + C$ . . . splits apart to give these two products.

An example of an elimination reaction is the reaction of an alkyl halide with base to yield an acid and an alkene:

This one reactant . . .

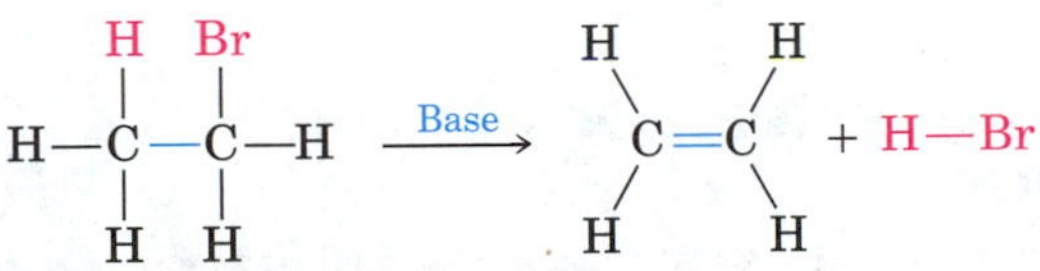

. . . gives these two products.

Bromoethane (an alkyl halide) **Ethylene (an alkene)**

Substitution reactions occur when two reactants exchange parts to give two new products:

These two reactants exchange parts . . . $A{-}B + C{-}D \longrightarrow A{-}C + B{-}D$ . . . to give these two new products.

An example of a substitution reaction is the reaction of an alkane with chlorine gas in the presence of ultraviolet light to yield an alkyl chloride. A –Cl group from chlorine substitutes for an –H group of the alkane, and two new products result:

These two reactants . . .

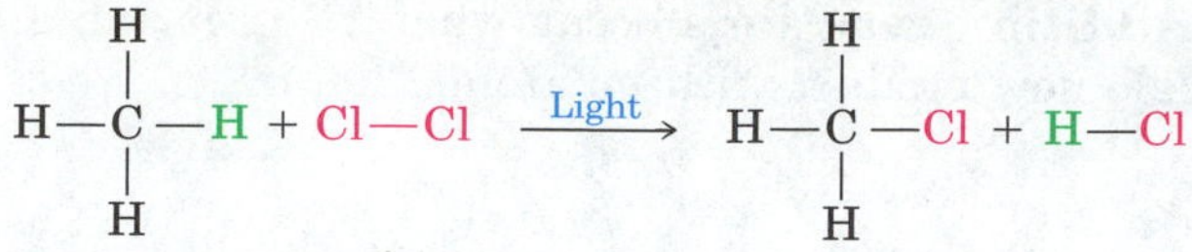

. . . give these two products.

Methane (an alkane)

Chloromethane (an alkyl halide)

Rearrangement reactions occur when a single reactant undergoes a reorganization of bonds and atoms to yield an isomeric product:

This single reactant . . . A $\longrightarrow$ B . . . gives this isomeric product.

An example of a rearrangement reaction is the conversion of the alkene 1-butene into its constitutional isomer 2-butene by treatment with an acid catalyst:

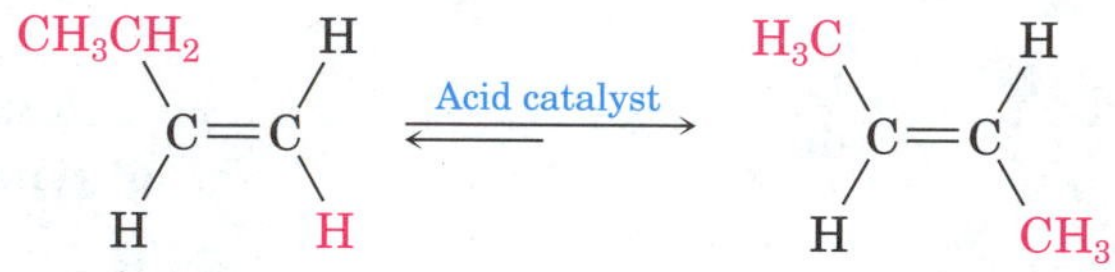

1-Butene **2-Butene**

PROBLEM

5.1 Classify these reactions as additions, eliminations, substitutions, or rearrangements:

(a) $CH_3Br + KOH \longrightarrow CH_3OH + KBr$

(b) $CH_3CH_2OH \longrightarrow H_2C{=}CH_2 + H_2O$

(c) $H_2C{=}CH_2 + H_2 \longrightarrow CH_3CH_3$

5.2 How Organic Reactions Occur: Mechanisms

Having looked at the kinds of reactions that take place, let's now see how reactions occur. An overall description of how a reaction occurs is called a **reaction mechanism**. A mechanism describes in detail exactly what takes place at each stage of a chemical transformation. It describes which bonds are broken and in what order, which bonds are formed and in what order, and what the relative rate of each step is. A complete mechanism must also account for all reactants used, all products formed, and the amount of each.

All chemical reactions involve bond breaking and bond making. When two reactants come together, react, and yield products, specific bonds in the reactants are broken, and specific bonds in the products are formed. Fundamentally, there are two ways in which a covalent two-electron bond can break: A bond can break in an electronically *symmetrical* way so that one electron remains with each product fragment, or a bond can break in an electronically *unsymmetrical* way so that both bonding electrons remain with one product fragment, leaving the other fragment with a vacant orbital. The symmetrical cleavage is said to be **homolytic**, and the unsymmetrical cleavage is said to be **heterolytic**.[1]

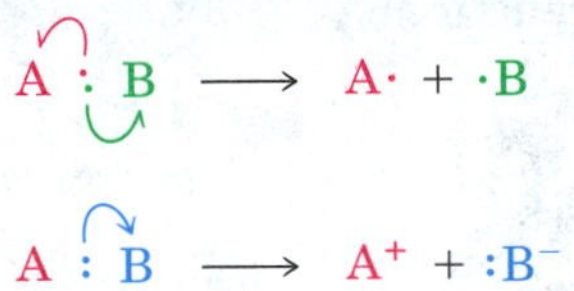

Homolytic bond breaking (radical)
(one electron stays with each fragment)

Heterolytic bond breaking (polar)
(two electrons stay with one fragment)

Conversely, there are two ways in which a covalent two-electron bond can form: A bond can form in an electronically symmetrical (**homogenic**) way when one electron is donated to the new bond by each reactant, or a bond can form in an electronically unsymmetrical (**heterogenic**) way when both bonding electrons are donated to the new bond by one reactant.

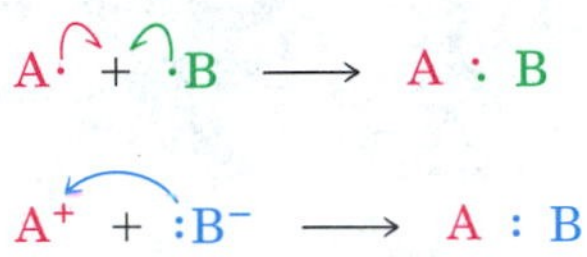

Homogenic bond making (radical)
(one electron donated by each fragment)

Heterogenic bond making (polar)
(two electrons donated by one fragment)

Processes that involve symmetrical bond breaking and bond making are called **radical reactions**. A **radical** (sometimes called a "free radical") is a chemical species that contains an odd number of valence electrons and thus has a single, unpaired electron in one of its orbitals. Processes that involve unsymmetrical bond breaking and bond making are called **polar reactions**. Polar reactions involve species that have an even number of valence electrons and thus have only electron pairs in their orbitals. Polar processes are the more common reaction type in organic chemistry, and a large part of this book is devoted to their description.

In addition to polar and radical reactions, there is a third, less commonly encountered type called *pericyclic reactions*. Rather than explain pericyclic reactions now, though, we'll study them in more detail in Chapter 31.

5.3 Radical Reactions and How They Occur

Radical reactions are not as common as polar reactions, but they're nevertheless important in organic chemistry, particularly in certain industrial processes. Let's see how they occur.

[1]We'll develop this point in more detail later, but you might notice that the movement of *one* electron in a homolytic process is indicated using a half-headed arrow, or "fishhook" (↷), whereas the movement of *two* electrons in a heterolytic process is indicated using a full-headed curved arrow (↷).

Although most radicals are electrically neutral, they are highly reactive because they contain an atom with an odd number of electrons (usually seven) in its valence shell, rather than a stable noble-gas octet. They can achieve a valence-shell octet in several ways. For example, a radical might abstract an atom from another molecule, leaving behind a new radical. The net result is a radical *substitution* reaction:

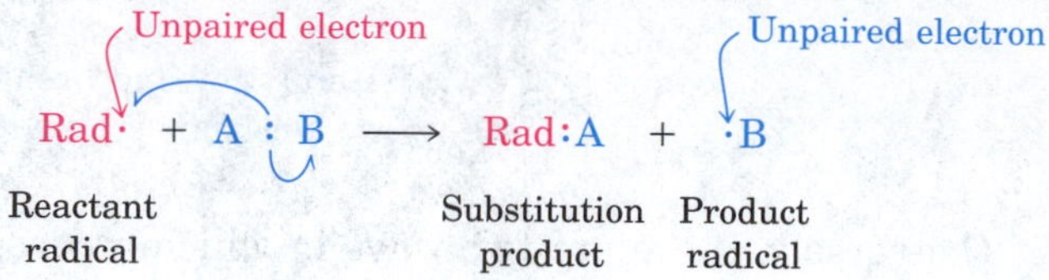

Alternatively, a reactant radical might add to an alkene, taking one electron from the alkene double bond and yielding a new radical. The net result is a radical *addition* reaction:

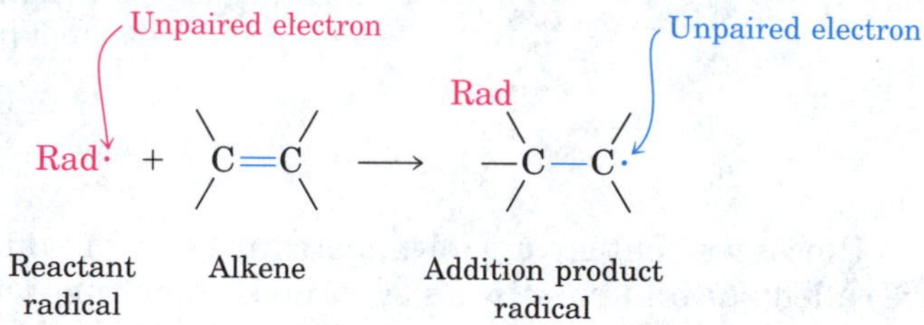

Let's look at a specific example of a radical reaction—the chlorination of methane—to see its characteristics. A more detailed discussion of this radical substitution reaction is given in Chapter 10. For the present, it's only necessary to know that methane chlorination is a multistep process involving radicals.

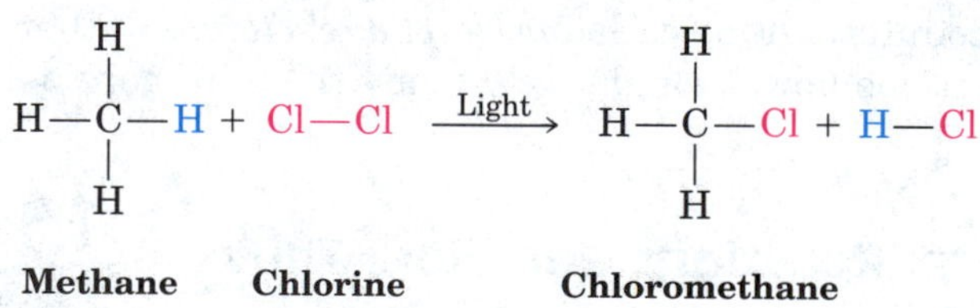

Radical substitution reactions normally require three kinds of steps: initiation, propagation, and termination.

1. *Initiation* The initiation step starts off the reaction by producing reactive radicals. In the present case, the relatively weak Cl–Cl bond is homolytically broken by irradiation with ultraviolet light. Two reactive chlorine radicals are produced, and further chemistry ensues.

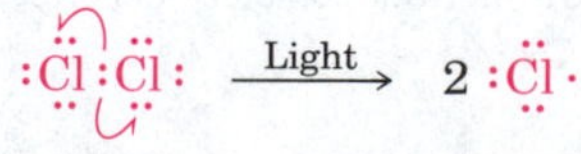

2. *Propagation steps* Once a small number of chlorine radicals have been produced, propagation steps take place. When a reactive chlorine radical collides with a methane molecule, it abstracts a hydrogen atom to produce HCl and a methyl radical ($\cdot CH_3$). This methyl radical reacts further with Cl_2 in a second propagation step to give the product chloromethane and a new chlorine radical, which cycles back into the first propagation step. Once the sequence has been initiated, it becomes a self-sustaining cycle of repeating steps (a) and (b), making the overall process a **chain reaction**.

   (a) $:\ddot{\underset{..}{Cl}}\cdot + H{:}CH_3 \longrightarrow H{:}\ddot{\underset{..}{Cl}}: + \cdot CH_3$

   (b) $\cdot CH_3 + :\ddot{\underset{..}{Cl}}:\ddot{\underset{..}{Cl}}: \longrightarrow :\ddot{\underset{..}{Cl}}:CH_3 + :\ddot{\underset{..}{Cl}}\cdot$

   (c) Repeat steps (a) and (b) over and over.

3. *Termination steps* Occasionally, two radicals might collide and combine to form a stable product. When this happens, the reaction cycle is broken and the chain is ended. Such termination steps occur infrequently, however, because the concentration of radicals in the reaction at any given moment is very small. Thus, the likelihood that two radicals will collide is also small.

$$\left.\begin{array}{l} :\ddot{\underset{..}{Cl}}\cdot + \cdot\ddot{\underset{..}{Cl}}: \longrightarrow :\ddot{\underset{..}{Cl}}:\ddot{\underset{..}{Cl}}: \\ :\ddot{\underset{..}{Cl}}\cdot + \cdot CH_3 \longrightarrow :\ddot{\underset{..}{Cl}}:CH_3 \\ H_3C\cdot + \cdot CH_3 \longrightarrow H_3C{:}CH_3 \end{array}\right\} \text{Possible termination steps}$$

The radical substitution reaction just discussed is only one of several different processes that radicals can undergo. The fundamental principle behind all radical reactions is the same, however: All bonds are broken and formed by reaction of odd-electron species.

PROBLEM

5.2 Alkane chlorination is not a generally useful reaction, because most alkanes have several different kinds of hydrogens, causing mixtures of chlorinated products to result. Draw and name all monochloro products you might obtain from reaction of 2-methylpentane with chlorine.

PROBLEM

5.3 Radical chlorination of pentane is a poor way to prepare 1-chloropentane, $CH_3CH_2CH_2CH_2CH_2Cl$, but radical chlorination of neopentane, $(CH_3)_4C$, is a good way to prepare neopentyl chloride, $(CH_3)_3CCH_2Cl$. Explain.

5.4 Polar Reactions and How They Occur

Polar reactions occur because of the attraction between positive and negative charges on molecules. To see how these reactions take place, we first need to recall the discussion of polar covalent bonds in Section 2.1 and then we need to look more deeply into the effects of bond polarity on organic molecules.

Most organic molecules are electrically neutral; they have no net charge, either positive or negative. We saw in Section 2.1, however, that certain bonds within a molecule, particularly the bonds in functional groups, are polar. Bond polarity is a consequence of an unsymmetrical electron distribution in a bond and is due to the difference in electronegativity of the bonded atoms. (See Table 5.1, which repeats part of Figure 2.2 for convenience.)

Table 5.1 Electronegativity of Some Common Elements

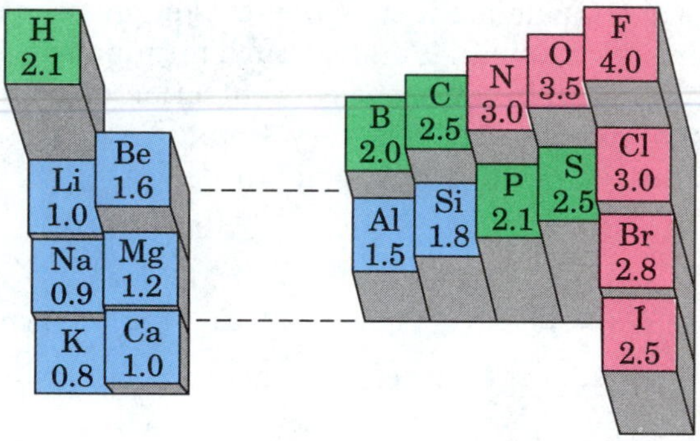

Elements such as oxygen, nitrogen, fluorine, chlorine, and bromine are more electronegative than carbon. Thus, a carbon atom bonded to one of these electronegative elements has a partial *positive* charge ($\delta+$), and the electronegative atom has a slight negative charge ($\delta-$). Conversely, metals are less electronegative than carbon. A carbon atom bonded to a metal therefore has a partial *negative* charge, and the metal has a partial positive charge.

where Y = O, N, Cl, Br, I   where M = a metal such as Mg or Li

The polarity patterns of some common functional groups are shown in Table 5.2.

This discussion of bond polarity is oversimplified in that we've considered only bonds that are inherently polar due to differences in electronegativity. Polar bonds can also result from the interaction of functional groups

Table 5.2 Polarity Patterns in Some Common Functional Groups

Compound type	*Functional group structure*	*Compound type*	*Functional group structure*
Alcohol	$-\overset{\delta+}{C}-\overset{\delta-}{O}H$	Carbonyl	$\overset{\delta+}{C}=\overset{\delta-}{O}$
Alkene	$C=C$ Symmetrical, nonpolar	Carboxylic acid	$-\overset{\delta+}{C}(=\overset{\delta-}{O})-\overset{\delta-}{O}H$
Alkyl halide	$-\overset{\delta+}{C}-\overset{\delta-}{X}$	Carboxylic acid chloride	$-\overset{\delta+}{C}(=\overset{\delta-}{O})-\overset{\delta-}{Cl}$
Amine	$-\overset{\delta+}{C}-\overset{\delta-}{N}H_2$	Aldehyde	$-\overset{\delta+}{C}(=\overset{\delta-}{O})-H$
Ether	$-\overset{\delta+}{C}-\overset{\delta-}{O}-\overset{\delta+}{C}-$	Ester	$-\overset{\delta+}{C}(=\overset{\delta-}{O})-\overset{\delta-}{O}-C$
Nitrile	$-\overset{\delta+}{C}\equiv\overset{\delta-}{N}$	Ketone	$-\overset{\delta+}{C}(=\overset{\delta-}{O})-C$
Grignard reagent	$-\overset{\delta-}{C}-\overset{\delta+}{Mg}Br$		
Alkyllithium	$-\overset{\delta-}{C}-\overset{\delta+}{Li}$		

with solvents and with Lewis acids or bases. For example, the polarity of the carbon–oxygen bond in methanol is greatly enhanced by protonation of the oxygen atom:

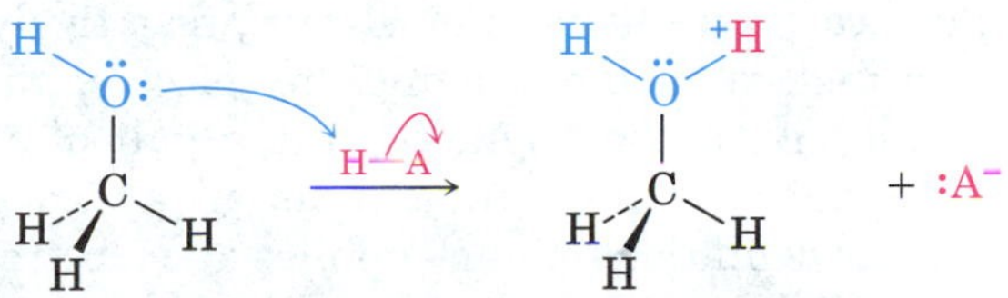

Methanol (weakly polar C–O bond)   **Protonated methanol** (strongly polar C–O bond)

In neutral methanol, the carbon atom is somewhat electron-poor because the electronegative oxygen attracts the electrons in the carbon–oxygen bond. In the protonated methanol cation, however, a full positive charge on oxygen *strongly* attracts the electrons in the carbon–oxygen bond and makes the carbon much more electron-poor.

Yet a further consideration is the *polarizability* (as opposed to polarity) of an atom. As the electric field around a given atom changes because of changing interactions with solvent or with other polar reagents, the electron distribution around that atom also changes. The measure of this response to an external influence is called the **polarizability** of the atom. Larger atoms with more loosely held electrons are more polarizable than smaller atoms with tightly held electrons. Thus, iodine is much more polarizable than fluorine. The effect of iodine's high polarizability is that the carbon–iodine bond, although electronically symmetrical according to the electronegativity table (Table 5.1), can nevertheless react as if it were polar.

What does functional-group polarity mean with respect to chemical reactivity? *Because unlike charges attract each other, the fundamental characteristic of all polar organic reactions is that electron-rich sites in one molecule react with electron-poor sites in another molecule*. Bonds are made when the electron-rich reagent donates a pair of electrons to the electron-poor reagent, and bonds are broken when one of the two product fragments leaves with the electron pair.

As we saw in Section 2.8, chemists normally indicate the movement of an electron pair during a polar reaction by using a curved arrow. *A curved arrow shows where electrons move when reactant bonds are broken and product bonds are formed*. It means that an electron pair moves *from* the atom at the tail of the arrow *to* the atom at the head of the arrow during the reaction.

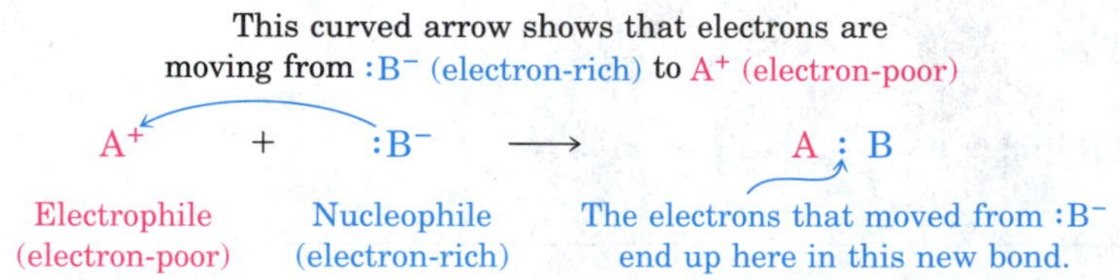

In referring to the species involved in a polar reaction, chemists have coined the words *nucleophile* and *electrophile*. A **nucleophile** is a substance that is "nucleus-loving"; a nucleophile has an electron-rich atom and can form a bond by donating a pair of electrons to an electron-poor atom. Nucleophiles are often, though not always, negatively charged. An **electrophile**, by contrast, is "electron-loving"; an electrophile has an electron-poor atom and can form a bond by accepting a pair of electrons from a nucleophile. Electrophiles are often, though not always, positively charged.

If the definitions of nucleophiles and electrophiles sound similar to those given in Section 2.8 for Lewis acids and Lewis bases, that's because there is indeed a correlation between electrophilicity/nucleophilicity and Lewis acidity/basicity. Lewis bases are electron donors and behave as nucleophiles, whereas Lewis acids are electron acceptors and behave as electrophiles. The main difference is that the terms *electrophile* and *nucleophile* are normally used when bonds to *carbon* are involved. We'll explore these ideas in more detail in Chapter 10.

PROBLEM

5.4 Identify the functional groups in the following molecules, and show the direction of polarity in each.

(a) Acetone, $CH_3\overset{O}{\overset{\|}{C}}CH_3$ (b) Ethyl propenoate, $H_2C{=}CH\overset{O}{\overset{\|}{C}}OCH_2CH_3$
(c) Chloroethylene, $H_2C{=}CHCl$
(d) Tetraethyllead, $(CH_3CH_2)_4Pb$ (the "lead" in gasoline)

PROBLEM

5.5 Which of the following species would you expect to behave as electrophiles, and which as nucleophiles?
(a) H^+ (b) OH^- (c) NH_3 (d) Mg^{2+}

5.5 An Example of a Polar Reaction: Addition of HBr to Ethylene

Let's look at a typical polar process—the addition reaction of ethylene with HBr. When ethylene is treated with hydrogen bromide at room temperature, bromoethane is produced. Overall, the reaction can be formulated as

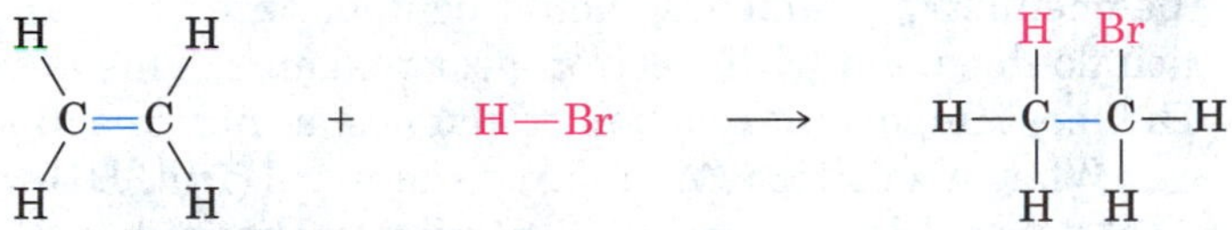

Ethylene (nucleophile) **Hydrogen bromide (electrophile)** **Bromoethane**

This reaction, an example of a polar reaction type known as an *electrophilic addition*, can be understood using the general concepts discussed in the previous section. We'll begin by looking at the nature of the two reactants.

What do we know about ethylene? We know from Section 1.10 that a carbon–carbon double bond results from orbital overlap of two sp^2-hybridized carbon atoms. The σ part of the double bond results from an sp^2–sp^2 overlap, and the π part results from a p–p overlap.

What kind of chemical reactivity might we expect of a carbon–carbon double bond? We know that *alkanes,* such as ethane, are relatively inert because all valence electrons are tied up in strong, nonpolar C–C and C–H bonds. Furthermore, the bonding electrons in alkanes are relatively inaccessible to approaching reagents because they are sheltered in σ bonds between nuclei. The electronic situation in *alkenes* is quite different, however. For one thing, double bonds have a greater electron density than single bonds—four electrons in a double bond versus only two electrons in a single bond. Equally important, the electrons in the π bond are accessible to approaching reagents because they are located above and below the plane of the double bond rather than being sheltered between the nuclei (Figure 5.1).

Carbon–carbon σ bond:
stronger; less accessible bonding electrons

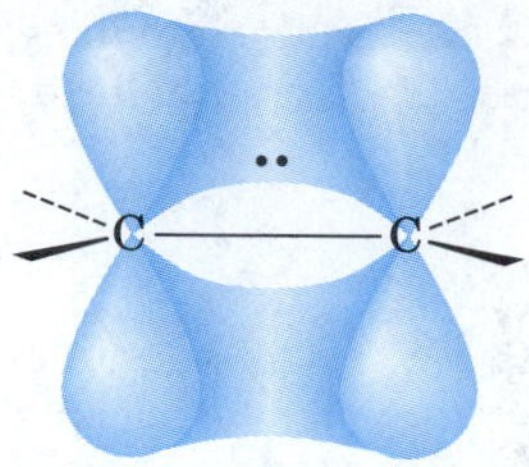

Carbon–carbon π bond:
weaker; more accessible electrons

Figure 5.1 A comparison of carbon–carbon single and double bonds. A double bond is both more electron-rich (nucleophilic) and more accessible to attack by approaching reagents than a single bond.

Both electron richness and electron accessibility lead to the prediction that carbon–carbon double bonds should behave as *nucleophiles*. That is, the chemistry of alkenes should be dominated by reaction of the electron-rich double bond with electron-poor reagents. This is exactly what we find: The most important reaction of alkenes is their reaction with electrophiles.

What about HBr? As a strong mineral acid, HBr is a powerful proton (H^+) donor. Since a proton is positively charged and electron-poor, it is a good electrophile. Thus, the reaction between H^+ and ethylene is a typical electrophile–nucleophile combination, characteristic of all polar reactions. (Although chemists often talk about "H^+" when referring to acids, remember that there's really no such species. As pointed out in Section 2.6, protons are always associated with another molecule for stability—for example, with water in H_3O^+.)

We'll see more details about alkene electrophilic addition reactions shortly, but for the present we can imagine the reaction as taking place by the pathway shown in Figure 5.2. The reaction begins when the alkene donates a pair of electrons from its C=C double bond to form a new single bond to H^+, as indicated by tracing the path of the curved arrow in the first step of Figure 5.2. One of the alkene carbon atoms bonds to the incoming hydrogen, but the other carbon atom, having lost its share of the double-bond electrons, now has only six valence electrons and is left with a positive charge. This positively charged species, a carbon cation or **carbocation**, is itself an electrophile that can accept an electron pair from nucleophilic bromide anion to form a C–Br bond, yielding the neutral addition product. Once again, a curved arrow in Figure 5.2 shows the electron-pair movement from bromide anion to carbon.

The electrophile H^+ is attacked by the π electrons of the double bond, and a new C–H σ bond is formed. This leaves the other carbon atom with a + charge and a vacant *p* orbital.

Br⁻ donates an electron pair to the positively charged carbon atom, forming a C–Br σ bond and yielding the neutral addition product.

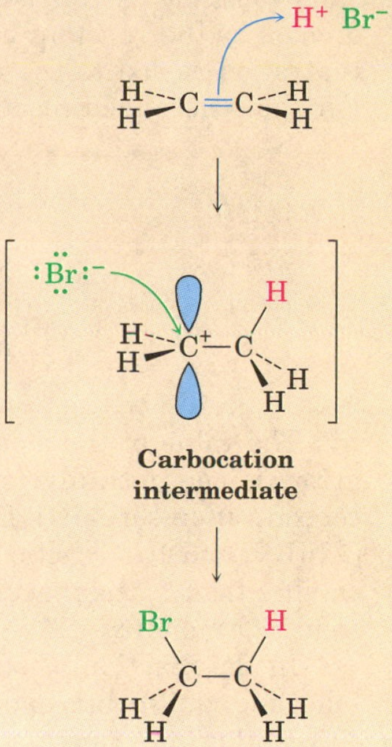

Figure 5.2 The electrophilic addition reaction of HBr and ethylene. The reaction takes place in two steps, both of which involve electrophile–nucleophile interactions.

The electrophilic addition of HBr to ethylene is only one example of a polar process; there are many other types that we'll study in detail in later chapters. Regardless of the details of individual reactions, all polar reactions take place between an electron-poor site and an electron-rich site and involve the sharing of an electron pair from a nucleophile with an electrophile.

PROBLEM

5.6 What product would you expect from reaction of HBr with cyclohexene?

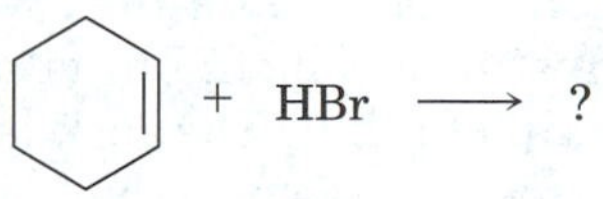

5.6 Describing a Reaction: Rates and Equilibria

Every chemical reaction can go in two directions. Reactants can give products, and products can revert to reactants. The position of the resulting chemical equilibrium is expressed by an equation in which K_{eq}, the equilibrium constant, is equal to the product concentrations multiplied together, divided by the reactant concentrations multiplied together, with each concentration raised to the power of its coefficient in the balanced equation. For the generalized reaction

$$a\mathrm{A} + b\mathrm{B} \rightleftharpoons c\mathrm{C} + d\mathrm{D}$$

we have

$$K_{eq} = \frac{[\text{Products}]}{[\text{Reactants}]} = \frac{[\mathrm{C}]^c[\mathrm{D}]^d}{[\mathrm{A}]^a[\mathrm{B}]^b}$$

The value of the equilibrium constant tells which side of the reaction arrow is energetically favored. If K_{eq} is larger than 1, then the product concentration term $[\mathrm{C}]^c[\mathrm{D}]^d$ is larger than the reactant concentration term $[\mathrm{A}]^a[\mathrm{B}]^b$, and the reaction proceeds as written from left to right. If K_{eq} is smaller than 1, the reaction does not take place as written but instead goes from right to left.

In the reaction of ethylene with HBr, for example, we can write the following equilibrium expression, and we can determine experimentally that the equilibrium constant at room temperature is approximately 7.5×10^7:

$$\mathrm{H_2C{=}CH_2} + \mathrm{HBr} \rightleftharpoons \mathrm{CH_3CH_2Br}$$

$$K_{eq} = \frac{[\mathrm{CH_3CH_2Br}]}{[\mathrm{HBr}][\mathrm{H_2C{=}CH_2}]} = 7.5 \times 10^7$$

Since K_{eq} is relatively large, the reaction proceeds as written, and greater than 99.999 99% of the ethylene is converted into bromoethane. Although we often speak of such reactions as "going to completion," this is imprecise terminology because in almost no reaction does *every* molecule react. For practical purposes, though, equilibrium constants greater than 10^3 can be considered to indicate "complete" reaction, because the amount of reactant left will be barely detectable (less than 0.1%).

What determines whether a reaction proceeds? *For a reaction to have a favorable equilibrium constant and proceed as written, the energy level of the products must be lower than the energy level of the reactants.* In other words, energy must be given off. The situation is analogous to that of a rock poised near the top of a hill. The rock is in a high-energy position because energy was required to raise it to the top. When it rolls downhill, it releases energy until it reaches a more stable low-energy position at the bottom.

The total energy change during a reaction is called the **Gibbs free-energy change, ΔG.**[2] (The Greek letter delta, Δ, is the mathematical symbol for the difference between two numbers, in this case the difference

[2]The free-energy change usually reported for a reaction is the *standard free-energy change*, symbolized $\Delta G°$, where the superscript ° means that the reaction is carried out with all substances in their standard states. For gases, the standard state is 1 atm pressure; for solutions, the standard state is 1 M concentration. The superscript is dropped and the free energy change is ΔG if a reaction is carried out under other than standard-state conditions.

between the free energy of the products and the free energy of the reactants.) For a favorable reaction, ΔG has a negative value, meaning that energy is released *to* the surroundings. For an unfavorable reaction, ΔG has a positive value, meaning that energy is absorbed *from* the surroundings.

Because the equilibrium constant, K_{eq}, and the standard free-energy change, $\Delta G°$, both measure whether a reaction is favored, they are mathematically related:

$$\Delta G° = G°_{\text{products}} - G°_{\text{reactants}}$$

$$\Delta G° = -RT \ln K_{eq} \qquad \text{or} \qquad K_{eq} = e^{-\Delta G°/RT}$$

where

$$R = 8.315 \text{ J/(K·mol)} = 1.987 \text{ cal/(K·mol)}$$

$$T = \text{Temperature in kelvins}$$

$$e = 2.718$$

$$\ln K_{eq} = \text{Natural logarithm of } K_{eq}$$

As an example of how this relationship can be used, the reaction of ethylene with HBr has $K_{eq} = 7.5 \times 10^7$. We can therefore calculate that $\Delta G° = -44.8$ kJ/mol (-10.7 kcal/mol) at room temperature (298 K):

$$K_{eq} = 7.5 \times 10^7 \qquad \text{and} \qquad \ln K_{eq} = 18.1$$

$$\Delta G° = -RT \ln K_{eq} = -[8.315 \text{ J/(K·mol)}] \times (298 \text{ K}) \times (18.1)$$

$$= -44{,}800 \text{ J/mol} = -44.8 \text{ kJ/mol}$$

To what is the free-energy change during a reaction due? The Gibbs free-energy change is made up of two terms, an **enthalpy** term, ΔH, and a temperature-dependent **entropy** term, $T\Delta S$, where T is the temperature in kelvins:

$$\Delta G° = \Delta H° - T\Delta S°$$

For the reaction of ethylene with HBr at room temperature (298 K), the values are $\Delta G° = -44.8$ kJ/mol, $\Delta H° = -84.1$ kJ/mol, and $\Delta S° = -132$ J/(K·mol).

$$H_2C{=}CH_2 + HBr \rightleftarrows CH_3CH_2Br \begin{cases} \Delta G° = -44.8 \text{ kJ/mol} \\ \Delta H° = -84.1 \text{ kJ/mol} \\ \Delta S° = -0.132 \text{ kJ(K·mol)} \\ T = 298 \text{ K} \end{cases}$$

The enthalpy change, $\Delta H°$, is called the **heat of reaction** and is a measure of the change in total bonding energy during a reaction. If $\Delta H°$ is negative, as in the reaction of HBr with ethylene, the bonds in the products are stronger (more stable) than the bonds in the reactants, heat is evolved, and the reaction is said to be **exothermic**. If $\Delta H°$ is positive, the bonds in the products are weaker (less stable) than the bonds in the reactants, heat is absorbed, and the reaction is said to be **endothermic**. For example, if a certain reaction breaks reactant bonds with a total strength of 380 kJ/mol and forms product bonds with a total strength of 400 kJ/mol, then $\Delta H°$ for the reaction is −20 kJ/mol and the reaction is exothermic. (*Remember*: Breaking bonds takes energy, and making bonds releases energy.)

Energy absorbed in breaking reactant bonds: $\Delta H°$ =	380 kJ/mol
Energy released in making product bonds: $\Delta H°$ =	−400 kJ/mol
Net change: $\Delta H°$ =	−20 kJ/mol

The entropy change, $\Delta S°$, is a measure of the change in the amount of molecular disorder, or freedom of motion, that accompanies a reaction. To illustrate with an elimination reaction of the type

$$A \longrightarrow B + C$$

there is more freedom of movement (disorder) in the products than in the reactant because one molecule has split into two. Thus, there is a net gain in entropy during the reaction, and $\Delta S°$ has a positive value.

On the other hand, for an addition reaction of the type

$$A + B \longrightarrow C$$

the opposite is true. Because such reactions restrict the freedom of movement of two molecules by joining them together, the product has *less* disorder than the reactants, and $\Delta S°$ has a negative value. The reaction of ethylene and HBr to yield bromoethane is an example [$\Delta S° = -132$ J/(K·mol)].

Of the two terms that make up $\Delta G°$, the enthalpy term ($\Delta H°$) is usually larger than the entropy term ($T\Delta S°$) at typical reaction temperatures. Furthermore, enthalpy changes during reactions are easily measured, and large compilations of data are available. For both reasons, the entropy contribution is sometimes ignored when making thermodynamic arguments, and chemists often make the simplifying assumption that $\Delta G° \approx \Delta H°$. Table 5.3 describes the thermodynamic terms more fully.

Table 5.3 Explanation of Thermodynamic Quantities: $\Delta G° = \Delta H° - T\Delta S°$

Term	*Name*	*Explanation*
$\Delta G°$	Gibbs free-energy change (kJ/mol)	Overall energy difference between reactants and products. When $\Delta G°$ is negative, a reaction can occur spontaneously. $\Delta G°$ is related to the equilibrium constant by the equation $\Delta G° = -RT \ln K_{eq}$
$\Delta H°$	Enthalpy change (kJ/mol)	Heat of reaction; the energy difference between strengths of bonds broken in a reaction and bonds formed
$\Delta S°$	Entropy change (J/K·mol)	Overall change in freedom of motion or "disorder" resulting from reaction; usually much smaller than $\Delta H°$

Knowing the value of K_{eq} for a reaction is extremely useful, but it's important to realize the limitations. An equilibrium constant tells only the *position* of the equilibrium, or how much product is ultimately formed. It does *not* tell the *rate* of reaction, or how fast the equilibrium is established. Some reactions are extremely slow even though they have favorable equilibrium constants. Gasoline is stable at room temperature, for example, because the rate of its reaction with oxygen is slow at 298 K. At higher temperatures, however, such as occur in contact with a lighted match, gasoline reacts rapidly with oxygen and undergoes complete conversion to the equilibrium products water and carbon dioxide. Rates (*how fast* a reaction occurs) and equilibria (*how much* a reaction occurs) are entirely different.

Rate ⟶ Is the reaction fast or slow?

Equilibrium ⟶ In what direction does the reaction proceed?

PROBLEM

5.7 Which reaction is more favored, one with $\Delta G = -44$ kJ/mol or one with $\Delta G = +44$ kJ/mol?

PROBLEM

5.8 Which reaction is more exothermic, one with $K_{eq} = 1000$ or one with $K_{eq} = 0.001$?

PROBLEM

5.9 What is the value of $\Delta G°$ at 298 K for reactions where $K_{eq} = 1000$, $K_{eq} = 1$, and $K_{eq} = 0.001$? What is the value of K_{eq} for reactions where $\Delta G° = -40$ kJ/mol, $\Delta G° =$ 0 kJ/mol, and $\Delta G° = +40$ kJ/mol?

5.7 Describing a Reaction: Bond Dissociation Energies

We've just seen that energy is released (negative $\Delta H°$) when a bond is made and is absorbed (positive $\Delta H°$) when a bond is broken. The measure of the energy change that occurs on bond breaking is a quantity called the **bond dissociation energy (*D*)**. Bond dissociation energy is defined as the amount of energy required to break a given bond to produce two radical fragments when the molecule is in the gas phase at 25°C.

$$\text{A}:\text{B} \xrightarrow[\text{energy}]{\text{Bond dissociation}} \text{A}\cdot + \cdot\text{B}$$

Each specific bond has its own characteristic strength, and extensive tables of data are available. For example, a C–H bond in methane has a bond dissociation energy $D = 438.4$ kJ/mol (104.8 kcal/mol), meaning that 438.4 kJ/mol is required to break a C–H bond of methane to give the two radical fragments $\cdot CH_3$ and ·H. Conversely, 438.4 kJ/mol of energy is *released* when a methyl radical and a hydrogen atom combine to form methane. Table 5.4 lists some other bond-strength data.

If enough bond dissociation energies were known, it would seem possible to calculate $\Delta H°$ for any reaction of interest, thereby avoiding a lot of time-consuming work in the laboratory. Unfortunately, there are two problems with this approach. First, the calculation says nothing about the rate of reaction, since a reaction may have a favorable $\Delta H°$ and still not take place on any reasonable time scale. Second, bond dissociation energies refer to molecules in the gas phase and aren't directly relevant to solution chemistry.

In practice, most organic reactions are carried out in solution, where solvent molecules can surround and interact with dissolved reactants, a phenomenon called *solvation*. Solvation can weaken bonds and cause large deviations from the gas-phase value of $\Delta H°$ for a reaction. In addition, the entropy term, $\Delta S°$, can also be different in solution because the solvation of a polar reactant by a polar solvent causes a certain amount of orientation in the solvent and thereby reduces the amount of disorder. Although we can often use bond-strength data to get a rough idea of how thermodynamically favorable a given reaction might be, we have to keep in mind that the answer is only approximate.

Table 5.4 Bond Dissociation Energies

Bond	*D (kJ/mol)*	*Bond*	*D (kJ/mol)*	*Bond*	*D (kJ/mol)*
H—H	436	$(CH_3)_3C$—Br	263	CH_3—CH_3	376
H—F	570	$(CH_3)_3C$—I	209	C_2H_5—CH_3	355
H—Cl	432	H_2C=CH—H	444	$(CH_3)_2CH$—CH_3	351
H—Br	366	H_2C=CH—Cl	368	$(CH_3)_3C$—CH_3	339
H—I	298	H_2C=$CHCH_2$—H	361	H_2C=CH—CH_3	406
Cl—Cl	243	H_2C=$CHCH_2$—Cl	289	H_2C=$CHCH_2$—CH_3	310
Br—Br	193	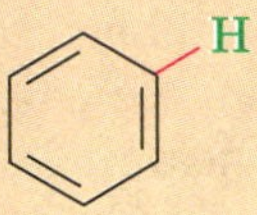	464	H_2C=CH_2	611
I—I	151			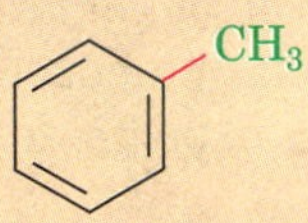	427
CH_3—H	438				
CH_3—Cl	351	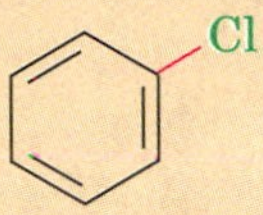	405		
CH_3—Br	293			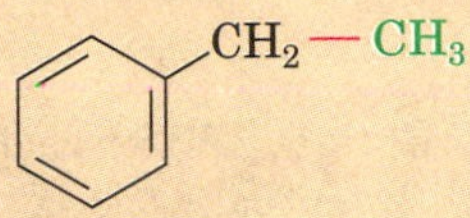	332
CH_3—I	234				
CH_3—OH	380	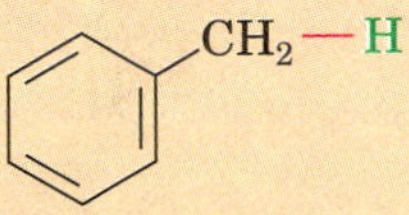	368		
CH_3—NH_2	335			$CH_3C(=O)$—H	368
C_2H_5—H	420	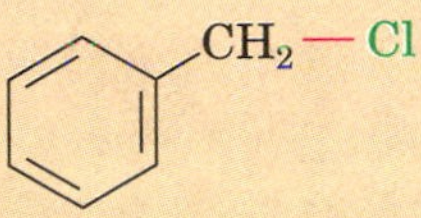	293	HO—H	498
C_2H_5—Cl	338			HO—OH	213
C_2H_5—Br	285			CH_3O—H	437
C_2H_5—I	222	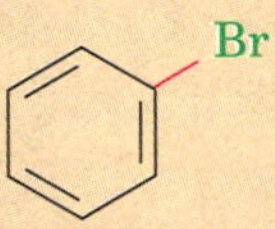	337	CH_3S—H	371
C_2H_5—OH	380			C_2H_5O—H	436
$(CH_3)_2CH$—H	401				
$(CH_3)_2CH$—Cl	339	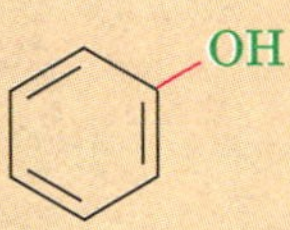	469	$CH_3C(=O)$—CH_3	322
$(CH_3)_2CH$—Br	274			CH_3CH_2O—CH_3	339
$(CH_3)_3C$—H	390	HC≡C—H	552	NH_2—H	449
$(CH_3)_3C$—Cl	330			H—CN	518

To take a familiar example, what do bond-dissociation data look like for the reaction of gaseous HBr with gaseous ethylene? By subtracting the dissociation energies of the bonds formed in the products from the dissociation energies of the bonds broken in the reactants, we can calculate an approximate $\Delta H°$ for the overall reaction.

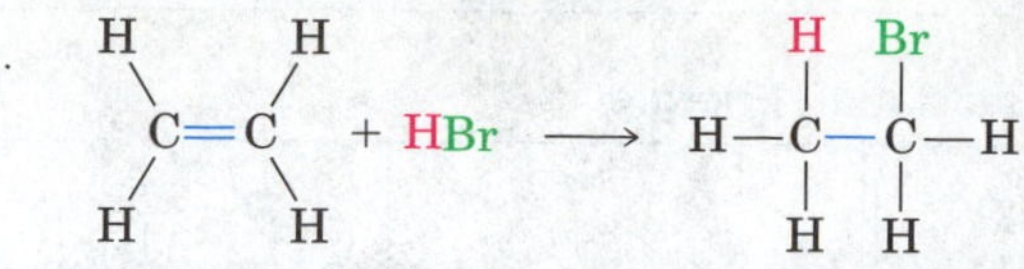

Bonds formed		*Bonds broken*	
C—H	D = 420 kJ/mol	H—Br	D = 366 kJ/mol
C—C	D = 376 kJ/mol	C═C	D = 611 kJ/mol
C—Br	D = 285 kJ/mol		
Total	D = 1081 kJ/mol	Total	D = 977 kJ/mol

$$\Delta H°: = 977\ \text{kJ/mol} - 1081\ \text{kJ/mol} = -104\ \text{kJ/mol}$$

Because the bonds formed in the gas-phase reaction of HBr with ethylene are stronger than the bonds broken, we calculate that the reaction is favorable by about −104 kJ/mol (−25 kcal/mol), in reasonable agreement with the experimental value of −84.1 kJ/mol.

PROBLEM.

5.10 Calculate $\Delta H°$ for each step in the gas-phase radical substitution reaction of chlorine with methane (Section 5.3):

(a) $Cl_2 \longrightarrow 2\ Cl\cdot$

(b) $CH_4 + Cl\cdot \longrightarrow \cdot CH_3 + HCl$

(c) $\cdot CH_3 + Cl_2 \longrightarrow CH_3Cl + Cl\cdot$

(d) What is the overall $\Delta H°$ for the reaction? Consider only the propagation steps (b) and (c) in deciding your answer.

PROBLEM.

5.11 Calculate $\Delta H°$ for these reactions:

(a) $CH_3CH_2OCH_3 + HI \longrightarrow CH_3CH_2OH + CH_3I$

(b) $CH_3Cl + NH_3 \longrightarrow CH_3NH_2 + HCl$

5.8 Describing a Reaction: Energy Diagrams and Transition States

For a reaction to take place, reactant molecules must collide, and reorganization of atoms and bonds must occur. Let's again look at the addition reaction of HBr with ethylene:

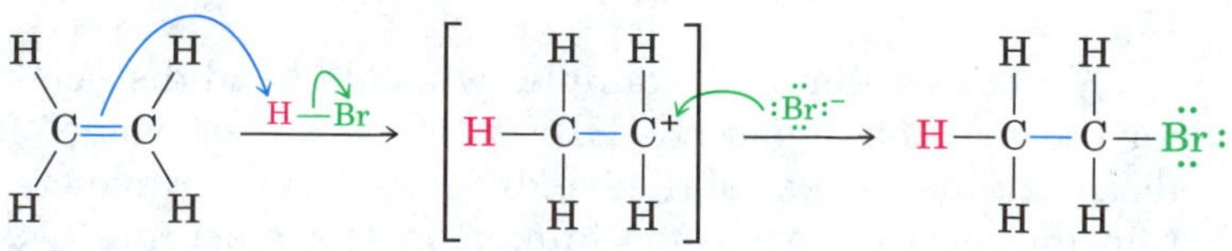

Carbocation

As the reaction proceeds, ethylene and HBr must approach each other, the ethylene π bond and H–Br bond must break, a new C–H bond must form in the first step, and a new C–Br bond must form in the second step.

Over the years, chemists have developed a method for graphically depicting the energy changes that occur during a reaction by using **reaction energy diagrams** as shown in Figure 5.3. The vertical axis represents the total energy of all reactants, and the horizontal axis, called the *reaction coordinate*, represents the progress of the reaction from beginning (left) to end (right). Let's see how the addition of HBr to ethylene can be described in a reaction energy diagram.

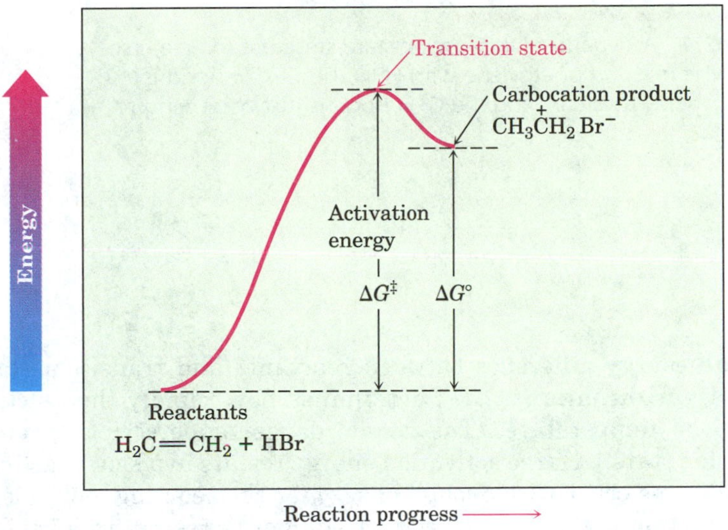

Figure 5.3 A reaction energy diagram for the first step in the reaction of ethylene with HBr. The energy difference between reactants and transition state, $\Delta G^{\ddagger}$, controls the reaction rate. The energy difference between reactants and carbocation product, ΔG°, controls the position of the equilibrium.

At the beginning of their reaction, ethylene and HBr have the total amount of energy indicated by the reactant level on the left side of the diagram in Figure 5.3. As the two molecules collide and reaction commences, their electron clouds repel each other, causing the energy level to rise. If the collision has occurred with sufficient force and proper orientation, the reactants continue to approach each other despite the rising repulsion until the new C–H bond starts to form. At some point, a structure of maximum energy is reached, a structure we call the *transition state*.

The **transition state** represents the highest-energy structure involved in the step. It is unstable and can't be isolated, but we can nevertheless imagine it to be an activated complex of the two reactants in which the carbon–carbon π bond is partially broken and the new carbon–hydrogen bond is partially formed (Figure 5.4, p. 166).

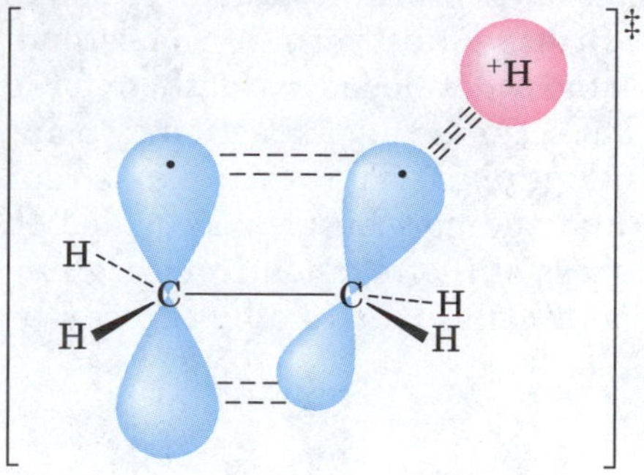

Figure 5.4 A hypothetical transition-state structure for the first step of the reaction of ethylene with HBr. The C–C π bond is just beginning to break, and the C–H bond is just beginning to form.

The energy difference between reactants and transition state, called the **activation energy, $\Delta G^{\ddagger}$**, determines how rapidly the reaction occurs at a given temperature. (The double dagger superscript, ‡, refers to the transition state.) A large activation energy results in a slow reaction because few collisions occur with enough energy for the reacting molecules to reach the transition state. A small activation energy results in a rapid reaction because almost all collisions occur with enough energy for the reacting molecules to reach the transition state.

The situation of reactants needing enough energy to climb the activation barrier from reactant to transition state is similar to the situation of hikers who need enough energy to climb over a mountain pass. If the pass is a high one, the hikers need a lot of energy and surmount the barrier slowly. If the pass is low, however, the hikers need less energy and reach the top quickly.

Although it's difficult to generalize, most organic reactions have activation energies in the range 40–150 kJ/mol (10–35 kcal/mol). The reaction of ethylene with HBr, for example, has an activation energy of approximately 140 kJ/mol (34 kcal/mol). Reactions with activation energies less than 80 kJ/mol take place at or below room temperature, whereas reactions with higher activation energies normally require a higher temperature. Heat provides the energy necessary for the reactants to climb the activation barrier.

Once the transition state is reached, the reaction can either continue on to give the carbocation product or revert back to reactants. Since both choices are energetically downhill from the high point of the transition state, both are equally likely. When reversion to reactants occurs, the transition-state structure comes apart and an amount of energy corresponding to $-\Delta G^\ddagger$ is released. When the reaction continues on to give carbocation, the new C–H bond forms fully and an amount of energy corresponding to the difference between transition state and carbocation product is released. The net change in energy for the step, $\Delta G°$, is represented in the energy diagram as the difference in level between reactant and product. Since the carbocation is higher in energy than the starting alkene, the step is endothermic, $\Delta G°$ has a positive value, and energy is absorbed.

Not all reaction energy diagrams are like the one shown for the reaction of ethylene and HBr. As illustrated in Figure 5.5, each specific reaction has

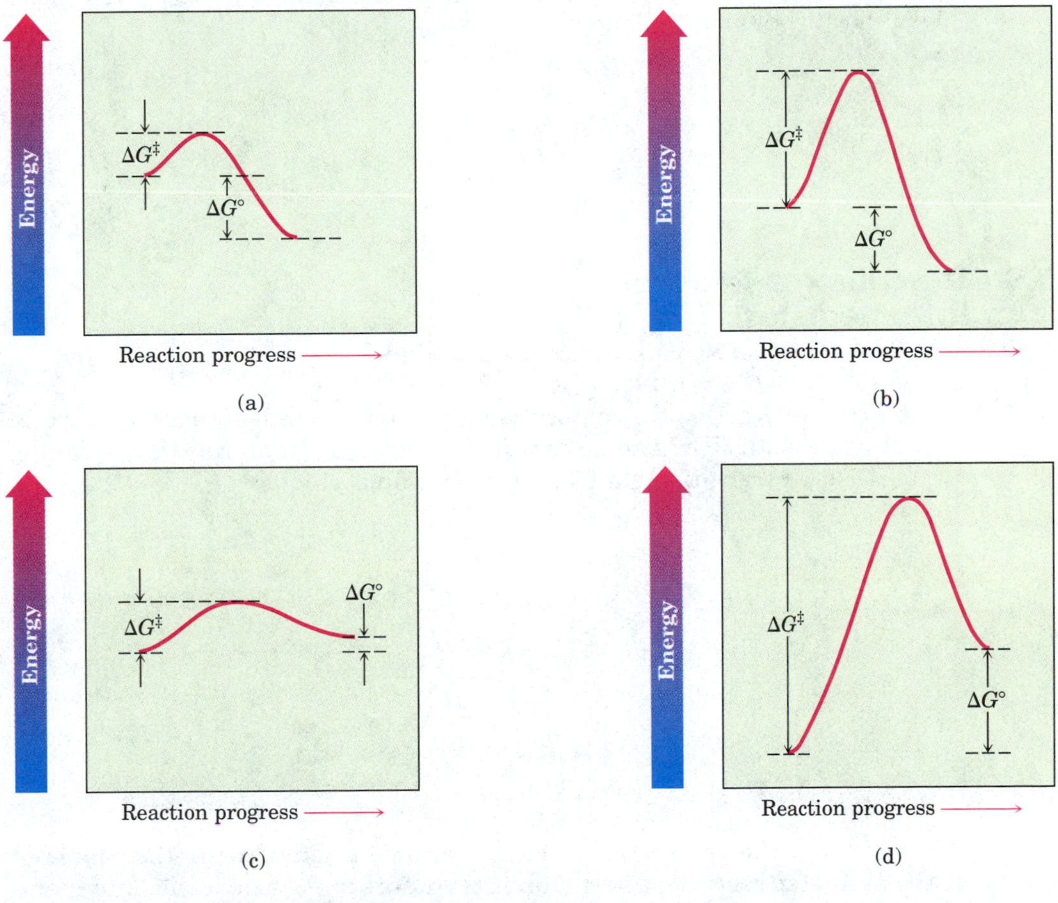

Figure 5.5 Some hypothetical reaction energy diagrams:
(a) a fast exothermic reaction (small $\Delta G^\ddagger$, negative $\Delta G°$);
(b) a slow exothermic reaction (large $\Delta G^\ddagger$, negative $\Delta G°$);
(c) a fast endothermic reaction (small $\Delta G^\ddagger$, small positive $\Delta G°$);
(d) a slow endothermic reaction (large $\Delta G^\ddagger$, positive $\Delta G°$).

its own specific energy profile. Some reactions are fast (small $\Delta G^{\ddagger}$) and some are slow (large $\Delta G^{\ddagger}$); some have a negative ΔG° and some have a positive ΔG°. Figure 5.5 illustrates some different possibilities for energy profiles. Note the use of the words *exothermic* and *endothermic* in this figure to refer to reactions in which ΔG° is negative and positive, respectively. This usage is not strictly correct because these words refer to ΔH°, not to ΔG°. As mentioned earlier, though, chemists often make the simplifying assumption that ΔG° and ΔH° are approximately equal.

PROBLEM.

5.12 Which reaction is faster, one with $\Delta G^{\ddagger} = 45$ kJ/mol or one with $\Delta G^{\ddagger} = 70$ kJ/mol? Which of the two has the larger K_{eq}?

5.9 Describing a Reaction: Intermediates

How can we describe the carbocation formed in the first step of the reaction of ethylene with HBr? The carbocation is clearly different from the reactants, yet it isn't a transition state and it isn't a final product.

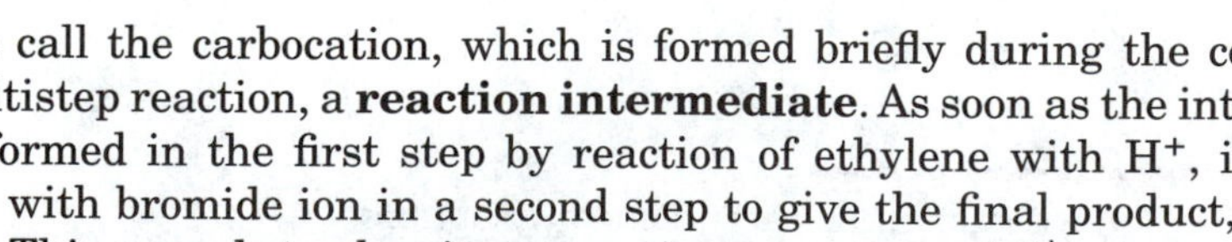

Reaction intermediate

We call the carbocation, which is formed briefly during the course of the multistep reaction, a **reaction intermediate**. As soon as the intermediate is formed in the first step by reaction of ethylene with H^+, it reacts further with bromide ion in a second step to give the final product, bromoethane. This second step has its own activation energy ($\Delta G^{\ddagger}$), its own transition state, and its own energy change (ΔG°). We can picture the second transition state as an activated complex between the electrophilic carbocation intermediate and the nucleophilic bromide anion, in which the bromide ion is donating a pair of electrons to the positively charged carbon atom and the new C–Br bond is just starting to form.

A complete energy diagram for the overall reaction of ethylene with HBr is shown in Figure 5.6. In essence, we draw a diagram for each of the individual steps and then join them in the middle so that the carbocation *product* of step 1 serves as the *reactant* for step 2. As indicated in Figure 5.6, the reaction intermediate lies at an energy minimum between steps 1 and 2. Since the energy level of this intermediate is higher than the level of either the initial reactants (ethylene + HBr) or the final product (bromoethane), the intermediate is reactive and can't be isolated. It is, however, more stable than either of the two transition states that neighbor it.

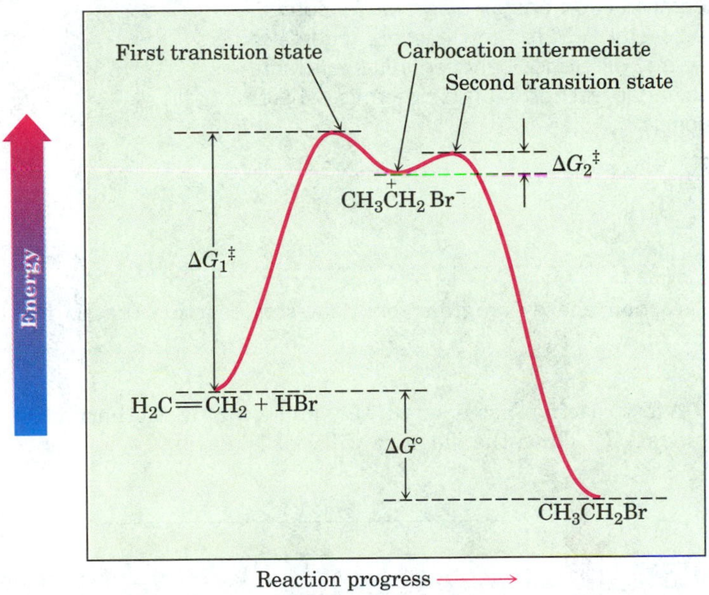

Figure 5.6 A reaction energy diagram for the overall reaction of ethylene with HBr. Two separate steps are involved, each with its own transition state. The energy minimum between the two steps represents the carbocation reaction intermediate.

Each step in a multistep process can always be considered separately. Each step has its own $\Delta G^\ddagger$ and its own $\Delta G°$. The *overall* $\Delta G°$ of the reaction, however, is the energy difference between initial reactants (far left) and final products (far right). Figure 5.7 (p. 170) illustrates some different possible cases.

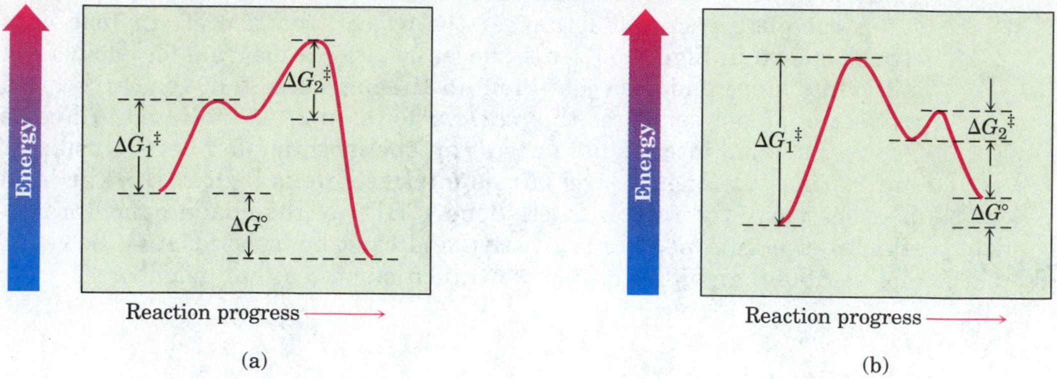

Figure 5.7 Hypothetical reaction energy diagrams for some two-step reactions. The overall $\Delta G°$ for any reaction, regardless of complexity, is the energy difference between initial reactants and final products. Note that reaction (a) is exothermic, whereas reaction (b) is endothermic.

PRACTICE PROBLEM

Sketch a reaction energy diagram for a one-step reaction that is fast and highly exothermic.

Solution A fast reaction has a small $\Delta G^{\ddagger}$, and a highly exothermic reaction has a large negative $\Delta G°$. Thus, the diagram will look like this:

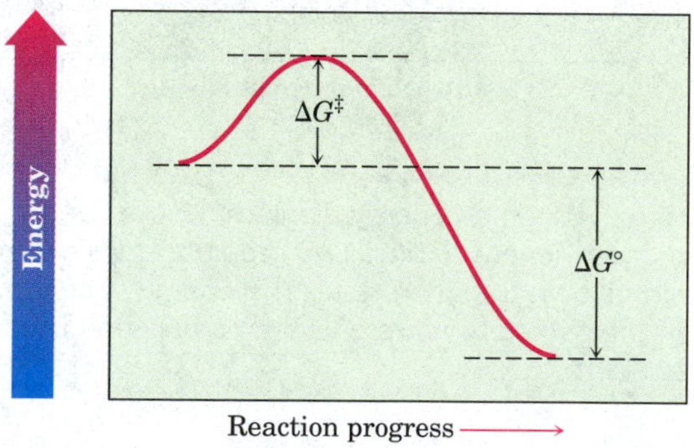

PROBLEM

5.13 Sketch a reaction energy diagram for a two-step reaction with an endothermic first step and an exothermic second step. Label the parts of the diagram corresponding to reactant, product, and intermediate.

PROBLEM

5.14 Sketch a reaction energy diagram showing both propagation steps in the radical reaction of chlorine with methane. Is the overall $\Delta G°$ for this reaction positive or negative? Label the parts of your diagram corresponding to $\Delta G°$ and $\Delta G^{\ddagger}$.

$$CH_4 + Cl_2 \xrightarrow{\text{Light}} CH_3Cl + HCl$$

INTERLUDE

Explosives

Demolition of the Alfred P. Murrah Federal Building in Oklahoma City, damaged by a car-bomb attack April 19, 1995

Most chemical reactions take place in one or more discrete steps, each of which has a rate, an equilibrium constant, and a well-defined mechanism. The steps can usually be studied, the rates and equilibrium constants can be measured, and the mechanism can be elucidated until the reaction is well understood. *Explosions*, however, are different. Their rates are so fast, and their mechanisms are so complex, that the details by which explosions occur defy a complete understanding.

Chemical explosions are characterized by the spontaneous breakdown of molecules into fragments that recombine to give the final products —usually stable gases such as N_2, H_2O, and CO_2. The result is a nearly instantaneous release of large quantities of hot gases, which set up a devastating shock wave as they expand. The shock wave can travel at speeds of up to 9000 m/s (approximately 20,000 mi/h) and generate a pressure of up to 700,000 atm, causing enormous physical devastation to the surroundings.

Explosives are categorized as either *primary* or *secondary*, depending on their sensitivity to shock. Primary explosives, such as lead azide, $Pb(N_3)_2$, are the most sensitive to heat and shock. They are used in detonators, blasting caps, and military fuses to initiate the explosion of a less sensitive, secondary explosive. Secondary explosives, or *high explosives*, are less sensitive to heat and shock than primary explosives

(continued)➤

and are therefore safer to manufacture and transport. Most secondary explosives simply burn rather than explode when ignited in air, and most can be detonated only by the nearby explosion of a primary initiator.

The first commercially important high explosive was nitroglycerin, prepared in 1847 by reaction of glycerin with nitric acid in the presence of sulfuric acid:

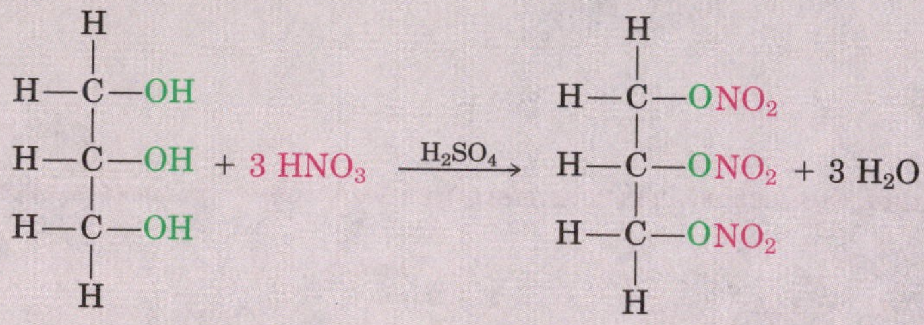

Glycerin Nitroglycerin

As you might expect, the reaction is extremely hazardous to carry out, and it was not until 1865 that the Swedish chemist Alfred Nobel succeeded in finding a reliable method of producing nitroglycerin and incorporating it into the commercial blasting product called *dynamite*. Modern industrial dynamite used for quarrying stone and blasting roadbeds is a mixture of ammonium nitrate and nitroglycerin absorbed onto diatomaceous earth.

The military explosives used as fillings for bombs or shells must have a low sensitivity to impact shock on firing, and must have good stability for long-term storage. TNT (trinitrotoluene), HMX (His Majesty's explosive), and RDX (research department explosive) are the most commonly used military high explosives. RDX is also compounded with waxes or synthetic polymers to make so-called plastic explosives frequently used by terrorist groups.

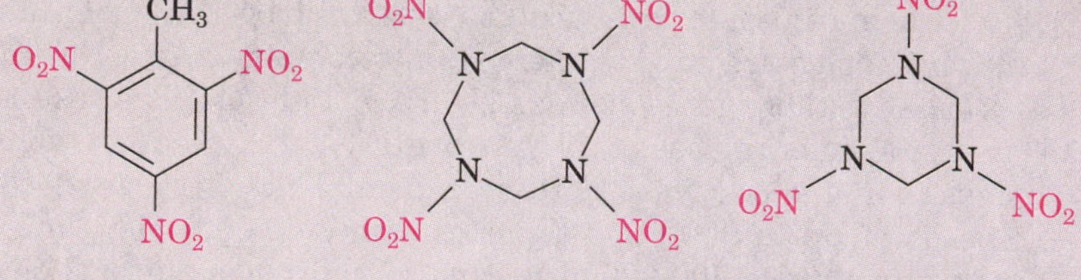

TNT (Trinitrotoluene) **HMX** **RDX**

Summary and Key Words

There are four main kinds of reactions: additions, eliminations, substitutions, and rearrangements. **Addition reactions** take place when two reactants add together to give a single product; **elimination reactions** take place when one reactant splits apart to give two products; **substitution reactions** take place when two reactants exchange parts to give two new products; and **rearrangement reactions** take place when one reactant undergoes a reorganization of bonds and atoms to give an isomeric product.

Addition	$A + B \longrightarrow C$
Elimination	$A \longrightarrow B + C$
Substitution	$A\text{–}B + C\text{–}D \longrightarrow A\text{–}C + B\text{–}D$
Rearrangement	$A \longrightarrow B$

A full description of how a reaction occurs is called its **mechanism**. There are two kinds of mechanisms by which reactions take place: **radical** mechanisms and **polar** mechanisms. Polar reactions, the most common type, occur because of an attractive interaction between an electron-rich (**nucleophilic**) site in one molecule and an electron-poor (**electrophilic**) site in another molecule. A bond is formed in a polar reaction when the nucleophile donates an electron pair to the electrophile. This movement of electrons is indicated by a curved arrow showing the direction of electron travel from the nucleophile to the electrophile. Radical reactions involve odd-electron species; a bond is formed when each reactant donates one electron.

Polar $B{:}^- + A^+ \longrightarrow A{:}B$

Nucleophile Electrophile

Radical $B\cdot + \cdot A \longrightarrow A{:}B$

The energy change that takes place during a reaction can be described by considering both rates (how fast reaction occurs) and equilibria (how much reaction occurs). The position of a chemical equilibrium is determined by the value of the **Gibbs free-energy change** ($\Delta G°$) that takes place during reaction. The free-energy change is composed of two parts, $\Delta G° = \Delta H° - T\Delta S°$. The **enthalpy** term ($\Delta H°$) corresponds to the net change in strength of chemical bonds broken and formed during reaction. The **entropy** term ($\Delta S°$) corresponds to the change in the amount of disorder during reaction. Since the enthalpy term is usually larger and more important than the entropy term, chemists often make the assumption that $\Delta G° \approx \Delta H°$.

A reaction can be described pictorially using a **reaction energy diagram**, which follows the reaction course from reactant through transition state to product. The **transition state** is an activated complex occurring at the highest-energy point of a reaction. The amount of energy needed by reactants to reach this high point is the **activation energy**, $\Delta G^{\ddagger}$. The higher the activation energy, the slower the reaction.

Many reactions take place in more than one step and involve the formation of a **reaction intermediate**. An intermediate is a species that lies at an energy minimum between steps on the reaction curve and is formed briefly during the course of a reaction.

ADDITIONAL PROBLEMS

5.15 Identify the functional groups in the following molecules:

(a) $CH_3CH_2C{\equiv}N$

(b) 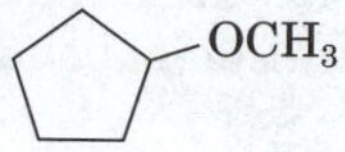

(c) $CH_3\overset{O}{\overset{\|}{C}}CH_2\overset{O}{\overset{\|}{C}}OCH_3$

(d) 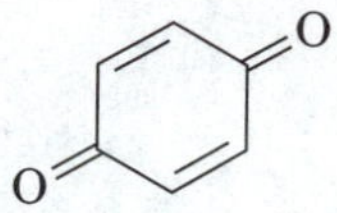

(e) 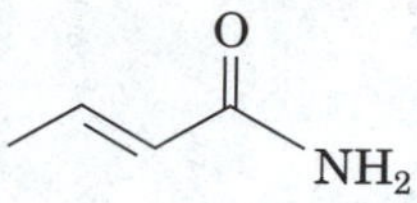

(f) 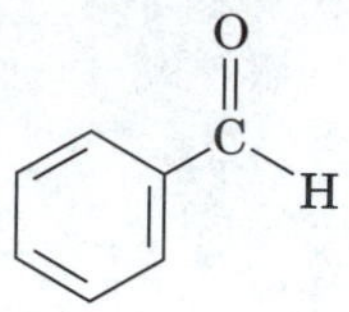

5.16 Predict the polarity of the functional groups you identified in Problem 5.15.

5.17 Identify these reactions as additions, eliminations, substitutions, or rearrangements:

(a) $CH_3CH_2Br + NaCN \longrightarrow CH_3CH_2CN\ (+ NaBr)$

(b) 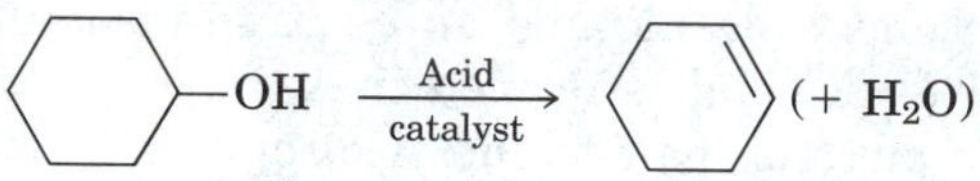

(c) 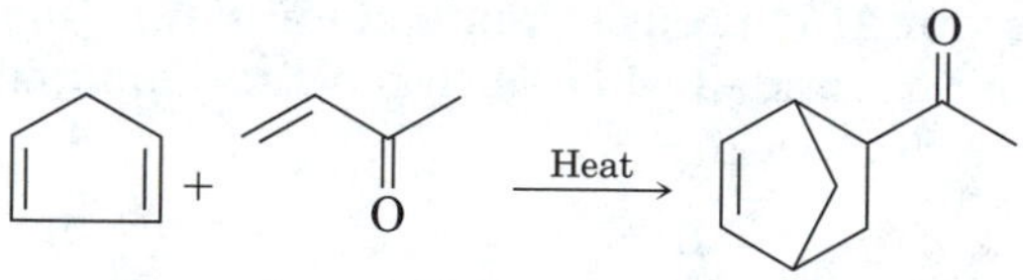

(d) 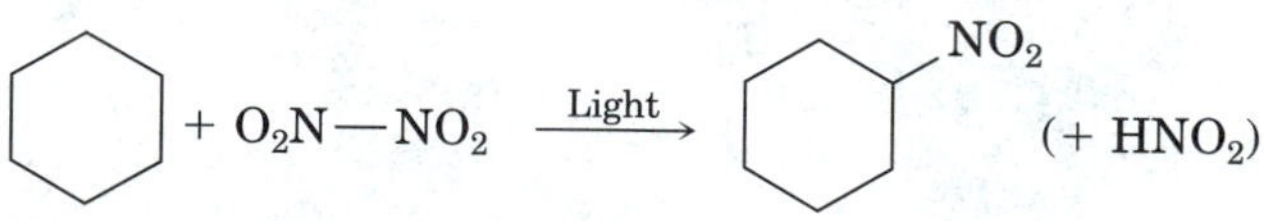

5.18 Explain the differences between addition, elimination, substitution, and rearrangement reactions.

5.19 Define the following:
(a) Polar reaction
(b) Heterolytic bond breakage
(c) Homolytic bond breakage
(d) Radical reaction
(e) Functional group
(f) Polarization

5.20 Give an example of each of the following:
(a) A nucleophile
(b) An electrophile
(c) A polar reaction
(d) A substitution reaction
(e) A heterolytic bond breakage
(f) A homolytic bond breakage

5.21 Classify each of the following as either nucleophiles or electrophiles:
(a) Cl^-
(b) BF_3
(c) CH_3NH_2

5.22 Draw a reaction energy diagram for a one-step endothermic reaction. Label the parts of the diagram corresponding to reactants, products, transition state, $\Delta G°$, and $\Delta G^{\ddagger}$. Is $\Delta G°$ positive or negative?

5.23 Draw a reaction energy diagram for a two-step exothermic reaction. Label the overall $\Delta G°$, transition states, and intermediate. Is $\Delta G°$ positive or negative?

5.24 What is the difference between a transition state and an intermediate?

5.25 Draw a reaction energy diagram for a two-step exothermic reaction whose second step is faster than its first step.

5.26 Draw a reaction energy diagram for a reaction with $K_{eq} = 1$. What is the value of $\Delta G°$ in this reaction?

5.27 Look at the reaction energy diagram shown here and answer the following questions:

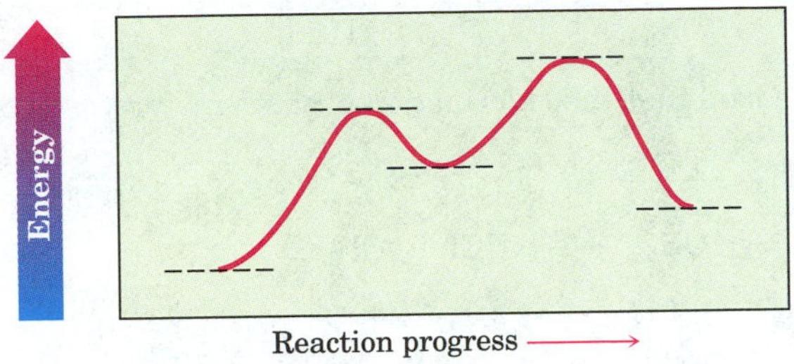

(a) Is $\Delta G°$ for the reaction positive or negative? Label it.
(b) How many steps are involved in the reaction?
(c) Which step is faster?
(d) How many transition states are there? Label them.

5.28 Use the information in Table 5.4 to calculate $\Delta H°$ for these reactions:
(a) $CH_3OH + HBr \longrightarrow CH_3Br + H_2O$
(b) $CH_3CH_2OH + CH_3Cl \longrightarrow CH_3CH_2OCH_3 + HCl$

5.29 Use the information in Table 5.4 to calculate $\Delta H°$ for the reaction of ethane with chlorine, bromine, and iodine:
(a) $CH_3CH_3 + Cl_2 \longrightarrow CH_3CH_2Cl + HCl$
(b) $CH_3CH_3 + Br_2 \longrightarrow CH_3CH_2Br + HBr$
(c) $CH_3CH_3 + I_2 \longrightarrow CH_3CH_2I + HI$
What can you conclude about the relative energetics of chlorination, bromination, and iodination?

5.30 An alternative course for the reaction of bromine with ethane could result in the formation of bromomethane:

$$H_3C{-}CH_3 + Br_2 \longrightarrow 2\ CH_3Br$$

Calculate $\Delta H°$ for this reaction, and comment on how it compares with the value you calculated in Problem 5.29 for the formation of bromoethane.

5.31 When a mixture of methane and chlorine is irradiated, reaction commences immediately. When irradiation is stopped, the reaction gradually slows down but does not stop immediately. How can you account for this behavior?

5.32 Radical chlorination of alkanes is not generally useful because mixtures of products often result when more than one kind of C–H bond is present in the substrate. Calculate approximate $\Delta H°$ values for the possible monochlorination reactions of 2-methylbutane. Use the bond dissociation energies measured for CH_3CH_2–H, H–$CH(CH_3)_2$, and H–$C(CH_3)_3$ as representative of typical primary, secondary, and tertiary C–H bonds.

5.33 Name each of the products formed in Problem 5.32.

5.34 Despite the limitations of radical chlorination of alkanes, the reaction is still useful for synthesizing certain halogenated compounds. For which of the following compounds does radical chlorination give a single monochloro product?

(a) C_2H_6 (b) $CH_3CH_2CH_3$ (c) 

(d) $(CH_3)_3CCH_2CH_3$ (e) 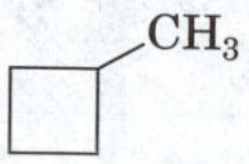

(f) $CH_3C{\equiv}CCH_3$

5.35 We've said that the chlorination of methane proceeds by the following steps:

(a) $Cl_2 \xrightarrow{\text{Light}} 2\ Cl\cdot$

(b) $Cl\cdot + CH_4 \longrightarrow HCl + \cdot CH_3$

(c) $\cdot CH_3 + Cl_2 \longrightarrow CH_3Cl + Cl\cdot$

Alternatively, one might propose a different series of steps:

(d) $Cl_2 \longrightarrow 2\ Cl\cdot$

(e) $Cl\cdot + CH_4 \longrightarrow CH_3Cl + H\cdot$

(f) $H\cdot + Cl_2 \longrightarrow HCl + Cl\cdot$

Calculate $\Delta H°$ for each step in both routes. What insight does this provide into the relative merits of each route?

5.36 When isopropylidenecyclohexane is treated with strong acid at room temperature, isomerization occurs by the mechanism shown below to yield 1-isopropylcyclohexene:

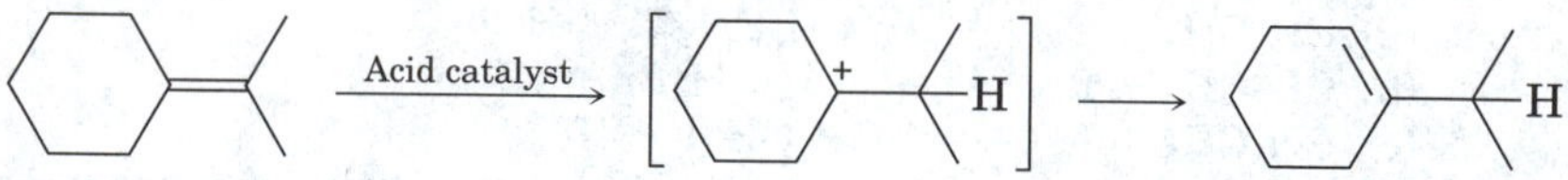

Isopropylidenecyclohexane **1-Isopropylcyclohexene**

At equilibrium, the product mixture contains about 30% isopropylidenecyclohexane and about 70% 1-isopropylcyclohexene.

(a) What kind of reaction is occurring? Is the mechanism polar or radical?
(b) Draw curved arrows to indicate electron movement in each step.
(c) Calculate K_{eq} for the reaction.
(d) Since the reaction occurs slowly at room temperature, what is its approximate $\Delta G^{\ddagger}$?
(e) Draw a quantitative reaction energy diagram for the reaction.

5.37 2-Chloro-2-methylpropane reacts with water in a two-step process to yield 2-methyl-2-propanol. The first step is slower than the second, the reaction takes place slowly at room temperature, and the equilibrium constant is near 1.

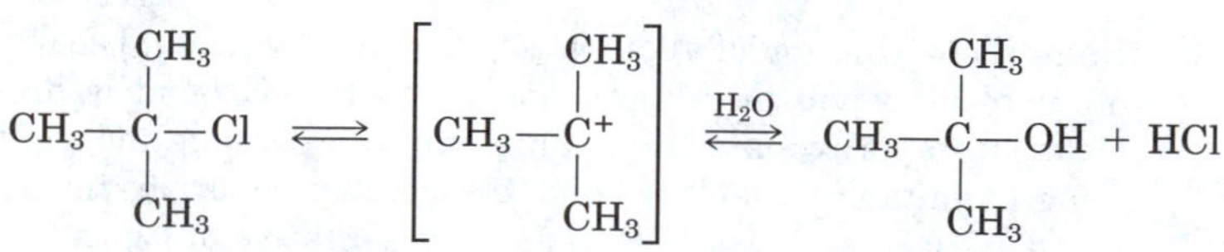

2-Chloro-2-methylpropane **2-Methyl-2-propanol**

(a) Give approximate values for $\Delta G^{\ddagger}$ and $\Delta G°$ that are consistent with the above information.

(b) Draw a reaction energy diagram, labeling all points of interest and making sure that the relative energy levels on the diagram correspond to the data.

5.38 The reaction of hydroxide ion with chloromethane to yield methanol and chloride ion is an example of a general reaction type called a nucleophilic substitution reaction:

$$HO^- + CH_3Cl \rightleftharpoons CH_3OH + Cl^-$$

The value of $\Delta H°$ for the reaction is -75 kJ/mol, and the value of $\Delta S°$ is $+54$ J/(K·mol). What is the value of $\Delta G°$ (in kJ/mol) at 25°C (298 K)? Is the reaction exothermic or endothermic?

5.39 Use the value of $\Delta G°$ you calculated in Problem 5.38 to find the equilibrium constant K_{eq} for the reaction of hydroxide ion with chloromethane.

A Look Ahead

5.40 Reaction of 2-methylpropene with HBr might, in principle, lead to a mixture of two bromoalkane addition products. Name them, and draw their structures.

5.41 Draw the structures of the two carbocation intermediates that might form during the reaction of 2-methylpropene with HBr (Problem 5.40). We'll see in the next chapter that the stability of carbocations depends on the number of alkyl substituents attached to the positively charged carbon—the more alkyl substituents there are, the more stable the cation. Which of the two carbocation intermediates you drew is more stable?

5.42 We'll see in Chapter 7 that alkenes can be converted into alcohols by acid-catalyzed addition of water. Review the mechanism of the addition of HBr to ethylene (Figure 5.2), and propose a mechanism for the corresponding addition of H_2O:

$$H_2C{=}CH_2 + H_2O \xrightarrow{H^+ \text{ catalyst}} CH_3CH_2OH$$

The isopropyl cation has a planar, trivalent carbon atom with a vacant *p* orbital.

6

Alkenes: Structure and Reactivity

Alkenes are hydrocarbons that contain a carbon–carbon double bond. The word *olefin* is often used as a synonym, but alkene is the generally preferred term. Alkenes occur abundantly in nature, and many have important biological roles. For example, ethylene is a plant hormone that induces ripening in fruit, and α-pinene is the major component of turpentine.

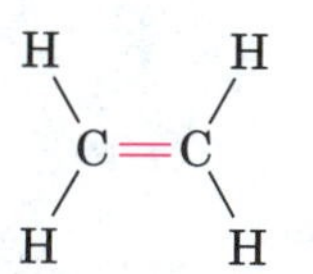

Ethylene

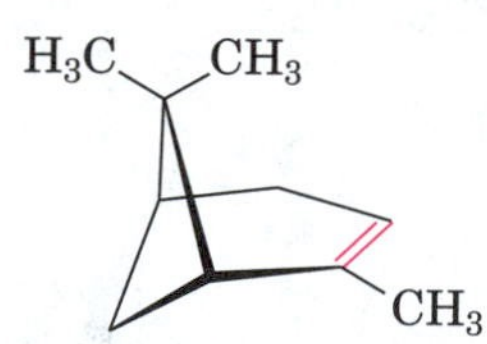

α-Pinene

Life itself would be impossible without such alkenes as β-carotene, a compound that contains 11 double bonds. β-Carotene, the orange pigment responsible for the color of carrots, serves as a valuable dietary source of vitamin A and is thought to offer some protection against certain types of cancer.

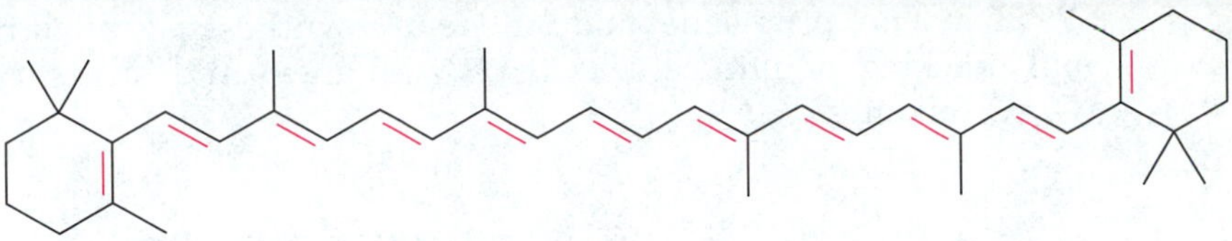

β-Carotene
(orange pigment and vitamin A precursor)

6.1 Industrial Preparation and Use of Alkenes

Ethylene (ethene) and propylene (propene), the simplest alkenes, are the two most important organic chemicals produced industrially. More than 48 billion pounds of ethylene and 28 billion pounds of propylene are produced each year in the United States for use in the synthesis of polyethylene, polypropylene, ethylene glycol, acetic acid, acetaldehyde, and a host of other substances (Figure 6.1).

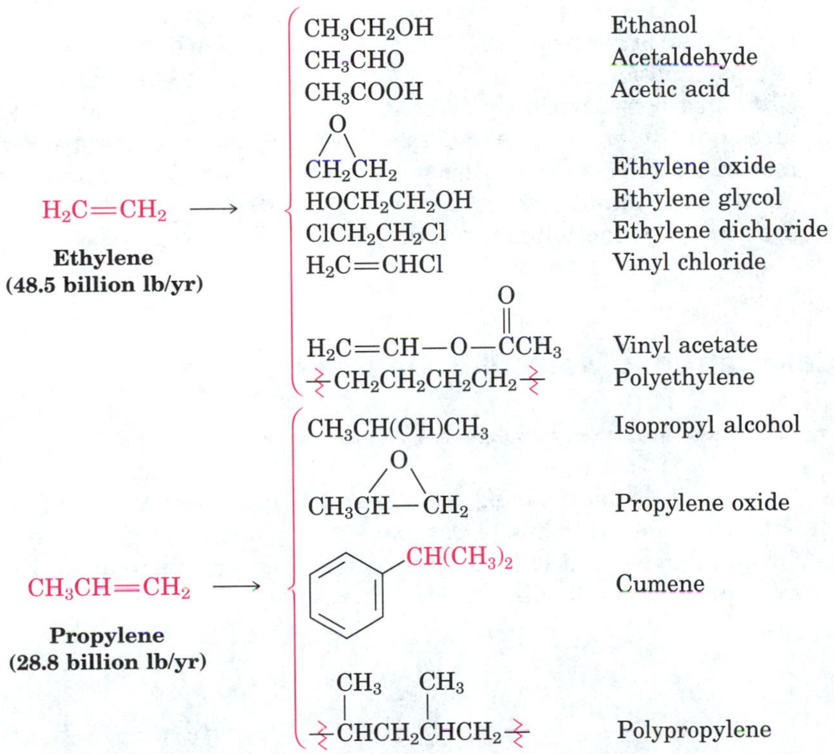

Figure 6.1 Compounds derived industrially from ethylene (ethene) and propylene (propene).

Ethylene, propylene, and butene are synthesized industrially by thermal cracking of natural gas (C_1–C_4 alkanes) and straight-run gasoline (C_4–C_8 alkanes):

$$CH_3(CH_2)_nCH_3 \xrightarrow[\text{Steam}]{850\text{–}900^\circ C} H_2 + CH_4 + H_2C{=}CH_2$$
$$n = 0\text{–}6 \qquad + CH_3CH{=}CH_2 + CH_3CH_2CH{=}CH_2$$

Thermal cracking, introduced in 1912, takes place in the absence of catalysts at temperatures up to 900°C. The exact processes are complex, although they undoubtedly involve radical reactions. The high-temperature reaction conditions cause spontaneous homolysis of C–C and C–H bonds, with resultant formation of smaller fragments. We might imagine, for instance, that a molecule of butane could split into two ethyl radicals, which could then each lose a hydrogen atom to generate two molecules of ethylene and H_2:

$$CH_3CH_2{-}CH_2CH_3 \xrightarrow{900^\circ C} 2\ \overset{\displaystyle H}{\overset{|}{C}}H_2{-}CH\cdot \longrightarrow 2\ H_2C{=}CH_2 + H_2$$

Thermal cracking is an example of a reaction whose energetics are dominated by entropy ($\Delta S°$) rather than by enthalpy ($\Delta H°$) in the free-energy equation $\Delta G° = \Delta H° - T\Delta S°$. Although the bond dissociation energy D for a carbon–carbon single bond is relatively high (about 375 kJ/mol), the large positive entropy change resulting from the fragmentation of one large molecule into several smaller pieces, together with the extremely high temperature, T, makes the $T\Delta S°$ term larger than the $\Delta H°$ term. As a result, the cracking reaction is favored.

6.2 Calculating a Molecule's Degree of Unsaturation

Because of its double bond, an alkene has fewer hydrogens than an alkane with the same number of carbons—C_nH_{2n} for an alkene versus C_nH_{2n+2} for an alkane—and is therefore referred to as **unsaturated**. Ethylene, for example, has the formula C_2H_4, whereas ethane has the formula C_2H_6.

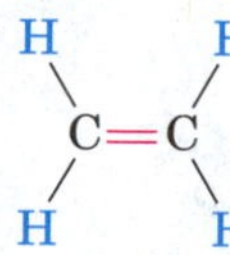

Ethylene: C_2H_4
(fewer hydrogens—*unsaturated*)

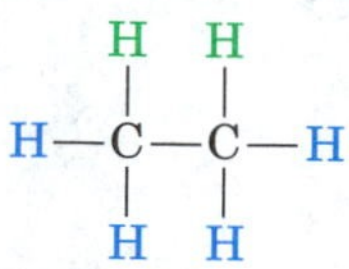

Ethane: C_2H_6
(more hydrogens—*saturated*)

In general, each ring or double bond in a molecule corresponds to a loss of two hydrogens from the alkane formula C_nH_{2n+2}. Knowing this relationship, it's possible to work backwards from a molecular formula to calculate a molecule's **degree of unsaturation**—the number of rings and/or multiple bonds present in the molecule.

Let's assume that we want to find the structure of an unknown hydrocarbon. A molecular weight determination on the unknown yields a value of 82, which corresponds to a molecular formula of C_6H_{10}. Since the saturated C_6 alkane (hexane) has the formula C_6H_{14}, the unknown compound has two fewer pairs of hydrogens ($H_{14} - H_{10} = H_4 = 2\,H_2$), and its degree of unsaturation is two. The unknown therefore contains two double bonds, one ring and one double bond, two rings, or one triple bond. There's still a long way to go to establish structure, but the simple calculation has told us a lot about the molecule.

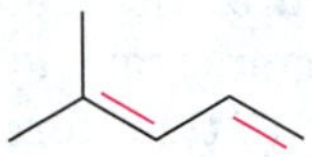

4-Methyl-1,3-pentadiene (two double bonds)

Cyclohexene (one ring, one double bond)

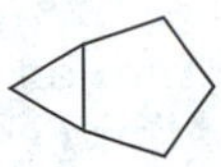

Bicyclo[3.1.0]hexane (two rings)

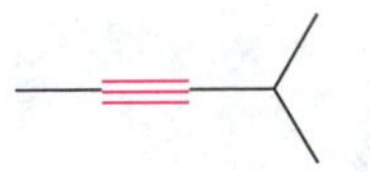

4-Methyl-2-pentyne (one triple bond)

C_6H_{10}

Similar calculations can be carried out for compounds containing elements other than just carbon and hydrogen.

1. *Organohalogen compounds, containing C, H, X, where X = F, Cl, Br, or I.* Because a halogen substituent is simply a replacement for hydrogen in an organic molecule (both are monovalent), we can *add* the number of halogens and hydrogens to arrive at an equivalent hydrocarbon formula from which the number of double bonds and/or rings can be found. For example, the alkyl halide formula $C_4H_6Br_2$ is equivalent to the hydrocarbon formula C_4H_8 and thus has one degree of unsaturation:

Replace 2 Br by 2 H

$$BrCH_2CH{=}CHCH_2Br = HCH_2CH{=}CHCH_2H$$

$$\underbrace{C_4H_6Br_2}_{\text{Add}} = \text{“}C_4H_8\text{”}$$

One unsaturation: one double bond

2. *Organooxygen compounds, containing C, H, O.* Because oxygen is divalent, it doesn't affect the formula of an equivalent hydrocarbon and can be ignored when calculating the degree of unsaturation. The easiest way to convince yourself of this is to see what happens when an oxygen atom is inserted into an alkane C–C or C–H bond:

There's no change in the number of hydrogen atoms. For example, the formula C_5H_8O is equivalent to the hydrocarbon formula C_5H_8 and thus has two degrees of unsaturation:

O removed from here

$$H_2C{=}CHCH{=}CHCH_2OH = H_2C{=}CHCH{=}CHCH_2{-}H$$

C_5H_8O = "C_5H_8" Two unsaturations: two double bonds

3. *Organonitrogen compounds, containing C, H, N.* Because nitrogen is trivalent, an organonitrogen compound has one more hydrogen than a related hydrocarbon has, and we therefore *subtract* the number of nitrogens from the number of hydrogens to arrive at the equivalent hydrocarbon formula. Again, the best way to convince yourself of this is to see what happens when a nitrogen atom is inserted into an alkane bond: Another hydrogen atom is required to fill the third valency on nitrogen, and we must therefore subtract this extra hydrogen atom to arrive at the equivalent hydrocarbon formula. For example, the formula C_5H_9N is equivalent to C_5H_8 and thus has two degrees of unsaturation:

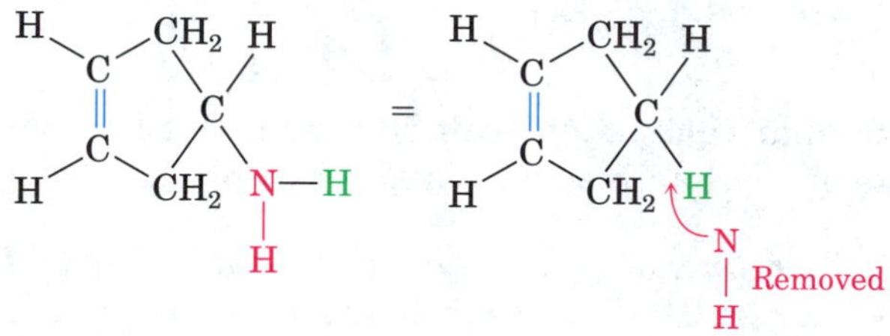

C_5H_9N = "C_5H_8" Two unsaturations: one ring and one double bond

To summarize:

Add the number of halogens to the number of hydrogens.

Ignore the number of oxygens.

Subtract the number of nitrogens from the number of hydrogens.

PROBLEM. . . . .

6.1 Calculate the degree of unsaturation in these hydrocarbons:
(a) C_8H_{14} (b) C_5H_6 (c) $C_{12}H_{20}$
(d) $C_{20}H_{32}$ (e) $C_{40}H_{56}$, β-carotene

PROBLEM. . . . .

6.2 Calculate the degree of unsaturation in the following formulas, and then draw as many structures as you can for each.
(a) C_4H_8 (b) C_4H_6 (c) C_3H_4

PROBLEM

6.3 Calculate the degree of unsaturation in these formulas:

(a) C_6H_5N (b) $C_6H_5NO_2$ (c) $C_8H_9Cl_3$

(d) $C_9H_{16}Br_2$ (e) $C_{10}H_{12}N_2O_3$ (f) $C_{20}H_{32}O_2$

6.3 Naming Alkenes

Alkenes are named using a series of rules similar to those developed for alkanes, with the suffix *-ene* used instead of *-ane* to identify the family. There are three steps:

1. Name the parent hydrocarbon. Find the longest carbon chain containing the double bond and name the compound accordingly, using the suffix *-ene*:

Named as a *pentene* *NOT* as a hexene, since the double bond is not contained in the six-carbon chain

2. Number the carbon atoms in the chain beginning at the end nearer the double bond. If the double bond is equidistant from the two ends, begin at the end nearer the first branch point. This rule ensures that the double-bond carbons receive the lowest possible numbers:

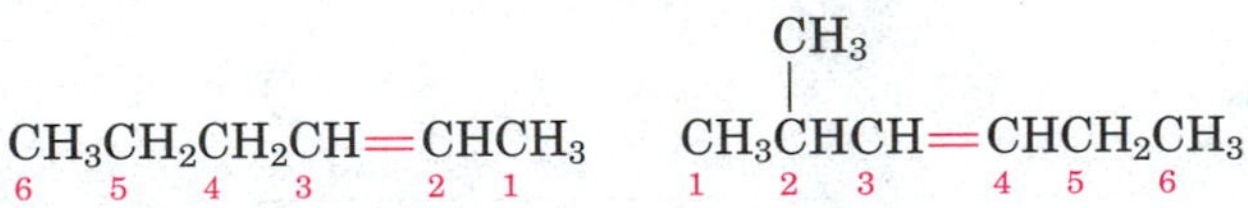

3. Write the full name, numbering the substituents according to their position in the chain and listing them alphabetically. Indicate the position of the double bond by giving the number of the *first* alkene carbon. If more than one double bond is present, indicate the position of each and use one of the suffixes *-diene, -triene*, and so on.

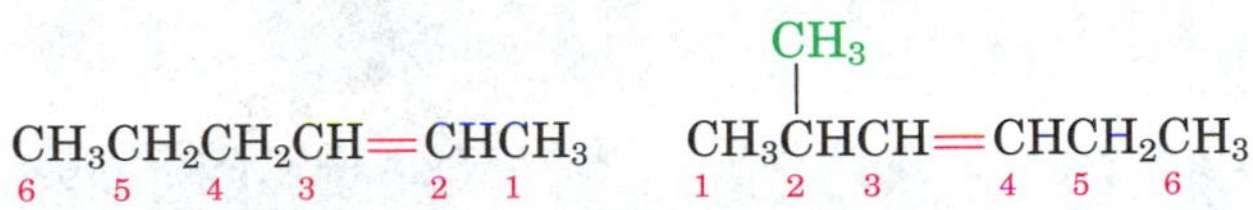

2-Hexene **2-Methyl-3-hexene**

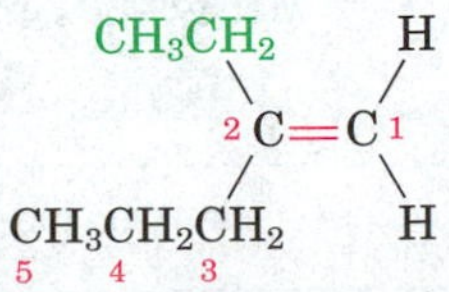

2-Ethyl-1-pentene

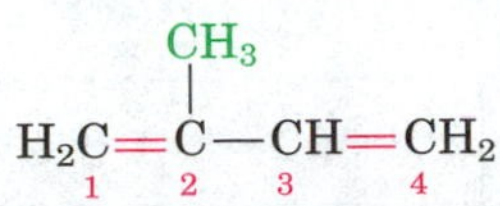

2-Methyl-1,3-butadiene

Cycloalkenes are named in a similar way, but because there is no chain end to begin from, we number the cycloalkene so that the double bond is between C1 and C2 and so that the first substituent has as low a number as possible. Note that it's not necessary to include the position of the double bond in the name because it is always between C1 and C2.

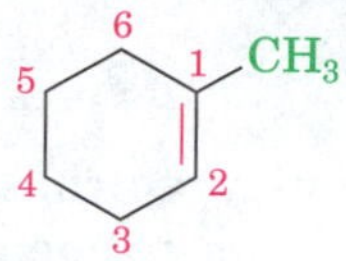

1-Methylcyclohexene

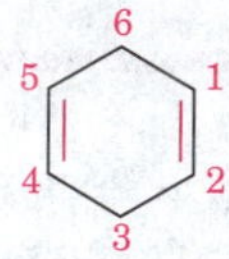

1,4-Cyclohexadiene

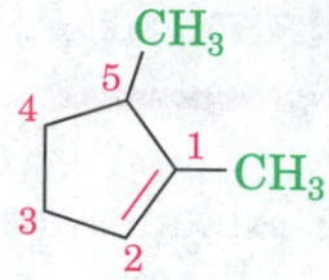

1,5-Dimethylcyclopentene

For historical reasons, there are a few alkenes whose names are firmly entrenched in common usage and don't conform to the rules. For example, the alkene derived from ethane should be called *ethene*, but the name *ethylene* has been used so long that it is accepted by IUPAC. Table 6.1 lists several other common names that are often used and are recognized by IUPAC. Note also that a $=CH_2$ substituent is called a **methylene group**, a $H_2C{=}CH-$ substituent is called a **vinyl group**, and a $H_2C{=}CHCH_2-$ substituent is called an **allyl group**:

$H_2C=$ — **A *methylene* group**

$H_2C{=}CH-$ — **A *vinyl* group**

$H_2C{=}CH{-}CH_2-$ — **An *allyl* group**

Table 6.1 Common Names of Some Alkenes[a]

Compound	*Systematic name*	*Common name*
$H_2C{=}CH_2$	Ethene	Ethylene
$CH_3CH{=}CH_2$	Propene	Propylene
$CH_3C(CH_3){=}CH_2$	2-Methylpropene	Isobutylene
$H_2C{=}C(CH_3){-}CH{=}CH_2$	2-Methyl-1,3-butadiene	Isoprene
$CH_3CH{=}CHCH{=}CH_2$	1,3-Pentadiene	Piperylene

[a]Both common and systematic names are recognized by IUPAC.

PROBLEM

6.4 Give IUPAC names for these compounds:

(a) 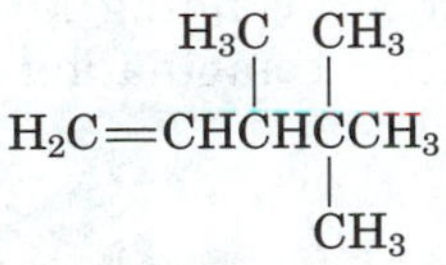

(b) 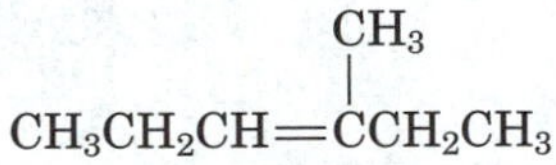

(c) 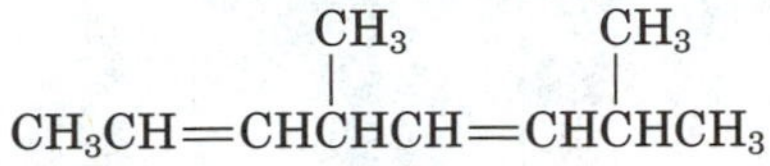

PROBLEM

6.5 Draw structures corresponding to the following IUPAC names:
(a) 2-Methyl-1,5-hexadiene
(b) 3-Ethyl-2,2-dimethyl-3-heptene
(c) 2,3,3-Trimethyl-1,4,6-octatriene
(d) 3,4-Diisopropyl-2,5-dimethyl-3-hexene
(e) 4-*tert*-Butyl-2-methylheptane

PROBLEM

6.6 Name the following cycloalkenes:

(a) 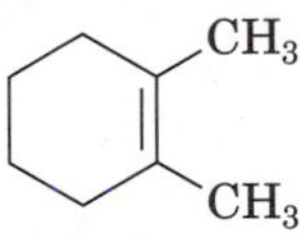
(b) 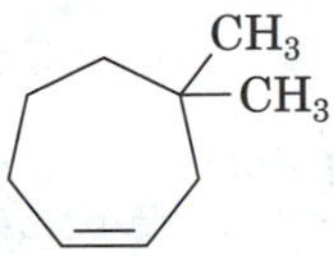
(c) 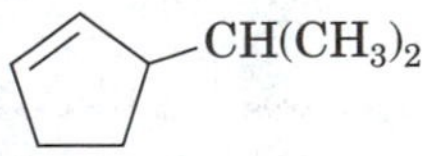

6.4 Electronic Structure of Alkenes

We saw in Section 1.10 that the carbon atoms in a double bond are sp^2-hybridized and have three equivalent orbitals, which lie in a plane at angles of 120° to one another. The fourth carbon orbital is an unhybridized p orbital perpendicular to the sp^2 plane. When two such carbon atoms approach each other, they form a σ bond by head-on overlap of sp^2 orbitals and a π bond by sideways overlap of p orbitals.

In molecular orbital terms, interaction of the *p* orbitals leads to one bonding and one antibonding π molecular orbital. The π bonding MO has no node between nuclei and results from an additive combination of *p* orbital lobes with the same algebraic sign. The π antibonding MO has a node between nuclei and results from a subtractive combination of lobes with different algebraic signs (Figure 6.2).

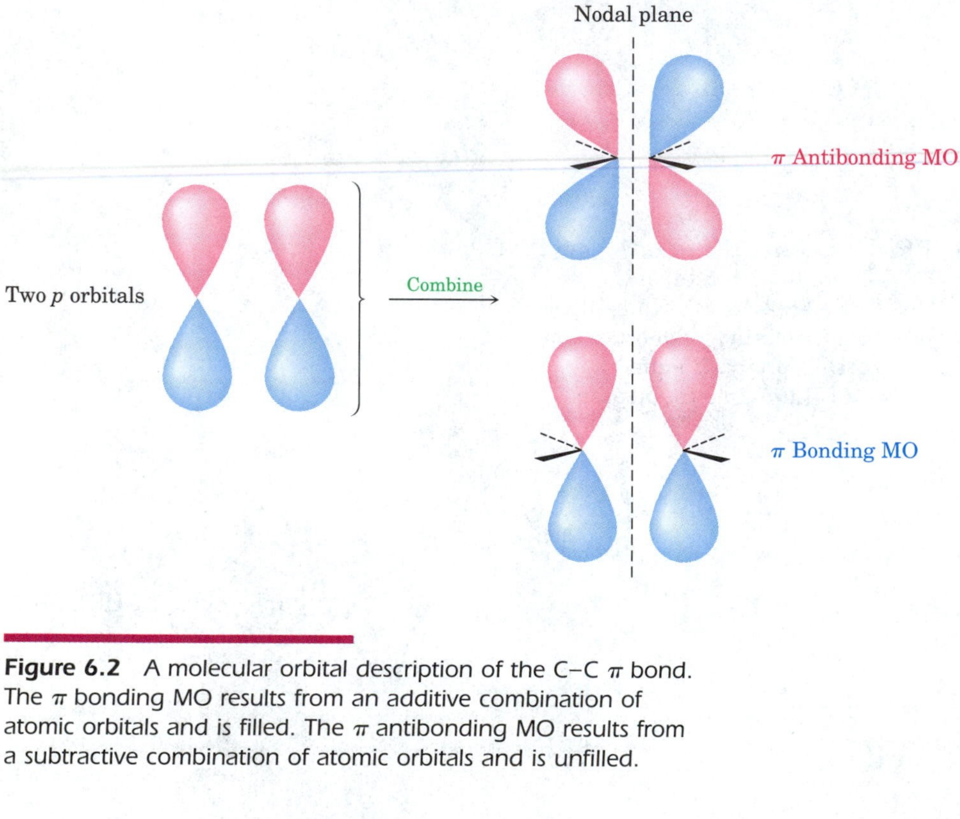

Figure 6.2 A molecular orbital description of the C–C π bond. The π bonding MO results from an additive combination of atomic orbitals and is filled. The π antibonding MO results from a subtractive combination of atomic orbitals and is unfilled.

We also know from Section 4.1 that relatively free rotation is possible around σ bonds and that open-chain alkanes like butane therefore have many rapidly interconverting conformations. The same is not true for double bonds, however. For rotation to occur around a double bond, the π bond must break temporarily (Figure 6.3). Thus, the barrier to double-bond rotation must be at least as great as the strength of the π bond itself.

A rough estimate of how much energy is required to break the π bond of an alkene can be made by subtracting the value for the strength of an average C–C σ bond (376 kJ/mol) from the total bond strength value for ethylene (611 kJ/mol). This calculation predicts an approximate bond strength of 235 kJ/mol (56 kcal/mol) for the ethylene π bond, and it's there-

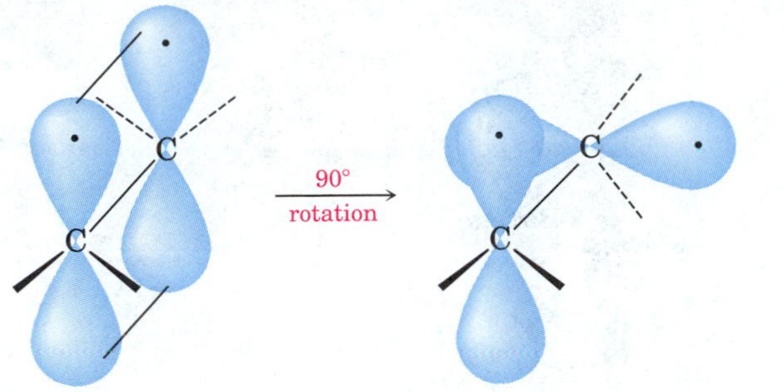

Figure 6.3 The π bond must break for rotation to take place around a carbon–carbon double bond.

fore clear why rotation does not occur. (Recall that the barrier to bond rotation in ethane is only 12 kJ/mol.)

Ethylene C=C bond strength ($\sigma + \pi$)	611 kJ/mol (146 kcal/mol)
Ethane C–C bond strength (σ only)	376 kJ/mol (90 kcal/mol)
Difference (π bond only)	235 kJ/mol (56 kcal/mol)

6.5 Cis–Trans Isomerism in Alkenes

The lack of rotation around the carbon–carbon double bond is of more than just theoretical interest; it also has chemical consequences. Imagine the situation for a disubstituted alkene such as 2-butene. (*Disubstituted* means that two substituents other than hydrogen are bonded to the double-bond carbons.) The two methyl groups in 2-butene can be either on the same side of the double bond or on opposite sides, a situation reminiscent of disubstituted cycloalkanes (Section 3.8). Figure 6.4 (p. 188) shows the two 2-butene isomers.

Since bond rotation can't occur, the two 2-butenes can't spontaneously interconvert; they are distinct, isolable compounds. As with disubstituted cycloalkanes (Section 3.8), we call such compounds *cis–trans stereoisomers*. The compound with substituents on the same side of the double bond is called *cis*-2-butene, and the isomer with substituents on opposite sides is *trans*-2-butene.

Cis–trans isomerism is not limited to *di*substituted alkenes. It can occur whenever both of the double-bond carbons are attached to two different groups. If one of the double-bond carbons is attached to two identical groups, however, then cis–trans isomerism is not possible (Figure 6.5).

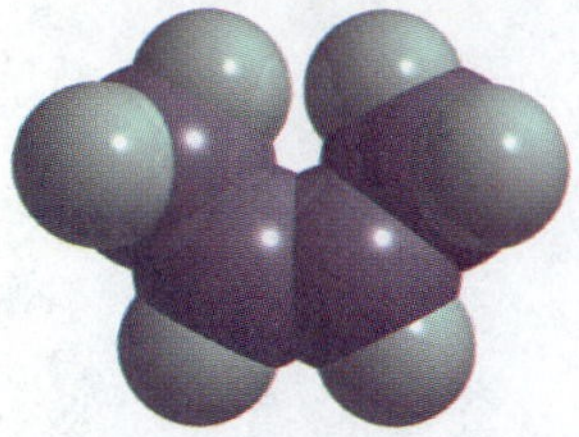

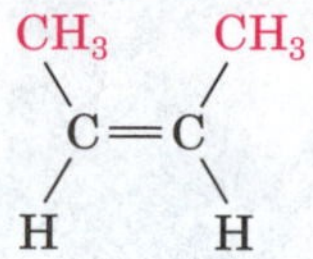

***cis*-2-Butene**

Stereo View

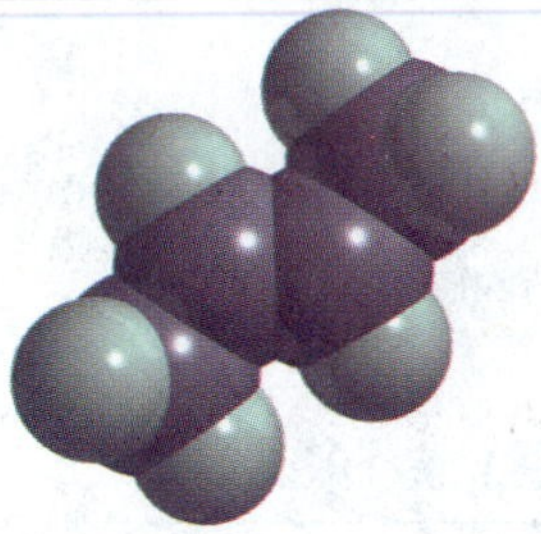

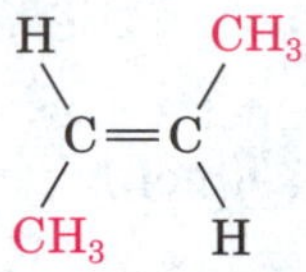

***trans*-2-Butene**

Stereo View

Figure 6.4 Cis and trans isomers of 2-butene. The cis isomer has the two methyl groups on the same side of the double bond, and the trans isomer has the methyl groups on opposite sides.

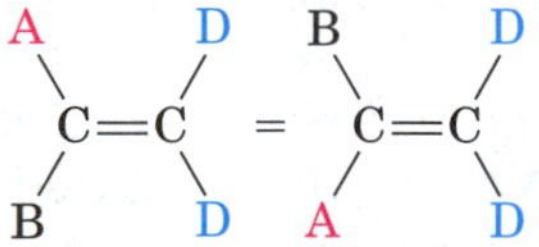

These two compounds are identical; they are not cis–trans isomers.

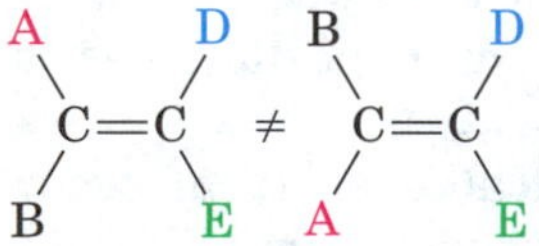

These two compounds are not identical; they are cis–trans isomers.

Figure 6.5 The requirement for cis–trans isomerism in alkenes. Compounds that have one of their carbons bonded to two identical groups can't exist as cis–trans isomers.

PROBLEM

6.7 Which of the following compounds can exist as pairs of cis–trans isomers? Draw each cis–trans pair, and indicate the geometry of each isomer.
(a) $CH_3CH{=}CH_2$ (b) $(CH_3)_2C{=}CHCH_3$ (c) $CH_3CH_2CH{=}CHCH_3$
(d) $(CH_3)_2C{=}C(CH_3)CH_2CH_3$ (e) $ClCH{=}CHCl$ (f) $BrCH{=}CHCl$

PROBLEM

6.8 Explain why cyclodecene can exist in both cis and trans forms but cyclohexene cannot. Making molecular models is helpful.

6.6 Sequence Rules: The *E,Z* Designation

The cis–trans nomenclature used in the previous section works well for describing the geometry of disubstituted double bonds, but fails with trisubstituted and tetrasubstituted double bonds. (*Trisubstituted* means three substituents other than hydrogen are attached to the double-bond carbons; *tetrasubstituted* means four substituents other than hydrogen.)

A more general method of describing double-bond geometry is provided by the *E,Z* system of nomenclature, which uses a series of **sequence rules** to assign priorities to the substituent groups on the double-bond carbons. Considering each end of the double-bond separately, the sequence rules are used to decide which of the groups attached to each carbon is higher in priority. If the higher-priority groups on each carbon are on the same side of the double bond, the alkene is designated *Z* (for the German *zusammen*, "together"). If the higher-priority groups are on opposite sides, the alkene is designated *E* (for the German *entgegen*, "opposite"). A simple way to remember which is which is to think with an accent: In the *Z* isomer, the groups are on "ze zame zide." The assignments are shown in Figure 6.6.

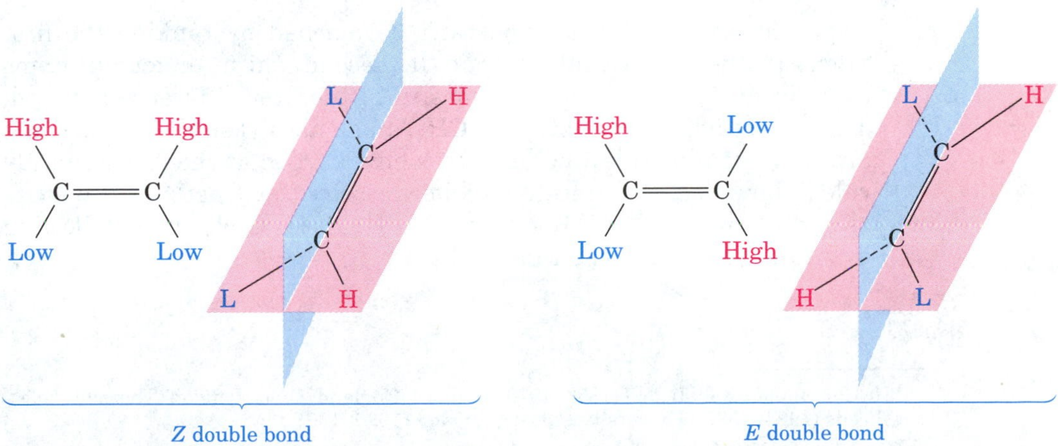

Figure 6.6 The *E,Z* system of nomenclature for substituted alkenes. The higher-priority groups on each carbon are on the same side in the *Z* isomer, but are on opposite sides in the *E* isomer.

Called the Cahn–Ingold[1]–Prelog[2] rules after the chemists who proposed them, the sequence rules are as follows:

Sequence rule 1 Considering each of the double-bond carbons separately, look at the two atoms directly attached and rank them according to atomic number. An atom with higher atomic number receives higher priority than an atom with lower atomic number. Thus, the atoms commonly found attached to a double bond are assigned the following order:

$$\overset{35}{Br} > \overset{17}{Cl} > \overset{8}{O} > \overset{7}{N} > \overset{6}{C} > \overset{1}{H}$$

For example:

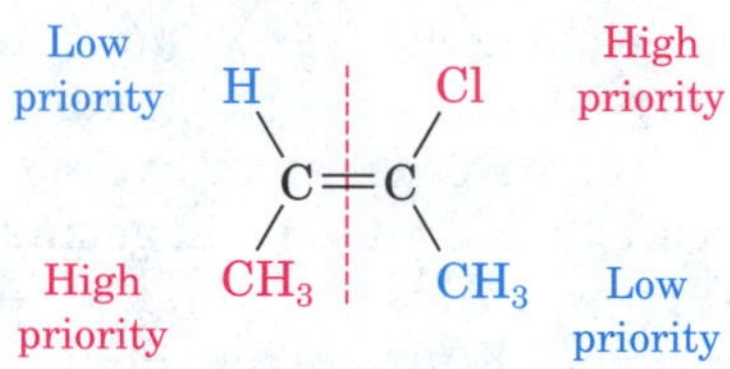

(a) (*E*)-2-Chloro-2-butene

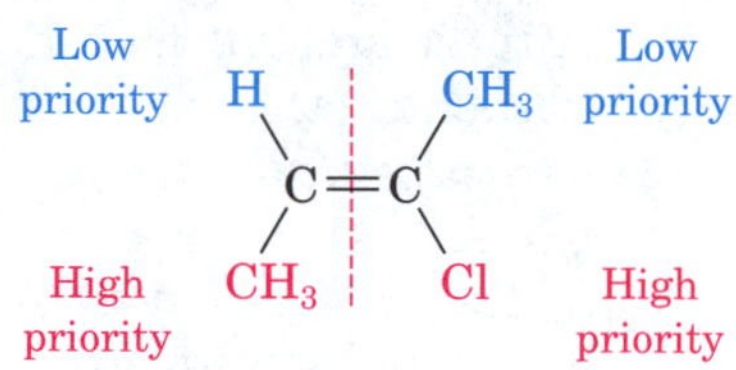

(b) (*Z*)-2-Chloro-2-butene

Because chlorine has a higher atomic number than carbon, a Cl substituent receives higher priority than a methyl (CH_3) group. Methyl receives higher priority than hydrogen, however, and isomer (a) is therefore assigned *E* geometry (high-priority groups on opposite sides of the double bond). Isomer (b) has *Z* geometry (high-priority groups on ze zame zide of the double bond).

Sequence rule 2 If a decision can't be reached by ranking the first atoms in the substituent, look at the second, third, or fourth atoms away from the double-bond carbons until the first difference is found. Thus, an ethyl substituent, $-CH_2CH_3$, and a methyl substituent, $-CH_3$, are equivalent by rule 1 because both have carbon as the first atom. By rule 2, however, ethyl receives higher priority than methyl because its *second* atoms are C, H, H rather than H, H, H. Look at the following examples to see how the rule works:

---

[1]Sir Christopher Kelk Ingold (1893–1970); b. Ilford, England; D.Sc., London (Thorpe); professor, Leeds (1924–1930), University College, London (1930–1961).

[2]Vladimir Prelog (1906–); b. Sarajevo (now in Bosnia-Herzegovina); Dr. Ing., Institute of Technology, Prague (Votocek); Professor, University of Zagreb, Federal Institute of Technology (ETH), Zürich (1941–1976); Nobel Prize (1975).

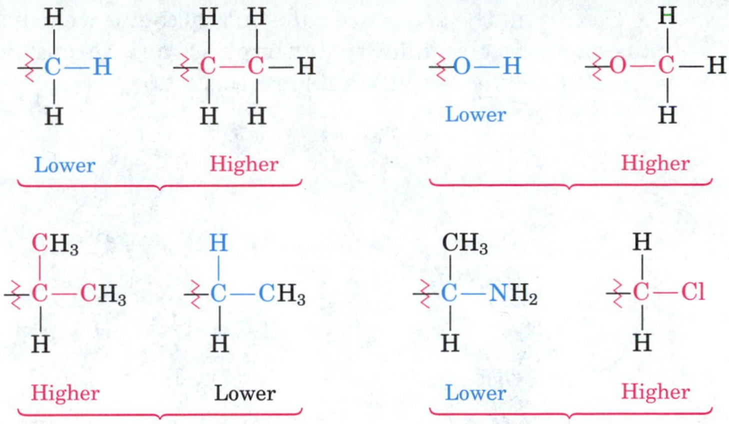

Sequence rule 3 Multiple-bonded atoms are equivalent to the same number of single-bonded atoms. For example, an aldehyde substituent (–CH=O), which has a carbon atom *doubly* bonded to *one* oxygen, is equivalent to a substituent with a carbon atom *singly* bonded to *two* oxygen atoms:

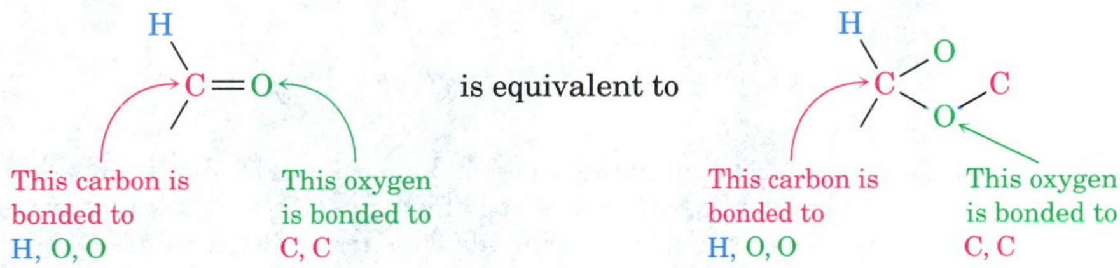

As further examples, the following pairs are equivalent:

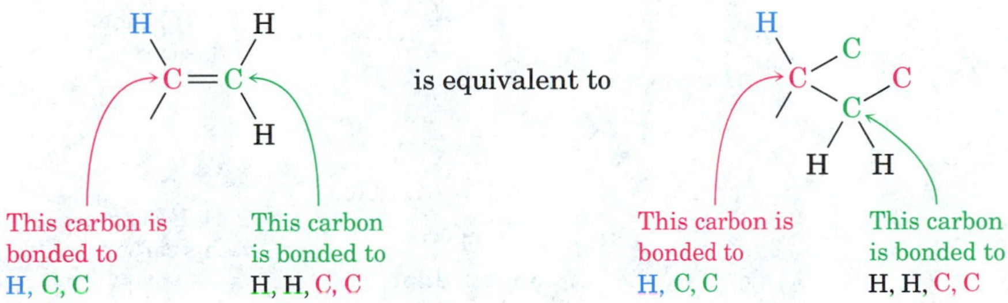

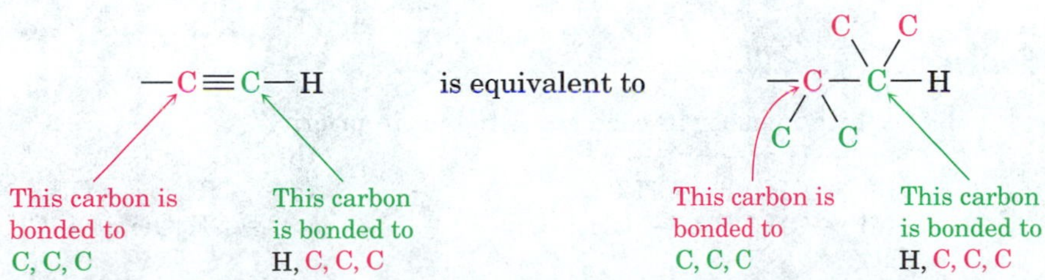

Taking all the sequence rules into account, we can assign the configurations shown in the following examples. Work through each one to convince yourself that the assignments are correct.

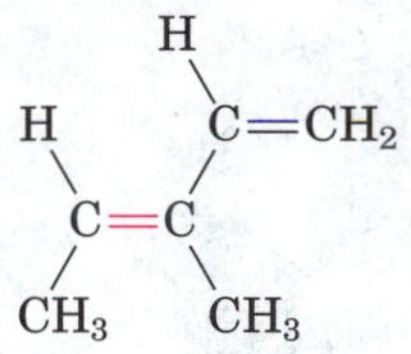

(*E*)-3-Methyl-1,3-pentadiene

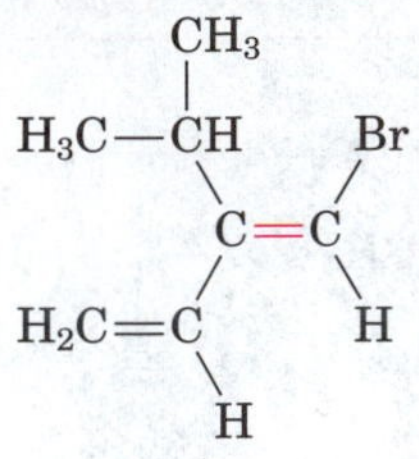

(*E*)-1-Bromo-2-isopropyl-1,3-butadiene

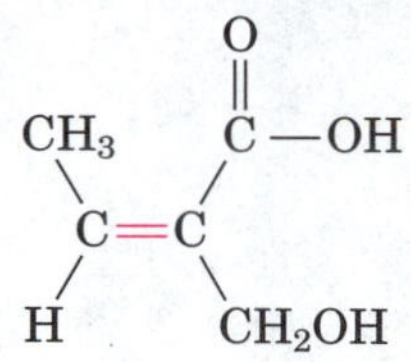

(*Z*)-2-Hydroxymethyl-2-butenoic acid

PRACTICE PROBLEM

Assign *E* or *Z* configuration to the double bond in this compound:

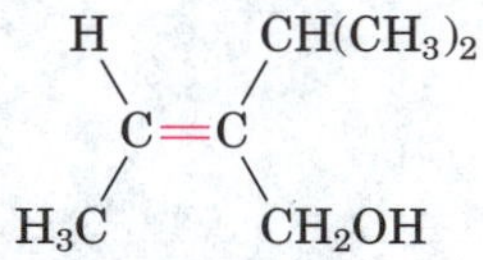

Solution Look at the double-bond carbons individually. The left-hand carbon has two substituents, $-H$ and $-CH_3$, of which $-CH_3$ receives higher priority by sequence rule 1. The right-hand carbon also has two substituents, $-CH(CH_3)_2$ and $-CH_2OH$, which are equivalent by rule 1. By rule 2, however, $-CH_2OH$ receives higher priority than $-CH(CH_3)_2$. The substituent $-CH_2OH$ has an *oxygen* and two hydrogens as its second atoms, but $-CH(CH_3)_2$ has two *carbons* and a hydrogen. The two high-priority groups are on the same side of the double bond, and we assign *Z* configuration.

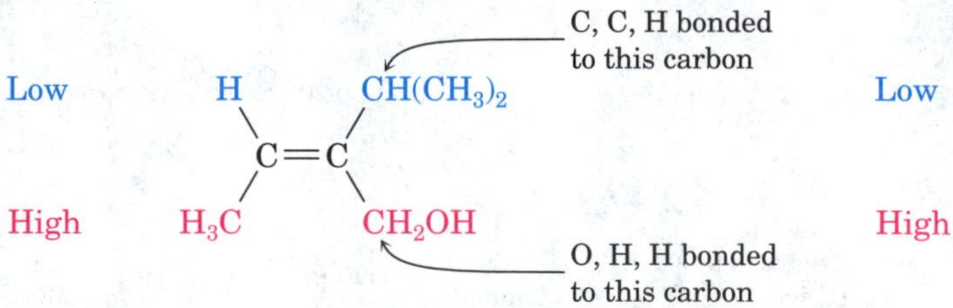

***Z* configuration**

PROBLEM

6.9 Which member in each set is higher in priority?

(a) $—H$ or $—Br$
(b) $—Cl$ or $—Br$
(c) $—CH_3$ or $—CH_2CH_3$
(d) $—NH_2$ or $—OH$
(e) $—CH_2OH$ or $—CH_3$
(f) $—CH_2OH$ or $—CH{=}O$

PROBLEM

6.10 Rank the following sets of substituents in order of Cahn–Ingold–Prelog priorities:
(a) $-CH_3$, $-OH$, $-H$, $-Cl$
(b) $-CH_3$, $-CH_2CH_3$, $-CH{=}CH_2$, $-CH_2OH$
(c) $-COOH$, $-CH_2OH$, $-C{\equiv}N$, $-CH_2NH_2$
(d) $-CH_2CH_3$, $-C{\equiv}CH$, $-C{\equiv}N$, $-CH_2OCH_3$

PROBLEM

6.11 Assign *E* or *Z* configuration to these alkenes:

(a)
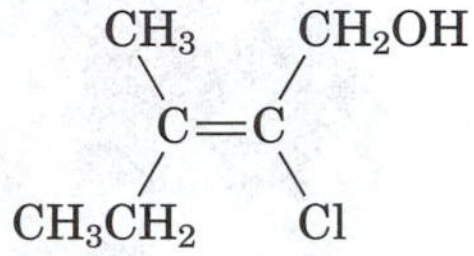

(b)
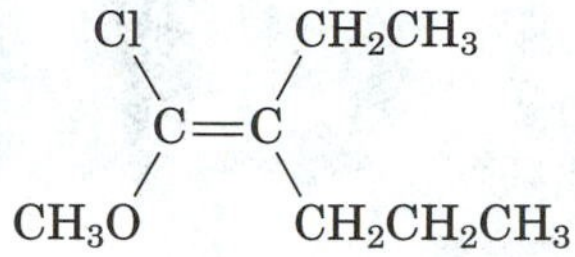

(c)
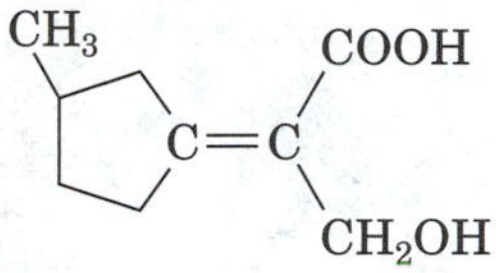

(d)
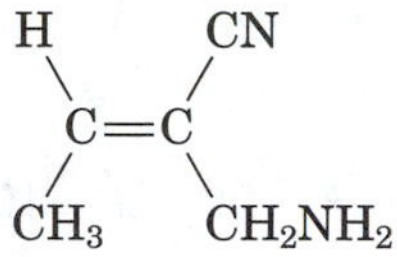

6.7 Alkene Stability

The cis–trans interconversion of alkene isomers does not occur spontaneously, but it can be made to happen by treating the alkene with a strong acid catalyst. If we interconvert *cis*-2-butene with *trans*-2-butene and allow them to reach equilibrium, we find that they aren't of equal stability. The trans isomer is more favored than the cis isomer by a ratio of 76% trans to 24% cis.

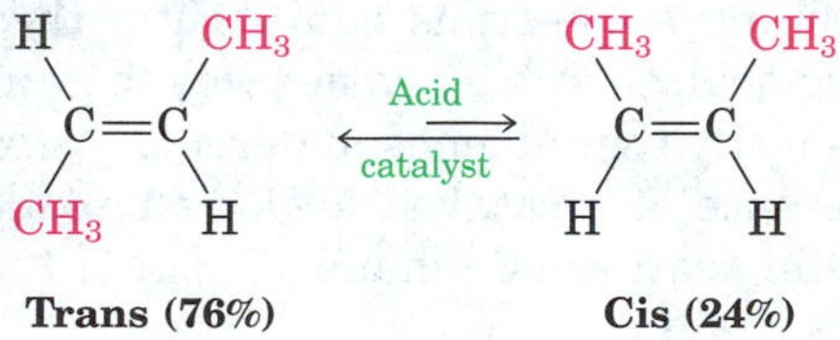

Using the relationship between equilibrium constant and free-energy differences shown previously in Figure 4.20, we can calculate that *cis*-2-butene is less stable than *trans*-2-butene by 2.8 kJ/mol (0.66 kcal/mol) at room temperature.

Cis alkenes are less stable than their trans isomers because of steric (spatial) strain between the two bulky substituents on the same side of the double bond. As shown in Figure 6.7, this is the same kind of steric interference that we saw previously in the axial conformation of methylcyclohexane (Section 4.12).

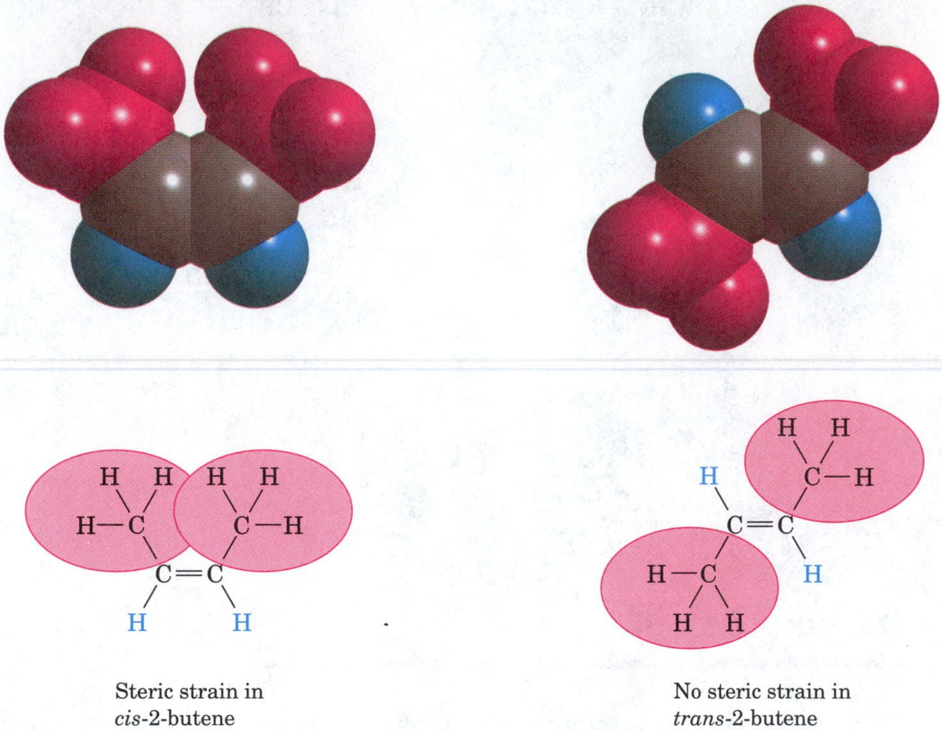

Figure 6.7 *cis*-2-Butene is less stable than its trans isomer because of the steric strain that occurs when the two methyl groups try to occupy the same space.

Although it's sometimes possible to find relative stabilities of alkenes by establishing a cis–trans equilibrium through treatment of the alkene with strong acid, there's an easier way to gain nearly the same information. We'll see shortly that alkenes undergo a *hydrogenation* reaction with H_2 gas in the presence of a catalyst such as palladium or platinum, to yield the corresponding saturated alkane.

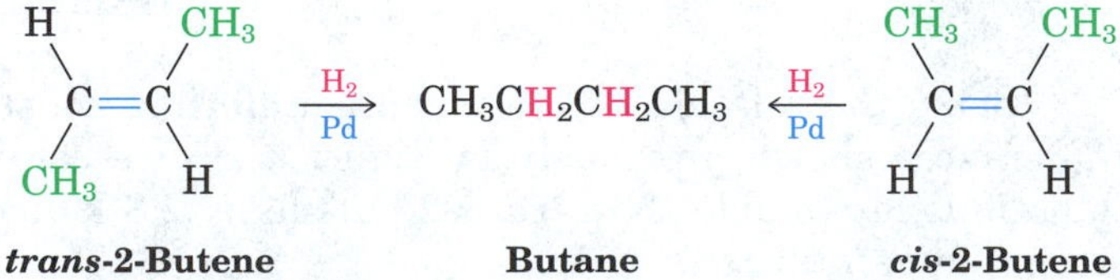

Look at the hydrogenations of *cis*- and *trans*-2-butene, both of which give the same product, butane. Energy diagrams for the two reactions are shown in Figure 6.8. Since *cis*-2-butene is less stable than *trans*-2-butene by 2.8 kJ/mol, the energy diagram shows the cis alkene at a higher energy level. After reaction, however, both curves are at the same energy level (butane). It therefore follows that $\Delta G°$ for reaction of the cis isomer must be larger than $\Delta G°$ for reaction of the trans isomer. In other words, more energy is evolved in the hydrogenation of the cis isomer than the trans isomer because the cis isomer has more energy to begin with.

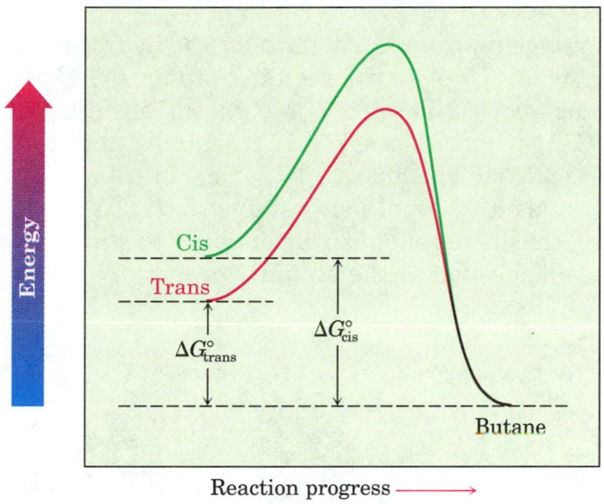

Figure 6.8 Reaction energy diagrams for hydrogenation of *cis*- and *trans*-2-butene. The cis isomer is higher in energy than the trans isomer by about 2.8 kJ/mol and therefore gives off more energy in the reaction.

If we were to measure the heats of reaction for the two hydrogenations and find their difference, we could determine the relative stabilities of cis and trans isomers without having to measure an equilibrium position. Many such **heats of hydrogenation** ($\Delta H°_{hydrog}$) have been measured, and the results bear out our expectation. For *cis*-2-butene, $\Delta H°_{hydrog} = -120$ kJ/mol (-28.6 kcal/mol); for the trans isomer, $\Delta H°_{hydrog} = -116$ kJ/mol (-27.6 kcal/mol).

Cis isomer
$\Delta H°_{hydrog} = -120$ kJ/mol

Trans isomer
$\Delta H°_{hydrog} = -116$ kJ/mol

Although the energy difference between the 2-butene isomers as calculated from heats of hydrogenation (4 kJ/mol) agrees reasonably well with the energy difference calculated from equilibrium data (2.8 kJ/mol), the two numbers aren't exactly the same for two reasons. The first is simply experimental error: Heats of hydrogenation require considerable expertise and specialized equipment to measure accurately, and we are looking at a small difference between two large numbers. The second reason is that heats of reaction and equilibrium constants don't measure exactly the same quantity. Heats of reaction measure enthalpy changes, $\Delta H°$, whereas equilibrium constants measure free-energy changes, $\Delta G°$ ($\Delta G° = \Delta H° - T\Delta S°$). We therefore expect a slight difference when comparing the two measurements.

Table 6.2 lists some representative data for the hydrogenation of different alkenes, and Figure 6.9 plots the results graphically. These data show that alkenes become more stable with increasing substitution. For example, ethylene has $\Delta H°_{hydrog} = -137$ kJ/mol (-32.8 kcal/mol), but when one alkyl substituent is attached to the double bond, as in 1-butene, the alkene becomes approximately 10 kJ/mol more stable ($\Delta H°_{hydrog} = -126$ kJ/mol). Further increasing the degree of substitution leads to still further stability. As a general rule, alkenes follow the stability order:

Tetrasubstituted > Trisubstituted > Disubstituted > Monosubstituted

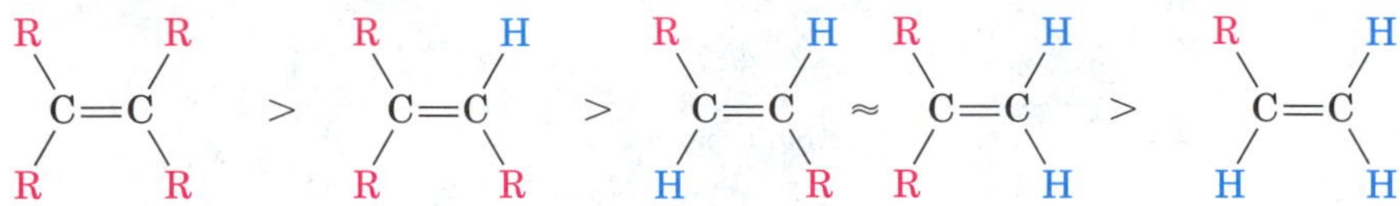

Table 6.2 Heats of Hydrogenation of Some Alkenes

Substitution	Alkene	$\Delta H°_{hydrog}$ (kJ/mol)	$\Delta H°_{hydrog}$ (kcal/mol)
	$H_2C{=}CH_2$	−137	−32.8
Monosubstituted	$CH_3CH{=}CH_2$	−126	−30.1
	$CH_3CH_2CH{=}CH_2$	−126	−30.1
	$(CH_3)_2CHCH{=}CH_2$	−127	−30.3
Disubstituted	$CH_3CH{=}CHCH_3$ (cis)	−120	−28.6
	$CH_3CH{=}CHCH_3$ (trans)	−115	−27.6
	$(CH_3)_2C{=}CH_2$	−119	−28.4
Trisubstituted	$(CH_3)_2C{=}CHCH_3$	−113	−26.9
Tetrasubstituted	$(CH_3)_2C{=}C(CH_3)_2$	−111	−26.6

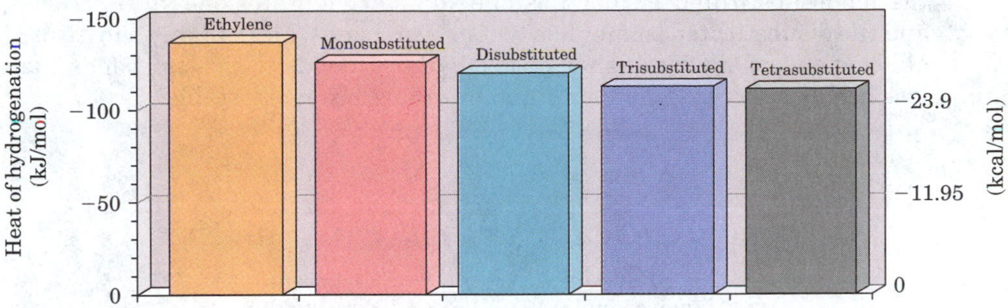

Figure 6.9 A plot of heat of hydrogenation versus substitution pattern for alkenes. Alkene stability increases with increasing substitution.

Two explanations have been advanced to account for the observed stability order of alkenes. Many chemists believe that the stability order is due primarily to **hyperconjugation** (Figure 6.10), a stabilizing effect resulting from interaction between the unfilled antibonding C=C π bond orbital (Section 1.10) and a filled C–H σ bond orbital on a neighboring substituent. The more substituents that are present, the more opportunities exist for hyperconjugation and the more stable the alkene.

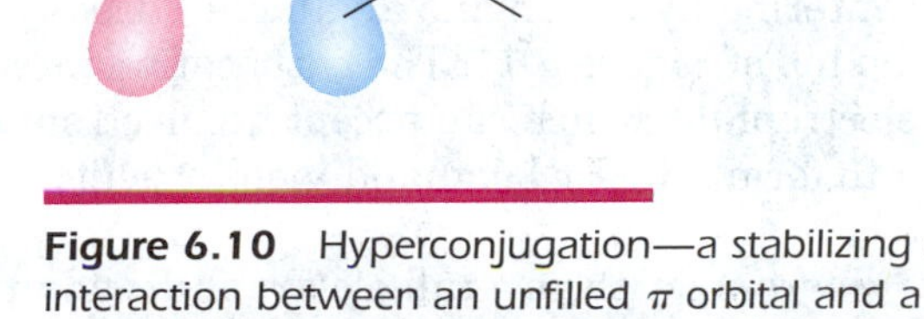

Figure 6.10 Hyperconjugation—a stabilizing effect due to interaction between an unfilled π orbital and a neighboring filled C–H σ bond orbital.

In addition to the effect of hyperconjugation, a simple bond-strength argument can also be used to explain the observed alkene stability order. A bond between an sp^2 carbon and an sp^3 carbon is somewhat stronger than a bond between two sp^3 carbons. Thus, in comparing 1-butene and 2-butene, the monosubstituted isomer has one sp^3–sp^3 bond and one sp^3–sp^2 bond, and the disubstituted isomer has two sp^3–sp^2 bonds. More highly substituted alkenes always have a higher ratio of sp^3–sp^2 bonds to sp^3–sp^3 bonds than less highly substituted alkenes and are therefore more stable.

sp^3–sp^2 ↓ sp^3–sp^2 ↓	sp^3–sp^3 ↓ sp^3–sp^2 ↓
CH_3—CH=CH—CH_3	CH_3—CH_2—CH=CH_2
2-Butene (more stable)	**1-Butene (less stable)**

PROBLEM

6.12 Which alkene in each of the following sets is more stable?

(a) 1-Butene or 2-methylpropene (b) (*Z*)-2-Hexene or (*E*)-2-hexene

(c) 1-Methylcyclohexene or 3-methylcyclohexene

6.8 Alkene Electrophilic Addition Reactions

Before beginning a detailed discussion of alkene reactions, let's review briefly some conclusions from Sections 5.5–5.9. We said there that alkenes behave as nucleophiles (Lewis bases) in polar reactions. The carbon–carbon double bond is electron-rich and can donate a pair of electrons to an electrophile (Lewis acid). For example, reaction of 2-methylpropene with HBr yields 2-bromo-2-methylpropane. Careful study of this and similar reactions by Christopher Ingold and others many years ago has led to the generally accepted mechanism shown in Figure 6.11 for what is called an **electrophilic addition reaction**.

The reaction begins with an attack on the electrophile, HBr, by the electrons of the nucleophilic π bond. Two electrons from the π bond form a new σ bond between the entering hydrogen and an alkene carbon, as shown by the curved arrow at the top of Figure 6.11. The carbocation intermediate that results is itself an electrophile, which can accept an electron pair from nucleophilic bromide ion to form a C–Br bond and yield a neutral addition product.

The reaction energy diagram for the overall electrophilic addition reaction (Figure 6.12) has two peaks (transition states) separated by a valley (carbocation intermediate). The energy level of the intermediate is higher than that of the starting alkene, but the reaction as a whole is exothermic (negative $\Delta G°$). The first step, protonation of the alkene to yield the intermediate cation, is relatively slow, but once formed, the cation intermediate rapidly reacts further to yield the final bromoalkane product. The relative rates of the two steps are indicated in Figure 6.12 by the fact that $\Delta G_1^‡$ is larger than $\Delta G_2^‡$.

The electrophile H^+ is attacked by the π electrons of the double bond, and a new C–H σ bond is formed. This leaves the other carbon atom with a + charge and a vacant p orbital.

Br$^-$ donates an electron pair to the positively charged carbon atom, forming a C–Br σ bond and yielding the neutral addition product.

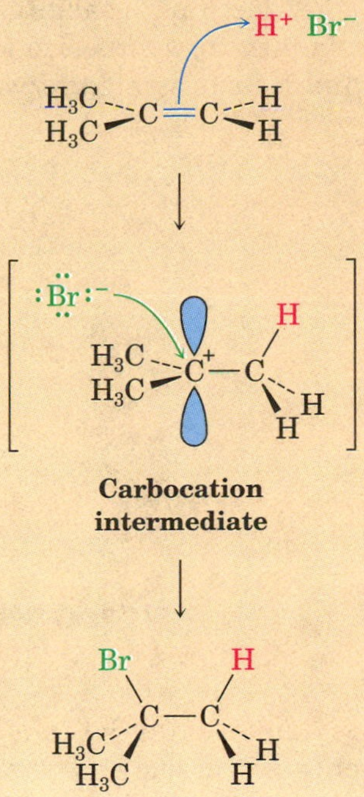

Figure 6.11 Mechanism of the electrophilic addition of HBr to 2-methylpropene. The reaction occurs in two steps and involves a carbocation intermediate.

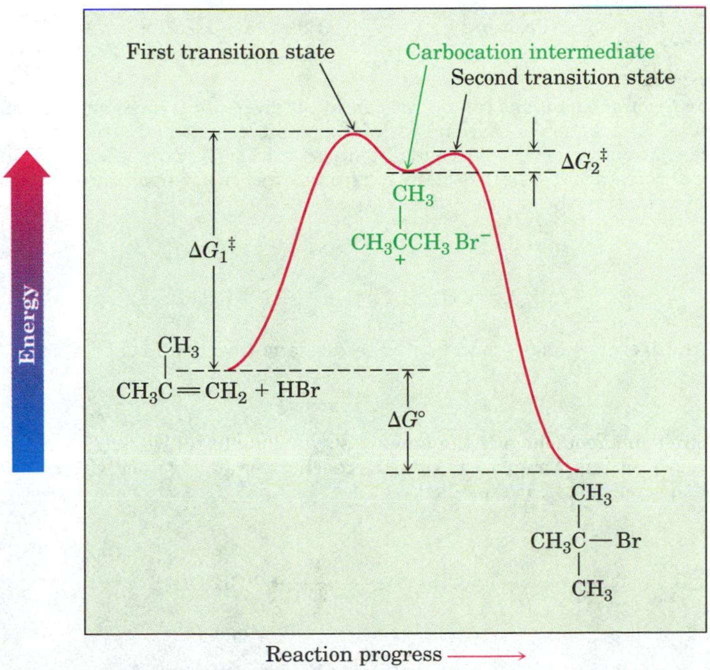

Figure 6.12 Reaction energy diagram for the two-step electrophilic addition of HBr to 2-methylpropene. The first step is slower than the second step.

Electrophilic addition of HX to alkenes is a general reaction that allows chemists to synthesize a variety of products. For example, addition of HCl and HBr is straightforward:[3]

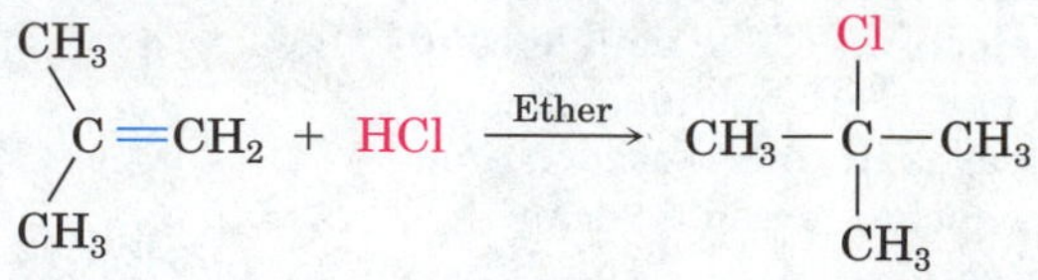

2-Methylpropene **2-Chloro-2-methylpropane (94%)**

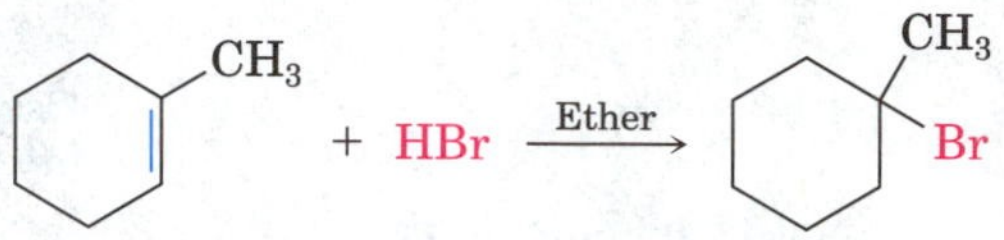

1-Methylcyclohexene **1-Bromo-1-methylcyclohexane (91%)**

Addition of HI to alkenes also occurs, but it's best to use a mixture of phosphoric acid and potassium iodide to generate HI in the reaction mixture, rather than to use HI directly. The mechanism is the same as for the other additions:

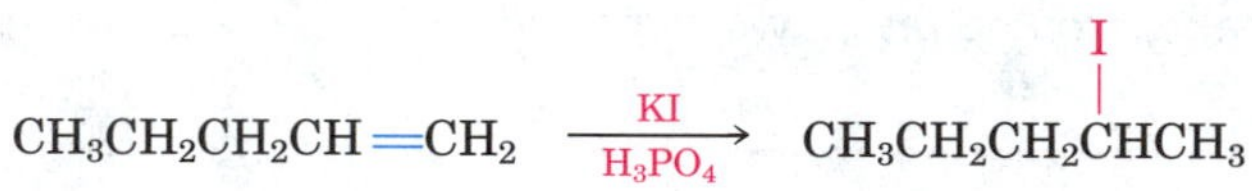

1-Pentene **(HI)** **2-Iodopentane**

[3]Organic reaction equations can be written in different ways to emphasize different points. For example, the reaction of ethylene with HBr might be written in the format A + B → C to emphasize that both reactants are equally important for the purposes of the discussion. The solvent and notes about other reaction conditions such as temperature are usually written either above or below the reaction arrow.

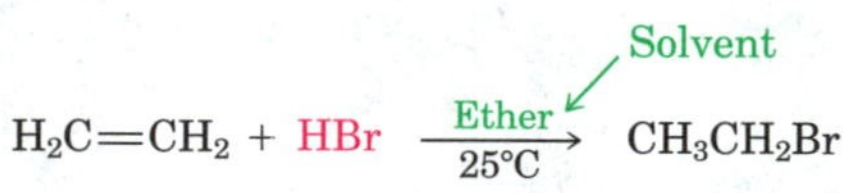

Alternatively, we might choose to write the same reaction in the format

$$A \xrightarrow{B} C$$

to emphasize that reactant A is the organic starting material whose chemistry is of greater interest. Reactant B is then placed above the reaction arrow, together with notes about solvent and reaction conditions. For example:

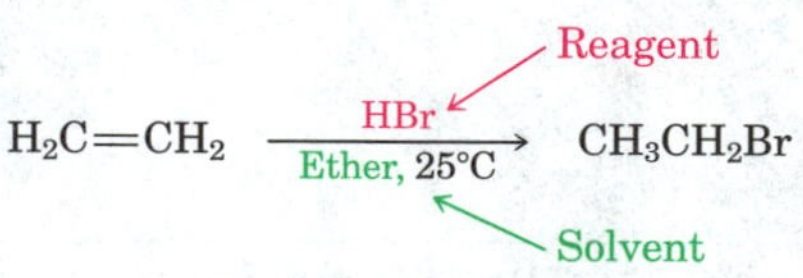

Both reaction formats are frequently used in chemistry, and you sometimes have to look carefully at the overall transformation to see what the different roles of the chemicals shown next to the reaction arrows are.

6.9 Orientation of Electrophilic Addition: Markovnikov's Rule

Look carefully at the reactions shown in the previous section. In every case, an unsymmetrically substituted alkene has given a single addition product, rather than the mixture that might have been expected. For example, 2-methylpropene *might* have reacted with HCl to give 1-chloro-2-methylpropane (isobutyl chloride) in addition to 2-chloro-2-methylpropane, but it didn't. We say that such reactions are **regiospecific** (**ree**-jee-oh-specific) when only one of two possible directions of addition occurs.

A regiospecific reaction:

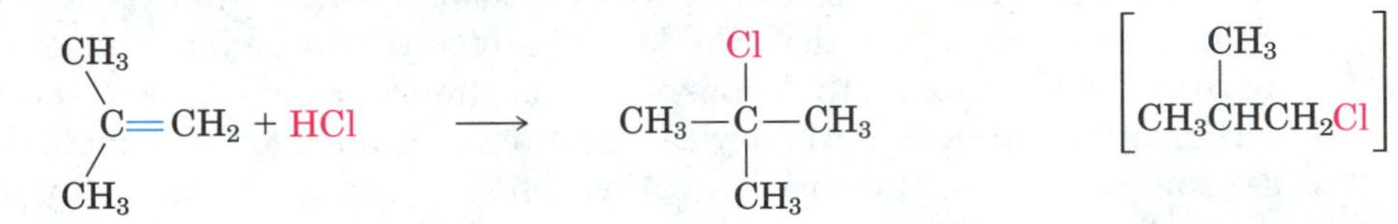

2-Methylpropene **2-Chloro-2-methylpropane** ***(sole product)*** **1-Chloro-2-methylpropane** **(*NOT* formed)**

After looking at the results of many such reactions, the Russian chemist Vladimir Markovnikov[4] proposed in 1869 what has become known as **Markovnikov's rule**:

Markovnikov's rule In the addition of HX to an alkene, the H attaches to the carbon with fewer alkyl substituents, and the X attaches to the carbon with more alkyl substituents.

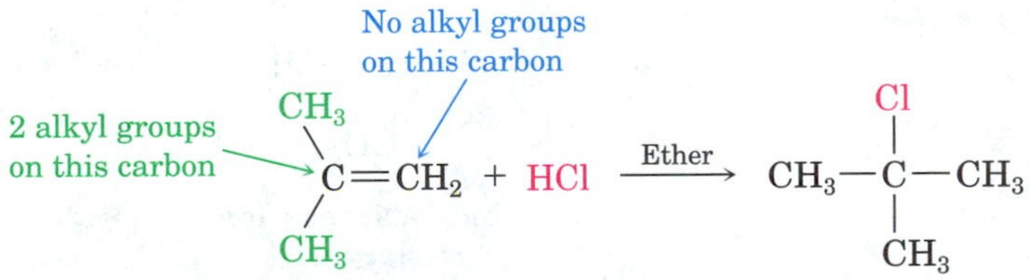

2-Methylpropene **2-Chloro-2-methylpropane**

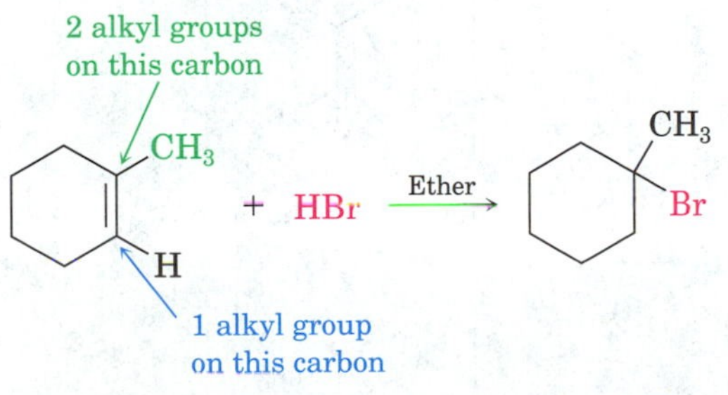

1-Methylcyclohexene **1-Bromo-1-methylcyclohexane**

---

[4]Vladimir Vassilyevich Markovnikov (1837–1904); b. Nijni-Novgorod, Russia; Ph.D. Kazan (A. M. Butlerov); professor in Kazan (1870), Odessa (1871), and Moscow (1873).

When both ends of the double bond have the same degree of substitution, a mixture of products results:

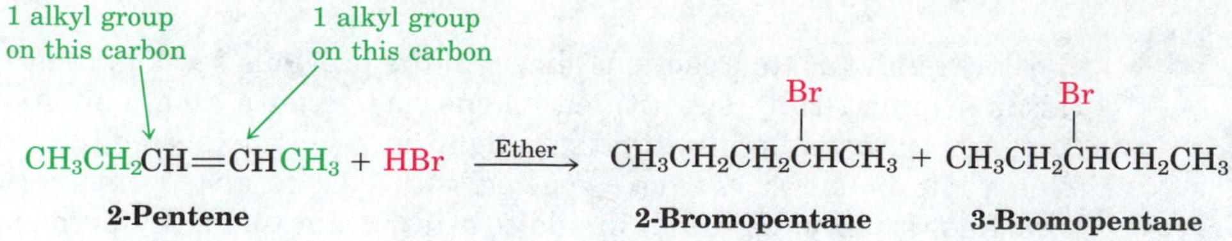

Since carbocations are involved as intermediates in these reactions, another way to express Markovnikov's rule is to say that, in the addition of HX to an alkene, the more highly substituted carbocation is formed as an intermediate rather than the less highly substituted one. For example, addition of H^+ to 2-methylpropene yields the intermediate *tertiary* carbocation rather than the primary carbocation, and addition to 1-methylcyclohexene yields a tertiary cation rather than a secondary one. Why should this be?

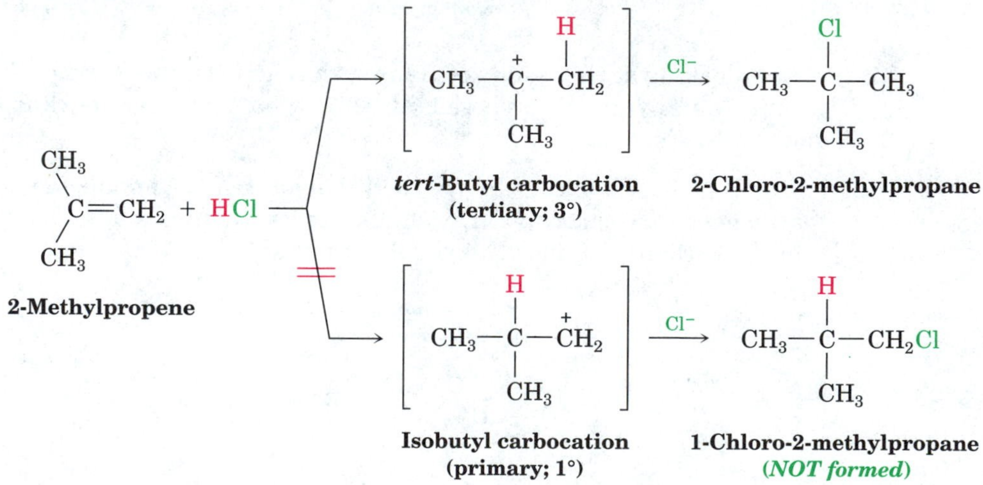

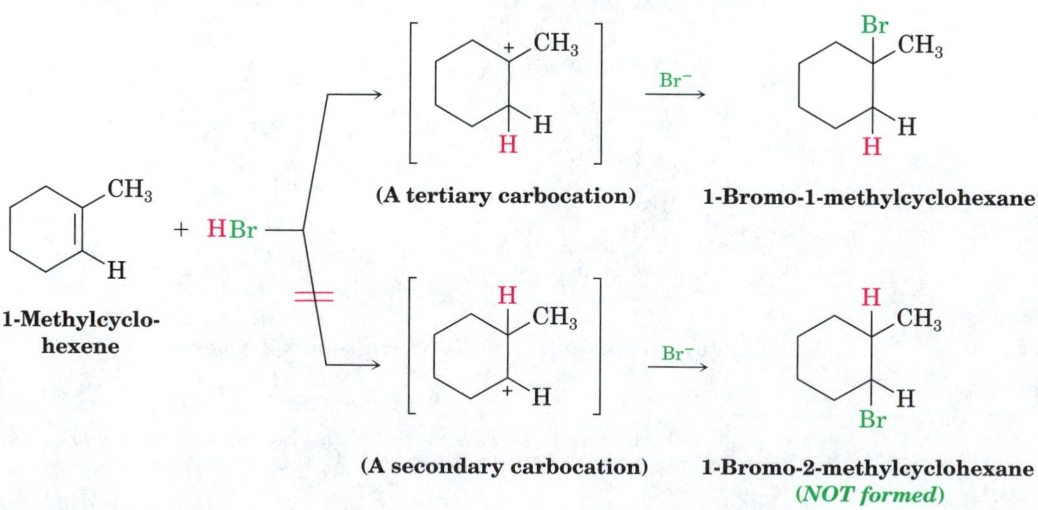

PRACTICE PROBLEM

What product would you expect from reaction of HCl with 1-ethylcyclopentene?

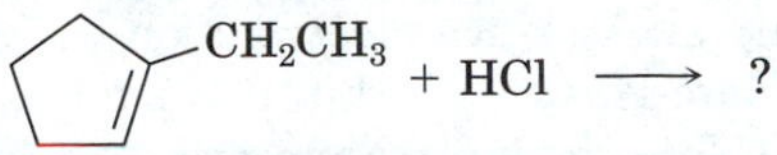

Solution Markovnikov's rule predicts that H will add to the double-bond carbon that has one alkyl group (C2 on the ring) and Cl will add to the double-bond carbon that has two alkyl groups (C1 on the ring). The expected product is 1-chloro-1-ethylcyclopentane.

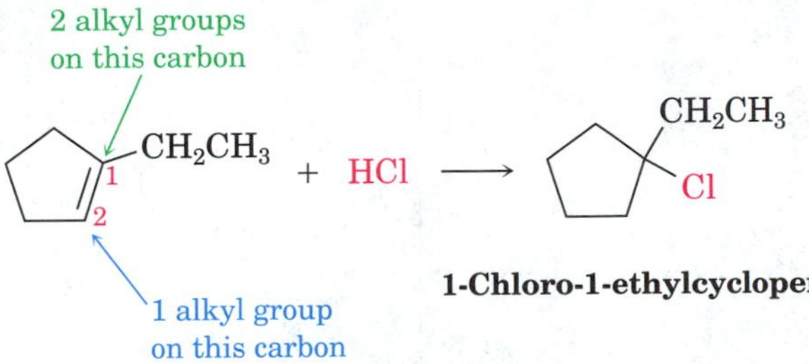

1-Chloro-1-ethylcyclopentane

PROBLEM

6.13 Predict the products of these reactions:

(a)  + HCl ⟶

(b) $(CH_3)_2C{=}CHCH_2CH_3 \xrightarrow{HBr}$

(c) $CH_3CH_2CH_2CH{=}CH_2 \xrightarrow[KI]{H_3PO_4}$

(d) 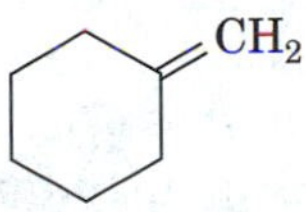

+ HBr ⟶

PROBLEM

6.14 What alkenes would you start with to prepare the following alkyl halides?

(a) Bromocyclopentane

(b) $CH_3CH_2CHBrCH_2CH_2CH_3$

(c) 1-Iodo-1-ethylcyclohexane

(d) 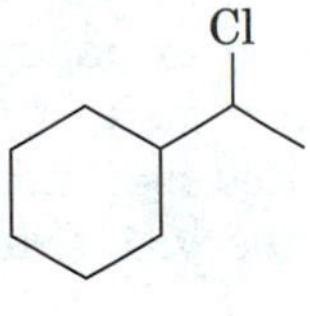

6.10 Carbocation Structure and Stability

To understand the reasons for the Markovnikov orientation of electrophilic addition reactions, we need to learn more about the structure and stability of carbocations and about the general nature of reactions and transition states. The first point to explore involves structure.

A great deal of evidence has shown that carbocations are *planar*. The trivalent carbon is sp^2-hybridized, and the three substituents are oriented to the corners of an equilateral triangle, as indicated in Figure 6.13. Since there are only six valence electrons on carbon, and since all six are used in the three σ bonds, the p orbital extending above and below the plane is unoccupied. (Note the electronic similarity of carbocations to trivalent boron compounds such as BF_3; see Section 1.12.)

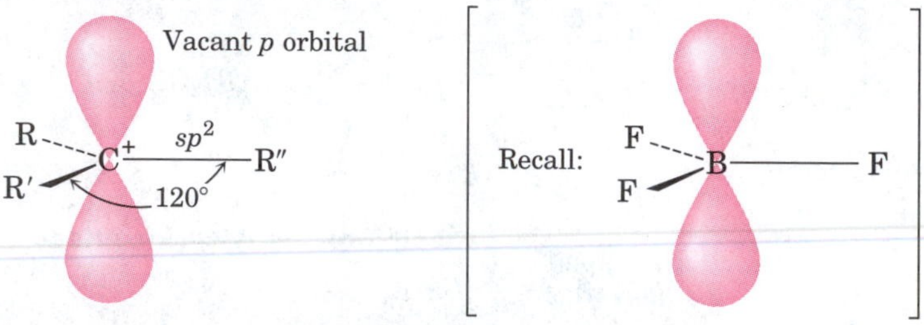

Figure 6.13 The electronic structure of a carbocation. The trivalent carbon is sp^2-hybridized and has a vacant p orbital.

The second point to explore involves carbocation stability. 2-Methylpropene might react with H^+ to form a carbocation having three alkyl substituents (a tertiary ion, 3°), or it might react to form a carbocation having one alkyl substituent (a primary ion, 1°). Since the tertiary chloride, 2-chloro-2-methylpropane, is the only product observed, formation of the tertiary cation is evidently favored over formation of the primary cation. Thermodynamic measurements show that, indeed, the stability of carbocations increases with increasing substitution: More highly substituted carbocations are more stable than less highly substituted ones.

One way of determining carbocation stabilities is to measure the amount of energy required to form the carbocation from its corresponding alkyl halide, $R–X \rightarrow R^+ + :X^-$. As shown in Figure 6.14, tertiary halides dissociate to give carbocations much more readily than secondary or primary halides. As a result, trisubstituted (tertiary, 3°) carbocations are more stable than disubstituted (secondary, 2°) ones, which are more stable than monosubstituted (primary, 1°) ones:

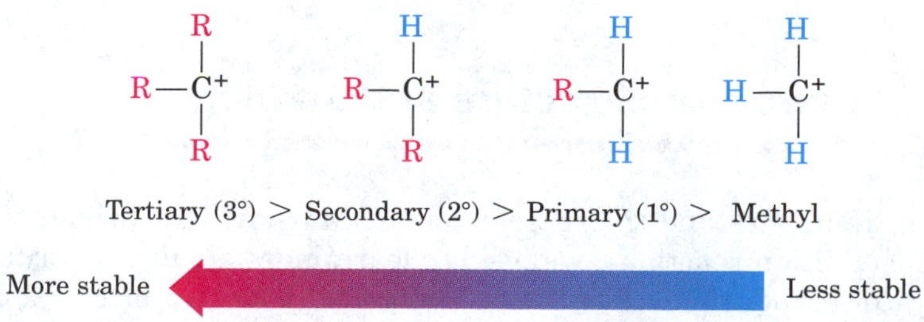

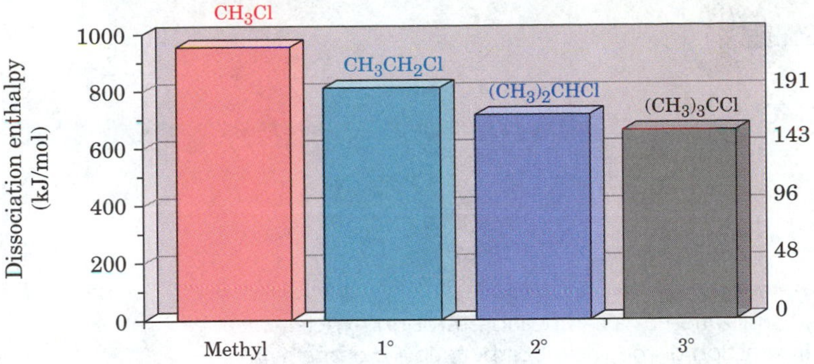

Figure 6.14 A plot of dissociation enthalpy versus substitution pattern for the gas-phase dissociation of alkyl chlorides to yield carbocations. More highly substituted alkyl halides dissociate more readily than less highly substituted ones. These enthalpies are calculated in the following way:

	$CH_3{-}Cl \rightarrow CH_3\cdot + Cl\cdot$	D =	351 kJ/mol	(Bond dissociation energy)
	$CH_3\cdot \rightarrow CH_3^+ + e^-$	E_i =	948 kJ/mol	(Ionization energy)
	$Cl\cdot + e^- \rightarrow Cl^-$	E_{ea} =	−348 kJ/mol	(Electron affinity)
Net:	$CH_3{-}Cl \rightarrow CH_3^+ + Cl^-$		951 kJ/mol	(Dissociation enthalpy)

The data in Figure 6.14 are taken from measurements made in the gas phase, but a similar stability order is found for carbocations in solution. The enthalpies for dissociation are much lower in solution because polar solvents can stabilize the ions, but the order of carbocation stability remains the same.

Why are more highly substituted carbocations more stable than less highly substituted ones? There are at least two reasons. Part of the answer has to do with hyperconjugation, and part has to do with inductive effects. Hyperconjugation, discussed in Section 6.7 in connection with the stability order of substituted alkenes, is the interaction of a vacant *p* orbital and a neighboring C–H σ orbital. This interaction stabilizes the carbocation and lowers its energy (Figure 6.15, p. 206). The more alkyl groups there are on the carbocation, the more possibilities there are for hyperconjugation and the more stable the carbocation—in other words: tertiary > secondary > primary > methyl.

Inductive effects, discussed in Section 2.1 in connection with polar covalent bonds, result from the shifting of electrons in a bond in response to the electronegativity of a nearby atom. In the present instance, electrons from a relatively large and polarizable alkyl group can shift toward a neighboring positive charge more easily than electrons from a hydrogen. Thus, the more

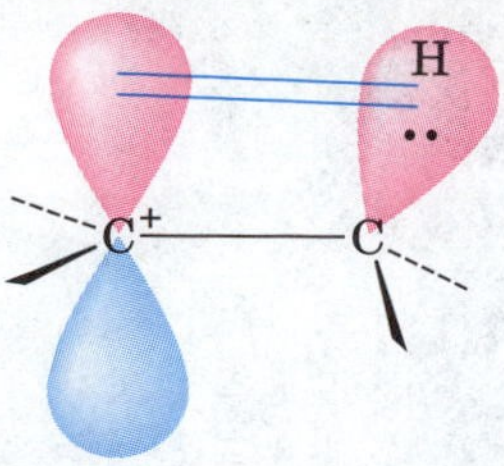

Figure 6.15 Stabilization of a carbocation through hyperconjugation. Interaction of the vacant carbocation *p* orbital with the neighboring C–H σ orbital stabilizes the cation and lowers its energy.

alkyl groups there are attached to the positively charged carbon, the more electron density shifts toward the charge and the more inductive stabilization of the cation occurs.

3°: Three alkyl groups donating electrons | 2°: Two alkyl groups donating electrons | 1°: One alkyl group donating electrons | Methyl: No alkyl groups donating electrons

PROBLEM

6.15 Show the structures of the carbocation intermediates you would expect in these reactions:

(a) 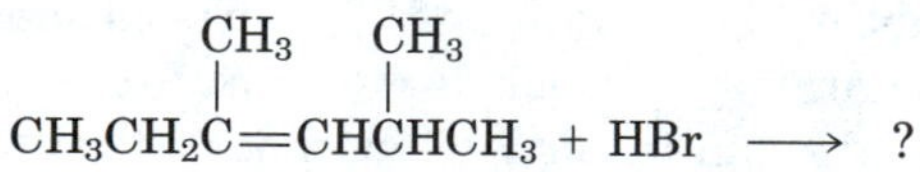

(b) 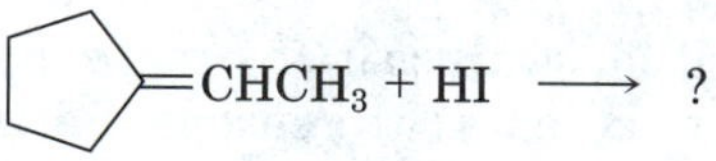

6.11 The Hammond Postulate

To summarize our knowledge of electrophilic addition reactions up to this point, we know that:

1. Electrophilic addition reactions to unsymmetrically substituted alkenes involve the more highly substituted carbocation. A more highly substituted carbocation forms faster than a less highly substituted one and, once formed, rapidly goes on to give the final product.

2. More highly substituted carbocations are more stable than less highly substituted ones. That is, the stability order of carbocations is tertiary > secondary > primary > methyl.

What we have not yet seen is how these two points are related. Why does the *stability* of the carbocation intermediate affect the *rate* at which it's formed and thereby determine the structure of the final product? After all, carbocation stability is determined by ΔG°, but reaction rate is determined by $\Delta G^\ddagger$ (activation energy). The two quantities aren't directly related.

Although there is no precise thermodynamic relationship between the stability of a high-energy carbocation intermediate and the rate of its formation, there *is* an intuitive relationship. It's generally true when comparing two similar reactions that the more stable intermediate forms faster than the less stable one. The situation is shown graphically in Figure 6.16, where the reaction energy profile in part (a) represents the usual situation rather than the profile in part (b). That is, the curves for two similar reactions don't cross one another.

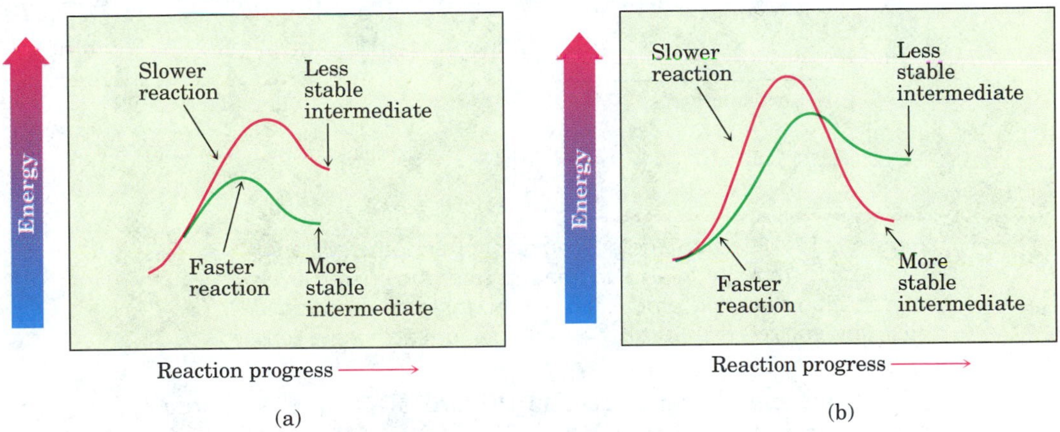

Figure 6.16 Reaction energy diagrams for two similar competing reactions. In (a), the faster reaction yields the more stable intermediate. In (b), the slower reaction yields the more stable intermediate. The curve shown in (a) represents the usual situation.

An explanation of the relationship between reaction rate and intermediate stability was first advanced in 1955. Known as the **Hammond postulate**,[5] this explanation is not a thermodynamic law; it is simply a reasonable account of observed facts. It intuitively links reaction rate and intermediate stability by looking at the energy level and structure of the transition state.

[5]George Simms Hammond (1921–); b. Auburn, Maine; Ph.D. (1947), Harvard University; professor, Iowa State University; California Institute of Technology; University of California, Santa Cruz; Allied Chemical Company.

Transition states represent energy maxima. They are high-energy activated complexes that occur transiently during the course of a reaction and immediately go on to a more stable species. Although we can't actually *observe* transition states, because they have no finite lifetime, the Hammond postulate says that we can get an *idea* of a particular transition state's structure by looking at the structure of the nearest stable species. Imagine the two cases shown in Figure 6.17, for example. The reaction profile in part (a) shows the energy curve for an endothermic reaction step, and the profile in part (b) shows the curve for an exothermic step.

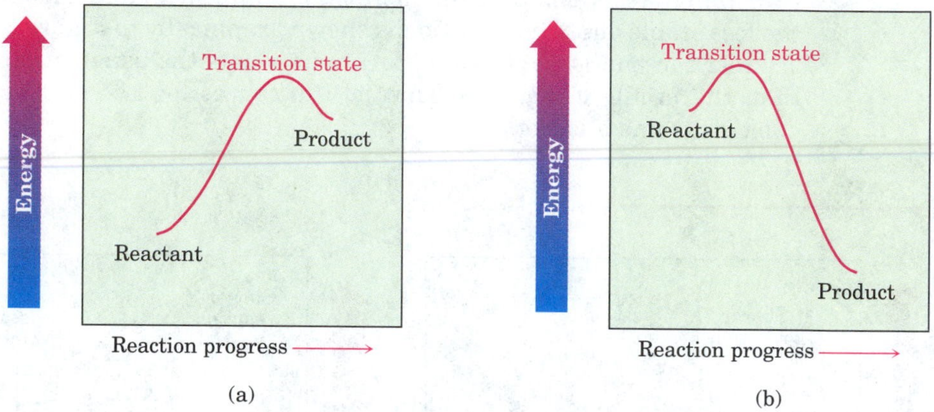

Figure 6.17 Reaction energy diagrams for endothermic and exothermic steps. (a) In an endothermic step, the energy levels of transition state and *product* are similar. (b) In an exothermic step, the energy levels of transition state and *reactant* are similar.

In an endothermic reaction [Figure 6.17(a)], the energy level of the transition state is closer to that of the product than to that of the reactant. Since the transition state is closer *energetically* to the product, we make the natural assumption that it's also closer *structurally.* In other words, we can say that *the transition state for an endothermic reaction step structurally resembles the product of that step*. Conversely, the transition state for an exothermic reaction [Figure 6.17(b)] is closer energetically to the reactant than to the product, and we say that *the transition state for an exothermic reaction step structurally resembles the reactant for that step*.

Hammond postulate The structure of a transition state resembles the structure of the nearest stable species. Transition states for endothermic steps structurally resemble products, and transition states for exothermic steps structurally resemble reactants.

How does the Hammond postulate apply to electrophilic addition reactions? We know that the formation of a carbocation by protonation of an alkene is an endothermic step. Therefore, the transition state for alkene protonation should structurally resemble the carbocation intermediate, and any factor that makes the carbocation product more stable should also make the nearby transition state more stable. Since increasing alkyl substitution stabilizes carbocations, it also stabilizes the transition states leading to those ions, thus resulting in faster reaction. *More highly substituted carbocations form faster because their stability is reflected in the transition state that forms them.* A hypothetical transition state for alkene protonation might be expected to look like that shown in Figure 6.18.

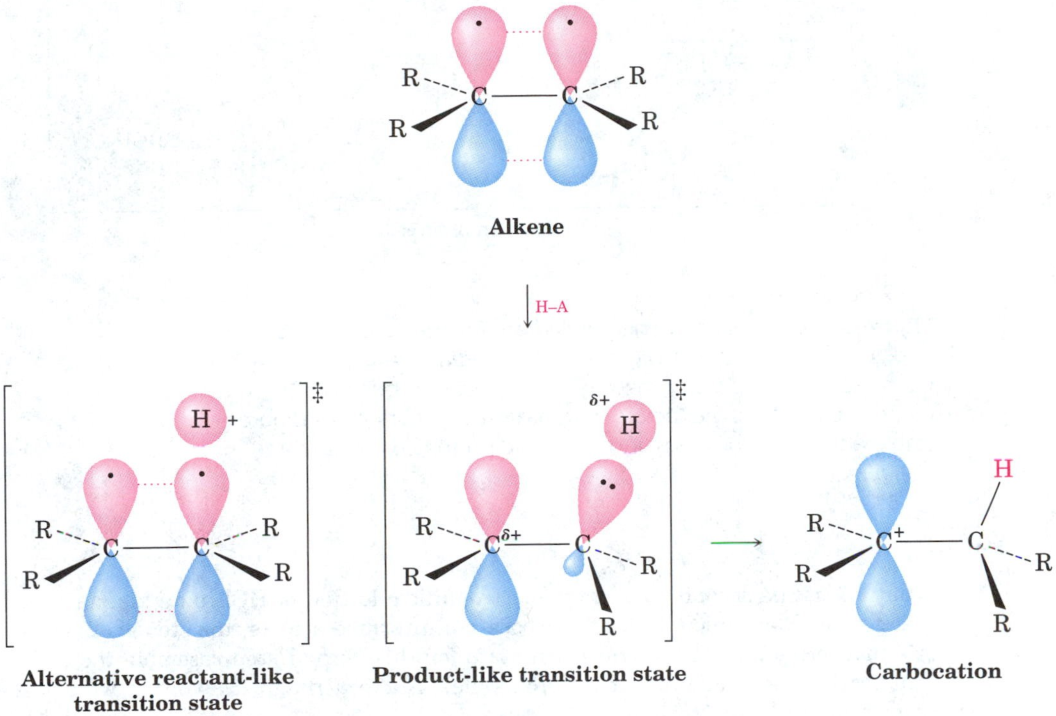

Figure 6.18 The structure of a hypothetical transition state for alkene protonation. The transition state is closer in both energy and structure to the carbocation than to the reactant. Thus, an increase in carbocation stability (lower $\Delta G°$) also causes an increase in transition-state stability (lower $\Delta G^{\ddagger}$).

Because the transition state for alkene protonation resembles the carbocation product, we can imagine it to be a structure in which one of the alkene carbon atoms has almost completely rehybridized from sp^2 to sp^3 and in which the remaining alkene carbon bears much of the positive charge. This transition state is stabilized by hyperconjugation in the same way as the product carbocation. The more alkyl groups that are present, the greater the extent of stabilization in the transition state and the faster the transition state forms. Figure 6.19 summarizes the situation by showing competing reaction energy profiles for the reaction of 2-methylpropene with HCl.

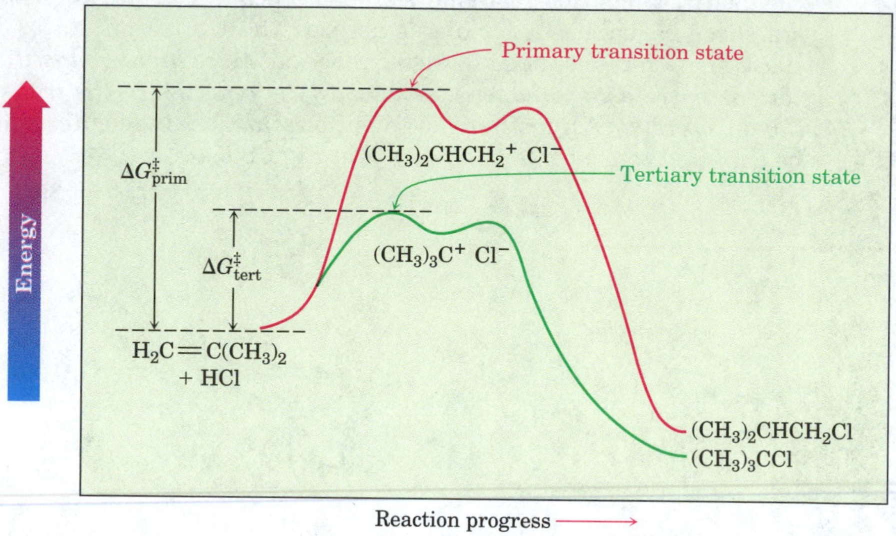

Figure 6.19 A reaction energy diagram for the electrophilic addition of HCl to 2-methylpropene. The tertiary cation intermediate forms faster than the primary cation because it's more stable. The same factors that make the tertiary cation more stable also make the transition state leading to it more stable.

PROBLEM.

6.16 What about the second step in the electrophilic addition of HCl to an alkene—the reaction of chloride ion with the carbocation intermediate? Is this step exothermic or endothermic? Does the transition state for this second step resemble the reactant (carbocation) or product (chloroalkane)? Make a rough drawing of what the transition-state structure might look like.

6.12 Mechanistic Evidence: Carbocation Rearrangements

How do we know that the carbocation mechanism for addition of HX to alkenes is correct? The answer is that we *don't* know it's correct, or at least we don't know with complete certainty. Although an *incorrect* reaction mechanism can be *disproved* by demonstrating that it doesn't satisfactorily account for observed data, a correct reaction mechanism can never be entirely proven. The best we can do is to show that a proposed mechanism is consistent with all known facts. If enough facts are satisfactorily accounted for, then the mechanism is probably correct.

What evidence is there to support the two-step, carbocation mechanism we've proposed for the reaction of HX with alkenes? How do we know that the two reactants, HX and alkene, don't simply come together in a single step to give the final product without going through a carbocation intermediate? One of the best pieces of evidence for a carbocation mechanism was discovered during the 1930s by F. C. Whitmore[6] at thc Pennsylvania State University, who found that structural rearrangements often occur during the reaction of HX with an alkene. For example, reaction of HCl with 3-methyl-1-butene yields a substantial amount of 2-chloro-2-methylbutane in addition to the "expected" product, 2-chloro-3-methylbutane:

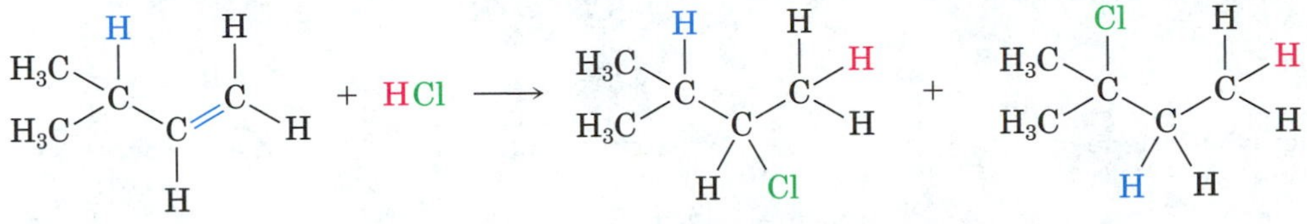

3-Methyl-1-butene **2-Chloro-3-methylbutane (approx. 50%)** **2-Chloro-2-methylbutane (approx. 50%)**

How can the formation of 2-chloro-2-methylbutane be explained? If the reaction takes place in a single step, it would be difficult to account for rearrangement, but if the reaction takes place in two steps, rearrangement is more easily explained. Whitmore suggested that it is a carbocation intermediate that is undergoing rearrangement. The secondary carbocation formed by protonation of 3-methyl-1-butene rearranges to a more stable tertiary carbocation by a **hydride shift**: the shift of a hydrogen atom and its electron pair (a hydride ion, $:H^-$).

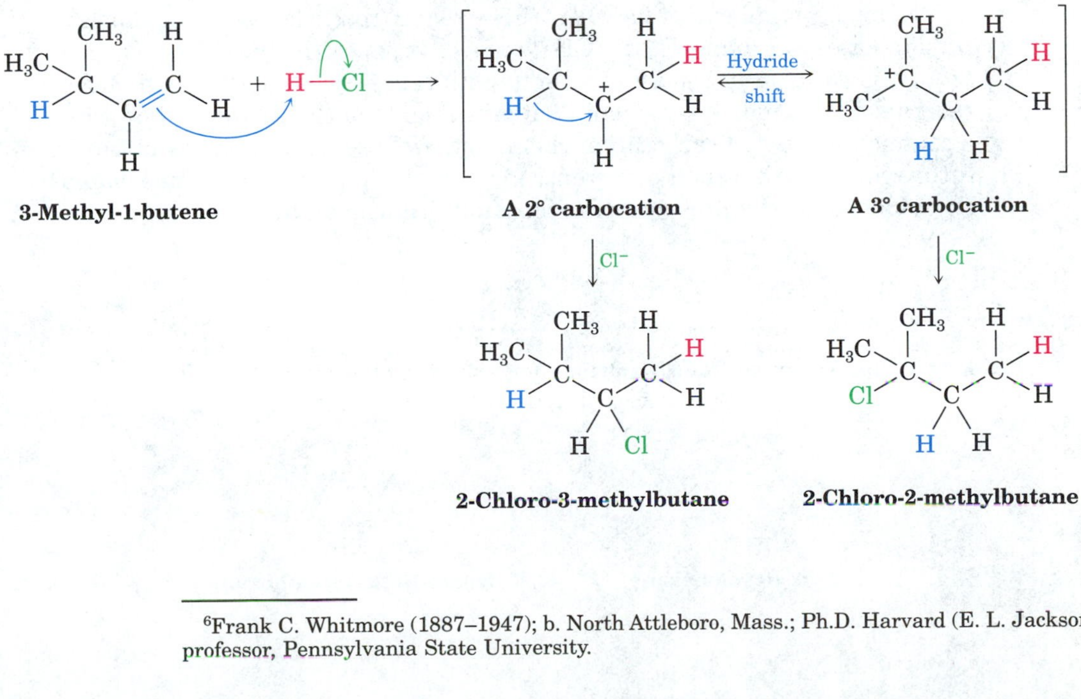

3-Methyl-1-butene **A 2° carbocation** **A 3° carbocation**

2-Chloro-3-methylbutane **2-Chloro-2-methylbutane**

[6]Frank C. Whitmore (1887–1947); b. North Attleboro, Mass.; Ph.D. Harvard (E. L. Jackson); professor, Pennsylvania State University.

Carbocation rearrangements can also occur by the shift of an *alkyl group* with its electron pair. For example, reaction of 3,3-dimethyl-1-butene with HCl leads to an equal mixture of unrearranged 2-chloro-3,3-dimethylbutane and rearranged 2-chloro-2,3-dimethylbutane. In this instance, a secondary carbocation rearranges to a more stable tertiary carbocation by the shift of a methyl group:

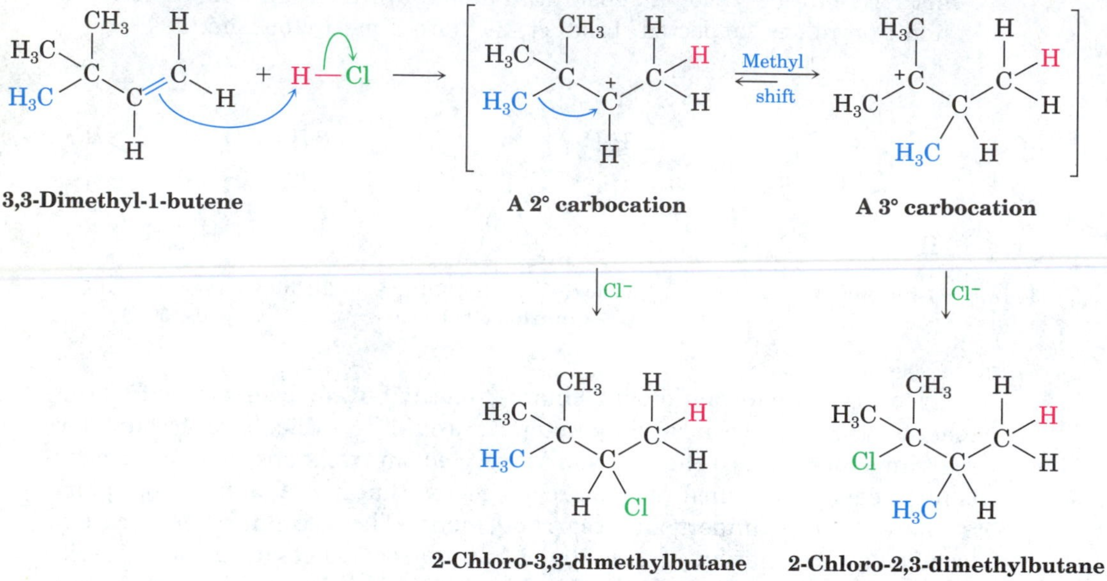

Note the similarities between these two carbocation rearrangements: In both cases, a group ($:H^-$ or $:CH_3^-$) moves to a positively charged carbon, *taking its electron pair with it*. Also in both cases, a less stable carbocation rearranges to a more stable ion. Rearrangements of the sort just shown are a common feature of carbocation chemistry. We'll see at numerous places in future chapters that their occurrence in a reaction provides strong mechanistic evidence for the presence of carbocation intermediates.

PROBLEM

6.17 Propose a mechanism to account for the formation of 1-bromo-1-ethylcyclohexane on reaction of vinylcyclohexane with HBr.

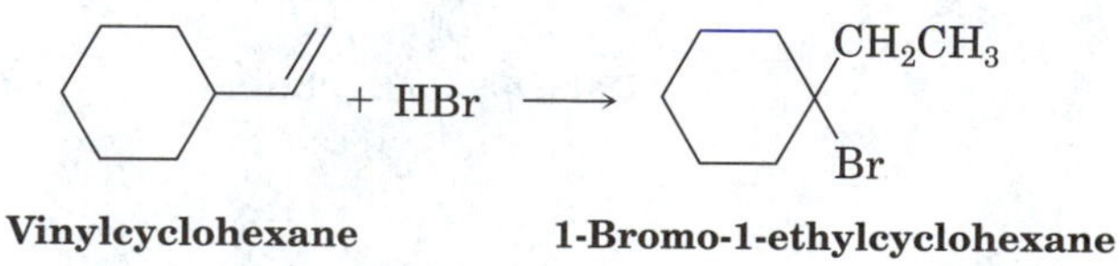

INTERLUDE

Carrots, Alkenes, and the Chemistry of Vision

The pigment responsible for the striking color of these flamingoes is derived from β-carotene.

Folk medicine has long held that eating carrots is good for your eyes. Although that's probably not true for healthy adults on a proper diet, there's no question that the chemistry of carrots and the chemistry of vision are related. Carrots are rich in β-carotene, a purple-orange alkene that is an excellent dietary source of vitamin A. β-Carotene is converted to vitamin A by enzymes in the liver, oxidized to an aldehyde called *all-trans-retinal*, and then isomerized by a change in geometry of the C11–C12 double bond to produce 11-*cis*-retinal, the light-sensitive pigment on which the visual systems of all living things are based.

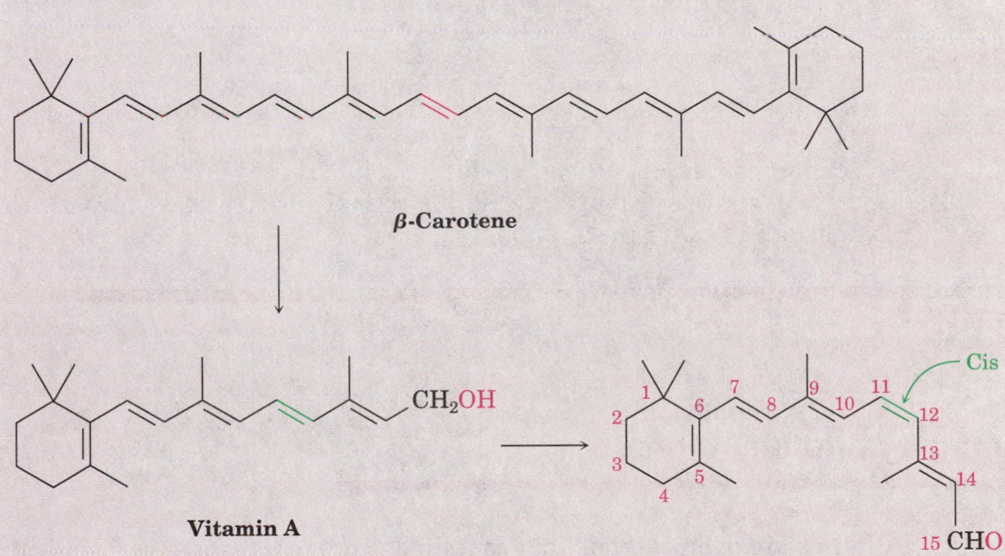

β-Carotene → **Vitamin A** → **11-*cis*-Retinal**

There are two types of light-sensitive receptor cells in the retina of the human eye, *rod* cells and *cone* cells. The three million or so rod cells are primarily responsible for seeing in dim light, whereas the hundred

(continued)➤

million cone cells are responsible for seeing in bright light and for the perception of bright colors. In the rod cells of the eye, 11-*cis*-retinal is converted into *rhodopsin*, a light-sensitive substance formed from the protein *opsin* and 11-*cis*-retinal. When light strikes the rod cell, isomerization of the C11–C12 double bond occurs and *trans*-rhodopsin, called *metarhodopsin II*, is produced. This cis–trans isomerization of rhodopsin is accompanied by a change in molecular geometry, which in turn causes a nerve impulse to be sent to the brain where it is perceived as vision.

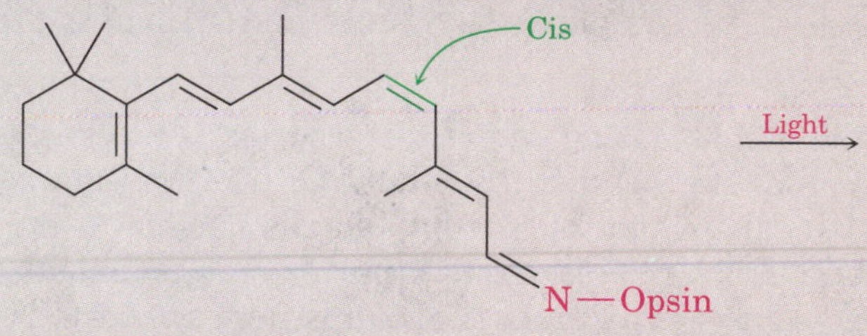

Rhodopsin

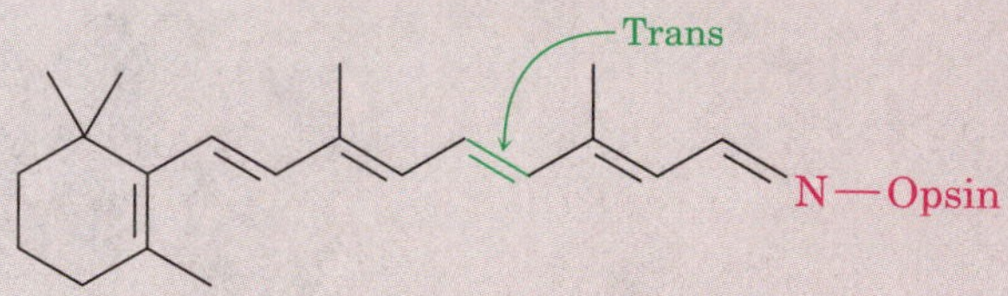

Metarhodopsin II

Metarhodopsin II is then recycled back into rhodopsin by a multistep sequence involving cleavage to all-*trans*-retinal and cis–trans isomerization back to 11-*cis*-retinal.

Summary and Key Words

Alkenes are hydrocarbons that contain one or more carbon–carbon double bonds. Because they contain fewer hydrogens than related alkanes, alkenes are often referred to as **unsaturated**. Alkenes are named by IUPAC rules using the suffix *-ene*.

Because rotation around the double bond is restricted, substituted alkenes can exist as cis–trans stereoisomers. The geometry of a double bond can be specified by application of the Cahn–Ingold–Prelog **sequence rules**, which assign priorities to double-bond substituents. If the high-priority groups on each carbon are on the same side of the double bond, the geometry is *Z* (*zusammen*, "together"); if the high-priority groups on each carbon are on opposite sides of the double bond, the geometry is *E* (*entgegen*, "apart"). The stability order of substituted alkenes is:

Tetrasubstituted	>	Trisubstituted	>	Disubstituted	>	Monosubstituted
$R_2C{=}CR_2$	>	$R_2C{=}CHR$	>	$RCH{=}CHR \approx R_2C{=}CH_2$	>	$RCH{=}CH_2$

Alkene chemistry is dominated by **electrophilic addition reactions**. When HX reacts with an unsymmetrically substituted alkene, **Markovnikov's rule** predicts that the H will add to the carbon having fewer alkyl substituents and the X group will add to the carbon having more alkyl substituents. Electrophilic additions to alkenes take place through carbocation intermediates formed by reaction of the nucleophilic alkene π bond with electrophilic H^+. Carbocation stability follows the order

Tertiary (3°)	>	Secondary (2°)	>	Primary (1°)	>	Methyl
R_3C^+	>	R_2CH^+	>	RCH_2^+	>	CH_3^+

Markovnikov's rule can be restated by saying that, in the addition of HX to an alkene, the more stable carbocation intermediate is formed. This result is explained by the **Hammond postulate**, which says that the transition state of an exothermic reaction step structurally resembles the reactant, whereas the transition state of an endothermic reaction step structurally resembles the product. Since an alkene protonation step is endothermic, the stability of the more highly substituted carbocation is reflected in the stability of the transition state leading to its formation.

One of the best pieces of evidence in support of a carbocation mechanism for electrophilic additions is the observation that structural rearrangements sometimes occur during reaction. Rearrangements occur by shift of either a hydride ion, $:H^-$ (a **hydride shift**), or an alkyl group anion, $:R^-$, from a neighboring carbon atom to the positively charged carbon. The result is isomerization of a less stable carbocation to a more stable one.

ADDITIONAL PROBLEMS

6.18 Calculate the number of double bonds and/or rings in these formulas:
(a) Benzene, C_6H_6
(b) Cyclohexene, C_6H_{10}
(c) Myrcene (bay oil), $C_{10}H_{16}$
(d) Lindane, $C_6H_6Cl_6$
(e) Pyridine, C_5H_5N
(f) Safrole, $C_{10}H_{10}O_2$

6.19 Calculate the number of multiple bonds and/or rings in the following formulas, and then draw five possible structures for each:
(a) $C_{10}H_{16}$
(b) C_8H_8O
(c) $C_7H_{10}Cl_2$
(d) $C_{10}H_{16}O_2$
(e) $C_5H_9NO_2$
(f) $C_8H_{10}ClNO$

6.20 A compound of formula $C_{10}H_{14}$ undergoes catalytic hydrogenation but absorbs only 2 equivalents of hydrogen. How many rings does the compound have?

6.21 A compound of formula $C_{12}H_{13}N$ contains two rings. How many equivalents of hydrogen does it absorb if all the remaining unsaturations are C–C double bonds?

6.22 Give IUPAC names for these alkenes (ignoring cis–trans isomers):

(a)
$$\begin{array}{c} CH_3 \\ | \\ CH_3CH{=}CHCHCH_2CH_3 \end{array}$$

(b)
$$\begin{array}{c} CH_2CH_2CH_2CH_3 \\ | \\ CH_3CH{=}CHCHCH_2CH_2CHCH_3 \\ | \\ CH_3 \end{array}$$

(c) $H_2C{=}C(CH_2CH_3)_2$

(d)
$$\begin{array}{c} CH_3 \\ | \\ H_2C{=}CHCHCHCH{=}CHCH_3 \\ | \\ CH_3 \end{array}$$

(e)
$$\begin{array}{c} CH_3 \\ | \\ CH_3CH_2C{=}CHCH{=}CH_2 \end{array}$$

(f) $H_2C{=}C{=}CHCH_3$

(g) $H_2C{=}CHC(CH_3)_3$

(h) $(CH_3)_3CCH{=}CHC(CH_3)_3$

6.23 Ocimene is a triene found in the essential oil of many plants. What is its IUPAC name?

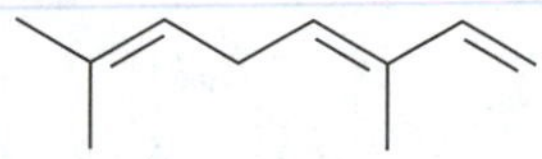

Ocimene

6.24 α-Farnesene is a constituent of the natural wax found on apples. What is its IUPAC name?

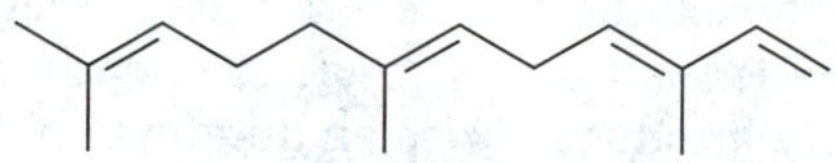

α-Farnesene

6.25 Indicate *E* or *Z* stereochemistry for each of the double bonds in α-farnesene (Problem 6.24).

6.26 Draw structures corresponding to these systematic names:
(a) (4*E*)-2,4-Dimethyl-1,4-hexadiene
(b) *cis*-3,3-Dimethyl-4-propyl-1,5-octadiene
(c) 4-Methyl-1,2-pentadiene
(d) (3*E*,5*Z*)-2,6-Dimethyl-1,3,5,7-octatetraene
(e) 3-Butyl-2-heptene
(f) *trans*-2,2,5,5-Tetramethyl-3-hexene

6.27 Menthene, a hydrocarbon found in mint plants, has the systematic name 1-isopropyl-4-methylcyclohexene. Draw its structure.

6.28 The following names are incorrect. Draw structures and give correct names.
(a) 2-Methyl-2,4-pentadiene
(b) 3-Methylene-1-pentene
(c) 3*E*,6*Z*-Octadiene
(d) *Z*-5-Ethyl-4-octene
(e) *E*-3-Propyl-3-heptene
(f) 3-Vinyl-1-propene

6.29 Draw and name the 5 pentene isomers, C_5H_{10}. Ignore *E*,*Z* isomers.

6.30 Draw and name the 13 hexene isomers, C_6H_{12}. Ignore *E*,*Z* isomers.

6.31 Which of the compounds you drew in Problems 6.29 and 6.30 show cis–trans isomerism?

6.32 According to data for heats of hydrogenation, *trans*-2-butene is more stable than its cis isomer by only 4 kJ/mol, but *trans*-2,2,5,5-tetramethyl-3-hexene is more stable than its cis isomer by 39 kJ/mol. (See the table at the top of the next page.) Explain.

Alkene	ΔH°_{hydrog} (*kJ/mol*)	ΔH°_{hydrog} (*kcal/mol*)
cis-2-Butene	−119.7	−28.6
trans-2-Butene	−115.5	−27.6
cis-2,2,5,5-Tetramethyl-3-hexene	−151.5	−36.2
trans-2,2,5,5-Tetramethyl-3-hexene	−112.6	−26.9

6.33 The double bonds in small-ring cycloalkenes must have cis geometry because a stable trans double bond is impossible in a five- or six-membered ring. At some point, however, a ring becomes large enough to accommodate a trans double bond. The following heats of hydrogenation have been measured:

Cycloalkene	ΔH°_{hydrog} (*kJ/mol*)	ΔH°_{hydrog} (*kcal/mol*)
cis-Cyclooctene	−96.2	−23.0
trans-Cyclooctene	−134.7	−32.2
cis-Cyclononene	−98.7	−23.6
trans-Cyclononene	−110.9	−26.5
cis-Cyclodecene	−86.6	−20.7
trans-Cyclodecene	−100.4	−24.0

How do you explain these data? Make molecular models of the trans cycloalkenes to see their conformations.

6.34 Allene (1,2-propadiene), $H_2C{=}C{=}CH_2$, has two adjacent double bonds. What kind of hybridization must the central carbon have? Sketch the bonding π orbitals in allene. What shape do you predict for allene?

6.35 1,4-Pentadiene, a compound with two nonadjacent double bonds, has $\Delta H^\circ_{hydrog} =$ −254 kJ/mol, which is approximately twice the value for 1-pentene ($\Delta H^\circ_{hydrog} = -126$ kJ/mol). 1,2-Pentadiene, however, has $\Delta H^\circ_{hydrog} = -298$ kJ/mol. What does this tell you about the stability of 1,2-pentadiene? Explain.

6.36 Predict the major product in each of the following reactions.

(a) $CH_3CH_2CH{=}C(CH_3)CH_2CH_3 + HCl \longrightarrow$

(b) 1-Ethylcyclopentene + HBr ⟶

(c) 2,2,4-Trimethyl-3-hexene + HI ⟶

(d) 1,6-Heptadiene + 2 HCl ⟶

(e) 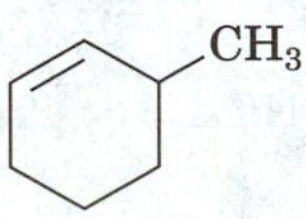 + HBr ⟶

6.37 Predict the major product from addition of HBr to each of the following alkenes:

(a) 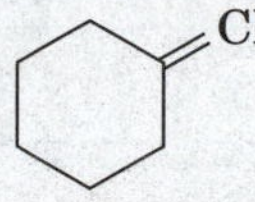 (b) 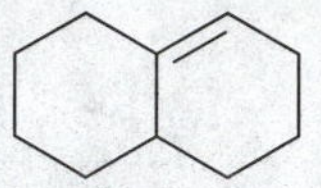 (c) 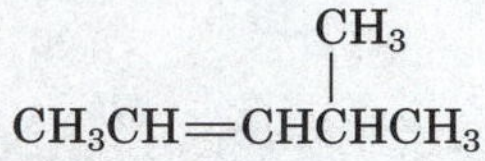

6.38 Rank the following sets of substituents in order of priority according to the Cahn–Ingold–Prelog sequence rules:

(a) $—CH_3$, $—Br$, $—H$, $—I$ (b) $—OH$, $—OCH_3$, $—H$, $—COOH$

(c) $—COOH$, $—COOCH_3$, $—CH_2OH$, $—CH_3$

(d) $—CH_3$, $—CH_2CH_3$, $—CH_2CH_2OH$, $—\overset{O}{\overset{\|}{C}}CH_3$

(e) $—CH{=}CH_2$, $—CN$, $—CH_2NH_2$, $—CH_2Br$

(f) $—CH{=}CH_2$, $—CH_2CH_3$, $—CH_2OCH_3$, $—CH_2OH$

6.39 Assign *E* or *Z* configuration to each of the following alkenes:

(a) 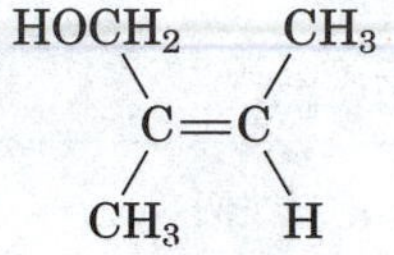 (b) 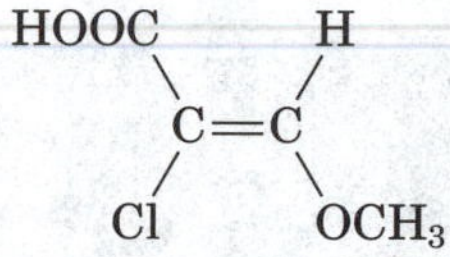

(c) 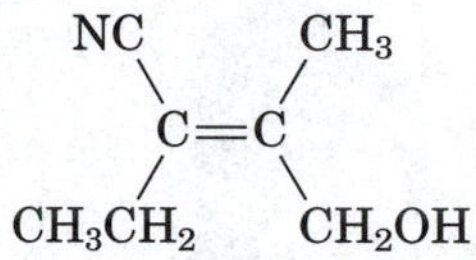 (d) 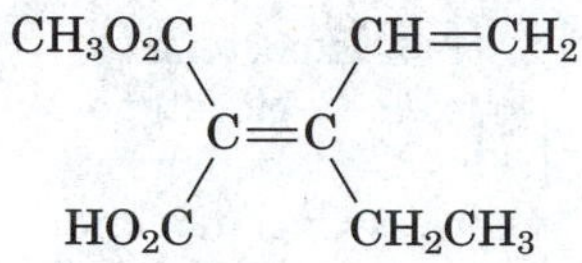

6.40 Name the following cycloalkenes according to IUPAC rules:

(a) 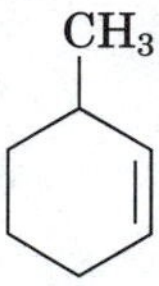 (b) 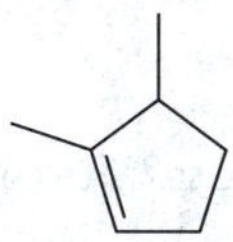 (c) 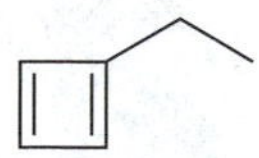

(d) 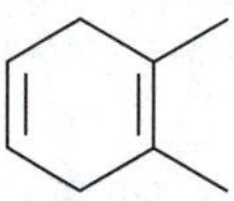 (e) 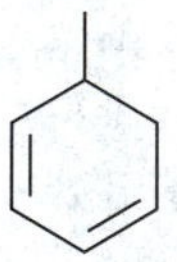 (f) 

6.41 Which of the following *E*,*Z* designations are correct, and which are incorrect?

(a) 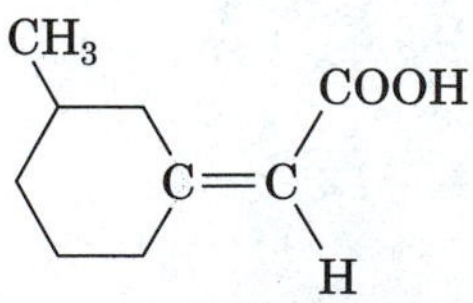

Z

(b) 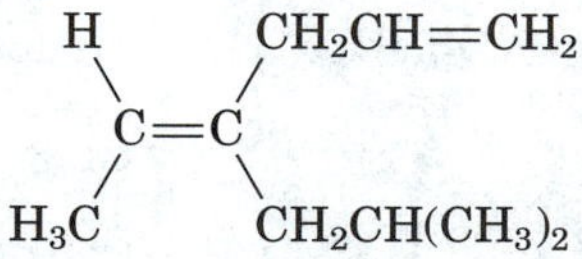

E

(c) 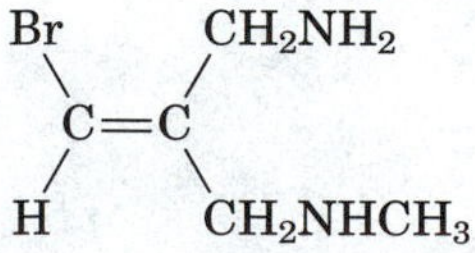

Z

(d) 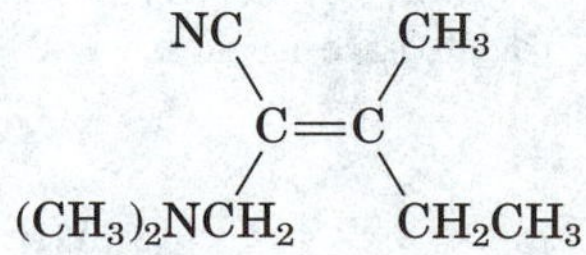

E

(e) 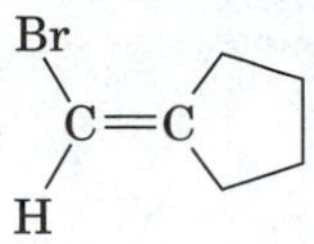

Z

(f) 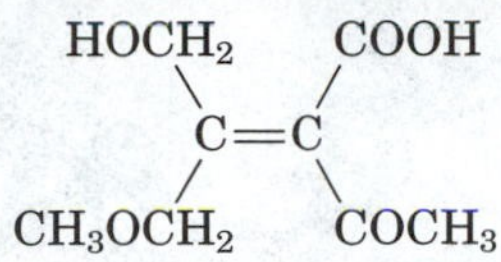

E

6.42 Use the bond dissociation energies in Table 5.4 to calculate $\Delta H°$ for the reaction of ethylene with HCl, HBr, and HI. Which reaction is most favorable?

6.43 Propose a mechanism to account for the following reaction. Show the structure of the intermediate(s), and use curved arrows to indicate electron flow in each step.

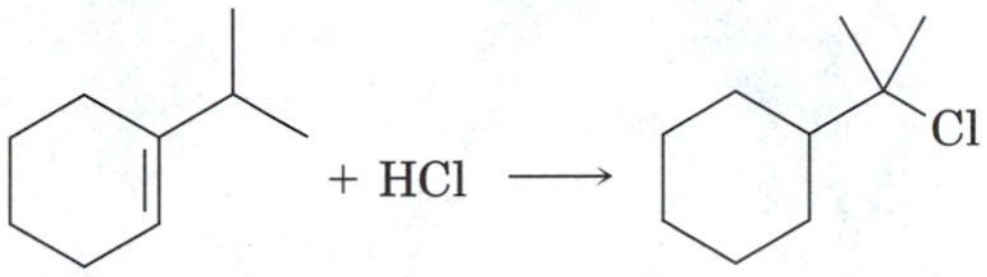

6.44 Repeat Problem 6.43 for the following reaction:

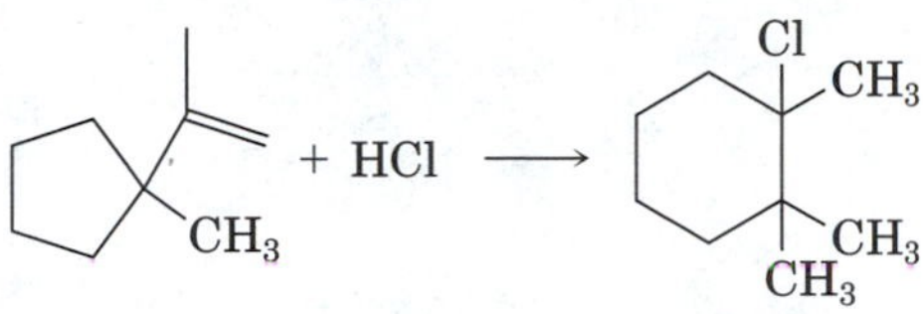

6.45 Calculate the degree of unsaturation in each of the following formulas:

(a) Cholesterol, $C_{27}H_{46}O$ (b) DDT, $C_{14}H_{9}Cl_{5}$
(c) Prostaglandin E_1, $C_{20}H_{34}O_{5}$ (d) Caffeine, $C_{8}H_{10}N_{4}O_{2}$
(e) Cortisone, $C_{21}H_{28}O_{5}$ (f) Atropine, $C_{17}H_{23}NO_{3}$

6.46 Draw a reaction energy diagram for the addition of HBr to 1-pentene. Let one curve on your diagram show the formation of 1-bromopentane product and another curve on the same diagram show the formation of 2-bromopentane product. Label the positions for all reactants, intermediates, and products. Which curve has the higher-energy carbocation intermediate? Which curve has the higher-energy first transition state?

6.47 Make sketches of the transition-state structures involved in the reaction of HBr with 1-pentene (Problem 6.46). Tell whether each structure resembles reactant or product.

A Look Ahead

6.48 We'll see in the next chapter that alkenes can be converted into alcohols by acid-catalyzed addition of water. Assuming that Markovnikov's rule is valid, predict the major alcohol product from each of the following alkenes:

(a) 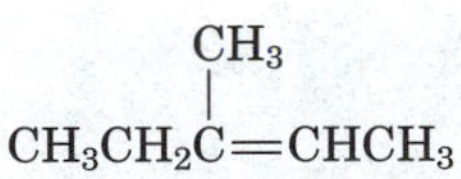 (b) 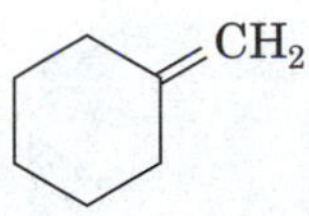 (c) 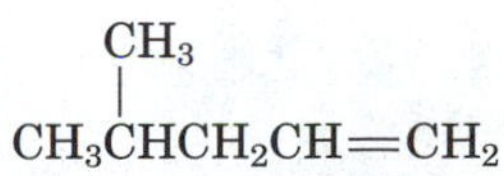

6.49 Reaction of 2,3-dimethyl-1-butene with HBr leads to a bromoalkane, $C_6H_{13}Br$. On treatment of this bromoalkane with KOH in methanol, a hydrocarbon that is isomeric with the starting alkene is formed. What is the structure of this hydrocarbon, and how do you think it is formed from the bromoalkane?

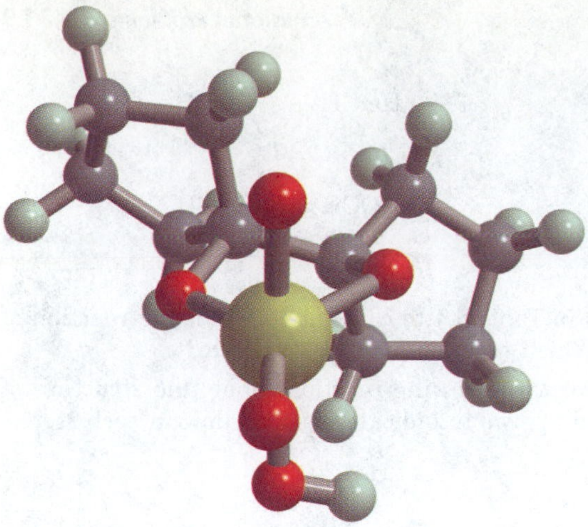

This cyclic periodate is an intermediate in the cleavage reaction of alkenes to yield carbonyl compounds.

7

Alkenes: Reactions and Synthesis

The addition of electrophiles to alkenes is a useful and general reaction that makes possible the synthesis of many different kinds of compounds. Although we've studied only the addition of HX thus far, many other electrophilic reagents also add to alkenes. In this chapter, we'll see how alkenes are prepared, we'll discuss many further examples of alkene addition reactions, and we'll review the wide variety of compounds that can be made from alkenes.

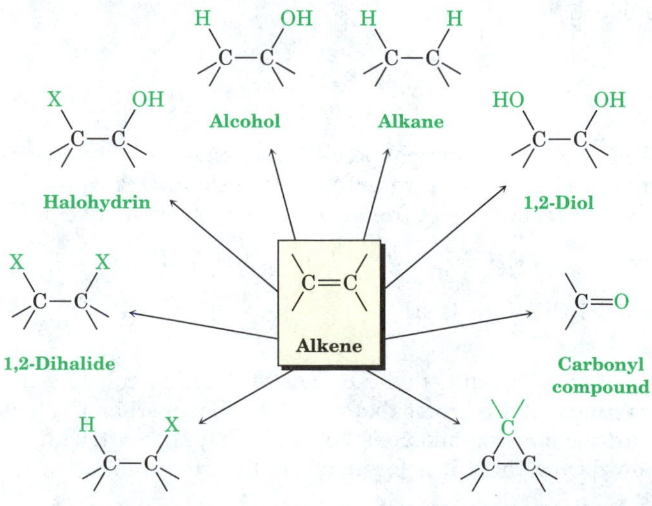

7.1 Preparation of Alkenes: A Preview of Elimination Reactions

Before getting to the main subject of this chapter—the reactions of alkenes—let's take a brief look at how alkenes are prepared. We'll return to this topic in Chapter 11 for a more detailed study.

Just as the chemistry of alkenes is dominated by addition reactions, the preparation of alkenes is dominated by elimination reactions. Additions and eliminations are in many respects two sides of the same coin. That is, an addition reaction might involve the *addition* of HBr or H_2O to an alkene to form an alkyl halide or alcohol, whereas an elimination reaction might involve the *loss* of HBr or H_2O from an alkyl halide or alcohol to form an alkene.

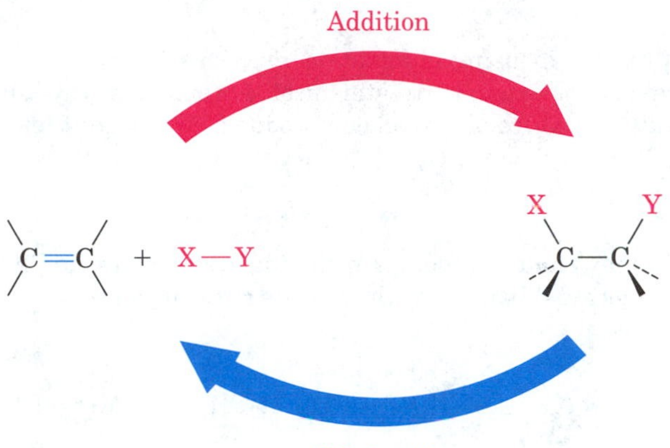

The two most common alkene-forming elimination reactions are **dehydrohalogenation**—the loss of HX from an alkyl halide—and **dehydration**—the loss of water from an alcohol. Dehydrohalogenation usually occurs by reaction of an alkyl halide with strong base such as potassium hydroxide. For example, bromocyclohexane yields cyclohexene when treated with KOH in ethanol solution:

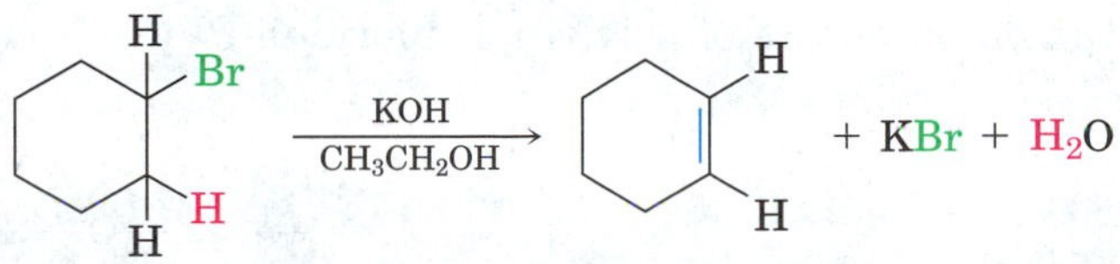

Bromocyclohexane **Cyclohexene (81%)**

Dehydration is often carried out by treatment of an alcohol with a strong acid. For example, loss of water occurs and 1-methylcyclohexene is formed when 1-methylcyclohexanol is warmed with aqueous sulfuric acid in tetrahydrofuran (THF) solvent:

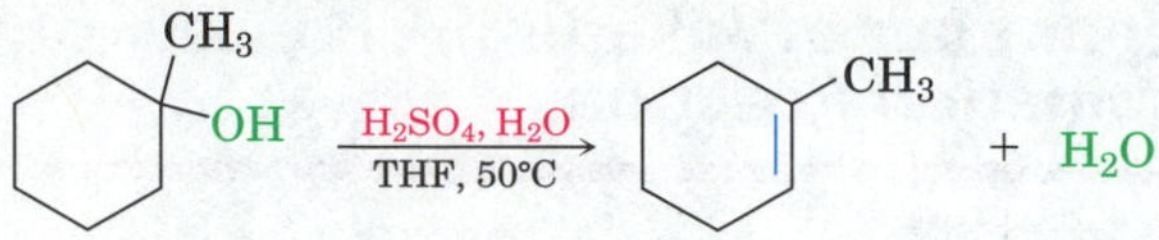

1-Methylcyclohexanol **1-Methylcyclohexene (91%)**

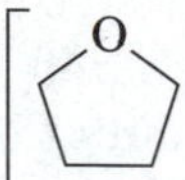

Tetrahydrofuran (THF)—a common solvent

Elimination reactions are sufficiently complex that it's best to defer a detailed discussion until Chapter 11. For the present, it's enough to realize that alkenes are readily available from simple precursors.

PROBLEM.

7.1 A problem in using elimination reactions is that mixtures of products are often formed. For example, treatment of 2-bromo-2-methylbutane with KOH in ethanol yields a mixture of two alkene products. What are their structures?

PROBLEM.

7.2 How many alkene products, including *E,Z* isomers, might be obtained by dehydration of 3-methyl-3-hexanol with aqueous sulfuric acid?

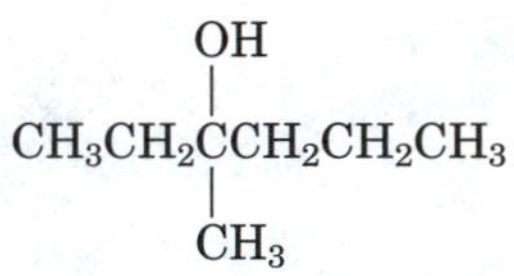

3-Methyl-3-hexanol

7.2 Addition of Halogens to Alkenes

Bromine and chlorine both add readily to alkenes to yield 1,2-dihaloalkanes. For example, nearly 7 million tons per year of 1,2-dichloroethane (ethylene dichloride) are synthesized industrially by the addition of Cl_2 to ethylene. The product is used both as a solvent and as starting material for use in the manufacture of poly(vinyl chloride), PVC.

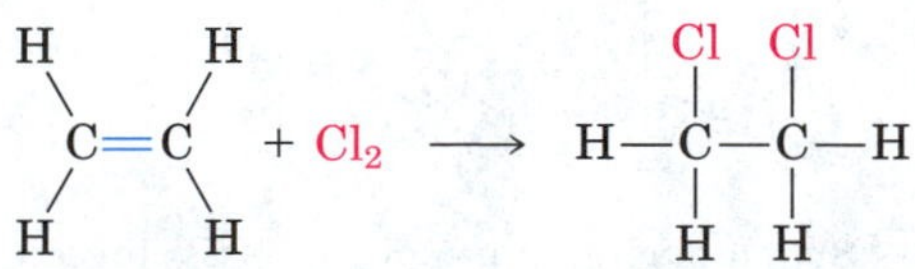

Ethylene **1,2-Dichloroethane (Ethylene dichloride)**

The addition of bromine to an alkene also serves as a simple and rapid laboratory test for unsaturation. A sample of unknown structure is dissolved in tetrachloromethane (carbon tetrachloride, CCl_4) and placed in a test tube to which several drops of bromine are added. Immediate disappearance of the reddish Br_2 color signals a positive test and indicates that the sample is an alkene.

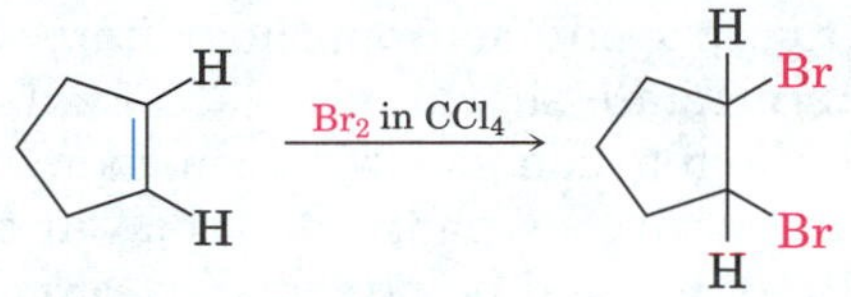

Cyclopentene **1,2-Dibromocyclopentane (95%)**

Fluorine is too reactive and difficult to control for most laboratory applications, and iodine does not react with most alkenes.

Based on what we've seen thus far, a possible mechanism for the reaction of halogens with alkenes is shown in Figure 7.1. As a Br_2 molecule approaches an alkene, the Br–Br bond becomes polarized (Section 5.5). The π electron pair of the alkene then attacks the positive end of the polarized bromine molecule, breaking the Br–Br bond and displacing bromide ion. The net result is that electrophilic Br^+ adds to the alkene in the same way that H^+ adds, giving a carbocation that undergoes further reaction with bromide ion and yields the dibromo addition product.

The electron pair from the double bond attacks the polarized bromine, forming a C–Br bond and causing the Br–Br bond to break. Bromide ion departs with both electrons from the former Br–Br bond.

Bromide ion then uses an electron pair to attack the carbocation intermediate, forming a C–Br bond and giving the neutral addition product.

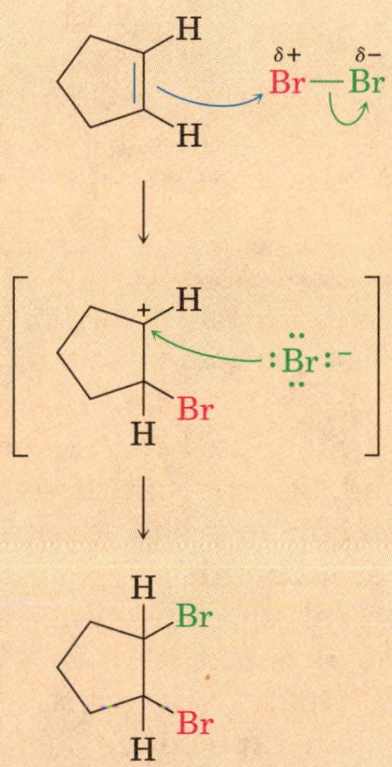

Figure 7.1 A possible mechanism for the electrophilic addition of Br_2 to an alkene.

Although the mechanistic description shown in Figure 7.1 for the addition of bromine to alkenes looks reasonable, it's not completely consistent with known facts. In particular, the proposed mechanism doesn't explain the *stereochemistry* of halogen addition. That is, the mechanism doesn't explain what product stereoisomers are formed in the reaction.

Let's look again at the reaction of Br_2 with cyclopentene and assume that Br^+ adds to cyclopentene from the bottom side of the ring to form the carbocation intermediate shown in Figure 7.2. (The addition could equally well occur from the top side, but we'll consider only one possibility for simplicity.) Since the positively charged carbon in the intermediate is planar and sp^2-hybridized, it could be attacked by bromide ion in the second step of the reaction from either the top or the bottom to give a *mixture* of products. One product has the two bromine atoms on the same side of the ring (cis), and the other has the bromines on opposite sides (trans). We find, however, that only *trans*-1,2-dibromocyclopentane is produced. None of the cis product is formed.

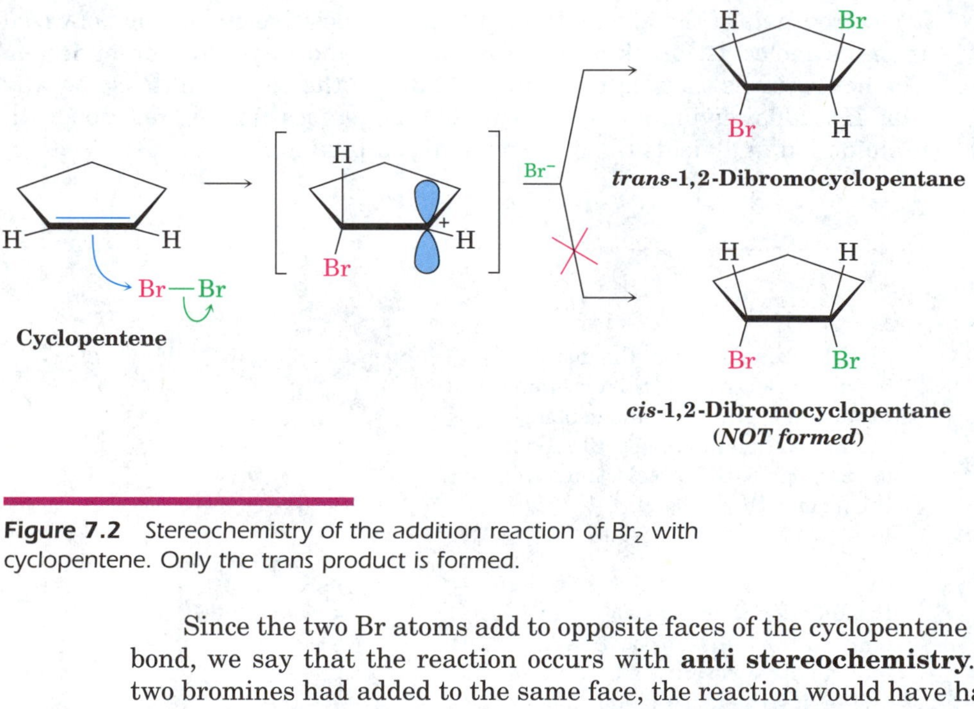

Figure 7.2 Stereochemistry of the addition reaction of Br_2 with cyclopentene. Only the trans product is formed.

Since the two Br atoms add to opposite faces of the cyclopentene double bond, we say that the reaction occurs with **anti stereochemistry**. If the two bromines had added to the same face, the reaction would have had **syn stereochemistry**. Note that the word *anti* has a similar meaning in the present stereochemical context to the meaning it has in a butane conformational context (Section 4.3). In both cases, the two groups of interest are 180° apart.

An explanation for anti addition was suggested in 1937 by George Kimball and Irving Roberts, who proposed that the true reaction intermediate is not a carbocation but is instead a **bromonium ion**, R_2Br^+. (A *chloronium ion,* similarly, contains a positively charged, divalent chlorine, R_2Cl^+.) In the present instance, the bromonium ion is in a three-membered ring and is formed by donation of bromine lone-pair electrons to the vacant *p* orbital of the neighboring carbocation (Figure 7.3). Although Figure 7.3 depicts bromonium ion formation as stepwise, this is done only for clarity. The bromonium ion is formed in a single step by interaction of the alkene with Br^+.

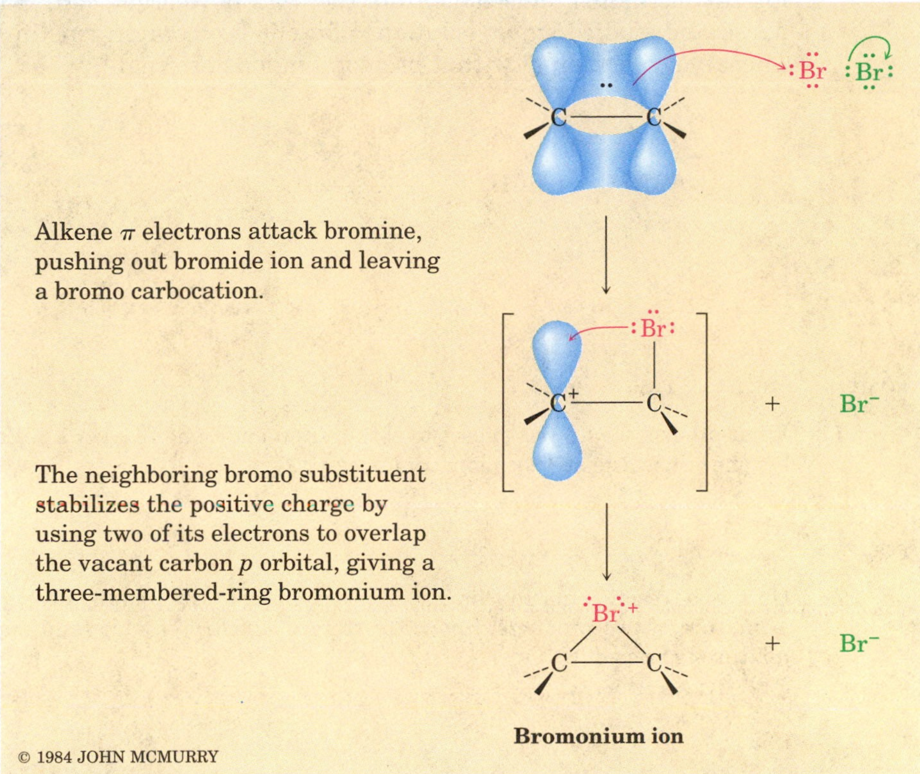

Figure 7.3 Formation of a bromonium ion intermediate by electrophilic addition of Br^+ to an alkene.

How does the formation of a bromonium ion account for the anti stereochemistry of addition to cyclopentene? If a bromonium ion is formed as an intermediate, we can imagine that the large bromine atom might "shield" one side of the molecule. Attack by bromide ion in the second step could then occur only from the opposite, unshielded side to give anti product.

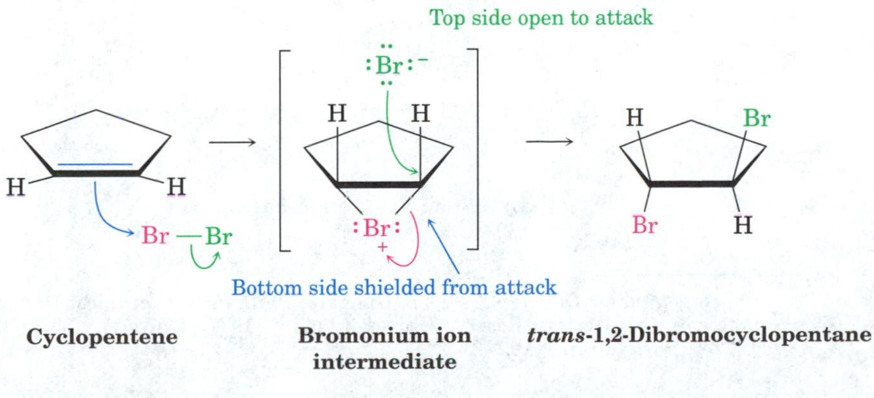

The halonium ion postulate, made nearly 60 years ago to explain the stereochemistry of halogen addition to alkenes, is a remarkable example of deductive logic in chemistry. Arguing from experimental results, chemists were able to make a hypothesis about the intimate mechanistic details of alkene electrophilic reactions. Much more recently, strong evidence supporting the mechanism has come from the work of George Olah,[1] who has prepared and studied *stable* solutions of cyclic bromonium ions in liquid SO_2. There's no question now that bromonium ions are real.

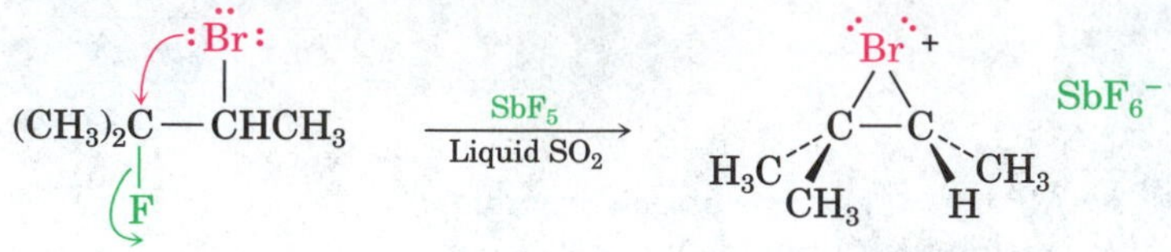

A bromonium ion stable in SO_2 solution

PROBLEM

7.3 What product would you expect to obtain from addition of Cl_2 to 1,2-dimethylcyclohexene? Show the stereochemistry of the product.

PROBLEM

7.4 Unlike the reaction in Problem 7.3, addition of HCl to 1,2-dimethylcyclohexene yields a mixture of two products. Show the stereochemistry of each, and explain why a mixture is formed.

7.3 Halohydrin Formation

Many different kinds of electrophilic additions to alkenes take place. For example, alkenes add HO–Cl or HO–Br under suitable conditions to yield 1,2-halo alcohols, or **halohydrins**. Halohydrin formation doesn't take place by direct reaction of an alkene with HOBr or HOCl, however. Rather, the addition is done indirectly by reaction of the alkene with either Br_2 or Cl_2 in the presence of water.

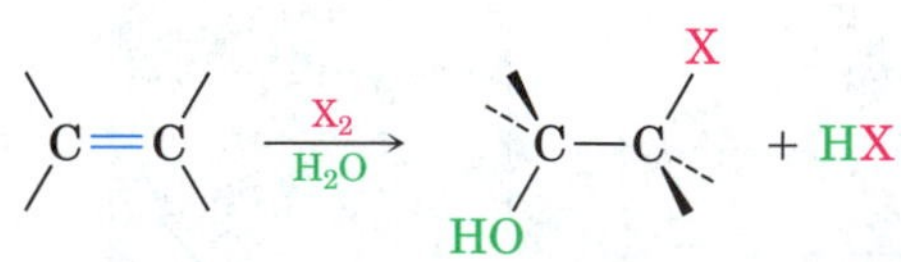

An alkene **A halohydrin**

[1]George Andrew Olah (1927–); b. Hungary; Ph.D. (1949) Technical University, Budapest; professor, Case-Western Reserve University (1965–1977), University of Southern California (1977–); Nobel Prize (1994).

We've seen that, when Br_2 reacts with an alkene in CCl_4 solution, the cyclic bromonium ion intermediate is trapped by the only nucleophile present, bromide ion. If the reaction is carried out *in the presence of an additional nucleophile*, however, the intermediate bromonium ion can be "intercepted" by the added nucleophile and diverted to a different product. When an alkene reacts with Br_2 in the presence of water, for example, water competes with bromide ion as nucleophile and reacts with the bromonium ion intermediate to yield a **bromohydrin**. The net effect is addition of HO–Br to the alkene. The reaction takes place by the pathway shown in Figure 7.4.

Reaction of the alkene with Br_2 yields a bromonium ion intermediate.

Water acts as a nucleophile, using a lone pair of electrons to open the bromonium ion ring and form a bond to carbon. Since oxygen donates its electrons in this step, it now has the positive charge.

Loss of a proton (H^+) from oxygen then gives HBr and the neutral bromohydrin addition product.

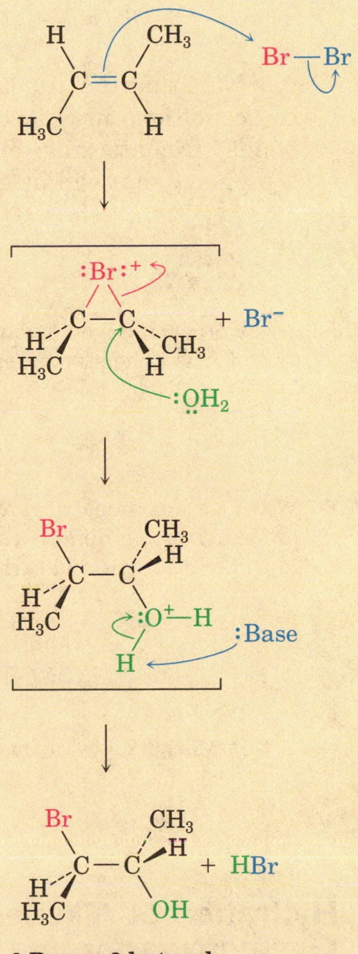

3-Bromo-2-butanol

Figure 7.4 Mechanism of bromohydrin formation by reaction of an alkene with bromine in the presence of water. Water acts as a nucleophile to react with the intermediate bromonium ion.

In practice, few alkenes are soluble in water, and bromohydrin formation is often carried out in a solvent such as aqueous dimethyl sulfoxide (DMSO), using a reagent called *N*-bromosuccinimide (NBS) as a source of Br_2. NBS is a stable, easily handled compound that slowly decomposes in water to yield Br_2 at a controlled rate. Bromine itself can also be used in the addition reaction, but it is more dangerous and more difficult to handle than NBS.

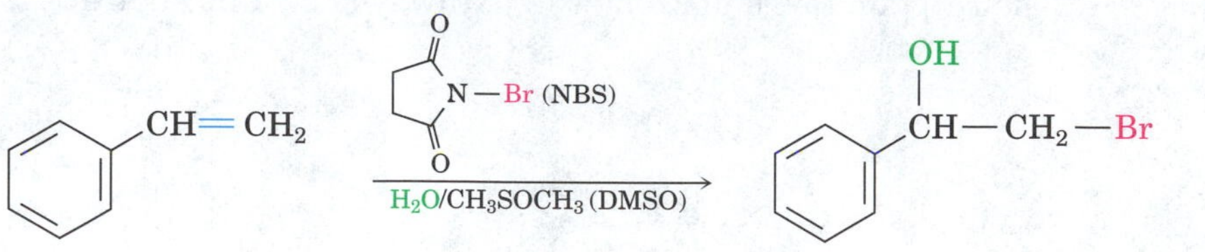

Styrene **2-Bromo-1-phenylethanol (76%)**

Note that the aromatic ring in the above example is inert to Br_2 under the conditions used, even though it contains three carbon–carbon double bonds. Aromatic rings are a good deal more stable than might be expected, a property that will be examined in Chapter 15.

PROBLEM...

7.5 What product(s) would you expect from the reaction of cyclopentene with NBS and water? Show the stereochemistry.

PROBLEM...

7.6 When an unsymmetrical alkene, such as propene, is treated with *N*-bromosuccinimide in aqueous dimethyl sulfoxide, the major product has the bromine atom bonded to the less substituted carbon atom:

$$CH_3CH{=}CH_2 \xrightarrow[\text{DMSO}]{\text{NBS, } H_2O} CH_3\overset{\overset{\displaystyle OH}{|}}{C}HCH_2Br$$

Is this Markovnikov or non-Markovnikov orientation? Explain.

....

7.4 Hydration of Alkenes: Oxymercuration

Water adds to simple alkenes such as ethylene and 2-methylpropene to yield alcohols, a process called **hydration**. The reaction takes place on treatment of the alkene with water and a strong acid catalyst (HA) by a mechanism similar to that of HX addition. Thus, protonation of an alkene double bond yields a carbocation intermediate, which reacts with water to yield a protonated alcohol product (ROH_2^+). Loss of H^+ from this protonated alcohol gives the neutral alcohol and regenerates the acid catalyst (Figure 7.5).

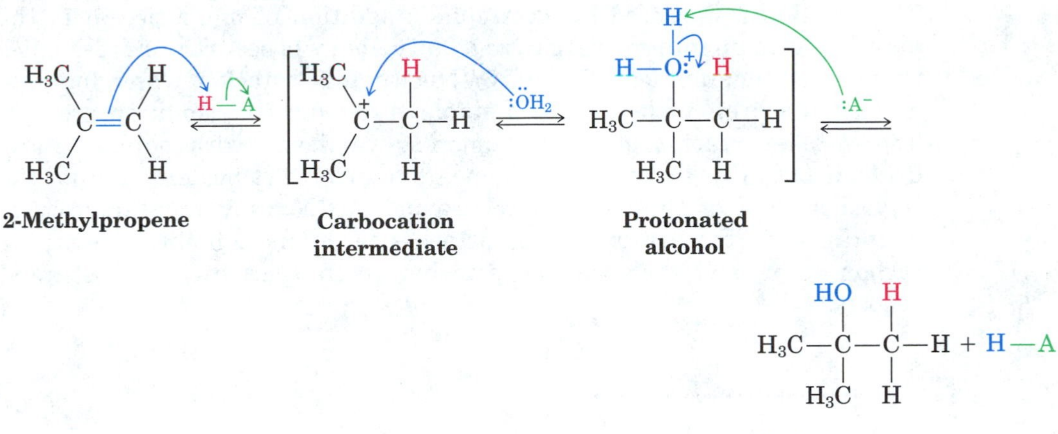

Figure 7.5 Mechanism of the acid-catalyzed hydration of an alkene. Protonation of the alkene gives a carbocation intermediate that reacts with water.

Acid-catalyzed alkene hydration is suitable for large-scale industrial procedures, and approximately 300,000 tons of ethanol are manufactured each year in the United States by hydration of ethylene. The reaction is of little value for most laboratory applications, though, because it requires high temperatures and strongly acidic conditions. The hydration of ethylene, for example, takes place at 250°C with sulfuric acid as catalyst.

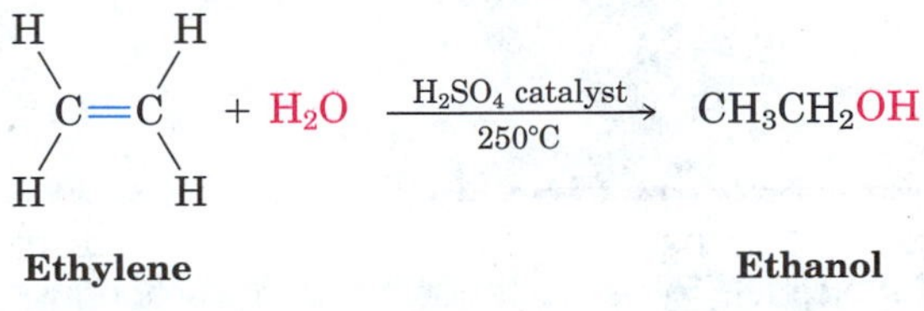

In the laboratory, most alkenes are best hydrated by the **oxymercuration** procedure. When an alkene is treated with mercuric acetate [$Hg(O_2CCH_3)_2$, usually abbreviated $Hg(OAc)_2$] in aqueous tetrahydrofuran (THF) solvent, electrophilic addition to the double bond rapidly occurs. The intermediate *organomercury* compound is then treated with sodium borohydride, $NaBH_4$, and an alcohol is produced. For example:

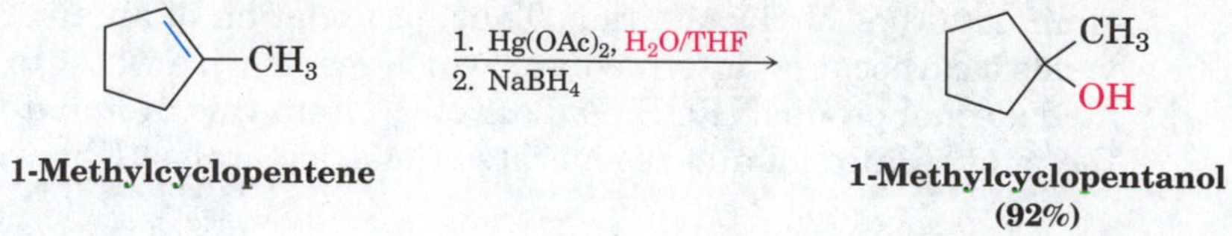

Alkene oxymercuration is closely analogous to halohydrin formation. The reaction is initiated by electrophilic addition of mercuric ion to the alkene to give an intermediate *mercurinium ion*, whose structure resembles that of a bromonium ion (Figure 7.6). Nucleophilic attack of water, followed by loss of a proton, then yields a stable organomercury addition product. The final step, reaction of the organomercury compound with sodium borohydride, is not fully understood but appears to involve radicals. Note that the regiochemistry of the reaction corresponds to Markovnikov addition of water; that is, the hydroxyl group attaches to the more highly substituted carbon atom, and the hydrogen attaches to the less highly substituted carbon.

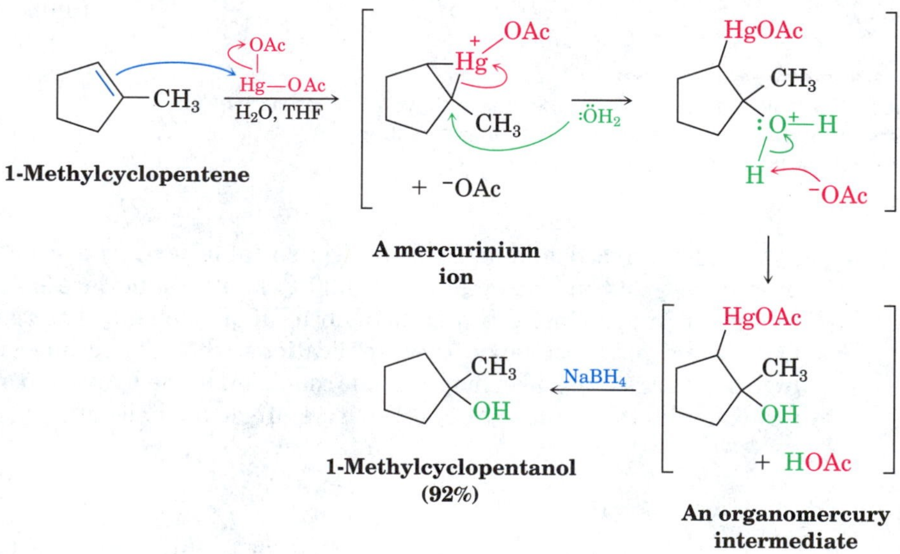

Figure 7.6 Mechanism of the oxymercuration of an alkene to yield an alcohol. This electrophilic addition reaction involves a mercurinium ion intermediate, and its mechanism is similar to that of halohydrin formation.

PROBLEM. ....

7.7 What products would you expect from oxymercuration of these alkenes?

(a) $CH_3CH_2CH_2CH{=}CH_2$ (b) 2-Methyl-2-pentene

PROBLEM

7.8 What alkenes might these alcohols have been prepared from?

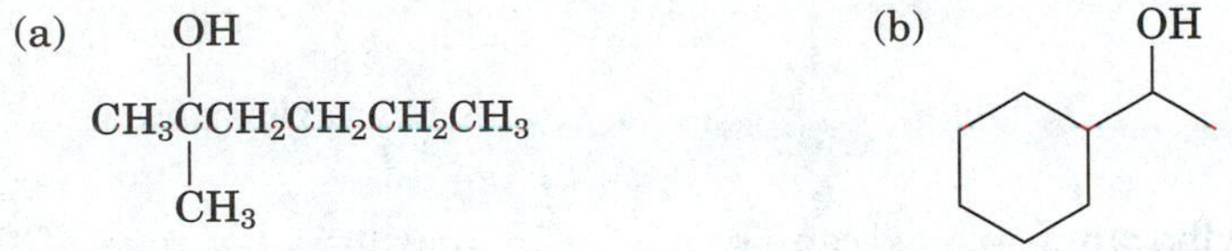

7.5 Hydration of Alkenes: Hydroboration

One of the most useful alkene additions is the **hydroboration** reaction reported in 1959 by H. C. Brown.[2] Hydroboration involves addition of a B–H bond of borane, BH_3, to an alkene to yield an organoborane intermediate, RBH_2:

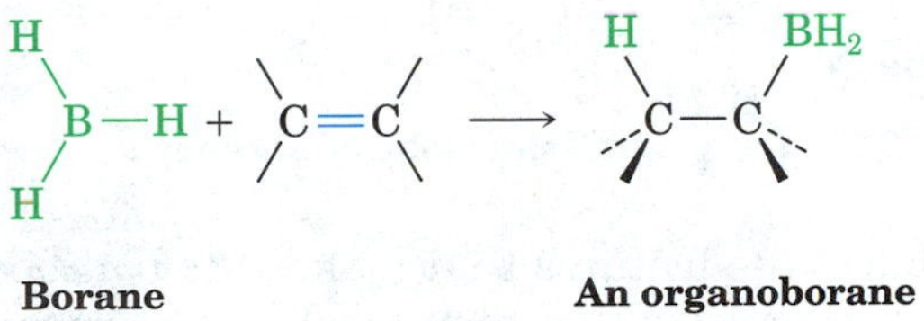

Borane is highly reactive since the boron atom has only six electrons in its valence shell. In tetrahydrofuran (THF) solution, however, BH_3 accepts an electron pair from a solvent molecule in a Lewis acid–base reaction to complete its octet and form a stable BH_3–THF complex:

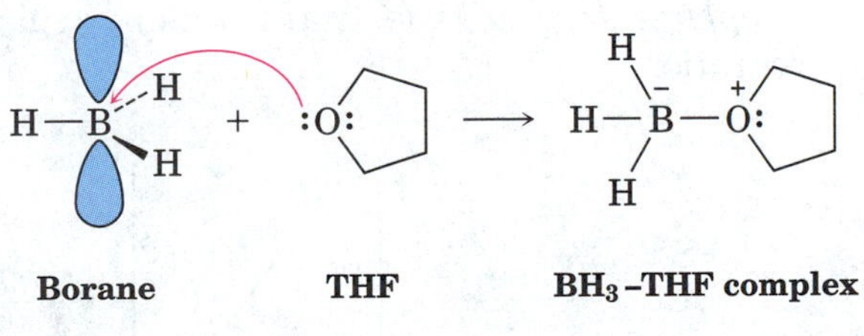

When an alkene reacts with BH_3 in THF solution, rapid addition to the double bond occurs. Since BH_3 has three hydrogens, addition occurs three times and a *trialkylborane*, R_3B, is formed. For example, 1 molar equivalent of BH_3 adds to 3 molar equivalents of ethylene to yield triethylborane:

[2]Herbert Charles Brown (1912–); b. London; Ph.D. (1938) University of Chicago (Schlesinger); professor, Purdue University (1947–); Nobel Prize (1979).

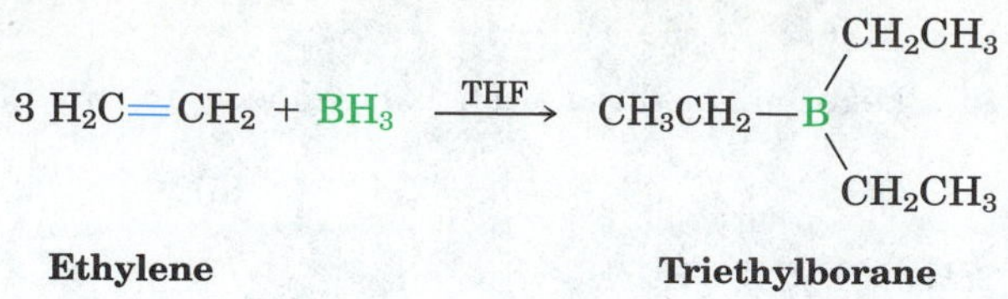

Alkylboranes are extremely useful in synthesis because of the further reactions they undergo. When tricyclohexylborane is treated with aqueous hydrogen peroxide (H_2O_2) in basic solution, for example, an oxidation takes place. The carbon–boron bond is broken, a hydroxyl group bonds to carbon, and 3 equivalents of cyclohexanol are produced. The net effect of the two-step hydroboration/oxidation sequence is hydration of the alkene double bond.

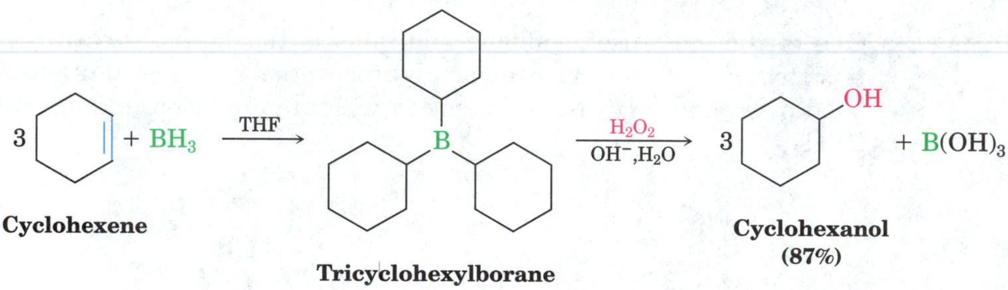

One of the features that makes the hydroboration reaction so useful is the regiochemistry that results when an unsymmetrical alkene is hydroborated. For example, hydroboration/oxidation of 1-methylcyclopentene yields *trans*-2-methylcyclopentanol. Boron and hydrogen add to the alkene with syn stereochemistry, with boron attaching to the less highly substituted carbon. During the oxidation step, the boron is replaced by a hydroxyl with the same stereochemistry, resulting in an overall syn non-Markovnikov addition of water. This stereochemical result is particularly useful because it is *complementary* to the Markovnikov regiochemistry observed for oxymercuration.

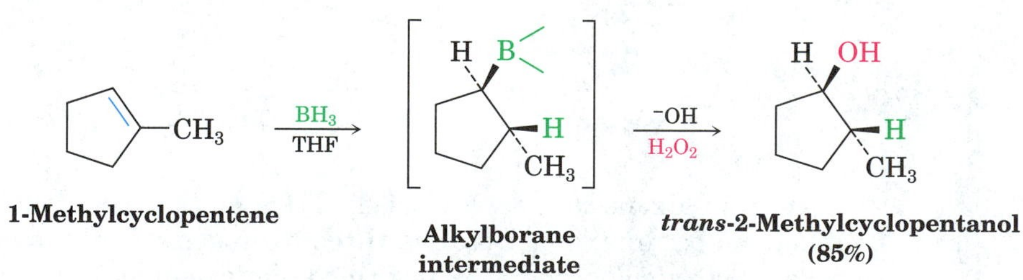

Why does alkene hydroboration take place with "non-Markovnikov" regiochemistry? Hydroboration differs from many other alkene addition reactions in that it occurs in a single step without a carbocation intermediate. We can view the reaction as taking place through a four-center, cyclic transition state, as shown in Figure 7.7. Since both C–H and C–B bonds form at the same time and from the same face of the alkene, syn stereochemistry is observed.

The mechanism shown in Figure 7.7 accounts for not only the reaction's stereochemistry but also its regiochemistry. Although hydroboration does not involve a carbocation intermediate as other alkene addition reactions do, the interaction of borane with an alkene nevertheless has a large amount of polar character to it. Borane, with only six valence electrons on boron, is a powerful Lewis acid and electrophile because of its vacant *p* orbital. Thus, the interaction of BH_3 with an alkene involves a partial transfer of electrons from the alkene to boron, with consequent buildup of polar character in the four-membered-ring transition state. Boron thus carries a partial negative charge ($\delta-$) because it has gained electrons, and the alkene carbons carry a partial positive charge ($\delta+$) because they have lost electrons.

Addition of borane to the alkene π bond occurs in a single step through a cyclic four-membered-ring transition state. The dotted lines indicate partial bonds that are breaking or forming.

A neutral alkylborane addition product is then formed when reaction is complete.

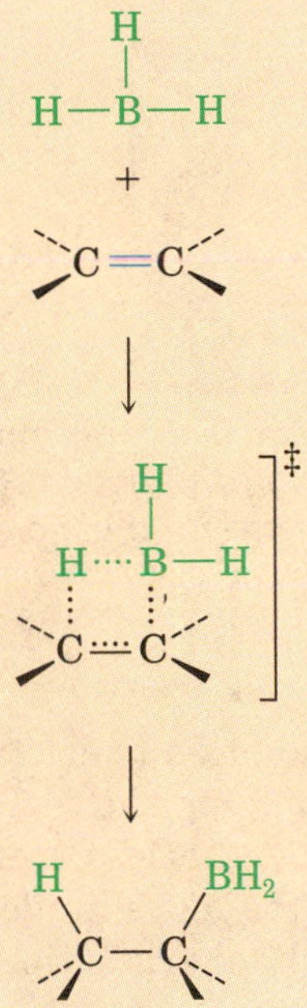

Figure 7.7 Mechanism of alkene hydroboration. The reaction occurs in a single step, in which both C–H and C–B bonds form at the same time and on the same face of the double bond.

In the addition of BH_3 to an unsymmetrically substituted alkene such as 1-methylcyclopentene, there are two possible four-center transition states. Since the transition state that places a partial positive charge on the more highly substituted carbon is favored over the alternative that places the charge on the less highly substituted carbon, boron tends to add to the less highly substituted carbon (Figure 7.8).

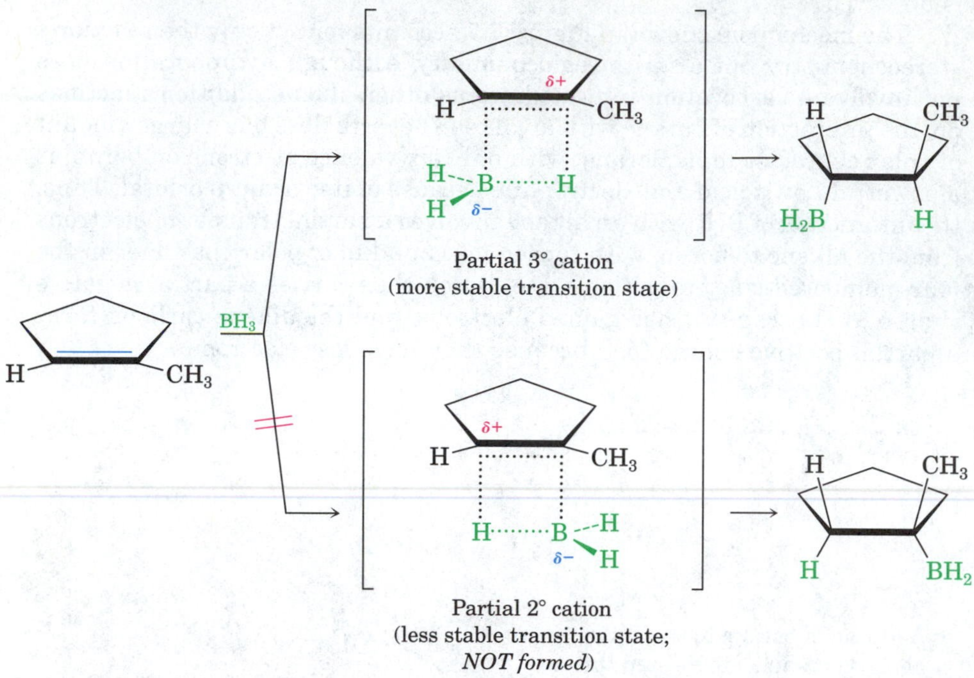

Figure 7.8 Mechanism of the hydroboration of 1-methylcyclopentene. The favored transition state is the one that places the partial positive charge on the more highly substituted carbon.

In addition to electronic factors, a steric factor is probably also involved in determining the regiochemistry of hydroboration. Attachment of boron is favored at the less sterically hindered carbon atom of the alkene, rather than at the more hindered carbon, because there is less steric crowding in the resultant transition state:

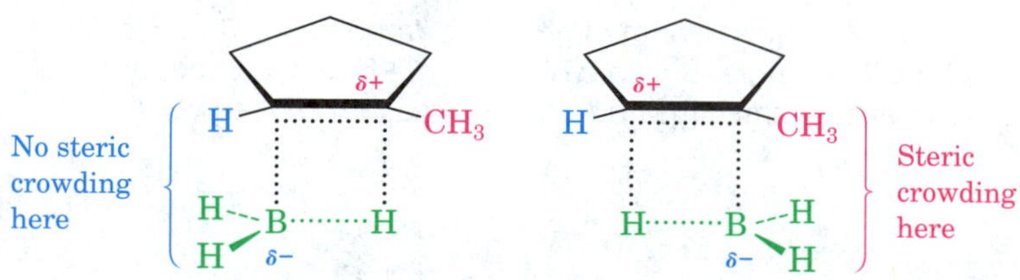

Both steric and electronic arguments predict the observed regiochemistry, but evidence suggests that steric factors probably have greater influence than electronic factors.

PROBLEM.

7.9 What product will result from hydroboration/oxidation of 1-methylcyclopentene with deuterated borane, BD_3? Show both the stereochemistry (spatial arrangement) and the regiochemistry (orientation) of the product.

PROBLEM.

7.10 What alkenes might be used to prepare the following alcohols by hydroboration/oxidation?

(a) $(CH_3)_2CHCH_2CH_2OH$

(b) $(CH_3)_2CHCH(OH)CH_3$

(c) 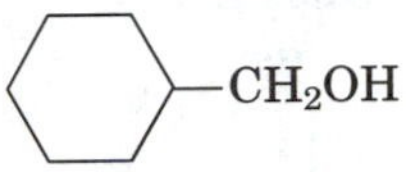

7.6 Addition of Carbenes to Alkenes: Cyclopropane Synthesis

Yet another kind of alkene addition is the reaction of a *carbene* with an alkene to yield a cyclopropane. A **carbene, R_2C:**, is a neutral molecule containing a divalent carbon that has only six electrons in its valence shell. It is therefore highly reactive and can be generated only as a reaction intermediate, rather than as an isolable molecule. Because they have only six valence electrons, carbenes are electron-deficient and behave as electrophiles. Thus, they react with nucleophilic carbon–carbon double bonds just as other electrophiles do. The reaction occurs in a single step without intermediates.

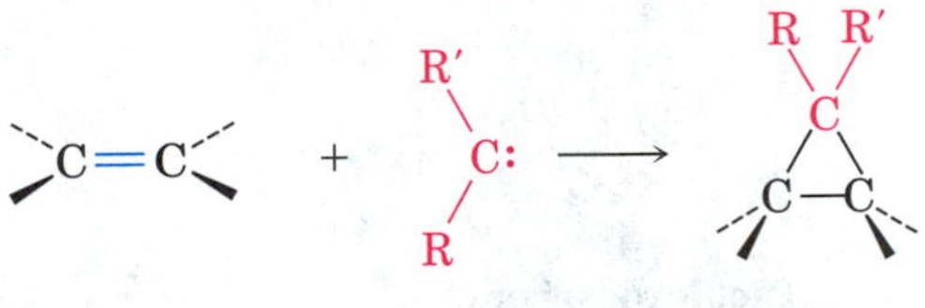

An alkene **A carbene** **A cyclopropane**

One of the simplest methods for generating a substituted carbene is by treatment of chloroform, $CHCl_3$, with a strong base such as potassium hydroxide. Loss of a proton from $CHCl_3$ generates the trichloromethanide anion, $^-:CCl_3$, which expels a chloride ion to give dichlorocarbene, $:CCl_2$ (Figure 7.9).

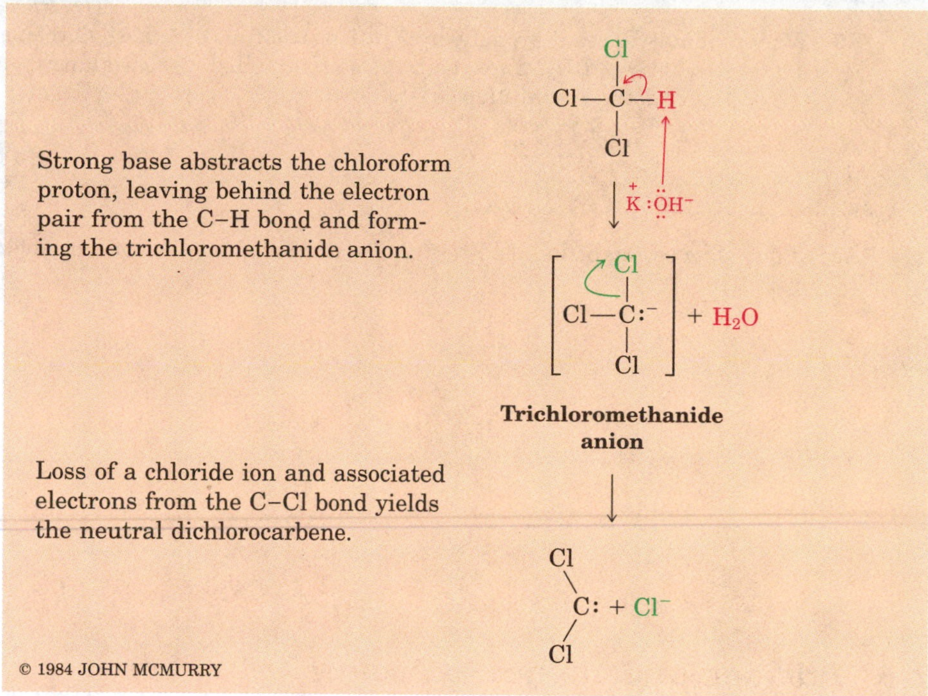

Figure 7.9 Mechanism of the formation of dichlorocarbene by reaction of chloroform with strong base.

The dichlorocarbene carbon atom is sp^2-hybridized, with a vacant p orbital extending above and below the plane of the three atoms and with an unshared pair of electrons occupying the third sp^2 lobe. Note that this electronic description of dichlorocarbene is similar to that for a carbocation (Section 6.10) with respect both to the sp^2 hybridization of carbon and to the vacant p orbital.

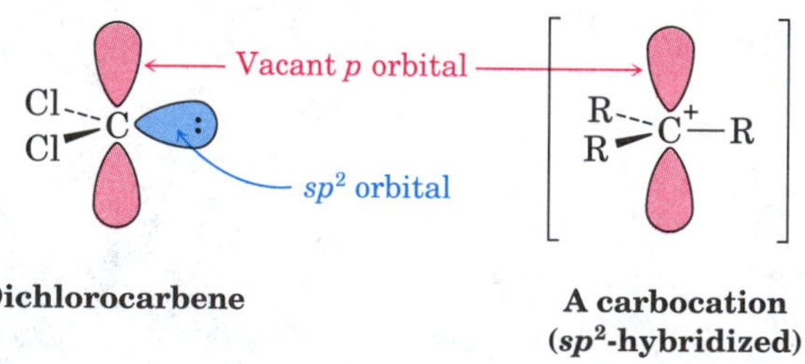

If dichlorocarbene is generated in the presence of an alkene, addition to the double bond occurs, and a dichlorocyclopropane is formed. As the reaction of dichlorocarbene with *cis*-2-pentene demonstrates, the addition is **stereospecific**, meaning that only a single stereoisomer is formed as product. Starting from the cis alkene, only cis-disubstituted cyclopropane is produced.

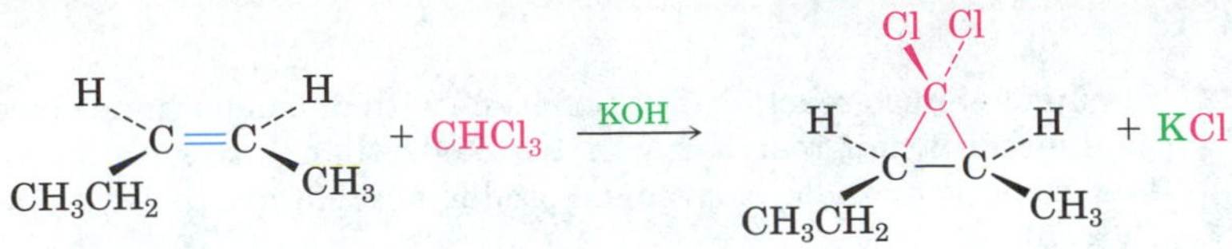

cis-2-Pentene **65%**

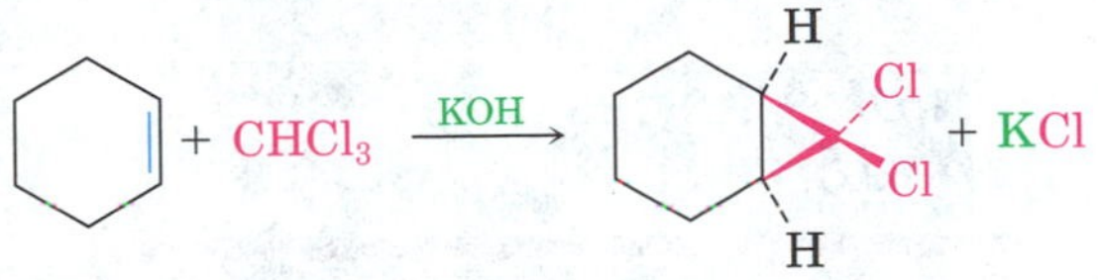

Cyclohexene **60%**

The best method for preparing nonhalogenated cyclopropanes is the **Simmons–Smith reaction**. This reaction, first investigated at the Du Pont company, does not involve a free carbene. Rather, it utilizes a *carbenoid*—a metal-complexed reagent with carbene-like reactivity. When diiodomethane is treated with a specially prepared zinc–copper alloy, (iodomethyl)zinc iodide, ICH_2ZnI, is formed. If an alkene is present, (iodomethyl)zinc iodide transfers a CH_2 group to the double bond and yields the cyclopropane. For example, cyclohexene reacts cleanly and in good yield to give the corresponding cyclopropane. Although we won't discuss the mechanistic details, carbene addition to an alkene is an example of a general class of reactions called *cycloadditions*, which we'll study more carefully in Chapter 31.

$$CH_2I_2 + Zn(Cu) \xrightarrow{\text{Ether}} I{-}CH_2{-}Zn{-}I = \text{“}{:}CH_2\text{”}$$

Diiodomethane **(Iodomethyl)zinc iodide (a carbenoid)**

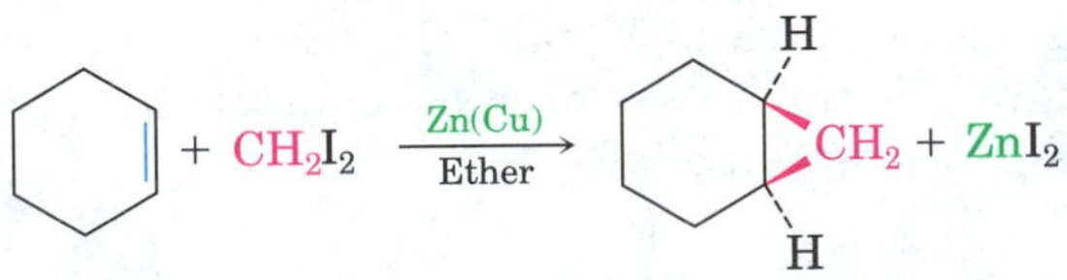

Cyclohexene **Bicyclo[4.1.0]heptane (92%)**

PROBLEM...

7.11 What products would you expect from these reactions?

(a) 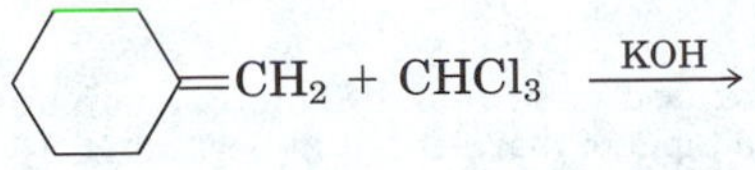

(b) $(CH_3)_2CHCH_2CH{=}CHCH_3 + CH_2I_2 \xrightarrow{Zn(Cu)}$

PROBLEM

7.12 Simmons–Smith reaction of cyclohexene with diiodomethane gives a single cyclopropane product, but reaction with 1,1-diiodoethane gives (in low yield) a mixture of two isomeric methylcyclopropane products. What are the two products, and how do they differ?

7.7 Reduction of Alkenes: Hydrogenation

Alkenes react with hydrogen in the presence of a catalyst to yield the corresponding saturated alkane addition products. We describe the result by saying that the double bond has been **hydrogenated**, or *reduced*.[3] Although we looked briefly at catalytic hydrogenation as a method of determining alkene stabilities in Section 6.7, the reaction also has enormous practical value.

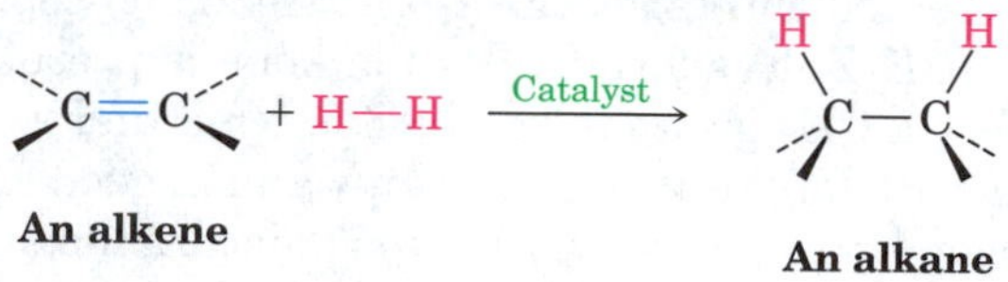

Platinum and palladium are the two catalysts most commonly used for alkene hydrogenations. Palladium is normally employed in a very finely divided state "supported" on an inert material such as charcoal to maximize surface area (Pd/C). Platinum is normally used as PtO_2, a reagent known as *Adams' catalyst* after its discoverer, Roger Adams.[4]

Catalytic hydrogenation, unlike most other organic reactions, is a *heterogeneous* process rather than a homogeneous one. That is, the hydrogenation reaction does not occur in solution but instead takes place on the surface of solid catalyst particles. Studies have shown that hydrogenation usually occurs with syn stereochemistry; both hydrogens add to the double bond from the same face.

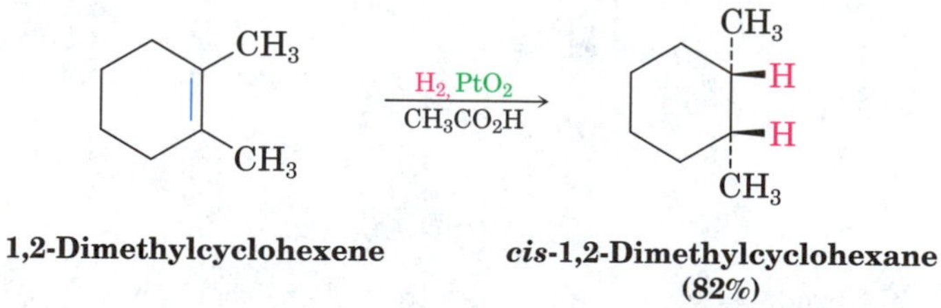

[3]The words *oxidation* and *reduction* are used somewhat differently in organic chemistry than in inorganic chemistry. We'll explore the topic in more detail in Section 10.10 but will note for the present that an organic oxidation often forms carbon–oxygen bonds, while a reduction often forms carbon–hydrogen bonds.

[4]Roger Adams (1889–1971); b. Boston; Ph.D. Harvard (Torrey, Richards) (1912); professor, University of Illinois (1916–1971).

The first step in the reaction is adsorption of hydrogen onto the catalyst surface. Complexation between catalyst and alkene then occurs as a vacant metal orbital interacts with the filled alkene π orbital. In the final steps, hydrogen is inserted into the double bond, and the saturated product diffuses away from the catalyst (Figure 7.10). The stereochemistry of hydrogenation is syn because both hydrogens add to the double bond from the same catalyst surface.

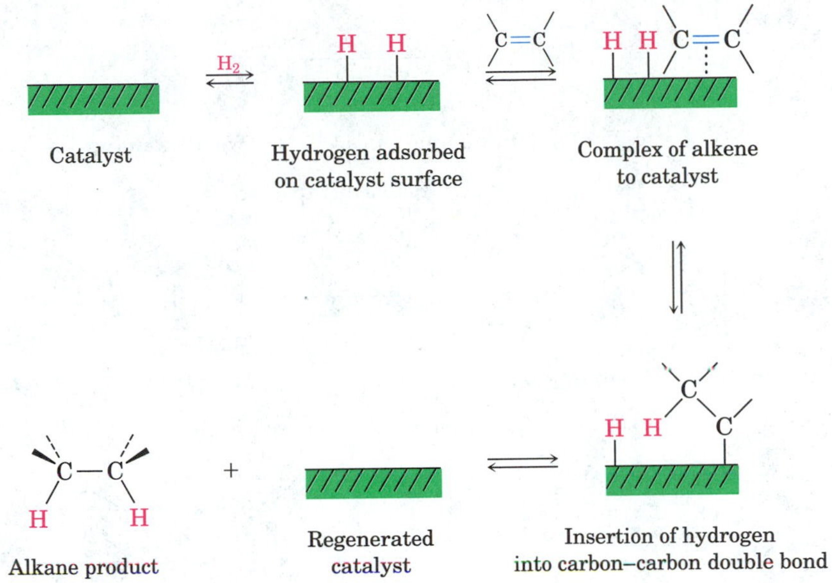

Figure 7.10 Mechanism of alkene hydrogenation. The reaction takes place with syn stereochemistry on the surface of catalyst particles.

An interesting stereochemical feature of catalytic hydrogenation is that the reaction is highly sensitive to the steric environment around the double bond. As a result, the catalyst often approaches only one face, giving rise to a single product. In α-pinene, for example, one of the methyl groups attached to the four-membered ring hangs over the top face of the double bond and blocks approach of the hydrogenation catalyst from that side. Reduction therefore occurs exclusively from the bottom face to yield the product shown.

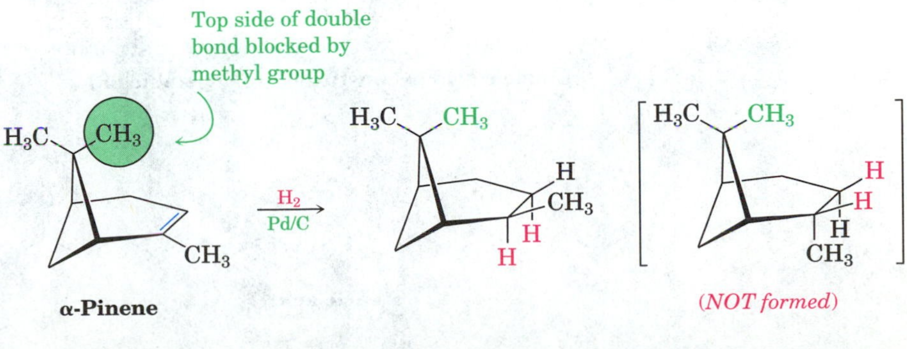

Alkenes are much more reactive than most other functional groups toward catalytic hydrogenation, and the reaction is therefore quite selective. Other functional groups such as ketones, esters, and nitriles survive normal alkene hydrogenation conditions unchanged, although reaction with these groups does occur under more vigorous conditions. Note that, in the hydrogenation of methyl 3-phenylpropenoate shown below, the aromatic ring is not reduced by hydrogen on palladium even though it contains three double bonds.

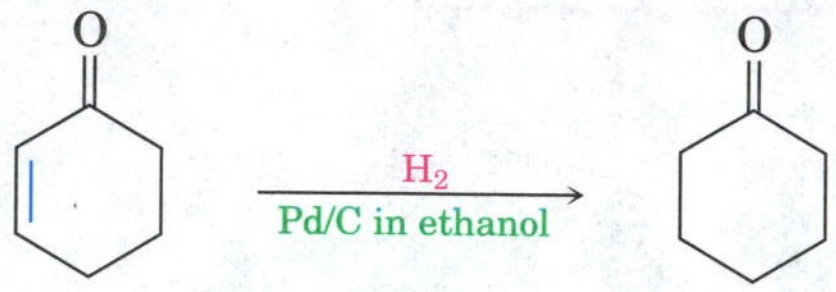

2-Cyclohexenone **Cyclohexanone**

Ketone not reduced

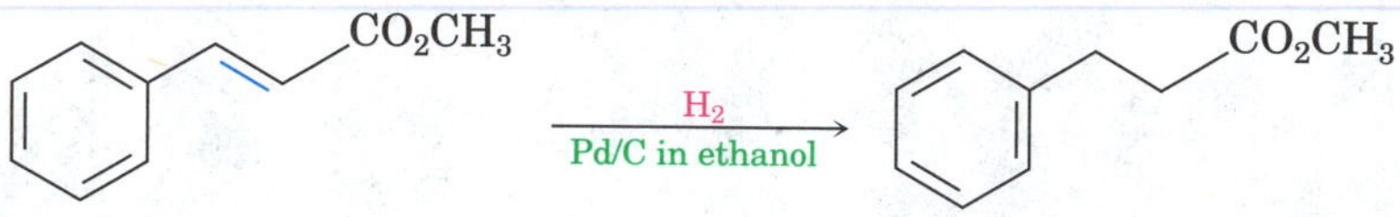

Methyl 3-phenylpropenoate **Methyl 3-phenylpropanoate**

Benzene ring, ester not reduced

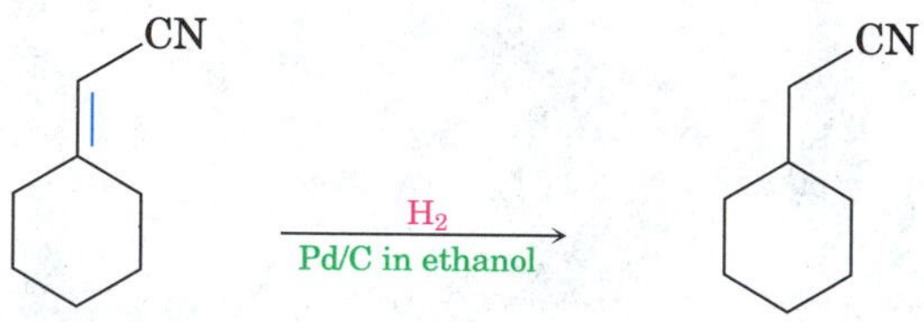

Cyclohexylideneacetonitrile **Cyclohexylacetonitrile**

Nitrile not reduced

In addition to its usefulness in the laboratory, catalytic hydrogenation is of great commercial value in the food industry. Unsaturated vegetable oils, which usually contain numerous double bonds, are catalytically hydrogenated on a vast scale to produce the saturated fats used in margarine and solid cooking fats.

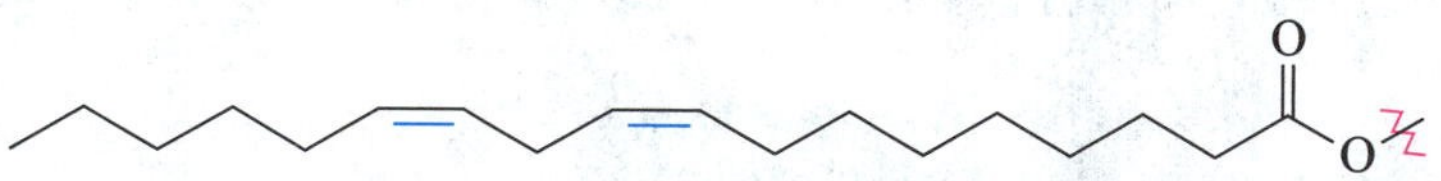

Linoleic acid (a constituent of vegetable oil)

$\downarrow$ 2 H_2 Pd/C

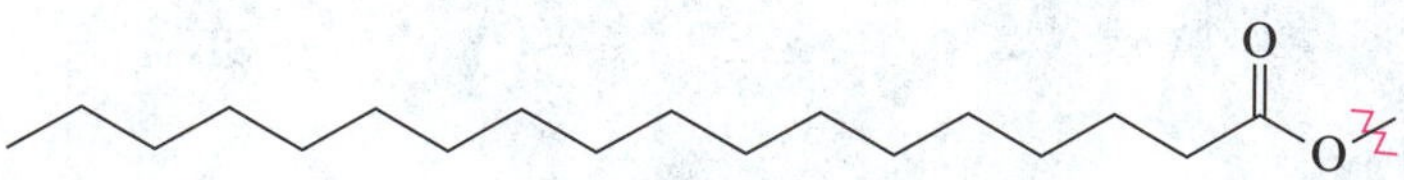

Stearic acid

PROBLEM

7.13 What product would you obtain from catalytic hydrogenation of the following alkenes?
(a) $(CH_3)_2C{=}CHCH_2CH_3$ (b) 3,3-Dimethylcyclopentene

7.8 Oxidation of Alkenes: Hydroxylation and Cleavage

Alkene Hydroxylation

Hydroxylation of an alkene—the addition of an –OH group to each of the two alkene carbons—can be carried out by reaction of the alkene with osmium tetroxide (OsO_4). The reaction occurs with syn stereochemistry and yields a 1,2-dialcohol, or **diol** product (also called a **glycol**).

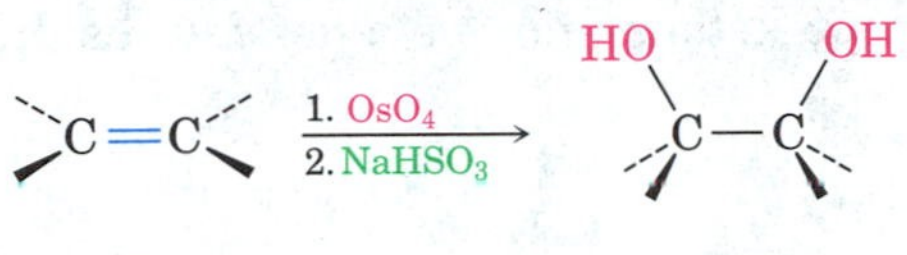

Alkene hydroxylation does not involve a carbocation intermediate but instead occurs through an intermediate cyclic *osmate* species, which is formed in a single step by addition of OsO_4 to the alkene. This cyclic osmate is cleaved in a separate step using aqueous sodium bisulfite, $NaHSO_3$.

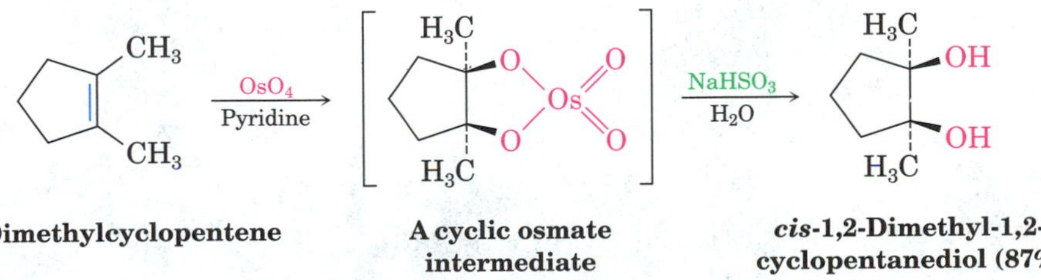

Alkene Cleavage

In all the alkene addition reactions we've seen thus far, the carbon–carbon double bond has been converted into a single bond and the carbon skeleton of the starting material has been left intact. There are, however, powerful oxidizing reagents that *cleave* C=C bonds and produce two fragments.

Ozone (O_3) is the most useful double-bond cleavage reagent. Prepared by passing a stream of oxygen through a high-voltage electrical discharge, ozone adds rapidly to alkenes at low temperature to give cyclic intermediates called **molozonides**. Once formed, molozonides then rapidly rearrange to form **ozonides**. Though we won't study the mechanism of this rearrangement in detail, it involves the molozonide coming apart into two fragments, which then recombine in a new way.

$$3\ O_2 \xrightarrow[\text{discharge}]{\text{Electric}} 2\ O_3$$

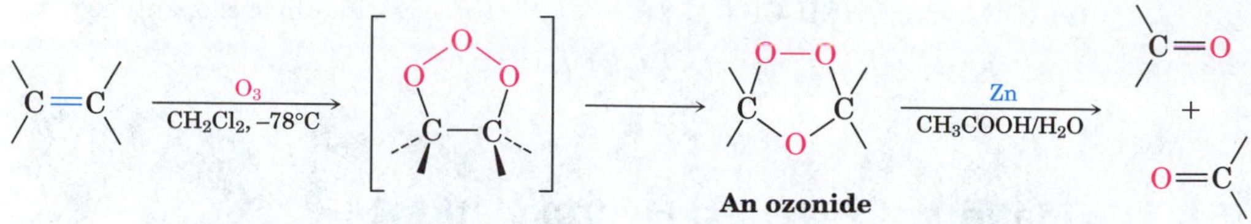

A molozonide

An ozonide

Low-molecular-weight ozonides are explosive and are therefore never isolated. Instead, ozonides are further treated with a reducing agent such as zinc metal in acetic acid to convert them to carbonyl compounds. The net result of the **ozonolysis**/zinc reduction sequence is that the C=C bond is cleaved, and oxygen becomes doubly bonded to each of the original alkene carbons. If an alkene with a tetrasubstituted double bond is ozonized, two ketone fragments result; if an alkene with a trisubstituted double bond is ozonized, one ketone and one aldehyde result, and so on.

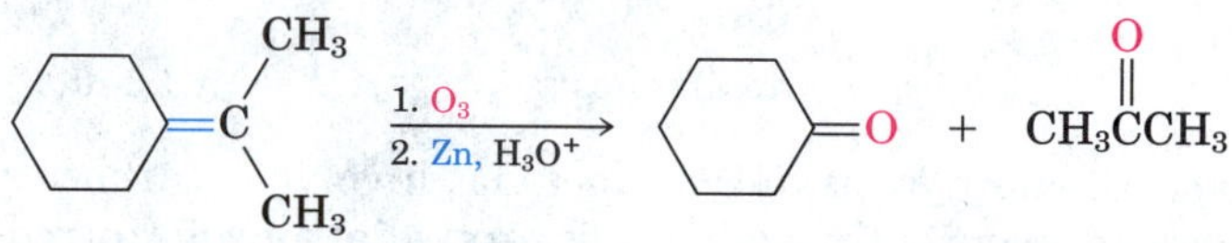

Isopropylidenecyclohexane (tetrasubstituted)

Cyclohexanone **Acetone**

84%; two ketones

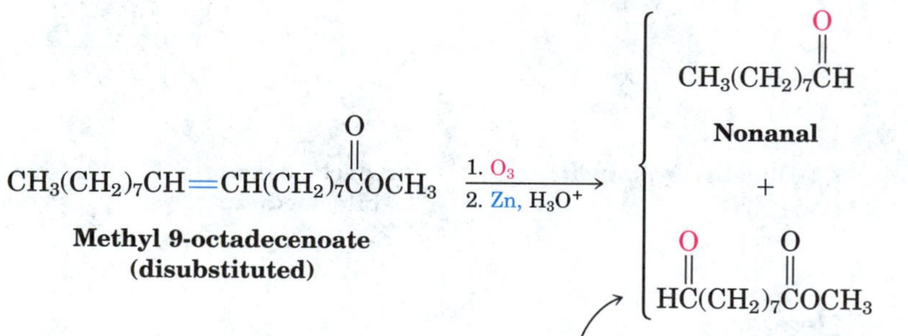

Methyl 9-octadecenoate (disubstituted)

Nonanal

Methyl 9-oxononanoate

78%; two aldehydes

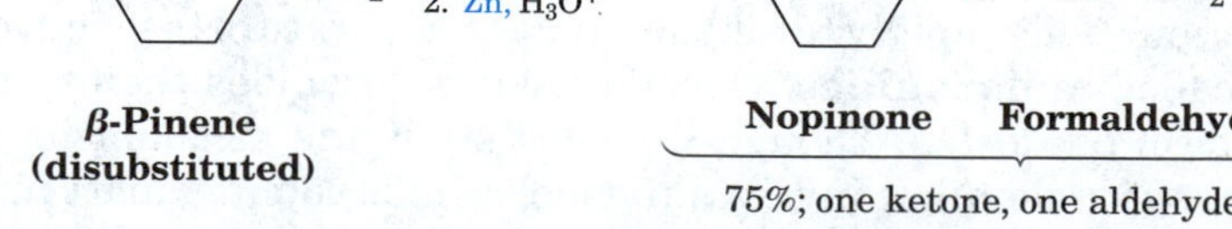

β-Pinene (disubstituted)

Nopinone **Formaldehyde**

75%; one ketone, one aldehyde

Several oxidizing reagents other than ozone also cause double-bond cleavage. For example, potassium permanganate in neutral or acidic solution cleaves alkenes, giving carbonyl-containing products in low to moderate yield. If hydrogens are present on the double bond, carboxylic acids are produced; if two hydrogens are present on one carbon, CO_2 is formed.

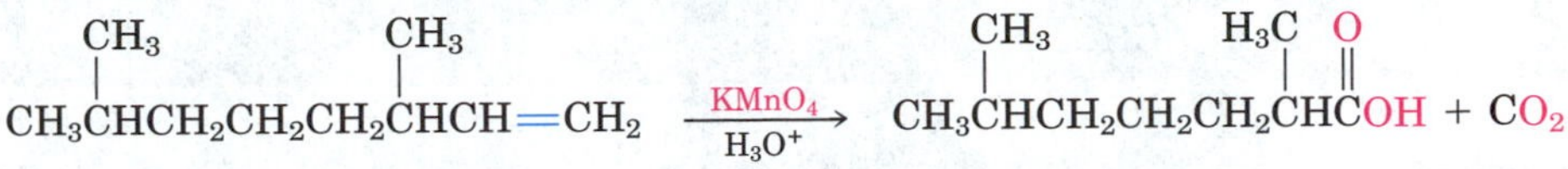

3,7-Dimethyl-1-octene **2,6-Dimethylheptanoic acid (45%)**

1,2-Diol Cleavage

1,2-Diols are oxidatively cleaved by reaction with periodic acid (HIO_4) to yield carbonyl compounds, a reaction similar to the potassium permanganate cleavage of alkenes just discussed. The sequence of (1) alkene hydroxylation with OsO_4 followed by (2) diol cleavage with HIO_4 is often an excellent alternative to direct alkene cleavage with ozone or potassium permanganate.

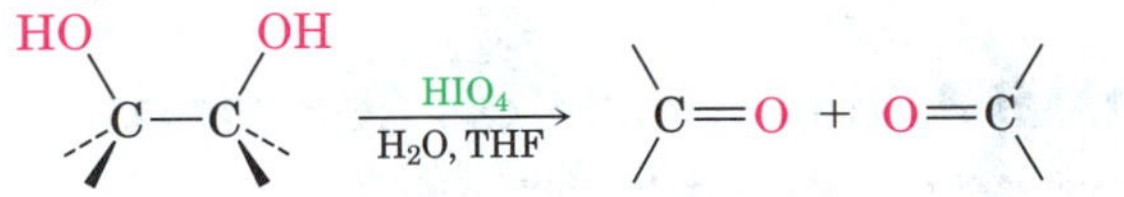

A 1,2-diol **Two carbonyl compounds**

If the two hydroxyls are on an open chain, two carbonyl compounds result. If the two hydroxyls are on a ring, a single dicarbonyl compound is formed. As indicated in the following examples, the cleavage reaction is believed to take place through a cyclic periodate intermediate.

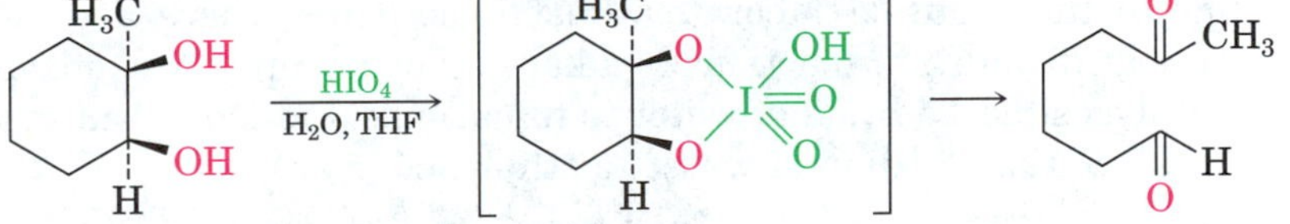

A 1,2-diol **Cyclic periodate intermediate** **6-Oxoheptanal (86%)**

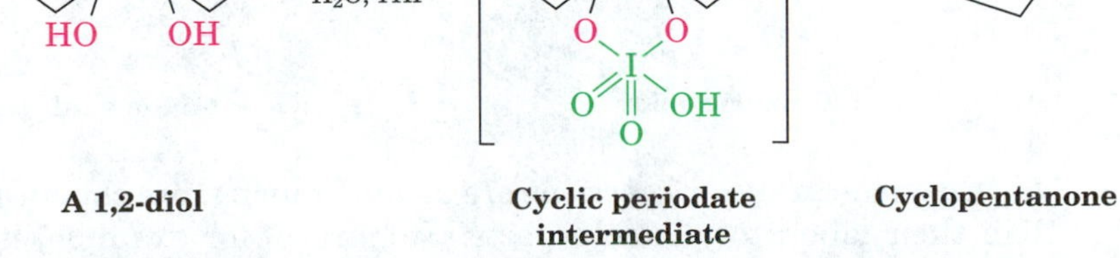

A 1,2-diol **Cyclic periodate intermediate** **Cyclopentanone (81%)**

PROBLEM.

7.14 What alkene would you start with to prepare each of the following compounds?

(a) 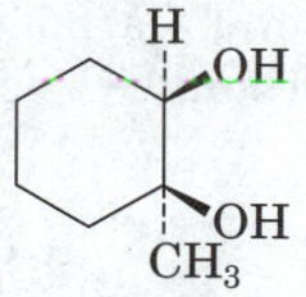

(b) 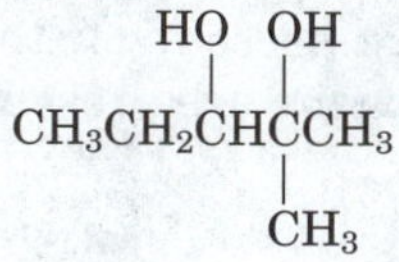

(c) 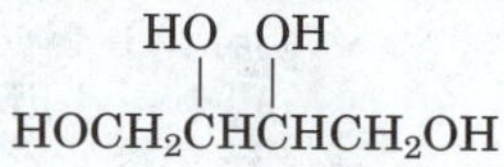

PROBLEM.

7.15 What products would you expect from reaction of 1-methylcyclohexene with these reagents?

(a) Aqueous acidic $KMnO_4$

(b) O_3, followed by Zn, CH_3COOH

PROBLEM.

7.16 Propose structures for alkenes that yield the following products on reaction with ozone followed by treatment with Zn.

(a) $(CH_3)_2C{=}O + H_2C{=}O$

(b) 2 equiv $CH_3CH_2CH{=}O$

7.9 Biological Alkene Addition Reactions

The chemistry of living organisms is a fascinating field of study—the simplest one-celled organism is capable of more complex organic syntheses than any human chemist. Yet as we learn more, it becomes clear that the same principles that apply to laboratory chemistry also apply to biological chemistry.

Biological organic chemistry takes place in the aqueous medium inside cells, rather than in organic solvents, and involves complex catalysts called *enzymes*. Nevertheless, the *kinds* of biological reactions are remarkably similar to the kinds of laboratory reactions. Thus, there are many cases of biological addition reactions to alkenes. For example, the enzyme fumarase catalyzes the addition of water to fumaric acid much as sulfuric acid might catalyze the addition of water to ethylene.

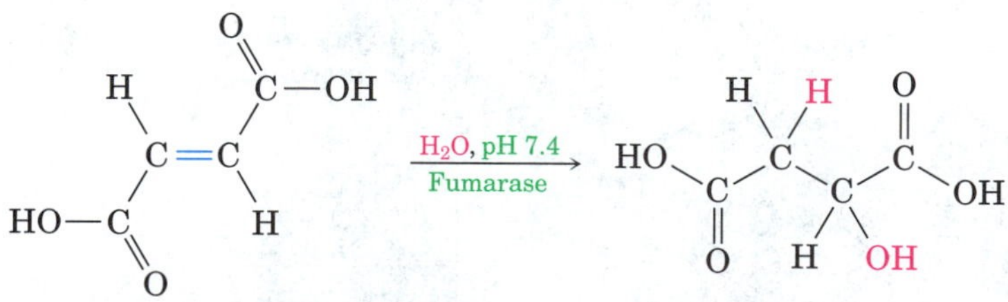

Fumaric acid **Malic acid**

Enzyme-catalyzed reactions are usually much more chemically selective than their laboratory counterparts. Fumarase, for example, is completely inert toward maleic acid, the cis isomer of fumaric acid. Nevertheless, the fundamental processes of organic chemistry are the same in the living cell and in the laboratory.

7.10 A Radical Addition to Alkenes: HBr/Peroxides

When HBr adds to an alkene under typical conditions, we now know that an intermediate carbocation is involved and that Markovnikov orientation will be observed. This was not the case prior to 1933, however. In the 1920s, the addition of HBr to 3-bromopropene had been studied several times by different workers, and conflicting results were reported. The usual rule of Markovnikov addition did not appear to hold; *both* possible addition products were usually obtained, but in widely different ratios by different workers.

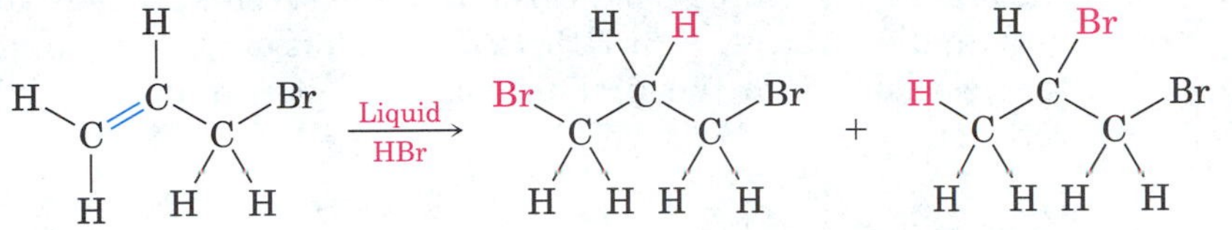

3-Bromopropene **1,3-Dibromopropane** **1,2-Dibromopropane**

Product mixture

Further work ultimately resolved the problem when it was found that HBr (but not HCl or HI) can add to alkenes by two entirely different mechanisms. It turns out that 3-bromopropene, the substrate many investigators had been using, is highly air-sensitive and absorbs oxygen to form *peroxides*—compounds with oxygen–oxygen bonds, R–O–O–R. Since the O–O bond is weak (about 155 kJ/mol; 37 kcal/mol), it is easily broken to give radicals that can catalyze a *radical addition* of HBr to the alkene, rather than a polar electrophilic addition. In subsequent work, it was found that non-Markovnikov radical addition of HBr occurs with many different alkenes. For example,

$$H_2C{=}CH(CH_2)_8COOH \xrightarrow[\text{Peroxides}]{\text{HBr}} BrCH_2CH_2(CH_2)_8COOH$$

10-Undecenoic acid **11-Bromoundecanoic acid**

Peroxide-catalyzed addition of HBr to alkenes proceeds by a radical chain mechanism involving addition of bromine radical, Br·, to the alkene. As with the radical chain process we saw earlier for the light-induced chlorination of methane (Section 5.3), both initiation steps and propagation steps are required. In writing the mechanism of this radical reaction, recall that a curved half-arrow, or "fishhook" ⌒, is used to show the movement of a single electron, as opposed to the full curved arrow used to show the movement of an electron pair in a polar reaction.

Initiation steps The reaction is initiated in two steps. In the first step, light-induced homolytic cleavage of the weak RO–OR peroxide bond generates two alkoxy radicals, RO·. An alkoxy radical then abstracts a hydrogen atom from HBr in the second initiation step to give a bromine radical, Br·.

1. 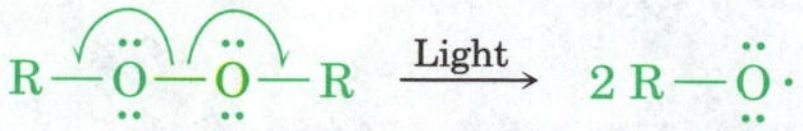

2. 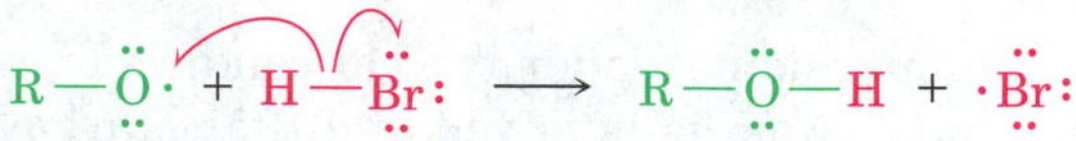

Propagation steps Once a bromine radical is formed in the initiation steps, a repeating cycle of two propagation steps begins. In the first propagation step, Br· adds to the alkene double bond, giving an alkyl radical. One electron from Br· and one electron from the double bond are used to form the new C–Br bond, leaving an unpaired electron on the remaining double-bond carbon. In the second propagation step, this alkyl radical reacts with HBr to yield addition product plus a new Br· to cycle back into the first propagation step and carry on the chain reaction.

3. 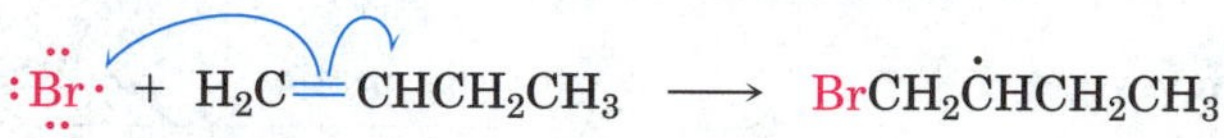

4. 

According to this radical chain mechanism, the regiochemistry of addition is determined in the first propagation step when Br· adds to the alkene. For an unsymmetrical alkene, such as 1-butene, this addition could take place at either of two carbons, to yield either a primary radical intermediate or a secondary radical. We find, however, that only the more highly substituted, secondary radical is formed. Why should this be?

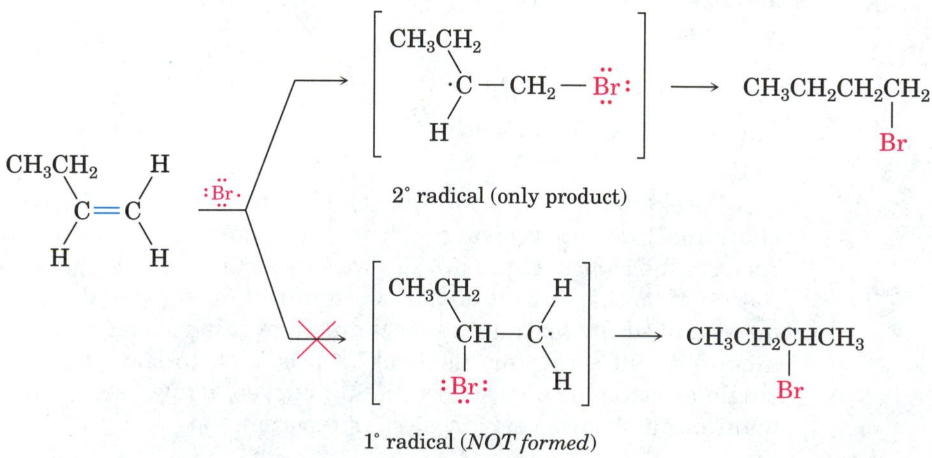

Stability of Radicals

In explaining the Markovnikov regiochemistry of polar electrophilic addition reactions in Section 6.11, we invoked the Hammond postulate to account for the fact that more stable carbocation intermediates form faster than less stable ones. In explaining the regiochemistry of the *radical* addition of HBr to alkenes, we need to invoke a similar argument. First, we need to compare the relative stabilities of substituted radicals by looking at the bond dissociation energies for different kinds of C–H bonds (Figure 7.11).

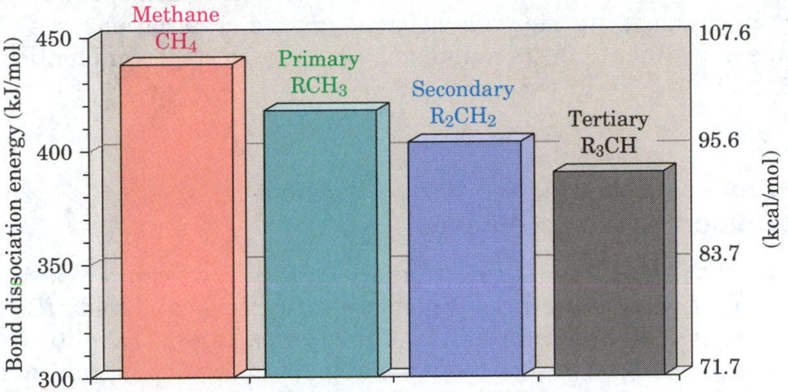

Figure 7.11 A plot of C–H bond dissociation energy versus substitution pattern. The order of bond strength is methane > 1° > 2° > 3°.

As we saw in Section 5.7, bond dissociation energy (D) is the amount of energy required to cleave a bond homolytically, yielding two radical fragments. When D is larger, the bond is stronger and there is a larger difference in stability between reactant and products. When D is smaller, the bond is weaker, and there is a smaller difference in stability between reactant and products. If the energy required to break a primary C–H bond of propane (418 kJ/mol; 100 kcal/mol) is compared with that required to break a secondary C–H bond of propane (401 kJ/mol; 96 kcal/mol), there is a difference of 17 kJ/mol. Since we are starting with the same reactant in both cases (propane), and since one product is the same in both cases (H·), the 17 kJ/mol energy difference is a direct measure of the difference in stability between the primary propyl radical ($CH_3CH_2CH_2\cdot$) and the secondary propyl radical [$(CH_3)_2CH\cdot$]. In other words, the secondary propyl radical is more stable than the primary propyl radical by 17 kJ/mol (4.1 kcal/mol).

A similar comparison between a primary C–H bond of 2-methylpropane (418 kJ/mol; 100 kcal/mol) and the tertiary C–H bond in the same molecule (390 kJ/mol; 93 kcal/mol) leads to the conclusion that the tertiary radical is more stable than the primary radical by 28 kJ/mol (6.7 kcal/mol). We thus find the following stability order of radicals:

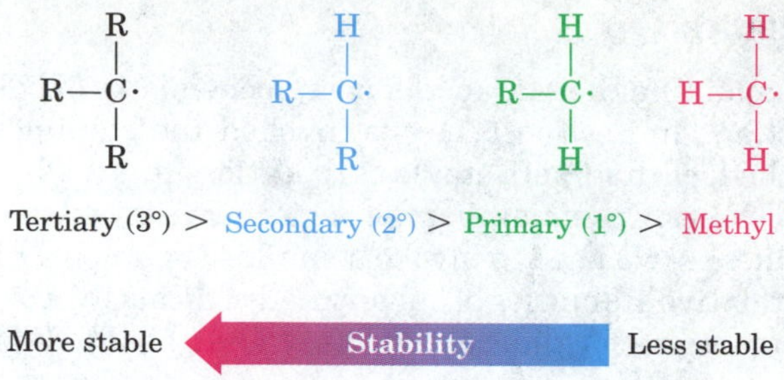

The stability order of radicals is identical to the stability order of carbocations, and for much the same reasons. Carbon radicals, even though they are uncharged, are electron-deficient just as carbocations are. Thus, the same substituent effects stabilize both through hyperconjugation (Section 6.10).

An Explanation of the Regiochemistry of Radical-Catalyzed Addition of HBr to Alkenes

Now that we know the stability order of radicals, we can complete an explanation for the observed regiochemistry of radical-catalyzed HBr additions to alkenes. Although the addition of Br· to an alkene is a slightly exothermic step, there is nevertheless a certain amount of developing radical character in the transition state. Thus, according to the Hammond postulate (Section 6.11), the same factors that make a secondary radical more stable than a primary one also make the transition state leading to it more stable. In other words, the more stable radical forms faster than the less stable one.

We conclude that, although peroxide-catalyzed radical addition of HBr to an alkene gives an *apparently* non-Markovnikov product, the reaction proceeds through the more stable intermediate just as a polar reaction does. A reaction energy diagram for the overall process is shown in Figure 7.12.

PROBLEM

7.17 How would you synthesize the following compounds? Identify the alkene starting material, and indicate what reagents you would use in each case.

(a) $CH_3CH_2C(CH_3)_2$ with Br on the C

(b) $CH_3CH_2CH_2CH_2Br$

(c) $CH_3CH_2CHCHCH_2CH_3$ with H_3C and Br on the two CH carbons

PROBLEM

7.18 Draw a reaction energy diagram for the propagation steps in the radical addition of HBr to 2-methyl-2-butene. Construct your diagram so that reactions leading to the two possible addition products are both shown. Which of the two curves has the lower activation energy for the first step? Which of the two curves has the lower $\Delta G°$ for the first step?

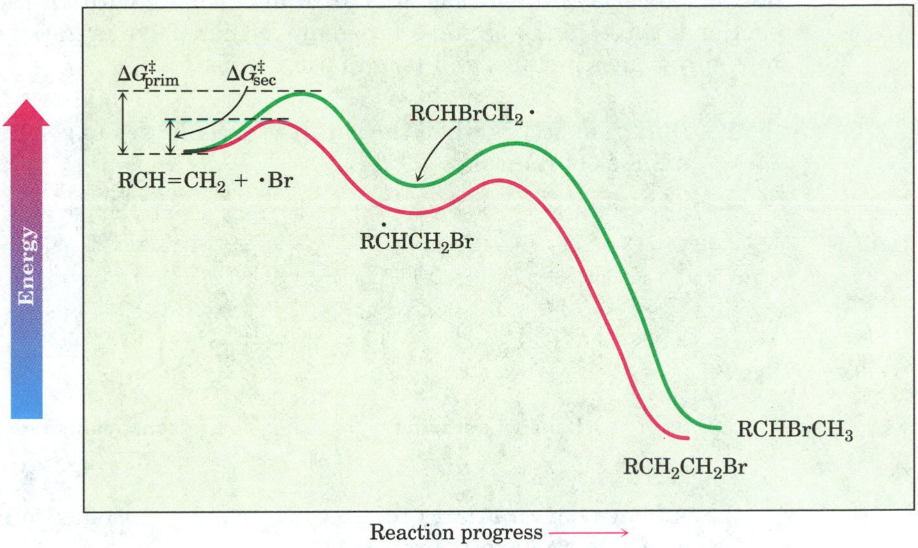

Figure 7.12 Reaction energy diagram for the addition of bromine radical to an alkene. The more stable secondary radical forms faster than the less stable primary one.

7.11 Radical Polymerization of Alkenes: Polyethylene

No other group of synthetic chemicals has had as great an impact on our day-to-day lives as the synthetic *polymers*. From carpets to clothes to foam coffee cups, it sometimes seems that we are surrounded by polymers.

A **polymer** is simply a large—sometimes *very* large—molecule built up by repetitive bonding together of many smaller molecules, called **monomers**. Polyethylene, for example, consists of enormous, long-chain alkane molecules prepared by bonding together of several thousand ethylene units. Nearly 12 million tons per year of polyethylene are manufactured in the United States alone.

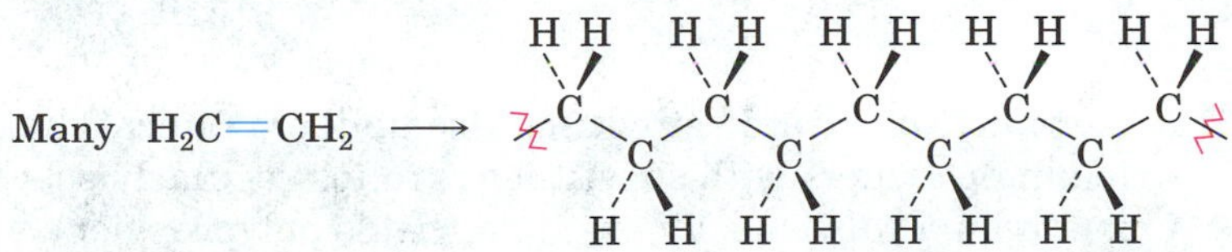

Ethylene polymerization is usually carried out at high pressure (1000–3000 atm) and high temperature (100–250°C) in the presence of a radical catalyst such as benzoyl peroxide. The key step is the addition of a radical to the ethylene double bond in a reaction similar to what takes place in the radical-catalyzed addition of HBr to alkenes. As with all radical chain reactions, three kinds of steps are required for polymerization: initiation steps, propagation steps, and termination steps.

1. *Initiation* occurs when the catalyst decomposes to generate a small number of radicals:

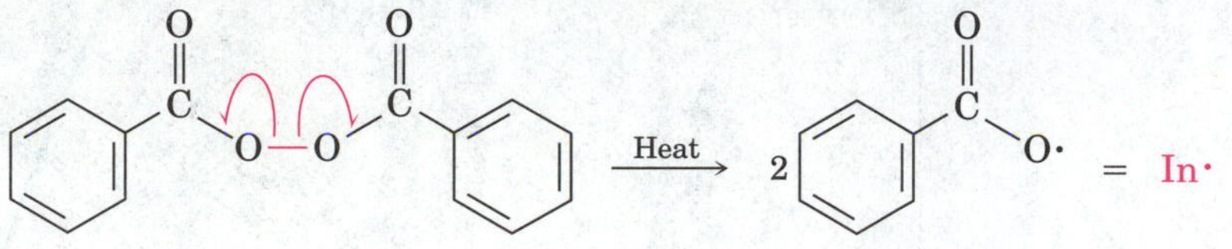

Benzoyl peroxide **Benzoyloxy radical**

2. A benzoyloxy radical produced in step 1 then adds to ethylene to generate a new alkyl radical:

$$In\cdot + H_2C{=}CH_2 \longrightarrow In{-}CH_2CH_2\cdot$$

3. *Propagation* of the polymerization process occurs when the alkyl radical produced in step 2 adds to another ethylene molecule. Repetition of this radical addition step for hundreds or thousands of times builds the polymer chain.

$$In{-}CH_2CH_2\cdot + H_2C{=}CH_2 \longrightarrow In{-}CH_2CH_2CH_2CH_2\cdot$$

$$\xrightarrow[\text{many times}]{\text{Repeat}} In{-}(CH_2CH_2)_nCH_2CH_2\cdot$$

4. *Termination* of the chain eventually occurs by a reaction that consumes the radical. Combination is one possible chain-terminating reaction:

$$2\,R\cdot \longrightarrow R{-}R$$

Many substituted ethylenes also undergo radical chain polymerization, yielding polymers with substituent groups regularly spaced along the polymer chain. Propylene, for example, yields polypropylene when polymerized, and styrene yields polystyrene.

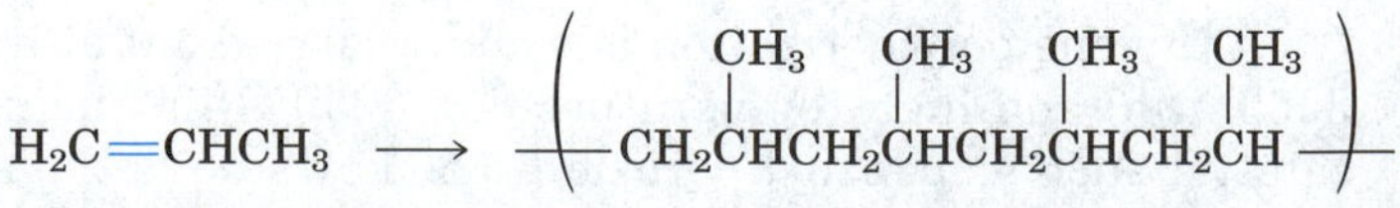

Propylene **Polypropylene**

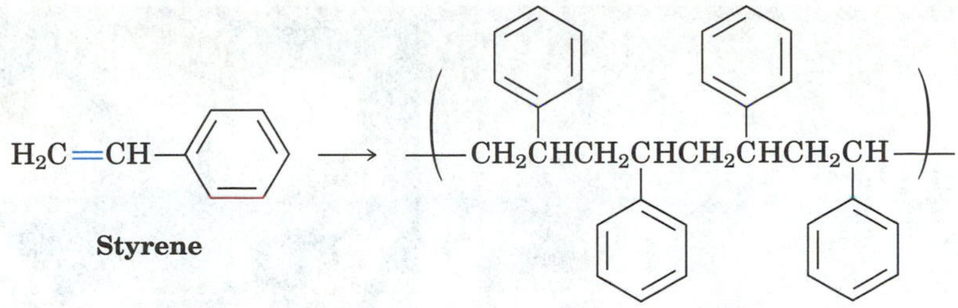

Styrene **Polystyrene**

Table 7.1 shows some of the more important monomers and lists the industrial uses of the polymers that result.

Table 7.1 Some Alkene Polymers and Their Uses

Monomer name	*Formula*	*Trade or common name of polymer*	*Uses*
Ethylene	$H_2C{=}CH_2$	Polyethylene	Packaging, bottles, cable insulation, films and sheets
Propene (propylene)	$H_2C{=}CHCH_3$	Polypropylene	Automotive moldings, rope, carpet fibers
Chloroethylene (vinyl chloride)	$H_2C{=}CHCl$	Poly(vinyl chloride), Tedlar	Insulation, films, pipes
Styrene	$H_2C{=}CHC_6H_5$	Polystyrene, Styron	Foam and molded articles
Tetrafluoroethylene	$F_2C{=}CF_2$	Teflon	Valves and gaskets, coatings
Acrylonitrile	$H_2C{=}CHCN$	Orlon, Acrilan	Fibers
Methyl methacrylate	$H_2C{=}C(CH_3)CO_2CH_3$	Plexiglas, Lucite	Molded articles, paints
Vinyl acetate	$H_2C{=}CHOCOCH_3$	Poly(vinyl acetate)	Paints, adhesives

PROBLEM

7.19 Show the monomer units you would use to prepare the following polymers.

(a) 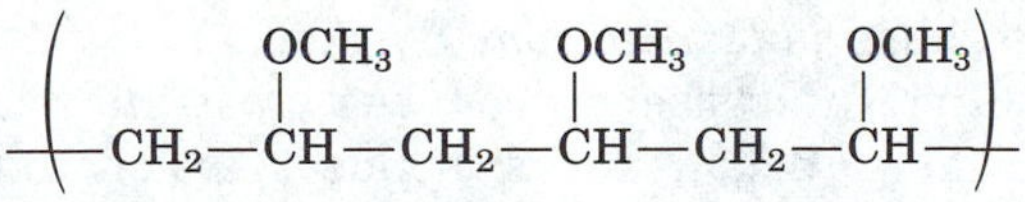

(b) 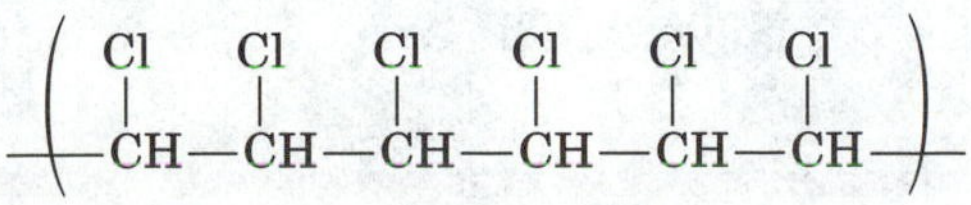

INTERLUDE

Natural Rubber

Toys, balloons, and a great many other items are made of rubber.

Rubber—an unusual name for a most unusual substance—is a naturally occurring alkene polymer produced by more than 400 different plants. The major source, however, is the so-called rubber tree, *Hevea brasiliensis*, from which the crude material is harvested as it drips from a slice made through the bark. The name *rubber* was coined by Joseph Priestley, the discoverer of oxygen and early researcher of rubber chemistry, for the simple reason that one of rubber's early uses was to rub out pencil marks on paper.

Unlike polyethylene and other simple alkene polymers, natural rubber is a polymer of a *diene*, isoprene (2-methyl-1,3-butadiene). The polymerization takes place by addition of each isoprene monomer unit to the growing chain, leading to formation of a polymer that still contains double bonds spaced regularly at four-carbon intervals. As the following structure shows, these double bonds have *Z* stereochemistry:

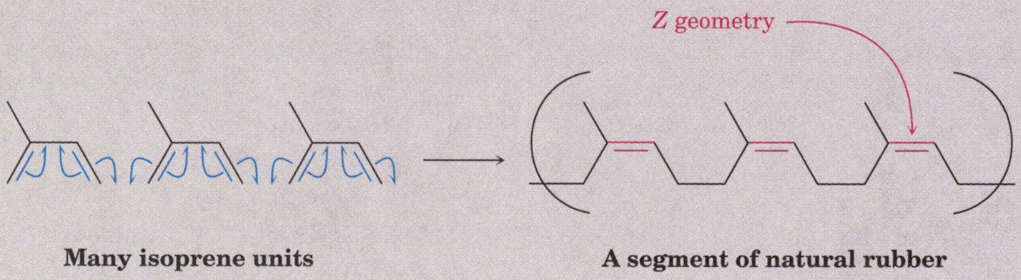

Crude rubber (latex) is collected from the tree as an aqueous dispersion that is washed, dried, and coagulated by warming in air to give a polymer with chains that average about 5000 monomer units in length and have molecular weights of 200,000–500,000. This crude coagulate is too soft and tacky to be useful until it is hardened by heating with elemental sulfur, a process called *vulcanization*. By mechanisms that are still not fully understood, vulcanization introduces cross-links between the rubber chains by forming carbon–sulfur bonds between them, thereby hardening and stiffening the polymer. The exact degree of hardening can

(continued)➤

be varied, yielding material soft enough for automobile tires or hard enough for bowling balls (*ebonite*).

The remarkable ability of rubber to stretch and then contract to its original shape is due to the irregular shapes of the polymer chains caused by the double bonds. These double bonds introduce bends and kinks into the polymer chains, thereby preventing neighboring chains from nestling together. When stretched, the randomly coiled chains straighten out and orient along the direction of the pull but are kept from sliding over each other by the cross-links. When the stretch is released, the polymer reverts to its original random state.

Summary and Key Words

Methods for the preparation of alkenes generally involve **elimination reactions**, such as **dehydrohalogenation**, the elimination of HX from an alkyl halide, and **dehydration**, the elimination of water from an alcohol.

HCl, HBr, and HI add to alkenes by a two-step electrophilic addition mechanism. Initial reaction of the nucleophilic double bond with H^+ gives a carbocation intermediate, which then reacts with halide ion. Bromine and chlorine add to alkenes via three-membered-ring **halonium ion** intermediates to give addition products having **anti stereochemistry**. If water is present during halogen addition reactions, a **halohydrin** is formed.

Hydration of alkenes (addition of water) is carried out by either of two procedures, depending on the product desired. **Oxymercuration** involves electrophilic addition of Hg^{2+} to an alkene, followed by trapping of the cation intermediate with water and subsequent treatment with $NaBH_4$. **Hydroboration** involves addition of borane (BH_3) followed by oxidation of the intermediate organoborane with alkaline H_2O_2. The two hydration methods are complementary: Oxymercuration gives the product of Markovnikov addition, whereas hydroboration/oxidation gives the product of non-Markovnikov syn addition.

Carbenes, R_2C:, are neutral molecules containing a divalent carbon with only six valence electrons. Carbenes are highly reactive toward alkenes, adding to give cyclopropanes. Dichlorocarbene, usually prepared from $CHCl_3$ by reaction with base, adds to alkenes to give 1,1-dichlorocyclopropanes. Nonhalogenated cyclopropanes are best prepared by treatment of the alkene with CH_2I_2 and zinc–copper alloy—the **Simmons–Smith reaction**.

Alkenes are reduced by addition of H_2 in the presence of a catalyst such as platinum or palladium to yield alkanes, a process called **catalytic hydrogenation**. Cis-1,2-diols can be made directly from alkenes by **hydroxylation** with OsO_4. Alkenes can also be cleaved to produce carbonyl compounds by reaction with ozone, followed by reduction with zinc metal.

HBr (but not HCl or HI) [illegible] alkenes by a radical chain pathway to give the non-Mark[illegible]. Radicals have the stability order:

Tertiary > Seco[illegible] > Methyl

$R_3C\cdot$ > R_2[illegible] > $\cdot CH_3$

Polymers—large molecule[illegible] repetitive bonding together of many hundreds or thousands [illegible]**mer** units—are formed by reaction of simple alkenes with a radical initiator at high temperature and pressure. Polyethylene, polypropylene, and polystyrene are common examples.

Learning Reactions

What's seven times nine? Sixty-three, of course. You didn't have to stop and figure it out; you knew the answer immediately because you long ago learned the multiplication tables. Learning the reactions of organic chemistry requires the same approach: Reactions have to be learned for immediate recall if they are to be useful.

Different people take different approaches to learning reactions. Some people make flashcards; others find studying with friends to be helpful. To help guide your study, most chapters in this book end with a summary of the reactions just presented. In addition, the accompanying *Study Guide and Solutions Manual* has several appendixes that organize organic reactions from other viewpoints. Fundamentally, though, there are no shortcuts. Learning organic chemistry takes effort.

Summary of Reactions

Note: No stereochemistry is implied unless specifically indicated with wedged, solid, and dashed lines.

1. Synthesis of alkenes
   (a) Dehydrohalogenation of alkyl halides (Section 7.1)

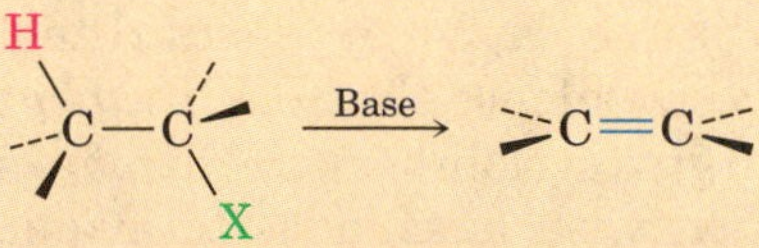

   (b) Dehydration of alcohols (Section 7.1)

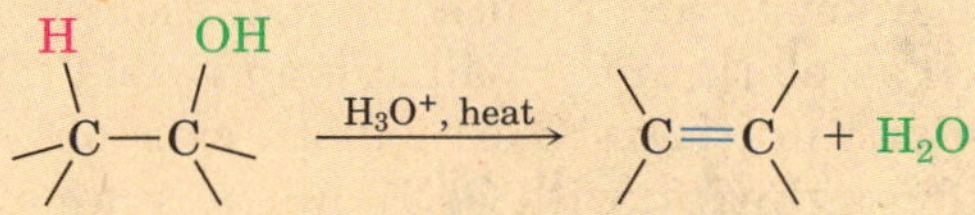

(continued)

2. Addition reactions of alkenes
   (a) Addition of HX, where X = Cl, Br, or I (Sections 6.8 and 6.9)

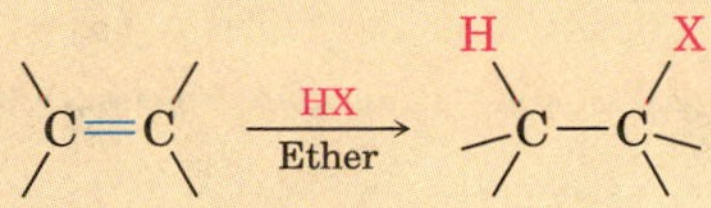

Markovnikov regiochemistry is observed: H adds to the less substituted carbon, and X adds to the more substituted carbon.

   (b) Addition of halogens, where X_2 = Cl_2 or Br_2 (Section 7.2)

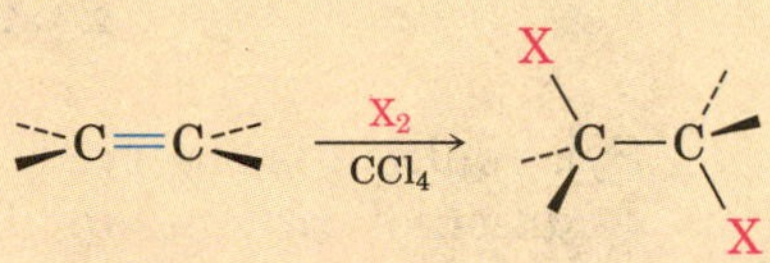

Anti addition is observed.

   (c) Halohydrin formation (Section 7.3)

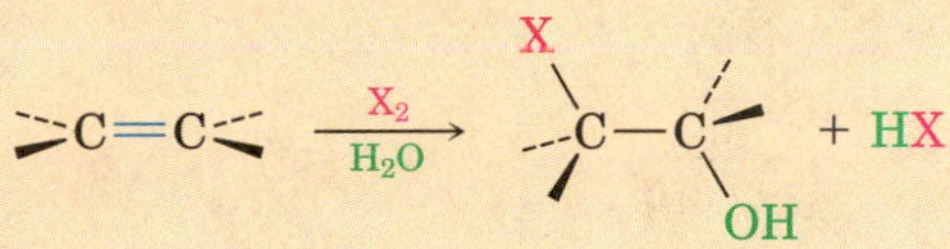

Markovnikov regiochemistry and anti stereochemistry are observed.

   (d) Addition of water by oxymercuration (Section 7.4)

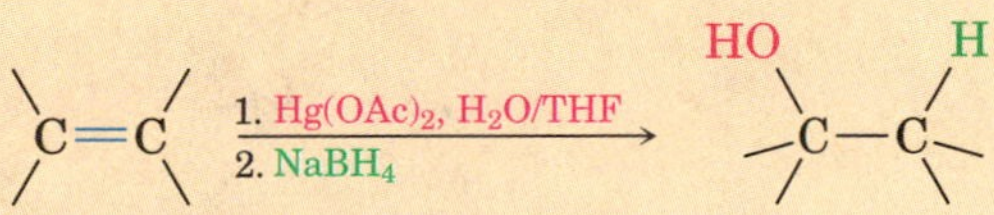

Markovnikov regiochemistry is observed, with the OH attaching to the more substituted carbon.

   (e) Addition of water by hydroboration/oxidation (Section 7.5)

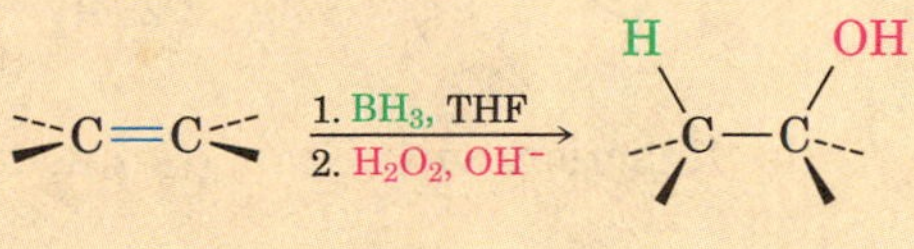

Non-Markovnikov syn addition is observed.

   (f) Radical addition of HBr to alkenes (Section 7.10)

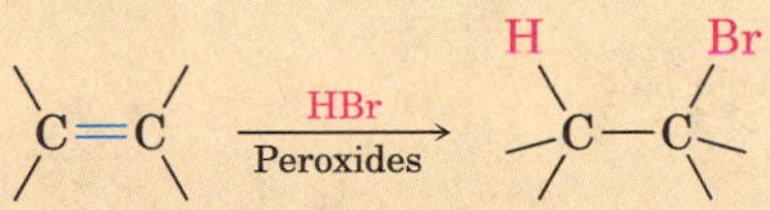

Non-Markovnikov addition is observed.

(continued)➤

(g) Hydrogenation of alkenes (Section 7.7)

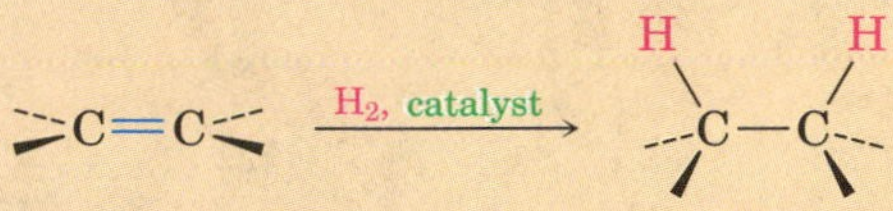

Syn addition is observed.

(h) Hydroxylation of alkenes (Section 7.8)

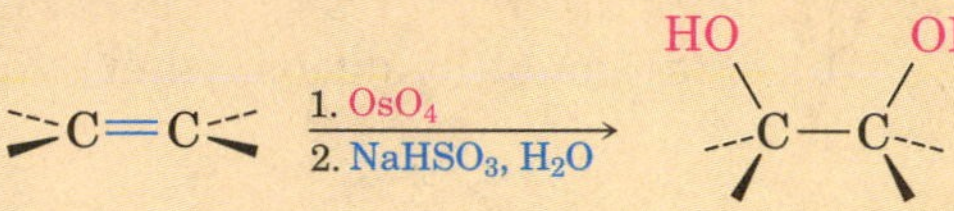

Syn addition is observed.

(i) Addition of carbenes to alkenes to yield cyclopropanes (Section 7.6)

(1) Dichlorocarbene addition

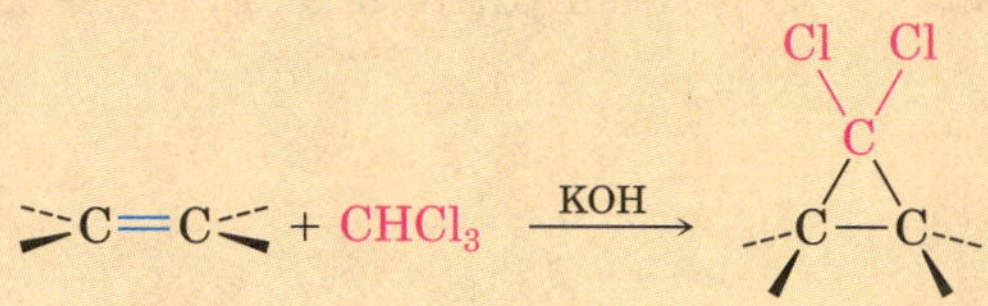

(2) Simmons–Smith reaction

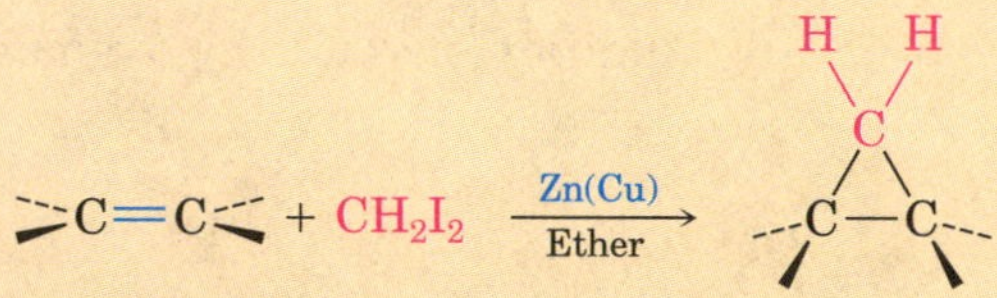

3. Oxidative cleavage of alkenes (Section 7.8)

(a) Treatment with ozone, followed by zinc in acetic acid

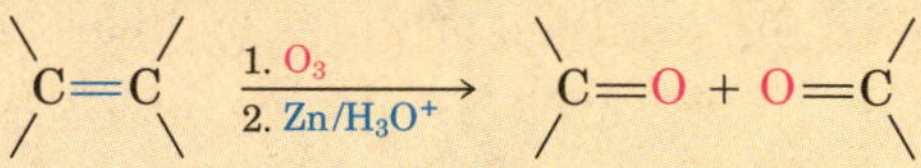

(b) Reaction with $KMnO_4$ in acidic solution

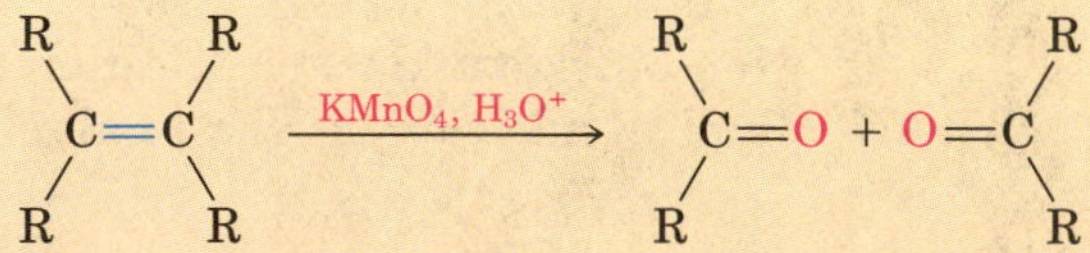

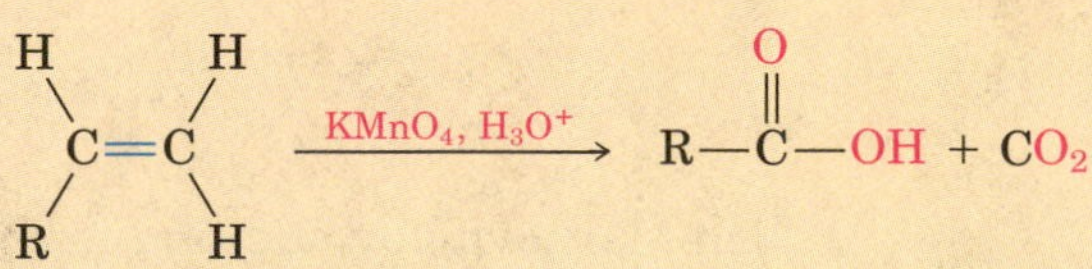

(continued)➤

4. Oxidative cleavage of 1,2-diols (Section 7.8)

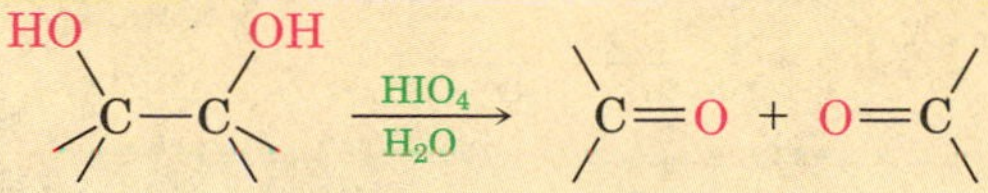

ADDITIONAL PROBLEMS

7.20 Iodine azide, IN_3, adds to alkenes by an electrophilic mechanism similar to that of bromine. If a monosubstituted alkene is used, only one product results:

$$CH_3CH_2CH{=}CH_2 + I{-}N_3 \longrightarrow CH_3CH_2CH(N_3)CH_2I$$

In light of this result, what is the polarity of the $I–N_3$ bond? Propose a mechanism for the reaction.

7.21 Predict the products of the following reactions (the aromatic ring is unreactive in all cases). Indicate regiochemistry when relevant.

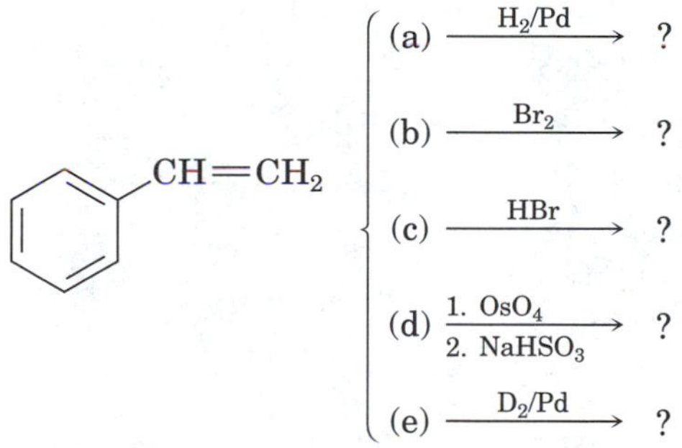

7.22 Suggest structures for alkenes that give the following reaction products. There may be more than one answer for some cases.

(a) ? $\xrightarrow{H_2/Pd}$ 2-Methylhexane

(b) ? $\xrightarrow{H_2/Pd}$ 1,1-Dimethylcyclohexane

(c) ? $\xrightarrow{Br_2/CCl_4}$ 2,3-Dibromo-5-methylhexane

(d) ? $\xrightarrow[2.\ NaBH_4]{1.\ Hg(OAc)_2,\ H_2O}$ $CH_3CH_2CH_2CH(OH)CH_3$

(e) ? $\xrightarrow{HBr/peroxides}$ 2-Bromo-3-methylheptane

(f) ? $\xrightarrow{HCl,\ ether}$ 2-Chloro-3-methylheptane

7.23 Predict the products of the following reactions, indicating both regiochemistry and stereochemistry where appropriate.

(a)
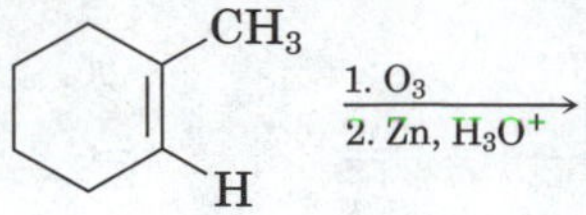

(b)
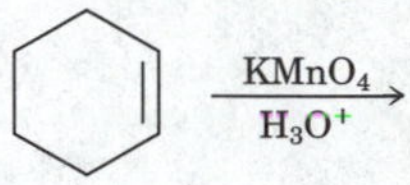

(c)
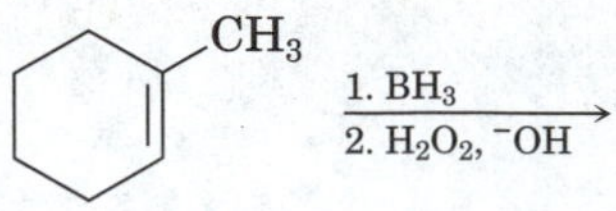

(d)
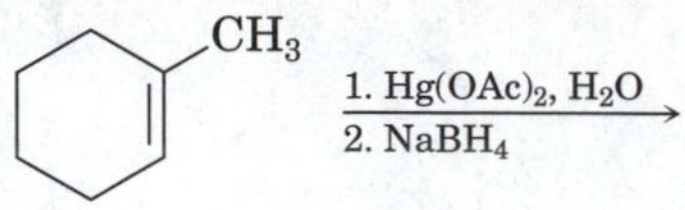

7.24 How would you carry out the following transformations? Indicate the proper reagents.

(a)
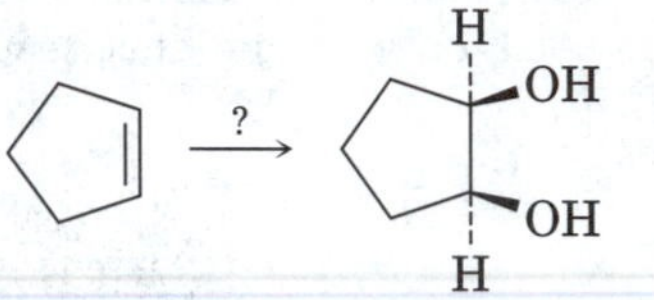

(b)
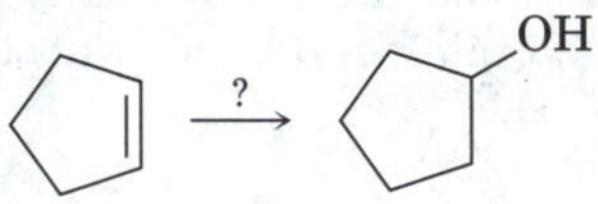

(c)
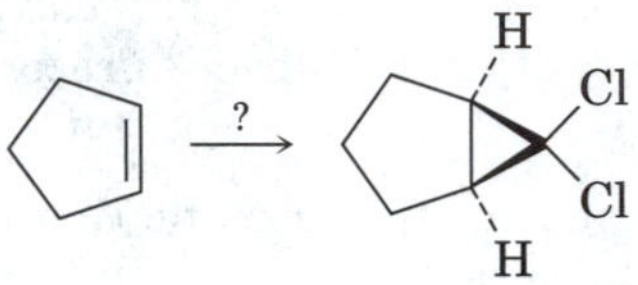

(d) $CH_3CH{=}CHCH(CH_3)CH_3 \xrightarrow{?} CH_3CHO + (CH_3)_2CHCHO$

(e) $CH_3C(CH_3){=}CH_2 \xrightarrow{?} CH_3CH(CH_3)CH_2OH$

(f)
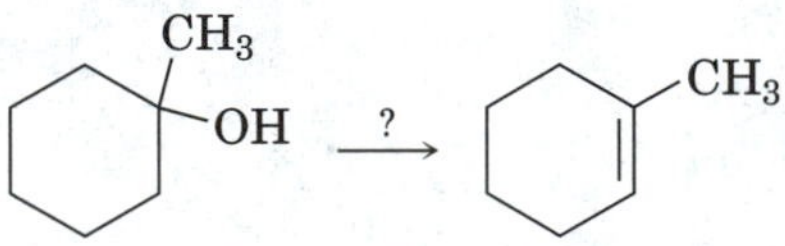

7.25 Draw the structure of an alkene that yields only $(CH_3)_2C{=}O$ on ozonolysis followed by treatment with zinc.

7.26 Draw the structure of a hydrocarbon that reacts with 1 mol equiv of hydrogen on catalytic hydrogenation and gives only pentanal, $CH_3CH_2CH_2CH_2CHO$, on ozonolysis. Write the reactions involved.

7.27 Show the structures of alkenes that give the following products on oxidative cleavage with $KMnO_4$ in acidic solution:

(a) $CH_3CH_2COOH + CO_2$

(b) $(CH_3)_2C{=}O + CH_3CH_2CH_2COOH$

(c) 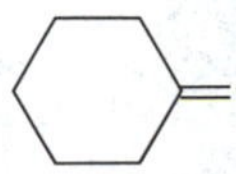 $=O + (CH_3)_2C{=}O$

7.28 Compound A has the formula $C_{10}H_{16}$. On catalytic hydrogenation over palladium, it reacts with only 1 equiv of hydrogen. Compound A undergoes reaction with ozone, followed by zinc treatment, to yield compound B, a symmetrical diketone ($C_{10}H_{16}O_2$).

(a) How many rings does A have?

(b) What are the structures of A and B?

(c) Write the reactions.

7.29 An unknown hydrocarbon, A, with formula C_6H_{12}, reacts with 1 equiv of hydrogen over a palladium catalyst. Hydrocarbon A also reacts with OsO_4 to give a diol, B. When oxidized with $KMnO_4$ in acidic solution, A gives two fragments. One fragment can be identified as propanoic acid, CH_3CH_2COOH, and the other fragment can be shown to be a ketone, C. What are the structures of A, B, and C? Write all reactions, and show your reasoning.

7.30 Using an oxidative cleavage reaction, explain how you would distinguish between the following two isomeric dienes:

 and 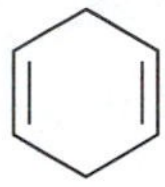

7.31 Compound A, $C_{10}H_{18}O$, undergoes reaction with dilute H_2SO_4 at 250°C to yield a mixture of two alkenes, $C_{10}H_{16}$. The major alkene product, B, gives only cyclopentanone after ozone treatment followed by reduction with zinc in acetic acid. Formulate the reactions involved, and identify A and B.

7.32 Which reaction would you expect to be faster, addition of HBr to cyclohexene or to 1-methylcyclohexene? Explain.

7.33 Predict the products of the following reactions, and indicate regiochemistry if relevant.

(a) $CII_3CII{-}CIICII_3 \xrightarrow{HBr}$

(b) $CH_3CH{=}CHCH_3 \xrightarrow{BH_3} A \xrightarrow[^{-}OH]{H_2O_2} B$

(c) $(CH_3)_2C{=}CH_2 \xrightarrow[\text{Peroxide}]{HBr}$

(d) $CH_3CH{=}C(CH_3)_2 \xrightarrow[\text{Peroxide}]{HI}$

7.34 Draw the structure of a hydrocarbon that absorbs 2 mol equiv of hydrogen on catalytic hydrogenation and gives only butanedial on ozonolysis.

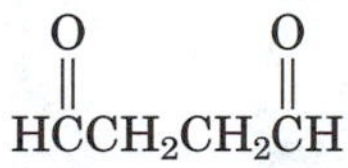 **Butanedial**

7.35 In planning the synthesis of one compound from another, it's just as important to know what *not* to do as to know what to do. The following proposed reactions all have serious drawbacks to them. Explain the potential problems of each.

(a) 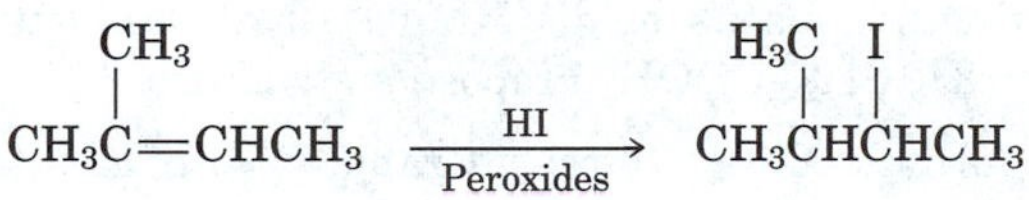

(b)

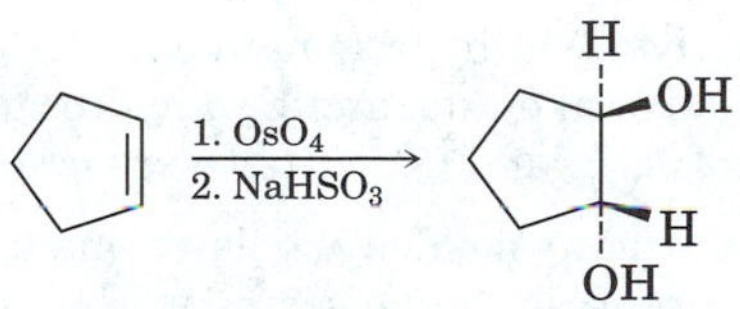

(c)

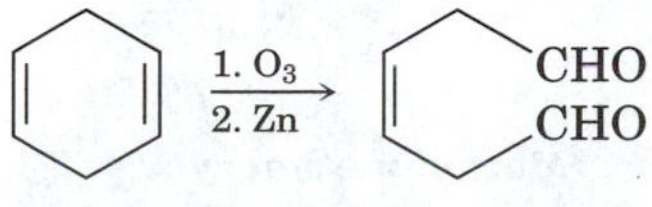

(d)

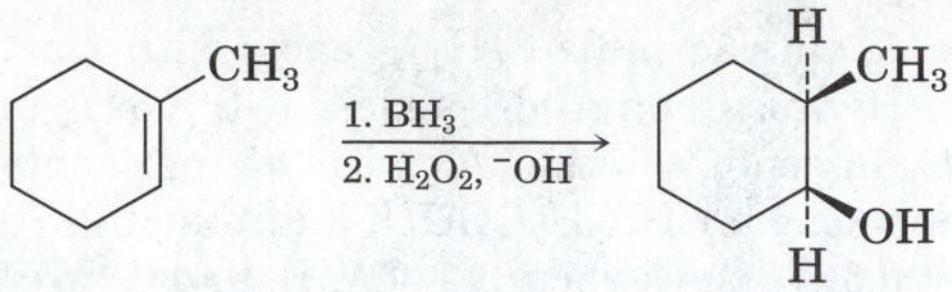

7.36 Which of the following alcohols could *not* be made selectively by hydroboration/oxidation of an alkene?

(a) $CH_3CH_2CH_2CH(OH)CH_3$

(b) $(CH_3)_2CHC(OH)(CH_3)_2$

(c)

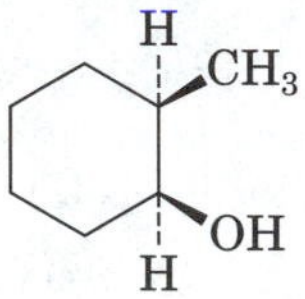

(d)

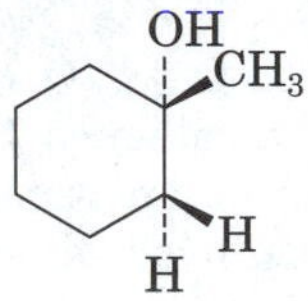

7.37 What alkenes might be used to prepare the following cyclopropanes?

(a)

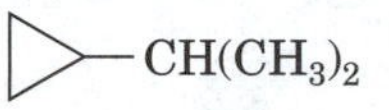

(b)

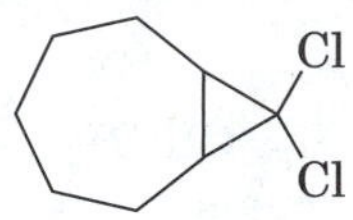

7.38 Predict the products of the following reactions. Don't worry about the size of the molecule; concentrate on the functional groups.

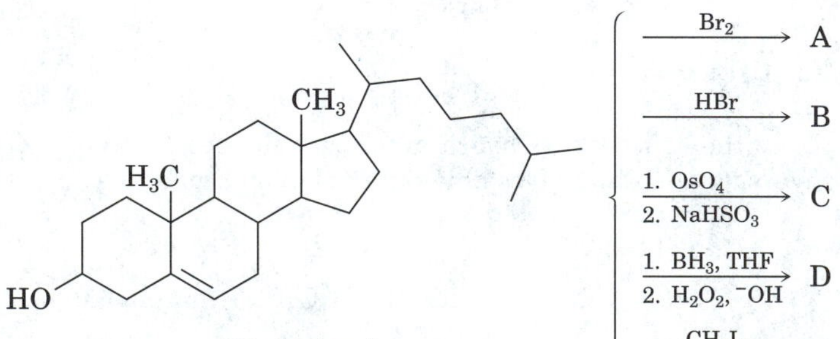

Cholesterol

7.39 The sex attractant of the common housefly is a hydrocarbon with the formula $C_{23}H_{46}$. On treatment with aqueous acidic $KMnO_4$, two products are obtained, $CH_3(CH_2)_{12}COOH$ and $CH_3(CH_2)_7COOH$. Propose a structure.

7.40 Compound A has the formula C_8H_8. It reacts rapidly with $KMnO_4$ to give CO_2 and B, a carboxylic acid ($C_7H_6O_2$), but reacts with only 1 equiv of H_2 on catalytic hydrogenation over a palladium catalyst. On hydrogenation under conditions that reduce aromatic rings, 4 equiv of H_2 are taken up, and C, a hydrocarbon (C_8H_{16}), is produced. What are the structures of A, B, and C? Write the reactions.

7.41 Plexiglas™, a clear plastic used to make many molded articles, is made by polymerization of methyl methacrylate. Draw a representative segment of Plexiglas.

$H_2C{=}C(CH_3){-}C(=O)OCH_3$ **Methyl methacrylate**

7.42 Draw representative segments of polymers made from the following monomers.
(a) Teflon, from $F_2C{=}CF_2$ (b) Poly(vinyl chloride), from $H_2C{=}CHCl$

7.43 Reaction of 2-methylpropene with methyl alcohol in the presence of sulfuric acid catalyst yields methyl *tert*-butyl ether, $CH_3OC(CH_3)_3$, by a mechanism analogous to that of acid-catalyzed alkene hydration. Write the mechanism in detail.

7.44 When 4-penten-1-ol is treated with aqueous bromine, a cyclic bromo ether is formed, rather than the expected bromohydrin. Propose a mechanism.

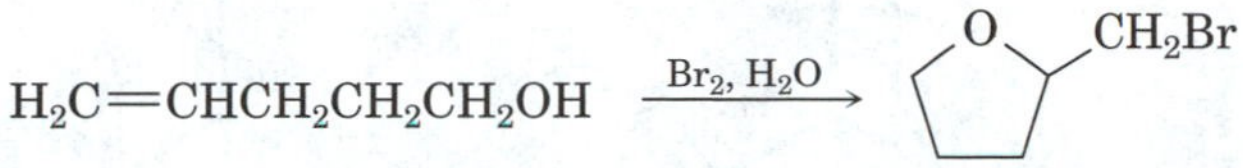

4-Penten-1-ol **2-(Bromomethyl)tetrahydrofuran**

7.45 How would you distinguish between the following pairs of compounds using simple chemical tests? Tell what you would do and what you would see.
(a) Cyclopentene and cyclopentane (b) 2-Hexene and benzene

7.46 Ethylidenecyclohexane, on treatment with a strong acid, isomerizes to yield 1-ethylcyclohexene. Propose a mechanism. Which alkene is more stable?

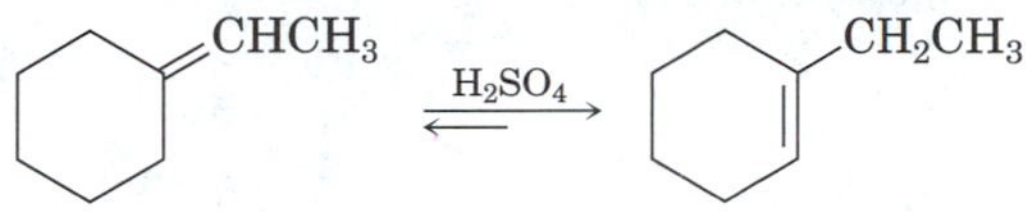

Ethylidenecyclohexane **1-Ethylcyclohexene**

7.47 Dichlorocarbene can be generated by heating sodium trichloroacetate:

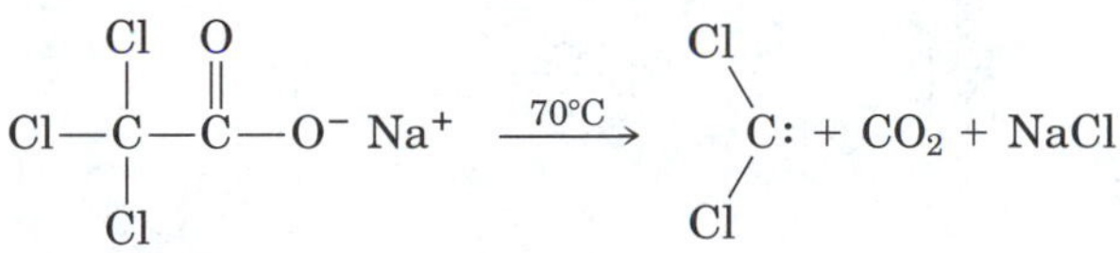

Propose a mechanism for the reaction, and use curved arrows to indicate the movement of electrons. What relation does your mechanism bear to the base-induced elimination of HCl from chloroform?

7.48 α-Terpinene, $C_{10}H_{16}$, is a pleasant-smelling hydrocarbon that has been isolated from oil of marjoram. On hydrogenation over a palladium catalyst, α-terpinene reacts with 2 mol equiv of hydrogen to yield a hydrocarbon, $C_{10}H_{20}$. On ozonolysis, followed by reduction with zinc and acetic acid, α-terpinene yields two products, glyoxal and 6-methyl-2,5-heptanedione.

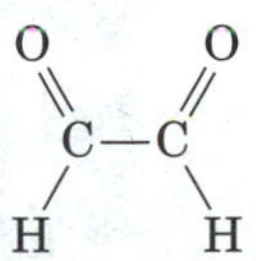

Glyoxal

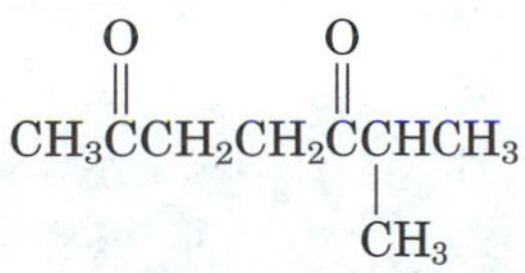

6-Methyl-2,5-heptanedione

(a) How many degrees of unsaturation does α-terpinene have?
(b) How many double bonds, and how many rings?
(c) Propose a structure for α-terpinene.

7.49 Evidence that cleavage of 1,2-diols by HIO_4 occurs through a five-membered cyclic periodate intermediate is based on *kinetic data*—the measurement of reaction rates. When diols A and B were prepared and the rates of their reaction with HIO_4 were measured, it was found that diol A cleaved approximately 1 million times faster than diol B. Make molecular models of A and B and of potential cyclic periodate intermediates, and then explain the kinetic results.

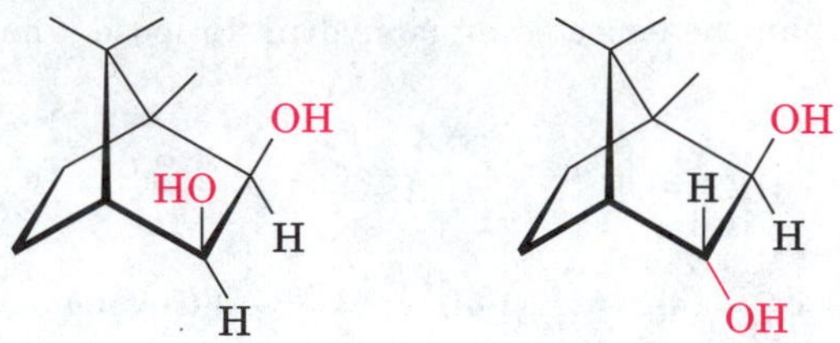

A (*cis* diol) B (*trans* diol)

7.50 Reaction of HBr with 3-bromocyclohexene yields *trans*-1,2-dibromocyclohexane as the sole product, but reaction of 3-methylcyclohexene with HBr yields a mixture of four products: *cis*- and *trans*-1-bromo-3-methylcyclohexane and *cis*- and *trans*-1-bromo-2-methylcyclohexane.

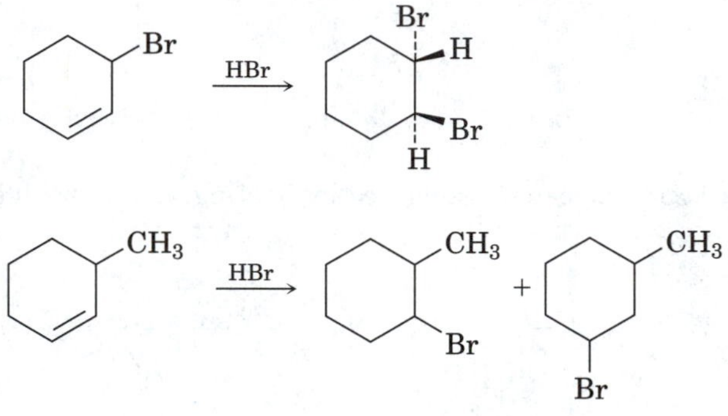

Draw structures of the possible intermediates, and then explain why only a single product is formed in the reaction of HBr with 3-bromocyclohexene.

7.51 The following reaction takes place in high yield:

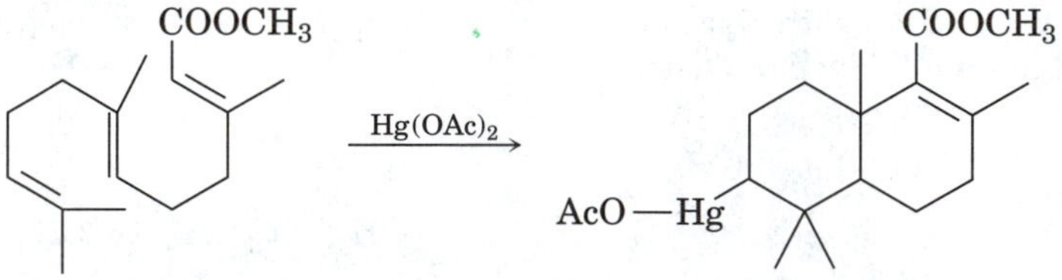

Even though you've never seen this reaction before, use your general knowledge of alkene chemistry to propose a mechanism.

7.52 Hydroboration of 2-methyl-2-pentene at 25°C followed by oxidation with alkaline H_2O_2 yields 2-methyl-3-pentanol, but hydroboration at 160°C followed by oxidation yields 4-methyl-1-pentanol. Explain.

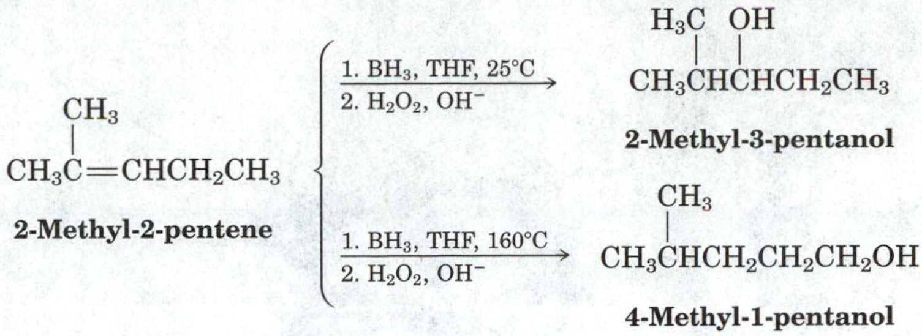

A Look Ahead

7.53 We'll see in the next chapter that alkynes undergo many of the same reactions that alkenes do. What product would you expect from each of the following reactions?

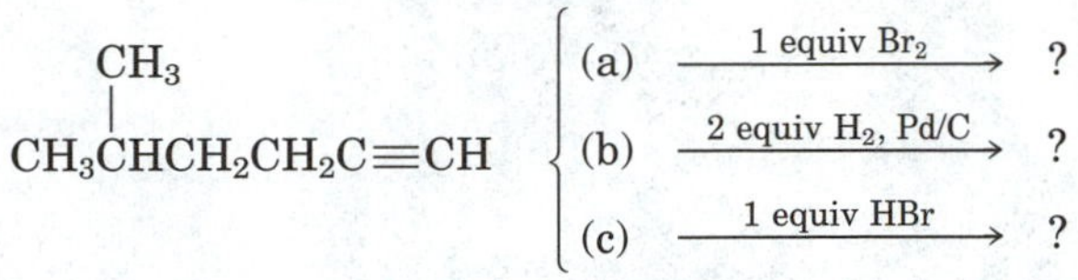

7.54 Explain the observation that hydroxylation of *cis*-2-butene with OsO_4 yields a different product than hydroxylation of *trans*-2-butene. First draw the structure and show the stereochemistry of each product, and then make molecular models. (We'll explore the stereochemistry of the products in more detail in Chapter 9.)

7.55 Reaction of cyclohexene with mercuric acetate in methyl alcohol rather than water, followed by treatment with $NaBH_4$, yields cyclohexyl methyl ether rather than cyclohexanol. Suggest a mechanism. (We'll look at this reaction again in Chapter 18.)

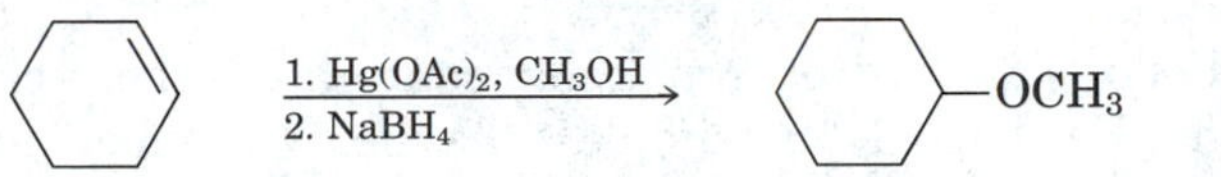

Cyclohexene **Cyclohexyl methyl ether**

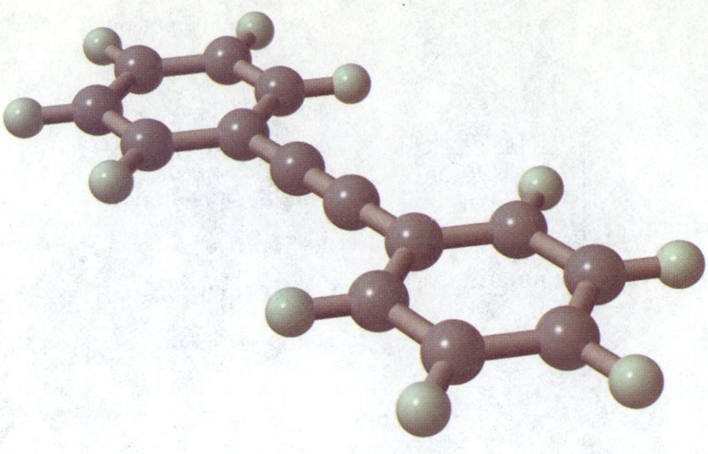

Diphenylacetylene—a typical alkyne

8

Alkynes

Alkynes are hydrocarbons that contain a carbon–carbon triple bond. Acetylene, H–C≡C–H, the simplest alkyne, was once widely used in industry as the starting material for the preparation of acetaldehyde, acetic acid, vinyl chloride, and other high-volume chemicals, but more efficient routes using ethylene as starting material are now more common. Acetylene is still used in the preparation of acrylic polymers, however, and is prepared industrially by high-temperature decomposition (*pyrolysis*) of methane. This method is not of general utility in the laboratory.

$$2\ CH_4 \xrightarrow[1200°C]{\text{Steam}} HC{\equiv}CH + 3\ H_2$$

Methane **Acetylene**

A large number of naturally occurring alkynes have been isolated from the plant kingdom. For example, the following *triyne* from the safflower, *Carthamus tinctorius* L., evidently forms part of the plant's chemical defenses against nematode infestation:

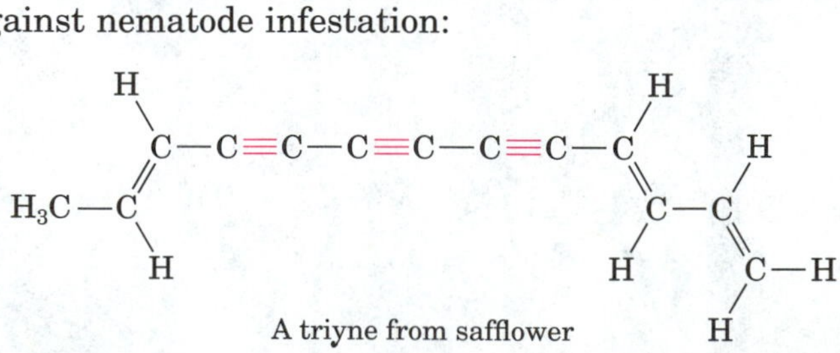

A triyne from safflower

8.1 Electronic Structure of Alkynes

A carbon–carbon triple bond results from the interaction of two *sp*-hybridized carbon atoms (Section 1.11). Recall that the two *sp* hybrid orbitals of carbon lie at an angle of 180° to each other along an axis perpendicular to the axes of the two unhybridized $2p_y$ and $2p_z$ orbitals. When two *sp*-hybridized carbons approach each other for bonding, the geometry is perfect for the formation of one *sp–sp* σ bond and two *p–p* π bonds—a net *triple* bond (Figure 8.1).

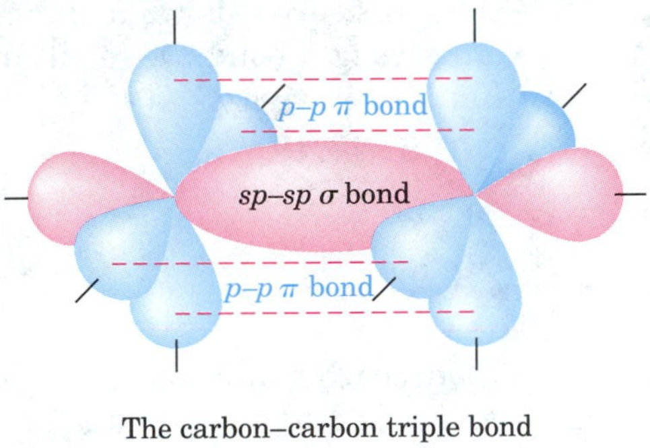

Figure 8.1 Formation of a carbon–carbon triple bond by interaction of two *sp*-hybridized carbons.

The two remaining *sp* orbitals form bonds to other atoms at an angle of 180° from the carbon–carbon bond. Thus, acetylene, C_2H_2, is a linear molecule with H-C-C bond angles of 180° (Figure 8.2).

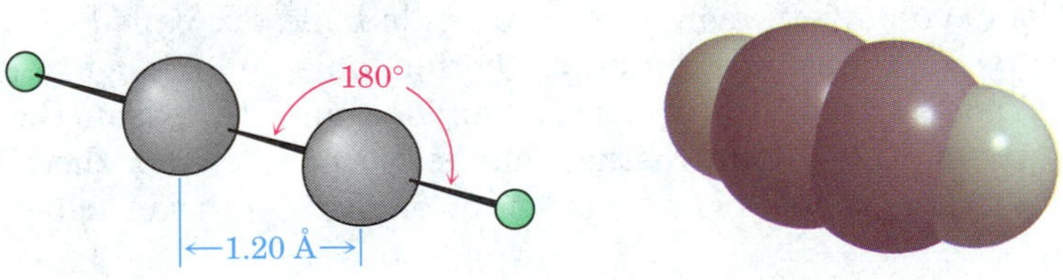

Figure 8.2 The structure of acetylene, H–C≡C–H. The H-C-C bond angles are 180°.

The length of the carbon–carbon triple bond in acetylene is 1.20 Å, and its strength is approximately 835 kJ/mol (200 kcal/mol), making it the shortest and strongest known carbon–carbon bond. In a purely bookkeeping sense, we can assign strengths to each of the three triple-bond "parts." Since we know that the carbon–carbon *single* bond in ethane has a strength of

376 kJ/mol (90 kcal/mol) and the carbon–carbon *double* bond of ethylene has a strength of 611 kJ/mol (146 kcal/mol), we can "dissect" the overall carbon–carbon triple bond:

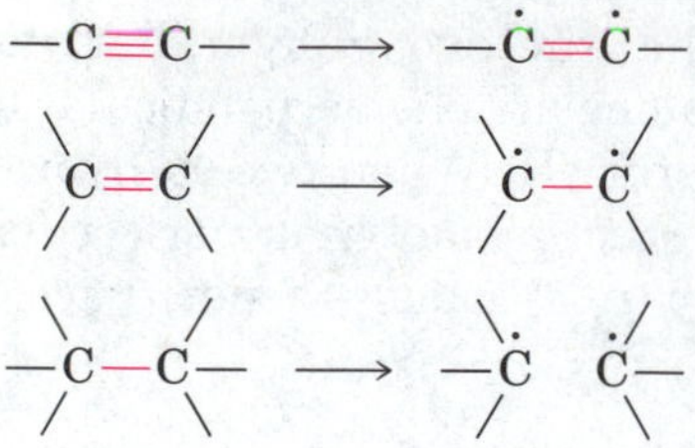

$D = (835 - 611)$ kJ/mol = 224 kJ/mol (54 kcal/mol)

$D = (611 - 376)$ kJ/mol = 235 kJ/mol (56 kcal/mol)

$D = 376$ kJ/mol (90 kcal/mol)

This crude calculation shows that approximately 224 kJ/mol (54 kcal/mol) is needed to break an alkyne π bond. Since this value is similar to the amount of energy needed to break an *alkene* π bond, we might predict that alkynes and alkenes should have similar reactivity.

8.2 Naming Alkynes

Alkynes follow the general rules of hydrocarbon nomenclature discussed in Sections 3.4 and 6.3. The suffix *-yne* is used to denote an alkyne, and the position of the triple bond is indicated by giving the number of the first alkyne carbon in the chain. Numbering the main chain begins at the end nearer the triple bond so that the triple bond receives as low a number as possible.

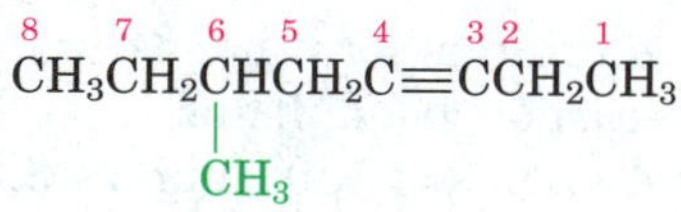

Begin numbering at the end nearer the triple bond.

6-Methyl-3-octyne

Compounds with more than one triple bond are called *diynes*, *triynes*, and so forth; compounds containing both double and triple bonds are called *enynes* (not *ynenes*). Numbering of an enyne chain starts from the end nearer the first multiple bond, whether double or triple. When there is a choice in numbering, double bonds receive lower numbers than triple bonds. For example:

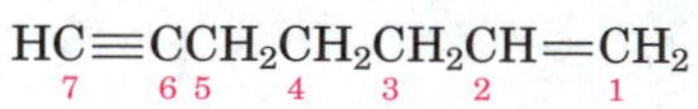

1-Hepten-6-yne

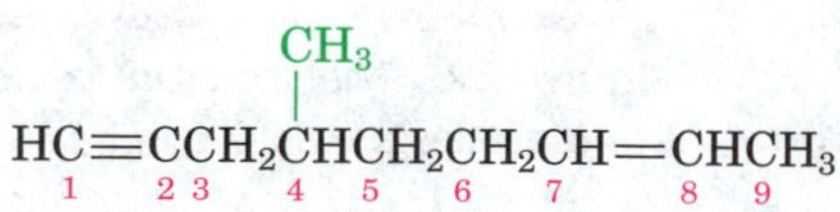

4-Methyl-7-nonen-1-yne

As with hydrocarbon substituents derived from alkanes and alkenes, *alkynyl* groups are also possible:

$CH_3CH_2CH_2CH_2$–	$CH_3CH_2CH{=}CH$–	$CH_3CH_2C{\equiv}C$–
Butyl (an alkyl group)	**1-Butenyl** (a vinylic group)	**1-Butynyl** (an alkynyl group)

PROBLEM

8.1 Give IUPAC names for the following compounds:

(a) 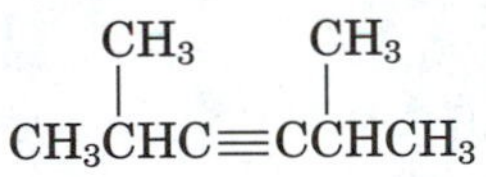

(b) 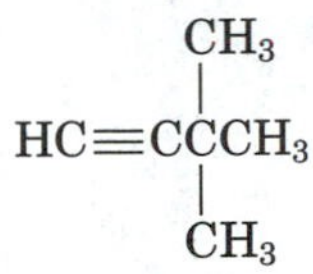

(c) $CH_3CH{=}CHCH{=}CHC{\equiv}CCH_3$

(d) 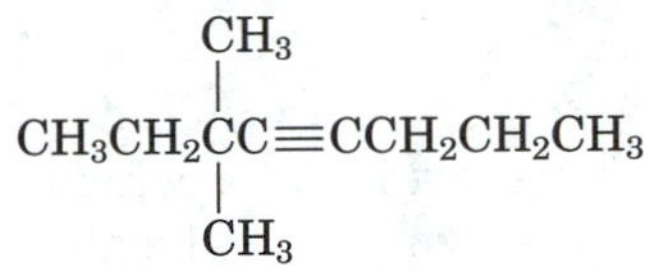

(e) 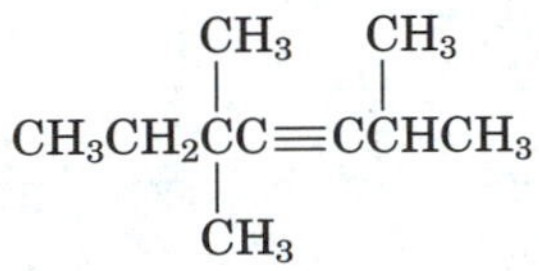

(f) 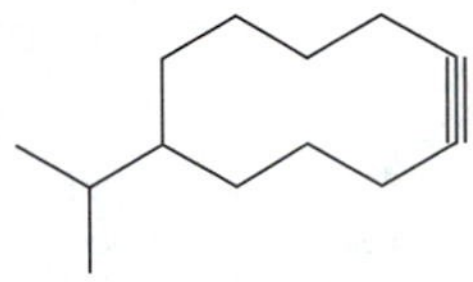

PROBLEM

8.2 There are seven isomeric alkynes with the formula C_6H_{10}. Draw them, and name them according to IUPAC rules.

8.3 Preparation of Alkynes: Elimination Reactions of Dihalides

Alkynes can be prepared by elimination of HX from alkyl halides in much the same manner as alkenes (Section 7.1). Treatment of a 1,2-dihalide (a **vicinal** dihalide) with excess strong base such as KOH or $NaNH_2$ results in a twofold elimination of HX and formation of an alkyne. As with the elimination of HX to form an alkene, we'll defer a discussion of the mechanism until Chapter 11.

The necessary vicinal dihalides are themselves readily available by addition of Br_2 or Cl_2 to alkenes. Thus, the overall halogenation/dehydrohalogenation sequence provides an excellent method for going from an alkene to an alkyne. For example, diphenylethylene is converted into diphenylacetylene by reaction with Br_2 and subsequent base treatment.

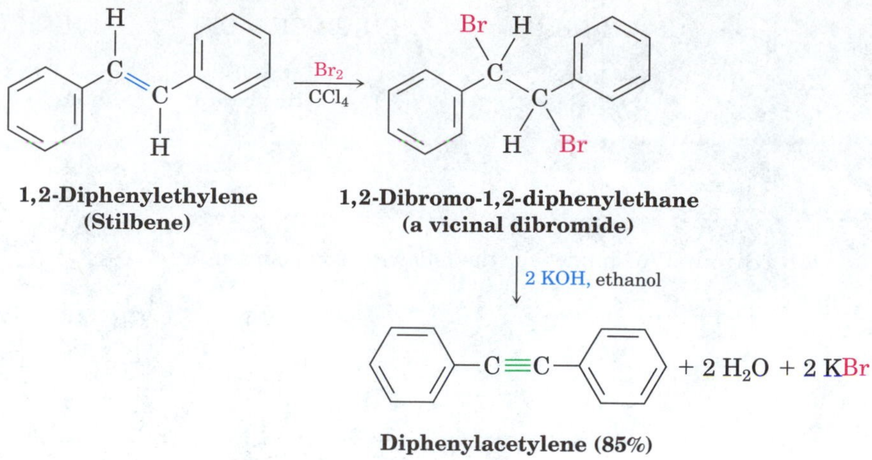

The twofold dehydrohalogenation takes place through a vinylic halide intermediate, which suggests that vinylic halides themselves should give alkynes when treated with strong base. This is indeed the case. For example,

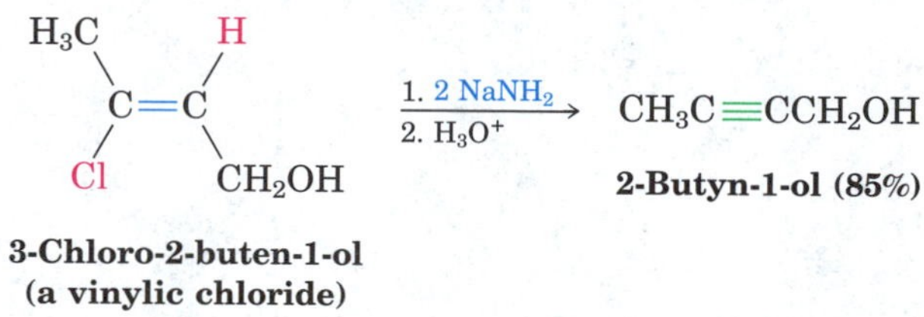

8.4 Reactions of Alkynes: Addition of HX and X_2

Based on the electronic similarity between alkenes and alkynes, you might expect that the chemical reactivity of the two functional groups should also be similar. The π part of the triple bond is relatively weak (224 kJ/mol), and the electrons are readily accessible to attacking reagents. Alkynes do indeed exhibit much chemistry similar to that of alkenes, but there are also significant differences.

As a general rule, electrophilic reagents add to alkynes in the same way that they add to alkenes. With HX, for example, alkynes give the expected addition products. Although the reactions usually can be stopped after addition of 1 equivalent of HX, an excess of acid leads to a dihalide product. For example, reaction of 1-hexyne with 2 equivalents of HBr yields 2,2-dibromohexane. As the following examples indicate, the regiochemistry of addition follows Markovnikov's rule. Halogen adds to the more highly substituted side of the alkyne bond, and hydrogen adds to the less highly substituted side. Trans stereochemistry of H and X is normally (though not always) found in the product.

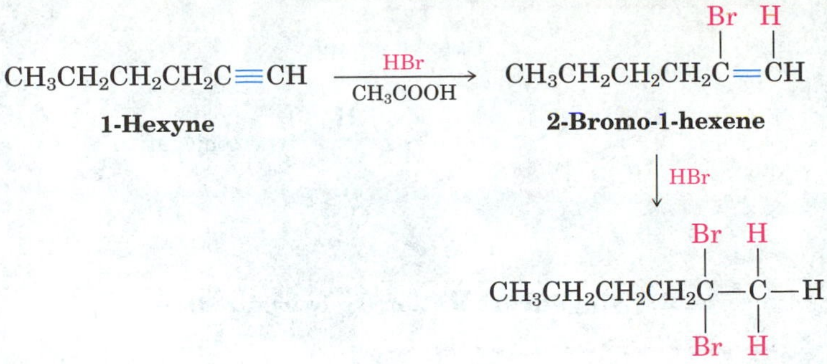

1-Hexyne → 2-Bromo-1-hexene → 2,2-Dibromohexane

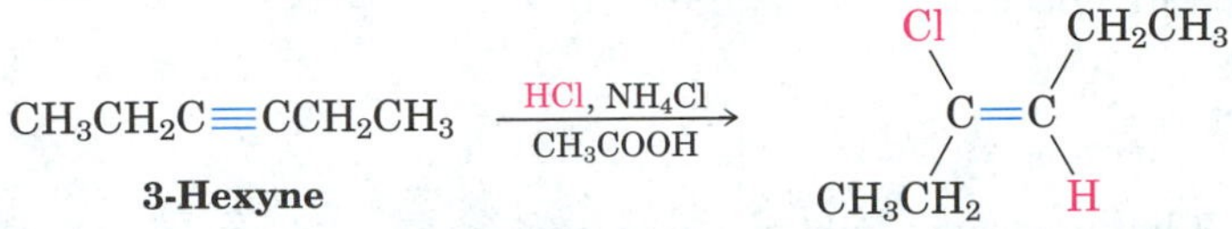

3-Hexyne → (*Z*)-3-Chloro-3-hexene (95%)

Bromine and chlorine also add to alkynes to give addition products, and trans stereochemistry again results:

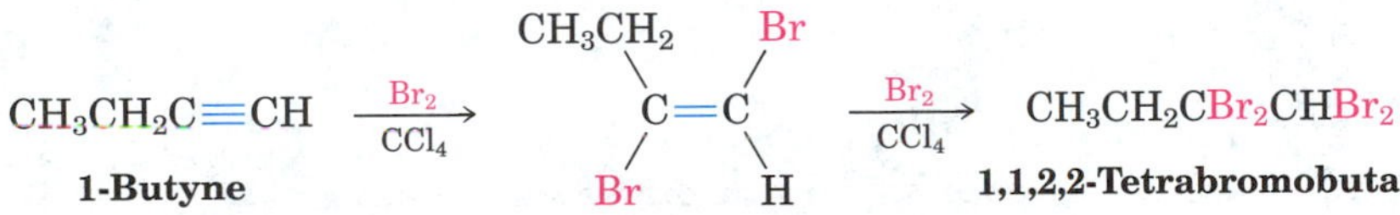

1-Butyne → (*E*)-1,2-Dibromo-1-butene → 1,1,2,2-Tetrabromobutane

The mechanism of electrophilic addition to an alkyne is similar to that of addition to an alkene. When an electrophile such as HBr adds to an *alkene* (Sections 6.8 and 6.9), the reaction takes place in two steps and involves an alkyl carbocation intermediate. When HBr adds to an *alkyne*, an analogous *vinylic carbocation* is formed as the intermediate. (*Remember: Vinylic* means "on a double bond.")

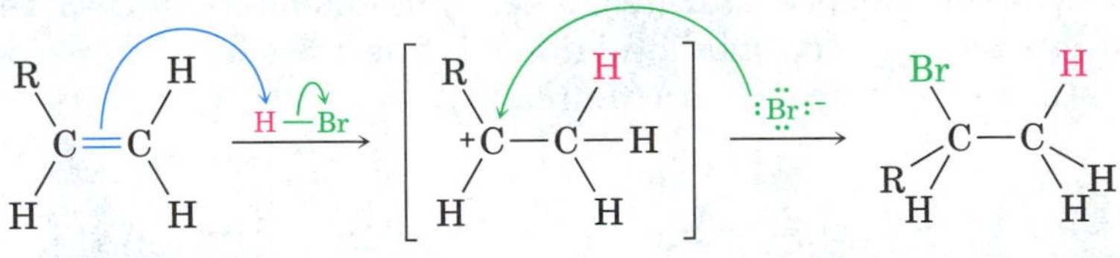

An alkene — An alkyl carbocation — An alkyl bromide

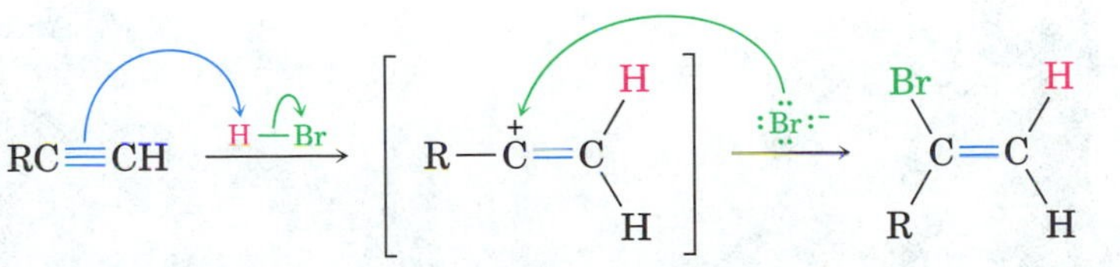

An alkyne — A vinylic carbocation — A vinylic bromide

The vinylic carbocation is *sp*-hybridized and is stabilized by hyperconjugation in the same way that an alkyl carbocation is (Figure 8.3).

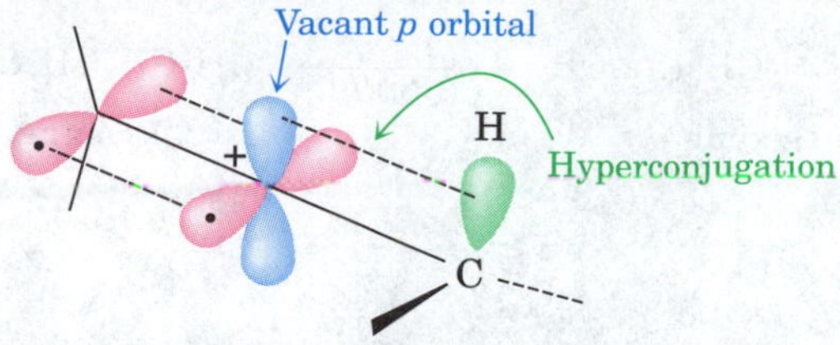

A secondary vinylic carbocation (*sp*-hybridized)

Figure 8.3 The electronic structure of a vinylic carbocation. The cationic carbon atom is *sp*-hybridized and is stabilized by hyperconjugation of its vacant *p* orbital with a neighboring C–H bond.

PROBLEM

8.3 What products would you expect from the following reactions?

(a) $CH_3CH_2CH_2C{\equiv}CH + 2\ Cl_2 \longrightarrow$

(b) 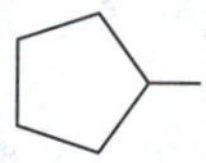 $-C{\equiv}CH + 1\ HBr \longrightarrow$

(c) $CH_3CH_2CH_2CH_2C{\equiv}CCH_3 + 1\ HBr \longrightarrow$

8.5 Hydration of Alkynes

Like alkenes (Sections 7.4 and 7.5), alkynes can be hydrated by either of two methods. Direct addition of water catalyzed by mercuric ion yields the Markovnikov product, and indirect addition of water by a hydroboration/oxidation sequence yields the non-Markovnikov product.

Mercuric Ion-Catalyzed Hydration of Alkynes

Alkynes can't be hydrated as easily as alkenes because of their lower reactivity toward electrophilic addition. In the presence of mercuric sulfate catalyst, however, hydration occurs readily.

$$\underset{\textbf{1-Hexyne}}{CH_3CH_2CH_2CH_2C{\equiv}CH} \xrightarrow[HgSO_4]{H_2O,\ H_2SO_4} \underset{\textbf{An enol}}{\left[CH_3CH_2CH_2CH_2\overset{\overset{OH}{|}}{C}{=}\underset{\underset{H}{|}}{CH}\right]}$$

$$\downarrow$$

$$\underset{\textbf{2-Hexanone (78\%)}}{CH_3CH_2CH_2CH_2\overset{\overset{O}{\|}}{C}-CH_3}$$

Alkyne hydration occurs with Markovnikov regiochemistry: The OH group adds to the more highly substituted carbon, and the H attaches to the less highly substituted one. Interestingly, though, the product is not the vinylic alcohol, or **enol** (*ene* + *ol*), but is instead a ketone. Although the enol may well be an intermediate in the reaction, it immediately rearranges to a ketone by a process called *keto–enol tautomerism*. The individual keto and enol forms are said to be **tautomers**, a word used to describe constitutional isomers that are rapidly interconverted. With few exceptions, the keto–enol tautomeric equilibrium lies on the side of the ketone; enols are almost never isolated. We'll look more closely at this equilibrium in Section 22.1.

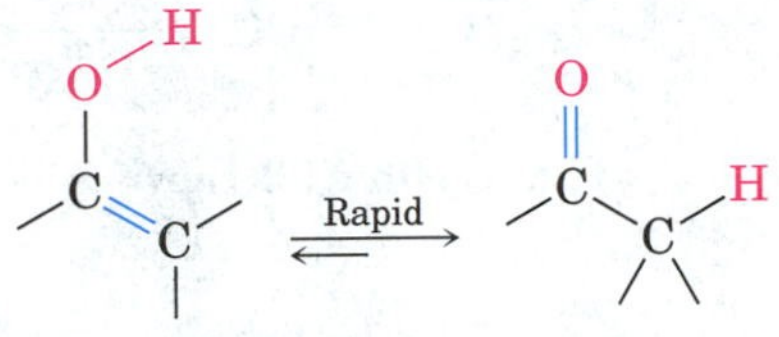

The mechanism of the mercuric ion-catalyzed alkyne hydration reaction is analogous to the oxymercuration reaction of alkenes (Section 7.4). Electrophilic addition of mercuric ion to the alkyne gives a vinylic cation, which reacts with water and loses a proton to yield an organomercury intermediate. In contrast to alkene oxymercuration, no treatment with $NaBH_4$ is necessary to remove the mercury; the acidic reaction conditions alone are sufficient to allow replacement of mercury by hydrogen (Figure 8.4).

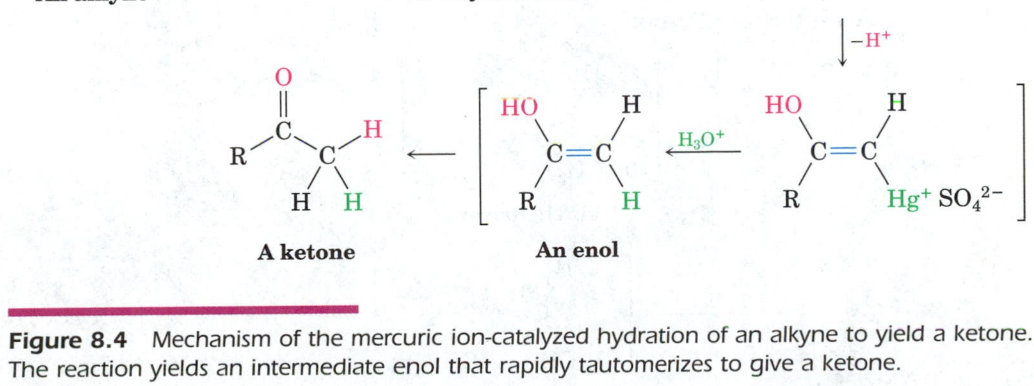

Figure 8.4 Mechanism of the mercuric ion-catalyzed hydration of an alkyne to yield a ketone. The reaction yields an intermediate enol that rapidly tautomerizes to give a ketone.

A mixture of both possible ketones results when an unsymmetrically substituted internal alkyne is hydrated. The reaction is therefore most useful when applied to a terminal alkyne (RC≡CH) because only a methyl ketone is formed.

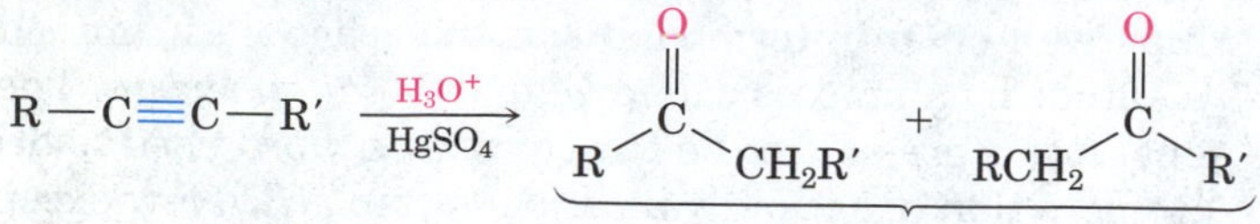

An internal alkyne **Mixture**

A terminal alkyne **A methyl ketone**

PROBLEM

8.4 What product would you obtain by hydration of 4-octyne? Of 2-methyl-4-octyne?

PROBLEM

8.5 What alkynes would you start with to prepare the following ketones?

(a) $CH_3CH_2CH_2\overset{\displaystyle O}{\overset{\|}{C}}CH_3$ (b) $CH_3CH_2\overset{\displaystyle O}{\overset{\|}{C}}CH_2CH_3$

Hydroboration/Oxidation of Alkynes

Borane adds rapidly to an alkyne, just as it does to an alkene, and the resulting vinylic borane can be oxidized by basic hydrogen peroxide to yield an enol. Tautomerization then gives either a ketone or an aldehyde, depending on the structure of the alkyne reactant. Hydroboration/oxidation of an internal alkyne such as 3-hexyne gives a ketone, and hydroboration/oxidation of a terminal alkyne gives an aldehyde. Note that the relatively unhindered terminal alkyne undergoes *two* additions, giving a doubly hydroborated intermediate. Oxidation with H_2O_2 at pH 8 then replaces both boron atoms by oxygen and generates the aldehyde.

An internal alkyne:

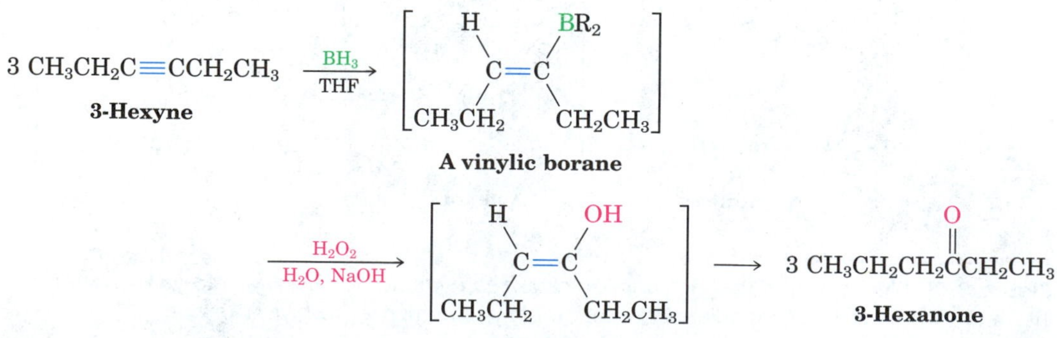

A terminal alkyne:

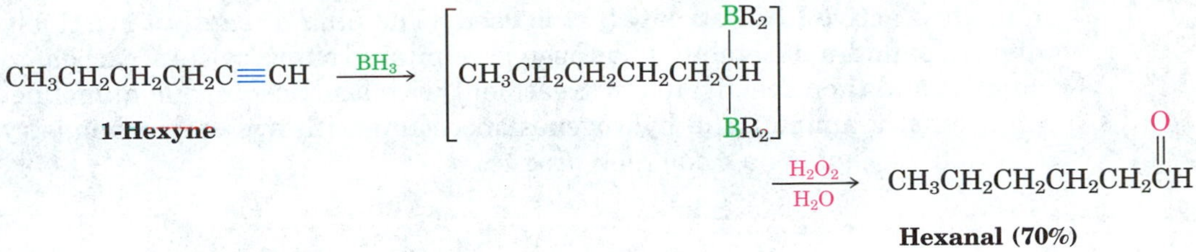

The hydroboration/oxidation sequence is *complementary* to the direct, mercuric ion-catalyzed hydration reaction of a terminal alkyne because different products result. Direct hydration with aqueous acid and mercuric sulfate leads to a methyl ketone, whereas hydroboration/oxidation of the same terminal alkyne leads to an aldehyde:

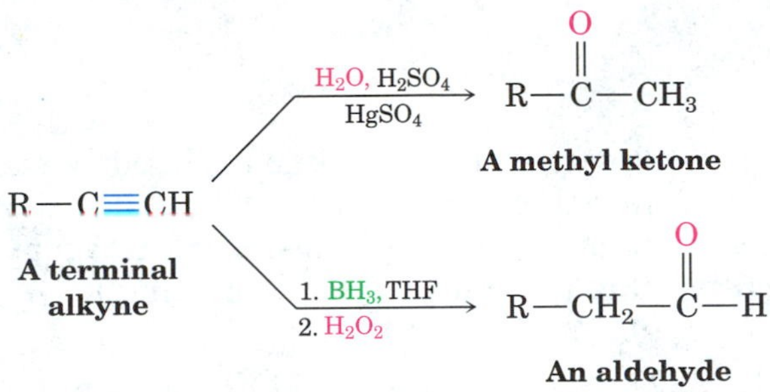

PROBLEM.

8.6 What alkynes would you start with to prepare the following compounds by a hydroboration/oxidation reaction?

(a) 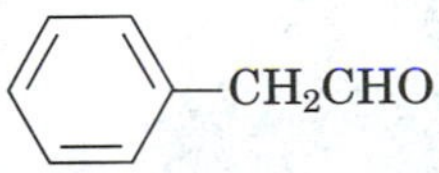

(b) $(CH_3)_2CHCH_2\overset{O}{\overset{\|}{C}}CH(CH_3)_2$

8.6 Reduction of Alkynes

Alkynes are easily reduced to alkanes by addition of H_2 over a metal catalyst. The reaction occurs in steps through an alkene intermediate, and measurements indicate that the first step in the reaction has a larger ΔH°_{hydrog} than the second step. As a result, alkynes reduce somewhat more readily than alkenes.

$$HC{\equiv}CH \xrightarrow[\text{Catalyst}]{H_2} H_2C{=}CH_2 \qquad \Delta H^\circ_{hydrog} = -176\text{ kJ/mol } (-42\text{ kcal/mol})$$

$$H_2C{=}CH_2 \xrightarrow[\text{Catalyst}]{H_2} CH_3{-}CH_3 \qquad \Delta H^\circ_{hydrog} = -137\text{ kJ/mol } (-33\text{ kcal/mol})$$

Complete reduction to the alkane occurs when palladium on carbon (Pd/C) is used as catalyst, but hydrogenation can be stopped at the alkene if the less active **Lindlar catalyst** is used. (The *Lindlar catalyst* is a finely divided palladium metal that has been precipitated onto a calcium carbonate support and then deactivated by treatment with lead acetate and quinoline, an aromatic amine.) The hydrogenation occurs with syn stereochemistry (Section 7.7), giving a cis alkene product.

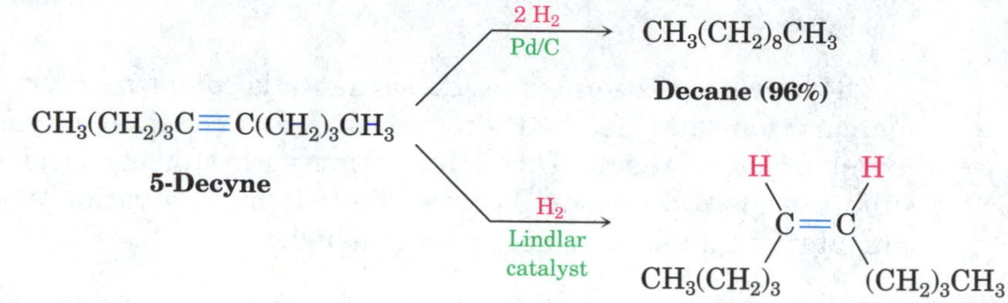

***cis*-5-Decene (96%)**

This alkyne reaction has been explored extensively by the Hoffmann–LaRoche pharmaceutical company, where it is used in the commercial synthesis of vitamin A. (The cis isomer of vitamin A produced on hydrogenation is converted to the trans isomer by heating.)

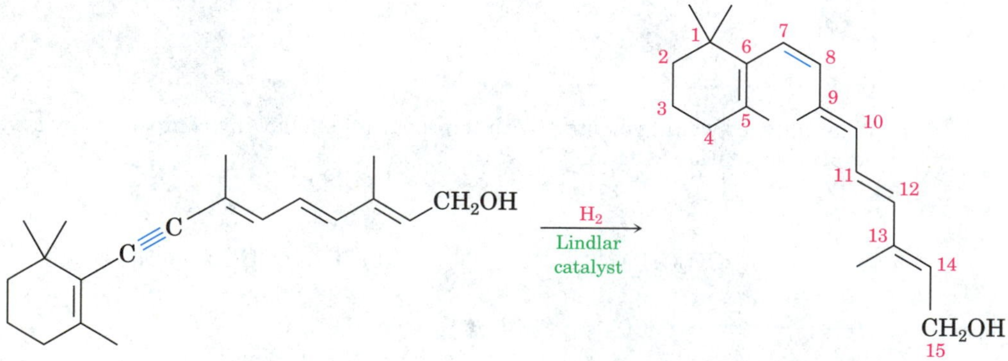

7-*cis*-Retinol
(7-*cis*-vitamin A; vitamin A has a trans double bond at C7)

Another method for the conversion of an alkyne to an alkene employs sodium or lithium metal in liquid ammonia as solvent. This method is complementary to the Lindlar reduction because it produces trans alkenes rather than cis. Alkali metals, such as lithium and sodium, dissolve in pure liquid ammonia at −33°C to produce a deep blue solution containing the metal cation and ammonia-solvated electrons. When an alkyne is added to this blue solution, reduction of the triple bond occurs and a trans alkene results. For example, 5-decyne gives *trans*-5-decene on treatment with lithium in liquid ammonia.

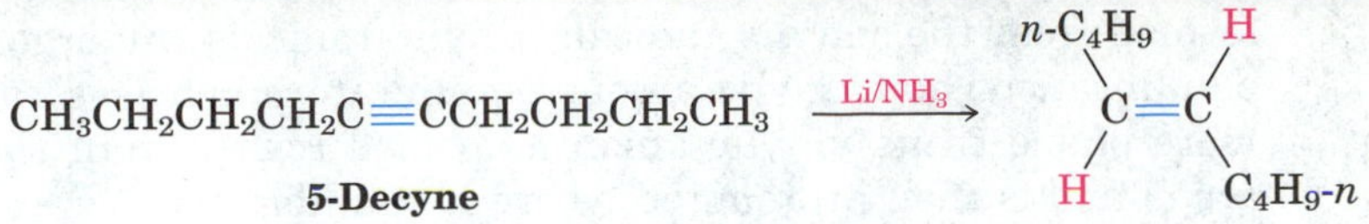

5-Decyne → *trans*-5-Decene (78%)

The mechanism of alkyne reduction by lithium in liquid ammonia involves addition of an electron to the triple bond to yield an intermediate *anion radical*—a species that is both a radical (has an odd number of electrons) and an anion (has a negative charge). This intermediate then removes a proton from ammonia to give a vinylic radical. Addition of a second electron to the vinylic radical gives a vinylic anion, which takes a second proton from ammonia to give trans alkene product (Figure 8.5).

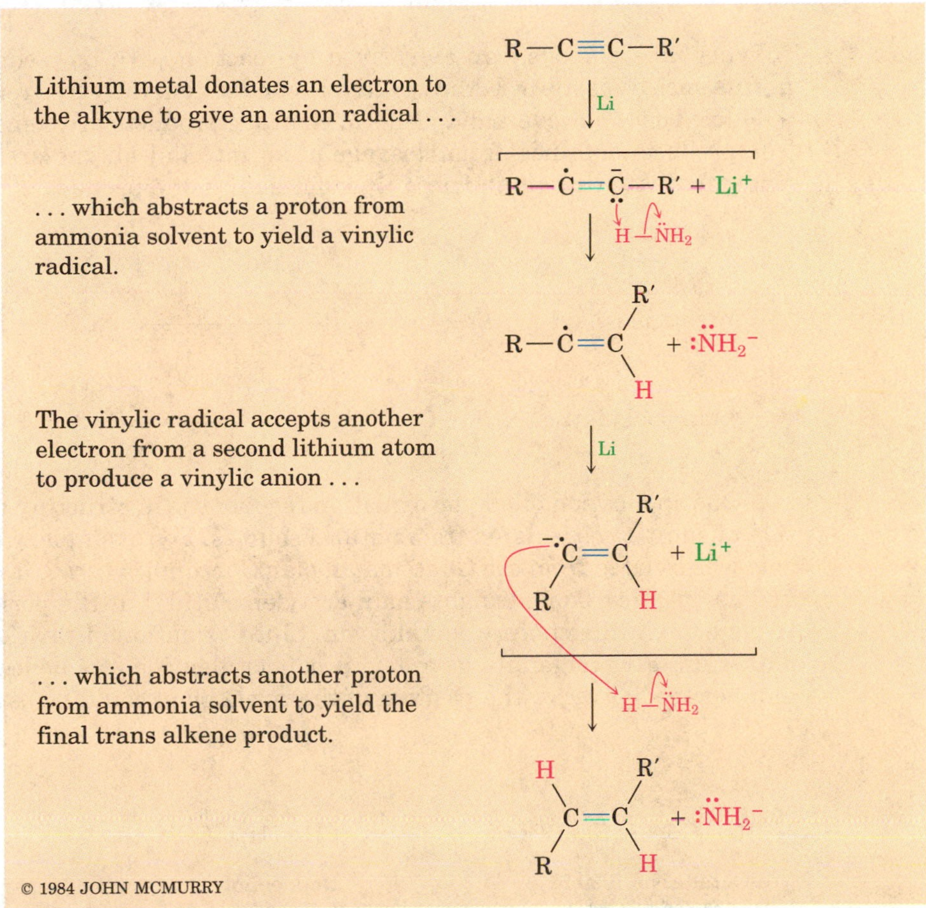

Figure 8.5 Mechanism of the lithium/ammonia reduction of an alkyne to produce a trans alkene.

The trans stereochemistry of the alkene product is established during the second reduction step when the less hindered, trans vinylic anion is formed from the vinylic radical. Vinylic radicals undergo rapid cis–trans equilibration, but vinylic anions equilibrate much less rapidly. Thus, the more stable trans vinylic anion is formed rather than the less stable cis anion and is then protonated before it equilibrates.

PROBLEM

8.7 Using any alkyne needed, how would you prepare these alkenes?
(a) *trans*-2-Octene (b) *cis*-3-Heptene (c) 3-Methyl-1-pentene

8.7 Oxidative Cleavage of Alkynes

Alkynes, like alkenes, can be cleaved by reaction with powerful oxidizing agents such as ozone or $KMnO_4$. A triple bond is generally less reactive than a double bond, however, and yields of cleavage products are sometimes low. The products obtained from cleavage of an internal alkyne are carboxylic acids; from a terminal alkyne, CO_2 is formed as one product.

An internal alkyne $$R{-}C{\equiv}C{-}R' \xrightarrow{KMnO_4 \text{ or } O_3} R\overset{O}{\overset{\|}{C}}OH + R'\overset{O}{\overset{\|}{C}}OH$$

A terminal alkyne $$R{-}C{\equiv}C{-}H \xrightarrow{KMnO_4 \text{ or } O_3} R\overset{O}{\overset{\|}{C}}OH + CO_2$$

One application of alkyne oxidation reactions is in structure determination of substances isolated from natural sources. For example, when tariric acid was isolated from the Guatemalan plant *Picramnia tariri*, it was identified as an 18-carbon, straight-chain acetylenic acid, but the position of the triple bond in the chain was unknown. Since oxidation of tariric acid with potassium permanganate gave two products identified as dodecanoic acid and hexanedioic acid, the position of the triple bond could be established.

$$CH_3(CH_2)_{10}C{\equiv}C(CH_2)_4COOH \xrightarrow[H_3O^+]{KMnO_4} CH_3(CH_2)_{10}\overset{O}{\overset{\|}{C}}OH + HO\overset{O}{\overset{\|}{C}}(CH_2)_4\overset{O}{\overset{\|}{C}}OH$$

6-Octadecynoic acid (Tariric acid) → **Dodecanoic acid (Lauric acid)** + **1,6-Hexanedioic acid (Adipic acid)**

PROBLEM

8.8 Propose structures for alkynes that give the following products on oxidative cleavage by $KMnO_4$.

(a) 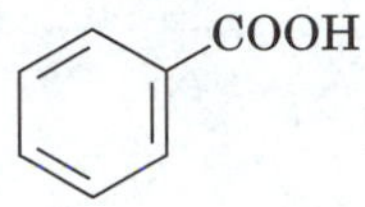 $+ CO_2$

(b) $2\ CH_3(CH_2)_7COOH + HO_2C(CH_2)_7COOH$

8.8 Alkyne Acidity: Formation of Acetylide Anions

The most striking difference between the chemistry of alkenes and alkynes is that terminal alkynes are weakly acidic. When a terminal alkyne is treated with a strong base, such as sodium amide, $NaNH_2$, the terminal hydrogen is removed and an **acetylide anion** is formed:

$$R{-}C{\equiv}C{-}H + :\ddot{N}H_2\ Na^+ \longrightarrow R{-}C{\equiv}C:^-\ Na^+ + :NH_3$$

Acetylide anion

According to the Brønsted–Lowry definition, an acid is any species that donates a proton (Section 2.6). Although we usually think of oxyacids (H_2SO_4, HNO_3) or halogen acids (HCl, HBr) in this context, *any* compound containing a hydrogen atom can be considered an acid under the right circumstances. By measuring dissociation constants of different acids and expressing the results as pK_a values, we can establish an acidity order. (Recall from Section 2.6 that a low pK_a corresponds to a strong acid, and a high pK_a corresponds to a weak acid.)

Since a stronger acid donates its proton to the anion of a weaker acid in an acid–base reaction, a rank-ordered list tells which bases are needed to deprotonate which acids. For example, since acetic acid ($pK_a = 4.75$) is a stronger acid than ethanol ($pK_a = 16$), we know that the anion of ethanol (ethoxide ion, $CH_3CH_2O^-$) will remove a proton from acetic acid. Similarly, amide ion ($^-NH_2$), the anion of ammonia ($pK_a = 35$), will remove a proton from ethanol ($pK_a = 16$).

$$CH_3CH_2O^- + CH_3\overset{\overset{O}{\|}}{C}OH \rightleftharpoons CH_3CH_2OH + CH_3\overset{\overset{O}{\|}}{C}O^-$$

Ethoxide ion **Acetic acid** **Ethanol** **Acetate ion**

$$H_2N^- + CH_3CH_2OH \rightleftharpoons H_2NH + CH_3CH_2O^-$$

Amide ion **Ethanol** **Ammonia** **Ethoxide ion**

Where do hydrocarbons lie on the acidity scale? As the data in Table 8.1 indicate, both methane ($pK_a \approx 60$) and ethylene ($pK_a = 44$) are very weak acids that do not react with bases for all practical purposes. Acetylene, however, has a pK_a of 25 and can thus be deprotonated by the conjugate base of any acid whose pK_a is greater than 25. Amide ion, NH_2^-, for example, is the conjugate base of ammonia ($pK_a = 35$) and is therefore able to abstract a proton from terminal alkynes.

Table 8.1 Acidity of Simple Hydrocarbons

Type	*Example*	K_a	pK_a	
				Stronger acid
Alkyne	$HC{\equiv}CH$	10^{-25}	25	↑
Alkene	$H_2C{=}CH_2$	10^{-44}	44	
Alkane	CH_4	$\sim 10^{-60}$	~ 60	
				Weaker acid

Why are terminal alkynes more acidic than alkenes or alkanes? In other words, why are acetylide anions more stable than vinylic or alkyl anions? The simplest explanation involves the hybridization of the negatively charged carbon atom. Acetylide anions are *sp*-hybridized, so the negative charge resides in an orbital that has $\frac{1}{2}$ "*s* character"; vinylic anions are sp^2-hybridized and therefore have $\frac{1}{3}$ *s* character in the relevant orbital; and alkyl anions (sp^3) have $\frac{1}{4}$ *s* character. Since *s* orbitals are lower in energy and nearer the positively charged nucleus than are *p* orbitals, a negative charge is stabilized to a greater extent in an orbital with high *s* character than in an orbital with low *s* character (Figure 8.6). Acetylide anions are therefore more stable than vinylic anions, which, in turn, are more stable than alkyl anions.

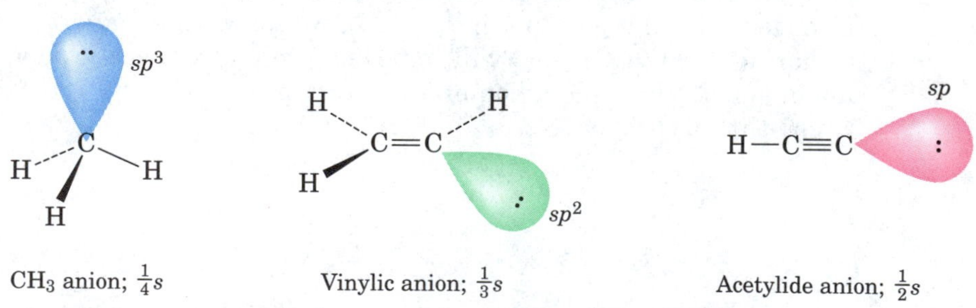

Figure 8.6 A comparison of alkyl, vinylic, and acetylide anions. The acetylide anion, with *sp* hybridization, has more *s* character and is more stable.

PROBLEM

8.9 The pK_a of acetone, CH_3COCH_3, is 19.3. Which of the following bases is strong enough to deprotonate acetone?

(a) KOH (pK_a of $H_2O = 15.7$)
(b) $Na^+ \ ^-C{\equiv}CH$ (pK_a of $C_2H_2 = 25$)
(c) $NaHCO_3$ (pK_a of $H_2CO_3 = 6.4$)
(d) $NaOCH_3$ (pK_a of $CH_3OH = 15.6$)

8.9 Alkylation of Acetylide Anions

The presence of a negative charge and an unshared electron pair on carbon makes acetylide anions strongly nucleophilic. As a result, acetylide anions can react with alkyl halides such as bromomethane to substitute for the halogen and yield a new alkyne product:

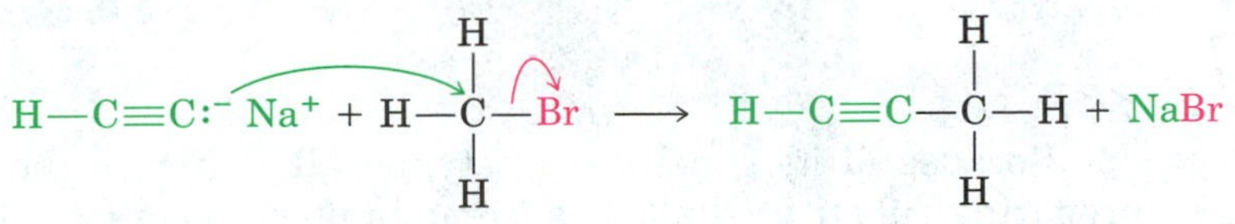

We won't study the details of this substitution reaction until Chapter 11, but we can picture it as happening by the pathway shown in Figure 8.7. The nucleophilic acetylide ion attacks the positively polarized (and therefore electrophilic) carbon atom of bromomethane and pushes out bromide ion with the electron pair from the former C–Br bond, yielding propyne as product. We call such a reaction an **alkylation** because a new alkyl group has become attached to the starting alkyne.

The nucleophilic acetylide anion uses its electron lone pair to form a bond to the positively polarized, electrophilic carbon atom of bromomethane. As the new C–C bond begins to form, the C–Br bond begins to break in the transition state.

The new C–C bond is fully formed and the old C–Br bond is fully broken at the end of the reaction.

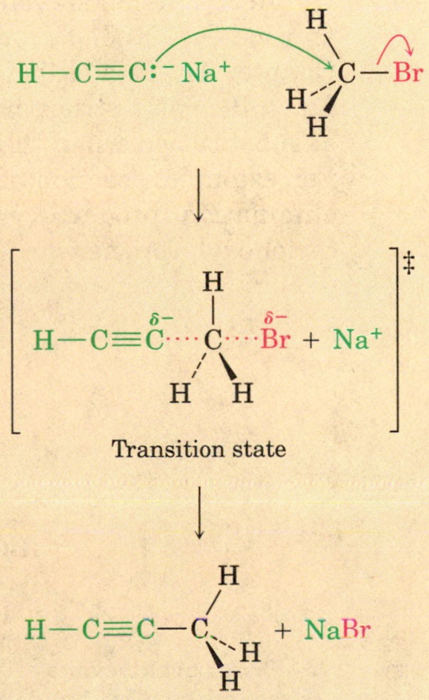

Figure 8.7 A mechanism for the alkylation reaction of acetylide anion with bromomethane.

Alkyne alkylation is not limited to acetylide ion. *Any* terminal alkyne can be converted into its corresponding anion and then alkylated by treatment with an alkyl halide. The product is an internal alkyne. For example, conversion of 1-hexyne into its anion, followed by reaction with 1-bromobutane, yields 5-decyne:

$$\underset{\textbf{1-Hexyne}}{CH_3CH_2CH_2CH_2C{\equiv}CH} \xrightarrow[\text{2. } CH_3CH_2CH_2CH_2Br\text{, THF}]{\text{1. } NaNH_2\text{, } NH_3} \underset{\textbf{5-Decyne (76\%)}}{CH_3CH_2CH_2CH_2C{\equiv}CCH_2CH_2CH_2CH_3}$$

Because of its generality, acetylide alkylation is the best method for preparing substituted alkynes from simpler precursors. Terminal alkynes can be prepared by alkylation of acetylene itself, and internal alkynes can be prepared by further alkylation of a terminal alkyne.

$$\underset{\textbf{Acetylene}}{HC{\equiv}CH} \xrightarrow{NaNH_2} HC{\equiv}C^-\ Na^+ \xrightarrow{RCH_2Br} \underset{\textbf{A terminal alkyne}}{HC{\equiv}CCH_2R'}$$

$$\underset{\textbf{A terminal alkyne}}{RC{\equiv}CH} \xrightarrow{NaNH_2} RC{\equiv}C^-\ Na^+ \xrightarrow{R'CH_2Br} \underset{\textbf{An internal alkyne}}{RC{\equiv}CCH_2R'}$$

Acetylide ion alkylation is limited to the use of primary alkyl bromides and iodides, RCH_2X, for reasons that will be discussed in more detail in Chapter 11. In addition to their reactivity as nucleophiles, acetylide ions are sufficiently strong bases that they cause dehydrohalogenation instead of substitution when they react with secondary and tertiary alkyl halides. For example, reaction of bromocyclohexane with propyne anion yields the elimination product cyclohexene rather than the substitution product cyclohexylpropyne.

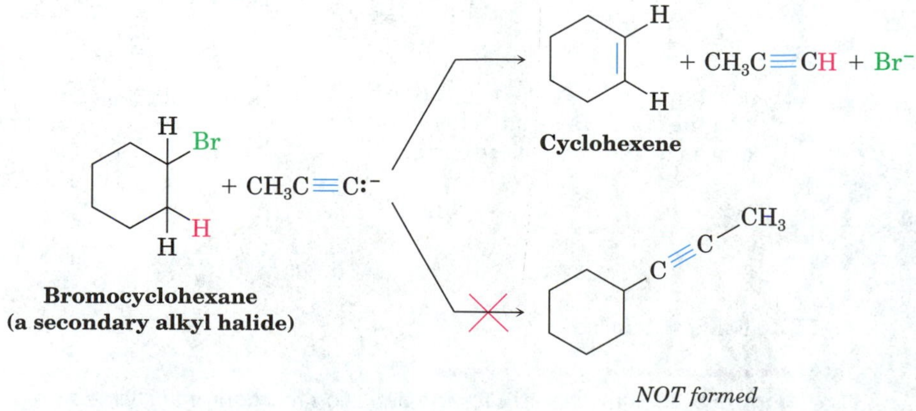

PROBLEM...

8.10 Show the terminal alkyne and alkyl halide from which the following products can be obtained. Where two routes look feasible, list both.

(a) $CH_3CH_2CH_2C{\equiv}CCH_3$ (b) $(CH_3)_2CHC{\equiv}CCH_2CH_3$

(c) 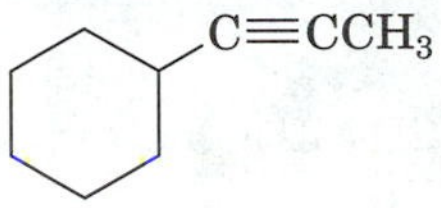

(d) 5-Methyl-2-hexyne

(e) 2,2-Dimethyl-3-hexyne

PROBLEM...

8.11 How would you prepare *cis*-2-butene starting from 1-propyne, an alkyl halide, and any other reagents needed? (This problem can't be worked in a single step. You'll have to carry out more than one reaction.)

....

8.10 Organic Synthesis

There are many possible reasons for carrying out the laboratory synthesis of an organic molecule from simple precursors. In the pharmaceutical industry, new organic molecules are designed and synthesized in the hope that some might be useful new drugs. In the chemical industry, synthesis is done to devise more economical routes to known compounds. In academic laboratories, the synthesis of complex molecules is sometimes done purely for the intellectual challenge involved in mastering so difficult a subject. The successful synthesis route is a highly creative work that is sometimes described by such subjective terms as *elegant* or *beautiful*.

In this book, too, we will often devise syntheses of molecules from simpler precursors. Our purpose, however, is pedagogical. The ability to plan a workable synthetic sequence demands knowledge of a wide variety of organic reactions. Furthermore, it requires a practical grasp for the proper fitting together of steps in a sequence such that each reaction does only what is desired. *Working synthesis problems is an excellent way to learn organic chemistry.*

Some of the syntheses we plan may appear trivial. Here's an example:

PRACTICE PROBLEM...

Prepare octane from 1-pentyne.

$$\underset{\textbf{1-Pentyne}}{CH_3CH_2CH_2C{\equiv}CH} \xrightarrow{?} \underset{\textbf{Octane}}{CH_3CH_2CH_2CH_2CH_2CH_2CH_2CH_3}$$

Solution First alkylate the acetylide anion of 1-pentyne with 1-bromopropane, and then reduce the product using catalytic hydrogenation:

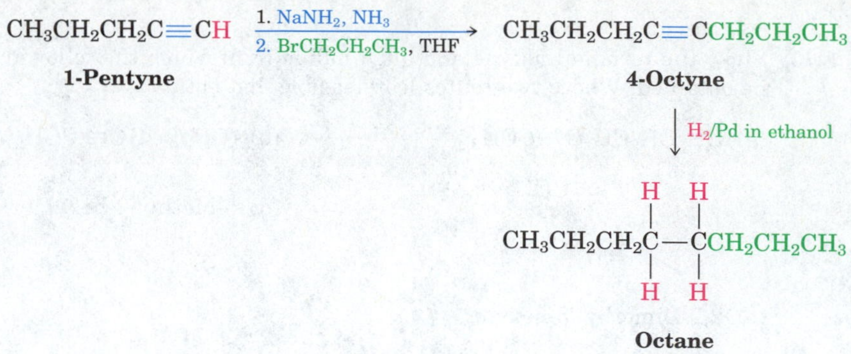

Although the synthesis route just presented will work perfectly well, it has little practical value because a chemist can simply *buy* octane from any of several dozen chemical supply companies. The value of working the problem is that it makes us approach a chemical problem in a logical way, draw on our knowledge of chemical reactions, and organize that knowledge into a workable plan—it helps us *learn* organic chemistry.

There's no secret to planning an organic synthesis. All it takes is a knowledge of the different reactions and a lot of practice. But here's a hint: *Work backward*. Look at the final product and ask, "What was the immediate precursor of that product?" For example, if the end product is an alkyl halide, the immediate precursor might be an alkene (via HX addition). Having found an immediate precursor, proceed backward again, one step at a time, until a suitable starting material is found.

Let's work some examples of increasing complexity.

PRACTICE PROBLEM

Starting from 1-pentyne and any alkyl halide needed, synthesize *cis*-2-hexene. More than one step is required.

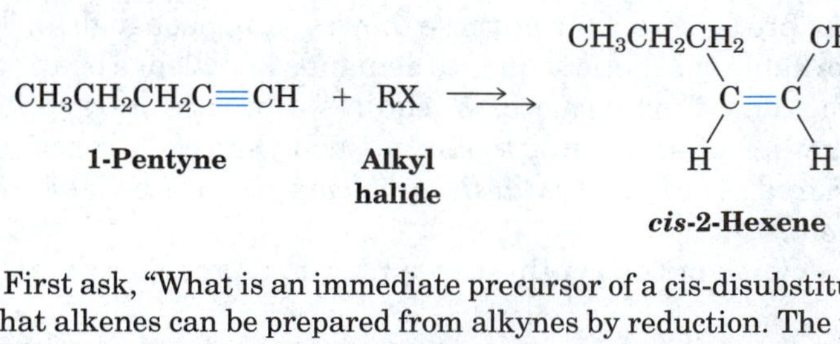

Solution First ask, "What is an immediate precursor of a cis-disubstituted alkene?" We know that alkenes can be prepared from alkynes by reduction. The proper choice of experimental conditions will allow us to prepare either a trans-disubstituted alkene (using lithium in liquid ammonia) or a cis-disubstituted alkene (using catalytic hydrogenation over the Lindlar catalyst). Thus, reduction of 2-hexyne by catalytic hydrogenation using the Lindlar catalyst should yield *cis*-2-hexene:

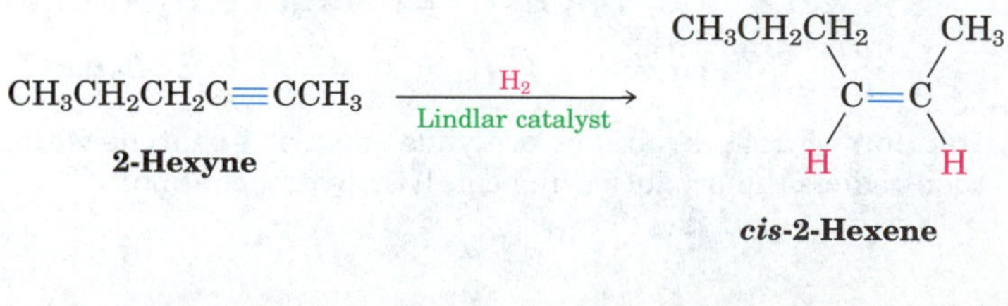

Next ask, "What is an immediate precursor of 2-hexyne?" We've seen that internal alkynes can be prepared by alkylation of terminal alkyne anions (acetylides). In the present instance, we're told to start with 1-pentyne. Thus, alkylation of the anion of 1-pentyne with iodomethane should yield 2-hexyne:

$$\underset{\textbf{1-Pentyne}}{CH_3CH_2CH_2C{\equiv}CH} + NaNH_2 \xrightarrow{\text{In } NH_3} CH_3CH_2CH_2C{\equiv}C{:}^-\ Na^+$$

$$CH_3CH_2CH_2C{\equiv}C{:}^-\ Na^+ + CH_3I \xrightarrow{\text{In THF}} \underset{\textbf{2-Hexyne}}{CH_3CH_2CH_2C{\equiv}CCH_3}$$

In three steps, we've synthesized *cis*-2-hexene from the given starting materials:

$$\underset{\textbf{1-Pentyne}}{CH_3CH_2CH_2C{\equiv}CH} \xrightarrow[\text{2. } CH_3I\text{, THF}]{\text{1. } NaNH_2\text{, } NH_3} \underset{\textbf{2-Hexyne}}{CH_3CH_2CH_2C{\equiv}CCH_3}$$

$$\downarrow H_2, \text{ Lindlar catalyst}$$

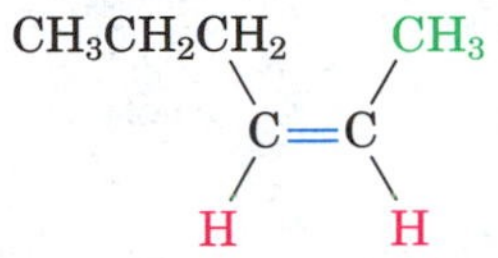

***cis*-2-Hexene**

PRACTICE PROBLEM

Starting from acetylene and any alkyl halide needed, synthesize 2-bromopentane. More than one step is required.

$$\underset{\textbf{Acetylene}}{HC{\equiv}CH} + \underset{\textbf{Alkyl halide}}{RX} \xrightarrow{?} \underset{\textbf{2-Bromopentane}}{CH_3CH_2CH_2\overset{Br}{\overset{|}{C}}HCH_3}$$

Solution "What is an immediate precursor of an alkyl halide?" Perhaps an alkene:

$$\begin{matrix} CH_3CH_2CH_2CH{=}CH_2 \\ \text{or} \\ CH_3CH_2CH{=}CHCH_3 \end{matrix} \xrightarrow[\text{Ether}]{HBr} CH_3CH_2CH_2\overset{Br}{\overset{|}{C}}HCH_3$$

Of the two possibilities, addition of HBr to 1-pentene looks like a better choice than addition to 2-pentene, because the latter reaction would give a mixture of isomers.

"What is an immediate precursor of an alkene?" Perhaps an alkyne, which could be reduced:

$$CH_3CH_2CH_2C{\equiv}CH \xrightarrow[\text{Lindlar catalyst}]{H_2} CH_3CH_2CH_2CH{=}CH_2$$

"What is an immediate precursor of a terminal alkyne?" Perhaps sodium acetylide and an alkyl halide:

$$Na^+ \ :\bar{C}{\equiv}CH + BrCH_2CH_2CH_3 \longrightarrow CH_3CH_2CH_2C{\equiv}CH$$

In four steps, we have synthesized the desired material from acetylene and 1-bromopropane.

$$\underset{\textbf{Acetylene}}{HC{\equiv}CH} \xrightarrow[\text{2. } CH_3CH_2CH_2Br\text{, THF}]{\text{1. } NaNH_2\text{, } NH_3} \underset{\textbf{1-Pentyne}}{CH_3CH_2CH_2C{\equiv}CH} \xrightarrow[\text{Lindlar catalyst}]{H_2} \underset{\textbf{1-Pentene}}{CH_3CH_2CH_2CH{=}CH_2}$$

$$\xrightarrow{HBr\text{, ether}} \underset{\textbf{2-Bromopentane}}{CH_3CH_2CH_2CH(Br)CH_3}$$

PRACTICE PROBLEM

Synthesize 1-hexanol from acetylene and an alkyl halide.

$$\underset{\textbf{Acetylene}}{HC{\equiv}CH} + \underset{\textbf{Alkyl halide}}{RX} \Longrightarrow \underset{\textbf{1-Hexanol}}{CH_3CH_2CH_2CH_2CH_2CH_2OH}$$

Solution "What is an immediate precursor of a primary alcohol?" Perhaps an alkene, which could be hydrated by reaction with borane followed by oxidation with H_2O_2:

$$CH_3CH_2CH_2CH_2CH{=}CH_2 \xrightarrow[\text{2. } H_2O_2\text{, NaOH}]{\text{1. } BH_3} CH_3CH_2CH_2CH_2CH_2CH_2OH$$

"What is an immediate precursor of a terminal alkene?" Perhaps a terminal alkyne, which could be reduced:

$$CH_3CH_2CH_2CH_2C{\equiv}CH \xrightarrow[\text{Lindlar catalyst}]{H_2} CH_3CH_2CH_2CH_2CH{=}CH_2$$

"What is an immediate precursor of 1-hexyne?" Perhaps acetylene and 1-bromobutane:

$$HC{\equiv}CH \xrightarrow{NaNH_2} Na^+ \ {}^-C{\equiv}CH \xrightarrow{CH_3CH_2CH_2CH_2Br} CH_3CH_2CH_2CH_2C{\equiv}CH$$

We have completed the synthesis in three steps by working backward.

PROBLEM

8.12 Beginning with 4-octyne as your only source of carbon, and using any inorganic reagents necessary, how would you synthesize the following compounds?
(a) Butanoic acid (b) *cis*-4-Octene (c) 4-Bromooctane
(d) 4-Octanol (4-hydroxyoctane) (e) 4,5-Dichlorooctane

PROBLEM

8.13 Beginning with acetylene and any alkyl halides needed, how would you synthesize the following compounds?
(a) Decane (b) 2,2-Dimethylhexane (c) Hexanal (d) 2-Heptanone

INTERLUDE

Polyynes: A New Form of Carbon

Elemental carbon appears in many structural forms, or *allotropes*. Three of these forms are crystalline—diamond, graphite, and the recently discovered fullerene—while more than 40 others, including coke and carbon black, are amorphous. Diamond, the rarest, hardest, and most valuable allotrope of carbon, is a covalent network solid composed of sp^3-hybridized atoms linked together in a vast, three-dimensional lattice of cyclohexane-like rings. Graphite, the most common carbon allotrope and the most stable one at room temperature, is composed of sp^2-hybridized atoms joined in two-dimensional sheets of benzene-like rings. Fullerene, a molecular allotrope of carbon first characterized in 1991, contains 60 sp^2-hybridized carbon atoms in 32 five- and six-membered rings that are joined into the spherical shape of a soccer ball (Figure 8.8).

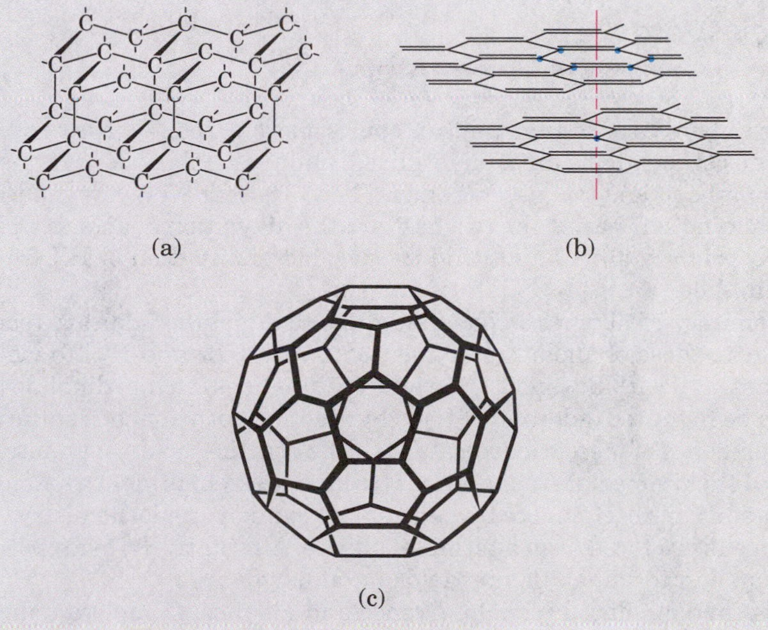

Figure 8.8 Three allotropes of carbon: (a) Diamond is a covalent network solid consisting of a three-dimensional lattice of sp^3-hybridized carbons in cyclohexane-like rings. (b) Graphite consists of two-dimensional sheets of sp^2-hybridized carbons in benzene-like rings. The atoms in each sheet are offset slightly from the atoms in the neighboring sheets. (c) Fullerene, C_{60}, is a molecular solid whose molecules have the shape of a soccer ball. The ball has 12 pentagonal and 20 hexagonal faces, and each carbon atom is sp^2-hybridized.

(continued)➤

Now there seems to be yet a fourth crystalline allotrope of carbon, the *polyyne* form. Reported in 1995 by Richard Lagow at The University of Texas, the polyyne allotrope consists of linear chains of up to 500 *sp*-hybridized carbons joined by alternating single and triple bonds. It is prepared by laser-induced vaporization of graphite in the presence of trifluoromethyl radicals ($\cdot CF_3$) to "cap" the ends of the long acetylenic chains and thereby stabilize the molecules.

$$\text{-}C{\equiv}C{-}C{\equiv}C{-}C{\equiv}C{-}C{\equiv}C{-}C{\equiv}C{-}C{\equiv}C{-}C{\equiv}C{-}C{\equiv}C{-}C{\equiv}C{-}C{\equiv}C\text{-}$$

Polyyne allotrope of carbon

Though yet too newly discovered to have been fully explored, polyyne carbon appears to have some remarkable properties. It is soluble in ordinary organic solvents like toluene; it appears to be electrically conducting along its long axis, perhaps allowing it to function as a "molecular wire" in molecule-sized electronic devices; and it is easily coated onto other substrates, allowing it to act as a precursor of graphite or diamond films.

Summary and Key Words

Alkynes are hydrocarbons that contain one or more carbon–carbon triple bonds. Alkyne carbon atoms are *sp*-hybridized, and the triple bond is formed by one *sp–sp* σ bond and two *p–p* π bonds. There are relatively few general methods of alkyne synthesis. The two best are the alkylation of an acetylide anion with a primary alkyl halide and the twofold elimination of HX from a **vicinal** dihalide.

The chemistry of alkynes is dominated by electrophilic addition reactions, similar to those of alkenes. Alkynes react with HBr and HCl to yield vinylic halides, and with Br_2 and Cl_2 to yield 1,2-dihalides (vicinal dihalides). Alkynes can be hydrated (addition of H_2O) by reaction with aqueous sulfuric acid in the presence of mercuric ion catalyst. The reaction leads to an intermediate **enol** that immediately isomerizes (**tautomerizes**) to yield a ketone. Since the addition reaction occurs with Markovnikov regiochemistry, a methyl ketone is produced from a terminal alkyne. Alternatively, hydroboration/oxidation of a terminal alkyne yields an aldehyde.

Alkynes can be reduced to yield alkenes and alkanes. Complete reduction of the triple bond over a normal palladium hydrogenation catalyst yields an alkane; partial reduction by catalytic hydrogenation over a **Lindlar catalyst** yields a cis alkene. Reduction of the alkyne with lithium in ammonia yields a trans alkene.

Terminal alkynes are weakly acidic. The alkyne hydrogen can be removed by a strong base to yield an **acetylide anion**. Acetylide anions act as nucleophiles and can displace halide ion from primary alkyl halides in an **alkylation** reaction. Acetylide anions are more stable than either alkyl anions or vinylic anions because their negative charge is in a hybrid orbital with $\frac{1}{2}$ *s* character, allowing the charge to be closer to the nucleus.

Summary of Reactions

1. Preparation of alkynes

(a) Dehydrohalogenation of vicinal dihalides (Section 8.

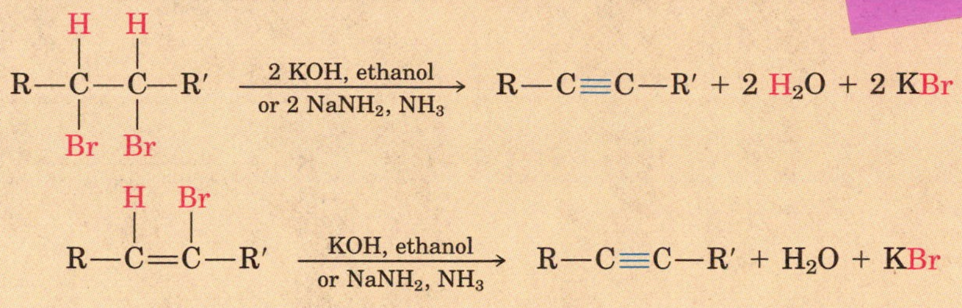

(b) Acetylide ion alkylation (Section 8.9)

$$HC{\equiv}CH \xrightarrow{NaNH_2} HC{\equiv}C^- \ Na^+ \xrightarrow{RCH_2Br} HC{\equiv}CCH_2R$$

Acetylene **A terminal alkyne**

$$RC{\equiv}CH \xrightarrow{NaNH_2} RC{\equiv}C^- \ Na^+ \xrightarrow{R'CH_2Br} RC{\equiv}CCH_2R'$$

A terminal alkyne **An internal alkyne**

2. Reactions of alkynes

(a) Addition of HX, where X = Br or Cl (Section 8.4)

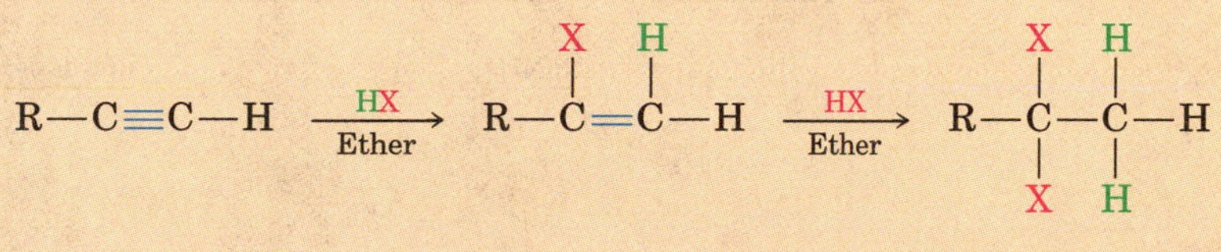

(b) Addition of X_2, where X = Br or Cl (Section 8.4)

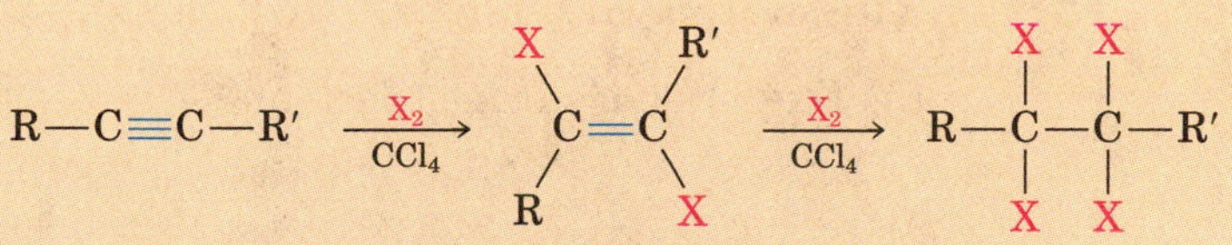

(c) Mercuric sulfate-catalyzed hydration (Section 8.5)

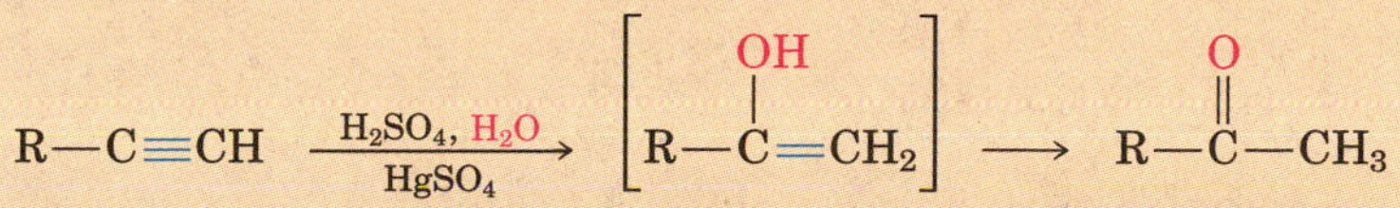

A methyl ketone

(d) Hydroboration/oxidation (Section 8.5)

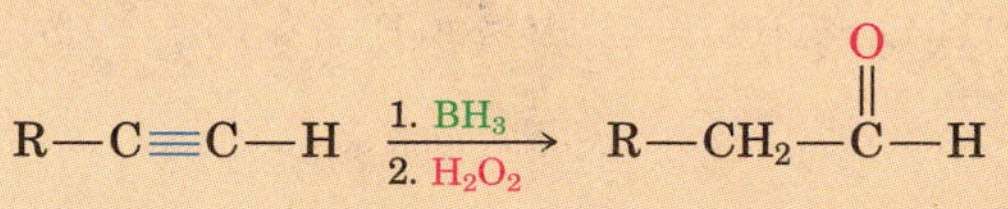

(continued)➤

(e) Reduction (Section 8.6)

(1) Catalytic hydrogenation

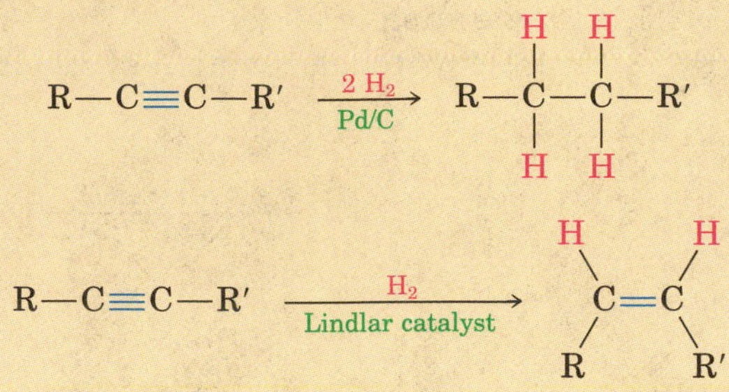

A cis alkene

(2) Lithium/ammonia

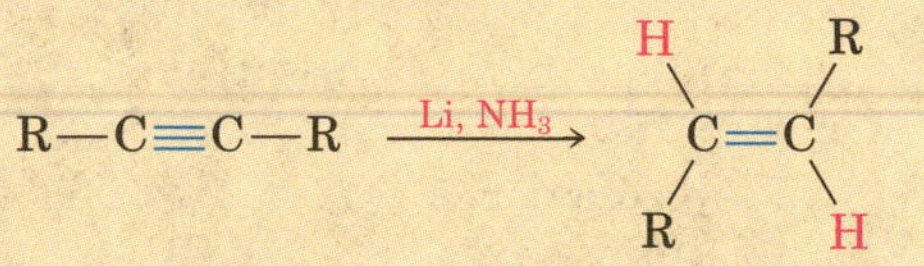

A trans alkene

(f) Acidity: conversion into acetylide anions (Section 8.8)

$$R{-}C{\equiv}C{-}H \xrightarrow[NH_3]{NaNH_2} R{-}C{\equiv}C{:}^- \; Na^+ + NH_3$$

(g) Acetylide ion alkylation (Section 8.9)

$$HC{\equiv}CH \xrightarrow{NaNH_2} HC{\equiv}C^- \; Na^+ \xrightarrow{RCH_2Br} HC{\equiv}CCH_2R$$

Acetylene → **A terminal alkyne**

$$RC{\equiv}CH \xrightarrow{NaNH_2} RC{\equiv}C^- \; Na^+ \xrightarrow{R'CH_2Br} RC{\equiv}CCH_2R'$$

A terminal alkyne → **An internal alkyne**

(h) Oxidative cleavage (Section 8.7)

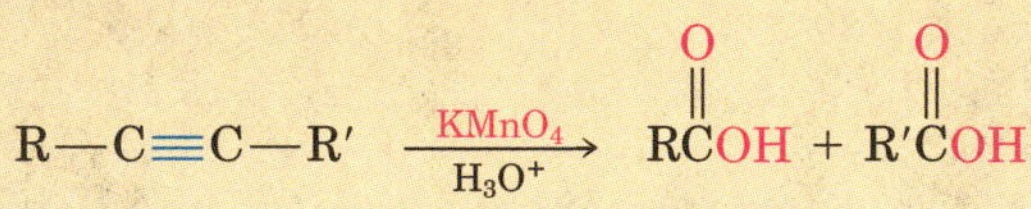

ADDITIONAL PROBLEMS

8.14 Give IUPAC names for the following compounds:

(a) $CH_3CH_2C{\equiv}CC(CH_3)_3$

(b) $CH_3C{\equiv}CCH_2C{\equiv}CCH_2CH_3$

(c) $CH_3CH{=}C(CH_3)C{\equiv}CCH(CH_3)_2$

(d) $HC{\equiv}CC(CH_3)_2CH_2C{\equiv}CH$

(e) $H_2C{=}CHCH{=}CHC{\equiv}CH$

(f)

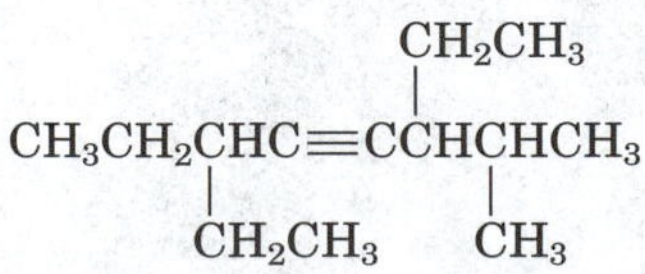

8.15 Draw structures corresponding to the following names:

(a) 3,3-Dimethyl-4-octyne
(b) 3-Ethyl-5-methyl-1,6,8-decatriyne
(c) 2,2,5,5-Tetramethyl-3-hexyne
(d) 3,4-Dimethylcyclodecyne
(e) 3,5-Heptadien-1-yne
(f) 3-Chloro-4,4-dimethyl-1-nonen-6-yne
(g) 3-*sec*-Butyl-1-heptyne
(h) 5-*tert*-Butyl-2-methyl-3-octyne

8.16 The following names are incorrect. Draw the structures and give the correct names.

(a) 1-Ethyl-5,5-dimethyl-1-hexyne
(b) 2,5,5-Trimethyl-6-heptyne
(c) 3-Methylhept-5-en-1-yne
(d) 2-Isopropyl-5-methyl-7-octyne
(e) 3-Hexen-5-yne
(f) 5-Ethynyl-1-methylcyclohexane

8.17 The following two hydrocarbons have been isolated from various plants in the sunflower family. Name them according to IUPAC rules.

(a) $CH_3CH{=}CHC{\equiv}CC{\equiv}CCH{=}CHCH{=}CHCH{=}CH_2$ (all trans)
(b) $CH_3C{\equiv}CC{\equiv}CC{\equiv}CC{\equiv}CC{\equiv}CCH{=}CH_2$

8.18 Predict the products of the following reactions.

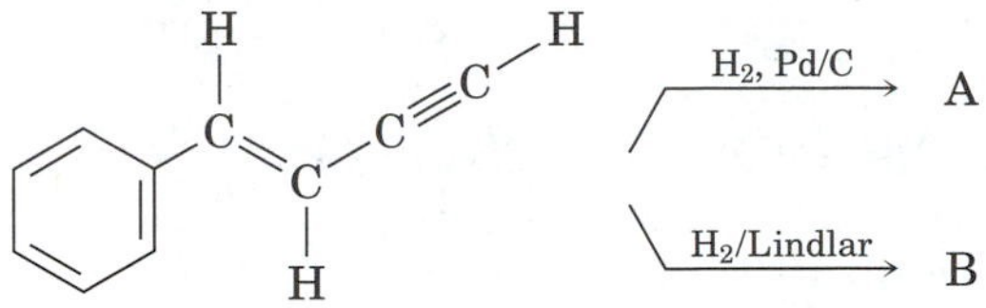

8.19 A hydrocarbon of unknown structure has the formula C_8H_{10}. On catalytic hydrogenation over the Lindlar catalyst, 1 equiv of H_2 is absorbed. On hydrogenation over a palladium catalyst, 3 equiv of H_2 are absorbed.

(a) How many rings/double bonds/triple bonds are present in the unknown?
(b) How many triple bonds are present?
(c) How many double bonds are present?
(d) How many rings are present?

Explain your answers, and draw a structure that fits the data.

8.20 Predict the products from reaction of 1-hexyne with the following reagents.

(a) 1 equiv HBr
(b) 1 equiv Cl_2
(c) H_2, Lindlar catalyst
(d) $NaNH_2$ in NH_3, then CH_3Br
(e) H_2O, H_2SO_4, $HgSO_4$
(f) 2 equiv HCl

8.21 Predict the products from reaction of 5-decyne with the following reagents.

(a) H_2, Lindlar catalyst
(b) Li in NH_3
(c) 1 equiv Br_2
(d) BH_3 in THF, then H_2O_2, OH^-
(e) H_2O, H_2SO_4, $HgSO_4$
(f) Excess H_2, Pd/C catalyst

8.22 Predict the products from reaction of 2-hexyne with the following reagents.

(a) 2 equiv Br_2
(b) 1 equiv HBr
(c) Excess HBr
(d) Li in NH_3
(e) H_2O, H_2SO_4, $HgSO_4$

8.23 Acetonitrile, CH_3CN, contains a carbon–nitrogen triple bond. Sketch the orbitals involved in the bonding in acetonitrile, and indicate the hybridization of each atom.

8.24 Hydrocarbon A has the formula C_9H_{12} and absorbs 3 equiv of H_2 to yield B, C_9H_{18}, when hydrogenated over a Pd/C catalyst. On treatment of A with aqueous H_2SO_4 in the presence of mercuric ion catalyst, two isomeric ketones, C and D, are produced. Oxidation of A with $KMnO_4$ gives a mixture of acetic acid (CH_3COOH) and the tricarboxylic acid E:

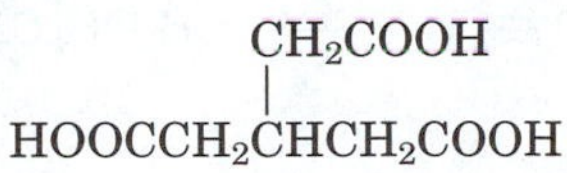

E

Propose structures for compounds A–D, and formulate the reactions.

8.25 How would you carry out the following reactions?

(a) $CH_3CH_2C{\equiv}CH \xrightarrow{?} CH_3CH_2\overset{\overset{O}{\|}}{C}CH_3$

(b) $CH_3CH_2C{\equiv}CH \xrightarrow{?} CH_3CH_2CH_2CHO$

(c) 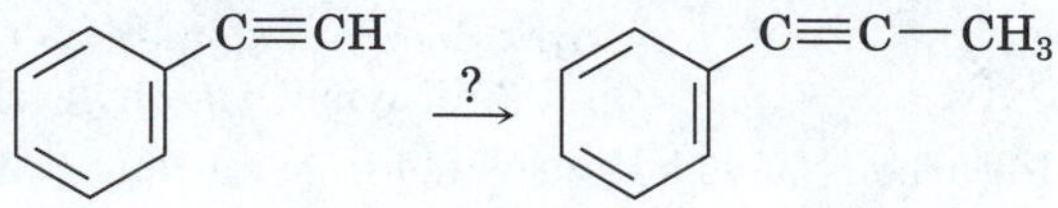

(d) 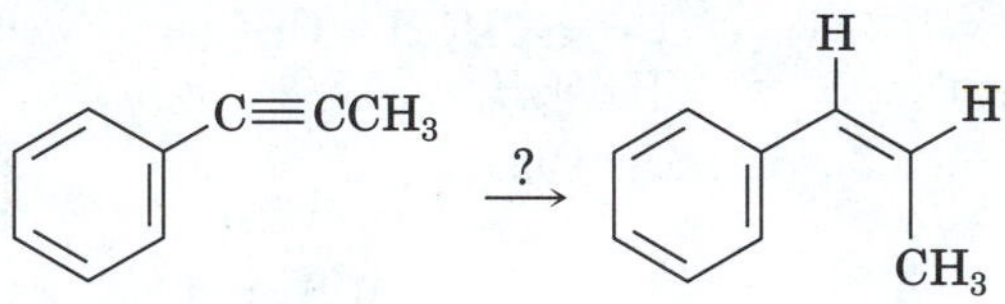

(e) $CH_3CH_2C{\equiv}CH \xrightarrow{?} CH_3CH_2COOH$

(f) $CH_3CH_2CH_2CH_2CH{=}CH_2 \xrightarrow{(2\ steps)} CH_3CH_2CH_2CH_2C{\equiv}CH$

8.26 Occasionally, chemists need to invert the stereochemistry of an alkene; that is, to convert a cis alkene to a trans alkene, or vice versa. There is no one-step method for doing this alkene inversion, but the transformation can be carried out by combining several reactions in the proper sequence. How would you carry out these reactions?

(a) *trans*-5-Decene $\xrightarrow{?}$ *cis*-5-Decene

(b) *cis*-5-Decene $\xrightarrow{?}$ *trans*-5-Decene

8.27 Propose structures for hydrocarbons that give the following products on oxidative cleavage by $KMnO_4$ or O_3.

(a) $CO_2 + CH_3(CH_2)_5COOH$

(b) CH_3COOH + 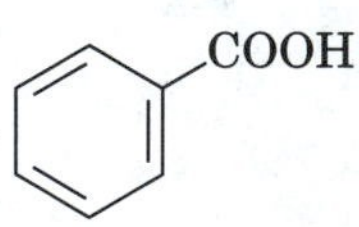

(c) $HOOC(CH_2)_8COOH$

(d) $CH_3CHO + CH_3\overset{\overset{O}{\|}}{C}CH_2CH_2COOH + CO_2$

(e) $OHCCH_2CH_2CH_2CH_2\overset{\overset{O}{\|}}{C}COOH + CO_2$

8.28 Each of the following syntheses requires more than one step. How would you carry them out?

(a) $CH_3CH_2CH_2C{\equiv}CH \xrightarrow{?} CH_3CH_2CH_2CHO$

(b) $(CH_3)_2CHCH_2C{\equiv}CH \xrightarrow{?}$ 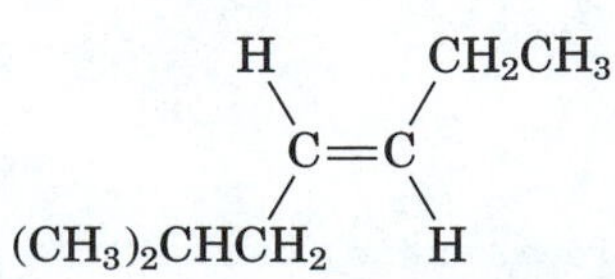

8.29 How would you carry out the following transformation? More than one step is required.

$CH_3CH_2CH_2CH_2C{\equiv}CH \xrightarrow{?}$

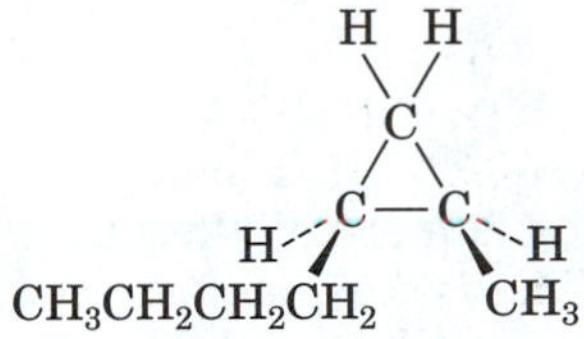

8.30 How would you carry out the following conversion? More than one step is needed.

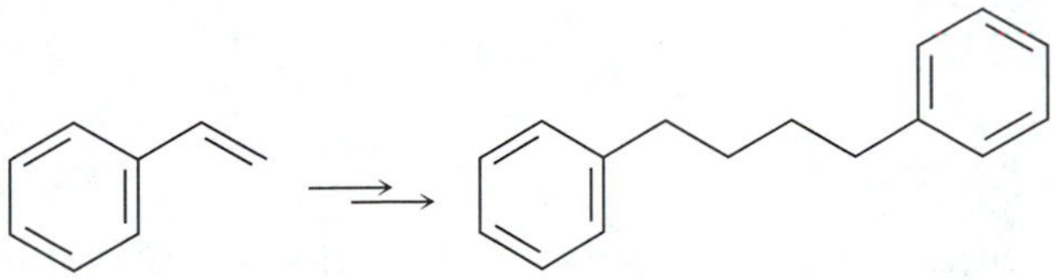

8.31 Predict the products of the following reactions:

(a) $CH_3CH_2CH_2CH_2CH_2C{\equiv}CH \xrightarrow[2.\ H_2O_2]{1.\ BH_3,\ THF}$

(b)

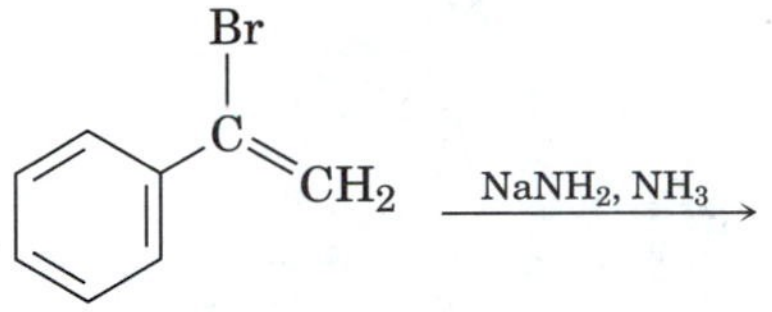

(c)

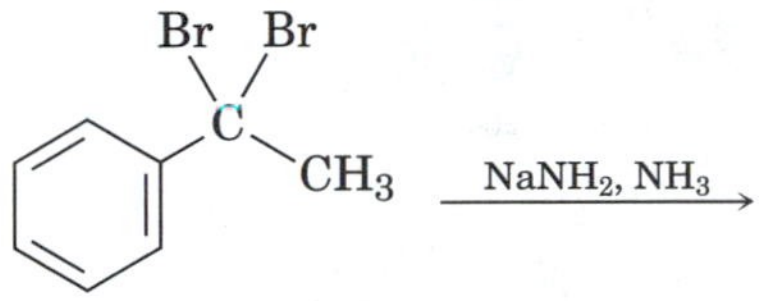

8.32 Using 1-butyne as the only source of carbon, along with any inorganic reagents you need, synthesize the following compounds. More than one step may be needed.
(a) 1,1,2,2-Tetrachlorobutane (b) Octane (c) Butanal

8.33 How would you synthesize the following compounds from acetylene and any alkyl halides with four or fewer carbons? More than one step may be required.

(a) $CH_3CH_2CH_2C{\equiv}CH$

(b) $CH_3CH_2C{\equiv}CCH_2CH_3$

(c) $(CH_3)_2CHCH_2CH{=}CH_2$

(d) $CH_3CH_2CH_2\overset{O}{\overset{\|}{C}}CH_2CH_2CH_2CH_3$

(e) $CH_3CH_2CH_2CH_2CH_2CHO$

8.34 How would you carry out the following reactions to introduce deuterium into organic molecules?

(a) $CH_3CH_2C{\equiv}CCH_2CH_3 \longrightarrow$

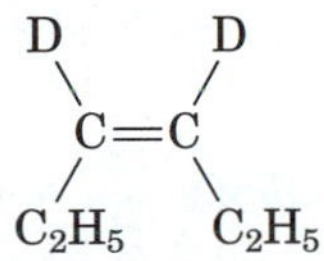

(b) $CH_3CH_2C{\equiv}CCH_2CH_3 \longrightarrow$

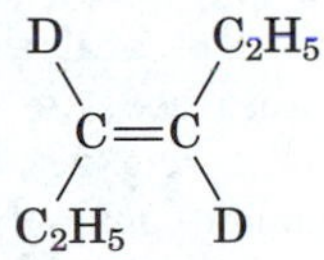

(c) $CH_3CH_2CH_2C{\equiv}CH \longrightarrow CH_3CH_2CH_2C{\equiv}CD$

(d) 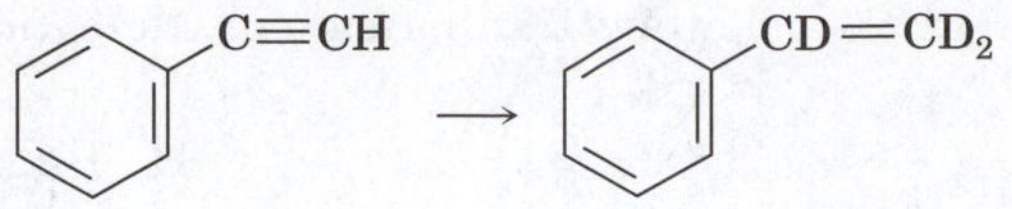

8.35 A *cumulene* is a compound with three adjacent double bonds. Draw an orbital picture of a cumulene. What kind of hybridization do the two central carbon atoms have? What is the geometric relationship of the substituents on one end to the substituents on the other end? What kind of isomerism is possible? Make a model to help see the answer.

$$R_2C{=}C{=}C{=}CR_2$$

A cumulene

8.36 Although it is geometrically impossible for a triple bond to exist in a small ring, large-ring cycloalkynes are quite stable. How would you prepare cyclodecyne starting from acetylene and any alkyl halide needed?

8.37 The sex attractant given off by the common housefly is an alkene named *muscalure*. Propose a synthesis of muscalure starting from acetylene and any alkyl halides. What is the IUPAC name for muscalure?

$$cis\text{-}CH_3(CH_2)_7CH{=}CH(CH_2)_{12}CH_3$$

Muscalure

8.38 Compound A (C_9H_{12}) absorbed 3 equiv of hydrogen on catalytic reduction over a palladium catalyst to give B (C_9H_{18}). On ozonolysis, compound A gave, among other things, a ketone that was identified as cyclohexanone. On treatment with $NaNH_2$ in NH_3, followed by addition of iodomethane, compound A gave C, a new hydrocarbon ($C_{10}H_{14}$). What are the structures of A, B, and C?

8.39 Hydrocarbon A has the formula $C_{12}H_8$. It absorbs 8 equiv of hydrogen on catalytic reduction over a palladium catalyst. On ozonolysis, only two products are formed: oxalic acid (HOOCCOOH) and succinic acid ($HOOCCH_2CH_2COOH$). Formulate these reactions and propose a structure for A.

8.40 Organometallic reagents such as sodium acetylide undergo an addition reaction with ketones, giving alcohols:

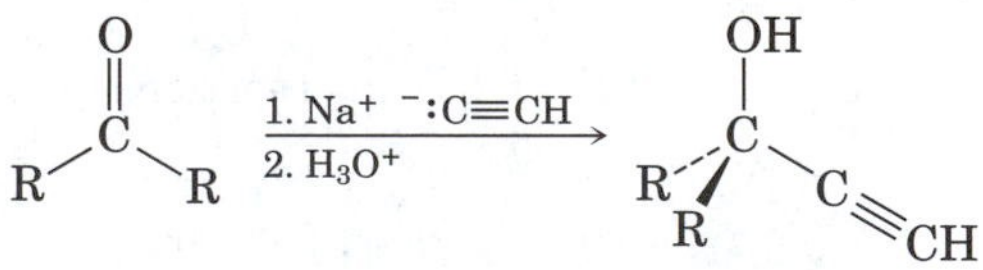

How might you use this reaction to prepare 2-methyl-1,3-butadiene, the starting material used in the manufacture of synthetic rubber?

8.41 Erythrogenic acid, $C_{18}H_{26}O_2$, is an interesting acetylenic fatty acid that turns a vivid red on exposure to light. On catalytic hydrogenation over a palladium catalyst, 5 equiv of hydrogen are absorbed, and stearic acid, $CH_3(CH_2)_{16}COOH$, is produced. Ozonolysis of erythrogenic acid gives four products: formaldehyde, CH_2O; oxalic acid, HOOCCOOH; azelaic acid, $HOOC(CH_2)_7COOH$; and the aldehyde acid $OHC(CH_2)_4COOH$. Draw two possible structures for erythrogenic acid, and suggest a way to tell them apart by carrying out some simple reactions.

8.42 Terminal alkynes react with Br_2 and water to yield bromo ketones. For example:

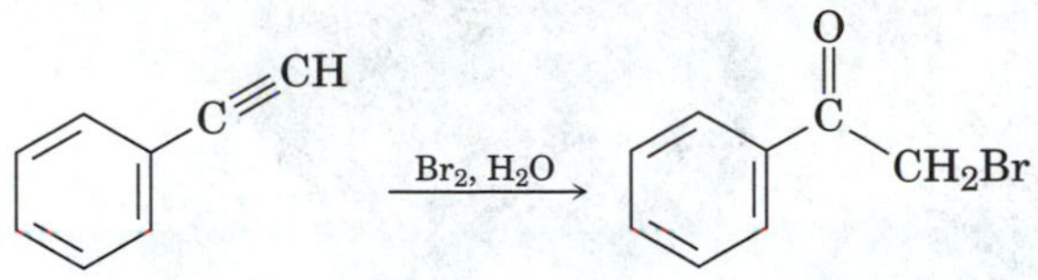

Propose a mechanism for the reaction. To what reaction of alkenes is the process analogous?

A Look Ahead

8.43 Reaction of acetone with deuterated aqueous acid yields hexadeuterioacetone. That is, all the hydrogens in acetone are exchanged for deuterium. Review the mechanism of alkyne hydration, and then propose a mechanism for this deuterium incorporation. We'll look at the process in more detail in Chapter 22.

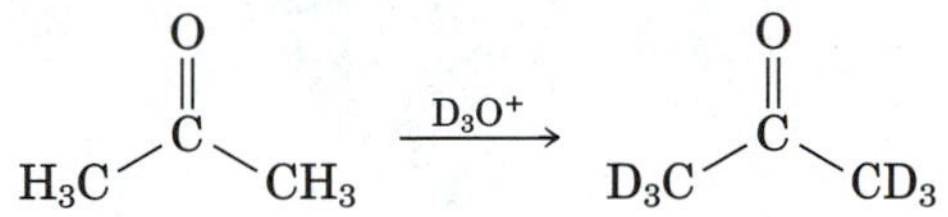

Acetone **Hexadeuterioacetone**

These two models of alanine [$CH_3CH(NH_2)COOH$] are *enantiomers,* or mirror images.

9

Stereochemistry

Are you right-handed or left-handed? Though most of us don't often think about it, handedness plays a surprisingly large role in our daily activities. Many musical instruments, such as oboes and clarinets, have a handedness to them; the last available softball glove always fits the wrong hand; left-handed people write in a "funny" way. The fundamental reason for these difficulties is that our hands aren't identical, they're *mirror images*. When you hold a *right* hand up to a mirror, the image you see looks like a *left* hand. Try it.

Handedness also plays a large role in organic chemistry as a direct consequence of the tetrahedral stereochemistry of sp^3-hybridized carbon. Let's see how handedness in organic molecules arises.

9.1 Enantiomers and the Tetrahedral Carbon

Look at the generalized molecules of the type CH_3X, CH_2XY, and CHXYZ shown in Figure 9.1. On the left are three molecules, and on the right are their images reflected in a mirror. The CH_3X and CH_2XY molecules are identical to their mirror images and thus are not handed. If you make a molecular model of each molecule and of its mirror image, you can superimpose one on the other.

Unlike the CH_3X and CH_2XY molecules, the CHXYZ molecule is *not* identical to its mirror image. You can't superimpose a model of the molecule on a model of its mirror image for the same reason that you can't superimpose a left hand on a right hand. You might get *two* of the substituents superimposed, X and Y for example, but H and Z would be reversed. If the H and Z substituents were superimposed, X and Y would be reversed.

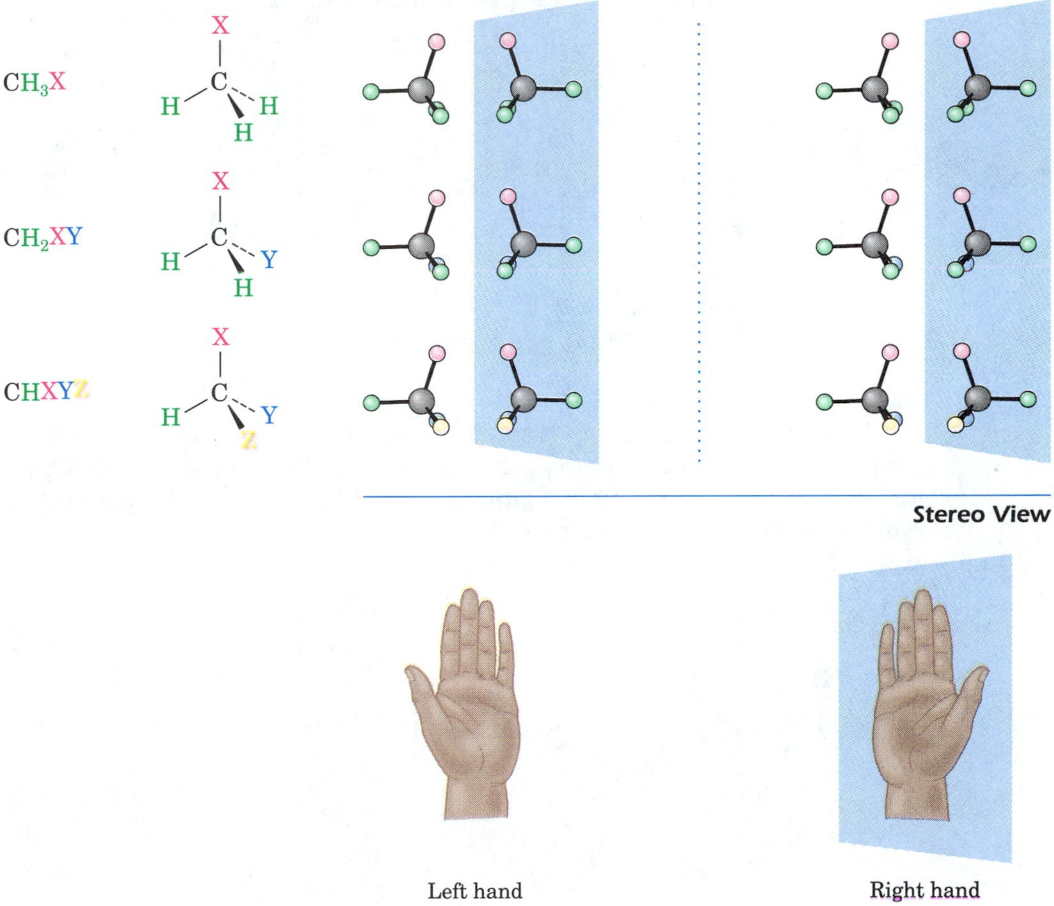

Figure 9.1 Tetrahedral carbon atoms and their mirror images. Molecules of the type CH_3X and CH_2XY are identical to their mirror images, but a molecule of the type CHXYZ is not. A CHXYZ molecule is related to its mirror image in the same way that a right hand is related to a left hand.

A molecule that is not identical to its mirror image is a special kind of stereoisomer called an **enantiomer** (e-**nan**-tee-o-mer; Greek *enantio*, "opposite"). Enantiomers are related to each other as a right hand is related to a left hand and result whenever a tetrahedral carbon is bonded to four different substituents (one need not be H). For example, lactic acid (2-hydroxypropanoic acid) exists as a pair of enantiomers because there are four different groups (–H, –OH, $-CH_3$, –COOH) bonded to the central carbon atom:

Lactic acid: a molecule of general formula CHXYZ

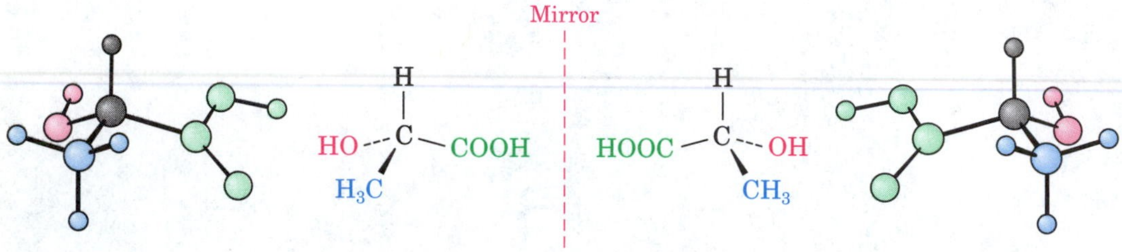

(+)-Lactic acid $[\alpha]_D = +3.82°$ **(−)-Lactic acid** $[\alpha]_D = -3.82°$

No matter how hard you try, you can't superimpose a molecule of (+)-lactic acid on a molecule of (−)-lactic acid; the two simply aren't identical, as Figure 9.2 shows. If any two groups, say –H and –COOH, match up, the remaining two groups don't match.

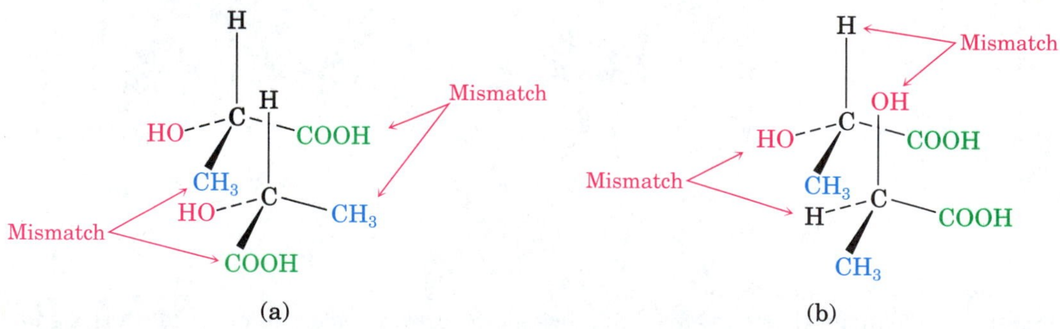

Figure 9.2 Attempts at superimposing the mirror-image forms of lactic acid: (a) When the –H and –OH substituents match up, the –COOH and $-CH_3$ substituents don't; (b) when –COOH and $-CH_3$ match up, –H and –OH don't. Regardless of how the molecules are oriented, they aren't identical.

9.2 The Reason for Handedness in Molecules: Chirality

Molecules that are not identical to their mirror images and thus exist in two enantiomeric forms are said to be **chiral** (**ky**-ral, from the Greek *cheir*, "hand"). You can't take a chiral molecule and its mirror image (enantiomer) and place one on the other so that all atoms coincide.

How can you predict whether a given molecule is or is not chiral? *A molecule can't be chiral if it contains a plane of symmetry.* A **plane of symmetry** is a plane that cuts through an object (or molecule) in such a way that one half of the object is an exact mirror image of the other half. For example, a laboratory flask has a plane of symmetry. If you were to cut the flask in half, one half would be an exact mirror image of the second half. A hand, however, has no plane of symmetry. One "half" of a hand is not a mirror image of the other half (Figure 9.3).

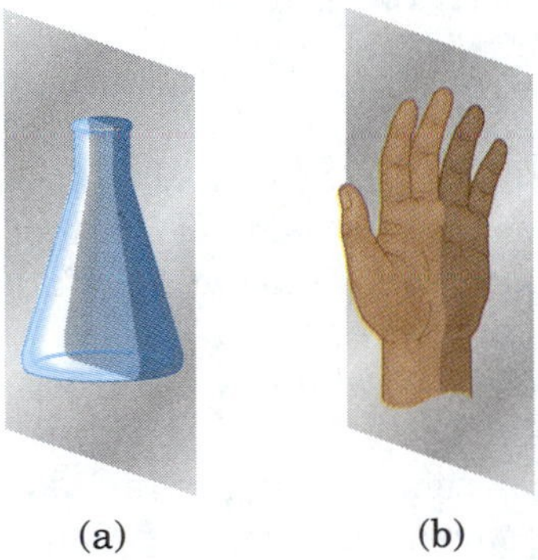

Figure 9.3 The meaning of *symmetry plane*. An object like the flask (a) has a symmetry plane cutting through it, making right and left halves mirror images. An object like a hand (b) has no symmetry plane; the right "half" of a hand is not a mirror image of the left half.

A molecule that has a plane of symmetry in any of its possible conformations must be identical to its mirror image and hence must be nonchiral, or **achiral** (a-**ky**-ral). Thus, propanoic acid has a plane of symmetry when it is lined up as shown in Figure 9.4 (p. 298) and is therefore achiral. Lactic acid (2-hydroxypropanoic acid), however, has no plane of symmetry and is thus chiral.

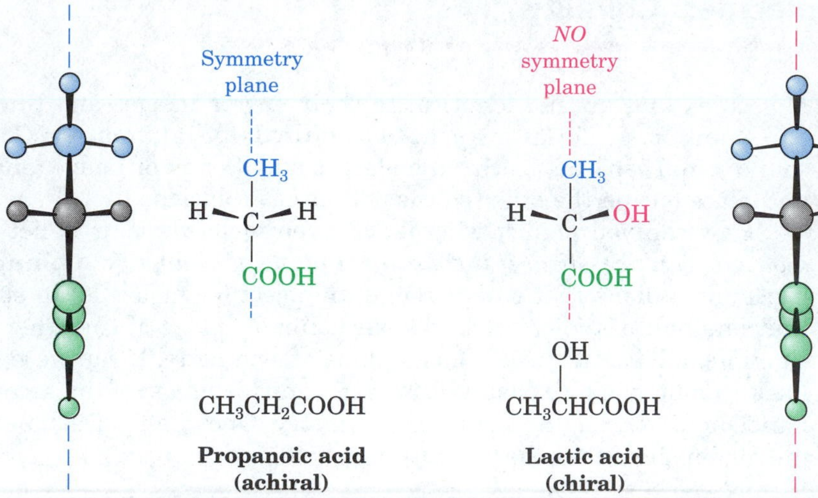

Figure 9.4 The achiral propanoic acid molecule versus the chiral lactic acid molecule. Propanoic acid has a plane of symmetry that makes one side of the molecule a mirror image of the other side. Lactic acid, however, has no such symmetry plane.

The most common (although not the only) cause of chirality in organic molecules is the presence of a carbon atom bonded to four different groups—for example, the central carbon atom in lactic acid. Such carbons are referred to as *asymmetric centers*, or **stereogenic centers**. Note that *chirality* is a property of the entire molecule, whereas a stereogenic center is the *cause* of chirality.

Detecting stereogenic centers in a complex molecule takes practice, because it's not always immediately apparent that four different groups are bonded to a given carbon. The differences don't necessarily appear right next to the stereogenic center. For example, 5-bromodecane is a chiral molecule because four different groups are bonded to C5, the stereogenic center (marked by an asterisk):

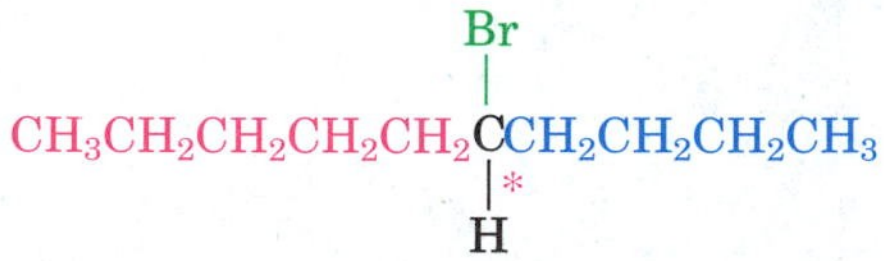

5-Bromodecane (chiral)

Substituents on carbon 5

—H

—Br

—$CH_2CH_2CH_2CH_3$ (butyl)

—$CH_2CH_2CH_2CH_2CH_3$ (pentyl)

A butyl substituent is *similar* to a pentyl substituent but is not identical. The difference isn't apparent until four carbon atoms away from the stereogenic center, but there's still a difference.

As other examples, look at methylcyclohexane and 2-methylcyclohexanone. Are either of these molecules chiral?

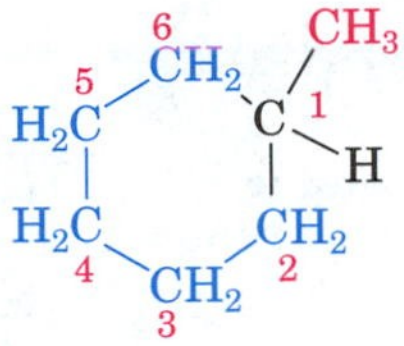

Methylcyclohexane (achiral)

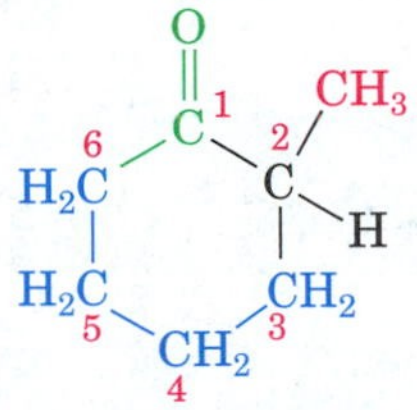

2-Methylcyclohexanone (chiral)

Methylcyclohexane is achiral because no carbon atom in the molecule is bonded to four different groups. You can immediately eliminate all $-CH_2-$ carbons and the $-CH_3$ carbon from consideration, but what about C1 on the ring? The C1 carbon atom is bonded to a $-CH_3$ group, to an –H atom, and to C2 and C6 of the ring. Carbons 2 and 6 are equivalent, however, as are carbons 3 and 5. Thus, the C6-C5-C4 "substituent" is equivalent to the C2-C3-C4 substituent, and methylcyclohexane is therefore achiral. Another way of reaching the same conclusion is to realize that methylcyclohexane has a symmetry plane passing through the methyl group and through carbons 1 and 4 of the ring. Make a molecular model to see this symmetry plane more clearly.

The situation is different for 2-methylcyclohexanone. 2-Methylcyclohexanone has no symmetry plane and is chiral because C2 is bonded to four different groups: a $-CH_3$ group, an –H atom, a $-COCH_2-$ ring bond (C1), and a $-CH_2CH_2-$ ring bond (C3). Additional examples of chiral molecules are shown below. Check for yourself that the labeled centers are stereogenic. (*Remember*: $-CH_2-$, $-CH_3$, C=C, and C≡C carbons *can't* be stereogenic centers.)

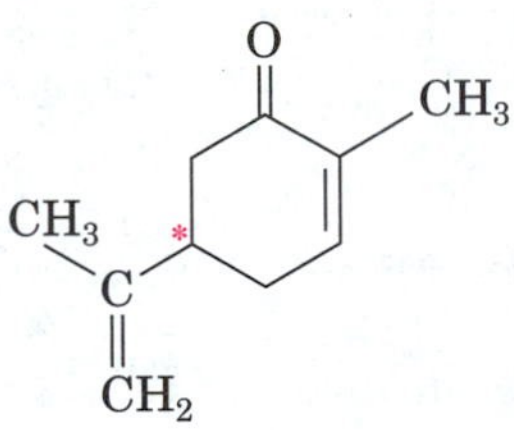

Carvone (spearmint oil)

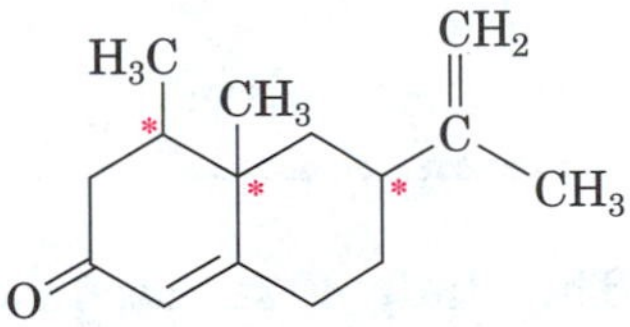

Nootkatone (grapefruit oil)

PROBLEM

9.1 Which of these objects are chiral?

(a) A screwdriver
(b) A screw
(c) A bean stalk
(d) A shoe
(e) A hammer

PROBLEM

9.2 Which of the following compounds are chiral? Build molecular models for help.

(a) 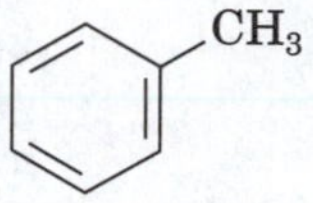

Toluene

(b) 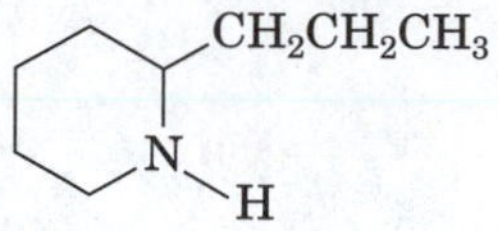

Coniine (from poison hemlock)

(c) 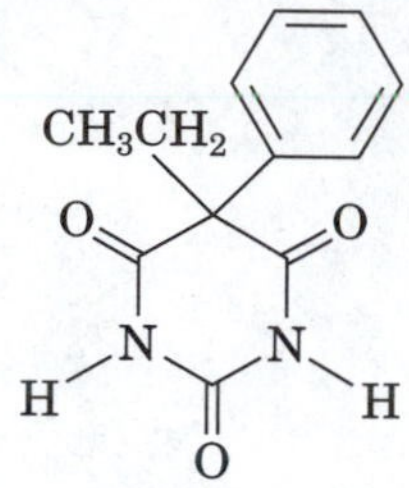

Phenobarbital (tranquilizer)

PROBLEM

9.3 Place asterisks at all stereogenic centers in these molecules:

(a) 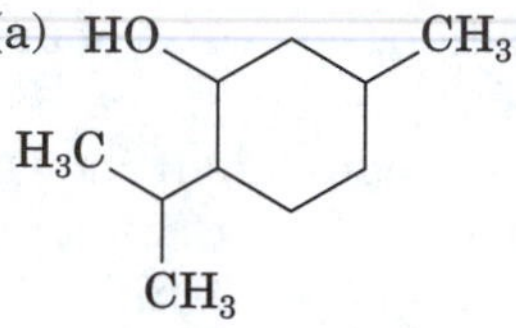

Menthol

(b) 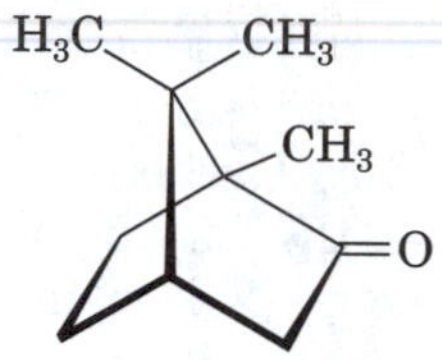

Camphor

(c) 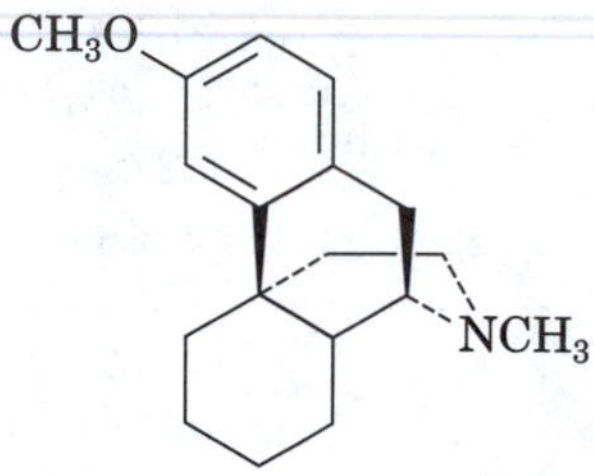

Dextromethorphan (a cough suppressant)

PROBLEM

9.4 Alanine, an amino acid found in proteins, is chiral. Draw the two enantiomers of alanine using the standard convention of wedged, solid, and dashed lines.

$$\begin{array}{c} \quad NH_2 \\ \quad | \\ CH_3CHCOOH \end{array} \quad \textbf{Alanine}$$

9.3 Optical Activity

The study of stereochemistry has its origins in the work of the French scientist Jean Baptiste Biot[1] in the early nineteenth century. Biot, a physicist, was investigating the nature of **plane-polarized light**. A beam of ordinary light consists of electromagnetic waves that oscillate in an infinite number of planes at right angles to the direction of light travel. When a beam of ordinary light is passed through a device called a *polarizer*, however, only the light waves oscillating in a *single* plane pass through—hence the name *plane-polarized light*. Light waves in all other planes are blocked out. The polarization process is represented in Figure 9.5.

[1]Jean Baptiste Biot (1774–1862); b. Paris; physicist, College de France.

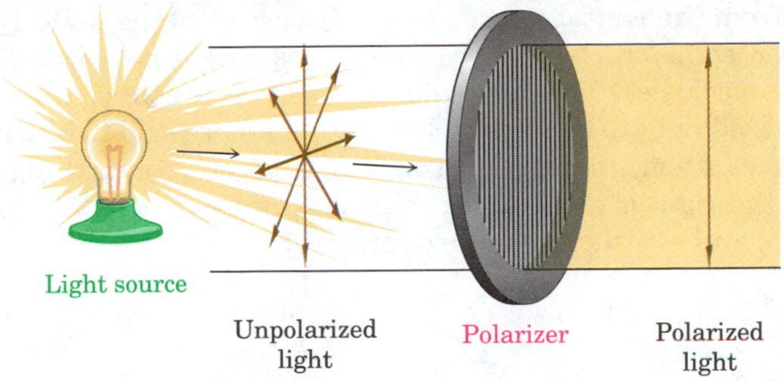

Figure 9.5 Plane-polarized light. Oscillation of the electromagnetic field occurs in a single plane.

Biot made the remarkable observation that, when a beam of plane-polarized light passes through a solution of certain organic molecules, such as sugar or camphor, the plane of polarization is *rotated*. Not all organic substances exhibit this property, but those that do are said to be **optically active**.

The amount of rotation can be measured with an instrument known as a *polarimeter*, represented schematically in Figure 9.6. A solution of optically active organic molecules is placed in a sample tube, plane-polarized light is passed through the tube, and rotation of the plane occurs. The light then goes through a second polarizer called the *analyzer*. By rotating the analyzer until the light passes through it, we can find the new plane of polarization and can tell to what extent rotation has occurred. The amount of rotation is denoted α (Greek alpha) and is expressed in degrees.

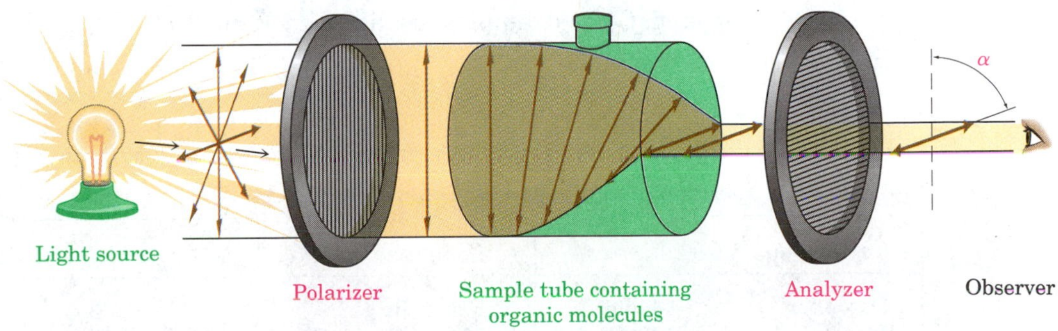

Figure 9.6 Schematic representation of a polarimeter. Plane-polarized light passes through a solution of optically active molecules, which rotate the plane of polarization.

In addition to determining the extent of rotation, we can also find the direction. From the vantage point of an observer looking directly end-on at the analyzer, some optically active molecules rotate polarized light to the left (counterclockwise) and are said to be **levorotatory**, whereas others rotate polarized light to the right (clockwise) and are said to be **dextrorotatory**. By convention, rotation to the left is given a minus sign (−), and rotation to the right is given a plus sign (+). For example, (−)-morphine is levorotatory, and (+)-sucrose is dextrorotatory.

9.4 Specific Rotation

The amount of rotation observed in a polarimetry experiment depends on the number of optically active molecules the light beam encounters. The more molecules the light encounters, the greater the observed rotation. Thus, the amount of rotation depends on both sample concentration and sample path length. If we double the concentration of sample, the observed rotation doubles. Similarly, if we keep the concentration constant but double the length of the sample tube, the observed rotation doubles. It also turns out that the amount of rotation depends on the wavelength of the light used.

To express optical rotation data in a meaningful way so that comparisons can be made, we have to choose standard conditions. The **specific rotation, $[\alpha]_D$**, of a compound is defined as the observed rotation when light of 589 nanometer (nm) wavelength is used with a sample path length l of 1 decimeter (1 dm = 10 cm) and a sample concentration C of 1 g/mL. (Light of 589 nm, the so-called sodium D line, is the yellow light emitted from common sodium lamps; 1 nm = 10^{-9} m.)

$$[\alpha]_D = \frac{\text{Observed rotation (degrees)}}{\text{Path length, } l \text{ (dm)} \times \text{Concentration, } C \text{ (g/mL)}} = \frac{\alpha}{l \times C}$$

When optical rotation data are expressed in this standard way, the specific rotation, $[\alpha]_D$, is a physical constant characteristic of a given optically active compound. Some examples are listed in Table 9.1.

Table 9.1 Specific Rotation of Some Organic Molecules

Compound	*$[\alpha]_D$ (degrees)*	*Compound*	*$[\alpha]_D$ (degrees)*
Camphor	+44.26	Penicillin V	+233
Morphine	−132	Monosodium glutamate	+25.5
Sucrose	+66.47	Benzene	0
Cholesterol	−31.5	Acetic acid	0

PROBLEM

9.5 A 1.50 g sample of coniine, the toxic extract of poison hemlock, was dissolved in 10.0 mL of ethanol and placed in a sample cell with a 5.00 cm path length. The observed rotation at the sodium D line was +1.21°. Calculate $[\alpha]_D$ for coniine.

9.5 Pasteur's Discovery of Enantiomers

Little was done after Biot's discovery of optical activity until Louis Pasteur[2] began work in 1849. Pasteur had received his formal training in chemistry but had become interested in the subject of crystallography. He began work on crystalline salts of tartaric acid derived from wine and was repeating some measurements published a few years earlier when he made a surprising observation. On recrystallizing a concentrated solution of sodium ammonium tartrate below 28°C, two distinct kinds of crystals precipitated. Furthermore, the two kinds of crystals were mirror images of each other. That is, the crystals were related to each other in the same way that a right hand is related to a left hand.

Working carefully with a pair of tweezers, Pasteur was able to separate the crystals into two piles, one of "right-handed" crystals and one of "left-handed" crystals like those shown in Figure 9.7. Although the original sample (a 50:50 mixture of right and left) was optically inactive, *solutions of the crystals from each of the sorted piles were optically active*, and their specific rotations were equal in amount but opposite in sign.

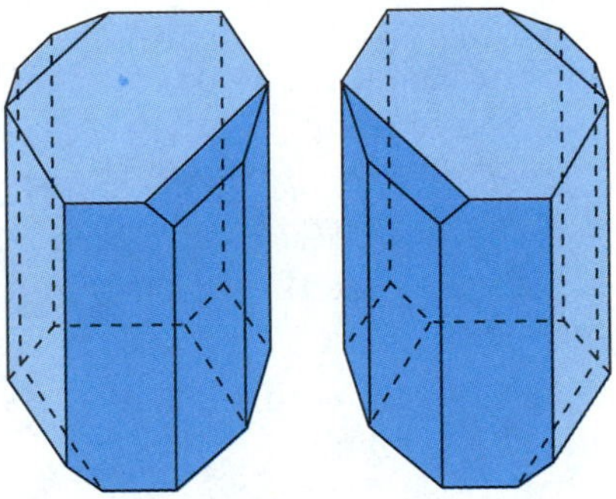

(a) (b)

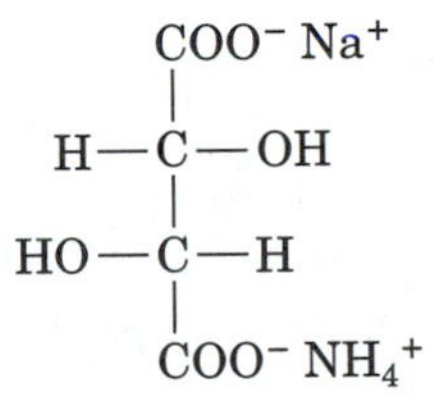

Sodium ammonium tartrate

Figure 9.7 Drawings of sodium ammonium tartrate crystals taken from Pasteur's original sketches. One of the crystals is "right-handed" and one is "left-handed."

[2]Louis Pasteur (1822–1895); b. Dôle, Jura, France; studied at Arbois, Besançon; professor, Dijon, Strasbourg (1849–1854), Lille (1854–1857), École Normale Supérieure (1857–1863).

Pasteur was far ahead of his time. Although the structural theory of Kekulé had not yet been proposed, Pasteur explained his results by speaking of the *molecules themselves*, saying, "There is no doubt that [in the *dextro* tartaric acid] there exists an asymmetric arrangement having a nonsuperimposable image. It is no less certain that the atoms of the *levo* acid possess precisely the inverse asymmetric arrangement." Pasteur's vision was extraordinary, for it was not until 25 years later that the theories of van't Hoff and Le Bel confirmed his ideas regarding the asymmetric carbon atom.

Today, we would describe Pasteur's work by saying that he had discovered the phenomenon of enantiomerism. The enantiomeric tartaric acid salts that Pasteur separated are physically identical in all respects except for their interaction with plane-polarized light. They have the same melting point, the same boiling point, the same solubilities, and the same spectroscopic properties.

9.6 Sequence Rules for Specification of Configuration

Although drawings provide a pictorial representation of stereochemistry, they are difficult to translate into words. Thus, a verbal method for indicating the three-dimensional arrangement of atoms (the **configuration**) at a stereogenic center is also necessary. The standard method employs the same Cahn–Ingold–Prelog sequence rules used for the specification of alkene geometry (*Z* versus *E*) in Section 6.6. Let's briefly review the sequence rules to see how they're used to specify the configuration of a stereogenic center. Refer to Section 6.6 for an explanation of each rule.

1. Look at the four atoms directly attached to the stereogenic center and assign priorities in order of decreasing atomic number. The atom with highest atomic number is ranked first; the atom with lowest atomic number is ranked fourth.
2. If a decision about priority can't be reached by applying rule 1, compare atomic numbers of the second atoms in each substituent, continuing on as necessary through the third or fourth atoms until a point of difference is reached.
3. Multiple-bonded atoms are considered equivalent to the same number of single-bonded atoms. For example:

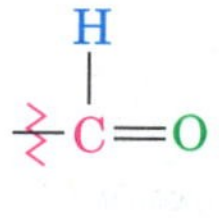

is equivalent to

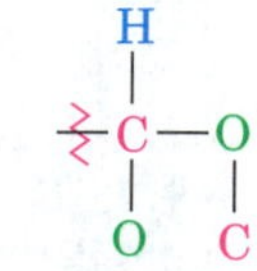

Having assigned priorities to the four groups attached to a stereogenic carbon, we describe the stereochemical configuration around the carbon by comparing the four groups to the thumb and fingers of a hand. Hold up one of your hands, with fingers curled and thumb outstretched, and orient it so that the thumb points from the carbon to the group of lowest priority (4). If you are using your *right* hand and the fingers curl in the direction of decreasing priority (1 → 2 → 3) for the remaining three groups, we say that the stereogenic center has the *R* configuration (Latin *rectus*, "right").[3] If it's the fingers of your *left* hand that curl in the 1 → 2 → 3 direction, the stereogenic center has the *S* configuration (Latin *sinister*, "left"). The assignments are shown in Figure 9.8.

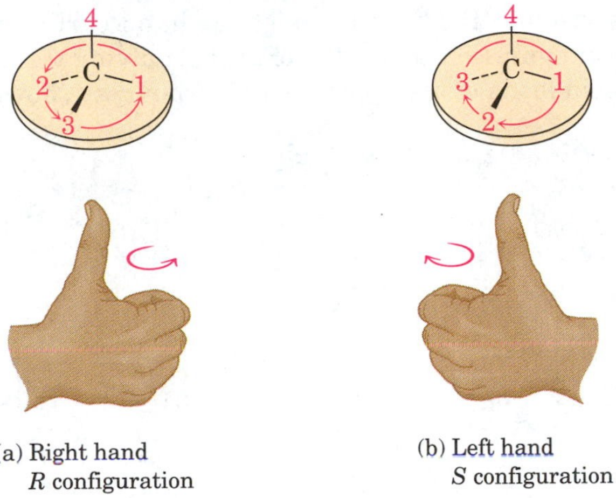

Figure 9.8 Assigning configuration to a stereogenic carbon. Priorities are assigned to the four groups, and the molecule is compared to a hand oriented so that the thumb points from the carbon to the group of lowest priority (4). (a) If the fingers of your *right* hand curl in the direction of decreasing priority (1 → 2 → 3) for the remaining three groups, the stereogenic center has the *R* configuration. (b) If the fingers of your *left* hand curl in the 1 → 2 → 3 direction, the stereogenic center has the *S* configuration.

Look at (−)-lactic acid to see how configuration can be assigned. Sequence rule 1 says that –OH has priority 1 and –H has priority 4, but it doesn't allow us to distinguish between $-CH_3$ and –COOH because both groups have carbon as their first atom. Sequence rule 2, however, says that –COOH is higher priority than $-CH_3$ because oxygen outranks hydrogen (the second atom in each group).

---

[3] *Rectus* actually means "right" in the sense of "correct," whereas *dexter* is the Latin word for "right" in the directional sense. In some European societies, however, left-handedness has historically been looked on with disfavor, and it has been considered "correct" to be right-handed.

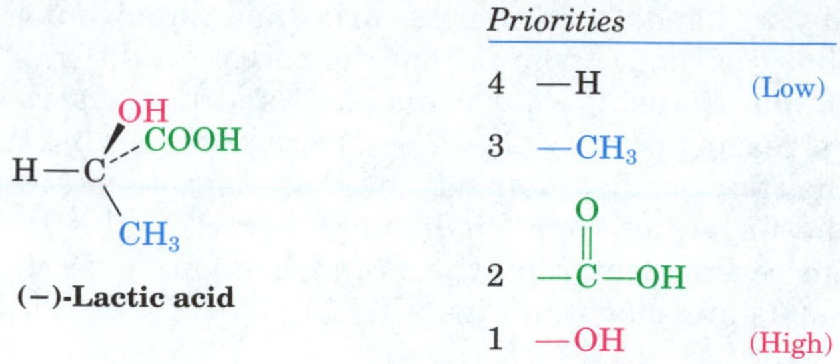

When a right hand is held so that the thumb points from the carbon to the fourth-priority group (–H), the fingers of the hand curl from 1 → 2 → 3, so we assign *R* configuration to (−)-lactic acid. Applying the same procedure to (+)-lactic acid leads to the opposite assignment, as shown in Figure 9.9.

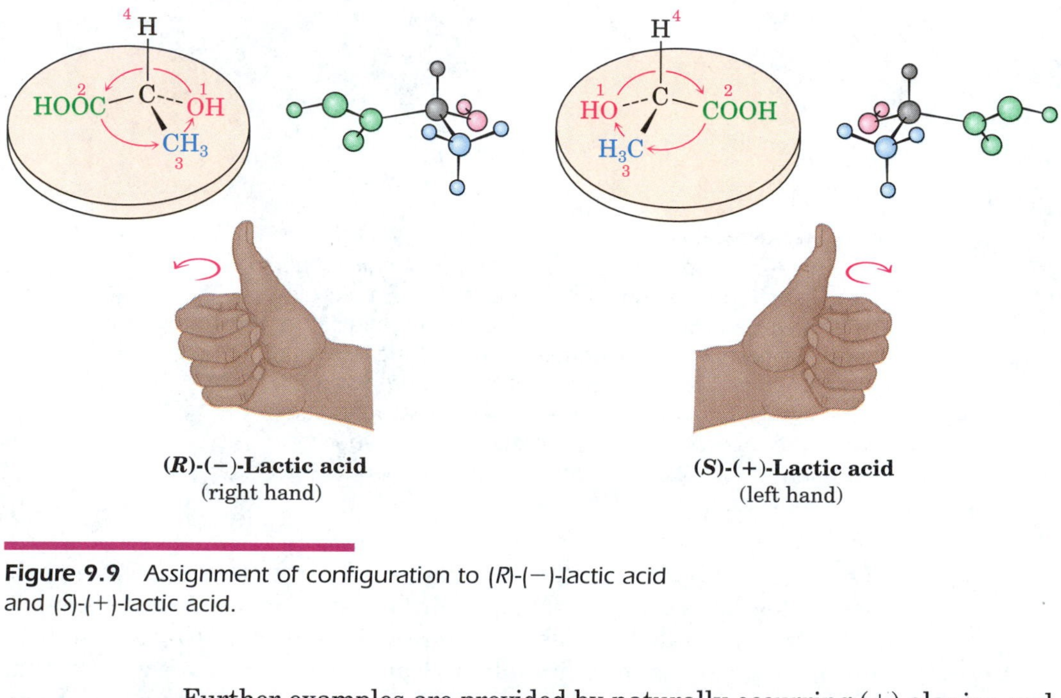

Figure 9.9 Assignment of configuration to (*R*)-(−)-lactic acid and (*S*)-(+)-lactic acid.

Further examples are provided by naturally occurring (+)-alanine and (−)-glyceraldehyde, which have the *S* configurations shown in Figure 9.10. *Note that the sign of optical rotation, (+) or (−), is not related to the R,S designation.* (*S*)-Alanine happens to be dextrorotatory (+), and (*S*)-glyceraldehyde happens to be levorotatory (−). There is no simple correlation between *R*,*S* configuration and direction or magnitude of optical rotation.

One further point needs mentioning: the matter of **absolute configuration**. How do we know that our assignments of *R*,*S* configuration are correct in an *absolute*, rather than a relative, sense? Since we can't see the

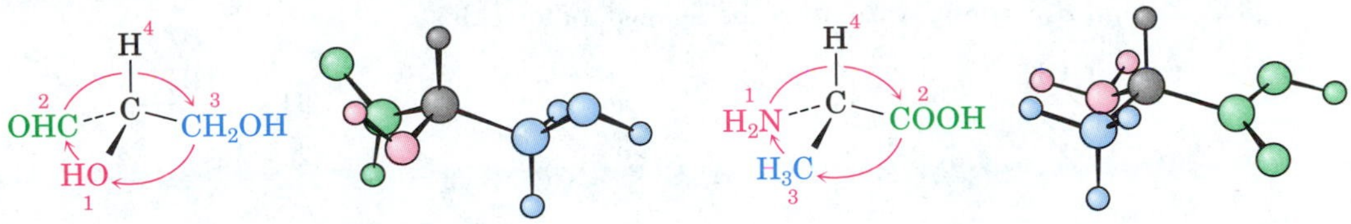

(S)-Glyceraldehyde
[(S)-(−)-2,3-Dihydroxypropanal]
$[\alpha]_D = -8.7°$

(S)-Alanine
[(S)-(+)-2-Aminopropanoic acid]
$[\alpha]_D = +8.5°$

Figure 9.10 Assignment of configuration to (−)-glyceraldehyde and (+)-alanine. Both happen to have the *S* configuration, although one is levorotatory and the other is dextrorotatory.

molecules themselves, how do we know for certain that it is the dextrorotatory enantiomer of lactic acid that has the *R* configuration? This difficult question was not solved until 1951 when J. M. Bijvoet of the University of Utrecht reported an X-ray spectroscopic method for determining the absolute spatial arrangement of atoms in a molecule. Based on his results, we can say with certainty that the *R*,*S* conventions are correct.

PRACTICE PROBLEM

Draw a tetrahedral representation of (*R*)-2-chlorobutane.

Solution The four substituents bonded to the stereogenic carbon of (*R*)-2-chlorobutane can be assigned the following priorities: (1) –Cl, (2) $-CH_2CH_3$, (3) $-CH_3$, (4) –H. To draw a tetrahedral representation of the molecule, it's easiest to orient the low-priority –H group toward the top and then arrange the other three groups so that the direction of travel from 1 → 2 → 3 corresponds to the curl of your right hand:

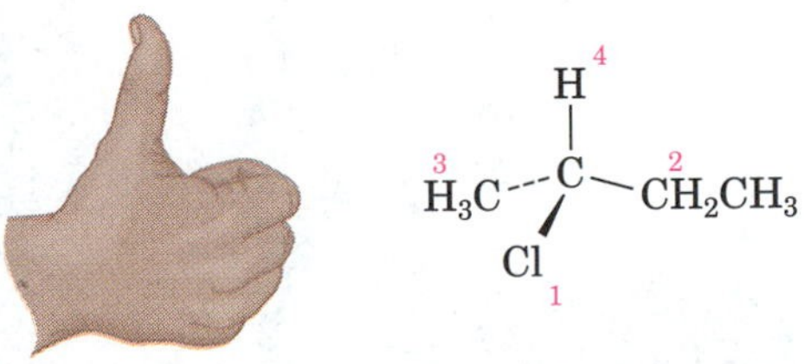

(R)-2-Chlorobutane

Using molecular models is a great help in working problems of this sort.

PROBLEM

9.6 Assign priorities to these sets of substituents:
(a) —H, —Br, $-CH_2CH_3$, $-CH_2CH_2OH$
(b) $-CO_2H$, $-CO_2CH_3$, $-CH_2OH$, —OH
(c) —CN, $-CH_2NH_2$, $-CH_2NHCH_3$, $-NH_2$
(d) —Br, $-CH_2Br$, —Cl, $-CH_2Cl$

PROBLEM.

9.7 Assign *R*,*S* configurations to the following molecules:

(a) 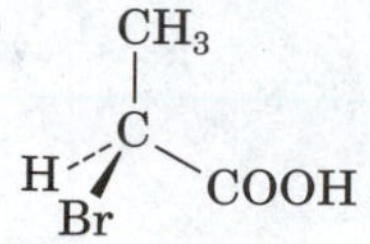

(b) 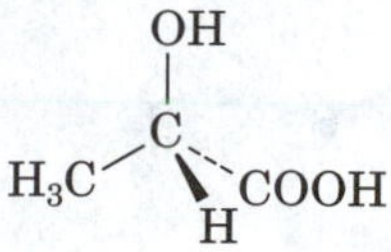

(c) 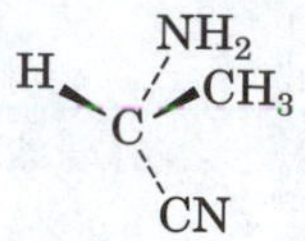

PROBLEM.

9.8 Draw a tetrahedral representation of (*S*)-2-pentanol (2-hydroxypentane).

9.7 Diastereomers

Molecules like lactic acid, alanine, and glyceraldehyde are relatively simple because each has only one stereogenic center and can exist in only two enantiomeric forms. The situation becomes more complex, however, with molecules that have more than one stereogenic center.

Look at the amino acid threonine (2-amino-3-hydroxybutanoic acid), for example. Since threonine has two stereogenic centers (C2 and C3), there are four possible stereoisomers, as shown in Figure 9.11. Check for yourself that the *R*,*S* configurations are correct as indicated.

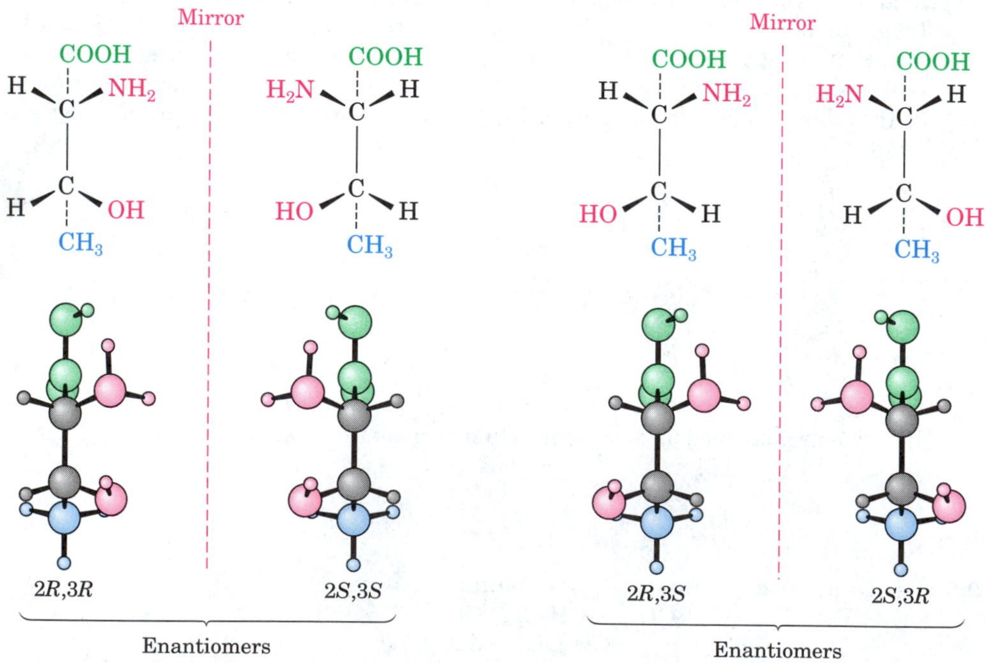

Figure 9.11 The four stereoisomers of 2-amino-3-hydroxybutanoic acid (threonine).

The four threonine stereoisomers can be classified into two pairs of enantiomers, or mirror images. The 2*R*,3*R* stereoisomer is the mirror image of 2*S*,3*S*, and the 2*R*,3*S* stereoisomer is the mirror image of 2*S*,3*R*. But what is the relationship between any two configurations that are not mirror images? What, for example, is the relationship between the 2*R*,3*R* isomer and the 2*R*,3*S* isomer? They are stereoisomers, yet they aren't enantiomers. To describe such a relationship, we need a new term—*diastereomer.*

Diastereomers are stereoisomers that are not mirror images of each other. Diastereomers have opposite configurations at *some* (one or more) stereogenic centers, but have the same configuration at others. Enantiomers, by contrast, have opposite configurations at *all* stereogenic centers. A full description of the four threonine stereoisomers is given in Table 9.2.

Table 9.2 Relationships Among Four Stereoisomers of Threonine

Stereoisomer	*Enantiomeric with*	*Diastereomeric with*
2*R*,3*R*	2*S*,3*S*	2*R*,3*S* and 2*S*,3*R*
2*S*,3*S*	2*R*,3*R*	2*R*,3*S* and 2*S*,3*R*
2*R*,3*S*	2*S*,3*R*	2*R*,3*R* and 2*S*,3*S*
2*S*,3*R*	2*R*,3*S*	2*R*,3*R* and 2*S*,3*S*

Of the four stereoisomers of threonine, only the 2*S*,3*R* isomer, $[\alpha]_D = -29.3°$, occurs naturally in plants and animals. This result is typical: Most biologically important molecules are chiral, and usually only a single stereoisomer is found in nature.

PROBLEM. . . . .

9.9 Assign *R,S* configurations to each stereogenic center in the following molecules. Which are enantiomers and which are diastereomers?

(a)
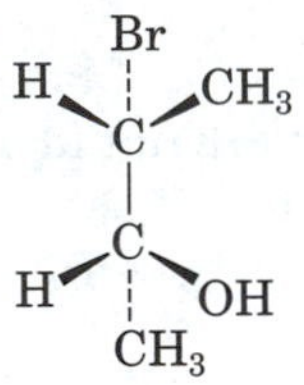

(b)
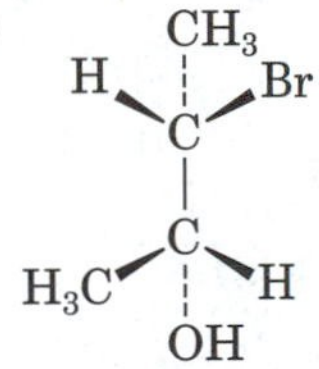

(c)
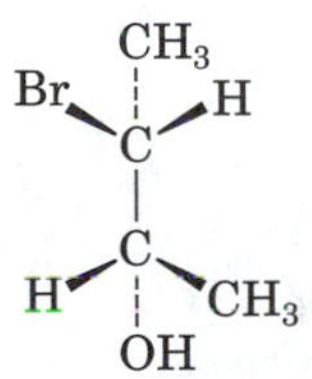

(d)
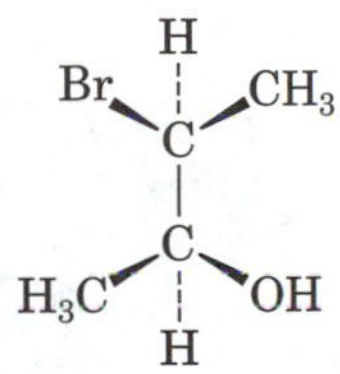

PROBLEM. . . . .

9.10 Chloramphenicol, a powerful antibiotic isolated in 1949 from the *Streptomyces venezuelae* bacterium, is active against a broad spectrum of bacterial infections and is particularly valuable against typhoid fever. Assign *R,S* configurations to the stereogenic centers in chloramphenicol.

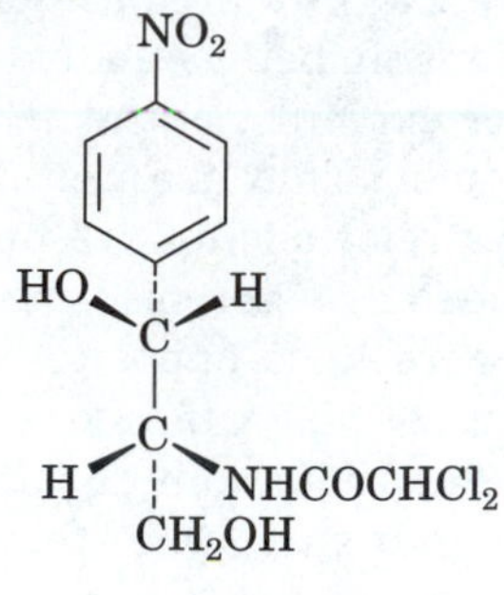

Chloramphenicol
$[\alpha]_D = +18.6°$

9.8 Meso Compounds

Let's look at one more example of a compound with two stereogenic centers: tartaric acid. We're already acquainted with tartaric acid because of its role in Pasteur's discovery of optical activity, and we can now draw the four stereoisomers:

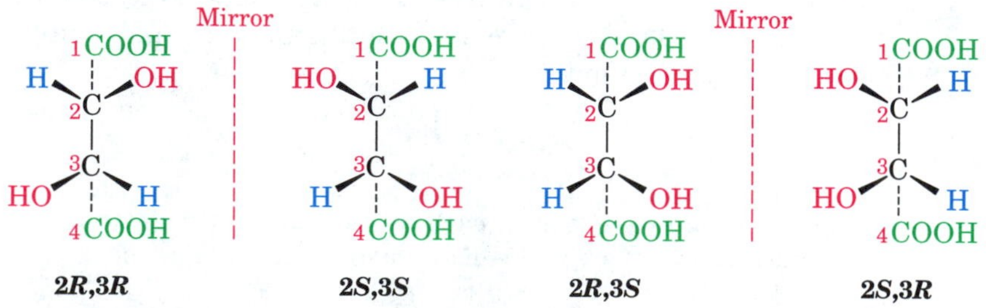

The mirror-image 2*R*,3*R* and 2*S*,3*S* structures are not identical and are therefore a pair of enantiomers. A careful look, however, shows that the 2*R*,3*S* and 2*S*,3*R* structures *are* identical, as can be seen by rotating one structure 180°:

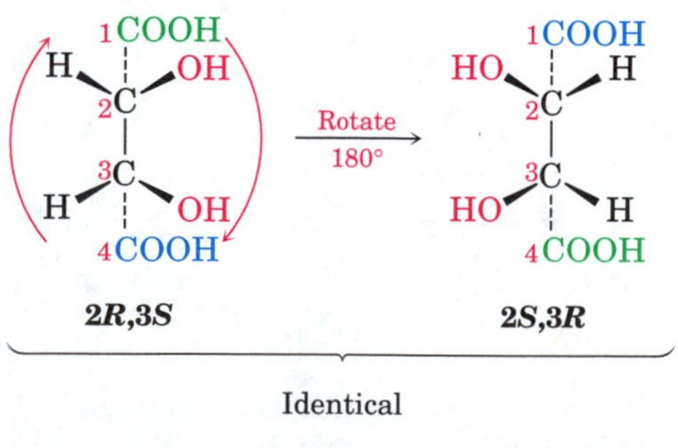

The 2*R*,3*S* and 2*S*,3*R* structures are identical because the molecule has a plane of symmetry and is therefore achiral. The symmetry plane cuts through the C2–C3 bond, making one half of the molecule a mirror image of the other half (Figure 9.12).

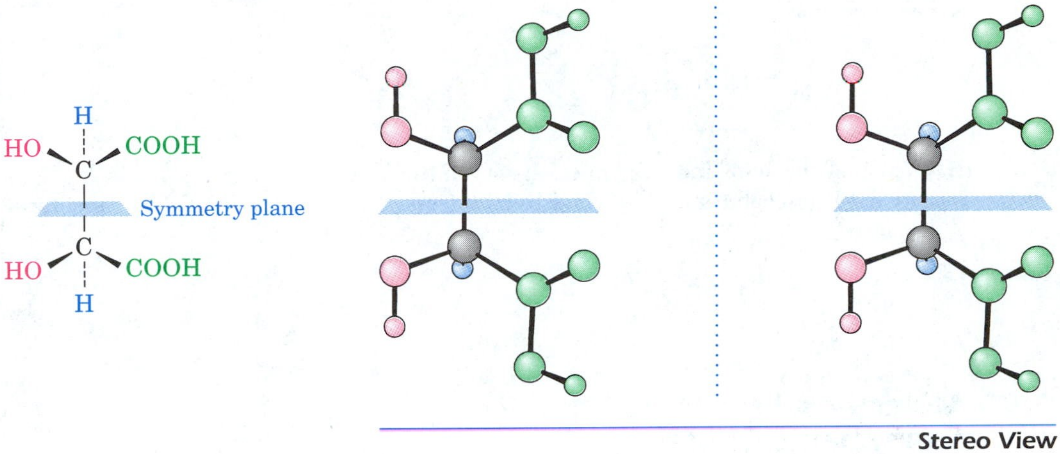

Figure 9.12 A symmetry plane through the C2–C3 bond of *meso*-tartaric acid makes the molecule achiral.

Because of the plane of symmetry, the tartaric acid stereoisomer shown in Figure 9.12 must be achiral, despite the fact that it has two stereogenic centers. Compounds that are achiral, yet contain stereogenic centers, are called **meso compounds**(**me**-zo). Thus, tartaric acid exists in three stereoisomeric forms: two enantiomers and one meso form.

PRACTICE PROBLEM

Does *cis*-1,2-dimethylcyclobutane have any stereogenic centers? Is it chiral?

Solution A look at the structure of *cis*-1,2-dimethylcyclobutane shows that both methyl-bearing ring carbons (C1 and C2) are stereogenic centers. Overall, though, the compound is achiral because there is a symmetry plane bisecting the ring between C1 and C2. Thus, the molecule is a meso compound.

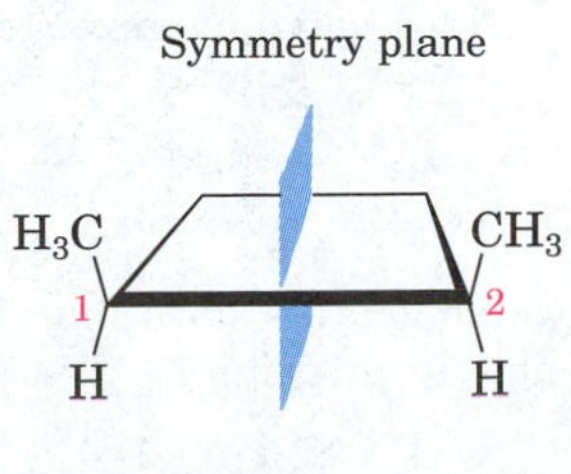

PROBLEM

9.11 Which of the following structures represent meso compounds?

(a)
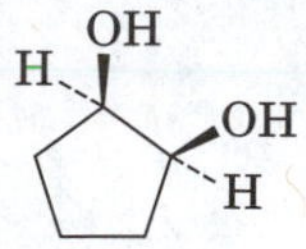

(b)
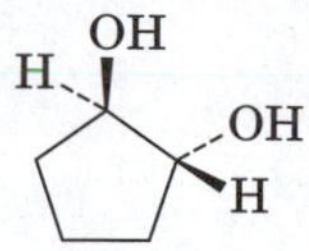

(c)
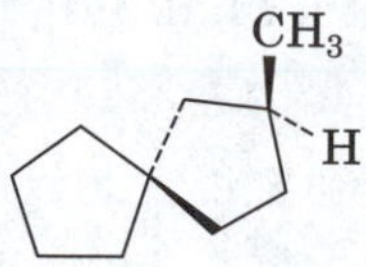

(d)
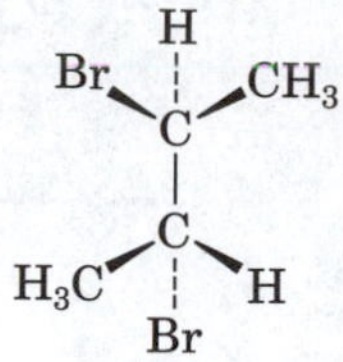

PROBLEM

9.12 Which of the following substances have a meso form?
(a) 2,3-Dibromobutane (b) 2,3-Dibromopentane (c) 2,4-Dibromopentane

9.9 Molecules with More Than Two Stereogenic Centers

We've seen now that a single stereogenic center in a molecule gives rise to two stereoisomers (one pair of enantiomers) and that two stereogenic centers in a molecule give rise to a maximum of four stereoisomers (two pairs of enantiomers). In general, a molecule with n stereogenic centers has a maximum of 2^n stereoisomers (2^{n-1} pairs of enantiomers). Cholesterol, for example, contains eight stereogenic centers, making possible $2^8 = 256$ stereoisomers (128 enantiomeric pairs), although many are too strained to exist. Only one, however, is produced in nature.

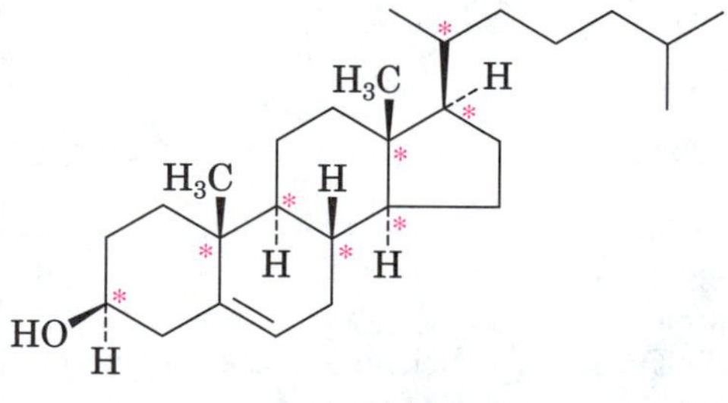

Cholesterol
(eight stereogenic centers)

PROBLEM

9.13 How many stereogenic centers does morphine have? How many stereoisomers of morphine are possible in principle?

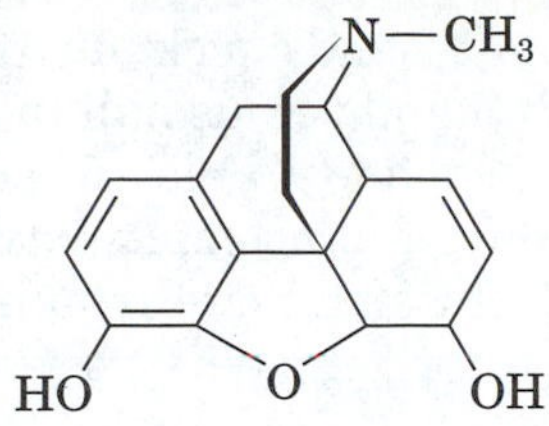

Morphine

9.10 Racemic Mixtures and Their Resolution

To conclude this discussion of stereoisomerism, let's return for a final look at Pasteur's pioneering work. Pasteur took an optically inactive tartaric acid salt and found that he could crystallize from it two optically active forms having the 2*R*,3*R* and 2*S*,3*S* configurations. But what was the optically inactive form he started with? It couldn't have been *meso*-tartaric acid, because *meso*-tartaric acid is a different chemical compound and can't interconvert with the two chiral enantiomers without breaking and re-forming chemical bonds.

The answer is that Pasteur started with a 50:50 *mixture* of the two chiral tartaric acid enantiomers. Such a mixture is called a **racemic** (ray-**see**-mic) **mixture**, or **racemate**, and is denoted either by the symbol (±) or by the prefix *d*,*l* to indicate a mixture of dextrorotatory and levorotatory forms. Racemic mixtures *must* show zero optical rotation because they contain equal amounts of (+) and (−) forms. The (+) rotation from one enantiomer exactly cancels the (−) rotation from the other. Through good fortune, Pasteur was able to separate, or **resolve**, racemic tartaric acid into its (+) and (−) enantiomers by fractional crystallization. Unfortunately, this method doesn't work for most racemic mixtures, so other techniques are required.

The most common method of resolution uses an acid–base reaction between a racemic mixture of chiral carboxylic acids (RCOOH) and an amine (RNH_2) to yield an ammonium salt:

$$\underset{\textbf{Carboxylic acid}}{R{-}\overset{O}{\overset{\|}{C}}{-}OH} + \underset{\textbf{Amine base}}{RNH_2} \xrightarrow{?} \underset{\textbf{Ammonium salt}}{R{-}\overset{O}{\overset{\|}{C}}{-}O^- \; RNH_3^+}$$

To understand how this method of resolution works, let's see what happens when a racemic mixture of chiral acids, such as (+)- and (−)-lactic acids, reacts with an achiral amine base, such as methylamine, CH_3NH_2, to yield the ammonium salt. Stereochemically, the situation is analogous to what happens when left and right hands (chiral) pick up a tennis ball (achiral).

Both left and right hands pick up the ball equally well, and the products—ball in right hand versus ball in left hand—are mirror images. In the same way, both (+)- and (−)-lactic acid react with methylamine equally well, and the product is a racemic mixture of two mirror-image salts, methylammonium (+)-lactate and methylammonium (−)-lactate (Figure 9.13).

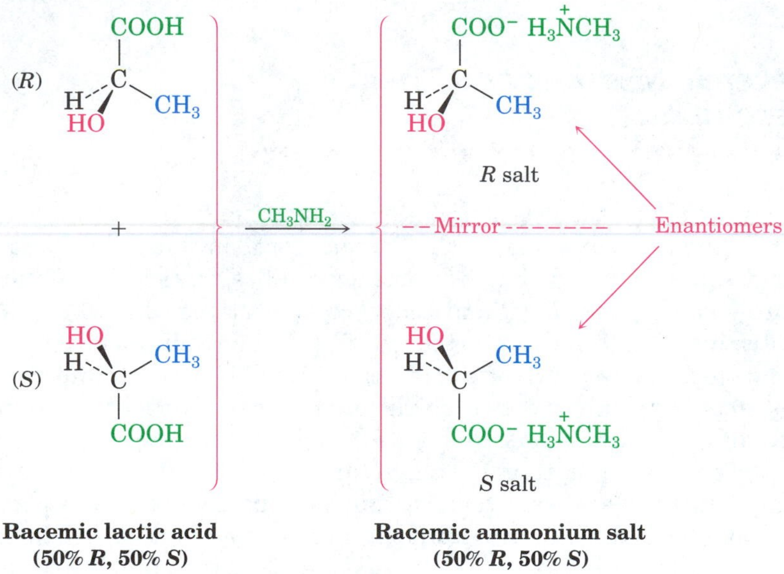

Figure 9.13 Reaction of racemic lactic acid with methylamine leads to a racemic mixture of ammonium salts.

Now let's see what happens when the racemic mixture of (+)- and (−)-lactic acids reacts with a single enantiomer of a chiral amine base, such as (*R*)-1-phenylethylamine (Figure 9.14). Stereochemically, the situation is analogous to what happens when a hand (a chiral reagent) puts on a right-handed glove (*also a chiral reagent). Left and right hands do not put on the same glove in the same way.* The products—right hand in right glove versus left hand in right glove—are not mirror images, they're altogether different.

In the same way, (+)- and (−)-lactic acids react with (*R*)-1-phenylethylamine to give two different products. (*R*)-Lactic acid reacts with (*R*)-1-phenylethylamine to give the *R,R* salt, and (*S*)-lactic acid reacts with *R* amine to give the *S,R* salt. *These two salts are diastereomers*; they are different compounds and therefore have different chemical and physical properties. It may therefore be possible to separate them by crystallization or some other means. Once separated, acidification of the two diastereomeric salts with mineral acid then allows us to isolate the two pure enantiomers of lactic acid and to recover the chiral amine for further use.

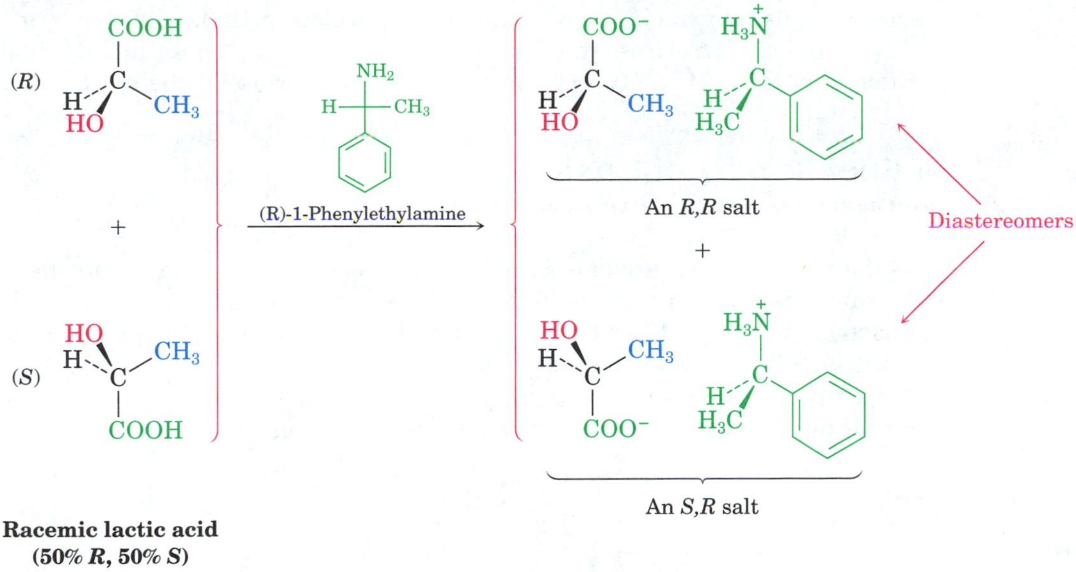

Figure 9.14 Reaction of racemic lactic acid with (*R*)-1-phenylethylamine yields a mixture of diastereomeric ammonium salts.

9.11 Physical Properties of Stereoisomers

Some physical properties of the three stereoisomers of tartaric acid and of the racemic mixture are listed in Table 9.3. As indicated, the (+)- and (−)-tartaric acids have identical melting points, solubilities, and densities. They differ only in the sign of their rotation of plane-polarized light. The meso isomer, by contrast, is diastereomeric with the (+) and (−) forms. As such, it has no mirror-image relationship to (+)- and (−)-tartaric acids, is a different compound altogether, and has different physical properties.

Table 9.3 Some Properties of the Stereoisomers of Tartaric Acid

Stereoisomer	*Melting point (°C)*	$[\alpha]_D$ *(degrees)*	*Density* (g/cm^3)	*Solubility at 20°C* ($g/100\ mL\ H_2O$)
(+)	168–170	+12	1.7598	139.0
(−)	168–170	−12	1.7598	139.0
Meso	146–148	0	1.6660	125.0
(±)	206	0	1.7880	20.6

The racemic mixture is different still. Though a mixture of enantiomers, racemates usually act as though they were pure compounds, different from either enantiomer. Thus, the physical properties of racemic tartaric acid differ from those of the two enantiomers and from those of the meso form.

9.12 A Brief Review of Isomerism

As noted on several previous occasions, isomers are compounds that have the same chemical formula but different structures. We've seen several kinds of isomers in the past few chapters, and it's a good idea at this point to see how they relate to one another by looking at the flowchart in Figure 9.15. There are two fundamental types of isomerism, both of which we've now encountered: constitutional isomerism and stereoisomerism.

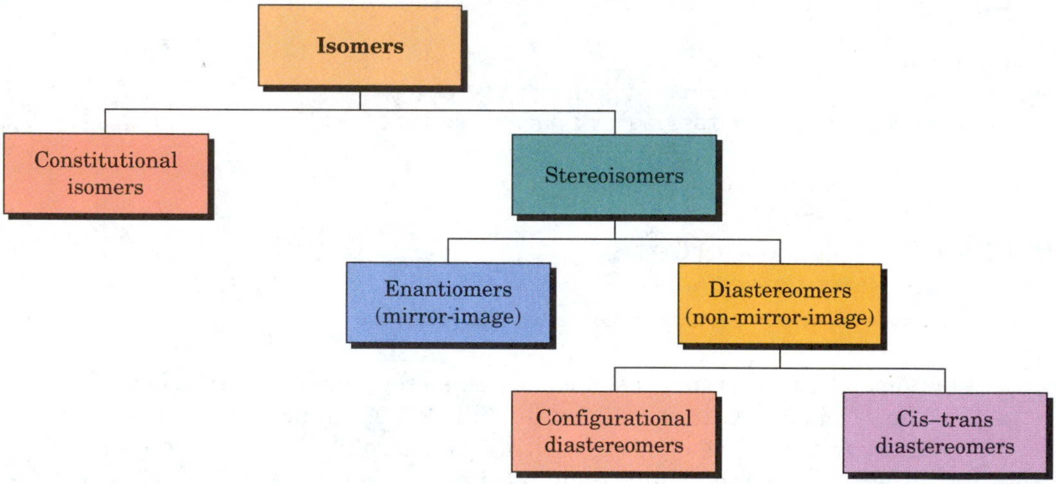

Figure 9.15 A flow diagram summarizing the different kinds of isomers.

Constitutional isomers (Section 3.2) are compounds whose atoms are connected differently. Among the kinds of constitutional isomers we've seen are skeletal, functional, and positional isomers.

Constitutional isomers—different connections among atoms:

Different carbon skeletons:	$CH_3CH(CH_3)CH_3$	and	$CH_3CH_2CH_2CH_3$
	Isobutane		**Butane**
Different functional groups:	CH_3CH_2OH	and	CH_3OCH_3
	Ethyl alcohol		**Dimethyl ether**

Different position of functional groups:

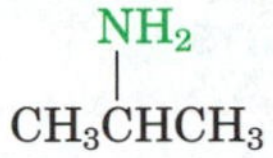

and $CH_3CH_2CH_2NH_2$

Isopropylamine **Propylamine**

Stereoisomers (Section 3.9) are compounds whose atoms are connected in the same order but with a different geometry. Among the kinds of stereoisomers we've seen are enantiomers, diastereomers, and cis–trans isomers (both in alkenes and in cycloalkanes). To be accurate, though, cis–trans isomers are really just another kind of diastereomers, because they are non-mirror-image stereoisomers.

Stereoisomers—same connections among atoms, but different geometry:

Enantiomers (nonsuperimposable mirror-image stereoisomers)

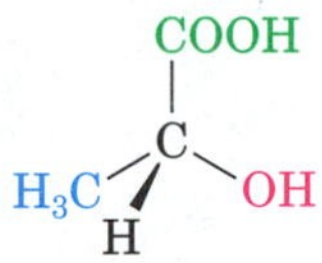

(*R*)-Lactic acid

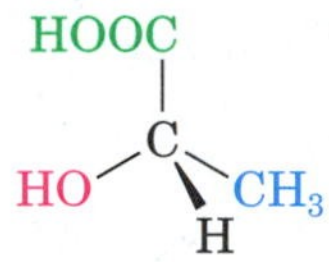

(*S*)-Lactic acid

Diastereomers (nonsuperimposable, non-mirror-image stereoisomers)

Configurational diastereomers

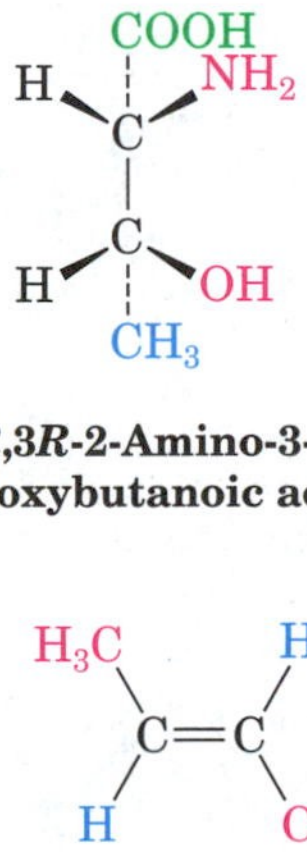

2*R*,3*R*-2-Amino-3-hydroxybutanoic acid

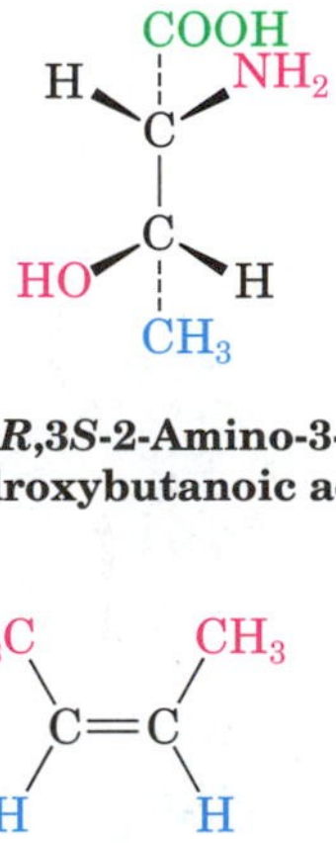

2*R*,3*S*-2-Amino-3-hydroxybutanoic acid

Cis–trans diastereomers (substituents on same side or opposite side of double bond or ring)

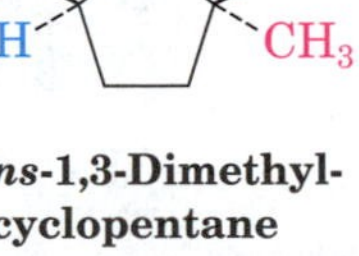

***trans*-2-Butene**

and

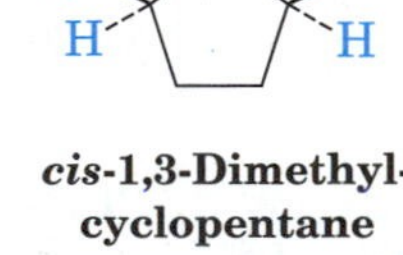

***cis*-2-Butene**

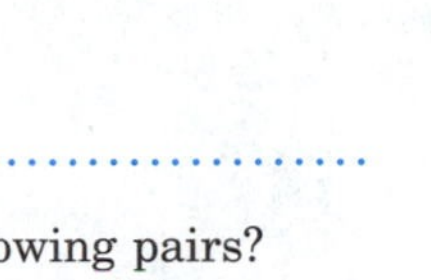

***trans*-1,3-Dimethylcyclopentane**

and

***cis*-1,3-Dimethylcyclopentane**

PROBLEM

9.14 What kinds of isomers are the following pairs?

(a) (*S*)-5-Chloro-2-hexene and chlorocyclohexane

(b) (2*R*,3*R*)-Dibromopentane and (2*S*,3*R*)-dibromopentane

9.13 Fischer Projections

When learning to visualize chiral molecules, it's best to begin by building molecular models. As more experience is gained, it becomes easier to draw pictures and work with mental images. To do this successfully, though, a standard method of representation is needed for depicting the three-dimensional arrangement of atoms on a page. In 1891, Emil Fischer suggested a method based on the projection of a tetrahedral carbon atom onto a flat surface. These **Fischer projections** were soon adopted and are now a standard means of depicting stereochemistry at stereogenic centers.

A tetrahedral carbon atom is represented in a Fischer projection by two perpendicular lines. The horizontal lines represent bonds coming out of the page, and the vertical lines represent bonds going into the page:

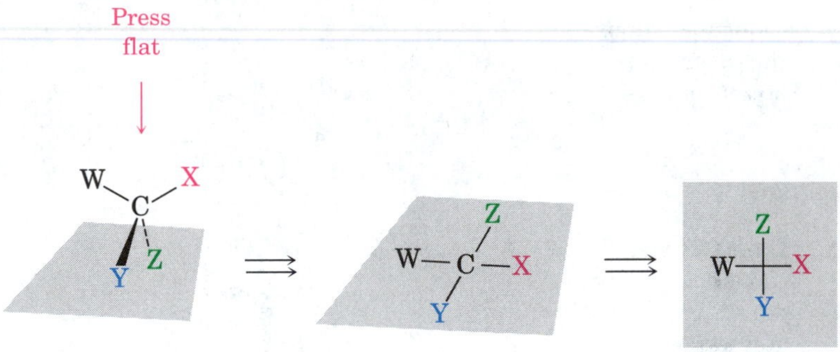

Fischer projection

For example, (*R*)-lactic acid can be drawn as follows:

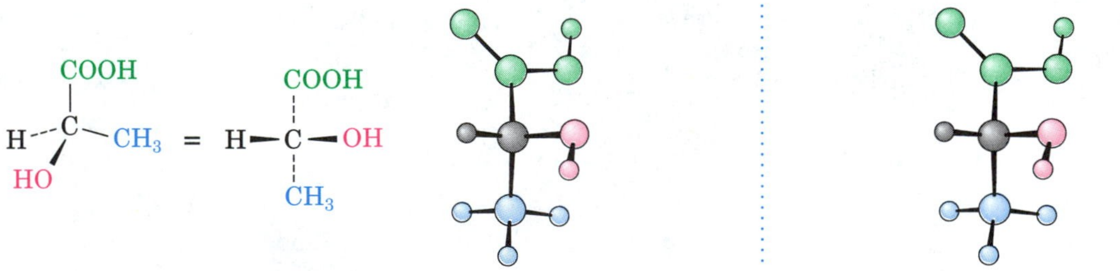

Stereo View

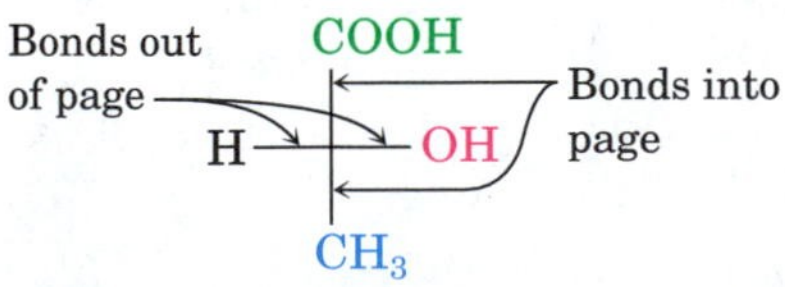

Fischer projection
(*R*)-Lactic acid

Because a given chiral molecule can be drawn in many different ways, it's often necessary to compare two projections to see if they represent the same or different enantiomers. To test for identity, Fischer projections can be moved around on the paper, but care must be taken not to change the meaning of the projection inadvertently. Only two kinds of motions are allowed:

1. A Fischer projection can be rotated on the page by 180°, but *not by 90° or 270°*. A 180° rotation maintains the Fischer convention by keeping the same substituent groups going into and coming out of the plane. In the following Fischer projection of (*R*)-lactic acid, for example, the –H and –OH groups come out of the plane both before and after a 180° rotation:

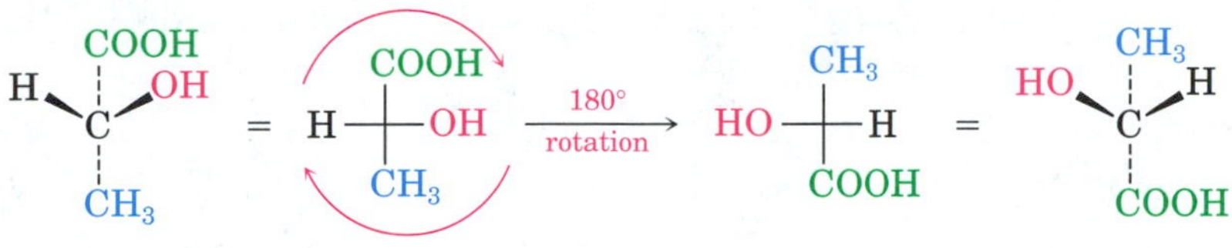

(*R*)-Lactic acid **(*R*)-Lactic acid**

A 90° rotation, however, breaks the Fischer convention by exchanging the groups that go into the plane and those that come out. In the following Fischer projection of (*R*)-lactic acid, for example, the –H and –OH groups come out of the plane before rotation but go into the plane after a 90° rotation. As a result, the rotated projection represents (*S*)-lactic acid:

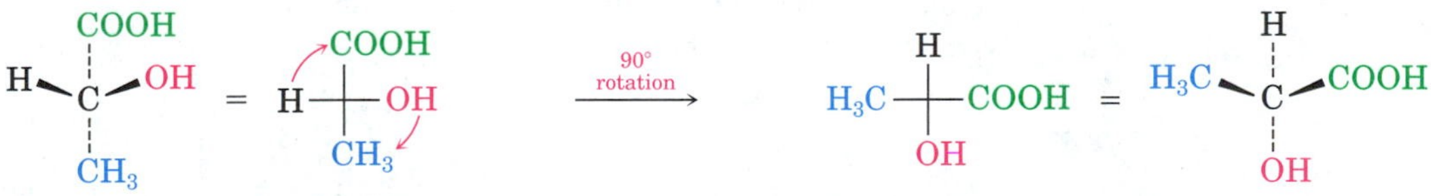

(*R*)-Lactic acid **(*S*)-Lactic acid**

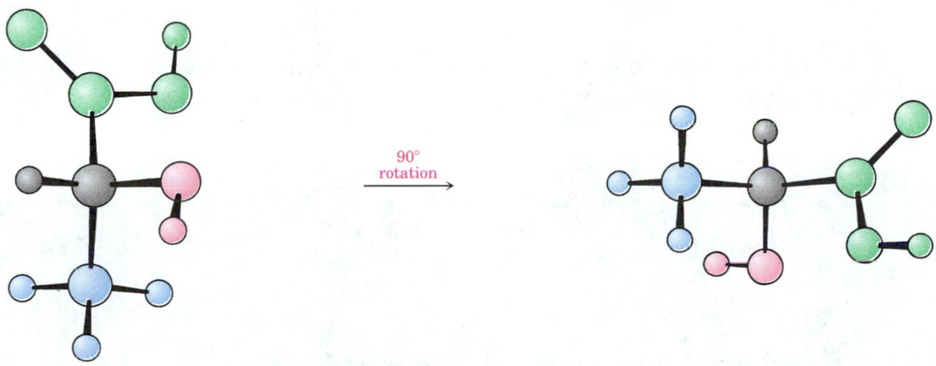

2. A Fischer projection can have one group held steady while the other three rotate in either a clockwise or a counterclockwise direction. For example:

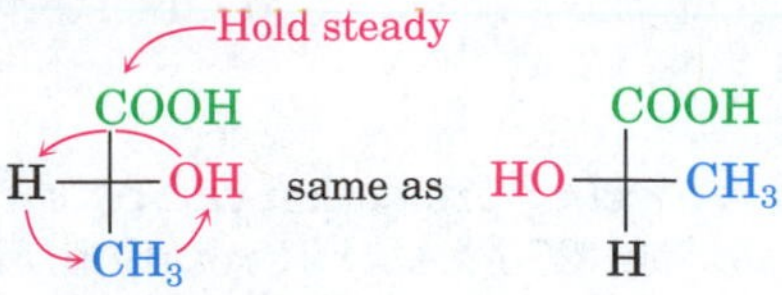

These are the only kinds of motion allowed. Moving a Fischer projection in any other way inverts its meaning.

Knowing the two rules provides a way to see if two projections represent the same or different enantiomers. For example, three different Fischer projections of 2-butanol follow. Do they all represent the same enantiomer, or is one different?

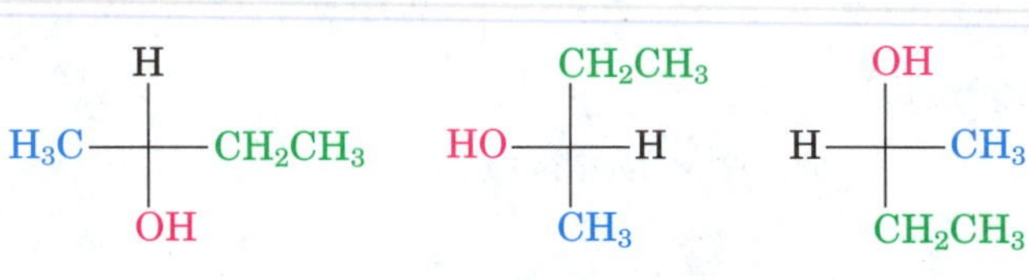

A B C

The simplest way to see if two Fischer projections represent the same enantiomer is to carry out allowed rotations until *two* groups are superimposed. If the other two groups are also superimposed, the Fischer projections are the same; if the other two groups are not superimposed, the Fischer projections are different.

Let's keep projection A unchanged and move B so that the $-CH_3$ and $-H$ substituents match up with those in A:

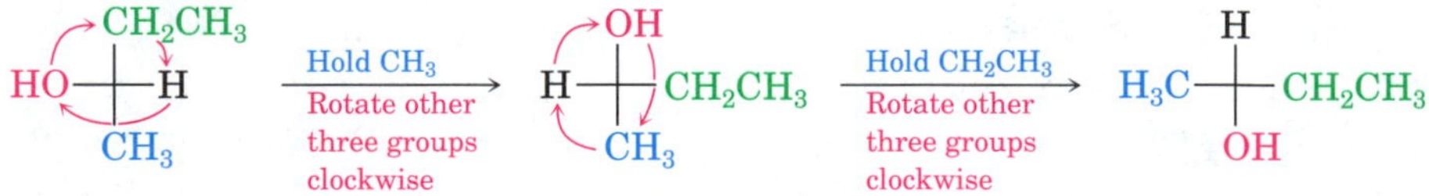

B A

By performing two allowed movements on B, we find that it is identical to A. Now let's do the same thing to C:

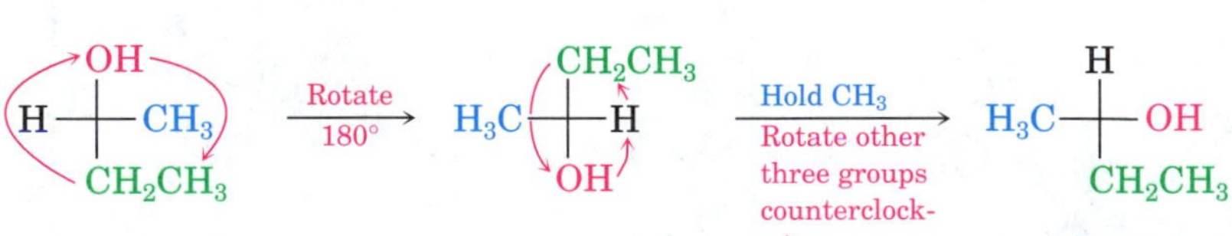

C Not A

By performing two allowed movements on C, we can match up the $-H$ and $-CH_3$ substituents with those in A. After doing so, however, the $-OH$ and

$-CH_2CH_3$ substituents *don't* match up. Thus, C is enantiomeric with A and B.

PRACTICE PROBLEM

Convert the following tetrahedral representation of (*R*)-2-butanol into a Fischer projection.

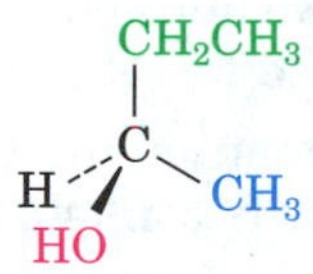

(*R*)-2-Butanol

Solution Orient the molecule so that two horizontal bonds are facing you and two vertical bonds are receding from you. Then press the molecule flat into the paper, indicating the stereogenic carbon as the intersection of two crossed lines.

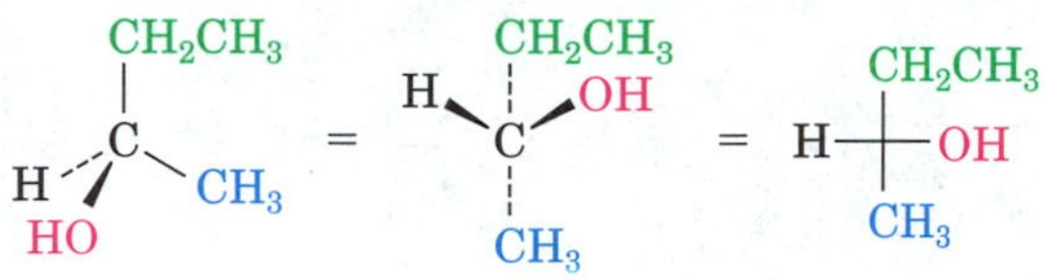

(*R*)-2-Butanol

PROBLEM

9.15 Which of the following Fischer projections represent the same enantiomer?

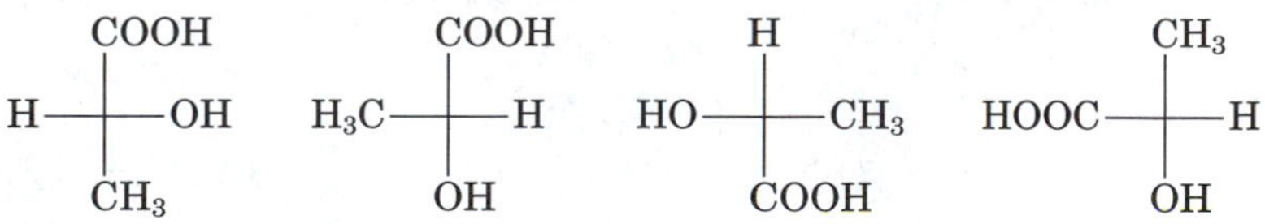

A B C D

PROBLEM

9.16 Are the following pairs of Fischer projections the same, or are they enantiomers?

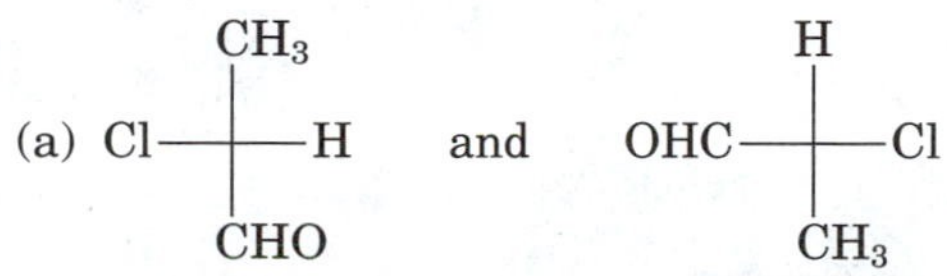

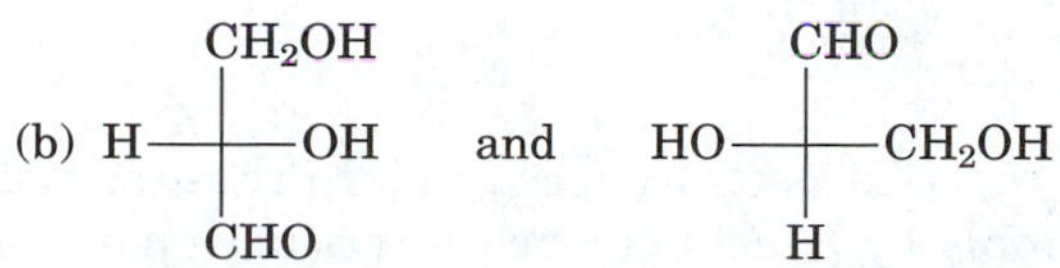

9.14 Assigning *R,S* Configurations to Fischer Projections

The *R*,*S* stereochemical designations (Section 9.6) can be assigned to Fischer projections by following three steps:

1. Assign priorities to the four substituents in the usual way.
2. Perform one of the two allowed motions to place the group of lowest (fourth) priority at the top of the Fischer projection.
3. Determine the direction of rotation in going from priority 1 to 2 to 3, and assign *R* or *S* configuration as in the following practice problem.

PRACTICE PROBLEM

Assign *R* or S configuration to this Fischer projection of alanine, an amino acid.

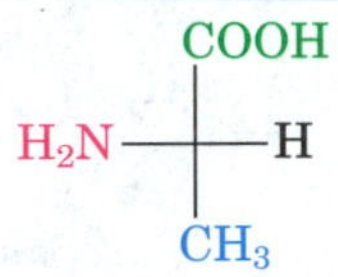

Alanine

Solution First, assign priorities to the four substituents on the stereogenic carbon. According to the sequence rules, the priorities are (1) $-NH_2$, (2) $-COOH$, (3) $-CH_3$, and (4) $-H$. Next, perform one of the allowed motions on the Fischer projection to bring the group of lowest priority ($-H$) to the top. In the present instance, we might want to hold the $-CH_3$ group steady while rotating the other three groups counterclockwise:

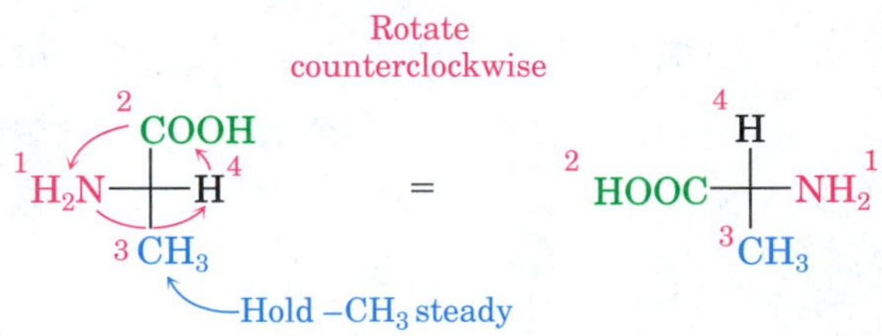

Going now from first to second to third highest priority requires a counterclockwise (left-hand) turn, corresponding to *S* stereochemistry.

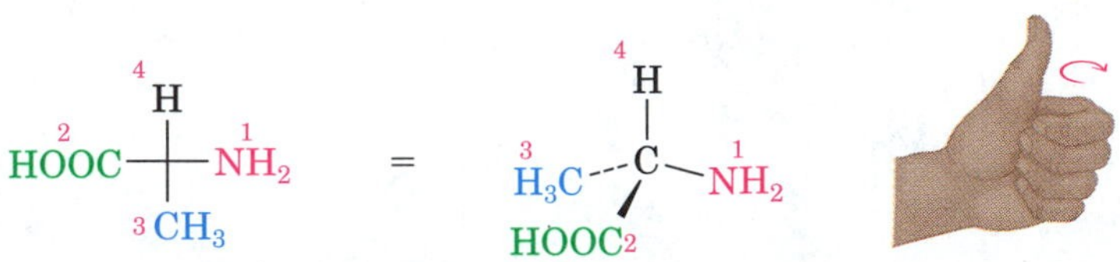

***S* stereochemistry**

Fischer projections can be used to specify more than one stereogenic center in a molecule simply by "stacking" the centers on top of one another. For example, threose, a simple four-carbon sugar, has the following 2*S*,3*R* configuration:

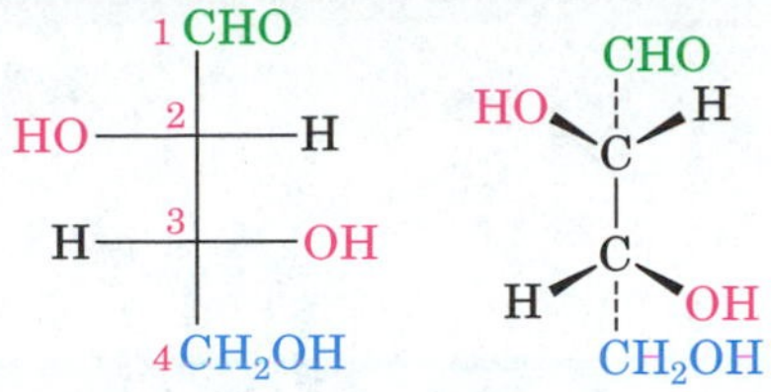

Threose [(2*S*,3*R*)-2,3,4-Trihydroxybutanal]

Molecular models are particularly helpful in visualizing these structures.

PROBLEM

9.17 Assign *R* or *S* configuration to the stereogenic centers in these molecules:

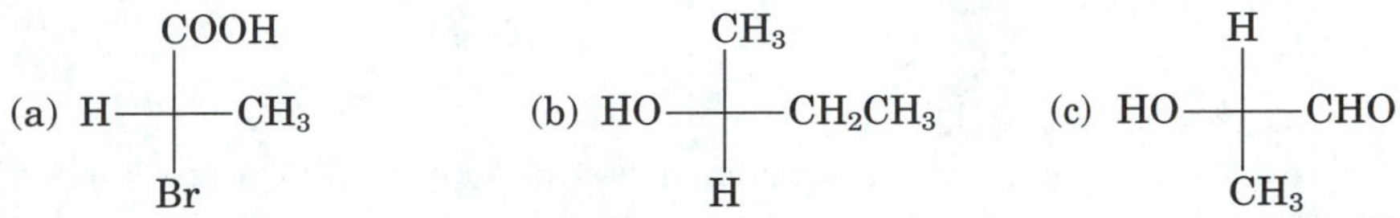

9.15 Stereochemistry of Reactions: Addition of HBr to Alkenes

Many organic reactions, including some that we've studied, yield products with stereogenic centers. For example, addition of HBr to 1-butene yields 2-bromobutane, a chiral molecule. What predictions can we make about the stereochemistry of this chiral product? If a single enantiomer is formed, is it *R* or *S*? If a mixture of enantiomers is formed, how much of each is present? In fact, the 2-bromobutane produced is a racemic mixture of *R* and *S* enantiomers. Let's see why.

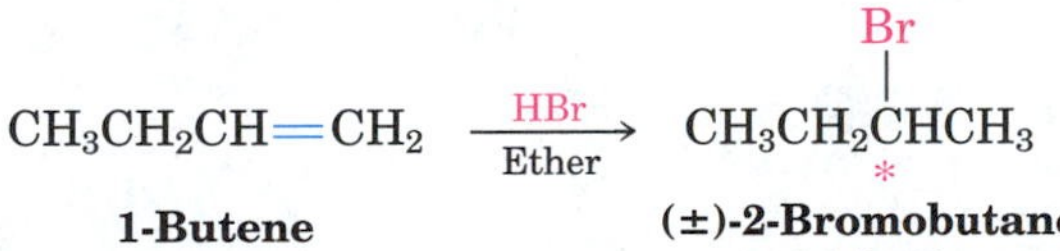

1-Butene (achiral) **(±)-2-Bromobutane (chiral)**

To understand why a racemic product results from the reaction of HBr with 1-butene, think about what happens during the reaction. 1-Butene is first protonated by acid to yield an intermediate secondary (2°) carbocation. Since the trivalent carbon is sp^2-hybridized, the cation has no stereogenic centers, has a plane of symmetry, and is achiral. As a result, it can be attacked by bromide ion equally well from either the top or the bottom. Attack from the top leads to (*S*)-2-bromobutane, and attack from the bottom leads to (*R*)-2-bromobutane. Since both pathways occur with equal probability, a racemic product mixture results (Figure 9.16).

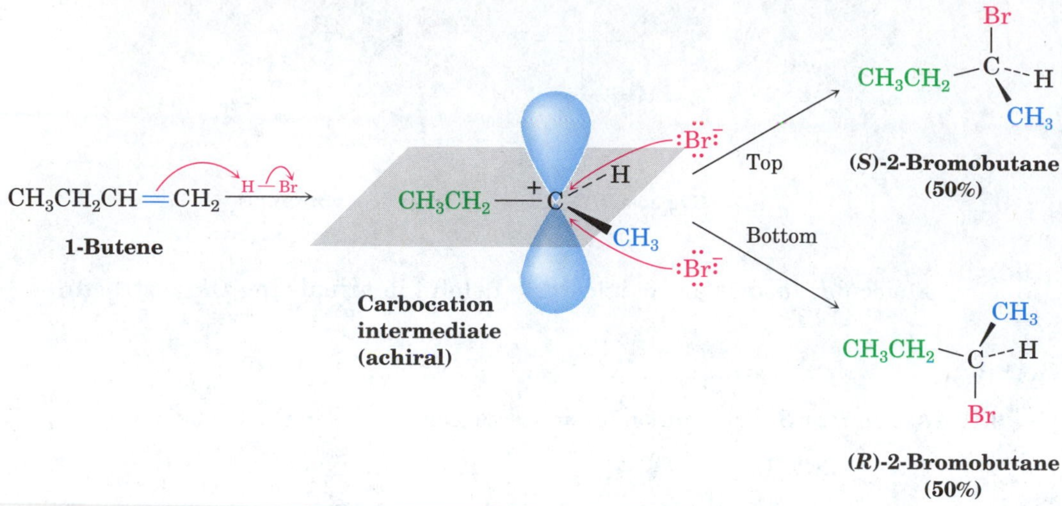

Figure 9.16 Stereochemistry of the addition of HBr to 1-butene. The achiral intermediate carbocation is attacked equally well from both top and bottom, giving a racemic product mixture.

Another way to think about the reaction is in terms of transition states. If the intermediate carbocation is attacked from the top, *S* product is formed through transition state 1 (TS 1) in Figure 9.17. If the cation is attacked from the bottom, *R* product is formed through TS 2. *The two transition states are mirror images*. They therefore have identical energies, form at identical rates, and are equally likely to occur.

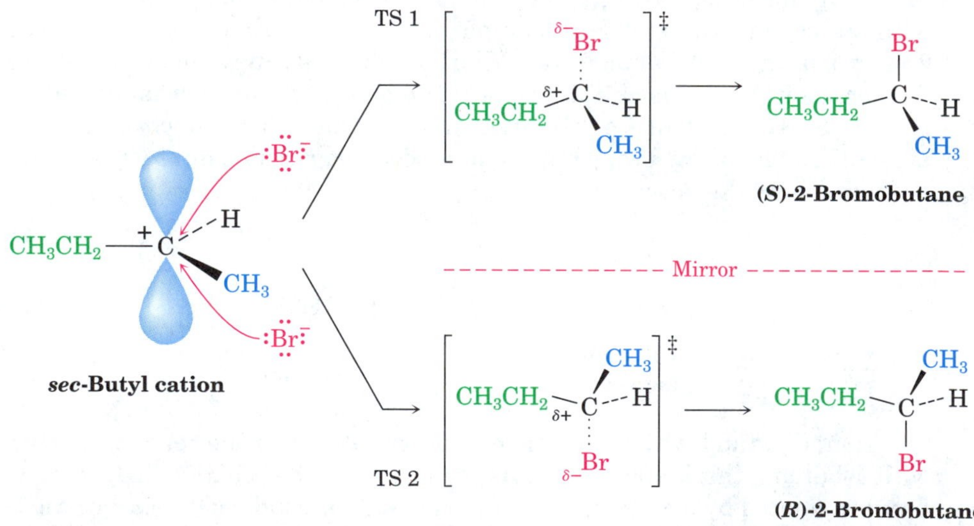

Figure 9.17 Attack of bromide ion on the *sec*-butyl carbocation. Attack from the top leads to *S* product and is the mirror image of attack from the bottom, which leads to *R* product. Since both are equally likely, racemic product is formed. The dotted C ··· Br bond in the transition state indicates partial bond formation.

9.16 Stereochemistry of Reactions: Addition of Br_2 to Alkenes

Addition of Br_2 to 2-butene leads to the formation of 2,3-dibromobutane and to the generation of two stereogenic centers. What stereochemistry should we predict for such a reaction? Starting with planar, achiral *cis*-2-butene, Br_2 can add to the double bond equally well from either the top or the bottom face to generate two intermediate bromonium ions. For the sake of simplicity, let's consider only the result of attack from the top face, keeping in mind that every structure we consider also has a mirror image.

The bromonium ion formed by addition to the top face of *cis*-2-butene can be attacked by bromide ion from either the right or the left side of the bottom face, as shown in Figure 9.18. Attack from the left (path a) leads to (2*S*,3*S*)-dibromobutane, and attack from the right (path b) leads to (2*R*,3*R*)-dibromobutane. Since both modes of attack on the achiral bromonium ion are equally likely, a 50:50 (racemic) mixture of the two enantiomeric products is formed. Thus, we obtain (±)-2,3-dibromobutane.

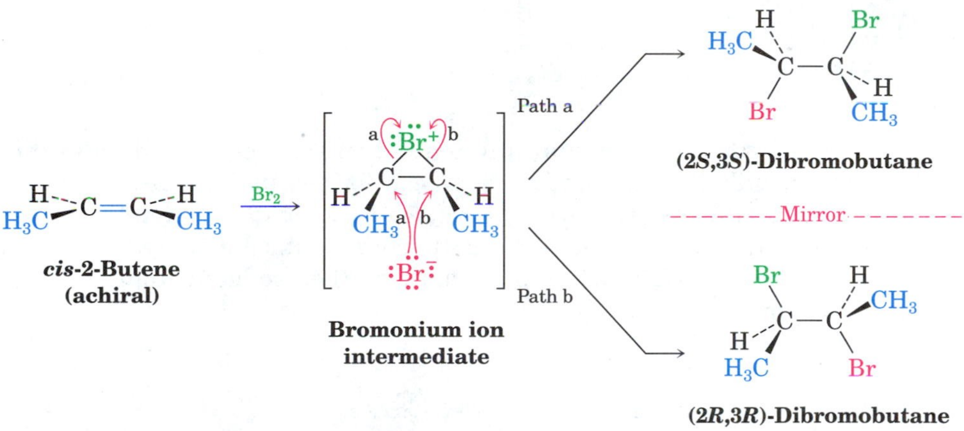

Figure 9.18 Stereochemistry of the addition of Br_2 to *cis*-2-butene. A racemic mixture of 2*S*,3*S* and 2*R*,3*R* products is formed.

What about the addition of Br_2 to *trans*-2-butene? Is the same racemic product mixture formed? Perhaps surprisingly at first glance, the answer is no. *trans*-2-Butene reacts with Br_2 to form a bromonium ion, and again we'll consider only top-face attack for simplicity. Attack of bromide ion on the bromonium ion intermediate takes place equally well from both right and left sides of the bottom face, leading to the formation of 2*R*,3*S* and 2*S*,3*R* products in equal amounts (Figure 9.19, p. 326). A close look at the two products, however, shows that they are *identical*. Both structures represent *meso*-2,3-dibromobutane.

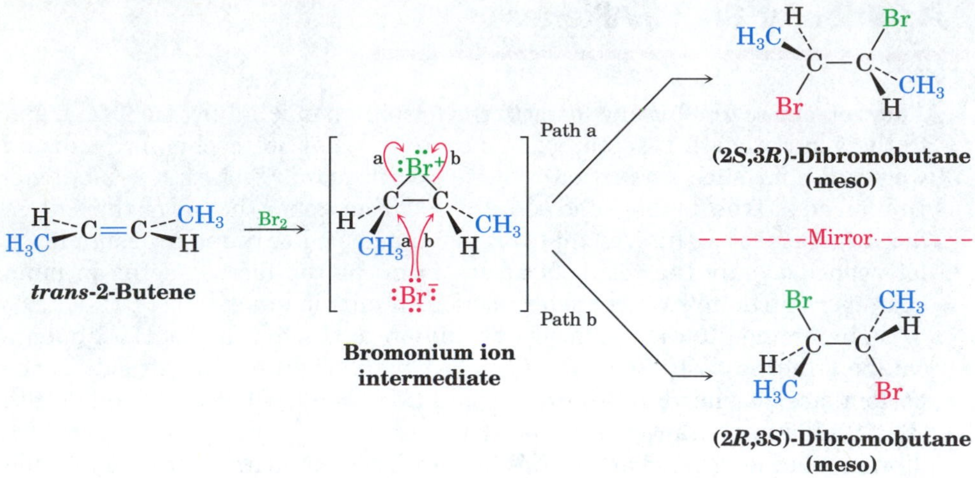

Figure 9.19 Stereochemistry of the addition of Br_2 to *trans*-2-butene. A meso product is formed.

The key conclusion from all three addition reactions just discussed is that an optically inactive product has been formed in each case. *Reaction between two achiral partners always leads to optically inactive products—either racemic or meso.* Put another way, optical activity can't come from nowhere; optically active products can't be produced from optically inactive reactants.

PROBLEM.

9.18 What is the stereochemistry of the product that results from addition of Br_2 to cyclohexene? Is the product optically active? Explain.

PROBLEM.

9.19 Addition of Br_2 to an unsymmetrical alkene such as *cis*-2-hexene leads to racemic product, even though attack of bromide ion on the unsymmetrical bromonium ion intermediate is not equally likely at both ends. Make drawings of the intermediate and the products, and explain the observed stereochemical result.

PROBLEM.

9.20 Predict the stereochemical outcome of the reaction of Br_2 with *trans*-2-hexene. Show your reasoning.

9.17 Stereochemistry of Reactions: Addition of HBr to a Chiral Alkene

The reactions considered in the previous two sections involve additions to achiral alkenes, and optically inactive products are formed in all cases. What would happen, though, if we were to carry out a reaction on a *single* enantiomer of a *chiral* reactant? For example, what stereochemical result would be obtained from addition of HBr to a chiral alkene, such as (*R*)-4-methyl-1-hexene? The product of the reaction, 2-bromo-4-methylhexane, has two stereogenic centers and four stereoisomeric configurations.

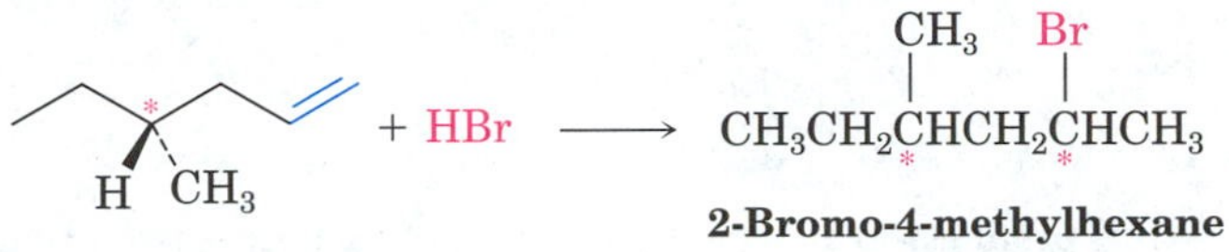

(*R*)-4-Methyl-1-hexene **2-Bromo-4-methylhexane**

Let's think about the two stereogenic centers separately. What about the configuration at C4, the methyl-bearing carbon atom? Since C4 has the *R* configuration in the starting material, and since this stereogenic center is unaffected by the reaction, its configuration remains unchanged. Thus, the configuration of C4 in the product remains *R*.

What about the configuration at C2, the newly formed stereogenic center? As illustrated in Figure 9.20, the stereochemistry at C2 is estab-

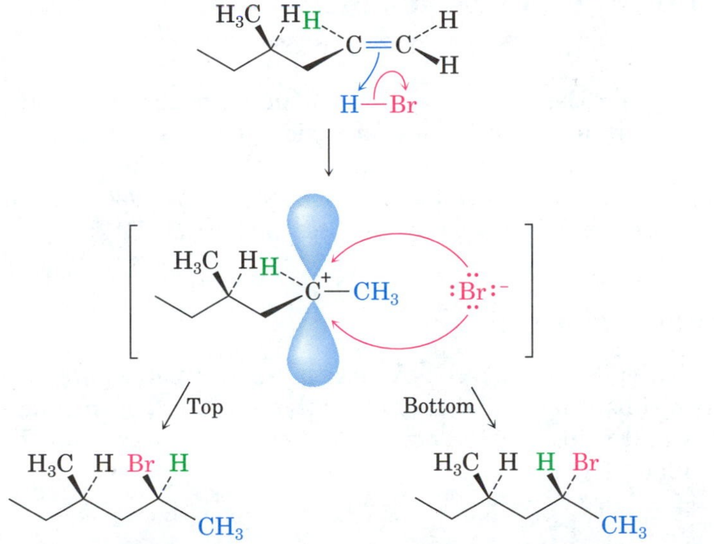

(2*S*,4*R*)-2-Bromo-4-methylhexane **(2*R*,4*R*)-2-Bromo-4-methylhexane**

Figure 9.20 Stereochemistry of the addition of HBr to the chiral alkene, (*R*)-4-methyl-1-hexene. A mixture of diastereomeric 2*R*,4*R* and 2*S*,4*R* products is formed in unequal amounts, because attack on the chiral carbocation intermediate is not equally likely from top and bottom.

lished by attack of bromide ion on a carbocation intermediate in the usual manner. *But this carbocation is not symmetrical; it is chiral because of the presence of the C4 center.* Since the carbocation has no plane of symmetry, it is not attacked equally well from top and bottom faces. One of the two faces is likely, for steric reasons, to be a bit more accessible than the other face, leading to a mixture of *R* and *S* products in some ratio other than 50:50. Thus, two diastereomeric products, (2*R*,4*R*)-2-bromo-4-methylhexane and (2*S*,4*R*)-2-bromo-4-methylhexane, are formed in unequal amounts.

As a general rule, *reaction between an achiral reactant* (such as HBr) *and a chiral reactant leads to unequal amounts of diastereomeric products.* If the chiral reactant is optically active because only one enantiomer is used, then the products are also optically active.

PROBLEM

9.21 What products are formed, and in what amounts, from reaction of HBr with racemic (±)-4-methyl-1-hexene? Is the product mixture optically active?

PROBLEM

9.22 What products are formed, and in what amounts, from reaction of HBr with 4-methylcyclopentene?

9.18 Stereoisomerism and Chirality in Substituted Cyclohexanes

We saw in Section 4.12 that substituted cyclohexane rings adopt a chair-like geometry and that the conformation of a given compound can be predicted by looking at steric interactions in the molecule. To complete a study of cyclohexane stereochemistry, we now need to look at the effect of conformation on stereoisomerism and chirality.

1,4-Disubstituted Cyclohexanes

1,4-Disubstituted cyclohexanes have a symmetry plane passing through the two substituents and through carbons 1 and 4 of the ring. As a result, both cis and trans diastereomers are achiral, as shown in Figure 9.21 for 1,4-dimethylcyclohexane.

1,3-Disubstituted Cyclohexanes

1,3-Disubstituted cyclohexanes have two stereogenic centers, and four stereoisomers are therefore possible. *trans*-1,3-Dimethylcyclohexane exists as a pair of enantiomers, but *cis*-1,3-dimethylcyclohexane has a symmetry plane and is thus a meso compound (Figure 9.22).

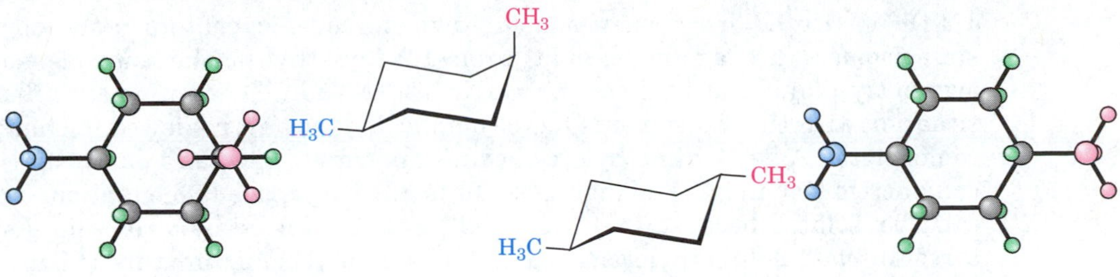

***cis*-1,4-Dimethylcyclohexane** ***trans*-1,4-Dimethylcyclohexane**

Diastereomers
(stereoisomers but not enantiomers)

Figure 9.21 The stereochemical relationships among 1,4-dimethylcyclohexanes. Both cis and trans isomers are achiral.

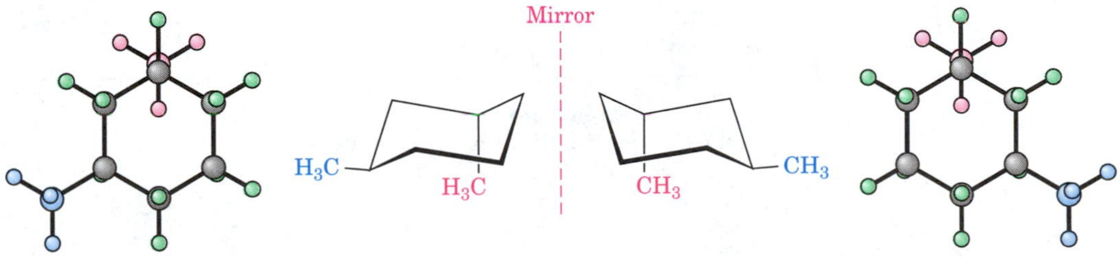

Enantiomers
(+) and (−)-*trans*-1,3-Dimethylcyclohexane

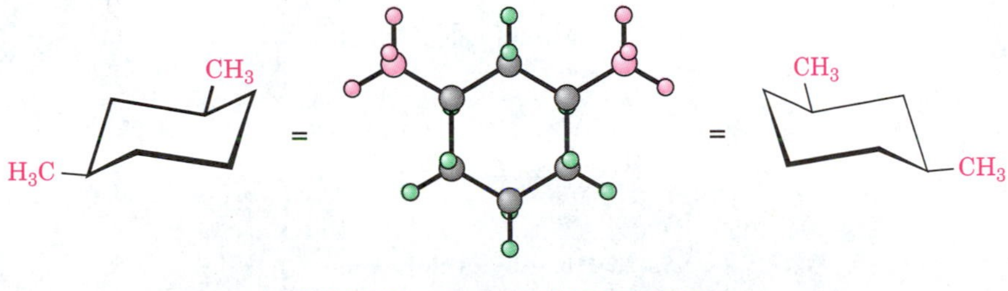

***cis*-1,3-Dimethylcyclohexane (meso)**

Figure 9.22 The stereochemical relationships among 1,3-dimethylcyclohexanes. The trans isomer is a pair of enantiomers, and the cis isomer is a meso compound.

1,2-Disubstituted Cyclohexanes

1,2-Disubstituted cyclohexanes have two stereogenic centers, and four stereoisomers are again possible. *trans*-1,2-Dimethylcyclohexane has no symmetry plane and is therefore a pair of (+) and (−) enantiomers. The situation with the cis isomer is more complicated, however. Viewed in chair conformation, *cis*-1,2-dimethylcyclohexane appears to exist as a pair of optically active (+) and (−) enantiomers. In fact, though, the two enantiomers can't be isolated because they are rapidly interconverted by a ring-flip. As a result, *cis*-1,2-dimethylcyclohexane exists as a racemic mixture (Figure 9.23).

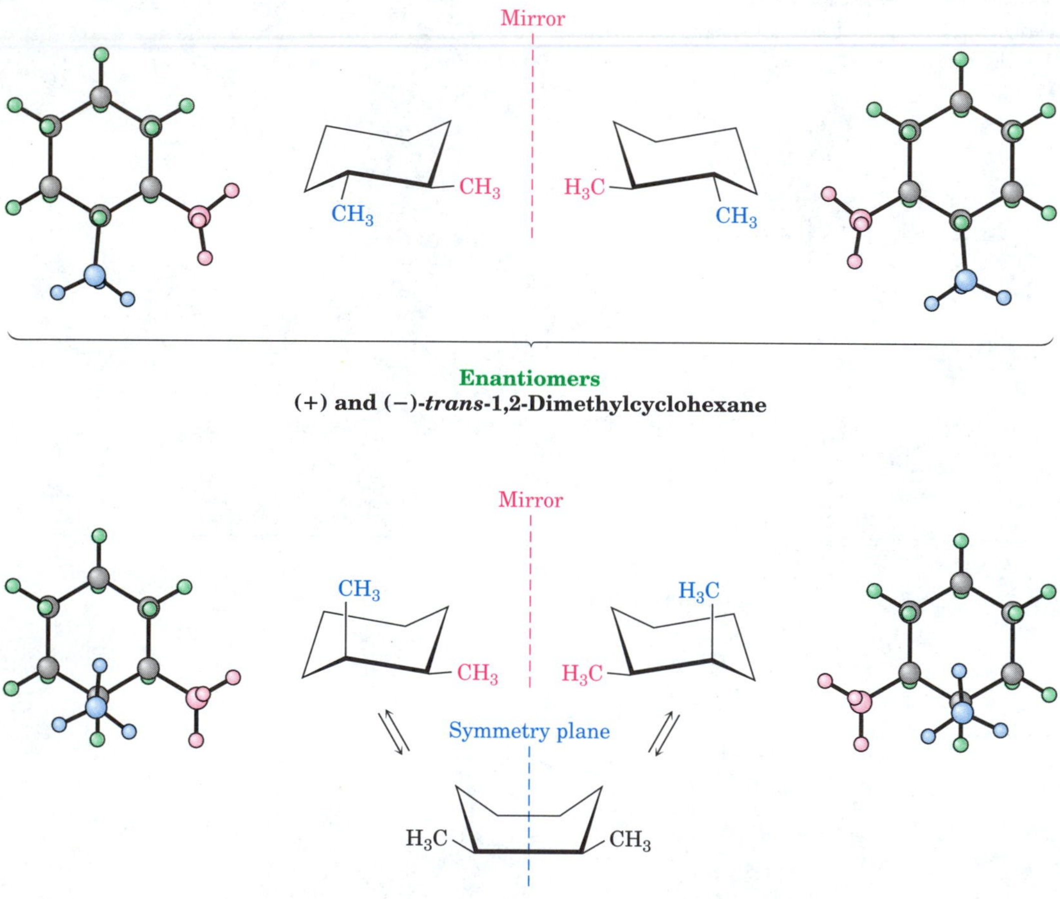

Figure 9.23 *trans*-1,2-Dimethylcyclohexane exists as a pair of (+) and (−) enantiomers. *cis*-1,2-Dimethylcyclohexane exists as a pair of enantiomers that rapidly interconvert by ring-flip. As a result, the compound is racemic.

PROBLEM.

9.23 How many stereoisomers of 1-chloro-3,5-dimethylcyclohexane are there? Draw the most stable conformation.

PROBLEM.

9.24 How many 1,2-dimethylcyclopentane stereoisomers are there? What are the stereochemical relationships among them?

9.19 Chirality at Atoms Other Than Carbon

Since the most common cause of chirality is the presence of four different substituents bonded to a tetrahedral atom, tetrahedral atoms other than carbon can also be stereogenic centers. Silicon, nitrogen, phosphorus, and sulfur are all commonly encountered in organic molecules, and all can be stereogenic centers under the proper circumstances. We know, for example, that trivalent nitrogen is tetrahedral, with its lone pair of electrons acting as the fourth "substituent" (Section 1.12). Is trivalent nitrogen chiral? Does a compound such as ethylmethylamine exist as a pair of enantiomers?

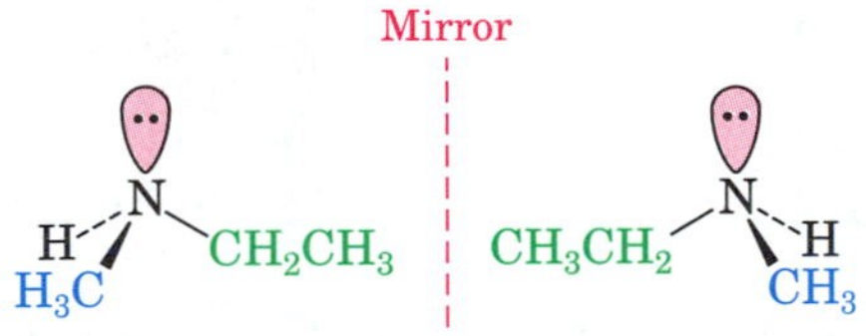

Ethylmethylamine

The answer is both yes and no. Yes in principle, but no in practice. Trivalent nitrogen compounds undergo a rapid umbrella-like inversion that interconverts enantiomers. We therefore can't isolate individual enantiomers except in special cases.

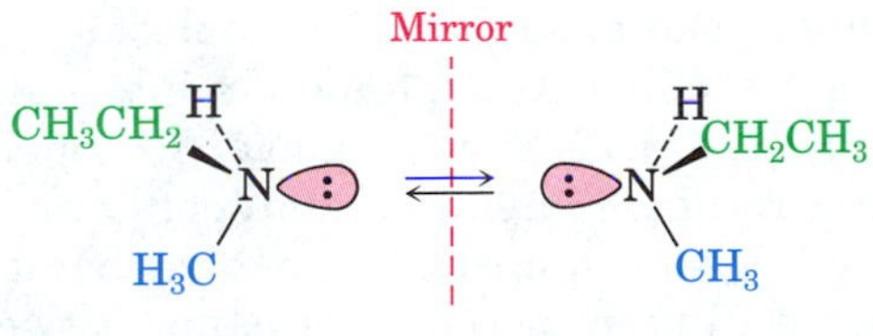

Rapid

9.20 Chirality in Nature

Just as different stereoisomeric forms of a chiral molecule have different physical properties, they usually have different biological properties as well. For example, the dextrorotatory enantiomer of carvone has the odor of caraway seeds, and the levorotatory enantiomer has the odor of spearmint.

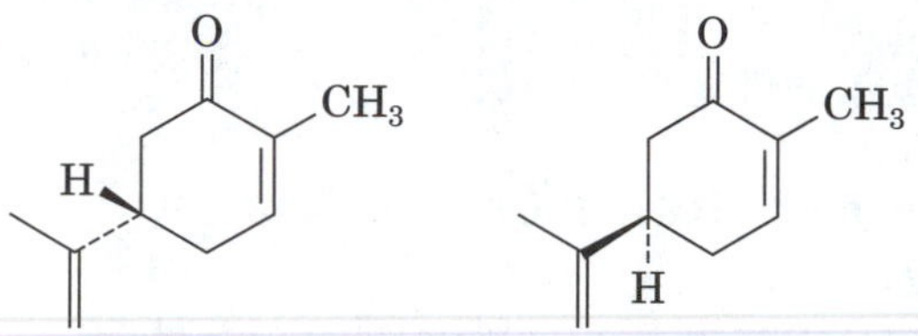

(+)-Carvone (in caraway seeds) **(−)-Carvone (in spearmint oil)**

Another example of how a change in chirality can affect the biological properties of a molecule is found in the amino acid, dopa. More properly named 2-amino-3-(3,4-dihydroxyphenyl)propanoic acid, dopa has a single stereogenic center and can thus exist in two stereoisomeric forms. The dextrorotatory enantiomer, D-dopa, has no physiological effect on humans, but the levorotatory enantiomer, L-dopa, is widely used for its potent activity against Parkinson's disease, a chronic malady of the central nervous system.

D-Dopa (no biological effect) **L-Dopa (anti-Parkinsonian agent)**

Why do different stereoisomers have different biological properties? To exert its biological action, a chiral molecule must fit into a chiral receptor at the target site, much as a hand fits into a glove. But just as a right hand can fit only into a right-hand glove, so a particular stereoisomer can fit only into a receptor having the proper complementary shape. Any other stereoisomer will be a misfit like a right hand in a left-hand glove. A schematic representation of the interaction between a chiral molecule and a chiral biological receptor is shown in Figure 9.24. One enantiomer fits the receptor perfectly, but the other does not.

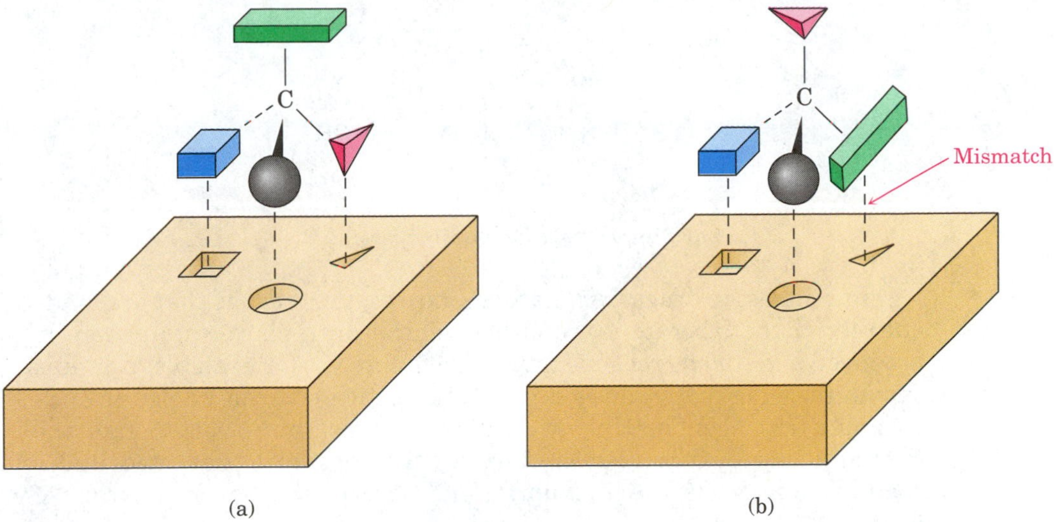

Figure 9.24 (a) One enantiomer fits easily into a chiral receptor site to exert its biological effect, but (b) the other enantiomer can't fit into the same receptor.

INTERLUDE

Chiral Drugs

The *S* enantiomer of ibuprofen soothes the aches and pains of athletic injuries.

The hundreds of different pharmaceutical agents approved for use by the U.S. Food and Drug Administration come from many sources. Many drugs are isolated directly from plants or bacteria, others are made by chemical modification of naturally occurring compounds, and still others are made entirely in the laboratory and have no relatives in nature.

Those drugs that come from natural sources, either directly or after chemical modification, are usually chiral and are generally found only as a single enantiomer rather than as a racemic mixture. Penicillin V, for example, an antibiotic isolated from the *Penicillium* mold, has the 2*S*,5*R*,6*R* configuration. Its enantiomer, which does not occur naturally but can be made in the laboratory, has essentially no biological activity.

(continued)➤

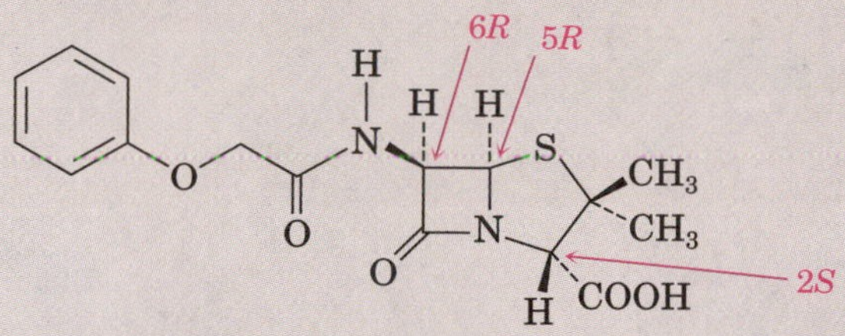

Penicillin V (2*S*,5*R*,6*R* configuration)

In contrast to drugs from natural sources, those drugs that are made entirely in the laboratory are either achiral or, if chiral, are generally produced and sold as racemic mixtures. Ibuprofen, for example, contains one stereogenic center, and only the *S* enantiomer is an analgesic/anti-inflammatory agent useful in treating rheumatism. Even though the *R* enantiomer of ibuprofen is inactive, the substance marketed under such trade names as Advil, Nuprin, and Motrin is a racemic mixture of *R* and *S*.

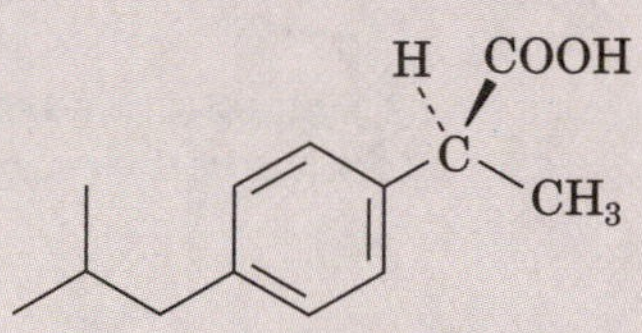

(*S*)-Ibuprofen
(an active analgesic agent)

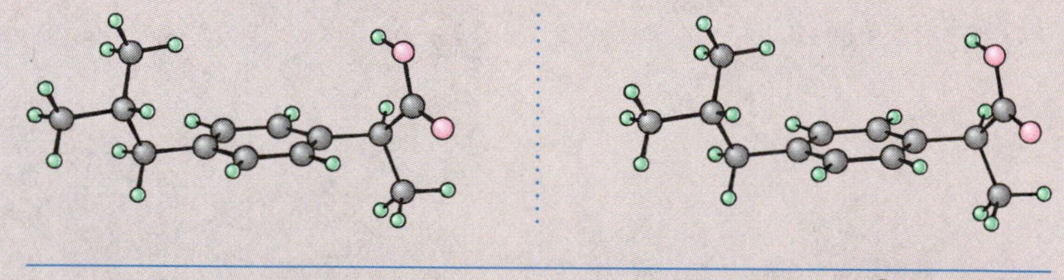

Stereo View

Not only is it chemically wasteful to synthesize and administer an enantiomer that does not serve the intended purpose, many examples are now known where the presence of the "wrong" enantiomer in a racemic mixture either affects the body's ability to utilize the "right" enantiomer or has unintended pharmacological effects of its own. The presence of (*R*)-ibuprofen in the racemic mixture, for instance, seems to slow down substantially the rate at which the *S* enantiomer takes effect.

To get around this problem, pharmaceutical companies are now devising methods of so-called *asymmetric synthesis*, which allows them to prepare only a single enantiomer rather than a racemic mixture. Viable methods have already been developed for the preparation of (*S*)-ibuprofen, and the time may not be far off when television ads show famous tennis players talking about the advantages of chiral drugs.

Summary and Key Words

When a beam of **plane-polarized light** is passed through a solution of certain organic molecules, the plane of polarization is rotated. Compounds that exhibit this behavior are called **optically active**. Optical activity is due to the asymmetric structure of the molecules themselves.

An object or molecule that is not superimposable on its mirror image is said to be **chiral**, meaning "handed." For example, a glove is chiral but a coffee cup is nonchiral, or **achiral**. A chiral molecule is one that does not contain a **plane of symmetry**—a plane that cuts through the molecule so that one half is a mirror image of the other half. The most common cause of chirality in organic molecules is the presence of a tetrahedral, sp^3-hybridized carbon atom bonded to four different groups. Compounds that contain such **stereogenic centers** exist as a pair of nonsuperimposable, mirror-image stereoisomers called **enantiomers.** Enantiomers are identical in all physical properties except for the direction in which they rotate plane-polarized light.

The stereochemical **configuration** of a carbon atom can be depicted using **Fischer projections**, in which horizontal lines (bonds) are understood to come out of the plane of the paper and vertical bonds are understood to go back into the plane of the paper. The configuration can be specified as either *R* (*rectus*) or *S* (*sinister*) by using the Cahn–Ingold–Prelog sequence rules. This is done by first assigning priorities to the four substituents on the stereogenic carbon atom and then comparing the atom with a hand, oriented so that the thumb points from the carbon to the group of lowest priority (4). If the fingers of a *right* hand curl in the direction of decreasing priority ($1 \rightarrow 2 \rightarrow 3$) for the remaining three groups, the stereogenic center has the *R* configuration. If the fingers of a *left* hand curl in the $1 \rightarrow 2 \rightarrow 3$ direction, the stereogenic center has the *S* configuration.

Some molecules have more than one stereogenic center. Enantiomers have opposite configuration at all stereogenic centers, whereas **diastereomers** have the same configuration in at least one center but opposite configurations at the others. A compound with n stereogenic centers can have a maximum of 2^n stereoisomers.

Meso compounds contain stereogenic centers, but are achiral overall because they have a plane of symmetry. **Racemic mixtures**, or **racemates**, are 50:50 mixtures of (+) and (−) enantiomers. Racemic mixtures and individual diastereomers differ in their physical properties, such as solubility, melting point, and boiling point.

Most reactions give chiral products. If the reactants are optically inactive, the products are also optically inactive—either meso or racemic. If one or both of the reactants is optically active, the product can also be optically active.

ADDITIONAL PROBLEMS

9.25 Cholic acid, the major steroid found in bile, was found to have a rotation of +2.22° when a 3.00 g sample was dissolved in 5.00 mL alcohol and the solution was placed in a sample tube with a 1.00 cm path length. Calculate $[\alpha]_D$ for cholic acid.

9.26 Polarimeters for measuring optical rotation are very sensitive and can measure rotations to 0.001°, an important fact when only small amounts of sample are available. Ecdysone, for example, is an insect hormone that controls molting in the silkworm moth. When 7.00 mg ecdysone was dissolved in 1.00 mL chloroform and the solution was placed in a 2.00 cm path-length cell, an observed rotation of +0.087° was found. Calculate $[\alpha]_D$ for ecdysone.

9.27 Suppose you have a sucrose solution that appears to rotate plane-polarized light 90° to the right (dextrorotatory). How do you know that the solution isn't rotating the plane of polarization to the *left* by 270° (levorotatory)? The analyzer would be in exactly the same position in either case.

9.28 What do these terms mean?
(a) Chirality (b) Stereogenic center (c) Optical activity
(d) Diastereomer (e) Enantiomer (f) Racemate

9.29 Which of the following compounds are chiral? Draw them and label the stereogenic centers.
(a) 2,4-Dimethylheptane (b) 3-Ethyl-5,5-dimethylheptane
(c) *cis*-1,4-Dichlorocyclohexane (d) 4,5-Dimethyl-2,6-octadiyne

9.30 Draw chiral molecules that meet these descriptions:
(a) A chloroalkane, $C_5H_{11}Cl$ (b) An alcohol, $C_6H_{14}O$
(c) An alkene, C_6H_{12} (d) An alkane, C_8H_{18}

9.31 Eight alcohols have the formula $C_5H_{12}O$. Draw them. Which are chiral?

9.32 Draw the nine chiral molecules that have the formula $C_6H_{13}Br$.

9.33 Draw compounds that fit these descriptions:
(a) A chiral alcohol with four carbons
(b) A chiral carboxylic acid with the formula $C_5H_{10}O_2$
(c) A compound with two stereogenic centers
(d) A chiral aldehyde with the formula C_3H_5BrO

9.34 Which of these objects are chiral?
(a) A basketball (b) A fork (c) A wine glass
(d) A golf club (e) A monkey wrench (f) A snowflake

9.35 Penicillin V is an important broad-spectrum antibiotic that contains three stereogenic centers. Identify them with asterisks.

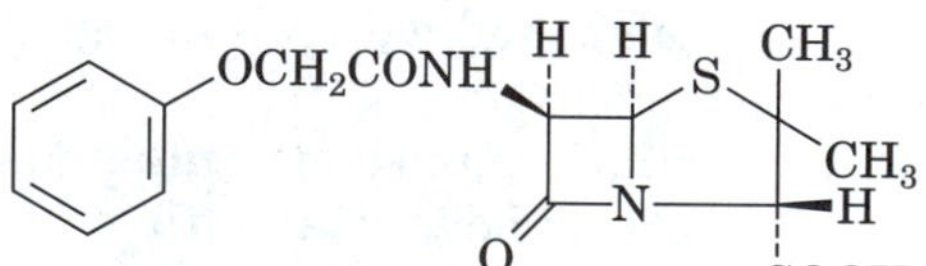

Penicillin V (antibiotic)

9.36 Draw examples of the following:
(a) A meso compound with the formula C_8H_{18}
(b) A meso compound with the formula C_9H_{20}
(c) A compound with two stereogenic centers, one *R* and the other *S*

9.37 What is the relationship between the specific rotations of (2*R*,3*R*)-dichloropentane and (2*S*,3*S*)-dichloropentane? Between (2*R*,3*S*)-dichloropentane and (2*R*,3*R*)-dichloropentane?

9.38 What is the stereochemical configuration of the enantiomer of (2*S*,4*R*)-dibromooctane?

9.39 What are the stereochemical configurations of the two diastereomers of (2*S*,4*R*)-dibromooctane?

9.40 Assign Cahn–Ingold–Prelog priorities to the following sets of substituents:

(a) $-CH{=}CH_2$, $-CH(CH_3)_2$, $-C(CH_3)_3$, $-CH_2CH_3$

(b) $-C{\equiv}CH$, $-CH{=}CH_2$, $-C(CH_3)_3$, 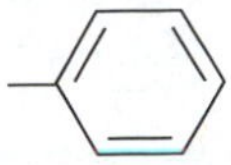

(c) $-CO_2CH_3$, $-COCH_3$, $-CH_2OCH_3$, $-CH_2CH_3$

(d) $-C{\equiv}N$, $-CH_2Br$, $-CH_2CH_2Br$, $-Br$

9.41 Assign *R* or *S* configurations to the stereogenic centers in the following molecules:

(a) 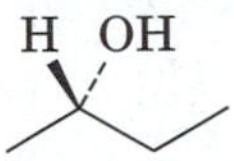

(b) 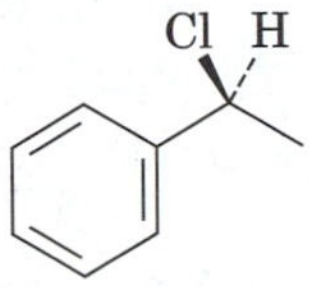

(c) 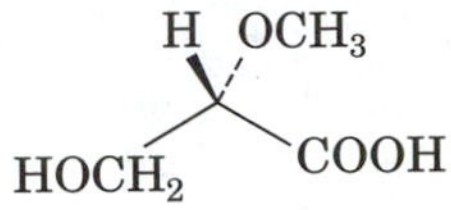

9.42 Assign *R* or *S* configurations to each stereogenic center in the following molecules:

(a) 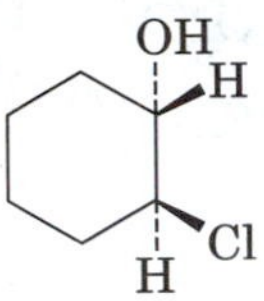

(b) 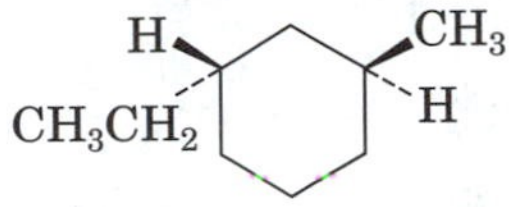

9.43 Draw tetrahedral representations of the following molecules.
(a) (*S*)-2-Butanol, $CH_3CH_2CH(OH)CH_3$ (b) (*R*)-3-Chloro-1-pentene

9.44 Draw tetrahedral representations of the two enantiomers of the amino acid cysteine, $HSCH_2CH(NH_2)COOH$, and identify each as *R* or *S*.

9.45 Which of the following pairs of Fischer projections represent the same enantiomer, and which represent different enantiomers?

(a) 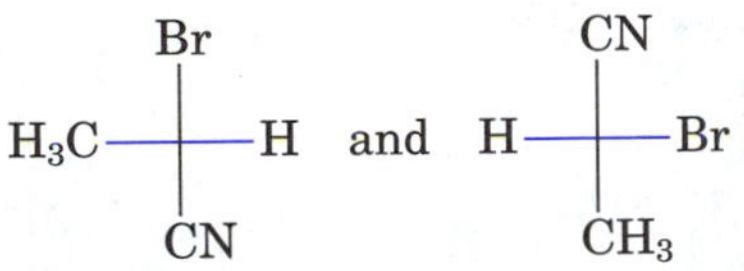

(b) 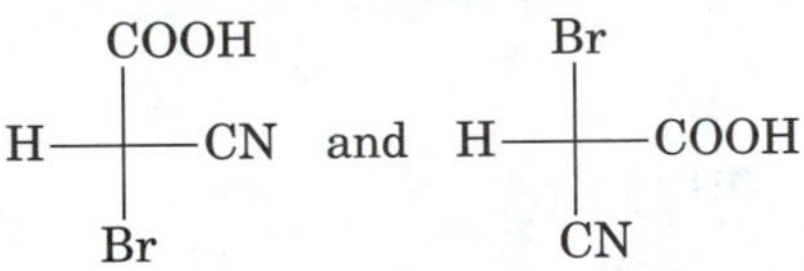

(c) 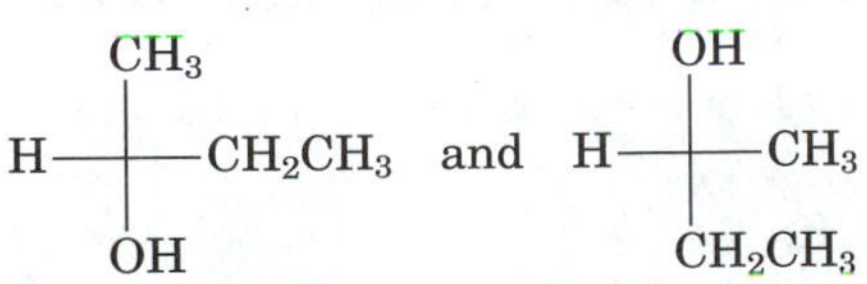

(d) 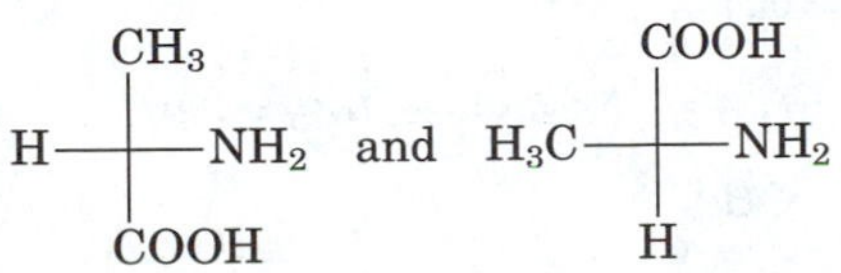

9.46 Assign *R* or *S* configurations to the following Fischer projections:

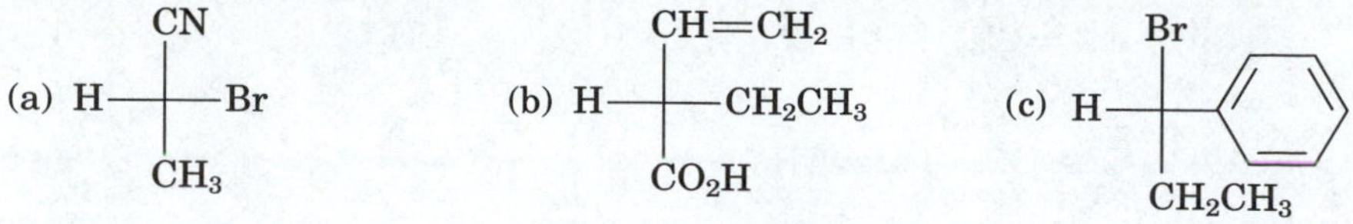

9.47 Assign *R* or *S* configurations to each stereogenic center in these molecules:

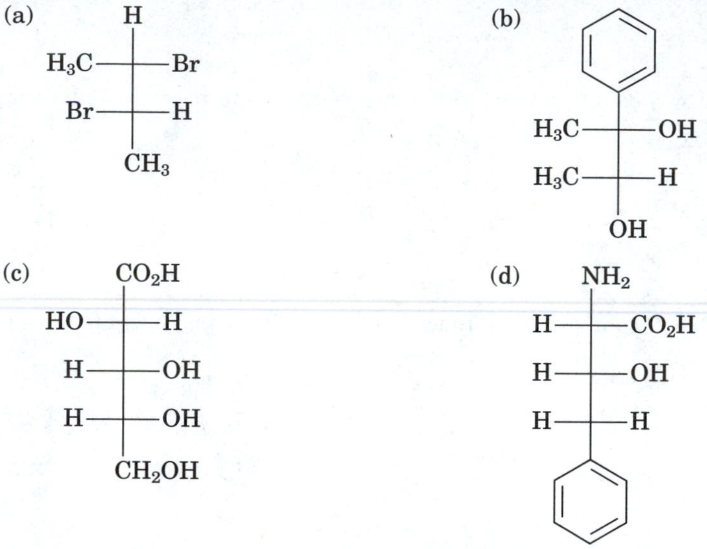

9.48 Draw Fischer projections that fit the following descriptions.
(a) The *S* enantiomer of 2-bromobutane
(b) The *R* enantiomer of alanine, $CH_3CH(NH_2)COOH$
(c) The *R* enantiomer of 2-hydroxypropanoic acid
(d) The *S* enantiomer of 3-methylhexane

9.49 Assign *R* or *S* configurations to each stereogenic center in ascorbic acid (vitamin C).

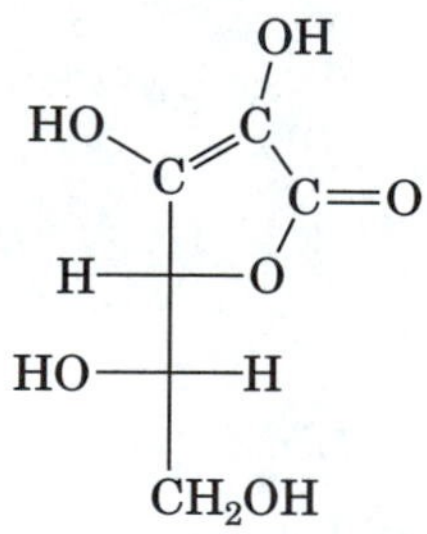

Ascorbic acid

9.50 Xylose is a common sugar found in many woods, including maple and cherry. Because it is much less prone to cause tooth decay than sucrose, xylose has been used in candy and chewing gum. Assign *R* or *S* configurations to the stereogenic centers in xylose.

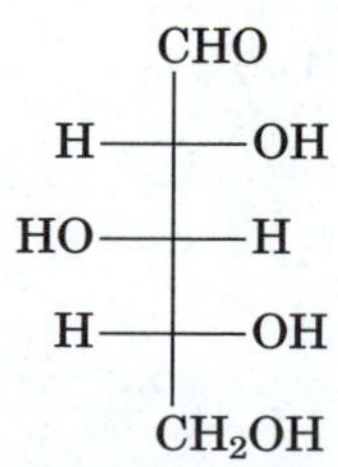

(+)-Xylose, $[\alpha]_D = +92°$

9.51 Hydroxylation of *cis*-2-butene with OsO_4 yields 2,3-butanediol. What stereochemistry do you expect for the product? (Review Section 7.8 if necessary.)

9.52 Hydroxylation of *trans*-2-butene with OsO_4 also yields 2,3-butanediol. What stereochemistry do you expect for the product?

9.53 We'll see in Chapter 18 that alkenes undergo reaction with peroxycarboxylic acids (RCO_3H) to give three-membered-ring cyclic ethers called *epoxides*. For example, 4-octene reacts with peroxyacids to yield 4,5-epoxyoctane:

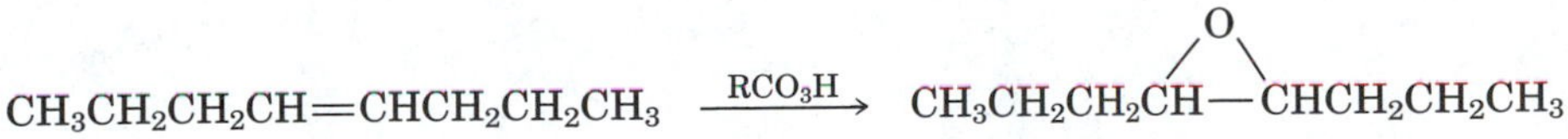

4-Octene **4,5-Epoxyoctane**

Assuming that this epoxidation reaction occurs with syn stereochemistry, draw the structure obtained from epoxidation of *cis*-4-octene. Is the product chiral? How many stereogenic centers does it have? How would you describe it stereochemically?

9.54 Answer Problem 9.53, assuming that the epoxidation reaction is carried out on *trans*-4-octene.

9.55 Write the products of the following reactions, and indicate the stereochemistry obtained in each instance.

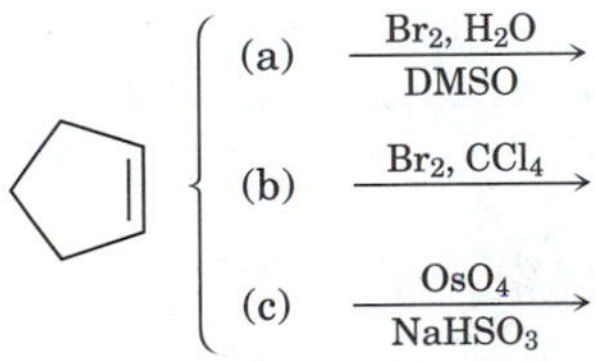

9.56 Draw all possible stereoisomers of cyclobutane-1,2-dicarboxylic acid, and indicate the interrelationships. Which, if any, are optically active? Do the same for cyclobutane-1,3-dicarboxylic acid.

9.57 Compound A, C_7H_{12}, was found to be optically active. On catalytic reduction over a palladium catalyst, 2 equiv of hydrogen were absorbed, yielding compound B, C_7H_{16}. On ozonolysis of A, two fragments were obtained. One fragment was identified as acetic acid. The other fragment, compound C, was an optically active carboxylic acid, $C_5H_{10}O_2$. Formulate the reactions, and draw structures for A, B, and C.

9.58 Compound A, $C_{11}H_{16}O$, was found to be an optically active alcohol. Despite its apparent unsaturation, no hydrogen was absorbed on catalytic reduction over a palladium catalyst. On treatment of A with dilute sulfuric acid, dehydration occurred, and B, an optically inactive alkene, $C_{11}H_{14}$, was produced as the major product. Alkene B, on ozonolysis, gave two products. One product was identified as propanal, CH_3CH_2CHO. Compound C, the other product, C_8H_8O, was shown to be a ketone. How many multiple bonds and/or rings does A have? Formulate the reactions and identify A, B, and C.

9.59 Draw the structure of (*R*)-2-methylcyclohexanone.

9.60 How many stereoisomers of 2,4-dibromo-3-chloropentane are there? Draw them and indicate which are optically active.

9.61 The so-called tetrahedranes are an interesting class of compounds, the first example of which was synthesized in 1979. Make a model of a substituted tetrahedrane with four different substituents. Is it chiral? Explain.

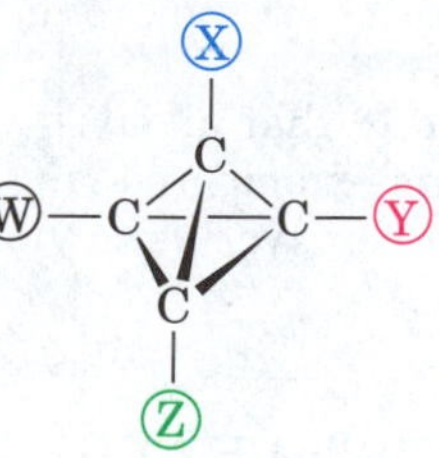

9.62 *Allenes* are compounds with adjacent carbon–carbon double bonds. Many allenes are chiral, even though they don't contain stereogenic centers. Mycomycin, for example, a naturally occurring antibiotic isolated from the bacterium *Nocardia acidophilus*, is chiral and has $[\alpha]_D = -130°$. Why is mycomycin chiral? Making a molecular model should be helpful.

$$HC{\equiv}C{-}C{\equiv}C{-}CH{=}C{=}CH{-}CH{=}CH{-}CH{=}CH{-}CH_2COOH$$

Mycomycin
(an allene)

9.63 Long before chiral allenes were known (Problem 9.62), the resolution of 4-methylcyclohexylideneacetic acid into two enantiomers had been carried out. Why is it chiral? What geometric relationship does it have to allenes?

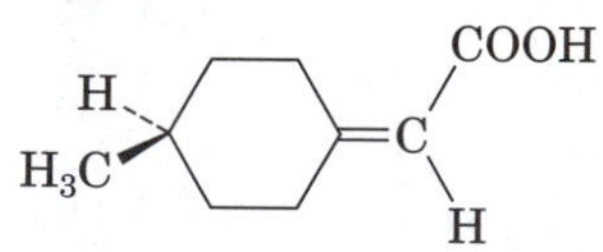

Methylcyclohexylideneacetic acid

9.64 Carboxylic acids react with alcohols to yield esters:

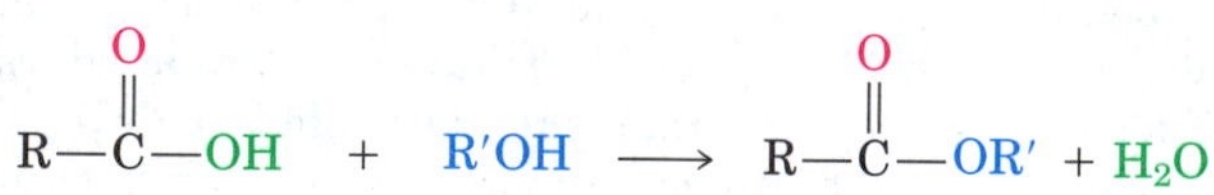

Carboxylic acid **Alcohol** **Ester**

Suppose that racemic lactic acid reacts with methanol, CH_3OH, to yield the ester methyl lactate. What stereochemistry would you expect the product(s) to have? What is the relationship of one product to another?

9.65 Suppose that (*S*)-lactic acid reacts with (*R*)-2-butanol to form an ester. What stereochemistry would you expect the product(s) to have? Draw the starting materials and product.

9.66 Suppose that racemic lactic acid reacts with (*S*)-2-butanol to form an ester. What stereochemistry does the product(s) have? What is the relationship of one product to another? Assuming that esters can be converted back into carboxylic acids, how might you use this reaction to resolve (±)-lactic acid?

A Look Ahead

9.67 We'll see in the next chapter that an alkyl halide reacts with a nucleophile to give a substitution product by a mechanism that involves *inversion* of stereochemistry at carbon:

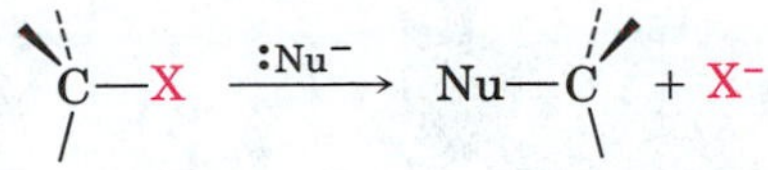

Formulate the reaction of (*S*)-2-bromobutane with HS^- ion to yield 2-butanethiol, $CH_3CH_2CH(SH)CH_3$. What is the stereochemistry of the product?

9.68 (*S*)-1-Chloro-2-methylbutane undergoes light-induced reaction with Cl_2 by a radical mechanism to yield a mixture of products (Section 10.4). Among the products are 1,4-dichloro-2-methylbutane and 1,2-dichloro-2-methylbutane.

(a) Formulate the reaction, showing the correct stereochemistry of the starting material.

(b) One of the two products is optically active, but the other is optically inactive. Which is which?

(c) What can you conclude about the stereochemistry of radical chlorination reactions?

9.69 Grignard reagents, RMgX, react with aldehydes to yield alcohols (Section 17.7). For example, the reaction of methylmagnesium bromide with propanal yields 2-butanol:

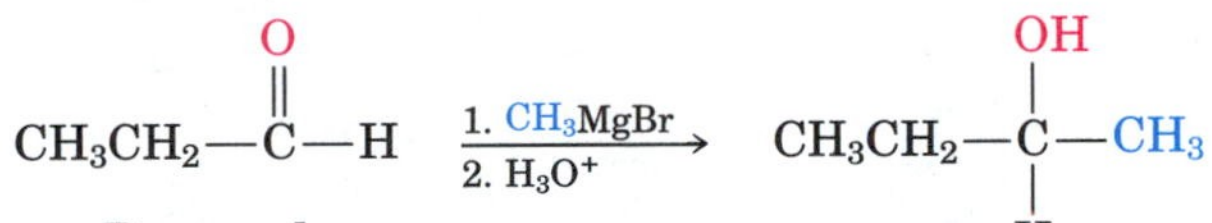

Propanal

2-Butanol

(a) Is the product chiral? Is it optically active?

(b) How many stereoisomers of butanol are formed, what are their stereochemical relationships, and what are their relative amounts?

9.70 Imagine that another Grignard reaction similar to that in Problem 9.69 is carried out between methylmagnesium bromide and (*R*)-2-phenylpropanal to yield 3-phenyl-2-butanol:

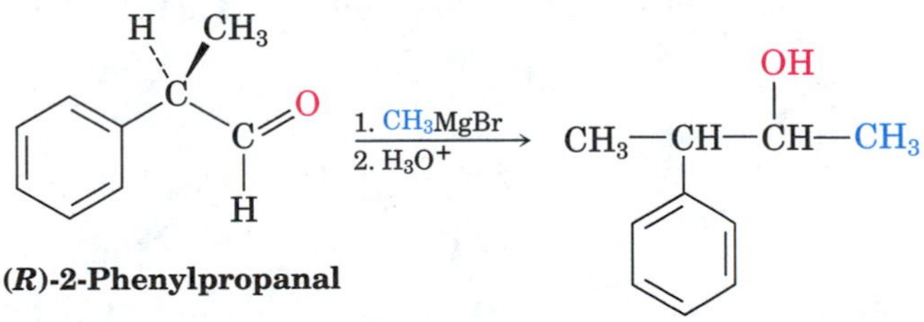

(*R*)-2-Phenylpropanal

3-Phenyl-2-butanol

(a) Is the product chiral? Is it optically active?

(b) How many stereoisomers of 3-phenyl-2-butanol are formed, what are their stereochemical relationships, and what are their relative amounts?

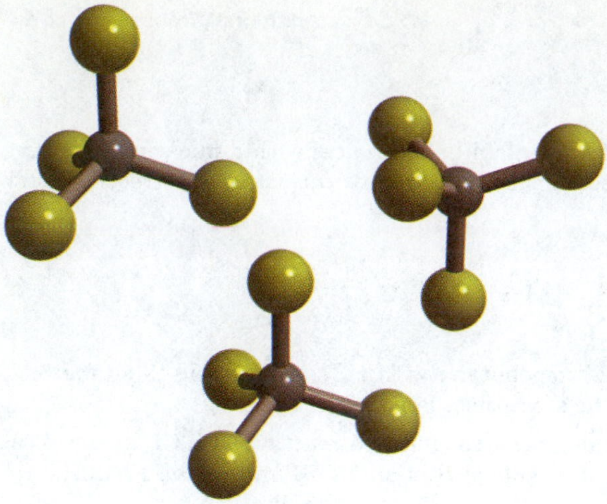

Tetrachloromethane is one of many alkyl halides produced by algae.

10

Alkyl Halides

It would be difficult to study organic chemistry for long without becoming aware of halo-substituted alkanes. Among their many uses, alkyl halides are employed as industrial solvents, as inhaled anesthetics in medicine, as refrigerants, and as pesticides and fumigating agents.

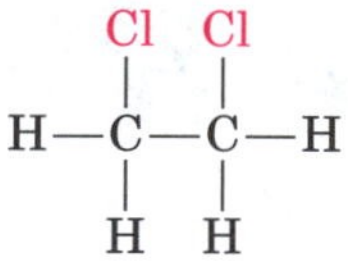

1,2-Dichloroethane (a solvent)

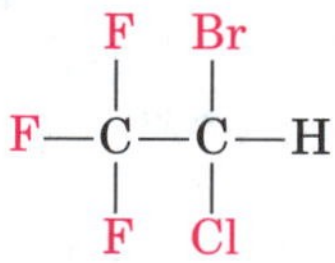

Halothane (an inhaled anesthetic)

Dichlorodifluoromethane (a refrigerant)

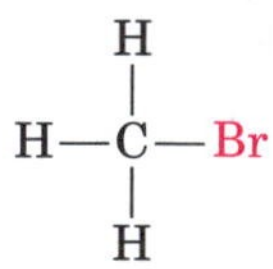

Bromomethane (a fumigant)

Alkyl halides occur widely in nature, though mostly in marine rather than terrestrial organisms. It has been estimated, for example, that the global emission rate of chloromethane, CH_3Cl, from natural sources is approximately 5 million tons per year. In addition, more than 100 different halogenated compounds have been found in the edible Hawaiian alga *Asparagopsis taxiformis*, and hundreds of others have been found in other marine sources. Many of these substances isolated from marine organisms exhibit interesting biological activity. For example, plocamene B, a trichlorocyclohexene derivative isolated from the red alga *Plocamium violaceum*, is similar in potency to DDT in its insecticidal activity against mosquito larvae.

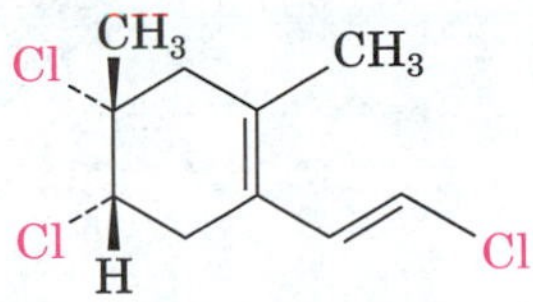

Plocamene B
(A trichloride)

In discussing the chemistry of alkyl halides in this chapter, we'll be talking primarily about compounds having halogen atoms bonded to saturated, sp^3-hybridized carbon atoms. Other classes of organohalides also exist, such as aromatic (*aryl*) and alkenyl (*vinylic*) halides, but much of their chemistry is different.

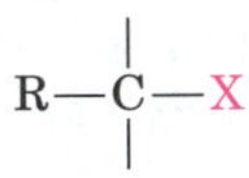

Alkyl halide

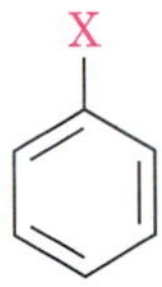

Aryl halide

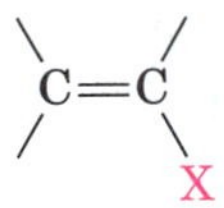

Vinylic halide

10.1 Naming Alkyl Halides

Alkyl halides are named in the same way as alkanes (Section 3.4), by treating the halogen as a substituent on a parent alkane chain. There are three rules:

1. Find and name the parent chain. As in naming alkanes, select the longest chain as the parent. If a double or triple bond is present, the parent chain must contain it.
2. Number the carbon atoms of the parent chain beginning at the end nearer the first substituent, regardless of whether it is alkyl or halo. Assign each substituent a number according to its position on the chain. For example:

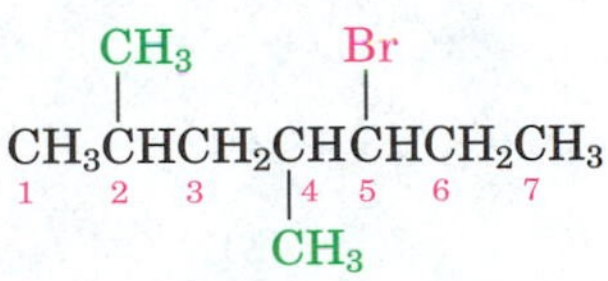

5-Bromo-2,4-dimethylheptane

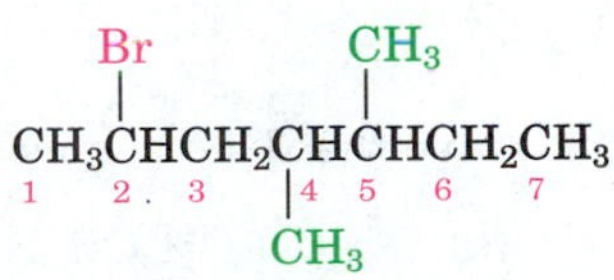

2-Bromo-4,5-dimethylheptane

(a) If more than one of the same kind of halogen is present, number each and use one of the prefixes *di-*, *tri-*, *tetra-*, and so on. For example:

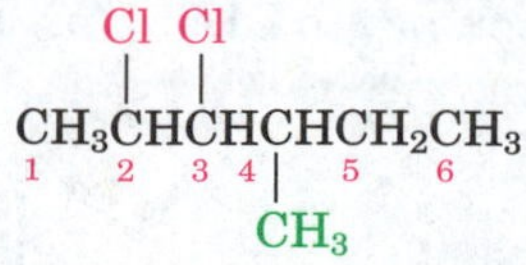

2,3-Dichloro-4-methylhexane

(b) If different halogens are present, number each according to its position on the chain and list them in alphabetical order when writing the name. For example:

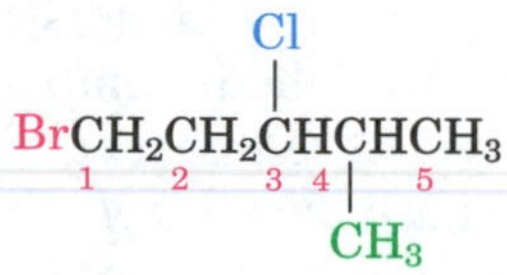

1-Bromo-3-chloro-4-methylpentane

3. If the parent chain can be properly numbered from either end by rule 2, begin at the end nearer the substituent (either alkyl or halo) that has alphabetical precedence. For example:

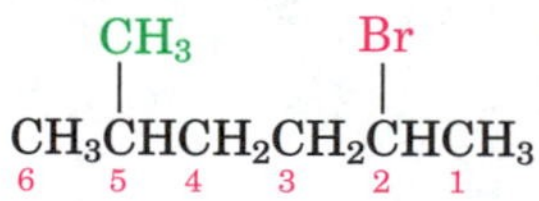

2-Bromo-5-methylhexane
(*NOT* 5-bromo-2-methylhexane)

In addition to their systematic names, many simple alkyl halides are also named by identifying first the alkyl group and then the halogen. For example, CH_3I can be called methyl iodide. Such names are well entrenched in the chemical literature and in daily usage, but they won't be used in this book.

CH_3I

Iodomethane
(or methyl iodide)

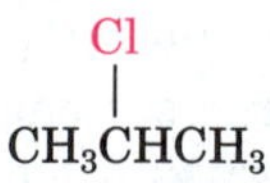

2-Chloropropane
(or isopropyl chloride)

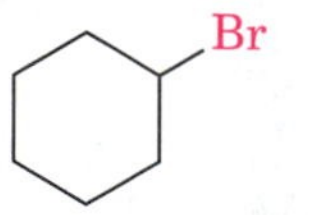

Bromocyclohexane
(or cyclohexyl bromide)

PROBLEM...

10.1 Give the IUPAC names of these alkyl halides:

(a) $CH_3CH_2CH_2CH_2I$

(b)

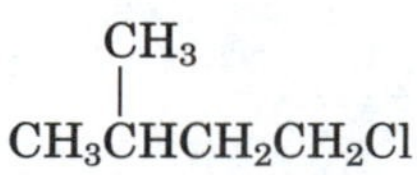

(c) 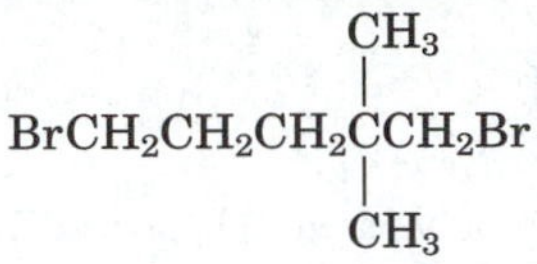

(d) 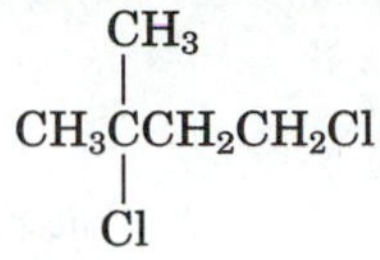

(e) 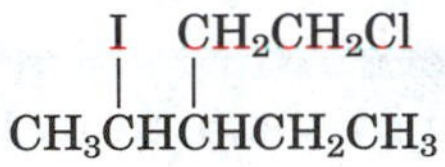

(f) 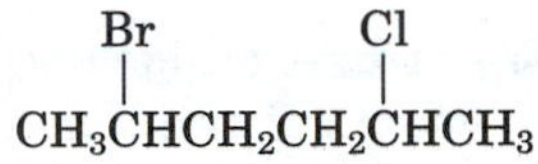

PROBLEM

10.2 Draw structures corresponding to these IUPAC names:

(a) 2-Chloro-3,3-dimethylhexane
(b) 3,3-Dichloro-2-methylhexane
(c) 3-Bromo-3-ethylpentane
(d) 1,1-Dibromo-4-isopropylcyclohexane
(e) 4-*sec*-Butyl-2-chlorononane
(f) 1,1-Dibromo-4-*tert*-butylcyclohexane

10.2 Structure of Alkyl Halides

The carbon–halogen bond in an alkyl halide results from the overlap of a carbon sp^3 hybrid orbital with a halogen orbital. Thus, alkyl halide carbon atoms have an approximately tetrahedral geometry with H-C-X bond angles near 109°. Halogens increase in size going down the periodic table, an increase that is reflected in the bond lengths of the halomethane series (Table 10.1). Table 10.1 also indicates that C–X bond strengths decrease going down the periodic table.

Table 10.1 A Comparison of the Halomethanes

Halomethane	*Bond length* (Å)	*Bond strength* (kJ/mol)	*Bond strength* (kcal/mol)	*Dipole moment* (D)
CH_3F	1.39	452	108	1.85
CH_3Cl	1.78	351	84	1.87
CH_3Br	1.93	293	70	1.81
CH_3I	2.14	234	56	1.62

In an earlier discussion of bond polarity in functional groups (Section 5.4), we noted that halogens are more electronegative than carbon. The C–X bond is therefore polar, with the carbon atom bearing a slight positive charge ($\delta+$) and the halogen a slight negative charge ($\delta-$):

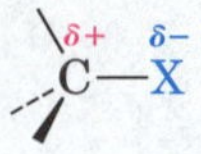

(X is the standard abbreviation for a halogen: F, Cl, Br, or I)

We can get an idea of the amount of bond polarity by measuring the dipole moments of the halomethanes. As indicated by the data in Table 10.1, all the halomethanes have substantial dipole moments; the specific identity of the halogen has a fairly small effect.

Since the carbon atom of alkyl halides is positively polarized, alkyl halides are good electrophiles (Section 5.4). We'll see in the next chapter that much of the chemistry alkyl halides undergo is dominated by their electrophilic behavior.

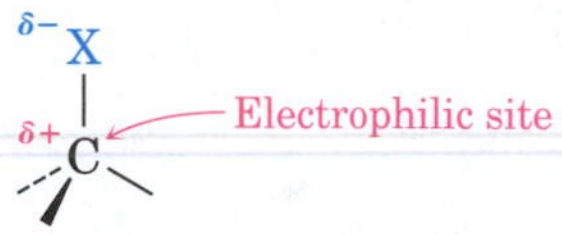

10.3 Preparation of Alkyl Halides

We've already seen several methods of alkyl halide preparation, including the reaction of both HX and X_2 with alkenes in electrophilic addition reactions (Sections 6.8 and 7.2). HCl, HBr, and HI react with alkenes by a polar mechanism to give the product of Markovnikov addition; HBr can also add by a radical pathway to give the non-Markovnikov product. Bromine and chlorine yield trans 1,2-dihalogenated addition products.

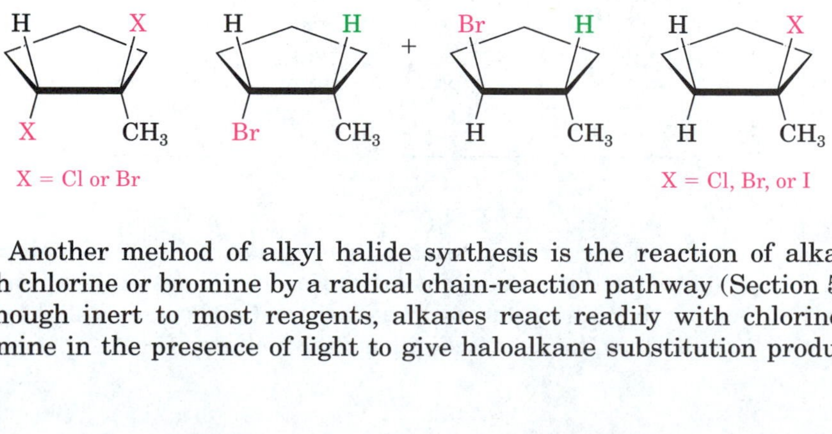

Another method of alkyl halide synthesis is the reaction of alkanes with chlorine or bromine by a radical chain-reaction pathway (Section 5.3). Although inert to most reagents, alkanes react readily with chlorine or bromine in the presence of light to give haloalkane substitution products.

The reaction occurs by the radical mechanism shown in Figure 10.1 for chlorination.

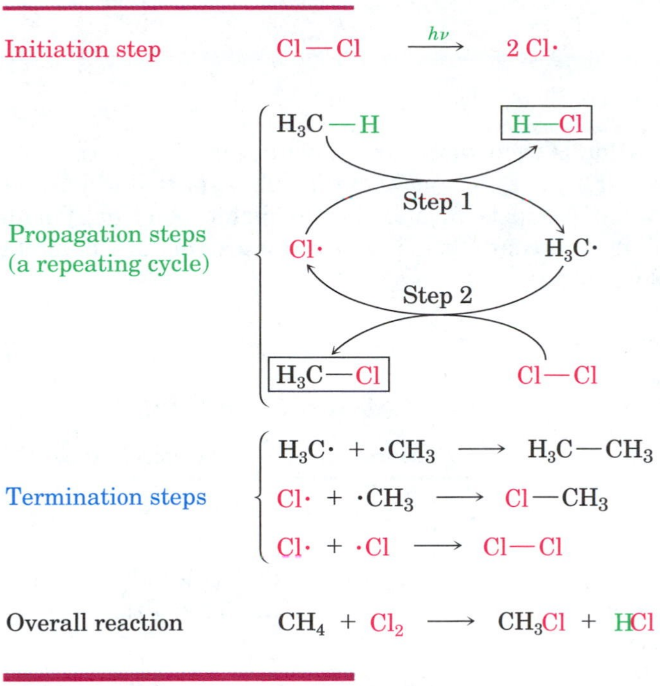

Figure 10.1 Mechanism of the radical chlorination of methane. Three kinds of steps are required: *initiation, propagation,* and *termination.* (The symbol $h\nu$ shown in the initiation step is the standard way of indicating irradiation with light.)

Recall from Section 5.3 that radical substitution reactions normally require three kinds of steps: *initiation*, *propagation*, and *termination*. Once an initiation step has started the process by producing radicals, the reaction continues in a self-sustaining cycle. The cycle requires two repeating propagation steps in which a radical, the halogen, and the alkane yield alkyl halide product plus more radical to carry on the chain. The chain is occasionally terminated by the combination of two radicals.

Though interesting from a mechanistic point of view, alkane halogenation is a poor synthetic method for preparing different haloalkanes. Let's see why.

10.4 Radical Halogenation of Alkanes

Alkane halogenation is a poor method of alkyl halide synthesis because mixtures of products invariably result. For example, chlorination of methane does not stop cleanly at the monochlorinated stage; the reaction continues on to give a mixture of dichloro, trichloro, and even tetrachloro products:

$$CH_4 + Cl_2 \longrightarrow CH_3Cl + HCl$$
$$CH_3Cl \xrightarrow{Cl_2} CH_2Cl_2 + HCl$$
$$CH_2Cl_2 \xrightarrow{Cl_2} CHCl_3 + HCl$$
$$CHCl_3 \xrightarrow{Cl_2} CCl_4 + HCl$$

The situation is even worse for chlorination of alkanes that have more than one type of hydrogen. For example, chlorination of butane gives two monochlorinated products in addition to dichlorobutane, trichlorobutane, and so on. Thirty percent of the monochloro product is 1-chlorobutane, and 70% is 2-chlorobutane:

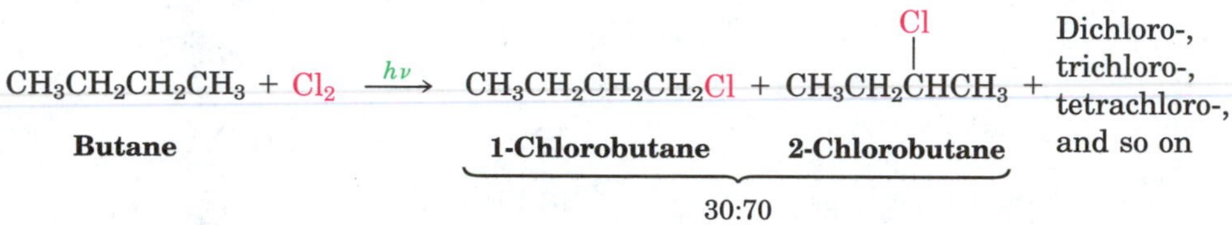

As another example, 2-methylpropane yields 2-chloro-2-methylpropane and 1-chloro-2-methylpropane in the ratio 35:65, along with more highly chlorinated products:

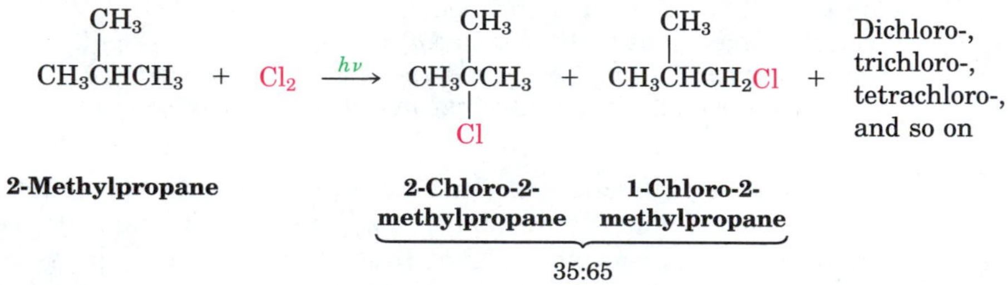

From these and similar reactions, it's possible to calculate a reactivity order toward chlorination for different types of hydrogen atoms in a molecule. Take the butane chlorination, for instance. Butane has six equivalent primary hydrogens ($-CH_3$) and four equivalent secondary hydrogens ($-CH_2-$). The fact that butane yields 30% of 1-chlorobutane product means that *each one* of the six primary hydrogens is responsible for 30%/6 = 5% of the product. Similarly, the fact that 70% of 2-chlorobutane is formed means that each of the four secondary hydrogens is responsible for 70%/4 = 17.5% of the product. Thus, reaction of a secondary hydrogen happens 17.5/5 = 3.5 times as often as reaction of a primary hydrogen.

A similar calculation for the chlorination of 2-methylpropane indicates that each of the nine primary hydrogens accounts for 65%/9 = 7.2% of the product, while the single tertiary hydrogen (R_3CH) accounts for 35% of the product. Thus, a tertiary hydrogen is 35/7.2 = 5 times as reactive as a primary hydrogen toward chlorination.

Relative reactivity toward chlorination: $R_3CH > R_2CH_2 > RCH_3$

5.0 3.5 1.0

What are the reasons for the observed reactivity order of alkane hydrogens toward radical chlorination? A look at the bond dissociation energies given previously in Table 5.4 hints at the answer. As the data in Table 5.4 indicate, tertiary C–H bonds (390 kJ/mol; 93 kcal/mol) are weaker than secondary C–H bonds (401 kJ/mol; 96 kcal/mol), which are in turn weaker than primary C–H bonds (420 kJ/mol; 100 kcal/mol). Since less energy is needed to break a tertiary C–H bond than to break a primary or secondary C–H bond, the resultant tertiary radical is more stable than a primary or secondary radical.

Relative stability of radicals: $R_3C\cdot > R_2\dot{C}H > R\dot{C}H_2$

An explanation of the relationship between reactivity and bond strength in radical chlorination reactions is similar to that invoked in Section 6.11 to explain why more stable carbocations form faster than less stable ones in alkene electrophilic addition reactions. A reaction energy diagram for the formation of an alkyl radical during alkane chlorination looks like that in Figure 10.2. Although the hydrogen-abstraction step is slightly exothermic, there is nevertheless a certain amount of developing radical character in the transition state. Thus, any factor (such as increased alkyl substitution) that stabilizes a radical intermediate also stabilizes the transition state leading to that intermediate (lowers $\Delta G^{\ddagger}$). In other words, a more stable radical forms faster than a less stable one. Tertiary radicals are more stable, and tertiary hydrogen atoms are therefore more easily removed.

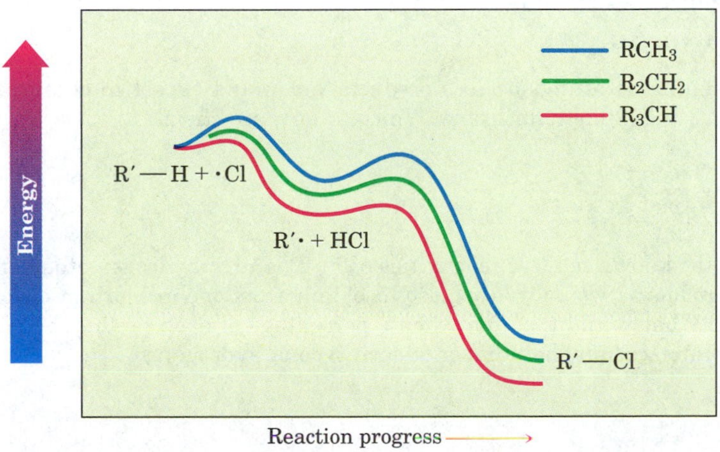

Figure 10.2 Reaction energy diagram for alkane chlorination. The stability order of tertiary, secondary, and primary radicals is the same as their relative rate of formation.

In contrast to alkane chlorination, alkane bromination is usually much more selective. In its reaction with 2-methylpropane, for example, bromine abstracts the tertiary hydrogen with greater than 99% selectivity, as opposed to the 35:65 mixture observed in the corresponding chlorination.

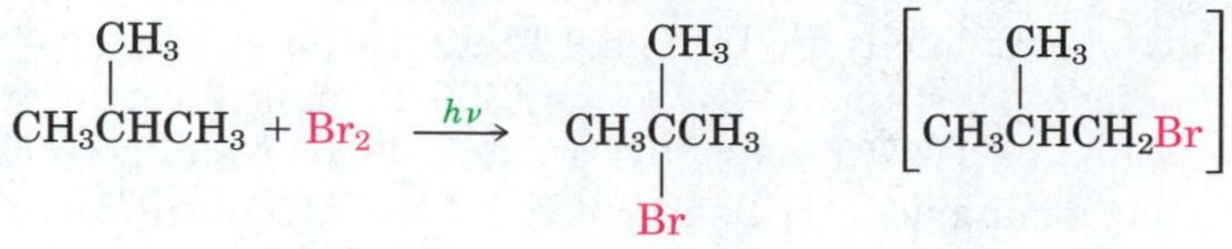

2-Methylpropane **2-Bromo-2-methylpropane (>99%)** **1-Bromo-2-methylpropane (<1%)**

The enhanced selectivity of alkane bromination over chlorination can be explained by turning once again to the Hammond postulate. In comparing the product-determining steps—abstraction of an alkane hydrogen by Cl· or Br· radical—a comparison of $\Delta H°$ values shows that reaction with Cl· is exothermic but that reaction with Br· is endothermic. As a result, the transition state for bromination resembles the alkyl radical more closely than the transition state for chlorination, and the stability of that radical is therefore more important for bromination than for chlorination.

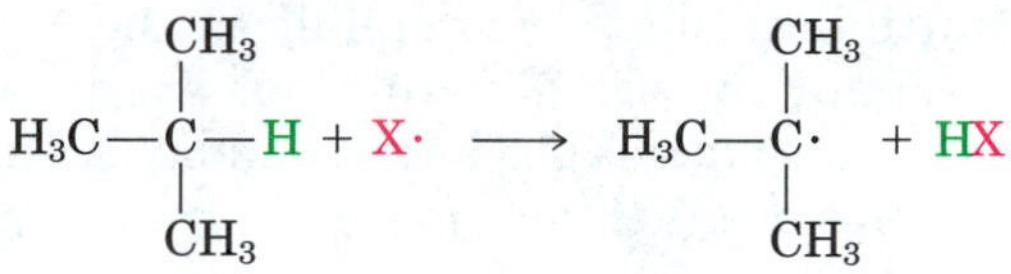

$\Delta H° = -50$ kJ for X = Cl
$\Delta H° = +13$ kJ for X = Br

2-Methylpropane

PROBLEM. . . . .

10.3 Draw and name all monochloro products you would expect to obtain from radical chlorination of 2-methylpentane. Which, if any, are chiral?

PROBLEM. . . . .

10.4 Taking the known relative reactivities of 1°, 2°, and 3° hydrogen atoms into account, what product(s) would you expect to obtain from monochlorination of 2-methylbutane? What would the approximate percentage of each product be? (Don't forget to take into account the number of each type of hydrogen.)

PROBLEM. . . . .

10.5 Use the data in Table 5.4 to calculate $\Delta H°$ for the reactions of Cl· and Br· with a secondary hydrogen atom of propane. Which reaction would you expect to be more selective?

10.5 Allylic Bromination of Alkenes

While repeating some earlier work of Wohl, Karl Ziegler[1] reported in 1942 that alkenes react with *N*-bromosuccinimide (abbreviated NBS) in the presence of light to give products resulting from substitution of hydrogen by bromine at the **allylic** position, the position *next to* the double bond. Cyclohexene, for example, gives 3-bromocyclohexene in 85% yield.

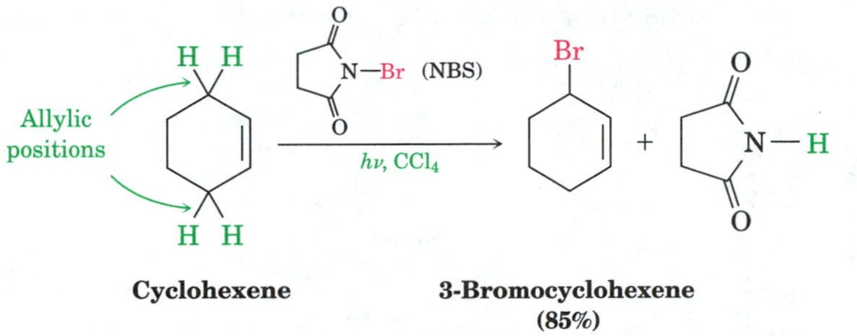

This allylic bromination with NBS looks very similar to the alkane halogenation reaction just discussed. In both cases, a C–H bond on a saturated carbon is broken and the hydrogen atom is replaced by halogen. The analogy is a good one, for studies have shown that allylic NBS brominations do in fact occur by a two-step, radical chain pathway. The product-determining step involves abstraction of an allylic hydrogen atom by Br· and formation of the corresponding allylic radical. This allylic radical then reacts with Br_2 to yield the product and a Br· radical, which cycles back into the first step to carry on the chain. (The Br_2 necessary for reaction with the allyl radical is produced by reaction of the HBr formed in the first step with NBS.)

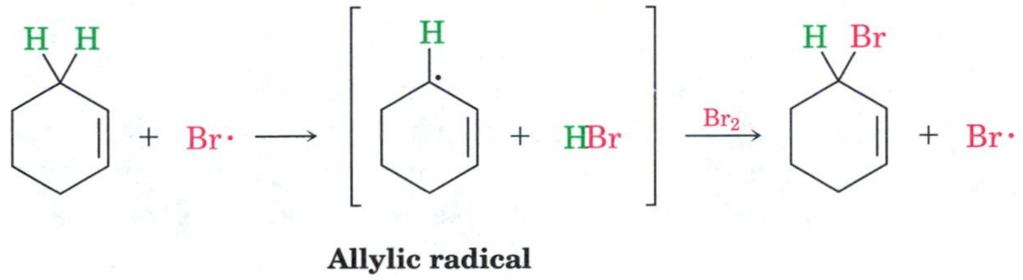

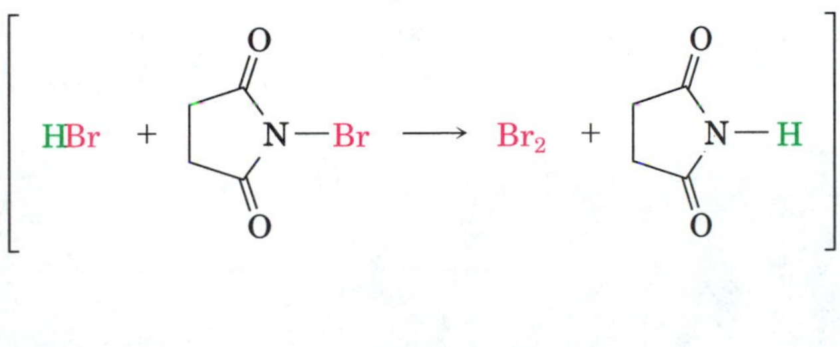

[1]Karl Ziegler (1889–1973); b. Helsa, near Kassel, Germany; Ph.D. University of Marburg (von Auwers); professor, Heidelberg (1927–1936); director, Kaiser Wilhelm Institute for Coal Research, Mülheim-Ruhr, Germany; Nobel Prize (1963).

Why does bromination occur exclusively at an allylic position rather than elsewhere in the molecule? The answer, once again, is found by looking at bond dissociation energies to see the relative stabilities of various kinds of radicals.

There are three types of C–H bonds in cyclohexene, and Table 5.4 gives an idea of their relative strengths. Although a typical secondary alkyl C–H bond has a strength of about 400 kJ/mol (96 kcal/mol), and a typical vinylic C–H bond has a strength of 445 kJ/mol (106 kcal/mol), an *allylic* C–H bond has a strength of only 360 kJ/mol (87 kcal/mol). An allylic radical is therefore more stable than a typical alkyl radical by about 40 kJ/mol (9 kcal/mol).

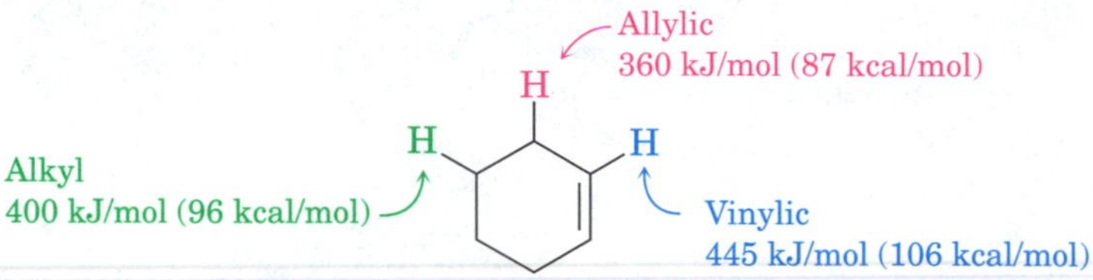

We can thus expand the stability ordering to include allylic and vinylic radicals:

$$\text{Allylic} > R_3C\cdot > R_2\dot{C}H > R\dot{C}H_2 > \cdot CH_3 > \text{Vinylic}$$

10.6 Stability of the Allyl Radical: Resonance Revisited

To see why allylic radicals are so stable, look at the orbital picture of the allyl radical in Figure 10.3. The radical carbon atom next to the double bond can adopt sp^2 hybridization, placing the unpaired electron in a *p* orbital and giving a structure that is electronically symmetrical. The *p* orbital on the central carbon can therefore overlap equally well with a *p* orbital on *either* of the two neighboring carbons.

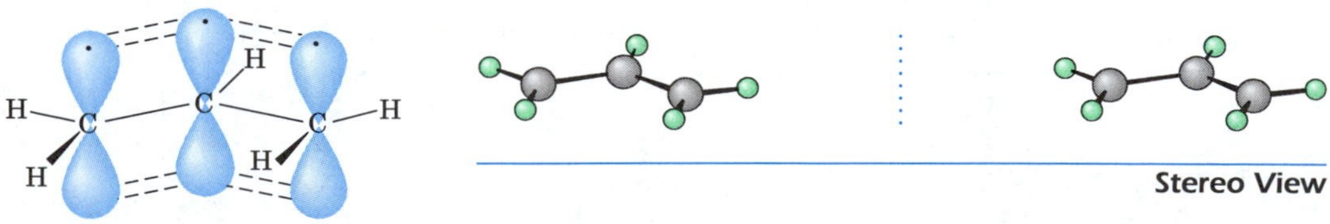

Figure 10.3 An orbital view of the allyl radical and a computer-generated model. The *p* orbital on the central carbon can overlap equally well with a *p* orbital on either neighboring carbon because the structure is electronically symmetrical.

Since the allyl radical is electronically symmetrical, there are two ways to draw it—with the unpaired electron on the left and the double bond on the right, or with the unpaired electron on the right and the double bond on the left. Neither structure is correct by itself; the true structure of the allyl radical is a resonance hybrid of the two. (You might want to review Section 2.4 if you need to brush up on resonance. In addition, the Practice Problem at the end of this section shows how resonance structures can be drawn.)

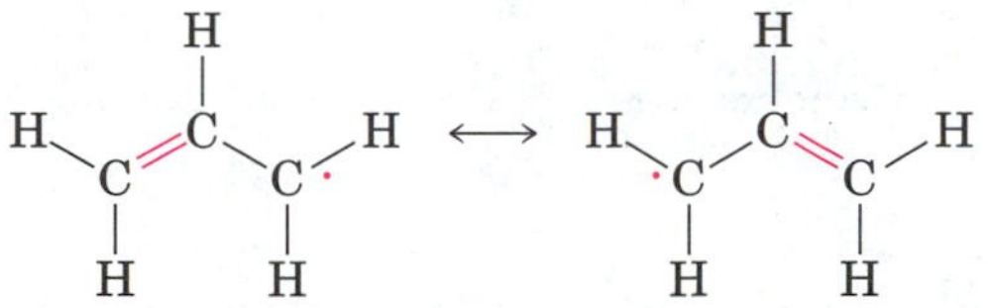

As noted in Section 2.5, the greater the number of possible resonance forms, the greater the stability of the compound. An allyl radical, with two resonance forms, is therefore more stable than a typical alkyl radical, which has only a single Kekulé structure. In molecular orbital terms, the stability of the allyl radical is due to the fact that the unpaired electron is **delocalized**, or spread out, over an extended π orbital network rather than localized at only one site.

$$H_2C{=}CH\dot{C}H_2 \longleftrightarrow \dot{C}H_2CH{=}CH_2 \qquad CH_3CH_2\dot{C}H_2$$

Allyl radical (delocalized, more stable) — **Propyl radical (localized, less stable)**

In addition to affecting stability, delocalization of the unpaired electron in the allyl radical has other chemical consequences. Since the unpaired electron can be found at either end of the π orbital system, bromination of the allyl radical can occur at either end. As a result, allylic bromination of an unsymmetrical alkene often leads to a mixture of products. For example, bromination of 1-octene gives a mixture of 3-bromo-1-octene and 1-bromo-2-octene. The two products are not formed in equal amounts, however, because the intermediate allylic radical is not symmetrical and reaction at the two ends is not equally likely. Reaction at the less hindered, primary end is favored.

$$CH_3CH_2CH_2CH_2CH_2CH_2CH{=}CH_2$$

1-Octene

$$\downarrow \text{NBS, } CCl_4$$

$$\left[CH_3CH_2CH_2CH_2CH_2\dot{C}HCH{=}CH_2 \longleftrightarrow CH_3CH_2CH_2CH_2CH_2CH{=}CH\dot{C}H_2\right]$$

$$\downarrow$$

$$CH_3CH_2CH_2CH_2CH_2CH(Br)CH{=}CH_2 + CH_3CH_2CH_2CH_2CH_2CH{=}CHCH_2Br$$

3-Bromo-1-octene (17%) — **1-Bromo-2-octene (83%) (53:47 trans:cis)**

The products of allylic bromination reactions are particularly useful for conversion into dienes by dehydrohalogenation with base. Cyclohexene can be converted into 1,3-cyclohexadiene, for example.

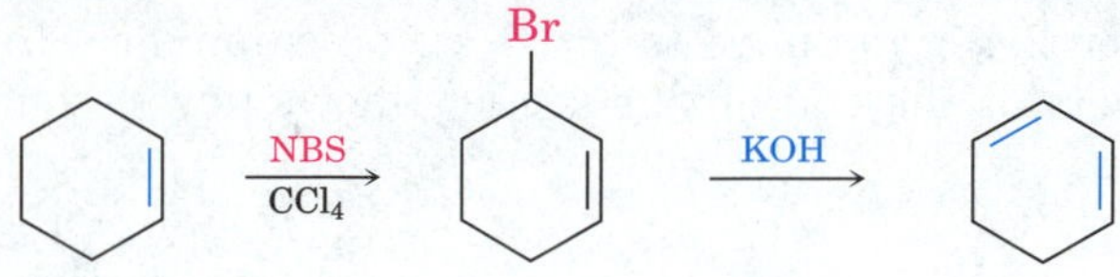

Cyclohexene **3-Bromocyclohexene** **1,3-Cyclohexadiene**

PRACTICE PROBLEM

1,3-Pentanedione reacts with sodium ethoxide (the sodium salt of ethyl alcohol, $Na^+\ {}^-OCH_2CH_3$) to yield an anion. Draw as many resonance forms as you can for this anion.

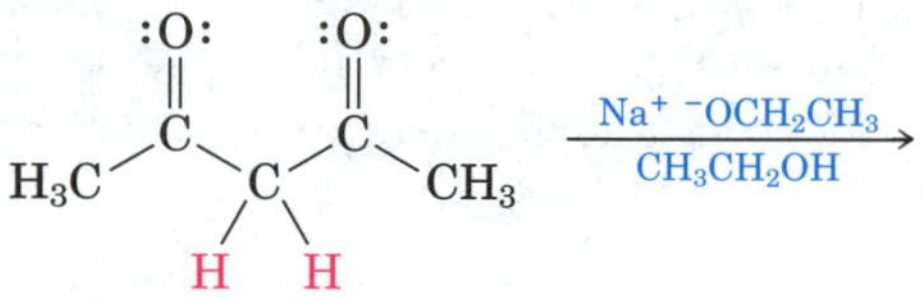

1,3-Pentanedione

Solution Just as an allylic radical has two resonance forms, so any analogous three-atom system has two resonance forms:

The atoms X, Y, and *Z* might be C, N, O, P, or S, and the asterisk (*) might represent a radical (·), a cation (+), or an anion (−). In the present instance, the anion of 1,3-pentanedione has a lone pair of electrons and a negative charge on the carbon atom next to a C=O bond. Thus, a resonance structure can be drawn in which the lone pair of electrons and the double bond move:

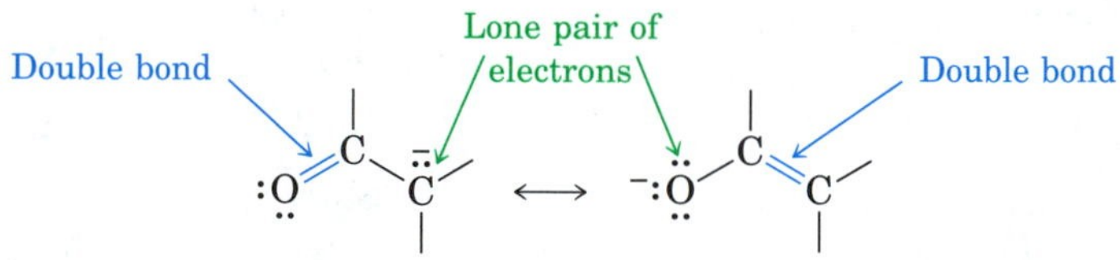

We can draw three resonance structures for the 1,3-pentanedione anion because there are two neighboring C=O groups over which to delocalize the negative charge.

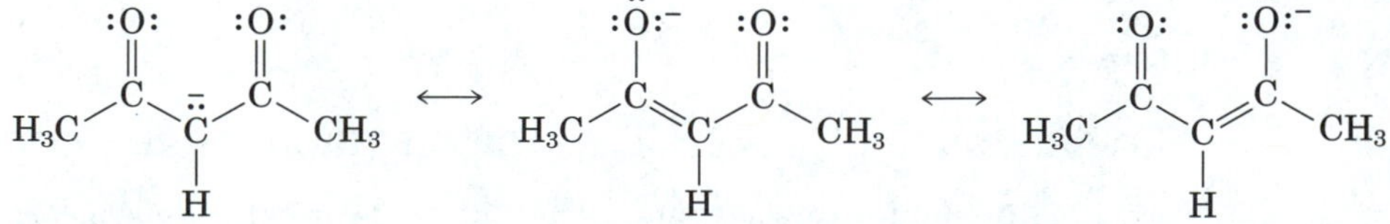

PROBLEM. . . . .

10.6 The major product of the reaction of methylenecyclohexane with *N*-bromosuccinimide is 1-(bromomethyl)cyclohexene. Explain.

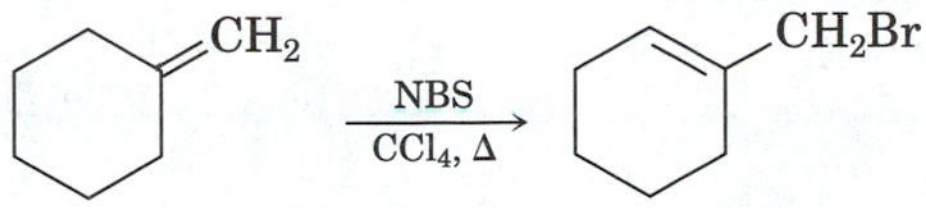

Major product

PROBLEM. . . . .

10.7 What products would you expect from reaction of the following alkenes with NBS? If more than one product is formed, show the structures of all.

(a) 5-Methylcycloheptene

(b) $CH_3CH(CH_3)CH{=}CHCH_2CH_3$

PROBLEM. . . . .

10.8 Draw as many resonance structures as you can for these species:

(a) 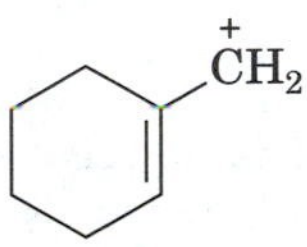

(b) 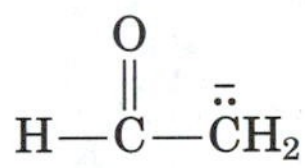

(c) 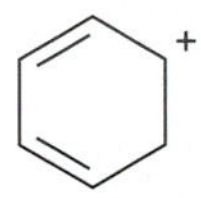

. . . .

10.7 Preparing Alkyl Halides from Alcohols

The most general method for preparing alkyl halides is to make them from alcohols. A great many alcohols are commercially available, and we'll see later that a great many more can be obtained from carbonyl compounds. Because of the importance of the reaction, many different reagents have been used for transforming alcohols into alkyl halides.

The simplest method for converting an alcohol to an alkyl halide involves treating the alcohol with HCl, HBr, or HI:

$$ROH + HX \longrightarrow RX + H_2O \qquad (X = Cl, Br, \text{ or } I)$$

For reasons that will be discussed in the next chapter (Section 11.16), the reaction works best when applied to tertiary alcohols, R_3COH. Primary and secondary alcohols also react, but at slower rates and at higher reaction temperatures. Although this is not a problem in simple cases, more complicated molecules are sometimes acid-sensitive and are destroyed by the reaction conditions.

$$\underset{3°}{R_3COH} > \underset{2°}{R_2CHOH} > \underset{1°}{RCH_2OH} > \underset{\text{Methanol}}{CH_3OH}$$

More reactive ← Reactivity — Less reactive

The reaction of HX with a tertiary alcohol is so rapid that it's often carried out simply by bubbling the pure HX gas into a cold ether solution of the alcohol. Reaction is usually complete within a few minutes.

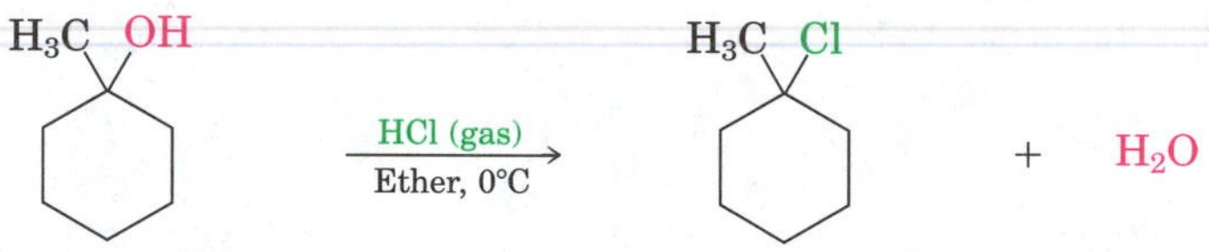

1-Methylcyclohexanol **1-Chloro-1-methylcyclohexane (90%)**

Primary and secondary alcohols are best converted into alkyl halides by treatment with such reagents as thionyl chloride ($SOCl_2$) or phosphorus tribromide (PBr_3). These reactions, which normally take place readily under mild conditions, are less acidic and less likely to cause acid-catalyzed rearrangements than the HX method.

$$3\ CH_3CH_2\overset{OH}{\overset{|}{C}}HCH_3 \xrightarrow[\text{Ether, 35°C}]{PBr_3} 3\ CH_3CH_2\overset{Br}{\overset{|}{C}}HCH_3 + P(OH)_3$$

2-Butanol **2-Bromobutane (86%)**

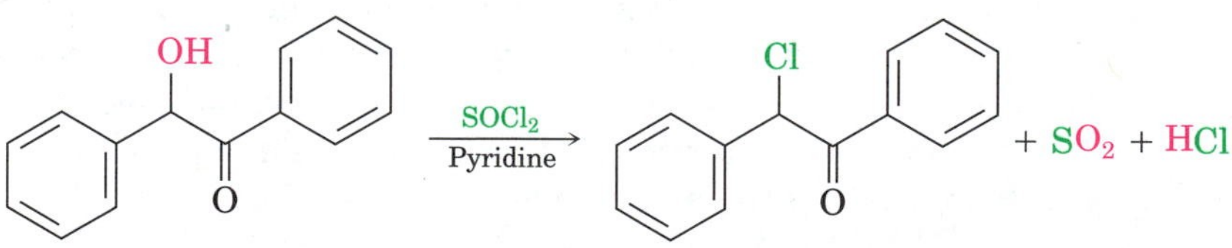

Benzoin **Desyl chloride (86%)**

As the preceding examples indicate, the yields of these PBr_3 and $SOCl_2$ reactions are generally high. Other functional groups such as ethers, carbonyls, and aromatic rings don't usually interfere. We'll look at the mechanisms of these substitution reactions in the next chapter.

PROBLEM

10.9 How would you prepare the following alkyl halides from the appropriate alcohols?
(a) 2-Chloro-2-methylpropane (b) 2-Bromo-4-methylpentane

(c) 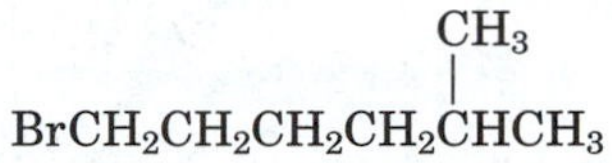

(d) 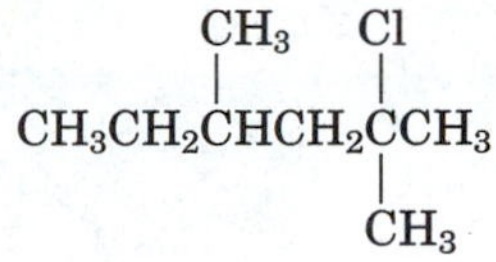

10.8 Reactions of Alkyl Halides: Grignard Reagents

Organic halides, RX, react with magnesium metal in ether or tetrahydrofuran (THF) solvent to yield organomagnesium halides, RMgX. The products, called **Grignard reagents** after their discoverer, Victor Grignard,[2] are *organometallic* compounds because they contain a carbon–metal bond.

$$R{-}X + Mg \xrightarrow[\text{or THF}]{\text{Ether}} R{-}Mg{-}X$$

where R = 1°, 2°, or 3° alkyl, aryl, or alkenyl

X = Cl, Br, or I

For example:

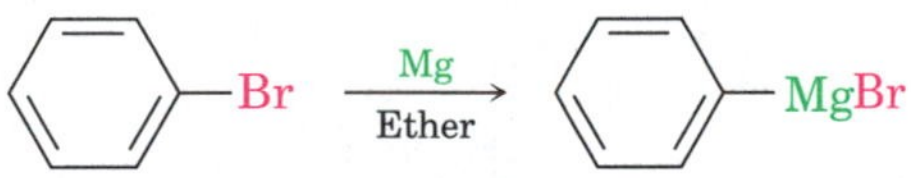

Bromobenzene **Phenylmagnesium bromide**

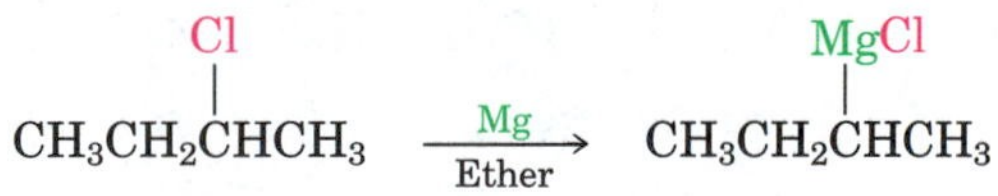

2-Chlorobutane ***sec*-Butylmagnesium chloride**

Many different kinds of organohalides form Grignard reagents. Steric hindrance in the halide is not a problem in the formation of Grignard reagents, since 1°, 2°, and 3° alkyl halides all react with similar ease. Aryl and alkenyl halides also react with magnesium, although it's best to use THF as solvent for these cases. The halogen may be Cl, Br, or I, although chlorides are less reactive than bromides and iodides. Organofluorides rarely react with magnesium.

[2]François Auguste Victor Grignard (1871–1935); b. Cherbourg, France; professor, University of Nancy, Lyons; Nobel Prize, 1912.

As you might expect from the discussion of electronegativity and bond polarity in Section 5.4, the carbon–magnesium bond is polarized, making the carbon atom both nucleophilic and basic:

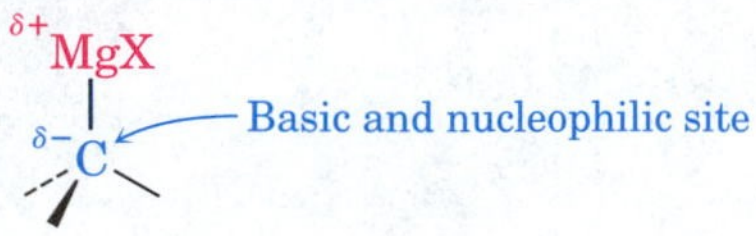

In a formal sense, a Grignard reagent can be thought of as the magnesium salt of a carbon anion, or **carbanion**. It's more accurate, though, to view Grignard reagents as containing a highly polar covalent C–Mg bond, rather than an ionic $R_3C^-\ ^+MgX$ bond.

Because of their basic character, Grignard reagents react with acids; because of their nucleophilic character, they react with a wide variety of electrophiles. For example, Grignard reagents react with proton donors, such as H_2O, ROH, RCOOH, or RNH_2, to yield hydrocarbons. The overall sequence of Grignard formation followed by acid treatment is a useful synthetic method for converting an organohalide into a hydrocarbon, R–X → R–H. For example,

$$CH_3(CH_2)_8CH_2Br \xrightarrow[2.\ H_2O]{1.\ Mg} CH_3(CH_2)_8CH_3$$

1-Bromodecane **Decane (85%)**

We'll see many more uses of Grignard reagents in later chapters.

PROBLEM...

10.10 How strong a base would you expect a Grignard reagent to be? Look at Table 8.1, and then predict whether the following reactions will occur. (The pK_a of NH_3 is 35.)

(a) $CH_3MgBr + H{-}C{\equiv}C{-}H \longrightarrow CH_4 + H{-}C{\equiv}C{-}MgBr$

(b) $CH_3MgBr + NH_3 \longrightarrow CH_4 + H_2N{-}MgBr$

PROBLEM...

10.11 An important advantage of alkyl halide reaction with Grignard reagents is that the sequence can be used to introduce deuterium into a specific site in a molecule. How might you do this?

$$CH_3\underset{}{\overset{Br}{CH}}CH_2CH_3 \xrightarrow{?} CH_3\overset{D}{CH}CH_2CH_3$$

PROBLEM...

10.12 Why do you suppose it's not possible to prepare a Grignard reagent from a bromo alcohol such as 4-bromo-1-pentanol?

$$\underset{}{CH_3\overset{Br}{\overset{|}{C}}HCH_2CH_2CH_2OH} \;\not\xrightarrow{Mg}\; CH_3\overset{MgBr}{\overset{|}{C}}HCH_2CH_2CH_2OH$$

Give another example of a molecule that is unlikely to form a Grignard reagent.

10.9 Organometallic Coupling Reactions

Several other kinds of organometallic reagents can be prepared in a manner similar to that of Grignard reagents. For example, alkyllithium reagents can be prepared by the reaction of an alkyl halide with lithium metal:

$$\underset{\textbf{1-Bromobutane}}{CH_3CH_2CH_2CH_2Br} \xrightarrow[\text{Pentane}]{2\ Li} \underset{\textbf{Butyllithium}}{CH_3CH_2CH_2\overset{\delta-}{C}H_2\overset{\delta+}{Li}} + LiBr$$

Alkyllithiums are both nucleophiles and bases, and their chemistry is similar in many respects to that of the alkylmagnesium halides. One of the most valuable reactions of alkyllithiums is their use in preparing lithium diorganocopper reagents, R_2CuLi, called **Gilman[3] reagents.**

$$CH_3Br + 2\ Li \xrightarrow{\text{Pentane}} \underset{\textbf{Methyllithium}}{CH_3Li} + LiBr$$

$$\underset{\textbf{Methyllithium}}{2\ CH_3Li} + CuI \xrightarrow{\text{Ether}} \underset{\substack{\textbf{Lithium dimethylcopper} \\ \textbf{(a Gilman reagent)}}}{(CH_3)_2Cu^-\ Li^+} + LiI$$

Gilman reagents are easily prepared by reaction of an alkyllithium with cuprous iodide, CuI, in ether solvent. Though rather unstable, they have the remarkable ability of undergoing organometallic *coupling* reactions with alkyl bromides and iodides (but not fluorides). One of the alkyl groups from the Gilman reagent replaces the halogen of the alkyl halide, forming a new carbon–carbon bond and yielding a hydrocarbon product. Lithium dimethylcopper, for example, reacts with 1-iododecane to give undecane in 90% yield.

$$\underset{\substack{\textbf{Lithium} \\ \textbf{dimethylcopper}}}{(CH_3)_2CuLi} + \underset{\textbf{1-Iododecane}}{CH_3(CH_2)_8CH_2I} \xrightarrow[0°C]{\text{Ether}} \underset{\textbf{Undecane (90\%)}}{CH_3(CH_2)_8CH_2CH_3} + LiI + CH_3Cu$$

[3]Henry Gilman (1893–1986); b. Boston; Ph.D. (1918) Harvard (Kohler); professor, Iowa State University (1923–1986).

This organometallic coupling reaction is extremely versatile and very useful in organic synthesis. As the following examples indicate, the coupling reaction can be carried out on aryl and vinylic halides as well as on alkyl halides.

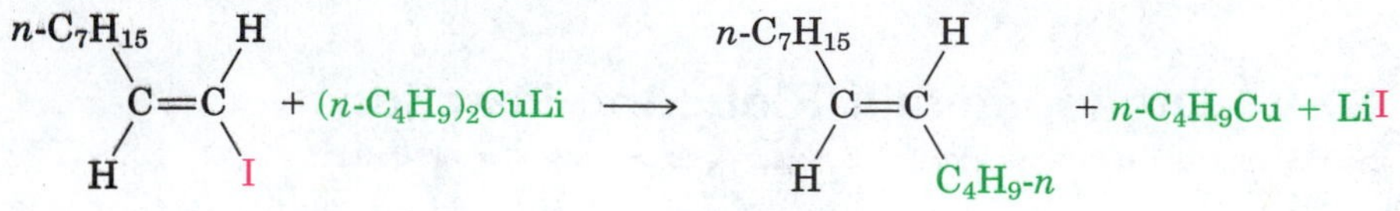

trans-1-Iodo-1-nonene **trans-5-Tridecene (71%)**

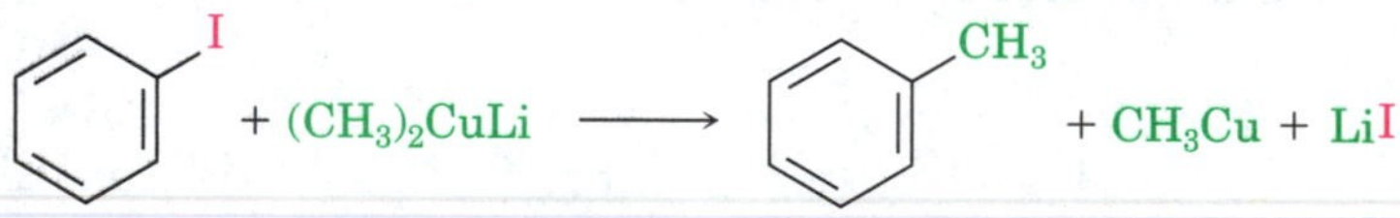

Iodobenzene **Toluene (91%)**

An organocopper coupling reaction is carried out commercially to synthesize *muscalure*, (9*Z*)-tricosene, the sex attractant of the common housefly. Tiny amounts of the insect hormone are used to lure flies to insecticide-treated traps, thereby providing a species-specific means of insect control. Coupling of *cis*-1-bromo-9-octadecene with lithium dipentylcopper produces muscalure in 100 lb batches.

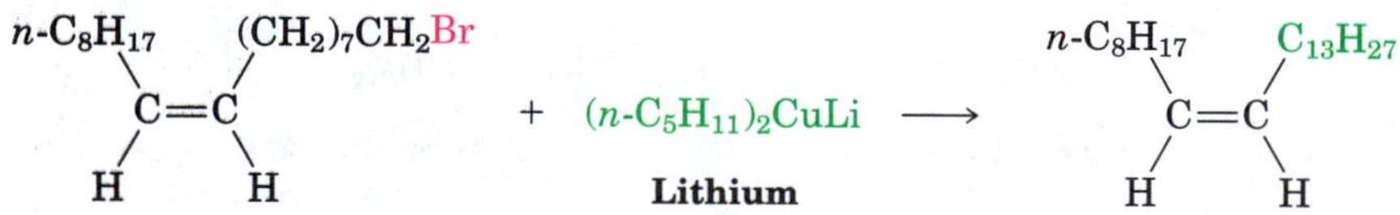

cis-1-Bromo-9-octadecene **Lithium dipentylcopper** **Muscalure [(9Z)-Tricosene] (99%)**

Although the details of the mechanism by which coupling occurs are not fully understood, radicals may well be involved. The coupling is not a typical polar nucleophilic substitution reaction of the sort considered in the next chapter.

PROBLEM

10.13 How would you prepare the following compounds using an organocopper coupling reaction? More than one step is required in each case.
(a) 3-Methylcyclohexene from cyclohexene (b) Octane from 1-bromobutane
(c) Decane from 1-pentene

10.10 Oxidation and Reduction in Organic Chemistry

Although we didn't point it out at the time, several of the reactions discussed earlier in this chapter can be thought of as *oxidation/reduction* processes. In inorganic chemistry, an oxidation is defined as the loss of one or more electrons by an atom and a reduction is defined as the gain of one or more electrons. In organic chemistry, however, the words are used somewhat differently.

In organic chemistry, an **oxidation** is a reaction that causes a decrease in electron ownership by carbon, either by bond *formation* between carbon and a more electronegative atom (usually oxygen, nitrogen, or a halogen) or by bond *breaking* between carbon and a less electronegative atom (usually hydrogen). Conversely, a **reduction** causes an increase of electron ownership by carbon, either by bond breaking between carbon and a more electronegative atom or by bond formation between carbon and a less electronegative atom. Note that an *oxidation* often adds *oxygen*, while a reduction usually adds hydrogen.

Oxidation	Decreasing electron density on carbon by forming one of these: C–O C–N C–Hal or breaking this: C–H
Reduction	Increasing electron density on carbon by forming this: C–H or breaking one of these: C–O C–N C–Hal

Based on these definitions, the chlorination reaction of methane to yield chloromethane is an oxidation because a C–H bond is replaced by a C–Cl bond. The conversion of an alkyl chloride to an alkane via a Grignard reagent followed by protonation is a reduction, however, because the overall result of the sequence is replacement of a C–Cl bond by a C–H bond.

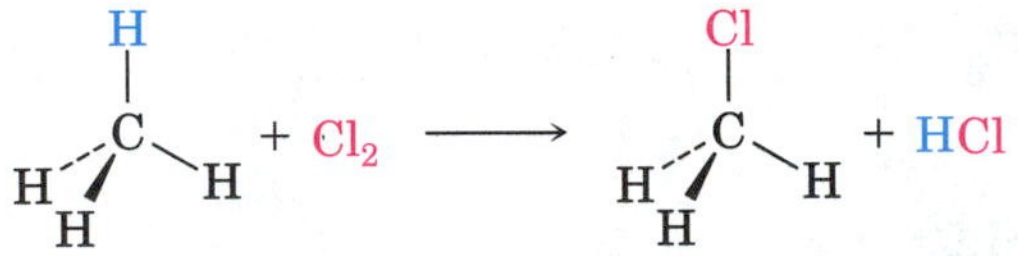

Methane **Chloromethane**

Oxidation: C–H bond broken and C–Cl bond formed

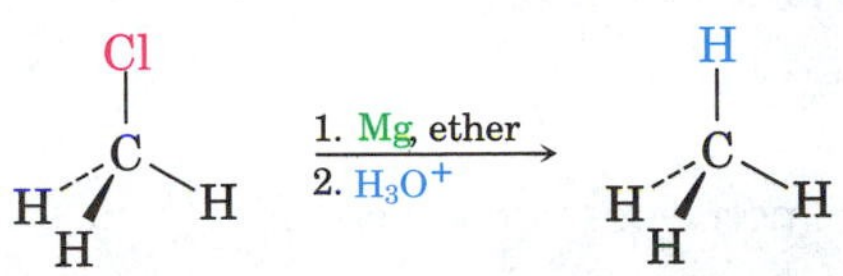

Chloromethane **Methane**

Reduction: C–Cl bond broken and C–H bond formed

As other examples, the reaction of ethylene with Br_2 to yield 1,2-dibromoethane is an oxidation because two C–Br bonds are formed, but the reaction of ethylene with H_2 to yield ethane is a reduction because two C–H bonds are formed. The reaction of ethylene with H_2O to yield ethyl alcohol is neither an oxidation nor a reduction, however, because a C–H and a C–O bond are both formed in the same reaction.

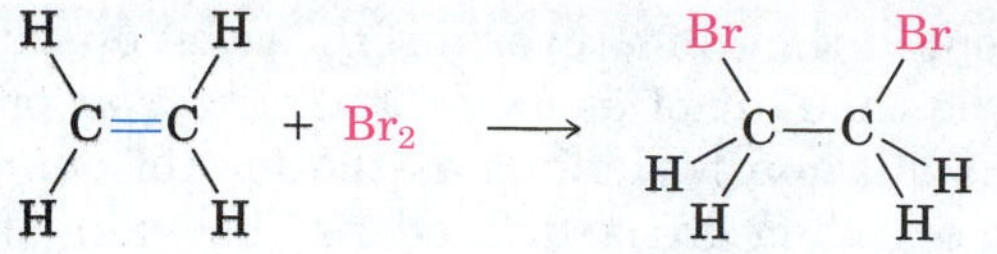

Ethylene **1,2-Dibromoethane**

Oxidation: Two new bonds formed between carbon and a more electronegative element

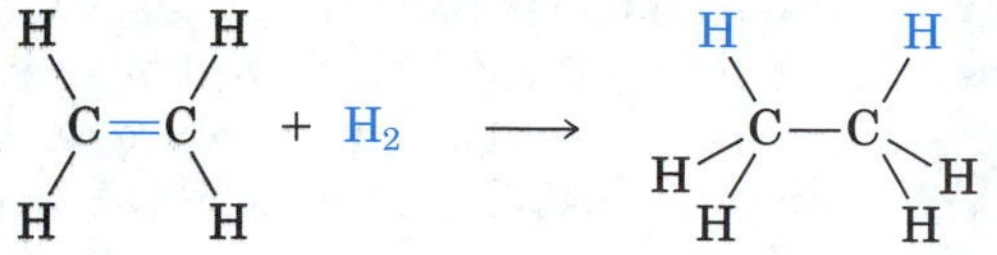

Ethylene **Ethane**

Reduction: Two new bonds formed between carbon and a less electronegative element

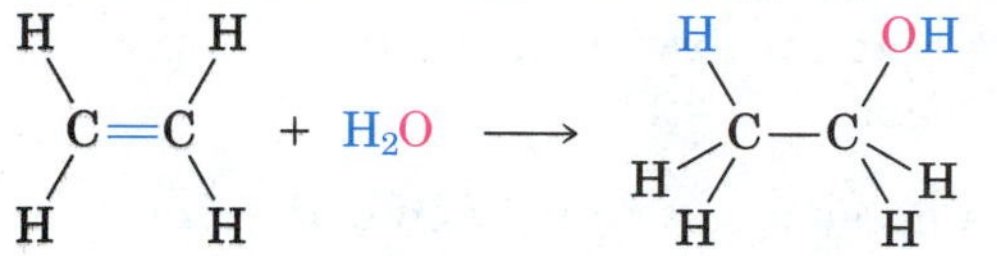

Ethylene **Ethyl alcohol**

Neither oxidation nor reduction: One new C–H bond and one new C–O bond formed

A list of compounds of increasing oxidation level is shown in Figure 10.4. Any reaction that converts a compound from a lower level to a higher level is an oxidation, any reaction that converts a compound from a higher level to a lower level is a reduction, and any reaction that doesn't change the level is neither an oxidation nor a reduction.

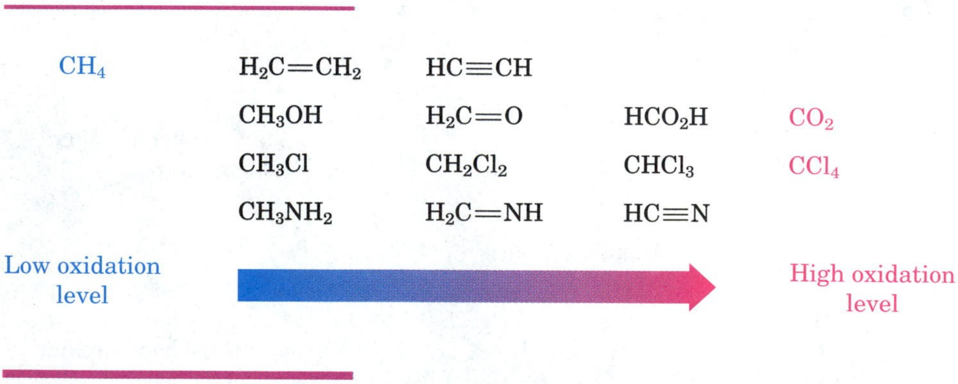

Figure 10.4 Oxidation levels of some common types of compounds.

PRACTICE PROBLEM

Rank the following compounds in order of increasing oxidation level:

(a) $CH_3CH{=}CH_2$ (b) $CH_3CH(OH)CH_3$ (c) $CH_3C(=O)CH_3$ (d) $CH_3CH_2CH_3$

Solution Count the number of C–O bonds in each compound and subtract from that value the number of C–H bonds. Compound (a) has six C–H bonds, giving an oxidation level of −6; (b) has one C–O bond and seven C–H bonds, giving an oxidation level of −6; (c) has two C–O bonds and six C–H bonds, giving an oxidation level of −4; and (d) has eight C–H bonds, giving an oxidation level of −8. Thus, the order of increasing oxidation level is (d) < (a) = (b) < (c).

PROBLEM

10.14 Rank each of the following series of compounds in order of increasing oxidation level:

(a)

(b) CH_3CN, $CH_3CH_2NH_2$, $H_2NCH_2CH_2NH_2$

INTERLUDE

Alkyl Halides and the Ozone Hole

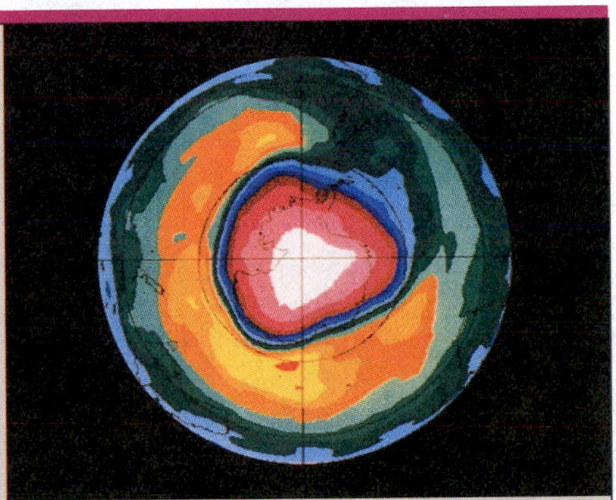

The Antarctic "ozone hole" in October 1993. Ozone values in the center are up to 50% lower than normal.

The aerosol can is a fixture of modern life—something we take for granted to spray our deodorants, paints, and insect repellents. In the early 1970s, though, it became apparent that the proliferation of aerosol sprays was leading to a serious environmental problem.

The volatile propellants used in aerosols at the time were various alkyl halides called *chlorofluorocarbons*, or *CFC's*, simple alkanes in which all the hydrogens have been replaced by either chlorine or fluorine. Fluorotrichloromethane (CCl_3F) and dichlorodifluoromethane (CCl_2F_2) are two of the most common CFC's. The advantage of using CFC's as aerosol propellants is that they are chemically inert and nonflammable. They don't react with the contents of the can, they leave no residue, they have no odor, and they're nontoxic. They do, however, escape into the atmosphere, where they ultimately find their way into the stratosphere.

The *ozone layer* is an atmospheric band extending from about 20 to

(continued)➤

40 km above the earth's surface. Ozone (O_3) is a severe pollutant at low altitudes but is critically important in the upper atmosphere, because it acts as a shield to protect the surface of the earth from intense solar ultraviolet radiation. If the ozone layer were depleted or destroyed, more ultraviolet radiation would reach the earth, causing an increased incidence of skin cancers and eye cataracts. Unfortunately, destruction of ozone is exactly what chlorofluorocarbons do. Beginning around 1976, a disturbing amount of ozone depletion, the so-called *ozone hole*, began showing up over the South Pole. More recently, a similar phenomenon has been found over the North Pole. Ozone levels drop to below 50% of normal in the polar spring before returning to near normal levels in the autumn.

The mechanism of ozone destruction by chlorofluorocarbons involves radical reactions of the same kind as occur in the radical chlorination of methane (Section 10.4). Ultraviolet light ($h\nu$) striking a CFC molecule breaks a C–Cl bond, producing a chlorine radical. This radical then reacts with ozone to yield O_2 and ClO:

$$CCl_3F \xrightarrow{h\nu} \cdot CCl_2F + \cdot Cl$$

$$\cdot Cl + O_3 \longrightarrow O_2 + ClO$$

Recognition of the problem led the U.S. government in 1980 to ban the use of CFC's for aerosol propellants, although they remained in use as refrigerants and solvents. Further action was taken in 1992 when an international agreement was reached calling for a total ban on all production and uses of CFC's by 1996. If nations comply with the agreement, the amount of CFC's in the stratosphere should peak around the year 2000 before slowly declining over the next century.

Summary and Key Words

Alkyl halides are compounds containing halogen bonded to a saturated, sp^3-hybridized carbon atom. The C–X bond is polar, and alkyl halides can therefore behave as electrophiles.

Alkyl halides can be prepared by **radical halogenation** of alkanes, but this method is of little general value since mixtures of products usually result. The reactivity order of alkanes toward halogenation is identical to the stability order of radicals: tertiary > secondary > primary. Alkyl halides can also be prepared from alkenes by reaction with *N*-bromosuccinimide (NBS) to give the product of **allylic** bromination. The NBS bromination of alkenes takes place through an intermediate allyl radical, which is stabilized by resonance.

Alcohols react with HX to form alkyl halides, but the reaction works well only for tertiary alcohols, R_3COH. Primary and secondary alkyl halides

...ly prepared from alcohols using either $SOCl_2$ or PBr_3. Alkyl ...ct with magnesium in ether solution to form organomagnesium ... **Grignard reagents**, RMgX. Since Grignard reagents are both ...ic and basic, they react with Brønsted–Lowry acids to yield hydro... ...he overall result of Grignard formation and protonation is the ... of an alkyl halide into an alkane (RX → RMgX → RH).

... halides also react with lithium metal to form **organolithium** ..., RLi. In the presence of CuI, these form diorganocoppers, or **Gilman reagents**, R_2CuLi. Gilman reagents react with alkyl halides to yield coupled hydrocarbon products.

In organic chemistry, an **oxidation** is a reaction that causes a decrease in electron ownership by carbon, either by bond *formation* between carbon and a more electronegative atom (usually oxygen, nitrogen, or a halogen) or by bond *breaking* between carbon and a less electronegative atom (usually hydrogen). Conversely, a **reduction** causes an increase of electron ownership by carbon, either by bond breaking between carbon and a more electronegative atom or by bond formation between carbon and a less electronegative atom. Thus, the halogenation of an alkane to yield an alkyl halide is an oxidation, while the conversion of an alkyl halide to an alkane by protonation of a Grignard reagent is a reduction.

Summary of Reactions

1. Preparation of alkyl halides
   (a) From alkenes by allylic bromination (Section 10.5)

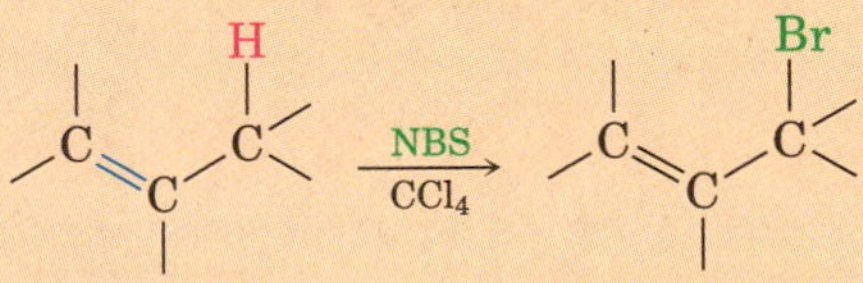

   (b) From alkenes by addition of HBr and HCl (Sections 6.8 and 6.9)

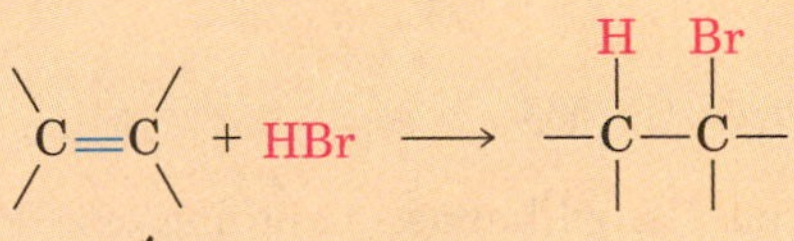

   (c) From alcohols
   (1) Reaction with HX, where X = Cl, Br, or I (Section 10.7)

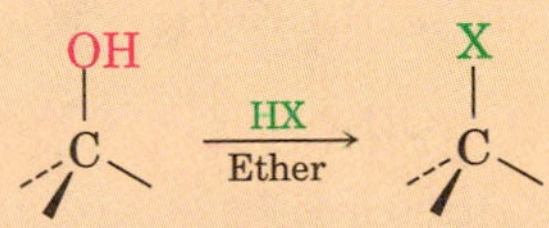

Reactivity order: 3° > 2° > 1°

(continued)➤

(2) Reaction of 1° and 2° alcohols with $SOCl_2$ (Section 10.7)

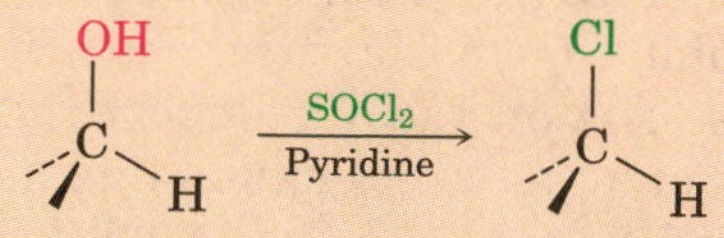

(3) Reaction of 1° and 2° alcohols with PBr_3 (Section 10.7)

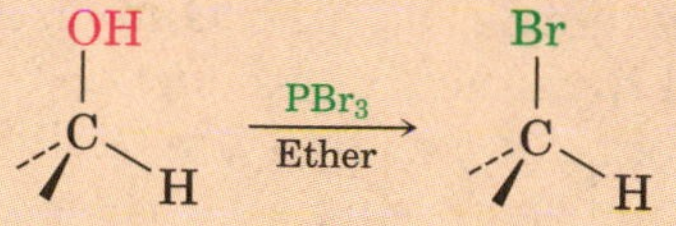

2. Reaction of alkyl halides

(a) Grignard reagent formation (Section 10.8)

$$R{-}X \xrightarrow[\text{Ether}]{\text{Mg}} R{-}Mg{-}X$$

where X = Br, Cl, or I

R = 1°, 2°, or 3° alkyl, aryl, or vinylic

(b) Diorganocopper (Gilman reagent) formation (Section 10.9)

$$R{-}X \xrightarrow[\text{Pentane}]{\text{2 Li}} R{-}Li + LiX$$

where R = 1°, 2°, or 3° alkyl, aryl, or vinylic

$$2\ R{-}Li + CuI \xrightarrow{\text{In ether}} [R{-}Cu{-}R]^- \ Li^+ + LiI$$

(c) Organometallic coupling (Section 10.9)

$$R_2CuLi + R'{-}X \xrightarrow{\text{In ether}} R{-}R' + RCu + LiX$$

(d) Conversion of alkyl halides to alkanes (Section 10.8)

$$R{-}X \xrightarrow[\text{Ether}]{\text{Mg}} R{-}Mg{-}X \xrightarrow{H_3O^+} R{-}H + HOMgX$$

ADDITIONAL PROBLEMS

10.15 Name these alkyl halides according to IUPAC rules:

(a)
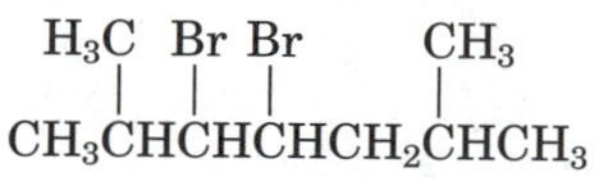

(b)
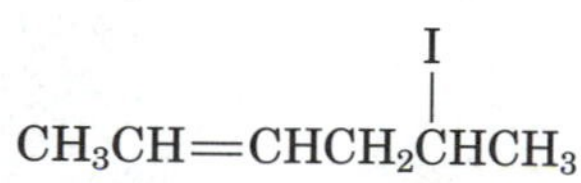

(c)
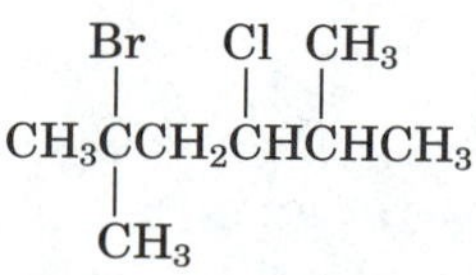

(d)
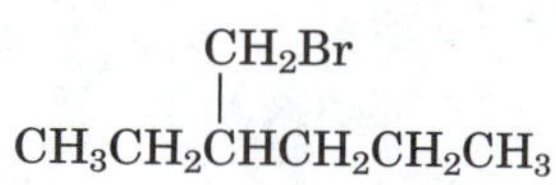

(e) $ClCH_2CH_2CH_2C{\equiv}CCH_2Br$

10.16 Draw structures corresponding to these IUPAC names:
(a) 2,3-Dichloro-4-methylhexane
(b) 4-Bromo-4-ethyl-2-methylhexane
(c) 3-Iodo-2,2,4,4-tetramethylpentane
(d) *cis*-1-Bromo-2-ethylcyclopentane

10.17 A chemist requires a large amount of 1-bromo-2-pentene as starting material for a synthesis. She finds a supply of 2-pentene in the stockroom and decides to carry out an NBS allylic bromination reaction:

$$CH_3CH_2CH{=}CHCH_3 \xrightarrow[CCl_4]{NBS} CH_3CH_2CH{=}CHCH_2Br$$

What is wrong with this synthesis plan? What side products would form in addition to the desired product?

10.18 What product(s) would you expect from the reaction of 1-methylcyclohexene with NBS? Would you use this reaction as part of a synthesis?

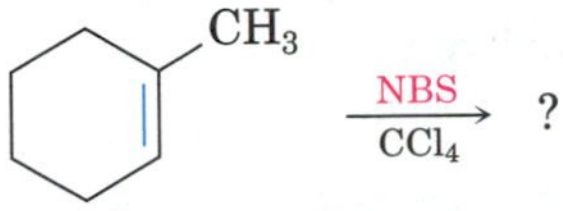

10.19 How would you prepare the following compounds, starting with cyclopentene and any other reagents needed?
(a) Chlorocyclopentane
(b) Methylcyclopentane
(c) 3-Bromocyclopentene
(d) Cyclopentanol
(e) Cyclopentylcyclopentane
(f) 1,3-Cyclopentadiene

10.20 Predict the product(s) of the following reactions:

(a) 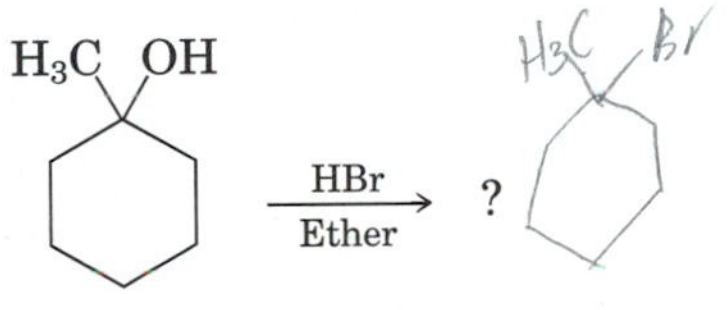

(b) $CH_3CH_2CH_2CH_2OH \xrightarrow{SOCl_2}$?

CH2CH2CH2CH2Cl

(c) 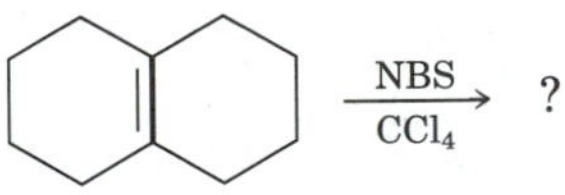

(d) 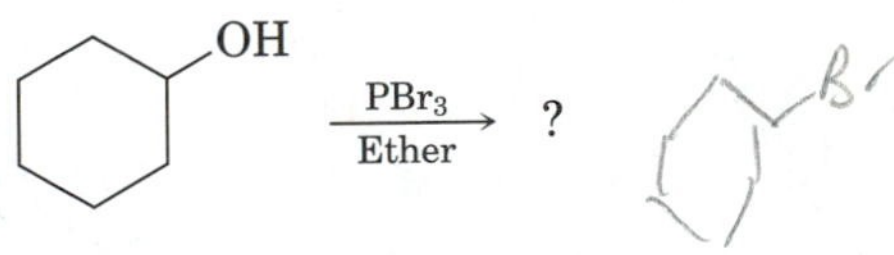

(e) $CH_3CH_2CHBrCH_3 \xrightarrow[Ether]{Mg} A \xrightarrow{H_2O} B$

(f) $CH_3CH_2CH_2CH_2Br \xrightarrow[Pentane]{Li} A \xrightarrow{CuI} B$

(g) $CH_3CH_2CH_2CH_2Br + (CH_3)_2CuLi \xrightarrow{Ether}$?

10.21 Draw and name the monochlorination products you might obtain by radical chlorination of 2-methylpentane. Which of the products are chiral? Are any of the products optically active?

10.22 (*S*)-3-Methylhexane undergoes radical bromination to yield optically inactive 3-bromo-3-methylhexane as the major product. Is the product chiral? What conclusions can you draw about the radical intermediate?

10.23 Assume that you have carried out a radical chlorination reaction on (*R*)-2-chloropentane and have isolated (in low yield) 2,4-dichloropentane. How many stereoisomers of the product are formed and in what ratio? Are any of the isomers optically active? (See Problem 10.22.)

10.24 Calculate $\Delta H°$ for the reactions of Cl· and Br· with CH_4, and then draw a reaction energy diagram showing both processes. Which reaction is likely to be faster?

10.25 What product(s) would you expect from the reaction of 1,4-hexadiene with NBS? What is the structure of the most stable radical intermediate?

10.26 We'll see in Chapter 16 that alkylbenzenes such as toluene (methylbenzene) react with NBS to give products in which bromine substitution has occurred at the position next to the aromatic ring (the *benzylic* position). Explain, based on the data in Table 5.4.

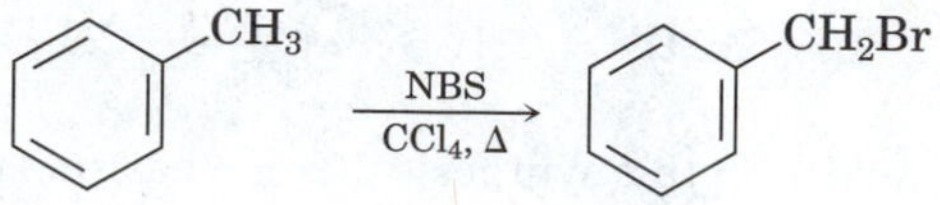

10.27 Draw as many resonance structures as you can for the benzyl radical, $C_6H_5CH_2\cdot$, the intermediate produced in the NBS bromination reaction of toluene (Problem 10.26).

10.28 What product would you expect from the reaction of 1-phenyl-2-butene with NBS? Explain.

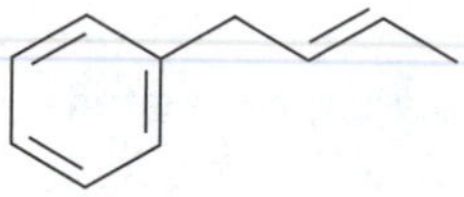

1-Phenyl-2-butene

10.29 Draw as many resonance structures as you can for the following species:

(a) $CH_3CH{=}CHCH{=}CHCH{=}CH\overset{+}{C}H_2$ (b)

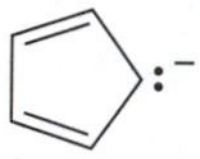

(c) $CH_3C{\equiv}\overset{+}{N}{-}\ddot{\underset{\cdot\cdot}{O}}{:}^-$

10.30 Rank the compounds in each of the following series in order of increasing oxidation level:

(a) $CH_3CH{=}CHCH_3$ $CH_3CH_2CH{=}CH_2$ $CH_3CH_2CH_2\overset{O}{\overset{\|}{C}}H$ $CH_3CH_2CH_2\overset{O}{\overset{\|}{C}}OH$

(b) $CH_3CH_2CH_2NH_2$ $CH_3CH_2CH_2Br$ $CH_3\overset{O}{\overset{\|}{C}}CH_2Cl$ $BrCH_2CH_2CH_2Cl$

10.31 Tell whether each of the following reactions is an oxidation or a reduction.

(a) $CH_3CH_2OH \xrightarrow{CrO_3} CH_3\overset{O}{\overset{\|}{C}}H$

(b) $H_2C{=}CH\overset{O}{\overset{\|}{C}}CH_3 + NH_3 \longrightarrow H_2NCH_2CH_2\overset{O}{\overset{\|}{C}}CH_3$

(c) $CH_3CH_2\overset{Br}{\overset{|}{C}}HCH_3 \xrightarrow[2.\ H_2O]{1.\ Mg} CH_3CH_2CH_2CH_3$

10.32 How would you carry out the following syntheses?
(a) Butylcyclohexane from cyclohexene (b) Butylcyclohexane from cyclohexanol
(c) Butylcyclohexane from cyclohexane

10.33 The syntheses shown here are unlikely to occur as written. What is wrong with each?

(a) $CH_3CH_2CH_2F \xrightarrow[\text{2. } H_3O^+]{\text{1. Mg}} CH_3CH_2CH_3$

(b) 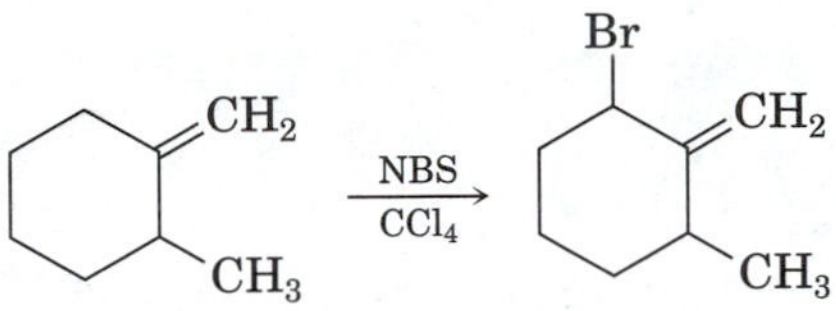

(c) 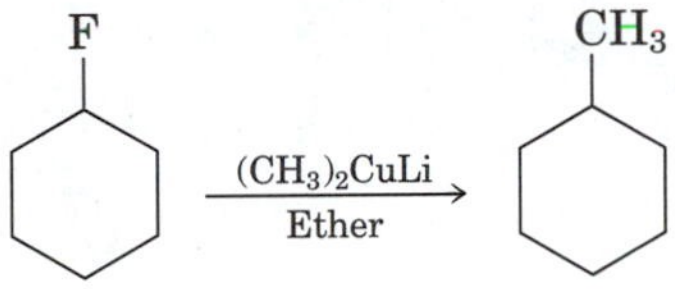

10.34 *Phenols*, compounds that have an –OH group bonded to a benzene ring, are relatively acidic because their anions are stabilized by resonance. Draw as many resonance structures as you can for the phenoxide ion.

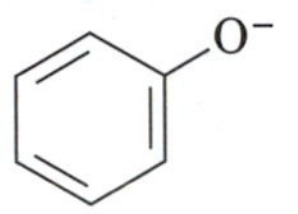

Phenoxide ion

10.35 Alkyl halides can be reduced to alkanes by a radical reaction with tributyltin hydride, $(C_4H_9)_3SnH$, in the presence of light ($h\nu$):

$$R{-}X + (C_4H_9)_3SnH \xrightarrow{h\nu} R{-}H + (C_4H_9)_3SnX$$

Propose a radical chain mechanism by which the reaction might occur. The initiation step is the light-induced homolytic cleavage of the Sn–H bond to yield a tributyltin radical.

A Look Ahead

10.36 We'll see in the next chapter that tertiary alkyl halides, R_3CX, undergo spontaneous dissociation to yield a carbocation, R_3C^+. Which do you think reacts faster, $(CH_3)_3CBr$ or $H_2C{=}CHC(CH_3)_2Br$? Explain.

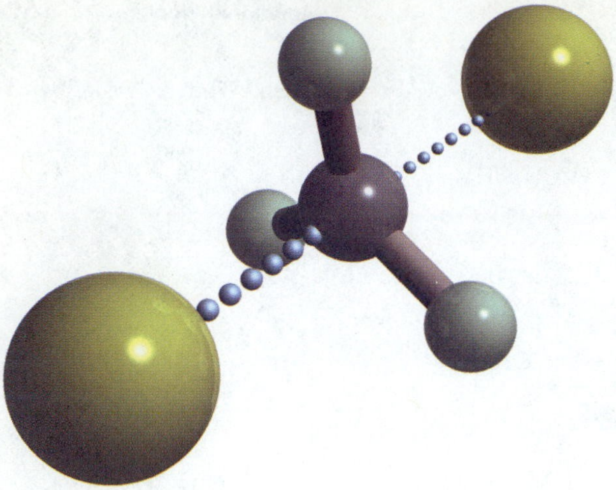

The S_N2 reaction of iodide ion on bromomethane occurs in a single step through a planar transition state.

11

Reactions of Alkyl Halides: Nucleophilic Substitutions and Eliminations

We saw in the preceding chapter that the carbon–halogen bond in alkyl halides is polar and that the carbon atom is electron-poor. Thus, alkyl halides are electrophiles, and much of their chemistry involves polar reactions with nucleophiles and bases.

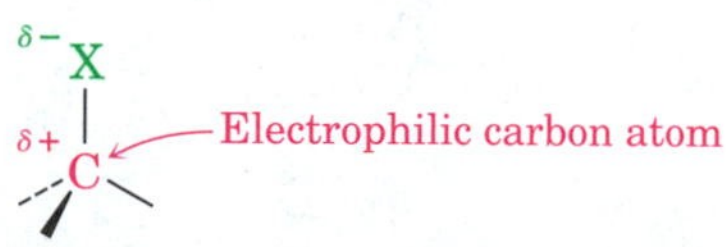

Alkyl halides do one of two things when they react with a nucleophile: Either they undergo *substitution* of the X group by the nucleophile, or they undergo *elimination* of HX to yield an alkene:

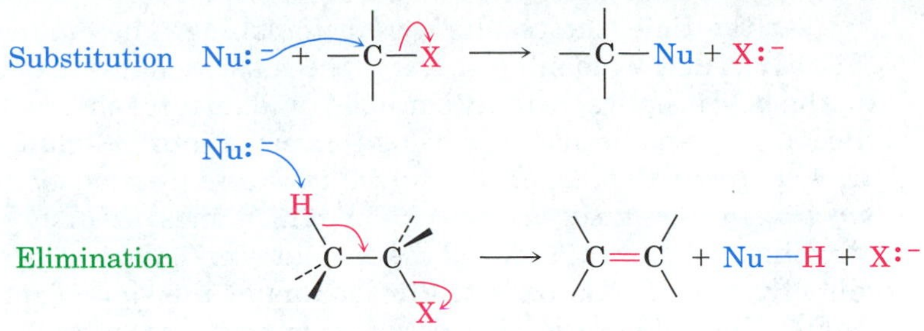

Nucleophilic substitution and base-induced elimination are two of the most widely occurring and versatile reactions in organic chemistry. We'll take a close look at both reactions in this chapter to see how they occur, what their characteristics are, and how they can be used to synthesize new molecules.

11.1 The Discovery of the Walden Inversion

In 1896 the German chemist Paul Walden[1] made a remarkable discovery. He found that the pure enantiomeric (+)- and (−)-malic acids could be interconverted by a series of simple substitution reactions. When Walden treated (−)-malic acid with PCl_5, he isolated (+)-chlorosuccinic acid. This, on treatment with wet Ag_2O, gave (+)-malic acid. Similarly, reaction of (+)-malic acid with PCl_5 gave (−)-chlorosuccinic acid, which was converted into (−)-malic acid when treated with wet Ag_2O. The full cycle of reactions reported by Walden is shown in Figure 11.1.

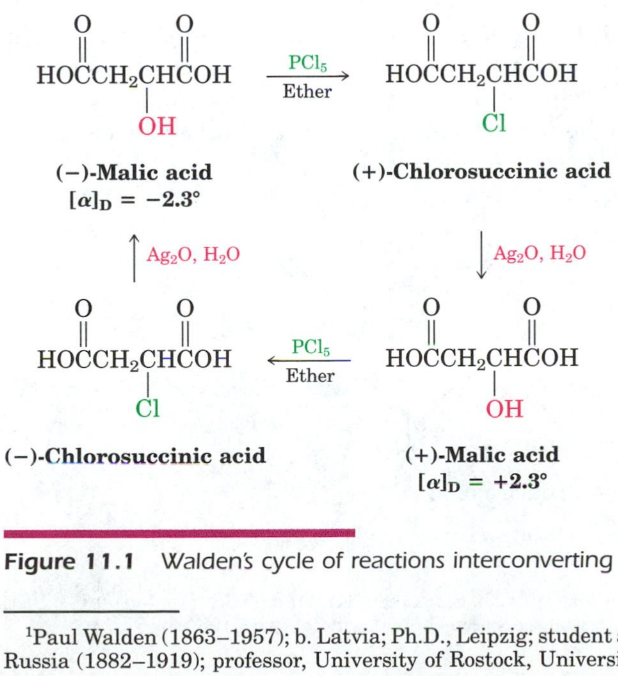

Figure 11.1 Walden's cycle of reactions interconverting (+)- and (−)-malic acids.

[1]Paul Walden (1863–1957); b. Latvia; Ph.D., Leipzig; student and professor, Riga Polytechnic, Russia (1882–1919); professor, University of Rostock, University of Tübingen, Germany.

At the time, the results were astonishing. The eminent chemist Emil Fischer called Walden's discovery "the most remarkable observation made in the field of optical activity since the fundamental observations of Pasteur." Because (−)-malic acid was being converted into (+)-malic acid, *some reactions in the cycle must have occurred with an inversion, or change, in configuration at the stereogenic center.* But which ones, and how? (Recall that the direction of light rotation and the absolute configuration of a molecule aren't directly related. You can't tell by looking at the sign of rotation whether a change in configuration has occurred during a reaction.)

Today, we refer to the transformations taking place in Walden's cycle as **nucleophilic substitution reactions**, because each step involves the substitution of one nucleophile (chloride ion, Cl^-, or hydroxide ion, HO^-) by another. Nucleophilic substitution reactions are one of the most common and versatile reaction types in organic chemistry.

11.2 Stereochemistry of Nucleophilic Substitution

Although Walden realized that changes in configuration had taken place during his reaction cycle, he didn't know at which steps the changes occurred. In the 1920s, Joseph Kenyon[2] and Henry Phillips began a series of investigations to elucidate the mechanism of nucleophilic substitution reactions and to find out how inversions of configuration occur. They recognized that the presence of the carboxylic acid group in Walden's work on malic acid may have led to complications, and they therefore carried out their own work on simpler cases. (In fact, the particular sequence of reactions studied by Walden *is* unusually complex for reasons we won't go into.)

Among the reaction series studied by Kenyon and Phillips was one that interconverted the two enantiomers of 1-phenyl-2-propanol (Figure 11.2). Although this particular series of reactions involves nucleophilic substitution of an alkyl toluenesulfonate (called a *tosylate*) rather than an alkyl halide, exactly the same type of reaction is involved as that studied by Walden. For all practical purposes, the *entire* tosylate group acts as if it were simply a halogen substituent:

$$\text{R—Y} + \text{Nu:}^- \longrightarrow \text{R—Nu} + \text{Y:}^-$$

where Y = Cl, Br, I, OTos

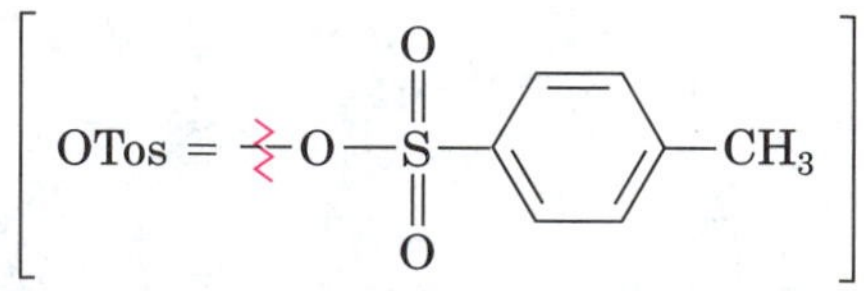

Nu = A nucleophile

[2]Joseph Kenyon (1885–1961); b. Blackburn, England; D.Sc. (London), 1914; British Dyestuffs Corp. (1916–1920); Battersea Polytechnic, London (1920–1950).

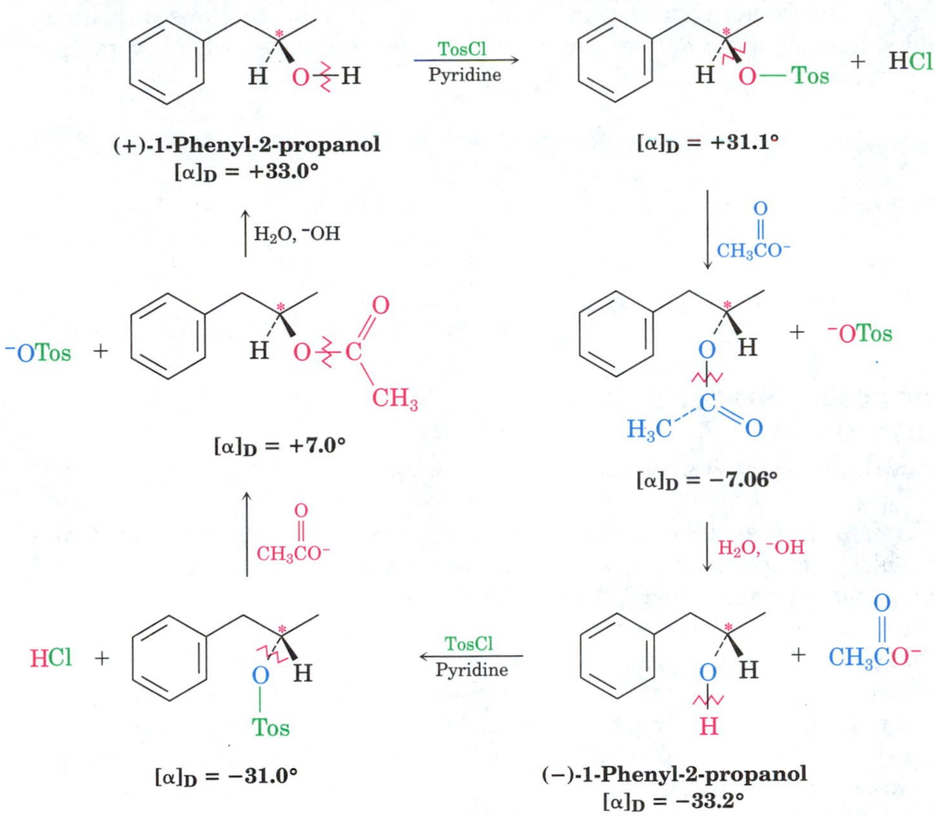

Figure 11.2 A Walden cycle interconverting (+) and (−) enantiomers of 1-phenyl-2-propanol. Stereogenic centers are marked by asterisks, and the bonds broken in each reaction are indicated by wavy lines.

In the three-step reaction sequence shown in Figure 11.2, (+)-1-phenyl-2-propanol is interconverted with its (−) enantiomer, and at least one of the three steps must therefore involve an inversion of configuration at the stereogenic center. The first step, formation of a toluenesulfonate, is known to occur by breaking the O–H bond of the alcohol rather than the C–O bond to the stereogenic carbon, and the configuration around carbon is therefore unchanged. Similarly, the third step, hydroxide ion cleavage of the acetate, also takes place without breaking the C–O bond at the stereogenic center. *The inversion of stereochemical configuration must therefore take place in the second step, the nucleophilic substitution of tosylate ion by acetate ion.*

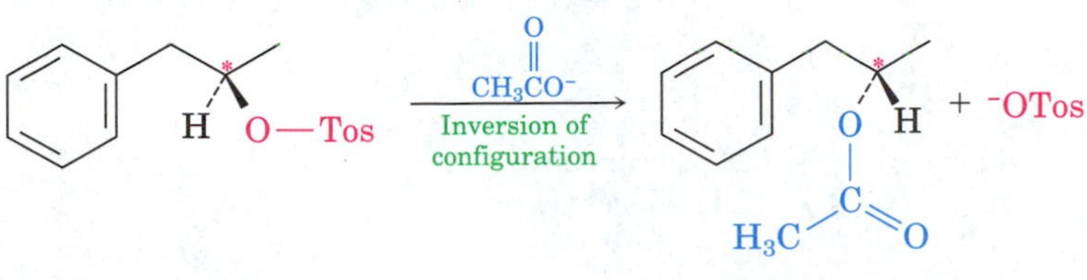

From this and nearly a dozen other series of reactions, Kenyon and Phillips concluded that the nucleophilic substitution reactions of primary and secondary alkyl halides and tosylates always proceed with inversion of configuration.

PROBLEM

11.1 What product would you expect to obtain from reaction of (*S*)-2-bromohexane with acetate ion, CH_3COO^-? Show the stereochemistry of both starting material and product.

11.3 Kinetics of Nucleophilic Substitution

Chemists often speak of a reaction as being "fast" or "slow." The exact rate at which a reactant is converted into product is called the **reaction rate** and can often be measured. The determination of reaction rates and of how those rates depend on reagent concentrations is a powerful tool for probing reaction mechanisms. Let's see what can be learned about the nucleophilic substitution reaction from a study of reaction rates.

In every chemical reaction, there is a direct relationship between reaction rate and reactant concentrations. When we measure this relationship, we measure the **kinetics** of the reaction. For example, let's look at the kinetics of a simple nucleophilic substitution—the reaction of CH_3Br with OH^- to yield CH_3OH.

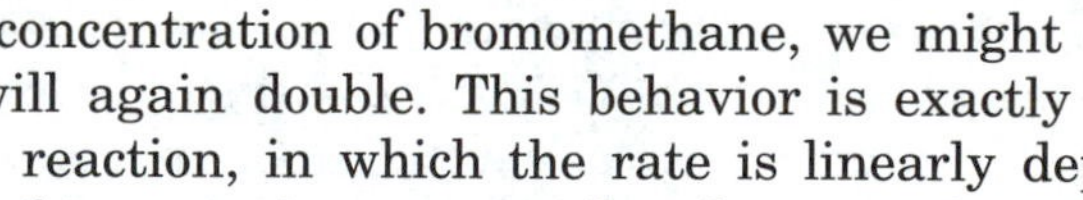

At a given temperature and concentration of reactants, the reaction occurs at a certain rate. If we double the concentration of hydroxide ion, the frequency of encounter between the reaction partners is also doubled, and we might therefore predict that the reaction rate will double. Similarly, if we double the concentration of bromomethane, we might expect that the reaction rate will again double. This behavior is exactly what is found. We call such a reaction, in which the rate is linearly dependent on the concentrations of two species, a **second-order reaction**. Mathematically, we can express this second-order dependence of the nucleophilic substitution reaction by setting up a **rate equation**:

$$\text{Reaction rate} = \text{Rate of disappearance of starting material}$$
$$= k \times [\text{RX}] \times [^-\text{OH}]$$

where $[\text{RX}] = CH_3Br$ concentration
$[^-\text{OH}] = {}^-OH$ concentration
k = A constant value

This equation says that the reaction rate can be expressed as the rate of disappearance of starting material and is equal to a coefficient, k, times the alkyl halide concentration times the hydroxide ion concentration. The constant k is called the **rate constant** for the reaction and has units of liters per mole second (L/mol·s). The rate equation says that as either [RX] or [$^-$OH] changes, the rate of the reaction changes proportionately. If the alkyl halide concentration is doubled, the reaction rate doubles; if the alkyl halide concentration is halved, the reaction rate is halved.

11.4 The S_N2 Reaction

At this point, we have two important pieces of information about the nature of nucleophilic substitution reactions on primary and secondary alkyl halides and tosylates.

1. The reactions occur with inversion of stereochemistry at the stereogenic carbon center.
2. The reactions show second-order kinetics and follow the rate law:

$$\text{Rate} = k \times [\text{RX}] \times [\text{Nu:}^-]$$

A mechanism that accounts for both the stereochemistry and the kinetics of nucleophilic substitution reactions was suggested in 1937 by E. D. Hughes[3] and C. Ingold, who formulated what they called the **S_N2 reaction**—short for *substitution, nucleophilic, bimolecular.* (**Bimolecular** means that two molecules, nucleophile and alkyl halide, take part in the step whose kinetics are measured.)

The essential feature of the S_N2 mechanism is that the reaction takes place in a single step without intermediates when the incoming nucleophile attacks the substrate from a direction directly opposite the departing group. As the nucleophile comes in on one side of the molecule and bonds to the carbon, the halide or tosylate departs from the other side, thereby inverting the molecule's stereochemical configuration. The process is shown in Figure 11.3 (page 376) for the reaction of (*S*)-2-bromobutane with hydroxide ion, leading to (*R*)-2-butanol.

We can picture an S_N2 reaction as occurring when an electron pair on the nucleophile, Nu:$^-$, forces out the leaving group, Y:$^-$, with the electron pair from the former C–Y bond. This occurs through a transition state in which the new Nu–C bond is partially forming at the same time that the old C–Y bond is partially breaking, and in which the negative charge is shared by both the incoming nucleophile and the outgoing leaving group. The transition state for this inversion must have the remaining three bonds to carbon in a planar arrangement, as shown in Figure 11.4 (page 377).

[3]Edward David Hughes (1906–1963); b. Criccieth, North Wales; Ph.D., Wales (Watson); D.Sc., London (Ingold); professor, University College, London (1930–1963).

The nucleophile ^{-}OH uses its lone-pair electrons to attack the alkyl halide carbon 180° away from the departing halogen. This leads to a transition state with a partially formed C–OH bond and a partially broken C–Br bond.

The stereochemistry at carbon is inverted as the C–OH bond forms fully and the bromide ion departs with the electron pair from the former C–Br bond.

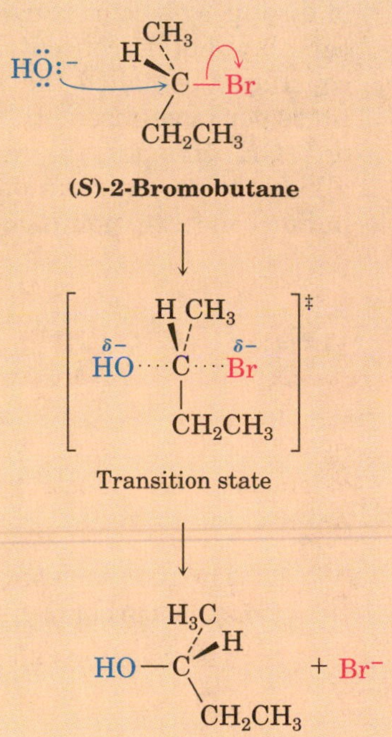

Figure 11.3 The mechanism of the S_N2 reaction. The reaction takes place in a single step when the incoming nucleophile approaches from a direction 180° away from the departing halide ion, thereby inverting the stereochemistry at carbon.

The mechanism proposed by Hughes and Ingold is fully consistent with experimental results, explaining both stereochemical and kinetic data. Thus, the requirement for back-side attack of the entering nucleophile from a direction 180° away from the departing Y group causes the stereochemistry of the substrate to invert, much like an umbrella turning inside out in the wind. The Hughes–Ingold mechanism also explains why second-order kinetics are found: The S_N2 reaction occurs in a single step that involves *both* alkyl halide and nucleophile. Two molecules are involved in the step whose rate is measured.

PROBLEM.

11.2 What product would you expect to obtain from reaction of OH^- with (*R*)-2-bromobutane? Show the stereochemistry of both starting material and product.

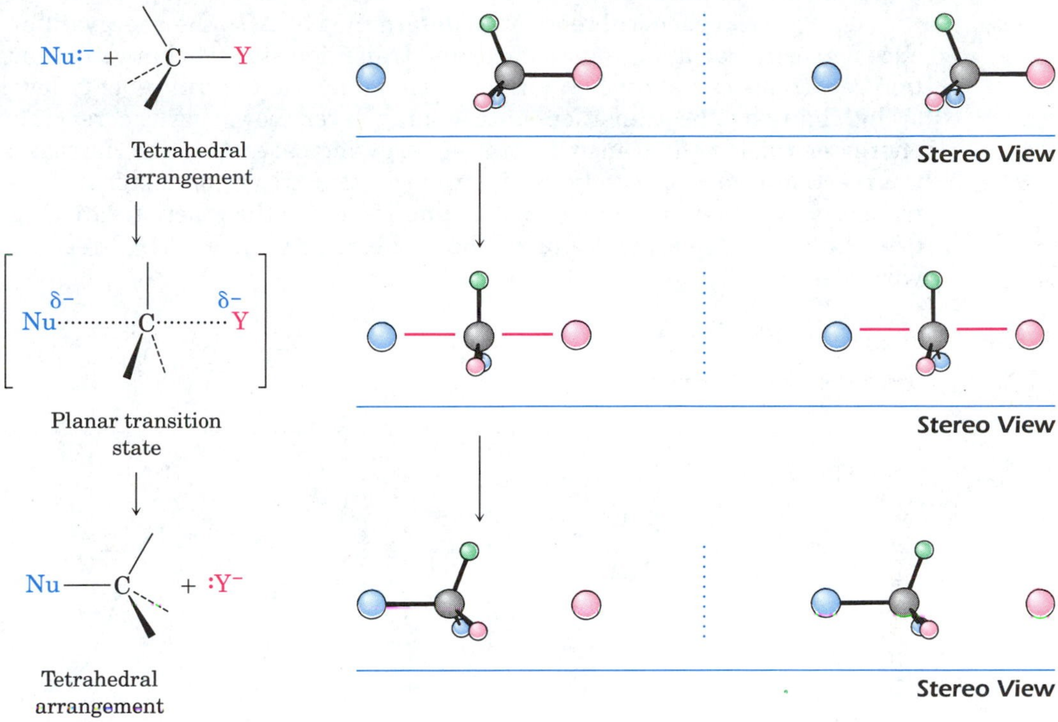

Figure 11.4 The transition state of an S_N2 reaction has a planar arrangement of the carbon atom and the remaining three groups.

PROBLEM.

11.3 A further piece of evidence in support of the requirement for back-side S_N2 displacement is the finding that the following alkyl bromide does not undergo a substitution reaction with hydroxide ion. Make a molecular model, and then suggest a reason for the lack of reactivity.

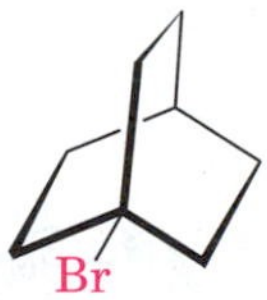

11.5 Characteristics of the S_N2 Reaction

We now have a good picture of how S_N2 reactions occur, but we also need to see how these substitutions can be used and what variables affect them. Some S_N2 reactions are fast and some are slow; some take place in high yield and others in low yield. Understanding the factors involved can be of tremendous value to chemists. Let's begin by reviewing what we know about reaction rates in general.

The rate of a chemical reaction is determined by $\Delta G^{\ddagger}$, the energy difference between reactant (ground state) and transition state. A change in reaction conditions can affect $\Delta G^{\ddagger}$ either by changing the reactant energy level or by changing the transition-state energy level. Lowering the reactant energy or raising the transition-state energy increases $\Delta G^{\ddagger}$ and decreases the reaction rate. Conversely, raising the reactant energy or decreasing the transition-state energy decreases $\Delta G^{\ddagger}$ and increases the reaction rate (Figure 11.5). We'll see examples of all these effects as we look at S_N2 reaction variables.

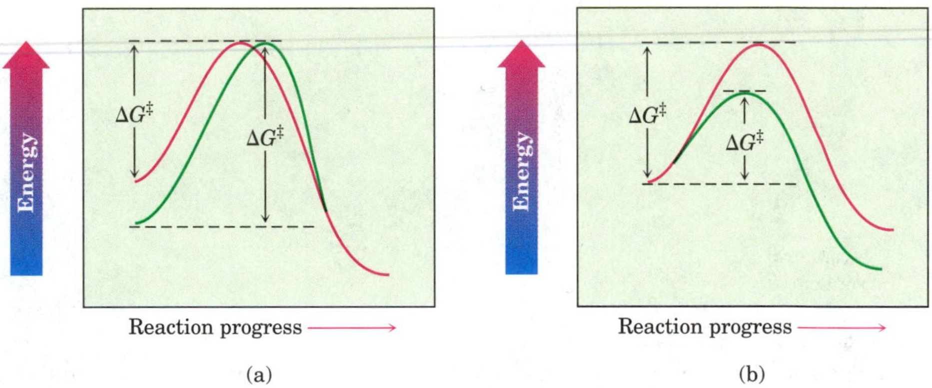

Figure 11.5 The effect of changes in reactant and transition-state energy levels on reaction rate. (a) A higher reactant energy level (red curve) corresponds to a faster reaction (smaller $\Delta G^{\ddagger}$). (b) A higher transition-state energy level (red curve) corresponds to a slower reaction (larger $\Delta G^{\ddagger}$).

The Substrate: Steric Effects in the S_N2 Reaction

The first S_N2 reaction variable we'll look at is the substitution pattern of the alkyl halide substrate. Since the S_N2 transition state involves partial bond formation between the incoming nucleophile and the alkyl halide, it seems reasonable that a hindered, bulky substrate should prevent easy approach of the nucleophile, making bond formation difficult. In other words, the transition state for reaction of a sterically hindered alkyl halide, whose carbon atom is "shielded" from attack of the incoming nucleophile, is higher in energy and forms more slowly than the corresponding transition state for a less hindered alkyl halide (Figure 11.6).

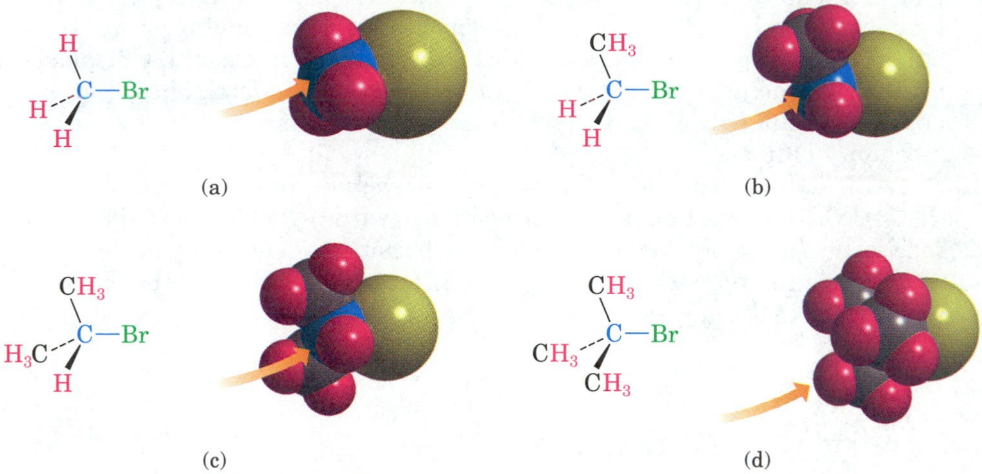

Figure 11.6 Steric hindrance to the S_N2 reaction. As the computer-generated models indicate, the carbon atom in (a) bromomethane is readily accessible, resulting in a fast S_N2 reaction. The carbon atoms in (b) bromoethane (primary), (c) 2-bromopropane (secondary), and (d) 2-bromo-2-methyl-propane (tertiary) are successively more hindered, resulting in successively slower S_N2 reactions.

As Figure 11.6 shows, the difficulty of nucleophilic attack increases as the three substituents bonded to the halo-substituted carbon atom increase in size. The relative reactivities for some different substrates are indicated below.

$$R\text{—}Br + Cl^- \longrightarrow R\text{—}Cl + Br^-$$

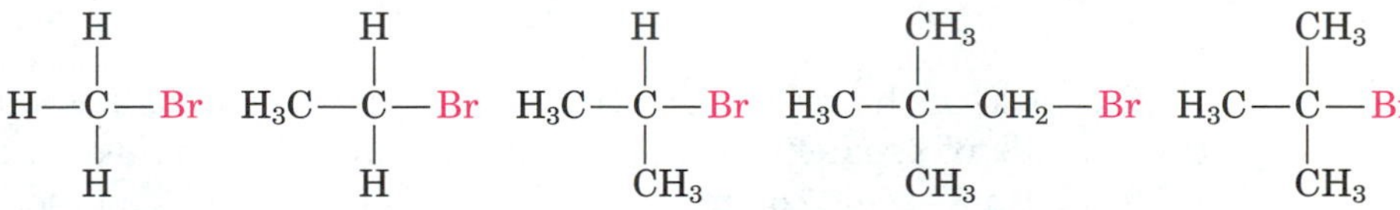

	(Methyl)	(Primary)	(Secondary)	(Neopentyl)	(Tertiary)
Relative reactivity	2,000,000	40,000	500	1	<1

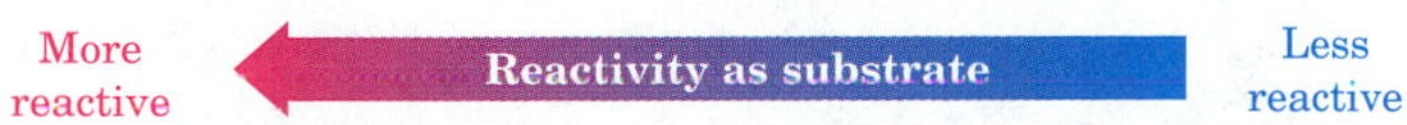

Methyl halides are by far the most reactive in S_N2 reactions, followed by primary alkyl halides such as ethyl and propyl. Alkyl branching next to the leaving group, as in isopropyl halides (2°), slows the reaction greatly, and further branching, as in *tert*-butyl halides (3°), effectively halts the reaction. Even branching one carbon removed from the leaving group, as in 2,2-dimethylpropyl (*neopentyl*) halides, greatly slows nucleophilic displacement. *S_N2 reactions can occur only at relatively unhindered sites*, and are normally useful only with methyl halides, primary halides, and a few simple secondary halides.

Although not shown in the preceding reactivity order, vinylic halides ($R_2C{=}CRX$) and aryl halides are unreactive toward S_N2 reaction. This lack of reactivity is probably due to steric factors, because the incoming nucleophile would have to approach in the plane of the carbon–carbon double bond to carry out a back-side displacement.

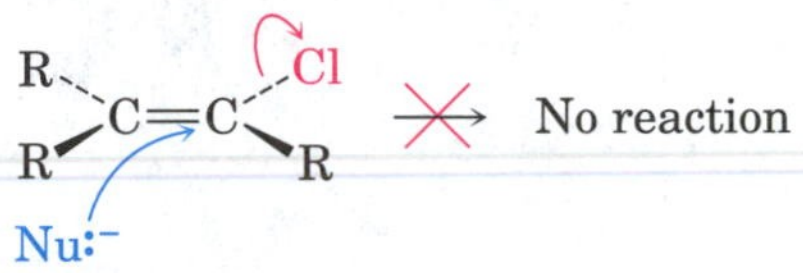

Vinylic halide

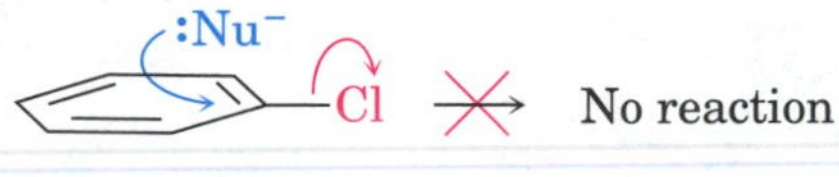

Aryl halide

The Attacking Nucleophile

The nature of the attacking nucleophile is another variable that has a major effect on the S_N2 reaction. Any species, either neutral or negatively charged, can act as a nucleophile as long as it has an unshared pair of electrons (that is, as long as it's a Lewis base). If the nucleophile is negatively charged, the product is neutral, but if the nucleophile is neutral, the product is positively charged.

Negatively charged $Nu:^-$ $\quad Nu:^- + R{-}Y \longrightarrow R{-}Nu + Y:^-$

Neutral Nu: $\quad Nu: + R{-}Y \longrightarrow R{-}Nu^+ + Y:^-$

Because of the great versatility of nucleophilic substitution reactions, many kinds of products can be prepared from alkyl halides and tosylates. Table 11.1 lists some common nucleophiles and shows the products of their reactions with bromomethane.

Although all the S_N2 reactions shown in Table 11.1 take place, some are much faster than others. What are the reasons for the reactivity differences? Why do some reagents appear to be much more "nucleophilic" than others?

The answers to these questions aren't straightforward. Part of the problem is that the term *nucleophilic* is imprecise. Although many chemists use the term *nucleophilicity* to mean a measure of the affinity of a species for a carbon atom in the S_N2 reaction, the reactivity of a given nucleophile can change from one reaction to the next.

Table 11.1 Some S_N2 Reactions with Bromomethane:
$Nu:^- + CH_3Br \longrightarrow NuCH_3 + Br^-$

Attacking nucleophile		*Product*	
Formula	*Name*	*Formula*	*Name*
$H:^-$	Hydride	CH_4	Methane
$CH_3\ddot{S}:^-$	Methanethiolate	CH_3SCH_3	Dimethyl sulfide
$H\ddot{S}:^-$	Hydrosulfide	$HSCH_3$	Methanethiol
$N{\equiv}C:^-$	Cyanide	$N{\equiv}CCH_3$	Acetonitrile
$:\ddot{I}:^-$	Iodide	ICH_3	Iodomethane
$H\ddot{O}:^-$	Hydroxide	$HOCH_3$	Methanol
$CH_3\ddot{O}:^-$	Methoxide	CH_3OCH_3	Dimethyl ether
$N{=}N{=}\ddot{N}:^-$	Azide	N_3CH_3	Azidomethane
$:\ddot{Cl}:^-$	Chloride	$ClCH_3$	Chloromethane
$CH_3CO_2:^-$	Acetate	$CH_3CO_2CH_3$	Methyl acetate
$H_3N:$	Ammonia	$H_3\overset{+}{N}CH_3\ Br^-$	Methylammonium bromide
$(CH_3)_3N:$	Trimethylamine	$(CH_3)_3\overset{+}{N}CH_3\ Br^-$	Tetramethylammonium bromide

The exact nucleophilicity of a species in a given reaction depends on the substrate, the solvent, and even the reactant concentrations. To speak with any precision, therefore, we must study the relative reactivity of various nucleophiles on a single substrate in a single solvent system. Much work has been carried out on the S_N2 reactions of bromomethane in aqueous ethanol, with the following results:

$$CH_3Br + Nu:^- \longrightarrow CH_3Nu + Br^-$$

Nu =	HS^-	CN^-	I^-	CH_3O^-	HO^-	Cl^-	NH_3	H_2O
Relative reactivity	125,000	125,000	100,000	25,000	16,000	1000	700	1

More reactive ⟵ Reactivity as nucleophile — Less reactive

Precise explanations for the observed nucleophilicities aren't known, but some trends can be detected in the data:

1. When comparing nucleophiles that have the same attacking atom, nucleophilicity roughly parallels basicity (Table 11.2). For example, OH^- is both more basic and more nucleophilic than H_2O. Since "nucleophilicity" measures the affinity of a Lewis base for a carbon

Table 11.2 Correlation of Basicity and Nucleophilicity

Nucleophile	CH_3O^-	HO^-	$CH_3CO_2^-$	H_2O
Rate of S_N2 reaction with CH_3Br	25	16	0.3	0.001
pK_a of conjugate acid	15.5	15.7	4.7	−1.7

atom in the S_N2 reaction, and "basicity" measures the affinity of a base for a proton, it's easy to see why there might be a rough correlation between the two kinds of behavior.

2. Nucleophilicity usually increases going down a column of the periodic table. Thus, HS^- is more nucleophilic than HO^-, and the halide reactivity order is $I^- > Br^- > Cl^-$. The matter is complex, though, and the nucleophilicity order can change depending on the solvent.

PROBLEM.

11.4 What products would you expect from reaction of 1-bromobutane with these reagents?
(a) NaI (b) KOH (c) H—C≡C—Li (d) NH_3

PROBLEM.

11.5 Which reagent in each of the following pairs is more nucleophilic? Explain.
(a) $(CH_3)_2N^-$ or $(CH_3)_2NH$ (b) $(CH_3)_3B$ or $(CH_3)_3N$ (c) H_2O or H_2S

The Leaving Group

Still another variable that can affect the S_N2 reaction is the nature of the group displaced by the attacking nucleophile—the **leaving group**. Because the leaving group is expelled with a negative charge in most S_N2 reactions, we might expect the best leaving groups to be those that best stabilize the negative charge. Furthermore, because the stability of an anion is related to basicity, the best leaving groups should be the weakest bases.

As indicated below, the weakest bases (anions derived from the strongest acids) are indeed the best leaving groups. The *p*-toluenesulfonate (tosylate) leaving group is very easily displaced, as are iodide and bromide ion, but chloride and fluoride ion are much less effective as leaving groups.

	$TosO^-$	I^-	Br^-	Cl^-	F^-	HO^-, H_2N^-, RO^-
Relative reactivity	60,000	30,000	10,000	200	1	~0

More reactive ← Reactivity as leaving group — Less reactive

The reason that stable anions make good leaving groups can be understood by looking at the transition state. In the transition state for an S_N2 reaction, the charge is distributed over both the attacking nucleophile and the leaving group. The greater the extent of charge stabilization by the leaving group, the lower the energy of the transition state and the more rapid the reaction.

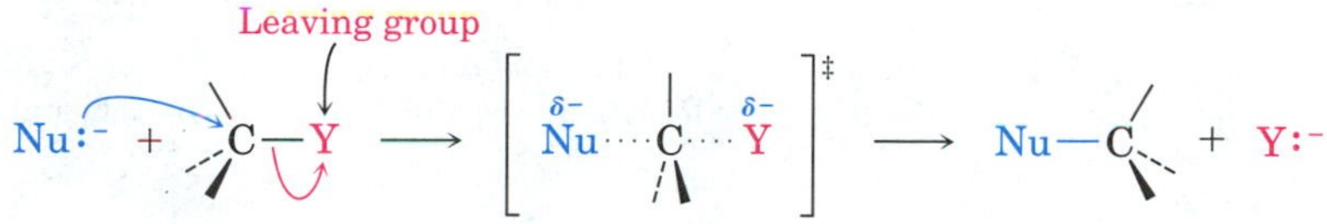

Transition state
(Negative charge is delocalized over both Nu: and Y)

It's just as important to know which are *poor* leaving groups as to know which are good, and the preceding data clearly indicate that F^-, HO^-, RO^-, and H_2N^- are not displaced by nucleophiles. In other words, alkyl fluorides, alcohols, ethers, and amines do not normally undergo S_N2 reactions.

R—F R—OH R—OR′ R—NH_2

These compounds do not undergo S_N2 reactions.

PROBLEM

11.6 Rank the following compounds in order of their expected reactivity toward S_N2 reaction:

CH_3Br CH_3OTos $(CH_3)_3CCl$ $(CH_3)_2CHCl$

The Solvent

The rates of many S_N2 reactions are affected by the solvent. Protic solvents, which contain –OH or –NH groups, are generally the worst solvents for S_N2 reactions; **polar aprotic solvents**, which have strong dipoles but don't have –OH or –NH groups, are the best.

Protic solvents, such as methanol and ethanol, slow down S_N2 reactions by clustering around the reactant nucleophile and lowering its energy, a process called **solvation**. Solvent molecules form hydrogen bonds to the nucleophile, orienting themselves into a "cage" around it and thereby stabilizing it.

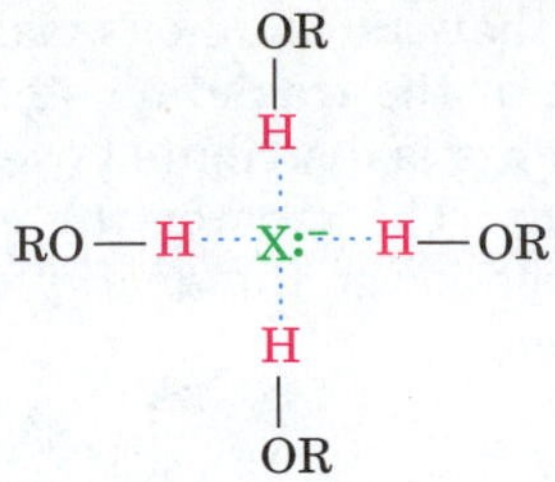

A solvated anion
(reduced nucleophilicity due to enhanced ground-state stability)

In contrast to protic solvents, polar aprotic solvents increase the rates of S_N2 reactions by raising the energy of the nucleophile. Particularly valuable are acetonitrile (CH_3CN), dimethylformamide [$(CH_3)_2NCHO$, abbreviated DMF], dimethyl sulfoxide [$(CH_3)_2SO$, abbreviated DMSO], and hexamethylphosphoramide {[$(CH_3)_2N]_3PO$, abbreviated HMPA}. These solvents can dissolve many salts because of their high polarity, but they tend to solvate metal cations rather than nucleophilic anions. As a result, the bare unsolvated anions have a greater nucleophilicity, and S_N2 reactions take place at correspondingly faster rates. A rate increase of 200,000 has been observed on changing from methanol to hexamethylphosphoramide for the reaction of azide ion with 1-bromobutane.

$$CH_3CH_2CH_2CH_2Br + N_3^- \xrightarrow{\text{Solvent}} CH_3CH_2CH_2CH_2N_3 + Br^-$$

Solvent =	HMPA	CH_3CN	DMF	DMSO	H_2O	CH_3OH
Relative reactivity	200,000	5000	2800	1300	7	1

PROBLEM.

11.7 Organic solvents such as benzene, ether, and chloroform are neither protic nor strongly polar. What effect would you expect these solvents to have on S_N2 reactions?

S_N2 Reaction Characteristics: A Summary

The effect on S_N2 reactions of the four variables—substrate structure, nucleophile, leaving group, and solvent—are summarized at the top of the next page and in the reaction energy diagrams of Figure 11.7.

Substrate	Steric hindrance raises the energy of the transition state, thus increasing $\Delta G^{\ddagger}$ and decreasing the reaction rate. As a result, S_N2 reactions are best for methyl and primary substrates.
Nucleophile	More reactive nucleophiles are higher in energy, thereby decreasing $\Delta G^{\ddagger}$ and increasing the reaction rate.
Leaving group	Good leaving groups (more stable anions) lower the energy of the transition state, thus decreasing $\Delta G^{\ddagger}$ and increasing the reaction rate.
Solvent	Protic solvents form hydrogen bonds to the nucleophile, thereby lowering the energy of the nucleophile, increasing $\Delta G^{\ddagger}$, and decreasing the reaction rate. Polar aprotic solvents surround the accompanying cation but not the nucleophilic anion, thereby raising the energy of the nucleophile, decreasing $\Delta G^{\ddagger}$, and increasing the reaction rate.

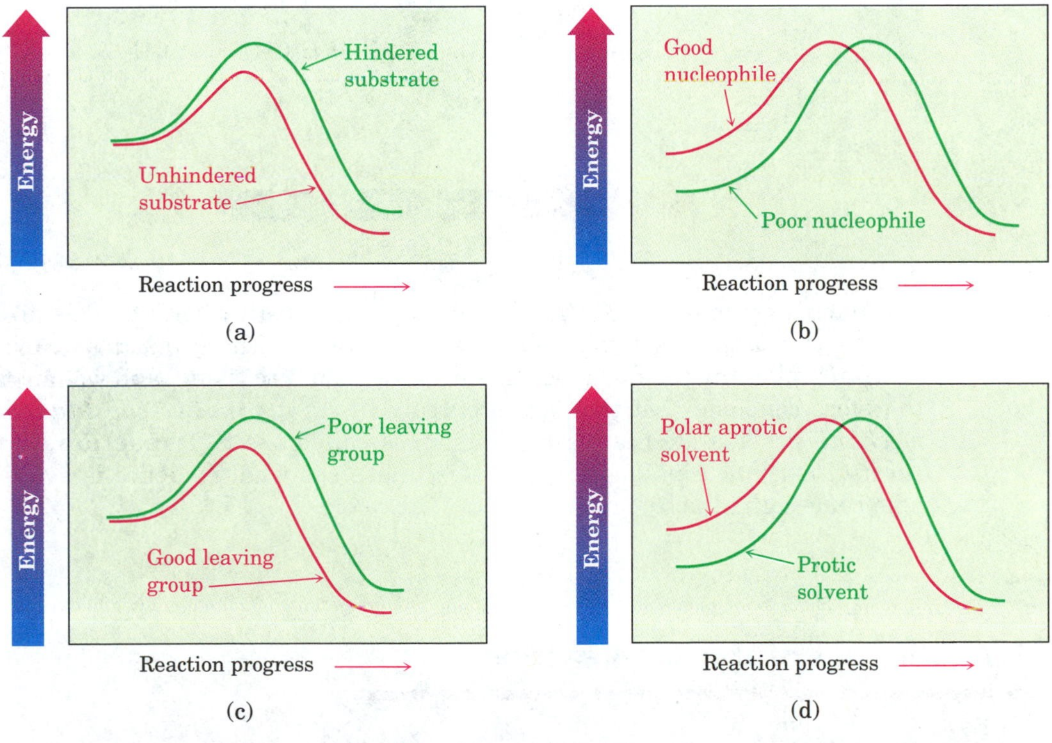

Figure 11.7 Reaction energy diagrams showing the effects of (a) substrate structure, (b) nucleophile, (c) leaving group, and (d) solvent on S_N2 reaction rates. Substrate and leaving group effects are felt primarily in the transition state. Nucleophile and solvent effects are felt primarily in the reactant ground state.

11.6 The S_N1 Reaction

As we've seen, the S_N2 reaction is disfavored by hindered substrates, nonbasic nucleophiles, and protic solvents. You might therefore expect the reaction of a tertiary substrate such as $(CH_3)_3CBr$ with neutral water to be among the slowest of substitution reactions. Remarkably, however, the opposite is true. The reaction is, in fact, quite rapid.

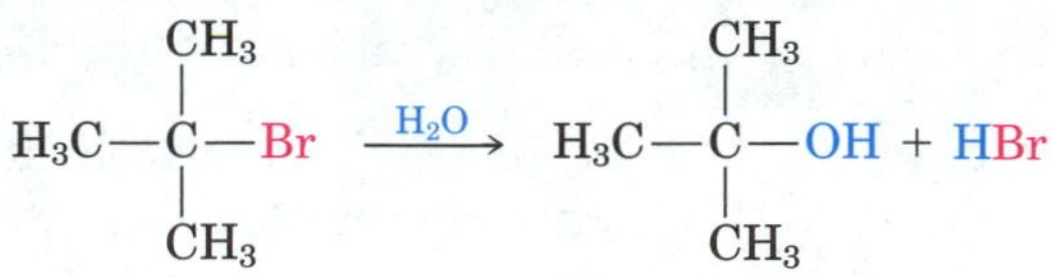

Perhaps even more surprising is that the reaction of the *tertiary* halide $(CH_3)_3CBr$ with neutral water is more than *1 million times* as fast as the corresponding reaction of the *methyl* halide CH_3Br.

$$RBr + H_2O \longrightarrow ROH + HBr$$

	$(CH_3)_3CBr$	$(CH_3)_2CHBr$	CH_3CH_2Br	CH_3Br
Relative reactivity	1,200,000	12	1	1

More reactive ⟵ **Reactivity** — Less reactive

What's going on here? Clearly, nucleophilic substitution reactions are occurring, yet the reactivity order seems backward. These reactions can't be taking place by the S_N2 mechanism we've been discussing, and we must therefore conclude that they are occurring by *an alternative substitution mechanism*. This alternative mechanism is called the **S_N1 reaction** (for *substitution, nucleophilic, unimolecular*). Let's see what evidence is available concerning the S_N1 reaction.

11.7 Kinetics of the S_N1 Reaction

The reaction of $(CH_3)_3CBr$ with H_2O looks analogous to the reaction of CH_3Br with OH^-, and we might therefore expect to observe second-order kinetics. In fact, we do not. We find instead that the reaction rate is dependent only on the alkyl halide concentration and is independent of the water concentration. In other words, the reaction is a **first-order process**; only one molecule is involved in the step whose kinetics are measured. *The concentration of the nucleophile does not appear in the rate expression.*

$$\text{Reaction rate} = \text{Rate of disappearance of alkyl halide}$$
$$= k \times [\text{RX}]$$

How can this result be explained? To answer this question, we must first learn more about kinetics measurements.

Many organic reactions are rather complex and occur in successive steps. One of these steps is usually slower than the others, and we call this the **rate-limiting step**. No reaction can proceed faster than its rate-limiting step, which acts as a kind of traffic jam, or bottleneck. The overall reaction rate that we actually measure in a kinetics experiment is determined by the height of the highest energy barrier between a low point and a subsequent high point in the energy diagram of the reaction. The reaction energy diagrams in Figure 11.8 illustrate the idea of the rate-limiting step.

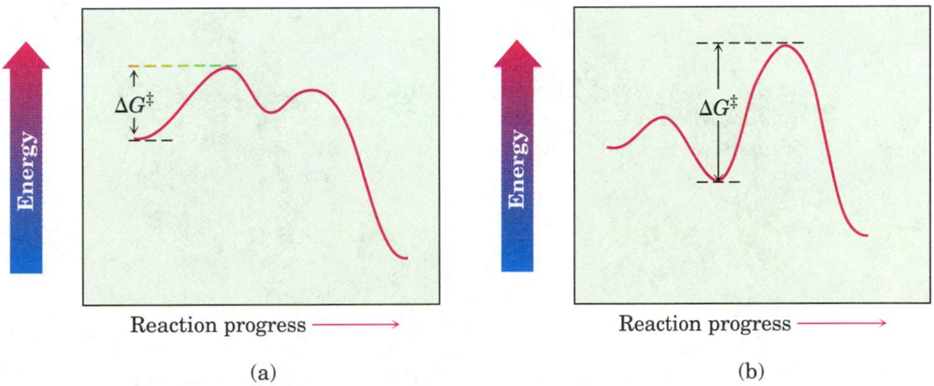

Figure 11.8 Reaction energy diagrams for two hypothetical reactions. The rate-limiting step in each is determined by the difference in height between a low point and a subsequent high point. In (a), the first step is rate-limiting; in (b), the second step is rate-limiting.

The observation of first-order kinetics for the S_N1 reaction of $(CH_3)_3CBr$ with water tells us that the alkyl halide is involved in a unimolecular rate-limiting step. In other words, 2-bromo-2-methylpropane undergoes some manner of spontaneous, rate-limiting reaction without involvement of the nucleophile. But because the nucleophile must be involved at *some* point, there must be at least two steps in the reaction. The mechanism shown in Figure 11.9 accounts for the kinetic observations.

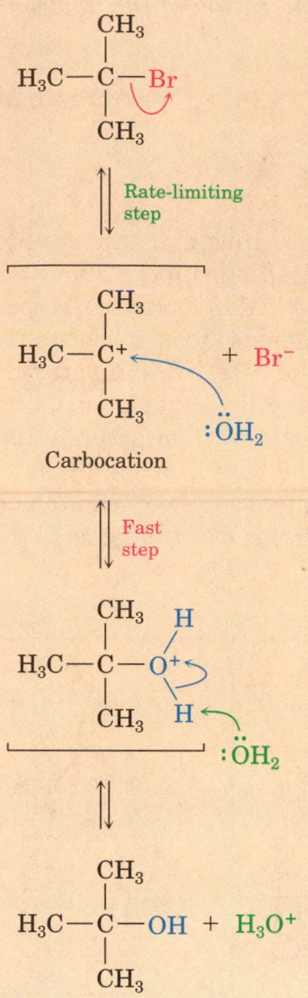

Figure 11.9 The mechanism of the S_N1 reaction of 2-bromo-2-methylpropane with water. The first of three steps—the spontaneous, unimolecular dissociation of the alkyl bromide to yield a carbocation—is rate-limiting.

Unlike what happens in an S_N2 reaction, where the leaving group is displaced at the same time that the incoming nucleophile is approaching, an S_N1 reaction takes place by loss of the leaving group *before* the incoming nucleophile approaches. 2-Bromo-2-methylpropane spontaneously dissociates to the *tert*-butyl carbocation plus bromide ion in a slow, rate-limiting step, and the intermediate ion is then immediately trapped by nucleophilic water in a fast step so that first-order kinetics are obtained. *Water is not a reactant in the step whose rate is measured by kinetics.* The reaction energy diagram is shown in Figure 11.10.

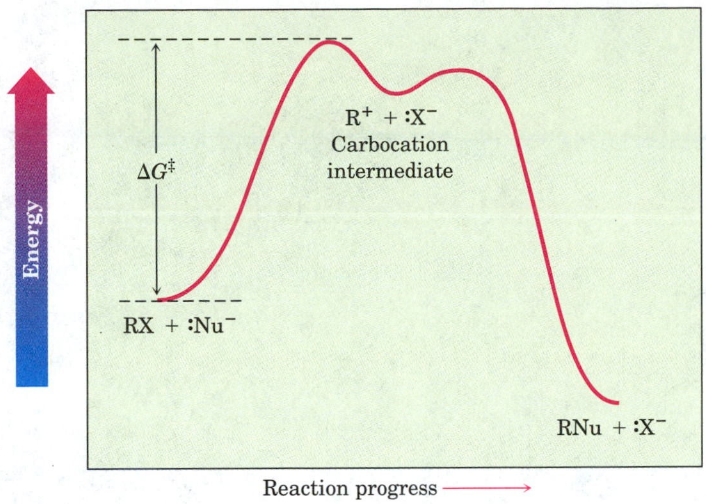

Figure 11.10 A reaction energy diagram for an S_N1 reaction. The rate-limiting step is the spontaneous dissociation of the alkyl halide.

11.8 Stereochemistry of the S_N1 Reaction

If S_N1 reactions occur through carbocation intermediates, the stereochemical consequences should be different from those for S_N2 reactions. Since carbocations are planar and sp^2-hybridized, they are achiral. Thus, if we carry out an S_N1 reaction on one enantiomer of a chiral reactant and go through an achiral carbocation intermediate, then the product must be optically inactive. The symmetrical intermediate carbocation can be attacked by a nucleophile equally well from either side, leading to a 50:50 mixture of enantiomers—a racemic mixture (Figure 11.11, p. 390).

The prediction that S_N1 reactions on optically active substrates should lead to racemic products is exactly what is observed. Surprisingly, though, few S_N1 displacements occur with complete racemization. Most give a minor (0–20%) excess of inversion. For example, the reaction of (*R*)-6-chloro-2,6-dimethyloctane with water leads to an alcohol product that is approximately 80% racemized and 20% inverted (80% *R*,*S* + 20% *S* is equivalent to 40% *R* + 60% *S*):

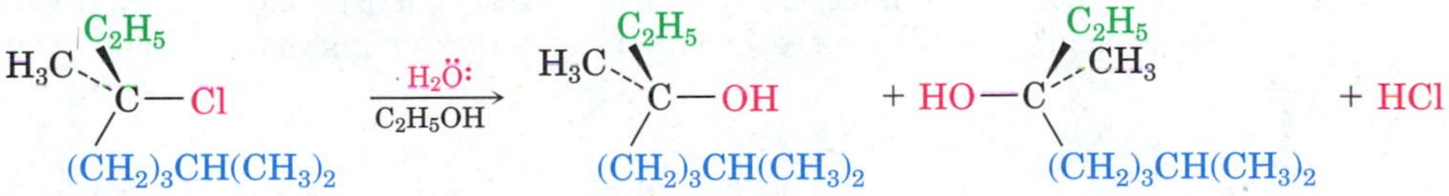

(*R*)-6-Chloro-2,6-dimethyloctane **40% *R* (retention)** **60% *S* (inversion)**

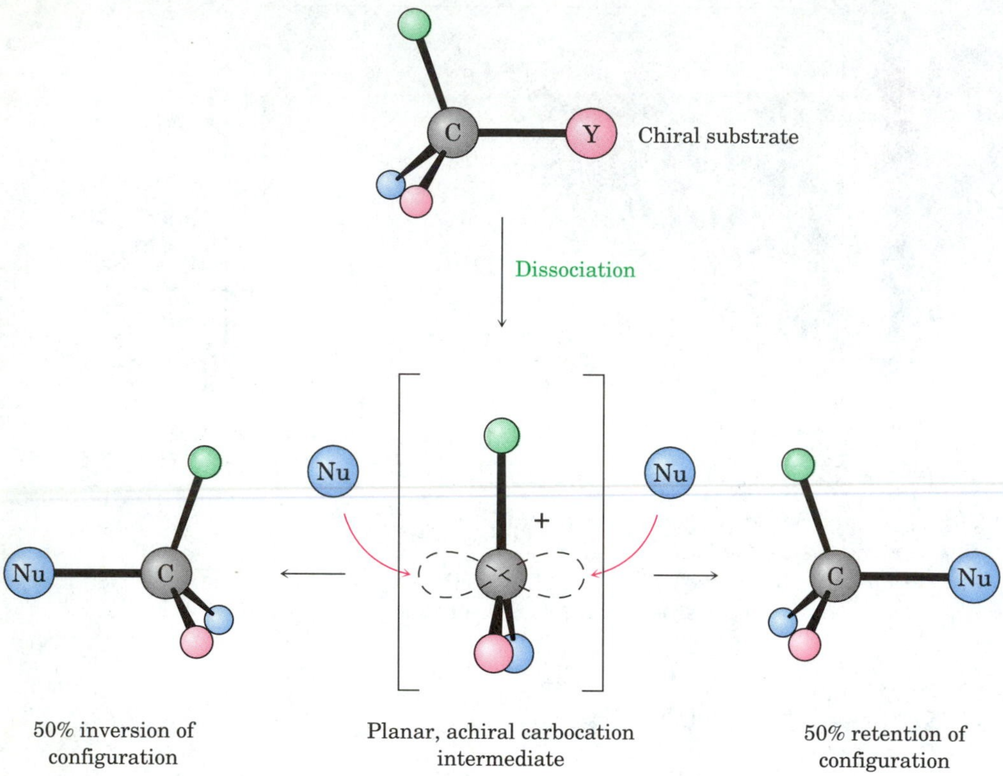

Figure 11.11 Stereochemistry of the S_N1 reaction. An optically active starting material must give a racemic product.

The reasons for the lack of complete racemization in most S_N1 reactions aren't completely clear, but an attractive suggestion first proposed by Saul Winstein[4] is that **ion pairs** are involved. According to this suggestion, dissociation of the substrate occurs to give a complex in which the two ions are still loosely associated and in which the carbocation is effectively shielded from nucleophilic attack on one side by the departing ion. If a certain amount of substitution occurs before the two ions fully diffuse away from each other, then a net inversion of configuration will be observed (Figure 11.12).

PROBLEM.

11.8 What product(s) would you expect from reaction of (*S*)-3-chloro-3-methyloctane with acetic acid? Show the stereochemistry of both starting material and product.

[4]Saul Winstein (1912–1969); b. Montreal; Ph.D., California Institute of Technology (Lucas); professor, University of California, Los Angeles (1942–1969).

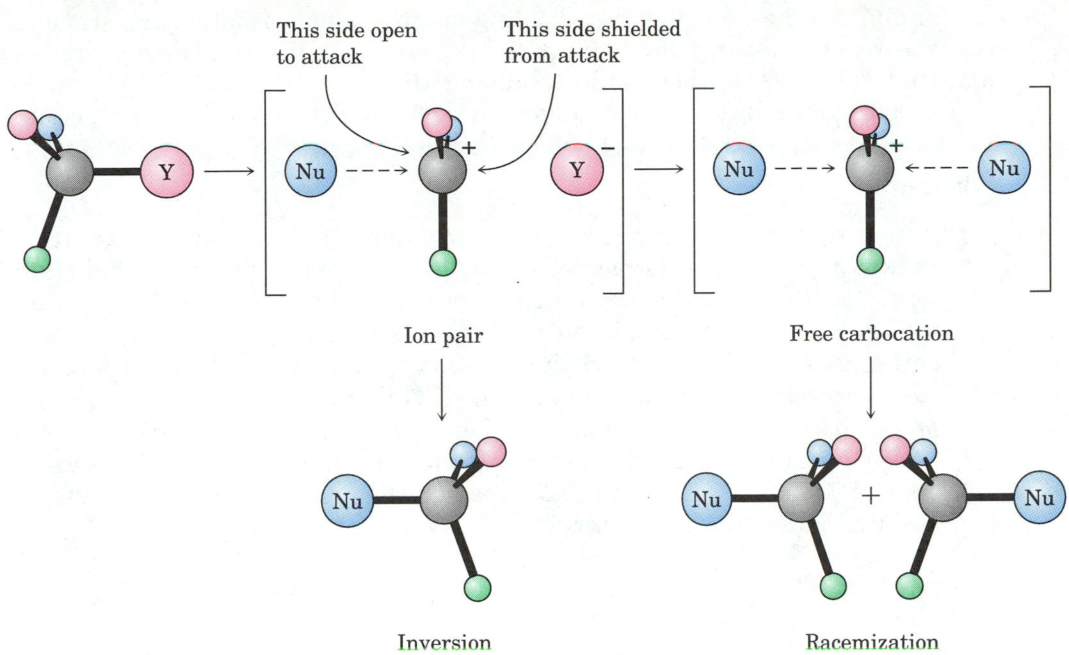

Figure 11.12 The ion-pair hypothesis in S_N1 reactions. The leaving group shields one side of the developing carbocation intermediate from attack by the nucleophile, thereby leading to some inversion of configuration rather than complete racemization.

PROBLEM.

11.9 Among the numerous examples of S_N1 reactions that occur with incomplete racemization is one reported by Winstein in 1952. The optically pure tosylate of 2,2-dimethyl-1-phenyl-1-propanol ($[\alpha]_D = -30.3°$) was heated in acetic acid to yield the corresponding acetate ($[\alpha]_D = +5.3°$). If complete inversion had occurred, the optically pure acetate would have had $[\alpha]_D = +53.6°$. What percentage racemization and what percentage inversion occurred in this reaction?

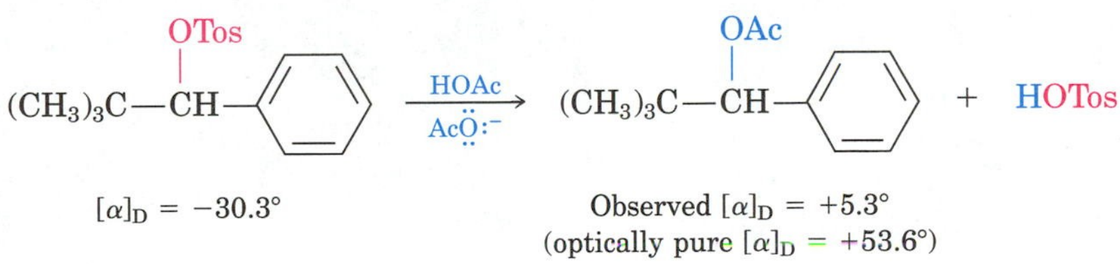

11.9 Characteristics of the S_N1 Reaction

Just as the S_N2 reaction is strongly influenced by such variables as solvent, leaving group, substrate structure, and nature of the attacking nucleophile, the S_N1 reaction is similarly influenced. Factors that lower $\Delta G^\ddagger$, either by stabilizing the transition state leading to carbocation formation or by raising the reactant energy level, favor faster S_N1 reactions. Conversely, factors that raise $\Delta G^\ddagger$, either by destabilizing the transition state leading to a carbocation or by lowering reactant energy level, slow down the S_N1 reaction.

The Substrate

According to the Hammond postulate (Section 6.11), any factor that stabilizes a high-energy intermediate should also stabilize the transition state leading to that intermediate. Since the rate-limiting step in the S_N1 reaction is the spontaneous, unimolecular dissociation of the substrate to yield a carbocation, we would expect the reaction to be favored whenever a stabilized carbocation intermediate is formed. This is exactly what is found: *The more stable the carbocation intermediate, the faster the S_N1 reaction.*

We've already seen (Section 6.10) that the stability order of alkyl carbocations is 3° > 2° > 1° > $-CH_3$. To this list we must also add the resonance-stabilized allyl and benzyl cations:

$$H_2C{=}CH{-}\overset{+}{C}H_2$$

Allyl carbocation

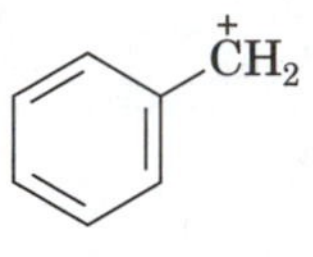

Benzyl carbocation

Just as allylic *radicals* are unusually stable because the unpaired electron can be delocalized over an extended π orbital system (Section 10.6), so allylic and benzylic *carbocations* are unusually stable. (The word **benzylic** means next to an aromatic ring.) As Figure 11.13 indicates, an allylic cation has two resonance forms. In one form the double bond is on the "left," and in the other form the double bond is on the "right." A benzylic cation, however, has *four* resonance forms, all of which make substantial contributions to the overall resonance hybrid.

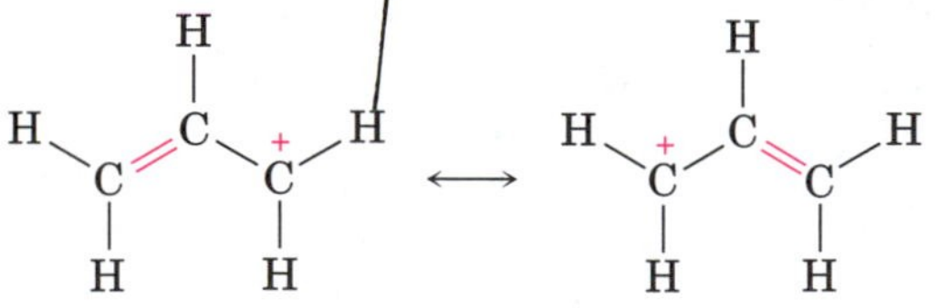

Allyl carbocation

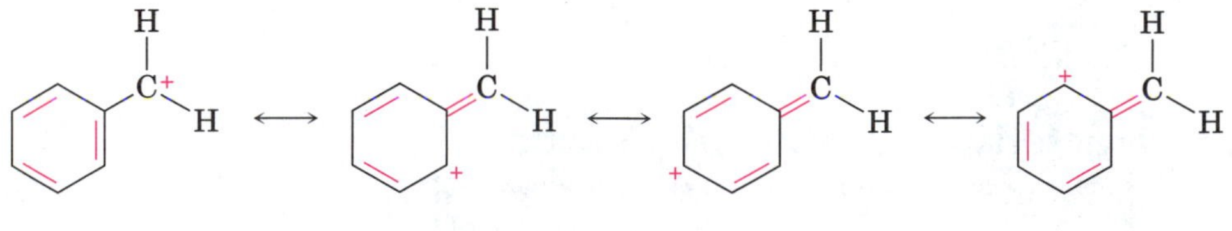

Benzyl carbocation

Figure 11.13 Resonance forms of allylic and benzylic carbocations.

Because of their resonance stabilization, a *primary* allylic or benzylic carbocation is as stable as a *secondary* alkyl carbocation. Similarly, a *secondary* allylic or benzylic carbocation is about as stable as a *tertiary* alkyl carbocation.

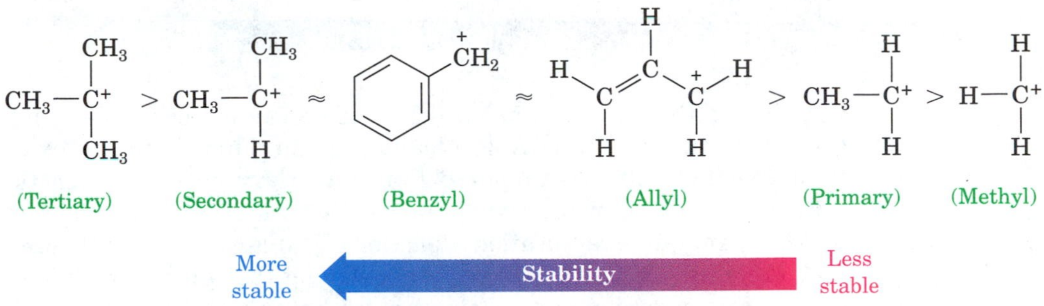

This stability order of carbocations is exactly the same as the order of S_N1 reactivity for alkyl halides and tosylates.

Parenthetically, it should also be noted that allylic and benzylic substrates are particularly reactive in S_N2 reactions as well as in S_N1 reactions. Allylic and benzylic C–X bonds are about 50 kJ/mol (12 kcal/mol) weaker than the corresponding saturated bonds and are therefore more easily broken.

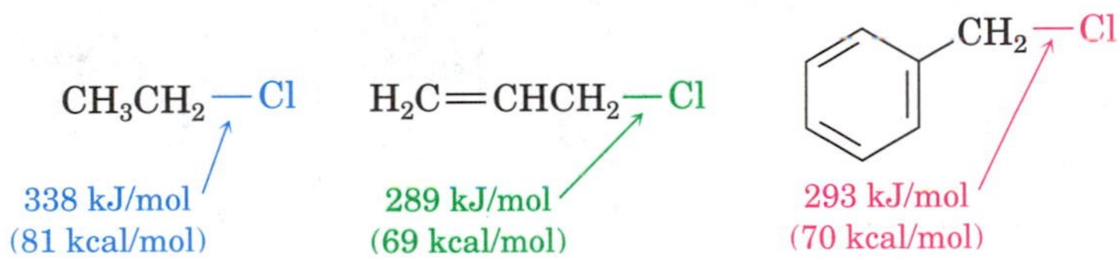

PROBLEM

11.10 Rank the following alkyl halides in order of their expected S_N1 reactivity:

CH_3CH_2Br $H_2C{=}CHCH(Br)CH_3$ $H_2C{=}CHBr$ $CH_3CH(Br)CH_3$

PROBLEM

11.11 How can you account for the fact that 3-bromo-1-butene and 1-bromo-2-butene undergo S_N1 reaction at nearly the same rate even though one is a secondary halide and the other is primary?

The Leaving Group

We reasoned during the discussion of S_N2 reactivity that the best leaving groups should be those that are most stable—that is, the conjugate bases of strong acids. An identical reactivity order is found for the S_N1 reaction, because the leaving group is directly involved in the rate-limiting step. Thus, we find the S_N1 reactivity order to be

$$TosO^- > I^- > Br^- > Cl^- \approx H_2O$$

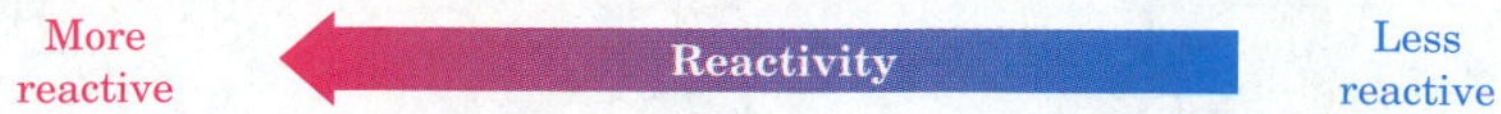

Note that in the S_N1 reaction, which is often carried out under acidic conditions, neutral water can act as a leaving group. This occurs, for example, when an alkyl halide is prepared from a tertiary alcohol by reaction with HBr or HCl (Section 10.7). The alcohol is first protonated and then loses water to generate a carbocation. Reaction of the carbocation with halide ion yields the alkyl halide (Figure 11.14). Knowing that an S_N1 reaction is involved in the conversion of alcohols to alkyl halides makes it clear why the reaction works well only for tertiary alcohols: Tertiary alcohols react fastest because they give the most stable carbocation intermediates.

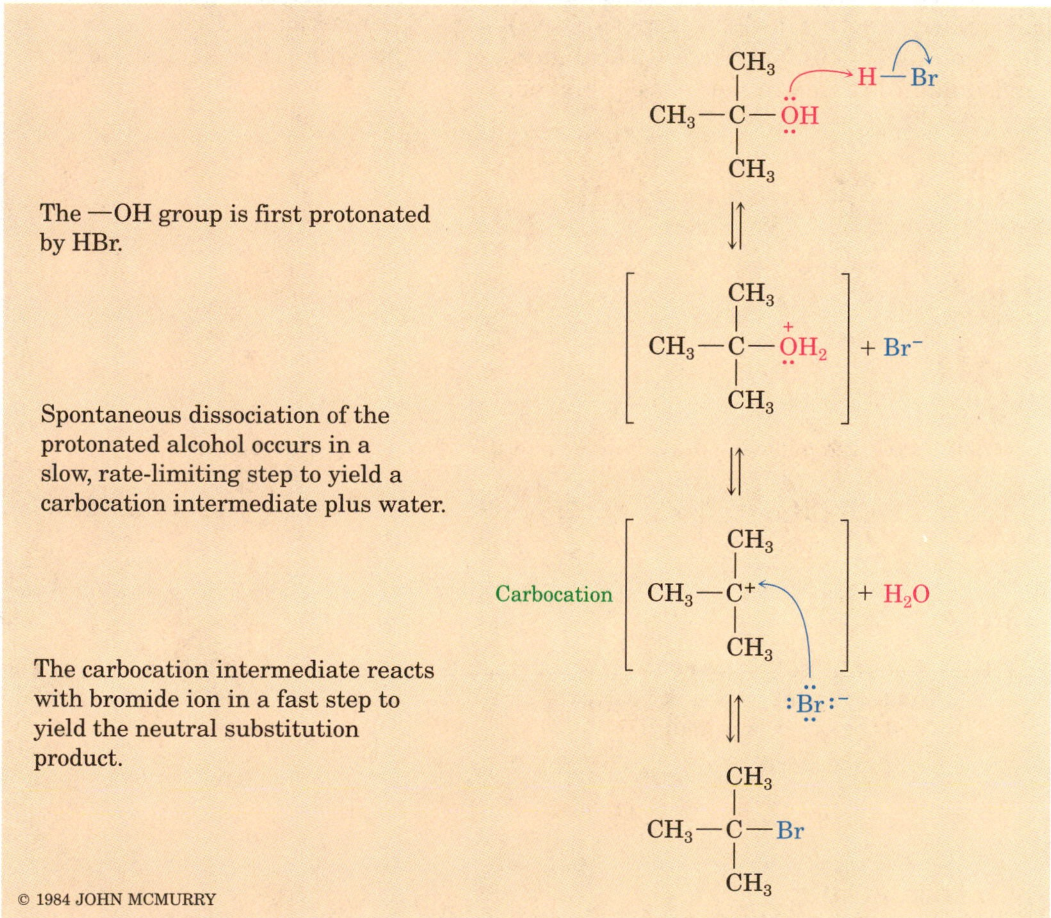

Figure 11.14 The mechanism of the S_N1 reaction of a tertiary alcohol with HBr to yield an alkyl halide. Neutral water is the leaving group.

The Nucleophile

The nature of the attacking nucleophile plays a major role in the S_N2 reaction. Is the nucleophile also important in determining the rate of an S_N1 reaction? The answer is no. The S_N1 reaction, by its very nature, occurs through a rate-limiting step in which the added nucleophile has no kinetic role. The nucleophile does not enter into the reaction until after rate-limiting dissociation has occurred and thus cannot affect the reaction rate. The reaction of 2-methyl-2-propanol with HX, for example, occurs at the same rate regardless of whether X is Cl, Br, or I:

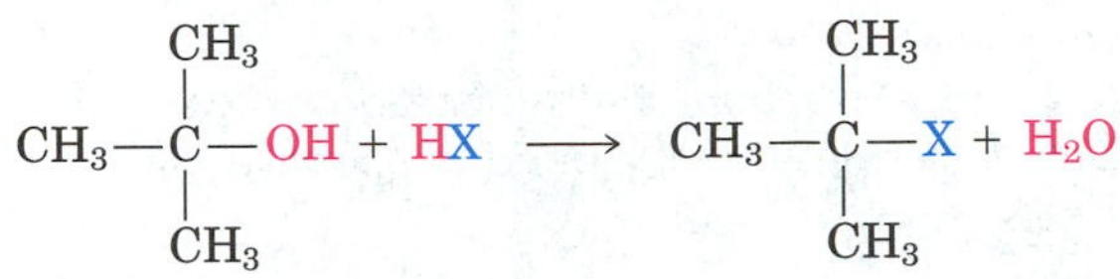

2-Methyl-2-propanol (same rate for X = Cl, Br, I)

PROBLEM

11.12 How do you account for the fact that 1-chloro-1,2-diphenylethane reacts with the nucleophiles fluoride ion and triethylamine at the same rate?

The Solvent

What about solvent? Do solvents have the same effect in S_N1 reactions that they have in S_N2 reactions? The answer is both yes and no. Yes, solvents have a large effect on S_N1 reactions, but no, the reasons for the effects are not the same. Solvent effects in the S_N2 reaction are due largely to stabilization or destabilization of the nucleophile *reactant*. Solvent effects in the S_N1 reaction, however, are due largely to stabilization or destabilization of the *transition state*. Let's look again at the Hammond postulate to see the effect of solvent on the transition state.

The Hammond postulate says that any factor stabilizing the intermediate carbocation should increase the rate of an S_N1 reaction. Solvation of the carbocation—the interaction of the ion with solvent molecules—has just such an effect. Solvent molecules orient around the cation so that the electron-rich ends of the solvent dipoles face the positive charge (Figure 11.15, p. 396), thereby stabilizing the ion.

The properties of a solvent that contribute to its ability to stabilize ions by solvation aren't fully understood but are related to the solvent's polarity. Polar solvents such as water, methanol, and dimethyl sulfoxide are good at solvating ions, but most ether and hydrocarbon solvents are very poor at solvating ions.

Solvent polarity is usually expressed in terms of **dielectric constants**, ϵ, which measure the ability of a solvent to act as an insulator of electric charges. Solvents of low dielectric constant, such as hydrocarbons, are nonpolar, whereas solvents of high dielectric constant, such as water, are polar. Table 11.3 lists the dielectric constants of some common solvents.

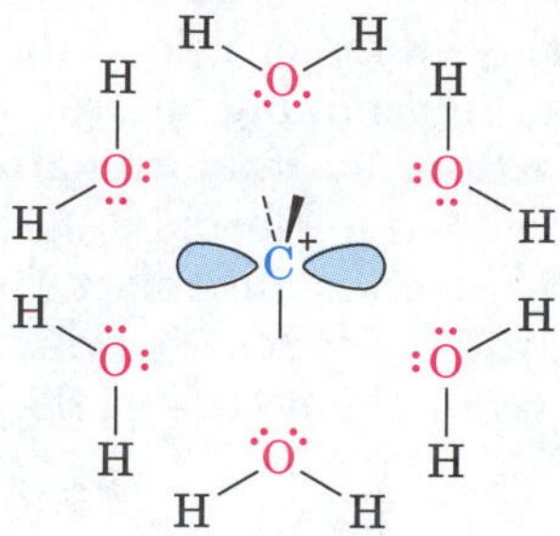

Figure 11.15 Solvation of a carbocation by water. The electron-rich oxygen atoms of solvent molecules orient around the positively charged carbocation and thereby stabilize it.

Table 11.3 Dielectric Constants of Some Common Solvents

Name	*Dielectric constant*	*Name*	*Dielectric constant*
Aprotic solvents		*Protic solvents*	
Hexane	1.9	Acetic acid	6.2
Benzene	2.3	Ethanol	24.3
Diethyl ether	4.3	Methanol	33.6
Chloroform	4.8	Formic acid	58.0
Hexamethylphosphoramide (HMPA)	30	Water	80.4
Dimethylformamide (DMF)	38		
Dimethyl sulfoxide (DMSO)	48		

S_N1 reactions take place much more rapidly in polar solvents than in nonpolar solvents. In the reaction of 2-chloro-2-methylpropane, for example, a rate increase of 100,000 is observed on going from the polar solvent ethanol to the even more polar solvent water. The rate increases on going from hydrocarbon solvents to water are too large to measure accurately.

$$(CH_3)_3CCl + ROH \longrightarrow (CH_3)_3COR + HCl$$

	Water	80% aqueous ethanol	40% aqueous ethanol	Ethanol
Relative reactivity	100,000	14,000	100	1

It should be emphasized again that both S_N1 and S_N2 reactions show large solvent effects, but that they do so for different reasons. S_N2 reactions are disfavored by protic solvents because the *ground-state energy* of the attacking nucleophile is lowered by solvation. S_N1 reactions are favored by protic solvents because the *transition-state energy* leading to carbocation intermediate is lowered by solvation. To see the difference, compare the S_N1 reaction energy diagram in Figure 11.16 to that in Figure 11.7(d), where the effect of solvent on the S_N2 reaction was illustrated.

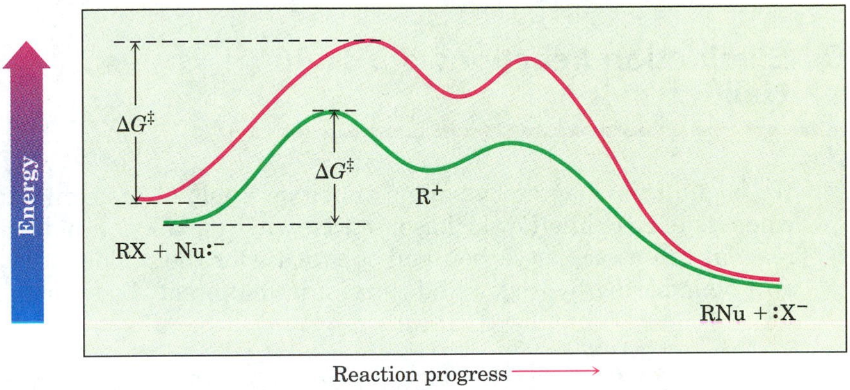

Figure 11.16 The effect of solvent on an S_N1 reaction. The energy level of the transition-state energy is lowered dramatically by solvation in a polar solvent. (Red curve, nonpolar solvent; green curve, polar solvent.)

PROBLEM.

11.13 As indicated in Problem 11.3, the halide shown below is inert to S_N2 displacement. Perhaps more surprisingly, it is also unreactive to S_N1 substitution even though it's tertiary. Explain.

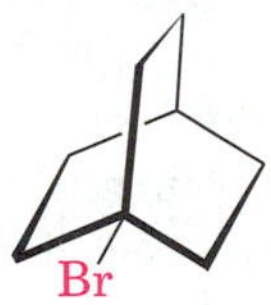

S_N1 Reaction Characteristics: A Summary

The effect on S_N1 reactions of the four variables—substrate structure, leaving group, nucleophile, and solvent—are summarized below.

Substrate	The best substrates are those that yield the most stable carbocations. As a result, S_N1 reactions are best for tertiary, allylic, and benzylic substrates.

Leaving group — Good leaving groups (more stable anions) lower the energy of the transition state leading to carbocation formation, thus increasing the reaction rate.

Nucleophile — The nucleophile must be nonbasic to prevent a competitive E2 elimination, but otherwise does not affect the reaction rate.

Solvent — Polar solvents, such as water, stabilize the carbocation intermediate by solvation, thereby increasing the reaction rate.

11.10 Elimination Reactions of Alkyl Halides

We began this chapter by saying that two kinds of reactions are possible when a nucleophile/Lewis base reacts with an alkyl halide. Either the reagent can attack at carbon and substitute for the halide, or it can attack at a neighboring hydrogen and cause elimination of HX to form an alkene.

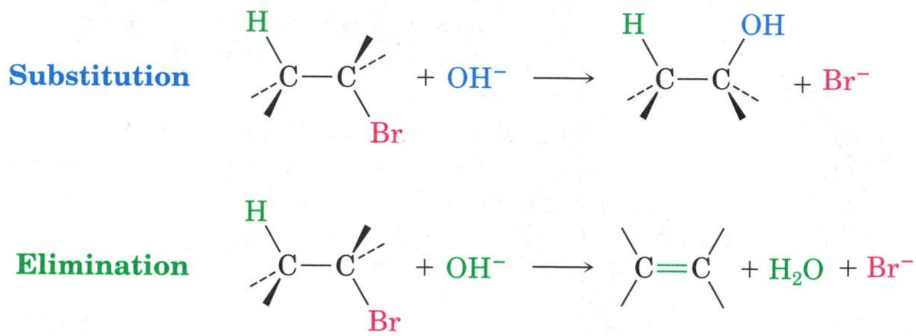

Elimination reactions are more complex than substitution reactions for several reasons. There is, for example, the problem of regiochemistry: What products result from dehydrohalogenation of an unsymmetrical halide? In fact, elimination reactions almost always give *mixtures* of alkene products, and the best we can usually do is to predict which will be the major product.

According to a rule formulated in 1875 by the Russian chemist Alexander Zaitsev,[5] base-induced elimination reactions generally give the more highly substituted (more stable) alkene product—that is, the alkene with more alkyl substituents on the double-bond carbons. In the following two cases, for example, **Zaitsev's rule** is clearly applicable. The more highly substituted alkene product predominates in both cases when sodium ethoxide in ethanol is used as the base.

[5]Alexander M. Zaitsev (1841–1910); b. Kazan, Russia; student of Butlerov and Kolbe; Ph.D., Leipzig (1866); professor, University of Kazan (1870–1903), Kiev University.

Zaitsev's rule In the elimination of HX from an alkyl halide, the more highly substituted alkene product predominates.

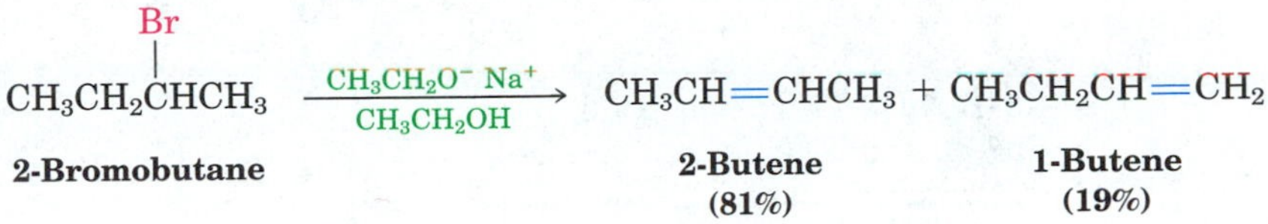

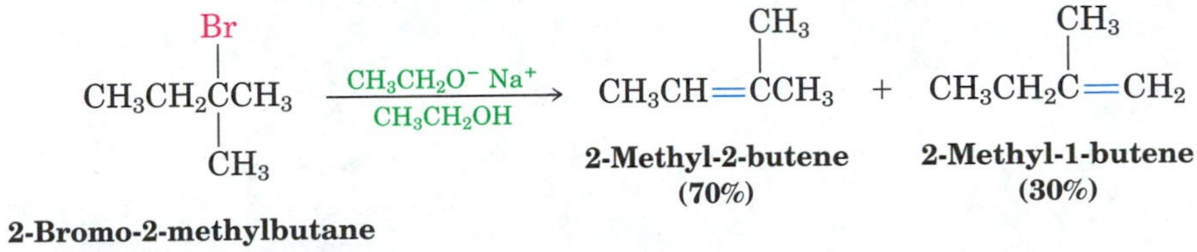

The elimination of HX from an alkyl halide is an excellent method for preparing alkenes, but the subject is complex because elimination reactions can take place through a variety of different mechanistic pathways, just as substitutions can. We'll consider two pathways: the E1 and E2 reactions.

PROBLEM

11.14 What products would you expect from elimination reactions of the following alkyl halides? Which product will be major in each case?

(a) 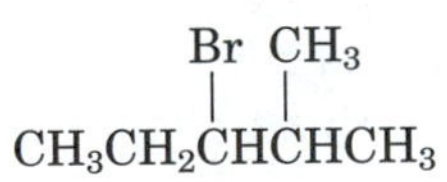

(b) 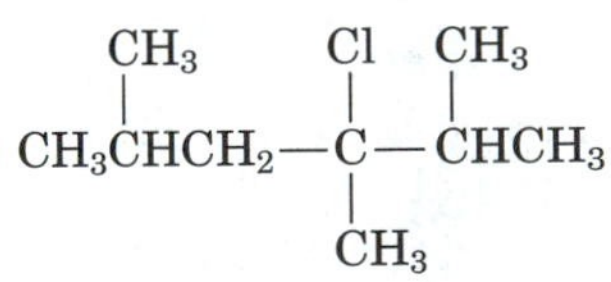

(c) 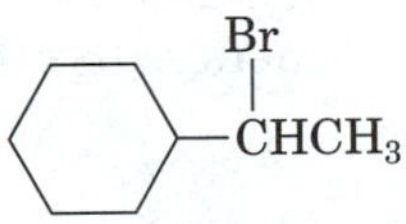

11.11 The E2 Reaction

The **E2 reaction** (for *elimination, bimolecular*) occurs when an alkyl halide is treated with a strong base, such as hydroxide ion or alkoxide ion (RO^-). It is the most commonly occurring pathway for elimination and can be formulated as shown in Figure 11.17.

Like the S_N2 reaction, the E2 reaction takes place in one step without intermediates. As the attacking base begins to abstract a proton from a

Base (B:) attacks a neighboring hydrogen and begins to remove the H at the same time as the alkene double bond starts to form and the X group starts to leave.

Neutral alkene is produced when the C–H bond is fully broken and the X group has departed with the C–X bond electron pair.

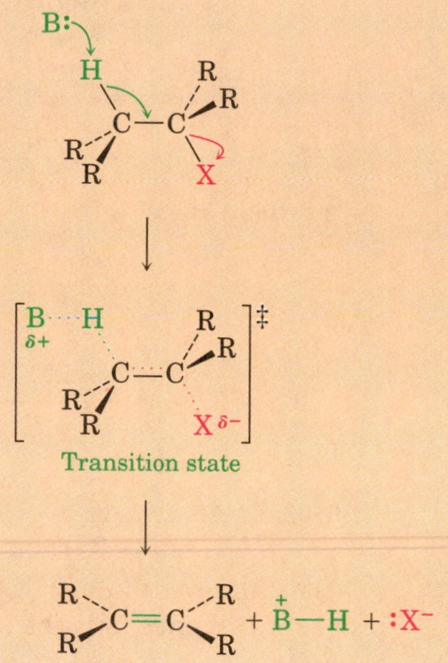

Figure 11.17 Mechanism of the E2 reaction of an alkyl halide. The reaction takes place in a single step through a transition state in which the double bond begins to form at the same time the H and X groups are leaving.

carbon next to the leaving group, the C–H bond begins to break, a C=C bond begins to form, and the leaving group begins to depart, taking with it the electron pair from the C–X bond.

Among the evidence supporting this mechanism is the measurement of reaction kinetics. Since both base and alkyl halide enter into the rate-limiting step, E2 reactions show second-order kinetics. In other words, E2 reactions follow the rate law

$$\text{Rate} = k \times [\text{RX}] \times [\text{Base}]$$

A second and more compelling piece of evidence involves the stereochemistry of E2 eliminations. As shown by a large number of experiments, E2 reactions always occur with a **periplanar** geometry, meaning that all four reacting atoms—the hydrogen, the two carbons, and the leaving group—lie in the same plane. Two such geometries are possible: **syn periplanar** geometry, in which the H and the X are on the same side of the molecule, and **anti periplanar** geometry, in which the H and the X are on opposite sides of the molecule. Of the two choices, anti periplanar geometry is energetically preferred because it allows the two carbon centers to adopt a staggered relationship, whereas syn geometry requires that the substituents on carbon be eclipsed.

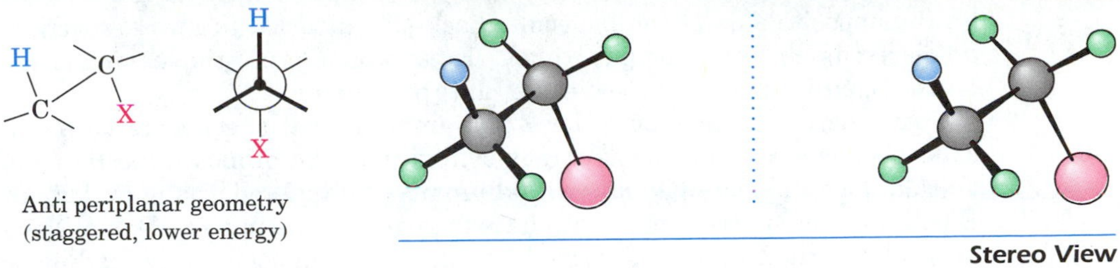

Anti periplanar geometry
(staggered, lower energy)

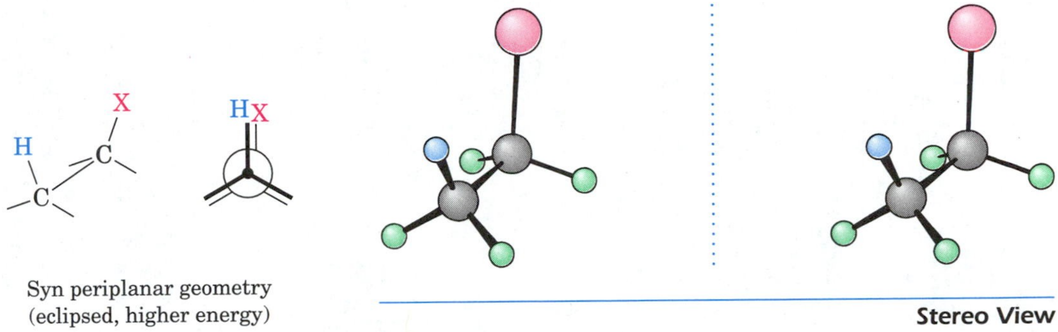

Syn periplanar geometry
(eclipsed, higher energy)

What's so special about periplanar geometry? Because the original C–H and C–X bond sp^3 σ orbitals in the reactant must overlap and become p π orbitals in the alkene product, there must also be some overlap in the transition state. This can occur only if all the orbitals are in the same plane to begin with—that is, if they're periplanar (Figure 11.18).

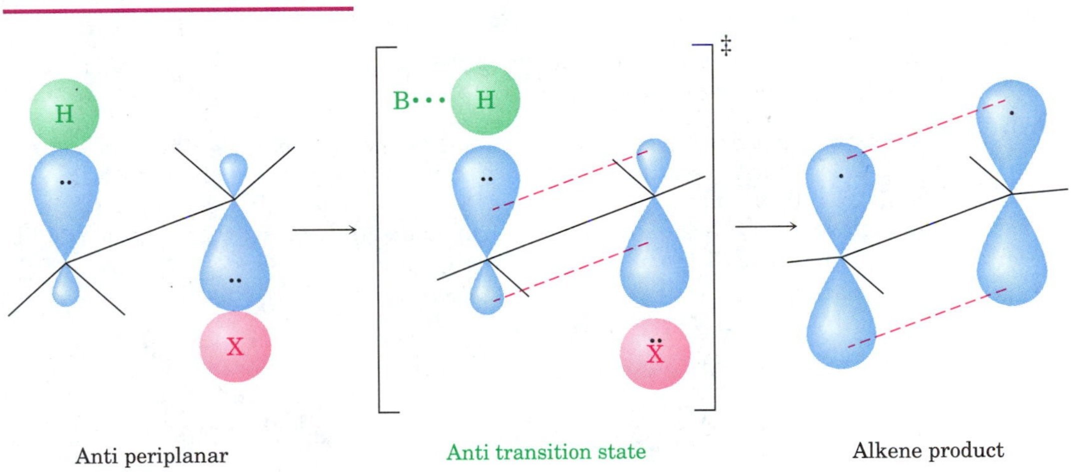

Figure 11.18 The transition state for the E2 reaction of an alkyl halide with base. Overlap of the developing p orbitals in the transition state requires periplanar geometry of the starting material.

It might help to think of E2 elimination reactions with periplanar geometry as similar to S_N2 reactions with 180° geometry. In an S_N2 reaction, an electron pair from the incoming nucleophile pushes out the leaving group on the opposite side of the molecule (back-side attack). In an E2 reaction, an electron pair from a neighboring C–H bond pushes out the leaving group on the opposite side of the molecule (anti periplanar).

Anti periplanar geometry for E2 eliminations has specific stereochemical consequences that provide strong evidence for the proposed mechanism. To take just one example, *meso*-1,2-dibromo-1,2-diphenylethane undergoes E2 elimination on treatment with base to give only the pure *E* alkene. None of the isomeric *Z* alkene is formed because the transition state leading to the *Z* alkene would have to have syn periplanar geometry.

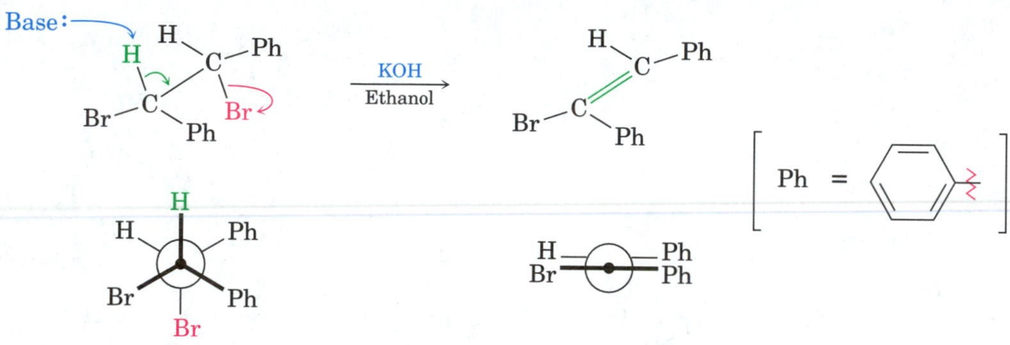

***meso*-1,2-Dibromo-1,2-diphenylethane**

(*E*)-1-Bromo-1,2-diphenylethylene

PRACTICE PROBLEM

What stereochemistry do you expect for the alkene obtained by E2 elimination of (1*S*,2*S*)-1,2-dibromo-1,2-diphenylethane?

Solution First draw (1*S*,2*S*)-1,2-dibromo-1,2-diphenylethane so that you can see its stereochemistry and so that the –H and –Br groups to be eliminated are anti periplanar (molecular models are extremely helpful here). Keeping all substituents in approximately their same positions, eliminate HBr and see what alkene results. The product is (*Z*)-1-bromo-1,2-diphenylethylene.

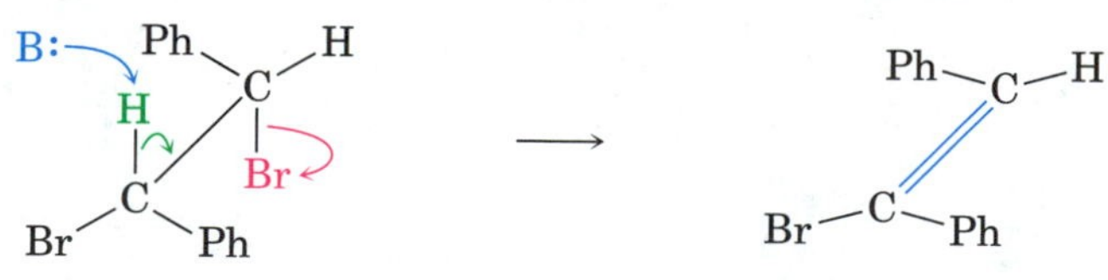

(1*S*,2*S*)-1,2-Dibromo-1,2-diphenylethane **(*Z*)-1-Bromo-1,2-diphenylethylene**

PROBLEM

11.15 What stereochemistry do you expect for the alkene obtained by E2 elimination of (1*R*,2*R*)-1,2-dibromo-1,2-diphenylethane? Draw a Newman projection of the reacting conformation.

11.12 Elimination Reactions and Cyclohexane Conformation

Anti periplanar geometry for E2 reactions is particularly important in cyclohexane rings where chair geometry forces a rigid relationship between the substituents on neighboring carbon atoms (Section 4.9). As pointed out by Derek Barton[6] in a landmark 1950 paper, much of the chemical reactivity of substituted cyclohexanes is controlled by their conformation. Let's look at the E2 dehydrohalogenation of chlorocyclohexanes to see an example of such conformational control.

The anti periplanar requirement for E2 reactions can be met in cyclohexanes only if the hydrogen and the leaving group are trans diaxial (Figure 11.19). If either the leaving group or the hydrogen is equatorial, E2 elimination can't occur.

Axial chlorine: H and Cl are anti periplanar

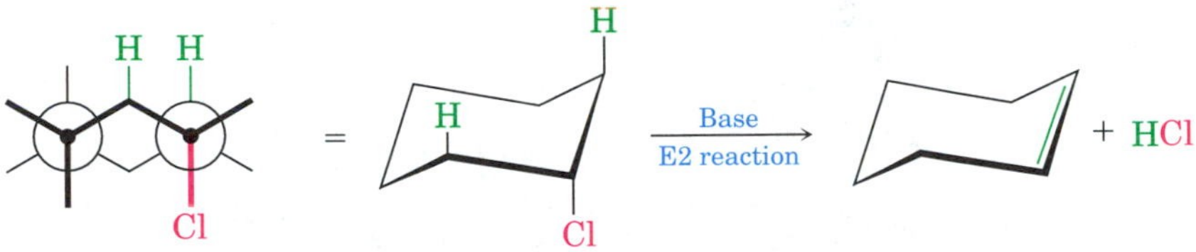

Equatorial chlorine: H and Cl are not anti periplanar

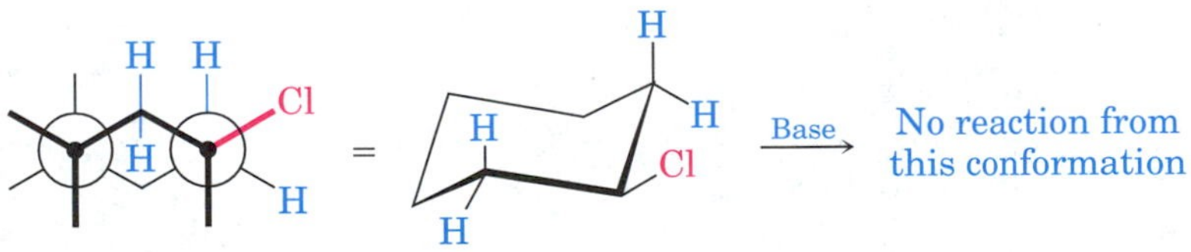

Figure 11.19 The geometric requirement for E2 reaction in a cyclohexane. The leaving group and the hydrogen must both be axial for anti periplanar elimination to be possible.

The elimination of HCl from the isomeric menthyl and neomenthyl chlorides shown in Figure 11.20 provides a good illustration of this trans-diaxial requirement. Neomenthyl chloride undergoes elimination of HCl on reaction with ethoxide ion 200 times as fast as menthyl chloride. Furthermore, neomenthyl chloride yields 3-menthene as the major alkene product, whereas menthyl chloride yields 2-menthene.

[6]Derek H. R. Barton (1918–); b. Gravesend, England; Ph.D. and D.Sc. London (Heilbron, E. R. H. Jones); professor, Birkbeck College, Harvard, Glasgow, Imperial College, Institut de Chimie des Substances Naturelles, Texas A and M; Nobel Prize (1969).

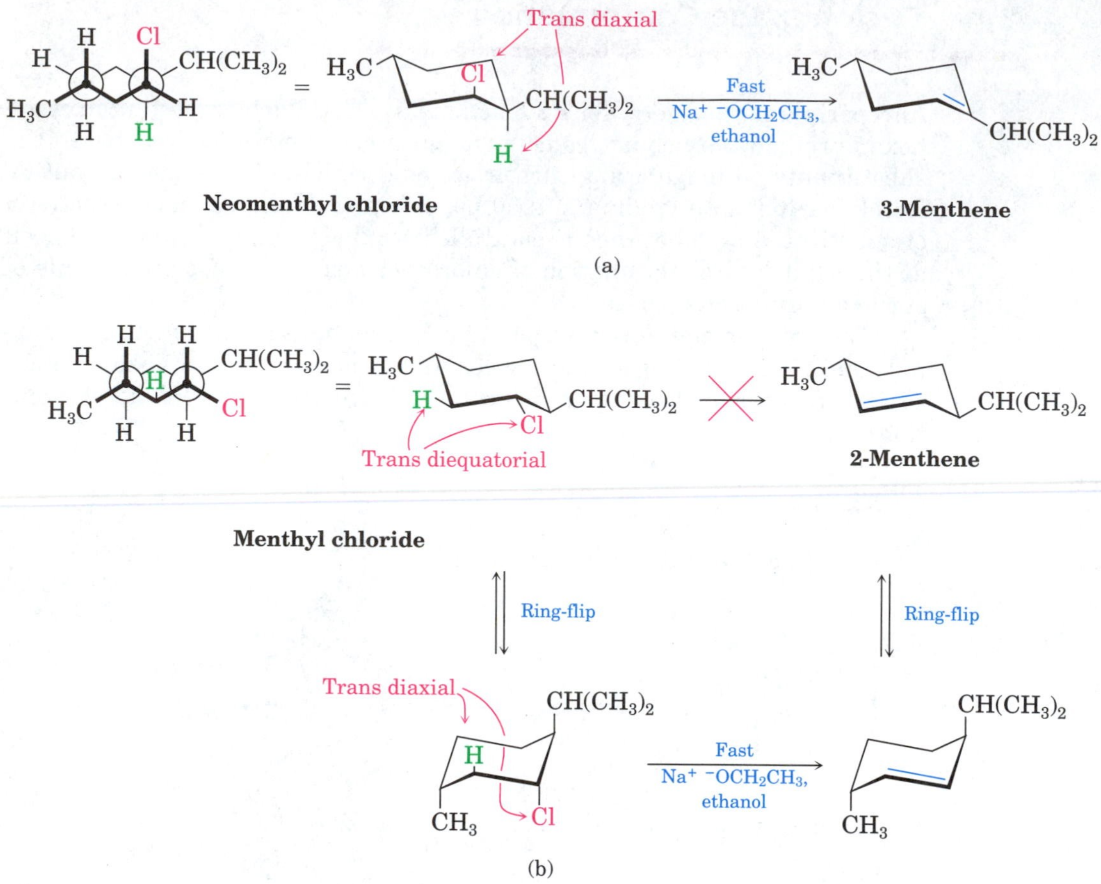

Figure 11.20 Dehydrochlorination of menthyl and neomenthyl chlorides. Neomenthyl chloride loses HCl from its more stable conformation, but menthyl chloride must first ring-flip before HCl loss can occur.

We can understand the difference in reactivity between the isomeric menthyl chlorides by looking at the more favorable chair conformations of the reactant molecules. Neomenthyl chloride has the conformation shown in Figure 11.20(a), with the methyl and isopropyl groups equatorial and the chlorine axial—a perfect geometry for E2 elimination. Loss of the hydrogen atom at C4 occurs easily to yield the more substituted alkene product 3-menthene, as predicted by Zaitsev's rule.

Menthyl chloride, by contrast, has a conformation in which all three substituents are equatorial [Figure 11.20(b)]. To achieve the necessary geometry for elimination, menthyl chloride must first ring-flip to a higher-energy chair conformation, in which all three substituents are axial. E2 elimination then occurs with loss of the only trans-diaxial hydrogen, leading to 2-menthene. The net effect of the simple change in chlorine stereochemistry is a 200-fold change in reaction rate and a complete change of product. The chemistry of the molecule is truly controlled by its conformation.

PROBLEM.

11.16 Which isomer would you expect to undergo E2 elimination faster, *trans*-1-bromo-4-*tert*-butylcyclohexane or *cis*-1-bromo-4-*tert*-butylcyclohexane? Draw each molecule in its more stable chair conformation, and explain your answer.

11.13 The Deuterium Isotope Effect

One final piece of evidence in support of the E2 mechanism is provided by a phenomenon known as the **deuterium isotope effect**. For reasons that we won't go into, a carbon–*hydrogen* bond is weaker by a small amount [about 5 kJ/mol (1.2 kcal/mol)] than a corresponding carbon–*deuterium* bond. Thus, a C–H bond is more easily broken than an equivalent C–D bond, and the rate of C–H bond cleavage is faster. As an example of how this effect can be used to obtain mechanistic information, the base-induced elimination of HBr from 1-bromo-2-phenylethane proceeds 7.11 times as fast as the corresponding elimination of DBr from 1-bromo-2,2-dideuterio-2-phenylethane:

Faster reaction

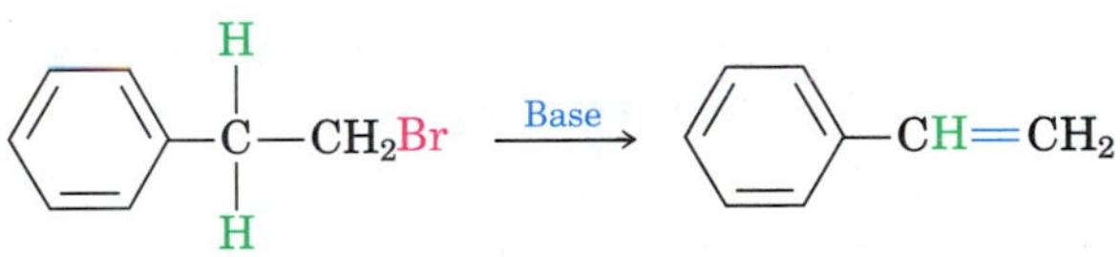

1-Bromo-2-phenylethane

Slower reaction

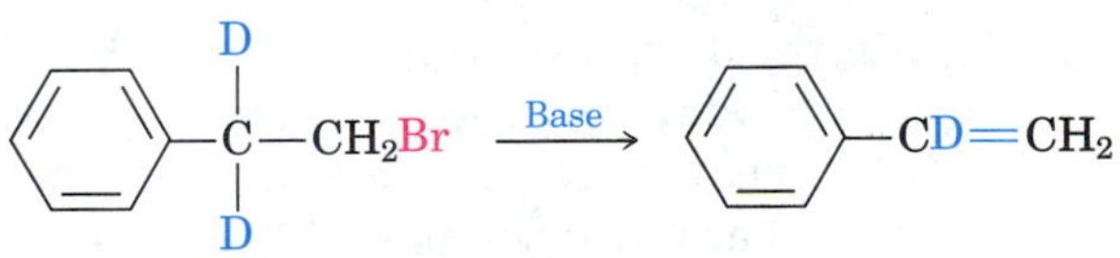

1-Bromo-2,2-dideuterio-2-phenylethane

This result tells us that the C–H (or C–D) bond is broken *in the rate-limiting step*, consistent with our picture of the E2 reaction as a one-step process. If it were otherwise, we couldn't measure a rate difference.

11.14 The E1 Reaction

Just as the S_N2 reaction is analogous to the E2 reaction, the S_N1 reaction has a close analog: the **E1 reaction** (for *elimination*, *unimolecular*). The E1 reaction can be formulated as shown in Figure 11.21 for the elimination of HCl from 2-chloro-2-methylpropane.

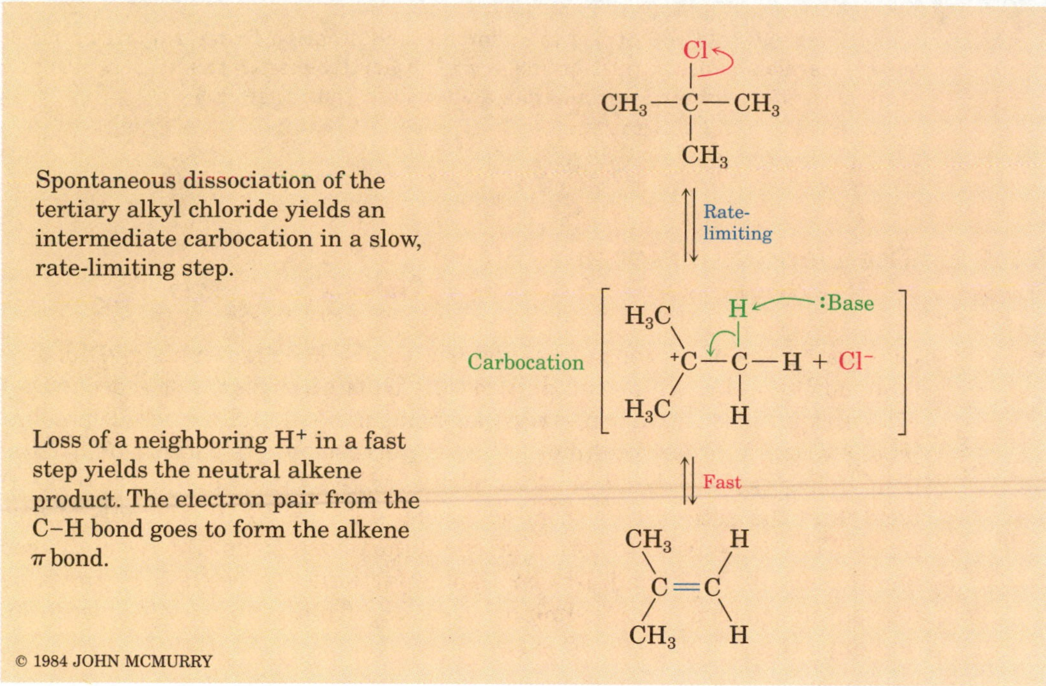

Figure 11.21 Mechanism of the E1 reaction. Two steps are involved, the first of which is rate-limiting, and a carbocation intermediate is present.

E1 eliminations begin with the same unimolecular dissociation we saw in the S_N1 reaction, but the dissociation is followed by loss of a proton from the intermediate carbocation rather than by substitution. In fact, the E1 and S_N1 reactions normally occur in competition whenever an alkyl halide is treated in a protic solvent with a nonbasic nucleophile. Thus, the best E1 substrates are also the best S_N1 substrates, and mixtures of products are usually obtained. For example, when 2-chloro-2-methylpropane is warmed to 65°C in 80% aqueous ethanol, a 64:36 mixture of 2-methyl-2-propanol (S_N1) and 2-methylpropene (E1) results:

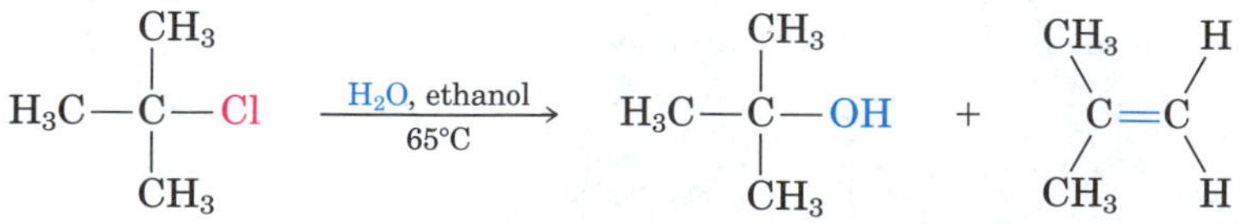

2-Chloro-2-methylpropane **2-Methyl-2-propanol (64%)** **2-Methylpropene (36%)**

Much evidence has been obtained in support of the E1 mechanism. As expected, E1 reactions show first-order kinetics, consistent with a spontaneous dissociation process:

$$\text{Rate} = k \times [\text{RX}]$$

Another piece of evidence involves the stereochemistry of elimination. Unlike the E2 reaction, where periplanar geometry is required, there is no geometric requirement on the E1 reaction because the halide and the hydrogen are lost in separate steps. We might therefore expect to obtain the more stable (Zaitsev's rule) product from E1 reaction, which is just what we find. To return to a familiar example, menthyl chloride loses HCl under E1 conditions in a polar solvent to give a mixture of alkenes in which the Zaitsev product, 3-menthene, predominates (Figure 11.22).

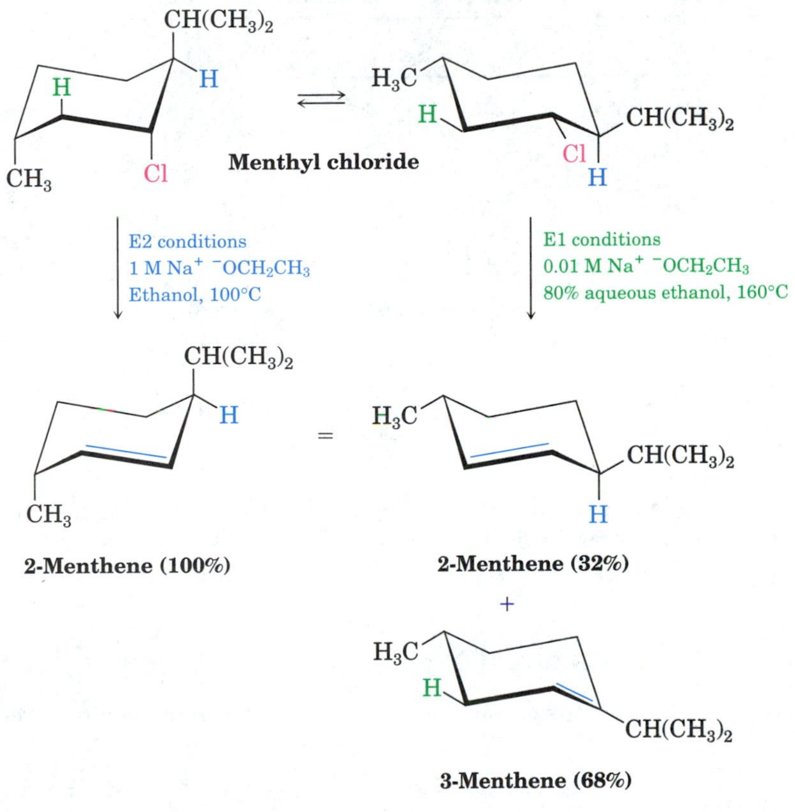

Figure 11.22 Elimination reactions of menthyl chloride. E2 conditions (strong base in pure ethanol) lead to 2-menthene, whereas E1 conditions (very dilute base in aqueous ethanol) lead to a mixture of 2-menthene and 3-menthene.

A final piece of evidence about E1 reactions is that they show no deuterium isotope effect. Because rupture of the C–H (or C–D) bond occurs *after* the rate-limiting step rather than during it, we can't measure a rate difference between a deuterated and nondeuterated substrate.

11.15 Summary of Reactivity: S_N1, S_N2, E1, E2

S_N1, S_N2, E1, E2: How can you keep it all straight? How can you predict what will happen in any given case? Will substitution or elimination occur? Will the reaction be bimolecular or unimolecular? There are no rigid answers to these questions, but it's possible to recognize some trends and make some generalizations (Table 11.4).

Table 11.4 A Summary of Substitution and Elimination Reactions

Halide type	S_N1	S_N2	E1	E2
RCH_2X (primary)	Does not occur	Highly favored	Does not occur	Occurs when strong bases are used
R_2CHX (secondary)	Can occur with benzylic and allylic halides	Occurs in competition with E2 reaction	Can occur with benzylic and allylic halides	Favored when strong bases are used
R_3CX (tertiary)	Favored in hydroxylic solvents	Does not occur	Occurs in competition with S_N1 reaction	Favored when bases are used

1. *Primary alkyl halides*: S_N2 substitution results if a good nucleophile such as RS^-, I^-, CN^-, NH_3, or Br^- is used. E2 elimination takes place if a strong, bulky base such as *tert*-butoxide is used.

$$\underset{\textbf{1-Bromobutane}}{CH_3CH_2CH_2CH_2Br} \xrightarrow[\text{THF–HMPA}]{Na^+\ ^-CN} \underset{\textbf{Pentanenitrile (90\%)}}{CH_3CH_2CH_2CH_2CN}$$

$$\underset{\textbf{1-Bromobutane}}{CH_3CH_2CH_2CH_2Br} \xrightarrow{(CH_3)_3CO^-\ K^+} \underset{\textbf{1-Butene (85\%)}}{CH_3CH_2CH{=}CH_2}$$

2. *Secondary alkyl halides*: S_N2 substitution and E2 elimination occur in competition, often leading to a mixture of products. If a weakly basic nucleophile is used in a polar aprotic solvent, S_N2 substitution predominates. If a strong base such as $CH_3CH_2O^-$, OH^-, or NH_2^- is used, E2 elimination predominates. For example, 2-bromopropane undergoes different reactions when treated with ethoxide ion (strong base; E2) and with acetate ion (weak base; S_N2):

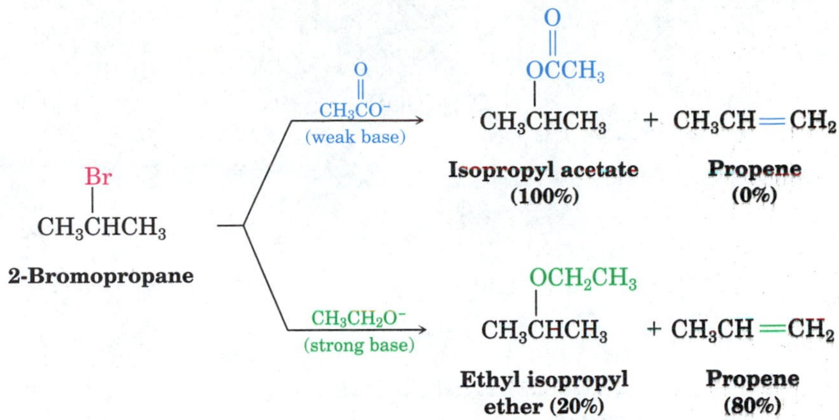

Secondary alkyl halides, particularly allylic and benzylic ones, can also undergo S_N1 and E1 reactions if weakly basic nucleophiles are used in protic solvents such as ethanol or acetic acid.

3. *Tertiary halides*: E2 elimination occurs when a base such as OH^- or RO^- is used. For example, 2-bromo-2-methylpropane gives 97% elimination product when treated with ethoxide ion in ethanol. By contrast, reaction under nonbasic conditions (heating in pure ethanol) leads to a mixture of products resulting from both S_N1 substitution and E1 elimination.

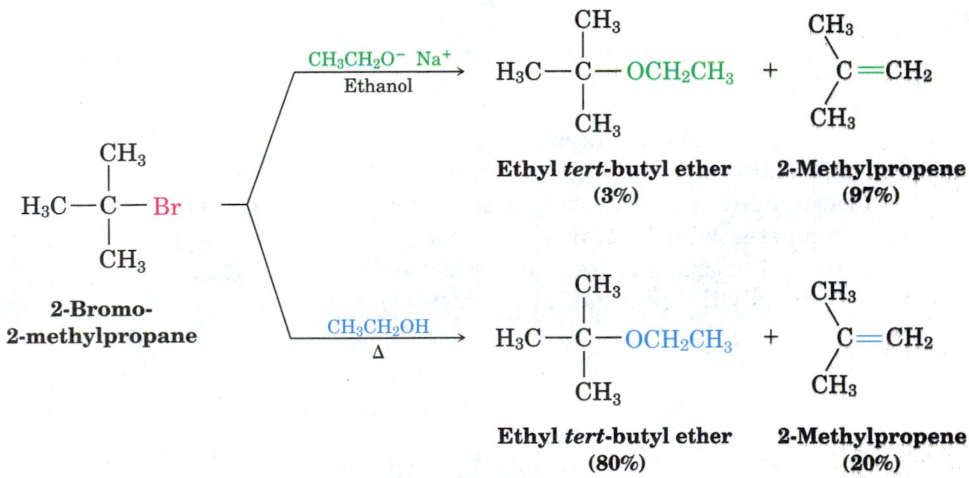

PROBLEM. . . . .

11.17 Tell whether the following reactions are S_N1, S_N2, E1, or E2.

(a) 1-Bromobutane + NaN_3 $\longrightarrow$ 1-Azidobutane

(b) $CH_3CH_2CH(Cl)CH_2CH_3 + KOH \longrightarrow CH_3CH_2CH{=}CHCH_3$

(c) 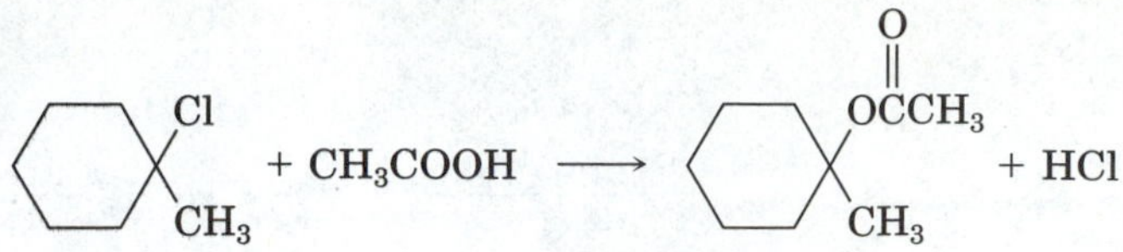

11.16 Substitution Reactions in Synthesis

The reason we've discussed nucleophilic substitution reactions in such detail is that they're so important in organic chemistry. In fact, we've already seen a number of substitution reactions in previous chapters, although they weren't identified as such at the time. An example is the alkylation of acetylide anions discussed in Section 8.9. We said that acetylide anions react well with primary alkyl bromides, iodides, and tosylates, to provide the alkyne product.

$$RC{\equiv}C{:}^-\ Na^+ + R'CH_2X \longrightarrow RC{\equiv}CCH_2R' + NaX$$

where X = Br, I, or OTos

Acetylide ion alkylation is an S_N2 reaction, and it's therefore understandable that only primary alkyl halides and tosylates react well. Since acetylide anion is a strong base as well as a good nucleophile, E2 elimination competes with S_N2 alkylation when a secondary or tertiary substrate is used. For example, reaction of sodio 1-hexyne with 2-bromopropane gives primarily the elimination product propene:

$$\underset{\textbf{Sodio 1-hexyne}}{CH_3(CH_2)_3C{\equiv}C{:}^-\ Na^+} + CH_3\overset{\overset{Br}{|}}{C}HCH_3 \longrightarrow \begin{cases} \underset{\textbf{7\% } S_N2}{CH_3(CH_2)_3C{\equiv}CCH(CH_3)_2} \\ + \\ \underset{\textbf{93\% E2}}{CH_3(CH_2)_3C{\equiv}CH + CH_3CH{=}CH_2} \end{cases}$$

Other substitution reactions we've seen include some of the reactions used for preparing alkyl halides from alcohols. We said in Section 10.7, for example, that alkyl halides can be prepared by treating alcohols with HX—reactions now recognizable as nucleophilic substitutions of halide ions on the protonated alcohols. Tertiary alcohols react by an S_N1 pathway involving unimolecular dissociation of the protonated alcohol to yield a carbocation, whereas primary alcohols react by an S_N2 pathway involving direct bimolecular displacement of water from the protonated alcohol (Figure 11.23).

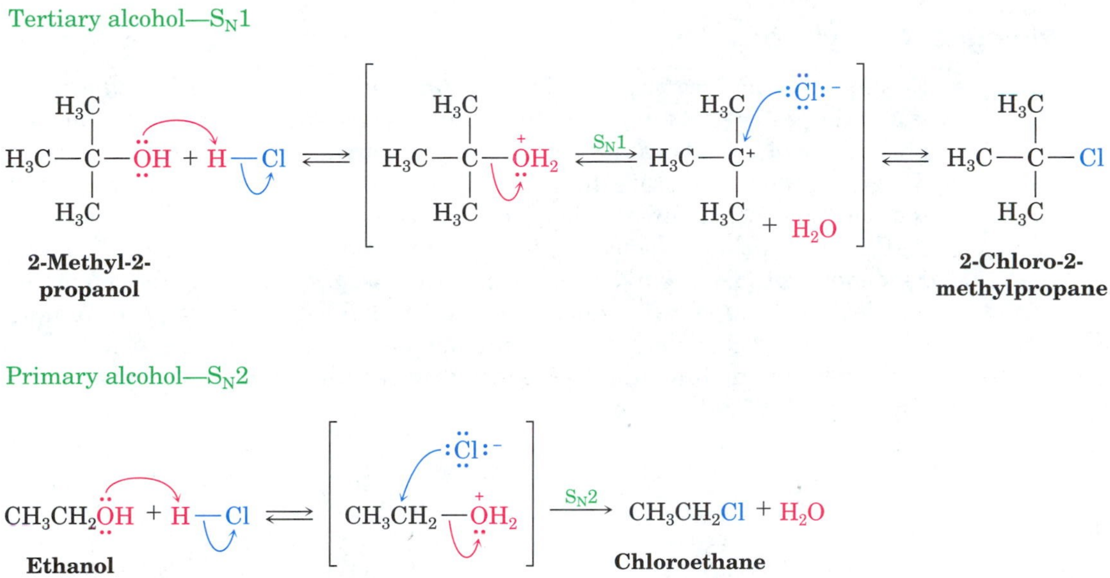

Figure 11.23 Mechanisms of reactions of HCl with alcohols. Tertiary alcohols react by an S_N1 mechanism; primary alcohols react by an S_N2 pathway.

Yet another substitution reaction we've seen is the conversion of primary and secondary alcohols into alkyl bromides by treatment with PBr_3 (Section 10.7). Although OH^- is a poor leaving group and can't be displaced directly by nucleophiles, reaction with PBr_3 transforms the hydroxyl into a better leaving group, thereby activating it for nucleophilic displacement. Alcohols react with PBr_3 to give dibromophosphites ($ROPBr_2$), which are highly reactive substrates in S_N2 reactions. Displacement by bromide ion then occurs rapidly on the primary carbon, and alkyl bromides are produced in good yield.

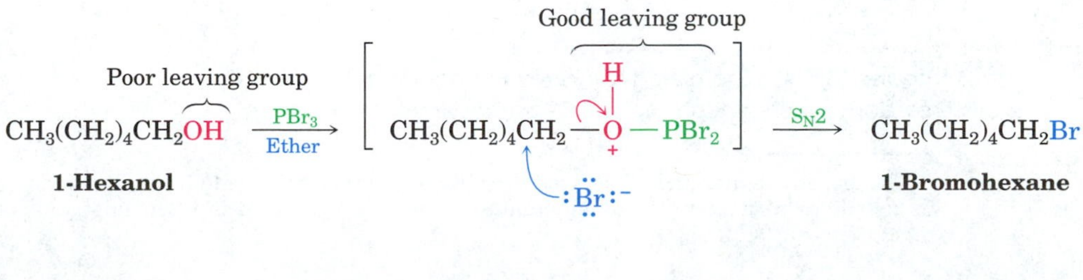

11.17 Substitution Reactions in Biological Systems

Most biological processes occur by reaction pathways analogous to those carried out in the laboratory. Thus, a number of reactions that occur in living organisms take place by nucleophilic substitution mechanisms.[7]

Biological Methylations

Perhaps the most common biological substitution is the *methylation* reaction—the transfer of a methyl group from an electrophilic donor to a nucleophile. Although a laboratory chemist might choose CH_3I for such a reaction, living organisms operate in a more subtle way because of solvent and temperature constraints. The large and complex molecule *S*-adenosylmethionine is the biological methyl group donor. Since the sulfur atom in *S*-adenosylmethionine has a positive charge (a *sulfonium* ion), it is an excellent leaving group for S_N2 displacements on the methyl carbon. An example of the action of *S*-adenosylmethionine in biological methylations takes place in the adrenal medulla during the formation of adrenaline from norepinephrine (Figure 11.24).

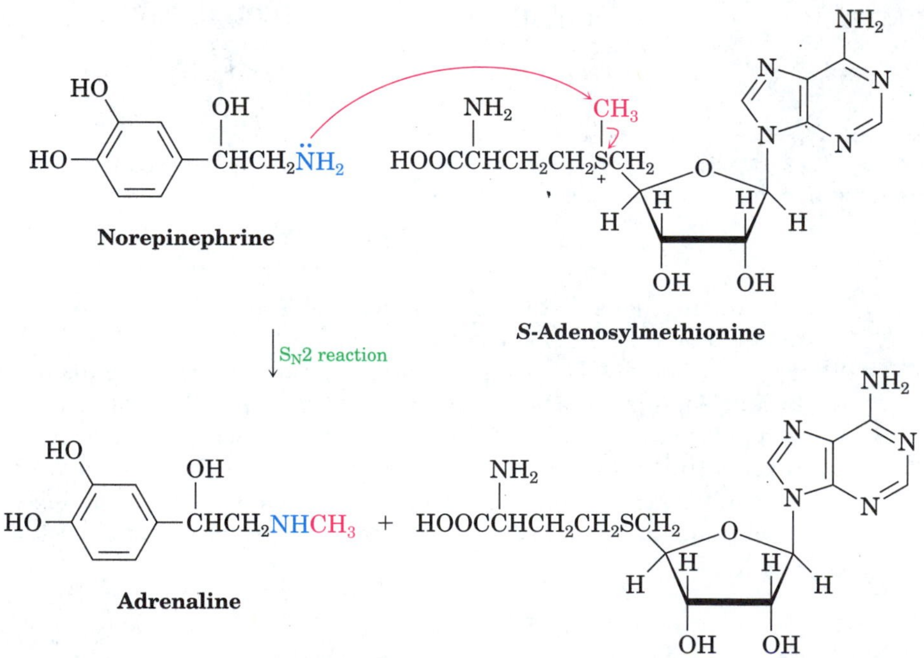

Figure 11.24 The biological formation of adrenaline by S_N2 reaction of norepinephrine with *S*-adenosylmethionine.

[7]The logo in the margin will be used throughout the text to indicate that the reactions being discussed occur commonly in living organisms and are significant in biochemistry.

After dealing only with simple halides such as iodomethane used for laboratory alkylations, it's something of a shock to encounter a molecule as complex as *S*-adenosylmethionine. From a chemical standpoint, however, CH_3I and *S*-adenosylmethionine do exactly the same thing: Both transfer a methyl group by an S_N2 reaction. The same principles of reactivity apply to both.

Other Biological Alkylations

Another example of a biological S_N2 reaction is involved in the response of organisms to certain toxic chemicals. Many reactive S_N2 substrates with deceptively simple structures are quite toxic to living organisms. Bromomethane, for example, was once widely used as a fumigant to kill termites. The toxicity of these chemicals derives from their ability to transfer alkyl groups to nucleophilic amino groups ($-NH_2$) and mercapto groups ($-SH$) on enzymes. With enzymes modified by alkylation, normal biological chemistry is altered.

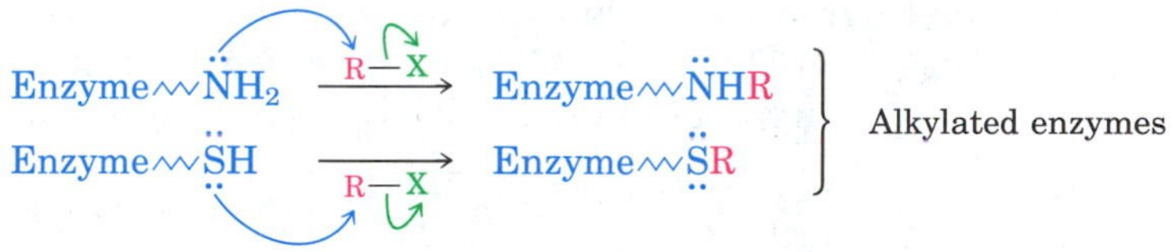

One of the best-known toxic alkylating agents is mustard gas, $ClCH_2CH_2SCH_2CH_2Cl$, which gained notoriety because of its use as a chemical warfare agent during World War I. An estimated 400,000 casualties resulted from its use. Mustard gas, a primary halide, is highly reactive toward S_N2 displacements by nucleophilic groups of proteins. It is thought to act through an intermediate sulfonium ion in much the same manner as *S*-adenosylmethionine.

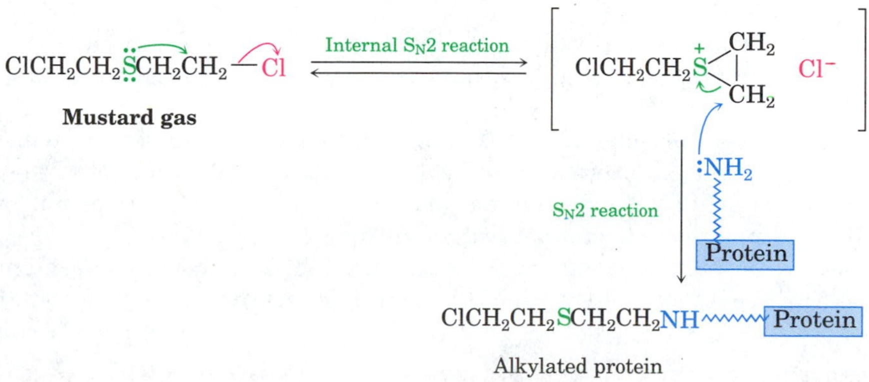

INTERLUDE

Naturally Occurring Organohalogen Compounds

Marine coral secretes organohalogen compounds that act as a feeding deterrent to starfish.

As recently as 1968, only about 30 naturally occurring organohalogen compounds were known. It was simply assumed that chloroform, halogenated phenols, chlorinated aromatic compounds called PCB's, and other such substances noted in the environment were synthetic "pollutants." Now, less than a quarter century later, the situation is quite different. More than 2000 naturally occurring organohalogen compounds have been identified, and many thousands more surely exist. From compounds as simple as chloromethane to those as complex as nostocyclophane D and chartelline A, an extraordinarily diverse range of organohalogen compounds has been found in plants, bacteria, and animals.

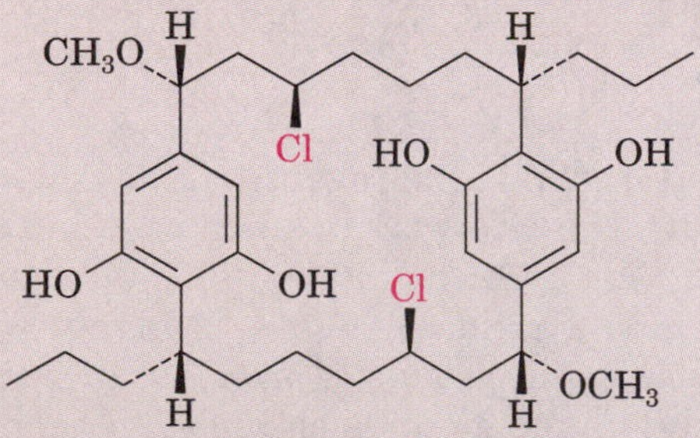

Nostocyclophane D
(from the blue-green alga *Nostoc linckia*)

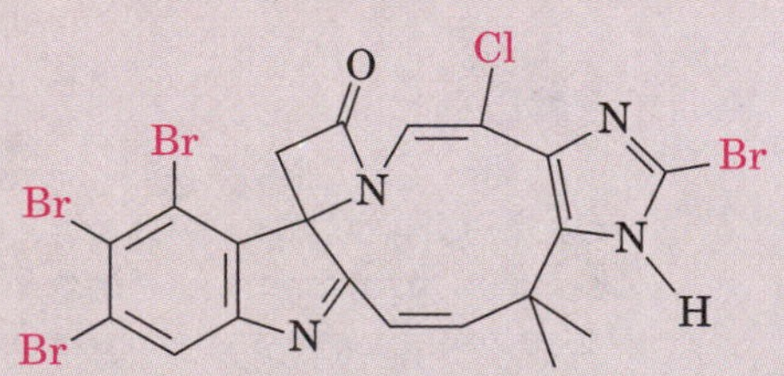

Chartelline A
(from the bryozoan *Chartella papyracea*)

Some naturally occurring organohalogens are produced in massive quantities. Forest fires, volcanoes, and marine kelp release up to 5 *million* tons of CH_3Cl per year, for example, while annual industrial emissions total only about 26,000 tons. A detailed examination of one species of Okinawan acorn worm in a 1 km^2 study area showed that they released nearly 100 pounds per day of halogenated phenols, compounds previously thought to be nonnatural pollutants.

Why do organisms produce organohalogen compounds, many of which are undoubtedly toxic? The answer seems to be that many organisms use

(continued)➤

organohalogen compounds for self-defense, either as feeding deterrents, as irritants to predators, or as natural pesticides. Marine sponges, coral, and sea hares, for example, release foul-tasting organohalogen compounds that deter fish, starfish, and other predators from eating them. More remarkably, even humans appear to produce halogenated compounds as part of their defense against infection. The human immune system contains a peroxidase enzyme capable of carrying out halogenation reactions on fungi and bacteria, thereby killing the pathogen.

Much remains to be learned—only a few hundred of the more than 500,000 known species of marine organisms have been examined—but it is already clear that organohalogen compounds are an integral part of the world around us.

Summary and Key Words

Reaction of an alkyl halide or tosylate with a nucleophile results either in **substitution** or in **elimination**. Nucleophilic substitutions are of two types: **S_N2 reactions** and **S_N1 reactions**. In the S_N2 reaction, the entering nucleophile attacks the halide from a direction 180° away from the leaving group, resulting in an umbrella-like Walden **inversion of configuration** at the carbon atom. The reaction shows **second-order kinetics** and is strongly inhibited by increasing steric bulk of the reagents. Thus, S_N2 reactions are favored for primary and secondary substrates.

The S_N1 reaction occurs when the substrate spontaneously dissociates to a carbocation in a slow **rate-limiting step**, followed by a rapid attack of nucleophile. As a result, S_N1 reactions show **first-order kinetics** and take place with racemization of configuration at the carbon atom. They are most favored for tertiary substrates.

Eliminations of alkyl halides to yield alkenes also occur by two different mechanisms: **E2 reaction** and **E1 reaction**. In the E2 reaction, a base abstracts a proton at the same time the leaving group departs. The reaction takes place preferentially through an **anti periplanar** transition state in which the four reacting atoms—hydrogen, two carbons, and leaving group—are in the same plane. The reaction shows second-order kinetics and a **deuterium isotope effect**, and occurs when a secondary or tertiary substrate is treated with a strong base. These elimination reactions usually give a mixture of alkene products in which the more highly substituted alkene predominates (**Zaitsev's rule**).

The E1 reaction takes place when the substrate spontaneously dissociates to yield a carbocation in the slow rate-limiting step before losing a proton from a neighboring carbon. The reaction shows first-order kinetics and no deuterium isotope effect, and occurs when a tertiary substrate reacts in polar, nonbasic solution.

In general, substrates react in the following way:

RCH_2X (primary)	⟶	Mostly S_N2 substitution
R_2CHX (secondary)	⟶	S_N2 substitution with nonbasic nucleop E2 elimination with strong bases
R_3CX (tertiary)	⟶	Mostly E2 elimination (S_N1 substitution and E1 elimination in nonbasic solvents)

Summary of Reactions

1. Nucleophilic substitutions
   (a) S_N1 reaction; carbocation intermediate is involved (Sections 11.6–11.9)

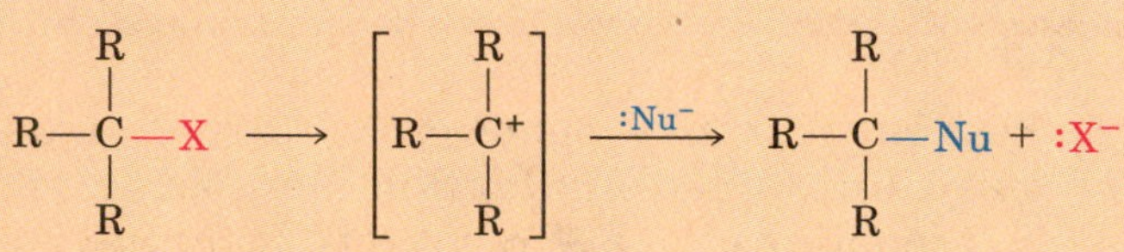

   Best for 3°, allylic, and benzylic halides and tosylates
   (b) S_N2 reaction; back-side attack of nucleophile occurs (Sections 11.4–11.5)

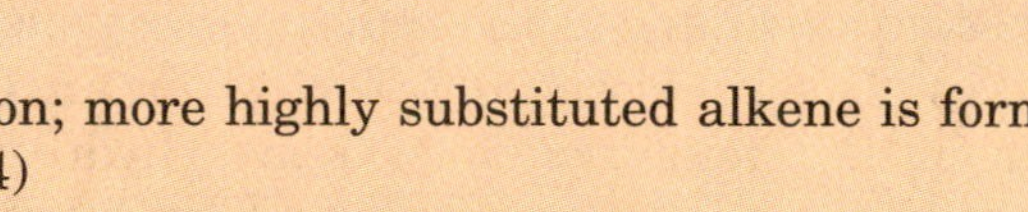

Best for 1° or 2° halides

$Nu:^- = H^-, {}^-CN, I^-, Br^-, Cl^-, {}^-OH, {}^-NH_2, CH_3O^-, CH_3CO_2^-, HS^-, H_2O, NH_3$, and so forth

2. Eliminations
   (a) E1 reaction; more highly substituted alkene is formed (Section 11.14)

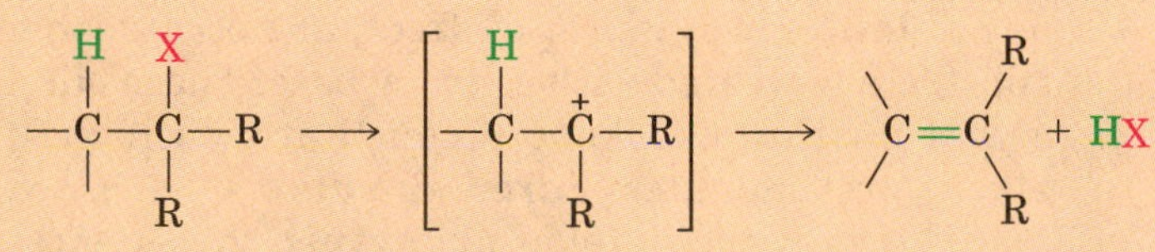

Best for 3° halides

(continued)➤

(b) E2 reaction; anti periplanar geometry is required (Section 11.11)

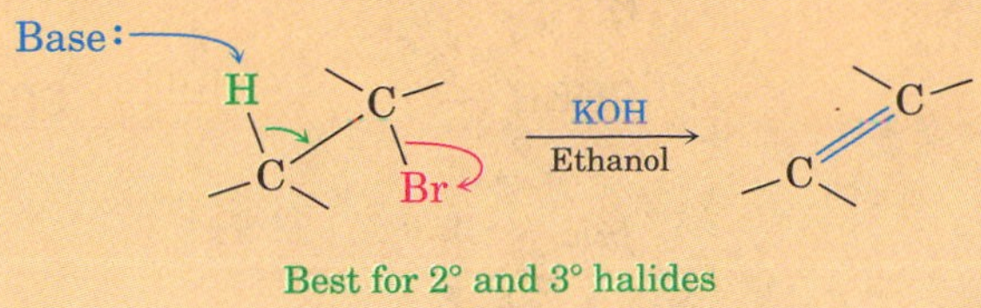

ADDITIONAL PROBLEMS

11.18 Which reagent in each pair will react faster in an S_N2 reaction with hydroxide ion?
(a) CH_3Br or CH_3I
(b) CH_3CH_2I in ethanol or dimethyl sulfoxide
(c) $(CH_3)_3CCl$ or CH_3Cl
(d) $H_2C{=}CHBr$ or $H_2C{=}CHCH_2Br$

11.19 What effect would you expect the following changes to have on the rate of the reaction of 1-iodo-2-methylbutane with cyanide ion?
(a) CN^- concentration is halved and 1-iodo-2-methylbutane concentration is doubled.
(b) Both CN^- and 1-iodo-2-methylbutane concentrations are tripled.

11.20 What effect would you expect on the rate of reaction of ethanol with 2-iodo-2-methylbutane if the concentration of the halide were tripled?

11.21 How might you prepare each of the following molecules using a nucleophilic substitution reaction at some step?
(a) $CH_3C{\equiv}CCH(CH_3)_2$
(b) $CH_3CH_2CH_2CH_2CN$
(c) $H_3C{-}O{-}C(CH_3)_3$
(d) $CH_3CH_2CH_2NH_2$

(e) 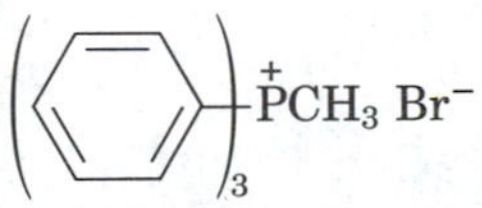

(f) 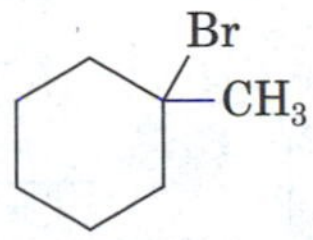

11.22 Which reaction in each of these pairs would you expect to be faster?
(a) The S_N2 displacement by iodide ion on CH_3Cl or on CH_3OTos
(b) The S_N2 displacement by acetate ion on bromoethane or on bromocyclohexane
(c) The S_N2 displacement on 2-bromopropane by ethoxide ion or by cyanide ion
(d) The S_N2 displacement by acetylide ion on bromomethane in benzene or in hexamethylphosphoramide

11.23 What products would you expect from the reaction of 1-bromopropane with each of the following?
(a) $NaNH_2$
(b) $K^+ \ {}^-OC(CH_3)_3$
(c) NaI
(d) NaCN
(e) $Na^+ \ {}^-C{\equiv}CH$
(f) Mg, then H_2O

11.24 Which reagent in each of the following pairs is more nucleophilic? Explain.
(a) ${}^-NH_2$ or NH_3
(b) H_2O or CH_3COO^-
(c) BF_3 or F^-
(d) $(CH_3)_3P$ or $(CH_3)_3N$
(e) I^- or Cl^-
(f) ${}^-C{\equiv}N$ or ${}^-OCH_3$

11.25 Among the Walden cycles carried out by Phillips and Kenyon is the following series of reactions reported in 1923. Explain the results, and indicate where Walden inversion is occurring.

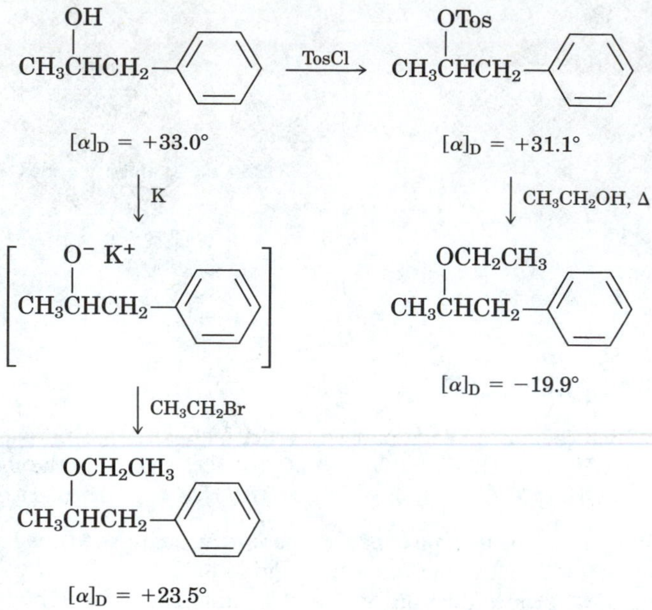

11.26 The synthetic sequences shown here are unlikely to occur as written. What is wrong with each?

(a)

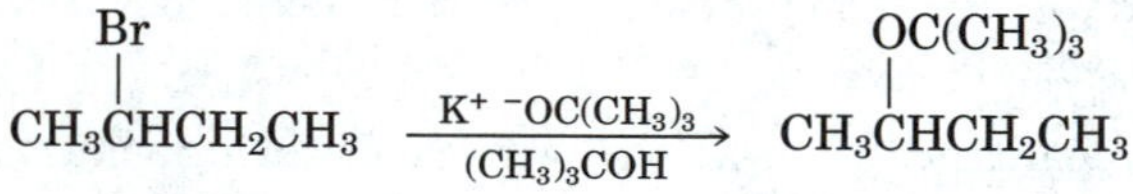

(b)

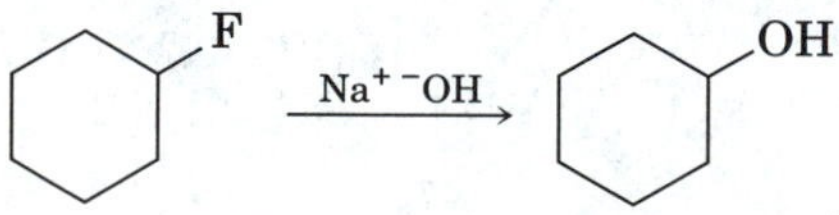

(c)

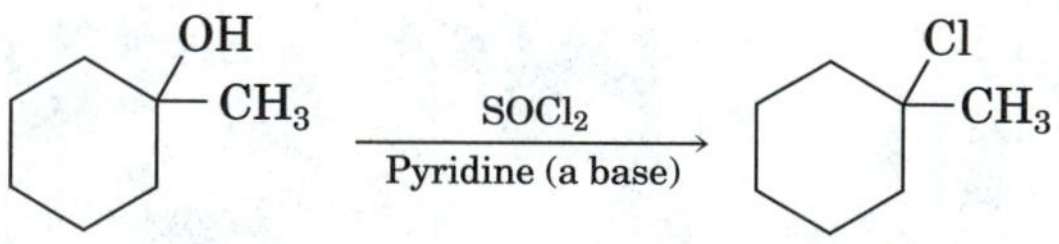

11.27 Order each set of compounds with respect to S_N1 reactivity.

(a)

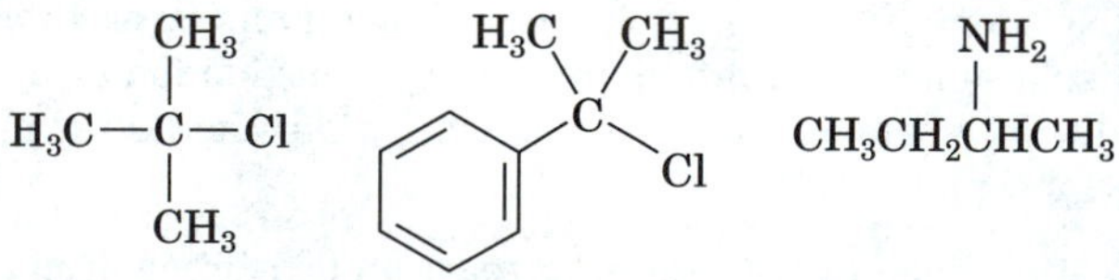

(b) $(CH_3)_3CF$ $(CH_3)_3CBr$ $(CH_3)_3COH$

(c)

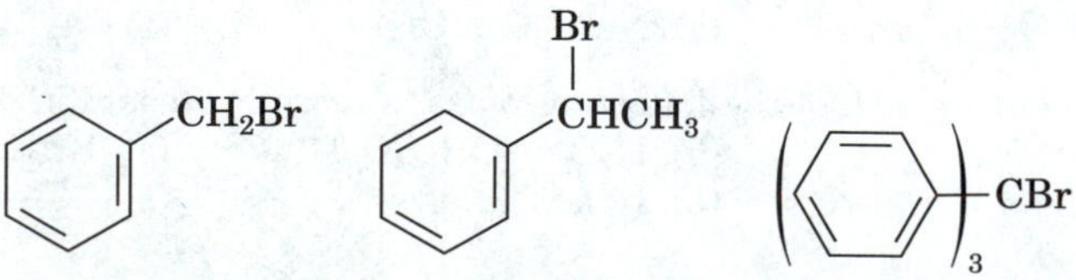

11.28 Order each set of compounds with respect to S_N2 reactivity.

(a) 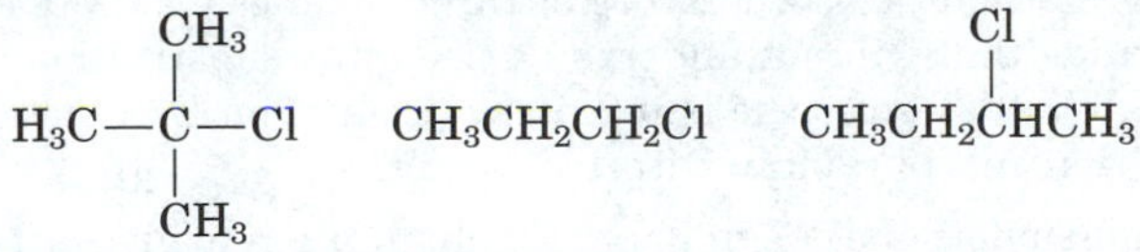

(b) 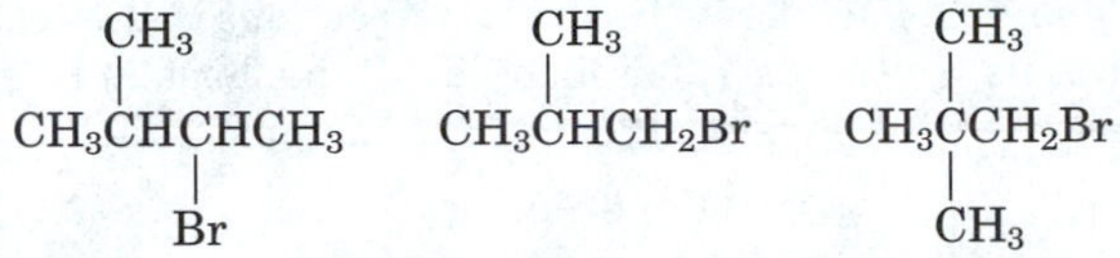

(c) $CH_3CH_2CH_2OCH_3$ $CH_3CH_2CH_2OTos$ $CH_3CH_2CH_2Br$

11.29 Predict the product and give the stereochemistry resulting from reaction of each of the following nucleophiles with (*R*)-2-bromooctane.
(a) ^-CN (b) $CH_3CO_2^-$ (c) CH_3S^- (d) Br^-

11.30 Describe the effects of each of the following variables on S_N2 and S_N1 reactions.
(a) Solvent (b) Leaving group
(c) Attacking nucleophile (d) Substrate structure

11.31 Ethers can often be prepared by S_N2 reaction of alkoxide ions with alkyl halides. Suppose you wanted to prepare cyclohexyl methyl ether. Which of the two possible routes shown here would you choose? Explain.

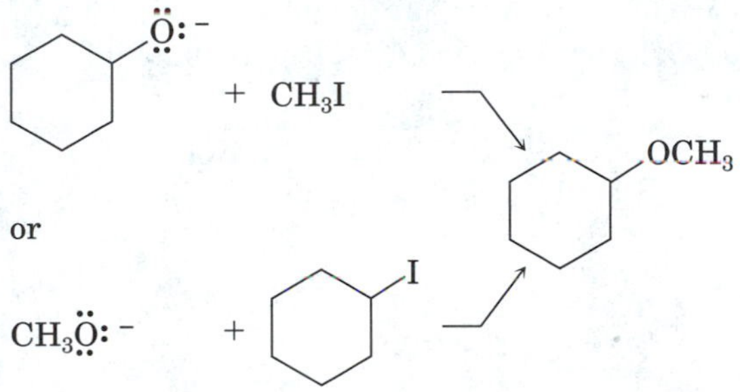

11.32 The S_N2 reaction can occur *intramolecularly* (within the same molecule). What product would you expect from treatment of 4-bromo-1-butanol with base?

$$BrCH_2CH_2CH_2CH_2OH \xrightarrow{Na^+ \ ^-OCH_3} CH_3OH + [BrCH_2CH_2CH_2CH_2O^- \ Na^+] \longrightarrow \ ?$$

11.33 In light of your answer to Problem 11.32, propose a synthesis of 1,4-dioxane starting only with 1,2-dibromoethane.

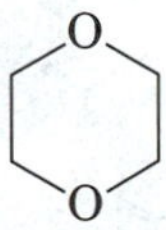

1,4-Dioxane

11.34 How can you explain the fact that *trans*-1-bromo-2-methylcyclohexane yields the non-Zaitsev elimination product 3-methylcyclohexene on treatment with base?

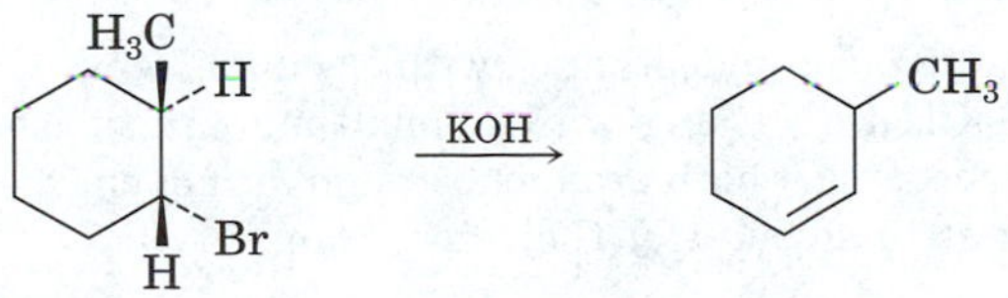

***trans*-1-Bromo-2-methylcyclohexane** **3-Methylcyclohexene**

11.35 Propose structures for compounds that fit the following descriptions:
(a) An alkyl halide that gives a mixture of three alkenes on E2 reaction
(b) An alkyl halide that will not undergo nucleophilic substitution
(c) An alkyl halide that gives the non-Zaitsev product on E2 reaction
(d) An alcohol that reacts rapidly with HCl at 0°C

11.36 1-Chloro-1,2-diphenylethane can undergo E2 elimination to give either *cis*- or *trans*-1,2-diphenylethylene (stilbene). Draw Newman projections of the reactive conformations leading to both *cis*- and *trans*-1,2-diphenylethylene. Look at both conformations and suggest a reason why the trans alkene is the major product.

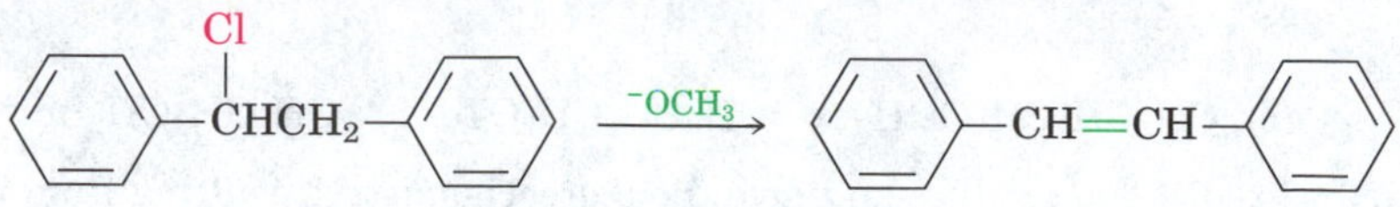

1-Chloro-1,2-diphenylethane **_trans_-1,2-Diphenylethylene**

11.37 Predict the major alkene product of the following E1 reaction:

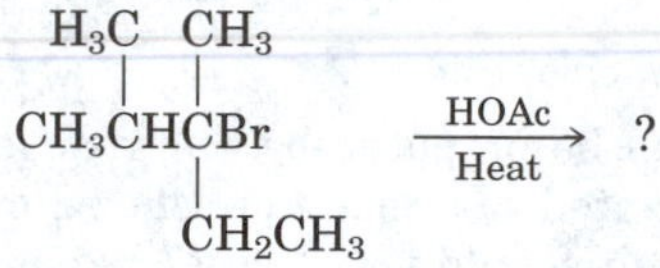

11.38 The tosylate of (2*R*,3*S*)-3-phenyl-2-butanol undergoes E2 elimination on treatment with ethoxide ion to yield (*Z*)-2-phenyl-2-butene:

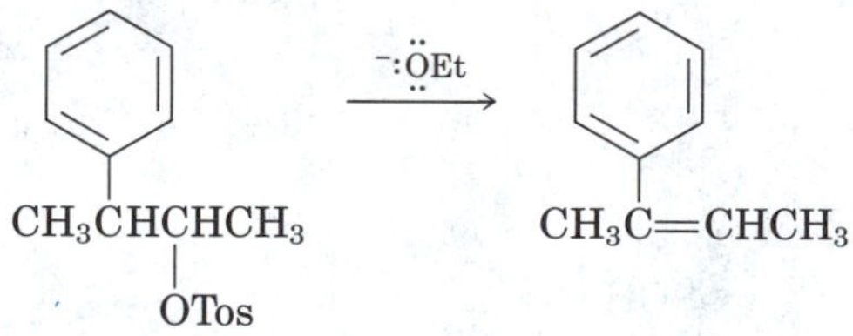

Explain the observed result using Newman projections.

11.39 In light of your answer to Problem 11.38, which alkene, *E* or *Z*, would you expect from an elimination reaction on the tosylate of (2*R*,3*R*)-3-phenyl-2-butanol? Which alkene would result from E2 reaction on the (2*S*,3*R*) and (2*S*,3*S*) tosylates? Explain.

11.40 Predict the product(s) of the following reaction, indicating stereochemistry where necessary.

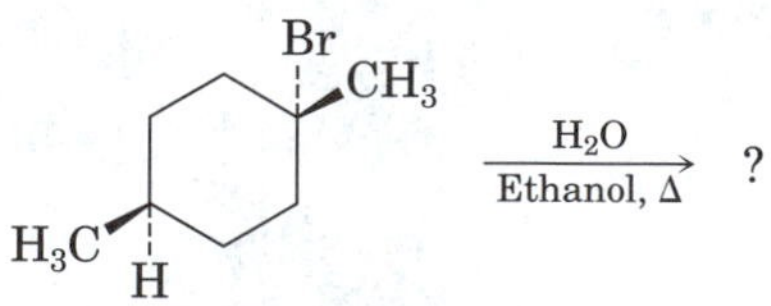

11.41 Draw all isomers of C_4H_9Br, name them, and arrange them in order of decreasing reactivity in the S_N2 reaction.

11.42 Reaction of iodoethane with cyanide ion, CN^-, yields primarily the nitrile $CH_3CH_2C{\equiv}N$ along with a small amount of isonitrile, $CH_3CH_2N{\equiv}C$. Write Lewis structures for both products, assign formal charges as necessary, and propose mechanisms to account for their formation.

11.43 Alkynes can be made by dehydrohalogenation of vinylic halides in a reaction that is essentially an E2 process. In studying the stereochemistry of this elimination, it was found that (*Z*)-2-chloro-2-butene-1,4-dioic acid reacts 50 times as fast as the corresponding *E* isomer. What conclusion can you draw about the stereochemistry of eliminations in vinylic halides? How does this result compare with eliminations of alkyl halides?

$$HOOC{-}C(H){=}C(Cl){-}COOH \xrightarrow[2.\ H_3O^+]{1.\ Na^+\ {}^-NH_2} HOOC{-}C{\equiv}C{-}COOH$$

11.44 (*S*)-2-butanol slowly racemizes on standing in dilute sulfuric acid. Propose a mechanism to account for this observation.

$CH_3CH_2CH(OH)CH_3$ **2-Butanol**

11.45 Suggest an explanation for the observation that reaction of HBr with (*R*)-3-methyl-3-hexanol leads to (±)-3-bromo-3-methylhexane.

$CH_3CH_2CH_2C(OH)(CH_3)CH_2CH_3$ **3-Methyl-3-hexanol**

11.46 Explain the fact that treatment of 1-bromo-2-deuterio-2-phenylethane with strong base leads to a mixture of deuterated and nondeuterated phenylethylenes in which the deuterated product predominates by approximately 7:1.

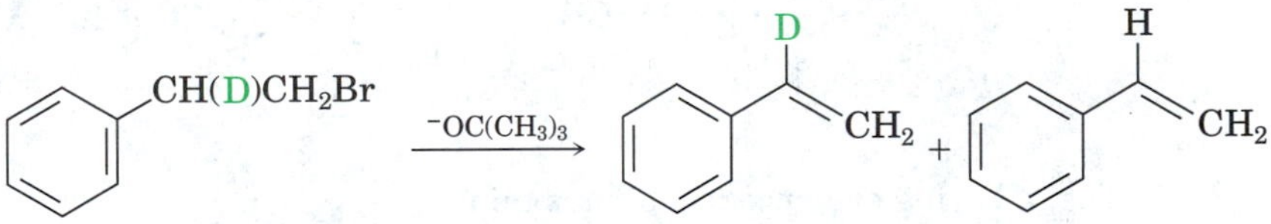

7:1 ratio

11.47 Although anti periplanar geometry is preferred for E2 reactions, it isn't absolutely necessary. The deuterated bromo compound shown here reacts with strong base to yield an undeuterated alkene. Clearly, a syn elimination has occurred. Make a molecular model of the reactant, and explain the result.

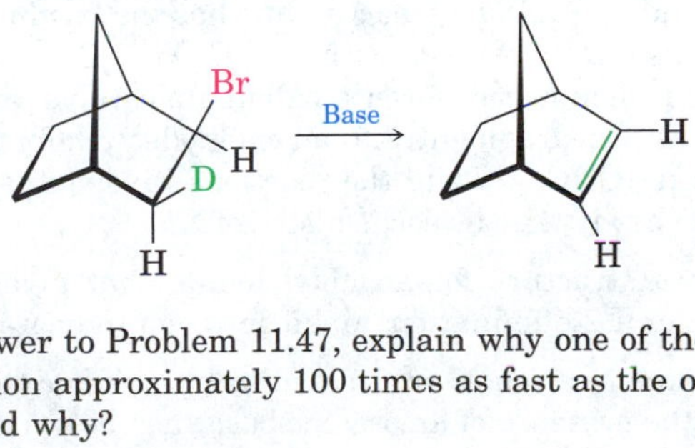

11.48 In light of your answer to Problem 11.47, explain why one of the following isomers undergoes E2 reaction approximately 100 times as fast as the other. Which isomer is more reactive, and why?

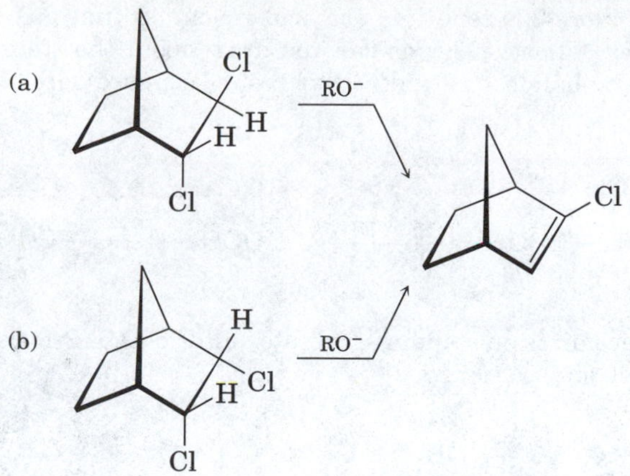

11.49 There are eight diastereomers of 1,2,3,4,5,6-hexachlorocyclohexane. Draw each in its more stable chair conformation. One isomer loses HCl in an E2 reaction nearly 1000 times more slowly than the others. Which isomer reacts so slowly, and why?

11.50 The tertiary amine quinuclidine reacts with CH_3I 50 times as fast as triethylamine, $(CH_3CH_2)_3N$. Explain.

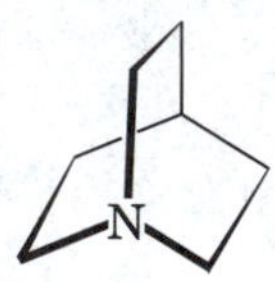

Quinuclidine

11.51 Consider the following methyl ester cleavage reaction:

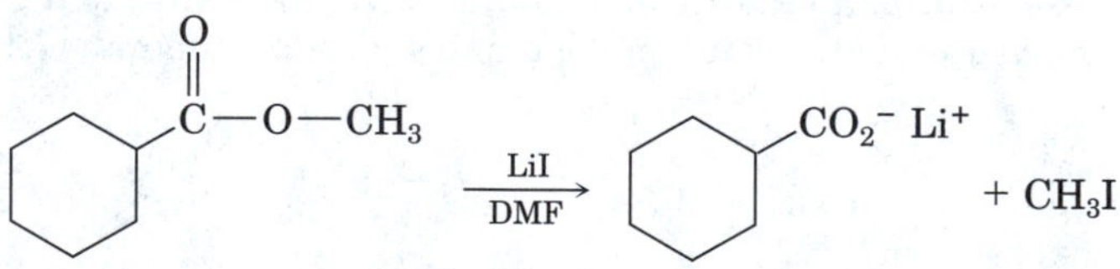

The following evidence has been obtained:
(a) The reaction occurs much faster in DMF than in ethanol.
(b) The corresponding ethyl ester cleaves approximately 10 times more slowly than the methyl ester.
Using this evidence, propose a mechanism for the reaction. What other kinds of experimental evidence could you gather to support your mechanistic hypothesis?

11.52 The reaction of 1-chlorooctane with acetate ion to give octyl acetate is greatly accelerated by the presence of a small quantity of iodide ion. Explain.

11.53 Compound X is optically inactive and has the formula $C_{16}H_{16}Br_2$. On treatment with strong base, X gives hydrocarbon Y, $C_{16}H_{14}$. Compound Y absorbs 2 equivalents of hydrogen when reduced over a palladium catalyst and reacts with ozone to give two fragments. One fragment, Z, is an aldehyde with formula C_7H_6O. The other fragment is glyoxal, $(CHO)_2$. Write the reactions involved, and suggest structures for X, Y, and Z. What is the stereochemistry of X?

11.54 Propose a structure for an alkyl halide that gives only (*E*)-3-methyl-2-phenyl-2-pentene on E2 elimination. Make sure you indicate the stereochemistry.

11.55 When primary alcohols are treated with *p*-toluenesulfonyl chloride at room temperature in the presence of an organic base such as pyridine, a tosylate is formed. When the same reaction is carried out at higher temperature, an alkyl chloride is often formed. Propose a mechanism.

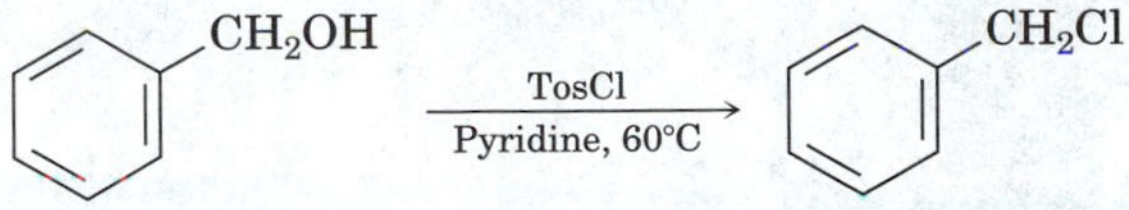

11.56 S_N2 reactions take place with inversion of configuration, and S_N1 reactions take place with racemization. The following substitution reaction, however, occurs with complete *retention* of configuration. Propose a mechanism.

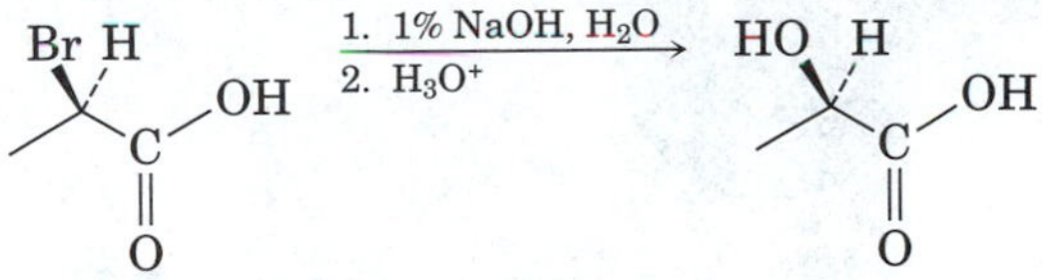

11.57 Propose a mechanism for the following reaction, an important step in the laboratory synthesis of proteins.

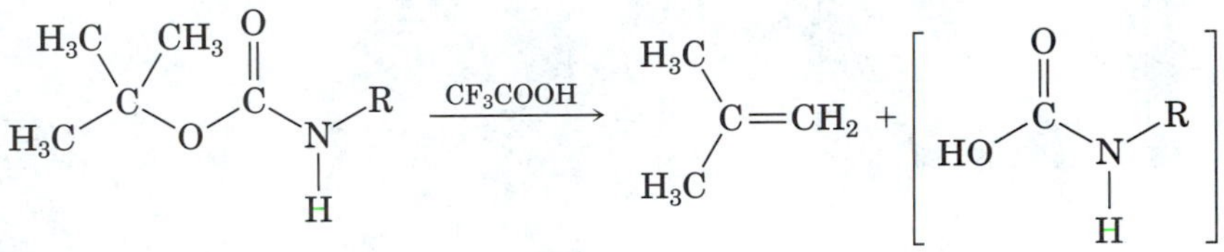

A Look Ahead

11.58 We'll see in Chapter 18 that bromohydrins (Section 7.3) are converted into cyclic ethers called *epoxides* when treated with base. Propose a mechanism.

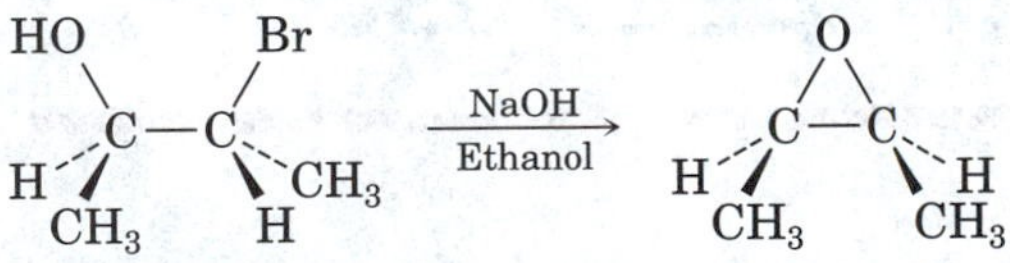

11.59 Show the stereochemistry of the epoxide (Problem 11.58) you would obtain by formation of a bromohydrin from *trans*-2-butene, followed by treatment with base.

11.60 We'll see in Chapter 24 that amines are converted into alkenes by a two-step process called the *Hofmann elimination*. Reaction of the amine with excess CH_3I in the first step yields an intermediate that undergoes E2 reaction when treated with basic silver oxide. Pentylamine, for example, yields 1-pentene.

$$CH_3CH_2CH_2CH_2CH_2NH_2 \xrightarrow[\text{2. } Ag_2O,\ H_2O]{\text{1. Excess } CH_3I} CH_3CH_2CH_2CH{=}CH_2$$

Propose a structure for the intermediate, and explain why it undergoes ready elimination.

The vibrational motions of methane are visible using IR spectroscopy.

12

Structure Determination: Mass Spectrometry and Infrared Spectroscopy

How do we know that Markovnikov's rule is followed in alkene electrophilic addition reactions and that treatment of 1-butene with HCl yields 2-chlorobutane rather than 1-chlorobutane? How do we know that Zaitsev's rule is followed in elimination reactions and that treatment of 2-chlorobutane with NaOH yields 2-butene rather than 1-butene? The answer to these and many thousands of similar questions is that the *structures* of the reaction products have been elucidated.

$$\underset{\textbf{1-Butene}}{CH_3CH_2CH{=}CH_2} + HCl \longrightarrow \underset{\textbf{2-Chlorobutane}}{CH_3CH_2\overset{\overset{Cl}{|}}{C}HCH_3} \quad \text{or} \quad \underset{\textbf{1-Chlorobutane}}{CH_3CH_2CH_2CH_2Cl}$$

$$\underset{\textbf{2-Chlorobutane}}{CH_3CH_2\overset{\overset{Cl}{|}}{C}HCH_3} \xrightarrow{NaOH} \underset{\textbf{2-Butene}}{CH_3CH{=}CHCH_3} \quad \text{or} \quad \underset{\textbf{1-Butene}}{CH_3CH_2CH{=}CH_2}$$

Determining the structure of an organic compound was a difficult and time-consuming process in the nineteenth and early twentieth centuries, but extraordinary advances have been made in the past few decades. Sophisticated techniques are now available that greatly simplify the problem of structure determination. In this and the next two chapters we'll look at four of the most useful such techniques—mass spectrometry (MS), infrared spectroscopy (IR), nuclear magnetic resonance spectroscopy (NMR), and ultraviolet spectroscopy (UV)—and we'll see the kind of information that can be obtained from each.

Mass spectrometry	What size and formula?
Infrared spectroscopy	What functional groups are present?
Ultraviolet spectroscopy	Is a conjugated π electron system present?
Nuclear magnetic resonance spectroscopy	What carbon–hydrogen framework is present?

12.1 Mass Spectrometry

At its simplest, **mass spectrometry** is a technique for measuring the mass, and therefore the molecular weight (MW), of a molecule. In addition, it's often possible to gain information about an unknown structure by measuring the masses of the fragments produced when high-energy molecules fly apart. There are several different kinds of mass spectrometers available, but one of the most common is the electron-ionization, magnetic-sector instrument shown schematically in Figure 12.1.

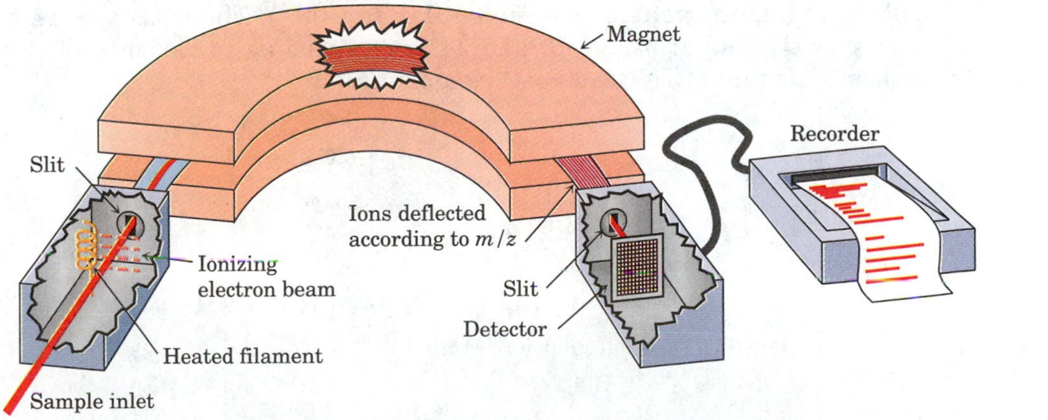

Figure 12.1 Schematic representation of an electron-ionization, magnetic-sector mass spectrometer. Molecules are ionized by collision with high-energy electrons, causing some of the molecules to fragment. Passage of the charged fragments through a magnetic field then sorts them according to their mass.

A small amount of sample is introduced into the mass spectrometer, where it is bombarded by a stream of high-energy electrons. The energy of the electron beam can be varied but is commonly around 70 electron volts (eV), or 6700 kJ/mol (1600 kcal/mol). When a high-energy electron strikes an organic molecule, it dislodges a valence electron from the molecule, producing a *cation radical—cation* because the molecule has lost an electron and now has a positive charge; *radical* because the molecule now has an odd number of electrons.

$$\underset{\text{Organic molecule}}{RH} \xrightarrow{e^-} \underset{\text{Cation radical}}{RH^{+\cdot}} + e^-$$

Electron bombardment transfers such a large amount of energy to the molecules that most of the cation radicals *fragment* after formation: They fly apart into smaller pieces, some of which retain the positive charge, and some of which are neutral. The fragments then flow through a curved pipe in a strong magnetic field, which deflects them according to their mass-to-charge ratio (*m/z*). Neutral fragments are not deflected by the magnetic field and are lost on the walls of the pipe, but positively charged fragments are sorted by the mass spectrometer onto a detector, which records them as peaks at the various *m/z* ratios. Since the number of charges *z* on each ion is usually 1, the value of *m/z* for each ion is simply its mass *m*.

The **mass spectrum** of a compound is usually presented as a bar graph with masses (*m/z* values) on the *x*-axis, and intensity (number of ions of a given *m/z* striking the detector) on the *y*-axis. The tallest peak, called the **base peak**, is arbitrarily assigned an intensity of 100%. Figure 12.2 shows mass spectra of methane and propane.

The mass spectrum of methane is relatively simple because few fragmentations are possible. As Figure 12.2(a) shows, the base peak has m/z = 16, which corresponds to the unfragmented methane cation radical, $CH_4^{+\cdot}$, called the **parent peak** or the **molecular ion (M⁺)**. The mass spectrum also shows peaks at m/z = 15 and 14, corresponding to cleavage of the molecular ion into CH_3^+ and $CH_2^{+\cdot}$ fragments.

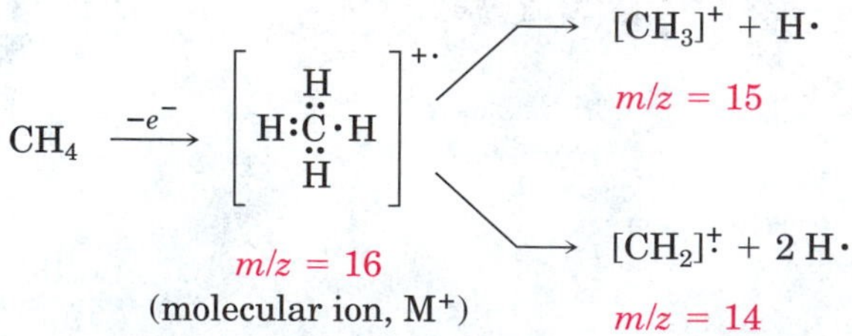

The mass spectral fragmentation patterns of larger molecules are usually complex, and the molecular ion is often not the highest (base) peak. For example, the mass spectrum of propane, shown in Figure 12.2(b), has a molecular ion at m/z = 44 that is only about 30% as high as the base peak at m/z = 29. In addition, many other fragment ions are observed.

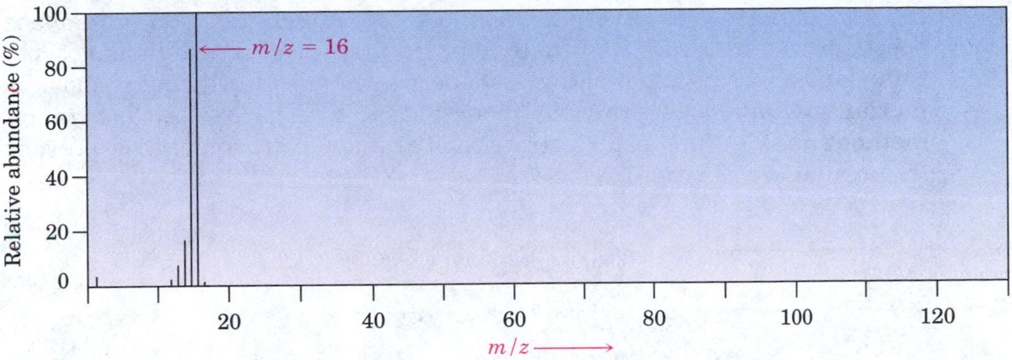

(a)

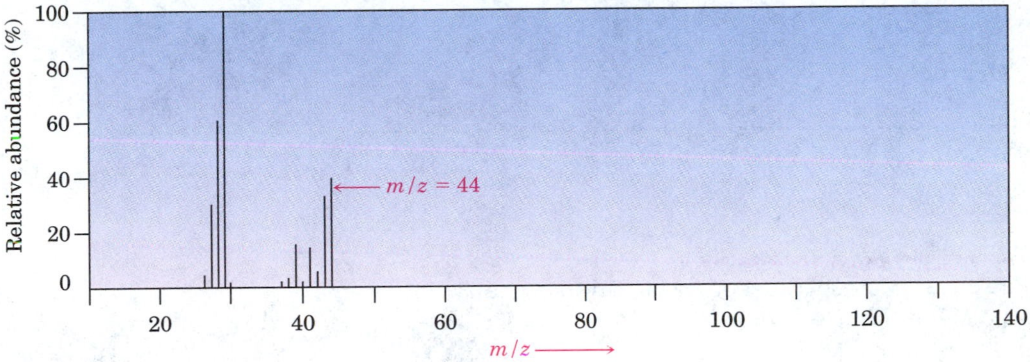

(b)

Figure 12.2 Mass spectra of (a) methane (CH_4; MW = 16) and (b) propane (C_3H_8; MW = 44).

12.2 Interpreting Mass Spectra

What kinds of information can we get from the mass spectrum of a compound? Certainly the most obvious information is the molecular weight, which in itself can be invaluable. For example, if we were given three unlabeled bottles containing hexane (MW = 86), 1-hexene (MW = 84), and 1-hexyne (MW = 82), mass spectrometry would easily distinguish among them.

Some instruments, called *double-focusing* mass spectrometers, are so precise that they provide mass measurements accurate to 0.0001 atomic mass units, making it possible to distinguish between two formulas with the same nominal mass. For example, both C_5H_{12} and C_4H_8O have MW = 72, but they differ slightly beyond the decimal point: C_5H_{12} has an exact mass of 72.0939, whereas C_4H_8O has an exact mass of 72.0575. A high-resolution instrument can easily distinguish between them.

Unfortunately, not every compound shows a molecular ion in its mass spectrum. Although M^+ is usually easy to identify if it's abundant, some compounds, such as 2,2-dimethylpropane, fragment so easily that no molecular ion is observed (Figure 12.3). In such cases, alternative "soft" ionization methods that do not use electron bombardment can sometimes prevent fragmentation.

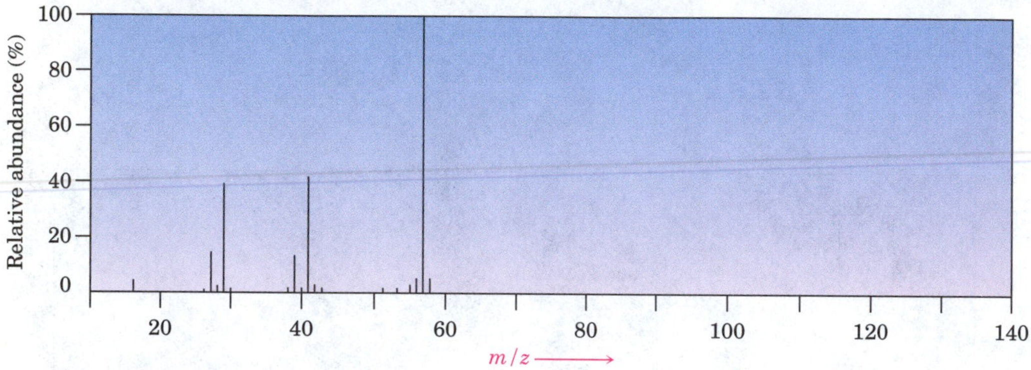

Figure 12.3 Mass spectrum of 2,2-dimethylpropane (C_5H_{12}; MW = 72). No molecular ion is observed.

Knowing the molecular weight makes it possible to narrow greatly the choices of molecular formula. For example, if the mass spectrum of an unknown compound shows a molecular ion at m/z = 110, the molecular formula is likely to be C_8H_{14}, $C_7H_{10}O$, $C_6H_6O_2$, or $C_6H_{10}N_2$. There are always a number of molecular formulas possible for all but the lowest molecular weights, and computer programs can easily generate a list of choices.

A further point about mass spectrometry is noticeable in the mass spectra of methane and propane in Figure 12.2. Perhaps surprisingly, the molecular ions are not the highest mass peaks in the two spectra; there is also a small peak in each spectrum at M+1 because of the presence in the samples of small amounts of isotopically substituted molecules. Although ^{12}C is the most abundant carbon isotope, a small amount (1.10% natural abundance) of ^{13}C is also present. Thus, a certain percentage of the molecules analyzed in the mass spectrometer are likely to contain a ^{13}C atom, giving rise to the observed M+1 peak. In addition, a small amount of ^2H (deuterium; 0.015% natural abundance) is present, making a further contribution to the M+1 peak.

PRACTICE PROBLEM

List the possible formulas of molecules with M^+ = 100. Assume that C, H, and O may be present.

Solution A good approach to this kind of problem is to begin by calculating the possible hydrocarbon formulas. First divide the molecular weight by 12 to find the maximum number of carbons possible. In this example, 100/12 = 8 (remainder 4), giving the formula C_8H_4. Each carbon is equal in mass to 12 hydrogens, so the next step is to replace 1 C by 12 H, giving the second possible formula C_7H_{16}.

Oxygen-containing formulas can be calculated by realizing that one oxygen is equal in mass to CH_4. Starting with C_8H_4 and replacing CH_4 by O gives C_7O as a possible (but unlikely) formula, and doing the same with C_7H_{16} gives $C_6H_{12}O$. Again replacing CH_4 by O gives $C_5H_8O_2$, and repeating the process a third time gives $C_4H_4O_3$. Thus, there are five likely formulas for a substance with MW = 100. A double-focusing instrument could distinguish among the five.

PROBLEM.

12.1 Write as many molecular formulas as you can for compounds that have the following molecular ions in their mass spectra. Assume that all the compounds contain C and H, and that O may or may not be present.
(a) $M^+ = 86$ (b) $M^+ = 128$ (c) $M^+ = 156$

PROBLEM.

12.2 Nootkatone, one of the chemicals responsible for the odor and taste of grapefruit, shows a molecular ion at $m/z = 218$ in its mass spectrum and contains C, H, and O. Suggest several possible molecular formulas for nootkatone.

PROBLEM.

12.3 By knowing the natural abundances of minor isotopes, it's possible to calculate the relative heights of M^+ and M+1 peaks. If ^{13}C has a natural abundance of 1.10%, what are the relative heights of the M^+ and M+1 peaks in the mass spectrum of benzene, C_6H_6?

12.3 Interpreting Mass Spectral Fragmentation Patterns

Mass spectrometry would be useful even if molecular weight and formula were the only information that could be obtained. In fact, though, we can get much more. For example, the mass spectrum of a compound serves as a kind of "molecular fingerprint." Each organic molecule fragments in a unique way depending on its structure, and the likelihood of two compounds having identical mass spectra is small. Thus, it's sometimes possible to identify an unknown by computer-based matching of its mass spectrum to one of the more than 220,000 mass spectra recorded in the computerized data base called the Registry of Mass Spectral Data.

It's also possible to derive structural information about a molecule by looking at its fragmentation pattern. Fragmentation occurs when the high-energy cation radical flies apart by spontaneous cleavage of a chemical bond.

One of the two fragments retains the positive charge (is a carbocation), and the other fragment is a neutral radical.

Not surprisingly, the positive charge often remains with the fragment that is better able to stabilize it. In other words, the more stable carbocation is often formed during fragmentation. For example, propane tends to fragment in such a way that the positive charge remains with the ethyl group rather than with the methyl, because an ethyl carbocation is more stable than a methyl carbocation (Section 6.10). Propane therefore has a base peak at m/z = 29 and a barely detectable peak at m/z = 15 [Figure 12.2(b)].

$$[CH_3CH_2CH_3]^{+\cdot} \longrightarrow CH_3CH_2^{+} + \cdot CH_3$$

M^+ = 44	m/z = 29	Neutral; not observed

Because mass-spectral fragmentation patterns are usually complex, it's often difficult to assign definite structures to fragment ions. Most hydrocarbons fragment in many ways, as the mass spectrum of hexane shown in Figure 12.4 demonstrates. The hexane spectrum shows a moderately abundant molecular ion at m/z = 86 and fragment ions at m/z = 71, 57, 43, and 29. Since all the carbon–carbon bonds of hexane are electronically similar, all break to a similar extent, giving rise to the observed ions.

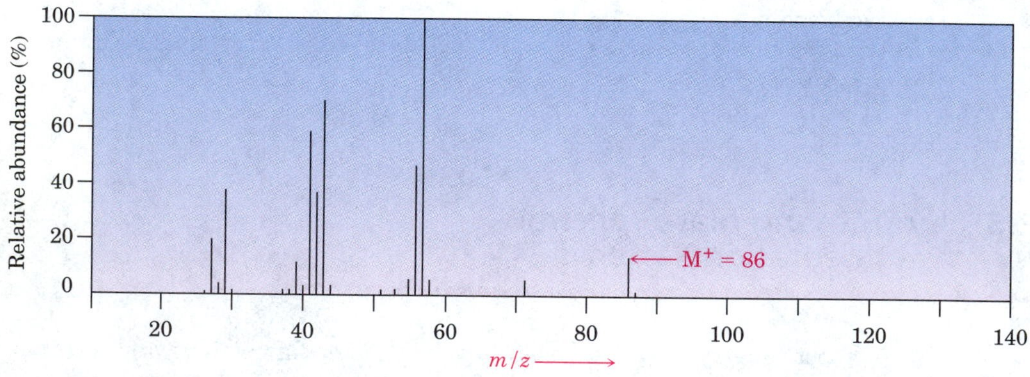

Figure 12.4 Mass spectrum of hexane (C_6H_{14}; MW = 86). The base peak is at m/z = 57, and numerous other ions are present.

Figure 12.5 shows how the hexane fragments might arise. The loss of a methyl radical from the hexane cation radical (M^+ = 86) gives rise to a fragment of mass 71; the loss of an ethyl radical accounts for a fragment of mass 57; the loss of a propyl radical accounts for a fragment of mass 43; and the loss of a butyl radical accounts for a fragment of mass 29. With skill and practice, chemists can learn to analyze the fragmentation patterns of unknown compounds and work backward to a structure that is compatible with the data.

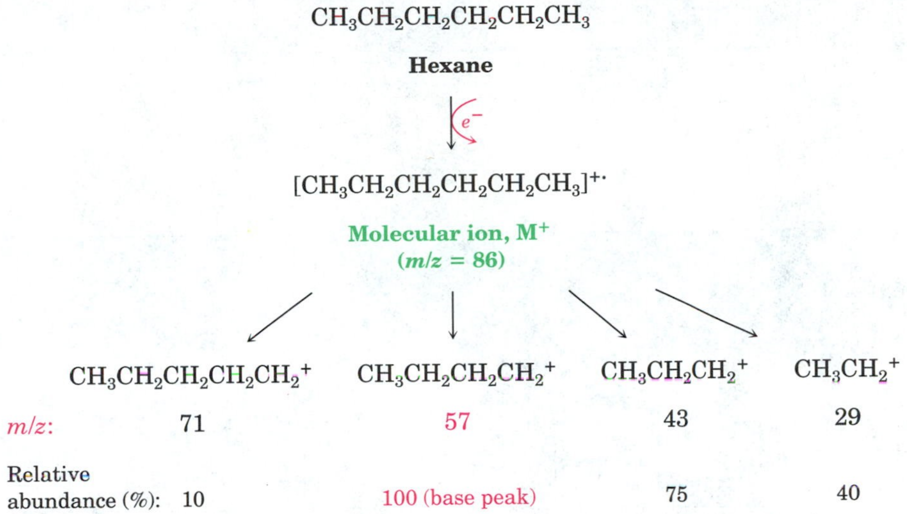

Figure 12.5 Fragmentation of hexane in a mass spectrometer.

An example of how information from fragmentation patterns can be used to solve structural problems is given in Figure 12.6 (p. 432). Assume we have two unlabeled bottles, A and B, one containing methylcyclohexane and the other containing ethylcyclopentane. The mass spectra of both samples show molecular ions at M^+ = 98, corresponding to C_7H_{14}, but the two spectra differ considerably in their fragmentation patterns. Sample B has its base peak at m/z = 83, corresponding to the loss of a CH_3 group (15 mass units) from the molecular ion, but sample A has only a small peak at m/z = 83. Conversely, A has its base peak at m/z = 69, corresponding to the loss of a CH_2CH_3 group (29 mass units), but B has a rather small peak at m/z = 69. We can therefore be reasonably certain that B is methylcyclohexane and A is ethylcyclopentane.

This example is a simple one, but the principles used are broadly applicable for organic structure determination by mass spectrometry. As we'll see in later chapters, specific functional groups such as alcohols, ketones, aldehydes, and amines often show specific kinds of mass spectral fragmentations that can be interpreted to provide structural information.

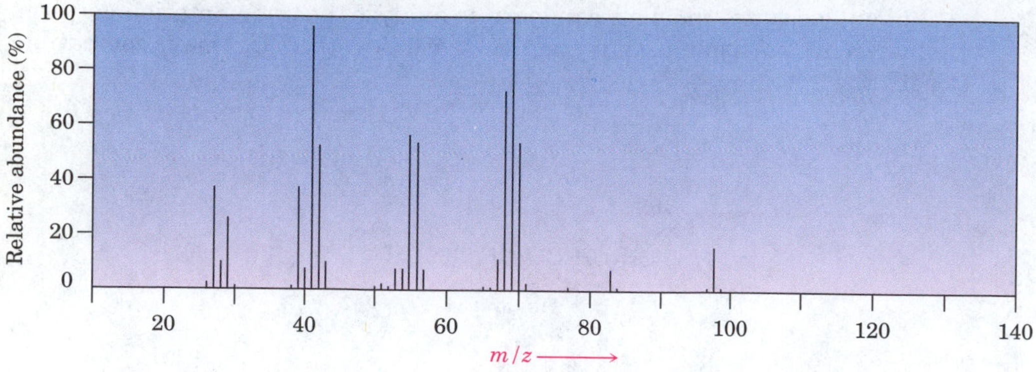

(a)

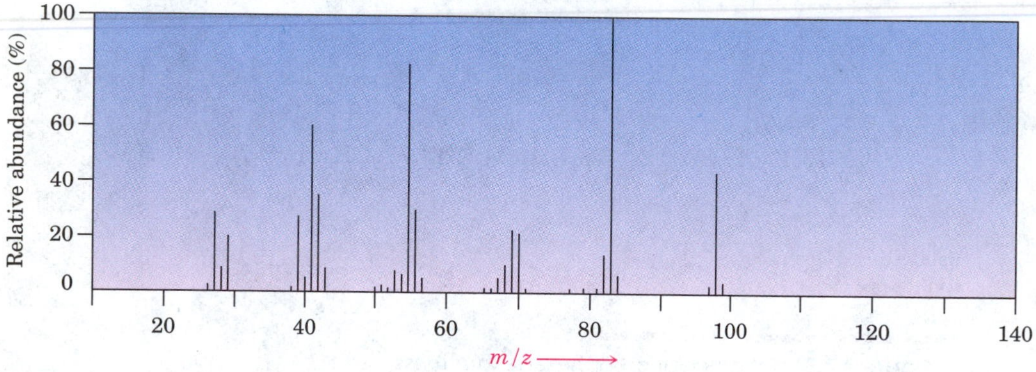

(b)

Figure 12.6 Mass spectra of unknown samples. The top spectrum (a) is that of ethylcyclopentane; the bottom spectrum (b) is that of methylcyclohexane.

PROBLEM

12.4 The mass spectrum of 2,2-dimethylpropane (Figure 12.3) shows a base peak at $m/z = 57$. What molecular formula does this correspond to? Suggest a structure for the $m/z = 57$ fragment ion.

PROBLEM

12.5 Two mass spectra are shown. One spectrum corresponds to 2-methyl-2-pentene; the other, to 2-hexene. Which is which? Explain.

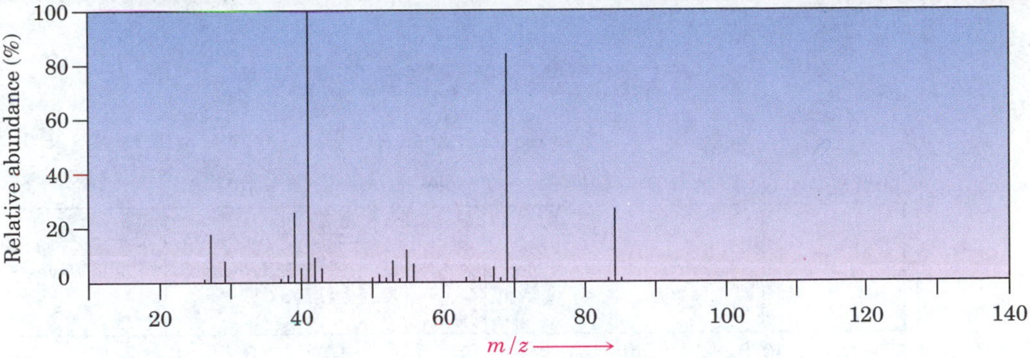

(a)

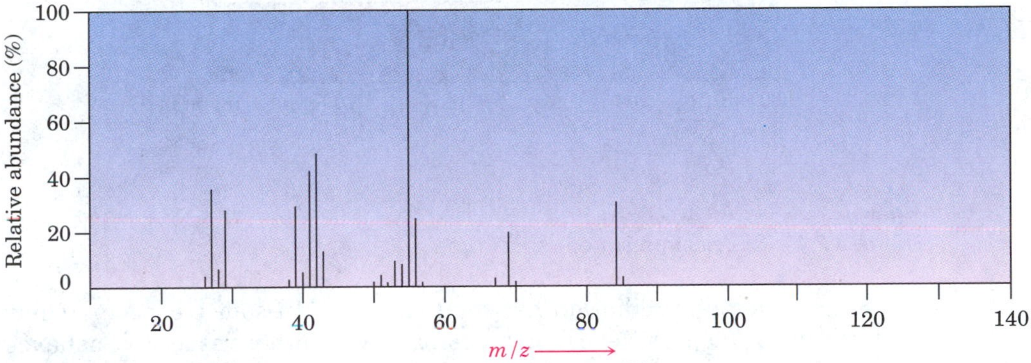

(b)

12.4 Spectroscopy and the Electromagnetic Spectrum

Infrared, ultraviolet, and nuclear magnetic resonance spectroscopies differ from mass spectrometry in that they involve the interaction of molecules with electromagnetic energy rather than with a high-energy electron beam. Before beginning a study of these techniques, we need to look into the nature of radiant energy and the electromagnetic spectrum.

Visible light, X rays, microwaves, radio waves, and so forth are all different kinds of **electromagnetic radiation**. Collectively, they make up the **electromagnetic spectrum**, shown in Figure 12.7 (p. 434). As indicated, the electromagnetic spectrum is arbitrarily divided into various regions, with the familiar visible region accounting for only a small portion of the overall spectrum, from 3.8×10^{-5} cm to 7.8×10^{-5} cm in wavelength. The visible region is flanked by the infrared and ultraviolet regions.

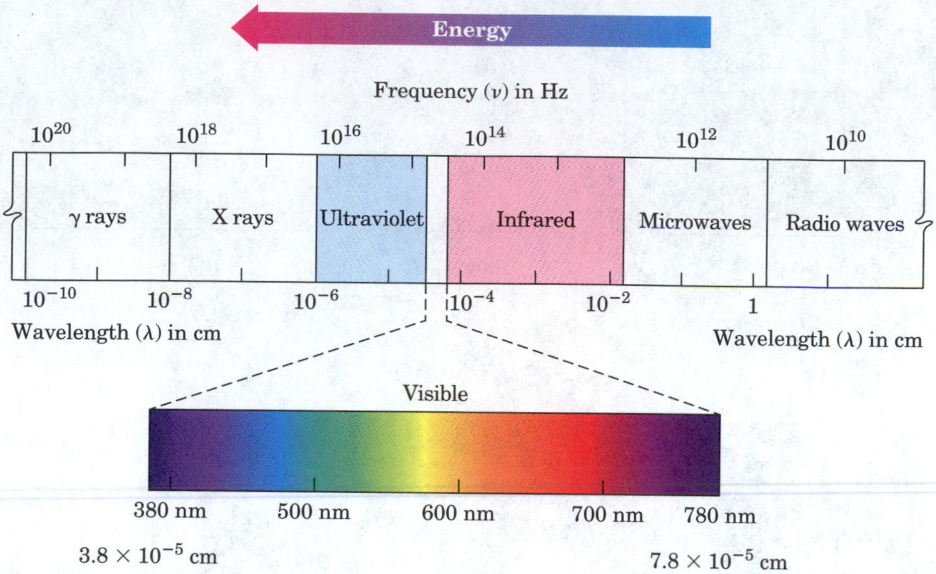

Figure 12.7 The electromagnetic spectrum.

Electromagnetic radiation has dual behavior. In some respects, it has the properties of a particle (called a *photon*), yet in other respects it behaves as an energy wave traveling at the speed of light. Like all waves, electromagnetic radiation is characterized by a *wavelength*, a *frequency*, and an *amplitude* (Figure 12.8). The **wavelength**, λ (Greek lambda), is simply the distance from one wave maximum to the next. The **frequency**, ν (Greek nu), is the number of wave maxima that pass by a fixed point per unit time, usually given in reciprocal seconds (s^{-1}) or **hertz, Hz** (1 Hz = 1 s^{-1}). The **amplitude** is the height of a wave, measured from the midpoint to the maximum. The intensity of radiant energy, whether a feeble beam or a blinding glare, is proportional to the square of the wave's amplitude.

Multiplying the wavelength of a wave in centimeters (cm) by its frequency in reciprocal seconds (s^{-1}) gives the speed of the wave in centimeters per second (cm/s). The rate of travel of all electromagnetic radiation in a vacuum is a constant value, commonly called the "speed of light" and abbreviated *c*. It is one of the most accurately known of all physical constants, with a numerical value of $2.997\,924\,58 \times 10^{10}$ cm/s, usually rounded off to 3.00×10^{10} cm/s.

$$\text{Wavelength} \times \text{Frequency} = \text{Speed}$$
$$\lambda\,(\text{cm}) \times \nu\,(\text{s}^{-1}) = c\,(\text{cm/s})$$

which can be rewritten as:

$$\lambda = \frac{c}{\nu} \qquad \text{or} \qquad \nu = \frac{c}{\lambda}$$

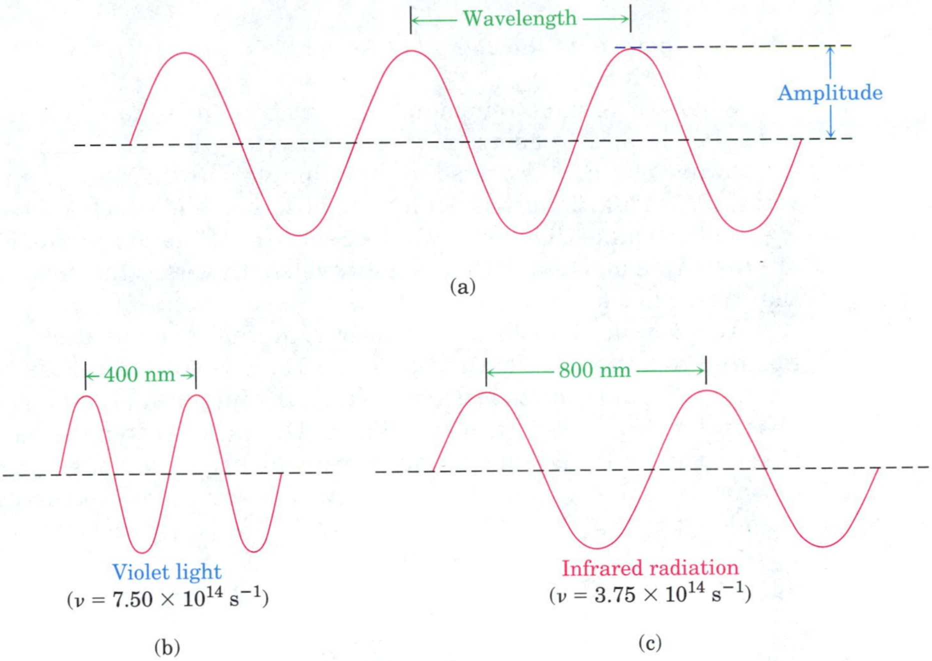

Figure 12.8 (a) Wavelength (λ) is the distance between two successive wave maxima. Amplitude is the height of the wave measured from the center. (b)–(c) What we perceive as different kinds of electromagnetic radiation are simply waves with different wavelengths and frequencies.

Electromagnetic energy is transmitted only in discrete energy bundles, called *quanta*. The amount of energy ε corresponding to 1 quantum of energy (or 1 photon) of a given frequency ν is expressed by the equation

$$\varepsilon = h\nu = \frac{hc}{\lambda}$$

where ε = Energy of 1 photon (1 quantum)
h = Planck's constant (6.62×10^{-34} J·s = 1.58×10^{-34} cal·s)
ν = Frequency (s^{-1})
λ = Wavelength (cm)
c = Speed of light (3.00×10^{10} cm/s)

This equation says that the energy of a given photon varies *directly* with its frequency ν but *inversely* with its wavelength λ. High frequencies and short wavelengths correspond to high-energy radiation such as gamma rays; low frequencies and long wavelengths correspond to low-energy radiation such as radio waves. If we multiply ε by Avogadro's number N_A, we arrive at the same equation expressed in units familiar to organic chemists:

$$E = \frac{N_A hc}{\lambda} = \frac{1.20 \times 10^{-2}\ \text{kJ/mol}}{\lambda\ (\text{cm})} = \frac{2.86 \times 10^{-3}\ \text{kcal/mol}}{\lambda\ (\text{cm})}$$

where E represents the energy of Avogadro's number (a "mole") of photons of wavelength λ.

When an organic compound is struck by a beam of electromagnetic radiation, it absorbs energy of certain wavelengths but transmits energy of other wavelengths. If we irradiate the sample with energy of many different wavelengths and determine which are absorbed and which are transmitted, we can determine the **absorption spectrum** of the compound. The results are displayed on a graph that plots wavelength versus the amount of radiation transmitted.

An example of an absorption spectrum, that of ethyl alcohol exposed to infrared radiation, is shown in Figure 12.9. The horizontal axis records the wavelength, and the vertical axis records the intensity of the various energy absorptions in percent transmittance. The baseline corresponding to 0% absorption (or 100% transmittance) runs along the top of the chart, and a downward spike means that energy absorption has occurred at that wavelength.

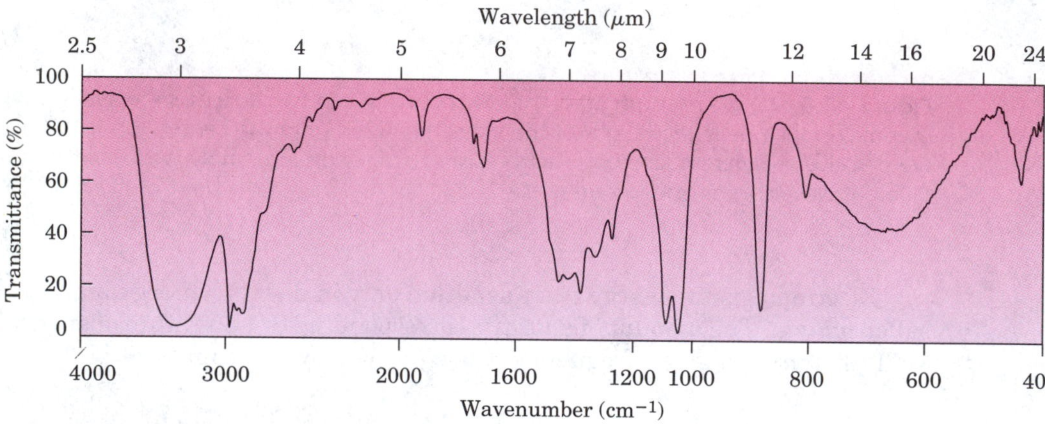

Figure 12.9 An infrared absorption spectrum of ethyl alcohol, CH_3CH_2OH. A transmittance of 100% means that all the energy is passing through the sample, whereas a lower transmittance means that some energy is being absorbed. Thus, each downward spike corresponds to an energy absorption.

The energy that a molecule gains when it absorbs radiation must be distributed over the molecule in some way. For example, absorption of radiation might increase a molecule's kinetic energy by causing bonds to stretch or bend more vigorously. Alternatively, absorption of radiation might cause an electron to jump from a lower-energy orbital to a higher one. Different radiation frequencies affect molecules in different ways, but each can provide structural information if the results are interpreted properly.

There are many kinds of spectroscopies, which differ according to which region of the electromagnetic spectrum is used. We'll look closely at two types—infrared spectroscopy and nuclear magnetic resonance spectroscopy—and have a brief introduction to a third—ultraviolet spectroscopy. Let's begin by seeing what happens when an organic sample absorbs infrared energy.

PRACTICE PROBLEM

Which is higher in energy, FM radio waves with a frequency of 1.015×10^8 Hz (101.5 MHz) or visible green light with a frequency of 5×10^{14} Hz?

Solution The equation $\varepsilon = h\nu$ says that energy increases as frequency increases. Thus, visible light is higher in energy than radio waves.

PROBLEM

12.6 Which has higher energy, infrared radiation with $\lambda = 1.0 \times 10^{-4}$ cm or an X ray with $\lambda = 3.0 \times 10^{-7}$ cm?

PROBLEM

12.7 Which is higher in energy, radiation with $\nu = 4.0 \times 10^9$ Hz or radiation with $\lambda = 9.0 \times 10^{-4}$ cm?

PROBLEM

12.8 It's useful to develop an intuitive feeling for the amounts of energy that correspond to different parts of the electromagnetic spectrum. Using the relationships

$$E = \frac{1.20 \times 10^{-2} \text{ kJ/mol}}{\lambda \text{ (cm)}} \qquad \text{and} \qquad \nu = \frac{c}{\lambda}$$

calculate the energies of each of the following kinds of radiation.
(a) A gamma ray with $\lambda = 5.0 \times 10^{-9}$ cm
(b) An X ray with $\lambda = 3.0 \times 10^{-7}$ cm
(c) Ultraviolet light with $\nu = 6.0 \times 10^{15}$ Hz
(d) Visible light with $\nu = 7.0 \times 10^{14}$ Hz
(e) Infrared radiation with $\lambda = 2.0 \times 10^{-3}$ cm
(f) Microwave radiation with $\nu = 1.0 \times 10^{11}$ Hz

12.5 Infrared Spectroscopy of Organic Molecules

The infrared (IR) region of the electromagnetic spectrum covers the range from just above the visible (7.8×10^{-5} cm) to approximately 10^{-2} cm, but only the midportion from 2.5×10^{-4} cm to 2.5×10^{-3} cm is used by organic chemists (Figure 12.10). Wavelengths within the IR region are usually given

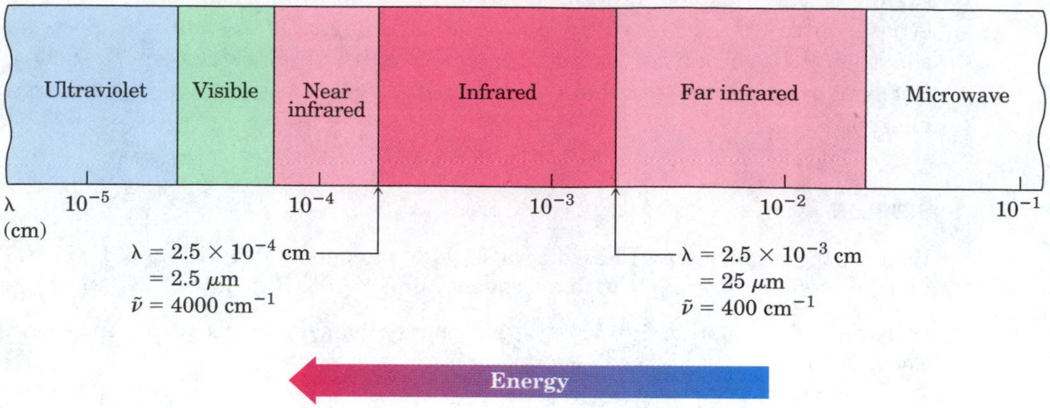

Figure 12.10 The infrared region of the electromagnetic spectrum.

in micrometers (1 μm = 10^{-4} cm), and frequencies are expressed in **wavenumbers** ($\tilde{\nu}$) rather than in hertz. The wavenumber, expressed in units of reciprocal centimeters (cm^{-1}), is simply the reciprocal of the wavelength:

$$\text{Wavenumber, } \tilde{\nu}\ (\text{cm}^{-1}) = \frac{1}{\lambda\ (\text{cm})}$$

Thus, the useful IR region is from 4000 cm^{-1} to 400 cm^{-1}. Using the equation $E = (1.20 \times 10^{-2}\ \text{kJ/mol})/\lambda$, we can calculate that the energy levels of IR radiation range from 48.0 kJ/mol to 4.80 kJ/mol (11.5–1.15 kcal/mol).

Why does an organic molecule absorb some wavelengths of IR radiation but not others? All molecules have a certain amount of energy distributed throughout their structure that causes bonds to stretch and bend, atoms to wag and rock, and other molecular vibrations to occur. Some of the kinds of allowed vibrations are shown:

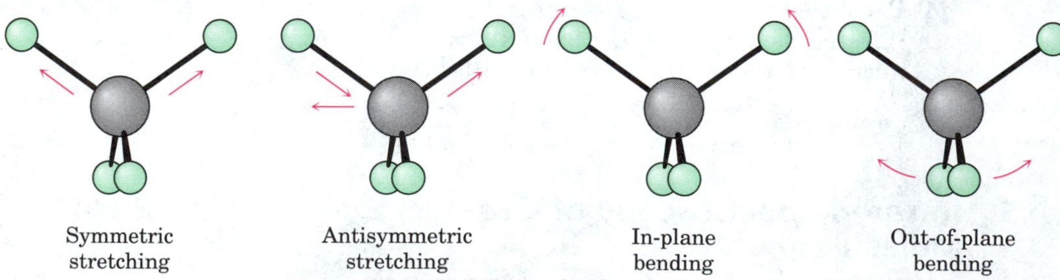

The amount of energy a molecule contains is not continuously variable but is *quantized*. That is, a molecule can stretch, bend, or vibrate only at specific frequencies corresponding to specific energy levels. Take bond stretching, for example. Although we usually speak of bond lengths as if they were fixed, the numbers given are actually averages. In reality, bonds are constantly stretching and bending, lengthening and contracting. Thus, a typical C–H bond with an average bond length of 1.10 Å is actually vibrating at a specific frequency, alternately stretching and compressing as if there were a spring connecting the two atoms. When the molecule is irradiated with electromagnetic radiation, *energy is absorbed when the frequency of the radiation matches the frequency of the vibration.*

When a molecule absorbs IR radiation, the molecular vibration with a frequency matching that of the radiation increases in amplitude. In other words, the "spring" connecting the two atoms stretches and compresses a bit further. Since each frequency absorbed by a molecule corresponds to a specific molecular motion, we can see what kinds of motions a molecule has by measuring its IR spectrum. By then interpreting those motions, we can find out what kinds of bonds (functional groups) are present in the molecule.

IR spectrum ⟶ **What molecular motions?** ⟶ **What functional groups?**

PROBLEM

12.9 Because IR absorptions can be expressed either in micrometers or in wavenumbers, it's useful to be able to interconvert between units. Do the following conversions:

(a) 3.10 μm to cm^{-1} (b) 5.85 μm to cm^{-1}

(c) 2250 cm^{-1} to μm (d) 970 cm^{-1} to μm

12.6 Interpreting Infrared Spectra

The full interpretation of an IR spectrum is difficult because most organic molecules are so large that they have dozens of different bond stretching and bending motions. Thus, an IR spectrum contains dozens of absorption bands. In one sense, this complexity is valuable because an IR spectrum serves as a unique fingerprint of a specific compound. In fact, the complex region of the IR spectrum from 1500 cm^{-1} to around 400 cm^{-1} is called the **fingerprint region**. If two compounds have identical IR spectra, they are almost certainly identical.

Fortunately, we don't need to interpret an IR spectrum fully to get useful structural information. *Most functional groups have characteristic IR absorption bands that don't change from one compound to another.* The C=O absorption of a ketone is almost always in the range 1680–1750 cm^{-1}; the O–H absorption of an alcohol is almost always in the range 3400–3650 cm^{-1}; the C=C absorption of an alkene is almost always in the range 1640–1680 cm^{-1}. By learning where characteristic functional-group absorptions occur, it's possible to get structural information from IR spectra. Table 12.1 lists the characteristic IR bands of some common functional groups.

Table 12.1 Characteristic IR Absorptions of Some Functional Groups

Functional group class	*Band position (cm^{-1})*	*Intensity of absorption*
Alkanes, alkyl groups		
C—H	2850–2960	Medium to strong
Alkenes		
=C—H	3020–3100	Medium
C=C	1640–1680	Medium
Alkynes		
≡C—H	3300	Strong
—C≡C—	2100–2260	Medium
Alkyl halides		
C—Cl	600–800	Strong
C—Br	500–600	Strong
C—I	500	Strong
Alcohols		
O—H	3400–3650	Strong, broad
C—O	1050–1150	Strong
Aromatics		
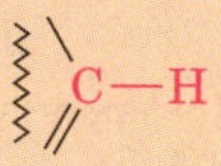	3030	Medium
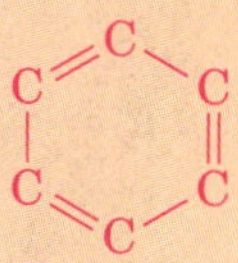	1600, 1500	Strong
Amines		
N—H	3300–3500	Medium
C—N	1030, 1230	Medium
Carbonyl compounds[a]		
C=O	1680–1750	Strong
Carboxylic acids		
O—H	2500–3100	Strong, very broad
Nitriles		
C≡N	2210–2260	Medium
Nitro compounds		
NO_2	1540	Strong

[a]Acids, esters, aldehydes, and ketones.

Look at the IR spectra of hexane, 1-hexene, and 1-hexyne in Figure 12.11 to see an example of how infrared spectroscopy can be used. Although all three IR spectra contain many peaks, there are characteristic absorptions of the C=C and C≡C functional groups that allow the three compounds to be distinguished. Thus, 1-hexene shows a characteristic C=C absorption at 1660 cm^{-1} and a vinylic =C–H absorption at 3100 cm^{-1}, whereas 1-hexyne has a C≡C absorption at 2100 cm^{-1} and a terminal alkyne ≡C–H absorption at 3300 cm^{-1}.

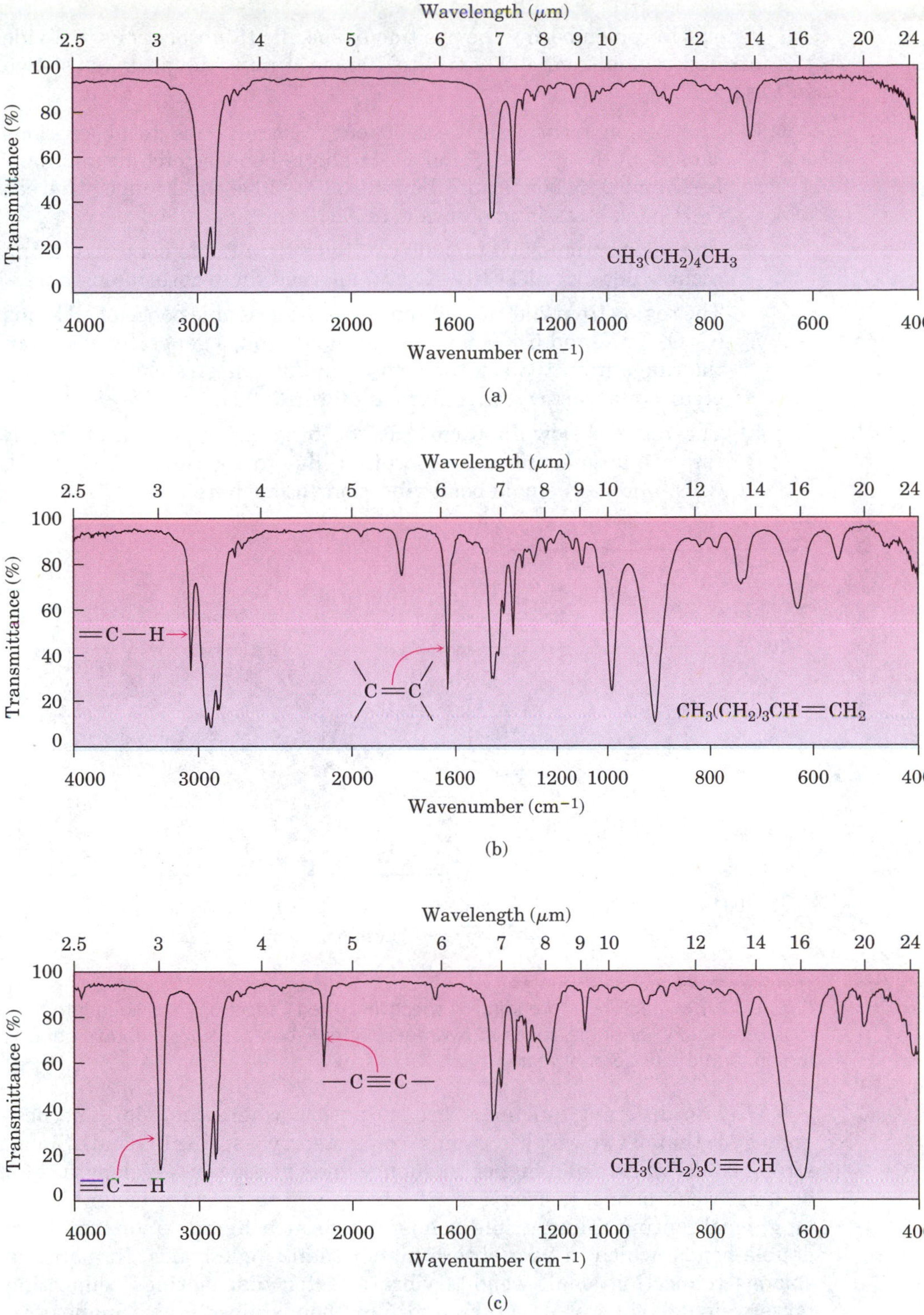

Figure 12.11 Infrared spectra of (a) hexane, (b) 1-hexene, and (c) 1-hexyne. Spectra like these are easily obtained on milligram amounts of material in a few minutes using commercially available instruments.

It helps in remembering the position of specific IR absorptions to divide the infrared region from 4000 cm^{-1} to 400 cm^{-1} into four parts, as shown in Figure 12.12:

1. The region from 4000 to 2500 cm^{-1} corresponds to absorptions caused by N–H, C–H, and O–H single-bond stretching motions. N–H and O–H bonds absorb in the 3300–3600 cm^{-1} range, whereas C–H bond stretching occurs near 3000 cm^{-1}.
2. The region from 2500 to 2000 cm^{-1} is where triple-bond stretching occurs. Both nitriles (RC≡N) and alkynes show absorptions here.
3. The region from 2000 to 1500 cm^{-1} is where double bonds of all kinds (C=O, C=N, and C=C) absorb. Carbonyl groups generally absorb in the range from 1680 to 1750 cm^{-1}, and alkene stretching normally occurs in a narrow range from 1640 to 1680 cm^{-1}.
4. The region below 1500 cm^{-1} is the fingerprint portion of the IR range. A large number of absorptions due to a variety of C–C, C–O, C–N, and C–X single-bond vibrations occur here.

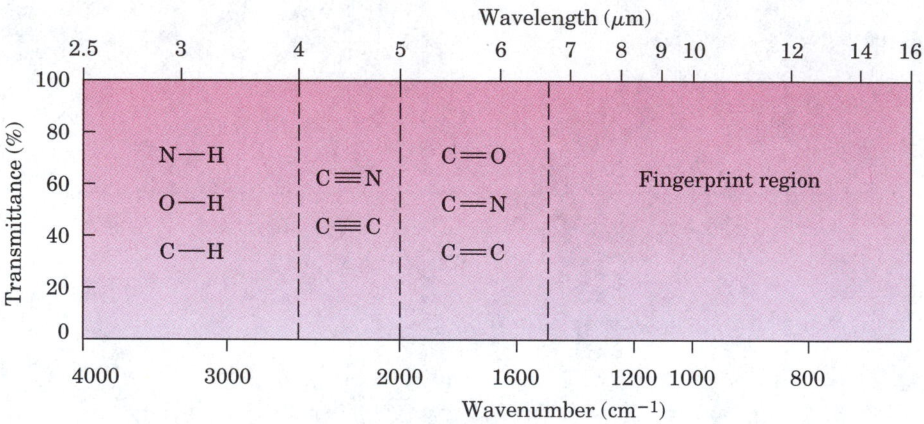

Figure 12.12 Regions in the infrared spectrum. The IR spectrum is divided into four regions: a region of single bonds to hydrogen, a triple-bond region, a double-bond region, and the fingerprint region.

Why do different functional groups absorb where they do? The best analogy is that of two weights (atoms) connected by a spring (a bond). Short, strong bonds vibrate at a higher frequency and with lower wavelength than do long, weak bonds, just as a short, strong spring vibrates faster than a long, weak spring. Thus, triple bonds absorb at a higher frequency than double bonds, which in turn absorb higher than single bonds. In addition, springs connecting small weights vibrate faster than springs connecting large weights. Thus, C–H, O–H, and N–H bonds vibrate at a higher frequency than bonds between heavier C, O, and N atoms.

PROBLEM

12.10 Refer to Table 12.1, and make educated guesses about which functional groups the following molecules might contain.

(a) A compound with a strong absorption at 1710 cm^{-1}
(b) A compound with a strong absorption at 1540 cm^{-1}
(c) A compound with strong absorptions at 1720 cm^{-1} and at 2500–3100 cm^{-1}

PROBLEM

12.11 How might you use IR spectroscopy to distinguish between the following pairs of isomers?
(a) CH_3CH_2OH and CH_3OCH_3
(b) Cyclohexane and 1-hexene
(c) CH_3CH_2COOH and $HOCH_2CH_2CHO$

12.7 Infrared Spectra of Hydrocarbons

Alkanes

The infrared spectrum of an alkane is fairly uninformative because no functional groups are present and all absorptions are due to C–H and C–C bond stretching and bending. Alkane C–H bonds always show a strong absorption from 2850 to 2960 cm^{-1}, and saturated C–C bonds show a number of bands in the 800–1300 cm^{-1} range. Since most organic compounds contain saturated alkane-like portions, most organic compounds have these characteristic IR absorptions. These C–H and C–C bands are clearly visible in the three spectra shown in Figure 12.11.

Alkanes	—C—H (saturated C)	2850–2960 cm^{-1}
	—C—C— (saturated C)	800–1300 cm^{-1}

Alkenes

Alkenes show several characteristic stretching absorptions. Vinylic =C–H bonds absorb from 3020 to 3100 cm^{-1}, and alkene C=C bonds usually absorb near 1650 cm^{-1}, although in some cases the peaks can be rather small and difficult to see clearly. Both absorptions are visible in the 1-hexene spectrum in Figure 12.11(b).

Mono- and disubstituted alkenes have characteristic =C–H out-of-plane bending absorptions in the 700–1000 cm^{-1} range, thereby allowing the substitution pattern on a double bond to be determined. Monosubstituted alkenes such as 1-hexene show strong characteristic bands at 910 and 990 cm^{-1}, and 2,2-disubstituted alkenes ($R_2C{=}CH_2$) have an intense band at 890 cm^{-1}.

Alkenes	=C—H	3020–3100 cm^{-1}
	C=C	1650–1670 cm^{-1}
	$RCH{=}CH_2$	910 and 990 cm^{-1}
	$R_2C{=}CH_2$	890 cm^{-1}

Alkynes

Alkynes show a C≡C stretching absorption at 2100–2260 cm^{-1}, an absorption that is much more intense for terminal alkynes than for internal alkynes. In fact, symmetrically substituted triple bonds like that in 3-hexyne show no absorption at all, for reasons we won't go into. Terminal alkynes such as 1-hexyne also have a characteristic ≡C–H stretch at 3300 cm^{-1}. This band is diagnostic for terminal alkynes because it is fairly intense and quite sharp.

Alkynes	—C≡C—	2100–2260 cm^{-1}
	≡C—H	3300 cm^{-1}

One other important point about IR spectroscopy: It's also possible to get structural information from an IR spectrum by noticing which absorptions are *not* present. If the spectrum of a compound has no absorptions at 3300 and 2150 cm^{-1}, the compound isn't a terminal alkyne; if the spectrum has no absorption near 3400 cm^{-1}, the compound isn't an alcohol; and so on.

PROBLEM

12.12 The infrared spectrum of phenylacetylene is shown. What absorption bands can you identify?

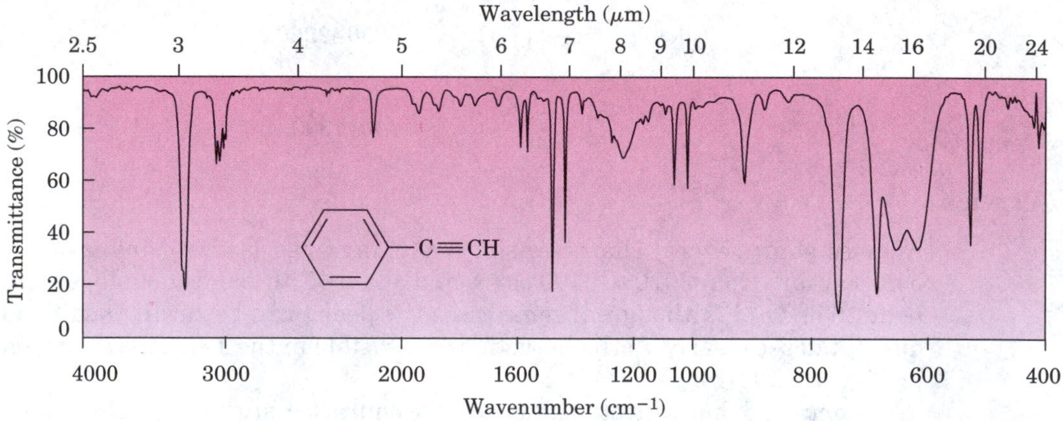

12.8 Infrared Spectra of Other Functional Groups

As each functional group is discussed in future chapters, the spectroscopic behavior of that group will be described. For the present, though, we'll simply point out some distinguishing features of the more important functional groups.

Alcohols

The O–H functional group of alcohols is easy to spot in the IR. Alcohols have a characteristic band in the range 3400–3650 cm^{-1} that is usually fairly broad and intense. If present, it's hard to miss this band or to confuse it with anything else.

Alcohols —O—H 3300–3650 cm^{-1} (broad, intense)

Amines

The N–H functional group of amines is also easy to spot in the IR, with a characteristic absorption in the 3300–3500 cm^{-1} range. Although alcohols absorb in the same range, an N–H absorption is much sharper and less intense than an O–H band.

Amines —N—H 3300–3500 cm^{-1} (sharp, medium intensity)

Carbonyl Compounds

Carbonyl functional groups are the easiest to identify of all IR absorptions because of their sharp, intense peak in the range 1680–1750 cm^{-1}. Most important, the exact position of absorption within the range can often be used to identify the exact kind of carbonyl functional group—aldehyde, ketone, ester, and so forth.

Aldehydes Saturated aldehydes absorb at 1730 cm^{-1}; aldehydes next to either a double bond or an aromatic ring absorb at 1705 cm^{-1}.

Aldehydes

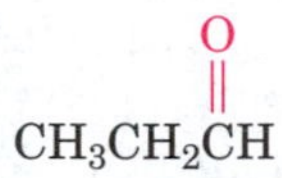

1730 cm^{-1}

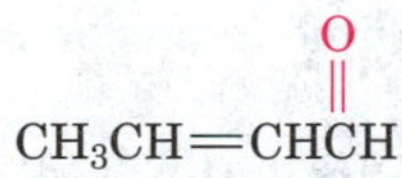

1705 cm^{-1}

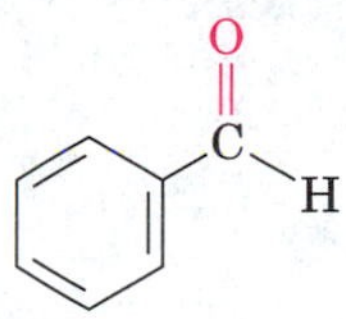

1705 cm^{-1}

Ketones Open-chain ketones and six-membered-ring cyclic ketones absorb at 1715 cm^{-1}, five-membered-ring ketones absorb at 1750 cm^{-1}, and ketones next to a double bond or an aromatic ring absorb at 1690 cm^{-1}.

Ketones

1715 cm^{-1}

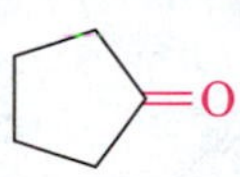

1750 cm^{-1}

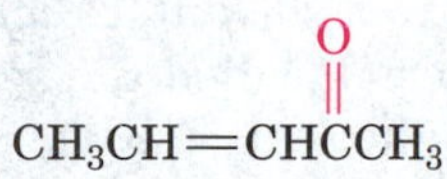

1690 cm^{-1}

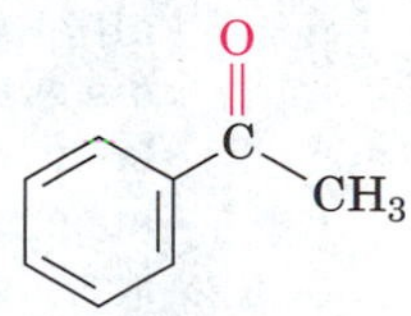

1690 cm^{-1}

Esters Saturated esters absorb at 1735 cm^{-1}; esters next to either an aromatic ring or a double bond absorb at 1715 cm^{-1}.

Esters

$CH_3C(=O)OCH_3$ — 1735 cm^{-1}

$CH_3CH{=}CHC(=O)OCH_3$ — 1715 cm^{-1}

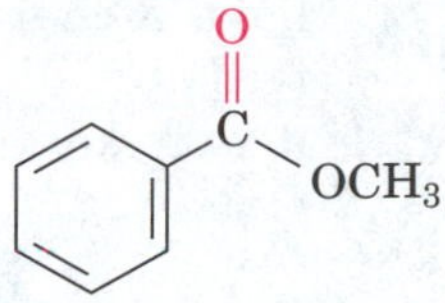

1715 cm^{-1}

PROBLEM

12.13 Where might the following compounds have IR absorptions?

(a) 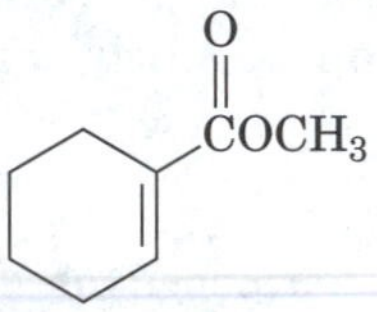

(b) $HC{\equiv}CCH_2CH_2C(=O)H$

INTERLUDE

Chromatography: Purifying Organic Compounds

Every time a new organic substance is isolated from a plant or animal, and every time a reaction is run, the target compound must be purified by separating it from all solvents and contaminants. Purification was an enormously time-consuming, hit-or-miss proposition in the nineteenth and early twentieth centuries, but the development of extraordinarily powerful instruments in the last few decades now simplifies the problem greatly.

Most organic purification is done by *chromatography* (literally, "color writing"), a separation technique that dates from the work of the Russian chemist Mikhail Tswett in 1903. Tswett described the separation of the pigments in green leaves by dissolving the leaf extract in an organic solvent and allowing the solution to run down through a vertical glass tube packed with chalk powder. Different pigments passed down the column at different rates, leaving a series of colored bands on the white chalk column.

There are a variety of chromatographic techniques in common use, all of which work on a similar principle: The mixture to be separated is dissolved in a solvent, called the *mobile phase*, and passed over an adsorbent material, called the *stationary phase*. Because different compounds adsorb to the stationary phase to different extents, they migrate through the phase at different rates and are separated as they emerge (*elute*) from the end of the chromatography column.

(continued)➤

Liquid chromatography, or *column chromatography*, is perhaps the most often used chromatographic method. As in Tswett's original experiments, a mixture of organic compounds is dissolved in a suitable solvent and adsorbed onto a stationary phase such as alumina (Al_2O_3) or silica gel (hydrated SiO_2) packed into a glass column. More solvent is then passed down the column, and different compounds are eluted at different times.

The time at which a compound is eluted is strongly influenced by its polarity. Molecules with polar functional groups are generally adsorbed more strongly and therefore migrate through the stationary phase more slowly than nonpolar molecules. A mixture of an alcohol and an alkene, for example, can be easily separated by liquid chromatography because the nonpolar alkene passes through the column much faster than the more polar alcohol.

High-performance liquid chromatography (HPLC) is a recent variant of the simple column technique, based on the discovery that chromatographic separations are vastly improved if the stationary phase is made up of very small, uniformly sized spherical particles. Small particle size ensures a large surface area for better adsorption, and a uniform spherical shape allows a tight, uniform packing. In practice, specially prepared and coated silica microspheres of 10–25 μm size are often used. Only 15 g of these microspheres have a surface area the size of a football field.

High-pressure pumps are required to force solvent through a tightly packed HPLC column, and sophisticated detectors are used for monitoring the appearance of material eluting from the column. Figure 12.13 shows the results of HPLC analysis of a mixture of 14 common pesticides, using coated silica microspheres as the stationary phase and acetonitrile/water as the mobile phase.

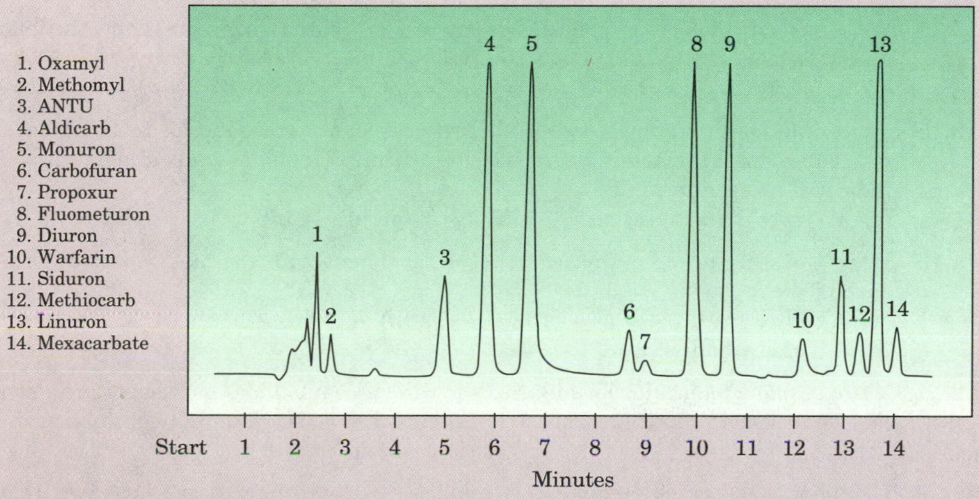

Figure 12.13 The HPLC analysis of a mixture of 14 agricultural pesticides. The structures of the pesticides can be found in the *Merck Index*.

Summary and Key Words

The structure of an organic molecule is usually determined using spectroscopic methods such as mass spectrometry and infrared spectroscopy. **Mass spectrometry (MS)** tells the molecular weight and formula of a molecule; **infrared (IR) spectroscopy** identifies the functional groups present in the molecule.

In mass spectrometry, molecules are first ionized by collision with a high-energy electron beam. The ions then fragment into smaller pieces, which are magnetically sorted according to their mass-to-charge ratio (m/z). The ionized sample molecule is called the **molecular ion, M^+**, and measurement of its mass gives the molecular weight of the sample. Structural clues about unknown samples can be obtained by interpreting the **fragmentation pattern** of the molecular ion. Mass spectral fragmentations are usually complex, however, and interpretation is often difficult.

Infrared spectroscopy involves the interaction of a molecule with **electromagnetic radiation**. When an organic molecule is irradiated with infrared energy, certain frequencies are absorbed by the molecule. The frequencies absorbed correspond to the amounts of energy needed to increase the amplitude of specific molecular vibrations such as bond stretchings and bendings. Since every functional group has a characteristic combination of bonds, every functional group has a characteristic set of infrared absorptions. For example, the terminal alkyne ≡C–H bond absorbs IR radiation of 3300 cm^{-1} frequency, and the alkene C=C bond absorbs in the range 1640–1680 cm^{-1}. By observing which frequencies of infrared radiation are absorbed by a molecule and which are not, it's possible to determine the functional groups a molecule contains.

ADDITIONAL PROBLEMS

12.14 Write as many molecular formulas as you can for hydrocarbons that show the following molecular ions in their mass spectra.
(a) $M^+ = 64$ (b) $M^+ = 186$ (c) $M^+ = 158$ (d) $M^+ = 220$

12.15 Write the molecular formulas of all hydrocarbons corresponding to the following molecular ions. How many degrees of unsaturation (double bonds and/or rings) are indicated by each formula?
(a) $M^+ = 86$ (b) $M^+ = 110$ (c) $M^+ = 146$ (d) $M^+ = 190$

12.16 Draw the structure of a molecule that is consistent with the mass spectral data in each of the following examples.
(a) A hydrocarbon with $M^+ = 132$ (b) A hydrocarbon with $M^+ = 166$
(c) A hydrocarbon with $M^+ = 84$

12.17 Write as many molecular formulas as you can for compounds that show the following molecular ions in their mass spectra. Assume that C, H, N, and O might be present.
(a) $M^+ = 74$ (b) $M^+ = 131$

12.18 Camphor, a saturated monoketone from the Asian camphor tree, is used as a moth repellent and as a constituent of embalming fluid, among other things. If camphor has $M^+ = 152$, what is a likely molecular formula? How many rings does camphor have?

12.19 The "nitrogen rule" of mass spectrometry says that a compound containing an odd number of nitrogens has an odd-numbered molecular ion. Conversely, a compound containing an even number of nitrogens has an even-numbered M^+ peak. Explain.

12.20 In light of the nitrogen rule mentioned in Problem 12.19, what is the molecular formula of pyridine, $M^+ = 79$?

12.21 Nicotine is a diamino compound that can be isolated from dried tobacco leaves. Nicotine has two rings and $M^+ = 162$ in its mass spectrum. Propose a molecular formula for nicotine, and calculate the number of double bonds. (There is no oxygen.)

12.22 Halogenated compounds are particularly easy to identify by their mass spectra because both chlorine and bromine occur naturally as mixtures of two abundant isotopes. Chlorine occurs as ^{35}Cl (75.8%) and ^{37}Cl (24.2%); bromine occurs as ^{79}Br (50.7%) and ^{81}Br (49.3%). At what masses do the molecular ion(s) occur for the following formulas? What are the relative percentages of each molecular ion?
(a) Bromomethane, CH_3Br (b) 1-Chlorohexane, $C_6H_{13}Cl$

12.23 Molecular ions can be particularly complex for polyhalogenated compounds. Taking the natural abundance of Cl into account (see Problem 12.22), calculate the masses of the molecular ions of the following formulas. What are the relative percentages of each ion?
(a) Chloroform, $CHCl_3$ (b) Freon 12, CF_2Cl_2 (Fluorine occurs only as ^{19}F.)

12.24 2-Methylpentane (C_6H_{14}) has the mass spectrum shown. Which peak represents M^+? Which is the base peak? Propose structures for fragment ions of m/z = 71, 57, 43, and 29. Why does the base peak have the mass it does?

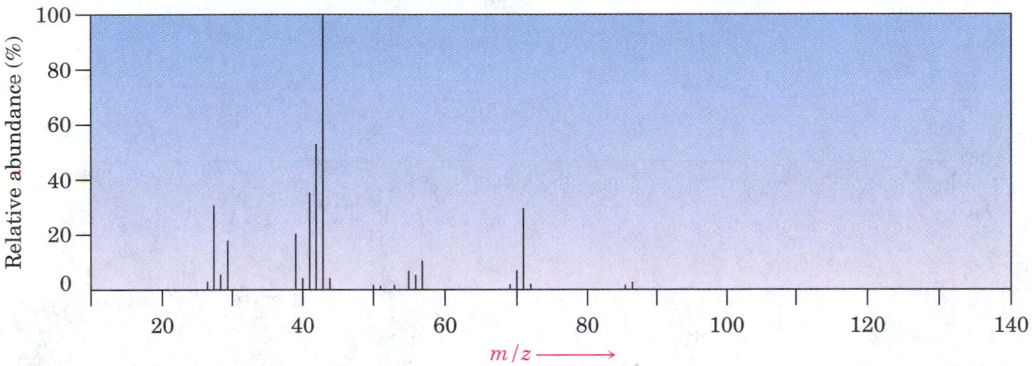

12.25 Assume that you are in a laboratory carrying out the catalytic hydrogenation of cyclohexene to cyclohexane. How could you use a mass spectrometer to determine when the reaction is finished?

12.26 Convert the following infrared absorption values from micrometers to wavenumbers:
(a) An alcohol, 2.98 μm (b) An ester, 5.81 μm (c) A nitrile, 4.93 μm

12.27 Convert the following infrared absorption values from wavenumbers to micrometers:
(a) A cyclopentanone, 1755 cm^{-1} (b) An amine, 3250 cm^{-1}
(c) An aldehyde, 1725 cm^{-1} (d) An acid chloride, 1780 cm^{-1}

12.28 How might you use IR spectroscopy to distinguish among the three isomers 1-butyne, 1,3-butadiene, and 2-butyne?

12.29 Would you expect two enantiomers such as (*R*)-2-bromobutane and (*S*)-2-bromobutane to have identical or different IR spectra? Explain.

12.30 Would you expect two diastereomers such as *meso*-2,3-dibromobutane and (2*R*,3*R*)-dibromobutane to have identical or different IR spectra? Explain.

12.31 Propose structures for compounds that meet the following descriptions:
(a) C_5H_8, with IR absorptions at 3300 and 2150 cm^{-1}

(b) C_4H_8O, with a strong IR absorption at 3400 cm^{-1}
(c) C_4H_8O, with a strong IR absorption at 1715 cm^{-1}
(d) C_8H_{10}, with IR absorptions at 1600 and 1500 cm^{-1}

12.32 How could you use infrared spectroscopy to distinguish between the following pairs of isomers?
(a) $HC{\equiv}CCH_2NH_2$ and $CH_3CH_2C{\equiv}N$
(b) CH_3COCH_3 and CH_3CH_2CHO

12.33 Two infrared spectra are shown. One is the spectrum of cyclohexane, and the other is the spectrum of cyclohexene. Identify them and explain your answer.

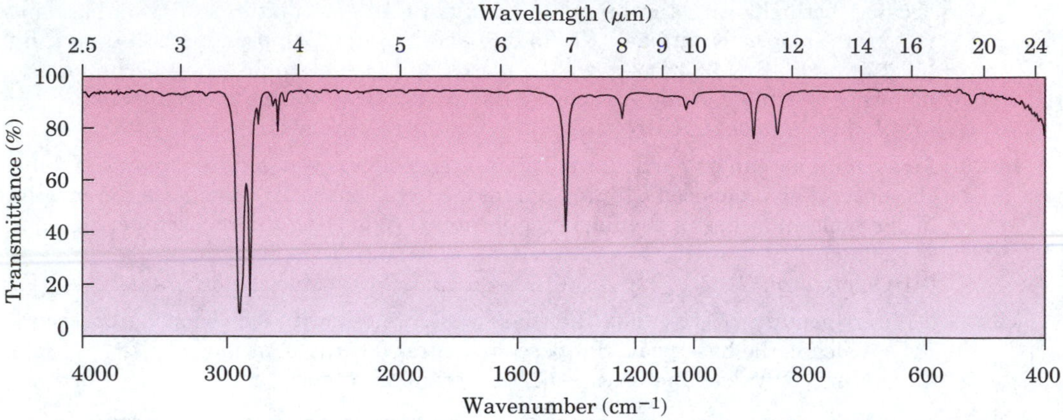

(a)

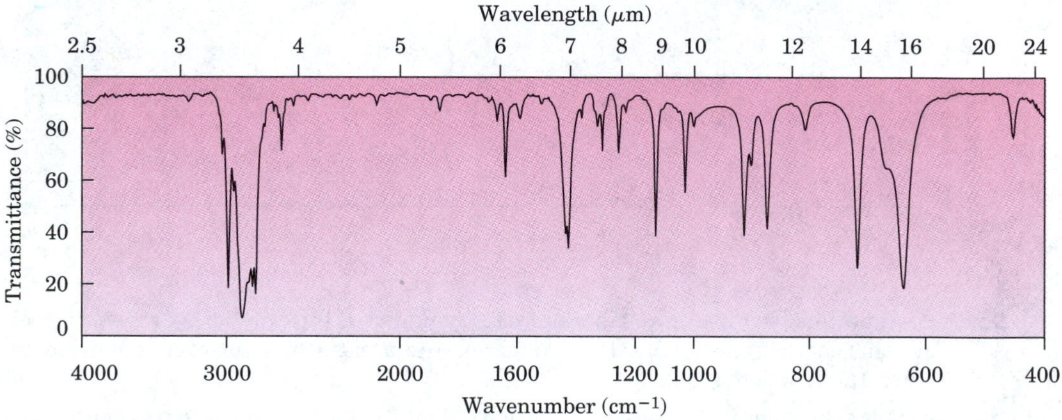

(b)

12.34 How would you use infrared spectroscopy to distinguish between the following pairs of constitutional isomers?

(a) $CH_3C{\equiv}CCH_3$ and $CH_3CH_2C{\equiv}CH$

(b) $CH_3\overset{O}{\overset{\|}{C}}CH{=}CHCH_3$ and $CH_3\overset{O}{\overset{\|}{C}}CH_2CH{=}CH_2$

(c) $H_2C{=}CHOCH_3$ and CH_3CH_2CHO

12.35 The hormone cortisone contains C, H, and O, and shows a molecular ion at M^+ = 360.1937 when analyzed by double-focusing mass spectrometry. What is the molecular formula of cortisone? (Isotopic masses are: ^{12}C, 12.0000 amu; ^{1}H, 1.007 83 amu; ^{16}O, 15.9949 amu. The degree of unsaturation of cortisone is 8.)

12.36 Assume you are carrying out the dehydration of 1-methylcyclohexanol to yield 1-methylcyclohexene. How could you use infrared spectroscopy to determine when the reaction is complete?

12.37 Assume that you are carrying out the base-induced dehydrobromination of 3-bromo-3-methylpentane (Section 11.10). How could you use IR spectroscopy to tell which of two possible elimination products is formed?

12.38 At what approximate positions might the following compounds show IR absorptions?

(a) $CH_3CH_2\overset{\overset{O}{\|}}{C}CH_3$ (b) $(CH_3)_2CHCH_2C{\equiv}CH$ (c) $(CH_3)_2CHCH_2CH{=}CH_2$

(d) $CH_3CH_2CH_2COOCH_3$ (e) 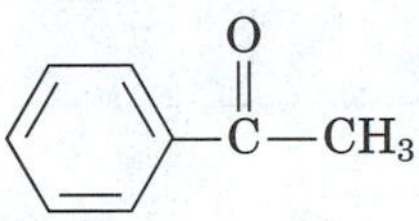

12.39 Which is stronger, the C=O bond in an ester (1735 cm^{-1}) or the C=O bond in a saturated ketone (1715 cm^{-1})? Explain.

12.40 Carvone is an unsaturated ketone responsible for the odor of spearmint. If carvone has $M^+ = 150$ in its mass spectrum, what molecular formulas are likely? If carvone has three double bonds and one ring, what molecular formula is correct?

12.41 Carvone (see Problem 12.40) has an intense infrared absorption at 1690 cm^{-1}. What kind of ketone does carvone contain?

12.42 The mass spectrum (a) and the infrared spectrum (b) of an unknown hydrocarbon are shown. Propose as many structures as you can.

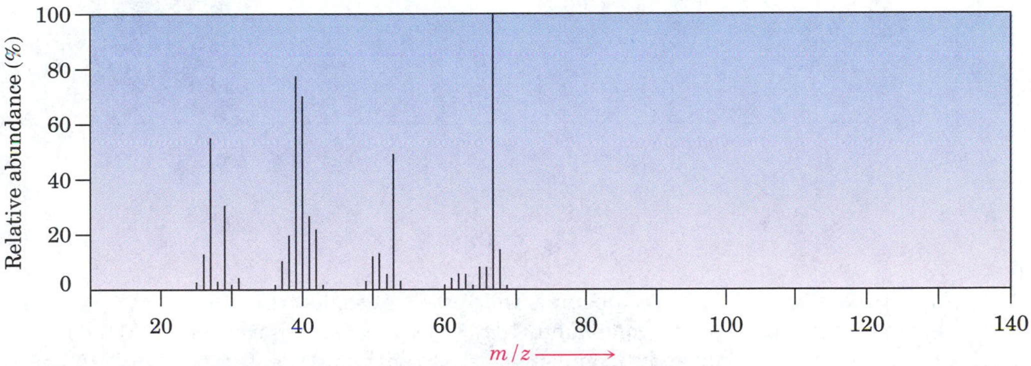

(a)

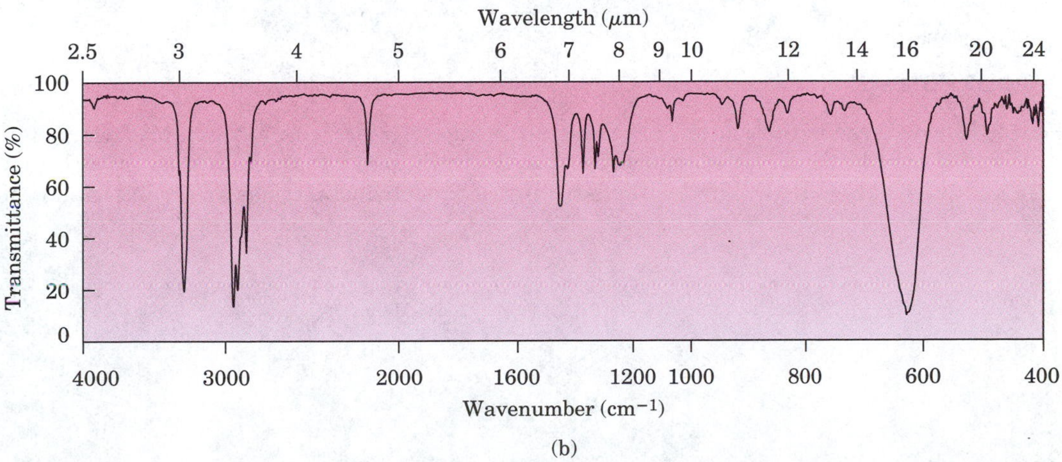

(b)

12.43 The mass spectrum (a) and the infrared spectrum (b) of another unknown hydrocarbon are shown. Propose as many structures as you can.

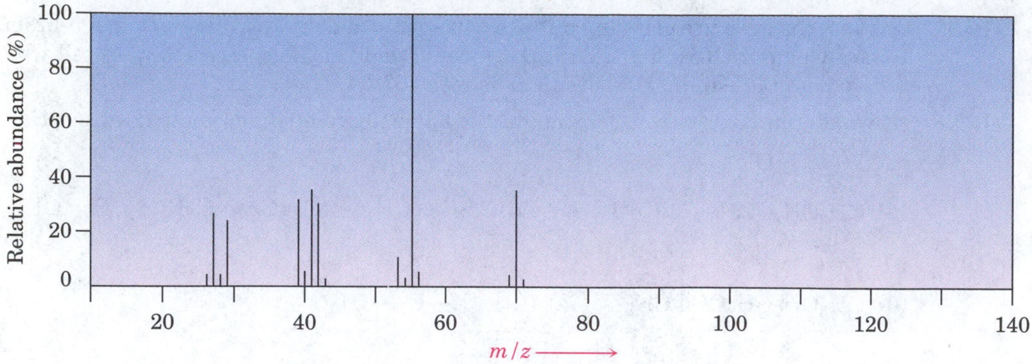

(a)

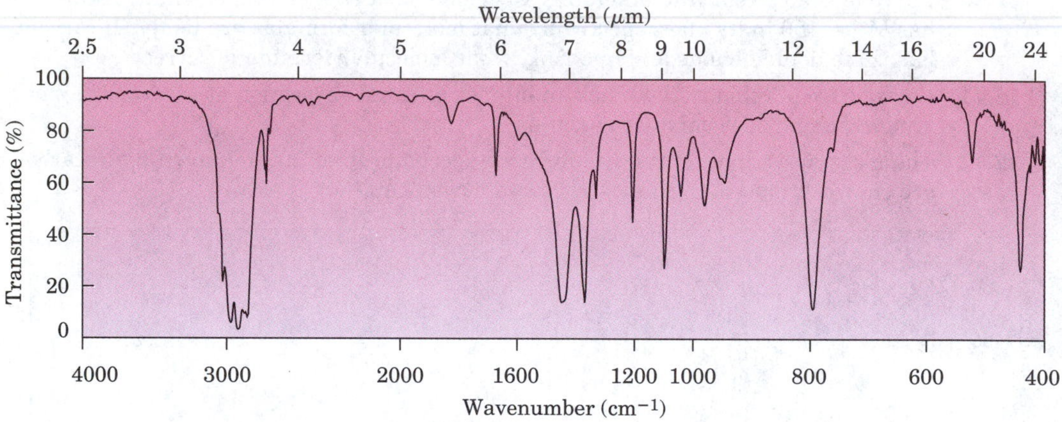

(b)

12.44 Propose structures for compounds that meet these descriptions:
(a) An optically active compound $C_5H_{10}O$ with an IR absorption at 1730 cm^{-1}.
(b) A non-optically-active compound C_5H_9N with an IR absorption at 2215 cm^{-1}.

A Look Ahead

12.45 We'll see in Section 17.7 that Grignard reagents react with ketones. Methylmagnesium bromide, for example, reacts with cyclohexanone to yield a product with the formula $C_7H_{14}O$. What is the structure of this product if it has an IR absorption at 3400 cm^{-1}?

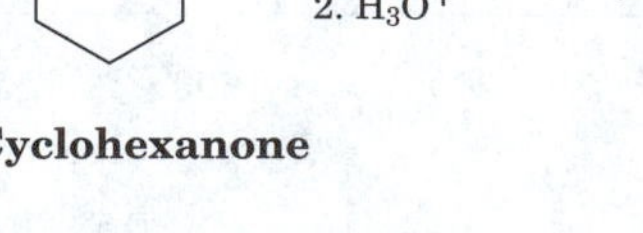

Cyclohexanone

12.46 Ketones undergo a reduction (Section 10.10) when treated with sodium borohydride, $NaBH_4$. What is the structure of the compound produced by reaction of 2-butanone with $NaBH_4$ if it has an IR absorption at 3400 cm^{-1} and $M^+ = 74$ in the mass spectrum?

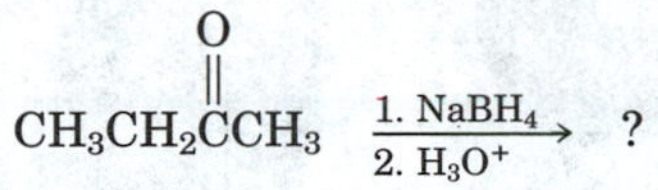

2-Butanone

12.47 Nitriles undergo a hydrolysis reaction when heated with aqueous acid (Section 21.9). What is the structure of the compound produced by hydrolysis of propanenitrile, $CH_3CH_2C{\equiv}N$, if it has IR absorptions at 2500 cm^{-1} and 1710 cm^{-1} and has $M^+ =$ 74?

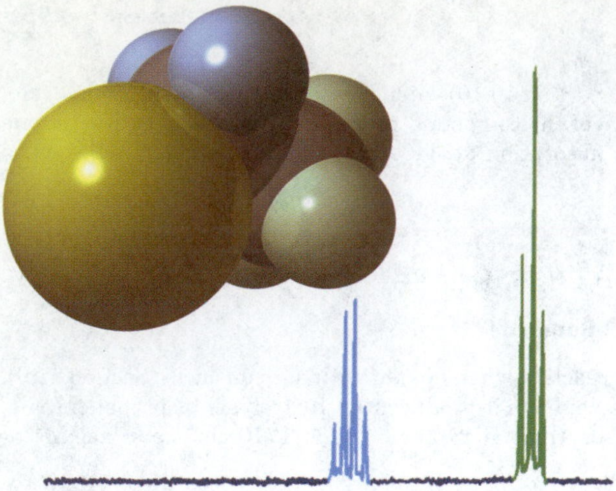

The different hydrogen atoms of chloroethane give rise to different absorption peaks in NMR spectroscopy.

13

Structure Determination: Nuclear Magnetic Resonance Spectroscopy

Nuclear magnetic resonance spectroscopy (NMR) is the most valuable spectroscopic technique available to organic chemists. It's the method of structure determination that organic chemists first turn to for information.

We saw in Chapter 12 that mass spectrometry provides information about a molecule's formula and that infrared spectroscopy provides information about a molecule's functional groups. Nuclear magnetic resonance spectroscopy does not replace either of these techniques; rather, it complements them by providing a "map" of the carbon–hydrogen framework of an organic molecule. Taken together, NMR, IR, and mass spectrometry often make it possible to determine the complete structures of even very complex molecules.

Mass spectrometry	Molecular size and formula
Infrared spectroscopy	Functional groups
NMR spectroscopy	**Map of carbon–hydrogen framework**

13.1 Nuclear Magnetic Resonance Spectroscopy

Many kinds of nuclei behave as if they were spinning about an axis, much as the earth spins daily. Since they're positively charged, these spinning nuclei act like tiny magnets and therefore interact with an external magnetic field (denoted $\boldsymbol{H}_0$). Not all nuclei act this way, but fortunately for organic chemists, both the proton (^{1}H) and the ^{13}C nucleus do have spins.[1] Let's see what the consequences of nuclear spin are and how we can use the results.

In the absence of an external magnetic field, the nuclear spins of magnetic nuclei are oriented randomly. When a sample containing these nuclei is placed between the poles of a strong magnet, however, the nuclei adopt specific orientations, much as a compass needle orients in the earth's magnetic field. A spinning ^{1}H or ^{13}C nucleus can orient so that its own tiny magnetic field is aligned either with (parallel to) or against (antiparallel to) the external field. The two orientations don't have the same energy and therefore aren't equally likely. The parallel orientation is slightly lower in energy, making this spin state slightly favored over the antiparallel orientation (Figure 13.1).

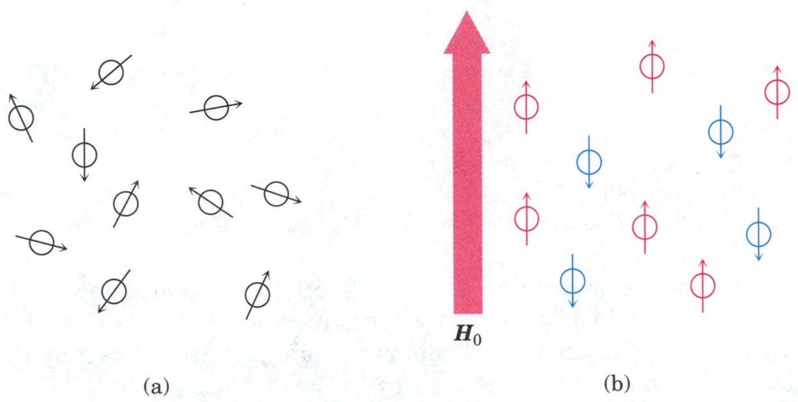

Figure 13.1 Nuclear spins are oriented randomly in the absence of an external magnetic field (a), but have a specific orientation in the presence of an external field, $\boldsymbol{H}_0$ (b). Note that some of the spins (red) are aligned parallel to the external field while others (blue) are antiparallel. The parallel spin state is lower in energy.

If the oriented nuclei are now irradiated with electromagnetic radiation of the right frequency, energy absorption occurs and the lower-energy state "spin-flips" to the higher-energy state. When this spin-flip occurs, the nuclei

[1]In speaking about ^{1}H NMR, the words *proton* and *hydrogen* are often used interchangeably.

are said to be in resonance with the applied radiation, hence the name *nuclear magnetic resonance*.

The exact frequency necessary for resonance depends both on the strength of the external magnetic field and on the identity of the nuclei. If a very strong magnetic field is applied, the energy difference between the two spin states is large, and higher-frequency (higher-energy) radiation is required for a spin-flip. If a weaker magnetic field is applied, less energy is required to effect the transition between nuclear spin states (Figure 13.2).

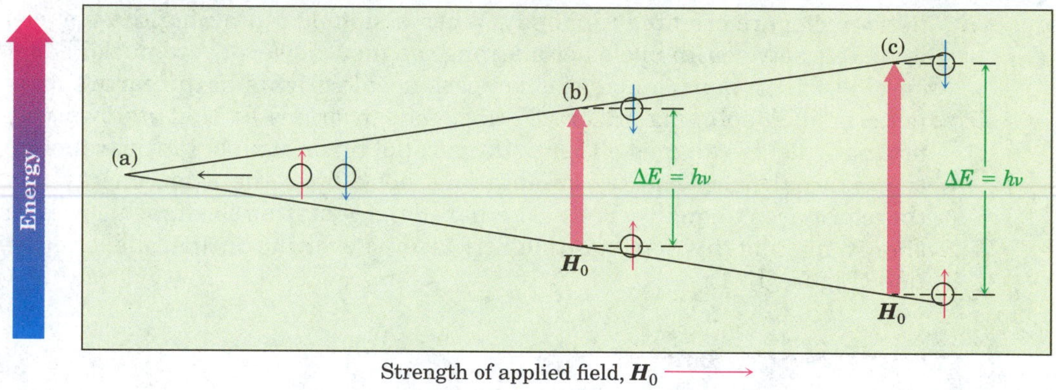

Figure 13.2 The energy difference ΔE between nuclear spin states is a function of applied magnetic field strength. Absorption of energy of frequency ν converts a nucleus from a lower spin state to a higher spin state. (a) Spin states have equal energies in the absence of an applied magnetic field, but (b) have unequal energies in the presence of a magnetic field. At $\nu = 60$ MHz, $\Delta E = 2.4 \times 10^{-5}$ kJ/mol (5.7×10^{-6} kcal/mol). (c) The energy difference between spin states is greater at larger applied fields. At $\nu = 100$ MHz, $\Delta E = 4.0 \times 10^{-5}$ kJ/mol.

In practice, superconducting magnets that produce enormously powerful fields up to 14.1 tesla (T) are sometimes used, but field strengths in the range 1.41–4.7 T are more common.[2] At a magnetic field strength of 1.41 T, so-called *radiofrequency* (rf) energy in the 60 MHz range (1 MHz = 10^6 Hz) is required to bring a ^{1}H nucleus into resonance, and rf energy of 15 MHz is required to bring a ^{13}C nucleus into resonance. These energy levels needed for NMR are much smaller than those required for infrared spectroscopy 60 MHz rf energy corresponds to only 2.4×10^{-5} kJ/mol (5.7×10^{-6} kcal/mol) versus 48–4.8 kJ/mol for IR spectroscopy.

^{1}H and ^{13}C nuclei are not unique in their ability to exhibit the nuclear magnetic resonance phenomenon. All nuclei with an odd number of protons (^{1}H, ^{2}H, ^{14}N, ^{19}F, ^{31}P, for example) and all nuclei with an odd number of neutrons (^{13}C, for example) show magnetic properties. Only nuclei with even numbers of both protons and neutrons (^{12}C, ^{16}O) do not give rise to magnetic phenomena (Table 13.1).

[2]The SI unit of magnetic flux density is the *tesla* (T), which has replaced the older unit *gauss* (G); 1 T = 10^4 G = 1 J/(A·m^2).

Table 13.1 The NMR Behavior of Some Common Nuclei

Magnetic nuclei		*Nonmagnetic nuclei*	
^{1}H	NMR observed	^{12}C	No NMR observed
^{13}C		^{16}O	
^{2}H		^{32}S	
^{14}N			
^{19}F			
^{31}P			

PROBLEM

13.1 The amount of energy required to spin-flip a nucleus depends both on the strength of the external magnetic field and on the nucleus. At a field strength of 1.41 T, rf energy of 60 MHz is required to bring a ^{1}H nucleus into resonance, but energy of only 56 MHz will bring a ^{19}F nucleus into resonance. Use the equation given in Problem 12.8 to calculate the amount of energy required to spin-flip a ^{19}F nucleus. Is this amount greater or less than that required to spin-flip a ^{1}H nucleus?

$E = 1.2 \times 10^{-2}$ kJ/mol　　$v = \frac{c}{\lambda}$

PROBLEM

13.2 Calculate the amount of energy required to spin-flip a proton in a spectrometer operating at 100 MHz. Does increasing the spectrometer frequency from 60 MHz to 100 MHz increase or decrease the amount of energy necessary for resonance?

13.2 The Nature of NMR Absorptions

From the description given thus far, you might expect all ^{1}H nuclei in a molecule to absorb rf energy at the same frequency and all ^{13}C nuclei to absorb at the same frequency. If this were true, we would observe only a single NMR absorption band in the ^{1}H or ^{13}C spectrum of a molecule, a situation that would be of little use for structure determination. In fact, the absorption frequency is not the same for all ^{1}H or all ^{13}C nuclei.

All nuclei in molecules are surrounded by electrons. When an external magnetic field is applied to a molecule, the electrons set up tiny local magnetic fields of their own. These local magnetic fields act in opposition to the applied field so that the *effective* field actually felt by the nucleus is a bit smaller than the applied field.

$$\boldsymbol{H}_{\text{effective}} = \boldsymbol{H}_{\text{applied}} - \boldsymbol{H}_{\text{local}}$$

In describing this effect, we say that nuclei are **shielded** from the full effect of the applied field by the circulating electrons that surround them. Since each specific nucleus in a molecule is in a slightly different electronic environment, each nucleus is shielded to a slightly different extent, and the effective magnetic field is not the same for each nucleus. If the NMR instrument is sensitive enough, the tiny differences in the effective magnetic fields experienced by different nuclei can be detected, and we can see a distinct NMR signal for each chemically distinct carbon or hydrogen nucleus in a molecule. Thus, the NMR spectrum of an organic compound effectively maps the carbon–hydrogen framework. With practice, it's possible to read the map and thereby derive structural information about an unknown molecule.

Figure 13.3 shows both the ^{1}H and the ^{13}C NMR spectra of methyl acetate, $CH_3CO_2CH_3$. The horizontal axis shows the effective field strength felt by the nuclei, and the vertical axis indicates intensity of absorption of rf energy. Each peak in the NMR spectrum corresponds to a chemically distinct nucleus in a molecule. Note, though, that ^{1}H and ^{13}C spectra can't

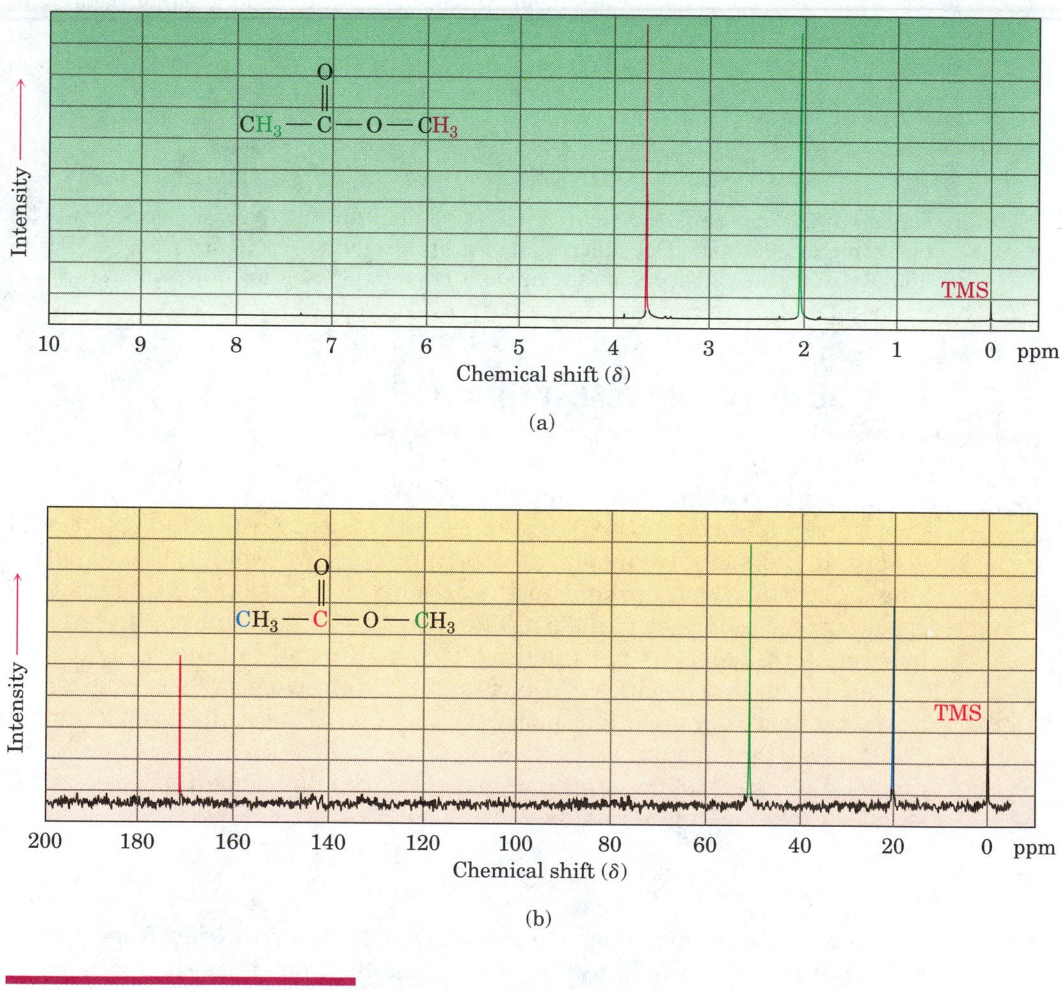

Figure 13.3 (a) The ^{1}H NMR spectrum and (b) the ^{13}C NMR spectrum of methyl acetate, $CH_3CO_2CH_3$.

both be observed at the same time on the same spectrometer because different amounts of energy are required to spin-flip the different kinds of nuclei. The two spectra must be recorded separately.

The ^{13}C spectrum of methyl acetate in Figure 13.3 shows three peaks, one for each of the three carbon atoms in the molecule. The ^{1}H NMR spectrum shows only *two* peaks, however, even though methyl acetate has *six* hydrogens. One peak is due to the CH_3CO hydrogens, and the other to the OCH_3 hydrogens. Because the three hydrogens of each methyl group have the same electronic (and magnetic) environment, they are shielded to the same extent and are said to be *equivalent*. Chemically equivalent nuclei always show a single absorption. The two methyl groups themselves, however, are nonequivalent and absorb at different positions.

The operation of a typical NMR spectrometer is illustrated schematically in Figure 13.4. An organic sample is dissolved in a suitable solvent (usually deuteriochloroform, $CDCl_3$) and placed in a thin glass tube between the poles of a magnet. The strong magnetic field causes the ^{1}H and ^{13}C nuclei in the molecule to align in one of the two possible orientations, and the sample is irradiated with rf energy. If the frequency of the rf irradiation is held constant and the strength of the applied magnetic field is changed, each nucleus comes into resonance at a slightly different field strength. A sensitive detector monitors the absorption of rf energy, and the electronic signal is then amplified and displayed as a peak on a recorder chart.

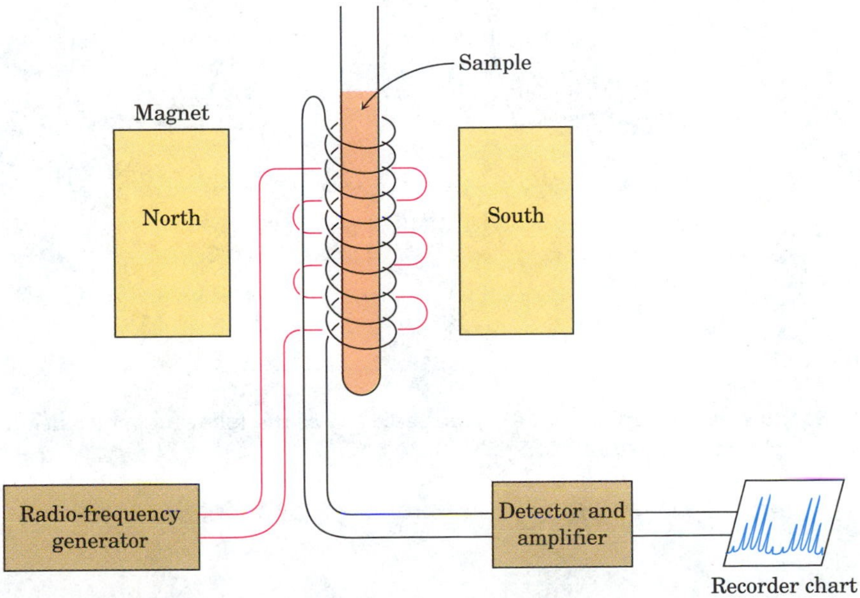

Figure 13.4 Schematic operation of an NMR spectrometer. A thin glass tube containing the sample solution is placed between the poles of a strong magnet and irradiated with rf energy.

NMR spectroscopy differs from IR spectroscopy (Sections 12.5–12.8) in that the time scales of the two techniques are quite different. The absorption of infrared energy by a molecule giving rise to a change in vibrational state is an essentially instantaneous process (about 10^{-13} s). The NMR process, however, requires much more time (about 10^{-3} s).

The difference in time scales between IR and NMR spectroscopy is comparable to the difference between a camera operating at a very fast shutter speed and a camera operating at a very slow shutter speed. The fast camera (IR) takes an instantaneous picture and "freezes" the action. If two rapidly interconverting species are present, IR spectroscopy records the spectrum of each. The slow camera (NMR), however, takes a blurred, "time-averaged" picture. If two species interconverting faster than 10^3 times per second are present in a sample, NMR records only a single, averaged spectrum, rather than separate spectra of the two discrete species.

Because of this "blurring" effect, NMR spectroscopy can be used to measure the rates and activation energies of very fast processes. In cyclohexane, for example, a ring-flip (Section 4.11) occurs so rapidly at room temperature that axial and equatorial hydrogens can't be distinguished by NMR; only a single ^1H NMR absorption is seen for cyclohexane at 25°C. At −90°C, however, the ring-flip is slow enough that two absorption peaks are seen, one for the six axial hydrogens and one for the six equatorial hydrogens. Knowing the temperature and the rate at which signal blurring begins to occur, it's possible to calculate that the activation energy for the cyclohexane ring-flip is 45 kJ/mol (10.8 kcal/mol).

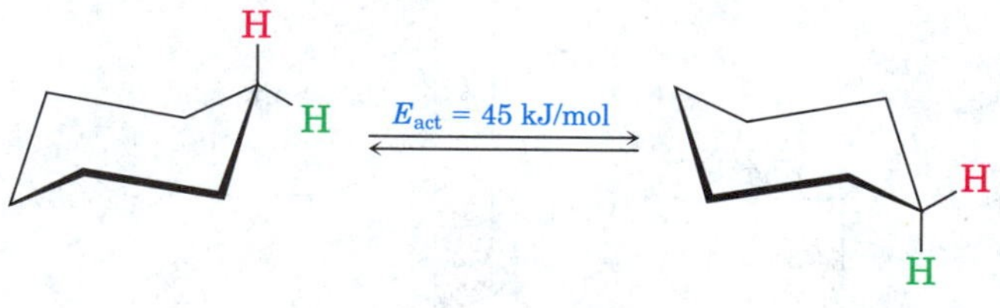

^1H NMR: 1 peak at 25°C
2 peaks at −90°C

PROBLEM

13.3 How many signals would you expect each of the following molecules to have in its ^1H and ^{13}C spectra at 25°C?

(a) $(CH_3)_2C{=}C(CH_3)_2$

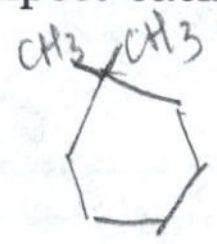

(b) 1,1-Dimethylcyclohexane

(c) $CH_3\overset{\overset{\displaystyle O}{\|}}{C}CH_3$

(d) $(CH_3)_3C\overset{\overset{\displaystyle O}{\|}}{C}OCH_3$

(e) 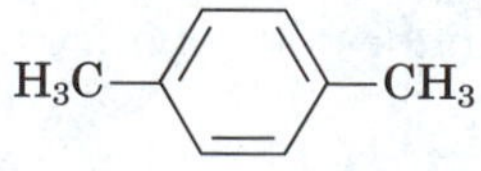

(f) 1,1-Dimethylcyclopropane

PROBLEM

13.4 2-Chloropropene shows signals for three kinds of protons in its ^1H NMR spectrum. Explain.

13.3 Chemical Shifts

NMR spectra are displayed on charts that show the applied field strength increasing from left to right (Figure 13.5). Thus, the left part of the chart is the low-field, or **downfield**, side, and the right part is the high-field, or **upfield**, side. To define the position of an absorption, the NMR chart is calibrated and a reference point is used. In practice, a small amount of tetramethylsilane [TMS, $(CH_3)_4Si$] is added to the sample so that a reference absorption is produced when the spectrum is run. TMS is used as reference for both ^1H and ^{13}C measurements because it produces in both kinds of spectra a single peak that occurs upfield of other absorptions normally found in organic compounds. The ^1H and ^{13}C spectra of methyl acetate in Figure 13.3 have the TMS reference peak indicated.

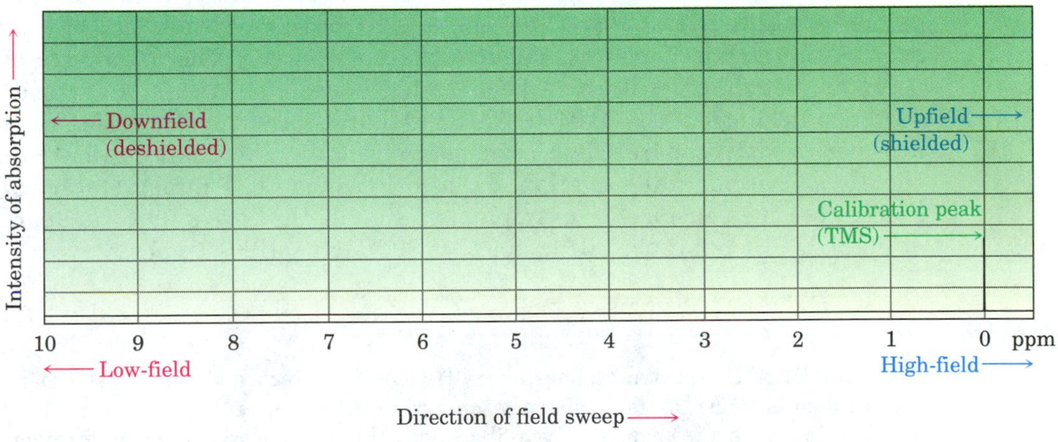

Figure 13.5 The NMR chart. The downfield side is on the left, and the upfield side is on the right. The tetramethylsilane (TMS) absorption is used as reference point.

The position on the chart at which a nucleus absorbs is called its **chemical shift**. By convention, the chemical shift of TMS is set as the zero point, and other absorptions normally occur downfield (to the left on the chart). NMR charts are calibrated using an arbitrary scale called the **delta scale**. One delta unit (δ) is equal to 1 part per million (ppm; one-millionth) of the spectrometer operating frequency. For example, if we were measuring the ^1H NMR spectrum of a sample using an instrument operating at 60 MHz,

1 δ would be 1 ppm of 60,000,000 Hz, or 60 Hz. Similarly, if we were measuring the spectrum using a 300 MHz instrument, then 1 δ = 300 Hz. The following equation can be used for any absorption:

$$\delta = \frac{\text{Observed chemical shift (number of Hz away from TMS)}}{\text{Spectrometer frequency in MHz}}$$

Although this method of calibrating NMR charts may seem needlessly complex, there's a good reason for it. As we saw earlier, the chemical shift of an NMR peak and the rf frequency required for resonance both depend on magnetic field strength. Since many different kinds of spectrometers operating at many different magnetic field strengths and rf frequencies are available, chemical shifts given in Hz vary greatly from one instrument to another. By using a system of measurement in which NMR absorptions are expressed in *relative* terms (ppm of spectrometer frequency) rather than in absolute terms (Hz), comparisons of spectra obtained on different instruments are possible. *The chemical shift of an NMR absorption given in δ units is constant, regardless of the operating frequency of the spectrometer.* A ^{1}H nucleus that absorbs at 2.0 δ on a 60 MHz instrument (2.0 ppm × 60 MHz = 120 Hz downfield from TMS) also absorbs at 2.0 δ on a 300 MHz instrument (2.0 ppm × 300 MHz = 600 Hz downfield from TMS).

The range in which most NMR absorptions occur is quite narrow. Almost all ^{1}H NMR absorptions occur 0–10 δ downfield from the proton absorption of TMS, and almost all ^{13}C absorptions occur 1–220 δ downfield from the carbon absorption of TMS. Thus, there is a considerable likelihood that accidental overlap of nonequivalent signals will occur. The advantage of using an instrument with high field strength (say, 300 MHz NMR) rather than low field strength (60 MHz NMR) is that different NMR absorptions are more widely separated at high field strength. The chances that two signals will accidentally overlap are also lessened, and interpretation of spectra becomes easier. For example, two signals that are only 6 Hz apart at 60 MHz (0.1 ppm) are 30 Hz apart at 300 MHz (still 0.1 ppm).

PROBLEM

13.5 When the ^{1}H NMR spectrum of acetone, CH_3COCH_3, is recorded on an instrument operating at 60 MHz, a single sharp resonance at 2.1 δ is seen.
(a) How many hertz downfield from TMS does the acetone resonance correspond to?
(b) If the ^{1}H NMR spectrum of acetone were recorded at 100 MHz, what would be the position of the absorption in δ units?
(c) How many hertz downfield from TMS does this 100 MHz resonance correspond to?

PROBLEM

13.6 The following ^{1}H NMR peaks were recorded on a spectrometer operating at 60 MHz. Convert each into δ units.
(a) $CHCl_3$; 436 Hz
(b) CH_3Cl; 183 Hz
(c) CH_3OH; 208 Hz
(d) CH_2Cl_2; 318 Hz

13.4 ¹H NMR Spectroscopy and Proton Equivalence

Everything we've said thus far about NMR spectroscopy applies to both ^1H and ^{13}C spectra, but let's now focus only on ^1H NMR spectroscopy. Since each chemically distinct hydrogen atom in a molecule normally has its own unique absorption, one use of NMR is to find out how many kinds of nonequivalent hydrogens are present. In the ^1H NMR spectrum of methyl acetate shown previously in Figure 13.3, for example, there are two signals, corresponding to the two nonequivalent kinds of protons present, CH_3CO– protons and –OCH_3 protons

A quick look at the structure is usually enough to decide how many kinds of nonequivalent protons are present in a molecule. If in doubt, though, the equivalence or nonequivalence of two protons can be determined by seeing whether the same structure would result if some group X were substituted for one of the protons. If the protons are chemically equivalent, the same product will be formed regardless of which proton is replaced. If the protons are not chemically equivalent, different products will be formed on substitution. In 2,3-dimethyl-2-butene, for example, all 12 protons are equivalent. No matter which proton we replace by an X group, we get the same structure. The 12 protons thus give rise to a single, sharp ^1H NMR peak (Figure 13.6).

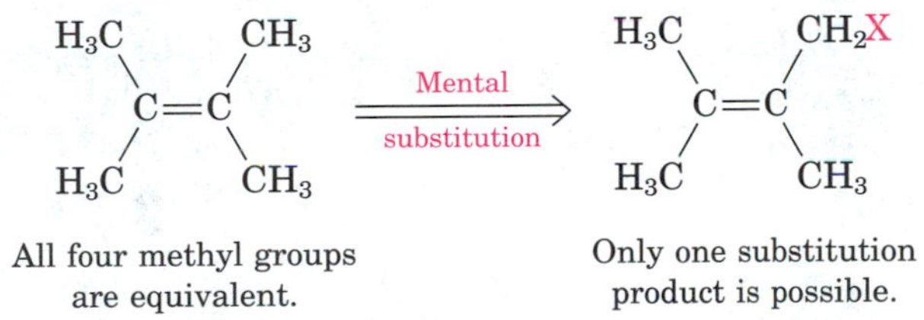

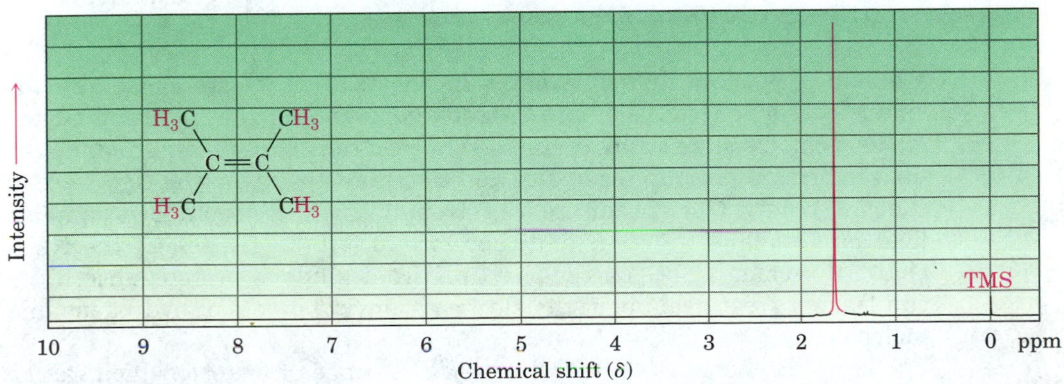

Figure 13.6 The ¹H NMR spectrum of 2,3-dimethyl-2-butene. Since all 12 protons in the molecule are equivalent, there is only one peak in the spectrum.

By contrast, the 10 protons of 2-methyl-2-butene are *not* all equivalent. There are three distinct kinds of methyl-group protons and one vinylic proton, leading to four different possible substitution products and four different signals in the ^{1}H NMR spectrum.

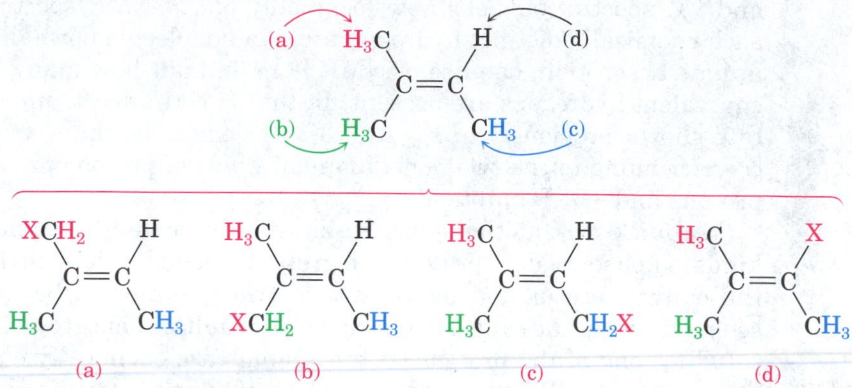

Four different substitution products

PROBLEM

13.7 How many kinds of nonequivalent protons are present in each of the following compounds?

(a) CH_3CH_2Br (b) $CH_3OCH_2CH(CH_3)_2$ (c) $CH_3CH_2CH_2NO_2$
(d) Methylbenzene (e) 2-Methyl-1-butene (f) *cis*-3-Hexene

13.5 Chemical Shifts in ^{1}H NMR Spectroscopy

We said previously that differences in chemical shifts are caused by the small local magnetic fields of electrons surrounding the different nuclei. Nuclei that are more strongly shielded by electrons require a higher applied field to bring them into resonance and therefore absorb on the right side of the NMR chart. Nuclei that are less strongly shielded need a lower applied field for resonance to occur and therefore absorb on the left of the NMR chart. By seeing where on the chart different kinds of protons absorb, we can learn a great deal about the electronic environments of hydrogens in a molecule.

Most ^{1}H chemical shifts fall within the range of 0–10 δ, which can be conveniently divided into the five regions shown in Table 13.2. By remembering the positions of these regions, it's often possible to tell at a glance what kinds of protons a molecule contains.

Table 13.2 Regions of the ¹H NMR Spectrum

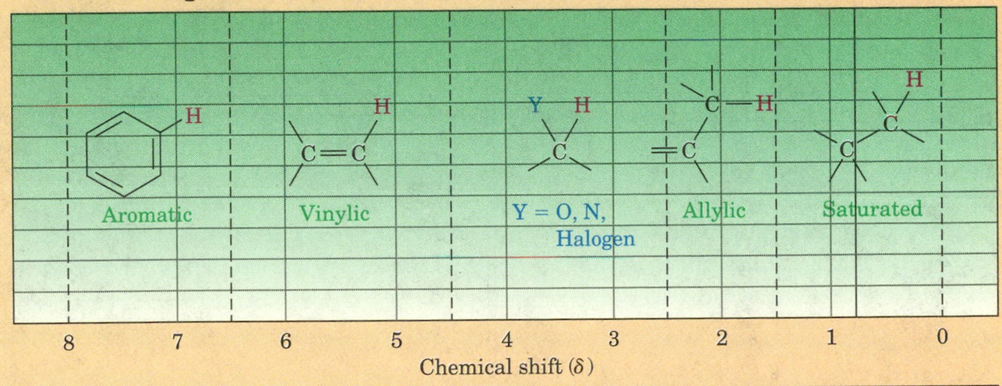

Region (δ)	*Proton type*	*Comments*
0–1.5	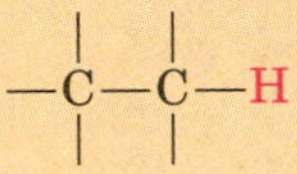 —C—C—H	Protons on carbon next to saturated centers absorb in this region. Thus, the alkane portions of most organic molecules show complex absorption here.
1.5–2.5	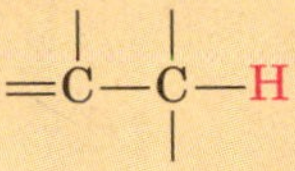 =C—C—H	Protons on carbon next to unsaturated centers (allylic, benzylic, next to carbonyl) show characteristic absorptions in this region, just downfield from other alkane resonance.
2.5–4.5	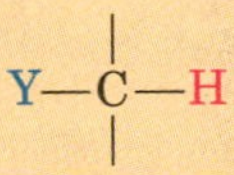	Protons on carbon next to electronegative atoms (halogen, O, N) are deshielded because of the electron-withdrawing ability of these atoms. Thus, the protons absorb in this midfield region.
4.5–6.5	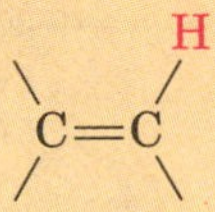	Protons on double-bond carbons (vinylic protons) are strongly deshielded by the neighboring π bond and therefore absorb in this characteristic downfield region.
6.5–8.0	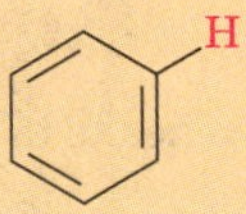	Protons on aromatic rings (aryl protons) are strongly deshielded by the π orbitals of the ring and absorb in this characteristic low-field range.

Table 13.3 shows the correlation of ¹H chemical shift with electronic environment in more detail. In general, protons bonded to saturated, sp^3-hybridized carbons absorb at higher fields, whereas protons bonded to sp^2-hybridized carbons absorb at lower fields. Protons on carbons that are bonded to electronegative atoms, such as N, O, or halogen, also absorb at lower fields.

Table 13.3 Correlation of ^1H Chemical Shift with Environment

Type of hydrogen		*Chemical shift (δ)*
Reference	$(CH_3)_4Si$	0
Saturated primary	$—CH_3$	0.7–1.3
Saturated secondary	$—CH_2—$	1.2–1.4
Saturated tertiary	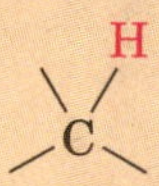	1.4–1.7
Allylic	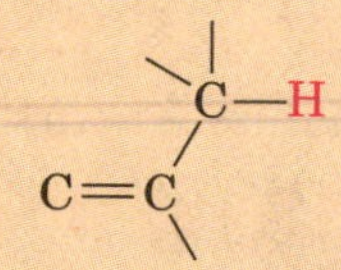	1.6–2.2
Methyl ketone	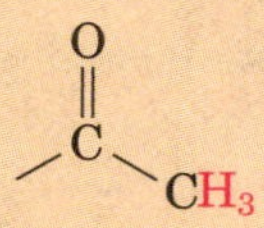	2.1–2.4
Aromatic methyl	$Ar—CH_3$	2.5–2.7
Alkyl halide X = Cl, Br, I	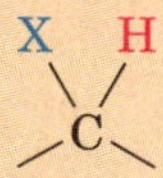	2.5–4.0
Alcohol, ether	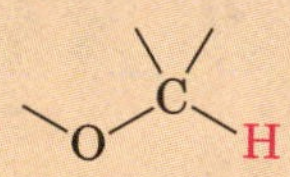	3.3–4.0
Alkynyl	C≡C—H	2.5–2.7
Vinylic	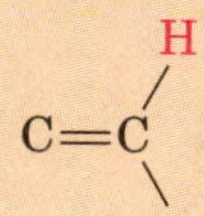	5.0–6.5
Aromatic	Ar—H	6.5–8.0
Aldehyde	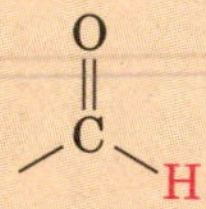	9.7–10.0
Carboxylic acid	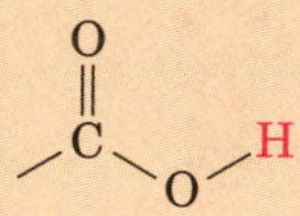	11.0–12.0
Alcohol	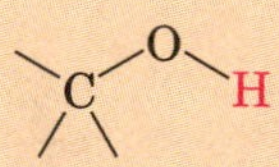	2.5–5.0 (Variable)

PRACTICE PROBLEM

Methyl 2,2-dimethylpropanoate $(CH_3)_3COOCH_3$ has two peaks in its ^1H NMR spectrum. What are their approximate chemical shifts?

Solution The $CH_3O–$ protons absorb around 3.5–4.0 δ because they are on carbon bonded to oxygen. The $(CH_3)_3C–$ protons absorb near 1.0 δ because they are typical alkane-like protons. (See Figure 13.7.)

PROBLEM

13.8 Each of the following compounds has a single ^1H NMR peak. Approximately where would you expect each compound to absorb?

(a) Cyclohexane (b) CH_3COCH_3 (c) Benzene

(d) Glyoxal, 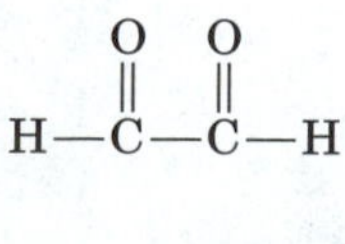 (e) CH_2Cl_2 (f) $(CH_3)_3N$

PROBLEM

13.9 Identify the different kinds of protons in the following molecule, and tell where you would expect each to absorb.

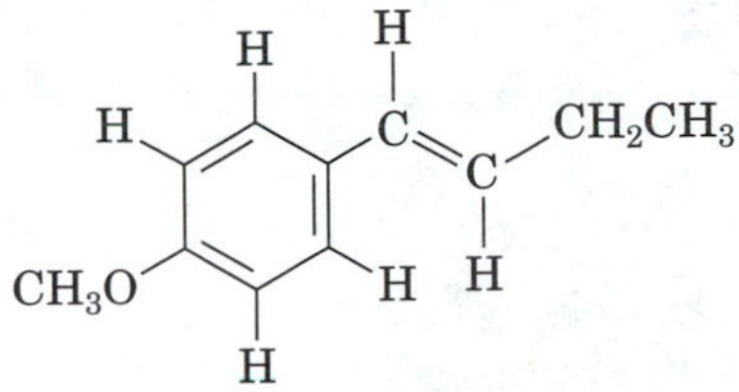

13.6 Integration of ¹H NMR Absorptions: Proton Counting

Look at the ¹H NMR spectrum of methyl 2,2-dimethylpropanoate in Figure 13.7. There are two peaks, corresponding to the two kinds of protons, but the peaks aren't the same size. The peak at 1.2 δ, due to the $(CH_3)_3C$– protons, is larger than the peak at 3.7 δ, due to the $-OCH_3$ protons.

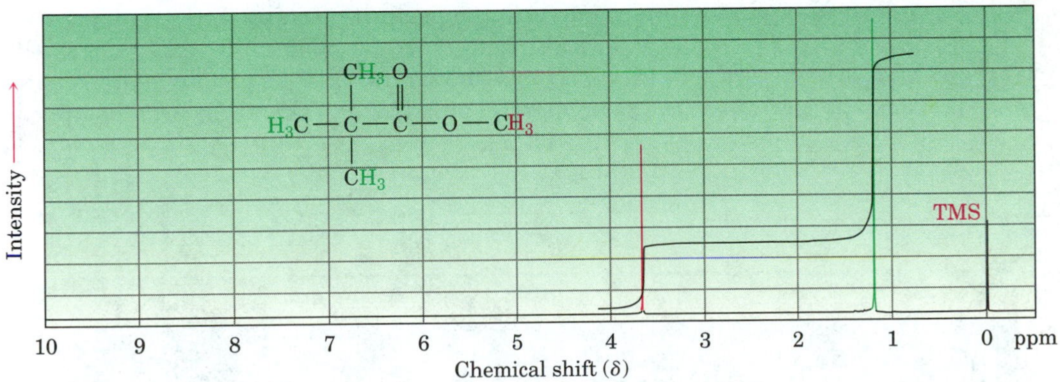

Figure 13.7 The ¹H NMR spectrum of methyl 2,2-dimethylpropanoate. Integrating the peaks in a "stair-step" manner shows that they have a 1:3 ratio, corresponding to the ratio of the numbers of protons (3:9) responsible for each peak.

The area under each peak is proportional to the number of protons causing that peak. By electronically measuring, or **integrating**, the area under each peak, it's possible to measure the relative number of each kind of proton in a molecule. Integrated peak areas are superimposed over the spectrum as a "stair-step" line, with the height of each step proportional to the area under the peak, and therefore proportional to the relative number of protons causing the peak. To compare the size of one peak against another, simply take out a ruler and measure the heights of the various steps. For example, the two peaks in methyl 2,2-dimethylpropanoate are found to have a 1:3 (or 3:9) ratio when integrated—exactly what we expect since the three $-OCH_3$ protons are equivalent and the nine $(CH_3)_3C-$ protons are equivalent.

PROBLEM

13.10 How many peaks would you expect in the ^{1}H NMR spectrum of 1,4-dimethylbenzene (*p*-xylene)? What ratio of peak areas would you expect on integration of the spectrum? Refer to Table 13.3 for approximate chemical shifts, and sketch what the spectrum would look like.

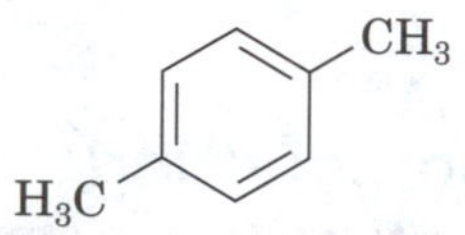

***p*-Xylene**

13.7 Spin–Spin Splitting in ^{1}H NMR Spectra

In the ^{1}H NMR spectra we've seen thus far, each different kind of proton in a molecule has given rise to a single peak. It often happens, though, that the absorption of a proton splits into *multiple* peaks. For example, in the ^{1}H NMR spectrum of bromoethane shown in Figure 13.8, the $-CH_2Br$ protons appear as four peaks (a *quartet*) at 3.42 δ and the $-CH_3$ protons appear as three peaks (a *triplet*) at 1.68 δ.

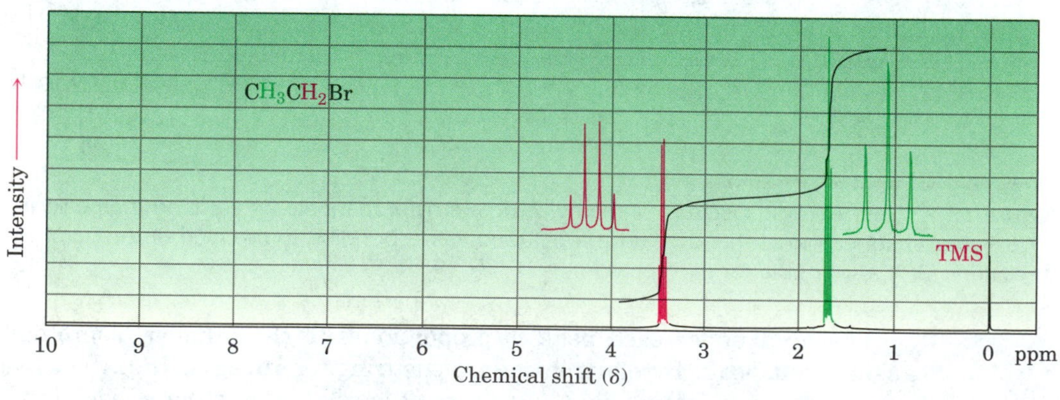

Figure 13.8 The ^{1}H NMR spectrum of bromoethane, CH_3CH_2Br. The $-CH_2Br$ protons appear as a quartet at 3.42 δ, and the $-CH_3$ protons appear as a triplet at 1.68 δ.

Called **spin–spin splitting**, the phenomenon of multiple absorptions is caused by the interaction, or **coupling**, of the nuclear spins of nearby atoms. In other words, the tiny magnetic field of one nucleus affects the magnetic field felt by neighboring nuclei.

How does spin–spin splitting arise? Let's review briefly. We've seen that, when a magnetic nucleus such as ^1H is placed in a magnetic field, the ^1H spin lines up either with or against the applied magnetic field. The applied magnetic field also causes electrons in the molecule to set up tiny local magnetic fields, which act in opposition to the applied field and therefore lessen the effective field actually felt by a nucleus. Differences in the extent of electron shielding at each nucleus account for the differences in chemical shifts between nuclei.

In addition to being affected by electron shielding, the magnetic field felt by a nucleus is also affected by neighboring magnetic nuclei. Look at the $-CH_3$ protons in bromoethane, for example. The three equivalent $-CH_3$ protons are neighbored by two other magnetic nuclei—the protons on the adjacent $-CH_2Br$ group. Each of the $-CH_2Br$ protons has its own nuclear spin, which can align either with or against the applied field, producing a tiny effect that is felt by the neighboring $-CH_3$ protons.

There are three ways in which the spins of the two $-CH_2Br$ protons can align, as shown schematically in Figure 13.9. If both proton spins align *with* the applied field, the total effective field felt by the neighboring $-CH_3$ protons is slightly larger than it would otherwise be. Consequently, the applied field necessary to cause resonance is slightly reduced. Alternatively, if one of the

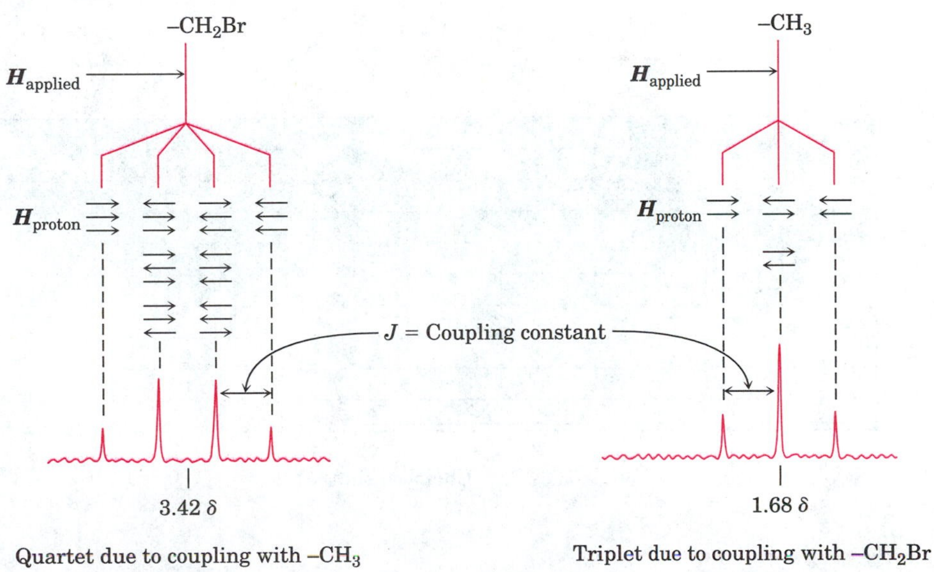

Figure 13.9 The origin of spin–spin splitting in bromoethane. The nuclear spins of neighboring protons, indicated by horizontal arrows, align either with or against the applied field, causing the splitting of absorptions into multiplets.

$-CH_2Br$ proton spins aligns *with* the field and one aligns *against* the field, there is no effect on the neighboring $-CH_3$ protons. (There are two ways this arrangement can occur, depending on which of the two proton spins aligns which way.) Finally, if both $-CH_2Br$ proton spins align *against* the applied field, the effective field felt by the $-CH_3$ protons is slightly smaller than it would otherwise be, and the applied field needed for resonance is slightly increased.

Any given molecule can adopt only one of the three possible alignments of $-CH_2Br$ spins, but in a large collection of molecules, all three spin states will be represented in a 1:2:1 statistical ratio. We therefore find that the neighboring $-CH_3$ protons come into resonance at three slightly different values of the applied field, and we see a 1:2:1 triplet in the NMR spectrum. One resonance is a little above where it would be without coupling, one is at the same place it would be without coupling, and the third resonance is a little below where it would be without coupling.

In the same way that the $-CH_3$ absorption of bromoethane is split into a triplet, the $-CH_2Br$ absorption is split into a quartet. The three spins of the neighboring $-CH_3$ protons can align in four possible combinations: all three with the applied field, two with and one against (three ways), one with and two against (three ways), or all three against. Thus, four peaks are produced for the $-CH_2Br$ protons in a 1:3:3:1 ratio.

As a general rule (called the ***n* + 1 rule**), protons that have n equivalent neighboring protons show $n + 1$ peaks in their NMR spectrum. For example, the spectrum of 2-bromopropane in Figure 13.10 shows a doublet at 1.71 δ and a seven-line multiplet, or *septet*, at 4.28 δ. The septet is caused by splitting of the –CHBr– proton signal by six equivalent neighboring protons

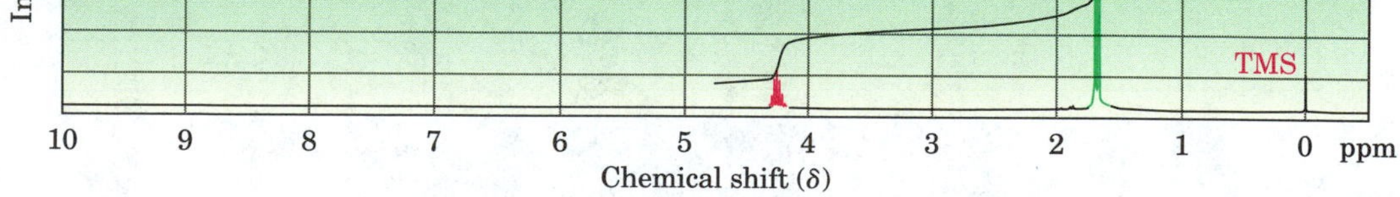

Figure 13.10 The ^1H NMR spectrum of 2-bromopropane. The $-CH_3$ proton signal at 1.71 δ is split into a doublet, and the –CHBr– proton signal at 4.28 δ is split into a septet.

on the two methyl groups (n = 6 leads to 6 + 1 = 7 peaks). The doublet is due to signal splitting of the six equivalent methyl protons by the single –CHBr– proton (n = 1 leads to 2 peaks). Integration confirms the expected 6:1 ratio.

The distance between peaks in a multiplet is called the **coupling constant**, denoted J. Coupling constants are measured in hertz and generally fall in the range 0–18 Hz. The exact value of the coupling constant between two neighboring protons depends on the geometry of the molecule, but a typical value for an open-chain alkane is J = 6–8 Hz. Note that the same coupling constant is shared by both groups of hydrogens whose spins are coupled and is independent of spectrometer field strength. In bromoethane, for instance, the $-CH_2Br$ protons are coupled to the $-CH_3$ protons and appear as a quartet with J = 7 Hz. The $-CH_3$ protons appear as a triplet with the same J = 7 Hz coupling constant.

Since coupling is a reciprocal interaction between two adjacent groups of protons, it's sometimes possible to tell which multiplets in a complex NMR spectrum are related to each other. If two multiplets have the same coupling constant, they are probably related, and the protons causing those multiplets are therefore adjacent in the molecule.

Spin–spin splitting in ^{1}H NMR can be summarized by three rules:

1. *Chemically equivalent protons do not show spin–spin splitting.* The equivalent protons may be on the same carbon or on different carbons, but their signals don't split.

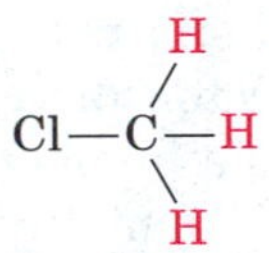

Three C–H protons are chemically equivalent; no splitting occurs.

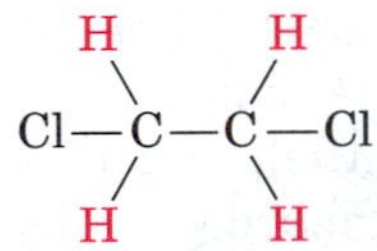

Four C–H protons are chemically equivalent; no splitting occurs.

2. *The signal of a proton that has n equivalent neighboring protons is split into a multiplet of n + 1 peaks with coupling constant J.* Protons that are farther than two carbon atoms apart don't usually couple, although they sometimes show small coupling when they are separated by a π bond.

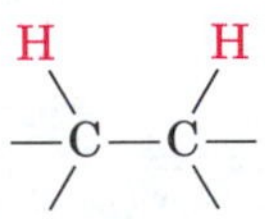

Splitting observed

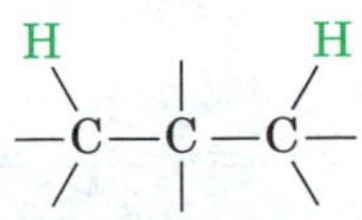

Splitting not usually observed

3. *Two groups of protons coupled to each other have the same coupling constant, J.*

The most commonly observed coupling patterns and the relative intensities of the multiplet signals are listed in Table 13.4.

Table 13.4 Some Common Spin Multiplicities

Number of equivalent adjacent protons	*Type of multiplet observed*	*Ratio of intensities*
0	Singlet	1
1	Doublet	1:1
2	Triplet	1:2:1
3	Quartet	1:3:3:1
4	Quintet	1:4:6:4:1
6	Septet	1:6:15:20:15:6:1

The spectrum of *para*-methoxypropiophenone in Figure 13.11 further illustrates the preceding three rules. The downfield absorptions at 6.91 and 7.93 δ are due to the four aromatic ring protons. There are two kinds of aromatic protons, each of which gives a signal that is split into a doublet by its neighbor. The $-OCH_3$ signal is unsplit and appears as a sharp singlet at 3.84 δ. The $-CH_2-$ protons next to the carbonyl group appear at 2.93 δ in the region expected for protons on carbon next to an unsaturated center, and their signal is split into a quartet by coupling with the protons of the neighboring methyl group. The methyl protons appear as a triplet at 1.20 δ in the usual upfield region.

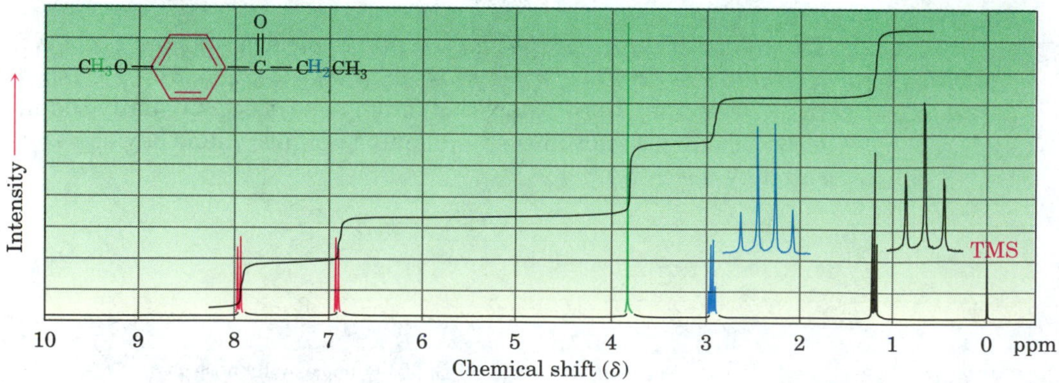

Figure 13.11 The ^{1}H NMR spectra of *para*-methoxypropiophenone.

PROBLEM.

13.11 Predict the splitting patterns you would expect for each proton in these molecules.

(a) $CHBr_2CH_3$ (b) $CH_3OCH_2CH_2Br$ (c) $ClCH_2CH_2CH_2Cl$

(d) $CH_3CH_2OC(=O)CH(CH_3)CH_3$

PROBLEM.

13.12 Draw structures for compounds that meet the following descriptions:

(a) C_2H_6O; one singlet (b) C_3H_7Cl; one doublet and one septet

(c) $C_4H_8Cl_2O$; two triplets (d) $C_4H_8O_2$; one singlet, one triplet, and one quartet

PROBLEM.

13.13 The integrated ^1H NMR spectrum of a compound of formula $C_4H_{10}O$ is shown. Propose a structure consistent with the data.

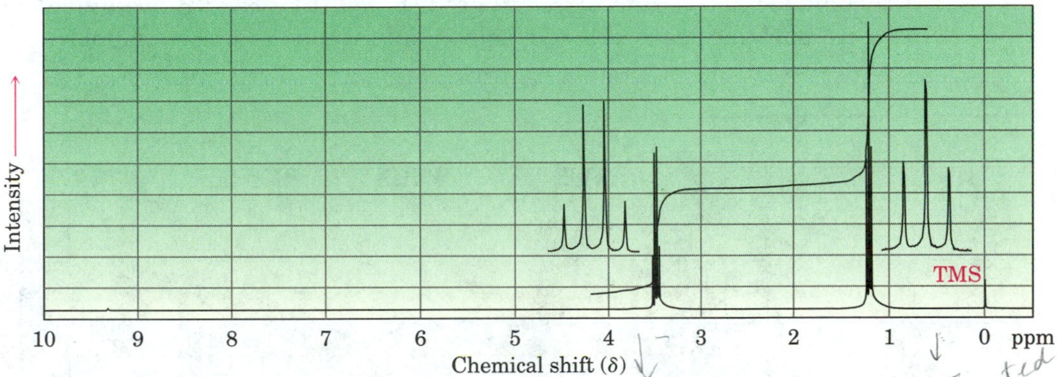

13.8 More Complex Spin–Spin Splitting Patterns

In all the NMR spectra we've seen so far, the chemical shifts of different protons have been distinct, and the spin–spin splitting patterns have been straightforward. It often happens, however, that different kinds of hydrogens in a molecule have *overlapping* signals. The spectrum of toluene (methylbenzene) in Figure 13.12, for example, shows that the five aromatic ring protons give a complex, overlapping pattern, even though they aren't all equivalent.

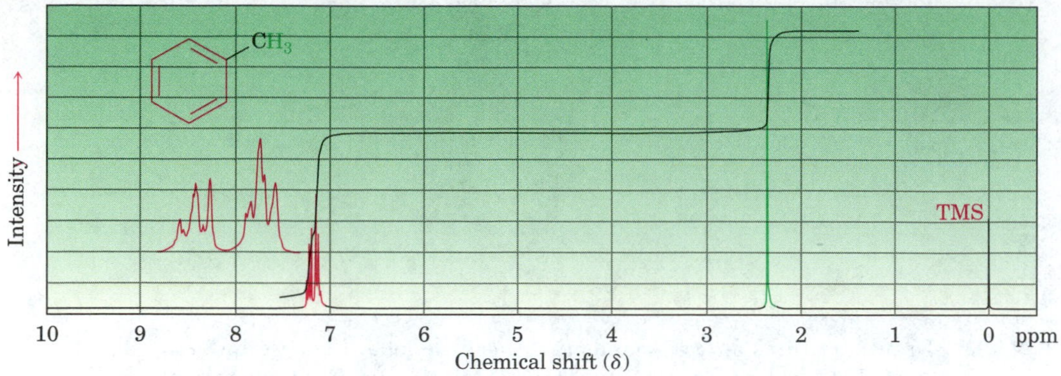

Figure 13.12 The ^{1}H NMR spectrum of toluene, showing the accidental overlap of the five nonequivalent aromatic ring protons.

Yet another complication in ^{1}H NMR spectroscopy arises when a signal is split by two or more *nonequivalent* kinds of protons, as is the case with *trans*-cinnamaldehyde, isolated from oil of cinnamon (Figure 13.13). Although the $n + 1$ rule correctly predicts splitting caused by *equivalent* protons, splittings caused by nonequivalent protons are more complex.

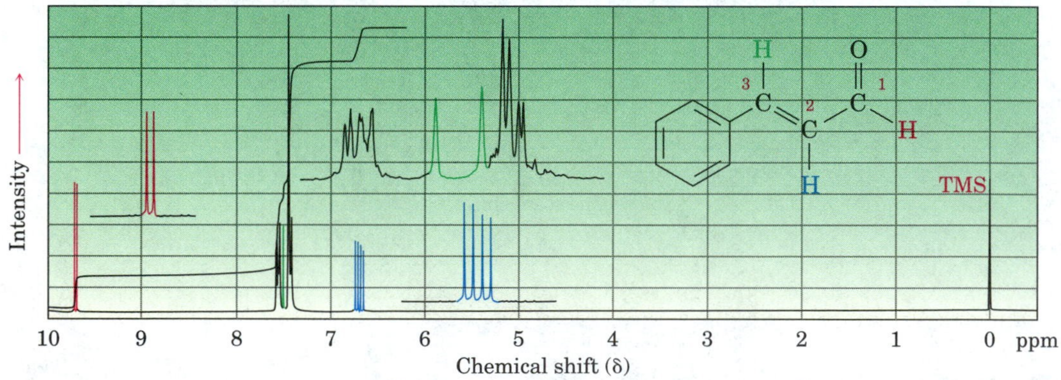

Figure 13.13 The ^{1}H NMR spectrum of *trans*-cinnamaldehyde. The signal of the proton at C2 is split into four peaks—a doublet of doublets—by the two nonequivalent neighboring protons.

To understand the ^{1}H NMR spectrum of *trans*-cinnamaldehyde, we have to isolate the different parts and look at the signal of each proton individually:

1. The five aromatic proton signals overlap into a complex pattern with a large peak at 7.42 δ and a broad absorption at 7.57 δ.

2. The aldehyde proton signal at C1 appears in the normal downfield position at 9.69 δ and is split into a doublet with J = 6 Hz by the adjacent proton at C2.
3. The vinylic proton at C3 is next to the aromatic ring and is therefore shifted downfield from the normal vinylic region. This C3 proton signal appears as a doublet centered at 7.49 δ. Because it has one neighbor proton at C2, its signal is split into a doublet, with J = 12 Hz.
4. The C2 vinylic proton signal appears at 6.73 δ and shows an interesting, four-line absorption pattern. It is coupled to the two nonequivalent protons at C1 and C3 with two different coupling constants: $J_{1\text{-}2}$ = 6 Hz and $J_{2\text{-}3}$ = 12 Hz.

The best way to understand the effect of multiple coupling such as occurs for the C2 proton of *trans*-cinnamaldehyde is to draw a *tree diagram*, like that in Figure 13.14. The diagram shows the individual effect of each coupling constant on the overall pattern. Coupling with the C3 proton splits the signal of the C2 proton in *trans*-cinnamaldehyde into a doublet with J =12 Hz. Further 6 Hz coupling with the aldehyde proton then splits each peak of the doublet into new doublets, and we therefore observe a four-line spectrum for the C2 proton.

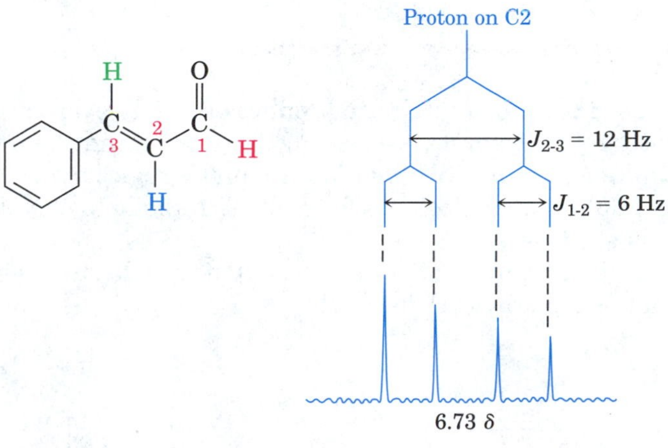

Figure 13.14 A tree diagram for *trans*-cinnamaldehyde shows how the C2 proton is coupled to the C1 and C3 protons with different coupling constants.

One further point about the cinnamaldehyde spectrum is that the four peaks of the C2 proton signal are not all the same size: The two left-hand peaks are somewhat larger than the two right-hand peaks. Such a size difference occurs whenever coupled nuclei have similar chemical shifts—in this case, 7.49 δ for the C3 proton and 6.73 δ for the C2 proton. The peaks nearer the signal of the coupled partner are always larger, and the peaks farther from the signal of the coupled partner are always smaller. Thus, the left-hand peaks of the C2 proton multiplet at 6.73 δ are closer to the C3 proton absorption at 7.49 δ and are larger than the right-hand peaks. (At the same time, the *right-hand* peak of the C3 proton doublet at 7.49 δ is larger than the left-hand peak because it is closer to the C2 proton multiplet at 6.73 δ.) This skewing effect on multiplets can often be useful because it tells where to look in the spectrum to find the coupled partner: Look toward the direction of the larger peaks.

PROBLEM

13.14 3-Bromo-1-phenyl-1-propene shows a complex NMR spectrum in which the vinylic proton at C2 is coupled with both the C1 vinylic proton (J = 16 Hz) and the C3 methylene protons (J = 8 Hz). Draw a tree diagram for the C2 proton signal and account for the fact that a five-line multiplet is observed.

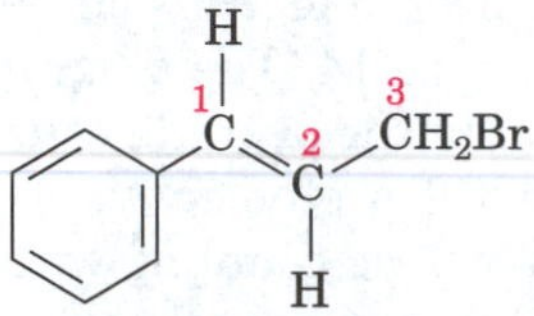

3-Bromo-1-phenyl-1-propene

13.9 Uses of ^{1}H NMR Spectroscopy

NMR can be used to help identify the product of nearly every reaction run in the laboratory. For example, we said in Section 7.5 that hydroboration/oxidation of alkenes occurs with non-Markovnikov regiochemistry; that is, the less highly substituted alcohol is formed. With the help of NMR, we can now prove this statement.

Does hydroboration/oxidation of methylenecyclohexane yield cyclohexylmethanol or 1-methylcyclohexanol?

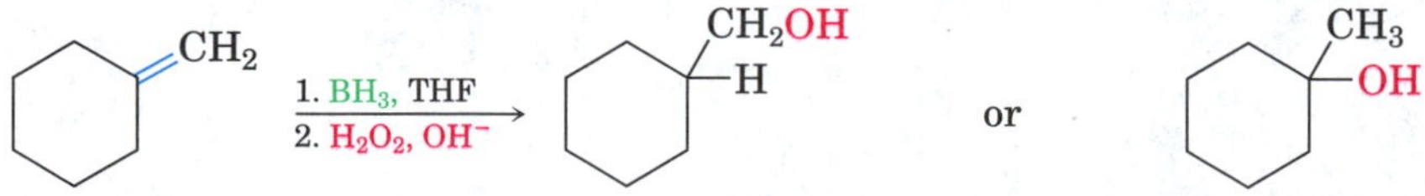

Methylenecyclohexane **Cyclohexylmethanol** **1-Methylcyclohexanol**

The ^{1}H NMR spectrum of the reaction product is shown in Figure 13.15(a). The spectrum shows a two-proton triplet absorption at 3.40 δ, indicating that the product has a $-CH_2-$ group bonded to an electronegative oxygen atom ($-CH_2OH$). Furthermore, the spectrum shows *no* large three-proton singlet absorption near 1 δ, where we would expect the signal of a quaternary $-CH_3$ group to occur. [Figure 13.15(b) gives the spectrum of 1-methylcyclohexanol, the alternative product.] Thus, it's clear that cyclohexylmethanol is the reaction product.

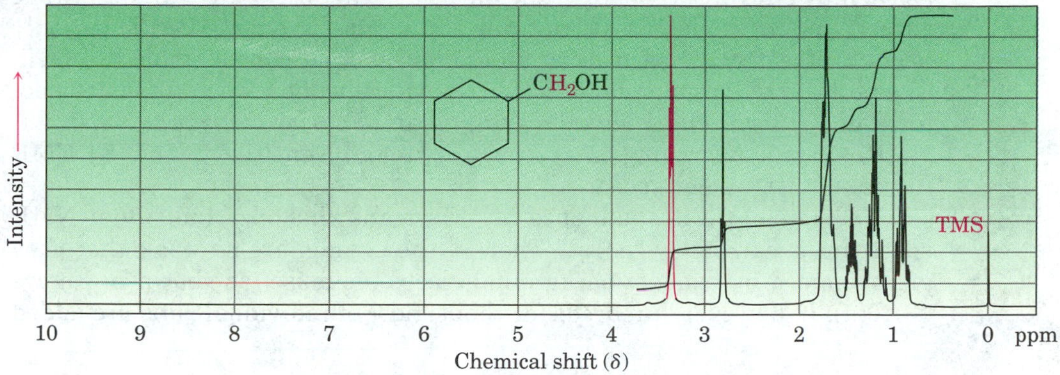

(a)

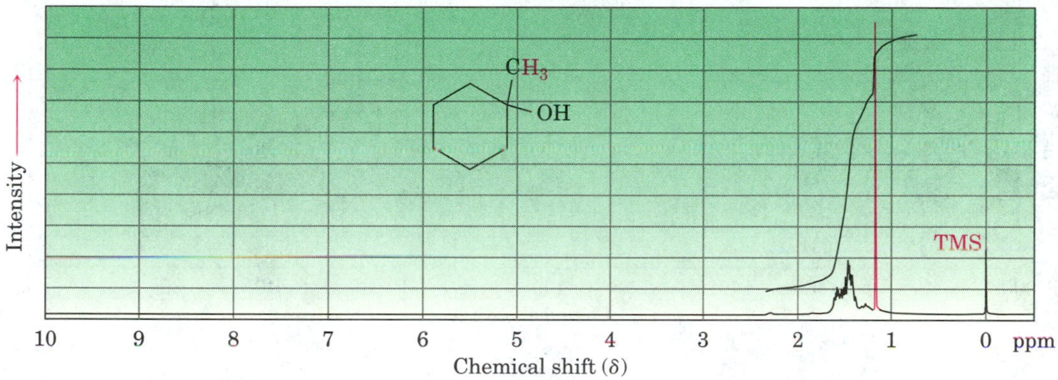

(b)

Figure 13.15 (a) The ^1H NMR spectrum of cyclohexylmethanol, the product from hydroboration/oxidation of methylenecyclohexane, and (b) the ^1H NMR spectrum of 1-methylcyclohexanol, the possible alternative reaction product.

PROBLEM.

13.15 How could you use ^1H NMR to help you determine the regiochemistry of electrophilic addition to alkenes? Does addition of HCl to 1-methylcyclohexene yield 1-chloro-1-methylcyclohexane or 1-chloro-2-methylcyclohexane?

13.10 ^{13}C NMR Spectroscopy: Signal Averaging and FT-NMR

Having looked in some detail at ^1H NMR, let's now do the same for ^{13}C NMR. In some ways, it's surprising that carbon NMR is even possible. After all, ^{12}C, the most abundant carbon isotope, has no nuclear spin and can't be seen by NMR. Carbon-13 is the only naturally occurring carbon isotope with a nuclear spin, but its natural abundance is only 1.1%. Thus, only about 1 of every 100 carbons in an organic sample is observable by NMR. The problem of low abundance has been overcome, however, by the development of two techniques: *signal averaging* and *Fourier-transform NMR* (FT-NMR). Signal averaging increases instrument sensitivity, and FT-NMR increases instrument speed.

The low natural abundance of ^{13}C means that any individual NMR spectrum is extremely "noisy." That is, the signals are so weak that they are swamped by random background electronic noise, as shown in Figure 13.16(a). If, however, hundreds (or thousands) of individual runs are added

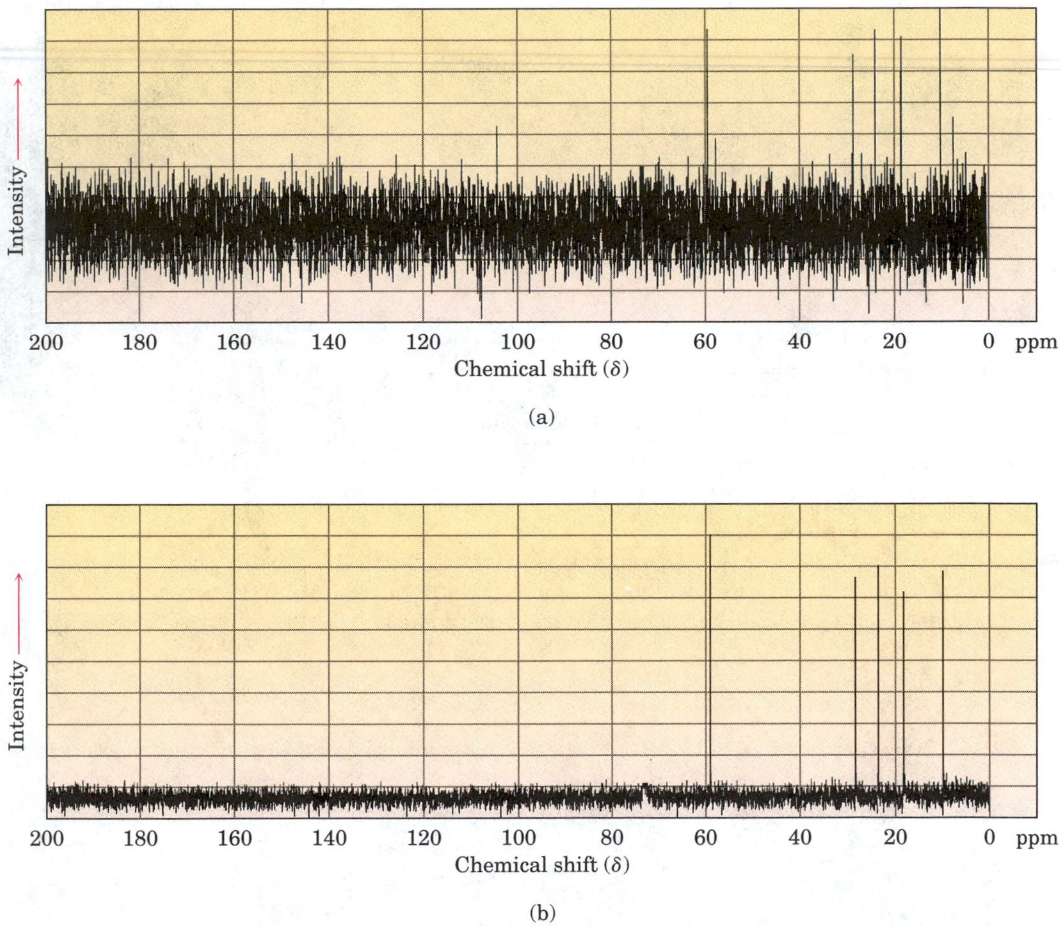

Figure 13.16 Carbon-13 NMR spectra of 1-pentanol, $CH_3CH_2CH_2CH_2CH_2OH$. Spectrum (a) is a single run, showing the large amount of background noise. Spectrum (b) is an average of 200 runs.

together by computer and then averaged, a greatly improved spectrum results [Figure 13.16(b)]. Background noise, because of its random nature, averages to zero, allowing the signals themselves to stand out clearly. Unfortunately, the value of signal averaging is limited when using the method of NMR spectrometer operation described at the beginning of this chapter, because it takes about 5–10 minutes to obtain a single spectrum. Thus, a faster way to obtain spectra is needed if signal averaging is to be used.

Recall that in the method of NMR spectrometer operation described in Section 13.2, either the rf frequency is held constant while the strength of the magnetic field is varied or the strength of the magnetic field is held constant while the rf frequency is varied. In either case, each signal in the spectrum is recorded sequentially. In the FT-NMR technique used by modern spectrometers, a sample is placed in a magnetic field of constant strength and is irradiated with a short burst, or "pulse," of rf energy that covers the entire range of useful frequencies at once. All ^{1}H or ^{13}C nuclei in the sample resonate at once, giving a complex, composite signal that must be mathematically manipulated using so-called *Fourier transforms* before it can be displayed in the usual way. Since all resonance signals are collected at once, it takes only a few seconds rather than a few minutes to record an entire spectrum.

Combining the speed of FT-NMR with the sensitivity enhancement of signal averaging is what gives modern NMR spectrometers their power. Literally thousands of spectra can be taken and averaged in a few hours, resulting in sensitivity so high that ^{13}C NMR spectra can be obtained with only a few milligrams of material, and ^{1}H spectra can be recorded with only a few *micro*grams.

13.11 Characteristics of ^{13}C NMR Spectroscopy

Look at the ^{13}C NMR spectra of methyl acetate and 1-pentanol shown previously in Figures 13.3(b) and 13.16(b). In each case, a single sharp resonance line is observed for each nonequivalent kind of carbon atom. As a result, ^{13}C NMR makes it possible to count the number of carbons in a molecule. Why, though, is there no splitting of carbon signals into multiplets, as occurs in ^{1}H NMR (Section 13.7)? You might expect that the spin of a given ^{13}C nucleus would couple with the spin of an adjacent magnetic nucleus (either ^{13}C or ^{1}H) leading to spin–spin splitting. In fact, the spins of nearby magnetic nuclei *do* couple, but the spectrometer operating conditions are such that the splitting of signals is suppressed.

No coupling of a ^{13}C nucleus with nearby *carbons* is seen because the low natural abundance of ^{13}C makes it unlikely that two such nuclei will be adjacent. No coupling of a ^{13}C nucleus with nearby *hydrogens* is seen because ^{13}C spectra are normally recorded using what is called *broadband decoupling*. At the same time that the sample is irradiated with a pulse of rf energy to cover the *carbon* resonance frequencies, it is also irradiated by a second broad band of rf energy covering all the *hydrogen* resonance frequencies. This second irradiation makes the hydrogens flip spins so rapidly that their local magnetic fields average to zero and no coupling with carbon spins occurs.

Most ^{13}C resonances are between 0 and 220 ppm downfield from the TMS reference line, with the exact chemical shift of each ^{13}C resonance dependent on that carbon's electronic environment within the molecule. Figure 13.17 shows the correlation of environment with chemical shift.

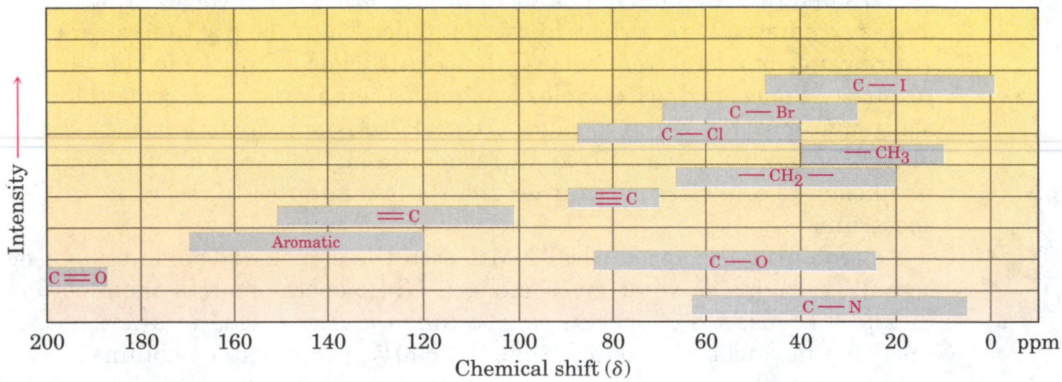

Figure 13.17 Chemical shift correlations for ^{13}C NMR.

The factors that determine chemical shifts are complex, but it's possible to make some generalizations from the data in Figure 13.17. One trend is that a carbon's chemical shift is affected by the electronegativity of nearby atoms: Carbons bonded to oxygen, nitrogen, or halogen absorb downfield (to the left) of typical alkane carbons. Since electronegative atoms attract electrons, they pull electrons away from neighboring carbon atoms, causing those carbons to be deshielded and to come into resonance at a lower field.

Another trend is that sp^3-hybridized carbons generally absorb in the range 0–90 δ, while sp^2 carbons absorb in the range 100–220 δ. Carbonyl carbons (C=O) are particularly distinct in ^{13}C NMR and are always found at the extreme low-field end of the spectrum, in the range 170–220 δ. Figure 13.18 shows the ^{13}C NMR spectra of 2-butanone and *para*-bromoacetophenone, and indicates the peak assignments. Note that the carbonyl carbons are at the left edge of the spectrum in each case.

The ^{13}C NMR spectrum of *para*-bromoacetophenone is interesting in several ways. Note particularly that only six carbon absorptions are observed even though the molecule contains eight carbons. *para*-Bromoacetophenone has a symmetry plane that makes ring carbons 4 and 4′, and carbons 5 and 5′, equivalent. Thus, the six ring carbons show only four absorptions in the range 128–137 δ.

A second interesting point about both spectra is that the peaks aren't uniform in size. Some peaks are larger than others even though all are one-carbon resonances (except for the two two-carbon peaks of *para*-bromoacetophenone). This difference in peak size is caused by several factors that we won't go into, but is a general feature of ^{13}C NMR spectra. (The size difference also makes it difficult, though not impossible, to integrate peaks in ^{13}C spectra. Integration is rarely necessary for ^{13}C spectra, however, because most peaks are caused by a single carbon.)

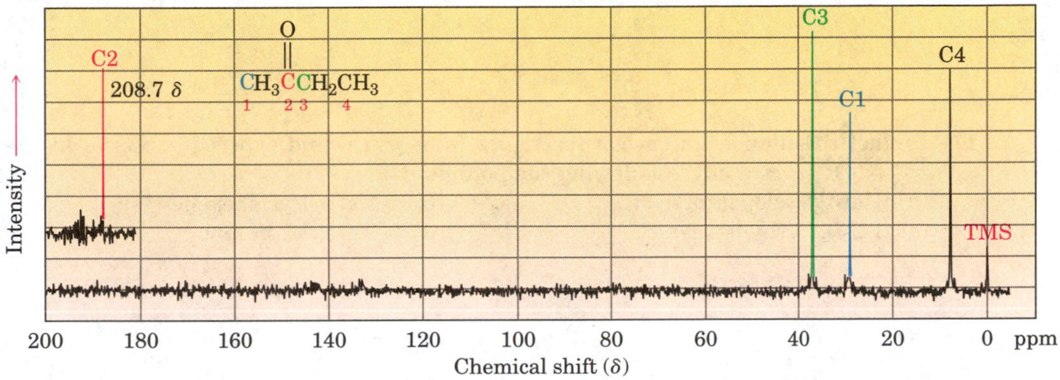

(a)

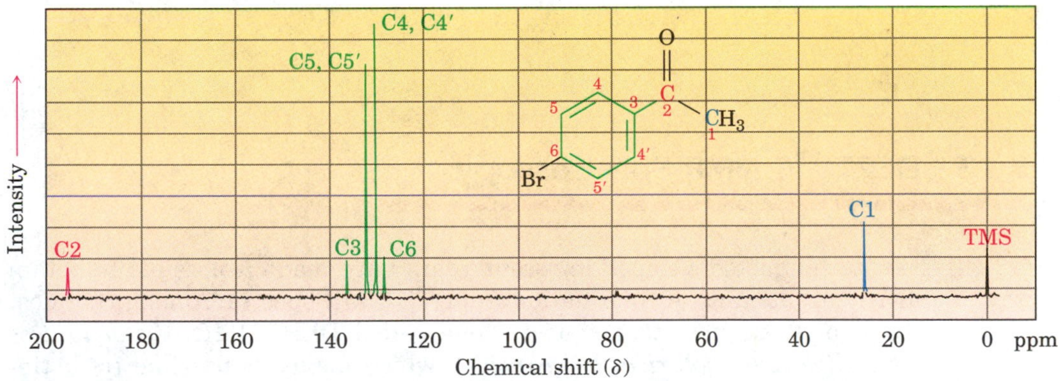

(b)

Figure 13.18 Carbon-13 NMR spectra of (a) 2-butanone and (b) *para*-bromoacetophenone.

PROBLEM

13.16 Assign the resonances in the following ^{13}C NMR spectrum of methyl propanoate, $CH_3CH_2CO_2CH_3$.

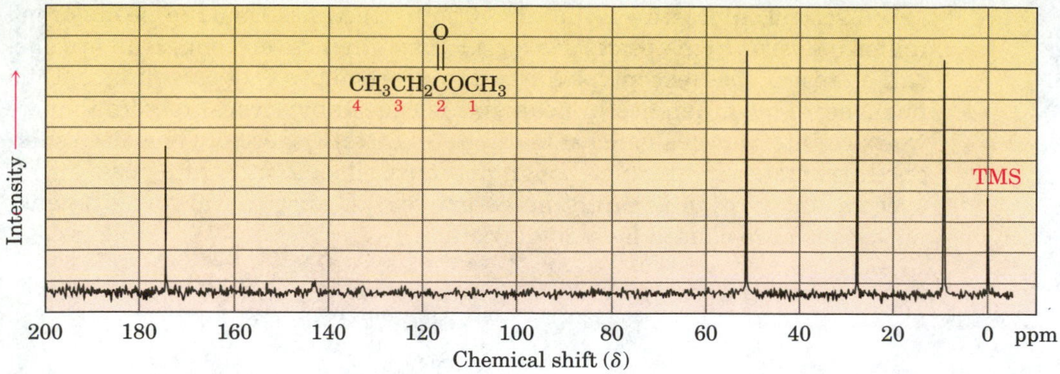

PROBLEM

13.17 Predict the number of carbon resonance lines you would expect to observe in the ^{13}C NMR spectra of the following compounds:

(a) Methylcyclopentane
(b) 1-Methylcyclohexene
(c) 1,2-Dimethylbenzene
(d) 2-Methyl-2-butene

PROBLEM

13.18 Propose structures for compounds that fit the following descriptions.

(a) A hydrocarbon with seven lines in its ^{13}C NMR spectrum
(b) A six-carbon compound with only five lines in its ^{13}C NMR spectrum
(c) A four-carbon compound with three lines in its ^{13}C NMR spectrum

13.12 DEPT ^{13}C NMR Spectroscopy

New techniques developed in recent years have made it possible to obtain enormous amounts of information from ^{13}C NMR spectra. Among the most useful of these new techniques is one called DEPT-NMR (*distortionless enhancement by polarization transfer*), which makes it possible to distinguish among signals due to CH_3, CH_2, CH, and quaternary carbons. That is, the number of hydrogens attached to each carbon can be determined.

A DEPT experiment is usually done in three stages, as shown in Figure 13.19 for 6-methyl-5-hepten-2-ol. The first stage is to run an ordinary broadband-decoupled spectrum to locate the chemical shifts of all carbons. Next, a second spectrum called a DEPT-90 is run, using special conditions under which *only signals due to CH carbons appear*. Signals due to CH_3, CH_2, and quaternary carbons are absent. Finally, a third spectrum called a DEPT-135 is run, using conditions under which CH_3 and CH resonances appear as positive signals, but CH_2 resonances appear as *negative* signals—that is, as peaks *below* the baseline.

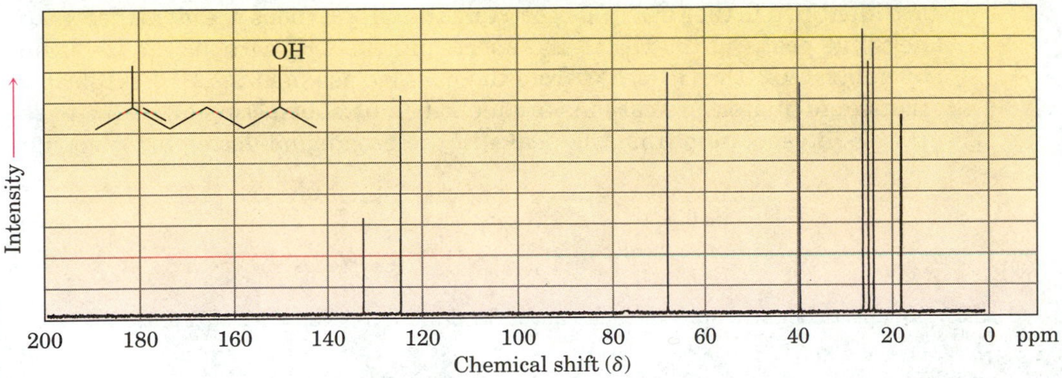

(a)

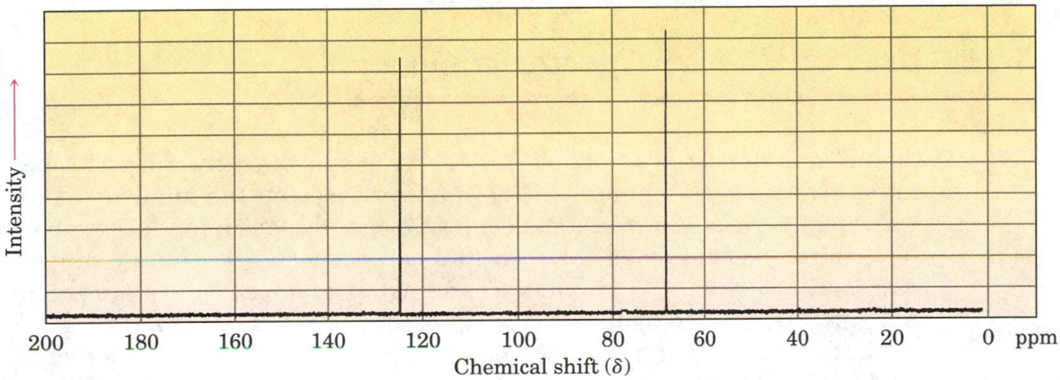

(b)

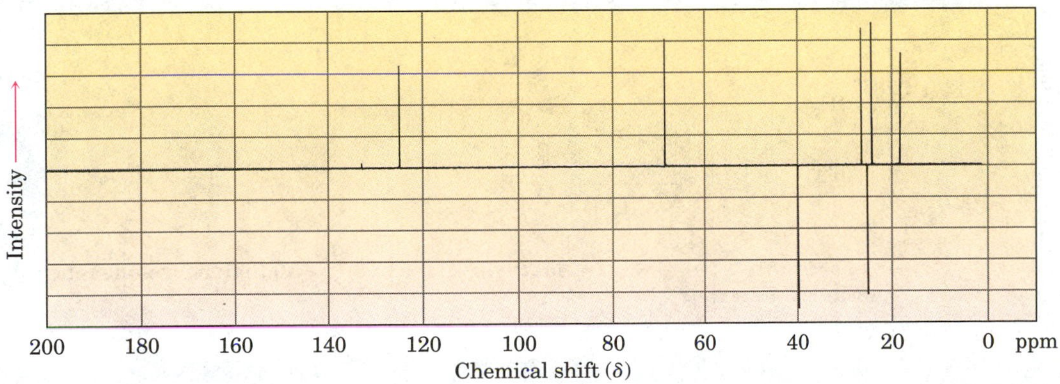

(c)

Figure 13.19 DEPT-NMR spectra for 6-methyl-5-hepten-2-ol. Part (a) is a broadband-decoupled spectrum, which shows signals for all eight carbons. Part (b) is a DEPT-90 spectrum, which shows only signals for the two CH carbons. Part (c) is a DEPT-135 spectrum, which shows positive signals for the two CH and three CH_3 carbons and negative signals for the two CH_2 carbons.

Putting together the information from all three spectra makes it possible to tell the number of hydrogens attached to each carbon. The CH carbons are identified in the DEPT-90 spectrum; the CH_2 carbons are identified as the negative peaks in the DEPT-135 spectrum; the CH_3 carbons are identified by subtracting the CH peaks from the positive peaks in the DEPT-135 spectrum; and quaternary carbons are identified by subtracting all peaks in the DEPT-135 spectrum from the peaks in the broadband-decoupled spectrum.

PROBLEM

13.19 Assign a chemical shift to each carbon in 6-methyl-5-hepten-2-ol (Figure 13.19).

13.13 Uses of ^{13}C NMR Spectroscopy

The information derived from ^{13}C NMR spectroscopy is extraordinarily useful for structure determination. Not only can we count the number of nonequivalent carbon atoms in a molecule, we can also get information about the electronic environment of each and can even find how many protons each is attached to. As a result, we can answer many structural questions that can't be handled by infrared spectroscopy or mass spectrometry.

Let's take an example. How might we prove that E2 elimination of an alkyl halide gives the more highly substituted alkene (Zaitsev's rule, Section 11.10)? Does reaction of 1-chloro-1-methylcyclohexane with strong base lead predominantly to 1-methylcyclohexene or to methylenecyclohexane?

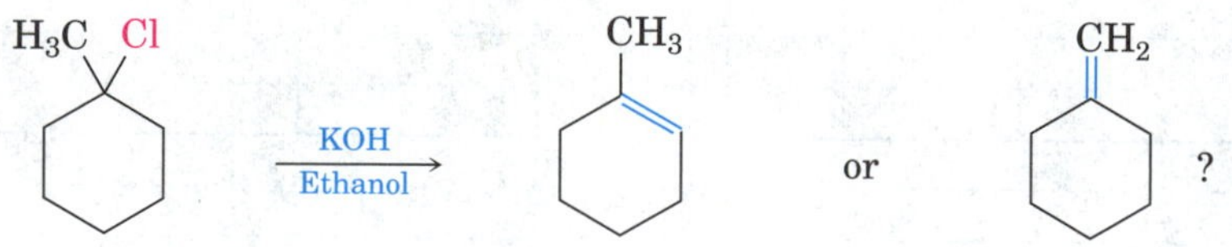

1-Chloro-1-methylcyclohexane **1-Methylcyclohexene** **Methylenecyclohexane**

1-Methylcyclohexene should have five sp^3-carbon resonances in the range 20–50 δ and two sp^2-carbon resonances in the range 100–150 δ. Methylenecyclohexane, however, because of its symmetry, should have only three sp^3-carbon resonance peaks and two sp^2-carbon peaks. The spectrum of the actual reaction product, shown in Figure 13.20, clearly identifies 1-methylcyclohexene as the substance formed in this E2 reaction.

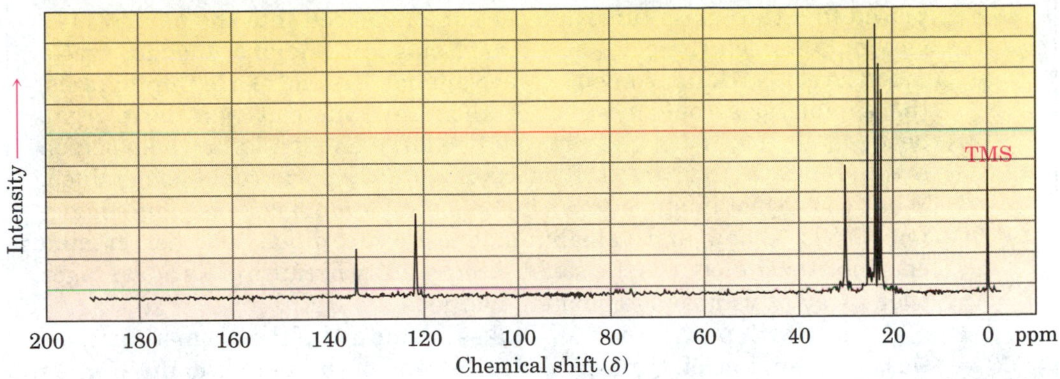

Figure 13.20 The ^{13}C NMR spectrum of 1-methylcyclohexene, the E2 reaction product from 1-chloro-1-methylcyclohexane.

PROBLEM

13.20 We saw in Section 7.10 that addition of HBr to alkenes under radical conditions leads to the non-Markovnikov product, with the bromine bonding to the less highly substituted carbon. How could you use ^{13}C NMR to identify the product of radical addition of HBr to 2-methylpropene?

INTERLUDE

Magnetic Resonance Imaging (MRI)

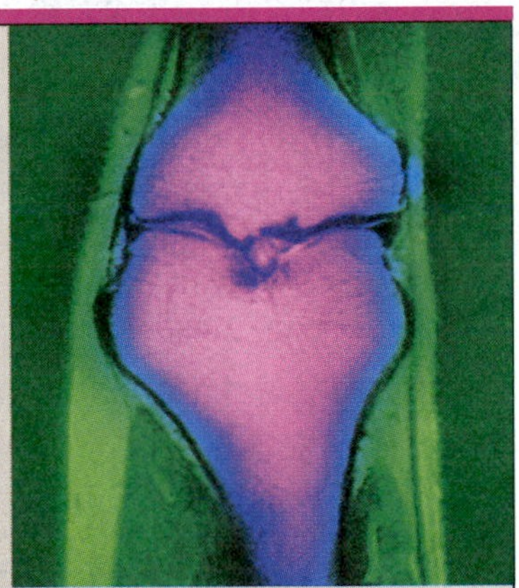

A false-color MRI image of a normal human knee, showing the femur (top) and tibia (bottom).

As practiced by organic chemists, NMR spectroscopy is a powerful method of structure determination. A small amount of sample, typically a few milligrams or less, is dissolved in 1 mL or so of solvent, the solution is

(continued)➤

placed in a thin glass tube, and the tube is placed into the narrow (1–2 cm) gap between the poles of a strong magnet. Imagine, though, that a much larger NMR instrument were available. Instead of a few milligrams, the sample size could be tens of kilograms; instead of a narrow gap between magnet poles, the gap could be large enough for a whole person to climb into so that an NMR spectrum of body parts could be obtained. What you've just imagined is an instrument for *magnetic resonance imaging* (*MRI*), a new and valuable diagnostic technique that has created enormous excitement in the medical community because of its advantages over X-ray or radioactive imaging methods.

Like NMR spectroscopy, MRI takes advantage of the magnetic properties of certain nuclei, typically hydrogen, and of the signals emitted when those nuclei are stimulated by radiofrequency energy. Unlike what happens in NMR spectroscopy, though, MRI instruments use powerful computers and data manipulation techniques to look at the three-dimensional *location* of magnetic nuclei in the body rather than at the chemical nature of the nuclei. As noted, most MRI instruments currently look at hydrogen, present in abundance wherever there is water or fat in the body.

The signals produced vary with the density of hydrogen atoms and with the nature of their surroundings, allowing identification of different types of tissue and even the visualization of motion. For example, the volume of blood leaving the heart in a single stroke can be measured, allowing observation of the heart in motion. Soft tissues that do not show up well on X rays can also be seen clearly, allowing diagnosis of brain tumors, strokes, and other conditions. The technique is also valuable in diagnosing damage to knees or other joints and is a painless alternative to arthroscopy, in which an endoscope is physically introduced into the knee joint.

Several types of atoms in addition to hydrogen can be detected by MRI, and the applications of images based on ^{31}P atoms are being explored. The technique holds great promise for studies of metabolism.

Summary and Key Words

When magnetic nuclei such as ^{1}H and ^{13}C are placed in a strong magnetic field, their spins orient either with or against the field. On irradiation with radiofrequency (rf) waves, energy is absorbed and the nuclei "spin-flip" from the lower energy state to the higher energy state. This absorption of rf energy is detected, amplified, and displayed as a **nuclear magnetic resonance (NMR) spectrum**.

A ^{1}H NMR spectrum is typically obtained by irradiating a sample with rf energy of constant frequency while slowly changing the value of the applied magnetic field. Each chemically distinct ^{1}H nucleus in a molecule comes into resonance at a slightly different value of the applied field, thereby producing a unique absorption signal. The exact position of each peak is called the **chemical shift**. Chemical shifts are caused by electrons setting up tiny local magnetic fields that **shield** a nearby nucleus from the applied field.

The NMR chart is calibrated in **delta units** (δ), where 1 δ = 1 part per million (ppm) of spectrometer frequency. Tetramethylsilane (TMS) is used as a reference point to which other peaks are compared because it shows both ^{1}H and ^{13}C absorptions at unusually high values of the applied magnetic field. The TMS absorption occurs at the right-hand (**upfield**) side of the chart and is arbitrarily assigned a value of 0 δ.

The area under each ^{1}H NMR absorption peak can be electronically **integrated** to determine the relative number of hydrogens responsible for each peak. In addition, neighboring nuclear spins can **couple**, splitting NMR peaks into **multiplets**. The NMR signal of a hydrogen neighbored by n equivalent adjacent hydrogens splits into $n + 1$ peaks (the $\boldsymbol{n + 1}$ **rule**) with **coupling constant** J.

Most ^{13}C spectra are run on Fourier-transform NMR (**FT-NMR**) spectrometers using broadband decoupling of proton spins so that each chemically distinct carbon shows a single unsplit resonance line. As with ^{1}H NMR, the chemical shift of each ^{13}C signal provides information about a carbon's chemical environment in the sample. In addition, the number of protons attached to each carbon can be determined using the **DEPT-NMR** technique.

ADDITIONAL PROBLEMS

13.21 The following ^{1}H NMR absorptions were obtained on a spectrometer operating at 60 MHz and are given in hertz downfield from the TMS standard. Convert the absorptions to δ units.
(a) 131 Hz (b) 287 Hz (c) 451 Hz (d) 543 Hz

13.22 The following ^{1}H NMR absorptions given in δ units were obtained on a spectrometer operating at 80 MHz. Convert the chemical shifts from δ units to hertz downfield from TMS.
(a) 2.1 δ (b) 3.45 δ (c) 6.30 δ (d) 7.70 δ

13.23 When measured on a spectrometer operating at 60 MHz, chloroform ($CHCl_3$) shows a single sharp absorption at 7.3 δ.
(a) How many parts per million downfield from TMS does chloroform absorb?
(b) How many hertz downfield from TMS would chloroform absorb if the measurement were carried out on a spectrometer operating at 360 MHz?
(c) What would be the position of the chloroform absorption in δ units when measured on a 360 MHz spectrometer?

13.24 How many absorptions would you expect to observe in the ^{13}C NMR spectra of the following compounds?
(a) 1,1-Dimethylcyclohexane
(b) $CH_3CH_2OCH_3$
(c) *tert*-Butylcyclohexane
(d) 3-Methyl-1-pentyne
(e) *cis*-1,2-Dimethylcyclohexane
(f) 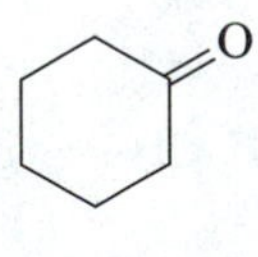

13.25 Suppose you ran a DEPT-135 spectrum for each substance in Problem 13.24. Indicate which carbon atoms in each molecule would show positive peaks and which would show negative peaks.

13.26 Why do you suppose accidental overlap of signals is much more common in ^{1}H NMR than in ^{13}C NMR?

13.27 What is meant by each of the following terms?
(a) Chemical shift (b) Spin–spin splitting
(c) Applied magnetic field (d) Spectrometer operating frequency
(e) Coupling constant (f) Upfield/downfield

13.28 How many types of nonequivalent protons are there in each of the following molecules?

(a) 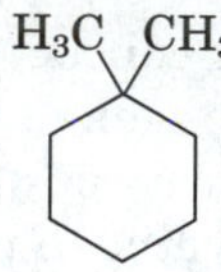

(b) $CH_3CH_2CH_2OCH_3$

(c) 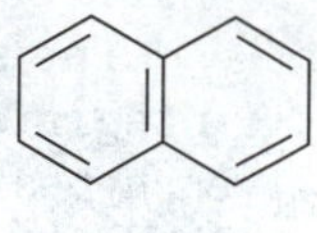

Naphthalene

(d) 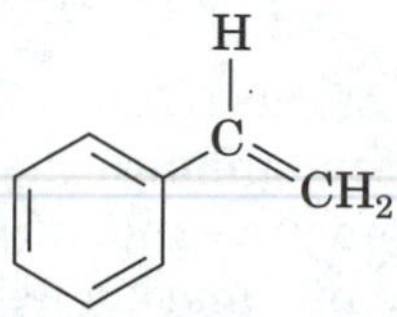

Styrene

(e) 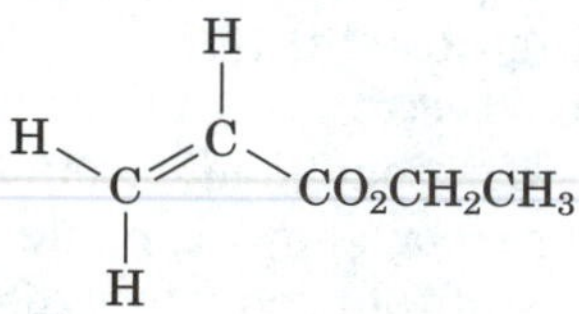

Ethyl acrylate

13.29 The following compounds all show a single line in their ^{1}H NMR spectra. List them in expected order of increasing chemical shift.

CH_4; CH_2Cl_2; cyclohexane; CH_3COCH_3; $H_2C{=}CH_2$; benzene

13.30 Predict the splitting pattern for each kind of hydrogen in these molecules.
(a) $(CH_3)_3CH$ (b) $CH_3CH_2COOCH_3$ (c) *trans*-2-Butene

13.31 Predict the splitting pattern for each kind of hydrogen in isopropyl propanoate, $CH_3CH_2COOCH(CH_3)_2$.

13.32 The acid-catalyzed dehydration of 1-methylcyclohexanol yields a mixture of two alkenes. How would you use ^{1}H NMR to help you decide which was which?

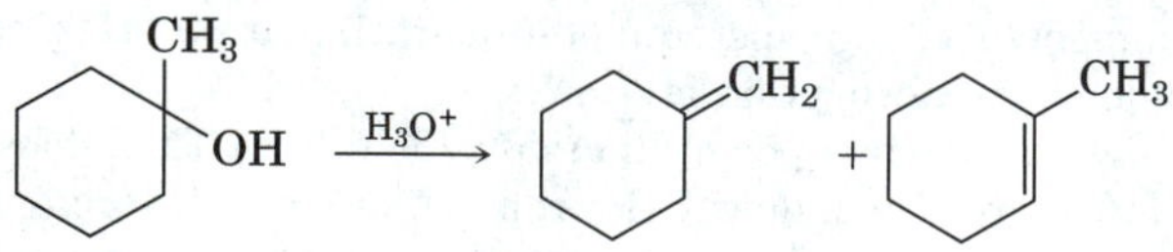

13.33 How would you use ^{1}H NMR to distinguish between the following pairs of isomers?

(a) $CH_3CH{=}CHCH_2CH_3$ and $H_2C{-}CHCH_2CH_3$ (with a CH_2 bridging the two ring carbons)

(b) $CH_3CH_2OCH_2CH_3$ and $CH_3OCH_2CH_2CH_3$

(c) $CH_3\overset{O}{\overset{\|}{C}}OCH_2CH_3$ and $CH_3CH_2\overset{O}{\overset{\|}{C}}CH_3$

(d) $H_2C{=}C(CH_3)\overset{O}{\overset{\|}{C}}CH_3$ and $CH_3CH{=}CH\overset{O}{\overset{\|}{C}}CH_3$

13.34 Propose structures for compounds with the following formulas that show only one peak in their ^{1}H NMR spectra.
(a) C_5H_{12} (b) C_5H_{10} (c) $C_4H_8O_2$

13.35 Assume that you have a compound with formula C_3H_6O.
(a) How many double bonds and/or rings does your compound contain?
(b) Propose as many structures as you can that fit the molecular formula.
(c) If your compound shows an infrared absorption peak at 1715 cm^{-1}, what functional group does it have?
(d) If your compound shows a single ^{1}H NMR absorption peak at 2.1 δ, what is its structure?

13.36 How would you use ^{1}H and ^{13}C NMR to help you distinguish among the following isomeric compounds of formula C_4H_8?

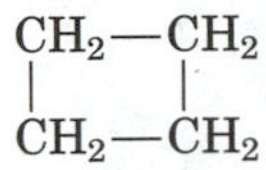

$H_2C{=}CHCH_2CH_3$ $CH_3CH{=}CHCH_3$ $(CH_3)_2C{=}CH_2$

13.37 How could you use ^{1}H and ^{13}C NMR to help you distinguish between these structures?

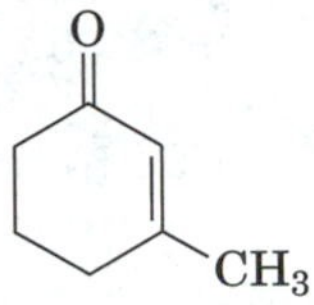

3-Methyl-2-cyclohexenone

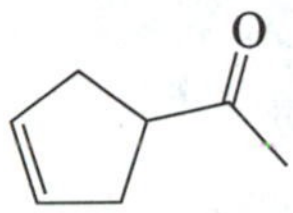

4-Cyclopentenyl methyl ketone

13.38 How could you use IR spectroscopy to help you distinguish between the two compounds shown in Problem 13.37?

13.39 The compound whose ^{1}H NMR spectrum is shown has the molecular formula $C_3H_6Br_2$. Propose a structure.

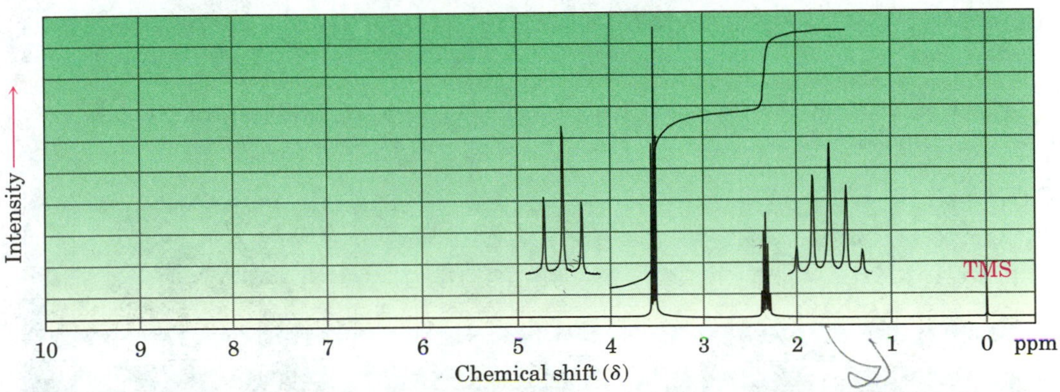

13.40 Propose structures for compounds that fit the following ^{1}H NMR data:
(a) $C_5H_{10}O$:
0.95 δ (6 H, doublet, J = 7 Hz)
2.10 δ (3 H, singlet)
2.43 δ (1 H, multiplet)

(b) C_3H_5Br:
2.32 δ (3 H, singlet)
5.35 δ (1 H, broad singlet)
5.54 δ (1 H, broad singlet)

13.41 The compound whose ^{1}H NMR spectrum is shown has the molecular formula $C_4H_7O_2Cl$ and has an infrared absorption peak at 1740 cm^{-1}. Propose a structure.

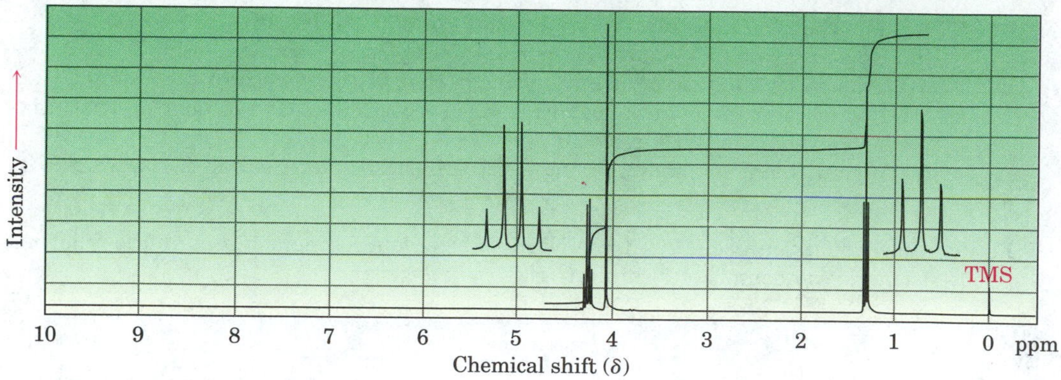

13.42 Propose structures for compounds that fit the following ^{1}H NMR data:

(a) $C_4H_6Cl_2$:
- 2.18 δ (3 H, singlet)
- 4.16 δ (2 H, doublet, J = 7 Hz)
- 5.71 δ (1 H, triplet, J = 7 Hz)

(b) $C_{10}H_{14}$:
- 1.30 δ (9 H, singlet)
- 7.30 δ (5 H, singlet)

(c) C_4H_7BrO:
- 2.11 δ (3 H, singlet)
- 3.52 δ (2 H, triplet, J = 6 Hz)
- 4.40 δ (2 H, triplet, J = 6 Hz)

(d) $C_9H_{11}Br$:
- 2.15 δ (2 H, quintet, J = 7 Hz)
- 2.75 δ (2 H, triplet, J = 7 Hz)
- 3.38 δ (2 H, triplet, J = 7 Hz)
- 7.22 δ (5 H, singlet)

13.43 How might you use NMR (either ^{1}H or ^{13}C) to differentiate between the following two isomeric structures? (You might want to build molecular models.)

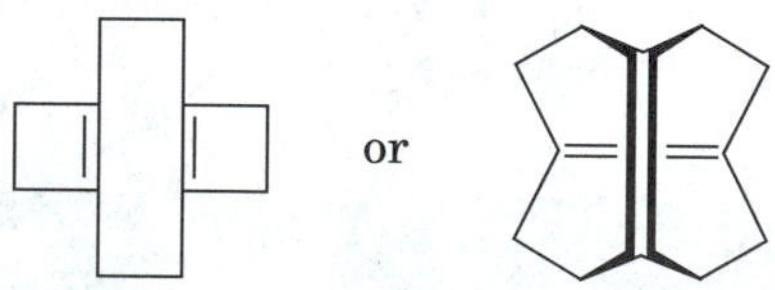

13.44 Propose structures for the two compounds whose ^{1}H NMR spectra are shown.
(a) C_4H_9Br

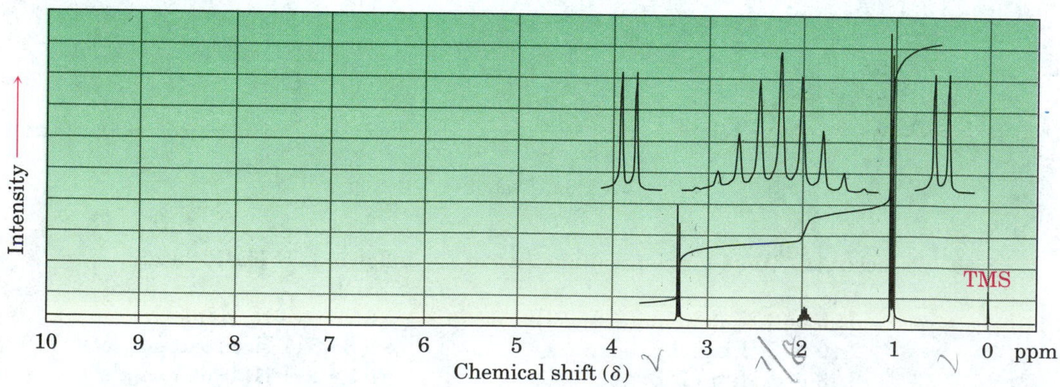

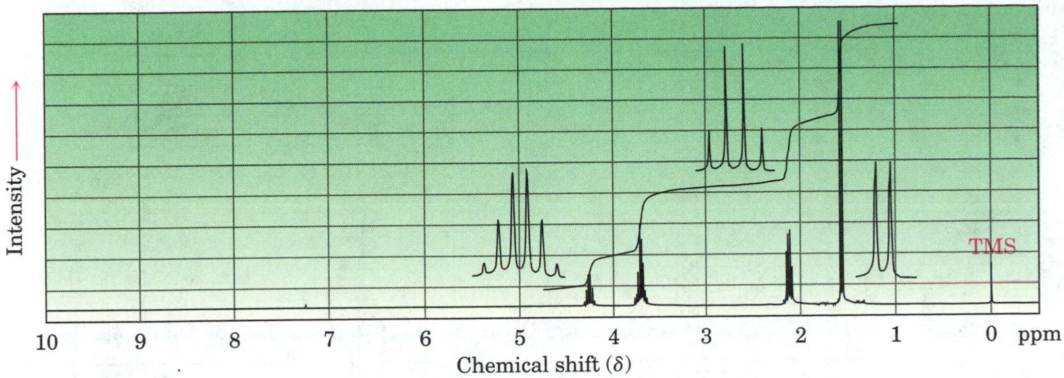

13.45 We saw earlier that long-range coupling between protons more than two carbon atoms apart is sometimes observed when π bonds intervene. One example is found in 1-methoxy-1-buten-3-yne, whose ^{1}H NMR spectrum is shown.

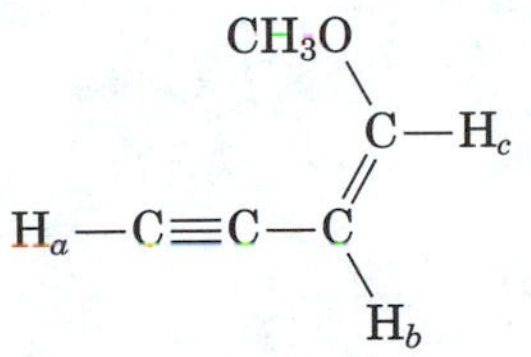

Not only does the acetylenic proton, H_a, couple with the vinylic proton H_b, it also couples with the vinylic proton H_c (four carbon atoms away). The following coupling constants are found:

$$J_{a\text{-}b} = 3 \text{ Hz} \qquad J_{a\text{-}c} = 1 \text{ Hz} \qquad J_{b\text{-}c} = 7 \text{ Hz}$$

Construct tree diagrams that account for the observed splitting patterns of H_a, H_b, and H_c.

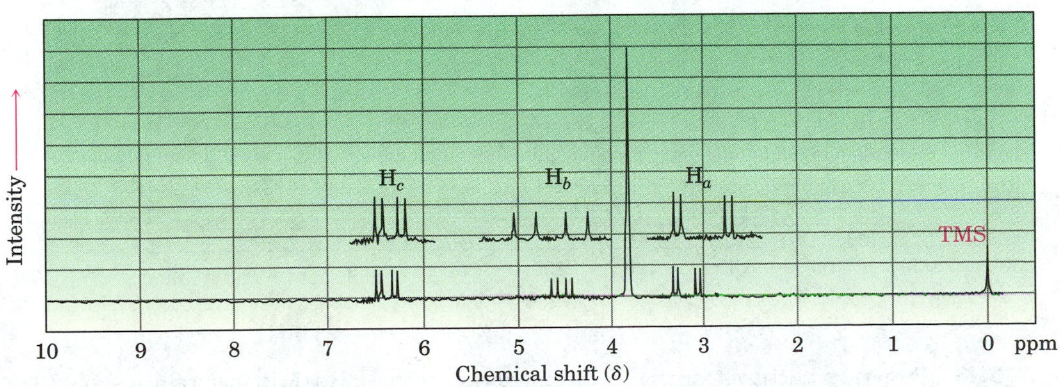

13.46 Assign as many of the resonances as you can to specific carbon atoms in the ^{13}C NMR spectrum of ethyl benzoate.

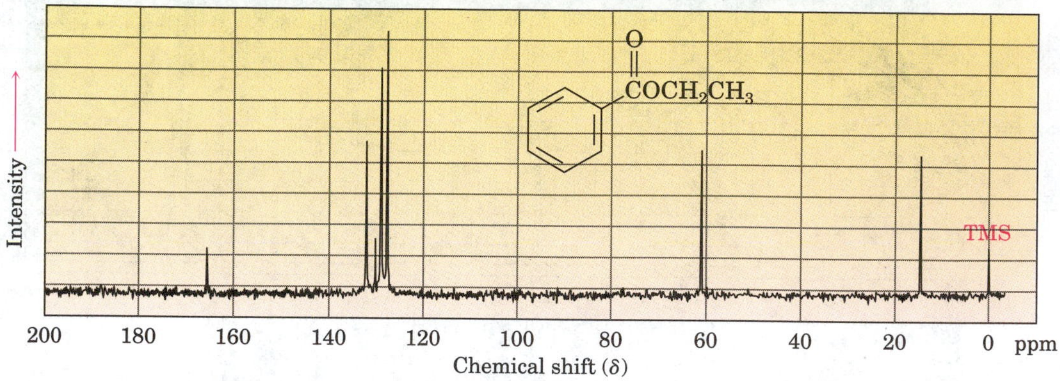

13.47 The ^{1}H and ^{13}C NMR spectra of compound A, C_8H_9Br, are shown. Propose a structure for A, and assign peaks in the spectra to your structure.

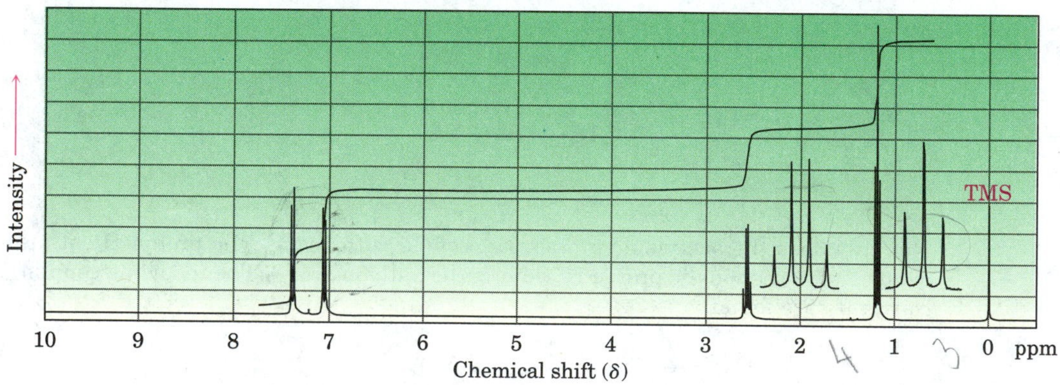

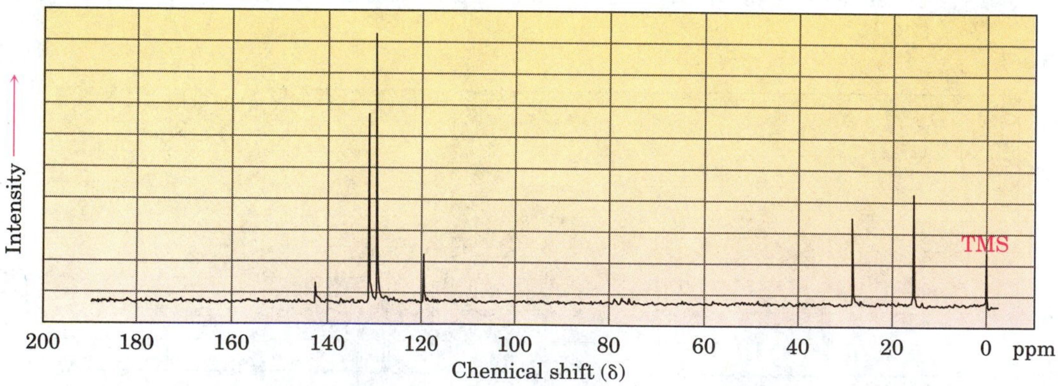

13.48 Propose structures for the three compounds whose ^{1}H NMR spectra are shown.

(a) $C_5H_{10}O$

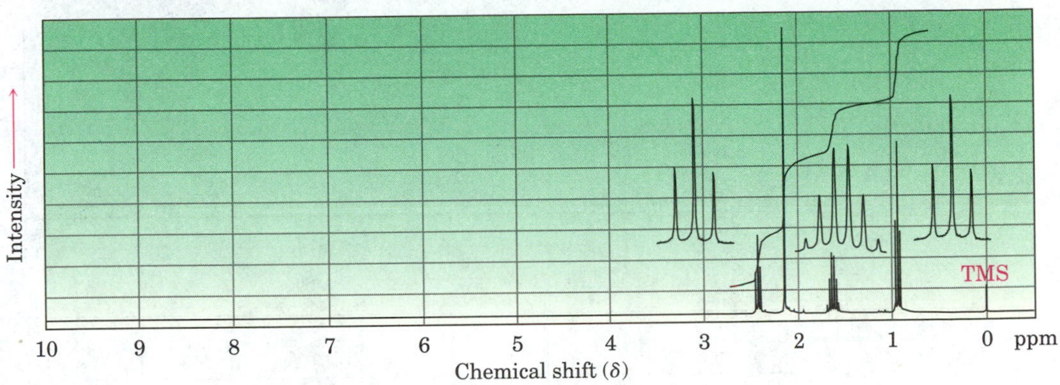

(b) C_7H_7Br

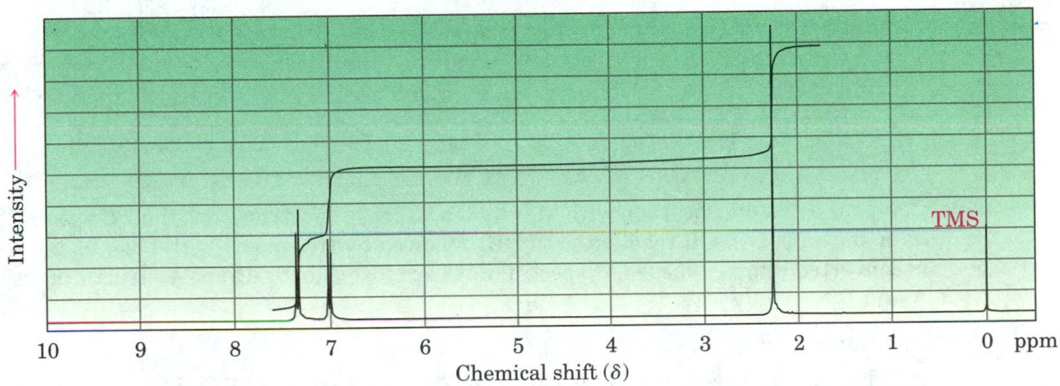

(c) C_8H_9Br

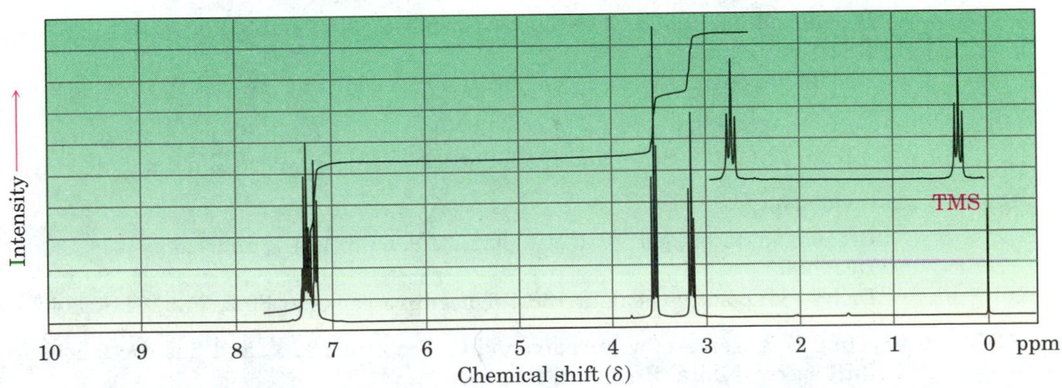

13.49 The mass spectrum and ^{13}C NMR spectrum of a hydrocarbon are shown at the top of page 494. Propose a structure for this hydrocarbon, and explain the spectral data.

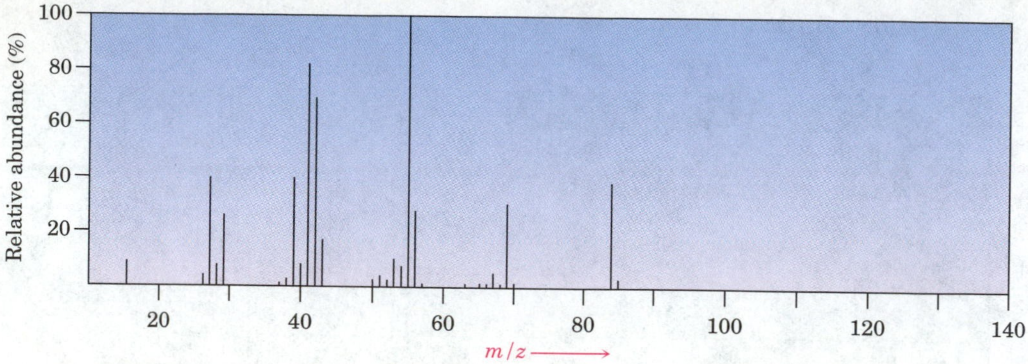

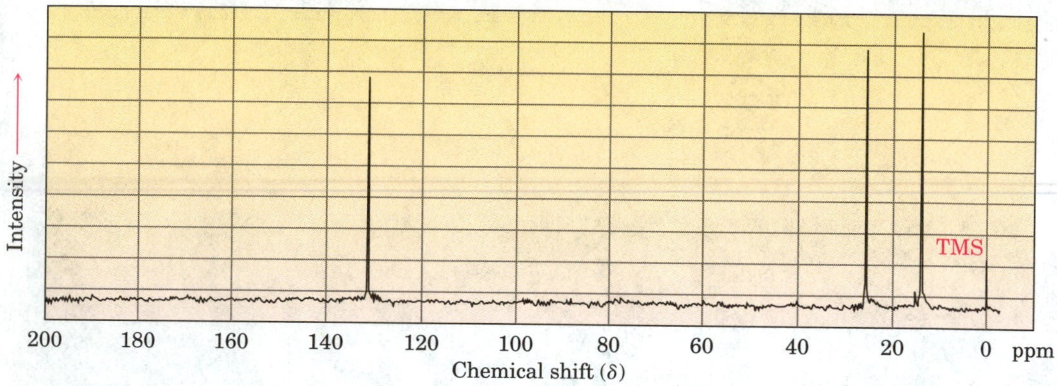

13.50 Compound A, a hydrocarbon with $M^+ = 96$ in its mass spectrum, has the ^{13}C spectral data shown below. On reaction with BH_3 followed by treatment with basic H_2O_2, A is converted into B, whose ^{13}C spectral data are also shown. Propose structures for A and B.

Compound A:

Broadband decoupled ^{13}C NMR: 26.8, 28.7, 35.7, 106.9, 149.7 δ
DEPT-90: no peaks
DEPT-135: no positive peaks; negative peaks at 26.8, 28.7, 35.7, 106.9 δ

Compound B:

Broadband decoupled ^{13}C NMR: 26.1, 26.9, 29.9, 40.5, 68.2 δ
DEPT-90: 40.5 δ
DEPT-135: positive peak at 40.5 δ; negative peaks at 26.1, 26.9, 29.9, 68.2 δ

13.51 Propose a structure for compound C, which has $M^+ = 86$ in its mass spectrum, an IR absorption at 3400 cm^{-1}, and the following ^{13}C NMR spectral data.

Compound C:

Broadband decoupled ^{13}C NMR: 30.2, 31.9, 61.8, 114.7, 138.4 δ
DEPT-90: 138.4 δ
DEPT-135: positive peak at 138.4 δ; negative peaks at 30.2, 31.9, 61.8, 114.7 δ

13.52 Compound D is isomeric with compound C (Problem 13.51) and has the following ^{13}C NMR spectral data. Propose a structure.

Compound D:

Broadband decoupled ^{13}C NMR: 9.7, 29.9, 74.4, 114.4, 141.4 δ
DEPT-90: 74.4, 141.4 δ
DEPT-135: positive peaks at 9.7, 74.4, 141.4 δ; negative peaks at 29.9, 114.4 δ

13.53 Propose a structure for compound E, $C_7H_{12}O_2$, which has the following ^{13}C NMR spectral data.

Compound E:

Broadband decoupled ^{13}C NMR: 19.1, 28.0, 70.5, 129.0, 129.8, 165.8 δ
DEPT-90: 28.0, 129.8 δ
DEPT-135: positive peaks at 19.1, 28.0, 129.8 δ; negative peaks at 70.5, 129.0 δ

13.54 Compound F, a hydrocarbon with $M^+ = 96$ in its mass spectrum, undergoes reaction with HBr to yield compound G. Propose structures for F and G, whose ^{13}C NMR spectral data are given below.

Compound F:

Broadband decoupled ^{13}C NMR: 27.6, 29.3, 32.2, 132.4 δ
DEPT-90: 132.4 δ
DEPT-135: positive peak at 132.4 δ; negative peaks at 27.6, 29.3, 32.2 δ

Compound G:

Broadband decoupled ^{13}C NMR: 25.1, 27.7, 39.9, 56.0 δ
DEPT-90: 56.0 δ
DEPT-135: positive peak at 56.0 δ; negative peaks at 25.1, 27.7, 39.9 δ

13.55 3-Methyl-2-butanol has five signals in its ^{13}C NMR spectrum at 17.90, 18.15, 20.00, 35.05, and 72.75 δ. Why are the two methyl groups attached to C3 nonequivalent? (Making a molecular model should be helpful.)

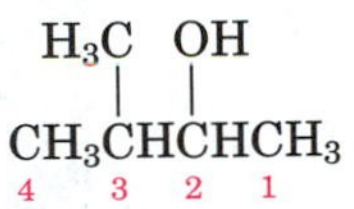

3-Methyl-2-butanol

13.56 A ^{13}C NMR spectrum of commercially available 2,4-pentanediol, shows five peaks at 23.3, 23.9, 46.5, 64.8, and 68.1 δ. Explain.

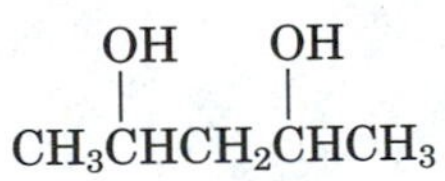

2,4-Pentanediol

A Look Ahead

13.57 We'll see in Chapter 21 that carboxylic acids (RCOOH) react with alcohols (R′OH) in the presence of an acid catalyst. The reaction product of propanoic acid with methanol has the following spectroscopic properties. Propose a structure.

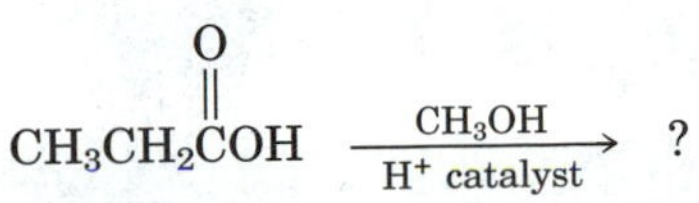

Propanoic acid

MS: $M^+ = 88$
IR: 1735 cm^{-1}
^1H NMR: 1.11 δ (3 H triplet, J = 7 Hz); 2.32 δ (2 H quartet, J = 7 Hz); 3.65 δ (1 H singlet)
^{13}C NMR: 9.3, 27.6, 51.4, 174.6 δ

13.58 We'll see in Chapter 21 that nitriles (RC≡N) react with Grignard reagents (R′MgBr). The reaction product from 2-methylpropanenitrile with methylmagnesium bromide has the following spectroscopic properties. Propose a structure.

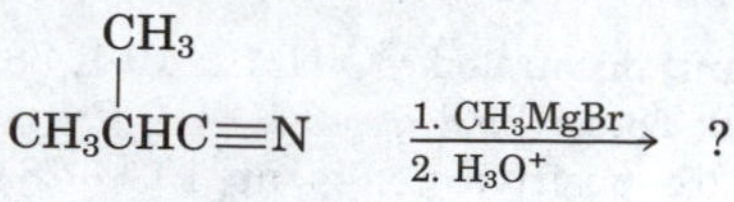

2-Methylpropanenitrile

MS: $M^+ = 86$
IR: 1715 cm^{-1}
^{1}H NMR: 1.05 δ (6 H doublet, $J = 7$ Hz); 2.12 δ (3 H singlet); 2.67 δ (1 H septet, $J = 7$ Hz)
^{13}C NMR: 18.2, 27.2, 41.6, 211.2 δ

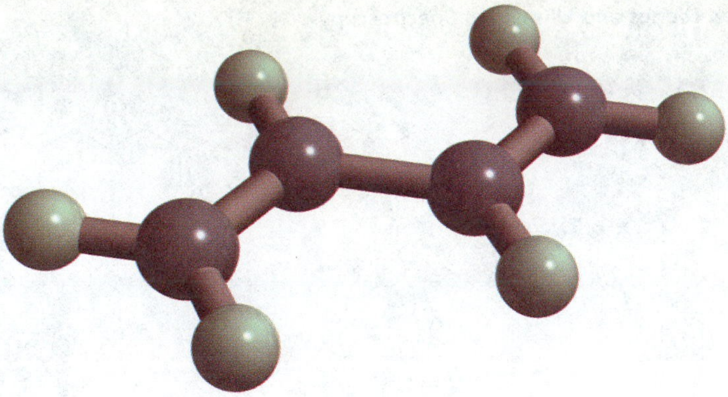

1,3-Butadiene, a *conjugated* diene, has an extended array of overlapping *p* orbitals.

14

Conjugated Dienes and Ultraviolet Spectroscopy

Multiple bonds that alternate with single bonds are said to be **conjugated**. Thus, 1,3-butadiene is a **conjugated diene**, whereas 1,4-pentadiene is a nonconjugated diene.

$$H_2C{=}CH{-}CH{=}CH_2 \qquad H_2C{=}CH{-}CH_2{-}CH{=}CH_2$$

1,3-Butadiene (conjugated; alternating double and single bonds)	**1,4-Pentadiene** (nonconjugated; nonalternating double and single bonds)

There are other types of conjugated systems besides dienes. For example, many of the pigments responsible for the brilliant reds and yellows of fruits and flowers are conjugated *polyenes*. Lycopene, the red pigment in tomatoes, is one such molecule. Conjugated *enones* (alkene + ketone) are common structural features of biologically important molecules such as progesterone, the so-called pregnancy hormone. Cyclic conjugated molecules such as benzene are a major field of study in themselves and will be considered in detail in the next chapter.

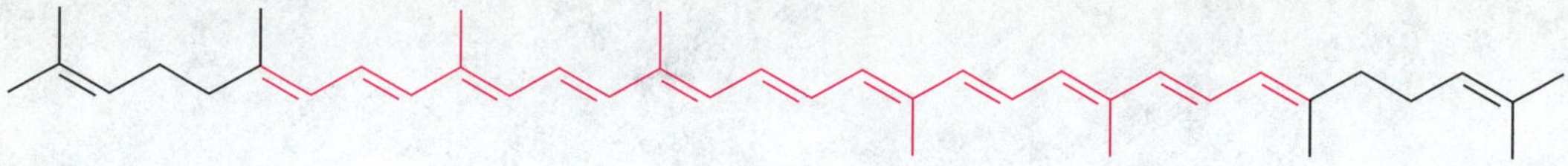

Lycopene

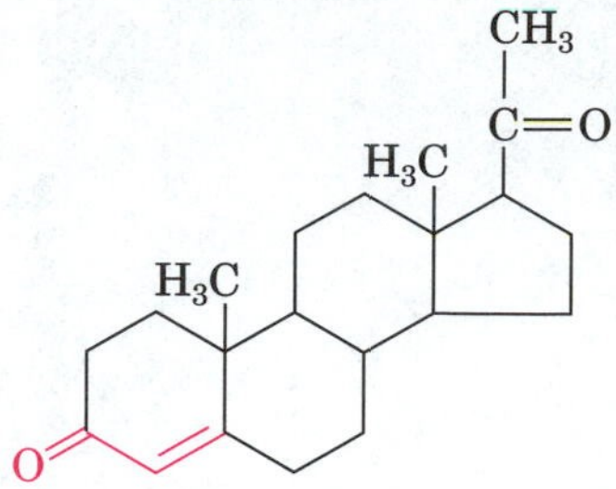

Progesterone

Benzene

PROBLEM. . . . .

14.1 Which of the following molecules contain conjugated systems? Circle the conjugated portion in each.

(a) 

(b) 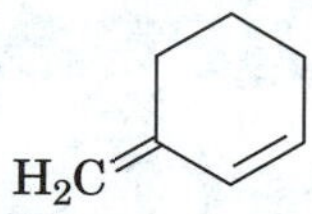

(c) $H_2C{=}CH{-}C{\equiv}N$

(d) 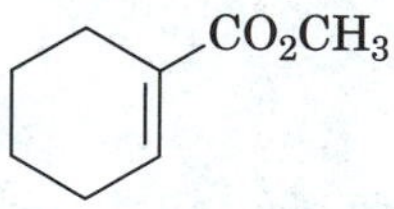

(e) 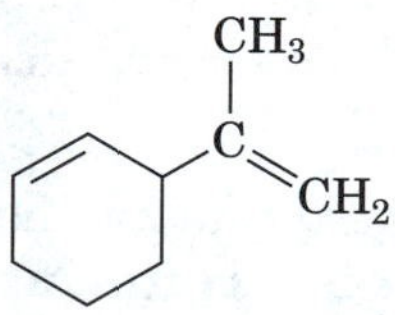

(f) 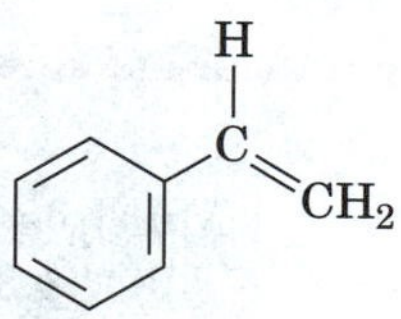

. . . .

14.1 Preparation of Conjugated Dienes

Conjugated dienes are generally prepared by the methods previously discussed for alkene synthesis. The base-induced elimination of HX from an allylic halide is one such method.

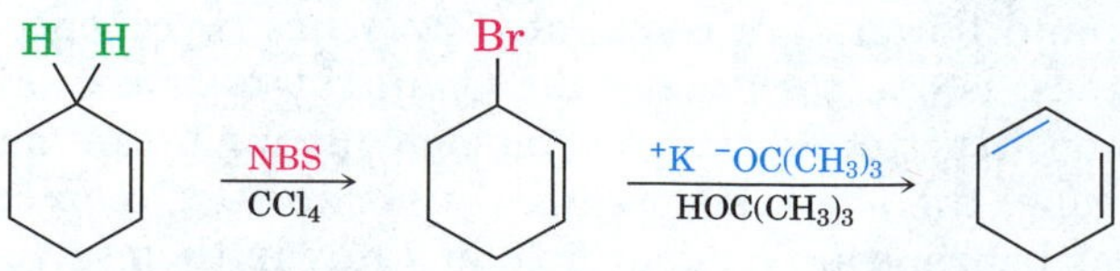

Cyclohexene **3-Bromocyclohexene** **1,3-Cyclohexadiene (76%)**

1,3-Butadiene is prepared industrially for use in polymer synthesis by thermal cracking of butane over a chromium oxide/aluminum oxide catalyst, but this procedure is of little use in the laboratory.

$$CH_3CH_2CH_2CH_3 \xrightarrow[\text{Catalyst}]{600^\circ C} H_2C{=}CHCH{=}CH_2 + 2\ H_2$$

Butane **1,3-Butadiene**

Other simple conjugated dienes used in polymer synthesis include chloroprene (2-chloro-1,3-butadiene) and isoprene (2-methyl-1,3-butadiene). Isoprene has been prepared industrially by a number of methods, including the acid-catalyzed double dehydration of 3-methyl-1,3-butanediol.

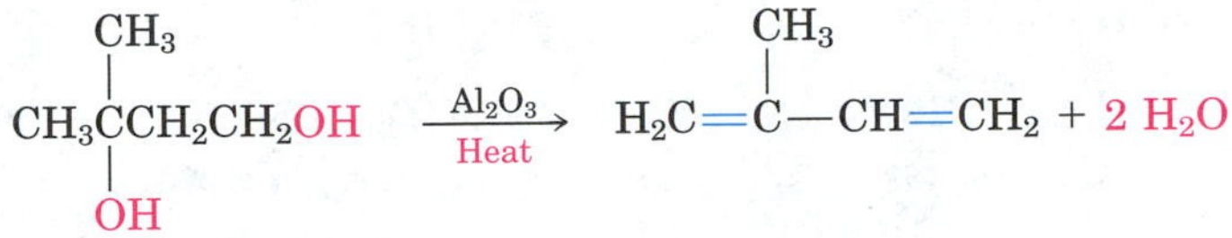

3-Methyl-1,3-butanediol **Isoprene (2-Methyl-1,3-butadiene)**

14.2 Stability of Conjugated Dienes

Conjugated dienes are similar to other alkenes in much of their chemistry, but there are also important differences. One such difference is *stability*: Conjugated dienes are somewhat more stable than nonconjugated dienes.

Evidence for the extra stability of conjugated dienes comes from measurements of heats of hydrogenation (Table 14.1, p. 500). We saw earlier in the discussion of alkene stabilities (Section 6.7) that alkenes of similar substitution pattern have remarkably similar $\Delta H^\circ_{\text{hydrog}}$ values. Monosubstituted alkenes such as 1-butene have values for $\Delta H^\circ_{\text{hydrog}}$ near -126 kJ/mol (-30 kcal/mol), whereas disubstituted alkenes such as 2-methyl-1-butene have $\Delta H^\circ_{\text{hydrog}}$ values approximately 7 kJ/mol less negative. We concluded from these data that highly substituted alkenes are more stable than less substituted ones. That is, more substituted alkenes release less heat on hydrogenation because they contain less energy to start with. A similar conclusion can be drawn for conjugated dienes.

Since a monosubstituted alkene such as 1-butene has $\Delta H^\circ_{\text{hydrog}} = -126$ kJ/mol, we might expect that a compound with two monosubstituted double bonds should have a $\Delta H^\circ_{\text{hydrog}}$ approximately twice this value, or -252 kJ/mol. Nonconjugated dienes, such as 1,4-pentadiene ($\Delta H^\circ_{\text{hydrog}} = -253$ kJ/mol), meet this expectation, but the conjugated diene 1,3-butadiene ($\Delta H^\circ_{\text{hydrog}} = -236$ kJ/mol) does not. 1,3-Butadiene is approximately 16 kJ/mol (3.8 kcal/mol) more stable than expected.

Table 14.1 Heats of Hydrogenation for Some Alkenes and Dienes

		ΔH°_{hydrog}	
Alkene	*Product*	*(kJ/mol)*	*(kcal/mol)*
$CH_3CH_2CH{=}CH_2$	$CH_3CH_2CH_2CH_3$	−126	−30.1
$CH_3CH_2C(CH_3){=}CH_2$	$CH_3CH_2CH(CH_3)CH_3$	−118	−28.2
$H_2C{=}CHCH{=}CH_2$	$CH_3CH_2CH{=}CH_2$	−110	−26.3
$H_2C{=}CHCH{=}CH_2$	$CH_3CH_2CH_2CH_3$	−236	−56.4
$H_2C{=}CHC(CH_3){=}CH_2$	$CH_3CH_2CH(CH_3)CH_3$	−229	−54.7
$H_2C{=}CHCH_2CH{=}CH_2$	$CH_3CH_2CH_2CH_2CH_3$	−253	−60.5

Confirmation of this unexpected stability comes from data on the *partial* hydrogenation of 1,3-butadiene. If 1,3-butadiene is partially hydrogenated to yield 1-butene, −110 kJ/mol (−26.3 kcal/mol) of energy is released, a value about 15 kJ/mol less than we might expect for an isolated monosubstituted double bond (−126 kJ/mol). The same is true of other conjugated dienes.

	ΔH°_{hydrog} *(kJ/mol)*	
$H_2C{=}CHCH_2CH{=}CH_2$ **1,4-Pentadiene**	−126 + (−126) = −252	Expected
	−253	Observed
	1	Difference
$H_2C{=}CHCH{=}CH_2$ **1,3-Butadiene**	−126 + (−126) = −252	Expected
	−236	Observed
	−16	Difference
$H_2C{=}CHC(CH_3){=}CH_2$ **2-Methyl-1,3-butadiene**	−126 + (−118) = −244	Expected
	−229	Observed
	−15	Difference

PROBLEM

14.2 Use the data in Table 14.1 to calculate an expected heat of hydrogenation for allene, $H_2C{=}C{=}CH_2$. The measured value is −298 kJ/mol (−71.3 kcal/mol). Rank a conjugated diene, a nonconjugated diene, and an allene in order of stability.

14.3 Molecular Orbital Description of 1,3-Butadiene

Why are conjugated dienes so stable? Two explanations have been advanced. One explanation says that the difference in stability between conjugated and nonconjugated dienes results from differences in orbital hybridization. In a nonconjugated diene, such as 1,4-pentadiene, the C–C single bonds result from σ overlap of an sp^2 orbital on one carbon with an sp^3 orbital on the neighboring carbon. In a conjugated diene, however, the C–C single bond results from σ overlap of sp^2 orbitals on both carbons. Since sp^2 orbitals have more *s* character than sp^3 orbitals, the electrons in sp^2 orbitals are closer to the nucleus, and the bonds they form are somewhat shorter and stronger. Thus, the "extra" stability of a conjugated diene results from the greater amount of *s* character in the orbitals forming the C–C single bond.

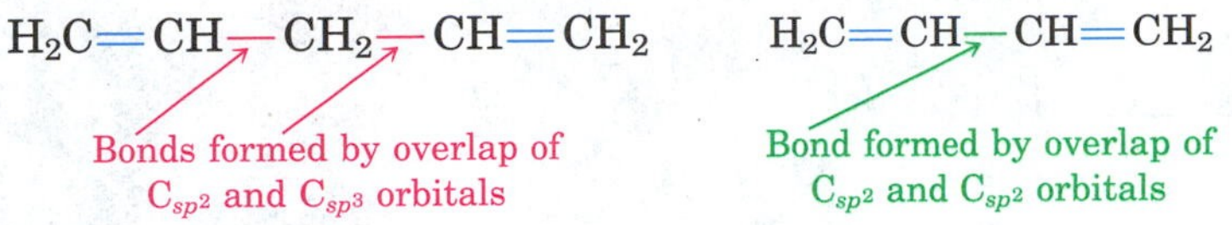

The second explanation for the stability of conjugated dienes focuses on the interaction between the π orbitals of the two double bonds. To see how this interaction arises, let's briefly review molecular orbital theory (Sections 1.7 and 1.10). When a covalent bond forms by combination of atomic orbitals, molecular orbitals result. For example, when two *p* atomic orbitals combine to form a π bond, two π molecular orbitals result. One is lower in energy than the starting *p* orbitals and is therefore bonding; the other is higher in energy, has a node between nuclei, and is antibonding. Both electrons occupy the low-energy, bonding orbital, resulting in formation of a stable bond between atoms (Figure 14.1).

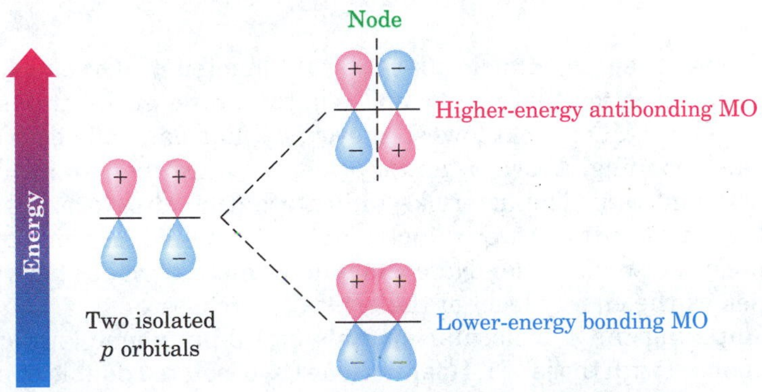

Figure 14.1 Two *p* orbitals combine to form two π molecular orbitals. When these orbitals are occupied by two electrons, both electrons occupy the low-energy, bonding orbital, leading to a net lowering of energy and formation of a stable bond.

Now let's combine four p atomic orbitals, as occurs in a conjugated diene. In so doing, we generate a set of four π molecular orbitals, two of which are bonding and two of which are antibonding (Figure 14.2). The four π electrons occupy the two bonding orbitals, leaving the antibonding orbitals vacant.

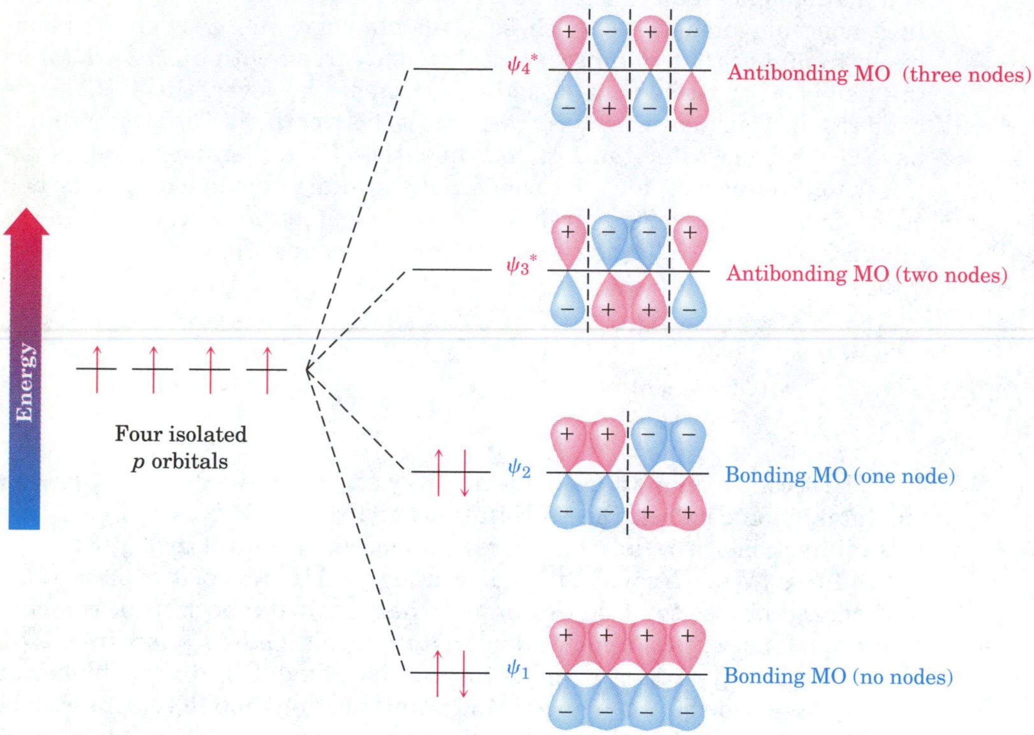

Figure 14.2 Four π molecular orbitals in 1,3-butadiene. The asterisks on ψ_3* and ψ_4* indicate antibonding orbitals. The number of nodes between nuclei increases as the energy level of the orbital increases.

The lowest-energy π molecular orbital (denoted ψ_1, Greek psi) is a fully additive combination that has no nodes between the nuclei and is therefore bonding. The π MO of next lowest energy, ψ_2, has one node between nuclei and is also bonding. Above ψ_1 and ψ_2 in energy are the two antibonding π MO's, ψ_3* and ψ_4*. (The asterisks indicate antibonding orbitals.) The ψ_3* orbital has two nodes between nuclei, and ψ_4*, the highest-energy MO, has three nodes between nuclei. Note that the number of nodes between nuclei increases as the energy level of the orbital increases.

Comparing the π molecular orbitals of 1,3-butadiene (two conjugated double bonds) with those of 1,4-pentadiene (two isolated double bonds) shows why the conjugated diene is more stable. In a conjugated diene, the lowest-energy π MO (ψ_1) has a favorable bonding interaction between C2 and C3 that is absent in a nonconjugated diene. As a result, there is a certain amount of double-bond character to the C2–C3 bond, making that bond stronger and stabilizing the molecule.

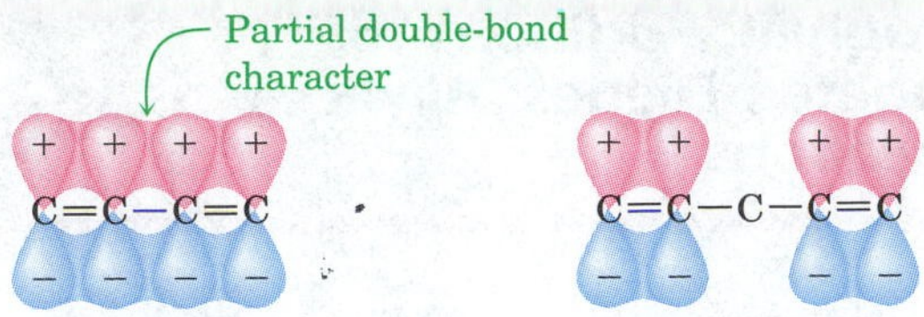

A conjugated diene

A nonconjugated diene

In describing the 1,3-butadiene molecular orbitals, we say that the π electrons are spread out, or *delocalized*, over the entire π framework rather than localized between two specific nuclei. Electron delocalization always leads to lower energy and greater stability of the molecule.

14.4 Bond Lengths in 1,3-Butadiene

Further evidence for the special nature of conjugated dienes comes from data on bond lengths (Table 14.2). If we compare the length of the carbon–carbon single bond of 1,3-butadiene (1.48 Å) to that of ethane (1.54 Å), we find that the 1,3-butadiene single bond is shorter by 0.06 Å.

Table 14.2 Some Carbon–Carbon Bond Lengths

Bond	*Bond length (Å)*	*Bond hybridization*
$CH_3—CH_3$	1.54	$C_{sp^3}–C_{sp^3}$
$H_2C{=}CH_2$	1.33	$C_{sp^2}–C_{sp^2}$
$H_2C{=}CH—CH_3$	1.49	$C_{sp^2}–C_{sp^3}$
$H_2C{=}CH—CH{=}CH_2$	1.48	$C_{sp^2}–C_{sp^2}$
$H_2C{=}CHCH{=}CH_2$	1.34	$C_{sp^2}–C_{sp^2}$

Both explanations advanced in the previous section to account for the stability of conjugated dienes also explain the bond shortening. According to the molecular orbital argument, the partial double-bond character of the C2–C3 bond gives the bond a length midway between a pure single bond and a pure double bond. Alternatively, it can be argued that the shortened 1,3-butadiene single bond is a consequence of orbital hybridization. The C2–C3 bond results from σ overlap of two carbon sp^2 orbitals, whereas a typical alkane C–C bond results from overlap of two carbon sp^3 orbitals. The greater amount of *s* character in the 1,3-butadiene single bond makes it a bit shorter and stronger than usual. Both explanations are valid, and both contribute to the bond shortening observed for 1,3-butadiene.

14.5 Electrophilic Additions to Conjugated Dienes: Allylic Carbocations

One of the most striking differences between conjugated dienes and typical alkenes is in their electrophilic addition reactions. To review briefly, the addition of an electrophile to a carbon–carbon double bond is a general reaction of alkenes (Section 6.8). Markovnikov regiochemistry is found because the more stable carbocation is involved as an intermediate. Thus, addition of HCl to 2-methylpropene yields 2-chloro-2-methylpropane rather than 1-chloro-2-methylpropane, and addition of 2 mol equiv of HCl to the nonconjugated diene 1,4-pentadiene yields 2,4-dichloropentane.

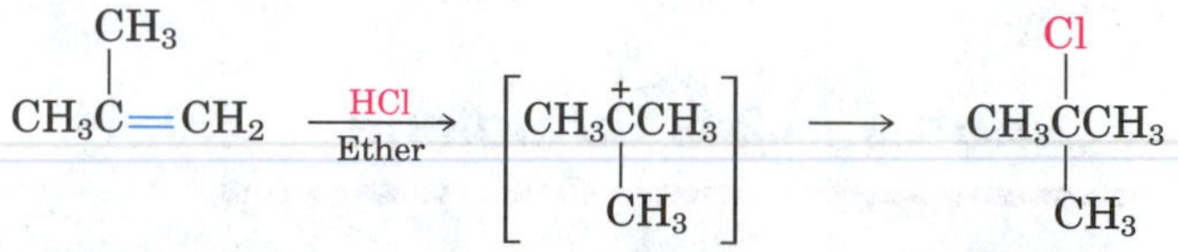

2-Methylpropene Tertiary carbocation **2-Chloro-2-methylpropane**

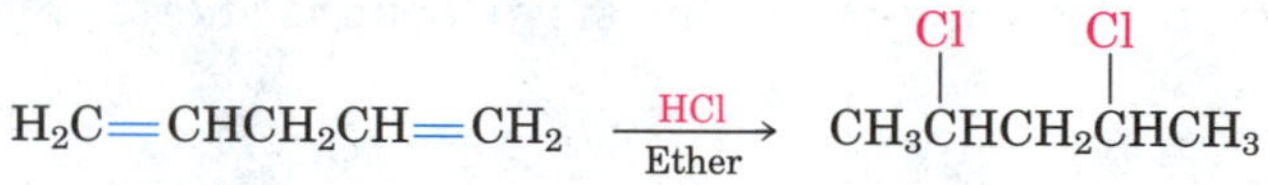

1,4-Pentadiene (nonconjugated) **2,4-Dichloropentane**

Conjugated dienes also undergo electrophilic addition reactions readily, but mixtures of products are invariably obtained. For example, addition of HBr to 1,3-butadiene yields a mixture of two products (not counting cis–trans isomers). 3-Bromo-1-butene is the typical Markovnikov product of **1,2 addition**, but 1-bromo-2-butene appears unusual. The double bond in this product has moved to a position between carbons 2 and 3, and HBr has added to carbons 1 and 4, a result described as **1,4 addition**.

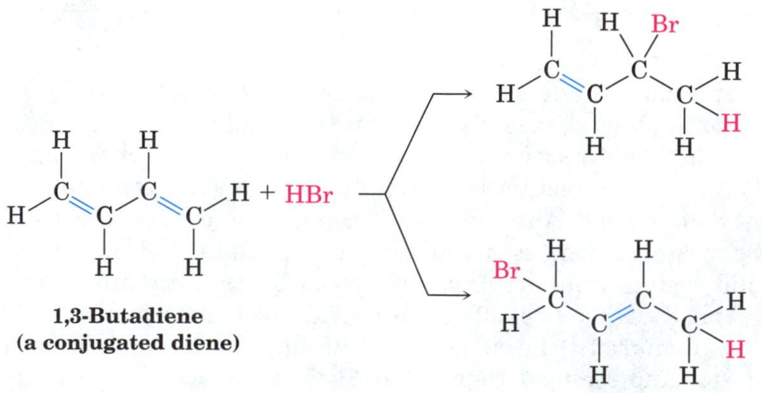

1,3-Butadiene (a conjugated diene)

3-Bromo-1-butene (71%; 1,2 addition)

1-Bromo-2-butene (29%; 1,4 addition)

Many other electrophiles besides HBr add to conjugated dienes, and mixtures of products are usually formed. For example, Br_2 adds to 1,3-butadiene to give a mixture of 1,4-dibromo-2-butene and 3,4-dibromo-1-butene.

$$H_2C{=}CHCH{=}CH_2 \xrightarrow[25°C]{Br_2} BrCH_2CH{=}CHCH_2Br \quad \text{1,4 addition}$$

1,3-Butadiene **1,4-Dibromo-2-butene (45%)**

+

$$BrCH_2CH(Br)CH{=}CH_2 \quad \text{1,2 addition}$$

3,4-Dibromo-1-butene (55%)

How can we account for the formation of 1,4-addition products? The answer is that *allylic carbocations* are involved as intermediates in the reactions. When 1,3-butadiene is protonated, two carbocation intermediates are possible: a primary carbocation and a secondary allylic cation (recall that *allylic* means "next to a double bond"). Since an allylic cation is stabilized by resonance (Section 11.9), it is more stable and forms faster than a less stable primary carbocation.

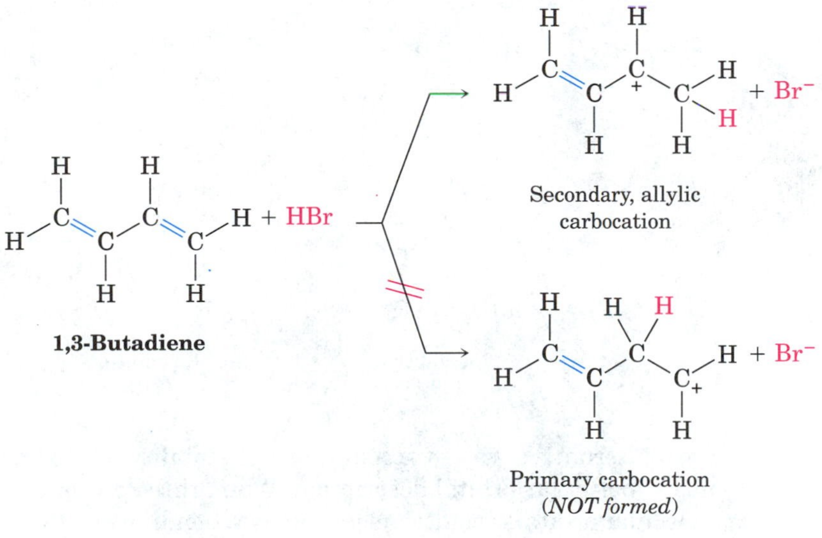

There are two ways to account for the stability of allylic carbocations. Resonance theory (Section 2.4) offers a pictorial representation of the situation through the use of different resonance forms. It says that the more resonance forms that are possible, the more stable the compound. Thus, an allylic cation has two resonance forms and is more stable than a nonallylic cation. Neither of the two resonance forms is correct by itself; the true structure of the allylic cation is a resonance hybrid of the two Kekulé structures.

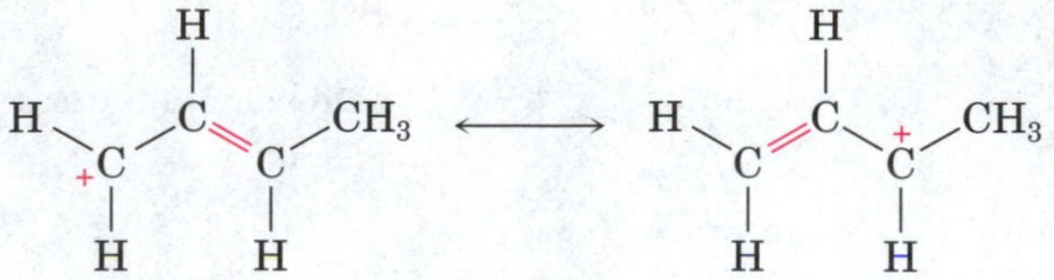

Two resonance forms
of an allylic carbocation

When the allylic cation reacts with Br^- to complete the electrophilic addition reaction, attack can occur either at C1 or at C3 because both carbons share the positive charge. Thus, a mixture of 1,2- and 1,4-addition products results. (Recall that a similar product mixture was found in Section 10.5 for NBS bromination of alkenes, a reaction that proceeds through an allylic *radical*.)

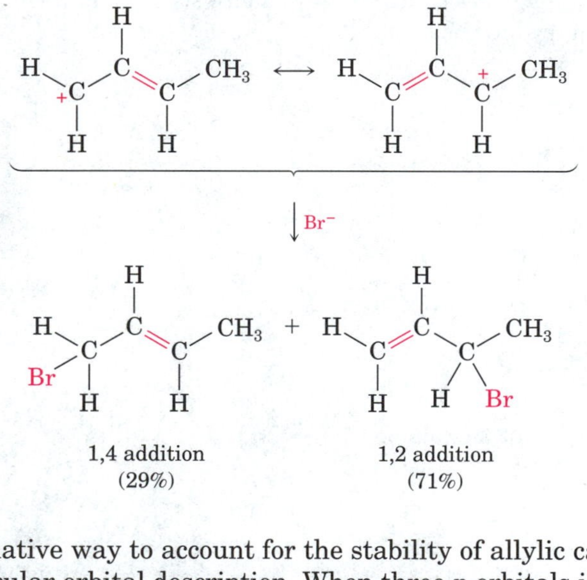

An alternative way to account for the stability of allylic carbocations is to use a molecular orbital description. When three *p* orbitals interact, three π molecular orbitals result: one low-energy, bonding orbital; one *nonbonding* orbital; and one high-energy, antibonding orbital (Figure 14.3). The two π electrons occupy the bonding orbital, indicating that a partial bond exists between carbons 2 and 3. Thus, the allylic cation is a conjugated system stabilized by electron delocalization in much the same way as 1,3-butadiene.

It's important to realize that the two approaches for describing allylic carbocations—the resonance approach and the molecular orbital approach —don't "compete" with each other. It's not that one approach is more "right" than the other; both are really just alternative ways of trying to visualize

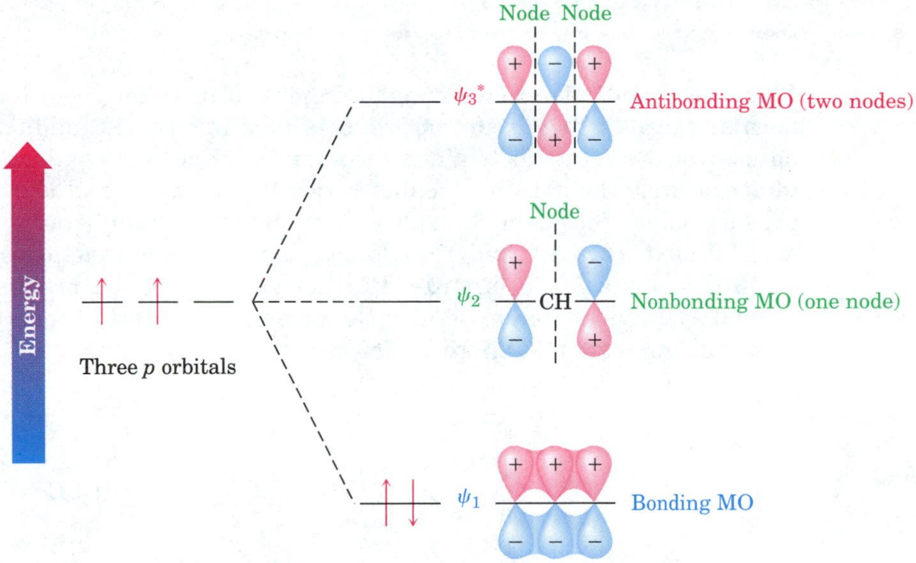

Figure 14.3 The π molecular orbitals of an allylic carbocation. The two electrons occupy the low-energy, bonding orbital, leading to a net stabilization.

a phenomenon that is best handled using quantum mechanics. Sometimes one approach is more convenient, and sometimes the other is more convenient. We'll use both at various times.

PROBLEM

14.3 Give the structures of the likely products from reaction of 1 equivalent of HCl with 1,3-pentadiene. Show both 1,2 and 1,4 adducts.

PROBLEM

14.4 Look at the possible carbocation intermediates produced during addition of HCl to 1,3-pentadiene (Problem 14.3), and predict which 1,2 adduct predominates. Which 1,4 adduct predominates?

PROBLEM

14.5 The π molecular orbital diagram for an allylic *radical* ($H_2C{=}CH{-}CH_2\cdot$) is similar to that for an allylic carbocation. Which orbitals do the three π electrons occupy?

14.6 Kinetic versus Thermodynamic Control of Reactions

Electrophilic addition to a conjugated diene at or below room temperature normally leads to a mixture of products in which the 1,2 adduct predominates over the 1,4 adduct. When the same reaction is carried out at higher temperatures, though, the product ratio often changes and the 1,4 adduct predominates. For example, addition of HBr to 1,3-butadiene at 0°C yields a 71:29 mixture of 1,2 and 1,4 adducts, but the same reaction carried out at 40°C yields a 15:85 mixture. Furthermore, when the product mixture formed at 0°C is heated to 40°C in the presence of HBr, the ratio of adducts slowly changes from 71:29 to 15:85. Why?

$$H_2C{=}CHCH{=}CH_2 + HBr \longrightarrow H_2C{=}CH\underset{}{\overset{Br}{\overset{|}{C}}}HCH_3 + CH_3CH{=}CHCH_2Br$$

	1,2 adduct	1,4 adduct
at 0°C:	71%	29%
at 40°C:	15%	85%

To understand the effect of temperature on product distribution, let's briefly review what we said in Section 5.6 about rates and equilibria. Imagine a reaction that can give either or both of two products, B and C:

$$A \longrightarrow B + C$$

Let's assume that B forms faster than C (in other words, $\Delta G_B^‡ < \Delta G_C^‡$) but that C is more stable than B (in other words, $\Delta G_C° > \Delta G_B°$). A reaction energy diagram for the two processes might look like that shown in Figure 14.4.

Let's first carry out the reaction under vigorous, high-temperature conditions so that both processes are reversible and an equilibrium is reached. Since C is more stable than B, C is the major product formed under the equilibrium conditions. It doesn't matter that C forms more slowly than B, because the reaction conditions have been chosen so that B and C interconvert rapidly. *The product of a reversible process depends only on thermodynamic stability.* Such reactions are said to be under equilibrium control, or **thermodynamic control**.

$$B \rightleftharpoons A \rightleftharpoons C$$

Thermodynamic control (vigorous conditions; reversible)

Now let's carry out the same reaction under milder, low-temperature conditions so that both processes are *irreversible* and no equilibrium is

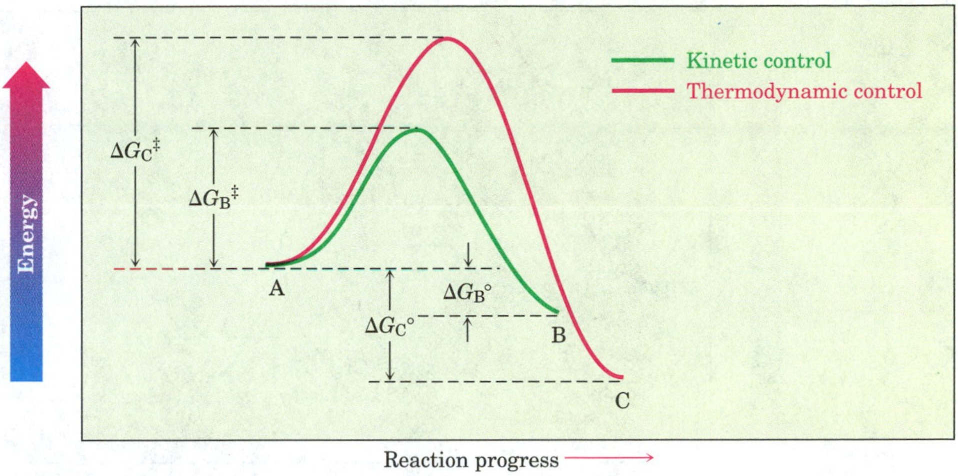

Figure 14.4 A reaction energy diagram for two competing reactions in which the less stable product forms faster than the more stable product.

reached. In other words, enough energy is present for reactant molecules to surmount the barriers separating A from B and C, but not for product B and C molecules to climb the higher barriers back to A. Since B forms faster than C, B is the major product. It doesn't matter that C is more stable than B, because the reaction conditions have been chosen so that B and C don't interconvert. *The product of an irreversible process depends only on the reaction rate.* Such reactions are said to be under **kinetic control**.

$$\mathrm{B} \longleftarrow \mathrm{A} \longrightarrow \mathrm{C}$$

Kinetic control
(mild conditions; irreversible)

We can now explain the effect of temperature on electrophilic addition reactions of conjugated dienes. Under mild conditions (0°C), HBr adds to 1,3-butadiene under kinetic control to give a 71:29 mixture of products with the more rapidly formed 1,2 adduct predominating. Since these mild conditions don't allow the reaction to reach equilibrium, the product that forms faster predominates. Under more vigorous conditions (40°C), however, the reaction occurs under thermodynamic control to give a 15:85 mixture of products, with the more stable 1,4 adduct predominating. The higher temperature makes the addition process reversible, and an equilibrium mixture of products therefore results. Figure 14.5 (p. 510) shows the situation in a reaction energy diagram.

The electrophilic addition of HBr to 1,3-butadiene is a good example of how a change in experimental conditions can change the product of a reaction. The concept of thermodynamic control versus kinetic control is a valuable one that we can often take advantage of in the laboratory.

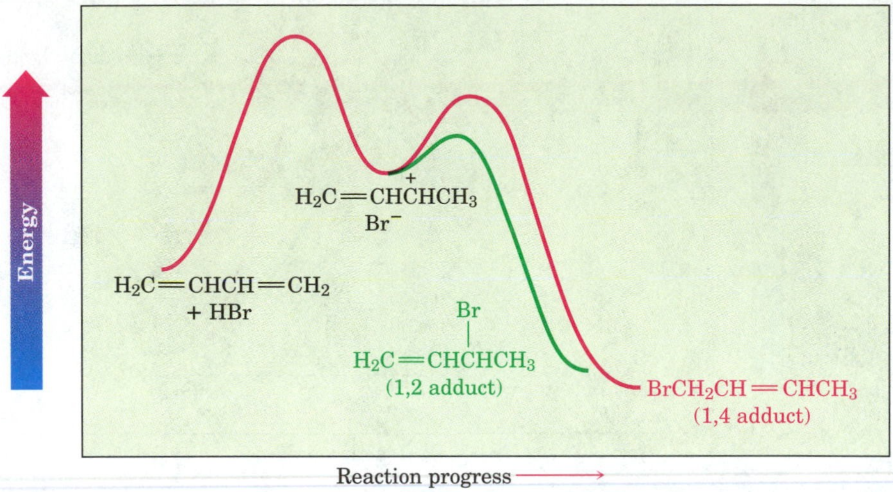

Figure 14.5 Reaction energy diagram for the electrophilic addition of HBr to 1,3-butadiene. The 1,2 adduct is the kinetic product because it forms faster, but the 1,4 adduct is the thermodynamic product because it is more stable.

PROBLEM. . . . .

14.6 The 1,2 adduct and the 1,4 adduct formed by reaction of HBr with 1,3-butadiene are in equilibrium at 40°C. Propose a mechanism by which the interconversion of products takes place. (See Section 11.6.)

PROBLEM. . . . .

14.7 Why do you suppose 1,4 adducts of 1,3-butadiene are generally more stable than 1,2 adducts?

14.7 Diene Polymers: Natural and Synthetic Rubbers

Conjugated dienes can be polymerized just as simple alkenes can (Section 7.11). Diene polymers are structurally more complex than simple alkene polymers, though, because double bonds remain every four carbon atoms along the chain, leading to the possibility of cis–trans isomers. The reaction can be initiated either by a radical, as occurs in ethylene polymerization, or by an acidic catalyst. Note that the polymerization is a 1,4 addition of the growing chain to a conjugated diene monomer.

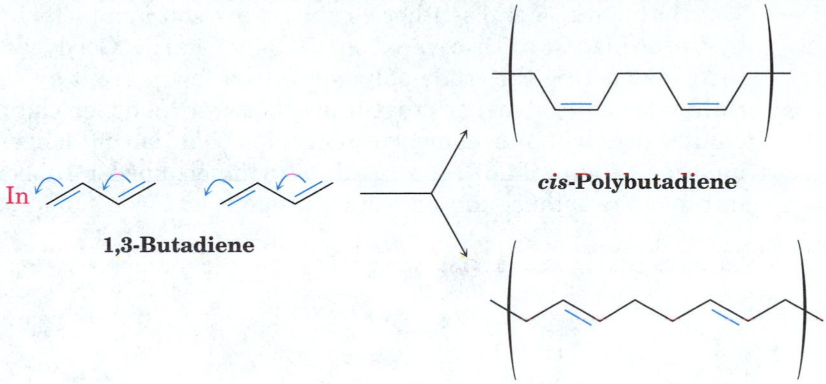

1,3-Butadiene

cis-Polybutadiene

trans-Polybutadiene

As noted in the Chapter 7 Interlude, rubber is a naturally occurring polymer of isoprene. The double bonds of rubber have *Z* stereochemistry, but *gutta-percha*, the *E* isomer of natural rubber, also occurs naturally. Harder and more brittle than rubber, gutta-percha has a variety of minor applications, including occasional use as the covering on golf balls.

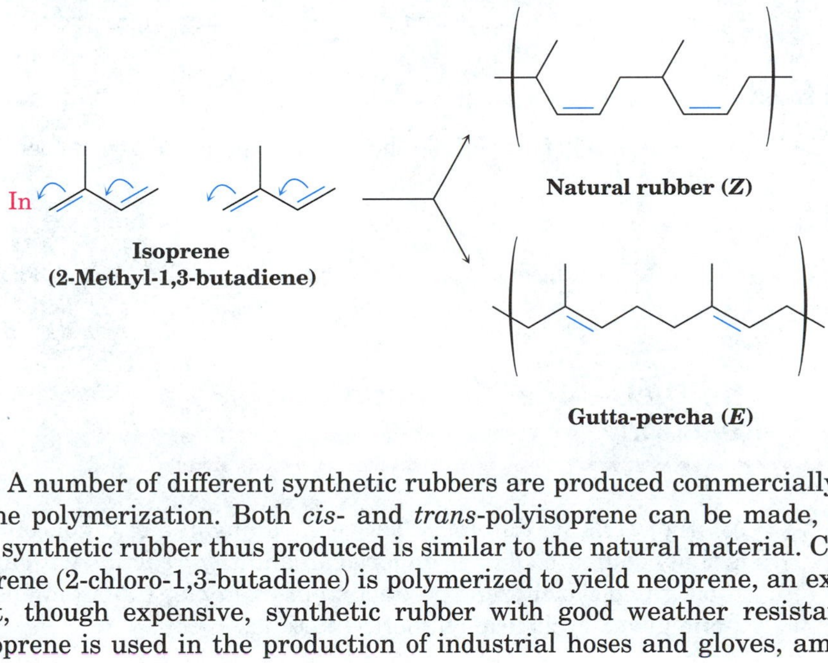

Isoprene
(2-Methyl-1,3-butadiene)

Natural rubber (*Z*)

Gutta-percha (*E*)

A number of different synthetic rubbers are produced commercially by diene polymerization. Both *cis*- and *trans*-polyisoprene can be made, and the synthetic rubber thus produced is similar to the natural material. Chloroprene (2-chloro-1,3-butadiene) is polymerized to yield neoprene, an excellent, though expensive, synthetic rubber with good weather resistance. Neoprene is used in the production of industrial hoses and gloves, among other things.

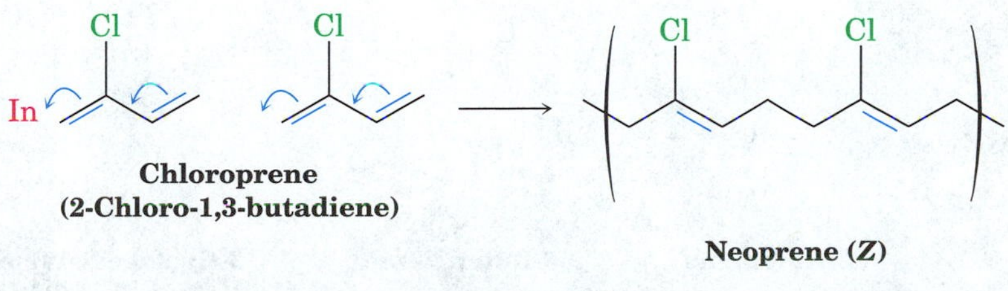

Chloroprene
(2-Chloro-1,3-butadiene)

Neoprene (*Z*)

Both natural and synthetic rubbers are soft and tacky unless hardened by **vulcanization**. Discovered in 1839 by Charles Goodyear, vulcanization involves heating the crude polymer with a few percent by weight of sulfur. Sulfur forms bridges, or cross-links, between polymer chains, locking the chains together into immense molecules that can no longer slip over one another (Figure 14.6). The result is a much harder rubber with greatly improved resistance to wear and abrasion.

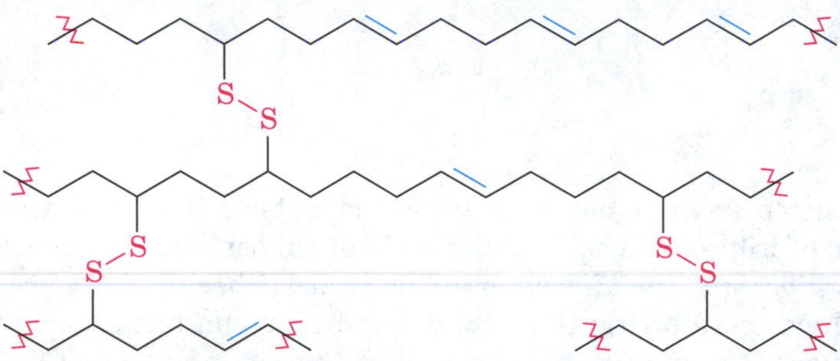

Figure 14.6 Sulfur cross-linked chains resulting from vulcanization of rubber.

PROBLEM

14.8 Draw a segment of the polymer that might be prepared from 2-phenyl-1,3-butadiene.

PROBLEM

14.9 Show the mechanism of the acid-catalyzed polymerization of 1,3-butadiene.

14.8 The Diels–Alder Cycloaddition Reaction

Another striking difference between conjugated and nonconjugated dienes is that conjugated dienes undergo an addition reaction with alkenes to yield substituted cyclohexene products. For example, 1,3-butadiene and 3-buten-2-one give 3-cyclohexenyl methyl ketone.

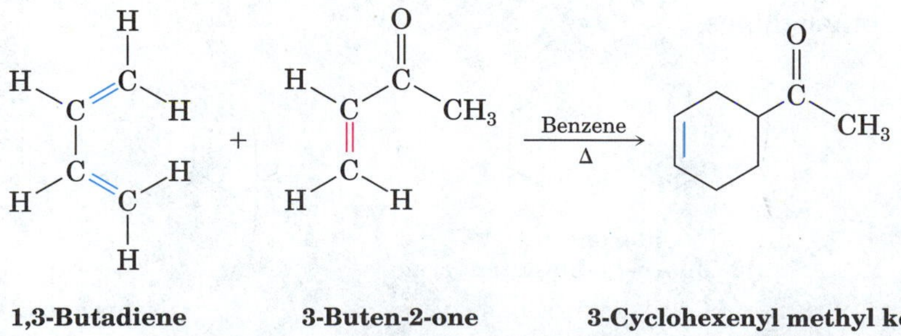

1,3-Butadiene **3-Buten-2-one** **3-Cyclohexenyl methyl ketone (96%)**

This process, named the **Diels–Alder cycloaddition reaction** after its discoverers,[1,2] is extremely useful in organic synthesis because it forms two carbon–carbon bonds in a single step and is one of the few methods available for making cyclic molecules. (As you might expect, a *cycloaddition* reaction is one in which two reactants *add* together to give a *cyclic* product.) The 1950 Nobel Prize in chemistry was awarded to Diels and Alder in recognition of the importance of their discovery.

The mechanism of the Diels–Alder cycloaddition is different from other reactions we've studied because it is neither polar nor radical. Rather, the Diels–Alder reaction is a *pericyclic* process. Pericyclic reactions, which we'll discuss in more detail in Chapter 31, take place in a single step by a cyclic redistribution of bonding electrons. The two reactants simply join together through a cyclic transition state in which both new carbon–carbon bonds form at the same time.

We can picture a Diels–Alder addition as occurring by head-on (σ) overlap of the two alkene *p* orbitals with the two *p* orbitals on carbons 1 and 4 of the diene (Figure 14.7). This is, of course, a *cyclic* orientation of the reactants.

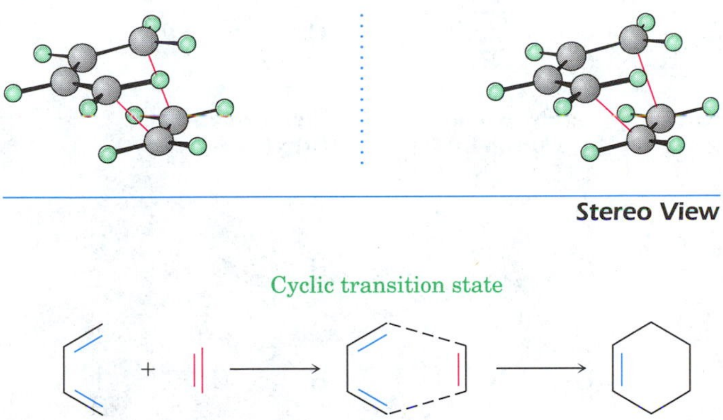

Figure 14.7 Mechanism of the Diels–Alder cycloaddition reaction. The reaction occurs in a single step through a cyclic transition state in which the two new carbon–carbon bonds form simultaneously.

In the Diels–Alder transition state, the two alkene carbons and carbons 1 and 4 of the diene rehybridize from sp^2 to sp^3 to form two new single bonds. Carbons 2 and 3 of the diene remain sp^2-hybridized to form the new double bond in the cyclohexene product. We'll study this mechanism at greater length in Chapter 31 and will concentrate for the present on learning more about the chemistry of the Diels–Alder reaction.

[1]Otto Diels (1876–1954); b. Hamburg; Ph.D. Berlin (E. Fischer); professor, University of Berlin (1906–1916), Kiel (1916–1948); Nobel Prize (1950).

[2]Kurt Alder (1902–1958); b. Königshütte; Ph.D. Kiel (Diels); professor, University of Cologne (1940–1958); Nobel Prize (1950).

14.9 Characteristics of the Diels–Alder Reaction

The Dienophile

The Diels–Alder cycloaddition reaction occurs most rapidly and in highest yield if the alkene component, or **dienophile** ("diene lover"), has an electron-withdrawing substituent group. Thus, ethylene itself reacts sluggishly, but propenal, ethyl propenoate, maleic anhydride, benzoquinone, propenenitrile, and others are highly reactive (Figure 14.8). Note also that alkynes, such as methyl propynoate, can act as Diels–Alder dienophiles. In all these cases, the dienophile double bond is next to the positively polarized carbon of a substituent that withdraws electrons.

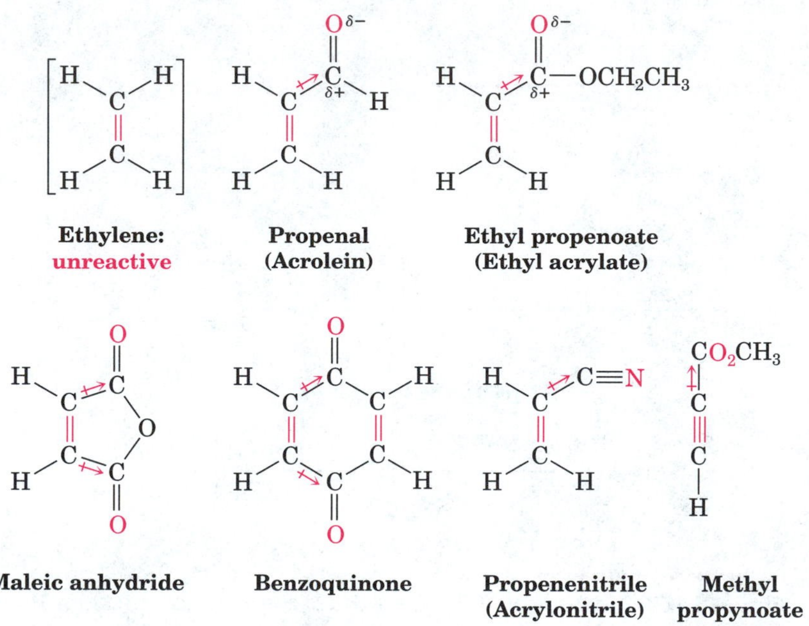

Figure 14.8 Some Diels–Alder dienophiles. All contain electron-withdrawing groups as substituents on the double bond.

One of the most useful features of the Diels–Alder reaction is that it is *stereospecific*: The stereochemistry of the starting dienophile is maintained during the reaction, and a single product stereoisomer results. If we carry out the cycloaddition with a cis alkene, such as methyl *cis*-2-butenoate, only the cis-substituted cyclohexene product is formed. Conversely, Diels–Alder reaction with methyl *trans*-2-butenoate yields only the trans-substituted cyclohexene product.

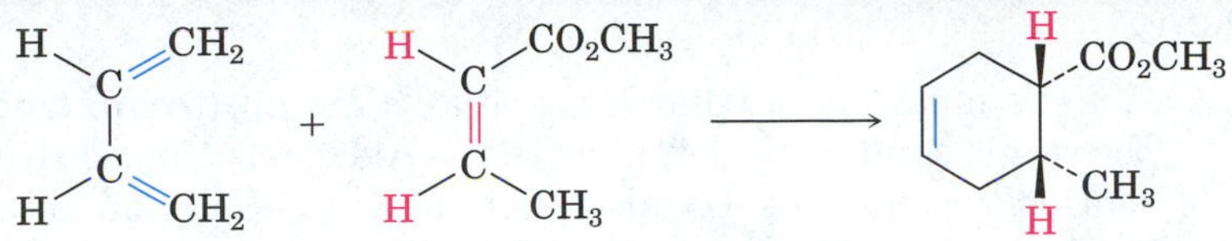

1,3-Butadiene **Methyl (*Z*)-2-butenoate** Cis product

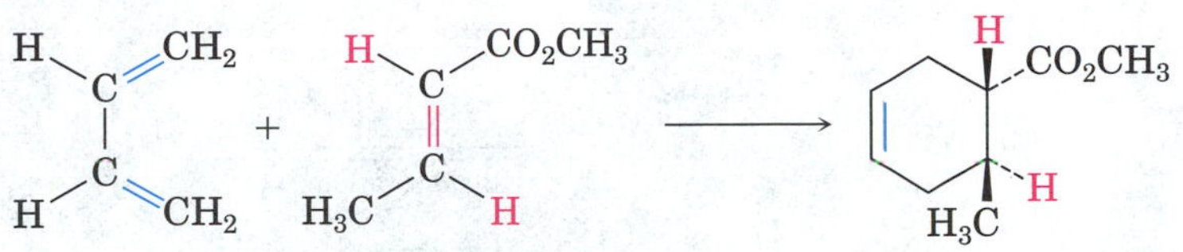

1,3-Butadiene **Methyl (*E*)-2-butenoate** Trans product

Another stereochemical feature of the Diels–Alder reaction is that the diene and dienophile partners line up so that the *endo* product rather than the alternative *exo* product is formed. The words *endo* and *exo* are used to indicate relative stereochemistry when referring to bicyclic structures like substituted norbornanes (Section 4.15). A substituent on one bridge is said to be *exo* if it is anti (trans) to the larger of the other two bridges, and is said to be *endo* if it is syn (cis) to the larger of the other two bridges.

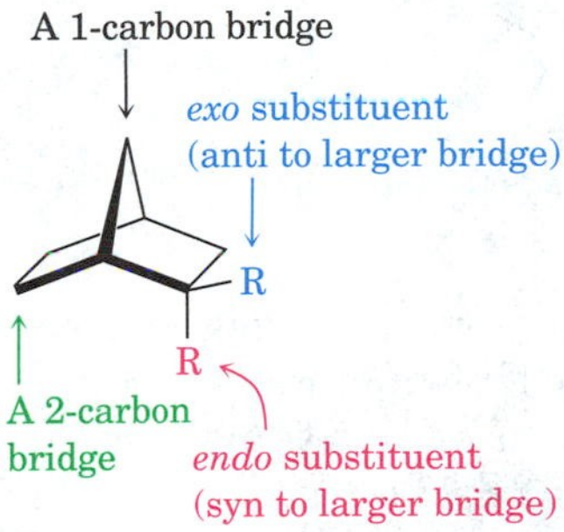

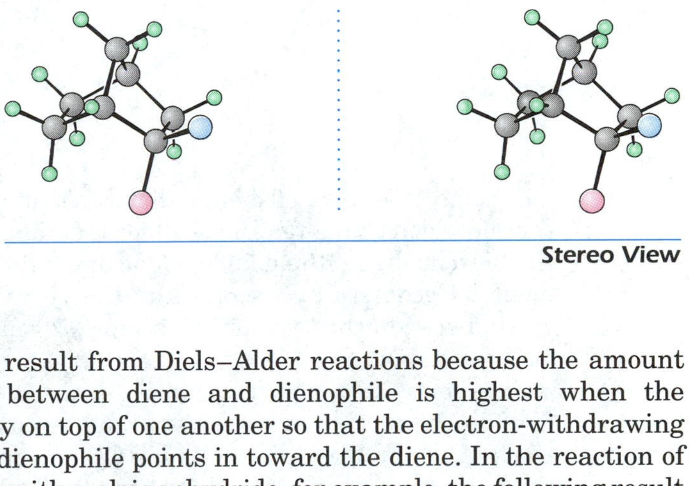

Stereo View

Endo products result from Diels–Alder reactions because the amount of orbital overlap between diene and dienophile is highest when the reactants lie directly on top of one another so that the electron-withdrawing substituent on the dienophile points in toward the diene. In the reaction of 1,3-cyclopentadiene with maleic anhydride, for example, the following result is obtained:

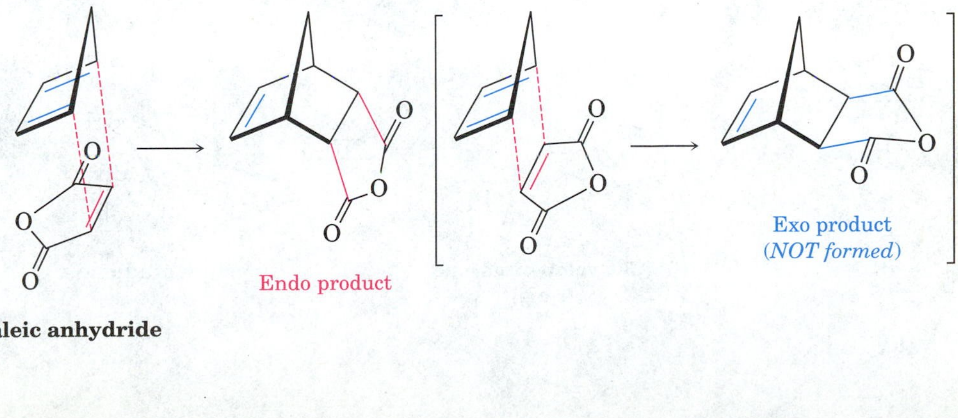

Maleic anhydride

The Diene

A diene must adopt what is called an **s-cis conformation** ("cis-like" about the *s*ingle bond) to undergo the Diels–Alder reaction. Only in the *s*-cis conformation are carbons 1 and 4 of the diene close enough to react through a cyclic transition state. In the alternative *s*-trans conformation, the ends of the diene partner are too far apart to overlap with the dienophile *p* orbitals.

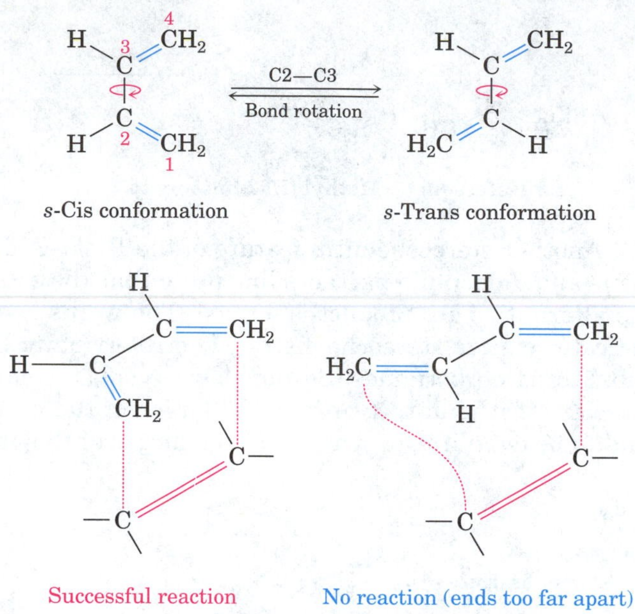

Some examples of dienes that can't adopt an *s*-cis conformation, and therefore don't undergo Diels–Alder reactions, are shown in Figure 14.9. In the bicyclic diene, the double bonds are rigidly fixed in an *s*-trans arrangement by geometric constraints of the rings. In (2*Z*,4*Z*)-hexadiene, steric strain between the two methyl groups prevents the molecule from adopting *s*-cis geometry.

In contrast to the unreactive *s*-trans dienes, other dienes are rigidly fixed in the correct *s*-cis geometry and are therefore highly reactive in the Diels–Alder cycloaddition reaction. Cyclopentadiene, for example, is so reactive that it reacts with itself. At room temperature, cyclopentadiene *dimerizes*: One molecule acts as diene and another acts as dienophile in a self Diels–Alder reaction.

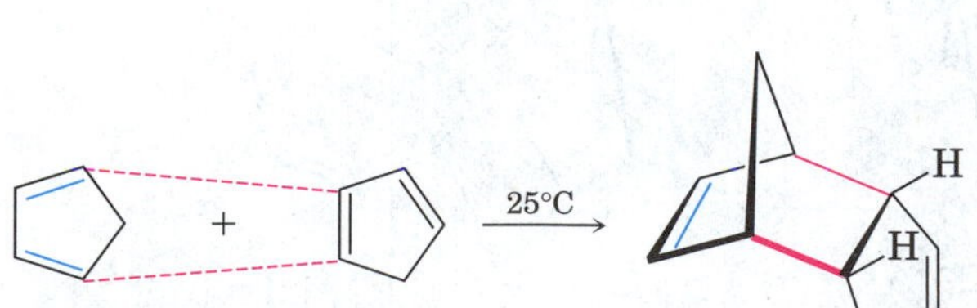

1,3-Cyclopentadiene (*s*-cis)

Bicyclopentadiene

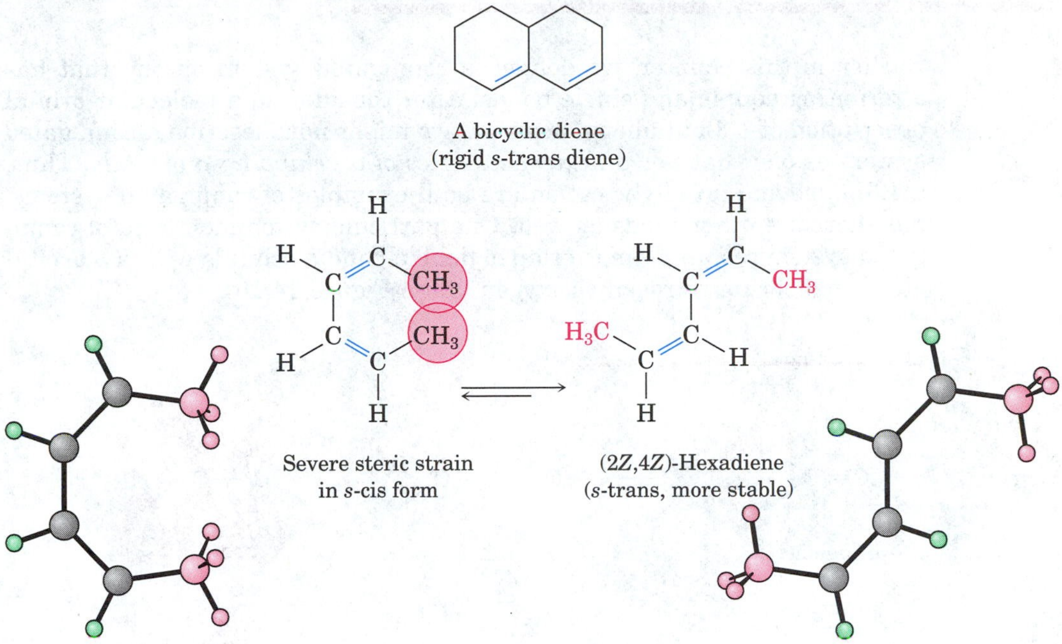

Figure 14.9 Two *s*-trans dienes that can't undergo Diels–Alder reactions.

PROBLEM. . . . .

14.10 Which of the following alkenes would you expect to be good Diels–Alder dienophiles?

(a) $H_2C{=}CH\overset{O}{\overset{\|}{C}}Cl$

(b) $H_2C{=}CHCH_2CH_2\overset{O}{\overset{\|}{C}}OCH_3$

(c) 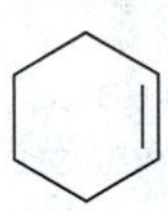

(d) 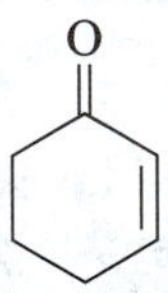

(e) 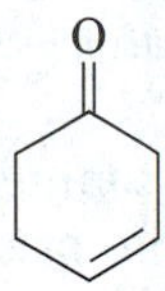

PROBLEM. . . . .

14.11 Which of the following dienes have an *s*-cis conformation, and which an *s*-trans conformation? Of the *s*-trans dienes, which can readily rotate to *s*-cis?

(a) 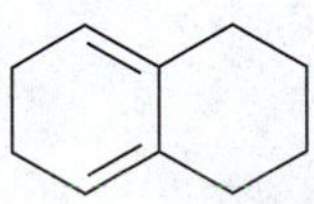

(b) 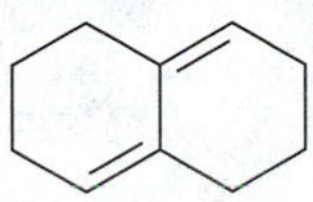

(c) 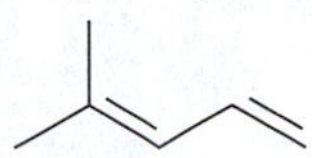

14.10 Other Conjugated Systems

Earlier in this chapter, we defined a conjugated system as one that has alternating double and single bonds. After considering a molecular orbital description of 1,3-butadiene, however, we might now describe a conjugated system as one that has *an extended series of overlapping p orbitals*. Thus, a 1,3-diene and an allylic cation are both examples of conjugated systems, but there are other kinds as well. One particularly common kind of conjugated system results from overlap of double-bond *p* orbitals with a *p* orbital on a neighboring nitrogen or oxygen atom (Figure 14.10).

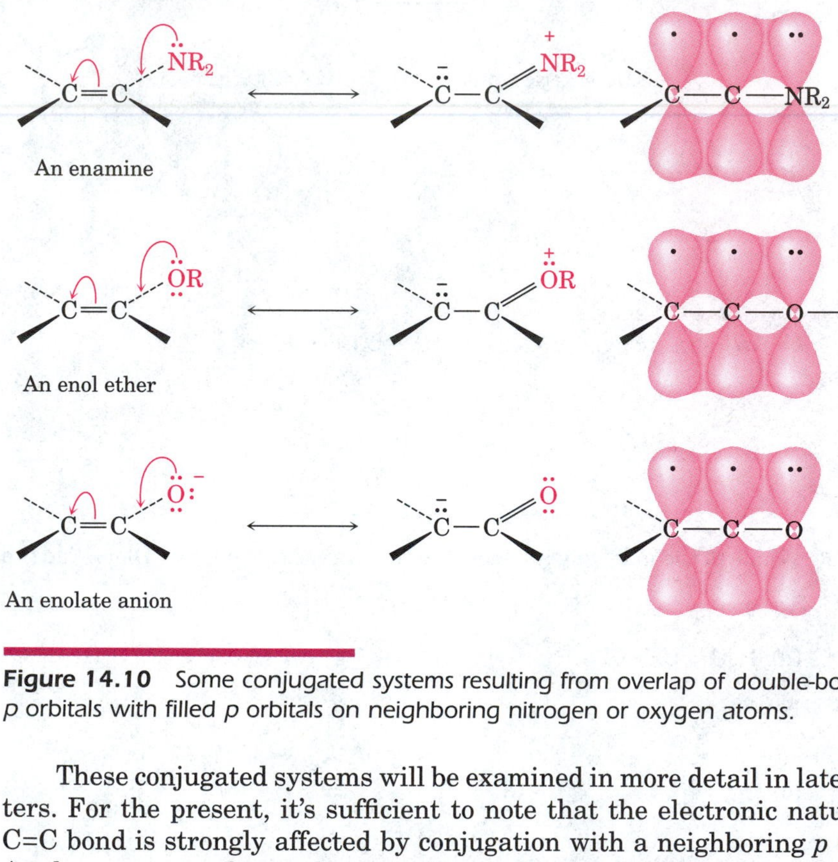

Figure 14.10 Some conjugated systems resulting from overlap of double-bond *p* orbitals with filled *p* orbitals on neighboring nitrogen or oxygen atoms.

These conjugated systems will be examined in more detail in later chapters. For the present, it's sufficient to note that the electronic nature of a C=C bond is strongly affected by conjugation with a neighboring *p* orbital. As the resonance forms shown in Figure 14.10 indicate, conjugation of the double bond with a filled neighboring orbital greatly increases the bond's electron density. Thus, enamines, enol ethers, and enolate ions are all much more strongly nucleophilic than alkenes.

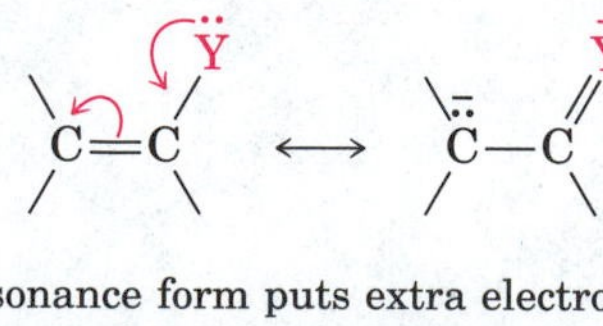

This resonance form puts extra electron density on *carbon*, making the carbon atom nucleophilic

14.11 Structure Determination in Conjugated Systems: Ultraviolet Spectroscopy

Mass spectrometry, infrared spectroscopy, and nuclear magnetic resonance spectroscopy are applicable to all organic structures. In addition to these three generally useful techniques of structure determination, there's a fourth—**ultraviolet (UV) spectroscopy**—that is applicable only to conjugated systems.

Mass spectrometry	Molecular size and formula
Infrared spectroscopy	Functional groups present
NMR spectroscopy	Carbon–hydrogen framework
Ultraviolet spectroscopy	**Nature of conjugated π electron system**

Ultraviolet spectroscopy is less commonly used than the other three spectroscopic techniques because of the specialized information it gives. We'll therefore study it only briefly.

The ultraviolet region of the electromagnetic spectrum extends from the low-wavelength end of the visible region (4×10^{-5} cm) to 10^{-6} cm, but the narrow range from 2×10^{-5} cm to 4×10^{-5} cm is the portion of greatest interest to organic chemists. Absorptions in this region are usually measured in nanometers (nm), where 1 nm = 10^{-9} m = 10^{-7} cm. Thus, the ultraviolet range of interest is from 200 to 400 nm (Figure 14.11).

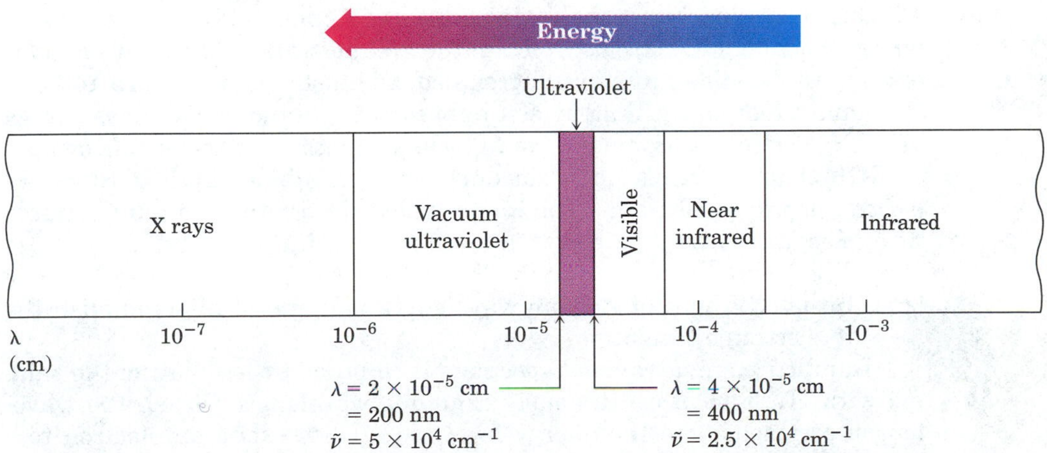

Figure 14.11 The ultraviolet (UV) region of the electromagnetic spectrum.

We saw in Section 12.4 that when an organic molecule is irradiated with electromagnetic energy, the radiation either is absorbed or not, depending on its energy. With IR irradiation, the energy absorbed corresponds to the amount necessary to increase molecular bending and stretching vibrations in functional groups. With UV radiation, the energy absorbed corresponds to the amount necessary to promote an electron from one orbital to another. Let's see what this means by looking first at 1,3-butadiene.

PROBLEM

14.12 Calculate the energy range of electromagnetic radiation in the UV region of the spectrum from 200 to 400 nm. Recall the equation

$$E = \frac{N_A hc}{\lambda} = \frac{1.20 \times 10^{-2}\ \text{kJ/mol}}{\lambda\ (\text{cm})}$$

PROBLEM

14.13 How does the energy you calculated in Problem 14.12 for UV radiation compare with the values calculated previously for IR and NMR spectroscopy?

14.12 Ultraviolet Spectrum of 1,3-Butadiene

1,3-Butadiene has four π molecular orbitals (Section 14.3). The two lower-energy, bonding MO's are occupied in the ground state, and the two higher-energy, antibonding MO's are unoccupied, as illustrated in Figure 14.12.

On irradiation with ultraviolet light ($h\nu$), 1,3-butadiene absorbs energy and a π electron is promoted from the highest occupied molecular orbital, or **HOMO**, to the lowest unoccupied molecular orbital, or **LUMO**. Since the electron is promoted from a bonding π molecular orbital to an antibonding π^* molecular orbital, we call this a $\pi \rightarrow \pi^*$ excitation (read as "pi to pi star"). The energy gap between the HOMO and the LUMO of 1,3-butadiene is such that UV light of 217 nm wavelength is required to accomplish the $\pi \rightarrow \pi^*$ electronic transition.

In practice, an ultraviolet spectrum is recorded by irradiating the sample with UV light of continuously changing wavelength. When the wavelength corresponds to the energy level required to excite an electron to a higher level, energy is absorbed. This absorption is detected and displayed on a chart that plots wavelength versus percent radiation absorbed (Figure 14.13). Note that UV spectra differ from IR spectra in the way they are

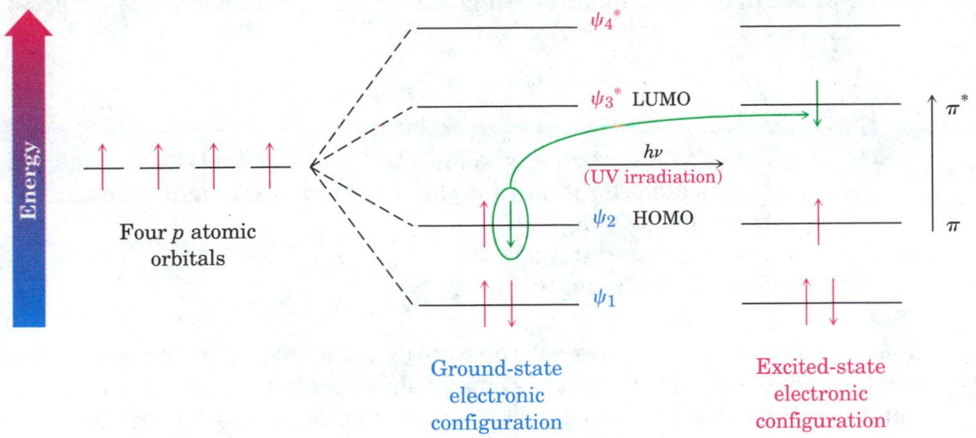

Figure 14.12 Ultraviolet excitation of 1,3-butadiene results in the promotion of an electron from ψ_2, the highest occupied molecular orbital (HOMO), to ψ_3^*, the lowest unoccupied molecular orbital (LUMO).

presented. IR spectra are usually displayed so that the baseline corresponding to zero absorption runs across the top of the chart and a valley indicates an absorption. UV spectra are displayed with the baseline at the bottom of the chart so that a peak indicates an absorption.

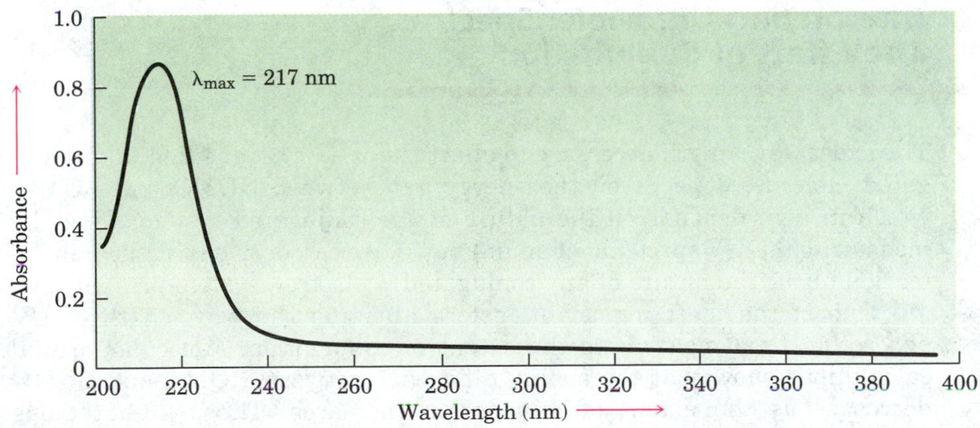

Figure 14.13 The ultraviolet spectrum of 1,3-butadiene, λ_{max} = 217 nm.

The exact amount of UV light absorbed is expressed as the sample's **molar absorptivity**, or **extinction coefficient**, ε, defined by the equation

$$\text{Molar absorptivity } \varepsilon = \frac{A}{C \times l}$$

where A = Absorbance, expressed as $\log(I_0/I)$, where I_0 is the intensity of the incident light and I is the intensity of the light transmitted through the sample
C = Concentration in mol/L
l = Sample path length in cm

Molar absorptivity is a physical constant, characteristic of the particular substance being observed and thus characteristic of the particular π electron system in the molecule. Typical values for conjugated dienes are in the range ε = 10,000–25,000.

Unlike IR and NMR spectra, which show many absorptions for a given molecule, UV spectra are usually quite simple—often only a single peak is seen. The peak is usually broad, however, and we identify its position by noting the wavelength at the very top of the peak (λ_{max}; read as "lambda max").

PROBLEM

14.14 A knowledge of molar absorptivities is particularly important in biochemistry where UV spectroscopy can provide an extremely sensitive method of analysis. For example, imagine that you wanted to determine the concentration of vitamin A in a sample. If pure vitamin A has λ_{max} = 325 (ε = 50,100) in a cell with a path length of 1.00 cm, what is the vitamin A concentration in a sample whose absorbance at 325 nm is A = 0.735?

14.13 Interpreting Ultraviolet Spectra: The Effect of Conjugation

The exact wavelength necessary to effect the $\pi \rightarrow \pi^*$ transition in a conjugated molecule depends on the energy gap between HOMO and LUMO, which in turn depends on the nature of the conjugated system. Thus, by measuring the UV spectrum of an unknown, we can derive structural information about the nature of any conjugated π electron system present.

One of the most important factors affecting the wavelength of UV absorption by a molecule is the extent of conjugation. Molecular orbital calculations show that the energy difference between HOMO and LUMO decreases as the extent of conjugation increases. Thus, 1,3-butadiene absorbs at λ_{max} = 217 nm, 1,3,5-hexatriene absorbs at λ_{max} = 258 nm, and 1,3,5,7-octatetraene absorbs at λ_{max} = 290 nm. (*Remember*: Longer wavelength means lower energy.)

Other kinds of conjugated systems, such as conjugated enones and aromatic rings, also exhibit characteristic UV absorptions that aid in structure determination. The UV absorption maxima of some representative conjugated molecules are given in Table 14.3.

Table 14.3 Ultraviolet Absorptions of Some Conjugated Molecules

Name	*Structure*	λ_{max} *(nm)*
2-Methyl-1,3-butadiene	$H_2C{=}C(CH_3){-}CH{=}CH_2$	220
1,3-Cyclohexadiene	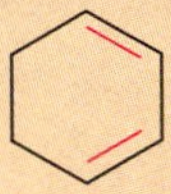	256
1,3,5-Hexatriene	$H_2C{=}CH{-}CH{=}CH{-}CH{=}CH_2$	258
1,3,5,7-Octatetraene	$H_2C{=}CH{-}CH{=}CH{-}CH{=}CH{-}CH{=}CH_2$	290
2,4-Cholestadiene	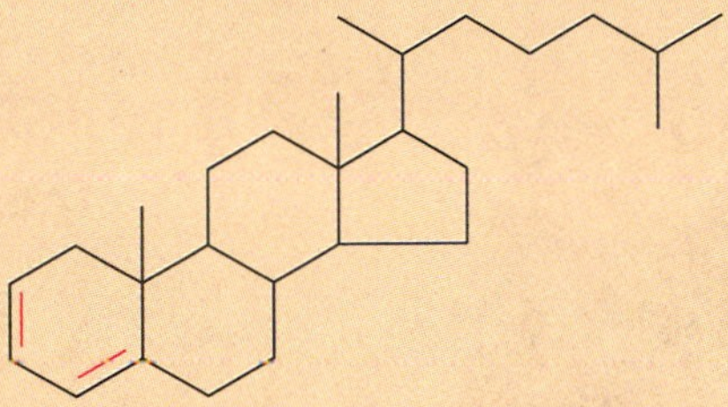	275
3-Buten-2-one	$H_2C{=}CH{-}C(CH_3){=}O$	219
Benzene	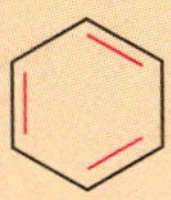	254
Naphthalene	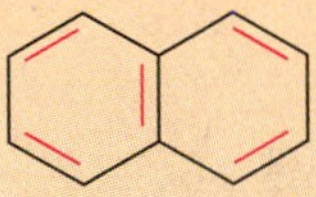	275

PROBLEM

14.15 Which of the following compounds would you expect to show ultraviolet absorptions in the 200–400 nm range?

(a) 1,4-Cyclohexadiene (b) 1,3-Cyclohexadiene (c) $H_2C{=}CH{-}C{\equiv}N$

(d)

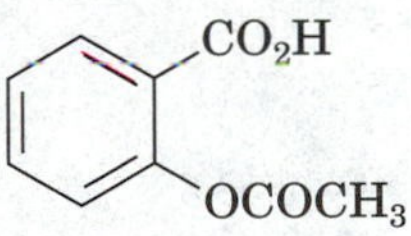

Aspirin

(e)

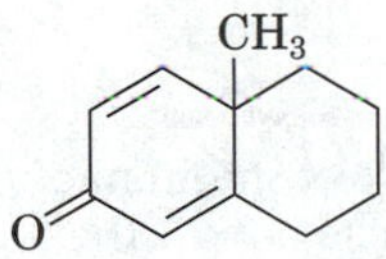

(f)

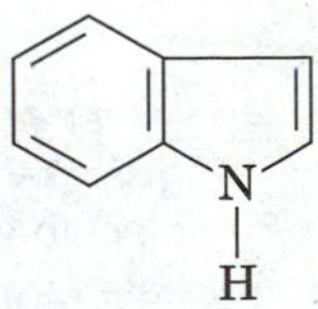

Indole

14.14 Colored Organic Compounds

Why are some organic compounds colored while others aren't? Why is β-carotene orange (the pigment in carrots), while benzene is colorless? The answers involve both the structures of colored molecules and the way we perceive light.

The visible region of the electromagnetic spectrum is adjacent to the ultraviolet region, extending from approximately 400 to 800 nm. Colored compounds have such extended systems of conjugation that their "UV" absorptions extend into the visible region. β-Carotene, for example, has 11 double bonds in conjugation, and its absorption occurs at λ_{max} = 455 nm (Figure 14.14).

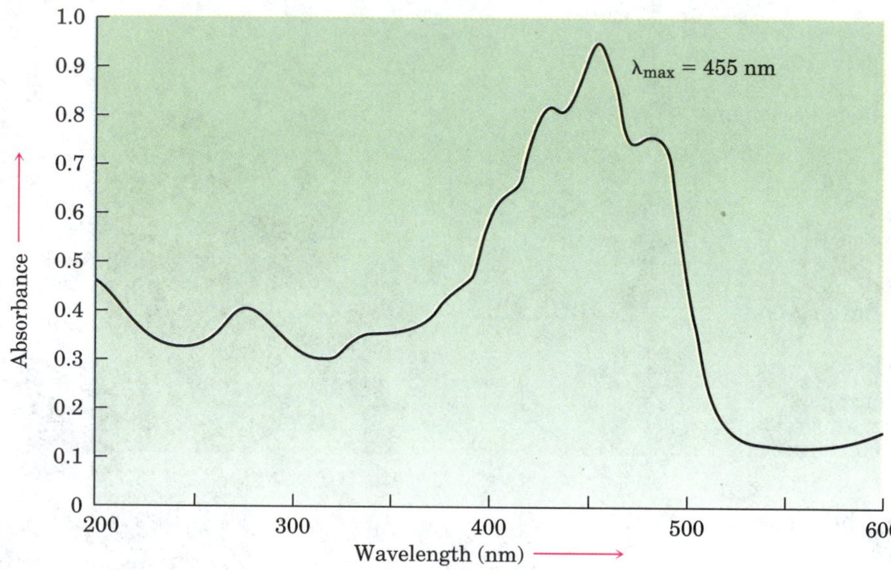

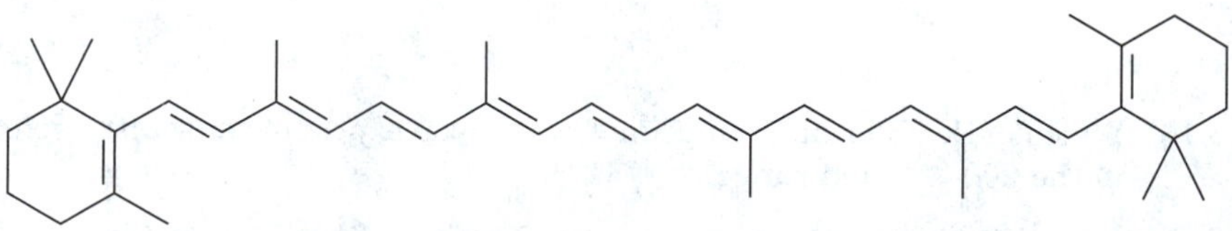

β-Carotene

Figure 14.14 Ultraviolet spectrum of β-carotene, a conjugated molecule with 11 double bonds. The absorption occurs in the visible region.

"White" light from the sun or from a lamp consists of all wavelengths in the visible region. When white light strikes β-carotene, the wavelengths from 400 to 500 nm (blue) are absorbed, while all other wavelengths are transmitted and can reach our eyes. We therefore see the white light with the blue removed, and we perceive a yellow-orange color for β-carotene.

What is true for β-carotene is also true for all other colored organic compounds: All have an extended system of π electron conjugation that gives rise to an absorption in the visible region of the electromagnetic spectrum.

INTERLUDE

Resists for Integrated Circuits

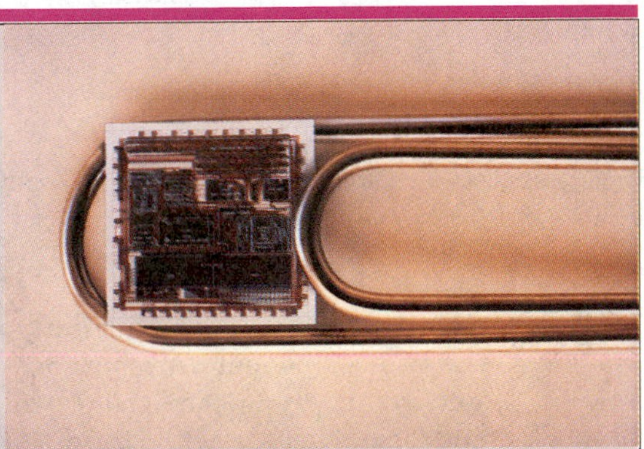

An integrated circuit chip next to a small paper clip

Twenty years ago, someone interested in owning a computer would have paid approximately $150,000 for 16 megabytes of random-access memory that would have occupied a volume the size of a small desk. Today, anyone can buy 16 MB of computer memory for under $1000 and can fit the chips into their shirt pocket. The difference between then and now is due to improvements in *photolithography*, the process by which integrated circuit chips are made.

Photolithography begins by coating a layer of SiO_2 onto a silicon wafer, further coating with a thin (0.5–1.0 μm thick) film of a light-sensitive organic polymer called a *resist*, and then irradiating the wafer with UV light through a stencil, or *mask*. The nonmasked, irradiated sections of the polymer undergo chemical changes that alter their solubility, making them more soluble than the masked, unirradiated sections. Washing with solvent selectively removes polymer from the irradiated areas, exposing a pattern of SiO_2 that can be chemically etched by reaction with hydrofluoric acid. Further washing removes the remaining polymer, leaving a positive image of the mask in the form of exposed ridges of SiO_2 (Figure 14.15). Additional cycles of coating, masking, and etching then produce the completed chips.

(continued)➤

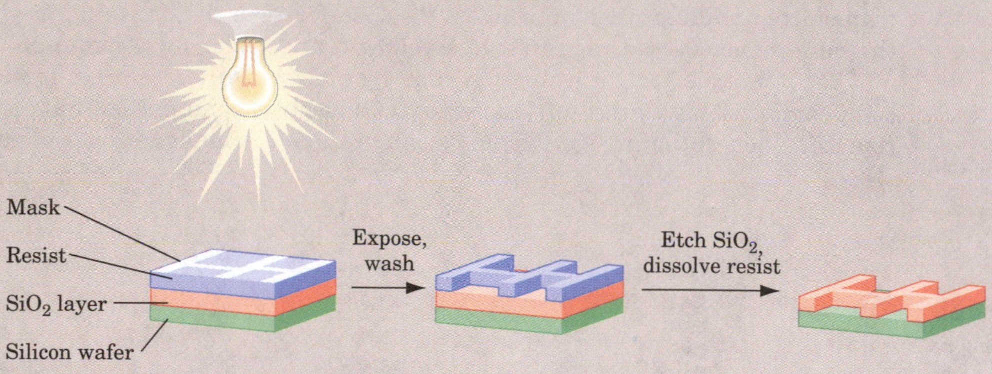

Figure 14.15 Outline of the photolithography process for producing integrated circuit chips.

The polymer resist currently used in chip manufacturing is based on the two-component *diazoquinone–novolac system*. The novolac resin is a soft, relatively low-molecular-weight polymer made from methylphenol and formaldehyde, while the diazoquinone is a bicyclic (two-ring) molecule containing a diazo group (=N_2) adjacent to a ketone carbonyl (C=O). The diazoquinone–novolac mix is relatively insoluble when fresh, but on exposure to ultraviolet light and water vapor, the diazoquinone component undergoes photodecomposition to yield a carboxylic acid plus N_2. On washing with dilute base, the carboxylic acid is converted into its soluble anion and removed. Novolac–diazoquinone technology is capable of producing features as small as 0.5 μm (5×10^{-7} m), but further improvements in miniaturization will have to come from newer resist materials currently being developed.

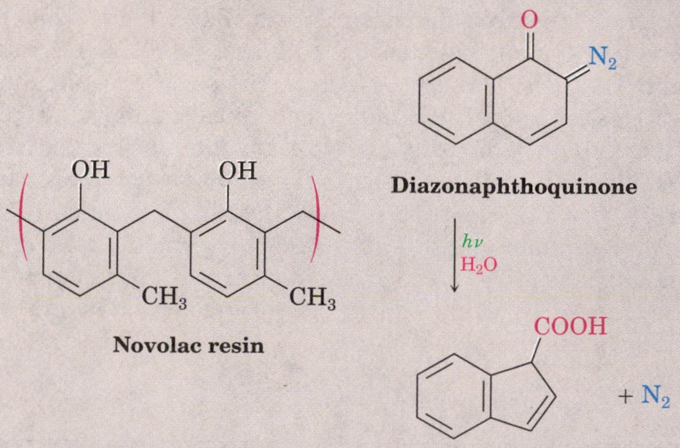

Summary and Key Words

A **conjugated diene** is one that contains alternating double and single bonds. One characteristic of conjugated dienes is that they are somewhat more stable than their nonconjugated counterparts. This unexpected stability can be explained by a molecular orbital description in which four *p* atomic orbitals combine to form four π molecular orbitals. Only the two bonding orbitals are occupied; the two antibonding orbitals are unoccupied. A π bonding interaction introduces some partial double-bond character between carbons 2 and 3, thereby strengthening the C2–C3 bond and stabilizing the molecule.

Conjugated dienes undergo two reactions not observed for nonconjugated dienes. The first is **1,4 addition** of electrophiles. When a conjugated diene is treated with an electrophile such as HCl, **1,2** and **1,4 adducts** are formed. Both products are formed from the same resonance-stabilized allylic carbocation intermediate and are produced in varying amounts depending on the reaction conditions. The 1,2 adduct is usually formed faster and is said to be the product of **kinetic control**. The 1,4 adduct usually predominates at equilibrium and is said to be the product of **thermodynamic control**.

The second reaction unique to conjugated dienes is **Diels–Alder cycloaddition**. Conjugated dienes react with electron-poor alkenes (**dienophiles**) in a single step through a cyclic transition state to yield a cyclohexene product. The reaction can occur only if the diene is able to adopt an ***s*-cis conformation**.

Ultraviolet (UV) spectroscopy is a method of structure determination applicable specifically to conjugated systems. When a conjugated molecule is irradiated with ultraviolet light, energy absorption occurs and a π electron is promoted from the highest occupied molecular orbital (**HOMO**) to the lowest unoccupied molecular orbital (**LUMO**). For 1,3-butadiene, radiation of λ_{max} = 217 nm is required. As a general rule, the greater the extent of conjugation, the less the energy (the longer the wavelength) needed.

ADDITIONAL PROBLEMS

14.16 Give IUPAC names for the following alkenes:

(a) $CH_3CH{=}\overset{\displaystyle CH_3}{\overset{|}{C}}CH{=}CHCH_3$

(b) $H_2C{=}CHCH{=}CHCH{=}CHCH_3$

(c) $CH_3CH{=}C{=}CHCH{=}CHCH_3$

(d) $CH_3CH{=}\overset{\displaystyle CH_2CH_2CH_3}{\overset{|}{C}}CH{=}CH_2$

14.17 Circle any conjugated portions of these molecules:

(a) $CH_3CH{=}C{=}CHCH{=}CHCH_3$

(b)

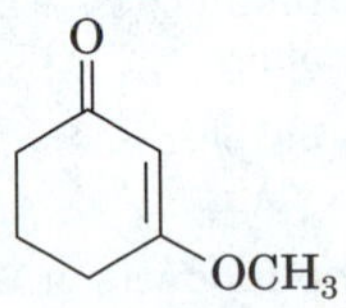

(c)

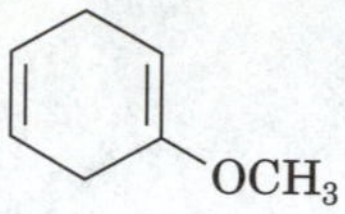

(d)

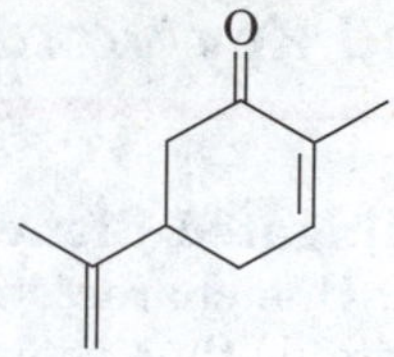

Carvone
(oil of spearmint)

(e)

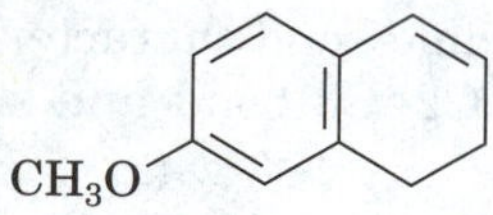

(f)

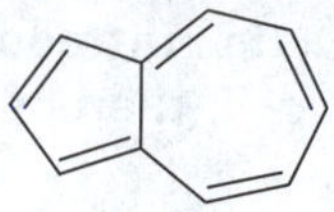

14.18 What product(s) would you expect to obtain from reaction of 1,3-cyclohexadiene with each of the following?

(a) 1 mol Br_2 in CCl_4
(b) O_3 followed by Zn
(c) 1 mol HCl in ether
(d) 1 mol DCl in ether
(e) 3-Buten-2-one ($H_2C{=}CHCOCH_3$)
(f) Excess OsO_4, followed by $NaHSO_3$

14.19 Draw and name the six possible diene isomers of formula C_5H_8. Which of the six are conjugated dienes?

14.20 Treatment of 3,4-dibromohexane with strong base leads to loss of 2 equivalents of HBr and formation of a product with formula C_6H_{10}. Three products are possible. Name each of the three, and tell how you would use ^{1}H and ^{13}C NMR spectroscopy to help you identify them. How would you use UV spectroscopy?

14.21 Electrophilic addition of Br_2 to isoprene yields the following product mixture:

$$H_2C{=}C(CH_3)CH{=}CH_2 \xrightarrow{Br_2} \underset{(3\%)}{H_2C{=}C(CH_3)CHBrCH_2Br} + \underset{(21\%)}{BrCH_2CBr(CH_3)CH{=}CH_2} + \underset{(76\%)}{BrCH_2C(CH_3){=}CHCH_2Br}$$

Of the 1,2-addition products, explain why 3,4-dibromo-3-methyl-1-butene (21%) predominates over 3,4-dibromo-2-methyl-1-butene (3%).

14.22 Diene polymers contain occasional vinyl branches along the chain. How do you think these branches might arise?

$$CH_2{=}CH{-}CH{=}CH_2 \longrightarrow \text{-}CH_2CH{=}CHCH_2CH_2\underset{\underset{\displaystyle CH{=}CH_2}{|}}{C}HCH_2CH{=}CHCH_2\text{-}$$

A vinyl branch

14.23 Tires whose sidewalls are made of natural rubber tend to crack and weather rapidly in areas around cities where high levels of ozone and other industrial pollutants are found. Explain.

14.24 Would you expect allene, $H_2C{=}C{=}CH_2$, to show a UV absorption in the 200–400 nm range? Explain.

14.25 Predict the products of the following Diels–Alder reactions:

(a) 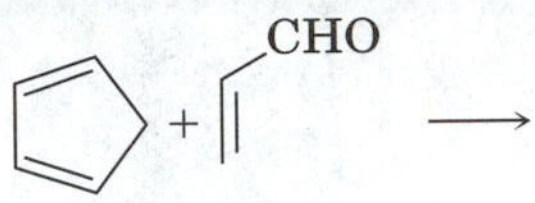

(b) 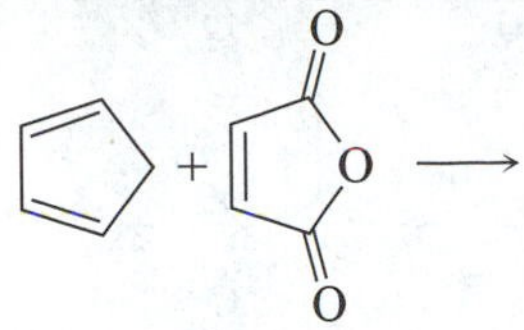

(c) 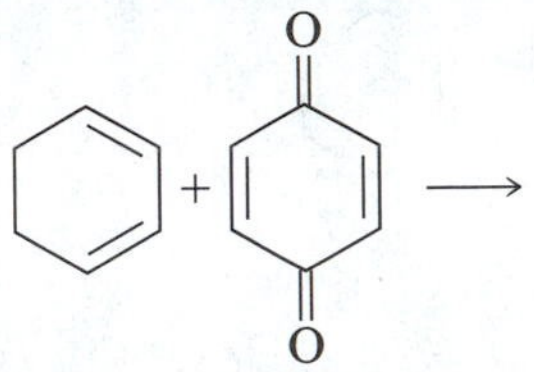

14.26 How can you account for the fact that *cis*-1,3-pentadiene is much less reactive than *trans*-1,3-pentadiene in the Diels–Alder reaction?

14.27 Which of the following compounds would you expect to have a $\pi \rightarrow \pi^*$ UV absorption in the 200–400 nm range?

(a) 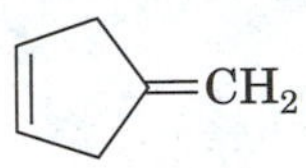

(b) 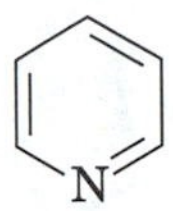

Pyridine

(c) $(CH_3)_2C{=}C{=}O$

A ketone

14.28 Would you expect a conjugated diyne such as 1,3-butadiyne to undergo Diels–Alder reaction with a dienophile? Explain.

14.29 Propose a structure for a conjugated diene that gives the same product from both 1,2 and 1,4 addition of HBr.

14.30 Draw the possible products resulting from addition of 1 equiv of HCl to 1-phenyl-1,3-butadiene. Which would you expect to predominate, and why?

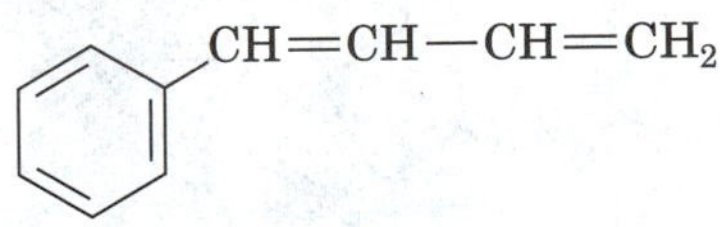

1-Phenyl-1,3-butadiene

14.31 Reaction of isoprene (2-methyl-1,3-butadiene) with ethyl propenoate gives a mixture of two Diels–Alder adducts. Show the structure of each product, and explain why a mixture is formed.

$$H_2C{=}C(CH_3){-}CH{=}CH_2 + H_2C{=}CH\overset{O}{\overset{\|}{C}}OCH_2CH_3 \longrightarrow ?$$

14.32 Rank the following dienophiles in order of their expected reactivity in the Diels–Alder reaction. Explain.

$H_2C{=}CHCH_3$, $H_2C{=}CHCHO$, $(N{\equiv}C)_2C{=}C(C{\equiv}N)_2$, $(CH_3)_2C{=}C(CH_3)_2$

14.33 Cyclopentadiene is highly reactive in Diels–Alder cycloaddition reactions, but 1,3-cyclohexadiene is less reactive, and 1,3-cycloheptadiene is nearly inert. Suggest a reason for this reactivity order. (Molecular models are helpful.)

14.34 How would you use Diels–Alder reactions to prepare the following products? Show the starting diene and dienophile in each case.

(a)

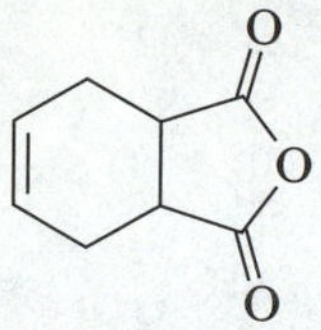

(b)

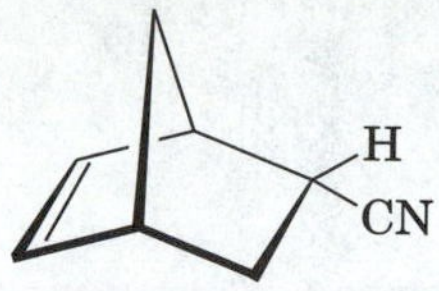

(c)

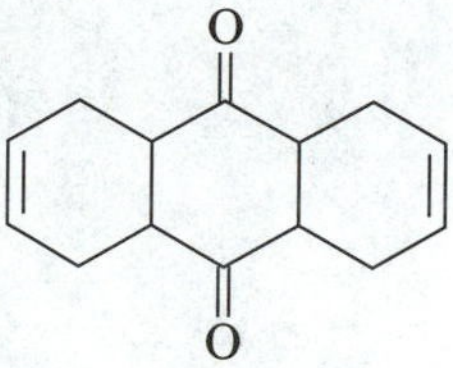

(d)

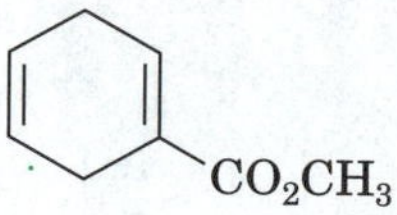

14.35 Aldrin, a chlorinated insecticide now banned for use in the United States, can be made by Diels–Alder reaction of hexachloro-1,3-cyclopentadiene with norbornadiene. What is the structure of aldrin?

Norbornadiene

14.36 Norbornadiene (Problem 14.35) can be prepared by reaction of chloroethylene with cyclopentadiene, followed by treatment of the product with sodium ethoxide. Formulate the overall scheme, and identify the two kinds of reactions.

14.37 We've seen that the Diels–Alder cycloaddition reaction is a one-step, pericyclic process that occurs through a cyclic transition state. Propose a mechanism for the following reaction:

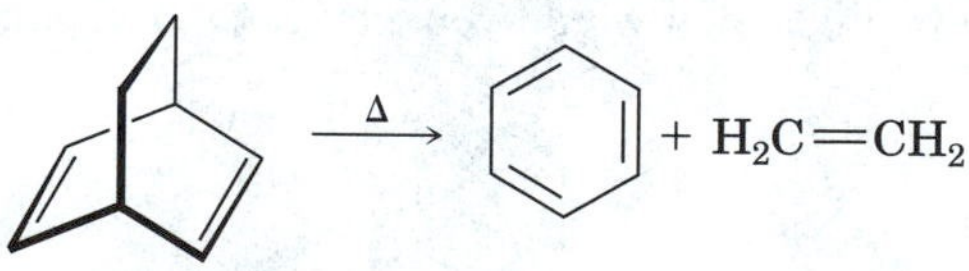

14.38 Propose a mechanism to explain the following reaction (see Problem 14.37):

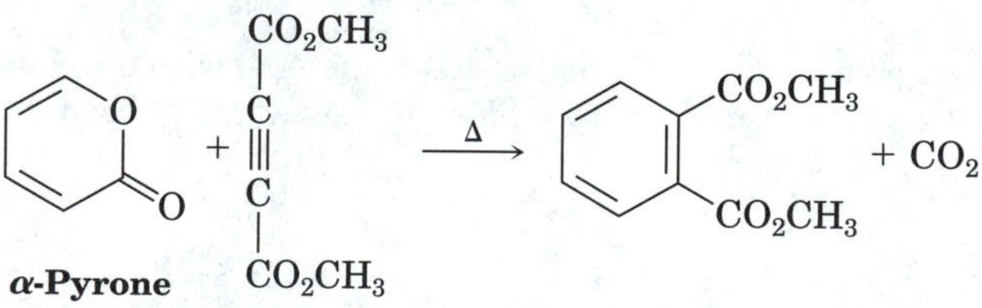

***α*-Pyrone**

14.39 The following ultraviolet absorption maxima have been measured:

	λ_{max} *(nm)*
1,3-Butadiene	217
2-Methyl-1,3-butadiene	220
1,3-Pentadiene	223
2,3-Dimethyl-1,3-butadiene	226
2,4-Hexadiene	227
2,4-Dimethyl-1,3-pentadiene	232
2,5-Dimethyl-2,4-hexadiene	240

What conclusion can you draw about the effect of alkyl substitution on UV absorption maxima? Approximately what effect does each added alkyl group have?

14.40 1,3,5-Hexatriene has λ_{max} = 258 nm. In light of your answer to Problem 14.39, approximately where would you expect 2,3-dimethyl-1,3,5-hexatriene to absorb? Explain.

14.41 β-Ocimene is a pleasant-smelling hydrocarbon found in the leaves of certain herbs. It has the molecular formula $C_{10}H_{16}$ and exhibits a UV absorption maximum at 232 nm. On hydrogenation with a palladium catalyst, 2,6-dimethyloctane is obtained. Ozonolysis of β-ocimene, followed by treatment with zinc and acetic acid, produces four fragments: acetone, formaldehyde, pyruvaldehyde, and malonaldehyde.

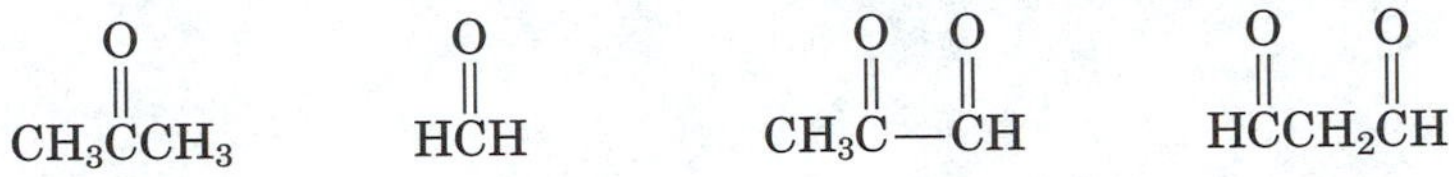

Acetone **Formaldehyde** **Pyruvaldehyde** **Malonaldehyde**

(a) How many double bonds does β-ocimene have?
(b) Is β-ocimene conjugated or nonconjugated?
(c) Propose a structure for β-ocimene.
(d) Formulate the reactions, showing starting material and products.

14.42 Myrcene, $C_{10}H_{16}$, is found in oil of bay leaves and is isomeric with β-ocimene (see Problem 14.41). It shows an ultraviolet absorption at 226 nm and can be catalytically hydrogenated to yield 2,6-dimethyloctane. On ozonolysis followed by zinc/acetic acid treatment, myrcene yields formaldehyde, acetone, and 2-oxopentanedial.

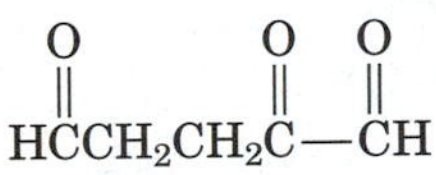

2-Oxopentanedial

Propose a structure for myrcene, and formulate the reactions, showing starting material and products.

14.43 Addition of HCl to 1-methoxycyclohexene yields 1-chloro-1-methoxycyclohexane as the sole product. Why is none of the other regioisomer formed?

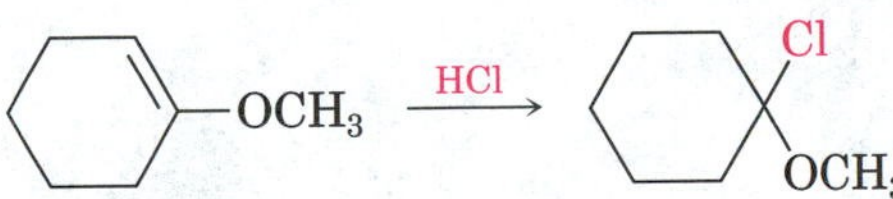

14.44 Hydrocarbon A, $C_{10}H_{14}$, has a UV absorption at λ_{max} = 236 nm and gives hydrocarbon B, $C_{10}H_{18}$, on catalytic hydrogenation. Ozonolysis of A followed by zinc/acetic acid treatment yields the following diketo dialdehyde:

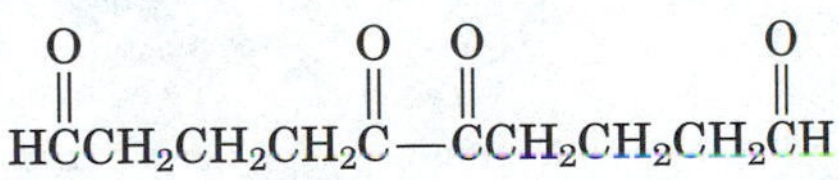

(a) Propose two possible structures for A.
(b) Hydrocarbon A reacts with maleic anhydride to yield a Diels–Alder adduct. Which of your structures for A is correct?
(c) Formulate the reactions, showing starting material and products.

14.45 Adiponitrile, a starting material used in the manufacture of nylon, can be prepared in three steps from 1,3-butadiene. How would you carry out this synthesis?

$$H_2C{=}CHCH{=}CH_2 \xrightarrow{\text{3 steps}} N{\equiv}CCH_2CH_2CH_2CH_2C{\equiv}N$$

Adiponitrile

14.46 Ergosterol, a precursor of vitamin D, has λ_{max} = 282 nm and molar absorptivity ε = 11,900. What is the concentration of ergosterol in a solution whose absorbance A = 0.065 with a sample path length l = 1.00 cm?

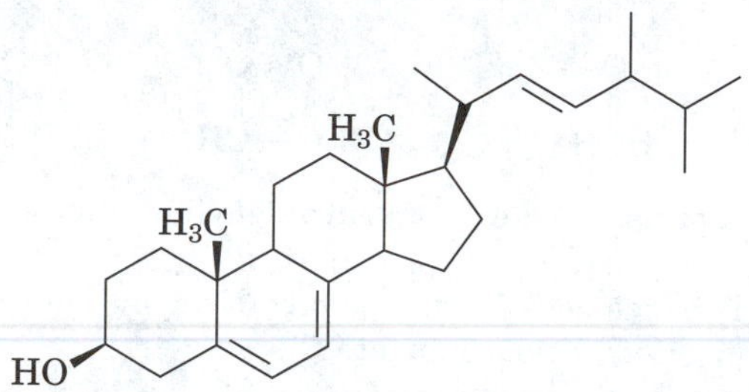

Ergosterol ($C_{28}H_{44}O$)

14.47 Cyclopentadiene polymerizes slowly at room temperature to yield a polymer that has no double bonds. On heating, the polymer breaks down to regenerate cyclopentadiene. Propose a structure for the product.

14.48 Dimethyl butynedioate undergoes a Diels–Alder reaction with (2*E*,4*E*)-hexadiene. Show the structure and stereochemistry of the product.

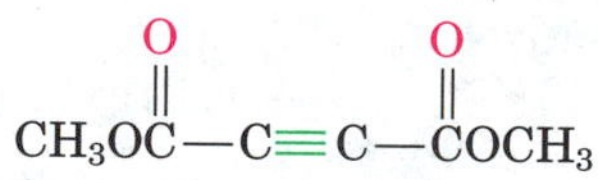

Dimethyl butynedioate

14.49 Dimethyl butynedioate also undergoes a Diels–Alder reaction with (2*E*,4*Z*)-hexadiene, but the stereochemistry of the product is different from that of the (2*E*,4*E*) isomer (Problem 14.48). Explain.

A Look Ahead

14.50 Benzene has an ultraviolet absorption at λ_{max} = 204 nm, and *para*-toluidine has λ_{max} = 235 nm. How do you account for this difference?

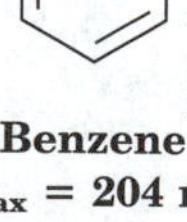

Benzene
(λ_{max} = 204 nm)

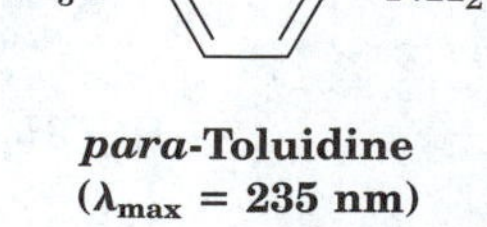

***para*-Toluidine**
(λ_{max} = 235 nm)

14.51 When the ultraviolet spectrum of *para*-toluidine (Problem 14.50) is measured in the presence of a small amount of HCl, the λ_{max} decreases to 207 nm, nearly the same value as benzene. Explain.

14.52 Phenol, a weak acid with pK_a = 10.0, has a UV absorption at λ_{max} = 210 nm in ethanol solution. When dilute NaOH is added, the absorption increases to λ_{max} = 235 nm. Explain.

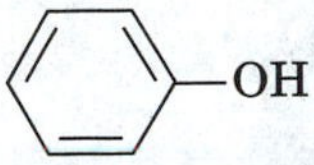

Phenol

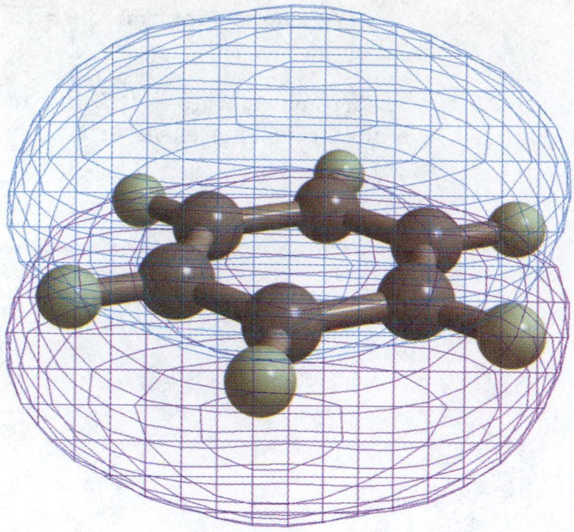

Benzene, a typical aromatic compound, is a cyclic conjugated molecule.

15

Benzene and Aromaticity

In the early days of organic chemistry, the word *aromatic* was used to describe such fragrant substances as benzaldehyde (from cherries, peaches, and almonds), toluene (from Tolu balsam), and benzene (from coal distillate). It was soon realized, however, that substances grouped as aromatic differed from most other organic compounds in their chemical behavior.

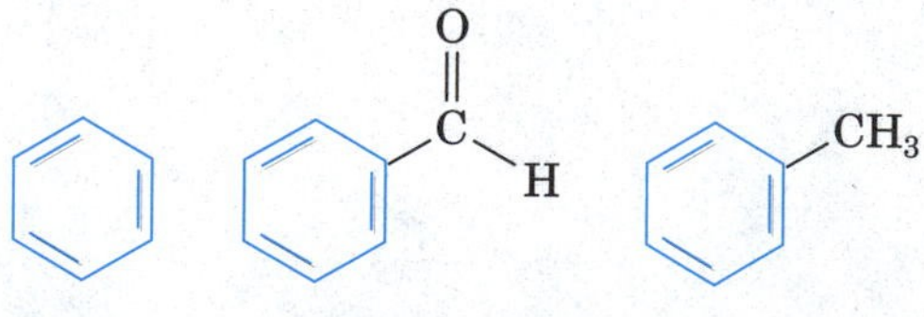

Benzene **Benzaldehyde** **Toluene**

Today, we use the word **aromatic** to refer to benzene and its structural relatives. We'll see in this and the next chapter that aromatic compounds show chemical behavior quite different from that of the aliphatic compounds we've studied to this point. Thus, chemists of the early nineteenth century were correct about there being a chemical difference between aromatic compounds and others, but the association of aromaticity with fragrance has long been lost.

Many compounds isolated from natural sources are aromatic in part. In addition to benzene, benzaldehyde, and toluene, such compounds as the steroidal hormone estrone and the well-known analgesic morphine have aromatic rings. Many synthetic drugs are also aromatic in part; the tranquilizer diazepam (Valium) is an example.

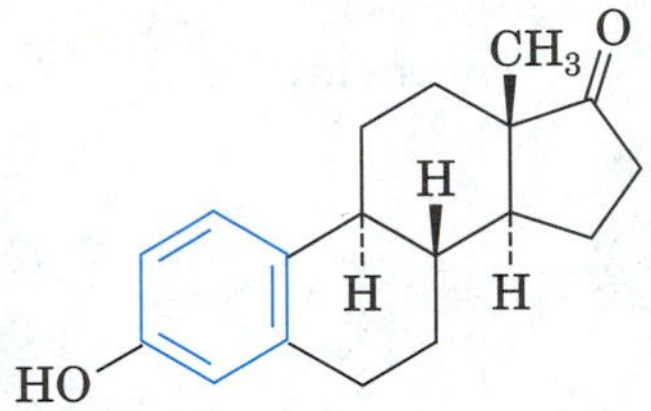

Estrone

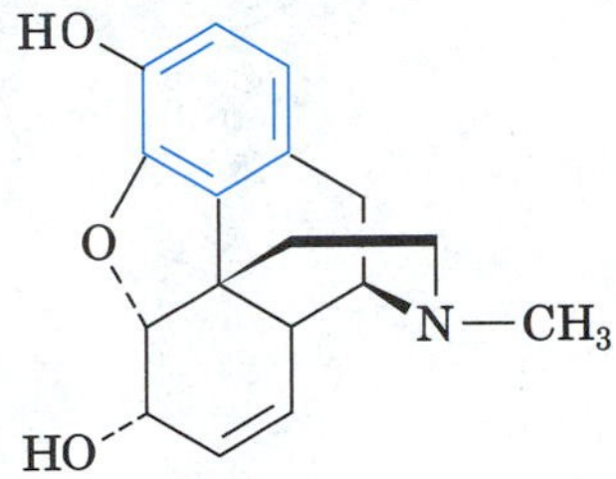

Morphine

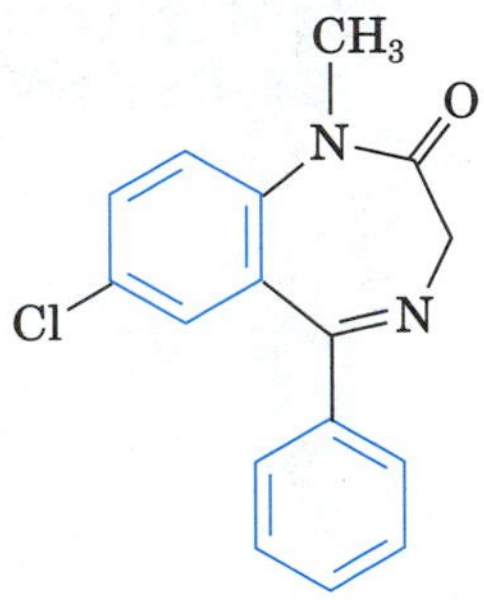

Diazepam (Valium)

Benzene itself has been found to cause bone-marrow depression and consequent leukopenia (depressed white blood cell count) on prolonged exposure. Benzene should therefore be used cautiously as a laboratory solvent.

15.1 Sources of Aromatic Hydrocarbons

Simple aromatic hydrocarbons come from two main sources: coal and petroleum. Coal is an enormously complex mixture made up primarily of large arrays of benzene-like rings linked together. When heated to 1000°C in the absence of air, thermal breakdown of coal occurs and a mixture of volatile products called *coal tar* boils off. Fractional distillation of coal tar yields benzene, toluene, xylene (dimethylbenzene), naphthalene, and a host of other aromatic compounds (Figure 15.1, p. 536).

Petroleum, unlike coal, contains few aromatic compounds and consists largely of alkanes (see the Chapter 3 Interlude). During petroleum refining, however, aromatic molecules are formed when alkanes are passed over a catalyst at about 500°C under high pressure. Heptane (C_7H_{16}), for example, is converted into toluene (C_7H_8) by dehydrogenation and cyclization.

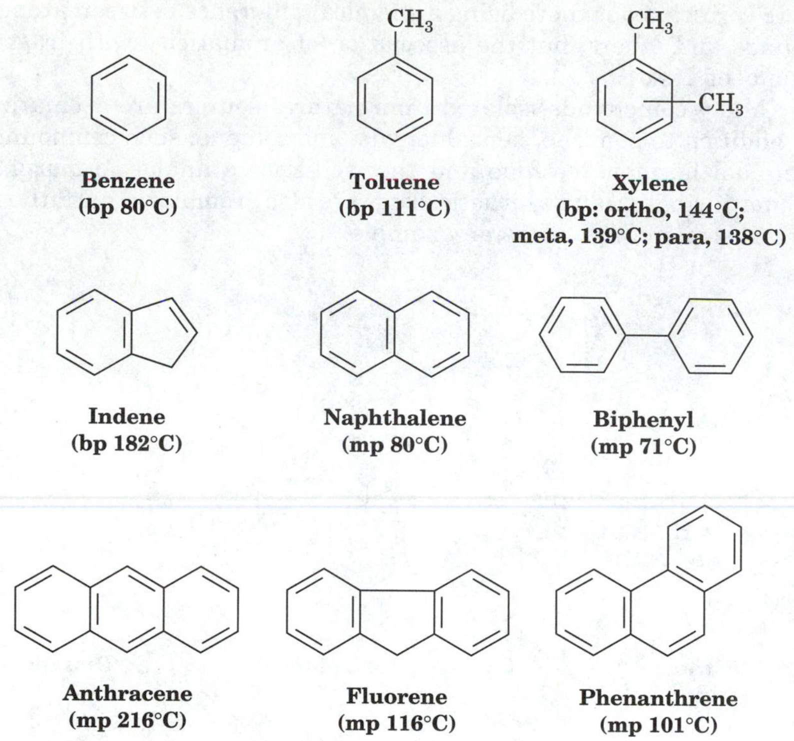

Figure 15.1 Some aromatic hydrocarbons found in coal tar.

15.2 Naming Aromatic Compounds

Aromatic substances, more than any other class of organic compounds, have acquired a large number of nonsystematic names. Although the use of such names is discouraged, IUPAC rules allow for some of the more widely used names to be retained (Table 15.1). Thus, methylbenzene is known commonly as *toluene*, hydroxybenzene as *phenol*, aminobenzene as *aniline*, and so on.

Monosubstituted benzene derivatives are systematically named in the same manner as other hydrocarbons, with *-benzene* as the parent name. Thus, C_6H_5Br is bromobenzene, $C_6H_5NO_2$ is nitrobenzene, and $C_6H_5CH_2CH_2CH_3$ is propylbenzene.

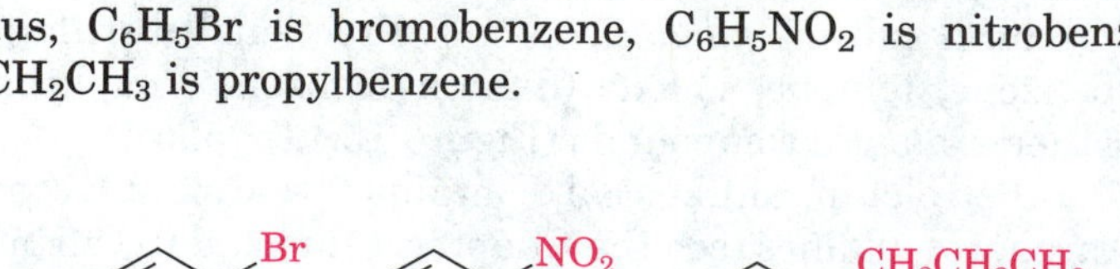

Table 15.1 Common Names of Some Aromatic Compounds

Formula	*Name*	*Formula*	*Name*
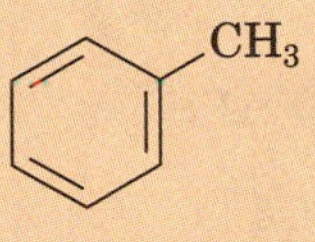	Toluene (bp 111°C)	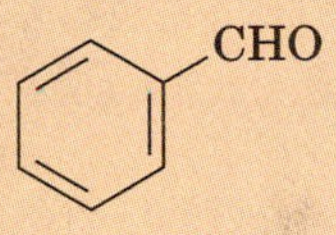	Benzaldehyde (bp 178°C)
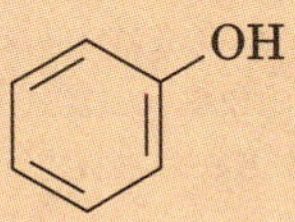	Phenol (mp 43°C)	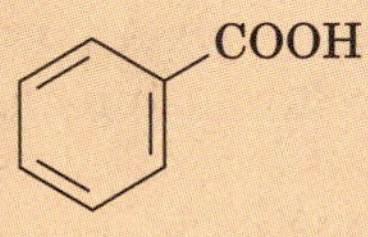	Benzoic acid (mp 122°C)
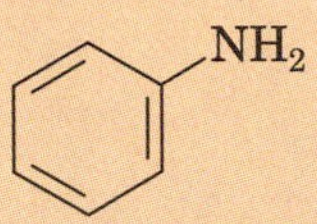	Aniline (bp 184°C)	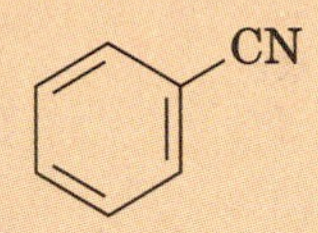	Benzonitrile (bp 191°C)
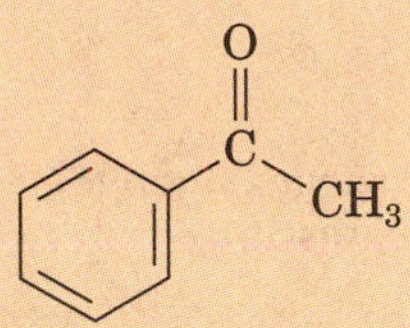	Acetophenone (mp 21°C)	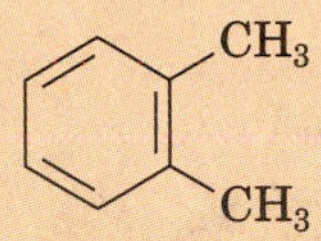	*ortho*-Xylene (bp 144°C)
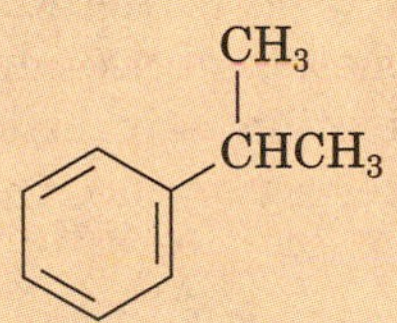	Cumene (bp 152°C)	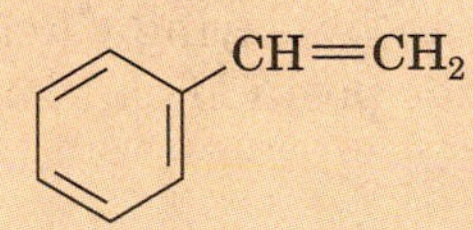	Styrene (bp 145°C)

Alkyl-substituted benzenes, sometimes referred to as **arenes**, are named in different ways depending on the size of the alkyl group. If the alkyl substituent has six or fewer carbons, the arene is named as an alkyl-substituted benzene. If the alkyl substituent has more than six carbons, the compound is named as a phenyl-substituted alkane. The name **phenyl**, pronounced **fen**-nil and often abbreviated as -Ph or Φ (Greek phi), is used for the $-C_6H_5$ unit when the benzene ring is considered as a substituent. The word is derived from the Greek *pheno* ("I bear light"), commemorating the fact that benzene was discovered by Michael Faraday[1] in 1825 from the oily residue left by the illuminating gas used in London street lamps. As mentioned previously (in Chapter 11), the $C_6H_5CH_2-$ group is called *benzyl*.

[1]Michael Faraday (1791–1867); b. Newington Butts, Surrey, England; assistant to Sir Humphry Davy (1813); director, laboratory of the Royal Institution (1825); Fullerian Professor of Chemistry, Royal Institution (1833).

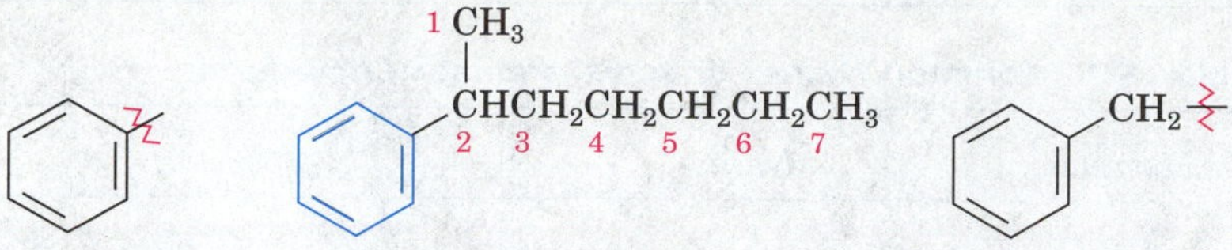

A phenyl group **2-Phenylheptane** **A benzyl group**

Disubstituted benzenes are named using one of the prefixes *ortho* (*o*), *meta* (*m*), or *para* (*p*). An ortho-disubstituted benzene has its two substituents in a 1,2 relationship on the ring; a meta-disubstituted benzene has its two substituents in a 1,3 relationship; and a para-disubstituted benzene has its substituents in a 1,4 relationship.

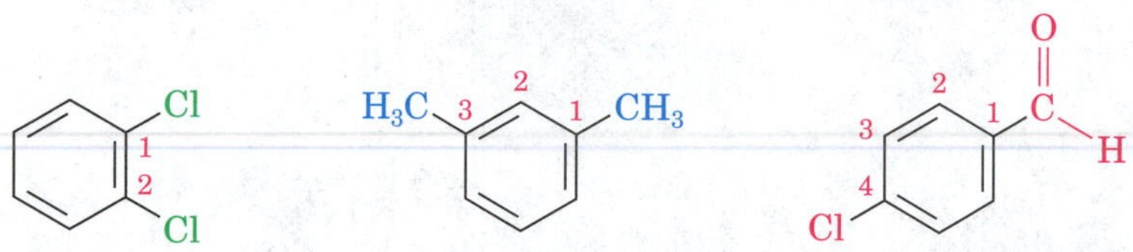

***ortho*-Dichlorobenzene** **1,2 disubstituted** — ***meta*-Xylene** **1,3 disubstituted** — ***para*-Chlorobenzaldehyde** **1,4 disubstituted**

The ortho, meta, para system of nomenclature is also useful when discussing reactions. For example, we might describe the reaction of bromine with toluene by saying, "Reaction occurs in the para position"—in other words, at the position para to the methyl group already present on the ring.

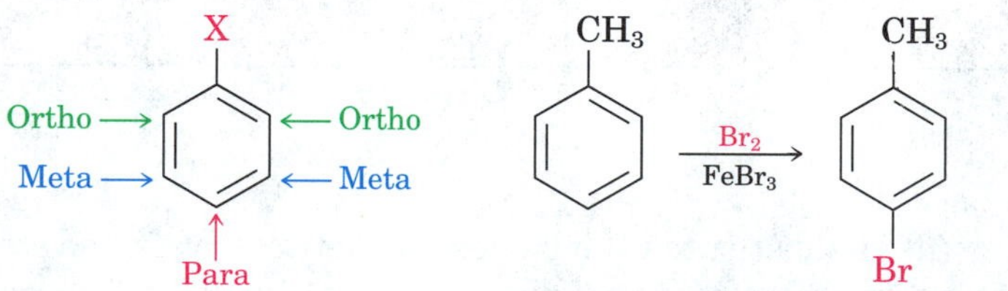

Toluene ***p*-Bromotoluene**

Benzenes with more than two substituents are named by numbering the position of each substituent so that the lowest possible numbers are used. The substituents are listed alphabetically when writing the name.

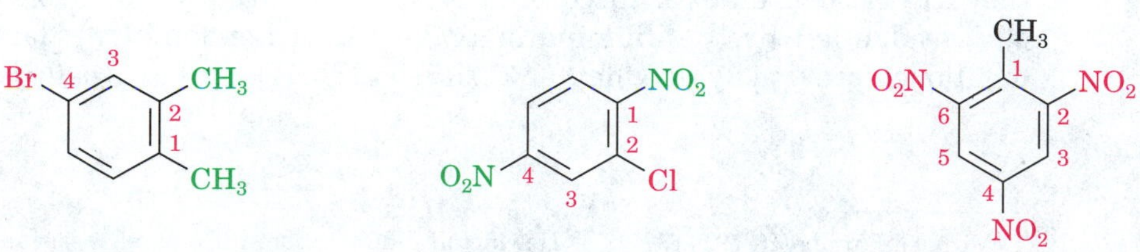

4-Bromo-1,2-dimethylbenzene **2-Chloro-1,4-dinitrobenzene** **2,4,6-Trinitrotoluene (TNT)**

Note in the third example shown that *-toluene* is used as the parent name rather than *-benzene*. Any of the monosubstituted aromatic compounds shown in Table 15.1 can serve as a parent name, with the principal substituent ($-CH_3$ in toluene) assumed to be on C1. The following two examples further illustrate this practice.

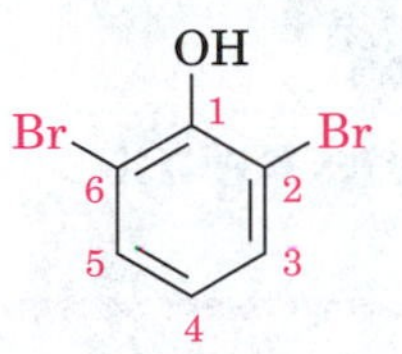

2,6-Dibromo*phenol*

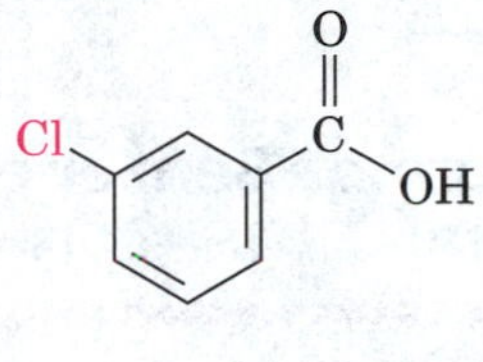

m*-Chloro*benzoic acid

PROBLEM.

15.1 Tell whether the following compounds are ortho, meta, or para disubstituted:

(a) 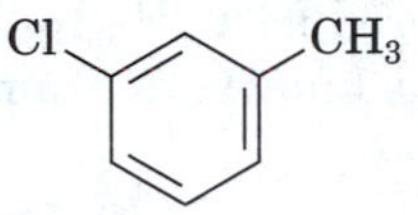

(b) 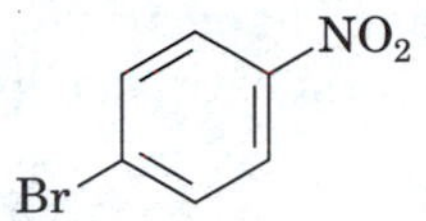

(c) 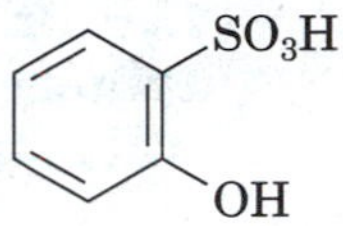

PROBLEM.

15.2 Give IUPAC names for the following compounds:

(a) 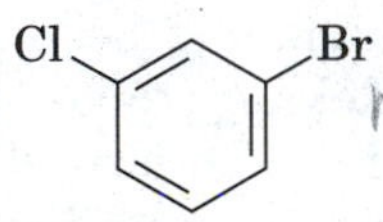

(b) 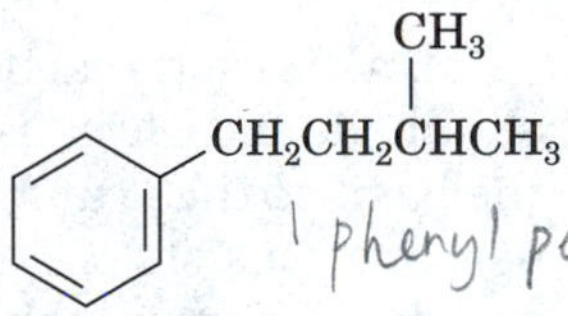

(c) 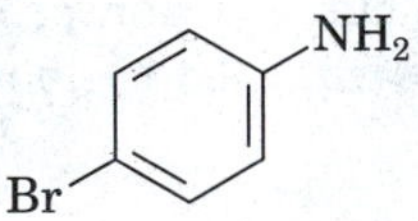

(d) 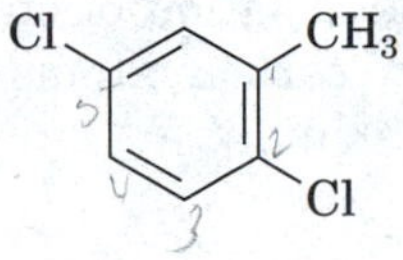

(e) 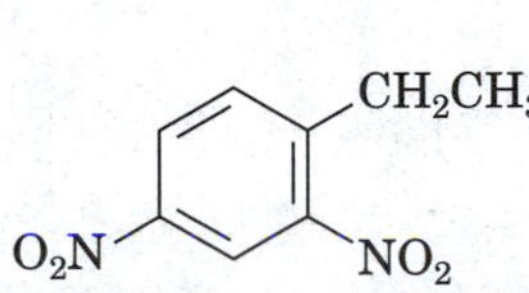

(f) 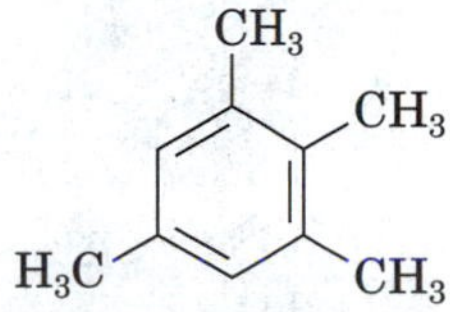

PROBLEM.

15.3 Draw structures corresponding to the following IUPAC names:

(a) *p*-Bromochlorobenzene
(b) *p*-Bromotoluene
(c) *m*-Chloroaniline
(d) 1-Chloro-3,5-dimethylbenzene

PROBLEM.

15.4 The following names are incorrect. Draw the structure represented by each, and give the correct IUPAC name.

(a) 2-Bromo-3-chlorobenzene (b) 4,6-Dinitrotoluene

(c) 4-Bromo-1-methylbenzene (d) 2-Chloro-*p*-xylene

15.3 Structure of Benzene: The Kekulé Proposal

By the mid-1800s, benzene was known to have the molecular formula C_6H_6, and its chemistry was being actively explored. The results, though, were puzzling. Although benzene is clearly "unsaturated"—the formula C_6H_6 requires a combination of four double bonds/rings—it nevertheless fails to undergo reactions characteristic of alkenes. For example, benzene reacts slowly with Br_2 in the presence of iron to give the *substitution* product C_6H_5Br, rather than the possible *addition* product $C_6H_6Br_2$. Furthermore, only one monobromo substitution product was known; no isomers of C_6H_5Br had been prepared.

$$C_6H_6 + Br_2 \xrightarrow{Fe} C_6H_5Br + HBr \quad \left[C_6H_6Br_2 \right]$$

Benzene — **Bromobenzene** **(substitution product)** — (Addition product —*NOT formed*)

On further reaction of C_6H_5Br with Br_2, disubstitution products are obtained, and three isomeric $C_6H_4Br_2$ compounds had been prepared. On the basis of these and similar results, August Kekulé proposed in 1865 that benzene consists of a ring of carbon atoms and can be formulated as 1,3,5-cyclohexatriene. Kekulé reasoned that this structure would readily account for the isolation of only a single monobromo substitution product, because all six carbon atoms and all six hydrogens in 1,3,5-cyclohexatriene are equivalent.

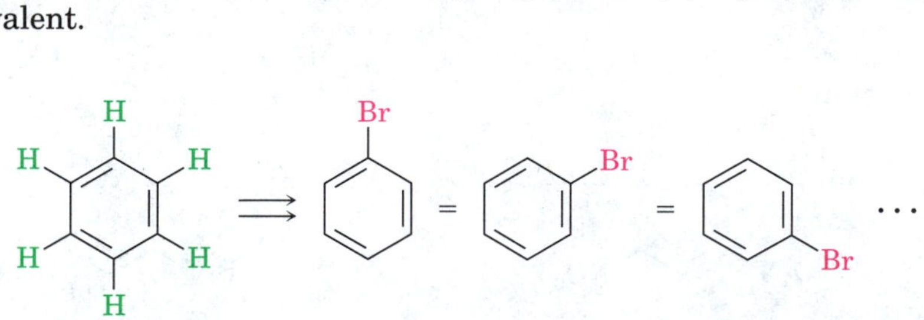

All six hydrogens are equivalent

Only one possible monobromo substitution product

The observation that only three isomeric dibromo substitution products were known was more difficult to explain because four structures can be written:

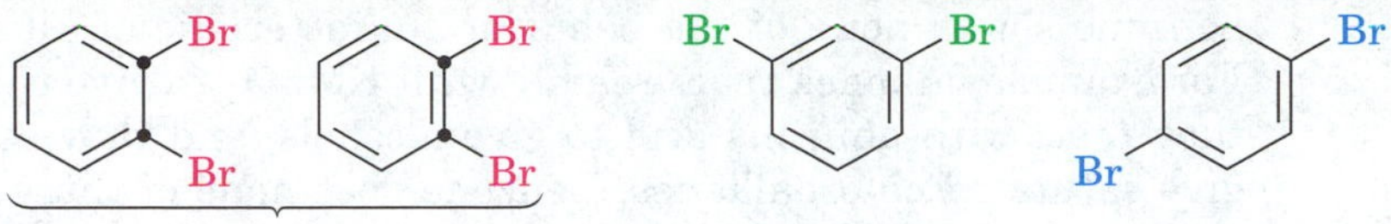

Two 1,2-dibromocyclohexatrienes | 1,3-Dibromocyclohexatriene | 1,4-Dibromocyclohexatriene

Although there is only one 1,3 derivative and one 1,4 derivative, there appear to be *two* 1,2-dibromo substitution products, which differ in the positions of the double bonds. One isomer has a single bond between the bromine-bearing carbons, and the other isomer has a double bond. Kekulé accounted for the formation of only three isomers by proposing that the double bonds in benzene "oscillate" rapidly between two positions. Thus, the two 1,2-dibromocyclohexatrienes can't be separated, according to Kekulé, because they interconvert too rapidly.

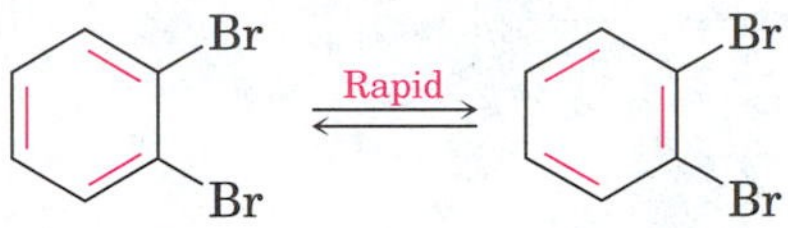

Kekulé's proposed structure for benzene was widely criticized at the time. Although it satisfactorily accounts for the correct number of mono- and disubstituted benzene isomers, it fails to answer two critical questions: Why is benzene unreactive compared with other alkenes, and why does benzene give a substitution product rather than an addition product on reaction with Br_2?

PROBLEM

15.5 How many tribromo benzene derivatives are possible according to Kekulé's theory? Draw and name them.

PROBLEM

15.6 The following structures with formula C_6H_6 were considered for benzene at one time. If we assume that bromine can be substituted for hydrogen, how many monobromo derivatives are possible for each? How many dibromo derivatives?

Ladenburg benzene **Dewar benzene**

15.4 Stability of Benzene

The unusual stability of benzene was a great puzzle to early chemists. Although its formula, C_6H_6, indicates that multiple bonds must be present, benzene shows none of the behavior characteristic of alkenes or alkynes. For example, alkenes react readily with $KMnO_4$ to give cleavage products, they react with aqueous acid to give alcohols, and they react with HCl to give saturated chloroalkanes. Benzene does none of these things. *Benzene does not undergo electrophilic addition reactions.*

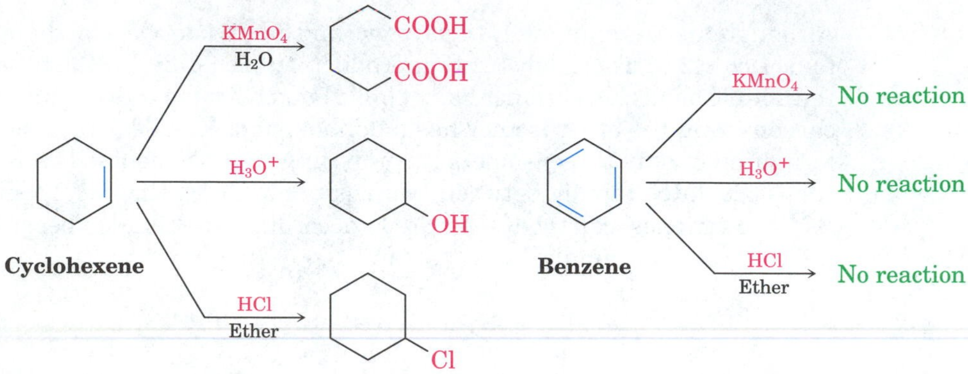

We can get a quantitative idea of benzene's unusual stability from data on heats of hydrogenation, shown in Table 15.2. Cyclohexene, an isolated alkene, has $\Delta H^\circ_{hydrog} = -118$ kJ/mol (-28.2 kcal/mol), and 1,3-cyclohexadiene, a conjugated diene, has $\Delta H^\circ_{hydrog} = -230$ kJ/mol (-55.0 kcal/mol). As expected, the value for 1,3-cyclohexadiene is a bit less than twice the cyclohexene value because conjugated dienes are unusually stable (Section 14.2).

Table 15.2 Heats of Hydrogenation of Cyclic Alkenes

		ΔH°_{hydrog}	
Reactant	*Product*	*(kJ/mol)*	*(kcal/mol)*
Cyclohexene	Cyclohexane	−118	−28.2
1,3-Cyclohexadiene	Cyclohexane	−230	−55.0
Benzene	Cyclohexane	−206	−49.2

Carrying the analogy one step further, we might expect the ΔH°_{hydrog} for "cyclohexatriene" (benzene) to be a bit less than -356 kJ/mol, or three times the cyclohexene value. *The actual value is −206 kJ/mol, some 150 kJ/mol (36 kcal/mol) less than expected.* Since 150 kJ/mol less heat than expected is released during hydrogenation of benzene, benzene must have 150 kJ/mol less energy than expected. In other words, benzene has 150 kJ/mol "extra" stability (Figure 15.2).

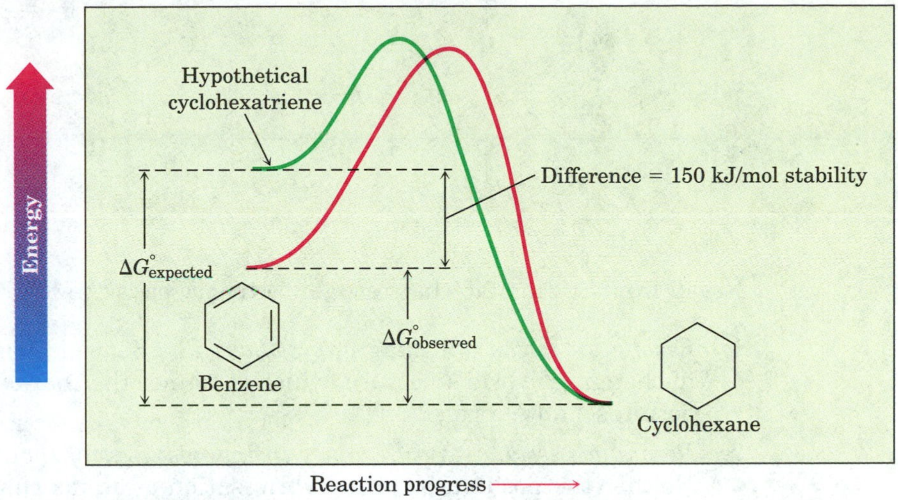

Figure 15.2 Reaction energy diagram for the hydrogenation of benzene compared with a hypothetical cyclohexatriene. Benzene is 150 kJ/mol (36 kcal/mol) more stable than "cyclohexatriene."

Further evidence for the unusual nature of benzene is that all carbon–carbon bonds in benzene have the same length, intermediate between typical single and double bonds. Most C–C single bonds are 1.54 Å long and most C=C double bonds are 1.34 Å long, but all C–C bonds in benzene are 1.39 Å long.

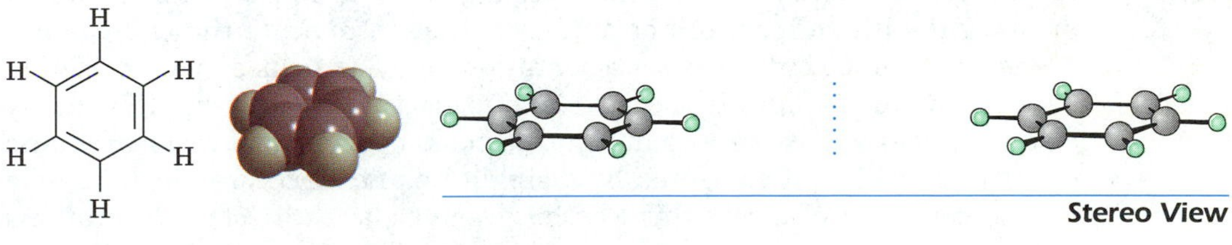

15.5 Representations of Benzene: The Resonance Approach

How can we account for benzene's properties, and how can we best represent its structure? Resonance theory answers this question by saying that benzene can be described as a hybrid of two equivalent Kekulé structures in which each C–C connection averages 1.5 bonds—midway between single and double.

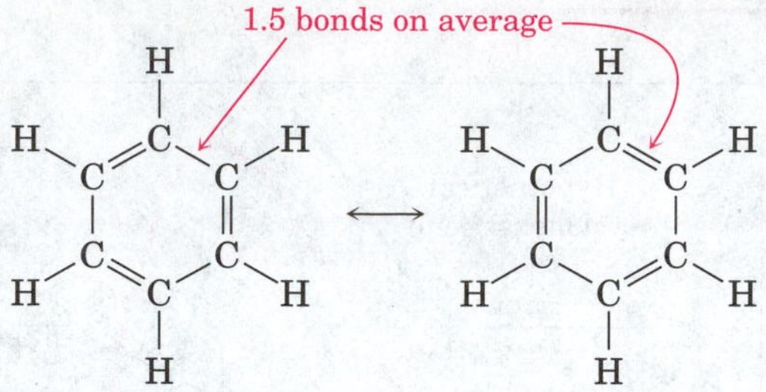

Recall from Section 2.5 that resonance theory says the following:

1. *Resonance forms are imaginary, not real.* Benzene has a single, unchanging hybrid structure, which combines the characteristics of both resonance forms.
2. *Resonance structures differ only in the positions of their electrons.* Neither the position nor the hybridization of atoms changes from one resonance structure to another. In benzene, the six carbon atoms form a regular hexagon with the π electrons shared equally between neighboring nuclei. Each C–C connection averages 1.5 bonds, and all bonds are equivalent.
3. *Different resonance forms don't have to be equivalent.* The more nearly equivalent the forms are, however, the more stable the molecule.
4. *The more resonance structures there are, the more stable the molecule.*

Benzene can't be represented accurately by either individual Kekulé structure and does not oscillate back and forth between the two. The true structure is somewhere in between the two extremes but is impossible to draw with our usual conventions. We might try to represent benzene by drawing it with either a full or a dotted circle to indicate the equivalence of the C–C bonds, but these representations have to be used very carefully because they don't indicate the number of π electrons in the ring: How many electrons does a circle represent? In this book, benzene and other aromatic compounds will be represented by a single Kekulé structure. We'll be able to keep count of π electrons this way, but we must be aware of the limitations of the drawings.

Some alternative representations of benzene. Such structures must be used carefully, since they don't indicate the number of π electrons.

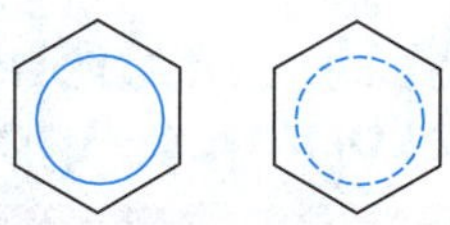

There is a subtle yet important difference between Kekulé's representation of benzene and the resonance representation. Kekulé considered benzene as rapidly oscillating back and forth between two cyclohexatriene structures, whereas resonance theory considers benzene to be a single "resonance hybrid" structure:

Kekulé benzene **Modern benzene**

At any given instant, Kekulé's oscillating structures have different carbon–carbon bond lengths—three bonds are short and three are long. This difference in bond length implies that the carbon atoms must change position in oscillating from one structure to another, and thus the two structures are not the same as resonance forms.

The same resonance argument that explains benzene's stability also explains why there is only one *o*-dibromobenzene rather than two. The two Kekulé structures of *o*-dibromobenzene are simply resonance forms of a single compound, whose true structure is intermediate between the forms.

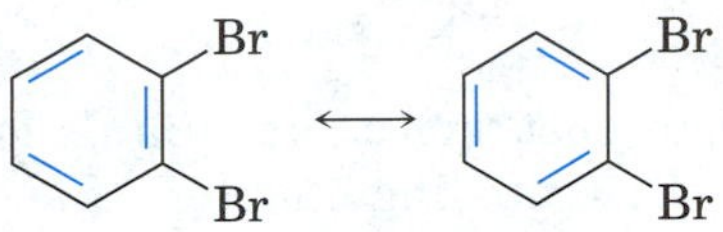

Two resonance forms of *o*-dibromobenzene

PROBLEM

15.7 In 1932, A. A. Levine and A. G. Cole studied the ozonolysis of *o*-xylene and isolated three products: glyoxal, 2,3-butanedione, and pyruvaldehyde.

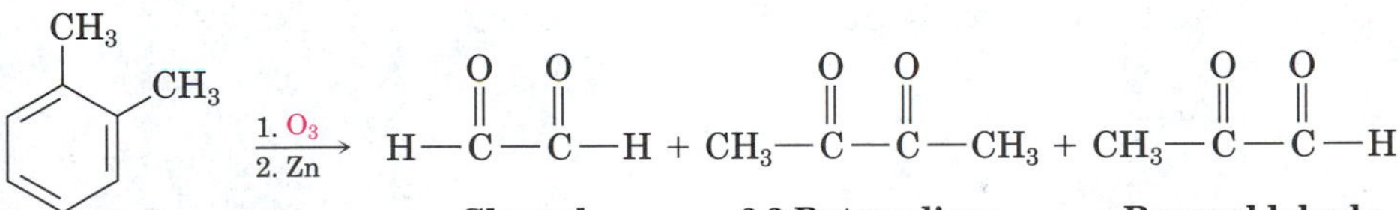

Glyoxal **2,3-Butanedione** **Pyruvaldehyde**

In what ratio would you expect the three products to be formed if *o*-xylene is a resonance hybrid of two Kekulé structures? The actual ratio found was 3 parts glyoxal, 1 part 2,3-butanedione, and 2 parts pyruvaldehyde. What conclusions can you draw about the structure of *o*-xylene?

15.6 Molecular Orbital Description of Benzene

Having just seen a resonance description of benzene, let's now see the alternative molecular orbital description. An orbital view of benzene emphasizes the cyclic conjugation of the benzene molecule and the equivalence of the six C–C bonds. Benzene is a planar molecule with the shape of a regular hexagon. All C-C-C bond angles are 120°, all six carbon atoms are sp^2-hybridized, and each carbon has a *p* orbital perpendicular to the plane of the six-membered ring.

Since all six carbon atoms and all six *p* orbitals in benzene are equivalent, it's impossible to define three localized π bonds in which a given *p* orbital overlaps only one neighboring *p* orbital. Rather, each *p* orbital overlaps equally well with *both* neighboring *p* orbitals, leading to a picture of benzene in which the six π electrons are completely delocalized around the ring. Benzene therefore has two doughnut-shaped clouds of electrons, one above and one below the ring:

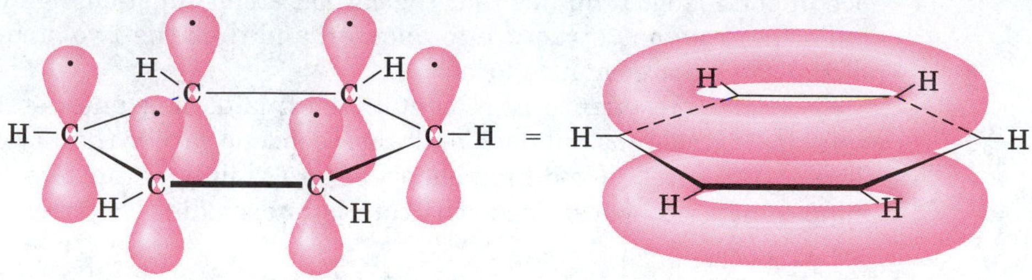

We can construct molecular orbitals for benzene just as we did for 1,3-butadiene in the preceding chapter. If six *p* atomic orbitals combine in a cyclic manner, six benzene molecular orbitals result, as shown in Figure 15.3. The three low-energy molecular orbitals, denoted ψ_1, ψ_2, and ψ_3, are bonding combinations, and the three high-energy orbitals are antibonding. Note that two of the bonding orbitals, ψ_2 and ψ_3, have the same energy, as do the antibonding orbitals ψ_4* and ψ_5*. Such orbitals are said to be **degenerate**. The six *p* electrons of benzene occupy the three bonding molecular orbitals and are delocalized over the entire conjugated system, leading to the observed 150 kJ/mol stabilization of benzene.

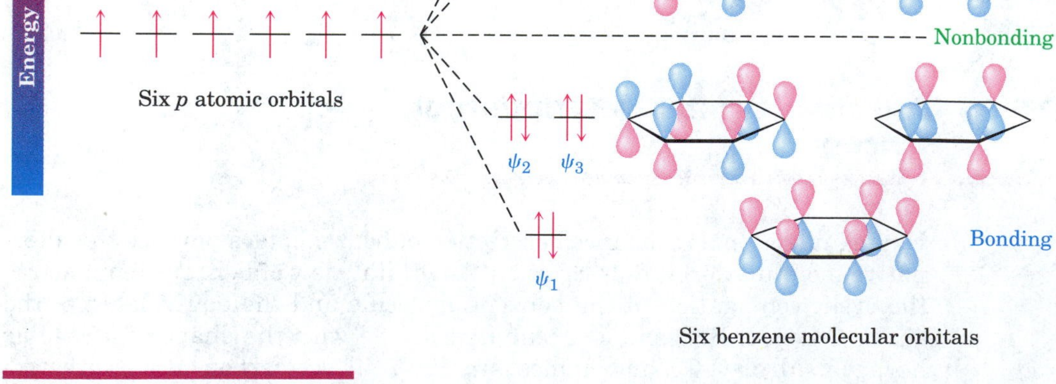

Figure 15.3 Energy levels of the six benzene molecular orbitals. The bonding orbitals ψ_2 and ψ_3 have the same energy and are said to be degenerate, as are the antibonding orbitals ψ_4* and ψ_5*.

PROBLEM.

15.8 Pyridine is a flat, hexagonal molecule with bond angles of 120°. It undergoes electrophilic substitution rather than addition and generally behaves like benzene. Draw an orbital picture of pyridine to explain its properties. Check your answer by looking ahead to Section 15.9.

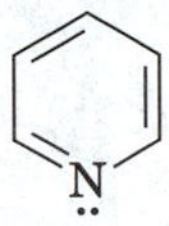

Pyridine

15.7 Aromaticity and the Hückel 4*n* + 2 Rule

Let's review what we've learned thus far about benzene and, by extension, about other benzene-like aromatic molecules:

1. Benzene is a cyclic conjugated molecule.
2. Benzene is unusually stable, having a heat of hydrogenation 150 kJ/mol less negative than we might expect for a cyclic triene.
3. Benzene is planar and has the shape of a regular hexagon. All bond angles are 120°, and all C–C bond lengths are 1.39 Å.
4. Benzene undergoes substitution reactions that retain the cyclic conjugation rather than electrophilic addition reactions that would destroy the conjugation.
5. Benzene is a resonance hybrid whose structure is intermediate between two Kekulé structures:

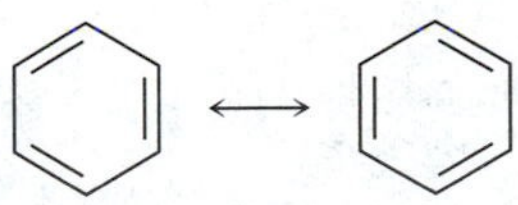

Although these facts would seem to provide a good description of benzene and other aromatic molecules, they aren't enough. Something else is needed to complete a description of aromaticity. According to a theory devised by the German physicist Erich Hückel[2] in 1931, a molecule is aromatic only if it has a planar, monocyclic system of conjugation with a *p* orbital on each atom and *only if the p orbital system contains* $4n + 2$ *π electrons*, where n is an integer ($n = 0, 1, 2, 3, \ldots$). In other words, only molecules with 2, 6, 10, 14, 18, . . . π electrons can be aromatic. Molecules with $4n$ π electrons (4, 8, 12, 16, . . .) *can't* be aromatic, even though they may be cyclic and

[2]Erich Hückel (1896–1980); b. Stuttgart, Germany; Ph.D. Göttingen (Debye); professor of physics at Stuttgart and Marburg (1937–1980).

apparently conjugated. In fact, planar, conjugated molecules with $4n$ π electrons are said to be **antiaromatic**, because delocalization of their π electrons leads to an *increase* in energy.

Let's look at some examples to see how the **Hückel $4n + 2$ rule** works.

1. *Cyclobutadiene* has four π electrons and is antiaromatic:

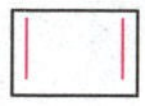

Two double bonds;
four π electrons

Cyclobutadiene

Stereo View

Cyclobutadiene is a highly reactive substance that shows none of the properties associated with aromaticity. A long history of attempts to synthesize the compound culminated in 1965 when Rowland Pettit[3] of the University of Texas prepared cyclobutadiene at low temperature but was unable to isolate it. Even at −78°C, cyclobutadiene dimerizes by a Diels–Alder reaction with itself. One molecule behaves as a diene and the other as a dienophile:

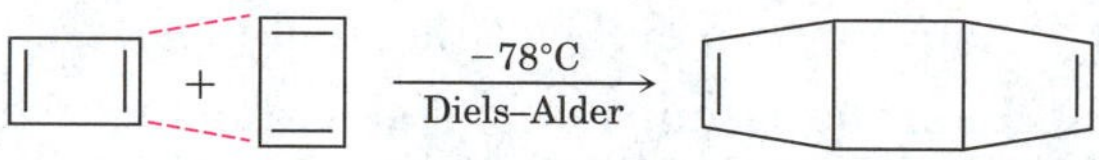

2. *Benzene* has six π electrons ($4n + 2 = 6$ when $n = 1$) and is aromatic:

Three double bonds;
six π electrons

Benzene

3. *Cyclooctatetraene* has eight π electrons and is not aromatic:

Four double bonds;
eight π electrons

Cyclooctatetraene

Chemists in the early 1900s believed that the only requirement for aromaticity was the presence of a cyclic conjugated system. It was therefore expected that cyclooctatetraene, as a close analog of benzene, would also prove to be unusually stable. The facts proved otherwise. When cyclooctatetraene was first prepared in 1911 by the German chemist Richard Willstätter,[4] it was found to resemble open-chain polyenes in its reactivity.

---

[3]Rowland Pettit (1927–1981); b. Port Lincoln, Australia; Ph.D. University of Adelaide (1952) and University of London (Dewar, 1956); University of Texas (1957–1981).

[4]Richard Willstätter (1872–1942); b. Karlsruhe, Germany; Technische Hochschule, Munich (Einhorn) (1895); professor, Zurich, Dahlem, Munich; Nobel Prize (1915).

Cyclooctatetraene reacts readily with Br_2, $KMnO_4$, and HCl, just as other alkenes do. We now know, in fact, that cyclooctatetraene is not even conjugated. It is tub-shaped rather than planar and has no cyclic conjugation because neighboring *p* orbitals don't have the proper alignment for overlap (Figure 15.4). The π electrons are localized in four discrete C=C bonds rather than delocalized as in benzene. X-ray studies show that the C–C single bonds are 1.47 Å long, and the double bonds are 1.34 Å long. In addition, the ^1H NMR spectrum shows a single sharp resonance line at 5.7 δ, a value characteristic of an alkene rather than an aromatic molecule (Section 15.12).

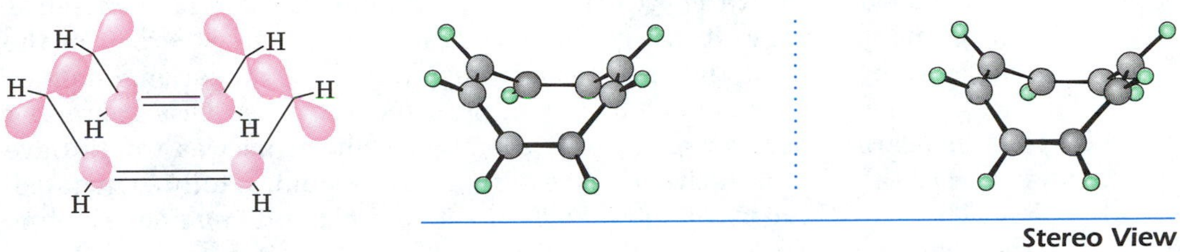

Figure 15.4 Cyclooctatetraene is a tub-shaped molecule that has no cyclic conjugation because its *p* orbitals are not aligned properly for overlap.

PROBLEM

15.9 To be aromatic, a molecule must be flat, so that *p*-orbital overlap can occur, and it must have $4n + 2$ π electrons. Cyclodecapentaene fulfills one of these criteria but not the other and is therefore nonaromatic. Explain. (Molecular models may be useful.)

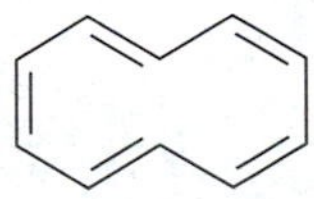

Cyclodecapentaene (not aromatic)

15.8 Aromatic Ions

Look back at the Hückel criteria for aromaticity in the preceding section. To be aromatic, a molecule must be cyclic, conjugated (that is, have a *p* orbital on each carbon), and have $4n + 2$ π electrons. Nothing in this definition says that the numbers of *p* orbitals and π electrons must be the same. In fact, they can be different. The $4n + 2$ rule is broadly applicable to many kinds of molecules, not just to neutral hydrocarbons. For example, both the cyclopentadienyl *anion* and the cycloheptatrienyl *cation* are aromatic.

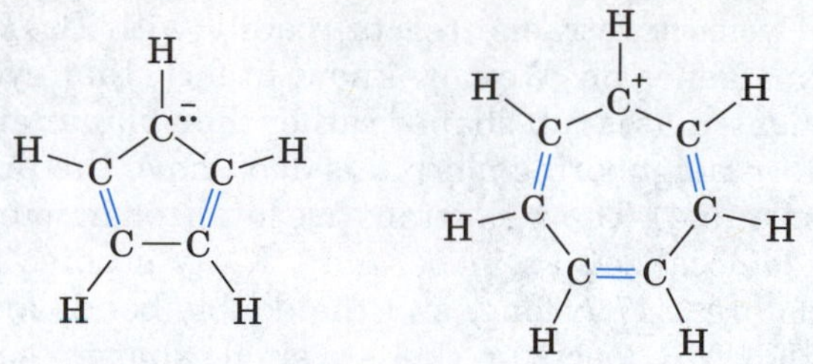

Cyclopentadienyl anion **Cycloheptatrienyl cation**

Six π electrons; aromatic ions

Let's look first at the cyclopentadienyl anion. Cyclopentadiene itself is not aromatic because it is not fully conjugated. The $-CH_2-$ carbon in the ring is sp^3-hybridized, thus preventing complete cyclic conjugation. Imagine, though, that we remove one hydrogen from the saturated CH_2 group and let that carbon become sp^2-hybridized. The resultant species would have five *p* orbitals, one on each of the five carbons, and would be fully conjugated.

There are three ways, shown in Figure 15.5, we might imagine removing the hydrogen:

1. We could remove the hydrogen atom and *both* electrons ($H:^-$) from the C–H bond. Since the hydrogen would have two electrons and a negative charge, the cyclopentadienyl group that remains is positively charged.
2. We could remove the hydrogen and *one* electron (H·) from the C–H bond, leaving a cyclopentadienyl radical.
3. We could remove a hydrogen ion with *no* electrons (H^+), leaving a cyclopentadienyl anion.

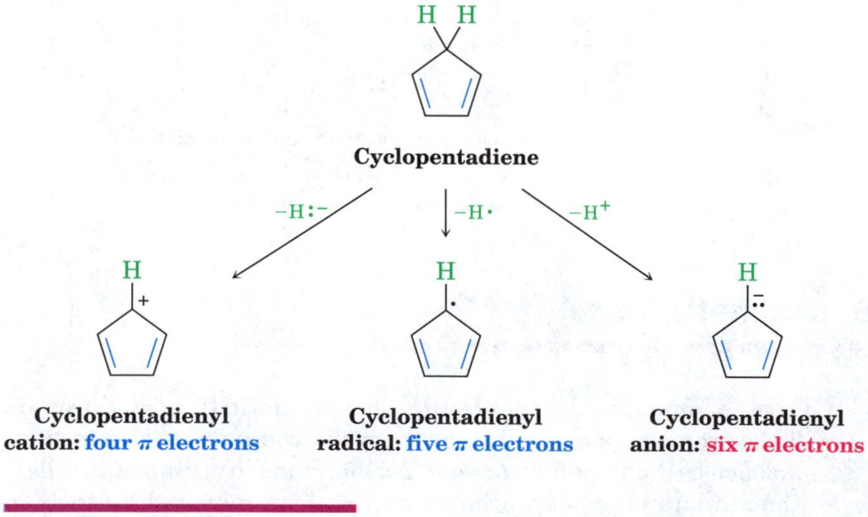

Figure 15.5 Generating the cyclopentadienyl cation, radical, and anion by removing a hydrogen from cyclopentadiene.

Although five equivalent resonance structures can be drawn for all three species, Hückel's rule predicts that *only the anion—with six π electrons—should be aromatic*. The cyclopentadienyl carbocation—with four π electrons—and the cyclopentadienyl radical—with five π electrons—are predicted to be unstable and antiaromatic.

In practice, both the cyclopentadienyl cation and the radical are highly reactive and difficult to prepare. Neither shows any sign of the unusual stability expected of an aromatic system. The cyclopentadienyl anion (with six π electrons), by contrast, is easily prepared and remarkably stable. In fact, cyclopentadiene is one of the most acidic hydrocarbons known. Although most hydrocarbons have a $pK_a > 45$, cyclopentadiene has $pK_a = 16$, a value comparable to that of water! Cyclopentadiene is acidic because the anion formed by dissociation is so stable. It doesn't matter that the cyclopentadienyl anion has only five *p* orbitals; all that matters is that there are six π electrons, a Hückel number (Figure 15.6).

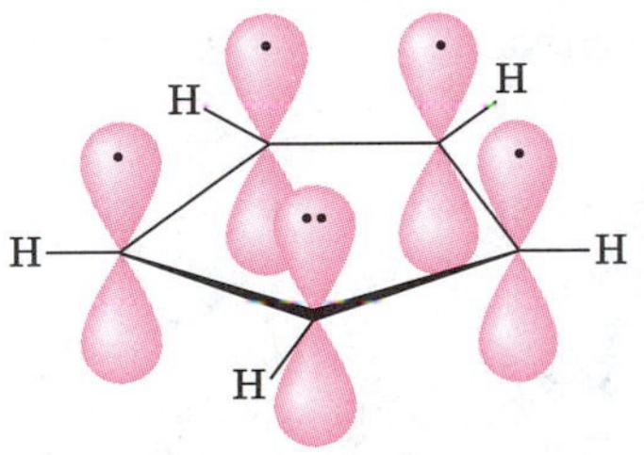

Aromatic cyclopentadienyl anion with six π electrons

Figure 15.6 An orbital view of the aromatic cyclopentadienyl anion, showing the cyclic conjugation and six π electrons in five *p* orbitals.

Similar arguments can be used to predict the relative stabilities of the cycloheptatrienyl cation, radical, and anion. Removal of a hydrogen from cycloheptatriene can generate the cation (six π electrons), the radical (seven π electrons), or the anion (eight π electrons), as shown in Figure 15.7 (p. 552). Once again, all three species have numerous resonance forms, but Hückel's rule predicts that only the cycloheptatrienyl cation (six π electrons) should be aromatic. The cycloheptatrienyl radical (seven π electrons) and the anion (eight π electrons) are antiaromatic.

Both the cycloheptatrienyl radical and the anion are reactive and difficult to prepare. The cation (with six π electrons), however, is extraordinarily stable. In fact, the cycloheptatrienyl cation was first prepared in 1891 by reaction of Br_2 with cycloheptatriene (Figure 15.8, p. 552), although its structure was not recognized at the time.

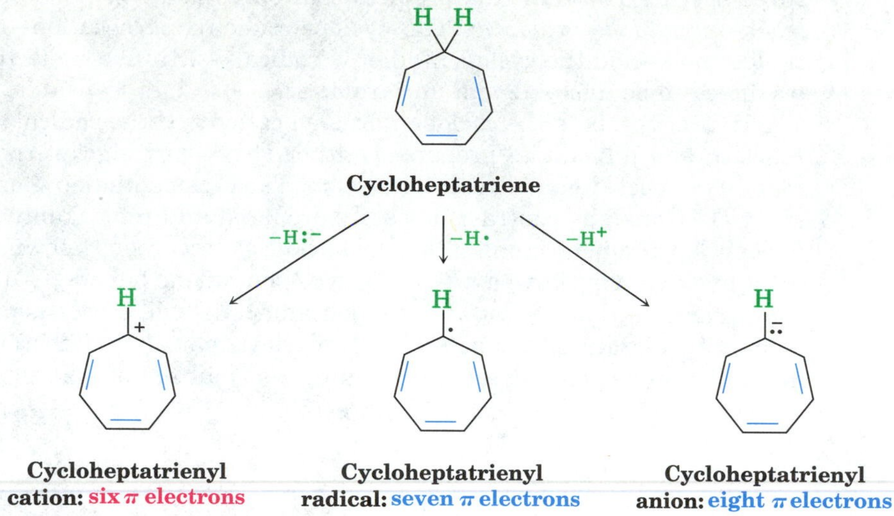

Figure 15.7 Generation of the cycloheptatrienyl cation, radical, and anion. Only the cation (six π electrons) is aromatic.

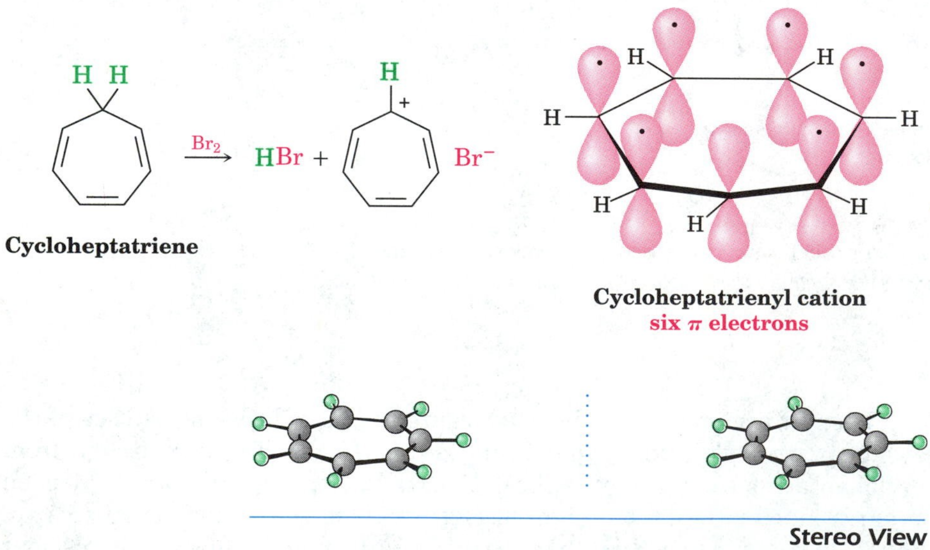

Figure 15.8 Reaction of cycloheptatriene with bromine yields cycloheptatrienylium bromide, a salt-like material containing the cycloheptatrienyl cation.

PROBLEM.

15.10 Draw the five resonance structures of the cyclopentadienyl anion. Are all carbon–carbon bonds equivalent? How many absorption lines would you expect to see in the ^{1}H NMR and ^{13}C NMR spectra of the anion?

PROBLEM

15.11 Cyclooctatetraene readily reacts with potassium metal to form the cyclooctatetraene dianion, $C_8H_8^{2-}$. Why do you suppose this reaction occurs so easily? What geometry do you expect for the cyclooctatetraene dianion?

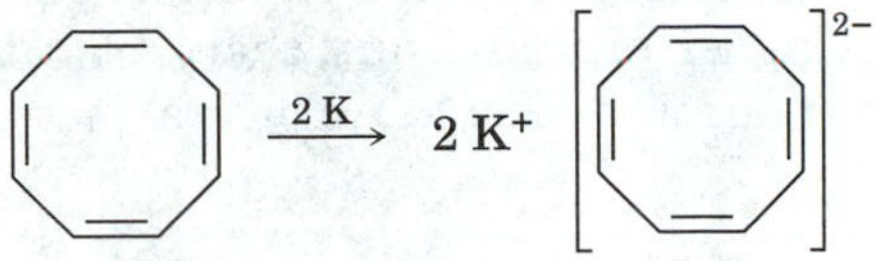

15.9 Pyridine and Pyrrole: Two Aromatic Heterocycles

Look back once again at the definition of aromaticity: . . . a cyclic conjugated molecule containing $4n + 2$ π electrons. Nothing in this definition says that the atoms in the ring must be *carbon*. In fact, *heterocyclic* compounds can also be aromatic. A **heterocycle** is a compound with a ring that has one or more atoms other than carbon. The heteroatom is often nitrogen or oxygen, but sulfur, phosphorus, and other elements are also found. Pyridine, for example, is a six-membered heterocycle with a nitrogen atom in its ring.

Pyridine is much like benzene in its π electron structure. Each of the five sp^2-hybridized carbons has a p orbital perpendicular to the plane of the ring, and each p orbital contains one π electron. The nitrogen atom is also sp^2-hybridized and has one electron in a p orbital, bringing the total to six π electrons. The nitrogen lone-pair electrons are in an sp^2 orbital in the plane of the ring perpendicular to the π system and are not involved with the aromatic π system because they don't have the correct alignment for overlap (Figure 15.9).

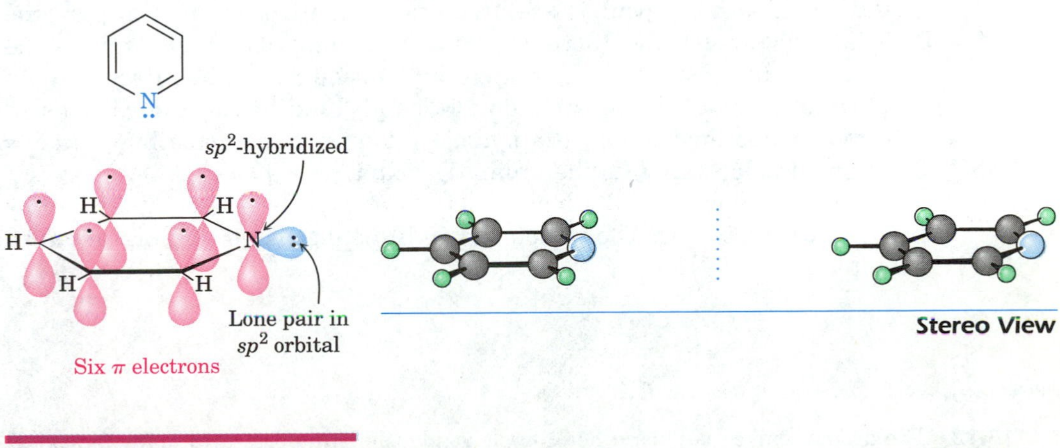

Figure 15.9 Pyridine, an aromatic heterocycle, has a π electron arrangement much like that of benzene.

The five-membered heterocycle, pyrrole (two *r*'s, one *l*), another example of an aromatic substance with six π electrons, has a π electron system similar to that of the cyclopentadienyl anion. Each of the four sp^2-hybridized carbons has a *p* orbital perpendicular to the ring, and each contributes one π electron. The nitrogen atom is sp^2-hybridized, with its lone pair of electrons also occupying a *p* orbital. Thus, there are a total of six π electrons, making pyrrole an aromatic molecule. An orbital picture of pyrrole is shown in Figure 15.10.

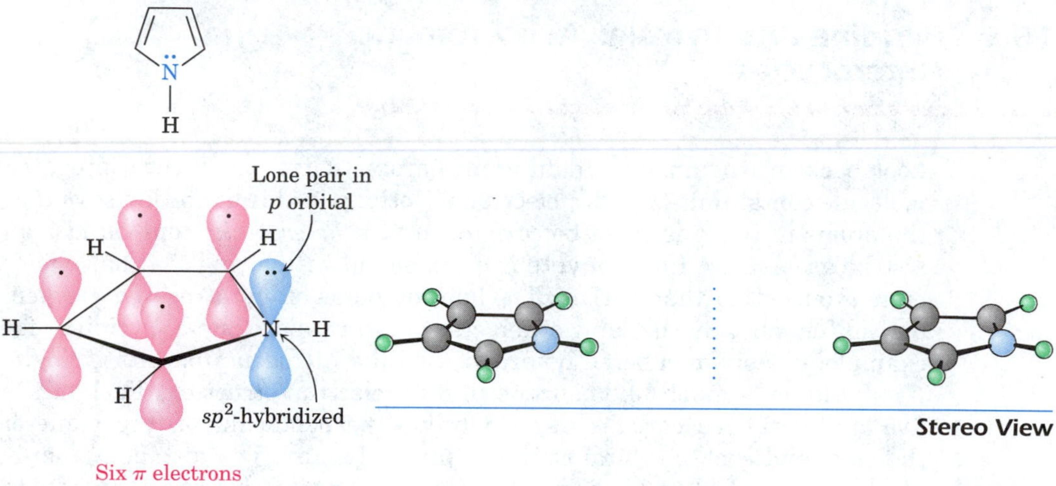

Figure 15.10 Pyrrole, a five-membered aromatic heterocycle, has a π electron arrangement much like that of the cyclopentadienyl anion.

Note that the nitrogen atoms have different roles in pyridine and pyrrole even though both compounds are aromatic. The nitrogen atom in pyridine is in a double bond and therefore contributes only *one* π electron to the aromatic sextet, just as a carbon atom in benzene does. The nitrogen atom in pyrrole, however, is not in a double bond. Like one of the carbons in the cyclopentadienyl anion, the pyrrole nitrogen atom contributes *two* π electrons (the lone pair) to the aromatic sextet.

Pyridine, pyrrole, and other aromatic heterocycles are crucial to many biochemical processes. Their chemistry will be discussed in more detail in Chapter 29.

PROBLEM.

15.12 The aromatic five-membered heterocycle imidazole is important in many biological processes. One of its nitrogen atoms is pyridine-like in that it contributes one π electron to the aromatic sextet, and the other nitrogen is pyrrole-like in that it contributes two π electrons. Draw an orbital picture of imidazole, and account for its aromaticity. Which atom is pyridine-like and which is pyrrole-like?

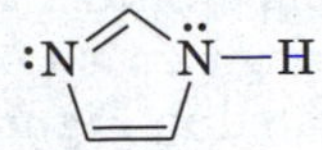

Imidazole

PROBLEM.

15.13 Assuming that the oxygen atom in furan is sp^2-hybridized, draw an orbital picture to show how the molecule is aromatic.

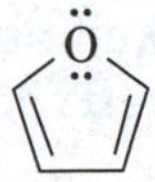

Furan

15.10 Why 4*n* + 2?

What's so special about $4n + 2$ π electrons? Why do 2, 6, 10, 14, . . . π electrons lead to aromatic stability, while other numbers of electrons do not? The answer to this question comes from molecular orbital theory and involves the relative energy levels of the π MO's.

When the energy levels of molecular orbitals for cyclic conjugated molecules are calculated, it turns out that there is always a *single* lowest-lying MO above which the MO's come in degenerate *pairs*. Thus, when electrons fill the various molecular orbitals, it takes two electrons (one pair) to fill the lowest-lying orbital and four electrons (two pairs) to fill each of *n* succeeding energy levels—a total of $4n + 2$. Any other number would leave an energy level partially filled.

The six π molecular orbitals of benzene were shown previously in Figure 15.3, and their relative energies are shown again in Figure 15.11. The lowest-energy MO, ψ_1, occurs singly and contains two electrons. The next two

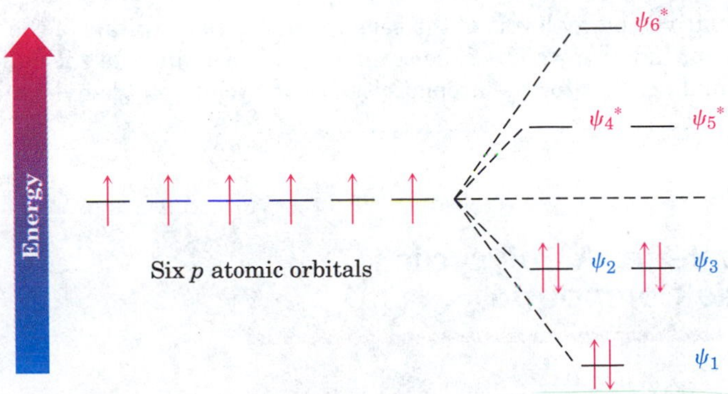

Figure 15.11 Energy levels of the six benzene π molecular orbitals.

lowest-energy orbitals, ψ_2 and ψ_3, are degenerate, and it therefore takes four electrons to fill them. The result is a stable aromatic molecule with six π electrons and filled bonding orbitals.

A similar line of reasoning carried out for the cyclopentadienyl cation, radical, and anion is illustrated in Figure 15.12. The five atomic *p* orbitals combine to give five π molecular orbitals in which there is a single lowest-energy orbital and higher-energy degenerate pairs of orbitals. In the cation (four π electrons), there are two electrons in ψ_1 but only one electron each in ψ_2 and ψ_3. Thus, the cation has two orbitals that are only partially filled, and it is therefore antiaromatic. In the radical (five π electrons), ψ_1 and ψ_2 are filled, but ψ_3 is still only half full. Only in the cyclopentadienyl anion (six π electrons) are all the bonding orbitals filled. Similar analyses can be carried out for all other aromatic species.

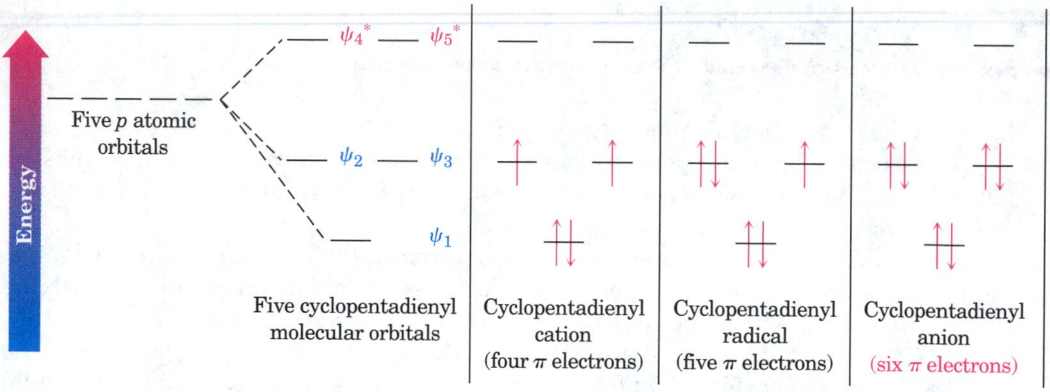

Figure 15.12 Energy levels of the five cyclopentadienyl molecular orbitals. Only the cyclopentadienyl anion (six π electrons) has a filled-shell configuration leading to aromaticity.

PROBLEM.

15.14 Show the relative energy levels of the seven π molecular orbitals of the cycloheptatrienyl system. Indicate which of the seven orbitals are filled in the cation, radical, and anion, and account for the aromaticity of the cycloheptatrienyl cation.

15.11 Naphthalene: A Polycyclic Aromatic Compound

The Hückel rule is strictly applicable only to *monocyclic* aromatic compounds, but the general concept of aromaticity can be extended beyond simple monocyclic compounds to include **polycyclic aromatic compounds**. Naphthalene, with two benzene-like rings fused together, anthracene, 1,2-benzpyrene, and coronene are all well known. Benzo[a]pyrene is particularly interesting because it is one of the cancer-causing substances that has been isolated from tobacco smoke.

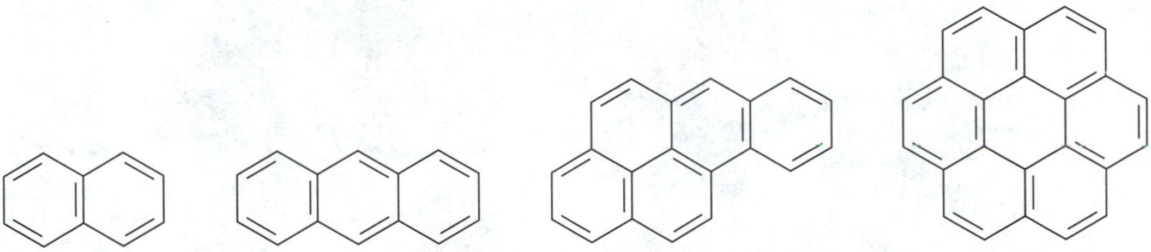

Naphthalene **Anthracene** **Benzo[a]pyrene** **Coronene**

All polycyclic aromatic hydrocarbons can be represented by a number of different resonance forms. Naphthalene, for instance, has three:

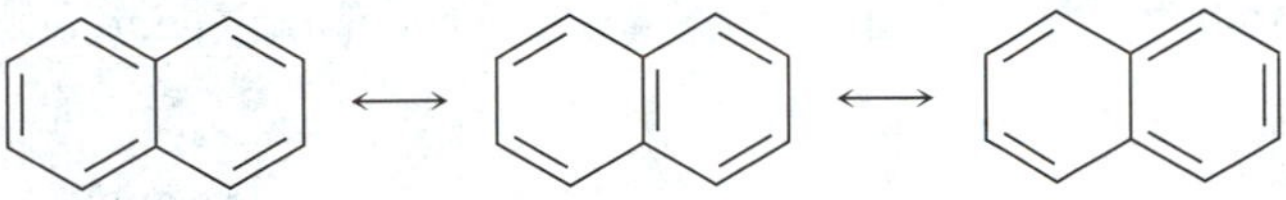

As was true for benzene with its two equivalent resonance forms, no individual Kekulé structure is a true representation of naphthalene. The true structure of naphthalene is a hybrid of the three resonance forms.

Naphthalene and other polycyclic aromatic hydrocarbons show many of the chemical properties associated with aromaticity. Thus, heat of hydrogenation measurements show an aromatic stabilization energy of approximately 250 kJ/mol (60 kcal/mol). Furthermore, naphthalene reacts slowly with electrophiles such as Br_2 to give substitution products rather than double-bond addition products.

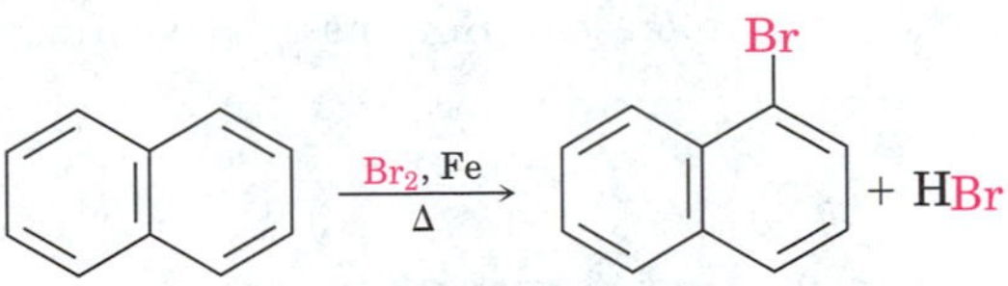

Naphthalene **1-Bromonaphthalene (75%)**

How can we explain the aromaticity of naphthalene? The orbital picture of naphthalene in Figure 15.13 shows a fully conjugated cyclic π electron system, with *p*-orbital overlap both around the ten-carbon periphery of the molecule and across the central bond. Since ten π electrons is a Hückel number, there is a high degree of π electron delocalization and consequent aromaticity in naphthalene.

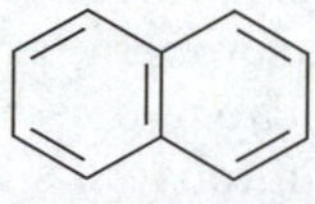

Naphthalene

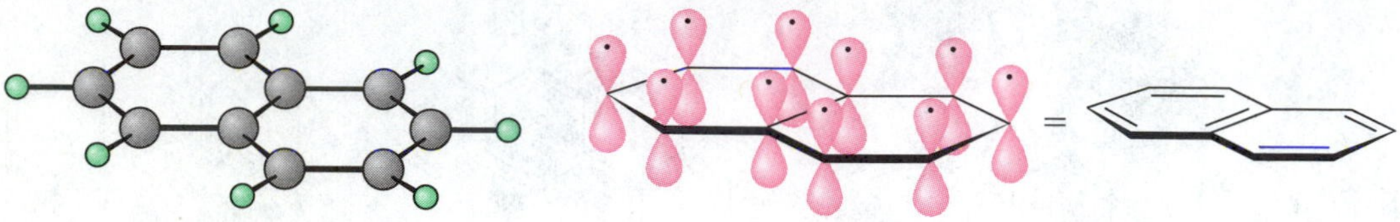

Figure 15.13 An orbital picture of naphthalene, showing that the ten π electrons are fully delocalized throughout both rings.

PROBLEM

15.15 Azulene, a beautiful blue hydrocarbon, is an isomer of naphthalene. Is azulene aromatic? Draw a second resonance form of azulene in addition to that shown.

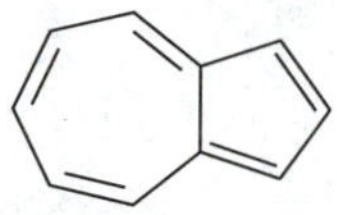

Azulene

PROBLEM

15.16 Naphthalene is sometimes represented with circles in each ring to represent aromaticity:

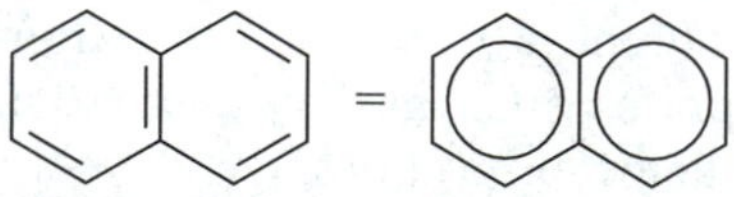

The difficulty with this representation is that it's not immediately apparent how many π electrons are present. How many π electrons are in each circle?

15.12 Spectroscopy of Aromatic Compounds

Infrared Spectroscopy

Aromatic rings show a characteristic C–H stretching absorption at 3030 cm^{-1} and a characteristic series of peaks in the 1450-1600 cm^{-1} range of the infrared spectrum. The aromatic C–H band at 3030 cm^{-1} generally has low intensity and occurs just to the left of a typical saturated C–H band.

As many as four absorptions are sometimes observed in the 1450–1600 cm^{-1} region because of complex molecular motions of the ring itself. Two bands, one at 1500 cm^{-1} and one at 1600 cm^{-1}, are usually the most intense. In addition, aromatic compounds show weak absorptions in the 1660–2000 cm^{-1} region and strong absorptions in the 690–900 cm^{-1} range due to C–H out-of-plane bending. The exact position of both sets of absorptions is diagnostic of the substitution pattern of the aromatic ring:

Monosubstituted:	690–710 cm^{-1} 730–770 cm^{-1}	*m*-Disubstituted:	690–710 cm^{-1} 810–850 cm^{-1}
o-Disubstituted:	735–770 cm^{-1}	*p*-Disubstituted:	810–840 cm^{-1}

The IR spectrum of toluene in Figure 15.14 shows these characteristic absorptions.

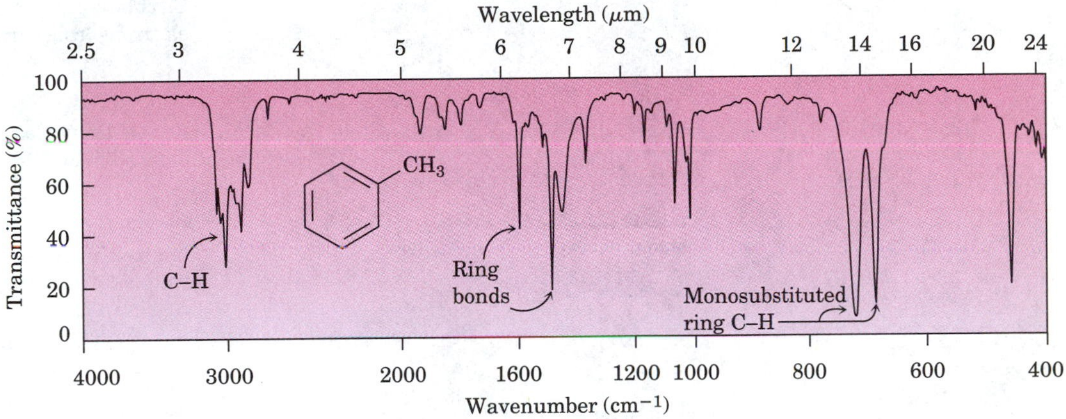

Figure 15.14 The infrared spectrum of toluene.

Ultraviolet Spectroscopy

Aromatic rings are detectable by ultraviolet spectroscopy because they contain a conjugated π electron system. In general, aromatic compounds show a series of bands, with a fairly intense absorption near 205 nm and a less intense absorption in the 255–275 nm range. The presence of these bands in the ultraviolet spectrum of a molecule of unknown structure is a sure indication of an aromatic ring.

Nuclear Magnetic Resonance Spectroscopy

Hydrogens directly bonded to an aromatic ring are easily identifiable in the ^{1}H NMR spectrum. Aromatic hydrogens are strongly deshielded by the ring and absorb between 6.5 and 8.0 δ. The spins of nonequivalent aromatic protons on substituted rings often couple with each other, giving rise to spin–spin splitting patterns that can give information about the substitution pattern of the ring.

Much of the difference in chemical shift between aromatic protons (6.5–8.0 δ) and vinylic protons (4.5–6.5 δ) is due to a property of aromatic rings called **ring current**. When an aromatic ring is oriented perpendicular to a strong magnetic field, the delocalized π electrons circulate around the ring, producing a small local magnetic field. This induced field *opposes* the applied field in the middle of the ring but *reinforces* the applied field outside the ring (Figure 15.15). Aromatic protons are therefore deshielded. They experience an effective magnetic field greater than the applied field and thus come into resonance at a lower applied field.

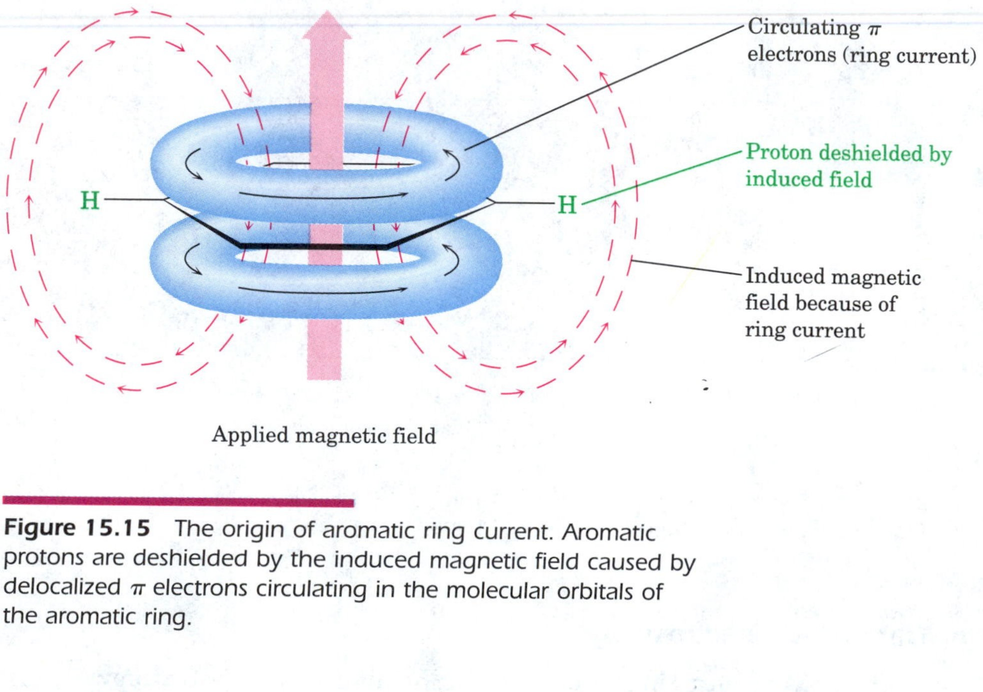

Figure 15.15 The origin of aromatic ring current. Aromatic protons are deshielded by the induced magnetic field caused by delocalized π electrons circulating in the molecular orbitals of the aromatic ring.

Note that the aromatic ring current produces different effects inside and outside the ring. If a ring were large enough to have both "inside" and "outside" protons, those protons on the outside should be deshielded and absorb at a field lower than normal, but those protons on the inside should be *shielded* and absorb at a field higher than normal. This prediction has been strikingly verified by studies on [18]annulene, a cyclic conjugated polyene with 18 π electrons—a Hückel number of electrons ($4n + 2 = 18$ when $n = 4$). The 6 inside protons of [18]annulene are strongly shielded by the aromatic ring current and absorb at −3.0 δ (that is, 3.0 ppm *upfield* from TMS), while the 12 outside protons are strongly deshielded and absorb in the typical aromatic region at 9.3 ppm downfield from TMS.

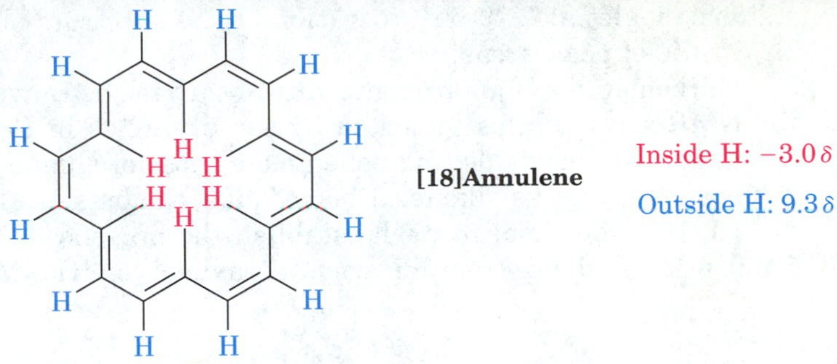

The presence of a ring current is characteristic of all Hückel aromatic molecules and is a good test of aromaticity. For example, benzene, an aromatic molecule with six π electrons, absorbs at 7.37 δ, but cyclooctatetraene, a nonaromatic molecule with eight π electrons, absorbs at 5.78 δ.

Hydrogens on carbon next to aromatic rings also show distinctive absorptions in the NMR spectrum. Benzylic protons normally absorb downfield from other alkane protons in the region from 2.3 to 3.0 δ.

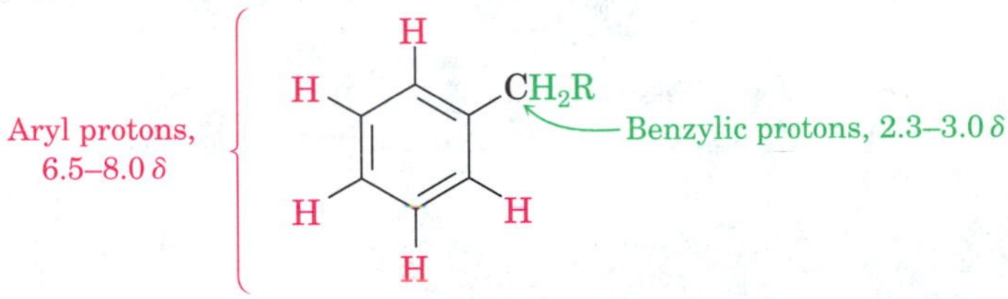

The ^{1}H NMR spectrum of *p*-bromotoluene shown in Figure 15.16 displays many of the features just discussed. The aromatic protons appear as two doublets at 7.02 and 7.45 δ, and the benzylic methyl protons absorb as

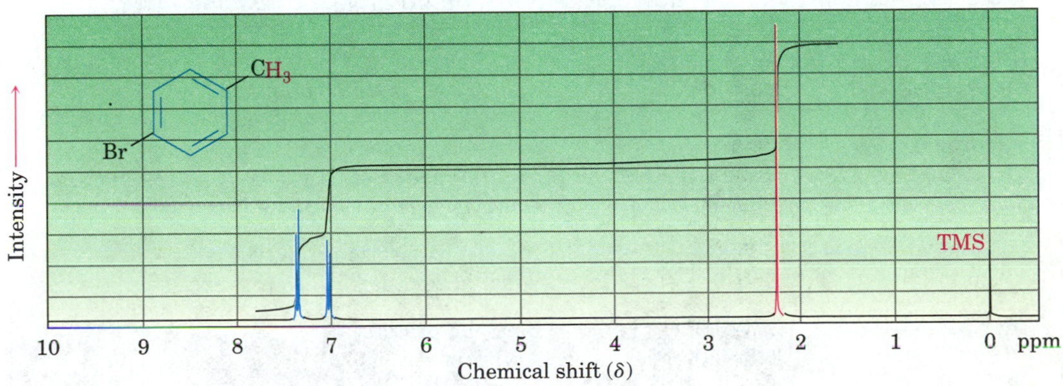

Figure 15.16 The ^{1}H NMR spectrum of *p*-bromotoluene.

a sharp singlet at 2.29 δ. Integration of the spectrum shows the expected 2:2:3 ratio of peak areas.

Carbon atoms of an aromatic ring absorb in the range 110–160 δ in the ^{13}C NMR spectrum, as indicated by the examples in Figure 15.17. These resonances are easily distinguished from those of alkane carbons but occur in the same range as alkene carbons. Thus, the presence of ^{13}C absorptions at 110–160 δ does not in itself establish the presence of an aromatic ring. Confirming evidence from infrared, ultraviolet, or ^{1}H NMR is needed.

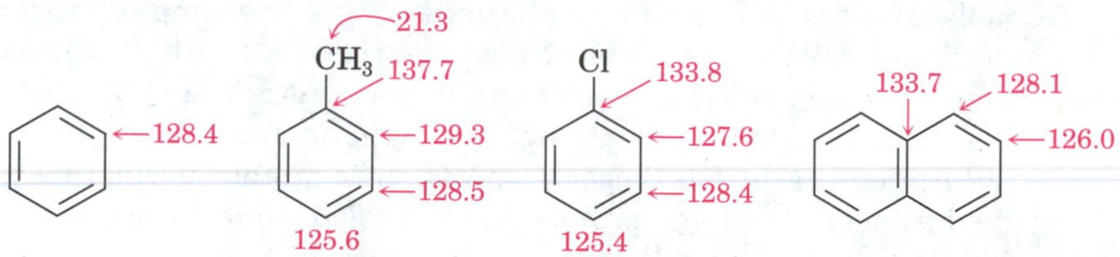

Benzene **Toluene** **Chlorobenzene** **Naphthalene**

Figure 15.17 Some ^{13}C NMR absorptions of aromatic compounds (δ units).

A summary of the kinds of information obtainable from different spectroscopic techniques is given in Table 15.3.

Table 15.3 Summary of Spectroscopic Information on Aromatic Compounds

Kind of spectroscopy	*Absorption position*	*Interpretation*
Infrared (cm^{-1})	3030	Aryl C–H stretch
	1500 and 1600	Two intense absorptions due to ring motions
	690–900	Intense C–H out-of-plane bending
Ultraviolet (nm)	205	Intense absorption
	260	Weak absorption
^{1}H NMR (δ)	2.3–3.0	Benzylic protons
	6.5–8.0	Aryl protons
^{13}C NMR (δ)	110–160	Aromatic ring carbons

INTERLUDE

Polycyclic Aromatic Hydrocarbons and Cancer

Cigarette smoke contains carcinogenic compounds called *polycyclic aromatic hydrocarbons* that react with cellular DNA.

In addition to naphthalene and anthracene, there are a great many more complex polycyclic aromatic hydrocarbons (PAH's). Benzo[a]pyrene is a particularly important and well-studied PAH because it is one of the cancer-causing (*carcinogenic*) substances found in chimney soot, barbecued meat, and cigarette smoke. Exposure to a tiny amount is sufficient to induce a skin tumor in susceptible mice.

Recent studies have given a clear picture of how these PAH's cause tumors. After a PAH is absorbed by eating or inhaling, the body attempts to rid itself of the foreign substance by converting it into a water-soluble metabolite that can be excreted. In the case of benzo[a]pyrene, oxidation in the liver converts it into an oxygenated product called a *diol epoxide*.

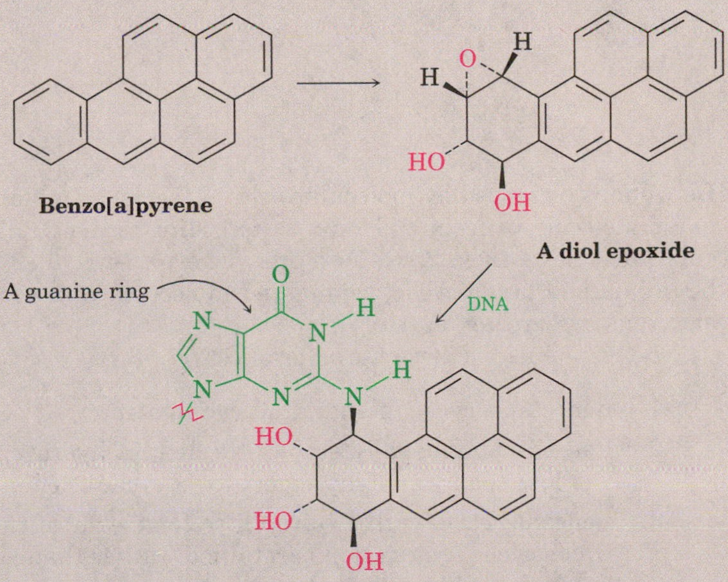

(continued)➤

Unfortunately, the diol epoxide metabolite reacts with and binds cellular DNA, thereby altering the DNA and leading to mutations or cancer. A guanine ring in DNA, acting as a nucleophile, attacks the three-membered, oxygen-containing ring of the diol epoxide in an S_N2 reaction, opening the ring and forming a C–N bond.

A number of other kinds of aromatic substances in addition to PAH's are carcinogenic. Even benzene can cause certain types of cancers, particularly leukemia, on prolonged exposure. Breathing the fumes of volatile aromatic hydrocarbons is therefore best avoided.

Summary and Key Words

The term **aromatic** is used for historical reasons to refer to the class of compounds related structurally to benzene. Aromatic compounds are systematically named according to IUPAC rules, but many common names are also used. Disubstituted benzenes are named as **ortho** (1,2 disubstituted), **meta** (1,3 disubstituted), or **para** (1,4 disubstituted) derivatives. The C_6H_5- unit itself is referred to as a **phenyl** group, and the $C_6H_5CH_2-$ unit is a **benzyl** group.

Benzene is described by resonance theory as a resonance hybrid of two equivalent Kekulé structures:

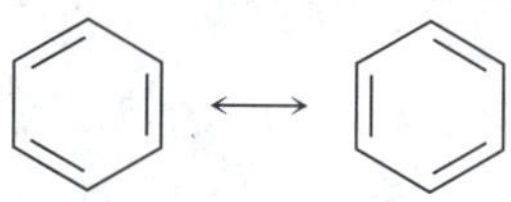

Benzene is described by molecular orbital theory as a planar, cyclic, conjugated molecule with six π electrons. According to the **Hückel rule**, a molecule must have **$4n + 2$ π electrons** (where $n = 0, 1, 2, 3$, and so on) to be aromatic. Planar, cyclic, conjugated molecules with other numbers of π electrons are **antiaromatic**.

Benzene has five key characteristics:

1. Benzene is cyclic, planar, and conjugated.
2. Benzene is unusually stable. Its heat of hydrogenation is 150 kJ/mol less than we might expect for a cyclic triene.
3. Benzene reacts with electrophiles to give substitution products, in which cyclic conjugation is retained, rather than addition products, in which conjugation is destroyed.
4. Benzene has six equivalent C–C bonds with a length of 1.39 Å, a value intermediate between single and double bonds.
5. Benzene has a Hückel number of π electrons, $4n + 2$ with $n = 1$. The π electrons are delocalized over all six carbons.

Other kinds of molecules besides benzene-like compounds can also be aromatic according to the Hückel $4n + 2$ definition. For example, the cyclopentadienyl anion and the cycloheptatrienyl cation are aromatic ions. **Heterocyclic compounds**, which have atoms other than carbon in the ring, can also be aromatic. Pyridine, a six-membered, nitrogen-containing heterocycle, resembles benzene electronically. Pyrrole, a five-membered heterocycle, resembles the cyclopentadienyl anion.

ADDITIONAL PROBLEMS

15.17 Give IUPAC names for the following compounds:

(a)

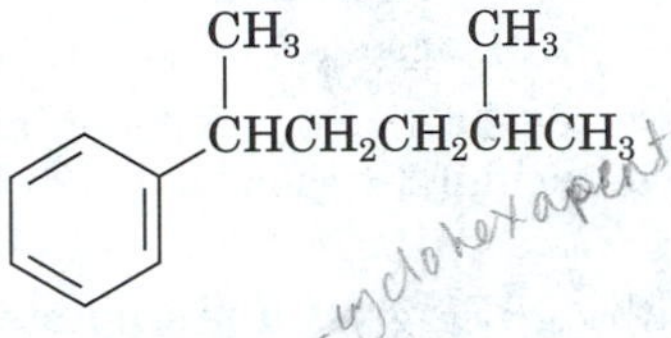

(b)

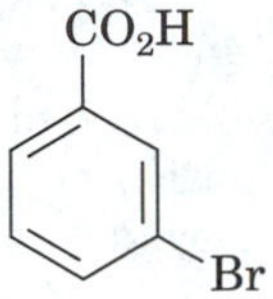

(c)

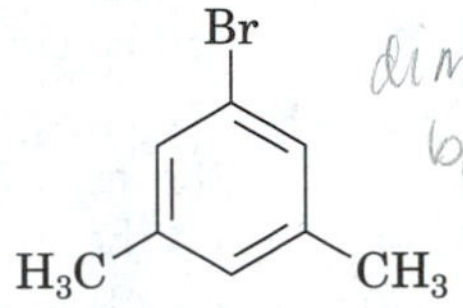

(d)

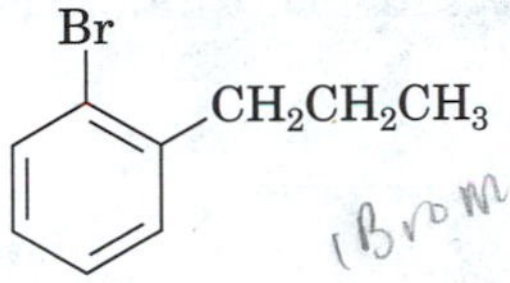

(e)

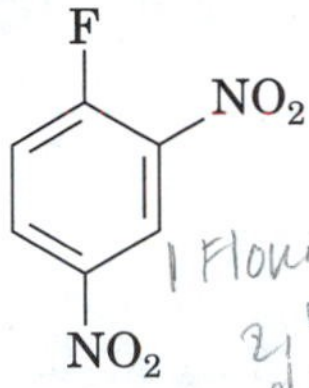

(f)

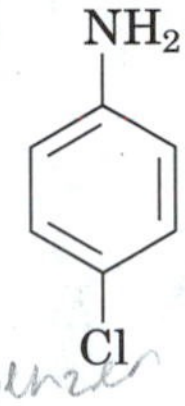

15.18 Draw structures corresponding to the following names:
(a) 3-Methyl-1,2-benzenediamine
(b) 1,3,5-Benzenetriol
(c) 3-Methyl-2-phenylhexane
(d) *o*-Aminobenzoic acid
(e) *m*-Bromophenol
(f) 2,4,6-Trinitrophenol (picric acid)
(g) *p*-Iodonitrobenzene

15.19 Draw and name all possible isomers of:
(a) Dinitrobenzene
(b) Bromodimethylbenzene
(c) Trinitrophenol

15.20 Draw and name all possible aromatic compounds with the formula C_7H_7Cl.

15.21 Draw and name all possible aromatic compounds with the formula C_8H_9Br. (There are 14.)

15.22 Propose structures for aromatic hydrocarbons that meet these descriptions:
(a) C_9H_{12}; gives only one $C_9H_{11}Br$ product on substitution with bromine
(b) $C_{10}H_{14}$; gives only one $C_{10}H_{13}Cl$ product on substitution with chlorine
(c) C_8H_{10}; gives three C_8H_9Br products on substitution with bromine
(d) $C_{10}H_{14}$; gives two $C_{10}H_{13}Cl$ products on substitution with chlorine

15.23 There are four resonance structures for anthracene, one of which is shown. Draw the other three.

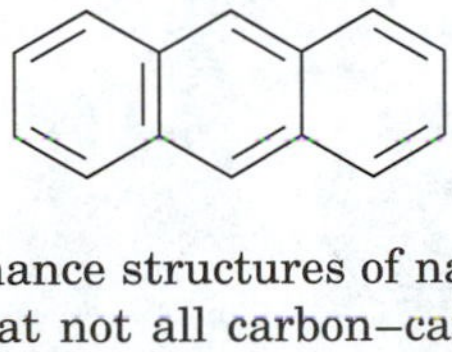

Anthracene

15.24 Look at the three resonance structures of naphthalene shown in Section 15.11, and account for the fact that not all carbon–carbon bonds have the same length. The C1–C2 bond is 1.36 Å long, whereas the C2–C3 bond is 1.39 Å long.

15.25 There are five resonance structures of phenanthrene, one of which is shown. Draw the other four.

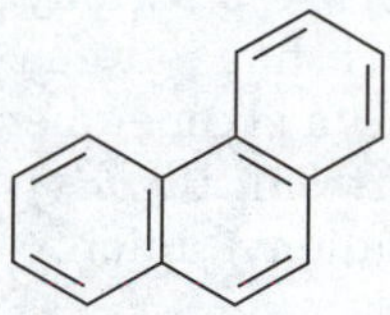

Phenanthrene

15.26 Look at the five resonance structures for phenanthrene (Problem 15.25) and predict which of its carbon–carbon bonds is shortest.

15.27 Define the following terms in your own words:
(a) Aromaticity (b) Conjugated
(c) Hückel $4n + 2$ rule (d) Resonance hybrid

15.28 Use the data in Table 15.2 to calculate the heat of hydrogenation, $\Delta H^\circ_{\text{hydrog}}$, for the partial hydrogenation of benzene to yield 1,3-cyclohexadiene. Is the reaction exothermic or endothermic?

15.29 3-Chlorocyclopropene, on treatment with $AgBF_4$, gives a precipitate of AgCl and a stable solution of a product that shows a single ^{1}H NMR absorption at 11.04 δ. What is a likely structure for the product, and what is its relation to Hückel's rule?

3-Chlorocyclopropene

15.30 Draw an energy diagram for the three molecular orbitals of the cyclopropenyl system (Problem 15.29). How are these three molecular orbitals occupied in the cyclopropenyl anion, cation, and radical? Which of the three substances is aromatic according to Hückel's rule?

15.31 If we were to use the "circle" notation for aromaticity, we would draw the cyclopropenyl cation as shown here. How many π electrons are represented by the circle?

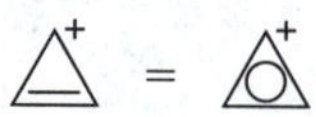

15.32 Cyclopropanone is highly reactive because of its large amount of angle strain. Methylcyclopropenone, although more strained than cyclopropanone, is nevertheless quite stable and can even be distilled. Explain. (*Hint*: Consider the polarity of the carbonyl group.)

Cyclopropanone **Methylcyclopropenone**

15.33 Cycloheptatrienone is stable, but cyclopentadienone is so reactive that it can't be isolated. Explain.

Cycloheptatrienone **Cyclopentadienone**

15.34 Which would you expect to be most stable, cyclononatetraenyl radical, cation, or anion?

15.35 How might you convert 1,3,5,7-cyclononatetraene to an aromatic substance?

15.36 Compound A, C_8H_{10}, yields three substitution products, C_8H_9Br, on reaction with Br_2. Propose two possible structures for A. The ^1H NMR spectrum of A shows a complex four-proton multiplet at 7.0 δ and a six-proton singlet at 2.30 δ. What is the correct structure of A?

15.37 Azulene, an isomer of naphthalene, has a large dipole moment (μ = 1.0 D). Explain.

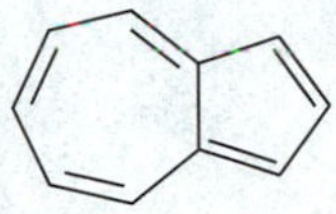

Azulene

15.38 What is the structure of a hydrocarbon that has M^+ = 120 in its mass spectrum and has the following ^1H NMR spectrum?

7.25 δ (5 H, broad singlet) — 2.90 δ (1 H, septet, J = 7 Hz)
1.22 δ (6 H, doublet, J = 7 Hz)

15.39 Propose structures for compounds that fit these descriptions:

(a) $C_{10}H_{14}$
^1H NMR: 7.18 δ (4 H, broad singlet) — IR: 745 cm^{-1}
2.70 δ (4 H, quartet, J = 7 Hz)
1.20 δ (6 H, triplet, J = 7 Hz)

(b) $C_{10}H_{14}$
^1H NMR: 7.0 δ (4 H, broad singlet) — IR: 825 cm^{-1}
2.85 δ (1 H, septet, J = 8 Hz)
2.28 δ (3 H, singlet)
1.20 δ (6 H, doublet, J = 8 Hz)

15.40 Indole is an aromatic heterocycle that has a benzene ring fused to a pyrrole ring. Draw an orbital picture of indole.

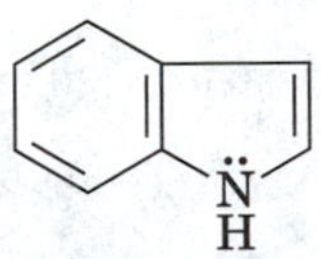

Indole

(a) How many π electrons does indole have?
(b) What is the electronic relationship of indole to naphthalene?

15.41 On reaction with acid, 4-pyrone is protonated on the carbonyl-group oxygen to give a stable cationic product. Explain the stability of the protonated product.

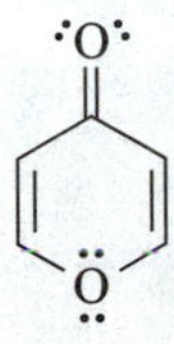

4-Pyrone

15.42 1-Phenyl-2-butene has an ultraviolet absorption at λ_{max} = 208 nm (ε = 8000). On treatment with a small amount of strong acid, isomerization occurs and a new substance with λ_{max} = 250 nm (ε = 15,800) is formed. Propose a structure for this isomer, and suggest a mechanism for its formation.

15.43 Propose structures for aromatic compounds that have the following ^1H NMR spectra.

(a) C_8H_9Br
IR: 820 cm^{-1}

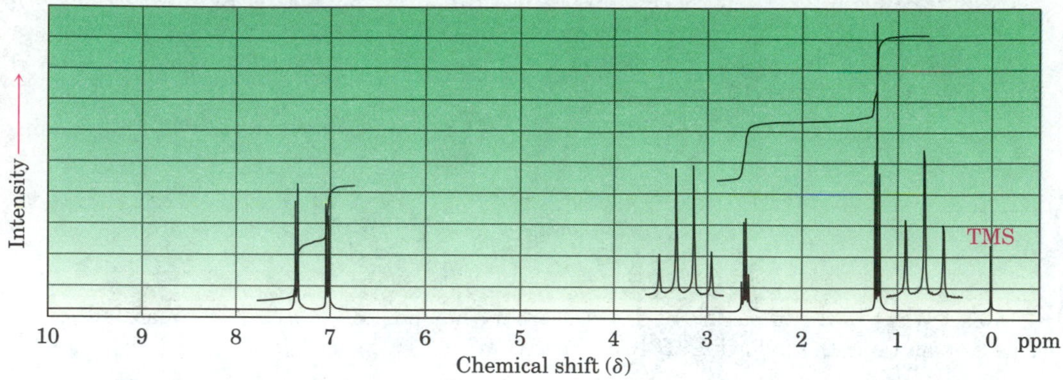

(b) C_9H_{12}
IR: 750 cm^{-1}

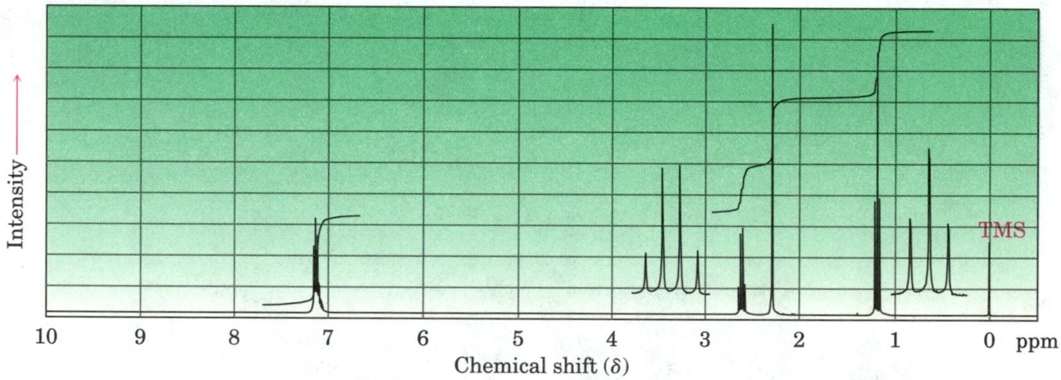

(c) $C_{11}H_{16}$
IR: 820 cm^{-1}

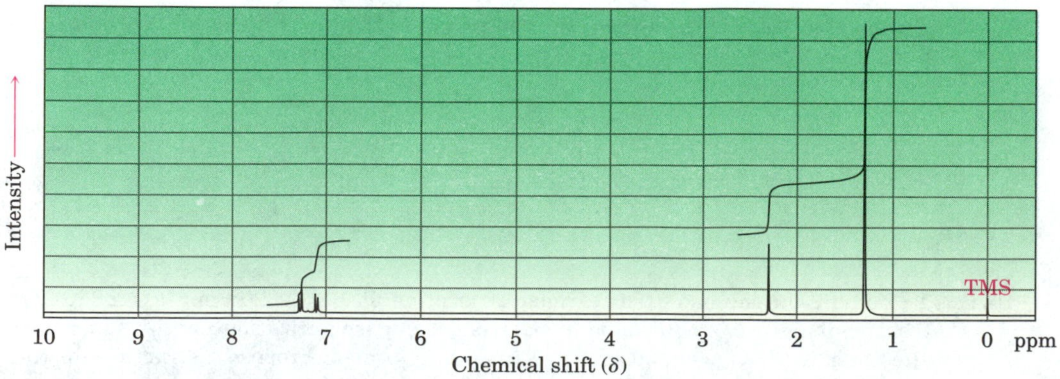

15.44 Propose a structure for a molecule $C_{14}H_{12}$ that has the following ^1H NMR spectrum and has IR absorptions at 700, 740, and 890 cm^{-1}.

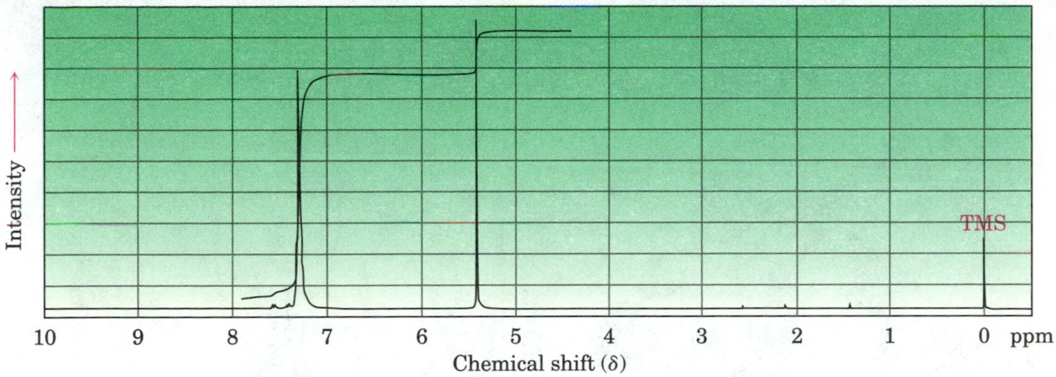

15.45 Pentalene is a most elusive molecule that has never been isolated. The pentalene dianion, however, is well known and quite stable. Explain.

Pentalene **Pentalene dianion**

15.46 Purine is a heterocyclic aromatic compound whose derivatives are constituents of DNA and RNA. Why is purine aromatic? How many *p* electrons does each nitrogen donate to the aromatic π system?

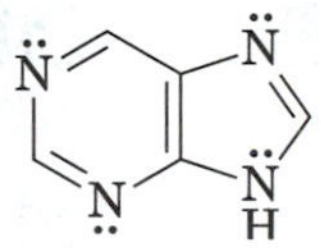

Purine

A Look Ahead

15.47 We'll see in the next chapter that aromatic substitution reactions occur by addition of an electrophile such as Br^+ to the aromatic ring to yield an allylic carbocation intermediate, followed by loss of H^+. Show the structure of the intermediate formed by reaction of benzene with Br^+.

15.48 The substitution reaction of toluene with Br_2 can, in principle, lead to the formation of three isomeric bromotoluene products. In practice, however, only *o*- and *p*-bromotoluene are formed in substantial amounts. The meta isomer is not formed. Draw the structures of the three possible carbocation intermediates (Problem 15.47), and explain why ortho and para products predominate over meta.

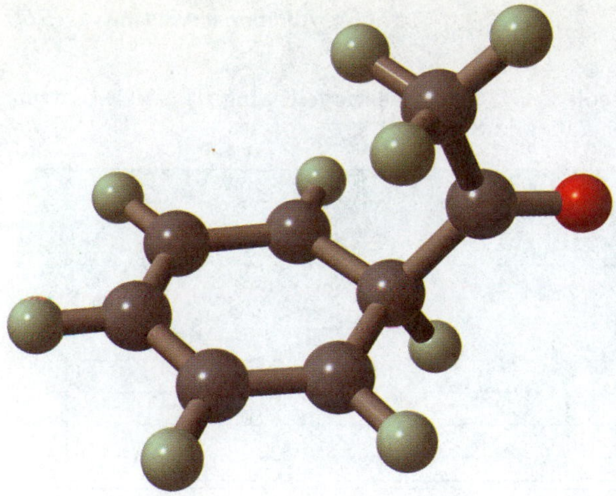

The Friedel–Crafts acylation reaction of benzene occurs via this carbocation intermediate.

16

Chemistry of Benzene: Electrophilic Aromatic Substitution

The most important reaction of aromatic compounds is **electrophilic aromatic substitution**. That is, an electrophile (E^+) reacts with an aromatic ring and substitutes for one of the hydrogens.

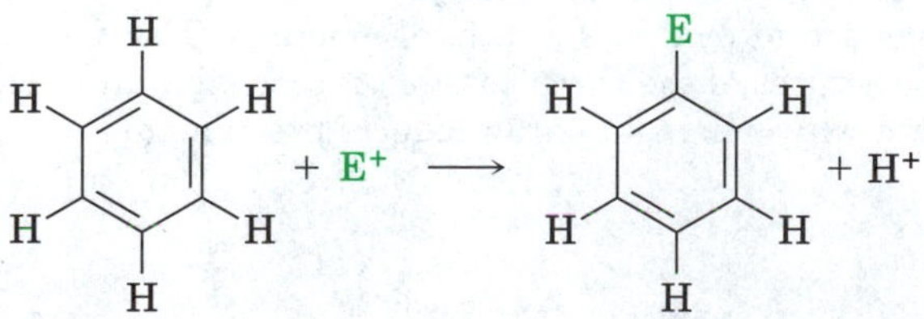

Many different substituents can be introduced onto the aromatic ring by electrophilic substitution reactions. By choosing the proper reagents, it's possible to **halogenate** the aromatic ring (substitute a halogen: –F, –Cl, –Br, or –I), **nitrate** it (substitute a nitro group: $-NO_2$), **sulfonate** it (substitute a sulfonic acid group: $-SO_3H$), **alkylate** it (substitute an alkyl group: –R), or **acylate** it (substitute an acyl group: –COR). Starting from only a few simple materials, we can prepare many thousands of substituted aromatic compounds (Figure 16.1).

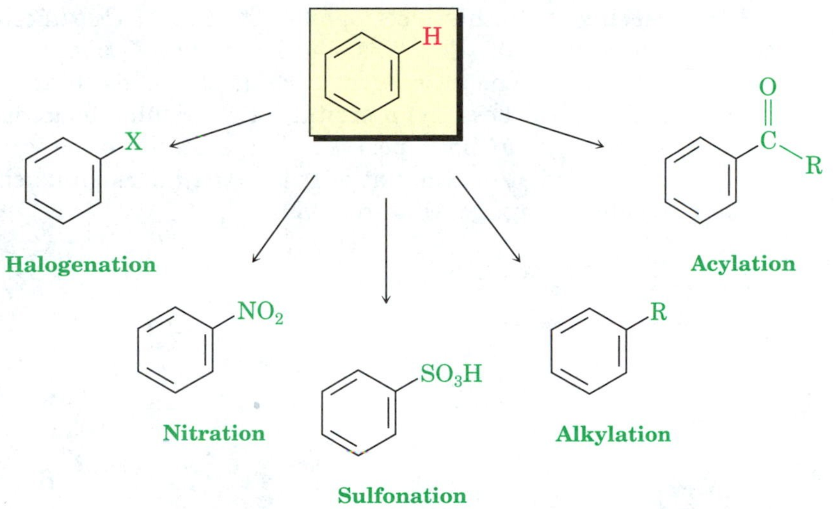

Figure 16.1 Some electrophilic aromatic substitution reactions.

All these reactions—and many more as well—take place by a similar mechanism. Let's begin a study of the process by looking at one reaction in detail, the bromination of benzene.

16.1 Bromination of Aromatic Rings

A benzene ring, with its six π electrons in a cyclic conjugated system, is a site of electron density. Furthermore, the benzene π electrons are sterically accessible to attacking reagents because of their location above and below the plane of the ring. Thus, benzene acts as an electron donor (a Lewis base, or nucleophile) in most of its chemistry, and most of its reactions take place with electron acceptors (Lewis acids, or electrophiles). For example, benzene reacts with Br_2 in the presence of $FeBr_3$ as catalyst to yield the substitution product, bromobenzene.

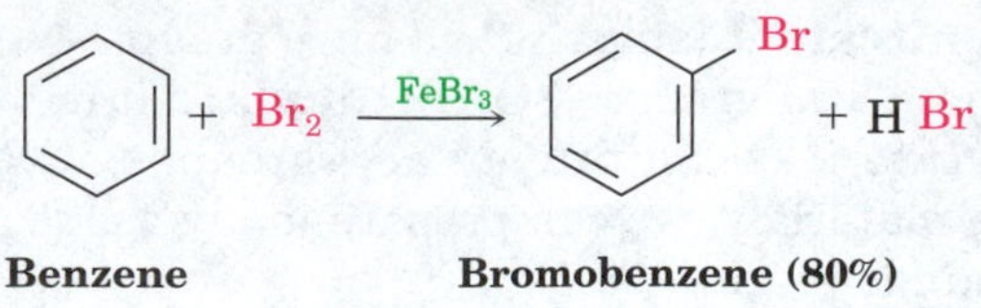

Electrophilic substitution reactions are characteristic of all aromatic rings, not just of benzene and substituted benzenes. Indeed, the ability of a compound to undergo electrophilic substitution is a good test of aromaticity.

Before seeing how this electrophilic aromatic *substitution* reaction occurs, let's briefly recall what was said in Chapter 6 about electrophilic *additions* to alkenes. When a reagent such as HCl adds to an alkene, the electrophilic H^+ approaches the *p* orbitals of the double bond and forms a bond to one carbon, leaving a positive charge at the other carbon. This carbocation intermediate is then attacked by the nucleophilic chloride ion to yield the addition product (Figure 16.2).

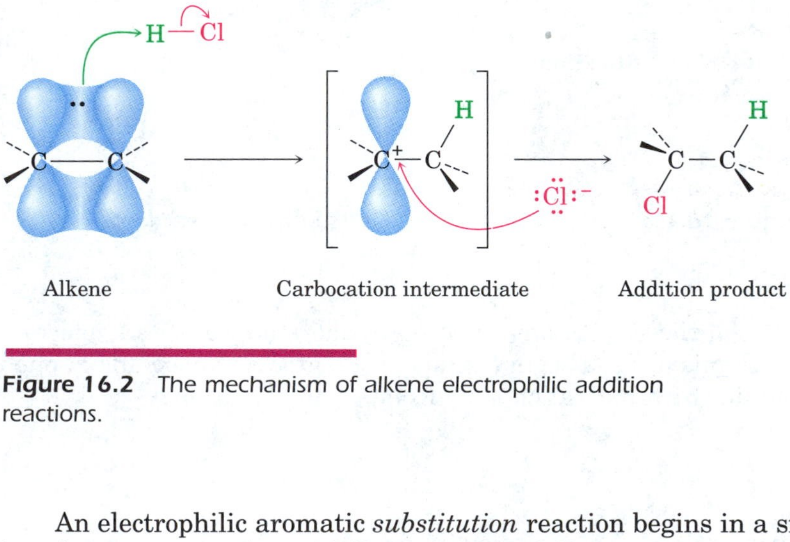

Figure 16.2 The mechanism of alkene electrophilic addition reactions.

An electrophilic aromatic *substitution* reaction begins in a similar way, but there are a number of differences. One difference is that aromatic rings are less reactive toward electrophiles than alkenes are. For example, Br_2 in CCl_4 solution reacts instantly with most alkenes but does not react with benzene. For bromination of benzene to take place, a catalyst such as $FeBr_3$ is needed. The catalyst makes the Br_2 molecule more electrophilic by complexing with it to give an $FeBr_4^-\ Br^+$ species that reacts as if it were Br^+.

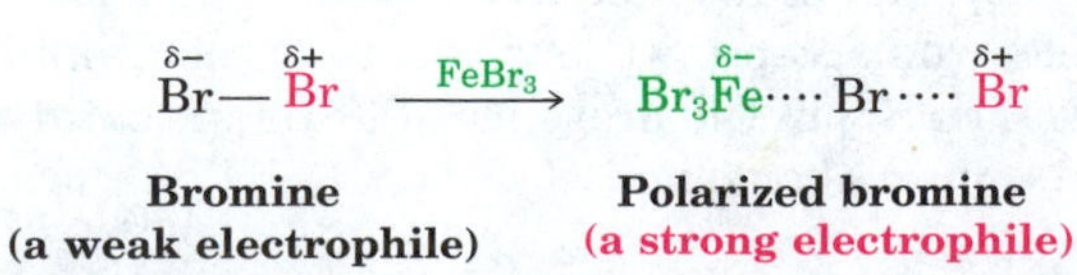

The complexed Br_2 molecule is then attacked by the π electron system of the nucleophilic benzene ring in a slow, rate-limiting step to yield a nonaromatic carbocation intermediate. This carbocation is doubly allylic (recall the allyl cation, Section 11.9) and can be written in three resonance forms:

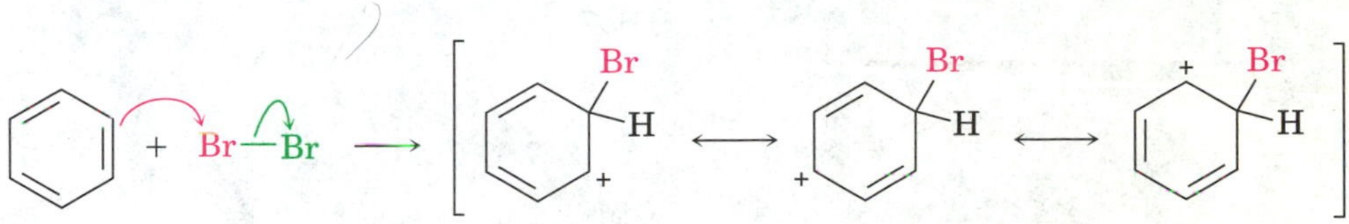

Although stable by comparison with typical alkyl carbocations, the intermediate in electrophilic aromatic substitution is nevertheless much less stable than the starting benzene ring itself with its 150 kJ/mol (36 kcal/mol) of aromatic stability. Thus, electrophilic attack on a benzene ring is endothermic, has a substantial activation energy, and is a rather slow reaction. Figure 16.3 gives reaction energy diagrams comparing the reaction of an electrophile with an alkene and with benzene. The benzene reaction is slower (higher $\Delta G^‡$) because the starting material is more stable.

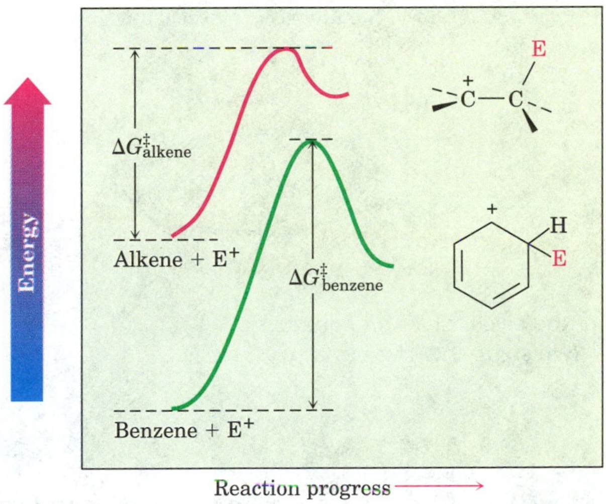

Figure 16.3 A comparison of the reactions of an electrophile (E^+) with an alkene and with benzene: $\Delta G^‡_{alkene} << \Delta G^‡_{benzene}$

A second difference between alkene addition and aromatic substitution occurs after the electrophile has added to the benzene ring to give the carbocation intermediate. Instead of Br^- adding to the carbocation to yield an addition product, a base present in solution abstracts H^+ from the bromine-bearing carbon to yield the neutral, aromatic substitution product. The net effect of reaction of Br_2 with benzene is the substitution of H^+ by Br^+ by the overall mechanism shown in Figure 16.4.

An electron pair from the benzene ring attacks Br_2, forming a new C–Br bond and leaving a carbocation intermediate.

The carbocation intermediate loses a proton and the neutral substitution product forms as two electrons from the C–H bond move into the new aromatic ring.

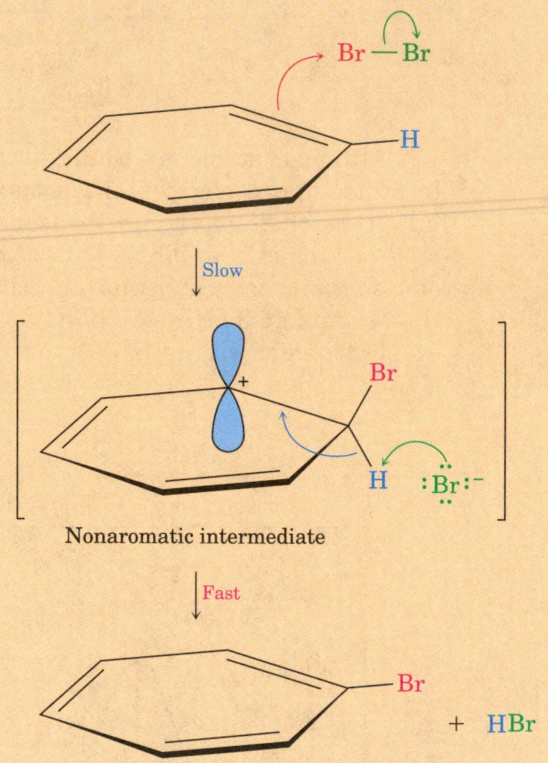

Figure 16.4 The mechanism of the electrophilic bromination of benzene. The reaction occurs in two steps and involves a carbocation intermediate.

Why does the reaction of Br_2 with benzene take a different course than its reaction with an alkene? The answer is simple: If *addition* occurred, the 150 kJ/mol stabilization energy of the aromatic ring would be lost, and the overall reaction would be endothermic. When *substitution* occurs, though, the stability of the aromatic ring is retained and the reaction is exothermic. A reaction energy diagram for the overall process is shown in Figure 16.5.

There are many other kinds of electrophilic aromatic substitutions besides bromination, and all are thought to occur by the same general mechanism. Let's look at some of these other reactions briefly.

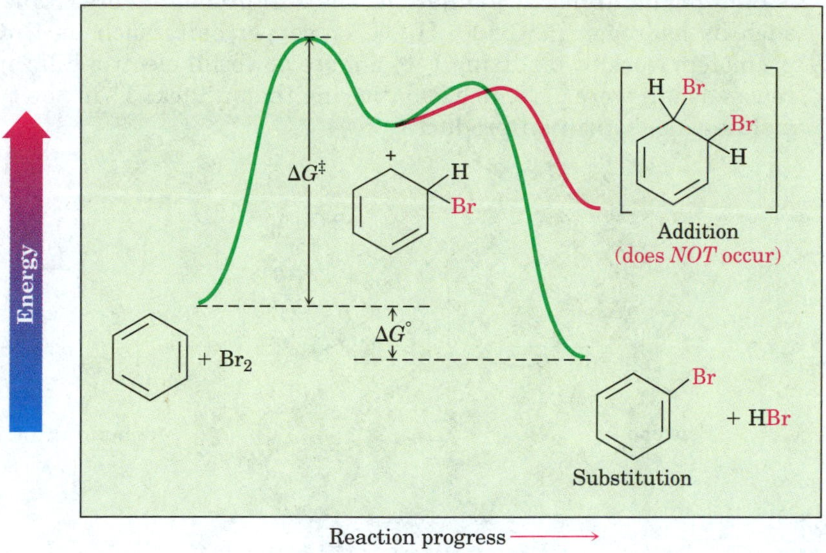

Figure 16.5 A reaction energy diagram for the electrophilic bromination of benzene.

16.2 Other Aromatic Substitutions

Aromatic Chlorination and Iodination

Chlorine and iodine can be introduced into aromatic rings by electrophilic substitution reactions, but fluorine is too reactive, and poor yields of monofluoroaromatic products are obtained. Aromatic rings react with Cl_2 in the presence of $FeCl_3$ catalyst to yield chlorobenzenes. This kind of reaction is used in the synthesis of numerous pharmaceutical agents, including the tranquilizer diazepam (Valium).

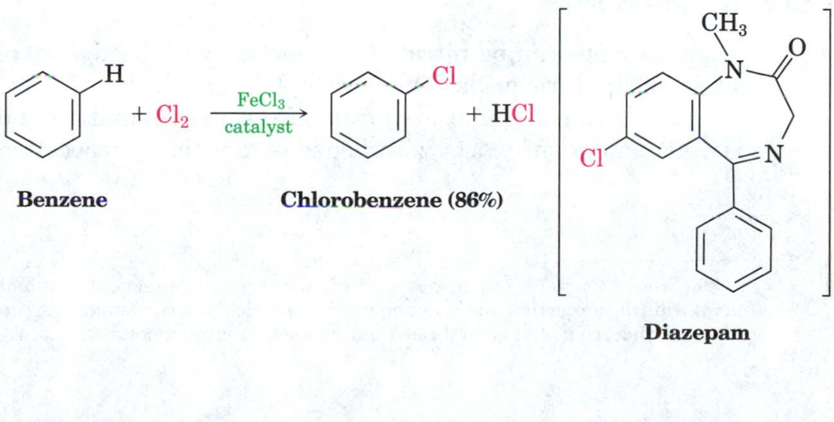

Iodine by itself is unreactive toward aromatic rings, and a promoter[1] is required for aromatic iodination. The best promoters are oxidizing agents such as hydrogen peroxide, H_2O_2, or copper salts such as $CuCl_2$. These promoters work by oxidizing I_2 to a more powerful electrophilic species that reacts as if it were I^+. The aromatic ring then attacks I^+ in the typical way, yielding a substitution product.

$$I_2 + 2\ Cu^{2+} \longrightarrow 2\ \text{“}I^{+}\text{”} + 2\ Cu^{+}$$

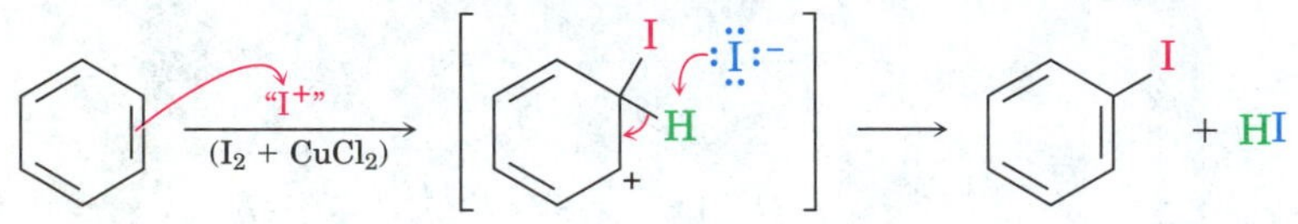

Benzene **Iodobenzene (65%)**

PROBLEM

16.1 Bromination of toluene gives a mixture of three bromotoluene products. Draw and name them.

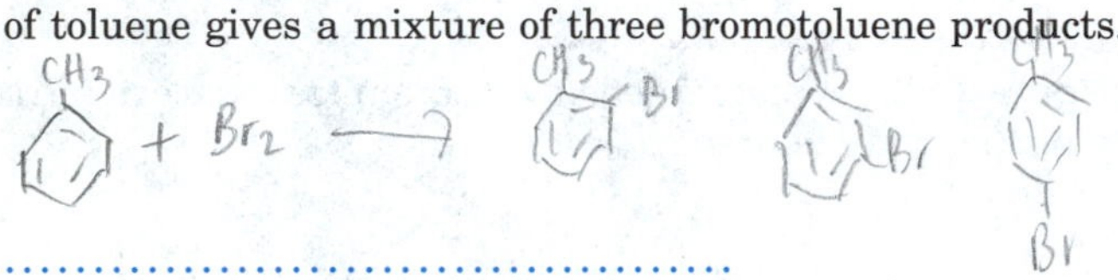

PROBLEM

16.2 Aromatic fluorides can be prepared by a two-step process involving electrophilic *thallation* of the aromatic ring (substitution of thallium):

$$ArH + Tl(OCOCF_3)_3 \longrightarrow Ar{-}Tl(OCOCF_3)_2 \xrightarrow{KF,\ BF_3} ArF$$

Thallium tris(trifluoroacetate)

Propose a mechanism for the thallation of benzene.

Aromatic Nitration

Aromatic rings can be nitrated by reaction with a mixture of concentrated nitric and sulfuric acids. The electrophile in this reaction is the nitronium ion, NO_2^+, which is generated from HNO_3 by protonation and loss of water. The nitronium ion reacts with benzene to yield a carbocation intermediate

[1]The word *promoter* is sometimes used when referring to a reagent that undergoes a chemical change during a reaction and must be used in a stoichiometric amount. A *catalyst*, by contrast, does not undergo a change and need not be used in large amounts.

in much the same way as Br^+. Loss of H^+ from this intermediate gives the neutral substitution product, nitrobenzene.

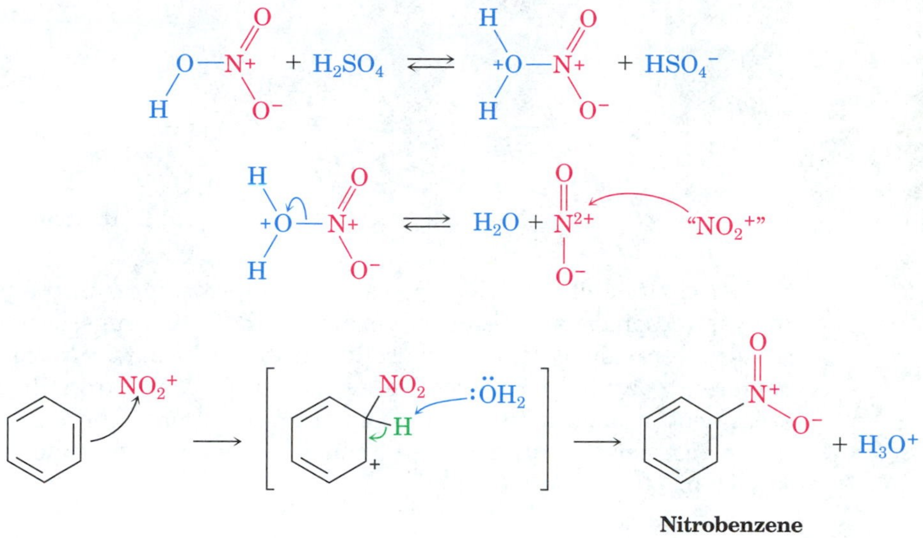

Nitration of an aromatic ring is a particularly important reaction because the nitro-substituted product can be reduced by reagents such as iron metal or $SnCl_2$ to yield an arylamine, $ArNH_2$. Attachment of a nitrogen to an aromatic ring by the two-step nitration/reduction sequence is a key part of the industrial synthesis of dyes and many pharmaceutical agents. We'll discuss this and other reactions of aromatic nitrogen compounds in Chapter 25.

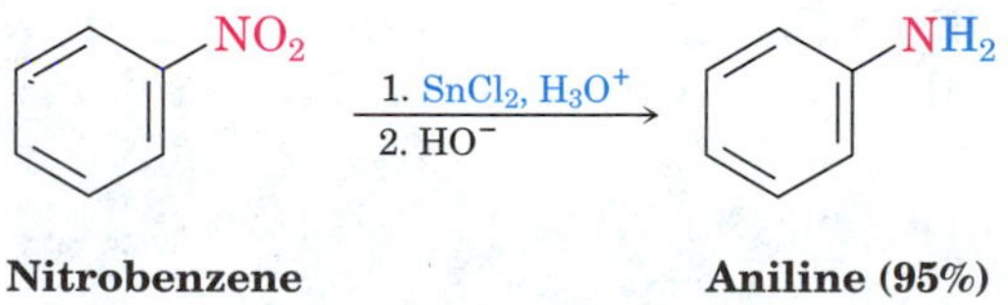

Nitrobenzene **Aniline (95%)**

Aromatic Sulfonation

Aromatic rings can be sulfonated by reaction with fuming sulfuric acid, a mixture of H_2SO_4 and SO_3. The reactive electrophile is either HSO_3^+ or neutral SO_3, depending on reaction conditions. Substitution occurs by the same two-step mechanism seen previously for bromination and nitration. Note, however, that the sulfonation reaction is readily reversible; it can occur either forward or backward, depending on the reaction conditions. Sulfonation is favored in strong acid, but desulfonation is favored in hot, dilute aqueous acid.

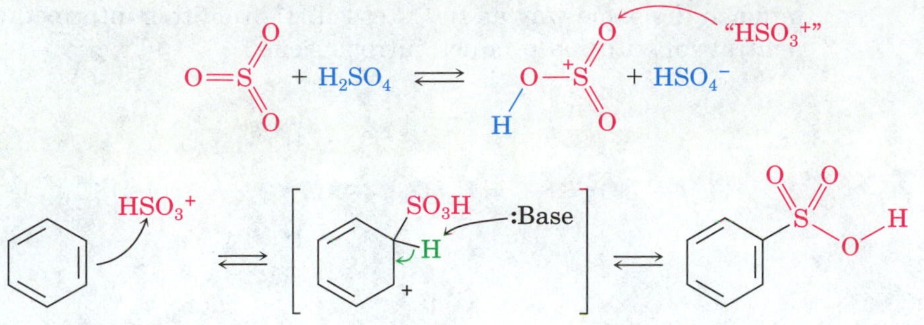

Benzenesulfonic acid

Aromatic sulfonic acids are valuable intermediates in the preparation of dyes and pharmaceuticals. For example, the sulfa drugs, such as sulfanilamide, were among the first useful antibiotics known. Although largely replaced today by more effective agents, sulfa drugs are still used in the treatment of meningitis and urinary-tract infections. These drugs are prepared commercially by a process that involves aromatic sulfonation as the key step.

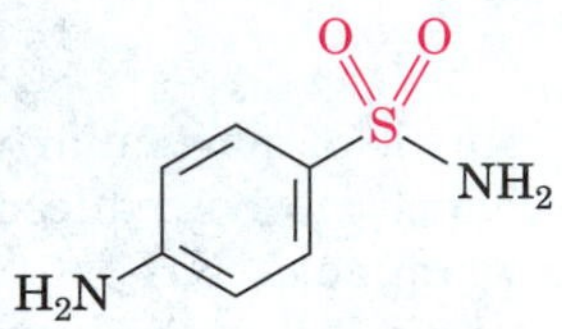

Sulfanilamide (an antibiotic)

Aromatic sulfonic acids are also useful because of the further chemistry they undergo. Heating an aromatic sulfonic acid with NaOH at 300°C in the absence of solvent effects a replacement of the $-SO_3H$ group by $-OH$ and gives a phenol. Yields in this so-called **alkali fusion** reaction are generally good, but the conditions are so vigorous that the reaction is not compatible with the presence of substituents other than alkyl groups on the aromatic ring.

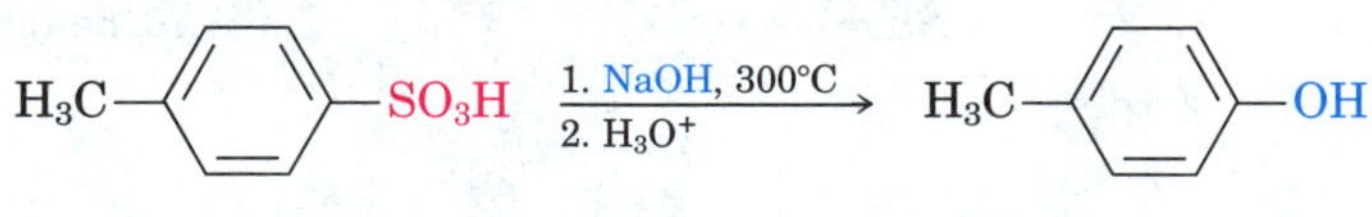

***p*-Toluenesulfonic acid**

***p*-Cresol (72%)**
(a phenol)

PROBLEM

16.3 Chlorination of *o*-xylene (dimethylbenzene) yields a mixture of two products, but chlorination of *p*-xylene yields a single product. Explain.

PROBLEM

16.4 How many products might be formed on chlorination of *m*-xylene?

PROBLEM

16.5 How can you account for the fact that deuterium slowly replaces hydrogen in the aromatic ring when benzene is treated with D_2SO_4?

16.3 Alkylation of Aromatic Rings: The Friedel–Crafts Reaction

One of the most useful of all electrophilic aromatic substitution reactions is *alkylation*, the attachment of an alkyl group to the benzene ring. Charles Friedel[2] and James Crafts[3] reported in 1877 that benzene rings can be alkylated by reaction with an alkyl chloride in the presence of aluminum chloride as catalyst. For example, benzene reacts with 2-chloropropane and $AlCl_3$ to yield isopropylbenzene, also called cumene.

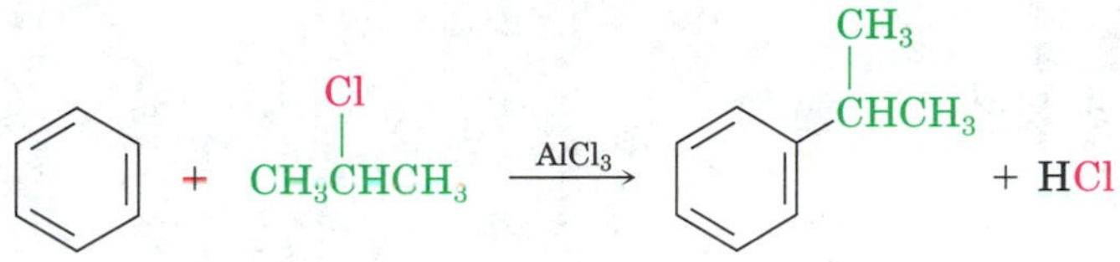

Benzene **2-Chloropropane** **Cumene (85%) (Isopropylbenzene)**

The **Friedel–Crafts alkylation reaction** is an electrophilic aromatic substitution in which the electrophile is a carbocation, R^+. Aluminum chloride catalyzes the reaction by helping the alkyl halide to ionize in much the same way that $FeBr_3$ catalyzes aromatic brominations by polarizing Br_2 (Section 16.1). Loss of a proton then completes the reaction, as shown in Figure 16.6 (p. 580).

Though broadly useful for the synthesis of alkylbenzenes, Friedel–Crafts alkylations nevertheless have strict limitations. One limitation is that only *alkyl* halides can be used. Alkyl fluorides, chlorides, bromides, and iodides all react well, but *aryl* halides and *vinylic* halides do not react. Aryl and vinylic carbocations are too high in energy to form under Friedel–Crafts conditions.

An aryl halide

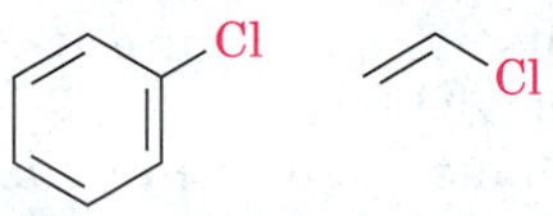

A vinylic halide

NOT reactive

[2]Charles Friedel (1832–1899); b. Strasbourg, France; studied at the Sorbonne; professor, École des Mines (1876–1884) and at Paris (1884–1899).

[3]James M. Crafts (1839–1917); b. Boston; L.L.D., Harvard (1898); professor, Cornell University (1868–1871); Massachusetts Institute of Technology (1871–1900).

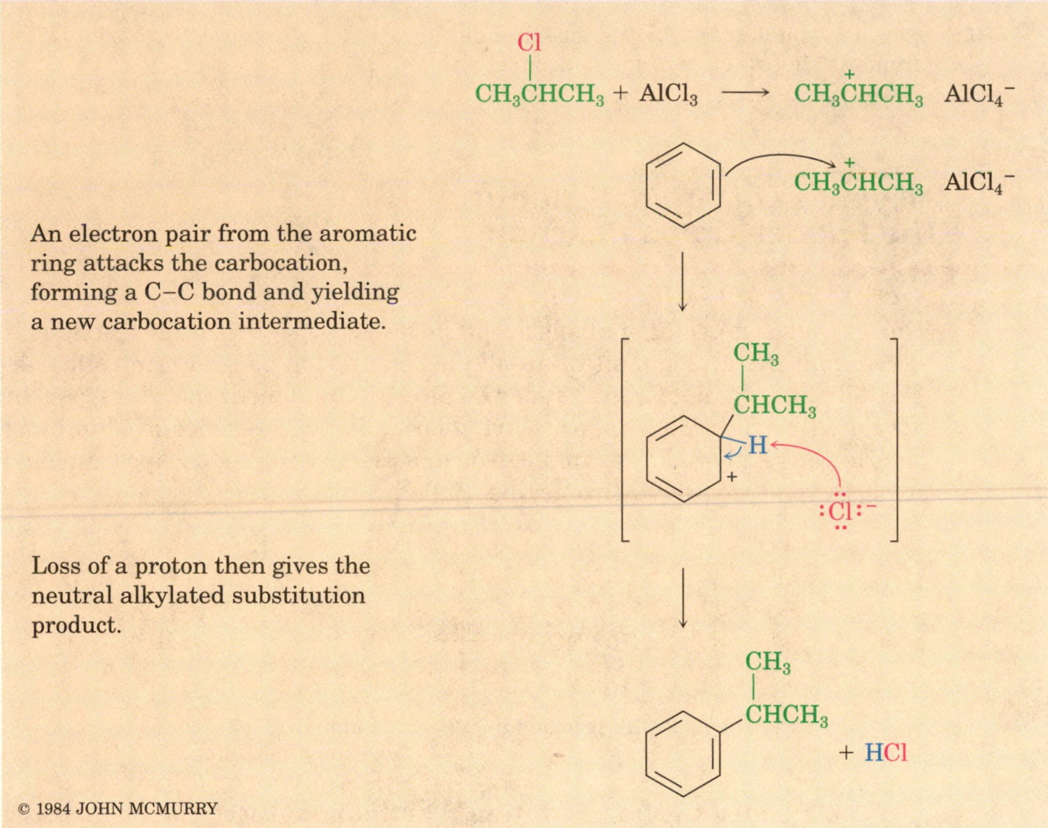

Figure 16.6 Mechanism of the Friedel–Crafts alkylation reaction. The electrophile is a carbocation, generated by $AlCl_3$-assisted ionization of an alkyl halide.

A second limitation is that Friedel–Crafts reactions don't succeed on aromatic rings that are substituted by strongly electron-withdrawing groups. We'll see in Section 16.5 that the presence of a substituent group already on a ring can have a dramatic effect on that ring's subsequent reactivity toward further electrophilic substitution. Rings that contain any of the substituents listed in Figure 16.7 are not reactive enough to undergo Friedel–Crafts alkylation.

Yet a third limitation of the Friedel–Crafts alkylation is that it's often difficult to stop the reaction after a single substitution. Once the first alkyl group is on the ring, a second substitution reaction is facilitated for reasons we'll discuss in the next section. Thus, we often observe *polyalkylation*. For example, reaction of benzene with 1 mol equiv of 2-chloro-2-methylpropane yields *p*-di-*tert*-butylbenzene as the major product, along with small amounts of *tert*-butylbenzene and unreacted starting material. High yields of monoalkylation product are obtained only when a large excess of benzene is used.

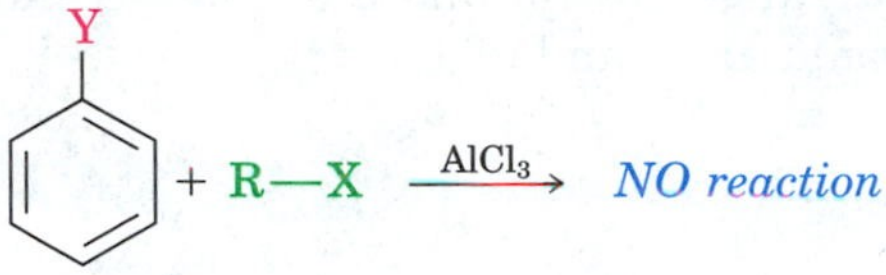

where $Y = -\overset{+}{N}R_3, -NO_2, -CN, -SO_3H, -CHO,$

$-COCH_3, -COOH, -COOCH_3$

$(-NH_2, -NHR, -NR_2)$

Figure 16.7 Limitations on the aromatic substrate in Friedel–Crafts reactions. No reaction occurs if the substrate has any of the electron-withdrawing substituents indicated.

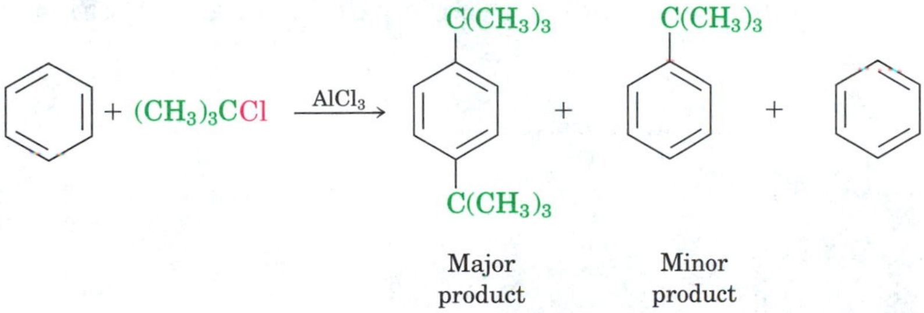

A final limitation to the Friedel–Crafts reaction is that skeletal rearrangement of the alkyl group sometimes occurs during reaction, particularly when primary alkyl halides are used. The amount of rearrangement depends on catalyst, reaction temperature, and reaction solvent. Less rearrangement is generally found at lower reaction temperatures, but mixtures of products are usually obtained. For example, treatment of benzene with 1-chlorobutane gives an approximately 2:1 ratio of rearranged (*sec*-butyl) to unrearranged (*n*-butyl) products when the reaction is carried out at 0°C using $AlCl_3$ as catalyst.

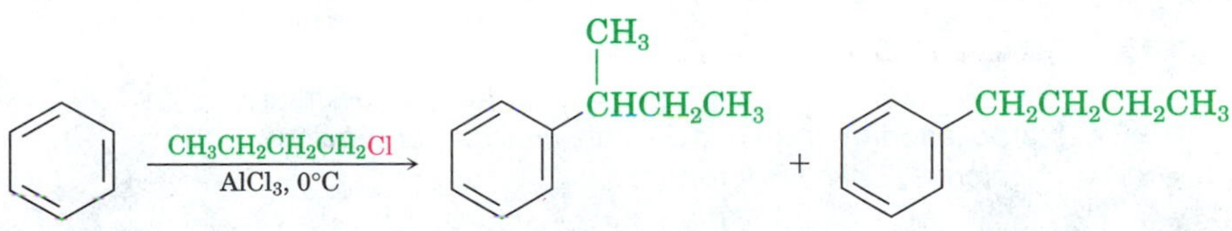

Benzene **sec-Butylbenzene (65%)** **Butylbenzene (35%)**

These carbocation rearrangements are similar to those we saw earlier during electrophilic additions to alkenes (Section 6.12). For example, the relatively unstable primary butyl carbocation produced from the reaction of 1-chlorobutane with $AlCl_3$ rearranges to the more stable secondary butyl carbocation by shift of a hydrogen atom and its electron pair (a hydride ion, $H:^-$) from C2 to C1.

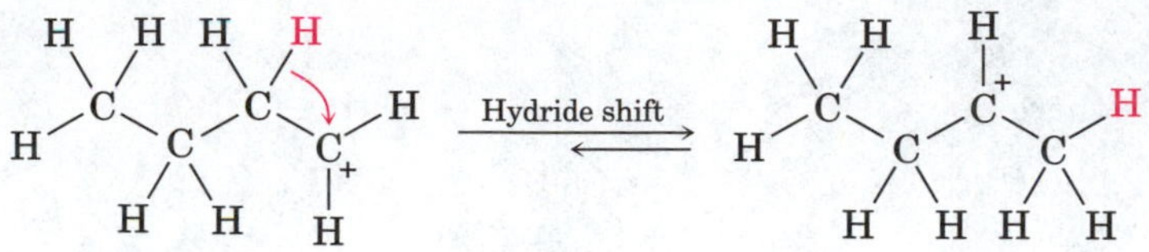

Primary butyl carbocation

Secondary butyl carbocation

Similarly, carbocation rearrangements can occur by *alkyl* shifts. For example, Friedel–Crafts alkylation of benzene with 1-chloro-2,2-dimethylpropane yields (1,1-dimethylpropyl)benzene as the sole product. The initially formed primary carbocation rearranges to a tertiary carbocation by shift of a methyl group and its electron pair from C2 to C1 (Figure 16.8).

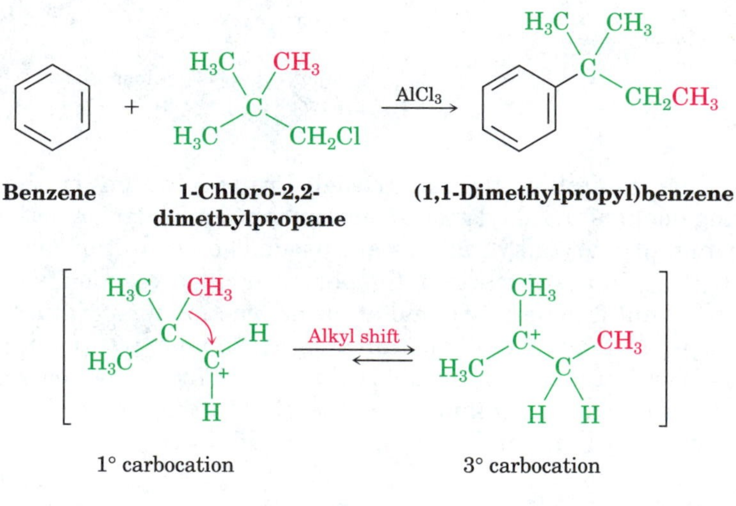

Figure 16.8 Rearrangement of a primary to a tertiary carbocation during Friedel–Crafts reaction of benzene with 1-chloro-2,2-dimethylpropane.

PROBLEM. . . . .

16.6 What is the major monosubstitution product from the Friedel–Crafts reaction of benzene with 1-chloro-2-methylpropane in the presence of $AlCl_3$?

PROBLEM

16.7 Which of the following alkyl halides would you expect to undergo Friedel–Crafts reaction *without* rearrangement? Explain.
(a) CH_3CH_2Cl (b) $CH_3CH_2CH(Cl)CH_3$ (c) $CH_3CH_2CH_2Cl$
(d) $(CH_3)_3CCH_2Cl$ (e) Chlorocyclohexane

16.4 Acylation of Aromatic Rings

An **acyl** group, –COR (pronounced **a**-sil), is introduced onto the ring when an aromatic compound reacts with a carboxylic acid chloride, RCOCl, in the presence of $AlCl_3$. For example, reaction of benzene with acetyl chloride yields the ketone, acetophenone.

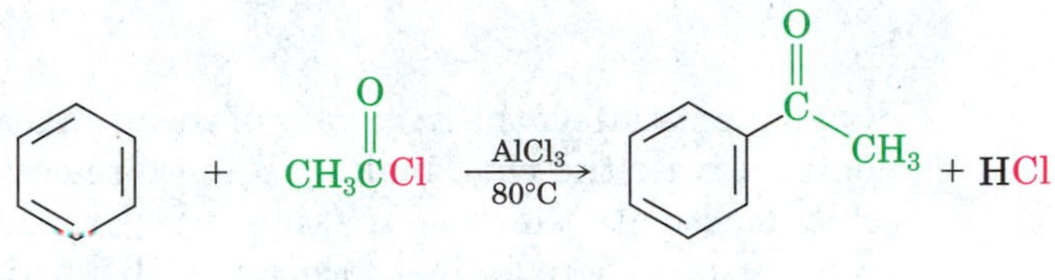

The mechanism of **Friedel–Crafts acylation** is similar to that of Friedel–Crafts alkylation. The reactive electrophile is a resonance-stabilized acyl cation, generated by reaction between the acyl chloride and $AlCl_3$ (Figure 16.9). As the resonance structures in Figure 16.9 indicate, an acyl cation is stabilized by interaction of the vacant orbital on carbon with a lone-pair orbital of the neighboring oxygen. Once formed, the acyl cation does not rearrange; rather, it is attacked by an aromatic ring to give unrearranged substitution product.

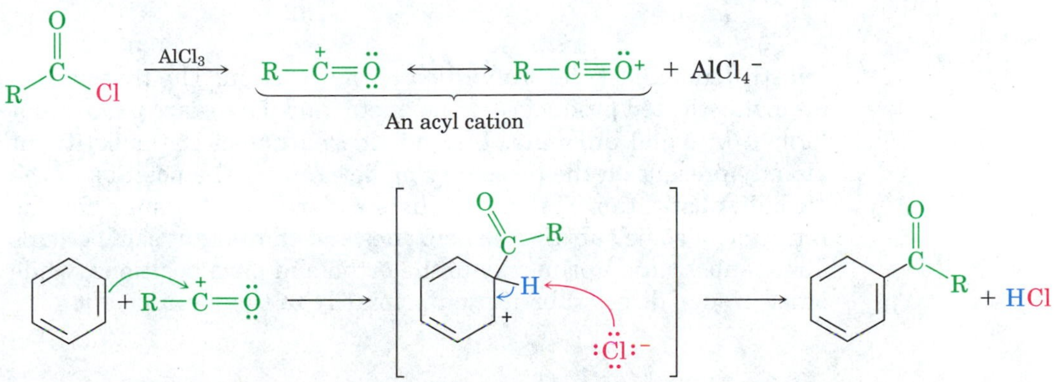

Figure 16.9 Mechanism of the Friedel–Crafts acylation reaction. The electrophile is a resonance-stabilized acyl cation.

Unlike the multiple substitutions that often occur in Friedel–Crafts alkylations, acylations never occur more than once on a ring because the product acylbenzene is always less reactive than the nonacylated starting material. We'll account for these reactivity differences in the next section.

16.5 Substituent Effects in Substituted Aromatic Rings

Only one product can form when an electrophilic substitution occurs on benzene, but what would happen if we were to carry out a reaction on an aromatic ring that already has a substituent? Substituents already present on the ring have two effects:

1. Substituents affect the *reactivity* of the aromatic ring. Some substituents activate the ring, making it more reactive than benzene, and some deactivate the ring, making it less reactive than benzene. In aromatic nitration, for instance, an –OH substituent makes the ring 1000 times more reactive than benzene, while an $-NO_2$ substituent makes the ring 2×10^7 times less reactive.

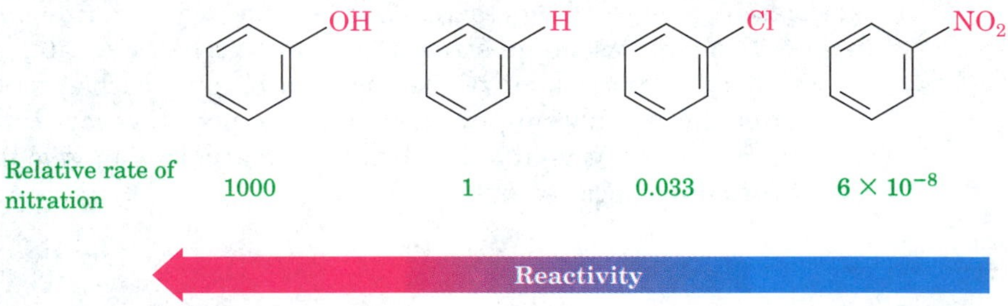

2. Substituents affect the *orientation* of the reaction. The three possible disubstituted products—ortho, meta, and para—are usually not formed in equal amounts. Instead, the nature of the substituent already present on the benzene ring determines the position of the second substitution. Table 16.1 lists experimental results for the nitration of some substituted benzenes and shows that some groups direct substitution primarily to the ortho and para positions, while other groups direct substitution primarily to the meta position.

Substituents can be classified into three groups: ortho- and para-directing activators, ortho- and para-directing deactivators, and meta-directing deactivators. No meta-directing activators are known. Figure 16.10 lists some groups in all three categories. Notice how the directing effects of

Table 16.1 Orientation of Nitration in Substituted Benzenes

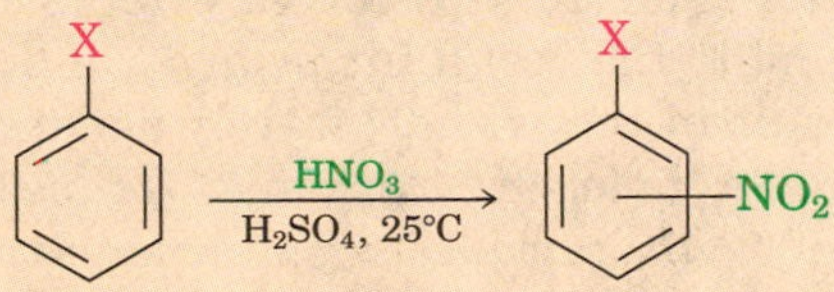

X	Product (%) *Ortho*	*Meta*	*Para*	X	Product (%) *Ortho*	*Meta*	*Para*
Meta-directing deactivators				**Ortho- and para-directing deactivators**			
$-\overset{+}{N}(CH_3)_3$	2	87	11	—F	13	1	86
$-NO_2$	7	91	2	—Cl	35	1	64
—COOH	22	76	2	—Br	43	1	56
—CN	17	81	2	—I	45	1	54
$-CO_2CH_2CH_3$	28	66	6	**Ortho- and para-directing activators**			
$-COCH_3$	26	72	2	$-CH_3$	63	3	34
—CHO	19	72	9	$-\ddot{O}H$	50	0	50
				$-\ddot{N}HCOCH_3$	19	2	79

the groups correlate with their reactivities. All meta-directing groups are strongly deactivating, and most ortho- and para-directing groups are activating. The halogens are unique in being ortho- and para-directing but weakly deactivating.

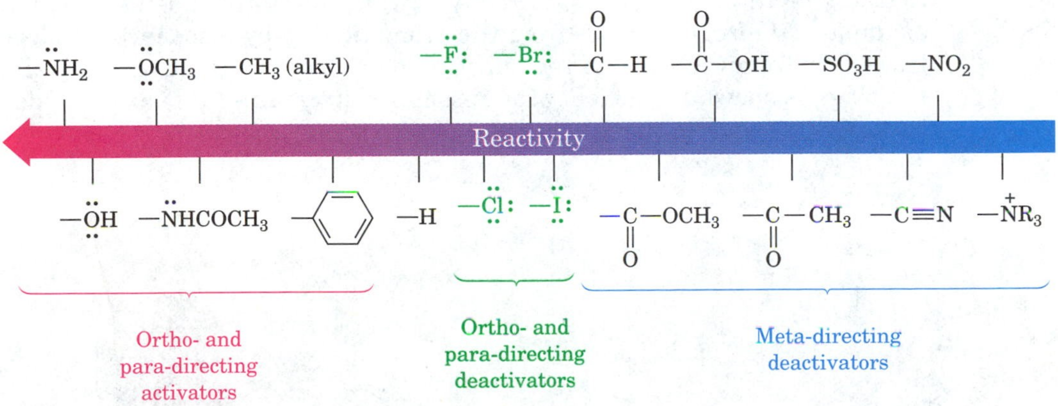

Figure 16.10 Classification of substituent effects in electrophilic aromatic substitution.

Reactivity and orientation in electrophilic aromatic substitutions are controlled by an interplay of *inductive effects* and *resonance effects*. As we saw in Sections 2.1 and 6.10, **inductive effects** are due to the electronegativity of atoms and to the resultant polarity of bonds in functional groups. These effects result in the withdrawal or donation of electrons through σ bonds. For example, halogens, carbonyl groups, cyano groups, and nitro groups inductively *withdraw* electrons through the σ bond linking the substituent to the ring.

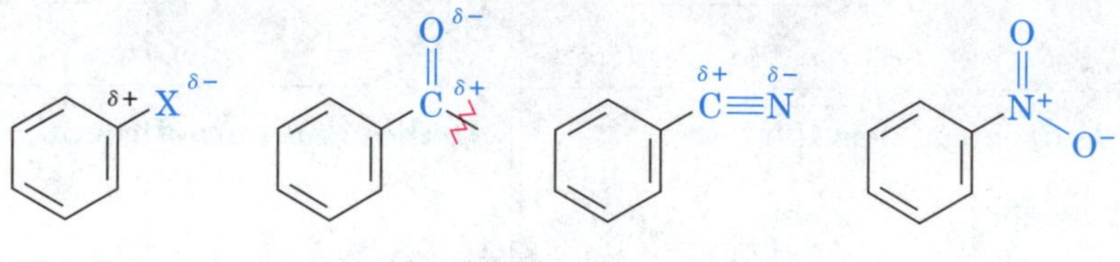

The groups attached to the aromatic rings are inductively electron-withdrawing because of the polarity of their bonds.

Alkyl groups, on the other hand, inductively *donate* electrons. The reasons for this behavior involve the same factors that cause alkyl substituents to stabilize alkenes (Section 6.7) and carbocations (Section 6.10).

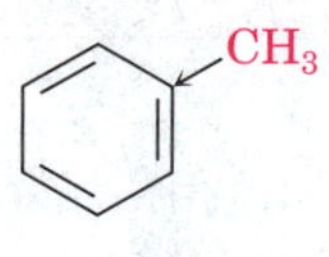

Alkyl group; inductively electron-donating

Resonance effects are due to the overlap of a *p* orbital on the substituent with a *p* orbital on the aromatic ring and result in withdrawal or donation of electrons through π bonds. Carbonyl, cyano, and nitro substituents, for example, *withdraw* electrons from the aromatic ring by resonance. Pi electrons flow from the rings to the substituents, placing a positive charge in the ring. As shown by the following resonance structures for benzaldehyde, the effect is greatest at the ortho and para positions.

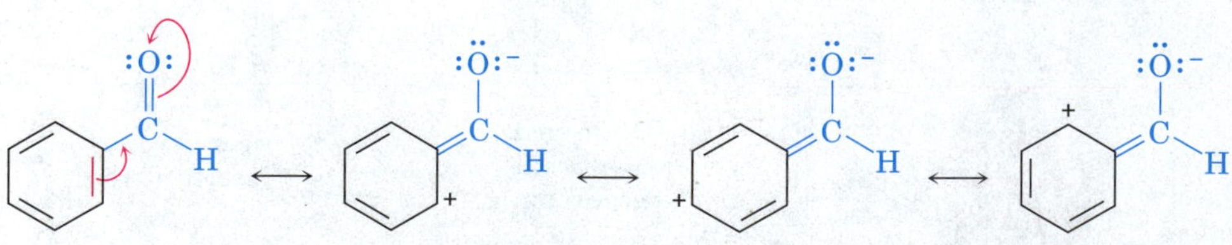

Benzaldehyde

Note that substituents with an electron-withdrawing resonance effect have the general structure –Y=Z, where the *Z* atom is more electronegative than Y:

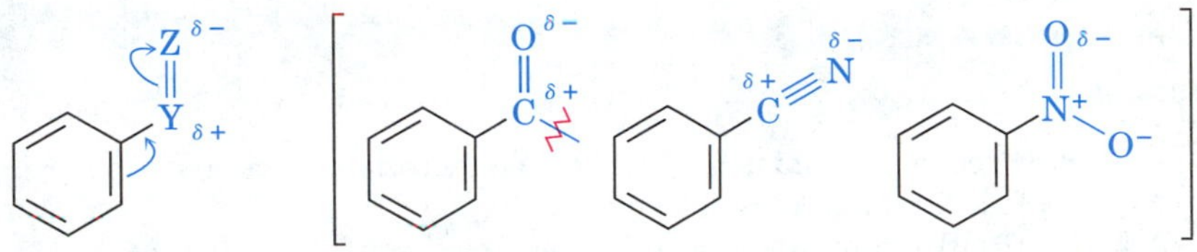

Rings substituted by a group with an electron-withdrawing resonance effect have this general structure.

Conversely, halogen, hydroxyl, alkoxyl (–OR), and amino substituents *donate* electrons to the aromatic ring by resonance. Pi electrons flow from the substituents to the ring, placing a negative charge in the ring as shown by the following resonance structures for chlorobenzene. (Recall the discussion of conjugation in Section 14.10 where a similar effect was noted.) Once again, the effect is greatest at the ortho and para positions.

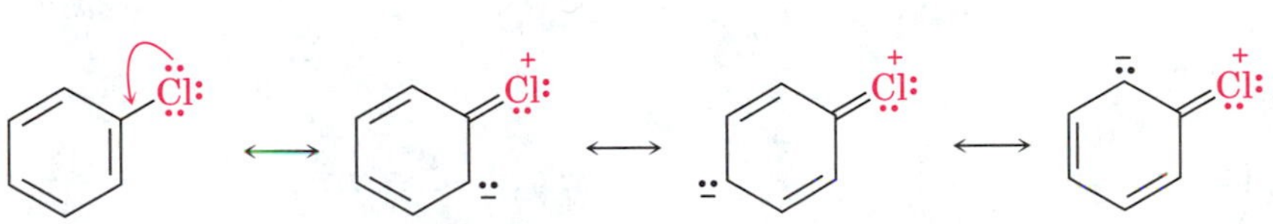

Chlorobenzene

Substituents with an electron-donating resonance effect have the general structure $-\ddot{Y}$, where the atom Y has a lone pair of electrons available for donation to the ring:

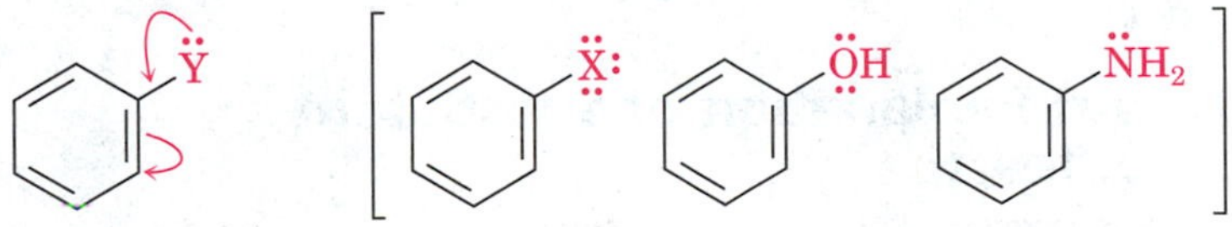

Rings substituted by a group with an electron-donating resonance effect have this general structure.

X = Halogen

One further point: It's not necessary for inductive effects and resonance effects to act in the same direction. Halogen, hydroxyl, alkoxyl, and amino substituents, for example, have electron-*withdrawing* inductive effects because of the electronegativity of the atom bonded to the aromatic ring but have electron-*donating* resonance effects because of the lone-pair electrons on the atom bonded to the ring.

PROBLEM...

16.8 Predict the major products of the following reactions:
(a) Nitration of bromobenzene
(b) Bromination of nitrobenzene
(c) Chlorination of phenol
(d) Bromination of aniline

PROBLEM...

16.9 Write resonance structures for nitrobenzene to show the electron-withdrawing resonance effect of the nitro group.

PROBLEM...

16.10 Write resonance structures for phenol to show the electron-donating resonance effect of the hydroxyl group.

PROBLEM...

16.11 Write as many resonance forms as you can for the carbocation intermediates formed by bromination of phenol at ortho, meta, and para positions. Be sure to consider the resonance effect of the hydroxyl substituent. Which of the three intermediates look most stable and which look least stable?

....

16.6 An Explanation of Substituent Effects

Activation and Deactivation of Aromatic Rings

How do inductive and resonance effects result in the activation or deactivation of an aromatic ring toward electrophilic substitution? *The common feature of all activating groups is that they donate electrons to the ring*, thereby stabilizing the carbocation intermediate from electrophilic addition and causing it to form faster. Hydroxyl, alkoxyl, and amino groups are activating because their stronger electron-donating resonance effect outweighs their weaker electron-withdrawing inductive effect. Alkyl groups are activating because of their electron-donating inductive effect.

The common feature of all deactivating groups is that they withdraw electrons from the ring, thereby destabilizing the carbocation intermediate and causing it to form more slowly. Carbonyl, cyano, and nitro groups are deactivating because of both electron-withdrawing resonance *and* inductive effects. Halogens are deactivating because their stronger electron-withdrawing inductive effect outweighs their weaker electron-donating resonance effect.

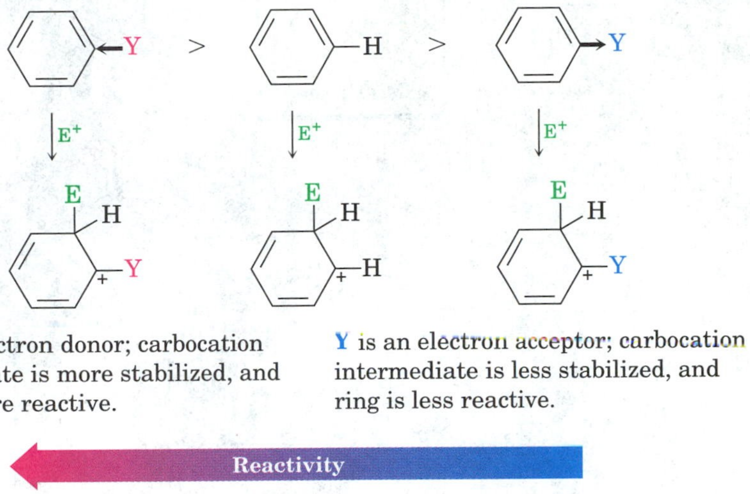

PROBLEM.

16.12 Rank the compounds in each group in order of their reactivity to electrophilic substitution.

(a) Nitrobenzene, phenol, toluene, benzene
(b) Phenol, benzene, chlorobenzene, benzoic acid
(c) Benzene, bromobenzene, benzaldehyde, aniline

PROBLEM.

16.13 Use Figure 16.10 to explain why Friedel–Crafts alkylations often give polysubstitution but Friedel–Crafts acylations do not.

Ortho- and Para-Directing Activators: Alkyl Groups

Inductive and resonance effects account for the directing ability of substituents as well as for their activating or deactivating ability. Take alkyl groups, for example, which have an electron-donating inductive effect and behave as ortho and para directors. The results of toluene nitration are shown in Figure 16.11.

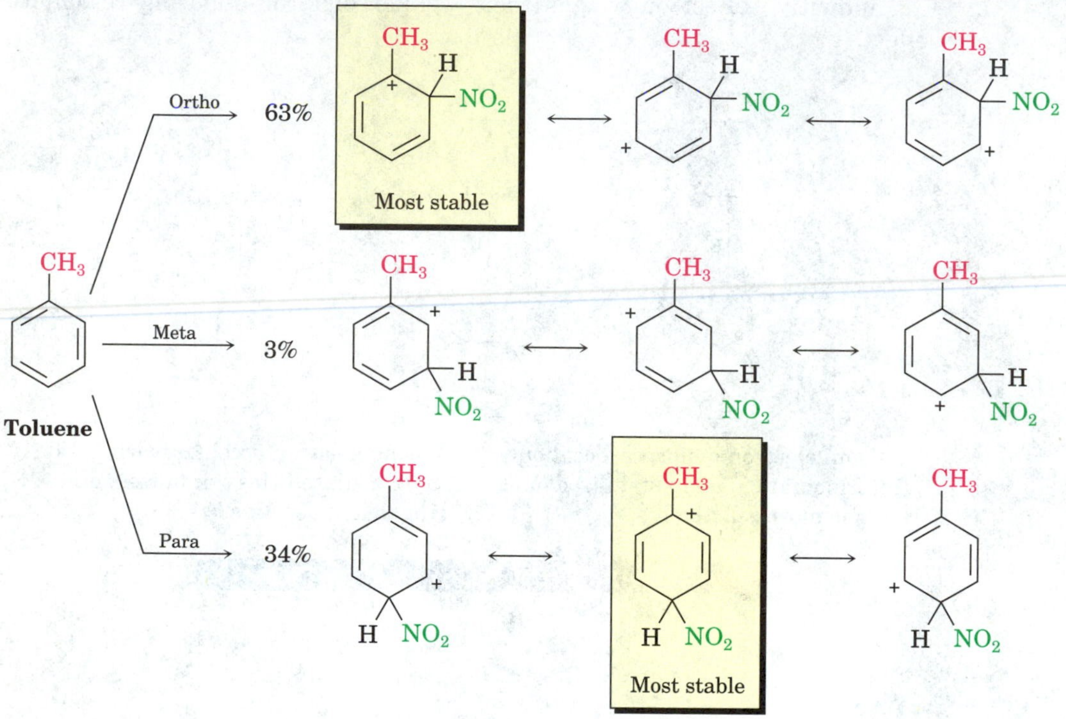

Figure 16.11 Carbocation intermediates in the nitration of toluene. Ortho and para intermediates are more stable than the meta intermediate.

Nitration of toluene might occur either ortho, meta, or para to the methyl group, giving the three carbocation intermediates shown in Figure 16.11. Although all three intermediates are resonance stabilized, the ortho and para intermediates are *most* stabilized. For both ortho and para attack, but not for meta, a resonance form places the positive charge directly on the methyl-substituted carbon, where it is in a tertiary position and can best be stabilized by the electron-donating inductive effect of the methyl group. The ortho and para intermediates are lower in energy than the meta intermediate and therefore form faster.

PROBLEM. . . . .

16.14 Which would you expect to be more reactive toward electrophilic substitution, toluene or (trifluoromethyl)benzene? Explain.

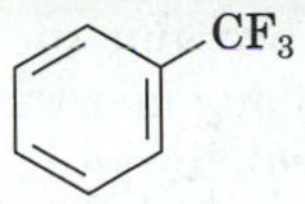

(Trifluoromethyl)benzene

Ortho- and Para-Directing Activators: OH and NH_2

Hydroxyl, alkoxyl, and amino groups are also ortho–para activators, but for a different reason than for alkyl groups. As mentioned in the previous section, hydroxyl, alkoxyl, and amino groups have a strong, electron-donating resonance effect that is most pronounced at the ortho and para positions and that outweighs a weaker electron-withdrawing inductive effect.

When phenol is nitrated, only ortho and para attack is observed, as shown in Figure 16.12. All three possible carbocation intermediates are

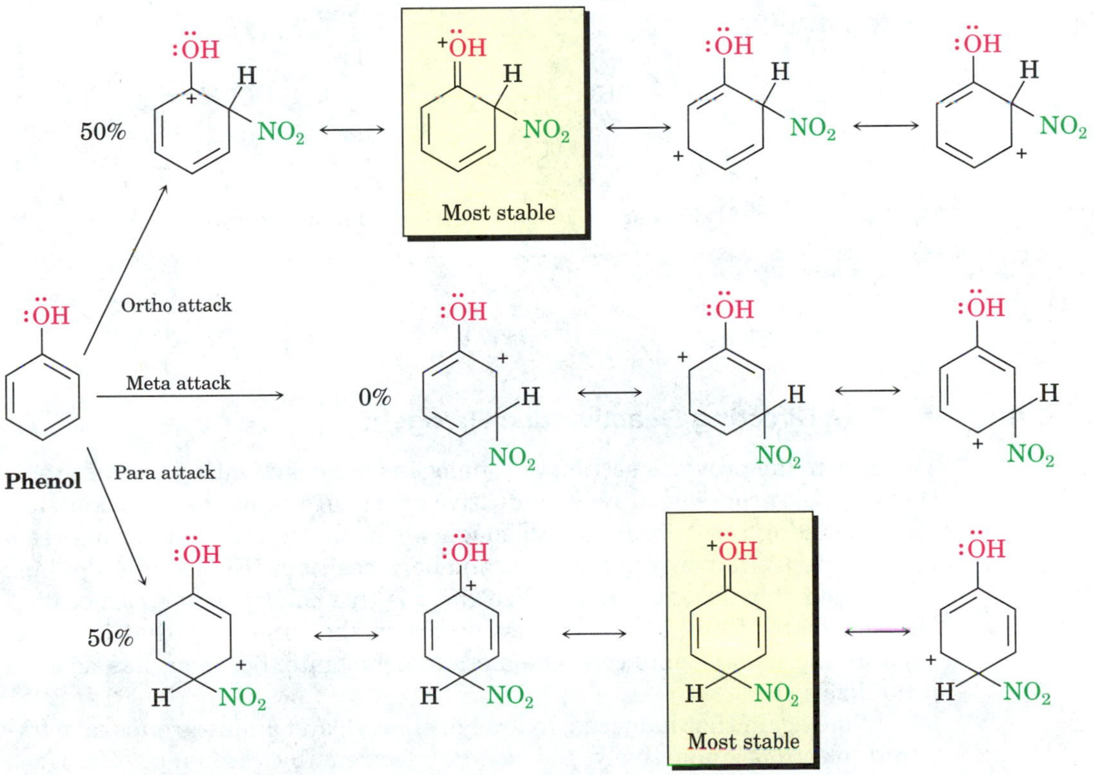

Figure 16.12 Carbocation intermediates in the nitration of phenol. The ortho and para intermediates are more stable than the meta intermediate because of resonance donation of electrons from oxygen.

stabilized by resonance, but the intermediates from ortho and para attack are stabilized most. Only in ortho and para attack are there resonance forms in which the positive charge is stabilized by donation of an electron pair from oxygen. The intermediate from meta attack has no such stabilization.

PROBLEM...

16.15 Acetanilide is less reactive than aniline toward electrophilic substitution. Explain.

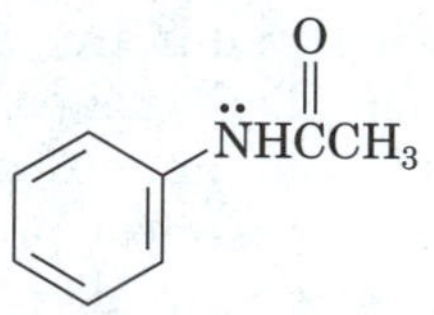

Acetanilide

PROBLEM...

16.16 What reactivity order toward electrophilic substitution would you expect for phenol, phenoxide ion, and phenyl acetate? Explain.

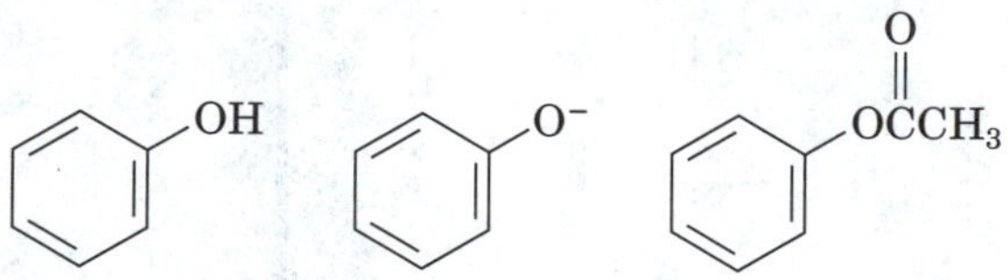

Phenol **Phenoxide ion** **Phenyl acetate**

....

Ortho- and Para-Directing Deactivators: Halogens

We saw in the previous section that halogens are deactivating because their stronger electron-withdrawing inductive effect outweighs their weaker electron-donating resonance effect. Though weak, that electron-donating resonance effect is felt only at the ortho and para positions (Figure 16.13). Thus, a halogen substituent can stabilize the positive charge of the carbocation intermediates from ortho and para attack in the same way that hydroxyl and amino substituents can. The meta intermediate, however, has no such stabilization.

Note again that halogens, hydroxyl, alkoxyl, and amino groups all *withdraw* electrons inductively and *donate* electrons by resonance. Halogens have a stronger electron-withdrawing inductive effect but a weaker electron-donating resonance effect and are thus deactivators. Hydroxyl, alkoxyl, and amino groups have a weaker electron-withdrawing inductive effect but a stronger electron-donating resonance effect and are thus activators. All are

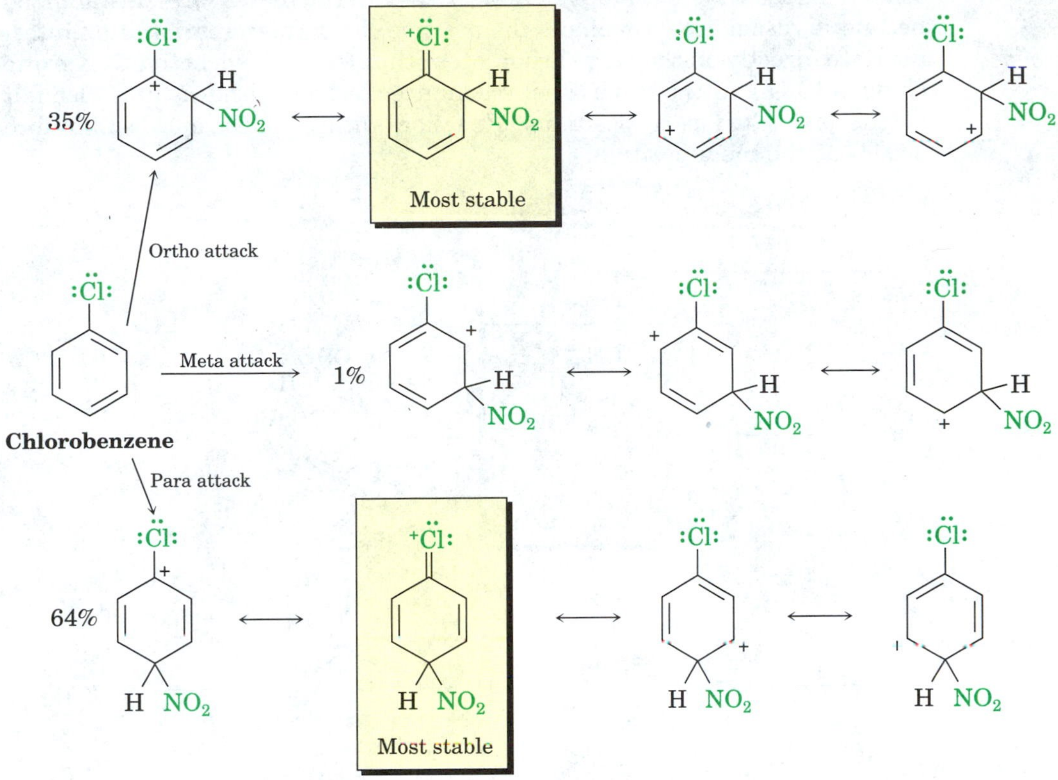

Figure 16.13 Carbocation intermediates in the nitration of chlorobenzene. The ortho and para intermediates are more stable than the meta intermediate because of electron donation of the halogen lone-pair electrons.

ortho and para directors because of the lone pair of electrons on the atom bonded to the aromatic ring.

PROBLEM.

16.17 The nitroso group, –N=O, is one of the few nonhalogens that is an ortho- and para-directing deactivator. Explain by drawing resonance structures of the carbocation intermediates in ortho, meta, and para electrophilic attack on nitrosobenzene, $C_6H_5N{=}O$.

Meta-Directing Deactivators

The influence of meta directors can be explained by the same kinds of arguments used for ortho and para directors. Meta-directing deactivators act through a combination of inductive and resonance effects that reinforce each other. Inductively, both ortho and para intermediates are destabilized because a resonance form places the positive charge of the carbocation intermediate directly on the ring carbon atom that bears the deactivating group (Figure 16.14). At the same time, resonance electron withdrawal is also felt at the ortho and para positions. Reaction with an electrophile therefore occurs at the meta position.

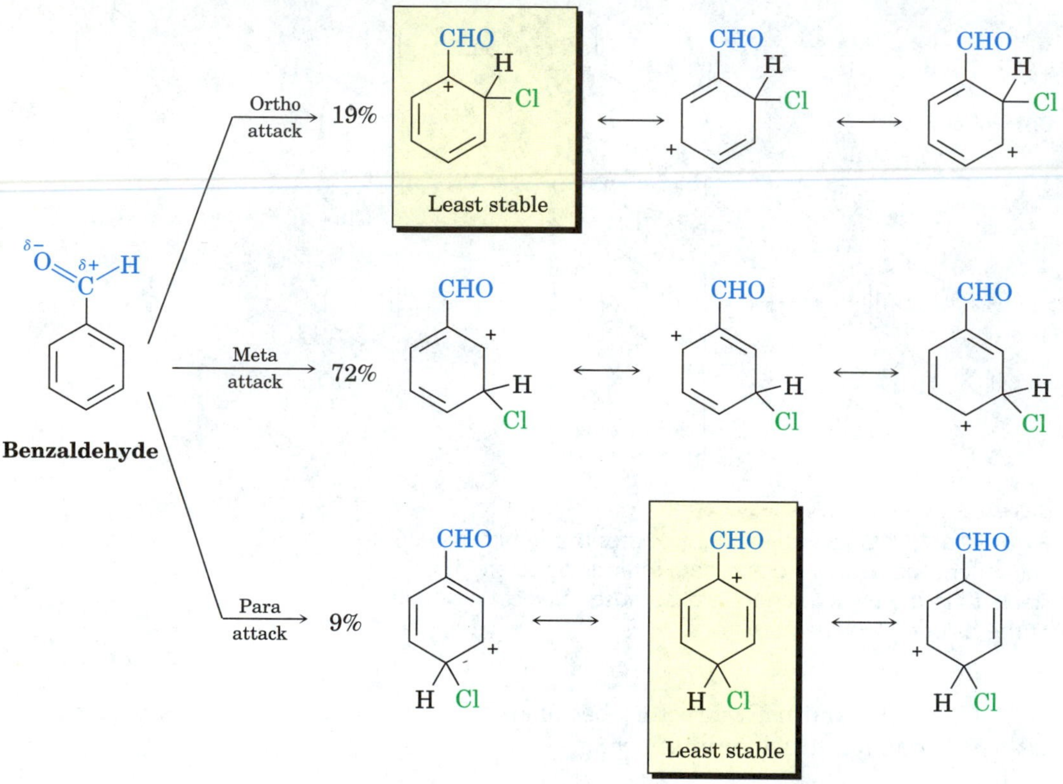

Figure 16.14 Carbocation intermediates in the chlorination of benzaldehyde. The meta intermediate is more stable than the ortho or para intermediate.

A Summary of Substituent Effects in Aromatic Substitution

A summary of the activating and directing effects of substituents in electrophilic aromatic substitution is shown in Table 16.2.

Table 16.2 Substituent Effects in Electrophilic Aromatic Substitution

Substituent	*Reactivity*	*Orientation*	*Inductive effect*	*Resonance effect*
$-CH_3$	Activating	Ortho, para	Weak; electron-donating	None
$-\ddot{O}H$, $-\ddot{N}H_2$	Activating	Ortho, para	Weak; electron-withdrawing	Strong; electron-donating
$-F$:, $-Cl$:, $-Br$:, $-I$:	Deactivating	Ortho, para	Strong; electron-withdrawing	Weak; electron-donating
$-\overset{+}{N}(CH_3)_3$	Deactivating	Meta	Strong; electron-withdrawing	None
$-NO_2$, $-CN$, $-CHO$, $-CO_2CH_3$, $-COCH_3$	Deactivating	Meta	Strong; electron-withdrawing	Strong; electron-withdrawing

PROBLEM

16.18 Draw resonance structures for the intermediates from attack of an electrophile at the ortho, meta, and para positions of nitrobenzene. Which intermediates are most favored?

16.7 Trisubstituted Benzenes: Additivity of Effects

Further electrophilic substitution of a disubstituted benzene is governed by the same resonance and inductive effects just discussed. The only difference is that it's now necessary to consider the additive effects of two different groups. In practice, this isn't as difficult as it sounds; three rules are usually sufficient:

1. If the directing effects of the two groups reinforce each other, there is no problem. In *p*-nitrotoluene, for example, both the methyl and the nitro group direct further substitution to the same position (ortho to the methyl = meta to the nitro). A single product is thus formed by electrophilic substitution.

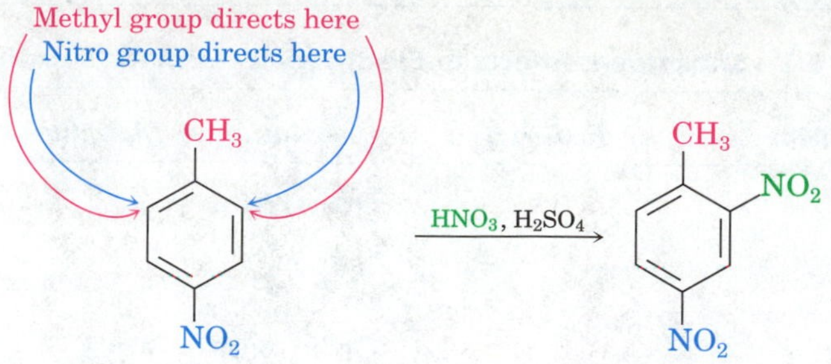

***p*-Nitrotoluene** **2,4-Dinitrotoluene**

2. If the directing effects of the two groups oppose each other, the more powerful activating group has the dominant influence, but mixtures of products often result. For example, bromination of *p*-methylphenol yields primarily 2-bromo-4-methylphenol because hydroxyl is a more powerful activator than methyl.

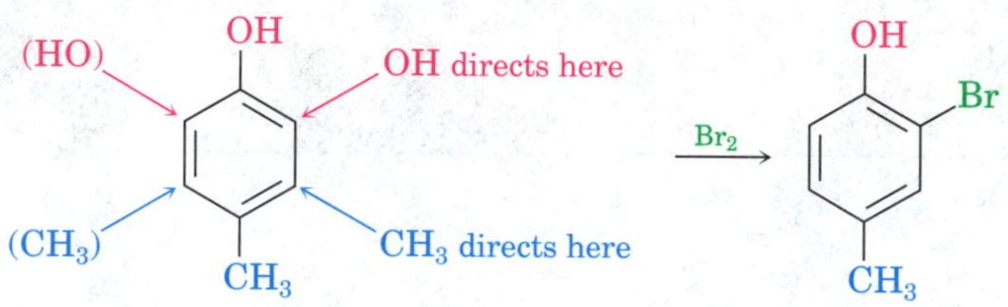

***p*-Methylphenol (*p*-Cresol)** **2-Bromo-4-methylphenol (major product)**

3. Further substitution rarely occurs between the two groups in a meta-disubstituted compound because this site is too hindered.

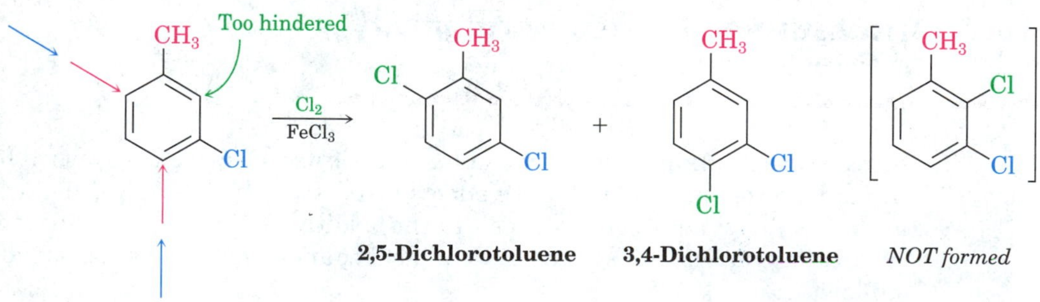

2,5-Dichlorotoluene **3,4-Dichlorotoluene** *NOT formed*

***m*-Chlorotoluene**

PRACTICE PROBLEM

What product would you expect from bromination of *p*-methylbenzoic acid?

Solution The carboxyl group (–COOH) is a meta director, and the methyl group is an ortho and para director. Both groups direct bromination to the position next to the methyl group, yielding 3-bromo-4-methylbenzoic acid.

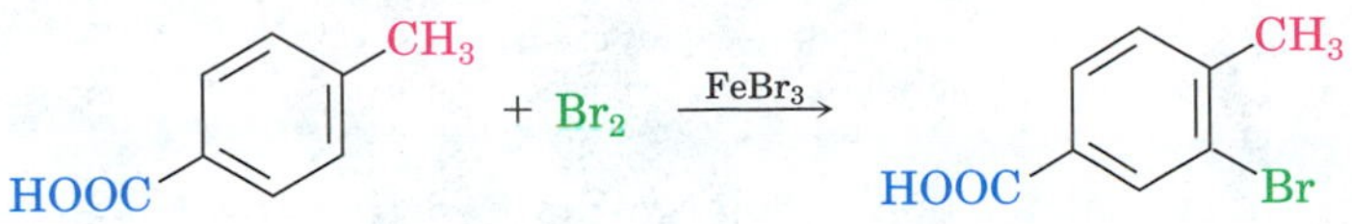

***p*-Methylbenzoic acid** **3-Bromo-4-methylbenzoic acid**

PROBLEM

16.19 Where would you expect electrophilic substitution to occur in these substances?

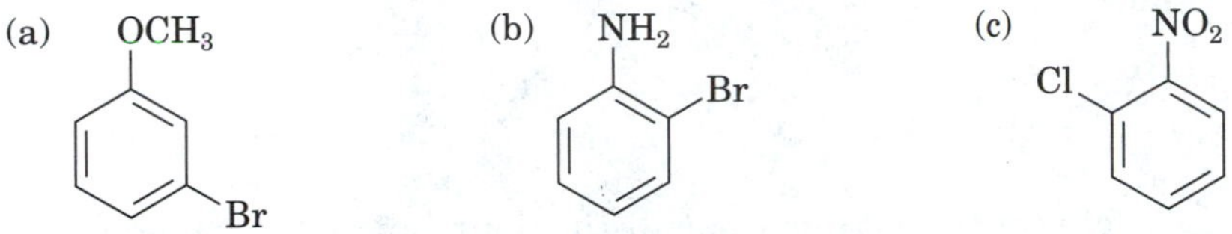

16.8 Nucleophilic Aromatic Substitution

Aromatic substitution reactions usually occur by an electrophilic mechanism. Aryl halides that have electron-withdrawing substituents, however, can also undergo *nucleophilic* aromatic substitution. For example, 2,4,6-trinitrochlorobenzene reacts with aqueous NaOH at room temperature to give 2,4,6-trinitrophenol in 100% yield. The nucleophile OH^- has substituted for Cl^-.

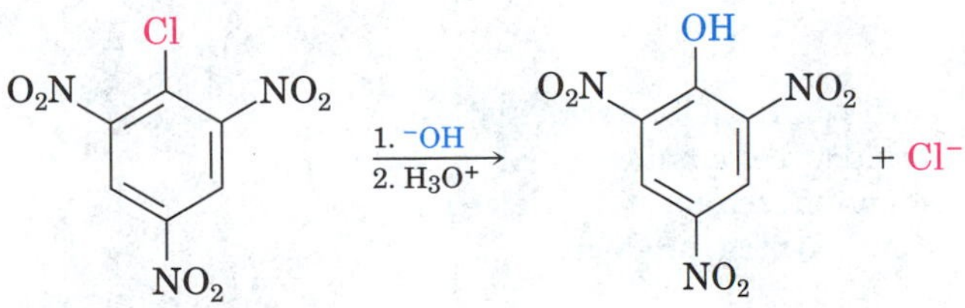

2,4,6-Trinitrochlorobenzene **2,4,6-Trinitrophenol (100%)**

How does this reaction take place? Although it appears superficially similar to the S_N1 and S_N2 nucleophilic substitution reactions of alkyl halides discussed in Chapter 11, it must somehow be different because aryl halides are inert to S_N1 and S_N2 conditions. Aryl halides don't undergo S_N1 reactions because aryl cations are relatively unstable. The dissociation of an aryl halide is therefore energetically unfavorable and does not occur easily.

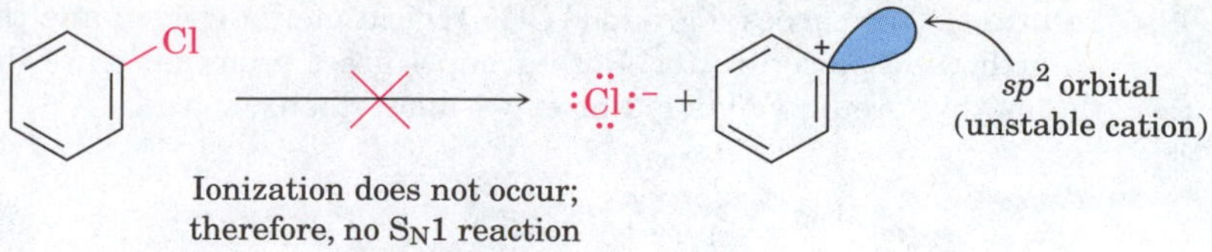

Aryl halides don't undergo S_N2 reactions because the halo-substituted carbon atom is sterically shielded from back-side attack by the aromatic ring. For a nucleophile to attack an aryl halide, it would have to approach directly through the aromatic ring and invert the stereochemistry of the aromatic ring—a geometric impossibility.

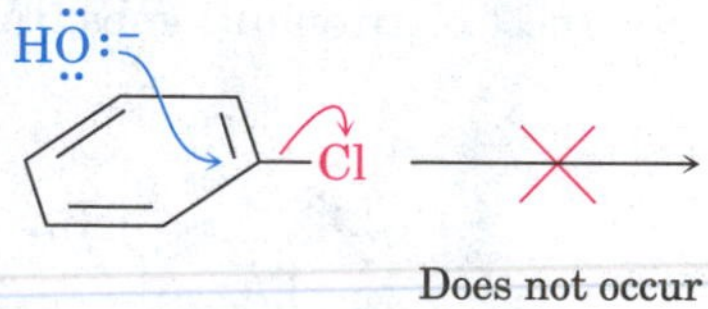

Nucleophilic aromatic substitutions proceed by the *addition/elimination* mechanism shown in Figure 16.15. The attacking nucleophile first adds to the electron-deficient aryl halide, forming a negatively charged intermediate called a *Meisenheimer*[4] *complex*. Halide ion is then eliminated in the second step.

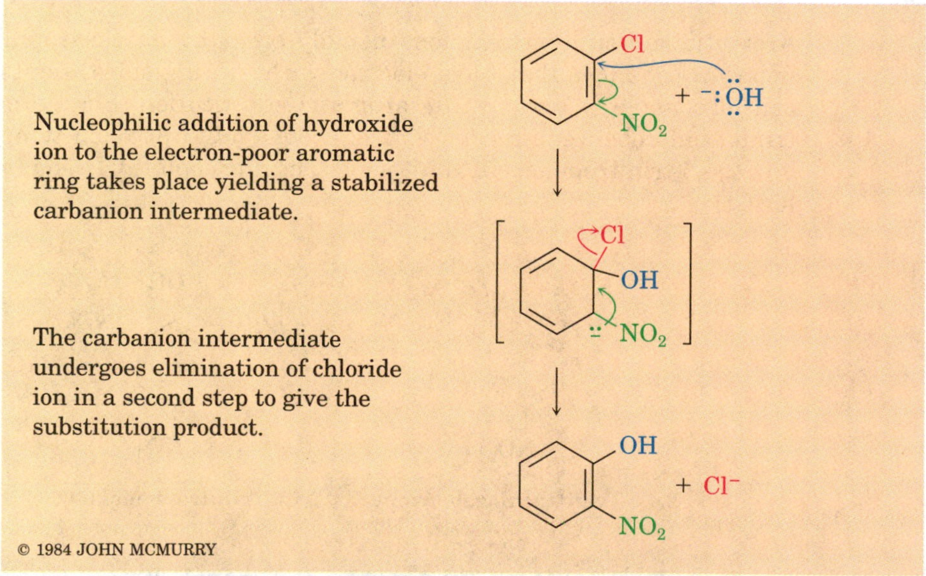

Figure 16.15 Mechanism of aromatic nucleophilic substitution. The reaction occurs in two steps and involves a stabilized carbanion intermediate.

[4]Jacob Meisenheimer (1876–1934); b. Greisheim; Ph.D. Munich; professor, universities of Berlin, Greifswald, Tübingen.

Nucleophilic aromatic substitution occurs only if the aromatic ring has an electron-withdrawing substituent in a position ortho or para to the halogen. The more such substituents there are, the faster the reaction goes. As shown in Figure 16.16, only ortho and para electron-withdrawing substituents can stabilize the anion intermediate through resonance; a meta substituent offers no such resonance stabilization. Thus, *p*-chloronitrobenzene and *o*-chloronitrobenzene react with hydroxide ion at 130°C to yield substitution products, but *m*-chloronitrobenzene is inert to OH^-.

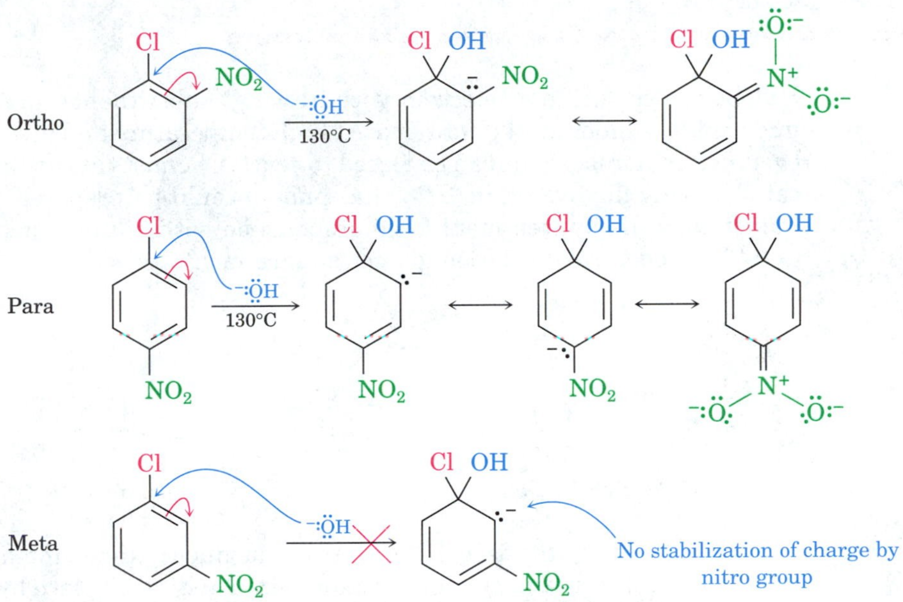

Figure 16.16 Nucleophilic aromatic substitution on nitrochlorobenzenes. Only the ortho and para isomers undergo reaction.

Note the different characteristics of electrophilic and nucleophilic aromatic substitutions: *Electrophilic* substitutions are favored by electron-*donating* substituents, while *nucleophilic* substitutions are favored by electron-*withdrawing* substituents. The electron-withdrawing groups that deactivate rings for electrophilic substitution (nitro, carbonyl, cyano, and so on) activate them for nucleophilic substitution. What's more, these groups are meta directors in electrophilic substitution, but are ortho–para directors in nucleophilic substitution.

PROBLEM.

16.20 Propose a mechanism for the reaction of 1-chloroanthraquinone with methoxide ion to give the substitution product 1-methoxyanthraquinone. (See the top of the next page.)

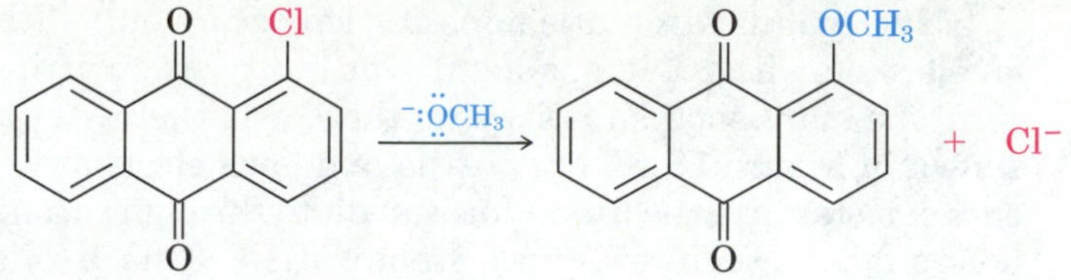

1-Chloroanthraquinone 1-Methoxyanthraquinone

16.9 Benzyne

Halobenzenes without electron-withdrawing substituents are inert to nucleophiles under most conditions. At high temperature and pressure, however, even chlorobenzene can be forced to react. Chemists at the Dow Chemical Company discovered in 1928 that phenol could be prepared on a large industrial scale by treatment of chlorobenzene with dilute aqueous NaOH at 340°C under 2500 psi (pounds per square inch) pressure.

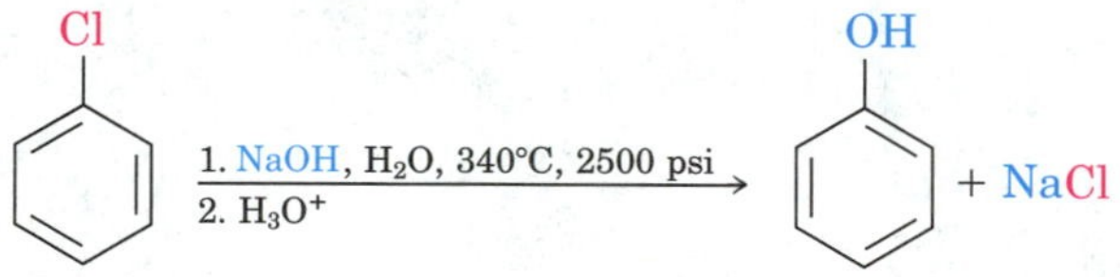

Chlorobenzene Phenol

This phenol synthesis is different from the nucleophilic aromatic substitutions discussed in the previous section because it takes place by an *elimination/addition* mechanism rather than an addition/elimination. Strong base first causes the elimination of HX from halobenzene in an E2 reaction, yielding a highly reactive **benzyne** intermediate, and a nucleophile then adds to benzyne in a second step to give the product. The two steps are similar to those in other nucleophilic aromatic substitutions, but their order is reversed: elimination before addition for the benzyne reaction rather than addition before elimination for the usual reaction.

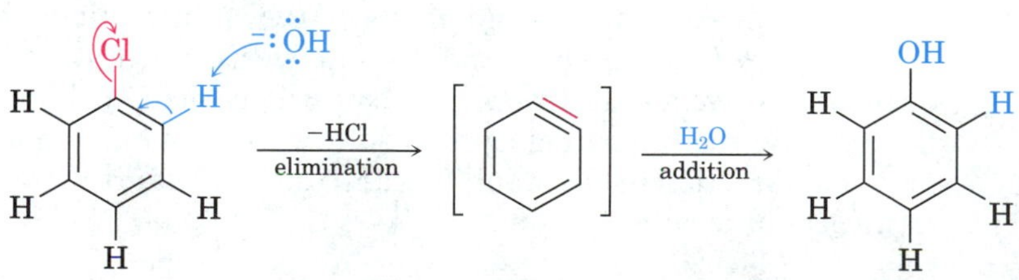

Chlorobenzene Benzyne Phenol

Evidence supporting the benzyne mechanism has been obtained by studying the reaction between bromobenzene and the strong base KNH_2 in NH_3 solvent. When bromobenzene labeled with radioactive ^{14}C at the C1 position is used, the substitution product has the label scrambled between C1 and C2. The reaction must therefore proceed through a symmetrical intermediate in which C1 and C2 are equivalent—a requirement that only benzyne can meet.

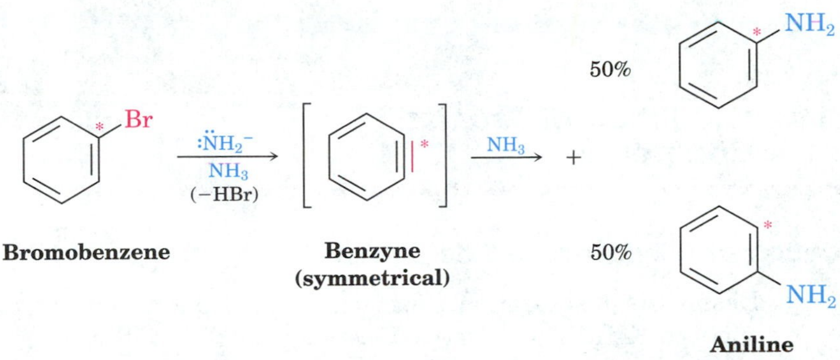

Further evidence for a benzyne intermediate comes from trapping experiments. Although benzyne is too reactive to be isolated as a pure compound, it can be intercepted in a Diels–Alder reaction if a diene such as furan is present when benzyne is generated.

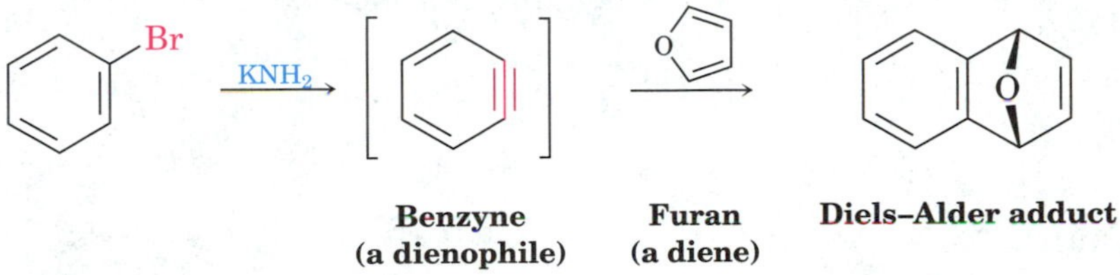

The electronic structure of benzyne, shown in Figure 16.17, is that of a highly distorted alkyne. Although a typical alkyne triple bond has two mutually perpendicular π bonds formed by p–p overlap, the benzyne triple bond has one π bond formed by p–p overlap and one π bond formed by sp^2–sp^2 overlap. The latter π bond is in the plane of the ring and is very weak.

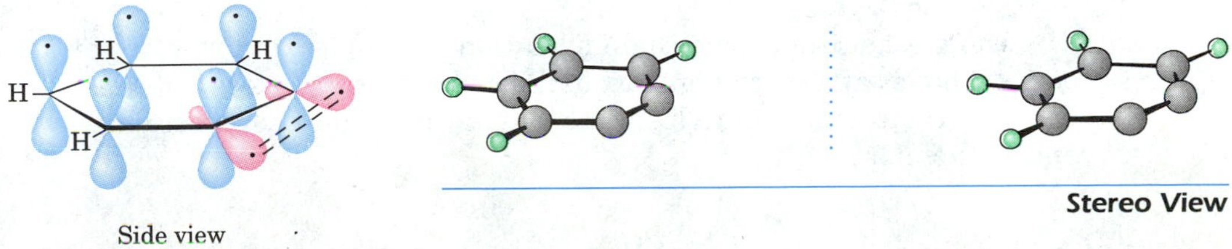

Figure 16.17 An orbital picture of benzyne. The benzyne carbons are sp^2-hybridized, and the "third" bond results from weak overlap of two adjacent sp^2 orbitals.

PROBLEM

16.21 Account for the fact that treatment of *p*-bromotoluene with NaOH at 300°C yields a mixture of two products, but treatment of *m*-bromotoluene with NaOH yields a mixture of three products.

16.10 Oxidation of Aromatic Compounds

Oxidation of Alkylbenzene Side Chains

Despite its unsaturation, the benzene ring is inert to strong oxidizing agents such as $KMnO_4$ and $Na_2Cr_2O_7$, reagents that will cleave alkene carbon–carbon bonds (Section 7.8). It turns out, however, that the presence of the aromatic ring has a dramatic effect on alkyl-group side chains. Alkyl side chains are readily attacked by oxidizing agents and are converted into carboxyl groups, –COOH. The net effect of side-chain oxidation is the conversion of an alkylbenzene into a benzoic acid, Ar–R → Ar–COOH. For example, *p*-nitrotoluene and butylbenzene are oxidized by aqueous $KMnO_4$ in high yield to give the corresponding benzoic acids.

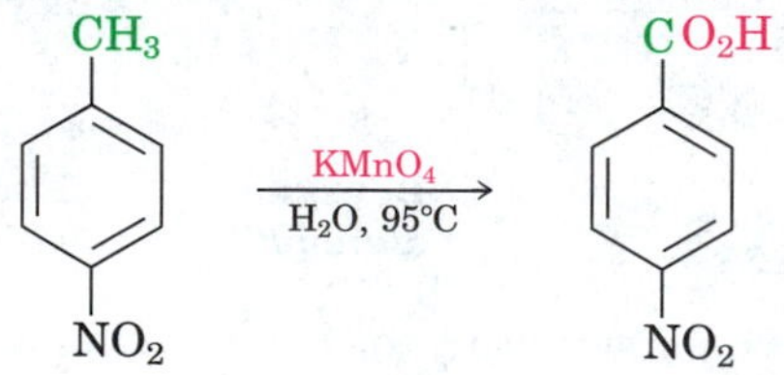

p-Nitrotoluene **p-Nitrobenzoic acid (88%)**

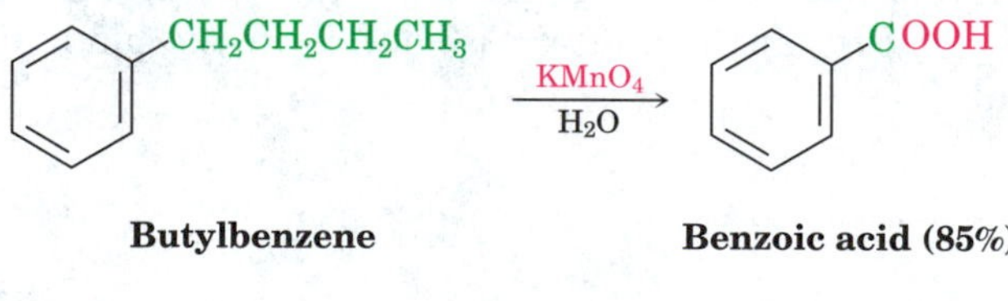

Butylbenzene **Benzoic acid (85%)**

The mechanism of side-chain oxidation is complex and involves attack on C–H bonds at the position next to the aromatic ring to form intermediate benzylic radicals. *tert*-Butylbenzene has no benzylic hydrogens, however, and is therefore inert.

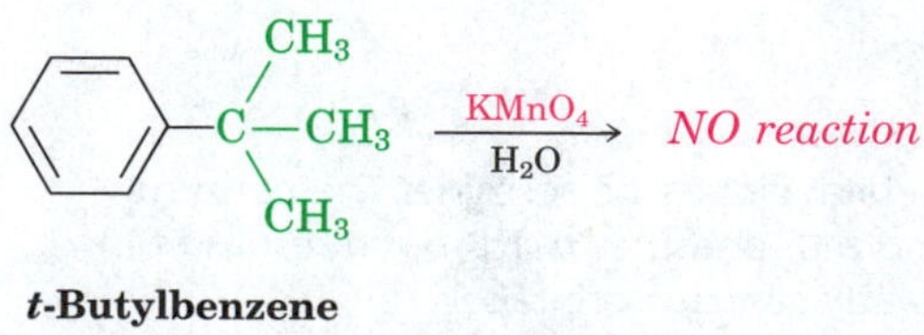

t-Butylbenzene

A similar oxidation is employed industrially for the preparation of terephthalic acid, used in the production of polyester fibers (Section 21.11). Approximately 5 million tons per year of *p*-xylene are oxidized, using air as the oxidant and Co(III) salts as catalyst.

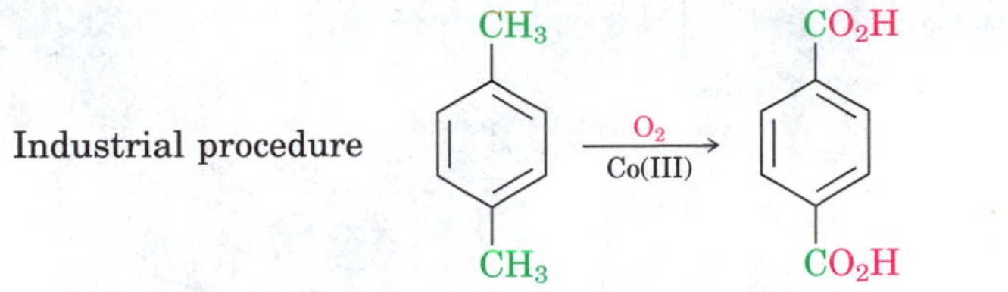

p-Xylene **Terephthalic acid**

PROBLEM.

16.22 What aromatic products would you obtain from the $KMnO_4$ oxidation of these substances?

(a) Tetralin, 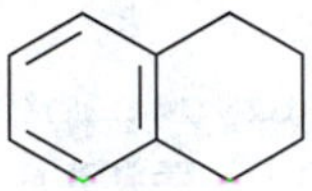 (b) *m*-Nitroisopropylbenzene

Bromination of Alkylbenzene Side Chains

Side-chain oxidation at the benzylic position also occurs when alkylbenzenes are treated with *N*-bromosuccinimide (NBS). For example, propylbenzene gives (1-bromopropyl)benzene in 97% yield on reaction with NBS in the presence of benzoyl peroxide, $(PhCO_2)_2$, as a radical initiator. Bromination occurs exclusively in the benzylic position and does not give a mixture of products.

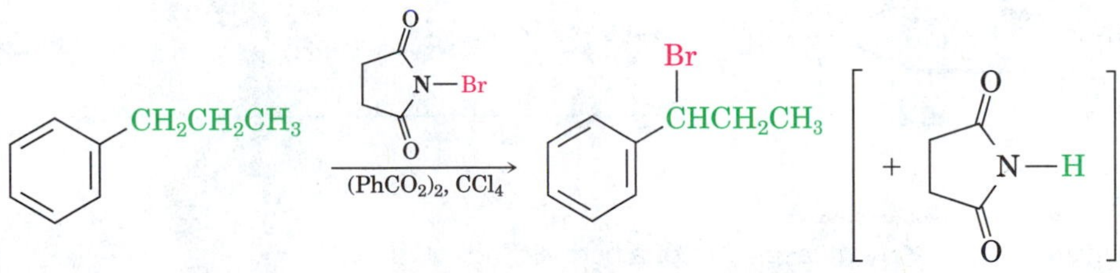

Propylbenzene **(1-Bromopropyl)benzene (97%)**

The mechanism of benzylic bromination is similar to that discussed in Section 10.5 for allylic bromination of alkenes and involves abstraction of a benzylic hydrogen atom to generate an intermediate benzyl radical. The radical then reacts with Br_2 to yield product and a Br· radical, which cycles back into the reaction to carry on the chain. The Br_2 necessary for reaction with the benzyl radical is produced by a concurrent reaction of HBr with NBS, as shown in Figure 16.18.

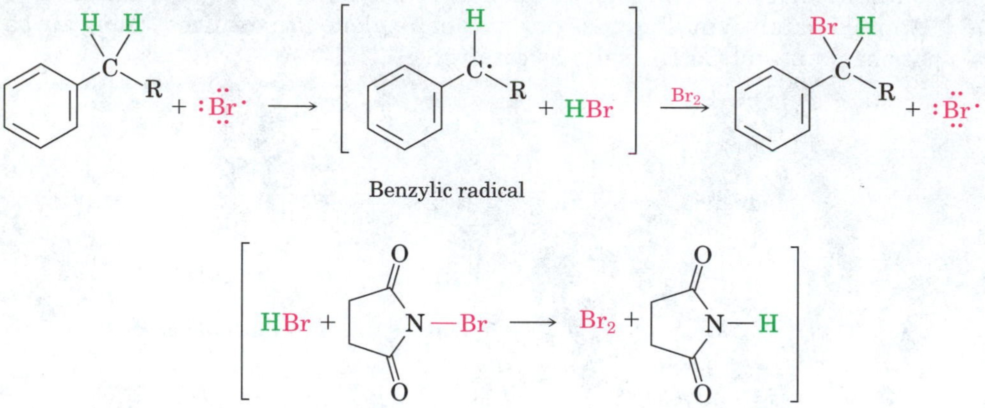

Figure 16.18 Mechanism of benzylic bromination with *N*-bromosuccinimide.

Reaction occurs exclusively at the benzylic position because the benzylic radical intermediate is highly stabilized by resonance. Figure 16.19 shows an orbital view of the benzyl radical, indicating how it is stabilized by overlap of its *p* orbital with the ring π electron system.

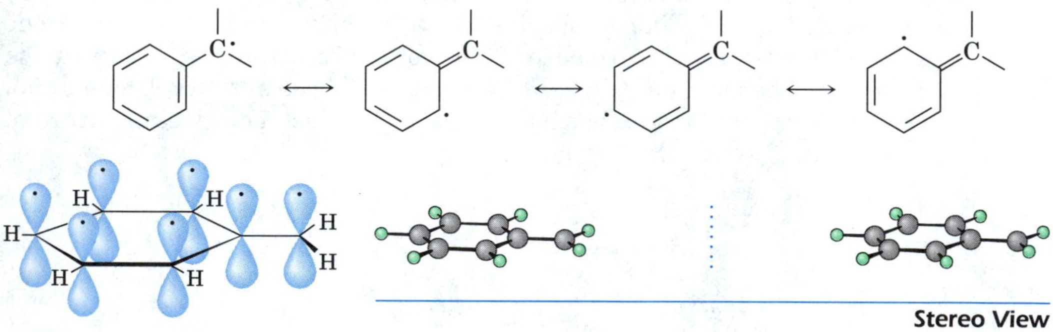

Figure 16.19 An orbital picture of a benzylic radical, showing the overlap of a side-chain *p* orbital with the aromatic ring orbitals.

PROBLEM. . . . .

16.23 Styrene, the simplest alkenylbenzene, is prepared commercially for use in plastics manufacture by catalytic dehydrogenation of ethylbenzene. How might you prepare styrene from benzene using reactions you've studied?

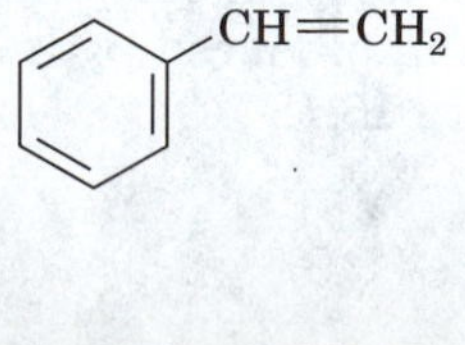

Styrene

PROBLEM

16.24 Refer to Table 5.4 for a quantitative idea of the stability of a benzyl radical. How much more stable (in kJ/mol) is the benzyl radical than a primary alkyl radical? How does a benzyl radical compare in stability to an allyl radical?

16.11 Reduction of Aromatic Compounds

Catalytic Hydrogenation of Aromatic Rings

Just as aromatic rings are inert to oxidation under most conditions, they're also inert to catalytic hydrogenation under conditions that reduce typical alkene double bonds. As a result, it's possible to selectively reduce alkene double bonds in the presence of aromatic rings. For example, 4-phenyl-3-buten-2-one is reduced to 4-phenyl-2-butanone when the reaction is carried out at room temperature and atmospheric pressure using a palladium catalyst. Neither the benzene ring nor the ketone carbonyl group is affected.

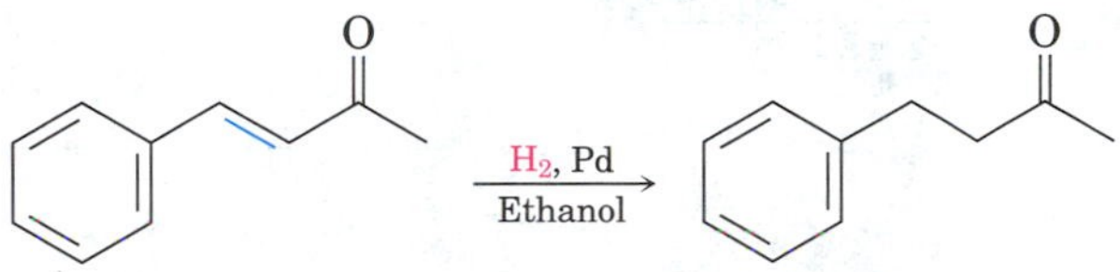

4-Phenyl-3-buten-2-one **4-Phenyl-2-butanone (100%)**

To hydrogenate an aromatic ring, it's necessary either to use a platinum catalyst with hydrogen gas at several hundred atmospheres pressure or to use a more powerful catalyst such as rhodium on carbon. Under these conditions, aromatic rings are readily reduced to cyclohexanes. For example, *o*-xylene yields 1,2-dimethylcyclohexane, and 4-*tert*-butylphenol gives 4-*tert*-butylcyclohexanol.

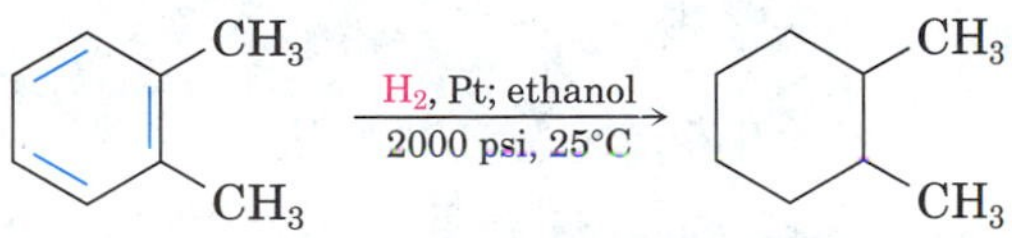

***o*-Xylene** **1,2-Dimethylcyclohexane (100%)**

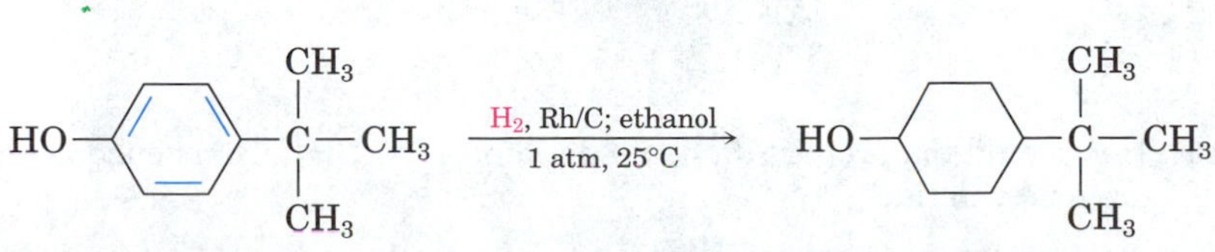

4-*tert*-Butylphenol **4-*tert*-Butylcyclohexanol (100%)**

Reduction of Aryl Alkyl Ketones

Just as an aromatic ring activates a neighboring (benzylic) C–H position toward oxidation, it also activates a neighboring carbonyl group toward reduction. Thus, an aryl alkyl ketone prepared by Friedel–Crafts acylation of an aromatic ring can be converted into an alkylbenzene by catalytic hydrogenation over a palladium catalyst. For example, propiophenone is reduced to propylbenzene in 100% yield by catalytic hydrogenation. Since the net effect of Friedel–Crafts acylation followed by reduction is the preparation of a primary alkylbenzene, this two-step sequence of reactions makes it possible to circumvent the carbocation rearrangement problems associated with direct Friedel–Crafts alkylation using primary alkyl halides.

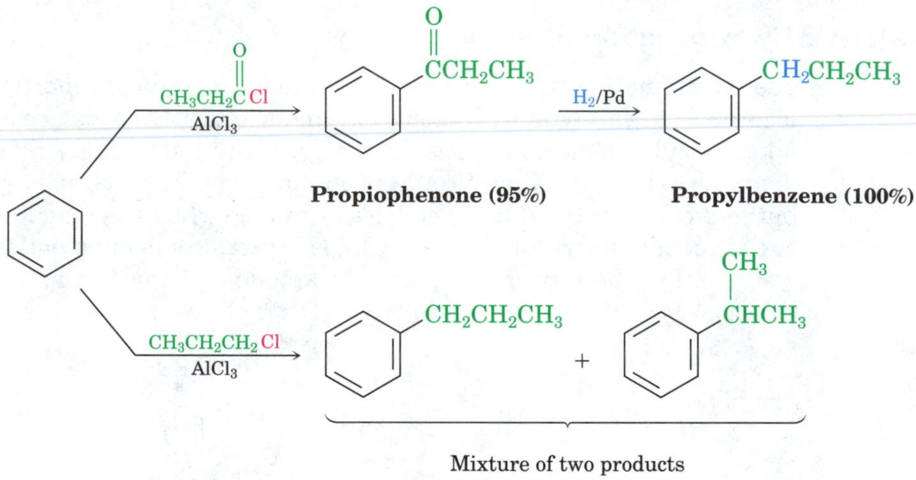

Note that the conversion of a carbonyl group into a methylene group ($C{=}O \rightarrow CH_2$) by catalytic hydrogenation is limited to *aryl* alkyl ketones; dialkyl ketones are not reduced under these conditions. It should also be pointed out that the catalytic reduction of aryl alkyl ketones is not compatible with the presence of a nitro substituent on the aromatic ring, because a nitro group is reduced to an amino group under the reaction conditions. We'll see a more general method for reducing all ketone carbonyl groups to yield alkanes in Section 19.13.

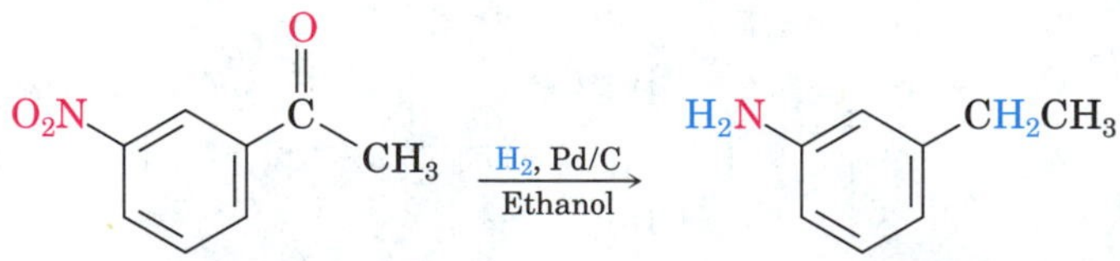

***m*-Nitroacetophenone** **_m_-Ethylaniline**

PROBLEM

16.25 How would you prepare diphenylmethane, $(Ph)_2CH_2$, from benzene and an appropriate acid chloride?

16.12 Synthesis of Substituted Benzenes

One of the surest ways to learn organic chemistry is to work synthesis problems. The ability to plan a successful multistep synthesis of a complex molecule requires a working knowledge of the uses and limitations of many hundreds of organic reactions. Not only must you know *which* reactions to use, you must also know *when* to use them. The order in which reactions are carried out is often critical to the success of the overall scheme.

The ability to plan a sequence of reactions in the right order is particularly valuable in the synthesis of substituted aromatic rings, where the introduction of a new substituent is strongly affected by the directing effects of other substituents. Planning syntheses of substituted aromatic compounds is therefore an excellent way to gain facility with the many reactions learned in the past few chapters.

During the previous discussion of strategies for working synthesis problems in Section 8.10, we said that it's usually best to work problems backward. Look at the target molecule and ask yourself, "What is an immediate precursor of this compound?" Choose a likely answer and continue working backward, one step at a time, until you arrive at a simple starting material. Let's try some examples.

PRACTICE PROBLEM

Synthesize *p*-bromobenzoic acid from benzene.

Solution Ask yourself, "What is an immediate precursor of *p*-bromobenzoic acid?"

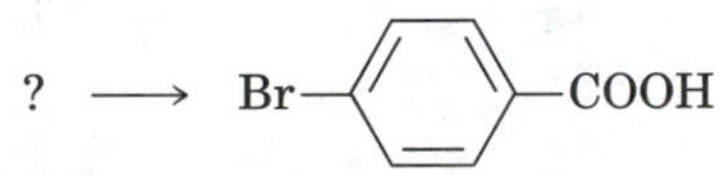

***p*-Bromobenzoic acid**

There are two substituents on the ring, a carboxyl group (COOH), which is meta-directing, and a bromine, which is ortho–para-directing. We can't brominate benzoic acid, because the wrong isomer (*m*-bromobenzoic acid) would be produced. We know, however, that oxidation of alkylbenzenes yields benzoic acids. Thus, an immediate precursor of our target molecule might be *p*-bromotoluene.

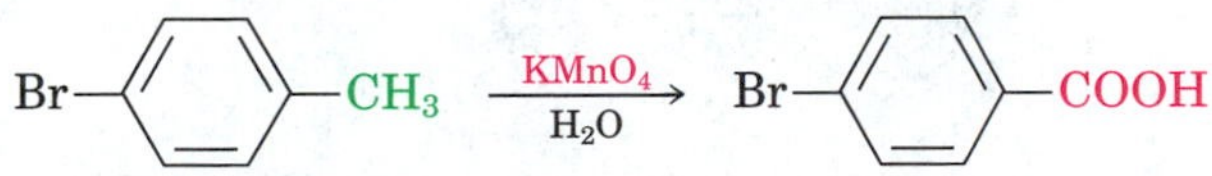

***p*-Bromotoluene** ***p*-Bromobenzoic acid**

Next ask yourself, "What is an immediate precursor of *p*-bromotoluene?" Perhaps toluene is an immediate precursor because the methyl group would direct bromination to the ortho and para positions, and the isomeric products could be separated. Alternatively, bromobenzene might be an immediate precursor because we could carry out a Friedel–Crafts methylation and obtain para product. Both answers are satisfactory, although, in view of the difficulties often observed with polyalkylation in Friedel–Crafts reactions, bromination of toluene may be the more efficient route.

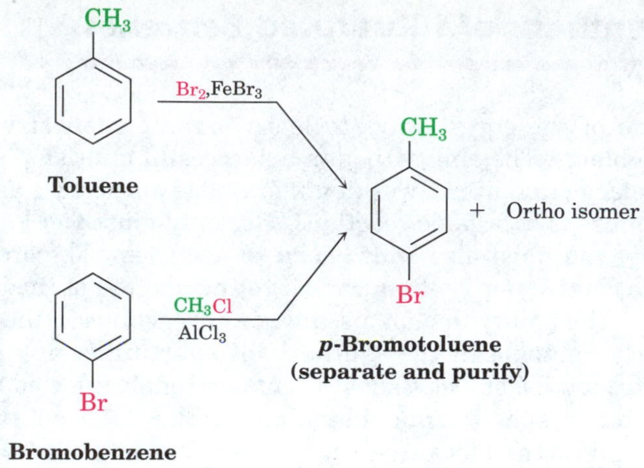

p-Bromotoluene (separate and purify)

"What is an immediate precursor of toluene?" Benzene, which could be methylated in a Friedel–Crafts reaction.

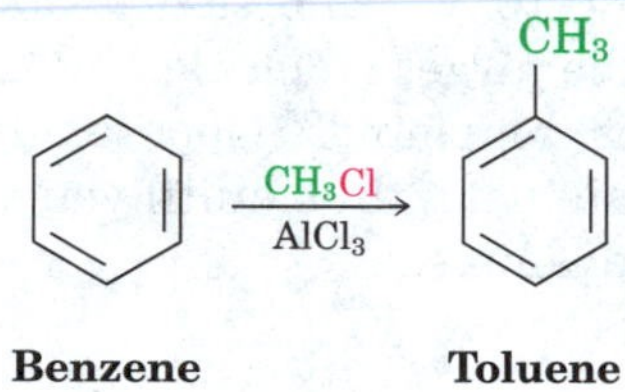

Alternatively, "What is an immediate precursor of bromobenzene?" Benzene, which could be brominated.

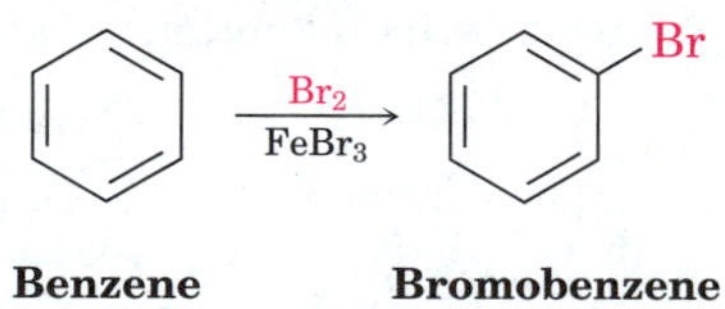

This backward synthetic (*retrosynthetic*) analysis has provided two valid routes from benzene to *p*-bromobenzoic acid (Figure 16.20).

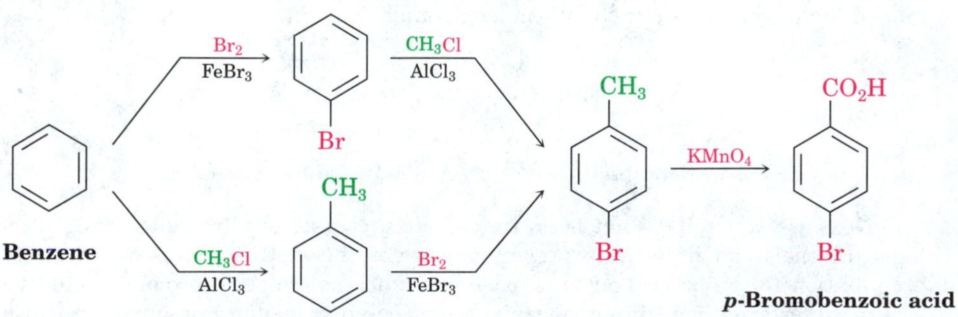

Figure 16.20 Two routes for the synthesis of *p*-bromobenzoic acid from benzene.

PRACTICE PROBLEM

Propose a synthesis of 4-chloro-1-nitro-2-propylbenzene from benzene.

Solution "What is an immediate precursor of the target?" Because the final step will involve introduction of one of three groups—chloro, nitro, or propyl—we have to consider three possibilities. Of the three, we know that chlorination of *o*-nitropropylbenzene can't be used because the reaction would occur at the wrong position. Similarly, a Friedel–Crafts reaction can't be used as the final step because these reactions don't work on nitro-substituted (deactivated) benzenes. Thus, the immediate precursor of our desired product is probably *m*-chloropropylbenzene, which can be nitrated. This nitration gives a mixture of product isomers, which must then be separated (Figure 16.21).

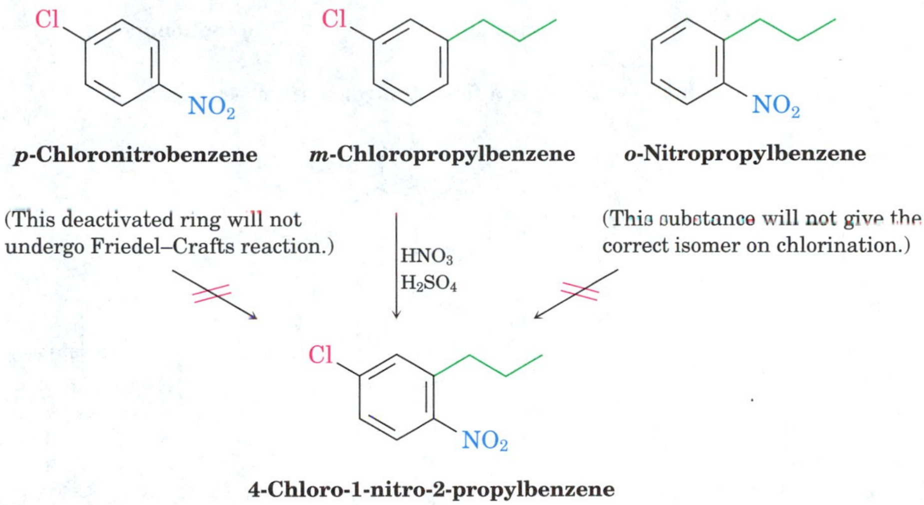

Figure 16.21 Possible routes for the synthesis of 4-chloro-1-nitro-2-propylbenzene.

"What is an immediate precursor of *m*-chloropropylbenzene?" Because the two substituents have a meta relationship, the first substituent placed on the ring must be a meta director so that the second substitution will take place at the proper position. Furthermore, because primary alkyl groups such as propyl can't be introduced directly by Friedel–Crafts alkylation, the precursor of *m*-chloropropylbenzene is probably *m*-chloropropiophenone, which could be catalytically reduced.

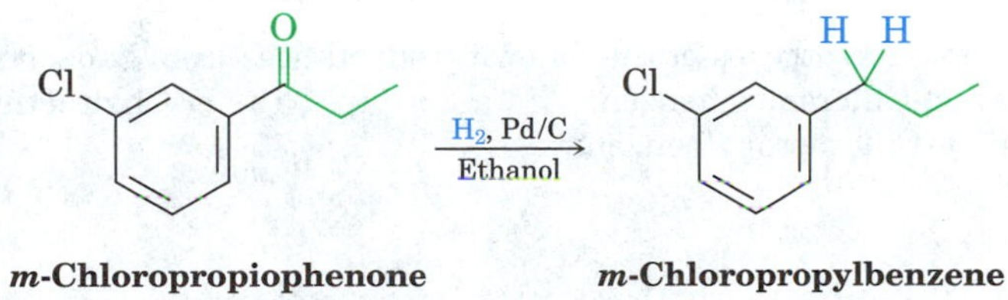

"What is an immediate precursor of *m*-chloropropiophenone?" Perhaps propiophenone, which could be chlorinated.

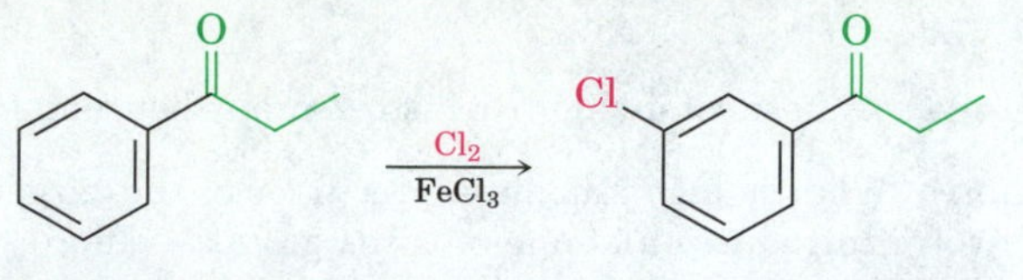

Propiophenone **m-Chloropropiophenone**

"What is an immediate precursor of propiophenone?" Benzene, which could undergo Friedel–Crafts acylation with propanoyl chloride and $AlCl_3$.

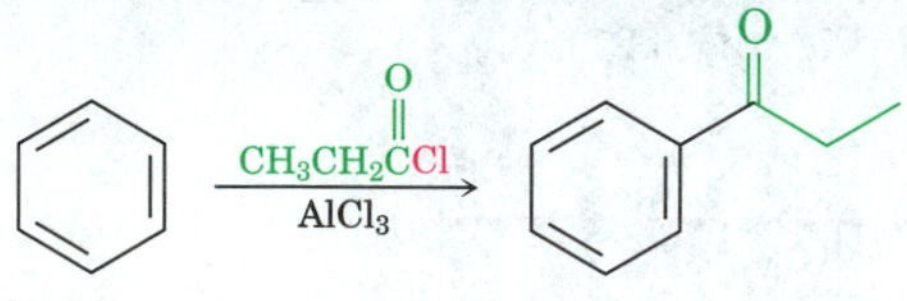

Benzene **Propiophenone**

The final synthesis is a four-step route from benzene:

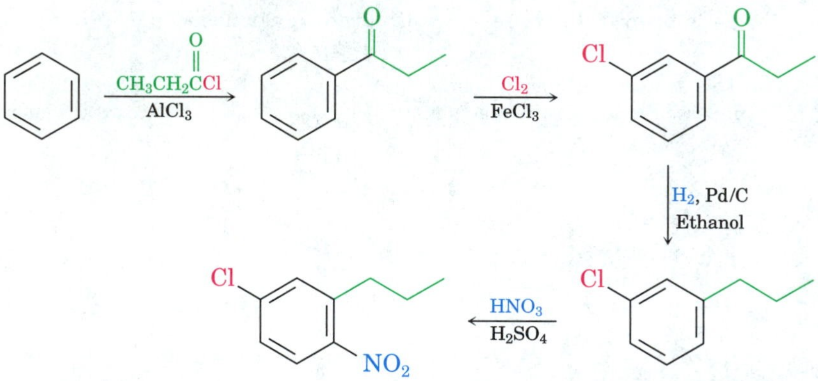

Planning organic syntheses has been compared to playing chess. There are no tricks; all that's required is a knowledge of the allowable moves (the organic reactions) and the discipline to evaluate carefully the consequences of each move. Practicing is not always easy, but there is no surer way to learn organic chemistry.

PROBLEM

16.26 Propose syntheses of the following substances from benzene.
(a) *m*-Chloronitrobenzene (b) *m*-Chloroethylbenzene
(c) *p*-Chloropropylbenzene

PROBLEM

16.27 In planning syntheses, it is as important to know what not to do as to know what to do. As written, the following reaction schemes have flaws in them. What is wrong with each?

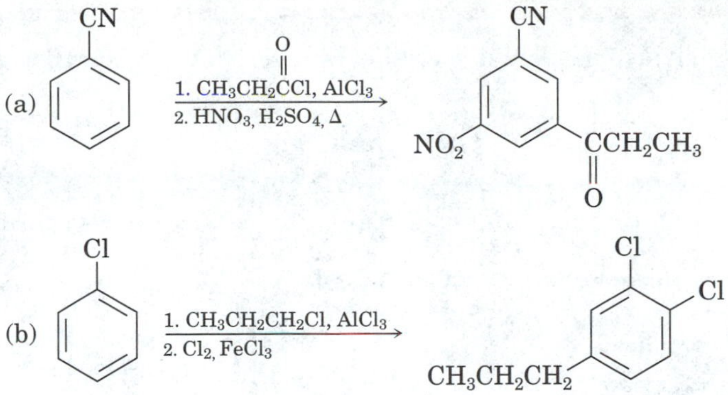

INTERLUDE

Aspirin and Other Aromatic NSAIDs

Long-distance runners sometimes call ibuprofen "the fifth basic food group" because of its usefulness in controlling aches and pains.

Whether from tennis elbow, a sprained ankle, or a wrenched knee, pain and inflammation seem to go together. They are, however, different in their origin, and powerful drugs are available for treating each separately. Codeine, for example, is a powerful *analgesic*, or pain reliever, used in the management of debilitating pain, while cortisone and related steroids are potent *anti-inflammatory* agents used for treating arthritis and other crippling inflammations. For minor pains and inflammation, both problems are often treated at the same time by using a common, over-the-counter medication called an *NSAID*, for *nonsteroidal anti-inflammatory drug*.

The most common NSAID is aspirin, or acetylsalicylic acid, whose use goes back to the late 1800s. It had been known from before the time of Hippocrates in 400 BC that fevers could be lowered by chewing the bark of willow trees. The active agent in willow bark was found in 1827 to be an aromatic compound called *salicin*, which could be converted by reaction with water (*hydrolysis*) to yield salicyl alcohol and then oxidized to give salicylic acid. Salicylic acid turned out to be even more effective than salicin for reducing fevers and to have analgesic and anti-inflammatory action as well. Unfortunately, it also turned out to be too corrosive to the walls of the stomach for everyday use. Conversion of the phenol –OH group into an acetate ester, however, yielded acetylsalicylic acid,

(continued)➤

which proved just as potent as salicylic acid but less corrosive to the stomach.

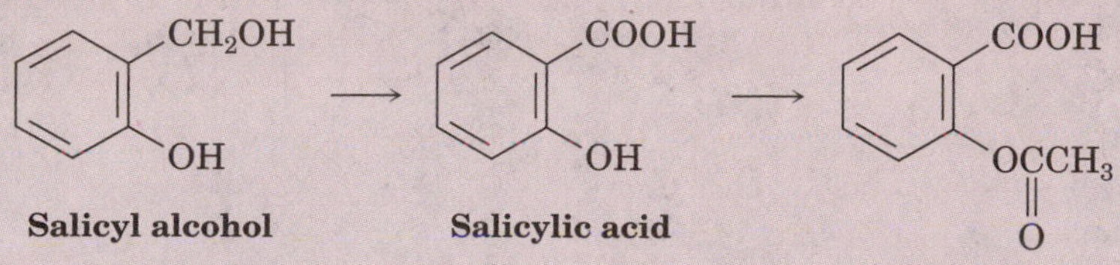

Salicyl alcohol **Salicylic acid** **Acetylsalicylic acid (Aspirin)**

Though extraordinary in its powers, aspirin is also more dangerous than commonly believed. Its toxicity is such that only about 15 g can be fatal to a small child, and it can cause stomach bleeding and allergic reactions in long-term users. Even more serious is a condition called *Reye's syndrome*, a potentially fatal reaction sometimes seen in children recovering from the flu. As a result of these problems, numerous other NSAIDs have been developed in the last two decades, most notably ibuprofen and naproxen. Like aspirin, both ibuprofen and naproxen are relatively simple aromatic compounds containing a side-chain carboxylic acid group. Ibuprofen, sold under the names Advil, Motrin, Nuprin, and others, has roughly the same potency as aspirin but is much less prone to cause stomach upset. Naproxen, sold under the names Naprosyn and Aleve, also has about the same potency as aspirin but remains active in the body six times longer.

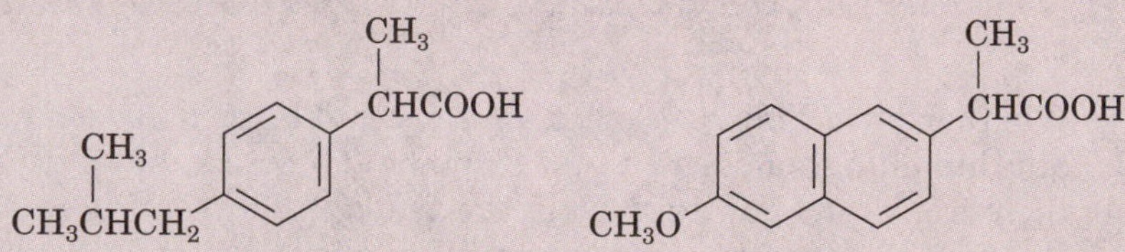

Ibuprofen (Advil, Motrin, Nuprin) **Naproxen (Naprosyn, Aleve)**

Summary and Key Words

Electrophilic aromatic substitution takes place in two steps—initial reaction of an electrophile, E^+, with the aromatic ring, followed by loss of H^+ from the resonance-stabilized carbocation intermediate to regenerate the aromatic ring:

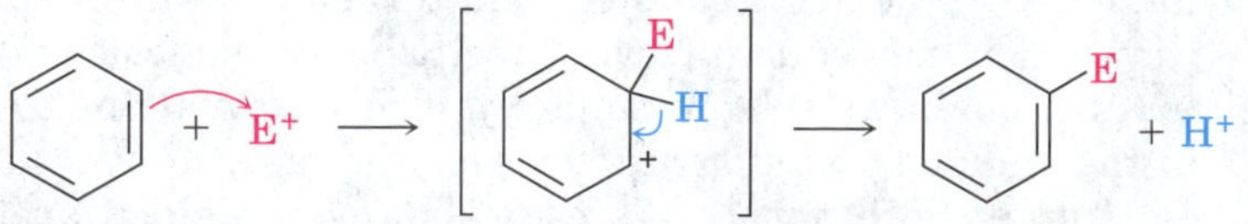

Many variations of the reaction can be carried out, including **bromination**, **chlorination**, **iodination**, **nitration**, and **sulfonation**. **Friedel–**

Crafts alkylation and **acylation**, which involve reaction of an aromatic ring with a carbocation electrophile, are particularly useful. Both are limited, however, by the fact that the aromatic ring must be at least as reactive as a halobenzene. In addition, polyalkylation and carbocation rearrangements often occur in Friedel–Crafts alkylation.

Substituents on the benzene ring affect both the reactivity of the ring toward further substitution and the orientation of that substitution. Groups can be classified as **ortho- and para-directing activators, ortho- and para-directing deactivators**, or **meta-directing deactivators**. Substituents influence aromatic rings by a combination of resonance and inductive effects. **Resonance effects** are transmitted through π bonds; **inductive effects** are transmitted through σ bonds.

Halobenzenes undergo **nucleophilic aromatic substitution** through either of two mechanisms. If the halobenzene has a strongly electron-withdrawing substituent in the ortho or para position, substitution occurs by addition of a nucleophile to the ring followed by elimination of halide from the intermediate anion. If the halobenzene is not activated by an electron-withdrawing substituent, substitution can occur by elimination of HX, followed by addition of a nucleophile to the intermediate **benzyne**.

The benzylic position of alkylbenzenes can be brominated by reaction with *N*-bromosuccinimide, and the entire side chain can be degraded to a carboxyl group by oxidation with aqueous $KMnO_4$. Although aromatic rings are less reactive than isolated alkene double bonds, they can be reduced to cyclohexanes by hydrogenation over a platinum or rhodium catalyst. In addition, aryl alkyl ketones are reduced to alkylbenzenes by hydrogenation over a platinum catalyst.

Summary of Reactions

1. Electrophilic aromatic substitution
   (a) Bromination (Section 16.1)

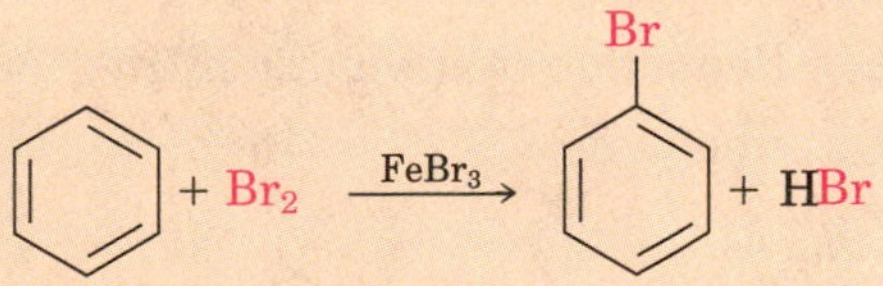

   (b) Chlorination (Section 16.2)

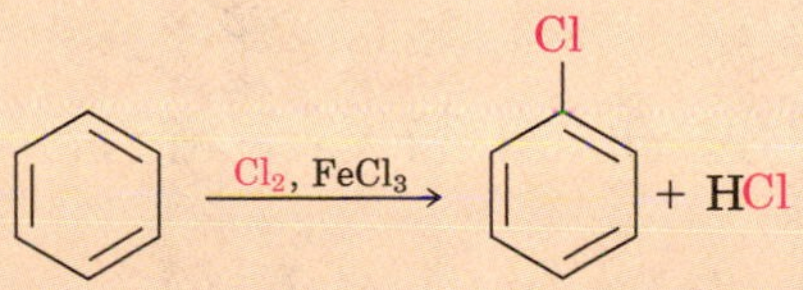

   (c) Iodination (Section 16.2)

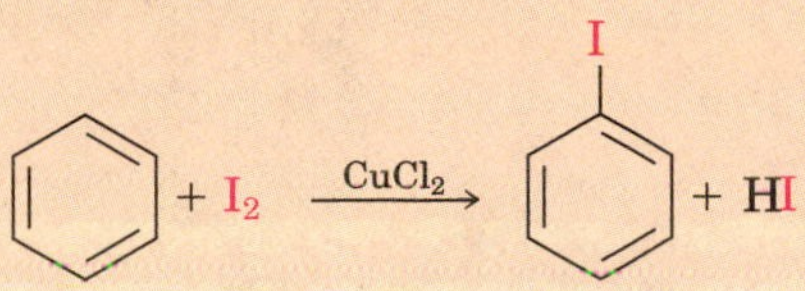

(continued)➤

(d) Nitration (Section 16.2)

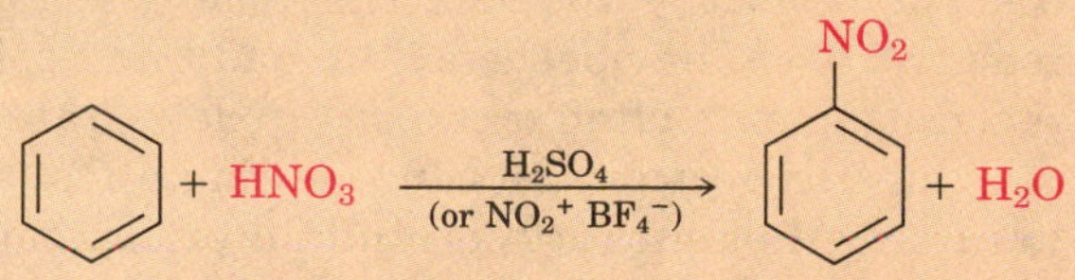

(e) Sulfonation (Section 16.2)

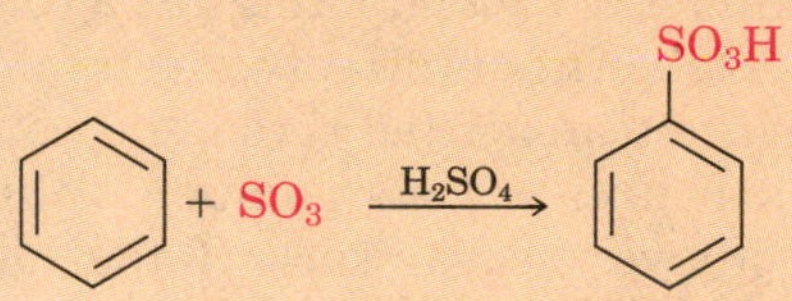

(f) Friedel–Crafts alkylation (Section 16.3)

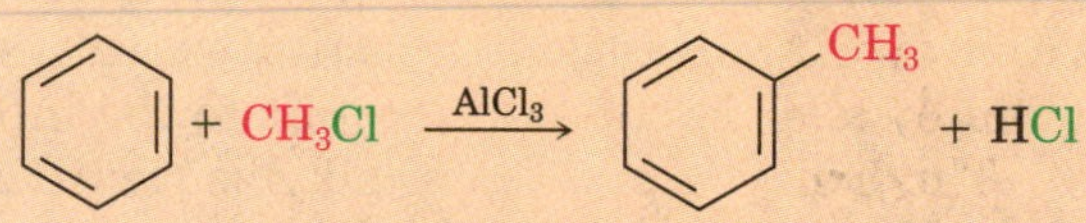

Aromatic ring: Must be at least as reactive as a halobenzene. Deactivated rings do not react.

Alkyl halide: Can be methyl, ethyl, 2°, or 3°; primary halides undergo carbocation rearrangement.

(g) Friedel–Crafts acylation (Section 16.4)

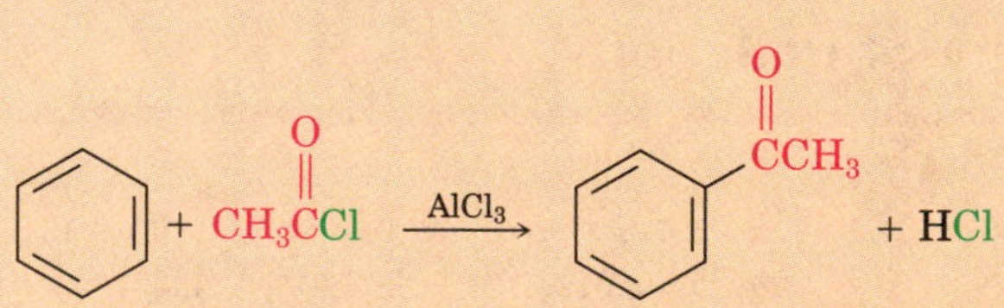

2. Reduction of aromatic nitro groups (Section 16.2)

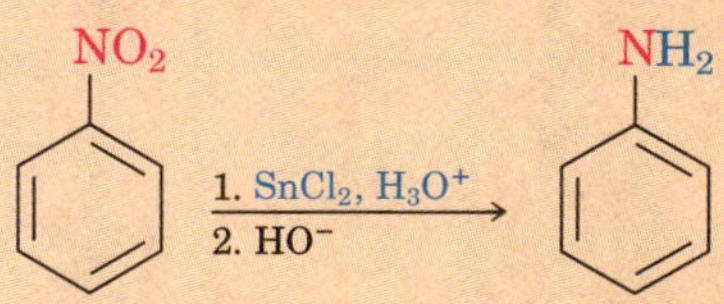

3. Alkali fusion of aromatic sulfonates (Section 16.2)

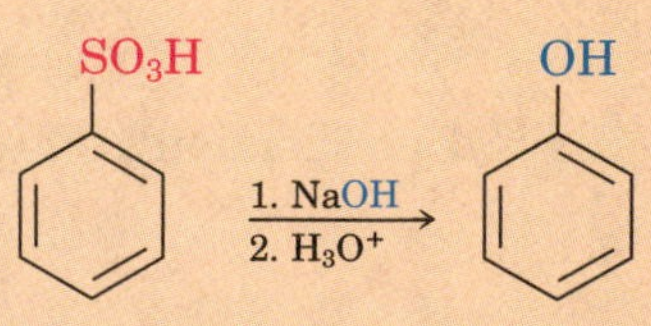

(continued)➤

4. Nucleophilic aromatic substitution
   (a) By addition/elimination to activated aryl halides (Section 16.8)

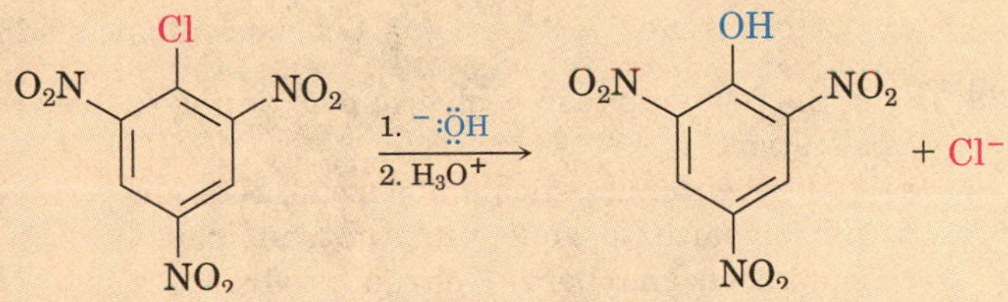

   (b) By benzyne intermediate for unactivated aryl halides (Section 16.9)

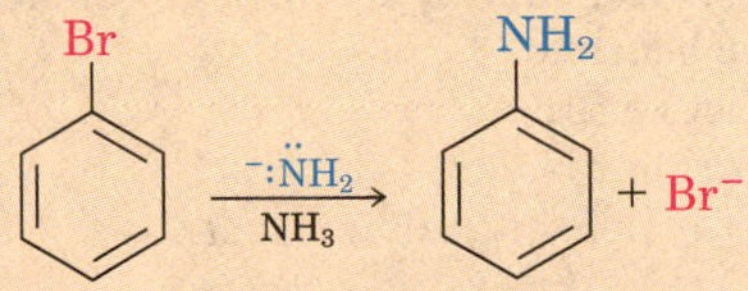

5. Oxidation of alkylbenzene side chains (Section 16.10)

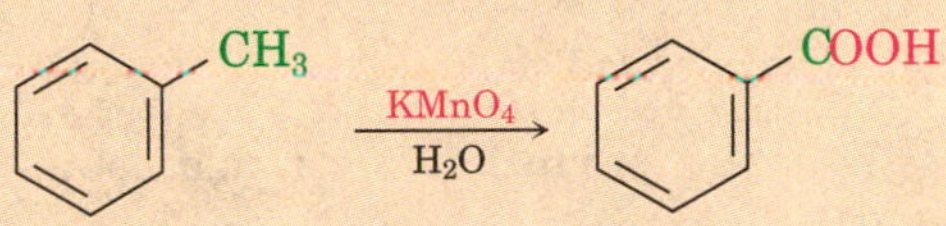

Reaction occurs with 1° and 2°, but not 3°, alkyl side chains.

6. Benzylic bromination of alkylbenzenes (Section 16.10)

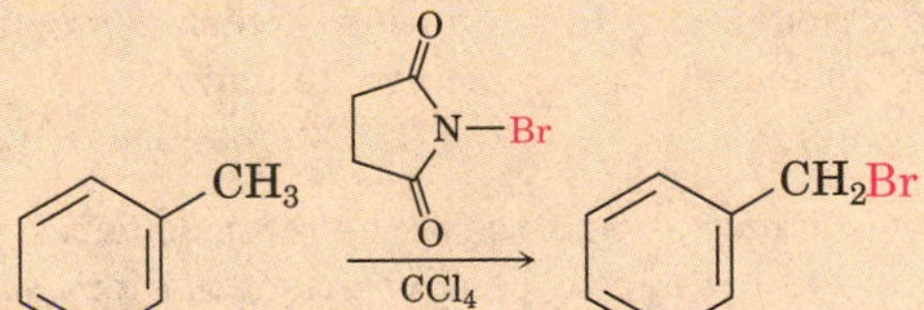

7. Catalytic hydrogenation of aromatic rings (Section 16.11)

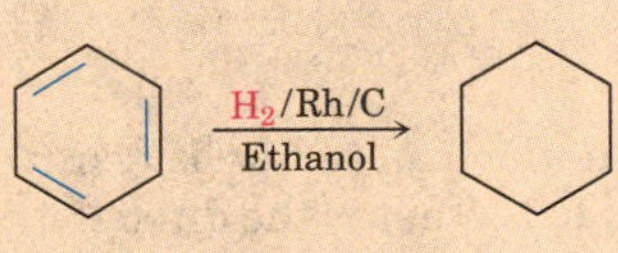

8. Reduction of aryl alkyl ketones (Section 16.11)

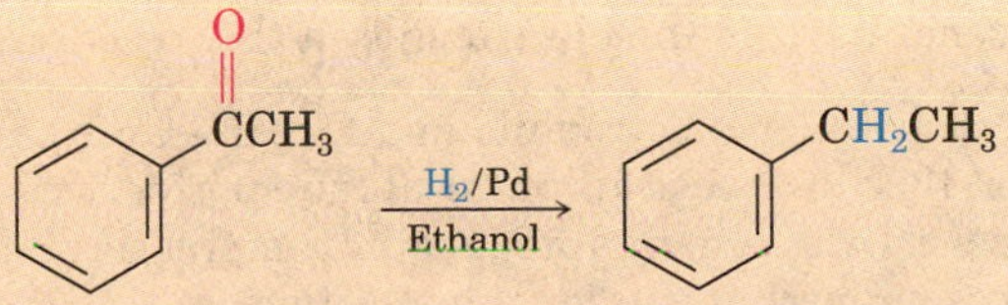

Reaction is specific for alkyl aryl ketones; dialkyl ketones are not affected.

ADDITIONAL PROBLEMS

16.28 Predict the major product(s) of mononitration of the following substances. Which react faster than benzene, and which slower?
(a) Bromobenzene (b) Benzonitrile (c) Benzoic acid
(d) Nitrobenzene (e) Benzenesulfonic acid (f) Methoxybenzene

16.29 Rank the compounds in each group according to their reactivity toward electrophilic substitution.
(a) Chlorobenzene, *o*-dichlorobenzene, benzene
(b) *p*-Bromonitrobenzene, nitrobenzene, phenol
(c) Fluorobenzene, benzaldehyde, *o*-xylene
(d) Benzonitrile, *p*-methylbenzonitrile, *p*-methoxybenzonitrile

16.30 Predict the major monoalkylation products you would expect to obtain from reaction of the following substances with chloromethane and $AlCl_3$.
(a) Bromobenzene (b) *m*-Bromophenol
(c) *p*-Chloroaniline (d) 2,4-Dichloronitrobenzene
(e) 2,4-Dichlorophenol (f) Benzoic acid
(g) *p*-Methylbenzenesulfonic acid (h) 2,5-Dibromotoluene

16.31 Name and draw the major product(s) of electrophilic chlorination of the following substances:
(a) *m*-Nitrophenol (b) *o*-Xylene
(c) *p*-Nitrobenzoic acid (d) *p*-Bromobenzenesulfonic acid

16.32 Predict the major product(s) you would obtain from sulfonation of the following compounds:
(a) Fluorobenzene (b) *m*-Bromophenol (c) *m*-Dichlorobenzene
(d) 2,4-Dibromophenol

16.33 Rank the following aromatic compounds in the expected order of their reactivity toward Friedel–Crafts alkylation. Which compounds are unreactive?
(a) Bromobenzene (b) Toluene (c) Phenol
(d) Aniline (e) Nitrobenzene (f) *p*-Bromotoluene

16.34 Using resonance structures of the intermediates, explain why bromination of biphenyl occurs at ortho and para positions rather than at meta.

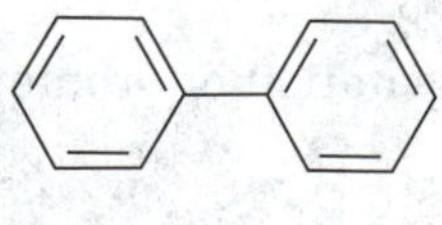

Biphenyl

16.35 Aromatic iodination can be carried out with a number of reagents, including iodine monochloride, ICl. What is the direction of polarization of ICl? Propose a mechanism to account for the iodination of an aromatic ring.

16.36 Show the mechanism of the desulfonation reaction of benzenesulfonic acid to yield benzene. What is the electrophile in this reaction?

16.37 The carbocation electrophile in a Friedel–Crafts reaction can be generated in ways other than by reaction of an alkyl chloride with $AlCl_3$. For example, reaction of benzene with 2-methylpropene in the presence of H_3PO_4 yields *tert*-butylbenzene. Propose a mechanism for this reaction.

16.38 The *N,N,N*-trimethylammonium group, $-\overset{+}{N}(CH_3)_3$, is one of the few groups that is a meta-directing deactivator yet has no electron-withdrawing resonance effect. Explain.

16.39 At what position and on what ring do you expect nitration of 4-bromobiphenyl to occur?

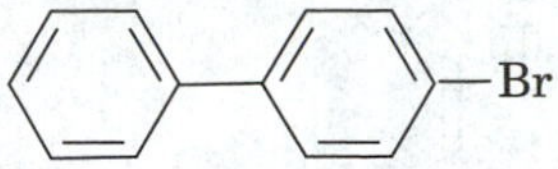

4-Bromobiphenyl

16.40 Use resonance structures of the intermediates to explain why electrophilic attack on 3-phenylpropanenitrile occurs at the ortho position and para position, whereas attack on 3-phenylpropenenitrile occurs at the meta position.

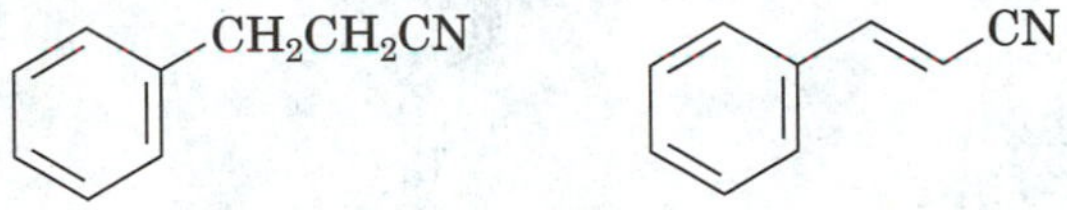

3-Phenylpropanenitrile **3-Phenylpropenenitrile**

16.41 Addition of HBr to 1-phenylpropene yields only (1-bromopropyl)benzene. Propose a mechanism for the reaction, and explain why none of the other regioisomer is produced.

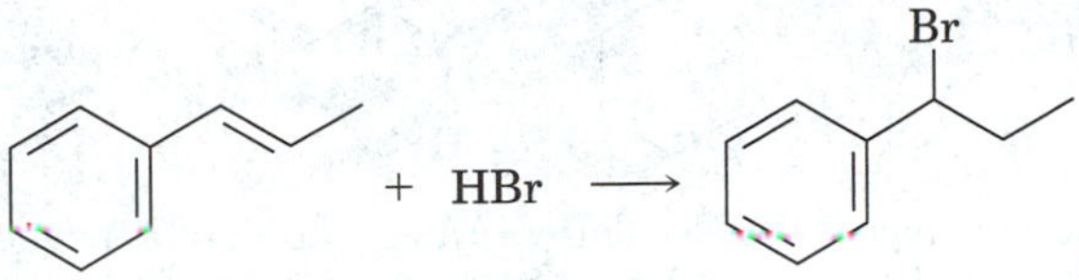

16.42 Triphenylmethane can be prepared by reaction of benzene and chloroform in the presence of $AlCl_3$. Propose a mechanism for the reaction.

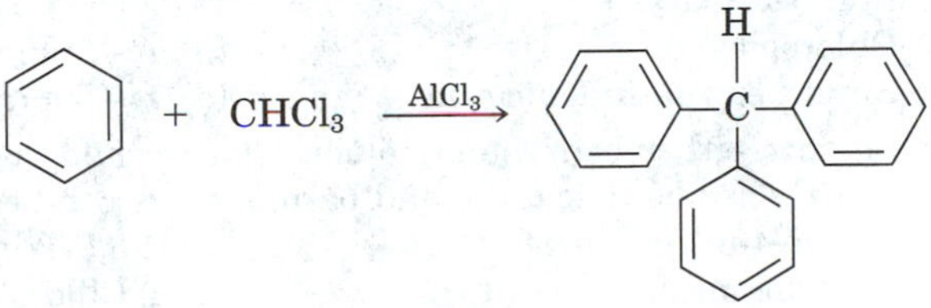

16.43 What product(s) would you expect to obtain from the following reactions?

(a) 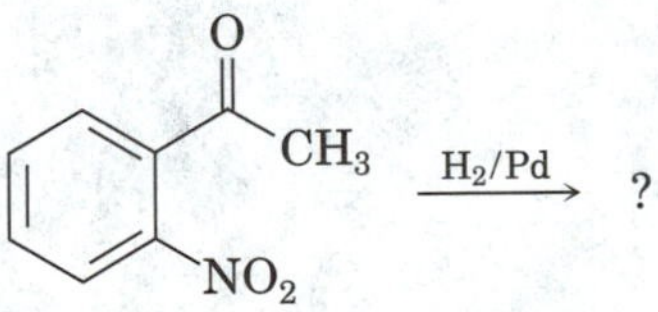

(b) 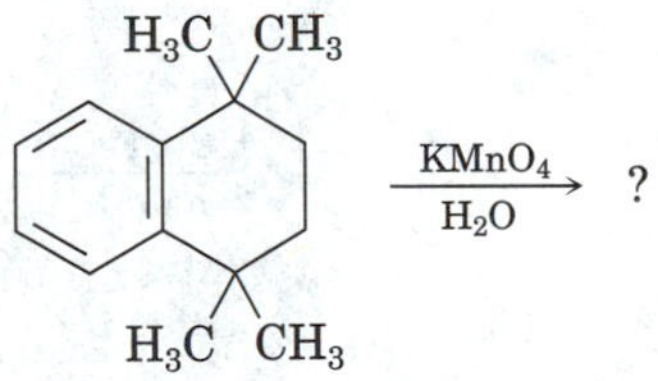

(c) 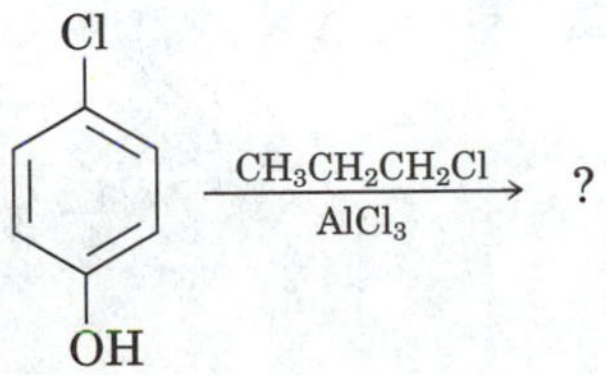

16.44 How would you synthesize the following substances starting from benzene? Assume that ortho- and para-substitution products can be separated.

(a) *o*-Methylphenol
(b) 2,4,6-Trinitrophenol
(c) 2,4,6-Trinitrobenzoic acid
(d) *m*-Bromoaniline

16.45 At what position, and on what ring, would you expect the following substances to undergo electrophilic substitution?

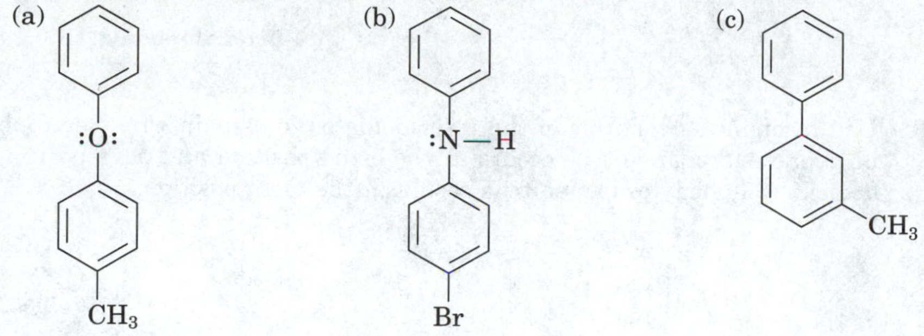

16.46 At what position, and on what ring, would you expect bromination of benzanilide to occur? Explain by drawing resonance structures of the intermediates.

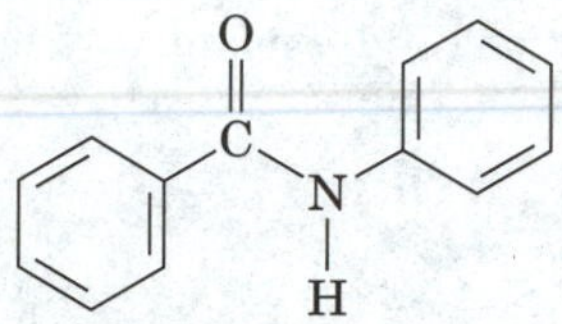

Benzanilide

16.47 Would you expect the Friedel–Crafts reaction of benzene with (*R*)-2-chlorobutane to yield optically active or racemic product? Explain.

16.48 Starting with benzene as your only source of aromatic compounds, how would you synthesize the following substances? Assume that you can separate ortho and para isomers if necessary.
(a) *p*-Chlorophenol (b) *m*-Bromonitrobenzene
(c) *o*-Bromobenzenesulfonic acid (d) *m*-Chlorobenzenesulfonic acid

16.49 Starting with either benzene or toluene, how would you synthesize the following substances? Assume that ortho and para isomers can be separated.
(a) 2-Bromo-4-nitrotoluene (b) 1,3,5-Trinitrobenzene
(c) 2,4,6-Tribromoaniline (d) 2-Chloro-4-methylphenol

16.50 As written, the following syntheses have flaws. What is wrong with each?

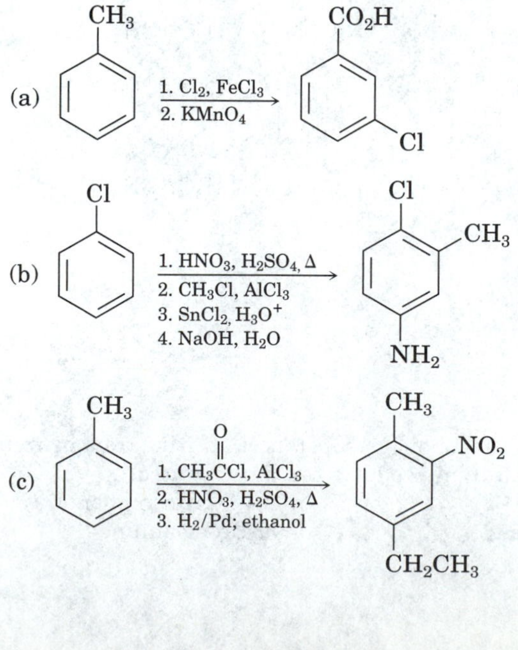

16.51 How would you synthesize the following substances starting from benzene?

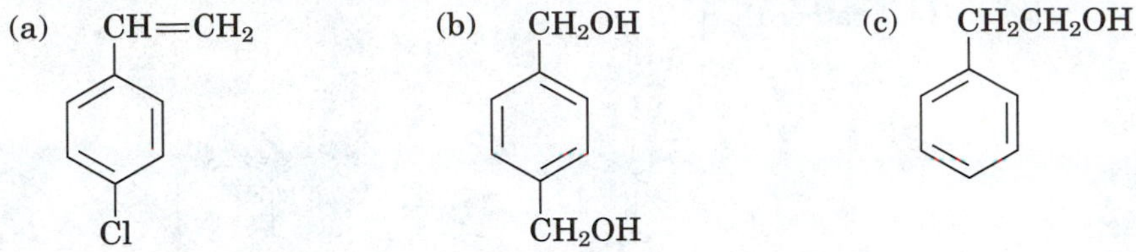

16.52 The compound MON-0585 is a nontoxic, biodegradable larvicide that is highly selective against mosquito larvae. Synthesize MON-0585 using only benzene as a source of the aromatic rings.

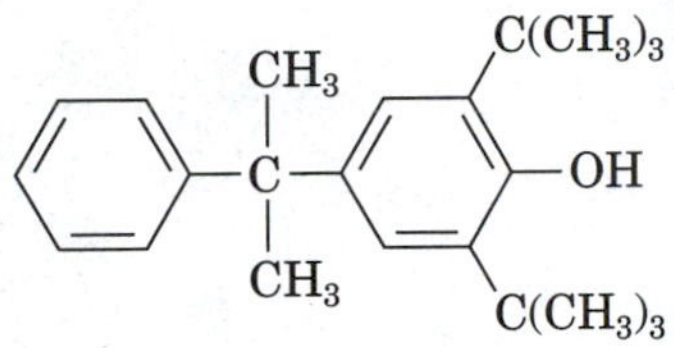

MON-0585

16.53 Hexachlorophene, a substance used in the manufacture of germicidal soaps, is prepared by reaction of 2,4,5-trichlorophenol with formaldehyde in the presence of concentrated sulfuric acid. Propose a mechanism for the reaction.

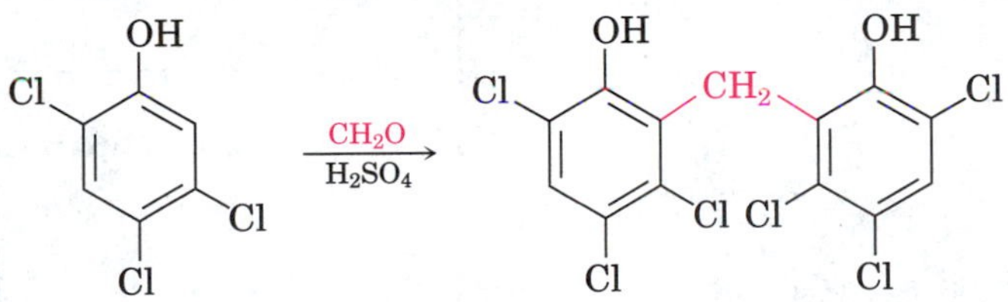

Hexachlorophene

16.54 Benzenediazonium carboxylate decomposes when heated to yield N_2, CO_2, and a reactive substance that can't be isolated. When benzenediazonium carboxylate is heated in the presence of furan, the following reaction is observed:

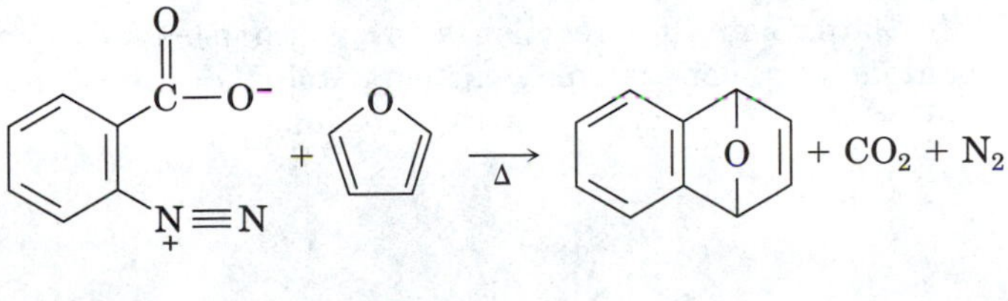

What intermediate is involved in this reaction? Propose a mechanism for its formation.

16.55 Phenylboronic acid, $C_6H_5B(OH)_2$, is nitrated to give 15% ortho-substitution product and 85% meta. Explain the meta-directing effect of the $-B(OH)_2$ group.

16.56 Draw resonance structures of the intermediate carbocations in the bromination of naphthalene, and account for the fact that naphthalene undergoes electrophilic attack at C1 rather than C2.

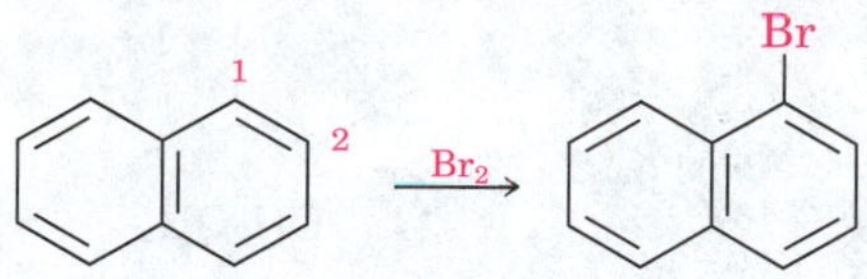

16.57 4-Chloropyridine undergoes reaction with dimethylamine to yield 4-dimethylaminopyridine. Propose a mechanism for the reaction.

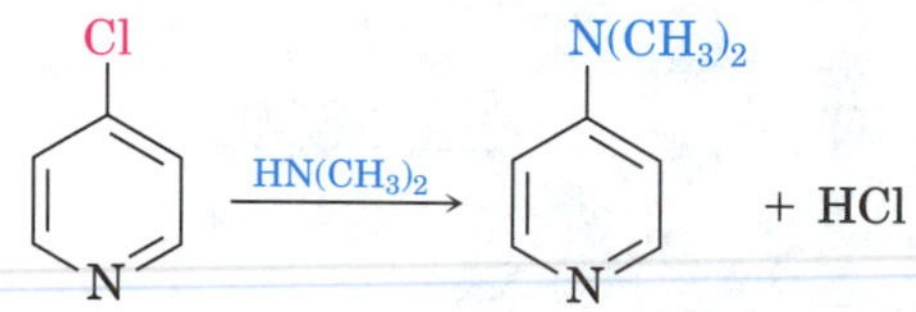

16.58 *p*-Bromotoluene reacts with potassium amide to give a mixture of *m*- and *p*-methylaniline. Explain.

16.59 Propose a synthesis of aspirin (acetylsalicylic acid) starting from benzene. You will need to use an acetylation reaction at some point in your scheme.

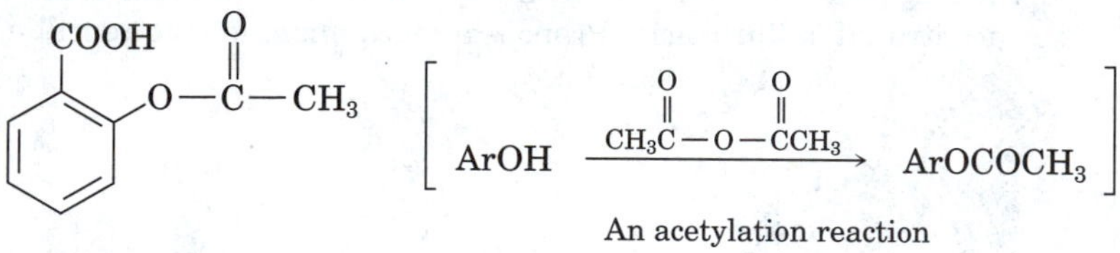

Aspirin

An acetylation reaction

16.60 Propose a mechanism to account for the reaction of benzene with 2,2,5,5-tetramethyltetrahydrofuran.

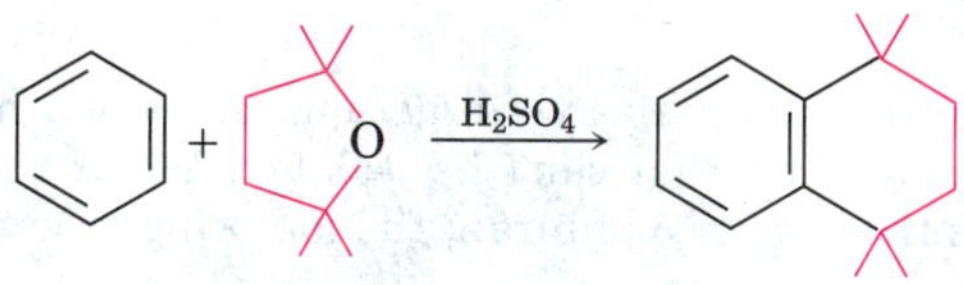

16.61 In the Gatterman–Koch reaction, a formyl group (–CHO) is introduced directly onto a benzene ring. For example, reaction of toluene with CO and HCl in the presence of mixed CuCl/$AlCl_3$ gives *p*-methylbenzaldehyde. Propose a mechanism.

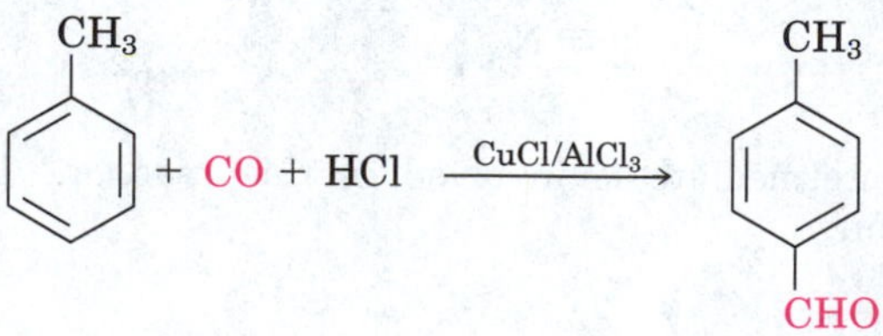

16.62 Triptycene is an unusual molecule that has been prepared by reaction of benzyne with anthracene. What kind of reaction is involved? Show the mechanism.

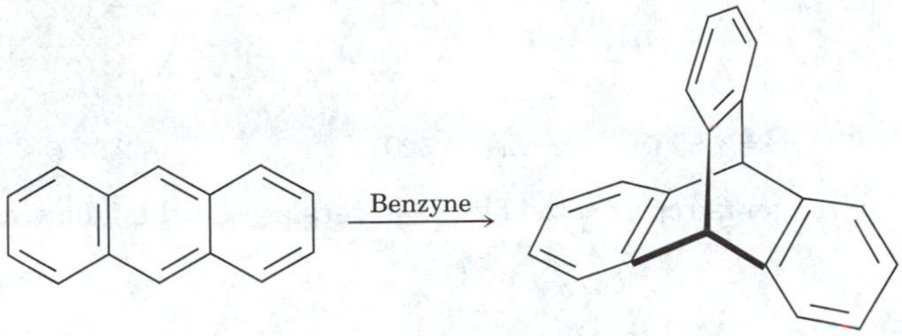

Triptycene

16.63 Treatment of *p-tert*-butylphenol with a strong acid such as H_2SO_4 yields phenol and 2-methylpropene. Propose a mechanism.

16.64 Benzene and alkyl-substituted benzenes can be hydroxylated by reaction with H_2O_2 in the presence of an acidic catalyst. What is the structure of the reactive electrophile? Propose a mechanism for the reaction.

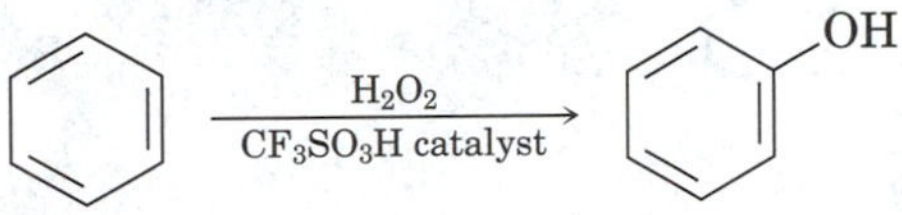

16.65 How would you synthesize the following compounds from benzene? Assume that ortho and para isomers can be separated.

(a)
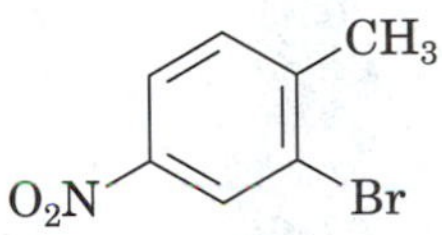

(b)
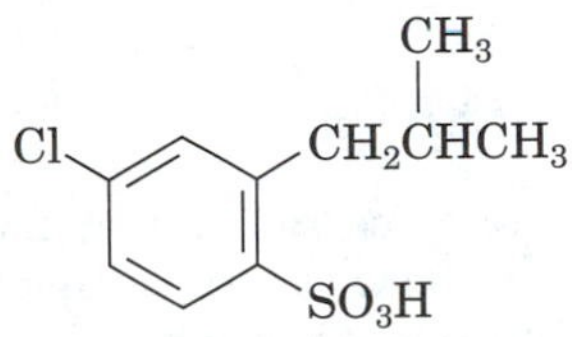

16.66 You know the mechanism of HBr addition to alkenes, and you know the effects of various substituent groups on aromatic substitution. Use this knowledge to predict which of the following two alkenes reacts faster with HBr. Explain your answer by drawing resonance structures of the carbocation intermediates.

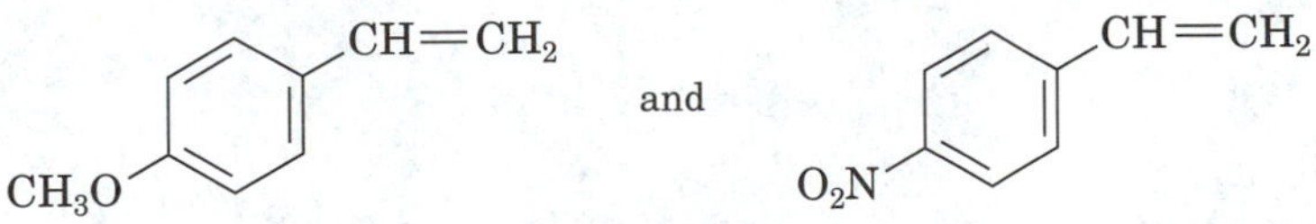

16.67 Draw a Fischer projection of (*R*)-2-phenylbutane, and predict the stereochemistry of its reaction with *N*-bromosuccinimide. Explain.

16.68 Benzyl bromide is converted into benzaldehyde by heating in dimethyl sulfoxide. Propose a structure for the intermediate, and show the mechanisms of the two steps in the reaction.

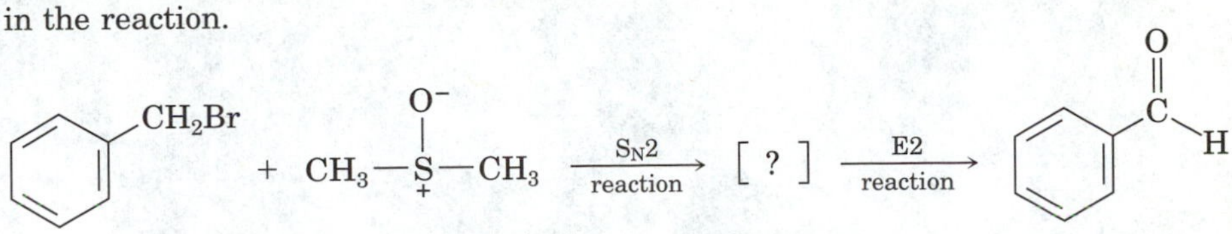

16.69 Use your knowledge of directing effects, along with the following data, to deduce the directions of the dipole moments in aniline and bromobenzene.

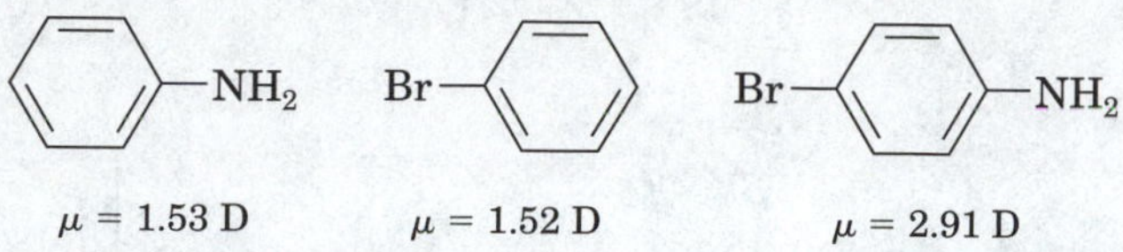

16.70 Identify the reagents represented by the letters a–e in the following scheme:

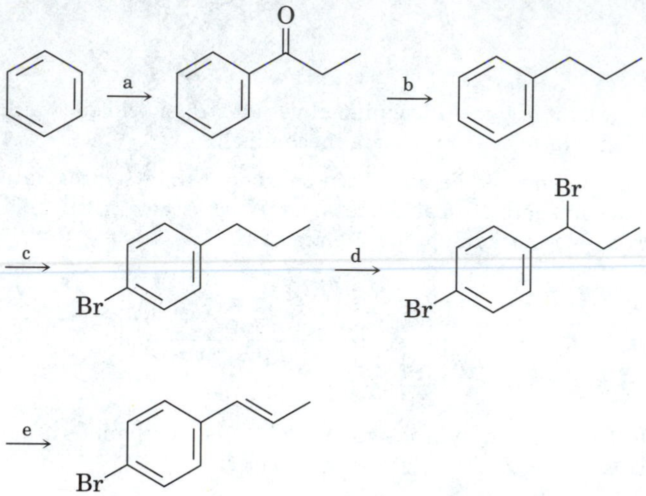

A Look Ahead

16.71 We'll see in Chapter 25 that phenols (ArOH) are relatively acidic and that the presence of a substituent group on the aromatic ring has a large effect. The pK_a of unsubstituted phenol, for example, is 10.00, while that of *p*-nitrophenol is 7.16. Explain.

16.72 Would you expect *p*-methylphenol to be more acidic or less acidic than unsubstituted phenol? Explain. (See Problem 16.71.)

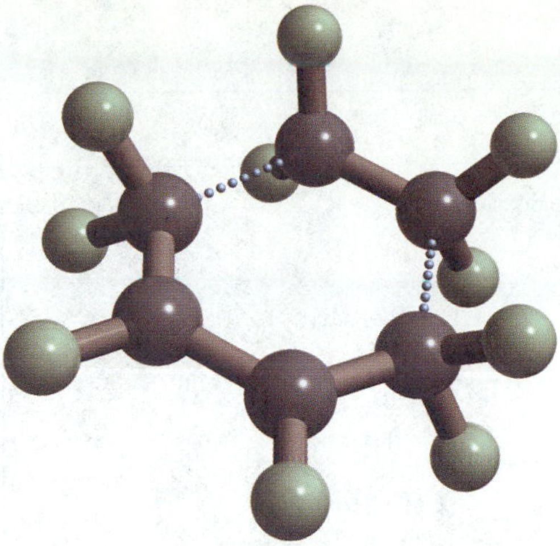

The Diels–Alder reaction of 1,3-butadiene yields a cyclohexene product.

Organic Reactions: A Brief Review

It's unavoidable; learning organic chemistry means knowing a large number of reactions. The way to simplify the job, of course, is to organize the material. We said in Chapter 5 that organic reactions can be organized in two ways: by what kinds of reactions occur and by how they occur. Let's briefly review both organizational methods in light of the past several chapters. You might find the following tables useful for seeing the overall picture when you study for exams.

I. A Summary of the Kinds of Organic Reactions

There are four important kinds of reactions: additions, eliminations, substitutions, and rearrangements. We've now seen examples of all four, as summarized in Review Tables 1–4.

Review Table 1 Some Addition Reactions

1. Additions to alkenes
   (a) Electrophilic addition of HX (X = Cl, Br, I; Sections 6.8 and 6.9)

$$CH_3CH{=}CH_2 \xrightarrow[\text{Ether}]{HCl} CH_3\overset{\overset{\displaystyle Cl}{|}}{C}HCH_3$$

(continued)➤

Review Table 1 *(continued)*

(b) Electrophilic addition of X_2 (X = Cl, Br; Section 7.2)

$$CH_3CH{=}CHCH_3 \xrightarrow[CCl_4]{Br_2} CH_3\overset{Br}{\overset{|}{C}}H{-}\overset{Br}{\overset{|}{C}}HCH_3$$

(c) Electrophilic addition of HO–X (X = Cl, Br, I; Section 7.3)

$$CH_3CH{=}CH_2 \xrightarrow[NaOH,\ H_2O]{Br_2} CH_3\overset{OH}{\overset{|}{C}}HCH_2Br$$

(d) Electrophilic addition of water by hydroxymercuration (Section 7.4)

$$CH_3CH{=}CH_2 \xrightarrow[H_2O]{Hg(OAc)_2} CH_3\overset{OH}{\overset{|}{C}}HCH_2HgOAc \xrightarrow{NaBH_4} CH_3\overset{OH}{\overset{|}{C}}HCH_3$$

(e) Addition of BH_3 (hydroboration; Section 7.5)

$$CH_3CH{=}CH_2 \xrightarrow{BH_3} CH_3\overset{H}{\overset{|}{C}}HCH_2BH_2 \xrightarrow[NaOH]{H_2O_2} CH_3CH_2CH_2OH$$

(f) Catalytic addition of H_2 (Section 7.7)

$$CH_3CH{=}CH_2 \xrightarrow[Pd/C\ catalyst]{H_2} CH_3\overset{H}{\overset{|}{C}}H{-}\overset{H}{\overset{|}{C}}H_2$$

(g) Hydroxylation with OsO_4 (Section 7.8)

$$CH_3CH{=}CH_2 \xrightarrow[2.\ NaHSO_3]{1.\ OsO_4} CH_3\overset{OH}{\overset{|}{C}}HCH_2OH$$

(h) Addition of carbenoids; cyclopropane formation (Section 7.6)

$$CH_3CH{=}CH_2 \xrightarrow{CH_2I_2,\ Zn(Cu)} CH_3\overset{\diagup CH_2 \diagdown}{CH{-}CH_2}$$

(i) Cycloaddition: Diels–Alder reaction (Sections 14.8 and 14.9)

$$H_2C{=}CH{-}CH{=}CH_2 + H_2C{=}CHCOCH_3 \longrightarrow$$

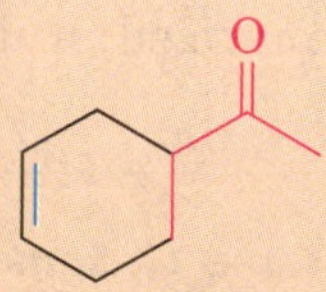

(j) Radical addition of HBr (Section 7.10)

$$CH_3CH{=}CH_2 \xrightarrow[Radicals]{HBr} CH_3\overset{H}{\overset{|}{C}}HCH_2Br$$

(continued)➤

Review Table 1 *(continued)*

2. Additions to alkynes
   (a) Electrophilic addition of HX (X = Cl, Br, I; Section 8.4)

$$CH_3C{\equiv}CH \xrightarrow[\text{Ether}]{HBr} CH_3\overset{\overset{Br}{|}}{C}{=}CH_2$$

   (b) Electrophilic addition of H_2O (Section 8.5)

$$CH_3C{\equiv}CH \xrightarrow[HgSO_4]{H_3O^+} CH_3\overset{\overset{OH}{|}}{C}{=}CH_2 \longrightarrow CH_3\overset{\overset{O}{\|}}{C}CH_3$$

   (c) Addition of H_2 (Section 8.6)

$$CH_3C{\equiv}CCH_3 \xrightarrow[\text{Lindlar catalyst}]{H_2} CH_3\overset{\overset{H}{|}}{C}{=}\overset{\overset{H}{|}}{C}CH_3$$

Review Table 2 Some Elimination Reactions

1. Dehydrohalogenation of alkyl halides (Section 11.10)

$$CH_3\overset{\overset{Br}{|}}{C}HCH_3 \xrightarrow{KOH} CH_3CH{=}CH_2 + KBr + H_2O$$

2. Dehydrohalogenation of vinylic halides (Section 8.3)

$$CH_3\overset{\overset{Br}{|}}{C}{=}CH_2 \xrightarrow{NaNH_2} CH_3C{\equiv}CH + NaBr + NH_3$$

3. Dehydrohalogenation of aryl halides: benzyne formation (Section 16.9)

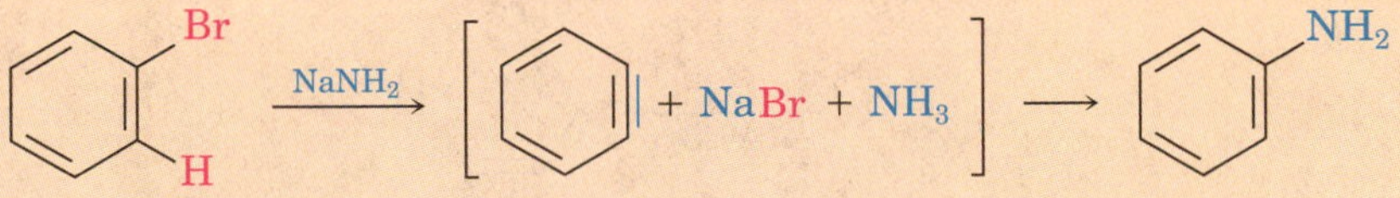

Review Table 3 Some Substitution Reactions

1. S_N2 reactions of primary alkyl halides (Sections 11.2–11.5)
   (a) General reaction

$$CH_3CH_2CH_2X + :Nu^- \longrightarrow CH_3CH_2CH_2Nu + :X^-$$

where X = Cl, Br, I, OTos

$:Nu^- = CH_3O^-, HO^-, CH_3S^-, HS^-, CN^-, CH_3COO^-, NH_3, (CH_3)_3N$, etc.

(continued)➤

Review Table 3 *(continued)*

(b) Alkyne alkylation (Section 8.9)

$$CH_3C{\equiv}C{:}^- \ Na^+ \xrightarrow{CH_3I} CH_3C{\equiv}CCH_3 + Na^+ \ I^-$$

2. S_N1 reactions of tertiary alkyl halides (Sections 11.8 and 11.9)
   (a) General reaction

$$(CH_3)_3CX \xrightarrow{:NuH} (CH_3)_3CNu + HX$$

   (b) Preparation of alkyl halides from alcohols (Section 10.7)

$$(CH_3)_3COH \xrightarrow{HBr} (CH_3)_3CBr + H_2O$$

3. Electrophilic aromatic substitution (Sections 16.1–16.4)
   (a) Halogenation of aromatic compounds (Section 16.1)

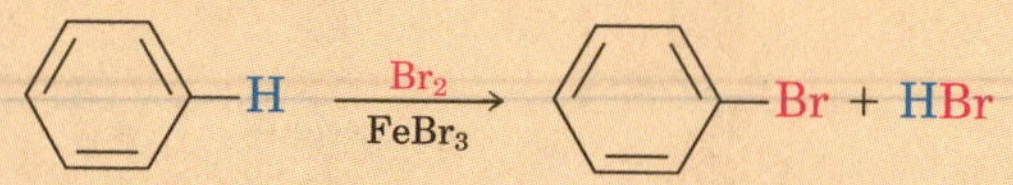

   (b) Nitration of aromatic compounds (Section 16.2)

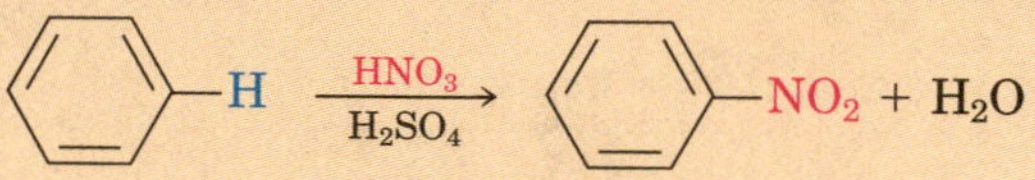

   (c) Sulfonation of aromatic compounds (Section 16.2)

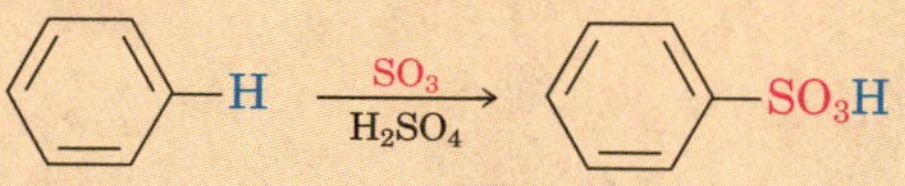

   (d) Alkylation of aromatic rings (Section 16.3)

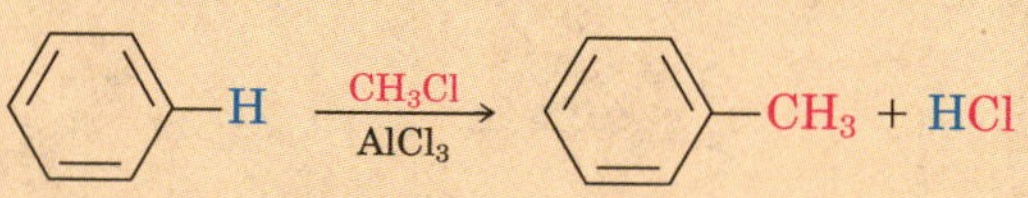

   (e) Acylation of aromatic rings (Section 16.4)

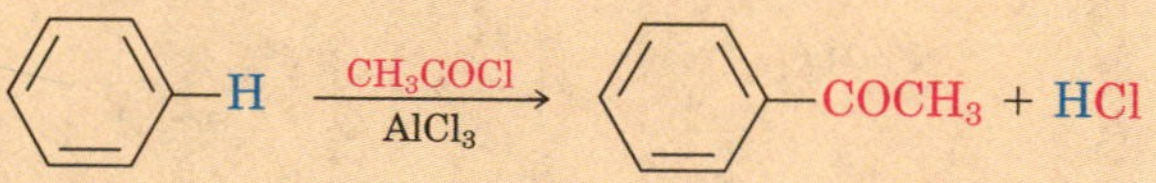

4. Nucleophilic aromatic substitution (Section 16.8)

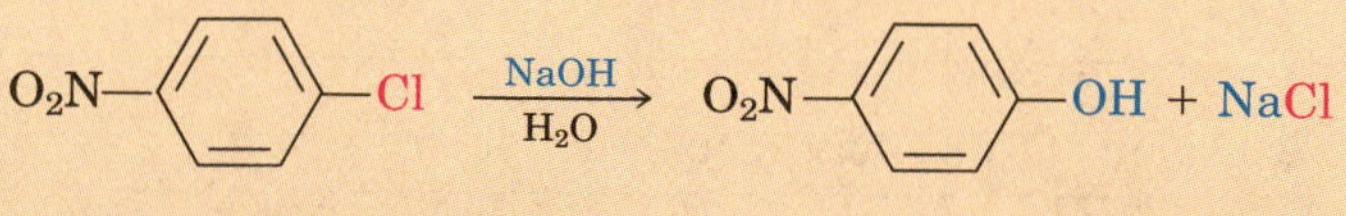

5. Radical substitution reactions
   (a) Chlorination of methane (Section 10.4)

$$CH_4 + Cl_2 \xrightarrow{Light} CH_3Cl + HCl$$

   (b) NBS allylic bromination of alkenes (Section 10.5)

$$CH_3CH{=}CH_2 \xrightarrow[Radicals]{NBS} BrCH_2CH{=}CH_2$$

Review Table 4 Some Rearrangement Reactions

1. Carbocation rearrangement during electrophilic addition to alkenes (Section 6.12)

$$(CH_3)_2\overset{\displaystyle H}{\overset{|}{C}}CH{=}CH_2 + HCl \longrightarrow (CH_3)_2\overset{\displaystyle Cl}{\overset{|}{C}}CH_2CH_3$$

2. Carbocation rearrangement during Friedel–Crafts alkylation (Section 16.3)

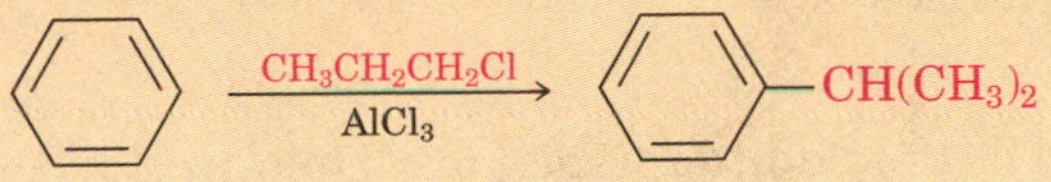

II. A Summary of How Reactions Occur

The second method of organizing reactions is by *how* they occur—that is, by their *mechanisms*. As we said in Chapter 5, there are three fundamental reaction types: polar reactions, radical reactions, and pericyclic reactions.

A. Polar Reactions

Polar reactions take place between electron-rich reagents (nucleophiles/Lewis bases) and electron-poor reagents (electrophiles/Lewis acids). These reactions are heterolytic processes and involve species with an even number of electrons. Bonds are made when a nucleophile donates an electron pair to an electrophile; bonds are broken when one product leaves with an electron pair.

Heterogenic bond formation $\quad A^+ + :B^- \longrightarrow A:B$

Electrophile Nucleophile

Heterolytic bond cleavage $\quad A:B \longrightarrow A^+ + :B^-$

The polar reactions we've studied can be grouped into five categories:

1. Electrophilic addition reactions
2. Elimination reactions
3. Nucleophilic alkyl substitution reactions
4. Electrophilic aromatic substitution reactions
5. Nucleophilic aromatic substitution reactions

1. Electrophilic Addition Reactions (Sections 6.8 and 6.9) Alkenes react with electrophiles such as HBr to yield saturated addition products. The reaction occurs in two steps: The electrophile first adds to the alkene double bond to yield a carbocation intermediate that reacts further to yield the addition product.

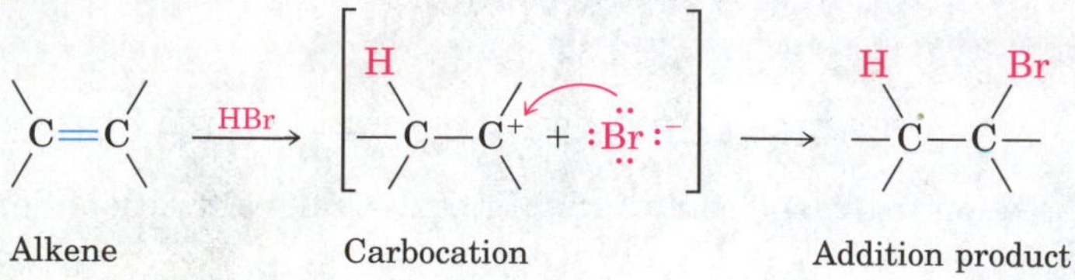

Many of the addition reactions listed in Review Table 1 take place by an electrophilic addition mechanism. The electrophile can be H^+, X^+, or Hg^{2+}, but the basic process is the same. The remaining addition reactions in Review Table 1 occur without carbocation intermediates, but it's still convenient to group them together.

2. Elimination Reactions

(a) *E2 Reaction (Sections 11.11–11.13)* Alkyl halides undergo elimination of HX to yield alkenes on treatment with base. When a strong base such as hydroxide ion (HO^-), alkoxide ion (RO^-), or amide ion (NH_2^-) is used, alkyl halides react by the E2 mechanism. E2 reactions occur in a single step involving removal by base of a neighboring hydrogen at the same time that the halide ion is leaving.

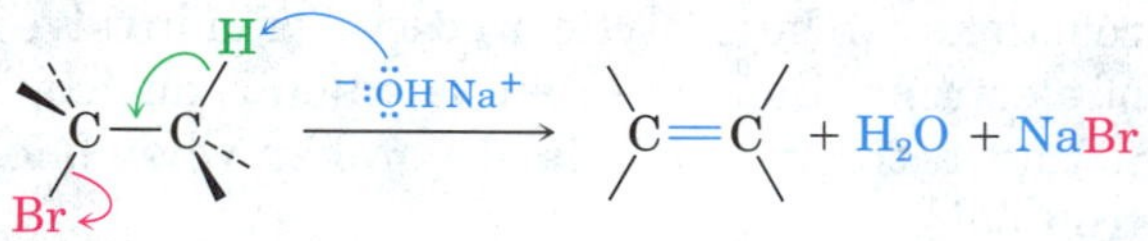

All the elimination reactions listed in Review Table 2 occur by the same E2 mechanism. Though they appear different, the elimination of an alkyl halide to yield an alkene (reaction 1), the elimination of a vinylic halide to yield an alkyne (reaction 2), and the elimination of an aryl halide to yield a benzyne (reaction 3) are all E2 reactions.

(b) *E1 Reaction (Section 11.14)* Tertiary alkyl halides undergo elimination by the E1 mechanism in competition with S_N1 substitution when a nonbasic nucleophile is used in a hydroxylic solvent. The reaction takes place in two steps: spontaneous dissociation of the alkyl halide, followed by loss of H^+ from the carbocation intermediate.

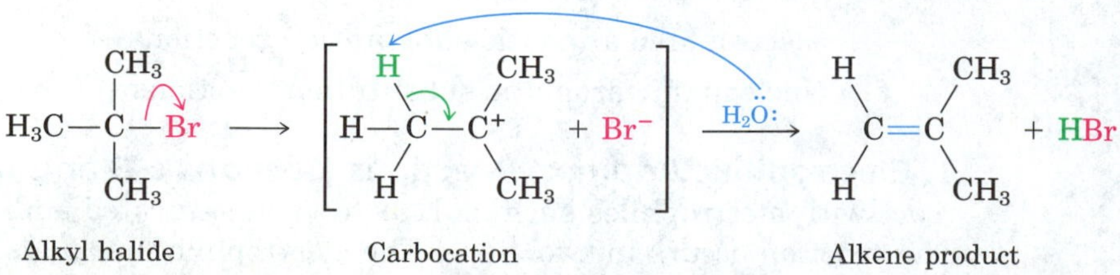

3. Nucleophilic Alkyl Substitution Reactions

(a) *S_N2 Reaction (Sections 11.2–11.5)* The nucleophilic alkyl substitution reaction is one of the most common reactions encountered in organic chemistry. As illustrated in Review Table 3 (reaction 1a), most primary halides and tosylates, and some secondary ones, undergo substitution reactions with a variety of different nucleophiles. One particularly useful S_N2 reaction is the alkylation of a terminal alkyne anion (reaction 1b) to yield an internal alkyne product.

Mechanistically, S_N2 reactions take place in a single step involving attack of the incoming nucleophile from a direction 180° away from the leaving group. This results in an umbrella-like inversion of stereochemistry (Walden inversion).

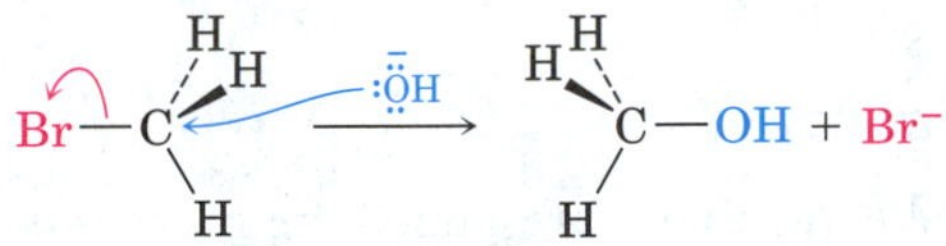

(b) *S_N1 Reaction (Sections 11.6–11.9)* Tertiary alkyl halides undergo nucleophilic substitution by the two-step S_N1 mechanism. Spontaneous dissociation of the alkyl halide into an anion and a carbocation intermediate takes place, followed by reaction of the carbocation with a nucleophile. The dissociation step is the slower of the two and is rate-limiting.

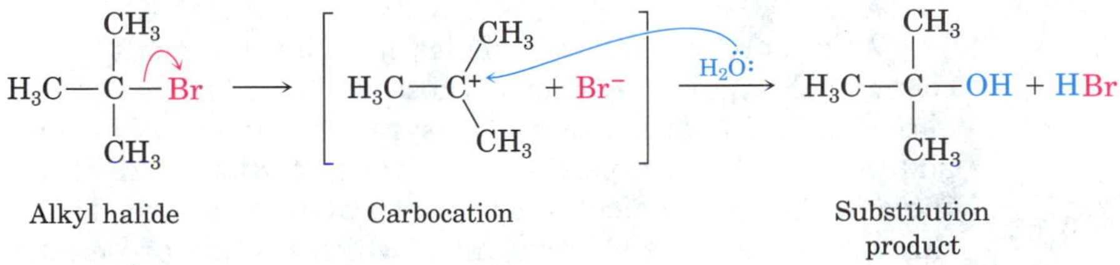

Among the more useful S_N1 reactions is the conversion of a secondary or tertiary alcohol into an alkyl halide by reaction with HX (reaction 2b, Review Table 3).

4. Electrophilic Aromatic Substitution Reactions (Sections 16.1–16.4)

All the electrophilic aromatic substitutions shown in Review Table 3 occur by the same two-step mechanism. The first step is similar to the first step in electrophilic addition to alkenes: An electron-poor reagent reacts with the electron-rich aromatic ring. The second step is identical to what happens during E2 elimination: A base abstracts a hydrogen atom next to the positively charged carbon, and elimination of the proton occurs.

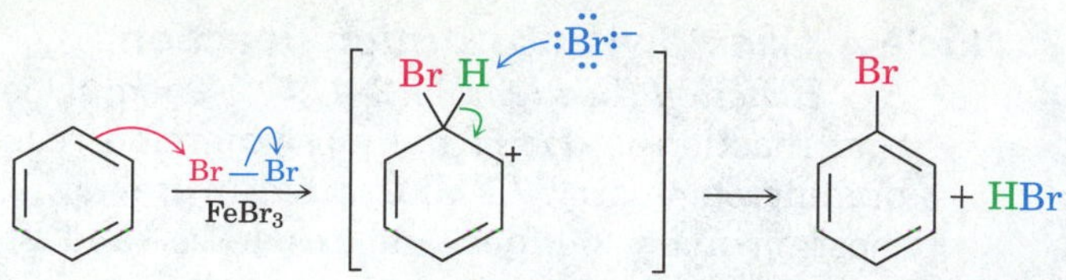

Benzene Carbocation **Bromobenzene**

5. Nucleophilic Aromatic Substitution Reactions (Section 16.8)

Nucleophilic aromatic substitution (reaction 4 in Review Table 3) occurs by addition of a nucleophile to an electrophilic aromatic ring, followed by elimination of the leaving group. The ring is made electrophilic, and hence reactive, only when substituted by strong electron-withdrawing groups such as nitro, cyano, and carbonyl.

B. Radical Reactions

Radical reactions are homolytic processes and involve species with an odd number of electrons. Bonds are made when each reactant donates one electron, and bonds are broken when each product fragment leaves with one electron.

Homogenic bond formation   A· + ·B ⟶ A:B

Homolytic bond cleavage   A:B ⟶ A· + ·B

We've only seen a few examples of radical reactions because they're less common than polar reactions. Those we have studied can be classified as either radical addition reactions or radical substitution reactions. Radical additions, such as the peroxide-catalyzed addition of HBr to alkenes, involve the addition of a radical to an unsaturated substrate. The reaction occurs through three kinds of steps, all of which involve odd-electron species: (1) initiation, (2) propagation, and (3) termination.

1. Initiation steps:
   (a) RO—OR ⟶ 2 RO·
   (b) RO· + H—Br ⟶ RO—H + ·Br
2. Propagation steps:

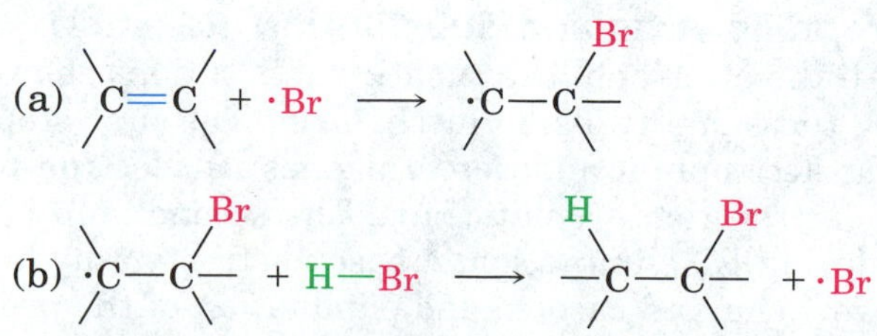

3. Termination steps:

(a) $Br\cdot + \cdot Br \longrightarrow Br{-}Br$

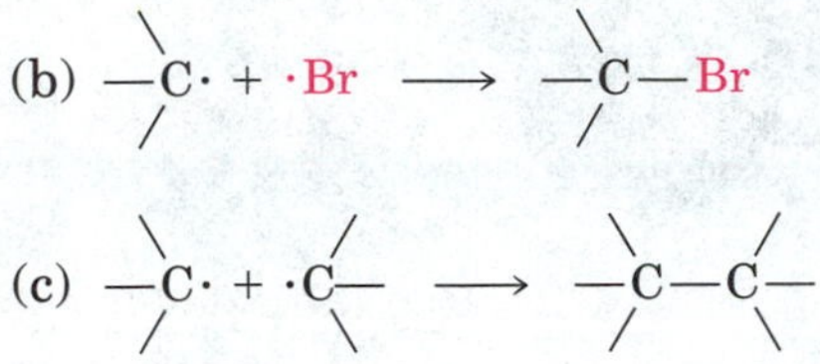

The reaction is initiated by homolytic cleavage of a peroxide, which forms two radicals. These radicals abstract H· from HBr, yielding a Br· radical that adds to the alkene, generating a new carbon radical and a C–Br bond. The reaction is completed by reaction of the carbon radical with HBr to yield neutral product and a Br· radical, which continues the chain.

Radical substitution reactions, such as the light-induced chlorination of methane and the allylic bromination of alkenes with *N*-bromosuccinimide (reactions 5a and 5b in Review Table 3), are also common. The key step in all these reactions is that a radical abstracts an atom from a neutral molecule, leaving a new radical.

C. Pericyclic Reactions

Pericyclic reactions, such as the addition of a carbene to an alkene and the Diels–Alder cycloaddition (reactions 1h and 1i, Review Table 1), involve neither radicals nor nucleophile–electrophile interactions. Rather, these processes take place in a single step by a reorganization of bonding electrons through a cyclic transition state. We'll look at these reactions more closely in Chapter 31.

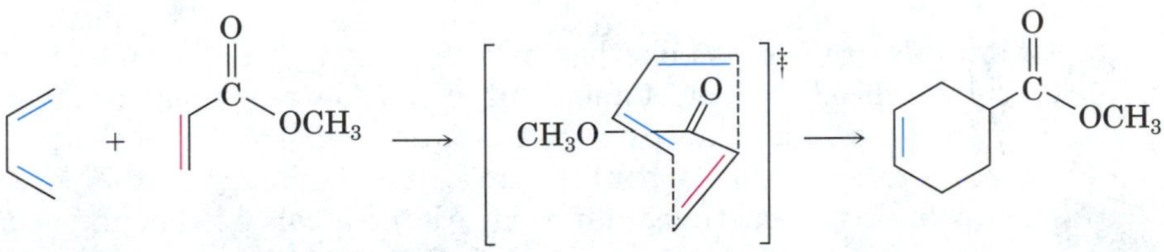

1,3-Butadiene **Methyl propenoate** Transition state **Methyl 3-cyclohexene-carboxylate**

Methanol and ethanol are the two most widely used alcohols.

17

Alcohols and Thiols

Alcohols are compounds that have hydroxyl groups bonded to saturated, sp^3-hybridized carbon atoms. This definition purposely excludes phenols (hydroxyl groups bonded to an aromatic ring) and enols (hydroxyl groups bonded to a vinylic carbon), because the chemistry of the three types of compounds is sometimes different. Alcohols can be thought of as organic derivatives of water in which one of the water hydrogens is replaced by an organic group: H–O–H versus R–O–H.

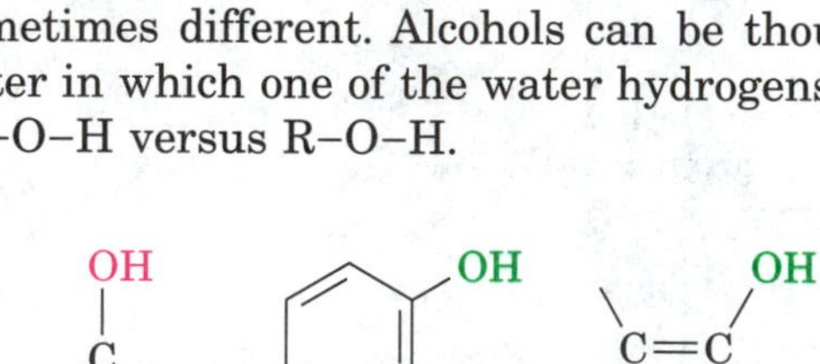

Alcohols occur widely in nature and have a great many industrial and pharmaceutical applications. Ethanol, for instance, is one of the simplest yet best known of all organic substances, finding use as a fuel additive, an industrial solvent, and a beverage. Menthol, an alcohol isolated from peppermint oil, is widely used as a flavoring and perfumery agent. Cholesterol, a complex-looking steroidal alcohol, plays a role in heart disease.

CH_3CH_2OH

Ethanol

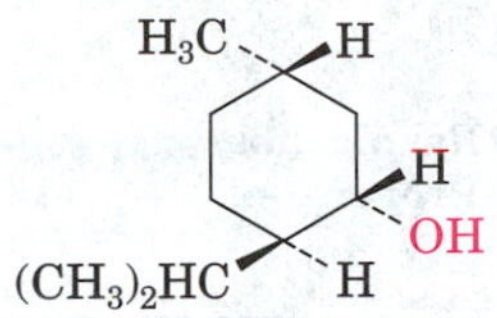

Menthol

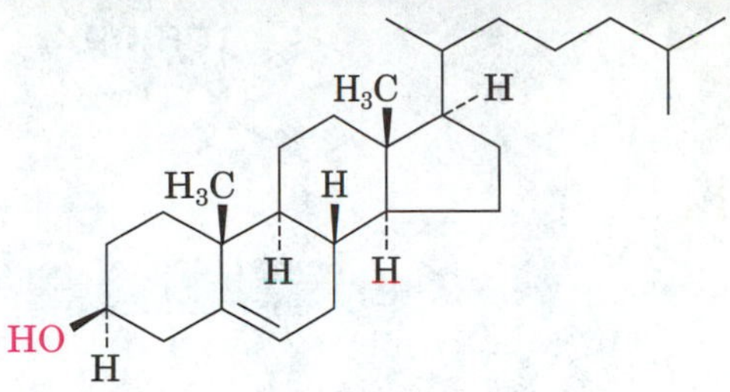

Cholesterol (a steroid)

17.1 Naming Alcohols

Alcohols are classified as primary (1°), secondary (2°), or tertiary (3°), depending on the number of organic groups bonded to the hydroxyl-bearing carbon.

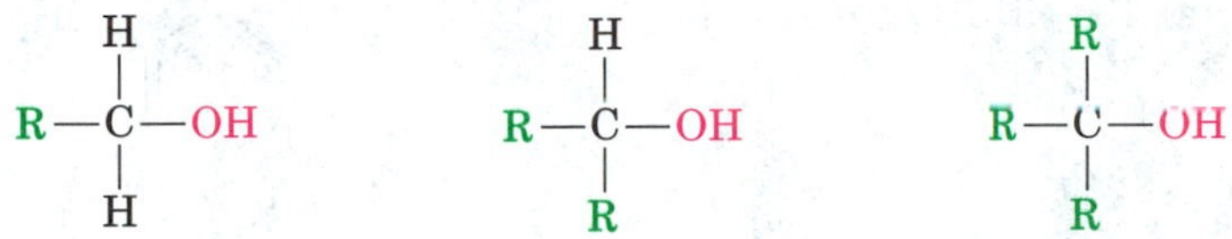

A primary alcohol (1°) A secondary alcohol (2°) A tertiary alcohol (3°)

Simple alcohols are named by the IUPAC system as derivatives of the parent alkane, using the suffix *-ol*:

1. Select the longest carbon chain *containing the hydroxyl group*, and derive the parent name by replacing the *-e* ending of the corresponding alkane with *-ol*.
2. Number the alkane chain beginning at the end nearer the hydroxyl group.
3. Number the substituents according to their position on the chain, and write the name listing the substituents in alphabetical order.

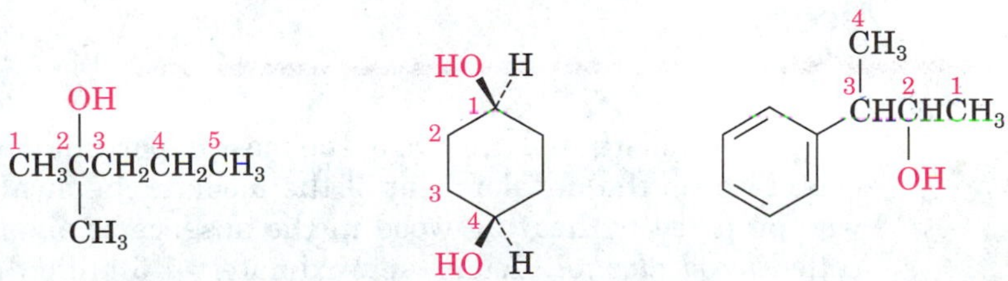

2-Methyl-2-pentanol ***cis*-1,4-Cyclohexanediol** **3-Phenyl-2-butanol**

Some simple and widely occurring alcohols have common names that are accepted by IUPAC. For example:

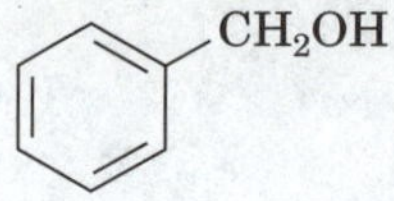

Benzyl alcohol
(Phenylmethanol)

$H_2C{=}CHCH_2OH$

Allyl alcohol
(2-Propen-1-ol)

$$\begin{array}{c} CH_3 \\ | \\ CH_3COH \\ | \\ CH_3 \end{array}$$

***tert*-Butyl alcohol**
(2-Methyl-2-propanol)

$HOCH_2CH_2OH$

Ethylene glycol
(1,2-Ethanediol)

$$\begin{array}{c} HOCH_2CHCH_2OH \\ | \\ OH \end{array}$$

Glycerol
(1,2,3-Propanetriol)

PROBLEM

17.1 Give IUPAC names for these compounds:

(a) $$\begin{array}{l} \quad\;\; OH \quad\;\; OH \\ \quad\;\;\, | \qquad\quad\; | \\ CH_3CHCH_2CHCH(CH_3)_2 \end{array}$$

(b)

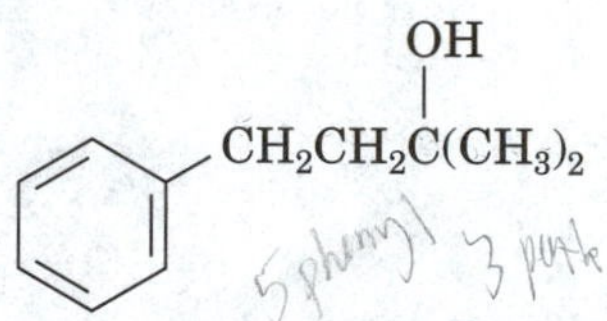

(c)

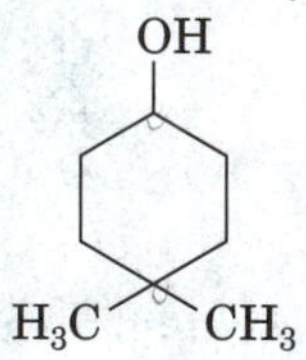

(d)

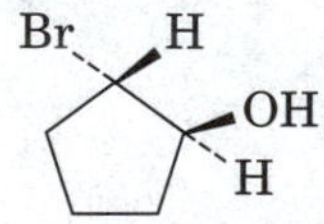

PROBLEM

17.2 Draw structures corresponding to the following IUPAC names:
(a) 2-Ethyl-2-buten-1-ol
(b) 3-Cyclohexen-1-ol
(c) *trans*-3-Chlorocycloheptanol
(d) 1,4-Pentanediol

17.2 Sources and Uses of Simple Alcohols

Methanol and ethanol are two of the most important of all industrial chemicals. Prior to the development of the modern chemical industry, methanol was prepared by heating wood in the absence of air and thus came to be called *wood alcohol*. Today, approximately 1.6 billion gallons of methanol are manufactured each year in the United States by catalytic reduction of carbon monoxide with hydrogen gas.

$$CO + 2\,H_2 \xrightarrow[\text{Zinc oxide/chromia}]{400°C} CH_3OH$$

Methanol is toxic to humans, causing blindness in low doses and death in larger amounts. Industrially, it is used both as a solvent and as a starting material for production of formaldehyde, CH_2O, and acetic acid, CH_3COOH.

Ethanol was one of the first organic chemicals to be prepared and purified. Its production by fermentation of grains and sugars has been carried out for millennia, and its purification by distillation goes back at least as far as the twelfth century AD. Only about 5% of the ethanol produced industrially comes from fermentation, although that figure may well change if demand for use in automobile fuel increases. Most ethanol is currently obtained by acid-catalyzed hydration of ethylene. Approximately 110 million gallons of ethanol a year are produced in the United States for use as a solvent or as a chemical intermediate in other industrial reactions.

$$H_2C{=}CH_2 + H_2O \xrightarrow{H_2SO_4} CH_3CH_2OH$$

17.3 Properties of Alcohols: Hydrogen Bonding

Alcohols have nearly the same geometry as water. The R-O-H bond angle has an approximately tetrahedral value (109° in methanol, for example), and the oxygen atom is sp^3-hybridized.

Methanol, CH_3OH

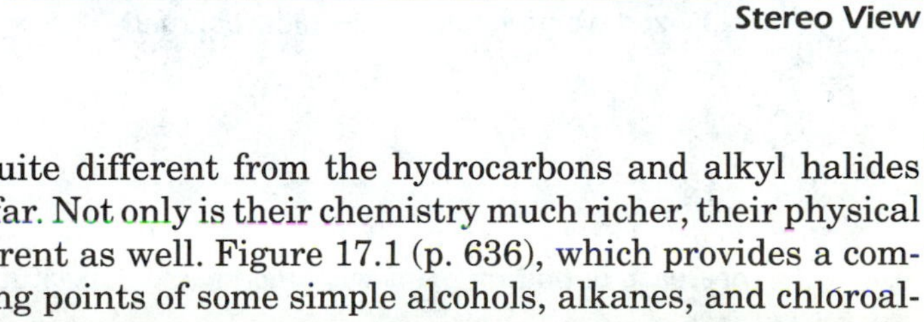

Stereo View

Alcohols are quite different from the hydrocarbons and alkyl halides we've studied thus far. Not only is their chemistry much richer, their physical properties are different as well. Figure 17.1 (p. 636), which provides a comparison of the boiling points of some simple alcohols, alkanes, and chloroalkanes, shows that alcohols have much higher boiling points. For example, 1-propanol (mol wt = 60), butane (mol wt = 58), and chloroethane (mol wt = 65) are close in molecular weight, yet 1-propanol boils at 97°C, compared to −0.5°C for the alkane and 12.5°C for the chloroalkane.

Alcohols have high boiling points because, like water, they form **hydrogen bonds** in the liquid state. A positively polarized –OH hydrogen atom from one molecule is attracted to a negatively polarized oxygen atom of

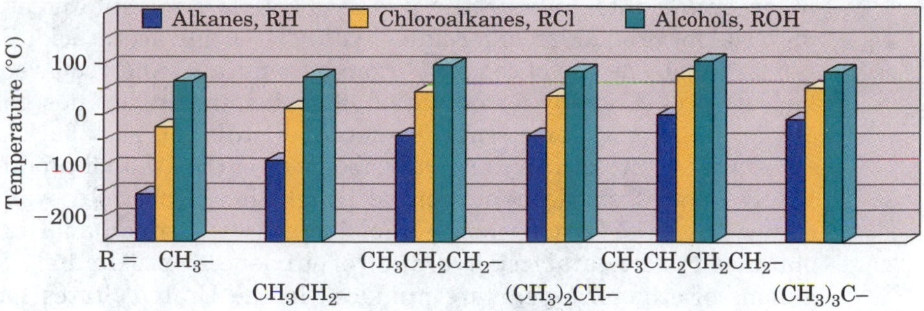

Figure 17.1 A comparison of boiling points for some alkanes, chloroalkanes, and alcohols. Alcohols generally have the highest boiling points.

another molecule (Figure 17.2), resulting in a weak force between the molecules that holds them together. These intermolecular attractions must be overcome for a molecule to break free from the liquid and enter the vapor state, so the boiling temperature is raised.

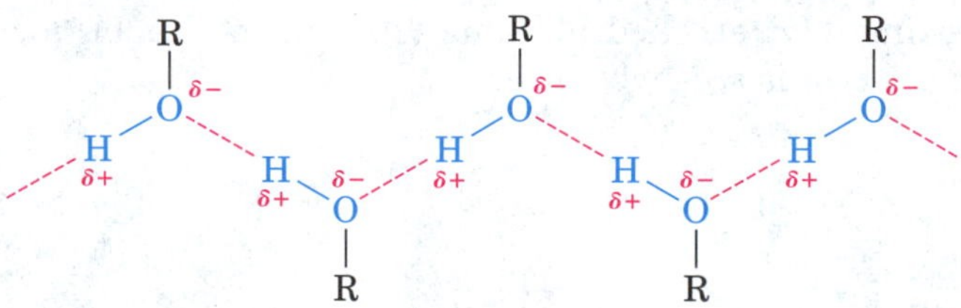

Figure 17.2 Hydrogen bonding in alcohols. A weak attraction between a positively polarized OH hydrogen and a negatively polarized oxygen holds molecules together.

PROBLEM.

17.3 The following data for three isomeric four-carbon alcohols show that there is a decrease in boiling point with increasing substitution:

1-Butanol, bp 117.5°C

2-Butanol, bp 99.5°C

2-Methyl-2-propanol, bp 82.2°C

Propose an explanation to account for this trend.

17.4 Properties of Alcohols: Acidity and Basicity

Like water, alcohols are both weakly basic and weakly acidic. As bases, alcohols are reversibly protonated by strong acids to yield oxonium ions, ROH_2^+:

$$R-\ddot{\underset{\cdot\cdot}{O}}-H + HX \rightleftharpoons R-\underset{\cdot\cdot}{\overset{H|}{O^+}}-H\ X^-$$

As acids, alcohols dissociate to a slight extent in dilute aqueous solution by donating a proton to water, generating H_3O^+ and an **alkoxide ion, RO^-**:

$$R\ddot{\underset{\cdot\cdot}{O}}-H + H_2\ddot{O}: \rightleftharpoons R\ddot{\underset{\cdot\cdot}{O}}:^- + H_3O:^+$$

In our earlier discussion of acidity (Sections 2.6–2.8), we said that the strength of any acid HA in water can be expressed by an *acidity constant*, K_a:

$$K_a = \frac{[A^-][H_3O^+]}{[HA]} \qquad pK_a = -\log K_a$$

Compounds with a smaller K_a (or larger pK_a) are weakly acidic, whereas compounds with a larger K_a (or smaller pK_a) are more strongly acidic. The data presented in Table 17.1 show that simple alcohols are about as acidic as water but that substituent groups can have significant effects. For example, methanol and ethanol are similar to water in acidity, but *tert*-butyl alcohol is a weaker acid and 2,2,2-trifluoroethanol is stronger.

Table 17.1 Acidity Constants of Some Alcohols

Alcohol	pK_a	
$(CH_3)_3COH$	18.00	Weaker acid
CH_3CH_2OH	16.00	
HOH (water)[a]	(15.74)	
CH_3OH	15.54	
CF_3CH_2OH	12.43	
$(CF_3)_3COH$	5.4	
HCl (hydrochloric acid)[a]	(−7.00)	Stronger acid

[a]Values for water and hydrochloric acid are shown for reference.

The effect of alkyl substitution on alcohol acidity is due primarily to solvent stabilization of the alkoxide ion that results from dissociation. The more easily the alkoxide ion is solvated by water, the more stable it is, the more its formation is energetically favored, and the greater the acidity of the parent alcohol. For example, the oxygen atom of an unhindered alkoxide ion, such as that from methanol, is sterically accessible and is easily solvated by water. The oxygen atom of a hindered alkoxide ion, however, such as that from *tert*-butyl alcohol, is less easily solvated and is therefore less stabilized.

Inductive effects (Section 16.6) are also important in determining alcohol acidities. Electron-withdrawing halogen substituents, for example, stabilize an alkoxide ion by spreading the charge over a larger area, thus making the alcohol more acidic. Compare, for example, the acidities of ethanol (pK_a = 16) and 2,2,2-trifluoroethanol (pK_a = 12.43), or of *tert*-butyl alcohol (pK_a = 18) and nonafluoro-*tert*-butyl alcohol (pK_a = 5.4).

Electron-withdrawing groups stabilize alkoxide and lower pK_a

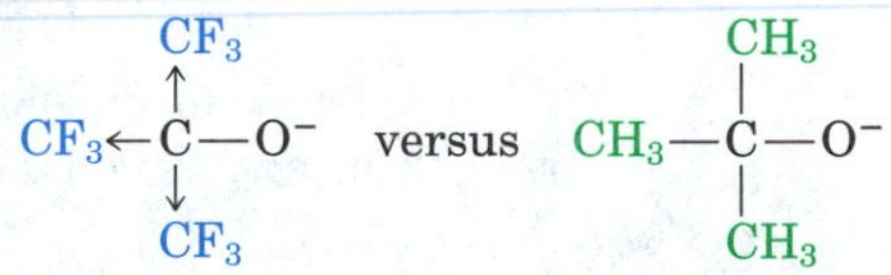

pK_a = 5.4 $\qquad$ pK_a = 18

Because alcohols are much less acidic than carboxylic acids or mineral acids, they don't react with weak bases such as amines or bicarbonate ion, and they react to only a limited extent with metal hydroxides. Alcohols do, however, react with alkali metals and with strong bases such as sodium hydride (NaH), sodium amide ($NaNH_2$), alkyllithium reagents (RLi), and Grignard reagents (RMgX). Metal alkoxides are themselves bases that are frequently used as reagents in organic chemistry.

$$CH_3OH + NaH \longrightarrow CH_3O^- \, Na^+ + H_2$$

Methanol $\qquad$ **Sodium methoxide**

$$CH_3CH_2OH + NaNH_2 \longrightarrow CH_3CH_2O^- \, Na^+ + NH_3$$

Ethanol $\qquad$ **Sodium ethoxide**

$$CH_3CH(OH)CH_3 + CH_3Li \longrightarrow CH_3CH(O^- \, Li^+)CH_3 + CH_4$$

Isopropyl alcohol $\qquad$ **Lithium isopropoxide**

Cyclohexanol $\qquad$ **Bromomagnesium cyclohexoxide**

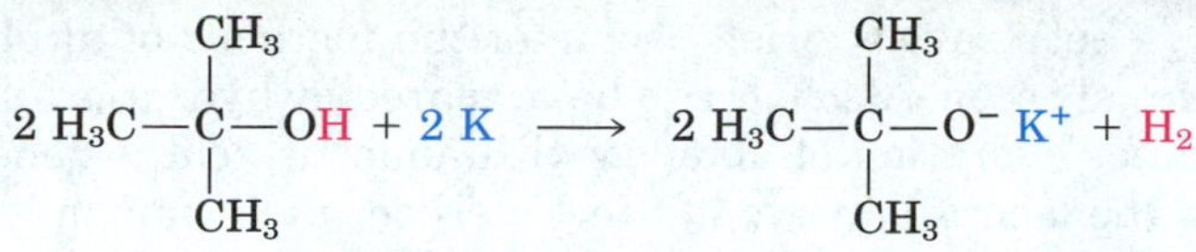

tert-Butyl alcohol — Potassium *tert*-butoxide

PROBLEM

17.4 Rank the following substances in order of increasing acidity.

$(CH_3)_2CHOH$ $\quad HC\equiv CH \quad$ $(CF_3)_2CHOH$ $\quad CH_3OH$

PROBLEM

17.5 *p*-Nitrobenzyl alcohol is more acidic than benzyl alcohol. Explain.

17.5 Preparation of Alcohols

Alcohols occupy a central position in organic chemistry. They can be prepared from many other kinds of compounds (alkenes, alkyl halides, ketones, esters, and aldehydes, among others), and they can be transformed into an equally wide assortment of compounds (Figure 17.3).

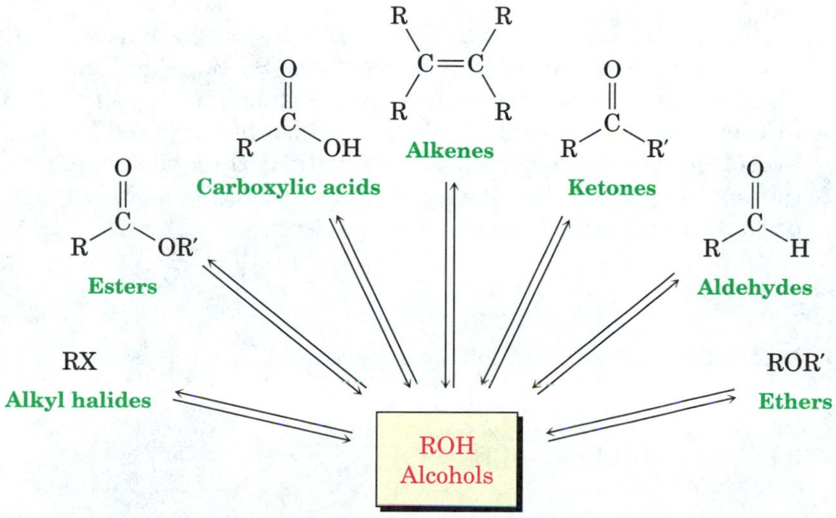

Figure 17.3 The central position of alcohols in organic chemistry. Alcohols can be prepared from, and converted into, many other kinds of compounds.

Let's review briefly some of the methods of alcohol preparation we've already seen. Alcohols can be prepared by hydration of alkenes. Because the direct hydration of alkenes with aqueous acid is generally a poor reaction in the laboratory, two indirect methods are commonly used. Hydroboration/oxidation (Section 7.5) yields the product of syn, non-Markovnikov hydration, whereas oxymercuration/reduction (Section 7.4) yields the product of Markovnikov hydration. Both reactions are generally applicable to most alkenes (Figure 17.4).

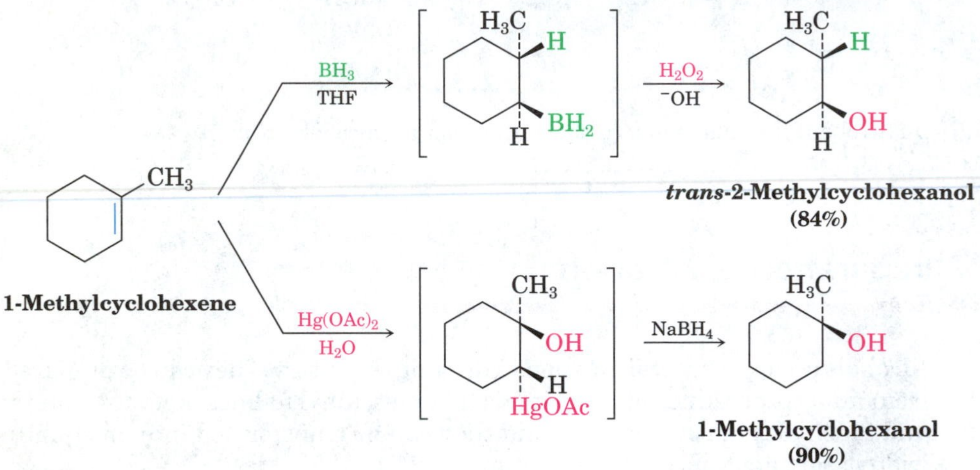

Figure 17.4 Two complementary methods for the hydration of an alkene to yield an alcohol.

1,2-Diols can be prepared by direct hydroxylation of an alkene with OsO_4 followed by reduction with $NaHSO_3$ (Section 7.8). The reaction takes place readily and occurs with syn stereochemistry (Figure 17.5). We'll see in the next chapter that 1,2-diols can also be prepared by acid-catalyzed hydrolysis of *epoxides*—compounds with a three-membered, oxygen-containing ring. Epoxide opening is complementary to direct hydroxylation because it occurs with anti stereochemistry.

PROBLEM

17.6 Predict the products of the following reactions:

(a)

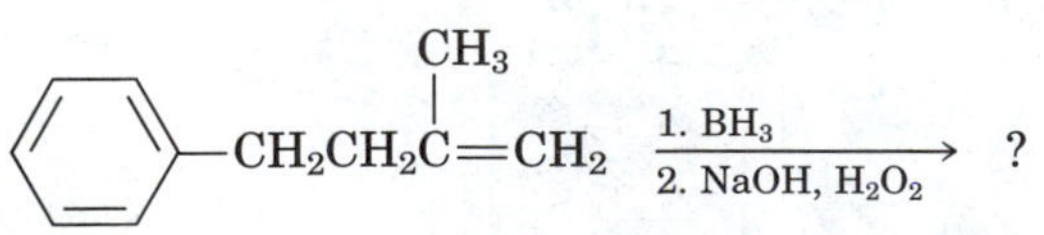

(b) $CH_3CH_2CH{=}C(CH_3)_2 \xrightarrow[\text{2. NaBH}_4]{\text{1. Hg(OAc)}_2\text{, H}_2\text{O}} ?$

(c) Reaction of *cis*-5-decene with OsO_4, followed by $NaHSO_3$ reduction. Be sure to indicate the stereochemistry of the product.

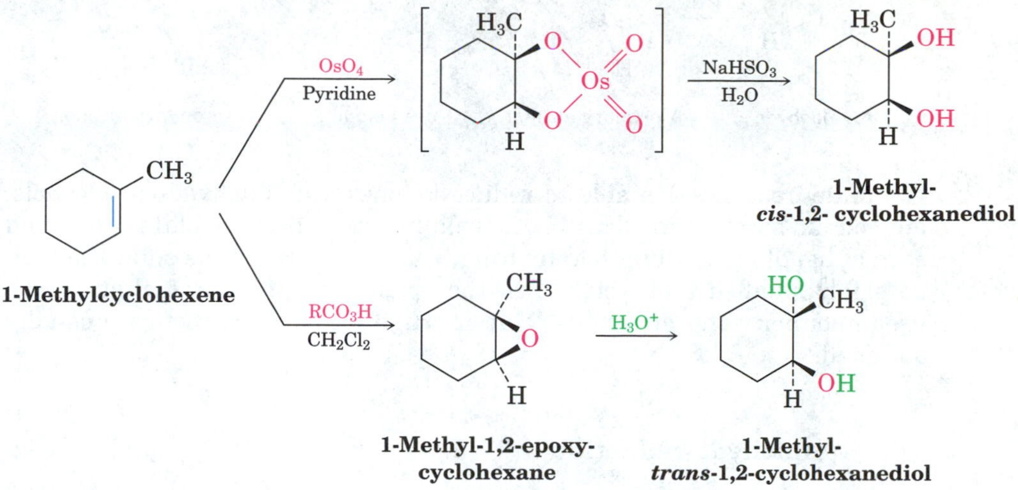

Figure 17.5 Two complementary methods for the preparation of 1,2-diols.

17.6 Alcohols from Reduction of Carbonyl Compounds

One of the most general methods for preparing alcohols is by *reduction* of a carbonyl compound. As we saw in Section 10.10, an organic reduction is a reaction that adds hydrogen to a molecule:

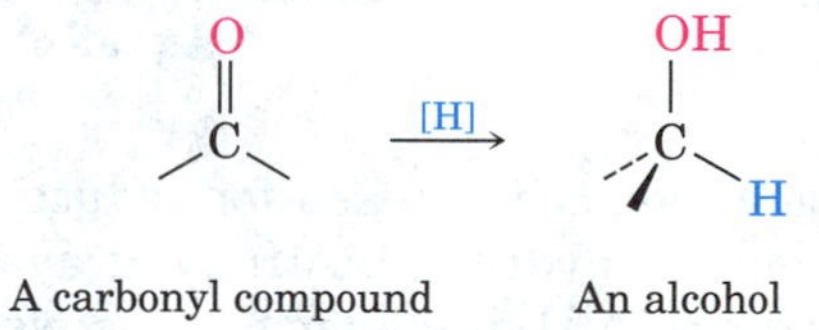

where [H] is a generalized reducing agent

All kinds of carbonyl compounds can be reduced, including aldehydes, ketones, carboxylic acids, and esters.

Reduction of Aldehydes and Ketones

Aldehydes and ketones are easily reduced to yield alcohols. Aldehydes are converted into primary alcohols, and ketones are converted into secondary alcohols.

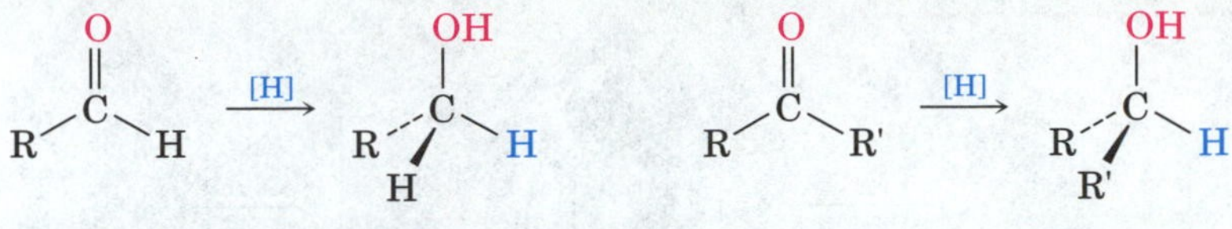

An aldehyde | A primary alcohol | A ketone | A secondary alcohol

Many reagents are able to reduce ketones and aldehydes to alcohols, but sodium borohydride, $NaBH_4$, is usually chosen because of its safety and ease of handling. Sodium borohydride is a white, crystalline solid that can be safely handled and weighed in the open atmosphere and that can be used in either water or alcohol solution. High yields of products are usually obtained.

Aldehyde reduction

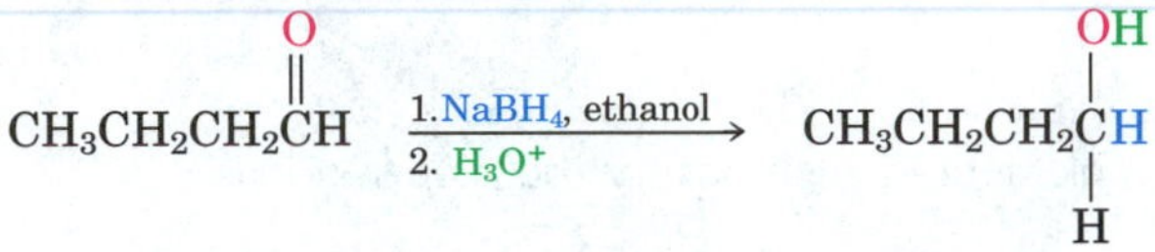

Butanal | **1-Butanol (85%) (a 1° alcohol)**

Ketone reduction

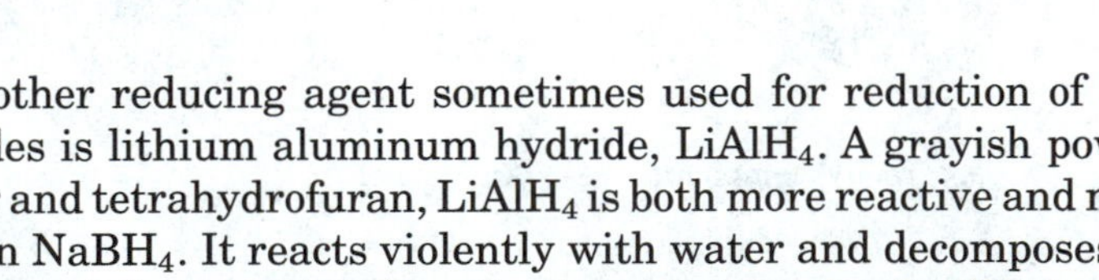

Dicyclohexyl ketone | **Dicyclohexylmethanol (88%) (a 2° alcohol)**

Another reducing agent sometimes used for reduction of ketones and aldehydes is lithium aluminum hydride, $LiAlH_4$. A grayish powder soluble in ether and tetrahydrofuran, $LiAlH_4$ is both more reactive and more dangerous than $NaBH_4$. It reacts violently with water and decomposes explosively when heated above 120°C. Despite these drawbacks, $LiAlH_4$ is an extremely valuable reagent that is used daily in thousands of laboratories.

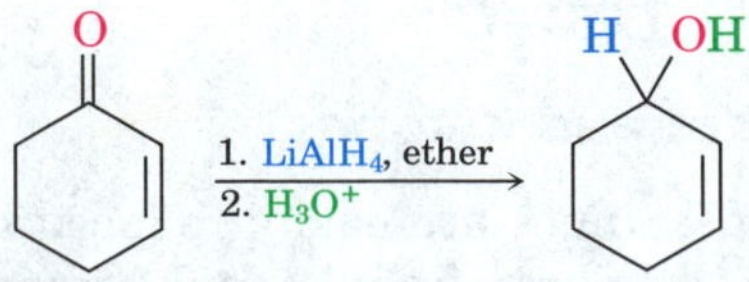

2-Cyclohexenone | **2-Cyclohexenol (94%)**

Reduction of Carboxylic Acids and Esters

Carboxylic acids and esters are reduced to give primary alcohols:

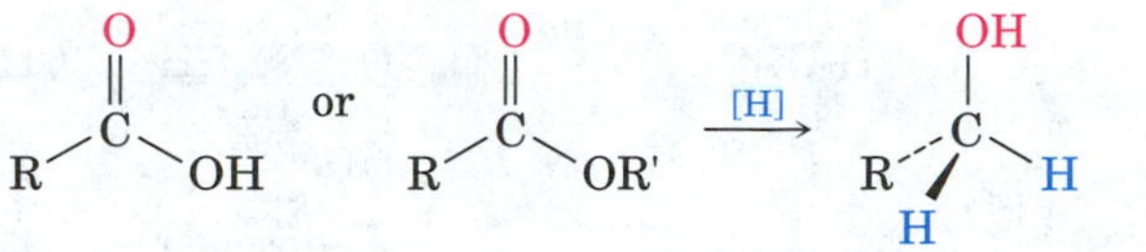

A carboxylic acid   An ester   A primary alcohol

These reactions are more difficult than the analogous reductions of aldehydes and ketones: $NaBH_4$ reduces esters slowly and does not reduce acids at all. Carboxylic acid and ester reductions are therefore usually carried out with $LiAlH_4$. All carbonyl groups, including acids, esters, ketones, and aldehydes, are rapidly reduced by $LiAlH_4$. Note that *one* hydrogen atom is delivered to the carbonyl carbon atom during ketone and aldehyde reductions, but that *two* hydrogens become bonded to the carbonyl carbon during carboxylic acid and ester reductions.

Carboxylic acid reduction

$$CH_3(CH_2)_7CH{=}CH(CH_2)_7\overset{\overset{O}{\|}}{C}OH \xrightarrow[\text{2. } H_3O^+]{\text{1. } LiAlH_4\text{, ether}} CH_3(CH_2)_7CH{=}CH(CH_2)_7CH_2OH$$

9-Octadecenoic acid (Oleic acid)   **9-Octadecen-1-ol (87%)**

Ester reduction

$$CH_3CH_2CH{=}CH\overset{\overset{O}{\|}}{C}OCH_3 \xrightarrow[\text{2. } H_3O^+]{\text{1. } LiAlH_4\text{, ether}} CH_3CH_2CH{=}CHCH_2OH + CH_3OH$$

Methyl 2-pentenoate   **2-Penten-1-ol (91%)**

We'll defer until Chapter 19 a detailed discussion of the mechanisms by which carbonyl compounds are reduced to give alcohols. For the moment, we'll simply note that these reactions involve the addition of nucleophilic hydride ion ($:H^-$) to the positively polarized carbon atom of the carbonyl group. The initial product is an alkoxide ion, which is protonated to yield the alcohol product by addition of H_3O^+ in a second step.

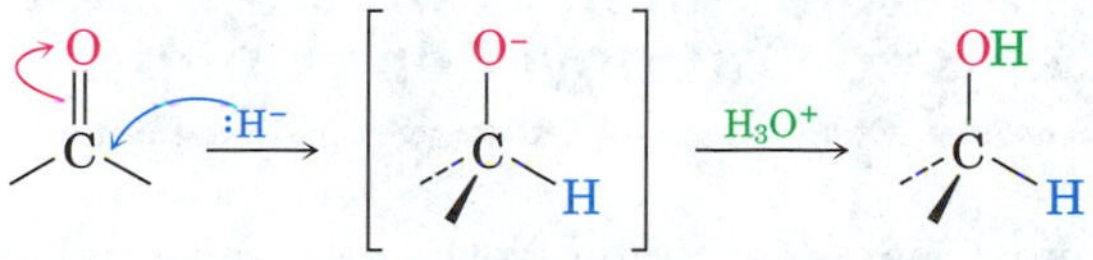

A carbonyl compound   An alkoxide ion intermediate   An alcohol

PROBLEM

17.7 What reagent would you use to accomplish each of the following reactions?

(a) 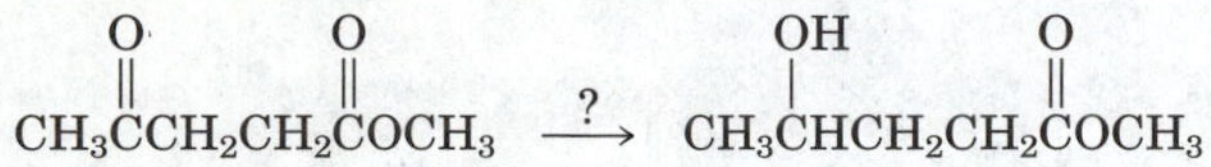

(b) 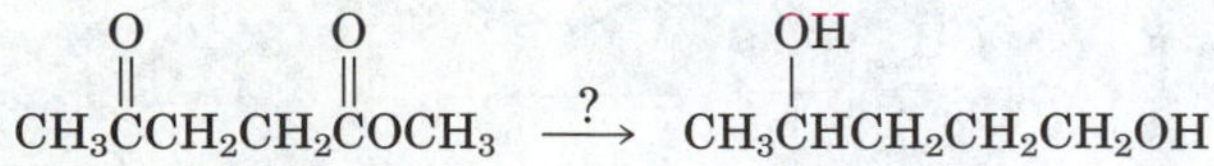

(c) 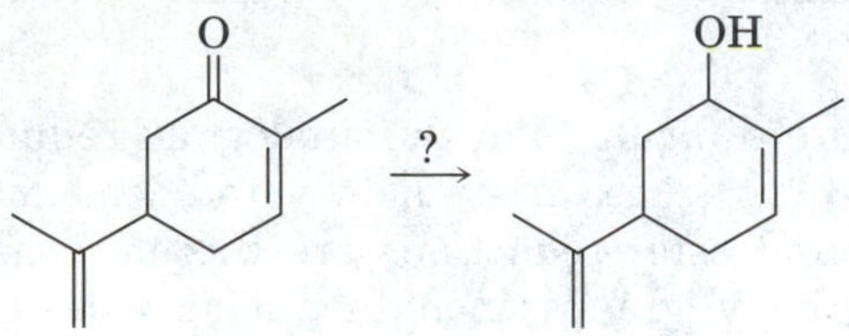

Carvone
(from spearmint oil)

(d) 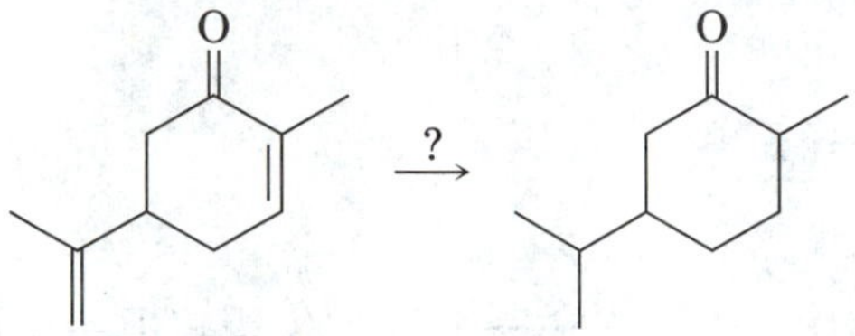

PROBLEM

17.8 What carbonyl compounds give the following alcohols on reduction with $LiAlH_4$? Show all possibilities.

(a) 

(b)

(c) 

(d) $(CH_3)_2CHCH_2OH$

17.7 Alcohols from Addition of Grignard Reagents to Carbonyl Compounds

Grignard reagents, RMgX, react with carbonyl compounds to yield alcohols in much the same way that hydride reagents do. The result is a useful and general method of alcohol synthesis.

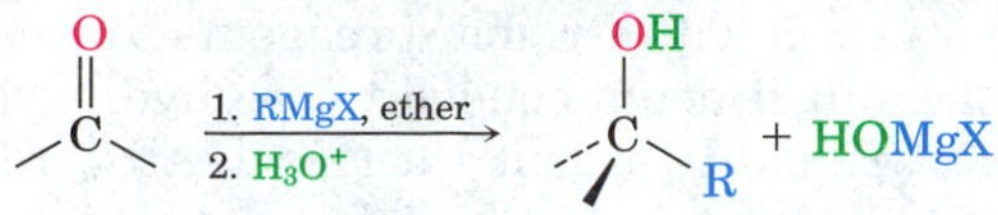

We saw in Section 10.8 that alkyl, aryl, and vinylic halides react with magnesium in ether or tetrahydrofuran solution to generate Grignard reagents, RMgX:

$$\text{R—X} + \text{Mg} \longrightarrow \overset{\delta-}{\text{R}}\text{—}\overset{\delta+}{\text{MgX}}$$

A Grignard reagent

where R = 1°, 2°, or 3° alkyl, aryl, or vinylic

X = Cl, Br, or I

A great many alcohols can be obtained from Grignard reactions, depending on the reagents used. For example, Grignard reagents react with formaldehyde, $H_2C{=}O$, to give primary alcohols, with aldehydes to give secondary alcohols, and with ketones to give tertiary alcohols:

Formaldehyde reaction

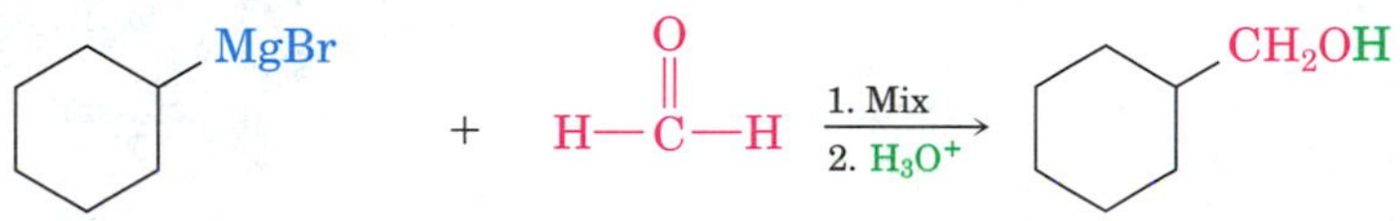

Cyclohexylmagnesium bromide | **Formaldehyde** | **Cyclohexylmethanol (65%) (a 1° alcohol)**

Aldehyde reaction

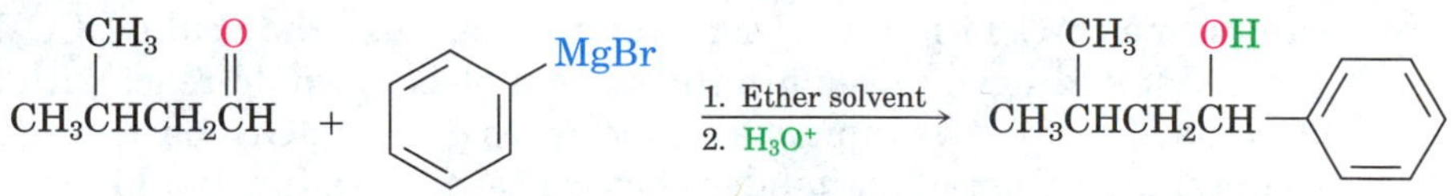

3-Methylbutanal | **Phenylmagnesium bromide** | **3-Methyl-1-phenyl-1-butanol (73%) (a 2° alcohol)**

Ketone reaction

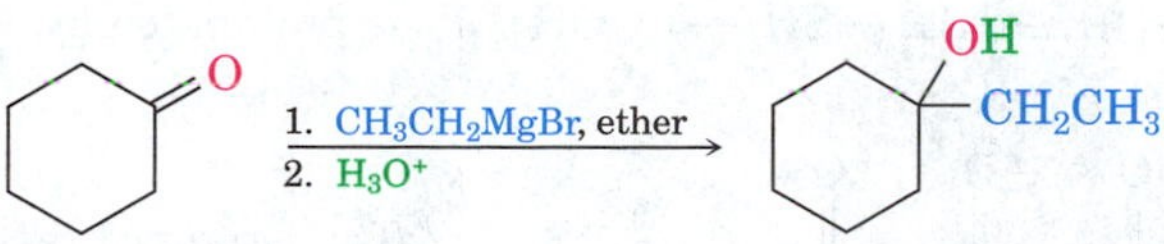

Cyclohexanone | **1-Ethylcyclohexanol (89%) (a 3° alcohol)**

Esters react with Grignard reagents to yield tertiary alcohols in which two of the substituents bonded to the hydroxyl-bearing carbon have come from the Grignard reagent (just as $LiAlH_4$ reduction of esters adds two hydrogens).

Ester reaction

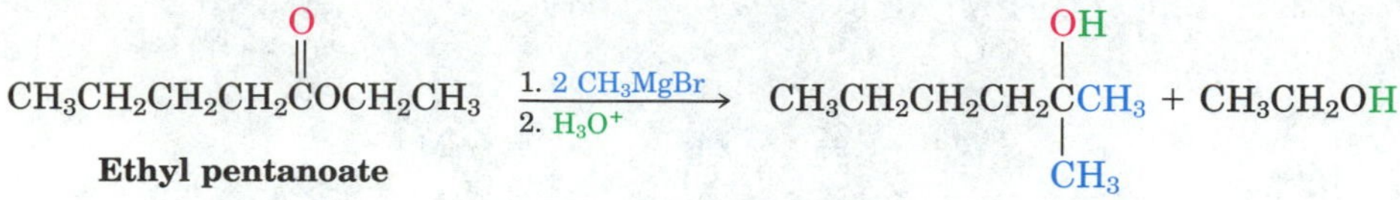

Ethyl pentanoate

2-Methyl-2-hexanol (85%)
(a 3° alcohol)

Carboxylic acids don't give addition products with Grignard reagents because the acidic carboxyl hydrogen reacts with the Grignard reagent to yield a hydrocarbon and the magnesium salt of the acid. We saw this reaction in Section 10.8 as a means of reducing alkyl halides to alkanes.

$$RBr + Mg \longrightarrow RMgBr$$

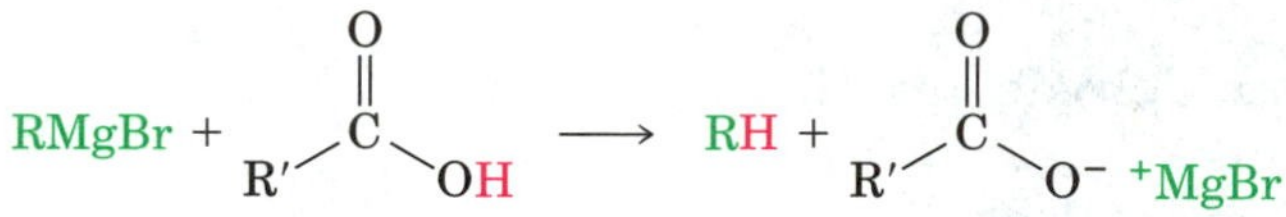

A carboxylic acid — A carboxylic acid salt

The Grignard reaction, though useful, has several limitations. One major problem is that a Grignard reagent can't be prepared from an organohalide if there are other reactive functional groups in the same molecule. For example, a compound that is both an alkyl halide and a ketone can't form a Grignard reagent because it would react with itself. Similarly, a compound that is both an alkyl halide and a carboxylic acid, an alcohol, or an amine can't form a Grignard reagent because the acidic $RCO_2\mathbf{H}$, $RO\mathbf{H}$, or $RN\mathbf{H}_2$ hydrogen present in the same molecule would react with the basic Grignard reagent as rapidly as it forms. In general, Grignard reagents can't be prepared from alkyl halides that contain these functional groups (FG):

where FG = —OH, —NH, —SH, —COOH } The Grignard reagent is protonated by these groups.

FG = —CHO, —COR, —CONR$_2$, —C≡N, —NO$_2$, —SO$_2$R } The Grignard reagent adds to these groups.

As with the reduction of carbonyl compounds discussed in the previous section, we'll defer a detailed treatment of the mechanism of Grignard reactions until Chapter 19. For the moment, it's sufficient to note that Grignard reagents act as nucleophilic carbon anions (carbanions, $:R^-$) and that the addition of a Grignard reagent to a carbonyl compound is analogous to the addition of hydride ion.

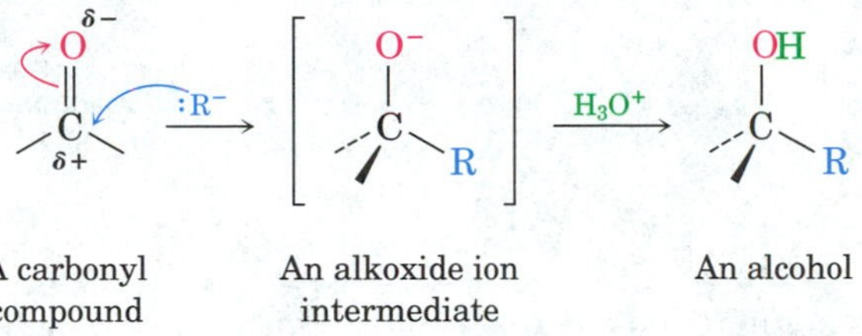

PRACTICE PROBLEM

How could you use the addition of a Grignard reagent to a ketone to synthesize 2-phenyl-2-propanol?

Solution First, draw the structure of the product and identify the groups bonded to the alcohol carbon atom. In the present instance, there are two methyl groups ($-CH_3$) and one phenyl ($-C_6H_5$). One of the three will have come from a Grignard reagent, and the remaining two will have come from a ketone. Thus, the possibilities are addition of methylmagnesium bromide to acetophenone and addition of phenylmagnesium bromide to acetone:

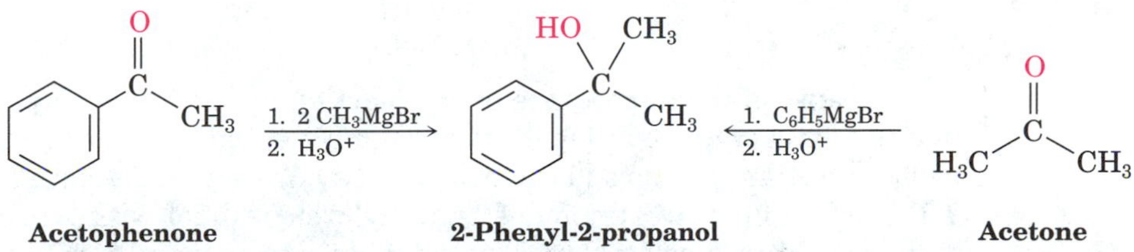

PROBLEM

17.9 Show the products obtained from addition of methylmagnesium bromide to the following compounds:
(a) Cyclopentanone (b) Benzophenone (diphenyl ketone) (c) 3-Hexanone

PROBLEM

17.10 How could you use a Grignard reaction to prepare the following alcohols?
(a) 2-Methyl-2-propanol (b) 1-Methylcyclohexanol
(c) 3-Methyl-3-pentanol (d) 2-Phenyl-2-butanol
(e) Benzyl alcohol

17.8 Reactions of Alcohols

Reactions of alcohols can be divided into two groups—those that occur at the C–O bond and those that occur at the O–H bond:

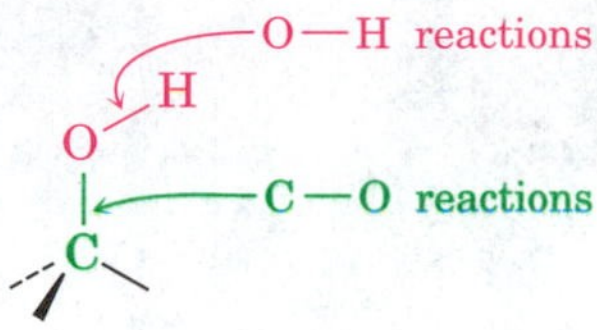

Let's begin to look at reactions of both types by reviewing some of the alcohol reactions seen in previous chapters.

Dehydration of Alcohols to Yield Alkenes

One of the most valuable C–O bond reactions of alcohols is dehydration to give alkenes. The C–O bond and a neighboring C–H are broken, and an alkene π bond is formed:

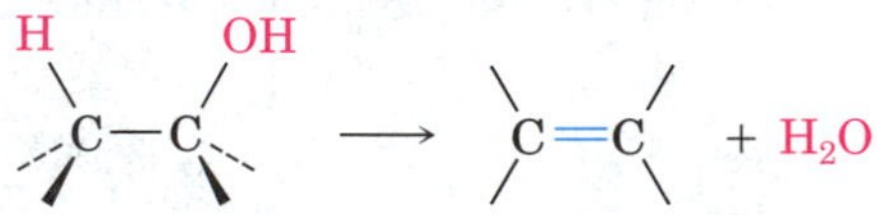

A dehydration reaction

Because of the importance of the reaction, a number of ways have been devised for carrying out dehydrations. One method that works particularly well for tertiary alcohols is the acid-catalyzed reaction discussed in Section 7.1. For example, treatment of 1-methylcyclohexanol with warm aqueous sulfuric acid in a solvent such as tetrahydrofuran results in loss of water and formation of 1-methylcyclohexene.

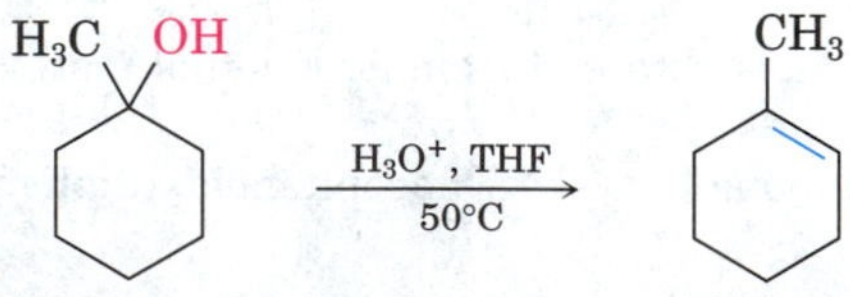

1-Methylcyclohexanol **1-Methylcyclohexene (91%)**

Acid-catalyzed dehydrations usually follow Zaitsev's rule (Section 11.10) and yield the more highly substituted alkene as the major product. Thus, 2-methyl-2-butanol gives primarily 2-methyl-2-butene (trisubstituted double bond) rather than 2-methyl-1-butene (disubstituted double bond).

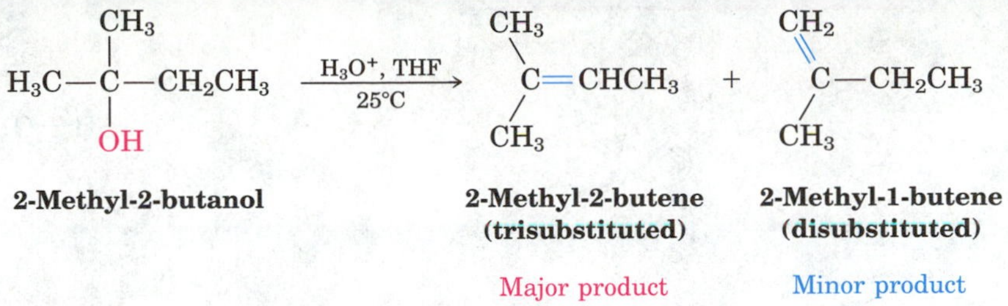

Only tertiary alcohols are normally dehydrated with acid. Secondary alcohols can be made to react, but the conditions are severe (75% H_2SO_4, 100°C) and sensitive molecules don't survive. Primary alcohols are even less reactive than secondary ones, and very harsh conditions are necessary to cause dehydration (95% H_2SO_4, 150°C). Thus, the reactivity order for acid-catalyzed dehydrations is

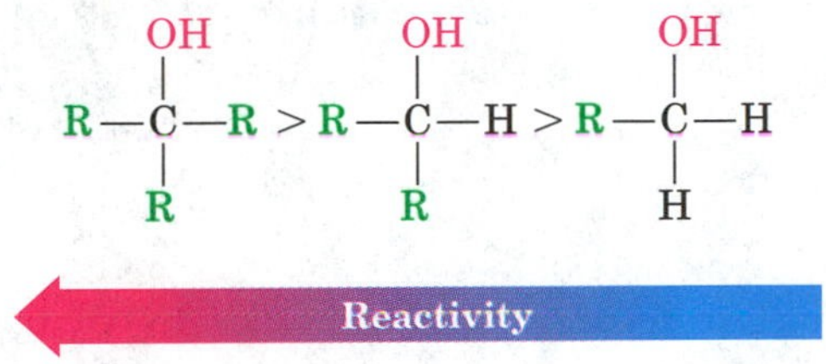

The reasons for the observed reactivity order are best understood by looking at the mechanism of the reaction (Figure 17.6, p. 650). Acid-catalyzed dehydrations are E1 reactions (Section 11.14), which occur by a three-step mechanism involving protonation of the alcohol oxygen, loss of water to generate a carbocation intermediate, and final loss of a proton (H^+) from the neighboring carbon atom. Tertiary substrates *always* react fastest in E1 reactions because they lead to highly stabilized, tertiary carbocation intermediates.

To circumvent the need for strong acid and allow the dehydration of secondary alcohols in a gentler way, reagents have been developed that are effective under mild, basic conditions. One such reagent, phosphorus oxychloride ($POCl_3$), is often able to effect the dehydration of secondary and tertiary alcohols at 0°C in the basic amine solvent pyridine:

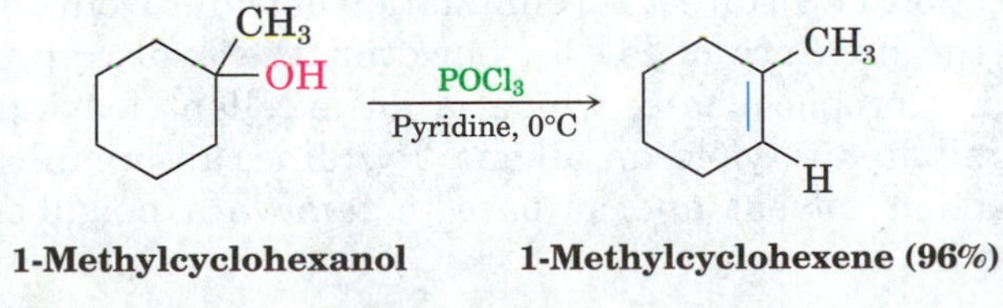

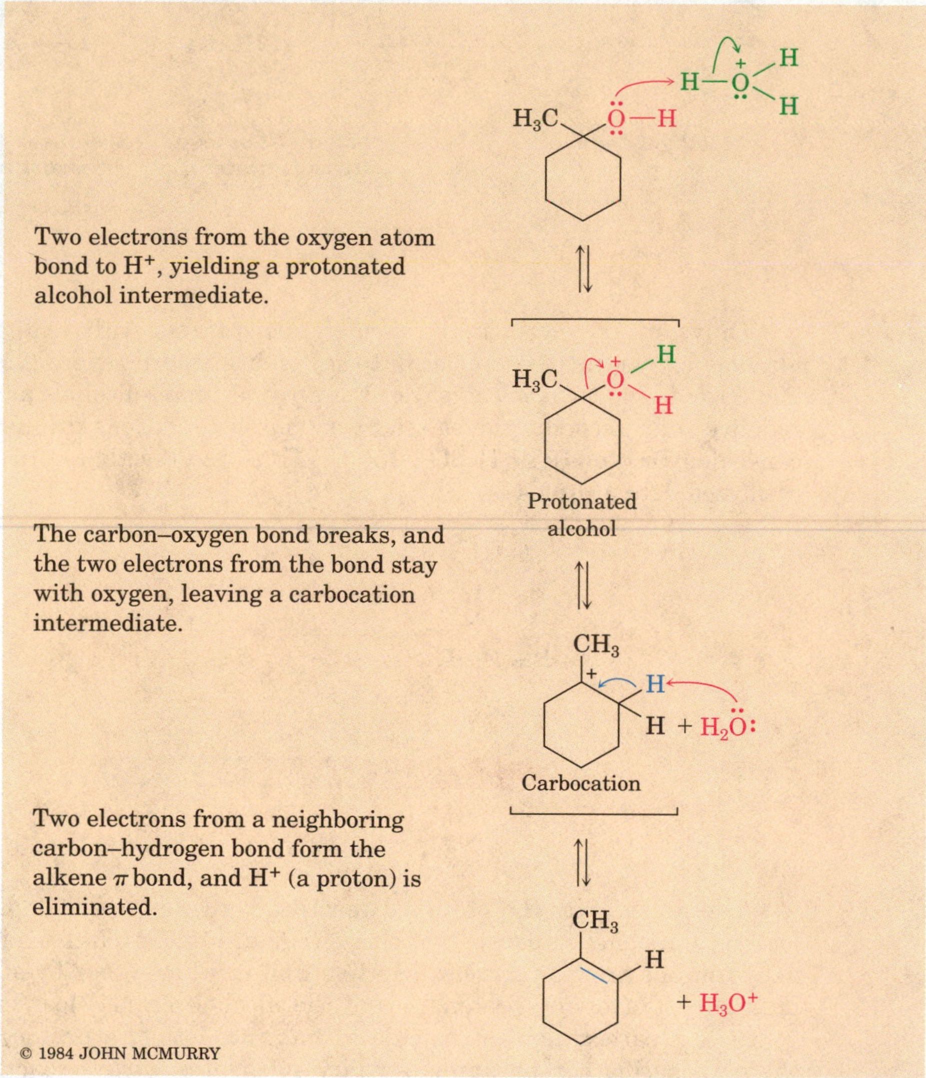

Figure 17.6 Mechanism of the acid-catalyzed dehydration of an alcohol to yield an alkene. The process is an E1 reaction and involves a carbocation intermediate.

Alcohol dehydrations carried out with $POCl_3$ take place by the E2 mechanism shown in Figure 17.7. Because hydroxide ion is a poor leaving group (Section 11.5), direct E2 elimination of water from an alcohol does not occur. In the presence of $POCl_3$, however, the hydroxyl group is converted into a dichlorophosphate, which is an excellent leaving group and is readily eliminated to yield an alkene. Pyridine, an organic amine, serves both as reaction solvent and as base to remove a neighboring proton in the E2 elimination step.

The alcohol hydroxyl group reacts with $POCl_3$ to form a dichlorophosphate intermediate.

E2 elimination then occurs by the usual one-step mechanism as the amine base pyridine abstracts a proton from the neighboring carbon at the same time that the dichlorophosphate group is leaving.

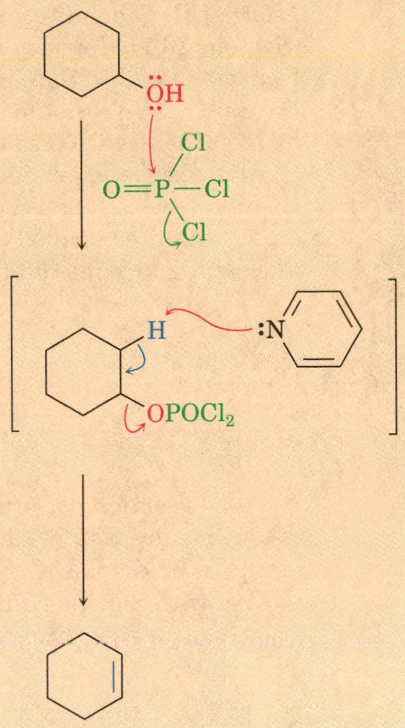

Figure 17.7 Mechanism of the dehydration of secondary and tertiary alcohols by reaction with $POCl_3$ in pyridine. The reaction is an E2 process.

PROBLEM. . . . .

17.11 What product(s) would you expect from dehydration of the following alcohols with $POCl_3$ in pyridine? Indicate the major product in each case.

(a) 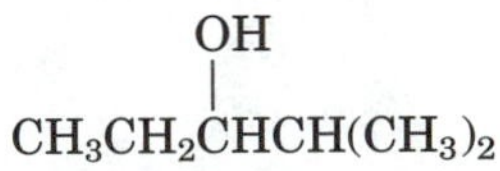

(b) *trans*-2-Methylcyclohexanol

(c) *cis*-2-Methylcyclohexanol

PROBLEM. . . . .

17.12 Good evidence for the intermediacy of carbocations in the acid-catalyzed dehydration of alcohols comes from the observation that rearrangements sometimes occur. Propose a mechanism to account for the formation of 2,3-dimethyl-2-butene from 3,3-dimethyl-2-butanol. (*Hint*: See Section 6.12.)

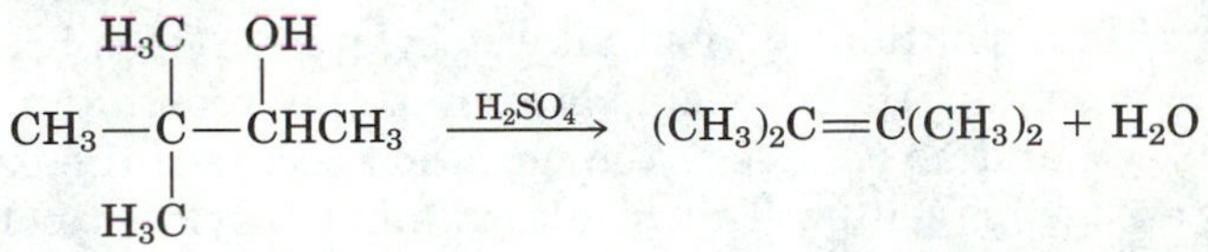

. . . .

Conversion of Alcohols into Alkyl Halides

Another C–O bond reaction of alcohols is their conversion into alkyl halides (Section 10.7). Tertiary alcohols are readily converted into alkyl halides by treatment with either HCl or HBr at 0°C. Primary and secondary alcohols are much more resistant to acid, however, and are best converted into halides by treatment with either $SOCl_2$ or PBr_3.

As discussed in Section 11.16, the reaction of a tertiary alcohol with HX takes place by an S_N1 mechanism. Acid protonates the hydroxyl oxygen atom, water is expelled to generate a carbocation, and the cation reacts with nucleophilic halide ion to give the alkyl halide product.

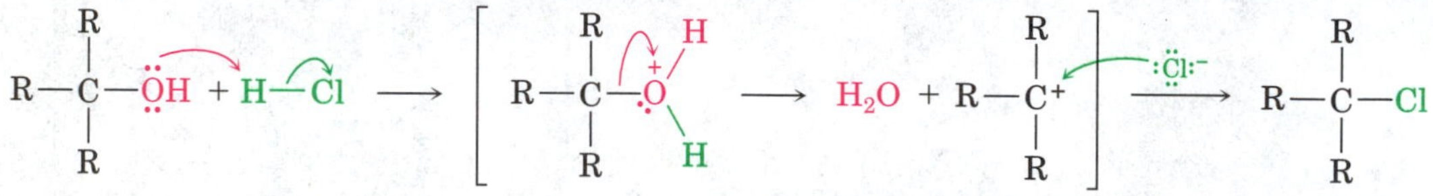

The reactions of primary and secondary alcohols with $SOCl_2$ and PBr_3 take place by S_N2 mechanisms. Hydroxide ion itself is too poor a leaving group to be displaced by nucleophiles in S_N2 reactions, but reaction of an alcohol with $SOCl_2$ or PBr_3 converts the hydroxyl into a much better leaving group that is readily expelled by back-side nucleophilic attack (Figure 17.8).

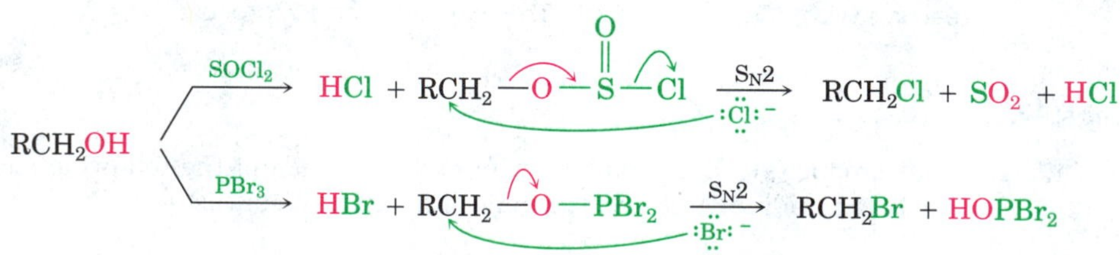

Figure 17.8 Conversion of a primary alcohol into alkyl halides by S_N2 reactions with $SOCl_2$ and PBr_3.

Conversion of Alcohols into Tosylates

Alcohols react with *p*-toluenesulfonyl chloride (tosyl chloride, *p*-TosCl) in pyridine solution to yield alkyl tosylates, ROTos (Section 11.2). Only the O–H bond of the alcohol is broken in this reaction; the C–O bond remains intact, and no change of configuration occurs if the oxygen is attached to a stereogenic center. The resultant alkyl tosylates behave much like alkyl halides in their chemistry, undergoing S_N1 and S_N2 substitution reactions with ease.

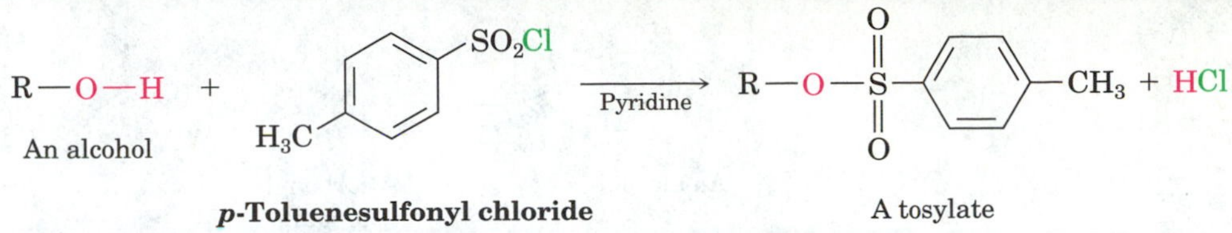

One of the most important reasons for using tosylates instead of halides in S_N2 reactions is stereochemical. The S_N2 reaction of an alcohol via an alkyl halide proceeds with *two* Walden inversions—one to make the halide from the alcohol and one to substitute the halide—and yields a product with the same absolute stereochemistry as the starting alcohol. The S_N2 reaction of an alcohol via a tosylate, however, proceeds with only *one* Walden inversion and yields a product of opposite absolute stereochemistry to the starting alcohol. Figure 17.9 shows a series of reactions on optically active 2-octanol that illustrates these stereochemical relationships.

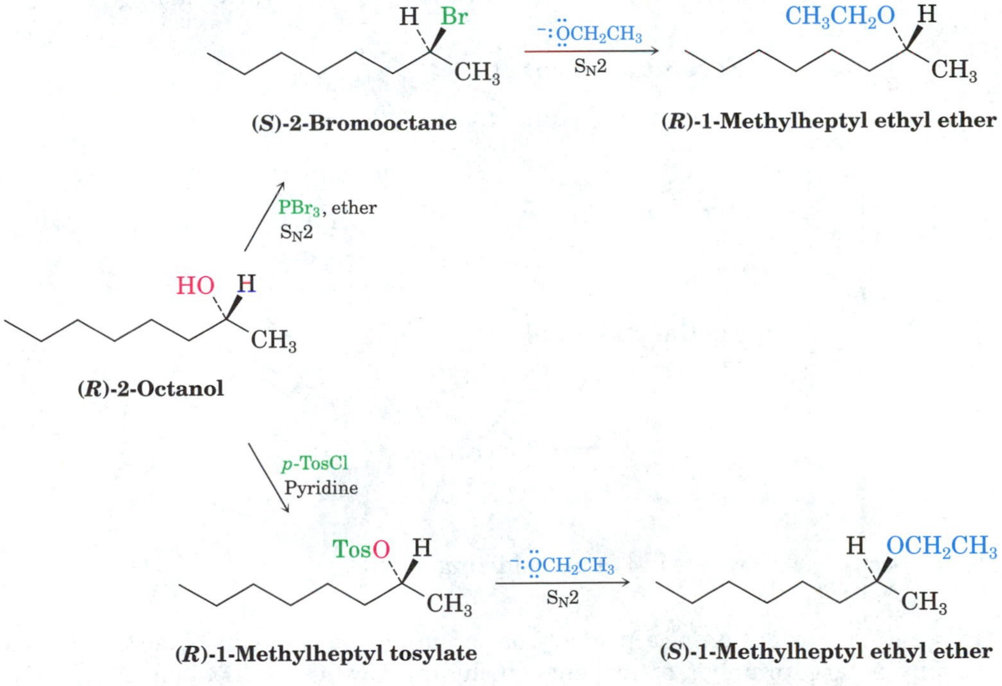

Figure 17.9 Stereochemical consequences of some S_N2 reactions on derivatives of (*R*)-2-octanol. Substitution via the halide is complementary to substitution via the tosylate.

17.9 Oxidation of Alcohols

One of the most valuable reactions of alcohols is their oxidation to yield carbonyl compounds—the opposite of the reduction of carbonyl compounds to yield alcohols:

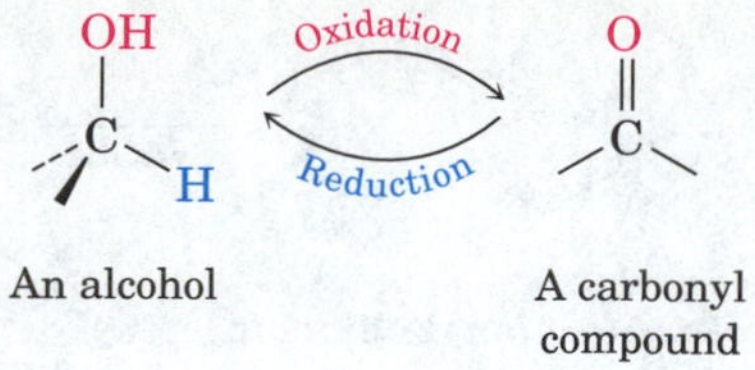

Primary alcohols yield aldehydes or carboxylic acids, secondary alcohols yield ketones, and tertiary alcohols don't normally react with most oxidizing agents.

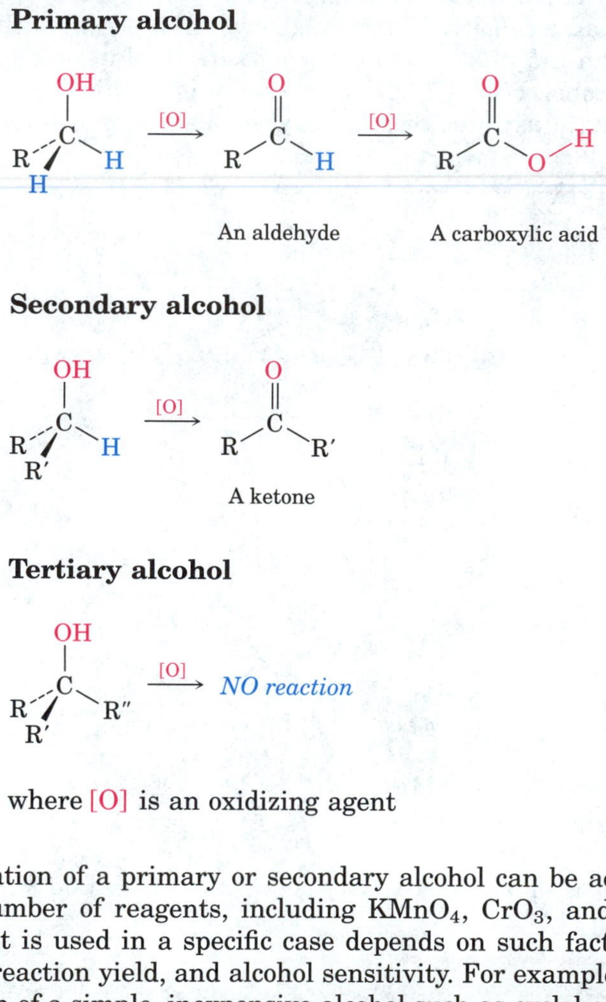

where [O] is an oxidizing agent

The oxidation of a primary or secondary alcohol can be accomplished by a large number of reagents, including $KMnO_4$, CrO_3, and $Na_2Cr_2O_7$. Which reagent is used in a specific case depends on such factors as cost, convenience, reaction yield, and alcohol sensitivity. For example, the large-scale oxidation of a simple, inexpensive alcohol such as cyclohexanol would best be done with a cheap oxidant such as $Na_2Cr_2O_7$. On the other hand, the small-scale oxidation of a delicate and expensive polyfunctional alcohol would best be done with a mild and high-yielding reagent, regardless of cost.

Primary alcohols are oxidized either to aldehydes or to carboxylic acids, depending on the reagents chosen and on the conditions used. Probably the best method for preparing aldehydes from primary alcohols on a laboratory scale (as opposed to an industrial scale) is by use of pyridinium chlorochromate (PCC, $C_5H_6NCrO_3Cl$) in dichloromethane solvent.

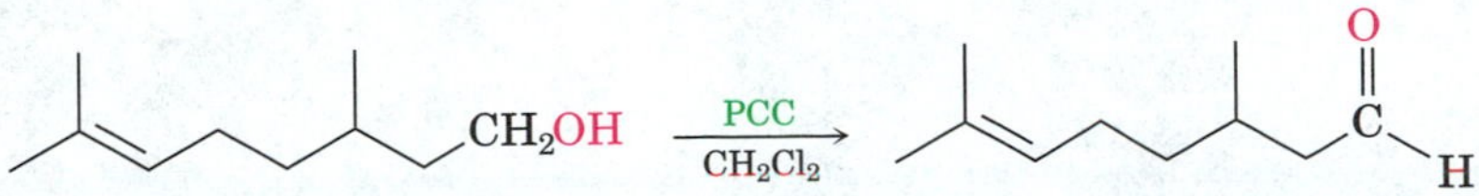

Citronellol (from rose oil) **Citronellal (82%)**

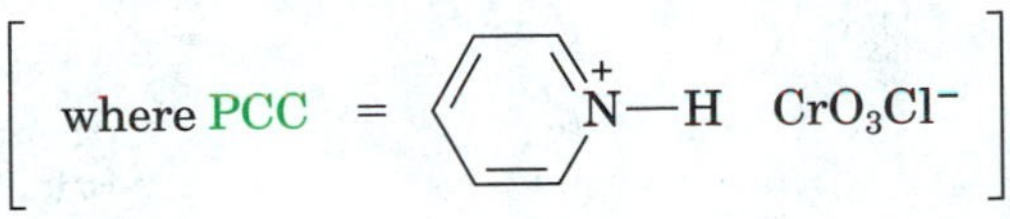

Most other oxidizing agents, such as chromium trioxide (CrO_3) in aqueous sulfuric acid (**Jones' reagent**), oxidize primary alcohols to carboxylic acids. Aldehydes are involved as intermediates in the Jones oxidation but can't usually be isolated because they are further oxidized too rapidly.

$$CH_3(CH_2)_8CH_2OH \xrightarrow[H_2O,\ acetone]{CrO_3,\ H_2SO_4} CH_3(CH_2)_8\overset{O}{\overset{\|}{C}}OH$$

1-Decanol **Decanoic acid (93%)**

Secondary alcohols are oxidized easily and in high yield to give ketones. For large-scale oxidations, an inexpensive reagent such as sodium dichromate in aqueous acetic acid is used.

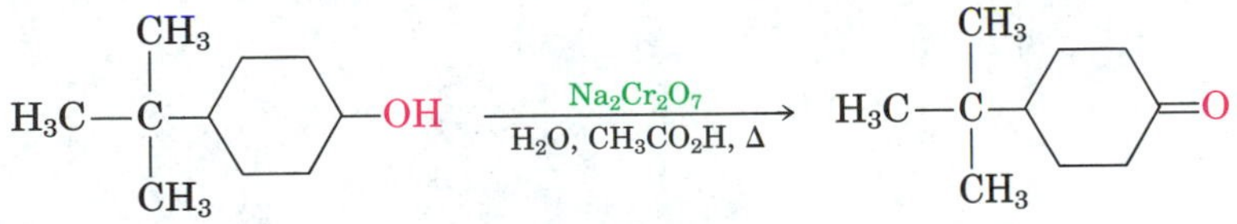

4-*tert*-Butylcyclohexanol **4-*tert*-Butylcyclohexanone (91%)**

For more sensitive alcohols, pyridinium chlorochromate is often used because the reaction is milder and occurs at lower temperatures.

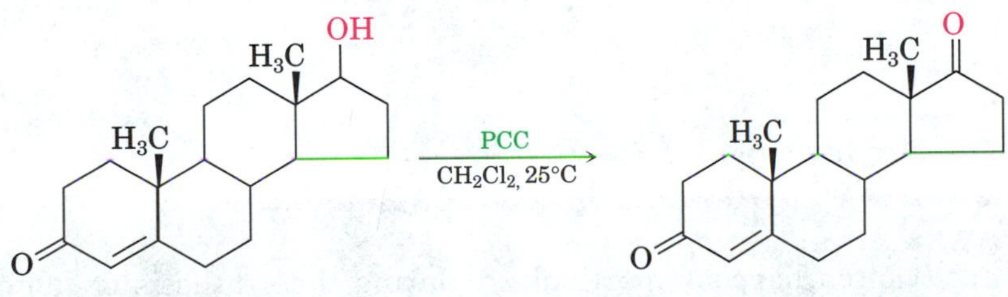

Testosterone (steroid; male sex hormone) **4-Androstene-3,17-dione (82%)**

All these oxidations occur by a pathway that is closely related to the E2 reaction. The first step involves reaction between the alcohol and a Cr(VI) reagent to form a *chromate* intermediate, which contains an O–Cr bond. Bimolecular elimination then yields the carbonyl product.

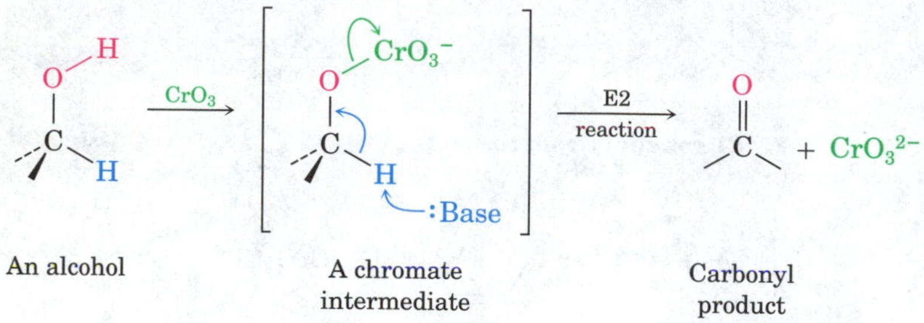

Although we usually think of the E2 reaction as a means of generating a carbon–*carbon* double bond by elimination of a halide leaving group, the reaction is also useful for generating a carbon–*oxygen* double bond by elimination of a high-valent metal as the leaving group. This is just one more example of how the same few fundamental mechanistic types keep reappearing in different variations.

PROBLEM

17.13 What alcohols would give these products on oxidation?

(a)

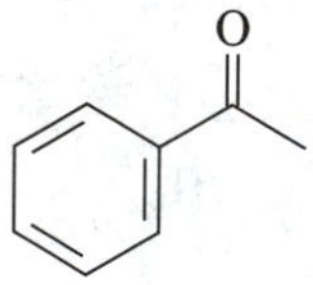

(b)

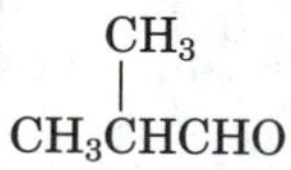

(c)

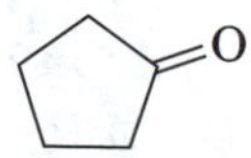

PROBLEM

17.14 What products would you expect from oxidation of the following compounds with Jones' reagent? With pyridinium chlorochromate?
(a) 1-Hexanol (b) 2-Hexanol (c) Hexanal

17.10 Protection of Alcohols

It often happens, particularly during the synthesis of complex molecules, that one functional group in a molecule interferes with an intended reaction on a second functional group elsewhere in the same molecule. For example, we saw earlier in this chapter that a Grignard reagent can't be prepared from a halo alcohol because the C–Mg bond is not compatible with the presence of an acidic –OH group in the same molecule.

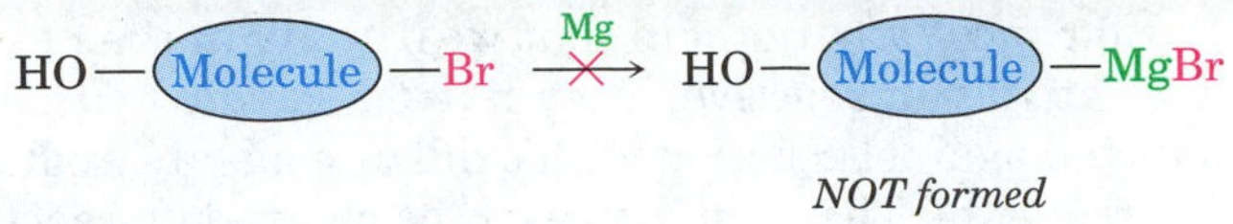

When this kind of incompatibility arises, it's sometimes possible to circumvent the problem by *protecting* the interfering functional group. Protection involves three steps: (1) introducing a **protecting group** to block the interfering function, (2) carrying out the desired reaction, and (3) removing the protecting group.

One of the most common methods of alcohol protection is reaction with chlorotrimethylsilane to yield a **trimethylsilyl (TMS) ether**. The reaction is carried out in the presence of a base, usually triethylamine, to help form the alkoxide anion from the alcohol and to scavenge the HCl by-product from the reaction.

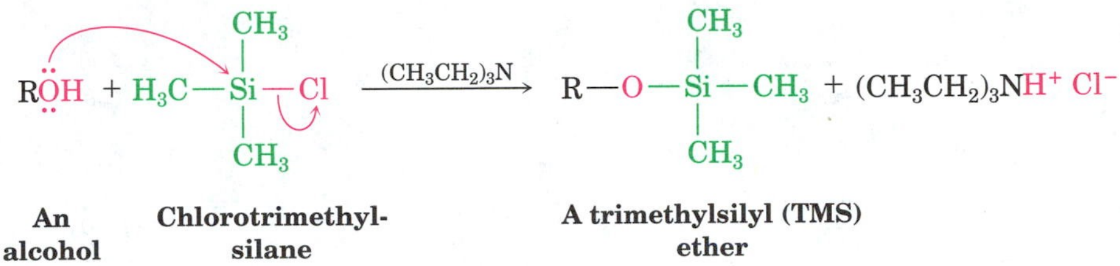

For example:

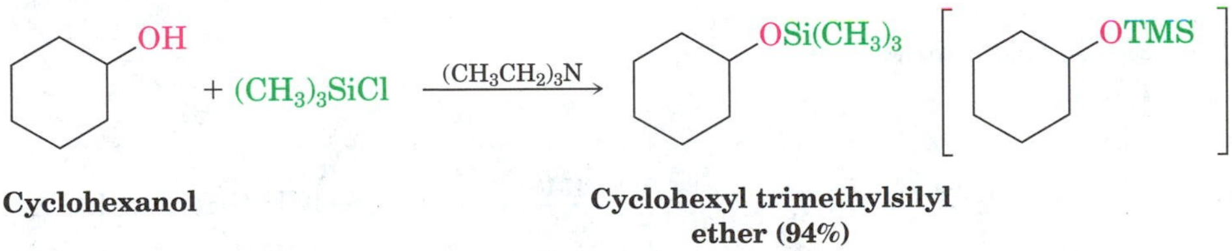

The ether-forming step is an attack of the alkoxide ion on the silicon atom, with concurrent loss of a leaving chloride anion. Unlike most S_N2 reactions, though, this reaction takes place at a *tertiary* center—a trialkyl-substituted silicon atom. The reaction occurs because silicon, a third-row atom, is larger than carbon and forms longer bonds. The three methyl substituents attached to silicon thus offer less steric hindrance to attack than they do in the analogous *tert*-butyl chloride.

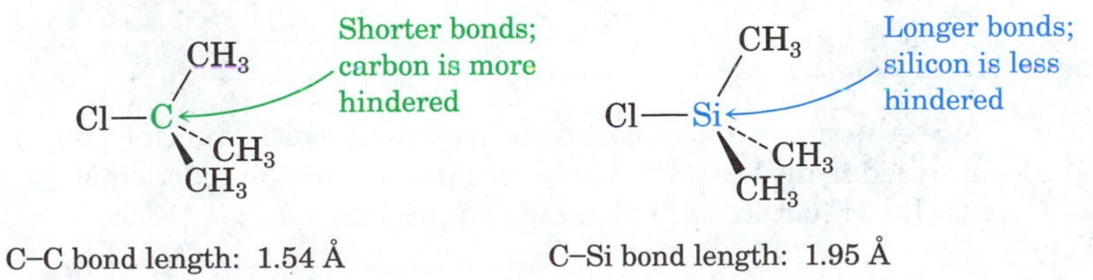

Like most other ethers that we'll study in the next chapter, TMS ethers are relatively unreactive. They have no acidic hydrogens and are therefore protected against reaction with oxidizing agents, reducing agents, and Grignard reagents. They can, however, be cleaved by reaction with aqueous acid or with fluoride ion to regenerate the alcohol.

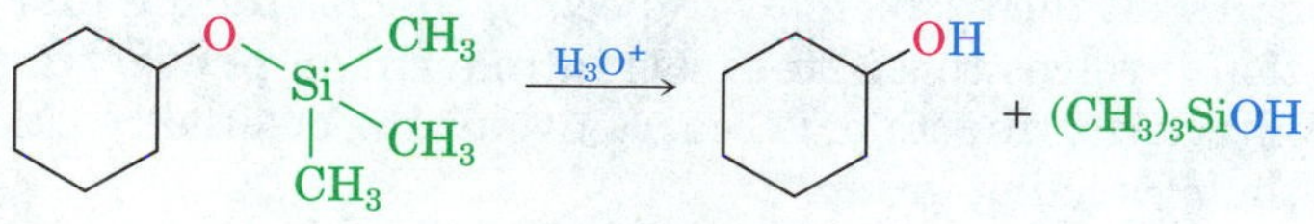

Cyclohexyl TMS ether **Cyclohexanol**

To complete the earlier example, it's possible to use a halo alcohol in a Grignard reaction by employing a protection sequence. For example, we can add 3-bromo-1-propanol to acetaldehyde by the route shown in Figure 17.10.

Step 1: Protect alcohol

$$HOCH_2CH_2CH_2Br + (CH_3)_3SiCl \xrightarrow{(CH_3CH_2)_3N} (CH_3)_3SiOCH_2CH_2CH_2Br$$

Step 2a: Form Grignard reagent

$$(CH_3)_3SiOCH_2CH_2CH_2Br \xrightarrow[\text{Ether}]{Mg} (CH_3)_3SiOCH_2CH_2CH_2MgBr$$

Step 2b: Do Grignard reaction

$$(CH_3)_3SiOCH_2CH_2CH_2MgBr \xrightarrow[2.\ H_3O^+]{1.\ CH_3CHO} (CH_3)_3SiOCH_2CH_2CH_2CH(OH)CH_3$$

Step 3: Remove protecting group

$$(CH_3)_3SiOCH_2CH_2CH_2CH(OH)CH_3 \xrightarrow{H_3O^+} HOCH_2CH_2CH_2CH(OH)CH_3 + (CH_3)_3SiOH$$

Figure 17.10 Use of a TMS-protected alcohol during a Grignard reaction.

PROBLEM

17.15 TMS ethers can be removed by treatment with fluoride ion as well as by acid-catalyzed hydrolysis. Propose a mechanism for the reaction of cyclohexyl TMS ether with LiF. Fluorotrimethylsilane is a product.

17.11 Spectroscopic Analysis of Alcohols

Infrared Spectroscopy

Alcohols show a characteristic O–H stretching absorption at 3300–3600 cm^{-1} in the infrared spectrum. The exact position of the absorption depends on the extent of hydrogen bonding in the sample. Unassociated alcohols show a fairly sharp absorption near 3600 cm^{-1}, whereas hydrogen-bonded alcohols show a broader absorption in the 3300–3400 cm^{-1} range. The hydrogen-bonded hydroxyl absorption is easily seen at 3350 cm^{-1} in the infrared spectrum of cyclohexanol (Figure 17.11). Alcohols also show a strong C–O stretching absorption near 1050 cm^{-1}.

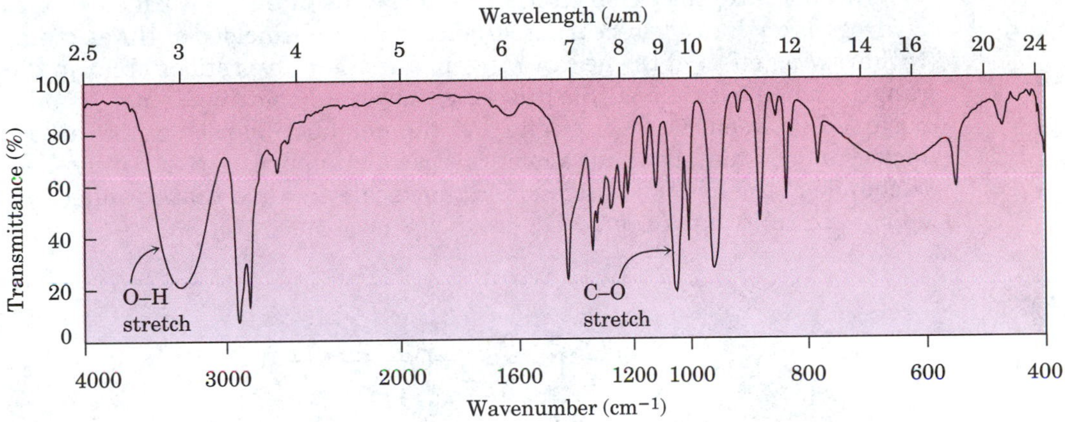

Figure 17.11 Infrared spectrum of cyclohexanol.

PROBLEM

17.16 Assume that you need to prepare 5-cholestene-3-one from cholesterol by Jones oxidation. How could you use infrared spectroscopy to tell if the reaction was successful? What differences would you look for in the infrared spectra of starting material and product?

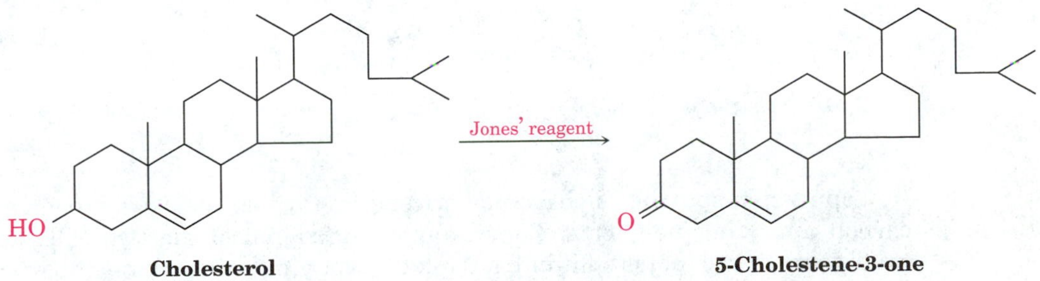

Nuclear Magnetic Resonance Spectroscopy

Carbon atoms bonded to electron-withdrawing –OH groups are deshielded and absorb at a lower field in the ^{13}C NMR spectrum than do normal alkane carbons. Most alcohol carbon absorptions fall in the range 50–80 δ, as the following data illustrate for cyclohexanol.

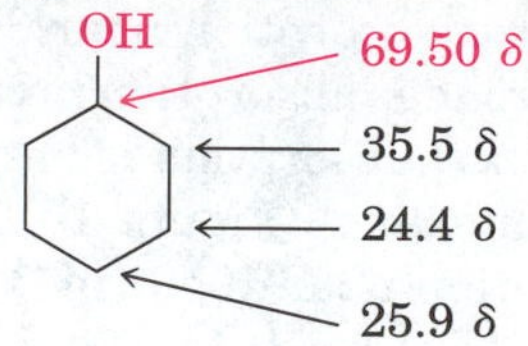

Alcohols also show characteristic absorptions in the ^{1}H NMR spectrum. Hydrogens on the oxygen-bearing carbon atom are deshielded by the electron-withdrawing effect of the nearby oxygen, and their absorptions occur in the range from 3.5 to 4.5 δ. Surprisingly, however, splitting is not usually observed between the O–H proton and the neighboring protons on carbon. Most samples contain small amounts of acidic impurities that catalyze an exchange of the hydroxyl proton on a time scale so rapid that the effects of spin–spin splitting are removed.

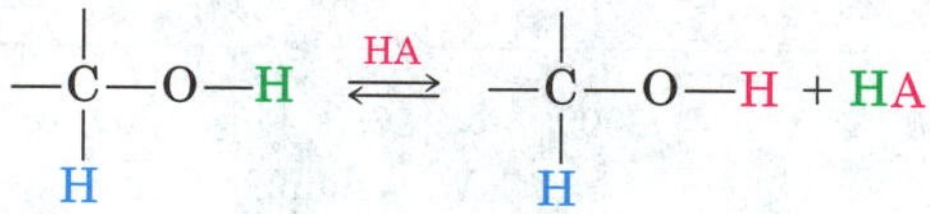

No NMR coupling observed

It's often possible to take advantage of this rapid proton exchange to identify the position of the O–H absorption. If a small amount of deuterated water, D_2O, is added to the NMR sample tube, the O–H proton is rapidly exchanged for deuterium, and the hydroxyl absorption disappears from the spectrum.

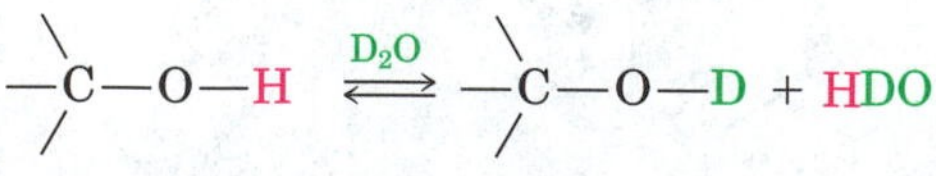

Spin–spin splitting *is* observed between protons on the oxygen-bearing carbon and other neighbors. For example, the signal of the two $-CH_2O-$ protons in 1-propanol is split into a triplet by coupling with the neighboring $-CH_2-$ protons (Figure 17.12).

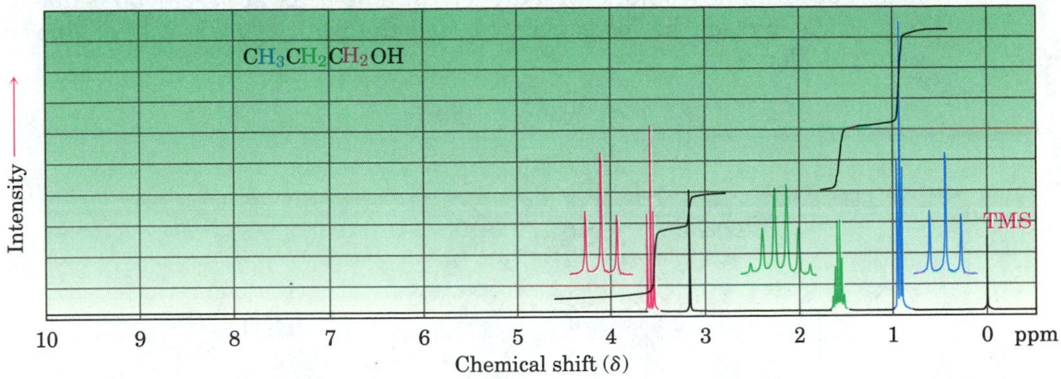

Figure 17.12 The ^{1}H NMR spectrum of 1-propanol. The protons on the oxygen-bearing carbon are split into a triplet at 3.58 δ.

PROBLEM

17.17 When the ^{1}H NMR spectra of an alcohol is run in dimethyl sulfoxide (DMSO) solvent, exchange of the O–H proton is slow, and spin–spin splitting is seen between the O–H proton and C–H protons on the adjacent carbon. What spin multiplicities would you expect for the hydroxyl protons in the following alcohols?

(a) 2-Methyl-2-propanol (b) Cyclohexanol (c) Ethanol
(d) 2-Propanol (e) Cholesterol (f) 1-Methylcyclohexanol

Mass Spectrometry

Alcohols undergo fragmentation in the mass spectrometer by two characteristic pathways, *alpha cleavage* and *dehydration*. In the alpha-cleavage pathway, a C–C bond nearest the hydroxyl group is broken, yielding a neutral radical plus a charged oxygen-containing fragment:

Alpha cleavage

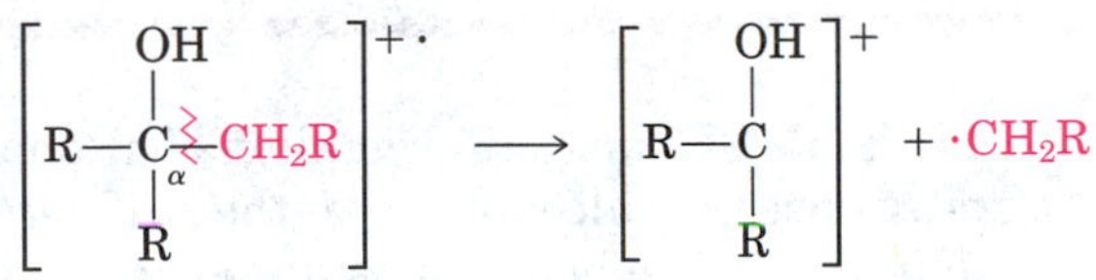

In the dehydration pathway, water is eliminated, yielding an alkene radical cation:

Dehydration

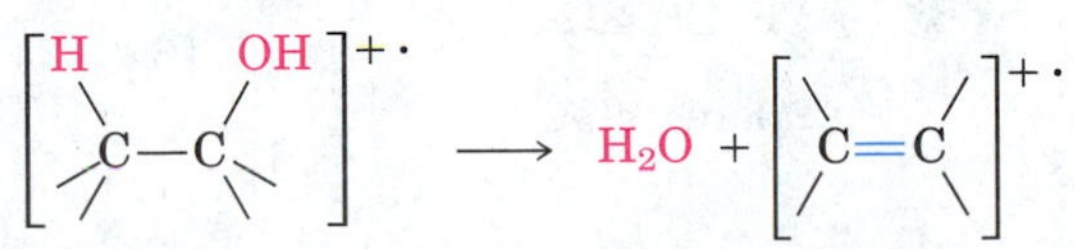

Both of these characteristic fragmentation modes are apparent in the mass spectrum of 1-butanol (Figure 17.13). The peak at $m/z = 56$ is due to loss of water from the molecular ion, and the peak at $m/z = 31$ is due to an alpha cleavage.

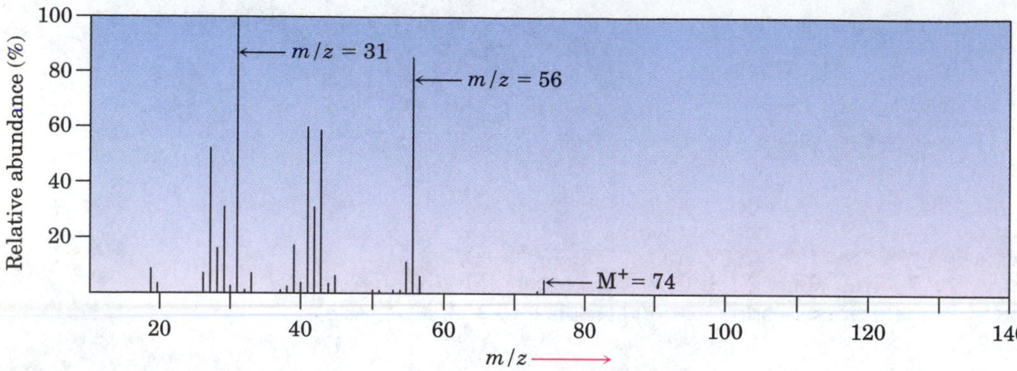

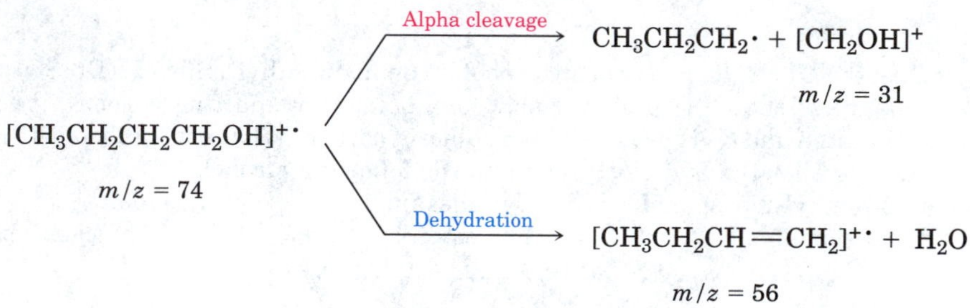

Figure 17.13 Mass spectrum of 1-butanol. Dehydration gives a peak at m/z = 56, and fragmentation by alpha cleavage gives a peak at m/z = 31.

17.12 Thiols

Thiols, R–SH, are sulfur analogs of alcohols. Thiols are named by the same system used for alcohols, with the suffix *-thiol* used in place of *-ol*. The –SH group itself is referred to as a **mercapto** group.

CH_3CH_2SH

Ethanethiol

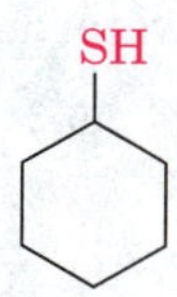

Cyclohexanethiol

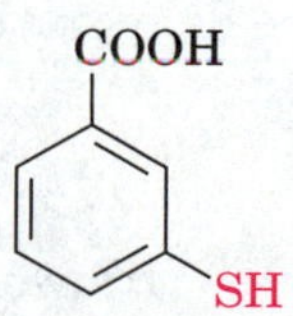

***m*-Mercaptobenzoic acid**

The most obvious characteristic of thiols is their appalling odor. Skunk scent, for example, is caused primarily by the simple thiols, 3-methyl-1-butanethiol and 2-butene-1-thiol. Volatile thiols are also added to natural gas to serve as an easily detectable warning in case of leaks.

Thiols are usually prepared from alkyl halides by S_N2 displacement with a sulfur nucleophile such as hydrosulfide anion, ^-SH:

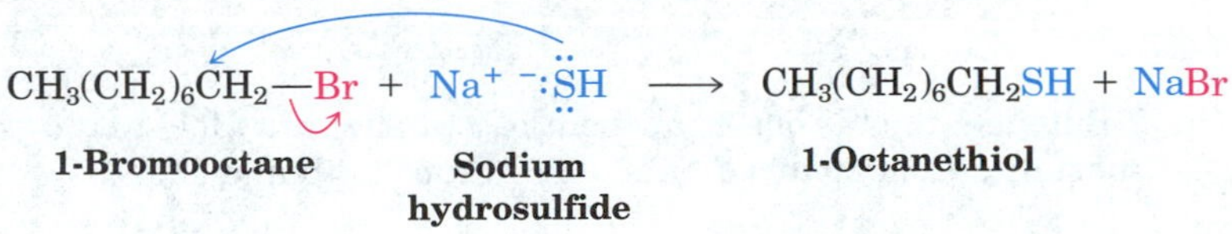

Yields are often poor in this reaction unless an excess of the nucleophile is used, because the product thiol can undergo further S_N2 reaction with alkyl halide, yielding a symmetrical **sulfide, RSR**, as a by-product. For this reason, thiourea, $(NH_2)_2C{=}S$, is often used as the nucleophile in the preparation of thiols from alkyl halides. The reaction occurs by displacement of the halide ion to yield an intermediate alkylisothiourea salt, which is hydrolyzed by subsequent reaction with aqueous base.

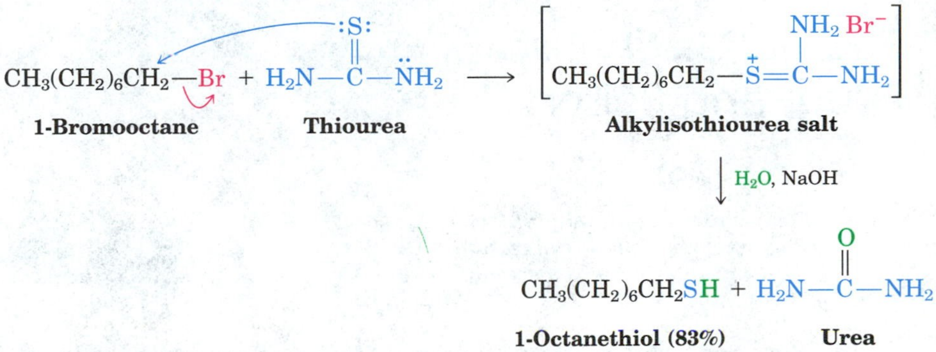

Thiols can be oxidized by Br_2 or I_2 to yield **disulfides, RSSR**. The reaction is easily reversed, and disulfides can be reduced back to thiols by treatment with zinc and acid:

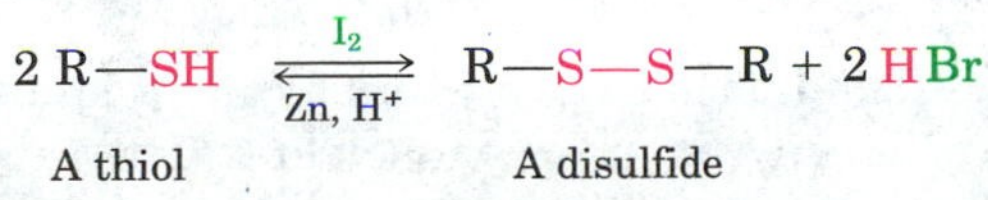

We'll see later that the thiol–disulfide interconversion is extremely important in biochemistry, where disulfide "bridges" form the cross-links between protein chains that help stabilize the three-dimensional conformations of proteins.

PROBLEM.

17.18 Name the following compounds:

(a) $CH_3CH_2CH(CH_3)SH$ (b) $(CH_3)_3CCH_2CH(SH)CH_2CH(CH_3)_2$ (c) 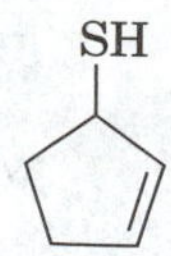

PROBLEM.

17.19 2-Butene-1-thiol is one component of skunk spray. How would you synthesize this substance from methyl 2-butenoate? From 1,3-butadiene?

$$CH_3CH{=}CHCOCH_3 \;(C{=}O) \xrightarrow{?} CH_3CH{=}CHCH_2SH$$

Methyl 2-butenoate → **2-Butene-1-thiol**

INTERLUDE

Ethanol as Chemical, Drug, and Poison

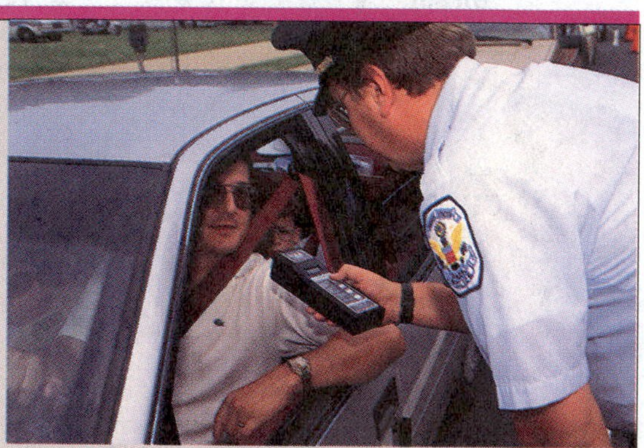

A Breathalyzer test showing a blood alcohol level above 0.10% will result in a charge of drunken driving.

The production of ethanol by fermentation of grains and sugars is one of the oldest known organic reactions, going back at least to the ancient Greeks. Fermentation is carried out by adding yeast to an aqueous sugar solution, where enzymes break down carbohydrates into ethanol and CO_2:

$$C_6H_{12}O_6 \xrightarrow{\text{Yeast}} 2\ CH_3CH_2OH + 2\ CO_2$$

A carbohydrate

Nearly 110 million gallons of ethanol are produced each year in the United States, primarily for use as a solvent. Only about 5% of this industrial ethanol comes from fermentation, though; most is obtained by acid-catalyzed hydration of ethylene.

$$H_2C{=}CH_2 + H_2O \xrightarrow[\text{catalyst}]{\text{Acid}} CH_3CH_2OH$$

(continued)➤

Ethanol is classified for medical purposes as a central nervous system (CNS) depressant. Its effects (that is, being drunk) resemble the human response to anesthetics. There is an initial excitability and increase in sociable behavior, but this results from depression of inhibition rather than from stimulation. At a blood alcohol concentration of 0.1–0.3%, motor coordination is affected, accompanied by loss of balance, slurred speech, and amnesia. When blood alcohol concentration rises to 0.3–0.4%, nausea and loss of consciousness occur. Above 0.6%, spontaneous respiration and cardiovascular regulation are affected, ultimately leading to death. The LD_{50} (see the Chapter 1 Interlude) of ethanol is 10.6 g/kg.

The passage of ethanol through the body begins with its absorption in the stomach and small intestine, followed by rapid distribution to all body fluids and organs. In the pituitary gland, ethanol inhibits the production of a hormone that regulates urine flow, causing increased urine production and dehydration. In the stomach, ethanol stimulates production of acid. Throughout the body, ethanol causes blood vessels to dilate, resulting in flushing of the skin and a sensation of warmth as blood moves into capillaries beneath the surface. The result is not a warming of the body, but an increased loss of heat at the surface.

The metabolism of ethanol occurs mainly in the liver and proceeds by oxidation in two steps, first to acetaldehyde (CH_3CHO) and then to acetic acid (CH_3COOH). In chronic alcoholics, ethanol and acetaldehyde are toxic, leading to devastating physical and metabolic deterioration. The liver usually suffers the worst damage since it is the major site of alcohol metabolism.

The quick and uniform distribution of ethanol in body fluids, the ease with which it crosses lung membranes, and its ready oxidizability provide the basis for simple tests for blood alcohol concentration. The *Breathalyzer test* measures alcohol concentration in expired air by the color change that occurs when the bright orange oxidizing agent potassium dichromate ($K_2Cr_2O_7$) is reduced to blue-green chromium(III). In most states, a blood alcohol level above 0.10% is sufficient for a charge of driving while intoxicated.

Summary and Key Words

Alcohols are among the most versatile of all organic compounds. They occur widely in nature, are important industrially, and have an unusually rich chemistry. The most important methods of alcohol synthesis involve carbonyl compounds. Aldehydes, ketones, esters, and carboxylic acids are reduced by reaction with either $NaBH_4$ or $LiAlH_4$. Aldehydes, esters, and carboxylic acids yield primary alcohols on reduction; ketones yield secondary alcohols.

The addition reaction of Grignard reagents with carbonyl compounds is another important method for preparing alcohols. Addition of a Grignard reagent to formaldehyde yields a primary alcohol, addition to an aldehyde yields a secondary alcohol, and addition to a ketone or an ester yields a tertiary alcohol. Carboxylic acids do not give Grignard addition products. The Grignard synthesis of alcohols is limited by the fact that Grignard reagents can't be prepared from alkyl halides that contain reactive functional groups in the same molecule. This problem can sometimes be avoided by **protecting** the interfering functional group. Alcohols are often protected by formation of **trimethylsilyl (TMS) ethers**.

Alcohols undergo a great many different reactions. They can be dehydrated by treatment with $POCl_3$ and can be transformed into alkyl halides by treatment with PBr_3 or $SOCl_2$. Furthermore, alcohols are weakly acidic ($pK_a \approx 16$–18). They react with strong bases and with alkali metals to form **alkoxide anions**, which are much used in organic synthesis.

One of the most important reactions of alcohols is their oxidation to carbonyl compounds. Primary alcohols yield either aldehydes or carboxylic acids, secondary alcohols yield ketones, and tertiary alcohols are not normally oxidized. **Pyridinium chlorochromate (PCC)** in dichloromethane is often used for oxidizing primary alcohols to aldehydes and secondary alcohols to ketones. The **Jones reagent**, a solution of CrO_3 in aqueous sulfuric acid, is frequently used for oxidizing primary alcohols to carboxylic acids and secondary alcohols to ketones.

Thiols, RSH, the sulfur analogs of alcohols, are usually prepared by S_N2 reaction of an alkyl halide with thiourea. Mild oxidation of a thiol yields a **disulfide, RSSR**, and mild reduction of a disulfide gives back the thiol.

Summary of Reactions

1. Synthesis of alcohols
   (a) Reduction of carbonyl compounds (Section 17.6)
      (1) Aldehydes

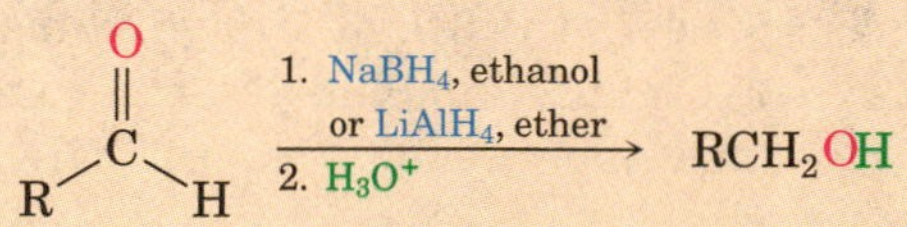

A primary alcohol

      (2) Ketones

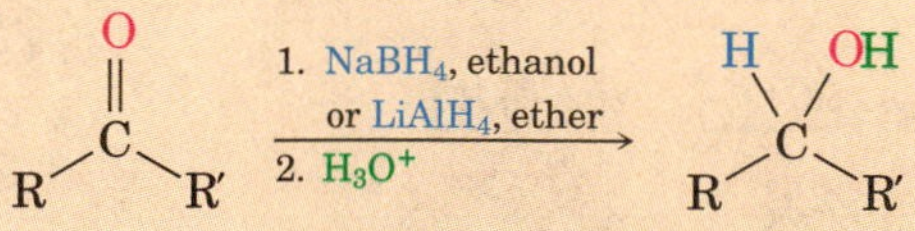

A secondary alcohol

      (3) Esters

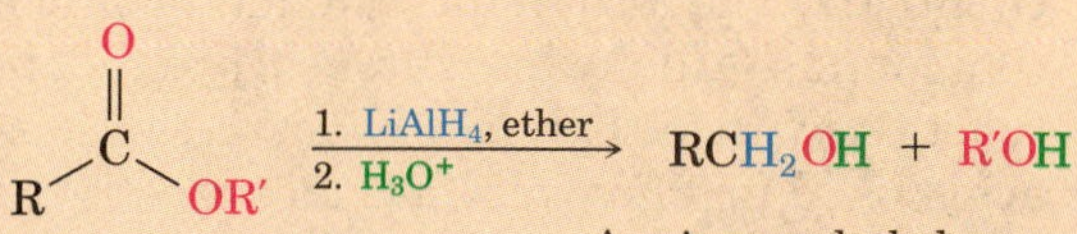

A primary alcohol

(continued)➤

(4) Carboxylic acids

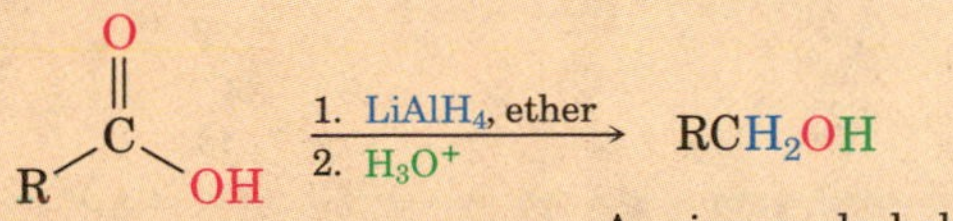

A primary alcohol

(b) Grignard addition to carbonyl compounds (Section 17.7)

(1) Formaldehyde

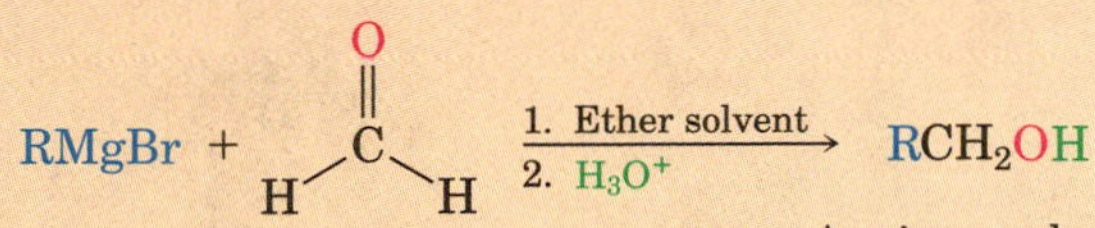

A primary alcohol

(2) Aldehydes

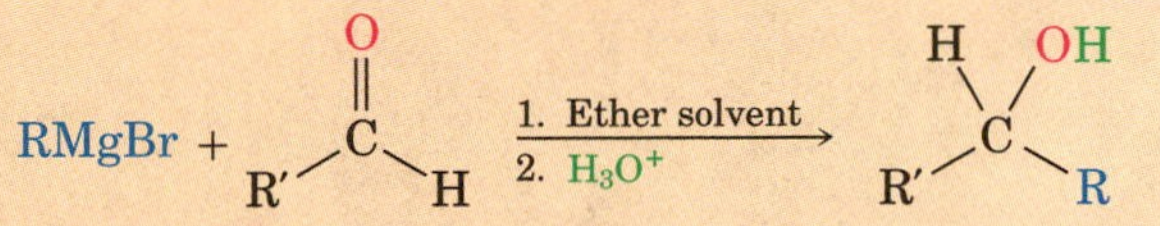

A secondary alcohol

(3) Ketones

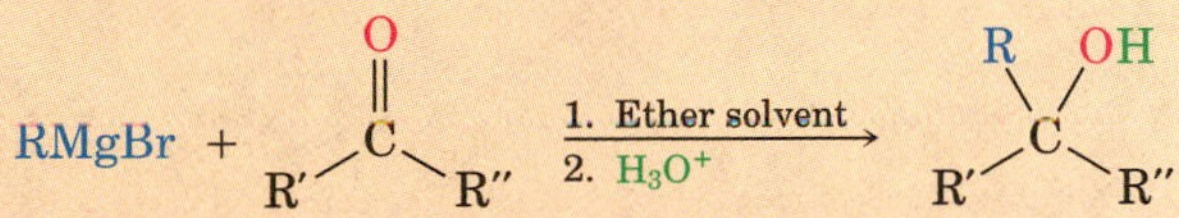

A tertiary alcohol

(4) Esters

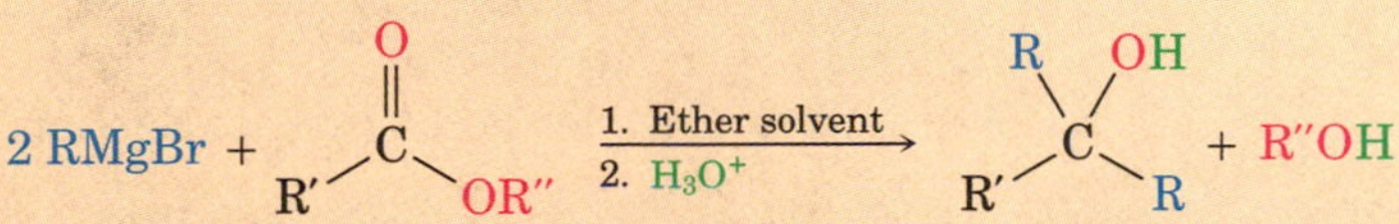

A tertiary alcohol

2. Reactions of alcohols

(a) Acidity (Section 17.4)

$$ROH + NaH \longrightarrow RO^- Na^+ + H_2$$
$$2\ ROH + 2\ Na \longrightarrow 2\ RO^- Na^+ + H_2$$

(b) Dehydration (Section 17.8)

(1) Tertiary alcohols

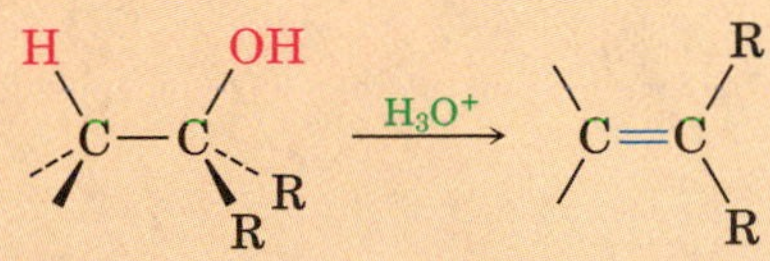

(continued)➤

(2) Secondary and tertiary alcohols

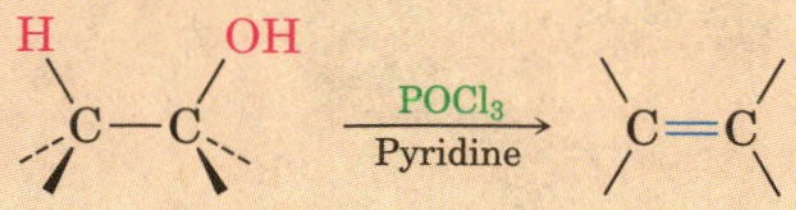

(c) Oxidation (Section 17.9)

(1) Primary alcohol

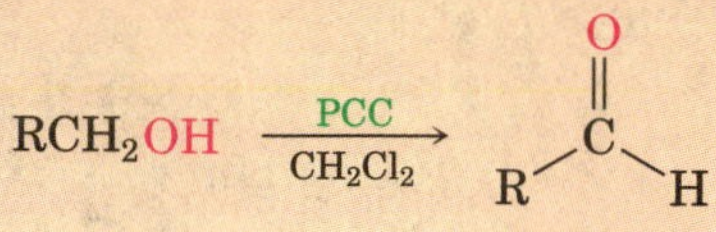

An aldehyde

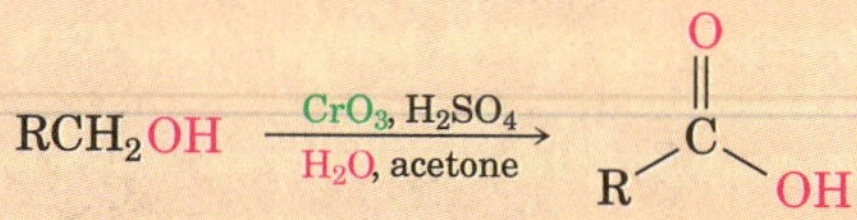

A carboxylic acid

(2) Secondary alcohol

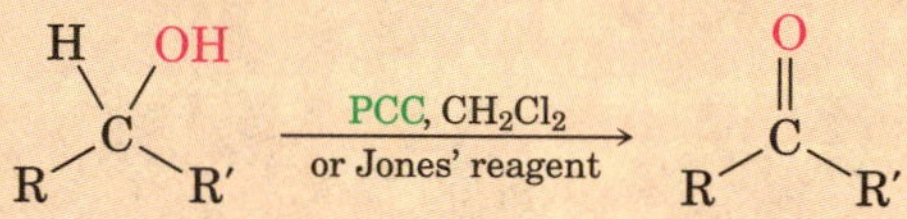

A ketone

3. Synthesis of thiols (Section 17.12)

$$RCH_2Br \xrightarrow[2.\ H_2O,\ NaOH]{1.\ (H_2N)_2C{=}S} RCH_2SH$$

4. Oxidation of thiols to disulfides (Section 17.12)

$$2\ RSH \xrightarrow{I_2,\ H_2O} RS{-}SR$$

ADDITIONAL PROBLEMS

17.20 Name the following compounds according to IUPAC rules:

(a) 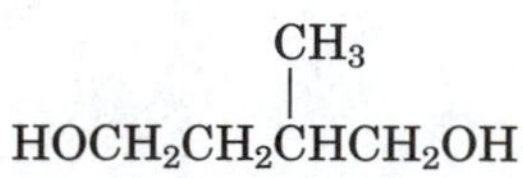

(b) 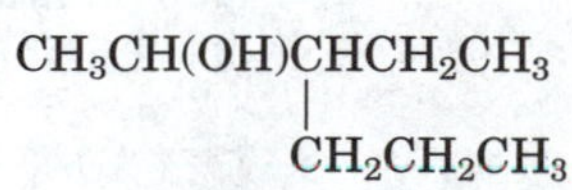

(c) 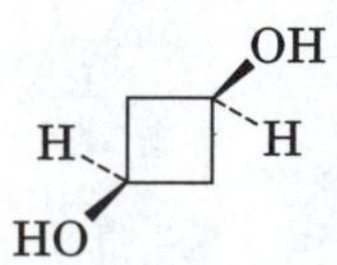

(d) 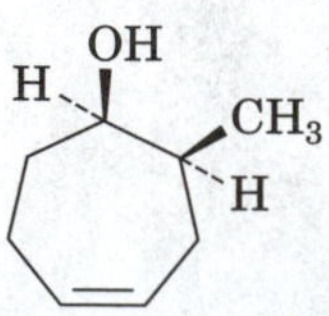

(e) 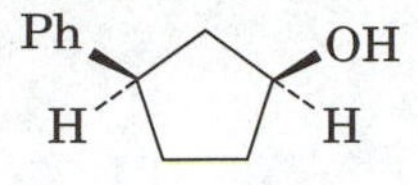

(f) $(CH_3)_2CHC(SH)(CH_3)CH_2CH_2CH_3$

17.21 Draw and name the eight isomeric alcohols with formula $C_5H_{12}O$.

17.22 Which of the eight alcohols you identified in Problem 17.21 react with Jones' reagent (CrO_3, H_2O, H_2SO_4)? Show the products you would expect from each reaction.

17.23 How would you prepare the following compounds from 2-phenylethanol?

(a) Styrene
(b) Phenylacetaldehyde($C_6H_5CH_2CHO$)
(c) Phenylacetic acid ($C_6H_5CH_2CO_2H$)
(d) Benzoic acid
(e) Ethylbenzene
(f) Benzaldehyde
(g) 1-Phenylethanol
(h) 1-Bromo-2-phenylethane

17.24 How would you carry out the following transformations?

(a) 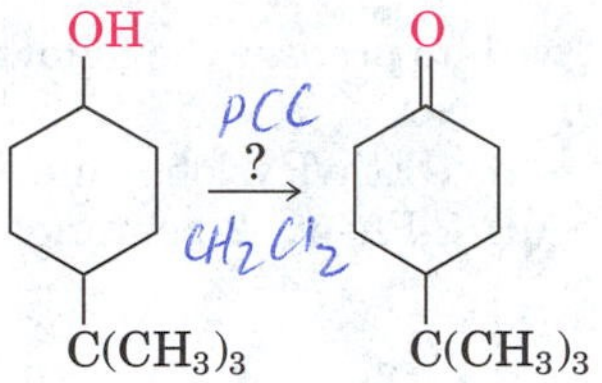

(b) 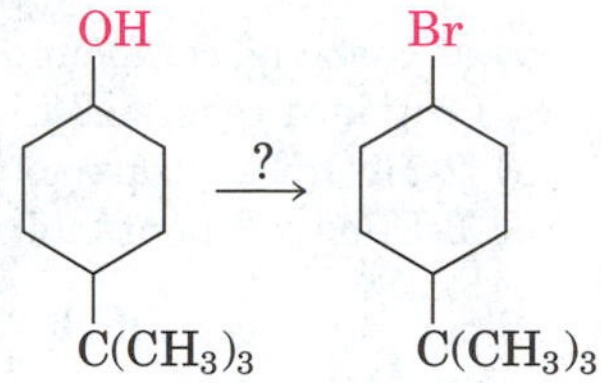

(c) 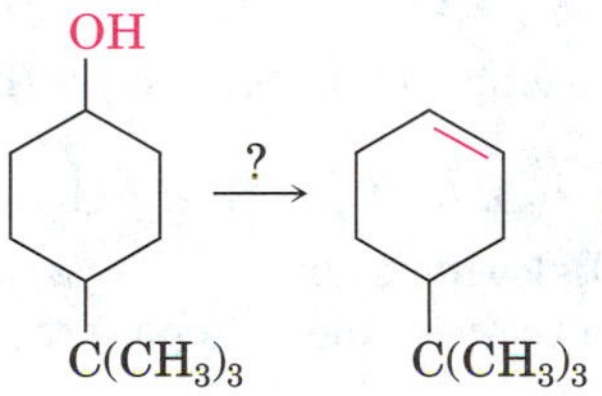

(d) $CH_3CH_2CH_2OH \xrightarrow{?} CH_3CH_2CHO$

(e) $CH_3CH_2CH_2OH \xrightarrow{?} CH_3CH_2CO_2H$

(f) $CH_3CH_2CH_2OH \xrightarrow{?} CH_3CH_2CH_2OTos$

(g) $CH_3CH_2CH_2OH \xrightarrow{?} CH_3CH_2CH_2Cl$

(h) $CH_3CH_2CH_2OH \xrightarrow{?} CH_3CH_2CH_2O^- \; Na^+$

17.25 What Grignard reagent and what carbonyl compound might you start with to prepare the following alcohols?

(a) $CH_3CH(OH)CH_2CH_3$

(b) 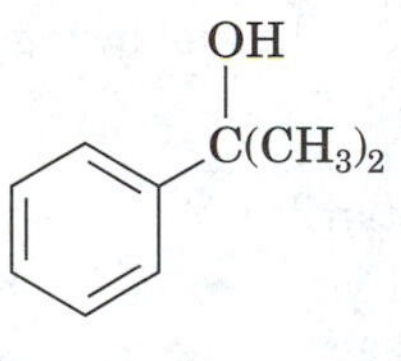

(c) $H_2C{=}C(CH_3){-}CH_2OH$

(d) $(Ph)_3COH$

(e) $CH_3CH(OH)CH_2CH_2CH_2Br$

17.26 Assume that you have been given a sample of (*S*)-2-octanol. How could you prepare (*R*)-2-chlorooctane? How could you prepare (*R*)-2-octanol?

17.27 When 4-chloro-1-butanol is treated with a strong base such as sodium hydride, NaH, tetrahydrofuran is produced. Suggest a mechanism.

$ClCH_2CH_2CH_2CH_2OH \xrightarrow[\text{Ether}]{\text{NaH}}$ 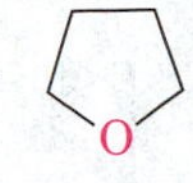 $+ \; H_2 + NaCl$

17.28 What carbonyl compounds would you reduce to prepare the following alcohols? List all possibilities.

(a) 2,2-Dimethyl-1-hexanol

(b) 3,3-Dimethyl-2-butanol

(c) 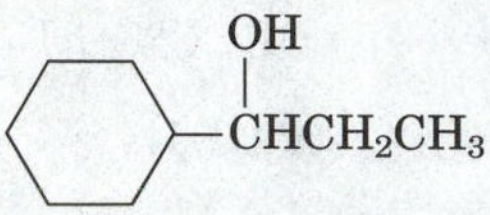

(d) 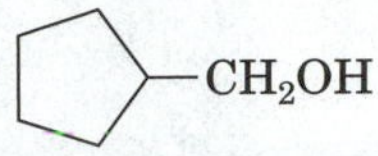

17.29 How would you carry out the following transformations?

(a) 

(b) 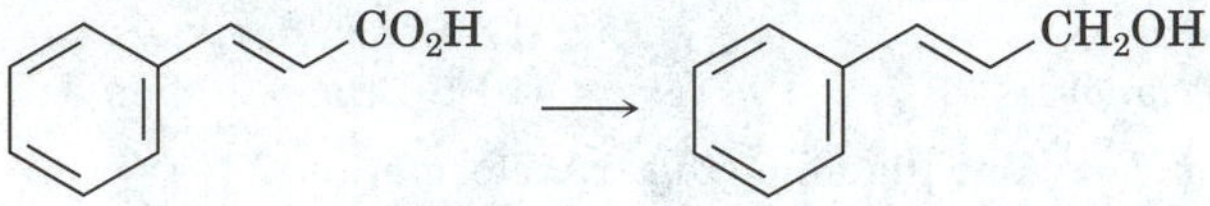

17.30 What carbonyl compounds might you start with to prepare the following compounds by Grignard reaction? List all possibilities.

(a) 2-Methyl-2-propanol

(b) 1-Ethylcyclohexanol

(c) 3-Phenyl-3-pentanol

(d) 2-Phenyl-2-pentanol

(e) 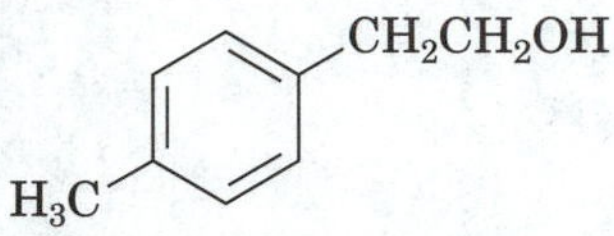

(f) 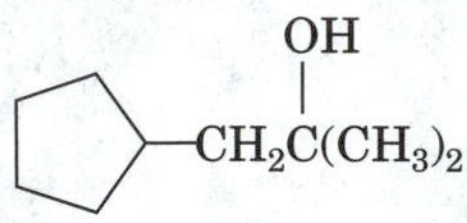

17.31 What products would you obtain from reaction of 1-pentanol with the following reagents?

(a) PBr_3 (b) $SOCl_2$ (c) CrO_3, H_2O, H_2SO_4 (d) PCC

17.32 Acid-catalyzed dehydration of 2,2-dimethylcyclohexanol yields a mixture of 1,2-dimethylcyclohexene and isopropylidenecyclopentane. Propose a mechanism to account for the formation of both products.

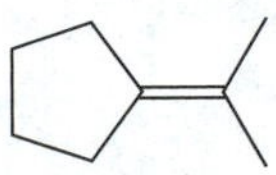

Isopropylidenecyclopentane

17.33 How would you prepare the following substances from cyclopentanol? More than one step may be required.

(a) Cyclopentanone

(b) Cyclopentene

(c) 1-Methylcyclopentanol

(d) *trans*-2-Methylcyclopentanol

17.34 What products would you expect to obtain from reaction of 1-methylcyclohexanol with the following reagents?

(a) HBr (b) NaH (c) H_2SO_4 (d) $Na_2Cr_2O_7$

17.35 Testosterone is one of the most important male steroid hormones. When testosterone is dehydrated by treatment with acid, rearrangement occurs to yield the product indicated. Propose a mechanism to account for this reaction. (*Hint*: See Section 6.12.)

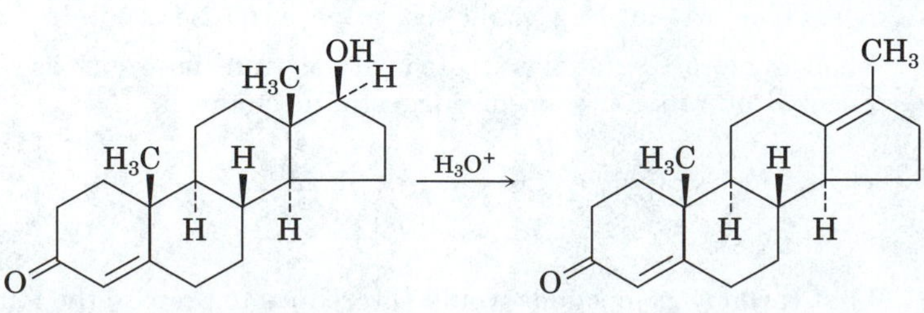

Testosterone

17.36 Starting from testosterone (Problem 17.35), how would you prepare the following substances?

(a) 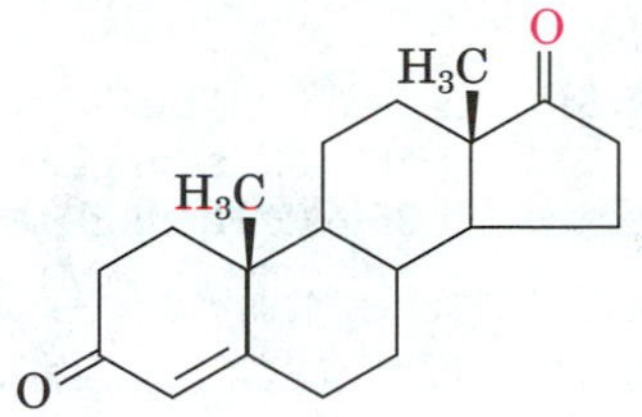

(b) 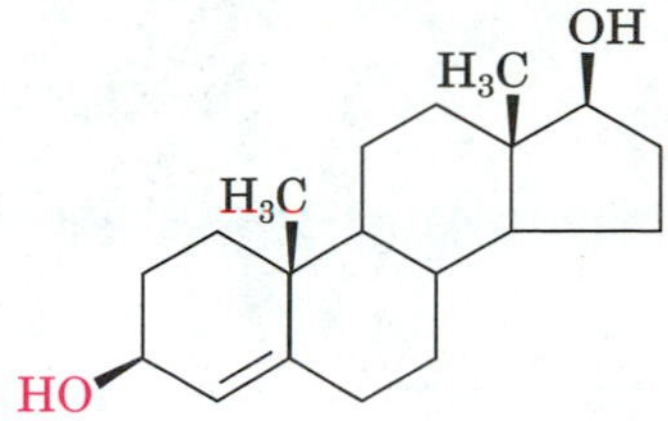

(c) 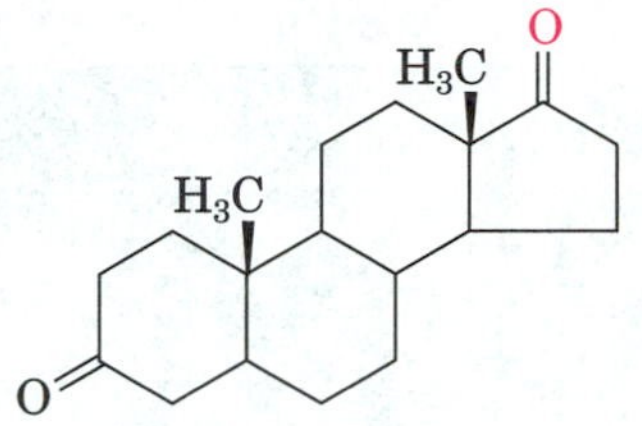

(d) 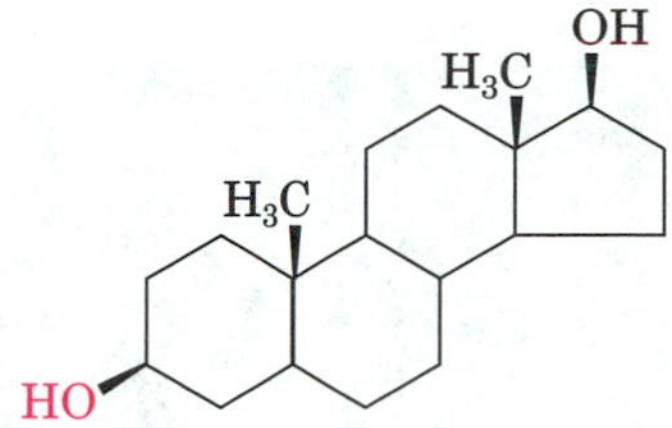

17.37 Compound A, $C_{10}H_{18}O$, undergoes reaction with dilute H_2SO_4 at 25°C to yield a mixture of two alkenes, $C_{10}H_{16}$. The major alkene product, B, gives only cyclopentanone after ozone treatment followed by reduction with zinc in acetic acid. Formulate the reactions involved, and identify A and B.

17.38 Dehydration of *trans*-2-methylcyclopentanol with $POCl_3$ in pyridine yields predominantly 3-methylcyclopentene. Is the stereochemistry of this dehydration syn or anti? Can you suggest a reason for formation of the observed product? (Make molecular models!)

17.39 How would you synthesize the following alcohols, starting with benzene and other alcohols of six or fewer carbons as your only organic reagents?

(a) 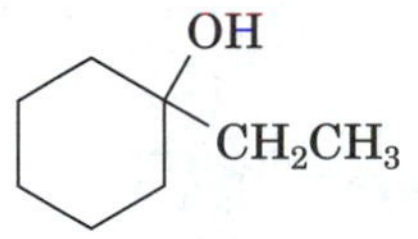

(b) $CH_3CH_2CH_2CH(CH_3)CH_2CH_2OH$

(c) 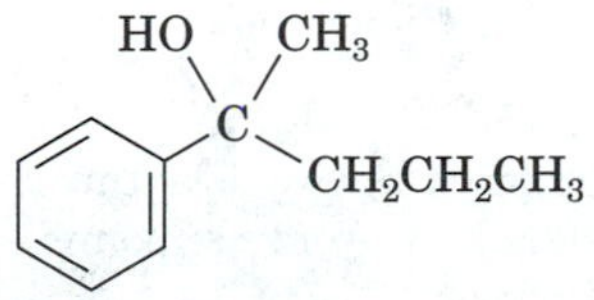

(d) $CH_3CH(CH_3)CH_2CH(OH)CH_2CH_3$

17.40 The ^1H NMR spectrum shown is that of 3-methyl-3-buten-1-ol. Assign all the observed resonance peaks to specific protons, and account for the splitting patterns.

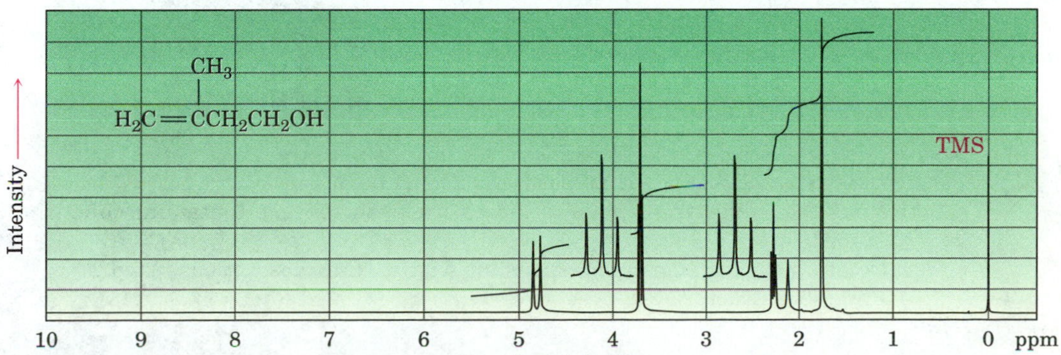

17.41 Propose a structure consistent with the following spectral data for a compound of formula $C_8H_{18}O_2$:

IR: 3350 cm^{-1} ^{1}H NMR: 1.24 δ (12 H, singlet)
1.56 δ (4 H, singlet)
1.95 δ (2 H, singlet)

17.42 The following ^{1}H NMR spectrum is of an alcohol, $C_8H_{10}O$. Propose a structure.

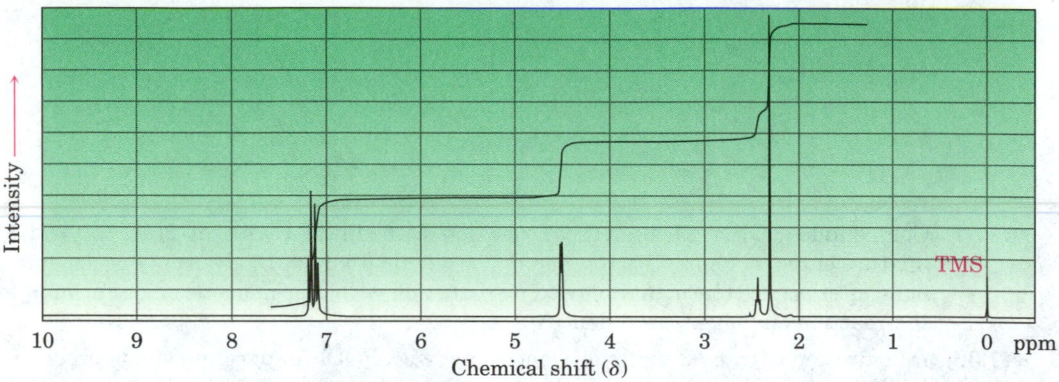

17.43 2,3-Dimethyl-2,3-butanediol has the common name *pinacol*. On heating with aqueous acid, pinacol rearranges to *pinacolone*, 3,3-dimethyl-2-butanone. Suggest a mechanism for this reaction. (*Hint*: See Section 6.12.)

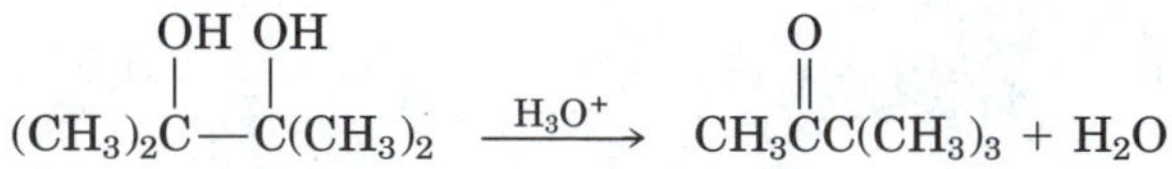

Pinacol **Pinacolone**

17.44 Compound A, $C_5H_{10}O$, is one of the basic building blocks of nature. All steroids and many other naturally occurring compounds are built from compound A. Spectroscopic analysis of A yields the following information:

IR: 3400 cm^{-1}; 1640 cm^{-1} ^{1}H NMR: 1.63 δ (3 H, singlet)
1.70 δ (3 H, singlet)
3.83 δ (1 H, broad singlet)
4.15 δ (2 H, doublet, J = 7 Hz)
5.70 δ (1 H, triplet, J = 7 Hz)

(a) How many double bonds and/or rings does A have?
(b) From the IR spectrum, what is the nature of the oxygen-containing functional group?
(c) What kinds of protons are responsible for the NMR absorptions listed?
(d) Propose a structure for A.

17.45 Propose structures for alcohols or thiols that have the following ^1H NMR spectra:

(a) $C_5H_{12}S$

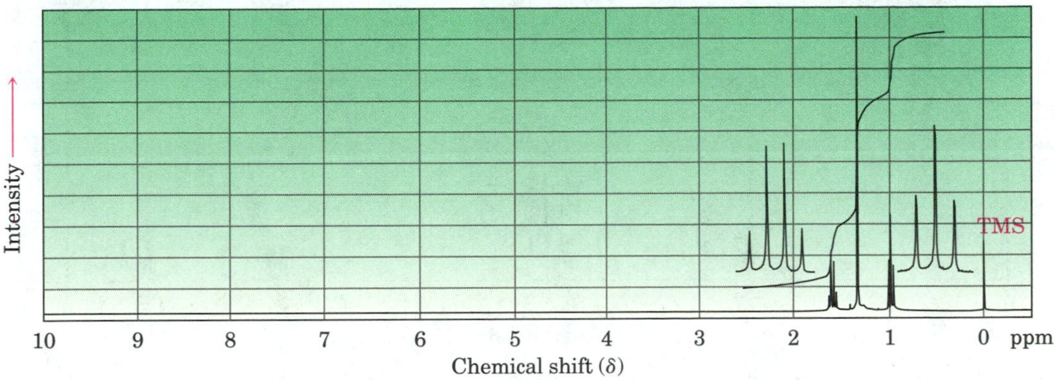

(b) $C_5H_{12}O$

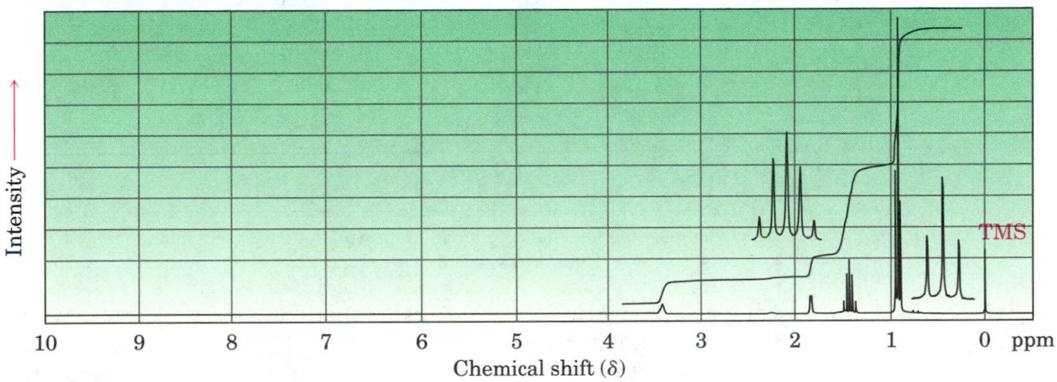

(c) $C_8H_{10}O$

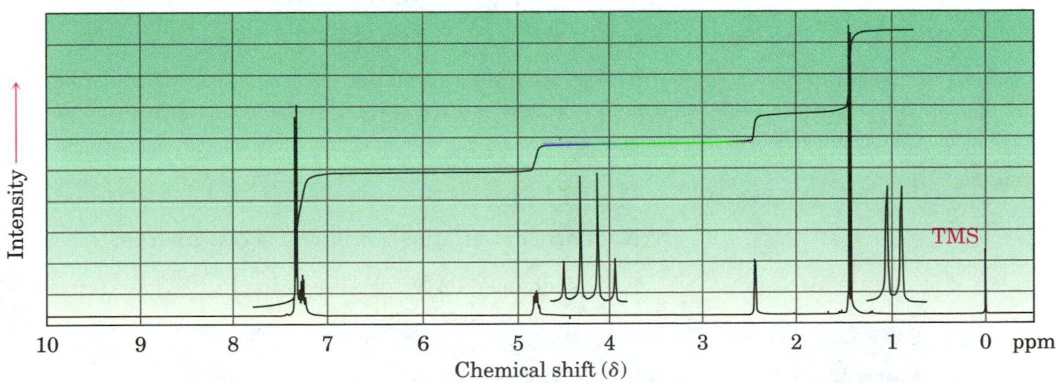

17.46 Propose structures for alcohols that have the following ^1H NMR spectra:
(a) $C_9H_{12}O$

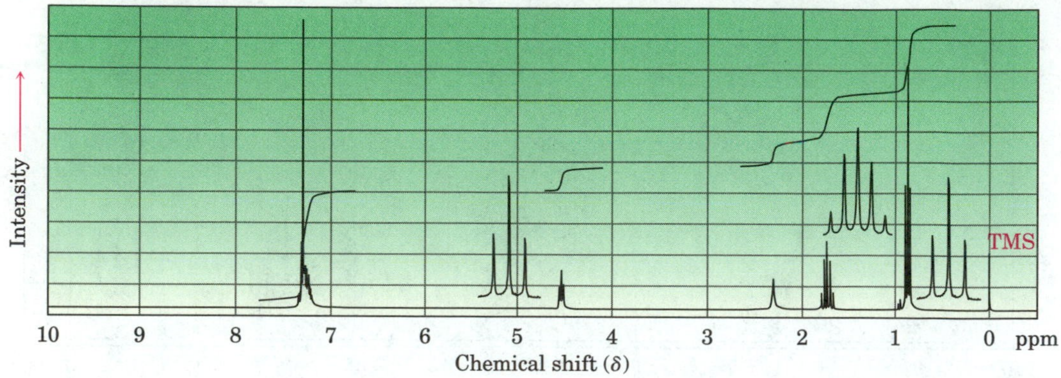

(b) $C_8H_{10}O_2$

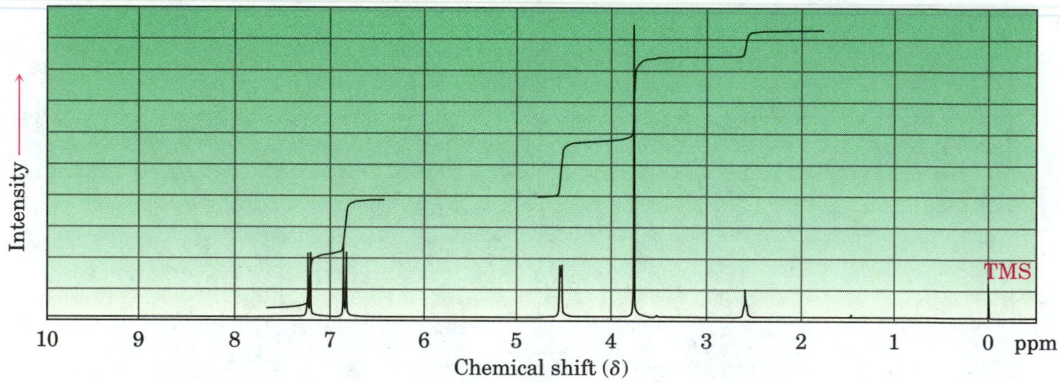

17.47 As a rule, axial alcohols oxidize somewhat faster than equatorial alcohols. Which would you expect to oxidize faster, *cis*-4-*tert*-butylcyclohexanol or *trans*-4-*tert*-butylcyclohexanol? Draw the more stable chair conformation of each molecule.

17.48 Propose a synthesis of bicyclohexylidene, starting from cyclohexanone as the only source of carbon.

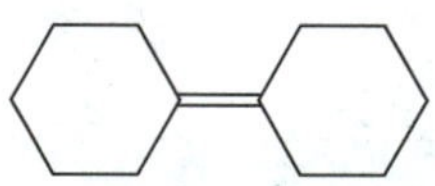

Bicyclohexylidene

17.49 Since all hamsters look pretty much alike, attraction between sexes is controlled by chemical secretions. The sex attractant exuded by the female hamster has the following spectral properties:

Mass spectrum:	$M^+ = 94$
IR:	No functional-group absorptions above 1500 cm^{-1}
^1H NMR:	2.10 δ (singlet); no other absorptions

Propose a structure for the hamster sex attractant.

17.50 A problem often encountered in the oxidation of primary alcohols to acids is that esters are sometimes produced as by-products. For example, oxidation of ethanol yields acetic acid and ethyl acetate:

$$\mathrm{CH_3CH_2OH} \xrightarrow[\text{reagent}]{\text{Jones'}} \mathrm{CH_3\overset{O}{\overset{\|}{C}}OH} + \mathrm{CH_3\overset{O}{\overset{\|}{C}}OCH_2CH_3}$$

Propose a mechanism to account for the formation of ethyl acetate. Take into account the reversible reaction between aldehydes and alcohols:

$$\mathrm{R\overset{O}{\overset{\|}{C}}H} + \mathrm{R'OH} \rightleftharpoons \mathrm{R\underset{OR'}{\underset{|}{\overset{OH}{\overset{|}{C}}}}H}$$

17.51 Treatment of 4-hydroxycyclohexanone with 1 equivalent of CH_3MgBr yields none of the expected addition product, whereas treatment with an excess of the Grignard reagent gives a good yield of 1-methyl-1,4-cyclohexanediol. Explain.

17.52 Identify the reagents a–f in the following scheme:

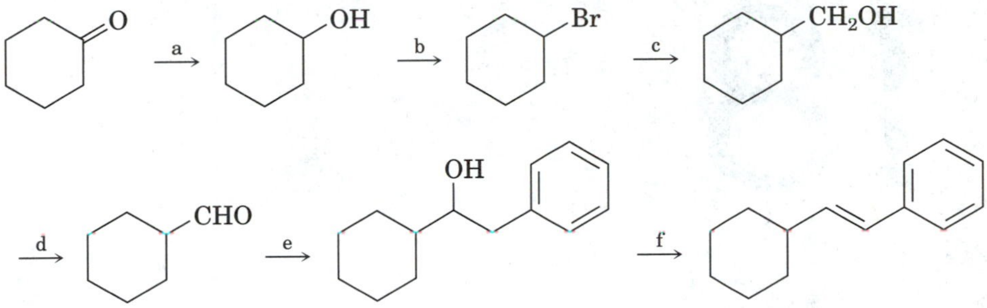

17.53 A compound of unknown structure gave the following spectroscopic data:

Mass spectrum:	$M^+ = 88.1$	^{1}H NMR:	1.4 δ (2 H, quartet, $J = 7$ Hz)
IR:	3600 cm^{-1}		1.2 δ (6 H, singlet)
			1.0 δ (1 H, singlet)
^{13}C NMR:	74, 35, 27, 25 δ		0.9 δ (3 H, triplet, $J = 7$ Hz)

(a) Assuming that the compound contains C and H, but may or may not contain O, give three possible molecular formulas.
(b) How many protons (H) does the compound contain?
(c) What functional group(s) does the compound contain?
(d) How many carbons does the compound contain?
(e) What is the molecular formula of the compound?
(f) What is the structure of the compound?
(g) Assign the peaks in the ^{1}H NMR spectrum of the molecule to specific protons.

A Look Ahead

17.54 The reduction of carbonyl compounds by reaction with hydride reagents (H:$^-$) and the Grignard addition by reaction with organomagnesium halides (R:$^-$ $^+$MgBr) are examples of *nucleophilic carbonyl addition reactions*, whose mechanism we'll study in Chapter 19. What analogous product do you think might result from reaction of HCN with a ketone?

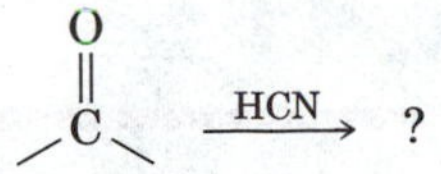

17.55 We'll see in Chapter 25 that phenol (C_6H_5OH, pK_a = 10.0) is a stronger acid than cyclohexanol (pK_a = 17). Explain, using your knowledge of resonance.

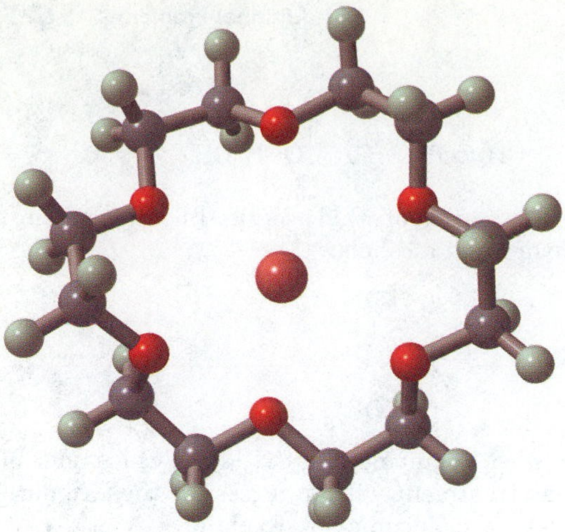

This cyclic "crown" ether has captured a potassium ion in its center.

18

Ethers, Epoxides, and Sulfides

An **ether** is a substance that has two organic groups bonded to the same oxygen atom, R–O–R′. The organic groups may be alkyl, aryl, or vinylic, and the oxygen atom can be in either an open chain or a ring. Perhaps the most well-known ether is diethyl ether, a familiar substance that has been used medicinally as an anesthetic and is used industrially as a solvent. Other useful ethers include anisole, a pleasant-smelling aromatic ether used in perfumery, and tetrahydrofuran (THF), a cyclic ether often used as a solvent.

$CH_3CH_2—O—CH_2CH_3$

Diethyl ether

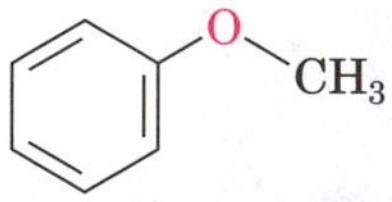

Anisole
(Methyl phenyl ether)

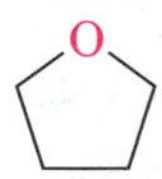

Tetrahydrofuran
(a cyclic ether)

18.1 Naming Ethers

Two systems for naming ethers are allowed by IUPAC rules. Simple ethers with no other functional groups are named by identifying the two organic substituents and adding the word *ether*.

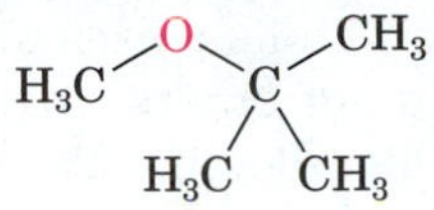

***tert*-Butyl methyl ether**

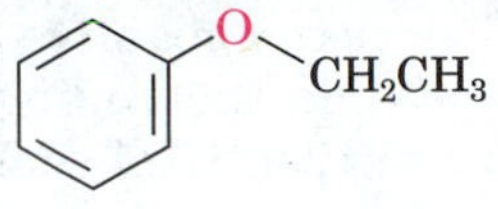

Ethyl phenyl ether

If other functional groups are present, the ether part is considered to be an alkoxy substituent. For example:

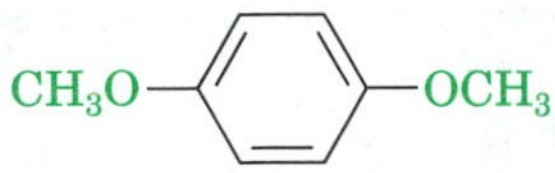

***p*-Dimethoxybenzene**

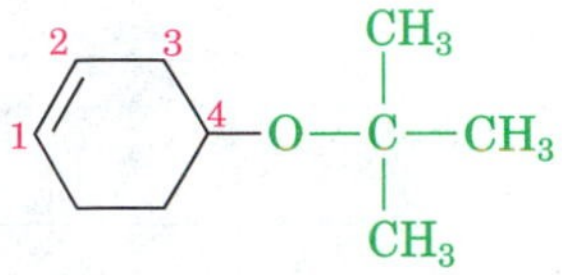

4-*tert*-Butoxy-1-cyclohexene

PROBLEM

18.1 Name these ethers according to IUPAC rules.

(a) 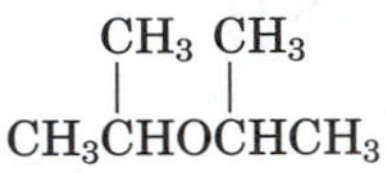

(b) 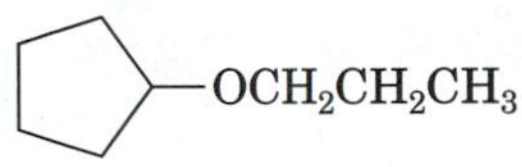

(c) 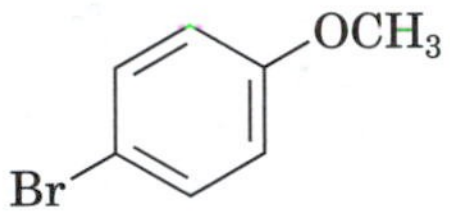

(d) 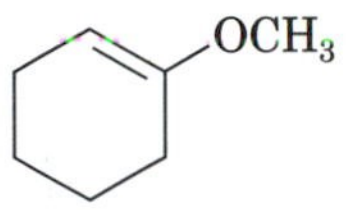

(e) 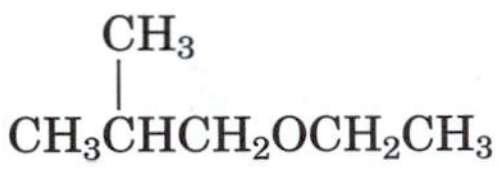

(f) $H_2C{=}CHCH_2OCH{=}CH_2$

18.2 Structure and Properties of Ethers

Ethers can be thought of as organic derivatives of water in which the hydrogen atoms have been replaced by organic fragments, H–O–H versus R–O–R. As such, ethers have nearly the same geometry as water. The R–O–R bonds have an approximately tetrahedral bond angle (112° in dimethyl ether), and the oxygen atom is sp^3-hybridized.

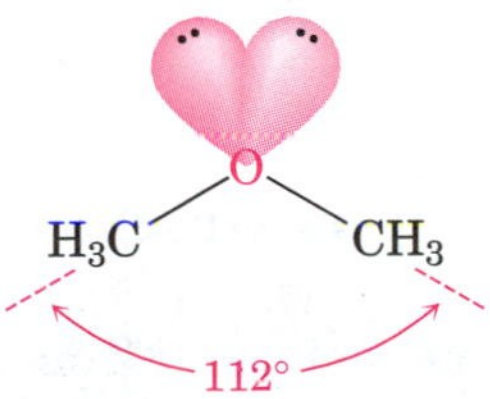

Stereo View

The electronegative oxygen atom causes ethers to have a slight dipole moment, and the boiling points of ethers are therefore often higher than the boiling points of comparable alkanes. Table 18.1 compares the boiling points of some common ethers with the corresponding hydrocarbons in which the ether oxygen atom has been replaced by a CH_2 group.

Table 18.1 Comparison of Boiling Points of Ethers and Hydrocarbons

Ether	*[Hydrocarbon]*	*Boiling point (°C)*	
CH_3OCH_3	$CH_3CH_2CH_3$	−25	−45
$CH_3CH_2OCH_2CH_3$	$CH_3CH_2CH_2CH_2CH_3$	34.6	36
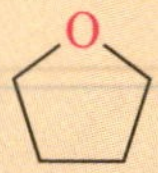	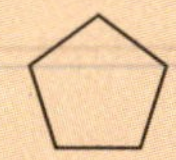	65	49
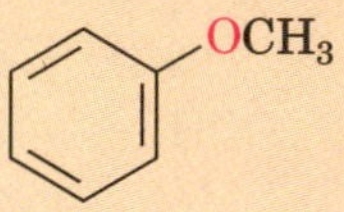	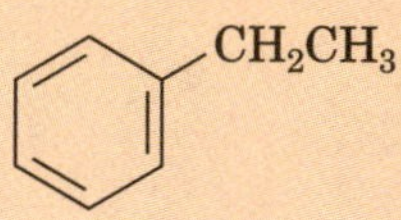	158	136

Although ethers are inert toward many reagents, certain ethers react slowly with air to give **peroxides**, compounds that contain O–O bonds. The peroxides from low-molecular-weight ethers such as diisopropyl ether and tetrahydrofuran are explosive and extremely dangerous, even in tiny amounts. Ether solvents are very useful in the laboratory, but they must always be treated with care.

18.3 Industrial Preparation of Ethers

Diethyl ether and other simple symmetrical ethers are prepared industrially by the sulfuric acid-catalyzed dehydration of alcohols:

$$2\ CH_3CH_2OH \xrightarrow{H_2SO_4} CH_3CH_2OCH_2CH_3 + H_2O$$

Ethanol **Diethyl ether**

The reaction occurs by S_N2 displacement of water from a protonated ethanol molecule by the oxygen atom of a second ethanol:

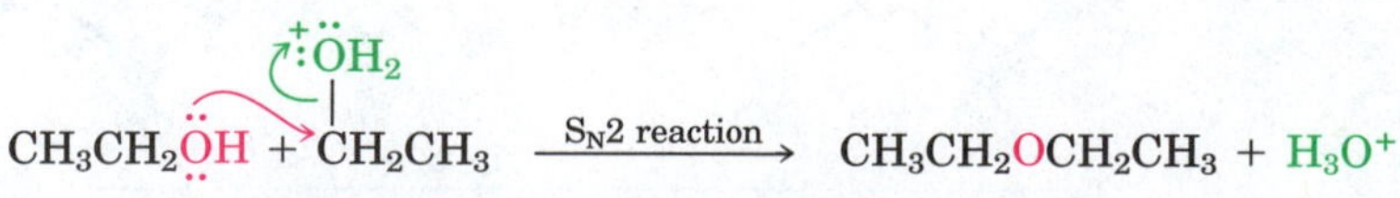

This acid-catalyzed method is limited to the production of symmetrical ethers from primary alcohols, because secondary and tertiary alcohols dehydrate to yield alkenes. The reaction conditions must be carefully controlled, and the method is of little practical value in the laboratory.

PROBLEM.

18.2 Why do you suppose only symmetrical ethers are prepared by the sulfuric acid-catalyzed dehydration procedure? What product(s) would you expect if ethanol and 1-propanol were allowed to react together? In what ratio would the products be formed if the two alcohols were of equal reactivity?

18.4 The Williamson Ether Synthesis

Metal alkoxides react with primary alkyl halides and tosylates by an S_N2 pathway to yield ethers, a process known as the **Williamson ether synthesis**. Discovered in 1850, the Williamson[1] synthesis is still the best method for the preparation of ethers, both symmetrical and unsymmetrical.

Cyclopentoxide ion **Cyclopentyl methyl ether (74%)**

The alkoxides needed in the Williamson reaction are normally prepared by reaction of an alcohol with a strong base such as sodium hydride, NaH (Section 17.4). An acid–base reaction occurs between the alcohol and sodium hydride to generate the sodium salt of the alcohol.

$$ROH + NaH \longrightarrow RO^- Na^+ + H_2$$

A useful variation of the Williamson synthesis involves silver oxide, Ag_2O, as base rather than NaH. Under these conditions, the free alcohol reacts directly with alkyl halide, and there is no need to preform the metal alkoxide. For example, glucose reacts with iodomethane in the presence of Ag_2O to generate a *pentaether* in 85% yield.

[1]Alexander W. Williamson (1824–1904); b. Wandsworth, England; Ph.D. Giessen (1846); professor, University College, London (1849–1904).

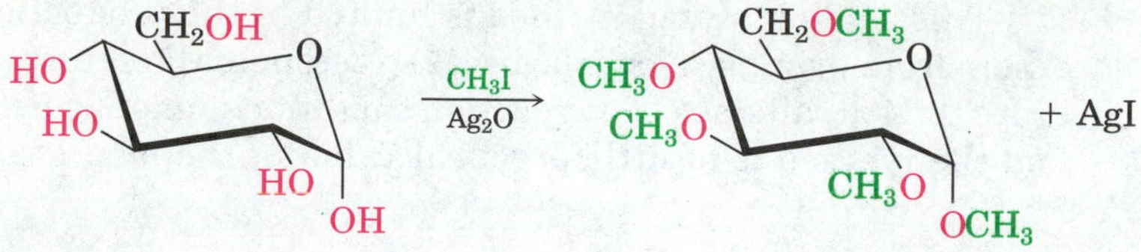

α-D-Glucose

α-D-Glucose pentamethyl ether (85%)

Mechanistically, the Williamson synthesis is simply an S_N2 displacement of halide ion by an alkoxide ion nucleophile. The Williamson synthesis is thus subject to all the usual constraints on S_N2 reactions discussed in Section 11.5. Primary halides and tosylates work best because competitive E2 elimination of HX can occur with more hindered substrates. Unsymmetrical ethers should therefore be synthesized by reaction between the more hindered alkoxide partner and less hindered halide partner rather than vice versa. For example, *tert*-butyl methyl ether, a substance used as an octane booster in gasoline, is best prepared by reaction of *tert*-butoxide ion with iodomethane rather than by reaction of methoxide ion with 2-chloro-2-methylpropane.

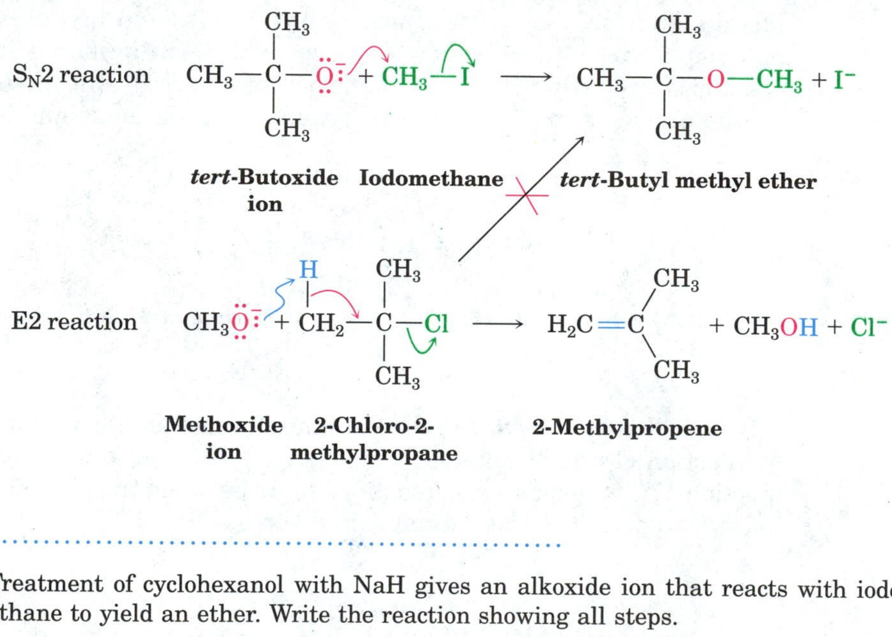

S_N2 reaction

tert-Butoxide ion — Iodomethane — *tert*-Butyl methyl ether

E2 reaction

Methoxide ion — 2-Chloro-2-methylpropane — 2-Methylpropene

PROBLEM

18.3 Treatment of cyclohexanol with NaH gives an alkoxide ion that reacts with iodoethane to yield an ether. Write the reaction showing all steps.

PROBLEM

18.4 How would you prepare the following compounds using a Williamson synthesis?
(a) Methyl propyl ether
(b) Anisole (methyl phenyl ether)
(c) Benzyl isopropyl ether
(d) Ethyl 2,2-dimethylpropyl ether

PROBLEM

18.5 Rank the following halides in order of their reactivity in the Williamson synthesis:
(a) Bromoethane, 2-bromopropane, bromobenzene
(b) Chloroethane, bromoethane, 1-iodopropene

18.5 Alkoxymercuration/Demercuration of Alkenes

We saw in Section 7.4 that alkenes react with water in the presence of mercuric acetate to yield a hydroxymercuration product. Subsequent treatment with $NaBH_4$ breaks the C–Hg bond and yields the alcohol. A similar **alkoxymercuration** reaction occurs when an alkene is treated with an *alcohol* in the presence of mercuric acetate. [Mercuric trifluoroacetate, $(CF_3CO_2)_2Hg$, works even better.] Demercuration by reaction with $NaBH_4$ then yields an ether. As indicated by the following examples, the net result is Markovnikov addition of alcohol to the alkene.

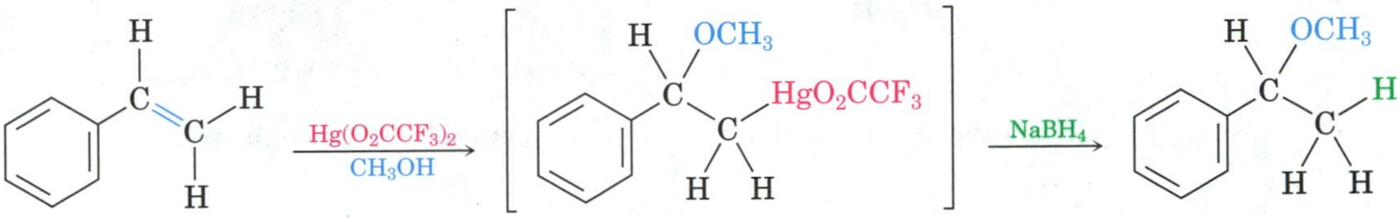

Styrene

1-Methoxy-1-phenylethane (97%)

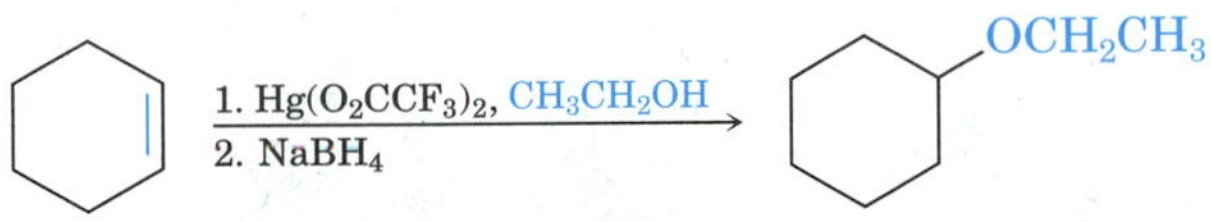

Cyclohexene

Cyclohexyl ethyl ether (100%)

The mechanism of the alkoxymercuration reaction is analogous to that described in Section 7.4 for hydroxymercuration. The reaction is initiated by electrophilic addition of Hg^{2+} to the alkene, followed by reaction of the intermediate cation with alcohol. Displacement of mercury by $NaBH_4$ completes the process.

A wide variety of alcohols and alkenes can be used in the alkoxymercuration reaction. Primary, secondary, and even tertiary alcohols react smoothly, but ditertiary ethers can't be prepared because of steric hindrance to reaction.

PROBLEM

18.6 Show in detail the mechanism of the reaction of 1-methylcyclopentene with ethanol and mercuric trifluoroacetate.

PROBLEM

18.7 How would you prepare the following ethers? Use whichever method you think is more appropriate, the Williamson synthesis or the alkoxymercuration reaction.
(a) Butyl cyclohexyl ether
(b) Benzyl ethyl ether ($C_6H_5CH_2OCH_2CH_3$)
(c) *tert*-Butyl *sec*-butyl ether
(d) Tetrahydrofuran

18.6 Reactions of Ethers: Acidic Cleavage

Ethers are unreactive to many reagents used in organic chemistry, a property that accounts for their wide use as reaction solvents. Halogens, dilute acids, bases, and nucleophiles have no effect on most ethers. In fact, ethers undergo only one reaction of general use—they are cleaved by strong acids.

The first example of acid-induced ether cleavage was observed in 1861 by Alexander Butleroff,[2] who found that 2-ethoxypropanoic acid reacts with aqueous HI at 100°C to yield iodoethane and lactic acid:

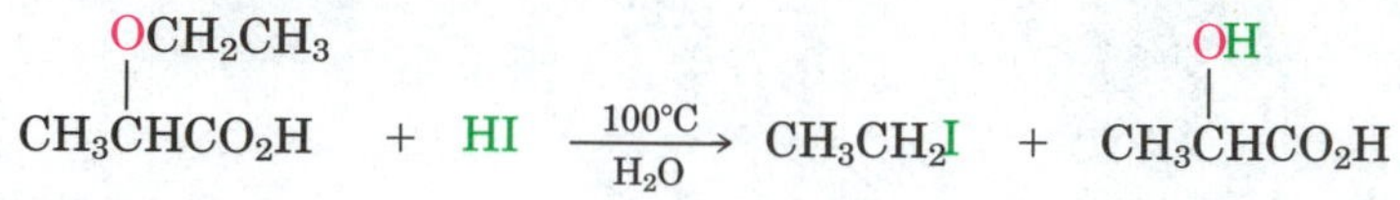

2-Ethoxypropanoic acid **Iodoethane** **Lactic acid**

Aqueous HI is still the preferred reagent for cleaving simple ethers, although concentrated aqueous HBr can also be used. HCl does not cleave ethers.

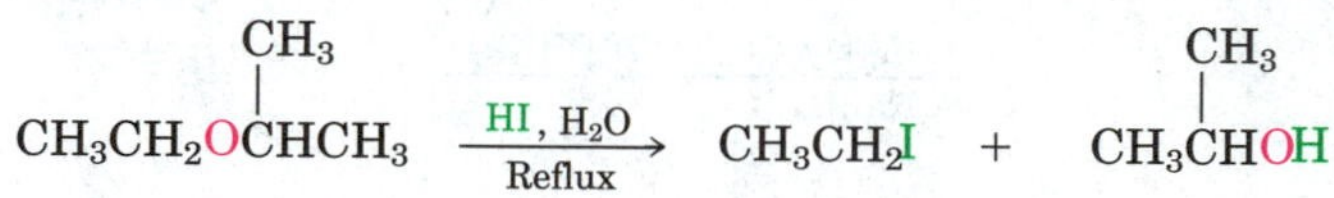

Ethyl isopropyl ether **Iodoethane** **Isopropyl alcohol**

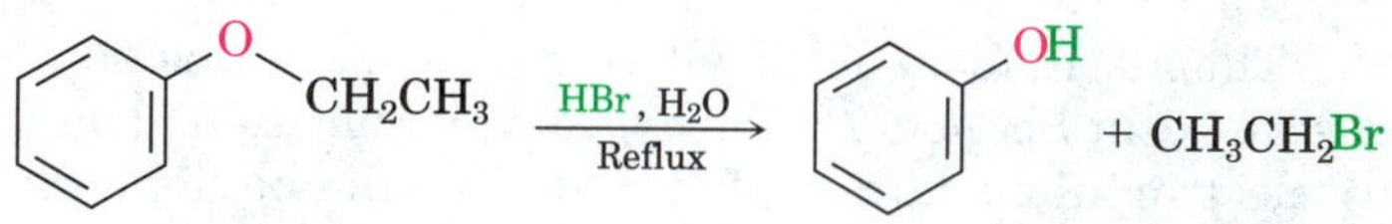

Ethyl phenyl ether **Phenol** **Bromoethane**

Acidic ether cleavages are typical nucleophilic substitution reactions that take place by either an S_N1 pathway or an S_N2 pathway, depending on the structure of the ether. Primary and secondary alkyl ethers react by an S_N2 pathway in which I^- or Br^- attacks the protonated ether at the less highly substituted site. This usually results in a selective cleavage into a single alcohol and a single alkyl halide. For example, butyl isopropyl ether yields exclusively isopropyl alcohol and 1-iodobutane on cleavage by HI, because nucleophilic attack by iodide ion occurs at the less hindered primary site rather than at the more hindered secondary site.

---

[2]Alexander M. Butleroff (1828–1886); b. Tschistopol, Russia; Ph.D. (1854), University of Moscow; professor, University of Kazan (1854–1867), University of St. Petersburg (1867–1880).

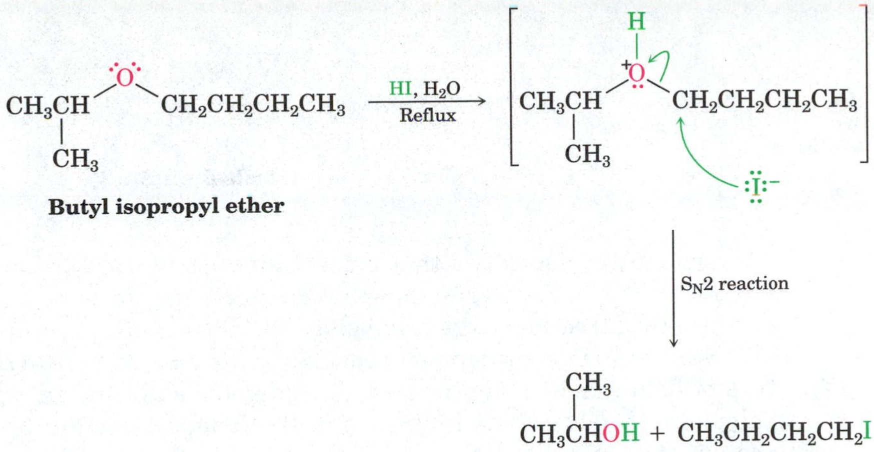

Butyl isopropyl ether

Isopropyl alcohol **1-Iodobutane**

Tertiary, benzylic, and allylic ethers cleave by either an S_N1 or an E1 mechanism because these substrates can produce stable intermediate carbocations. These reactions are often fast and take place at moderate temperatures. *tert*-Butyl ethers, for example, are often cleaved at room temperature or below using trifluoroacetic acid, although HBr and HI also work.

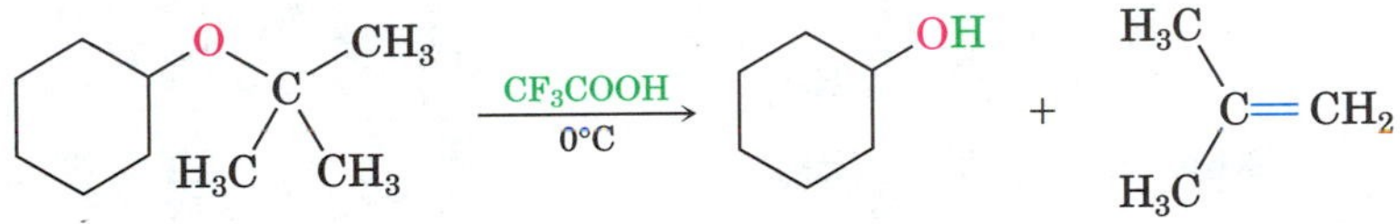

***tert*-Butyl cyclohexyl ether** **Cyclohexanol (90%)** **2-Methylpropene**

PROBLEM

18.8 Write the mechanism of the acid-catalyzed cleavage of a *tert*-butyl ether. Account for the fact that 2-methylpropene is formed.

PROBLEM

18.9 Suggest an explanation for the observation that HI and HBr are more effective than HCl in cleaving ethers. (*Hint*: See Section 11.5.)

18.7 Cyclic Ethers: Epoxides

For the most part, cyclic ethers behave like acyclic ethers. The chemistry of the ether functional group is the same, whether it's in an open chain or in a ring. Common cyclic ethers such as tetrahydrofuran and dioxane, for example, are often used as solvents because of their inertness, yet they can be cleaved by strong acids.

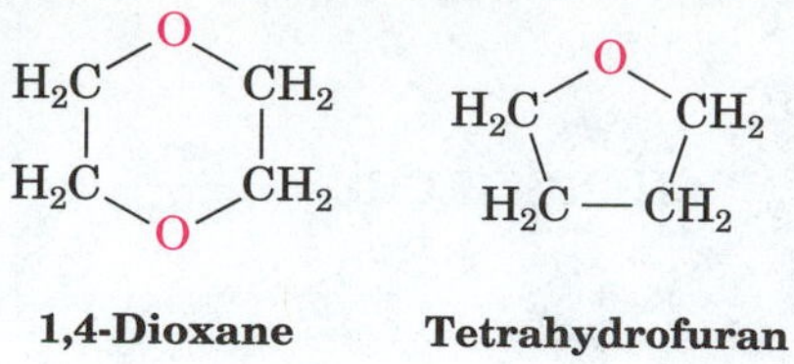

The only cyclic ethers that behave differently from open-chain ethers are the three-membered ring compounds called **epoxides**, or **oxiranes**. The strain of the three-membered ring gives epoxides unique chemical reactivity.

Ethylene oxide, the simplest epoxide, is an intermediate in the manufacture of both ethylene glycol, used for automobile antifreeze, and polyester polymers. More than 3.4 million tons of ethylene oxide are produced each year in the United States by air oxidation of ethylene over a silver oxide catalyst at 300°C. This process is not useful for other epoxides, however, and is of little value in the laboratory. (Note that the name *ethylene oxide* is not a systematic one because the *-ene* ending implies the presence of a double bond in the molecule. The name is frequently used, however, because ethylene oxide is derived from ethylene by addition of an oxygen atom. The systematic name for ethylene oxide is 1,2-epoxyethane.)

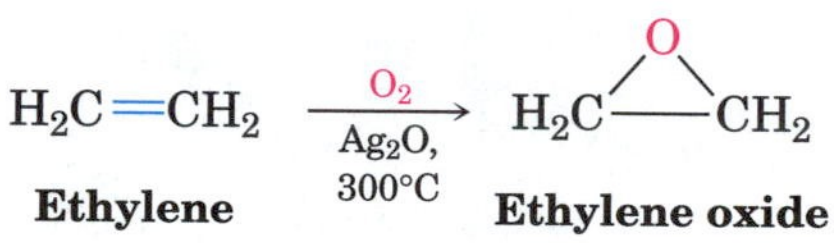

In the laboratory, epoxides are normally prepared by treatment of an alkene with a **peroxyacid, RCO_3H**:

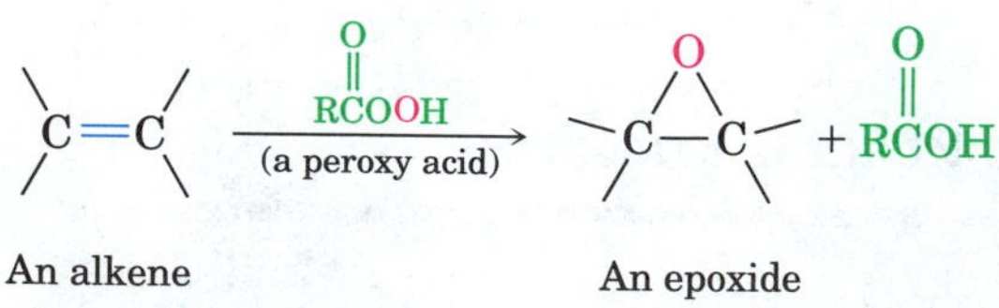

Many different peroxyacids can be used to accomplish epoxidation, but *m*-chloroperoxybenzoic acid is the most common choice.

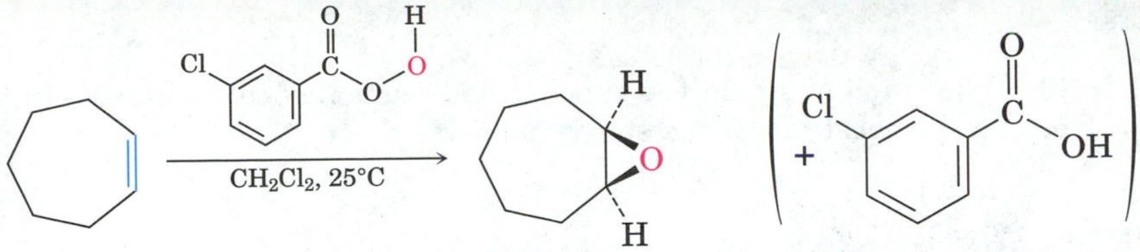

Cycloheptene 1,2-Epoxycycloheptane (78%)

Peroxyacids transfer oxygen to an alkene with syn stereochemistry through a one-step mechanism without intermediates. Studies have shown that the oxygen atom farthest from the carbonyl group is the one transferred.

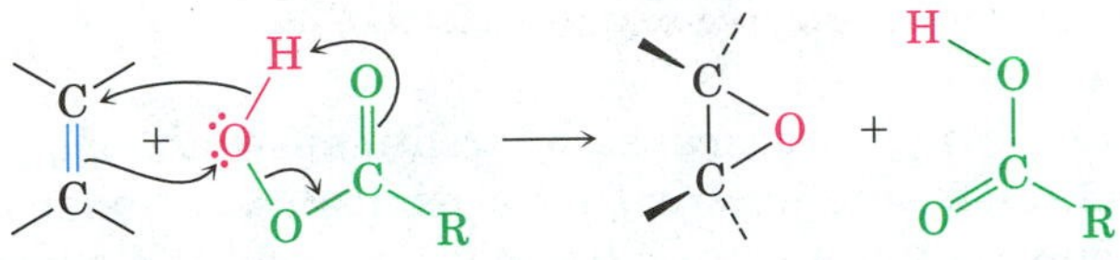

Alkene Peroxyacid Epoxide Acid

Another method for the synthesis of epoxides is through the use of halohydrins, prepared by electrophilic addition of HO–X to alkenes (Section 7.3). When halohydrins are treated with base, HX is eliminated, and an epoxide is produced.

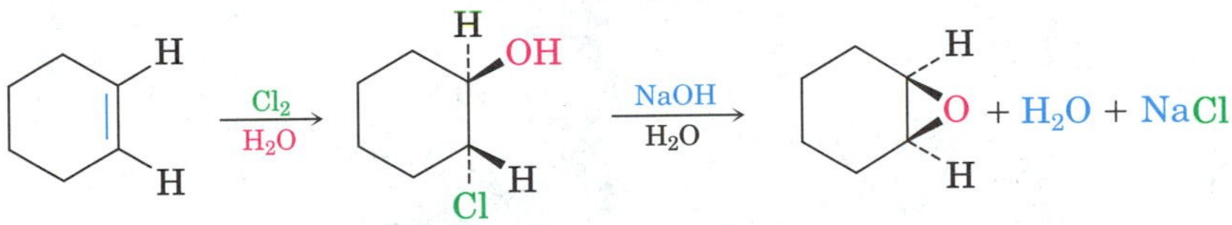

Cyclohexene *trans*-2-Chloro-cyclohexanol 1,2-Epoxycyclohexane (73%)

Note that the formation of an epoxide by treatment of a halohydrin with base is just an *intramolecular* Williamson ether synthesis. The nucleophilic alkoxide and the electrophilic alkyl halide are in the same molecule.

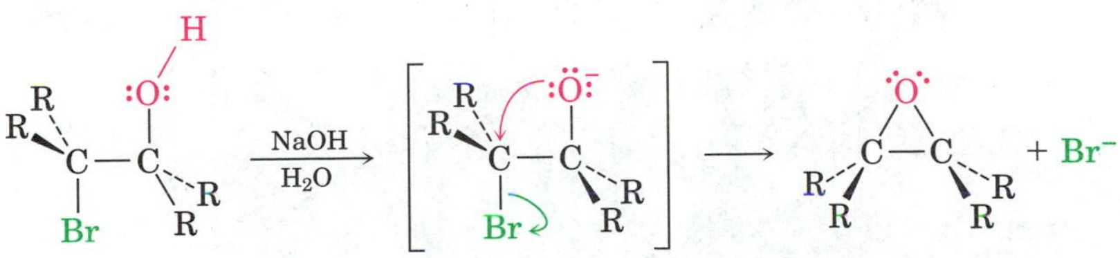

Bromohydrin Intramolecular substitution (within the same molecule) Epoxide

PROBLEM.

18.10 What product would you expect from reaction of *cis*-2-butene with *m*-chloroperoxybenzoic acid? Show the stereochemistry.

PROBLEM.

18.11 Reaction of *trans*-2-butene with *m*-chloroperoxybenzoic acid yields an epoxide different from that obtained by reaction of the cis isomer (Problem 18.10). Explain.

18.8 Ring-Opening Reactions of Epoxides

Epoxide rings are cleaved by treatment with acid just as other ethers are. The major difference is that epoxides react under much milder conditions because of ring strain. Dilute aqueous acid at room temperature is sufficient to cause the hydrolysis of epoxides to 1,2-diols, also called *vicinal glycols*. (The word *vicinal* means "adjacent," and a *glycol* is a diol.) More than 2.8 million tons of ethylene glycol, most of it used for automobile antifreeze, are produced each year in the United States by acid-catalyzed hydration of ethylene oxide. (Note that the name *ethylene glycol* refers to the glycol derived from ethylene just as ethylene oxide refers to the epoxide derived from ethylene.)

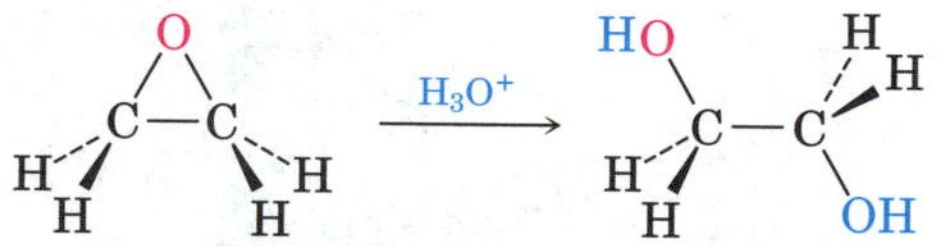

Ethylene oxide **Ethylene glycol (1,2-Ethanediol)**

Acid-catalyzed epoxide cleavage takes place by S_N2-like attack of a nucleophile on the protonated epoxide, in a manner analogous to the final step of alkene bromination in which a cyclic bromonium ion is opened by nucleophilic attack (Section 7.2). When a cycloalkane epoxide is opened by aqueous acid, a *trans*-1,2-diol results, just as a *trans*-1,2-dibromide results from alkene bromination.

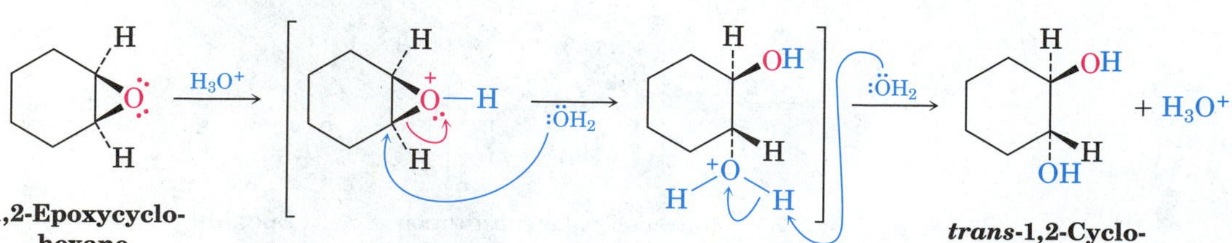

1,2-Epoxycyclohexane **_trans_-1,2-Cyclohexanediol (86%)**

Recall the following:

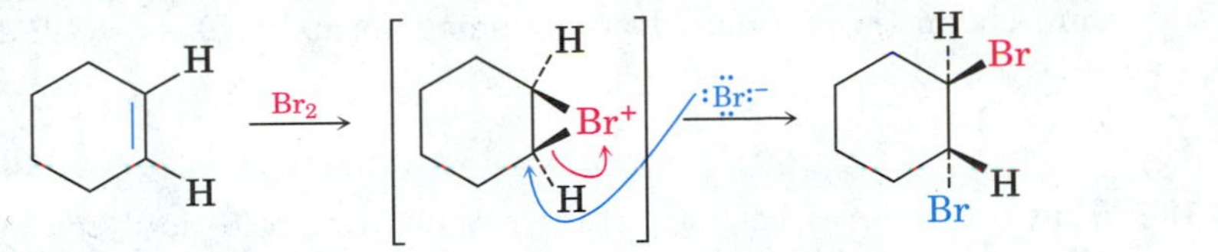

Epoxides can also be opened by reaction with nucleophiles other than water. For example, if anhydrous HX is used, epoxides are converted into trans halohydrins:

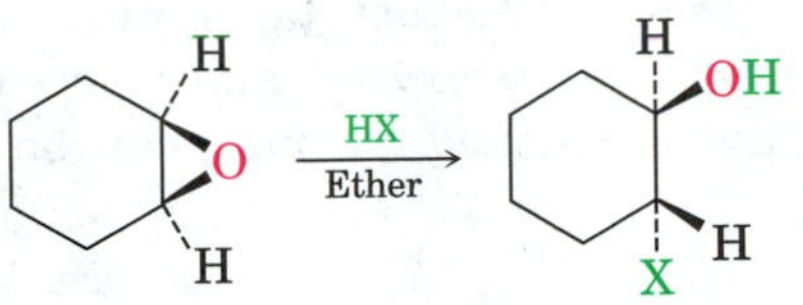

A trans 2-halocyclohexanol

where X = F, Br, Cl, or I

Unlike other ethers, epoxides can be cleaved by base as well as by acid. Although an ether oxygen is normally a poor leaving group in an S_N2 reaction (Section 11.5), the reactivity of the three-membered ring is sufficient to allow epoxides to react with hydroxide ion at elevated temperatures.

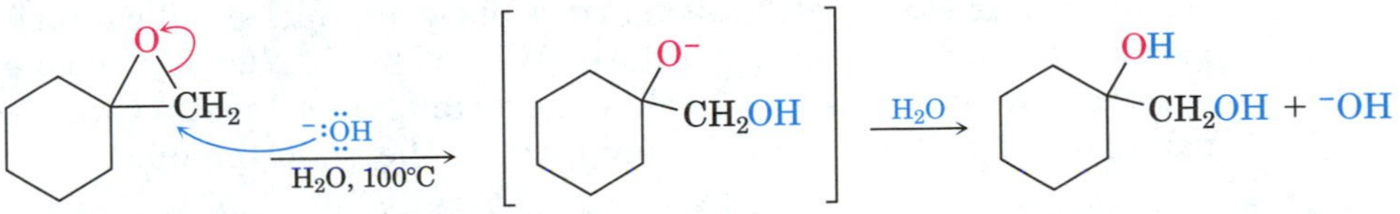

Methylenecyclohexane oxide **1-Hydroxymethylcyclohexanol (70%)**

A similar nucleophilic ring opening occurs when epoxides are treated with Grignard reagents. Ethylene oxide is frequently used, allowing the conversion of a Grignard reagent into a primary alcohol having two more carbons than the starting alkyl halide. 1-Bromobutane, for example, is converted into 1-hexanol by reaction of its Grignard reagent with ethylene oxide.

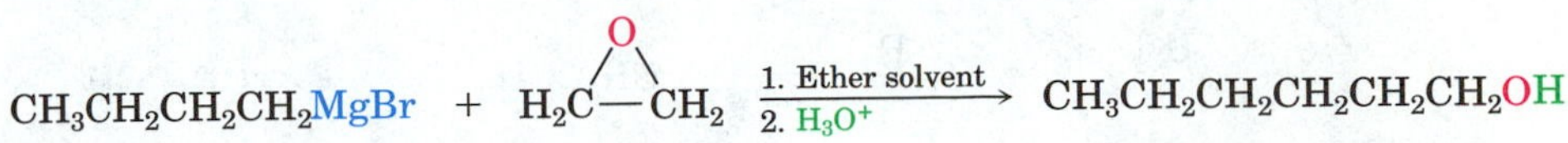

Butylmagnesium bromide **Ethylene oxide** **1-Hexanol (62%)**

PROBLEM.

18.12 Show the steps involved in the acidic hydrolysis of *cis*-5,6-epoxydecane. What is the stereochemistry of the product, assuming normal back-side S_N2 attack?

PROBLEM.

18.13 What is the stereochemistry of the product from acid-catalyzed hydrolysis of *trans*-5,6-epoxydecane? How does the product differ from that formed in Problem 18.12?

Regiochemistry of Epoxide Ring Opening

The direction in which an unsymmetrical epoxide ring is opened depends on its structure and on the reaction conditions. If a basic nucleophile is used in a typical S_N2 reaction, attack takes place at the less hindered epoxide carbon. For example, 1,2-epoxypropane reacts with ethoxide ion exclusively at the less highly substituted, primary carbon to give 1-ethoxy-2-propanol.

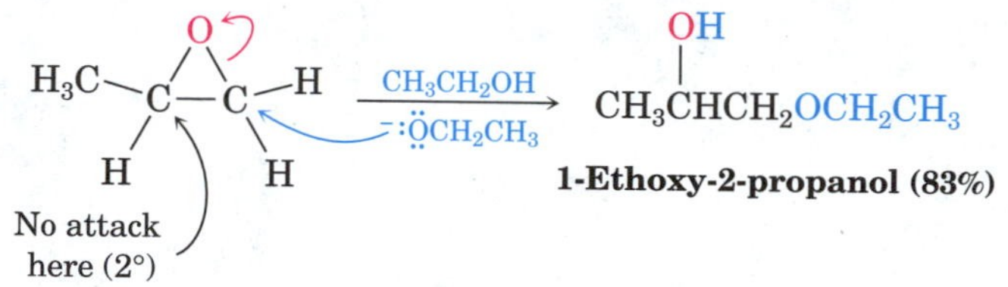

If acidic conditions are used, the regiochemistry of ring opening depends on the epoxide's structure, and a mixture of products is usually formed. When both epoxide carbon atoms are either primary or secondary, attack of the nucleophile occurs mostly at the *less* highly substituted site. When one of the epoxide carbon atoms is tertiary, however, nucleophilic attack occurs mostly at the *more* highly substituted site. Thus, 1,2-epoxypropane reacts with HCl to give primarily 1-chloro-2-propanol, but 2-methyl-1,2-epoxypropane gives 2-chloro-2-methyl-1-propanol as the major product.

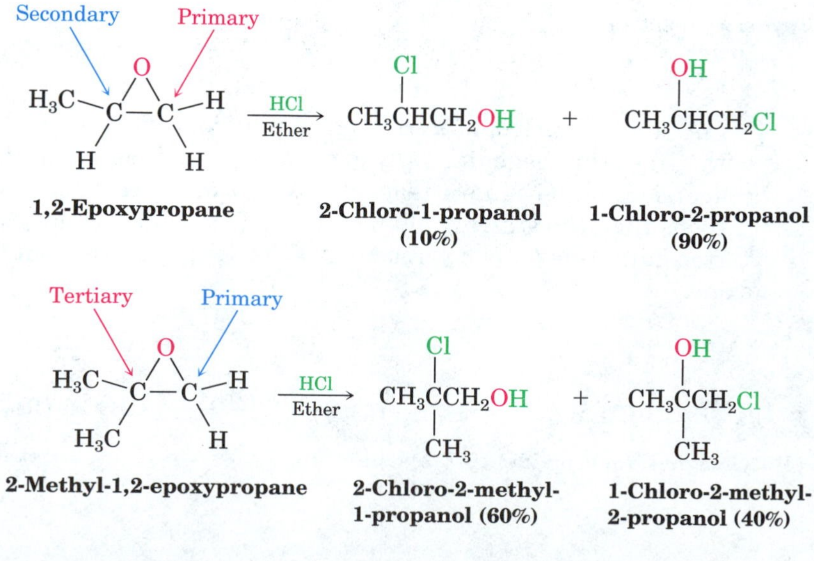

The mechanism of these acid-catalyzed epoxide openings is interesting because it appears to be *midway* between typical S_N1 and S_N2 pathways and has characteristics of both. Take the reaction of 1,2-epoxy-1-methylcyclohexane with HBr, for example. This reaction yields a single isomer of 2-bromo-2-methylcyclohexanol in which the –Br and –OH groups are trans. The fact that the product has the entering bromine and the leaving oxygen on opposite sides of the ring is an S_N2-like result, with back-side displacement of the leaving group. The fact that Br^- attacks the more hindered tertiary side of the epoxide rather than the less hindered secondary side is an S_N1-like result.

Evidently, the transition state for acid-induced epoxide opening has an S_N2-like geometry and also has a large amount of S_N1-like carbocationic character. Since the positive charge in the protonated epoxide is shared by the more highly substituted carbon atom, back-side attack of Br^- occurs at the more highly substituted site (Figure 18.1).

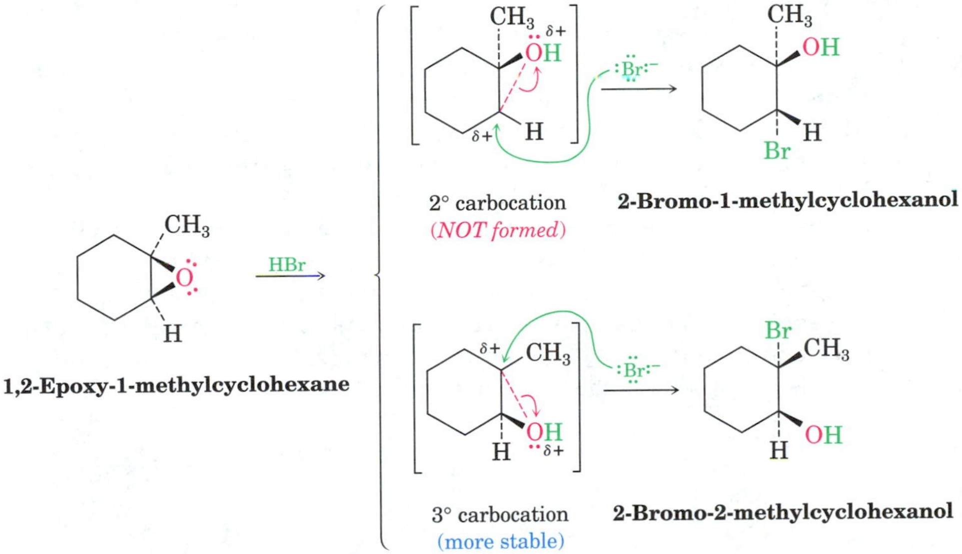

Figure 18.1 Acid-induced ring opening of 1,2-epoxy-1-methylcyclohexane. There is a high degree of S_N1-like carbocationic character in the transition state, which leads to back-side attack of the nucleophile at the tertiary center and to formation of the isomer of 2-bromo-2-methylcyclohexanol that has –Br and –OH groups trans.

PROBLEM

18.14 Predict the major product of the following reactions:

(a) 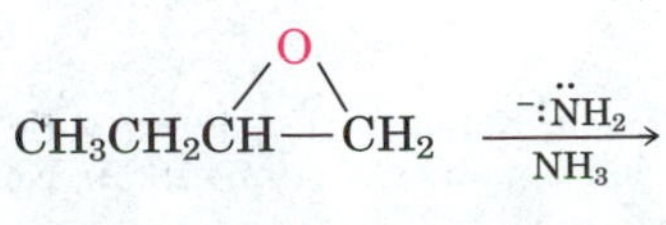

(b) 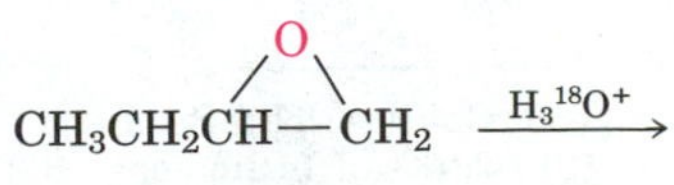

18.9 Crown Ethers

Discovered in the early 1960s by Charles Pedersen[3] at the Du Pont Company, **crown ethers** are a relatively recent addition to the ether family. Crown ethers are named according to the general format *x*-crown-*y*, where *x* is the total number of atoms in the ring and *y* is the number of oxygen atoms. Thus, 18-crown-6 ether is an 18-membered ring containing 6 ether oxygen atoms.

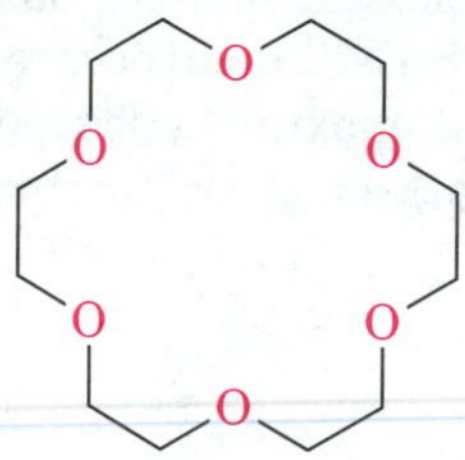

18-Crown-6 ether

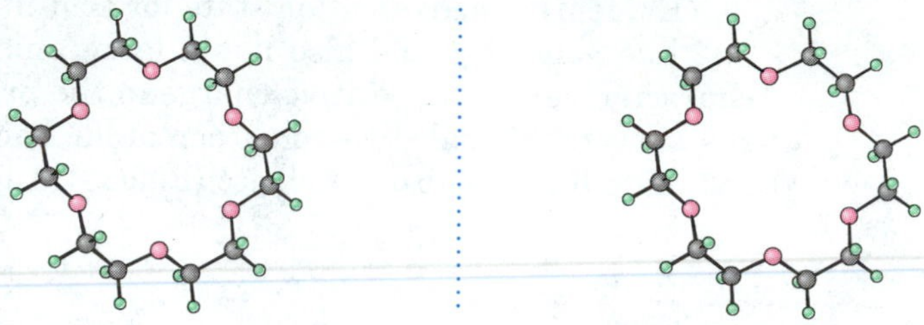

Stereo View

The importance of crown ethers derives from their extraordinary ability to solvate metal cations by sequestering the metal in the center of the polyether cavity. Different crown ethers solvate different metal cations, depending on the match between ion size and cavity size. For example, 18-crown-6 complexes strongly with potassium ion.

Complexes between crown ethers and inorganic salts are soluble in nonpolar organic solvents, thus allowing many reactions to be carried out under aprotic conditions that would otherwise have to be carried out in aqueous solution. For example, the inorganic salt $KMnO_4$ actually dissolves in benzene in the presence of 18-crown-6. The resulting solution of "purple benzene" is a valuable reagent for oxidizing alkenes.

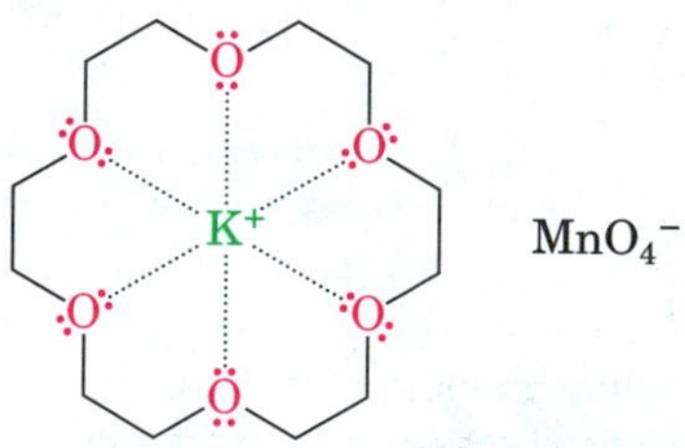

$KMnO_4$ solvated by 18-crown-6
(this solvate is soluble in benzene)

Many other inorganic salts, including KF, KCN, and NaN_3, can be dissolved in organic solvents with the help of crown ethers. The effect of using a crown ether to dissolve a salt in a hydrocarbon or ether solvent is similar to the effect of dissolving the salt in a polar aprotic solvent such as DMSO,

---

[3]Charles John Pedersen (1904–1989); b. Pusan, Korea (U.S. citizen); M.Sc., Massachusetts Institute of Technology (1927); E.I. Du Pont de Nemours Co. (1927–1969); Nobel Prize (1987).

DMF, or HMPA (Section 11.5). In both cases, the metal cation is strongly solvated, leaving the anion bare. Thus, the S_N2 reactivity of an anion is tremendously enhanced in the presence of a crown ether.

PROBLEM

18.15 15-Crown-5 and 12-crown-4 ethers complex Na^+ and Li^+, respectively. Make models of these crown ethers, and compare the sizes of the cavities.

18.10 Spectroscopic Analysis of Ethers

Infrared Spectroscopy

Ethers are difficult to distinguish by IR spectroscopy. Although they show an absorption due to carbon–oxygen single-bond stretching in the range 1050–1150 cm^{-1}, many other kinds of absorptions occur in the same range. Figure 18.2 shows the IR spectrum of diethyl ether and identifies the C–O stretch.

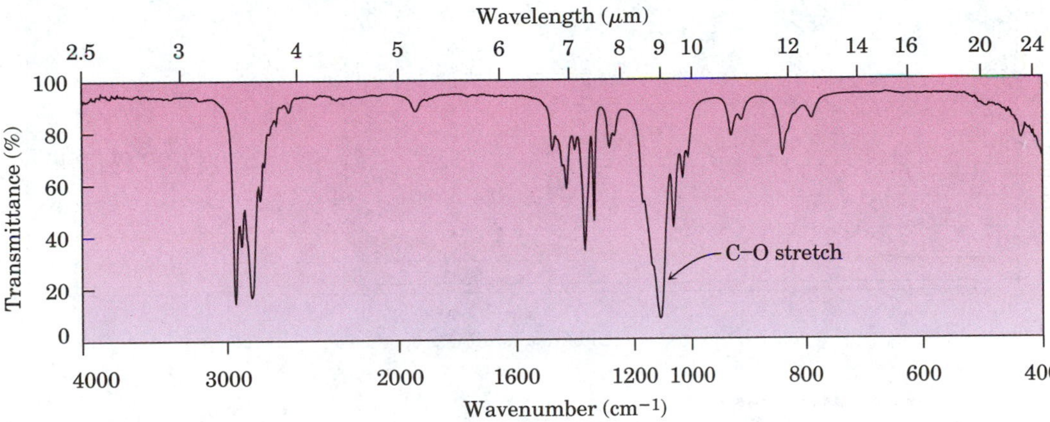

Figure 18.2 The infrared spectrum of diethyl ether, $CH_3CH_2OCH_2CH_3$.

Nuclear Magnetic Resonance Spectroscopy

Hydrogens on carbon next to an ether oxygen are shifted downfield from the normal alkane resonance and show ^1H NMR absorptions in the region 3.4–4.5 δ. This downfield shift is clearly seen in the spectrum of dipropyl ether shown in Figure 18.3.

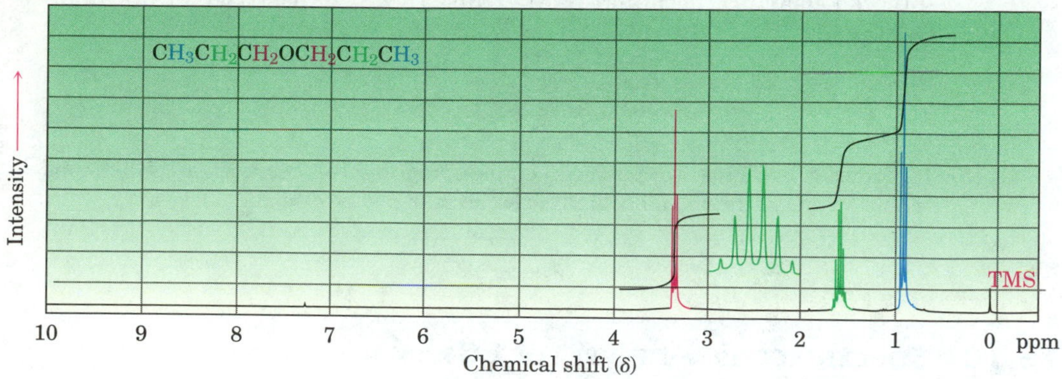

Figure 18.3 The ^{1}H NMR spectrum of dipropyl ether. Protons on carbon next to oxygen are shifted downfield to 3.4 δ.

Epoxides absorb at a slightly higher field than other ethers and show characteristic resonances at 2.5–3.5 δ in their ^{1}H NMR spectrum, as indicated for 1,2-epoxypropane in Figure 18.4.

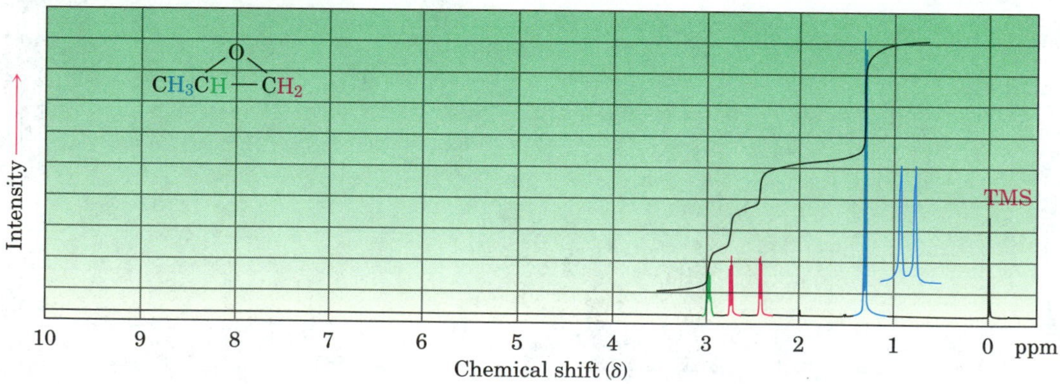

Figure 18.4 The ^{1}H NMR spectrum of 1,2-epoxypropane.

Ether carbon atoms also exhibit a downfield shift in the ^{13}C NMR spectrum, where they usually absorb in the range 50–80 δ. For example, the carbon atoms next to oxygen in methyl propyl ether absorb at 58.5 and 74.8 δ. Similarly, the methyl carbon in anisole absorbs at 54.8 δ.

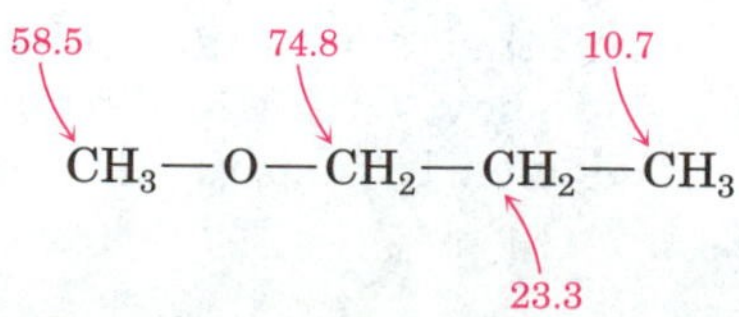

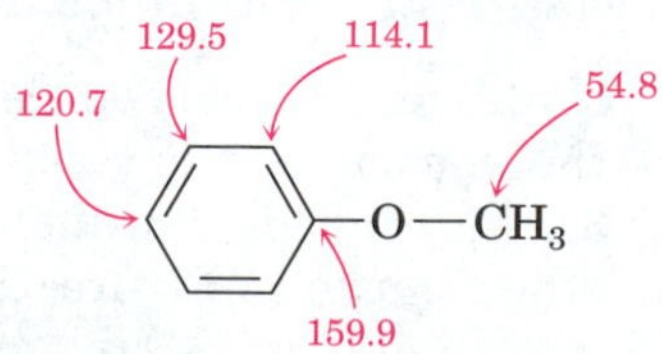

PROBLEM

18.16 The ^{1}H NMR spectrum shown is that of a substance with the formula C_4H_8O. Propose a structure.

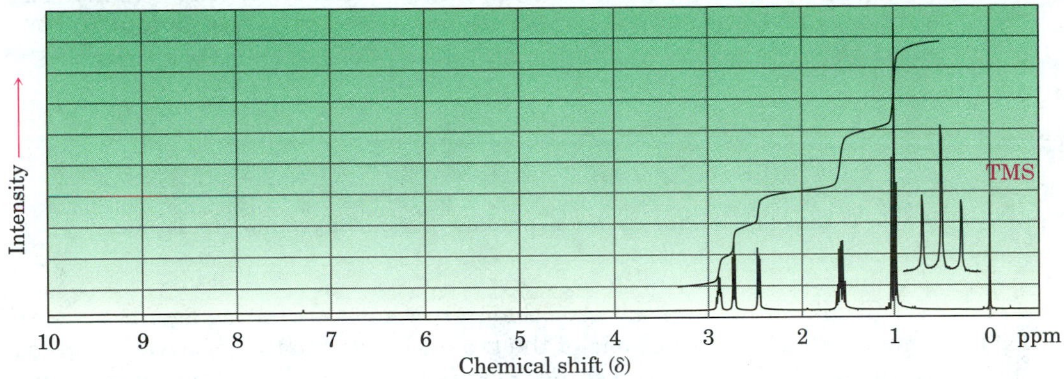

18.11 Sulfides

Sulfides, RSR′, are sulfur analogs of ethers in the same way that thiols are sulfur analogs of alcohols (Section 17.12). They are named by following the same rules used for ethers, with *sulfide* used in place of *ether* for simple compounds and *alkylthio* used in place of *alkoxy* for more complex substances.

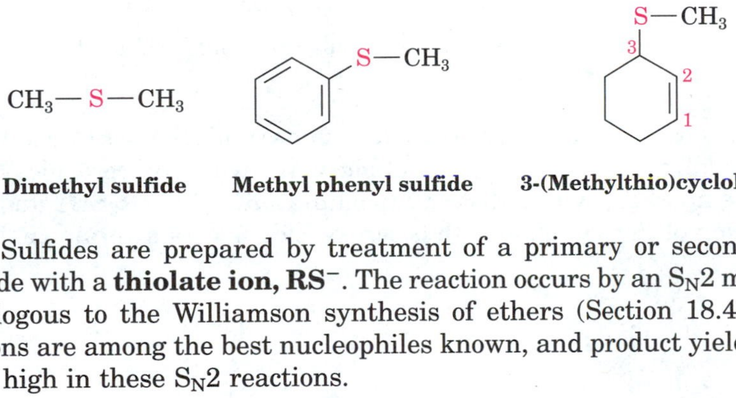

Dimethyl sulfide **Methyl phenyl sulfide** **3-(Methylthio)cyclohexene**

Sulfides are prepared by treatment of a primary or secondary alkyl halide with a **thiolate ion, RS$^-$**. The reaction occurs by an S_N2 mechanism, analogous to the Williamson synthesis of ethers (Section 18.4). Thiolate anions are among the best nucleophiles known, and product yields are usually high in these S_N2 reactions.

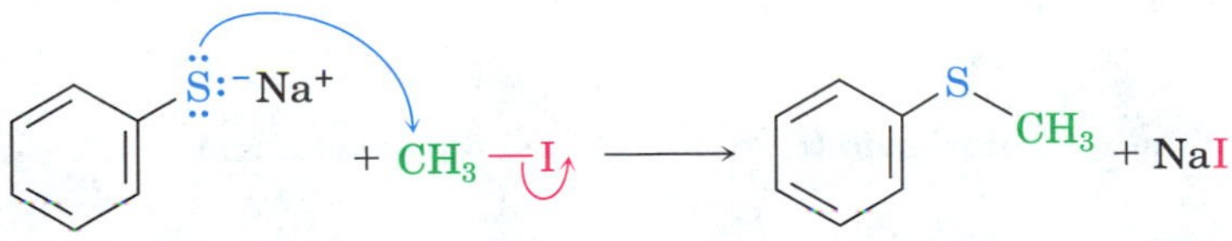

Sodium benzenethiolate **Methyl phenyl sulfide (96%)**

Since the valence electrons on sulfur are farther from the nucleus and are less tightly held than those on oxygen (3*p* electrons versus 2*p* electrons), there are some important differences between the chemistries of ethers and sulfides. For example, sulfur is more polarizable than oxygen, and sulfur compounds are thus more nucleophilic than their oxygen analogs. Unlike dialkyl ethers, dialkyl sulfides are good nucleophiles that react rapidly with primary alkyl halides by an S_N2 mechanism to give **trialkylsulfonium salts (R_3S^+).**

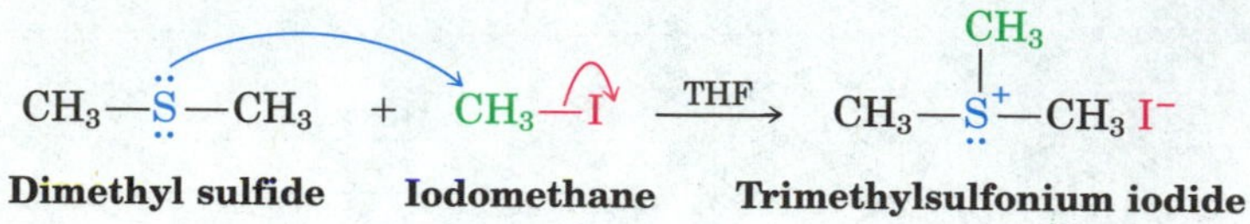

Trialkylsulfonium salts are themselves good alkylating agents because a nucleophile can attack one of the groups bonded to the positively charged sulfur, displacing a neutral sulfide as leaving group. Nature makes extensive use of the trialkylsulfonium salt *S*-adenosylmethionine as a biological methylating agent (Section 11.17).

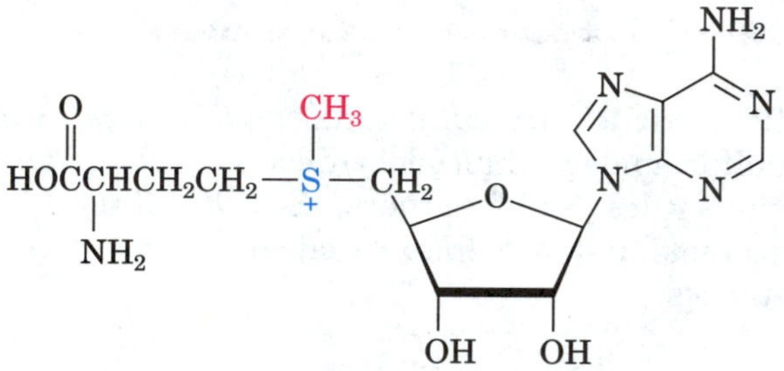

***S*-Adenosylmethionine (a sulfonium salt)**

Another difference between sulfides and ethers is that sulfides are easily oxidized. Treatment of a sulfide with hydrogen peroxide, H_2O_2, at room temperature yields the corresponding **sulfoxide (R_2SO)**, and further oxidation of the sulfoxide with a peroxyacid yields a **sulfone (R_2SO_2).**

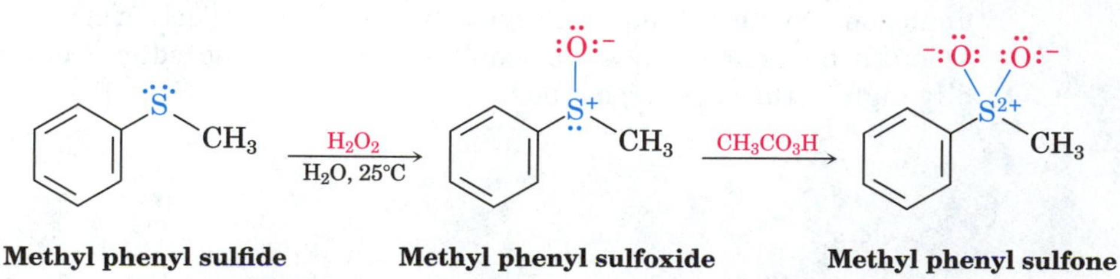

Dimethyl sulfoxide (DMSO) is a particularly well-known sulfoxide that is often used as a polar aprotic solvent. It must be handled with care, however, because it has a remarkable ability to penetrate the skin, carrying along whatever is dissolved in it.

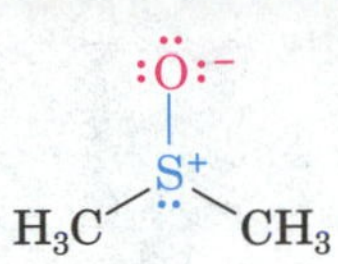

Dimethyl sulfoxide (a polar aprotic solvent)

PROBLEM

18.17 Name the following compounds by IUPAC rules:

(a) $CH_3CH_2SCH_3$

(b) $(CH_3)_3CSCH_2CH_3$

(c)

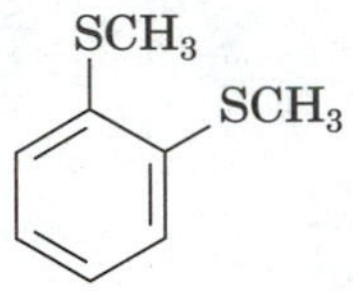

(d)

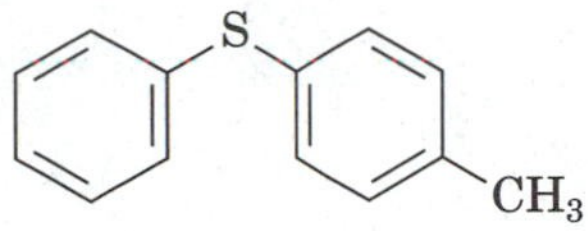

PROBLEM

18.18 How can you account for the fact that dimethyl sulfoxide has a boiling point of 189°C and is miscible with water, whereas dimethyl sulfide has a boiling point of 37°C and is immiscible with water?

INTERLUDE

Epoxy Resins and Adhesives

Racing canoes are often made of a high-strength polymer such as Kevlar coated with epoxy resin.

Few people know what an epoxide is, but practically everyone has used an "epoxy glue" for household repairs or an epoxy resin for a protective coating. Epoxy resins and adhesives generally consist of two components that must be mixed prior to use. One component is a liquid "prepolymer," and the second is a "curing agent" that reacts with the prepolymer and causes it to solidify.

The most widely used epoxy resins and adhesives are based on a prepolymer made from bisphenol A and epichlorohydrin. On treatment with base under carefully controlled conditions, bisphenol A is converted into its anion, which acts as a nucleophile in an S_N2 reaction with epichlorohydrin. Each epichlorohydrin molecule can react with two molecules of

(continued)➤

bisphenol A, once by S_N2 displacement of chloride ion and once by opening of the epoxide ring. At the same time, each bisphenol A molecule can react with two epichlorohydrins, leading to a long polymer chain. Each end of a prepolymer chain has an unreacted epoxy group, and each chain has numerous secondary alcohol groups in its middle.

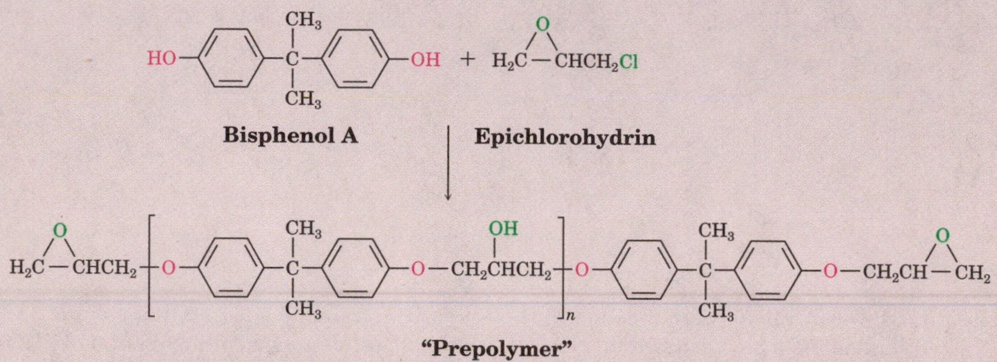

When the epoxide is to be used, a basic curing agent such as an amine, R_3N, is added to cause the individual prepolymer chains to link together. This "cross-linking" of chains is simply a base-catalyzed epoxide ring-opening of an –OH group in the middle of one chain with an epoxide group on the end of another chain. The result of such cross-linking is formation of a vast, three-dimensional tangle that has enormous strength and chemical resistance.

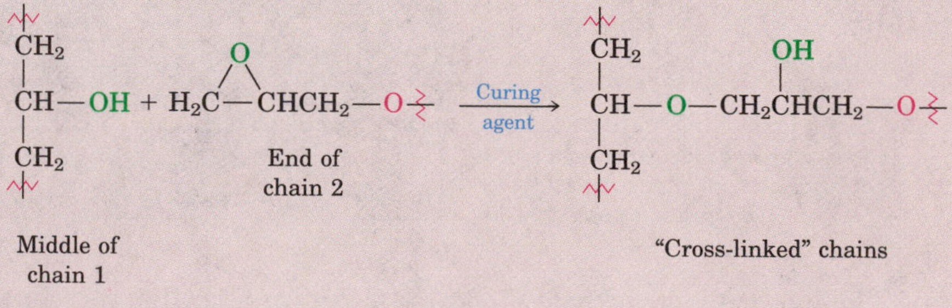

Summary and Key Words

Ethers are compounds that have two organic groups bonded to the same oxygen atom, ROR′. The organic groups can be alkyl, vinylic, or aryl, and the oxygen atom can be in a ring or in an open chain.

Ethers are prepared either by a Williamson synthesis or by an alkoxymercuration/demercuration sequence. The **Williamson ether synthesis** involves S_N2 attack of an alkoxide ion on a primary alkyl halide. The **alkoxymercuration/demercuration** sequence involves the formation of an intermediate organomercury compound, followed by $NaBH_4$ reduction of the C–Hg bond. The net result is Markovnikov addition of an alcohol to an alkene.

Ethers are inert to most reagents but are attacked by strong acids to give cleavage products. Both HI and HBr are often used. The cleavage reaction takes place by an S_N2 mechanism if primary and secondary alkyl groups are bonded to the ether oxygen, but by an S_N1 or E1 mechanism if one of the alkyl groups bonded to oxygen is tertiary.

Epoxides, cyclic ethers with a three-membered, oxygen-containing ring, differ from other ethers in their ease of cleavage. The high reactivity of the strained three-membered ether ring allows epoxide rings to be opened by nucleophilic attack of bases as well as acids. Base-catalyzed epoxide ring opening occurs by S_N2 attack of a nucleophile at the less hindered epoxide carbon, whereas acid-induced epoxide ring opening occurs by S_N1-like attack at the more highly substituted epoxide carbon.

Sulfides, RSR′, are sulfur analogs of ethers. They are prepared by a Williamson-type S_N2 reaction between a thiolate anion and a primary or secondary alkyl halide. Sulfides are much more nucleophilic than ethers and can be oxidized to **sulfoxides (R_2SO)** and to **sulfones (R_2SO_2)**. Sulfides can also be alkylated by reaction with a primary alkyl halide to yield **sulfonium salts, R_3S^+**.

Summary of Reactions

1. Preparation of ethers
   (a) Williamson synthesis (Section 18.4)

$$RO^- + R'CH_2X \longrightarrow ROR' + X^-$$

   Alkyl halide should be primary.
   (b) Alkoxymercuration/demercuration (Section 18.5)

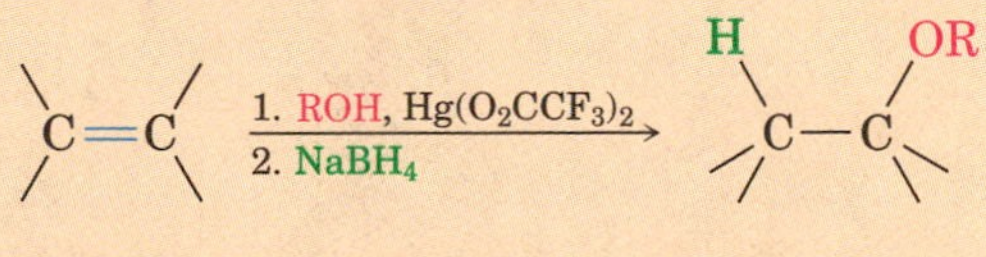

   Markovnikov orientation is observed.
   (c) Epoxidation of alkenes with peroxyacids (Section 18.7)

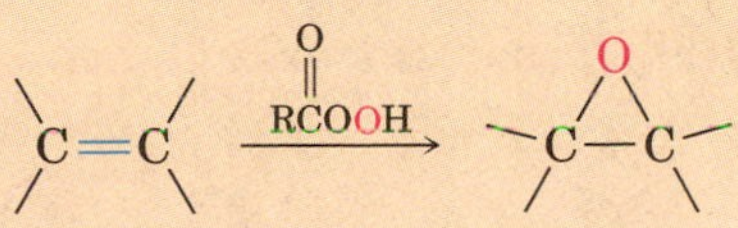

(continued)➤

2. Reaction of ethers
   (a) Cleavage by HX (Section 18.6)

$$R{-}O{-}R' \xrightarrow[H_2O]{HX} RX + R'OH$$

   where HX = HBr or HI

   (b) Base-catalyzed epoxide ring opening (Section 18.8)

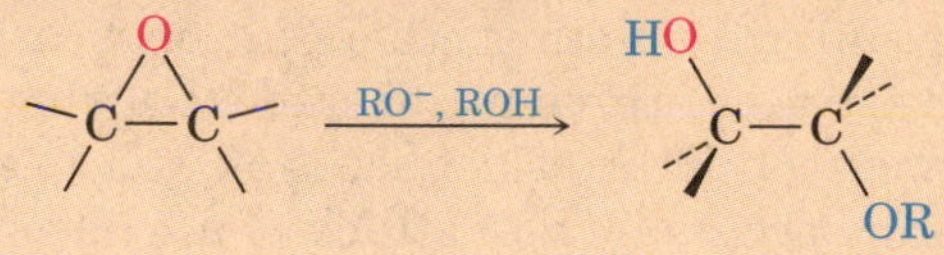

   Reaction occurs at least hindered site.

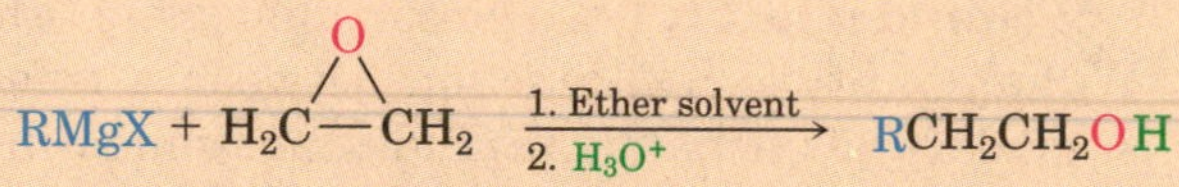

   (c) Acid-catalyzed hydrolysis of epoxides (Section 18.8)

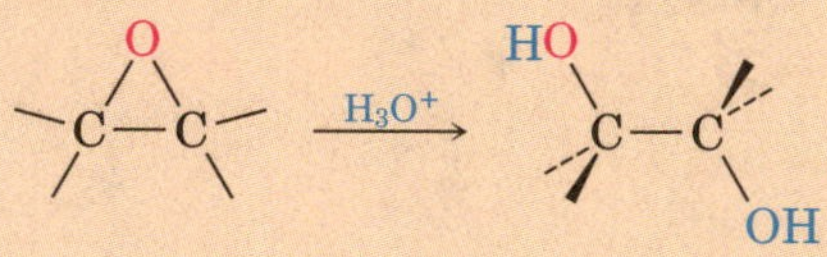

   Trans 1,2-diols are produced from cyclic epoxides.

   (d) Acid-induced epoxide ring opening (Section 18.8)

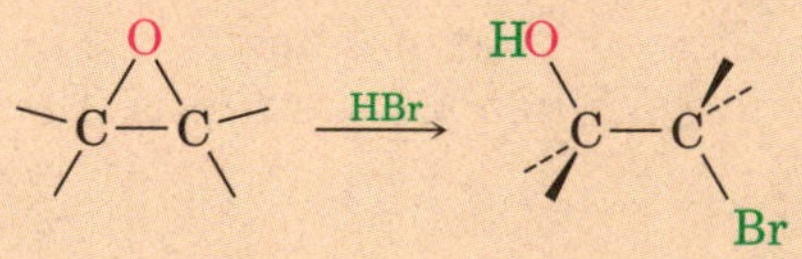

3. Preparation of sulfides (Section 18.11)

$$RS^- + R'CH_2Br \longrightarrow RSCH_2R' + Br^-$$

4. Oxidation of sulfides (Section 18.11)
   (a) Preparation of sulfoxides

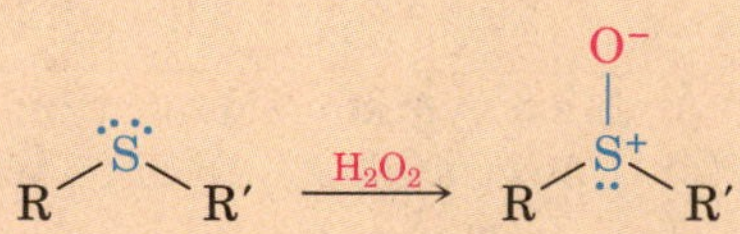

   (b) Preparation of sulfones

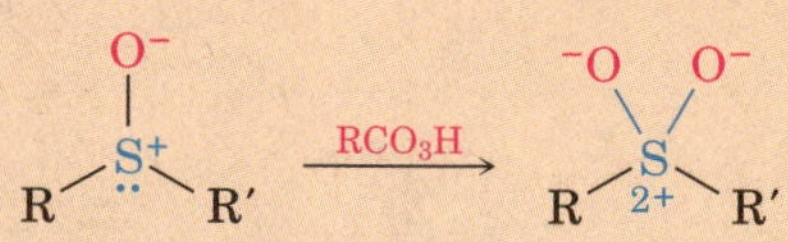

ADDITIONAL PROBLEMS

18.19 Draw structures corresponding to the following IUPAC names:
(a) Ethyl 1-ethylpropyl ether
(b) Di(*p*-chlorophenyl) ether
(c) 3,4-Dimethoxybenzoic acid
(d) Cyclopentyloxycyclohexane
(e) 4-Allyl-2-methoxyphenol (eugenol; from oil of cloves)

18.20 Give IUPAC names for the following structures:

(a) 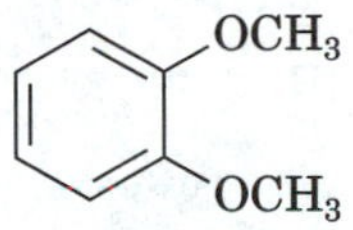

(b) 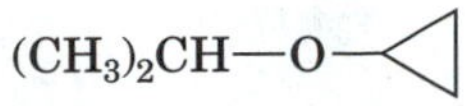

(c) 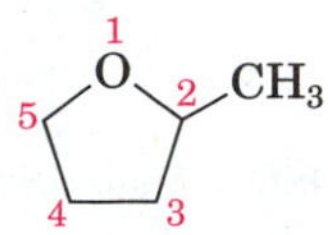

(d) 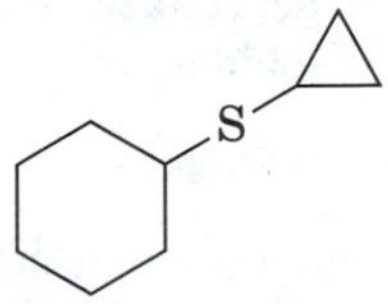

(e) $(CH_3)_2C(OCH_3)_2$

(f) 

(g) p-$NO_2C_6H_4OCH_2CH_3$

(h) 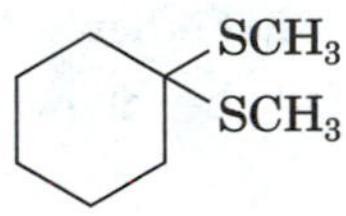

18.21 When 2-methylpentane-2,5-diol is treated with sulfuric acid, dehydration occurs and 2,2-dimethyltetrahydrofuran is formed. Suggest a mechanism for this reaction. Which of the two oxygen atoms is most likely to be eliminated, and why?

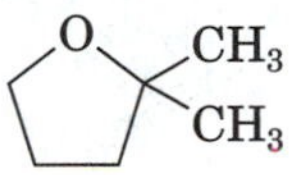

2,2-Dimethyltetrahydrofuran

18.22 Predict the products of the following ether cleavage reactions.

(a) 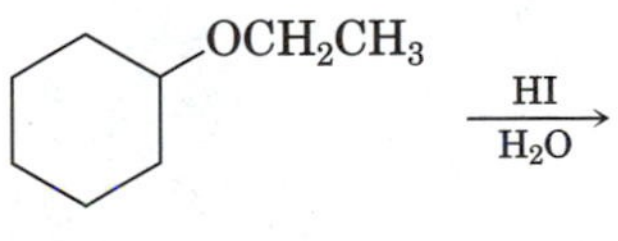

(b) 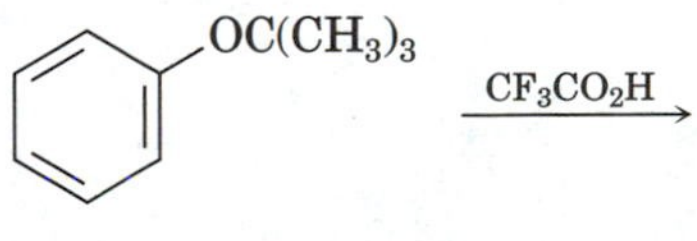

(c) $H_2C{=}CHOCH_2CH_3 \xrightarrow[H_2O]{HI}$

(d) $(CH_3)_3CCH_2OCH_2CH_3 \xrightarrow[H_2O]{HI}$

18.23 The *Zeisel method* is an analytical procedure for determining the number of methoxyl groups in a compound. A weighed amount of the compound is heated with concentrated HI, ether cleavage occurs, and the iodomethane product is distilled off and passed into an alcohol solution of $AgNO_3$, where it reacts to form a precipitate of silver iodide. The AgI is then collected and weighed, and the percentage of methoxyl groups in the sample is thereby determined. For example, 1.06 g of vanillin, the material responsible for the characteristic odor of vanilla, yields 1.60 g of AgI. If vanillin has a molecular weight of 152, how many methoxyls does it contain?

18.24 How would you prepare the following ethers?

(a) 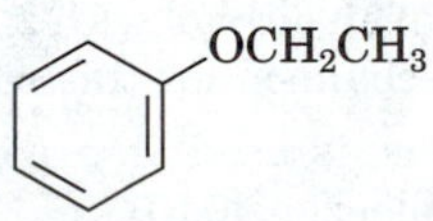

(b) 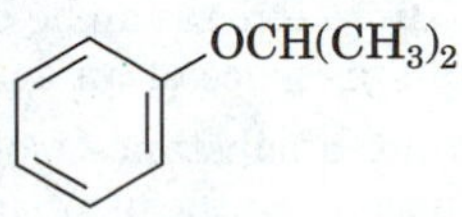

(c) 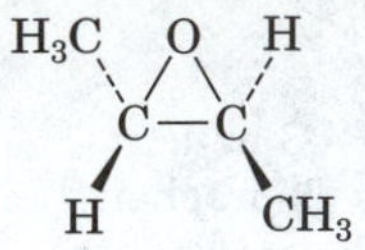

(d) 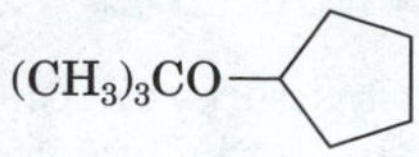

(e) 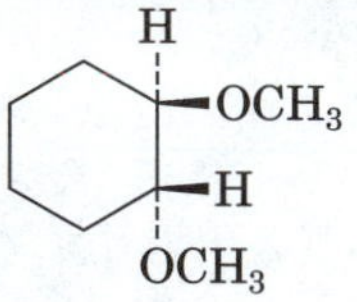

(f) 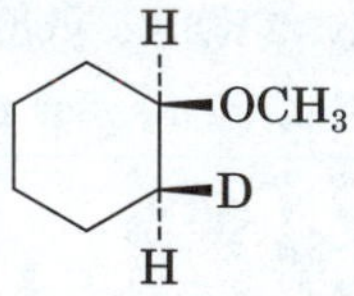

18.25 *Meerwein's reagent*, triethyloxonium tetrafluoroborate, is a powerful ethylating agent that converts alcohols into ethyl ethers at neutral pH. Show the reaction of Meerwein's reagent with cyclohexanol, and account for the fact that trialkyloxonium salts are much more reactive alkylating agents than alkyl iodides.

$(CH_3CH_2)_3O^+\ BF_4^-$ **Meerwein's reagent**

18.26 How would you prepare the following compounds from 1-phenylethanol?
(a) Methyl 1-phenylethyl ether
(b) Phenylepoxyethane
(c) *tert*-Butyl 1-phenylethyl ether

18.27 How would you carry out the following transformations? More than one step may be required.

(a) 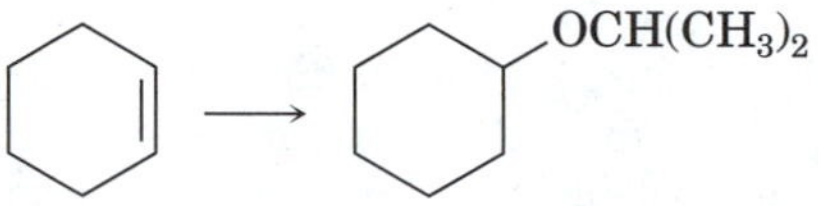

(b) 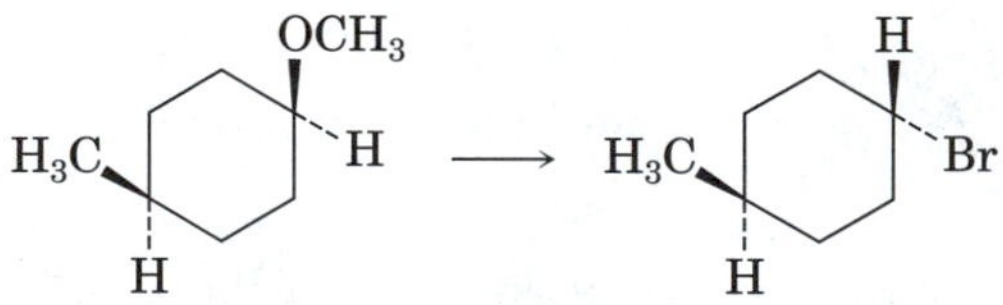

(c) 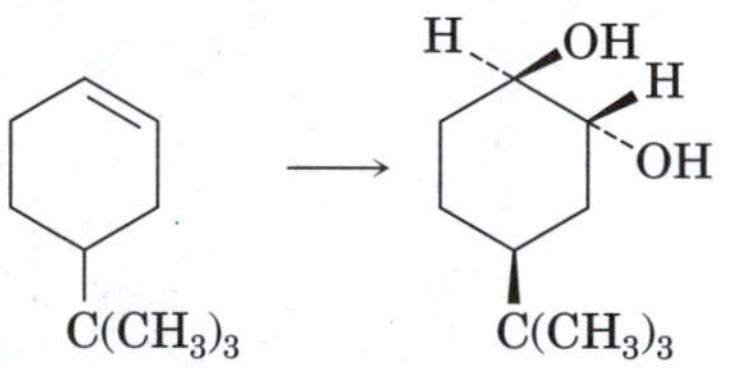

(d) $CH_3CH_2CH_2CH_2C{\equiv}CH \longrightarrow CH_3CH_2CH_2CH_2CH_2CH_2OCH_3$

(e) $CH_3CH_2CH_2CH_2C{\equiv}CH \longrightarrow CH_3CH_2CH_2CH_2\overset{\displaystyle OCH_3}{\overset{|}{C}}HCH_3$

18.28 What product would you expect from cleavage of tetrahydrofuran with HI?

18.29 How could you prepare benzyl phenyl ether from benzene? More than one step is required.

18.30 Methyl aryl ethers, such as anisole, are cleaved to iodomethane and a phenoxide ion by treatment with LiI in hot DMF. Propose a mechanism for this reaction.

18.31 *tert*-Butyl ethers can be prepared by the reaction of an alcohol with 2-methylpropene in the presence of an acid catalyst. Propose a mechanism for this reaction.

18.32 Safrole, a substance isolated from oil of sassafras, is used as a perfumery agent. Propose a synthesis of safrole from catechol (1,2-benzenediol).

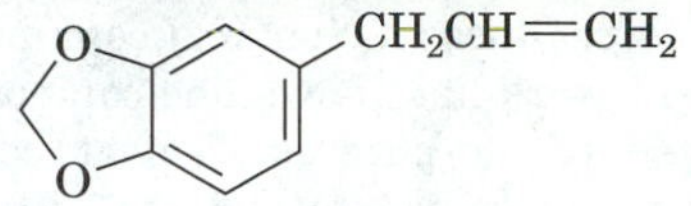

Safrole

18.33 Epoxides are reduced by treatment with lithium aluminum hydride to yield alcohols. Propose a mechanism for this reaction.

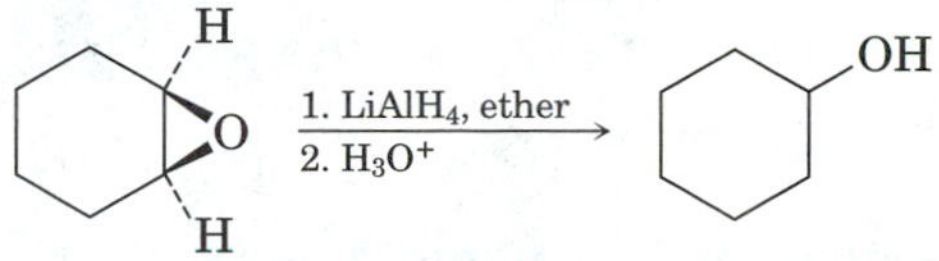

18.34 Show the structure and stereochemistry of the alcohol that would result if 1,2-epoxycyclohexane (Problem 18.33) were reduced with lithium aluminum deuteride, $LiAlD_4$.

18.35 Acid-induced hydrolysis of a 1,2-epoxycyclohexane produces a trans-diaxial 1,2-diol. What product would you expect to obtain from acidic hydrolysis of *cis*-3-*tert*-butyl-1,2-epoxycyclohexane? (Recall that the bulky *tert*-butyl group locks the cyclohexane ring into a specific conformation.)

18.36 Grignard reagents react with oxetane, a four-membered cyclic ether, to yield primary alcohols, but the reaction is much slower than the corresponding reaction with ethylene oxide. Suggest a reason for the difference in reactivity between oxetane and ethylene oxide.

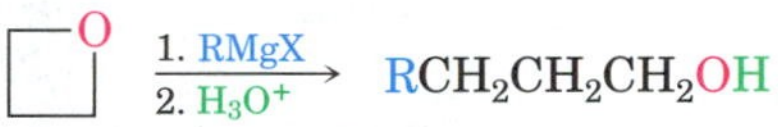

Oxetane

18.37 Treatment of *trans*-2-chlorocyclohexanol with NaOH yields 1,2-epoxycyclohexane, but reaction of the cis isomer under the same conditions yields cyclohexanone. Propose mechanisms for both reactions, and explain why different results are obtained.

18.38 Ethyl vinyl ether reacts with ethanol in the presence of an acid catalyst to yield 1,1-diethoxyethane, rather than 1,2-diethoxyethane. How can you account for the observed regioselectivity of addition?

18.39 Anethole, $C_{10}H_{12}O$, a major constituent of the oil of anise, has the ^1H NMR spectrum shown. On oxidation with $Na_2Cr_2O_7$, anethole yields *p*-methoxybenzoic acid. What is the structure of anethole? Assign all peaks in the NMR spectrum, and account for the observed splitting patterns.

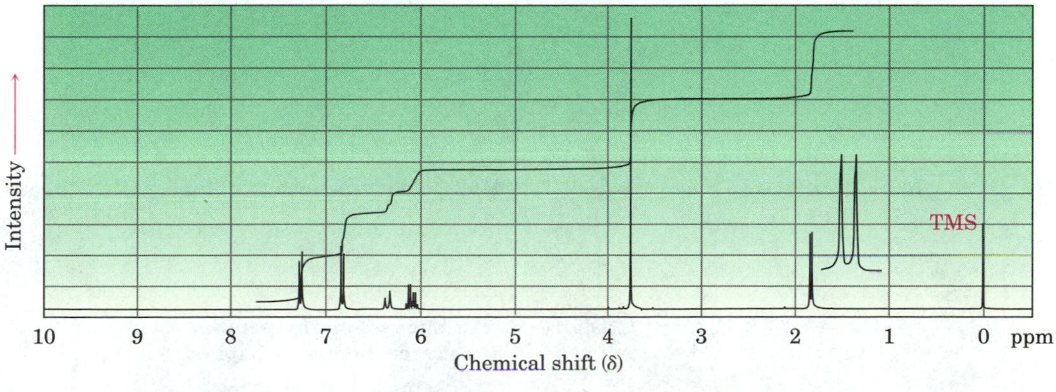

18.40 How would you synthesize anethole (Problem 18.39) from benzene?

18.41 The red fox (*Vulpes vulpes*) uses a chemical communication system based on scent marks in urine. Recent work has shown one component of fox urine to be a sulfide. Mass spectral analysis of the pure scent-mark component shows $M^+ = 116$. IR spectroscopy shows an intense band at 890 cm^{-1}, and ^1H NMR spectroscopy reveals the following peaks:

1.74 δ (3 H, singlet) — 2.11 δ (3 H, singlet)
2.27 δ (2 H, triplet, J = 4.2 Hz) — 2.57 δ (2 H, triplet, J = 4.2 Hz)
4.73 δ (2 H, broad)

Propose a structure consistent with these data. [*Note*: $(CH_3)_2S$ absorbs at 2.1 δ.]

18.42 Treatment of bornene with a peroxyacid yields a different epoxide from that obtained by reaction of bornene with aqueous bromine followed by base treatment. Propose structures for epoxides A and B. (Build molecular models of bornene.)

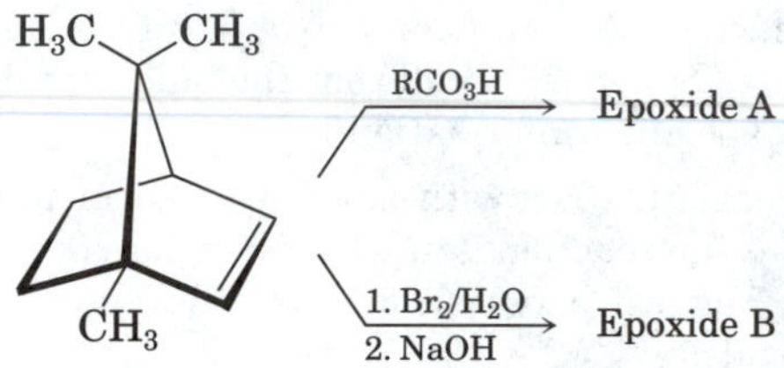

Bornene

18.43 Disparlure, $C_{19}H_{38}O$, is a sex attractant released by the female gypsy moth, *Lymantria dispar*. The ^1H NMR spectrum of disparlure shows a large absorption in the alkane region, 1–2 δ, and a triplet at 2.8 δ. Treatment of disparlure, first with aqueous acid and then with $KMnO_4$, yields two carboxylic acids identified as undecanoic acid and 6-methylheptanoic acid. ($KMnO_4$ cleaves 1,2-diols to yield carboxylic acids.) Neglecting stereochemistry, propose a structure for disparlure. The actual compound is a chiral molecule with 7*R*,8*S* stereochemistry. Draw disparlure, showing the correct stereochemistry.

18.44 How would you synthesize racemic disparlure (Problem 18.43) from compounds having ten or fewer carbons?

18.45 Treatment of 1,1-diphenyl-1,2-epoxyethane with aqueous acid yields diphenylacetaldehyde as the major product. Propose a mechanism for the reaction.

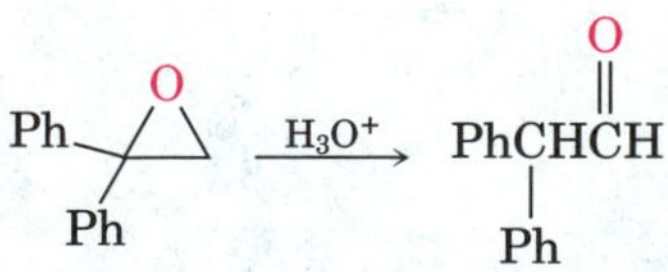

18.46 Imagine that you have treated (2*R*,3*R*)-2,3-epoxy-3-methylpentane with aqueous acid to carry out a ring-opening reaction.

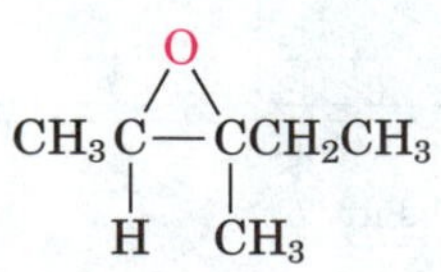

2,3-Epoxy-3-methylpentane (no stereochemistry implied)

(a) Draw the epoxide, showing stereochemistry.
(b) Draw and name the product, showing stereochemistry.
(c) Is the product chiral? Explain.
(d) Is the product optically active? Explain.

18.47 Identify the reagents a–e in the following scheme:

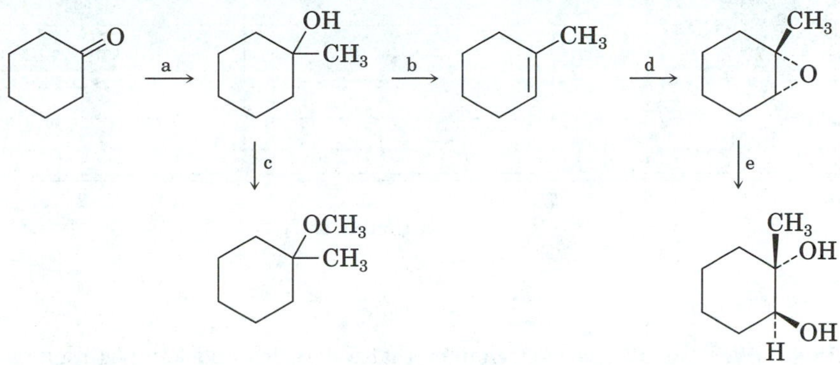

18.48 Propose structures for ethers that have the following ^{1}H NMR spectra:
(a) $C_9H_{11}BrO$

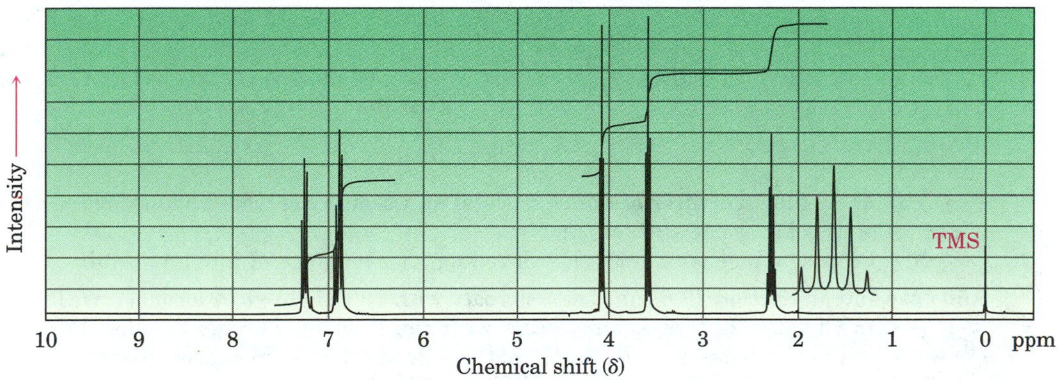

(b) $C_4H_{10}O_2$

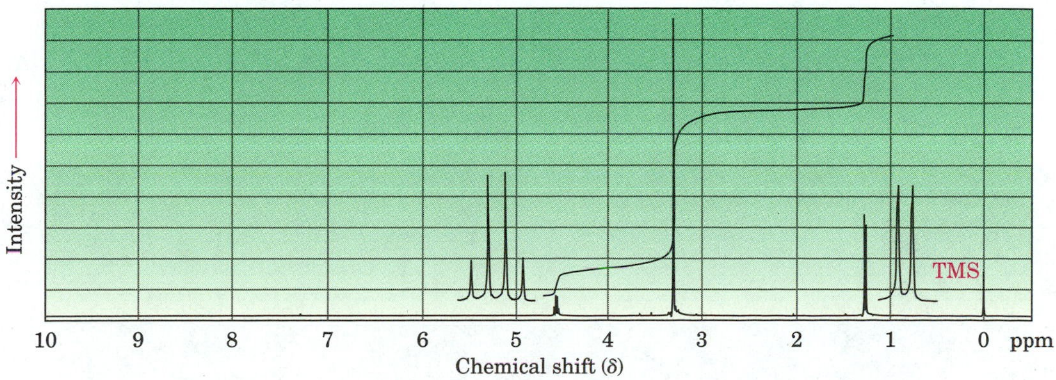

(c) $C_9H_{10}O$

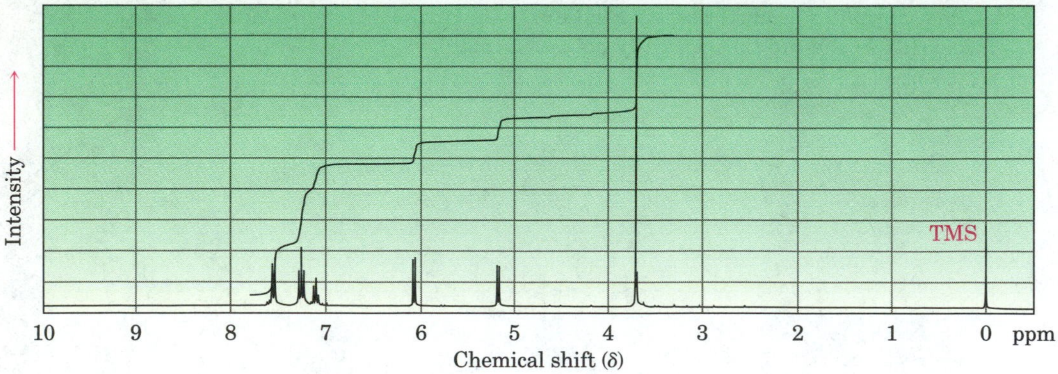

A Look Ahead

18.49 We'll see in the next chapter that aldehydes and ketones undergo acid-catalyzed reaction with alcohols to yield *hemiacetals*, compounds that have one alcohol-like oxygen and one ether-like oxygen bonded to the same carbon. Further reaction of a hemiacetal with alcohol then yields an *acetal*, a compound that has two ether-like oxygens bonded to the same carbon.

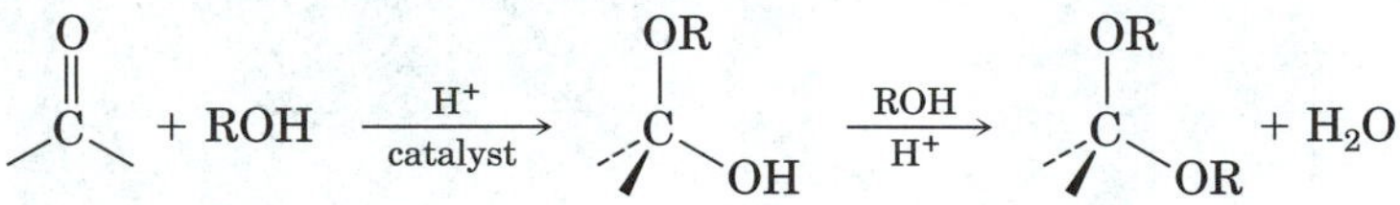

A hemiacetal An acetal

(a) Show the structures of the hemiacetal and acetal you would obtain by reaction of cyclohexanone with ethanol.

(b) Propose a mechanism for the conversion of a hemiacetal into an acetal.

18.50 We saw in Section 17.6 that ketones react with $NaBH_4$ to yield alcohols. We'll also see in Chapter 22 that ketones react with Br_2 to yield α-bromo ketones. Perhaps surprisingly, treatment with $NaBH_4$ of the α-bromo ketone from acetophenone yields an epoxide rather than a bromo alcohol. Show the structure of the epoxide, and explain its formation.

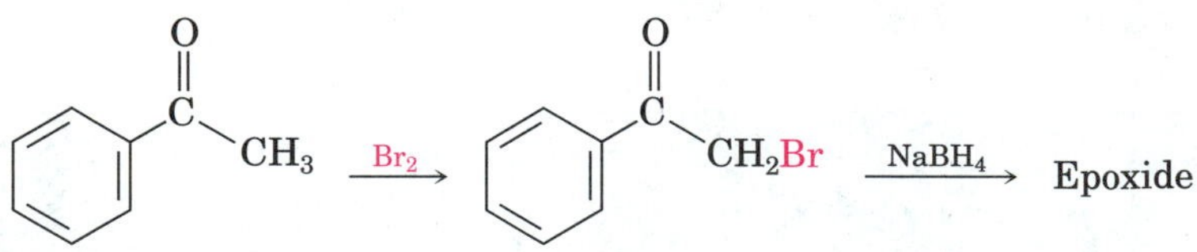

Acetophenone An α-bromo ketone

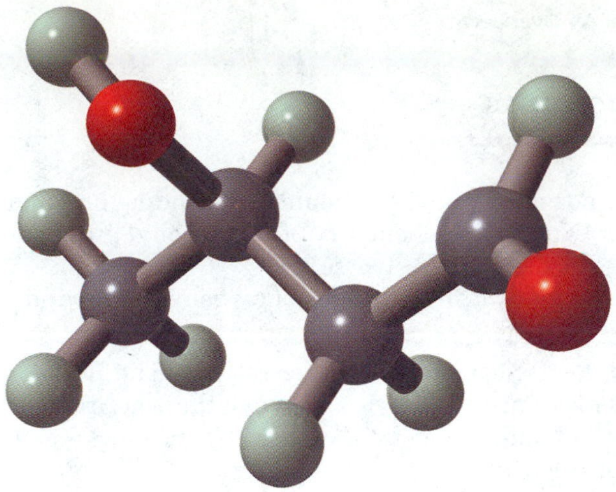

This β-hydroxy ketone, known commonly as *aldol*, is formed by a fundamental carbonyl-group reaction.

Chemistry of Carbonyl Compounds: An Overview

In this and the next five chapters, we'll discuss the most important functional group in organic chemistry—the **carbonyl group, C=O** (pronounced car-bo-**neel**). Although there are many different kinds of carbonyl compounds and many different reactions, there are only a few fundamental principles that tie the entire field together. The point of this brief overview is to make you aware that these principles exist and to provide a framework for learning carbonyl-group chemistry. Read through this overview now, and return to it regularly to remind yourself of the larger picture.

Carbonyl compounds are everywhere in nature. The majority of biologically important molecules contain carbonyl groups, as do most pharmaceutical agents and many of the synthetic chemicals that touch our everyday lives. Acetic acid, the chief component of vinegar; acetaminophen, the active ingredient in many over-the-counter headache remedies; and Dacron, the polyester material used in clothing, all contain different kinds of carbonyl groups.

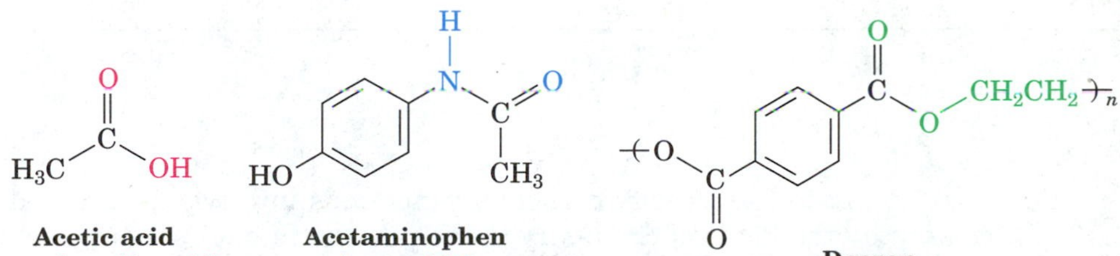

Acetic acid (a carboxylic acid) **Acetaminophen** (an amide) **Dacron** (a polyester)

I. Kinds of Carbonyl Compounds

There are many different kinds of carbonyl compounds, depending on what groups are bonded to the C=O unit. The chemistry of all carbonyl groups is similar, however, regardless of their exact structure.

Table 1 shows some of the many different kinds of carbonyl compounds. All contain an **acyl group**, $\mathrm{R{-}\overset{\displaystyle O}{\overset{\|}{C}}{-}}$, bonded to another residue. The R group of the acyl fragment may be alkyl, aryl, alkenyl, or alkynyl; the other residue to which the acyl fragment is bonded may be a carbon, hydrogen, oxygen, halogen, nitrogen, sulfur, or other atom.

Table 1 Some Types of Carbonyl Compounds

Name	General formula	Name ending	Name	General formula	Name ending
Aldehyde	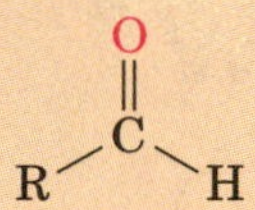	*-al*	Ester	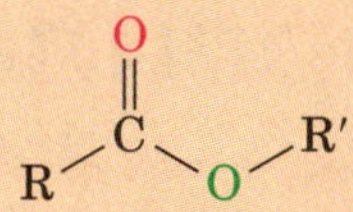	*-oate*
Ketone	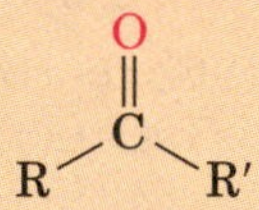	*-one*	Lactone (cyclic ester)	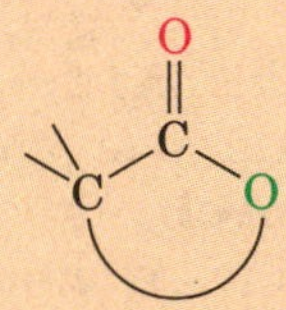	None
Carboxylic acid	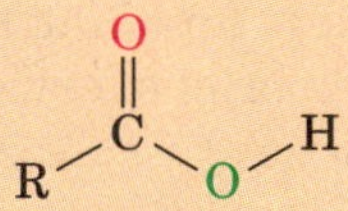	*-oic acid*	Amide	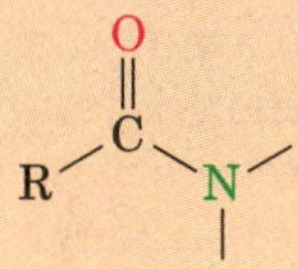	*-amide*
Acid halide	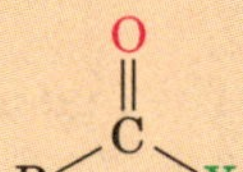 (X = halogen)	*-yl* or *-oyl halide*	Lactam (cyclic amide)	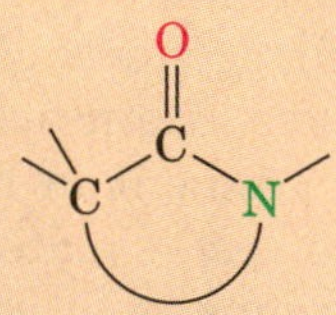	None
Acid anhydride	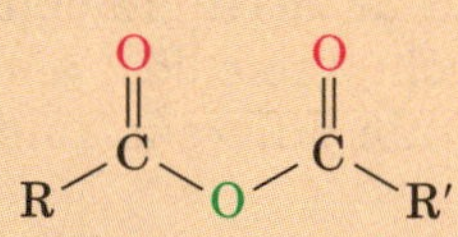	*-oic anhydride*			

It's useful to classify carbonyl compounds into two general categories based on the kinds of chemistry they undergo. In one category are ketones and aldehydes; in the other are carboxylic acids and their derivatives:

Aldehydes (RCHO) Ketones (R_2CO)	The acyl groups in these two families are bonded to substituents (–H and –R, respectively) that *can't stabilize a negative charge and therefore can't act as leaving groups.* Aldehydes and ketones behave similarly and undergo many of the same reactions.
Carboxylic acids (RCOOH) Esters (RCOOR′) Acid chlorides (RCOCl) Acid anhydrides (RCOOCOR′) Amides ($RCONH_2$)	The acyl groups in carboxylic acids and their derivatives are bonded to substituents (oxygen, halogen, nitrogen) that *can stabilize a negative charge and can serve as leaving groups in substitution reactions.* The chemistry of these compounds is therefore similar.

II. Nature of the Carbonyl Group

The carbon–oxygen double bond of carbonyl groups is similar in many respects to the carbon–carbon double bond of alkenes (Figure 1). The carbonyl carbon atom is sp^2-hybridized and forms three σ bonds. The fourth valence electron remains in a carbon p orbital and forms a π bond to oxygen by overlap with an oxygen p orbital. The oxygen atom also has two nonbonding pairs of electrons, which occupy its remaining two orbitals.

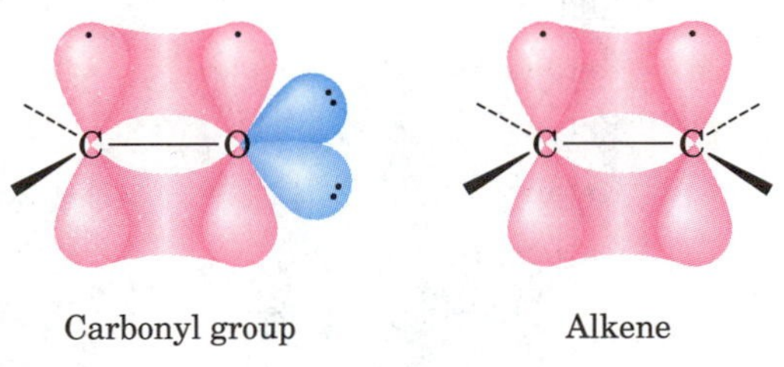

Figure 1 Electronic structure of the carbonyl group.

Like alkenes, carbonyl compounds are planar about the double bond and have bond angles of approximately 120°. Figure 2 (p. 708) shows the structure of acetaldehyde and indicates the experimentally determined bond lengths and angles. As you might expect, the C=O double bond is both shorter (1.22 Å versus 1.43 Å) and stronger [732 kJ/mol (175 kcal/mol) versus 385 kJ/mol (92 kcal/mol)] than a C–O single bond.

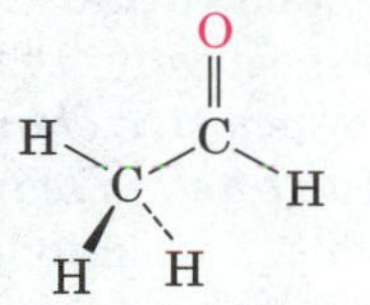

Bond angle (°)		*Bond length (Å)*	
H-C-C	118	C═O	1.22
C-C-O	121	C—C	1.50
H-C-O	121	OC—H	1.09

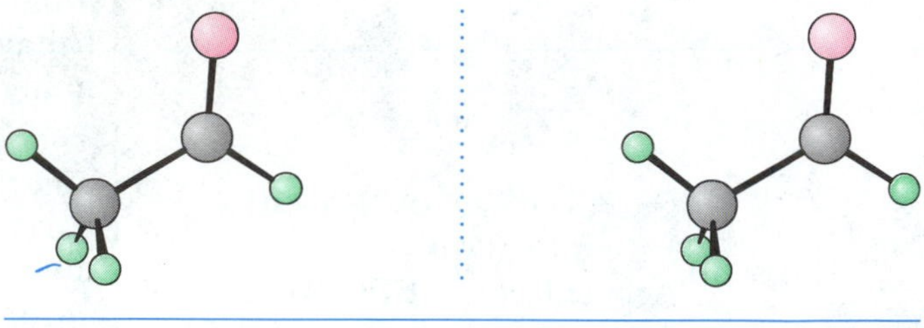

Figure 2 Structure of acetaldehyde.

Carbon–oxygen double bonds are polarized because of the high electronegativity of oxygen relative to carbon. Thus, all types of carbonyl compounds have substantial dipole moments, as listed in Table 2.

Table 2 Dipole Moments of Some Carbonyl Compounds, R_2CO

Carbonyl compound	*Type of carbonyl compound*	*Observed dipole moment (D)*
HCHO	Aldehyde	2.33
CH_3CHO	Aldehyde	2.72
$(CH_3)_2CO$	Ketone	2.88
$PhCOCH_3$	Ketone	3.02
Cyclobutanone	Ketone	2.99
CH_3COOH	Carboxylic acid	1.74
CH_3COCl	Acid chloride	2.72
$CH_3CO_2CH_3$	Ester	1.72
CH_3CONH_2	Amide	3.76
$CH_3CON(CH_3)_2$	Amide	3.81

The most important effect of carbonyl-group polarization is on the chemical reactivity of the C=O double bond. Because the carbonyl carbon carries a partial positive charge, it is an electrophilic (Lewis acidic) site and reacts with nucleophiles. Conversely, the carbonyl oxygen carries a partial negative charge, is a nucleophilic (Lewis basic) site, and reacts with electrophiles.

We'll see in the next five chapters that the majority of carbonyl-group reactions can be rationalized by simple bond-polarization arguments.

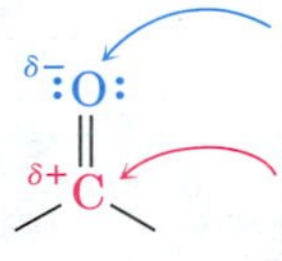

III. General Reactions of Carbonyl Compounds

Most reactions of carbonyl groups occur by one of four general mechanisms: *nucleophilic addition, nucleophilic acyl substitution, alpha substitution*, and *carbonyl condensation*. These mechanisms have many variations, just as alkene electrophilic addition reactions and S_N2 reactions do, but the variations are much easier to learn when the fundamental features of the mechanisms are understood. Let's see what the four mechanisms are and what kinds of chemistry carbonyl groups undergo.

Nucleophilic Addition Reactions of Ketones and Aldehydes (Chapter 19)

The most common reaction of ketones and aldehydes is the **nucleophilic addition reaction**, in which a nucleophile adds to the electrophilic carbon of the carbonyl group. Since the nucleophile uses an electron pair to form a new bond to carbon, two electrons from the carbon–oxygen double bond must move toward the electronegative oxygen atom, where they can be stabilized on an alkoxide anion. The carbonyl carbon rehybridizes from sp^2 to sp^3 during the reaction, and the alkoxide ion therefore has tetrahedral geometry.

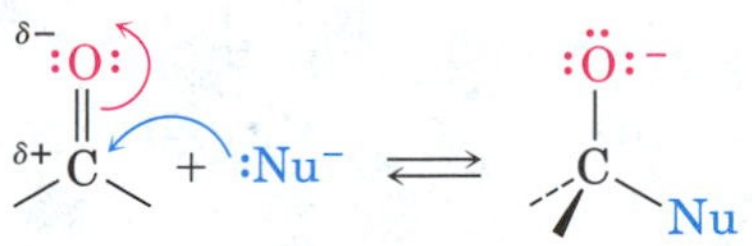

A carbonyl compound (sp^2-hybridized carbon)

A tetrahedral intermediate (sp^3-hybridized carbon)

Once formed, and depending on the nature of the nucleophile, the tetrahedral intermediate can undergo either of the reactions shown in Figure 3 (p. 710). Often, the tetrahedral alkoxide intermediate is simply protonated by water or acid to form an alcohol product. Alternatively, the tetrahedral intermediate can expel the oxygen to form a new double bond between the carbonyl-group carbon and the nucleophile. We'll study both processes in detail in Chapter 19.

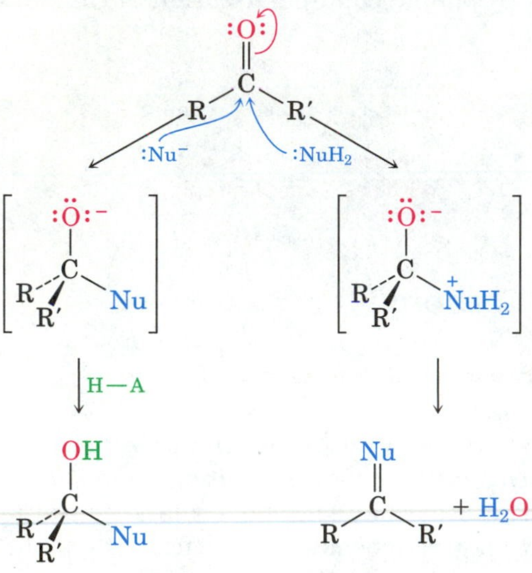

Figure 3 The addition reaction of a nucleophile with a ketone or an aldehyde. Depending on the nucleophile, either an alcohol or a compound with a C=Nu double bond is formed.

Formation of an Alcohol The simplest reaction of a tetrahedral intermediate is protonation to yield an alcohol. We've already seen two examples of this kind of process during reduction of ketones and aldehydes with hydride reagents such as $NaBH_4$ and $LiAlH_4$ (Section 17.6), and during Grignard reactions (Section 17.7). In the case of reduction, the nucleophile that adds to the carbonyl group is a hydride ion, $H:^-$, while in the case of Grignard reaction, the nucleophile is a carbanion, $R_3C:^-$.

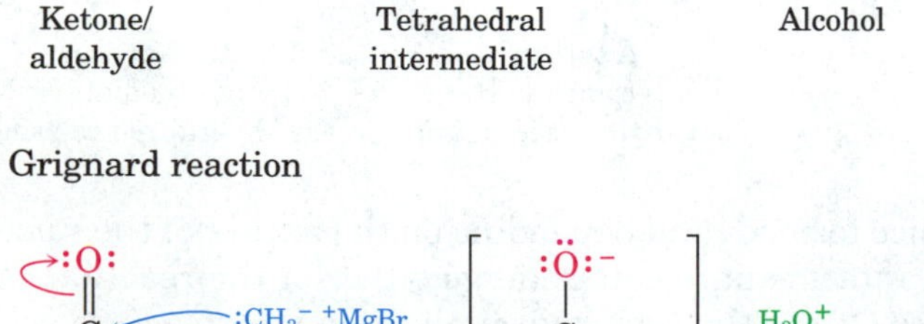

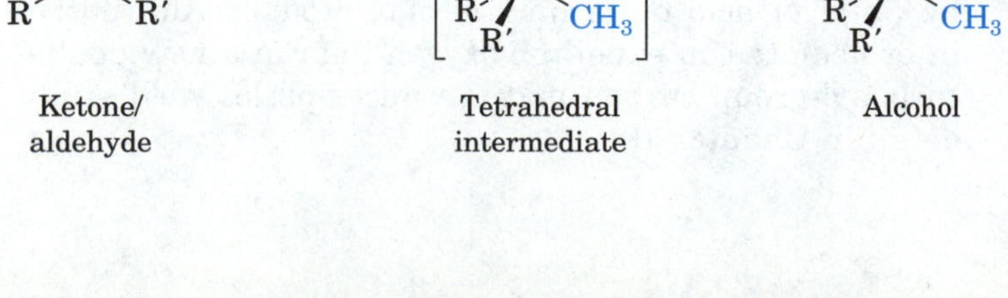

Formation of C=Nu The second mode of nucleophilic addition, which often happens with amine nucleophiles, involves elimination of oxygen and formation of a C=Nu double bond. For example, ketones and aldehydes react with primary amines, RNH_2, to form *imines*, $R_2C{=}NR'$. These reactions proceed through exactly the same kind of tetrahedral intermediate as that formed during hydride reduction and Grignard reaction, but the initially formed alkoxide ion is not isolated. Instead, it loses water to form an imine, as shown in Figure 4.

Addition to the ketone or aldehyde carbonyl group by the neutral amine nucleophile gives a dipolar tetrahedral intermediate.

Transfer of a proton from nitrogen to oxygen then yields a nonpolar amino alcohol intermediate.

Dehydration of the amino alcohol intermediate gives neutral imine plus water as final products.

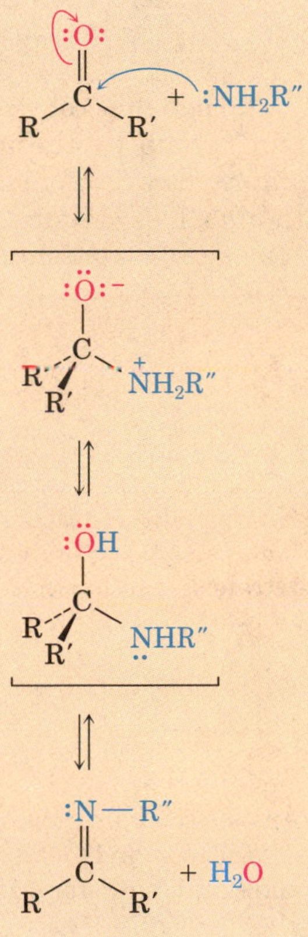

Figure 4 Mechanism of imine formation by reaction of an amine with a ketone or an aldehyde.

Nucleophilic Acyl Substitution Reactions of Carboxylic Acid Derivatives (Chapter 21)

The second fundamental reaction of carbonyl compounds, **nucleophilic acyl substitution**, is related to the nucleophilic addition reaction just discussed but occurs only with carboxylic acid derivatives rather than with ketones and aldehydes. When the carbonyl group of a carboxylic acid derivative reacts with a nucleophile, addition occurs, but the initially formed tetrahedral intermediate is not isolated. Because carboxylic acid derivatives have a leaving group bonded to the carbonyl-group carbon, the tetrahedral intermediate can react further by expelling the leaving group and forming a new carbonyl compound:

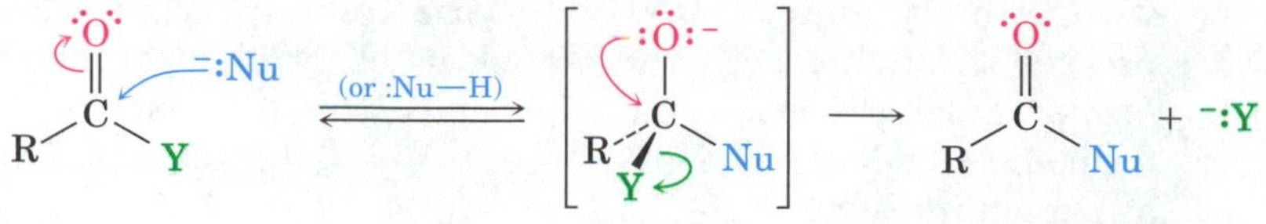

where Y = –OR (ester), –Cl (acid chloride), $-NH_2$ (amide), or –OCOR′ (acid anhydride)

The net effect of nucleophilic acyl substitution is the replacement of the leaving group by the attacking nucleophile. We'll see in Chapter 21, for example, that acid chlorides are rapidly converted into esters by treatment with alkoxides (Figure 5).

Nucleophilic addition of alkoxide ion to an acid chloride yields a tetrahedral intermediate.

An electron pair from oxygen expels chloride ion and yields the substitution product, an ester.

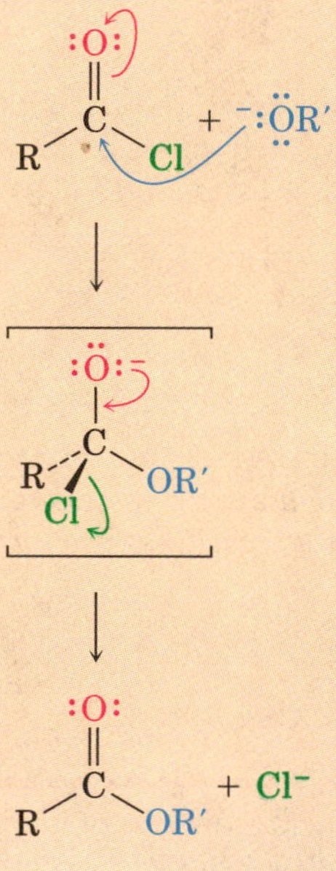

Figure 5 Mechanism of the nucleophilic acyl substitution reaction of an acid chloride with an alkoxide ion to yield an ester.

Alpha-Substitution Reactions (Chapter 22)

The third major reaction of carbonyl compounds, **alpha substitution**, occurs at the position *next to* the carbonyl group—the alpha (α) position. This reaction, which takes place with all carbonyl compounds regardless of structure, results in the substitution of an α hydrogen by another group and involves the formation of an intermediate *enol* or *enolate ion*:

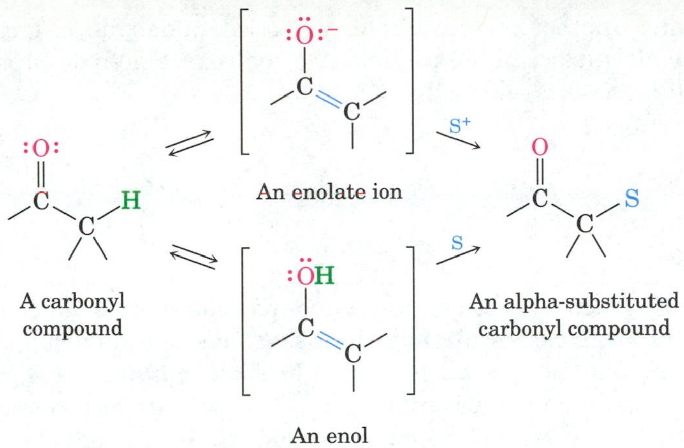

For reasons that we'll explore in Chapter 22, the presence of a carbonyl group renders the hydrogens on the α carbon acidic. Carbonyl compounds therefore react with strong base to yield enolate ions.

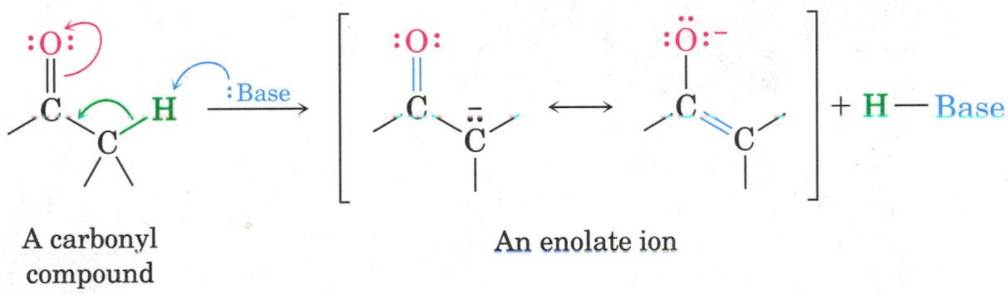

What chemistry might we expect of enolate ions? Since they are negatively charged, enolate ions behave as nucleophiles and undergo many of the reactions we've already studied. For example, enolates react with primary alkyl halides in the S_N2 reaction. The nucleophilic enolate ion displaces halide ion and a new C–C bond forms:

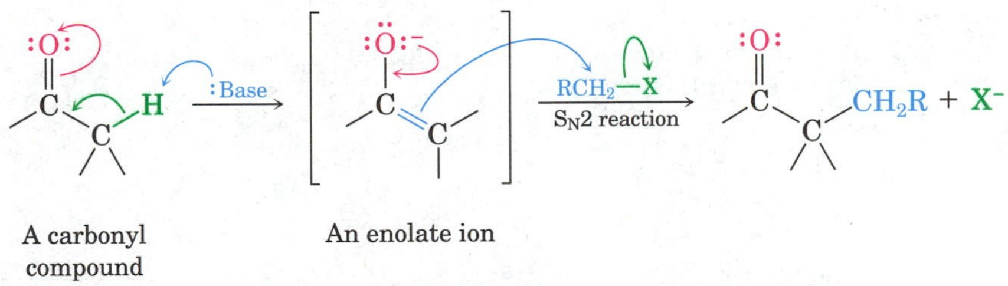

The S_N2 alkylation reaction between an enolate ion and an alkyl halide is one of the most powerful methods available for making C–C bonds, thereby building up larger molecules from smaller precursors. We'll study the alkylation of many kinds of carbonyl groups in Chapter 22.

Carbonyl Condensation Reactions (Chapter 23)

The fourth and last fundamental reaction of carbonyl groups, **carbonyl condensation**, takes place when two carbonyl compounds react with one another. For example, when acetaldehyde is treated with base, two molecules combine to yield the hydroxy aldehyde product known as *aldol* (*ald*ehyde + alcoh*ol*):

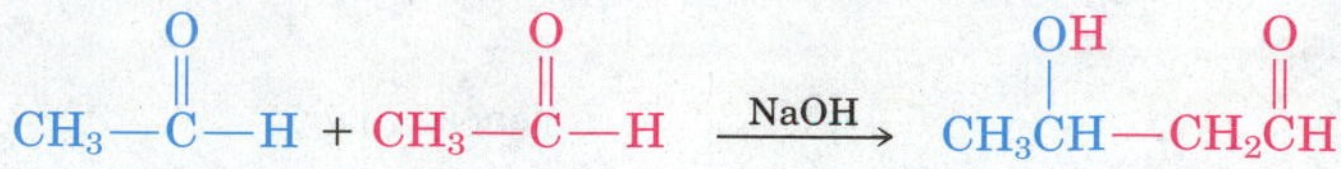

Two acetaldehydes

Aldol

Although the carbonyl condensation reaction appears different from the three processes already discussed, it's actually quite similar. A carbonyl condensation reaction is simply a *combination* of a nucleophilic addition step and an α-substitution step. The initially formed enolate ion of acetaldehyde acts as a nucleophile and adds to the carbonyl group of another acetaldehyde molecule. Reaction occurs by the pathway shown in Figure 6.

Hydroxide ion abstracts an acidic alpha proton from one molecule of acetaldehyde, yielding an enolate ion.

The enolate ion adds as a nucleophile to the carbonyl group of a second molecule of acetaldehyde, producing a tetrahedral intermediate.

The intermediate is protonated by water solvent to yield the neutral aldol product and regenerate hydroxide ion.

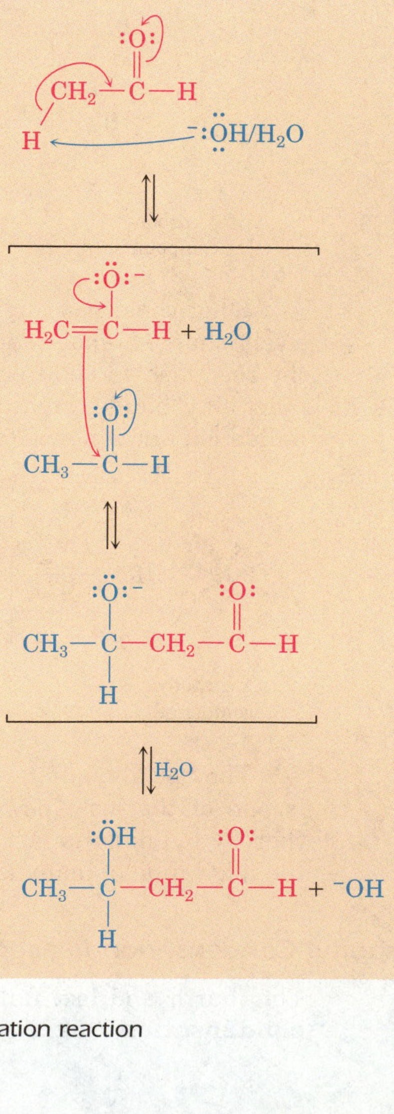

Figure 6 Mechanism of a carbonyl condensation reaction between two molecules of acetaldehyde.

IV. Summary

The purpose of this short overview of carbonyl-group reactions is not to show details of specific reactions but rather to lay the groundwork for the next five chapters. Although some of the reactions we'll be seeing appear unrelated, each is one of the four fundamental carbonyl-group processes. Knowing where we'll be heading should help you to keep matters straight in understanding this most important of all functional groups.

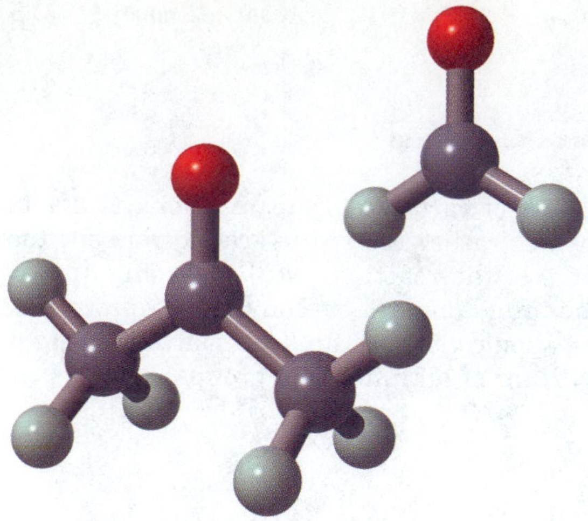

Acetone and formaldehyde are typical carbonyl compounds.

19

Aldehydes and Ketones: Nucleophilic Addition Reactions

Aldehydes and ketones are among the most widely occurring of all compounds, both in nature and in the chemical industry. In nature, many substances required by living organisms are aldehydes or ketones. In the chemical industry, simple aldehydes and ketones are produced in large quantities for use both as solvents and as starting materials for a host of other compounds. For example, approximately 1.4 million tons per year of formaldehyde, $H_2C{=}O$, are produced in the United States for use in building insulation materials and in the adhesive resins that bind particle board and plywood. Acetone, $(CH_3)_2C{=}O$, is widely used as an industrial solvent; some 1.2 million tons per year are produced in the United States.

Formaldehyde is synthesized industrially by catalytic oxidation of methanol, and one method of acetone preparation involves oxidation of 2-propanol.

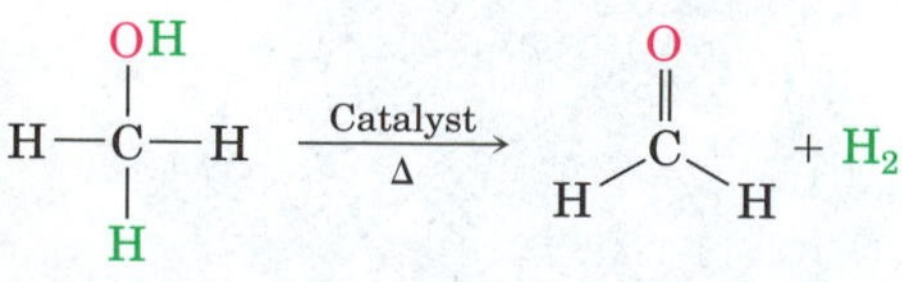

Methanol **Formaldehyde**

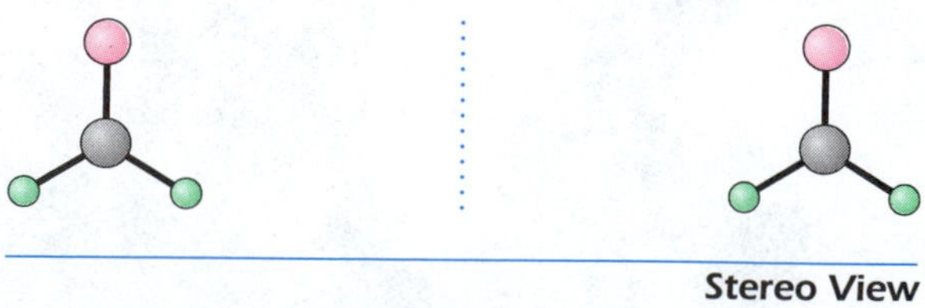

Stereo View

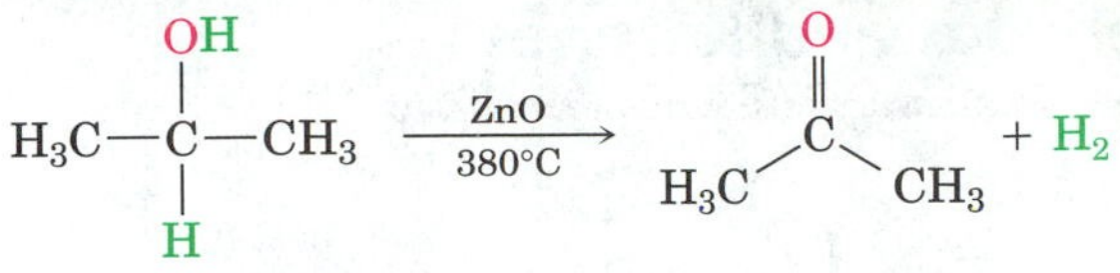

2-Propanol Acetone

Stereo View

19.1 Properties of Aldehydes and Ketones

As we saw in the previous overview of carbonyl-group chemistry, a carbonyl group is planar, with bond angles of approximately 120°, and the C=O double bond is polar.

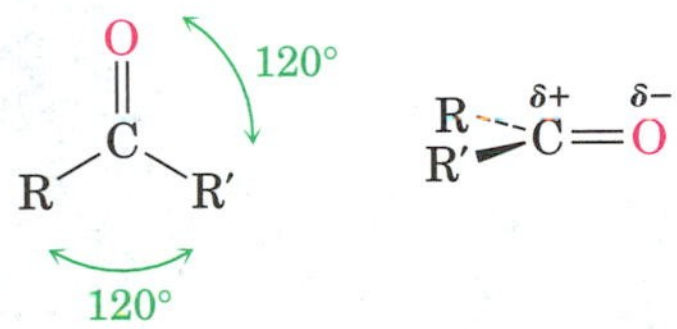

One consequence of carbonyl polarity is that aldehydes and ketones have stronger intermolecular forces and higher boiling points than alkanes of similar molecular weight. Because they can't form hydrogen bonds, however, ketones and aldehydes have lower boiling points than the corresponding alcohols. Formaldehyde, the simplest aldehyde, is a gas at room temperature, but other simple aldehydes and ketones are liquid (Table 19.1).

Table 19.1 Physical Properties of Some Aldehydes and Ketones

Name	*Structure*	*Melting point (°C)*	*Boiling point (°C)*
Formaldehyde	HCHO	−92	−21
Acetaldehyde	CH_3CHO	−121	21
Benzaldehyde	C_6H_5CHO	−26	178
Acetone	CH_3COCH_3	−95	56
2-Butanone	$CH_3CH_2COCH_3$	−86	80
Cyclohexanone	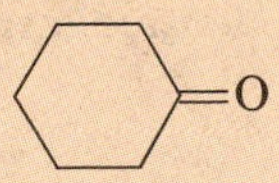	−16	156

19.2 Naming Aldehydes and Ketones

Aldehydes

Aldehydes are named by replacing the terminal *-e* of the corresponding alkane name with *-al*. The parent chain must contain the –CHO group, and the –CHO carbon is numbered as carbon 1. For example:

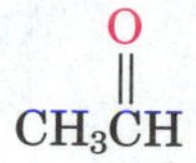

Ethanal (Acetaldehyde)

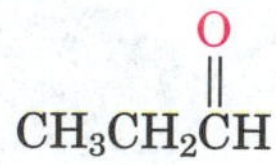

Propanal (Propionaldehyde)

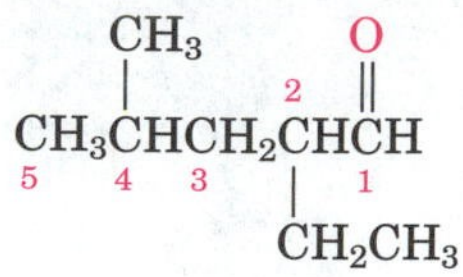

2-Ethyl-4-methylpentanal

Note that the longest chain in 2-ethyl-4-methylpentanal is a hexane, but this chain does not include the –CHO group and thus is not considered the parent.

For more complex aldehydes in which the –CHO group is attached to a ring, the suffix *-carbaldehyde* is used:

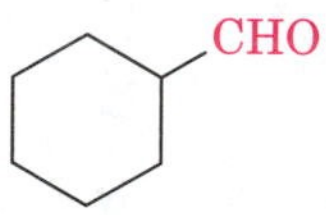

Cyclohexanecarbaldehyde

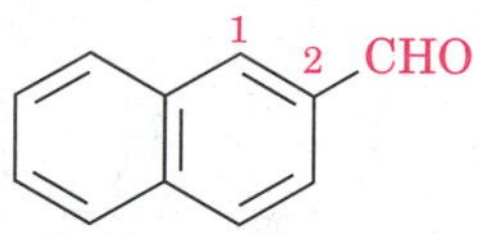

2-Naphthalenecarbaldehyde

Certain simple and well-known aldehydes have common names that are recognized by IUPAC. Some of the more important common names are given in Table 19.2.

Table 19.2 Common Names of Some Simple Aldehydes

Formula	*Common name*	*Systematic name*
HCHO	Formaldehyde	Methanal
CH_3CHO	Acetaldehyde	Ethanal
CH_3CH_2CHO	Propionaldehyde	Propanal
$CH_3CH_2CH_2CHO$	Butyraldehyde	Butanal
$CH_3CH_2CH_2CH_2CHO$	Valeraldehyde	Pentanal
$H_2C{=}CHCHO$	Acrolein	2-Propenal
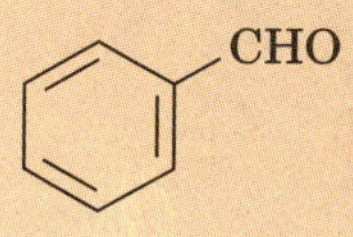	Benzaldehyde	Benzenecarbaldehyde

Ketones

Ketones are named by replacing the terminal *-e* of the corresponding alkane name with *-one* (pronounced oan). The parent chain is the longest one that contains the ketone group, and the numbering begins at the end nearer the carbonyl carbon. For example:

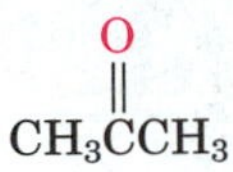

Propanone (Acetone)

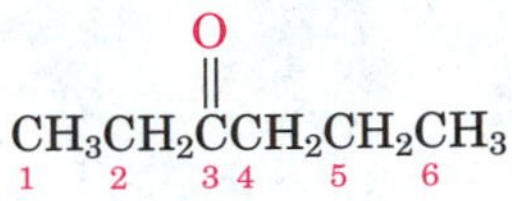

3-Hexanone

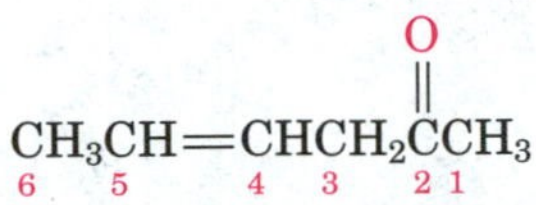

4-Hexen-2-one

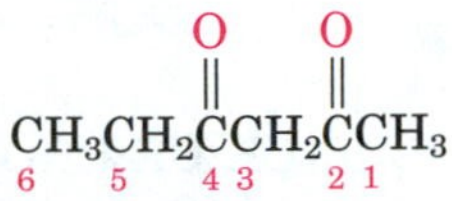

2,4-Hexanedione

A few ketones are allowed by IUPAC to retain their common names:

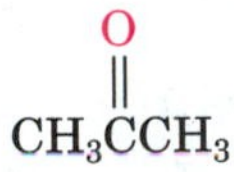

Acetone

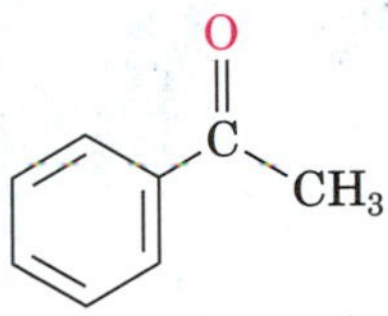

Acetophenone

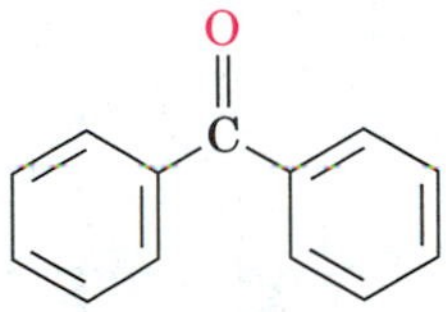

Benzophenone

When it is necessary to refer to the RCO– group as a substituent, the term **acyl** (**a**-sil) is used and the name ending *-yl* is attached. For example, CH_3CO– is an **acetyl** group, –CHO is a **formyl** group, and C_6H_5CO– is a **benzoyl** group.

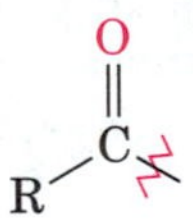

An acyl group

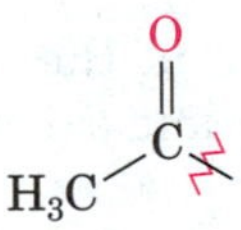

Acetyl

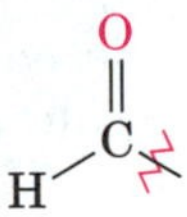

Formyl

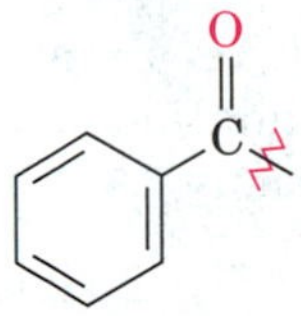

Benzoyl

If other functional groups are present and the doubly bonded oxygen is considered a substituent, the prefix *oxo-* is used. For example:

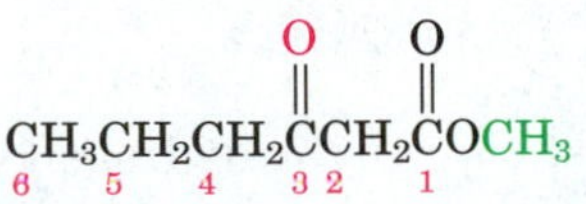

Methyl 3-oxohexanoate

PROBLEM

19.1 Name the following aldehydes and ketones according to IUPAC rules:

(a) $CH_3CH_2C(=O)CH(CH_3)_2$

(b) 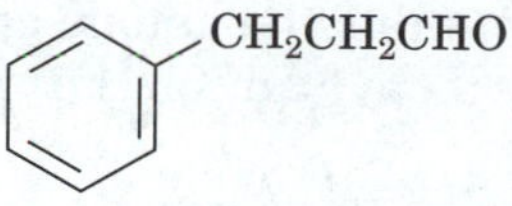

(c) $CH_3C(=O)CH_2CH_2CH_2C(=O)CH_2CH_3$

(d) 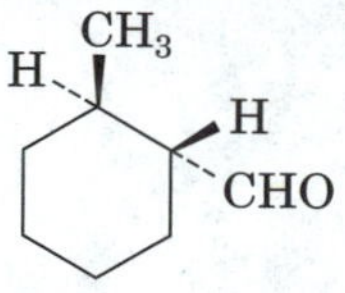

(e) $OHCCH_2CH_2CH_2CHO$

(f) 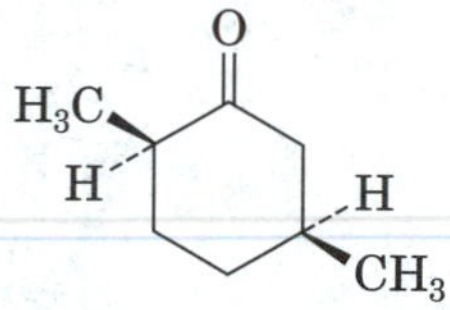

(g) $CH_3CH_2CH(CH_3)CH(CH_2CH_2CH_3)C(=O)CH_3$

(h) $CH_3CH{=}CHCH_2CH_2CHO$

PROBLEM

19.2 Draw structures corresponding to the following names:
(a) 3-Methylbutanal
(b) 4-Chloro-2-pentanone
(c) Phenylacetaldehyde
(d) *cis*-3-*tert*-Butylcyclohexanecarbaldehyde
(e) 3-Methyl-3-butenal
(f) 2-(1-Chloroethyl)-5-methylheptanal

19.3 Preparation of Aldehydes

We've already discussed two of the best methods of aldehyde synthesis: oxidation of primary alcohols and oxidative cleavage of alkenes. Let's review briefly.

1. Primary alcohols can be oxidized to give aldehydes (Section 17.9). The reaction is often carried out using pyridinium chlorochromate (PCC) in dichloromethane solvent at room temperature.

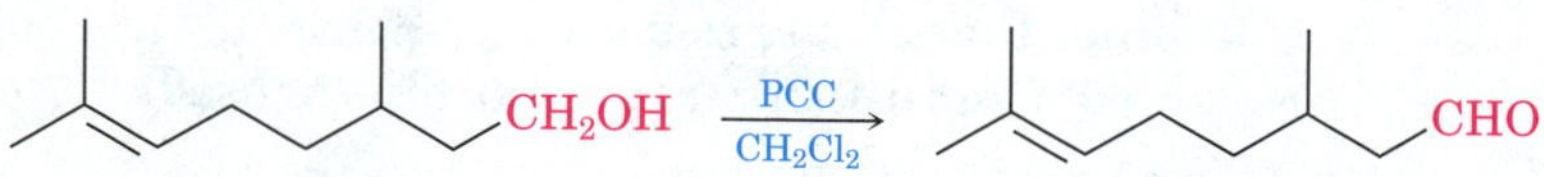

Citronellol **Citronellal (82%)**

2. Alkenes with at least one vinylic hydrogen undergo oxidative cleavage when treated with ozone to yield aldehydes (Section 7.8). If

the ozonolysis reaction is carried out on cyclic alkenes, dicarbonyl compounds result.

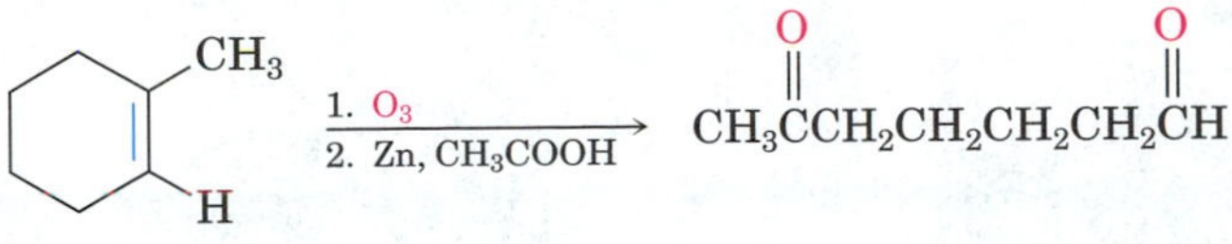

1-Methylcyclohexene **6-Oxoheptanal (86%)**

Yet a third method of aldehyde synthesis is one that we'll mention here just briefly and then return to for a more detailed explanation in Section 21.7. Certain carboxylic acid derivatives can be partially reduced to yield aldehydes.

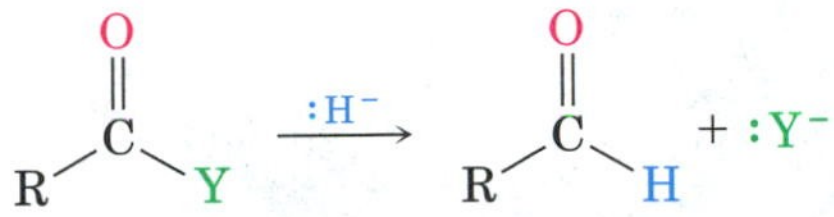

For example, the partial reduction of esters by diisobutylaluminum hydride (DIBAH) is an important laboratory-scale method of aldehyde synthesis. The reaction is normally carried out at −78°C (dry-ice temperature) in toluene solution, and yields are often excellent.

$$CH_3(CH_2)_{10}\overset{O}{\overset{\|}{C}}OCH_3 \xrightarrow[2.\ H_3O^+]{1.\ DIBAH,\ toluene,\ -78°C} CH_3(CH_2)_{10}\overset{O}{\overset{\|}{C}}H$$

Methyl dodecanoate **Dodecanal (88%)**

$$\text{where DIBAH} = (CH_3)_2CHCH_2{-}\overset{H}{\overset{|}{Al}}{-}CH_2CH(CH_3)_2$$

PROBLEM.

19.3 How would you prepare pentanal from these starting materials?
(a) 1-Pentanol (b) 1-Hexene (c) $CH_3CH_2CH_2CH_2COOCH_3$

19.4 Preparation of Ketones

For the most part, methods of ketone synthesis are analogous to those for aldehydes:

1. Secondary alcohols are oxidized by a variety of reagents to give ketones (Section 17.9). The choice of oxidant depends on such factors as reaction scale, cost, and acid or base sensitivity of the alcohol.

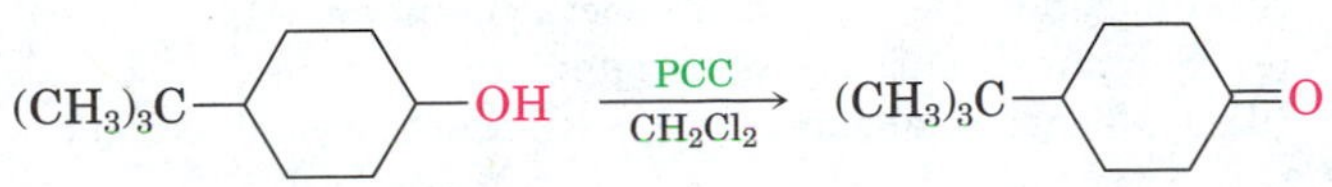

4-*tert*-Butylcyclohexanol **4-*tert*-Butylcyclohexanone (90%)**

2. Ozonolysis of alkenes yields ketones if one of the unsaturated carbon atoms is disubstituted (Section 7.8):

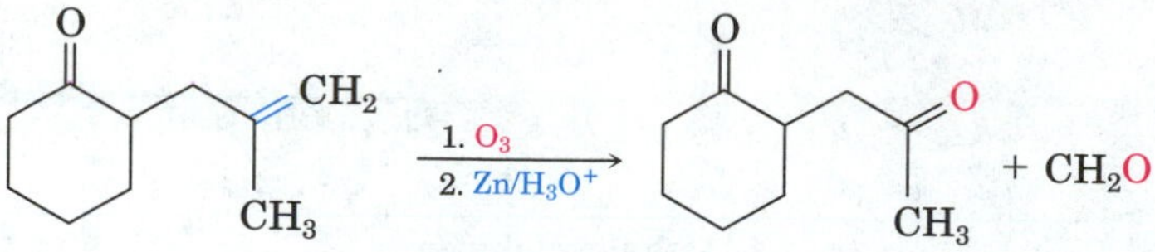

70%

3. Aryl ketones are prepared by Friedel–Crafts acylation of an aromatic ring with an acid chloride in the presence of $AlCl_3$ catalyst (Section 16.4):

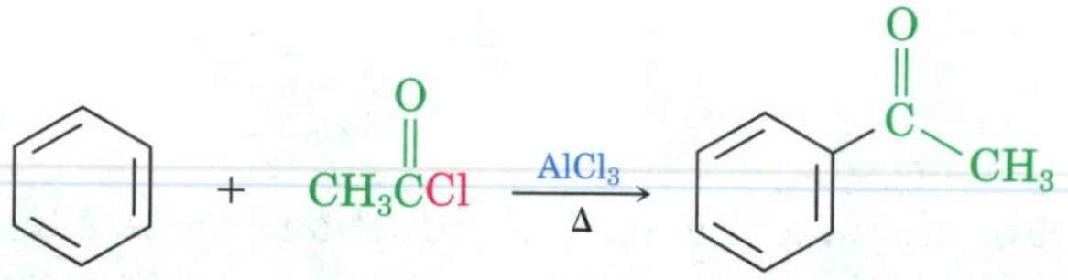

Benzene **Acetyl chloride** **Acetophenone (95%)**

4. Methyl ketones are prepared by hydration of terminal alkynes in the presence of Hg^{2+} catalyst (Section 8.5):

$$CH_3(CH_2)_3C{\equiv}CH \xrightarrow[Hg(OAc)_2]{H_3O^+} CH_3(CH_2)_3\overset{O}{\overset{\|}{C}}{-}CH_3$$

1-Hexyne **2-Hexanone (78%)**

5. Ketones can be prepared from certain carboxylic acid derivatives, just as aldehydes can. Among the most useful reactions of this type is that between an acid chloride and a diorganocopper reagent. We'll discuss this subject in more detail in Section 21.5.

$$CH_3(CH_2)_4\overset{O}{\overset{\|}{C}}Cl \; + \; (CH_3)_2\bar{C}u\ Li^+ \longrightarrow CH_3(CH_2)_4\overset{O}{\overset{\|}{C}}CH_3$$

Hexanoyl chloride **Dimethylcopper lithium** **2-Heptanone (81%)**

PROBLEM

19.4 How would you carry out the following reactions? More than one step may be required.

(a) 3-Hexyne ⟶ 3-Hexanone

(b) Benzene ⟶ *m*-Bromoacetophenone

(c) Bromobenzene ⟶ Acetophenone

(d) 1-Methylcyclohexene ⟶ 2-Methylcyclohexanone

19.5 Oxidation of Aldehydes and Ketones

Aldehydes are readily oxidized to yield carboxylic acids, but ketones are generally inert toward oxidation. The difference is a consequence of structure: Aldehydes have a –CHO proton that can be abstracted during oxidation, but ketones do not.

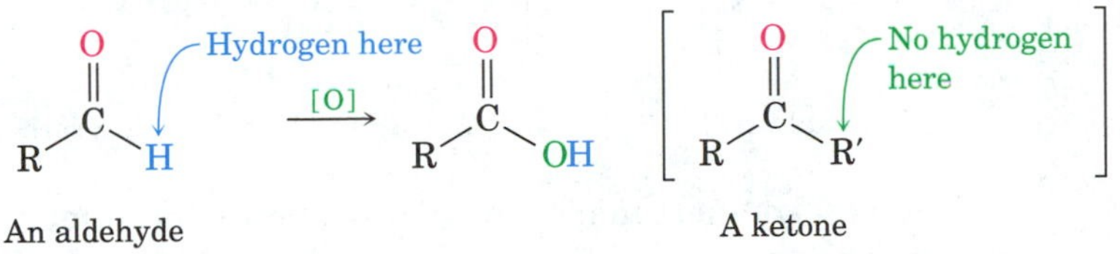

Many oxidizing agents, including hot HNO_3 and $KMnO_4$, convert aldehydes into carboxylic acids, but the Jones reagent, CrO_3 in aqueous sulfuric acid, is a more common choice in the laboratory (Section 17.9). Jones oxidations occur rapidly at room temperature and give good yields of product.

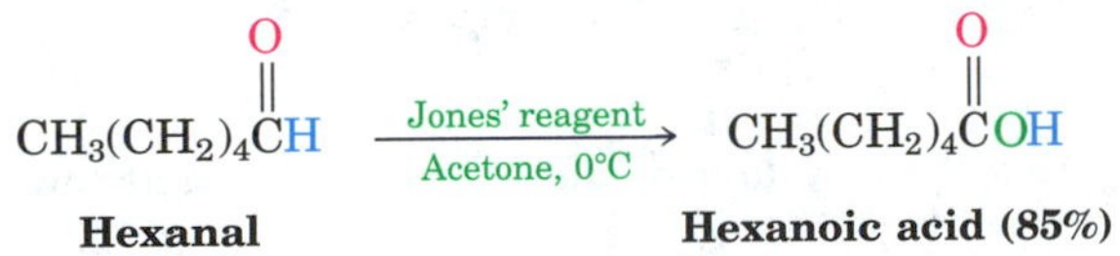

One drawback to the Jones oxidation is that it takes place under acidic conditions, and sensitive molecules sometimes undergo side reactions. In such cases, aldehyde oxidations are often carried out using a solution of silver oxide, Ag_2O, in aqueous ammonia, the so-called **Tollens[1] reagent**. Aldehydes are oxidized by the Tollens reagent in high yield without harming carbon–carbon double bonds or other functional groups in the molecule.

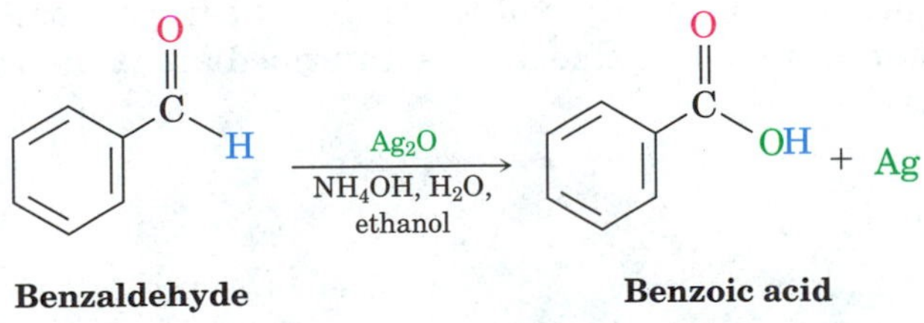

A shiny mirror of metallic silver is deposited on the walls of the reaction flask during a Tollens oxidation, forming the basis of a qualitative test for the presence of an aldehyde functional group in a molecule of unknown structure. A small sample of the unknown is dissolved in ethanol in a test tube, and a few drops of Tollens' reagent are added. If the test tube becomes silvery, the unknown is presumed to be an aldehyde.

---

[1]Bernhard Tollens (1841–1918); b. Hamburg, Germany; Ph.D. Göttingen; professor, University of Göttingen.

Aldehyde oxidations occur through intermediate 1,1-diols, or *hydrates*, which are formed by a reversible nucleophilic addition of water to the carbonyl group. Even though formed to only a small extent at equilibrium, the hydrate reacts like any typical primary or secondary alcohol and is oxidized to a carbonyl compound (Section 17.9).

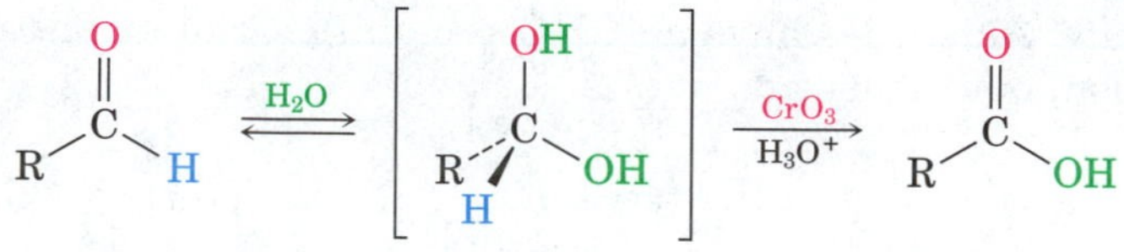

Ketones are inert to most oxidizing agents but undergo a slow cleavage reaction when treated with hot alkaline $KMnO_4$. The C–C bond next to the carbonyl group is broken, and carboxylic acids are produced. The reaction is useful only for symmetrical ketones such as cyclohexanone because product mixtures are formed from unsymmetrical ketones.

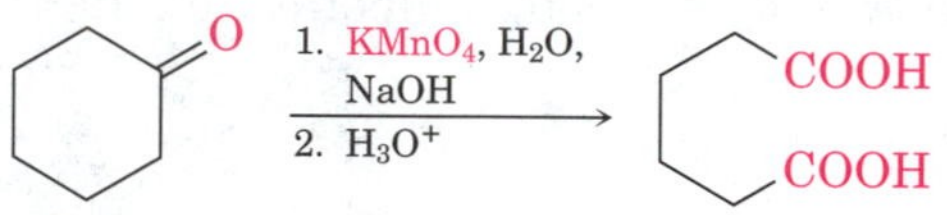

Cyclohexanone **Hexanedioic acid (79%)**

19.6 Nucleophilic Addition Reactions of Aldehydes and Ketones

The most general reaction of ketones and aldehydes is the **nucleophilic addition reaction**. A nucleophile attacks the electrophilic C=O carbon atom from a direction approximately perpendicular to the plane of the carbonyl group. Rehybridization of the carbonyl carbon from sp^2 to sp^3 then occurs, and a tetrahedral alkoxide ion intermediate is produced (Figure 19.1).

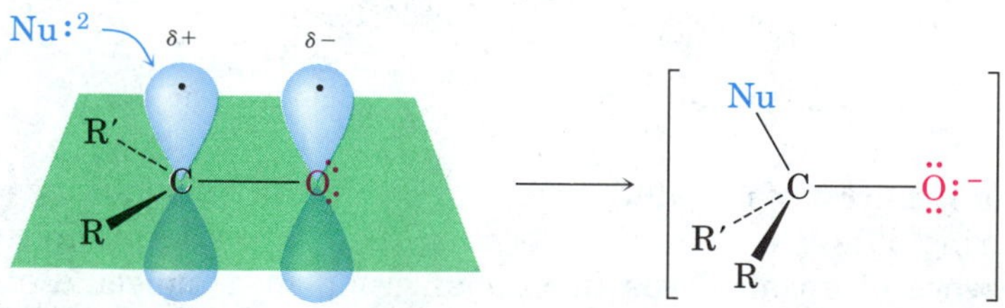

Figure 19.1 A nucleophilic addition reaction to a ketone or aldehyde. The attacking nucleophile approaches the carbonyl group from a direction approximately perpendicular to the plane of the sp^2 orbitals.

The attacking nucleophile can be either negatively charged (Nu:⁻) or neutral (:Nu–H). If it's neutral, however, the nucleophile usually carries a hydrogen atom that can subsequently be eliminated. For example:

Some negatively charged nucleophiles
- $HO:^-$ (hydroxide ion)
- $H:^-$ (hydride ion)
- $R_3C:^-$ (a carbanion)
- $RO:^-$ (an alkoxide ion)
- $N{\equiv}C:^-$ (cyanide ion)

Some neutral nucleophiles
- HOH (water)
- ROH (an alcohol)
- $H_3N:$ (ammonia)
- RNH_2 (an amine)

Nucleophilic additions to ketones and aldehydes have two general variations, as shown in Figure 19.2: (1) The tetrahedral intermediate can be protonated by water or acid to give an alcohol, or (2) the carbonyl oxygen atom can be eliminated as HO^- or H_2O to give a product with a C=Nu double bond.

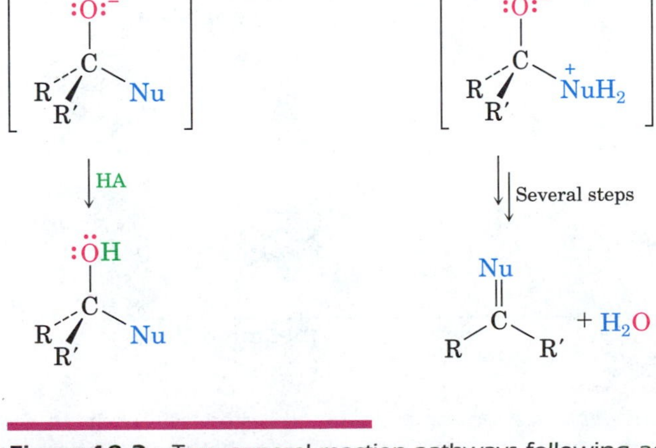

Figure 19.2 Two general reaction pathways following addition of a nucleophile to a ketone or aldehyde.

In the remainder of this chapter, we'll look at specific examples of nucleophilic addition reactions. In so doing, we'll be concerned both with the *reversibility* of a given reaction and with the acid or base *catalysis* of that reaction. Some nucleophilic addition reactions take place without catalysis, but many others require acid or base to proceed.

PROBLEM.

19.5 What product would you obtain by adding cyanide ion ($^{-}$:C≡N) to acetone and then protonating the tetrahedral intermediate?

19.7 Relative Reactivity of Aldehydes and Ketones

Aldehydes are generally more reactive than ketones in nucleophilic addition reactions for both steric and electronic reasons. Sterically, the presence of two relatively large substituents in ketones versus only one large substituent in aldehydes means that attacking nucleophiles are able to approach aldehydes more readily. Thus, the transition state leading to the tetrahedral intermediate is less crowded and lower in energy for aldehydes than for ketones (Figure 19.3).

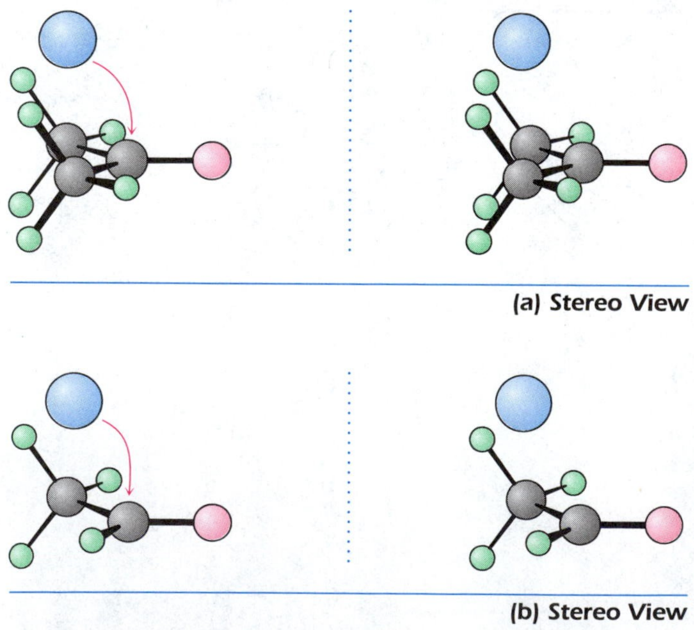

Figure 19.3 Nucleophilic attack on a ketone (a) is sterically hindered because of the two relatively large substituents attached to the carbonyl-group carbon. An aldehyde (b) has only one large substituent and is less hindered.

Electronically, aldehydes are more reactive than ketones because of the greater degree of polarity of aldehyde carbonyl groups. To see this polarity difference, recall the stability order of carbocations (Section 6.10). Primary carbocations are less stable than secondary ones because there is only one alkyl group inductively stabilizing the positive charge rather than two. For similar reasons, aldehydes are more reactive than ketones because there is only one alkyl group inductively stabilizing the partial positive charge on the carbonyl carbon rather than two.

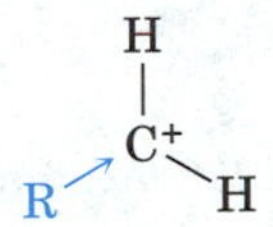

1° carbocation
(less stable, more reactive)

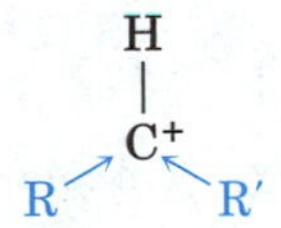

2° carbocation
(more stable, less reactive)

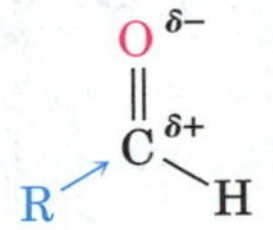

Aldehyde
(less stabilization of $\delta+$, more reactive)

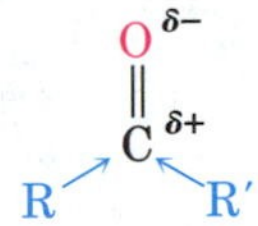

Ketone
(more stabilization of $\delta+$, less reactive)

PROBLEM

19.6 Aromatic aldehydes, such as benzaldehyde, are less reactive in nucleophilic addition reactions than aliphatic aldehydes. Explain.

PROBLEM

19.7 Which would you expect to be more reactive toward nucleophilic additions, *p*-methoxybenzaldehyde or *p*-nitrobenzaldehyde? Explain.

19.8 Nucleophilic Addition of H_2O: Hydration

Aldehydes and ketones undergo reaction with water to yield 1,1-diols, or **geminal (gem) diols**. The hydration reaction is reversible, and gem diols can eliminate water to regenerate ketones or aldehydes.

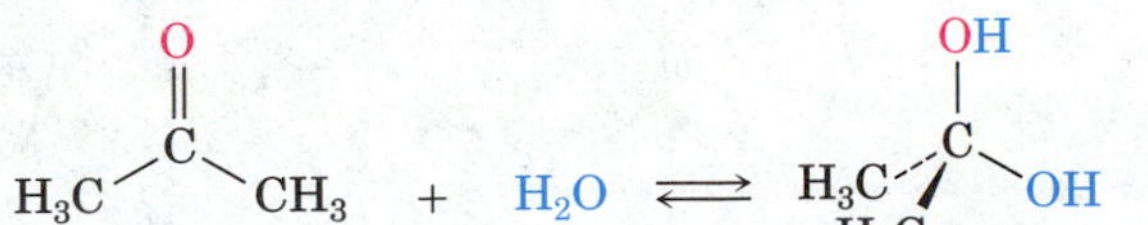

Acetone

Acetone hydrate (a gem diol)

The exact position of the equilibrium between a gem diol and a ketone or aldehyde depends on the structure of the carbonyl compound. Although the equilibrium strongly favors the carbonyl compound in most cases, the gem diol is favored for a few simple aldehydes. For example, an aqueous solution of formaldehyde consists of 99.9% gem diol and 0.1% aldehyde, whereas an aqueous solution of acetone consists of only about 0.1% gem diol and 99.9% ketone.

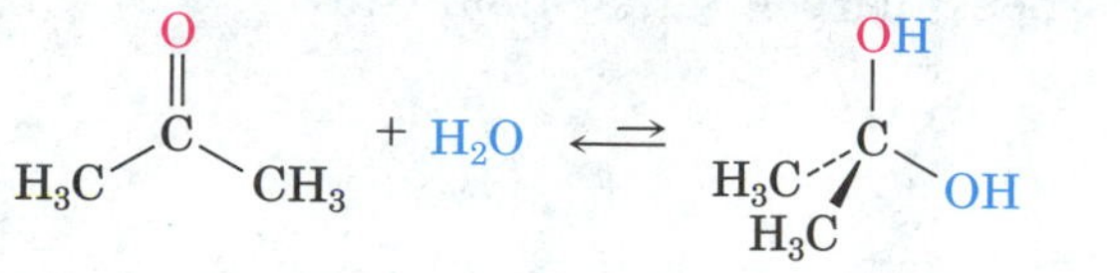

Acetone (99.9%) **Acetone hydrate (0.1%)**

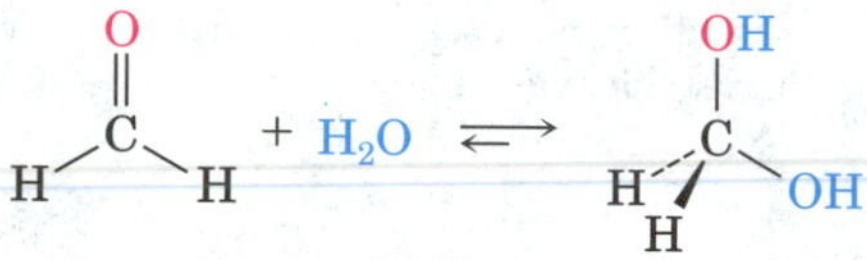

Formaldehyde (0.1%) **Formaldehyde hydrate (99.9%)**

The nucleophilic addition of water to a ketone or aldehyde is slow in pure water but is catalyzed by both acid and base. Like all catalysts, acids and bases don't change the *position* of the equilibrium; they affect only the rate at which the hydration reaction occurs.

The base-catalyzed hydration reaction takes place as shown in Figure 19.4. The attacking nucleophile is the negatively charged hydroxide ion.

Hydroxide ion nucleophile adds to the ketone or aldehyde carbonyl group to yield an alkoxide ion intermediate.

The basic alkoxide ion intermediate abstracts a proton (H^+) from water to yield gem diol product and regenerate hydroxide ion catalyst.

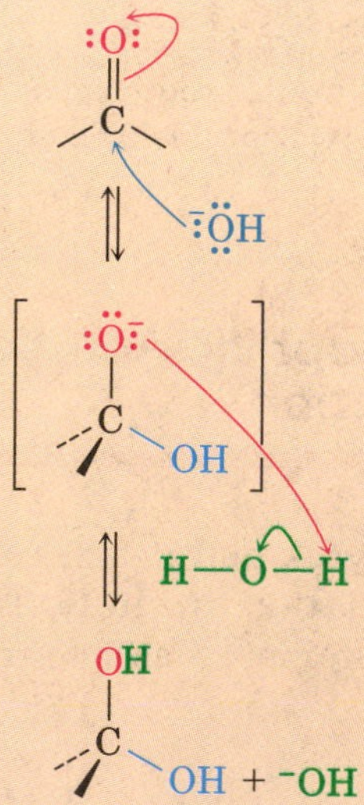

Figure 19.4 Mechanism of base-catalyzed hydration of a ketone or aldehyde. Hydroxide ion is a more reactive nucleophile than neutral water.

The acid-catalyzed hydration reaction also takes place in several steps. An acid catalyst first protonates the oxygen atom of the carbonyl group, placing a positive charge on oxygen and making the carbonyl group more electrophilic. Subsequent nucleophilic addition of water to the protonated ketone or aldehyde then yields a protonated gem diol, which loses H^+ to give the neutral product (Figure 19.5).

Acid catalyst protonates the basic carbonyl oxygen atom, making the ketone or aldehyde a much better acceptor of nucleophiles.

Nucleophilic addition of neutral water yields a protonated gem diol.

Loss of a proton regenerates the acid catalyst and gives neutral gem diol product.

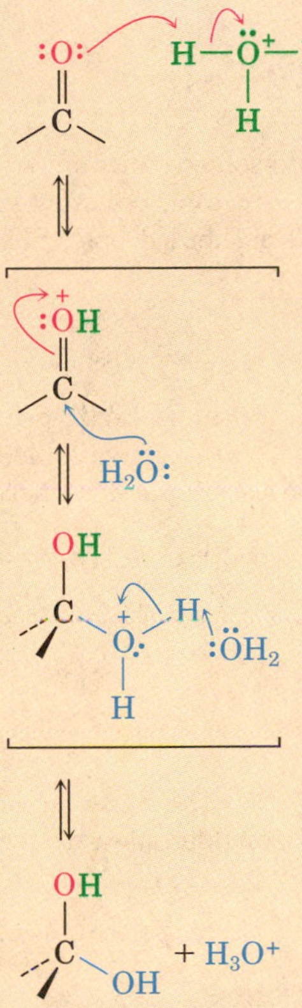

Figure 19.5 Mechanism of acid-catalyzed hydration of a ketone or aldehyde. Acid protonates the carbonyl group, thus making it more electrophilic and more reactive.

Note the difference between the base-catalyzed and the acid-catalyzed processes. The base-catalyzed reaction takes place rapidly because water is converted into hydroxide ion, a much better nucleophilic *donor*. The acid-catalyzed reaction takes place rapidly because the carbonyl compound is converted by protonation into a much better electrophilic *acceptor*.

The hydration reaction just described is typical of what happens when a ketone or aldehyde is treated with a nucleophile of the type H–Y, where the Y atom is electronegative and can stabilize a negative charge (oxygen, halogen, or sulfur, for example). In such reactions, nucleophilic addition is reversible, with the equilibrium favoring the carbonyl starting material rather than the tetrahedral adduct. In other words, treatment of ketones and aldehydes with reagents such as H_2O, HCl, HBr, or H_2SO_4 does not normally lead to addition products.

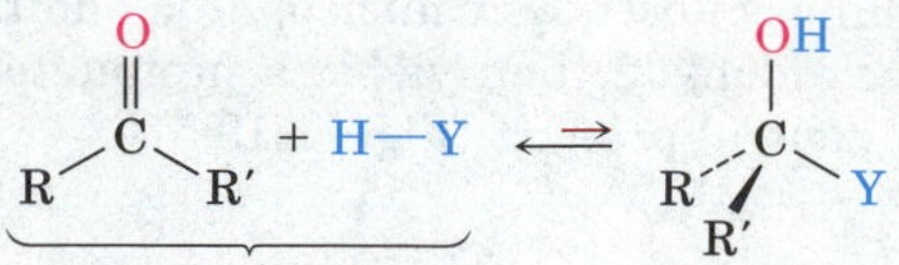

Favored when
Y = —OH, —Br, —Cl, HSO_4^-

PROBLEM.

19.8 When dissolved in water, trichloroacetaldehyde (chloral, CCl_3CHO) exists primarily as the gem diol, chloral hydrate, $CCl_3CH(OH)_2$, better known by the non-IUPAC name "knockout drops." Show the structure of chloral hydrate.

PROBLEM.

19.9 The oxygen in water is primarily (99.8%) ^{16}O, but water enriched with the heavy isotope ^{18}O is also available. When a ketone or aldehyde is dissolved in ^{18}O-enriched water, the isotopic label becomes incorporated into the carbonyl group:

$$R_2C{=}O + H_2O^* \longrightarrow R_2C{=}O^* + H_2O \qquad \text{(where } O^* = {}^{18}O\text{)}$$

Explain.

PROBLEM.

19.10 Explain why the S_N2 reaction of (dibromomethyl)benzene, $C_6H_5CHBr_2$, with NaOH yields benzaldehyde rather than (dihydroxymethyl)benzene, $C_6H_5CH(OH)_2$.

19.9 Nucleophilic Addition of HCN: Cyanohydrins

Aldehydes and unhindered ketones react with HCN to yield **cyanohydrins, RCH(OH)C≡N**. For example, benzaldehyde gives the cyanohydrin mandelonitrile in 88% yield on treatment with HCN:

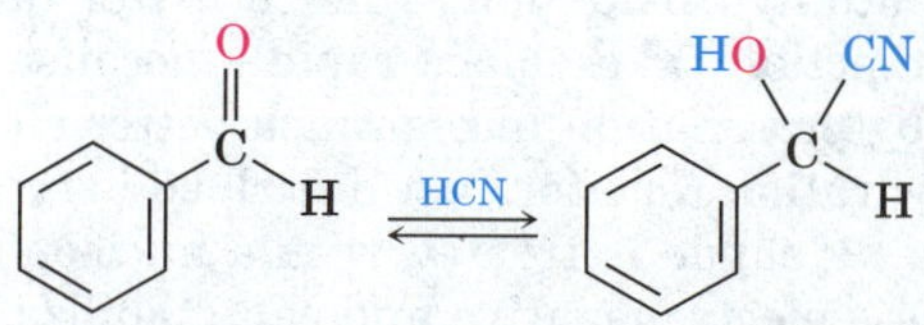

Benzaldehyde **Mandelonitrile (88%) (a cyanohydrin)**

Detailed studies carried out in the early 1900s by Arthur Lapworth[2] showed that cyanohydrin formation is reversible and base-catalyzed. Reaction occurs slowly with pure HCN but rapidly when a small amount of base is added to generate the strongly nucleophilic cyanide ion, CN^-. Addition of CN^- to ketones and aldehydes occurs by a typical nucleophilic addition pathway, yielding a tetrahedral intermediate that is protonated by HCN to give cyanohydrin product plus regenerated CN^-.

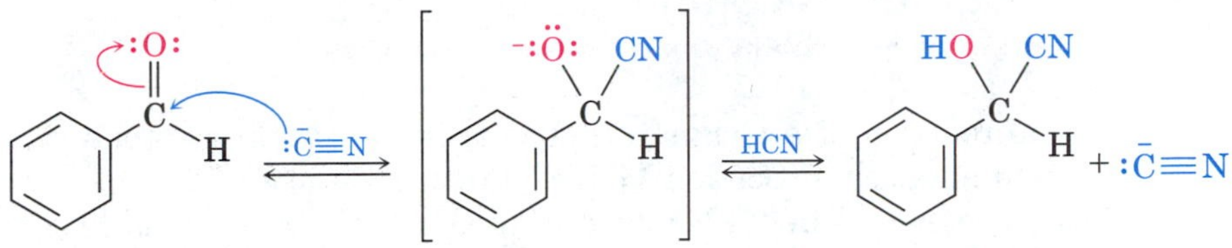

Benzaldehyde Tetrahedral intermediate **Mandelonitrile (88%)**

Cyanohydrin formation is particularly interesting because it is one of the few examples of the addition of an acid to a carbonyl group. As noted in the previous section, acids such as HBr, HCl, H_2SO_4, and CH_3COOH don't form carbonyl adducts because the equilibrium constant for reaction is unfavorable. With HCN, however, the equilibrium favors the adduct.

Cyanohydrin formation is useful because of the further chemistry that can be carried out. For example, nitriles (RCN) can be reduced with $LiAlH_4$ to yield primary amines (RCH_2NH_2) and can be hydrolyzed by hot aqueous acid to yield carboxylic acids. Thus, cyanohydrin formation provides a method for transforming a ketone or aldehyde into a different functional group.

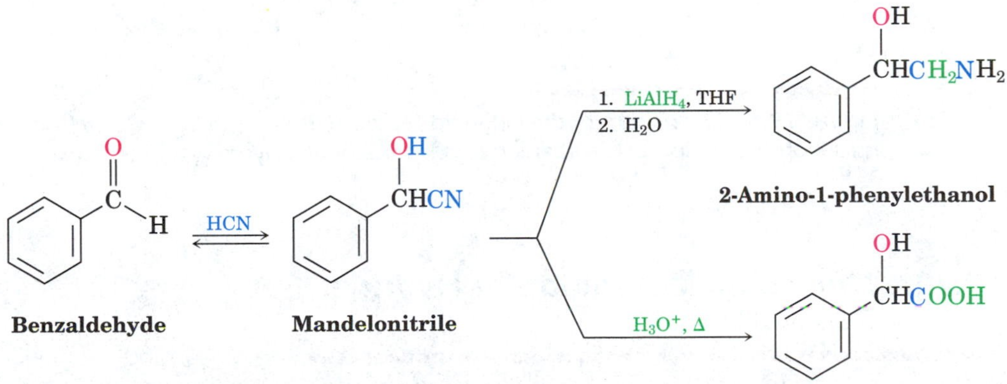

Benzaldehyde **Mandelonitrile** **2-Amino-1-phenylethanol** **Mandelic acid (90%)**

---

[2]Arthur Lapworth (1872–1941); b. Galashiels, Scotland; D.Sc., City and Guilds Institute, London; professor, University of Manchester (1909–1941).

PROBLEM.

19.11 Explain the observation that cyclohexanone forms a cyanohydrin in good yield but that 2,2,6-trimethylcyclohexanone is unreactive to HCN/KCN.

19.10 Nucleophilic Addition of Grignard Reagents: Alcohol Formation

The reaction of a Grignard reagent, RMgX, with a ketone or aldehyde to yield an alcohol (Section 17.7) is a nucleophilic addition of a *carbon* anion, or carbanion. The C–Mg bond in the Grignard reagent is so strongly polarized that Grignard reagents act for all practical purposes as $R:^- \, ^+MgX$.

Acid–base complexation of Mg^{2+} with the carbonyl oxygen atom first serves to make the carbonyl group a better acceptor, and nucleophilic addition of $R:^-$ then produces a tetrahedral magnesium alkoxide intermediate. Protonation by addition of dilute aqueous acid yields the neutral alcohol (Figure 19.6). Unlike the nucleophilic additions of water and HCN, Grignard additions are irreversible because a carbanion is too poor a leaving group to be expelled in a reversal step.

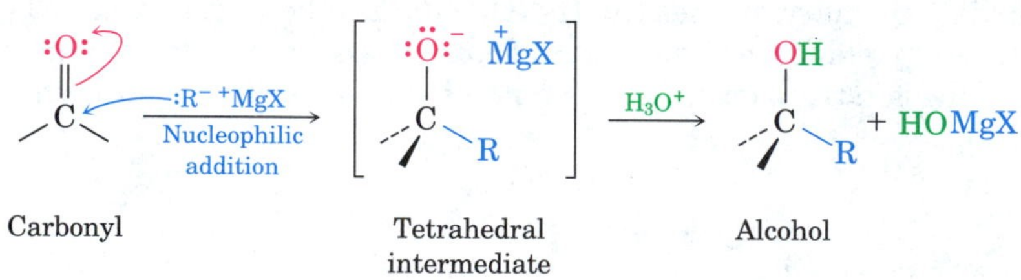

Figure 19.6 Mechanism of the Grignard reaction, the nucleophilic addition of a carbanion to a ketone or aldehyde.

19.11 Nucleophilic Addition of Hydride: Reduction

The reduction of a ketone or aldehyde to yield an alcohol (Section 17.6) is a nucleophilic addition reaction of a hydride ion, $:H^-$. The exact details of carbonyl-group reduction are complex, but the common reducing agents $LiAlH_4$ and $NaBH_4$ act as if they were hydride-ion donors (Figure 19.7). Addition of water or aqueous acid after the hydride addition step protonates the tetrahedral alkoxide intermediate and gives the alcohol product.

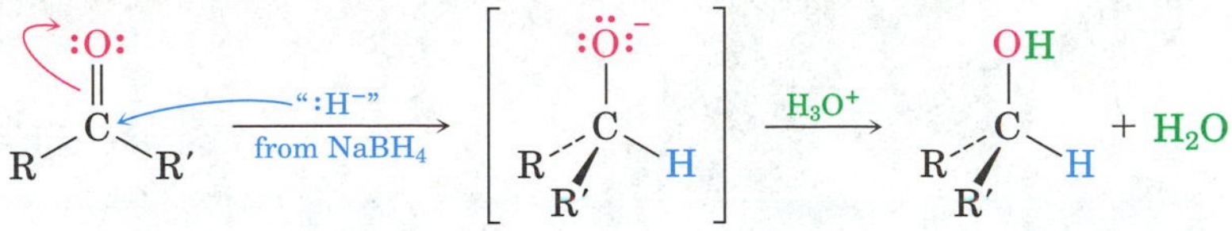

Figure 19.7 Mechanism of carbonyl-group reduction by nucleophilic addition of "hydride ion" from $NaBH_4$ or $LiAlH_4$.

19.12 Nucleophilic Addition of Amines: Imine and Enamine Formation

Primary amines, RNH_2, add to aldehydes and ketones to yield **imines, $R_2C{=}NR$**. Secondary amines add in a similar manner to yield **enamines, $R_2N{-}CR{=}CR_2$** (*ene* + *amine* = unsaturated amine). Imines are important intermediates in many metabolic pathways, and we'll see frequent examples of their occurrence in Chapter 30.

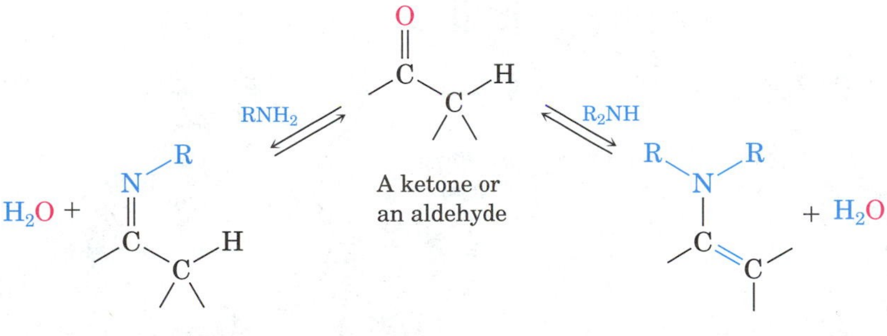

Imine formation and enamine formation appear different because one leads to a product with a C=N double bond and the other leads to a product with a C=C double bond, but they're actually quite similar. Both are typical examples of nucleophilic addition reactions in which the initially formed tetrahedral intermediate is not isolated. Instead, water is eliminated and a new C=Nu double bond is formed.

Imines are formed in a reversible, acid-catalyzed process involving nucleophilic attack on the carbonyl group by the primary amine, followed by transfer of a proton from nitrogen to oxygen to yield a neutral amino alcohol, or **carbinolamine**. Protonation of the carbinolamine oxygen by the acid catalyst then converts the hydroxyl into a good leaving group, and E1-like loss of water produces an iminium ion. Loss of a proton gives the final product and regenerates acid catalyst (Figure 19.8).

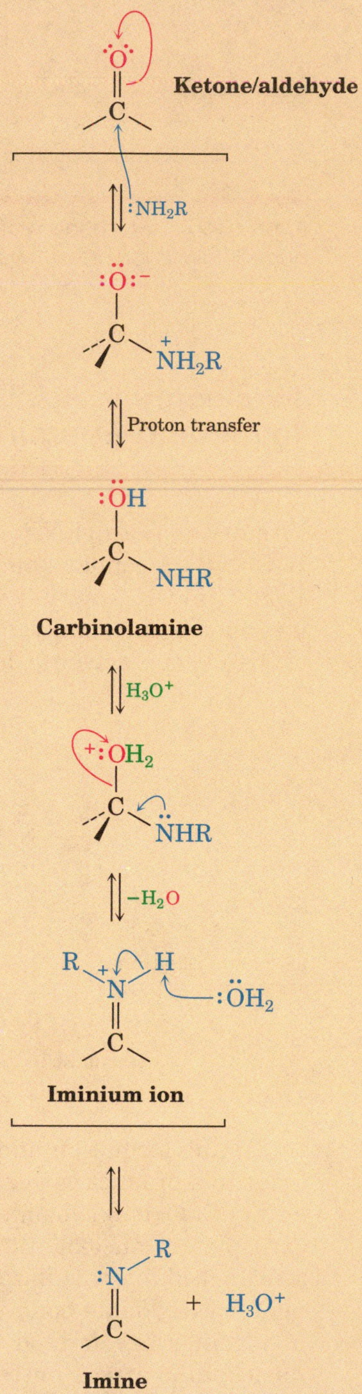

Nucleophilic attack on the ketone or aldehyde by the lone-pair electrons of an amine leads to a dipolar tetrahedral intermediate.

A proton is then transferred from nitrogen to oxygen, yielding a neutral carbinolamine.

Acid catalyst protonates the hydroxyl oxygen.

The nitrogen lone-pair electrons expel water, giving an iminium ion.

Loss of H^+ from nitrogen then gives the neutral imine product.

Figure 19.8 Mechanism of imine formation by reaction of a ketone or aldehyde with a primary amine.

Imine formation is slow at both high pH and low pH but reaches a maximum rate at weakly acidic pH. For example, the profile of pH versus rate obtained for the reaction between acetone and hydroxylamine, H_2NOH, shows that the maximum reaction rate is obtained at pH 4.5 (Figure 19.9).

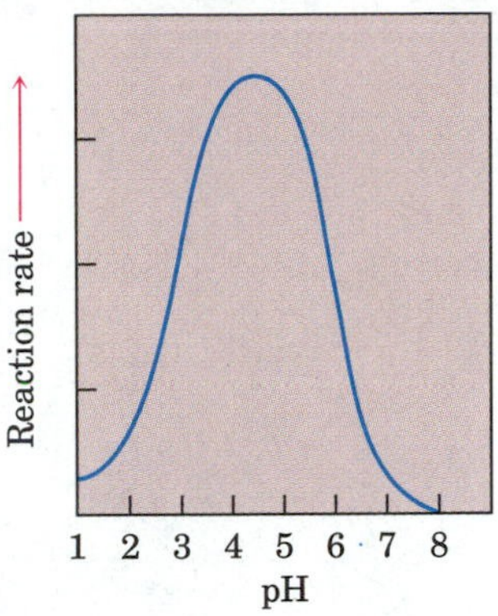

Figure 19.9 Dependence on pH of the rate of reaction between acetone and hydroxylamine: $(CH_3)_2C{=}O + H_2NOH \rightarrow (CH_3)_2C{=}NOH + H_2O$.

We can account for the observed rate maximum at pH 4.5 by looking at each individual step in the mechanism. As indicated in Figure 19.8, an acid catalyst is required to protonate the intermediate carbinolamine, thereby converting the –OH into a better leaving group. Thus, reaction can't occur if there is not enough acid present (that is, at high pH). On the other hand, if *too much* acid is present (low pH), the attacking amine nucleophile is completely protonated and the initial nucleophilic addition step can't occur.

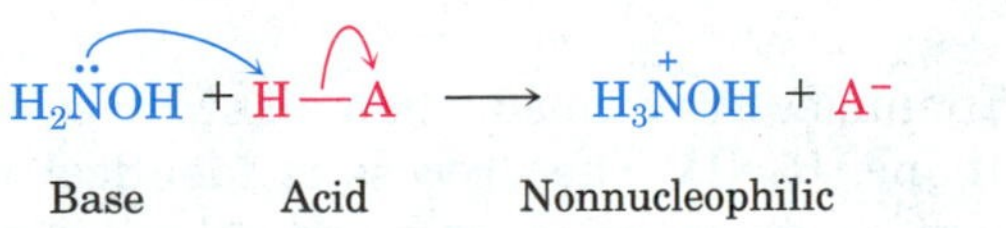

Evidently, pH 4.5 represents a compromise between the need for *some* acid to catalyze the rate-limiting dehydration step but *not too much* acid so as to avoid complete protonation of the amine. Each individual nucleophilic addition reaction has its own specific requirements, and reaction conditions must be controlled to obtain maximum reaction rates.

Imine formation from such reagents as hydroxylamine (NH_2OH), semicarbazide ($NH_2NHCONH_2$), and 2,4-dinitrophenylhydrazine are particularly useful because the products of these reactions—**oximes, semicarbazones**, and **2,4-dinitrophenylhydrazones (2,4-DNP's)**—are often crystalline and easy to handle. Such crystalline derivatives are sometimes prepared as a means of identifying liquid ketones or aldehydes.

Oxime

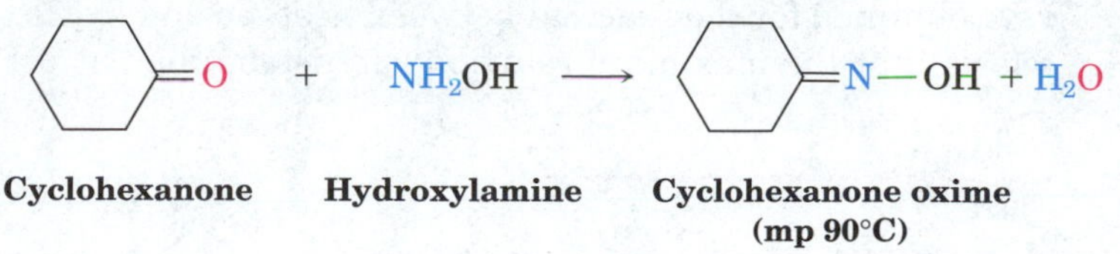

Semicarbazone

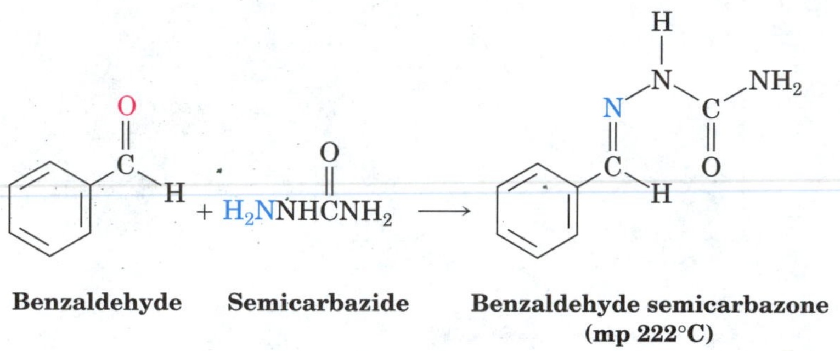

2,4-Dinitrophenylhydrazone

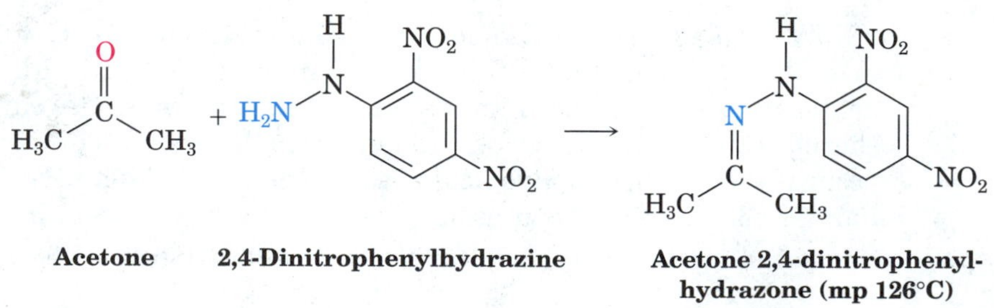

Enamines are formed when a ketone or aldehyde reacts with a secondary amine, R_2NH. The process is identical to imine formation up to the iminium ion stage, but at this point there is no proton on nitrogen that can be lost to yield neutral product. Instead, a proton is lost from the α carbon atom, yielding an enamine (Figure 19.10).

PROBLEM.

19.12 Show the products you would obtain by reaction of cyclohexanone with ethylamine, $CH_3CH_2NH_2$, and with diethylamine, $(CH_3CH_2)_2NH$.

PROBLEM.

19.13 Imine formation is reversible. Show all the steps involved in the hydrolysis of an imine to yield a ketone or aldehyde plus primary amine.

Nucleophilic addition of a secondary amine to the ketone or aldehyde, followed by proton transfer from nitrogen to oxygen, yields an intermediate carbinolamine in the normal way.

Protonation of the hydroxyl by acid catalyst converts it into a better leaving group.

Elimination of water by the lone-pair electrons on nitrogen then yields an intermediate iminium ion.

Loss of a proton from the alpha-carbon atom yields the enamine product and regenerates the acid catalyst.

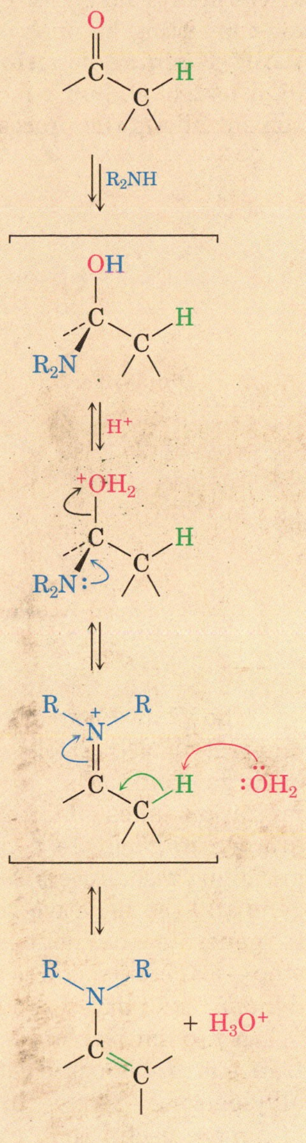

An enamine

Figure 19.10 Mechanism of enamine formation by reaction of a ketone or aldehyde with a secondary amine, R_2NH.

19.13 Nucleophilic Addition of Hydrazine: The Wolff–Kishner Reaction

A useful variant of the imine-forming reaction just discussed involves the treatment of a ketone or aldehyde with hydrazine, H_2NNH_2, in the presence of KOH. This reaction, discovered independently in 1911 by Ludwig Wolff[3] in Germany and N. M. Kishner[4] in Russia, is a valuable synthetic method for converting ketones or aldehydes into alkanes, $R_2C{=}O \rightarrow R_2CH_2$. The **Wolff–Kishner reaction** was originally carried out at temperatures as high as 240°C, but a modification in which dimethyl sulfoxide is used as solvent allows the process to take place near room temperature.

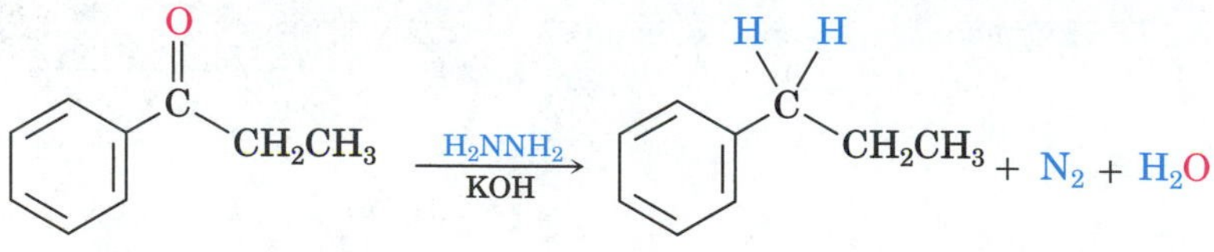

Propiophenone **Propylbenzene (82%)**

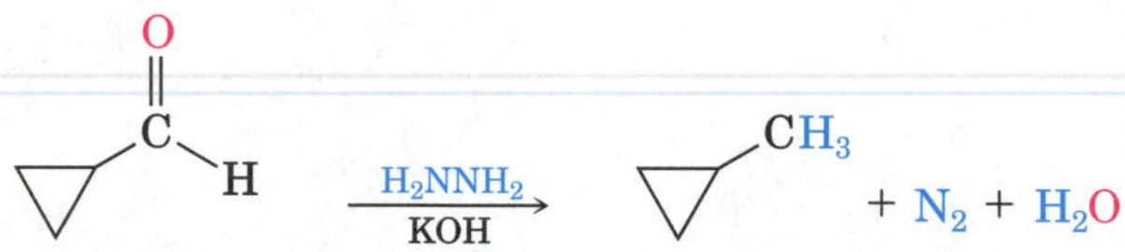

Cyclopropanecarbaldehyde **Methylcyclopropane (72%)**

The Wolff–Kishner reaction involves formation of a hydrazone intermediate, $R_2C{=}NNH_2$, followed by base-catalyzed double-bond migration, loss of N_2 gas, and formation of alkane product (Figure 19.11). The double-bond migration takes place when base removes one of the weakly acidic NH protons to generate a hydrazone anion. Since the hydrazone anion has an allylic resonance structure that places the double bond between nitrogens and the negative charge on carbon, protonation can occur on carbon to generate the double-bond rearrangement product. The next-to-last step—loss of nitrogen to give an alkyl anion—is driven by the large thermodynamic stability of the N_2 molecule.

Note that the Wolff–Kishner reduction accomplishes the same overall transformation as the catalytic hydrogenation of an acylbenzene to yield an alkylbenzene (Section 16.11). The Wolff–Kishner reduction is more general and more useful than catalytic hydrogenation, however, because it works well with both alkyl and aryl ketones.

In addition to the Wolff–Kishner reaction, another process called the **Clemmensen[5] reduction** also accomplishes the conversion of a ketone or

[3]Ludwig Wolff (1857–1919); b. Neustadt/Hardt; Ph.D. Strasbourg (Fittig); professor, University of Jena.

[4]N. M. Kishner (1867–1935); b. Moscow; Ph.D. Moscow (Markovnikov); professor, universities of Tomsk and Moscow.

[5]E. C. Clemmensen (1876–1941); b. Odense, Denmark; Ph.D. Copenhagen; Clemmensen Chemical Corp., Newark, NY.

Reaction of the ketone or aldehyde with hydrazine yields a hydrazone in the normal way.

Base then abstracts one of the weakly acidic protons from $-NH_2$, yielding a hydrazone anion. This anion has an "allylic" resonance form that places the negative charge on carbon and the double bond between nitrogens.

Protonation of the hydrazone anion takes place on carbon to yield a neutral intermediate.

Base-induced loss of nitrogen then gives a carbanion . . .

. . . that is protonated to yield neutral alkane product.

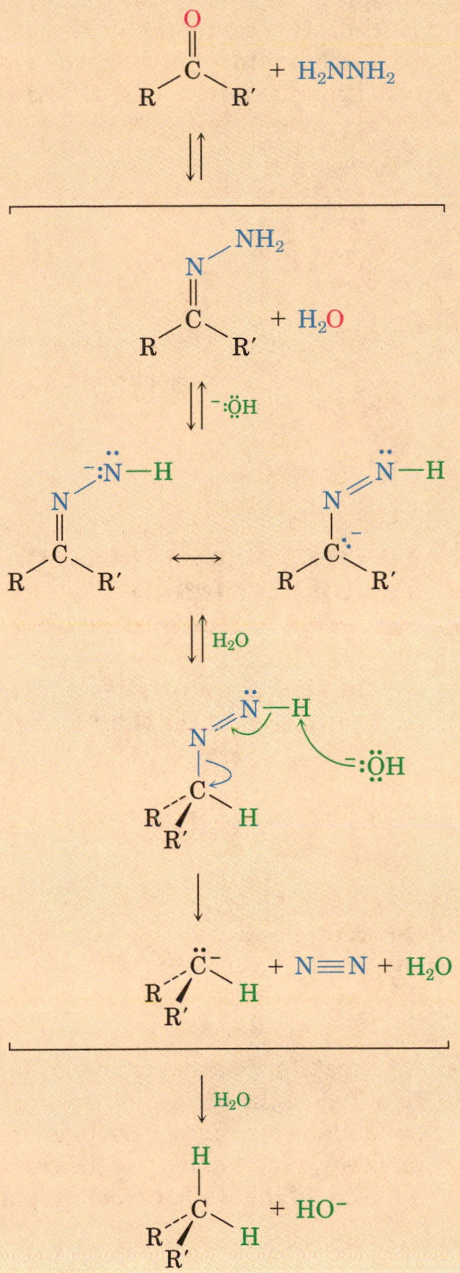

Figure 19.11 Mechanism of the Wolff–Kishner reduction of a ketone or aldehyde to yield an alkane.

aldehyde to an alkane. The Clemmensen reduction, whose mechanism is complex, involves treatment of the carbonyl compound with amalgamated zinc, Zn(Hg), and concentrated aqueous HCl. The reaction is used primarily when the ketone or aldehyde reactant is sensitive to the strongly basic conditions required by Wolff–Kishner reduction.

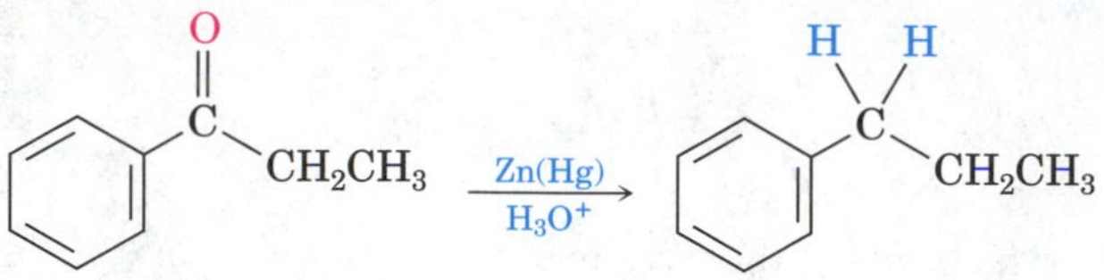

Propiophenone **Propylbenzene (86%)**

19.14 Nucleophilic Addition of Alcohols: Acetal Formation

Ketones and aldehydes react reversibly with alcohols in the presence of an acid catalyst to yield **acetals, $R_2C(OR')_2$** (formerly also called *ketals*).

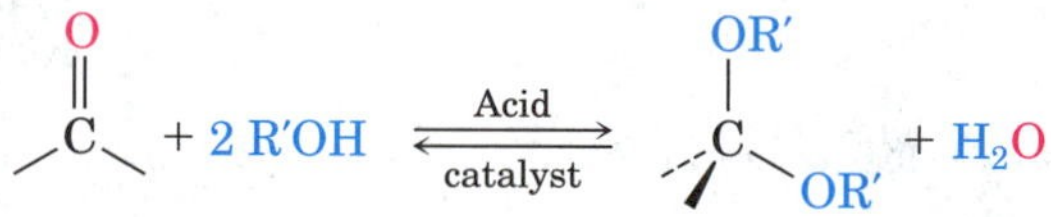

Ketone/aldehyde An acetal

Acetal formation is similar to the hydration reaction discussed in Section 19.8. Like water, alcohols are weak nucleophiles that add to ketones and aldehydes only slowly under neutral conditions. Under acidic conditions, however, the reactivity of the carbonyl group is increased by protonation so addition of an alcohol occurs rapidly.

Neutral carbonyl group (moderately nucleophilic)

Protonated carbonyl group (strongly electrophilic and highly reactive toward nucleophiles)

Nucleophilic addition of alcohol to the carbonyl group initially yields a hydroxy ether called a **hemiacetal**, analogous to the gem diol formed by addition of water (Section 19.8). Hemiacetals are formed reversibly, with the equilibrium normally favoring the carbonyl compound. In the presence of acid, however, a further reaction can occur. Protonation of the hydroxyl group followed by an E1-like loss of water leads to an **oxonium ion, $R_2C{=}OR^+$**, which reacts with a second equivalent of alcohol to yield the acetal. For example, reaction of cyclohexanone with methanol yields the dimethyl acetal. The mechanism is shown in Figure 19.12 (p. 742).

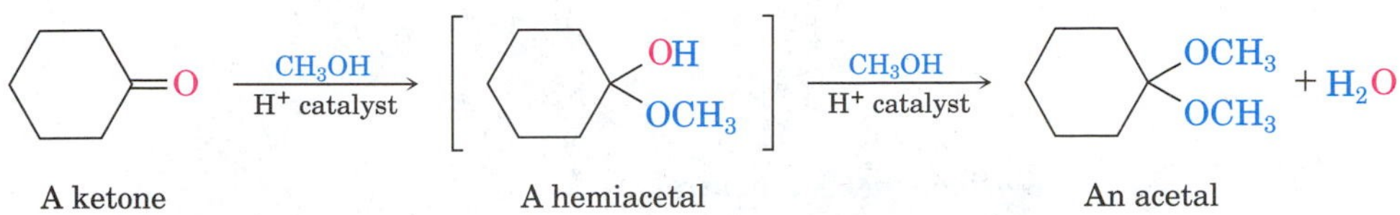

Because all the steps in acetal formation are reversible, the reaction can be driven either forward (from carbonyl compound to acetal) or backward (from acetal to carbonyl compound), depending on the conditions. The forward reaction is favored by conditions that remove water from the medium and thus drive the equilibrium to the right. In practice, this is often done by distilling off water as it forms. The reverse reaction is favored by treating the acetal with a large excess of aqueous acid to drive the equilibrium to the left.

Acetals are useful because they can serve as protecting groups for ketones and aldehydes in the same way that trimethylsilyl ethers serve as protecting groups for alcohols (Section 17.10). As we saw previously, it sometimes happens that one functional group interferes with intended chemistry elsewhere in a complex molecule. For example, if we wanted to reduce only the ester group of ethyl 4-oxopentanoate, the ketone would interfere. Treatment of the starting keto ester with $LiAlH_4$ would reduce both the keto and the ester groups to give a diol product.

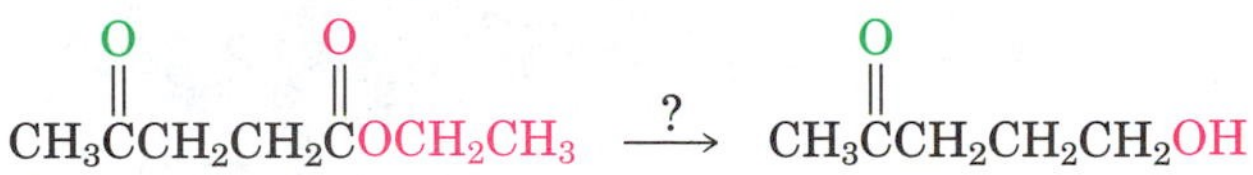

Ethyl 4-oxopentanoate **5-Hydroxy-2-pentanone**

By protecting the keto group as an acetal, however, the problem can be circumvented. Like other ethers, acetals are inert to bases, hydride reducing agents, Grignard reagents, and catalytic reducing conditions. Thus, we can

Protonation of the carbonyl oxygen strongly polarizes the carbonyl group and . . .

. . . activates the carbonyl group for nucleophilic attack by oxygen lone-pair electrons from alcohol.

Loss of a proton yields a neutral hemiacetal tetrahedral intermediate.

Protonation of the hemiacetal hydroxyl converts it into a good leaving group.

Dehydration yields an intermediate oxonium ion.

Addition of a second equivalent of alcohol gives protonated acetal.

Loss of a proton yields neutral acetal product.

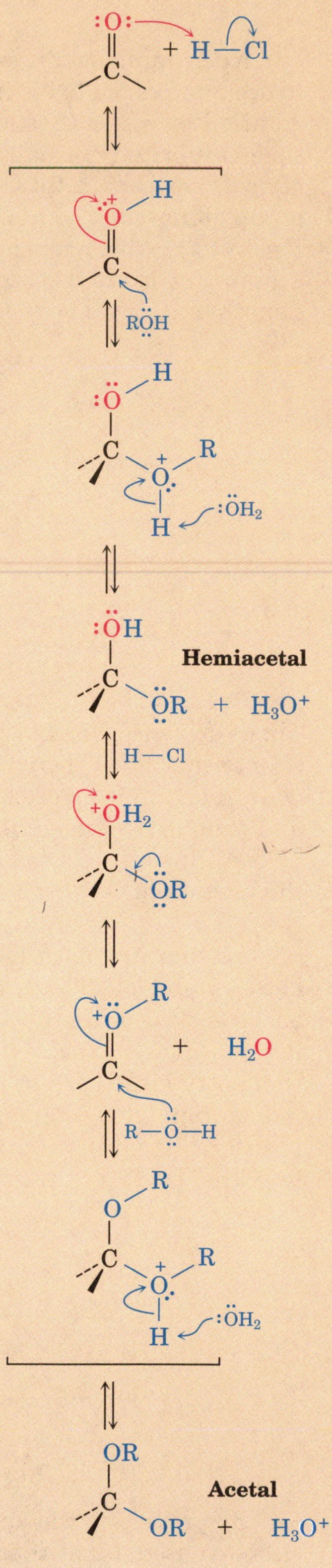

Figure 19.12 Mechanism of acid-catalyzed acetal formation by reaction of a ketone or aldehyde with an alcohol.

accomplish the selective reduction of the ester group in ethyl 4-oxopentanoate by first converting the keto group to an acetal, then reducing the ester with $LiAlH_4$, and then removing the acetal by treatment with aqueous acid:

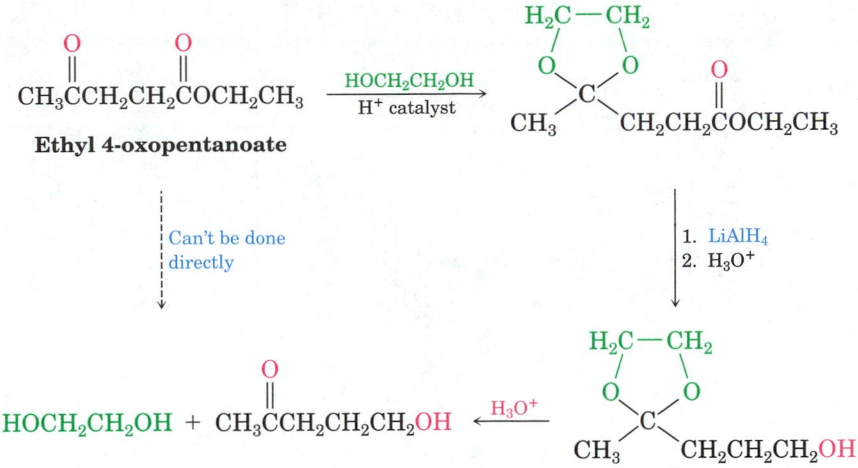

In practice, it's convenient to use ethylene glycol as the alcohol and to form a *cyclic* acetal. The mechanism of cyclic acetal formation using 1 equivalent of ethylene glycol is exactly the same as that using 2 equivalents of methanol or other monoalcohol. The only difference is that both alcohol groups are now in the same molecule.

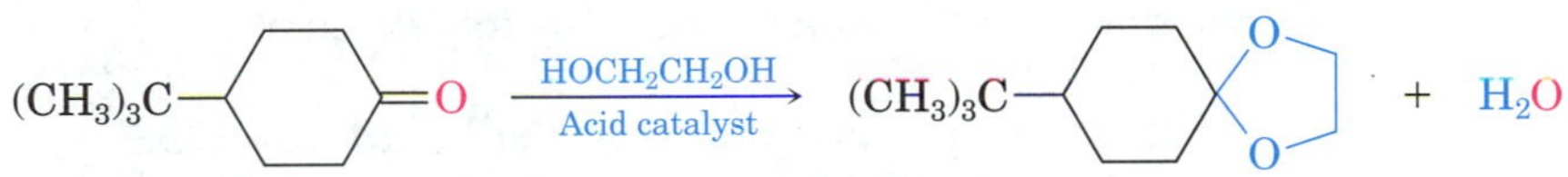

4-*tert*-Butylcyclohexanone **4-*tert*-Butylcyclohexanone ethylene acetal (88%)**

PROBLEM

19.14 Show all the steps in the acid-catalyzed formation of a cyclic acetal from ethylene glycol and a ketone or aldehyde.

19.15 Nucleophilic Addition of Phosphorus Ylides: The Wittig Reaction

Ketones and aldehydes are converted into alkenes by means of the **Wittig**[6] **reaction**. In this process, a phosphorus **ylide**, $\mathbf{R_2\overset{-}{C}{-}\overset{+}{P}(C_6H_5)_3}$ (also called

[6]Georg Wittig (1897–1987); b. Berlin; Ph.D. Marburg (von Auwers); professor, universities of Freiburg, Tübingen, and Heidelberg; Nobel Prize (1979).

a **phosphorane**), adds to a ketone or aldehyde to yield a dipolar intermediate called a **betaine**.[7] The betaine intermediate is not isolated; rather, it immediately decomposes to yield alkene and triphenylphosphine oxide. The net result is replacement of carbonyl oxygen by the organic group originally bonded to phosphorus (Figure 19.13).

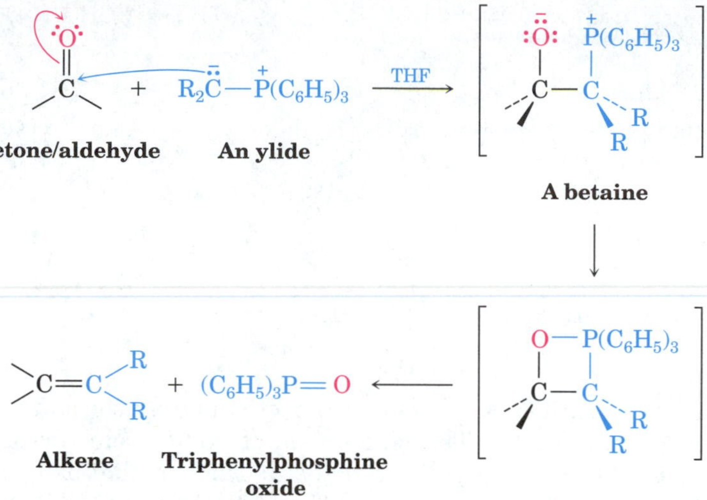

Figure 19.13 The mechanism of the Wittig reaction between a phosphorus ylide and a ketone or aldehyde to yield an alkene.

The phosphorus ylides necessary for Wittig reaction are easily prepared by S_N2 reaction of primary (and some secondary) alkyl halides with triphenylphosphine, followed by treatment with base. Triorganophosphines, R_3P, are good nucleophiles in S_N2 reactions, and yields of crystalline tetraorganophosphonium salts are high. The hydrogen on the carbon next to the positively charged phosphorus is weakly acidic and can be removed by a base such as sodium hydride or butyllithium (BuLi) to generate the neutral ylide. For example:

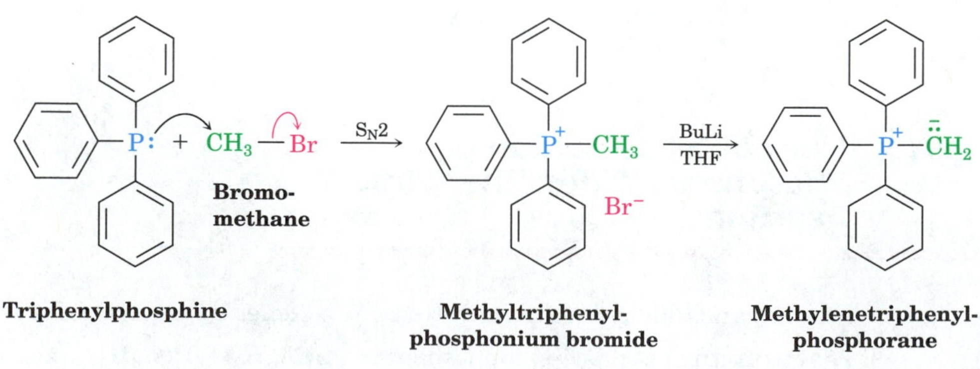

[7]An ylide—pronounced **ill**-id—is a neutral, dipolar compound with adjacent plus and minus charges. A betaine—pronounced **bay**-ta-een—is a neutral, dipolar compound with nonadjacent charges.

The Wittig reaction is extremely general, and a great many mono-, di-, and trisubstituted alkenes can be prepared from the appropriate combination of phosphorane and ketone or aldehyde. Tetrasubstituted alkenes can't be prepared, however, because of steric hindrance during the reaction.

The real value of the Wittig reaction is that pure alkenes of known structure are prepared. The alkene double bond is always exactly where the carbonyl group was in the precursor, and no product mixtures (other than *E,Z* isomers) are formed. For example, Wittig reaction of cyclohexanone with methylenetriphenylphosphorane yields only the single alkene product, methylenecyclohexane. By contrast, addition of methylmagnesium bromide to cyclohexanone, followed by dehydration with $POCl_3$, yields a mixture of two alkenes:

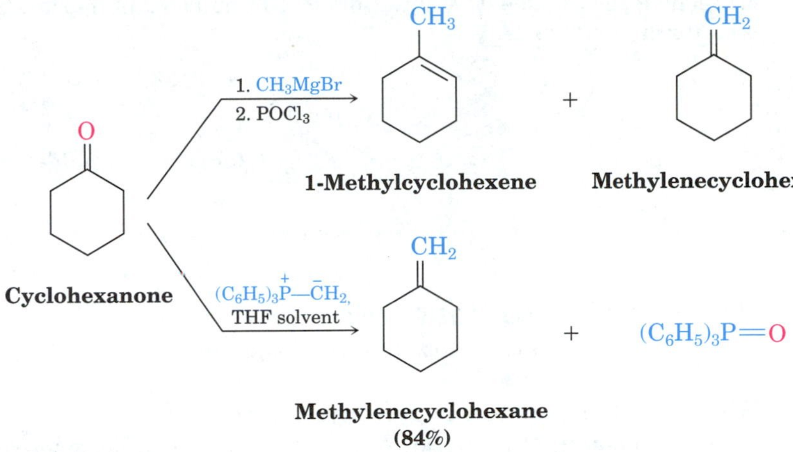

Wittig reactions are used commercially in a number of pharmaceutical applications. For example, the Swiss chemical firm of Hoffmann-LaRoche prepares β-carotene, a yellow food-coloring agent and dietary source of vitamin A, by Wittig reaction between retinal and retinylidenetriphenylphosphorane.

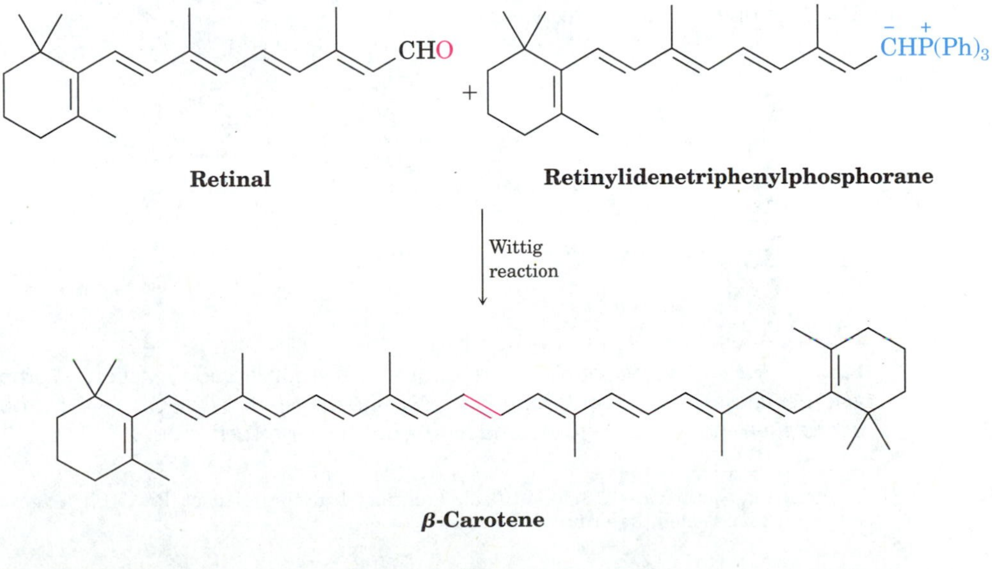

PROBLEM

19.15 What carbonyl compounds and what phosphorus ylides might you use to prepare the following compounds?

(a) 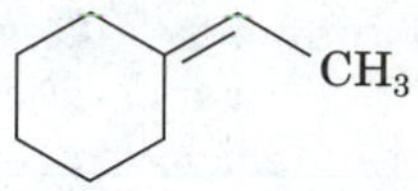
(b) 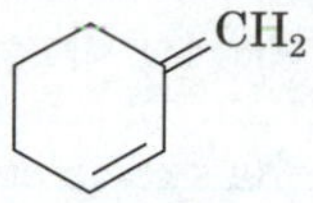
(c) 2-Methyl-2-hexene

(d) $C_6H_5CH{=}C(CH_3)_2$
(e) 1,2-Diphenylethylene

PROBLEM

19.16 Another route to β-carotene involves a double Wittig reaction between 2 equivalents of β-ionylideneacetaldehyde and a diylide. Formulate the reaction, and show the structure of the diylide.

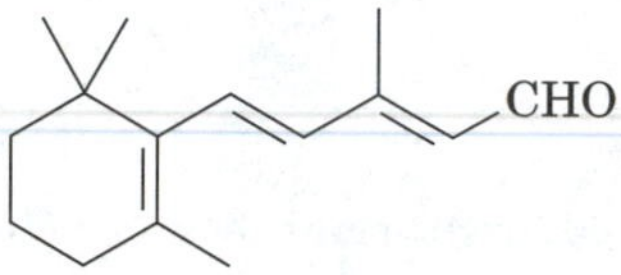

β-Ionylideneacetaldehyde

19.16 The Cannizzaro Reaction

We said in the overview of carbonyl chemistry preceding this chapter that nucleophilic addition reactions are characteristic of ketones and aldehydes but not of carboxylic acid derivatives. The reason for the difference is structural. As shown in Figure 19.14, tetrahedral intermediates produced by addition of a nucleophile to a carboxylic acid derivative can eliminate a leaving group, leading to a net nucleophilic acyl substitution reaction. Tetrahedral intermediates produced by addition of a nucleophile to a ketone or aldehyde, however, have only alkyl or hydrogen substituents and thus can't usually expel a leaving group. The **Cannizzaro[8] reaction**, discovered in 1853, is one exception to this rule.

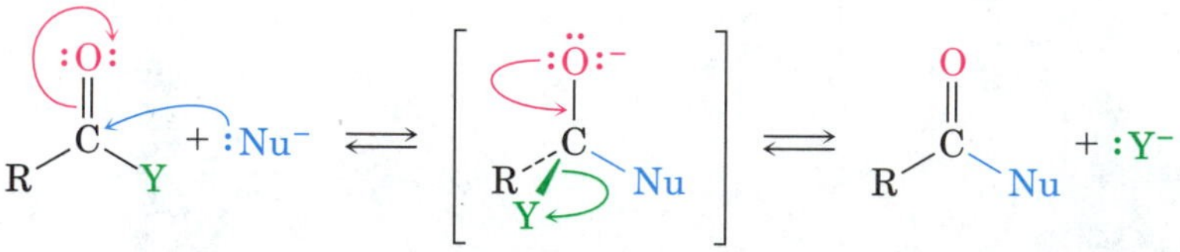

Figure 19.14 The mechanism of nucleophilic acyl substitution reactions. Carboxylic acid derivatives have an electronegative substituent Y (–Br, –Cl, –OR′, –NH_2) that can act as a leaving group. Ketones and aldehydes have no such group.

[8]Stanislao Cannizzaro (1826–1910); b. Palermo, Italy; studied at Pisa (Piria); professor, universities of Genoa, Palermo, and Rome.

When an aldehyde, such as benzaldehyde, that has no hydrogens on the carbon next to the –CHO group is heated with hydroxide ion, a reaction occurs yielding 1 equivalent of carboxylic acid and 1 equivalent of alcohol:

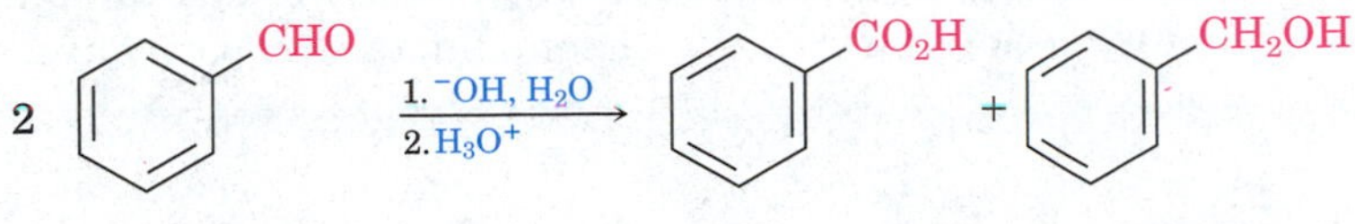

The Cannizzaro reaction takes place by nucleophilic addition of OH^- to an aldehyde to give a tetrahedral intermediate, *which expels hydride ion as a leaving group*. A second aldehyde molecule accepts the hydride ion in another nucleophilic addition step, resulting in a *disproportionation*. One molecule of aldehyde undergoes a substitution of H^- by OH^- and is thereby oxidized to an acid, while a second molecule of aldehyde undergoes an addition of H^- and is thereby reduced to an alcohol.

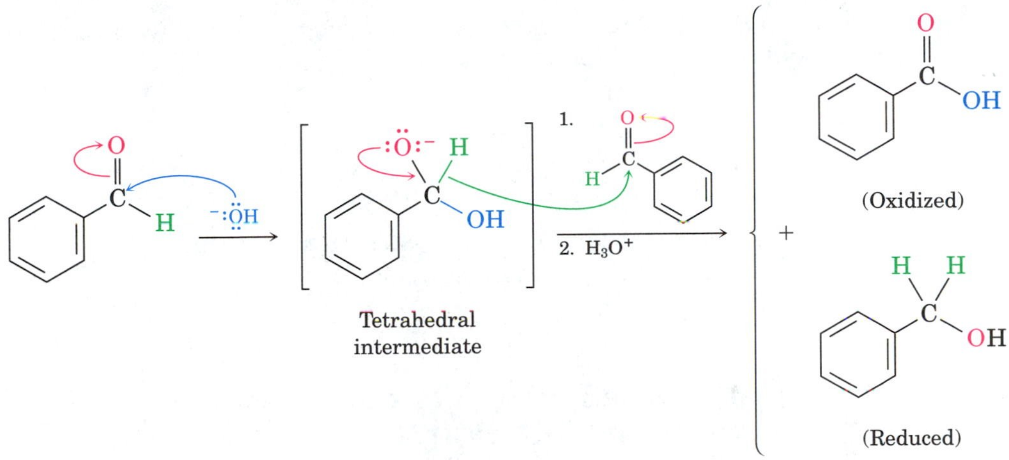

The Cannizzaro reaction has few practical applications and is effectively limited to formaldehyde and substituted benzaldehydes. It is interesting mechanistically, however, because it serves as a simple model for an important biological pathway by which reductions occur in living organisms. In nature, the most important reducing agent is a complex molecule named *reduced nicotinamide adenine dinucleotide*, abbreviated NADH.

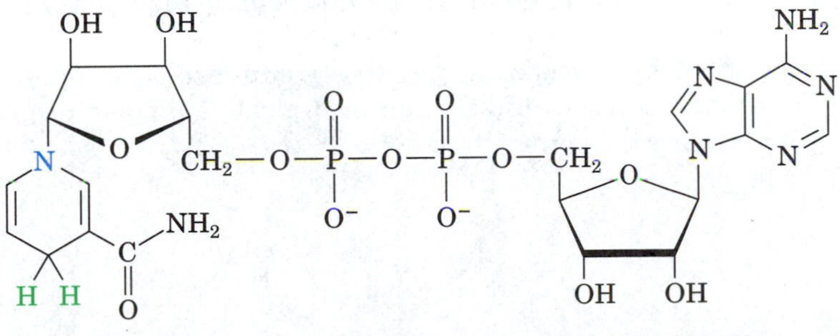

Reduced nicotinamide adenine dinucleotide (NADH)

NADH functions as a hydride donor in much the same way as the tetrahedral intermediate in a Cannizzaro reaction. The electron lone pair on a nitrogen atom expels H:⁻, which adds to a carbonyl group in another molecule to effect a reduction. We'll see this reaction again in Chapter 30 when we look at the details of some metabolic pathways.

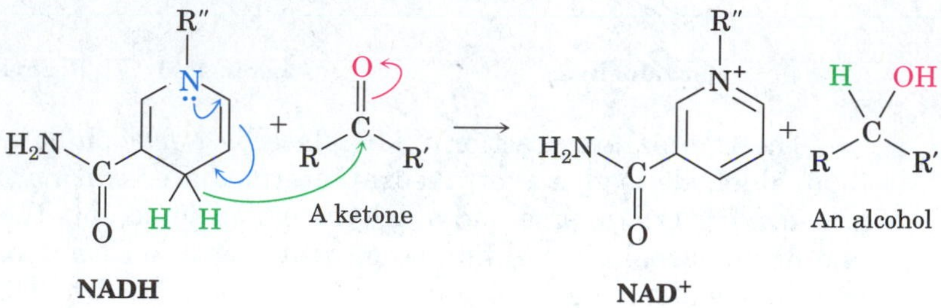

PROBLEM

19.17 When *o*-phthalaldehyde is treated with base, *o*-(hydroxymethyl)benzoic acid is formed. Show the mechanism of this reaction.

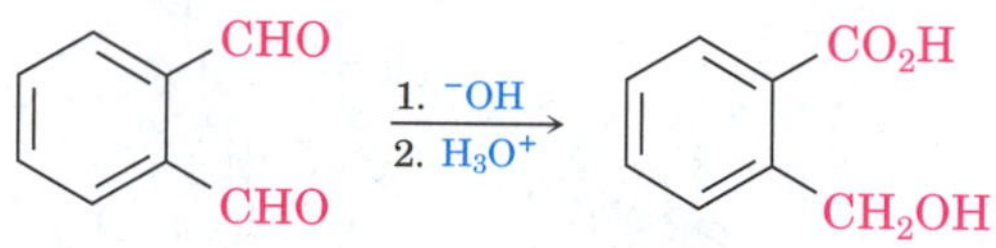

***o*-Phthalaldehyde** ***o*-(Hydroxymethyl)benzoic acid**

19.17 Conjugate Nucleophilic Addition to α,β-Unsaturated Carbonyl Groups

The reactions we've been discussing have all involved addition of a nucleophile *directly* to the carbonyl group. Closely related to this direct addition is the **conjugate addition** of a nucleophile to the C=C double bond of an α,β-unsaturated ketone or aldehyde (Figure 19.15). The two processes are often called *1,2 addition* and *1,4 addition*, respectively.

The conjugate addition of a nucleophile to an α,β-unsaturated carbonyl compound is due to the same electronic factors that are responsible for direct addition. We've seen that carbonyl groups are polarized so that the carbonyl carbon is positive, and we can even draw a dipolar resonance structure to underscore the point:

Carbonyl group

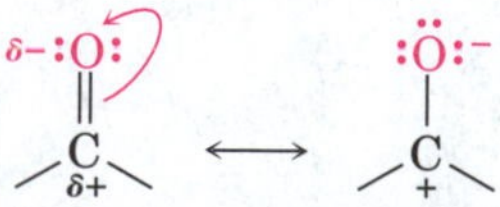

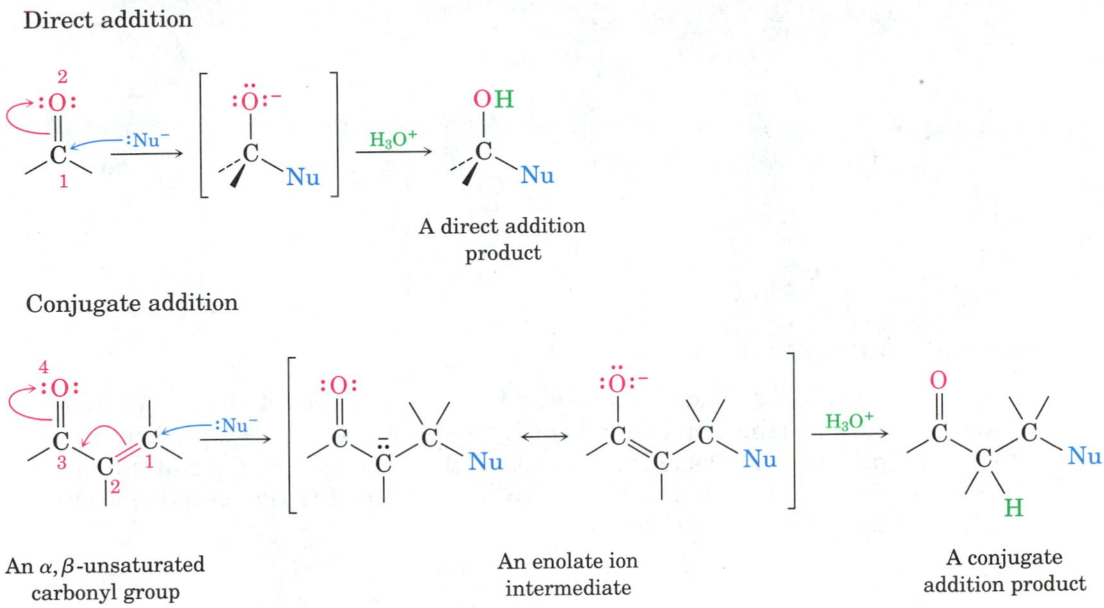

Figure 19.15 A comparison of direct (1,2) and conjugate (1,4) nucleophilic addition reactions.

When a similar resonance structure is drawn for an α,β-unsaturated carbonyl compound, however, the positive charge is allylic and is shared by the β carbon. In other words, the β carbon of an α,β-unsaturated carbonyl compound is an electrophilic site and can react with nucleophiles.

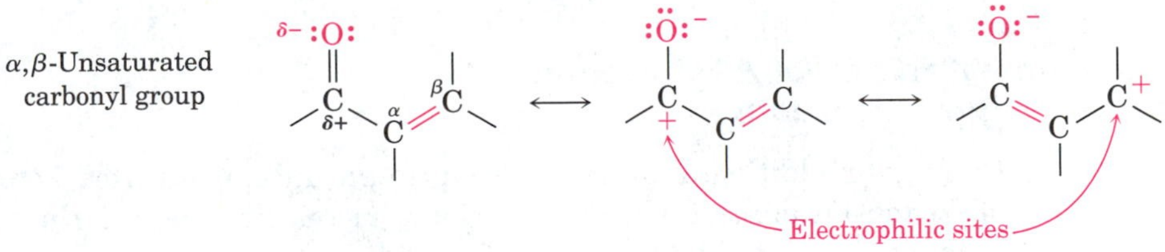

Conjugate addition of a nucleophile to the β carbon of an α,β-unsaturated carbonyl compound leads to an enolate ion intermediate, which is protonated on the α carbon to give the saturated product (Figure 19.15). The net effect is addition of the nucleophile to the C=C double bond, with the carbonyl group itself unchanged. In fact, of course, the carbonyl group is crucial to the success of the reaction. The C=C double bond would not be polarized, and no reaction would occur, without the carbonyl group.

Activated double bond

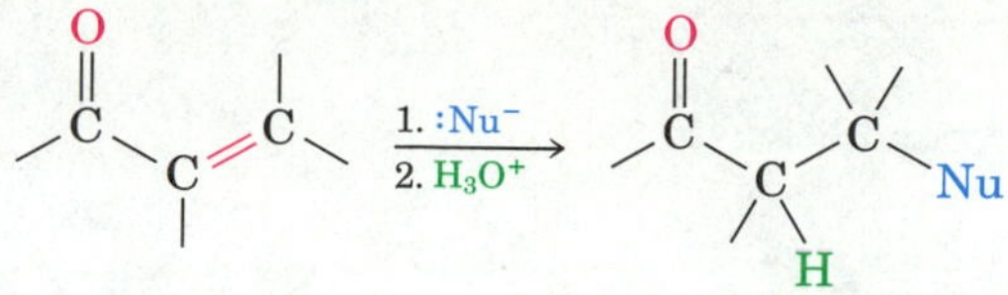

Unactivated double bond

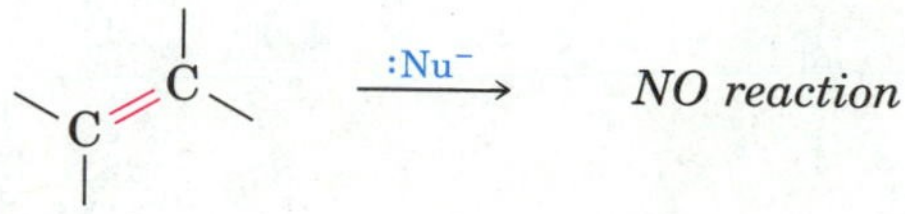

Conjugate Addition of Amines

Primary and secondary amines add to α,β-unsaturated carbonyl compounds to yield β-amino ketones and aldehydes. Reaction occurs rapidly under mild conditions, and yields are good. Note that, if only 1 equivalent of amine is used, the conjugate addition product is obtained to the complete exclusion of the direct addition product.

$$CH_3\overset{O}{\overset{\|}{C}}CH{=}CH_2 + H\ddot{N}(CH_2CH_3)_2 \xrightarrow{\text{Ethanol}} CH_3\overset{O}{\overset{\|}{C}}CH_2CH_2N(CH_2CH_3)_2$$

3-Buten-2-one **Diethylamine** **4-*N,N*-Diethylamino-2-butanone (92%)**

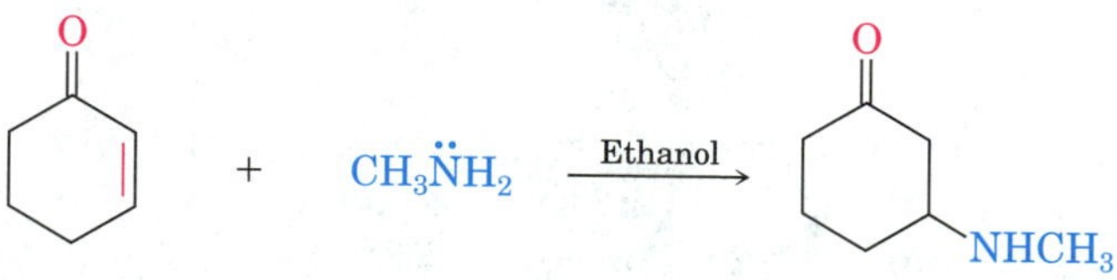

2-Cyclohexenone **Methylamine** **3-(*N*-methylamino)cyclohexanone**

Conjugate Addition of Alkyl Groups: Organocopper Reactions

Conjugate addition of an alkyl group to an α,β-unsaturated ketone is one of the most useful 1,4-addition reactions, just as direct addition of a Grignard reagent is one of the most useful 1,2 additions.

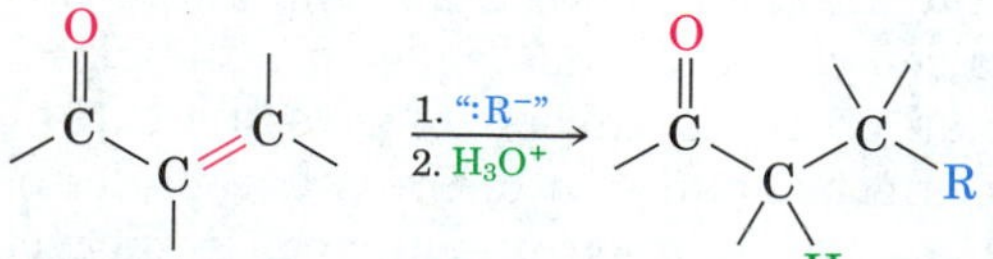

α,β-Unsaturated ketone/aldehyde

Conjugate addition of an alkyl group is carried out by treating the α,β-unsaturated ketone with a lithium diorganocopper reagent. As we saw in Section 10.9, diorganocopper reagents can be prepared by reaction between 1 equivalent of cuprous iodide and 2 equivalents of organolithium:

$$RX \xrightarrow[\text{Pentane}]{\text{2 Li}} RLi + Li^+ X^-$$

$$2\ RLi \xrightarrow[\text{Ether}]{\text{CuI}} Li^+(R\overset{-}{Cu}R) + Li^+ I^-$$

A lithium diorganocopper (Gilman reagent)

Primary, secondary, and even tertiary alkyl groups undergo the addition reaction, as do aryl and alkenyl groups. Alkynyl groups, however, react poorly in the conjugate addition process.

$$CH_3\overset{O}{\overset{\|}{C}}CH{=}CH_2 \xrightarrow[\text{2. } H_3O^+]{\text{1. } Li(CH_3)_2Cu\text{, ether}} CH_3\overset{O}{\overset{\|}{C}}CH_2CH_2CH_3$$

3-Buten-2-one **2-Pentanone (97%)**

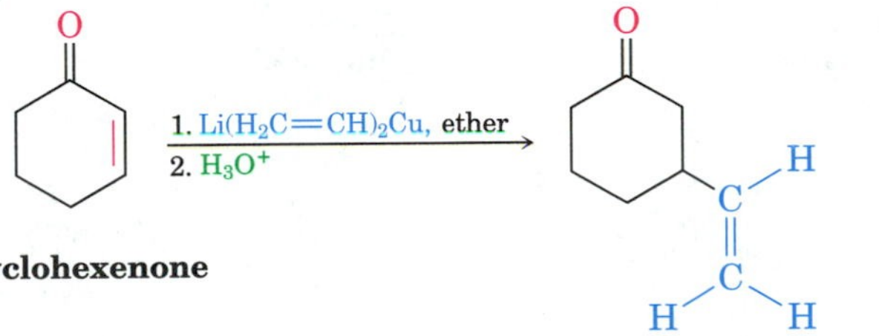

2-Cyclohexenone

3-Vinylcyclohexanone (65%)

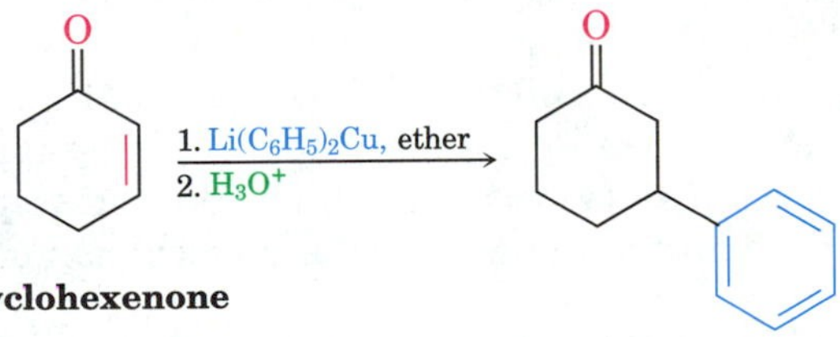

2-Cyclohexenone

3-Phenylcyclohexanone (70%)

Diorganocopper reagents are unique in their ability to give high yields of conjugate addition products. Other organometallic reagents, such as organomagnesiums (Grignard reagents) and organolithiums, normally give direct carbonyl addition on reaction with α,β-unsaturated ketones.

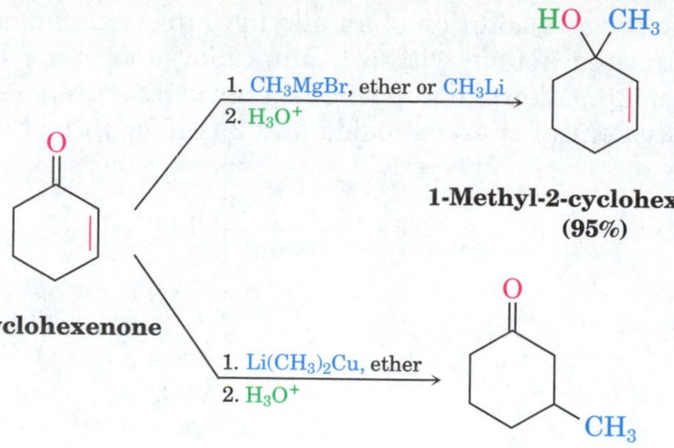

2-Cyclohexenone

1-Methyl-2-cyclohexen-1-ol (95%)

3-Methylcyclohexanone (97%)

The mechanism of diorganocopper addition is not fully understood but probably involves radicals. The reaction is not a typical polar process like other nucleophilic additions.

PROBLEM.

19.18 How might conjugate addition reactions of lithium diorganocopper reagents be used to synthesize the following compounds?

(a) 2-Heptanone

(b) 3,3-Dimethylcyclohexanone

(c) 4-*tert*-Butyl-3-ethylcyclohexanone

(d)

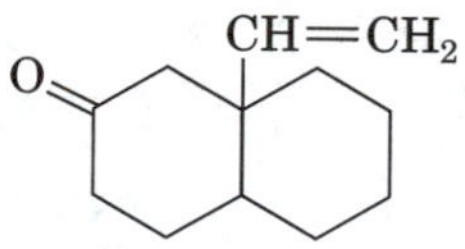

19.18 Some Biological Nucleophilic Addition Reactions

We'll see in Chapter 30 that living organisms use many of the same reactions that chemists use in the laboratory. This is particularly true of carbonyl-group reactions, where nucleophilic addition steps play a critical role in the biological synthesis of many vital molecules. For example, one of the pathways by which amino acids are made involves nucleophilic addition of ammonia to α-keto acids. To choose a specific example, the bacterium *Bacillus subtilis* synthesizes alanine from pyruvic acid and ammonia:

$$CH_3\overset{\overset{\Large O}{\|}}{C}COOH + :NH_3 \xrightarrow{\textit{B. subtilis}} CH_3\overset{\overset{\Large NH_2}{|}}{C}HCOOH$$

Pyruvic acid (an α-keto acid)

Alanine (an amino acid)

The key step in this biological transformation is the nucleophilic addition of ammonia to the ketone carbonyl group of pyruvic acid. The tetrahedral intermediate loses water to yield an imine, which is further reduced in a second nucleophilic addition step to yield alanine.

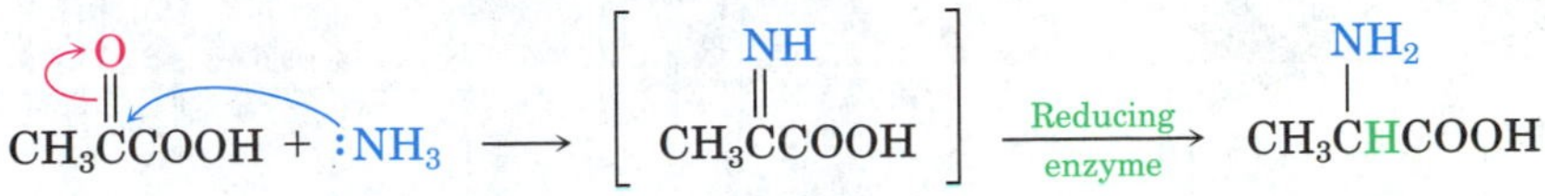

Pyruvic acid An imine **Alanine**

Another nucleophilic addition reaction—this time in reverse—plays an interesting role in the chemical defense mechanisms used by many plants and insects to protect themselves against predators. When the millipede *Apheloria corrugata* is attacked by ants, it secretes the cyanohydrin mandelonitrile and an enzyme that catalyzes the decomposition of mandelonitrile into benzaldehyde and HCN. The millipede actually protects itself by discharging poisonous HCN at would-be attackers.

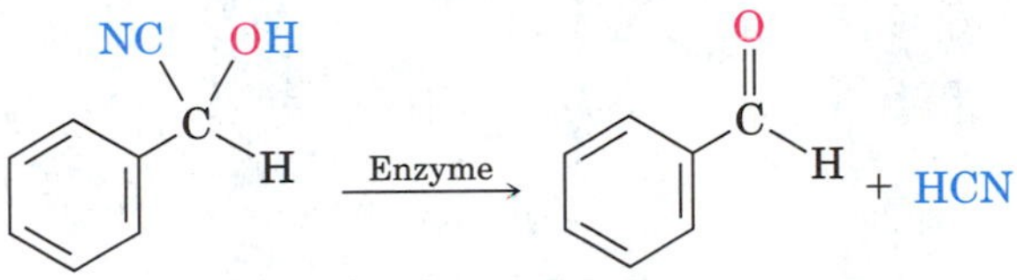

Mandelonitrile
(from *Apheloria corrugata*)

19.19 Spectroscopic Analysis of Ketones and Aldehydes

Infrared Spectroscopy

Ketones and aldehydes show a strong C=O bond absorption in the infrared region from 1660 to 1770 cm^{-1}, as the spectra of benzaldehyde and cyclohexanone demonstrate (Figure 19.16, p. 754). In addition, aldehydes show two characteristic C–H absorptions in the range 2720–2820 cm^{-1}. The exact position of the C=O absorption varies slightly from compound to compound but is highly diagnostic of the exact nature of the carbonyl group. Table 19.3 (p. 755) shows the correlation between the IR absorption maximum and carbonyl-group structure.

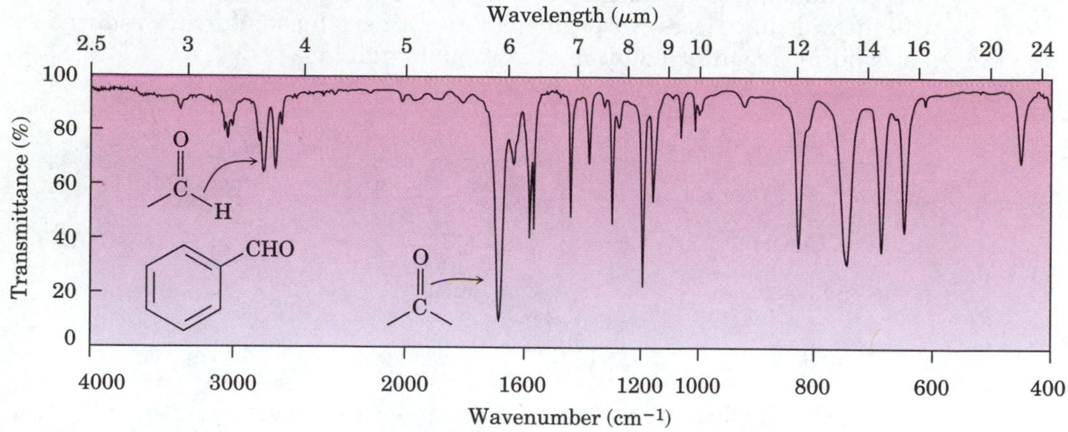

(a)

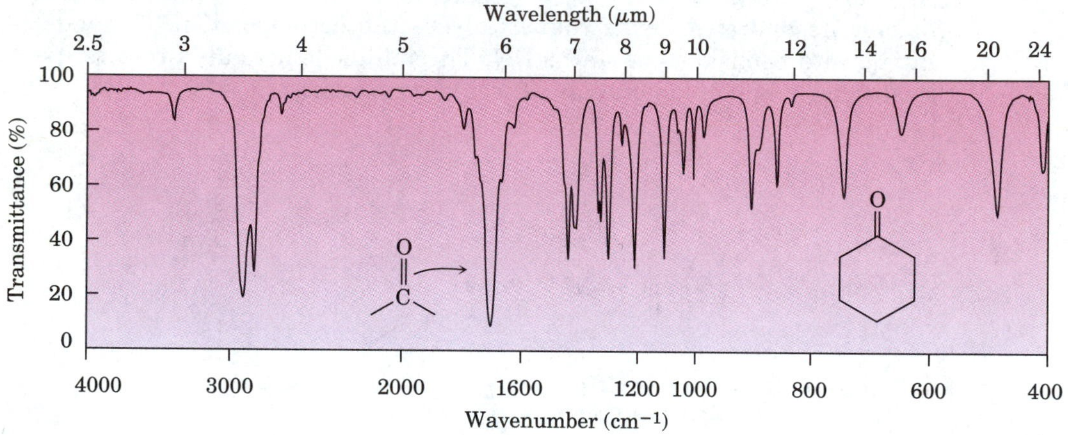

(b)

Figure 19.16 Infrared spectra of (a) benzaldehyde and (b) cyclohexanone.

As the data in Table 19.3 indicate, saturated aldehydes usually show carbonyl absorptions near 1730 cm^{-1} in the IR spectrum, but conjugation of the aldehyde to an aromatic ring or a double bond lowers the absorption by 25 cm^{-1} to near 1705 cm^{-1}. Saturated aliphatic ketones and cyclohexanones both absorb near 1715 cm^{-1}, and conjugation with a double bond or an aromatic ring again lowers the absorption by 30 cm^{-1} to 1685–1690 cm^{-1}. Angle strain in the carbonyl group caused by reducing the ring size of cyclic ketones to four or five raises the absorption position.

The values given in Table 19.3 are remarkably constant from one ketone to another or from one aldehyde to another. As a result, IR spectroscopy is

Table 19.3 Infrared Absorptions of Some Ketones and Aldehydes

Carbonyl type	*Example*	*Infrared absorption (cm^{-1})*
Aliphatic aldehyde	Acetaldehyde	1730
Aromatic aldehyde	Benzaldehyde	1705
α,β-Unsaturated aldehyde	$H_2C{=}CH{-}CHO$	1705
Aliphatic ketone	Acetone	1715
Six-membered-ring ketone	Cyclohexanone	1715
Five-membered-ring ketone	Cyclopentanone	1750
Four-membered-ring ketone	Cyclobutanone	1785
Aromatic ketone	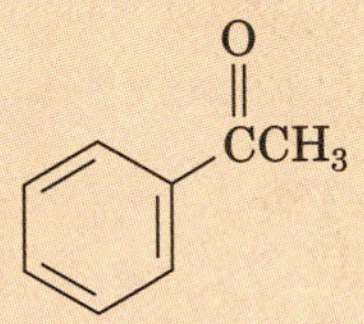	1690
α,β-Unsaturated ketone	$H_2C{=}CH\overset{\overset{O}{\|}}{C}CH_3$	1685

a powerful tool for diagnosing the nature and chemical environment of a carbonyl group in a molecule of unknown structure. An unknown that shows an IR absorption at 1730 cm^{-1} is almost certainly an aldehyde rather than a ketone; an unknown that shows an IR absorption at 1750 cm^{-1} is almost certainly a cyclopentanone, and so on.

PROBLEM. . . . .

19.19 How might you use IR spectroscopy to determine whether reaction between 2-cyclohexenone and lithium dimethylcopper gives the direct addition product or the conjugate addition product?

PROBLEM. . . . .

19.20 Where would you expect each of the following compounds to absorb in the IR spectrum?

(a) 4-Penten-2-one
(b) 3-Penten-2-one
(c) 2,2-Dimethylcyclopentanone
(d) *m*-Chlorobenzaldehyde
(e) 3-Cyclohexenone
(f) 2-Hexenal

PROBLEM

19.21 Acid-catalyzed dehydration of 3-hydroxy-3-phenylcyclohexanone leads to an unsaturated ketone. What possible structures are there for the product? At what position in the IR spectrum would you expect each to absorb? If the actual product has an absorption at 1670 cm^{-1}, what is its structure?

Nuclear Magnetic Resonance Spectroscopy

Proton (^1H) NMR is useful for analyzing aldehydes, though less so for ketones. Aldehyde protons (RCHO) absorb near 10 δ in the ^1H NMR spectrum and are very distinctive, since no other absorptions occur in this region. The aldehyde proton usually shows spin–spin coupling to neighbor protons, with coupling constant $J \approx 3$ Hz. Acetaldehyde, for example, shows a quartet at 9.8 δ for the aldehyde proton, indicating that there are three protons neighboring the –CHO group (Figure 19.17).

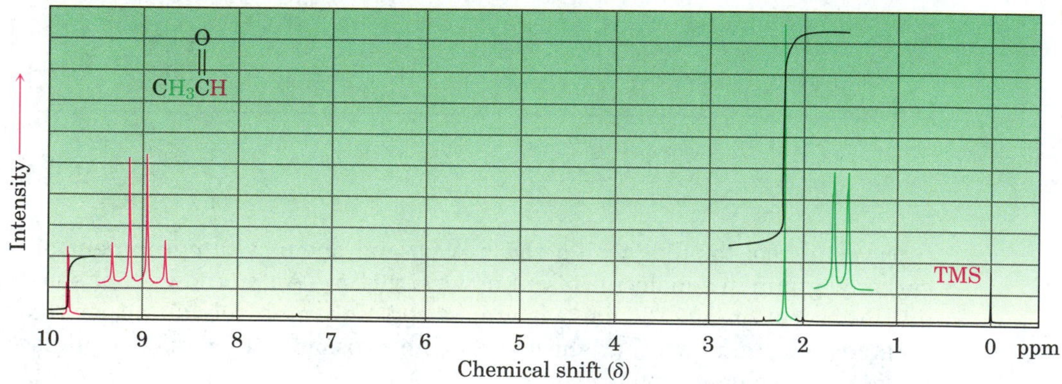

Figure 19.17 The ^1H NMR spectrum of acetaldehyde.

Hydrogens on the carbon next to a carbonyl group are slightly deshielded and normally absorb near 2.0–2.3 δ. (Note that the acetaldehyde methyl group in Figure 19.17 absorbs at 2.20 δ.) Methyl ketones are particularly distinctive because they show a large sharp three-proton singlet near 2.1 δ. Complex spin–spin splittings often obscure the absorption patterns of other ketones, however, and reduce the diagnostic usefulness of ^1H NMR.

Carbonyl-group carbon atoms show readily identifiable and highly characteristic ^{13}C NMR resonances in the range 190–215 δ. Since no other kinds of carbons absorb in this range, the presence of an NMR absorption near 200 δ is clear evidence for a carbonyl group. Saturated ketone or aldehyde carbons usually absorb in the region from 200 to 215 δ, while α,β-unsaturated carbonyl carbons absorb in the 190–200 δ region. Table 19.4 gives some examples.

Table 19.4 The ^{13}C NMR Absorptions of Some Ketones and Aldehydes

Carbonyl compound	Carbon-13 NMR absorption of >C=O (δ)
Acetaldehyde	201
Benzaldehyde	192
2-Butanone	207
Cyclohexanone	211
Acetophenone	196

Mass Spectrometry

Aliphatic ketones and aldehydes that have hydrogens on their gamma (γ) carbon atoms undergo a characteristic mass spectral cleavage called the **McLafferty[9] rearrangement**. A hydrogen atom is transferred from the γ carbon to the carbonyl oxygen, the bond between the α and β carbons is broken, and a neutral alkene fragment is produced. The charge remains with the oxygen-containing fragment.

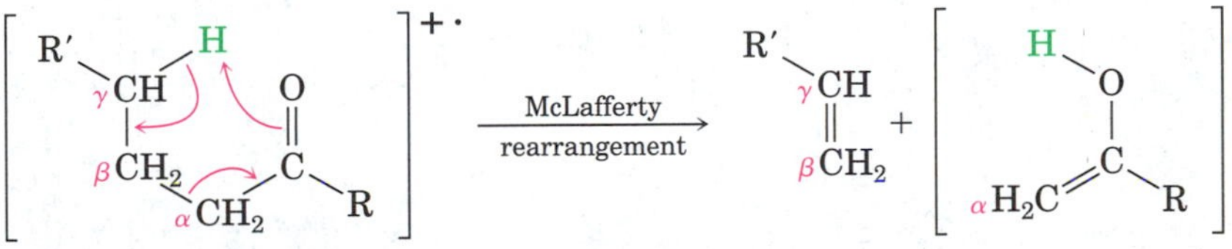

In addition to fragmentation by the McLafferty rearrangement, ketones and aldehydes also undergo cleavage of the bond between the carbonyl group and the α carbon, a so-called *α cleavage*. Alpha cleavage yields a neutral radical and an oxygen-containing cation.

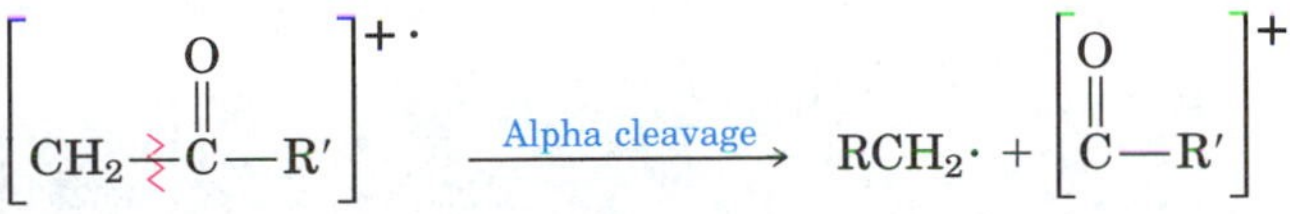

[9]Fred Warren McLafferty (1923–); b. Evanston, Ill.; Ph.D. (1950), Cornell University; Dow Chemical (1950–1964), professor, Purdue University (1964–1968), Cornell University (1968–).

Fragment ions from both α cleavage and McLafferty rearrangement are visible in the mass spectrum of 5-methyl-2-hexanone shown in Figure 19.18. Alpha cleavage occurs primarily at the more substituted side of the carbonyl group, leading to a $[CH_3CO]^+$ fragment with $m/z = 43$. McLafferty rearrangement and loss of 2-methylpropene yields a fragment with $m/z = 58$.

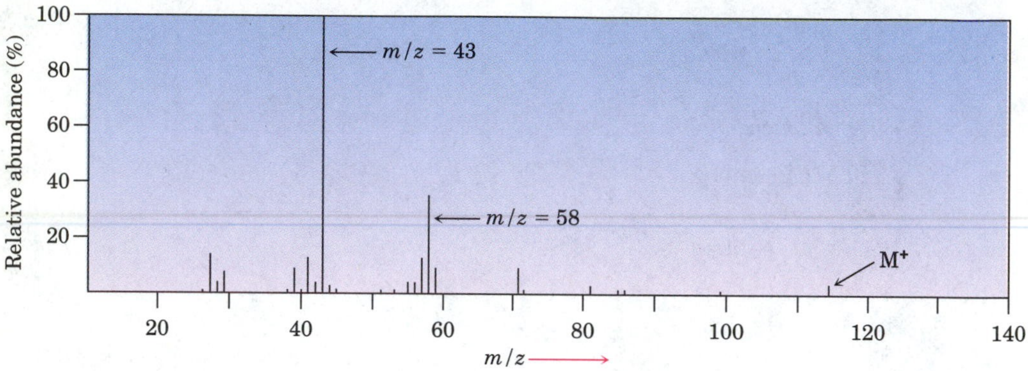

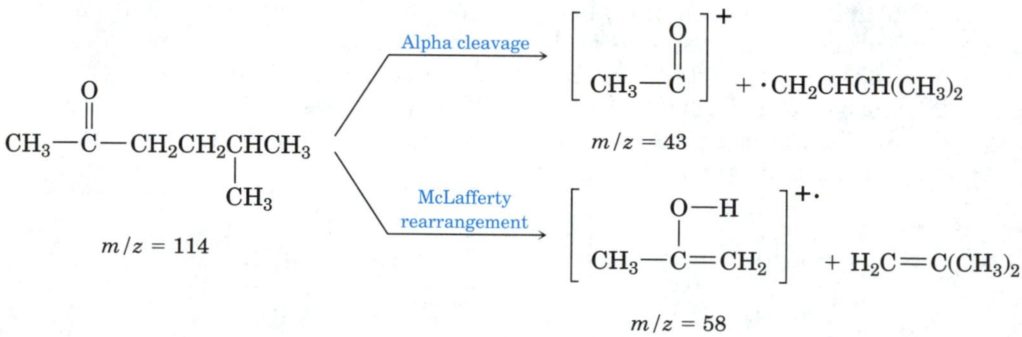

Figure 19.18 Mass spectrum of 5-methyl-2-hexanone. The abundant peak at $m/z = 43$ is due to α cleavage at the more highly substituted side of the carbonyl group. The peak at $m/z = 58$ is due to McLafferty rearrangement. Note that the peak due to the molecular ion is very small.

PROBLEM

19.22 How might you use mass spectrometry to distinguish between the following pairs of isomers?

(a) 3-Methyl-2-hexanone and 4-methyl-2-hexanone
(b) 3-Heptanone and 4-heptanone
(c) 2-Methylpentanal and 3-methylpentanal

INTERLUDE

Insect Antifeedants

Approximately 15% of the world's crops are lost each year to insects.

It has been estimated by the World Health Organization that approximately 15% of the world's crops are lost to insects each year and that more than 2 billion dollars is spent each year on crop insecticides. Though remarkably effective, the powerful, broad-spectrum insecticides in current use are far from an ideal solution for insect control because of their potential toxicity to animals and their lack of selectivity. Beneficial as well as harmful insects are killed indiscriminately.

One new approach to insect control, which promises to be more selective and less ecologically harmful than present-day insecticides, is through the use of *insect antifeedants*, substances that prevent an insect from eating but do not kill it directly. The idea of using antifeedants came from the observations of chemical ecologists that many plants have evolved elaborate and sophisticated chemical defenses against insects. Among these defenses, some plants contain chemicals that appear to block the ability or desire of an insect to feed. The insect often remains nearby, where it dies of starvation.

Most naturally occurring antifeedants are relatively complex molecules. Polygodial, for instance, is a dialdehyde active against African army worms, and ajugarin I shows activity against locusts. Both polygodial and ajugarin I are probably too complex to be manufactured economically, but recent research has discovered a number of substances that are structurally simpler than naturally occurring antifeedants yet retain potent activity. Among these substances synthesized in the laboratory is the tricyclic acetal shown below.

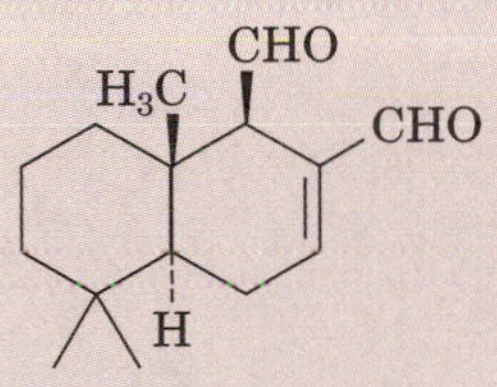

Polygodial

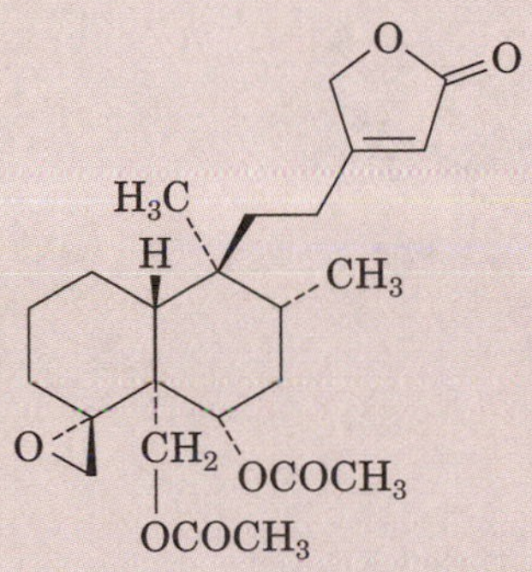

Ajugarin I

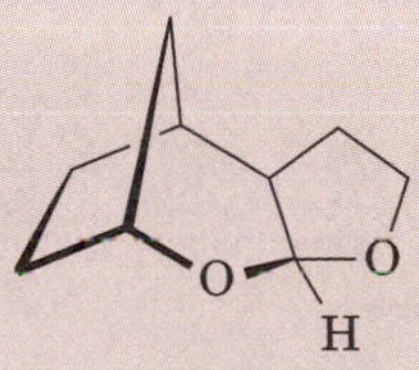

Synthetic compound

Summary and Key Words

Aldehydes and ketones are among the most important of all compounds, both in biochemistry and in the chemical industry. Aldehydes are normally prepared in the laboratory by oxidative cleavage of alkenes, by oxidation of primary alcohols, or by partial reduction of esters. Ketones are similarly prepared by oxidative cleavage of alkenes, by oxidation of secondary alcohols, or by addition of diorganocopper reagents to acid chlorides.

The **nucleophilic addition reaction** is the most common reaction of aldehydes and ketones. As shown in Figure 19.19, many different kinds of products can be prepared by nucleophilic additions. Ketones and aldehydes are reduced by $NaBH_4$ or $LiAlH_4$ to yield secondary and primary alcohols, respectively. Addition of Grignard reagents to ketones and aldehydes also gives alcohols (tertiary and secondary, respectively), and addition of HCN yields cyanohydrins. Primary amines add to carbonyl compounds yielding

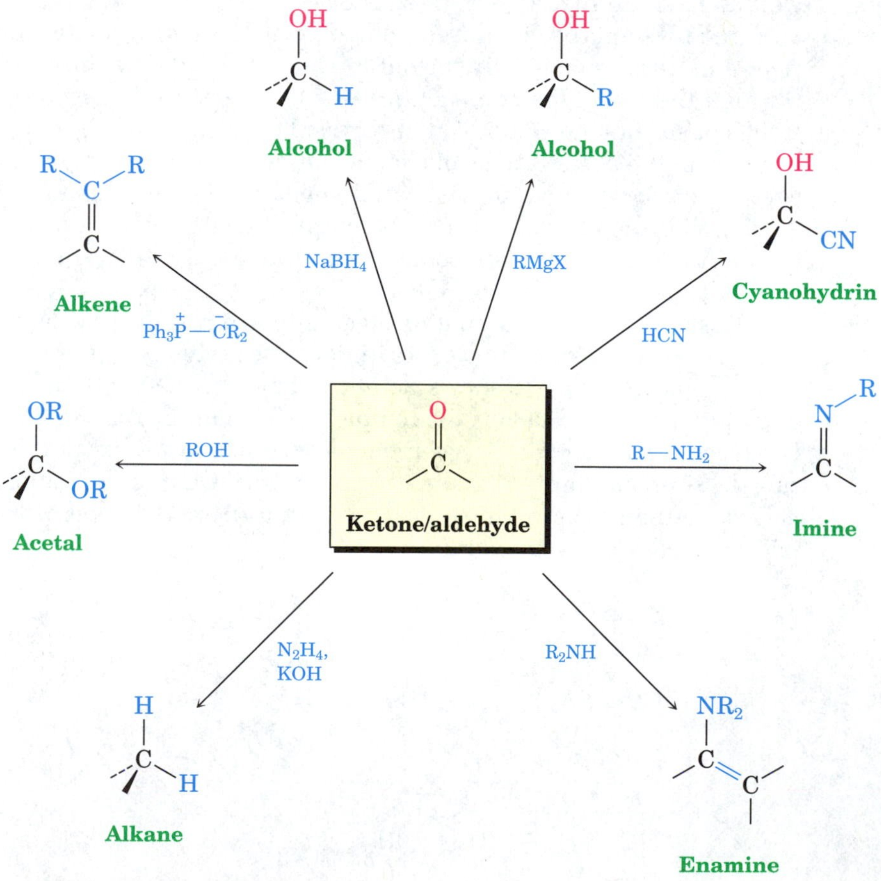

Figure 19.19 Some nucleophilic addition reactions of ketones and aldehydes.

imines, and secondary amines yield **enamines**. Reaction of a ketone or aldehyde with hydrazine and base yields an alkane (the **Wolff–Kishner reaction**). Alcohols add to carbonyl groups to yield **acetals**, which are valuable as protecting groups. Phosphoranes add to ketones and aldehydes, giving alkenes (the **Wittig reaction**) in which the new C=C– in the product is exactly where the C=O bond was in the starting material.

α,β-Unsaturated ketones and aldehydes often react with nucleophiles to give the product of **conjugate addition**, or **1,4 addition**. Particularly useful is the reaction with diorganocopper reagents, which results in the addition of an alkyl, aryl, or alkenyl group.

Ketones and aldehydes can be analyzed by IR and NMR spectroscopy. Carbonyl groups absorb in the IR range 1660–1770 cm^{-1}, with the exact position highly diagnostic of the kind of carbonyl group present in the molecule. Carbon-13 NMR spectroscopy is also useful for aldehydes and ketones because carbonyl carbons show resonances in the 190–215 δ range. Proton (^1H) NMR is useful largely for aldehydes because aldehyde –CHO protons absorb near 10 δ. Ketones and aldehydes undergo two characteristic kinds of fragmentation in the mass spectrometer: ***α* cleavage** and **McLafferty rearrangement**.

Summary of Reactions

1. Preparation of aldehydes (Section 19.3)
   (a) Oxidation of primary alcohols (Section 17.9)

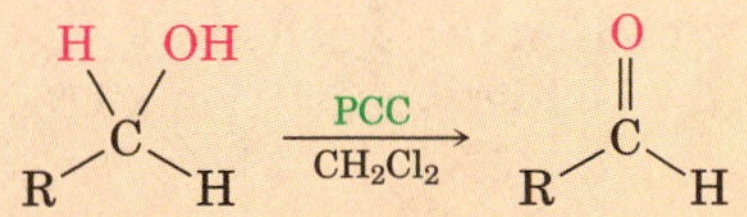

   (b) Ozonolysis of alkenes (Section 7.8)

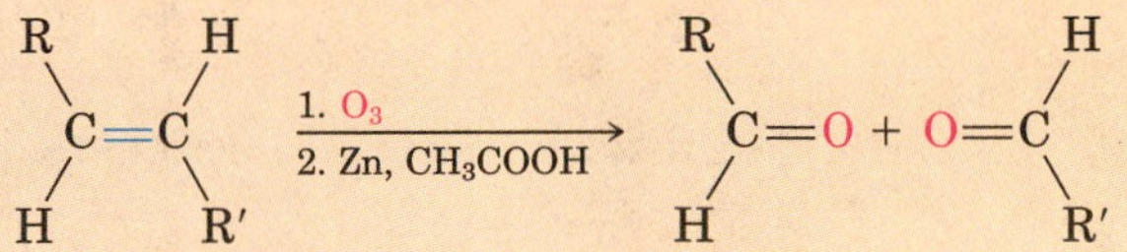

   (c) Partial reduction of esters (Section 19.3)

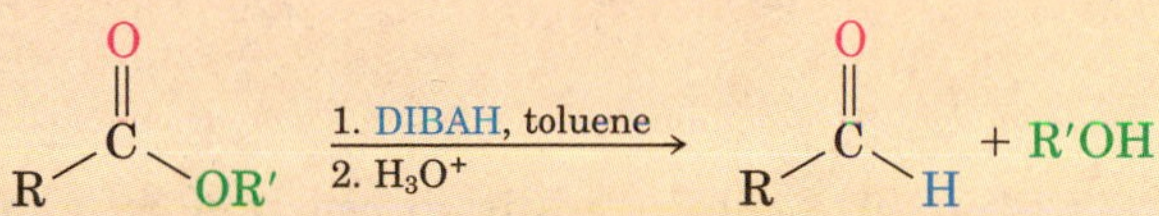

2. Preparation of ketones (Section 19.4)
   (a) Oxidation of secondary alcohols (Section 17.9)

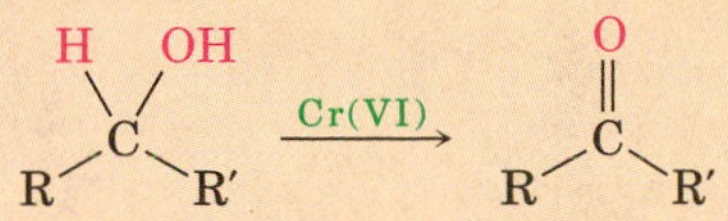

(continued)➤

(b) Ozonolysis of alkenes (Section 7.8)

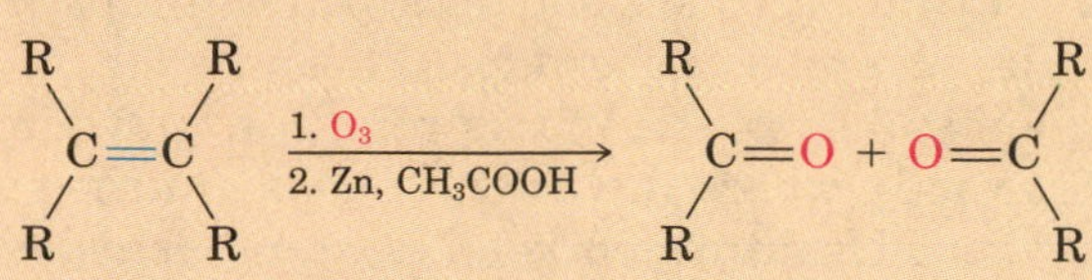

(c) Friedel–Crafts acylation (Section 16.4)

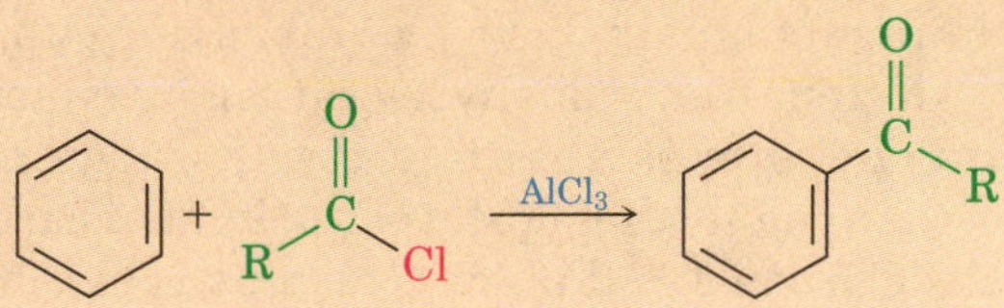

(d) Alkyne hydration (Section 8.5)

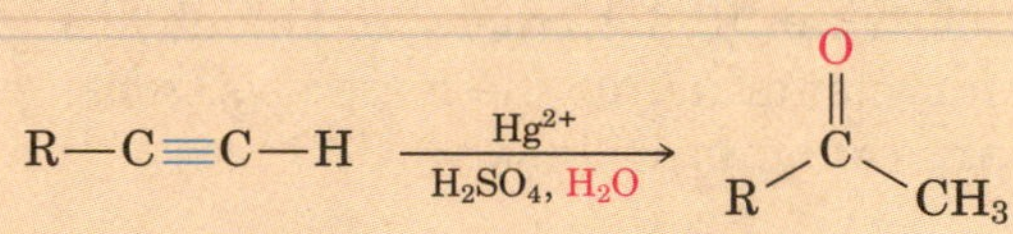

(e) Diorganocopper reaction with acid chlorides (Section 19.4)

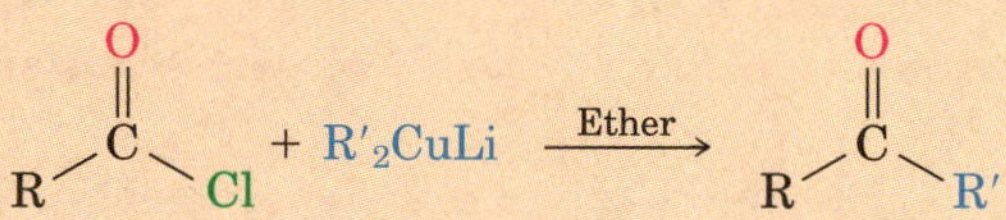

3. Reactions of aldehydes
   (a) Oxidation (Section 19.5)

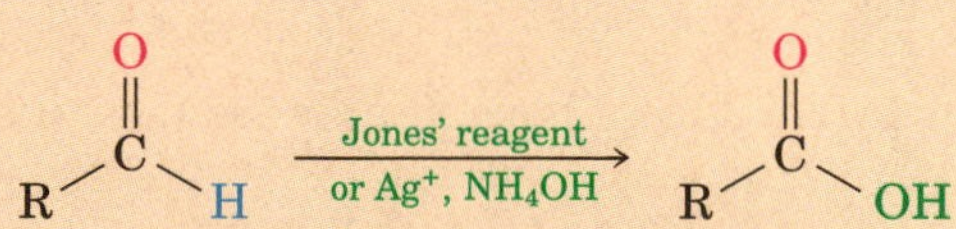

   (b) Cannizzaro reaction (Section 19.16)

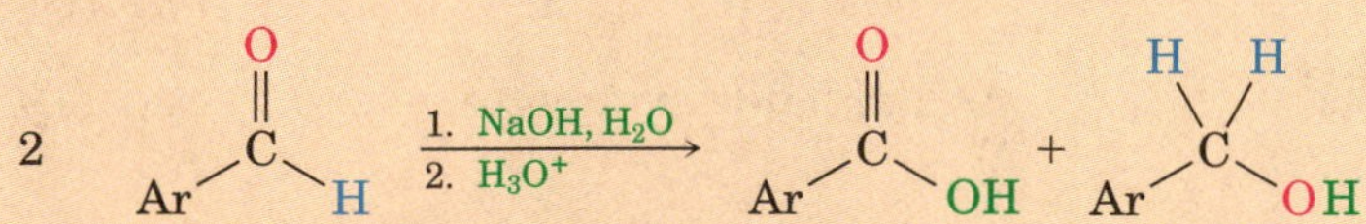

4. Nucleophilic addition reactions of aldehydes and ketones
   (a) Addition of hydride: reduction (Section 19.11)

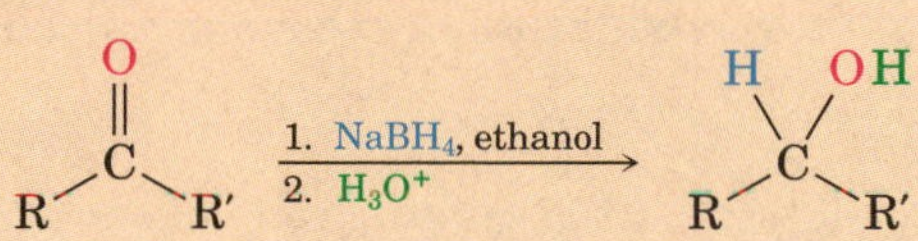

(continued)➤

(b) Addition of Grignard reagents (Section 19.10)

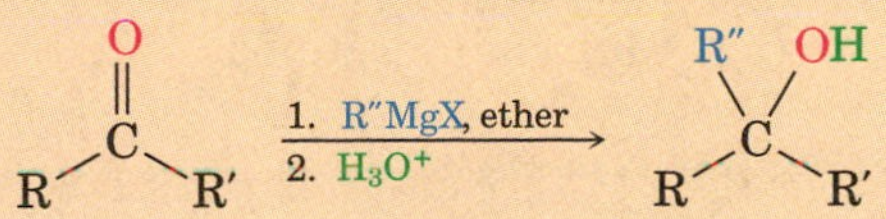

(c) Addition of HCN: Cyanohydrins (Section 19.9)

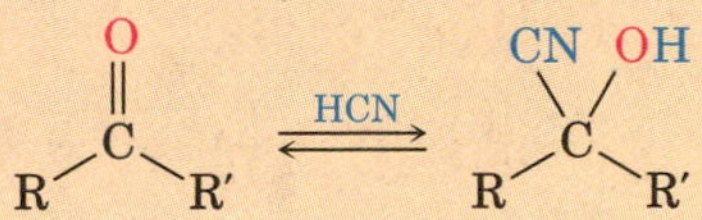

(d) Addition of primary amines: Imines (Section 19.12)

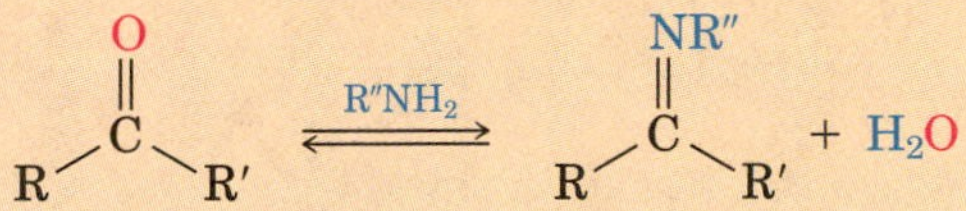

For example: oximes, $R_2C{=}N{-}OH$
semicarbazones, $R_2C{=}N{-}NHCONH_2$
2,4-dinitrophenylhydrazones,
$R_2C{=}N{-}NH{-}C_6H_4(NO_2)_2$

(e) Addition of secondary amines: Enamines (Section 19.12)

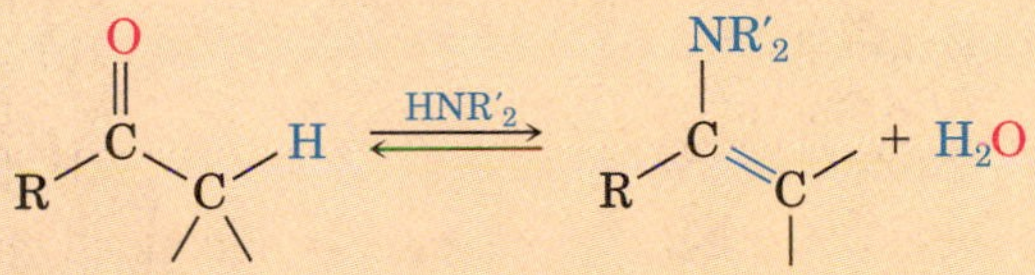

(f) Wolff–Kishner reaction (hydrazine addition) (Section 19.13)

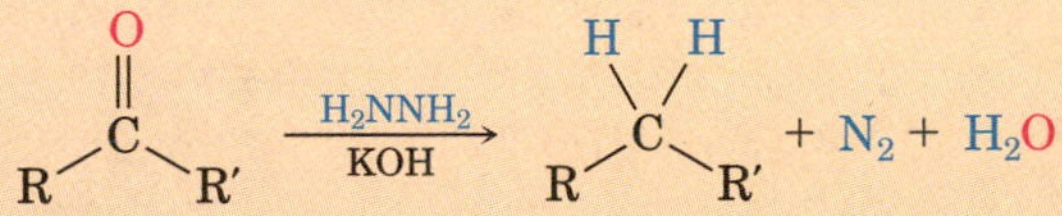

(g) Addition of alcohols: Acetals (Section 19.14)

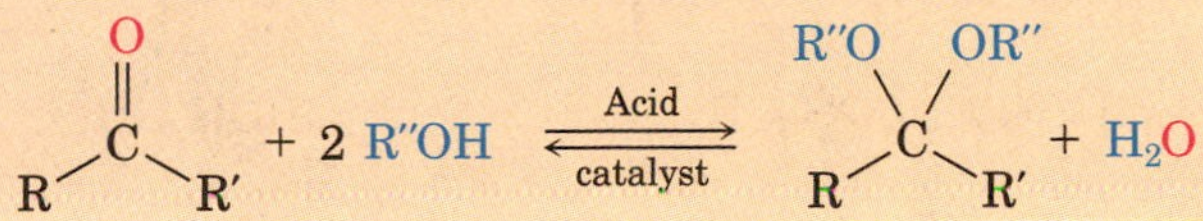

(h) Addition of phosphorus ylides: Wittig reaction (Section 19.15)

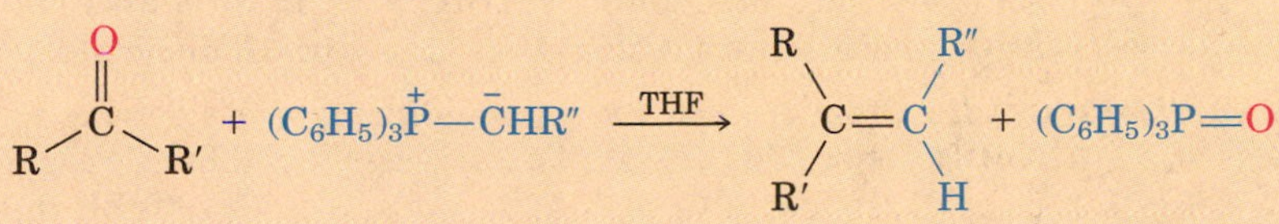

(continued)➤

5. Conjugate additions to α,β-unsaturated ketones and aldehydes (Section 19.17)
   (a) Addition of amines

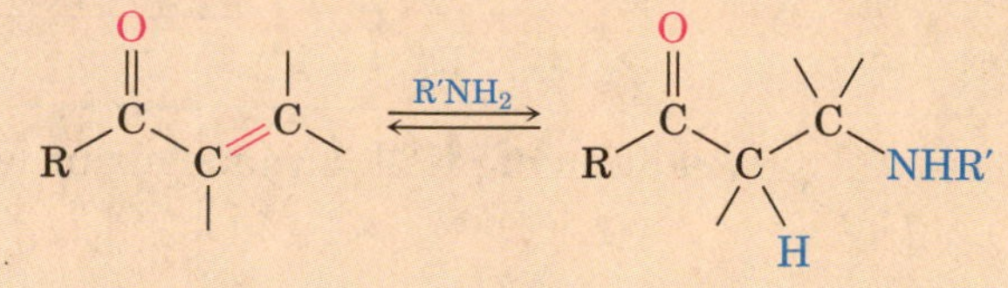

   (b) Addition of alkyl groups: Diorganocopper reaction

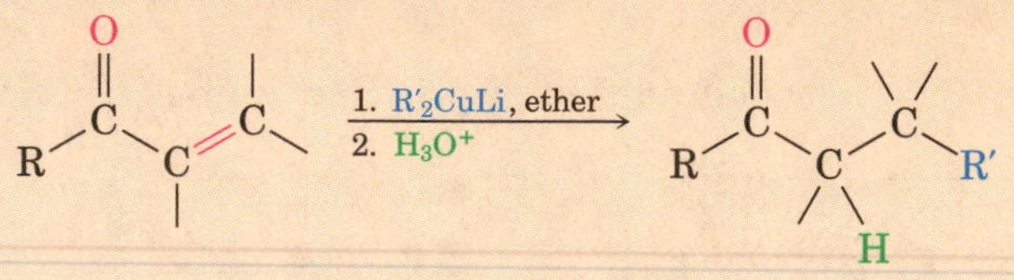

ADDITIONAL PROBLEMS

19.23 Draw structures corresponding to the following names:
(a) Bromoacetone
(b) 3,5-Dinitrobenzenecarbaldehyde
(c) 2-Methyl-3-heptanone
(d) 3,5-Dimethylcyclohexanone
(e) 2,2,4,4-Tetramethyl-3-pentanone
(f) 4-Methyl-3-penten-2-one
(g) Butanedial
(h) 3-Phenyl-2-propenal
(i) 6,6-Dimethyl-2,4-cyclohexadienone
(j) *p*-Nitroacetophenone
(k) (*S*)-2-Hydroxypropanal
(l) (2*S*,3*R*)-2,3,4-Trihydroxybutanal

19.24 Draw and name the seven ketones and aldehydes with the formula $C_5H_{10}O$.

19.25 Give IUPAC names for the following structures:

(a)
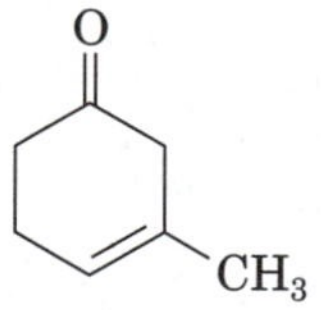

(b)
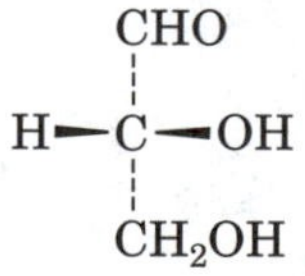

(c)
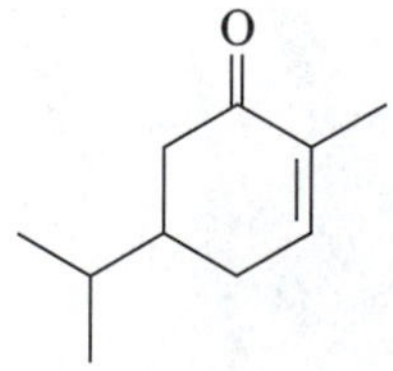

(d)
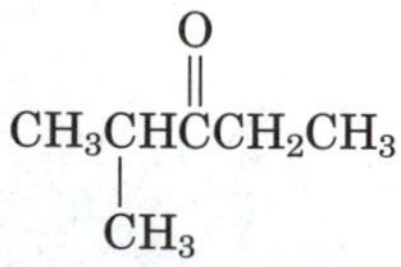

(e)
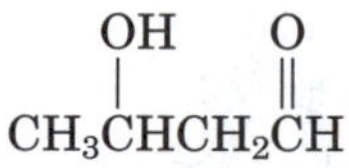

(f)
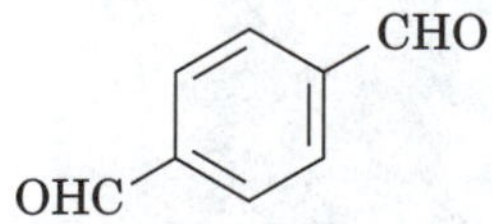

19.26 Give structures that fit the following descriptions:
(a) An α,β-unsaturated ketone, C_6H_8O
(b) An α-diketone
(c) An aromatic ketone, $C_9H_{10}O$
(d) A diene aldehyde, C_7H_8O

19.27 Predict the products of the reaction of phenylacetaldehyde with these reagents:
(a) $NaBH_4$, then H_3O^+
(b) Tollens' reagent
(c) NH_2OH, HCl catalyst
(d) CH_3MgBr, then H_3O^+
(e) CH_3OH, HCl catalyst
(f) H_2NNH_2, KOH
(g) $(C_6H_5)_3P{=}CH_2$
(h) HCN, KCN

19.28 Answer Problem 19.27 for reaction with acetophenone.

19.29 How would you prepare the following substances from 2-cyclohexenone? More than one step may be required.

(a) 

(b) 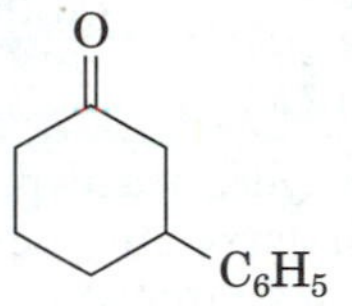

(c) 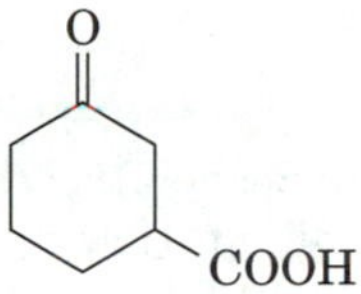

(d) 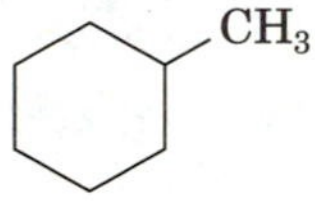
(Two ways)

19.30 How can you account for the fact that glucose reacts with the Tollens reagent to give a silver mirror, but glucose α-methyl glycoside does not?

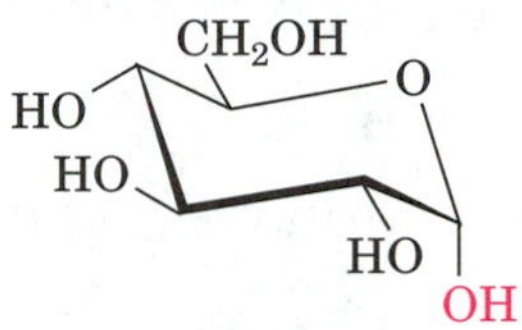

Glucose

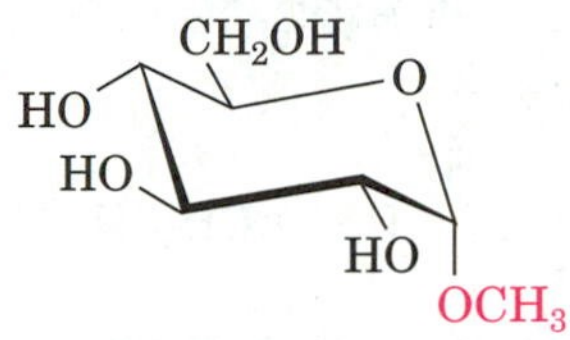

Glucose α-methyl glycoside

19.31 Reaction of 2-butanone with $NaBH_4$ yields a chiral product. What stereochemistry does the product have? Is it optically active?

19.32 Show how the Wittig reaction might be used to prepare the following alkenes. Identify the alkyl halide and the carbonyl components that would be used.

(a) $C_6H_5CH{=}CH{-}CH{=}CHC_6H_5$

(b) 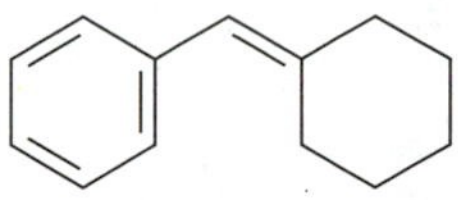

(c) 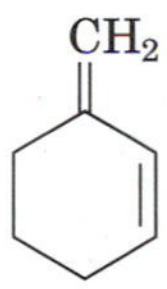

(d) 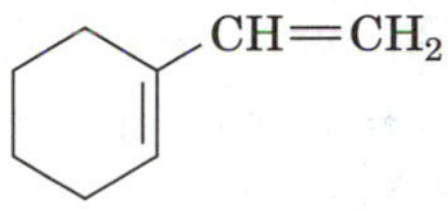

19.33 Why do you suppose tri*phenyl*phosphine rather than, say, tri*methyl*phosphine, is used to prepare Wittig reagents? What problems might you run into if trimethylphosphine were used?

19.34 How would you use a Grignard reaction on a ketone or aldehyde to synthesize the following compounds?

(a) 2-Pentanol
(b) 1-Butanol
(c) 1-Phenylcyclohexanol
(d) Diphenylmethanol

19.35 Aldehydes can be prepared by the Wittig reaction using (methoxymethylene)triphenylphosphorane as the Wittig reagent and then hydrolyzing the product with acid. For example,

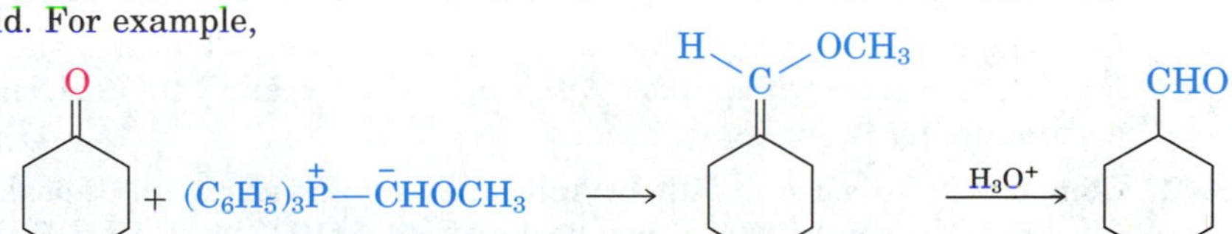

(Methoxymethylene)-triphenylphosphorane

(a) How would you prepare the required phosphorane?
(b) Propose a mechanism for the hydrolysis step.

19.36 When 4-hydroxybutanal is treated with methanol in the presence of an acid catalyst, 2-methoxytetrahydrofuran is formed. Explain.

$$HOCH_2CH_2CH_2CHO \xrightarrow[H^+]{CH_3OH}$$

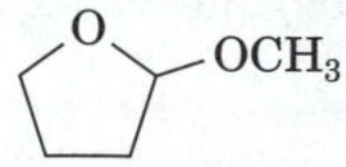

19.37 We saw in Section 19.7 that aldehydes are more reactive than ketones toward nucleophilic addition. How might you carry out the following selective transformations? One of the two schemes requires a protection step.

(a) $CH_3\overset{O}{\overset{\|}{C}}CH_2CH_2CH_2\overset{O}{\overset{\|}{C}}H \longrightarrow CH_3\overset{O}{\overset{\|}{C}}CH_2CH_2CH_2CH_2OH$

(b) $CH_3\overset{O}{\overset{\|}{C}}CH_2CH_2CH_2\overset{O}{\overset{\|}{C}}H \longrightarrow CH_3\overset{OH}{\overset{|}{C}}HCH_2CH_2CH_2\overset{O}{\overset{\|}{C}}H$

19.38 How would you synthesize the following substances from benzaldehyde and any other reagents needed?

(a)

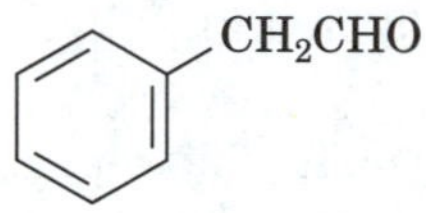

(b)

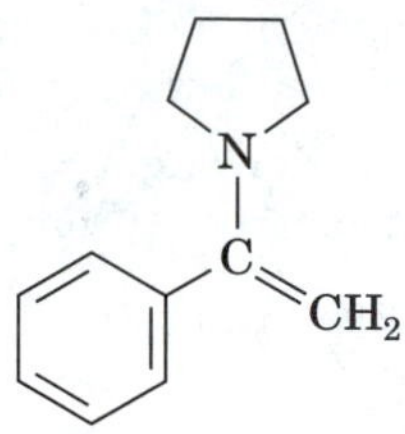

(c)

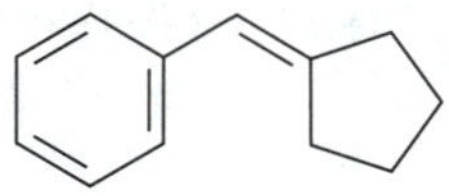

19.39 Give three methods for reducing a carbonyl group to a methylene group, $R_2C{=}O \rightarrow R_2CH_2$. What are the advantages and disadvantages of each?

19.40 Carvone is the major constituent of spearmint oil. What products would you expect from reaction of carvone with the following reagents?

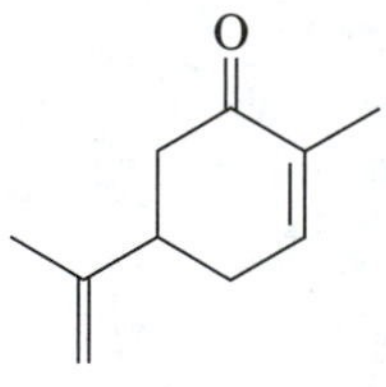

Carvone

(a) $(CH_3)_2Cu^+\ Li^-$, then H_3O^+
(b) $LiAlH_4$, then H_3O^+
(c) CH_3NH_2
(d) C_6H_5MgBr, then H_3O^+
(e) H_2/Pd
(f) Jones' reagent
(g) $(C_6H_5)_3\overset{+}{P}—\overset{-}{C}HCH_3$
(h) $HOCH_2CH_2OH$, HCl

19.41 Compound A, MW = 86, shows an IR absorption at 1730 cm^{-1} and a very simple ^1H NMR spectrum with peaks at 9.7 δ (1 H, singlet) and 1.2 δ (9 H, singlet). Propose a structure for A.

19.42 Compound B is isomeric with A (Problem 19.41) and shows an IR peak at 1715 cm^{-1}. The ^1H NMR spectrum of B has peaks at 2.4 δ (1 H, septet, J = 7 Hz), 2.1 δ (3 H, singlet), and 1.2 δ (6 H, doublet, J = 7 Hz). What is the structure of B?

19.43 How would you synthesize the following compounds from cyclohexanone?
(a) 1-Methylcyclohexene
(b) 2-Phenylcyclohexanone
(c) *cis*-1,2-Cyclohexanediol
(d) 1-Cyclohexylcyclohexanol

19.44 At what position would you expect to observe IR absorptions for the following molecules?

(a)

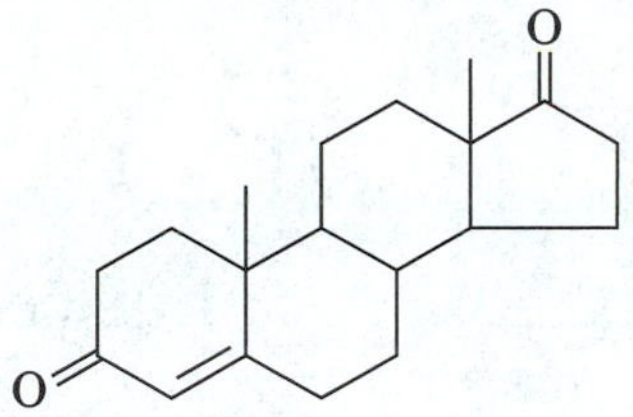

4-Androstene-3,17-dione

(b)

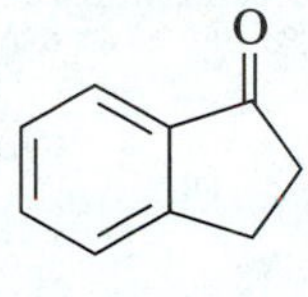

1-Indanone

(c)

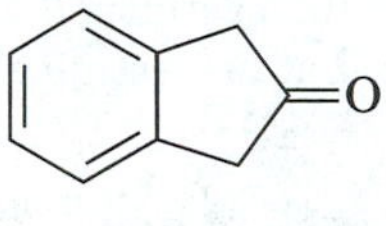

2-Indanone

(d)

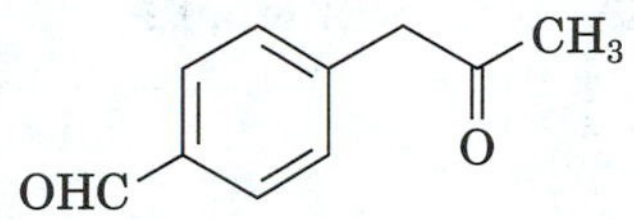

19.45 Each of the following reaction schemes contains one or more flaws. What is wrong in each case? How would you correct each scheme?

(a)

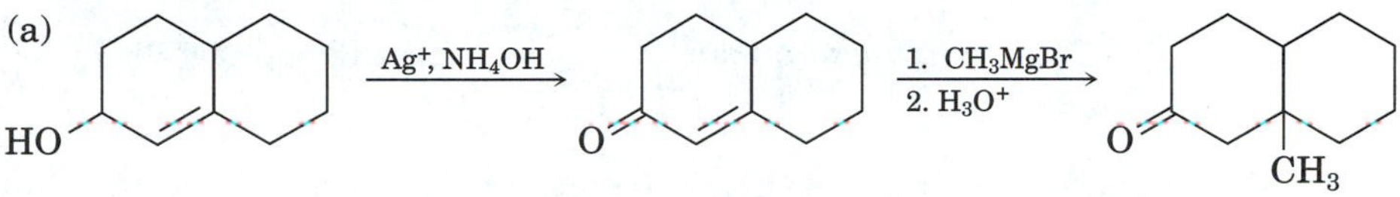

(b) $C_6H_5CH{=}CHCH_2OH \xrightarrow[\text{reagent}]{\text{Jones'}} C_6H_5CH{=}CHCHO \xrightarrow{H^+,\ CH_3OH} C_6H_5CH{=}CHCH(OCH_3)_2$

(c)

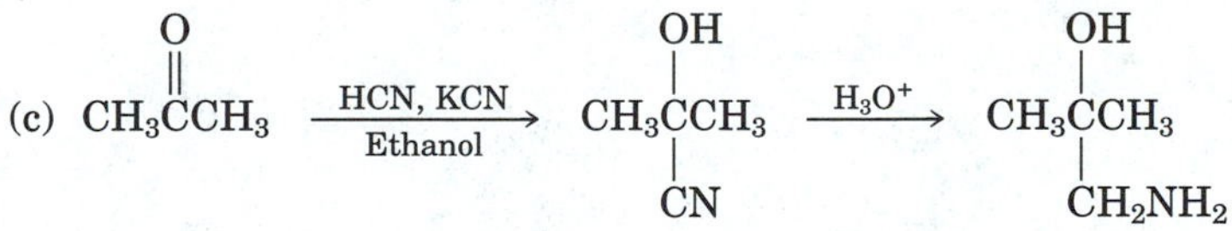

19.46 6-Methyl-5-hepten-2-one is a constituent of many essential oils, particularly the lemongrass species. How could you synthesize this natural product from methyl 4-oxopentanoate?

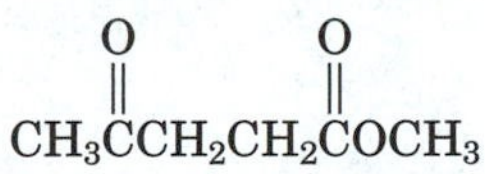

Methyl 4-oxopentanoate

19.47 Ketones react with dimethylsulfonium methylide to yield epoxides. Suggest a mechanism for the reaction.

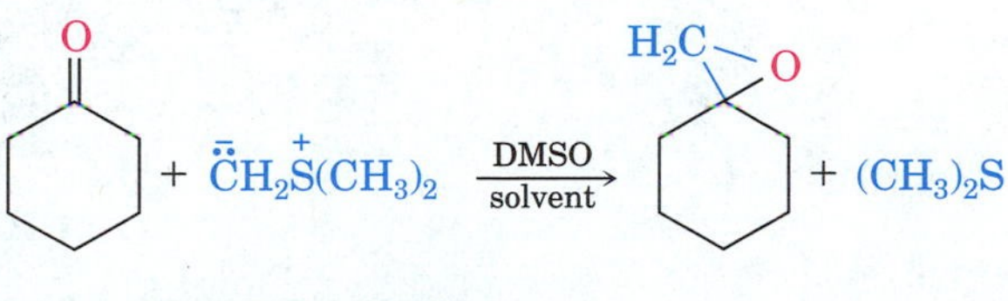

Dimethylsulfonium methylide

19.48 When cyclohexanone is heated in the presence of a large amount of acetone cyanohydrin and a small amount of base, cyclohexanone cyanohydrin and acetone are formed. Propose a mechanism.

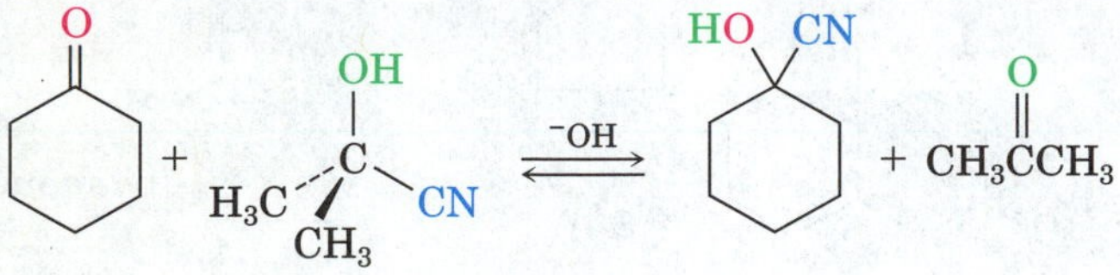

19.49 Treatment of an alcohol with dihydropyran yields an acetal called a tetrahydropyranyl ether, a reaction that can be used as a method of protecting alcohols (Section 17.10). Show the mechanism of the reaction.

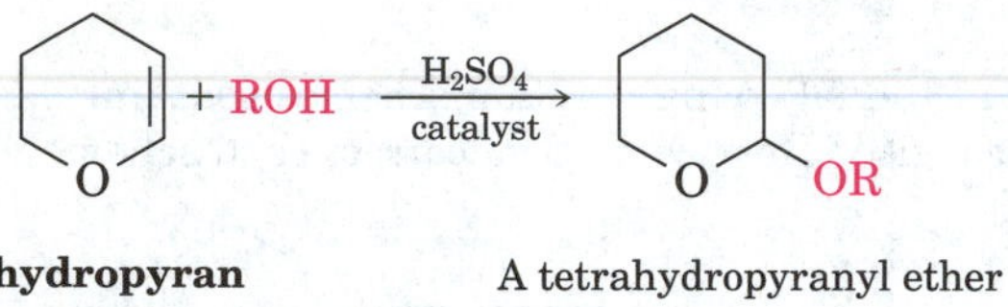

19.50 Tamoxifen is a drug used in the treatment of breast cancer. How would you prepare tamoxifen from benzene, the following ketone, and any other reagents needed?

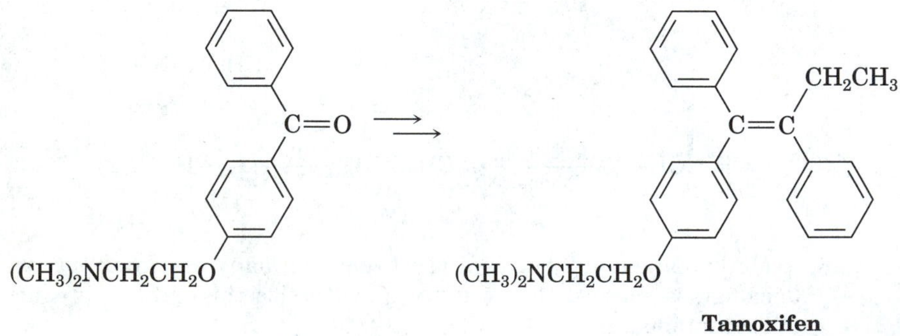

19.51 Paraldehyde, a sedative and hypnotic agent, is prepared by treatment of acetaldehyde with an acidic catalyst. Propose a mechanism for the reaction.

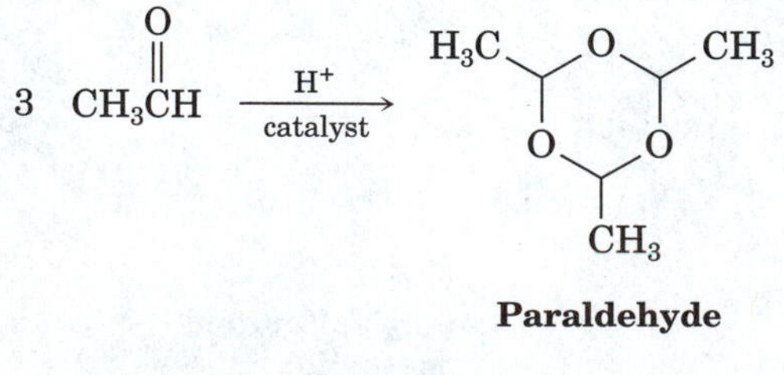

19.52 The ¹H NMR spectrum shown is that of a compound with formula $C_9H_{10}O$. How many double bonds and/or rings does this compound contain? If the unknown has an IR absorption at 1690 cm^{-1}, what is a likely structure?

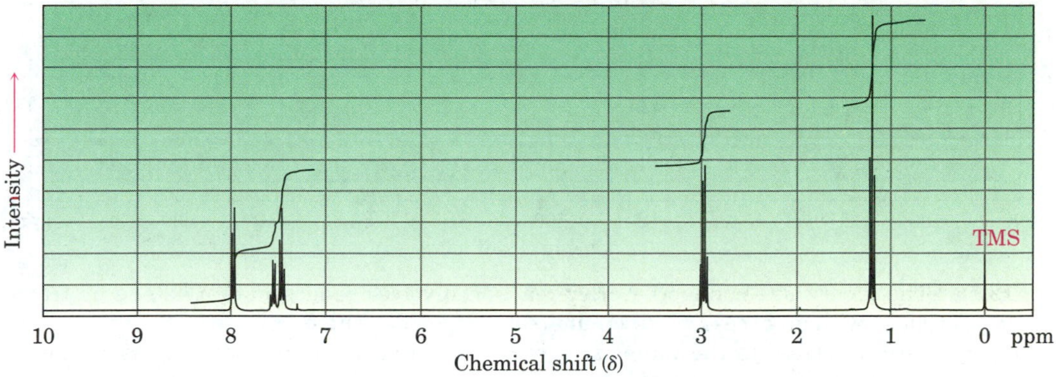

19.53 The ¹H NMR spectrum shown is that of a compound isomeric with the one in Problem 19.52. This isomer has an IR absorption at 1725 cm^{-1}. Propose a structure.

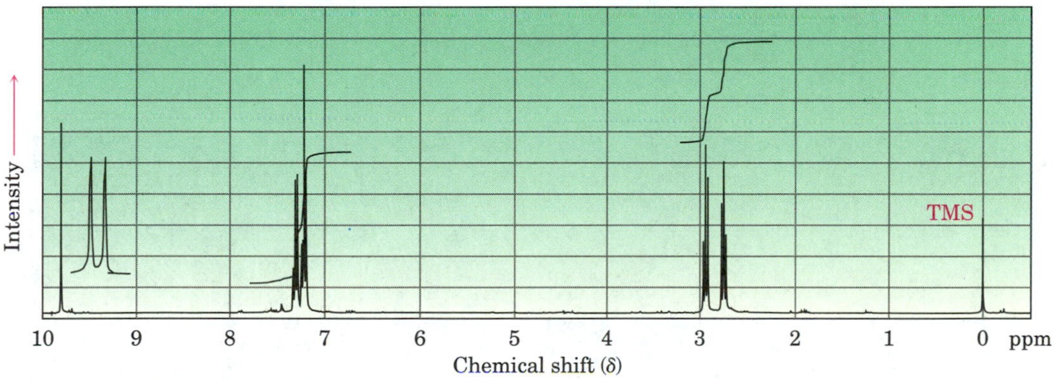

19.54 Propose structures for molecules that meet the following descriptions. Assume that the kinds of carbons (1°, 2°, 3°, or 4°) have been assigned by DEPT-NMR.

(a) $C_6H_{12}O$:
IR: 1715 cm^{-1}
^{13}C NMR: 8.0 δ (1°), 18.5 δ (1°), 33.5 δ (2°), 40.6 δ (3°), 214.0 δ (4°)

(b) $C_5H_{10}O$:
IR: 1725 cm^{-1}
^{13}C NMR: 22.6 δ (1°), 23.6 δ (3°), 52.8 δ (2°), 202.4 δ (3°)

(c) C_6H_8O:
IR: 1680 cm^{-1}
^{13}C NMR: 22.9 δ (2°), 25.8 δ (2°), 38.2 δ (2°), 129.8 δ (3°), 150.6 δ (3°), 198.7 δ (4°)

19.55 Compound A, $C_8H_{10}O_2$, has an intense IR absorption at 1750 cm^{-1} and gives the ^{13}C NMR spectrum shown at the top of the next page. Propose a structure for A.

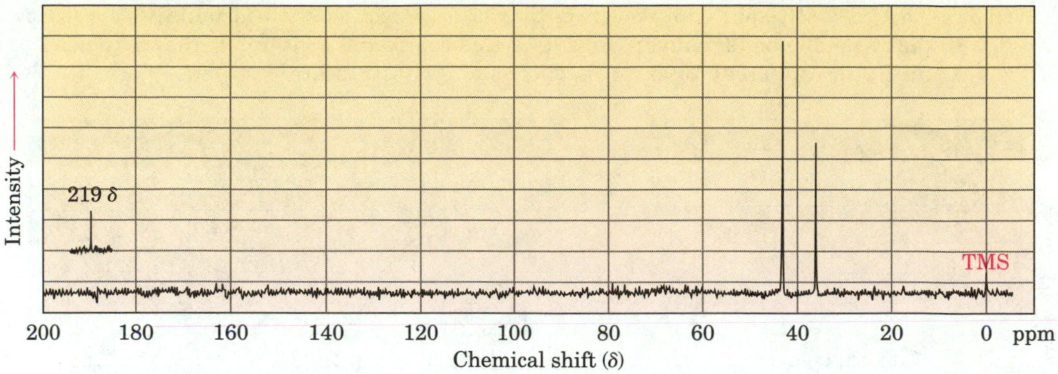

19.56 The Meerwein–Ponndorf–Verley reaction involves reduction of a ketone by treatment with an excess of aluminum triisopropoxide. The mechanism of the process is closely related to the Cannizzaro reaction in that a hydride ion acts as a leaving group. Propose a reasonable mechanism.

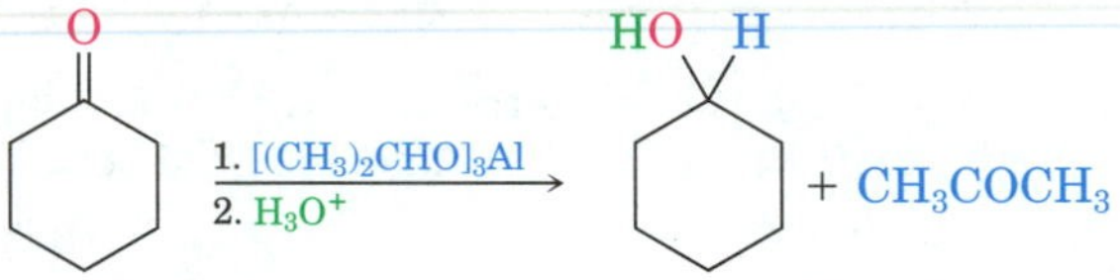

19.57 Propose structures for ketones or aldehydes that have the given ^{1}H NMR spectra.

(a) C_4H_7ClO
IR: 1715 cm^{-1}

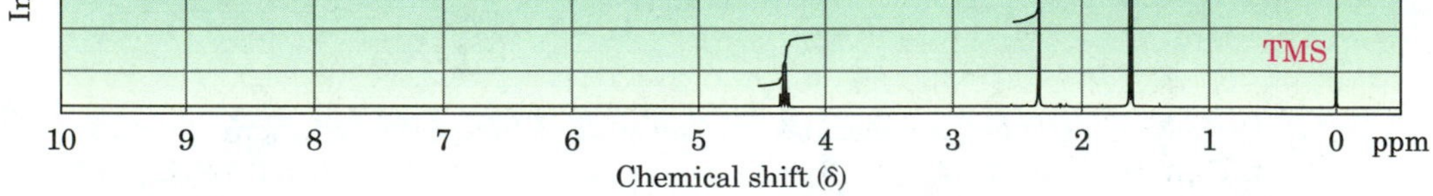

(b) $C_7H_{14}O$
IR: 1710 cm^{-1}

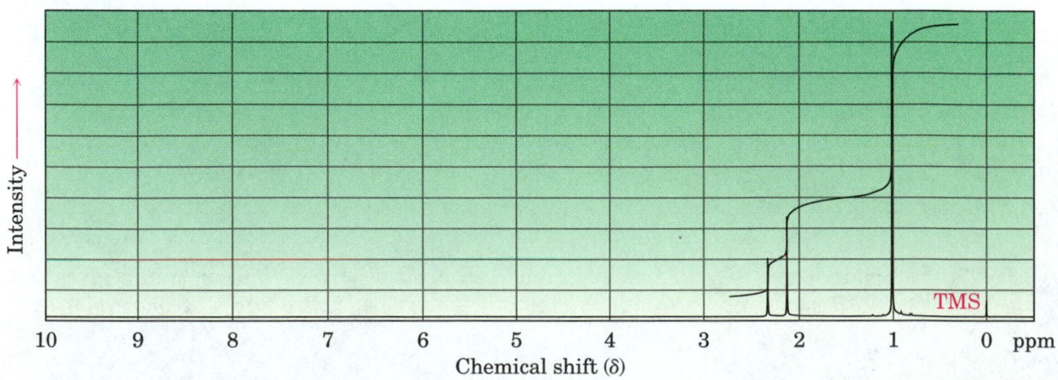

(c) $C_9H_{10}O_2$
IR: 1695 cm^{-1}

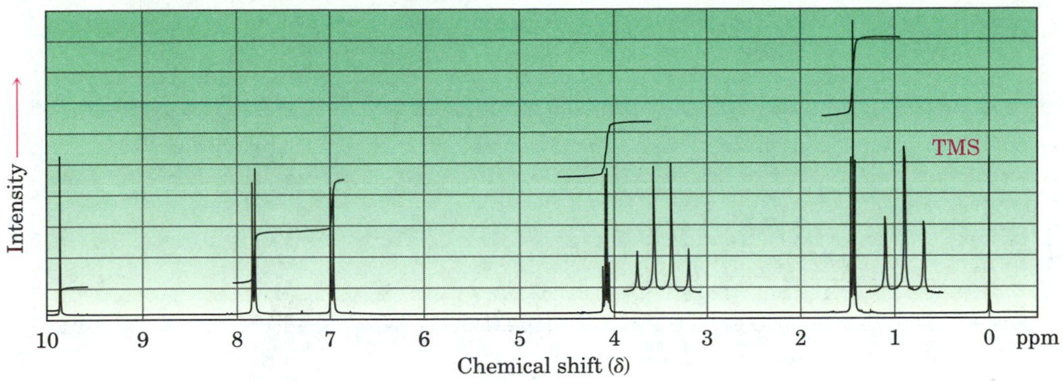

19.58 Propose structures for ketones or aldehydes that have the given ¹H NMR spectra.
(a) $C_{10}H_{12}O$
IR: 1710 cm^{-1}

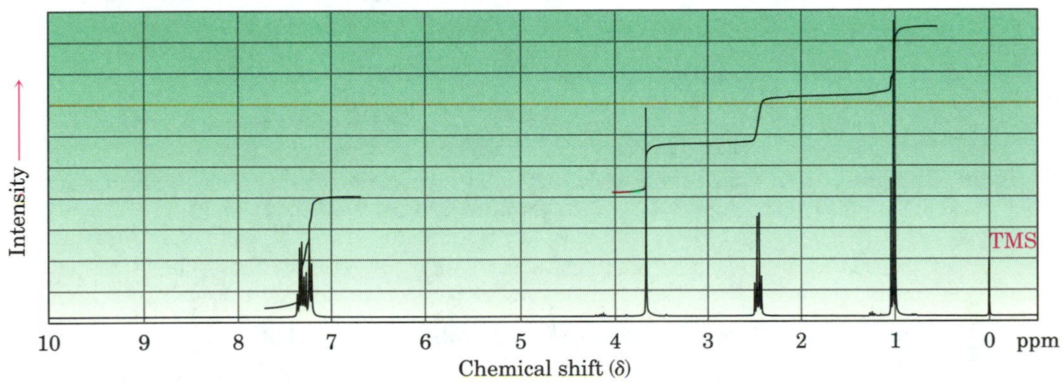

(b) $C_6H_{12}O_3$
IR: 1715 cm^{-1}

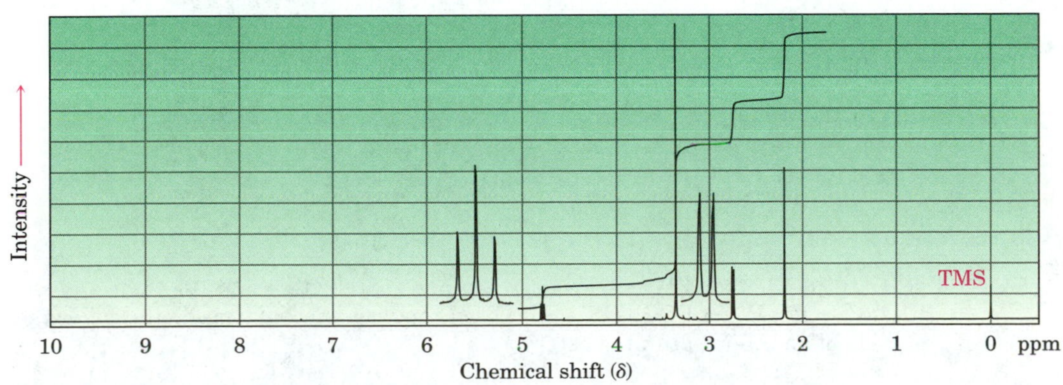

(c) C_4H_6O
IR: 1690 cm^{-1}

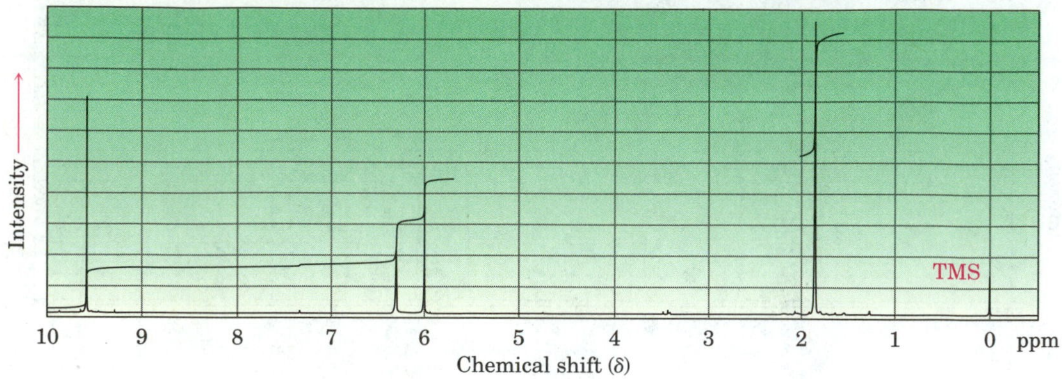

19.59 Propose a mechanism to account for the formation of 3,5-dimethylpyrazole from hydrazine and 2,4-pentanedione. Look carefully to see what has happened to each carbonyl carbon in going from starting material to product.

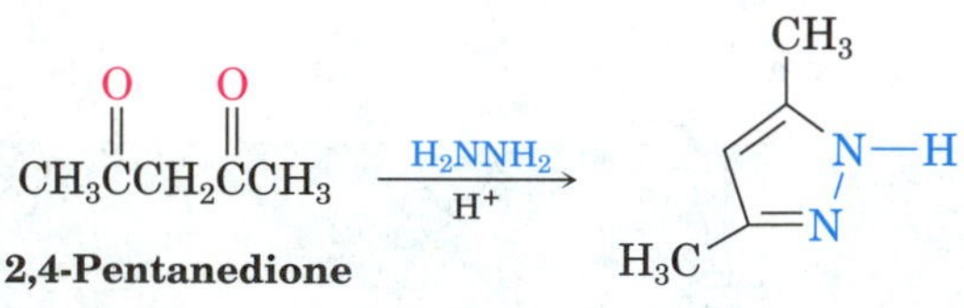

19.60 In light of your answer to Problem 19.59, propose a mechanism for the formation of 3,5-dimethylisoxazole from hydroxylamine and 2,4-pentanedione.

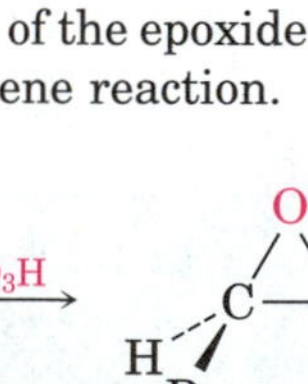

3,5-Dimethylisoxazole

19.61 Trans alkenes are converted into their cis isomers and vice versa on epoxidation followed by treatment of the epoxide with triphenylphosphine. Propose a mechanism for the epoxide → alkene reaction.

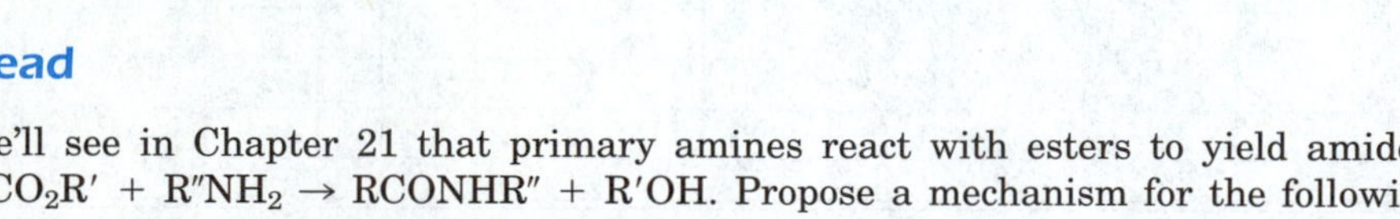

A Look Ahead

19.62 We'll see in Chapter 21 that primary amines react with esters to yield amides: $RCO_2R' + R''NH_2 \rightarrow RCONHR'' + R'OH$. Propose a mechanism for the following reaction of an α,β-unsaturated ester:

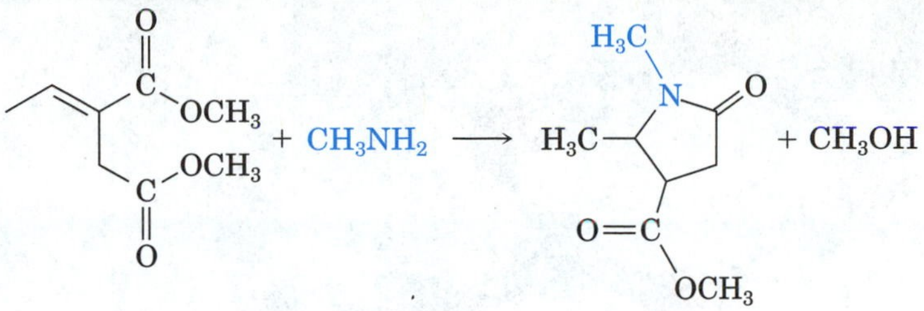

19.63 When crystals of pure α-glucose are dissolved in water, isomerization slowly occurs to produce β-glucose. Propose a mechanism for the isomerization.

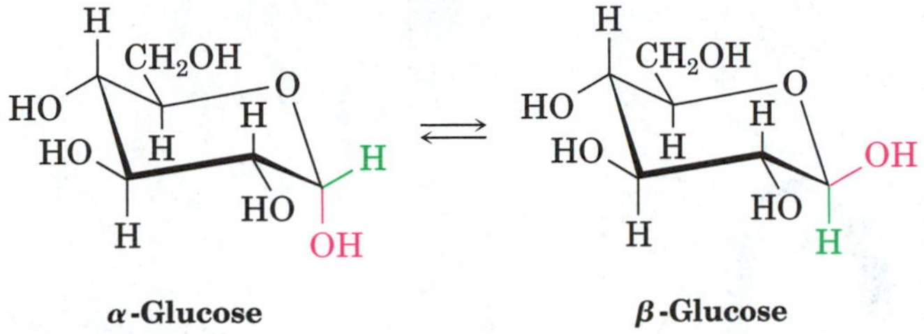

19.64 When glucose (Problem 19.63) is treated with $NaBH_4$, reaction occurs to yield *sorbitol*, a polyalcohol commonly used as a food additive. Show how this reduction occurs.

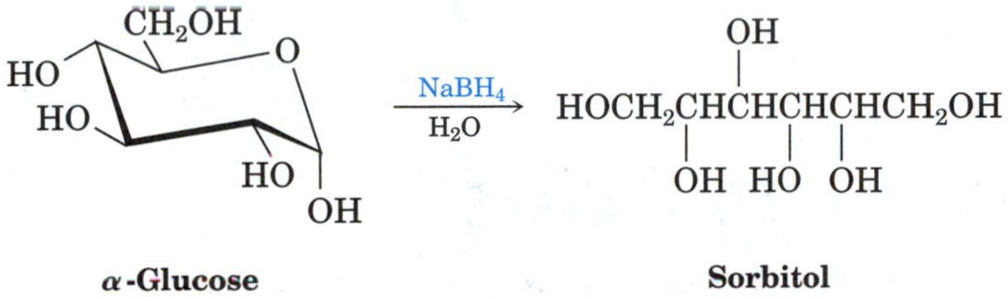

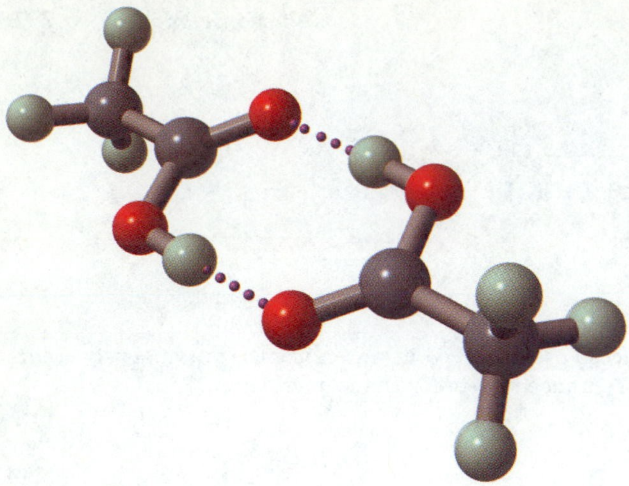

Carboxylic acids form dimers that are held together by hydrogen bonds.

20

Carboxylic Acids

Carboxylic acids occupy a central place among acyl derivatives. Not only are they important compounds themselves, they also serve as building blocks for preparing derivatives such as esters and amides. A great many carboxylic acids are found in nature. For example, acetic acid, CH_3COOH, is the chief organic component of vinegar; butanoic acid, $CH_3CH_2CH_2COOH$, is responsible for the rancid odor of sour butter; and hexanoic acid (caproic acid), $CH_3(CH_2)_4COOH$, is responsible for the unmistakable aroma of goats and dirty gym socks (Latin *caper*, "goat"). Other examples are cholic acid, a major component of human bile, and long-chain aliphatic acids such as oleic acid and linoleic acid, which are biological precursors of fats and other lipids.

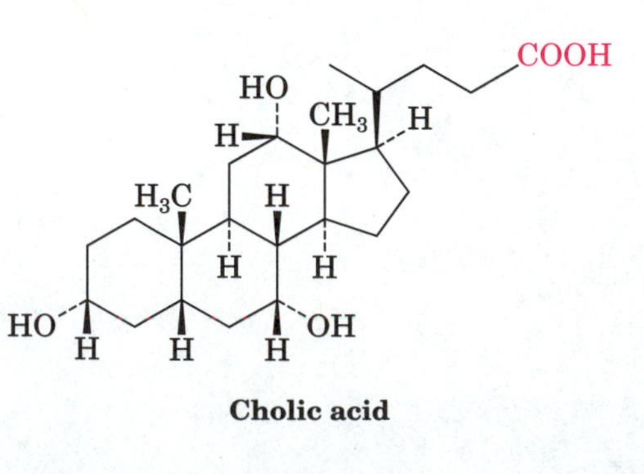

Cholic acid

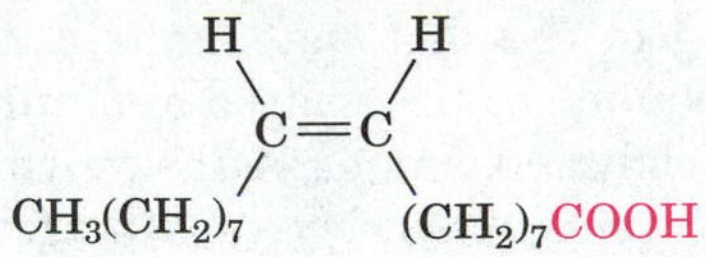

Oleic acid

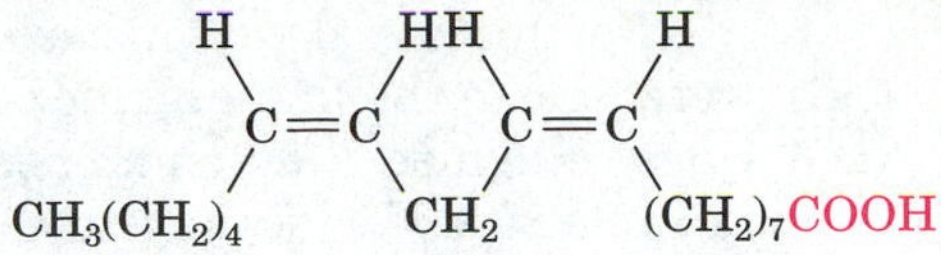

Linoleic acid

Approximately 1.9 million tons of acetic acid are produced each year in the United States for a variety of purposes, including preparation of the vinyl acetate polymer used in paints and adhesives. The industrial method of acetic acid synthesis involves a cobalt acetate-catalyzed air oxidation of acetaldehyde, but this method is not used in the laboratory.

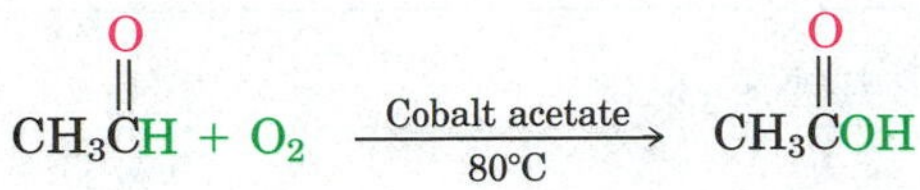

The Monsanto Company has developed an even more efficient synthesis based on the rhodium-catalyzed reaction of methanol with carbon monoxide:

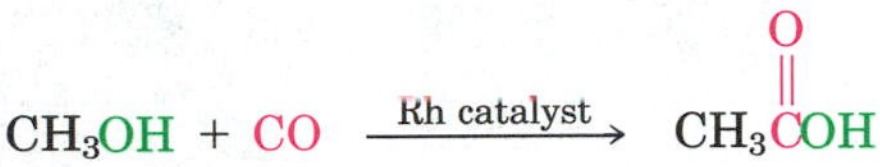

20.1 Naming Carboxylic Acids

IUPAC rules allow for two systems of nomenclature, depending on the complexity of the acid molecule. Carboxylic acids that are derived from open-chain alkanes are systematically named by replacing the terminal *-e* of the corresponding alkane name with *-oic acid*. The carboxyl carbon atom is numbered C1 in this system.

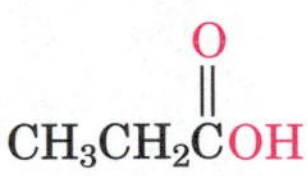

Propanoic acid

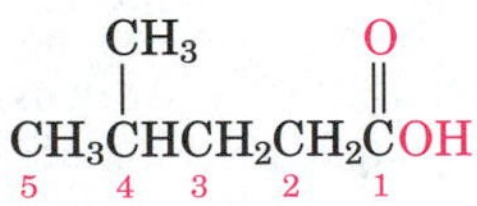

4-Methylpentanoic acid

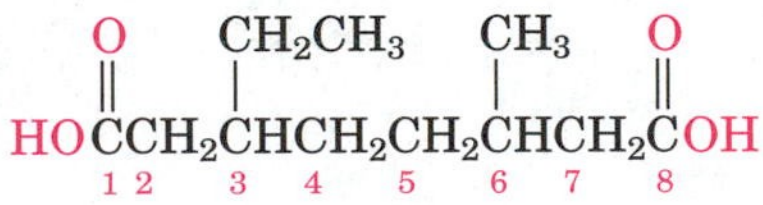

3-Ethyl-6-methyloctanedioic acid

Alternatively, compounds that have a –COOH group bonded to a ring are named using the suffix *-carboxylic acid*. The COOH carbon is attached to C1 and is not itself numbered in this system.

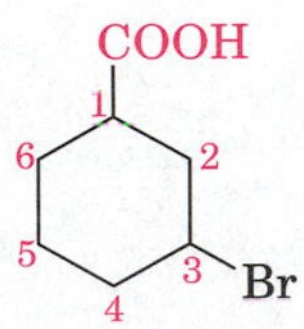

3-Bromocyclohexanecarboxylic acid

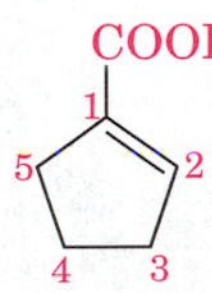

1-Cyclopentenecarboxylic acid

For historical reasons relating to the fact that many carboxylic acids were among the first organic compounds to be isolated and purified, IUPAC rules recognize a large number of common names, some of which are given in Table 20.1. We'll use systematic names in this book, with a few exceptions, such as formic (methanoic) acid and acetic (ethanoic) acid, whose names are so well known that it makes little sense to refer to them any other way. Also listed in Table 20.1 are the common names used for acyl groups derived from the parent acids.

Table 20.1 Common Names of Some Carboxylic Acids and Acyl Groups

Carboxylic acid		*Acyl group*	
Structure	*Name*	*Name*	*Structure*
$HCOOH$	Formic	Formyl	$HCO—$
CH_3COOH	Acetic	Acetyl	$CH_3CO—$
CH_3CH_2COOH	Propionic	Propionyl	$CH_3CH_2CO—$
$CH_3CH_2CH_2COOH$	Butyric	Butyryl	$CH_3(CH_2)_2CO—$
$(CH_3)_3CCOOH$	Pivalic	Pivaloyl	$(CH_3)_3CCO—$
$HOOCCOOH$	Oxalic	Oxalyl	$—OCCO—$
$HOOCCH_2COOH$	Malonic	Malonyl	$—OCCH_2CO—$
$HOOCCH_2CH_2COOH$	Succinic	Succinyl	$—OC(CH_2)_2CO—$
$HOOCCH_2CH_2CH_2COOH$	Glutaric	Glutaryl	$—OC(CH_2)_3CO—$
$HOOCCH_2CH_2CH_2CH_2COOH$	Adipic	Adipoyl	$—OC(CH_2)_4CO—$
$H_2C{=}CHCOOH$	Acrylic	Acryloyl	$H_2C{=}CHCO—$
$H_2C{=}C(CH_3)COOH$	Methacrylic	Methacryloyl	$H_2C{=}C(CH_3)CO—$
$HOOCCH{=}CHCOOH$	{ *cis*-Maleic { *trans*-Fumaric	Maleoyl } Fumaroyl }	$—OCCH{=}CHCO—$
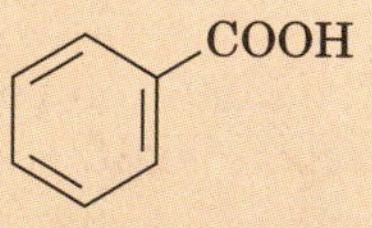	Benzoic	Benzoyl	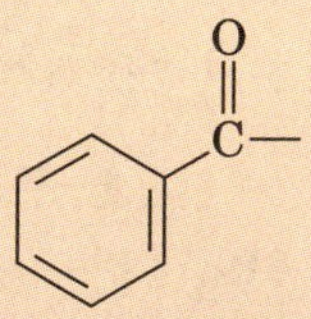
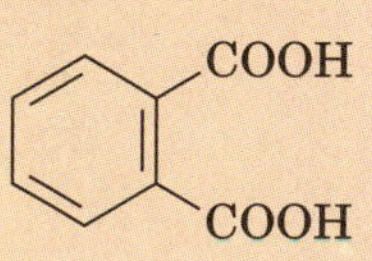	Phthalic	Phthaloyl	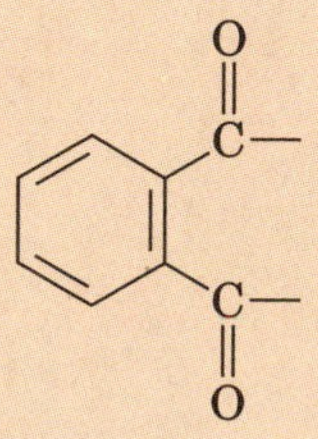

PROBLEM

20.1 Give IUPAC names for the following compounds:

(a) $(CH_3)_2CHCH_2COOH$

(b) $CH_3CHBrCH_2CH_2COOH$

(c) $CH_3CH{=}CHCH{=}CHCOOH$

(d) $CH_3CH_2CH(COOH)CH_2CH_2CH_3$

(e)

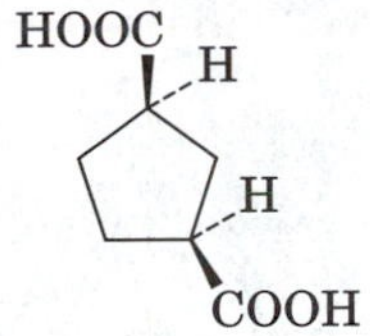

(f)

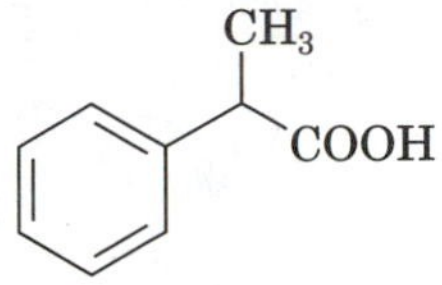

PROBLEM

20.2 Draw structures corresponding to the following IUPAC names:

(a) 2,3-Dimethylhexanoic acid

(b) 4-Methylpentanoic acid

(c) *trans*-1,2-Cyclobutanedicarboxylic acid

(d) *o*-Hydroxybenzoic acid

(e) (9*Z*,12*Z*)-9,12-Octadecadienoic acid

20.2 Structure and Physical Properties of Carboxylic Acids

Since the carboxylic acid functional group is structurally related to both ketones and alcohols, we might expect to see some familiar properties. Like ketones, the carboxyl carbon has sp^2 hybridization, and carboxylic acid groups are therefore planar with C-C-O and O-C-O bond angles of approximately 120°. The physical parameters of acetic acid are given in Table 20.2.

Table 20.2 Physical Parameters for Acetic Acid

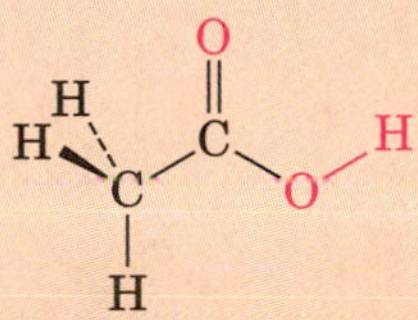

Bond angle (degrees)		*Bond length (Å)*	
C—C=O	119	C—C	1.52
C—C—OH	119	C=O	1.25
O=C—OH	122	C—OH	1.31

Like alcohols, carboxylic acids are strongly associated because of hydrogen bonding. Most carboxylic acids exist as cyclic dimers held together by two hydrogen bonds:

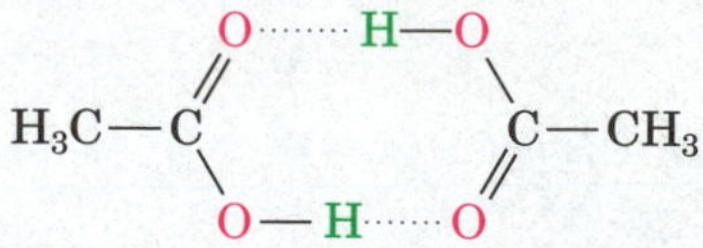

Acetic acid dimer

Stereo View

This strong hydrogen bonding has a noticeable effect on boiling points, making carboxylic acids much higher boiling than the corresponding alcohols. Table 20.3 lists the properties of some common acids.

Table 20.3 Physical Constants of Some Carboxylic Acids

Name	*Structure*	*Melting point (°C)*	*Boiling point (°C)*
Formic	HCOOH	8.4	100.7
Acetic	CH_3COOH	16.6	117.9
Propanoic	CH_3CH_2COOH	−20.8	141
Propenoic	$H_2C{=}CHCOOH$	13	141.6
Benzoic	C_6H_5COOH	122.1	249
Oxalic	$(COOH)_2$	189.5	Decomposes
Malonic	$CH_2(COOH)_2$	135.6	Decomposes
Maleic	(*Z*)-HOOCCH=CHCOOH	139	Decomposes

20.3 Dissociation of Carboxylic Acids

As their name implies, carboxylic acids are *acidic*. They therefore react with bases such as sodium hydroxide and sodium bicarbonate to give metal carboxylate salts. Carboxylic acids with more than six carbons are only slightly soluble in water, but alkali metal salts of carboxylic acids are generally quite water-soluble because of their ionic nature. It's often possible to purify acids by extracting their salts into aqueous base, then reacidifying and extracting the pure acid back into an organic solvent.

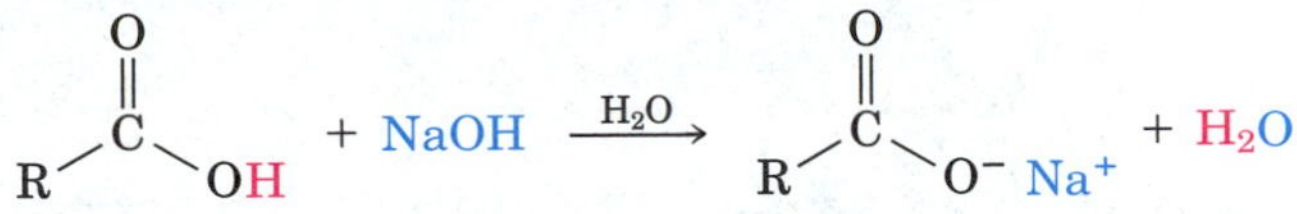

A carboxylic acid (water-insoluble) — A carboxylic acid salt (water-soluble)

Like other Brønsted–Lowry acids discussed in Section 2.6, carboxylic acids dissociate slightly in dilute aqueous solution to give H_3O^+ and carboxylate anions, $RCOO^-$.

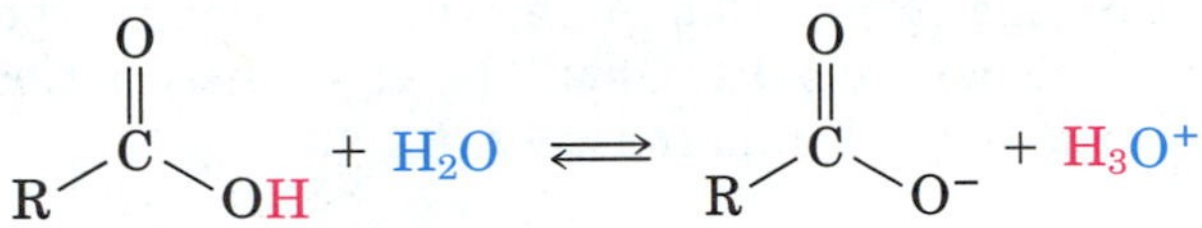

As with all acids, we can define an acidity constant, K_a:

$$K_a = \frac{[RCOO^-][H_3O^+]}{[RCOOH]} \quad \text{and} \quad pK_a = -\log K_a$$

For most carboxylic acids, K_a is approximately 10^{-5}. Acetic acid, for example, has $K_a = 1.76 \times 10^{-5}$, which corresponds to a pK_a of 4.75. In practical terms, K_a values near 10^{-5} mean that only about 0.1% of the molecules in a 0.1 M solution are dissociated, as opposed to the 100% dissociation found with strong mineral acids such as HCl and H_2SO_4.

Although much weaker than mineral acids, carboxylic acids are nevertheless much stronger acids than alcohols. The K_a of ethanol, for example, is approximately 10^{-16}, making ethanol a weaker acid than acetic acid by a factor of 10^{11}.

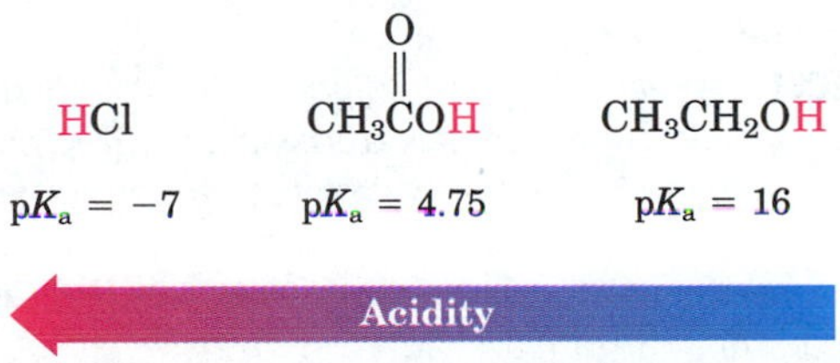

Why are carboxylic acids so much more acidic than alcohols, even though both contain –OH groups? The answer to this question involves the relative stability of carboxylate anions versus alkoxide anions. Alkoxides are oxygen anions in which the negative charge is localized on a single electronegative atom. Carboxylates, by contrast, have the negative charge *delocalized* over both oxygen atoms. In resonance terms (Section 2.4), a carboxylate ion is a stabilized resonance hybrid of two equivalent Kekulé structures.

$$CH_3CH_2\ddot{\underset{\cdot\cdot}{O}}H + H_2O \rightleftharpoons CH_3CH_2\ddot{\underset{\cdot\cdot}{O}}:^- + H_3O^+$$

Alcohol — Unstabilized alkoxide ion

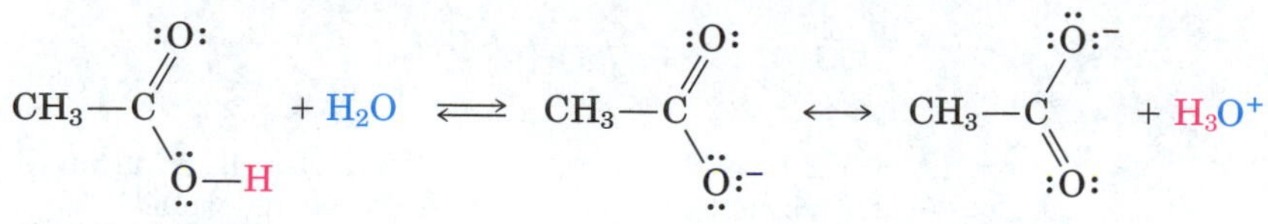

Carboxylic acid — Resonance-stabilized carboxylate ion (two equivalent resonance forms)

Since a carboxylate ion is more stable than an alkoxide ion, it is lower in energy and is present in greater amount at equilibrium, as shown in the reaction energy diagram in Figure 20.1. Put another way, dissociation of a carboxylic acid has a smaller $\Delta G°$ than dissociation of an alcohol, leading to a larger equilibrium constant, K_a.

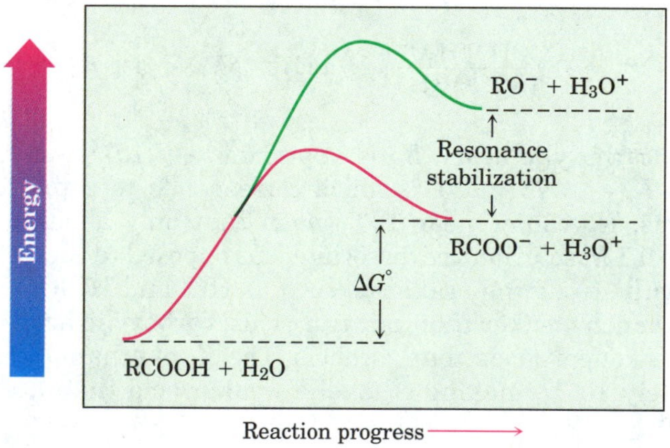

Figure 20.1 A reaction energy diagram for the dissociation of an alcohol (green curve) and a carboxylic acid (red curve). Resonance stabilization of the carboxylate anion lowers $\Delta G°$ for dissociation of the acid, leading to a larger K_a. (The starting energy levels of alcohol and acid are shown at the same point for ease of comparison.)

We can't really draw an accurate representation of the carboxylate resonance hybrid using Kekulé structures, but an orbital picture of acetate ion makes it clear that the carbon–oxygen bonds are equivalent and that each is intermediate between a single and a double bond (Figure 20.2). The *p* orbital on the carboxylate carbon atom overlaps equally well with *p* orbitals from *both* oxygens, and the four *p* electrons are delocalized throughout the three-atom π electron system.

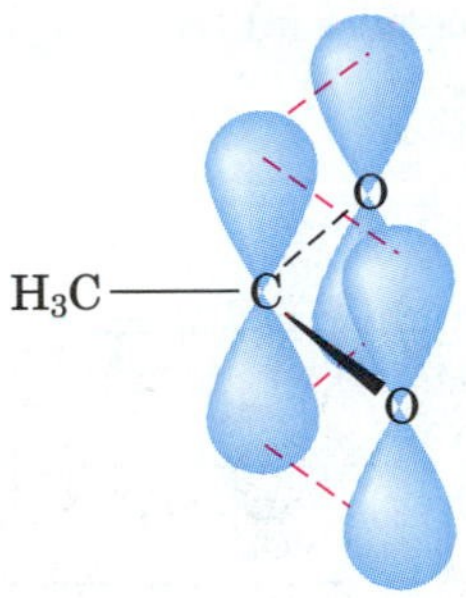

Figure 20.2 An orbital picture of acetate ion, showing the equivalence of the two oxygen atoms.

Evidence for the equivalence of the two carboxylate oxygens comes from X-ray studies on sodium formate. Both carbon–oxygen bonds are 1.27 Å in length, midway between the C=O double bond (1.20 Å) and C–O single bond (1.34 Å) of formic acid.

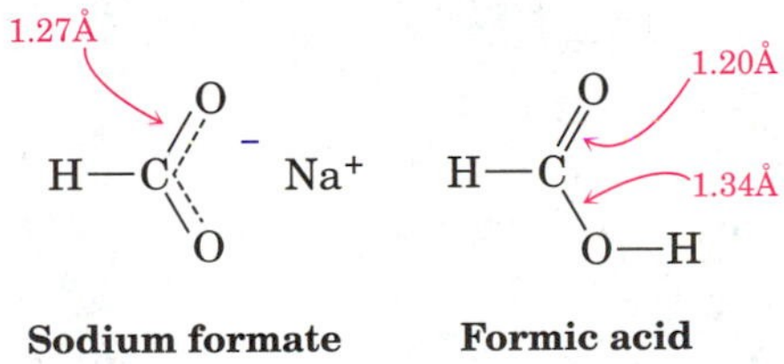

PROBLEM...

20.3 Use the equation $\Delta G° = -2.303RT \log K_a$ to calculate values of $\Delta G°$ for the dissociation of ethanol (pK_a = 16.0) and acetic acid (pK_a = 4.75) at 300 K (27°C). The gas constant R has the value 8.314 J/K·mol.

PROBLEM...

20.4 Assume you have a mixture of naphthalene and benzoic acid that you want to separate. How might you take advantage of the acidity of one component in the mixture to effect a separation?

PROBLEM...

20.5 The K_a for dichloroacetic acid is 3.32×10^{-2}. Approximately what percentage of the acid is dissociated in a 0.10 M aqueous solution?

....

20.4 Substituent Effects on Acidity

The pK_a values listed in Table 20.4 indicate a substantial difference in acidities for different carboxylic acids. For example, trifluoroacetic acid (K_a = 0.59) is more than 32,000 times as strong as acetic acid ($K_a = 1.76 \times 10^{-5}$). How can we account for such differences?

Table 20.4 Acidity of Some Carboxylic Acids

Structure	K_a	pK_a	
HCl (hydrochloric acid)[a]	(10^7)	(−7)	
F_3CCOOH	0.59	0.23	
Cl_3CCOOH	0.23	0.64	Stronger acid
$Cl_2CHCOOH$	3.3×10^{-2}	1.48	↑
FCH_2COOH	2.6×10^{-3}	2.59	
$ClCH_2COOH$	1.4×10^{-3}	2.85	
$BrCH_2COOH$	2.1×10^{-3}	2.68	
ICH_2COOH	7.5×10^{-4}	3.12	
HCOOH	1.77×10^{-4}	3.75	
$HOCH_2COOH$	1.5×10^{-4}	3.83	
C_6H_5COOH	6.46×10^{-5}	4.19	
$H_2C{=}CHCOOH$	5.6×10^{-5}	4.25	
$C_6H_5CH_2COOH$	5.2×10^{-5}	4.28	
CH_3COOH	1.76×10^{-5}	4.75	Weaker acid
CH_3CH_2COOH	1.34×10^{-5}	4.87	
CH_3CH_2OH (ethanol)[a]	(10^{-16})	(16)	

[a]Values for hydrochloric acid and ethanol are shown for reference.

Since the dissociation of a carboxylic acid is an equilibrium reaction, any factor that stabilizes the carboxylate anion relative to undissociated carboxylic acid will drive the equilibrium toward increased dissociation and result in increased acidity. Conversely, any factor that destabilizes carboxylate relative to undissociated acid will result in decreased acidity. For example, an electron-*withdrawing* group attached to the carboxyl should inductively withdraw electron density, thereby stabilizing the carboxylate anion and increasing acidity. An electron-*donating* group should have exactly the opposite effect, destabilizing the carboxylate anion and decreasing acidity.

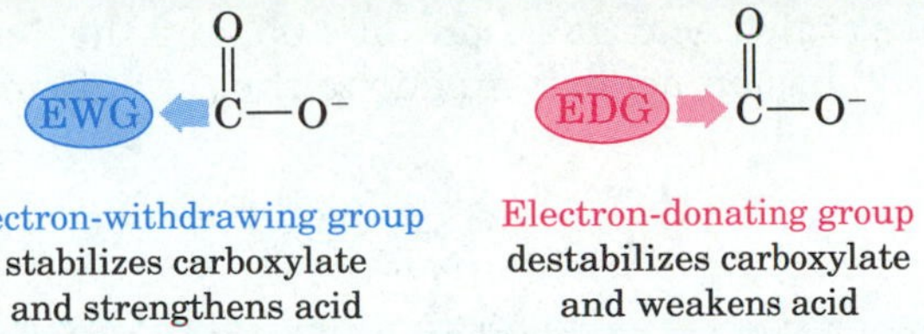

The pK_a data in Table 20.4 show exactly the expected effect. Electronegative substituents, such as the halogens, make the carboxylate anion more stable by inductively withdrawing electrons. Fluoroacetic, chloroacetic, bromoacetic, and iodoacetic acids are all stronger than acetic acid by factors of 50–150. Introduction of two electronegative substituents makes dichloroacetic acid some 3000 times as strong as acetic acid, and introduction of three substituents makes trichloroacetic acid more than 12,000 times as strong (Figure 20.3).

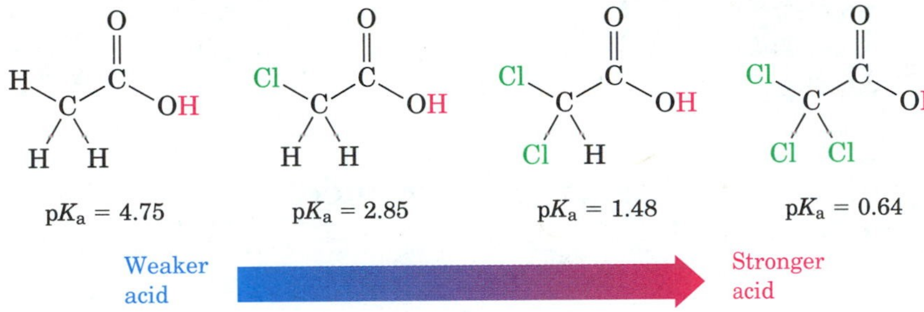

Figure 20.3 Relative strengths of chlorosubstituted acetic acids.

Because inductive effects operate through σ bonds and are dependent on distance, the effect of halogen substitution decreases as the substituent is moved farther from the carboxyl. The chlorobutanoic acids show this clearly (Table 20.5). 2-Chlorobutanoic acid has a pK_a of 2.86, the

Table 20.5 Acidity of Chlorosubstituted Butanoic Acids

Structure	K_a	pK_a
$CH_3CH_2CH(Cl)COOH$	1.39×10^{-3}	2.86
$CH_3CH(Cl)CH_2COOH$	8.9×10^{-5}	4.05
$ClCH_2CH_2CH_2COOH$	3.0×10^{-5}	4.52
$CH_3CH_2CH_2COOH$	1.5×10^{-5}	4.82

3-substituted acid has a pK_a of 4.05, and the 4-substituted acid, with a pK_a of 4.52, has an acidity similar to that of butanoic acid itself.

PROBLEM

20.6 Without looking at a table of pK_a values, rank the substances in each of the following groups in order of increasing acidity.
(a) CH_3CH_2COOH, $BrCH_2COOH$, FCH_2COOH
(b) Benzoic acid, *p*-nitrobenzoic acid, *p*-methoxybenzoic acid
(c) CH_3CH_2OH, $CH_3CH_2NH_2$, CH_3CH_2COOH

PROBLEM

20.7 Dicarboxylic acids have two dissociation constants, one for the initial dissociation into a monoanion and one for the second dissociation into a dianion. For oxalic acid, HOOC–COOH, the first ionization constant has a pK_1 of 1.2 and the second ionization constant has a pK_2 of 4.2. Why is the second carboxyl group so much less acidic than the first?

20.5 Substituent Effects in Substituted Benzoic Acids

We saw during the discussion of electrophilic aromatic substitution in Section 16.5 that substituents on the aromatic ring dramatically affect reactivity. Aromatic rings with electron-donating groups are activated toward further electrophilic substitution, and aromatic rings with electron-withdrawing groups are deactivated. Exactly the same effects are noticed on the acidity of substituted benzoic acids (Table 20.6).

As Table 20.6 shows, electron-withdrawing (deactivating) groups increase acidity by stabilizing the carboxylate anion, and electron-donating (activating) groups decrease acidity by destabilizing the carboxylate anion. Thus, an activating group such as methoxy decreases the acidity of benzoic acid, but a deactivating group such as nitro increases the acidity.

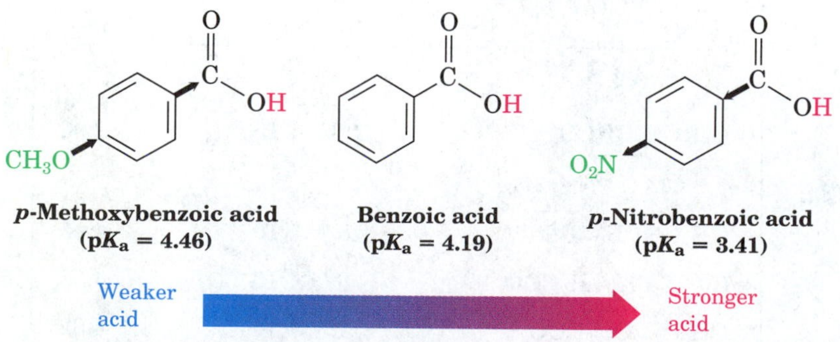

Table 20.6 Substituent Effects on Acidity of *p*-Substituted Benzoic Acids

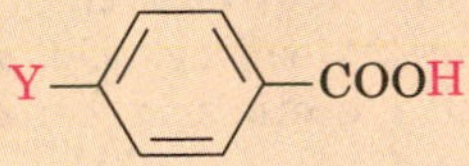

	Y	K_a	pK_a	
	—OH	3.3×10^{-5}	4.48	Activating groups
Weaker acid	—OCH_3	3.5×10^{-5}	4.46	
	—CH_3	4.3×10^{-5}	4.34	
↓	—H	6.46×10^{-5}	4.19	
	—Cl	1.0×10^{-4}	4.0	Deactivating groups
	—Br	1.1×10^{-4}	3.96	
Stronger acid	—CHO	1.8×10^{-4}	3.75	
	—CN	2.8×10^{-4}	3.55	
	—NO_2	3.9×10^{-4}	3.41	

Since it's much easier to measure the acidity of a substituted benzoic acid than to determine the relative electrophilic reactivity of a substituted benzene, the correlation between the two effects is useful for predicting reactivity. If we want to know the effect of a certain substituent on electrophilic reactivity, we can simply find the acidity of the corresponding benzoic acid.

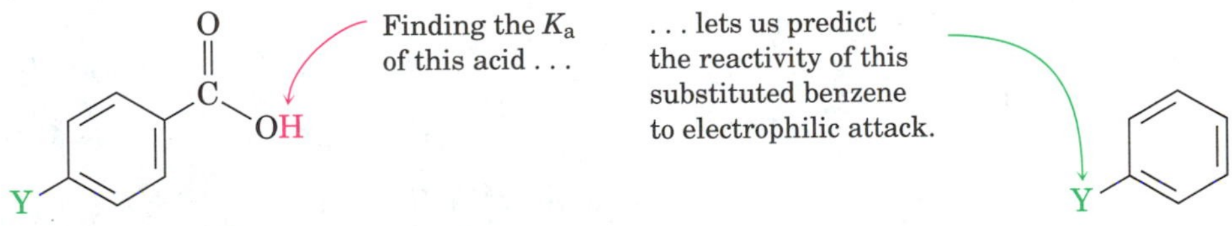

PRACTICE PROBLEM

The pK_a of *p*-(trifluoromethyl)benzoic acid is 3.6. Would you expect the trifluoromethyl substituent to be an activating or deactivating group in the Friedel–Crafts reaction?

Solution A pK_a of 3.6 means that *p*-(trifluoromethyl)benzoic acid is stronger than benzoic acid, whose pK_a is 4.19. Thus, the trifluoromethyl substituent favors dissociation by helping to stabilize negative charge. Trifluoromethyl must therefore be an electron-withdrawing, deactivating group.

PROBLEM

20.8 The pK_a of *p*-cyclopropylbenzoic acid is 4.45. Is cyclopropylbenzene likely to be more reactive or less reactive than benzene toward electrophilic bromination? Explain.

PROBLEM.

20.9 Rank the following compounds in order of increasing acidity. Don't look at a table of pK_a data to help with your answer.
(a) Benzoic acid, *p*-methylbenzoic acid, *p*-chlorobenzoic acid
(b) *p*-Nitrobenzoic acid, acetic acid, benzoic acid

20.6 Preparation of Carboxylic Acids

We've already seen most of the common methods for preparing carboxylic acids, but let's review them briefly:

1. Oxidation of a substituted alkylbenzene with $KMnO_4$ or $Na_2Cr_2O_7$ gives a substituted benzoic acid (Section 16.10). Both primary and secondary alkyl groups can be oxidized, but tertiary groups are not affected.

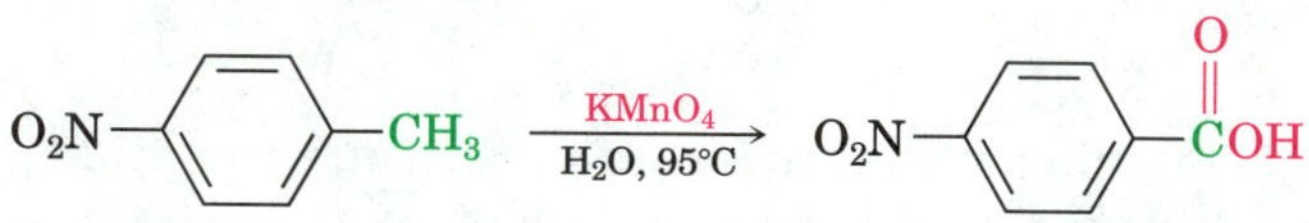

p-Nitrotoluene **p-Nitrobenzoic acid (88%)**

2. Oxidative cleavage of an alkene with $KMnO_4$ gives a carboxylic acid if the alkene has at least one vinylic hydrogen (Section 7.8).

$$CH_3(CH_2)_7CH{=}CH(CH_2)_7COOH \xrightarrow[H_3O^+]{KMnO_4} \begin{cases} CH_3(CH_2)_7COOH \\ + \\ HOOC(CH_2)_7COOH \end{cases}$$

Oleic acid → **Nonanoic acid** + **Nonanedioic acid**

3. Oxidation of primary alcohols and aldehydes yields carboxylic acids (Sections 17.9 and 19.5). Primary alcohols are often oxidized with Jones' reagent (CrO_3, H_2O, H_2SO_4), and aldehydes are oxidized with either Jones' reagent or basic silver oxide (Tollens' reagent).

$$CH_3(CH_2)_8CH_2OH \xrightarrow[H_2O,\ H_2SO_4]{CrO_3} CH_3(CH_2)_8\overset{O}{\overset{\|}{C}}OH$$

1-Decanol **Decanoic acid (93%)**

$$\underset{\textbf{Hexanal}}{CH_3CH_2CH_2CH_2CH_2\overset{\overset{\displaystyle O}{\parallel}}{C}H} \xrightarrow[NH_4OH]{AgNO_3} \underset{\textbf{Hexanoic acid (85\%)}}{CH_3CH_2CH_2CH_2CH_2\overset{\overset{\displaystyle O}{\parallel}}{C}OH}$$

Hydrolysis of Nitriles

Nitriles, R–C≡N, can be hydrolyzed by strong, hot, aqueous acid or base to yield carboxylic acids. Since nitriles themselves are usually prepared by S_N2 reaction of an alkyl halide with cyanide ion, the two-step sequence of cyanide displacement followed by nitrile hydrolysis is an excellent method for preparing carboxylic acids from alkyl halides (RBr → RC≡N → RCOOH). Note that the product acid has one more carbon than the starting alkyl halide.

$$RCH_2Br \xrightarrow[(S_N2)]{Na^+\ ^-CN} RCH_2C{\equiv}N \xrightarrow{H_3O^+} RCH_2COOH + NH_3$$

The method works best with primary halides because a competitive E2 elimination reaction can occur when a secondary or tertiary alkyl halide is used (Section 11.15). Nevertheless, some unhindered secondary halides react well. A good example occurs in the commercial synthesis of fenoprofen, a nonsteroidal anti-inflammatory drug, or NSAID (see the Chapter 16 Interlude), marketed under the trade name Mylan.

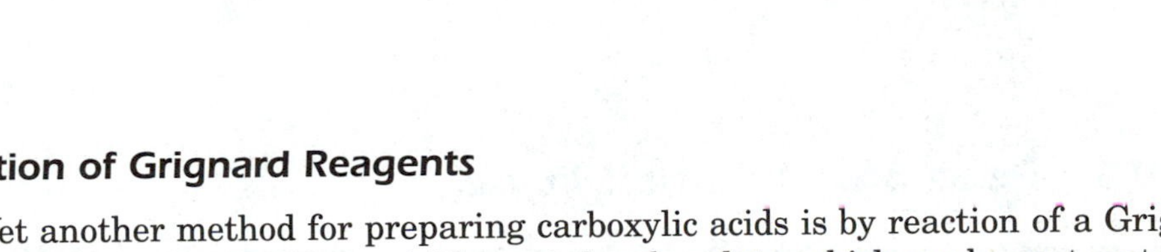

Fenoprofen
(an antiarthritic agent)

Carboxylation of Grignard Reagents

Yet another method for preparing carboxylic acids is by reaction of a Grignard reagent with CO_2 to yield a metal carboxylate, which can be protonated to give a carboxylic acid. This **carboxylation** reaction is carried out either by pouring the Grignard reagent over dry ice (solid CO_2) or by bubbling a stream of dry CO_2 through the Grignard reagent solution. Grignard carboxylation generally gives good yields of acids from alkyl halides, but is obviously limited to those alkyl halides that can form Grignard reagents in the first place (Section 17.7).

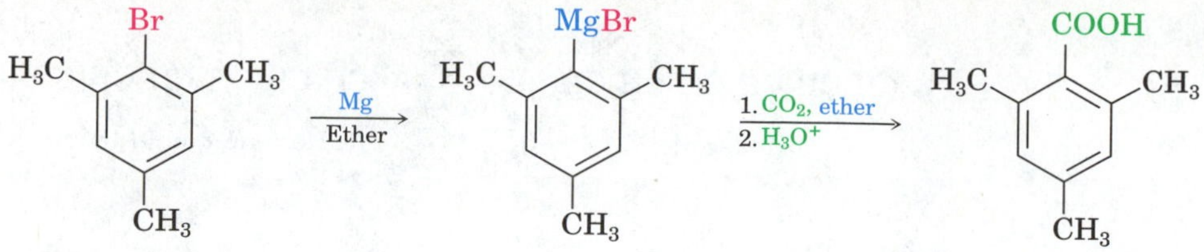

1-Bromo-2,4,6-trimethylbenzene **2,4,6-Trimethylbenzoic acid (87%)**

The mechanism of Grignard carboxylation is similar to that of other Grignard reactions (Section 19.10). The organomagnesium halide adds to a C=O bond of carbon dioxide in a typical nucleophilic addition reaction. Protonation of the carboxylate by addition of aqueous HCl then gives the free carboxylic acid product.

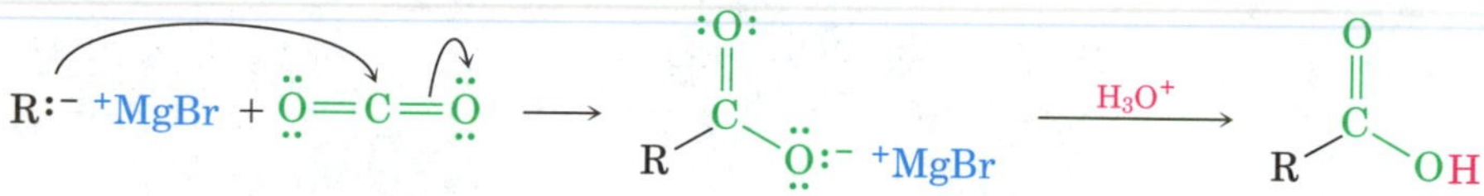

PROBLEM

20.10 We've seen two methods for converting an alkyl halide into a carboxylic acid having one more carbon atom: (1) substitution with CN^- followed by hydrolysis, and (2) formation of a Grignard reagent followed by carboxylation. What are the strengths and weaknesses of the two methods? Under what circumstances might one method be better than the other?

PROBLEM

20.11 In light of your answer to Problem 20.10, what methods would you use to prepare the following carboxylic acids from organohalides?
(a) Benzoic acid from bromobenzene
(b) $(CH_3)_3CCOOH$ from $(CH_3)_3CCl$
(c) $CH_3CH_2CH_2COOH$ from $CH_3CH_2CH_2Br$

20.7 Reactions of Carboxylic Acids

We commented earlier in this chapter that carboxylic acids are similar in some respects to both alcohols and ketones. Like alcohols, carboxylic acids can be deprotonated to give anions, which are good nucleophiles in S_N2 reactions. Like ketones, carboxylic acids undergo attack by nucleophiles on the carbonyl group. In addition, carboxylic acids undergo other reactions characteristic neither of alcohols nor ketones. Figure 20.4 shows some of the general reactions of carboxylic acids.

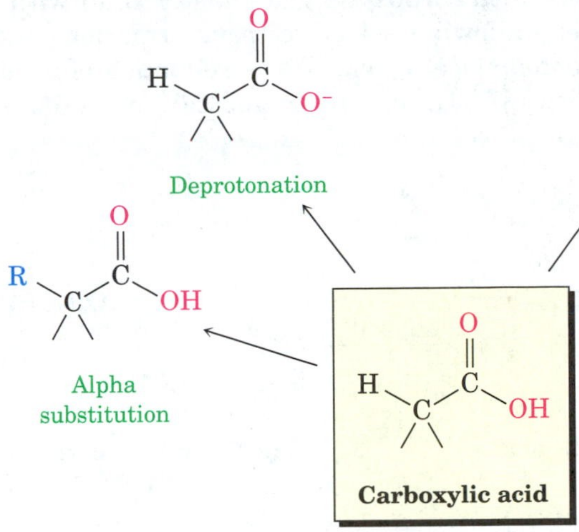

Figure 20.4 Some general reactions of carboxylic acids.

Reactions of carboxylic acids can be grouped into the four categories indicated in Figure 20.4. Of the four, we've already discussed the acidic behavior of carboxylic acids in Sections 20.3–20.5, and we'll discuss reduction in the next section. The remaining two categories are examples of fundamental carbonyl-group reaction mechanisms—nucleophilic acyl substitution and α substitution—and will be discussed in detail in Chapters 21–23.

20.8 Reduction of Carboxylic Acids

Carboxylic acids are reduced by $LiAlH_4$, but not by $NaBH_4$, to yield primary alcohols (Section 17.6). The reaction is difficult, however, and often requires heating in tetrahydrofuran solvent to go to completion.

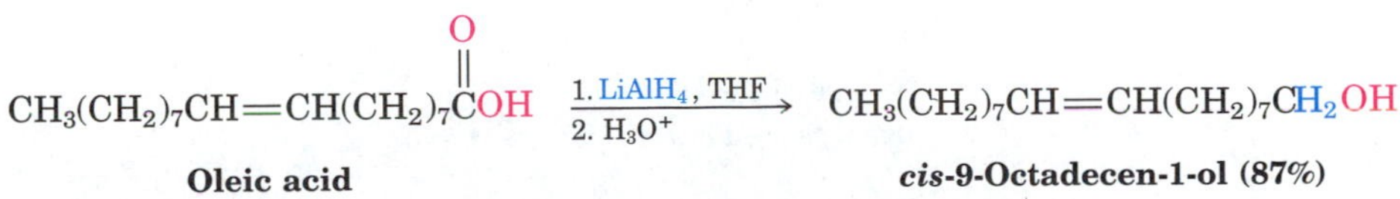

Alternatively, borane in tetrahydrofuran (BH_3/THF) is a useful reagent for reducing carboxylic acids to primary alcohols. Reaction of an acid with BH_3/THF occurs rapidly at room temperature, and the procedure is often preferred to reduction with $LiAlH_4$ because of its relative ease, safety, and specificity. Borane reacts with carboxylic acids faster than with any other functional group, thereby allowing selective transformations such as that shown below on *p*-nitrophenylacetic acid. If the reduction of *p*-nitrophenylacetic acid was done with $LiAlH_4$, both nitro and carboxyl groups would be reduced.

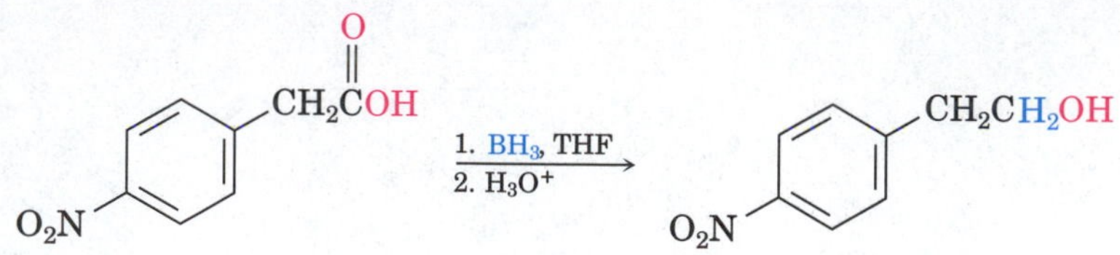

p-Nitrophenylacetic acid — **2-(p-Nitrophenyl)ethanol (94%)**

20.9 Spectroscopic Analysis of Carboxylic Acids

Infrared Spectroscopy

Carboxylic acids show two characteristic IR absorptions that make the –COOH group easily identifiable. The O–H bond of the carboxyl group gives rise to a very broad absorption over the range 2500–3300 cm^{-1}, and the C=O bond shows an absorption between 1710 cm^{-1} and 1760 cm^{-1}. The exact position of C=O absorption depends both on the structure of the molecule and on whether the acid is free (monomeric) or hydrogen-bonded (dimeric). Free carboxyl groups absorb at 1760 cm^{-1}, but the more commonly encountered dimeric carboxyl groups absorb in a broad band centered around 1710 cm^{-1}.

Free carboxyl (uncommon), 1760 cm^{-1}

Associated carboxyl (usual case), 1710 cm^{-1}

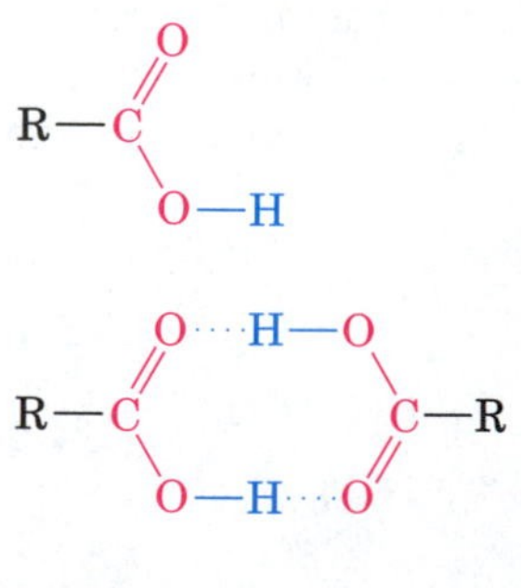

The IR spectrum of butanoic acid shown in Figure 20.5 has both the broad O–H absorption and the C=O absorption at 1710 cm^{-1} (dimeric) identified.

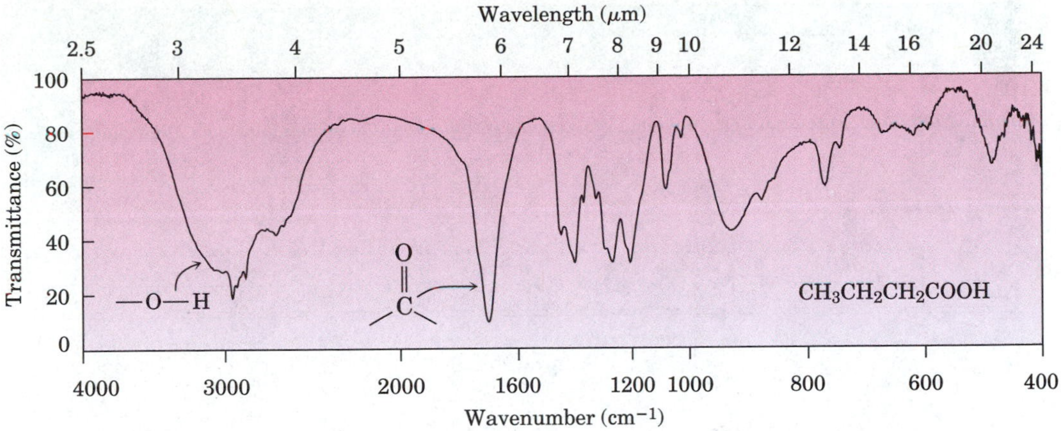

Figure 20.5 Infrared spectrum of butanoic acid, $CH_3CH_2CH_2COOH$.

Nuclear Magnetic Resonance Spectroscopy

Carboxylic acid groups can be detected by both ^{1}H and ^{13}C NMR spectroscopy. Carboxyl carbon atoms absorb in the range 165–185 δ in the ^{13}C NMR spectrum, with aromatic and α,β-unsaturated acids near the upfield end of the range (~165 δ) and saturated aliphatic acids near the downfield end (~185 δ). The acidic –COO**H** proton normally absorbs as a singlet near 12 δ in the ^{1}H NMR spectrum. As with alcohols (Section 17.11), the –COOH proton can be replaced by deuterium when D_2O is added to the sample tube, causing the –COOH absorption to disappear from the NMR spectrum.

Figure 20.6 indicates the positions of the ^{13}C NMR absorptions for several carboxylic acids, and Figure 20.7 shows the ^{1}H NMR spectrum of phenylacetic acid. Note that the carboxyl proton occurs at 12.0 δ.

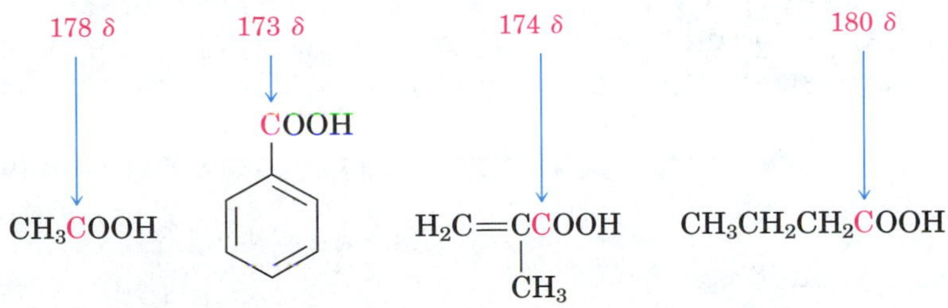

Figure 20.6 Carbon-13 NMR absorptions of some carboxylic acids.

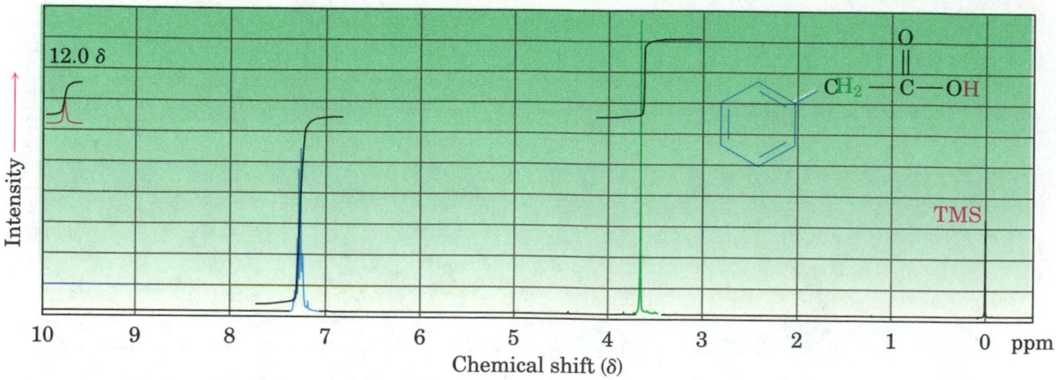

Figure 20.7 Proton NMR spectrum of phenylacetic acid.

INTERLUDE

Vitamin C

Sailors in the Middle Ages had a high death toll from scurvy, caused by a dietary vitamin C deficiency.

Vitamin C, or ascorbic acid, is surely the best known of all vitamins. It was the first vitamin to be discovered (1928), the first to be structurally characterized (1933), and the first to be synthesized in the laboratory (1933). Over 80 million pounds of vitamin C are now synthesized worldwide each year, more than the total amount of all other vitamins combined. In addition to its use as a vitamin supplement, vitamin C is used as a food preservative, a "flour improver" in bakeries, and an animal food additive.

Vitamin C is perhaps most famous for its antiscorbutic properties, meaning that it prevents the onset of scurvy, a bleeding disease affecting those with a deficiency of fresh vegetables and citrus fruits in their diet. Seamen in the Age of Exploration were particularly susceptible to scurvy, and the death toll among sailors was high in the Middle Ages. Vasco da

(continued)➤

Gama, for instance, lost more than half his crew to scurvy during his 2 year voyage around the Cape of Good Hope in 1497–1499.

In more recent times, large doses of vitamin C have been claimed to prevent the common cold, cure infertility, delay the onset of symptoms in AIDS, and inhibit the development of gastric and cervical cancers. Proof is still lacking for most of these claims, but a recent study in Europe did find statistical evidence for an inhibitory effect against gastric cancers. Although large daily doses of vitamin C are probably not warranted, the harmful side effects of vitamin C appear minimal, and many people have adopted a better-safe-than-sorry approach.

Chemically, vitamin C is interesting because its industrial preparation involves an unusual blend of biological and laboratory organic chemistry. The Hoffmann-LaRoche Company synthesizes ascorbic acid from glucose through the five-step route shown in Figure 20.8. Glucose, a pentahydroxy aldehyde, is first reduced to sorbitol, which is then oxidized by the microorganism *Acetobacter suboxydans*. No chemical reagent exists that is selective enough to oxidize only one of the six alcohol groups in sorbitol, so an enzymatic reaction is used. Treatment with acetone and an acid catalyst then protects four of the remaining hydroxyl groups in acetal linkages, and the unprotected hydroxyl group is chemically oxidized to the carboxylic acid by reaction with aqueous NaOCl (household bleach). Hydrolysis with acid then removes the two acetal groups, causes an internal ester-forming reaction to take place, and catalyzes a keto–enol tautomerization to give ascorbic acid. Each of the five steps takes place in better than 90% yield.

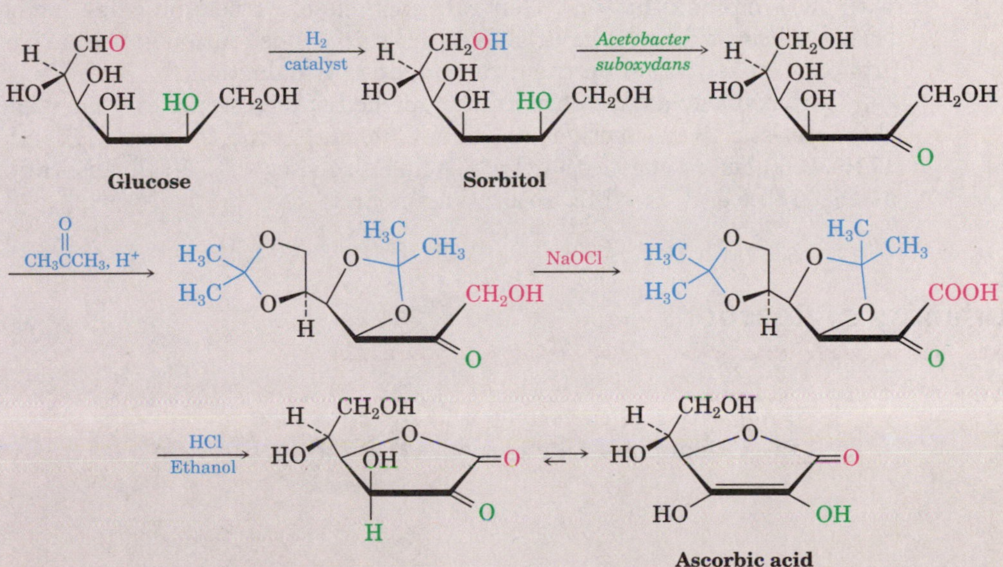

Figure 20.8 The industrial synthesis of ascorbic acid from glucose.

Summary and Key Words

Carboxylic acids are among the most useful building blocks for synthesizing other molecules, both in nature and in the chemical laboratory. They are named systematically by replacing the terminal *-e* of the corresponding alkane name with *-oic acid*. Like ketones and aldehydes, the carbonyl carbon atom is sp^2-hybridized; like alcohols, carboxylic acids are associated through hydrogen bonding and therefore have high boiling points.

The distinguishing characteristic of carboxylic acids is their acidity. Although weaker than mineral acids such as HCl, carboxylic acids dissociate much more readily than alcohols because the resultant carboxylate ions are stabilized by resonance between two equivalent forms:

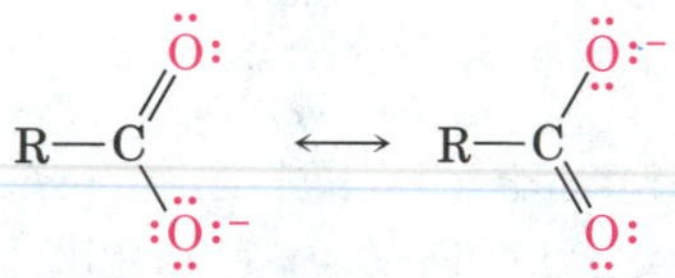

Most alkanoic acids have pK_a values near 5, but the exact pK_a of a given acid depends on structure. Carboxylic acids substituted by electron-withdrawing groups are more acidic (have a lower pK_a) because their carboxylate ions are stabilized. Carboxylic acids substituted by electron-donating groups are less acidic (have a higher pK_a) because their carboxylate ions are destabilized.

Methods of synthesis for carboxylic acids include: (1) oxidation of alkylbenzenes, (2) oxidative cleavage of alkenes, (3) oxidation of primary alcohols or aldehydes, (4) hydrolysis of nitriles, and (5) reaction of Grignard reagents with CO_2 (**carboxylation**). General reactions of carboxylic acids include: (1) loss of the acidic proton, (2) nucleophilic acyl substitution at the carbonyl group, (3) substitution on the α carbon, and (4) reduction.

Carboxylic acids are easily distinguished spectroscopically. They show characteristic IR absorptions at 2500–3300 cm^{-1} (due to the O–H) and at 1710–1760 cm^{-1} (due to the C=O). Acids also show ^{13}C NMR absorptions at 165–185 δ and ^1H NMR absorptions near 12 δ.

Summary of Reactions

1. Preparation of carboxylic acids (Section 20.6)
   (a) Oxidation of alkylbenzenes (Section 16.10)

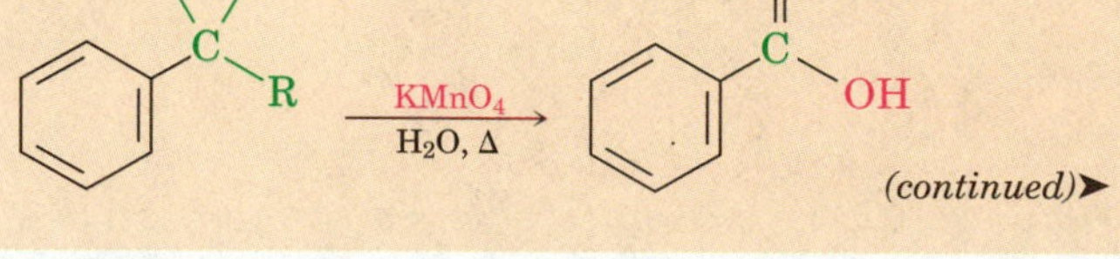

(continued)➤

(b) Oxidative cleavage of alkenes (Section 7.8)

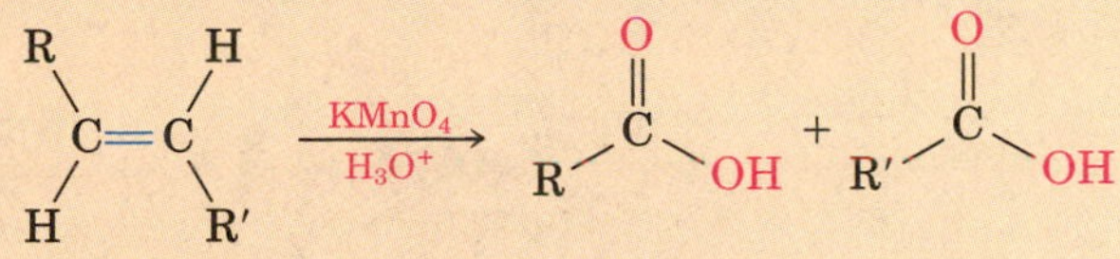

(c) Oxidation of primary alcohols (Section 17.9)

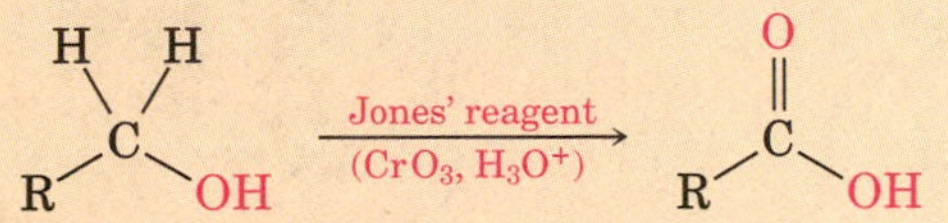

(d) Oxidation of aldehydes (Section 19.5)

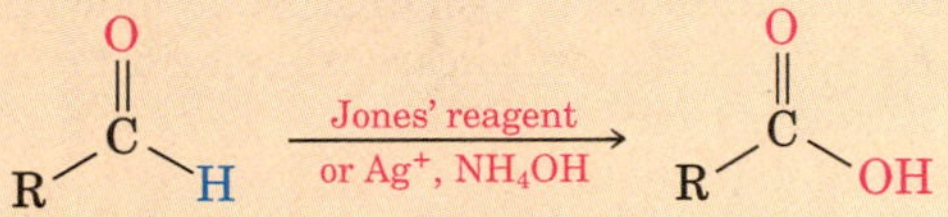

(e) Hydrolysis of nitriles (Section 20.6)

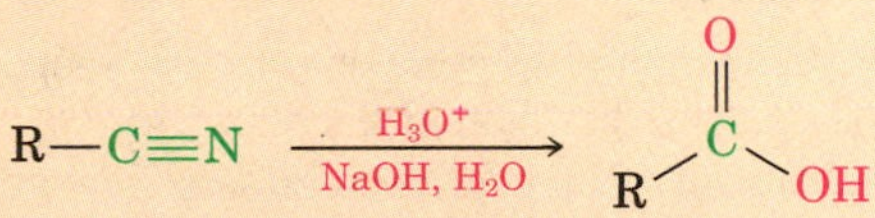

(f) Carboxylation of Grignard reagents (Section 20.6)

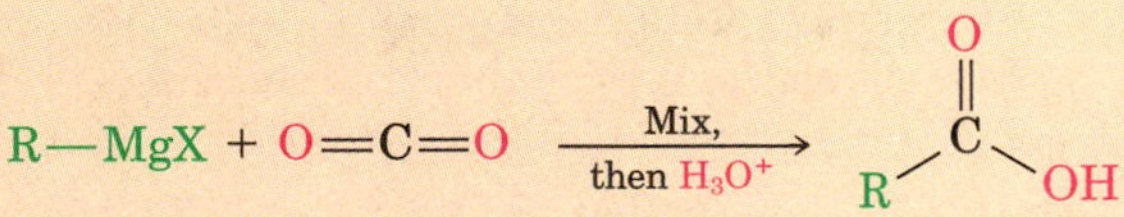

2. Reactions of carboxylic acids

(a) Deprotonation (Section 20.3)

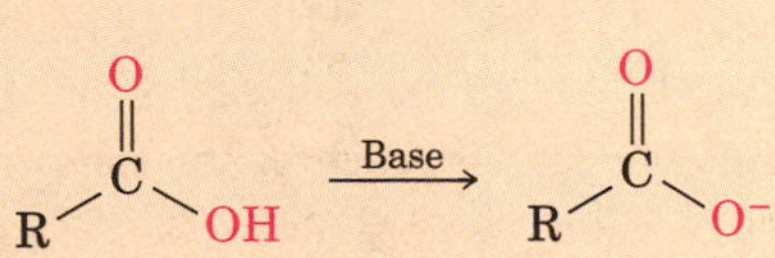

(b) Reduction to primary alcohols (Section 20.8)

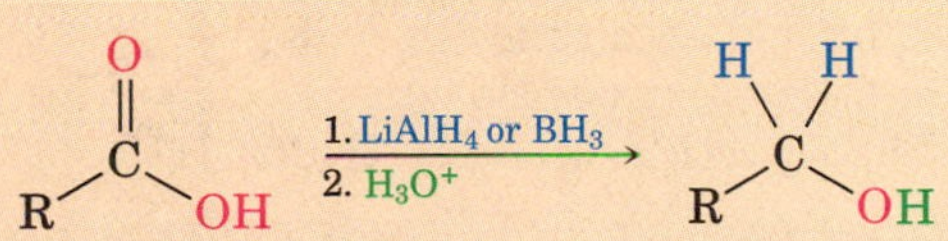

ADDITIONAL PROBLEMS

20.12 Give IUPAC names for the following compounds:

(a) 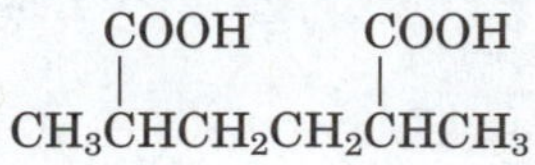

(b) $(CH_3)_3CCOOH$

(c) 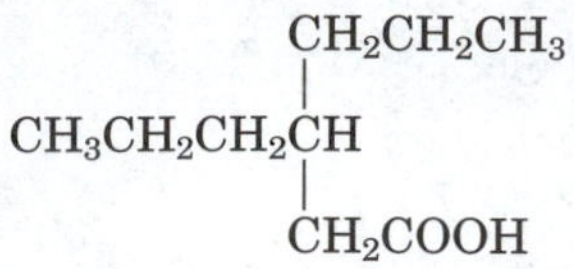

(d) 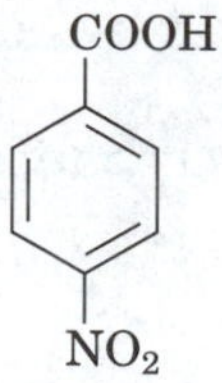

(e) 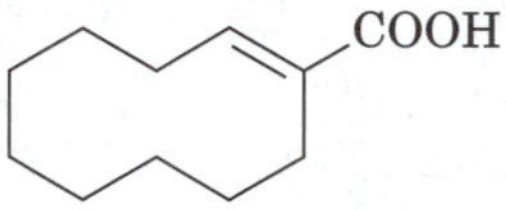

(f) $BrCH_2CHBrCH_2CH_2COOH$

20.13 Draw structures corresponding to the following IUPAC names:
(a) *cis*-1,2-Cyclohexanedicarboxylic acid
(b) Heptanedioic acid
(c) 2-Hexen-4-ynoic acid
(d) 4-Ethyl-2-propyloctanoic acid
(e) 3-Chlorophthalic acid
(f) Triphenylacetic acid

20.14 Acetic acid boils at 118°C, but its ethyl ester boils at 77°C. Why is the boiling point of the acid so much higher even though it has the lower molecular weight?

20.15 Draw and name the eight carboxylic acid isomers with the formula $C_6H_{12}O_2$.

20.16 Order the compounds in each set with respect to increasing acidity:
(a) Acetic acid, oxalic acid, formic acid
(b) *p*-Bromobenzoic acid, *p*-nitrobenzoic acid, 2,4-dinitrobenzoic acid
(c) Phenylacetic acid, diphenylacetic acid, 3-phenylpropanoic acid
(d) Fluoroacetic acid, 3-fluoropropanoic acid, iodoacetic acid

20.17 Arrange the compounds in each set in order of increasing basicity:
(a) Magnesium acetate, magnesium hydroxide, methylmagnesium bromide
(b) Sodium benzoate, sodium *p*-nitrobenzoate, sodium acetylide
(c) Lithium hydroxide, lithium ethoxide, lithium formate

20.18 Account for the fact that phthalic acid (1,2-benzenedicarboxylic acid) has $pK_2 = 5.4$ but terephthalic acid (1,4-benzenedicarboxylic acid) has $pK_2 = 4.8$.

20.19 How could you convert butanoic acid into the following compounds? Write each step showing the reagents needed.
(a) 1-Butanol
(b) 1-Bromobutane
(c) Pentanoic acid
(d) 1-Butene
(e) Octane

20.20 How could you convert each of the following compounds into butanoic acid? Write each step showing all reagents.
(a) 1-Butanol
(b) 1-Bromobutane
(c) 1-Butene
(d) 1-Bromopropane
(e) 4-Octene

20.21 How would you prepare the following compounds from benzene? More than one step is required in each case.
(a) *m*-Chlorobenzoic acid
(b) *p*-Bromobenzoic acid
(c) Phenylacetic acid, $C_6H_5CH_2COOH$

20.22 Calculate pK_a's for the following acids:
(a) Lactic acid, $K_a = 8.4 \times 10^{-4}$
(b) Acrylic acid, $K_a = 5.6 \times 10^{-6}$

20.23 Calculate K_a's for the following acids:
(a) Citric acid, $pK_a = 3.14$
(b) Tartaric acid, $pK_a = 2.98$

20.24 Shown here are some pK_a data for simple dibasic acids. How do you account for the fact that the difference between the first and second ionization constants decreases with increasing distance between the carboxyl groups?

Name	*Structure*	pK_1	pK_2
Oxalic	HOOCCOOH	1.2	4.2
Succinic	$HOOCCH_2CH_2COOH$	4.2	5.6
Adipic	$HOOC(CH_2)_4COOH$	4.4	4.4

20.25 Predict the product of the reaction of *p*-methylbenzoic acid with each of the following reagents:
(a) BH_3, then H_3O^+
(b) *N*-Bromosuccinimide in CCl_4
(c) CH_3MgBr in ether, then H_3O^+
(d) $KMnO_4$, H_3O^+
(e) $LiAlH_4$, then H_3O^+

20.26 Using $^{13}CO_2$ as your only source of labeled carbon, along with any other compounds needed, how would you synthesize the following compounds?
(a) $CH_3CH_2{}^{13}COOH$
(b) $CH_3{}^{13}CH_2COOH$

20.27 Propose a structure for an organic compound, $C_6H_{12}O_2$, that dissolves in dilute NaOH and shows the following ^{1}H NMR spectrum: 1.08 δ (9 H, singlet), 2.2 δ (2 H, singlet), and 11.2 δ (1 H, singlet).

20.28 How would you carry out the following transformations?

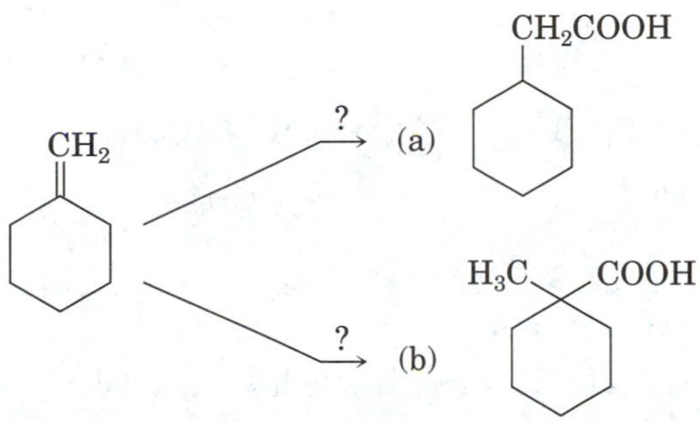

20.29 What spectroscopic method could you use to distinguish among the following three isomeric acids? Tell what characteristic features you would expect for each acid.

$CH_3(CH_2)_3COOH$	$(CH_3)_2CHCH_2COOH$	$(CH_3)_3CCOOH$
Pentanoic acid	**3-Methylbutanoic acid**	**2,2-Dimethylpropanoic acid**

20.30 Which method—Grignard carboxylation or nitrile hydrolysis—would you use for each of the following reactions? Explain.

(a)
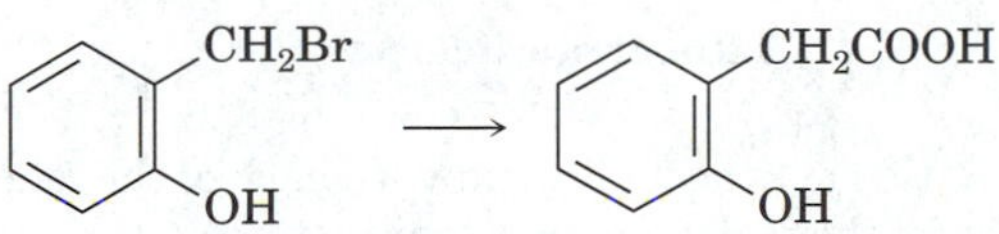

(b)
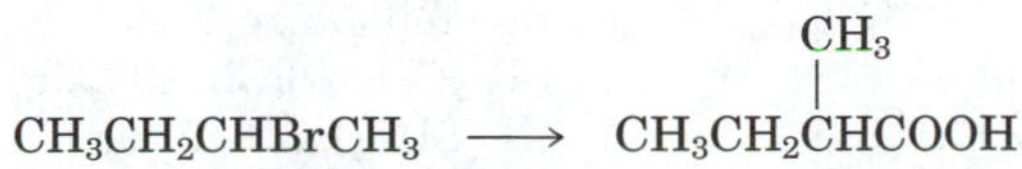

(c) $CH_3C(=O)CH_2CH_2CH_2I \longrightarrow CH_3C(=O)CH_2CH_2CH_2COOH$

(d) $HOCH_2CH_2CH_2Br \longrightarrow HOCH_2CH_2CH_2COOH$

20.31 A chemist in need of 2,2-dimethylpentanoic acid decided to synthesize some by reaction of 2-chloro-2-methylpentane with NaCN, followed by hydrolysis of the product. After carrying out the reaction sequence, however, none of the desired product could be found. What do you suppose went wrong?

20.32 The following synthetic schemes all have at least one flaw in them. What is wrong with each?

(a) 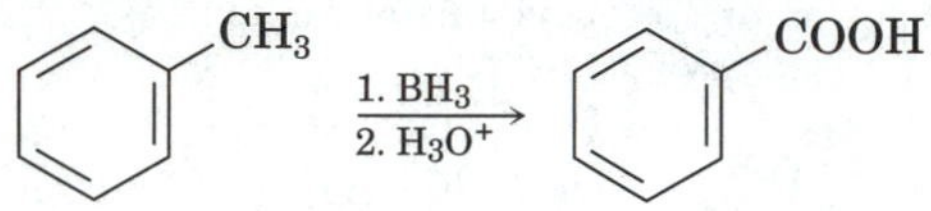

(b) $CH_3CH_2CHBrCH_2CH_3 \xrightarrow[\text{2. NaCN; 3. } H_3O^+]{\text{1. Mg}} CH_3CH_2CH(COOH)CH_2CH_3$

(c) 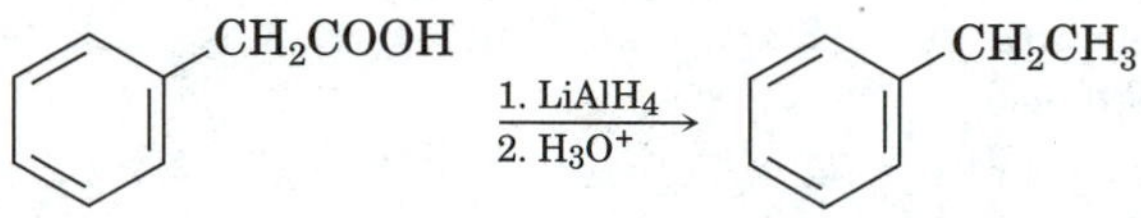

(d) $CH_3C(OH)(CH_3)CH_2CH_2Cl \xrightarrow[\text{2. } H_3O^+]{\text{1. NaCN}} CH_3C(OH)(CH_3)CH_2CH_2C(=O)OH$

20.33 *p*-Aminobenzoic acid (PABA) is widely used as a sunscreen agent. Propose a synthesis of PABA starting from toluene.

20.34 Lithocholic acid is a steroid found in human bile:

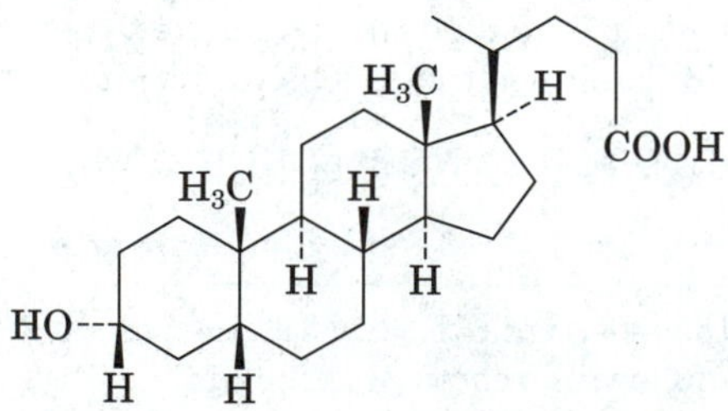

Lithocholic acid

Predict the product of reaction of lithocholic acid with each of the following reagents. Don't worry about the size of the molecule; just concentrate on the functional groups.

(a) Jones' reagent
(b) Tollens' reagent
(c) BH_3, then H_3O^+
(d) $(CH_3)_3SiCl$, $(CH_3CH_2)_3N$
(e) CH_3MgBr, then H_3O^+
(f) $LiAlH_4$, then H_3O^+

20.35 Propose a synthesis of the anti-inflammatory drug Fenclorac from phenylcyclohexane.

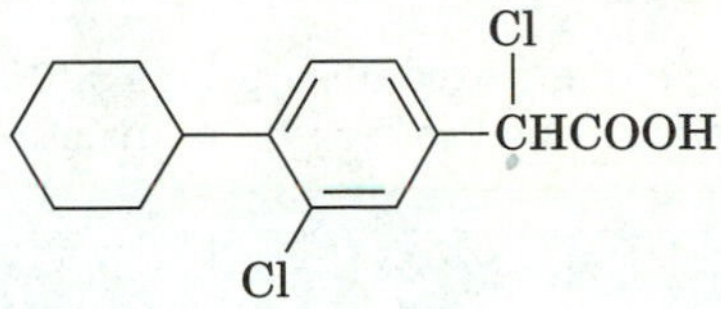

Fenclorac

20.36 How would you use NMR (either ^{13}C or ^{1}H) to distinguish between the following isomeric pairs?

(a)

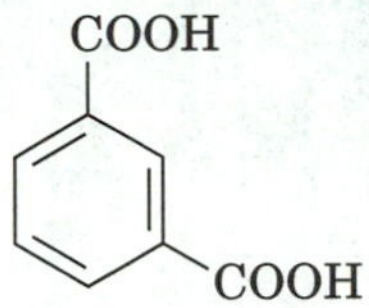

and

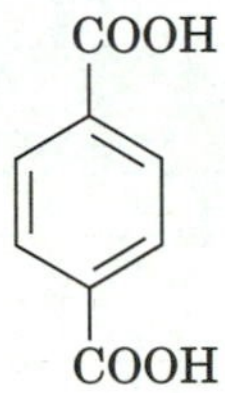

(b) $HOOCCH_2CH_2COOH$ and $CH_3CH(COOH)_2$

(c) $CH_3CH_2CH_2COOH$ and $HOCH_2CH_2CH_2CHO$

(d) $(CH_3)_2C{=}CHCH_2COOH$ and

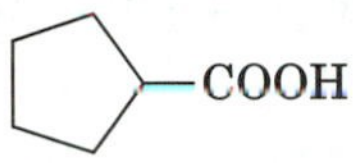

20.37 The pK_a's of five *p*-substituted benzoic acids (YC_6H_4COOH) are given in the following table. Rank the corresponding substituted benzenes (YC_6H_5) in order of their increasing reactivity toward electrophilic aromatic substitution. If benzoic acid has pK_a = 4.19, which of the substituents are activators and which are deactivators?

Substituent Y	pK_a *of* 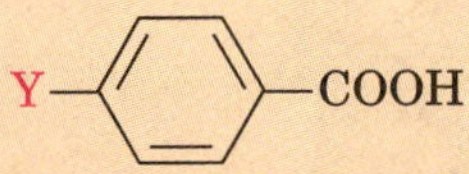
—$Si(CH_3)_3$	4.27
—$CH{=}CHC{\equiv}N$	4.03
—$HgCH_3$	4.10
—OSO_2CH_3	3.84
—PCl_2	3.59

20.38 How would you carry out the following transformations? More than one step is required in each case.

(a)

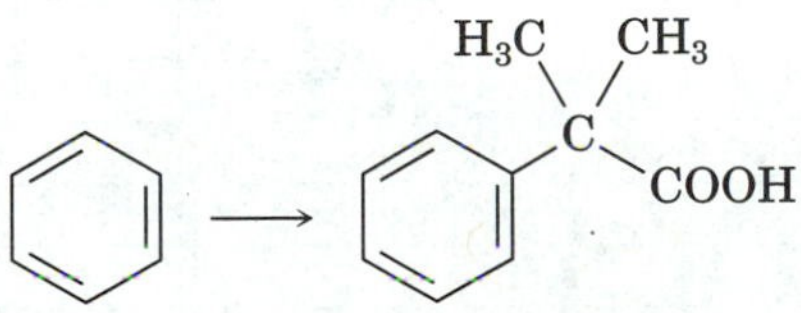

(b)

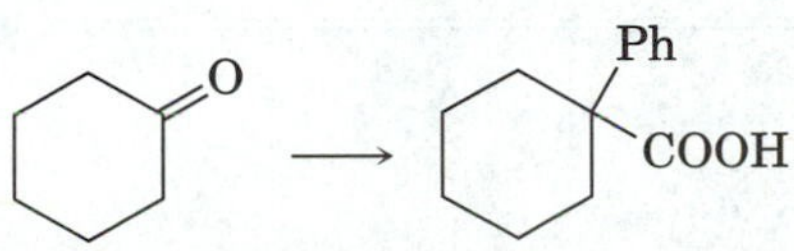

20.39 How can you explain the observation that hydroxyacetic acid ($pK_a = 3.83$) is *stronger* than acetic acid ($pK_a = 4.75$), yet *p*-hydroxybenzoic acid ($pK_a = 4.48$) is *weaker* than benzoic acid ($pK_a = 4.19$)?

20.40 Compound A, $C_4H_8O_3$, has infrared absorptions at 1710 and 2500–3100 cm^{-1}, and has the ^1H NMR spectrum shown. Propose a structure for A.

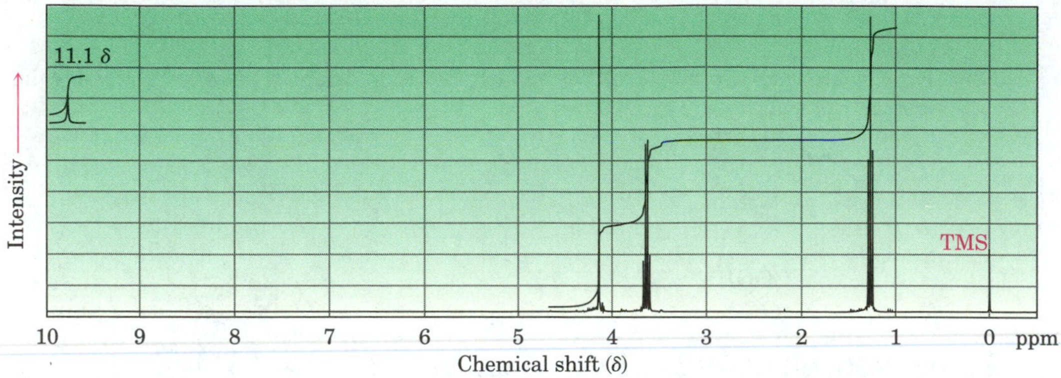

20.41 The two ^1H NMR spectra shown here belong to crotonic acid (*trans*-$CH_3CH{=}CHCOOH$) and methacrylic acid [$H_2C{=}C(CH_3)COOH$]. Which spectrum corresponds to which acid? Explain.

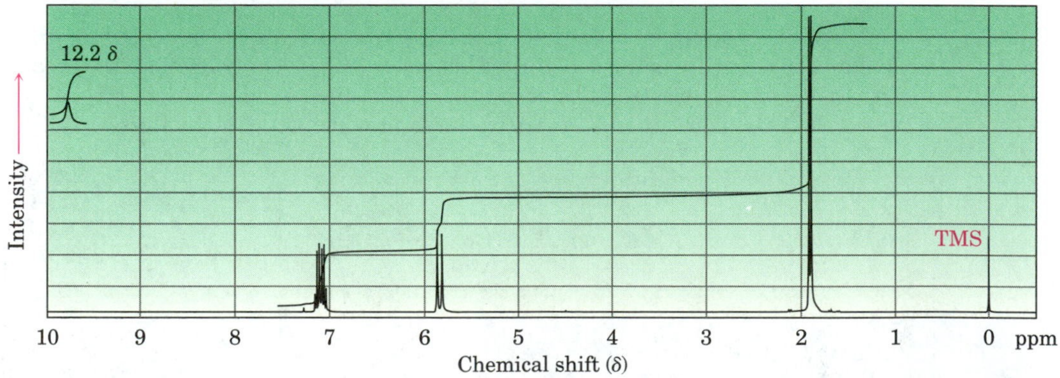

(a)

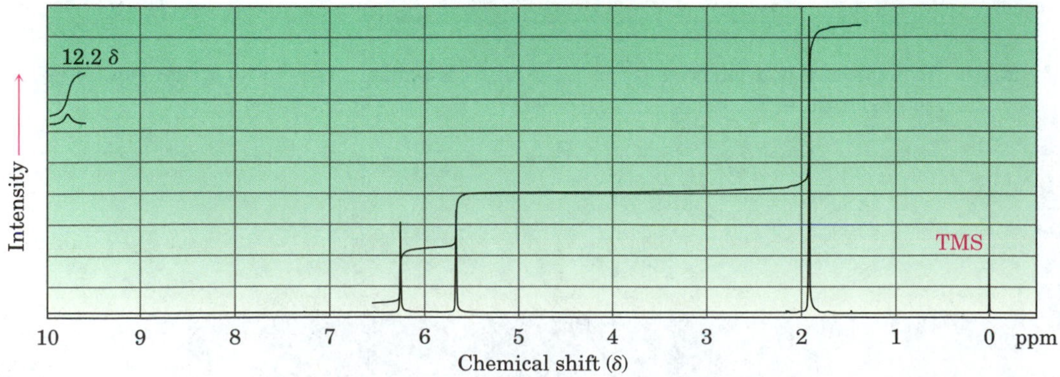

(b)

20.42 Propose structures for carboxylic acids that show the following peaks in their ^{13}C NMR spectra. Assume that the kinds of carbons (1°, 2°, 3°, or 4°) have been assigned by DEPT-NMR.

(a) $C_7H_{12}O_2$: 25.5 δ (2°), 25.9 δ (2°), 29.0 δ (2°), 43.1 δ (3°), 183.0 δ (4°)

(b) $C_8H_8O_2$: 21.4 δ (1°), 128.3 δ (4°), 129.0 δ (3°), 129.7 δ (3°), 143.1 δ (4°), 168.2 δ (4°)

20.43 3-Methyl-2-hexenoic acid (mixture of *E* and *Z* isomers) has been identified as the substance responsible for the odor of human sweat. Devise a synthesis of the compound from starting materials having five or fewer carbons.

20.44 Treatment of an α,β-unsaturated ketone with basic aqueous hydrogen peroxide yields an epoxy ketone. The reaction is specific to unsaturated ketones; isolated alkene double bonds do not react. Propose a mechanism. (The pK_a of H_2O_2 is 11.6.)

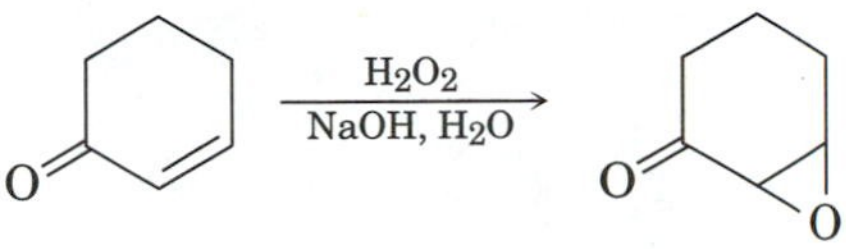

20.45 Identify the missing reagents a–f in the following scheme:

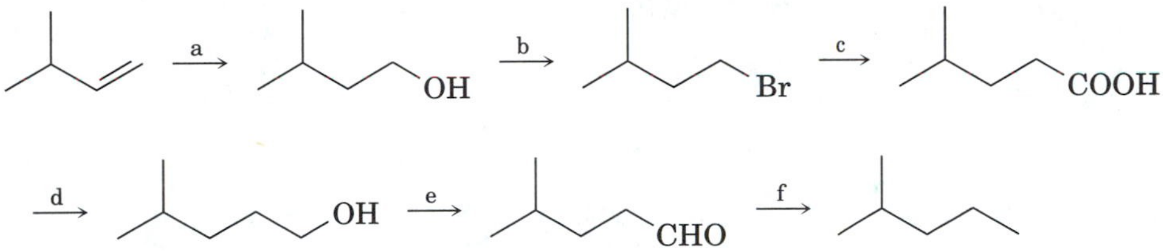

A Look Ahead

20.46 We'll see in the next chapter that carboxylic acids react with alcohols to yield esters: $RCO_2H + R'OH \rightarrow RCO_2R'$. Propose a mechanism for the following reaction. (*Hint*: See Section 19.14.)

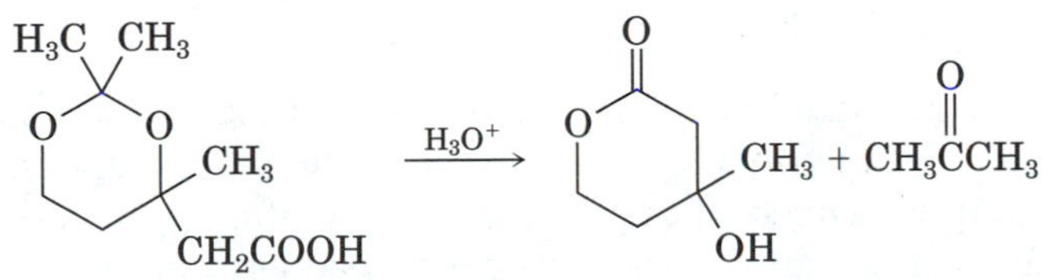

20.47 Look at the overview of carbonyl chemistry preceding Chapter 19 and then write a mechanism for the formation of ethyl acetate by acid-catalyzed reaction of acetic acid with ethanol. Check your answer in Figure 21.5.

$$CH_3\overset{O}{\overset{\|}{C}}OH + CH_3CH_2OH \xrightleftharpoons[\text{catalyst}]{H_2SO_4} CH_3\overset{O}{\overset{\|}{C}}OCH_2CH_3 + H_2O$$

Ethyl acetate

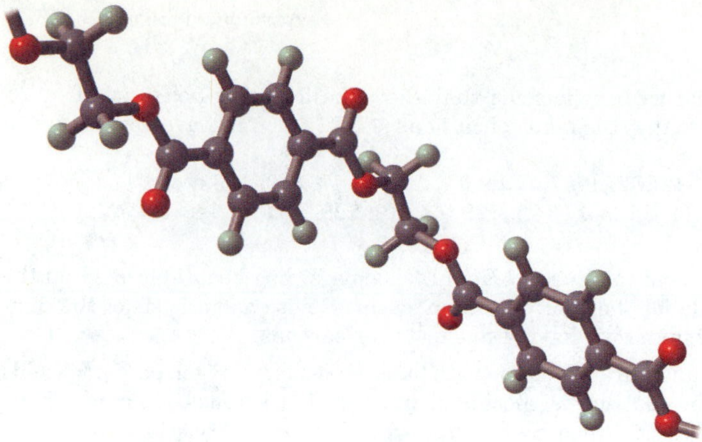

Polyester polymers are used in a great many consumer applications.

21

Carboxylic Acid Derivatives and Nucleophilic Acyl Substitution Reactions

Closely related to the carboxylic acids discussed in the previous chapter are the **carboxylic acid derivatives, RCOY**, compounds in which the acyl group is bonded to an electronegative atom or substituent –Y that can act as a leaving group in substitution reactions. Many kinds of acid derivatives are known, but we'll be concerned only with four of the more common ones: acid halides, acid anhydrides, esters, and amides. Also in this chapter, we'll discuss *nitriles*, another class of compounds closely related to carboxylic acids.

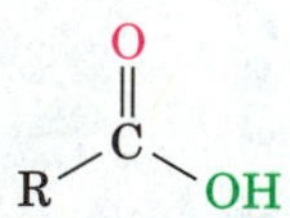

Carboxylic acid

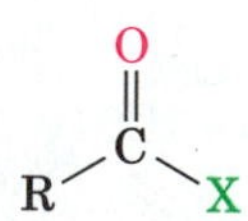

Acid halide (X=F, Cl, Br, I)

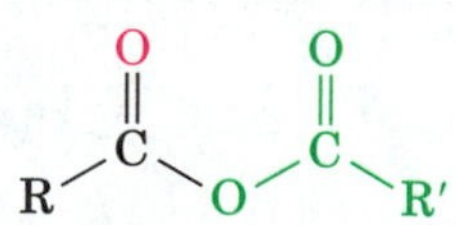

Acid anhydride

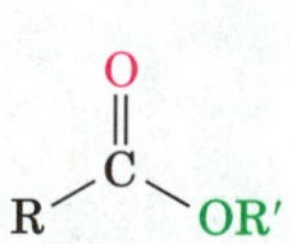

Ester

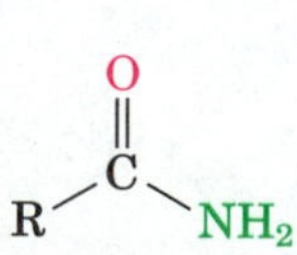

Amide

R—C≡N

Nitrile

The chemistry of all acid derivatives is similar and is dominated by a single general reaction—the **nucleophilic acyl substitution reaction**—that we saw briefly in the overview of carbonyl chemistry:

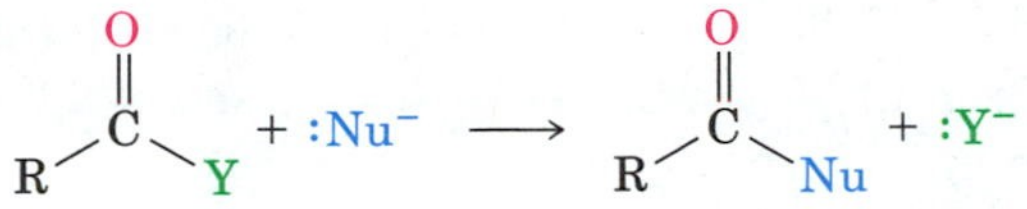

Let's first learn more about acid derivatives and then explore the chemistry of acyl substitution reactions.

21.1 Naming Carboxylic Acid Derivatives

Acid Halides: RCOX

Acid halides are named by identifying first the acyl group and then the halide. The acyl group name is derived from the carboxylic acid name by replacing the *-ic acid* ending with *-yl* or the *-carboxylic acid* ending with *-carbonyl*. For example:

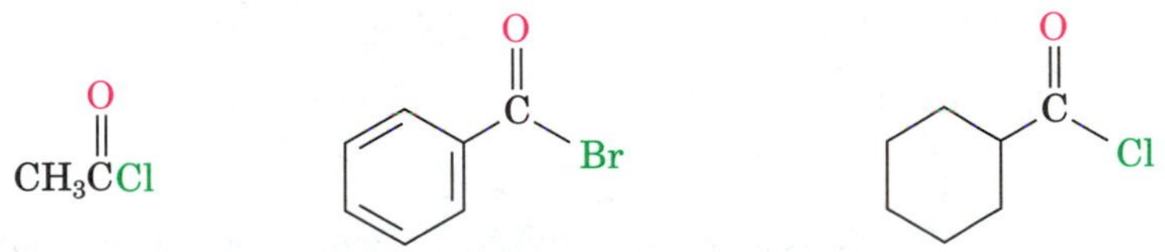

Acetyl chloride (from acetic acid) **Benzoyl bromide (from benzoic acid)** **Cyclohexanecarbonyl chloride (from cyclohexanecarboxylic acid)**

Acid Anhydrides: RCO_2COR'

Symmetrical anhydrides of unsubstituted monocarboxylic acids and cyclic anhydrides of dicarboxylic acids are named by replacing the word *acid* with *anhydride*:

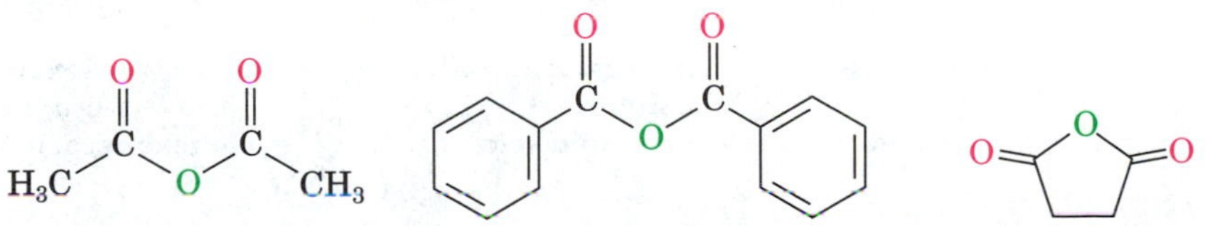

Acetic anhydride **Benzoic anhydride** **Succinic anhydride**

If the anhydride is derived from a substituted monocarboxylic acid, it is named by adding the prefix *bis-* (meaning two) to the acid name:

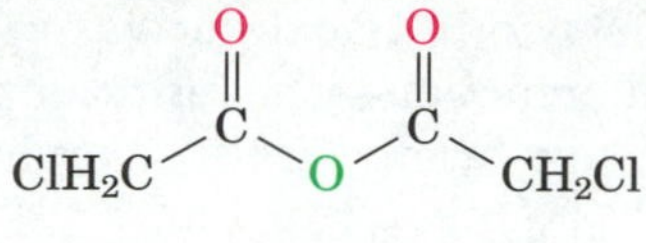

bis(chloroacetic) anhydride

Amides: $RCONH_2$

Amides with an unsubstituted $-NH_2$ group are named by replacing the *-oic acid* or *-ic acid* ending with *-amide*, or by replacing the *-carboxylic acid* ending with *-carboxamide*. For example:

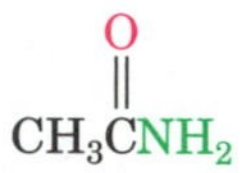

Acetamide (from acetic acid)

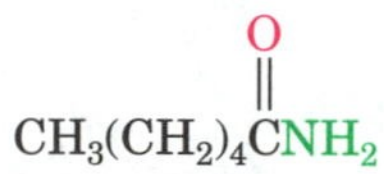

Hexanamide (from hexanoic acid)

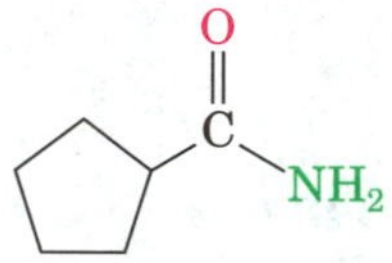

Cyclopentanecarboxamide (from cyclopentanecarboxylic acid)

If the nitrogen atom is further substituted, the compound is named by first identifying the substituent groups and then the parent amide name. The substituents are preceded by the letter *N* to identify them as being directly attached to nitrogen.

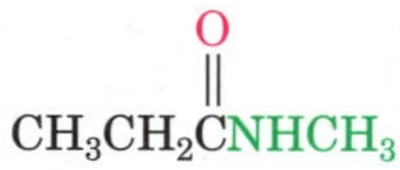

***N*-Methylpropanamide**

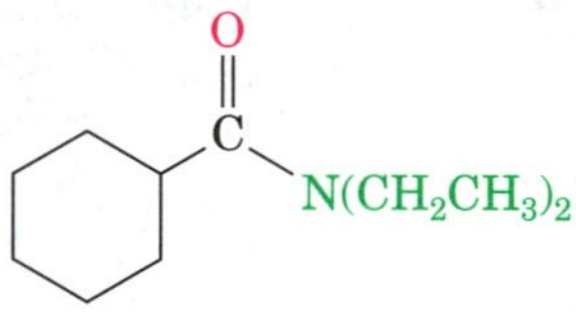

***N,N*-Diethylcyclohexanecarboxamide**

Esters: RCO_2R'

Esters are named by first identifying the alkyl group attached to oxygen and then the carboxylic acid, with the *-ic acid* ending replaced by *-ate*:

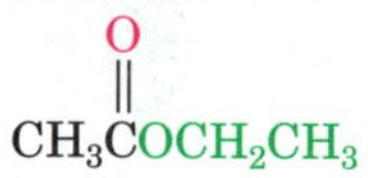

Ethyl acetate (the ethyl ester of acetic acid)

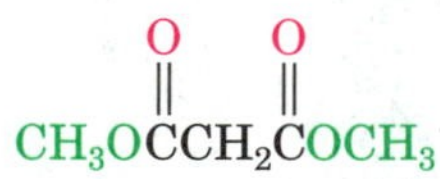

Dimethyl malonate (the dimethyl ester of malonic acid)

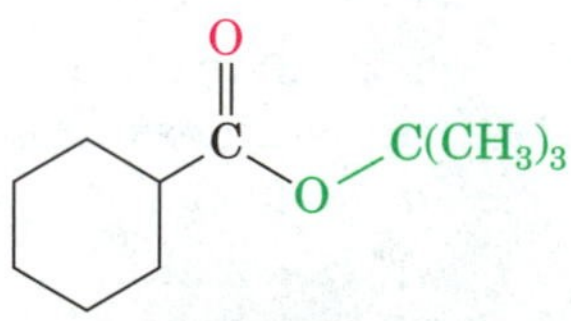

***tert*-Butylcyclohexanecarboxylate (the *tert*-butyl ester of cyclohexanecarboxylic acid)**

Nitriles: $RC{\equiv}N$

Compounds containing the $-C{\equiv}N$ functional group are called **nitriles**. Simple acyclic nitriles are named by adding *-nitrile* as a suffix to the alkane name, with the nitrile carbon numbered C1:

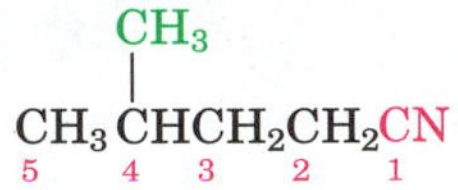

4-Methylpentanenitrile

More complex nitriles are named as derivatives of carboxylic acids by replacing the *-ic acid* or *-oic acid* ending with *-onitrile*, or by replacing the *-carboxylic acid* ending with *-carbonitrile*. The nitrile carbon atom is attached to C1 but is not itself numbered in this system.

$CH_3C{\equiv}N$

Acetonitrile
(from acetic acid)

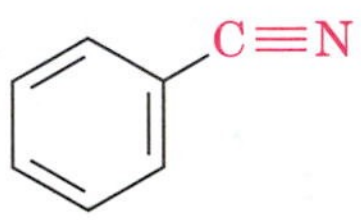

Benzonitrile
(from benzoic acid)

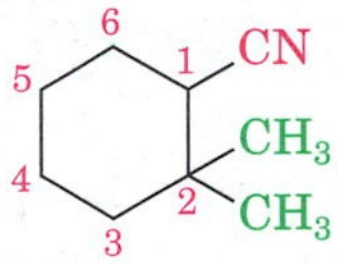

2,2-Dimethylcyclohexanecarbonitrile
(from 2,2-dimethylcyclohexane-carboxylic acid)

A summary of nomenclature rules for carboxylic acid derivatives is given in Table 21.1.

Table 21.1 Nomenclature of Carboxylic Acid Derivatives

Functional group	*Structure*	*Name ending*
Carboxylic acid	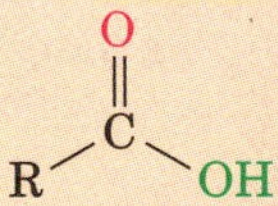	*-ic acid* (*-carboxylic acid*)
Acid halide	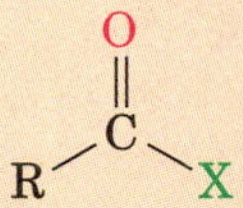	*-yl halide* (*-carbonyl halide*)
Acid anhydride	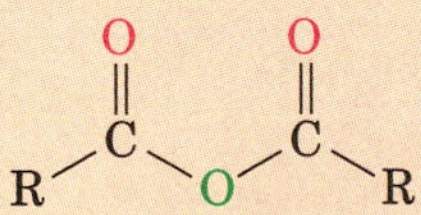	*anhydride*
Amide	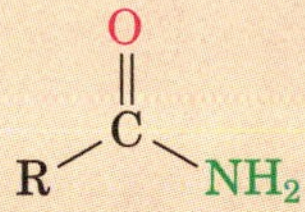	*-amide* (*-carboxamide*)
Ester	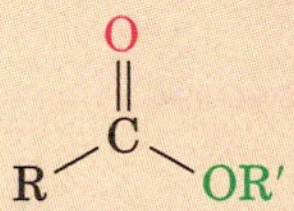	*-ate* (*-carboxylate*)
Nitrile	R—C≡N	*-onitrile* (*-carbonitrile*)

PROBLEM.

21.1 Give IUPAC names for the following substances:

(a) $CH_3\overset{CH_3}{\overset{|}{C}}HCH_2CH_2\overset{O}{\overset{\|}{C}}Cl$

(b) 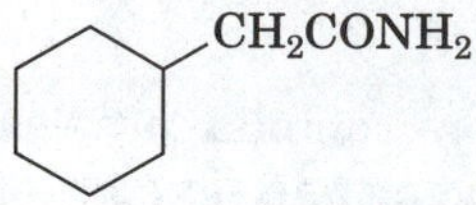

(c) $CH_3CH_2\overset{CH_3}{\overset{|}{C}}HCN$

(d) 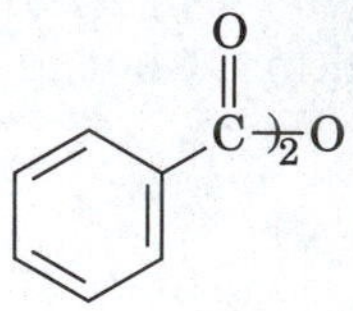

(e) 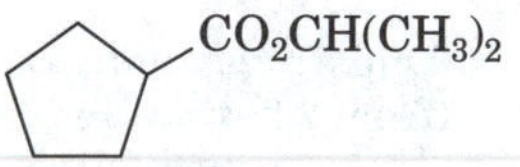

(f) 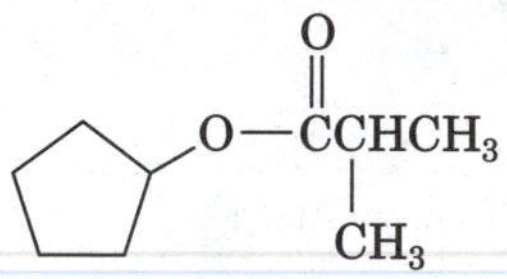

(g) $H_2C{=}CHCH_2CH_2\overset{O}{\overset{\|}{C}}NH_2$

(h) $CH_3CH_2\overset{CN}{\overset{|}{C}}HCH_2CH_3$

(i) 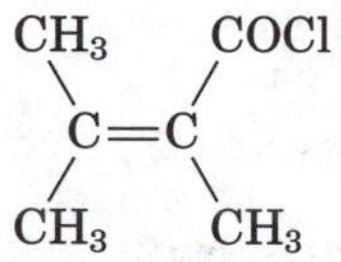

(j) $CF_3\overset{O}{\overset{\|}{C}}O\overset{O}{\overset{\|}{C}}CF_3$

PROBLEM.

21.2 The names shown for the following compounds are incorrect. Give the correct names.

(a) $\overset{4}{C}H_3\overset{3}{C}H_2\overset{2}{C}H{=}\overset{1}{C}HCN$

1-Pentenenitrile

(b) $CH_3CH_2CH_2CONHCH_3$

Methylbutanamide

(c) $(CH_3)_2CHCH_2\overset{CH_3}{\overset{|}{C}}HCOCl$

2,4-Methylpentanoyl chloride

(d) 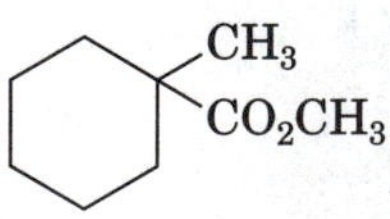

Methyl-2-methylcyclohexane carboxylate

21.2 Nucleophilic Acyl Substitution Reactions

The addition of a nucleophile to the polar C=O bond is the key step in three of the four major carbonyl-group reactions. When nucleophiles add to aldehydes and ketones, we saw in Chapter 19 that the initially formed tetrahedral intermediate either can be protonated to yield an alcohol or can eliminate the carbonyl oxygen, leading to a new C=Nu bond. When nucleophiles add to carboxylic acid derivatives, however, a different reaction course is followed. The initially formed tetrahedral intermediate eliminates one of the two substituents originally bonded to the carbonyl carbon, leading to a net nucleophilic acyl substitution (Figure 21.1).

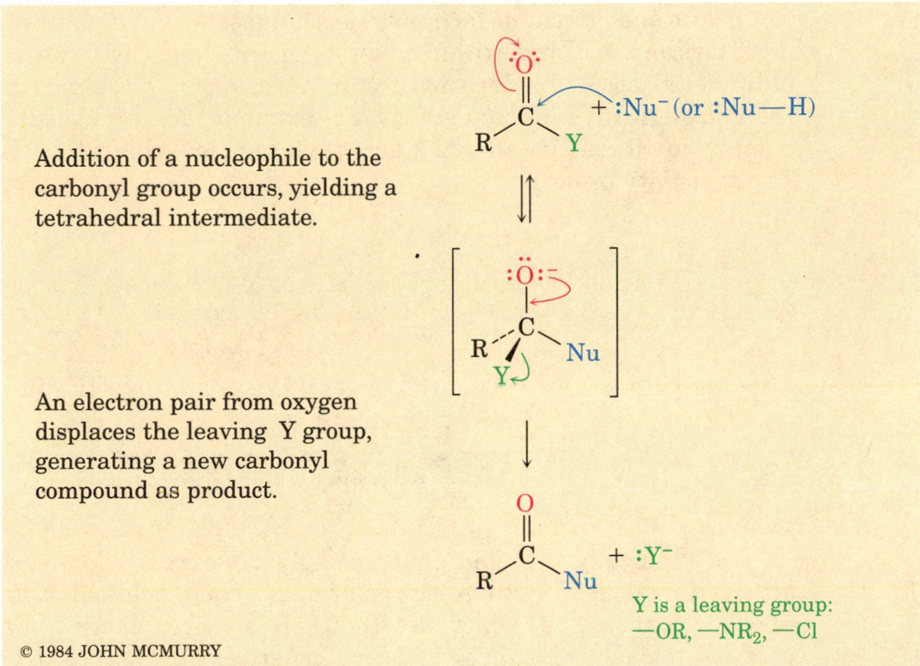

Figure 21.1 General mechanism of nucleophilic acyl substitution reactions.

The difference in behavior of ketones/aldehydes and carboxylic acid derivatives is a consequence of structure. Carboxylic acid derivatives have an acyl function bonded to a potential leaving group, Y. As soon as the tetrahedral intermediate is formed, the leaving group is expelled to generate a new carbonyl compound. Ketones and aldehydes have no such leaving group, however, and therefore don't undergo substitution.

As shown in Figure 21.1, the net effect of the addition/elimination sequence is a substitution by the attacking nucleophile for the Y group originally bonded to the acyl carbon. Thus, the overall reaction is superficially similar to the kind of nucleophilic substitution that occurs during S_N2 reactions (Section 11.5), but the *mechanisms* of the two reactions are completely different. S_N2 reactions occur in a single step by back-side displacement of the leaving group; nucleophilic acyl substitutions take place in two steps and involve a tetrahedral intermediate.

21.3 Relative Reactivity of Carboxylic Acid Derivatives

Nucleophilic acyl substitution reactions take place in two steps: addition of the nucleophile, and elimination of a leaving group. Although both steps can affect the overall rate of reaction, it's generally the first step that is rate-limiting. Thus, any factor that makes the carbonyl group more easily attacked by nucleophiles favors the reaction.

Steric and electronic factors are both important in determining reactivity. Sterically, we find within a series of the same acid derivatives that unhindered, accessible carbonyl groups react with nucleophiles more readily than do sterically hindered groups. For example, acetyl chloride, CH_3COCl, is much more reactive than 2,2-dimethylpropanoyl chloride, $(CH_3)_3CCOCl$. The reactivity order is:

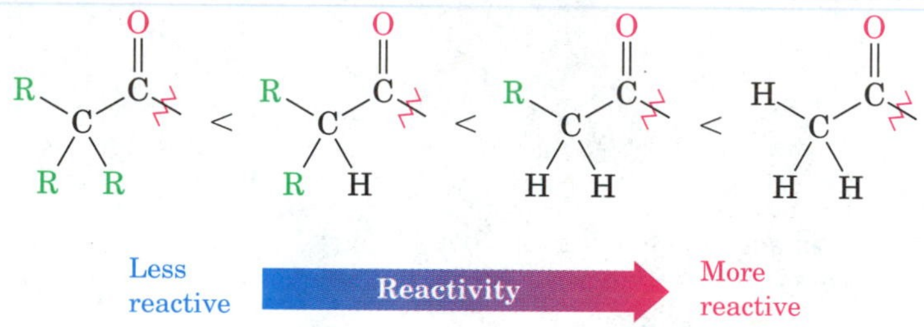

Electronically, we find that strongly polarized acid derivatives are attacked more readily than less polar ones. Thus, acid chlorides are more reactive than esters, which are themselves more reactive than amides, because the electronegative chlorine polarizes the carbonyl group more strongly than does an alkoxy group or an amino group. The reactivity order is:

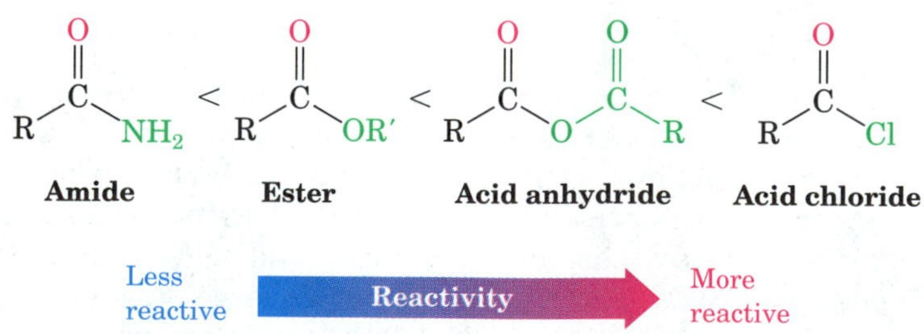

The way that various substituents affect the polarization of a carbonyl group is similar to the way they affect the reactivity of an aromatic ring toward electrophilic substitution (Section 16.6). A chlorine substituent, for example, inductively withdraws electrons from an acyl group in the same way that it withdraws electrons from an aromatic ring. Similarly, amino and methoxyl substituents donate electrons to acyl groups by resonance in the same way that they donate electrons to aromatic rings.

An important consequence of the observed reactivity order is that *it's usually possible to transform a more reactive acid derivative into a less reactive one*. As we'll see in the next few sections, acid chlorides can be converted into anhydrides, esters, and amides, but amides can't readily be converted into esters, anhydrides, or acid chlorides. Remembering the reactivity order is therefore a way to keep track of a large number of reactions. Figure 21.2 shows the transformations that can be carried out.

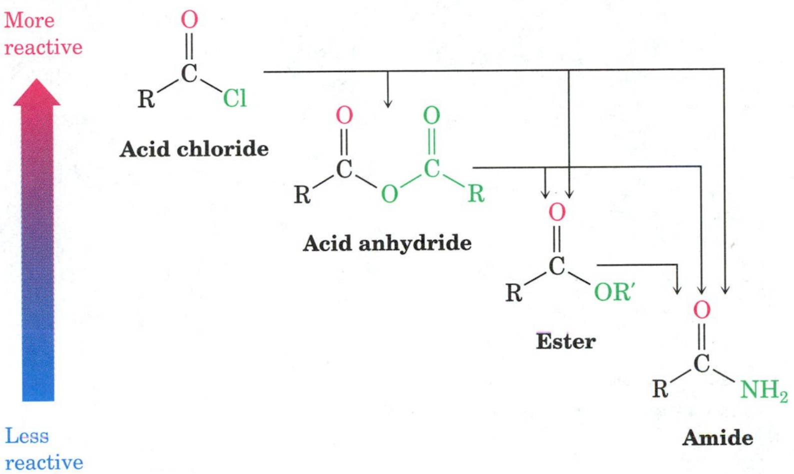

Figure 21.2 Interconversions of carboxylic acid derivatives.

Another consequence of the reactivity differences among carboxylic acid derivatives is that only esters and amides are commonly found in nature. Acid halides and acid anhydrides undergo rapid nucleophilic attack by water and are too reactive to exist in living organisms. Esters and amides, however, are stable enough to occur widely and to be vitally important in many life processes.

In studying the chemistry of acid derivatives in the next few sections, we'll find that there are striking similarities among the various types of compounds. We'll be concerned largely with the reactions of just a few nucleophiles and will see that the same kinds of reactions keep occurring (Figure 21.3, p. 810):

Hydrolysis: Reaction with water to yield a carboxylic acid

Alcoholysis: Reaction with an alcohol to yield an ester

Aminolysis: Reaction with ammonia or an amine to yield an amide

Reduction: Reaction with a hydride reducing agent to yield an aldehyde or an alcohol

Grignard reaction: Reaction with an organometallic reagent to yield a ketone or an alcohol

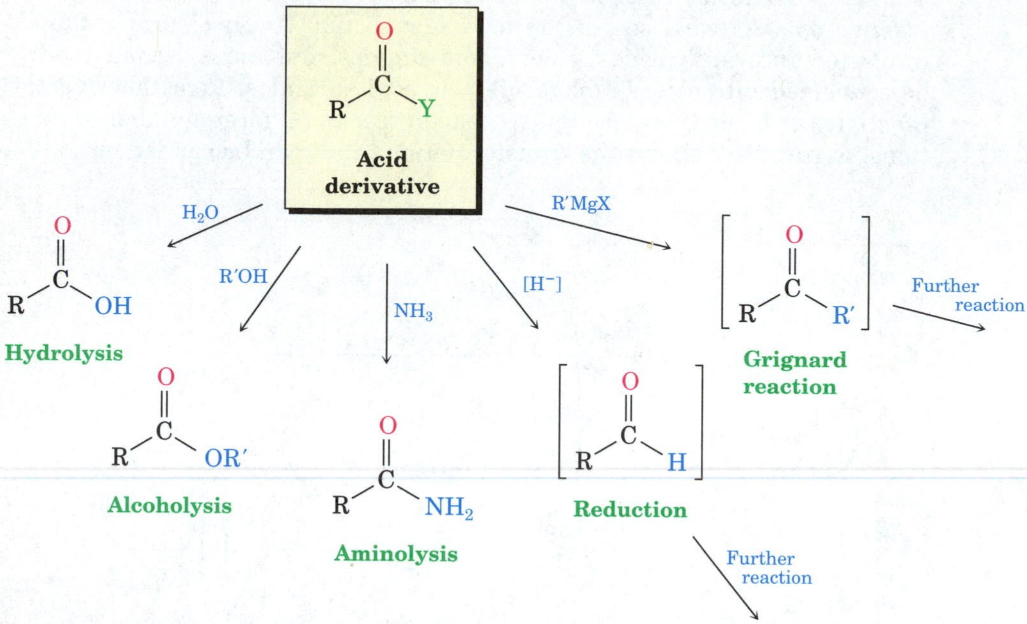

Figure 21.3 Some general reactions of carboxylic acid derivatives.

PROBLEM. . . . .

21.3 Rank the compounds in each of the following sets with regard to their expected reactivity toward nucleophilic acyl substitution.

(a) $CH_3\overset{\overset{O}{\|}}{C}Cl$, $CH_3\overset{\overset{O}{\|}}{C}OCH_3$, $CH_3\overset{\overset{O}{\|}}{C}NH_2$

(b) $CH_3\overset{\overset{O}{\|}}{C}OCH_3$, $CH_3\overset{\overset{O}{\|}}{C}OCH_2CCl_3$, $CH_3\overset{\overset{O}{\|}}{C}OCH(CF_3)_2$

PROBLEM. . . . .

21.4 How can you account for the fact that methyl trifluoroacetate, CF_3COOCH_3, is more reactive than methyl acetate, CH_3COOCH_3, in nucleophilic acyl substitution reactions?

. . . .

21.4 Nucleophilic Acyl Substitution Reactions of Carboxylic Acids

Among the most important reactions of carboxylic acids are those that convert the carboxyl group into another acid derivative by a nucleophilic acyl substitution, RCOOH → RCOY. Acid chlorides, anhydrides, esters, and amides can all be prepared from carboxylic acids (Figure 21.4).

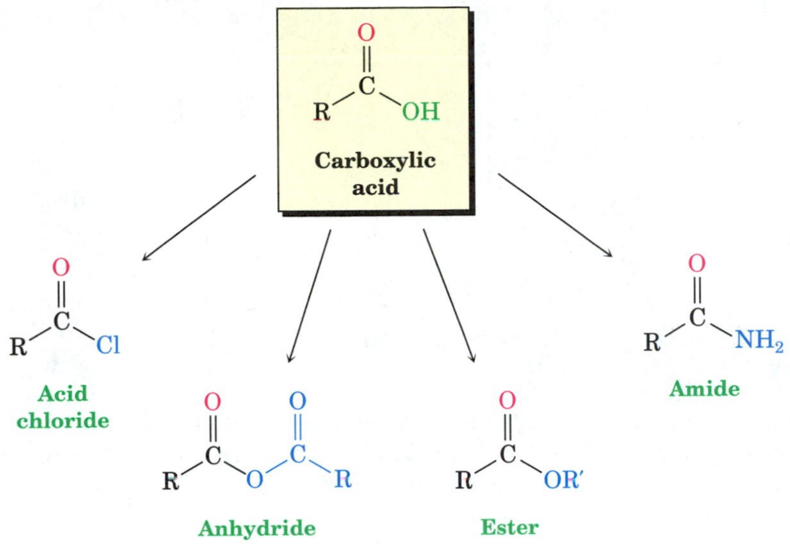

Figure 21.4 Some nucleophilic acyl substitution reactions of carboxylic acids.

Conversion of Carboxylic Acids into Acid Chlorides

Carboxylic acids are converted into acid chlorides by treatment with thionyl chloride ($SOCl_2$):

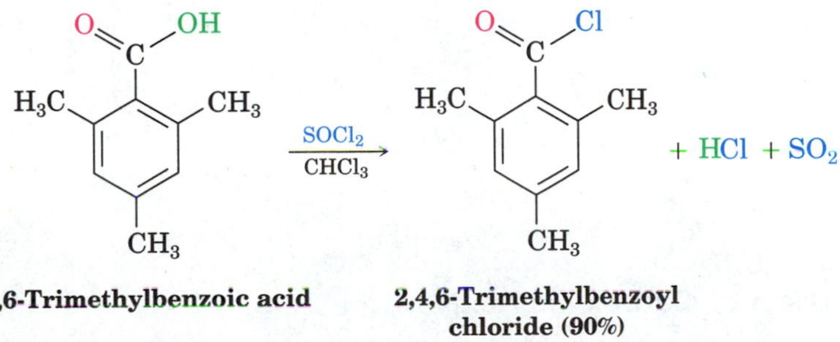

The reaction occurs by a nucleophilic acyl substitution pathway in which the carboxylic acid is first converted into a reactive *chlorosulfite* intermediate, which is then attacked by a nucleophilic chloride ion.

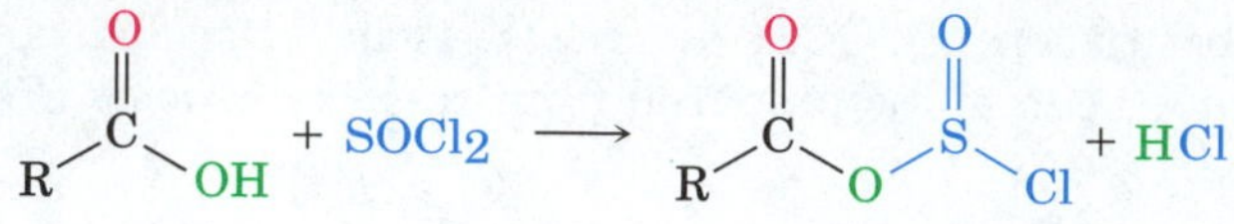

Carboxylic acid **A chlorosulfite**

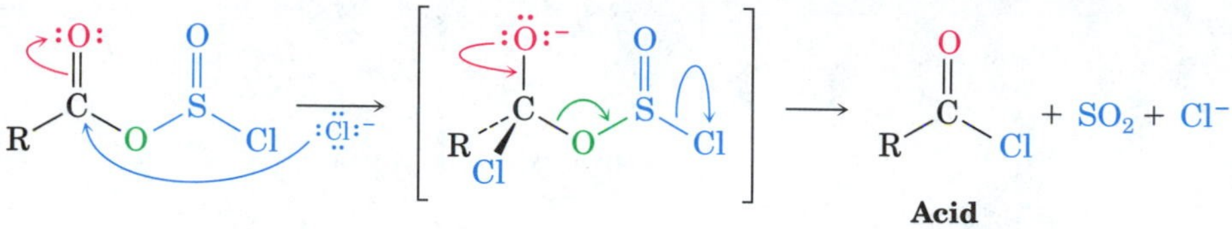

Acid chloride

Conversion of Carboxylic Acids into Acid Anhydrides

Acid anhydrides are derived from two molecules of carboxylic acid by removing 1 equivalent of water. Acyclic anhydrides are difficult to prepare directly from the corresponding acids, however, and only acetic anhydride is commonly used.

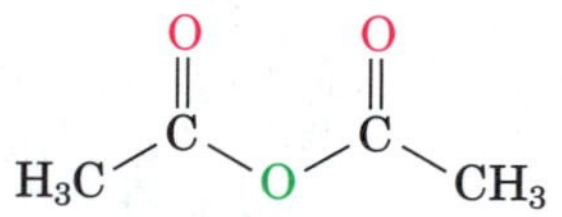

Acetic anhydride

Cyclic anhydrides with five- or six-membered rings are obtained by high-temperature dehydration of the diacids.

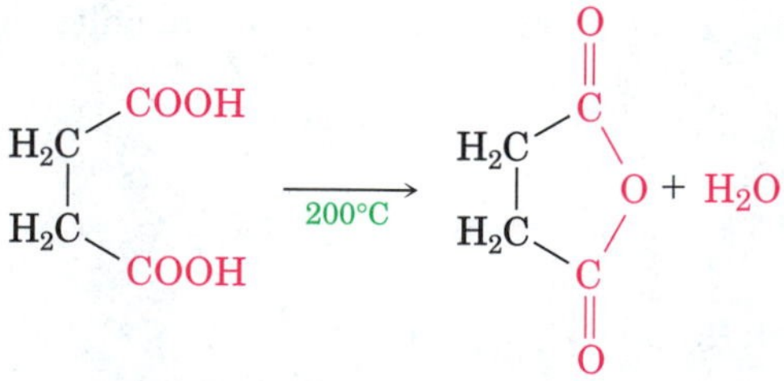

Succinic acid **Succinic anhydride**

Conversion of Carboxylic Acids into Esters

Perhaps the most important reaction of carboxylic acids is their conversion into esters, RCOOH → RCOOR′. There are many methods for accomplishing the transformation, including the S_N2 reaction between a carboxylate anion and a primary alkyl halide that was studied in Section 11.5.

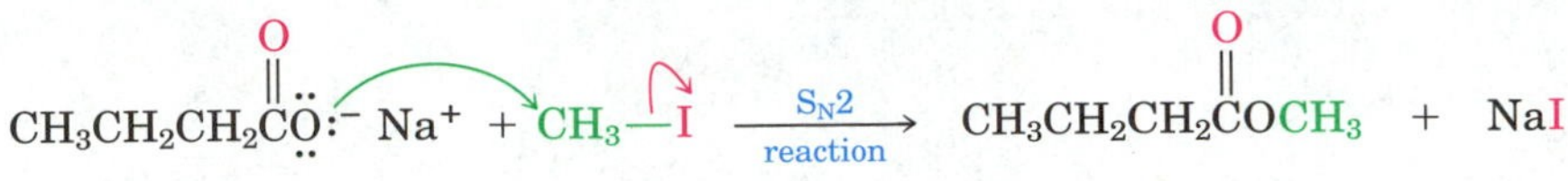

Sodium butanoate **Methyl butanoate, an ester (97%)**

Esters can also be synthesized by a nucleophilic acyl substitution reaction between a carboxylic acid and an alcohol. Fischer[1] and Speier discovered in 1895 that esters result simply from heating a carboxylic acid in alcohol solution containing a small amount of strong acid catalyst. Yields are good in this **Fischer esterification reaction**, but the need to use excess alcohol as solvent effectively limits the method to the synthesis of methyl, ethyl, and propyl esters.

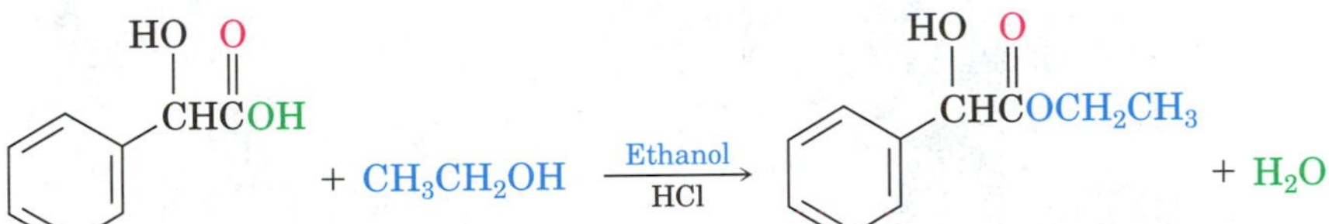

Mandelic acid **Ethyl mandelate (86%)**

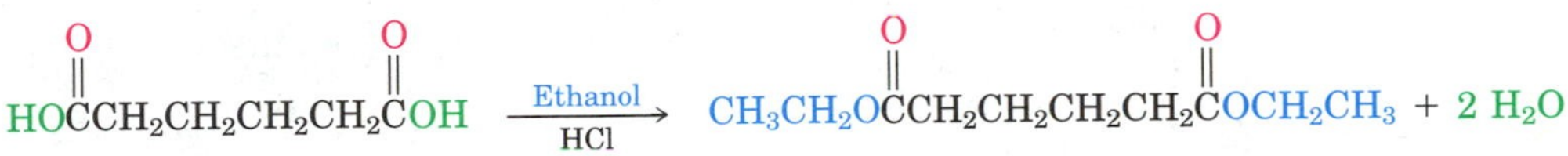

Hexanedioic acid (Adipic acid) **Diethyl hexanedioate (95%)**

The Fischer esterification reaction is a nucleophilic acyl substitution reaction carried out under acidic conditions, as shown in Figure 21.5 (p. 814). Carboxylic acids are not reactive enough to be attacked by neutral alcohols, but they can be made much more reactive in the presence of a strong mineral acid such as HCl or H_2SO_4. The mineral acid protonates the carbonyl-group oxygen atom, thereby giving the carboxylic acid a positive charge and rendering it much more reactive toward nucleophilic attack by alcohol. Subsequent loss of water from the tetrahedral intermediate yields the ester product.

The net effect of Fischer esterification is substitution of an –OH group by –OR′. All steps are reversible, and the reaction can be driven in either direction by choice of reaction conditions. Ester formation is favored when a large excess of alcohol is used as solvent, but carboxylic acid formation is favored when a large excess of water is present.

[1]Emil Fischer (1852–1919); b. Euskirchen, Germany; Ph.D. Strasbourg (Baeyer); professor, universities of Erlangen, Würzburg, and Berlin; Nobel Prize (1902).

Protonation of the carbonyl oxygen activates the carboxylic acid . . .

. . . toward nucleophilic attack by alcohol, yielding a tetrahedral intermediate.

Transfer of a proton from one oxygen atom to another yields a second tetrahedral intermediate and converts the OH group into a good leaving group.

Loss of a proton regenerates the acid catalyst and gives the ester product.

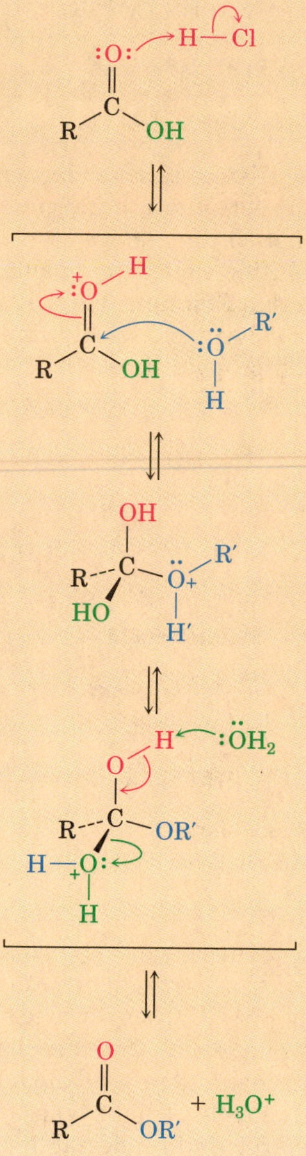

Figure 21.5 Mechanism of Fischer esterification. The reaction is an acid-catalyzed, nucleophilic acyl substitution of a carboxylic acid.

One of the best pieces of evidence in support of the mechanism shown in Figure 21.5 comes from isotope-labeling experiments. When ^{18}O-labeled methanol reacts with benzoic acid, the methyl benzoate produced is found to be ^{18}O-labeled, but the water produced is unlabeled. Thus, it is the CO–OH bond of the carboxylic acid that is broken during the reaction rather than the COO–H bond, and the RO–H bond of the alcohol that is broken rather than the R–OH bond.

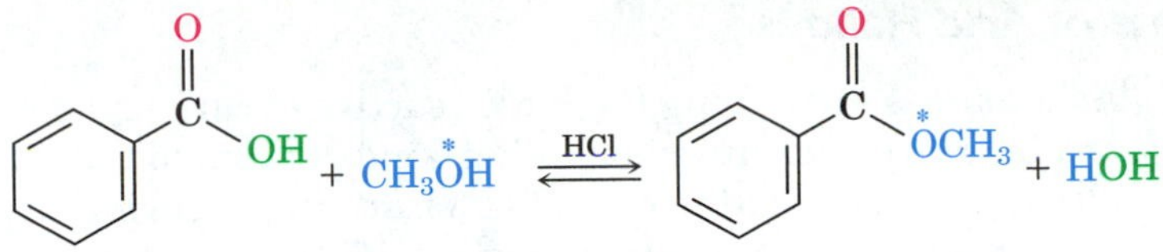

PROBLEM...

21.5 How would you prepare the following esters?
(a) Butyl acetate (b) Methyl butanoate

PROBLEM...

21.6 If 5-hydroxypentanoic acid is treated with acid catalyst, an intramolecular esterification reaction occurs. What is the structure of the product? (*Intramolecular* means within the same molecule.)

Conversion of Carboxylic Acids into Amides

Amides are difficult to prepare by direct reaction of carboxylic acids with amines because amines are bases that convert acidic carboxyl groups into their carboxylate anions. Since the carboxylate anion has a negative charge, it is no longer electrophilic and no longer likely to be attacked by nucleophiles except at a high temperature. We'll see a better method for making amides from acids in Section 27.11 in connection with the synthesis of proteins from amino acids.

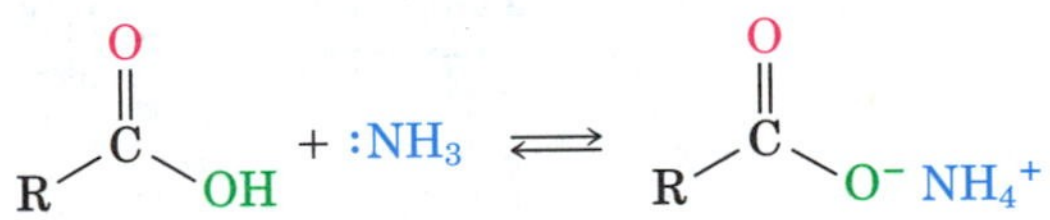

21.5 Chemistry of Acid Halides

Preparation of Acid Halides

Acid chlorides are prepared from carboxylic acids by reaction with thionyl chloride ($SOCl_2$), as we saw in the previous section. Reaction of a carboxylic acid with phosphorus tribromide (PBr_3) yields the acid bromide.

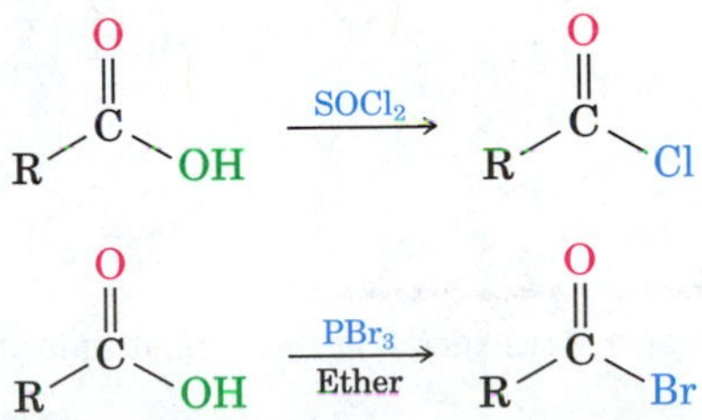

Reactions of Acid Halides

Acid halides are among the most reactive of carboxylic acid derivatives and can be converted into many other kinds of compounds. For example, we've already seen the value of acid chlorides in preparing aryl alkyl ketones by the Friedel–Crafts reaction (Section 16.3).

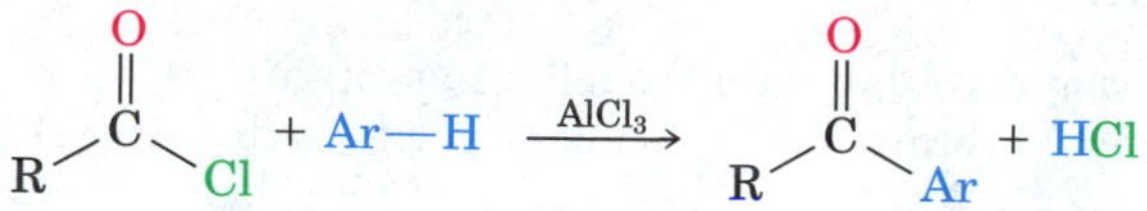

Most acid halide reactions occur by a nucleophilic acyl substitution mechanism. As shown in Figure 21.6, the halogen can be replaced by –OH to yield an acid, by –OR to yield an ester, or by $-NH_2$ to yield an amide. In addition, the reduction of an acid halide yields a primary alcohol, and reaction with a Grignard reagent yields a tertiary alcohol. Although the reactions we'll be discussing in this section are illustrated only for acid chlorides, they also occur with other acid halides.

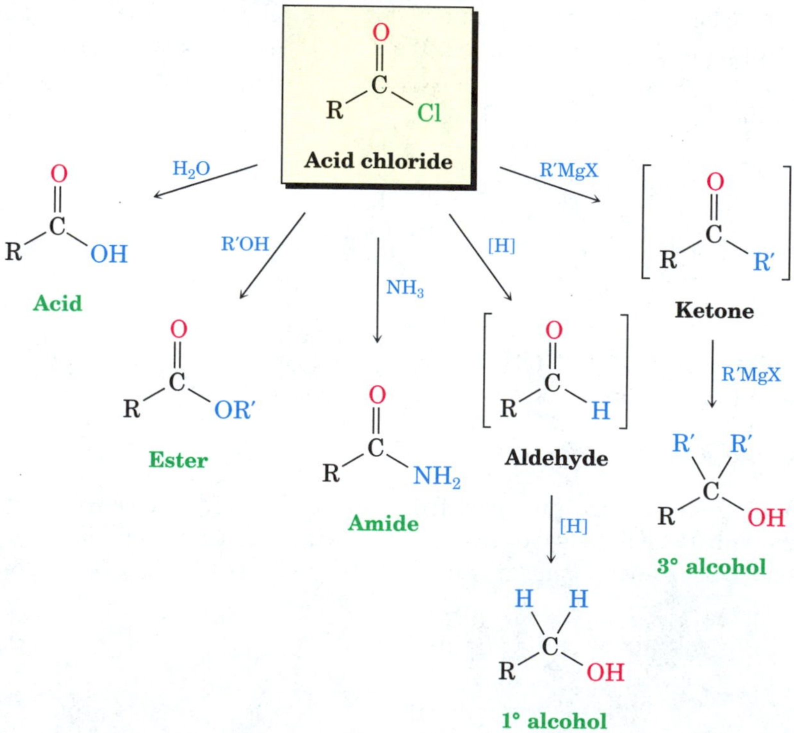

Figure 21.6 Some nucleophilic acyl substitution reactions of acid chlorides.

Hydrolysis: Conversion of Acid Halides into Acids Acid chlorides react with water to yield carboxylic acids. This hydrolysis reaction is a typical nucleophilic acyl substitution process, initiated by attack of water on the acid chloride carbonyl group. The tetrahedral intermediate undergoes elimination of Cl^- and loss of H^+ to give the product carboxylic acid plus HCl.

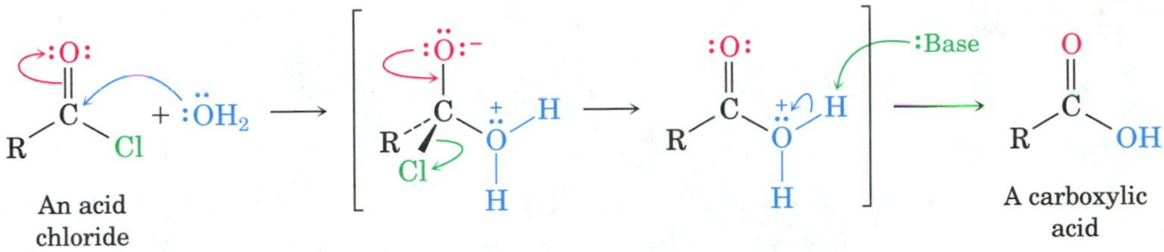

Since HCl is generated during the hydrolysis, the reaction is often carried out in the presence of a base such as pyridine or NaOH to scavenge the HCl and prevent it from causing side reactions.

Alcoholysis: Conversion of Acid Halides into Esters Acid chlorides react with alcohols to yield esters in a process analogous to their reaction with water to yield acids.

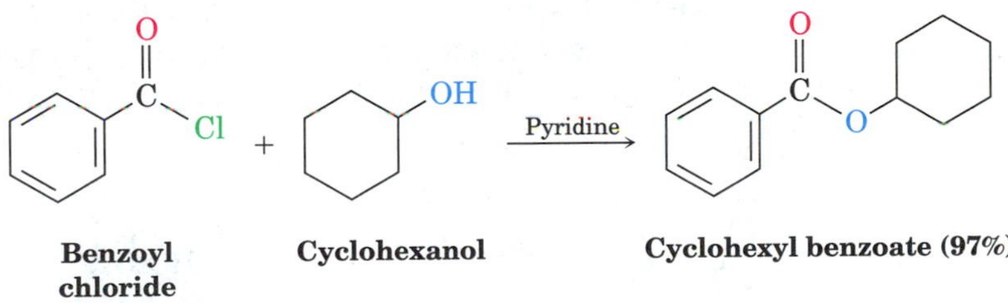

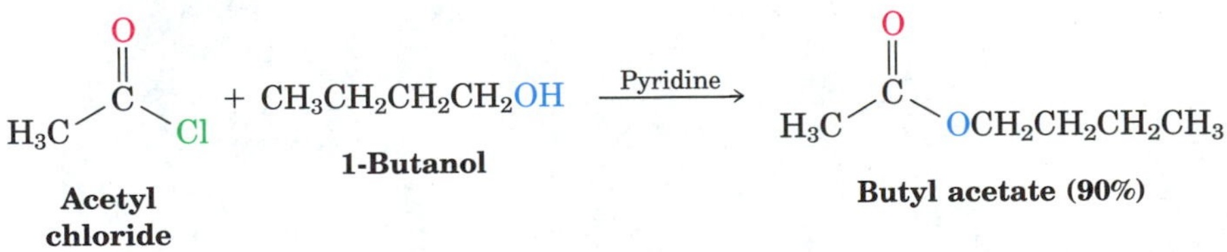

As with hydrolysis, alcoholysis reactions are usually carried out in the presence of pyridine or NaOH to react with the HCl formed and prevent it from causing side reactions.

The esterification reaction of an alcohol with an acid chloride is strongly affected by steric hindrance. Bulky groups on either partner slow down the reaction considerably, resulting in a reactivity order among alcohols of primary > secondary > tertiary. As a result, it's often possible to esterify an unhindered alcohol selectively in the presence of a more hindered one. This can be important in complex syntheses, where it is often necessary to distinguish between similar functional groups. For example:

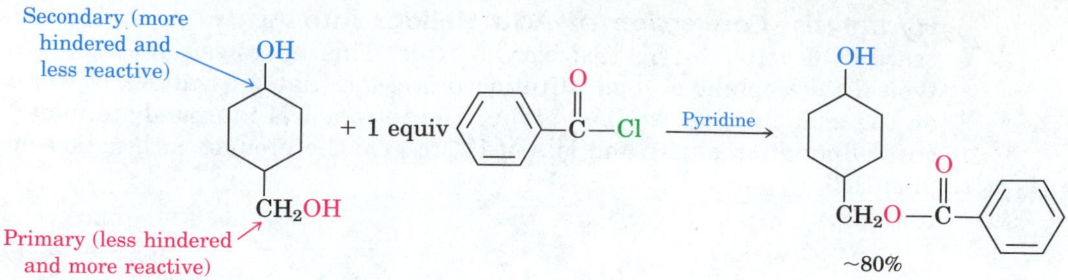

PROBLEM. . . . .

21.7 How might you prepare the following esters using a nucleophilic acyl substitution reaction of an acid chloride?
(a) $CH_3CH_2COOCH_3$ (b) $CH_3COOCH_2CH_3$ (c) Ethyl benzoate

PROBLEM. . . . .

21.8 Which method would you choose if you wanted to prepare cyclohexyl benzoate, Fischer esterification or reaction of an acid chloride with an alcohol? Explain.

. . . .

Aminolysis: Conversion of Acid Halides into Amides Acid chlorides react rapidly with ammonia and amines to give amides in good yield. Both mono- and disubstituted amines can be used, but not trisubstituted amines.

$$(CH_3)_2CH\overset{O}{\overset{\|}{C}}Cl + 2\ {:}NH_3 \longrightarrow (CH_3)_2CH\overset{O}{\overset{\|}{C}}NH_2 + \overset{+}{N}H_4\overset{-}{Cl}$$

2-Methylpropanoyl chloride **2-Methylpropanamide (83%)**

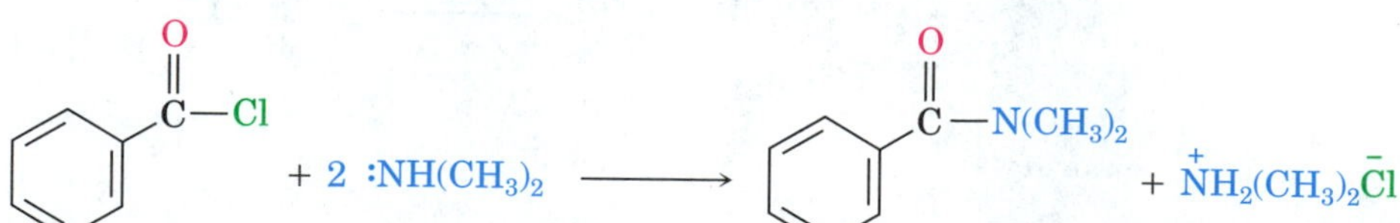

Benzoyl chloride ***N,N*-Dimethylbenzamide (92%)**

Since HCl is formed during the reaction, 2 equivalents of the amine must be used. One equivalent reacts with the acid chloride, and 1 equivalent reacts with the HCl by-product to form an ammonium chloride salt. If, however, the amine component is valuable, amide synthesis is often carried out using 1 equivalent of the desired amine plus 1 equivalent of an inexpensive base such as NaOH.

Aminolysis reactions carried out with NaOH present are sometimes referred to as *Schotten–Baumann reactions*, after their discoverers.[2] For example, the medically useful sedative trimetozine is prepared by reaction of 3,4,5-trimethoxybenzoyl chloride with the amine morpholine in the presence of 1 equivalent of NaOH.

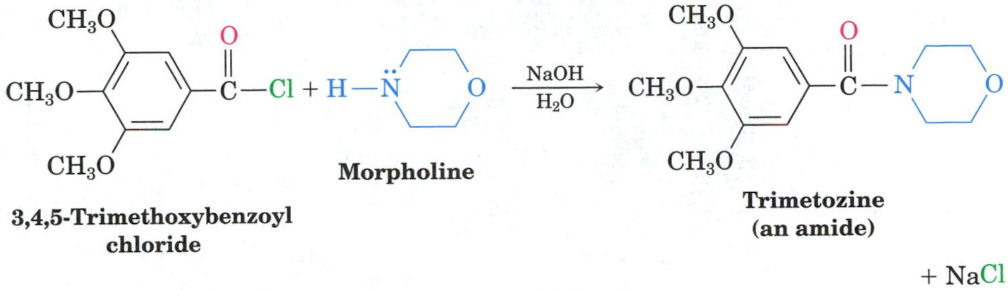

3,4,5-Trimethoxybenzoyl chloride **Morpholine** **Trimetozine (an amide)**

PROBLEM

21.9 Write the mechanism of the reaction between 3,4,5-trimethoxybenzoyl chloride and morpholine to form trimetozine.

PROBLEM

21.10 Trisubstituted amines such as triethylamine can be used in place of NaOH to scavenge HCl during aminolysis reactions. Why doesn't triethylamine react with acid chlorides to yield amides?

PROBLEM

21.11 How could you prepare the following amides using an acid chloride and an amine or ammonia?
(a) $CH_3CH_2CONHCH_3$ (b) *N,N*-Diethylbenzamide (c) Propanamide

Reduction: Conversion of Acid Chlorides into Alcohols Acid chlorides are reduced by $LiAlH_4$ to yield primary alcohols. The reaction is of little practical value, however, because the parent carboxylic acids are generally more readily available and are themselves reduced by $LiAlH_4$ to yield alcohols.

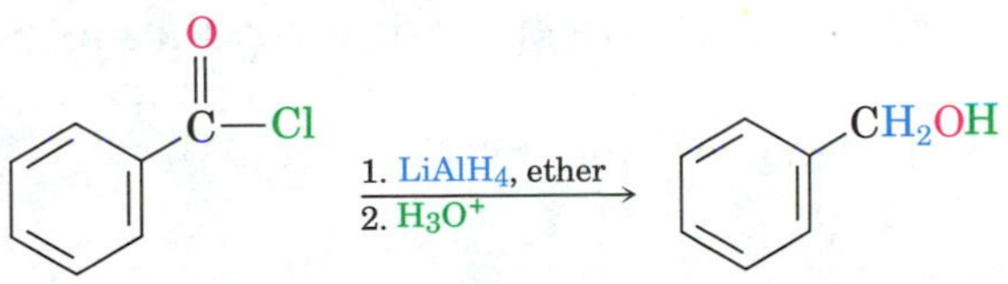

Benzoyl chloride **Benzyl alcohol (96%)**

[2]Carl Schotten (1853–1910); b. Marburg, Germany; Ph.D. Berlin (Hofmann); professor, University of Berlin.

Reduction occurs via a typical nucleophilic acyl substitution mechanism in which a hydride ion (H:⁻) attacks the carbonyl group, yielding a tetrahedral intermediate that expels Cl^-. The net effect is a substitution of –Cl by –H to yield an aldehyde, which is then immediately reduced by $LiAlH_4$ in a second step to yield the primary alcohol.

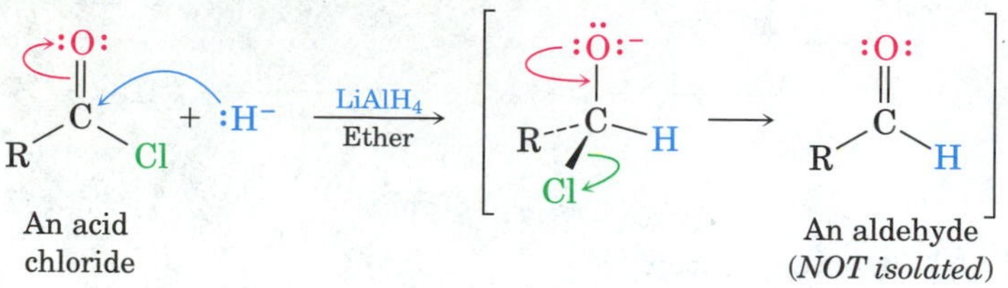

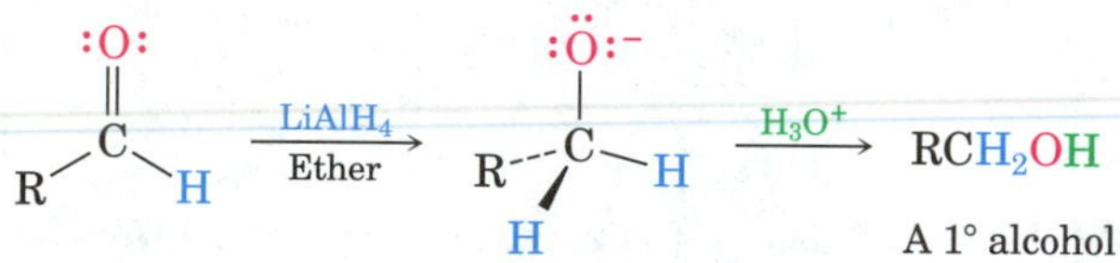

The aldehyde intermediate can be isolated if the less powerful hydride reducing agent lithium tri-*tert*-butoxyaluminum hydride is used in place of $LiAlH_4$. This reagent, which is obtained by reaction of $LiAlH_4$ with 3 equivalents of *tert*-butyl alcohol, is particularly effective for carrying out the partial reduction of acid chlorides to aldehydes (Section 19.3).

$$3\ (CH_3)_3COH + LiAlH_4 \longrightarrow Li^+\ {}^-AlH[OC(CH_3)_3]_3 + 3\ H_2$$

Lithium tri-*tert*-butoxyaluminum hydride

***p*-Nitrobenzoyl chloride** ***p*-Nitrobenzaldehyde (81%)**

Reaction of Acid Chlorides with Organometallic Reagents Grignard reagents react with acid chlorides to yield tertiary alcohols in which two of the substituents are the same:

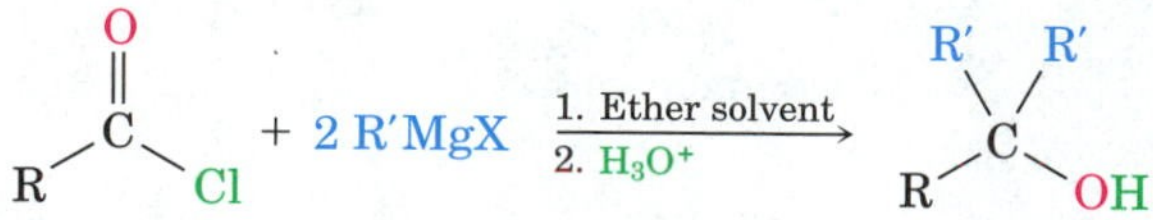

The mechanism for this Grignard reaction is similar to that of $LiAlH_4$ reduction. The first equivalent of Grignard reagent adds to the acid chloride, loss of Cl^- from the tetrahedral intermediate yields a ketone, and a second equivalent of Grignard reagent immediately adds to the ketone to produce an alcohol.

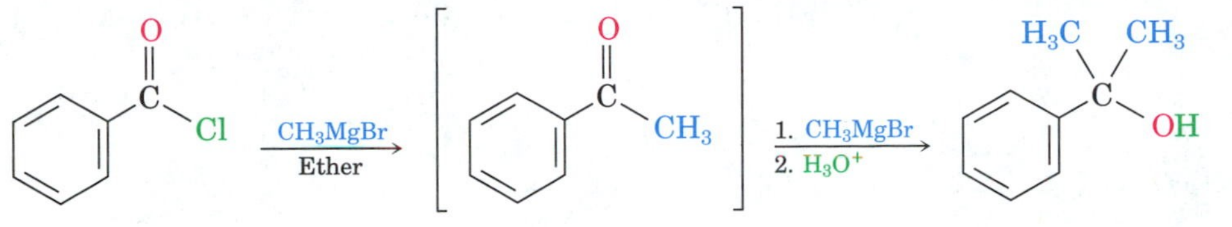

Benzoyl chloride **Acetophenone (*NOT isolated*)** **2-Phenyl-2-propanol (92%)**

The ketone intermediate formed during the Grignard reaction of an acid chloride can't usually be isolated because addition of the second equivalent of organomagnesium reagent occurs so rapidly. Ketones can, however, be isolated from the reaction of acid chlorides with diorganocopper (Gilman) reagents (Section 19.4):

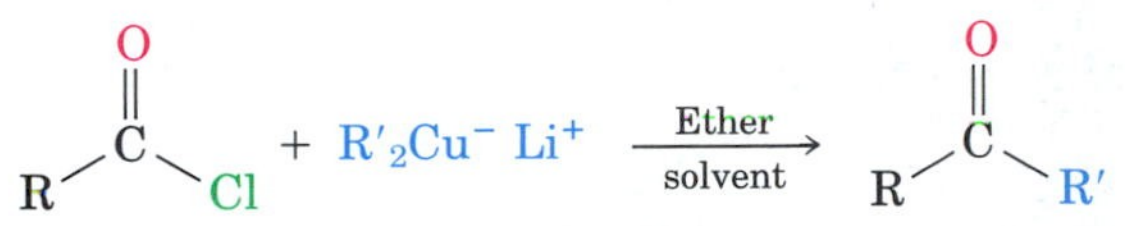

An acid chloride — A Gilman reagent — A ketone

Despite their apparent similarity to Grignard reactions, these diorganocopper reactions are probably not typical nucleophilic acyl substitution processes. Rather, it's thought that diorganocopper reactions occur by a radical pathway. The reactions are generally carried out at −78°C in ether solution, and yields are often excellent. For example, manicone, a substance secreted by male ants to coordinate ant pairing and mating, has been synthesized by reaction of lithium diethylcopper with (*E*)-2,4-dimethyl-2-hexenoyl chloride:

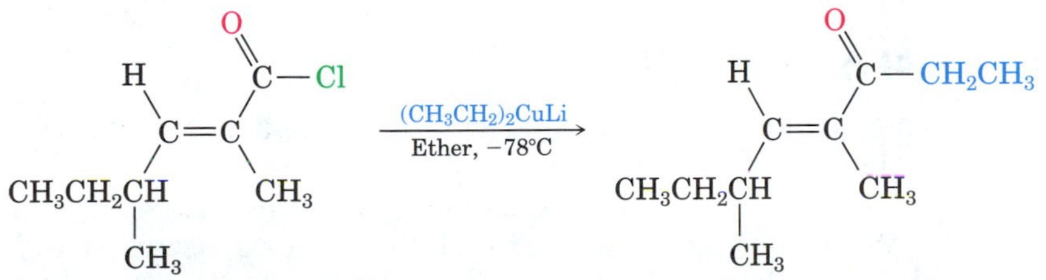

2,4-Dimethyl-2-hexenoyl chloride **Manicone (92%)**

Note that lithium diorganocopper reagents react only with acid chlorides. Acids, esters, anhydrides, and amides are inert to diorganocopper reagents.

PROBLEM.

21.12 Show how the following ketones might be prepared by reaction of an acid chloride with a lithium diorganocopper reagent.

(a) 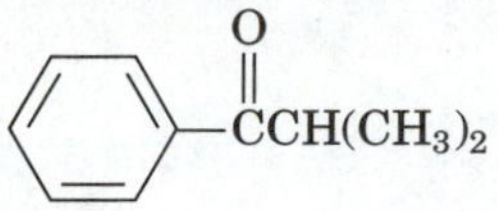

(b) 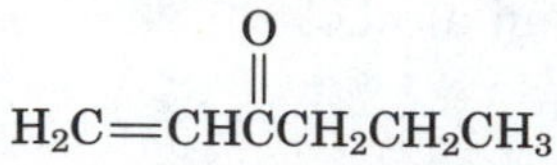

PROBLEM.

21.13 Ketones can sometimes be prepared by reaction of acid chlorides with 1 equivalent of Grignard reagent if the Grignard reagent is slowly added at low temperature to a solution of the acid chloride, rather than vice versa. Explain.

21.6 Chemistry of Acid Anhydrides

Preparation of Acid Anhydrides

The most general method for preparing acid anhydrides is by nucleophilic acyl substitution reaction of an acid chloride with a carboxylate anion. Both symmetrical and unsymmetrical acid anhydrides can be prepared in this way.

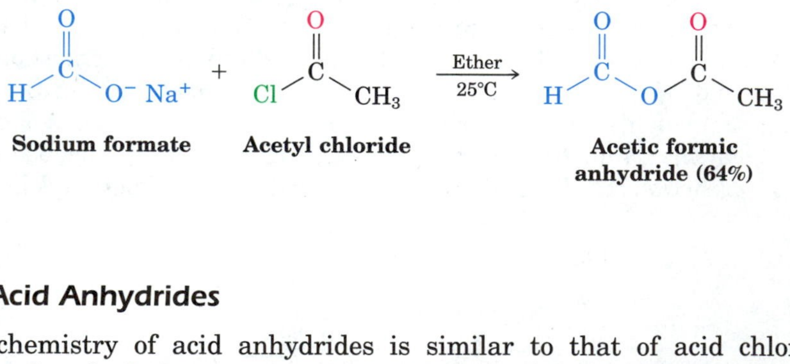

Sodium formate **Acetyl chloride** **Acetic formic anhydride (64%)**

Reactions of Acid Anhydrides

The chemistry of acid anhydrides is similar to that of acid chlorides. Although anhydrides react more slowly than acid chlorides, the kinds of reactions the two groups undergo are the same. Thus, acid anhydrides react with water to form acids, with alcohols to form esters, with amines to form amides, and with $LiAlH_4$ to form primary alcohols (Figure 21.7).

Acetic anhydride is often used to prepare acetate esters from alcohols and *N*-substituted acetamides from amines. For example, acetaminophen, a drug used in over-the-counter headache remedies, is prepared by reaction of *p*-hydroxyaniline with acetic anhydride. Aspirin (acetylsalicylic acid) is prepared similarly by the acetylation of *o*-hydroxybenzoic acid (salicylic acid) with acetic anhydride.

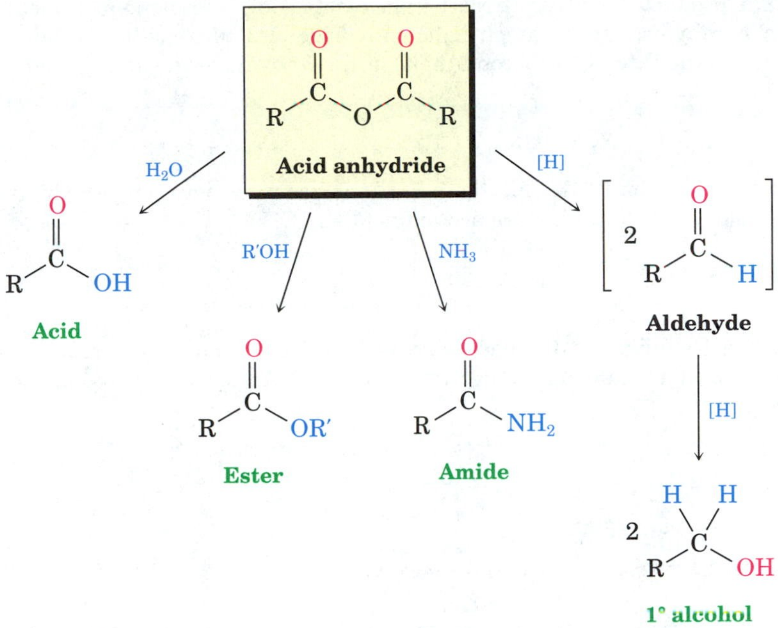

Figure 21.7 Some reactions of acid anhydrides.

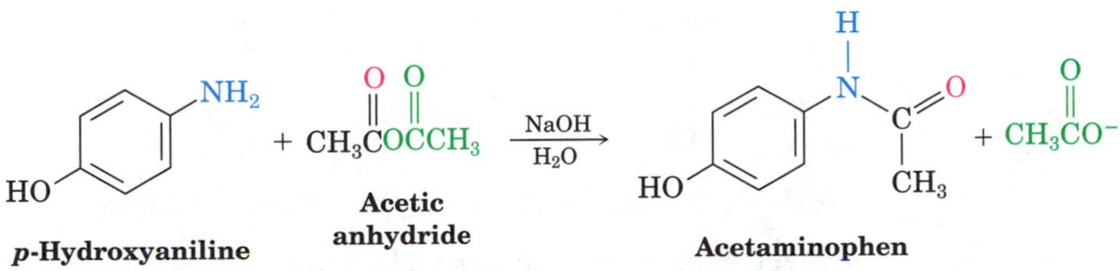

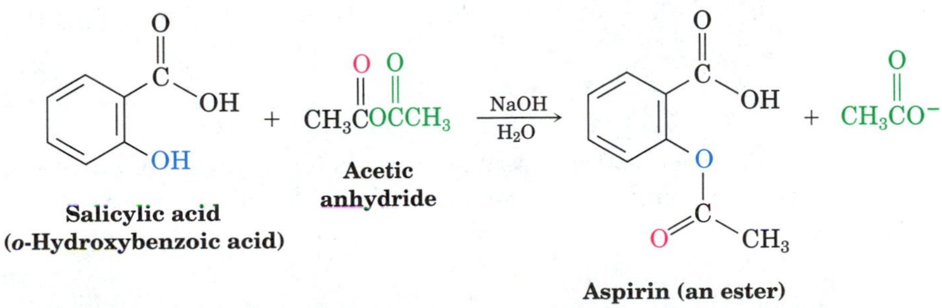

Notice in both these examples that only "half" of the anhydride molecule is used; the other half acts as the leaving group during the nucleophilic acyl substitution step and produces acetate ion as a by-product. Thus, anhydrides are inefficient to use, and acid chlorides are normally preferred for introducing acyl substituents other than acetyl groups.

PROBLEM

21.14 What product would you expect from reaction of 1 equivalent of methanol with a cyclic anhydride, such as phthalic anhydride (1,2-benzenedicarboxylic anhydride)? What is the fate of the second "half" of the anhydride in such a case?

PROBLEM

21.15 Write the steps in the mechanism of the reaction between *p*-hydroxyaniline and acetic anhydride to prepare acetaminophen.

PROBLEM

21.16 Why is 1 equivalent of a base such as NaOH required for the reaction between an amine and an anhydride to go to completion? What would happen if no base were present?

21.7 Chemistry of Esters

Esters are among the most widespread of all naturally occurring compounds. Many simple esters are pleasant-smelling liquids that are responsible for the fragrant odors of fruits and flowers. For example, methyl butanoate is found in pineapple oil, and isopentyl acetate is a constituent of banana oil. The ester linkage is also present in animal fats and in many biologically important molecules.

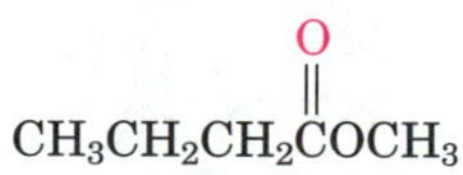

Methyl butanoate (from pineapples)

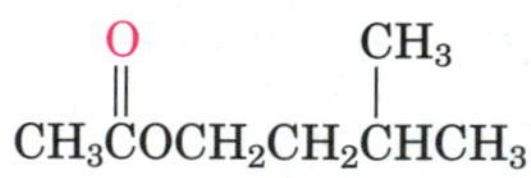

Isopentyl acetate (from bananas)

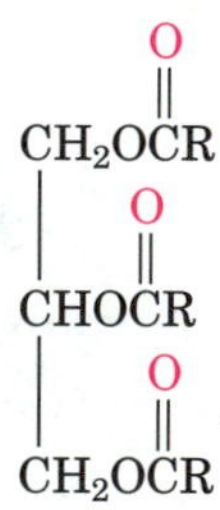

A fat ($R = C_{11-17}$ chains)

The chemical industry uses esters for a variety of purposes. Ethyl acetate, for example, is a common solvent found in nail-polish remover, and dialkyl phthalates are used as plasticizers to keep plastics from becoming brittle.

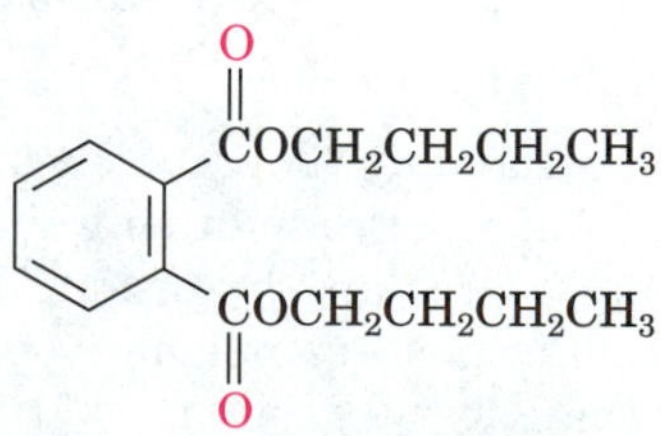

Dibutyl phthalate (a plasticizer)

Preparation of Esters

Esters are usually prepared from acids by the methods already discussed. Thus, carboxylic acids are converted directly into esters by S_N2 reaction of a carboxylate ion with a primary alkyl halide or by Fischer esterification of a carboxylic acid with an alcohol in the presence of a mineral acid catalyst. In addition, acid chlorides are converted into esters by treatment with an alcohol in the presence of base (Section 21.5).

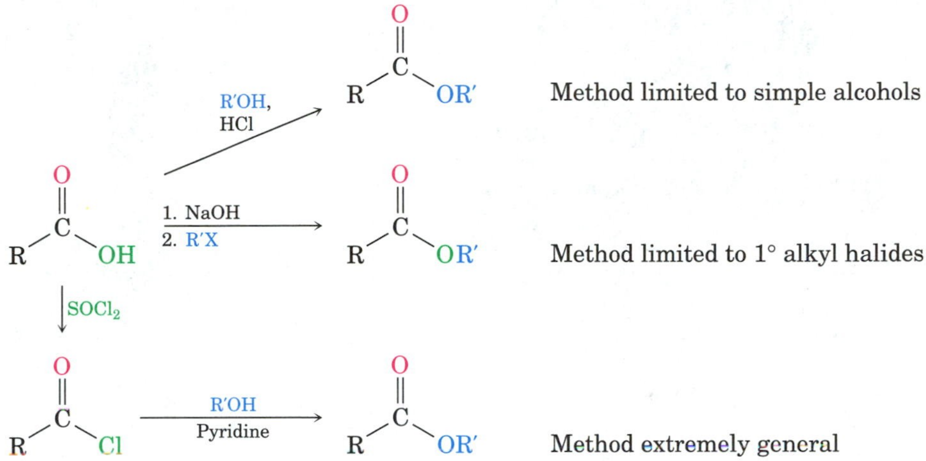

Reactions of Esters

Esters show the same kinds of chemistry that we've seen for other acid derivatives, but they are less reactive toward nucleophiles than either acid chlorides or anhydrides. Figure 21.8 (p. 826) shows some general reactions of esters, all of which are equally applicable to both acyclic and cyclic esters (**lactones**).

Hydrolysis: Conversion of Esters into Carboxylic Acids

Esters are hydrolyzed, either by aqueous base or by aqueous acid, to yield carboxylic acids plus alcohols:

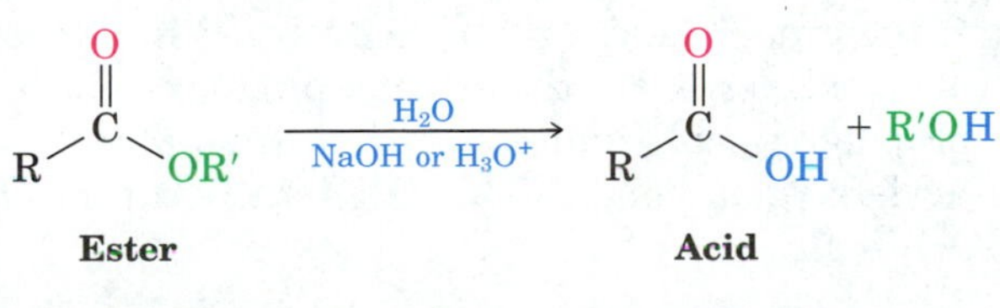

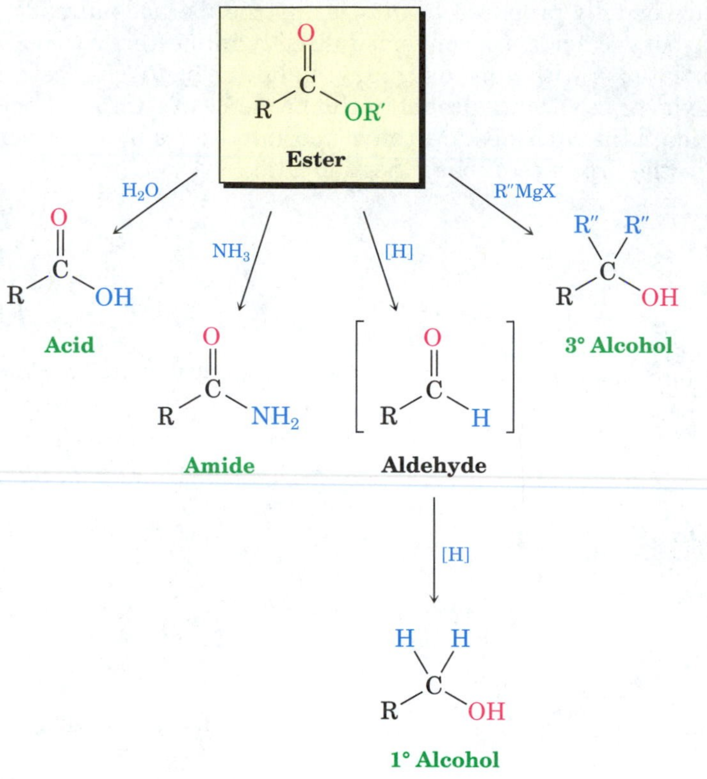

Figure 21.8 Some reactions of esters.

Ester hydrolysis in basic solution is called **saponification**, after the Latin *sapo*, meaning "soap." As we'll see in Section 28.2, the boiling of animal fat with extract of wood ash to make soap is indeed a saponification, because wood ash contains alkali and fats have ester linkages.

Ester hydrolysis occurs through the typical nucleophilic acyl substitution pathway shown in Figure 21.9, in which hydroxide ion is the nucleophile that adds to the ester carbonyl group to give a tetrahedral intermediate. Loss of alkoxide ion then gives a carboxylic acid, which is deprotonated to give the carboxylate ion. The free acid is then obtained by addition of aqueous HCl to protonate the carboxylate after the saponification is complete.

One of the most elegant experiments in support of the mechanism shown in Figure 21.9 involves isotope labeling. When ethyl propanoate labeled with ^{18}O in the ether-like oxygen is hydrolyzed in aqueous NaOH, the ^{18}O label shows up exclusively in the ethanol product. None of the label remains with the propanoic acid, indicating that saponification occurs by cleavage of the acyl–oxygen bond (RCO–OR′) rather than the alkyl–oxygen bond (RCOO–R′).

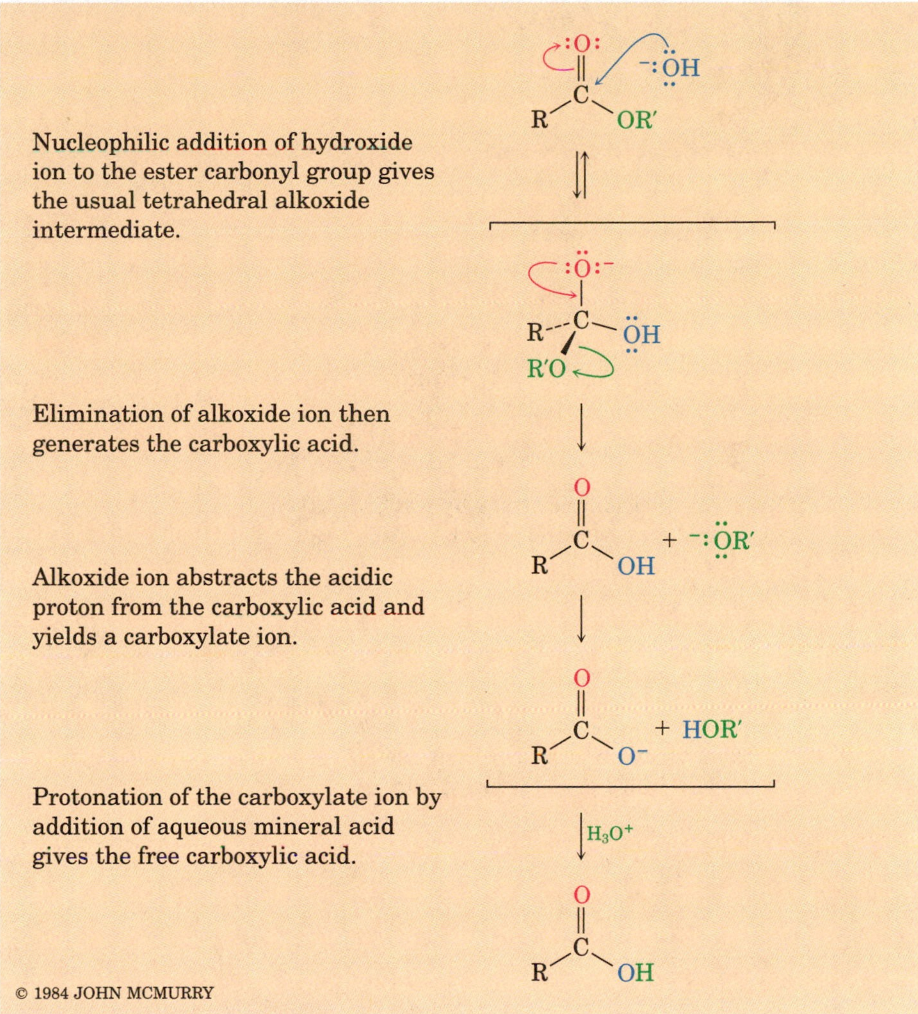

Figure 21.9 Mechanism of base-induced ester hydrolysis (saponification).

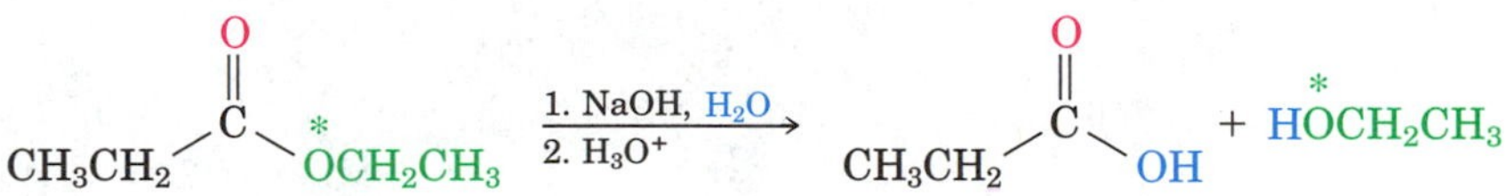

Acidic hydrolysis of esters can occur by more than one mechanism, depending on the structure of substrate. The usual pathway, however, is just the reverse of the Fischer esterification reaction (Section 21.4). The ester is first activated toward nucleophilic attack by protonation of the carboxyl oxygen atom, and nucleophilic attack by water occurs. Transfer of a proton and elimination of alcohol then yields the carboxylic acid (Figure 21.10).

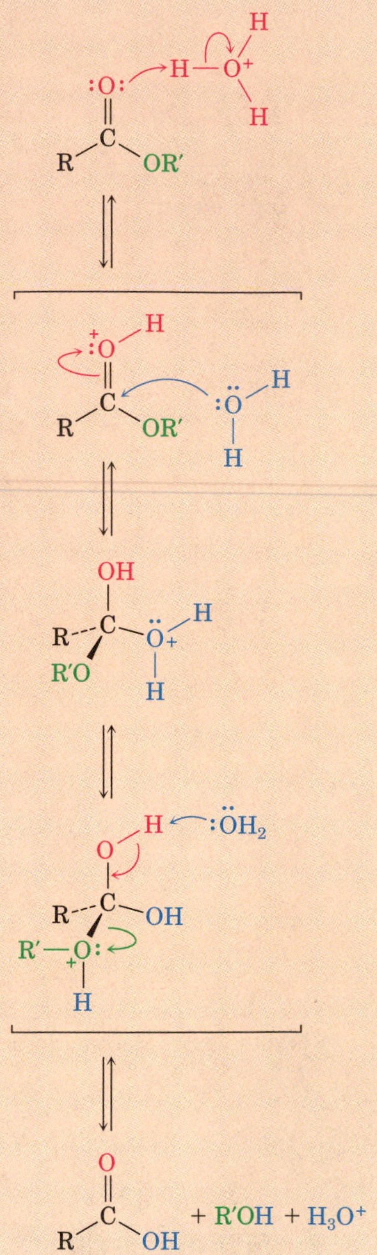

Protonation of the carbonyl group activates it . . .

. . . for nucleophilic attack by water to yield a tetrahedral intermediate.

Transfer of a proton then converts the OR′ into a good leaving group.

Expulsion of alcohol yields the free carboxylic acid product and regenerates the acid catalyst.

Figure 21.10 Mechanism of acid-catalyzed ester hydrolysis. The forward reaction is a hydrolysis; the back-reaction is a Fischer esterification.

PROBLEM. . . . .

21.17 How would you synthesize the ^{18}O-labeled ethyl propanoate used in mechanistic studies? Assume that ^{18}O-labeled acetic acid is your only source of isotopic oxygen.

PROBLEM

21.18 Why is saponification of esters irreversible? In other words, why doesn't treatment of a carboxylic acid with alkoxide ion lead to ester formation?

Aminolysis: Conversion of Esters into Amides Esters react with ammonia and amines to yield amides. The reaction is not often used, however, because higher yields are normally obtained by aminolysis of acid chlorides (Section 21.5).

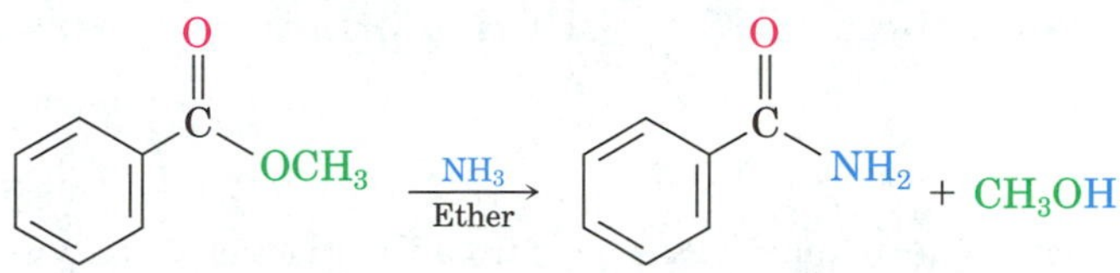

Methyl benzoate **Benzamide**

Reduction: Conversion of Esters into Alcohols Esters are easily reduced by treatment with $LiAlH_4$ to yield primary alcohols (Section 17.6).

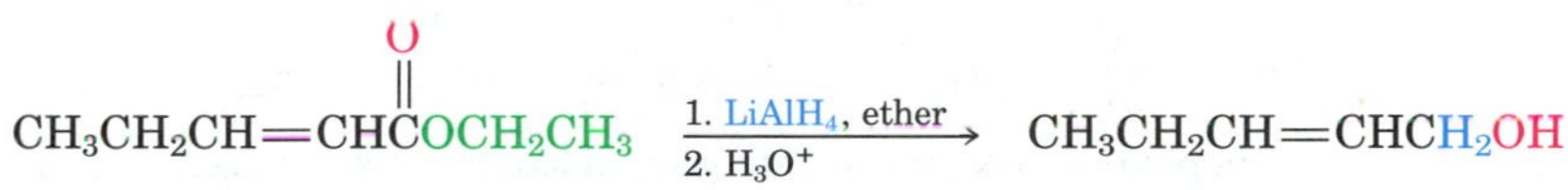

Ethyl 2-pentenoate **2-Penten-1-ol (91%)**

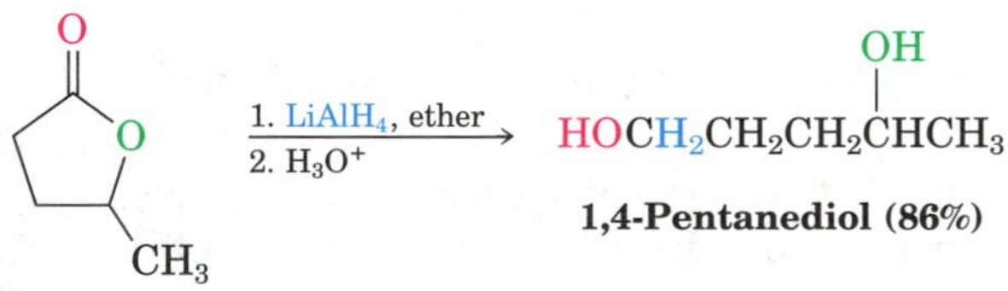

1,4-Pentanediol (86%)

A lactone

The mechanism of ester (and lactone) reductions is similar to that of acid chloride reduction. A hydride ion first adds to the carbonyl group, followed by elimination of alkoxide ion to yield an aldehyde. Further addition of H^- to the aldehyde gives the primary alcohol.

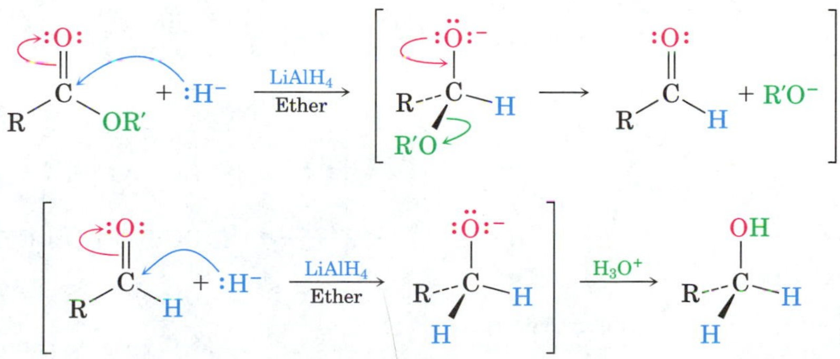

The aldehyde intermediate can be isolated if diisobutylaluminum hydride (DIBAH) is used as the reducing agent instead of $LiAlH_4$. Exactly 1 equivalent of hydride reagent must be used, and the reaction must be carried out at −78°C:

$$CH_3(CH_2)_{10}\overset{O}{\overset{\|}{C}}OCH_2CH_3 \xrightarrow[\text{2. } H_3O^+]{\text{1. DIBAH in toluene}} CH_3(CH_2)_{10}\overset{O}{\overset{\|}{C}}H + CH_3CH_2OH$$

Ethyl dodecanoate **Dodecanal (88%)**

where DIBAH = $[(CH_3)_2CHCH_2]_2AlH$.

PROBLEM

21.19 What product would you expect from the reaction of butyrolactone with DIBAH?

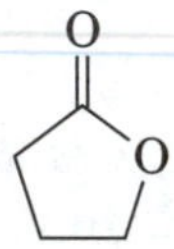

Butyrolactone

PROBLEM

21.20 Show the products you would obtain by reduction of the following esters with $LiAlH_4$.
(a) $CH_3CH_2CH_2CH(CH_3)COOCH_3$ (b) Phenyl benzoate

Reaction of Esters with Grignard Reagents Esters and lactones react with 2 equivalents of Grignard reagent or organolithium reagent to yield tertiary alcohols in which two of the substituents are identical (Section 17.7). The reaction occurs by the usual nucleophilic substitution mechanism to give an intermediate ketone, which reacts further with Grignard reagent to yield a tertiary alcohol.

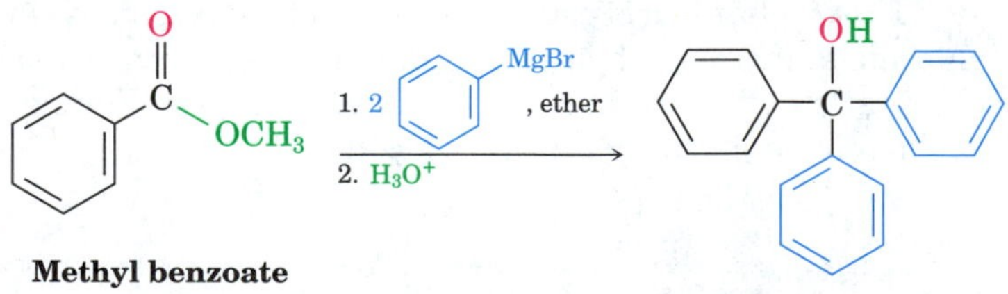

Methyl benzoate

Triphenylmethanol (96%)

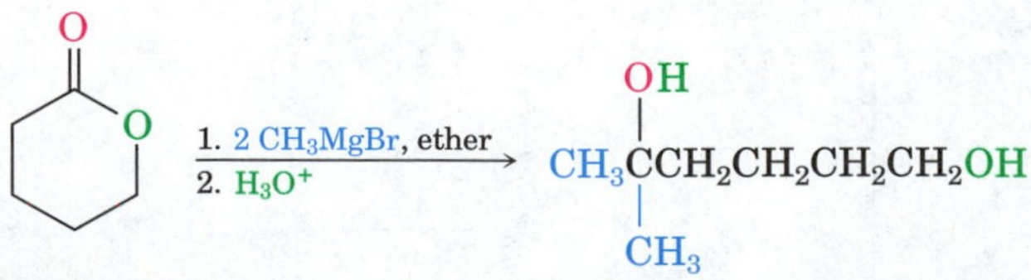

Valerolactone

5-Methyl-1,5-hexanediol

PROBLEM

21.21 What ester and what Grignard reagent might you start with to prepare the following alcohols?
(a) 2-Phenyl-2-propanol (b) 1,1-Diphenylethanol (c) 3-Ethyl-3-heptanol

21.8 Chemistry of Amides

Preparation of Amides

Amides are usually prepared by reaction of an acid chloride with an amine (Section 21.5). Ammonia, monosubstituted amines, and disubstituted amines all undergo the reaction.

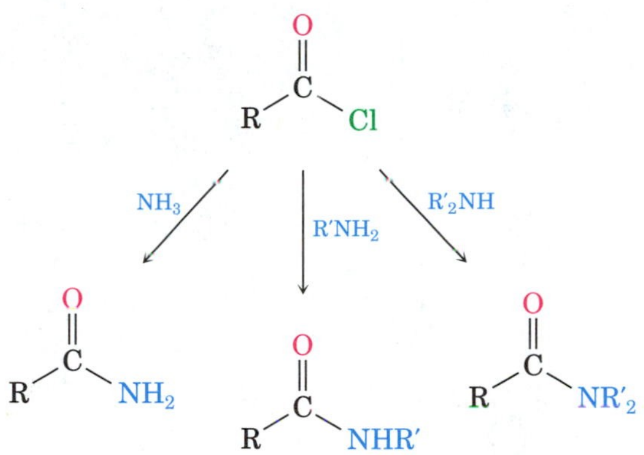

Reactions of Amides

Amides are much less reactive than acid chlorides, acid anhydrides, or esters. We'll see in Chapter 27, for example, that the amide linkage is stable enough to serve as the basic unit from which proteins are made.

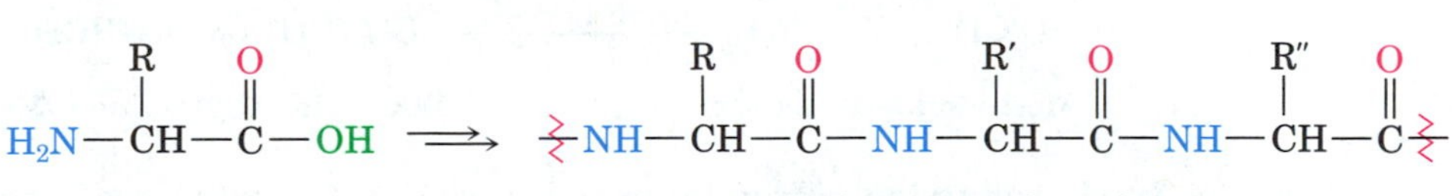

Amino acids

A protein (polyamide)

Amides undergo hydrolysis to yield carboxylic acids plus amine on heating in either aqueous acid or aqueous base. The conditions required for amide hydrolysis are more severe than those required for the hydrolysis of acid chlorides or esters, but the mechanisms are similar. The basic hydrolysis of amides occurs by nucleophilic addition of OH^- to the amide carbonyl group, followed by elimination of amide ion ($^-NH_2$). The acidic hydrolysis reaction occurs by nucleophilic addition of water to the protonated amide, followed by loss of ammonia.

Acidic hydrolysis

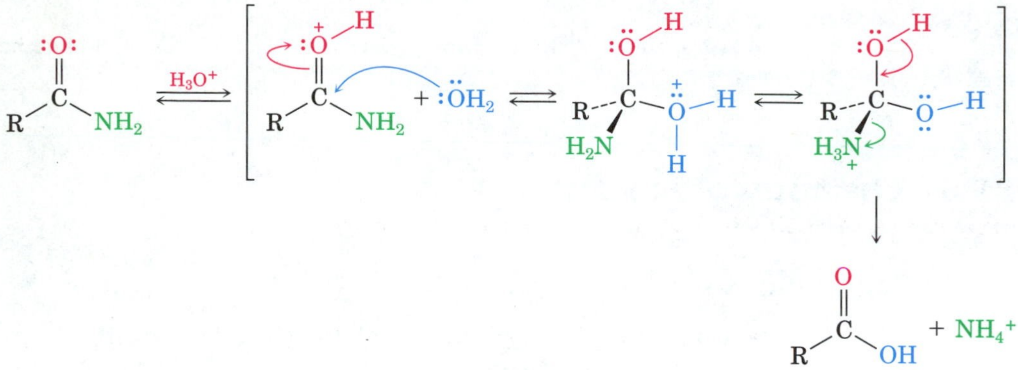

Basic hydrolysis

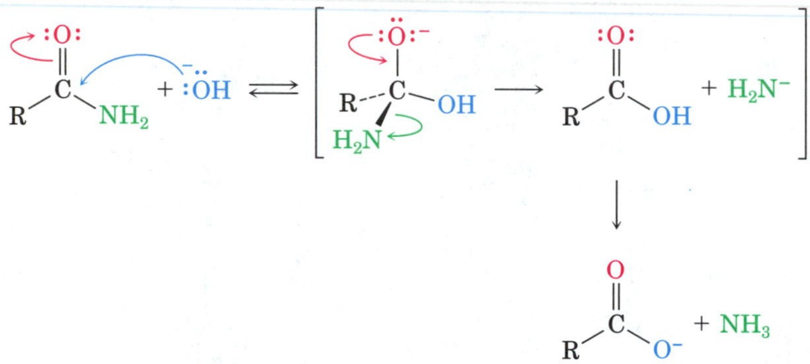

Like other carboxylic acid derivatives, amides can be reduced by $LiAlH_4$. The product of the reduction, however, is an *amine* rather than an alcohol. The net effect of an amide reduction reaction is thus to convert the amide carbonyl group into a methylene group ($C{=}O \rightarrow CH_2$). This kind of reaction is specific for amides and does not occur with other carboxylic acid derivatives.

$$CH_3(CH_2)_{10}\overset{\displaystyle O}{\overset{\|}{C}}NHCH_3 \xrightarrow[\text{2. } H_2O]{\text{1. } LiAlH_4\text{, ether}} CH_3(CH_2)_{10}CH_2NHCH_3$$

N-Methyldodecanamide **Dodecylmethylamine (95%)**

Amide reduction occurs by initial nucleophilic addition of hydride ion to the amide carbonyl group, followed by expulsion of the oxygen atom as an aluminate anion to give an iminium ion intermediate. The intermediate iminium ion is then further reduced by $LiAlH_4$ to yield the amine.

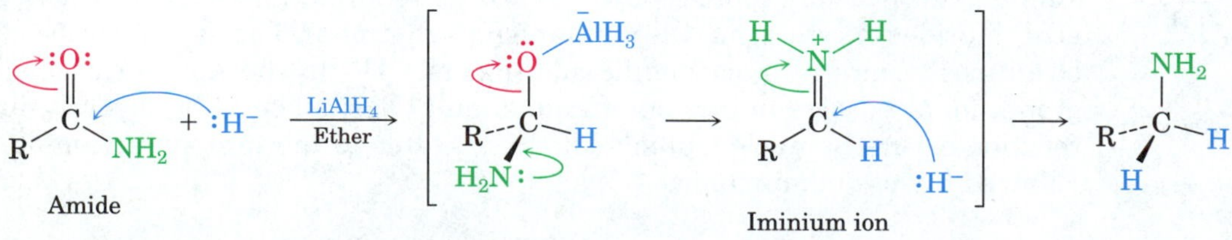

Lithium aluminum hydride reduction is equally effective with both acyclic and cyclic amides, or **lactams**. Lactam reduction is a valuable method for preparing cyclic amines.

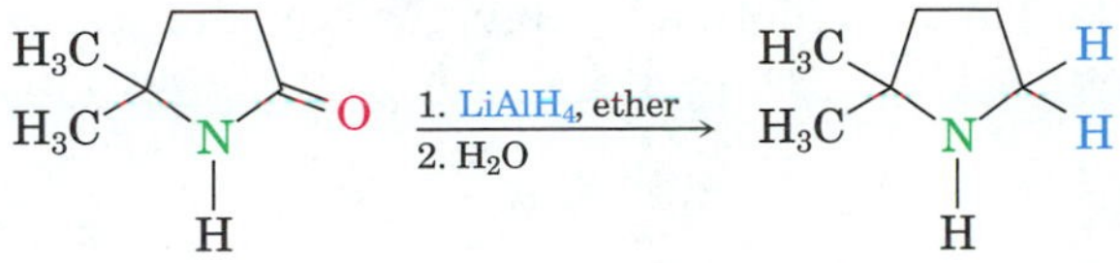

A lactam

A cyclic amine (80%)

PROBLEM.

21.22 How would you convert *N*-ethylbenzamide to the following products?
(a) Benzoic acid (b) Benzyl alcohol (c) $C_6H_5CH_2NHCH_2CH_3$

PROBLEM.

21.23 How would you use the reaction between an amide and $LiAlH_4$ as the key step in going from bromocyclohexane to (dimethylaminomethyl)cyclohexane? Formulate all steps involved in the reaction sequence.

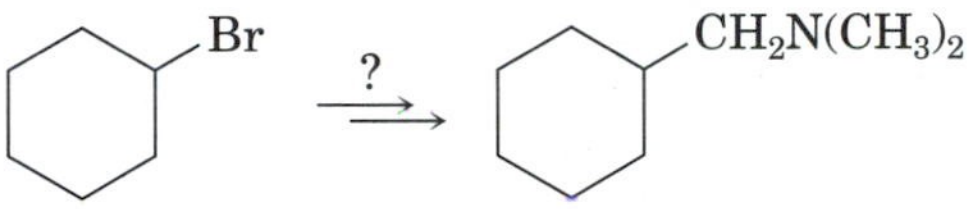

(Dimethylaminomethyl)cyclohexane

21.9 Chemistry of Nitriles

Nitriles are not related to carboxylic acids in the same sense that acid derivatives are, but the structures and reactions of nitriles and carboxylic acids are nevertheless similar. Both kinds of compounds have a carbon atom with three bonds to an electronegative atom, and both contain a π bond.

A nitrile—three bonds to nitrogen

An acid—three bonds to two oxygens

Preparation of Nitriles

The simplest method of nitrile preparation is S_N2 reaction of cyanide ion with a primary alkyl halide, a reaction discussed in Section 20.6. This method is limited by the usual S_N2 steric constraints to the synthesis of α-unsubstituted nitriles, RCH_2CN.

$$RCH_2Br + Na^+\ CN^- \xrightarrow[\text{reaction}]{S_N2} RCH_2CN + NaBr$$

Another method for preparing nitriles is by dehydration of a primary amide. Thionyl chloride is often used for the reaction, although other dehydrating agents such as P_2O_5, $POCl_3$, and acetic anhydride also work well.

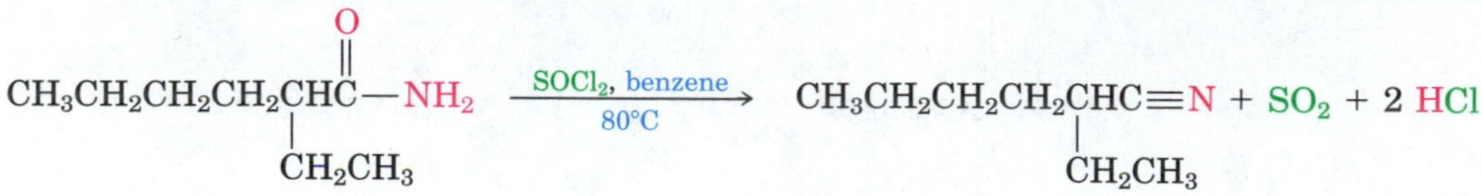

2-Ethylhexanamide **2-Ethylhexanenitrile (94%)**

Amide dehydration occurs by initial reaction of $SOCl_2$ on the amide oxygen atom, followed by an elimination reaction.

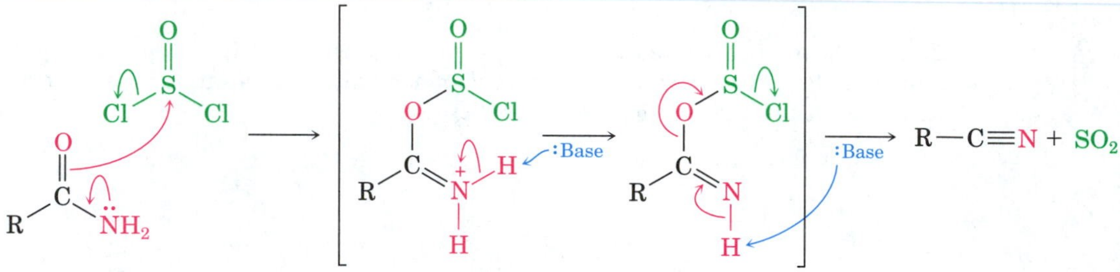

Both methods of nitrile synthesis—S_N2 displacement by CN^- on an alkyl halide and amide dehydration—are useful, but the synthesis from amides is more general because it is not limited by steric hindrance.

Reactions of Nitriles

The chemistry of nitriles is similar in many respects to the chemistry of carbonyl compounds. Like carbonyl groups, a nitrile group is strongly polarized, making the carbon atom electrophilic. Nitriles are therefore attacked by nucleophiles to yield sp^2-hybridized imine anions in a reaction analogous to the formation of an sp^3-hybridized alkoxide ion by nucleophilic addition to a carbonyl group.

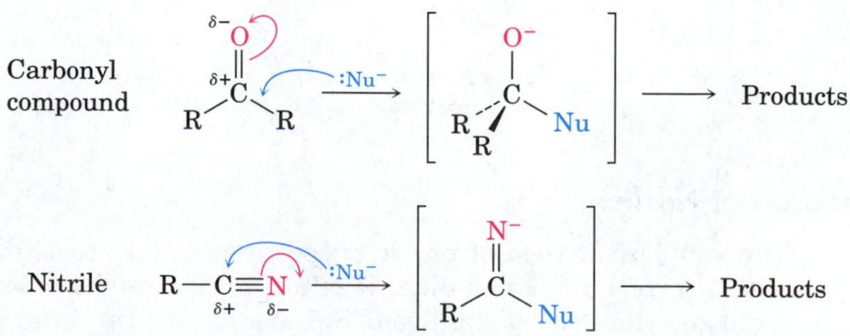

Among the important reactions of nitriles are hydrolysis and reduction. In addition, nitriles can be partially reduced and hydrolyzed to yield aldehydes, and can be treated with Grignard reagents to yield ketones (Figure 21.11).

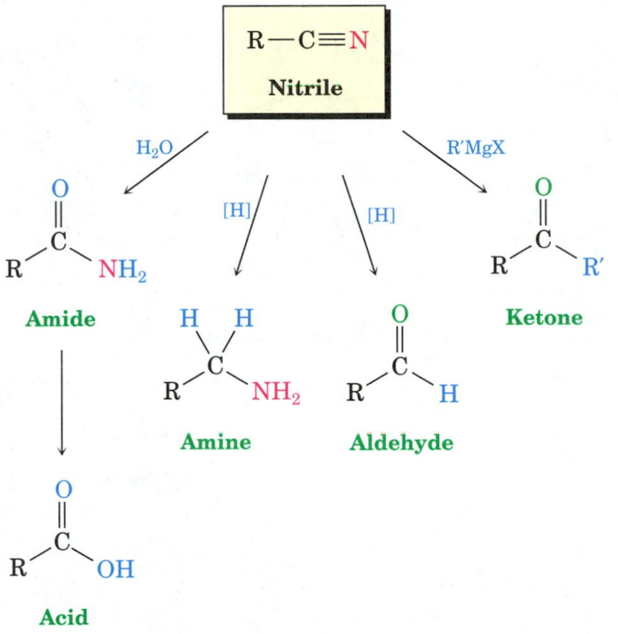

Figure 21.11 Some reactions of nitriles.

Hydrolysis: Conversion of Nitriles into Acids Nitriles are hydrolyzed in either acidic or basic aqueous solution to yield carboxylic acids plus ammonia or an amine:

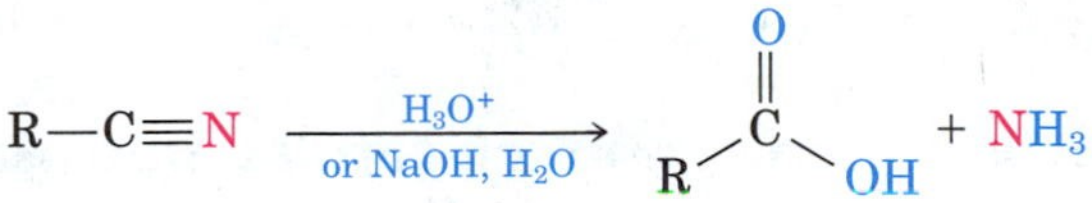

The mechanism of the basic hydrolysis involves nucleophilic addition of hydroxide ion to the polar C≡N bond in a manner analogous to that of nucleophilic addition to a polar carbonyl C=O bond. The initially formed hydroxy imine is rapidly isomerized to an amide in a step similar to the isomerization of an enol to a ketone (Section 8.5). Further hydrolysis of the amide, as discussed in the previous section, then yields the carboxylic acid (Figure 21.12).

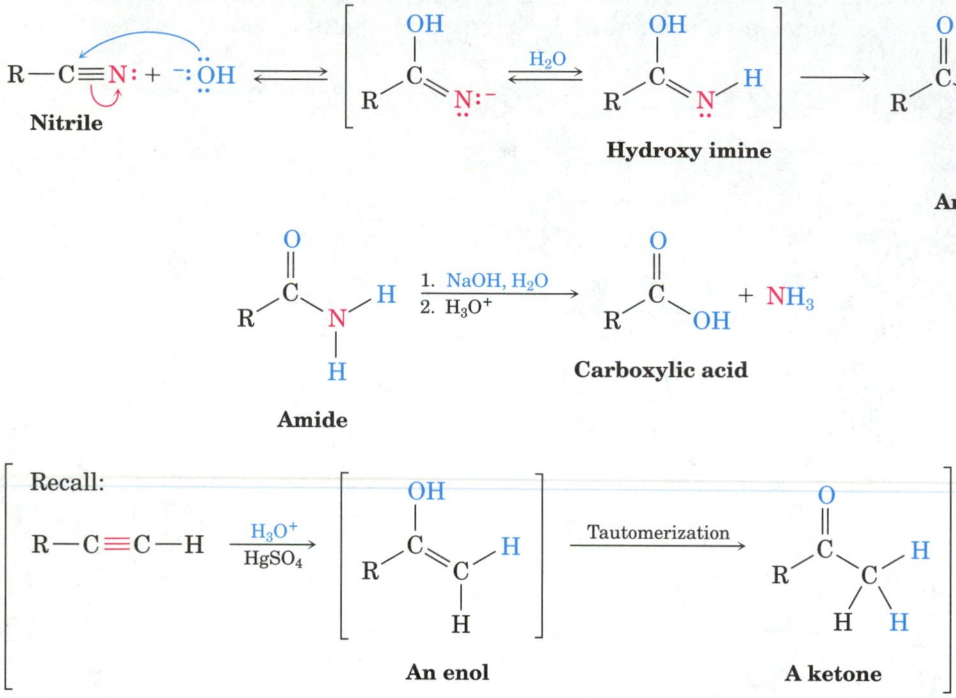

Figure 21.12 Mechanism of basic amide hydrolysis to yield a carboxylic acid.

The conditions required for nitrile alkaline hydrolysis are severe (KOH, 200°C), and the amide intermediate can sometimes be isolated if milder conditions are used.

PROBLEM. . . . .

21.24 Acid-catalyzed nitrile hydrolysis occurs by initial protonation of the nitrogen atom, followed by nucleophilic addition of water. Show all the steps involved in the acidic hydrolysis of a nitrile to yield a carboxylic acid.

. . . .

Reduction: Conversion of Nitriles into Amines and Aldehydes

Reduction of nitriles with $LiAlH_4$ gives primary amines in high yields. For example:

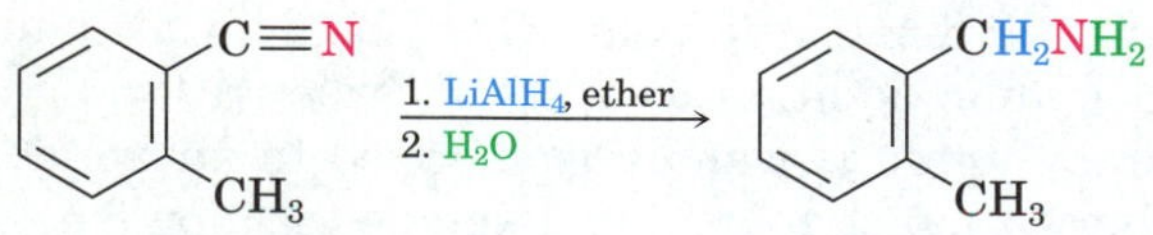

o-Methylbenzonitrile

o-Methylbenzylamine (88%)

The reaction occurs by nucleophilic addition of hydride ion to the polar C≡N bond, yielding an imine anion that undergoes further addition of a second equivalent of hydride. If, however, a less powerful reducing agent such as DIBAH is used, the second addition of hydride does not occur, and the imine intermediate can be hydrolyzed to yield an aldehyde.

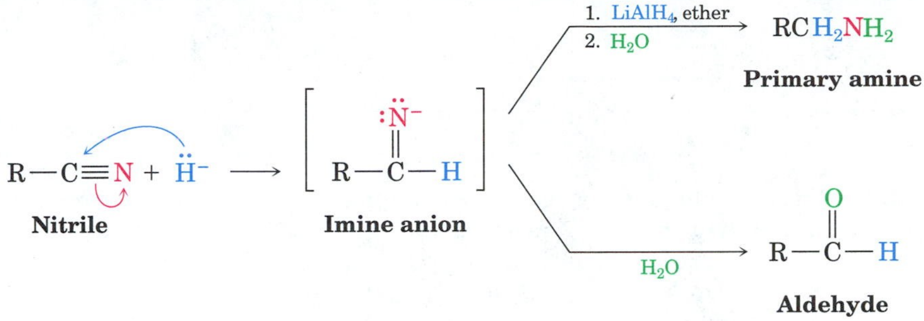

For example:

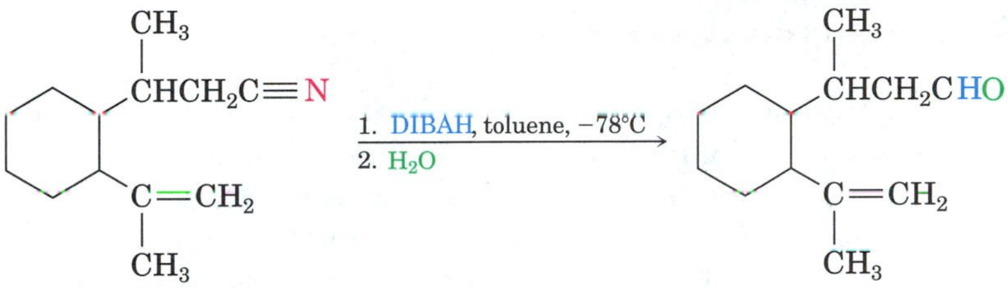

96%

Reaction of Nitriles with Organometallic Reagents Grignard reagents add to nitriles giving intermediate imine anions that can be hydrolyzed to yield ketones:

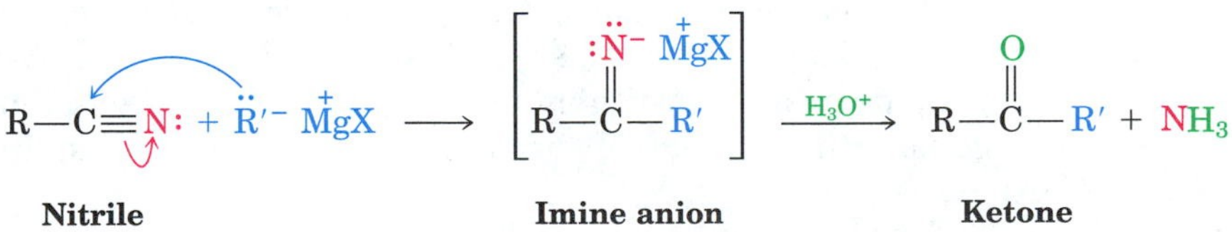

The reaction is similar to the DIBAH reduction of nitriles, except that the attacking nucleophile is a carbanion ($:R^-$) rather than a hydride ion. Yields are generally high. For example:

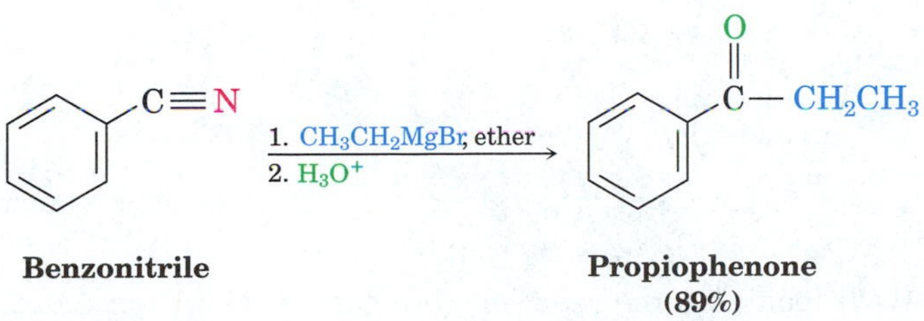

PROBLEM.

21.25 How would you prepare the following carbonyl compounds from a nitrile?
(a) $CH_3CH_2COCH_2CH_3$ (b) $CH_3CH_2COCH(CH_3)_2$ (c) $(CH_3)_2CHCHO$
(d) Acetophenone (e)

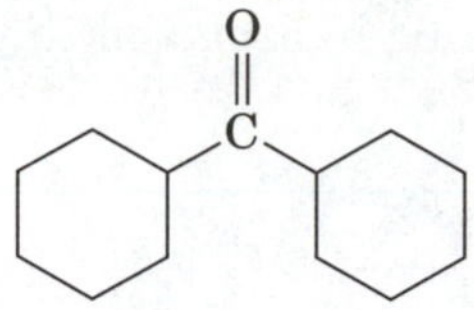

PROBLEM.

21.26 How would you prepare 1-phenyl-2-butanone from benzyl bromide, $C_6H_5CH_2Br$? More than one step is required.

21.10 Thiol Esters: Biological Carboxylic Acid Derivatives

Nucleophilic acyl substitution reactions take place in living organisms just as they take place in the chemical laboratory. The same principles apply in both cases. Nature, however, uses **thiol esters, RCOSR′**, as reactive acid derivatives because they are intermediate in reactivity between acid anhydrides and esters. Thiol esters aren't so reactive that they hydrolyze as rapidly as anhydrides, yet they're more reactive than typical esters toward nucleophilic attack.

Acetyl coenzyme A (usually abbreviated **acetyl CoA**; Figure 21.13) is the most common thiol ester found in nature. Acetyl CoA is a much more complex molecule than acetyl chloride or acetic anhydride, yet it serves exactly the same purpose as these simpler reagents. Nature uses acetyl CoA as a reactive acylating agent in nucleophilic acyl substitution reactions.

As an example of how acetyl CoA is used in nature, *N*-acetylglucosamine, an important constituent of surface membranes in mammalian cells, is synthesized by an aminolysis reaction between glucosamine and acetyl CoA:

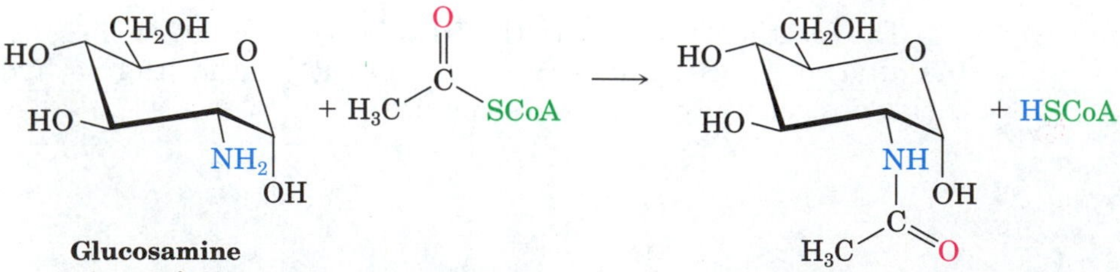

Glucosamine (an amine)

***N*-Acetylglucosamine (an amide)**

We'll look at some reactions of acetyl CoA in more detail in Chapter 30.

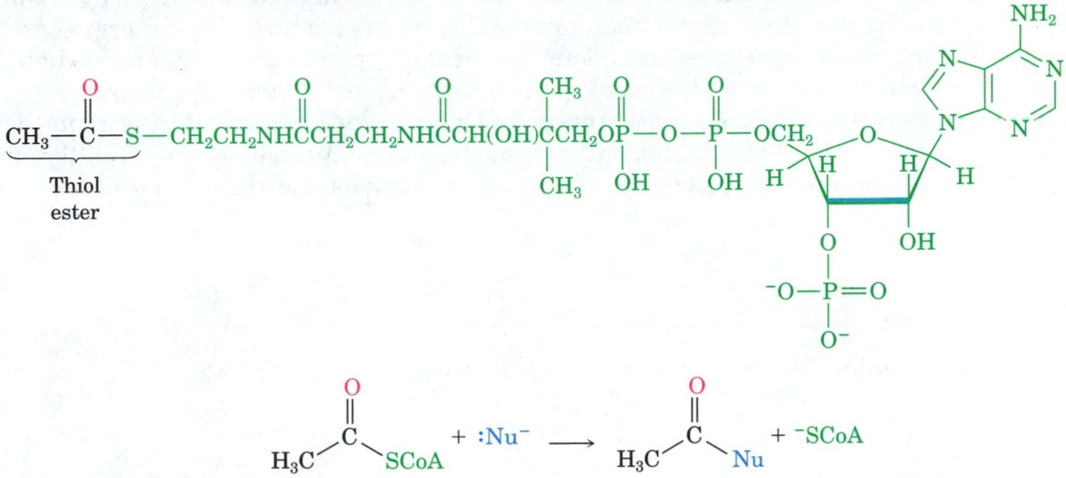

Figure 21.13 The structure of acetyl coenzyme A, abbreviated acetyl CoA or $CH_3COSCoA$.

21.11 Polyamides and Polyesters: Step-Growth Polymers

When an amine reacts with an acid chloride, an amide results (Section 21.5). What would happen, though, if a *diamine* and a *diacid chloride* were allowed to react? Each partner could form *two* amide bonds, linking more and more molecules together until a giant **polyamide** resulted. In the same way, reaction of a diol with a diacid would lead to a **polyester**.

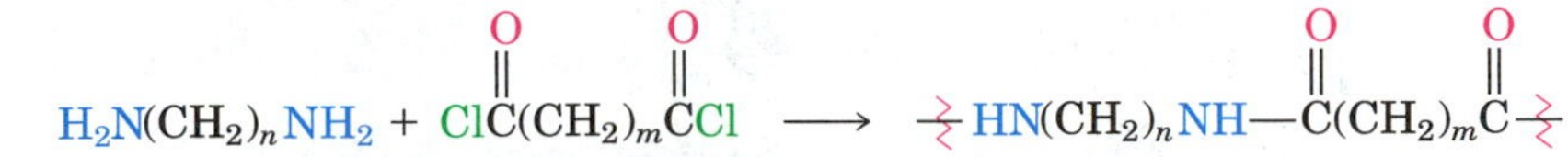

A diamine **A diacid chloride** **A polyamide (nylon)**

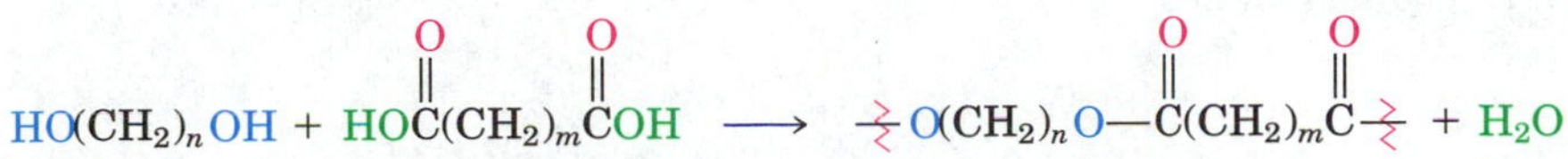

A diol **A diacid** **A polyester**

The alkene and diene polymers discussed in Sections 7.11 and 14.7 are called **chain-growth polymers** because they are produced by chain reactions. An initiator adds to a C=C bond to give a reactive intermediate, which adds to a second alkene molecule to produce a new intermediate, which adds to a third molecule, and so on. By contrast, polyamides and polyesters are said to be **step-growth polymers** because each bond in the polymer is formed independently of the others. A large number of different step-growth polymers have been made; some of the more important ones are shown in Table 21.2.

Table 21.2 Some Common Step-Growth Polymers and Their Uses

Monomer name	*Formula*	*Trade or common name of polymer*	*Uses*
Hexamethylene diamine	$H_2N(CH_2)_6NH_2$	Nylon 66	Fibers, clothing, tire cord, bearings
Adipic acid	$HOOC(CH_2)_4COOH$		
Ethylene glycol	$HOCH_2CH_2OH$	Dacron, Terylene, Mylar	Fibers, clothing, tire cord, film
Dimethyl terephthalate	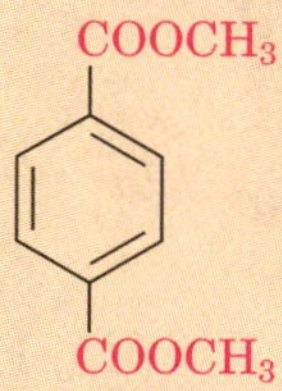		
Caprolactam	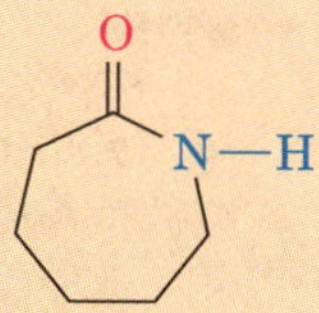	Nylon 6, Perlon	Fibers, large cast articles
Bisphenol A	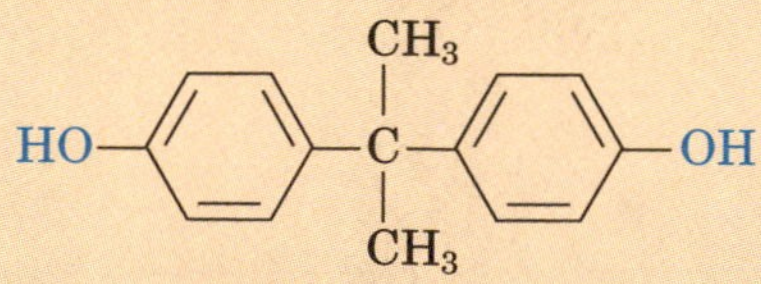	Lexan, polycarbonate	Molded articles, machine housings
Diphenyl carbonate	$C_6H_5OCOOC_6H_5$		
Poly(2-butene-1,4-diol)	$HO{+}CH_2CH{=}CHCH_2{+}_nOH$	Polyurethane, Spandex	Foams, fibers, coatings
Tolylene diisocyanate	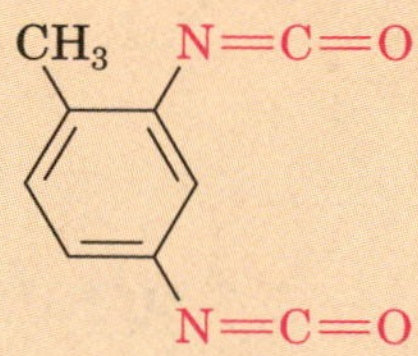		

Nylons

The best-known step-growth polymers are the polyamides, or **nylons**, first prepared by Wallace Carothers[3] at the Du Pont Company by heating diamines with diacids. For example, nylon 66 is prepared by reaction of the six-carbon adipic acid with the six-carbon hexamethylenediamine at 280°C.

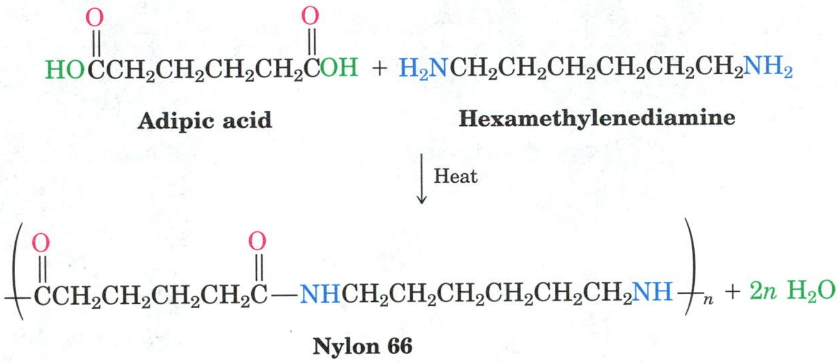

Nylons are used both in engineering applications and in making fibers. A combination of high impact strength and abrasion resistance makes nylon an excellent metal substitute for bearings and gears. As fiber, nylon is used in a wide variety of applications, from clothing to tire cord to Perlon mountaineering ropes.

Polyesters

The most generally useful polyester is made by an ester exchange reaction between dimethyl terephthalate and ethylene glycol. The product is used under the trade name Dacron to make clothing fiber and tire cord, and under the name Mylar to make recording tape. The tensile strength of poly(ethylene terephthalate) film is nearly equal to that of steel.

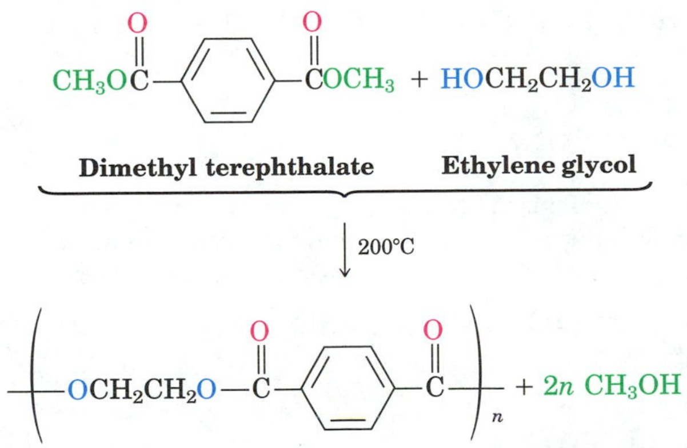

Polyester, Dacron, Mylar

[3]Wallace H. Carothers (1896–1937); b. Burlington, Iowa; Ph.D. Illinois (Adams); Du Pont Company.

Lexan, a polycarbonate prepared from diphenyl carbonate and bisphenol A, is another commercially valuable polyester. Lexan has an unusually high impact strength, making it valuable for use in machinery housings, telephones, and bicycle safety helmets.

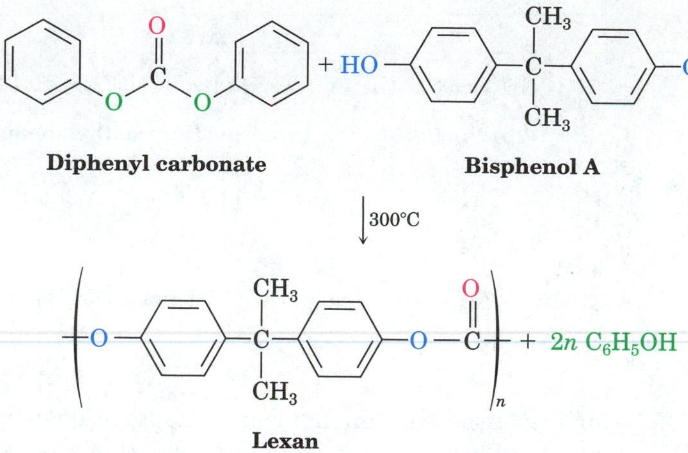

PROBLEM.

21.27 Draw structures of the step-growth polymers you would expect to obtain from the following reactions:

(a) $BrCH_2CH_2CH_2Br + HOCH_2CH_2CH_2OH \xrightarrow{\text{Base}}$

(b) $HOCH_2CH_2OH + HOOC(CH_2)_6COOH \xrightarrow{H_2SO_4 \text{ catalyst}}$

(c) $H_2N(CH_2)_6NH_2 + ClOC(CH_2)_4COCl \longrightarrow$

PROBLEM.

21.28 Kevlar, a nylon polymer prepared by reaction of 1,4-benzenedicarboxylic acid (terephthalic acid) with 1,4-diaminobenzene (*p*-phenylenediamine), is so strong that it's used to make bulletproof vests. Draw the structure of a segment of Kevlar.

PROBLEM.

21.29 Draw the structure of the polymer you would expect to obtain from reaction of dimethyl terephthalate with a triol such as glycerol. What structural feature would this new polymer have that was not present in Dacron? How do you think this new feature might affect the properties of the polymer?

21.12 Spectroscopy of Carboxylic Acid Derivatives

Infrared Spectroscopy

All carbonyl-containing compounds have intense IR absorptions in the range 1650–1850 cm^{-1}. As shown in Table 21.3, the exact position of the absorption provides information about the specific kind of carbonyl group. For comparison, the IR absorptions of ketones, aldehydes, and acids are included in the table, along with values for carboxylic acid derivatives.

Table 21.3 Infrared Absorptions of Some Carbonyl Compounds

Carbonyl type	*Example*	*Infrared absorption (cm^{-1})*
Aliphatic acid chloride	Acetyl chloride	1810
Aromatic acid chloride	Benzoyl chloride	1770
Aliphatic acid anhydride	Acetic anhydride	1820, 1760
Aliphatic ester	Ethyl acetate	1735
Aromatic ester	Ethyl benzoate	1720
Six-membered-ring lactone	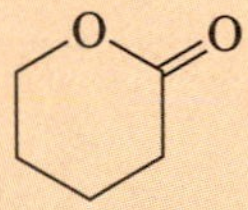	1735
Aliphatic amide	Acetamide	1690
Aromatic amide	Benzamide	1675
N-Substituted amide	*N*-Methylacetamide	1680
N,N-Disubstituted amide	*N,N*-Dimethylacetamide	1650
Aliphatic nitrile	Acetonitrile	2250
Aromatic nitrile	Benzonitrile	2230
Aliphatic aldehyde	Acetaldehyde	1730
Aliphatic ketone	Acetone	1715
Aliphatic carboxylic acid	Acetic acid	1710

As the data in the table indicate, acid chlorides are readily detected by their characteristic carbonyl-group absorption near 1800 cm^{-1}. Acid anhydrides can be identified by the fact that they show two absorptions in the carbonyl region, one at 1820 cm^{-1} and another at 1760 cm^{-1}. Esters are detected by their absorption at 1735 cm^{-1}, a position somewhat higher than for either ketones or aldehydes. Amides, by contrast, absorb near the low end of the carbonyl region, with the degree of substitution on nitrogen affecting the exact position of the IR band. Nitriles are easily recognized by the presence of an intense absorption near 2250 cm^{-1}. Since few other functional groups absorb in this region, IR spectroscopy is highly diagnostic for nitriles.

PROBLEM.

21.30 What kinds of functional groups might compounds have if they show the following IR absorptions?

(a) Absorption at 1735 cm^{-1}
(b) Absorption at 1810 cm^{-1}
(c) Absorptions at 2500–3300 cm^{-1} and 1710 cm^{-1}
(d) Absorption at 2250 cm^{-1}
(e) Absorption at 1715 cm^{-1}

PROBLEM.

21.31 Propose structures for compounds that have the following formulas and IR absorptions.

(a) C_3H_5N, 2250 cm^{-1}
(b) $C_6H_{12}O_2$, 1735 cm^{-1}
(c) C_4H_9NO, 1650 cm^{-1}
(d) C_4H_5ClO, 1780 cm^{-1}

Nuclear Magnetic Resonance Spectroscopy

Hydrogens on the carbon next to a carbonyl group are slightly deshielded and absorb near 2 δ in the ^1H NMR spectrum. The exact nature of the carbonyl group can't be distinguished by ^1H NMR, however, because all acid derivatives absorb in the same range. Figure 21.14 shows the ^1H NMR spectrum of ethyl acetate.

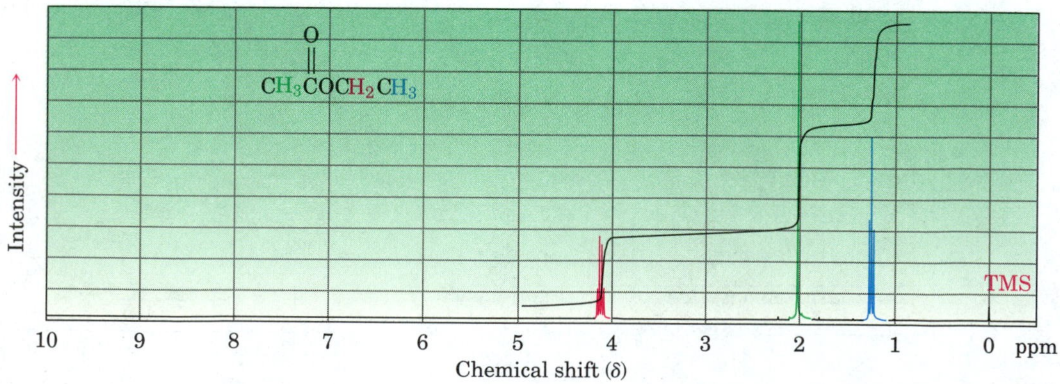

Figure 21.14 Proton NMR spectrum of ethyl acetate.

Although ^{13}C NMR is useful for determining the presence or absence of a carbonyl group in a molecule of unknown structure, precise information about the nature of the carbonyl group is difficult to obtain. Ketones and aldehydes have resonances near 200 δ, while the carbonyl carbon atoms of various acid derivatives have resonances in the range 160–180 δ (Table 21.4).

Table 21.4 Positions of ^{13}C NMR Absorptions in Some Carbonyl Compounds

Compound	*Absorption (δ)*	*Compound*	*Absorption (δ)*
Acetic acid	177.3	Acetic anhydride	166.9
Ethyl acetate	170.7	Acetonitrile	117.4
Acetyl chloride	170.3	Acetone	205.6
Acetamide	172.6	Acetaldehyde	201.0

INTERLUDE

β-Lactam Antibiotics

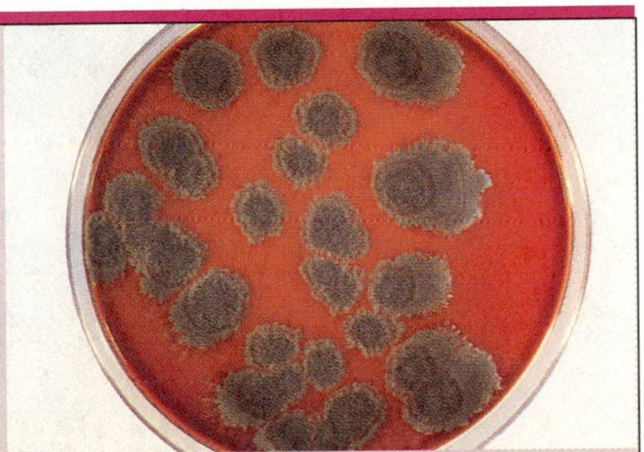

***Penicillium* mold growing in a Petri dish.**

The value of hard work and logical thinking shouldn't be underestimated, but sheer good luck also plays a role in most real scientific breakthroughs. What has been called "the supreme example [of luck] in all scientific history" occurred in the late summer of 1928 when the Scottish bacteriologist Alexander Fleming went on vacation, leaving in his lab a culture plate recently inoculated with the bacterium *Staphylococcus aureus*.

While Fleming was away, an extraordinary chain of events occurred. First, a 9 day cold spell lowered the laboratory temperature to a point where the *Staphylococcus* on the plate could not grow. During this time, spores from a colony of the mold *Penicillium notatum* being grown on the floor below wafted up into Fleming's lab and landed in the culture plate. The temperature then rose, and both *Staphylococcus* and *Penicillium* began to grow. On returning from vacation, Fleming discarded the plate into a tray of antiseptic, intending to sterilize it. Evidently, though, the plate did not sink deeply enough into the antiseptic, because when Fleming happened to glance at it a few days later, what he saw changed the course of human history: He noticed that the growing *Penicillium* mold appeared to dissolve the colonies of staphylococci.

(continued)➤

Fleming realized that the mold must be producing a chemical that killed bacteria, and he spent several years trying to isolate the substance. Finally, in 1939, the Australian pathologist Howard Florey and the German refugee Ernst Chain managed to isolate the active substance, called *penicillin*. The dramatic ability of penicillin to cure infections in mice was soon demonstrated, and successful tests in humans followed shortly thereafter. By 1943, penicillin was being produced on a large scale for military use, and by 1944 it was being used on civilians. Fleming, Florey, and Chain shared the 1945 Nobel Prize in Medicine.

Now called benzylpenicillin, or penicillin G, the substance first discovered by Fleming is but one member of a large class of so-called *β-lactam antibiotics*, compounds with a four-membered lactam (cyclic amide) ring. The four-membered lactam ring is fused to a five-membered, sulfur-containing ring, and the carbon atom next to the lactam carbonyl group is bonded to an acylamino substituent, RCONH−. This acylamino side chain can be varied in the laboratory to provide literally hundreds of penicillin analogs with different biological activity profiles. Ampicillin, for instance, has an α-aminophenylacetamido substituent [$PhCH(NH_2)CONH-$].

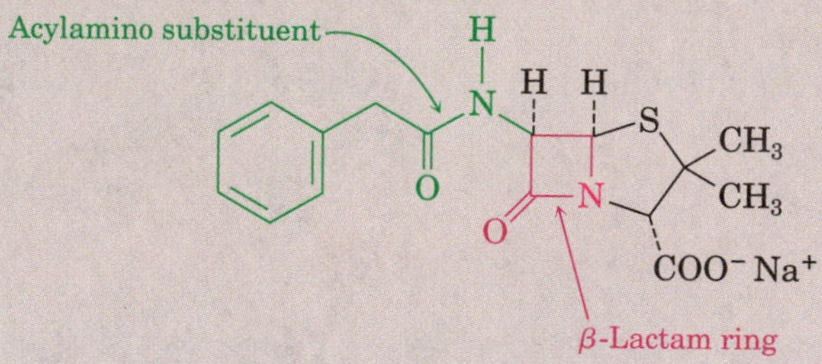

Benzylpenicillin (Penicillin G)

Closely related to the penicillins are the *cephalosporins*, a group of β-lactam antibiotics that contain an unsaturated six-membered, sulfur-containing ring. Cephalexin, marketed under the trade name Keflex, is an example. Cephalosporins generally have much greater antibacterial activity than penicillins, particularly against resistant strains of bacteria.

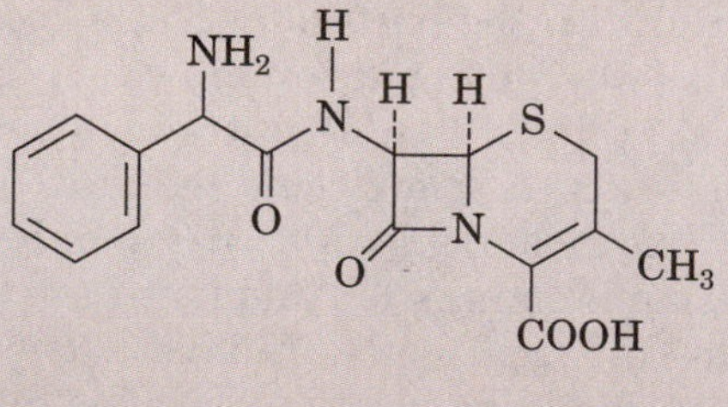

Cephalexin (a cephalosporin)

The biological activity of penicillins and cephalosporins is due to the presence of the strained β-lactam ring, which reacts with and deactivates the *transpeptidase* enzyme needed to synthesize and repair bacterial cell walls. With the wall either incomplete or weakened, the bacterial cell ruptures and dies.

Summary and Key Words

Carboxylic acids can be transformed into a variety of acid derivatives in which the carboxyl –OH group has been replaced by another substituent. **Acid halides, acid anhydrides, esters**, and **amides** are the most important such derivatives.

The chemistry of carboxylic acid derivatives is dominated by the **nucleophilic acyl substitution reaction**. Mechanistically, these substitutions take place by addition of a nucleophile to the polar carbonyl group of the acid derivative, followed by expulsion of a leaving group from the tetrahedral intermediate.

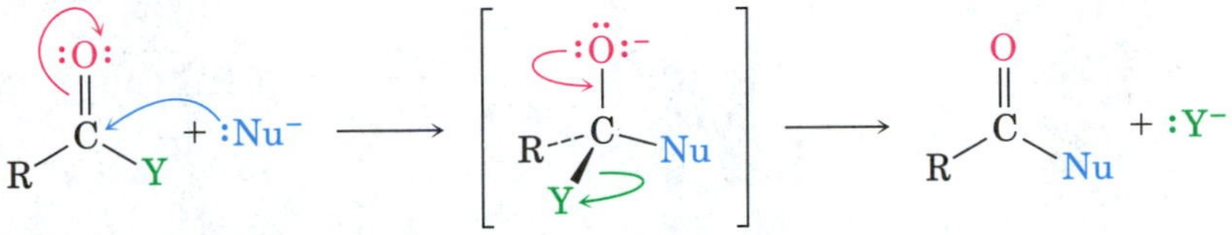

where Y = Cl, Br, I (acid halide); OR (ester); OCOR (anhydride); or NH_2 (amide).

The reactivity of an acid derivative toward substitution depends both on the steric environment near the carbonyl group and on the electronic nature of the substituent, Y. The reactivity order is:

Acid halide > Acid anhydride > Ester > Amide

The most common reactions of carboxylic acid derivatives are substitution by water (**hydrolysis**) to yield an acid, by an alcohol (**alcoholysis**) to yield an ester, by an amine (**aminolysis**) to yield an amide, by hydride (**reduction**) to yield an alcohol, and by an organometallic reagent (**Grignard reaction**) to yield an alcohol.

Nitriles undergo nucleophilic addition to the polar C≡N bond in the same way that carbonyl compounds do. The most important reactions of nitriles are their hydrolysis to carboxylic acids, reduction to primary amines, partial reduction to aldehydes, and reaction with organometallic reagents to yield ketones.

Nature employs nucleophilic acyl substitution reactions in the biosynthesis of many molecules, using **thiol esters** for the purpose. Acetyl coenzyme A (**acetyl CoA**) is a complex thiol ester that is employed in living systems to acetylate amines and alcohols.

Step-growth polymers, such as polyamides and polyesters, are prepared by reactions between difunctional molecules. **Polyamides (nylons)** are formed by step-growth polymerization between a diacid and a diamine; **polyesters** are formed from a diacid and a diol.

Infrared spectroscopy is a valuable tool for the structural analysis of acid derivatives. Acid chlorides, anhydrides, esters, amides, and nitriles all show characteristic infrared absorptions that can be used to identify these functional groups in unknowns.

Summary of Reactions

1. Reactions of carboxylic acids
   (a) Conversion into acid chlorides (Section 21.4)

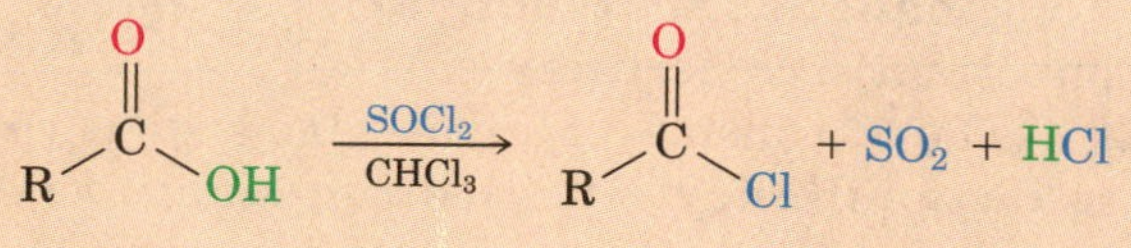

   (b) Conversion into cyclic acid anhydrides (Section 21.4)

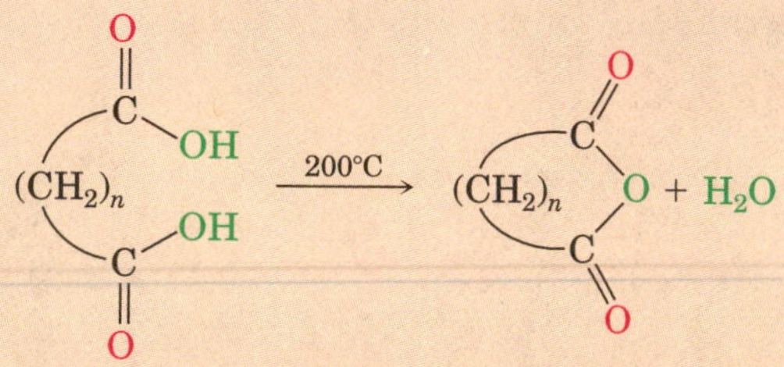

where $n = 2$ or 3

   (c) Conversion into esters (Section 21.4)

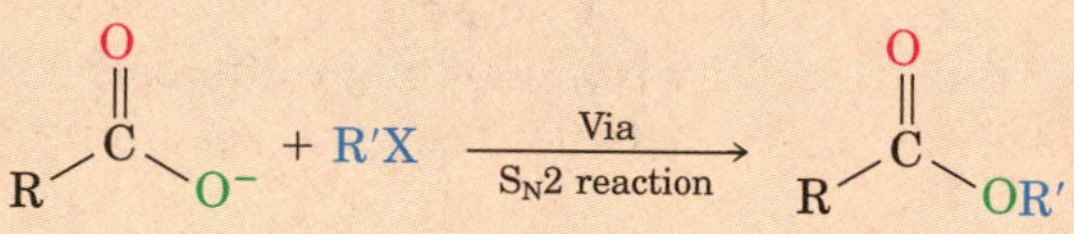

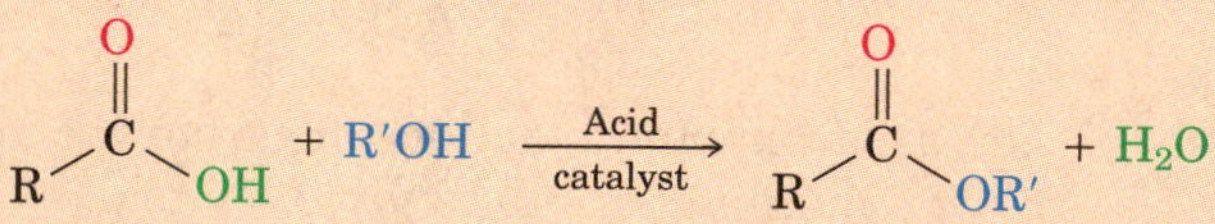

2. Reactions of acid chlorides
   (a) Hydrolysis to yield acids (Section 21.5)

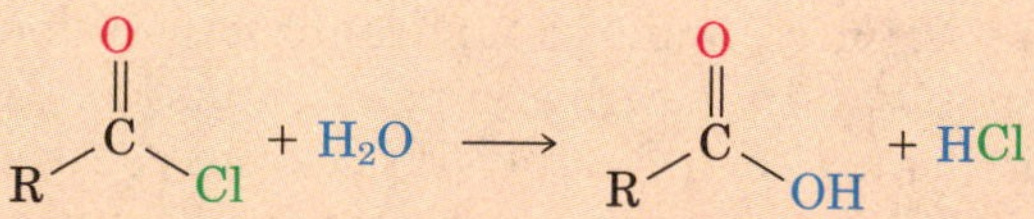

   (b) Alcoholysis to yield esters (Section 21.5)

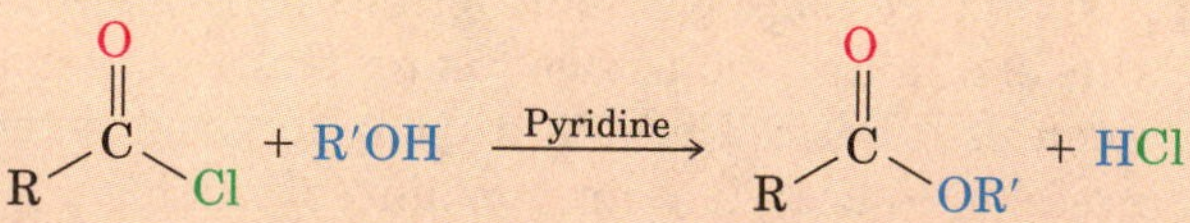

   (c) Aminolysis to yield amides (Section 21.5)

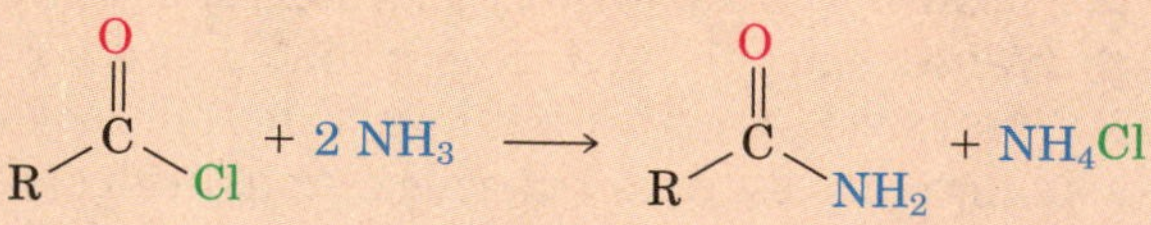

(continued)➤

(d) Reduction to yield primary alcohols (Section 21.5)

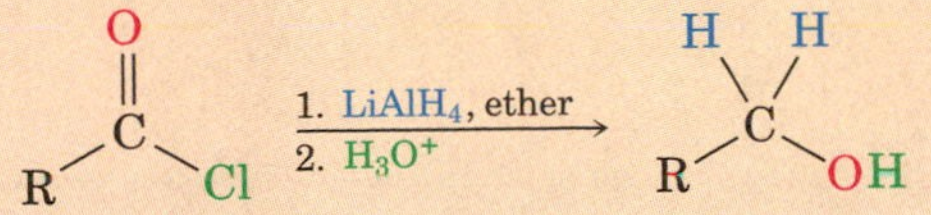

(e) Partial reduction to yield aldehydes (Section 21.5)

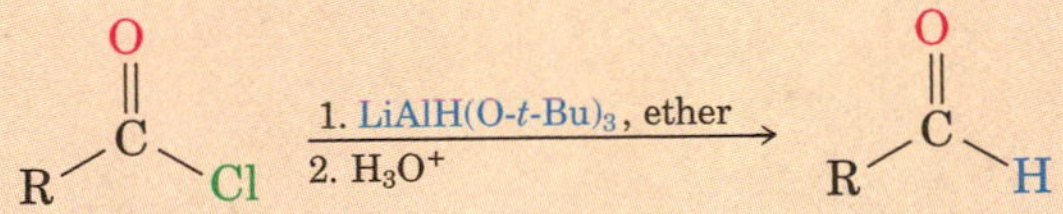

(f) Grignard reaction to yield tertiary alcohols (Section 21.5)

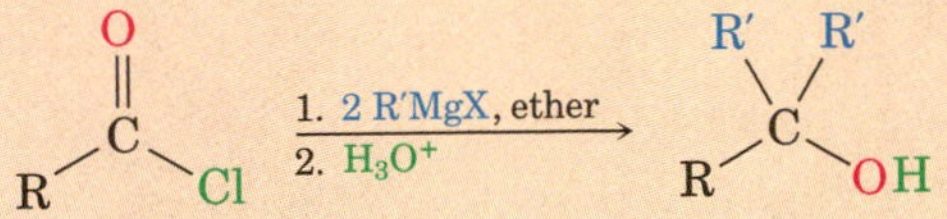

(g) Diorganocopper reaction to yield ketones (Section 21.5)

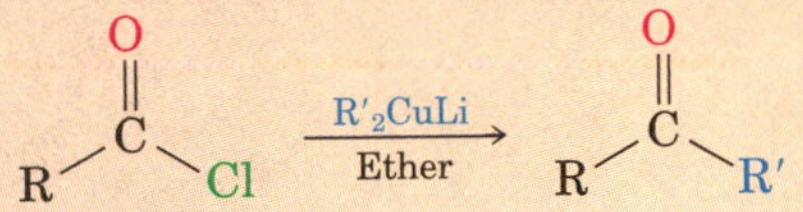

3. Reactions of acid anhydrides
   (a) Hydrolysis to yield acids (Section 21.6)

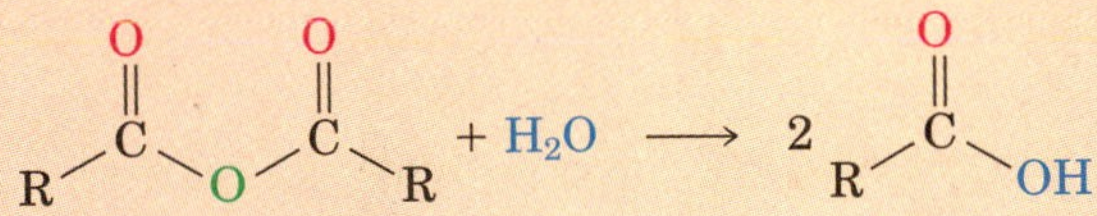

(b) Alcoholysis to yield esters (Section 21.6)

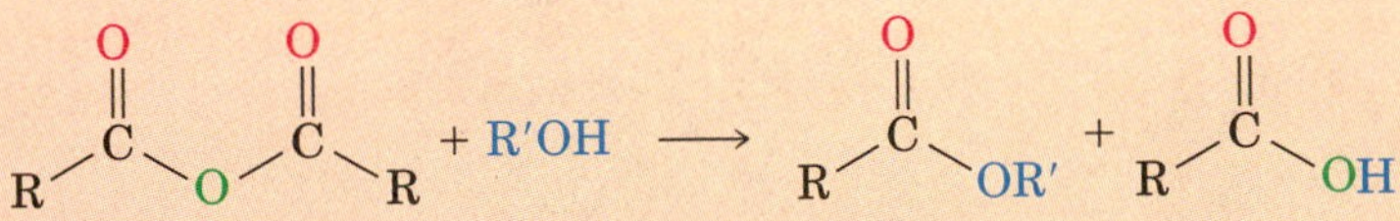

(c) Aminolysis to yield amides (Section 21.6)

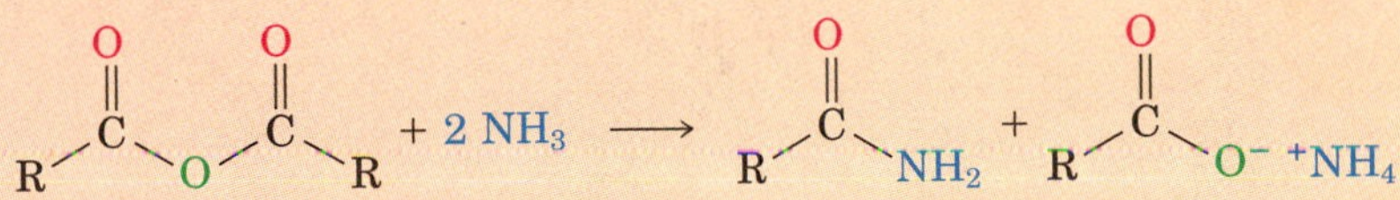

(d) Reduction to yield primary alcohols (Section 21.6)

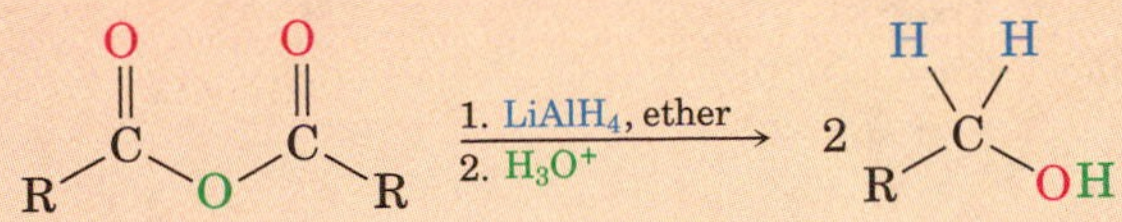

(continued)➤

4. Reactions of amides and lactams
   (a) Hydrolysis to yield acids (Section 21.8)

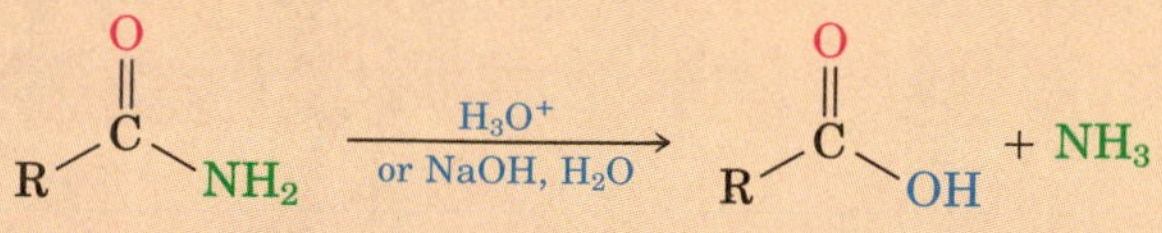

   (b) Reduction to yield amines (Section 21.8)

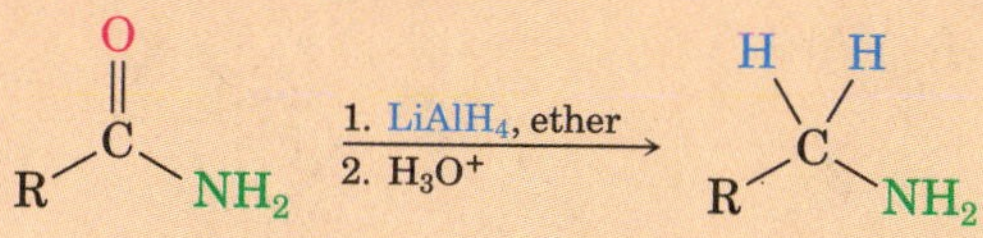

   (c) Dehydration of primary amides to yield nitriles (Section 21.9)

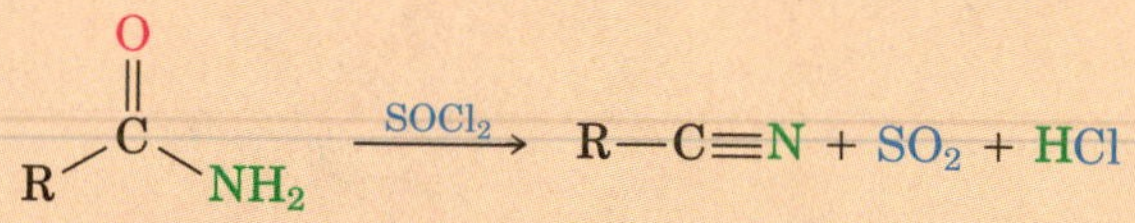

5. Reactions of esters and lactones
   (a) Hydrolysis to yield acids (Section 21.7)

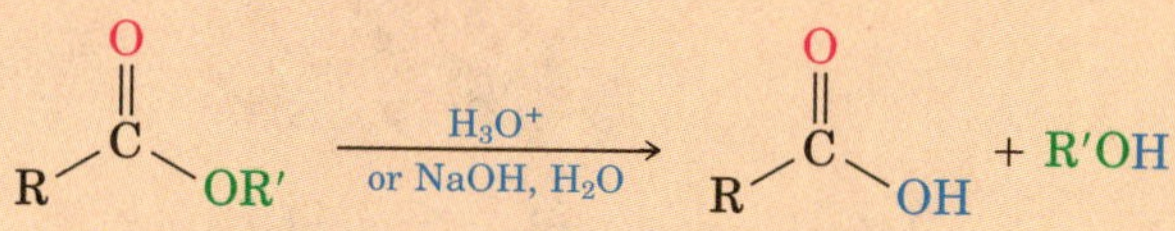

   (b) Aminolysis to yield amides (Section 21.7)

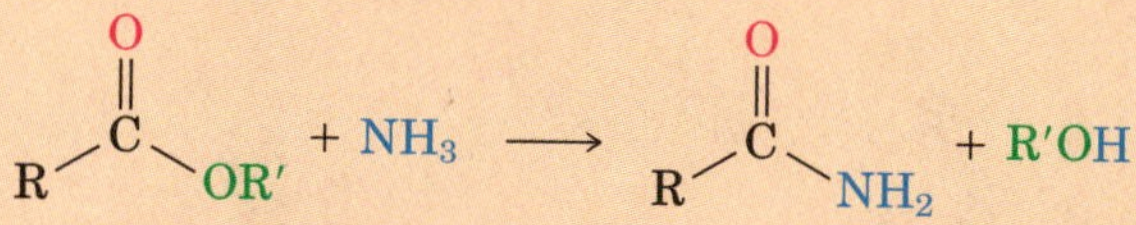

   (c) Reduction to yield primary alcohols (Section 21.7)

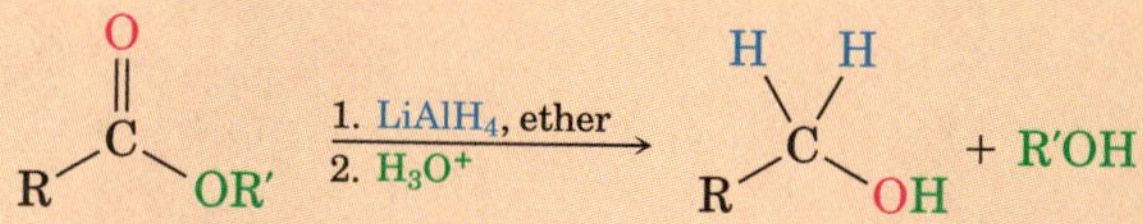

   (d) Partial reduction to yield aldehydes (Section 21.7)

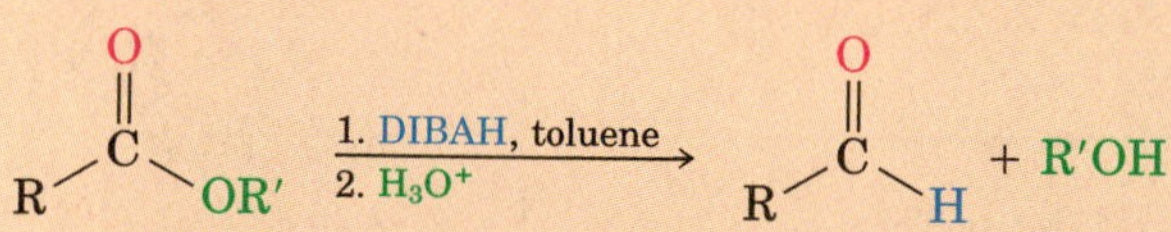

   (e) Grignard reaction to yield tertiary alcohols (Section 21.7)

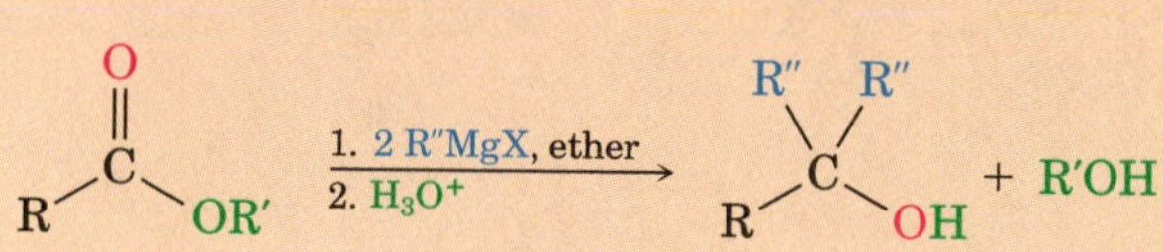

(continued)➤

6. Reactions of nitriles

(a) Hydrolysis to yield carboxylic acids (Section 21.9)

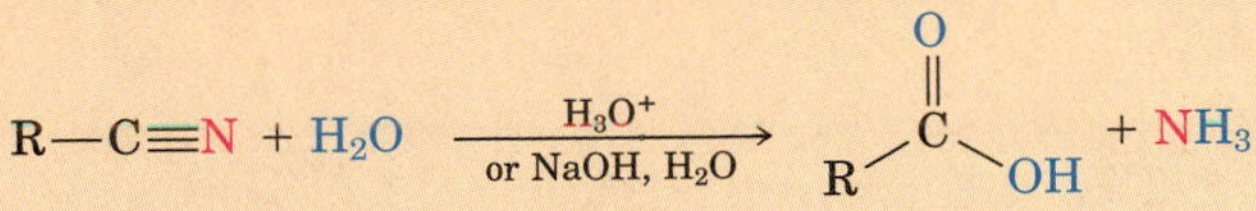

(b) Partial hydrolysis to yield amides (Section 21.9)

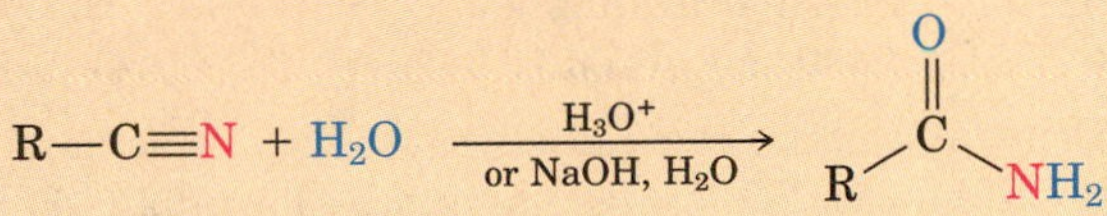

(c) Reduction to yield primary amines (Section 21.9)

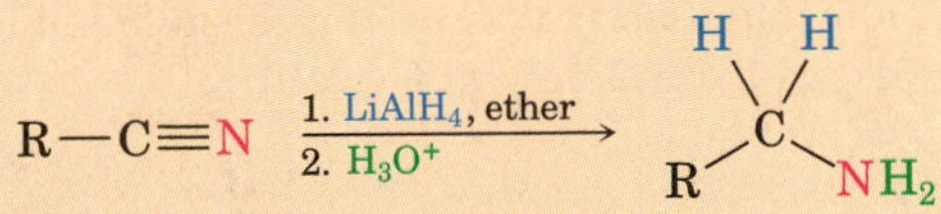

(d) Partial reduction to yield aldehydes (Section 21.9)

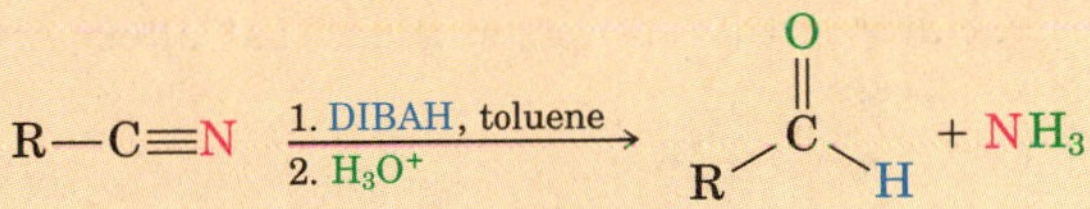

(e) Reaction with Grignard reagents to yield ketones (Section 21.9)

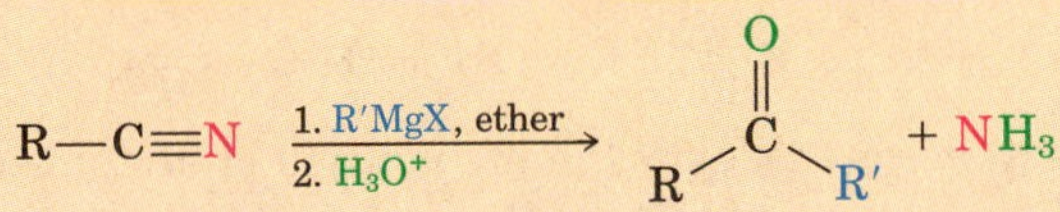

ADDITIONAL PROBLEMS

21.32 Give IUPAC names for the following compounds:

(a) 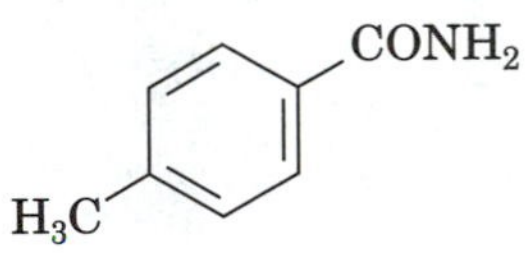

(b) $(CH_3CH_2)_2CHCH{=}CHCN$

(c) $CH_3O_2CCH_2CH_2CO_2CH_3$

(d) 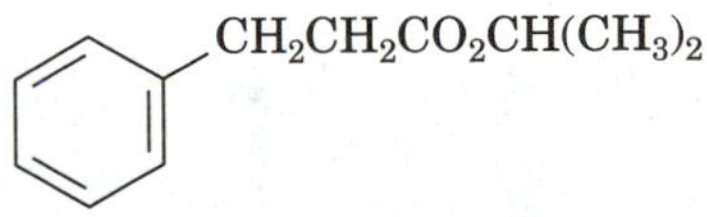

(e) 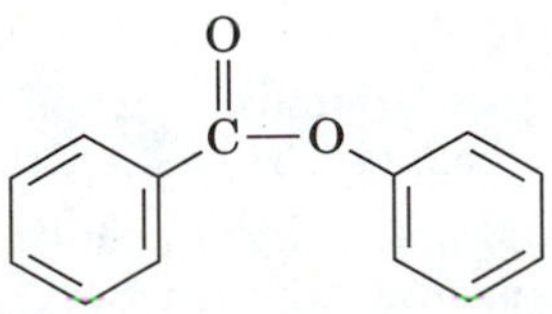

(f) $CH_3CHBrCH_2CONHCH_3$

(g) 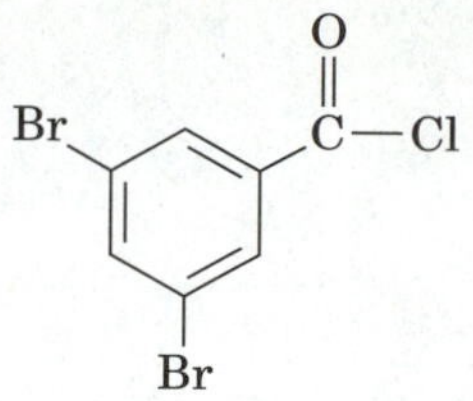

(h) 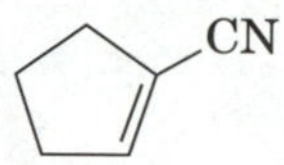

21.33 Draw structures corresponding to the following names:

(a) *p*-Bromophenylacetamide
(b) *m*-Benzoylbenzonitrile
(c) 2,2-Dimethylhexanamide
(d) Cyclohexyl cyclohexanecarboxylate
(e) 2-Cyclobutenecarbonitrile
(f) 2-Propylbutanedioyl dichloride

21.34 Draw and name compounds that meet these descriptions:

(a) Three acid chlorides having the formula C_6H_9ClO
(b) Three amides having the formula $C_7H_{11}NO$
(c) Three nitriles having the formula C_5H_7N

21.35 The following reactivity order has been found for the saponification of alkyl acetates by aqueous hydroxide ion. Explain.

$$CH_3CO_2CH_3 > CH_3CO_2CH_2CH_3 > CH_3CO_2CH(CH_3)_2 > CH_3CO_2C(CH_3)_3$$

21.36 How can you explain the observation that attempted Fischer esterification of 2,4,6-trimethylbenzoic acid with methanol/HCl is unsuccessful? No ester is obtained, and the acid is recovered unchanged. What alternative method of esterification might be successful?

21.37 How can you account for the fact that, when a carboxylic acid is dissolved in isotopically labeled water, the label rapidly becomes incorporated into *both* oxygen atoms of the carboxylic acid?

21.38 Outline methods for the preparation of acetophenone (phenyl methyl ketone) from the following starting materials.

(a) Benzene
(b) Bromobenzene
(c) Methyl benzoate
(d) Benzonitrile
(e) Styrene

21.39 In the basic hydrolysis of *p*-substituted methyl benzoates, the following reactivity order has been found for Y: $NO_2 > Br > H > CH_3 > OCH_3$. How can you explain this reactivity order? Where would you expect Y = C≡N, Y = CHO, and Y = NH_2 to be in the reactivity list?

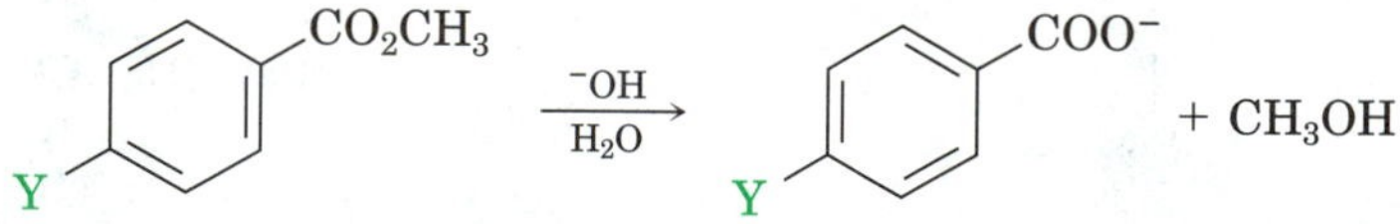

21.40 How might you prepare the following compounds from butanoic acid?

(a) 1-Butanol
(b) Butanal
(c) 1-Bromobutane
(d) Pentanenitrile
(e) 1-Butene
(f) *N*-Methylpentanamide
(g) 2-Hexanone
(h) Butylbenzene

21.41 What product would you expect to obtain from Grignard reaction of an excess of phenylmagnesium bromide with dimethyl carbonate, $CH_3OCOOCH_3$?

21.42 When ethyl benzoate is heated in methanol containing a small amount of HCl, methyl benzoate is formed. Propose a mechanism for the reaction.

21.43 *Qiana*, a polyamide fiber with a silk-like feel, has the following structure. What are the monomer units used in the synthesis of Qiana?

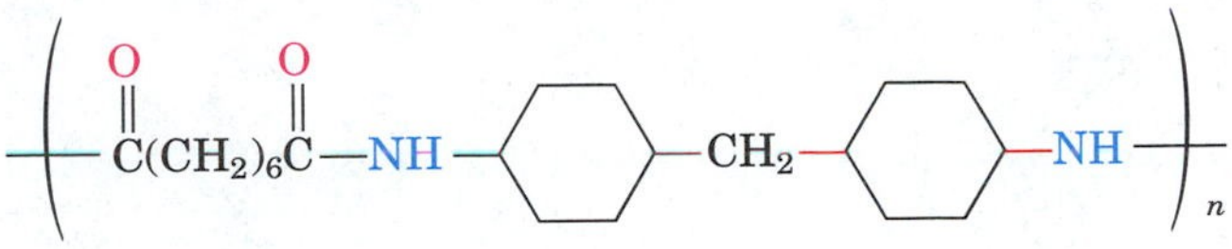

Qiana

21.44 What is the structure of the polymer produced by treatment of β-propiolactone with a small amount of hydroxide ion?

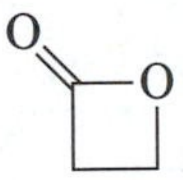

β-Propiolactone

21.45 *tert*-Butoxycarbonyl azide, a reagent used in protein synthesis, is prepared by treating *tert*-butoxycarbonyl chloride with sodium azide. Propose a mechanism for this reaction.

$$(CH_3)_3COCOCl + NaN_3 \longrightarrow (CH_3)_3COCON_3 + NaCl$$

21.46 Predict the product, if any, of reaction between propanoyl chloride and the following reagents:

(a) $(Ph)_2CuLi$ in ether
(b) $LiAlH_4$, then H_3O^+
(c) CH_3MgBr, then H_3O^+
(d) $Li(O\text{-}t\text{-}Bu)_3AlH$
(e) H_3O^+
(f) Cyclohexanol
(g) Aniline
(h) $CH_3COO^-\ {}^+Na$

21.47 Answer Problem 21.46 for reaction of the listed reagents with methyl propanoate.

21.48 Answer Problem 21.46 for reaction of the listed reagents with propanamide and with propanenitrile.

21.49 Treatment of a carboxylic acid with trifluoroacetic anhydride leads to a mixed anhydride that rapidly reacts with alcohol to give an ester:

$$RCOOH + (CF_3CO)_2O \longrightarrow R{-}\overset{O}{\overset{\|}{C}}{-}O{-}\overset{O}{\overset{\|}{C}}{-}CF_3 \xrightarrow{R'OH} R{-}\overset{O}{\overset{\|}{C}}{-}OR' + CF_3COOH$$

(a) Propose a mechanism for formation of the mixed anhydride.
(b) Why is the mixed anhydride unusually reactive?
(c) Why does the mixed anhydride react as indicated, rather than giving trifluoroacetate esters plus carboxylic acid?

21.50 How would you accomplish the following transformations? More than one step may be required.

(a) $CH_3CH_2CH_2CH_2CN \longrightarrow CH_3CH_2CH_2CH_2CH_2NH_2$
(b) $CH_3CH_2CH_2CH_2CN \longrightarrow CH_3CH_2CH_2CH_2CH_2N(CH_3)_2$
(c) $CH_3CH_2CH_2CH_2CN \longrightarrow CH_3CH_2CH_2CH_2C(CH_3)_2OH$
(d) $CH_3CH_2CH_2CH_2CN \longrightarrow CH_3CH_2CH_2CH_2CH(OH)CH_3$
(e) $CH_3CH_2CH_2CH_2CN \longrightarrow CH_3CH_2CH_2CH_2CHO$

21.51 List as many ways as you can think of for transforming cyclohexanol into cyclohexanecarbaldehyde (try to get at least four).

21.52 One method for preparing the 1,6-hexanediamine needed in nylon production starts with 1,3-butadiene. How would you accomplish this synthesis?

$$H_2C{=}CHCH{=}CH_2 \xrightarrow{?} H_2NCH_2CH_2CH_2CH_2CH_2CH_2NH_2$$

21.53 Succinic anhydride yields succinimide when heated with ammonium chloride at 200°C. Propose a mechanism for this reaction. Why do you suppose such a high reaction temperature is required?

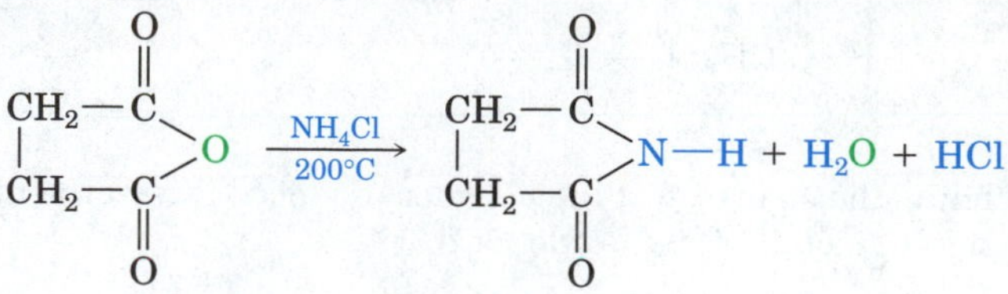

21.54 Polyimides having the structure shown are used as coatings on glass and plastics to improve scratch resistance. How would you synthesize a polyimide? (See Problem 21.53.)

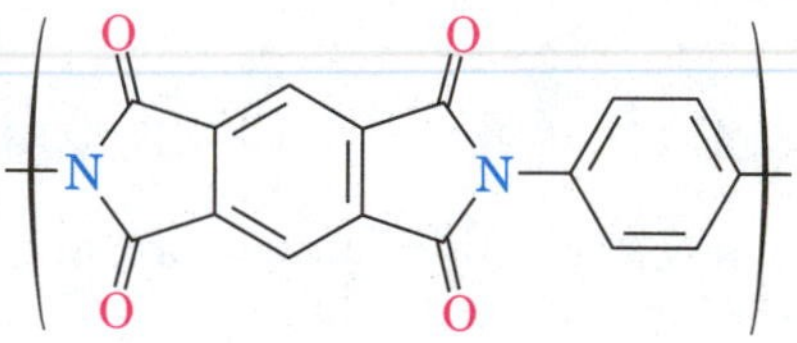

A polyimide

21.55 Butacetin is an analgesic (pain-killing) agent that is synthesized commercially from *p*-fluoronitrobenzene. Propose a likely synthesis route.

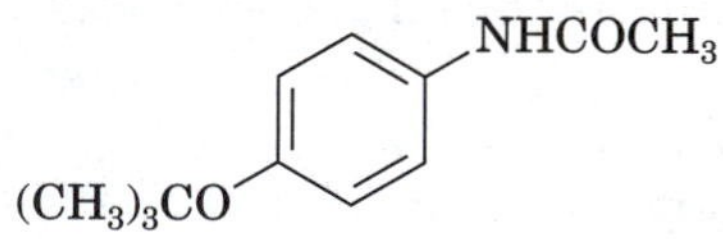

Butacetin

21.56 Phenyl 4-aminosalicylate is a drug used in the treatment of tuberculosis. Propose a synthesis of this compound starting from 4-nitrosalicylic acid.

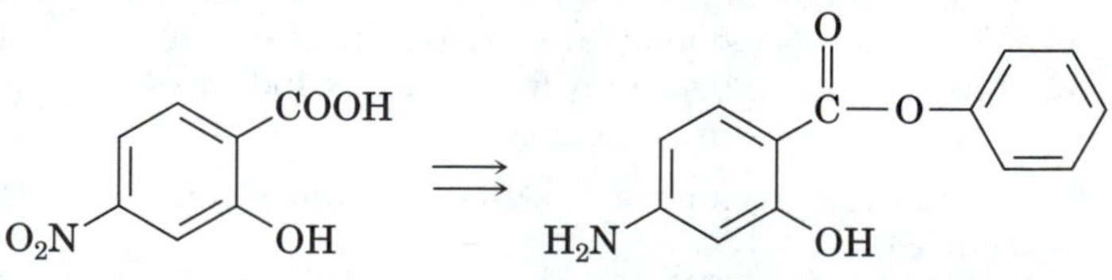

4-Nitrosalicylic acid **Phenyl 4-aminosalicylate**

21.57 How would you distinguish spectroscopically between the following isomer pairs? Tell what differences you would expect to see.
(a) *N*-Methylpropanamide and *N*,*N*-dimethylacetamide
(b) 5-Hydroxypentanenitrile and cyclobutanecarboxamide
(c) 4-Chlorobutanoic acid and 3-methoxypropanoyl chloride
(d) Ethyl propanoate and propyl acetate

21.58 *N*,*N*-Diethyl-*m*-toluamide (DEET) is the active ingredient in many insect-repellent preparations. How might you synthesize this substance from *m*-bromotoluene?

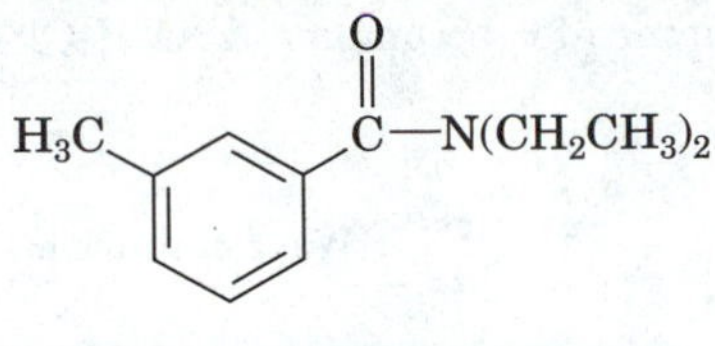

N,N-Diethyl-m-toluamide

21.59 Tranexamic acid, a drug useful against blood clotting, is prepared commercially from *p*-methylbenzonitrile. Formulate the steps likely to be used in the synthesis. (Don't worry about cis–trans isomers. Heating to 300°C interconverts the isomers.)

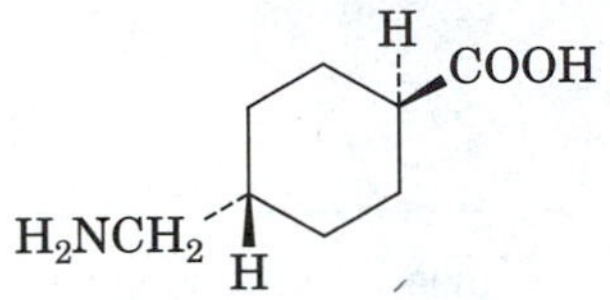

Tranexamic acid

21.60 One frequently used method for preparing methyl esters is by reaction of carboxylic acids with diazomethane, CH_2N_2:

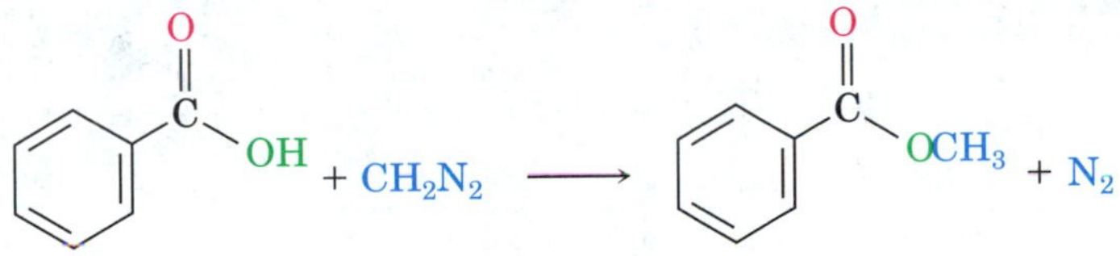

Benzoic acid **Diazomethane** **Methyl benzoate (100%)**

The reaction occurs in two steps: (1) protonation of diazomethane by the carboxylic acid to yield methyldiazonium ion, $CH_3N_2^+$, plus a carboxylate ion; and (2) reaction of the carboxylate ion with $CH_3N_2^+$.

(a) Draw two resonance structures of diazomethane and account for step (1).

(b) What kind of reaction occurs in step (2)?

21.61 The following reaction is called the benzilic acid rearrangement. You haven't seen it before, but it takes place by typical carbonyl-group mechanisms of the sort you've seen dozens of times. Propose a mechanism (Ph = phenyl).

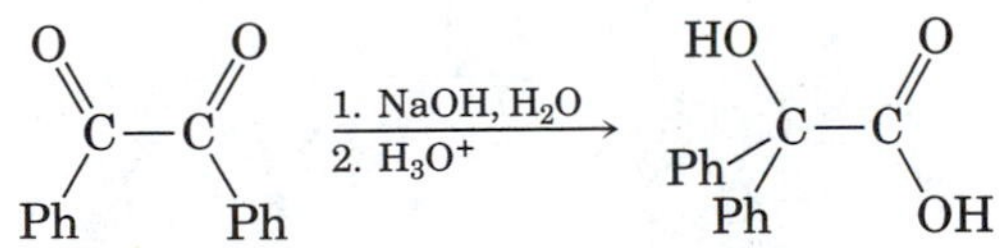

Benzil **Benzylic acid**

21.62 In light of your answer to Problem 21.61, what product is likely to result from the following reaction?

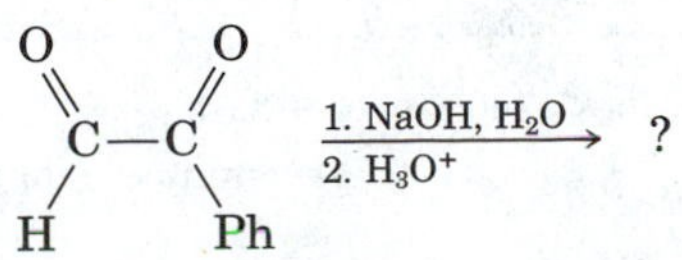

21.63 Propose a structure for a compound, $C_4H_7ClO_2$, that has the following IR and ^1H NMR spectra.

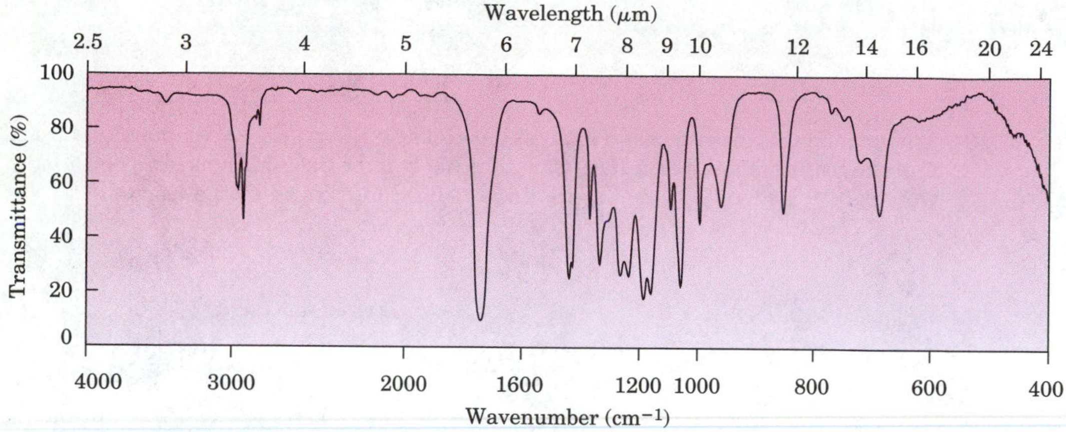

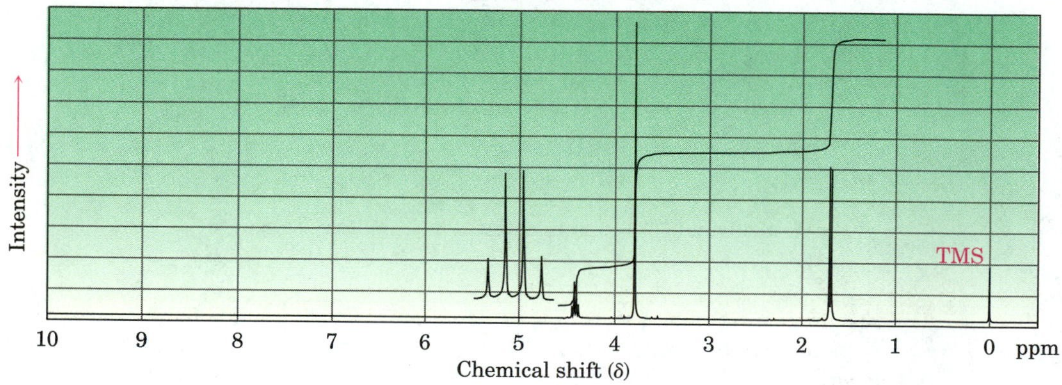

21.64 Propose a structure for a compound, C_4H_7N, that has the following IR and ^1H NMR spectra.

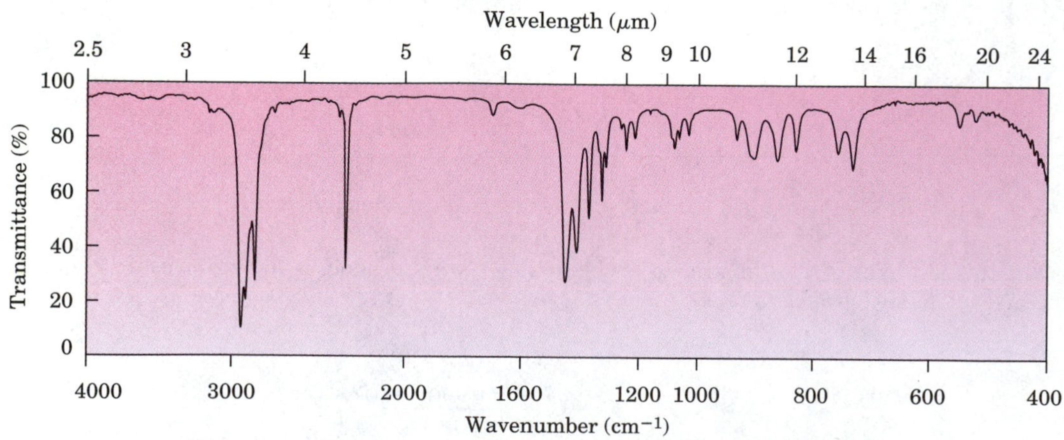

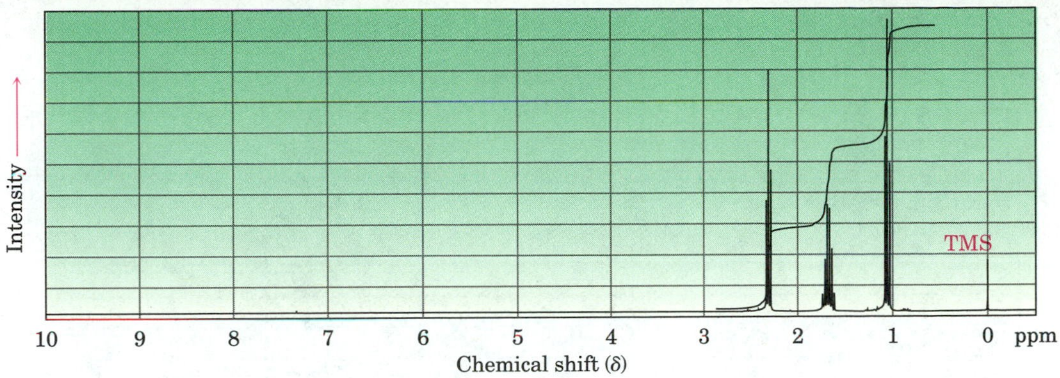

21.65 Assign structures to compounds with the given ^{1}H NMR spectra.

(a) C_4H_7ClO
IR: 1810 cm^{-1}

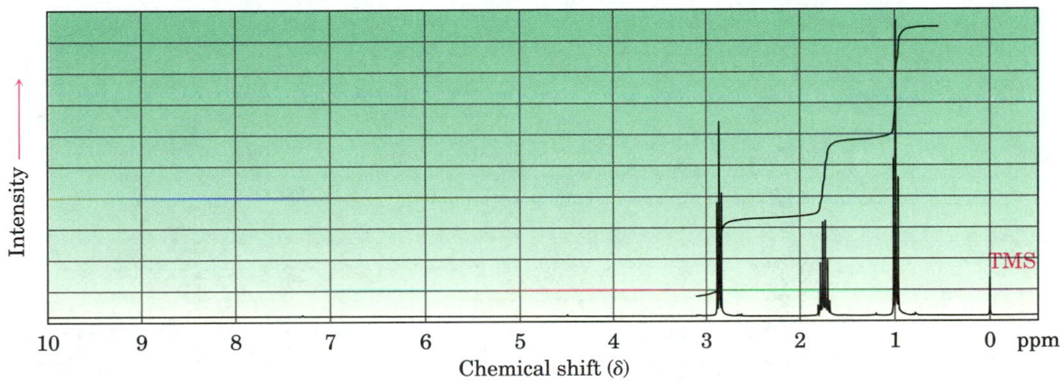

(b) $C_5H_7NO_2$
IR: 2250, 1735 cm^{-1}

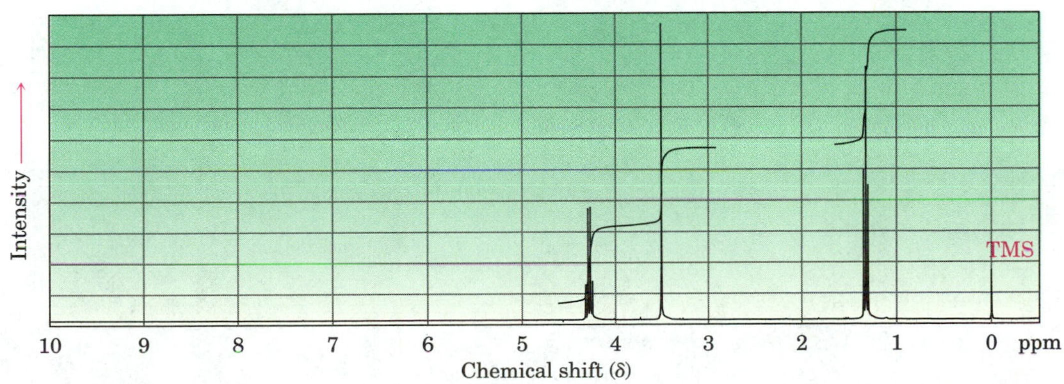

(c) $C_5H_{10}O_2$
IR: 1735 cm^{-1}

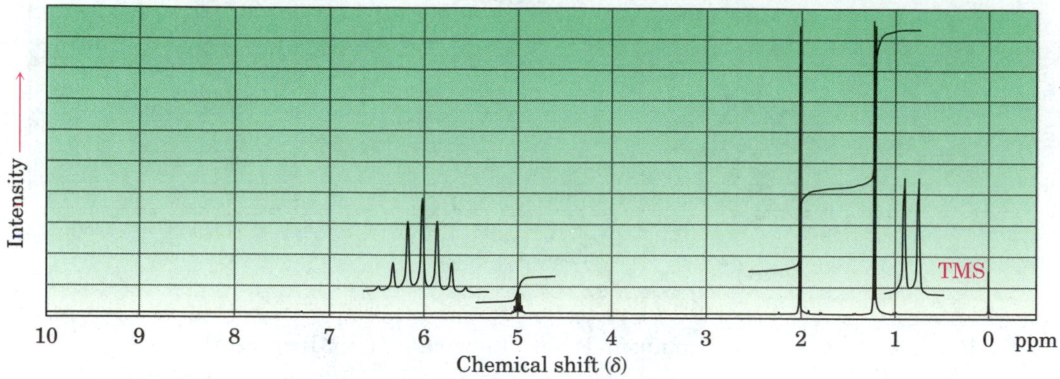

21.66 Propose structures for compounds with the given ^{1}H NMR spectra.
(a) $C_5H_9ClO_2$
IR: 1735 cm^{-1}

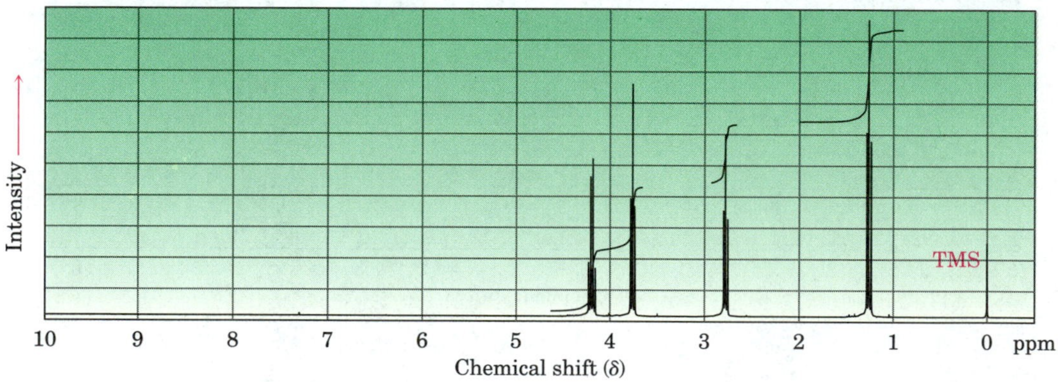

(b) $C_7H_{12}O_4$
IR: 1735 cm^{-1}

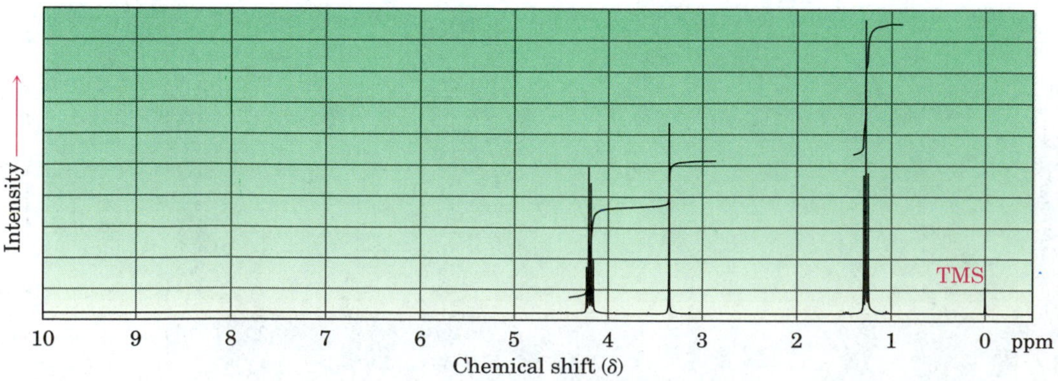

(c) $C_{11}H_{12}O_2$
IR: 1710 cm^{-1}

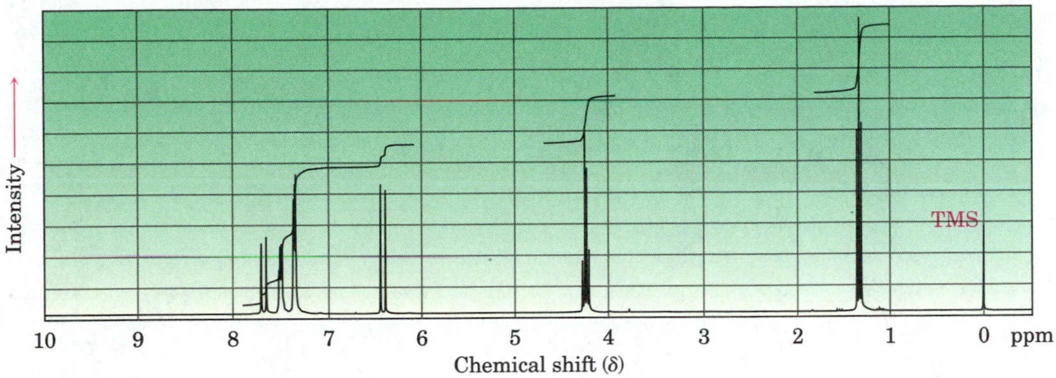

A Look Ahead

21.67 Epoxy adhesives are prepared in two steps. S_N2 reaction of the disodium salt of bisphenol A with epichlorohydrin forms a "prepolymer," which is then "cured" by treatment with a triamine such as $H_2NCH_2CH_2NHCH_2CH_2NH_2$.

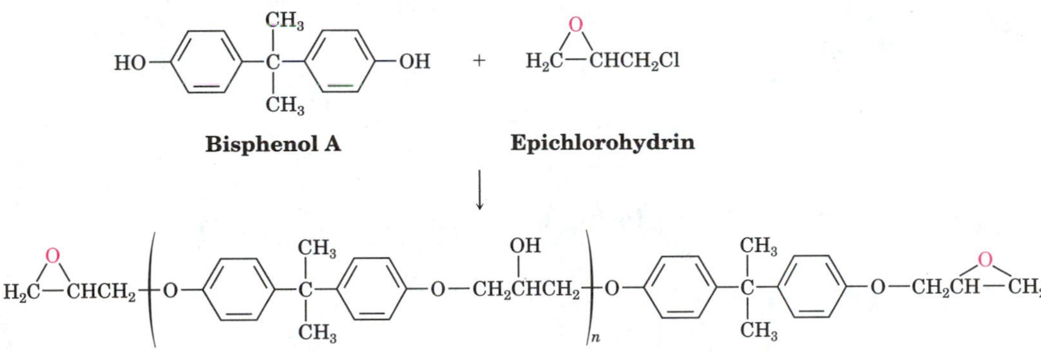

"Prepolymer"

Draw structures to show how addition of the triamine results in strengthening the polymer. We'll see in Chapter 24 that amines are good nucleophiles and can open epoxide rings in the same way other bases can (Section 18.8).

21.68 We'll see in the next chapter a process called the iodoform reaction, in which a triiodomethyl ketone reacts with aqueous base to yield a carboxylate ion and iodoform (triiodomethane). Propose a mechanism for this reaction.

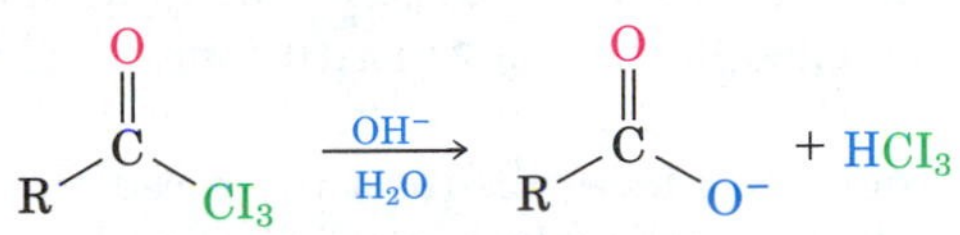

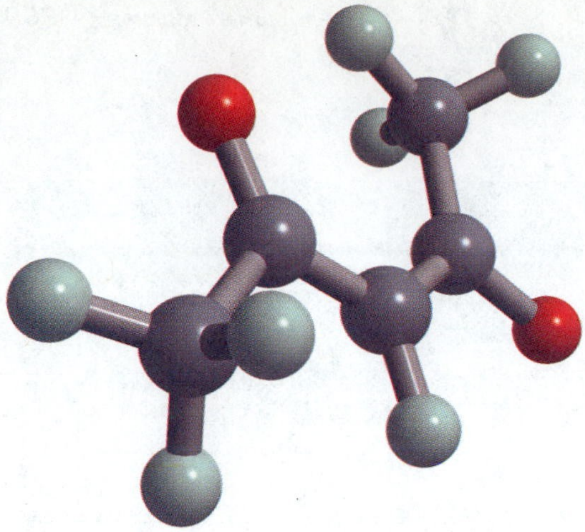

Enolate ions are reactive intermediates in many carbonyl-group reactions.

22

Carbonyl Alpha-Substitution Reactions

We said in the overview of carbonyl chemistry that most carbonyl-group reactions can be explained by just four fundamental processes: nucleophilic additions, nucleophilic acyl substitutions, α substitutions, and carbonyl condensations. Having looked at the chemistry of nucleophilic addition reactions and nucleophilic acyl substitution reactions in the past three chapters, we'll now look at the chemistry of the third major carbonyl-group process—the **α-substitution reaction**.

Alpha-substitution reactions occur at the position *next to* the carbonyl group—the **α position**—and involve the substitution of an α hydrogen atom by some other group through either an enol or enolate ion intermediate. Let's begin by learning more about these two species.

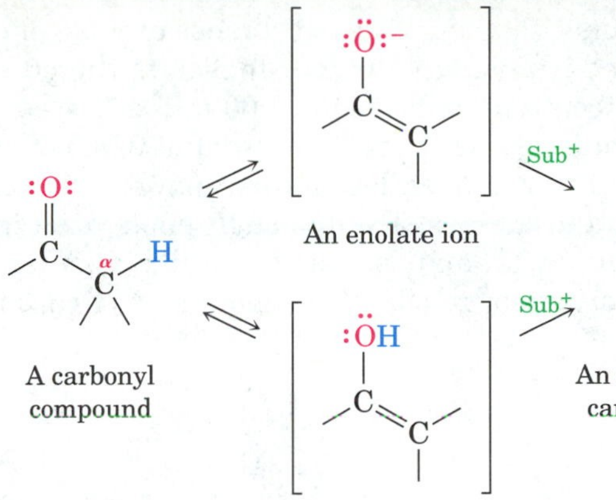

An enolate ion

A carbonyl compound

An alpha-substituted carbonyl compound

An enol

22.1 Keto–Enol Tautomerism

Carbonyl compounds with one or more hydrogen atoms on their α carbons rapidly interconvert with their corresponding **enols** (Section 8.5). This rapid interconversion between two substances is a special kind of isomerism known as **tautomerism** (taw-**tom**-er-ism; from the Greek *tauto*, "the same," and *meros*, "part"). Individual isomers are called **tautomers** (**taw**-toe-mers).

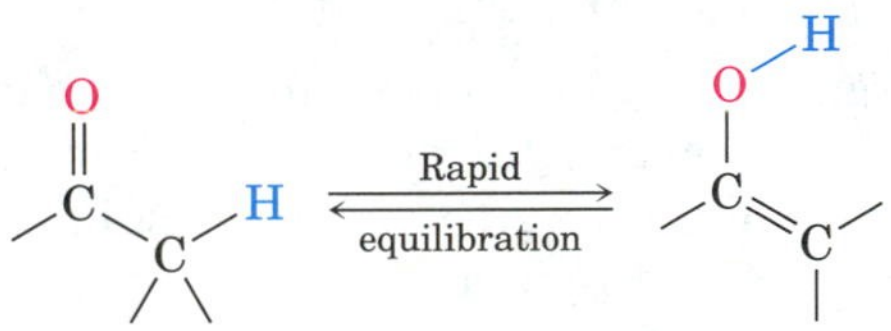

Keto tautomer

Enol tautomer

Note the difference between tautomers and resonance forms: Tautomers are different compounds (isomers) with different structures, while resonance forms are different representations of a single structure. Tautomers have their *atoms* arranged differently, while resonance forms differ only in the position of their *electrons*.

Note also that tautomers are *rapidly* interconvertible. Thus, keto and enol carbonyl isomers are tautomers, but two isomeric alkenes such as 1-butene and 2-butene are not, because they don't interconvert rapidly under most circumstances.

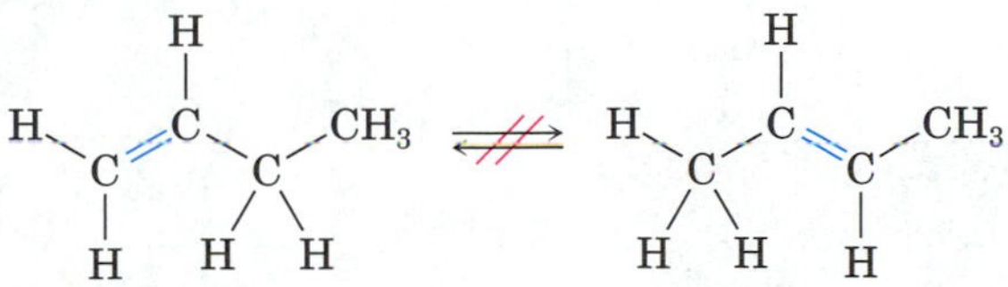

1-Butene

2-Butene

Most carbonyl compounds exist almost exclusively in the keto form at equilibrium, and it's usually difficult to isolate the pure enol. For example, cyclohexanone contains only about 0.0001% of its enol tautomer at room temperature, and acetone contains only about 0.000 001% enol. The percentage of enol tautomer is even less for carboxylic acids and acid derivatives, such as esters and amides. Even though enols are difficult to isolate and are present only to a small extent at equilibrium, they are nevertheless extremely important in much of the chemistry of carbonyl compounds.

Cyclohexanone

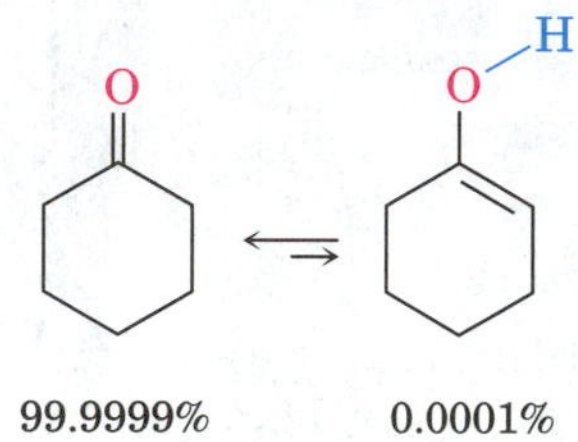

Acetone

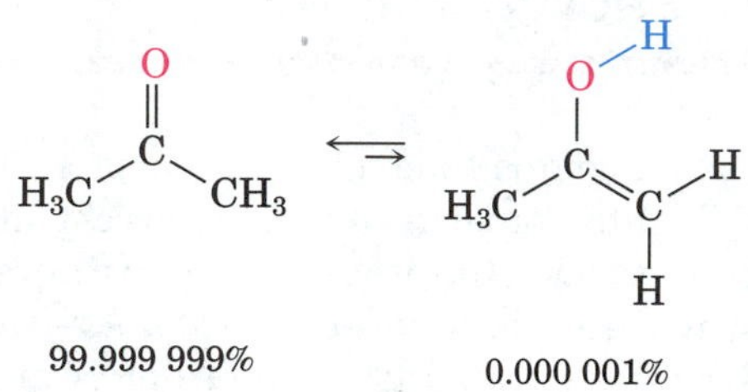

Keto–enol tautomerism of carbonyl compounds is catalyzed by both acids and bases. Acid catalysis involves protonation of the carbonyl oxygen atom (a Lewis base) to give an intermediate cation that can lose a proton from the α carbon to yield a neutral enol (Figure 22.1). This proton loss from the cation intermediate is analogous to what occurs during an E1 reaction when a carbocation loses a proton to form an alkene (Section 11.14).

Base-catalyzed enol formation occurs by an acid–base reaction between catalyst and carbonyl compound. The carbonyl compound acts as a weak protic acid and donates one of its α hydrogens to the base. The resultant anion—an **enolate ion**—is then reprotonated to yield a neutral compound. Since the enolate ion is a resonance hybrid of two forms, it can be protonated either on carbon to regenerate the keto tautomer or on oxygen to give an enol tautomer (Figure 22.2).

Note that only the hydrogens on the α positions of carbonyl compounds are acidic. Hydrogens at β, γ, δ, and so on, are not acidic and can't be removed by base. We'll account for this unique behavior of α hydrogens shortly.

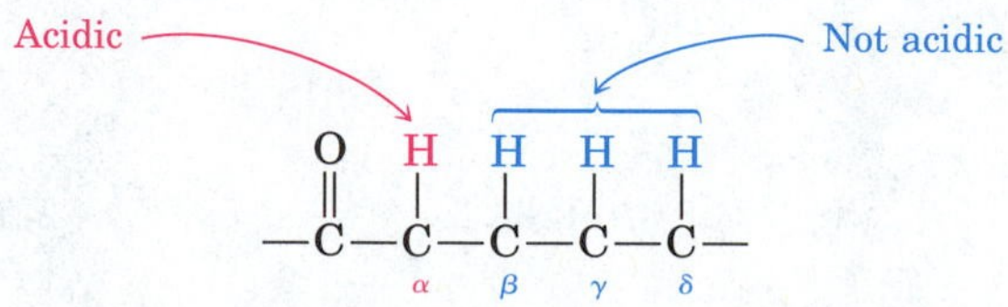

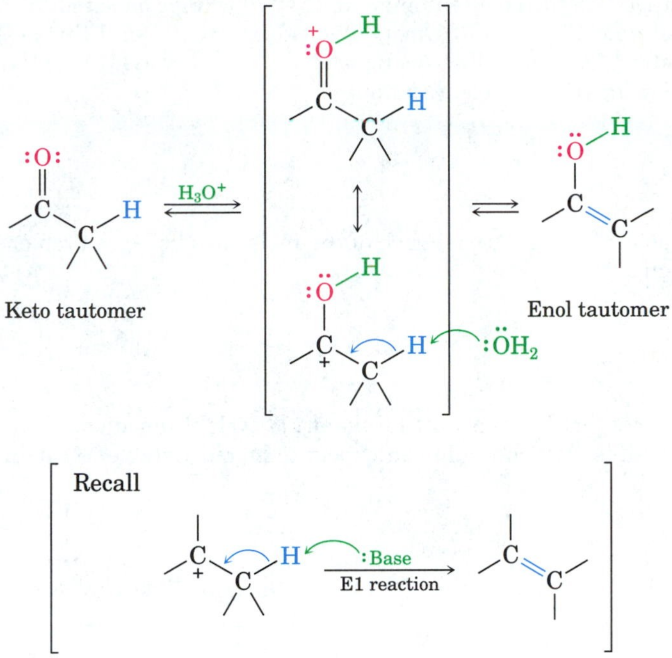

Figure 22.1 Mechanism of acid-catalyzed enol formation.

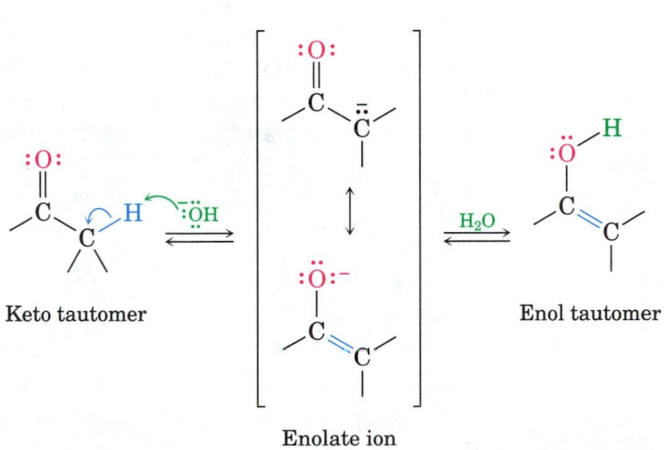

Figure 22.2 Mechanism of base-catalyzed enol formation. The intermediate enolate ion, a resonance hybrid of two forms, can be protonated either on carbon to regenerate starting ketone or on oxygen to give an enol.

PROBLEM

22.1 Draw structures for the enol tautomers of the following compounds:
(a) Cyclopentanone (b) Acetyl chloride (c) Ethyl acetate
(d) Propanal (e) Acetic acid (f) Phenylacetone
(g) Acetophenone (methyl phenyl ketone)

PROBLEM

22.2 How many acidic hydrogens does each of the molecules listed in Problem 22.1 have? Identify them.

PROBLEM

22.3 Draw structures for the monoenol forms of 1,3-cyclohexanedione. How many enol forms are possible? Which would you expect to be most stable? Explain.

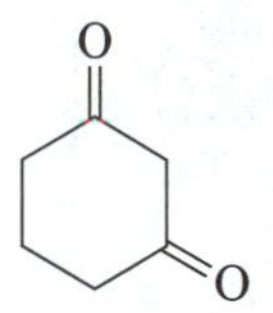

1,3-Cyclohexanedione

22.2 Reactivity of Enols: The Mechanism of Alpha-Substitution Reactions

What kind of chemistry do enols have? Since their double bonds are electron-rich, enols behave as nucleophiles and react with electrophiles in much the same way that alkenes do. Because of resonance electron donation of the lone-pair electrons on oxygen, though, enols are more reactive than alkenes (Section 14.10).

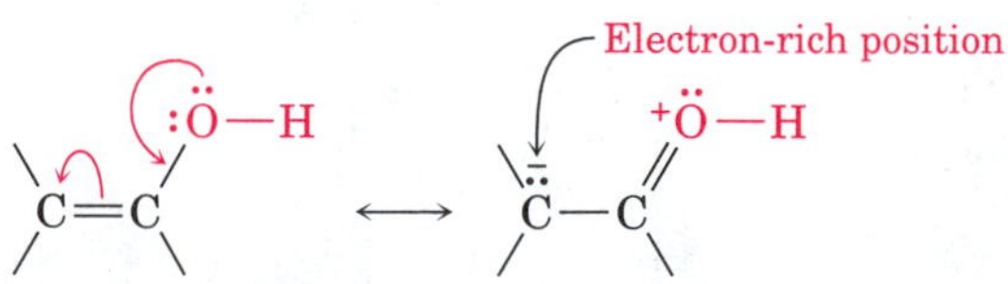

When an *alkene* reacts with an electrophile, such as Br_2, addition of Br^+ occurs to give an intermediate cation that reacts with Br^-. When an *enol* reacts with an electrophile, however, the initial addition step is the same but the intermediate cation loses the –OH proton to regenerate a carbonyl compound. The net result of the reaction of an enol with an electrophile is α substitution by the general mechanism shown in Figure 22.3.

Acid-catalyzed enol formation occurs by the usual mechanism.

An electron pair from the enol oxygen attacks an electrophile, forming a new bond and leaving a cation intermediate that is stabilized by resonance between two forms.

Loss of a proton from oxygen yields the neutral alpha-substitution product as a new C=O bond is formed.

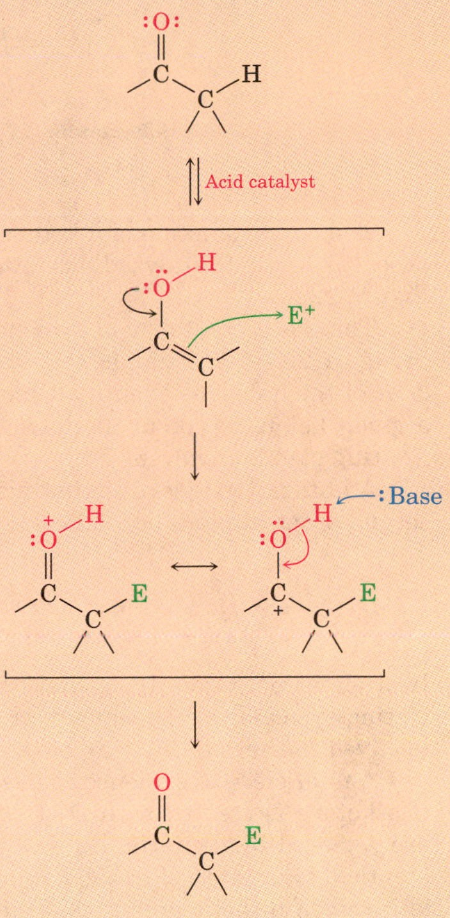

Figure 22.3 General mechanism of a carbonyl α-substitution reaction.

22.3 Alpha Halogenation of Ketones and Aldehydes

Ketones and aldehydes can be halogenated at their α positions by reaction with Cl_2, Br_2, or I_2 in acidic solution. Bromine is most often used, and acetic acid is often employed as solvent.

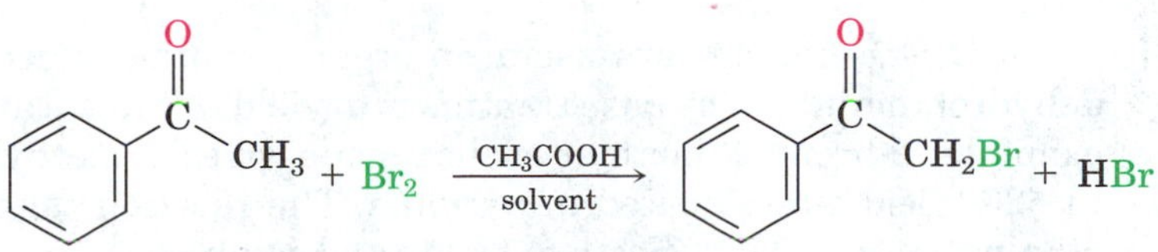

Acetophenone **α-Bromoacetophenone (72%)**

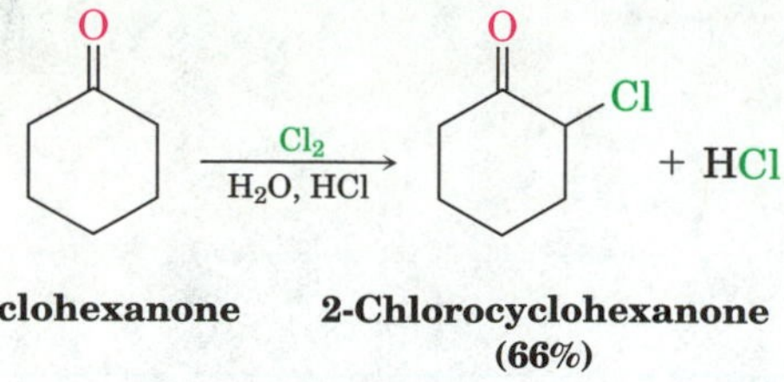

The α halogenation of a ketone is a typical α-substitution reaction that proceeds by acid-catalyzed formation of an enol intermediate, as shown in Figure 22.4.

There is a great deal of evidence to substantiate the mechanism shown in Figure 22.4. For example, the rate of the halogenation reaction is independent of the halogen's identity. Chlorination, bromination, and iodination of a given ketone all occur at the same rate, indicating that the same rate-limiting step is involved.

Additional evidence is that acid-catalyzed ketone halogenations show second-order kinetics and follow the rate law

$$\text{Reaction rate} = k[\text{Ketone}][\text{H}^+]$$

In other words, ketone halogenations depend only on ketone and acid concentrations, and are independent of halogen concentration. Halogen is not involved in the rate-limiting step.

A final piece of evidence comes from deuterium-exchange experiments. If a ketone is treated with D_3O^+ instead of H_3O^+, the acidic α hydrogens are replaced by deuterium. For a given ketone, the rate of deuteration is identical to the rate of halogenation, indicating that the same intermediate is involved in both processes. That common intermediate can only be an enol.

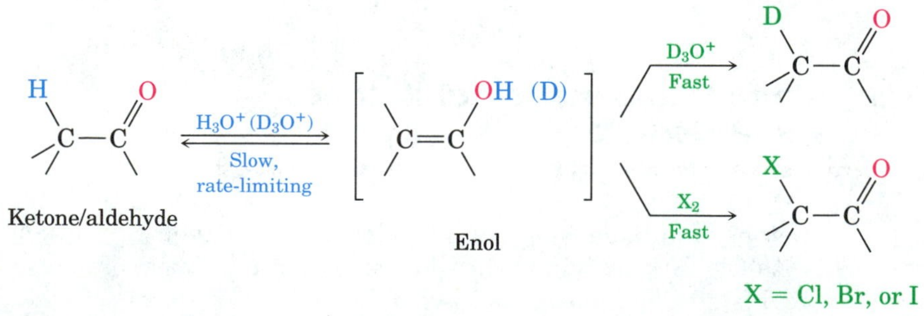

α-Bromo ketones are useful in organic synthesis because they can be dehydrobrominated by base treatment to yield α,β-unsaturated ketones. For example, 2-bromo-2-methylcyclohexanone gives 2-methyl-2-cyclohexenone in 62% yield when heated in pyridine. The reaction takes place by an E2 elimination pathway (Section 11.11) and is an excellent method for introducing C=C bonds into molecules.

The carbonyl oxygen atom is protonated by acid catalyst.

Loss of an acidic proton from the alpha carbon takes place in the normal way to yield an enol intermediate.

An electron pair from the enol attacks bromine, giving an intermediate cation that is stabilized by resonance between two forms.

Loss of the –OH proton then gives the alpha-halogenated product and generates more acid catalyst.

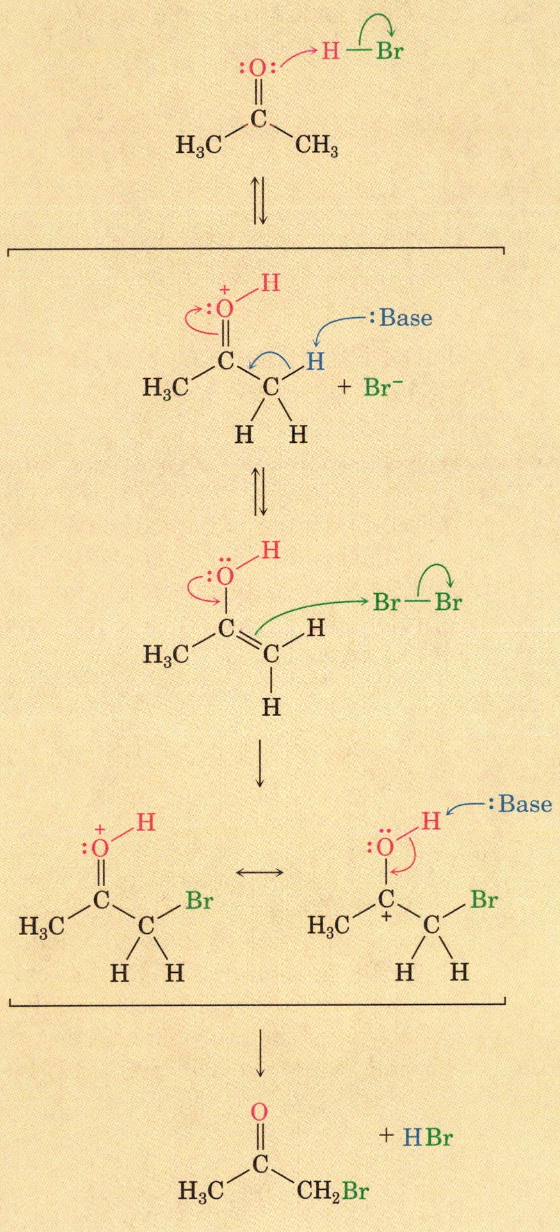

Figure 22.4 Mechanism of the acid-catalyzed bromination of ketones and aldehydes.

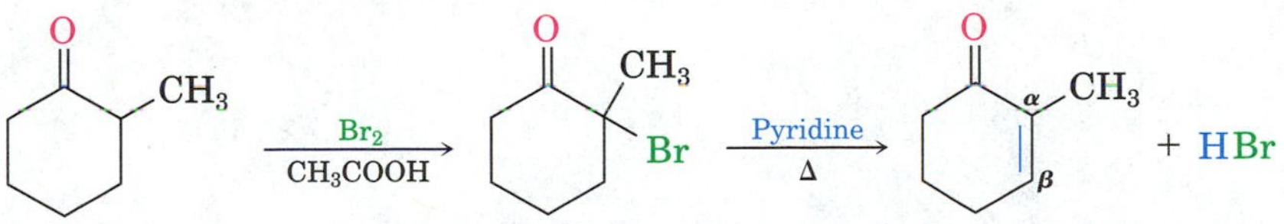

2-Methylcyclohexanone | **2-Bromo-2-methylcyclohexanone** | **2-Methyl-2-cyclohexenone (62%)**

PROBLEM

22.4 Show the mechanism of the deuteration of acetone on treatment with D_3O^+.

$$CH_3\overset{O}{\overset{\|}{C}}CH_3 \xrightarrow{D_3O^+} CH_3\overset{O}{\overset{\|}{C}}CH_2D$$

PROBLEM

22.5 How might you prepare 1-penten-3-one from 3-pentanone?

22.4 Alpha Bromination of Carboxylic Acids: The Hell–Volhard–Zelinskii Reaction

Direct α bromination of carbonyl compounds by Br_2 in acetic acid is limited to ketones and aldehydes because acids, esters, and amides don't enolize sufficiently for halogenation to take place. Carboxylic acids, however, can be α brominated by a mixture of Br_2 and PBr_3—the **Hell–Volhard–Zelinskii (HVZ) reaction**.

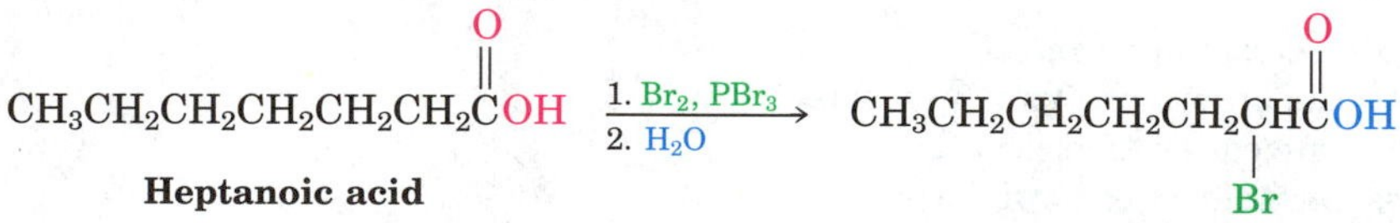

Heptanoic acid **2-Bromoheptanoic acid (90%)**

The first step in the Hell–Volhard–Zelinskii reaction takes place between PBr_3 and a carboxylic acid to yield an intermediate acid bromide plus HBr (Section 21.5). The HBr catalyzes enolization of the acid bromide, and the resultant enol reacts rapidly with Br_2 in an α-substitution reaction. Hydrolysis of the α-bromo acid bromide by addition of water then gives the α-bromo carboxylic acid product.

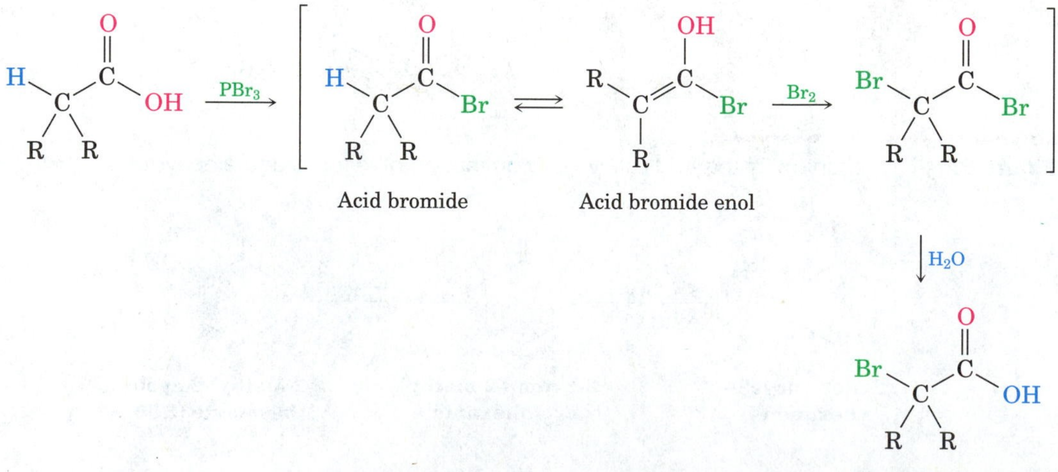

The overall result of the Hell–Volhard–Zelinskii reaction is the transformation of an acid into an α-bromo acid. Note, though, that the key step involves α substitution of an acid bromide enol rather than a carboxylic acid enol. The reaction is analogous in all respects to what occurs during ketone bromination.

PROBLEM...

22.6 If an optically active carboxylic acid such as (*R*)-2-phenylpropanoic acid were brominated under Hell–Volhard–Zelinskii conditions, would you expect the product to be optically active or racemic? Explain.

PROBLEM...

22.7 If methanol rather than water is added at the end of the Hell–Volhard–Zelinskii reaction, an ester rather than an acid is produced. Propose a mechanism for this transformation.

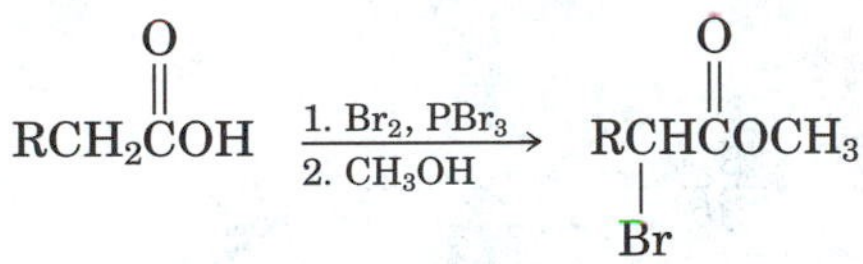

....

22.5 Acidity of Alpha Hydrogen Atoms: Enolate Ion Formation

During the discussion of base-catalyzed enol formation in Section 22.1, we said that carbonyl compounds act as weak protic acids. Strong bases can abstract acidic α hydrogens from carbonyl compounds to yield enolate ions:

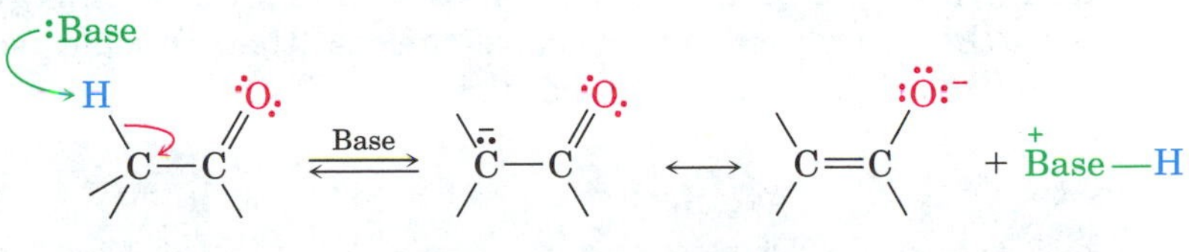

An enolate anion

Why are carbonyl compounds weakly acidic? If we compare acetone (pK_a = 19.3) with ethane ($pK_a \approx 60$), we find that the presence of a neighboring carbonyl group increases the acidity of a ketone over an alkane by a factor of 10^{40}.

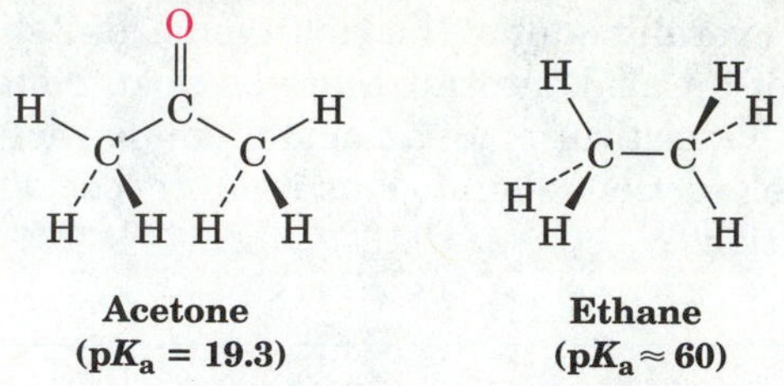

The reason for this increased acidity is best seen by looking at an orbital picture of the enolate ion (Figure 22.5). Proton abstraction from a carbonyl compound occurs when the α C–H bond is oriented roughly parallel to the carbonyl-group p orbitals. The α carbon atom of the enolate ion product is sp^2-hybridized and has a p orbital that overlaps the neighboring carbonyl-group p orbitals. Thus, the negative charge is shared by the electronegative oxygen atom, and the enolate ion is stabilized by resonance between two forms.

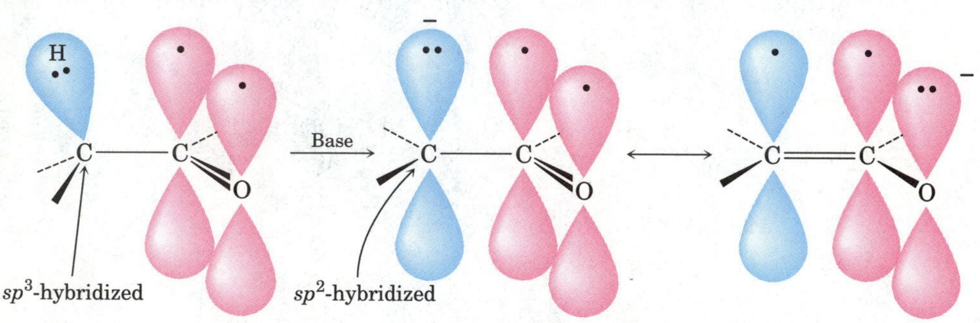

Figure 22.5 Mechanism of enolate ion formation by abstraction of an α proton from a carbonyl compound.

Carbonyl compounds are more acidic than alkanes for the same reason that carboxylic acids are more acidic than alcohols (Section 20.3). In both cases, the anions are stabilized by resonance. Enolate ions differ from carboxylate ions in that their two resonance forms are not equivalent—the form with the negative charge on oxygen is lower in energy than the form with the charge on carbon. Nevertheless, the principle behind resonance stabilization is the same in both cases.

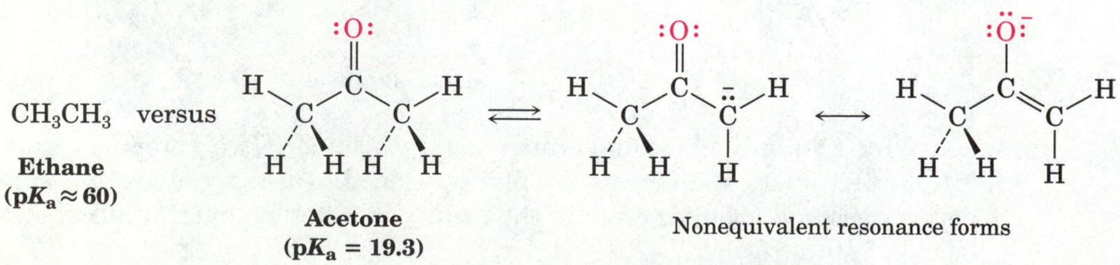

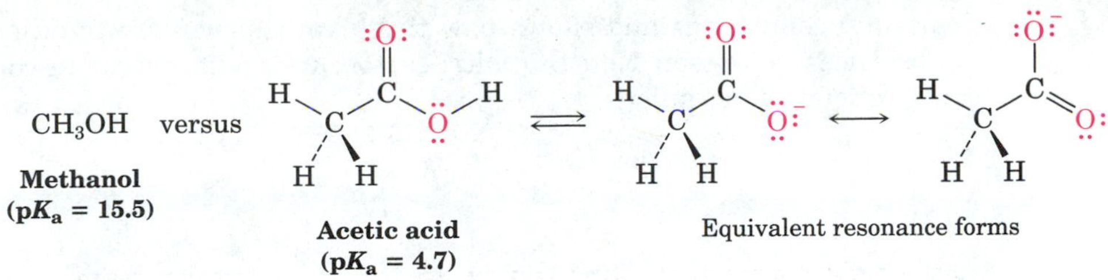

Since α hydrogen atoms of carbonyl compounds are only weakly acidic, strong bases are needed for enolate ion formation. If an alkoxide ion such as sodium ethoxide is used as base, ionization takes place only to the extent of about 0.1%, because ethanol ($pK_a = 16$) is a stronger acid than acetone. If, however, a more powerful base such as sodium hydride (NaH, the sodium "salt" of H_2) or lithium diisopropylamide [$LiN(i\text{-}C_3H_7)_2$, the lithium salt of diisopropylamine] is used, a carbonyl compound can be completely converted into its enolate ion.

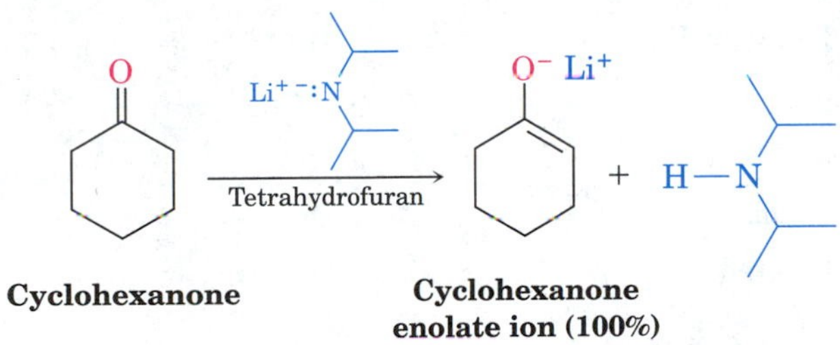

Lithium diisopropylamide, usually abbreviated LDA, is easily prepared by reaction between butyllithium and diisopropylamine and is widely used as a base for preparing enolate ions from carbonyl compounds. LDA has nearly ideal properties: It is a very strong base because diisopropylamine has $pK_a \approx 40$, it is soluble in organic solvents such as THF, it is sufficiently hindered so that it doesn't add to carbonyl groups in nucleophilic addition reactions, and it is reactive even at −78°C.

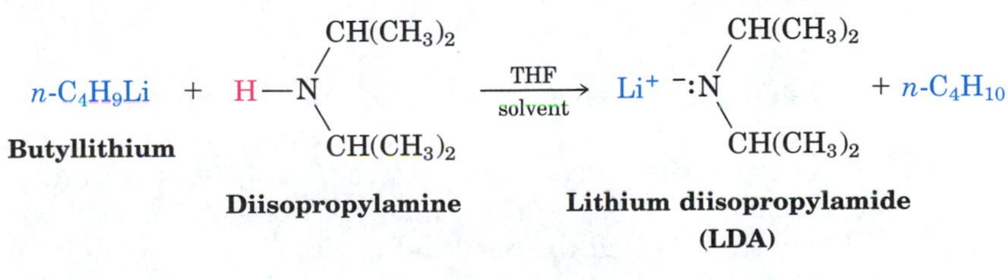

Many types of carbonyl compounds, including aldehydes, ketones, esters, acids, and amides, can be converted into enolate ions by reaction with LDA. Table 22.1 lists the approximate pK_a values of different types of carbonyl compounds and shows how these values compare to other acidic substances we've seen. Note that nitriles, too, are acidic and can be converted into "enolate-like" anions.

Table 22.1 Acidity Constants for Some Organic Compounds

Compound type	*Compound*	pK_a
Carboxylic acid	CH_3COOH	5
1,3-Diketone	$CH_2(COCH_3)_2$	9
1,3-Keto ester	$CH_3COCH_2CO_2C_2H_5$	11
1,3-Dinitrile	$CH_2(CN)_2$	11
1,3-Diester	$CH_2(CO_2C_2H_5)_2$	13
Water	HOH	16
Primary alcohol	CH_3CH_2OH	16
Acid chloride	CH_3COCl	16
Aldehyde	CH_3CHO	17
Ketone	CH_3COCH_3	19
Ester	$CH_3CO_2C_2H_5$	25
Nitrile	CH_3CN	25
Dialkylamide	$CH_3CON(CH_3)_2$	30
Ammonia	NH_3	35
Dialkylamine	$HN(i\text{-}C_3H_7)_2$	40
Alkyne	$HC{\equiv}CH$	25
Alkene	$CH_2{=}CH_2$	44
Alkane	CH_3CH_3	60

When a hydrogen atom is flanked by two carbonyl groups, its acidity is enhanced even more. Thus, Table 22.1 shows that such compounds as 1,3-diketones (**β-diketones**), 3-oxo esters (**β-keto esters**), and 1,3-diesters are more acidic than water. This enhanced acidity of β-dicarbonyl compounds is due to the fact that the resultant enolate ions are stabilized by delocalization of the negative charge over *both* neighboring carbonyl groups. The enolate ion of 2,4-pentanedione, for example, has three resonance forms. Similar resonance forms can be drawn for other doubly stabilized enolate ions.

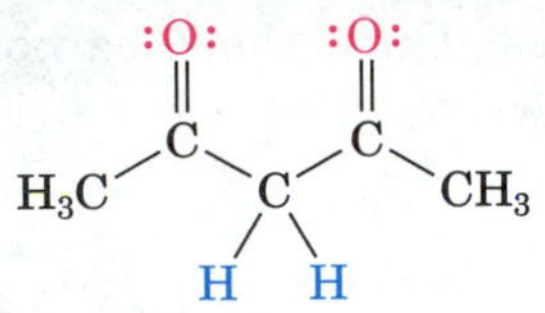

2,4-Pentanedione (pK_a = 9)

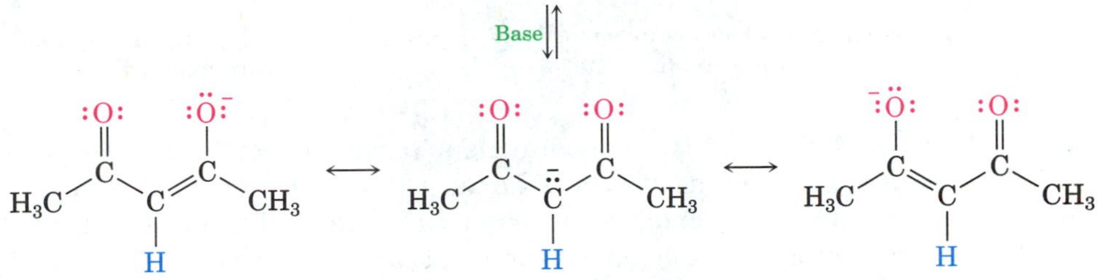

PROBLEM

22.8 Identify the most acidic hydrogens in the following molecules:

(a) CH_3CH_2CHO (b) $(CH_3)_3CCOCH_3$ (c) CH_3COOH
(d) Benzamide (e) $CH_3CH_2CH_2CN$ (f) $CH_3CON(CH_3)_2$
(g) 1,3-Cyclohexanedione

PROBLEM

22.9 Draw a resonance structure of the acetonitrile anion, $^-:CH_2C\equiv N$, to account for the acidity of nitriles.

PROBLEM

22.10 When optically active (*R*)-2-methylcyclohexanone is treated with aqueous NaOH, racemization occurs. Explain.

PROBLEM

22.11 Would you expect optically active (*S*)-3-methylcyclohexanone to be racemized on base treatment in the same way as 2-methylcyclohexanone (Problem 22.10)? Explain.

22.6 Reactivity of Enolate Ions

Enolate ions are more useful than enols for two reasons. First, pure enols can't normally be isolated. They are usually generated only as short-lived intermediates in low concentration. By contrast, stable solutions of pure enolate ions are easily prepared from most carbonyl compounds by reaction with a strong base. Second and more important, enolate ions are much more reactive than enols and undergo many reactions that enols don't. Whereas enols are neutral, enolate ions are negatively charged, making them much better nucleophiles. Thus, the α carbon atom of an enolate ion is highly reactive toward electrophiles.

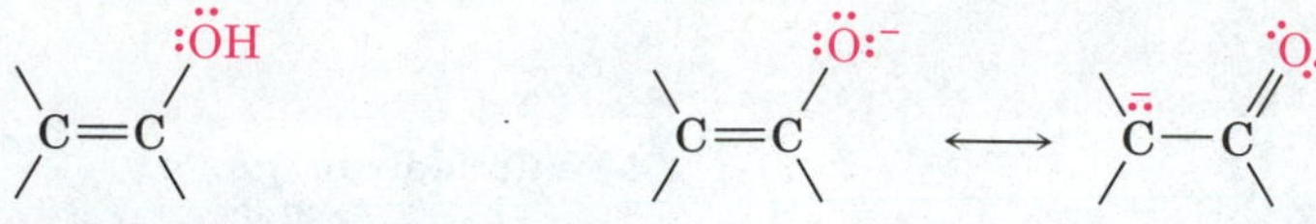

Enol: neutral, moderately reactive, very difficult to isolate

Enolate: negatively charged, very reactive, easily prepared

As hybrids of two nonequivalent resonance forms, enolate ions can be looked at either as vinylic alkoxides (C=C–O$^-$) or as α-keto carbanions ($^-$C–C=O). Thus, enolate ions can react with electrophiles either on oxygen or on carbon. Reaction on oxygen yields an enol derivative, whereas reaction on carbon yields an α-substituted carbonyl compound (Figure 22.6). Both kinds of reactivity are known, but reaction on carbon is more common.

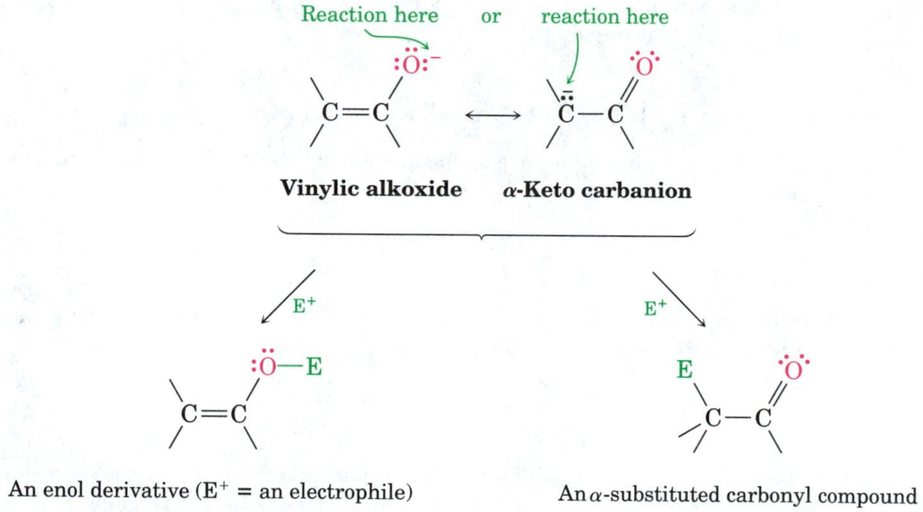

Figure 22.6 Two modes of enolate ion reactivity. Reaction on carbon to yield an α-substituted carbonyl product is the more commonly followed path.

22.7 Halogenation of Enolate Ions: The Haloform Reaction

Halogenation of ketones is promoted by base as well as by acid. As you might expect, the base-promoted reaction occurs through an enolate ion intermediate. Even relatively weak bases such as hydroxide ion are effective for halogenation because it's not necessary to convert the ketone completely into its enolate ion. Only a small amount of enolate need be generated at any one time because the reaction with halogen occurs as soon as the enolate ion is formed.

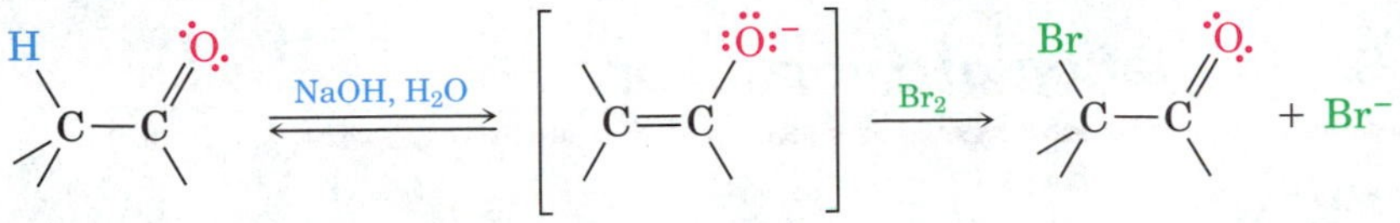

Base-promoted halogenation of ketones is little used in practice because it's difficult to stop the reaction at the monosubstituted product. An α-halogenated ketone is generally more acidic than the starting unsubstituted ketone because of the electron-withdrawing inductive effect of the halogen atom. Thus, monohalogenated products are themselves rapidly turned into enolate ions and further halogenated.

If excess base and halogen are used, methyl ketones are triply halogenated and then cleaved by base in the **haloform reaction**:

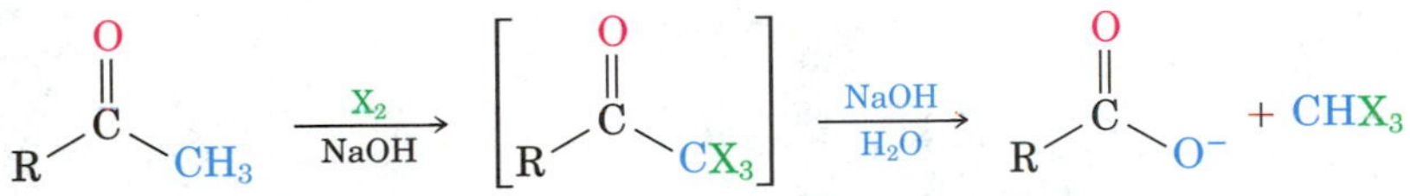

A methyl ketone

where X = Cl, Br, or I.

The haloform reaction, which converts a methyl ketone into a carboxylic acid plus a **haloform** (chloroform, $CHCl_3$; bromoform, $CHBr_3$; or iodoform, CHI_3), is the basis for a qualitative test for methyl ketones. A sample of unknown structure is dissolved in THF or ether, and dilute solutions of aqueous NaOH and I_2 are added. Formation of a yellow precipitate of solid iodoform signals a positive test and indicates that the sample is a methyl ketone.

PROBLEM. . . . .

22.12 Base-promoted chlorination and bromination of a given ketone occur at the same rate. Explain.

PROBLEM. . . . .

22.13 Why do you suppose that ketone halogenations in acidic media are referred to as being acid-*catalyzed*, whereas halogenations in basic media are base-*promoted*? Why is a full equivalent of base required for halogenation?

PROBLEM

22.14 The second step of the haloform reaction is a nucleophilic acyl substitution of $^{-}CX_3$ by ^{-}OH. Show the mechanism of this step.

22.8 Alkylation of Enolate Ions

Perhaps the single most important reaction of enolate ions is their **alkylation** by treatment with an alkyl halide. The alkylation reaction is useful because it forms a new C–C bond, thereby joining two smaller pieces into one larger molecule. Alkylation occurs when the nucleophilic enolate ion reacts with the electrophilic alkyl halide in an S_N2 reaction and displaces the leaving group by back-side attack.

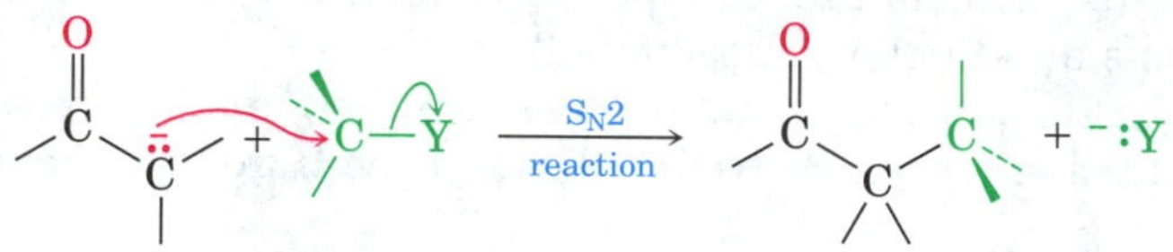

Enolate ion   Alkyl halide

Alkylation reactions are subject to the same constraints that affect all S_N2 reactions (Section 11.4). Thus, the leaving group Y in the alkylating agent can be chloride, bromide, iodide, or tosylate. The alkyl group R must be primary or methyl, and preferably should be allylic or benzylic. Secondary halides react poorly, and tertiary halides don't react at all because a competing E2 elimination of HY occurs instead. Vinylic and aryl halides are also unreactive, because back-side attack is sterically prevented.

$$\text{R—Y}\begin{cases}\text{—Y:} & \text{Tosylate} > \text{—I} > \text{—Br} > \text{—Cl}\\ \text{R—:} & \text{Allylic} \approx \text{Benzylic} > \text{H}_3\text{C—} > \text{RCH}_2\text{—}\end{cases}$$

The Malonic Ester Synthesis

The **malonic ester synthesis**, one of the oldest and best-known carbonyl alkylation reactions, is an excellent method for preparing a substituted acetic acid from an alkyl halide:

$$\underset{\text{Alkyl halide}}{\text{R—X}} \xrightarrow[\text{ester synthesis}]{\text{Via malonic}} \underset{\alpha\text{-Substituted acetic acid}}{\text{R—CH}_2\overset{\text{O}}{\overset{\|}{\text{C}}}\text{OH}}$$

Diethyl propanedioate, commonly called diethyl malonate or *malonic ester*, is more acidic than monocarbonyl compounds (pK_a = 13) because its α hydrogens are flanked by two carbonyl groups. Thus, malonic ester is easily converted into its enolate ion by reaction with sodium ethoxide in ethanol. The enolate ion, in turn, is a good nucleophile that reacts rapidly with an alkyl halide, yielding an α-substituted malonic ester.

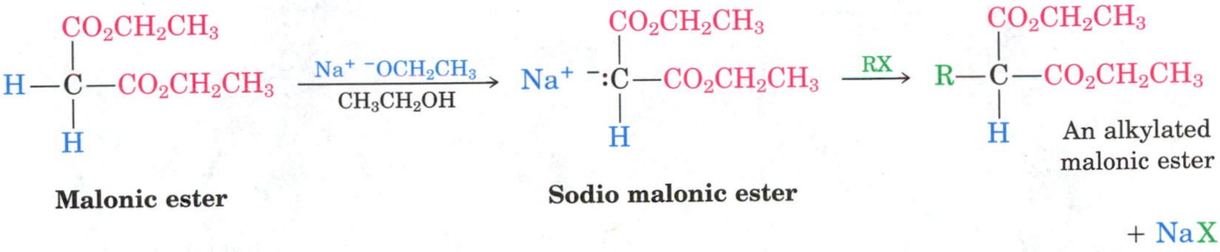

The product of malonic ester alkylation has one acidic α hydrogen atom left, and the alkylation process can therefore be repeated a second time to yield a dialkylated malonic ester:

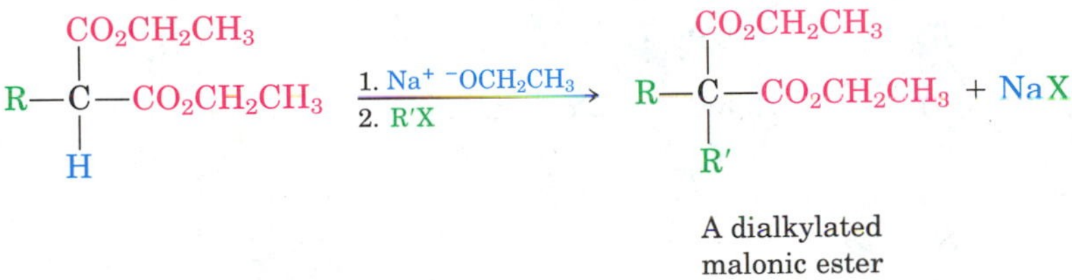

The alkylated malonic ester can be hydrolyzed and *decarboxylated* (lose CO_2) when heated with aqueous acid to yield a substituted monoacid:

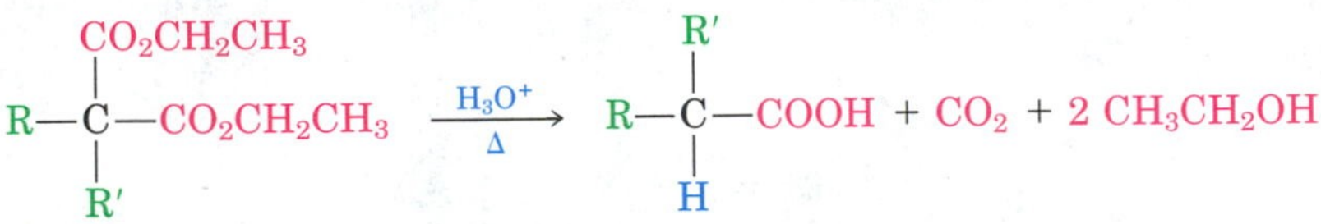

Note that decarboxylation is not a general reaction of carboxylic acids. It is a unique feature of compounds like malonic acid that have a *second* carbonyl group two atoms away from the –COOH. That is, only β-keto acids and substituted malonic acids undergo loss of CO_2 on heating. The decarboxylation reaction occurs in two steps and involves initial acid-catalyzed hydrolysis of the diester to a diacid. The diacid then loses CO_2 by a cyclic mechanism.

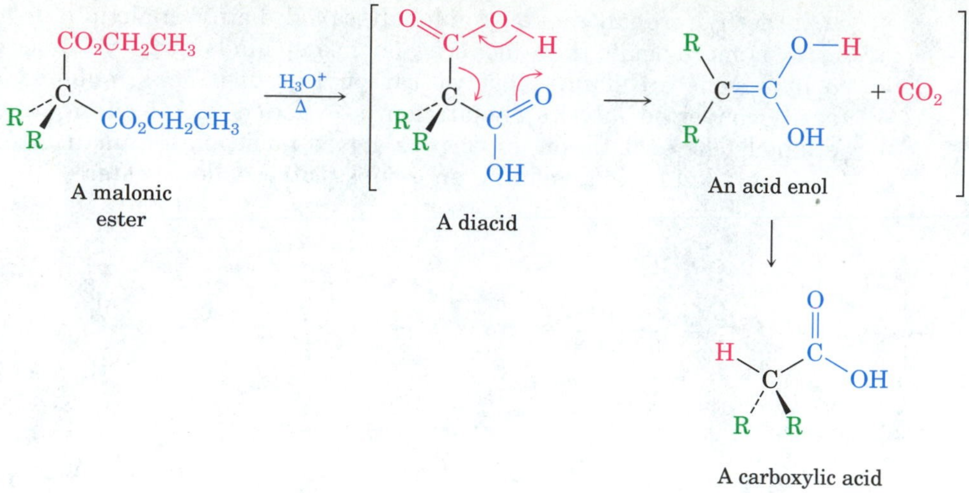

The overall effect of the malonic ester synthesis is to convert an alkyl halide into a carboxylic acid while lengthening the carbon chain by two atoms ($RX \rightarrow RCH_2COOH$). Note in the following example that the abbreviation "Et" is used as a space-saving way of indicating an ethyl group.

$$\underset{\textbf{1-Bromobutane}}{CH_3CH_2CH_2CH_2Br} + \underset{\textbf{Sodio diethylmalonate}}{Na^+ \ ^-{:}CH(CO_2Et)_2} \longrightarrow \underset{(84\%)}{CH_3CH_2CH_2CH_2CH(CO_2Et)_2}$$

$$\xrightarrow{H_3O^+,\ \Delta} CH_3CH_2CH_2CH_2CH_2COOH$$

Hexanoic acid (75%)

$$CH_3CH_2CH_2CH_2CH(CO_2Et)_2 \xrightarrow[2.\ CH_3I]{1.\ Na^+\ ^-OEt} CH_3CH_2CH_2CH_2C(CH_3)(CO_2Et)_2$$

$$\xrightarrow{H_3O^+,\ \Delta} CH_3CH_2CH_2CH_2CH(CH_3)COOH$$

2-Methylhexanoic acid (74%)

The malonic ester synthesis can also be used to prepare *cyclo*alkanecarboxylic acids. For example, if 1,4-dibromobutane is treated with diethyl malonate in the presence of 2 equivalents of sodium ethoxide base, the first alkylation occurs as expected, but the second alkylation step occurs *intramolecularly* to yield a five-membered cyclic product. Hydrolysis and decarboxylation then lead to cyclopentanecarboxylic acid (Figure 22.7). Three-, four-, five-, and six-membered rings can be prepared in this way, but yields decrease for larger ring sizes.

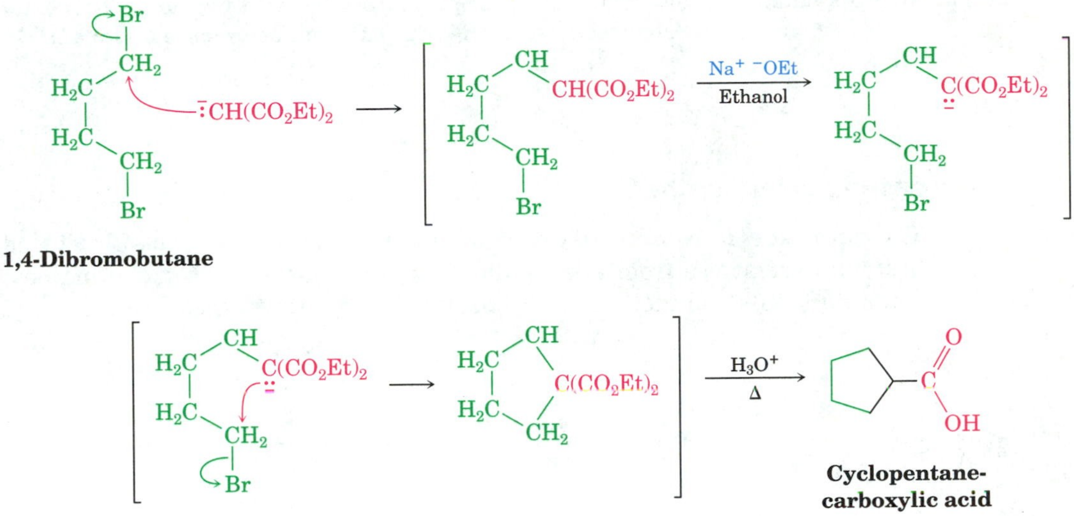

Figure 22.7 Malonic ester synthesis of cyclopentanecarboxylic acid. (The abbreviation "Et" means an ethyl group.)

PRACTICE PROBLEM

How would you prepare heptanoic acid using a malonic ester synthesis?

Solution The malonic ester synthesis converts an alkyl halide into a carboxylic acid having two more carbons. Thus, a *seven*-carbon acid chain must be derived from the *five*-carbon alkyl halide 1-bromopentane.

$$CH_3CH_2CH_2CH_2CH_2Br + CH_2(CO_2Et)_2 \xrightarrow[2.\ H_3O^+,\ \Delta]{1.\ Na^+\ ^-OEt} CH_3CH_2CH_2CH_2CH_2CH_2COOH$$

PROBLEM

22.15 What alkyl halide would you use to prepare the following compounds by a malonic ester synthesis?
(a) Butanoic acid (b) 5-Methylhexanoic acid

PROBLEM

22.16 How could you use a malonic ester synthesis to prepare the following compounds? Show all steps.

(a) 3-Phenylpropanoic acid
(b) 2-Methylpentanoic acid
(c) 4-Methylpentanoic acid
(d) Ethyl cyclobutanecarboxylate

PROBLEM

22.17 Monoalkylated and dialkylated acetic acids can be prepared by the malonic ester synthesis, but trialkylated acetic acids (R_3CCOOH) can't be prepared. Explain.

The Acetoacetic Ester Synthesis

The **acetoacetic ester synthesis** is a method for preparing α-substituted acetone derivatives from alkyl halides in the same way that the malonic ester synthesis is a method for preparing α-substituted acetic acids.

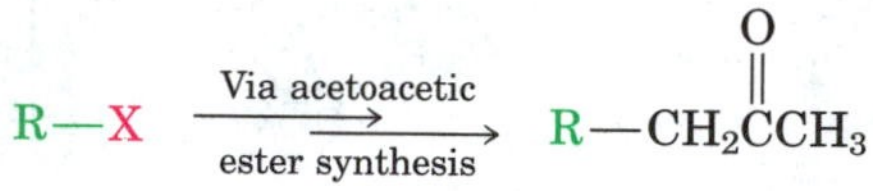

α-Substituted acetone

Ethyl 3-oxobutanoate, commonly called ethyl acetoacetate or *acetoacetic ester*, is much like malonic ester in that its α hydrogens are flanked by two carbonyl groups. It is therefore readily converted into its enolate ion, which can be alkylated by reaction with an alkyl halide. A second alkylation can also be carried out if desired, since acetoacetic ester has two acidic α hydrogens that can be replaced.

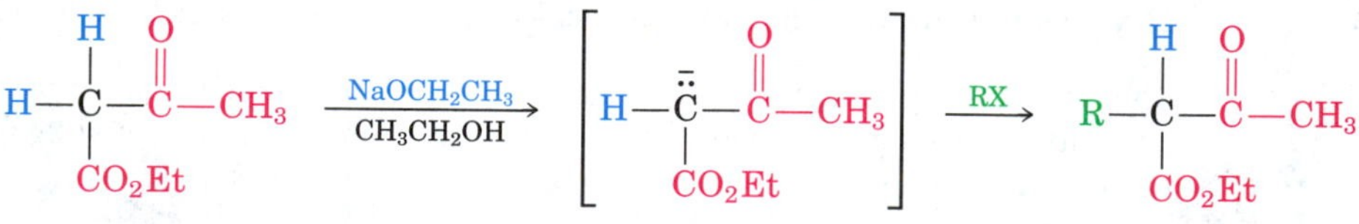

Acetoacetic ester

A monoalkylated acetoacetic ester

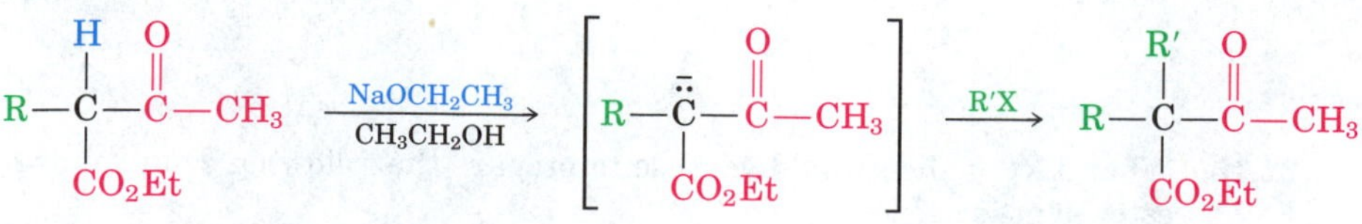

A monoalkylated acetoacetic ester

A dialkylated acetoacetic ester

On heating with aqueous HCl, the alkylated acetoacetic ester is hydrolyzed and decarboxylated through a β-keto acid intermediate to yield an α-substituted acetone product. If a monoalkylated acetoacetic ester is hydrolyzed and decarboxylated, an α-monosubstituted acetone is formed; if a dialkylated acetoacetic ester is hydrolyzed and decarboxylated, an α,α-disubstituted acetone is formed.

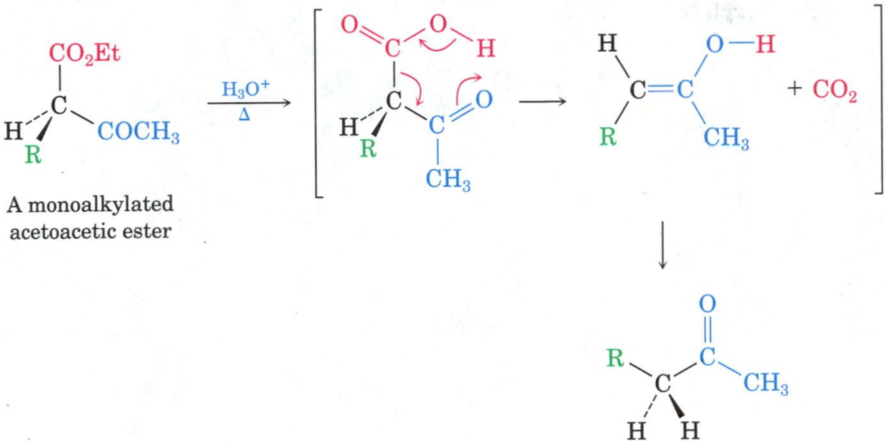

A monoalkylated acetoacetic ester

An α-monosubstituted acetone

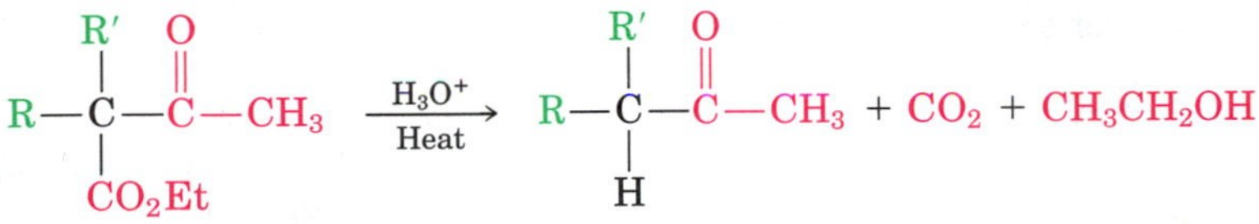

A dialkylated acetoacetic ester

A disubstituted acetone

For example:

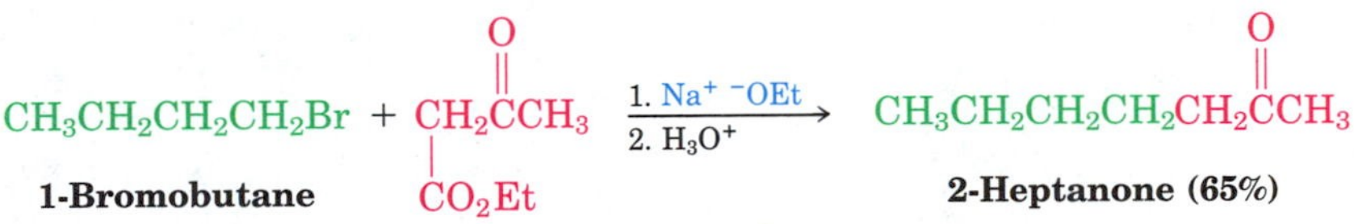

1-Bromobutane

Acetoacetic ester

2-Heptanone (65%)

The three-step sequence of (1) enolate ion formation, (2) alkylation, and (3) hydrolysis/decarboxylation is applicable to all β-keto esters with acidic α hydrogens, not just to acetoacetic ester itself. Thus, cyclic β-keto esters such as ethyl 2-oxocyclohexanecarboxylate can be alkylated and decarboxylated to give 2-substituted cyclohexanones in high yield. For example:

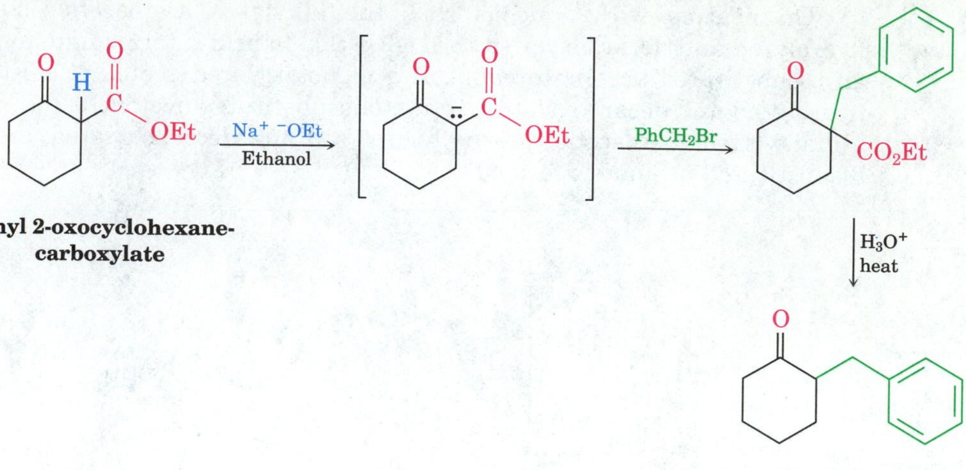

Ethyl 2-oxocyclohexane-carboxylate

2-Benzylcyclohexanone (77%)

PRACTICE PROBLEM. . . . .

How would you prepare 2-pentanone by an acetoacetic ester synthesis?

Solution The acetoacetic ester synthesis yields a ketone product by adding three carbons to an alkyl halide:

$$\underbrace{CH_3\overset{O}{\overset{\|}{C}}CH_2}_{\text{These three carbons from acetoacetic acid}}\!-\!\underbrace{CH_2CH_3}_{\text{These carbons from alkyl halide}}$$

This bond formed

Thus, the acetoacetic ester synthesis of 2-pentanone would involve reaction of bromoethane:

$$CH_3\overset{O}{\overset{\|}{C}}CH_2\overset{O}{\overset{\|}{C}}OCH_2CH_3 + CH_3CH_2Br \xrightarrow[\text{2. } H_3O^+]{\text{1. } Na^+\ ^-OCH_2CH_3} CH_3\overset{O}{\overset{\|}{C}}CH_2CH_2CH_3$$

PROBLEM. . . . .

22.18 What alkyl halides would you use to prepare the following ketones by an acetoacetic ester synthesis?

(a) 5-Methyl-2-hexanone

(b) 5-Phenyl-2-pentanone

PROBLEM

22.19 How would you prepare methyl cyclopentyl ketone using an acetoacetic ester synthesis?

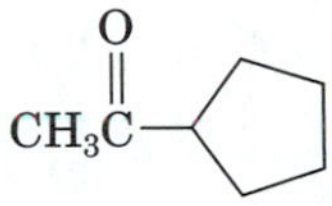

Methyl cyclopentyl ketone

PROBLEM

22.20 Which of the following compounds can't be prepared by an acetoacetic ester synthesis? Explain.

(a) 2-Butanone
(b) Phenylacetone
(c) Acetophenone
(d) 3,3-Dimethyl-2-butanone

Direct Alkylation of Ketones, Esters, and Nitriles

The malonic ester synthesis and the acetoacetic ester synthesis are relatively easy to carry out because both involve reactions that take place at doubly carbonyl-activated centers. As a result, relatively mild bases like sodium ethoxide in an alcohol solvent can be used. By contrast, it's also possible in many cases to alkylate directly at the singly activated α position of *mono*ketones, *mono*esters, or nitriles. A strong, sterically hindered base is needed, so that complete conversion to the enolate ion takes place rather than a nucleophilic addition, and a nonprotic solvent must be used.

Ketones, esters, and nitriles can all be alkylated by using LDA or related dialkylamide bases in THF. (Aldehydes rarely give high yields of pure products because their enolate ions undergo carbonyl condensation reactions instead of alkylation. We'll study this condensation reaction in the next chapter.) Some specific examples of alkylation reactions are shown below.

Lactone

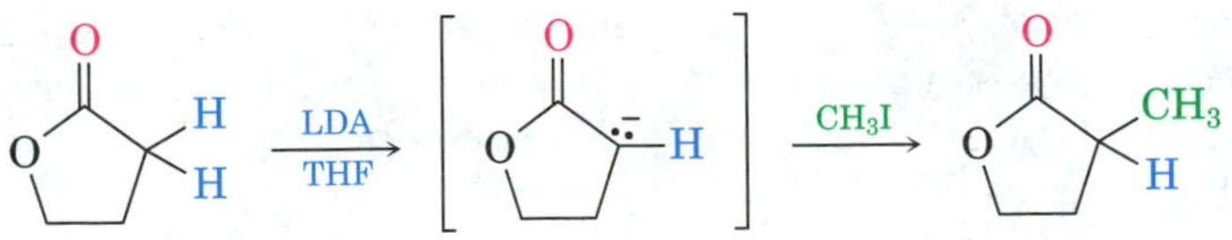

Butyrolactone **2-Methylbutyrolactone (88%)**

Ester

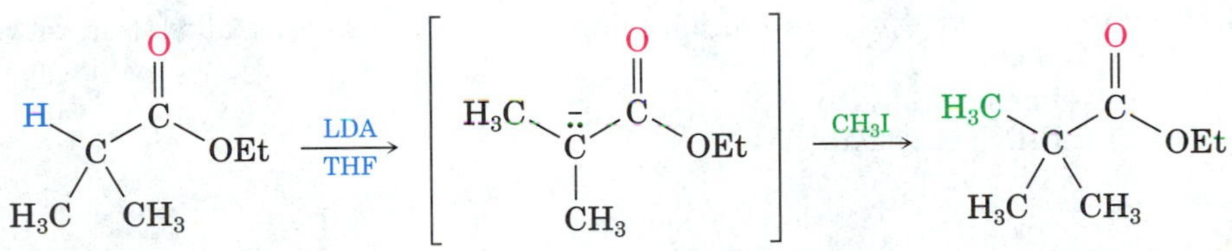

Ethyl 2-methylpropanoate **Ethyl 2,2-dimethylpropanoate (87%)**

Nitrile

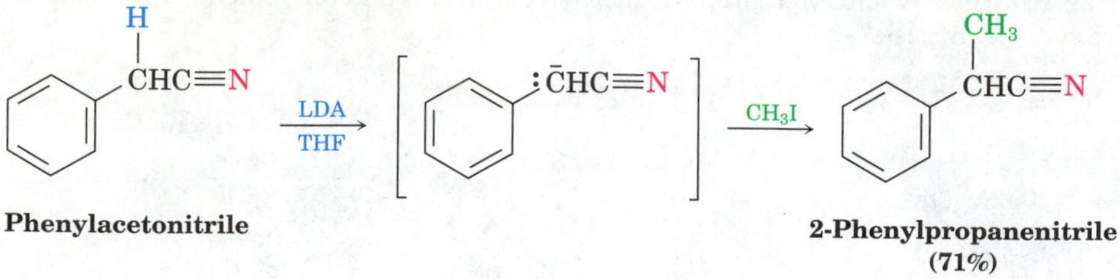

Ketone

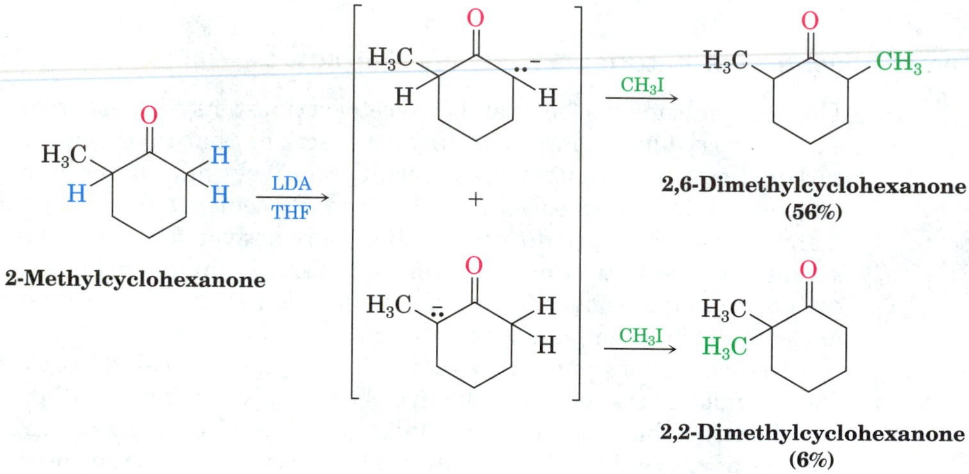

Note in the previous examples that alkylation of the unsymmetrically substituted ketone 2-methylcyclohexanone leads to a mixture of products because both possible enolate ions are formed. In general, the major product in such cases occurs by alkylation at the less hindered, more accessible position. Thus, alkylation of 2-methylcyclohexanone occurs primarily at C6 (secondary) rather than at C2 (tertiary).

PROBLEM. . . . .

22.21 How might you prepare the following compounds using an alkylation reaction as the key step?

(a) 3-Phenyl-2-butanone
(b) 2-Ethylpentanenitrile
(c) 2-Allylcyclohexanone
(d) 2,2,6,6-Tetramethylcyclohexanone

INTERLUDE

Unusual Elements in Organic Chemistry

Carbon, hydrogen, oxygen, nitrogen, sulfur, the halogens, and a few others—these are the elements usually thought of in connection with organic chemistry. In recent years, though, the list has grown to include such elements as boron, copper, and mercury among the more familiar elements, and also selenium, samarium, zirconium, titanium, thallium, cerium, and others among the less familiar elements. Modern organic chemistry, in fact, makes use of almost all the 90 naturally occurring elements except the noble gases (Figure 22.8).

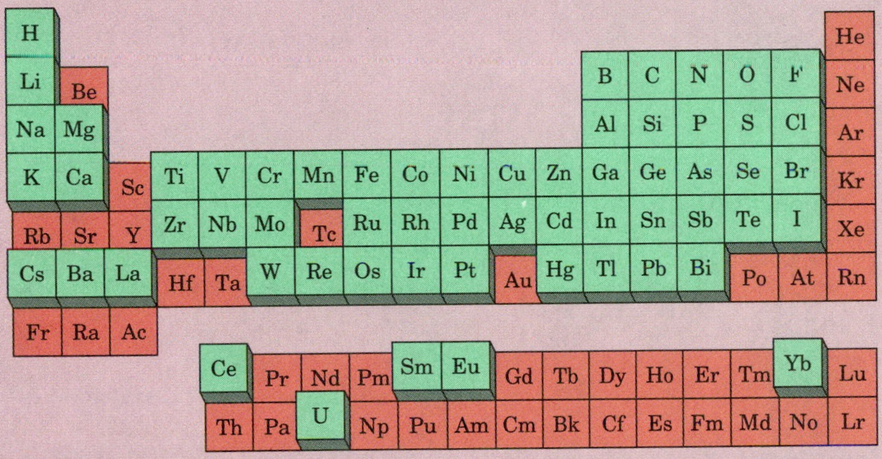

Figure 22.8 The elements highlighted in green are all used in modern organic chemistry.

One of the more useful recent advances in enolate ion chemistry is the finding that carbonyl compounds can be *selenenylated*. That is, a selenium atom can be introduced onto the α position of a carbonyl compound. Selenenylation is accomplished by allowing the carbonyl compound to react with LDA to generate an enolate ion, followed by addition of benzeneselenenyl bromide, C_6H_5SeBr. Immediate α-substitution reaction yields an α-phenylseleno-substituted product.

(continued)➤

The value of the selenenylation reaction is that the product can be easily converted into an α,β-unsaturated carbonyl compound. On treatment with dilute H_2O_2 at room temperature, the selenium atom is oxidized, elimination occurs, and an α,β-unsaturated carbonyl compound is formed. The net result is introduction of a C=C bond into the α,β position of the carbonyl starting material. Yields are usually excellent, and the method is often superior to the alternative α-bromination/dehydrobromination route (Section 22.3). No added base is required (as in dehydrobromination), and the reaction occurs quickly at room temperature.

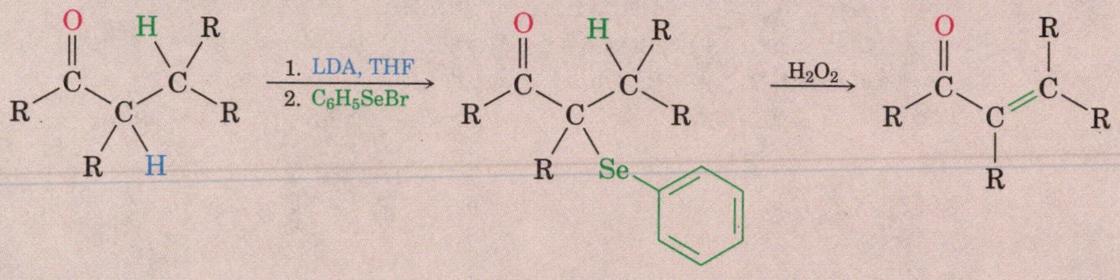

An α-phenylseleno ketone

For example:

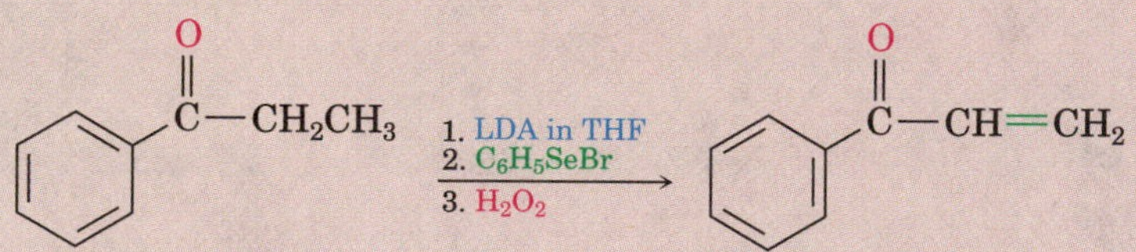

Propiophenone

Phenyl vinyl ketone (80%)

What is true of selenenylation is also true of many other newly discovered reactions. The elements involved may be unusual, but the chemistry is often mild, selective, and much superior to older, more classical methods.

Summary and Key Words

The ***α* substitution** of a carbonyl compound through an **enol** or **enolate ion** intermediate is one of the four fundamental reaction types in carbonyl-group chemistry.

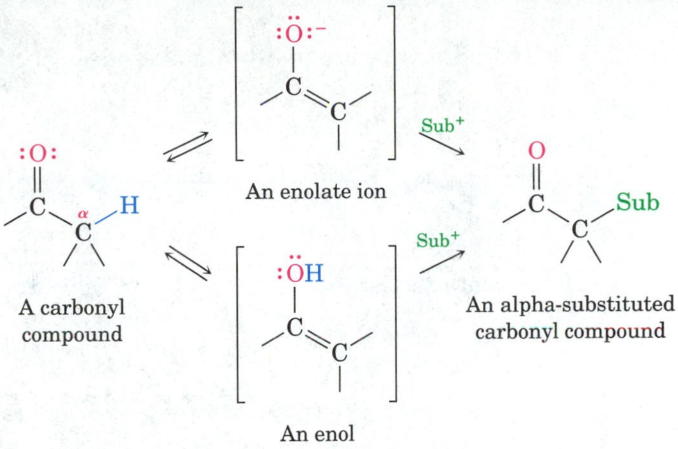

All carbonyl compounds rapidly equilibrate with their enols, a process called **tautomerism**. Although enol tautomers are normally present to only a small extent at equilibrium and can't usually be isolated in pure form, they nevertheless contain a highly nucleophilic double bond and react rapidly with electrophiles. For example, ketones and aldehydes are rapidly halogenated at the α position by reaction with Cl_2, Br_2, or I_2 in acetic acid solution. Alpha bromination of carboxylic acids can be similarly accomplished by the **Hell–Volhard–Zelinskii reaction**, in which an acid is treated with Br_2/PBr_3. The α-halogenated products can then undergo base-induced E2 elimination to yield α,β-unsaturated carbonyl products.

Alpha hydrogen atoms of carbonyl compounds are acidic and can be removed by strong bases, such as lithium diisopropylamide (LDA), to yield highly nucleophilic enolate ions. The most important reaction of enolate ions is their S_N2 **alkylation** with alkyl halides. The **malonic ester synthesis** provides a method for preparing monoalkylated or dialkylated acetic acids. Similarly, the **acetoacetic ester synthesis** provides a method for preparing monoalkylated or dialkylated acetone derivatives. In addition, many carbonyl compounds, including ketones, esters, and nitriles, can be directly alkylated by treatment with LDA and an alkyl halide.

Summary of Reactions

1. Ketone/aldehyde halogenation, where X = Cl, Br, or I (Section 22.3)

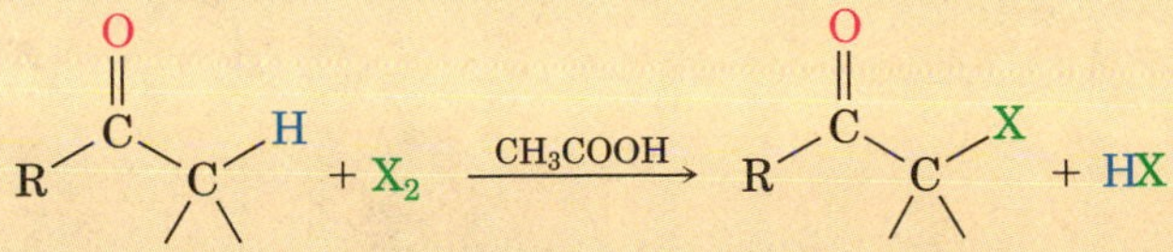

2. Hell–Volhard–Zelinskii bromination of acids (Section 22.4)

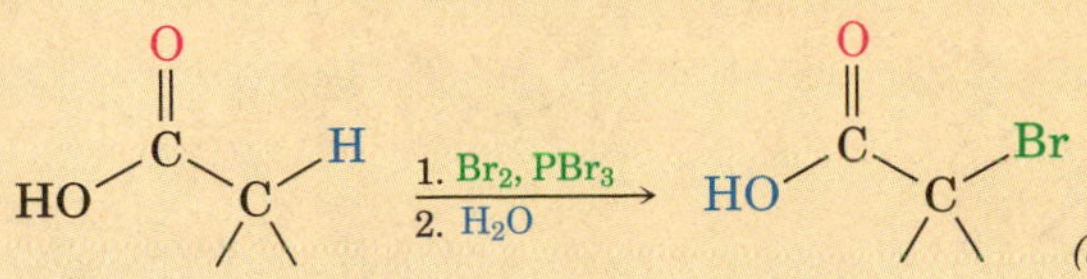

(continued)➤

3. Dehydrobromination of α-bromo ketones (Section 22.3)

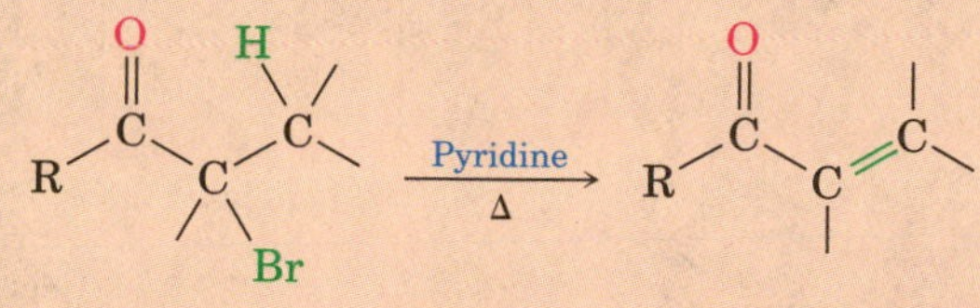

4. Haloform reaction, where X = Cl, Br, or I (Section 22.7)

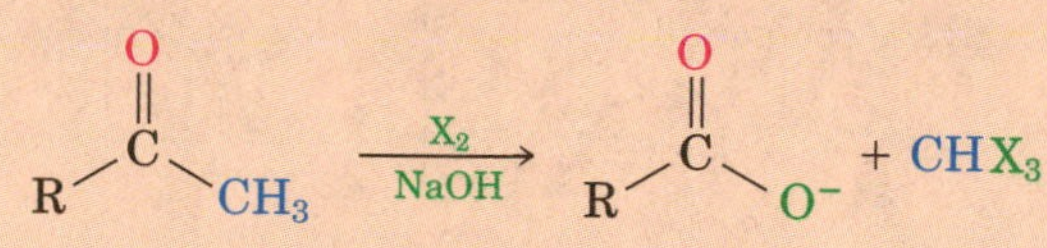

5. Alkylation of enolate ions (Section 22.8)
   (a) Malonic ester synthesis

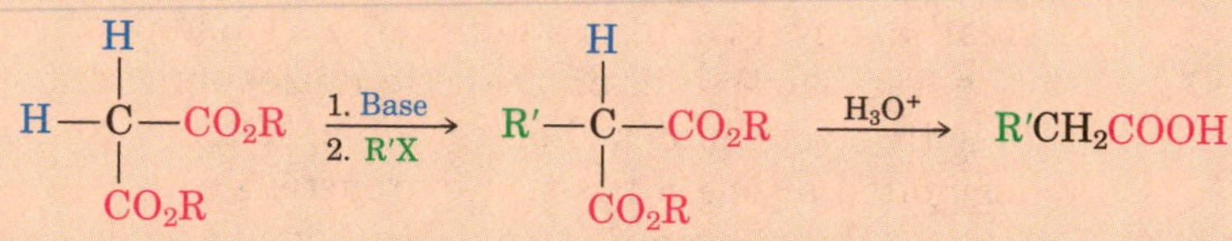

   (b) Acetoacetic ester synthesis

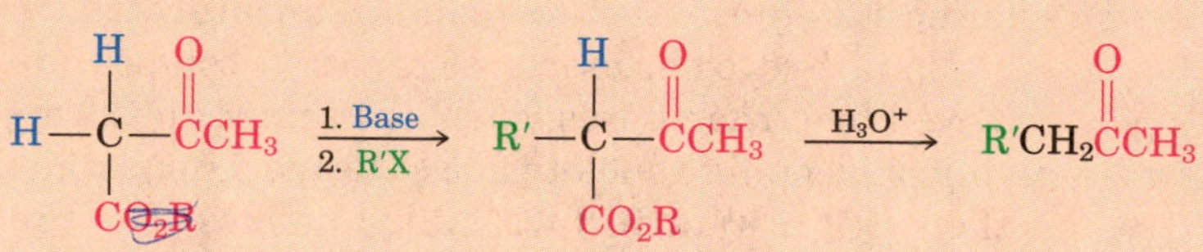

   (c) Alkylation of ketones

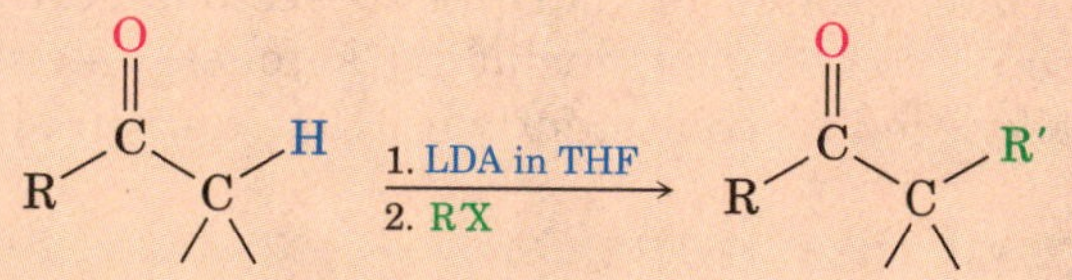

   (d) Alkylation of esters

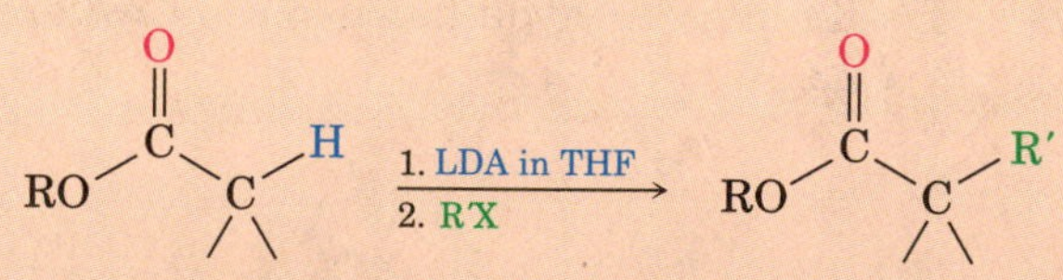

   (e) Alkylation of nitriles

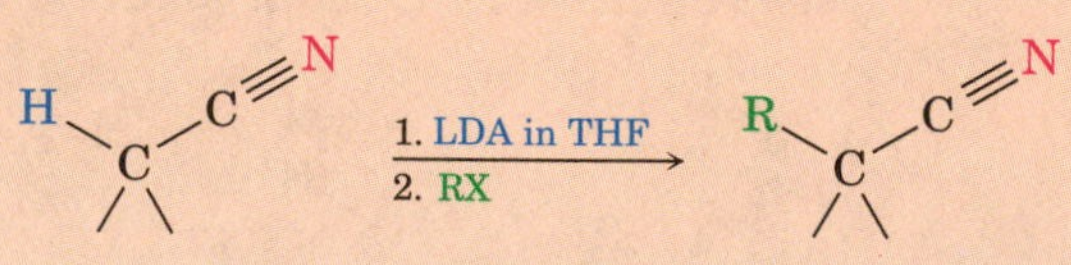

ADDITIONAL PROBLEMS

22.22 Acetone is enolized only to the extent of about 0.000 001% at equilibrium, whereas 2,4-pentanedione is 76% enolized. Explain.

22.23 Write resonance structures for the following anions:

(a) $CH_3C(=O)\ddot{C}^{-}HC(=O)CH_3$

(b) $CH_3CH{=}CH\ddot{C}^{-}HC(=O)CH_3$

(c) $N{\equiv}C\ddot{C}^{-}HCO_2C_2H_5$

(d) 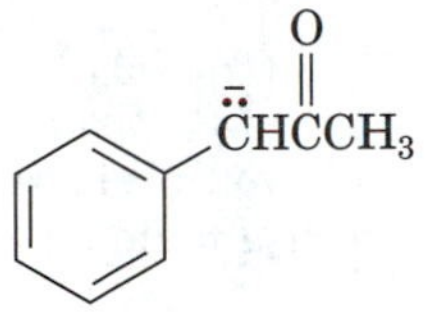

22.24 Indicate all the acidic hydrogen atoms in the following structures:

(a) $HOCH_2CH_2C(=O)CH_3$

(b) $HOCH_2CH_2C(=O)C(CH_3)_3$

(c) 1,3-Cyclopentanedione

22.25 Treatment of an α,β-unsaturated carbonyl compound with base yields an anion by removal of H^+ from the γ carbon. Why are hydrogens on the γ carbon atom acidic?

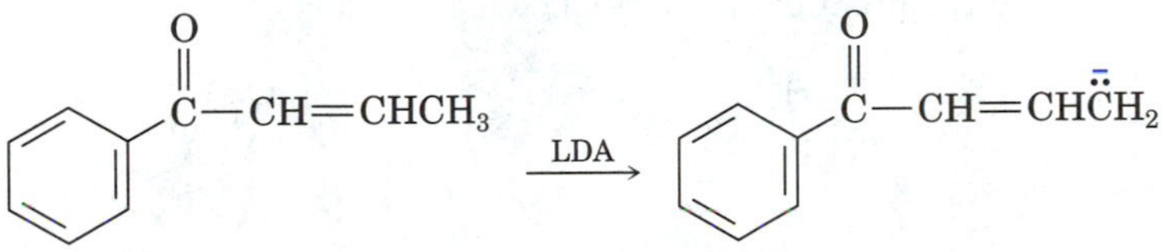

22.26 One way to determine the number of acidic hydrogens in a molecule is to treat the compound with NaOD in D_2O, isolate the product, and determine its molecular weight by mass spectroscopy. For example, if cyclohexanone is treated with NaOD in D_2O, the product has a molecular weight of 102. Explain how this method works.

22.27 2-Methylcycloheptanone and 3-methylcycloheptanone are nearly indistinguishable by spectroscopic techniques. How could you use the method outlined in Problem 22.26 to differentiate them?

22.28 Rank the following compounds in order of increasing acidity:

(a) CH_3CH_2COOH (b) CH_3CH_2OH (c) $(CH_3CH_2)_2NH$

(d) CH_3COCH_3 (e) $CH_3C(=O)CH_2C(=O)CH_3$ (f) CCl_3COOH

22.29 All attempts to isolate primary and secondary nitroso compounds result only in the formation of oximes. Tertiary nitroso compounds, however, are stable. Explain.

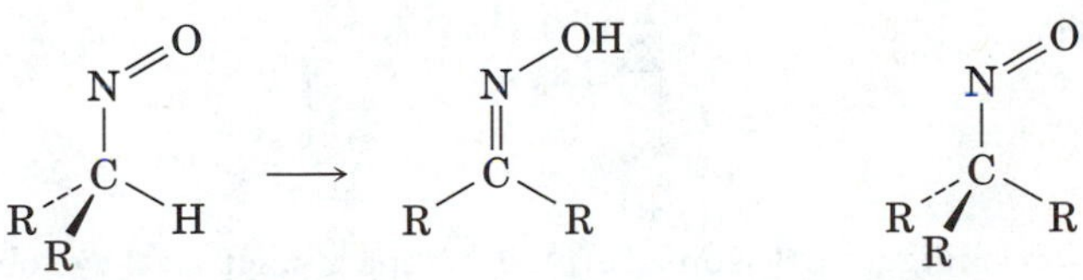

A 1° or 2° nitroso compound (unstable) An oxime A 3° nitroso compound (stable)

22.30 Which of the following compounds can be prepared by a malonic ester synthesis? Show the alkyl halide you would use in each case.

(a) Ethyl pentanoate
(b) Ethyl 3-methylbutanoate
(c) Ethyl 2-methylbutanoate
(d) Ethyl 2,2-dimethylpropanoate

22.31 How would you prepare the following ketones using an acetoacetic ester synthesis?

(a) $CH_3CH_2CH(CH_2CH_3)C(=O)CH_3$

(b) $CH_3CH_2CH_2CH(CH_3)C(=O)CH_3$

22.32 How would you prepare the following compounds using either an acetoacetic ester synthesis or a malonic ester synthesis?

(a) $CH_3C(CH_3)(CO_2Et)CO_2Et$

(b)

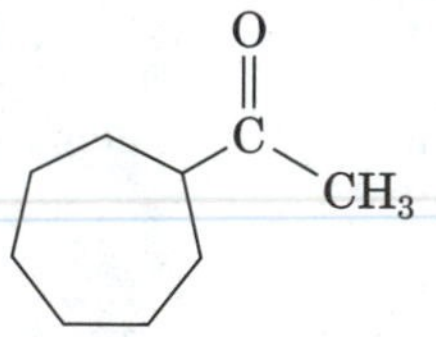

(c)

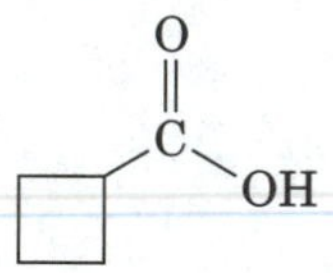

(d) $H_2C{=}CHCH_2CH_2C(=O)CH_3$

22.33 Predict the product(s) of the following reactions:

(a)

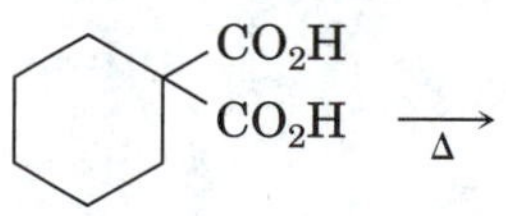

(b)

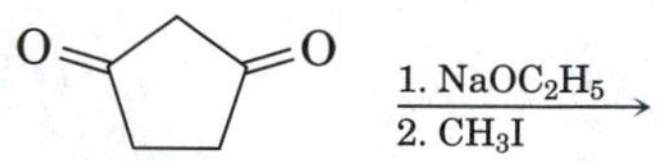

(c) $CH_3CH_2CH_2COOH \xrightarrow{Br_2,\ PBr_3} A \xrightarrow{H_2O} B$

(d)

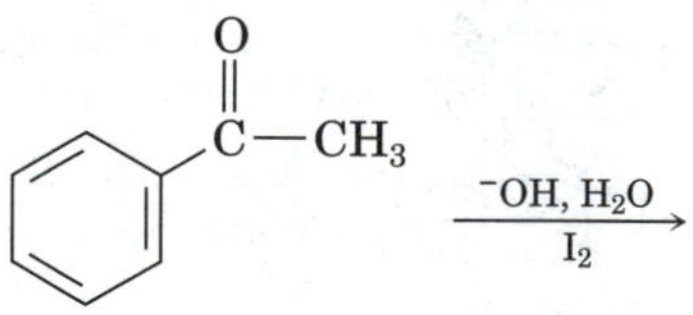

22.34 When optically active (*R*)-3-phenyl-2-butanone is exposed to aqueous acid, a loss of optical activity occurs and racemic 3-phenyl-2-butanone is produced. Explain.

22.35 In light of your answer to Problem 22.34, would you expect optically active (*R*)-3-methyl-3-phenyl-2-pentanone to be racemized by acid treatment? Explain.

22.36 Fill in the reagents a–c that are missing from the following scheme:

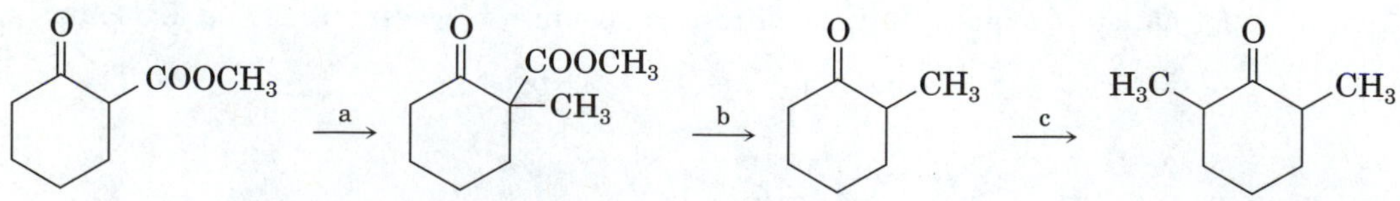

22.37 Nonconjugated β,γ-unsaturated ketones, such as 3-cyclohexenone, are in an acid-catalyzed equilibrium with their conjugated α,β-unsaturated isomers. Propose a mechanism for the acid-catalyzed isomerization.

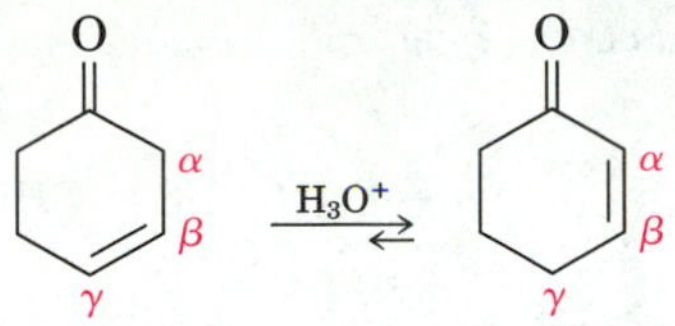

22.38 The β,γ to α,β interconversion of unsaturated ketones described in Problem 22.37 is also catalyzed by base. Explain.

22.39 One interesting consequence of the base-catalyzed β,γ to α,β isomerization of unsaturated ketones (Problem 22.38) is that 2-substituted 2-cyclopentenones can be interconverted with 5-substituted 2-cyclopentenones. Propose a mechanism for this isomerization.

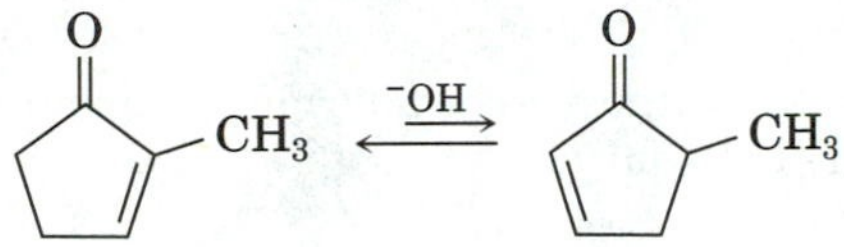

22.40 Although 2-substituted 2-cyclopentenones are in a base-catalyzed equilibrium with their 5-substituted 2-cyclopentenone isomers (Problem 22.39), the analogous isomerization is not observed for 2-substituted 2-cyclohexenones. Explain.

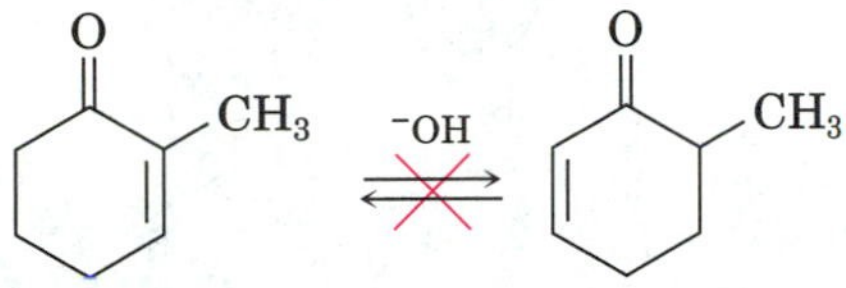

22.41 At least as far back as the sixteenth century, the Incas chewed the leaves of the coca bush, *Erythroxylon coca*, to combat fatigue. Chemical studies of *Erythroxylon coca* by Friedrich Wöhler in 1862 resulted in the discovery of *cocaine* as the active component. It was soon found that basic hydrolysis of cocaine led to methanol, benzoic acid, and another compound called *ecgonine*. Oxidation of ecgonine with CrO_3 led to a keto acid that readily lost CO_2 on heating, giving tropinone.

$$\underset{\textbf{Cocaine}}{C_{17}H_{21}NO_4} \xrightarrow[H_2O]{^-OH} CH_3OH + C_6H_5COOH + \underset{\textbf{Ecgonine}}{C_9H_{15}NO_3}$$

$$\text{Ecgonine} \xrightarrow{CrO_3} \underset{\textbf{Keto acid}}{C_9H_{13}NO_3} \xrightarrow[\Delta]{} CO_2 +$$

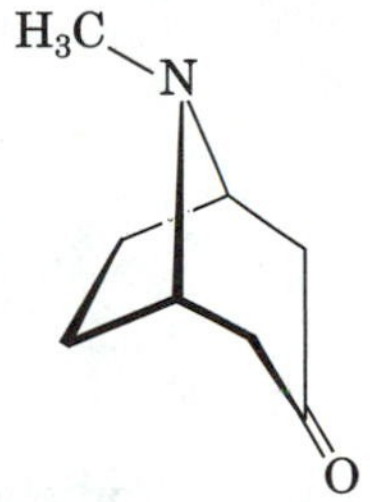

Tropinone

(a) What is a likely structure for the keto acid?
(b) What is a likely structure for ecgonine?
(c) What is a likely structure for cocaine?
(d) Formulate the reactions.

22.42 Which of the following substances would give a positive haloform reaction?

(a) CH_3COCH_3 (b) Acetophenone (c) CH_3CH_2CHO
(d) CH_3COOH (e) $CH_3C{\equiv}N$

22.43 How might you convert geraniol into either ethyl geranylacetate or geranylacetone?

$$(CH_3)_2C{=}CHCH_2CH_2C(CH_3){=}CHCH_2OH \xrightarrow{?} (CH_3)_2C{=}CHCH_2CH_2C(CH_3){=}CHCH_2CH_2CO_2C_2H_5$$

Geraniol → **Ethyl geranylacetate**

$$(CH_3)_2C{=}CHCH_2CH_2C(CH_3){=}CHCH_2OH \xrightarrow{?} (CH_3)_2C{=}CHCH_2CH_2C(CH_3){=}CHCH_2CH_2COCH_3$$

Geraniol → **Geranylacetone**

22.44 How would you synthesize the following compounds from cyclohexanone? More than one step may be required.

(a) 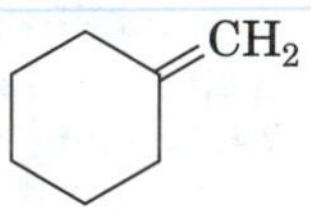

(b) 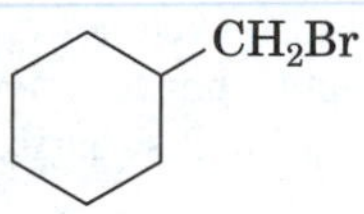

(c) 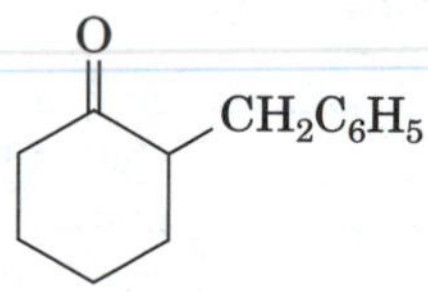

(d) 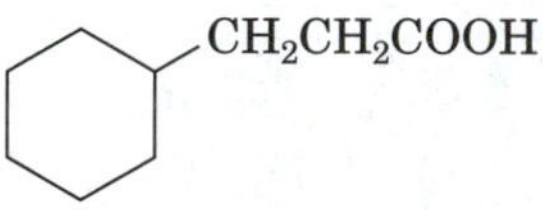

(e) 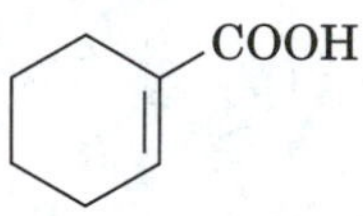

(f) 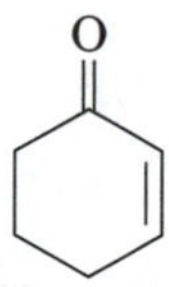

(g) 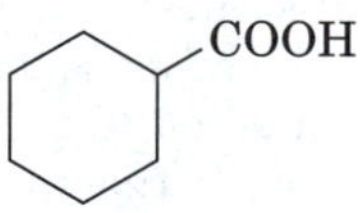

22.45 How can you account for the fact that *cis*- and *trans*-4-*tert*-butyl-2-methyl-cyclohexanone are interconverted by base treatment?

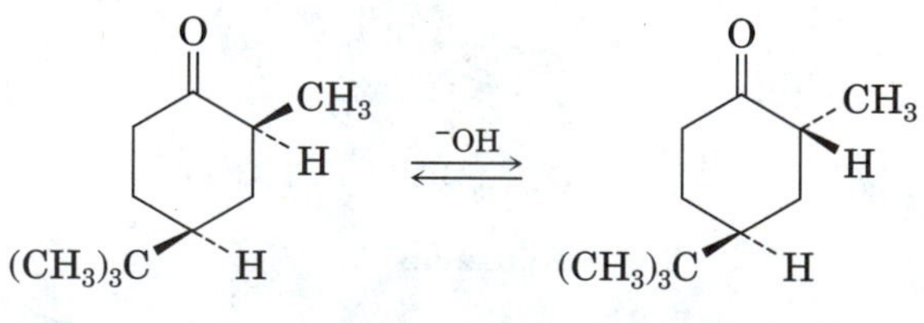

Which of the two isomers do you think is more stable, and why?

22.46 The following synthetic routes are incorrect. What is wrong with each?

(a) $CH_3CH_2CH_2CH_2CO_2CH_3 \xrightarrow[\text{2. }\Delta\text{, pyridine}]{\text{1. }Br_2,\ CH_3COOH} CH_3CH_2CH{=}CHCO_2CH_3$

(b) $CH_3CH(CO_2CH_2CH_3)_2 \xrightarrow[\text{2. }C_6H_5Br]{\text{1. }^-OC_2H_5/C_2H_5OH}$ 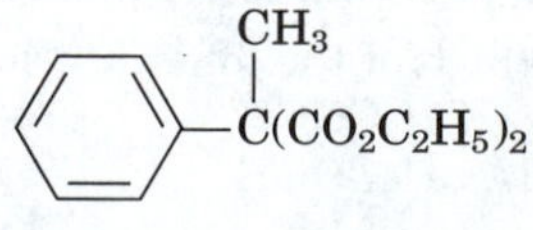

(c) $CH_3COCH_2CO_2C_2H_5 \xrightarrow[\text{2. } H_2C{=}CHCH_2Br \text{; 3. } H_3O^+]{\text{1. } ^-OC_2H_5/C_2H_5OH} H_2C{=}CHCH_2CH_2COOH$

(d) 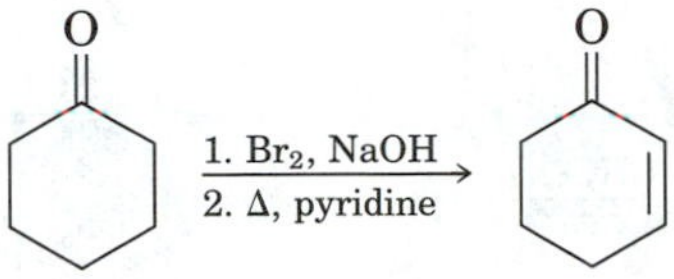

22.47 Unlike most β-diketones, the β-diketone shown below has no detectable enol content and is only about as acidic as acetone. Explain. Molecular models should be helpful.

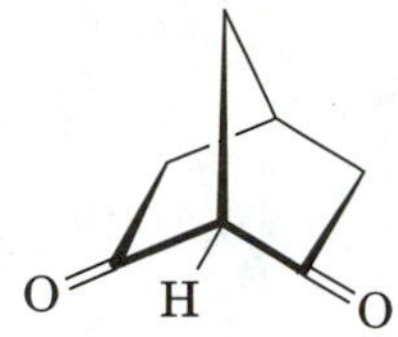

22.48 Methylmagnesium bromide adds to cyclohexanone to give the expected tertiary alcohol product in high yield. *tert*-Butylmagnesium bromide, however, gives only about a 1% yield of the addition product, along with 99% recovered starting material. Furthermore, if D_3O^+ is added to the reaction mixture after a suitable period, one deuterium atom is incorporated into the recovered cyclohexanone. Explain.

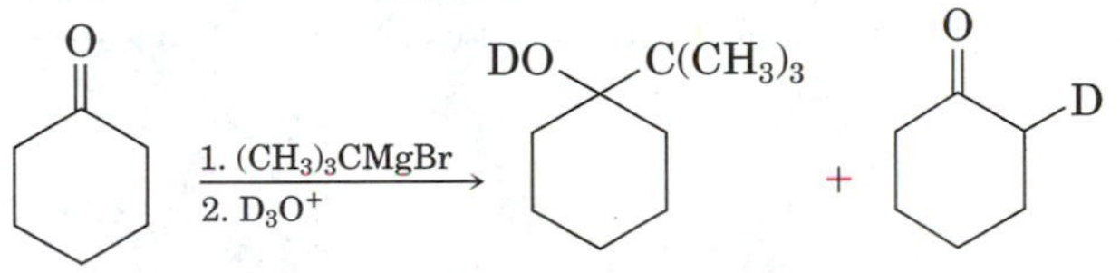

22.49 The final step in an attempted synthesis of laurene, a hydrocarbon isolated from the marine alga *Laurencia glandulifera*, involved the Wittig reaction shown. The product obtained, however, was not laurene but an isomer. Propose a mechanism to account for these unexpected results.

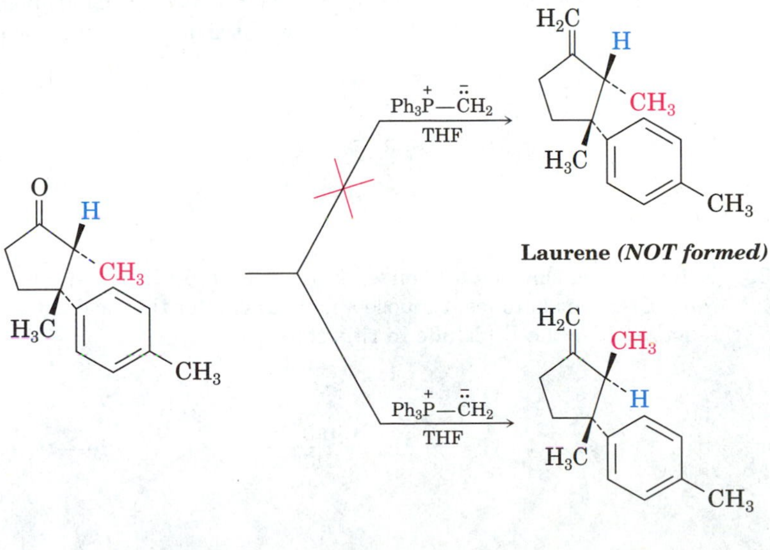

22.50 The key step in a reported laboratory synthesis of sativene, a hydrocarbon isolated from the mold *Helminthosporium sativum*, is shown. What kind of reaction is occurring? How would you complete the synthesis?

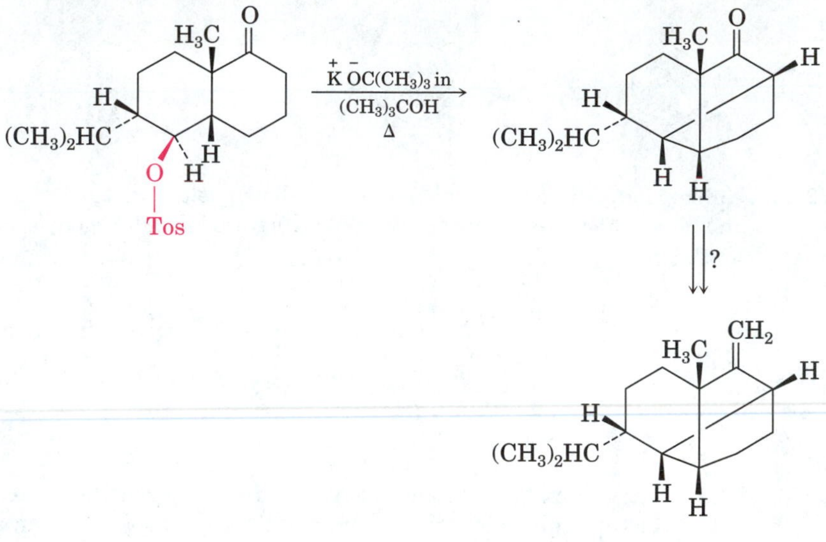

Sativene

22.51 The *Favorskii reaction* involves treatment of an α-bromo ketone with base to yield a ring-contracted product. For example, reaction of 2-bromocyclohexanone with aqueous NaOH yields cyclopentanecarboxylic acid. Propose a mechanism to account for this reaction.

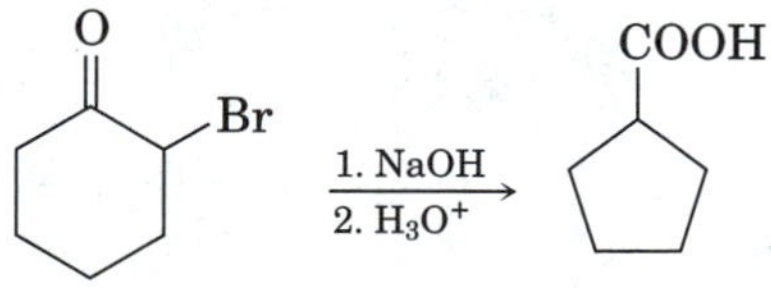

22.52 Treatment of a cyclic ketone with diazomethane is a method for accomplishing a *ring-expansion reaction*. For example, treatment of cyclohexanone with diazomethane yields cycloheptanone. Propose a mechanism to account for this reaction.

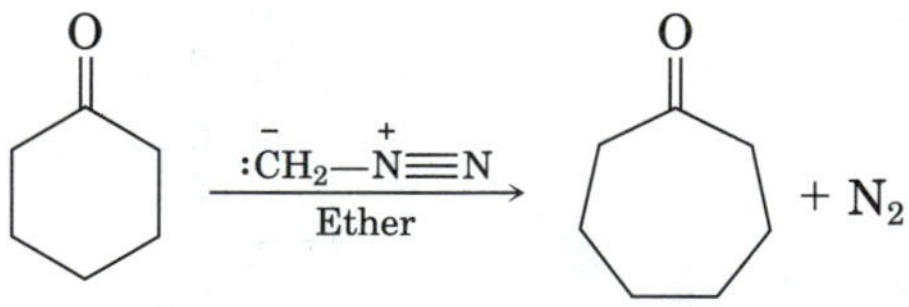

22.53 Ketones react slowly with benzeneselenenyl chloride in the presence of HCl to yield α-phenylseleno ketones. Propose a mechanism for this acid-catalyzed α-substitution reaction. (See the Interlude in this chapter.)

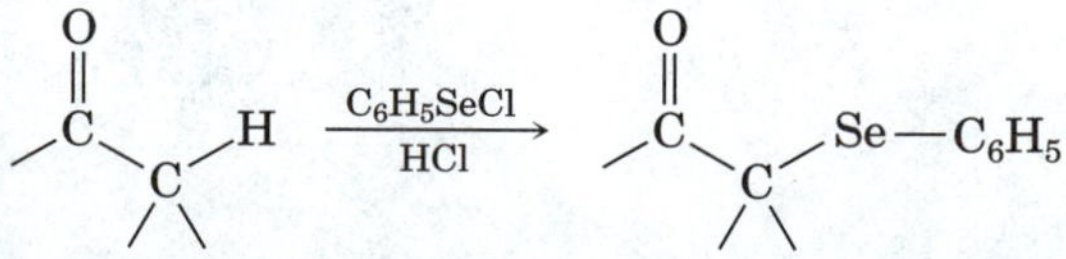

A Look Ahead

22.54 We'll see in Chapter 27 that amino acids can be prepared by reaction of alkyl halides with diethyl acetamidomalonate, followed by heating the initial alkylation product with aqueous HCl. Show how you would prepare alanine, $CH_3CH(NH_2)COOH$, one of the 20 amino acids found in proteins.

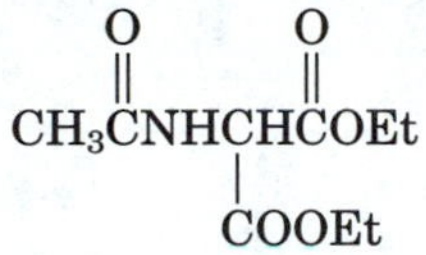

Diethyl acetamidomalonate

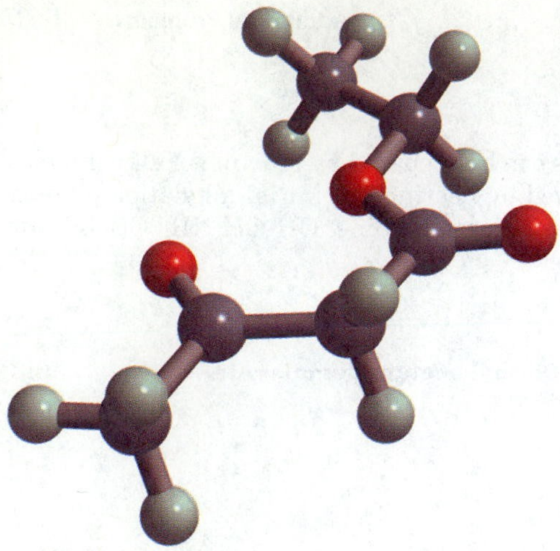

This *β*-keto ester is formed by the condensation reaction of a simple ester.

23

Carbonyl Condensation Reactions

We've seen three general kinds of carbonyl-group reactions in the past four chapters and have studied two general kinds of behavior. In nucleophilic addition and nucleophilic acyl substitution reactions, the carbonyl group behaves as an electrophile by accepting electrons from an attacking nucleophile. In α-substitution reactions, the carbonyl compound behaves as a nucleophile when it is converted into its enolate ion or enol tautomer. The **carbonyl condensation reactions** that we'll study in the present chapter involve *both* types of reactivity.

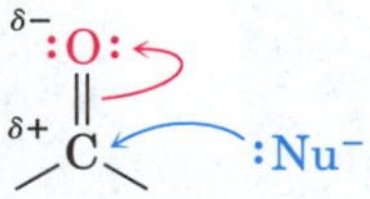

Electrophilic carbonyl group is attacked by nucleophiles

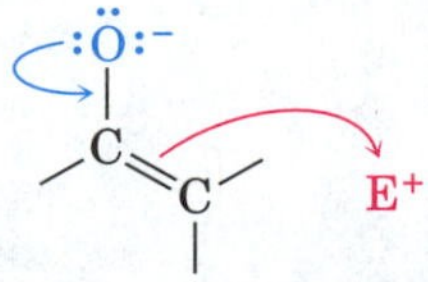

Nucleophilic enolate ion attacks electrophiles

23.1 Mechanism of Carbonyl Condensation Reactions

Carbonyl condensation reactions take place between two carbonyl components and involve a *combination* of nucleophilic addition and α-substitution steps. One component (the nucleophilic donor) is converted into its enolate ion and undergoes an α-substitution reaction by reacting with the second component (the electrophilic acceptor), which undergoes an addition. The general mechanism of a carbonyl condensation reaction is shown in Figure 23.1.

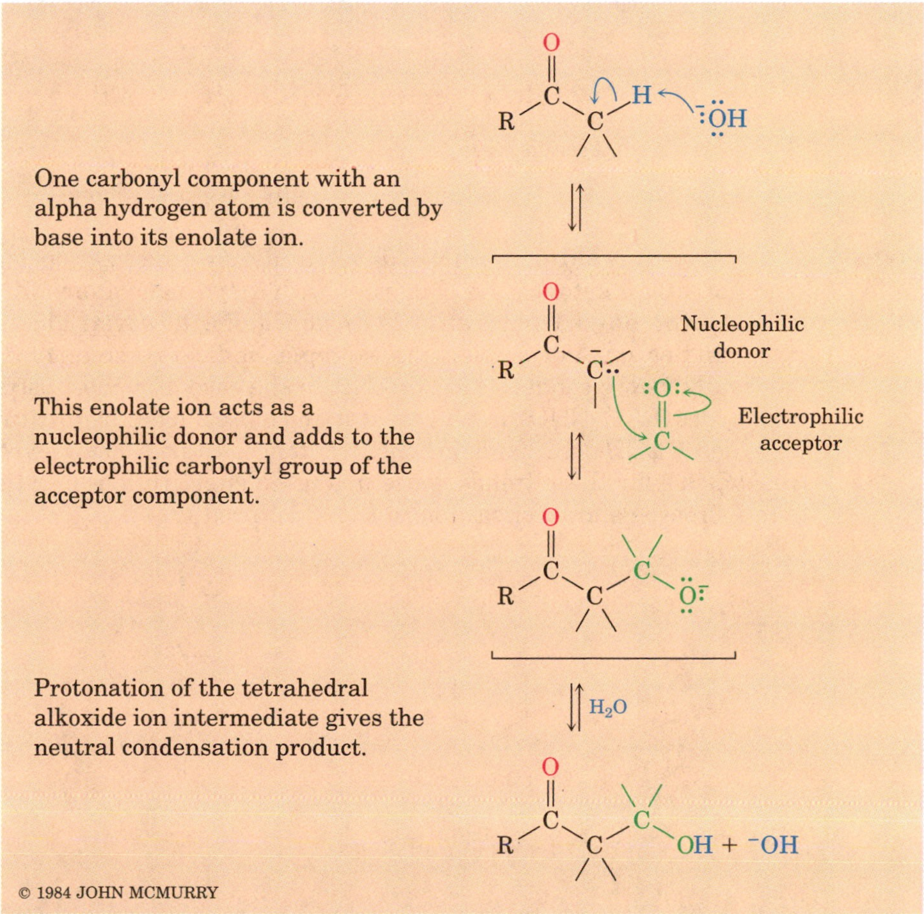

Figure 23.1 The general mechanism of a carbonyl condensation reaction. One component (the donor) acts as a nucleophile, while the other component (the acceptor) acts as an electrophile.

All kinds of carbonyl compounds, including aldehydes, ketones, esters, amides, acid anhydrides, thiol esters, and nitriles, enter into condensation reactions. Nature uses carbonyl condensation reactions as key steps in the biosynthesis of many naturally occurring compounds, and chemists use the same reactions in the laboratory.

23.2 Condensations of Aldehydes and Ketones: The Aldol Reaction

When acetaldehyde is treated with a base, such as sodium ethoxide or sodium hydroxide, a rapid and reversible condensation reaction occurs. The product is the β-hydroxy aldehyde known commonly as *aldol* (*ald*ehyde + alcoh*ol*).

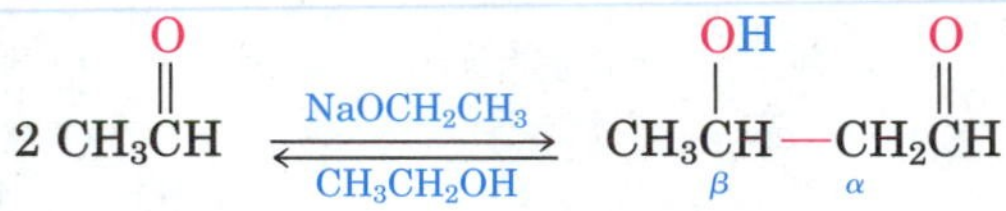

Acetaldehyde

Aldol
(a β-hydroxy aldehyde)

Called the **aldol reaction**, base-catalyzed dimerization is a general reaction for all ketones and aldehydes with α hydrogen atoms. If the ketone or aldehyde doesn't have an α hydrogen atom, however, aldol condensation can't occur. As the following examples indicate, the aldol equilibrium generally favors condensation product in the case of monosubstituted acetaldehydes (RCH_2CHO), but favors starting material for disubstituted acetaldehydes (R_2CHCHO) and for most ketones. Steric factors are probably responsible for these trends, since increased substitution near the reaction site increases steric congestion in the aldol product.

Ketones

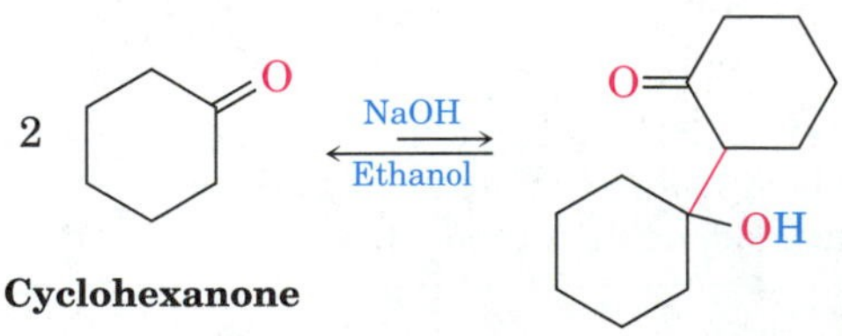

Cyclohexanone

22%

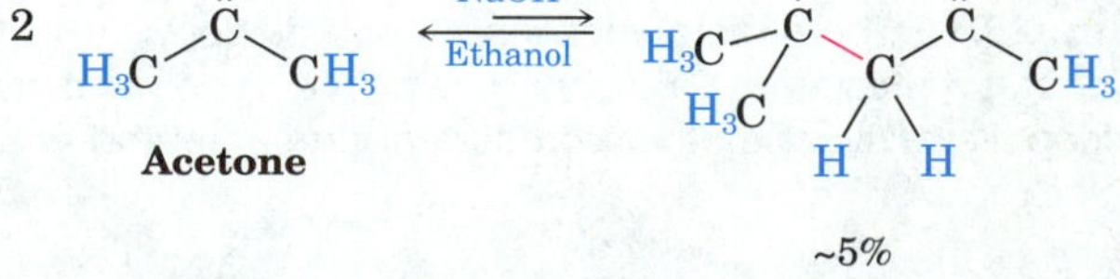

Acetone

~5%

Aldehydes

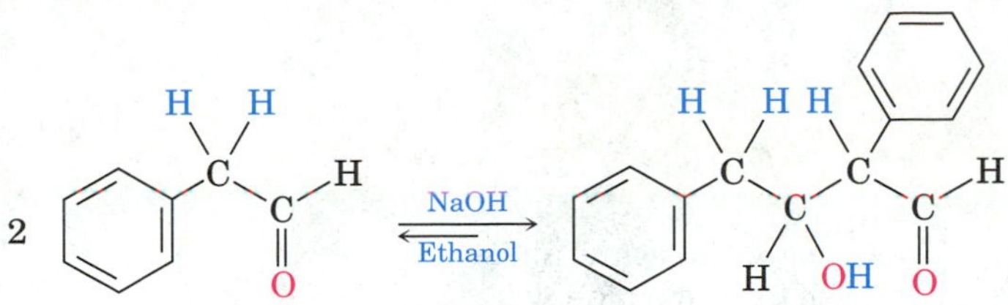

Phenylacetaldehyde 90%

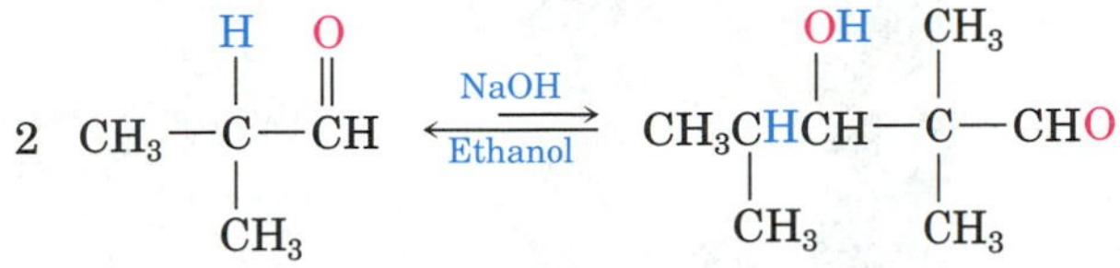

2-Methylpropanal Low yield

Aldol reactions are typical carbonyl condensations. They occur by nucleophilic addition of the enolate ion of the donor molecule to the carbonyl group of the acceptor molecule, yielding a tetrahedral intermediate that is protonated to give an alcohol product (Figure 23.2, p. 900). The reverse process occurs in exactly the opposite manner: Base abstracts the –OH hydrogen from the aldol to yield a β-keto alkoxide ion, which fragments to give one molecule of enolate ion and one molecule of neutral carbonyl compound.

PRACTICE PROBLEM

What is the structure of the aldol product from propanal?

Solution An aldol reaction combines two molecules of reactant, forming a bond between the α carbon of one partner and the carbonyl carbon of the second partner:

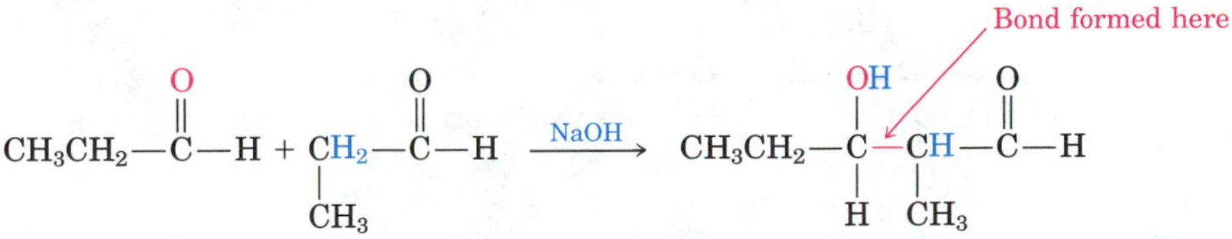

PROBLEM

23.1 Predict the product of aldol reaction of the following compounds:
(a) Butanal (b) 2-Butanone (c) Cyclopentanone

PROBLEM

23.2 The aldol reaction is catalyzed by acid as well as by base. What is the reactive nucleophile in the acid-catalyzed aldol reaction? Propose a possible mechanism.

PROBLEM

23.3 Using curved arrows, show how the base-catalyzed reverse aldol reaction of 4-methyl-4-hydroxy-2-pentanone takes place to yield 2 equivalents of acetone.

Base removes an acidic alpha hydrogen from one aldehyde molecule, yielding a resonance-stabilized enolate ion.

The enolate ion attacks a second aldehyde molecule in a nucleophilic addition reaction to give a tetrahedral alkoxide ion intermediate.

Protonation of the alkoxide ion intermediate yields neutral aldol product and regenerates the base catalyst.

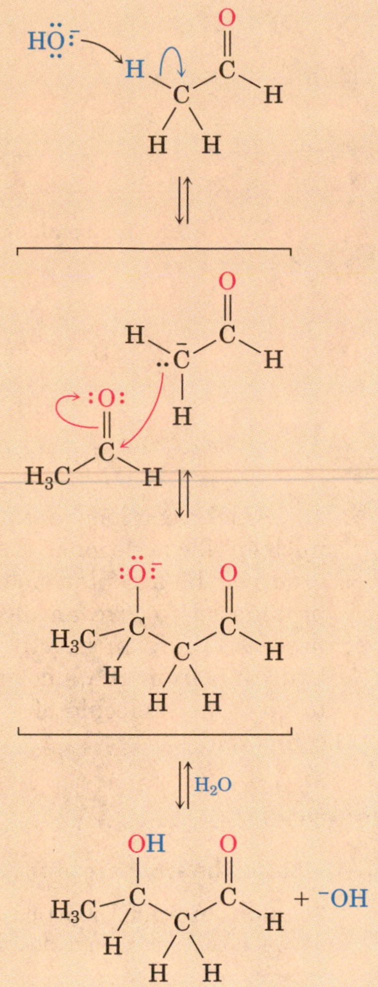

Figure 23.2 Mechanism of the aldol reaction.

23.3 Carbonyl Condensation Reactions versus Alpha-Substitution Reactions

Two of the four general carbonyl-group reactions—carbonyl condensation and α substitution—take place under basic conditions and involve enolate ion intermediates. Since reaction conditions for both processes are similar, how can we predict which of the two will occur in a given case? When we generate an enolate ion with the intention of carrying out an α alkylation, how can we be sure that a carbonyl condensation reaction won't occur instead?

Although there is no simple answer to this question, the experimental conditions usually have much to do with the result. Alpha-substitution reactions require a full equivalent of strong base and are normally carried out so that the carbonyl compound is rapidly and completely converted into its enolate ion at a low temperature. An electrophile is then added rapidly to ensure that the reactive enolate ion is quenched quickly. In a ketone alkylation reaction, for instance, we might use 1 equivalent of lithium diisopropylamide (LDA) in tetrahydrofuran solution at dry-ice temperature (−78°C). Rapid and complete generation of the ketone enolate ion would occur, and no reactant would be left so that no condensation reaction could take place. We would then immediately add an alkyl halide to complete the alkylation reaction.

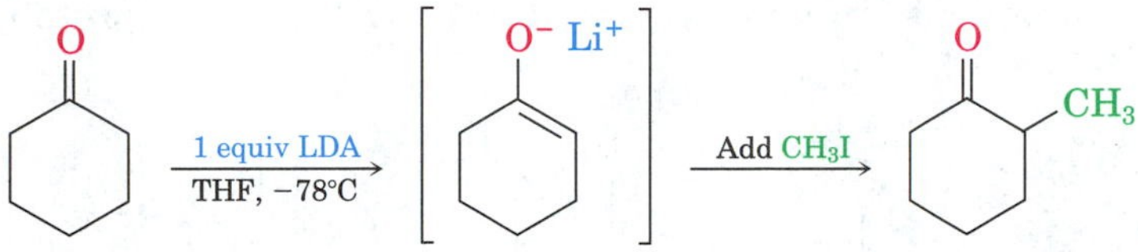

On the other hand, we might want to carry out a carbonyl condensation reaction. Since we need to generate only a small amount of the enolate ion in the presence of unreacted carbonyl compound, the aldol reaction requires only a *catalytic* amount of a weaker base, rather than a full equivalent. Once a condensation has occurred, the basic catalyst is regenerated. To carry out an aldol reaction on propanal, for example, we might dissolve the aldehyde in methanol, add 0.05 equivalent of sodium methoxide, and then warm the mixture. A high yield of aldol product would result.

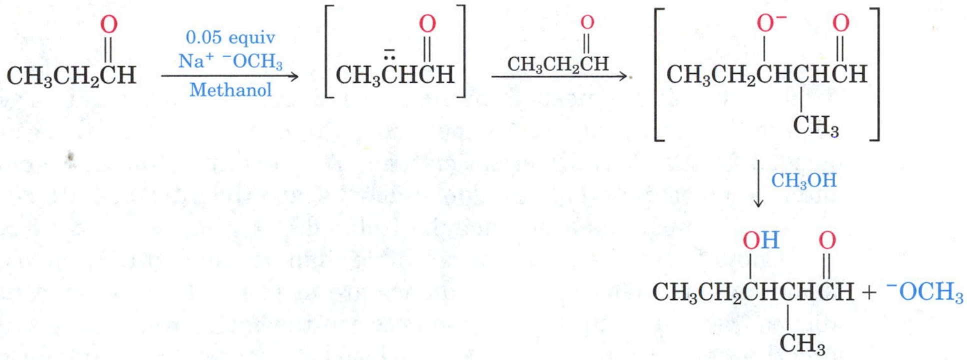

23.4 Dehydration of Aldol Products: Synthesis of Enones

The β-hydroxy ketones and β-hydroxy aldehydes formed in aldol reactions can be easily dehydrated to yield conjugated enones. In fact, it's this loss of water that gives the aldol *condensation* its name, since water condenses out of the reaction when the enone product forms.

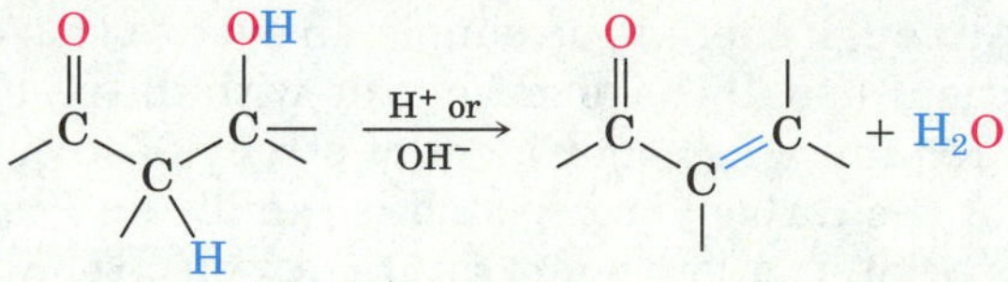

A β-hydroxy ketone or aldehyde — A conjugated enone

Most alcohols are resistant to dehydration by dilute acid or base, and reagents such as $POCl_3$ must therefore be used (Section 17.8). Hydroxyl groups β to a carbonyl group are special, however, because of the nearby carbonyl group. Under *basic* conditions, an acidic α hydrogen is abstracted, yielding an enolate ion that expels the ^{-}OH leaving group. Under *acidic* conditions, an enol is formed, the –OH group is protonated, and water is expelled.

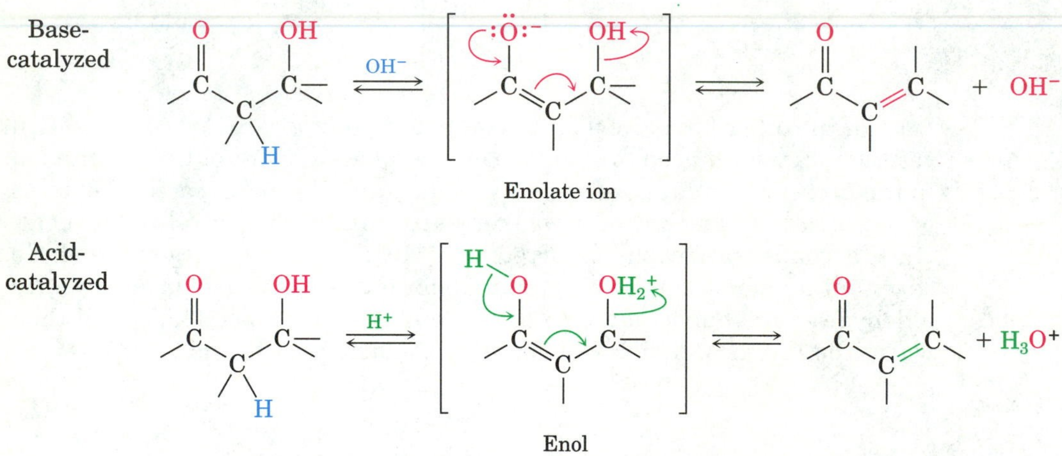

The conditions needed for aldol dehydration are often only a bit more vigorous (slightly higher temperature, for instance) than the conditions needed for the aldol dimerization itself. As a result, conjugated enones are often obtained directly from aldol reactions, and the intermediate β-hydroxy carbonyl compounds aren't usually isolated.

Conjugated enones are more stable than nonconjugated enones for the same reasons that conjugated dienes are more stable than nonconjugated dienes (Section 14.2). Interaction between the π electrons of the C=C bond and the π electrons of the C=O group leads to a molecular orbital description of conjugated enones that shows a partial delocalization of the π electrons over all four atomic centers.

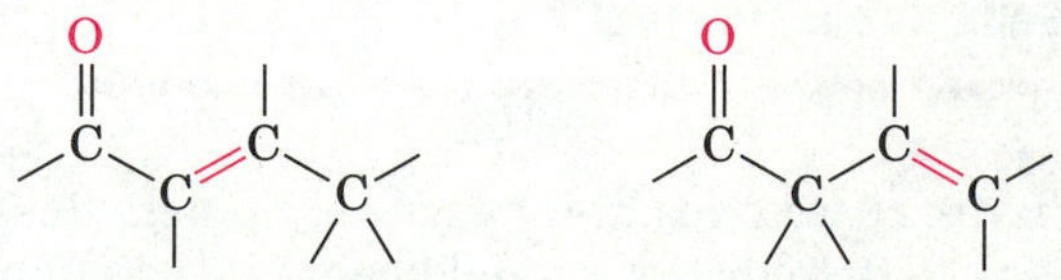

Conjugated enone (more stable) — Nonconjugated enone (less stable)

The real value of aldol dehydration is that removal of water from the reaction mixture drives the aldol equilibrium toward product formation. Even though the initial aldol step itself may be unfavorable (as it usually is for ketones), the subsequent dehydration step nevertheless allows most aldol condensations to be carried out in good yield. Cyclohexanone, for example, gives cyclohexylidenecyclohexanone in 92% yield even though the initial equilibrium is unfavorable.

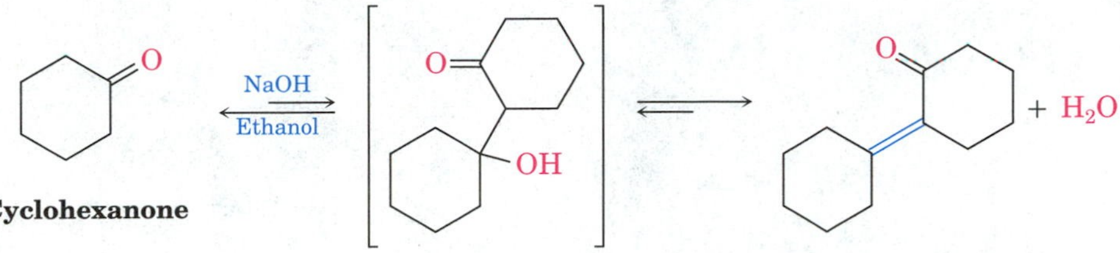

Cyclohexanone

Cyclohexylidenecyclohexanone (92%)

PRACTICE PROBLEM.

What is the structure of the enone obtained from aldol condensation of acetaldehyde?

Solution In the aldol reaction, H_2O is eliminated by removing two hydrogens from the acidic α position of one partner and the carbonyl oxygen from the second partner.

$$H_3C{-}\overset{\displaystyle H}{\overset{|}{C}}{=}O + H_2\overset{\displaystyle H}{\overset{|}{C}}{-}CHO \xrightarrow{NaOH} H_3C{-}\overset{\displaystyle H}{\overset{|}{C}}{=}\overset{\displaystyle H}{\overset{|}{C}}{-}CHO + H_2O$$

2-Butenal

PROBLEM.

23.4 What enone products would you expect from aldol condensation of the following compounds?
(a) Cyclopentanone (b) Acetophenone (c) 3-Methylbutanal

PROBLEM.

23.5 Aldol condensation of 3-methylcyclohexanone leads to a mixture of two products, not counting double-bond isomers. Draw them.

23.5 Recognizing Aldol Products

The aldol condensation reaction yields either a β-hydroxy ketone/aldehyde or an α,β-unsaturated ketone/aldehyde, depending on the specific case and on the reaction conditions. By learning how to think *backward*, it's possible to predict when the aldol reaction might be useful in synthesis. Any time the target molecule contains either a β-hydroxy ketone/aldehyde or a conjugated enone functional group, it might come from an aldol reaction.

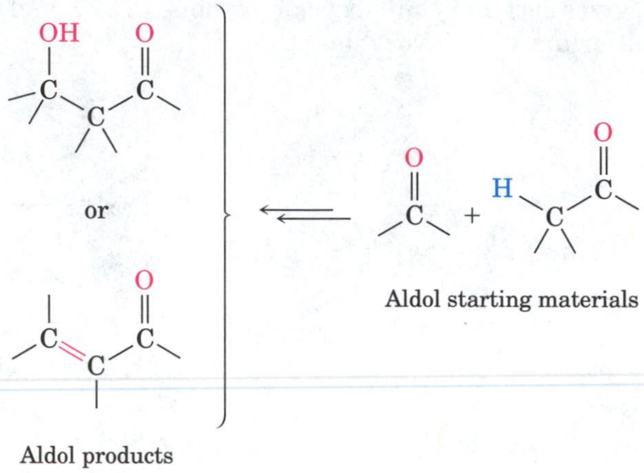

We can extend this kind of reasoning even further by considering that subsequent transformations might be carried out on the aldol products. For example, a saturated ketone might be prepared by catalytic hydrogenation of an enone product. A good example can be found in the industrial preparation of 2-ethyl-1-hexanol, an alcohol used in the synthesis of plasticizers for polymers. Although 2-ethyl-1-hexanol bears little resemblance to an aldol product at first glance, it is prepared commercially from butanal by an aldol reaction. Working backward, we can reason that 2-ethyl-1-hexanol might come from 2-ethylhexanal by a reduction. 2-Ethylhexanal, in turn, might be prepared by catalytic reduction of 2-ethyl-2-hexenal, which is the aldol condensation product of butanal. The reactions that follow show the sequence in reverse order.

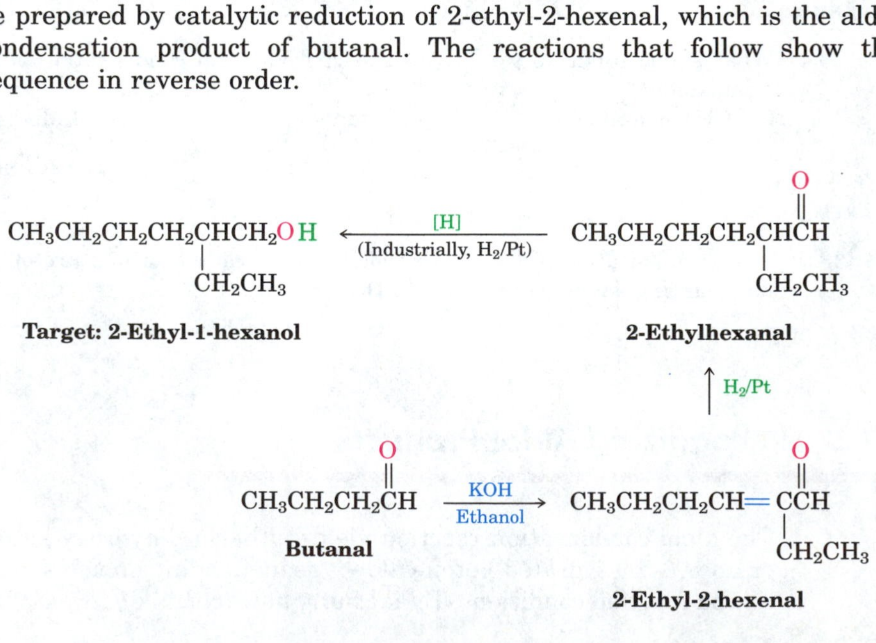

PROBLEM.

23.6 Which of the following compounds are aldol condensation products? What is the ketone or aldehyde precursor of each?
(a) 2,2,3-Trimethyl-3-hydroxybutanal (b) 2-Methyl-2-hydroxypentanal
(c) 5-Ethyl-4-methyl-4-hepten-3-one

PROBLEM.

23.7 1-Butanol is prepared commercially by a route that begins with the aldol reaction of acetaldehyde. Show the steps that are likely to be involved.

23.6 Mixed Aldol Reactions

Until now, we've considered only symmetrical aldol reactions, in which the two carbonyl components have been the same. What would happen, though, if a *mixed* aldol reaction were carried out between two different carbonyl partners?

In general, a mixed aldol reaction between two similar ketone or aldehyde components leads to a mixture of four possible products. For example, base treatment of a mixture of acetaldehyde and propanal gives a complex product mixture containing two "symmetrical" aldol products and two "mixed" aldol products. Clearly, such a reaction is of little practical value.

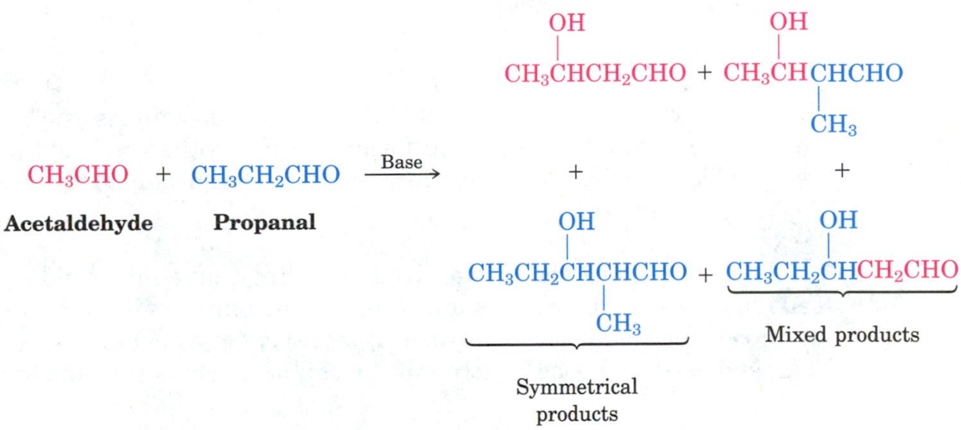

On the other hand, mixed aldol reactions *can* lead cleanly to a single product, if one of two conditions is met:

1. If one of the carbonyl components contains no α hydrogens (and thus can't form an enolate ion to become a donor) but does contain a reactive carbonyl group that is a good acceptor of nucleophiles, then a mixed aldol reaction is likely to be successful. This is the case, for example, when benzaldehyde or formaldehyde is used as one of the carbonyl components:

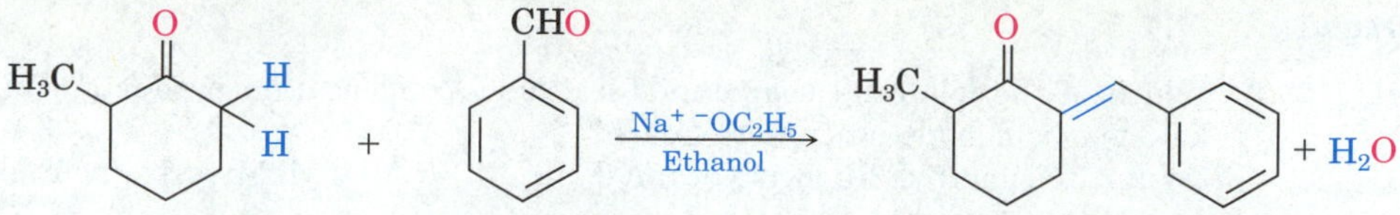

2-Methylcyclohexanone (donor) **Benzaldehyde (acceptor)** 78%

Neither benzaldehyde nor formaldehyde can form an enolate ion to condense with itself or with another partner, yet both carbonyl groups are unhindered and reactive. Thus, in the presence of a ketone such as 2-methylcyclohexanone, the ketone enolate adds preferentially to benzaldehyde, giving the mixed aldol product.

2. If one of the carbonyl components is much more acidic than the other and is easily transformed into its enolate ion, then a mixed aldol reaction is likely to be successful. This is the case, for example, when ethyl acetoacetate is used as the donor carbonyl component:

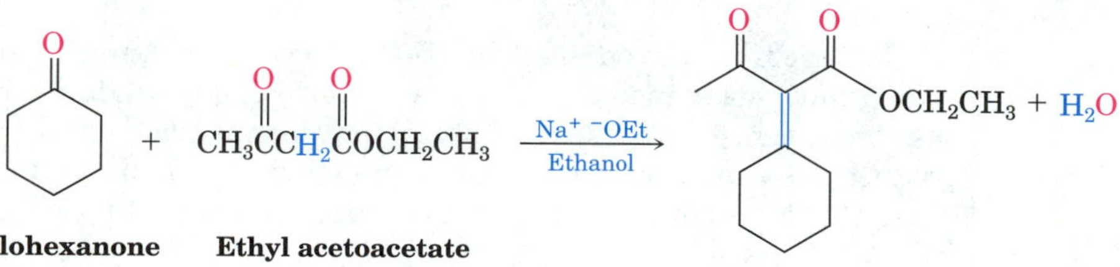

Cyclohexanone (acceptor) **Ethyl acetoacetate (donor)** 80%

Ethyl acetoacetate is completely converted into its enolate ion in preference to enolate ion formation from other carbonyl partners. Aldol condensation therefore occurs preferentially to give the mixed product.

The situation can be summarized by saying that a mixed aldol reaction leads to a mixture of products unless one of the partners is either an unusually good nucleophilic donor (such as acetoacetic ester) or else has no α hydrogens and is a good electrophilic acceptor (such as benzaldehyde).

PROBLEM.

23.8 Which of the following compounds can probably be prepared by a mixed aldol reaction?

(a) $C_6H_5CH{=}CH\overset{O}{\overset{\|}{C}}CH_3$

(b) $C_6H_5\underset{CH_3}{\underset{|}{C}}{=}CH\overset{O}{\overset{\|}{C}}CH_3$

(c)

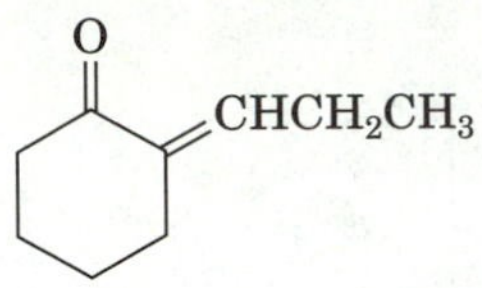

(d)

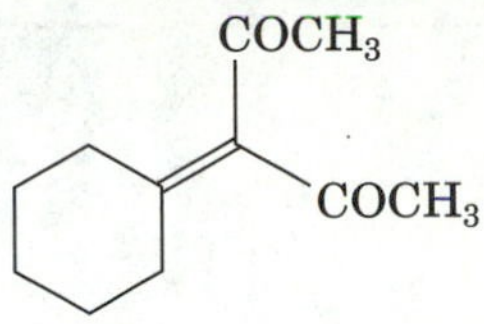

23.7 Intramolecular Aldol Reactions

The aldol reactions we've seen up to this point have all been *inter*molecular. That is, they have taken place between two different molecules. When certain *di*carbonyl compounds are treated with base, however, *intra*molecular aldol reactions can occur, leading to the formation of cyclic products. For example, base treatment of a 1,4-diketone such as 2,5-hexanedione yields a cyclopentenone product, and base treatment of a 1,5-diketone such as 2,6-heptanedione yields a cyclohexenone.

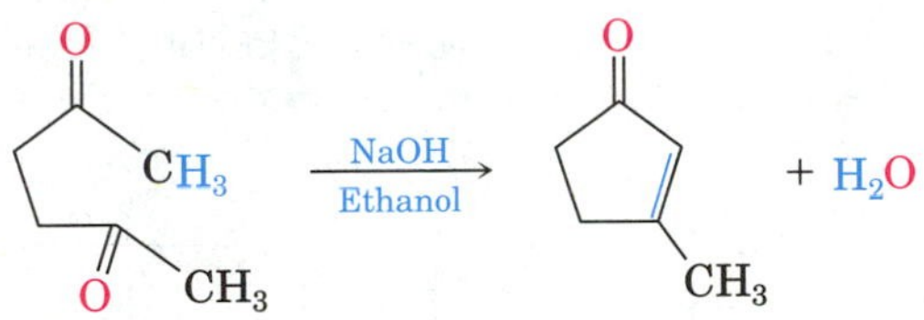

2,5-Hexanedione (a 1,4-diketone) **3-Methyl-2-cyclopentenone**

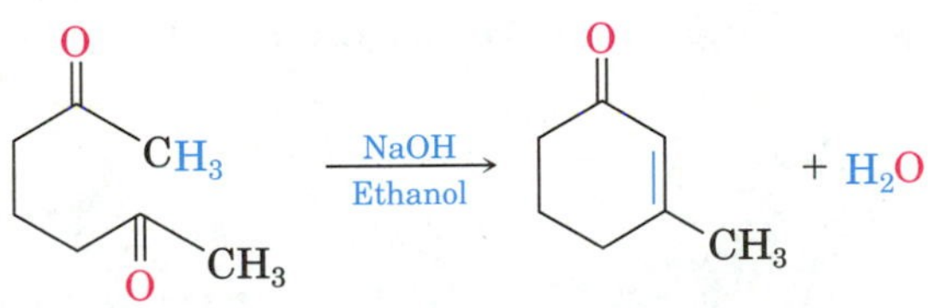

2,6-Heptanedione (a 1,5-diketone) **3-Methyl-2-cyclohexenone**

The mechanism of these intramolecular aldol reactions is similar to the mechanism of the corresponding intermolecular reactions. The only difference is that both the nucleophilic carbonyl anion donor and the electrophilic carbonyl acceptor are in the same molecule.

In principle, many intramolecular aldol reactions can lead to a mixture of products, depending on which enolate ion is formed. For example, 2,5-hexanedione might yield either the five-membered-ring product 3-methyl-2-cyclopentenone or the three-membered-ring product (2-methylcyclopropenyl)ethanone (Figure 23.3). In practice, though, only the cyclopentenone is formed.

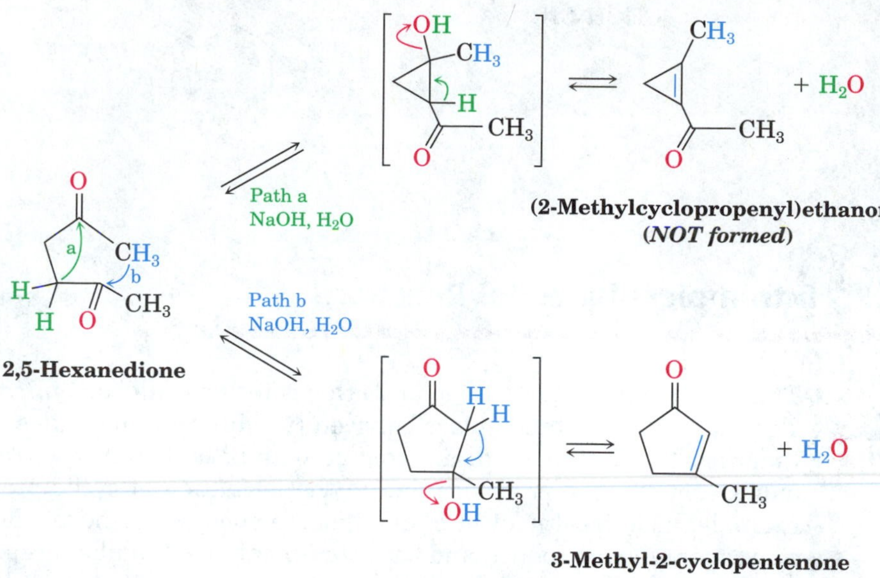

Figure 23.3 Intramolecular aldol reaction of 2,5-hexanedione yields 3-methyl-2-cyclopentenone rather than the alternative acetylcyclopropene.

The selectivity observed in intramolecular aldol reactions is due to the fact that all steps in the mechanism are reversible and that an equilibrium is reached. Thus, the relatively strain-free cyclopentenone product is considerably more stable than the highly strained cyclopropene alternative. For similar reasons, intramolecular aldol reactions of 1,5-diketones lead only to cyclohexenone products rather than to cyclobutenes.

PROBLEM.

23.9 Why do you suppose that 1,3-diketones do not undergo internal aldol condensation to yield cyclobutenones?

PROBLEM.

23.10 What product would you expect to obtain from base treatment of 1,6-cyclodecanedione?

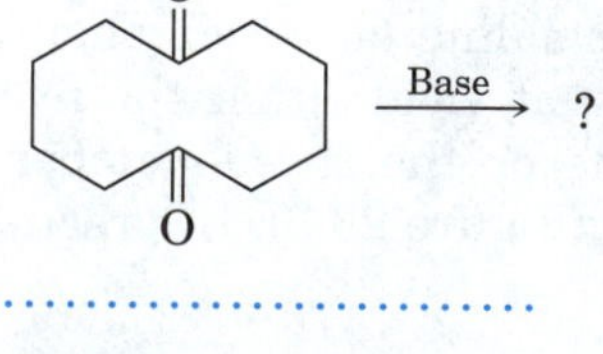

23.8 The Claisen Condensation Reaction

Esters, like aldehydes and ketones, are weakly acidic. When an ester with an α hydrogen is treated with 1 equivalent of a base such as sodium ethoxide, a reversible condensation reaction occurs to yield a β-keto ester. For example, ethyl acetate yields ethyl acetoacetate on treatment with base. This reaction between two ester components is known as the **Claisen condensation reaction.**[1]

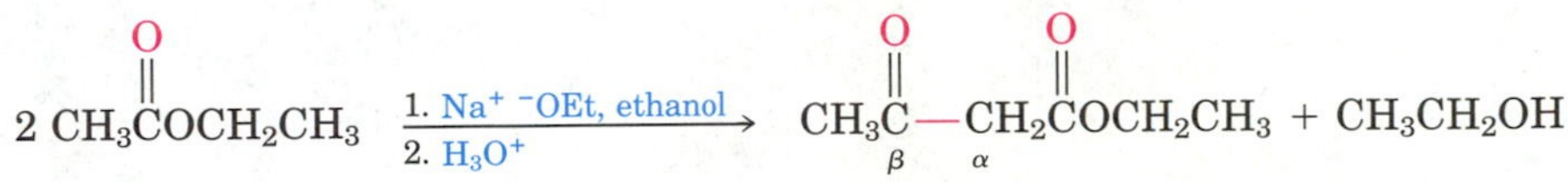

Ethyl acetate **Ethyl acetoacetate, a β-keto ester (75%)**

The mechanism of the Claisen condensation is similar to that of the aldol reaction. As shown in Figure 23.4 (p. 910), the Claisen condensation involves the nucleophilic addition of an ester enolate ion to the carbonyl group of a second ester molecule.

The only difference between an aldol condensation of a ketone/aldehyde and a Claisen condensation of an ester involves the fate of the initially formed tetrahedral intermediate. The tetrahedral intermediate in the aldol reaction is protonated to give an alcohol product—exactly the behavior previously seen for ketones and aldehydes (Section 19.6). The tetrahedral intermediate in the Claisen reaction expels an alkoxide leaving group to yield an acyl substitution product—exactly the behavior previously seen for esters (Section 21.7).

If the starting ester has more than one acidic α hydrogen, the product β-keto ester has a highly acidic, doubly activated hydrogen atom that can be abstracted by base. This deprotonation of the product requires that a full equivalent of base rather than a catalytic amount be used in the reaction. Furthermore, the deprotonation serves to drive the Claisen equilibrium completely to the product side so that high yields are often obtained.

PRACTICE PROBLEM

What product would you obtain from Claisen condensation of methyl propanoate?

Solution The Claisen condensation of an ester results in loss of one molecule of alcohol and formation of a product in which an acyl group of one reactant bonds to the α carbon of the second reactant:

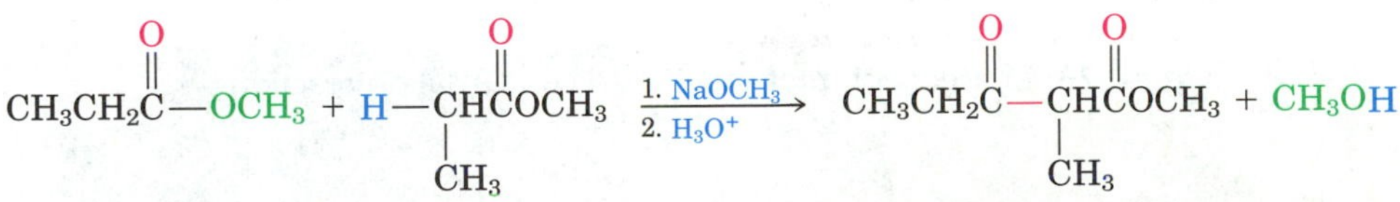

2 Methyl propanoate **Methyl 2-methyl-3-oxopentanoate**

[1]Ludwig Claisen (1851–1930); b. Cologne; Ph.D. Bonn (Kekulé); professor, University of Bonn, Owens College (Manchester), universities of Munich, Aachen, Kiel, and Berlin; Godesberg (private laboratory).

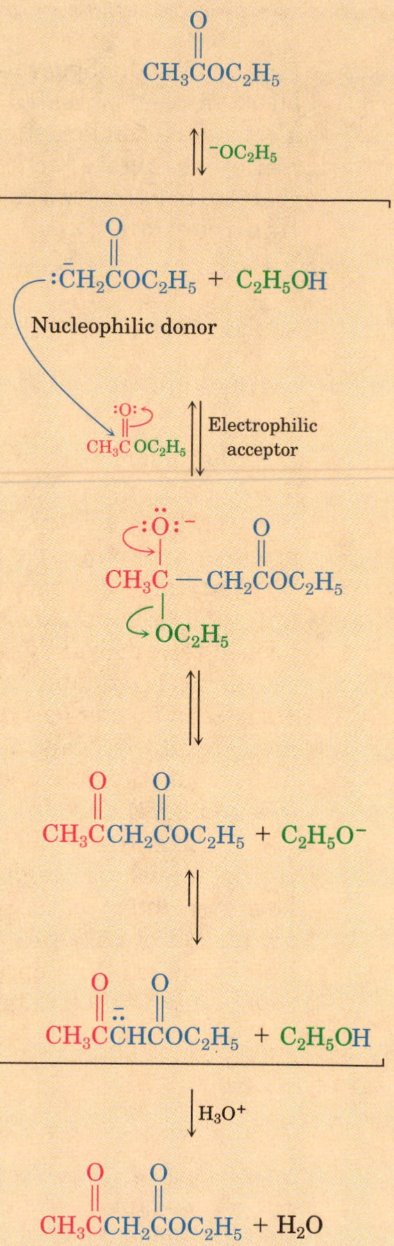

Ethoxide base abstracts an acidic alpha hydrogen atom from an ester molecule, yielding an ester enolate ion.

This ion does a nucleophilic addition to a second ester molecule, giving a tetrahedral intermediate.

The tetrahedral intermediate is not stable. It expels ethoxide ion to yield the new carbonyl compound, ethyl acetoacetate.

But ethoxide ion is a base. It therefore converts the β-keto ester product into its enolate, thus shifting the equilibrium and driving the reaction to completion.

Protonation by addition of acid yields the final product.

Figure 23.4 Mechanism of the Claisen condensation reaction.

PROBLEM. . . . .

23.11 Show the products you would expect to obtain by Claisen condensation of these esters.

(a) $(CH_3)_2CHCH_2COOCH_3$ (b) Methyl phenylacetate

(c) Methyl cyclohexylacetate

PROBLEM.

23.12 As shown in Figure 23.4, the Claisen reaction is reversible. That is, a β-keto ester can be cleaved by base into two fragments. Show the mechanism by which this cleavage occurs.

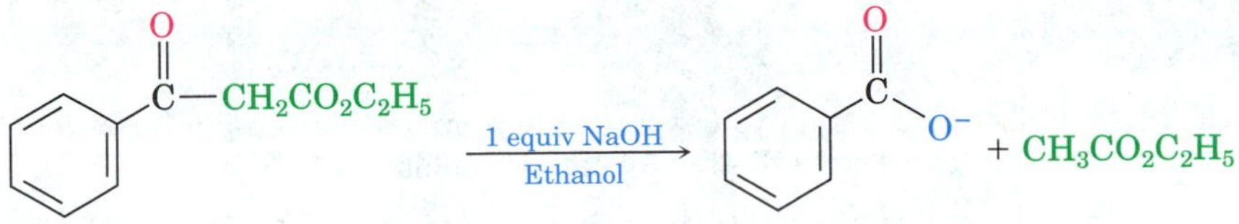

23.9 Mixed Claisen Condensations

The mixed Claisen condensation of two different esters is similar to the mixed aldol condensation of two different aldehydes or ketones (Section 23.6). Mixed Claisen reactions are generally successful only when one of the two ester components has no α hydrogens and thus can't form an enolate ion. For example, ethyl benzoate and ethyl formate can't form enolate ions and thus can't serve as donors. They can, however, act as the electrophilic acceptor components in reactions with other ester anions to give good yields of mixed β-keto ester products.

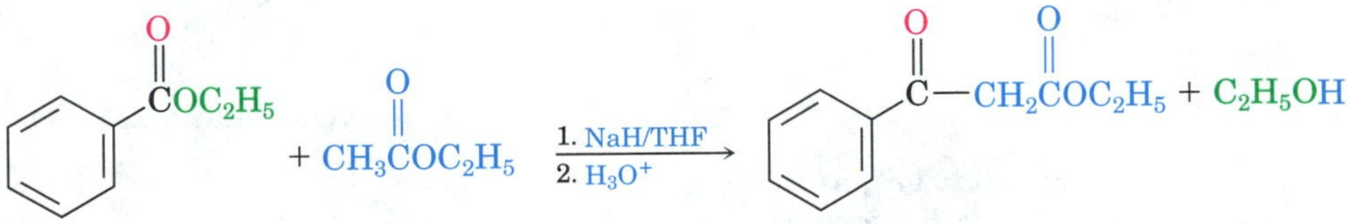

Ethyl benzoate (acceptor) **Ethyl acetate (donor)** **Ethyl benzoylacetate**

Mixed Claisen-like reactions can also be carried out between esters and ketones. The result is an excellent synthesis of β-diketones. The reaction works best when the ester component has no α hydrogens and thus can't act as the nucleophilic donor. For example, ethyl formate gives particularly high yields in mixed Claisen condensations with ketones.

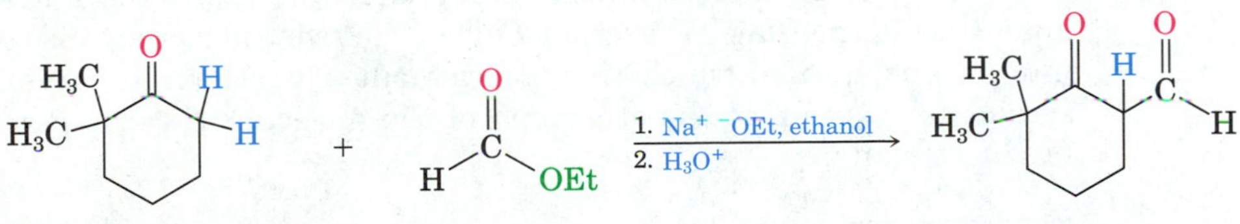

2,2-Dimethylcyclohexanone (donor) **Ethyl formate (acceptor)** **A β-keto aldehyde (91%)**

PROBLEM

23.13 Would you expect diethyl oxalate, $(CO_2Et)_2$, to give good yields in mixed Claisen reactions? What product would you expect to obtain from a mixed Claisen reaction of ethyl acetate with diethyl oxalate?

PROBLEM

23.14 What product would you expect from a mixed Claisen-like reaction of 2,2-dimethylcyclohexanone with diethyl oxalate (Problem 23.13)?

23.10 Intramolecular Claisen Condensations: The Dieckmann Cyclization

Intramolecular Claisen condensations can be carried out with diesters, just as intramolecular aldol condensations can be carried out with diketones (Section 23.7). Called the **Dieckmann[2] cyclization**, the reaction works best on 1,6-diesters and 1,7-diesters. Five-membered cyclic β-keto esters result from Dieckmann cyclization of 1,6-diesters, and six-membered cyclic β-keto esters result from cyclization of 1,7-diesters.

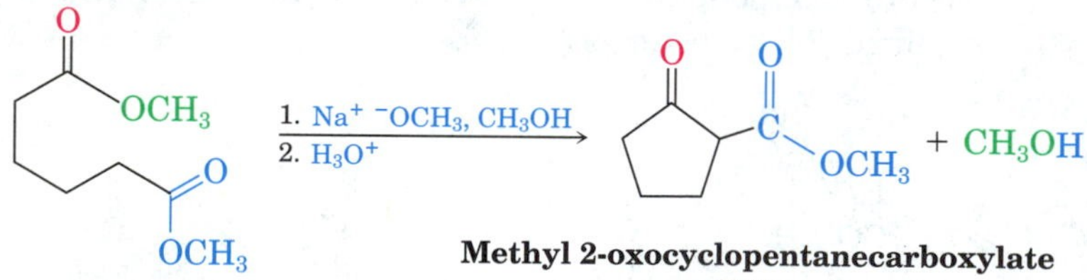

Dimethyl hexanedioate (a 1,6-diester)

Methyl 2-oxocyclopentanecarboxylate (82%)

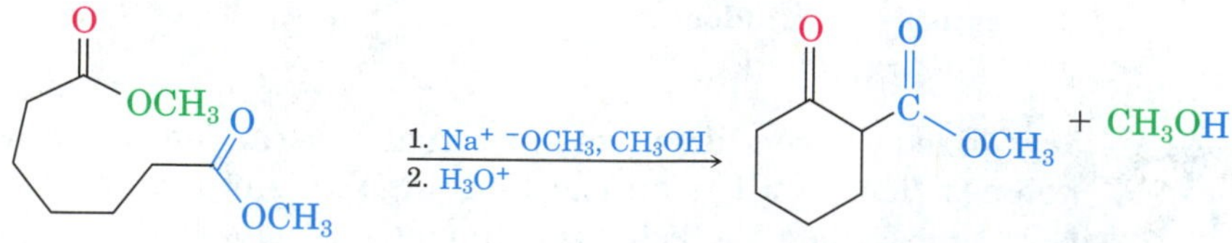

Dimethyl heptanedioate (a 1,7-diester)

Methyl 2-oxocyclohexanecarboxylate

The mechanism of the Dieckmann cyclization, shown in Figure 23.5, is analogous to the Claisen reaction. One of the two ester groups is converted into an enolate ion, which then carries out a nucleophilic attack on the second ester group at the other end of the molecule. A cyclic β-keto ester product results.

[2]Walter Dieckmann (1869–1925); b. Hamburg, Germany; Ph.D. Munich (Bamberger); professor, University of Munich.

Base abstracts an acidic α proton from the carbon atom next to one of the ester groups, yielding an enolate ion.

Intramolecular nucleophilic addition of the ester enolate ion to the carbonyl group of the second ester group at the other end of the chain then gives a cyclic tetrahedral intermediate.

Loss of alkoxide ion from the tetrahedral intermediate forms a cyclic β-keto ester.

Deprotonation of the acidic β-keto ester gives an enolate ion . . .

. . . which is protonated by addition of aqueous acid at the end of the reaction to generate the neutral β-keto ester product.

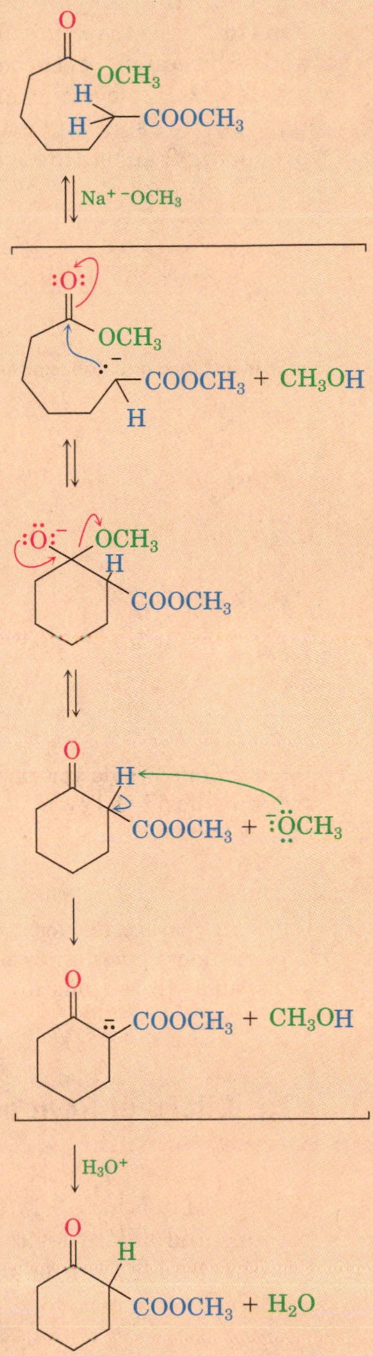

Figure 23.5 Mechanism of the Dieckmann cyclization of a diester to yield a cyclic β-keto ester product.

The products of the Dieckmann cyclization are cyclic β-keto esters that can be further alkylated and decarboxylated by a series of reactions analogous to the acetoacetic ester synthesis (Section 22.8). For example, alkylation and subsequent decarboxylation of methyl 2-oxocyclohexanecarboxylate yields a 2-alkylcyclohexanone. The overall sequence of Dieckmann cyclization, β-keto ester alkylation, and decarboxylation is an excellent method for preparing 2-substituted cyclohexanones and cyclopentanones.

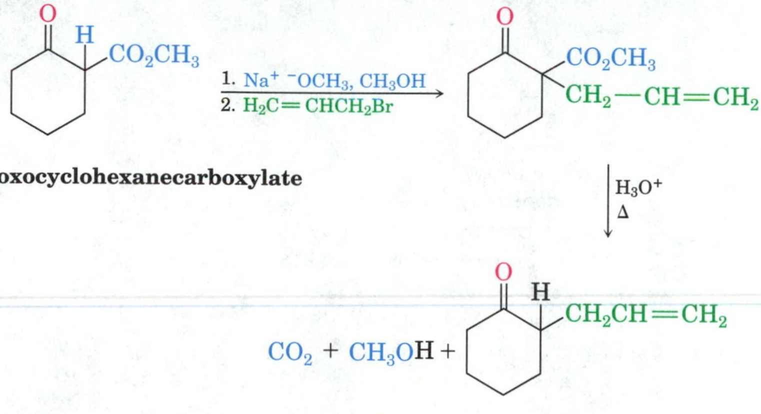

Methyl 2-oxocyclohexanecarboxylate

2-Allylcyclohexanone (83%)

PROBLEM.

23.15 What product would you expect on treatment of diethyl 4-methylheptanedioate with sodium ethoxide followed by acidification?

PROBLEM.

23.16 How can you account for the fact that Dieckmann cyclization of diethyl 3-methylheptanedioate gives a mixture of two β-keto ester products? What are their structures, and why is a mixture formed?

23.11 The Michael Reaction

We saw in Section 19.17 that nucleophiles can react with α,β-unsaturated ketones and aldehydes to give the conjugate addition product, rather than the direct addition product:

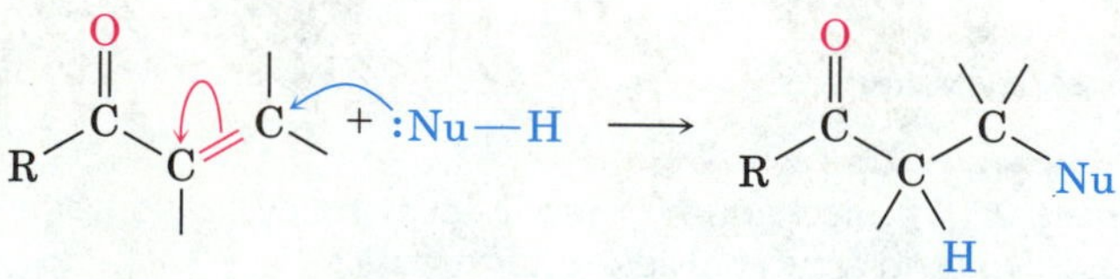

Conjugate addition product

Exactly the same kind of conjugate addition can occur when a nucleophilic enolate ion reacts with an α,β-unsaturated carbonyl compound—a process known as the **Michael[3] reaction**.

The best Michael reactions take place when particularly stable enolate ions, such as those derived from β-keto esters or malonic esters, add to unhindered α,β-unsaturated ketones. For example, ethyl acetoacetate reacts with 3-buten-2-one in the presence of sodium ethoxide catalyst to yield the conjugate addition product.

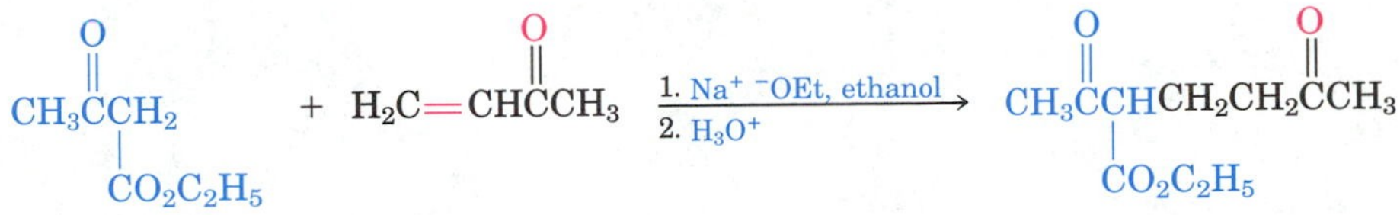

Ethyl acetoacetate **3-Buten-2-one** 94%

Michael reactions take place by addition of a nucleophilic enolate ion donor to the β carbon of an α,β-unsaturated carbonyl acceptor, according to the mechanism shown in Figure 23.6 (p. 916).

The Michael reaction occurs with a wide variety of α,β-unsaturated carbonyl compounds, not just conjugated enones. Conjugated aldehydes, esters, nitriles, amides, and nitro compounds can all act as the electrophilic acceptor component in the Michael reaction (Table 23.1). Similarly, a variety of different donors can be used, including β-diketones, β-keto esters, malonic esters, β-keto nitriles, and nitro compounds.

Table 23.1 Some Michael Acceptors and Michael Donors

Michael acceptors		*Michael donors*	
$H_2C{=}CHCHO$	Propenal	$RCOCH_2COR'$	β-Diketone
$H_2C{=}CHCO_2CH_3$	Methyl propenoate	$RCOCH_2CO_2CH_3$	β-Keto ester
$H_2C{=}CHC{\equiv}N$	Propenenitrile	$CH_3O_2CCH_2CO_2CH_3$	Malonic ester
$H_2C{=}CHCOCH_3$	3-Buten-2-one	$RCOCH_2C{\equiv}N$	β-Keto nitrile
$H_2C{=}CHNO_2$	Nitroethylene	RCH_2NO_2	Nitro compound
$H_2C{=}CHCONH_2$	Propenamide		

[3]Arthur Michael (1853–1942); b. Buffalo, New York; studied Heidelberg, Berlin, École de Médecine, Paris; professor, Tufts University (1882–1889 and 1894–1907), Harvard University (1912–1936).

The base catalyst removes an acidic alpha proton from the starting β-keto ester to generate a stabilized enolate ion nucleophile.

The nucleophile adds to the α,β-unsaturated ketone electrophile in a Michael reaction to generate a new enolate as product.

The enolate product abstracts an acidic proton, either from solvent or from starting keto ester, to yield the final addition product.

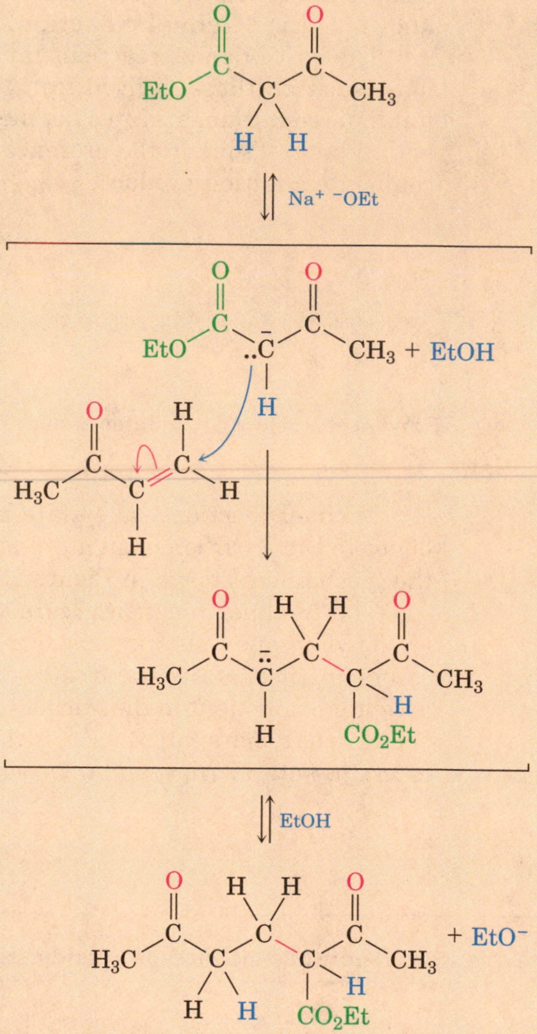

Figure 23.6 Mechanism of the Michael reaction between a β-keto ester and an α,β-unsaturated ketone.

PROBLEM.

23.17 What products would you obtain from base-catalyzed Michael reaction of 2,4-pentanedione with the following α,β-unsaturated acceptors?

(a) 2-Cyclohexenone (b) Propenenitrile (c) Methyl 2-butenoate

PROBLEM.

23.18 What products would you obtain from base-catalyzed Michael reaction of 3-buten-2-one with the following nucleophilic donors?

(a) Diethyl malonate (b) 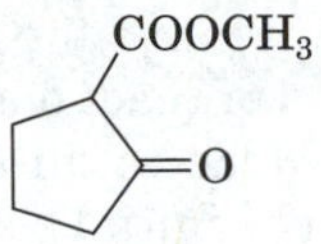 (c) Nitromethane

23.12 The Stork Enamine Reaction

Other kinds of carbon nucleophiles besides enolate ions add to α,β-unsaturated acceptors in the Michael reaction, greatly extending the usefulness and versatility of the process. Among the most important such nucleophiles are *enamines*. Recall from Section 19.12 that enamines are readily prepared by reaction between a ketone and a secondary amine.

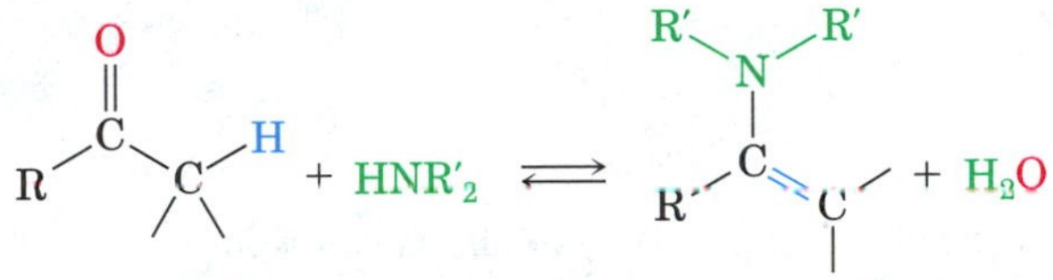

For example:

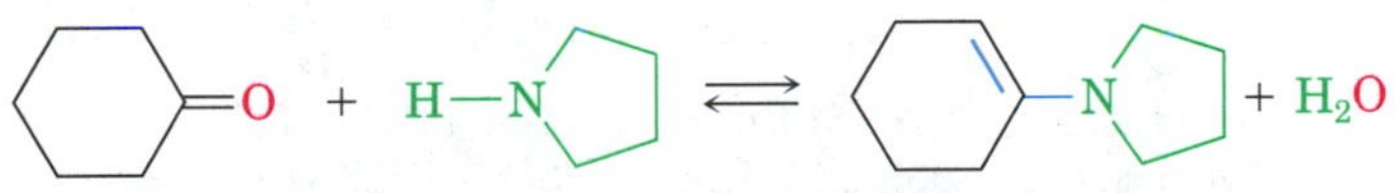

As the following resonance structures indicate, enamines are electronically similar to enolate ions. Overlap of the nitrogen lone-pair orbital with the double-bond p orbitals leads to an increase in electron density on the α carbon atom, making that carbon strongly nucleophilic.

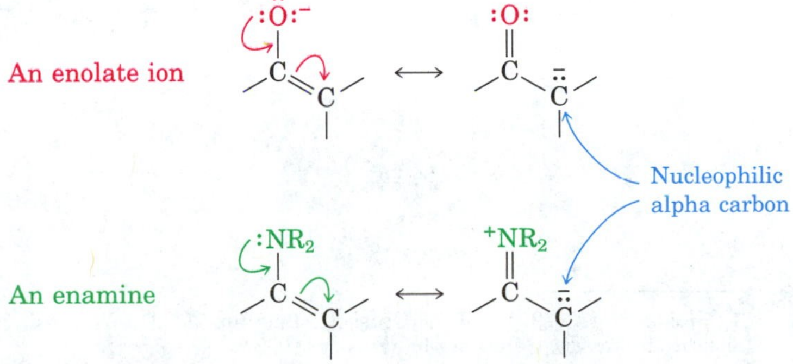

Enamines behave in much the same way as enolate ions and enter into many of the same kinds of reactions. In the **Stork[4] enamine reaction**, for example, an enamine adds to an α,β-unsaturated carbonyl acceptor in a Michael-type process. The initial product is then hydrolyzed by aqueous acid to yield a 1,5-dicarbonyl compound. The overall Stork enamine reaction is a three-step sequence:

1. Enamine formation from a ketone
2. Michael-type addition to an α,β-unsaturated carbonyl compound
3. Enamine hydrolysis back to a ketone

The net effect of the Stork enamine sequence is to carry out a Michael addition of a ketone to an α,β-unsaturated carbonyl compound. For example, cyclohexanone reacts with the cyclic amine pyrrolidine to yield an enamine; further reaction with an enone such as 3-buten-2-one yields a Michael-type adduct; and aqueous hydrolysis completes the sequence to provide a 1,5-diketone product, as shown in Figure 23.7.

PROBLEM.

23.19 Draw the structures of the enamines you would obtain from reaction of pyrrolidine with the following ketones:
(a) Cyclopentanone (b) 2,2-Dimethylcyclohexanone

PROBLEM.

23.20 What products would result from reaction of each enamine you prepared in Problem 23.19 with the following α,β-unsaturated acceptors? (Assume that the initial product is hydrolyzed.)
(a) Ethyl propenoate (b) Propenal (acrolein)

PROBLEM.

23.21 Show how you might use an enamine reaction to prepare the following compounds:

(a)

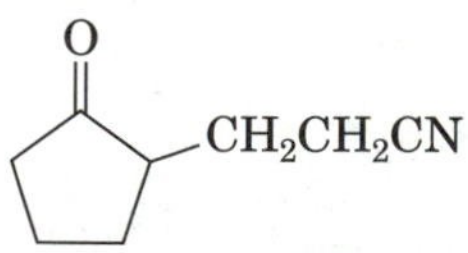

(b)

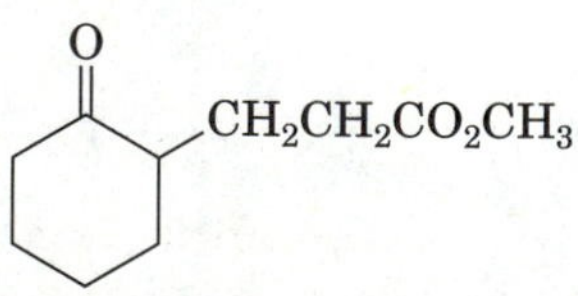

[4] Gilbert Stork (1921–); b. Brussels, Belgium; Ph.D. Wisconsin (McElvain); professor, Harvard University, Columbia University (1953–).

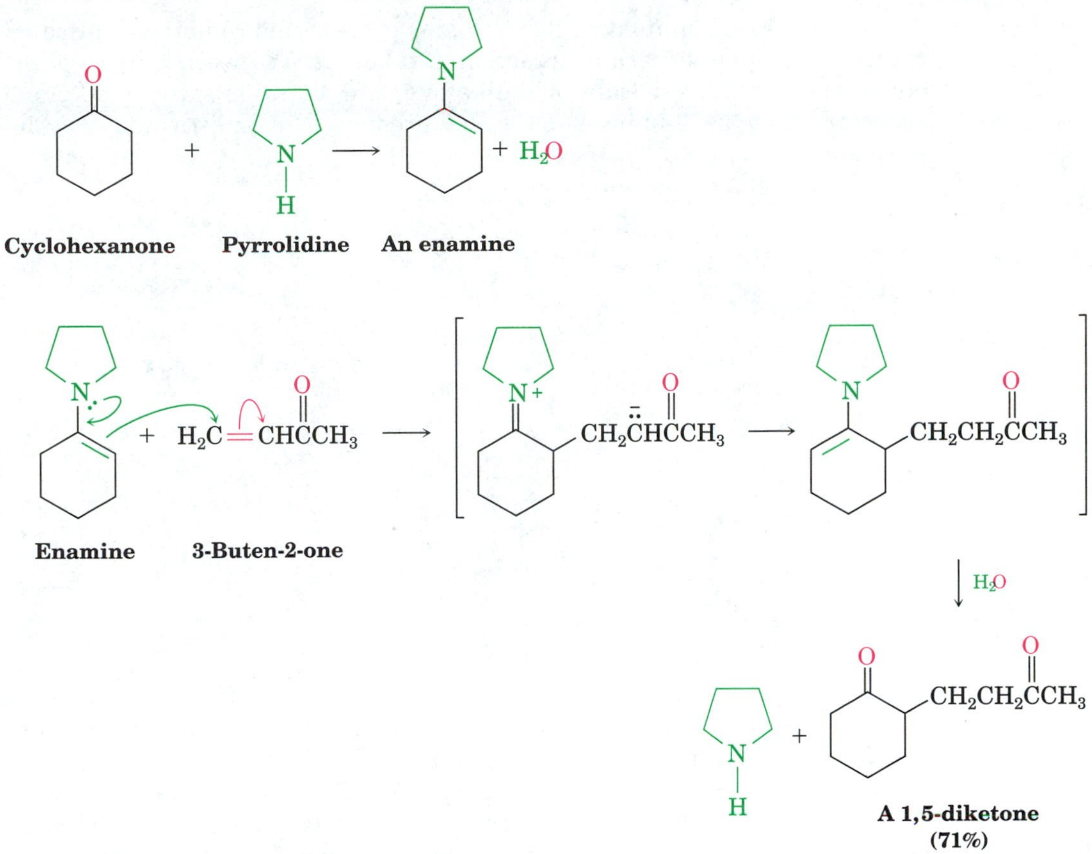

Figure 23.7 A Stork enamine reaction between cyclohexanone and 3-buten-2-one. Cyclohexanone is first converted into an enamine; the enamine then adds to the α,β-unsaturated ketone in a Michael reaction; and the initial product is hydrolyzed to yield a 1,5-diketone.

23.13 Carbonyl Condensation Reactions in Synthesis: The Robinson Annulation Reaction

Carbonyl condensation reactions are among the most valuable methods available for synthesizing complex molecules. By putting a few fundamental reactions together in the proper sequence, some remarkably useful transformations can be carried out. One such example is the **Robinson[5] annulation reaction**, used for the synthesis of polycyclic molecules. (An *annulation*

[5]Sir Robert Robinson (1886–1975); b. Rufford/Chesterfield, England; D.Sc. Manchester (Perkin); professor, Sydney, Liverpool, Manchester (1922–1928), University College London, Oxford (1930–1955); Nobel Prize (1947).

reaction, from the Latin *annulus*, meaning "ring," is one that builds a new ring onto a molecule.)

The Robinson annulation is a two-step process that combines a Michael reaction with an internal aldol reaction. It takes place between a nucleophilic donor, such as a β-keto ester or β-diketone, and an α,β-unsaturated ketone acceptor, such as 3-buten-2-one. The product is a substituted 2-cyclohexenone.

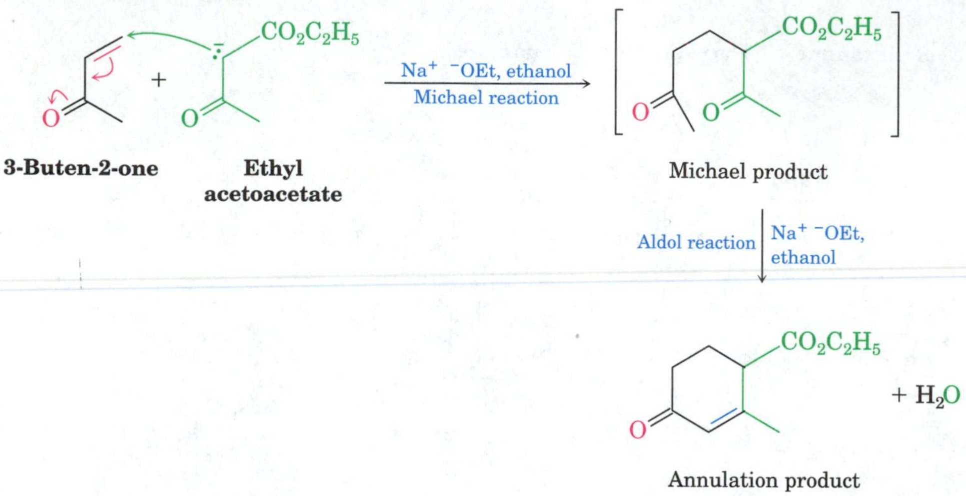

The first step of the Robinson annulation is simply a Michael reaction: An enolate ion from a β-diketone or β-keto ester effects a conjugate addition to an α,β-unsaturated ketone, yielding a 1,5-diketone. But as we saw in Section 23.7, 1,5-diketones undergo intramolecular aldol condensation to yield cyclohexenones when treated with base. Thus, if base treatment of the initial Michael addition product is continued at a higher temperature, the final product contains a six-membered ring, and an annulation has been accomplished. An excellent example occurs during the commercial synthesis of the female steroid hormone estrone (Figure 23.8).

In this example, 2-methyl-1,3-cyclopentanedione (a β-diketone) is used to generate the enolate ion required for Michael reaction, and an aryl-substituted α,β-unsaturated ketone is used as the acceptor. Base-catalyzed Michael reaction between the two yields an intermediate triketone, which cyclizes in an internal aldol condensation to give a Robinson annulation product. Several further transformations are then required to complete the synthesis of estrone.

PROBLEM

23.22 What product would you expect from a Robinson annulation reaction of 2-methyl-1,3-cyclopentanedione with 3-buten-2-one?

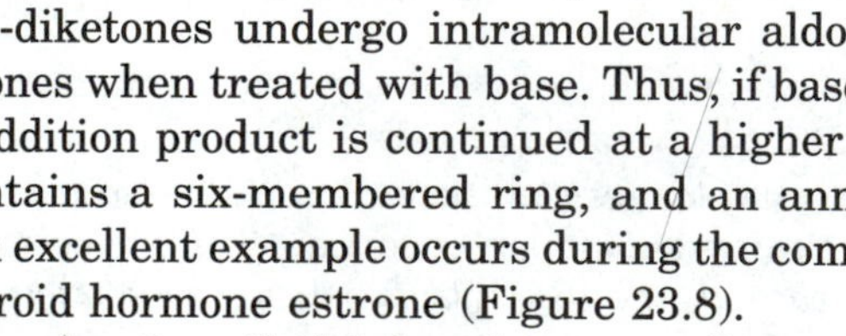

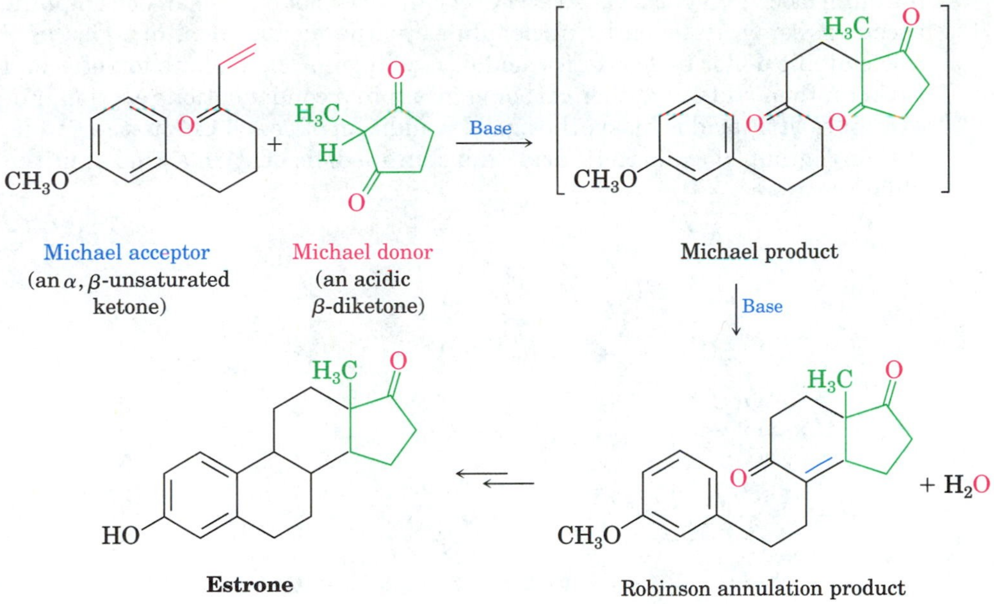

Figure 23.8 A Robinson annulation reaction, used in the commercial synthesis of the steroid hormone estrone.

PROBLEM

23.23 How would you prepare the following compound using a Robinson annulation reaction between a β-diketone and an α,β-unsaturated ketone? Draw the structures of both reactants and the intermediate Michael addition product.

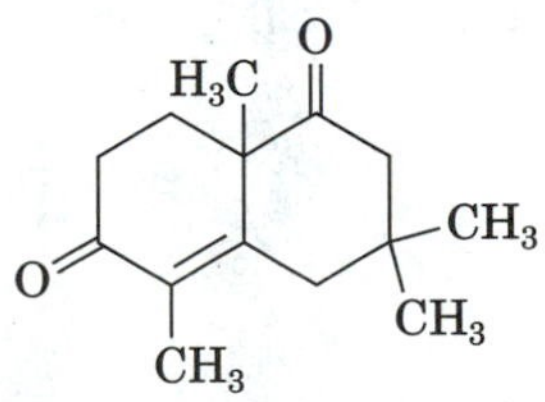

23.14 Biological Carbonyl Condensation Reactions

Carbonyl condensation reactions are used in nature for the biological synthesis of a great many different molecules. Fats, amino acids, steroid hormones, and many other kinds of compounds are synthesized by plants and animals using carbonyl condensation reactions as the key step.

Nature uses the two-carbon acetate fragment of acetyl CoA as the major building block for synthesis. Acetyl CoA can serve not only as an electrophilic acceptor, being attacked by nucleophiles at the acyl carbon, but also as a nucleophilic donor by loss of its acidic α hydrogen. The enolate ion of acetyl CoA can then add to another carbonyl group in a condensation reaction. For example, citric acid is biosynthesized by addition of acetyl CoA to the ketone carbonyl group of oxaloacetic acid (2-oxobutanedioic acid) in a kind of mixed aldol reaction.

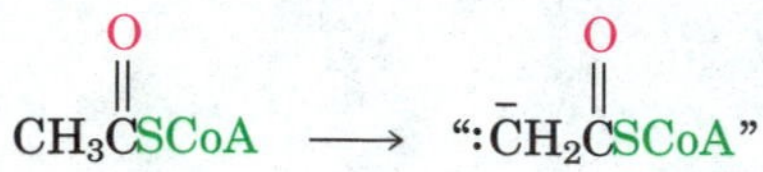

Acetyl CoA, a thiol ester

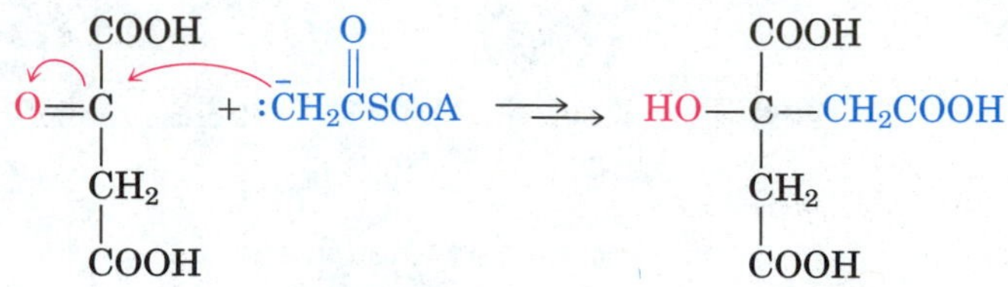

Oxaloacetic acid **Citric acid**

Acetyl CoA is also involved as a primary building block in the biosynthesis of steroids, fats, and other lipids, where the key step is a Claisen-like condensation reaction. We'll go into more of the details of this process in Section 30.7.

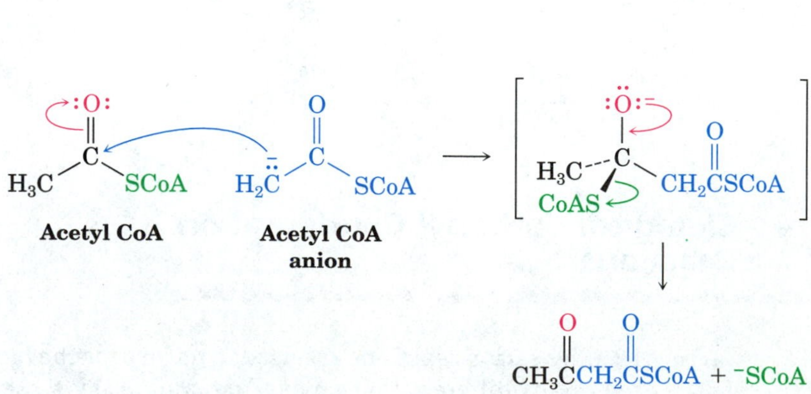

Acetyl CoA **Acetyl CoA anion**

Acetoacetyl CoA

INTERLUDE

A Prologue to Metabolism

You are what you eat. All classes of food molecules are metabolized by pathways that involve the four major carbonyl-group reactions.

Biochemistry *is* carbonyl chemistry. Almost all metabolic processes used by all living organisms involve one or more of the four fundamental carbonyl-group reactions. The digestion and metabolic breakdown of all the major classes of food molecules—fats, carbohydrates, and proteins—take place by nucleophilic addition reactions, nucleophilic acyl substitutions, α substitutions, and carbonyl condensations. Similarly, hormones and other crucial biological molecules are built up from smaller precursors by these same carbonyl-group reactions.

Take *glycolysis*, for example, the metabolic pathway by which organisms convert glucose to pyruvate as the first step in extracting energy from carbohydrates.

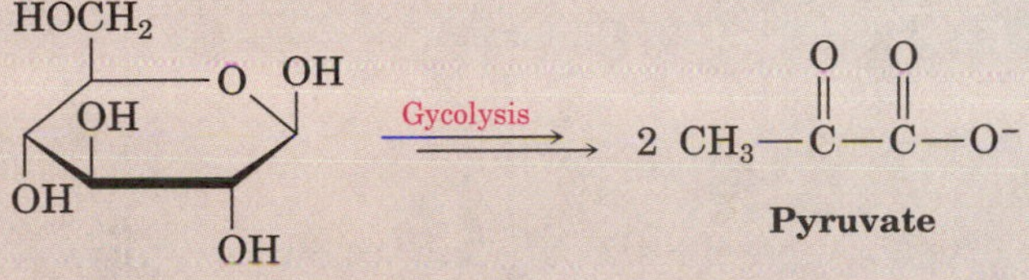

Glucose

(continued)➤

Glycolysis is a ten-step process that begins with conversion of glucose from its cyclic hemiacetal form to its open-chain aldehyde form—a reverse nucleophilic addition reaction. The aldehyde then undergoes tautomerism to yield an enol, which undergoes yet another tautomerism to give the ketone fructose.

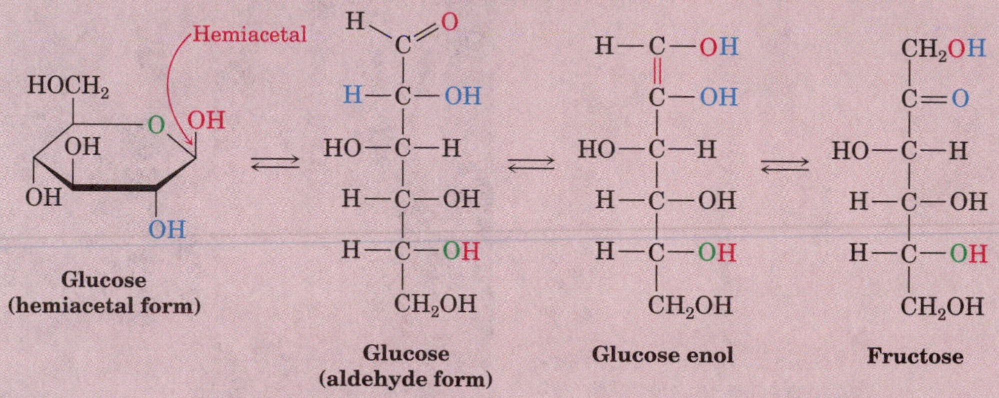

Fructose, a β-hydroxy ketone, is then cleaved into two three-carbon molecules—one ketone and one aldehyde—by a reverse aldol reaction. Still further carbonyl-group reactions then occur until pyruvate results.

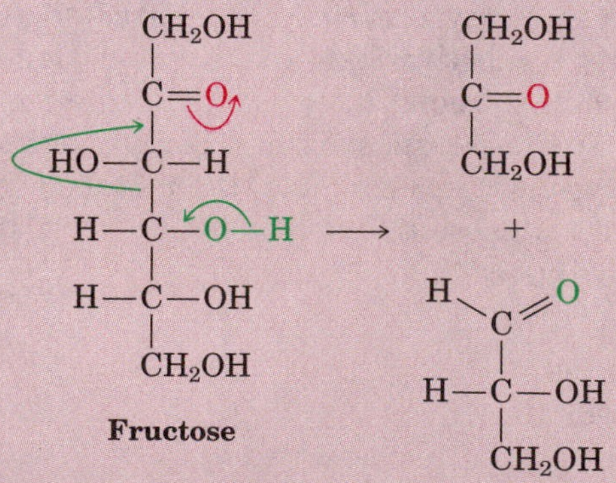

The few examples just given are only an introduction; we'll look at several of the major metabolic pathways in much more detail in Chapter 30. You haven't seen the end of carbonyl-group chemistry, though. A good grasp of carbonyl-group reactions is crucial to an understanding of biochemistry.

Summary and Key Words

A **carbonyl condensation reaction** takes place between two carbonyl components and involves a combination of nucleophilic addition and α-substitution steps. One carbonyl component (the donor) is converted by base into a nucleophilic enolate ion, which adds to the electrophilic carbonyl group of the second component (the acceptor). The donor molecule undergoes an α substitution, while the acceptor molecule undergoes a nucleophilic addition.

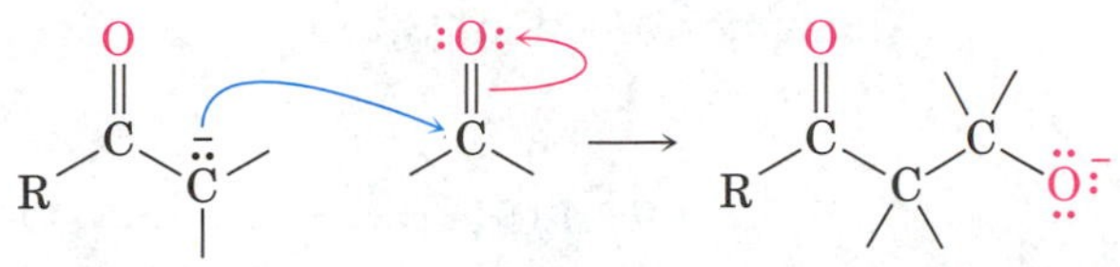

The **aldol reaction** is a carbonyl condensation that occurs between two ketones or aldehydes. Aldol reactions are reversible, leading first to β-hydroxy ketones/aldehydes and then to α,β-unsaturated ketones. Mixed aldol condensations between two different ketones generally give a mixture of all four possible products. A mixed reaction can be successful, however, if one of the two components is an unusually good donor (as with ethyl acetoacetate) or if it can act only as an acceptor (as with formaldehyde and benzaldehyde). Internal aldol condensations of 1,4- and 1,5-diketones are also successful and provide a good way to make five- and six-membered rings.

The **Claisen reaction** is a carbonyl condensation that occurs between two ester components and leads to a β-keto ester product. Mixed Claisen condensations between two different esters are successful only when one of the two components has no acidic α hydrogens (as with ethyl benzoate and ethyl formate) and thus can function only as the acceptor component. Internal Claisen condensations, called **Dieckmann cyclization reactions**, provide excellent syntheses of five- and six-membered cyclic β-keto esters starting from 1,6- and 1,7-diesters.

The conjugate addition of a carbon nucleophile to an α,β-unsaturated acceptor is known as the **Michael reaction**. The best Michael reactions take place between unusually acidic donors (β-keto esters or β-diketones) and unhindered α,β-unsaturated acceptors. Enamines, prepared by reaction of a ketone with a disubstituted amine, are also good Michael donors.

Carbonyl condensation reactions are widely used in synthesis. One example of their versatility is the **Robinson annulation reaction**, which leads to the formation of a substituted cyclohexenone. Treatment of a β-diketone or β-keto ester with an α,β-unsaturated ketone leads first to a Michael addition, which is followed by internal aldol cyclization. Condensation reactions are also used widely in nature for the biosynthesis of such molecules as fats and steroids.

Summary of Reactions

1. Aldol reaction: a condensation between two ketones, two aldehydes, or one ketone and one aldehyde

(a) Ketones (Section 23.2)

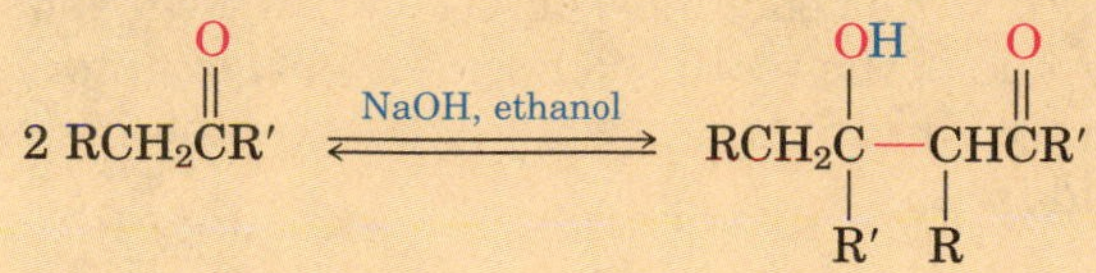

(b) Aldehydes (Section 23.2)

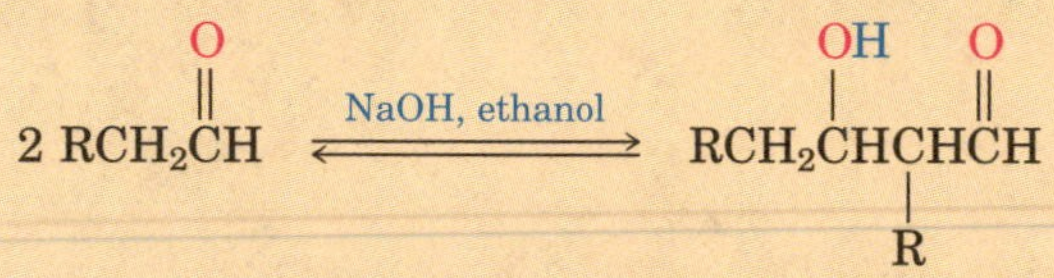

(c) Mixed aldol reaction (Section 23.6)

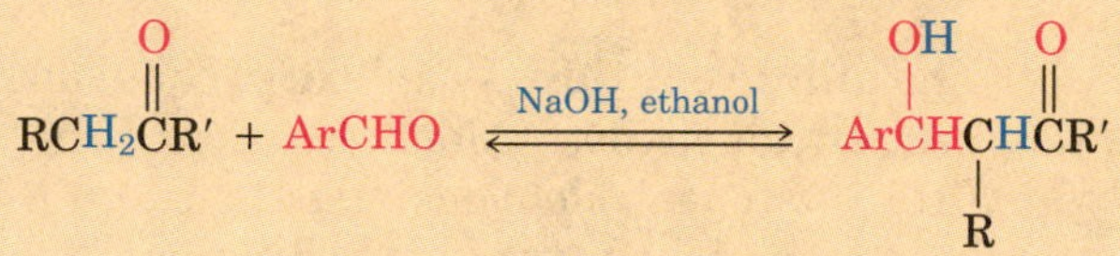

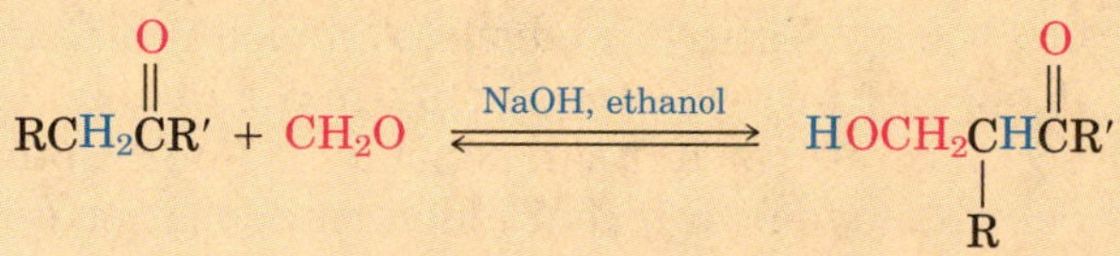

(d) Intramolecular aldol reaction (Section 23.7)

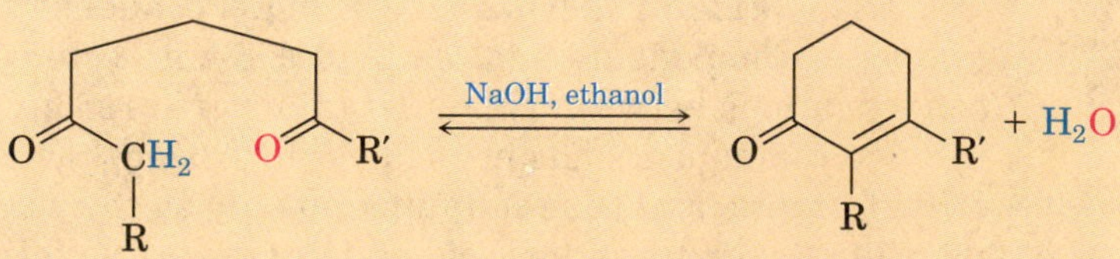

2. Dehydration of aldol products (Section 23.4)

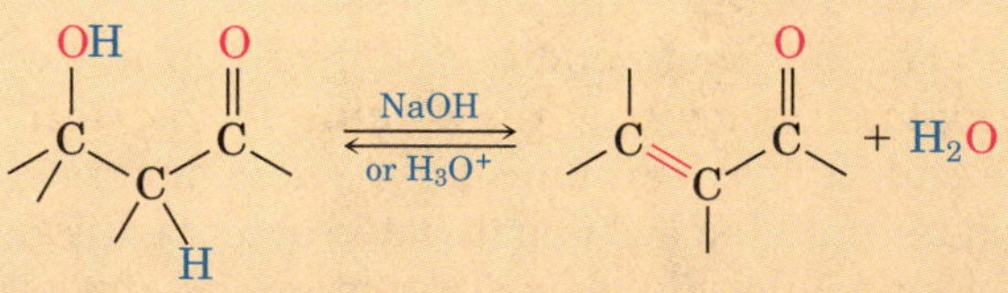

3. Claisen reaction: the condensation between two esters or between one ester and one ketone (Section 23.8)

(a) 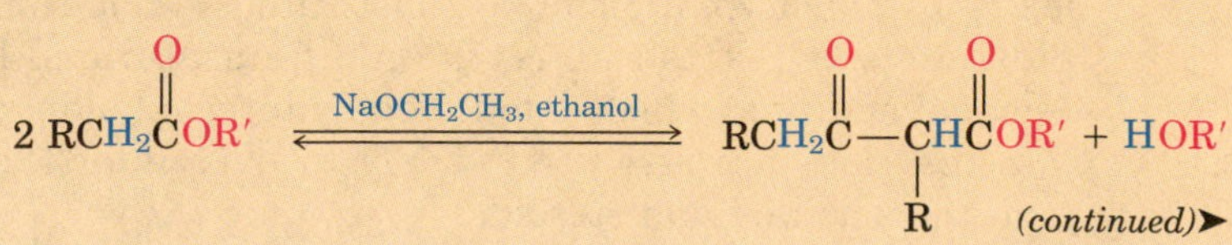

(continued)➤

(b) Mixed Claisen reaction (Section 23.9)

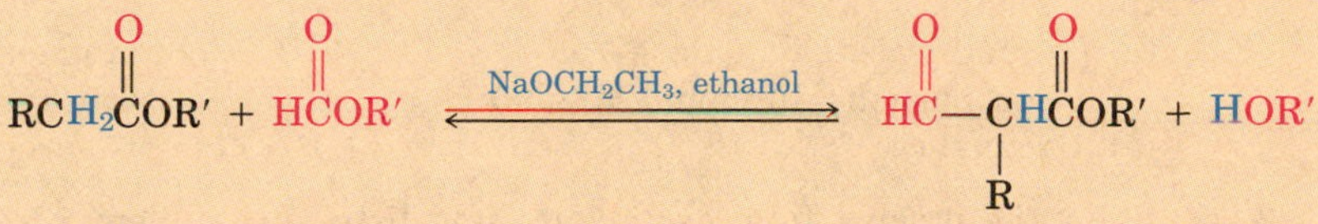

4. Dieckmann cyclization; internal Claisen condensation (Section 23.10)

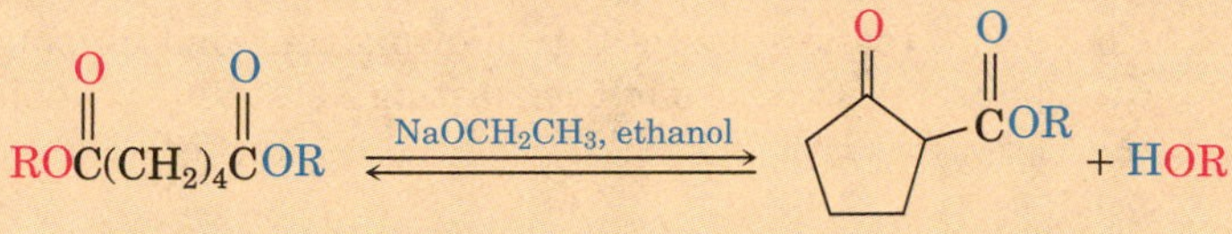

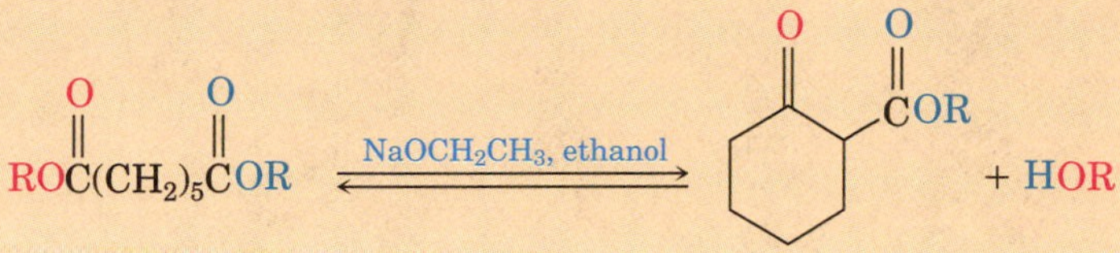

5. Michael reaction (Section 23.11)

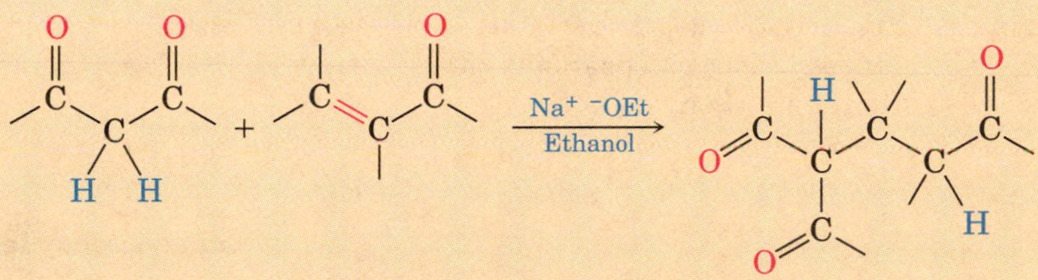

6. Enamine reaction (Section 23.12)

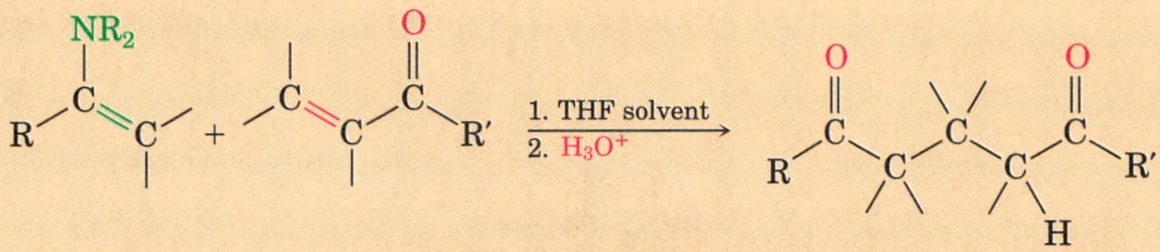

7. Robinson annulation reaction (Section 23.13)

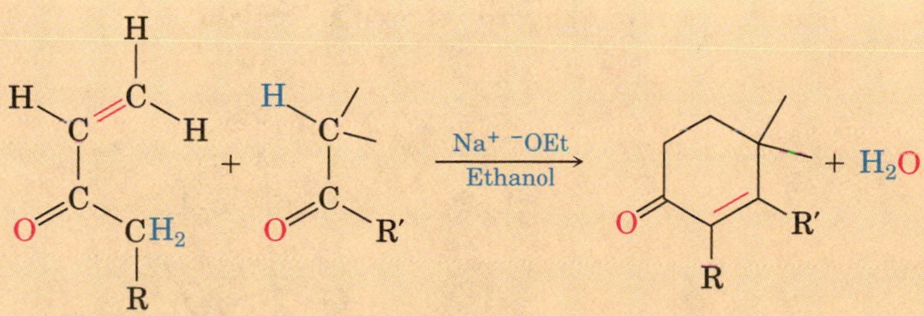

ADDITIONAL PROBLEMS

23.24 Which of the following compounds would you expect to undergo aldol self-condensation?
(a) Trimethylacetaldehyde (b) Cyclobutanone
(c) Benzophenone (diphenyl ketone) (d) 3-Pentanone
(e) Decanal (f) 3-Phenyl-2-propenal

23.25 Show the product from each compound listed in Problem 23.24 that is capable of undergoing the aldol reaction.

23.26 What product would you expect to obtain from aldol cyclization of hexanedial ($OHCCH_2CH_2CH_2CH_2CHO$)?

23.27 How might you synthesize the following compounds using aldol reactions? In each case, show the structure of the starting ketone(s) or aldehyde(s) you would use.
(a) $C_6H_5CH{=}CHCOC_6H_5$ (b) 2-Cyclohexenone

(c) 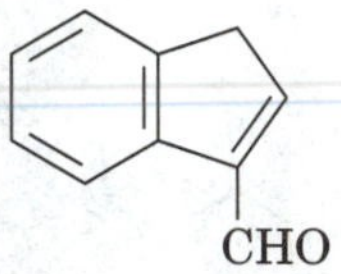

(d) 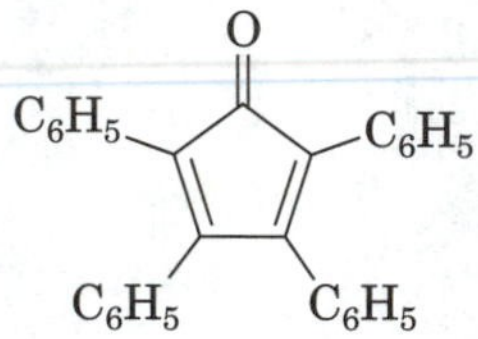

23.28 How can you account for the fact that 2,2,6-trimethylcyclohexanone yields no detectable aldol product even though it has an acidic α hydrogen?

23.29 Cinnamaldehyde, the aromatic constituent of cinnamon oil, can be synthesized by a mixed aldol condensation. Show the starting materials you would use, and formulate the reaction.

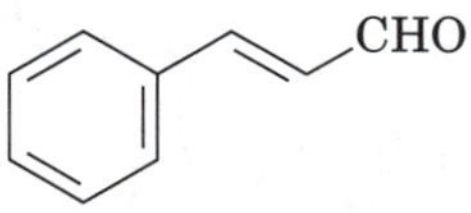

Cinnamaldehyde

23.30 The so-called Wieland–Miescher ketone is a valuable starting material used in the synthesis of steroid hormones. How might you prepare it from 1,3-cyclohexanedione?

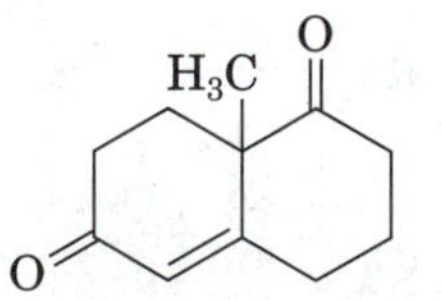

Wieland–Miescher ketone

23.31 The bicyclic ketone shown below does not undergo aldol self-condensation even though it has two α hydrogen atoms. Explain.

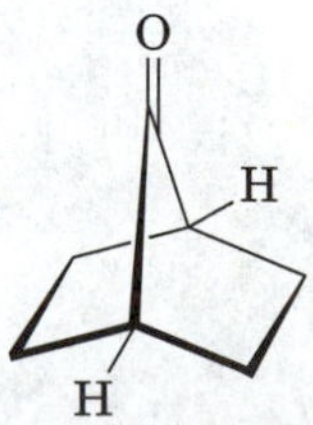

23.32 What condensation products would you expect to obtain by treatment of the following substances with sodium ethoxide in ethanol?

(a) Ethyl butanoate
(b) Cycloheptanone
(c) 3,7-Nonanedione
(d) 3-Phenylpropanal

23.33 Give the structures of the possible Claisen condensation products from the following reactions. Tell which, if any, you would expect to predominate in each case.

(a) $CH_3CO_2CH_3 + CH_3CH_2CO_2CH_3$
(b) $C_6H_5CO_2CH_3 + C_6H_5CH_2CO_2CH_3$
(c) $CH_3OCO_2CH_3$ + Cyclohexanone
(d) $C_6H_5CHO + CH_3CO_2CH_3$

23.34 The following reactions are unlikely to provide the desired product in high yield. What is wrong with each?

(a) $CH_3CHO + CH_3COCH_3 \xrightarrow[\text{Ethanol}]{Na^+\ ^-OC_2H_5} CH_3CH(OH)CH_2COCH_3$

(b) $CH_2(CO_2C_2H_5)_2 + H_2C{=}CHCOCH_3 \xrightarrow{\text{Base}} H_2C{=}CHC(OH)(CH_3){-}CH(CO_2C_2H_5)_2$

(c)

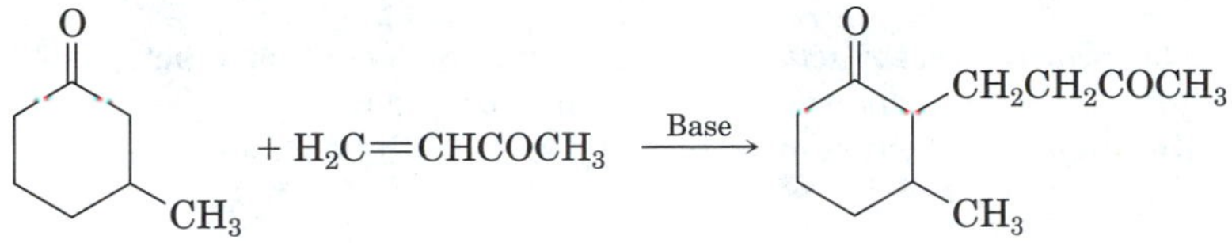

(d)

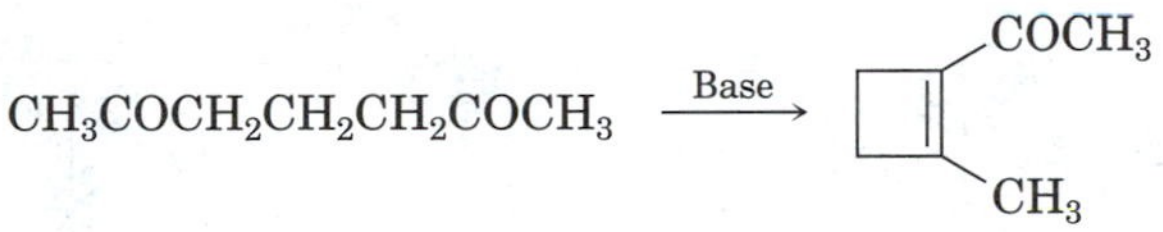

(e)

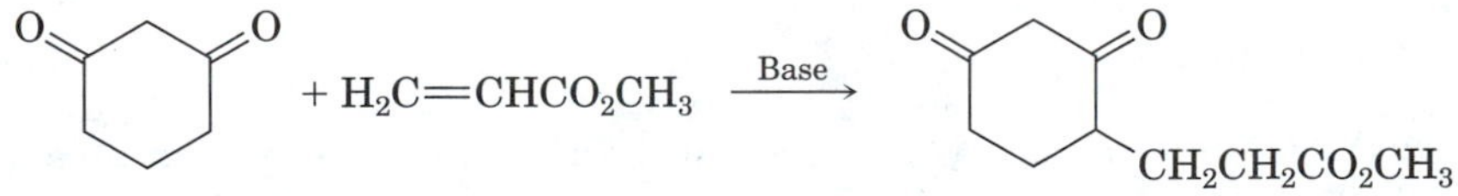

23.35 How might the following compounds be prepared using Michael reactions? Show the nucleophilic donor and the electrophilic acceptor in each case.

(a) $CH_3COCH(CO_2CH_3)CH_2CH_2COC_6H_5$

(b) $CH_3COCH_2CH_2CH_2COCH_3$

(c) $(CH_3O_2C)_2CHCH_2CH_2CN$

(d) $CH_3CH(NO_2)CH_2CH_2CO_2CH_3$

(e) $(CH_3O_2C)_2CHCH_2CH_2NO_2$

23.36 Fill in the missing reagents a–h in the following scheme:

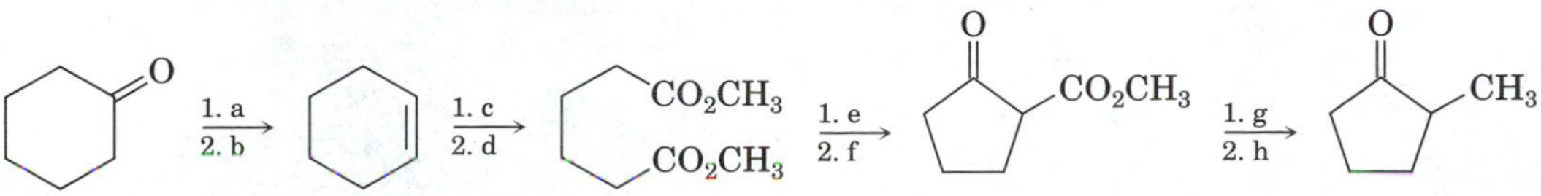

23.37 In the mixed Claisen reaction of cyclopentanone with ethyl formate, a much higher yield of the desired product is obtained by first mixing the two carbonyl components and then adding base, rather than by first mixing base with cyclopentanone and then adding ethyl formate. Explain.

23.38 Ethyl dimethylacetoacetate reacts instantly at room temperature when treated with ethoxide ion to yield two products, ethyl acetate and ethyl 2-methylpropanoate. Propose a mechanism for this cleavage reaction.

$$CH_3\overset{O}{\overset{\|}{C}}C(CH_3)_2\overset{O}{\overset{\|}{C}}OCH_2CH_3 \xrightarrow[\text{Ethanol, 25°C}]{Na^+\ ^-OEt} CH_3COOCH_2CH_3 + (CH_3)_2CH\overset{O}{\overset{\|}{C}}OCH_2CH_3$$

23.39 In contrast to the rapid reaction shown in Problem 23.38, ethyl acetoacetate requires temperatures over 150°C to undergo the same kind of cleavage reaction. How can you explain the difference in reactivity?

$$CH_3\overset{O}{\overset{\|}{C}}CH_2\overset{O}{\overset{\|}{C}}OCH_2CH_3 \xrightarrow[\text{Ethanol, 150°C}]{Na^+\ ^-OEt} 2\ CH_3\overset{O}{\overset{\|}{C}}OCH_2CH_3$$

23.40 The *Knoevenagel reaction* is a carbonyl condensation reaction of an ester with a ketone or aldehyde to yield an α,β-unsaturated product. Show the mechanism of the Knoevenagel reaction of diethyl malonate with benzaldehyde.

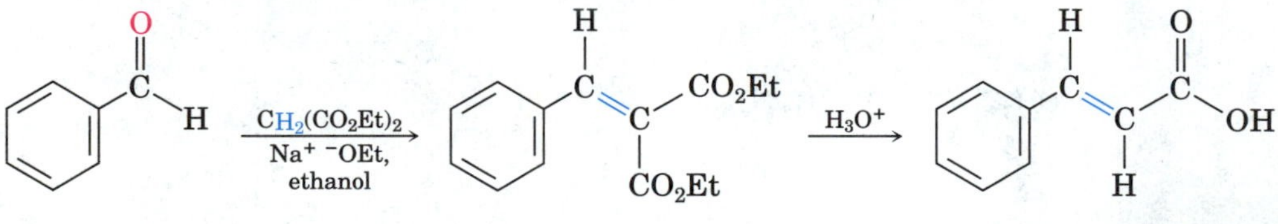

Benzaldehyde **Cinnamic acid (91%)**

23.41 In the *Perkin reaction*, acetic anhydride condenses with an aromatic aldehyde to yield a cinnamic acid. The reaction takes place by a mixed carbonyl condensation of the anhydride with the aldehyde to yield an α,β-unsaturated intermediate that undergoes hydrolysis to yield the cinnamic acid. What is the structure of the unsaturated intermediate?

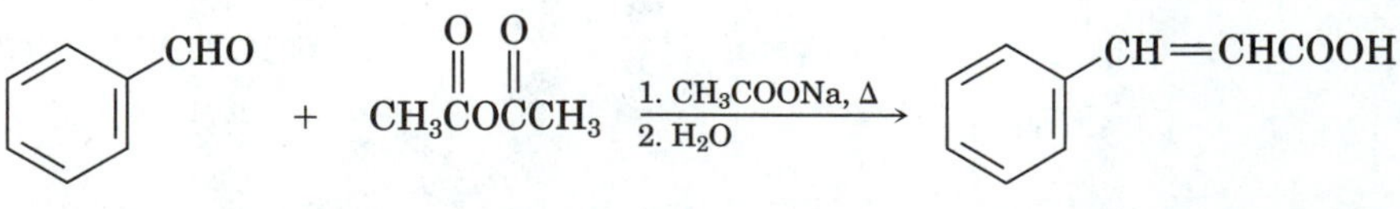

Benzaldehyde **Acetic anhydride** **Cinnamic acid (64%)**

23.42 The *Darzens reaction* involves a two-step, base-catalyzed condensation of ethyl chloroacetate with a ketone to yield an epoxy ester. The first step is a carbonyl condensation reaction, and the second step is an S_N2 reaction. Show the mechanism by which this reaction occurs.

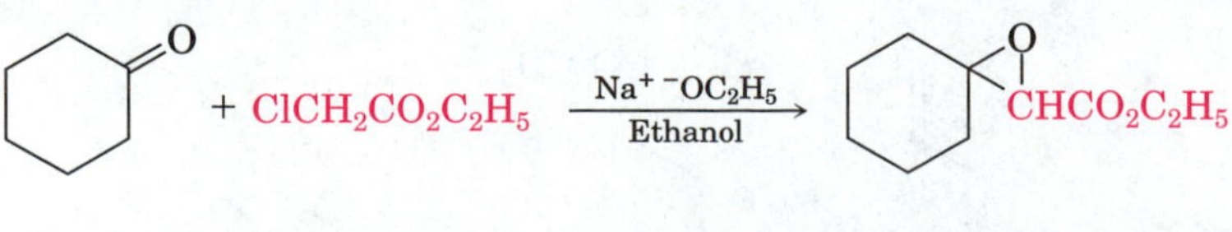

23.43 How would you prepare the following compounds from cyclohexanone?

(a)
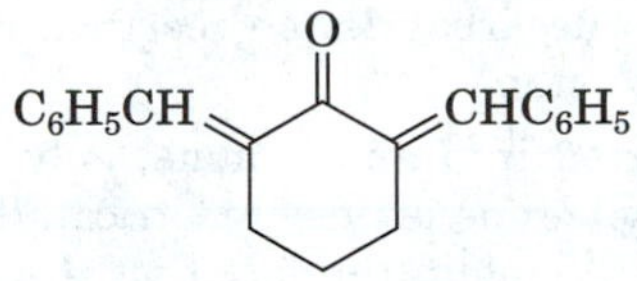

(b)
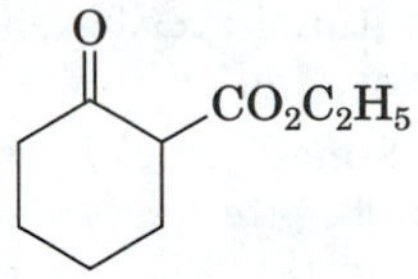

(c)
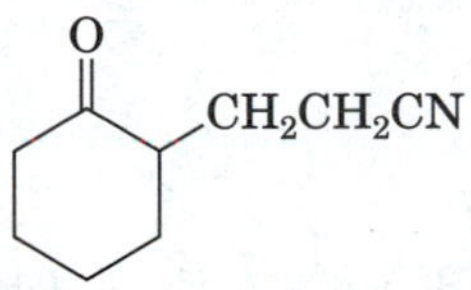

(d)
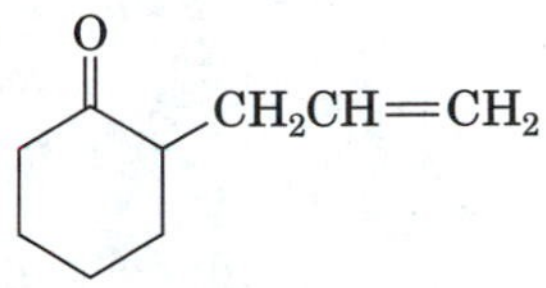

23.44 Griseofulvin, an antibiotic produced by the mold *Penicillium griseofulvum* (Dierckx), has been synthesized by a route that employs a twofold Michael reaction as the key step. Propose a mechanism for this transformation.

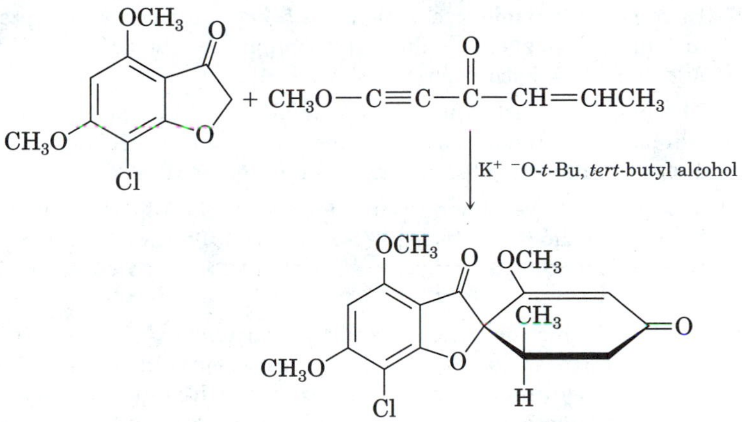

Griseofulvin (10%)

23.45 The compound known as *Hagemann's ester* is prepared by treatment of a mixture of formaldehyde and ethyl acetoacetate with base, followed by acid-catalyzed decarboxylation.

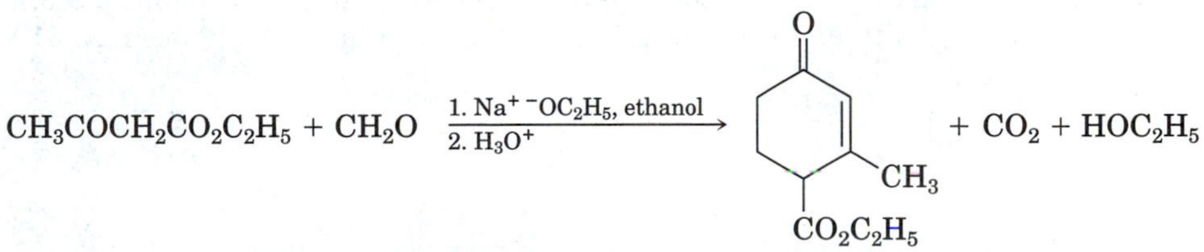

Hagemann's ester

(a) The first step is an aldol-like condensation between ethyl acetoacetate and formaldehyde to yield an α,β-unsaturated product. Write the reaction, and show the structure of the product.

(b) The second step is a Michael reaction between ethyl acetoacetate and the unsaturated product of the first step. Show the structure of the product.

23.46 The third and fourth steps in the synthesis of Hagemann's ester from ethyl acetoacetate and formaldehyde (Problem 23.45) are an internal aldol cyclization to yield a substituted cyclohexenone ring, and a decarboxylation reaction. Formulate both reactions, and write the products of each step.

23.47 When 2-methylcyclohexanone is converted into an enamine, only one product is formed despite the fact that the starting ketone is unsymmetrical. Build molecular models of the two possible products, and explain the fact that the sole product is the one with the double bond away from the methyl-substituted carbon.

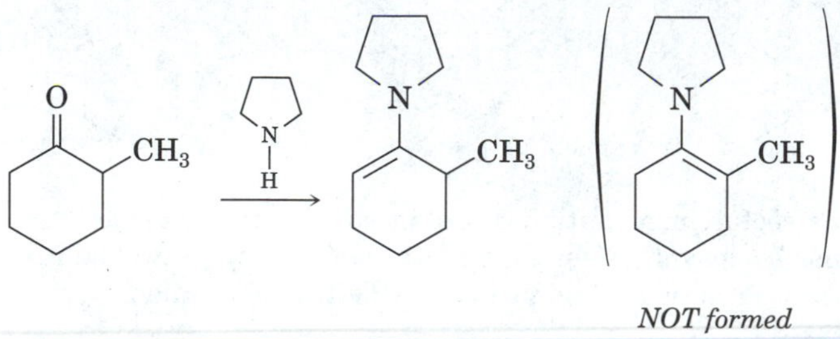

23.48 Intramolecular aldol cyclization of 2,5-heptanedione with aqueous NaOH yields a mixture of two enone products in the approximate ratio 9:1. Write their structures, and show how each is formed.

23.49 The major product formed by intramolecular aldol cyclization of 2,5-heptanedione (Problem 23.48) has two singlet absorptions in the ^{1}H NMR spectrum at 1.65 δ and 1.90 δ, and has no absorptions in the range 3–10 δ. What is the structure?

23.50 Treatment of the minor product formed in the intramolecular aldol cyclization of 2,5-heptanedione (Problems 23.48 and 23.49) with aqueous NaOH converts it into the major product. Propose a mechanism to account for this base-catalyzed isomerization.

23.51 The Stork enamine reaction and the intramolecular aldol reaction can be carried out in sequence to allow the synthesis of cyclohexenone rings. For example, reaction of the pyrrolidine enamine of cyclohexanone with 3-buten-2-one, followed by enamine hydrolysis and base treatment, yields the product indicated. Show the mechanisms of the three steps.

23.52 How could you prepare the following cyclohexenones by combining a Stork enamine reaction with an intramolecular aldol condensation? (See Problem 23.51.)

(a)

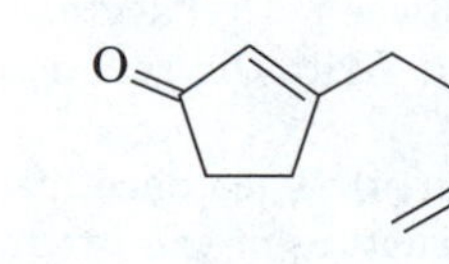

(b)

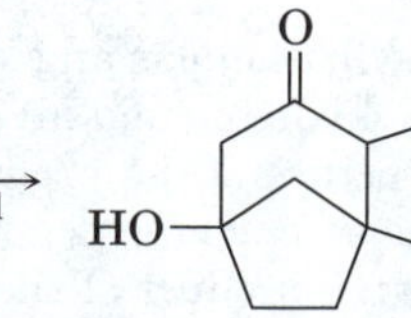

23.53 Propose a mechanism to account for the following reaction:

23.54 Propose a mechanism to account for the following reaction:

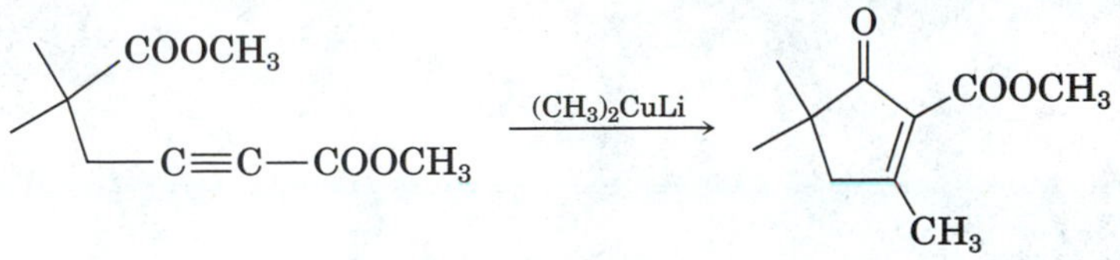

23.55 Propose a mechanism to account for the following reaction:

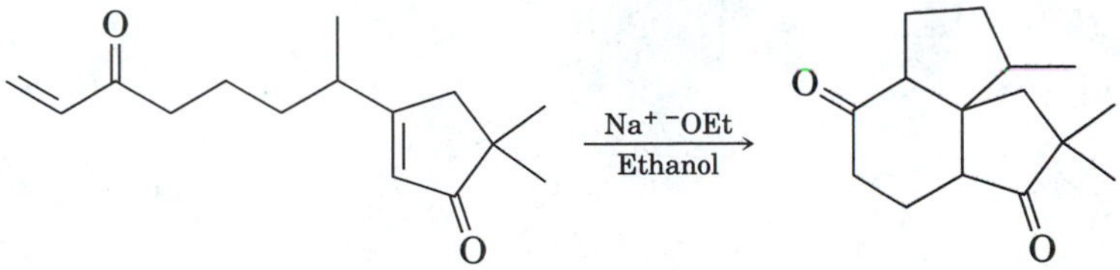

23.56 Propose a mechanism to account for the following reaction:

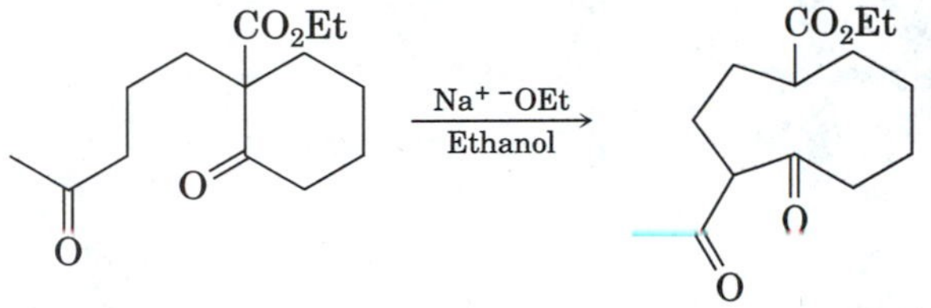

A Look Ahead

23.57 The *Mannich reaction* of a ketone, an amine, and an aldehyde is one of the few three-component reactions in organic chemistry. Cyclohexanone, for example, reacts with dimethylamine and acetaldehyde to yield an amino ketone:

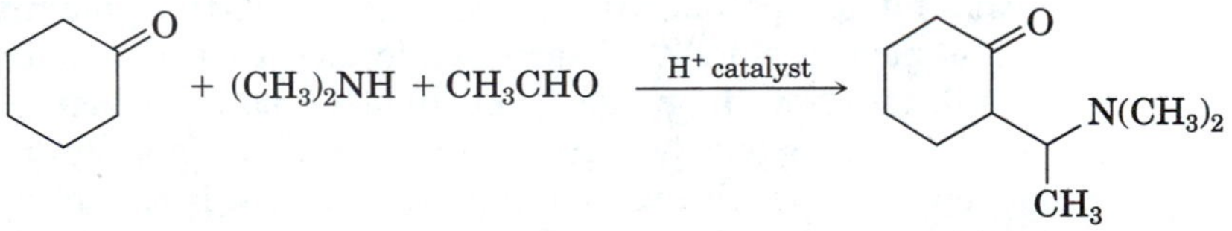

The reaction takes place in two steps, both of which are typical carbonyl-group reactions. The first step is the reaction between the aldehyde and the amine to yield an intermediate iminium ion ($R_2C{=}NR_2^+$) plus water. The second step is the reaction between the iminium ion intermediate and the ketone to yield the final product. Propose mechanisms for both steps, and show the structure of the intermediate iminium ion.

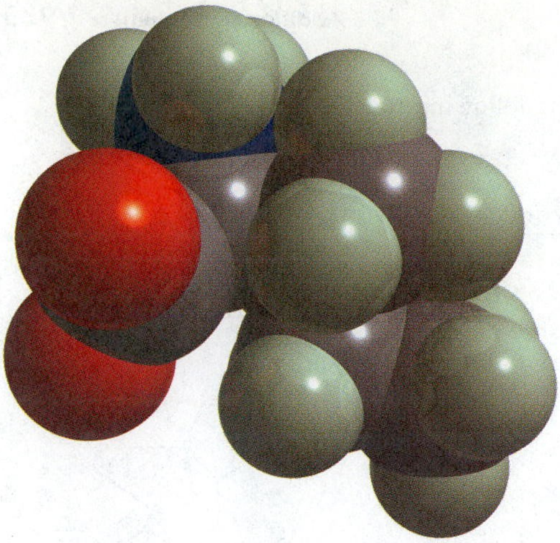

Amino acids, such as isoleucine, contain a basic amine functional group.

24

Aliphatic Amines

Amines are organic derivatives of ammonia, NH_3, in the same way that alcohols and ethers are organic derivatives of water. Like ammonia, amines contain a nitrogen atom with a lone pair of electrons, making amines both basic and nucleophilic. We'll soon see, in fact, that most of the chemistry of amines depends on the presence of this lone pair of electrons.

Amines occur widely throughout both plants and animals. Trimethylamine, for instance, occurs in animal tissues and is partially responsible for the distinctive odor of many fish; quinine is an important antimalarial drug isolated from the bark of the South American *Cinchona* tree; and codeine is an analgesic (painkiller) found in the opium poppy.

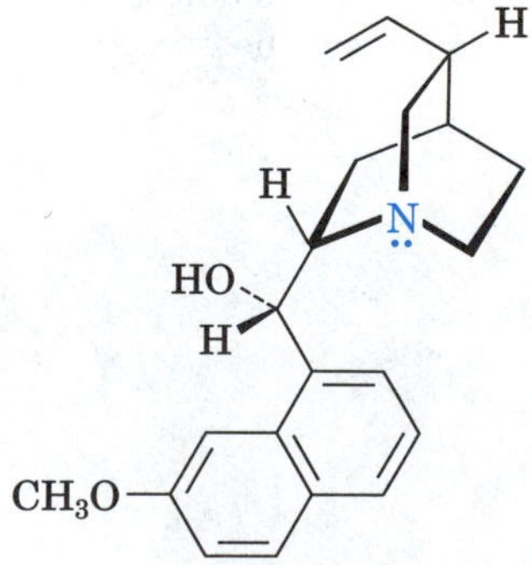

Quinine—an antimalarial

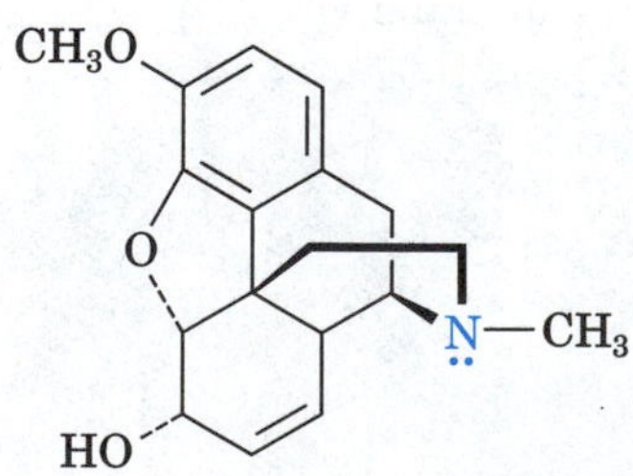

Codeine—an analgesic

24.1 Naming Amines

Amines are classified as **primary** (RNH_2), **secondary** (R_2NH), or **tertiary** (R_3N), depending on the number of organic substituents attached to nitrogen. For example, methylamine (CH_3NH_2) is a primary amine, dimethylamine [$(CH_3)_2NH$] is a secondary amine, and trimethylamine [$(CH_3)_3N$] is a tertiary amine. Note that this usage of the terms *primary*, *secondary*, and *tertiary* is different from our previous usage. When we speak of a tertiary alcohol or alkyl halide, we refer to the degree of substitution at the alkyl carbon atom, but when we speak of a tertiary amine, we refer to the degree of substitution at the nitrogen atom.

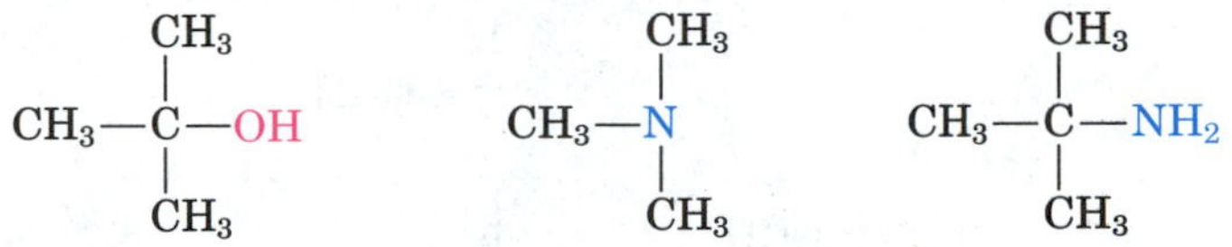

***tert*-Butyl alcohol (a tertiary alcohol)** **Trimethylamine (a tertiary amine)** ***tert*-Butylamine (a primary amine)**

Compounds containing a nitrogen atom with four attached groups are also possible, but the nitrogen atom must carry a positive charge. Such compounds are called **quaternary ammonium salts.**

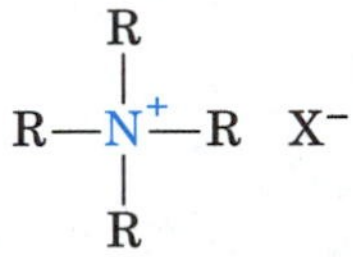

A quaternary ammonium salt

Amines can be either alkyl-substituted or aryl-substituted. Much of the chemistry of the two classes is similar, but there are sufficient differences that we'll consider them separately. Arylamines will be discussed in Chapter 25.

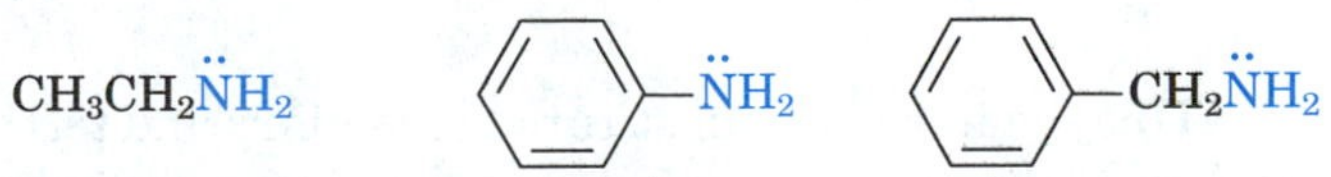

Ethylamine (an aliphatic amine) **Aniline (an arylamine)** **Benzylamine (an aliphatic amine)**

Primary amines (RNH_2) are named in the IUPAC system in several ways, depending on their structure. For simple amines, the suffix *-amine* is added to the name of the alkyl substituent:

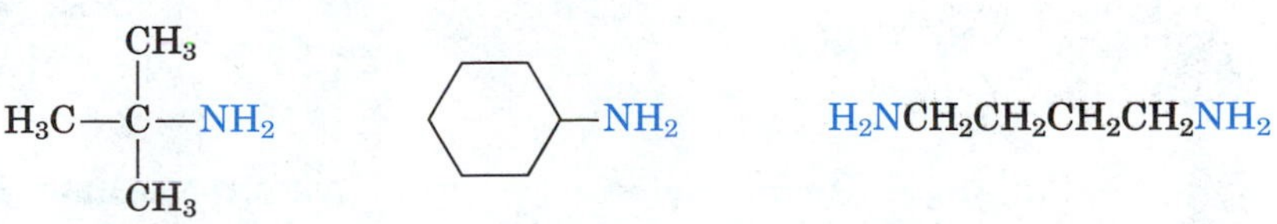

***tert*-Butylamine** **Cyclohexylamine** **1,4-Butanediamine**

Alternatively, the suffix *-amine* can be used in place of the final *-e* in the name of the parent compound:

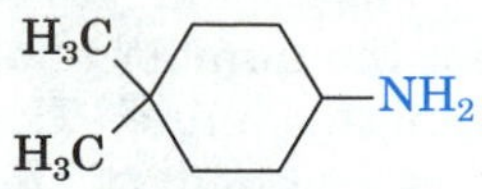

4,4-Dimethylcyclohexanamine

Amines with more than one functional group are named by considering the $-NH_2$ as an *amino* substituent on the parent molecule:

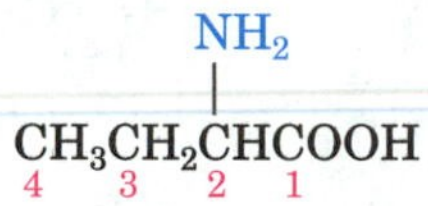

2-Aminobutanoic acid

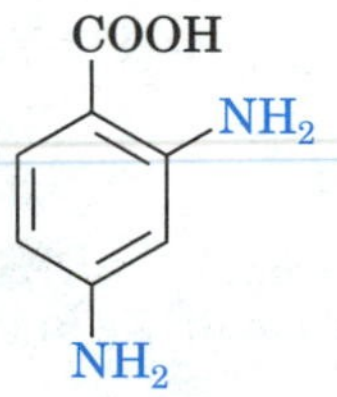

2,4-Diaminobenzoic acid

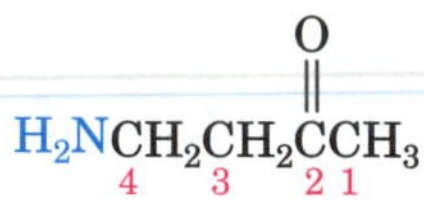

4-Amino-2-butanone

Symmetrical secondary and tertiary amines are named by adding the prefix *di-* or *tri-* to the alkyl group:

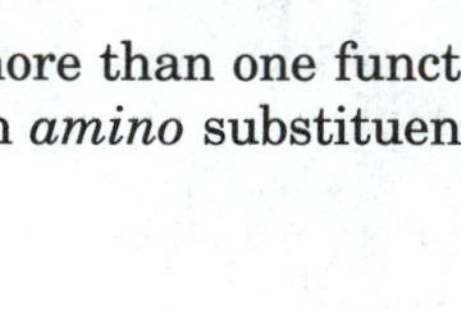

Diphenylamine

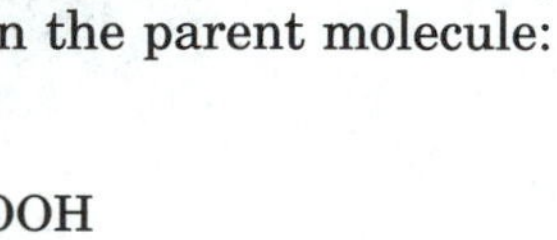

Triethylamine

Unsymmetrically substituted secondary and tertiary amines are named as *N*-substituted primary amines. The largest alkyl group is chosen as the parent name, and the other alkyl groups are considered *N*-substituents on the parent (*N* because they're attached to nitrogen).

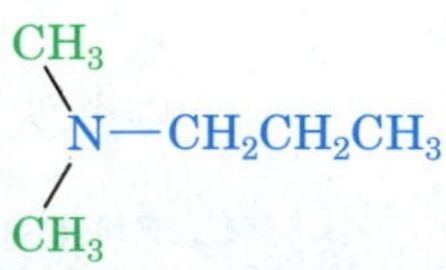

***N,N*-Dimethylpropylamine**
(propylamine is the parent name; the two methyl groups are substituents on nitrogen)

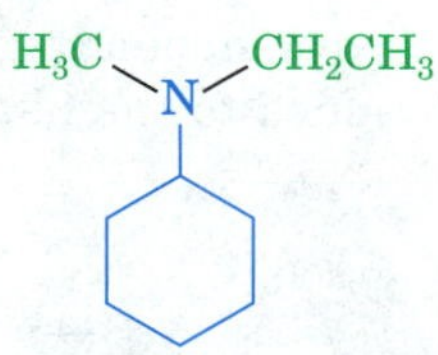

***N*-Ethyl-*N*-methylcyclohexylamine**
(cyclohexylamine is the parent name; methyl and ethyl are *N*-substituents)

There are relatively few common names for simple amines, but IUPAC rules recognize the names *aniline* for aminobenzene and *toluidine* for aminotoluene.

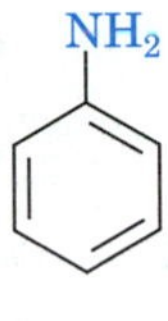

Aniline

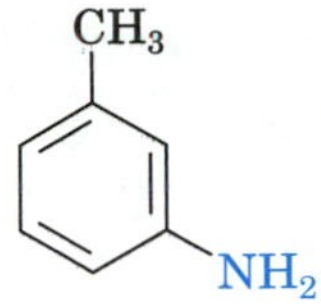

***m*-Toluidine**

Heterocyclic amines, compounds in which the nitrogen atom occurs as part of a ring, are also common, and each different heterocyclic ring system has its own parent name. The heterocyclic nitrogen atom is always numbered as position 1.

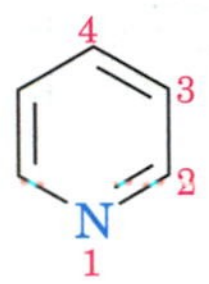

Pyridine

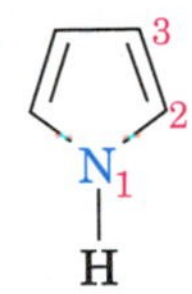

Pyrrole

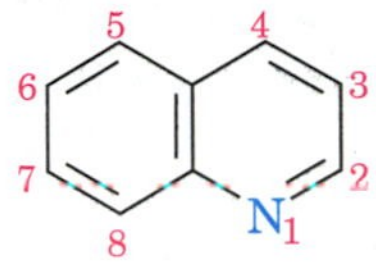

Quinoline

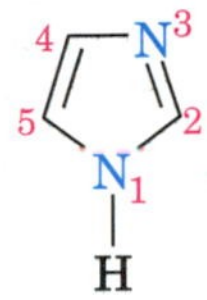

Imidazole

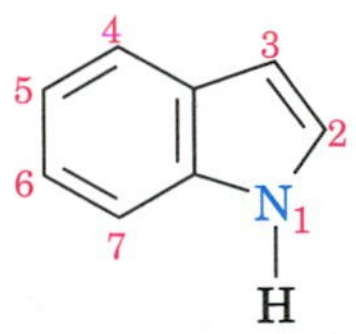

Indole

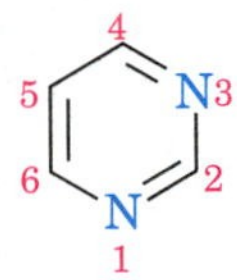

Pyrimidine

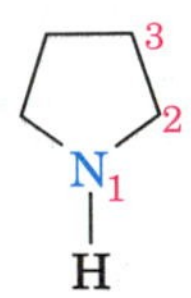

Pyrrolidine

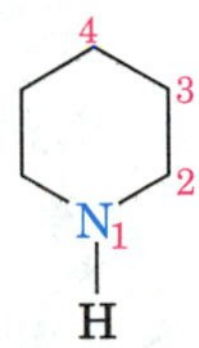

Piperidine

PROBLEM. . . . .

24.1 Classify the following compounds as primary, secondary, or tertiary amines, or as quaternary ammonium salts:

(a) 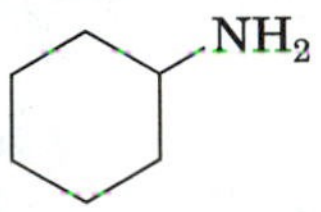

(b) 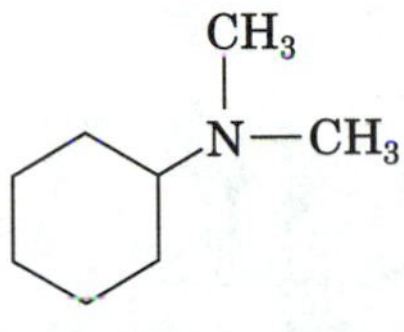

(c) 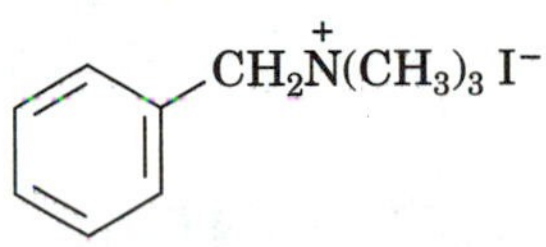

(d) $[(CH_3)_2CH]_2NH$

(e) 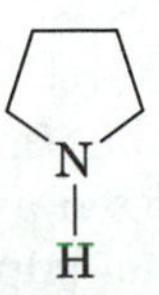

PROBLEM

24.2 Name the following compounds by IUPAC rules:

(a) $CH_3NHCH_2CH_3$

(b) 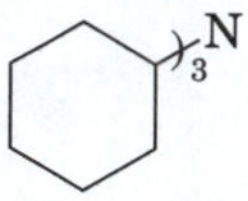

(c) 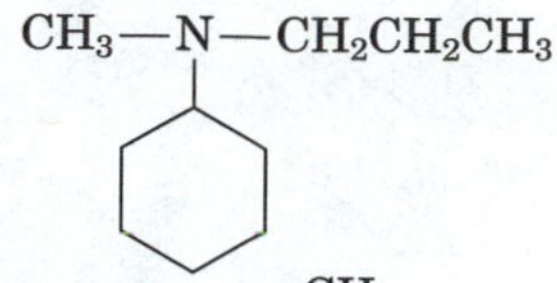

(d) 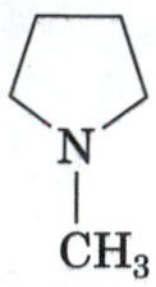

(e) $[(CH_3)_2CH]_2NH$

(f) 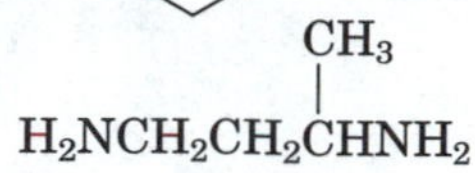

PROBLEM

24.3 Draw structures corresponding to the following IUPAC names:

(a) Triethylamine
(b) Triallylamine
(c) *N*-Methylaniline
(d) *N*-Ethyl-*N*-methylcyclopentylamine
(e) *N*-Isopropylcyclohexylamine
(f) *N*-Ethylpyrrole

PROBLEM

24.4 Draw structures for the following heterocyclic amines:

(a) 5-Methoxyindole
(b) 1,3-Dimethylpyrrole
(c) 4-(*N*,*N*-Dimethylamino)pyridine
(d) 5-Aminopyrimidine

24.2 Structure and Bonding in Amines

The bonding in amines is similar to the bonding in ammonia. The nitrogen atom is sp^3-hybridized, with the three substituents occupying three corners of a tetrahedron and the lone pair of electrons occupying the fourth corner. As you might expect, the C-N-C bond angles are close to the 109° tetrahedral value. For trimethylamine, the C-N-C bond angle is 108°, and the C–N bond length is 1.47 Å.

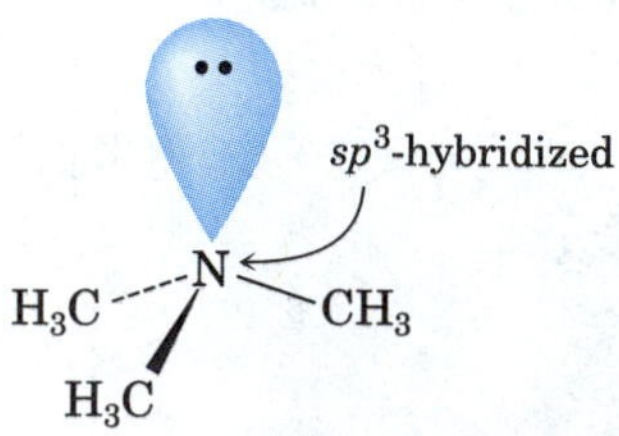

Trimethylamine

Stereo View

One consequence of tetrahedral geometry is that amines with three different substituents on nitrogen are chiral. Such an amine has no plane of symmetry and therefore is not superimposable on its mirror image. If we consider the lone pair of electrons to be the fourth substituent on nitrogen, these chiral amines are analogous to chiral alkanes with four different substituents attached to carbon:

A chiral amine

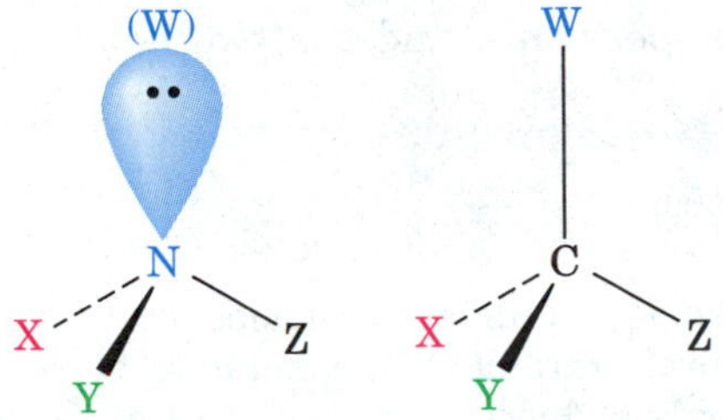

A chiral alkane

Unlike chiral hydrocarbons, most chiral amines can't be resolved into their two enantiomers because the two enantiomeric forms rapidly interconvert by a *pyramidal inversion*, much as an alkyl halide inverts in an S_N2 reaction. Pyramidal inversion occurs by a momentary rehybridization of the nitrogen atom to planar, sp^2 geometry, followed by rehybridization of the planar intermediate to tetrahedral, sp^3 geometry (Figure 24.1).

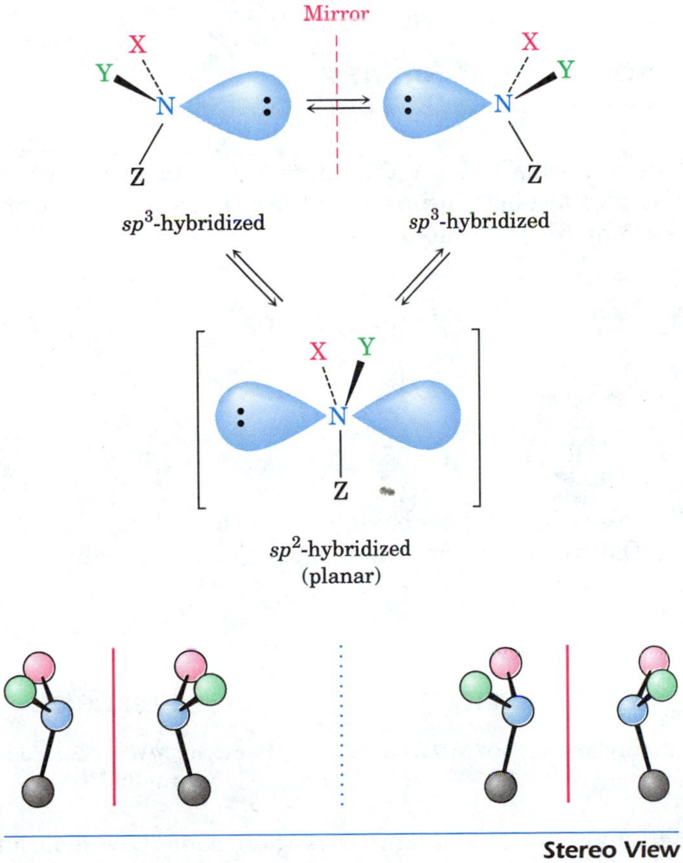

Figure 24.1 Pyramidal inversion of amines interconverts the two mirror-image (enantiomeric) forms.

Spectroscopic studies have shown that the barrier to nitrogen inversion is about 25 kJ/mol (6 kcal/mol), a figure only twice as large as the barrier to rotation about a C–C single bond. Pyramidal inversion is therefore rapid at room temperature, and the two enantiomeric forms can't normally be isolated.

PROBLEM.

24.5 Although rapid pyramidal inversions prevent chiral trialkylamines from being resolved, chiral tetraalkylammonium salts such as *N*-ethyl-*N*-methyl-*N*-propylbenzylammonium chloride are configurationally stable. Draw both the *R* and the *S* enantiomers of this chiral ammonium salt, and explain why they don't interconvert by pyramidal inversion.

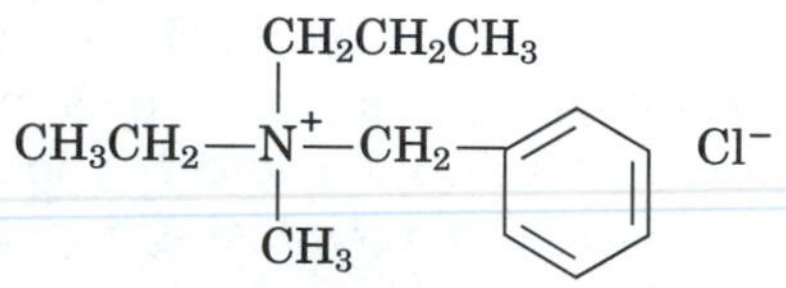

***N*-Ethyl-*N*-methyl-*N*-propylbenzylammonium chloride**

24.3 Physical Properties of Amines

Like alcohols, amines with fewer than five carbon atoms are generally water-soluble. Also like alcohols, primary and secondary amines form hydrogen bonds and are highly associated:

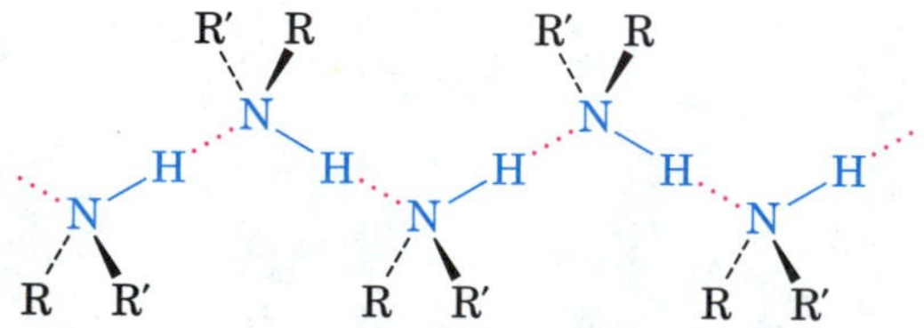

As a result, amines have higher boiling points than alkanes of similar molecular weight. Diethylamine, for example, boils at 56.3°C, while pentane boils at 36.1°C.

$CH_3CH_2N(H)CH_2CH_3$ | $CH_3CH_2CH_2CH_2CH_3$

Diethylamine, MW = 71.1 amu
bp = 56.3°C

Pentane, MW = 72.1 amu
bp = 36.1°C

Another characteristic of amines is their *odor*. Low-molecular-weight amines such as trimethylamine have a distinctive fish-like aroma, while diamines such as putrescine (1,4-butanediamine) have names that are self-explanatory.

24.4 Amine Basicity

As mentioned at the beginning of this chapter, the chemistry of amines is dominated by the lone pair of electrons on nitrogen. Because of this lone pair, amines are both basic and nucleophilic. They react with acids to form acid–base salts, and they react with electrophiles in many of the polar reactions seen in past chapters.

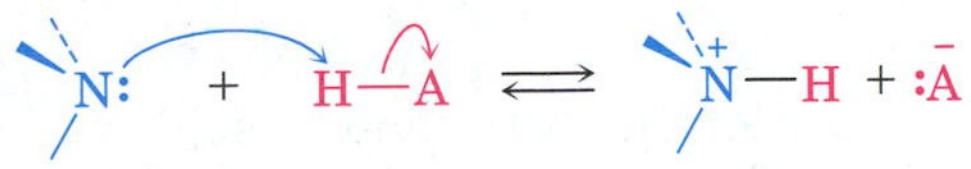

An amine (a Lewis base) An acid A salt

Amines are much more basic than alcohols, ethers, or water. When an amine is dissolved in water, an equilibrium is established in which water acts as a protic acid and transfers a proton to the amine. Just as we were able to measure the acid strength of a carboxylic acid by defining an acidity constant K_a (Section 2.6), we can measure the base strength of an amine by defining an analogous *basicity constant* K_b.

$$\text{For the reaction:}\quad RNH_2 + H_2O \rightleftharpoons RNH_3^+ + OH^-$$

$$K_b = \frac{[RNH_3^+][OH^-]}{[RNH_2]}$$

$$pK_b = -\log K_b$$

The larger the value of K_b (and the smaller the value of pK_b), the more favorable the proton-transfer equilibrium and the stronger the base.

In practice, however, K_b values (or pK_b values) are not often used. Instead, the most convenient way to measure the *basicity* of an amine (RNH_2) is to look at the *acidity* (pK_a) of the corresponding ammonium ion (RNH_3^+).

$$\text{For the reaction:}\quad RNH_3^+ + H_2O \rightleftharpoons RNH_2 + H_3O^+$$

$$K_a = \frac{[RNH_2][H_3O^+]}{[RNH_3^+]}$$

$$\text{so:}\quad K_a \cdot K_b = \left[\frac{[RNH_2][H_3O^+]}{[RNH_3^+]}\right]\left[\frac{[RNH_3^+][OH^-]}{[RNH_2]}\right]$$

$$= [H_3O^+][OH^-] = K_w = 1.00 \times 10^{-14}$$

$$\text{Thus:}\quad K_a = \frac{K_w}{K_b} \quad \text{and} \quad K_b = \frac{K_w}{K_a}$$

$$\text{and:}\quad pK_a + pK_b = 14$$

These equations say that the K_b of an amine multiplied by the K_a of the corresponding ammonium ion is equal to K_w, the ion-product constant for water (1.00×10^{-14}). Thus, if we know K_a for an ammonium ion, we also know K_b for the corresponding amine base because $K_b = K_w/K_a$.

The more acidic the ammonium ion (larger K_a or smaller pK_a), the weaker the base. Thus, a weaker base has an ammonium ion with a smaller pK_a, and a stronger base has an ammonium ion with a larger pK_a.

Weaker base: Smaller pK_a for ammonium ion
Stronger base: Larger pK_a for ammonium ion

This relationship between the acidity of a conjugate acid (RNH_3^+) and the basicity of its conjugate base (RNH_2) is just another example of the general relationship we saw in Section 2.7. A more strongly basic amine holds a proton more tightly, so its corresponding ammonium ion is less acidic. Conversely, a more weakly basic amine holds a proton less tightly, so its corresponding ammonium ion is more acidic.

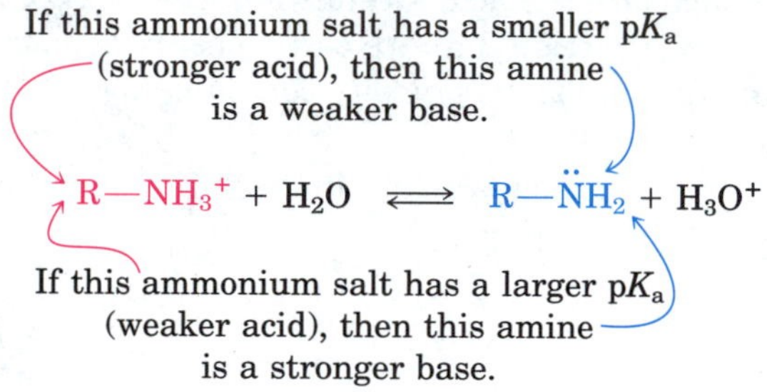

Table 24.1 lists the pK_a's of some ammonium ions and indicates that there is little effect of substitution on alkylamine basicity. The ions from most simple alkylamines have pK_a's in the narrow range 10–11, regardless of their exact structure. Arylamines, however, are considerably less basic than alkylamines, as is pyridine.

In contrast to amines, *amides* ($RCONH_2$) are completely nonbasic. Amides don't form ions when treated with aqueous acids, and they are poor nucleophiles. There are two reasons for the difference in basicity between amines and amides. First, the ground state of an amide is stabilized by delocalization of the nitrogen lone-pair electrons through orbital overlap with the carbonyl group. In resonance language, we can draw two contributing forms:

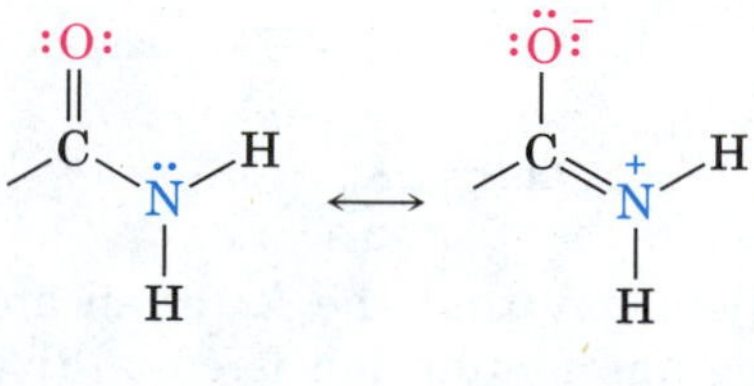

Table 24.1 Basicity of Some Common Amines

Name	*Structure*	pK_a *of ammonium ion*
Ammonia	NH_3	9.26
Primary alkylamine		
Methylamine	CH_3NH_2	10.66
Ethylamine	$CH_3CH_2NH_2$	10.81
Secondary alkylamine		
Dimethylamine	$(CH_3)_2NH$	10.73
Diethylamine	$(CH_3CH_2)_2NH$	10.49
Pyrrolidine	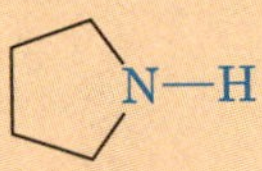	11.27
Tertiary alkylamine		
Trimethylamine	$(CH_3)_3N$	9.81
Triethylamine	$(CH_3CH_2)_3N$	11.01
Pyridine	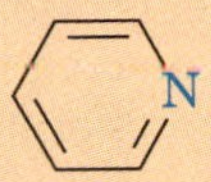	5.25
Arylamine		
Aniline	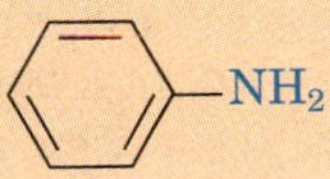	4.63

Since this amide resonance stabilization is lost in the protonated product, protonation is disfavored. Second, a protonated amide is higher in energy than a protonated amine because the electron-withdrawing carbonyl group inductively destabilizes the neighboring positive charge.

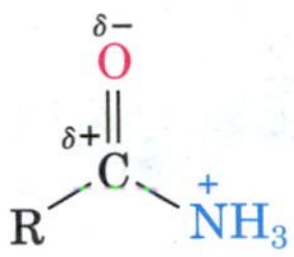

A protonated amide has no resonance stabilization and has inductive destabilization of the positive charge.

Both factors—increased stability of an amide versus an amine and decreased stability of a protonated amide versus a protonated amine—lead to a substantial difference in $\Delta G°$ and a resultant difference in basicity for amines and amides. Figure 24.2 shows these relationships on a reaction energy diagram.

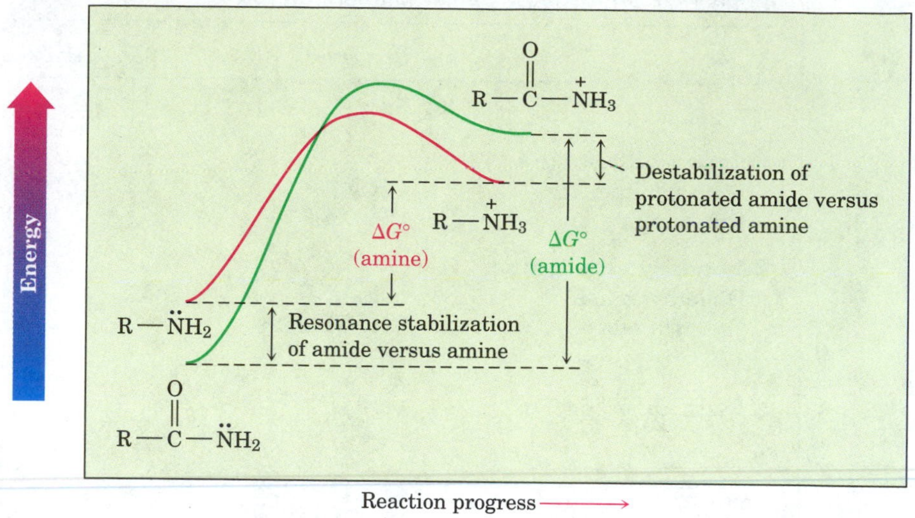

Figure 24.2 A reaction energy diagram comparing the protonation of amines and amides. The $\Delta G°$ for amide protonation is larger because of increased reactant stability and decreased product stability.

It's often possible to take advantage of their basicity to purify amines. For example, if a mixture of a basic amine and a neutral compound such as a ketone, an alcohol, or an ether is dissolved in an organic solvent and extracted with aqueous acid, the basic amine dissolves in the water layer as its protonated salt, while the neutral compound remains in the organic solvent layer. Separation, basification, and extraction of the aqueous layer with organic solvent then provides the pure amine (Figure 24.3).

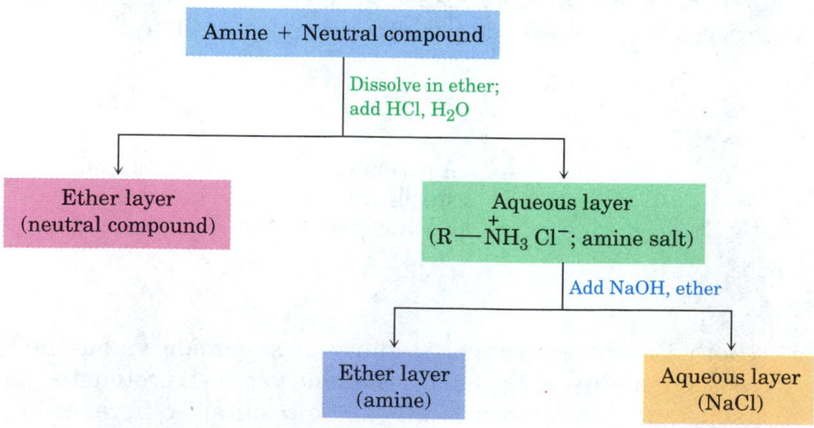

Figure 24.3 Separation and purification of an amine component from a mixture.

In addition to their behavior as bases, primary and secondary amines can also act as very weak *acids* because an N–H proton can be removed by a sufficiently strong base. We've already seen, for example, how diisopropylamine ($pK_a \approx 40$) reacts with butyllithium to yield lithium diisopropylamide (LDA; Section 22.5).

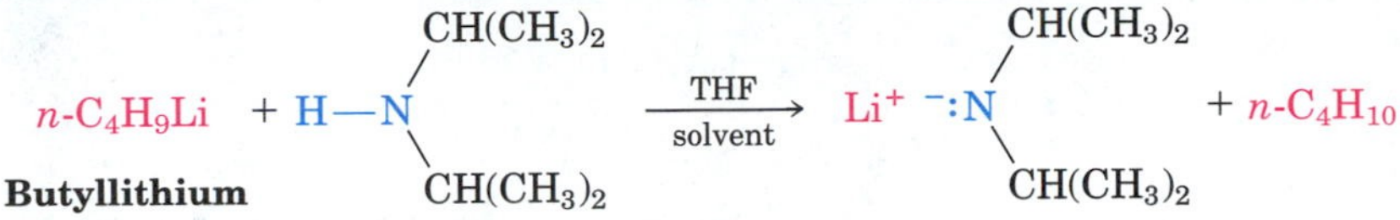

Butyllithium **Diisopropylamine** **Lithium diisopropylamide (LDA)**

Dialkylamide anions like LDA are extremely powerful bases that are much used in organic chemistry, particularly for the generation of enolate ions from carbonyl compounds (Section 22.8).

PROBLEM

24.6 Which compound in each of the following pairs is more basic?
(a) $CH_3CH_2NH_2$ or $CH_3CH_2CONH_2$ (b) NaOH or CH_3NH_2
(c) CH_3NHCH_3 or CH_3OCH_3

PROBLEM

24.7 The benzylammonium ion ($C_6H_5CH_2NH_3^+$) has $pK_a = 9.33$, and the propylammonium ion has $pK_a = 10.71$. Which is the stronger base, benzylamine or propylamine? Explain.

PROBLEM

24.8 What are the pK_b's of benzylamine and propylamine (see Problem 24.7)?

PROBLEM

24.9 Protonation of an amide occurs on oxygen rather than on nitrogen. Suggest a reason for this behavior.

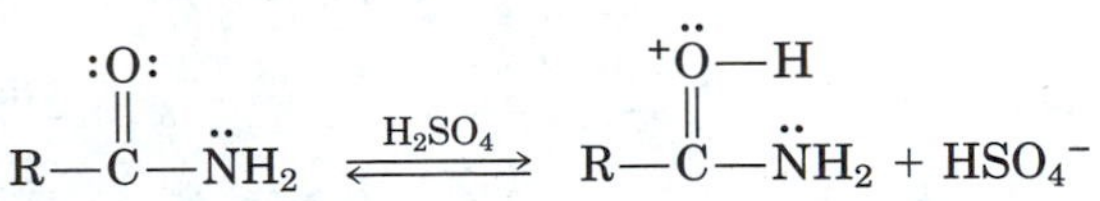

24.5 Industrial Sources and Uses of Alkylamines

Alkylamines have a variety of minor applications in the chemical industry as starting materials for the preparation of insecticides and pharmaceuticals. For example, propranolol, a heart stimulant used in the control of cardiac arrhythmia, is prepared by S_N2 reaction of an epoxide with isopropylamine.

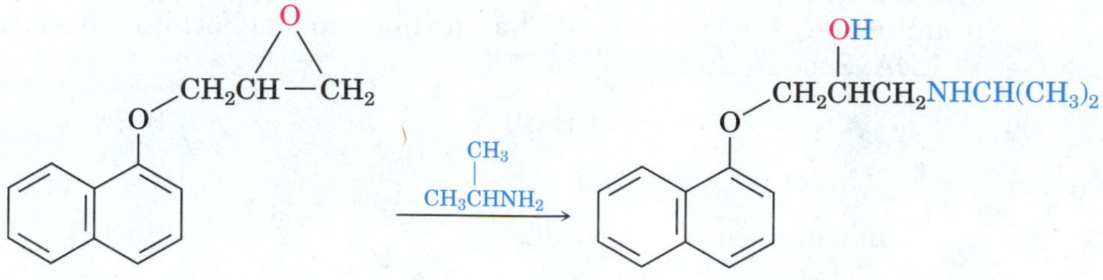

Propranolol (a heart stimulant)

Simple methylated amines are prepared by reaction of ammonia with methanol in the presence of an alumina catalyst. The reaction yields a mixture of mono-, di-, and trimethylated products, but is nonetheless useful industrially because the separation of the three products by distillation is easy.

$$NH_3 + CH_3OH \xrightarrow[450°C]{Al_2O_3} CH_3NH_2 + CH_3NHCH_3 + CH_3N(CH_3)CH_3$$

24.6 Synthesis of Amines

S_N2 Reactions of Alkyl Halides

Ammonia and other alkylamines are good nucleophiles in S_N2 reactions. As a result, the simplest method of amine synthesis is by S_N2 alkylation of ammonia or an alkylamine with an alkyl halide. If ammonia is used, a primary amine results; if a primary amine is used, a secondary amine results; and so on. Even tertiary amines react rapidly with alkyl halides to yield quaternary ammonium salts, $R_4N^+ X^-$.

		S_N2 reaction			
Ammonia	$\ddot{N}H_3 + R-X$	$\longrightarrow$	$RNH_3^+ X^-$	$\xrightarrow{NaOH}$ RNH_2	Primary
Primary	$R\ddot{N}H_2 + R-X$	$\longrightarrow$	$R_2NH_2^+ X^-$	$\xrightarrow{NaOH}$ R_2NH	Secondary
Secondary	$R_2\ddot{N}H + R-X$	$\longrightarrow$	$R_3NH^+ X^-$	$\xrightarrow{NaOH}$ R_3N	Tertiary
Tertiary	$R_3\ddot{N} + R-X$	$\longrightarrow$	$R_4N^+ X^-$	Quaternary ammonium salt	

Unfortunately, these reactions don't stop cleanly after a single alkylation has occurred. Because primary, secondary, and tertiary amines all have similar reactivity, the initially formed monoalkylated substance often undergoes further reaction to yield a mixture of products. For example, treatment of 1-bromooctane with a twofold excess of ammonia leads to a mixture containing only 45% of octylamine. A nearly equal amount of dioctylamine is produced by double alkylation, along with smaller amounts of trioctylamine and tetraoctylammonium bromide. Higher yields of monoalkylated product can sometimes be obtained by using a large excess of the starting amine, but even so the reaction is a poor one.

$$CH_3(CH_2)_6CH_2Br + :NH_3 \longrightarrow CH_3(CH_2)_6CH_2\ddot{N}H_2 + [CH_3(CH_2)_6CH_2]_2\ddot{N}H$$

1-Bromooctane → **Octylamine (45%)** + **Dioctylamine (43%)**

$$+ [CH_3(CH_2)_6CH_2]_3N: + [CH_3(CH_2)_6CH_2]_4\overset{+}{N}\ \overset{-}{Br}$$

Trace Trace

A better method for preparing primary amines is to use the **azide synthesis**, in which azide ion, N_3^-, is used to displace halide ion from a primary or secondary alkyl halide to give an alkyl azide, RN_3. Since alkyl azides are not nucleophilic, overalkylation can't occur. Reduction of the alkyl azide, either by catalytic hydrogenation over a palladium catalyst or by reaction with $LiAlH_4$, leads to the desired primary amine. Although the method works well, low-molecular-weight alkyl azides are explosive and must be handled carefully.

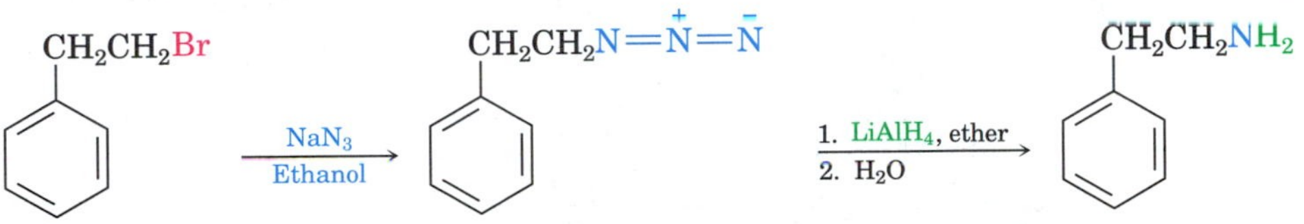

1-Bromo-2-phenylethane **2-Phenylethyl azide** **2-Phenylethylamine (89%)**

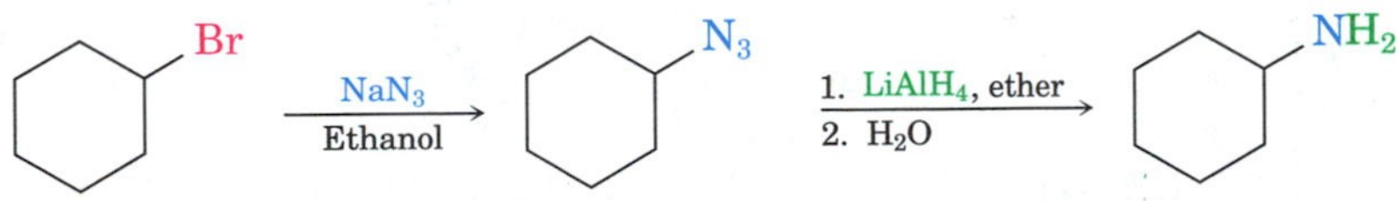

Bromocyclohexane **Cyclohexyl azide** **Cyclohexylamine (54%)**

An alternative to the azide synthesis is the **Gabriel[1] amine synthesis**, which uses a *phthalimide* alkylation for preparing a primary amine from an alkyl halide. **Imides** (–CONHCO–) are similar to ethyl acetoacetate in that the N–H hydrogen is flanked by two carbonyl groups. Thus, imides are deprotonated by such bases as KOH, and the resultant anions are readily alkylated in a reaction similar to the acetoacetic ester synthesis (Section 22.8). Basic hydrolysis of the *N*-alkylated imide then yields a primary amine product. Note that the imide hydrolysis step is closely analogous to the hydrolysis of an amide (Section 21.8).

[1] Siegmund Gabriel (1851–1924); b. Berlin; Ph.D. University of Berlin (1874); assistant to A. W. von Hofmann; professor, University of Berlin.

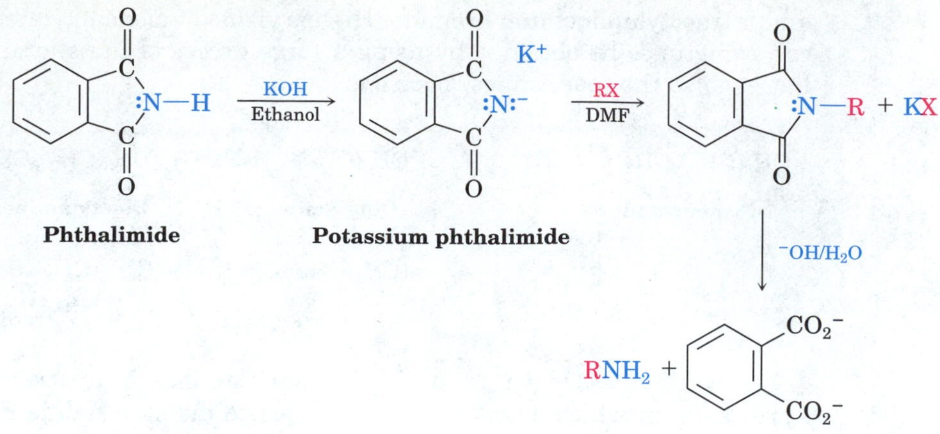

For example,

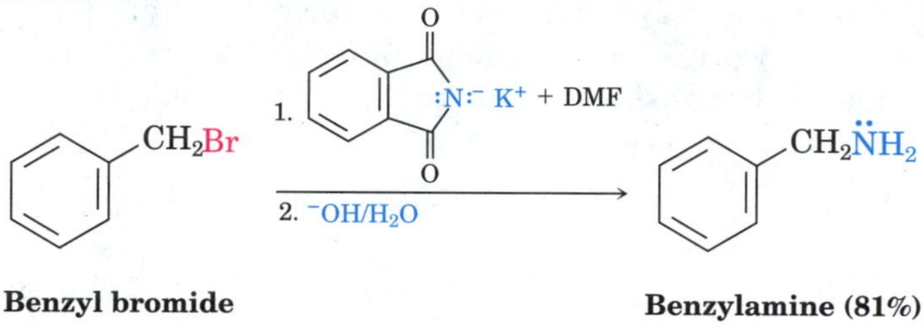

Recall the acetoacetic ester synthesis:

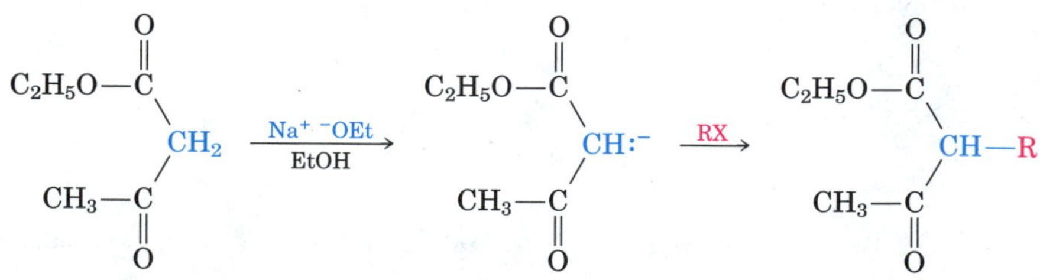

PROBLEM

24.10 Show the mechanism of the last step in the Gabriel amine synthesis, the base-promoted hydrolysis of a phthalimide to yield an amine plus phthalate ion.

PROBLEM

24.11 Show two methods for the synthesis of dopamine, a neurotransmitter involved in regulation of the central nervous system. Use any alkyl halide needed.

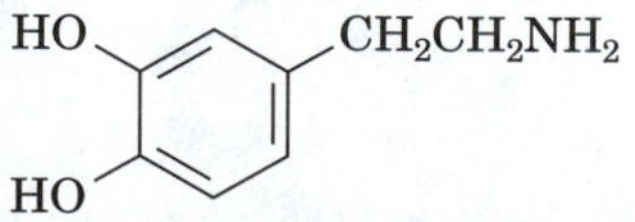

Dopamine

Reduction of Nitriles and Amides

We've already seen in Sections 21.8 and 21.9 how amines can be prepared by reduction of amides and nitriles with $LiAlH_4$. The two-step sequence of S_N2 displacement with CN^- followed by reduction is an excellent method for converting an alkyl halide into a primary amine having one more carbon atom. Amide reduction provides an excellent method for converting carboxylic acids and their derivatives into amines with the same number of carbon atoms.

$$\underset{\textbf{Alkyl halide}}{RX} \xrightarrow{NaCN} RCN \xrightarrow[2.\ H_2O]{1.\ LiAlH_4,\ ether} \underset{\textbf{1° amine}}{RCH_2NH_2}$$

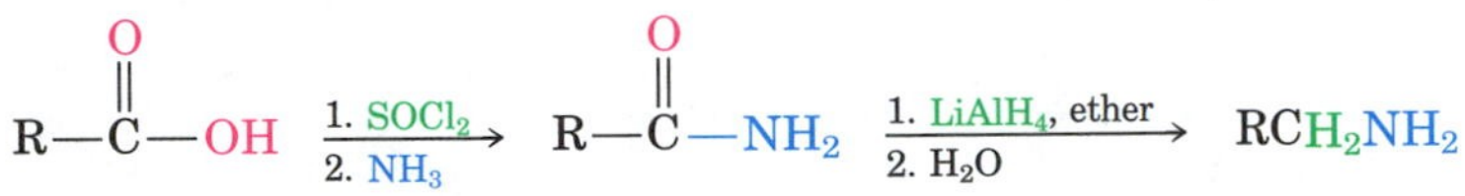

Carboxylic acid **1° amine**

PROBLEM

24.12 Propose structures for either a nitrile or an amide that might be a precursor of each of these amines.
(a) Propylamine
(b) Dipropylamine
(c) Benzylamine, $C_6H_5CH_2NH_2$
(d) *N*-Ethylaniline

Reductive Amination of Ketones and Aldehydes

Amines can be synthesized in a single step by treatment of a ketone or aldehyde with ammonia or an amine in the presence of a reducing agent, a process called **reductive amination**. For example, amphetamine, a central nervous system stimulant, is prepared commercially by reductive amination of phenyl-2-propanone with ammonia, using hydrogen gas over a nickel catalyst as the reducing agent.

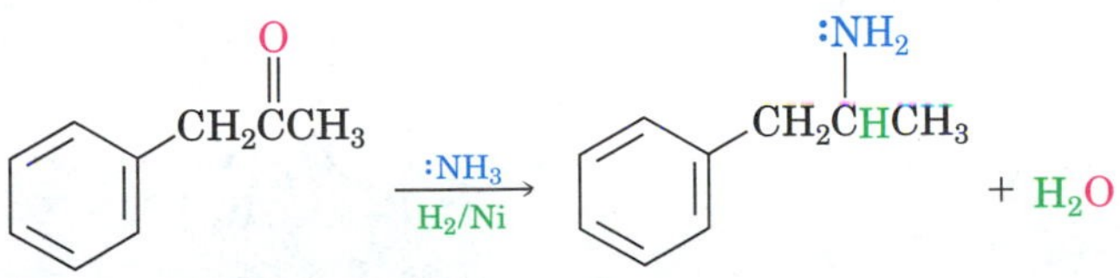

Phenyl-2-propanone **Amphetamine**

Reductive amination takes place by the pathway shown in Figure 24.4. An imine intermediate is first formed by a nucleophilic addition reaction (Section 19.12), and the imine is then reduced.

Ammonia attacks the carbonyl group in a nucleophilic addition reaction to yield an intermediate carbinolamine.

The intermediate loses water to give an imine.

The imine is reduced catalytically over nickel to yield the amine product.

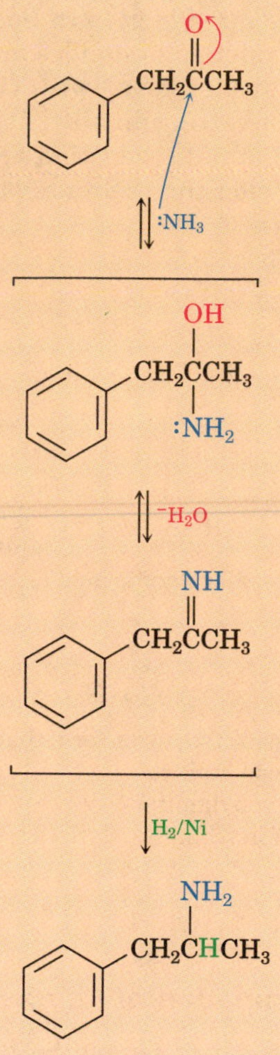

Figure 24.4 Mechanism of reductive amination of a ketone to yield an amine. (Details of the imine-forming step are shown in Figure 19.8.)

Ammonia, primary amines, and secondary amines can all be used in the reductive amination reaction to yield primary, secondary, and tertiary amines, respectively.

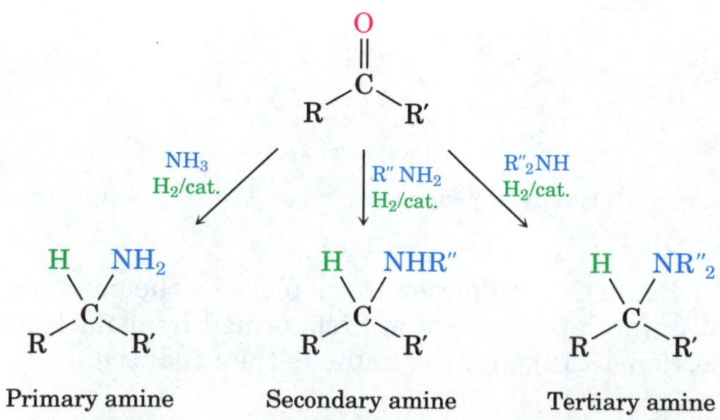

Many different reducing agents are effective, but the most common choice in the laboratory is sodium cyanoborohydride, $NaBH_3CN$, a relative of $NaBH_4$.

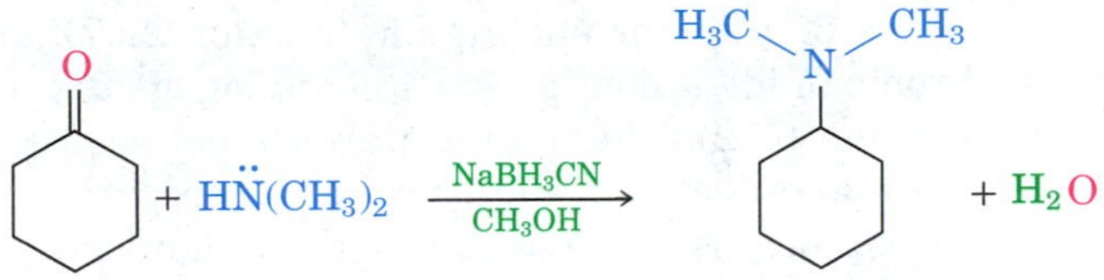

Cyclohexanone **N,N-Dimethylcyclohexylamine (85%)**

We'll see in Section 30.6 that a process closely related to reductive amination occurs frequently in the biological pathways by which amino acids are synthesized.

PROBLEM

24.13 How might the following amines be prepared by reductive amination of a ketone or aldehyde? Show all precursors if more than one is possible.
(a) $CH_3CH_2NHCH(CH_3)_2$ (b) *N*-Ethylaniline (c) *N*-Methylcyclopentylamine

PROBLEM

24.14 Show the mechanism of reductive amination of cyclohexanone and dimethylamine with $NaBH_3CN$.

Hofmann and Curtius Rearrangements

Carboxylic acid derivatives can be converted into primary amines with loss of one carbon atom by both the **Hofmann[2] rearrangement** and the **Curtius[3] rearrangement**. Although the Hofmann rearrangement involves a primary amide and the Curtius rearrangement involves an acyl azide, both proceed through similar mechanisms:

Hofmann rearrangement

Curtius rearrangement

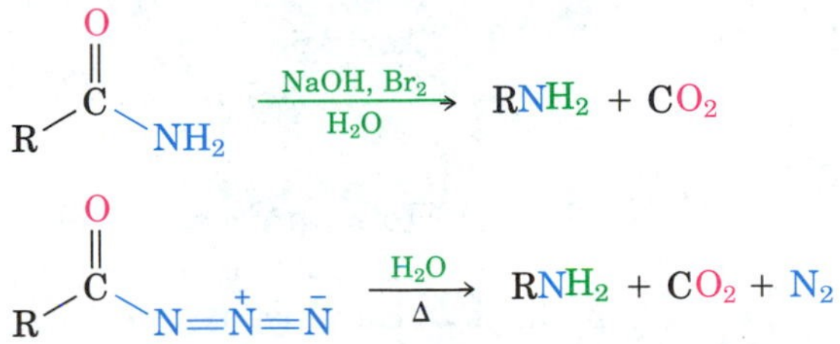

[2]August Wilhelm von Hofmann (1818–1892); b. Giessen, Germany; professor, Bonn, the Royal College of Chemistry, London (1845–1864), Berlin (1865–1892).

[3]Theodor Curtius (1857–1928); b. Duisberg; Ph.D. Leipzig; professor, universities of Kiel, Bonn, and Heidelberg (1898–1926).

Hofmann rearrangement occurs when a primary amide, $RCONH_2$, is treated with Br_2 and base (Figure 24.5). The overall mechanism is lengthy, but most of the individual steps have been encountered before. Thus, the bromination of an amide in steps 1 and 2 is analogous to the base-promoted bromination of a ketone enolate ion (Section 22.7), and the rearrangement of the bromoamide anion in step 4 is analogous to a carbocation rearrangement (Section 6.12). The main difference between the migration step in a Hofmann rearrangement and that in a carbocation rearrangement is that the –R group begins its migration to the neighboring atom *at the same time* the bromide ion is leaving, rather than after it has left. Nucleophilic addition of water to the isocyanate carbonyl group in step 5 is a typical carbonyl-group process (Section 19.8), as is the final decarboxylation step (Section 22.8).

Despite its mechanistic complexity, the Hofmann rearrangement often gives high yields of both aryl- and alkylamines. For example, the appetite-suppressing drug phentermine is prepared commercially by Hofmann rearrangement of a primary amide.

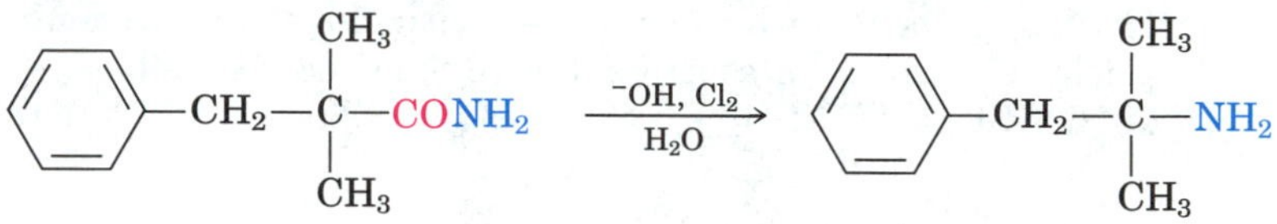

2,2-Dimethyl-3-phenylpropanamide **Phentermine**

The Curtius rearrangement, like the Hofmann rearrangement, involves migration of an –R group from the C=O carbon atom to the neighboring nitrogen with simultaneous loss of a leaving group. The reaction takes place on heating an acyl azide that is itself prepared by nucleophilic acyl substitution of an acid chloride.

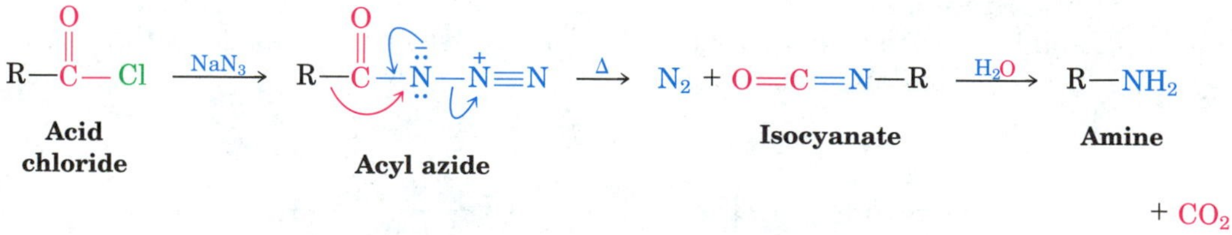

Acid chloride **Acyl azide** **Isocyanate** **Amine**

Like the Hofmann rearrangement, the Curtius rearrangement is often used commercially. For example, the antidepressant drug tranylcypromine is made by Curtius rearrangement of 2-phenylcyclopropanecarbonyl chloride.

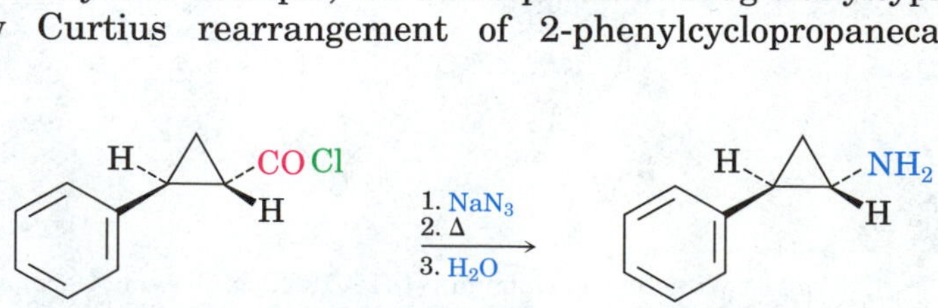

***trans*-2-Phenylcyclopropanecarbonyl chloride** **Tranylcypromine**

Base abstracts an acidic N–H proton, yielding an anion.

The anion reacts with bromine in an alpha-substitution reaction to give an *N*-bromoamide.

Base abstraction of the remaining amide proton gives a bromoamide anion.

The bromoamide anion rearranges as the R group attached to the carbonyl carbon migrates to nitrogen at the same time the bromide ion leaves, giving an isocyanate.

The isocyanate adds water in a nucleophilic addition step to yield a carbamic acid.

The carbamic acid spontaneously loses CO_2, yielding the amine product.

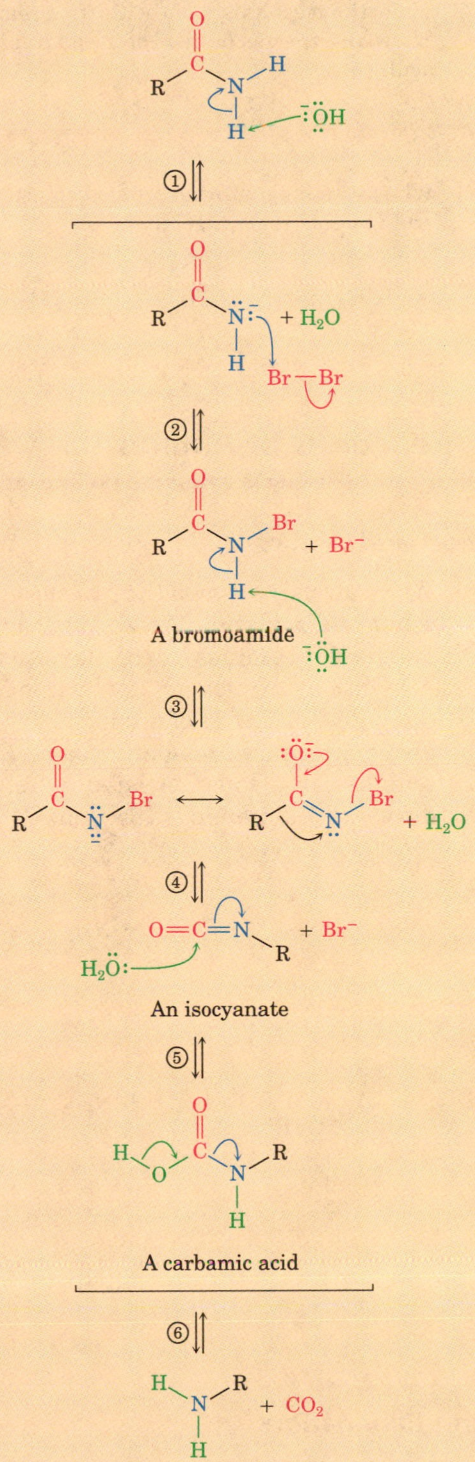

Figure 24.5 Mechanism of the Hofmann rearrangement of an amide to an amine. Each step is analogous to a reaction studied previously.

PROBLEM. . . . .

24.15 Show the mechanism of the Curtius rearrangement of an acyl azide to an isocyanate. Show also the mechanism of the addition of water to an isocyanate to yield a carbamic acid.

PROBLEM. . . . .

24.16 What starting materials would you use to prepare the following amines by Hofmann and Curtius rearrangements?

(a) $(CH_3)_3CCH_2CH_2NH_2$

(b) 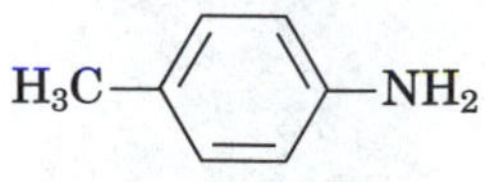

24.7 Reactions of Amines

Alkylation and Acylation

We've already studied the two most general reactions of alkylamines—alkylation and acylation. As we saw earlier in this chapter, primary, secondary, and tertiary amines can be alkylated by reaction with a primary alkyl halide. Alkylations of primary and secondary amines are difficult to control and often give mixtures of products, but tertiary amines are cleanly alkylated to give quaternary ammonium salts. Primary and secondary (but not tertiary) amines can also be acylated by reaction with acid chlorides or acid anhydrides to yield amides (Sections 21.5 and 21.6).

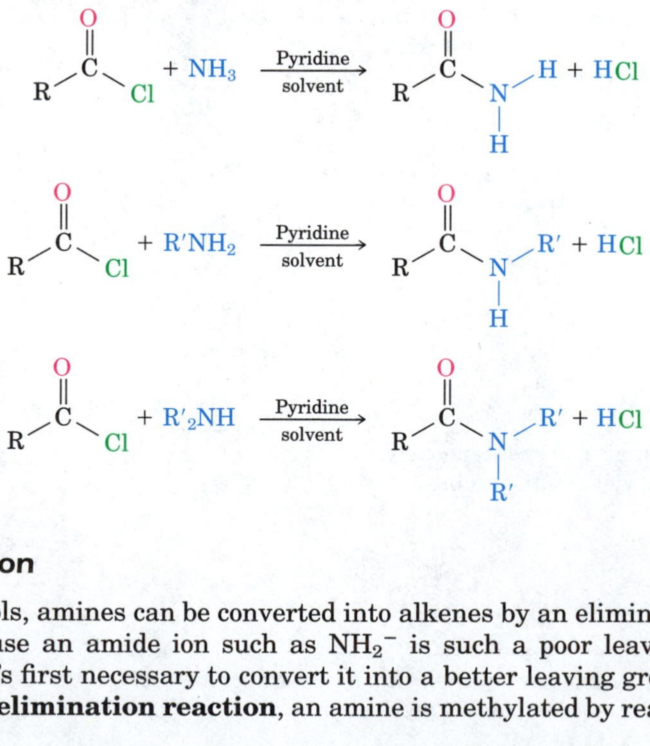

Hofmann Elimination

Like alcohols, amines can be converted into alkenes by an elimination reaction. Because an amide ion such as NH_2^- is such a poor leaving group, however, it's first necessary to convert it into a better leaving group. In the **Hofmann elimination reaction**, an amine is methylated by reaction with excess iodomethane to produce a quaternary ammonium salt, which then undergoes elimination to give an alkene on heating with silver oxide, Ag_2O. For example, hexylamine is converted into 1-hexene in 60% yield.

$$CH_3CH_2CH_2CH_2CH_2CH_2NH_2 \xrightarrow[\text{(excess)}]{CH_3I} CH_3CH_2CH_2CH_2CH_2CH_2\overset{+}{N}(CH_3)_3\,I^-$$

Hexylamine **Hexyltrimethylammonium iodide**

$$\xrightarrow{Ag_2O,\ H_2O,\ \Delta} CH_3CH_2CH_2CH_2CH{=}CH_2 + N(CH_3)_3$$

1-Hexene (60%)

Silver oxide functions by exchanging hydroxide ion for iodide ion in the quaternary salt, thus providing the base necessary to cause elimination. The actual elimination step is an E2 reaction (Section 11.11) in which hydroxide ion removes a proton and the positively charged nitrogen atom acts as the leaving group.

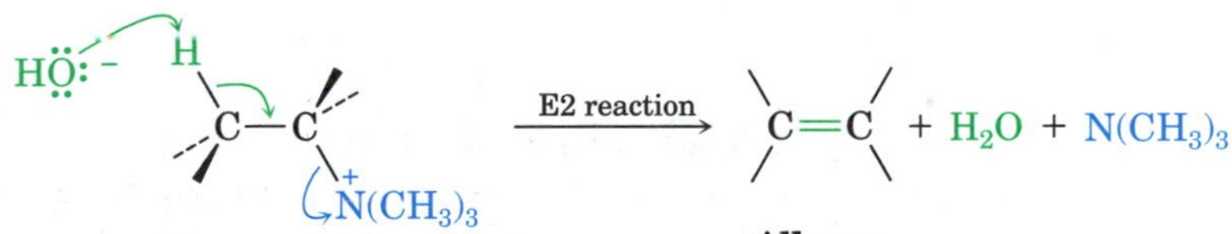

Quaternary ammonium salt **Alkene**

An interesting feature of the Hofmann elimination is that it gives products different from those of most other E2 reactions. Whereas the *more* highly substituted alkene product generally predominates in the E2 reaction of an alkyl halide (Zaitsev's rule; Section 11.10), the *less* highly substituted alkene predominates in the Hofmann elimination of a quaternary ammonium salt. For example, (1-methylbutyl)trimethylammonium hydroxide yields 1-pentene and 2-pentene in a 94:6 ratio. The reason for this selectivity is probably steric. Because of the large size of the trialkylamine leaving group, the attacking base abstracts a hydrogen from the most sterically accessible, least hindered position.

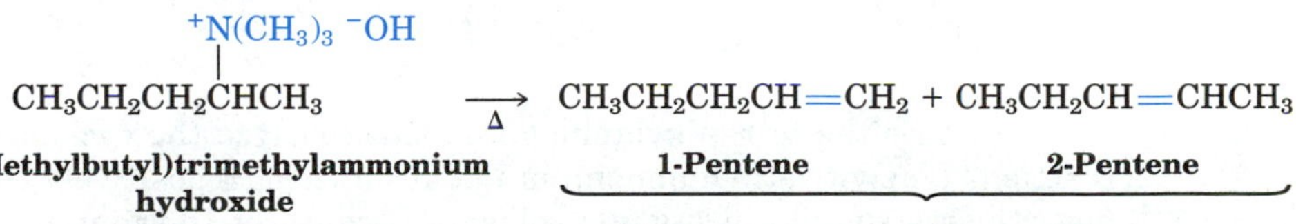

(1-Methylbutyl)trimethylammonium hydroxide **1-Pentene** **2-Pentene**

94:6 ratio

The Hofmann elimination reaction is important primarily because of its historical use as a degradative tool in the structure determination of many complex naturally occurring amines. The reaction is not often used because the product alkenes can usually be made more easily in other ways.

PROBLEM

24.17 What products would you expect to obtain from Hofmann elimination of the following amines? If more than one product is formed, indicate which is major.

(a) $CH_3CH_2CH_2CH(NH_2)CH_2CH_2CH_2CH_3$ (b) Cyclohexylamine

(c) $CH_3CH_2CH_2CH(NH_2)CH_2CH_2CH_3$ (d) *N*-Ethylcyclohexylamine

PROBLEM

24.18 What product would you expect from Hofmann elimination of a cyclic amine such as piperidine? Formulate all the steps.

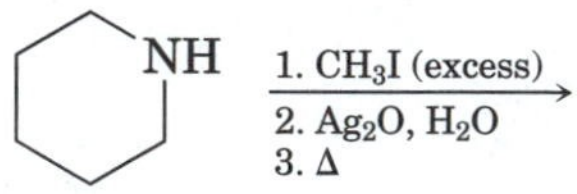

24.8 Tetraalkylammonium Salts as Phase-Transfer Agents

Tetraalkylammonium salts, $R_4N^+ X^-$, are widely used as catalysts for many different kinds of organic reactions. As an example, imagine an experiment in which cyclohexene is dissolved in chloroform and treated with aqueous NaOH. Since the organic layer and the water layer are immiscible, the base in the aqueous phase does not come into contact with chloroform in the organic phase, and there is no reaction. If, however, a small amount of benzyltriethylammonium chloride is added, an immediate reaction occurs. The chloroform reacts with NaOH to generate dichlorocarbene, which adds to the cyclohexene double bond to give a dichlorocyclopropane in 77% yield (Section 7.6).

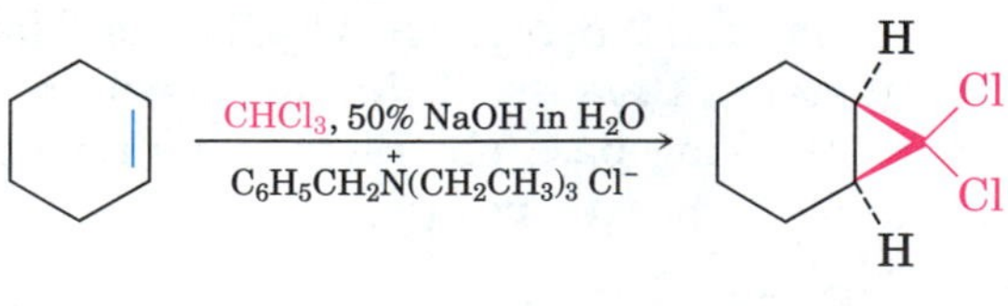

How does the tetraalkylammonium salt catalyze the cyclopropanation reaction? Benzyltriethylammonium ion, even though positively charged, is nevertheless soluble in organic solvents because of the four hydrocarbon substituents on nitrogen. But when the *positively* charged tetraalkylammonium ion goes into the organic layer, a *negatively* charged counter-ion must follow to preserve charge neutrality. Hydroxide ion, present in far greater amount than chloride ion, is thus transferred into the organic phase where reaction with chloroform immediately occurs (Figure 24.6).

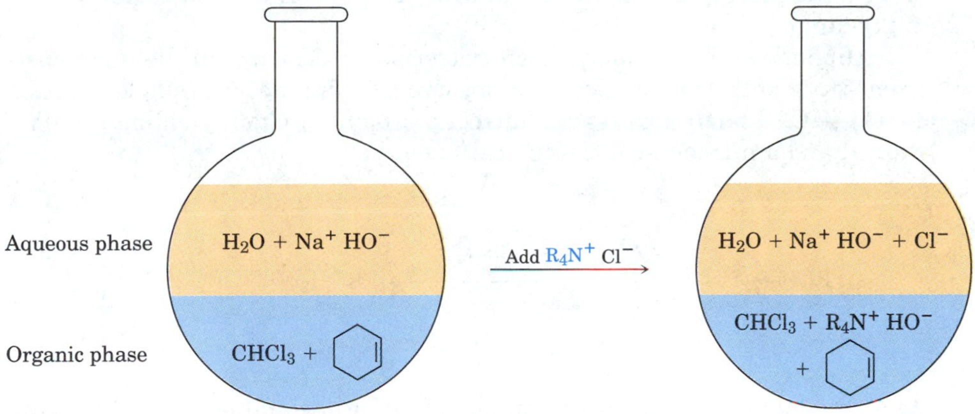

Figure 24.6 Phase-transfer catalysis. Addition of a small amount of a tetraalkylammonium salt to a two-phase mixture allows an inorganic anion to be transferred from the aqueous phase into the organic phase, where a reaction can occur.

The transfer of an inorganic ion such as OH^- from one phase to another is called **phase transfer**, and the tetraalkylammonium salt is referred to as a *phase-transfer catalyst*. Many different kinds of organic reactions, including oxidations, reductions, carbonyl-group alkylations, and S_N2 reactions, are subject to phase-transfer catalysis, often with considerable improvements in yield. S_N2 reactions are particularly good candidates for phase-transfer catalysis because inorganic nucleophiles can be transferred from an aqueous (protic) phase to an organic (aprotic) phase, where they are much more reactive. For example:

$$CH_3(CH_2)_6CH_2Br + NaCN \xrightarrow[C_6H_5CH_2\overset{+}{N}(CH_2CH_3)_3\ Cl^-]{H_2O,\ benzene} CH_3(CH_2)_6CH_2CN + NaBr$$

1-Bromooctane **Nonanenitrile (92%)**

24.9 Spectroscopy of Amines

Mass Spectrometry

The **nitrogen rule** of mass spectrometry says that a compound with an odd number of nitrogen atoms has an odd-numbered molecular weight. Thus, the presence of nitrogen in a molecule is detected simply by observing its mass spectrum. An odd-numbered molecular ion usually means that the unknown compound has one or three nitrogen atoms, and an even-numbered molecular ion usually means that a compound has either zero or two nitrogen atoms. The logic behind the rule derives from the fact that nitrogen is trivalent, thus requiring an odd number of hydrogen atoms in a molecule.

For example, methylamine has the formula CH_5N and a molecular weight of 31 amu; morphine has the formula $C_{17}H_{19}NO_3$ and a molecular weight of 285 amu.

Aliphatic amines undergo a characteristic α cleavage in the mass spectrometer, similar to the cleavage observed for aliphatic alcohols (Section 17.11). A C–C bond nearest the nitrogen atom is broken, yielding an alkyl radical and a nitrogen-containing cation:

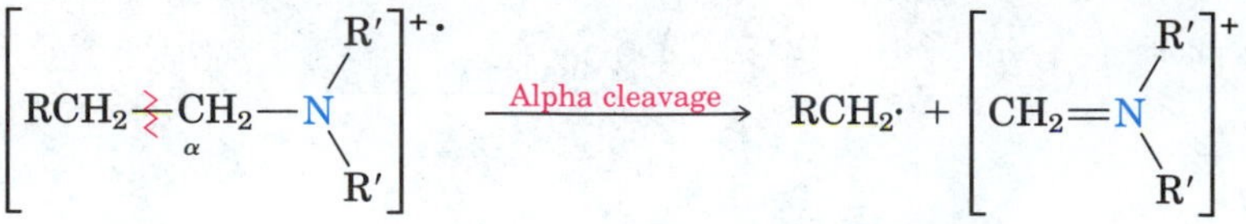

As an example, the mass spectrum of *N*-ethylpropylamine shown in Figure 24.7 has peaks at $m/z = 58$ and $m/z = 72$, corresponding to the two possible modes of α cleavage.

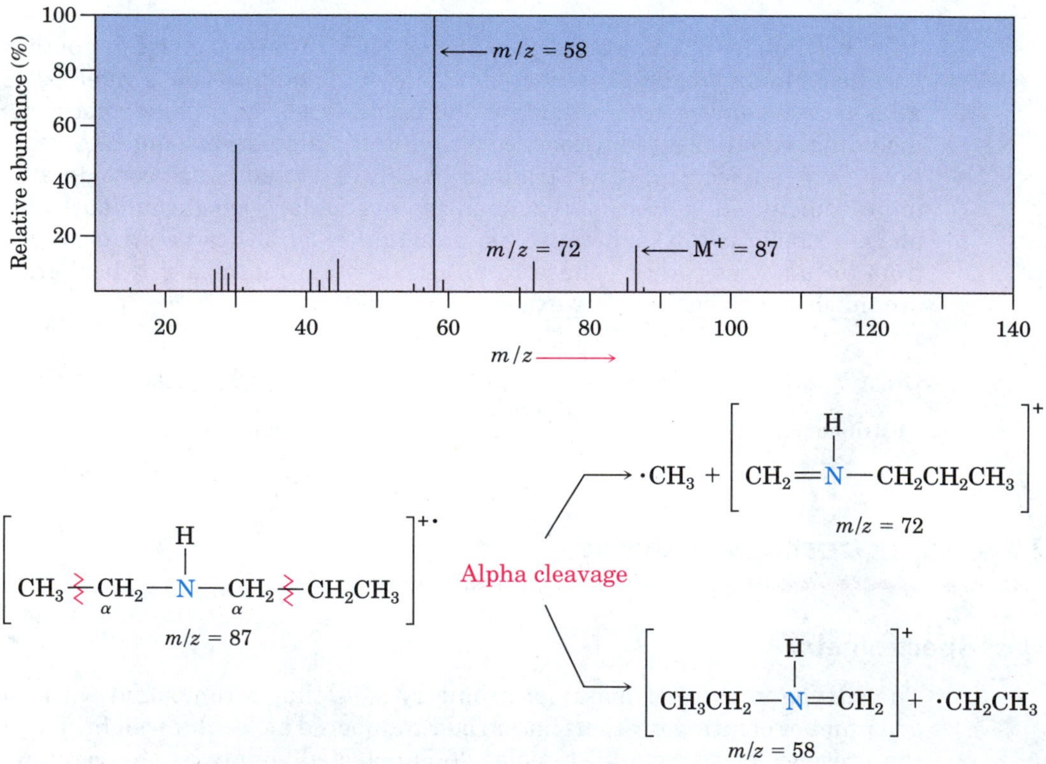

Figure 24.7 Mass spectrum of *N*-ethylpropylamine. The two possible modes of α cleavage lead to the observed fragment ions at $m/z = 58$ and $m/z = 72$.

Infrared Spectroscopy

Primary and secondary amines can be identified by characteristic N–H stretching absorptions in the 3300–3500 cm^{-1} range of the IR spectrum. Alcohols also absorb in this range (Section 17.11), but amine absorption bands are generally sharper and less intense than hydroxyl bands. Primary amines show a pair of bands at about 3350 and 3450 cm^{-1}, and secondary amines show a single band at 3350 cm^{-1}. Tertiary amines show no absorption in this region because they have no N–H bonds. Representative IR spectra of both primary and secondary amines are shown in Figure 24.8.

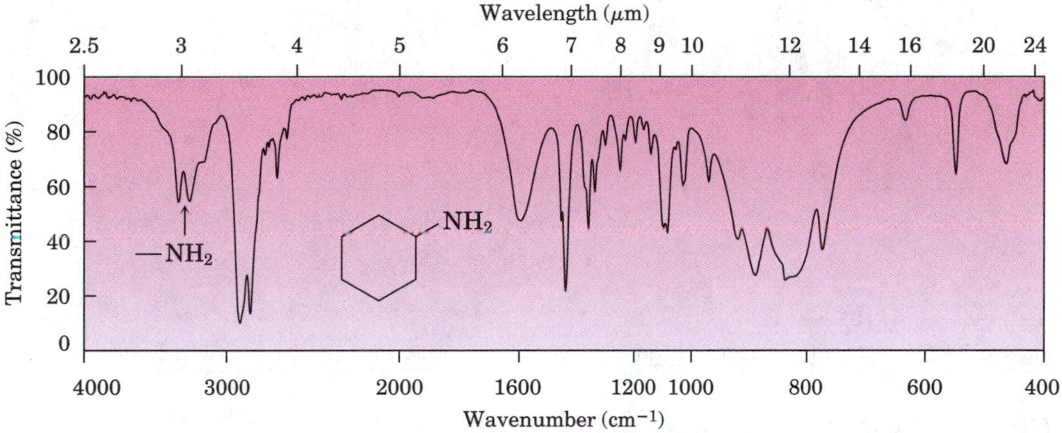

(a)

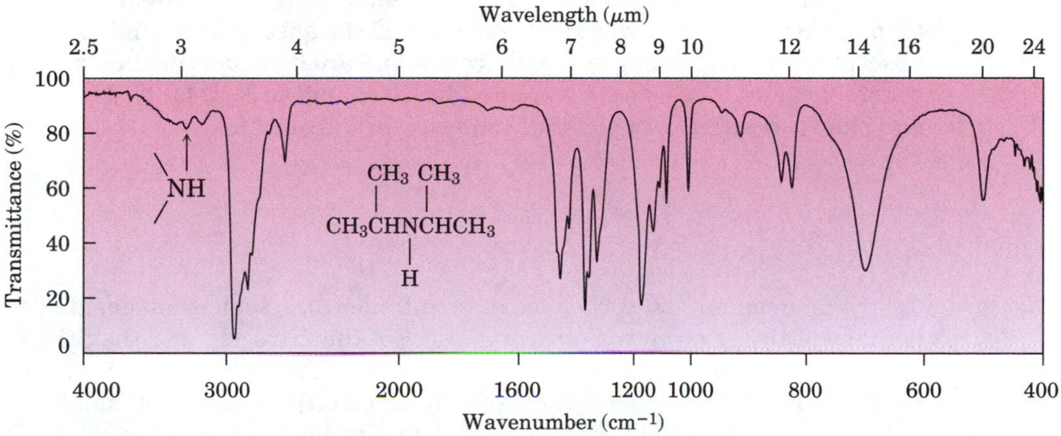

(b)

Figure 24.8 Infrared spectra of (a) cyclohexylamine and (b) diisopropylamine.

In addition to looking for characteristic N–H absorptions, there's also a simple trick for telling whether a compound is an amine. Addition of a small amount of HCl produces a broad and strong ammonium band in the 2200–3000 cm^{-1} range if the sample contains an amino group. All protonated amines show this readily observable absorption caused by the ammonium $R_3N–H^+$ bond. Figure 24.9 gives an example.

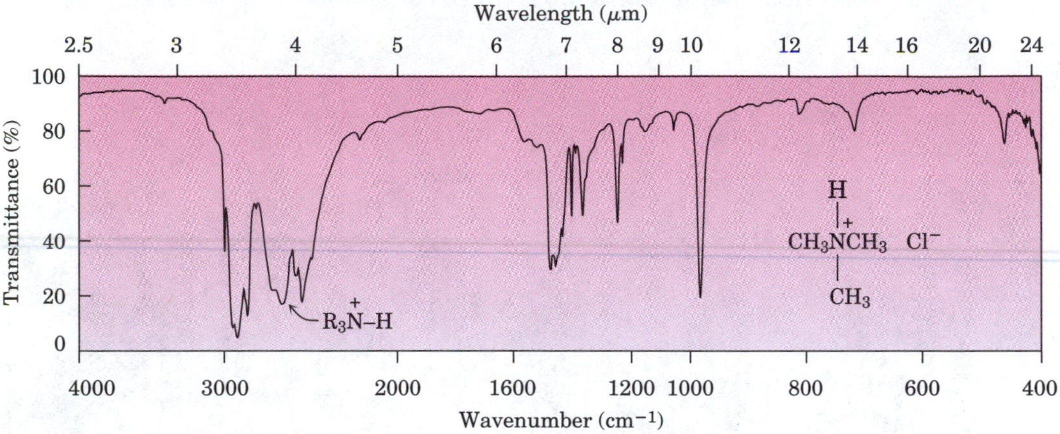

Figure 24.9 Infrared spectrum of trimethylammonium chloride.

Nuclear Magnetic Resonance Spectroscopy

Amines are often difficult to identify solely by ^1H NMR spectroscopy because N–H hydrogens tend to appear as broad signals without clear-cut coupling to neighboring C–H hydrogens. As with O–H absorptions, amine N–H absorptions can appear over a wide range and are best identified by adding a small amount of D_2O to the sample tube. Exchange of N–D for N–H occurs, and the N–H signal disappears from the NMR spectrum.

$$\rangle N{-}H \xrightleftharpoons{D_2O} \rangle N{-}D + HDO$$

Hydrogens on the carbon next to nitrogen are somewhat deshielded because of the electron-withdrawing effect of the nitrogen, and they therefore absorb at lower field than alkane hydrogens. *N*-Methyl groups are particularly distinctive because they absorb as a sharp three-proton singlet at 2.2–2.6 δ. The *N*-methyl resonance at 2.42 δ is easily seen in the ^1H NMR spectrum of *N*-methylcyclohexylamine (Figure 24.10).

Carbons next to amine nitrogens are slightly deshielded in the ^{13}C NMR spectrum and absorb about 20 ppm downfield from where they would absorb in an alkane of similar structure. In *N*-methylcyclohexylamine, for example, the ring carbon to which nitrogen is attached absorbs at a position 24 ppm lower than that of any other ring carbon (Figure 24.11).

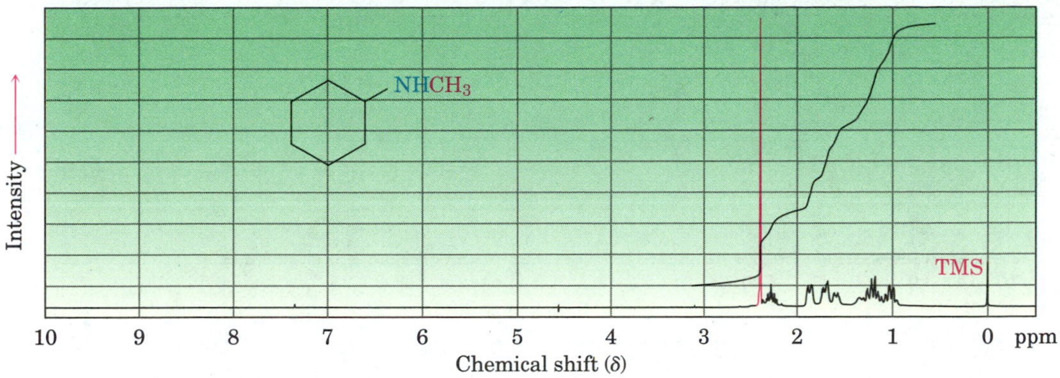

Figure 24.10 Proton NMR spectrum of *N*-methylcyclohexylamine.

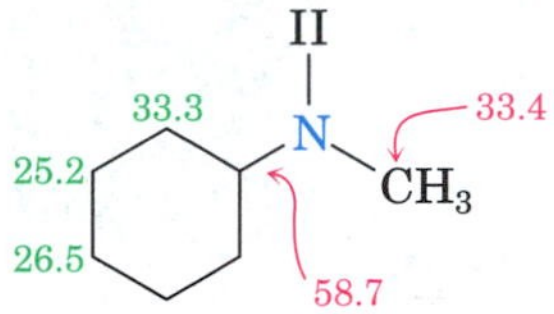

Figure 24.11 The ^{13}C NMR absorptions for *N*-methylcyclohexylamine.

PROBLEM

24.19 Compound A, $C_6H_{12}O$, has an IR absorption at 1715 cm^{-1} and gives compound B, $C_6H_{15}N$, when treated with ammonia and $NaBH_3CN$. The IR and ^1H NMR spectra of B are shown. What are the structures of A and B?

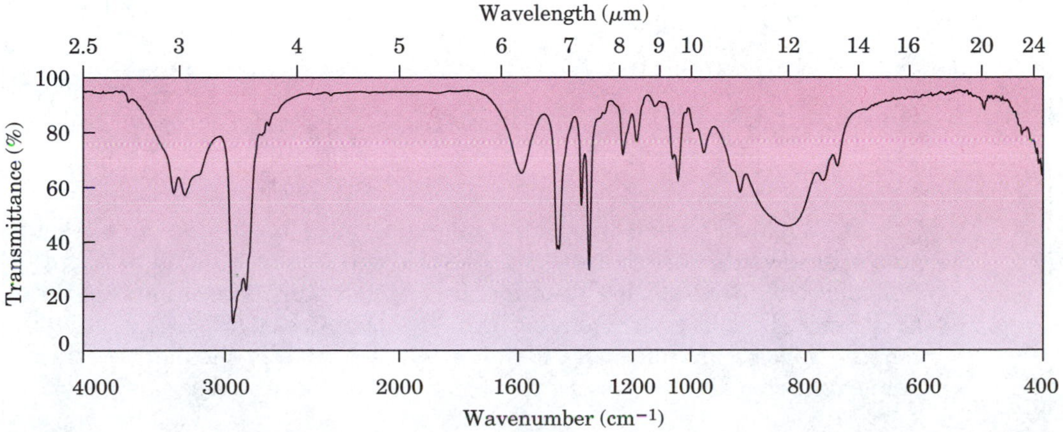

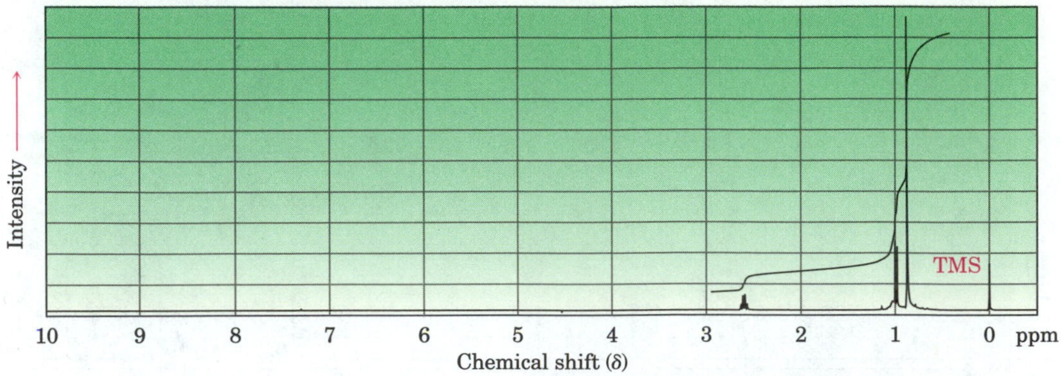

INTERLUDE

Morphine Alkaloids: Naturally Occurring Amines

Morphine and several of its relatives are isolated from the opium poppy, *Papaver somniferum*.

Naturally occurring amines derived from plant sources were once known as "vegetable alkali" because their aqueous solutions are slightly basic, but they are now referred to as **alkaloids**. The study of alkaloids provided much of the impetus for the growth of organic chemistry in the nineteenth century and remains today a fascinating area of research. Let's look briefly at one particular group, the morphine alkaloids.

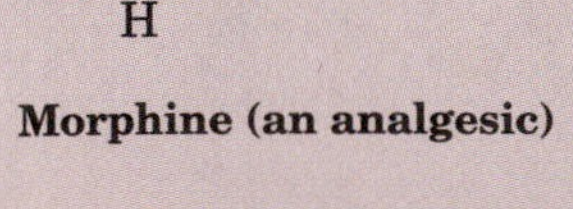

Morphine (an analgesic)

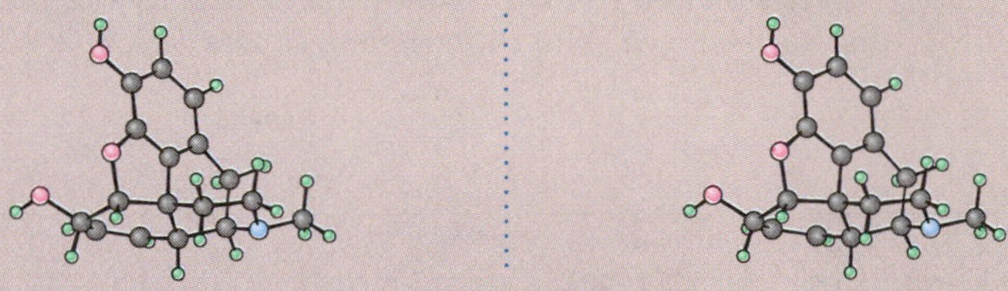

Stereo View

The medical uses of morphine alkaloids have been known at least since the seventeenth century, when crude extracts of the opium poppy, *Papaver somniferum*, were used for the relief of pain. Morphine was the first pure alkaloid to be isolated from the poppy, but its close relative, codeine, also occurs naturally. Codeine, which is simply the methyl ether of morphine, is used in prescription cough medicines and as an analgesic.

(continued)➤

Heroin, another close relative of morphine, does not occur naturally but is synthesized by diacetylation of morphine.

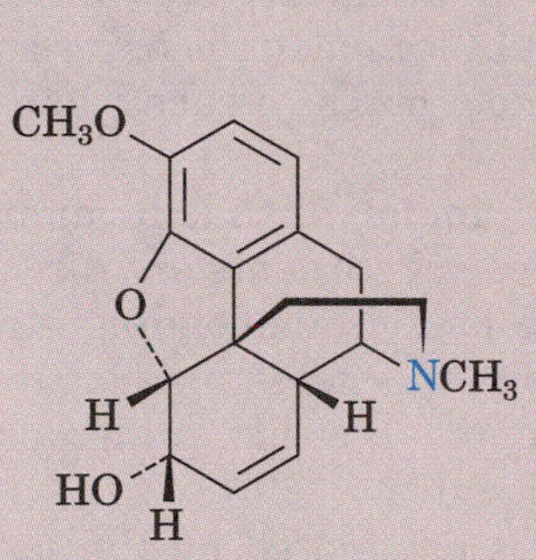

Codeine

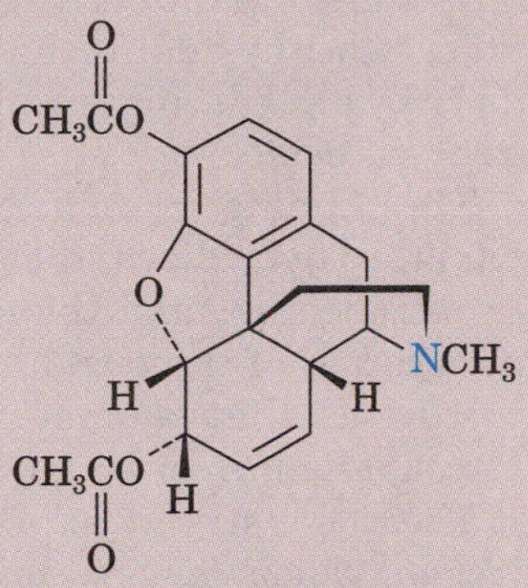

Heroin

Chemical investigations into the structure of morphine occupied some of the finest chemical minds of the nineteenth and early twentieth centuries, and it was not until 1924 that the puzzle was finally solved by Robert Robinson. The key reaction used to establish structure was the Hofmann elimination.

Morphine and its relatives are extremely useful pharmaceutical agents, yet they also pose an enormous social problem because of their addictive properties. Much effort has therefore gone into understanding how morphine works and into developing modified morphine analogs that retain the analgesic activity but don't cause physical dependence. Our present understanding is that morphine binds to opiate receptor sites in the brain. It doesn't interfere with the transmission of a pain signal to the brain but rather changes the brain's reception of the signal.

Hundreds of morphine-like molecules have been synthesized and tested for their analgesic properties. Research has shown that not all the complex framework of morphine is necessary for biological activity. According to the "morphine rule," biological activity requires: (1) an aromatic ring attached to (2) a quaternary carbon atom and (3) a tertiary amine situated (4) two carbon atoms farther away. Meperidine (Demerol), a widely used analgesic, and methadone, a substance used in the treatment of heroin addiction, are two compounds that fit the morphine rule.

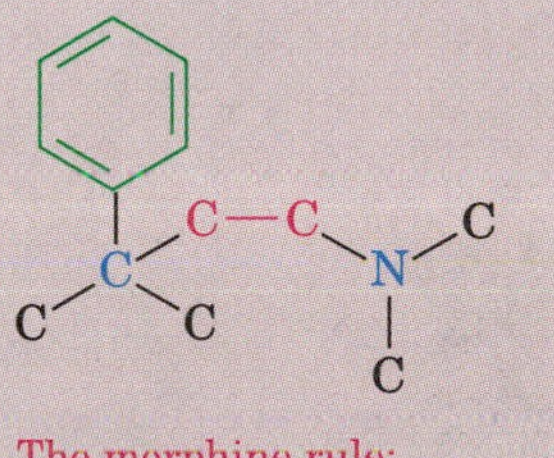

The morphine rule: an aromatic ring, attached to a quaternary carbon, attached to two more carbons, attached to a tertiary amine

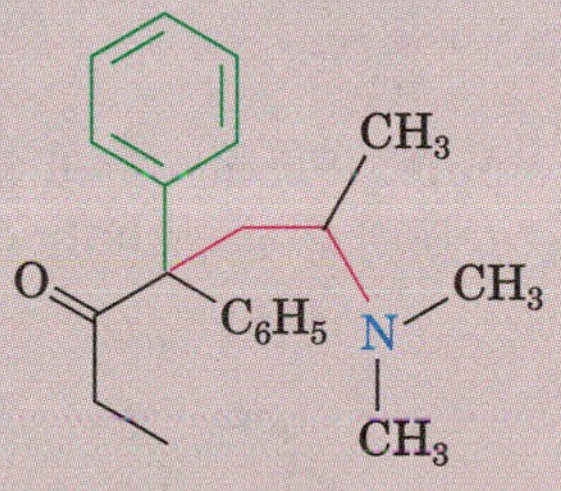

Methadone

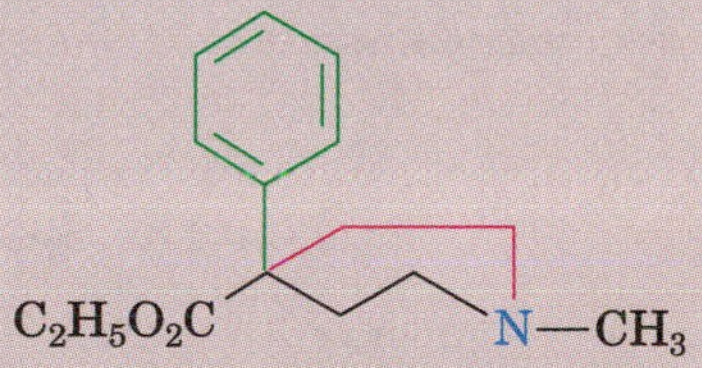

Meperidine

Summary and Key Words

Amines are organic derivatives of ammonia. They are named in the IUPAC system either by adding the suffix *-amine* to the names of the alkyl substituents or by considering the amino group as a substituent on a more complex parent molecule.

The bonding in amines is similar to that in ammonia. The nitrogen atom is sp^3-hybridized, the three substituents are directed to three corners of a tetrahedron, and the lone pair of nonbonding electrons occupies the fourth corner of the tetrahedron. An interesting feature of this tetrahedral structure is that amines undergo a rapid pyramidal inversion, which interconverts mirror-image structures.

The chemistry of amines is dominated by the lone-pair electrons on nitrogen, which makes amines both basic and nucleophilic. The simplest method of amine synthesis involves S_N2 reaction of ammonia or an amine with an alkyl halide. Alkylation of ammonia yields a primary amine; alkylation of a primary amine yields a secondary amine; and so on. This method often gives poor yields, however, and an alternative such as the **Gabriel amine synthesis** is often preferred.

Amines can also be prepared by a number of reductive methods, including $LiAlH_4$ reduction of amides, nitriles, and azides. Even more important is the **reductive amination** reaction in which a ketone or an aldehyde is treated with an amine in the presence of a reducing agent such as $NaBH_3CN$. An intermediate imine is formed and then immediately reduced by the hydride reagent present.

A final method of amine synthesis involves the **Hofmann** and **Curtius rearrangements** of carboxylic acid derivatives. Both methods involve migration of the –R group bonded to the carbonyl carbon and yield a product that has one less carbon atom than the starting material.

Many of the reactions of amines are familiar from past chapters. Thus, amines react with alkyl halides in S_N2 reactions and with acid chlorides in nucleophilic acyl substitution reactions. Amines also undergo E2 elimination to yield alkenes if they are first quaternized by treatment with methyl iodide and then heated with silver oxide (the **Hofmann elimination**).

Summary of Reactions

1. Preparation of amines (Section 24.6)
   (a) The S_N2 alkylation of alkyl halides

Ammonia	$:NH_3 + RX \longrightarrow$	$RNH_3^+ X^-$	$\xrightarrow{NaOH}$ RNH_2	Primary
Primary	$:NH_2R + RX \longrightarrow$	$R_2NH_2^+ X^-$	$\xrightarrow{NaOH}$ R_2NH	Secondary
Secondary	$:NHR_2 + RX \longrightarrow$	$R_3NH^+ X^-$	$\xrightarrow{NaOH}$ R_3N	Tertiary
Tertiary	$:NR_3 + RX \longrightarrow$	$R_4N^+ X^-$	Quaternary ammonium salt	

(continued)➤

(b) Gabriel amine synthesis

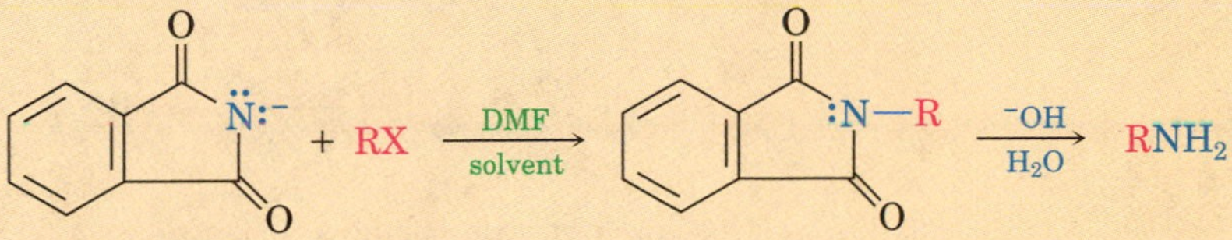

(c) Reduction of azides

$$RCH_2X + Na^+ \ {}^-N_3 \xrightarrow[\text{solvent}]{\text{Ethanol}} RCH_2{-}N{=}\overset{+}{N}{=}\overset{-}{N} \xrightarrow[\text{2. } H_2O]{\text{1. } LiAlH_4\text{, ether}} RCH_2NH_2$$

(d) Reduction of nitriles

$$RCH_2X + Na^+ \ {}^-CN \xrightarrow[\text{solvent}]{\text{DMF}} RCH_2C{\equiv}N \xrightarrow[\text{2. } H_2O]{\text{1. } LiAlH_4\text{, ether}} RCH_2CH_2NH_2$$

(e) Reduction of amides

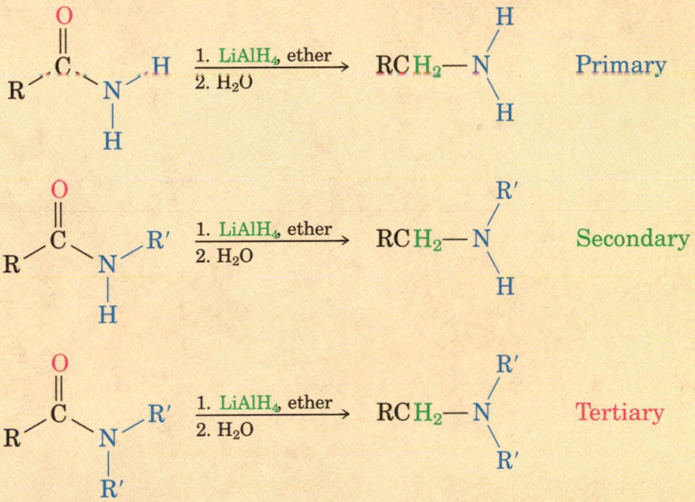

(f) Reductive amination of ketones/aldehydes

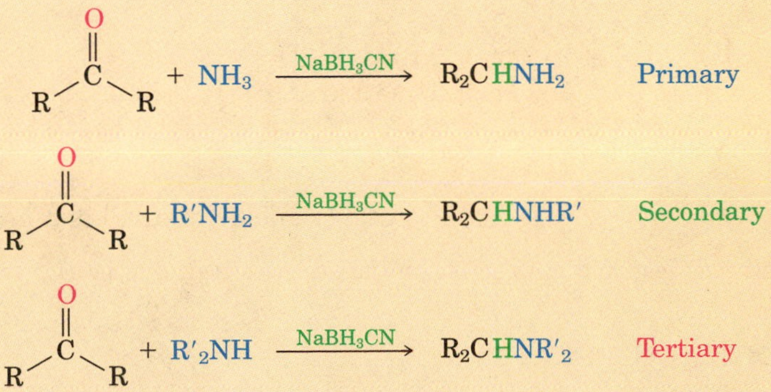

(continued)➤

(g) Hofmann rearrangement of amides

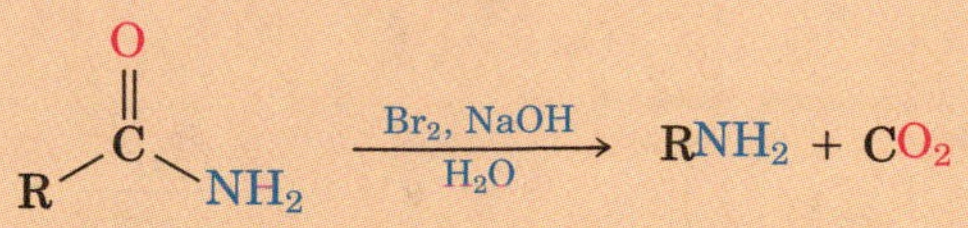

(h) Curtius rearrangement of acyl azides

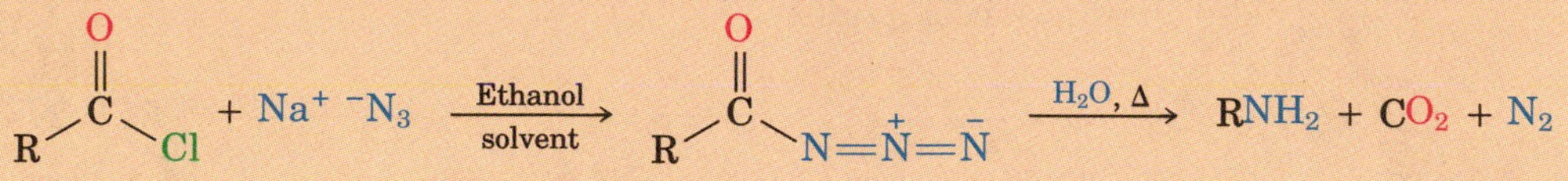

2. Reactions of amines (Section 24.7)
   (a) Alkylation of alkyl halides [see reaction 1(a)]
   (b) Nucleophilic acyl substitution (see also Section 21.5)

Ammonia

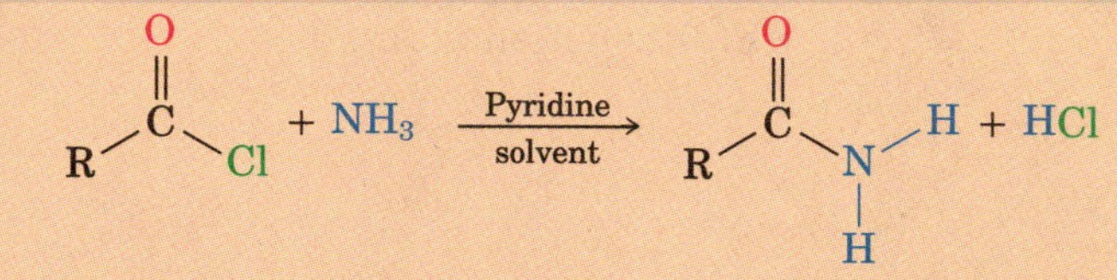

Primary

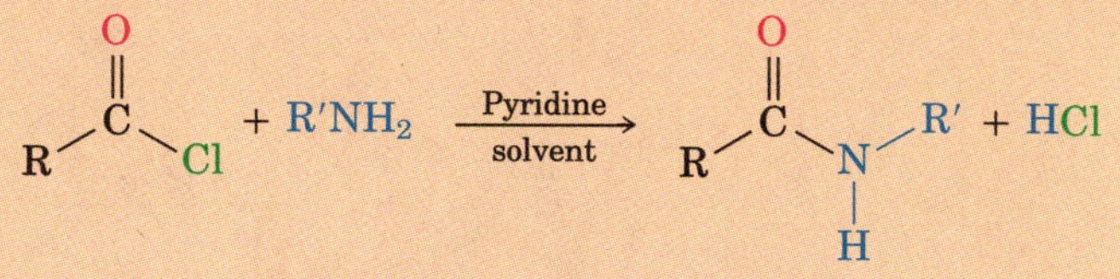

Secondary

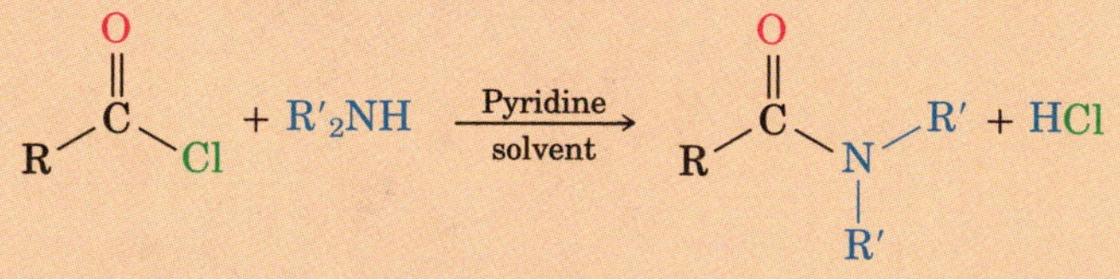

(c) Hofmann elimination

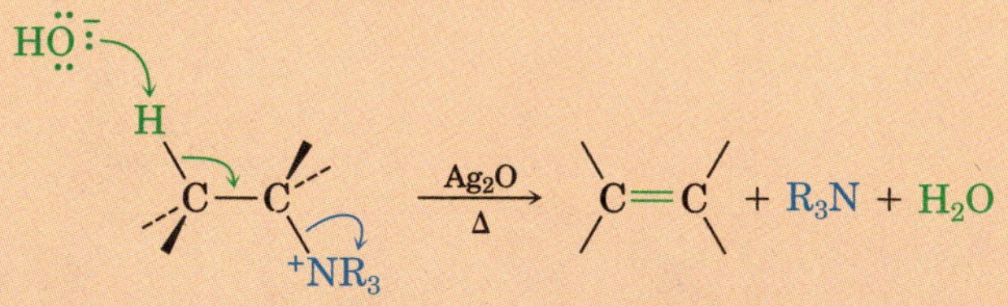

The less highly substituted alkene product is favored.

ADDITIONAL PROBLEMS

24.20 Classify each of the amine nitrogen atoms in the following substances as either primary, secondary, or tertiary.

(a)

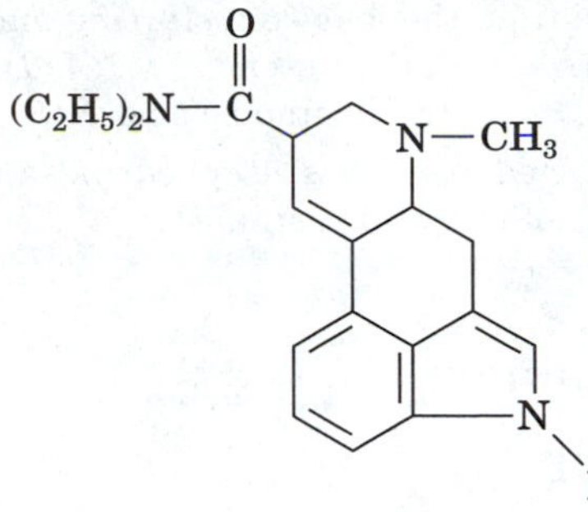

Lysergic acid diethylamide

(b)

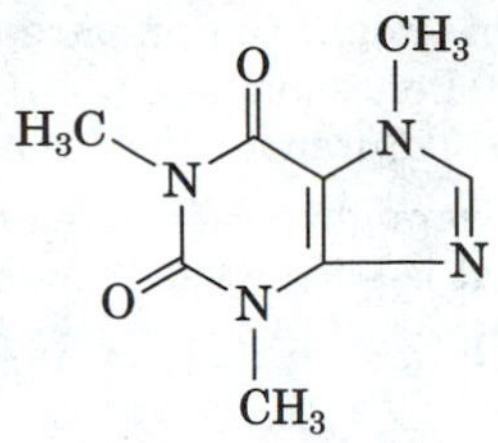

Caffeine

24.21 Draw structures corresponding to the following IUPAC names:
(a) *N,N*-Dimethylaniline
(b) (Cyclohexylmethyl)amine
(c) *N*-Methylcyclohexylamine
(d) (2-Methylcyclohexyl)amine
(e) 3-(*N,N*-Dimethylamino)propanoic acid
(f) *N*-Isopropyl-*N*-methylcyclohexylamine

24.22 Name the following compounds by IUPAC rules:

(a) 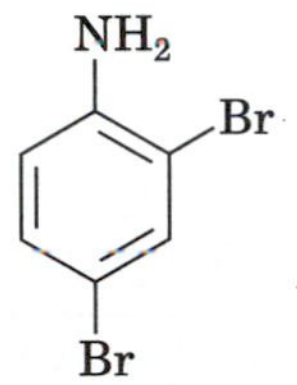

(b) 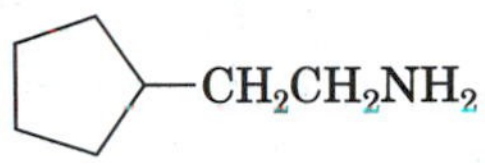

(c) 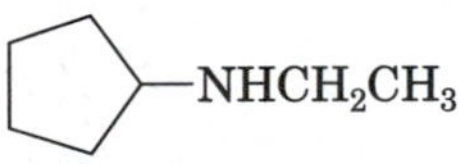

(d) 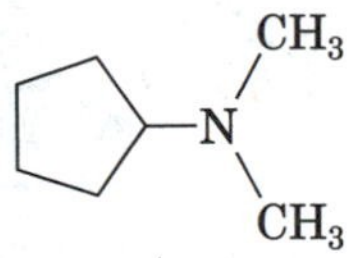

(e) 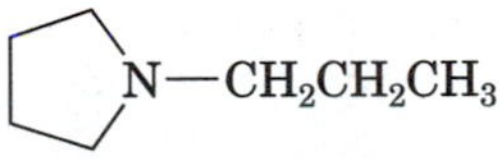

(f) $H_2NCH_2CH_2CH_2CN$

24.23 How can you explain the fact that trimethylamine (bp 3°C) boils lower than dimethylamine (bp 7°C), even though it has a higher molecular weight?

24.24 How would you prepare the following substances from 1-butanol?
(a) Butylamine
(b) Dibutylamine
(c) Propylamine
(d) Pentylamine
(e) *N,N*-Dimethylbutylamine
(f) Propene

24.25 How would you prepare the following substances from pentanoic acid?
(a) Pentanamide
(b) Butylamine
(c) Pentylamine
(d) 2-Bromopentanoic acid
(e) Hexanenitrile
(f) Hexylamine

24.26 Treatment of bromoacetone with ammonia yields a compound having the formula $C_6H_{10}N_2$ rather than the expected 1-amino-2-propanone. What is a likely structure for the product?

24.27 Propose structures for substances that fit the following descriptions:
(a) A chiral quaternary ammonium salt
(b) A five-membered heterocyclic amine
(c) A secondary amine, $C_6H_{11}N$

24.28 How might you prepare pentylamine from the following starting materials?
(a) Pentanamide (b) Pentanenitrile (c) 1-Butene (d) Hexanamide
(e) 1-Butanol (f) 5-Decene (g) Pentanoic acid

24.29 How might a reductive amination be used to synthesize ephedrine, an amino alcohol that is widely used for the treatment of bronchial asthma?

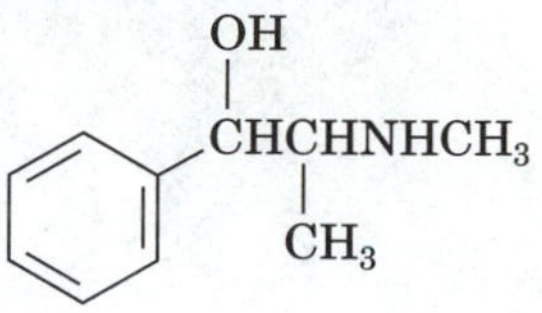

Ephedrine

24.30 Most chiral trisubstituted amines can't be resolved into enantiomers because nitrogen pyramidal inversion occurs too rapidly, but the substance known as *Tröger's base* is an exception. Make molecular models of Tröger's base, and then explain why it is resolvable into enantiomers.

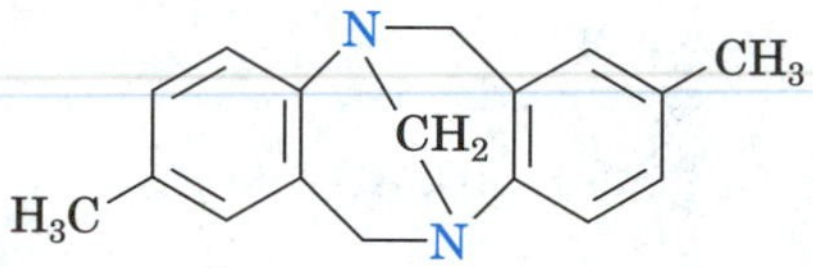

Tröger's base

24.31 Predict the product(s) of the following reactions. If more than one product is formed, tell which is major.

(a) 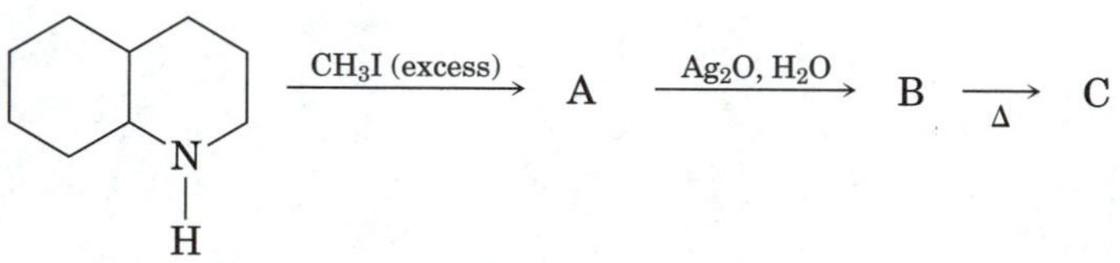

(b) 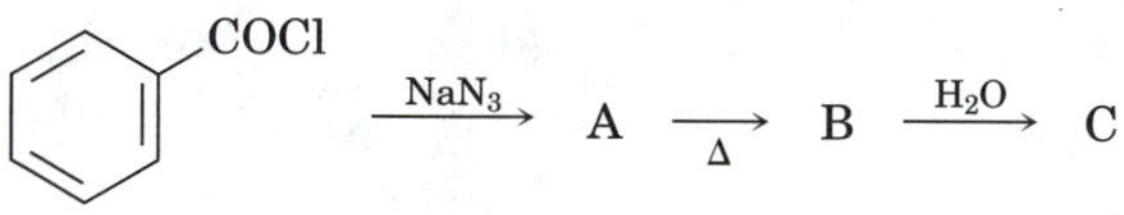

(c) 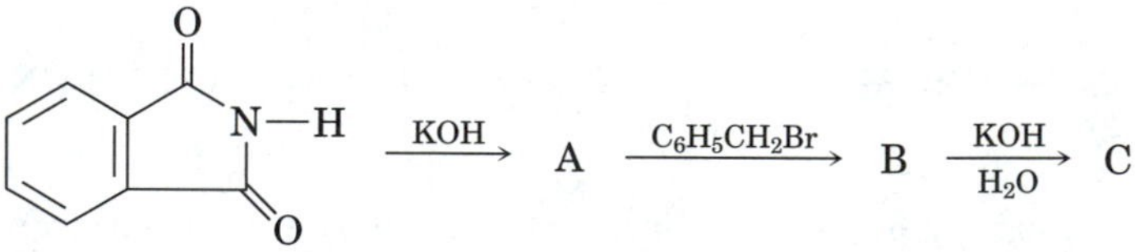

(d) $BrCH_2CH_2CH_2CH_2Br$ + 1 equiv CH_3NH_2 $\xrightarrow[H_2O]{NaOH}$?

24.32 Fill in the missing reagents a–e in the following scheme:

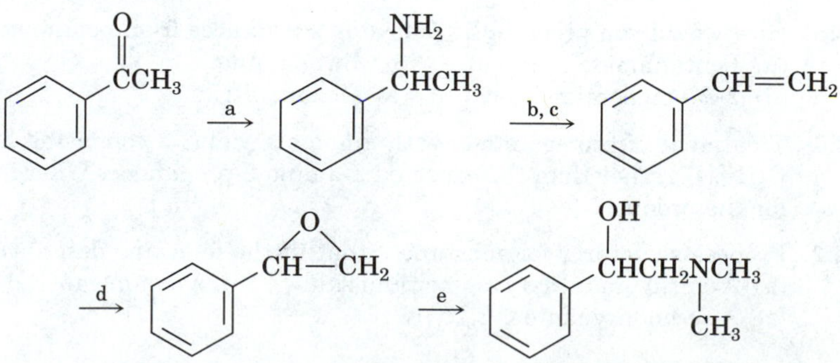

24.33 The following syntheses are incorrect. What is wrong with each?

(a) $CH_3CH_2CONH_2 \xrightarrow{Br_2,\ NaOH,\ H_2O} CH_3CH_2CH_2NH_2$

(b) 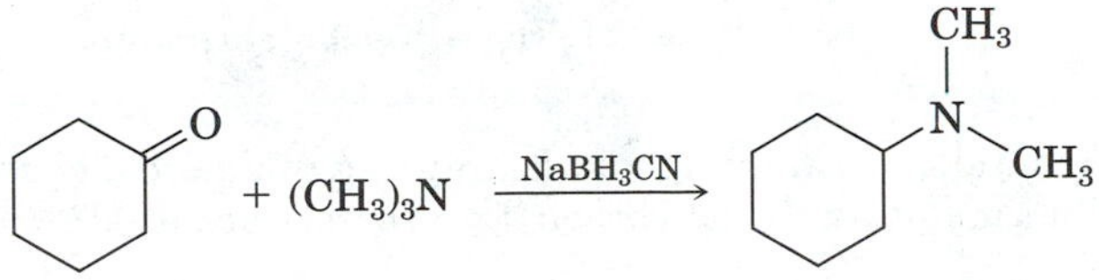

(c) $(CH_3)_3C{-}Br + NH_3 \longrightarrow (CH_3)_3C{-}NH_2$

(d) 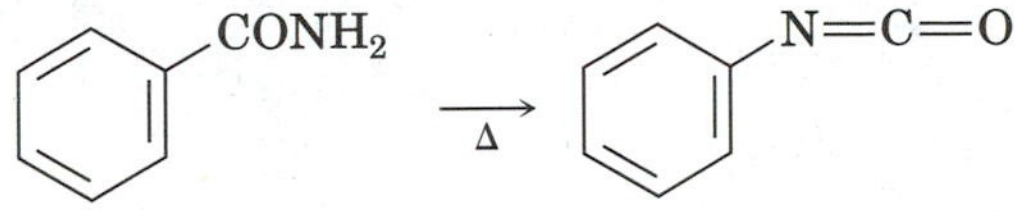

(e) $CH_3CH_2CH_2CH(NH_2)CH_3 \xrightarrow[\text{2. } Ag_2O \quad \text{3. } \Delta]{\text{1. } CH_3I \text{ (excess)}} CH_3CH_2CH{=}CHCH_3$

24.34 Phthalimide, used in the Gabriel synthesis, is prepared by reaction of ammonia with phthalic anhydride (1,2-benzenedicarboxylic anhydride). Propose a mechanism for the reaction.

24.35 Coniine, $C_8H_{17}N$, is the toxic principle of poison hemlock, drunk by Socrates. When subjected to Hofmann elimination, coniine yields 5-(*N,N*-dimethylamino)-1-octene. If coniine is a secondary amine, what is its structure?

24.36 Atropine, $C_{17}H_{23}NO_3$, is a poisonous alkaloid isolated from the leaves and roots of *Atropa belladonna*, the deadly nightshade. In low doses, atropine acts as a muscle relaxant; 0.5 ng (nanogram, 10^{-9} g) is sufficient to cause pupil dilation. On basic hydrolysis, atropine yields tropic acid, $C_6H_5CH(CH_2OH)COOH$, and tropine, $C_8H_{15}NO$. Tropine is an optically inactive alcohol that yields tropidene on dehydration with H_2SO_4. Propose a structure for atropine.

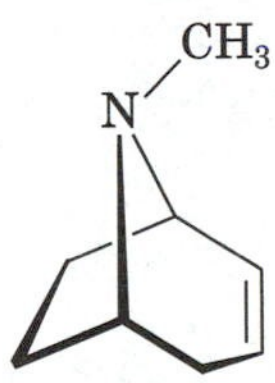

Tropidene

24.37 Tropidene (Problem 24.36) can be converted by a series of steps into tropilidene (1,3,5-cycloheptatriene). How would you accomplish this conversion?

24.38 One problem with reductive amination as a method of amine synthesis is that by-products are sometimes obtained. For example, reductive amination of benzaldehyde with methylamine leads to a mixture of methylbenzylamine and methyldibenzylamine. How do you suppose the tertiary amine by-product is formed? Propose a mechanism.

24.39 What are the major products you would expect from Hofmann elimination of the following amines?

(a) *N*-Methylcyclopentylamine
(b) $(CH_3)_2CHCH(NH_2)CH_2CH_2CH_3$
(c) *N*-Phenyl-*N*-(1-methyl)pentylamine

24.40 Cyclopentamine is an amphetamine-like central nervous system stimulant. Propose a synthesis of cyclopentamine from materials of five carbons or less.

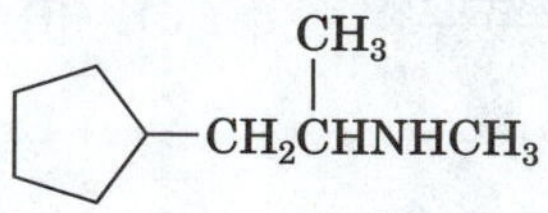

Cyclopentamine

24.41 Prolitane is an antidepressant drug that is prepared commercially using a reductive amination. What amine and what carbonyl precursors are used?

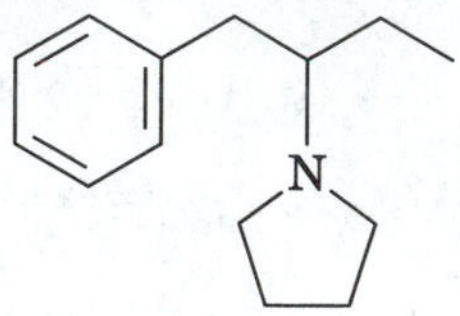

Prolitane

24.42 Tetracaine is a substance used medicinally as a spinal anesthetic during lumbar punctures (spinal taps).

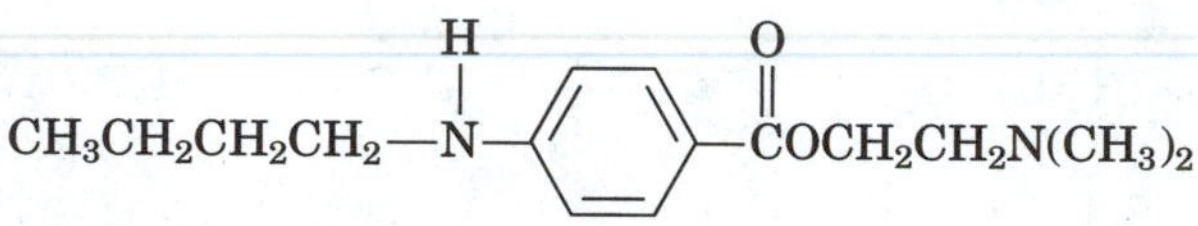

Tetracaine

(a) How would you prepare tetracaine from the corresponding aniline derivative, $ArNH_2$?
(b) How would you prepare tetracaine from *p*-nitrobenzoic acid?
(c) How would you prepare tetracaine from benzene?

24.43 Propose a structure for the product with formula $C_9H_{17}N$ that results when 2-(2-cyanoethyl)cyclohexanone is reduced catalytically.

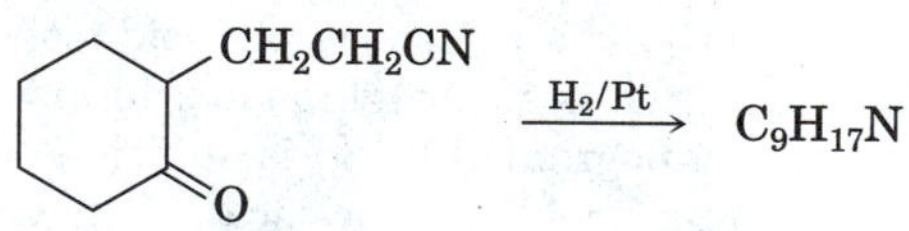

24.44 How would you synthesize coniine (Problem 24.35) from acrylonitrile ($H_2C{=}CHCN$) and ethyl 3-oxohexanoate ($CH_3CH_2CH_2COCH_2CO_2C_2H_5$)? (*Hint*: See Problem 24.43.)

24.45 How would you synthesize the heart stimulant propranolol starting from 1-hydroxynaphthalene and any other reagents needed?

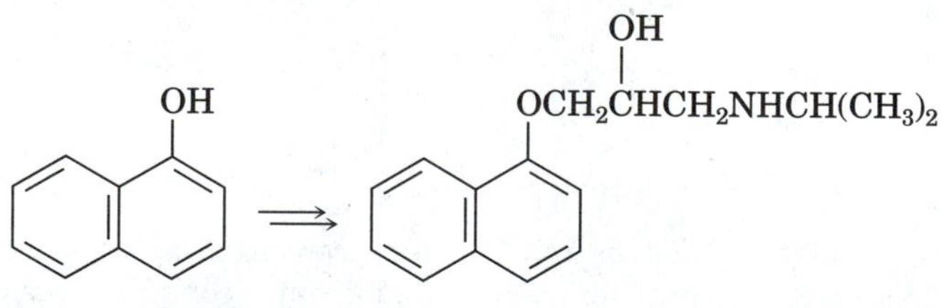

1-Hydroxynaphthalene **Propranolol**

24.46 Although the barrier to nitrogen inversion is normally too low to allow isolation of one enantiomer of a chiral amine, (+)-1-chloro-2,2-diphenylaziridine has been prepared optically pure and has been shown to be stable for several hours at 0°C. Explain.

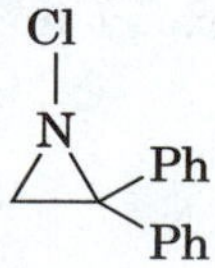

1-Chloro-2,2-diphenylaziridine

24.47 The hydrolysis of phthalimides is often slow, and an alternative method is sometimes needed to liberate the primary amine in the last stage of a Gabriel synthesis. In the Ing–Manske modification, reaction of an *N*-alkylphthalimide with hydrazine is used. Propose a mechanism for this *hydrazinolysis*.

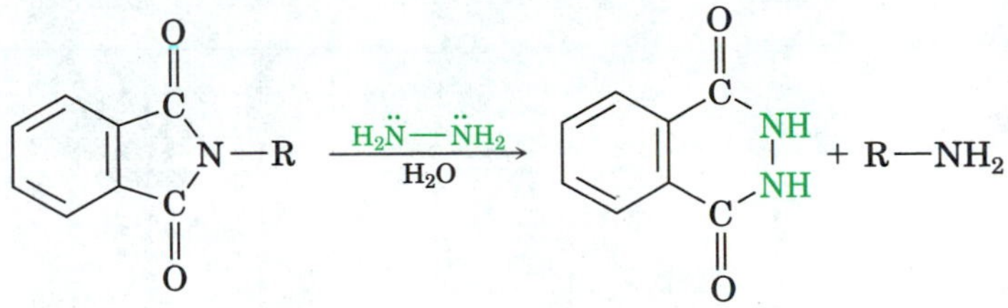

Phthalhydrazide

24.48 Cyclooctatetraene was first synthesized in 1911 by a route that involved the following transformation:

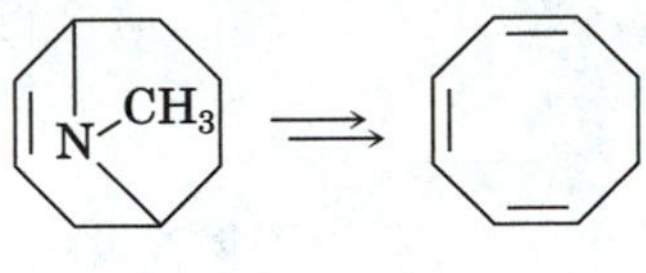

How might you use the Hofmann elimination to accomplish this reaction? How would you finish the synthesis by converting cyclooctatriene into cyclooctatetraene?

24.49 When an α-hydroxy amide is treated with Br_2 in aqueous NaOH under Hofmann rearrangement conditions, loss of CO_2 occurs and a chain-shortened aldehyde is formed. Propose a mechanism.

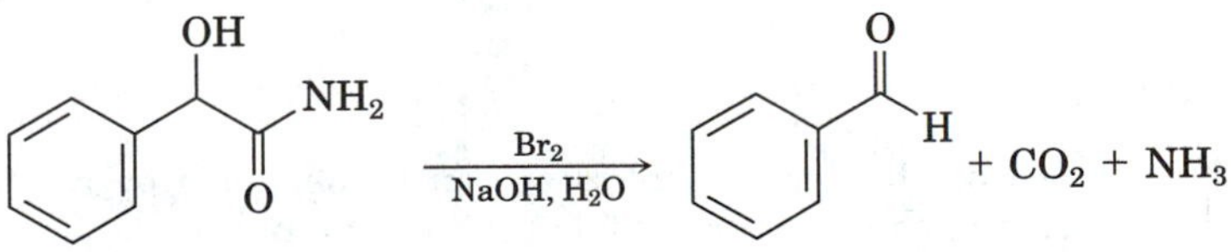

24.50 Propose a mechanism for the following reaction:

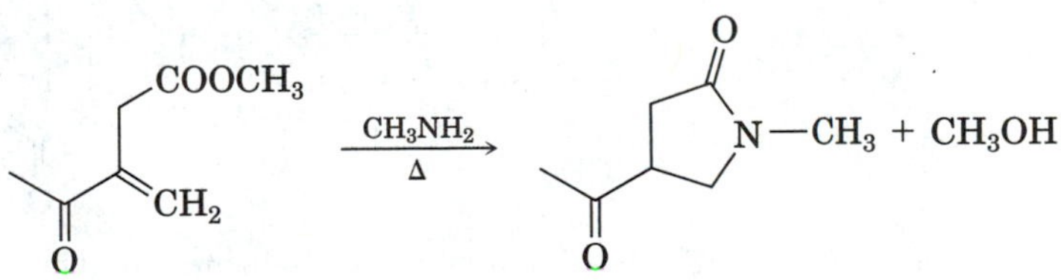

24.51 Propose a mechanism for the following reaction:

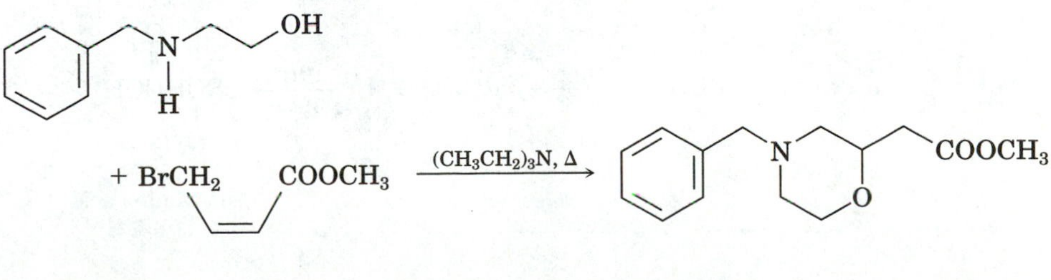

24.52 Propose structures for amines with the following ^1H NMR spectra.

(a) C_3H_9NO

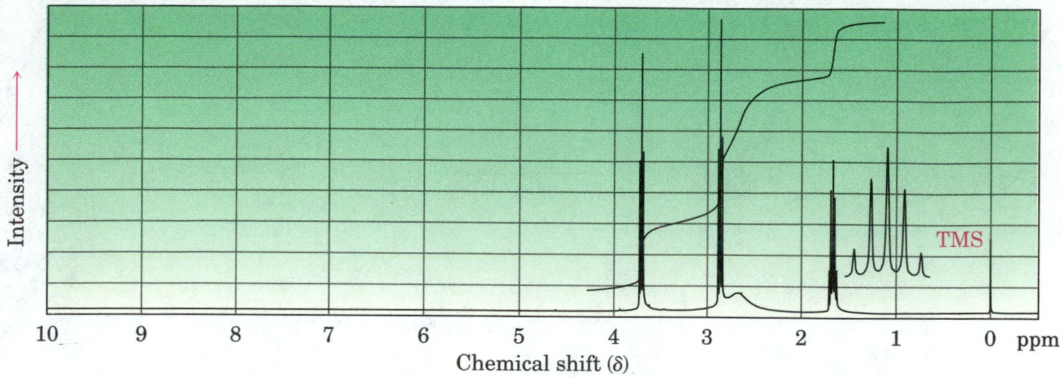

(b) $C_4H_{11}NO_2$

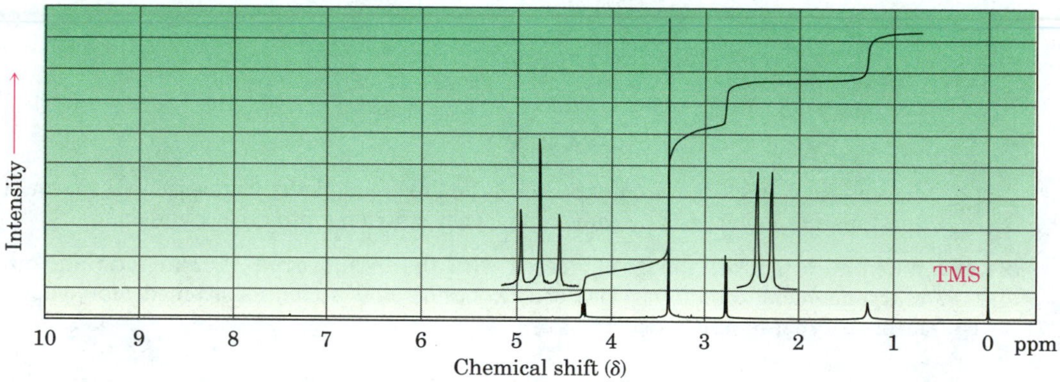

A Look Ahead

24.53 We'll see in Chapter 29 that DNA can be degraded by reaction with hydrazine. Propose a mechanism for the following reaction:

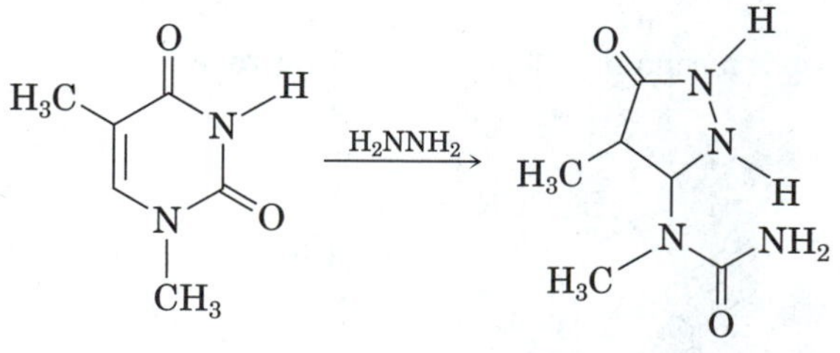

24.54 We'll see in Chapter 27 that α-amino acids can be prepared by the *Strecker synthesis*, a two-step process in which an aldehyde is treated with ammonium cyanide followed by hydrolysis of the amino nitrile intermediate with aqueous acid. Propose a mechanism for the reaction.

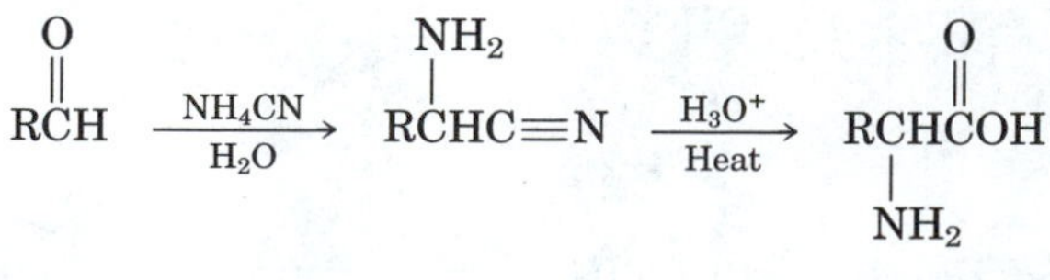

An α-amino acid

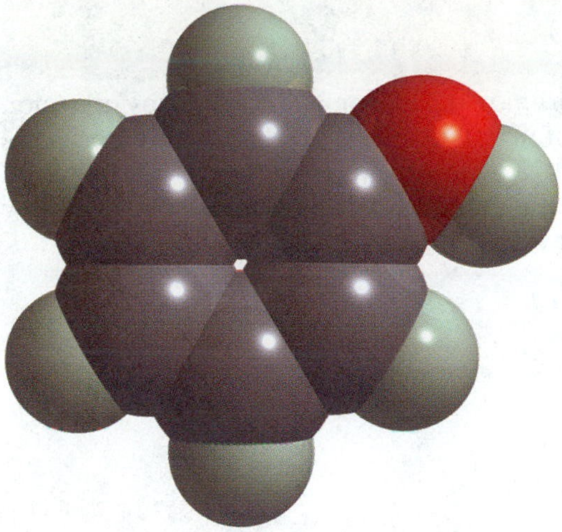

Phenol is a frequently used antiseptic agent.

25

Arylamines and Phenols

The founding of the modern organic chemical industry can be traced to the need for a single compound—aniline—and to the activities of one person—William Henry Perkin.[1] Perkin, a student of Hofmann's at the Royal College of Chemistry in London, worked during the day on problems assigned him by Hofmann but spent his free time working on his own in an improvised home laboratory. One day during Easter vacation in 1856, he decided to examine the oxidation of aniline with potassium dichromate. Although the reaction appeared unpromising at first, yielding a tarry black product, Perkin was able by careful extraction with methanol to isolate a small amount of a beautiful purple pigment that could dye silk.

Since the only dyes known at the time were naturally occurring vegetable dyes such as indigo, Perkin's synthetic purple dye, which he named *mauve*, created a sensation. Realizing the possibilities, Perkin resigned his post with Hofmann and, at the age of 18, formed a company to exploit his remarkable discovery.

No chemical industry existed at the time, since there had never before been a need for synthetic chemicals. Large-scale chemical manufacture was unknown, and Perkin's first task was to devise a procedure for preparing the needed quantities of aniline. He therefore worked out the techniques of

[1]Sir William Henry Perkin (1838–1907); b. London; studied with Hofmann at the Royal College of Chemistry, London; industrial consultant, London.

manufacture and soon learned to prepare aniline on a large scale by nitration of benzene, followed by reduction of nitrobenzene with iron and hydrochloric acid. A similar procedure is used today to prepare some 600,000 tons of aniline annually in the United States, although the reduction step is now carried out by catalytic hydrogenation.

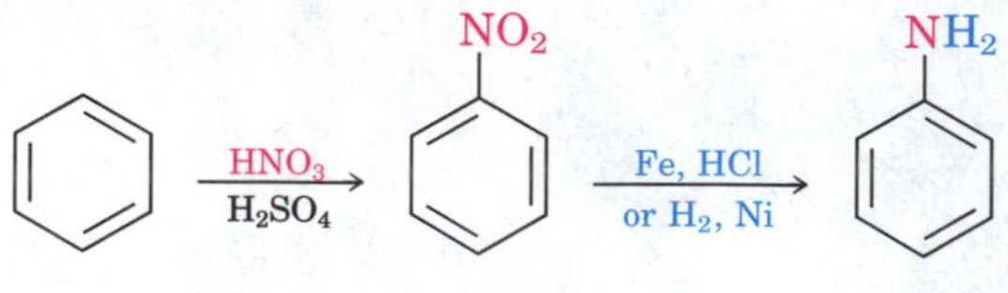

Subsequent work showed that Perkin's original mauve was not derived from aniline but from a toluidine (methylaniline) impurity in his starting material. Pure aniline yields a similar dye, however, which came to be marketed under the name *pseudomauveine*.

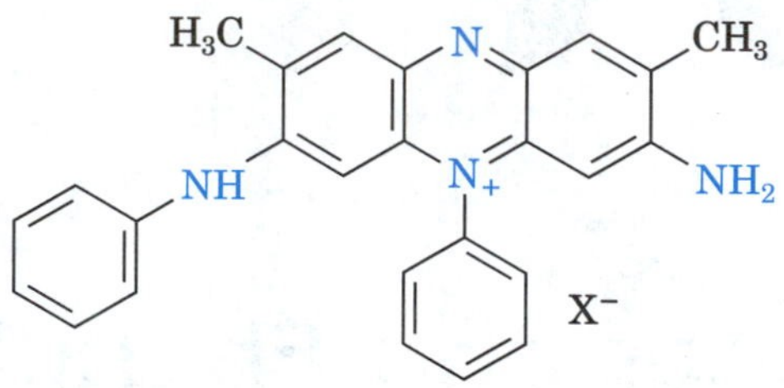

Perkin's mauve
(pseudomauveine has no methyl groups)

Dyestuff manufacture is today a thriving and important part of the chemical industry, and many commonly used pigments are derived from aniline. Although aniline itself and several substituted anilines are available naturally from coal tar, synthesis from benzene is the major source.

25.1 Basicity of Arylamines

Arylamines, like their aliphatic counterparts, are basic. The lone pair of nonbonding electrons on nitrogen can bond to acids, yielding an arylammonium salt. The base strength of arylamines is generally lower than that of aliphatic amines, however. Thus, methylammonium ion has $pK_a = 10.66$, whereas anilinium ion has $pK_a = 4.63$. [Remember from Section 24.4 that the base strength of an amine is inversely related to the acid strength of its corresponding ammonium ion. A stronger base corresponds to a less acidic ammonium ion (higher pK_a), and a weaker base corresponds to a more acidic ammonium ion (lower pK_a).]

Arylamines are less basic than alkylamines because the nitrogen lone-pair electrons are delocalized by interaction with the aromatic ring π electron system and are less available for bonding. In resonance terms, arylamines are stabilized relative to alkylamines because of the five resonance structures:

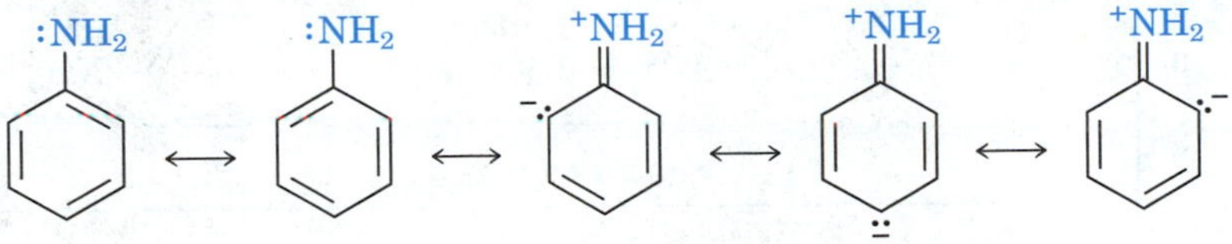

Resonance stabilization is lost on protonation, though, because only two resonance structures are possible for the arylammonium ion:

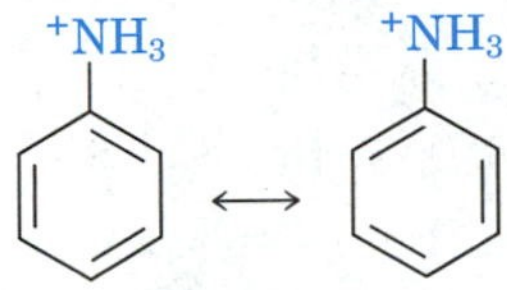

As a result, the energy difference ($\Delta G°$) between protonated and nonprotonated forms is higher for arylamines than it is for alkylamines, and arylamines are therefore less basic. Figure 25.1, which compares reaction energy diagrams for protonation of alkylamines and arylamines, illustrates the difference in $\Delta G°$ for the two reactions.

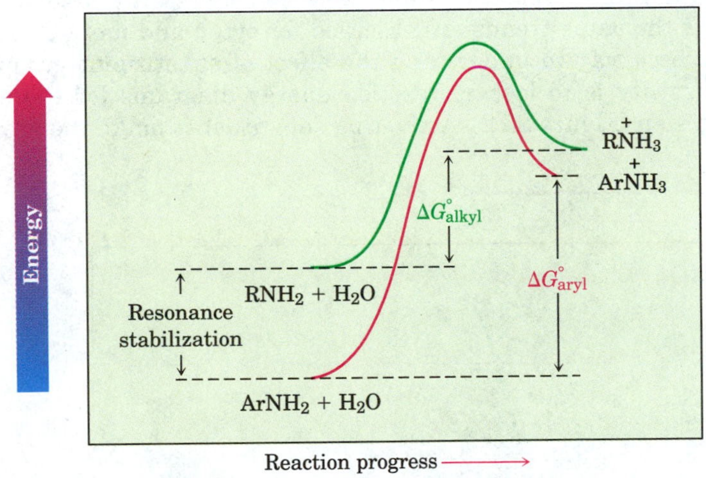

Figure 25.1 Reaction energy diagrams for the protonation of alkylamines (green curve) and arylamines (red curve). Arylamines have a larger positive $\Delta G°$ and are therefore less basic than alkylamines, primarily because of resonance stabilization of their ground state.

Substituted arylamines can be either more basic or less basic than aniline, depending on the substituent. Table 25.1 gives data for a variety of *p*-substituted anilines.

Table 25.1 Base Strength of *p*-Substituted Anilines

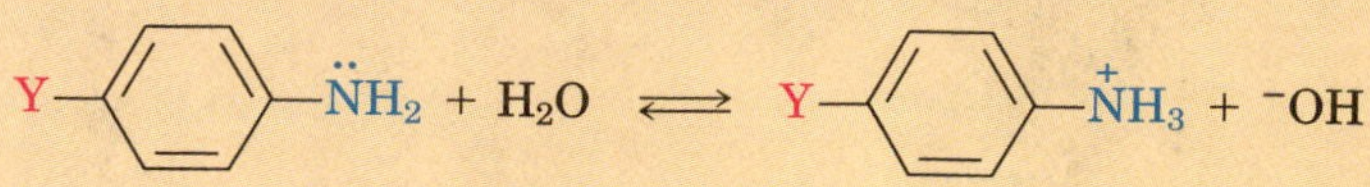

	Substituent, Y	pK_a	
Stronger base	$-NH_2$	6.15	Activating groups
	$-OCH_3$	5.34	Activating groups
	$-CH_3$	5.08	Activating groups
	—H	4.63	
	—Cl	3.98	Deactivating groups
	—Br	3.86	Deactivating groups
Weaker base	—C≡N	1.74	Deactivating groups
	$-NO_2$	1.00	Deactivating groups

Electron-donating substituents such as $-CH_3$, $-NH_2$, and $-OCH_3$, which increase the reactivity of an aromatic ring toward electrophilic substitution (Section 16.5), also increase the basicity of the corresponding arylamine. Conversely, electron-withdrawing substituents such as $-Cl$, $-NO_2$, and $-CN$, which decrease ring reactivity toward electrophilic substitution, also decrease arylamine basicity. Table 25.1 considers only *p*-substituted anilines, but the same trends are observed for ortho and meta derivatives.

The best way to understand the effect of substituent groups on arylamine basicity is to look at reaction energy diagrams for the amine protonation step (Figure 25.2). Activating substituents make the aromatic ring

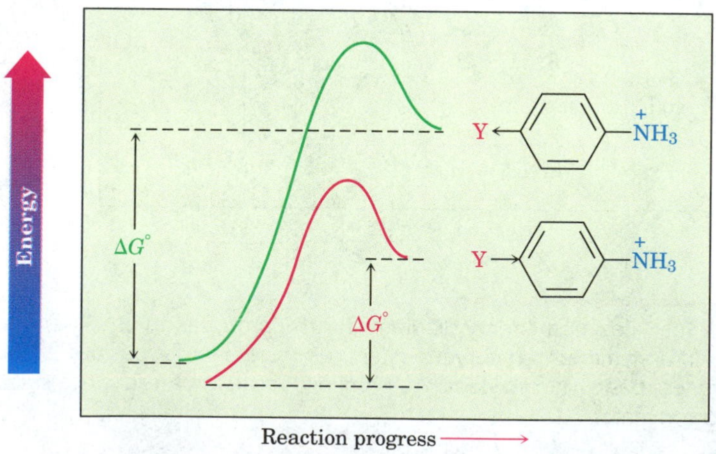

Figure 25.2 Reaction energy diagrams for the protonation of substituted arylamines. An electron-donating substituent (red curve) stabilizes the ammonium salt much more than an electron-withdrawing substituent does (green curve).

electron-rich, thereby increasing the stability of the positively charged ammonium ion. Deactivating substituents make the aromatic ring electron-poor, thereby decreasing the stability of the ammonium ion. We therefore find a lower positive $\Delta G°$ for protonation of an activated arylamine than for protonation of a deactivated arylamine.

PROBLEM.

25.1 Account for the fact that *p*-nitroaniline ($pK_a = 1.0$) is less basic than *m*-nitroaniline ($pK_a = 2.5$) by a factor of 30. Draw resonance structures to support your argument. (The pK_a values refer to the corresponding ammonium ions.)

PROBLEM.

25.2 Rank the following compounds in order of ascending basicity. Don't look at Table 25.1.

(a) *p*-Nitroaniline, *p*-aminobenzaldehyde, *p*-bromoaniline
(b) *p*-Chloroaniline, *p*-aminoacetophenone, *p*-methylaniline
(c) *p*-(Trifluoromethyl)aniline, *p*-methylaniline, *p*-(fluoromethyl)aniline

25.2 Preparation of Arylamines

Arylamines are usually prepared by nitration of an aromatic starting material, followed by reduction of the nitro group. No other method of synthesis approaches this nitration/reduction route for versatility and generality.

The reduction step can be carried out in many different ways, depending on the circumstances. Catalytic hydrogenation over platinum is clean and gives high yields, but is often incompatible with the presence elsewhere in the molecule of other reducible groups, such as C=C bonds or carbonyl groups. Iron, zinc, tin, and stannous chloride ($SnCl_2$) are also effective when used in acidic aqueous solution. Stannous chloride is particularly mild and is often used when other reducible functional groups are present.

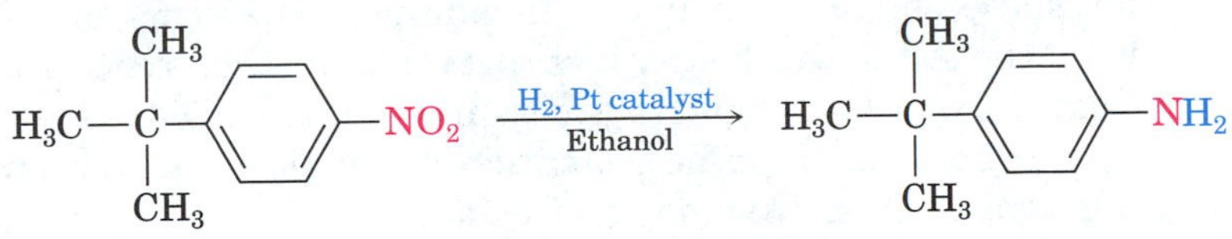

***p-tert*-Butylnitrobenzene** **p-tert-Butylaniline (100%)**

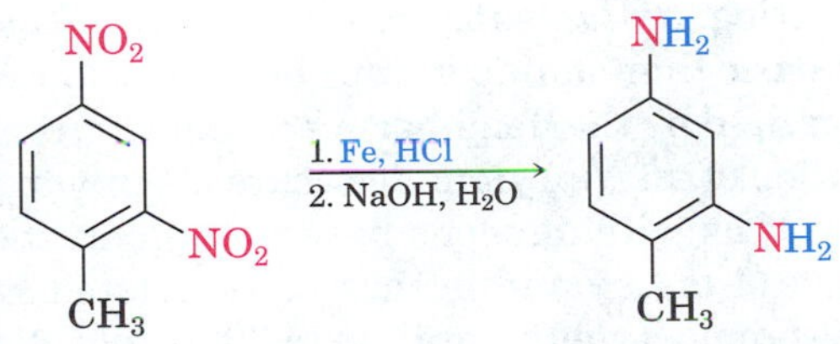

2,4-Dinitrotoluene **Toluene-2,4-diamine (74%)**

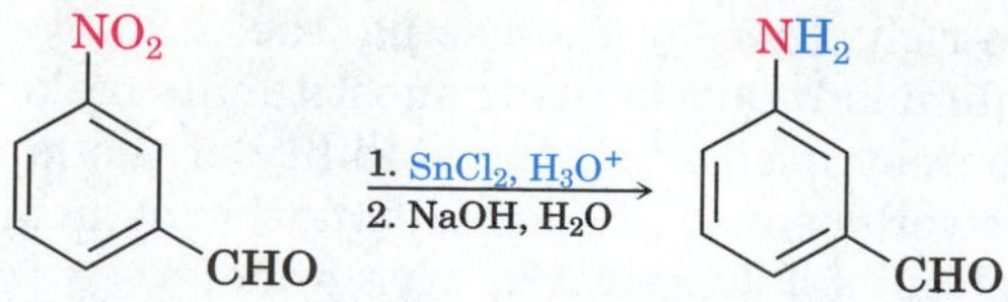

***m*-Nitrobenzaldehyde** ***m*-Aminobenzaldehyde (90%)**

25.3 Reactions of Arylamines

Electrophilic Aromatic Substitution

Amino substituents are strongly activating, ortho- and para-directing groups in electrophilic aromatic substitution reactions (Section 16.5). The high reactivity of amino-substituted benzenes can be a drawback at times because it's sometimes difficult to prevent polysubstitution. For example, reaction of aniline with Br_2 takes place rapidly and yields the 2,4,6-tribrominated product. The amino group is so strongly activating that it's not possible to stop at the monobromo stage.

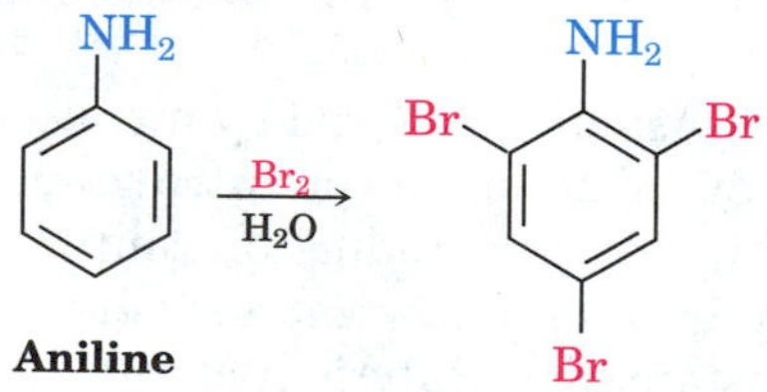

Aniline **2,4,6-Tribromoaniline (100%)**

Another drawback to the use of amino-substituted benzenes in electrophilic aromatic substitution reactions is that Friedel–Crafts reactions are not successful (Section 16.3). The amino group forms an acid–base complex with the $AlCl_3$ catalyst, which prevents further reaction from occurring. Both drawbacks—high reactivity and amine basicity—can be overcome by carrying out electrophilic aromatic substitution reactions on the corresponding amide rather than on the free amine.

As we saw in Section 21.6, treatment of an arylamine with acetic anhydride yields an *N*-acetylated product. Though still activating and ortho-, para-directing, *amido* substituents (–NHCOR) are less strongly activating and less basic than amino groups because their nitrogen lone-pair electrons are delocalized by the neighboring carbonyl group (Section 24.4). As a result, bromination of an *N*-arylamide occurs cleanly to give a monobromo product, and hydrolysis with aqueous base then gives the free amine. For example, *p*-toluidine (4-methylaniline) can be acetylated, brominated, and hydrolyzed to yield 2-bromo-4-methylaniline in 79% yield. None of the 2,6-dibrominated product is obtained.

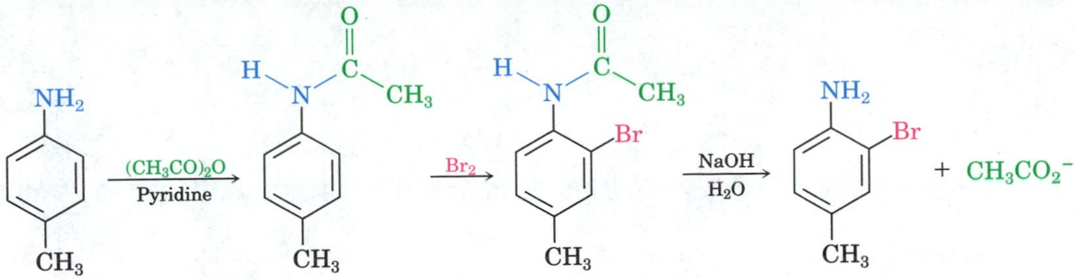

p-Toluidine

2-Bromo-4-methylaniline (79%)

Friedel–Crafts alkylations and acylations of *N*-arylamides also proceed normally. For example, benzoylation of acetanilide (*N*-acetylaniline) under Friedel–Crafts conditions gives 4-aminobenzophenone in 80% yield after hydrolysis:

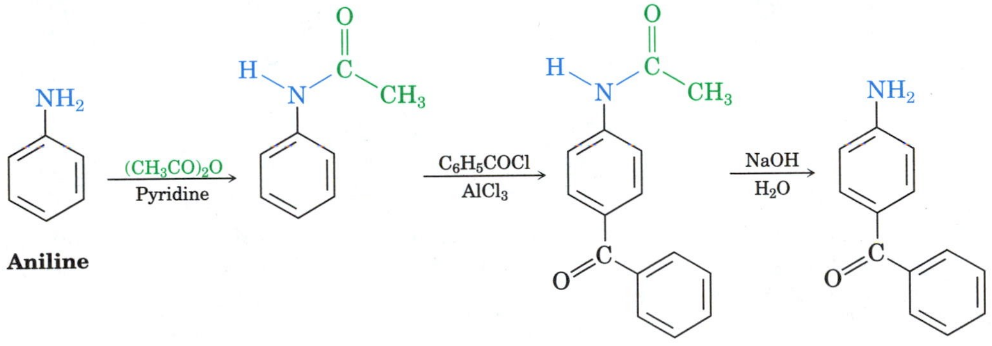

Aniline

4-Aminobenzophenone (80%)

Modulating the reactivity of an amino-substituted benzene by forming an amide is a useful trick that allows many kinds of electrophilic aromatic substitutions to be carried out that would otherwise be impossible. A good example is the preparation of *sulfa drugs*.

Sulfa drugs, such as sulfanilamide, were among the first pharmaceutical agents to be used clinically against infection. Although they have largely been replaced by safer and more powerful antibiotics, sulfa drugs were widely used in the 1940s and were credited with saving the lives of thousands of wounded during World War II. They are prepared by chlorosulfonation of acetanilide, followed by reaction of *p*-(*N*-acetylamino)benzenesulfonyl chloride with ammonia or some other amine to give a sulfonamide. Hydrolysis of the amide then yields the sulfa drug. Note that this amide hydrolysis can be carried out in the presence of the sulfonamide group because sulfonamides hydrolyze very slowly.

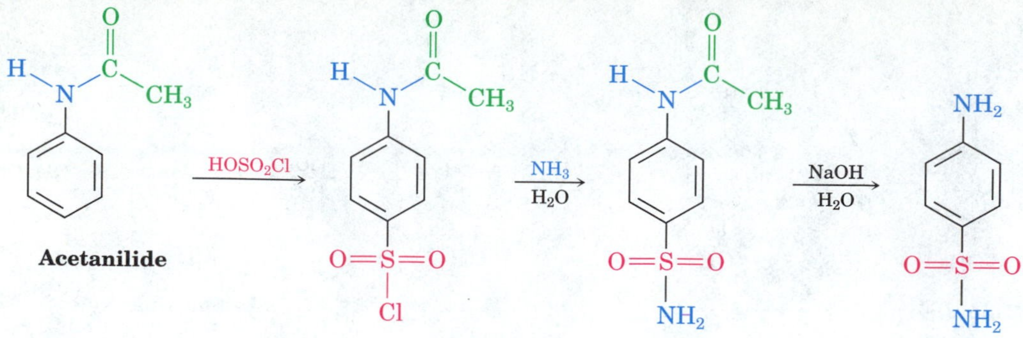

Sulfanilamide (a sulfa drug)

PROBLEM.

25.3 Propose a synthesis of sulfathiazole from benzene and any necessary amine.

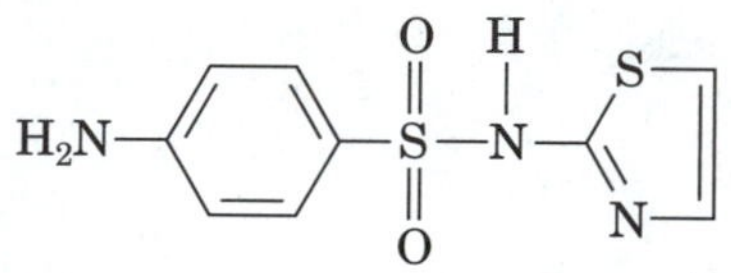

Sulfathiazole

PROBLEM.

25.4 Account for the fact that an amido substituent (–NHCOR) is ortho- and para-directing by drawing resonance structures that share the nitrogen lone-pair electrons with the aromatic ring.

PROBLEM.

25.5 Propose syntheses of the following compounds from benzene:
(a) *N,N*-Dimethylaniline
(b) *p*-Chloroaniline
(c) *m*-Chloroaniline
(d) 2,4-Dimethylaniline

Diazonium Salts: The Sandmeyer Reaction

Primary arylamines react with nitrous acid, HNO_2, to yield stable **arenediazonium salts**, $Ar{-}\overset{+}{N}{\equiv}N\ X^-$. This *diazotization* reaction is compatible with the presence of a wide variety of substituents on the aromatic ring.

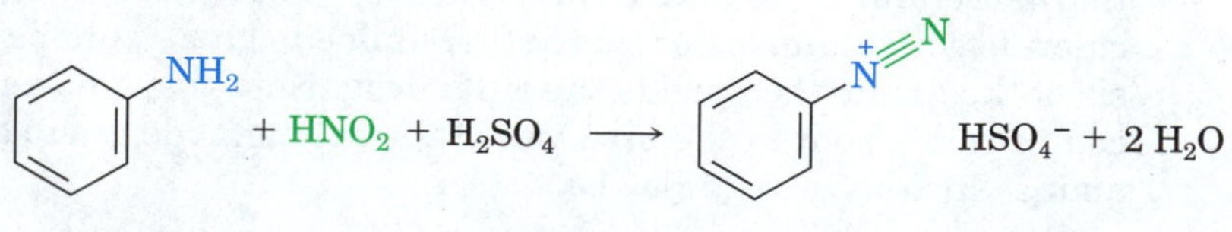

*Alkyl*amines also react with nitrous acid, but the alkanediazonium products of these reactions can't be isolated. Instead, they lose nitrogen instantly to yield carbocations. (The analogous loss of N_2 from an arenediazonium ion to yield an aryl cation is disfavored by the instability of the cation.)

$$RNH_2 + HNO_2 + H_2SO_4 \longrightarrow [R\overset{+}{N}{\equiv}N\ {}^{-}HSO_4] \longrightarrow R^+\ {}^{-}HSO_4 + N_2$$

Arenediazonium salts are extremely useful in synthesis because the diazonio group (N_2^+) can be replaced by nucleophiles in a substitution reaction:

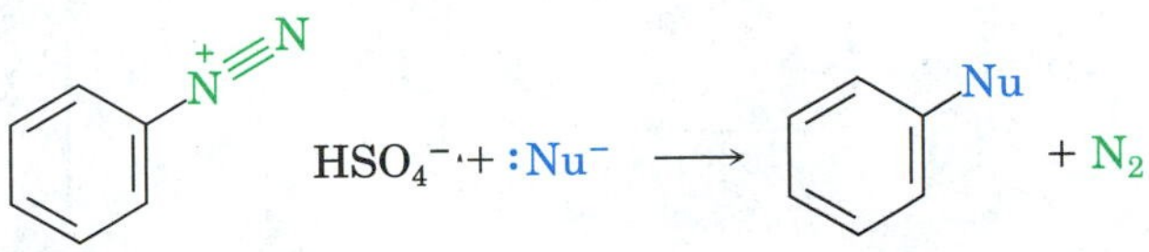

Many different nucleophiles react with arenediazonium salts yielding many different kinds of substituted benzenes. The overall sequence of (1) nitration, (2) reduction, (3) diazotization, and (4) nucleophile replacement is probably the single most versatile method of aromatic substitution (Figure 25.3). Although the details aren't fully understood, the mechanism by which these substitutions occur involves radicals.

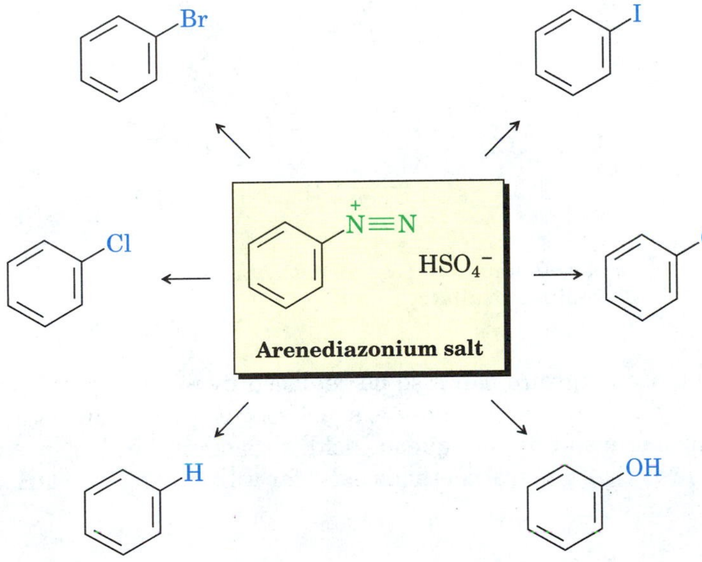

Figure 25.3 Preparation of substituted aromatic compounds by diazonio replacement reactions.

Aryl chlorides and bromides are prepared by reaction of an arenediazonium salt with the corresponding cuprous halide, CuX, a process called the **Sandmeyer[2] reaction.** Aryl iodides can be prepared by direct reaction with NaI without using a cuprous salt. Yields generally fall in the range 60–80%.

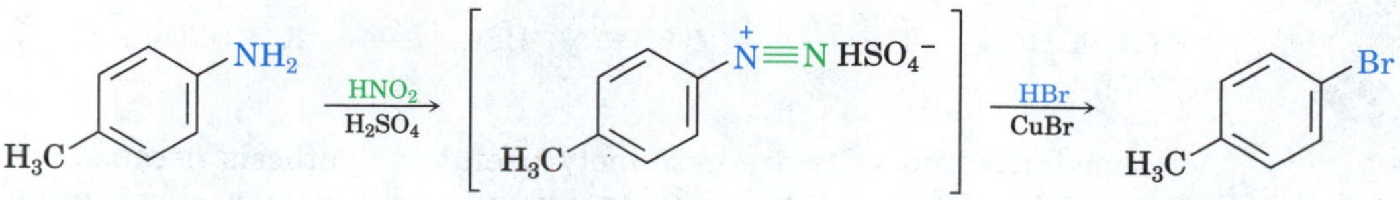

p-Methylaniline **p-Bromotoluene (73%)**

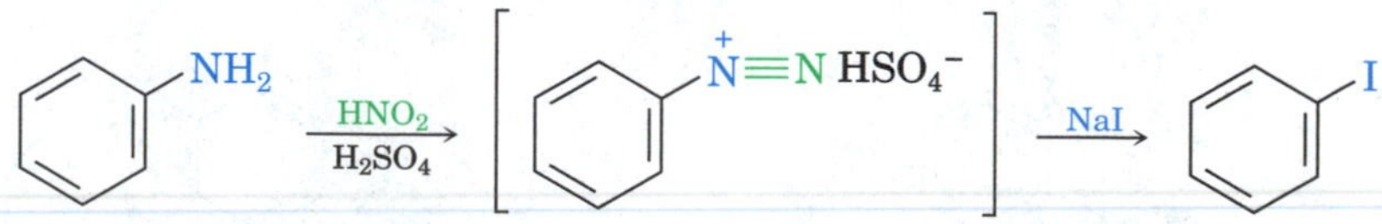

Aniline **Iodobenzene (67%)**

Similar treatment of an arenediazonium salt with CuCN yields the arenenitrile, ArCN. This reaction is particularly useful because it allows the replacement of a nitrogen substituent by a carbon substituent. The nitrile can then be further converted into other functional groups such as carboxyl. For example, Sandmeyer reaction of *o*-methylbenzenediazonium bisulfate with CuCN yields *o*-methylbenzonitrile, which can be hydrolyzed to give *o*-methylbenzoic acid. This product could not be prepared from *o*-xylene by the usual side-chain oxidation route because both methyl groups would be oxidized.

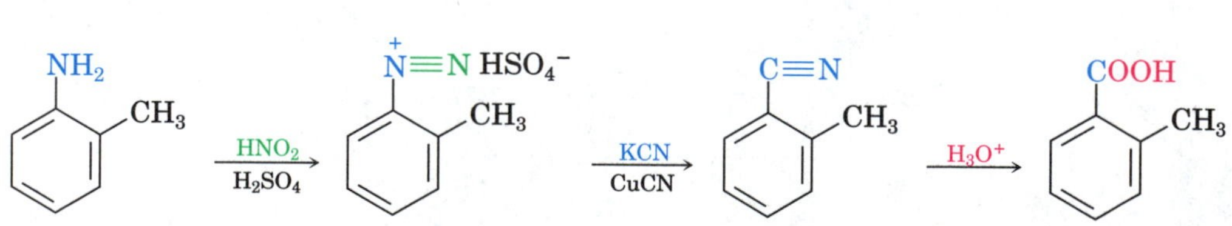

o-Methylaniline **o-Methylbenzene-diazonium bisulfate** **o-Methylbenzonitrile** **o-Methylbenzoic acid**

The diazonio group can also be replaced by –OH to yield phenols and by –H to yield arenes. Phenols are usually prepared by addition of the arenediazonium salt to hot aqueous acid, a reaction that is especially useful because few other general methods exist for introducing an –OH group onto an aromatic ring.

[2]Traugott Sandmeyer (1854–1922); b. Wettingen, Switzerland; Ph.D. Heidelberg (Gattermann); Geigy Company, Basel, Switzerland.

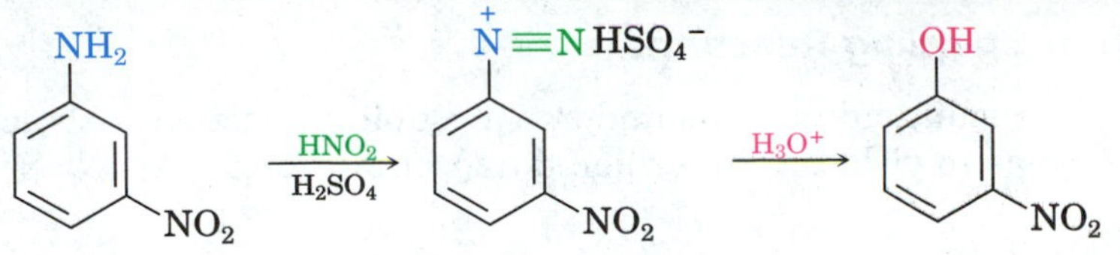

m-Nitroaniline

m-Nitrophenol (86%)

Arenes are produced by reduction of a diazonium salt with hypophosphorous acid, H_3PO_2. This reaction is used primarily when there is a need for temporarily introducing an amino substituent onto a ring to take advantage of its directing effect. Suppose, for example, that you needed to make 3,5-dibromotoluene. The product can't be made by direct bromination of toluene because reaction would occur at positions 2 and 4. Starting with *p*-methylaniline (*p*-toluidine), however, dibromination occurs ortho to the strongly directing amino substituent, and diazotization followed by treatment with H_3PO_2 yields the desired product.

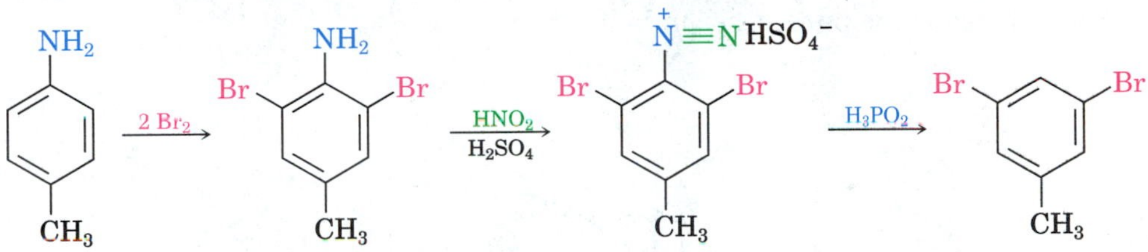

p-Methylaniline

3,5-Dibromotoluene

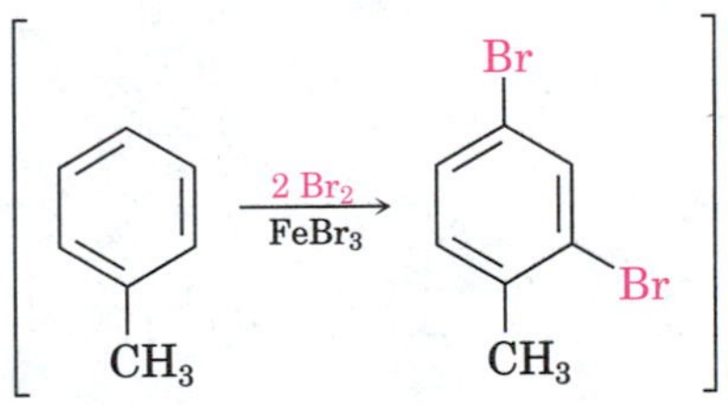

Toluene 2,4-Dibromotoluene

PROBLEM

25.6 How might you convert aniline ($C_6H_5NH_2$) into benzylamine ($C_6H_5CH_2NH_2$) using a diazonium replacement reaction?

PROBLEM

25.7 How would you prepare the following compounds from benzene using a diazonium replacement reaction at some point?

(a) *p*-Bromobenzoic acid
(b) *m*-Bromobenzoic acid
(c) *m*-Bromochlorobenzene
(d) *p*-Methylbenzoic acid
(e) 1,2,4-Tribromobenzene

Diazonium Coupling Reactions

Arenediazonium salts undergo a coupling reaction with activated aromatic rings to yield brightly colored **azo compounds**, Ar–N=N–Ar′:

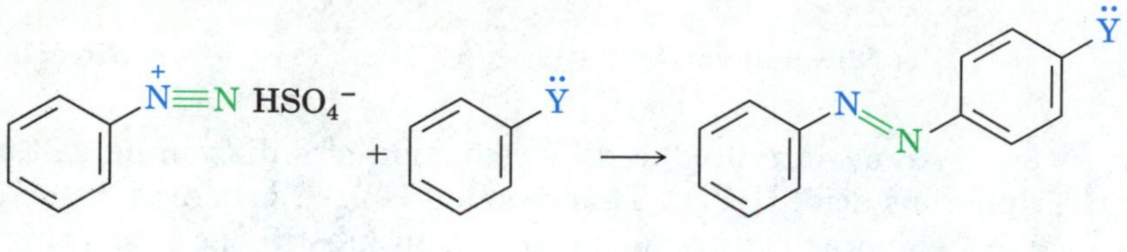

An azo compound

where Y = –OH or $-NR_2$

Diazonium coupling reactions are typical electrophilic aromatic substitutions (Section 16.2) in which the positively charged diazonium ion is the electrophile that reacts with the electron-rich ring of a phenol or arylamine. Reaction usually occurs at the para position, although ortho attack can take place if the para position is blocked.

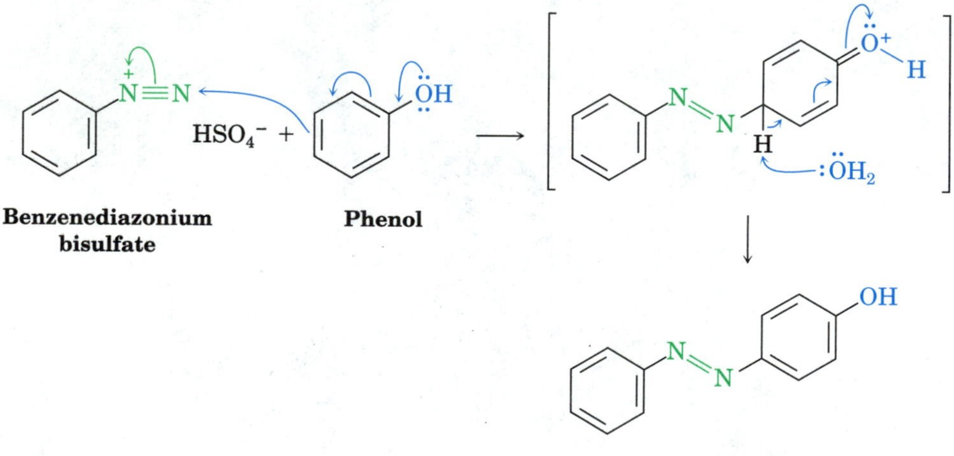

Benzenediazonium bisulfate

Phenol

***p*-Hydroxyazobenzene (orange crystals, mp 152°C)**

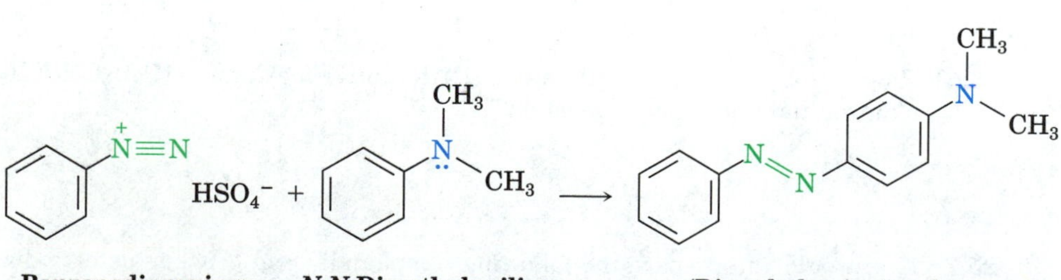

Benzenediazonium bisulfate

***N,N*-Dimethylaniline**

***p*-(Dimethylamino)azobenzene (yellow crystals, mp 127°C)**

Azo-coupled products are widely used as dyes because their extended conjugated π electron system causes them to absorb in the visible region of the electromagnetic spectrum (Section 14.14). *p*-(Dimethylamino)azobenzene, for example, is bright yellow and was at one time used as a coloring agent in margarine. Another azo compound, Alizarin Yellow R, is used for dyeing wool.

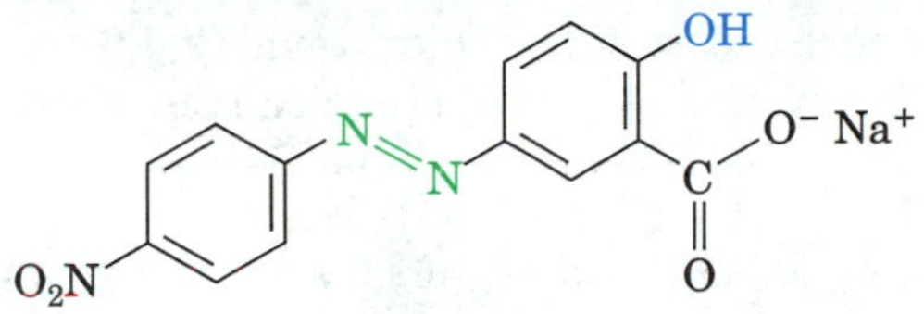

Alizarin Yellow R

PROBLEM

25.8 Propose a synthesis of *p*-(dimethylamino)azobenzene from benzene.

PROBLEM

25.9 Methyl orange is an azo dye that is widely used as a pH indicator. How would you synthesize methyl orange from benzene?

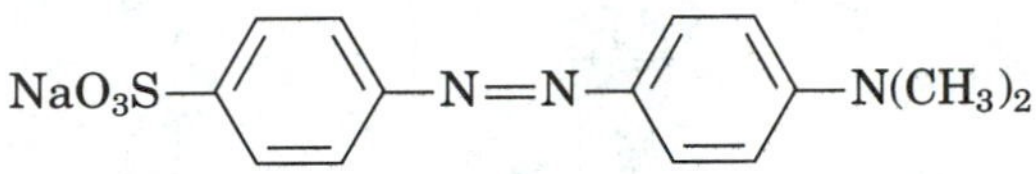

Methyl orange

25.4 Phenols

Phenols are compounds with an –OH group directly bonded to an aromatic ring, ArOH. They occur widely throughout nature, and they serve as intermediates in the industrial synthesis of products as diverse as adhesives and antiseptics. Phenol itself is a general disinfectant found in coal tar; methyl salicylate is a flavoring agent found in oil of wintergreen; and the urushiols are the allergenic constituents of poison oak and poison ivy. Note that the word *phenol* is the name both of a specific compound and of a class of compounds.

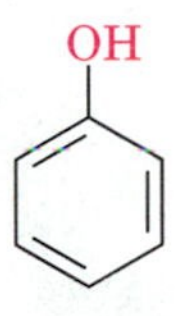

Phenol (also known as carbolic acid)

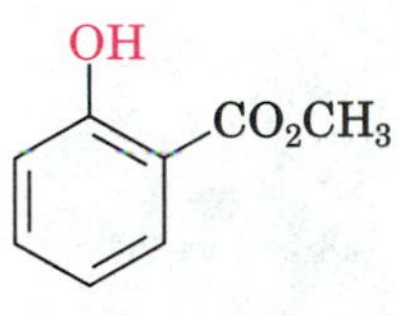

Methyl salicylate

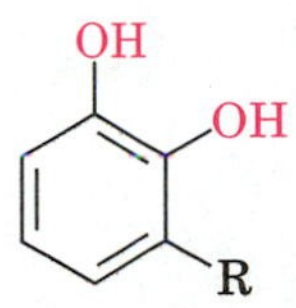

Urushiols (R = different C_{15} alkyl and alkenyl chains)

25.5 Industrial Uses of Phenols

The outbreak of World War I provided a stimulus for industrial preparation of large amounts of synthetic phenol, which was needed as raw material to manufacture the explosive, picric acid (2,4,6-trinitrophenol). Today, approximately 1.8 million tons of phenol are manufactured each year in the United States for use in such products as Bakelite resin and adhesives for binding plywood.

For many years, phenol was manufactured by the Dow process in which chlorobenzene reacts with NaOH at high temperature and pressure (Section 16.9). Now, however, an alternative synthesis from isopropylbenzene (cumene) is used. Cumene reacts with air at high temperature by a radical mechanism to form cumene hydroperoxide, which is converted into phenol and acetone by treatment with acid. This is a particularly efficient process because two valuable chemicals are prepared at the same time.

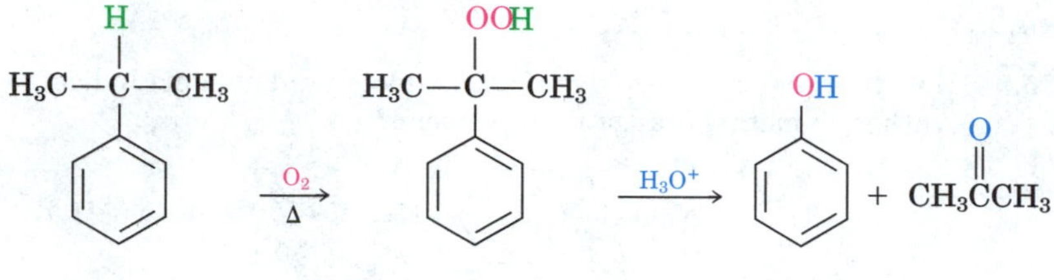

The reaction occurs by protonation of oxygen, followed by rearrangement of the phenyl group from carbon to oxygen with simultaneous loss of water. Readdition of water then yields a hemiacetal, which breaks down to phenol and acetone (Figure 25.4).

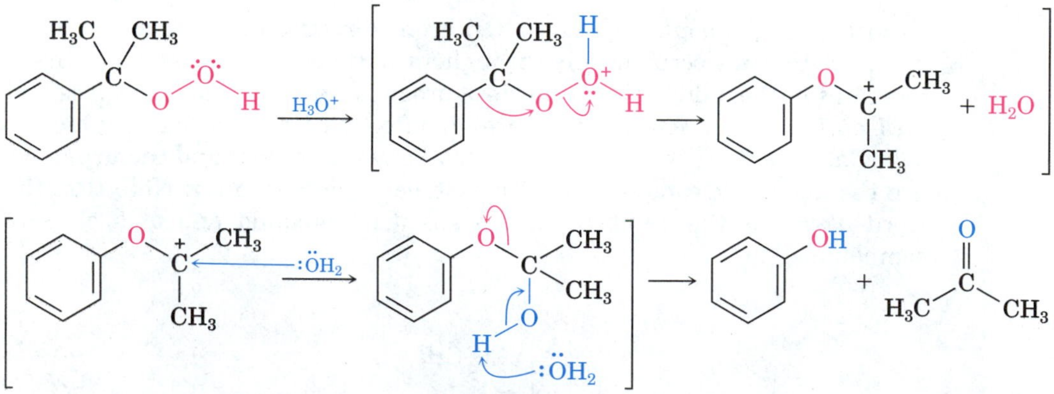

Figure 25.4 Mechanism of the formation of phenol by acid-catalyzed reaction of cumene hydroperoxide.

In addition to its use in resins and adhesives, phenol is also the starting material for the synthesis of chlorinated phenols and the food preservatives BHT (butylated hydroxytoluene) and BHA (butylated hydroxyanisole). Pentachlorophenol, a widely used wood preservative, is prepared by reaction of phenol with excess Cl_2. The herbicide 2,4-D (2,4-dichlorophenoxyacetic acid) is prepared from 2,4-dichlorophenol, and the hospital antiseptic agent hexachlorophene is prepared from 2,4,5-trichlorophenol.

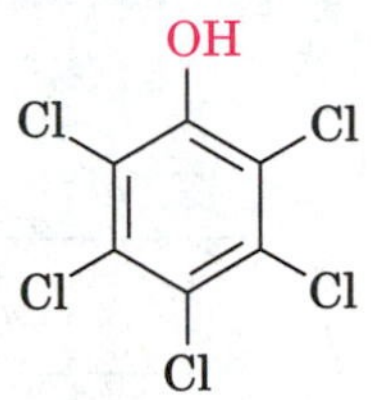

Pentachlorophenol (wood preservative)

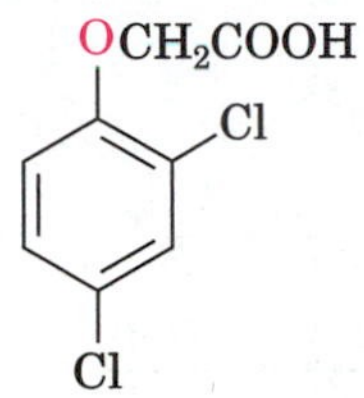

2,4-Dichlorophenoxyacetic acid, 2,4-D (herbicide)

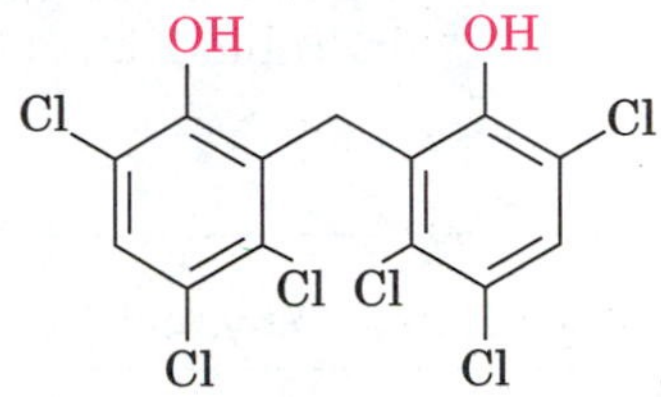

Hexachlorophene (antiseptic)

The food preservative BHT is prepared by Friedel–Crafts alkylation of *p*-methylphenol (*p*-cresol) with 2-methylpropene in the presence of acid; BHA is prepared similarly by alkylation of *p*-methoxyphenol.

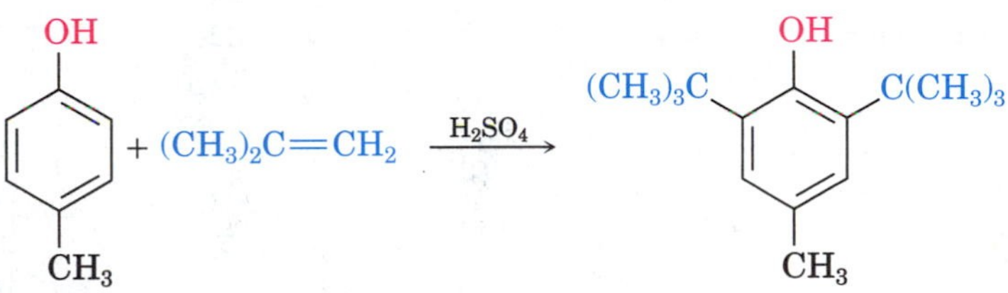

***p*-Methylphenol** **BHT**

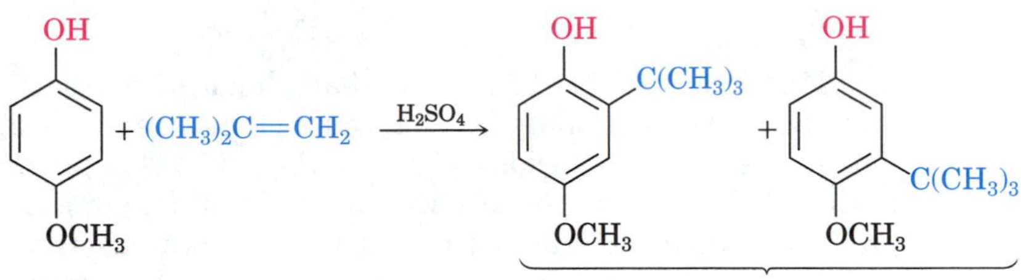

***p*-Methoxyphenol** **BHA**

PROBLEM. . . . .

25.10 2,4-Dichlorophenoxyacetic acid is prepared from phenol by a two-step sequence involving an electrophilic aromatic substitution followed by a Williamson ether synthesis (Section 18.4). Formulate the reactions.

25.6 Properties of Phenols: Acidity

Phenols are similar in many respects to alcohols. Like alcohols, low-molecular-weight phenols are generally somewhat water-soluble and are high boiling because of intermolecular hydrogen bonding. Perhaps the most important property of phenols, however, is their acidity. Phenols are weak acids that dissociate to a small extent in aqueous solution to give H_3O^+ and a phenoxide anion, ArO^-. Acidity values for some common phenols are given in Table 25.2.

Table 25.2 Physical Properties of Some Phenols

Phenol	*Melting point (°C)*	*Boiling point (°C)*	pK_a	
Acetic acid[a]			4.75	Stronger acid
2,4,6-Trinitrophenol	122	—	0.60	
p-Nitrophenol	115	—	7.15	
o-Nitrophenol	97	—	7.17	
m-Nitrophenol	45	216	8.28	
p-Iodophenol	94	—	9.30	
p-Bromophenol	66	238	9.35	
p-Chlorophenol	43	220	9.38	
Phenol	43	182	9.89	
p-Methylphenol (*p*-cresol)	35	202	10.17	
p-Methoxyphenol	57	243	10.21	
p-Aminophenol	186	—	10.46	
Ethanol[a]			16.00	Weaker acid

[a]Values for acetic acid and ethanol are given for reference.

The data in Table 25.2 show that phenols are about a million times more acidic than alcohols. Indeed, some phenols, such as 2,4,6-trinitrophenol, even surpass the acidity of carboxylic acids. One practical consequence of this acidity is that phenols are soluble in dilute aqueous NaOH. Thus, a phenolic component can often be separated from a mixture of compounds simply by basic extraction into aqueous solution, followed by reacidification.

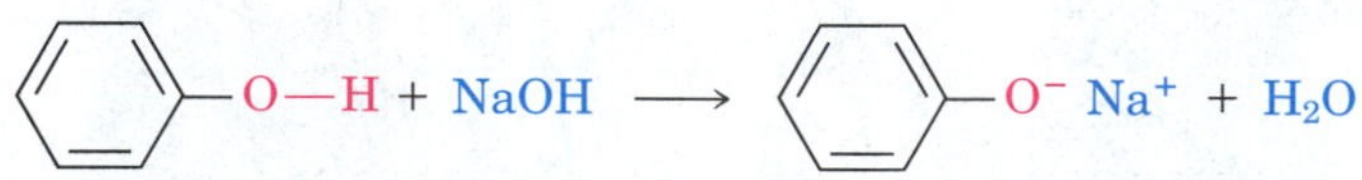

Phenol **Sodium phenoxide**

Phenols are more acidic than alcohols because the phenoxide anion is resonance stabilized. Delocalization of the negative charge over the ortho and para positions of the aromatic ring results in increased stability of the phenoxide anion relative to undissociated phenol and in a consequently lower $\Delta G°$ for the dissociation reaction. Figure 25.5 compares the acidity of phenols and alcohols in reaction energy diagrams.

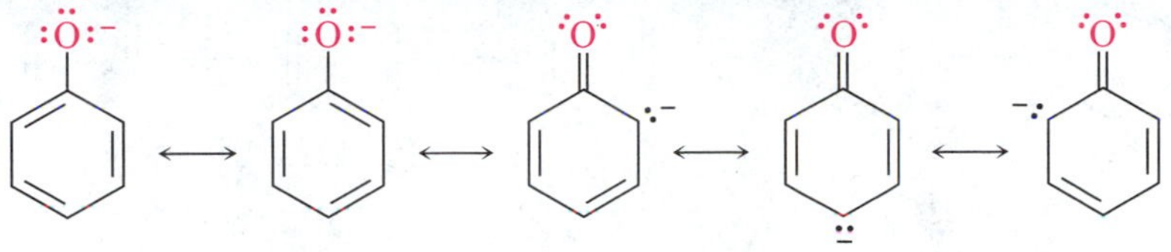

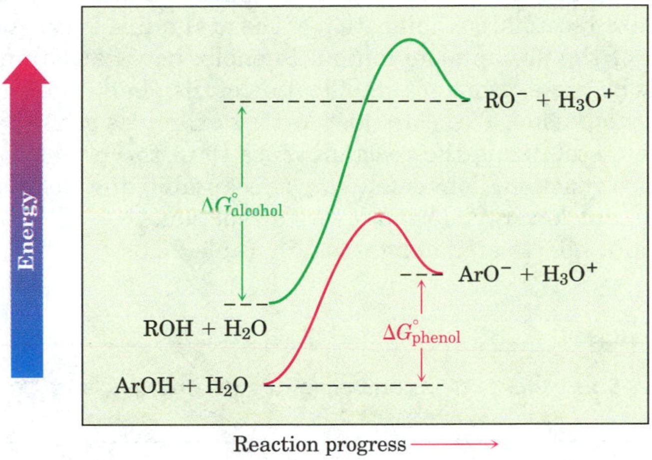

Figure 25.5 A comparison of phenol and alcohol acidities. Phenols (red curve) are more acidic than alcohols (green curve) because a phenoxide ion is stabilized relative to free phenol more than an alkoxide ion is stabilized relative to free alcohol.

Substituted phenols can be either more acidic or less acidic than phenol itself. Phenols with an electron-withdrawing substituent are generally more acidic because these substituents stabilize the phenoxide ion by delocalizing the negative charge. Phenols with an electron-donating substituent are less acidic because these substituents destabilize the phenoxide ion by localizing the charge.

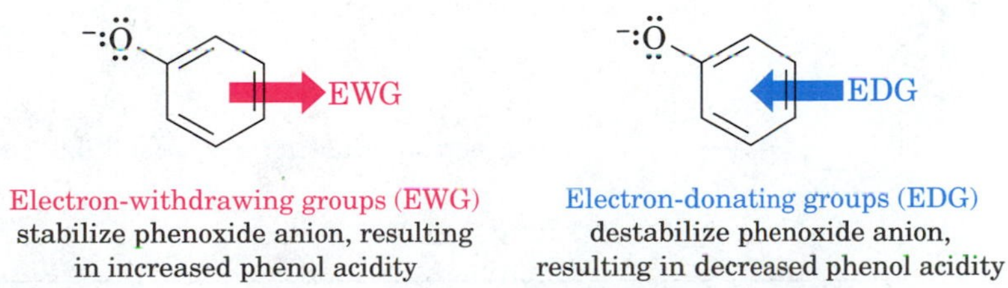

The acidifying effect of an electron-withdrawing substituent is particularly noticeable for phenols having a nitro group at the ortho or para position.

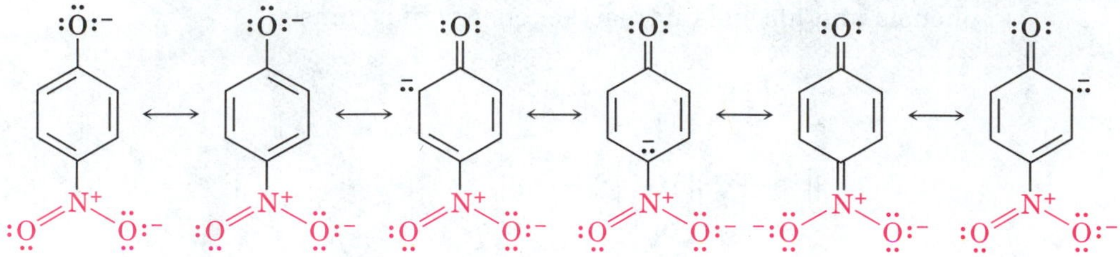

Note that the effect of substituents on phenol acidity is the same as their effect on benzoic acid acidity (Section 20.5) but opposite to their effect on arylamine basicity (Section 25.1). Arylamines are less basic than alkylamines because resonance stabilization of the arylamine is greater than the stabilization of the arylammonium ion. Phenols, however, are more acidic than alcohols because resonance stabilization of the phenoxide ion is greater than that of the phenol. These are just further examples of the general rule that any effect stabilizing the reactant more than the product raises $\Delta G°$ and disfavors a reaction. Conversely, any effect stabilizing the product more than the reactant lowers $\Delta G°$ and favors a reaction.

Substituent effects are summarized in Table 25.3.

Table 25.3 Substituent Effects on Benzoic Acids, Phenols, and Arylamines

Type of compound	*Effect of group* Y
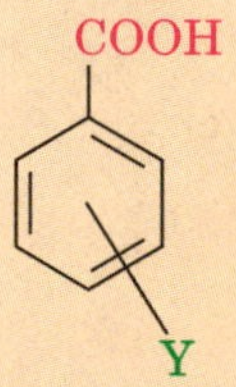	Electron-withdrawing; increased acidity Electron-donating; decreased acidity
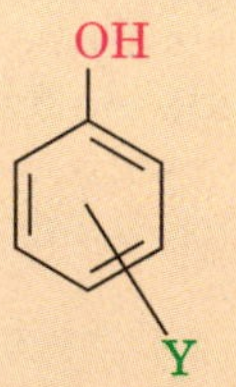	Electron-withdrawing; increased acidity Electron-donating; decreased acidity
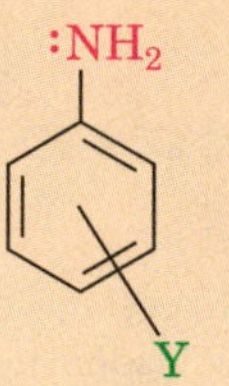	Electron-withdrawing; decreased basicity Electron-donating; increased basicity

PROBLEM.

25.11 Rank the compounds in each group in order of increasing acidity.
(a) Phenol, *p*-methylphenol, *p*-(trifluoromethyl)phenol
(b) Benzyl alcohol, phenol, *p*-hydroxybenzoic acid
(c) *p*-Bromophenol, 2,4-dibromophenol, 2,4,6-tribromophenol

25.7 Preparation of Phenols

We've already seen the two best methods of phenol synthesis. To review briefly:

1. Alkali fusion of aromatic sulfonates takes place when an arenesulfonic acid is melted with NaOH at high temperature (Section 16.2). Few functional groups can survive such harsh conditions, and the reaction is therefore limited to the preparation of alkyl-substituted phenols.

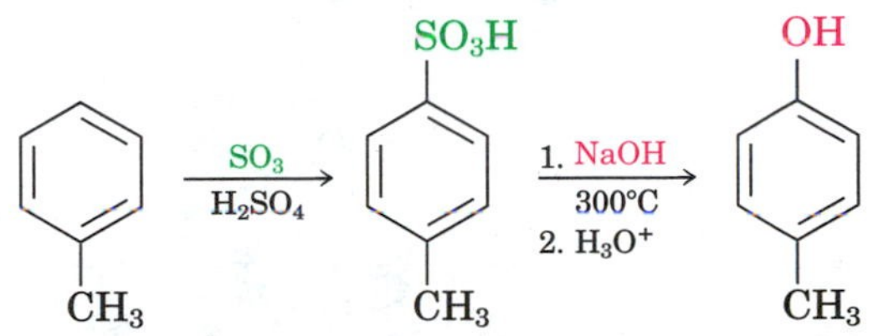

Toluene **p-Toluenesulfonic acid** **p-Methylphenol (72%)**

2. Hydrolysis of arenediazonium salts in a Sandmeyer-type substitution reaction is the most versatile and widely used laboratory method of phenol synthesis (Section 25.3). Most functional groups are compatible with the reaction conditions, and yields are generally good.

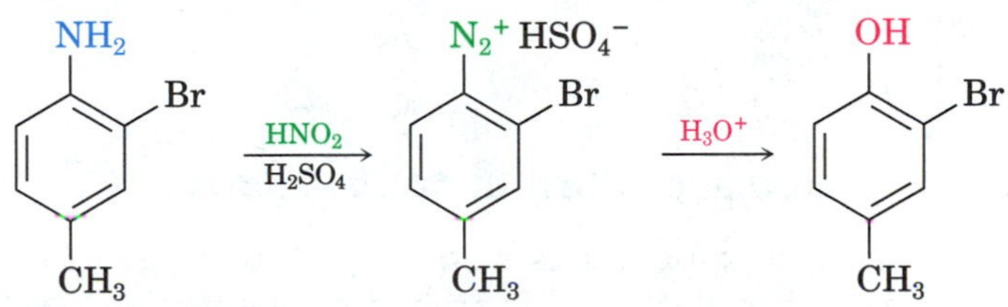

2-Bromo-4-methylaniline **2-Bromo-4-methylphenol (92%)**

PROBLEM.

25.12 *p*-Cresol (*p*-methylphenol) is used both as an antiseptic and as a starting material to prepare the food additive BHT. How would you prepare *p*-cresol from benzene?

PROBLEM

25.13 Carvacrol (5-isopropyl-2-methylphenol) is a naturally occurring substance isolated from oregano, thyme, and marjoram. Propose two different syntheses of carvacrol from benzene.

25.8 Reactions of Phenols

Alcohol-like Reactions

Much of the chemistry of phenols is similar to that of alcohols. Thus, phenols can be converted into esters by reaction with acid chlorides or acid anhydrides, and into ethers by reaction with alkyl halides in the presence of base (the Williamson synthesis; Section 18.4). Both reactions occur easily because the reactive phenoxide ion intermediate is formed more readily than an alkoxide ion. Direct Fischer esterification by acid-catalyzed reaction between a phenol and a carboxylic acid, however, is not usually successful.

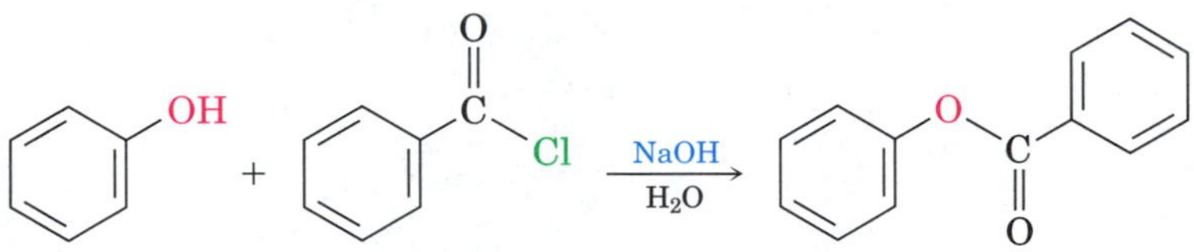

Phenol **Benzoyl chloride** **Phenyl benzoate (96%)**

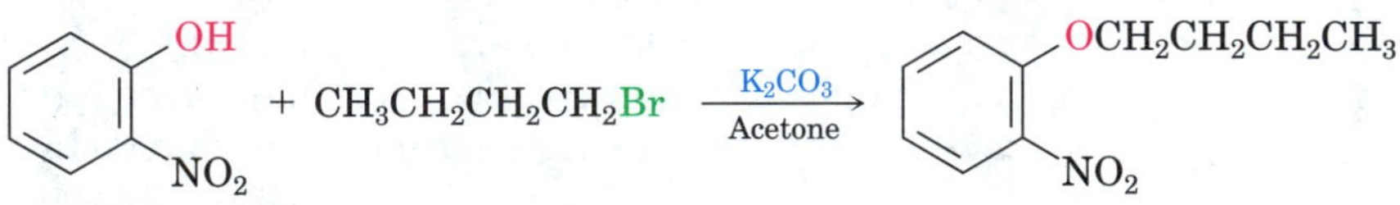

***o*-Nitrophenol** **1-Bromobutane** **Butyl *o*-nitrophenyl ether (80%)**

Electrophilic Aromatic Substitution Reactions

The hydroxyl group is a strongly activating, ortho- and para-directing substituent in electrophilic aromatic substitution reactions (Section 16.5). As a result, phenols are highly reactive substrates for electrophilic halogenation, nitration, and sulfonation, as well as for coupling with diazonium salts to produce azo dyes.

Not surprisingly, phenoxide *ions* are even more reactive toward electrophilic aromatic substitution than neutral phenols because they have a full negative charge:

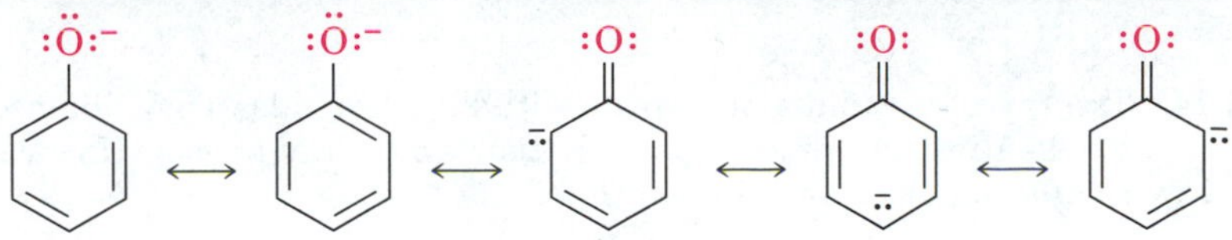

Phenoxide ion

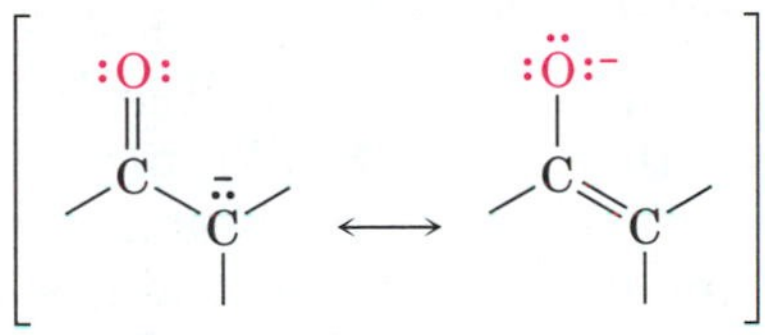

An enolate ion

Resonance structures of a phenoxide anion show a similarity to the resonance structures of an enolate anion, which suggests the possibility that phenoxides might undergo α-substitution reactions similar to those of ketones (Section 22.6). In practice, phenoxide ions are less reactive than enolate ions because of the stability of the benzene ring. Nevertheless, there are a number of examples of enolate-like reactivity of phenoxide ions. For example, in the **Kolbe–Schmitt carboxylation reaction**,[3,4] a phenoxide ion adds to CO_2 under pressure to yield an intermediate keto acid anion that enolizes to give an *o*-hydroxybenzoic acid (salicylic acid). This reaction is the key step in the industrial synthesis of aspirin (acetylsalicylic acid).

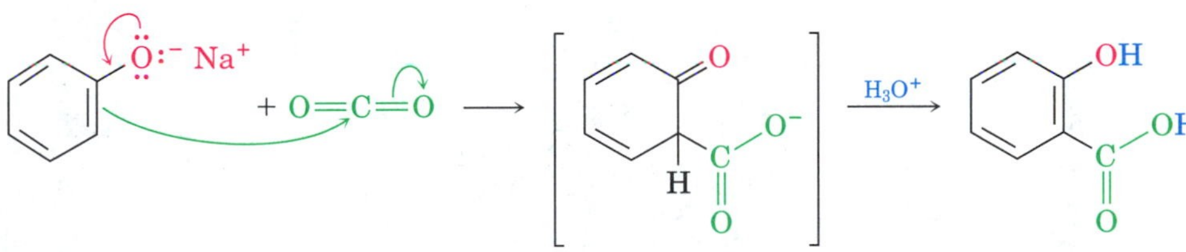

Sodium phenoxide **Salicylic acid**

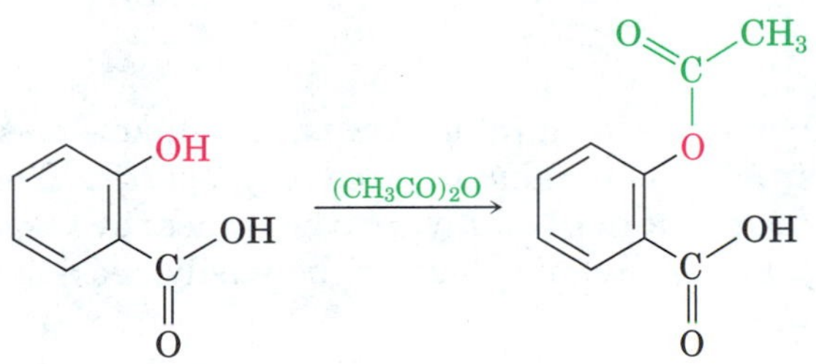

Salicylic acid **Acetylsalicylic acid (Aspirin)**

[3]Herman Kolbe (1818–1884); b. Germany; Ph.D. Göttingen; professor, universities of Marburg and Leipzig.

[4]Rudolf Schmitt (1830–1898); b. Wippershain, Germany; Ph.D. Marburg; professor, University of Dresden.

PROBLEM.

25.14 Treatment of sodium phenoxide with allyl bromide in a Williamson ether synthesis yields a mixture of allyl phenyl ether and *o*-allylphenol. How can you account for the formation of *o*-allylphenol?

Oxidation of Phenols: Quinones

The susceptibility of phenols to electrophilic aromatic substitution is one consequence of the electron-rich nature of the phenol ring. Another consequence is the susceptibility of phenols to oxidation. Treatment of a phenol with any of a number of strong oxidizing agents yields a 2,5-cyclohexadiene-1,4-dione, or **quinone**. Older procedures employed $Na_2Cr_2O_7$ as oxidant, but Fremy's salt [potassium nitrosodisulfonate, $(KSO_3)_2NO$] is now preferred. The reaction takes place under mild conditions through a radical mechanism, and good yields are normally obtained. Arylamines are similarly oxidized to quinones.

Phenol

Benzoquinone (79%)

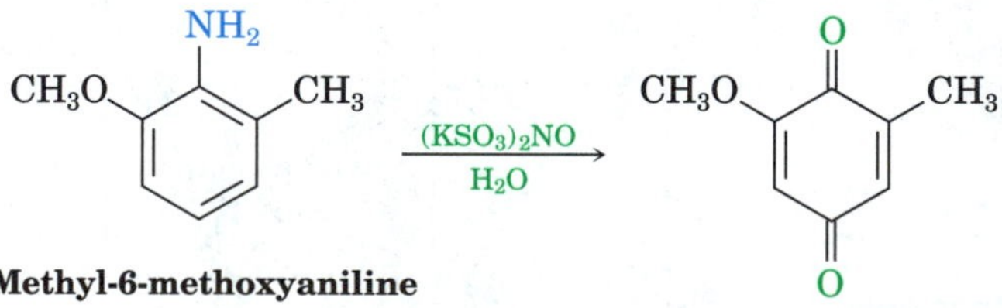

2-Methyl-6-methoxyaniline

2-Methyl-6-methoxybenzoquinone (96%)

Quinones are an interesting and valuable class of compounds because of their oxidation–reduction (redox) properties. They can be easily reduced to **hydroquinones** (*p*-dihydroxybenzenes) by reagents such as $NaBH_4$ and $SnCl_2$, and hydroquinones can be easily reoxidized back to quinones by Fremy's salt.

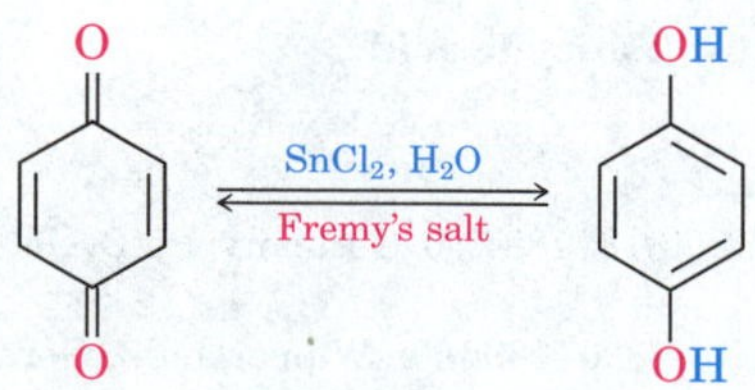

Benzoquinone **Hydroquinone**

The redox properties of quinones are important to the functioning of living cells, where compounds called *ubiquinones* act as biochemical oxidizing agents to mediate the electron-transfer processes involved in energy production. Ubiquinones, also called *coenzymes Q*, are components of the cells of all aerobic organisms, from the simplest bacterium to humans. They are so named because of their ubiquitous occurrence in nature.

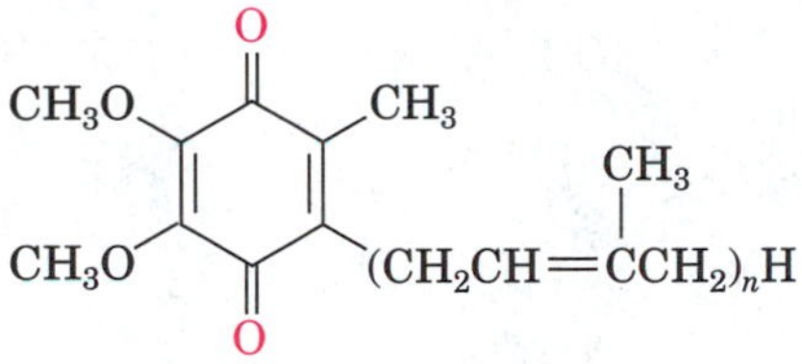

Ubiquinones (*n* = 1–10)

Ubiquinones function within the mitochondria of cells to mediate the respiration process in which electrons are transported from the biological reducing agent NADH to molecular oxygen. Although a complex series of steps is involved in the overall process, the ultimate result is a cycle whereby NADH is oxidized to NAD^+, O_2 is reduced to water, and energy is produced. Ubiquinone acts only as an intermediary and is itself unchanged.

Step 1:

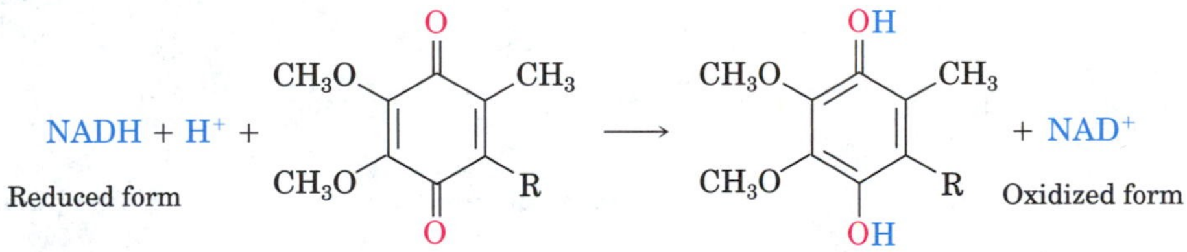

Step 2:

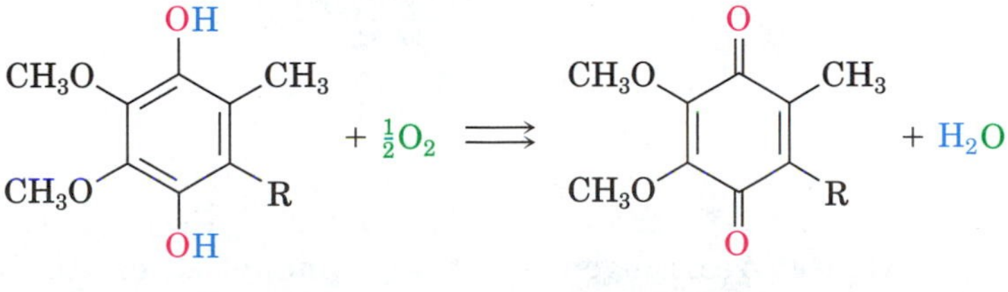

Net change:

$$NADH + \tfrac{1}{2}O_2 + H^+ \rightarrow NAD^+ + H_2O$$

PROBLEM

25.15 Early work on the structural elucidation of ubiquinones was complicated by the fact that extraction of the compounds from cells was carried out using basic ethanol solution. Under these conditions, the ubiquinone $-OCH_3$ groups became exchanged for $-OCH_2CH_3$. Propose a mechanism to account for this exchange. (See Section 19.17.)

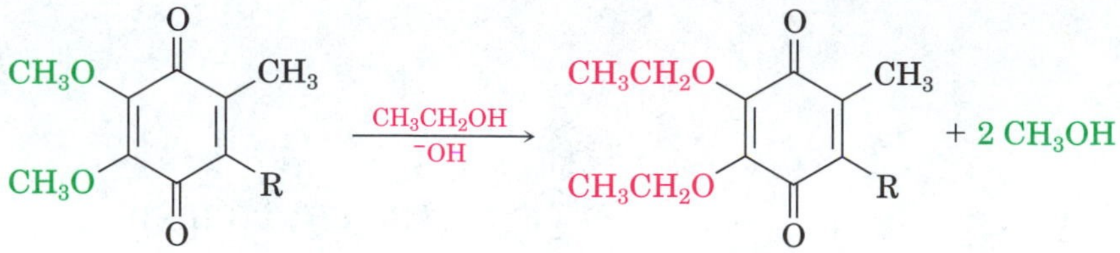

Claisen Rearrangement

Treatment of a phenoxide ion with 3-bromopropene (allyl bromide) results in a Williamson ether synthesis and production of an allyl phenyl ether. Heating the allyl phenyl ether to 200–250°C then effects **Claisen rearrangement**, leading to an *o*-allylphenol. The net effect is alkylation of the ortho position of the phenol.

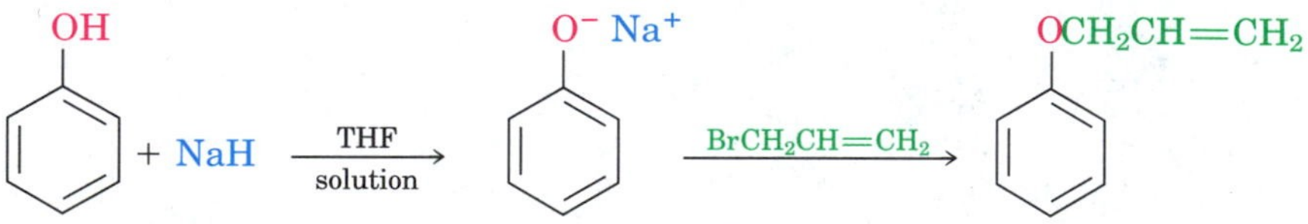

Phenol **Sodium phenoxide** **Allyl phenyl ether**

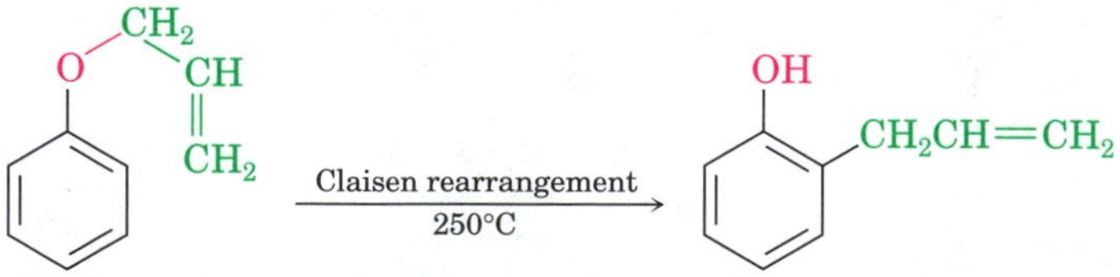

Allyl phenyl ether ***o*-Allylphenol**

Claisen rearrangement of allyl phenyl ethers to yield *o*-allylphenols is a general reaction that is compatible with the presence of many other substituents on the benzene ring. The reaction proceeds through a pericyclic mechanism in which a concerted reorganization of bonding electrons occurs by a six-membered, cyclic transition state. The 2-allylcyclohexadienone intermediate then tautomerizes to *o*-allylphenol.

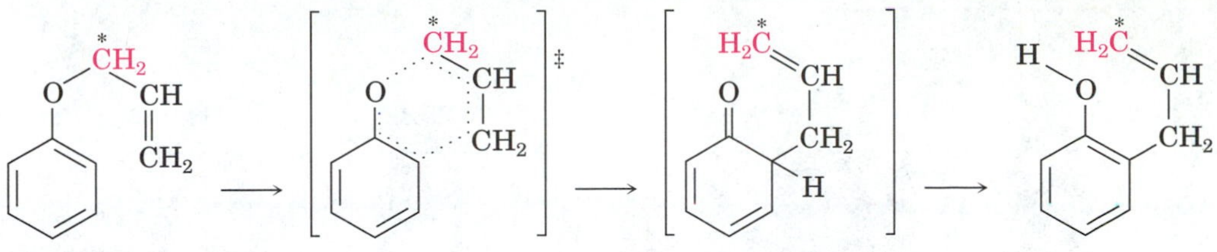

Allyl phenyl ether **Transition state** **Intermediate (6-Allyl-2,4-cyclohexadienone)** ***o*-Allylphenol**

Good evidence for this mechanism comes from the observation that the rearrangement takes place with an *inversion* of the allyl unit. That is, allyl phenyl ether containing a ^{14}C label on the allyl ether carbon atom yields *o*-allylphenol in which the label is on the terminal carbon. It would be very difficult to explain this result by any mechanism other than a pericyclic one. We'll look at more details in Section 31.10.

PROBLEM.

25.16 What product would you expect to obtain from Claisen rearrangement of 2-butenyl phenyl ether?

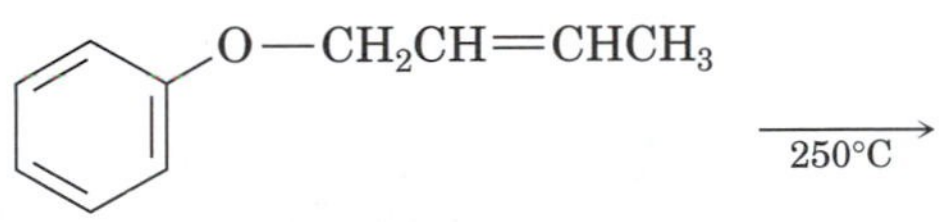

2-Butenyl phenyl ether

25.9 Spectroscopy of Arylamines and Phenols

Infrared Spectroscopy

The infrared spectra of arylamines and phenols are similar to those of aliphatic amines and alcohols. Aniline, for instance, shows the usual N–H IR absorptions at 3350 and 3450 cm^{-1} characteristic of a primary amine, as well as a pair of bands at 1500 and 1600 cm^{-1} characteristic of aromatic rings (Figure 25.6). Note that the IR spectrum of aniline also shows the typical monosubstituted aromatic-ring peaks at 690 and 760 cm^{-1}.

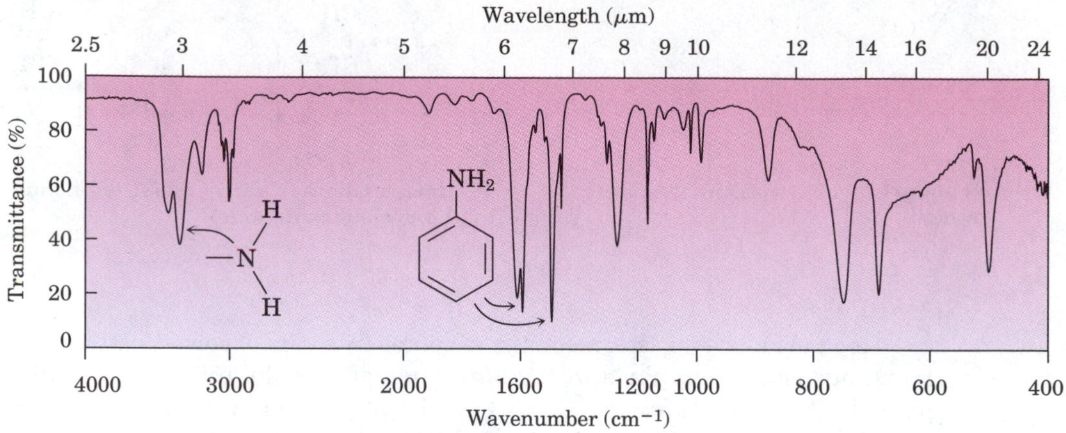

Figure 25.6 Infrared spectrum of aniline.

Phenol shows a characteristic broad IR absorption at 3500 cm^{-1} due to the –OH group, as well as the usual 1500 and 1600 cm^{-1} aromatic bands (Figure 25.7). Here, too, the monosubstituted aromatic-ring peaks at 690 and 760 cm^{-1} are visible.

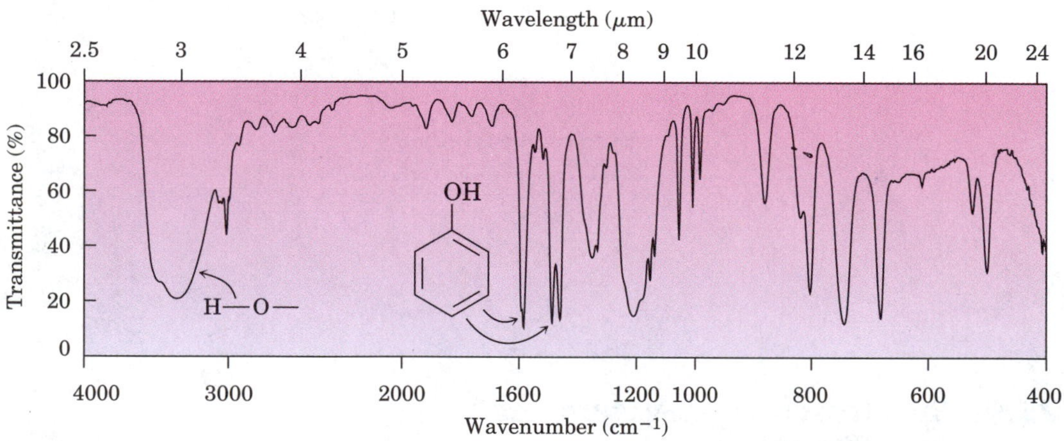

Figure 25.7 Infrared spectrum of phenol.

Nuclear Magnetic Resonance Spectroscopy

Arylamines and phenols, like all aromatic compounds, show ^{1}H NMR absorptions near 7–8 δ, the expected position for aromatic-ring protons. In addition, amine N–H protons usually absorb in the 2–3 δ range, and phenol O–H protons absorb at 3–8 δ. In neither case are these absorptions uniquely diagnostic for arylamines or phenols, since other kinds of protons absorb in the same range. As a result, a combination of NMR and IR evidence is usually needed to assign structure.

As was true for alcohols (Section 17.11), the identity of the NMR peak due to N–H or O–H protons is easily determined by adding a small amount of D_2O to the sample tube. The O–H and N–H protons rapidly exchange with added D_2O, and their peaks disappear from the spectrum.

INTERLUDE

Ubiquinones and Metabolic Energy

We eat and breathe, the food we eat is oxidized in our bodies to yield CO_2 and H_2O, and the energy released during that oxidation sustains life:

$$\text{Food (C, H)} + O_2 \longrightarrow CO_2 + H_2O + \text{Energy}$$

This overall picture of metabolism is simple enough—it's the details that are complex. One of the more intricate parts of the metabolism picture involves the way in which oxygen is used. Unlike what happens when a fuel is burned in a furnace or automobile engine, oxygen does not combine directly with food molecules. Instead, a much more elaborate process occurs.

Much of the body's chemistry involves the substances *nicotinamide adenine dinucleotide* (NAD^+) and *flavin adenine dinucleotide* (FAD), which are converted into their reduction products NADH and $FADH_2$ during the course of their reactions. For reuse, NADH and $FADH_2$ must undergo oxidation back to their oxidized forms, giving up two hydrogen atoms in the process. In biochemistry, however, the removal of a hydrogen *atom* is actually accomplished by the loss of a *proton* (H^+) and an electron (e^-). The electrons then travel through a series of intermediates until they ultimately combine with O_2 and H^+ to give water.

Crucial to this electron-transport process are a group of compounds called *ubiquinones*. To take $FADH_2$ as the example, two electrons and two H^+ ions are transferred from $FADH_2$ to a group of substances called *iron–sulfur proteins*, reducing two of the iron atoms in the protein from Fe(III) to Fe(II). The iron–sulfur protein, in turn, passes on the two electrons to ubiquinone, a substituted quinone containing a large (up to 50 carbons), nonpolar appendage that allows it to carry the electrons through the nonpolar membranes surrounding many subcellular structures. Ubiquinone then passes the electrons on to still further substances, and ultimately to oxygen.

Quinones are unusual among organic molecules in that they can act as single-electron carriers to facilitate oxidation–reduction reactions. Addition of H^+ and one electron to a quinone yields a stable radical called a *semiquinone*, and addition of a second electron plus H^+ yields a

(continued)➤

hydroquinone. In reverse, a hydroquinone readily *loses* one hydrogen atom to give the semiquinone and loses two hydrogen atoms to give a quinone. As a result, they are uniquely capable of shuttling electrons from a source (reducing agent) to a sink (oxidizing agent).

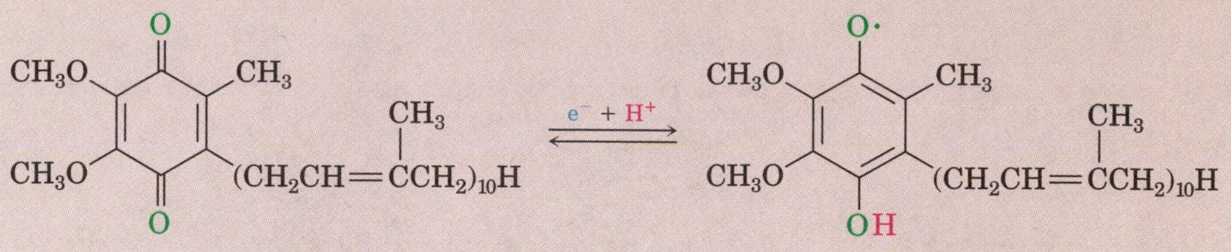

Ubiquinone **A semiquinone**

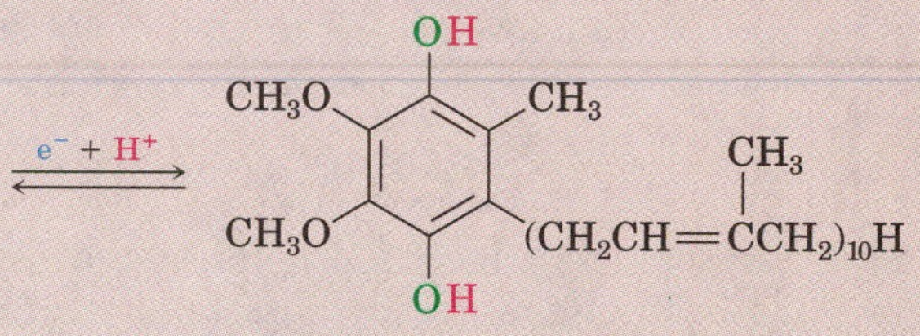

Ubiquinol (a hydroquinone)

In plant tissues, a group of compounds called *plastoquinones* behave in the same way as the ubiquinones, but in reverse. During photosynthesis, plastoquinones use energy from sunlight to shuttle electrons from water to CO_2, forming O_2 in the process.

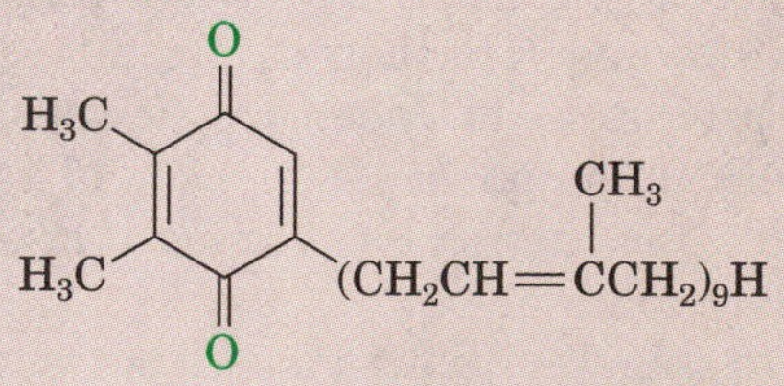

Plastoquinone

Summary and Key Words

Arylamines, like their aliphatic counterparts, are basic. The base strength of arylamines is generally lower than that of aliphatic amines, though, because the nitrogen lone-pair electrons are delocalized by interaction with the aromatic π system. Electron-withdrawing substituents on the ring further weaken the basicity of a substituted aniline, while electron-donating substituents increase basicity.

Substituted anilines are prepared by nitration of an aromatic ring followed by reduction. An amine group is a strongly activating, ortho- and para-directing substituent in electrophilic aromatic substitution reactions. If the amine group makes the ring too reactive, however, its reactivity can be modulated by converting it into an amide.

The most generally useful reaction of arylamines is conversion by diazotization with nitrous acid into **arenediazonium salts**, $ArN_2^+ X^-$. The diazonio group can then be replaced by many other substituents in the **Sandmeyer reaction** to give a wide variety of substituted aromatic compounds. Aryl chlorides, bromides, iodides, and nitriles can be prepared from arenediazonium salts, as can arenes and phenols. In addition to their reactivity toward substitution reactions, diazonium salts undergo coupling with phenols and arylamines to give brightly colored azo dyes.

Phenols are aromatic counterparts of alcohols but are much more acidic because phenoxide anions are stabilized by delocalization of the negative charge into the aromatic ring. Substitution of the aromatic ring by an electron-withdrawing group increases phenol acidity, and substitution by an electron-donating group decreases acidity. Phenols are generally prepared either by alkali fusion of an aromatic sulfonate or by hydrolysis of an arenediazonium salt.

Reactions of phenols can occur either at the hydroxyl group or on the aromatic ring. For example, a phenol can be converted into an ester or an ether. In addition, phenols can be oxidized to **quinones** by reaction with Fremy's salt, potassium nitrosodisulfonate. Phenyl allyl ethers are particularly interesting because they undergo a **Claisen rearrangement** to give *o*-allylphenols when heated to 250°C.

Summary of Reactions

1. Preparation of arylamines: Reduction of nitrobenzenes (Section 25.2)

$$ArNO_2 + H_2 \xrightarrow[\text{Ethanol}]{\text{Pt}} ArNH_2$$

$$ArNO_2 + Fe \xrightarrow[\text{2. } HO^-]{\text{1. } H_3O^+} ArNH_2$$

$$ArNO_2 + SnCl_2 \xrightarrow[\text{2. } HO^-]{\text{1. } H_3O^+} ArNH_2$$

2. Reactions of arylamines
   (a) Electrophilic aromatic substitution (Sections 16.2 and 25.3)

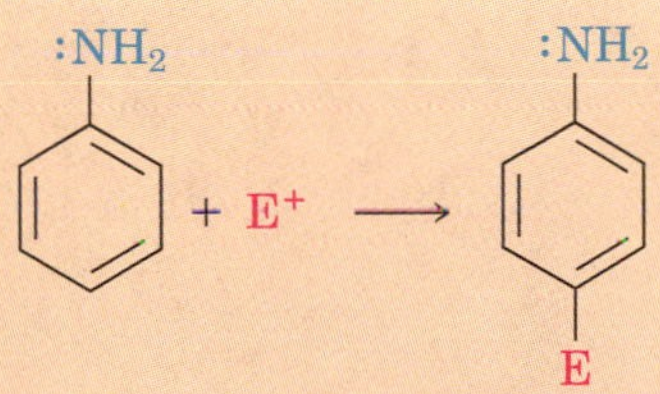

Ortho- and para-directing

(continued)➤

(b) Formation of arenediazonium salts (Section 25.3)

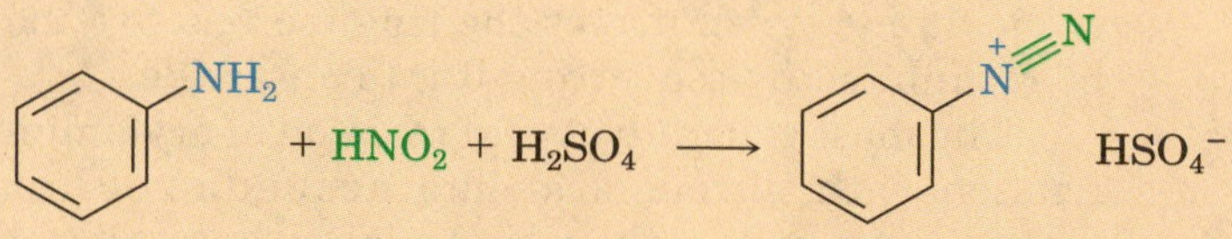

(c) Reaction of arenediazonium salts (Section 25.3)
(1) Sandmeyer-type reactions

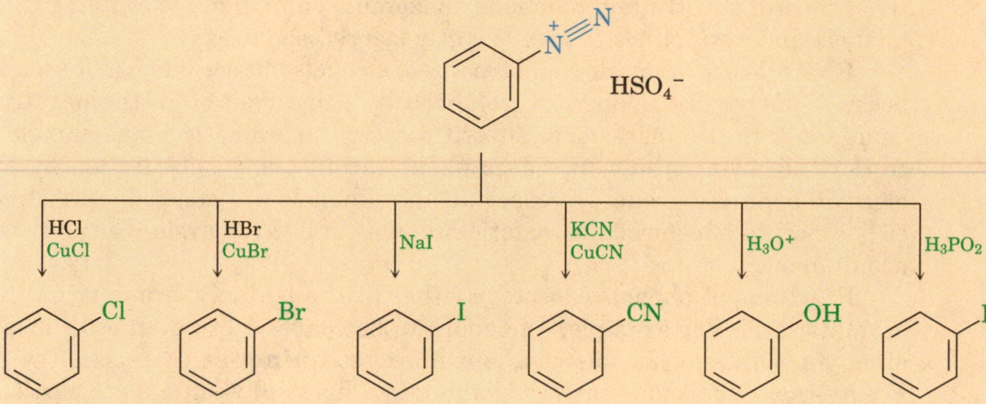

(2) Diazonium coupling

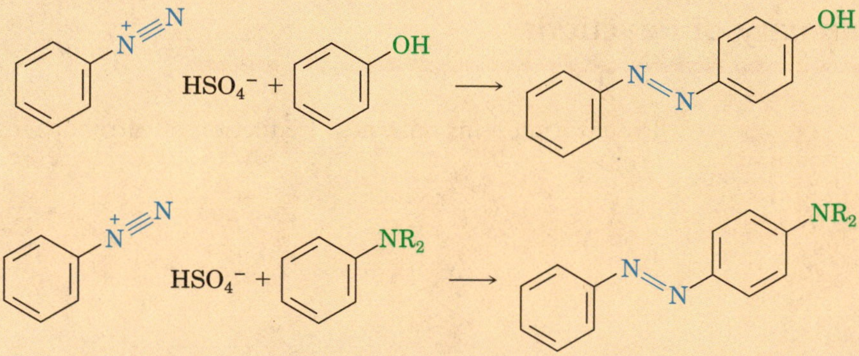

(d) Oxidation to quinones (Section 25.8)

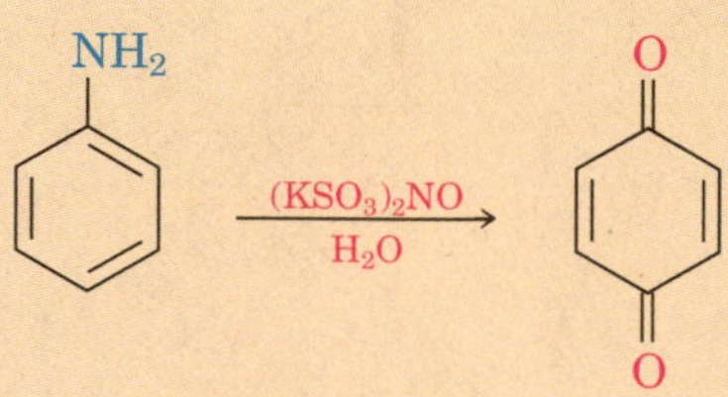

(continued)➤

3. Preparation of phenols
   (a) Alkali fusion of aryl sulfonates (Sections 16.2 and 25.7)

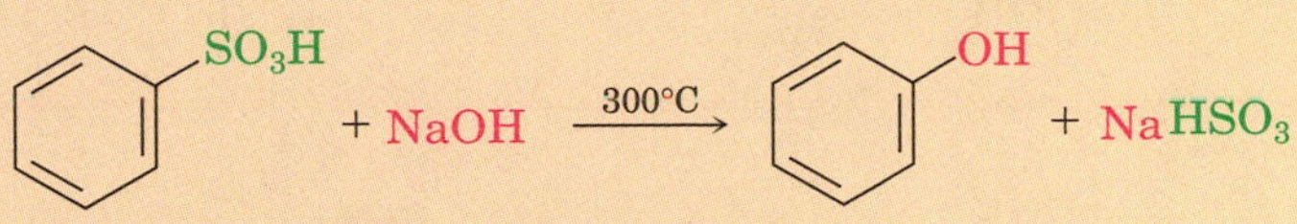

   (b) Hydrolysis of arenediazonium salts (Section 25.3)

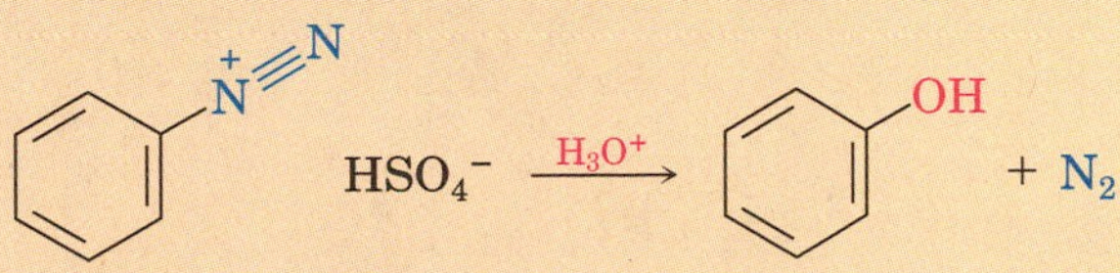

4. Reactions of phenols
   (a) Ester formation (Section 25.8)

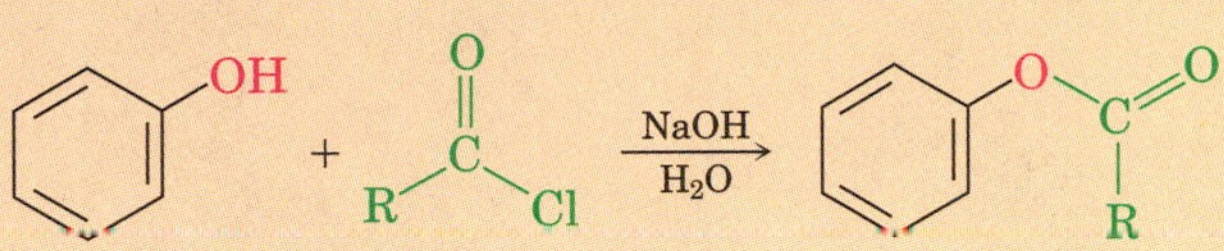

   (b) Williamson ether synthesis (Sections 18.4 and 25.8)

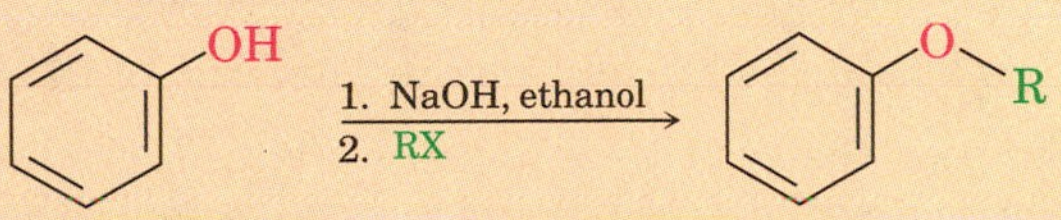

   (c) Kolbe–Schmitt carboxylation (Section 25.8)

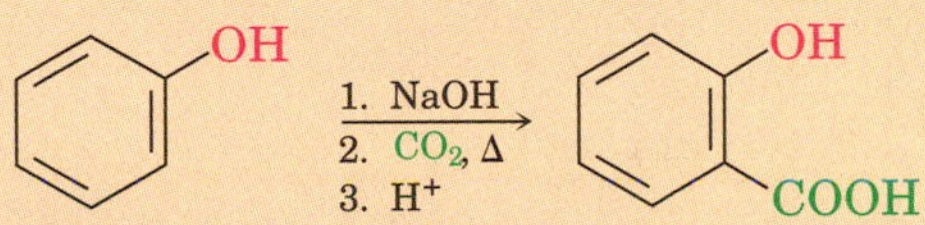

   (d) Oxidation to quinones (Section 25.8)

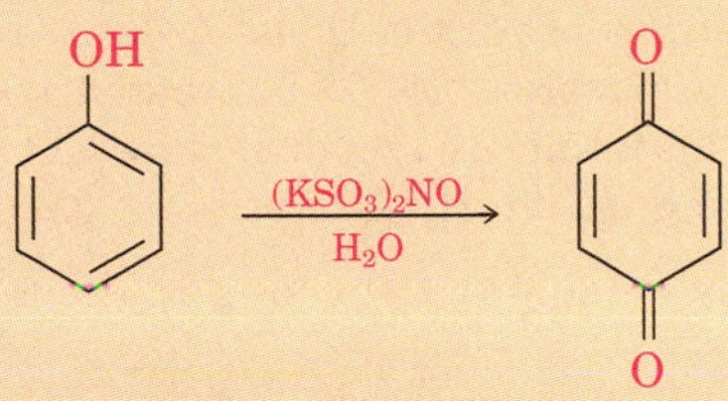

   (e) Claisen rearrangement (Section 25.8)

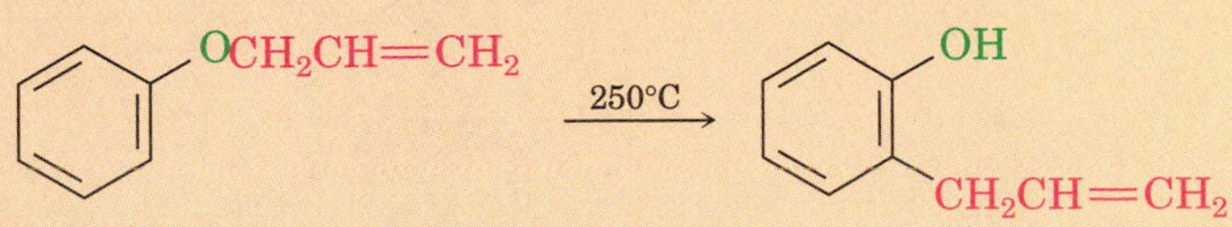

ADDITIONAL PROBLEMS

25.17 Give IUPAC names for the following compounds:

(a)
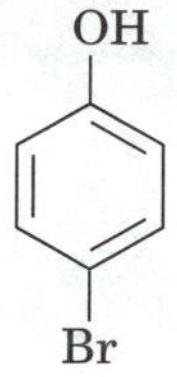

(b)
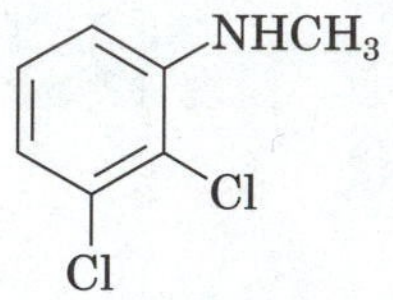

(c)
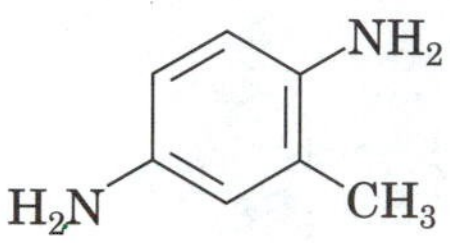

(d)
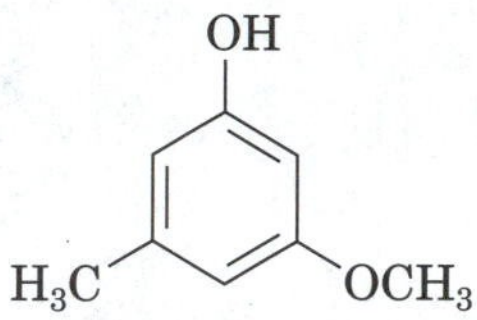

(e)
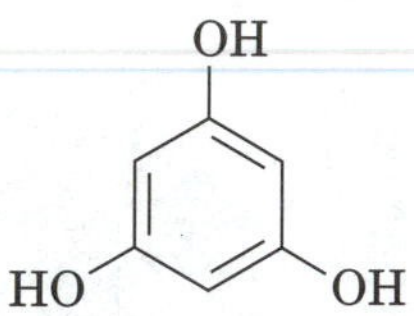

(f)
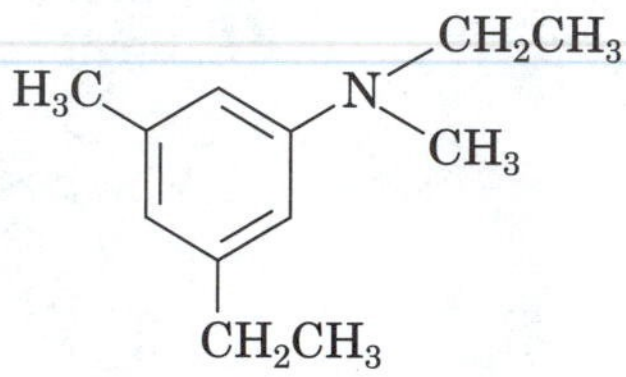

25.18 How would you prepare aniline from the following starting materials?
(a) Benzene (b) Benzamide (c) Toluene

25.19 How would you convert aniline into each of the products listed in Problem 25.18?

25.20 Suppose that you were given a mixture of toluene, aniline, and phenol. Describe how you would separate the mixture into its three pure components.

25.21 Give the structures of the major organic products you would expect from reaction of *m*-toluidine (*m*-methylaniline) with the following reagents.
(a) Br_2 (1 equiv)
(b) $(KSO_3)_2NO$
(c) CH_3I (excess)
(d) $CH_3Cl + AlCl_3$
(e) CH_3COCl in pyridine
(f) The product of (e), then HSO_3Cl

25.22 Benzoquinone is an excellent dienophile in the Diels–Alder reaction. What product would you expect from reaction of benzoquinone with 1 equivalent of butadiene? From reaction with 2 equivalents of butadiene?

25.23 When the product of the Diels–Alder reaction of benzoquinone and 1 equivalent of butadiene (Problem 25.22) is treated with dilute acid or base, an isomerization occurs. This isomerized product shows a two-proton singlet in the ^1H NMR spectrum at 6.7 δ and has an IR absorption at 3500 cm^{-1}. Propose a structure.

25.24 Fill in the missing reagents a–g in the following scheme:

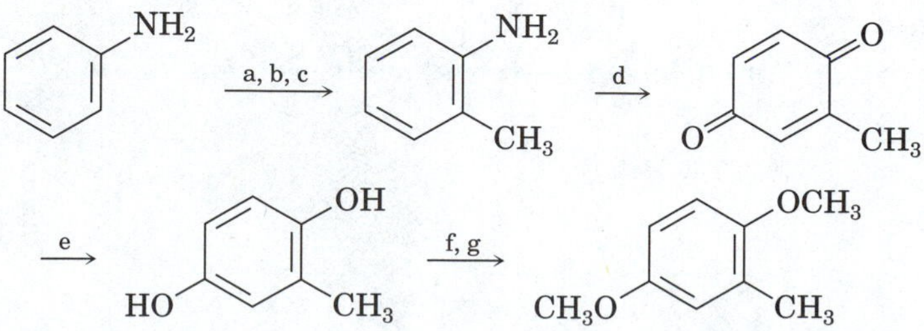

25.25 Tyramine is an alkaloid found, among other places, in mistletoe and ripe cheese. How would you synthesize tyramine from benzene? From toluene?

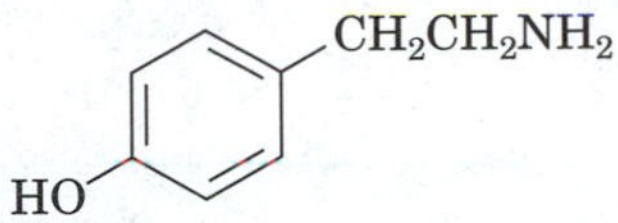

Tyramine

25.26 How can you account for the fact that diphenylamine does not dissolve in dilute aqueous HCl and appears to be nonbasic?

25.27 How would you prepare the following compounds from toluene? A diazonio replacement reaction is needed in some instances.

(a)

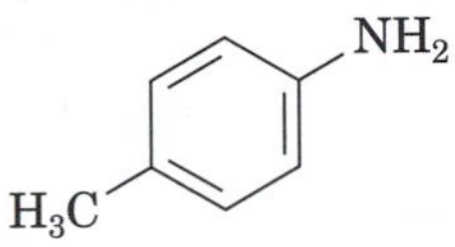

(b)

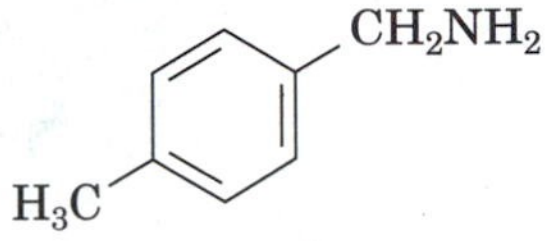

(c)

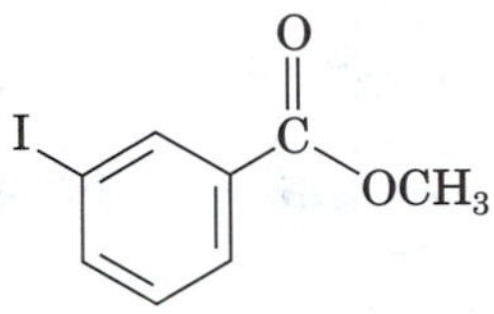

25.28 Show the products from reaction of *p*-bromoaniline with the following reagents:
(a) Excess CH_3I (b) HCl (c) $NaNO_2$, H_2SO_4
(d) CH_3COCl (e) CH_3MgBr (f) CH_3CH_2Cl, $AlCl_3$
(g) Product of (c) with CuCl (h) Product of (d) with CH_3CH_2Cl, $AlCl_3$

25.29 Show the products from reaction of *o*-chlorophenol with the following reagents:
(a) NaOH, then CH_3I (b) CH_3COCl, pyridine
(c) Fremy's salt (d) $CH_3CH_2CH_2Cl$, $AlCl_3$

25.30 Reaction of anthranilic acid (*o*-aminobenzoic acid) with $NaNO_2$ and H_2SO_4 yields a diazonium salt that can be treated with base to yield a neutral diazonium carboxylate.
(a) What is the structure of the neutral diazonium carboxylate?
(b) Heating the diazonium carboxylate results in the formation of CO_2, N_2, and an intermediate that reacts with 1,3-cyclopentadiene to yield the following product:

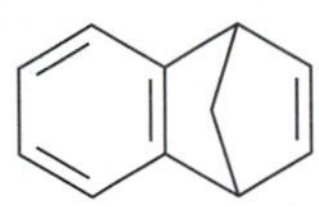

What is the structure of the intermediate, and what kind of reaction does it undergo with cyclopentadiene?

25.31 Mephenesin is a drug used as a muscle relaxant and sedative. Propose a synthesis of mephenesin from benzene and any other reagents needed.

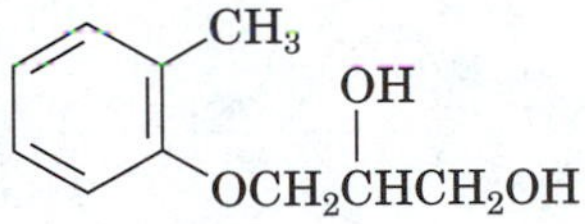

Mephenesin

25.32 How would you prepare the following substances?
(a) *p*-Iodobenzoic acid from aniline (b) *o*-Iodobromobenzene from benzene

25.33 Gentisic acid is a naturally occurring hydroquinone found in gentian. Its sodium salt is used medicinally as an antirheumatic agent. How would you prepare gentisic acid from benzene?

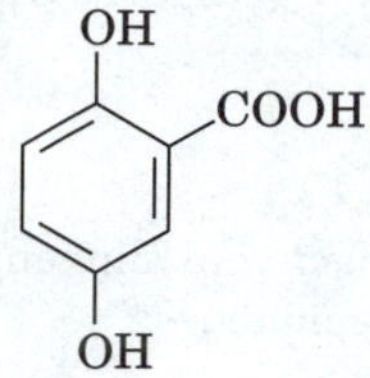

Gentisic acid

25.34 Prontosil is an antibacterial azo dye that was once used for urinary tract infections. How would you prepare prontosil from benzene?

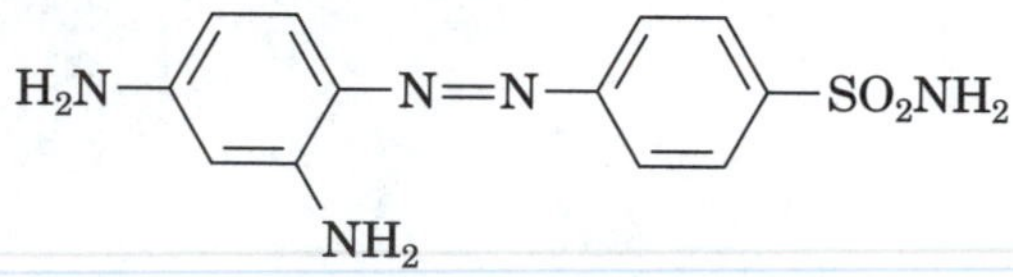

Prontosil

25.35 How would you synthesize the dye Orange II from benzene and β-naphthol?

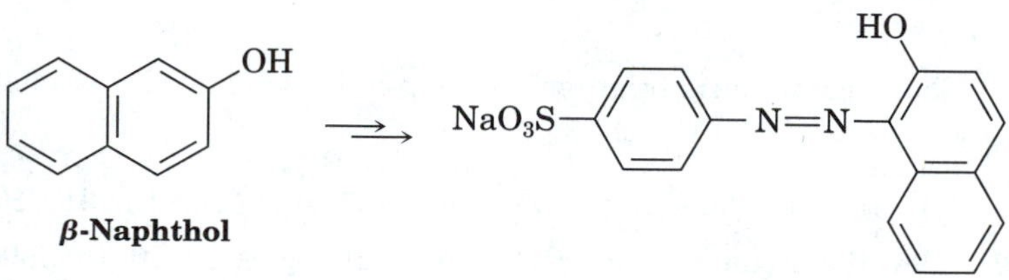

Orange II

25.36 2-Nitro-3,4,6-trichlorophenol is used as a lampricide (a compound toxic to lampreys) to combat the intrusion of sea lampreys into the Great Lakes. How would you synthesize this material from benzene?

25.37 The germicidal agent hexachlorophene is prepared by condensation of 2,4,5-trichlorophenol with formaldehyde in the presence of H_2SO_4. Propose a mechanism for the reaction.

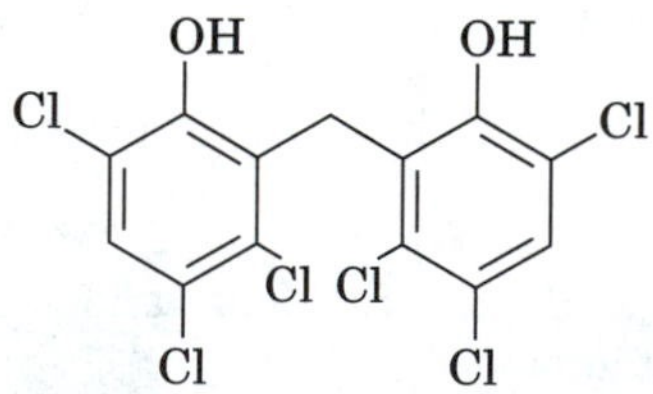

Hexachlorophene

25.38 Propose a route from benzene for the synthesis of the antiseptic agent trichlorosalicylanilide.

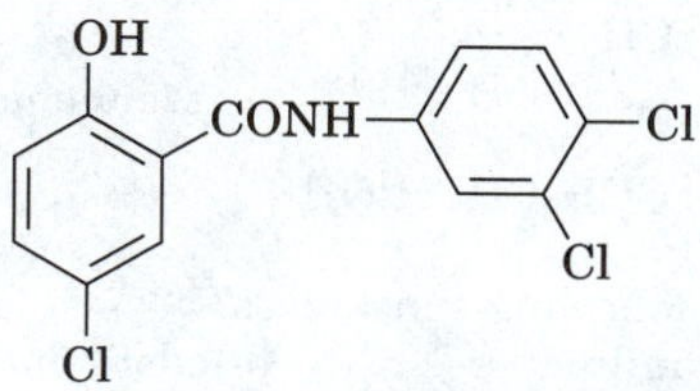

Trichlorosalicylanilide

25.39 Compound A, $C_8H_{10}O$, has the IR and ^1H NMR spectra shown. Propose a structure consistent with the observed spectra, and assign each peak in the NMR spectrum. Note that the absorption at 5.5 δ disappears when D_2O is added.

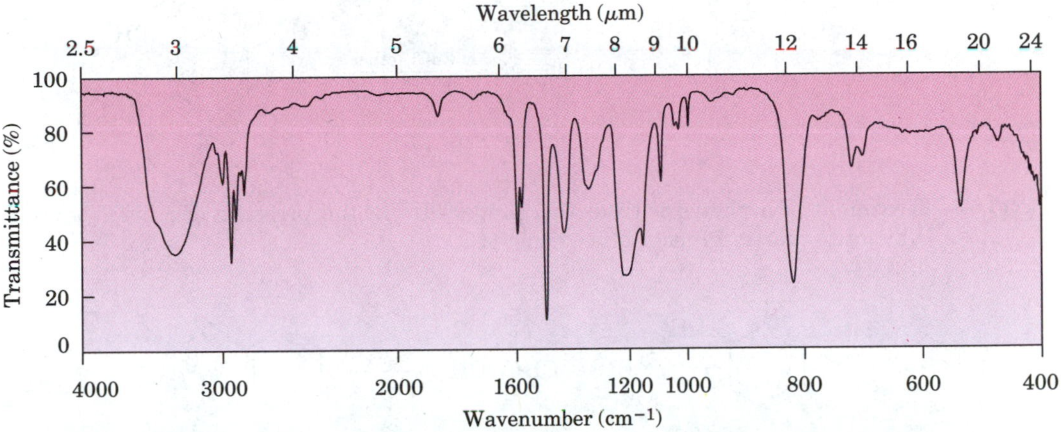

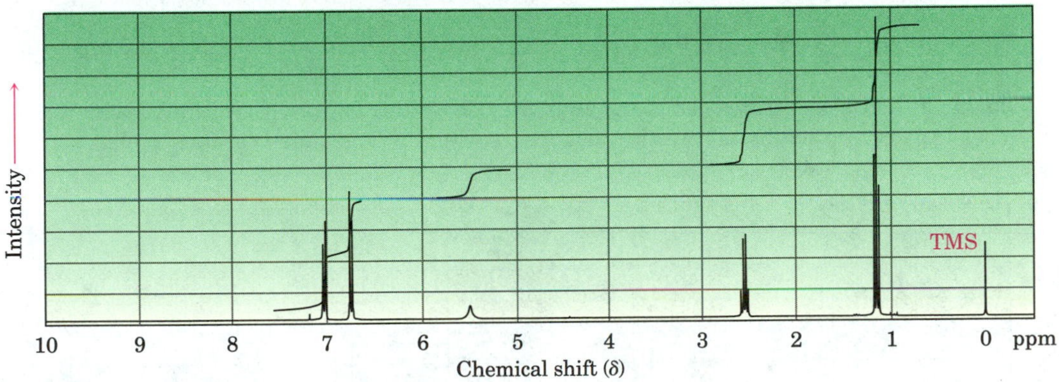

25.40 Phenacetin, a substance formerly used in over-the-counter headache remedies, has the formula $C_{10}H_{13}NO_2$. Phenacetin is neutral and does not dissolve in either acid or base. When warmed with aqueous NaOH, phenacetin yields an amine, $C_8H_{11}NO$, whose ^1H NMR spectrum is shown. When heated with HI, the amine is cleaved to an aminophenol, C_6H_7NO, which, on treatment with Fremy's salt, yields benzoquinone. What is the structure of phenacetin, and what are the structures of the amine and the aminophenol?

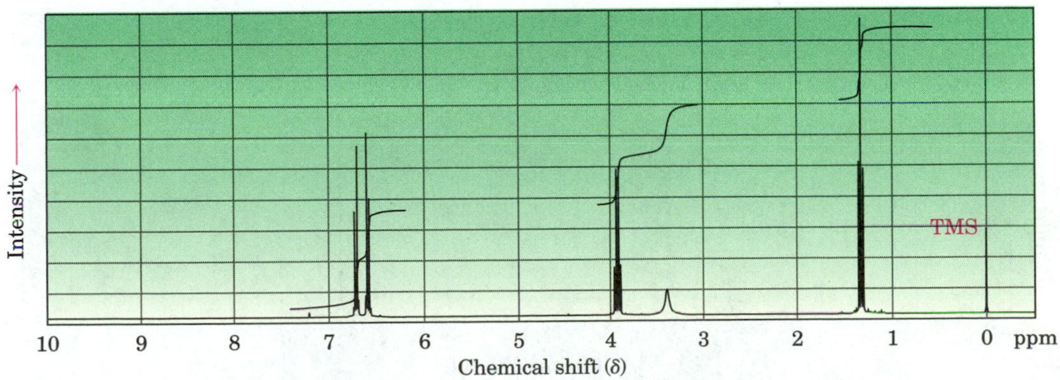

25.41 In the *Hoesch reaction*, resorcinol (*m*-dihydroxybenzene) is treated with a nitrile in the presence of a Lewis acid catalyst. After hydrolysis, an acyl resorcinol is isolated. Propose a mechanism for the Hoesch reaction. What other well-known reaction is similar to this reaction?

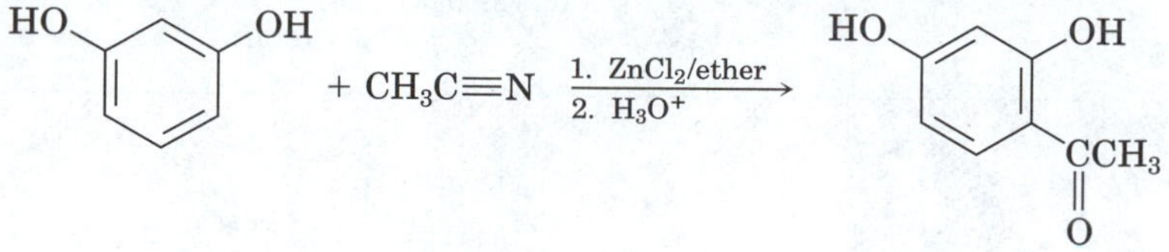

25.42 Treatment of a phenol with methyl propenoate in the presence of a base yields a dihydrocoumarin. Propose a mechanism.

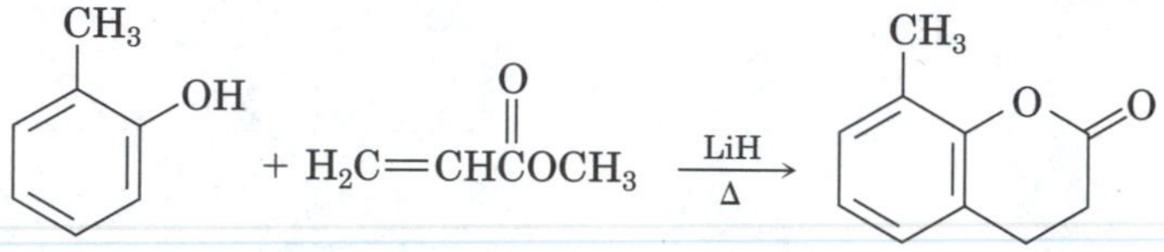

A dihydrocoumarin

25.43 Propose a route for the synthesis of 1-bromo-2,4-dimethylbenzene starting from benzene.

25.44 Propose structures for compounds that show the following ^{1}H NMR spectra. The peak marked by an asterisk disappears when D_2O is added to the sample.

(a) $C_{15}H_{24}O$

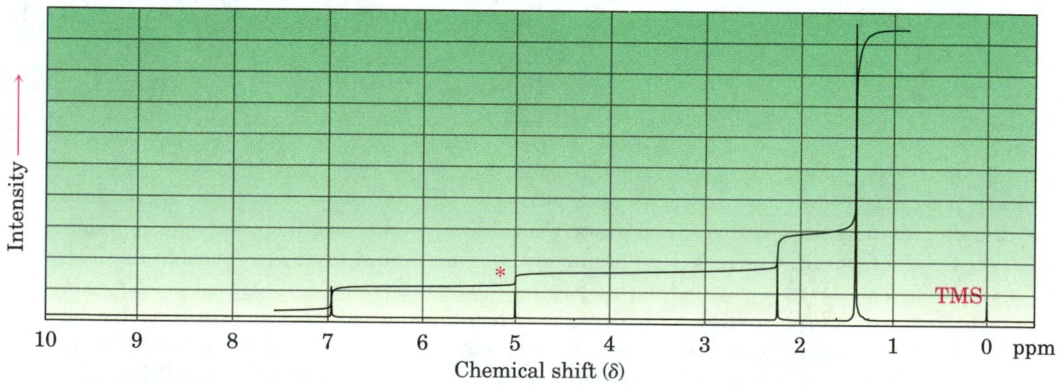

(b) $C_9H_{13}N$

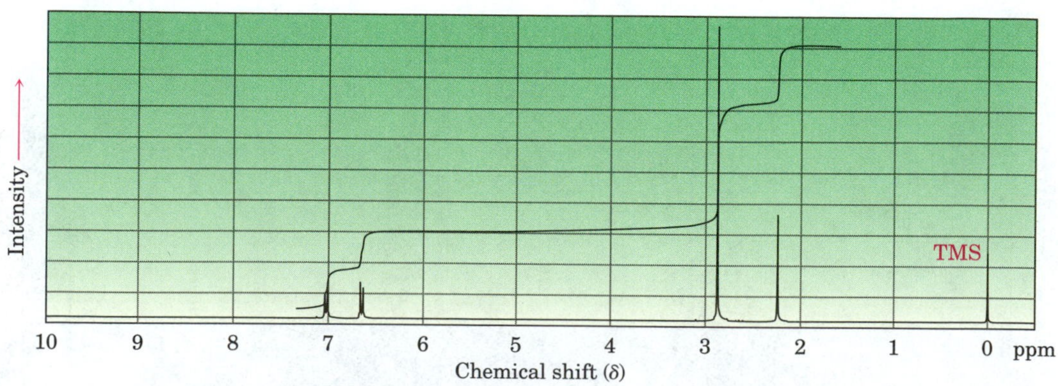

25.45 Propose structures for compounds that show the following ^1H NMR spectra.

(a) $C_{15}H_{17}N$

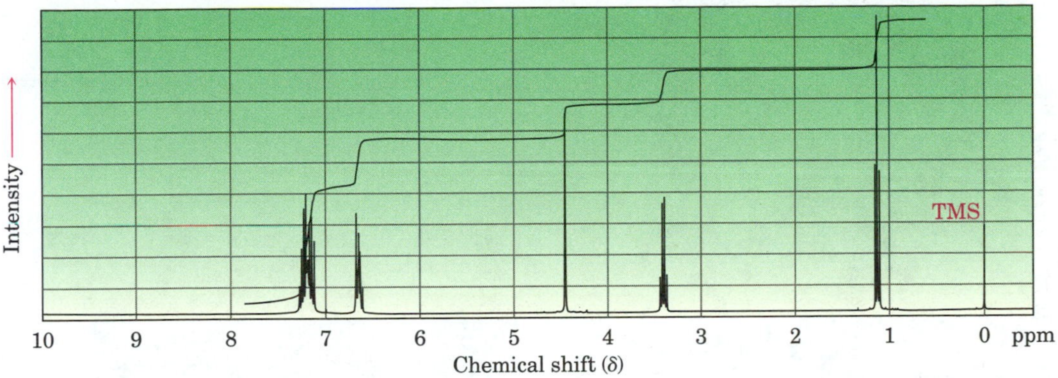

(b) $C_{10}H_{11}ClO_2$

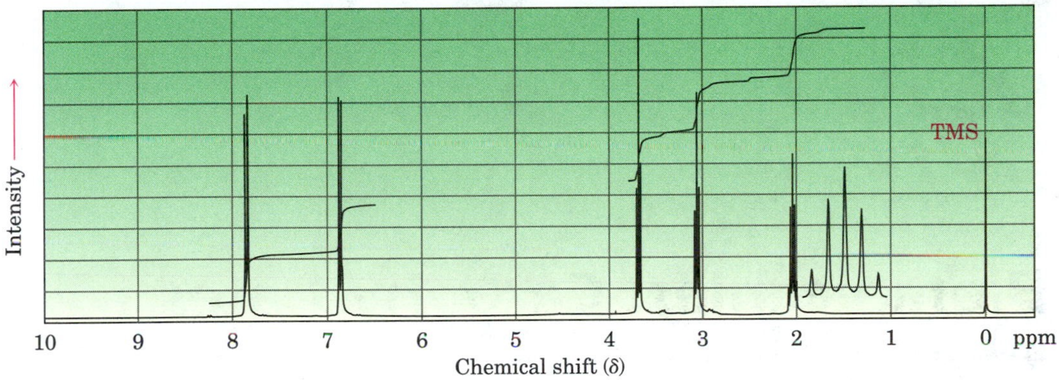

A Look Ahead

25.46 We'll see in Chapter 29 that certain amino groups in DNA are more basic than others and can be alkylated more easily. Deoxyguanosine, for example, reacts with the alkylating agent dimethyl sulfate in the following way:

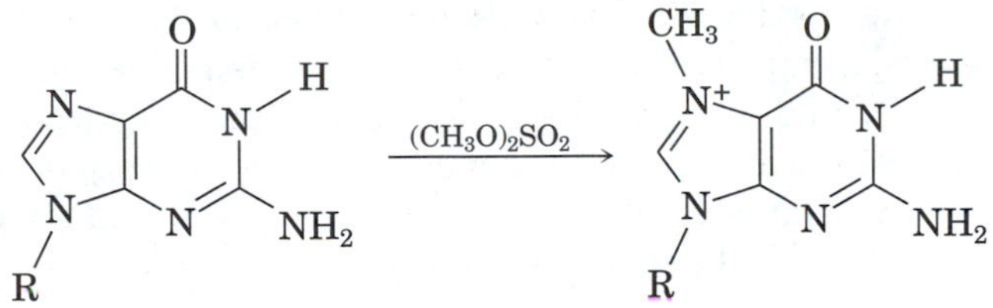

Deoxyguanosine

Look back at Sections 15.9 and 24.4, and explain why the doubly bonded nitrogen atom in the five-membered ring is the most basic and nucleophilic.

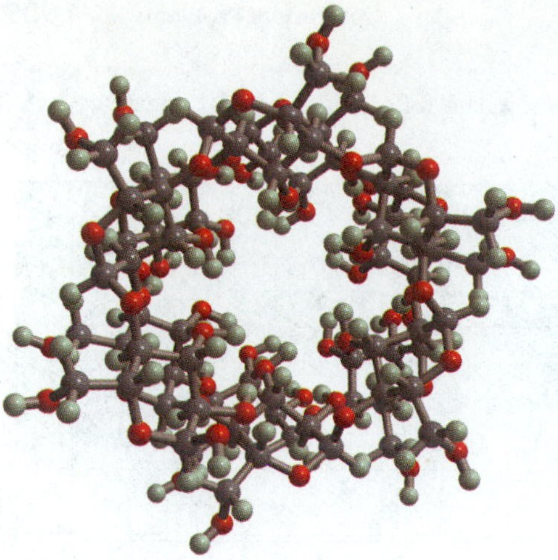

Amylose, a constituent of starch, has a complex, helical structure.

26

Biomolecules: Carbohydrates

Carbohydrates occur in every living organism. The sugar and starch in food, and the cellulose in wood, paper, and cotton, are nearly pure carbohydrate. Modified carbohydrates form part of the coating around living cells, other carbohydrates are found in the DNA that carries genetic information, and still others are used as medicines.

The word **carbohydrate** derives historically from the fact that glucose, the first simple carbohydrate to be obtained pure, has the molecular formula $C_6H_{12}O_6$ and was originally thought to be a "hydrate of carbon, $C_6(H_2O)_6$." This view was soon abandoned, but the name persisted. Today, the term *carbohydrate* is used to refer loosely to the broad class of polyhydroxylated aldehydes and ketones commonly called *sugars*.

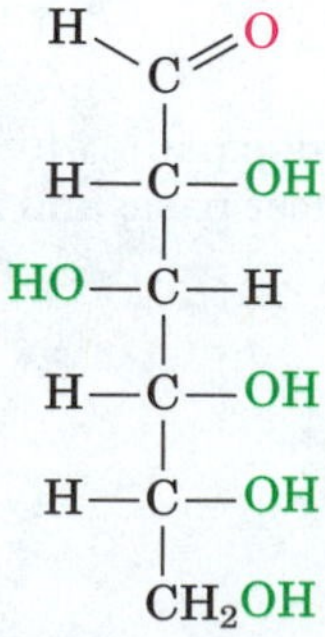

Glucose (also called dextrose), a pentahydroxyhexanal

Carbohydrates are synthesized by green plants during photosynthesis, a complex process in which sunlight provides the energy to convert CO_2 into glucose. Many molecules of glucose are then chemically linked for storage by the plant in the form of either cellulose or starch. It has been estimated that more than 50% of the dry weight of the earth's biomass—all plants and animals—consists of glucose polymers. When eaten and metabolized, carbohydrates provide the major source of energy required by organisms. Thus, carbohydrates act as the chemical intermediaries by which solar energy is stored and used to support life.

$$6\ CO_2 + 6\ H_2O \xrightarrow{\text{Sunlight}} 6\ O_2 + \underset{\textbf{Glucose}}{C_6H_{12}O_6} \longrightarrow \text{Cellulose, starch}$$

When eaten, glucose can be either metabolized in the body to provide immediate energy or stored in the form of glycogen for later use. Since humans and most other mammals lack the enzymes needed for digestion of cellulose, they require starch as their dietary source of carbohydrates. Grazing animals such as cows, however, contain in their rumen microorganisms that are able to digest cellulose. The energy stored in cellulose is thus moved up the biological food chain when these animals are used for food.

26.1 Classification of Carbohydrates

Carbohydrates are generally classed into two groups, *simple* and *complex*. **Simple sugars**, or **monosaccharides**, are carbohydrates like glucose and fructose that can't be hydrolyzed into smaller molecules. **Complex carbohydrates** are made of two or more simple sugars linked together. Sucrose (table sugar), for example, is a **disaccharide** made up of one glucose molecule linked to one fructose molecule. Similarly, cellulose is a **polysaccharide** made up of several thousand glucose molecules linked together. Hydrolysis of these polysaccharides breaks them down into their constituent monosaccharide units.

$$1\ \text{Sucrose} \xrightarrow{H_3O^+} 1\ \text{Glucose} + 1\ \text{Fructose}$$

$$\text{Cellulose} \xrightarrow{H_3O^+} {\sim}3000\ \text{Glucose}$$

Monosaccharides can be further classified as either **aldoses** or **ketoses**. The *-ose* suffix is used to designate a carbohydrate, and the *aldo-* and *keto-* prefixes identify the nature of the carbonyl group. The number of carbon atoms in the monosaccharide is indicated by using *tri-*, *tetr-*, *pent-*, *hex-*, and so forth in the name. For example, glucose is an *aldohexose*, a six-carbon aldehydo sugar; fructose is a *ketohexose*, a six-carbon keto sugar; and ribose is an *aldopentose*, a five-carbon aldehydo sugar. Most of the commonly occurring sugars are either aldopentoses or aldohexoses.

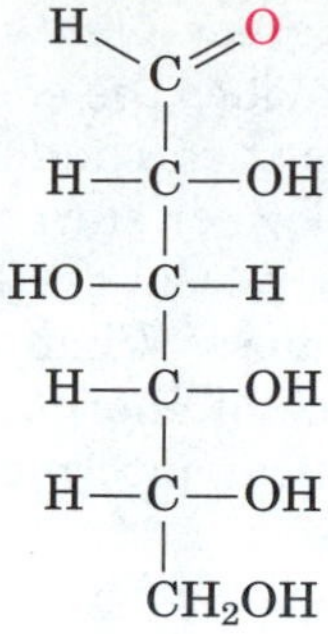

Glucose
(an aldohexose)

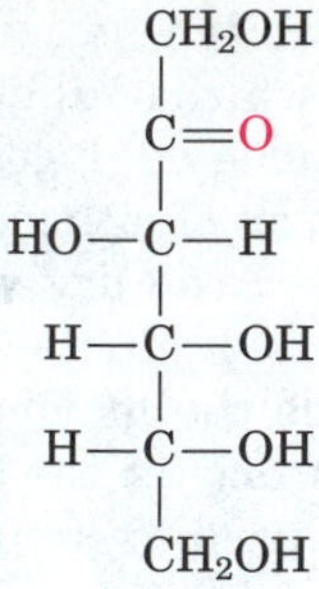

Fructose
(a ketohexose)

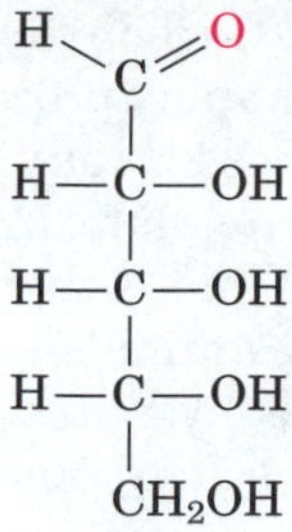

Ribose
(an aldopentose)

PROBLEM.

26.1 Classify each of the following monosaccharides:

(a) 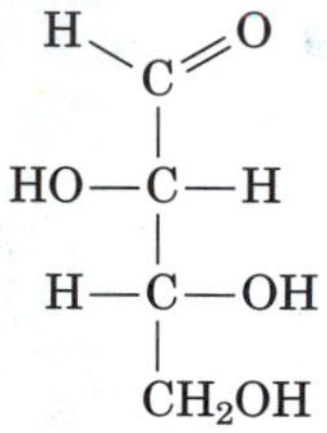

Threose

(b) 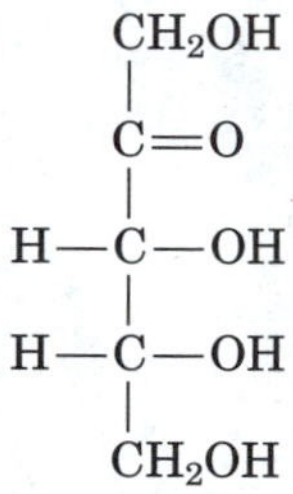

Ribulose

(c) 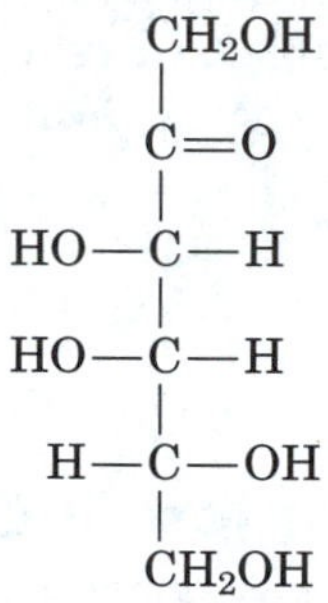

Tagatose

(d) 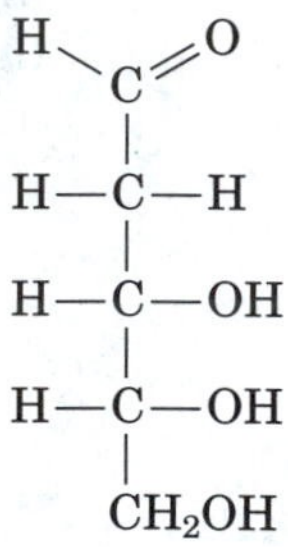

2-Deoxyribose

26.2 Configurations of Monosaccharides: Fischer Projections

Since all carbohydrates have stereogenic carbon atoms, it was recognized long ago that a standard method of representation is needed to describe carbohydrate stereochemistry. The method most commonly used employs **Fischer projections** for depicting stereogenic centers on a flat page.

Recall from Section 9.13 that a tetrahedral carbon atom is represented in a Fischer projection by two crossed lines. The horizontal lines represent bonds coming out of the page, and the vertical lines represent bonds going into the page. By convention, the carbonyl carbon is placed at or near the top in Fischer projections. Thus, (*R*)-glyceraldehyde, the simplest monosaccharide, is drawn as shown in Figure 26.1.

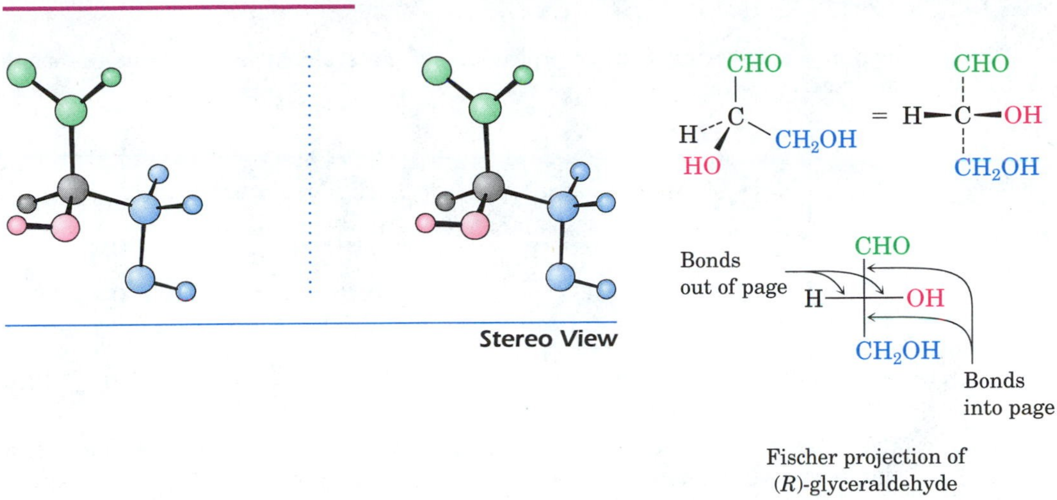

Figure 26.1 A Fischer projection of (*R*)-glyceraldehyde.

Recall also that Fischer projections can be rotated on the page by 180° without changing their meaning, but not by 90° or 270°.

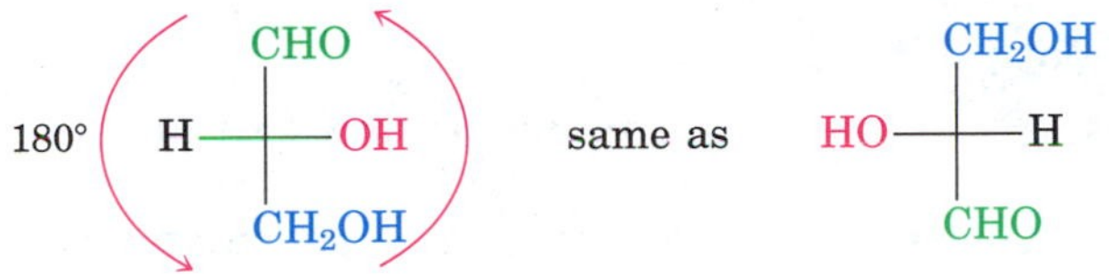

(*R*)-Glyceraldehyde

Carbohydrates with more than one stereogenic center are shown by stacking the centers on top of one another, with the carbonyl carbon again placed either at or near the top. Glucose, for example, has four stereogenic centers stacked on top of one another in Fischer projection:

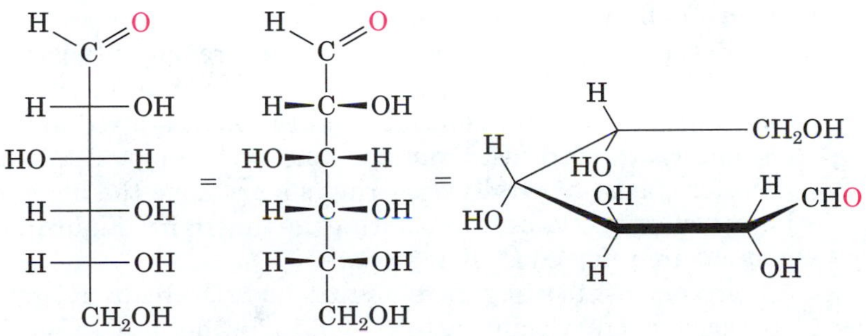

Glucose
(carbonyl group at top)

PROBLEM

26.2 Which of the following Fischer projections of glyceraldehyde represent the same enantiomer?

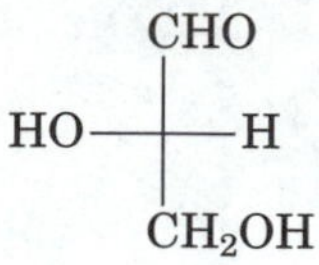
A

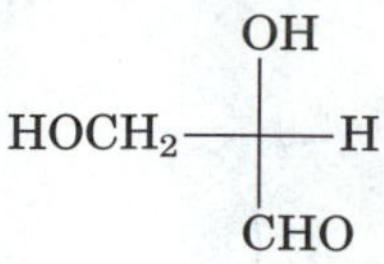
B

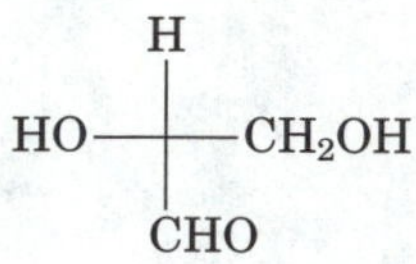
C

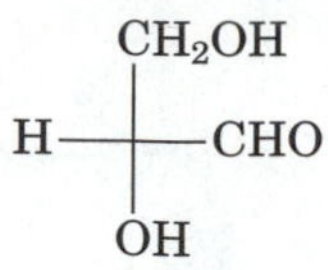
D

PROBLEM

26.3 Convert the following Fischer projections into tetrahedral representations, and assign *R* or *S* stereochemistry to each.

(a)
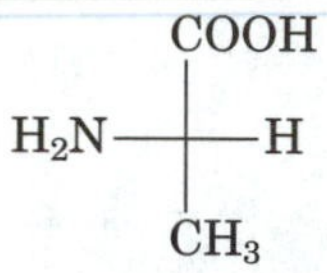

(b)
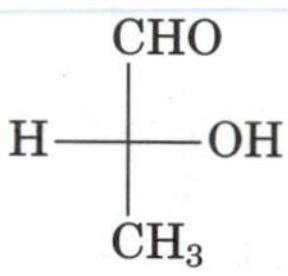

(c)
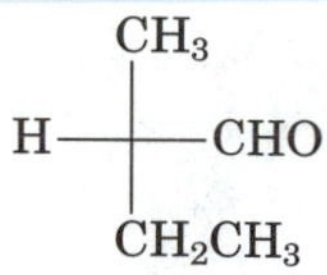

26.3 D,L Sugars

Glyceraldehyde has only one stereogenic center and therefore has two enantiomeric (mirror-image) forms. Only the dextrorotatory enantiomer occurs naturally, however. That is, a sample of naturally occurring glyceraldehyde placed in a polarimeter rotates plane-polarized light in a clockwise direction, denoted (+).

Since (+)-glyceraldehyde is known to have the *R* configuration at C2, it can be represented in Fischer projection as shown in Figure 26.1. For historical reasons dating back long before the adoption of the *R*,*S* system, (*R*)-(+)-glyceraldehyde is also referred to as D-*glyceraldehyde* (D for dextrorotatory). The other enantiomer, (*S*)-(−)-glyceraldehyde, is known as L-*glyceraldehyde* (L for levorotatory).

Because of the way monosaccharides are biosynthesized in nature, glucose, fructose, and almost all other naturally occurring monosaccharides have the same stereochemical configuration as D-glyceraldehyde at the stereogenic center farthest from the carbonyl group. In Fischer projections, therefore, most naturally occurring sugars have the hydroxyl group at the lowest stereogenic center pointing to the right (Figure 26.2). Such compounds are referred to as **D sugars**.

In contrast to D sugars, **L-sugars** have the hydroxyl group at the lowest stereogenic center pointing to the left in Fischer projection. Thus, an L sugar is the mirror image (enantiomer) of the corresponding D sugar and differs from the D sugar at all stereogenic centers. Note that the D and L notations

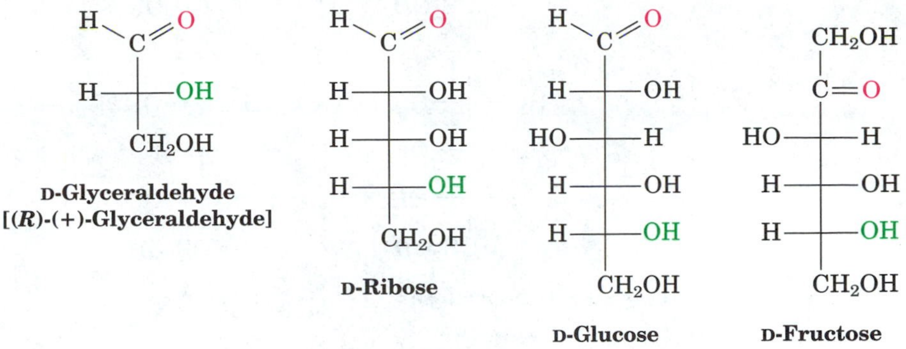

Figure 26.2 Some naturally occurring D sugars. The hydroxyl group at the stereogenic center farthest from the carbonyl group is on the right when the molecule is drawn in Fischer projection.

have no relation to the direction in which a given sugar rotates plane-polarized light; a D sugar can be either dextrorotatory or levorotatory. The prefix D indicates only that the stereochemistry of the lowest stereogenic carbon atom is to the right in Fischer projection when the molecule is drawn in the standard way with the carbonyl group at or near the top.

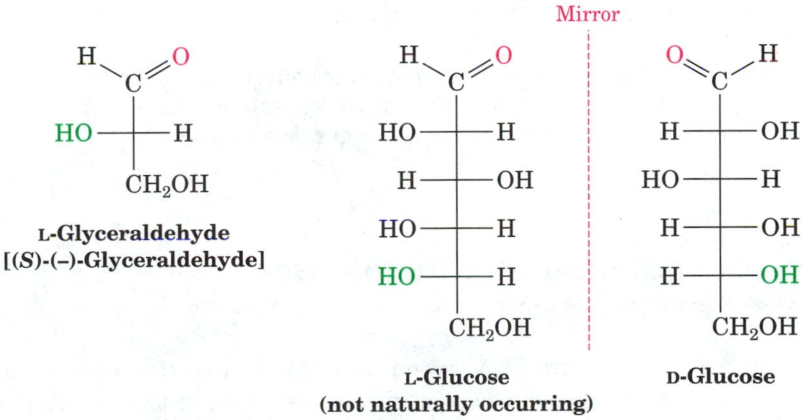

The D,L system of carbohydrate nomenclature describes the configuration at only one stereogenic center and says nothing about other stereogenic centers that may be present. The advantage of the system, though, is that it allows us to relate one sugar to another rapidly and visually.

PRACTICE PROBLEM

Draw a Fischer projection of L-fructose.

Solution Since L-fructose is the enantiomer of D-fructose, simply look at the structure of D-fructose and then reverse the configuration at each stereogenic center.

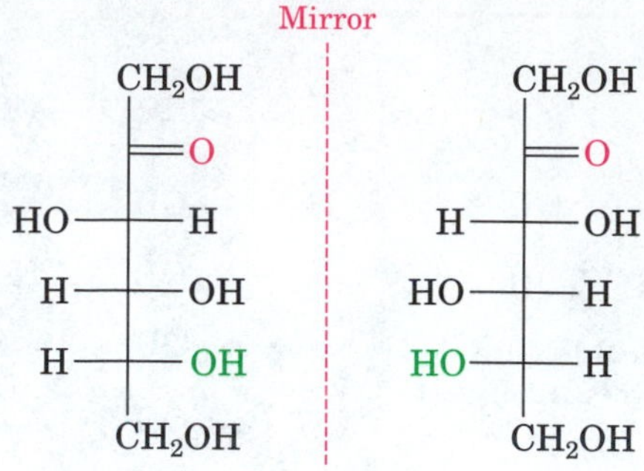

D-Fructose L-Fructose

PROBLEM

26.4 Assign *R* or *S* configuration to each stereogenic center in the following sugars, and tell whether each is a D sugar or an L sugar.

(a)

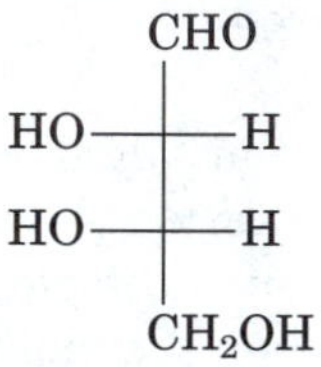

(b)

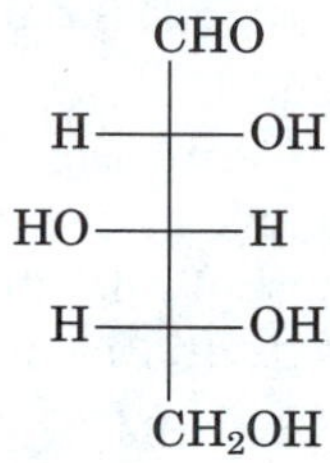

(c)

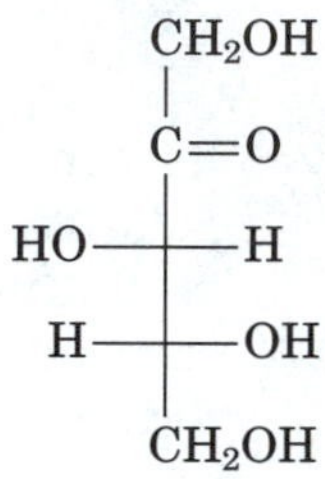

PROBLEM

26.5 (+)-Arabinose, an aldopentose that is widely distributed in plants, is systematically named (2*R*,3*S*,4*S*)-2,3,4,5-tetrahydroxypentanal. Draw a Fischer projection of (+)-arabinose and identify it as a D sugar or an L sugar.

26.4 Configurations of the Aldoses

Aldotetroses are four-carbon sugars with two stereogenic centers. There are $2^2 = 4$ possible stereoisomeric aldotetroses, or two D,L pairs of enantiomers, called *erythrose* and *threose*.

Aldopentoses have three stereogenic centers, leading to a total of $2^3 = 8$ possible stereoisomers, or four D,L pairs of enantiomers. These four pairs are called *ribose*, *arabinose*, *xylose*, and *lyxose*. All except lyxose occur widely. D-Ribose is an important constituent of RNA (ribonucleic acid), L-arabinose is found in many plants, and D-xylose is found in wood.

Aldohexoses have four stereogenic centers and a total of $2^4 = 16$ possible stereoisomers, or eight D,L pairs of enantiomers. The names of the eight are *allose*, *altrose*, *glucose*, *mannose*, *gulose*, *idose*, *galactose*, and *talose*. Of the eight, only D-glucose (from starch and cellulose) and D-galactose (from gums and fruit pectins) are found widely in nature. D-Mannose and D-talose also occur naturally, but in lesser abundance.

Fischer projections of the four-, five-, and six-carbon D aldoses are shown in Figure 26.3. Starting from D-glyceraldehyde, we can imagine constructing

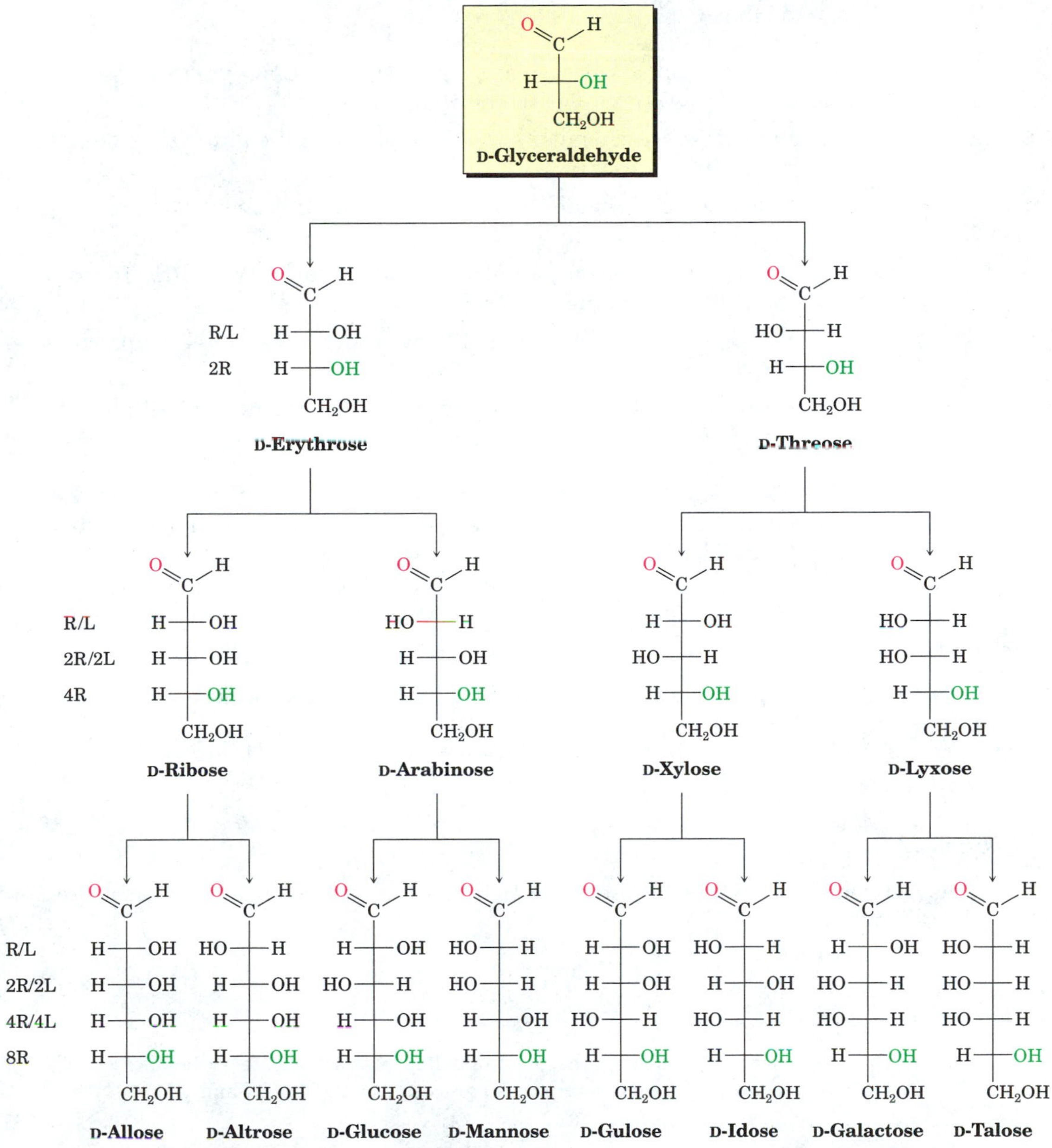

Figure 26.3 Configurations of D aldoses. The structures are arranged in order from left to right so that the hydroxyl groups on C2 alternate right/left (R/L) in going across a series. Similarly, the hydroxyl groups at C3 alternate two right/two left (2R/2L); the hydroxyl groups at C4 alternate 4R/4L; and the hydroxyl groups at C5 are to the right in all eight (8R).

the two D-aldotetroses by inserting a new stereogenic center just below the aldehyde carbon. Each of the two D-aldotetroses leads to two D-aldopentoses (four total), and each of the four D-aldopentoses leads to two D-aldohexoses (eight total).

Louis Fieser[1] of Harvard University suggested the following procedure, shown in Figure 26.3, for remembering the names and structures of the eight D-aldohexoses:

1. Set up eight Fischer projections with the –CHO group on top and the $-CH_2OH$ group at the bottom.
2. Indicate stereochemistry at C5 by placing all eight –OH groups to the right (D series).
3. Indicate stereochemistry at C4 by alternating four –OH groups to the right and four to the left.
4. Indicate stereochemistry at C3 by alternating two –OH groups to the right, two to the left, and so on.
5. Indicate stereochemistry at C2 by alternating –OH groups right, left, right, left, and so on.
6. Name the eight isomers according to the mnemonic "All altruists gladly make gum in gallon tanks."

The four D-aldopentose structures can be generated in a similar way and can be named by the mnemonic suggested by a Cornell undergraduate: "Ribs are extra lean."

PROBLEM

26.6 Only the D sugars are shown in Figure 26.3. Draw Fischer projections for the following L sugars:
(a) L-Xylose (b) L-Galactose (c) L-Glucose

PROBLEM

26.7 How many aldoheptoses are there? How many are D sugars, and how many are L sugars?

PROBLEM

26.8 Draw Fischer projections for the two D-aldoheptoses whose stereochemistry at C3, C4, C5, and C6 is the same as that of D-glucose at C2, C3, C4, and C5.

[1]Louis F. Fieser (1899–1977); b. Columbus, Ohio; Ph.D. Harvard (Conant); professor, Bryn Mawr College, Harvard University.

26.5 Cyclic Structures of Monosaccharides: Hemiacetal Formation

We said in Section 19.14 that alcohols undergo a rapid and reversible nucleophilic addition reaction with ketones and aldehydes to form hemiacetals:

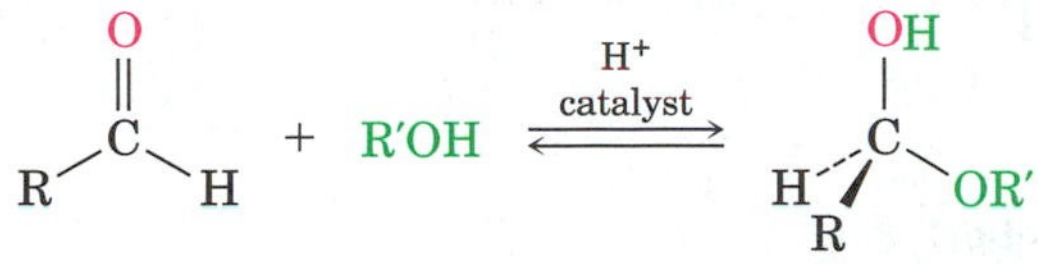

If both the hydroxyl and the carbonyl group are in the same molecule, an *intramolecular* nucleophilic addition can take place, leading to the formation of a *cyclic* hemiacetal. Five- and six-membered cyclic hemiacetals are particularly stable, and many carbohydrates therefore exist in an equilibrium between open-chain and cyclic forms. Glucose, for example, exists in aqueous solution primarily as the six-membered **pyranose** form resulting from intramolecular nucleophilic addition of the –OH group at C5 to the C1 carbonyl group. Fructose, on the other hand, exists to the extent of about 80% in the pyranose form and about 20% as the five-membered **furanose** form resulting from addition of the –OH group at C5 to the C2 carbonyl. The words *pyranose* for a six-membered ring and *furanose* for a five-membered ring are derived from the names of the simple cyclic ethers pyran and furan. The cyclic forms of glucose and fructose are shown in Figure 26.4 (p. 1020).

Pyranose and furanose rings are often represented using the **Haworth**[2] **projections** shown in Figure 26.4, rather than Fischer projections. In a Haworth projection, the hemiacetal ring is drawn as if it were flat and is viewed edge-on with the oxygen atom at the upper right. Though convenient, this view is not really accurate because pyranose rings are actually chair-shaped like cyclohexane (Section 4.9) rather than flat. Nevertheless, Haworth projections are widely used because they make it possible to see at a glance the cis–trans relationships among hydroxyl groups on the ring.

When converting from one kind of projection to the other, remember that an –OH on the *right* in a Fischer projection is *down* in a Haworth projection. Conversely, an –OH on the *left* in a Fischer projection is *up* in a Haworth projection. For D sugars, the terminal $-CH_2OH$ group is always up in Haworth projections, whereas for L sugars the $-CH_2OH$ group is down. Figure 26.5 (p. 1021) illustrates the conversion for D-glucose.

---

[2]Sir Walter Norman Haworth (1883–1950); b. Chorley, Lancashire; Ph.D. Göttingen; D.Sc. Manchester; professor, University of Birmingham; Nobel Prize (1937).

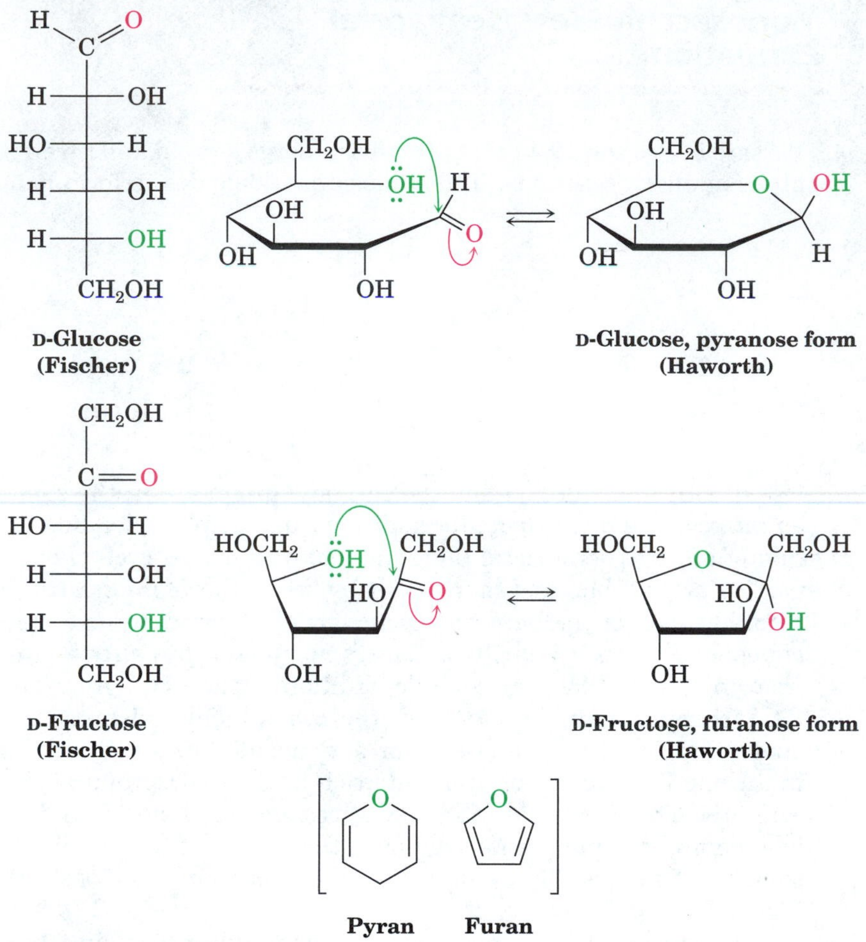

Figure 26.4 Glucose and fructose in their cyclic furanose and pyranose forms.

PRACTICE PROBLEM

D-Mannose differs from D-glucose in its stereochemistry at C2. Draw a Haworth projection of D-mannose in its pyranose form.

Solution First draw a Fischer projection of D-mannose. Then lay it on its side, and curl it around so that the –CHO group (C1) is toward the front and the $-CH_2OH$ group (C6) is toward the rear. Now connect the –OH at C5 to the C1 carbonyl group to form a pyranose ring.

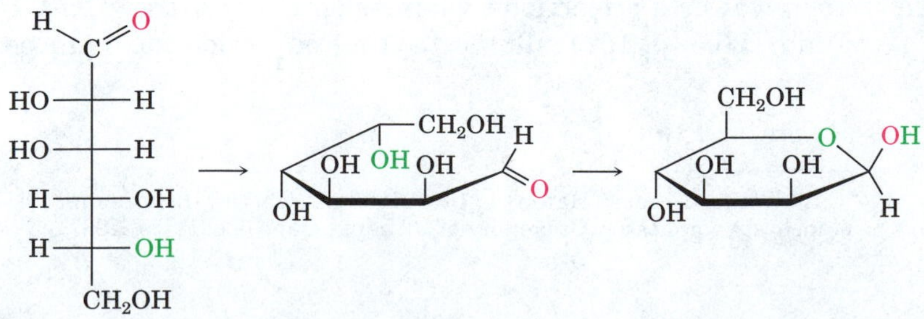

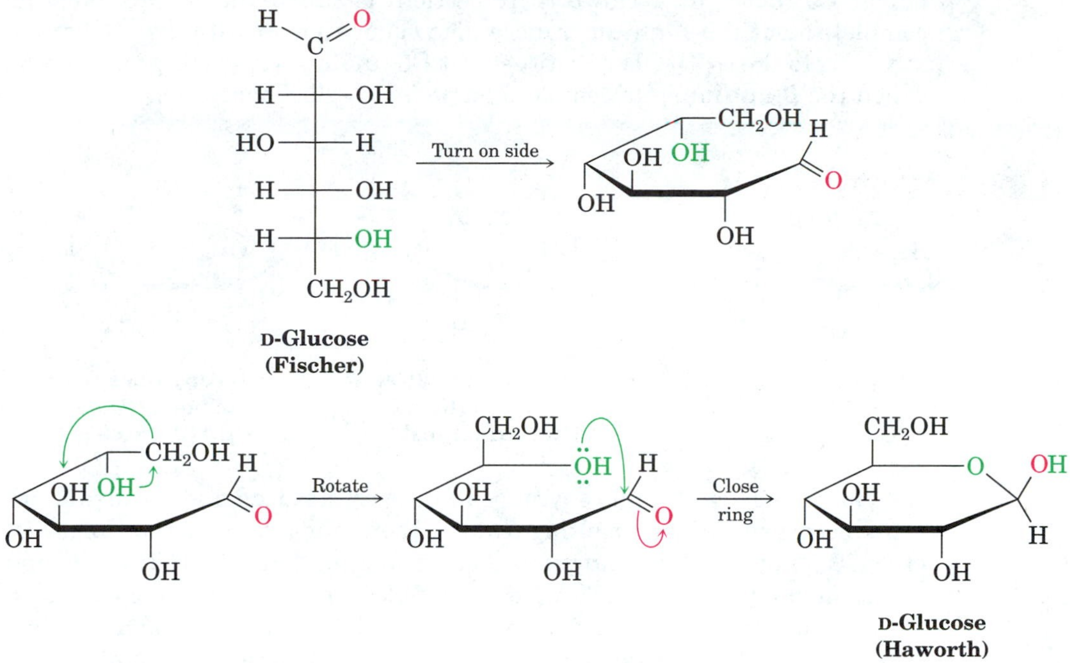

Figure 26.5 Interconversion of Fischer and Haworth projections of D-glucose.

PROBLEM

26.9 D-Galactose differs from D-glucose in its stereochemistry at C4. Draw a Haworth projection of D-galactose in its pyranose form.

PROBLEM

26.10 Draw Haworth projections of L-glucose in its pyranose form and D-ribose in its furanose form.

26.6 Monosaccharide Anomers: Mutarotation

When an open-chain monosaccharide cyclizes to a furanose or pyranose form, a new stereogenic center is generated at the former carbonyl carbon. The two diastereomers produced are called **anomers**, and the hemiacetal carbon atom is referred to as the **anomeric center**. For example, glucose cyclizes reversibly in aqueous solution to a 36:64 mixture of two anomers.

The minor anomer with the $-OH$ group at C1 trans to the $-CH_2OH$ substituent at C5 (down in a Haworth projection) is called the ***α* anomer**;[3] its complete name is α-D-glucopyranose. The major anomer with the $-OH$ group at C1 cis to the $-CH_2OH$ substituent at C5 (up in a Haworth projection) is called the ***β* anomer**; its complete name is β-D-glucopyranose.

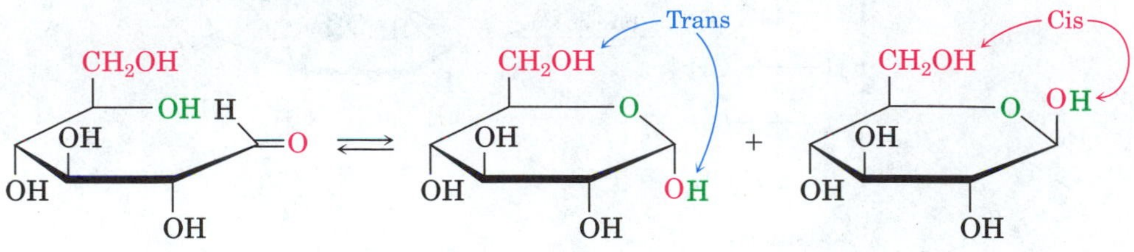

D-Glucose **α-D-Glucopyranose (36%) (α anomer: OH and CH_2OH are trans)** **β-D-Glucopyranose (64%) (β anomer: OH and CH_2OH are cis)**

Both anomers of D-glucopyranose can be crystallized and purified. Pure α-D-glucopyranose has a melting point of 146°C and a specific rotation, $[\alpha]_D$, of +112.2°; pure β-D-glucopyranose has a melting point of 148–155°C and a specific rotation of +18.7°. When a sample of either pure anomer is dissolved in water, however, the optical rotation slowly changes and ultimately reaches a constant value of +52.6°. The specific rotation of the α-anomer solution decreases from +112.2° to +52.6°, and the specific rotation of the β-anomer solution increases from +18.7° to +52.6°. Known as **mutarotation**, this phenomenon is due to the slow conversion of the pure α and β anomers into the 36:64 equilibrium mixture.

Mutarotation occurs by a reversible ring-opening of each anomer to the open-chain aldehyde, followed by reclosure. Although equilibration is slow at neutral pH, it is catalyzed by both acid and base.

D-Glucose

α-D-Glucopyranose (36%)
$[\alpha]_D$ = +112.2°

β-D-Glucopyranose (64%)
$[\alpha]_D$ = +18.7°

---

[3]An α anomer is formally defined as having the same configuration at the anomeric center and at the last stereogenic center. A β anomer has opposite configurations at the anomeric center and the last stereogenic center.

PROBLEM

26.11 Draw the two anomers of D-fructose in their furanose forms.

PROBLEM

26.12 Knowing that the specific rotation of pure α-D-glucopyranose is $+112.2°$ and that the specific rotation of pure β-D-glucopyranose is $+18.7°$, show how the equilibrium percentages of α and β anomers can be calculated from the equilibrium specific rotation of $+52.6°$.

26.7 Conformations of Monosaccharides

Haworth projections are easy to draw, but they don't give an accurate three-dimensional picture of molecular conformation. Pyranose rings, like cyclohexane rings, have a chair-like geometry with axial and equatorial substituents. Any substituent that is up in a Haworth projection is also up in a chair conformation, and any substituent that is down in a Haworth projection is down in the chair conformation. Haworth projections can be converted into chair representations by following three steps:

1. Draw the Haworth projection with the ring oxygen atom at the upper right.
2. Raise the leftmost carbon atom (C4) *above* the ring plane.
3. Lower the anomeric carbon atom (C1) *below* the ring plane.

Figure 26.6 (p. 1024) shows how this conversion is done for α-D-glucopyranose and β-D-glucopyranose. Make molecular models to see the process more clearly.

Note that in β-D-glucopyranose, all the substituents on the ring are equatorial. Thus, β-D-glucopyranose is the least sterically crowded and most stable of the eight D-aldohexoses.

PROBLEM

26.13 Draw chair conformations of β-D-galactopyranose and β-D-mannopyranose. Label the ring substituents as either axial or equatorial. Which would you expect to be more stable, galactose or mannose?

PROBLEM

26.14 Draw a chair conformation of β-L-glucopyranose, and label the substituents as either axial or equatorial.

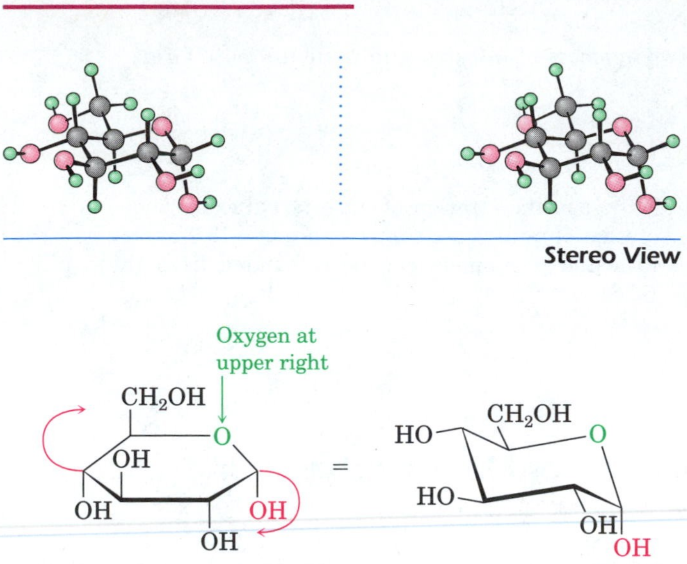

α-D-Glucopyranose

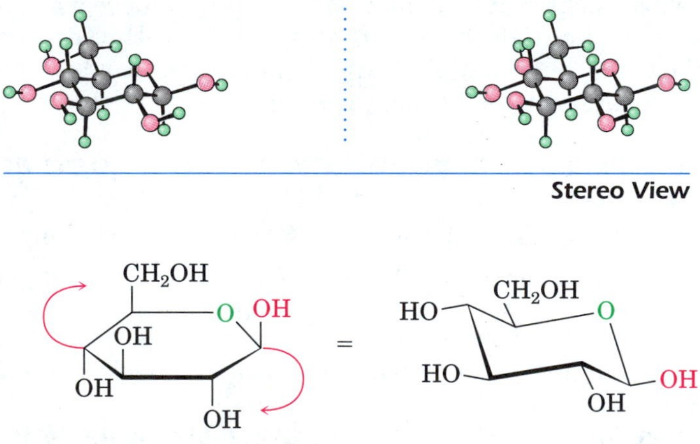

β-D-Glucopyranose

Figure 26.6 Chair representations of α-D-glucopyranose and β-D-glucopyranose.

26.8 Reactions of Monosaccharides

Since monosaccharides contain only two kinds of functional groups, carbonyls and hydroxyls, most of the chemistry of monosaccharides is the now familiar chemistry of these two groups.

Ester and Ether Formation

Monosaccharides behave as simple alcohols in much of their chemistry. For example, carbohydrate hydroxyl groups can be converted into esters and ethers, which are often easier to work with than the free sugars. Because of their many hydroxyl groups, monosaccharides are usually soluble in water but insoluble in organic solvents such as ether. They are also difficult to purify and have a tendency to form syrups rather than crystals when water is removed. Ester and ether derivatives, however, are soluble in organic solvents and are easily purified and crystallized.

Esterification is normally carried out by treating the carbohydrate with an acid chloride or acid anhydride in the presence of a base. All the hydroxyl groups react, including the anomeric one. For example, β-D-glucopyranose is converted into its pentaacetate by treatment with acetic anhydride in pyridine solution.

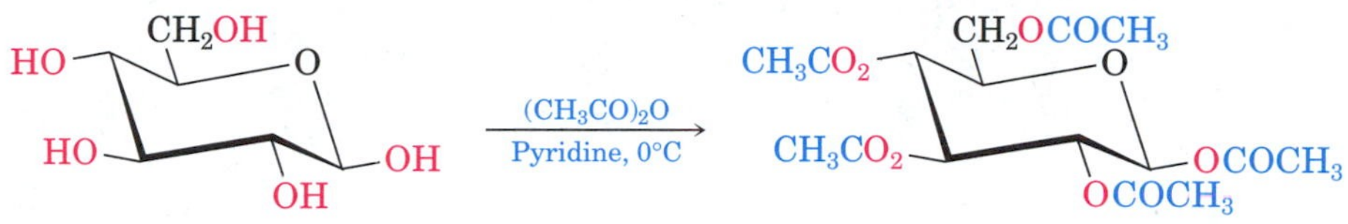

β-D-Glucopyranose **Penta-*O*-acetyl-β-D-glucopyranose (91%)**

Carbohydrates are converted into ethers by treatment with an alkyl halide in the presence of base (the Williamson ether synthesis, Section 18.4). Standard Williamson conditions using a strong base tend to degrade the sensitive sugar molecules, but Purdie[4] showed in 1903 that silver oxide works particularly well and that high yields of ethers are obtained. For example, α-D-glucopyranose is converted into its pentamethyl ether in 85% yield on reaction with iodomethane and Ag_2O.

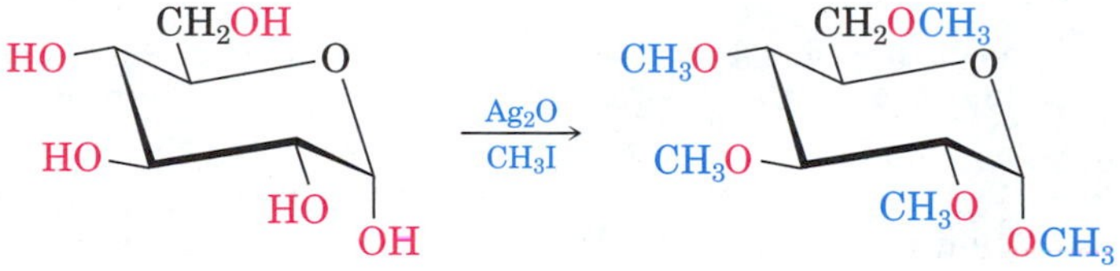

α-D-Glucopyranose **α-D-Glucopyranose pentamethyl ether (85%)**

[4]Thomas Purdie (1843–1916); b. Biggar, Scotland; Ph.D. Würzburg; professor, St. Andrews University.

PROBLEM

26.15 Draw the products you would obtain by reaction of β-D-ribofuranose with:
(a) CH_3I, Ag_2O (b) $(CH_3CO)_2O$, pyridine

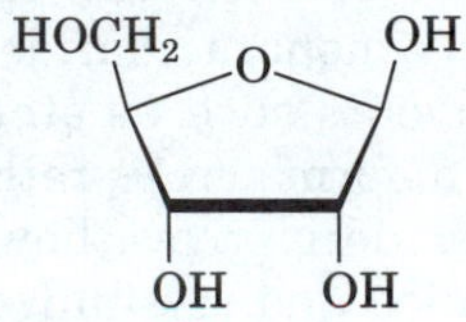

β-D-Ribofuranose

Glycoside Formation

We saw in Section 19.14 that treatment of a hemiacetal with an alcohol and an acid catalyst yields an acetal:

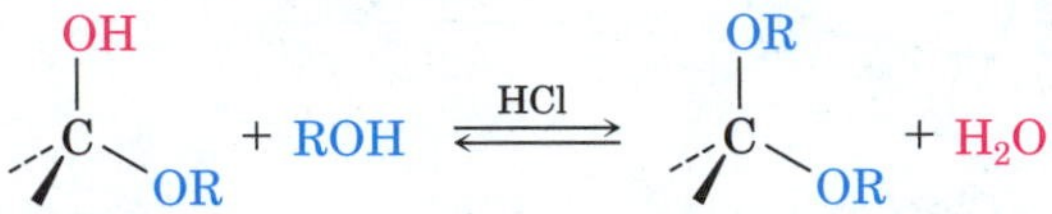

In the same way, treatment of a monosaccharide hemiacetal with an alcohol and an acid catalyst yields an acetal in which the anomeric –OH has been replaced by an –OR group. For example, reaction of glucose with methanol gives methyl β-D-glucopyranoside:

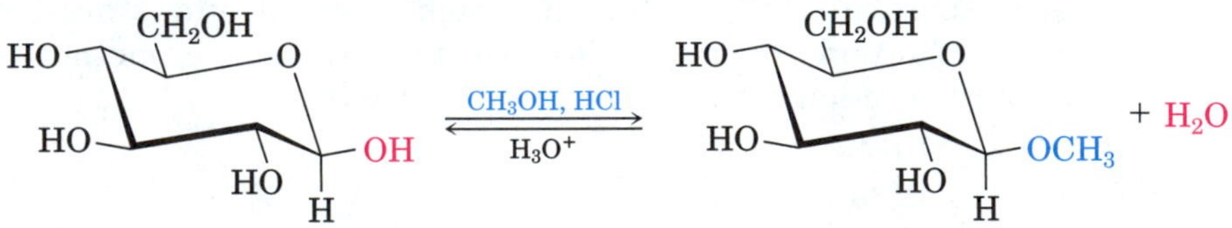

β-D-Glucopyranose (a hemiacetal)

Methyl β-D-glucopyranoside (an acetal)

Called **glycosides**, carbohydrate acetals are named by first citing the alkyl group and replacing the *-ose* ending of the sugar with *-oside*. Note that glycosides, like all acetals, are stable to neutral water. They aren't in equilibrium with an open-chain form, and they don't show mutarotation. They can, however, be converted back to the free monosaccharide by hydrolysis with aqueous acid.

Glycosides are widespread in nature, and a great many biologically important molecules contain glycosidic linkages. For example, digitoxin, the active component of the digitalis preparations used for treatment of heart disease, is a glycoside consisting of a complex steroid alcohol linked to a trisaccharide. Note also that the three sugars are linked to each other by glycosidic bonds.

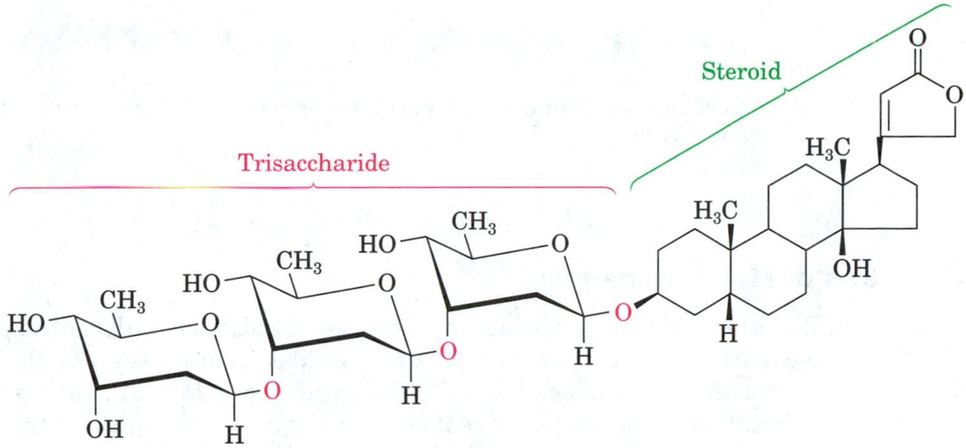

Digitoxin, a complex glycoside

Reduction of Monosaccharides

Treatment of an aldose or a ketose with $NaBH_4$ reduces it to a polyalcohol called an **alditol**. The reduction occurs by interception of the open-chain form present in the aldehyde/ketone ⇄ hemiacetal equilibrium. Although only a small amount of open-chain form is present at any given time, that small amount is reduced; then more is produced by opening of the pyranose form, that additional amount is reduced; and so on, until the entire sample has undergone reaction.

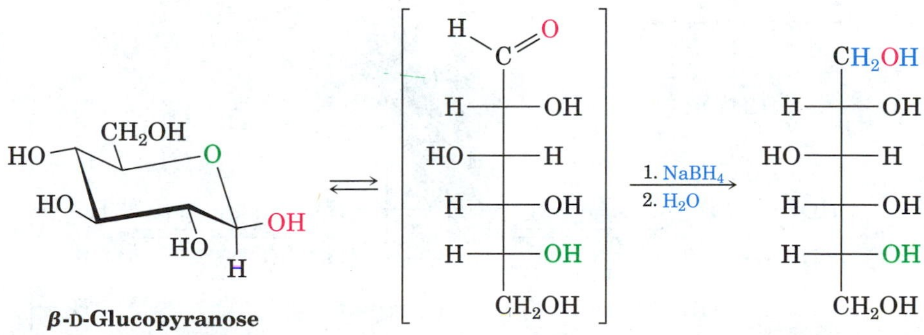

β-D-Glucopyranose **D-Glucose** **D-Glucitol (D-Sorbitol), an alditol**

D-Glucitol, the alditol produced by reduction of D-glucose, is itself a naturally occurring substance present in many fruits and berries. It is used under its alternative name D-sorbitol as an artificial sweetener and sugar substitute in foods.

PROBLEM.

26.16 How can you account for the fact that reduction of D-galactose (Figure 26.3) with $NaBH_4$ leads to an alditol that is optically inactive?

PROBLEM.

26.17 Reduction of L-gulose with $NaBH_4$ leads to the same alditol (D-glucitol) as reduction of D-glucose. Explain.

Oxidation of Monosaccharides

Like other aldehydes, aldoses are easily oxidized to yield the corresponding monocarboxylic acids, called **aldonic acids**. Aldoses react with Tollens' reagent (Ag^+ in aqueous NH_3), Fehling's reagent (Cu^{2+} in aqueous sodium tartrate), and Benedict's reagent (Cu^{2+} in aqueous sodium citrate) to yield the oxidized sugar and a reduced metallic species. All three reactions serve as simple chemical tests for what are called **reducing sugars** (*reducing* because the sugar reduces the oxidizing agent).

If Tollens' reagent is used, metallic silver is produced as a shiny mirror on the walls of the reaction flask or test tube. If Fehling's or Benedict's reagent is used, a reddish precipitate of Cu_2O signals a positive result. Some diabetes self-test kits sold for home use employ the Benedict test. As little as 0.1% glucose in urine gives a positive test.

All aldoses are reducing sugars because they contain an aldehyde carbonyl group, but some ketoses are reducing sugars as well. Fructose reduces Tollens' reagent, for example, even though it contains no aldehyde group. This occurs because fructose is readily isomerized to an aldose in basic solution by a series of keto–enol tautomeric shifts (Figure 26.7). Glycosides,

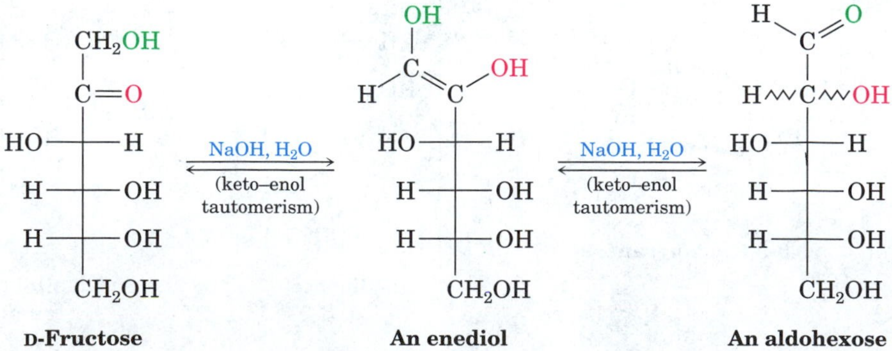

Figure 26.7 Fructose is a reducing sugar because it undergoes base-catalyzed keto–enol tautomerism that results in its conversion to an aldohexose.

however, are nonreducing. They don't react with Tollens' reagent because the acetal group can't open to an aldehyde under basic conditions.

Although the Tollens and Fehling reactions serve as useful tests for reducing sugars, they don't give good yields of aldonic acid products because the alkaline conditions cause decomposition of the carbohydrate. For preparative purposes, a buffered solution of aqueous Br_2 is the best oxidant. The reaction is specific for aldoses; ketoses are not oxidized by aqueous Br_2.

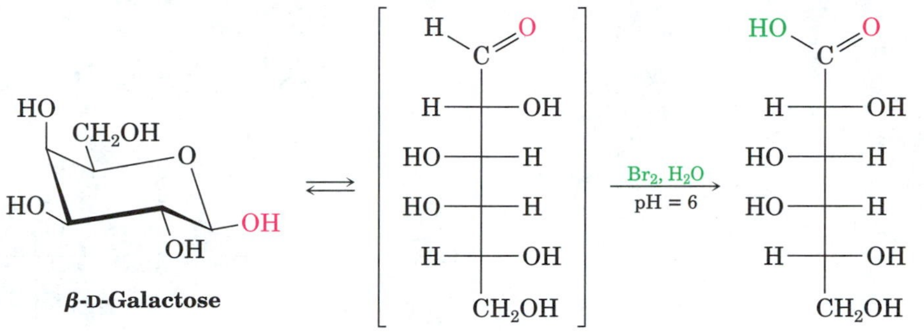

D-Galactonic acid (an aldonic acid)

If a more powerful oxidizing agent such as warm dilute HNO_3 is used, aldoses are oxidized to dicarboxylic acids called **aldaric acids**. Both the $-CHO$ group at C1 and the terminal $-CH_2OH$ group are oxidized in this reaction.

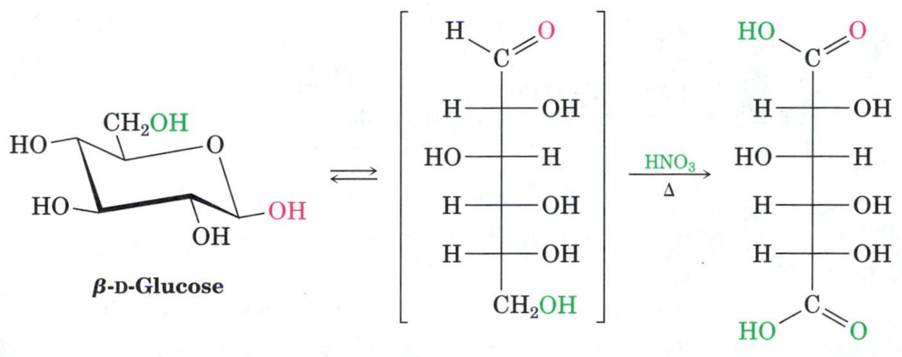

D-Glucaric acid (an aldaric acid)

A summary of the various carbohydrate derivatives is shown in Figure 26.8.

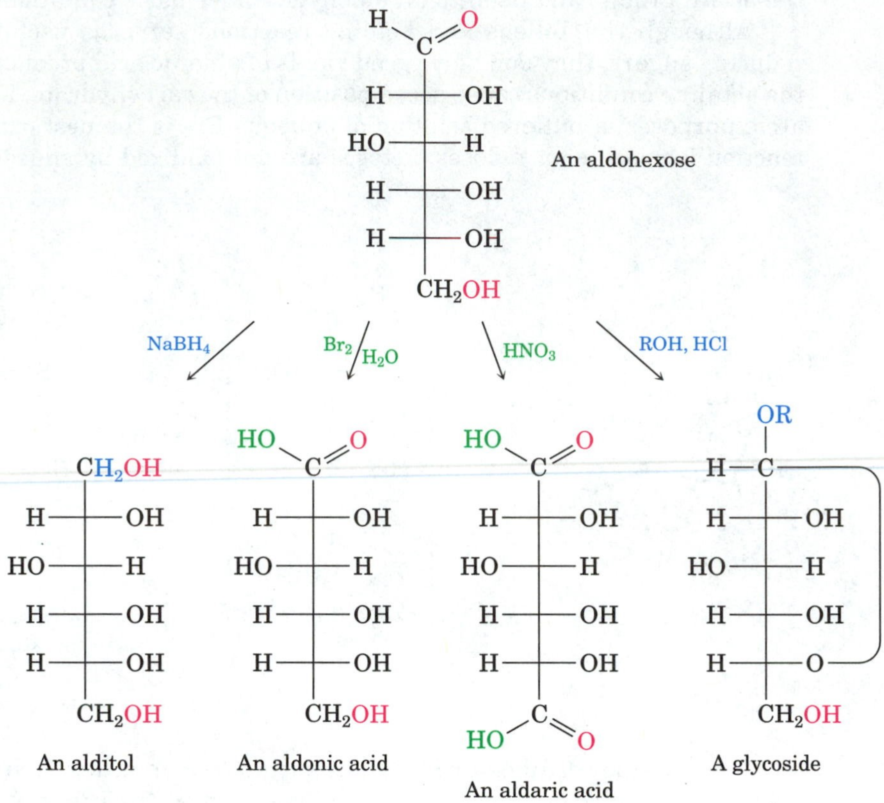

Figure 26.8 A summary of carbohydrate derivatives.

PROBLEM.

26.18 D-Glucose yields an optically active aldaric acid on treatment with HNO_3, but D-allose yields an optically inactive aldaric acid. Explain.

PROBLEM.

26.19 Which of the other six D-aldohexoses yield optically active aldaric acids on oxidation, and which yield meso aldaric acids? (See Problem 26.18.)

Chain Lengthening: The Kiliani-Fischer Synthesis

Much early activity in carbohydrate chemistry was devoted to unraveling the various stereochemical relationships among monosaccharides. One of the most important methods used was the **Kiliani–Fischer synthesis**, which results in the lengthening of an aldose chain by one carbon atom. The C1 aldehyde group of the starting sugar becomes C2 of the chain-lengthened sugar, and a new C1 carbon is added. For example, an aldo*pent*ose is converted by the Kiliani–Fischer synthesis into an aldo*hex*ose.

Discovery of the chain-lengthening sequence was initiated by the observation of Heinrich Kiliani[5] in 1886 that aldoses react with HCN to form cyanohydrins (Section 19.9). Emil Fischer immediately realized the importance of Kiliani's discovery and in 1890 published a method for converting the cyanohydrin nitrile group into an aldehyde.

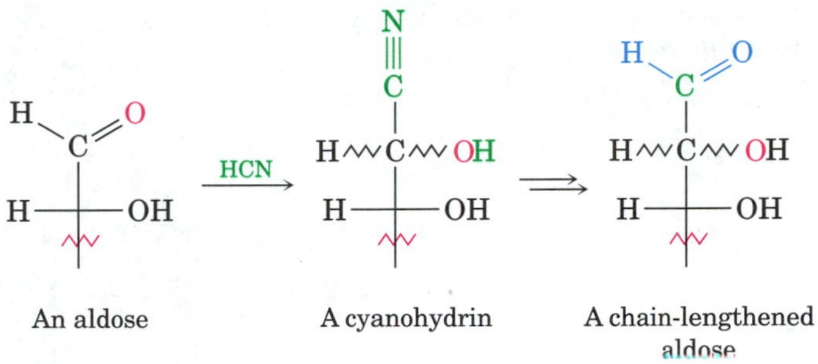

Fischer's original method for conversion of the nitrile into an aldehyde involved hydrolysis to a carboxylic acid, ring closure to a cyclic ester (lactone), and subsequent reduction. A modern improvement is to catalytically reduce the nitrile over a palladium catalyst to yield an imine intermediate, followed by hydrolysis. Note that the initial cyanohydrin is a mixture of stereoisomers at the new stereogenic center. Thus, *two* new aldoses, differing only in their stereochemistry at C2, result from Kiliani–Fischer synthesis. Chain extension of D-arabinose, for example, yields a mixture of D-glucose and D-mannose (Figure 26.9, p. 1032).

PROBLEM

26.20 What product(s) would you expect from Kiliani–Fischer reaction of D-ribose?

PROBLEM

26.21 What aldopentose would give a mixture of L-gulose and L-idose on Kiliani–Fischer chain extension?

[5]Heinrich Kiliani (1855–1945); b. Würzburg, Germany; Ph.D. Munich (Erlenmeyer); professor, University of Freiburg.

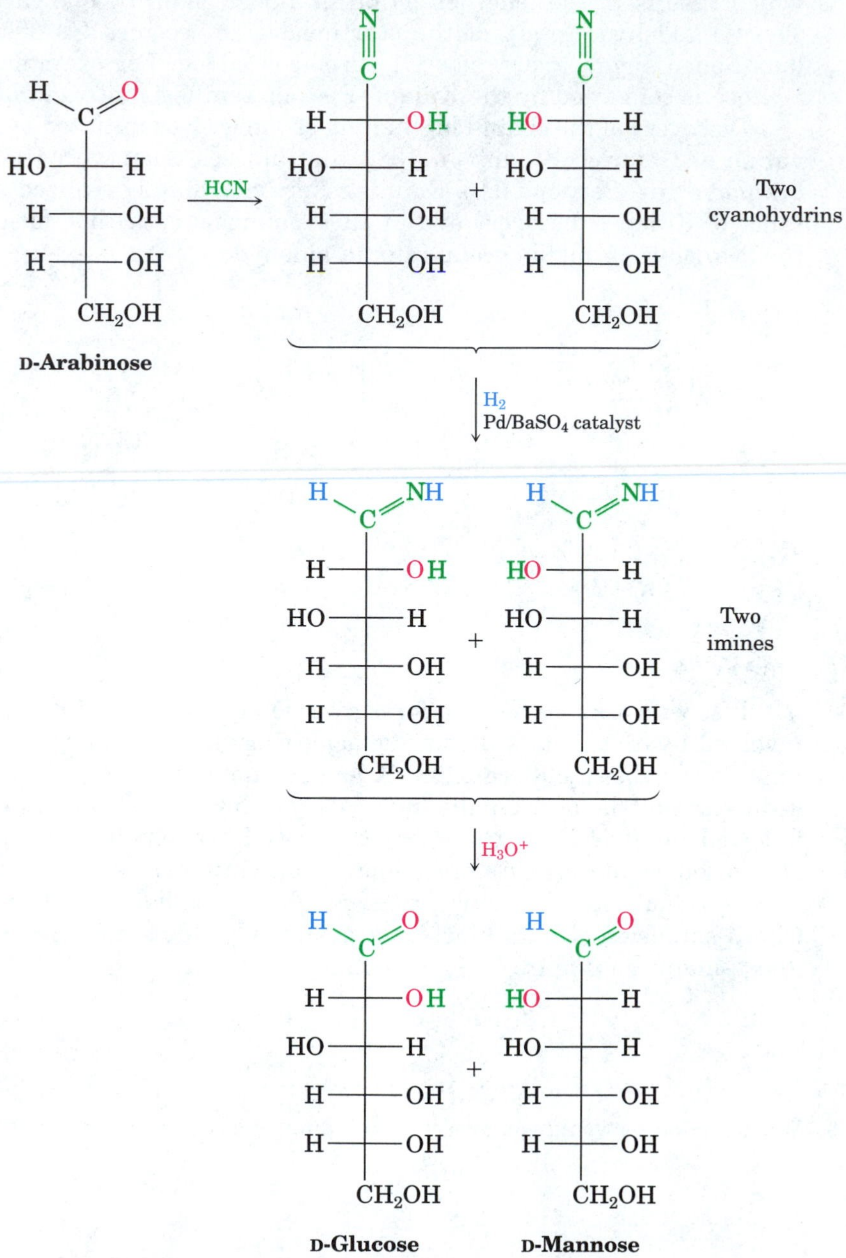

Figure 26.9 Kiliani–Fischer chain lengthening of D-arabinose leads to a mixture of D-glucose and D-mannose.

Chain Shortening: The Wohl Degradation

Just as the Kiliani–Fischer synthesis lengthens an aldose chain by one carbon, the **Wohl[6] degradation** shortens an aldose chain by one carbon. The Wohl degradation is almost exactly the opposite of the Kiliani–Fischer sequence: The aldose aldehyde carbonyl group is first converted into a nitrile, and the resulting cyanohydrin loses HCN under basic conditions—the reverse of a nucleophilic addition reaction.

Conversion of the aldehyde into a nitrile is accomplished by treatment of an aldose with hydroxylamine, followed by dehydration of the oxime product with acetic anhydride. The Wohl degradation does not give particularly high yields of chain-shortened aldoses, but the reaction is general for all aldopentoses and aldohexoses. For example, D-galactose is converted by Wohl degradation into D-lyxose:

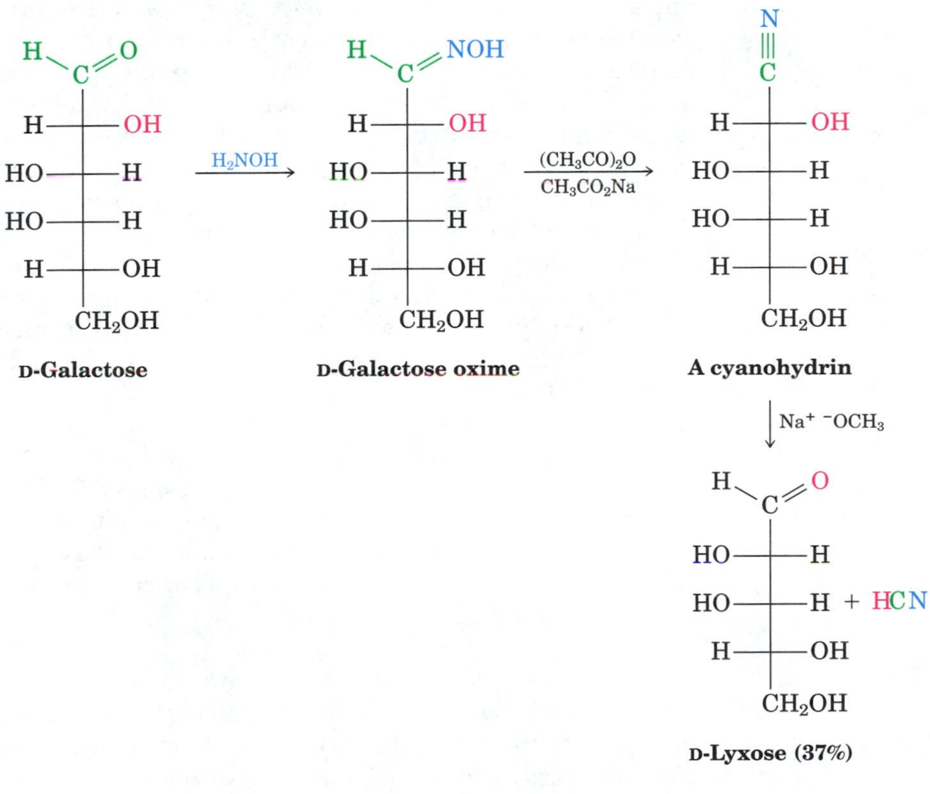

PROBLEM

26.22 What two D-aldopentoses yield D-threose on Wohl degradation?

[6]Alfred Wohl (1863–1933); b. Graudenz; Ph.D. Berlin (Hofmann); professor, University of Danzig.

26.9 Stereochemistry of Glucose: The Fischer Proof

In the late 1800s, the stereochemical theories of van't Hoff and Le Bel on the tetrahedral geometry of carbon were barely a decade old, modern methods of product purification were unknown, and modern spectroscopic techniques of structure determination were undreamed of. Despite these obstacles, Emil Fischer published in 1891 what remains today perhaps the finest use of chemical logic ever recorded—a structure proof of the stereochemistry of glucose. Let's follow Fischer's logic and see how he arrived at his conclusions.

1. *(+)-Glucose is an aldohexose.* (+)-Glucose has four stereogenic centers and can therefore be any one of $2^4 = 16$ possible stereoisomers. Since no method was available at the time for determining the *absolute* three-dimensional stereochemistry of a molecule, Fischer decided to simplify matters by considering only the eight enantiomers having the C5 hydroxyl group on the right in Fischer projections—what we now call the D series. Fischer was well aware that this arbitrary choice of D-series stereochemistry had only a 50:50 chance of being right, but it was finally shown some 60 years later (by the use of sophisticated X-ray techniques) that the choice was indeed correct.

   The four possible D-aldopentoses and the eight possible D-aldohexoses derived from them by Kiliani–Fischer synthesis are shown in Figure 26.10. One of the eight aldohexoses is glucose, but which one?

2. *Arabinose, an aldopentose, is converted by Kiliani–Fischer chain extension into a mixture of glucose and mannose.* This means that glucose and mannose have the same stereochemistry at C3, C4, and C5, and differ only at C2. Glucose and mannose are therefore represented by one of the pairs of structures 1 and 2, 3 and 4, 5 and 6, or 7 and 8 in Figure 26.10.

3. *Arabinose is converted by treatment with warm HNO_3 into an optically active aldaric acid.* Of the four aldopentoses (A, B, C, and D in Figure 26.10), A and C give optically inactive meso aldaric acids when oxidized, whereas B and D give optically active products. Thus, arabinose must be either B or D, and mannose and glucose must be either 3 and 4 or 7 and 8 (Figure 26.11, p. 1036).

4. *Both glucose and mannose are oxidized by warm HNO_3 to optically active aldaric acids.* Of the possibilities left at this point, the pair represented by structures 3 and 4 would both be oxidized to optically active aldaric acids, but the pair represented by 7 and 8 would not *both* give optically active products. Compound 7 would give an optically inactive meso aldaric acid (Figure 26.12, p. 1036). Thus, glucose and mannose must be 3 and 4, though we can't yet tell which is which.

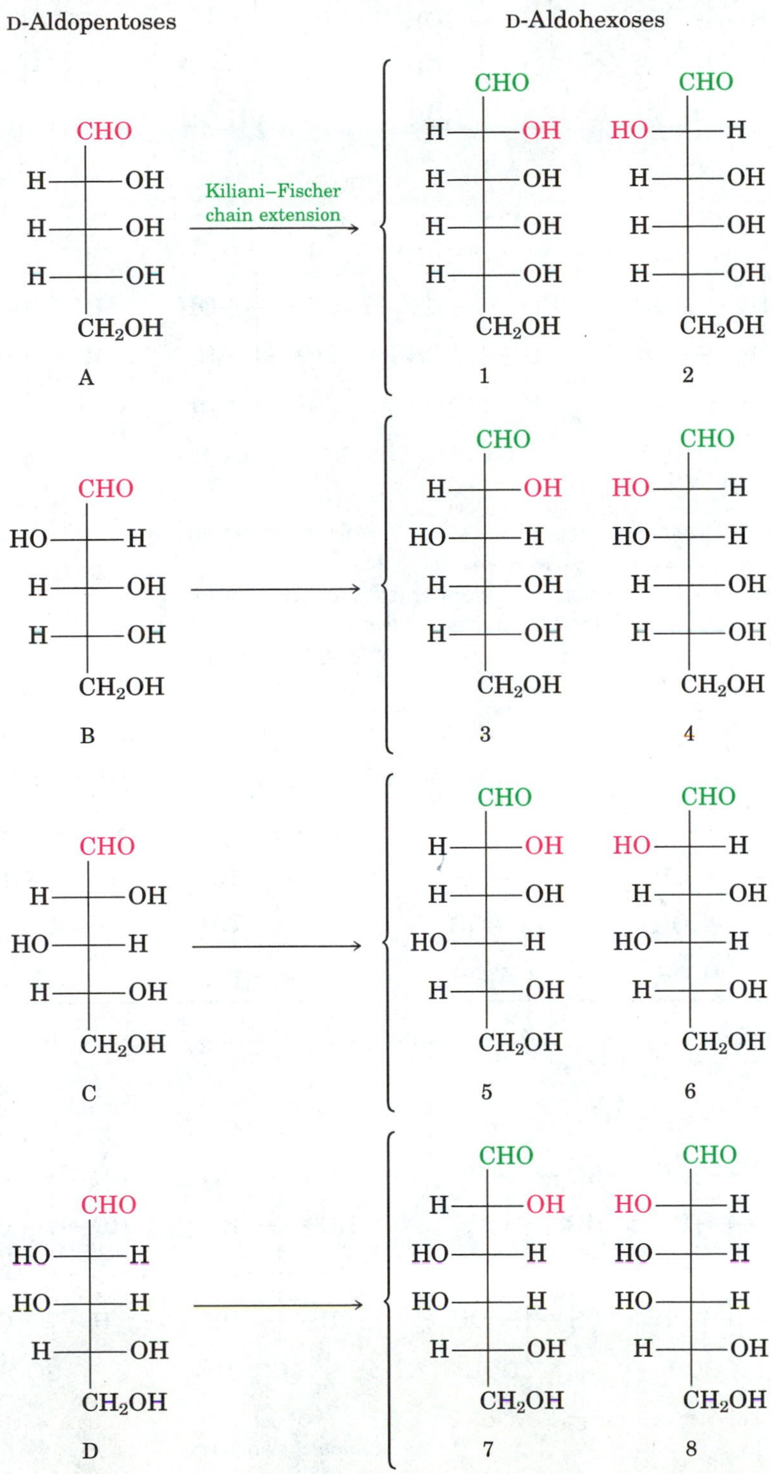

Figure 26.10 The four D-aldopentoses and the eight D-aldohexoses derived from them by Kiliani–Fischer chain extension.

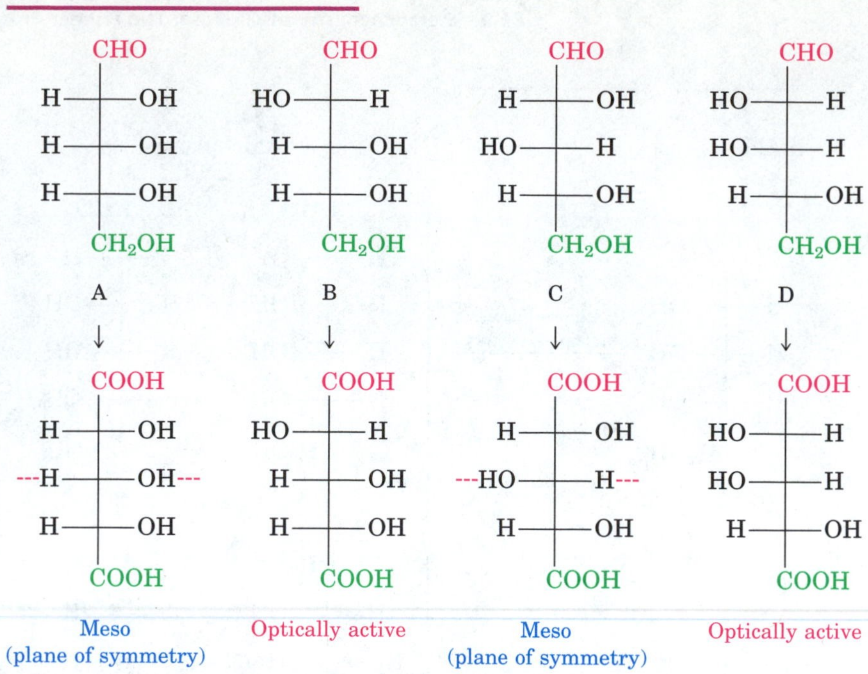

Figure 26.11 Oxidation of aldopentoses to aldaric acids. Only structures B and D lead to optically active products.

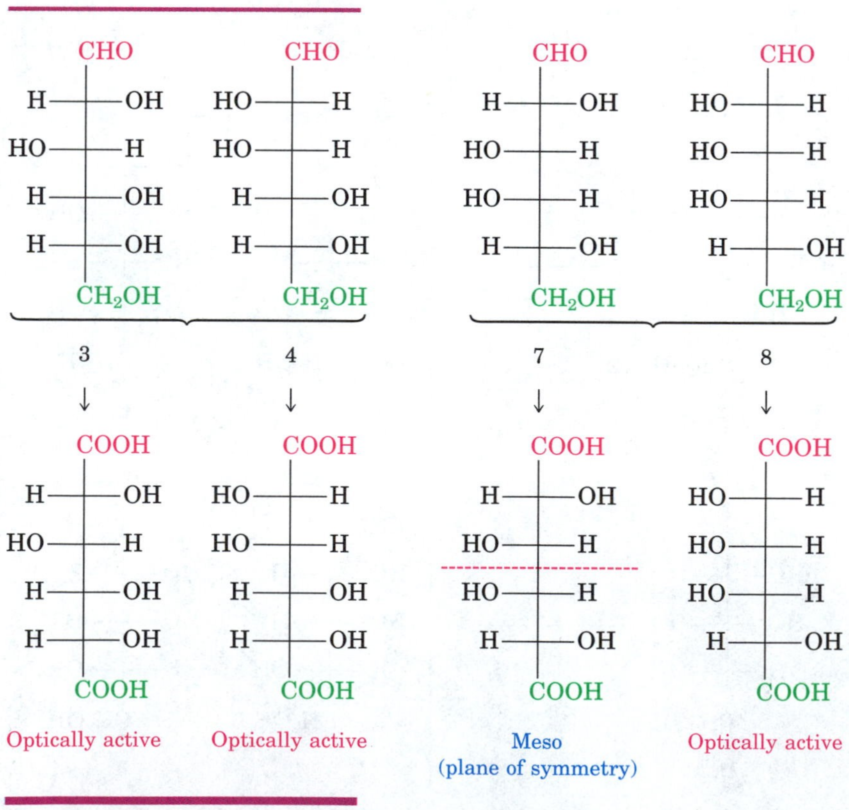

Figure 26.12 Oxidation of aldohexoses to aldaric acids. Only the pair of structures 3 and 4 *both* give optically active products.

5. *One of the other 15 aldohexoses is converted by nitric acid oxidation to the same aldaric acid as that derived from glucose.* How can two different aldohexoses give the same aldaric acid? Since aldaric acids have –COOH groups at both ends of the carbon chain, there is no way to tell which was originally the –CHO end and which was the $-CH_2OH$ end. Thus, a given aldaric acid has two different precursors. The aldaric acid from compound 3, for example, might also come from oxidation of a second aldohexose, and the aldaric acid from compound 4 might come from oxidation of a second aldohexose (Figure 26.13).

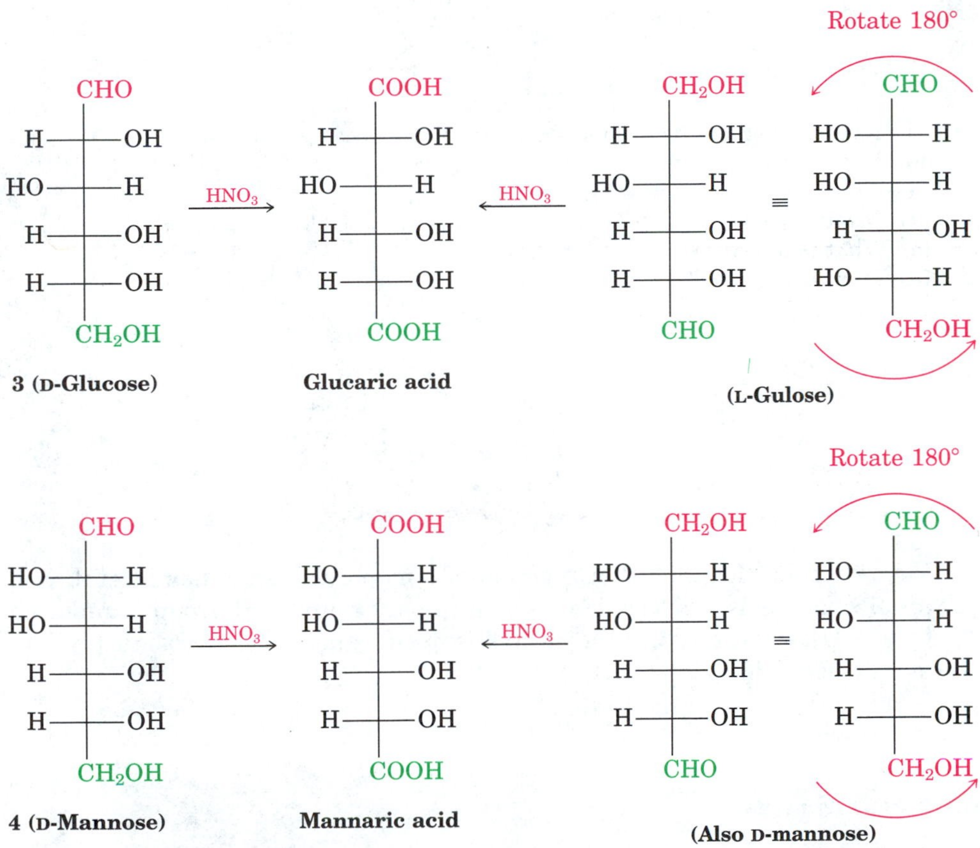

Figure 26.13 There is another aldohexose (L-gulose) that can produce the same aldaric acid as compound 3, but there is no other aldohexose that can produce the same aldaric acid as compound 4. Thus, glucose has structure 3.

If we look carefully at the aldaric acids derived from compounds 3 and 4, we find that the aldaric acid derived from 3 could also come from oxidation of another aldohexose (L-gulose), but that the aldaric acid derived from 4 could not. The "other" aldohexose that could produce the same aldaric acid as that from compound 4 is in fact identical to 4. Thus, glucose must have structure 3 and mannose must have structure 4 (Figure 26.13).

Reasoning similar to that just described for glucose allowed Fischer to determine the stereochemistry of 12 of the 16 aldohexoses. For this remarkable achievement, he was awarded the 1902 Nobel Prize in chemistry.

PROBLEM.

26.23 The structures of the four aldopentoses, A, B, C, and D, are shown in Figure 26.10. In light of point 2 presented by Fischer, what is the structure of arabinose? In light of point 3, what is the structure of lyxose, another aldopentose that yields an optically active aldaric acid?

PROBLEM.

26.24 The aldotetrose D-erythrose yields a mixture of D-ribose and D-arabinose on Kiliani–Fischer chain extension.
(a) What is the structure of D-ribose?
(b) What is the structure of D-xylose, the fourth possible aldopentose?
(c) What is the structure of D-erythrose?
(d) What is the structure of D-threose, the other possible aldotetrose?

26.10 Disaccharides

We saw in Section 26.8 that reaction of a monosaccharide hemiacetal with an alcohol yields a glycoside in which the anomeric –OH group is replaced by an –OR substituent. If the alcohol is itself a sugar, the glycosidic product is a disaccharide.

Cellobiose and Maltose

Disaccharides are compounds that contain a glycosidic acetal bond between C1 of one sugar and an –OH group at *any* position on the other sugar. A glycosidic bond between C1 of the first sugar and the –OH at C4 of the second sugar is particularly common. Such a bond is called a **1,4′ link** (read as "one, four-prime"). The prime superscript indicates that the 4′ position is on a different sugar than the nonprime 1 position.

A glycosidic bond to the anomeric carbon can be either α or β. Maltose, the disaccharide obtained by enzyme-catalyzed hydrolysis of starch, consists of two D-glucopyranoses joined by a 1,4′-α-glycoside bond. Cellobiose, the disaccharide obtained by partial hydrolysis of cellulose, consists of two D-glucopyranoses joined by a 1,4′-β-glycoside bond.

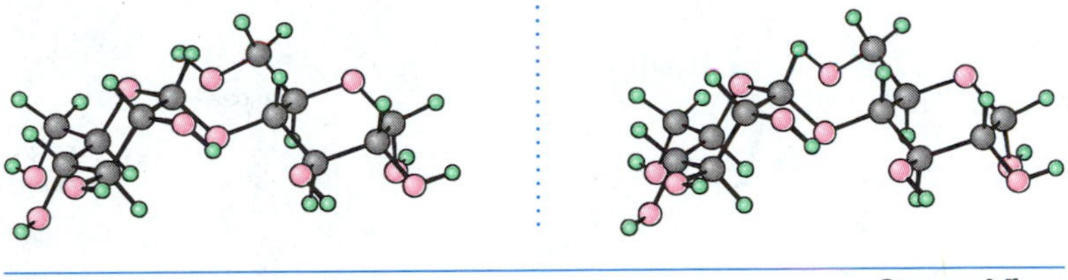

Stereo View

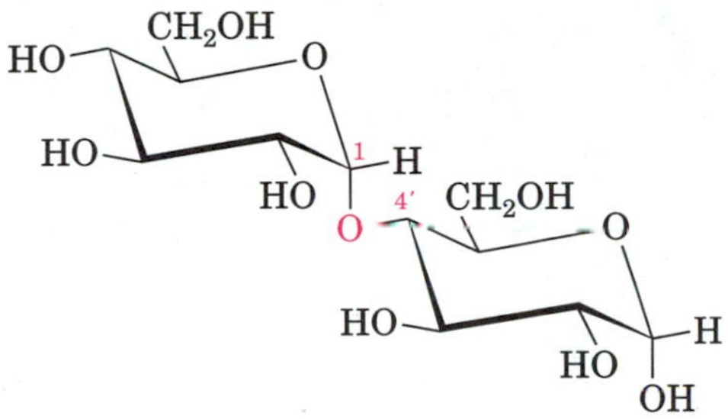

Maltose, a 1,4′-α-glycoside
[4-*O*-(α-D-Glucopyranosyl)-α-D-glucopyranose]

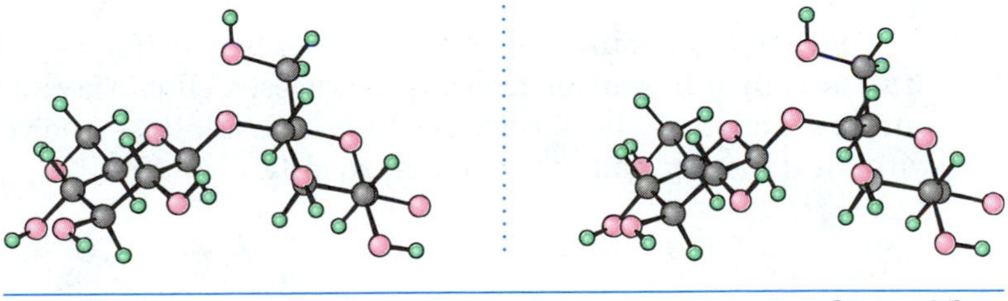

Stereo View

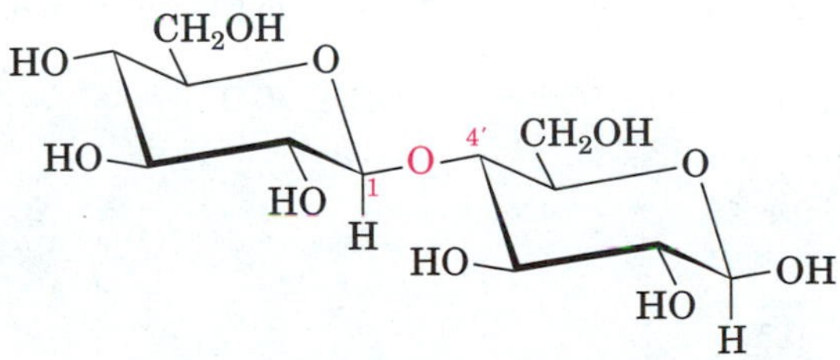

Cellobiose, a 1,4′-β-glycoside
[4-*O*-(β-D-Glucopyranosyl)-β-D-glucopyranose]

Maltose and cellobiose are both reducing sugars because the anomeric carbons on the right-hand glucopyranose units have hemiacetal groups. Both are therefore in equilibrium with aldehyde forms, which can reduce Tollens' or Fehling's reagent. For a similar reason, both maltose and cellobiose exhibit mutarotation of α and β anomers of the glucopyranose unit on the right (Figure 26.14).

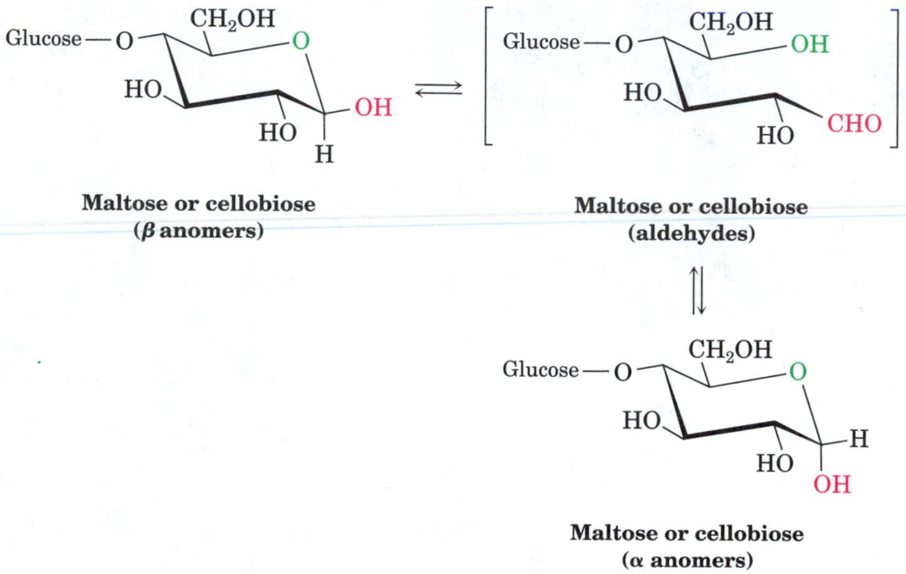

Figure 26.14 Mutarotation of maltose and cellobiose.

Despite the similarities of their structures, cellobiose and maltose have dramatically different biological properties. Cellobiose can't be digested by humans and can't be fermented by yeast. Maltose, however, is digested without difficulty and is fermented readily.

PROBLEM

26.25 Show the product you would obtain from the reaction of cellobiose with the following reagents:
(a) $NaBH_4$ (b) Br_2, H_2O (c) CH_3COCl, pyridine

Lactose

Lactose is a disaccharide that occurs naturally in both human and cow's milk. It is widely used in baking and in commercial milk formulas for infants.

Like cellobiose and maltose, lactose is a reducing sugar. It exhibits mutarotation and is a 1,4′-β-linked glycoside. Unlike cellobiose and maltose, however, lactose contains two *different* monosaccharide units—D-glucose and D-galactose—joined by a β-glycosidic bond between C1 of galactose and C4 of glucose.

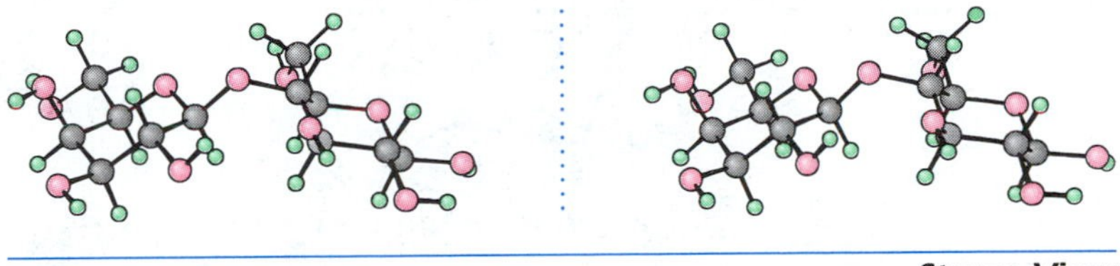

Stereo View

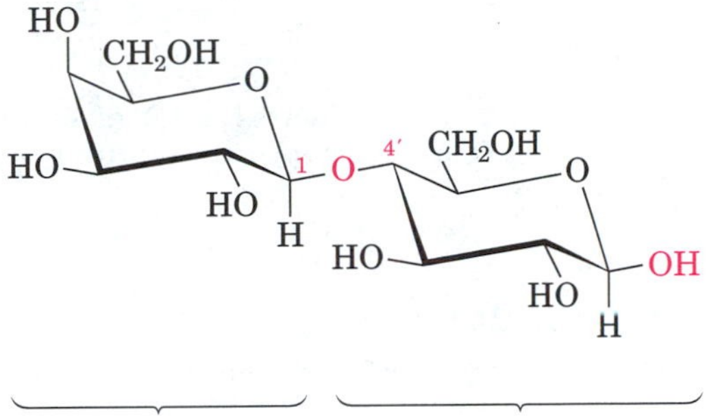

β-Galactopyranoside **β-Glucopyranose**

Lactose, a 1,4′-β-glycoside
[4-*O*-(β-D-Galactopyranosyl)-β-D-glucopyranose]

Sucrose

Sucrose, or ordinary table sugar, is probably the single most abundant pure organic chemical in the world and the one most widely known to nonchemists. Whether from sugar cane (20% by weight) or sugar beets (15% by weight), and whether raw or refined, all table sugar is sucrose.

Sucrose is a disaccharide that yields 1 equivalent of glucose and 1 equivalent of fructose on hydrolysis. This 1:1 mixture of glucose and fructose is often referred to as *invert sugar,* because the sign of optical rotation changes (inverts) during the hydrolysis from sucrose ($[\alpha]_D = +66.5°$) to a glucose/fructose mixture ($[\alpha]_D \approx -22.0°$). Insects such as honeybees have enzymes called *invertases* that catalyze the hydrolysis of sucrose to a glucose + fructose mixture. Honey, in fact, is primarily a mixture of glucose, fructose, and sucrose.

Unlike most other disaccharides, sucrose is not a reducing sugar and does not exhibit mutarotation. These observations imply that sucrose has no hemiacetal groups and suggest that glucose and fructose must both be glycosides. This can happen only if the two sugars are joined by a glycoside link between C1 of glucose and C2 of fructose.

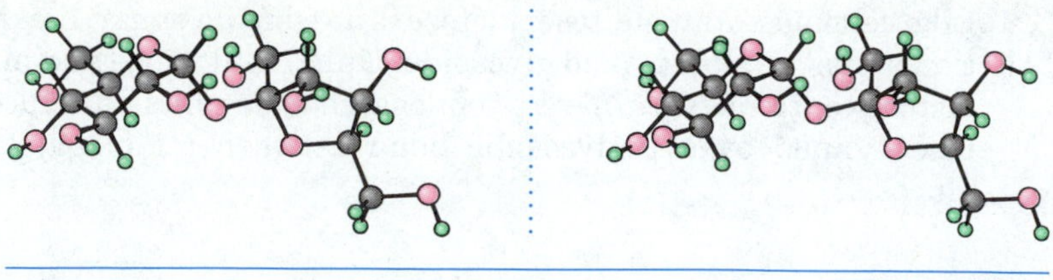

Stereo View

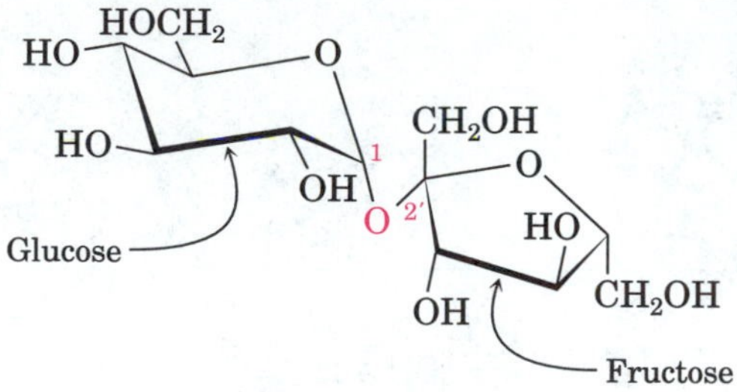

Sucrose, a 1,2′-glycoside
[2-*O*-(α-D-Glucopyranosyl)-β-D-fructofuranoside]

26.11 Polysaccharides

Polysaccharides are carbohydrates in which tens, hundreds, or even thousands of simple sugars are linked together through glycoside bonds. Since they have no free anomeric hydroxyls (except for one at the end of the chain), polysaccharides aren't reducing sugars and don't show mutarotation. Cellulose and starch are the two most widely occurring polysaccharides.

Cellulose

Cellulose consists of D-glucose units linked by 1,4′-β-glycoside bonds like those in cellobiose. Several thousand glucose units are linked to form one large molecule, and different molecules then interact to form a large aggregate structure held together by hydrogen bonds.

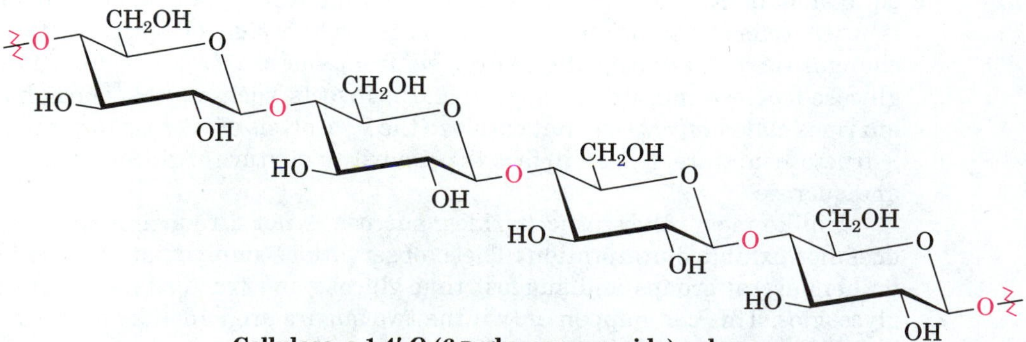

Cellulose, a 1,4′-*O*-(β-D-glucopyranoside) polymer

Nature uses cellulose primarily as a structural material to impart strength and rigidity to plants. Leaves, grasses, and cotton are primarily cellulose. Cellulose also serves as raw material for the manufacture of cellulose acetate, known commercially as acetate rayon.

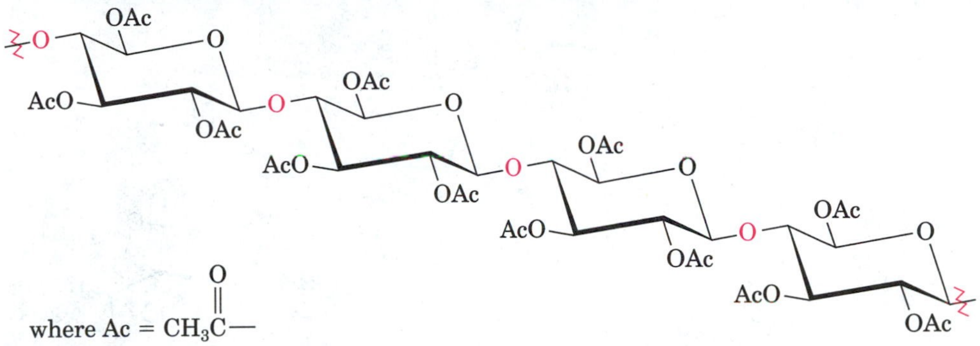

A segment of cellulose acetate (acetate rayon)

Starch and Glycogen

Potatoes, corn, and cereal grains contain large amounts of *starch*, a polymer of glucose in which the monosaccharide units are linked by 1,4′-α-glycoside bonds like those in maltose. Starch can be separated into two fractions: a fraction that is insoluble in cold water, called *amylose*, and a fraction that is soluble in cold water, called *amylopectin*. Amylose accounts for about 20% by weight of starch and consists of several hundred glucose molecules linked together by 1,4′-α-glycoside bonds.

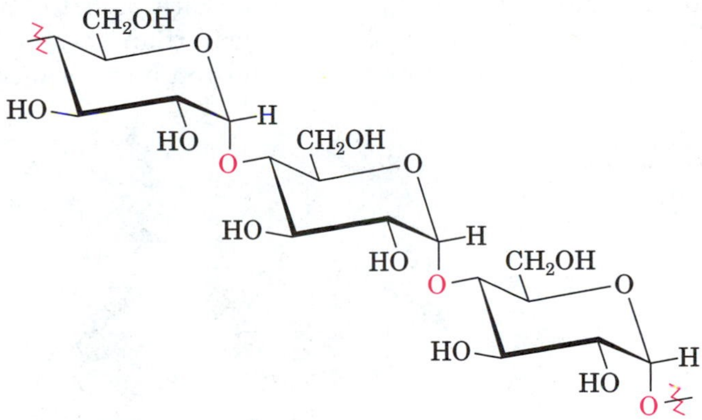

Amylose, a 1,4′-*O*-(α-D-glucopyranoside) polymer

Amylopectin accounts for the remaining 80% of starch and is more complex in structure than amylose. Unlike cellulose and amylose, which are linear polymers, amylopectin contains 1,6′-α-glycoside *branches* approximately every 25 glucose units. As a result, amylopectin has an exceedingly complex three-dimensional structure.

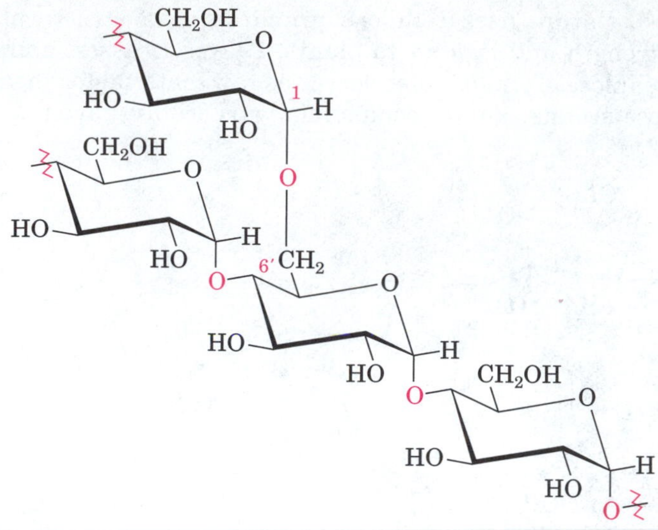

Amylopectin

When eaten, starch is digested in the mouth and stomach by *glycosidase* enzymes, which catalyze the hydrolysis of glycoside bonds and release individual molecules of glucose. Like most enzymes, glycosidases are highly selective in their action. They hydrolyze only the α-glycoside links in starch and leave the β-glycoside links in cellulose untouched. Thus, humans can eat potatoes and grains but not grass and leaves.

Glycogen is a polysaccharide that serves the same purpose of energy storage in animals that starch serves in plants. Dietary carbohydrate not needed for immediate energy is converted by the body to glycogen for long-term storage. Like the amylopectin found in starch, glycogen contains a complex three-dimensional structure with both 1,4′ and 1,6′ links (Figure 26.15). Glycogen molecules are larger than those of amylopectin—up to 100,000 glucose units—and contain even more branches.

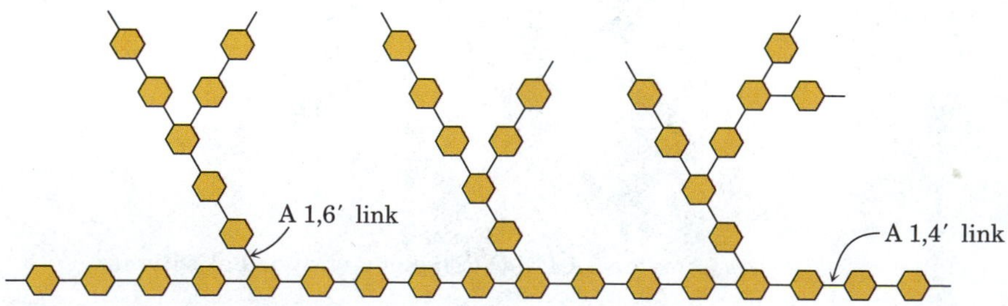

Figure 26.15 A representation of the structure of glycogen. The hexagons represent glucose units linked by 1,4′ and 1,6′ acetal bonds.

26.12 Other Important Carbohydrates

In addition to the common carbohydrates mentioned in previous sections, there are a variety of important carbohydrate-derived materials. Their structural resemblance to sugars is clear, but they aren't simple aldoses or ketoses.

Deoxy sugars differ from other sugars by having one of their oxygen atoms "missing." In other words, an –OH group is replaced by an –H. 2-Deoxyribose, a sugar found in DNA (deoxyribonucleic acid), is the most common deoxy sugar. Note that 2-deoxyribose adopts a furanose (five-membered) form.

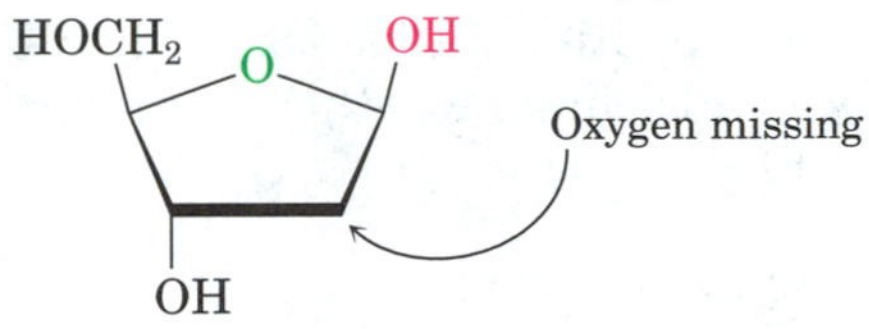

2-Deoxyribose

Amino sugars, such as D-glucosamine, have one of their –OH groups replaced by an $-NH_2$. The *N*-acetyl amide derived from D-glucosamine is the monosaccharide unit from which *chitin*, the hard crust that protects insects and shellfish, is built. Still other amino sugars are found in antibiotics such as streptomycin and gentamicin.

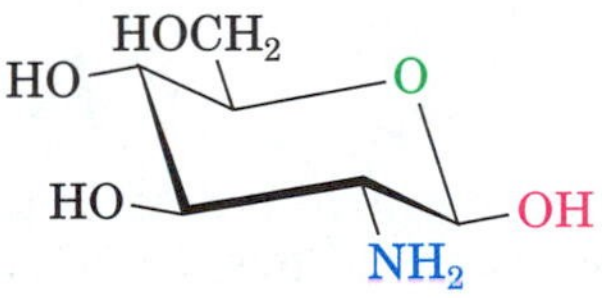

β-D-Glucosamine (an amino sugar)

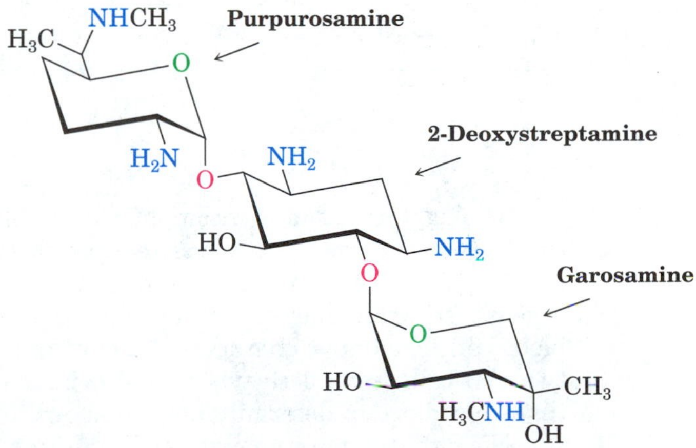

Gentamicin (an antibiotic)

26.13 Cell-Surface Carbohydrates

Carbohydrates were once thought to be dull compounds whose only biological purposes were to serve as structural materials and as energy sources. Although carbohydrates do indeed fill these two roles, it's now known that they perform many other important biochemical functions as well. For example, polysaccharides are centrally involved in the critical process by which one cell type recognizes another. Small polysaccharide chains, covalently bound by glycosidic links to hydroxyl groups on proteins (*glycoproteins*), act as biochemical labels on cell surfaces, as illustrated by the human blood-group antigens.

It has been known for a century that human blood can be classified into four blood-group types (A, B, AB, and O), and that blood from a donor of one type can't be transfused into a recipient with another type unless the two types are compatible (Table 26.1). Should an incompatible mix be made, the red blood cells clump together, or *agglutinate*.

Table 26.1 Human Blood-Group Compatibilities

Donor blood type	*Acceptor blood type*			
	A	B	AB	O
A	○	×	○	×
B	×	○	○	×
AB	×	×	○	×
O	○	○	○	○

○ = Compatible; × = Incompatible.

The agglutination of incompatible red blood cells, which indicates that the body's immune system has recognized the presence of foreign cells in the body and has formed antibodies against them, results from the presence of polysaccharide markers on the surface of the cells. Types A, B, and O red blood cells each have characteristic markers, called *antigenic determinants*; type AB cells have both type A and type B markers. The structures of all three blood-group determinants are shown in Figure 26.16.

Note that some unusual carbohydrates are involved. All three blood-group antigenic determinants contain *N*-acetylamino sugars as well as the unusual monosaccharide L-fucose.

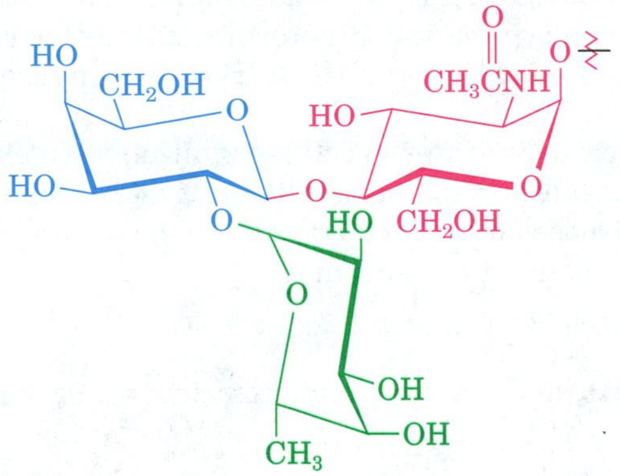

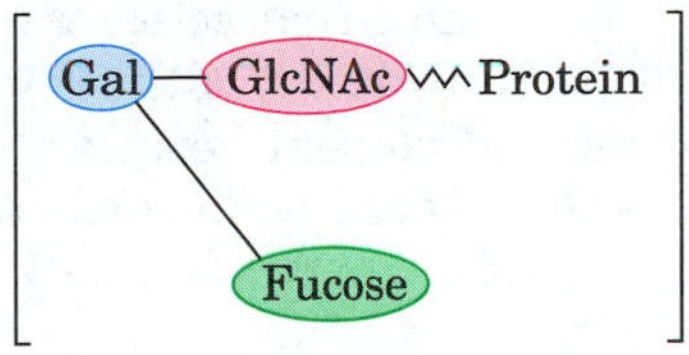

Blood group O

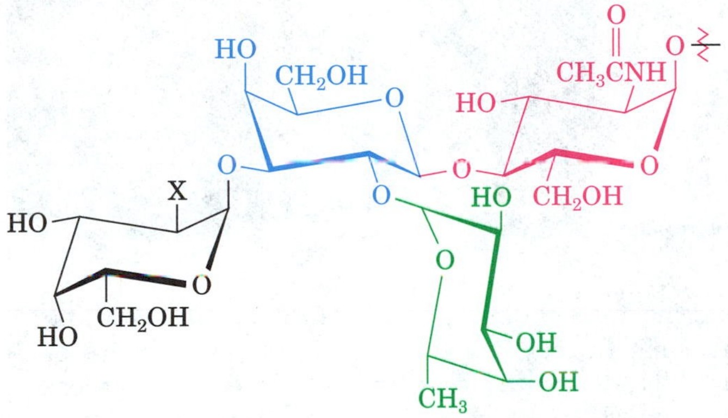

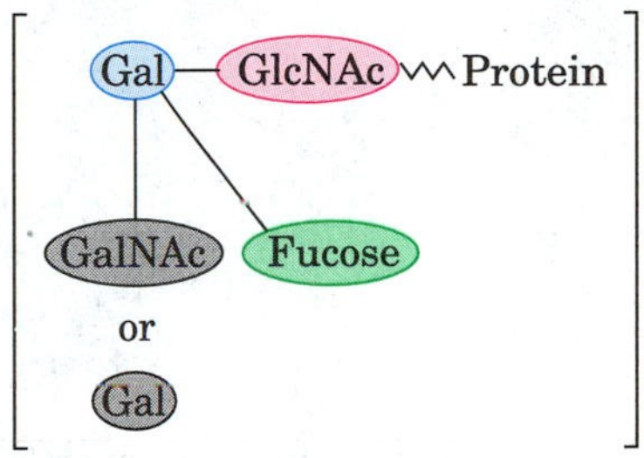

Blood group A, X = $NHCOCH_3$
Blood group B, X = OH

Figure 26.16 Structures of the A, B, and O blood-group antigenic determinants (Gal = D-galactose; GlcNAc = *N*-acetylglucosamine; GalNAc = *N*-acetylgalactosamine).

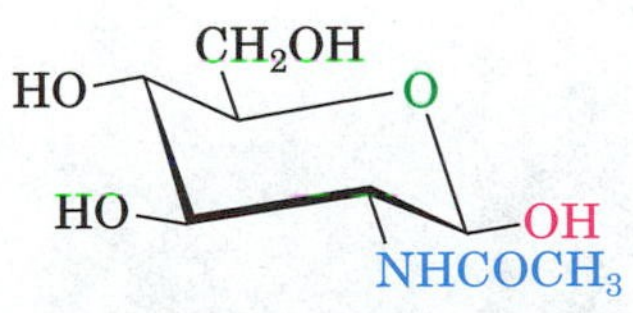

β-D-*N*-Acetylglucosamine (D-2-Acetamino-2-deoxyglucose)

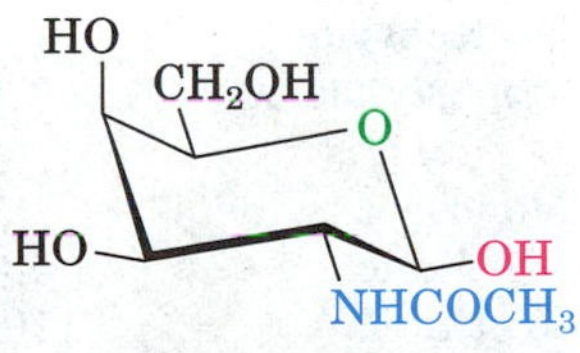

β-D-*N*-Acetylgalactosamine (D-2-Acetamino-2-deoxygalactose)

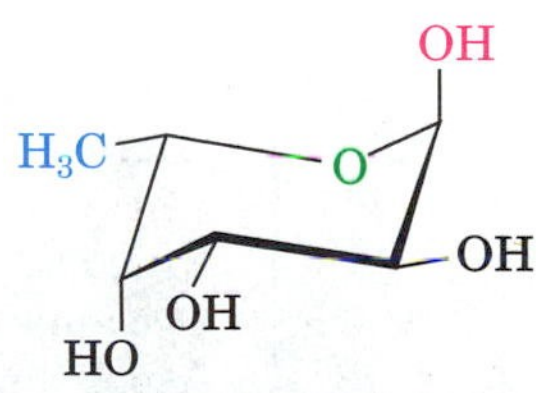

α-L-Fucose (L-6-Deoxygalactose)

The antigenic determinant of blood group O is a trisaccharide, whereas the determinants of blood groups A and B have an additional sugar attached at C3 of the galactose unit. The type A and B determinants differ only in the substitution of an acetylamino group ($-NHCOCH_3$) for a hydroxyl in the terminal galactose residue.

Elucidation of the role of carbohydrates in cell recognition is an exciting area of current research that offers hope of breakthroughs in the understanding of a wide range of diseases from bacterial infections to cancer. The potential benefits of work in this field are enormous.

INTERLUDE

Sweetness

A sugar overload!

Say the word *sugar* and most people immediately think of sweet-tasting candies, desserts, and such. In fact, most of the simple carbohydrates we've discussed in this chapter *do* taste sweet, but the degree of sweetness varies greatly from one sugar to another. With sucrose (table sugar) as a reference point, fructose is nearly twice as sweet, but lactose is only about one-sixth as sweet. Comparisons are difficult, though, because sweetness is simply a matter of taste, and the ranking of sugars is a matter of personal opinion. Nevertheless, the ordering in Table 26.2 is generally accepted.

Table 26.2 Sweetness of Some Sugars and Sugar Substitutes

Name	*Type*	*Sweetness*
Lactose	Disaccharide	0.16
Glucose	Monosaccharide	0.75
Sucrose	**Disaccharide**	**1.00**
Fructose	Monosaccharide	1.75
Cyclamate	Synthetic	300
Aspartame	Synthetic	1500
Saccharin	Synthetic	3500

(continued)➤

The desire of many people to cut their caloric intake has led to the development of synthetic sweeteners such as aspartame, saccharin, and cyclamate. All are far sweeter than natural sugars, but doubts have been raised as to their long-term safety. Cyclamate was briefly banned in the United States (but not in Canada), and saccharin has been banned in Canada (but not in the United States). None of the three has any structural resemblance to carbohydrates.

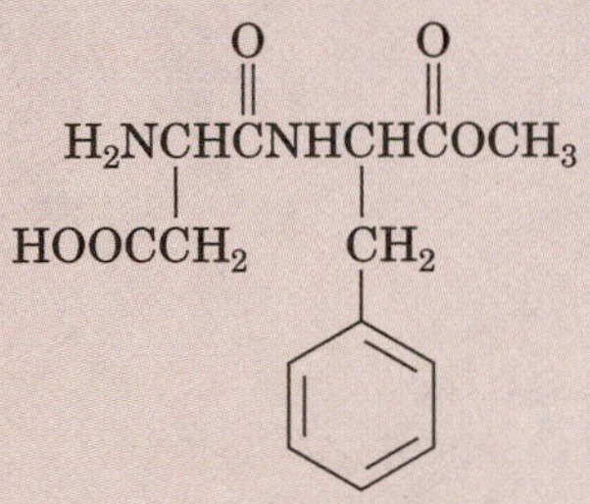

Aspartame

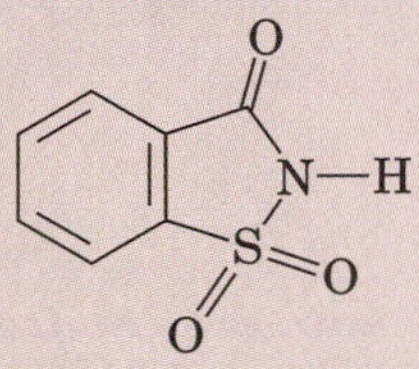

Saccharin

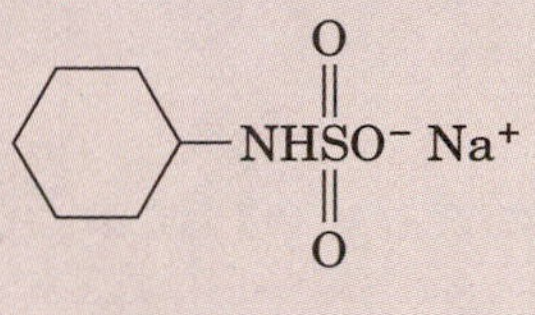

Sodium cyclamate

Summary and Key Words

Carbohydrates are polyhydroxy aldehydes and ketones. They are classified according to the number of carbon atoms and the kind of carbonyl group they contain. Glucose, for example, is an **aldohexose**, a six-carbon aldehydo sugar. Monosaccharides are further classified as either **D** or **L sugars**, depending on the stereochemistry of the stereogenic center farthest from the carbonyl group.

Monosaccharides normally exist as cyclic hemiacetals rather than as open-chain aldehydes or ketones. The hemiacetal linkage results from reaction of the carbonyl group with an –OH group three or four carbon atoms away. A five-membered cyclic hemiacetal is called a **furanose**, and a six-membered cyclic hemiacetal is called a **pyranose**. Cyclization leads to the formation of a new stereogenic center and production of two diastereomeric hemiacetals, called α **and** β **anomers**.

Stereochemical relationships among monosaccharides are portrayed in several ways. **Fischer projections** display stereogenic carbon atoms as a pair of crossed lines; cyclic **Haworth projections** provide a more accurate view.

Much of the chemistry of monosaccharides is the familiar chemistry of alcohols and aldehydes/ketones. Thus, the hydroxyl groups of carbohydrates form esters and ethers. The carbonyl group of a monosaccharide can be reduced with $NaBH_4$ to form an **alditol**, oxidized with aqueous Br_2 to form an **aldonic acid**, oxidized with HNO_3 to form an **aldaric acid**, or treated with an alcohol in the presence of acid to form a **glycoside**. Monosaccharides can also be chain-lengthened by the multistep **Kiliani–Fischer synthesis**, and can be chain-shortened by the **Wohl degradation**.

Disaccharides are complex carbohydrates in which two simple sugars are linked by a glycoside bond between the anomeric carbon of one unit and a hydroxyl of the second unit. The two sugars can be the same, as in maltose and cellobiose, or different, as in lactose and sucrose. The glycosidic bond can be either α (maltose) or β (cellobiose, lactose), and can involve any hydroxyl of the second sugar. A 1,4′ link is most common (cellobiose, maltose), but others such as 1,2′ (sucrose) are also known.

Summary of Reactions

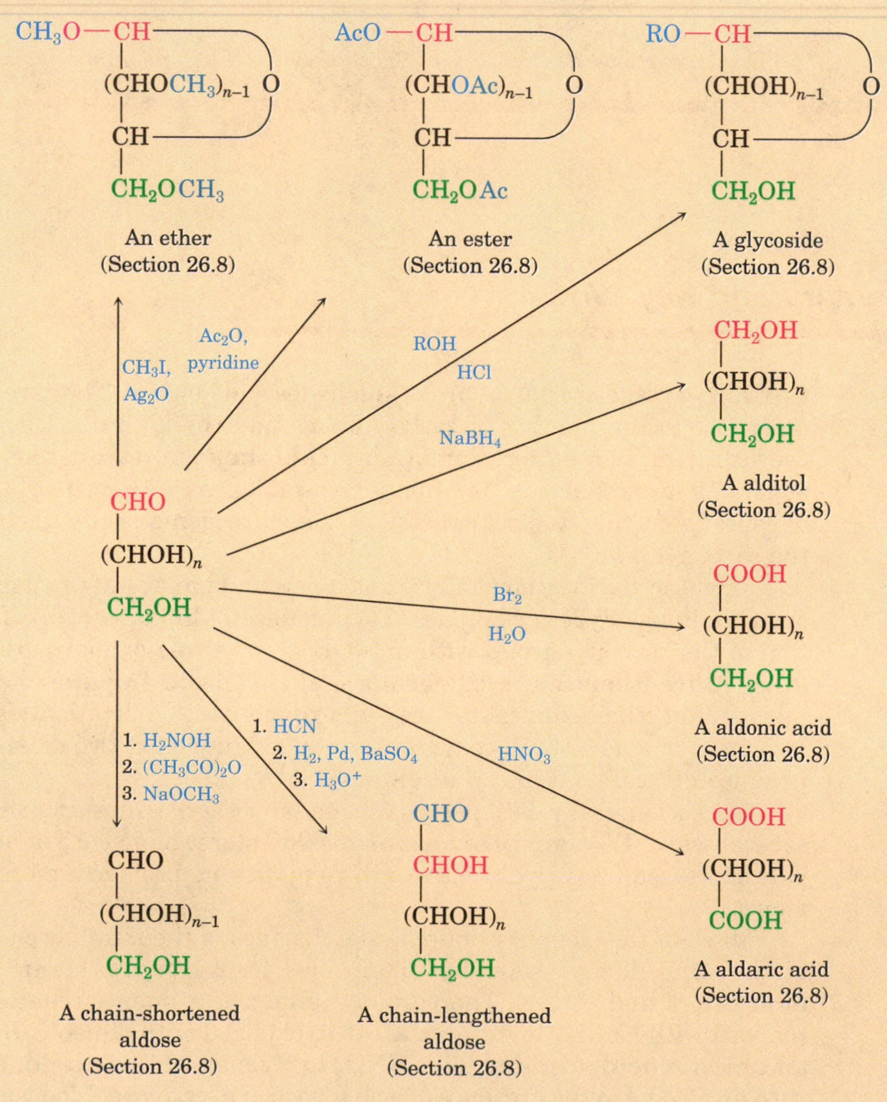

ADDITIONAL PROBLEMS

26.26 Classify the following sugars. (For example, glucose is an aldohexose.)

(a)

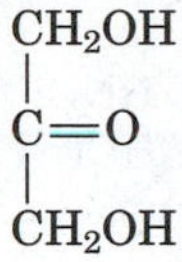

(b)

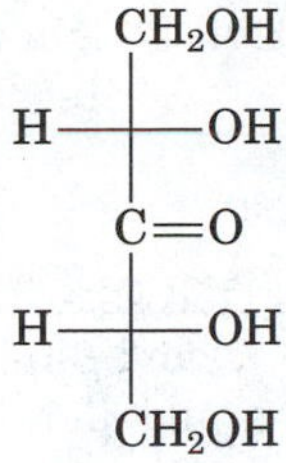

(c)

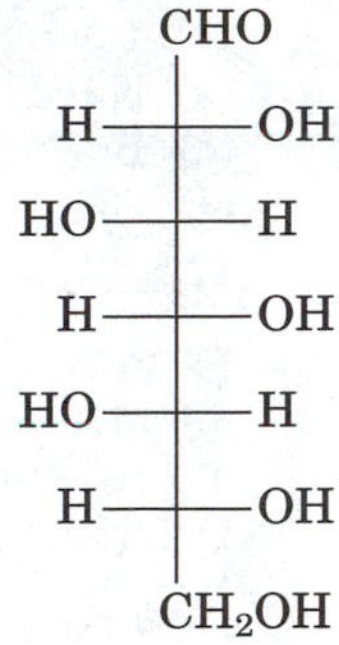

26.27 Write open-chain structures for a ketotetrose and a ketopentose.

26.28 Write an open-chain structure for a deoxyaldohexose.

26.29 Write an open-chain structure for a five-carbon amino sugar.

26.30 Does ascorbic acid (vitamin C) have a D or L configuration?

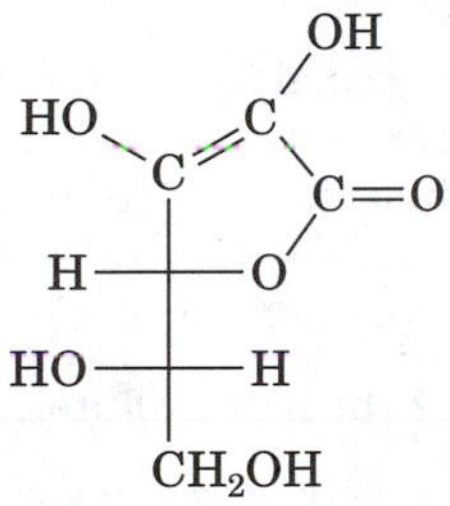

Ascorbic acid

26.31 Draw a Haworth projection of ascorbic acid (Problem 26.30).

26.32 Define the following terms, and give an example of each:

(a) Monosaccharide (b) Anomeric center (c) Haworth projection
(d) Fischer projection (e) Glycoside (f) Reducing sugar
(g) Pyranose form (h) 1,4′ Link (i) D Sugar

26.33 The following cyclic structure is that of gulose. Is this a furanose or pyranose form? Is it an α or β anomer? Is it a D or L sugar?

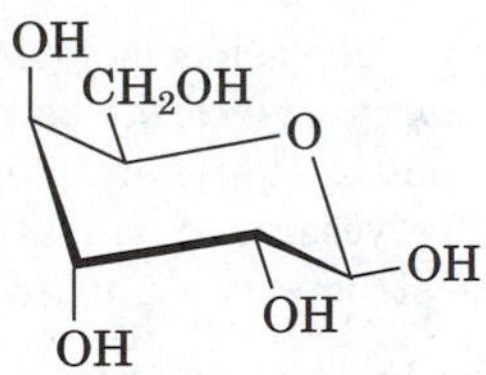

Gulose

26.34 Uncoil gulose (Problem 26.33), and write it in its open-chain form.

26.35 Draw D-ribulose in its five-membered cyclic β-hemiacetal form.

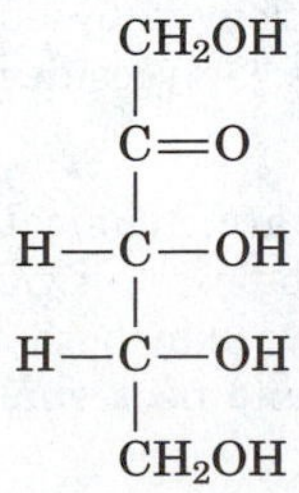

Ribulose

26.36 Look up the structure of D-talose in Figure 26.3, and draw the β anomer in its pyranose form. Identify the ring substituents as axial or equatorial.

26.37 Draw structures for the products you would expect to obtain from reaction of β-D-talopyranose with each of the following reagents:
(a) $NaBH_4$ in H_2O
(b) Warm dilute HNO_3
(c) Br_2, H_2O
(d) CH_3CH_2OH, HCl
(e) CH_3I, Ag_2O
(f) $(CH_3CO)_2O$, pyridine

26.38 Many other sugars besides glucose exhibit mutarotation. For example, α-D-galactopyranose has $[\alpha]_D = +150.7°$, and β-D-galactopyranose has $[\alpha]_D = +52.8°$. If either anomer is dissolved in water and allowed to reach equilibrium, the specific rotation of the solution is $+80.2°$. What are the percentages of each anomer at equilibrium? Draw the pyranose forms of both anomers using Haworth projections.

26.39 How many D-2-ketohexoses are possible? Draw them.

26.40 One of the D-2-ketohexoses (Problem 26.39) is called *sorbose*. On treatment with $NaBH_4$, sorbose yields a mixture of gulitol and iditol. What is the structure of sorbose?

26.41 Another D-2-ketohexose, *psicose*, yields a mixture of allitol and altritol when reduced with $NaBH_4$. What is the structure of psicose?

26.42 Fischer prepared the L-gulose needed for his structure proof of glucose in the following way. D-Glucose was oxidized to D-glucaric acid, which can form two six-membered-ring lactones. These were separated and reduced with sodium amalgam to give D-glucose and L-gulose. What are the structures of the two lactones, and which one is reduced to L-gulose?

26.43 What other D-aldohexose gives the same alditol as D-talose?

26.44 Which of the eight D-aldohexoses give the same aldaric acids as their L enantiomers?

26.45 Which of the other three D-aldopentoses gives the same aldaric acid as D-lyxose?

26.46 Gentiobiose, a rare disaccharide found in saffron and gentian, is a reducing sugar and forms only D-glucose on hydrolysis with aqueous acid. Reaction of gentiobiose with iodomethane and Ag_2O yields an octamethyl derivative, which can be hydrolyzed with aqueous acid to give 1 equiv of 2,3,4,6-tetra-*O*-methyl-D-glucopyranose and 1 equiv of 2,3,4-tri-*O*-methyl-D-glucopyranose. If gentiobiose contains a β-glycoside link, what is its structure?

26.47 Amygdalin, or Laetrile, is a glycoside isolated in 1830 from almond and apricot seeds. It is known as a *cyanogenic glycoside* because acidic hydrolysis liberates HCN, along with benzaldehyde and 2 equivalents of D-glucose. Structural studies have shown amygdalin to be a β-glycoside of benzaldehyde cyanohydrin with gentiobiose (Problem 26.46). Draw the structure of amygdalin.

26.48 Trehalose is a nonreducing disaccharide that is hydrolyzed by aqueous acid to yield 2 equiv of D-glucose. Methylation followed by hydrolysis yields 2 equiv of 2,3,4,6-tetra-*O*-methylglucose. How many structures are possible for trehalose?

26.49 Trehalose (Problem 26.48) is cleaved by enzymes that hydrolyze α-glycosides but not by enzymes that hydrolyze β-glycosides. What is the structure and systematic name of trehalose?

26.50 Isotrehalose and neotrehalose are chemically similar to trehalose (Problems 26.48 and 26.49) except that neotrehalose is hydrolyzed only by β-glycosidase enzymes, whereas isotrehalose is hydrolyzed by both α- and β-glycosidase enzymes. What are the structures of isotrehalose and neotrehalose?

26.51 D-Glucose reacts with acetone in the presence of acid to yield the nonreducing 1,2:5,6-diisopropylidene-D-glucofuranose. Propose a mechanism.

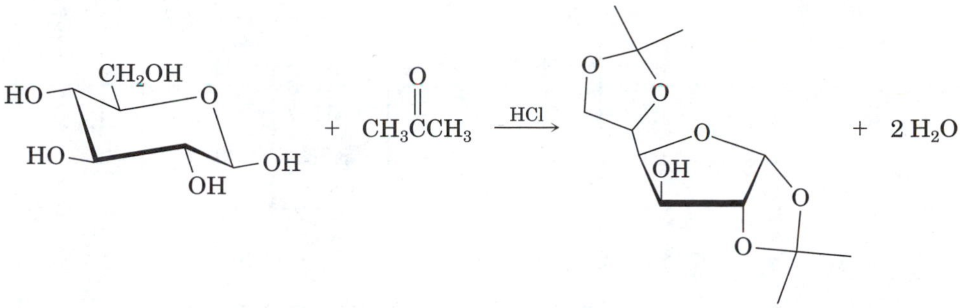

1,2:5,6-Diisopropylidene-D-glucofuranose

26.52 D-Mannose reacts with acetone to give a diisopropylidene derivative that is still reducing toward Tollens' reagent. Propose a likely structure for this derivative.

26.53 Propose a mechanism to account for the fact that D-gluconic acid and D-mannonic acid are interconverted when either is heated in pyridine solvent.

26.54 The *cyclitols* are a group of carbocyclic sugar derivatives having the general formulation 1,2,3,4,5,6-cyclohexanehexol. How many stereoisomeric cyclitols are possible? Draw them in Haworth projection.

26.55 Compound A is a D-aldopentose that can be oxidized to an optically inactive aldaric acid, B. On Kiliani–Fischer chain extension, A is converted into C and D; C can be oxidized to an optically active aldaric acid E, but D is oxidized to an optically inactive aldaric acid, F. What are the structures of A–F?

26.56 Simple sugars undergo reaction with phenylhydrazine, $PhNHNH_2$, to yield crystalline derivatives called *osazones*. The reaction is a bit complex, however, as shown by the fact that glucose and fructose yield the same osazone.

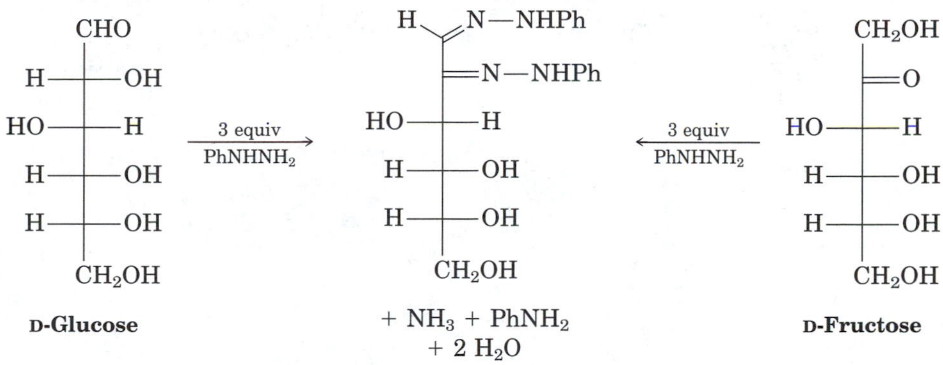

(a) Draw the structure of a third sugar that yields the same osazone as glucose and fructose.

(b) Using glucose as the example, the first step in osazone formation is reaction of the sugar with phenylhydrazine to yield an imine called a *phenylhydrazone*. Draw the structure of the product.

(c) The second and third steps in osazone formation are tautomerization of the phenylhydrazone to give an enol, followed by elimination of aniline to give a keto imine. Draw the structures of both the enol tautomer and the keto imine.

(d) The final step is reaction of the keto imine with 2 equivalents of phenylhydrazine to yield the osazone plus ammonia. Propose a mechanism for this step.

26.57 When heated to 100°C, D-idose undergoes a reversible loss of water and exists primarily as 1,6-anhydro-D-idopyranose.

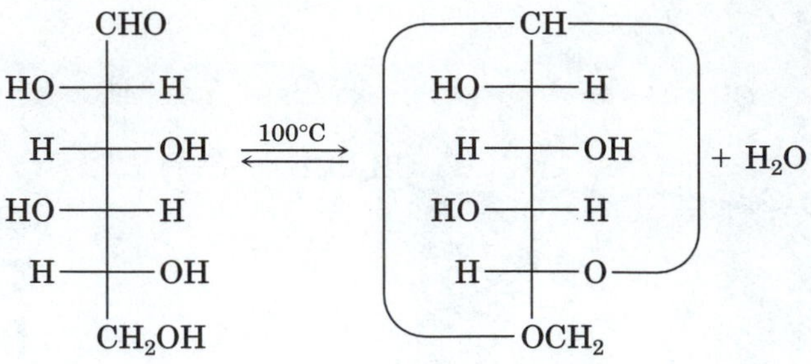

D-Idose **1,6-Anhydro-D-idopyranose**

(a) Draw D-idose in its pyranose form, showing the more stable chair conformation of the ring.
(b) Which is more stable, α-D-idopyranose or β-D-idopyranose? Explain.
(c) Draw 1,6-anhydro-D-idopyranose in its most stable conformation.
(d) When heated to 100°C under the same conditions as those used for D-idose, D-glucose does not lose water and does not exist in a 1,6-anhydro form. Explain.

A Look Ahead

26.58 We'll see in Chapter 30 that acetyl coenzyme A (acetyl CoA) is the key intermediate in food metabolism. What sugar is present in acetyl CoA?

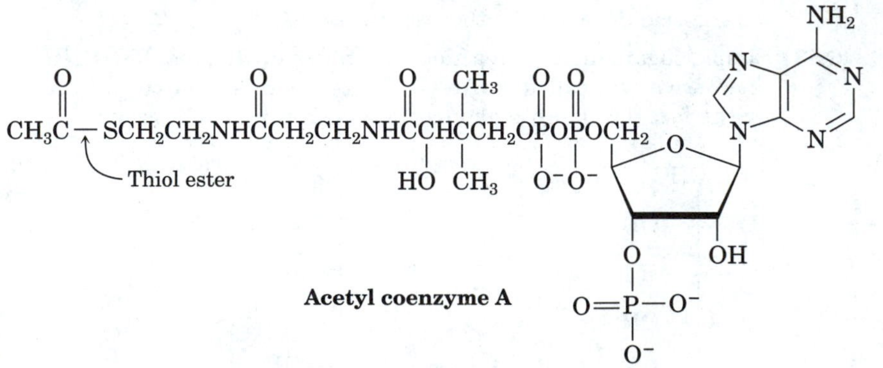

Acetyl coenzyme A

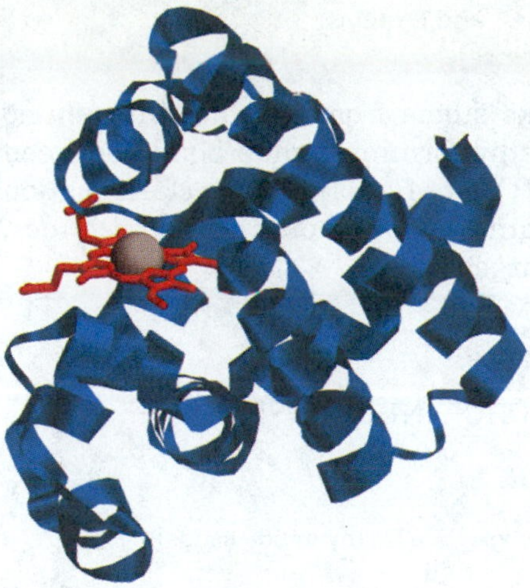

Enzymes, such as this myoglobin molecule, are frequently displayed as "ribbon" models that show their helical structures.

27

Biomolecules: Amino Acids, Peptides, and Proteins

Proteins are large biomolecules that occur in every living organism. They are of many different types and have many different biological functions. The keratin of skin and fingernails, the fibroin of silk and spider webs, and the DNA polymerase that catalyzes the synthesis of DNA in cells are all proteins. Regardless of their appearance or function, all proteins are made up of many *amino acid* units linked together into a long chain.

Amino acids, as their name implies, are difunctional. They contain both a basic amino group and an acidic carboxyl group:

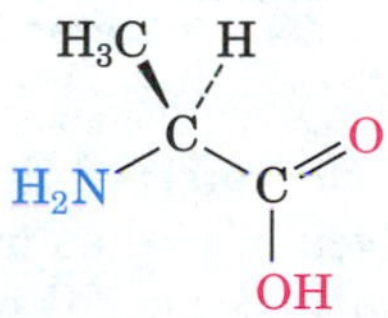

Alanine, an amino acid

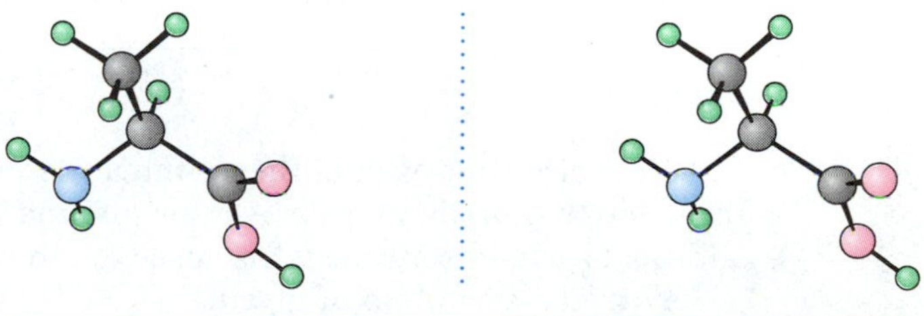

Stereo View

Their value as biological building blocks stems from the fact that amino acids can join together into long chains by forming amide bonds between the $-NH_2$ of one amino acid and the $-COOH$ of another. For classification purposes, chains with fewer than 50 amino acids are often called *peptides*, while the term *protein* is reserved for larger chains.

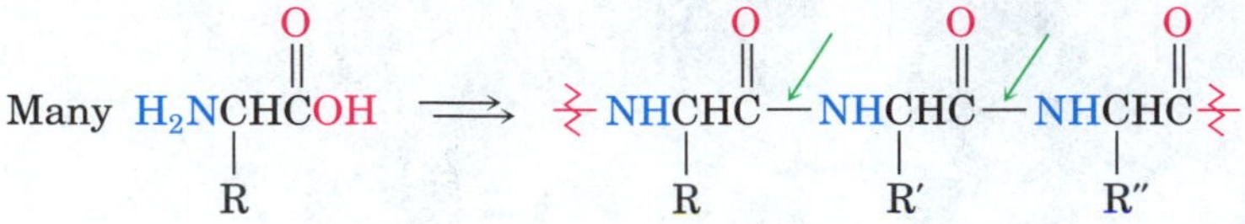

A polypeptide (many amide bonds)

27.1 Structures of Amino Acids

The structures of the 20 amino acids commonly found in proteins are shown in Table 27.1 (pp. 1058–1059). All are **α-amino acids**, meaning that the amino group in each is a substituent on the α carbon atom—the one next to the carbonyl group. Note that 19 of the 20 amino acids listed are primary amines, RNH_2, and differ only in the nature of the side-chain substituents. Proline, however, is a secondary amine whose nitrogen and α carbon atoms are part of a pyrrolidine ring.

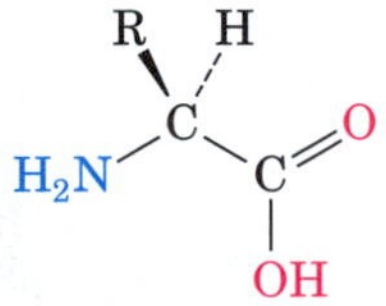

A primary α-amino acid
(R = a side chain)

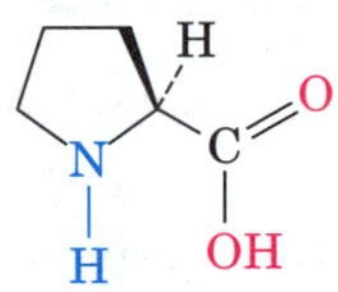

Proline, a secondary α-amino acid

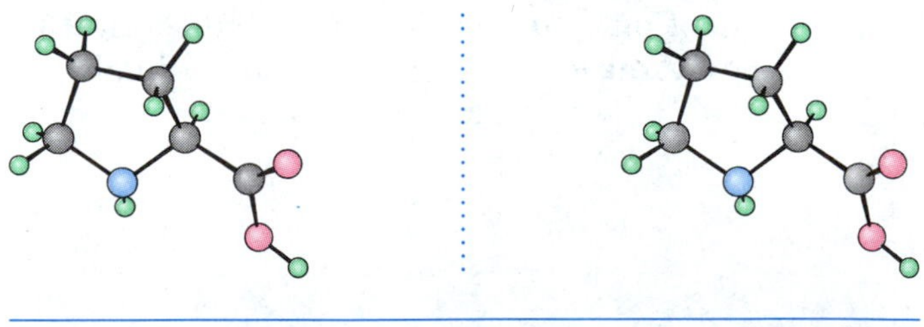

Stereo View

Note also that each of the amino acids in Table 27.1 is referred to by a three-letter shorthand code: Ala for alanine, Gly for glycine, and so on. In addition, a one-letter code is also used, as shown in parentheses in the table.

With the exception of glycine, H_2NCH_2COOH, the α carbons of amino acids are stereogenic centers. Two enantiomeric forms of each amino acid are therefore possible, but nature uses only a single enantiomer to build proteins. In Fischer projections, naturally occurring amino acids are represented by placing the –COOH group at the top as if drawing a carbohydrate (Section 26.2) and then placing the $-NH_2$ group on the left. Because of their stereochemical similarity to L sugars (Section 26.3), the naturally occurring α-amino acids are often referred to as L-amino acids.

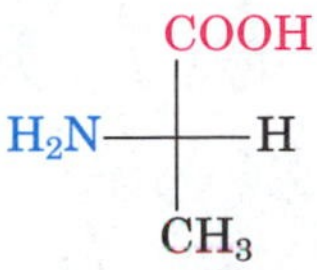

(*S*)-Alanine
(L-Alanine)

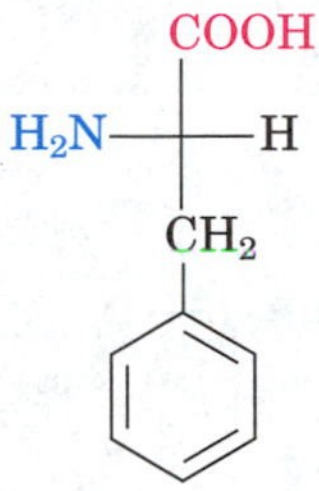

(*S*)-Phenylalanine
(L-Phenylalanine)

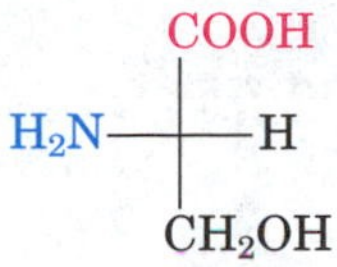

(*S*)-Serine
(L-Serine)

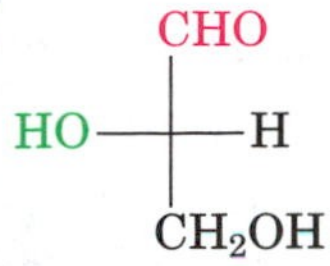

(Stereochemically similar to L-glyceraldehyde)

The 20 common amino acids can be further classified as either neutral, acidic, or basic, depending on the structure of their side chain. Fifteen of the 20 have neutral side chains, two (aspartic acid and glutamic acid) have an extra carboxylic acid function in their side chains, and three (lysine, arginine, and histidine) have basic amino groups in their side chains.

All 20 of the amino acids are necessary for protein synthesis, but humans can synthesize only 10 of the 20. The remaining 10 are called *essential amino acids* because they must be obtained from dietary sources. Failure to include an adequate dietary supply of these essential amino acids can lead to deficiency diseases.

PROBLEM. . . . .

27.1 Look carefully at the α-amino acids shown in Table 27.1. How many contain aromatic rings? How many contain sulfur? How many contain alcohols? How many contain hydrocarbon side chains?

PROBLEM. . . . .

27.2 Eighteen of the 19 L-amino acids have the *S* configuration at the α carbon. Cysteine is the only L-amino acid that has an *R* configuration. Explain.

PROBLEM. . . . .

27.3 The amino acid threonine, (2*S*,3*R*)-2-amino-3-hydroxybutanoic acid, has two stereogenic centers. Draw a Fischer projection of threonine.

PROBLEM. . . . .

27.4 Draw the Fischer projection of a threonine diastereomer, and label its stereogenic centers as *R* or *S* (Problem 27.3).

. . . .

Table 27.1 The 20 Common Amino Acids in Proteins
(Names of the amino acids essential to the human diet are shown in red.)

Name	*Abbreviations*		*Molecular weight*	*Structure*	*Isoelectric point*	pK_{a1} α-COOH	pK_{a2} α-NH_3^+
Neutral Amino Acids							
Alanine	Ala	(A)	89	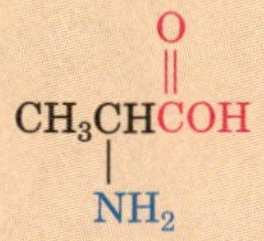	6.00	2.34	9.69
Asparagine	Asn	(N)	132	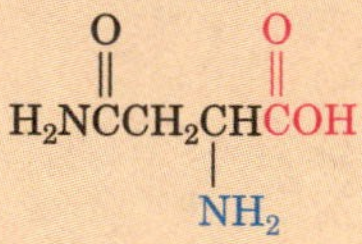	5.41	2.02	8.80
Cysteine	Cys	(C)	121	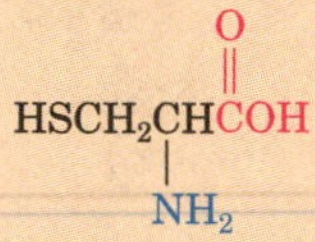	5.07	1.96	10.28
Glutamine	Gln	(Q)	146	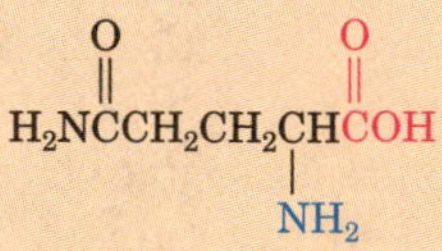	5.65	2.17	9.13
Glycine	Gly	(G)	75	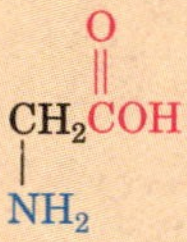	5.97	2.34	9.60
Isoleucine	Ile	(I)	131	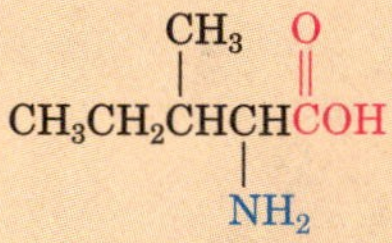	6.02	2.36	9.60
Leucine	Leu	(L)	131	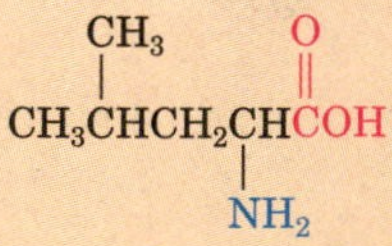	5.98	2.36	9.60
Methionine	Met	(M)	149	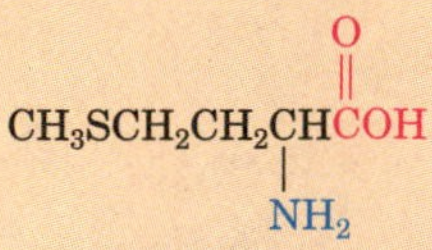	5.74	2.28	9.21
Phenylalanine	Phe	(F)	165	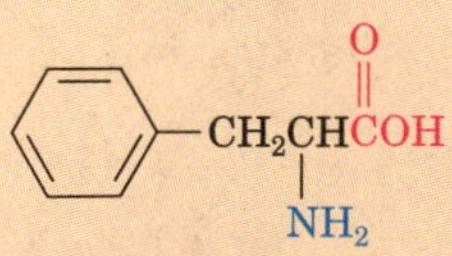	5.48	1.83	9.13
Proline	Pro	(P)	115	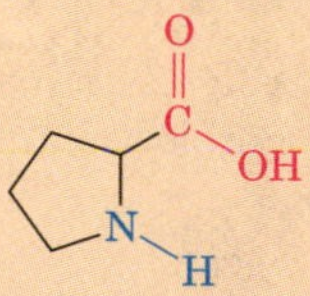	6.30	1.99	10.60
Serine	Ser	(S)	105	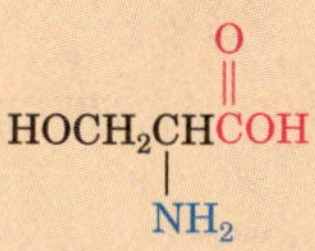	5.68	2.21	9.15

Table 27.1 (continued)

Name	*Abbreviations*		*Molecular weight*	*Structure*	*Isoelectric point*	pK_{a1} α-COOH	pK_{a2} α-NH_3^+
Threonine	Thr	(T)	119	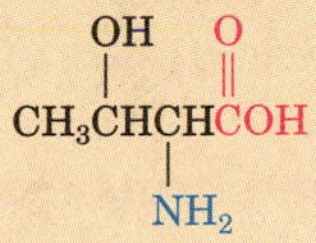	5.60	2.09	9.10
Tryptophan	Trp	(W)	204	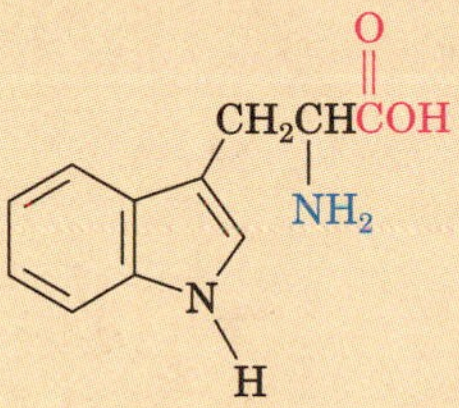	5.89	2.83	9.39
Tyrosine	Tyr	(Y)	181	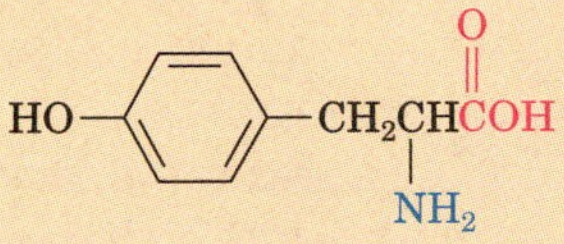	5.66	2.20	9.11
Valine	Val	(V)	117	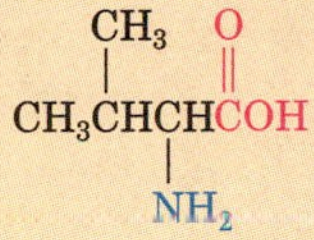	5.96	2.32	9.62
Acidic Amino Acids							
Aspartic acid	Asp	(D)	133	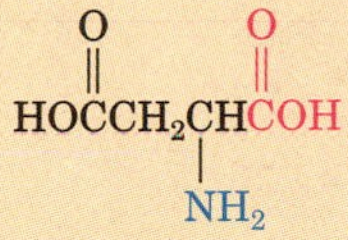	2.77	1.88	9.60
Glutamic acid	Glu	(E)	147	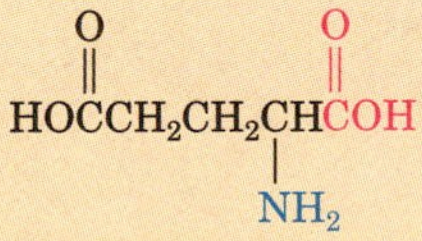	3.22	2.19	9.67
Basic Amino Acids							
Arginine	Arg	(R)	174	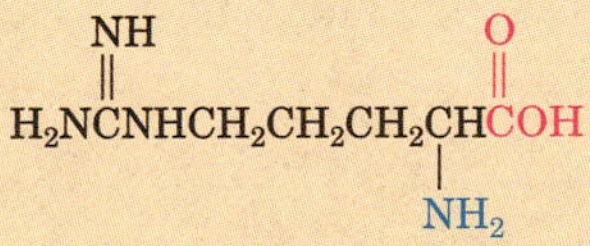	10.76	2.17	9.04
Histidine	His	(H)	155	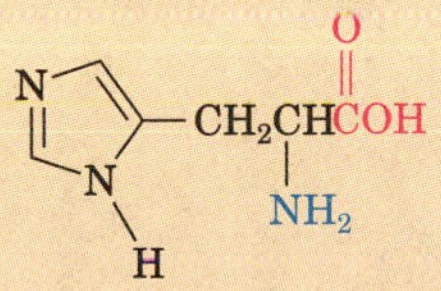	7.59	1.82	9.17
Lysine	Lys	(K)	146	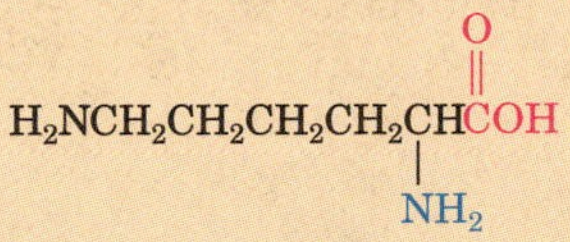	9.74	2.18	8.95

27.2 Dipolar Structure of Amino Acids

Since amino acids contain both an acidic and a basic group, they undergo an intramolecular acid–base reaction and exist primarily in the form of a dipolar ion, or **zwitterion** (German *zwitter*, "hybrid"):

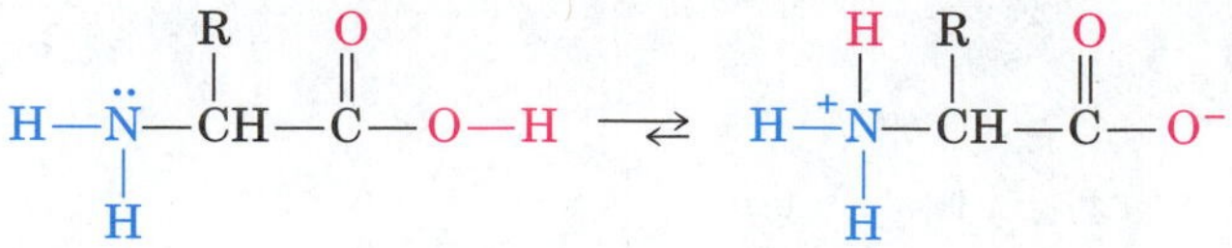

A zwitterion

Amino acid zwitterions are a kind of internal salt and therefore have many of the physical properties associated with salts. They have large dipole moments, are soluble in water but insoluble in hydrocarbons, and are crystalline substances with high melting points. In addition, amino acids are *amphoteric*: They can react either as acids or as bases, depending on the circumstances. In aqueous acid solution, an amino acid zwitterion *accepts* a proton to yield a cation; in aqueous basic solution, the zwitterion *loses* a proton to form an anion.

In acid solution

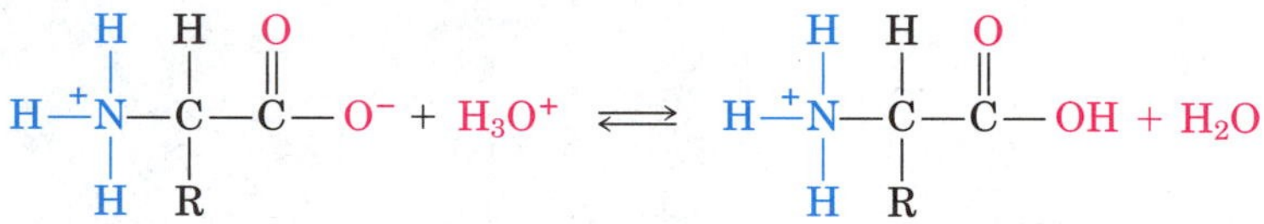

In base solution

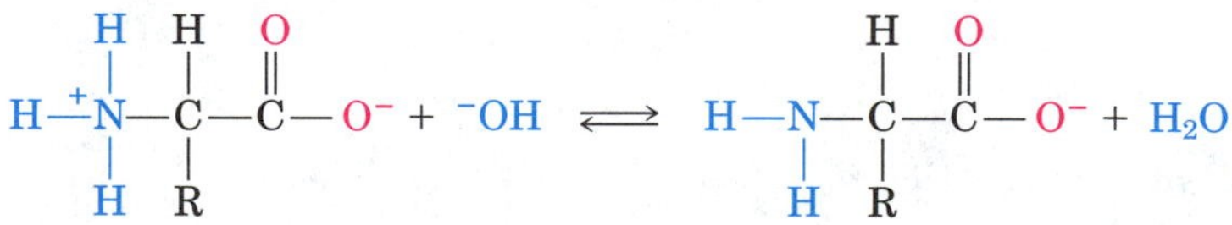

Note that it's the carboxylate anion, $-COO^-$, rather than the amino group that acts as the basic site and accepts the proton in acid solution. Similarly, it's the ammonium cation, $-NH_4^+$, rather than the carboxyl group that acts as the acidic site and donates a proton in base solution.

PROBLEM

27.5 Draw the following amino acids in their zwitterionic form:
(a) Phenylalanine (b) Serine (c) Proline

PROBLEM

27.6 Write structural formulas for the following equations:
(a) Phenylalanine + 1 equiv NaOH ⟶ ?
(b) Product of (a) + 1 equiv HCl ⟶ ?
(c) Product of (a) + 2 equiv HCl ⟶ ?

27.3 Isoelectric Points

In acid solution at low pH, an amino acid is protonated and exists primarily as a cation. In basic solution at high pH, an amino acid is deprotonated and exists primarily as an anion. Thus, there must be some intermediate pH at which the amino acid is exactly balanced between anionic and cationic forms and exists primarily as the neutral, dipolar zwitterion. This pH is called the amino acid's **isoelectric point**.

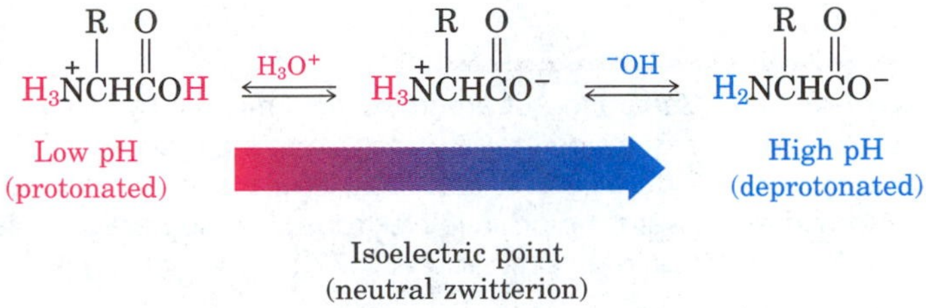

The isoelectric point of an amino acid depends on its structure, with values for the 20 common amino acids given in Table 27.1. The 15 amino acids with neutral side chains have isoelectric points near neutrality, in the pH range 5.0–6.5. (These values aren't exactly at neutral pH 7, because carboxyl groups are stronger acids in aqueous solution than amino groups are bases.) The two amino acids with acidic side chains have isoelectric points at lower pH, which suppresses dissociation of the extra –COOH in the side chain, and the three amino acids with basic side chains have isoelectric points at higher pH, which suppresses protonation of the extra amino group.

We can take advantage of the differences in isoelectric points to separate a mixture of amino acids (or a mixture of proteins) into its pure constituents. Using a technique known as **electrophoresis**, a solution of different amino acids is placed near the center of a strip of paper or gel. The paper or gel is moistened with an aqueous buffer of a given pH, and electrodes are connected to the ends of the strip. When an electric potential is applied, those amino acids with negative charges (those that are deprotonated because their isoelectric points are below the pH of the buffer) migrate slowly toward the positive electrode. At the same time, those amino acids with positive charges (those that are protonated because their isoelectric points are above the pH of the buffer) migrate toward the negative electrode.

Different amino acids migrate at different rates, depending on their isoelectric point and on the pH of the aqueous buffer. Thus, the different amino acids can be separated. Figure 27.1 (p. 1062) illustrates this separation for a mixture of lysine (basic), glycine (neutral), and aspartic acid (acidic).

If exact pK_a values for the acidic sites of an amino acid are known (Table 27.1), the percentages of protonated, neutral, and deprotonated forms at any pH can be calculated using the **Henderson–Hasselbalch equation**, a simple expression derived by rearranging the standard equation for pK_a.

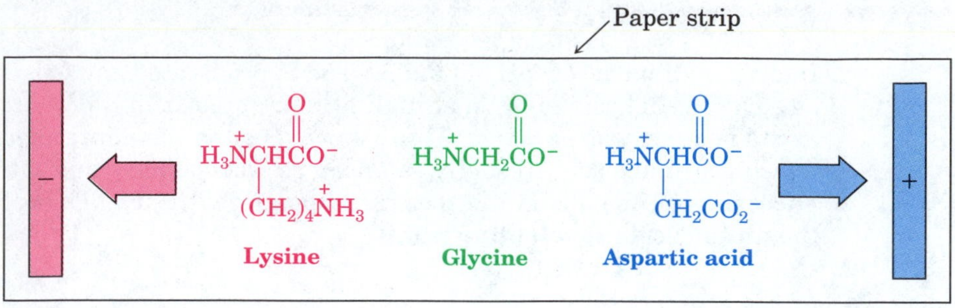

Figure 27.1 Separation of an amino acid mixture by electrophoresis. At pH = 5.97, glycine molecules are neutral and don't migrate; lysine molecules are protonated and migrate toward the negative electrode; and aspartic acid molecules are deprotonated and migrate toward the positive electrode. (Lysine has its isoelectric point at 9.74, glycine at 5.97, and aspartic acid at 2.77.)

Since

$$pK_a = -\log \frac{[H_3O^+][A^-]}{[HA]} = -\log[H_3O^+] - \log \frac{[A^-]}{[HA]} = pH - \log \frac{[A^-]}{[HA]}$$

then

$$pH = pK_a + \log \frac{[A^-]}{[HA]}$$

or $$\log \frac{[A^-]}{[HA]} = pH - pK_a$$

Henderson–Hasselbalch equation

In other words, the logarithm of the conjugate base concentration divided by the weak acid concentration is equal to the pH minus the pK_a of the acid. The protonated form of alanine, for example, is a weak acid with pK_{a1} = 2.34, and the neutral form of alanine is a still weaker acid with pK_{a2} = 9.69:

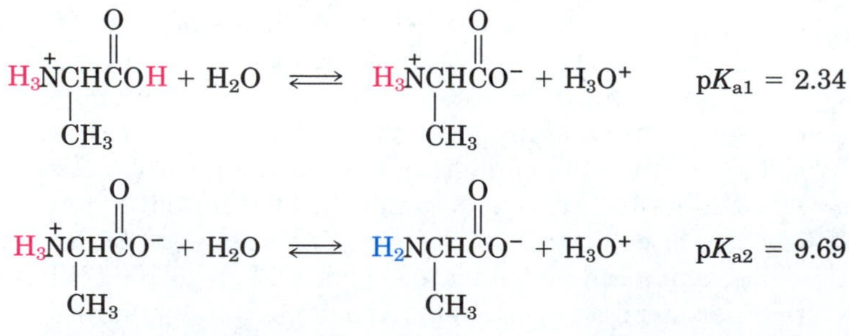

To see how to use the Henderson–Hasselbalch equation, let's find out what species are present in a 1.00 M solution of alanine at pH 1.00. Since this is a strongly acidic pH, the equilibrium we need to consider is that between protonated alanine, H_2A^+, and neutral, zwitterionic alanine, HA. According to Table 27.1, the pK_a of protonated alanine is 2.34. Thus,

$$\log \frac{[HA]}{[H_2A^+]} = pH - pK_a = 1.00 - 2.34 = -1.34$$

so

$$\frac{[HA]}{[H_2A^+]} = \text{antilog}(-1.34) = 0.046$$

But since

$$[HA] = 1.00\ M - [H_2A^+]$$

then

$$\frac{1.00 - [H_2A^+]}{[H_2A^+]} = 0.046$$

Thus,

$$[H_2A^+] = 0.956 \qquad \text{and} \qquad [HA] = 0.044$$

At pH 1.00, 4.4% of alanine molecules in a 1.00 M solution are neutral (zwitterionic) and 95.6% are protonated. Similar calculations can be done at any other pH, leading to the *titration curve* shown in Figure 27.2.

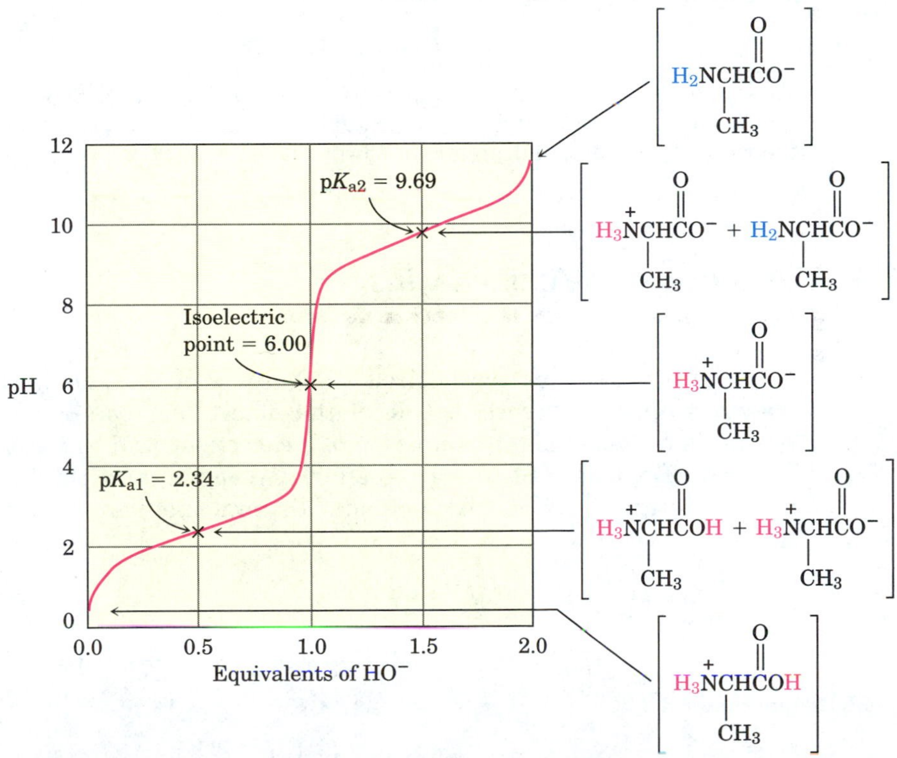

Figure 27.2 A titration curve for alanine plotted using the Henderson–Hasselbalch equation. Each of the two legs is plotted separately. At pH < 1, alanine is entirely protonated; at pH = 2.34, alanine is a 50:50 mix of protonated and neutral forms; at pH 6.00, alanine is entirely neutral; at pH = 9.69, alanine is a 50:50 mix of neutral and deprotonated forms; at pH > 11, alanine is entirely deprotonated.

Each "leg" of the curve—the first leg from pH 1 to 6 and the second leg from pH 6 to 11—is calculated separately, corresponding to the dissociations of protonated alanine and zwitterionic alanine. Exactly halfway between the two legs is the isoelectric point at 6.00. In essence, it's as if we started with protonated alanine at low pH and then titrated with NaOH until complete deprotonation occurred. When 0.5 equiv of NaOH is added, the first deprotonation is 50% done; when 1.0 equiv of NaOH is added, the first deprotonation is complete and the isoelectric point is reached; when 1.5 equiv of NaOH is added, the second deprotonation is 50% done; and when 2.0 equiv of NaOH is added, the second deprotonation is complete.

PROBLEM.

27.7 Look up the values of pK_{a1}, pK_{a2}, and the isoelectric point for glycine in Table 27.1. Draw structures of the predominant forms of glycine at pH = 2.0, 6.0, and 10.0.

PROBLEM.

27.8 For the mixtures of amino acids indicated, predict the direction of migration of each component (toward anode or cathode) and relative rate of migration.
(a) Valine, glutamic acid, and histidine at pH = 7.6
(b) Glycine, phenylalanine, and serine at pH = 5.7
(c) Glycine, phenylalanine, and serine at pH = 5.5
(d) Glycine, phenylalanine, and serine at pH = 6.0

PROBLEM.

27.9 Threonine has $pK_{a1} = 2.09$ and $pK_{a2} = 9.10$. Use the Henderson–Hasselbalch equation to calculate the ratio of protonated and neutral forms at pH = 1.50. Calculate the ratio of neutral and deprotonated forms at pH = 10.00.

27.4 Synthesis of α-Amino Acids

α-Amino acids can be synthesized using some of the standard chemical reactions already discussed. One of the oldest methods of α-amino acid synthesis begins with α bromination of a carboxylic acid by treatment with Br_2 and PBr_3 (the Hell–Volhard–Zelinskii reaction, Section 22.4). Nucleophilic substitution of the α-bromo acid with ammonia then yields an α-amino acid.

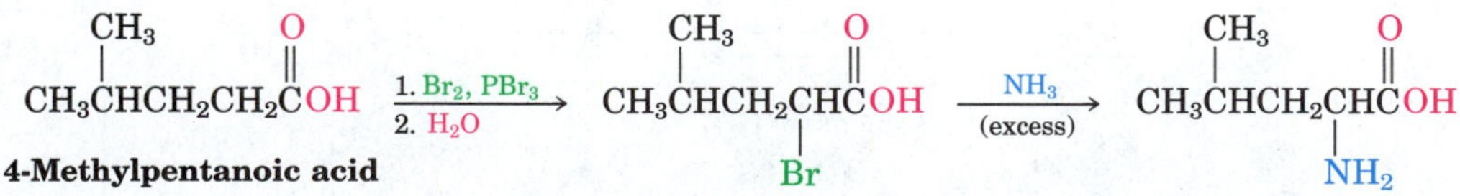

4-Methylpentanoic acid

2-Bromo-4-methylpentanoic acid

(*R*,*S*)-Leucine (45%)

Alternatively, higher product yields are obtained when the bromide displacement reaction is carried out by the Gabriel phthalimide method (Section 24.6) rather than by the ammonia method.

PROBLEM.

27.10 Show how you could prepare the following α-amino acids from the appropriate carboxylic acids.
(a) Phenylalanine (b) Valine

PROBLEM.

27.11 Show how a Gabriel amine synthesis (Section 24.6) might be used to prepare isoleucine.

The Strecker Synthesis

Another method for preparing racemic α-amino acids is the **Strecker[1] synthesis.** Developed in 1850, this versatile two-step process involves treatment of an aldehyde with KCN and aqueous ammonia to yield an intermediate α-amino nitrile. Hydrolysis of the nitrile then gives an α-amino acid:

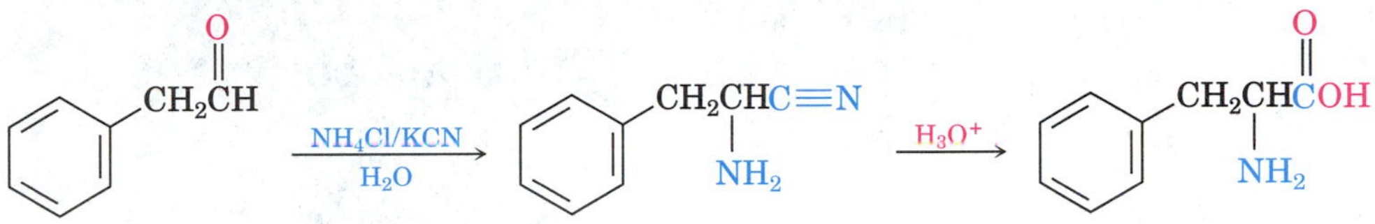

Phenylacetaldehyde **An α-amino nitrile** **(*R,S*)-Phenylalanine (53%)**

The first step of the Strecker synthesis, formation of an α-amino nitrile by reaction of an aldehyde with ammonia and KCN, is simply a combination of two carbonyl-group reactions seen earlier in Chapter 19. Reaction of the aldehyde with ammonia yields an imine (Section 19.12), which then adds HCN in a nucleophilic addition step similar to that involved in cyanohydrin formation (Section 19.9). The mechanism is shown in Figure 27.3.

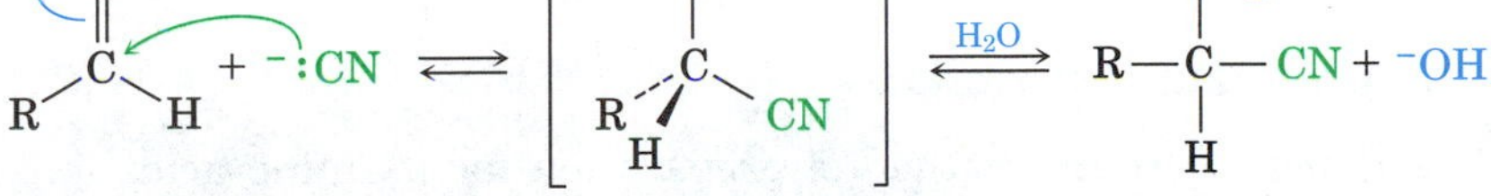

Figure 27.3 Mechanism of α-amino nitrile formation in the Strecker amino acid synthesis.

[1]Adolph Friedrich Ludwig Strecker (1822–1871); Ph.D. Giessen (1842); assistant to Liebig at Tübingen.

PROBLEM.

27.12 The rare amino acid L-dopa (3,4-dihydroxyphenylalanine) is useful as a drug against Parkinson's disease. Show how (±)-dopa might be synthesized from 3,4-dihydroxyphenylacetaldehyde.

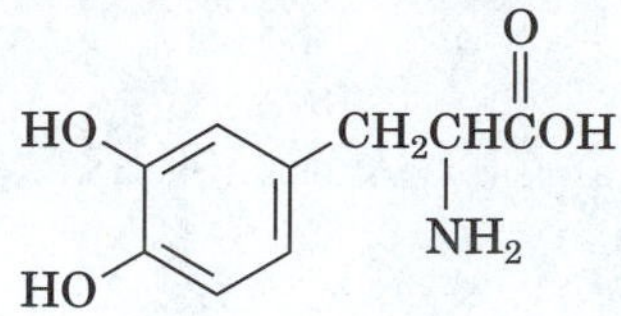

Dopa

Reductive Amination of α-Keto Acids: Biosynthesis

Yet a third method for the synthesis of α-amino acids is by reductive amination of an α-keto acid with ammonia and a reducing agent (Section 24.6):

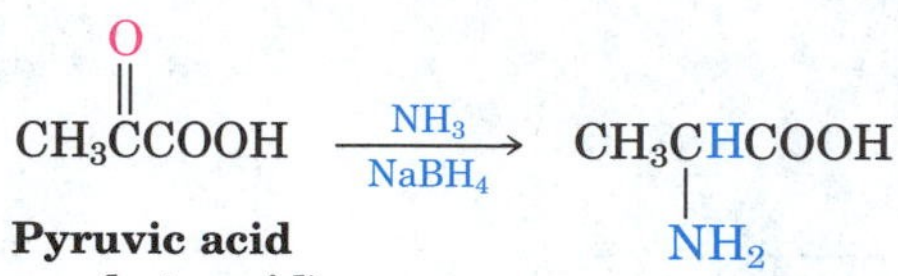

Pyruvic acid (an α-keto acid)

(*R*,*S*)-Alanine

This reductive amination method is particularly interesting because it is a close laboratory analogy of a pathway by which some amino acids are biosynthesized in nature. For example, the major route for glutamic acid synthesis in most organisms is reductive amination of α-ketoglutaric acid. The biological reducing agent is the rather complex molecule nicotinamide adenine dinucleotide (NADH), and the reaction is catalyzed by an enzyme, L-glutamate dehydrogenase. Nevertheless, the fundamental chemical principles of this biosynthetic reaction are identical to those of the laboratory reaction. We'll look at a related process in more detail in Section 30.6.

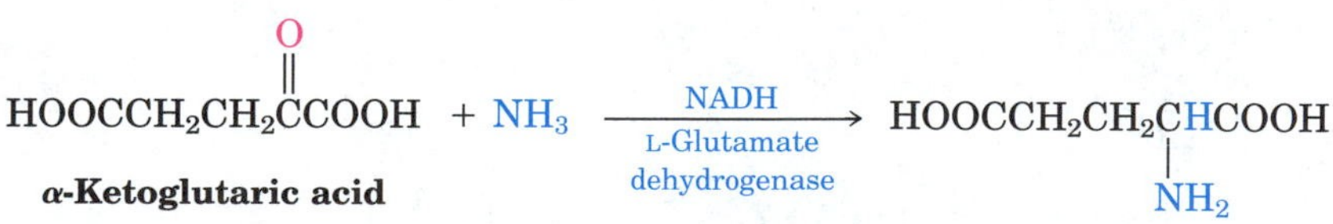

α-Ketoglutaric acid

(*S*)-Glutamic acid

The Amidomalonate Synthesis

The most general method of preparation for α-amino acids is the **amidomalonate synthesis**. This route, a straightforward extension of the malonic ester synthesis (Section 22.8), involves initial conversion of diethyl acetamidomalonate into its enolate anion by treatment with base, followed by S_N2 reaction with a primary alkyl halide. Hydrolysis and decarboxylation occur when the alkylated product is warmed with aqueous acid, and a racemic α-amino acid results. For example, aspartic acid is prepared in good yield when diethyl acetamidomalonate is alkylated with ethyl bromoacetate, followed by hydrolysis and decarboxylation:

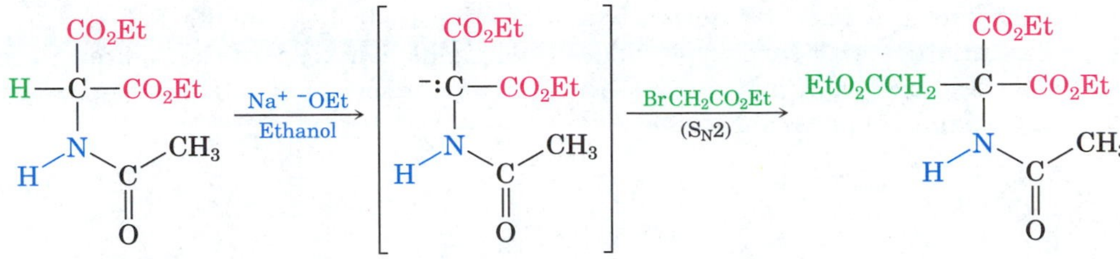

Diethyl acetamidomalonate

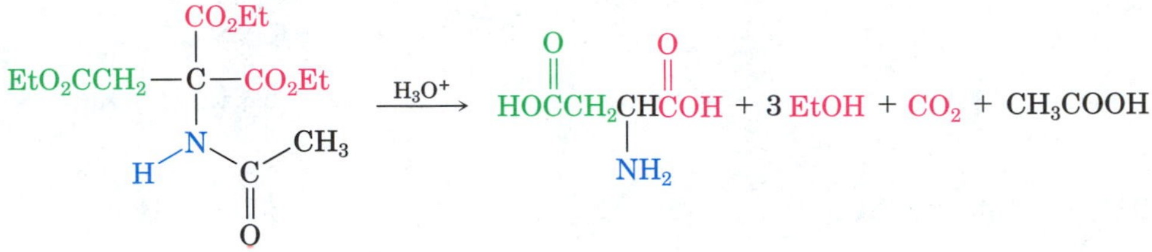

(*R,S*)-Aspartic acid (55%)

PROBLEM.

27.13 Show the alkyl halides you would use to prepare the following α-amino acids by the amidomalonate method.
(a) Leucine (b) Histidine (c) Tryptophan (d) Methionine

PROBLEM.

27.14 Serine can be synthesized by a simple variation of the amidomalonate method using formaldehyde rather than an alkyl halide. How might this be done?

27.5 Resolution of *R,S* Amino Acids

The synthesis of chiral amino acids from achiral precursors by any one of the methods just described yields a racemic mixture—an equal mixture of *S* and *R* products. To use these synthetic amino acids for the laboratory synthesis of naturally occurring proteins, the racemic mixture must first be resolved into pure enantiomers. Often this resolution can be done by allowing the racemic amino acid to react with a chiral acid or base to give a mixture of diastereomeric salts, which are then separated by fractional crystallization (Section 9.10).

Alternatively, biological methods of resolution can be used. For example, the enzyme carboxypeptidase selectively catalyzes the hydrolysis of *S* amido acids but not *R* amido acids. We can therefore resolve an *R*,*S* mixture of amino acids by first allowing the mixture to react with acetic anhydride to form the *N*-acetyl derivatives. Selective hydrolysis of the *R*,*S* amido acid mixture with carboxypeptidase then yields a mixture of the desired *S* amino acid and the unchanged *N*-acetyl *R* amido acid, which can be separated by usual chemical techniques.

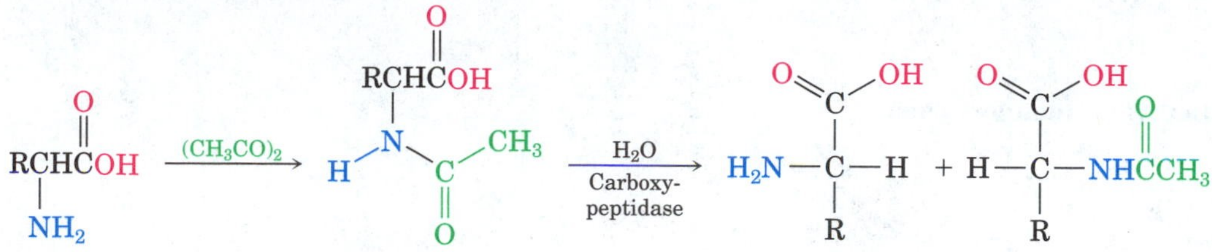

An *R*,*S* mixture — An *R*,*S* mixture — An *S* amino acid — An *R* amido acid

27.6 Peptides

Peptides are amino acid polymers in which the individual amino acid units, called **residues**, are linked together by amide bonds. An amino group from one residue forms an amide bond with the carboxyl of a second residue; the amino group of the second forms an amide bond with the carboxyl of a third, and so on. For example, alanylserine is the *dipeptide* that results when an amide bond is formed between the alanine carboxyl and the serine amino group:

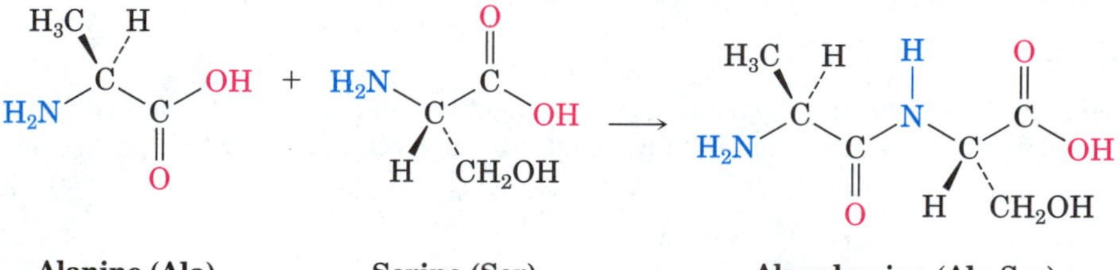

Alanine (Ala) — **Serine (Ser)** — **Alanylserine (Ala-Ser)**

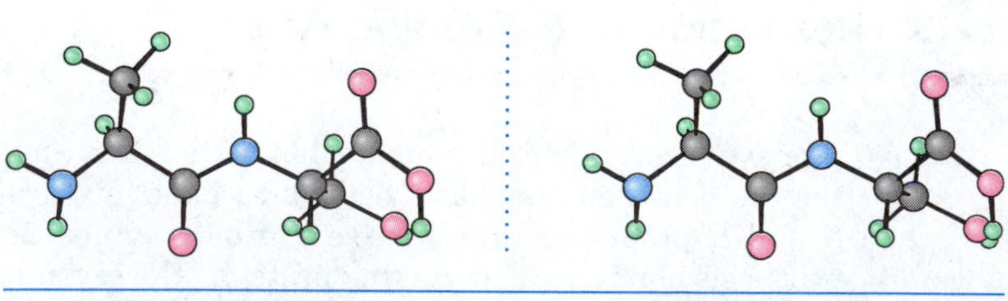

Stereo View

Note that two dipeptides can result from reaction between alanine and serine, depending on which carboxyl group reacts with which amino group.

If the alanine amino group reacts with the serine carboxyl, serylalanine results:

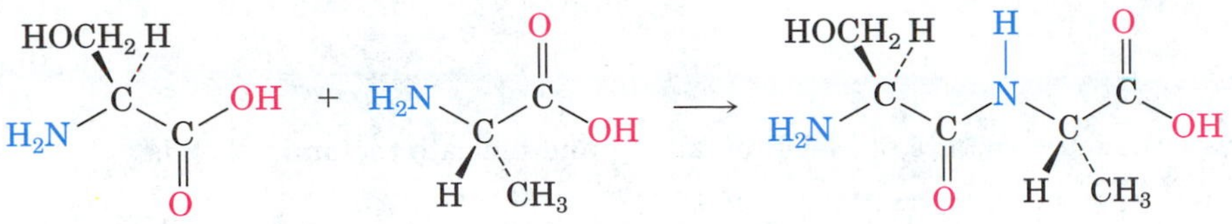

Serine (Ser) Alanine (Ala) Serylalanine (Ser-Ala)

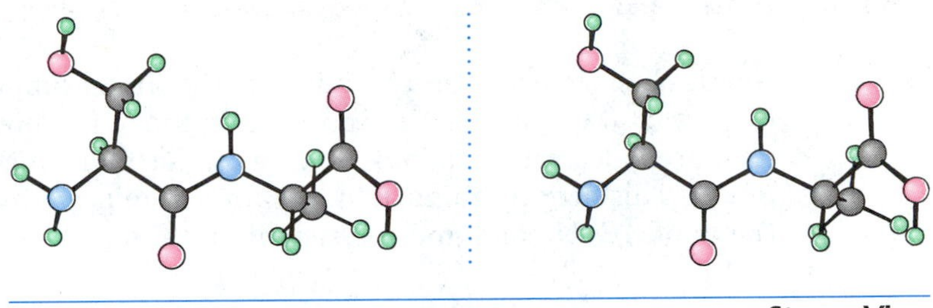

Stereo View

By convention, peptides are always written with the ***N*-terminal amino acid** (the one with the free $-NH_2$ group) on the left and the **C–terminal amino acid** (the one with the free –COOH group) on the right. The name of the peptide is usually indicated by using the three-letter abbreviations listed in Table 27.1 for each amino acid. Thus, alanylserine is abbreviated Ala-Ser and serylalanine is abbreviated Ser-Ala.

The number of possible isomeric peptides increases rapidly as the number of amino acid units increases. There are six ways in which three amino acids can be joined and more than 40,000 ways in which the eight amino acids present in the blood-pressure-regulating hormone angiotensin II can be joined (Figure 27.4).

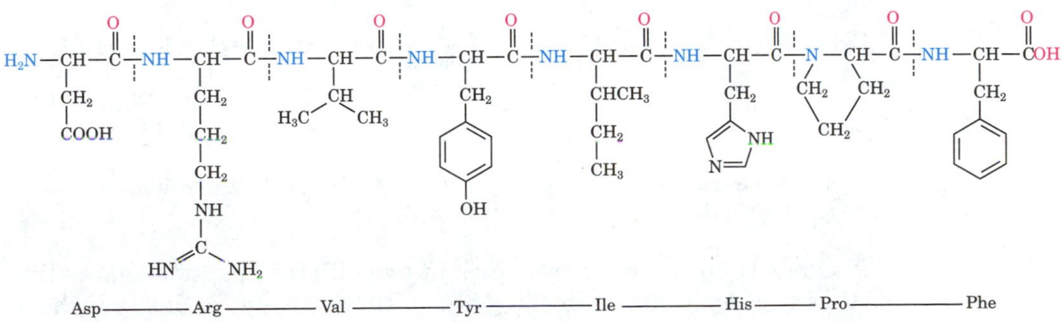

Figure 27.4 The structure of angiotensin II, a blood-pressure-regulating hormone present in blood plasma.

PROBLEM

27.15 Name the six possible isomeric tripeptides that contain valine, tyrosine, and glycine. Use the three-letter shorthand notation for each amino acid.

PROBLEM

27.16 Draw the full structure of Met-Pro-Val-Gly, and indicate the amide bonds.

27.7 Covalent Bonding in Peptides

Amide bonds in peptides are similar to the simple amide bonds we've already discussed (Section 24.4). Amide nitrogens are nonbasic because their unshared electron pair is delocalized by orbital overlap with the carbonyl group. This overlap imparts a certain amount of double-bond character to the amide C–N bond and restricts its rotation.

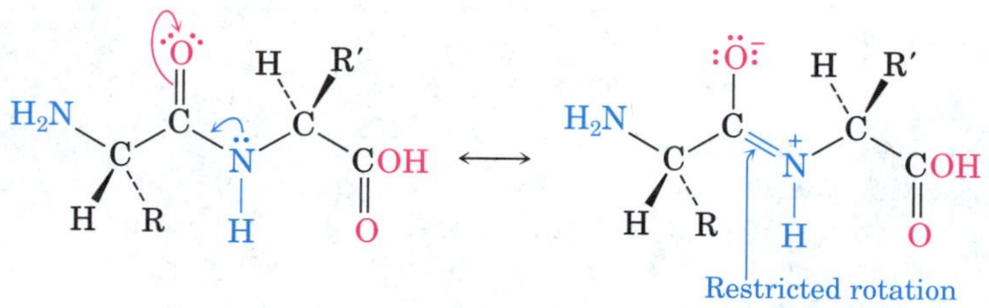

A second kind of covalent bonding in peptides occurs when a disulfide linkage, RS–SR, is formed between two cysteines. The linkage is sometimes represented by writing CyS (with a capital "S" for sulfur) and then drawing a line from one CyS to the other: CyS CyS. As we saw in Section 17.12, disulfide bonds are easily formed by mild oxidation of thiols, RSH, and are easily cleaved by mild reduction.

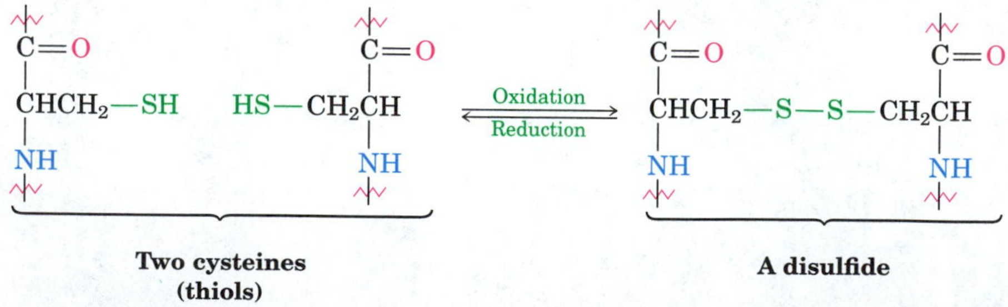

Disulfide bonds between cysteines in two different peptide chains link the otherwise separate chains together. Alternatively, a disulfide bond between two cysteines within the same chain causes a loop in the chain. Such is the case with the nonapeptide vasopressin, an antidiuretic hormone involved in controlling water balance in the body. Note that the C-terminal end of vasopressin occurs as the primary amide, $-CONH_2$, rather than as the free acid.

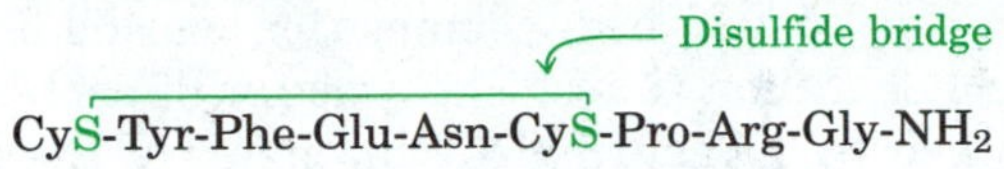

Vasopressin

27.8 Peptide Structure Determination: Amino Acid Analysis

Determining the structure of a peptide requires answering three questions: What amino acids are present? How much of each is present? In what sequence do the amino acids occur in the peptide chain? The answers to the first two questions are provided by an instrument called an *amino acid analyzer*.

An amino acid analyzer is an automated instrument based on analytical techniques worked out in the 1950s at the Rockefeller University in New York by William Stein[2] and Stanford Moore.[3] In preparation for analysis, the peptide is broken into its constituent amino acids by reducing all disulfide bonds and hydrolyzing all amide bonds with aqueous HCl. The resultant amino acid mixture is then analyzed by placing it at the top of a glass column (a *chromatography* column) filled with a special adsorbent material and pumping a series of aqueous buffers through the column. The various amino acids migrate down the column at different rates depending on their structures and are separated as they exit (*elute* from) the end of the column.

As each amino acid elutes from the end of the chromatography column, it mixes with a solution of *ninhydrin*, a reagent that reacts with α-amino acids to form an intense purple color. The purple color is detected by a spectrometer, and a plot of elution time versus spectrometer absorbance is obtained.

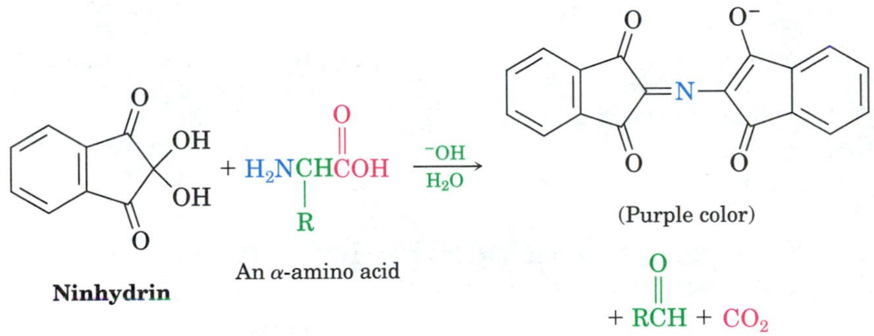

Since the amount of time required for a given amino acid to elute from the chromatography column is reproducible, the identity of the amino acids

[2]William H. Stein (1911–1980); b. New York; Ph.D. Columbia; professor, Rockefeller Institute; Nobel Prize (1972).

[3]Stanford Moore (1913–1982); b. Chicago; Ph.D. Wisconsin; professor, Rockefeller Institute; Nobel Prize (1972).

in a peptide of unknown composition can be determined simply by noting the various elution times. The amount of each amino acid in the sample is determined by measuring the intensity of the purple color resulting from its reaction with ninhydrin. Thus, the identity and amount of each amino acid in a peptide are found. Figure 27.5 shows the results of amino acid analysis of a standard equimolar mixture of 17 α-amino acids.

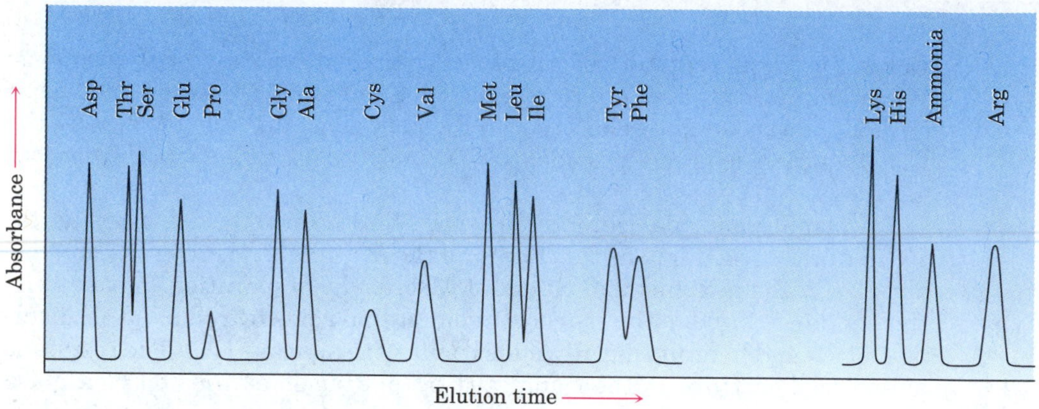

Figure 27.5 Amino acid analysis of an equimolar mixture of 17 amino acids.

PROBLEM.

27.17 Show the structures of the products obtained on reaction of valine with ninhydrin.

PROBLEM.

27.18 The data for amino acid analysis in Figure 27.5 indicate that proline is not easily detected by reaction with ninhydrin; only a very small peak is seen on the chromatogram. Why might this be?

27.9 Peptide Sequencing: The Edman Degradation

With the identity and amount of each amino acid known, the final task of structure determination is to *sequence* the peptide—that is, to find out in what order the amino acids are linked together. The general idea of peptide sequencing is to cleave one amino acid at a time from the end of the peptide chain (either N terminus or C terminus). That terminal amino acid is then separated and identified, and the cleavage reactions are repeated on the chain-shortened peptide until the entire peptide sequence is determined.

Most peptide sequencing is now done by **Edman[4] degradation**, an efficient method of *N*-terminal analysis. Automated Edman protein sequenators are available that allow a series of 50 or so repetitive sequencing steps to be carried out before a buildup of unwanted by-products begins to interfere with the results.

Edman degradation involves treatment of a peptide with phenyl isothiocyanate, $C_6H_5{-}N{=}C{=}S$, followed by mild acid hydrolysis, as shown in Figure 27.6 (p. 1074). The first step attaches a marker to the $-NH_2$ group of the *N*-terminal amino acid, and the second step splits the *N*-terminal residue from the chain, yielding a *phenylthiohydantoin* derivative plus the chain-shortened peptide. The phenylthiohydantoin is then identified chromatographically by comparison with known derivatives of the common amino acids, and the chain-shortened peptide is resubmitted to another round of Edman degradation.

Complete sequencing of large peptides and proteins by Edman degradation is impractical because buildup of unwanted by-products limits the method to about 25 cycles. Instead, a large peptide chain is first cleaved by partial hydrolysis into a number of smaller fragments, the sequence of each fragment is determined, and the individual fragments are fitted together like pieces in a jigsaw puzzle.

Partial hydrolysis of a peptide can be carried out either chemically with aqueous acid, or enzymatically with enzymes such as trypsin and chymotrypsin. Acidic hydrolysis is unselective and leads to a more or less random mixture of small fragments. Enzymic hydrolysis, however, is quite specific. Trypsin catalyzes hydrolysis only at the carboxyl side of the basic amino acids arginine and lysine; chymotrypsin cleaves only at the carboxyl side of the aryl-substituted amino acids phenylalanine, tyrosine, and tryptophan.

Val-Phe-Leu-Met-Tyr-Pro-Gly-Trp-Cys-Glu-Asp-Ile-Lys-Ser-Arg-His

Chymotrypsin cleaves these bonds. (after Phe, Tyr, Trp)

Trypsin cleaves these bonds. (after Lys, Arg)

To take an example of peptide sequencing, let's look at a hypothetical structure determination of angiotensin II, a hormonal octapeptide involved in controlling blood pressure by regulating the sodium–potassium salt balance in the body.

1. Amino acid analysis of angiotensin II shows the presence of eight different amino acids: Arg, Asp, His, Ile, Phe, Pro, Tyr, and Val in equimolar amounts.
2. An *N*-terminal analysis by the Edman method shows that angiotensin II has an aspartic acid at the N terminus.

[4]Pehr Edman (1916–); b. Stockholm; M.D. Karolinska Institute (E. Jorpes); professor, University of Lund.

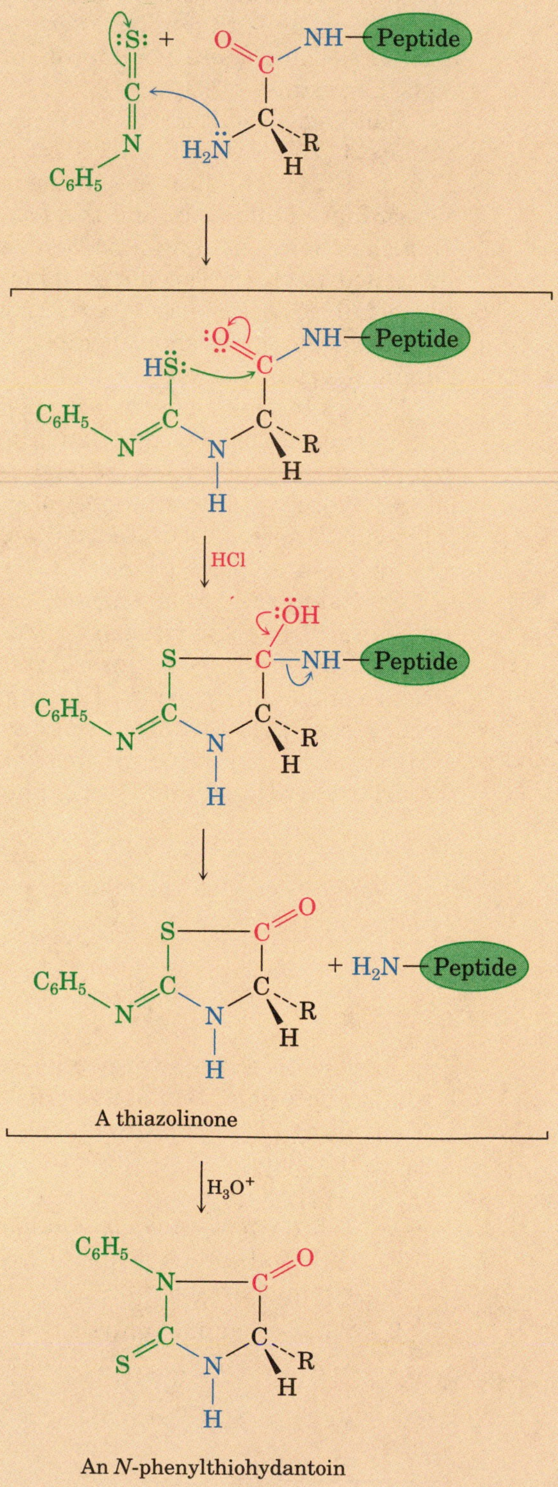

Nucleophilic addition of the peptide terminal amino group to phenyl isothiocyanate yields an *N*-phenylthiourea derivative.

Acid-catalyzed cyclization then yields a tetrahedral intermediate . . .

. . . which expels the chain-shortened peptide and forms a thiazolinone.

The thiazolinone rearranges in the presence of aqueous acid to yield the final *N*-phenylthiohydantoin derivative.

Figure 27.6 Mechanism of the Edman degradation for N-terminal analysis of peptides.

3. Partial hydrolysis of angiotensin II with dilute hydrochloric acid might yield the following fragments, whose sequences could be determined by Edman degradation:
   (a) Asp-Arg-Val (b) Ile-His-Pro (c) Arg-Val-Tyr
   (d) Pro-Phe (e) Val-Tyr-Ile
4. Matching the overlapping regions of the various fragments provides the full sequence of angiotensin II:

```
Asp-Arg-Val
    :   :
    Arg-Val-Tyr
        :   :
        Val-Tyr-Ile
                :
                Ile-His-Pro
                        :
                        Pro-Phe
Asp-Arg-Val-Tyr-Ile-His-Pro-Phe
```

Angiotensin II

The structure of angiotensin II is relatively simple—the entire sequence could easily be done by a protein sequenator—but the methods and logic used here are the same as those used to solve more complex structures. Indeed, protein chains with more than 400 amino acids have been sequenced by these methods.

PROBLEM

27.19 What fragments would result if angiotensin II were cleaved with trypsin? With chymotrypsin?

PROBLEM

27.20 Give the amino acid sequence of hexapeptides that produce these fragments on partial acid hydrolysis:
(a) Arg, Gly, Ile, Leu, Pro, Val gives Pro-Leu-Gly, Arg-Pro, Gly-Ile-Val
(b) Asp, Leu, Met, Trp, Val_2 gives Val-Leu, Val-Met-Trp, Trp-Asp-Val

27.10 Peptide Sequencing: C-Terminal Residue Determination

The Edman degradation is an excellent method of analysis for the N-terminal residue, but a complementary method of analysis for the C-terminal residue is also valuable. The best method currently available makes use of the enzyme carboxypeptidase to cleave the C-terminal amide bond in a peptide chain.

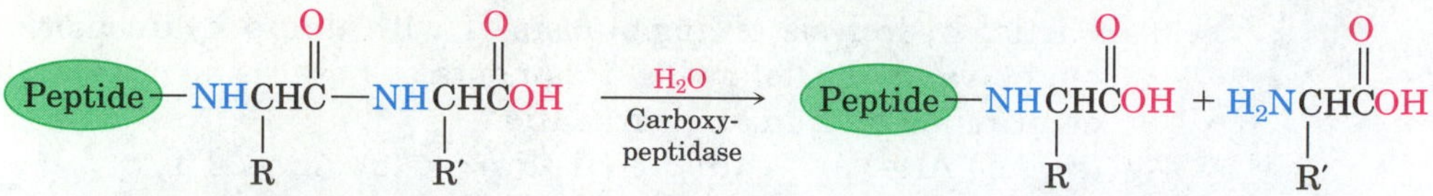

The analysis is carried out by incubating the polypeptide with carboxypeptidase and watching for the appearance of the first free amino acid that appears in solution. Of course, further degradation also occurs, since a new C–terminus is produced when the first amino acid is cleaved off. Ultimately the entire peptide is hydrolyzed.

Figure 27.7 shows an X-ray structure of carboxypeptidase A with a glycyltyrosine dipeptide held in its catalytic pocket, where it will be hydrolyzed.

Figure 27.7 An X-ray structure of carboxypeptidase A with a glycyltyrosine dipeptide held in the active site, where it can be hydrolyzed. The backbone of the peptide chain is displayed as a ribbon, and the α helical portions are clearly visible.

PROBLEM.

27.21 A hexapeptide with the composition Arg, Gly, Leu, Pro_3 is found to have proline at both C-terminal and N-terminal positions. Partial hydrolysis gives the following fragments:

Gly-Pro-Arg, Arg-Pro, Pro-Leu-Gly

What is the structure of the hexapeptide?

PROBLEM.

27.22 Propose two structures for a tripeptide that gives Leu, Ala, and Phe on hydrolysis but does not react with carboxypeptidase and does not react with phenyl isothiocyanate.

27.11 Peptide Synthesis

Once a peptide's structure has been determined, synthesis is sometimes the next goal, in order to obtain larger amounts for biological evaluation.

Although simple amides are usually formed by reaction between amines and acid chlorides (Section 21.8), peptide synthesis is more difficult because of the requirement for specificity. Many different amide bonds must be formed in a specific order rather than at random.

The solution to the specificity problem is *protection* (Section 17.10). We can force a reaction to take only the desired course by protecting all the amine and acid functional groups except for those we want to have react. For example, if we wanted to couple alanine with leucine to synthesize Ala-Leu, we could protect the $-NH_2$ group of alanine and the $-COOH$ group of leucine to render them unreactive, then form the desired amide bond, and finally, remove the protecting groups.

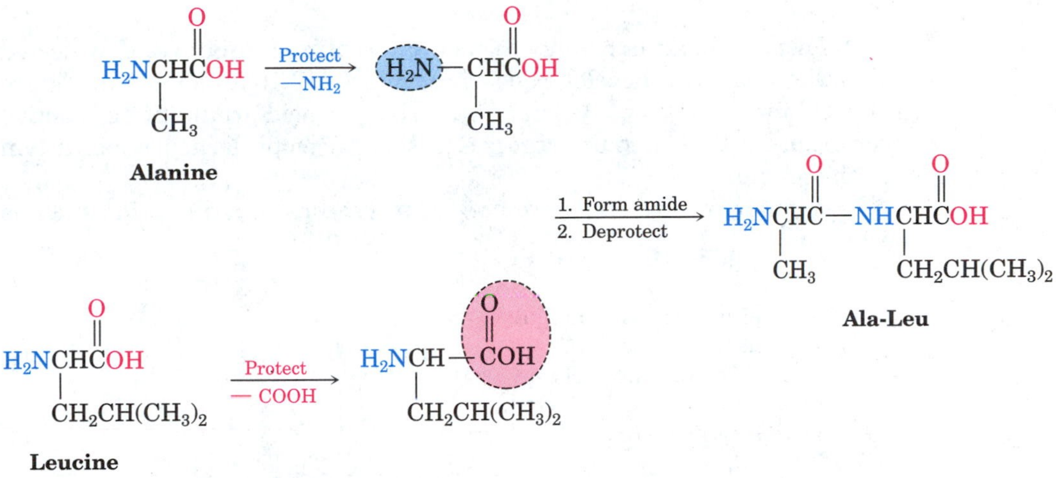

Many different amino- and carboxyl-protecting groups have been devised, but only a few are widely used. Carboxyls are often protected simply by converting them into methyl or benzyl esters. Both groups are easily introduced by standard methods of ester formation and are easily removed by mild hydrolysis with aqueous NaOH. Benzyl esters can also be cleaved by catalytic hydrogenolysis of the weak benzylic C–O bond ($ArCH_2–OCOR + H_2 \rightarrow ArCH_3 + RCOOH$).

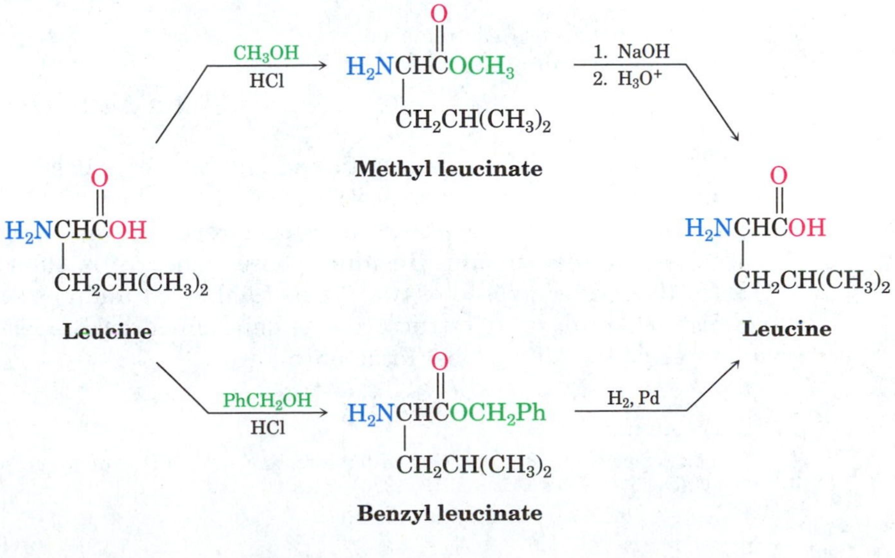

Amino groups are often protected as their *tert*-butoxycarbonyl amide (BOC) derivatives. The BOC protecting group is easily introduced by reaction of the amino acid with di-*tert*-butyl dicarbonate in a nucleophilic acyl substitution reaction (Section 21.6) and is removed by brief treatment with a strong acid such as trifluoroacetic acid, CF_3COOH.

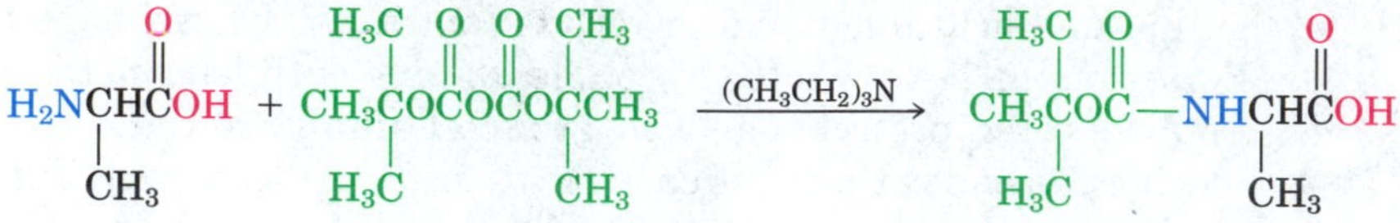

Alanine **Di-*tert*-butyl dicarbonate** **BOC–Ala**

The peptide bond is usually formed by treating a mixture of protected acid and amine with dicyclohexylcarbodiimide (DCC). As shown in Figure 27.8, DCC functions by converting the carboxylic acid group into a reactive acylating agent that then undergoes a further nucleophilic acyl substitution with the amine.

To summarize, five steps are needed to synthesize a dipeptide such as Ala-Leu:

1.–2. The amino group of alanine is protected as the BOC derivative, and the carboxyl group of leucine is protected as the methyl ester.

3. The two protected amino acids are coupled using DCC.

4. The BOC protecting group is removed by acid treatment.

5. The methyl ester is removed by basic hydrolysis.

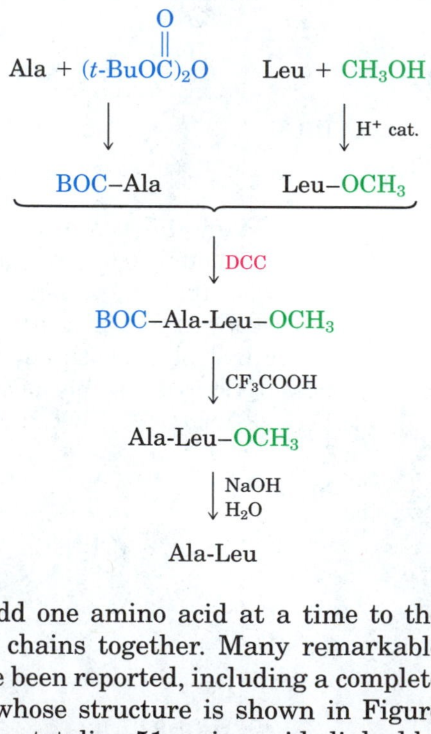

These steps can be repeated to add one amino acid at a time to the growing chain or to link two peptide chains together. Many remarkable achievements in peptide synthesis have been reported, including a complete synthesis of human insulin. Insulin, whose structure is shown in Figure 27.9 (p. 1080), is composed of two chains totaling 51 amino acids linked by two disulfide bridges. Its structure was determined by Frederick Sanger,[5] who received the 1958 Nobel Prize for his work.

---

[5]Frederick Sanger (1918–); b. Gloucestershire, England; Ph.D. Cambridge; professor, Cambridge University; Nobel Prize (1958, 1980).

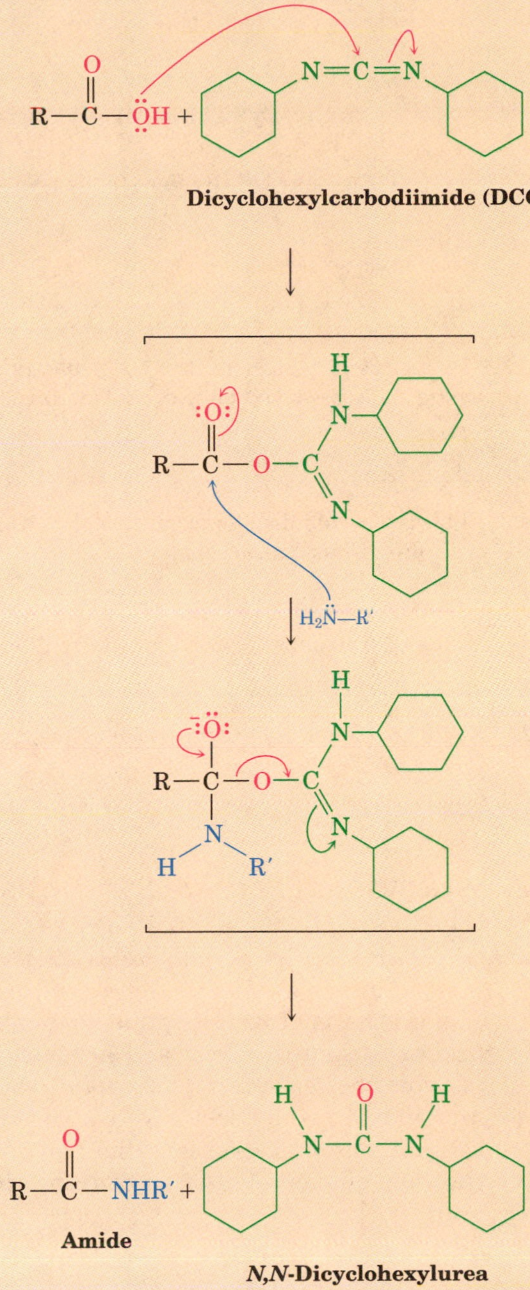

The carboxylic acid first adds to the carbodiimide reagent to yield a reactive acylating agent.

Nucleophilic attack of the amine on the acylating agent gives a tetrahedral intermediate.

The intermediate loses dicyclohexylurea and yields the desired amide.

Figure 27.8 The mechanism of amide formation by reaction of a carboxylic acid and an amine with DCC (dicyclohexylcarbodiimide).

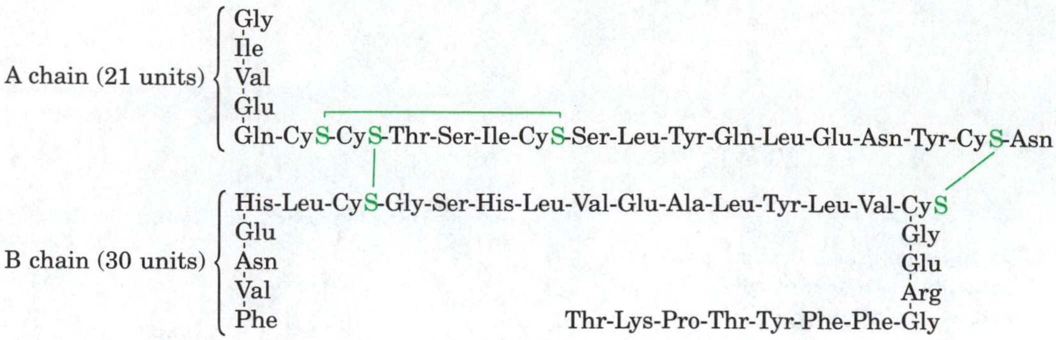

Figure 27.9 Structure of human insulin. Two separate chains totaling 51 amino acids are linked by two disulfide bridges.

PROBLEM.

27.23 Show the mechanism for formation of a BOC derivative by reaction of an amino acid with di-*tert*-butyl dicarbonate.

PROBLEM.

27.24 Write all five steps required for the synthesis of Leu-Ala from alanine and leucine.

PROBLEM.

27.25 Show how you would prepare the following tripeptides:
(a) Leu-Ala-Gly (b) Gly-Leu-Ala

27.12 Automated Peptide Synthesis: The Merrifield Solid-Phase Technique

The synthesis of large peptide chains by sequential addition of one amino acid at a time is a long and arduous task. An immense simplification is possible, however, using the *solid-phase* method introduced by R. Bruce Merrifield[6] at the Rockefeller University. In the Merrifield method, peptide synthesis is carried out on solid polymer beads of polystyrene, prepared so that one of every 100 or so benzene rings bears a chloromethyl ($-CH_2Cl$) group:

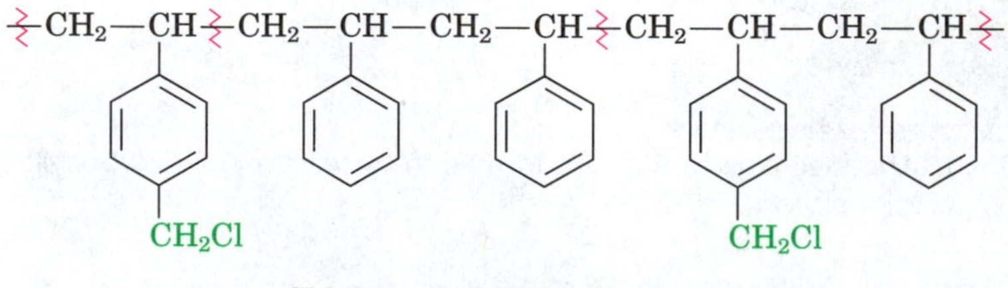

Chloromethylated polystyrene

[6]Robert Bruce Merrifield (1921–); Ph.D. University of California, Los Angeles (1949); professor, Rockefeller Institute; Nobel Prize (1984).

In the standard solution-phase method discussed in the previous section, a methyl ester was used to protect the carboxyl group during formation of the amide bond. In the solid-phase method, however, a solid *polymer* particle is the ester protecting group. Four steps are required in solid-phase peptide synthesis:

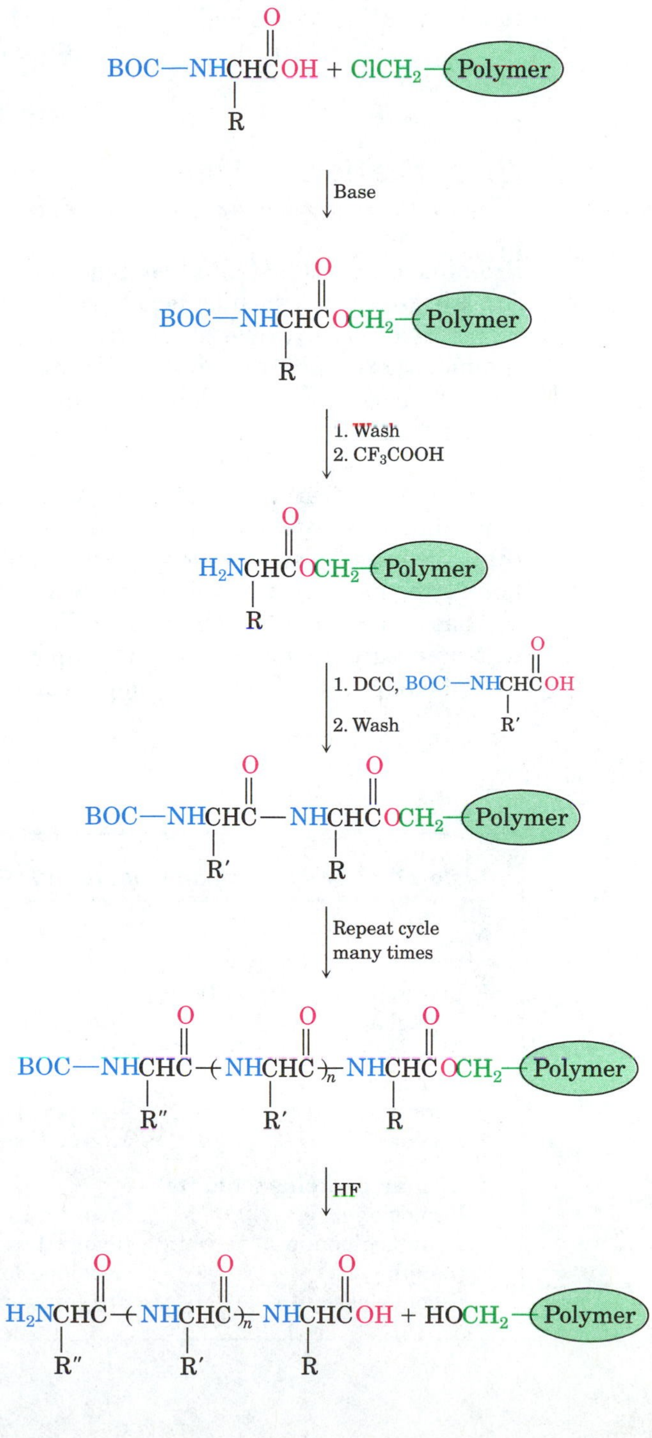

1. A BOC-protected amino acid is covalently linked to the polystyrene polymer by formation of an ester bond (S_N2 reaction).

2. The polymer-bonded amino acid is washed free of excess reagent and then treated with trifluoroacetic acid to remove the BOC group.

3. A second BOC-protected amino acid is coupled to the first by reaction with DCC. Excess reagents are removed by washing them from the insoluble polymer.

The cycle of deprotection, coupling, and washing is repeated as many times as desired to add amino acid units to the growing chain.

4. After the desired peptide has been made, treatment with anhydrous HF removes the final BOC group and cleaves the ester bond to the polymer, yielding the free peptide.

The solid-phase technique has been automated, and computer-controlled peptide synthesizers are available for automatically repeating the coupling and deprotection steps with different amino acids as many times as desired. Each step occurs in extremely high yield, and mechanical losses are minimized because the peptide intermediates are never removed from the insoluble polymer until the final step. Among the many remarkable achievements recorded by Merrifield is the synthesis of bovine pancreatic ribonuclease, a protein containing 124 amino acid units. The entire synthesis required only 6 weeks and took place in 17% overall yield.

27.13 Classification of Proteins

Proteins are classified into two major types according to their composition. **Simple proteins**, such as blood serum albumin, are those that yield only amino acids on hydrolysis. **Conjugated proteins**, which are much more common than simple proteins, yield other compounds such as carbohydrates, fats, or nucleic acids in addition to amino acids on hydrolysis.

Another way to classify proteins is as either *fibrous* or *globular*, according to their three-dimensional shape. **Fibrous proteins**, such as collagen and keratin, consist of polypeptide chains arranged side by side in long filaments. Because these proteins are tough and insoluble in water, they are used in nature for structural materials such as tendons, hooves, horns, and muscles. **Globular proteins**, by contrast, are usually coiled into compact, roughly spherical shapes. These proteins are generally soluble in water and are mobile within cells. Most of the 2000 or so known enzymes are globular. Table 27.2 lists some common examples of both kinds.

Table 27.2 Some Common Fibrous and Globular Proteins

Name	*Occurrence and use*
Fibrous proteins (insoluble)	
Collagens	Animal hide, tendons, connective tissues
Elastins	Blood vessels, ligaments
Fibrinogen	Necessary for blood clotting
Keratins	Skin, wool, feathers, hooves, silk, fingernails
Myosins	Muscle tissue
Globular proteins (soluble)	
Hemoglobin	Involved in oxygen transport
Immunoglobulins	Involved in immune response
Insulin	Hormone for controlling glucose metabolism
Ribonuclease	Enzyme controlling RNA synthesis

27.14 Protein Structure

Proteins are so large that the word *structure* takes on a broader meaning than it does with most other organic compounds. In fact, chemists speak of four different levels of structure when describing proteins. At its simplest, the structure of a protein is the sequence in which amino acids are bound together. Called the **primary structure** of a protein, this is the most fundamental structural level.

There is much more to protein structure than just amino acid sequence. The chemical properties of a protein are also dependent on higher levels of structure—on exactly how the peptide backbone folds to give the molecule a specific three-dimensional shape. Thus, the term **secondary structure** refers to the way in which *segments* of the peptide backbone orient into a regular pattern; **tertiary structure** refers to the way in which the entire protein molecule coils into an overall three-dimensional shape; and **quaternary structure** refers to the way in which several protein molecules come together to yield large aggregate structures.

Let's look at three examples—α-keratin (fibrous), fibroin (fibrous), and myoglobin (globular)—to see how higher structure affects protein properties.

α-Keratin

α-Keratin is the fibrous structural protein found in wool, hair, nails, and feathers. Studies show that α-keratin is coiled into a right-handed helical secondary structure like that of a telephone cord. Illustrated in Figure 27.10 (p. 1084), this so-called **α-helix** is stabilized by hydrogen bonding between amide N–H groups and C=O groups four residues away. Although the strength of a typical hydrogen bond (about 20 kJ/mol) is only about 5% the strength of a C–C or C–H covalent bond, the large number of hydrogen bonds made possible by helical winding imparts a great deal of stability to the α-helical structure. Each coil of the helix contains 3.6 amino acids, and the distance between coils (the *repeat distance*) is 540 pm, or 5.40 Å.

Although α-keratin is the best example of an almost entirely helical protein, most globular proteins contain α-helical *segments*. Both hemoglobin and myoglobin, for example, contain many short helical sections in their chains.

Fibroin

Fibroin, the fibrous protein found in silk, has a secondary structure called a **β-pleated sheet**. In this pleated-sheet structure, polypeptide chains line up in a parallel arrangement held together by hydrogen bonds between chains (Figure 27.11, p. 1085). Although not as common as the α-helix, small β-pleated-sheet regions are often found in proteins where sections of peptide chains double back on themselves.

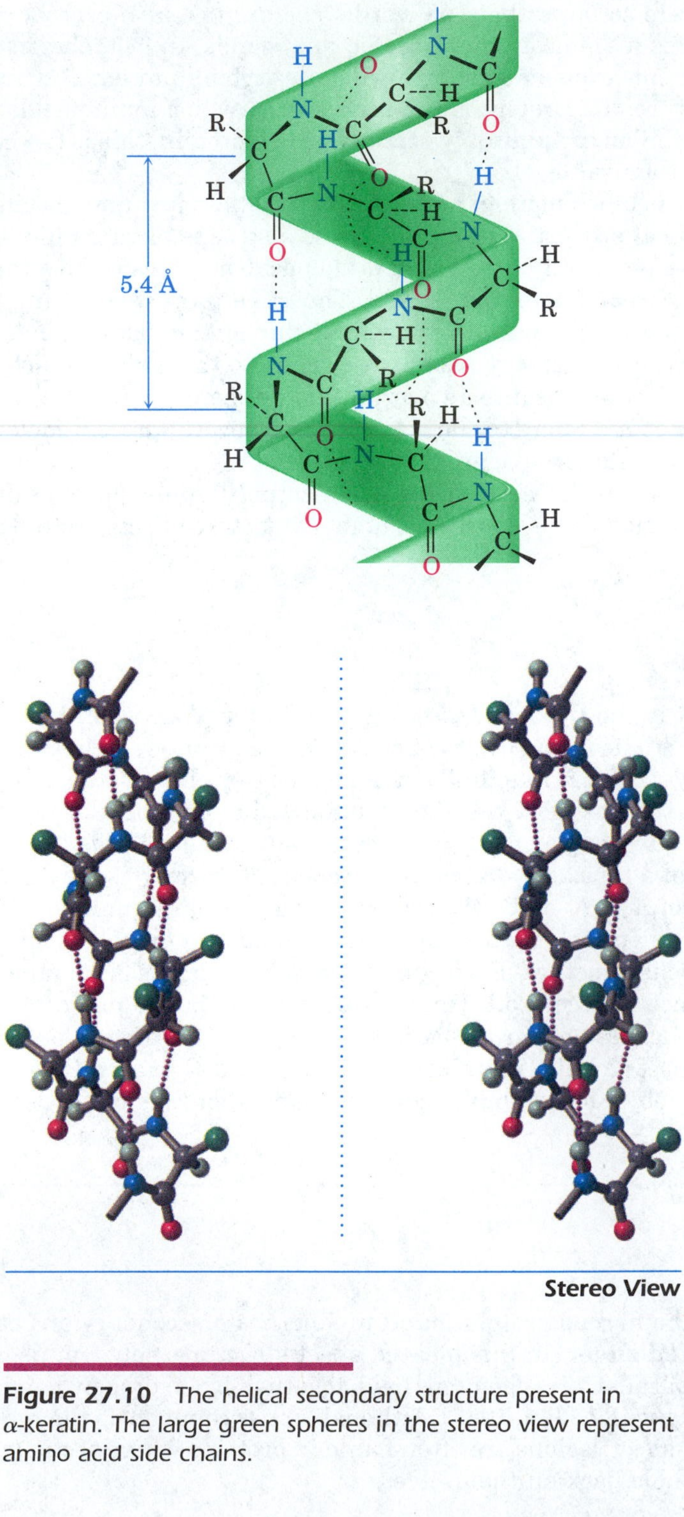

Figure 27.10 The helical secondary structure present in α-keratin. The large green spheres in the stereo view represent amino acid side chains.

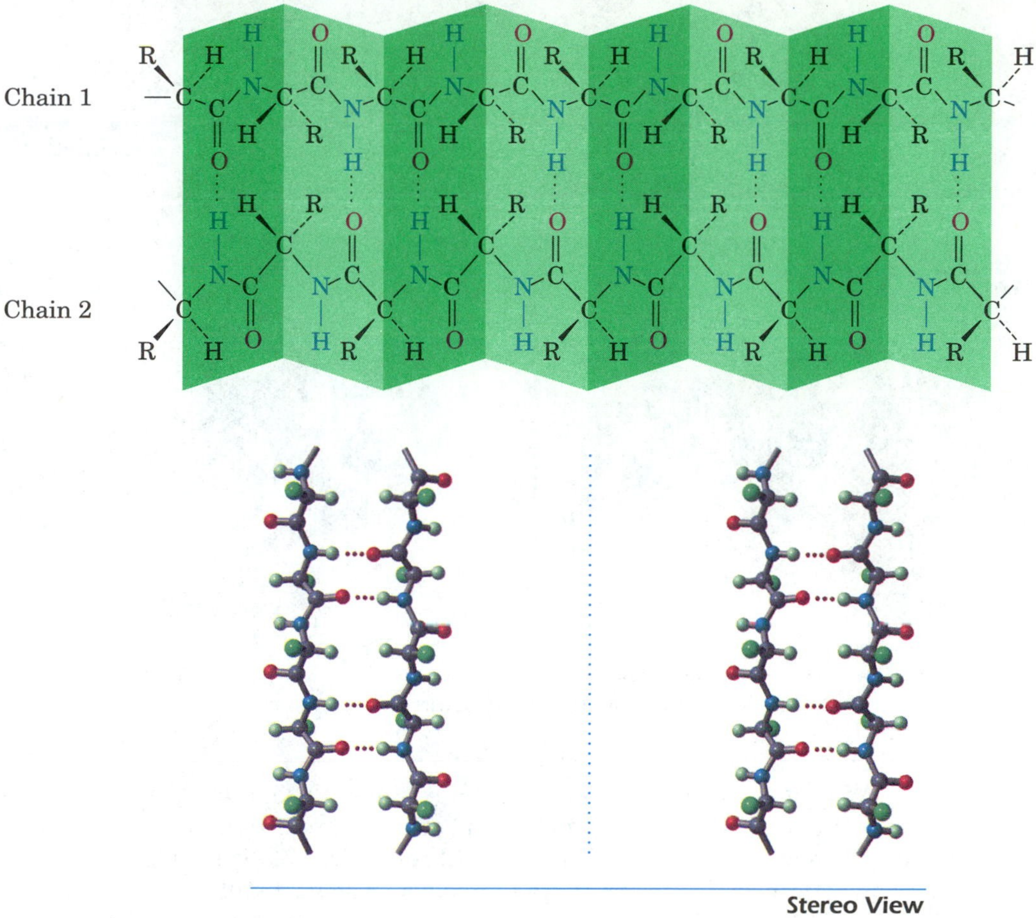

Figure 27.11 The β-pleated-sheet structure present in silk fibroin. The large green spheres in the stereo view represent amino acid side chains.

Myoglobin

Myoglobin is a small globular protein containing 153 amino acids in a single chain. A relative of hemoglobin, myoglobin is found in the skeletal muscles of sea mammals, where it stores oxygen needed to sustain the animals during long dives. X-ray evidence obtained by Sir John Kendrew[7] and Max Perutz[8] has shown that myoglobin consists of eight helical segments connected by bends to form a compact, nearly spherical, tertiary structure (Figure 27.12, p. 1086).

[7]Sir John C. Kendrew (1917–); b. Oxford, England; Ph.D. Cambridge; professor, Cambridge University; Nobel Prize (1962).

[8]Max Ferdinand Perutz (1914–); b. Vienna; universities of Vienna and Cambridge; professor, Cambridge University; Nobel Prize (1962).

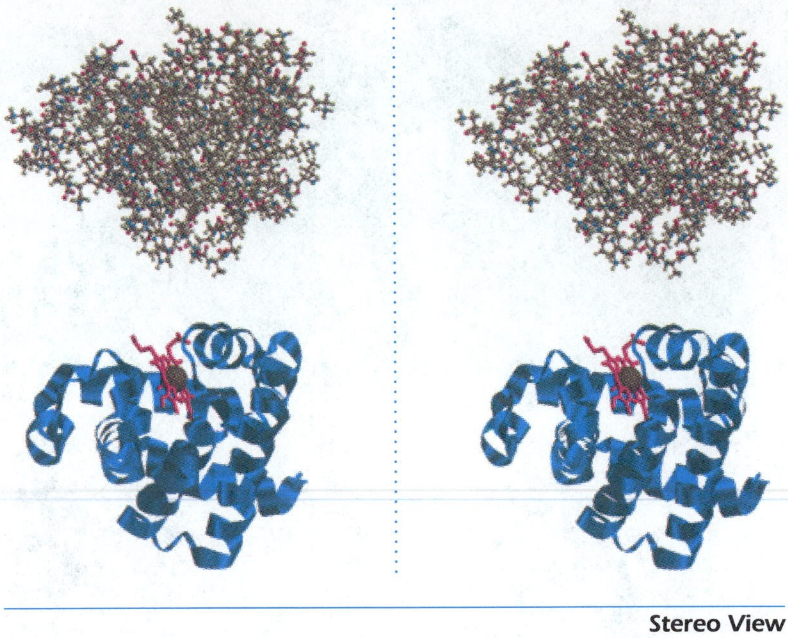

Figure 27.12 Secondary and tertiary structure of myoglobin, a globular protein.

Why does myoglobin adopt the shape it does? The forces that determine the tertiary structure of myoglobin and other globular proteins are the same simple forces that act on all molecules, regardless of size. By bending and twisting in exactly the right way, myoglobin achieves maximum stability. Although the bends appear irregular and the three-dimensional structure appears random, this isn't the case. All myoglobin molecules adopt this same shape because it is the most stable.

Particularly important among the forces stabilizing a protein's tertiary structure are the hydrophobic (water-repelling) interactions of hydrocarbon side chains on neutral amino acids. Those amino acids with neutral, nonpolar side chains have a strong tendency to congregate on the hydrocarbon-like interior of a protein molecule, away from the aqueous medium. Those acidic or basic amino acids with charged side chains, by contrast, tend to congregate on the exterior of the protein where they can be solvated by water.

Also important for stabilizing a protein's tertiary structure are the formation of disulfide bridges between cysteine residues, the formation of hydrogen bonds between nearby amino acids, and the development of ionic attractions, called *salt bridges*, between positively and negatively charged sites on various amino acid side chains within the protein.

Note that myoglobin is a conjugated protein that contains a covalently bound organic group (a **prosthetic group**) called *heme*. A great many proteins contain such prosthetic groups, which are crucial to their mechanism of action.

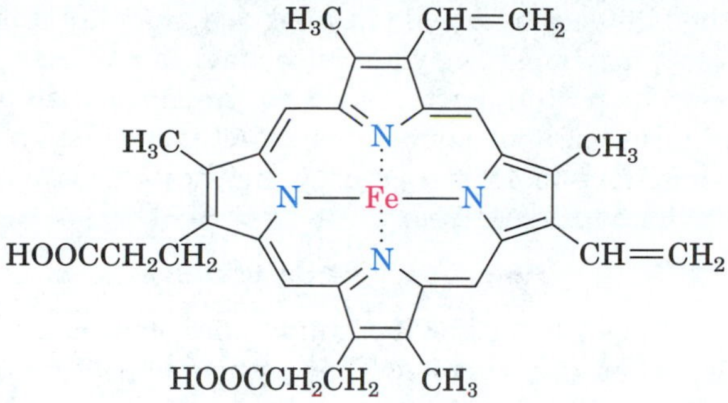

Heme

PROBLEM.

27.26 How can you account for the fact that proline is never present in a protein α-helix? The α-helical parts of myoglobin and other proteins stop whenever a proline residue is encountered in the chain.

27.15 Enzymes

Enzymes are large proteins that act as catalysts for biological reactions. Unlike many of the simple catalysts that chemists use in the laboratory, enzymes are usually specific in their action. Often, in fact, an enzyme can catalyze only a single reaction of a single compound, called the enzyme's *substrate*. For example, the enzyme amylase found in the human digestive tract catalyzes only the hydrolysis of starch to yield glucose; cellulose and other polysaccharides are untouched by amylase.

Different enzymes have different specificities. Some, such as amylase, are specific for a single substrate, but others operate on a range of substrates. Papain, for instance, a globular protein of 212 amino acids isolated from papaya fruit, catalyzes the hydrolysis of many kinds of peptide bonds. In fact, it's this ability to hydrolyze peptide bonds that makes papain useful as a meat tenderizer and a cleaner for contact lenses.

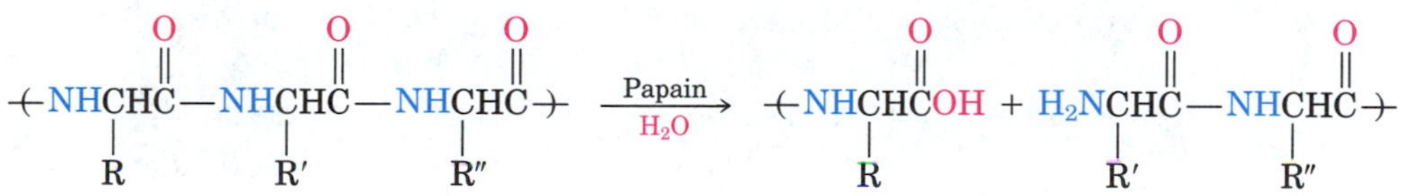

Like all catalysts, enzymes don't affect the equilibrium constant of a reaction and can't bring about chemical changes that are otherwise unfavorable. Enzymes act only to lower the activation energy for a reaction, thereby making the reaction take place more rapidly. Starch and water, for example, react very slowly in the absence of a catalyst because the activation energy is too high. When amylase is present, however, the energy barrier is lowered, and the hydrolysis reaction occurs rapidly.

All the 2000 or so known enzymes are globular proteins. In addition to the protein part, most enzymes also have small nonprotein parts called **cofactors**. The protein part in such enzymes is called an **apoenzyme**, and the combination of apoenzyme plus cofactor is called a **holoenzyme**. Only holoenzymes have biological activity; neither cofactor nor apoenzyme catalyze reactions by themselves.

Holoenzyme = Cofactor + Apoenzyme

Cofactors can be either inorganic ions, such as Zn^{2+}, or small organic molecules, called **coenzymes**. The requirement of many enzymes for inorganic cofactors is the main reason for our dietary need of trace minerals. Iron, zinc, copper, manganese, and numerous other metal ions are all essential minerals that act as enzyme cofactors, though the exact biological role isn't known in all cases.

A variety of organic molecules act as coenzymes. Many, though not all, coenzymes are **vitamins**, small organic molecules that must be obtained in the diet and are required in trace amounts for proper growth. Table 27.3 lists the 13 known vitamins required in the human diet and their enzyme functions.

Table 27.3 Vitamins and Their Enzyme Functions

Vitamin	*Enzyme function*	*Deficiency symptom*
Water-Soluble Vitamins		
Ascorbic acid (Vitamin C)	Hydroxylases	Bleeding gums, bruising
Thiamin (Vitamin B_1)	Reductases	Fatigue, depression
Riboflavin (Vitamin B_2)	Reductases	Cracked lips, scaly skin
Pyridoxine (Vitamin B_6)	Aminotransferases	Anemia, irritability
Niacin	Reductases	Dermatitis, dementia
Folic acid (Vitamin M)	Methyltransferases	Megaloblastic anemia
Vitamin B_{12}	Isomerases	Megaloblastic anemia, neurodegeneration
Pantothenic acid	Acyltransferases	Weight loss, irritability
Biotin (Vitamin H)	Carboxylases	Dermatitis, anorexia, depression
Fat-Soluble Vitamins		
Vitamin A	Visual system	Night blindness, dry skin
Vitamin D	Calcium metabolism	Rickets, osteomalacia
Vitamin E	Antioxidant	Hemolysis of red blood cells
Vitamin K	Blood clotting	Hemorrhage, delayed blood clotting

Enzymes are grouped into six classes according to the kind of reaction they catalyze (Table 27.4). *Hydrolases* catalyze hydrolysis reactions; *isomerases* catalyze isomerizations; *ligases* catalyze the bonding together of two molecules; *lyases* catalyze the breaking away of a small molecule such as H_2O from a substrate; *oxidoreductases* catalyze oxidations and reductions; and *transferases* catalyze the transfer of a group from one substrate to another.

Table 27.4 Classification of Enzymes

Main class	*Some subclasses*	*Type of reaction catalyzed*
Hydrolases	Lipases	Hydrolysis of an ester group
	Nucleases	Hydrolysis of a phosphate group
	Proteases	Hydrolysis of an amide group
Isomerases	Epimerases	Isomerization of stereogenic center
Ligases	Carboxylases	Addition of CO_2
	Synthetases	Formation of new bond
Lyases	Decarboxylases	Loss of CO_2
	Dehydrases	Loss of H_2O
Oxidoreductases	Dehydrogenases	Introduction of double bond by removal of H_2
	Oxidases	Oxidation
	Reductases	Reduction
Transferases	Kinases	Transfer of a phosphate group
	Transaminases	Transfer of an amino group

Although some enzymes, like papain and trypsin, have uninformative common names, the systematic name of an enzyme has two parts, ending with *-ase*. The first part identifies the enzyme's substrate, and the second part identifies its class. For example, *hexose kinase* is an enzyme that catalyzes the transfer of a phosphate group from adenosine triphosphate (ATP) to glucose.

PROBLEM

27.27 To what classes do the following enzymes belong?
(a) Pyruvate decarboxylase
(b) Chymotrypsin
(c) Alcohol dehydrogenase

27.16 Protein Denaturation

The tertiary structure of a globular protein is delicately held together by weak intramolecular attractions. Often, a modest change in temperature or pH will disrupt the tertiary structure and cause the protein to become **denatured**. Denaturation occurs under such mild conditions that covalent bonds aren't affected; the polypeptide primary structure remains intact, but the tertiary structure unfolds from a well-defined spherical shape to a randomly looped chain (Figure 27.13).

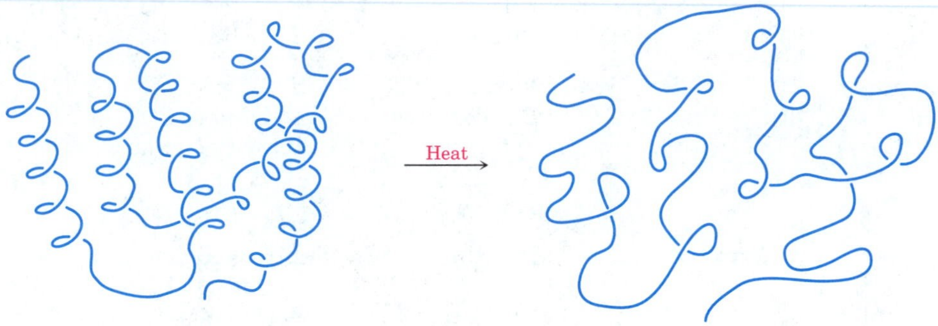

Figure 27.13 Schematic representation of protein denaturation. A globular protein loses its specific three-dimensional shape and becomes randomly looped.

Denaturation is accompanied by changes in both physical and biological properties. Solubility is drastically decreased, as occurs when egg white is cooked and the albumins unfold and coagulate to an insoluble white mass. Most enzymes also lose all catalytic activity when denatured, since a precisely defined tertiary structure is required for their action.

Most, but not all, denaturation is irreversible. Eggs don't become uncooked when their temperature is lowered, and curdled milk doesn't become homogeneous. Many cases have now been found, however, where spontaneous **renaturation** of an unfolded protein occurs. Renaturation is accompanied by a full recovery of biological activity in the case of enzymes, indicating that the protein has completely returned to its stable tertiary structure.

INTERLUDE

Protein and Nutrition

Dietary protein is necessary to build the body mass of these athletes.

Everyone—from infants to weightlifters—needs protein. Children need large amounts of protein for proper growth, and adults need protein to replace what is lost each day by the body's normal biochemical reactions. Dietary protein is necessary because our bodies can synthesize only 10 of the 20 common amino acids from simple precursor molecules; the other 10 amino acids must be obtained from food by digestion of edible proteins. Table 27.5 shows the estimated essential amino acid requirements of an infant and an adult.

Table 27.5 Estimated Essential Amino Acid Requirements

	Daily requirement (mg/kg body weight)	
Amino acid	*Infant*	*Adult*
Arginine	?	?
Histidine	33	?
Isoleucine	83	12
Leucine	35	16
Lysine	99	12
Methionine (+ Cysteine)	49	10
Phenylalanine (+ Tyrosine)	141	16
Threonine	68	8
Tryptophan	21	3
Valine	92	14

Not all foods provide sufficient amounts of the 10 essential amino acids to meet our minimum daily needs. Most meat and dairy products are satisfactory, but many vegetable sources, such as wheat and corn, are

(continued)➤

incomplete; that is, many vegetable proteins contain too little of one or more essential amino acids to sustain the growth of laboratory animals. Wheat is low in lysine, for example, and corn is low in both lysine and tryptophan.

Using an incomplete food as the sole source of protein can cause nutritional deficiencies, particularly in children. Vegetarians must therefore be careful to adopt a varied diet that provides proteins from several sources. Legumes and nuts, for example, are useful for overcoming the deficiencies of wheat and grains. Some of the limiting amino acids found in various foods are listed in Table 27.6.

Table 27.6 Limiting Amino Acids in Some Foods

Food	*Limiting amino acid*
Wheat, grains	Lysine, threonine
Peas, beans, legumes	Methionine, tryptophan
Nuts, seeds	Lysine
Leafy green vegetables	Methionine

Summary and Key Words

Proteins are large biomolecules made up of **α-amino acid residues** linked together by amide, or **peptide**, bonds. Twenty amino acids are commonly found in proteins; all are α-amino acids and all except glycine have stereochemistry similar to that of L sugars.

Amino acids can be synthesized by several methods, including ammonolysis of α-bromo acids, reductive amination of α-keto acids, **Strecker reaction** of aldehydes with KCN/NH_4Cl followed by hydrolysis, and alkylation of diethyl acetamidomalonate. Resolution of the synthetic racemate then provides the optically active amino acid.

Determining the structure of a large polypeptide or protein is carried out in several steps. The identity and amount of each amino acid present in a peptide is determined by **amino acid analysis**. The peptide is first hydrolyzed to its constituent α-amino acids, which are then separated and identified. Next, the peptide is **sequenced**. **Edman degradation** by treatment with phenyl isothiocyanate cleaves one residue from the N terminus of the peptide and forms an easily identifiable derivative of the N-terminal amino acid. A series of sequential Edman degradations allows the sequencing of a peptide chain up to 25 residues in length.

Peptide synthesis is made possible by the use of selective protecting groups. An N-protected amino acid with a free carboxyl group is coupled to an O-protected amino acid with a free amino group in the presence of dicyclohexylcarbodiimide (DCC). Amide formation occurs, the protecting groups are removed, and the sequence is repeated. Amines are usually protected as their *tert*-butoxycarbonyl (BOC) derivatives, and acids are protected as esters. This synthetic sequence is often carried out by the **Merrifield solid-phase technique**, in which the peptide is esterified to an insoluble polymeric support.

Proteins are classified as either globular or fibrous, depending on their **secondary** and **tertiary structures**. **Fibrous proteins** such as α-keratin are tough, rigid, and water-insoluble; **globular proteins** such as myoglobin are water-soluble and roughly spherical in shape. Most of the 2000 or so known enzymes are globular proteins.

Summary of Reactions

1. Amino acid synthesis
   (a) From α-bromo acids (Section 27.4)

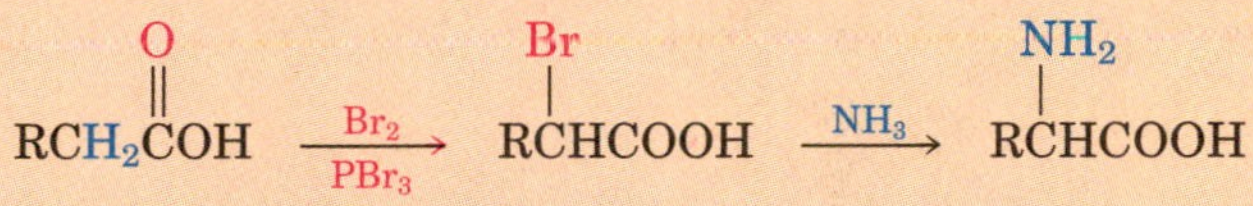

   (b) Strecker synthesis (Section 27.4)

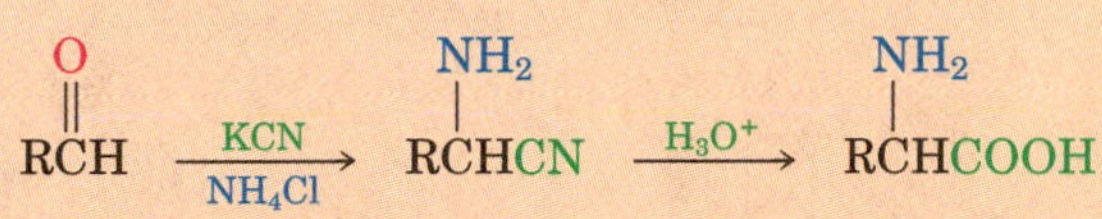

   (c) Reductive amination (Sections 24.6 and 27.4)

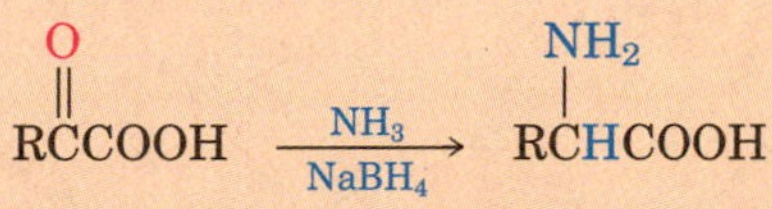

   (d) Diethyl acetamidomalonate synthesis (Sections 22.8 and 27.4)

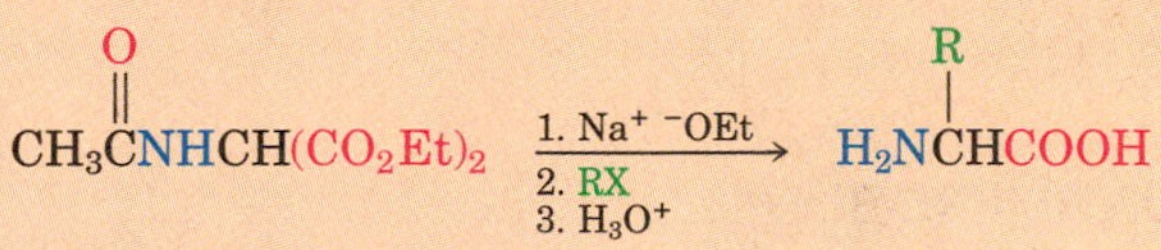

2. Peptide synthesis
   (a) Nitrogen protection (Section 27.11)

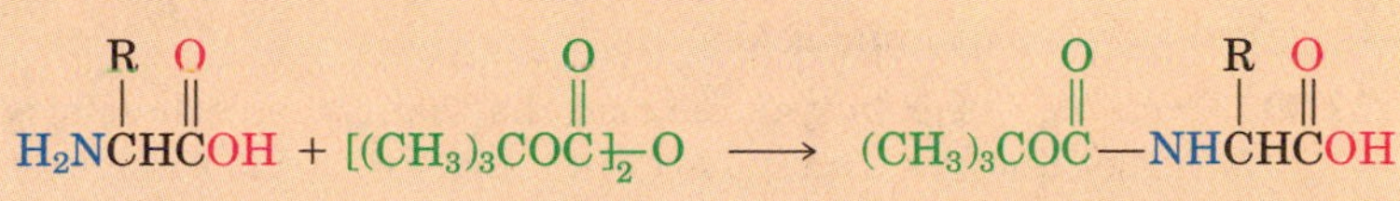

BOC-protected amino acid

(continued)➤

The BOC protecting group can be removed by acid treatment:

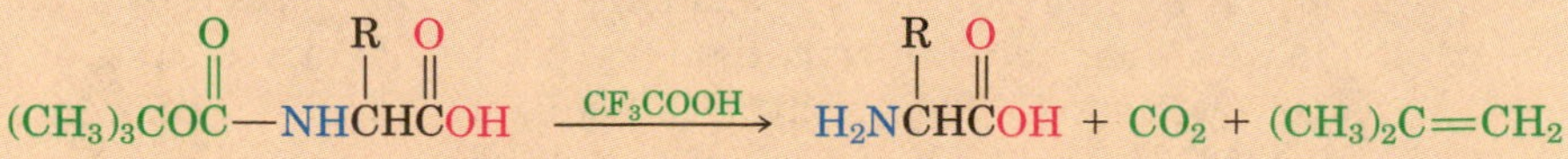

(b) Oxygen protection (Section 27.11)

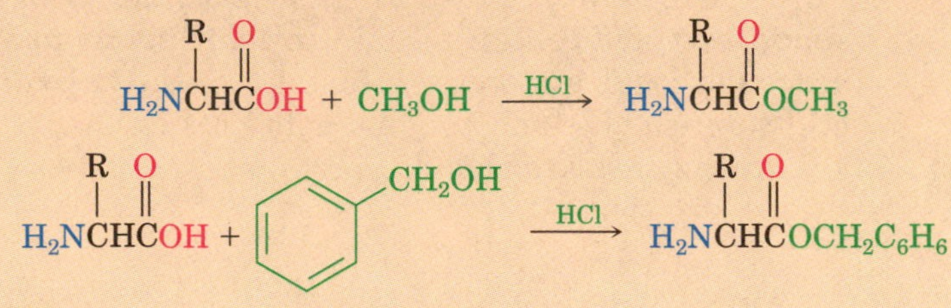

The ester protecting group can be removed by base hydrolysis:

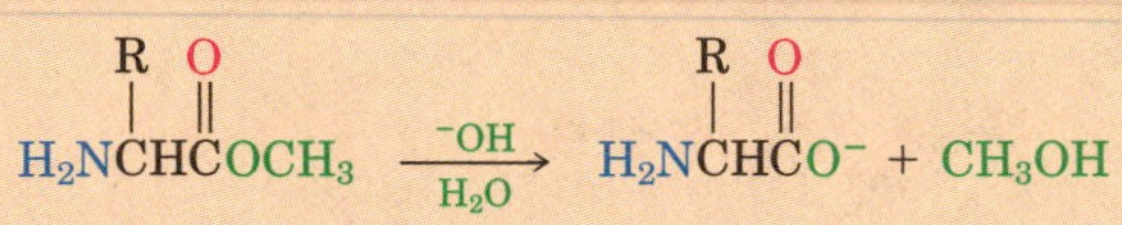

(c) Amide bond formation (Section 27.11)

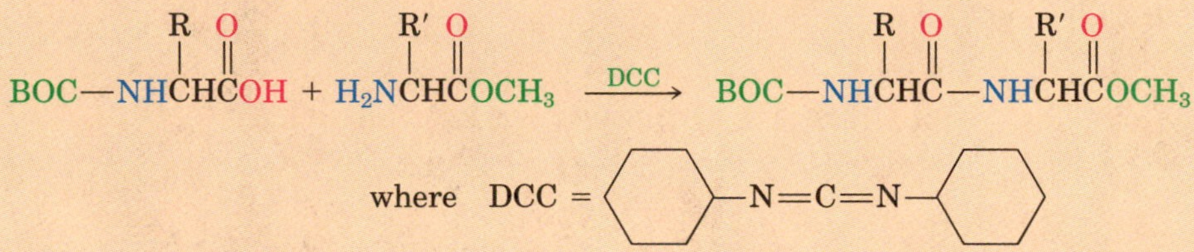

3. Peptide sequencing: Edman degradation (Section 27.9)

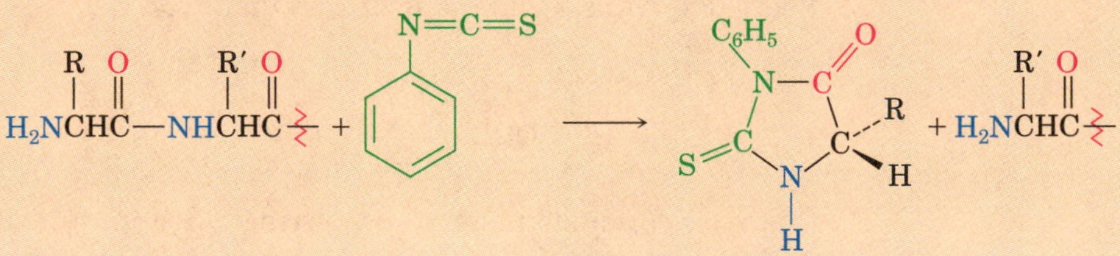

ADDITIONAL PROBLEMS

27.28 Except for cysteine, only *S* amino acids occur in proteins. Several *R* amino acids are also found in nature, however. (*R*)-Serine is found in earthworms, and (*R*)-alanine is found in insect larvae. Draw Fischer projections of (*R*)-serine and (*R*)-alanine. Are these D or L amino acids?

27.29 Cysteine is the only amino acid that has L stereochemistry but an *R* configuration. Design another L amino acid that also has an *R* configuration.

27.30 Draw a Fischer projection of (*S*)-proline.

27.31 Show the structures of the following amino acids:
(a) Trp (b) Ile (c) Cys (d) His

27.32 Draw the following amino acids in their zwitterionic forms:
(a) Tyrosine (b) Threonine

27.33 Explain the observation that amino acids exist as dipolar zwitterions in aqueous solution but exist largely as amino carboxylic acids in chloroform solution.

$$\overset{+}{H_3N}CH(R)CO_2^- \qquad H_2NCH(R)COOH$$

In H_2O In $CHCl_3$

27.34 At what pH would you carry out an electrophoresis experiment if you wanted to separate a mixture of histidine, serine, and glutamic acid? Explain.

27.35 Define the following terms:
(a) Amphoteric (b) Isoelectric point (c) Zwitterion

27.36 Proline has $pK_{a1} = 1.99$ and $pK_{a2} = 10.60$. Use the Henderson–Hasselbalch equation to calculate the ratio of protonated and neutral forms at pH = 2.50. Calculate the ratio of neutral and deprotonated forms at pH = 9.70.

27.37 Using the three-letter code names for amino acids, write the structures of all possible peptides containing the following amino acids:
(a) Val, Ser, Leu (b) Ser, Leu_2, Pro

27.38 Cytochrome *c* is an enzyme found in the cells of all aerobic organisms. Elemental analysis of cytochrome *c* shows that it contains 0.43% iron. What is the minimum molecular weight of this enzyme?

27.39 Predict the product of the reaction of valine with the following reagents:
(a) CH_3CH_2OH, acid (b) Di-*tert*-butyl dicarbonate
(c) KOH, H_2O (d) CH_3COCl, pyridine; then H_2O

27.40 Show how you could use the Strecker synthesis to prepare these amino acids:
(a) Glycine (b) Valine

27.41 Show how you could use the acetamidomalonate method to prepare these amino acids:
(a) Leucine (b) Tryptophan

27.42 Show how you could prepare the following amino acids using a reductive amination:
(a) Methionine (b) Isoleucine

27.43 Write full structures for the following peptides:
(a) Val-Phe-Cys-Ala (b) Glu-Pro-Ile-Leu

27.44 Show the steps involved in a synthesis of Phe-Ala-Val using the Merrifield procedure.

27.45 Draw the structure of the phenylthiohydantoin product you would obtain by Edman degradation of the following peptides.
(a) Ile-Leu-Pro-Phe (b) Asp-Thr-Ser-Gly-Ala

27.46 The chloromethylated polystyrene resin used for Merrifield solid-phase peptide synthesis is prepared by treatment of polystyrene with chloromethyl methyl ether and a Lewis acid catalyst. Propose a mechanism for the reaction.

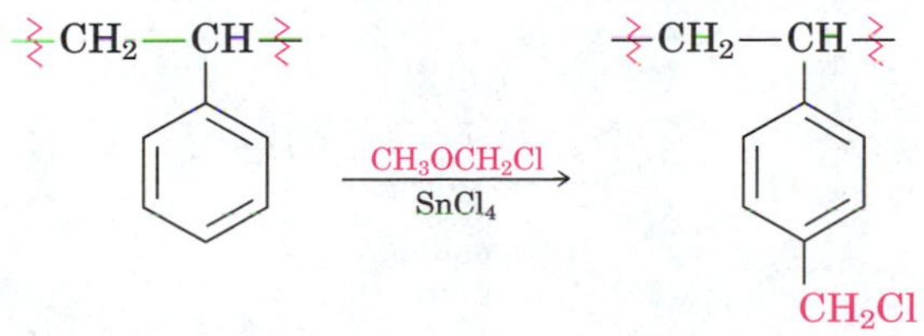

27.47 Which amide bonds in the following polypeptide are cleaved by trypsin? By chymotrypsin?

Phe-Leu-Met-Lys-Tyr-Asp-Gly-Gly-Arg-Val-Ile-Pro-Tyr

27.48 How can you account for the fact that tryptophan has a lower isoelectric point than histidine, even though both have amine nitrogen atoms in five-membered rings? Which nitrogen in the five-membered ring of histidine is more basic? Explain. (See Section 15.9.)

27.49 The *Sanger end-group determination* is sometimes used as an alternative to the Edman degradation. In the Sanger method, a peptide is allowed to react with 2,4-dinitrofluorobenzene, the peptide is hydrolyzed, and the N-terminal amino acid is identified by separation as its *N*-2,4-dinitrophenyl derivative. Propose a mechanism to account for the initial reaction between peptide and dinitrofluorobenzene.

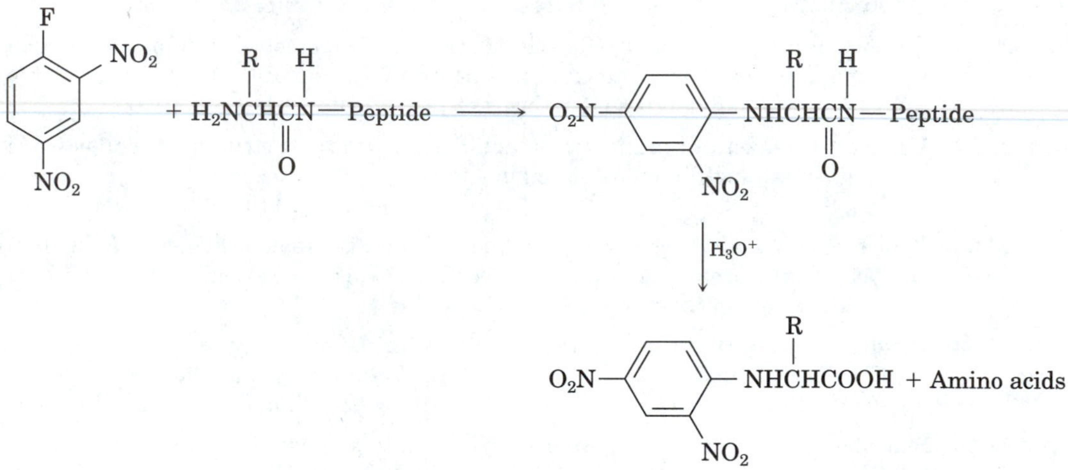

27.50 Do you foresee any problems in using the Sanger end-group determination method (Problem 27.49) on a peptide such as Gly-Pro-Lys-Ile? Explain.

27.51 When α-amino acids are treated with dicyclohexylcarbodiimide (DCC), 2,5-diketopiperazines result. Propose a mechanism.

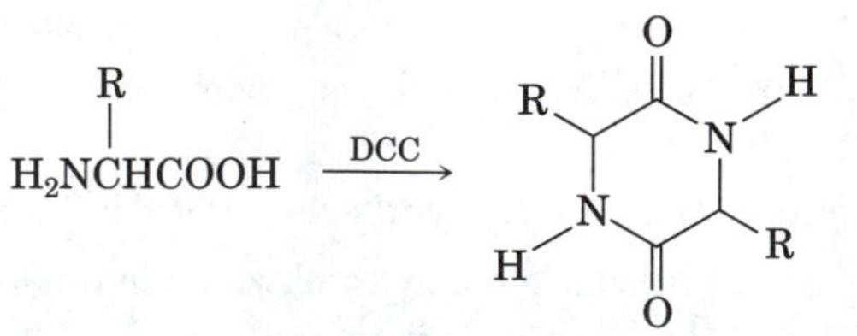

A 2,5-diketopiperazine

27.52 Arginine, which contains a guanidine functional group in its side chain, is the most basic of the 20 common amino acids. Explain using resonance structures to show how the protonated guanidino group is stabilized.

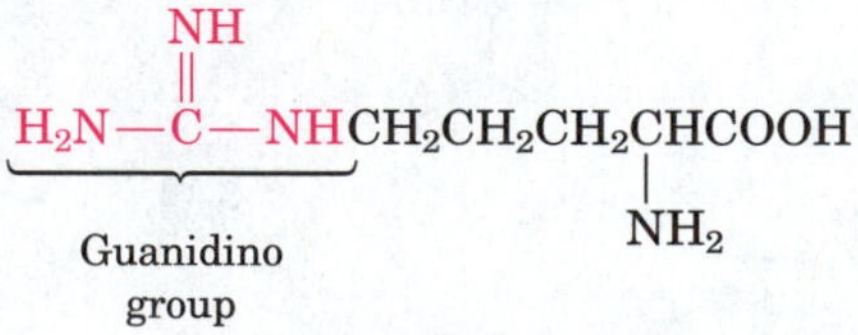

Arginine

27.53 Evidence for restricted rotation around amide CO–N bonds comes from NMR studies. At room temperature, the ^1H NMR spectrum of *N,N*-dimethylformamide shows three peaks: 2.9 δ (singlet, 3 H), 3.0 δ (singlet, 3 H), 8.0 δ (singlet, 1 H). As the temperature is raised, however, the two singlets at 2.9 δ and 3.0 δ slowly merge. At 180°C, the ^1H NMR spectrum shows only two peaks: 2.95 δ (singlet, 6 H) and 8.0 δ (singlet, 1 H). Explain this temperature-dependent behavior.

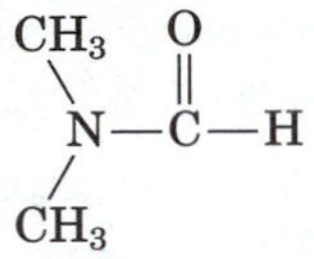

***N,N*-Dimethylformamide**

27.54 An octapeptide shows the composition Asp, Gly_2, Leu, Phe, Pro_2, Val on amino acid analysis. Edman analysis shows a glycine N-terminal group, and carboxypeptidase cleavage produces leucine as the first amino acid to appear. Acidic hydrolysis gives the following fragments:

Val-Pro-Leu, Gly, Gly-Asp-Phe-Pro, Phe-Pro-Val

Propose a structure for the starting octapeptide.

27.55 Propose a mechanism to account for the reaction of ninhydrin with an α-amino acid.

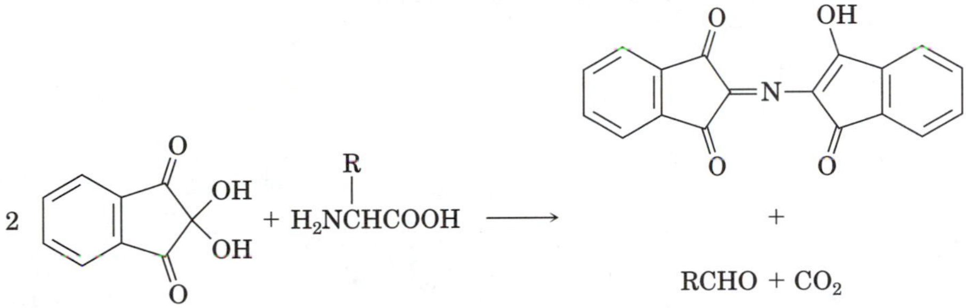

27.56 Draw as many resonance forms as you can for the purple anion obtained by reaction of ninhydrin with an α-amino acid (Problem 27.55).

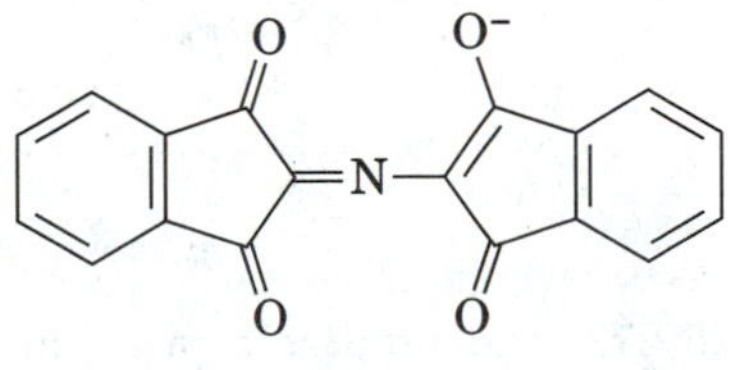

27.57 Look up the structure of human insulin (Figure 27.9), and indicate where in each chain the molecule is cleaved by trypsin and chymotrypsin.

27.58 What is the structure of a nonapeptide that gives the following fragments when cleaved?

Trypsin cleavage: Val-Val-Pro-Tyr-Leu-Arg, Ser-Ile-Arg
Chymotrypsin cleavage: Leu-Arg, Ser-Ile-Arg-Val-Val-Pro-Tyr

27.59 Oxytocin, a nonapeptide hormone secreted by the pituitary gland, functions by stimulating uterine contraction and lactation during childbirth. Its sequence was determined from the following evidence:

1. Oxytocin is a cyclic compound containing a disulfide bridge between two cysteine residues.

2. When the disulfide bridge is reduced, oxytocin has the constitution Asn, Cys_2, Gln, Gly, Ile, Leu, Pro, Tyr.
3. Partial hydrolysis of reduced oxytocin yields seven fragments:

Asp-Cys, Ile-Glu, Cys-Tyr, Leu-Gly, Tyr-Ile-Glu, Glu-Asp-Cys, Cys-Pro-Leu

4. Gly is the C-terminal group.
5. Both Glu and Asp are present as their side-chain amides (Gln and Asn) rather than as free side-chain acids.

What is the amino acid sequence of reduced oxytocin? What is the structure of oxytocin itself?

27.60 *Aspartame*, a nonnutritive sweetener marketed under the trade name Nutra-Sweet, is the methyl ester of a simple dipeptide, Asp-Phe–OCH_3.
(a) Draw the structure of aspartame.
(b) The isoelectric point of aspartame is 5.9. Draw the principal structure present in aqueous solution at this pH.
(c) Draw the principal form of aspartame present at physiological pH = 7.6.

27.61 The following drawing shows a titration curve for an amino acid.
(a) What are the approximate values of pK_{a1}, pK_{a2}, and the isoelectric point for this amino acid?
(b) Is this a neutral, acidic, or basic amino acid?

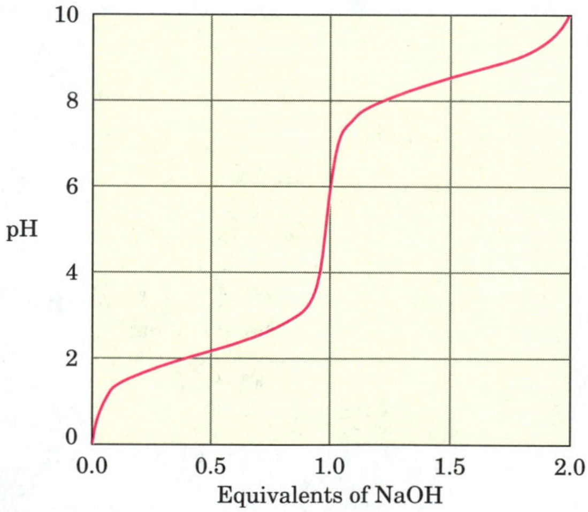

27.62 Refer to Figure 27.6 and propose a mechanism for the final step in the Edman degradation—the acid-catalyzed rearrangement of the thiazolinone to the *N*-phenylthiohydantoin.

A Look Ahead

27.63 We'll see in Chapter 30 that amino acids are metabolized by a transamination reaction in which the $-NH_2$ group of the amino acid changes places with the keto group of an α-keto acid. The products are a new amino acid and a new α-keto acid. Show the product from transamination of isoleucine.

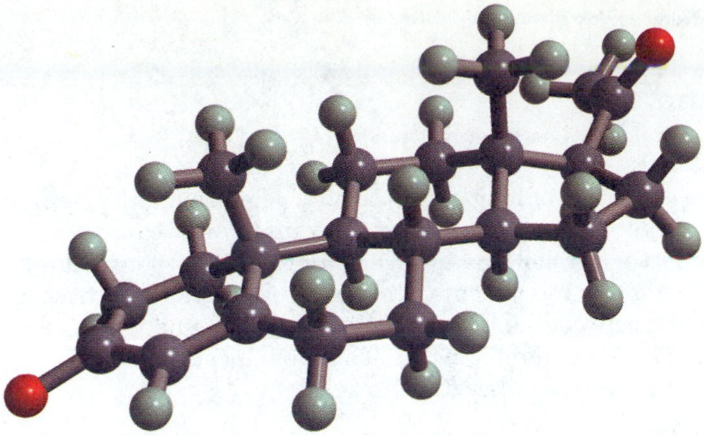

Progesterone is a steroid hormone that mediates many of the changes taking place during pregnancy.

28

Biomolecules: Lipids

Lipids are the naturally occurring organic molecules isolated from cells and tissues by extraction with nonpolar organic solvents. Note that this definition differs from the sort used for carbohydrates and proteins in that lipids are defined by physical property (solubility) rather than by structure.

Lipids are classified into two general types: those like fats and waxes, which contain ester linkages and can be hydrolyzed, and those like cholesterol and other steroids, which don't have ester linkages and can't be hydrolyzed.

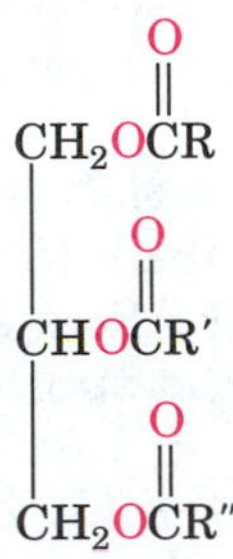

Animal fat, an ester
(R, R′, R″ = C_{11}–C_{19} chains)

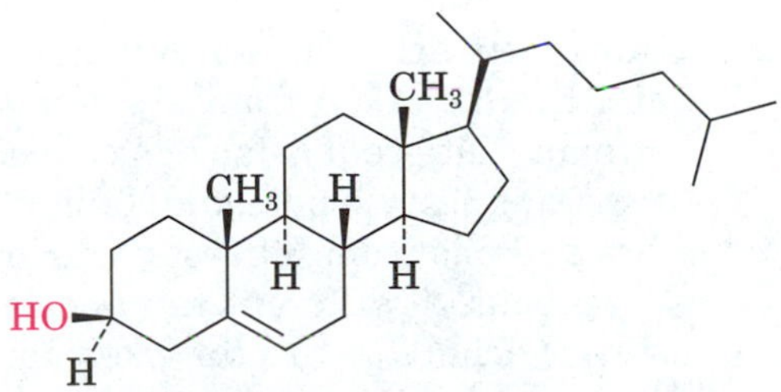

Cholesterol

28.1 Waxes, Fats, and Oils

Waxes are mixtures of esters of long-chain carboxylic acids and long-chain alcohols. The carboxylic acid usually has an even number of carbons from 16 through 36, while the alcohol has an even number of carbons from 24 through 36. One of the major components of beeswax, for instance, is triacontyl hexadecanoate, the ester of the C_{30} alcohol triacontanol and the C_{16} acid hexadecanoic acid. The waxy protective coatings on most fruits, berries, leaves, and animal furs have similar structures.

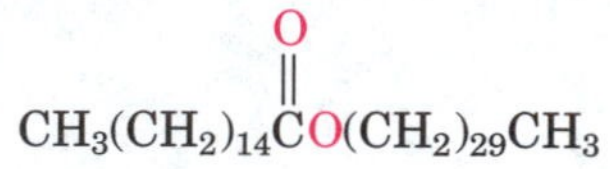

Triacontyl hexadecanoate (from beeswax)

Animal fats and **vegetable oils** are the most widely occurring lipids. Although they appear different—animal fats like butter and lard are solids, whereas vegetable oils like corn and peanut oil are liquid—their structures are closely related. Chemically, fats and oils are **triacylglycerols** (TAG's, also called *triglycerides*), triesters of glycerol with three long-chain carboxylic acids. Hydrolysis of a fat or oil with aqueous NaOH yields glycerol and three **fatty acids:**

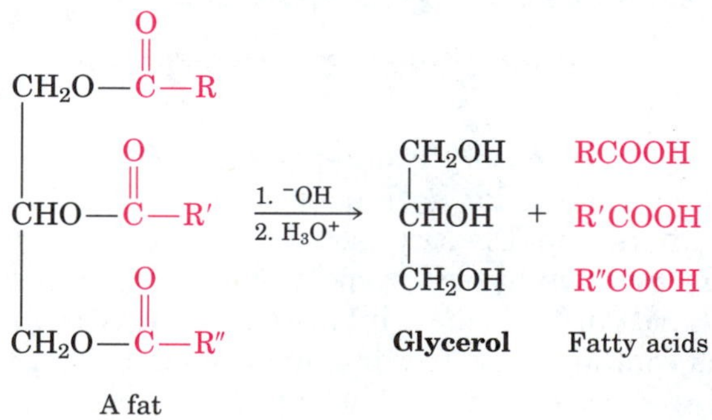

The fatty acids obtained by hydrolysis of triacylglycerols are generally unbranched and contain an even number of carbon atoms between 12 and 20. If double bonds are present, they usually have *Z* (cis) geometry. The three fatty acids of a specific triacylglycerol molecule need not be the same, and the fat or oil from a given source is likely to be a complex mixture of many different triacylglycerols. Table 28.1 lists some of the commonly occurring fatty acids, and Table 28.2 lists the approximate composition of fats and oils from different sources.

About 40 different fatty acids occur naturally. Palmitic acid (C_{16}) and stearic acid (C_{18}) are the most abundant saturated acids; oleic and linoleic acids (both C_{18}) are the most abundant unsaturated ones. Oleic acid is

Table 28.1 Structures of Some Common Fatty Acids

Name	*Carbons*	*Structure*	*Melting point (°C)*
Saturated			
Lauric	12	$CH_3(CH_2)_{10}COOH$	44
Myristic	14	$CH_3(CH_2)_{12}COOH$	58
Palmitic	16	$CH_3(CH_2)_{14}COOH$	63
Stearic	18	$CH_3(CH_2)_{16}COOH$	70
Arachidic	20	$CH_3(CH_2)_{18}COOH$	75
Unsaturated			
Palmitoleic	16	$CH_3(CH_2)_5CH{=}CH(CH_2)_7COOH$ (cis)	32
Oleic	18	$CH_3(CH_2)_7CH{=}CH(CH_2)_7COOH$ (cis)	16
Ricinoleic	18	$CH_3(CH_2)_5CH(OH)CH_2CH{=}CH(CH_2)_7COOH$ (cis)	5
Linoleic	18	$CH_3(CH_2)_4CH{=}CHCH_2CH{=}CH(CH_2)_7COOH$ (cis,cis)	−5
Arachidonic	20	$CH_3(CH_2)_4(CH{=}CHCH_2)_4CH_2CH_2COOH$ (all cis)	−50

Table 28.2 Approximate Composition of Some Fats and Oils

	Saturated fatty acids (%)				*Unsaturated fatty acids (%)*		
Source	C_{12} *Lauric*	C_{14} *Myristic*	C_{16} *Palmitic*	C_{18} *Stearic*	C_{18} *Oleic*	C_{18} *Ricinoleic*	C_{18} *Linoleic*
Animal fat							
Lard	—	1	25	15	50	—	6
Butter	2	10	25	10	25	—	5
Human fat	1	3	25	8	46	—	10
Whale blubber	—	8	12	3	35	—	10
Vegetable oil							
Coconut	50	18	8	2	6	—	1
Corn	—	1	10	4	35	—	45
Olive	—	1	5	5	80	—	7
Peanut	—	—	7	5	60	—	20
Linseed	—	—	5	3	20	—	20
Castor bean	—	—	—	1	8	85	4

monounsaturated since it has only one double bond, whereas linoleic, linolenic, and arachidonic acid are **polyunsaturated fatty acids**, or **PUFA's**, because they have more than one double bond. Linoleic and linolenic acids occur in cream and are essential in the human diet; infants grow poorly and develop skin lesions if fed a diet of nonfat milk for prolonged periods.

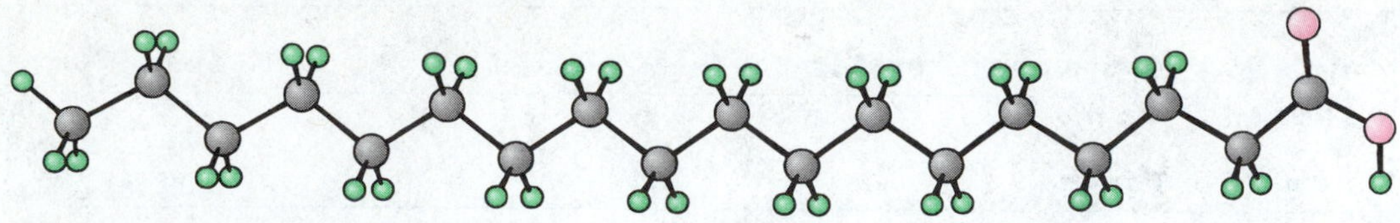

$$CH_3CH_2CH_2CH_2CH_2CH_2CH_2CH_2CH_2CH_2CH_2CH_2CH_2CH_2CH_2CH_2CH_2\overset{\overset{\displaystyle O}{\|}}{C}OH$$

Stearic acid

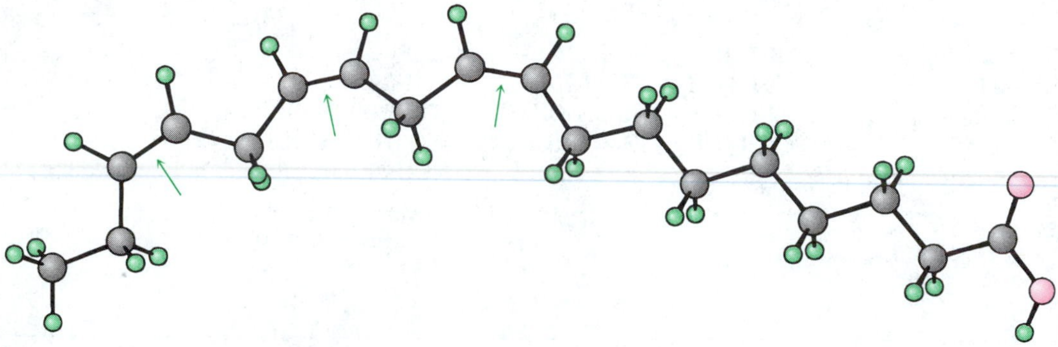

$$CH_3CH_2CH{=}CHCH_2CH{=}CHCH_2CH{=}CHCH_2CH_2CH_2CH_2CH_2CH_2CH_2\overset{\overset{\displaystyle O}{\|}}{C}OH$$

Linolenic acid, a polyunsaturated fatty acid (PUFA)

The data in Table 28.1 show that unsaturated fatty acids generally have lower melting points than their saturated counterparts, a trend that also holds true for triacylglycerols. Since vegetable oils generally have a higher proportion of unsaturated to saturated fatty acids than animal fats (Table 28.2), they have lower melting points. The difference is a consequence of structure. Saturated fats have a uniform shape that allows them to pack together easily in a crystal lattice. In unsaturated vegetable oils, however, the C=C bonds introduce bends and kinks into the hydrocarbon chains, making crystal formation difficult. The more double bonds there are, the harder it is for the molecules to crystallize and the lower the melting point of the oil.

The C=C bonds in vegetable oils can be reduced by catalytic hydrogenation (Section 7.7) to produce saturated solid or semisolid fats. Margarine and solid cooking fats, such as Crisco, are produced by hydrogenating soybean, peanut, or cottonseed oil until the proper consistency is obtained.

PROBLEM

28.1 Carnauba wax, used in floor and furniture polishes, contains an ester of a C_{32} straight-chain alcohol with a C_{20} straight-chain carboxylic acid. Draw its structure.

PROBLEM

28.2 Draw structures of the following molecules. Which would you expect to have a higher melting point?
(a) Glyceryl tripalmitate (b) Glyceryl trioleate

PROBLEM

28.3 Stearolic acid, $C_{18}H_{32}O_2$, yields stearic acid on catalytic hydrogenation, and undergoes oxidative cleavage with ozone to yield nonanoic acid and nonanedioic acid. What is the structure of stearolic acid?

28.2 Soap

Soap has been known since at least 600 BC, when the Phoenicians prepared a curdy material by boiling goat fat with extracts of wood ash. The cleansing properties of soap weren't generally recognized, however, and the use of soap did not become widespread until the eighteenth century. Chemically, soap is a mixture of the sodium or potassium salts of the long-chain fatty acids produced by hydrolysis (*saponification*) of animal fat with alkali. Wood ash was used as a source of alkali until the mid-1800s, when NaOH became commercially available:

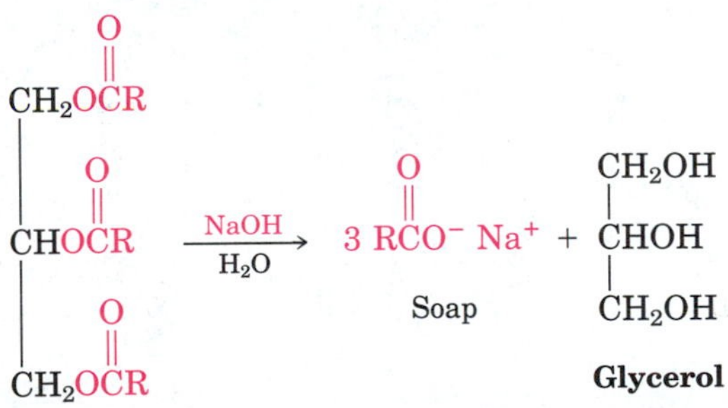

A fat
(R = C_{11}–C_{19} aliphatic chains)

Crude soap curds contain glycerol and excess alkali as well as soap but can be purified by boiling with water and adding NaCl to precipitate the pure sodium carboxylate salts. The smooth soap that precipitates is dried, perfumed, and pressed into bars for household use. Dyes are added to make colored soaps, antiseptics are added for medicated soaps, pumice is added for scouring soaps, and air is blown in for soaps that float. Regardless of these extra treatments and regardless of price, though, all soaps are basically the same.

Soaps act as cleansers because the two ends of a soap molecule are so different. The sodium carboxylate end of the long-chain molecule is ionic and therefore *hydrophilic* (water-loving). As a result, it tries to dissolve in water. The long hydrocarbon portion of the molecule, however, is nonpolar and *hydrophobic* (water-fearing). It therefore tries to avoid water and to dissolve in grease. The net effect of these two opposing tendencies is that soaps are attracted to both grease and water and are therefore valuable as cleansers.

When soaps are dispersed in water, the long hydrocarbon tails cluster together in a tangled hydrophobic ball, while the ionic heads on the surface of the cluster stick out into the water layer. These spherical clusters, called **micelles**, are shown schematically in Figure 28.1. Grease and oil droplets are solubilized in water when they are coated by the nonpolar tails of soap molecules in the center of micelles. Once solubilized, the grease and dirt can be rinsed away.

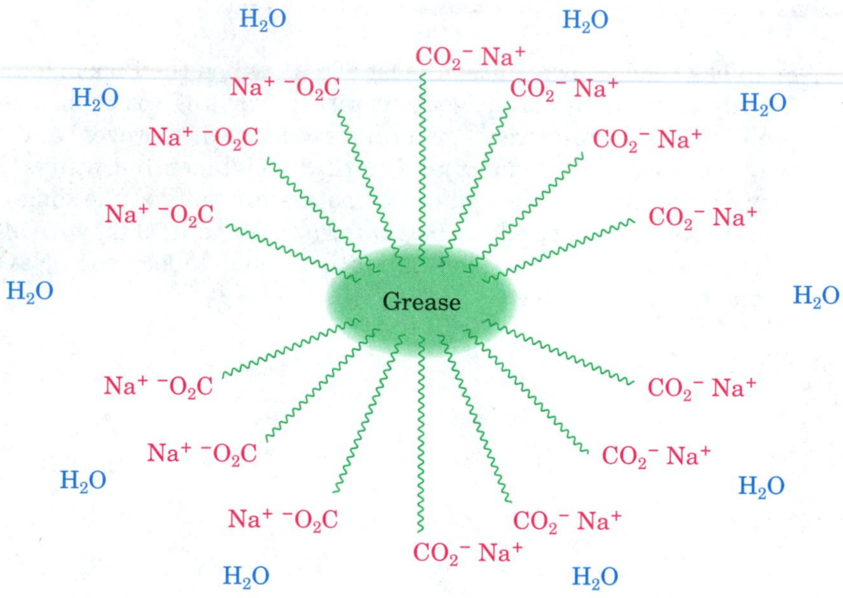

Figure 28.1 A soap micelle solubilizing a grease particle in water.

Soaps make life more pleasant than it might otherwise be, but they also have drawbacks. In hard water, which contains metal ions, soluble sodium carboxylates are converted into insoluble magnesium and calcium salts, leaving the familiar ring of scum around bathtubs and the gray tinge on white clothes. Chemists have circumvented these problems by synthesizing a class of synthetic detergents based on salts of long-chain alkylbenzenesulfonic acids. The principle of synthetic detergents is the same as that of soaps: The alkylbenzene end of the molecule is attracted to grease, while the charged sulfonate end is attracted to water. Unlike soaps, though, sulfonate detergents don't form insoluble metal salts in hard water and don't leave an unpleasant scum.

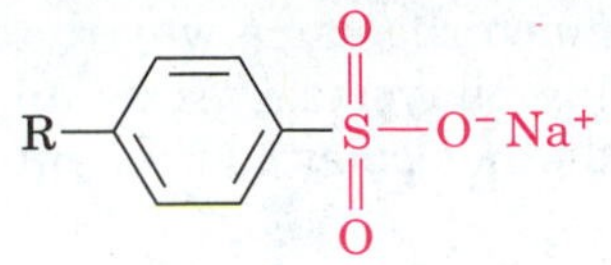

A synthetic detergent
(R = a mixture of C_{12} aliphatic chains)

PROBLEM.

28.4 Draw the structure of magnesium oleate, a component of bathtub scum.

28.3 Phospholipids

Just as waxes, fats, and oils are esters of carboxylic acids, **phospholipids** are esters of phosphoric acid, H_3PO_4. There are two main kinds of phospholipids: *phosphoglycerides* and *sphingolipids*.

Phosphoglycerides are closely related to fats and oils in that they contain a glycerol backbone linked by ester bonds to two fatty acids and one phosphoric acid. Although the fatty acid residues can be any of the C_{12}–C_{20} units typically present in fats, the acyl group at C1 is usually saturated and that at C2 is usually unsaturated. The phosphate group at C3 is also bound by a separate ester link to an amino alcohol such as choline, $HOCH_2CH_2\overset{+}{N}(CH_3)_3$, or ethanolamine, $HOCH_2CH_2NH_2$.

The most important phosphoglycerides are the *lecithins* and *cephalins*. Note that these compounds are chiral and that they have the L, or *R*, configuration at C2:

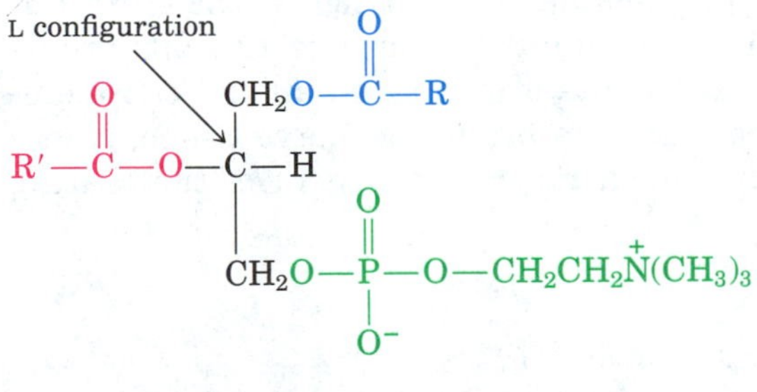

Phosphatidycholine, a lecithin

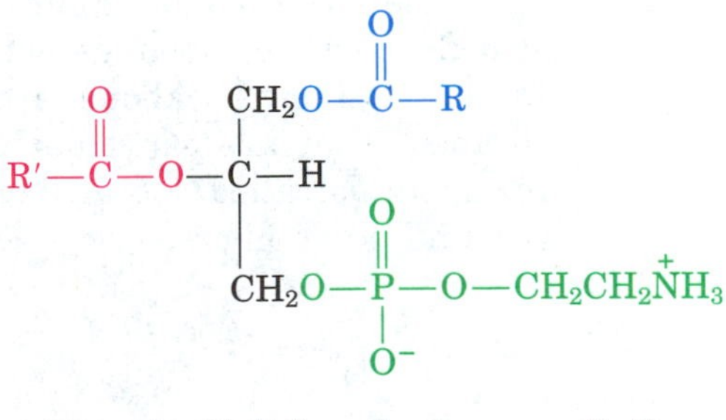

Phosphatidylethanolamine, a cephalin

where R is saturated and R′ is unsaturated

Found widely in both plant and animal tissues, phosphoglycerides are the major lipid component of cell membranes (approximately 40%). Like soaps, phosphoglycerides have a long, nonpolar hydrocarbon tail bound to a polar ionic head (the phosphate group). Cell membranes are composed mostly of phosphoglycerides oriented into a **lipid bilayer** about 50 Å thick. As shown in Figure 28.2, the hydrophobic tails aggregate in the center of the bilayer in much the same way that soap tails aggregate in the center of a micelle. This bilayer serves as an effective barrier to the passage of water, ions, and other polar components into and out of the cell.

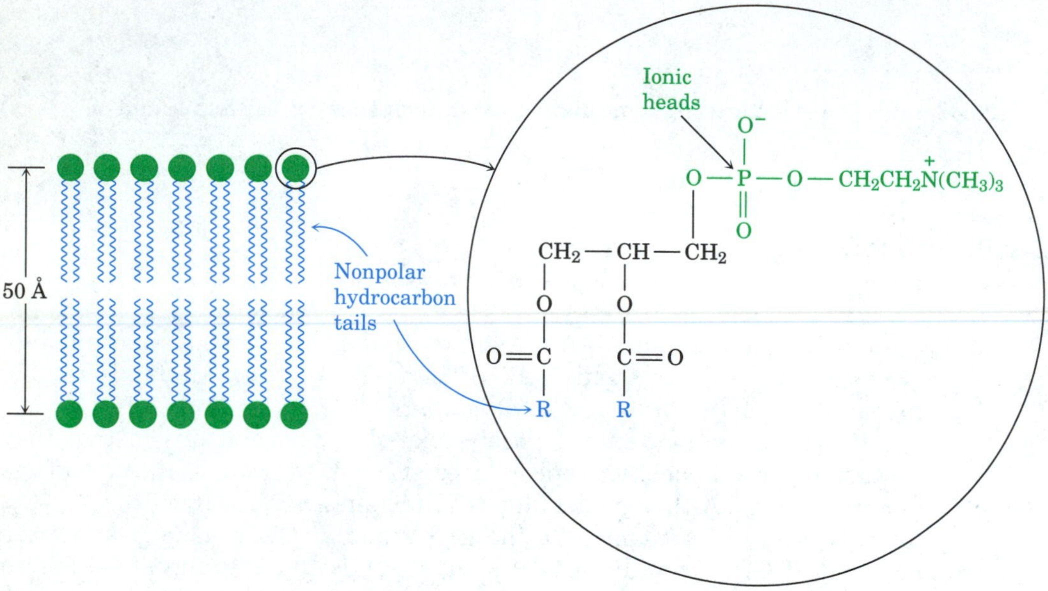

Figure 28.2 Aggregation of phosphoglycerides into the lipid bilayer that composes cell membranes.

The second major group of phospholipids is comprised of the **sphingolipids**. These compounds, which have sphingosine or a related dihydroxyamine as their backbone, are constituents of plant and animal cell membranes. They are particularly abundant in brain and nerve tissue, where compounds called *sphingomyelins* are a major constituent of the coating around nerve fibers.

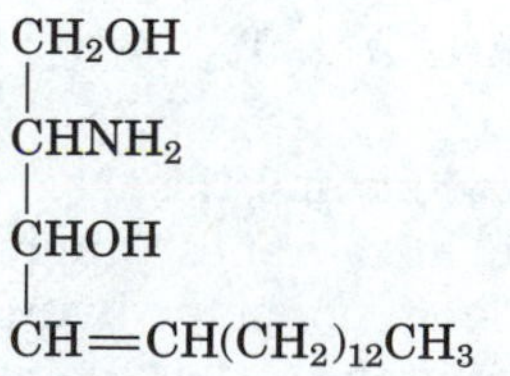

Sphingosine

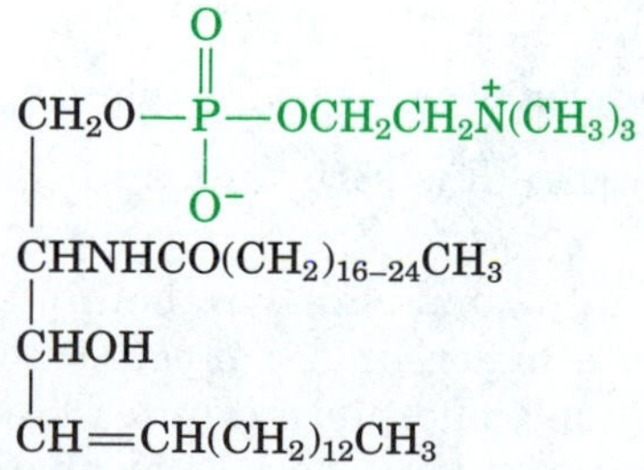

Sphingomyelin, a sphingolipid

28.4 Prostaglandins

The **prostaglandins** are a group of C_{20} carboxylic acids that contain a five-membered ring with two long side chains. First isolated by Sune Bergstrom,[1] Bengt Samuelsson,[2] and their collaborators at the Karolinska Institute in Sweden, these lipids are synthesized in nature from the C_{20} fatty acid, arachidonic acid. The name *prostaglandin* derives from the fact that these compounds were first isolated from sheep prostate glands, but they have subsequently been shown to be present in small amounts in all body tissues and fluids. Prostaglandin E_1 (PGE_1) and prostaglandin $F_{2\alpha}$ ($PGF_{2\alpha}$) are representative structures:

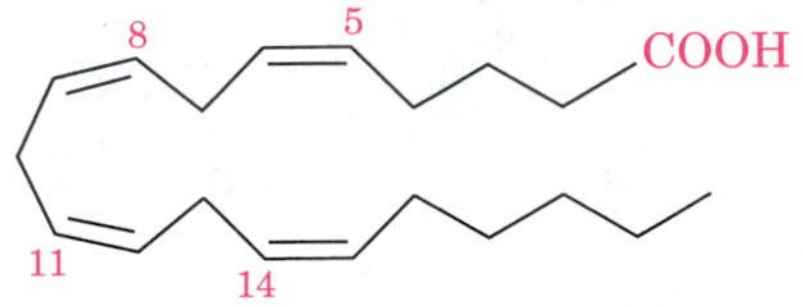

Arachidonic acid
[(5Z,8Z,11Z,14Z)-Eicosatetraenoic acid]

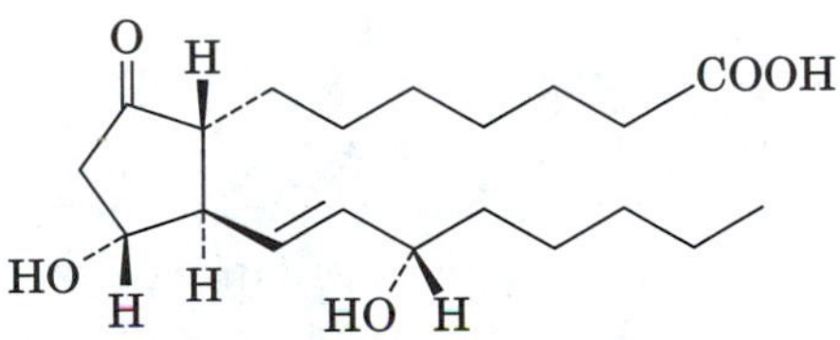

Prostaglandin E_1

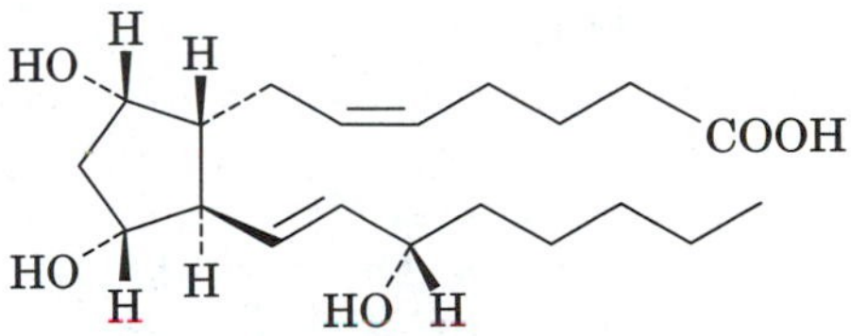

Prostaglandin $F_{2\alpha}$

The several dozen known prostaglandins have an extraordinarily wide range of biological effects. They can lower blood pressure, affect blood-platelet aggregation during clotting, lower gastric secretions, control inflammation, affect kidney function, affect reproductive systems, and stimulate uterine contractions during childbirth. In addition, compounds that are closely related to the prostaglandins have still other effects. Interest has centered particularly on the thromboxanes, on prostacyclin, and on the leukotrienes, compounds whose release in the body appears to trigger the asthmatic response.

---

[1]Sune Bergstrom (1916–); M.D. Karolinska Institute; professor, Karolinska Institute; Nobel Prize (1982).

[2]Bengt Samuelsson (1934–); M.D. Lund; professor, Karolinska Institute; Nobel Prize (1982).

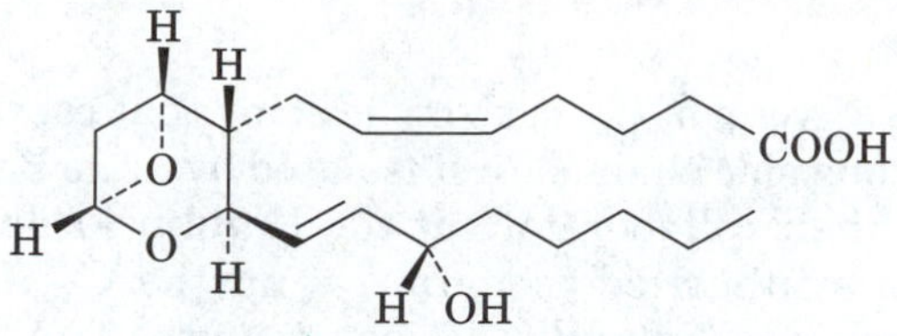

Thromboxane A_2

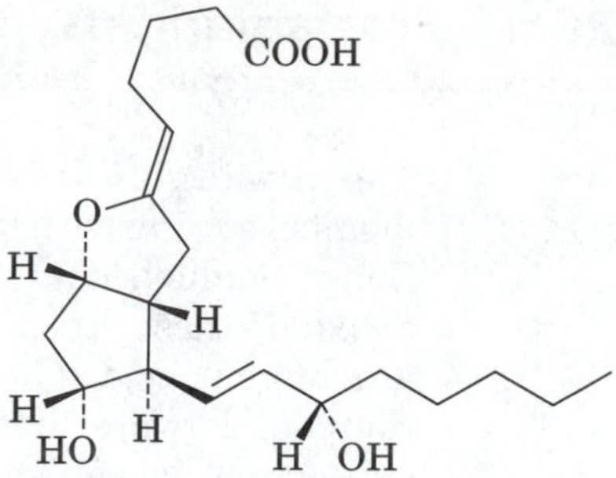

Prostacyclin

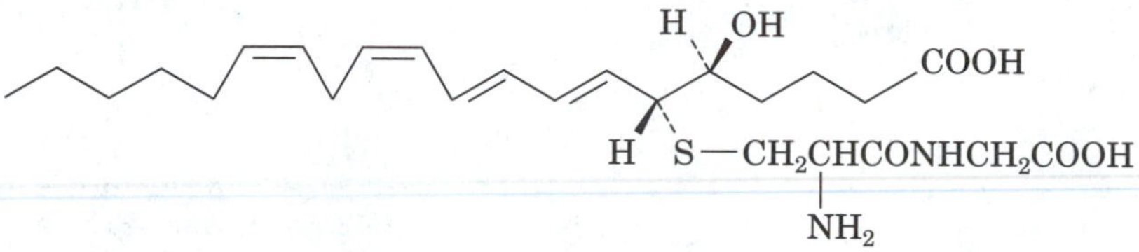

Leukotriene D_4

28.5 Terpenes

It has been known for centuries that codistillation of many plant materials with steam, a technique called *steam distillation*, produces a fragrant mixture of liquids known as plant **essential oils**. For thousands of years, such plant extracts have been used as medicines, spices, and perfumes. The investigation of essential oils also played a major role in the emergence of organic chemistry as a science during the nineteenth century.

Chemically, plant essential oils consist largely of mixtures of lipids called *terpenes*. **Terpenes** are small organic molecules that have an immense diversity of structure. Thousands of different terpenes are known. Some are hydrocarbons, and others contain oxygen; some are open-chain molecules, and others contain rings. Figure 28.3 gives some examples.

All terpenes are related, regardless of their apparent structural differences. According to the **isoprene rule** proposed by Leopold Ruzicka,[3] terpenes can be thought of as arising from head-to-tail joining of five-carbon isoprene (2-methyl-1,3-butadiene) units. Carbon 1 is called the head of the isoprene unit, and carbon 4 is the tail. For example, myrcene contains two isoprene units joined head to tail, forming an eight-carbon chain with two one-carbon branches. α-Pinene similarly contains two isoprene units assembled into a more complex cyclic structure.

---

[3]Leopold Ruzicka (1887–1976); b. Vukovar (now Yugoslavia); Ph.D., 1910, Karlsruhe (Staudinger); professor, Swiss Federal Institute (E.T.H.), Zürich (1923–1926 and 1929–1957); Nobel Prize, 1939.

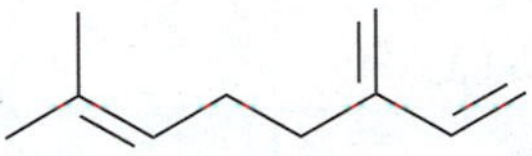

Myrcene (oil of bay)

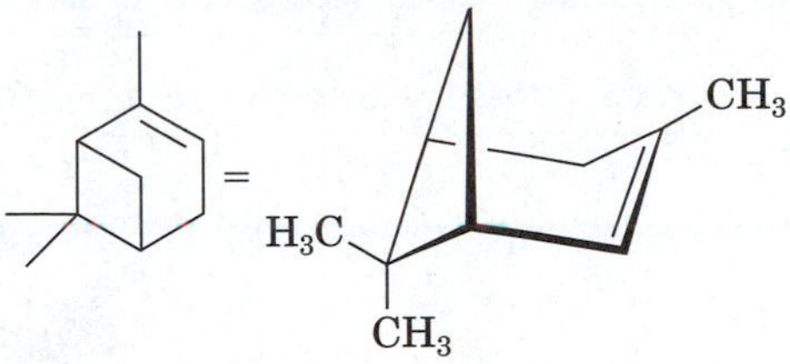

α-Pinene (oil of turpentine)

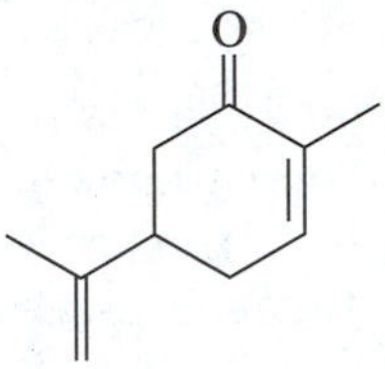

Carvone (oil of spearmint)

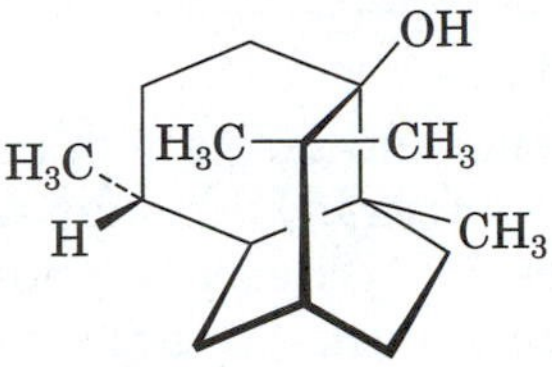

Patchouli alcohol (patchouli oil)

Figure 28.3 The structures of some terpenes isolated from plant essential oils.

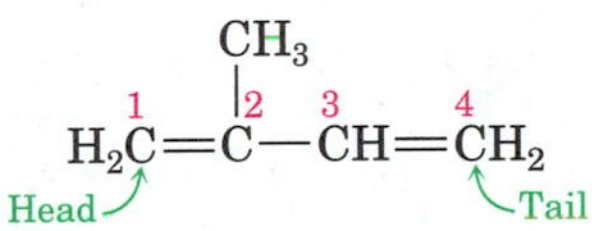

Isoprene (2-methyl-1,3-butadiene)

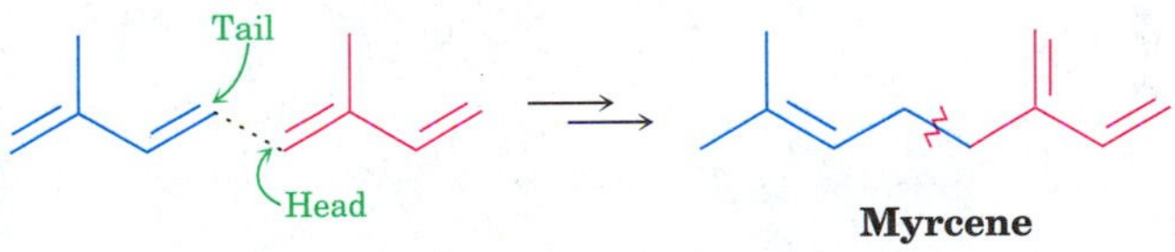

Myrcene

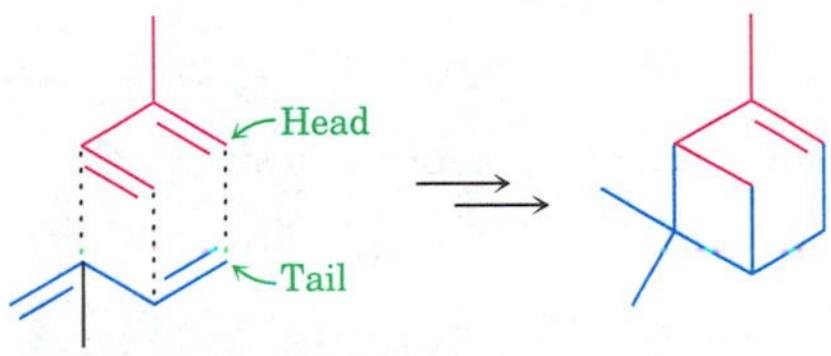

Two isoprenes **α-Pinene**

Terpenes are classified according to the number of isoprene units they contain. Thus, **monoterpenes** are 10-carbon substances biosynthesized from two isoprene units, **sesquiterpenes** are 15-carbon molecules from three isoprene units, and so on (Table 28.3).

Table 28.3 Classification of Terpenes

Carbon atoms	*Isoprene units*	*Classification*
10	2	Monoterpene
15	3	Sesquiterpene
20	4	Diterpene
25	5	Sesterterpene
30	6	Triterpene
40	8	Tetraterpene

Mono- and sesquiterpenes are found primarily in plants, but the higher terpenes occur in both plants and animals, and many have important biological roles. The triterpene lanosterol, for example, is the precursor from which all steroid hormones are made, and the tetraterpene β-carotene is a major dietary source of vitamin A.

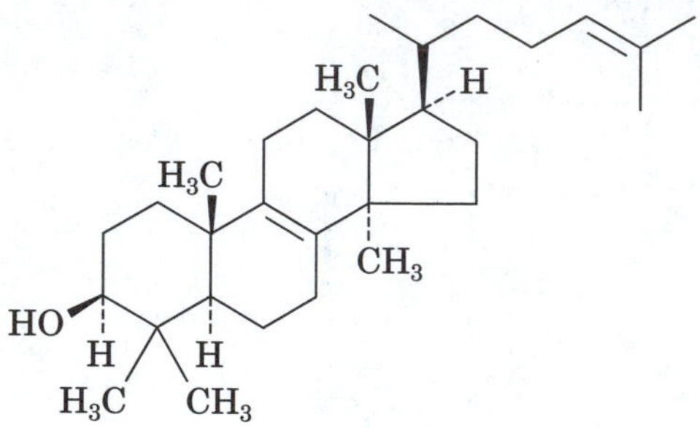

Lanosterol, a triterpene (C_{30})

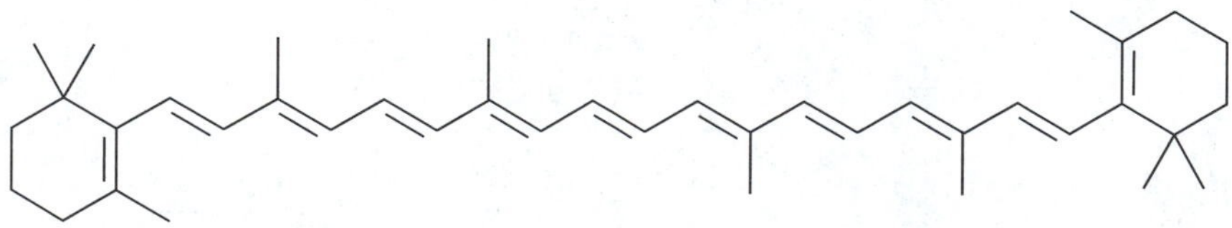

β-Carotene, a tetraterpene (C_{40})

PROBLEM.

28.5 Show the positions of the isoprene units in the following terpenes:

(a)

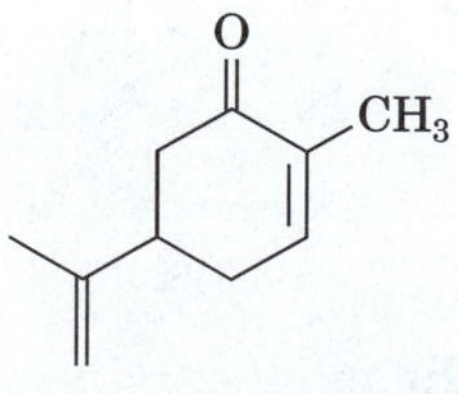

Carvone (spearmint oil)

(b)

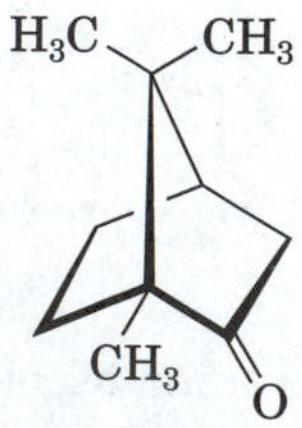

Camphor

(c)

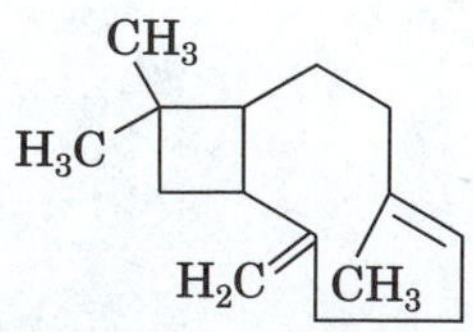

Caryophyllene (cloves)

28.6 Biosynthesis of Terpenes

The isoprene rule is a convenient formalism, but isoprene itself is not the biological precursor of terpenes. Nature instead uses two "isoprene equivalents"—isopentenyl pyrophosphate and dimethylallyl pyrophosphate. These five-carbon molecules are themselves made from condensation of three acetyl CoA units (Section 21.10).

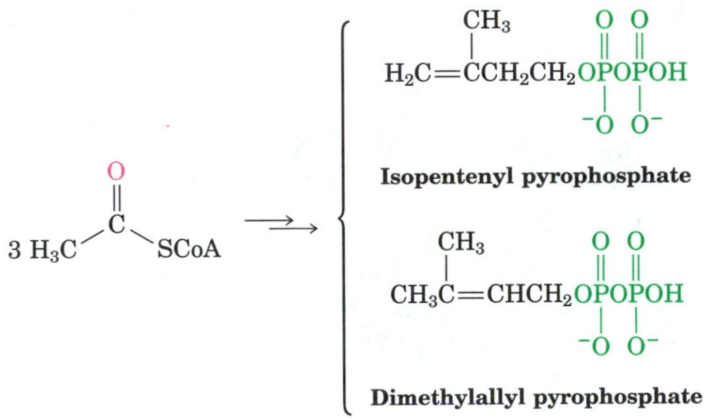

Dimethylallyl pyrophosphate is an effective alkylating agent in S_N2-like reactions because the primary, allylic pyrophosphate group (abbreviated OPP) is a good leaving group. Thus, displacement of this leaving group by the nucleophilic C=C bond of isopentenyl pyrophosphate, followed by loss of a proton from the carbocationic reaction intermediate, leads to the head-to-tail coupled 10-carbon unit, geranyl pyrophosphate. The corresponding alcohol, geraniol, is itself a fragrant terpene that occurs in rose oil.

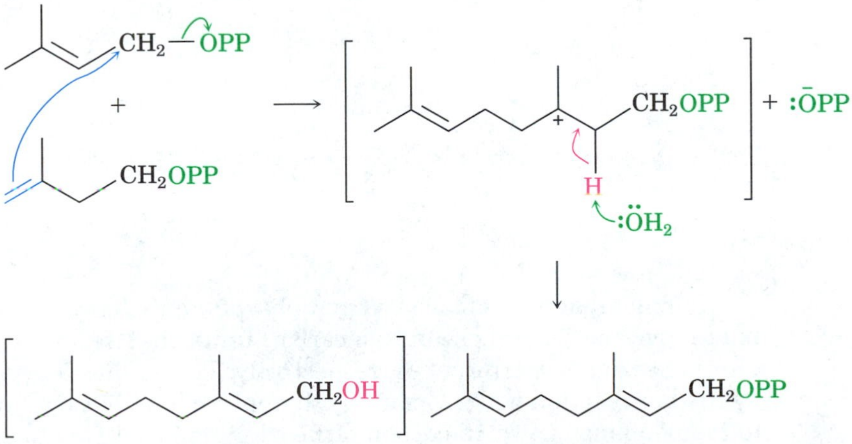

Geranyl pyrophosphate is the precursor of all monoterpenes. Limonene, for example, a monoterpene found in many citrus oils, arises from geranyl pyrophosphate by a cis–trans double-bond isomerization followed by internal nucleophilic displacement of the pyrophosphate group and subsequent loss of a proton.

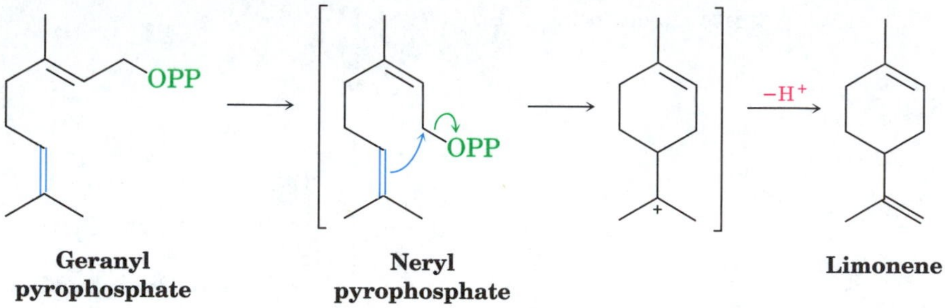

Further reaction of geranyl pyrophosphate with isopentenyl pyrophosphate yields the 15-carbon farnesyl pyrophosphate, the precursor of all sesquiterpenes. Farnesol, the corresponding alcohol, is found in citronella oil and lemon oil.

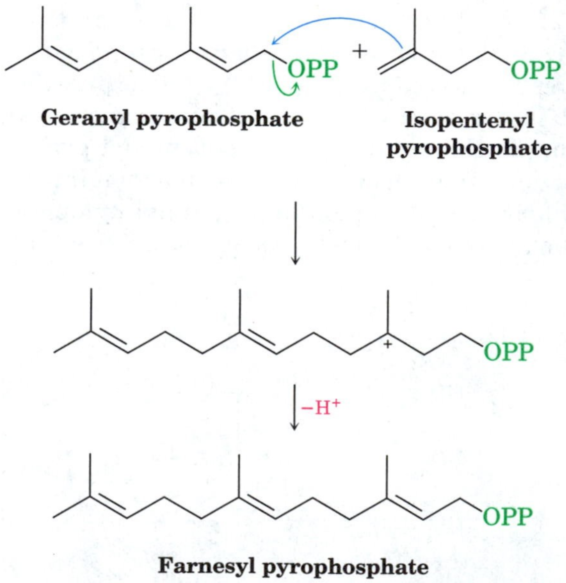

Further reactions of farnesyl pyrophosphate with isopentenyl pyrophosphate give the 20-carbon and 25-carbon units that serve as precursors of diterpenes and sesterterpenes, respectively. Triterpenes, however, arise not by further reaction with isopentenyl pyrophosphate but by a reductive tail-to-tail coupling of two 15-carbon farnesyl pyrophosphates to give squalene, a 30-carbon hexaene. Squalene, a major constituent of shark oil, is the precursor from which all triterpenes and steroids arise.

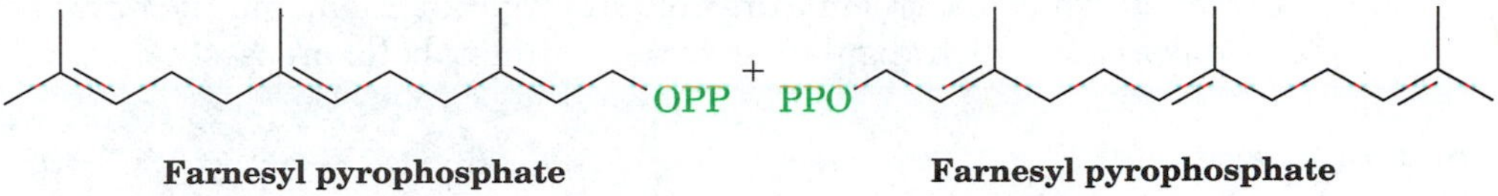

Squalene

PROBLEM

28.6 Propose a mechanistic pathway for the biosynthetic formation of γ-bisabolene from farnesyl pyrophosphate.

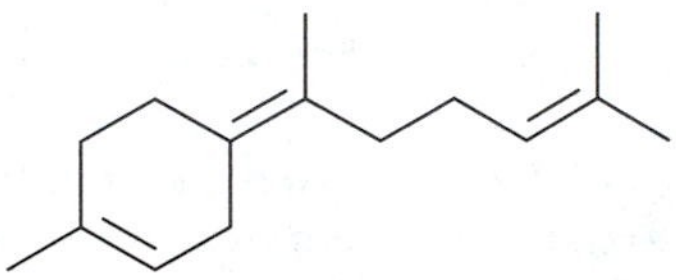

γ-Bisabolene

28.7 Steroids

In addition to fats, phospholipids, and terpenes, the lipid extracts of plants and animals also contain **steroids**, molecules whose structures are based on the tetracyclic ring system shown below. The four rings are designated A, B, C, and D, beginning at the lower left, and the carbon atoms are numbered beginning in the A ring. The three six-membered rings (A, B, and C) adopt chair conformations but are prevented by their rigid geometry from undergoing the usual cyclohexane ring-flips (Section 4.11).

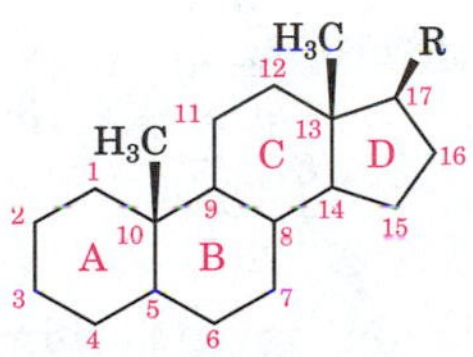

A steroid
(R = various side chains)

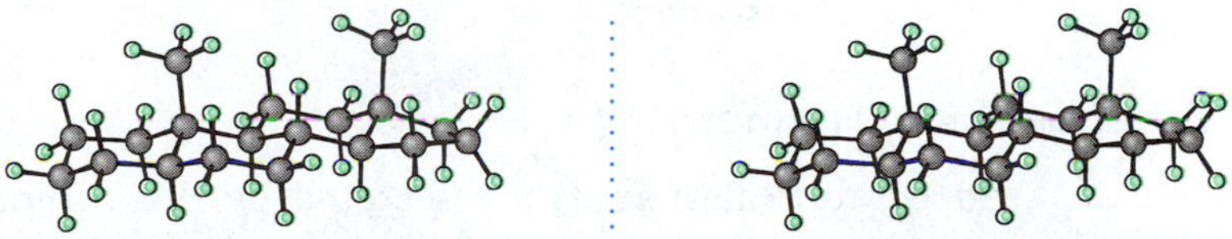

Stereo View

In humans, most steroids function as **hormones**, chemical messengers that are secreted by glands and are carried through the bloodstream to target tissues. There are two main classes of steroid hormones: the *sex hormones*, which control maturation and reproduction, and the *adrenocortical hormones*, which regulate a variety of metabolic processes.

Sex Hormones

Testosterone and androsterone are the two most important male sex hormones, or **androgens**. Androgens are responsible for the development of male secondary sex characteristics during puberty and for promoting tissue and muscle growth. Both are synthesized in the testes from cholesterol.

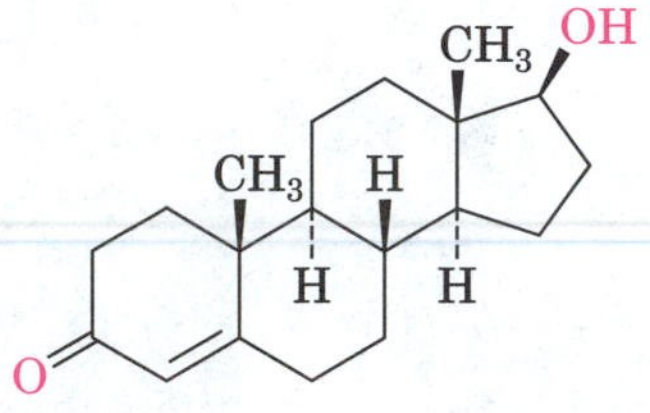

Testosterone

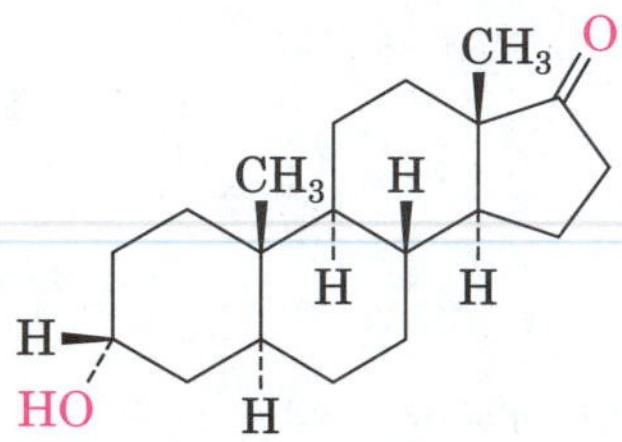

Androsterone

(Androgens)

Estrone and estradiol are the two most important female sex hormones, or **estrogens**. Synthesized in the ovaries from testosterone, estrogenic hormones are responsible for the development of female secondary sex characteristics and for regulation of the menstrual cycle. Note that both have a benzene-like aromatic A ring. In addition, another kind of sex hormone called a *progestin* is essential for preparing the uterus for implantation of a fertilized ovum during pregnancy. Progesterone is the most important progestin.

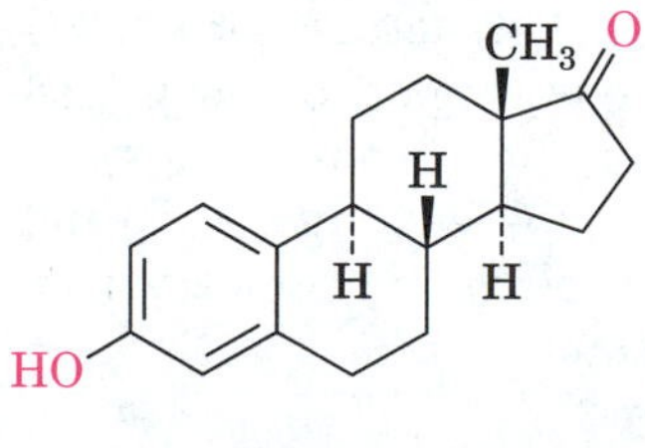

Estrone

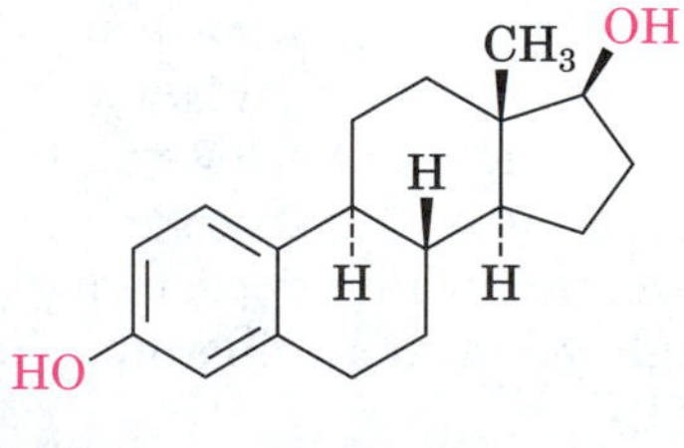

Estradiol

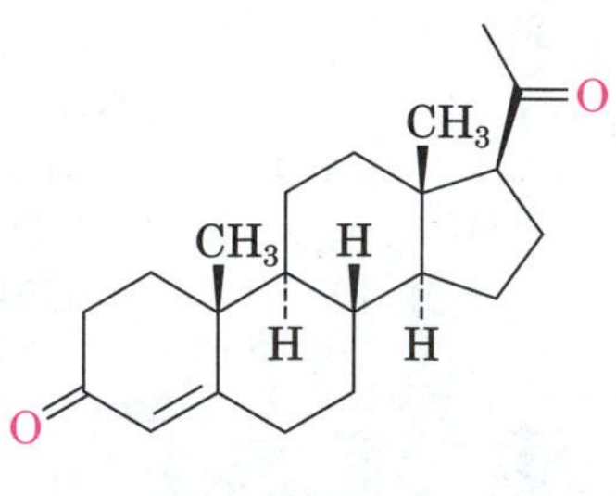

Progesterone
(a progestin)

(Estrogens)

Adrenocortical Hormones

Adrenocortical steroids are secreted by the adrenal glands, small organs located near the upper end of each kidney. There are two types of adrenocortical steroids, called *mineralocorticoids* and *glucocorticoids*. Mineralocorticoids, such as aldosterone, control tissue swelling by regulating cellular salt balance between Na^+ and K^+. Glucocorticoids, such as hydrocortisone, are involved in the regulation of glucose metabolism and in the control of inflammation. Glucocorticoid ointments are widely used to bring down the swelling from exposure to poison oak or poison ivy.

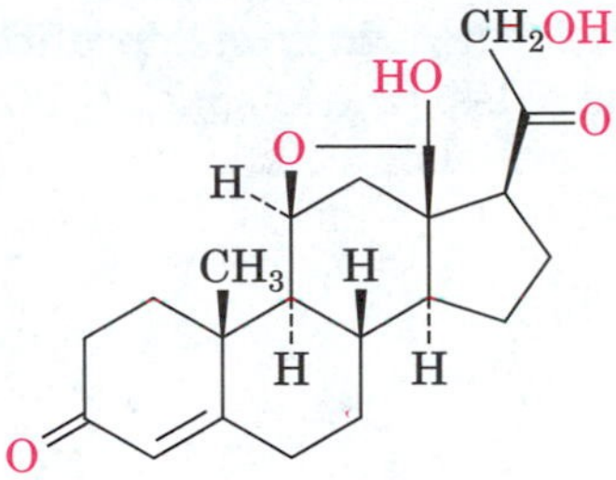

Aldosterone
(a mineralocorticoid)

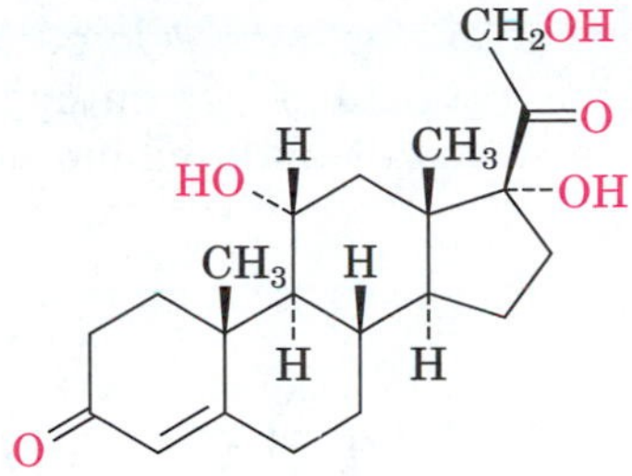

Hydrocortisone
(a glucocorticoid)

Synthetic Steroids

In addition to the many hundreds of steroids isolated from plants and animals, thousands more have been synthesized in pharmaceutical laboratories in a search for new drugs. The idea is to start with a natural hormone, carry out a chemical modification of the structure, and then see what biological properties the modified steroid has.

Among the best-known synthetic steroids are the oral contraceptives and anabolic agents. Most birth-control pills are a mixture of two compounds, a synthetic estrogen, such as ethynylestradiol, and a synthetic progestin, such as norethindrone. Anabolic steroids, such as stanozolol, detected in several athletes during the 1988 Olympics, are synthetic androgens that mimic the tissue-building effects of natural testosterone.

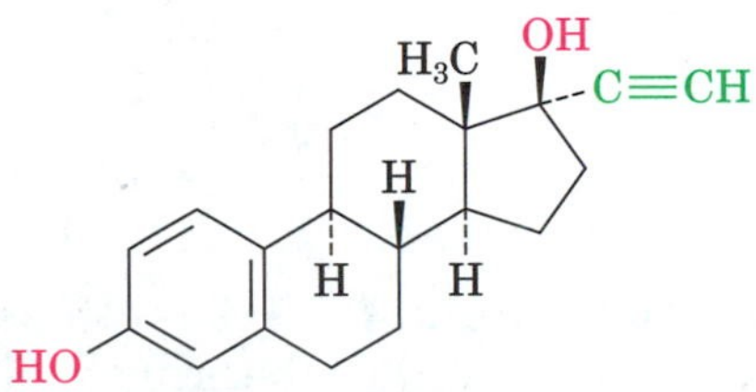

Ethynylestradiol
(a synthetic estrogen)

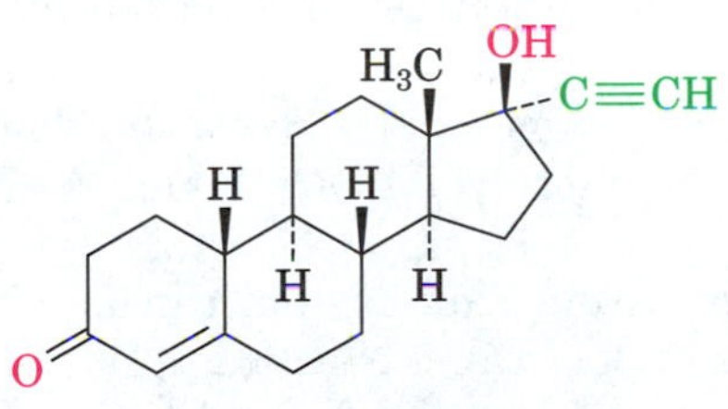

Norethindrone
(a synthetic progestin)

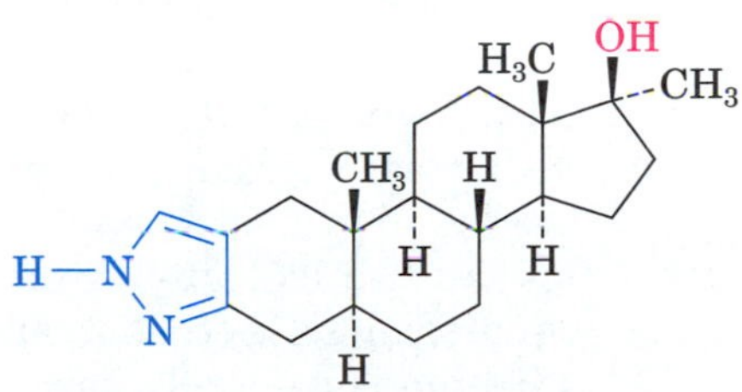

Stanozolol
(an anabolic agent)

28.8 Stereochemistry of Steroids

Two cyclohexane rings can be joined in either a cis or a trans manner. In *cis*-decalin, both groups at the ring-junction positions (the *angular* groups) are on the same side of the two rings. In *trans*-decalin, the groups at the ring junctions are on opposite sides. These spatial relationships are best grasped by building molecular models.

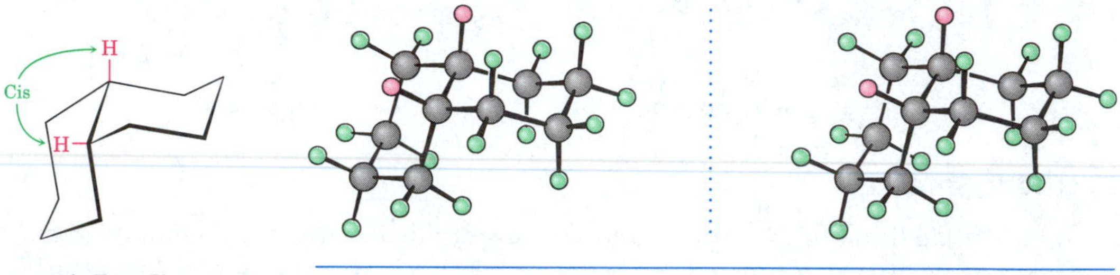

***cis*-Decalin** **Stereo View**

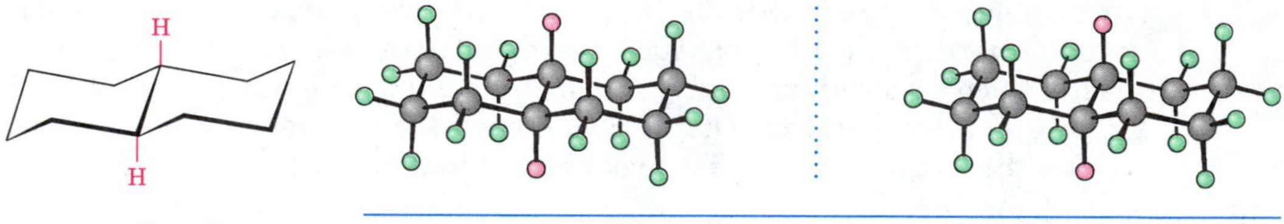

***trans*-Decalin** **Stereo View**

As shown in Figure 28.4, steroids can have either a cis or a trans fusion of the A and B rings, but the other ring fusions (B–C and C–D) are usually trans. A–B trans steroids have the C19 angular methyl group "up" (denoted β) and the hydrogen atom at C5 "down" (denoted α) on opposite sides of the molecule. A–B cis steroids, by contrast, have both the C19 angular methyl group and the C5 hydrogen atom on the same side (β) of the molecule. Both kinds of steroids are relatively long, flat molecules that have their two methyl groups (C18 and C19) protruding axially above the ring system. The A–B trans steroids are by far the more common, though A–B cis steroids are found in liver bile.

Substituent groups on the steroid ring system can be either axial or equatorial. As is true for simple cyclohexanes (Section 4.12), equatorial substitution is generally more favorable than axial substitution for steric reasons. The hydroxyl group at C3 of cholesterol, for example, has the more stable equatorial orientation (Figure 28.5).

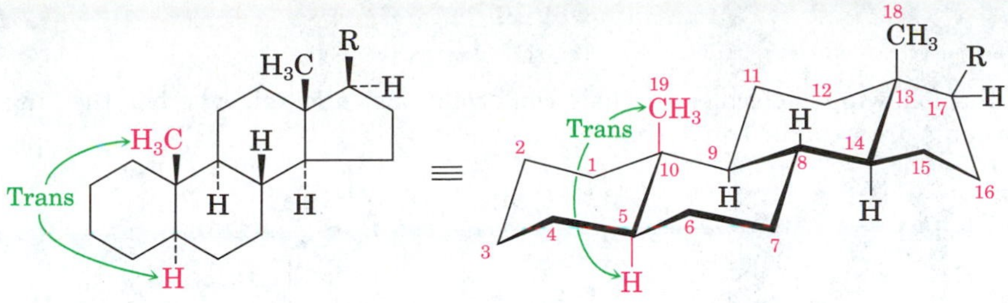

An A–B trans steroid

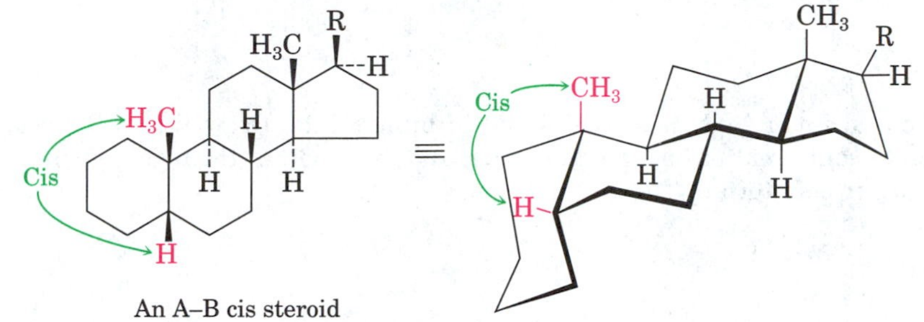

An A–B cis steroid

Figure 28.4 Steroid conformations. The three six-membered rings have chair conformations but are unable to undergo ring-flips. The A and B rings can be either cis-fused or trans-fused.

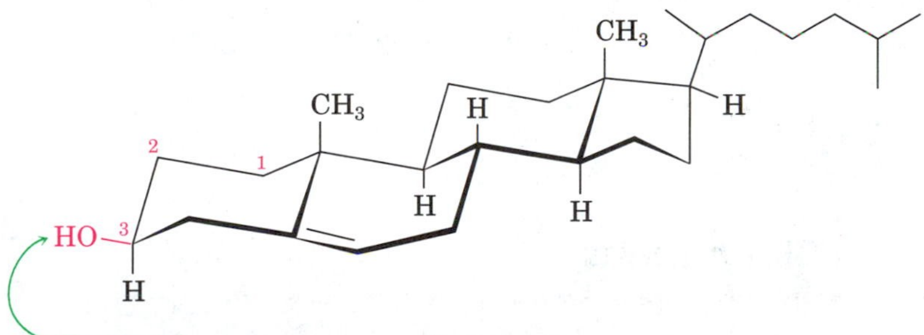

Cholesterol

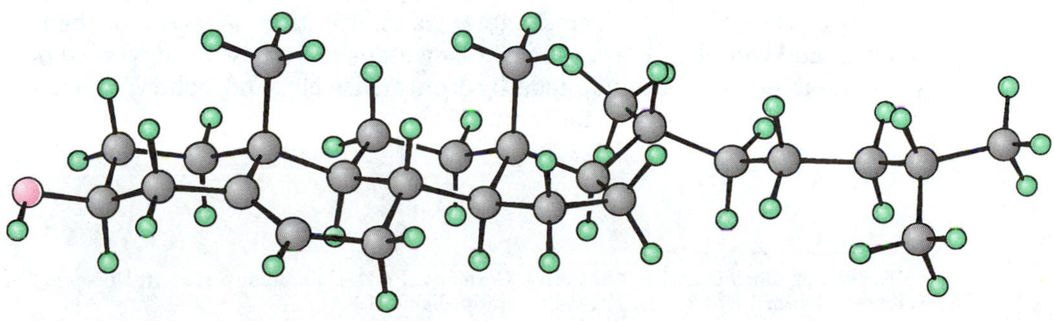

Figure 28.5 The stereochemistry of cholesterol.

PROBLEM

28.7 Draw the following molecules in chair conformations, and tell whether the ring substituents are axial or equatorial.

PROBLEM

28.8 Lithocholic acid is an A–B cis steroid found in human bile. Draw lithocholic acid showing chair conformations as in Figure 28.4, and tell whether the hydroxyl group at C_3 is axial or equatorial.

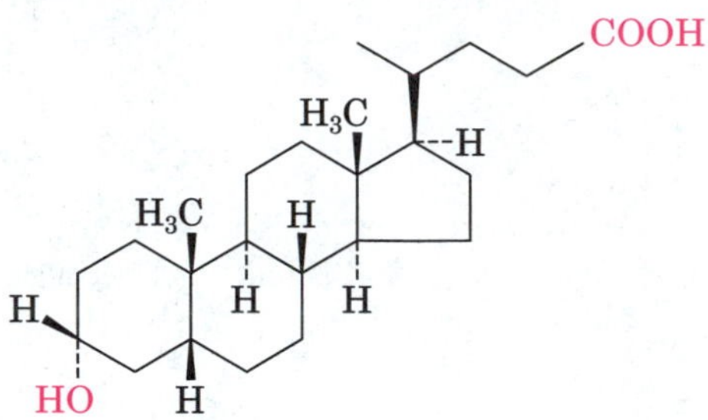

Lithocholic acid

28.9 Steroid Biosynthesis

Steroids are heavily modified triterpenes that are biosynthesized in living organisms from the acyclic hydrocarbon squalene (Section 28.6). The exact pathway by which this remarkable transformation is accomplished is lengthy and complex, but most of the key steps have now been worked out, with notable contributions made by Konrad Bloch[4] and John Cornforth,[5] who received Nobel Prizes for their efforts.

[4]Konrad E. Bloch (1912–); b. Neisse, Germany; Ph.D. Columbia; professor, University of Chicago, Harvard University; Nobel Prize in medicine (1964).

[5]John Warcup Cornforth (1917–); b. Australia; Ph.D. Oxford (Robinson); National Institute of Medical Research (Great Britain); Nobel Prize (1975).

Steroid biosynthesis occurs by enzyme-catalyzed epoxidation of squalene to yield squalene oxide, followed by acid-catalyzed cyclization and an extraordinary cascade of carbocation rearrangements to yield lanosterol (Figure 28.6, p. 1120). Lanosterol is then degraded by other enzymes to produce cholesterol, which is itself converted by enzymes to produce a host of different steroids. The exact series of carbocation rearrangements involved in the biosynthetic conversion of squalene to lanosterol involves the following steps:

1. The enzyme *squalene oxidase* selectively epoxidizes a terminal double bond of squalene to yield squalene oxide.
2. Squalene oxide is protonated, and the epoxide ring is opened by nucleophilic attack of the double bond six carbons away, to yield a cyclic carbocation intermediate.
3. The tertiary carbocation intermediate produced in step 2 is attacked by another double bond six carbons away to yield a second carbocation intermediate.
4. A third cyclization occurs by attack of an appropriately positioned double bond on a carbocation.
5. A fourth and last cyclization takes place, this one giving a five-membered ring.
6. Carbocation rearrangement occurs by a hydride shift.
7. A second hydride shift gives still another carbocation.
8. Carbocation rearrangement occurs by shift of a methyl group.
9. A second methyl-group shift gives a final carbocation intermediate.
10. Loss of a proton (E1 reaction) from the carbon next to the cationic center gives lanosterol.

Although written in a stepwise format in Figure 28.6 for convenience, it's thought that the entire cyclization sequence (steps 2–5) takes place at one time without intermediates. Similarly, the carbocation rearrangements and proton loss (steps 6–10) take place at essentially the same time without intermediates.

PROBLEM

28.9 Look at the structures of lanosterol and cholesterol, and catalog the changes that have occurred in the transformation.

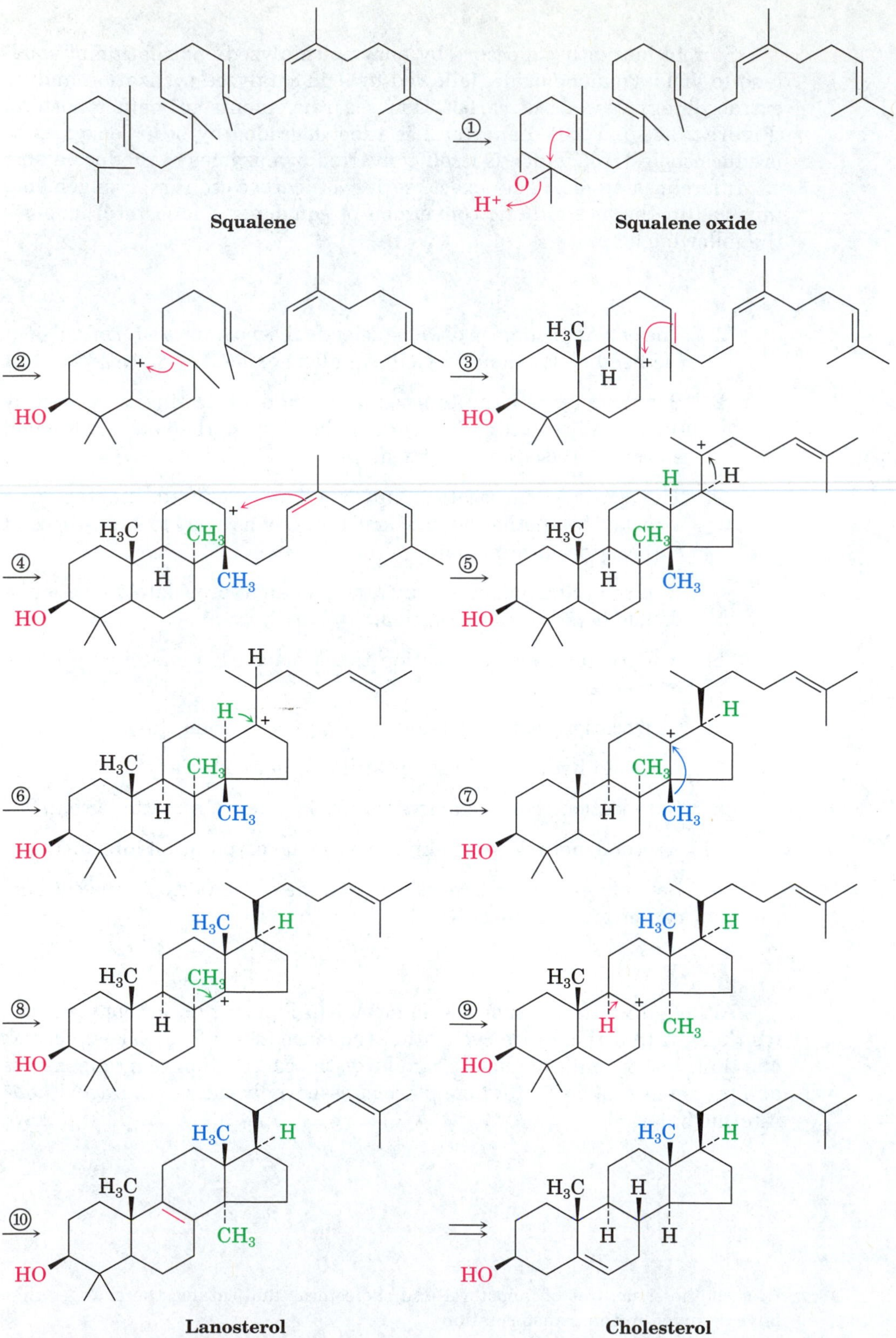

Figure 28.6 Biosynthesis of cholesterol from squalene.

INTERLUDE

Cholesterol and Heart Disease

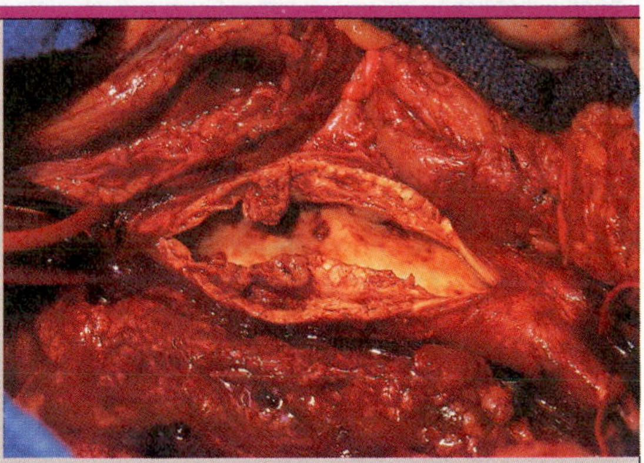

This coronary artery is partially blocked by deposits of cholesterol.

We read a lot about the relationship between cholesterol and heart disease. What are the facts? It's well established that a diet rich in saturated animal fats often leads to an increase in blood serum cholesterol, at least in sedentary, overweight people. Conversely, a diet lower in saturated fats and higher in polyunsaturated fats (PUFA's) leads to a lower serum cholesterol level. Studies have shown that a serum cholesterol level greater than 300 mg/dL (a normal value is 150–200 mg/dL) is weakly correlated with an increased incidence of *atherosclerosis*, a form of heart disease in which cholesterol deposits build up on the inner walls of coronary arteries, blocking the flow of blood to the heart muscles.

A better indication of a person's risk of heart disease comes from a measurement of blood lipoprotein levels. **Lipoproteins** are complex molecules with both lipid and protein parts that transport lipids through the body. They can be divided into four types according to density, as shown in Table 28.4. People with a high serum level of high-density lipoproteins (HDL's) seem to have a decreased risk of heart disease. As a rule of thumb, a person's risk drops about 25% for each increase of 5 mg/dL in HDL concentration. Normal values are about 45 mg/dL for men and 55 mg/dL for women, perhaps explaining why women are generally less susceptible than men to heart disease.

Table 28.4 Serum Lipoproteins

Name	*Density (g/mL)*	*% Lipid*	*% Protein*
Chylomicrons	<0.94	98	2
VLDL's (very-low-density lipoproteins)	0.940–1.006	90	10
LDL's (low-density lipoproteins)	1.006–1.063	75	25
HDL's (high-density lipoproteins)	1.063–1.210	60	40

(continued)➤

Chylomicrons and very-low-density lipoproteins (VLDL's) act primarily as carriers of triglycerides from the intestines to peripheral tissues, whereas LDL's and HDL's act as carriers of cholesterol to and from the liver. Present evidence suggests that LDL's transport cholesterol as its fatty acid ester *to* peripheral tissues, whereas HDL's remove cholesterol as its stearate ester *from* dying cells and transport it back to the liver. If LDL's deliver more cholesterol than is needed, and if insufficient HDL's are present to remove it, the excess is deposited in arteries. The higher the HDL level, the less the likelihood of deposits and the lower the risk of heart disease.

Not surprisingly, the most important factor in gaining high HDL levels is a generally healthy lifestyle. Obesity, smoking, and lack of exercise lead to low HDL levels, whereas regular exercise and a sensible diet lead to high HDL levels. Distance runners and other endurance athletes have HDL levels nearly 50% higher than the general population.

Summary and Key Words

Lipids are the naturally occurring materials isolated from plants and animals by extraction with organic solvents. **Animal fats** and **vegetable oils** are the most widely occurring lipids. Both are triesters of glycerol with long-chain **fatty acids**, but animal fats are usually saturated, whereas vegetable oils usually have unsaturated fatty acid residues.

Phosphoglycerides such as lecithin and cephalin are closely related to fats. The glycerol backbone in these molecules is esterified to two fatty acids (one saturated and one unsaturated) and to one phosphate ester. **Sphingolipids**, another major class of phospholipids, have an amino alcohol such as sphingosine for their backbone. These compounds are important constituents of cell membranes.

Prostaglandins and **terpenes** are still other classes of lipids. Prostaglandins, which are found in all body tissues, have a wide range of physiological actions. Terpenes are often isolated from the essential oils of plants. They have an immense diversity of structure and are produced biosynthetically by head-to-tail coupling of two five-carbon "isoprene equivalents"—isopentenyl pyrophosphate and dimethylallyl pyrophosphate.

Steroids are plant and animal lipids with a characteristic tetracyclic carbon skeleton. Like the prostaglandins, steroids occur widely in body tissues and have a large variety of physiological activities. Steroids are closely related to terpenes and arise biosynthetically from the triterpene precursor lanosterol. Lanosterol, in turn, arises from cyclization of the acyclic hydrocarbon squalene.

ADDITIONAL PROBLEMS

28.10 Fats can be either optically active or optically inactive, depending on their structure. Draw the structure of an optically active fat that yields 2 equivalents of stearic acid and 1 equivalent of oleic acid on hydrolysis. Draw the structure of an optically inactive fat that yields the same products.

28.11 Spermaceti, a fragrant substance from sperm whales, was much used in cosmetics until it was banned in 1976 to protect the whales from extinction. Chemically, spermaceti is cetyl palmitate, the ester of cetyl alcohol (n-$C_{16}H_{33}OH$) with palmitic acid. Draw its structure.

28.12 The *plasmalogens* are a group of lipids found in nerve and muscle cells. How do plasmalogens differ from fats, lecithins, and cephalins?

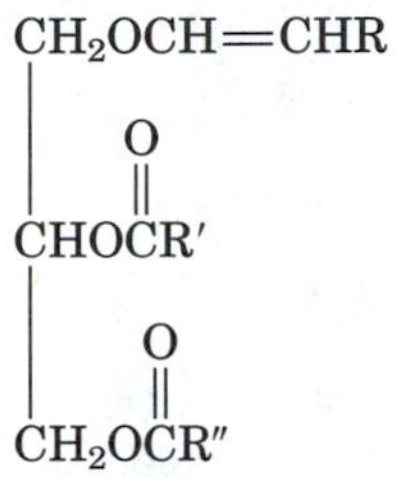

A plasmalogen

28.13 What product would you obtain from hydrolysis of a plasmalogen (Problem 28.12) with aqueous NaOH? With H_3O^+?

28.14 *Cardiolipins* are a group of lipids found in heart muscles. What products would be formed if all ester bonds were saponified by treatment with aqueous NaOH?

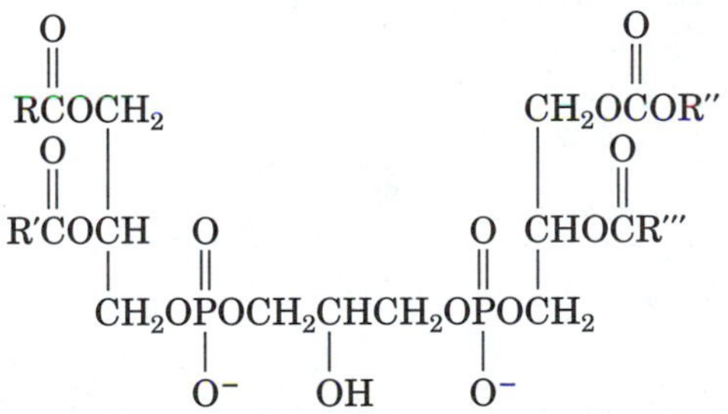

A cardiolipin

28.15 Show the products you would expect to obtain from reaction of glyceryl trioleate with the following reagents:
(a) Excess Br_2 in CCl_4
(b) H_2/Pd
(c) NaOH/H_2O
(d) O_3, then Zn/CH_3COOH
(e) $LiAlH_4$, then H_3O^+
(f) CH_3MgBr, then H_3O^+

28.16 Eleostearic acid, $C_{18}H_{30}O_2$, is a rare fatty acid found in the tung oil used for finishing furniture. On ozonolysis followed by treatment with zinc, eleostearic acid furnishes one part pentanal, two parts glyoxal (OHC–CHO), and one part 9-oxononanoic acid [$OHC(CH_2)_7COOH$]. What is the structure of eleostearic acid?

28.17 Draw representative structures for:
(a) A fat
(b) A prostaglandin
(c) A steroid

28.18 How would you convert oleic acid into the following substances?
(a) Methyl oleate
(b) Methyl stearate
(c) Nonanal
(d) Nonanedioic acid
(e) 9-Octadecynoic acid (stearolic acid)
(f) 2-Bromostearic acid
(g) 18-Pentatriacontanone, $CH_3(CH_2)_{16}CO(CH_2)_{16}CH_3$

28.19 How would you synthesize stearolic acid (Problem 28.18e) from 1-decyne and 1-chloro-7-iodoheptane?

28.20 Vaccenic acid, $C_{18}H_{34}O_2$, is a rare fatty acid that gives heptanal and 11-oxoundecanoic acid [$OHC(CH_2)_9COOH$] on ozonolysis followed by zinc treatment. When allowed to react with CH_2I_2/Zn–Cu, vaccenic acid is converted into lactobacillic acid. What are the structures of vaccenic and lactobacillic acids?

28.21 Show the location of the isoprene units in the following terpenes:

(a)

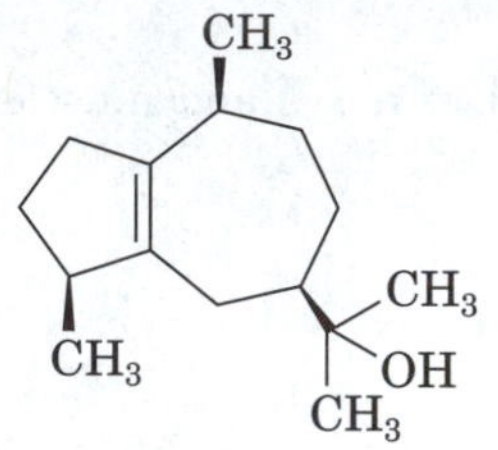

Guaiol

(b)

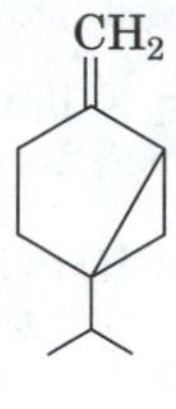

Sabinene

(c)

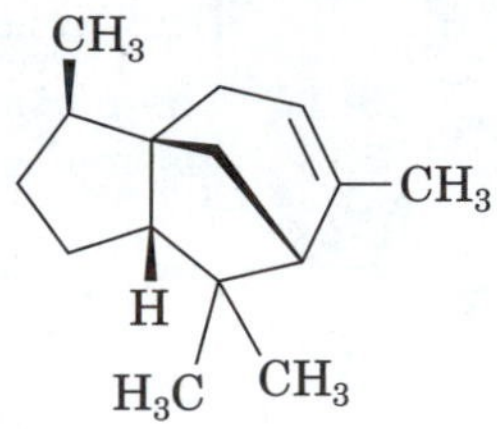

Cedrene

28.22 Indicate by asterisks the stereogenic centers present in each of the terpenes shown in Problem 28.21. How many stereoisomers of each are possible?

28.23 Assume that the three terpenes in Problem 28.21 are derived biosynthetically from isopentenyl pyrophosphate and dimethylallyl pyrophosphate, each of which was isotopically labeled at the pyrophosphate-bearing carbon atom (C1). At what positions would the terpenes be isotopically labeled?

28.24 Suggest a mechanistic pathway by which α-pinene might arise biosynthetically from geranyl pyrophosphate.

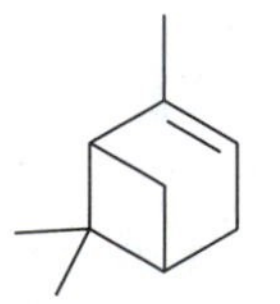

α-Pinene

28.25 Suggest a mechanism by which ψ-ionone is transformed into β-ionone on treatment with acid.

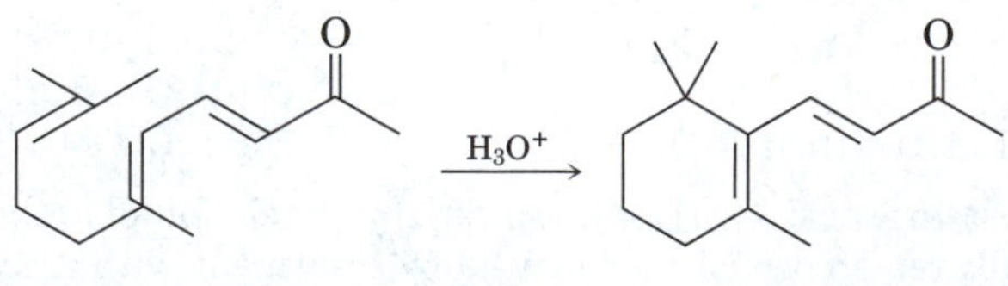

ψ-Ionone **β-Ionone**

28.26 Which isomer would you expect to be more stable, *cis*-decalin or *trans*-decalin? Explain. (Review Section 4.15.)

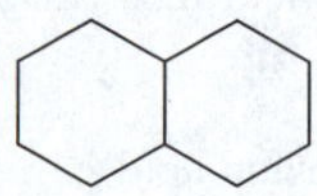

Decalin

28.27 Draw the most stable chair conformation of dihydrocarvone.

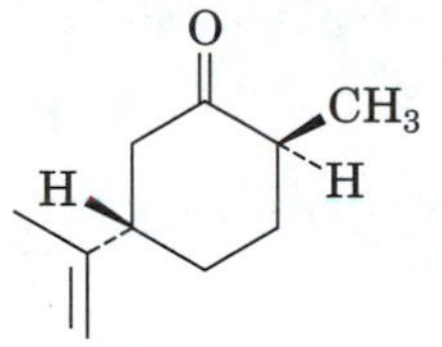

Dihydrocarvone

28.28 Draw the most stable chair conformation of menthol, and label each substituent as axial or equatorial.

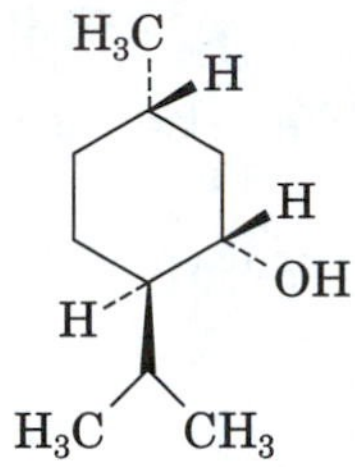

Menthol (from peppermint oil)

28.29 Cholic acid, a major steroidal constituent of human bile, has the structure shown. Draw a conformational structure of cholic acid, and label the three hydroxyl groups as axial or equatorial.

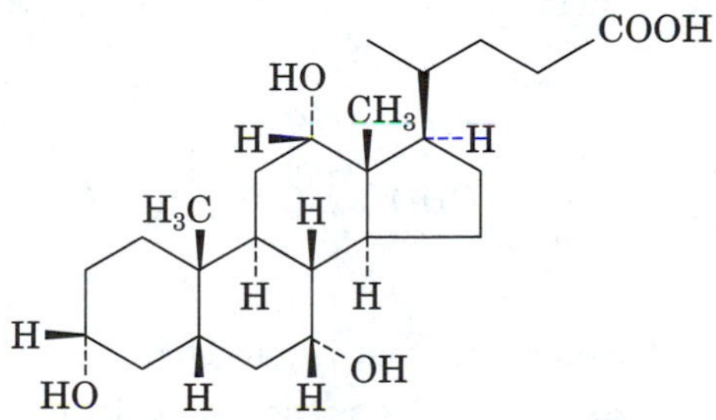

Cholic acid

28.30 How many stereogenic centers does cholic acid have (Problem 28.29)? How many stereoisomers are possible?

28.31 Show the products you would expect to obtain from reaction of cholic acid with these reagents:
(a) C_2H_5OH, HCl
(b) Excess pyridinium chlorochromate in CH_2Cl_2
(c) BH_3 in THF, then H_3O^+

28.32 As a general rule, equatorial alcohols are esterified more readily than axial alcohols. What product would you expect to obtain from reaction of the following two compounds with 1 equivalent of acetic anhydride?

(a)

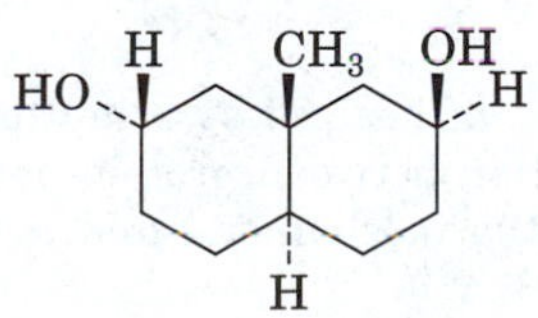

(b)

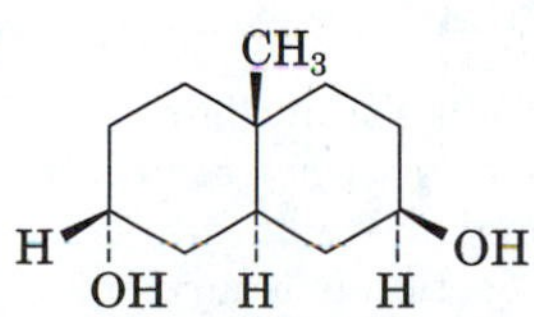

28.33 Diethylstilbestrol (DES) has estrogenic activity even though it is structurally unrelated to steroids. Once used as an additive in animal feed, DES has been implicated as a causative agent in several types of cancer. Look up the structure of estradiol (Section 28.7), and show how DES can be drawn so that it is sterically similar to estradiol.

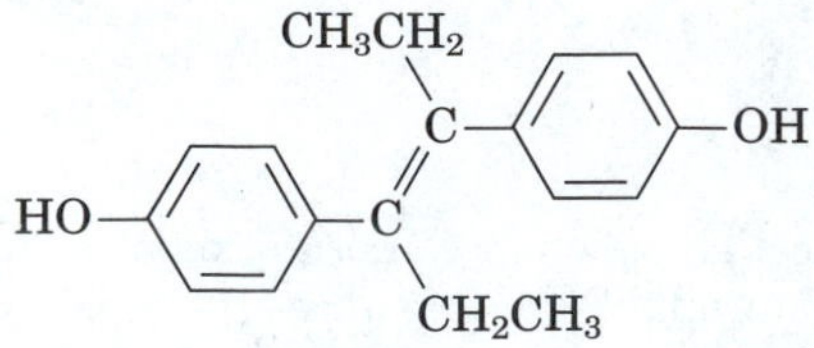

Diethylstilbestrol

28.34 Propose a synthesis of diethylstilbestrol (Problem 28.33) from phenol and any other organic compound required.

28.35 What products would you expect from reaction of estradiol (Section 28.7) with the following reagents?
(a) NaH, then CH_3I
(b) CH_3COCl, pyridine
(c) Br_2, $FeBr_3$
(d) Pyridinium chlorochromate in CH_2Cl_2

28.36 Cembrene, $C_{20}H_{32}$, is a diterpene hydrocarbon isolated from pine resin. Cembrene has a UV absorption at 245 nm, but dihydrocembrene ($C_{20}H_{34}$), the product of hydrogenation with 1 equiv H_2, has no UV absorption. On exhaustive hydrogenation, 4 equiv H_2 react, and octahydrocembrene, $C_{20}H_{40}$, is produced. On ozonolysis of cembrene, followed by treatment of the ozonide with zinc, four carbonyl-containing products are obtained:

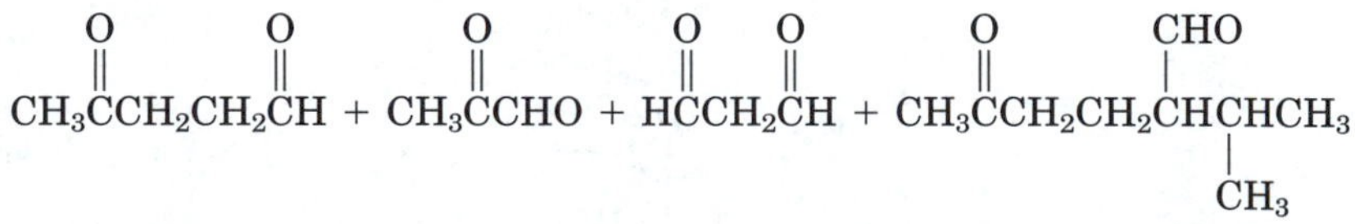

Propose a structure for cembrene that is consistent with the isoprene rule.

28.37 α-Fenchone is a pleasant-smelling terpene isolated from oil of lavender. Propose a pathway for the formation of α-fenchone from geranyl pyrophosphate. (*Hint*: A carbocation rearrangement is required.)

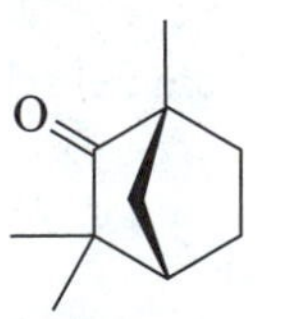

α-Fenchone

A Look Ahead

28.38 We'll see in Chapter 30 that fatty acids are synthesized by a multistep route that starts with acetate. The first step is a reaction between protein-bound acetyl and malonyl units to give 3-ketobutyryl. Show the mechanism, and tell what kind of reaction is occurring.

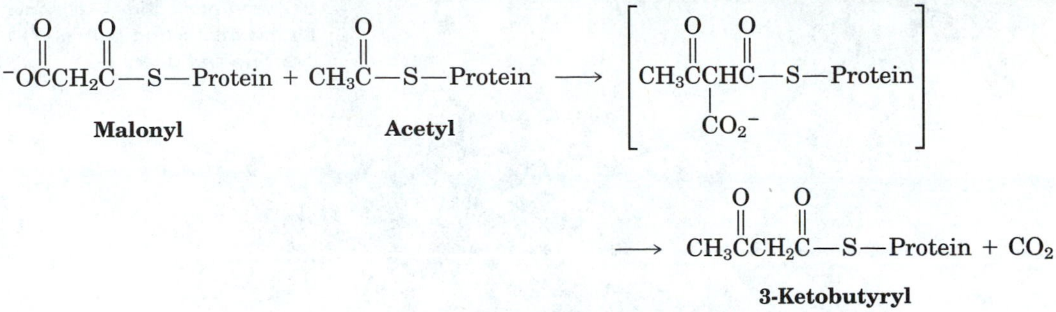

28.39 Isopentenyl pyrophosphate arises biosynthetically by loss of CO_2 from mevalonate, which itself results from a reaction of acetate with 3-ketobutyrate. Show the mechanism, and tell what kind of reaction is occurring in the formation of mevalonate.

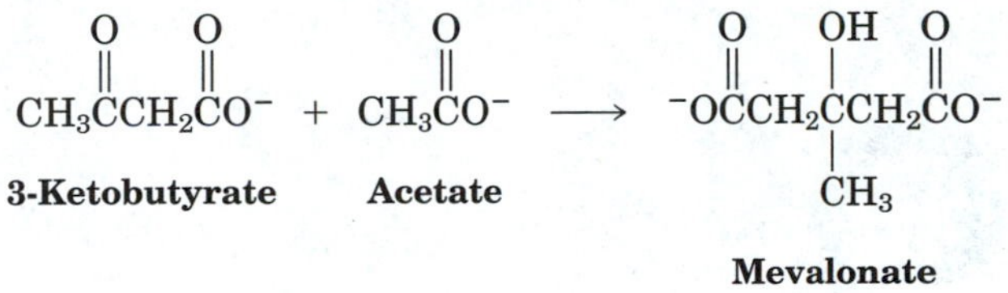

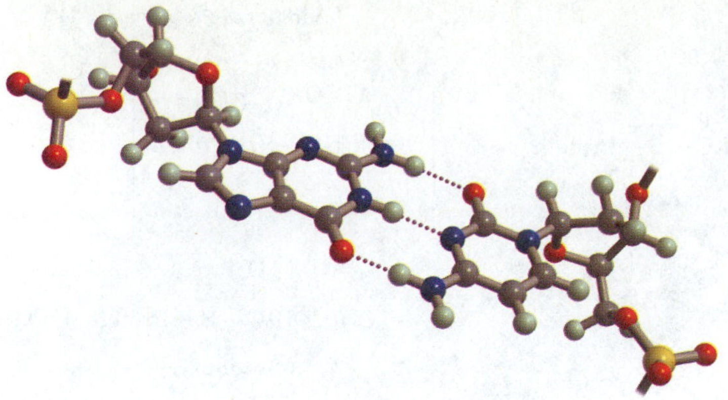

Strands of DNA are held together by hydrogen bonds between nucleotides, as shown by this interaction between cytosine and guanine.

29

Biomolecules: Heterocycles and Nucleic Acids

Cyclic organic compounds are classified either as **carbocycles** or as **heterocycles**. Carbocyclic rings contain only carbon atoms, but heterocyclic rings contain one or more different atoms in addition to carbon. Nitrogen, oxygen, and sulfur are the most common heteroatoms.

Heterocyclic compounds are common in organic chemistry, and many have important biological properties. For example, the antibiotic penicillin, the antiulcer agent cimetidine, the sedative phenobarbital, and the non-nutritive sweetener saccharin all have heterocyclic rings.

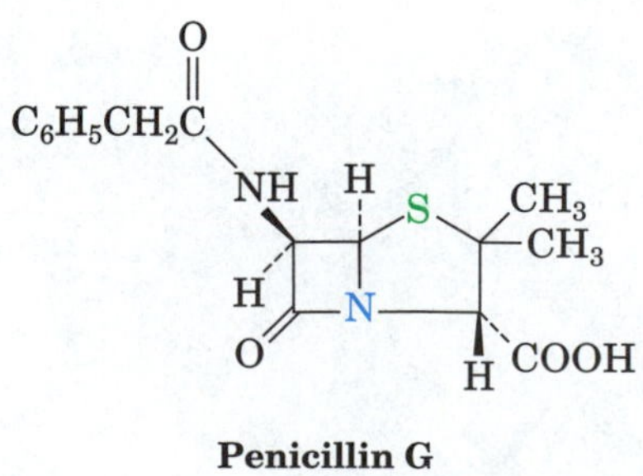

Penicillin G

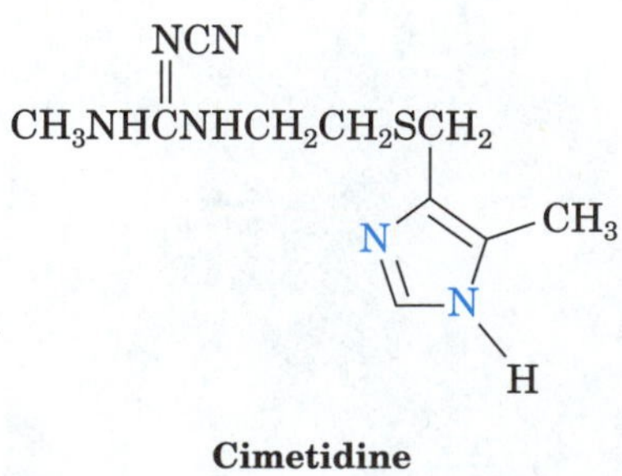

Cimetidine

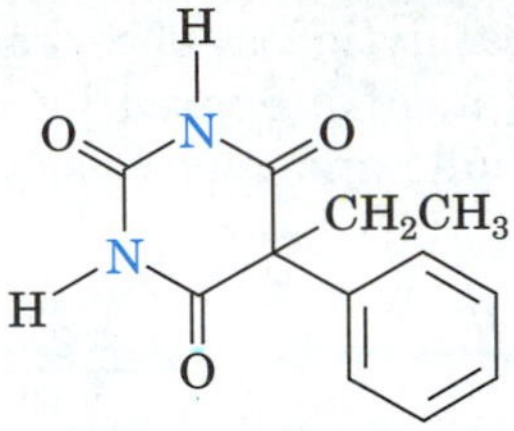

Phenobarbital

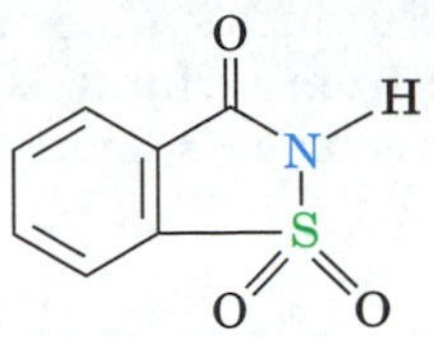

Saccharin

Heterocycles aren't new at this point; we've encountered them many times in previous chapters, usually without comment. Thus, epoxides (three-membered cyclic ethers), lactones (cyclic esters), and lactams (cyclic amides) are heterocycles, as are the solvents tetrahydrofuran (a cyclic ether) and pyridine (an unsaturated cyclic amine). In addition, carbohydrates exist as heterocyclic hemiacetals (Section 26.5).

Most heterocycles have the same chemistry as their open-chain counterparts. Lactones and acyclic esters behave similarly, lactams and acyclic amides behave similarly, and cyclic and acyclic ethers behave similarly. In certain cases, however, particularly when the ring is unsaturated, heterocycles have unique and interesting properties. Let's look first at the five-membered unsaturated heterocycles.

29.1 Five-Membered Unsaturated Heterocycles

Pyrrole, furan, and thiophene are the most common five-membered unsaturated heterocycles. Each has two double bonds and one heteroatom (N, O, or S).

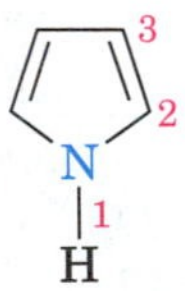

Pyrrole

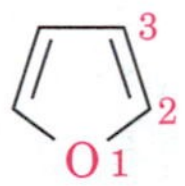

Furan

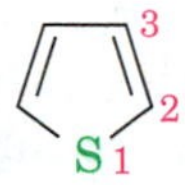

Thiophene

Pyrrole is obtained commercially either directly from coal tar or by treatment of furan with ammonia over an alumina catalyst at 400°C.

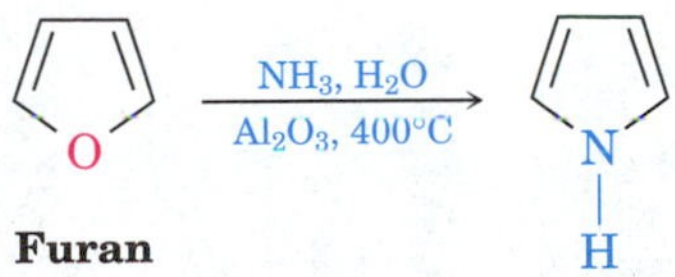

Furan

Pyrrole

Furan is synthesized by catalytic loss of carbon monoxide (decarbonylation) from furfural, which is itself prepared by acidic dehydration of the pentose sugars found in oat hulls and corncobs.

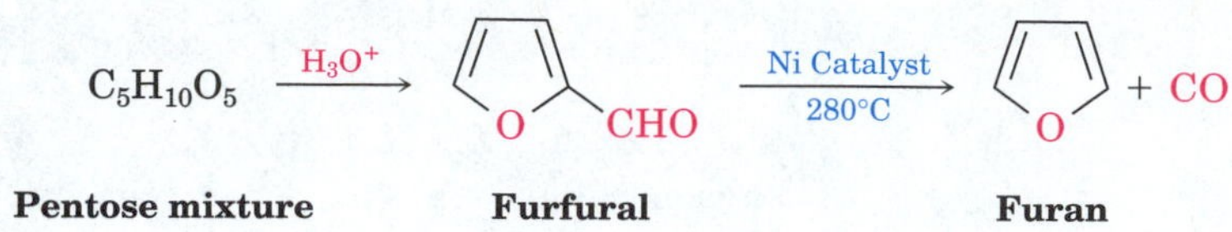

Thiophene is found in small amounts in coal tar and is synthesized industrially by cyclization of butane or butadiene with sulfur at 600°C.

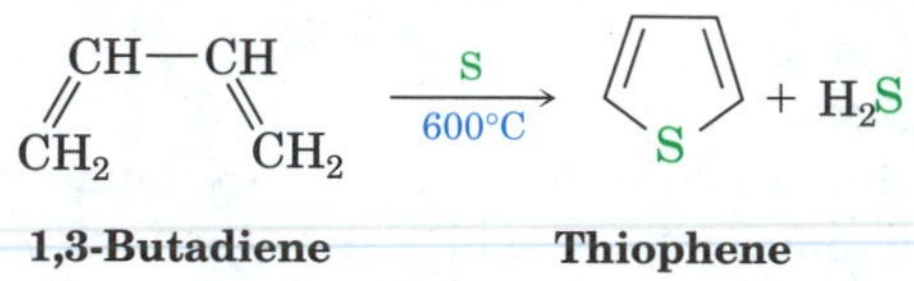

The chemistry of all three heterocyclic ring systems contains some surprises. Pyrrole, for example, is both an amine and a conjugated diene, yet its chemical properties are not consistent with either of these structural features. Unlike most other amines, pyrrole is not basic; unlike most other conjugated dienes, pyrrole undergoes electrophilic substitution rather than addition reactions. The same is true of furan and thiophene: Both react with electrophiles to give substitution products.

29.2 Structures of Pyrrole, Furan, and Thiophene

Pyrrole, furan, and thiophene give electrophilic substitution products because they're *aromatic*, as we saw in Section 15.9. Each has six π electrons in a cyclic conjugated system of overlapping *p* orbitals. Taking pyrrole as an example, each of the four carbon atoms of pyrrole contributes one π electron, and the sp^2-hybridized nitrogen atom contributes two (its lone pair). The six π electrons occupy *p* orbitals, with lobes above and below the plane of the ring, as shown in Figure 29.1. Overlap of the five *p* orbitals forms aromatic molecular orbitals just as in benzene.

Note that the pyrrole nitrogen atom uses all five valence electrons in bonding. Three electrons are used in forming three σ bonds (two to carbon and one to hydrogen), and the two lone-pair electrons are involved in aromatic π bonding.

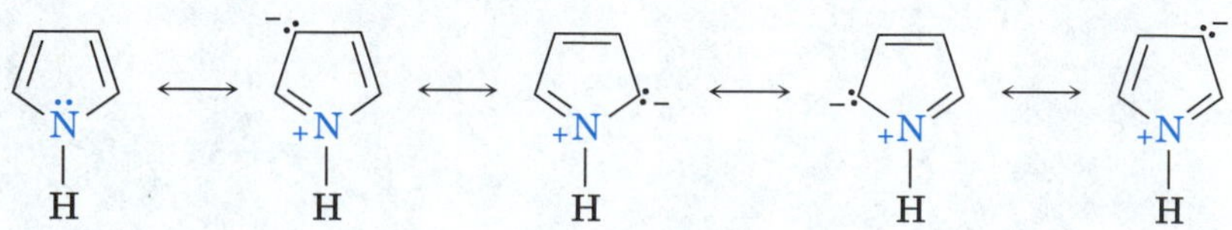

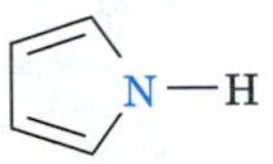

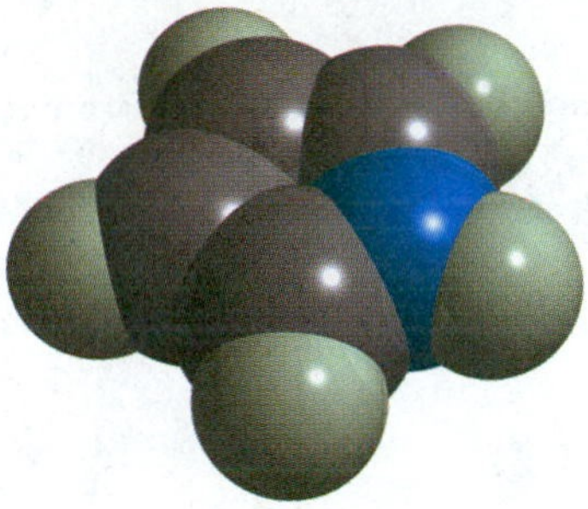

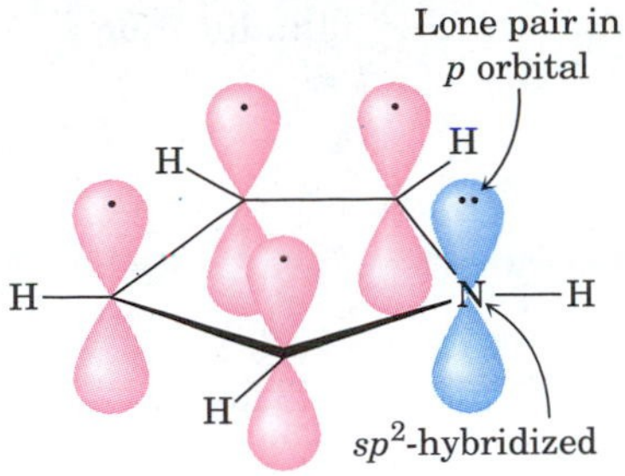

Figure 29.1 Pyrrole, a six-π-electron aromatic heterocycle.

Because the nitrogen lone pair is a part of the aromatic sextet, it is less available for bonding to acids. Pyrrole is therefore less basic and less nucleophilic than aliphatic amines (pK_a of pyrrolinium ion = 0.4). By the same token, however, the carbon atoms of pyrrole are *more* electron-rich and more nucleophilic than typical double-bond carbon atoms. The pyrrole ring is therefore reactive toward electrophiles in the same way that activated benzene rings are reactive.

PROBLEM.

29.1 Draw an orbital picture of furan. Assume that the oxygen atom is sp^2-hybridized, and show the orbitals that the two oxygen lone pairs occupy.

PROBLEM.

29.2 Pyrrole has a dipole moment μ = 1.8 D. Look at the resonance structures of pyrrole, and predict which is the positive end and which is the negative end of the dipole.

29.3 Electrophilic Substitution Reactions of Pyrrole, Furan, and Thiophene

The chemistry of pyrrole, furan, and thiophene is similar to that of activated benzene rings. In general, however, the heterocycles are more reactive toward electrophiles than benzene rings are, and low temperatures are often necessary to control the reactions. Halogenation, nitration, sulfonation, and Friedel–Crafts acylation can all be accomplished if the proper reaction conditions are chosen. A reactivity order of furan > pyrrole > thiophene is normally found.

Bromination

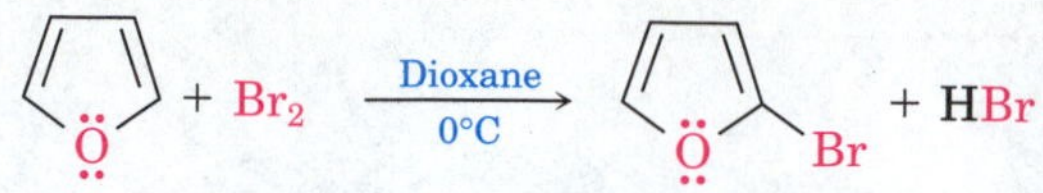

Furan **2-Bromofuran (90%)**

Nitration

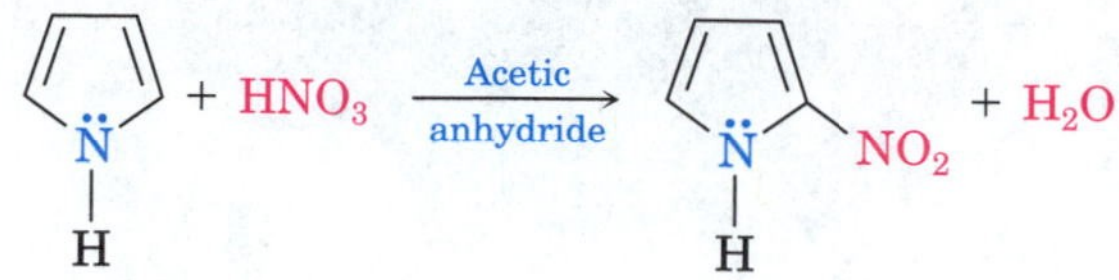

Pyrrole **2-Nitropyrrole (83%)**

Friedel–Crafts acylation

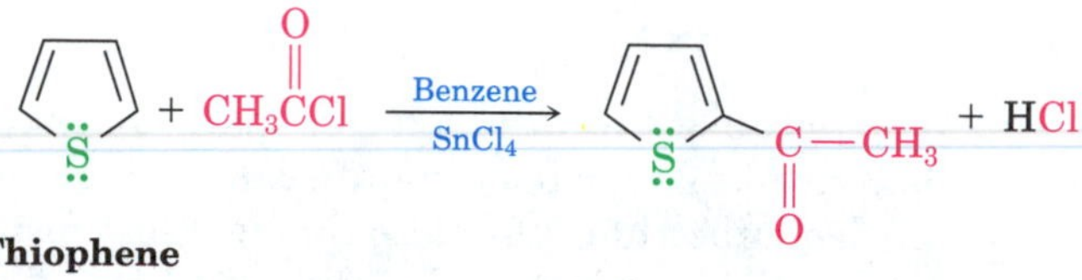

Thiophene **2-Acetylthiophene (83%)**

Electrophilic substitution normally occurs at C2, the position next to the heteroatom, because C2 is the most electron-rich position on the ring. Another way of saying the same thing is to note that electrophilic attack at C2 leads to a more stable intermediate cation having three resonance forms, while attack at C3 gives a less stable cation with only two resonance forms (Figure 29.2).

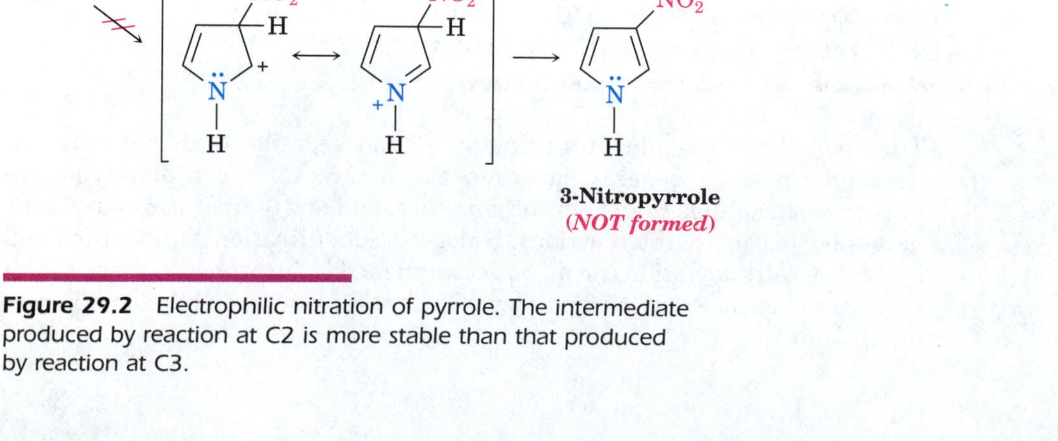

Figure 29.2 Electrophilic nitration of pyrrole. The intermediate produced by reaction at C2 is more stable than that produced by reaction at C3.

PROBLEM

29.3 Propose a mechanism to account for the fact that treatment of pyrrole with deuteriosulfuric acid, D_2SO_4, leads to formation of 2-deuteriopyrrole.

29.4 Pyridine, a Six-Membered Heterocycle

Pyridine, obtained commercially by distillation of coal tar, is the nitrogen-containing heterocyclic analog of benzene. Like benzene, pyridine is a flat, aromatic molecule with bond angles of 120° and C–C bond lengths of 1.39 Å, intermediate between typical single and double bonds. The five carbon atoms and the sp^2-hybridized nitrogen atom each contribute one π electron to the aromatic sextet. Unlike the situation in pyrrole, the lone pair of electrons on the pyridine nitrogen atom occupies an sp^2 orbital in the plane of the ring and is not involved in bonding (Figure 29.3).

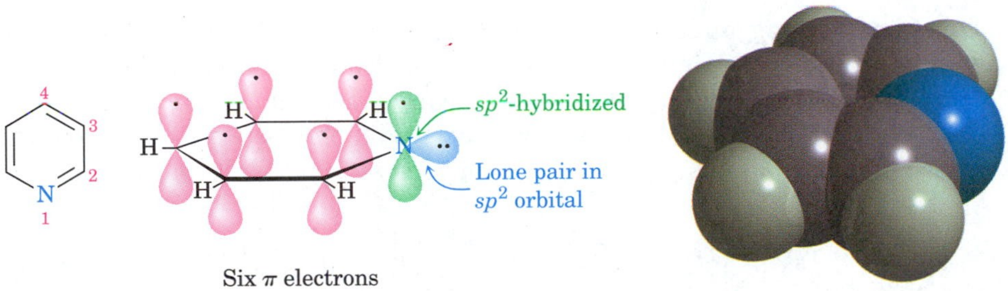

Figure 29.3 Electronic structure of pyridine, a six-π-electron, nitrogen-containing analog of benzene.

Because of its electronic structure, pyridine is a stronger base than pyrrole (pK_a of pyridinium ion = 5.25). The pyridine lone-pair electrons are not involved in aromatic π bonding and are thus available for donation to a Lewis acid. Also because of its electronic structure, however, pyridine is a *weaker* base than alkylamines. The lone-pair electrons on the nitrogen atom in pyridine are in an sp^2 orbital, while those in an alkylamine are in an sp^3 orbital. Because *s* orbitals have their maximum electron density at the nucleus but *p* orbitals have a node at the nucleus (Section 1.2), electrons in an orbital with more *s* character are held more closely to the positively charged nucleus and are less available for bonding. As a result, the sp^2-hybridized nitrogen atom (33% *s* character) in pyridine is less basic than the sp^3-hybridized nitrogen in an alkylamine (25% *s* character).

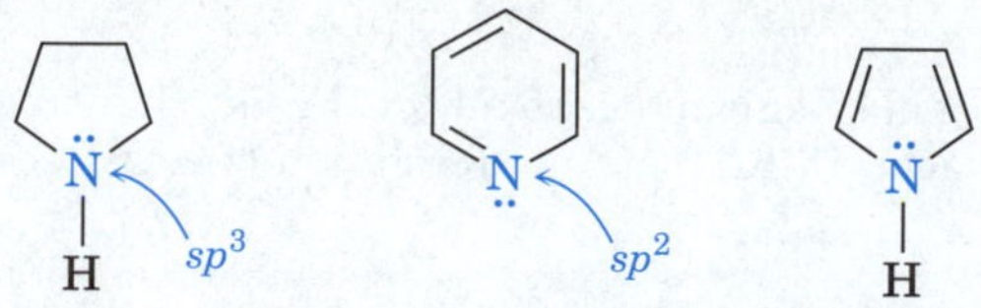

Pyrrolidinium ion (pK_a = 11.27) **Pyridinium ion** (pK_a = 5.25) **Pyrrolinium ion** (pK_a = 0.4)

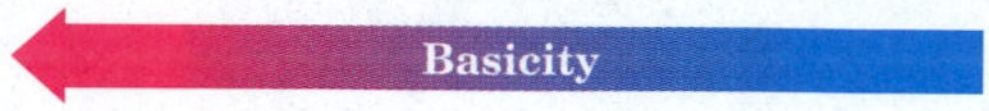

PROBLEM

29.4 Imidazolium ion has $pK_a = 6.95$. Draw an orbital structure of imidazole, and indicate which nitrogen is more basic.

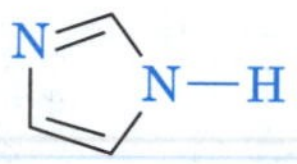

Imidazole

29.5 Electrophilic Substitution of Pyridine

The pyridine ring undergoes electrophilic aromatic substitution reactions with great difficulty. Halogenation and sulfonation can be carried out under drastic conditions, but nitration occurs in very low yield, and Friedel–Crafts reactions aren't successful. Reactions usually give the 3-substituted product.

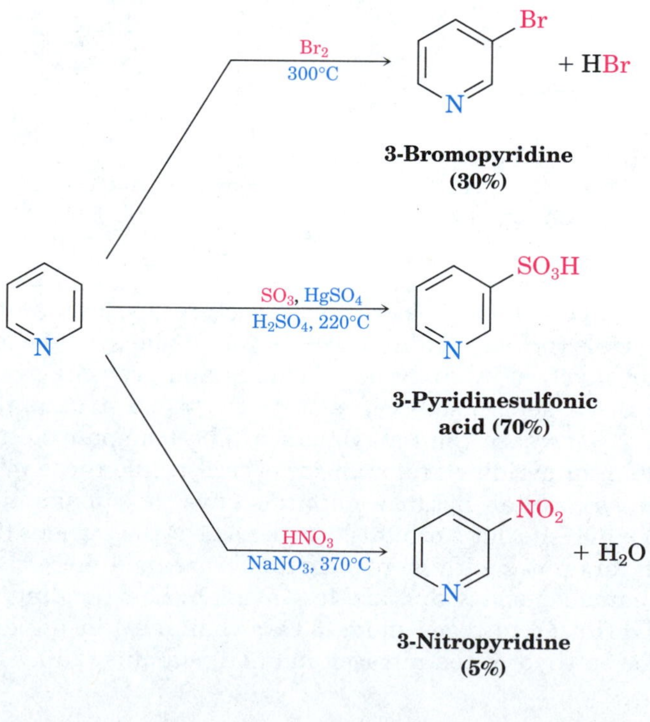

The low reactivity of pyridine toward electrophilic aromatic substitution is due to a combination of factors. Most important is that the electron density of the ring is decreased by the electron-withdrawing inductive effect of the electronegative nitrogen atom. Thus, pyridine has a substantial dipole moment $\mu = 2.26$ D, with the ring carbons acting as the positive end of the dipole. Electrophilic attack on the positively polarized carbon atoms is therefore difficult.

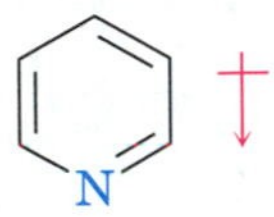

$\mu = 2.26$ D

A second factor decreasing the reactivity of the pyridine ring toward electrophilic attack is that acid–base complexation between the basic ring nitrogen atom and the attacking electrophile places a positive charge on the ring, further deactivating it.

PROBLEM.

29.5 Electrophilic aromatic substitution reactions of pyridine normally occur at C3. Draw the carbocation intermediates resulting from electrophilic attack at all possible positions, and explain the observed result.

29.6 Nucleophilic Substitution of Pyridine

In contrast to their lack of reactivity toward electrophilic substitution, 2- and 4-substituted (but not 3-substituted) halopyridines undergo *nucleophilic* aromatic substitution with relative ease.

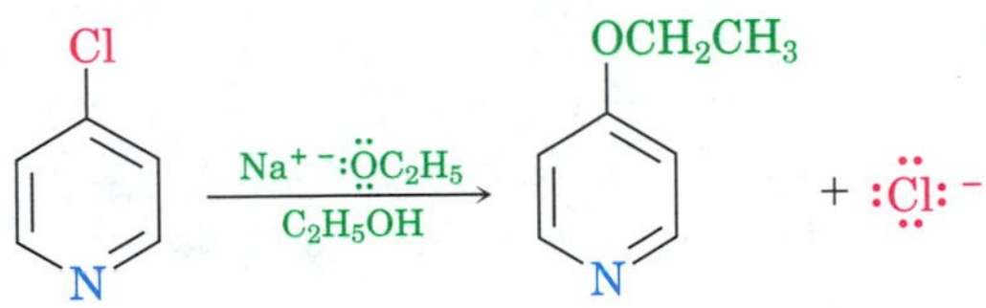

4-Chloropyridine

4-Ethoxypyridine (75%)

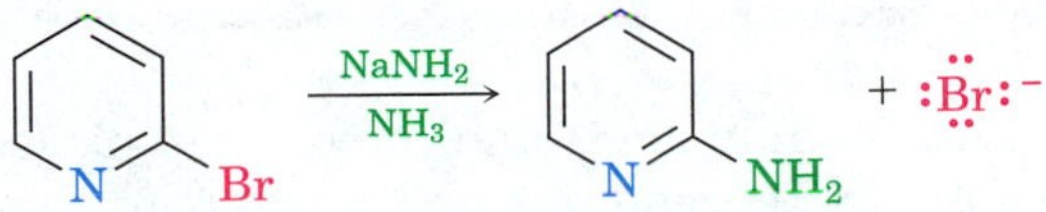

2-Bromopyridine

2-Aminopyridine (67%)

These reactions are typical nucleophilic aromatic substitutions, similar to those we saw earlier for halobenzenes (Section 16.8). Although a benzene ring needs to be further activated by the presence of an electron-withdrawing substituent for nucleophilic substitution to occur, pyridines are already sufficiently activated. Reaction occurs by addition of the nucleophile to the C=N bond, followed by loss of halide ion from the anion intermediate.

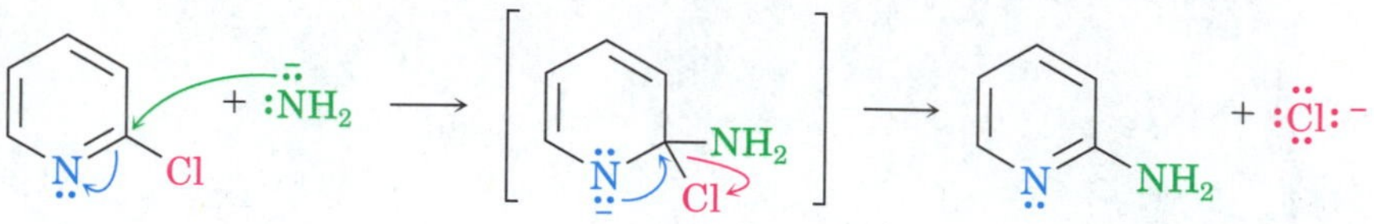

This nucleophilic aromatic substitution is in some ways analogous to the nucleophilic acyl substitution of acid chlorides (Section 21.5). In both cases, the initial addition step is favored by the ability of the electronegative atom (nitrogen or oxygen) to stabilize the anion intermediate. The intermediate then expels chloride ion to yield the substitution product.

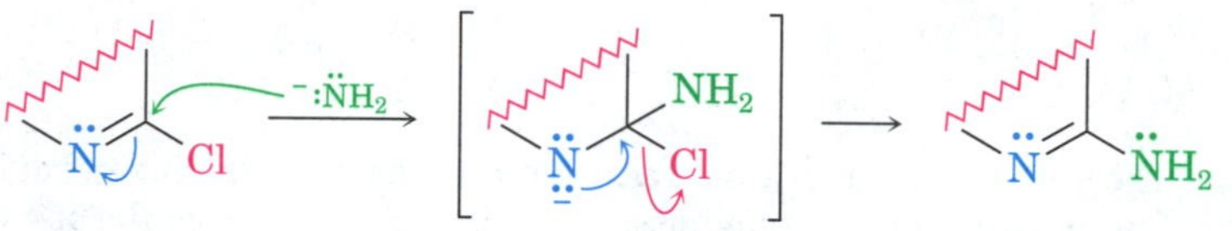

2-Chloropyridine **Stabilized anion** **2-Aminopyridine**

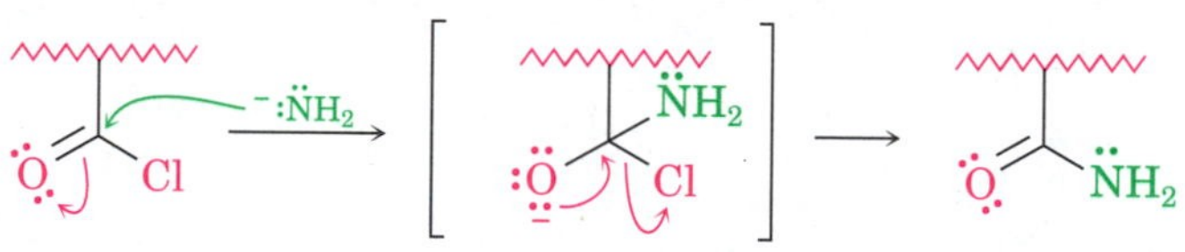

Acid chloride **Stabilized anion** **Amide**

PROBLEM

29.6 Draw the anion intermediates from nucleophilic attack at C4 of a 4-halopyridine and at C3 of a 3-halopyridine. Why does substitution of the 4-halopyridine occur so much more easily?

PROBLEM

29.7 If 3-bromopyridine is heated with $NaNH_2$, a mixture of 3- and 4-aminopyridine is obtained. Explain. (*Hint*: See Section 16.9.)

29.7 Fused-Ring Heterocycles

Quinoline, isoquinoline, and indole are **fused-ring heterocycles** containing both a benzene ring and a heterocyclic aromatic ring. All three ring systems occur commonly in nature, and many members of the class have pronounced biological activity. Thus, the quinoline alkaloid quinine is widely used as an antimalarial drug, and the indole alkaloid *N,N*-dimethyltryptamine is a powerful hallucinogen.

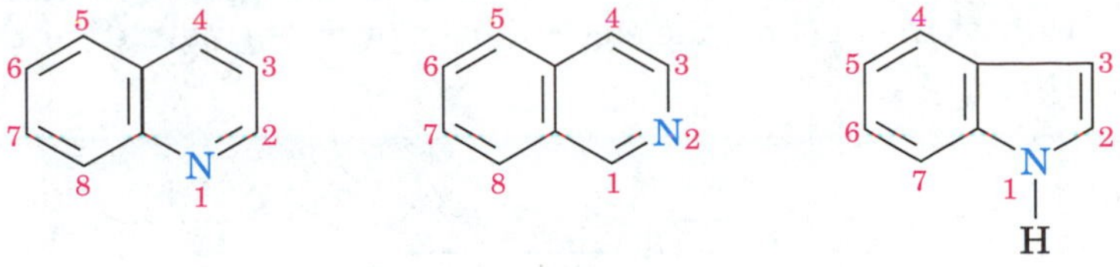

Quinoline **Isoquinoline** **Indole**

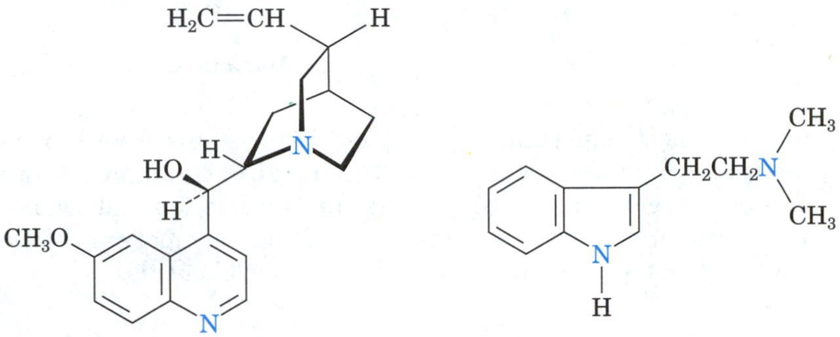

Quinine, an antimalarial drug (a quinoline alkaloid)

_N,N_-Dimethyltryptamine, a hallucinogen (an indole alkaloid)

The chemistry of these fused-ring heterocycles is just what you might expect from a knowledge of the simpler heterocycles pyridine and pyrrole. Quinoline and isoquinoline both undergo electrophilic substitution more easily than pyridine but less easily than benzene, consistent with the previous observation that pyridine rings are deactivated compared to benzene. Note that reaction occurs on the benzene ring rather than on the pyridine ring and that a mixture of C5 and C8 substitution products is obtained.

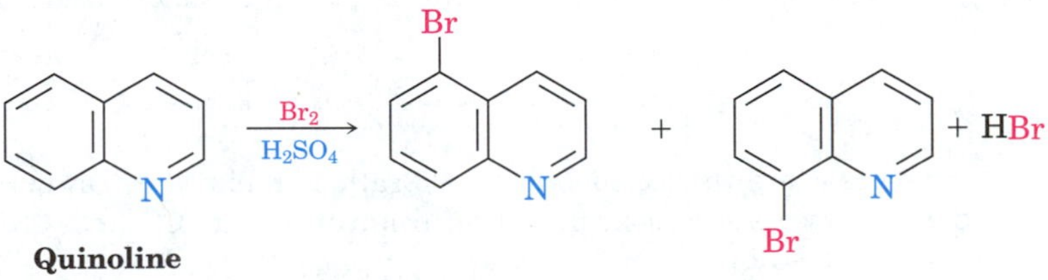

Quinoline

5-Bromoquinoline **8-Bromoquinoline**

A 51:49 ratio

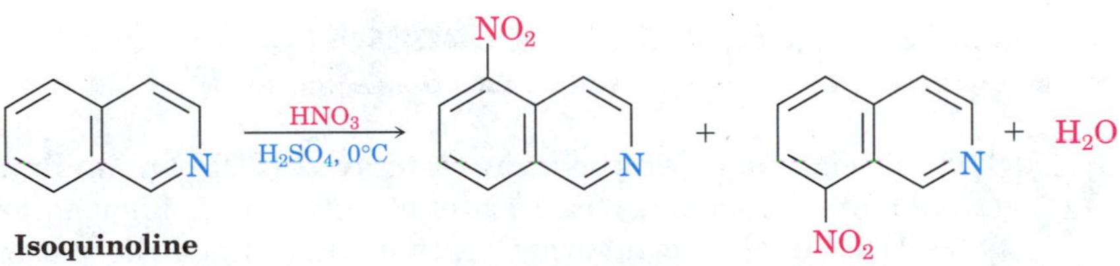

Isoquinoline

5-Nitroisoquinoline **8-Nitroisoquinoline**

A 90:10 ratio

Indole undergoes electrophilic substitution more easily than benzene but less easily than pyrrole. Again, this is consistent with the observation that pyrrole rings are more strongly activated than benzene rings. Substitution occurs at C3 of the electron-rich pyrrole ring, rather than on the benzene ring.

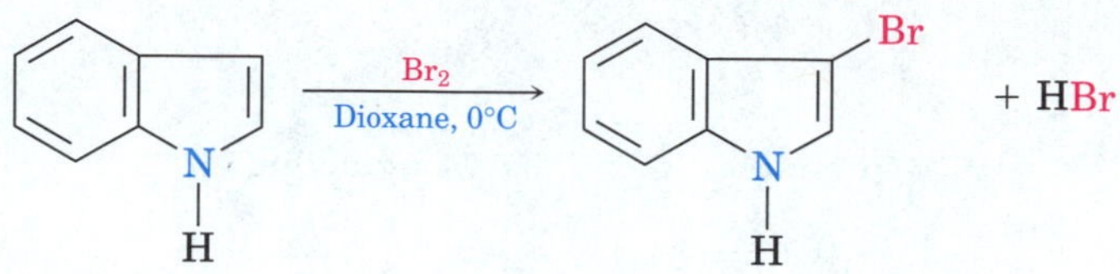

Indole **3-Bromoindole**

The most important heterocyclic ring systems from a biological viewpoint are *pyrimidine* and *purine*. **Pyrimidine** contains two nitrogens in a six-membered aromatic ring, while **purine** has four nitrogens in a fused-ring structure. Both heterocycles are essential components of the last major class of biomolecules we'll consider—the nucleic acids.

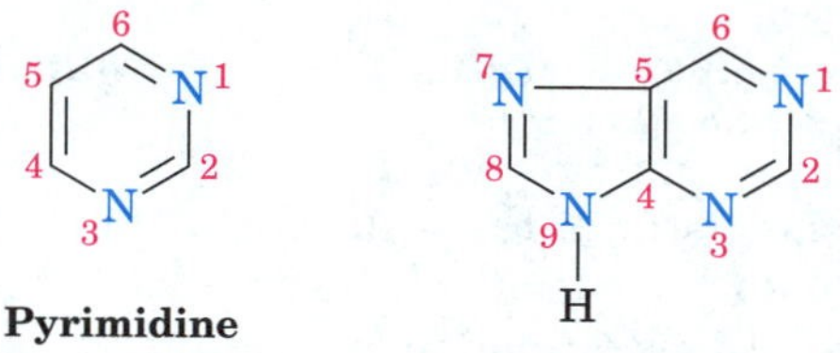

Pyrimidine **Purine**

PROBLEM

29.8 Which nitrogen atom in *N,N*-dimethyltryptamine is more basic? Explain.

PROBLEM

29.9 Indole reacts with electrophiles at C3 rather than at C2. Draw resonance forms of the intermediate cations resulting from attack at C2 and C3, and explain the observed results.

29.8 Nucleic Acids and Nucleotides

The nucleic acids, **deoxyribonucleic acid (DNA)** and **ribonucleic acid (RNA)**, are the chemical carriers of a cell's genetic information. Coded in a cell's DNA is all the information that determines the nature of the cell, controls cell growth and division, and directs biosynthesis of the enzymes and other proteins required for all cellular functions.

Just as proteins are biopolymers made of amino acid units, nucleic acids are biopolymers made of **nucleotides** joined together to form a long chain. Each nucleotide is composed of a **nucleoside** bonded to a phosphate group, and each nucleoside is composed of an aldopentose sugar linked to a heterocyclic purine or pyrimidine base.

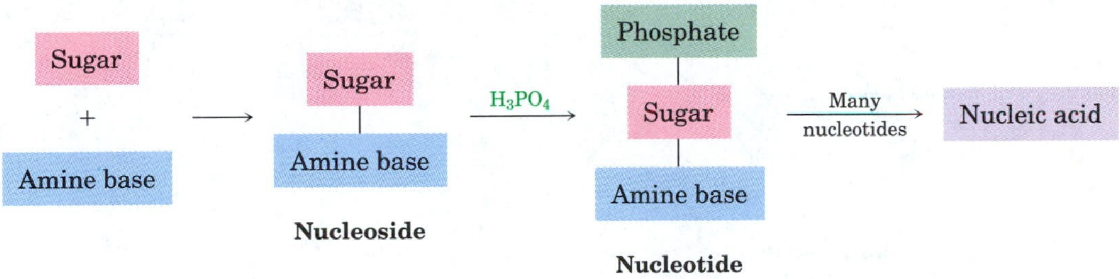

The sugar component in RNA is ribose, and the sugar in DNA is 2′-deoxyribose. (The prefix *2′-deoxy* indicates that oxygen is missing from the 2′ position of ribose. Numbers with a prime superscript refer to positions on the sugar component of a nucleotide, and numbers without a prime refer to positions on the heterocyclic amine base.)

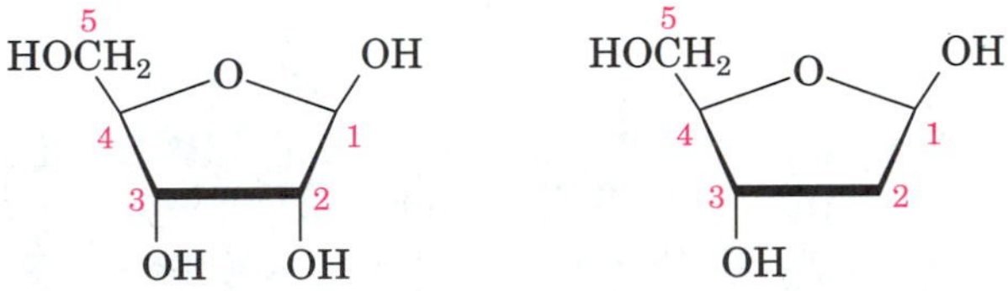

Ribose **2-Deoxyribose**

There are four different heterocyclic amine bases in deoxyribonucleotides. Two are substituted purines (**adenine** and **guanine**), and two are substituted pyrimidines (**cytosine** and **thymine**). Adenine, guanine, and cytosine also occur in RNA, but thymine is replaced in RNA by a different pyrimidine base called **uracil**.

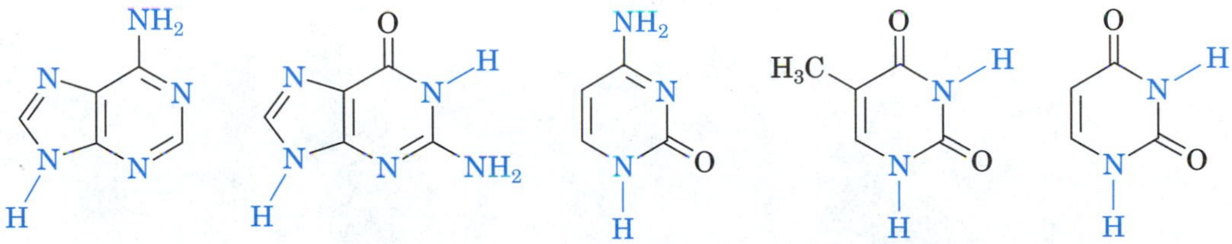

Adenine (A) **DNA** **RNA**

Guanine (G) **DNA** **RNA**

Cytosine (C) **DNA** **RNA**

Thymine (T) **DNA**

Uracil (U) **RNA**

In both DNA and RNA, the heterocyclic amine base is bonded to C1′ of the sugar, and the phosphoric acid is bonded by a phosphate ester linkage to the C5′ sugar position. Thus, nucleosides and nucleotides have the general structure shown in Figure 29.4. The full structures of all four deoxyribonucleotides and all four ribonucleotides are shown in Figure 29.5.

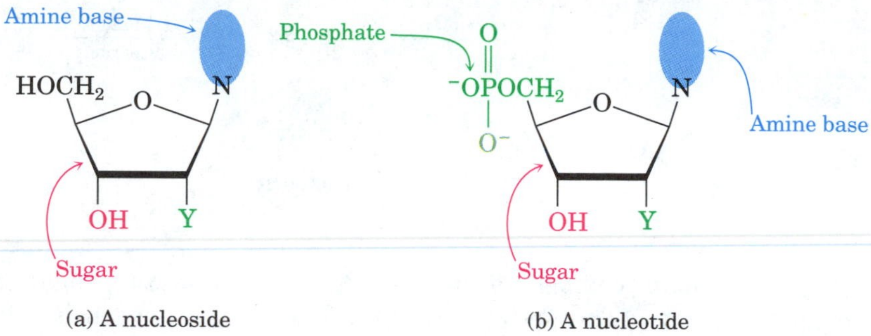

Figure 29.4 (a) A nucleoside and (b) a nucleotide. When Y = H, the sugar is deoxyribose. When Y = OH, the sugar is ribose.

Though chemically similar, DNA and RNA differ in size and have different roles within the cell. Molecules of DNA are enormous. They have molecular weights of up to 150 *billion* and lengths of up to 12 cm, and they are found mostly in the nucleus of cells. Molecules of RNA, by contrast, are much smaller (as low as 35,000 MW) and are found mostly outside the cell nucleus. We'll consider the two kinds of nucleic acids separately, beginning with DNA.

PROBLEM.

29.10 2′-Deoxythymidine exists largely in the lactam form rather than in the tautomeric lactim form. Explain.

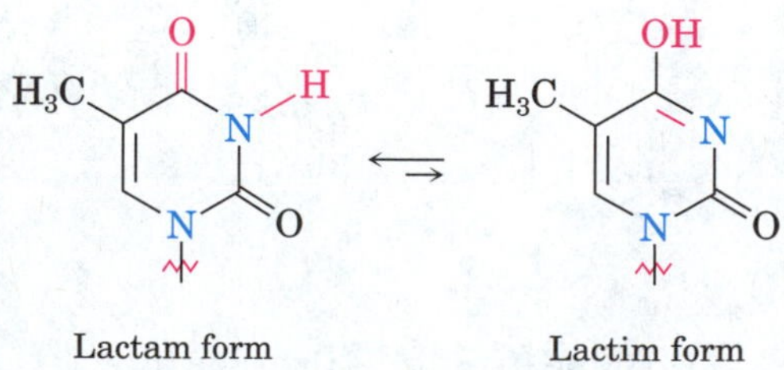

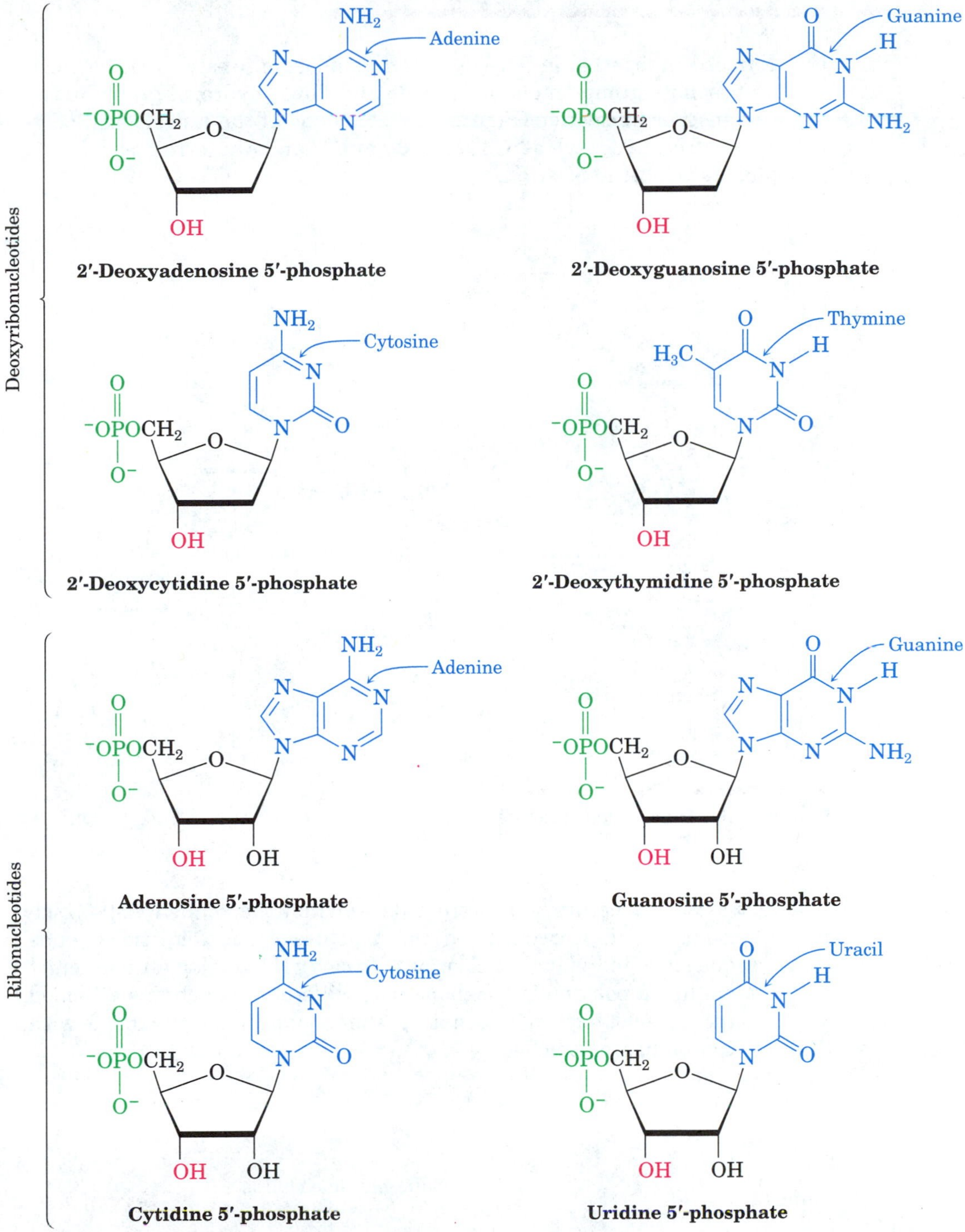

Figure 29.5 Names and structures of the four deoxyribonucleotides and the four ribonucleotides.

29.9 Structure of DNA

Nucleotides join together in DNA by forming a phosphate ester bond between the 5′-phosphate group on one nucleotide and the 3′-hydroxyl group on the sugar of another nucleotide (Figure 29.6). One end of the nucleic acid polymer has a free hydroxyl at C3′ (the **3′ end**), and the other end has a phosphate at C5′ (the **5′ end**).

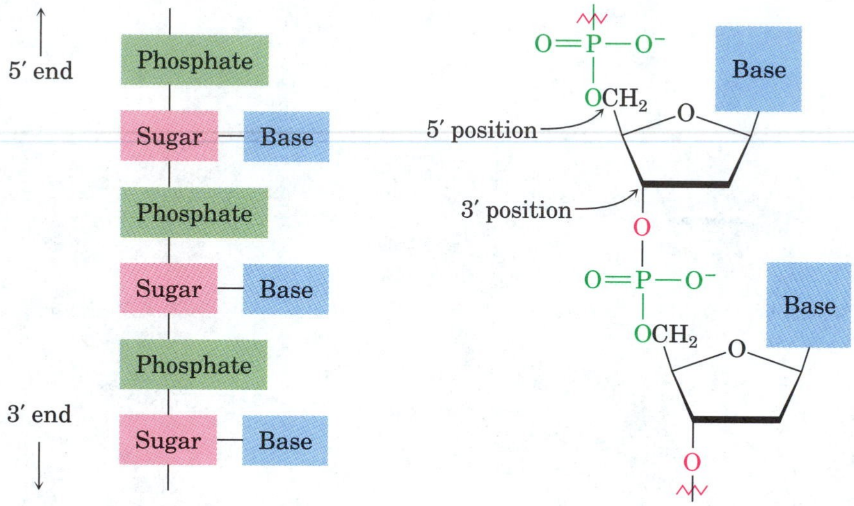

Figure 29.6 Generalized structure of DNA.

Just as the structure of a protein depends on the sequence in which individual amino acids are connected, the structure of a nucleic acid depends on the sequence of individual nucleotides. To carry the analogy further, just as a protein has a polyamide backbone with different side chains attached to it, a nucleic acid has an alternating sugar–phosphate backbone with different amine bases attached.

A protein

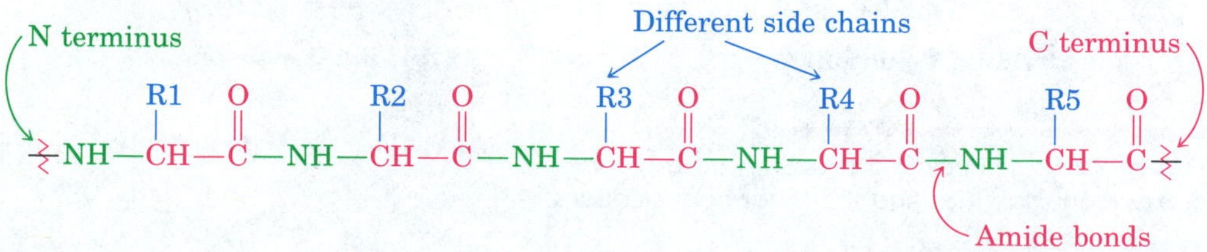

A nucleic acid

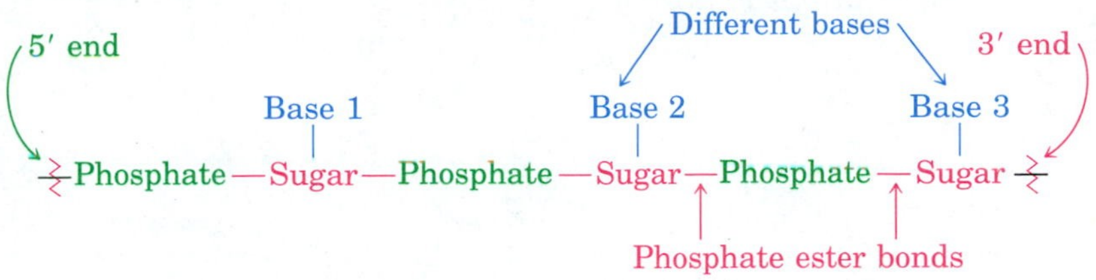

The sequence of nucleotides in a chain is described by starting at the 5′ end and identifying the bases in order of occurrence. Rather than write the full name of each nucleotide, though, it's more convenient to use abbreviations: A for adenosine, T for thymine, G for guanosine, and C for cytidine. Thus, a typical DNA sequence might be written as T-A-G-G-C-T.

PROBLEM

29.11 Draw the full structure of the DNA dinucleotide A-G.

PROBLEM

29.12 Draw the full structure of the RNA dinucleotide U-A.

29.10 Base Pairing in DNA: The Watson–Crick Model

Samples of DNA isolated from different tissues of the same species have the same proportions of heterocyclic bases, but samples from different species can have greatly differing proportions of bases. Human DNA, for example, contains about 30% each of adenine and thymine and about 20% each of guanine and cytosine. The bacterium *Clostridium perfringens*, however, contains about 37% each of adenine and thymine and only 13% each of guanine and cytosine. Note that in both examples the bases occur in pairs. Adenine and thymine are usually present in equal amounts, as are cytosine and guanine. Why should this be?

In 1953, James Watson[1] and Francis Crick[2] made their now-classic proposal for the secondary structure of DNA. According to the Watson–Crick model, DNA consists of two polynucleotide strands coiled around each other in a **double helix**. The two strands run in opposite directions and are held together by hydrogen bonds between specific pairs of bases. Adenine (A) and thymine (T) form strong hydrogen bonds to each other but not to C or G.

[1]James Dewey Watson (1928–); b. Chicago; Ph.D. Indiana; professor, Harvard University; Nobel Prize in medicine (1960).

[2]Francis H. C. Crick (1916–); b. England; Ph.D. Cambridge; professor, Cambridge University; Nobel Prize in medicine (1960).

Similarly, guanine (G) and cytosine (C) form strong hydrogen bonds to each other but not to A or T.

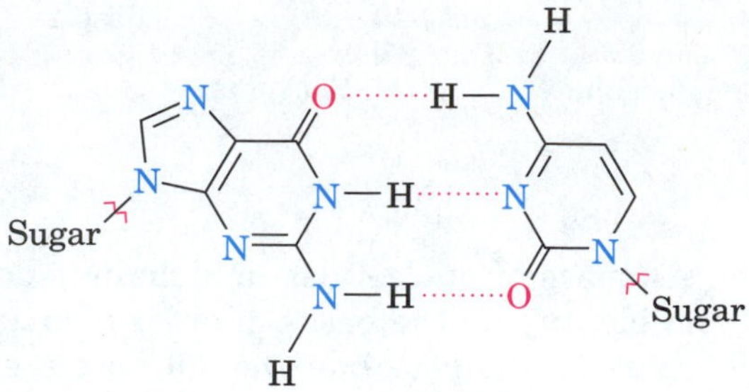

(Guanine) G ⋮⋮⋮⋮⋮⋮ C (Cytosine)

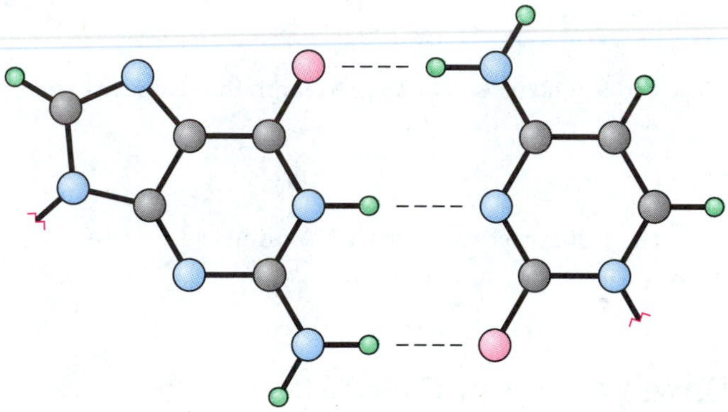

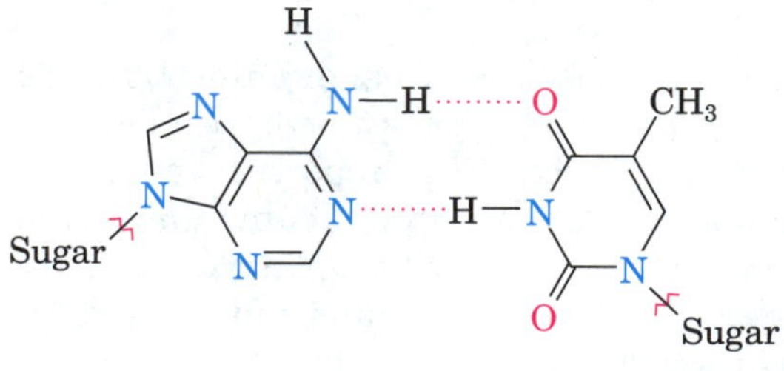

(Adenine) A : : : : : : T (Thymine)

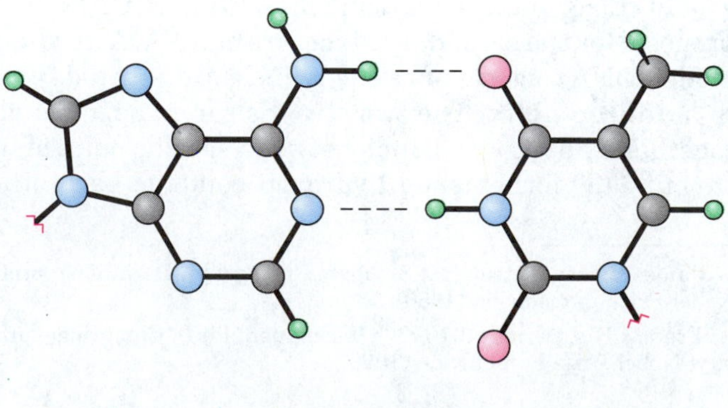

The two strands of the DNA double helix aren't identical; rather, they're complementary. Whenever a C base occurs in one strand, a G base occurs opposite it in the other strand. When an A base occurs in one strand, a T appears opposite it in the other strand. This complementary pairing of bases explains why A and T are always found in equal amounts, as are C and G. Figure 29.7 illustrates this base pairing, showing how the two complementary strands are coiled into the double helix. X-ray measurements show that the DNA double helix is 20 Å wide, that there are 10 base pairs in each full turn, and that each turn is 34 Å in height.

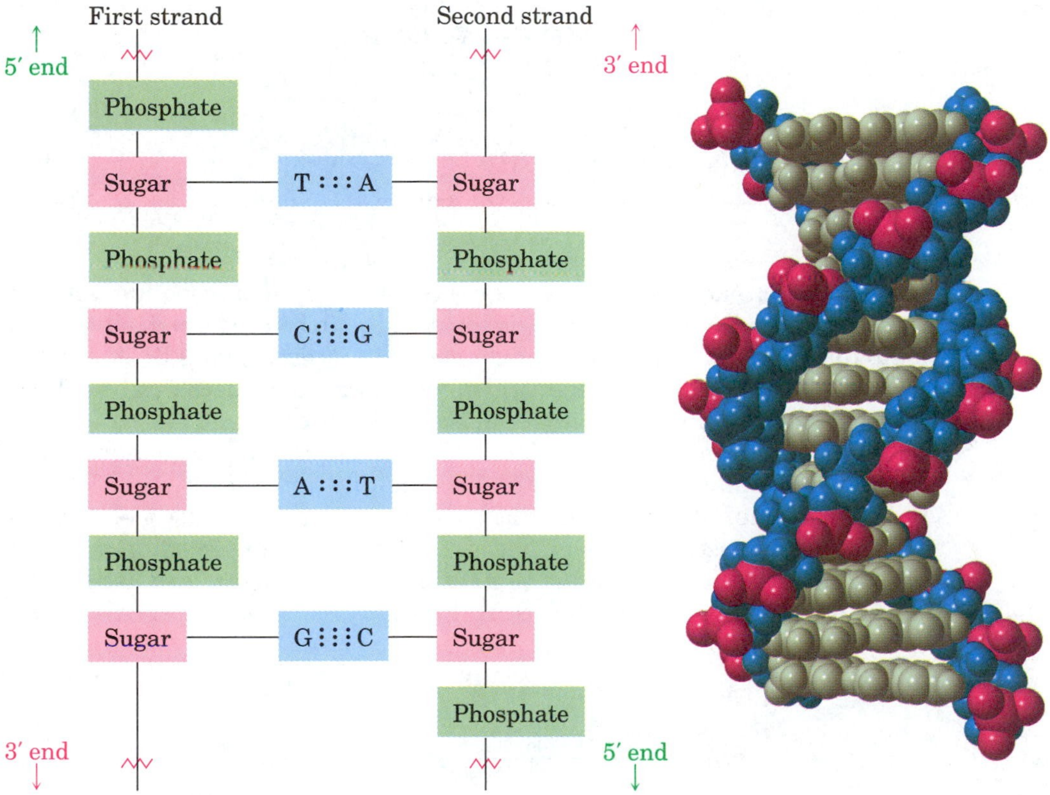

Figure 29.7 Complementarity in base pairing in the DNA double helix as shown by this computer-generated structure. The sugar–phosphate backbone runs along the outside of the helix, while the amine bases hydrogen-bond to one another on the inside.

A helpful mnemonic device to remember the nature of the hydrogen bonding between the four DNA bases is the simple phrase "Pure silver taxi."

Pure	Silver	Taxi
Pur	Ag	TC
The purine bases,	A and G,	hydrogen bond to T and C.

Notice in Figure 29.7 that the two strands of the double helix coil in such a way that two kinds of "grooves" result, a *major groove* that's 12 Å wide and a *minor groove* that's 6 Å wide. Interestingly, a variety of flat, polycyclic aromatic molecules are able to fit sideways into one of the grooves between the strands and *intercalate*, or insert themselves, between the stacked base pairs. Many cancer-causing and cancer-preventing agents are thought to function by intercalating with DNA in this way.

PROBLEM

29.13 What sequence of bases on one strand of DNA is complementary to the following sequence on another strand?

G-G-C-T-A-A-T-C-C-G-T

29.11 Nucleic Acids and Heredity

A DNA molecule is the chemical repository of an organism's genetic information, which is stored as a sequence of deoxyribonucleotides strung together in the DNA chain. For this information to be preserved and passed on to future generations, a mechanism must exist for copying DNA. For the information to be used, a mechanism must exist for decoding the DNA message and for implementing the instructions it contains.

What Crick has termed the "central dogma of molecular genetics," says that the function of DNA is to store information and pass it on to RNA at the proper time. The function of RNA, in turn, is to read, decode, and use the information received from DNA to make proteins. By decoding the right bit of DNA at the right time in the right place, an organism can use genetic information to synthesize the many thousands of proteins necessary for carrying out its biochemical reactions.

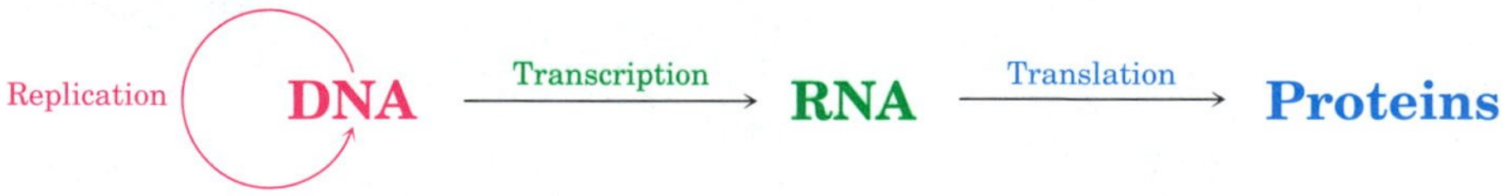

Three fundamental processes take place in the transfer of genetic information:

1. **Replication** is the process by which identical copies of DNA are made so that information can be preserved and handed down to offspring.
2. **Transcription** is the process by which the genetic messages contained in DNA are read and carried out of the nucleus to parts of the cell called ribosomes where protein synthesis occurs.
3. **Translation** is the process by which the genetic messages are decoded and used to build proteins.

29.12 Replication of DNA

Replication of DNA is an enzyme-catalyzed process that begins by a partial unwinding of the double helix. As the strands separate and bases are exposed, new nucleotides line up on each strand in a complementary manner, A to T and C to G, and two new strands begin to grow. Each new strand is complementary to its old template strand, and two new identical DNA double helices are produced (Figure 29.8). Crick described the process best when he used the analogy of the two DNA strands fitting together like a hand in a glove. The hand and glove separate, a new hand forms inside the glove, and a new glove forms around the hand. Two identical copies now exist where only one existed before.

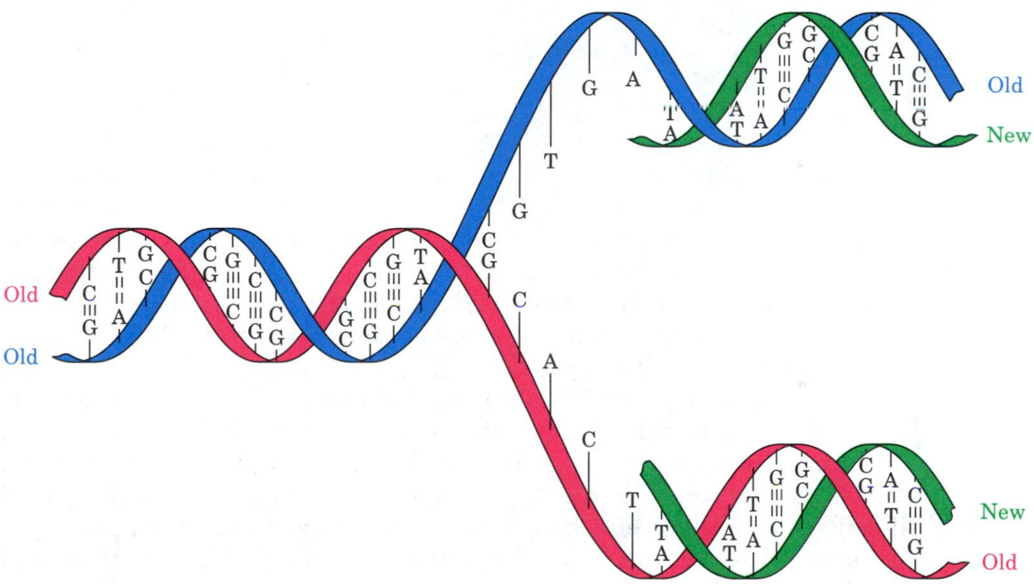

Figure 29.8 Schematic representation of DNA replication. The original double-stranded DNA partially unwinds, bases are exposed, nucleotides line up on each strand in a complementary manner, and two new strands begin to grow.

The process by which the individual nucleotides are joined to create new DNA strands involves many steps and many different enzymes. Addition of new nucleotide units to the growing chain is catalyzed by the enzyme DNA polymerase and occurs by addition of a 5′-mononucleotide triphosphate to the free 3′-hydroxyl group of the growing chain.

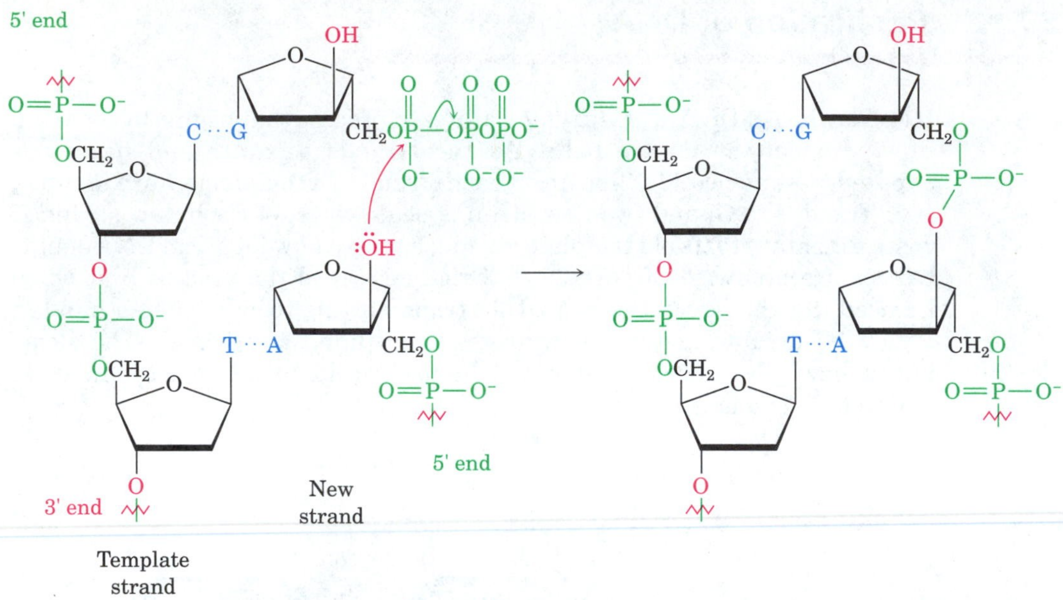

Both new DNA strands are synthesized in the same 5′ → 3′ direction, which implies that they can't be made in exactly the same way. Since the two complementary DNA strands are lined up in opposite directions, one strand must have its 3′ end near the point of unraveling (the **replication fork**) while the other strand has its 5′ end near the replication fork. What evidently happens is that the complement of the original 3′ → 5′ strand is synthesized continuously in a single piece, but the complement of the original 5′ → 3′ strand is synthesized discontinuously in small pieces that are then linked by DNA ligase enzymes.

The magnitude of the replication process is truly staggering. The nucleus of every human cell contains 46 chromosomes (23 pairs), each of which consists of one very large DNA molecule. Each chromosome, in turn, is made up of several thousand DNA segments called *genes*, and the sum of all genes in a human cell (the *genome*) is estimated to be approximately 3 billion base pairs. A single DNA chain might have a length of over 12 cm and contain up to 250 million pairs of bases. Despite the size of these enormous molecules, their base sequence is faithfully copied during replication. The copying process takes only minutes, and an error occurs only about once each 10–100 billion bases.

29.13 Structure and Synthesis of RNA: Transcription

As noted previously, RNA is structurally similar to DNA but contains ribose rather than deoxyribose and uracil rather than thymine. There are three major kinds of RNA, each of which serves a specific function. All three are much smaller molecules than DNA and all remain single-stranded rather than double-stranded.

> **Messenger RNA (mRNA)** carries genetic messages from DNA to *ribosomes*, small granular particles in the cytoplasm of a cell where protein synthesis takes place.
>
> **Ribosomal RNA (rRNA)** complexed with protein provides the physical makeup of the ribosomes.
>
> **Transfer RNA (tRNA)** transports amino acids to the ribosomes where they are joined together to make proteins.

The conversion of the information in DNA into proteins begins in the nucleus of cells with the synthesis of mRNA by **transcription** of DNA. Several turns of the DNA double helix unwind, forming a "bubble" and exposing the bases of the two strands. Ribonucleotides line up in the proper order by hydrogen bonding to their complementary bases on DNA, bond formation occurs in the $5' \rightarrow 3'$ direction, and the growing RNA molecule unwinds from DNA (Figure 29.9).

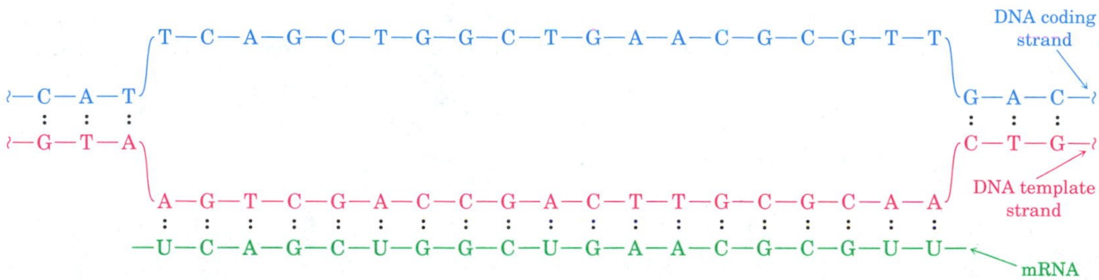

Figure 29.9 Biosynthesis of RNA using a DNA segment as template.

Unlike what happens in DNA replication, where both strands are copied, only one of the two DNA strands is transcribed into mRNA. The strand that contains the gene is called the **coding strand**, or **sense strand**, and the strand that gets transcribed is called the **template strand**, or **antisense strand**. Since the template strand and the coding strand are complementary, and since the template strand and the RNA molecule are also complementary, it follows that *the RNA molecule produced during transcription is a copy of the coding strand*. The only difference is that the RNA molecule has a U everywhere the DNA coding strand has a T.

Transcription of DNA by the process just discussed raises many questions. How does the DNA know where to unwind? Where along the chain does one gene stop and the next one start? How do the ribonucleotides know the right place along the template strand to begin lining up and the right place to stop? The picture that has emerged in the last decade is that a DNA chain contains specific base sequences called *promoter sites* that lie at positions 10 base pairs and 35 base pairs upstream from the beginning of the coding region and signal the beginning of a gene. Similarly, there are other base sequences near the end of the gene that signal a stop.

Another part of the picture that has recently emerged is that genes are not necessarily continuous segments of the DNA chain. Often a gene will begin in one small section of DNA called an **exon**, then be interrupted by a seemingly nonsensical section called an **intron**, and then take up again further down the chain in another exon. The final mRNA molecule results only after the nonsense sections are cut out and the remaining pieces are spliced together. Current evidence is that up to 90% of human DNA is made up of introns and only about 10% of DNA actually contains genetic instructions.

PROBLEM.

29.14 Show how uracil can form strong hydrogen bonds to adenine.

PROBLEM.

29.15 What RNA base sequence is complementary to the following DNA base sequence?

G-A-T-T-A-C-C-G-T-A

29.14 RNA and Protein Biosynthesis: Translation

The primary cellular function of RNA is to direct biosynthesis of the thousands of diverse peptides and proteins required by an organism. The mechanics of protein biosynthesis are directed by mRNA and take place on ribosomes, small granular particles in the cytoplasm of a cell that consist of about 60% ribosomal RNA and 40% protein. On the ribosome, mRNA serves as a template to pass on the genetic information it has transcribed from DNA.

The specific ribonucleotide sequence in mRNA forms a message that determines the order in which different amino acid residues are to be joined.

Each "word," or **codon**, along the mRNA chain consists of a sequence of three ribonucleotides that is specific for a given amino acid. For example, the series U-U-C on mRNA is a codon directing incorporation of the amino acid phenylalanine into the growing protein. Of the $4^3 = 64$ possible triplets of the four bases in RNA, 61 code for specific amino acids, and 3 code for chain termination. Table 29.1 shows the meaning of each codon.

Table 29.1 Codon Assignments of Base Triplets

First base (5′ end)	*Second base*	*Third base (3′ end)*			
		U	*C*	*A*	*G*
U	U	Phe	Phe	Leu	Leu
	C	Ser	Ser	Ser	Ser
	A	Tyr	Tyr	Stop	Stop
	G	Cys	Cys	Stop	Trp
C	U	Leu	Leu	Leu	Leu
	C	Pro	Pro	Pro	Pro
	A	His	His	Gln	Gln
	G	Arg	Arg	Arg	Arg
A	U	Ile	Ile	Ile	Met
	C	Thr	Thr	Thr	Thr
	A	Asn	Asn	Lys	Lys
	G	Ser	Ser	Arg	Arg
G	U	Val	Val	Val	Val
	C	Ala	Ala	Ala	Ala
	A	Asp	Asp	Glu	Glu
	G	Gly	Gly	Gly	Gly

The message carried by mRNA is read by transfer RNA (tRNA) in a process called **translation**. There are 61 different tRNA's, one for each of the 61 codons in Table 29.1 that specifies an amino acid. A typical tRNA is roughly the shape of a cloverleaf, as shown in Figure 29.10 (p. 1152). It consists of about 70–100 ribonucleotides and is bonded to a specific amino acid by an ester linkage through the 3′ hydroxyl on ribose at the 3′ end of the tRNA. Each tRNA also contains in its structure a segment called an **anticodon**, a sequence of three ribonucleotides complementary to the codon sequence. For example, the codon sequence U-U-C present on mRNA is read by a phenylalanine-bearing tRNA having the complementary anticodon base sequence A-A-G.

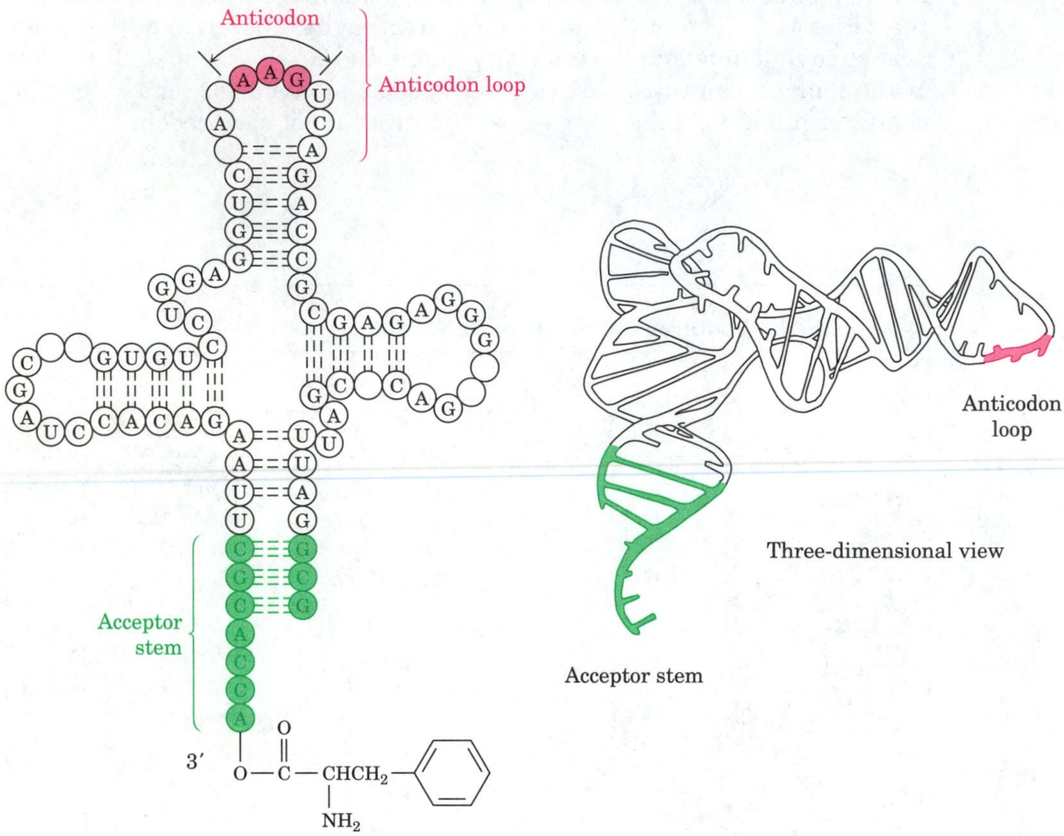

Figure 29.10 Structure of a tRNA molecule. The tRNA is a roughly cloverleaf-shaped molecule containing an anticodon triplet on one "leaf" and a covalently attached amino acid unit at its 3′ end. The example shown is a yeast tRNA that codes for phenylalanine. The nucleotides not specifically identified are chemically modified analogs of the four standard nucleotides.

As each successive codon on mRNA is read, different tRNA's bring the correct amino acids into position for enzyme-mediated transfer to the growing peptide. When synthesis of the proper protein is completed, a "stop" codon signals the end, and the protein is released from the ribosome. The process is illustrated schematically in Figure 29.11.

PROBLEM

29.16 List codon sequences for the following amino acids:
(a) Ala (b) Phe (c) Leu (d) Tyr

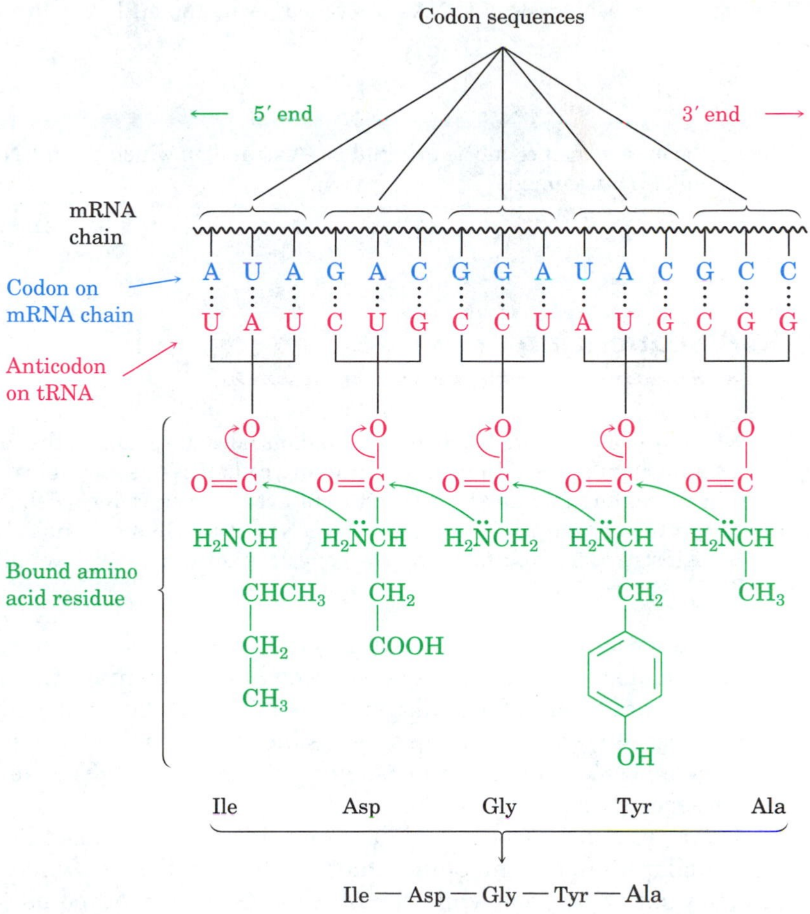

Figure 29.11 A schematic representation of protein biosynthesis. The codon base sequences on mRNA are read by tRNA's containing complementary anticodon base sequences. Transfer RNA's assemble the proper amino acids into position for incorporation into the growing peptide.

PROBLEM. . . . .

29.17 List anticodon sequences on the tRNA's carrying the amino acids shown in Problem 29.16.

PROBLEM. . . . .

29.18 What amino acid sequence is coded for by the following mRNA base sequence?

CUU-AUG-GCU-UGG-CCC-UAA

PROBLEM.

29.19 What anticodon sequences of tRNA's are coded for by the mRNA in Problem 29.18?

PROBLEM.

29.20 What is the base sequence in the original DNA strand on which the mRNA sequence in Problem 29.18 was made?

29.15 DNA Sequencing

> When we work out the structure of DNA molecules, we examine the fundamental level that underlies all processes in living cells. DNA is the information store that ultimately dictates the structure of every gene product, delineates every part of the organism. The order of the bases along DNA contains the complete set of instructions that make up the genetic inheritance. (Walter Gilbert, Nobel Prize Lecture, 1980)

One of the greatest scientific revolutions in history is now under way in molecular biology as scientists are learning how to manipulate and harness the genetic machinery of organisms. None of the extraordinary advances of the past decade would have been possible, however, were it not for the discovery in 1977 of methods for sequencing immense DNA chains to find the messages therein.

There are two methods of DNA sequencing in general use. Both operate along similar lines, but the Sanger method, often called the *dideoxy method*, uses enzymatic reactions while the Maxam–Gilbert[3] method uses chemical techniques. There are five steps to the Maxam–Gilbert method:

Step 1 The first problem in DNA sequencing is to cleave the enormous DNA chain at specific points to produce smaller, more manageable pieces, a task accomplished by the use of enzymes called **restriction endonucleases**. Each different restriction enzyme, of which more than 200 are available, cleaves a DNA molecule at well-defined points in the chain where a specific base sequence occurs. For example, the restriction enzyme *Alu I* cleaves between G and C in the four-base sequence AG–CT (Figure 29.12). If the original DNA molecule is cut with another restriction enzyme having a different specificity for cleavage, still other segments are produced whose sequences partially overlap those produced by the first enzyme. Sequencing of all the segments, followed by identification of the overlapping sequences, then allows complete DNA sequencing.

[3]Walter Gilbert (1932–); b. Boston; Ph.D. Cambridge University (1957); professor, Harvard University (1958–); Nobel Prize (1980).

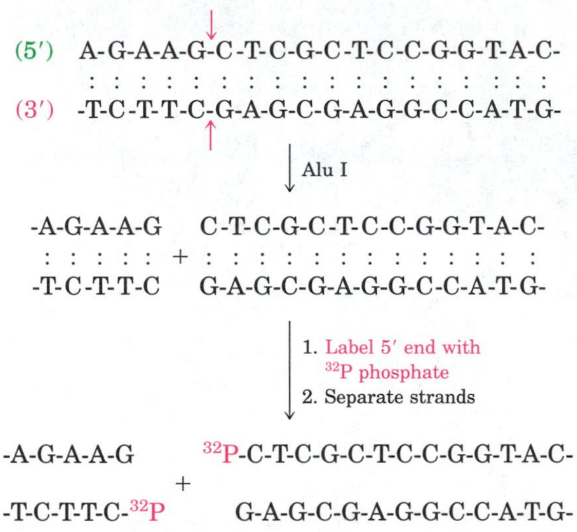

Figure 29.12 Cleavage of a double-stranded DNA molecule with the restriction enzyme Alu I cleaves at the sequence AG-CT. After cleavage, the fragments are isolated and each is radioactively labeled at its 5′ end by enzyme-catalyzed formation of a ^{32}P-containing phosphate ester. The strands are then separated.

Step 2 After cleavage of the DNA into smaller, more manageable pieces called *restriction fragments*, each piece is radioactively tagged by enzymatically incorporating a labeled ^{32}P phosphate group onto the 5′-hydroxyl group of the terminal nucleotide. The double-stranded restriction fragments are then separated into single strands by heating, and the strands are isolated. Imagine, for example, that we now have a single-stranded DNA fragment with the following partial structure:

(5′ end) ^{32}P-G-A-T-C-A-G-C-G-A-T---- (3′ end)

Step 3 The labeled DNA sample is divided into four subsamples and subjected to four parallel sets of chemical reactions under conditions that cause:

(a) splitting of the DNA chain next to A
(b) splitting of the DNA chain next to G
(c) splitting of the DNA chain next to C
(d) splitting of the DNA chain next to *both* T and C

Mild reaction conditions are used so that *only a few of the many possible splittings occur*. In our example, the pieces shown in Table 29.2 would be produced.

Table 29.2 Splitting of a DNA Fragment Under Four Conditions

Cleavage conditions	*Labeled DNA pieces produced*
Original DNA fragment	^{32}P-G-A-T-C-A-G-C-G-A-T-
next to A	^{32}P-G ^{32}P-G-A-T-C ^{32}P-G-A-T-C-A-G-C-G + larger pieces
next to G	^{32}P-G-A-T-C-A ^{32}P-G-A-T-C-A-G-C + larger pieces
next to C	^{32}P-G-A-T ^{32}P-G-A-T-C-A-G + larger pieces
next to C + T	^{32}P-G-A ^{32}P-G-A-T ^{32}P-G-A-T-C-A-G ^{32}P-G-A-T-C-A-G-C-G-A + larger pieces

Cleavages next to A and G are accomplished by treating a restriction fragment with dimethyl sulfate [$(CH_3O)_2SO_2$]. Deoxyadenosine (A) is methylated at N3 (S_N2 reaction), and deoxyguanosine (G) is methylated at N7, but T and C aren't affected.

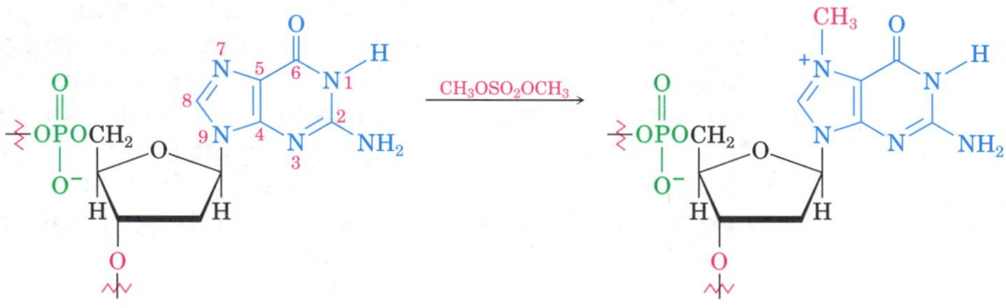

Deoxyguanosine

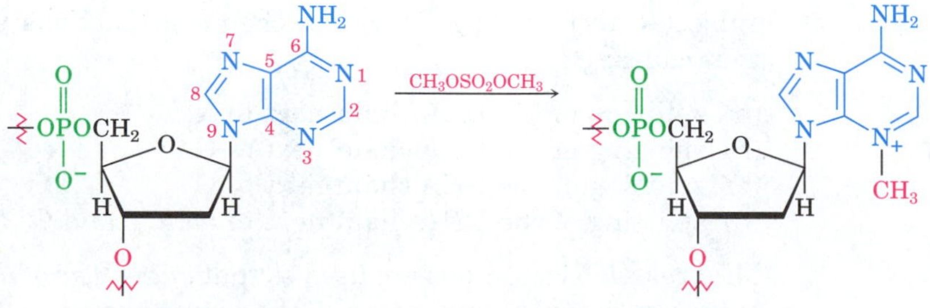

Deoxyadenosine

Treatment of methylated DNA with an aqueous solution of the secondary amine piperidine then brings about destruction of the methylated nucleotides and opening of the DNA chain at both the 3′ and 5′ positions next to the methylated bases. The mechanism of the cleavage process is shown in Figure 29.13 for deoxyguanosine.

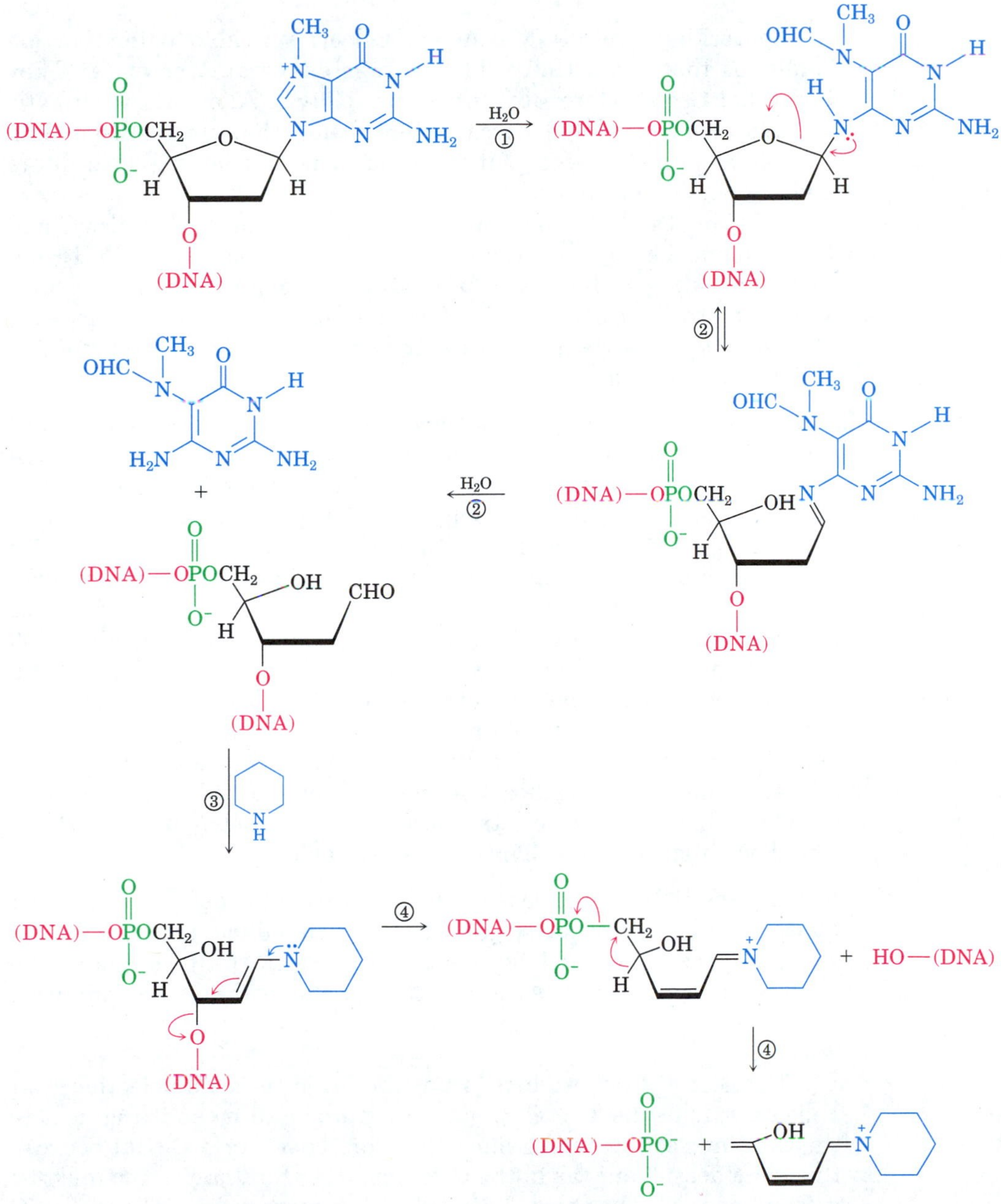

Figure 29.13 Mechanism of DNA cleavage at deoxyguanosine (G).

1. Hydrolysis occurs, opening the five-membered heterocycle.
2. Cleavage of the aminoglycoside linkage opens the sugar and yields an imine, which hydrolyzes to give the corresponding aldehyde.
3. Formation of an enamine between piperidine and the 2-deoxyribose aldehyde group occurs.
4. Two E2-like eliminations of the 2-deoxyribose oxygen substituents at C3 and C5 take place, breaking open the DNA chain.

By working carefully, Maxam and Gilbert were able to find reaction conditions that are selective for cleavage either at A or at G. (They found that G methylates five times as rapidly as A, but the hydrolytic breakdown of methylated A occurs more rapidly than the corresponding breakdown of methylated G if the product is first heated with dilute acid prior to base treatment.)

Breaking the DNA chain next to the pyrimidine nucleotides C and T is accomplished by treatment of DNA with hydrazine, H_2NNH_2, followed by heating with aqueous piperidine. Although no conditions have been found that are selective for cleavage next to T, a selective cleavage next to C can be accomplished by carrying out the hydrazine reaction in 5 M NaCl solution.

Step 4 Each of the product mixtures from the four cleavage reactions is separated by electrophoresis (Section 27.3). When a mixture of DNA cleavage products is placed at one end of a strip of buffered polyacrylamide gel and a voltage is applied, electrically charged pieces move along the gel. Each piece moves at a rate that depends both on its size and on the number of negatively charged phosphate groups (the number of nucleotides) it contains. Smaller pieces move rapidly, and larger pieces move more slowly. The technique is so sensitive that up to 600 DNA pieces, differing in size by only one nucleotide, can be separated.

Once separated, the locations of the DNA cleavage products are detected by exposing the gel to a photographic plate. Each radioactive end piece containing a ^{32}P label appears as a dark band on the photographic plate, but nonradioactive pieces from the middle of the chain aren't visualized. The gel electrophoresis pattern shown in Figure 29.14 would be obtained in our hypothetical example.

Step 5 The DNA sequence is read directly from the gel. The band that appears farthest from the origin is the 5′ terminal mononucleotide (the smallest piece) and can't be identified. Because the terminal mononucleotide appears in the A column, though, it must have been produced by splitting *next to* an A. Thus, the *second* nucleotide in the sequence is an A.

The second farthest band from the origin is a dinucleotide that appears only in the T + C column. It is produced by splitting next to the third nucleotide, which must therefore be a T or a C. But because this piece doesn't appear in the C column, the third nucleotide must be a T and not a C. The third farthest band appears in both C and T + C columns, meaning that the fourth nucleotide is a C. Continuing in this manner, the entire sequence of the DNA can be read from the gel simply by noting in what columns the successively larger labeled polynucleotide pieces appear. Once read, the entire sequence can be checked by determining the sequence of the complementary strand.

The Maxam–Gilbert method of DNA sequencing is so efficient that a single worker can sequence up to 2000 base pairs per day. The genome of the *Hemophilus influenzae* bacterium containing 1.8 million base pairs (1743 genes) has been sequenced, and work is now under way to sequence the 3 billion base pairs of the entire human genome. At least 10 years and several billion dollars will be needed.

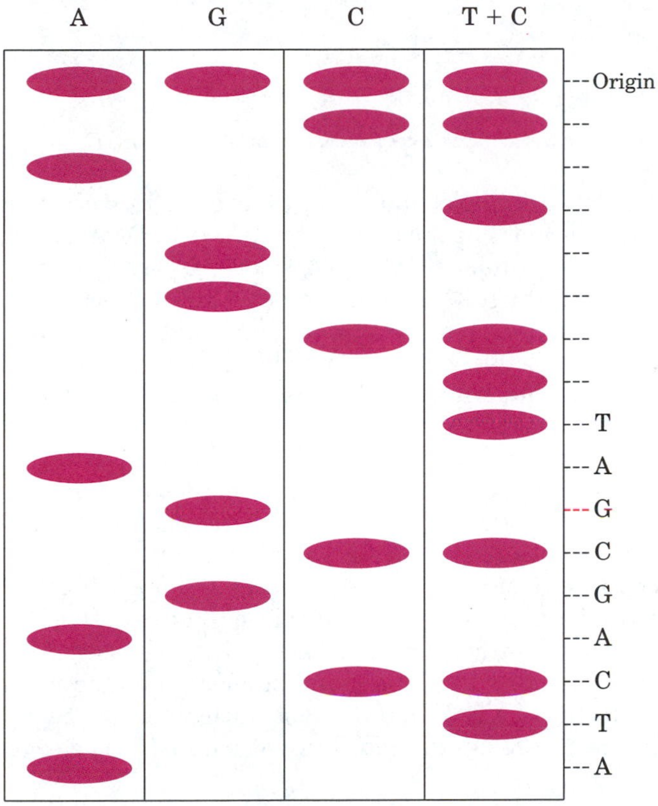

Figure 29.14 Representation of a gel electrophoresis pattern. The products of the four cleavage experiments are placed at the top of the gel, and a voltage is applied between top and bottom. Smaller products migrate along the gel at a faster rate and thus appear at the bottom. The DNA sequence can be read from the positions of the radioactive spots.

PROBLEM..

29.21 Show the labeled cleavage products you would expect to obtain if the following DNA segment were subjected to each of the four cleavage reactions:

^{32}P-A-A-C-A-T-G-G-C-G-C-T-T-A-T-G-A-C-G-A

PROBLEM.

29.22 Sketch what you would expect the gel electrophoresis pattern to look like if the DNA segment in Problem 29.21 were sequenced.

PROBLEM.

29.23 Finish assigning the sequence to the gel electrophoresis pattern shown in Figure 29.14.

29.16 DNA Synthesis

The recent revolution in molecular biology has brought with it an increased demand for the efficient chemical synthesis of short DNA segments, called *oligonucleotides*. The problems of DNA synthesis are similar to those of protein synthesis (Section 27.11) but are more difficult because of the complexity of the nucleotide monomers. Each nucleotide has multiple reactive sites that must be selectively protected and deprotected at the proper times, and coupling of the four nucleotides must be carried out in the proper sequence. Despite these difficulties, some impressive early achievements were recorded, most notably the synthesis by Khorana[4] in 1979 of the tyrosine suppressor tRNA gene from the bacterium *E. coli*. Some 207 base pairs were assembled in an effort that required 10 years of work. More recently, automated DNA synthesizers have become available, which allow the fast and reliable synthesis of DNA segments up to 200 nucleotides in length.

DNA synthesizers operate on a principle similar to that of the Merrifield solid-phase peptide synthesizer (Section 27.12). In essence, a protected nucleotide is covalently bound to a solid support, and one nucleotide at a time is added to the chain. When the last nucleotide has been added, the protecting groups are removed, and the synthetic DNA is cleaved from the solid support.

The first step in DNA synthesis involves attachment of a protected deoxynucleoside to a silica (SiO_2) support by an ester linkage to the 3′ –OH group of the deoxynucleoside. Both the 5′ –OH group on the sugar and free $-NH_2$ groups on the heterocyclic bases must be protected. Adenine and cytosine bases are protected by benzoyl groups, guanine is protected by an isobutyryl group, and thymine requires no protection. The deoxyribose 5′ –OH is protected as its *p*-dimethoxytrityl (DMT) ether.

[4]Har Gobind Khorana (1922–); b. Raipur, India; Ph.D. University of Liverpool; professor, Massachusetts Institute of Technology; Nobel Prize in medicine (1968).

Step 1

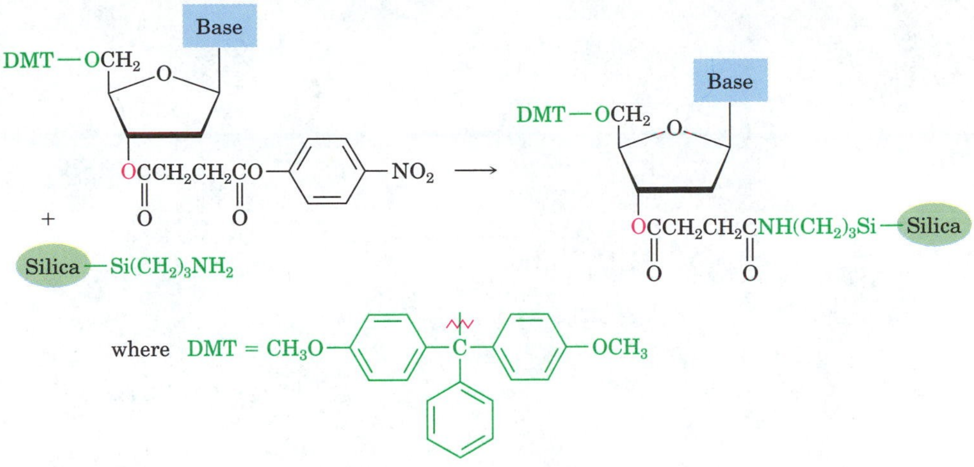

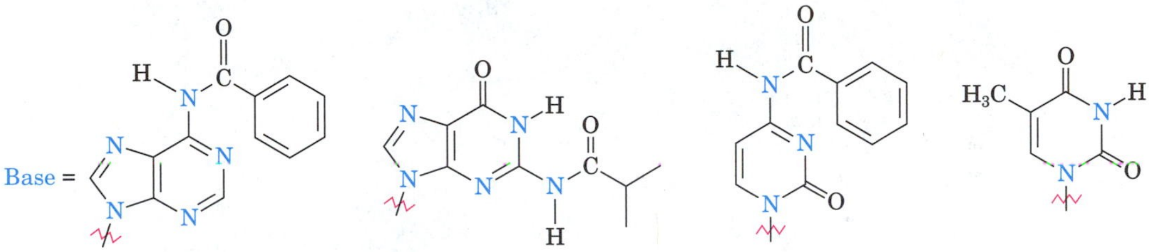

N-protected adenine **N-protected guanine** **N-protected cytosine** **Thymine**

The second step involves removal of the DMT protecting group by treatment with dichloroacetic acid in CH_2Cl_2. The reaction occurs by an S_N1 mechanism and proceeds rapidly because of the stability of the tertiary, benzylic dimethoxytrityl cation.

Step 2

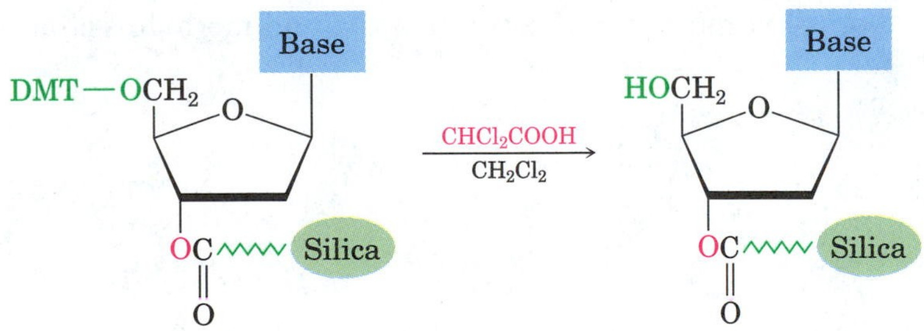

The third step involves coupling of the polymer-bonded deoxynucleoside with a protected deoxynucleoside containing a *phosphoramidite* group at its 3′ position. [A phosphoramidite has the structure $R_2NP(OR)_2$.] The coupling reaction takes place in the polar aprotic solvent acetonitrile, requires catalysis by the heterocyclic amine tetrazole, and yields a *phosphite*, $P(OR)_3$, as product. Note that one of the phosphorus oxygen atoms is protected by a β-cyanoethyl group. The coupling step takes place in better than 99% yield.

Step 3

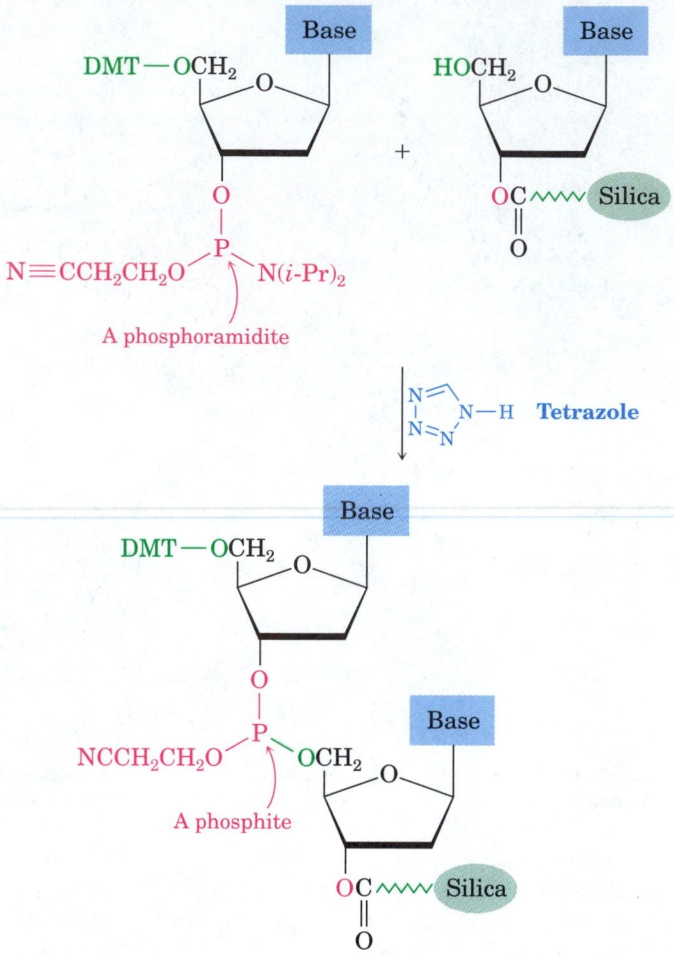

With the coupling completed, the phosphite product is oxidized to a phosphate by treatment with iodine. The reaction is carried out in aqueous tetrahydrofuran in the presence of 2,6-dimethylpyridine.

Step 4

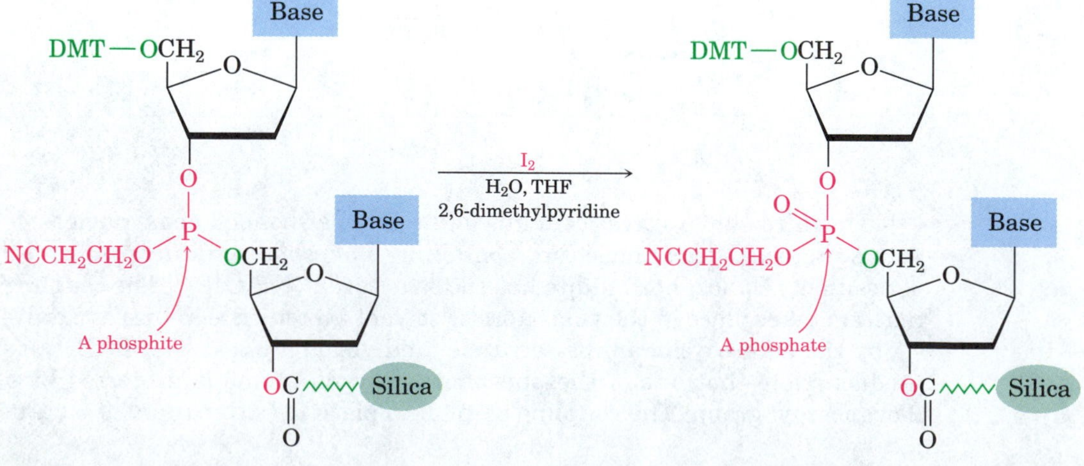

The cycle (1) deprotection, (2) coupling, and (3) oxidation is then repeated until an oligonucleotide chain of the desired sequence has been built. The final step is to remove all protecting groups from the heterocyclic bases and from the phosphates and to cleave the ester bond holding the DNA to the silica. All these reactions are done at the same time by treatment with aqueous NH_3. Purification by electrophoresis then yields the synthetic DNA.

Step 5

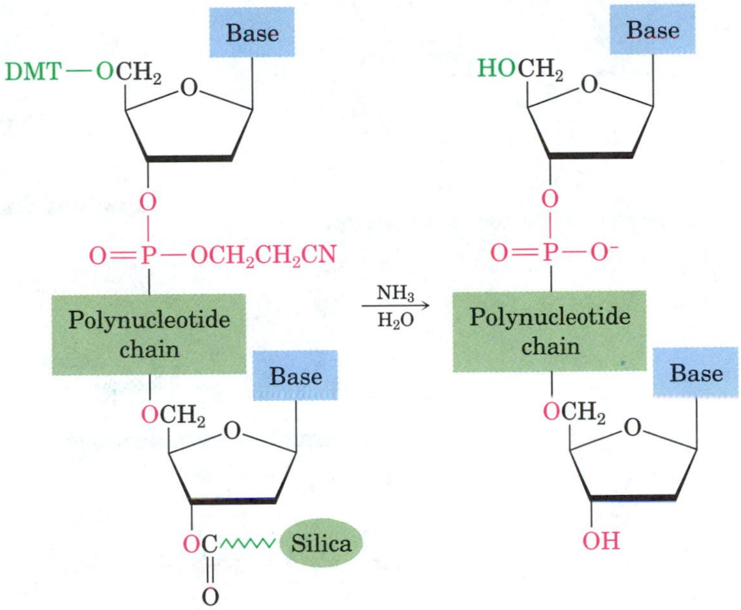

PROBLEM

29.24 *p*-Dimethoxytrityl (DMT) ethers are easily cleaved by mild acid treatment. Show the mechanism of the cleavage reaction in detail.

PROBLEM

29.25 Propose a mechanism to account for cleavage of the β-cyanoethyl protecting group from the phosphate groups on treatment with aqueous ammonia. What kind of reaction is occurring?

29.17 The Polymerase Chain Reaction

The invention of the **polymerase chain reaction (PCR)** by Kary Mullis in 1986 has been described as being to genes what Gutenberg's invention of the printing press was to the written word. Just as the printing press produces multiple copies of a book, PCR produces multiple copies of a given DNA sequence. Starting from less than 1 *picogram* of DNA with a chain length of 10,000 nucleotides (1 pg = 10^{-12} g; about 100,000 molecules), PCR makes it possible to obtain several micrograms (1 $\mu g = 10^{-6}$ g; about 10^{11} molecules) in a few hours time.

The key to the polymerase chain reaction is *Taq* DNA polymerase, a heat-stable enzyme isolated from the thermophilic bacterium *Thermus aquaticus* found in a hot spring in Yellowstone National Park.[5] *Taq* polymerase is able to take a single strand of DNA that has a short, primer segment of complementary chain at one end and then finish constructing the entire complementary strand. The overall process takes three steps, as shown schematically in Figure 29.15.

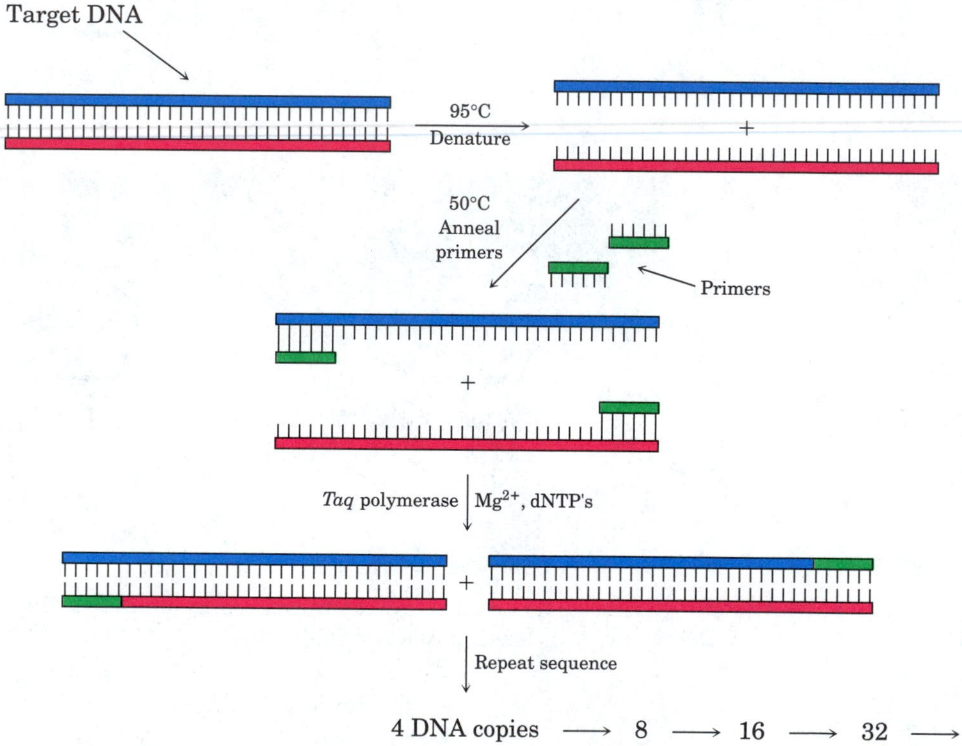

Figure 29.15 The polymerase chain reaction. Double-stranded DNA is heated to 95°C in the presence of two short oligonucleotide primer sequences, each of which is complementary to the end of one of the strands. After the DNA denatures, the temperature is lowered and the primer sequences anneal to the strand ends. Raising the temperature in the presence of *Taq* polymerase, Mg^{2+}, and a mixture of the four deoxynucleotide triphosphates (dNTP's) effects strand replication, producing two DNA copies. Each further repetition of the sequence again doubles the number of copies.

[5]Two new heat-stable DNA polymerase enzymes have recently become available, Vent polymerase and *Pfu* polymerase, both isolated from bacteria growing near geothermal vents in the ocean floor. The error rate of both enzymes is substantially less than that of *Taq*.

1. The double-stranded DNA to be amplified is first heated to 95°C in the presence of *Taq* polymerase, Mg^{2+} ion, the four deoxynucleotide triphosphate monomers (dNTP's), and a large excess of two short oligonucleotide primers of about 20 bases each. Each primer is complementary to the sequence at the end of one of the target DNA segments. At a temperature of 95°C, double-stranded DNA *denatures*, spontaneously breaking apart into two single strands.
2. The temperature is lowered to between 37°C and 50°C, allowing the primers, because of their relatively high concentration, to anneal to their complementary sequence at the end of each target strand.
3. The temperature is then raised to 72°C, and *Taq* polymerase catalyzes the addition of further nucleotides to the two primed DNA strands. When replication of each strand is finished, *two* copies of the original DNA now exist. Repeating the denature–anneal–synthesize cycle a second time yields four DNA copies, repeating a third time yields eight copies, and so on, in an exponential series.

PCR has been automated, and 30 or so cycles can be carried out in an hour, resulting in a theoretical amplification factor of 2^{30} ($\sim 10^9$). In practice, however, the efficiency of each cycle is less than 100%, and an experimental amplification of about 10^6–10^8 is routinely achieved for 30 cycles.

INTERLUDE

DNA-Based Computing?

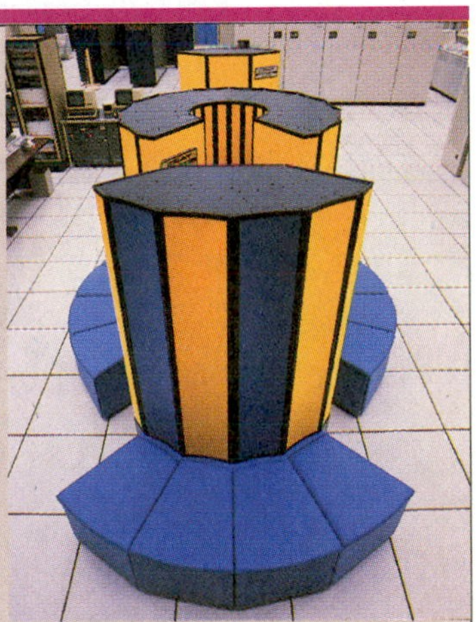

Silicon-based supercomputers like this Cray X-MP/48 may ultimately be replaced in certain applications by DNA-based machines.

Modern electronic computers are wondrous machines, masterpieces of scientific, engineering, and manufacturing genius. A typical desktop computer now executes approximately 10^6 mathematical operations per second, and the fastest supercomputers can execute approximately 10^{12} operations per second. There are, however, limits. As components get ever smaller, and as more and more transistors are packed on a silicon chip, the point must come when electronic miniaturization can go no further. At that point, new computing techniques will be necessary.

(continued)➤

The ultimate limit to miniaturization may well be a *molecular* computer, and DNA may be the molecule used. Looking at a string of A's, T's, G's, and C's in DNA as "bits" of information analogous to the 0's and 1's of binary computer code, it has been estimated that 1 pound of DNA can hold more information than all the memories of all the electronic computers ever made.

The first experiment in what may lead ultimately to DNA-based computers was reported in 1994 when techniques of molecular biology were used to solve a specialized kind of combinatorial mathematics problem called a *directed-path graph*. Given a start, a stop, a number of other points, and an assortment of allowed moves between points, the idea is to determine whether a path exists in which all points are passed through only once. For example, imagine the following graph containing seven points and a number of allowed moves as shown by the arrows. Starting at point 1 and ending at point 7, is there a path that passes through all the other points only once? (The answer in this simple case is the path $1 \rightarrow 2 \rightarrow 3 \rightarrow 4 \rightarrow 5 \rightarrow 6 \rightarrow 7$.)

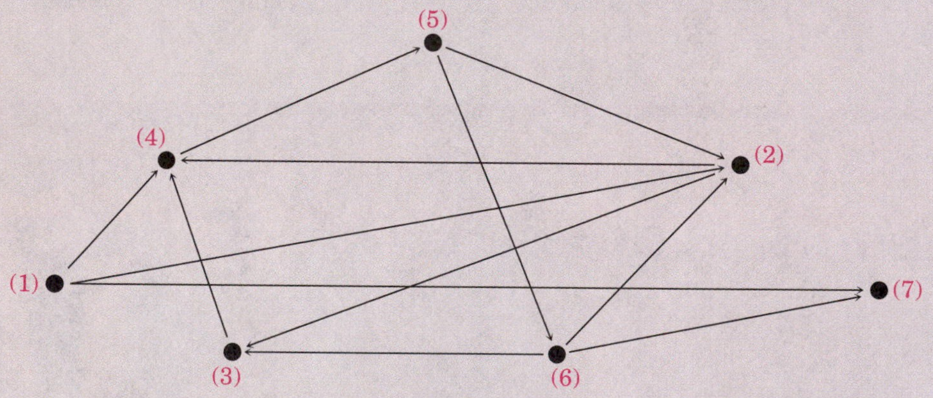

To solve this problem, each of the seven points was represented as a 20-unit oligonucleotide, and the allowed moves between any two points were represented as 20-unit oligonucleotides whose first 10 units were complementary to half of one point and whose second 10 units were complementary to half of the other point. Thus, if two oligonucleotides representing points 1 and 2 line up next to each other, the oligonucleotide representing the move $1 \rightarrow 2$ will act as a splint to hold them together by hydrogen bonding to 10 units of each. If now, a DNA polymerase enzyme is introduced, the two oligonucleotides representing points 1 and 2 will be joined, or *ligated*. When a grand mix of oligonucleotides representing all seven points and all possible moves was ligated with DNA polymerase, DNA pieces representing all possible paths were produced.

(continued)➤

This oligonucleotide represents move 1 ⟶ 2.

G-T-C-G-C-C-T-A-C-G-C-T-T-G-C-A-T-A-A-C

-A-T-C-A-G-C-G-G-A-T-G-C G-A-A-C-G-T-A-T-T-G-C-G-C-

This oligonucleotide represents point 1.

Ligate here

This oligonucleotide represents point 2.

To find the piece of DNA representing all possible paths starting at 1 and ending at 7, a polymerase chain reaction (Section 29.17) was carried out using primers whose sequences corresponded to the end sequences of the oligonucleotides representing 1 and 7. To find only those DNA pieces representing the correct number of other points (5), electrophoresis was carried out and all DNA pieces with lengths of 140 nucleotides were isolated. To make sure that points 2, 3, 4, 5, and 6 had each been passed through, the binding ability of the DNA pieces with a complementary oligonucleotide representing each point was checked. At the end of it all, only a single piece of DNA representing the path 1 → 2 → 3 → 4 → 5 → 6 → 7 remained.

This preliminary example is simple, yet its implications are profound. Computing experiments are now being devised in which DNA-based methods may well be *superior* to those of electronic computers. The future of computing will be interesting.

Summary and Key Words

A **heterocycle** is a compound containing a ring that has more than one kind of atom. Nitrogen, oxygen, and sulfur are often found along with carbon in heterocyclic rings. Saturated heterocyclic amines, ethers, and sulfides usually have the same chemistry as their open-chain analogs, but unsaturated heterocycles such as pyrrole, furan, and thiophene are aromatic. All three are unusually stable, and all three undergo aromatic substitution on reaction with electrophiles.

Pyridine is the six-membered-ring, nitrogen-containing heterocyclic analog of benzene. The pyridine ring is electron-poor and undergoes electrophilic aromatic substitution reactions with difficulty. Nucleophilic aromatic substitutions of 2- or 4-halopyridines take place readily, however.

The nucleic acids **DNA (deoxyribonucleic acid)** and **RNA (ribonucleic acid)** are biological polymers that act as chemical carriers of an organism's genetic information. Enzyme-catalyzed hydrolysis of nucleic acids yields **nucleotides**, the monomer units from which RNA and DNA are constructed. Each nucleotide consists of a **purine** or **pyrimidine base** linked to C1′ of an aldopentose sugar (ribose in RNA and 2′-deoxyribose in

DNA), with the sugar in turn linked through its C5′ hydroxyl to a phosphate group. The nucleotides are joined by phosphate links between the phosphate of one nucleotide and the 3′-hydroxyl on the sugar of another nucleotide.

Molecules of DNA consist of two polynucleotide strands held together by hydrogen bonds between heterocyclic bases on the different strands and coiled into a **double helix**. Adenine and thymine form hydrogen bonds to each other, as do cytosine and guanine. The two strands of DNA are not identical but are complementary.

Three processes take place in deciphering the genetic information of DNA:

1. **Replication** of DNA is the process by which identical DNA copies are made and genetic information is preserved. This occurs when the DNA double helix unwinds, complementary deoxyribonucleotides line up in order, and two new DNA molecules are produced.
2. **Transcription** is the process by which RNA is produced in order to carry genetic information from the nucleus to the ribosomes. This occurs when a short segment of the DNA double helix unwinds and complementary ribonucleotides line up to produce **messenger RNA (mRNA)**.
3. **Translation** is the process by which mRNA directs protein synthesis. Each mRNA is divided into **codons**, ribonucleotide triplets that are recognized by small amino acid-carrying molecules of **transfer RNA (tRNA)**, which deliver the appropriate amino acids needed for protein synthesis.

Small DNA segments can be synthesized in the laboratory, and commercial instruments are available for automating the work. Sequencing of DNA can be carried out rapidly and efficiently by the **Maxam–Gilbert method**. Small amounts of DNA can be amplified by factors of 10^8 using the **polymerase chain reaction (PCR)**.

Summary of Reactions

1. Electrophilic aromatic substitution (Section 29.3)
   (a) Bromination

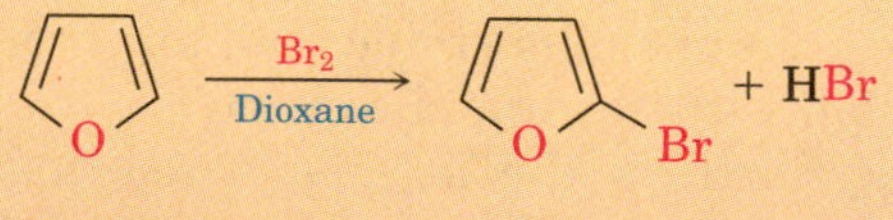

Furan

(b) Nitration

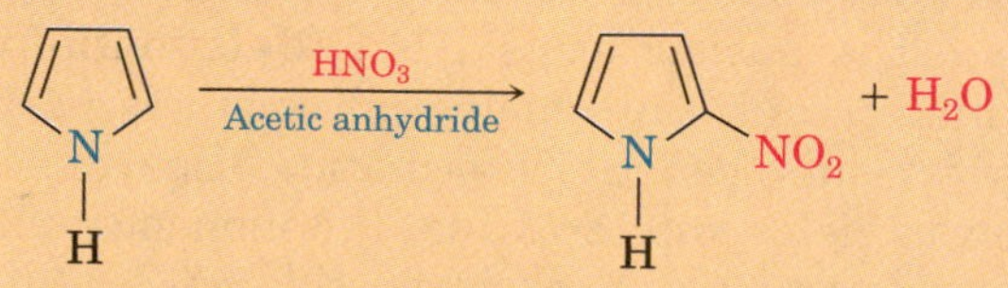

Pyrrole

(continued)

(c) Friedel–Crafts acylation

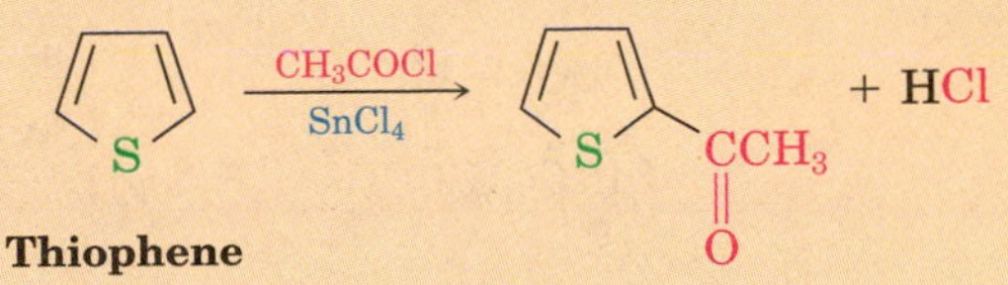

Thiophene

2. Nucleophilic aromatic substitution of halopyridines (Section 29.6)

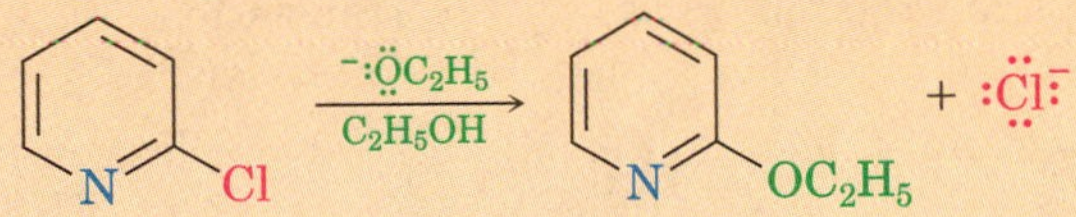

ADDITIONAL PROBLEMS

29.26 Although pyrrole is a much weaker base than most other amines, it is a much stronger acid (pK_a ≈ 15 for pyrrole versus 35 for diethylamine). The N–H proton is readily abstracted by base to yield the pyrrole anion, $C_4H_4N^-$. Explain.

29.27 Oxazole is a five-membered aromatic heterocycle. Draw an orbital picture of oxazole, showing all *p* orbitals and all lone-pair orbitals. Would you expect oxazole to be more basic or less basic than pyrrole? Explain.

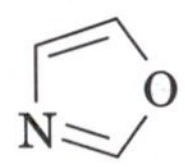

Oxazole

29.28 Write the products of the reaction of furan with each of the following reagents:
(a) Br_2, dioxane, 0°C
(b) HNO_3, acetic anhydride
(c) CH_3COCl, $SnCl_4$
(d) H_2/Pd
(e) SO_3, pyridine

29.29 Nitrofuroxime is a pharmaceutical agent used in the treatment of urinary tract infections. Propose a synthesis of nitrofuroxime from furfural.

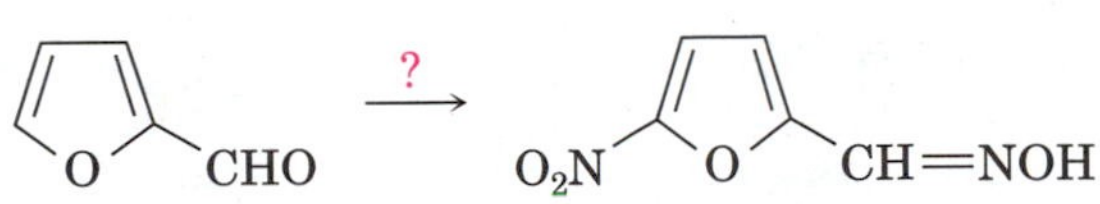

Furfural **Nitrofuroxime**

29.30 Substituted pyrroles can be prepared by treatment of a 1,4-diketone with ammonia. Suggest a mechanism. (Review Section 19.12.)

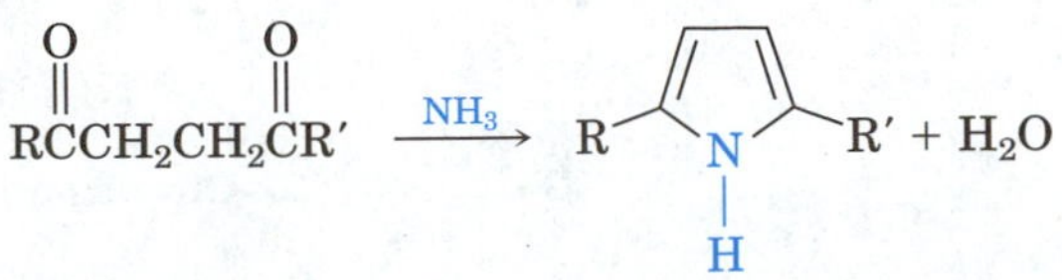

29.31 3,5-Dimethylisoxazole is prepared by reaction of 2,4-pentanedione with hydroxylamine. Propose a mechanism.

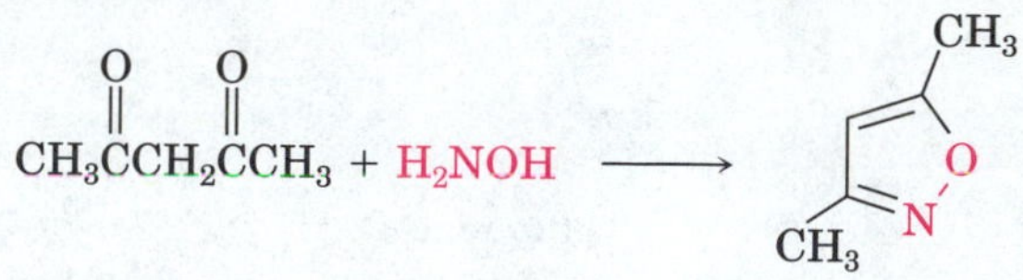

29.32 Define the following terms:
(a) Heterocycle (b) DNA (c) Base pair
(d) Transcription (e) Translation (f) Replication (of DNA)
(g) Codon (h) Anticodon

29.33 What amino acids do the following ribonucleotide triplets code for?
(a) AAU (b) GAG (c) UCC (d) CAU

29.34 From what DNA sequences were each of the mRNA codons in Problem 29.33 transcribed?

29.35 What anticodon sequences of tRNA's are coded for by the codons in Problem 29.33?

29.36 Draw the complete structure of the ribonucleotide codon U-A-C. For what amino acid does this sequence code?

29.37 Draw the complete structure of the deoxyribonucleotide sequence from which the mRNA codon in Problem 29.36 was transcribed.

29.38 Give an mRNA sequence that would code for synthesis of metenkephalin:

Tyr-Gly-Gly-Phe-Met

29.39 Look up the structure of angiotensin II (Figure 27.4), and give an mRNA sequence that would code for its synthesis.

29.40 What amino acid sequence is coded for by this mRNA base sequence?

CUA-GAC-CGU-UCC-AAG-UGA

29.41 Isoquinolines are often synthesized by the *Bischler–Napieralski* cyclization of an *N*-acyl-2-phenylethyl amine with strong acid and P_2O_5, followed by oxidation of the initially formed dihydroisoquinoline. Suggest a mechanism for the cyclization.

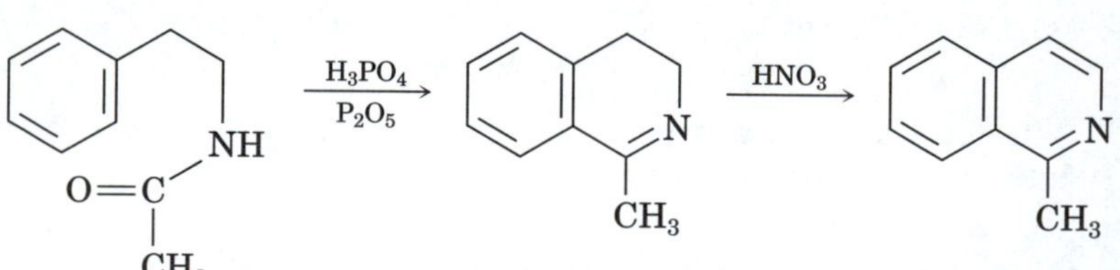

A dihydroisoquinoline **1-Methylisoquinoline**

29.42 Quinolines are often prepared by the *Skraup synthesis*, in which an aniline reacts with an α,β-unsaturated aldehyde and the dihydroquinoline product is oxidized. Suggest a mechanism for the Skraup reaction.

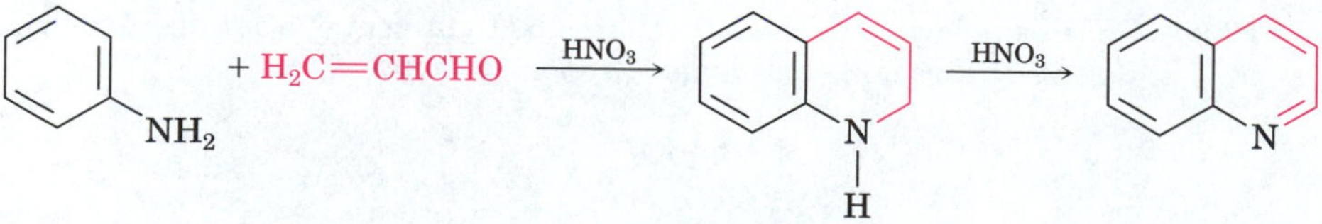

1,2-Dihydroquinoline **Quinoline**

29.43 Show the steps involved in a laboratory synthesis of the DNA fragment with the sequence C-T-A-G.

29.44 Sodium nitrite, a food preservative used in meats, causes the mutation of cytosine into uracil under acidic conditions. Propose a mechanism. (See Section 25.3.)

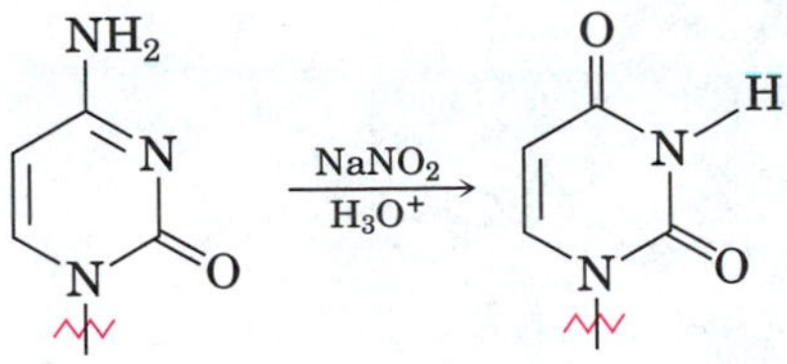

29.45 Review the mechanism shown in Figure 29.13 for the cleavage of deoxyguanosine residues, and propose a mechanism to account for the similar cleavage of deoxyadenosine residues in a DNA chain. Recall that deoxyadenosine is first methylated at N3 prior to hydrolysis.

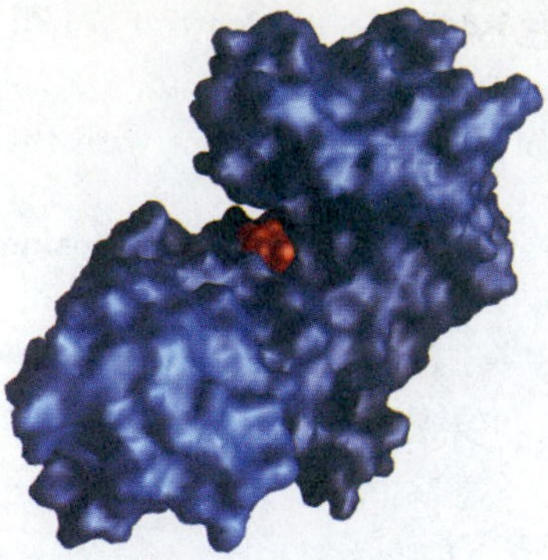

Glucose (in red) is held inside the active site of this hexokinase enzyme during glycolysis.

30

The Organic Chemistry of Metabolic Pathways

The organic chemical reactions that take place in the smallest and simplest living organism are more complex than those carried out in any laboratory. Yet regardless of their complexity, the reactions in living organisms follow the same rules of reactivity and proceed by the same mechanisms that we've seen in the preceding chapters.

In this chapter, we'll look at some of the pathways by which organisms carry out their chemistry, focusing primarily on how they metabolize fats and carbohydrates. Though nearly all the reactions in living organisms are catalyzed by enzymes (Section 27.15), our emphasis will be on the organic chemistry of the various pathways. We'll pay particular attention to recognizing similarities between the mechanisms of biological reactions and the mechanisms of analogous laboratory reactions.

30.1 An Overview of Metabolism and Biochemical Energy

The many reactions that go on in the cells of living organisms are collectively called **metabolism**. The group of pathways that break down larger molecules into smaller ones are called **catabolism**, while the pathways that synthesize larger biomolecules from smaller ones are known as **anabolism**. Catabolic reaction pathways usually release energy, while anabolic reaction pathways often absorb energy. Catabolism can be divided into the four stages shown in Figure 30.1.

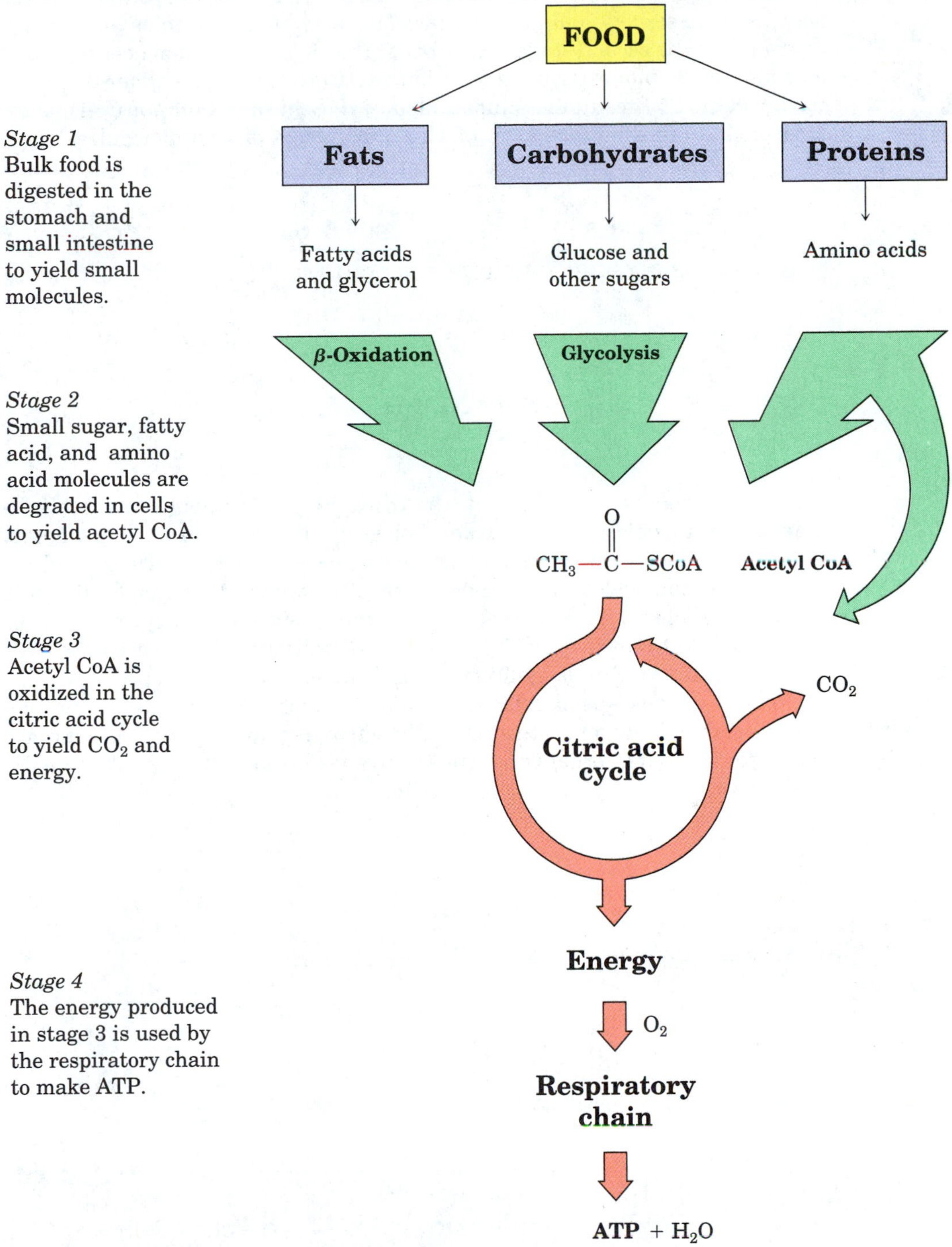

Figure 30.1 An overview of catabolic pathways for the degradation of food and the production of biochemical energy. The ultimate products of food catabolism are CO_2, H_2O, and adenosine triphosphate (ATP).

In the first catabolic stage, **digestion**, food is broken down by hydrolysis of ester, glycoside (acetal), and peptide (amide) bonds to yield primarily fatty acids, simple sugars, and amino acids. These smaller molecules are further degraded in the second stage of catabolism to yield two-carbon acetyl groups attached by a thiol ester bond (Section 21.10) to the large carrier molecule *coenzyme A*. The resultant compound, *acetyl coenzyme A* (*acetyl CoA*), is an intermediate in the breakdown of all main classes of food molecules.

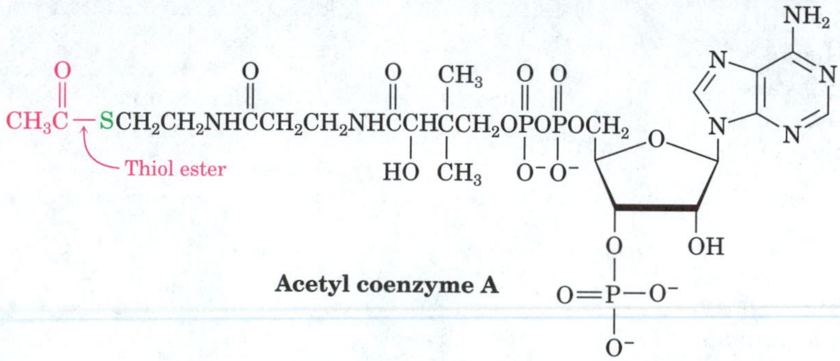

Acetyl coenzyme A

Acetyl groups are oxidized in the third stage of catabolism, the **citric acid cycle**, to yield carbon dioxide. This stage also releases a large amount of energy that is used in the fourth stage, the **respiratory chain**, to produce molecules of the nucleotide **adenosine triphosphate, ATP**. The final result of food catabolism, ATP has been called the "energy currency" of the cell. Catabolic reactions "pay off" in ATP by synthesizing it from **adenosine diphosphate, ADP**, plus hydrogen phosphate ion, HPO_4^{2-} (abbreviated P_i). Anabolic reactions "spend" ATP by transferring a phosphate group to another molecule, thereby regenerating ADP. The entire process of energy production and use in living organisms thus revolves around the ATP ⇄ ADP interconversion.

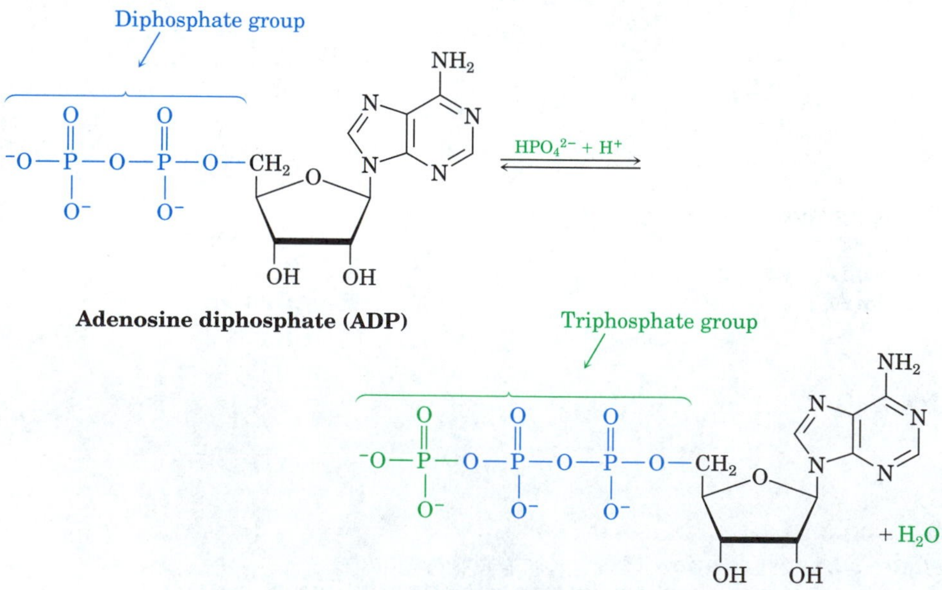

Adenosine diphosphate (ADP)

Adenosine triphosphate (ATP)

Note that both ADP and ATP are **phosphoric acid anhydrides**, which contain $-\overset{\overset{\large O}{\|}}{P}-O-\overset{\overset{\large O}{\|}}{P}-$ linkages analogous to the $-\overset{\overset{\large O}{\|}}{C}-O-\overset{\overset{\large O}{\|}}{C}-$linkage in carboxylic acid anhydrides. Just as carboxylic anhydrides react with alcohols by breaking a C–O bond and forming a carboxylic ester (Section 21.6), phosphoric anhydrides react with alcohols by breaking a P–O bond and forming a phosphate ester, $ROPO_3^{2-}$.

How does the body use ATP? Recall from Section 5.6 that ΔG, the free-energy change, must be negative for a reaction to occur spontaneously. That is, energy must be released. If ΔG is positive, then the reaction is unfavorable and the process can't occur spontaneously.[1] What normally happens for an energetically unfavorable reaction to occur is that it is "coupled" to an energetically favorable reaction so that the *overall* free-energy change for the two reactions together is favorable. Take the phosphorylation reaction of glucose to yield glucose 6-phosphate, an important step in the breakdown of dietary carbohydrates. The reaction of glucose with hydrogen phosphate ion does not occur directly because it is energetically unfavorable, with $\Delta G°' = +13.8$ kJ (3.3 kcal).[2]

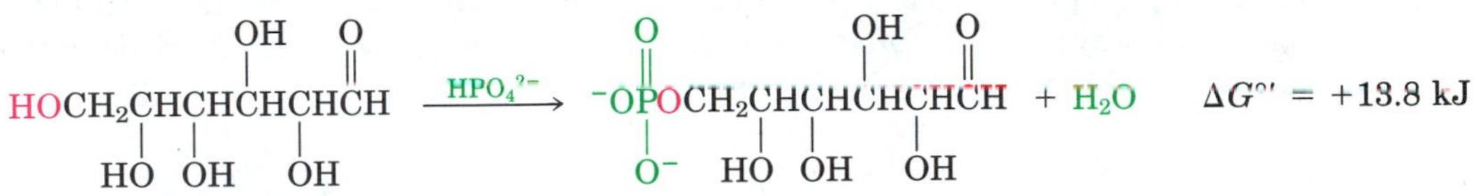

Glucose **Glucose 6-phosphate**

With ATP, however, glucose undergoes an energetically favorable reaction to yield glucose 6-phosphate plus ADP. The effect is as if the ATP reacted with the water produced in the unfavorable HPO_4^{2-} reaction, making the overall process favorable by about 16.7 kJ (4.0 kcal). We therefore say that ATP "drives" the phosphorylation reaction of glucose:

$$\begin{array}{lcll} \text{Glucose} + HPO_4^{2-} & \longrightarrow & \text{Glucose 6-phosphate} + H_2O & \Delta G°' = +13.8 \text{ kJ} \\ ATP + H_2O & \longrightarrow & ADP + HPO_4^{2-} + H^+ & \Delta G°' = -30.5 \text{ kJ} \\ \hline \textit{Net:}\ \text{Glucose} + ATP & \longrightarrow & \text{Glucose 6-phosphate} + ADP + H^+ & \Delta G°' = -16.7 \text{ kJ} \end{array}$$

It's this ability to drive otherwise unfavorable phosphorylation reactions that makes ATP so useful. The resultant phosphates are much more reactive molecules than the corresponding compounds they are derived from.

---

[1]As noted in Section 5.6, the free-energy change (ΔG) is the sum of two terms, an enthalpy term (ΔH), which measures the heat change in a reaction, and a temperature-dependent entropy term ($T\Delta S$), which measures the change in disorder: $\Delta G = \Delta H - T\Delta S$. In laboratory chemistry, ΔH is often large enough to dwarf $T\Delta S$, and chemists make the simplifying assumption that $\Delta G \approx \Delta H$. In biological chemistry, however, ΔH is often small, and the simplifying assumption is no longer valid.

[2]The standard free-energy change for a biological reaction, denoted $\Delta G°'$, refers to a process in which reactants and products are in their biological standard states, defined as having a concentration of 1.0 M and pH = 7.00.

PROBLEM

30.1 One of the steps in fat metabolism is the reaction of glycerol (1,2,3-propanetriol) with ATP to yield glycerol 1-phosphate. Write the reaction, and draw the structure of glycerol 1-phosphate.

PROBLEM

30.2 The reaction of ATP with an alcohol to yield ADP and an alkyl phosphate is analogous to that of a carboxylic acid anhydride with an alcohol to yield a carboxylate ion and an ester (Section 21.6). Propose a mechanism for the reaction of ATP with methanol to yield methyl phosphate, $CH_3OPO_3^{2-}$.

30.2 Catabolism of Fats: β-Oxidation

The catabolism of fats and oils (triacylglycerols) begins with their hydrolysis in the stomach and small intestine to yield glycerol plus fatty acids. Glycerol is first phosphorylated by reaction with ATP and is then oxidized to yield glyceraldehyde 3-phosphate, which enters the carbohydrate catabolic pathway. (We'll discuss this in detail in Section 30.3.)

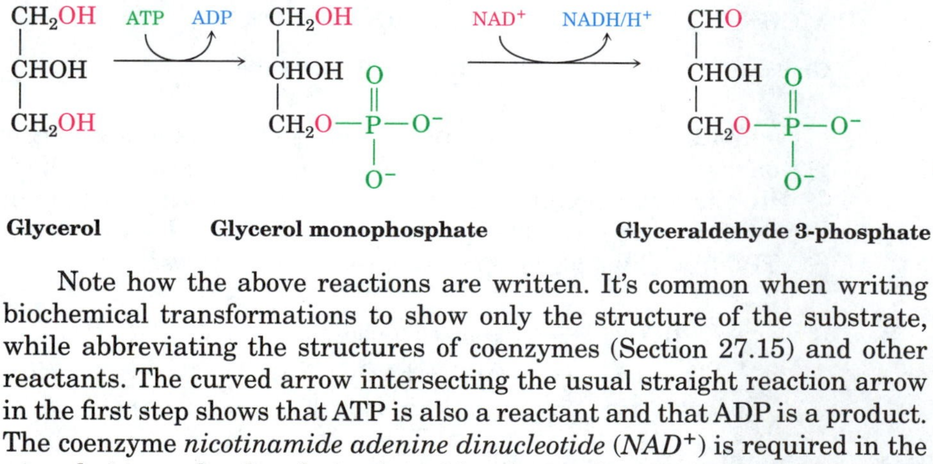

Glycerol **Glycerol monophosphate** **Glyceraldehyde 3-phosphate**

Note how the above reactions are written. It's common when writing biochemical transformations to show only the structure of the substrate, while abbreviating the structures of coenzymes (Section 27.15) and other reactants. The curved arrow intersecting the usual straight reaction arrow in the first step shows that ATP is also a reactant and that ADP is a product. The coenzyme *nicotinamide adenine dinucleotide* (*NAD*$^+$) is required in the second step, and *reduced nicotinamide adenine dinucleotide* (*NADH*) plus a proton are products. We'll see shortly that NAD$^+$ is often involved as a biochemical oxidizing agent for converting alcohols to ketones or aldehydes.

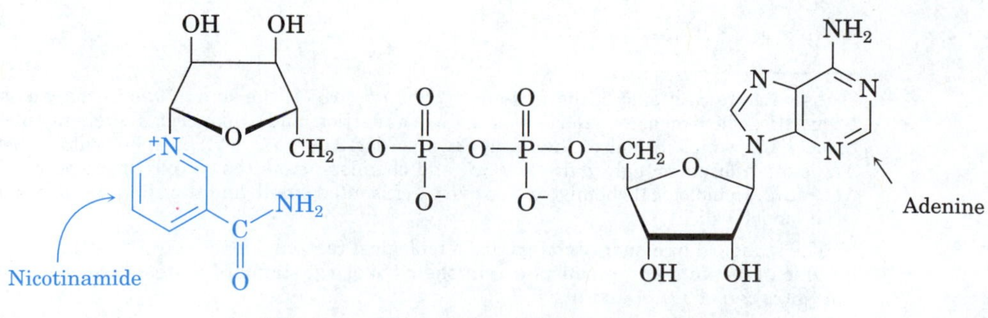

Nicotinamide adenine dinucleotide (NAD$^+$)

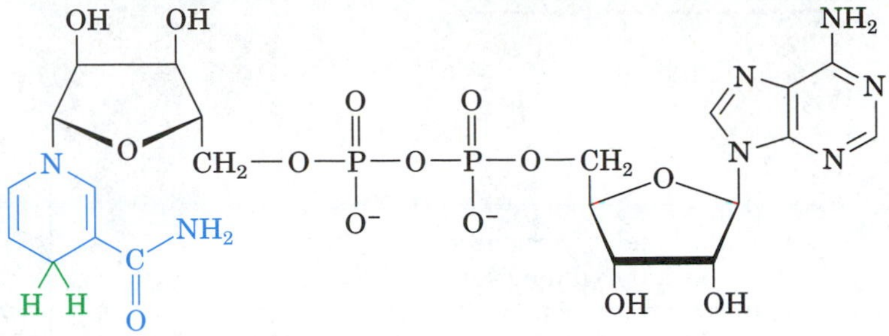

Reduced nicotinamide adenine dinucleotide (NADH)

Note also that glyceraldehyde 3-phosphate is written with its phosphate group dissociated—that is, $-OPO_3^{2-}$ rather than $-OPO_3H_2$. It's standard practice in writing biochemical structures to show carboxylic acids and phosphoric acids as their anions because they exist in this form at the physiological pH found in the cells of organisms.

Fatty acids are catabolized by a repetitive four-step sequence of enzyme-catalyzed reactions called the *fatty acid spiral*, or the **β-oxidation pathway**, shown in Figure 30.2 (p. 1178). Each passage along the pathway results in the cleavage of a two-carbon acetyl group from the end of the fatty acid chain, until the entire molecule is ultimately degraded. As each acetyl group is produced, it enters the citric acid cycle and is further catabolized, as we'll see in Section 30.5.

Step 1: Introduction of a double bond The β-oxidation pathway begins when a fatty acid forms a thiol ester with coenzyme A to give a fatty acyl CoA, and two hydrogen atoms from carbons 2 and 3 are removed by an acyl CoA dehydrogenase enzyme to yield an α,β-unsaturated acyl CoA. This kind of oxidation—the introduction of a conjugated double bond into a molecule—occurs frequently in biochemical pathways and is usually carried out by the coenzyme *flavin adenine dinucleotide* (*FAD*). Reduced $FADH_2$ is the by-product.

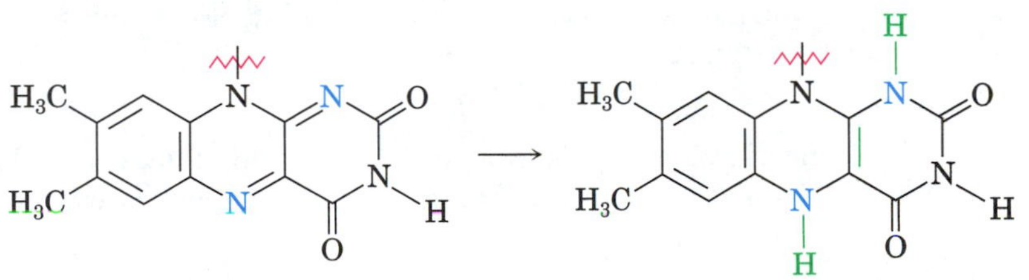

FAD **$FADH_2$**

Step 2: Conjugate addition of water The α,β-unsaturated acyl CoA produced in step 1 reacts with water by a conjugate addition pathway to yield a β-hydroxy acyl CoA in a process catalyzed by the enzyme enoyl CoA hydratase. The reaction is closely analogous to the conjugate nucleophilic addition of an amine to a ketone to yield a β-amino ketone

Step 1
A double bond is introduced by enzyme-catalyzed removal of hydrogens from C2 and C3.

Step 2
Water adds to the double bond in a conjugate addition reaction to yield an alcohol.

Step 3
The alcohol is oxidized by NAD^+ to give a β-keto thiol ester.

Step 4
The bond between C2 and C3 is broken by nucleophilic attack of coenzyme A on the C3 carbonyl group in a retro-Claisen reaction to yield acetyl CoA and a chain-shortened fatty acid.

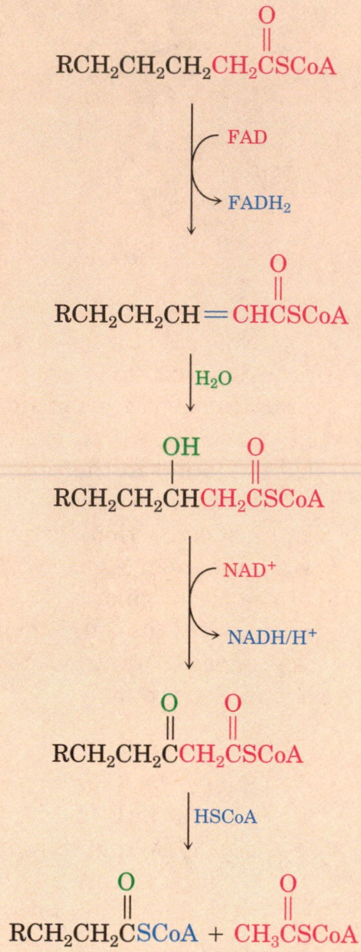

Figure 30.2 The four steps of the β-oxidation pathway, resulting in the cleavage of a two-carbon acetyl group from the end of the fatty acid chain. The key, chain-shortening step is a retro-Claisen reaction of a β-keto thiol ester.

(Section 19.17). Water as nucleophile adds to the double bond conjugated with the C=O group, yielding an enolate ion intermediate, which is then protonated.

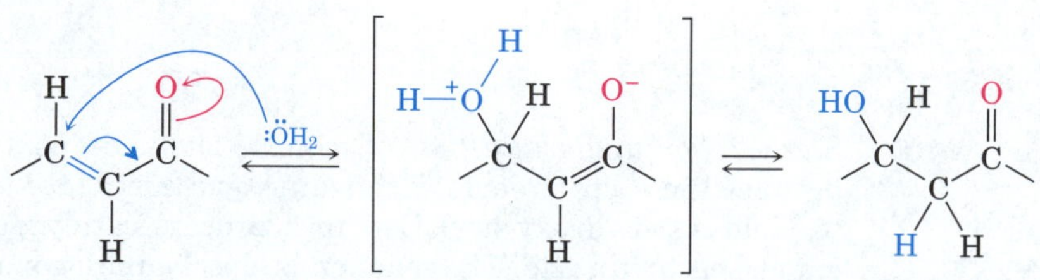

α,β-Unsaturated carbonyl **β-Hydroxy carbonyl**

Recall:

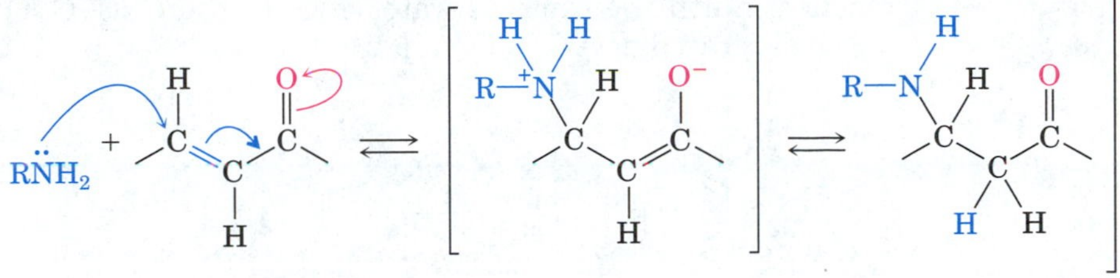

Step 3: Alcohol oxidation The β-hydroxy acyl CoA that results from step 2 is oxidized to a β-keto acyl CoA in a reaction catalyzed by the enzyme L-3-hydroxyacyl CoA dehydrogenase. As in the oxidation of glycerol 1-phosphate to glyceraldehyde 3-phosphate mentioned earlier, alcohol oxidation in the β-oxidation pathway requires NAD^+ as a coenzyme and yields reduced $NADH/H^+$ as by-product.

As noted in the Chapter 25 Interlude, it's useful when thinking about enzyme-catalyzed oxidation and reduction reactions to recognize that a hydrogen *atom* is equivalent to a hydrogen *ion*, H^+, plus an electron, e^-. Thus, for the two hydrogen atoms removed in the oxidation of an alcohol, 2 H atoms = $2\ H^+ + 2\ e^-$. When NAD^+ is involved, both electrons accompany one H^+, in effect adding a hydride ion, $H:^-$, to NAD^+ to give NADH. The second hydrogen removed from the oxidized substrate enters the solution as H^+.

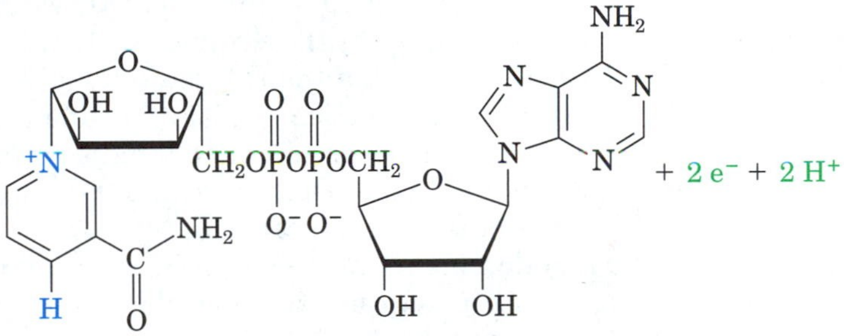

NAD^+

↓

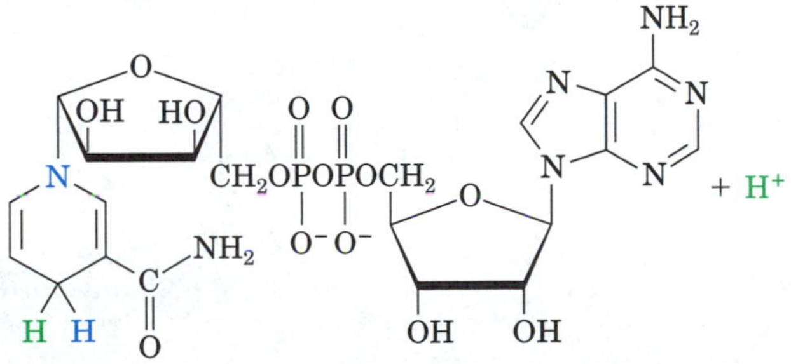

$NADH/H^+$

The mechanism of alcohol oxidation with NAD^+ has several analogies in the laboratory. A base removes the O–H proton from the alcohol and generates an alkoxide ion, which expels a hydride ion leaving group as in the Cannizzaro reaction (Section 19.16). The nucleophilic hydride ion then adds to the C=C–C=N$^+$ part of NAD$^+$ in a conjugate addition reaction, much the same as water adds to the C=C–C=O part of the α,β-unsaturated acyl CoA in step 2.

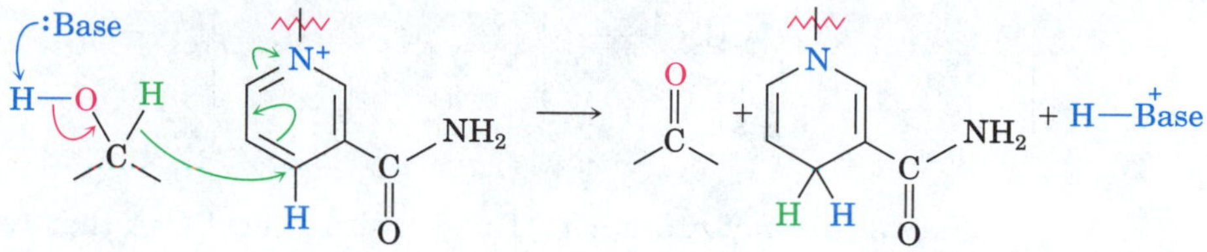

Alcohol NAD$^+$ Ketone NADH

Recall the Cannizzaro reaction:

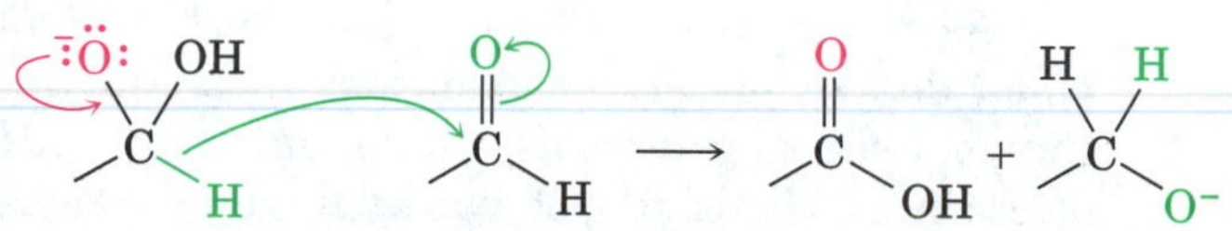

Step 4: Chain cleavage An acetyl group is split off from the acyl chain in the final step of β-oxidation and is attached to a new coenzyme A molecule, leaving behind an acyl CoA that is two carbon atoms shorter. The reaction is catalyzed by the enzyme β-keto thiolase and is mechanistically the exact reverse of the Claisen condensation reaction (Section 23.8). In the forward direction, a Claisen condensation joins two esters together to form a β-keto ester product. In the reverse direction, a retro-Claisen reaction splits a β-keto ester (or β-keto thiol ester) apart to form two esters (or two thiol esters).

The reaction occurs by nucleophilic addition of coenzyme A to the keto group of the β-keto acyl CoA to yield an alkoxide ion intermediate, followed by cleavage of the C2–C3 bond with expulsion of an acetyl CoA enolate ion. Protonation of the enolate ion gives acetyl CoA, and the chain-shortened acyl CoA enters another round of the β-oxidation pathway for further degradation.

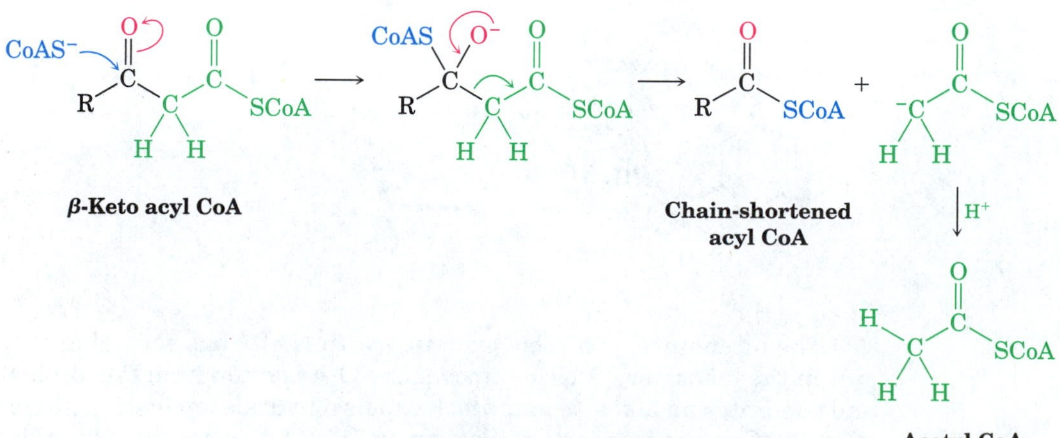

β-Keto acyl CoA

Chain-shortened acyl CoA

Acetyl CoA

Look at the catabolism of myristic acid shown in Figure 30.3 to see the overall results of the β-oxidation pathway. The first passage along the pathway converts the 14-carbon myristyl CoA into the 12-carbon lauryl CoA plus acetyl CoA; the second passage converts lauryl CoA into the 10-carbon capryl CoA plus acetyl CoA; the third passage converts capryl CoA into the 8-carbon caprylyl CoA; and so on. Note that the last passage produces *two* molecules of acetyl CoA because the precursor has four carbons.

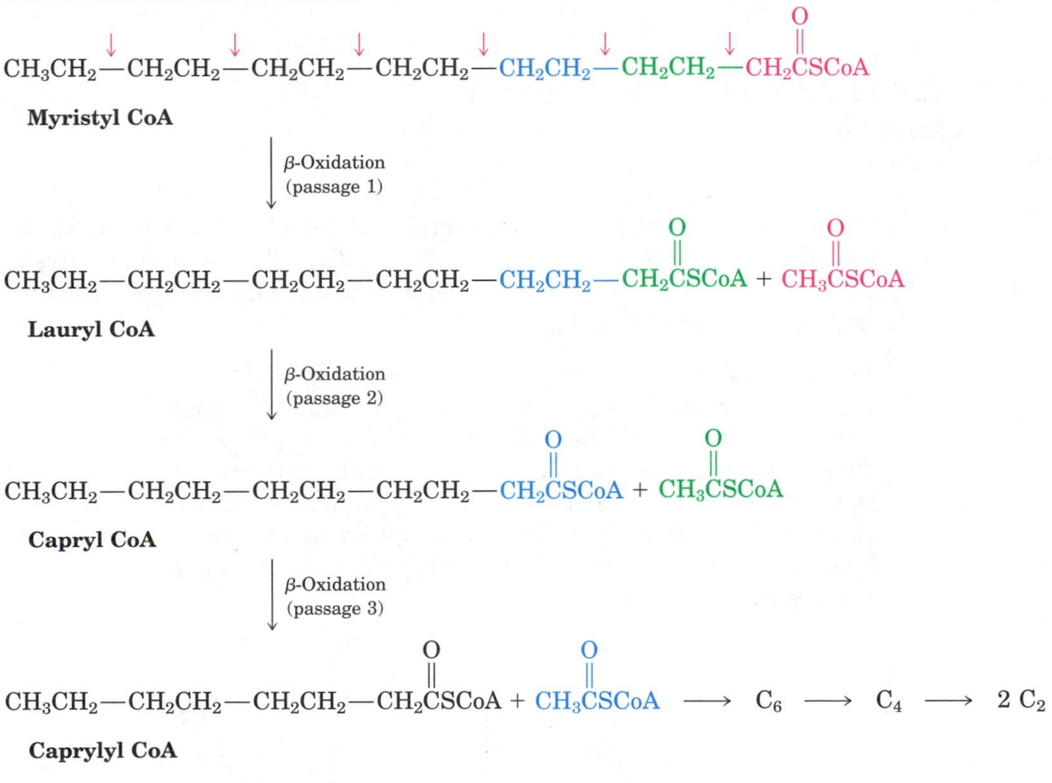

Figure 30.3 Catabolism of the 14-carbon myristic acid by the β-oxidation pathway yields seven molecules of acetyl CoA after six passages.

You can predict how many molecules of acetyl CoA will be obtained from a given fatty acid simply by counting the number of carbon atoms and dividing by two. For example, the 14-carbon myristic acid yields seven molecules of acetyl CoA after six passages through the β-oxidation pathway. The number of passages is always one less than the number of acetyl CoA molecules produced because the last passage cleaves a four-carbon chain into two acetyl CoA's.

Most fatty acids have an even number of carbon atoms, so that none are left over after β-oxidation. Those fatty acids with an odd number of carbon atoms or with double bonds require additional steps for degradation, but all carbon atoms are ultimately released for further oxidation in the citric acid cycle.

PROBLEM.

30.3 Write the equations for the remaining passages of the β-oxidation pathway following those shown in Figure 30.3.

PROBLEM.

30.4 How many molecules of acetyl CoA are produced by catabolism of the following fatty acids, and how many passages of the β-oxidation pathway are needed?
(a) Palmitic acid, $CH_3(CH_2)_{14}COOH$ (b) Arachidic acid, $CH_3(CH_2)_{18}COOH$

30.3 Catabolism of Carbohydrates: Glycolysis

Glycolysis is a series of 10 enzyme-catalyzed reactions that break down glucose into 2 equivalents of pyruvate, $CH_3COCO_2^-$. The steps of glycolysis, also called the *Embden–Meyerhof pathway* after its discoverers,[3,4] are summarized in Figure 30.4 (pp. 1184–1185).

Steps 1–3: Phosphorylation and isomerization Glucose, produced by the digestion of dietary carbohydrates, is first phosphorylated at the hydroxyl group on C6 by reaction with ATP in a step catalyzed by the enzyme hexokinase. The glucose 6-phosphate that results is isomerized by phosphoglucose isomerase to fructose 6-phosphate. As the open-chain structures in Figure 30.4 show, this isomerization reaction takes place by keto–enol tautomerism (Section 22.1), since both glucose and fructose share a common enol:

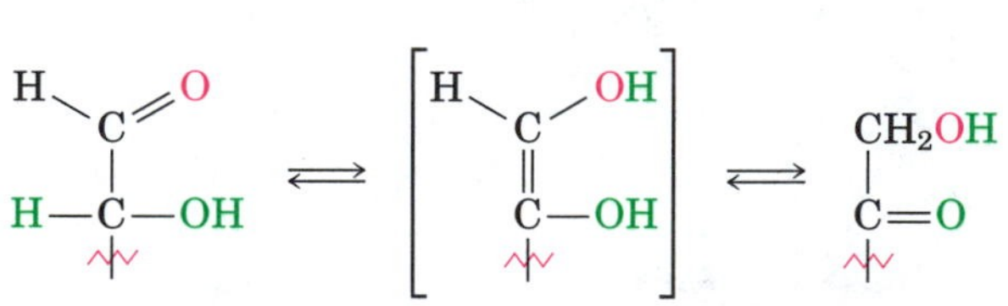

Glucose **Glucose/fructose enol** **Fructose**

Fructose 6-phosphate is then converted to fructose 1,6-bisphosphate by phosphofructokinase-catalyzed reaction with ATP (the prefix "bis" means two). The result is a molecule ready to be split into the two three-carbon intermediates that will ultimately become two molecules of pyruvate.

[3]Gustav George Embden (1874–1933); b. Hamburg, Germany; educated at Freiburg, Munich, Berlin, Strasbourg; professor, Frankfurt (1904–1933).

[4]Otto Fritz Meyerhof (1884–1951); b. Hanover, Germany; professor, Kiel, Heidelberg, Berlin, Paris, University of Pennsylvania; Nobel Prize (1922).

Steps 4–5: Cleavage and isomerization Fructose 1,6-bisphosphate is cleaved in step 4 into two, three-carbon monophosphates, one an aldose and one a ketose. The bond between carbons 3 and 4 of fructose 1,6-bisphosphate breaks, and a C=O group is formed. Mechanistically, the cleavage is the reverse of an aldol reaction (Section 23.2) and is carried out by an aldolase enzyme. (A *forward* aldol reaction joins two ketones/aldehydes to give a β-hydroxy ketone or aldehyde; a *retro* aldol reaction cleaves a β-hydroxy ketone or aldehyde into two ketones/aldehydes.) The reaction can be imagined to occur by attack of a base on the O–H group at C4, followed by loss of an enolate ion that is protonated to yield dihydroxyacetone phosphate.

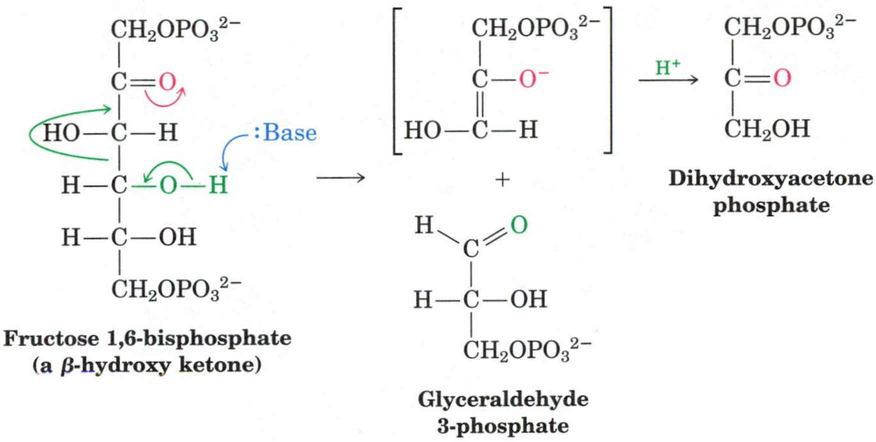

Actually, the reaction is a bit more complex than shown above because it does not take place on the free ketone. Instead, fructose 1,6-bisphosphate undergoes reaction with the side-chain $-NH_2$ group of a lysine residue on the aldolase enzyme to yield an imine (Section 19.12), also called a *Schiff base*. Protonation of the imine makes it more reactive; a retro aldol-like reaction ensues, giving glyceraldehyde 3-phosphate and the imine of dihydroxyacetone phosphate; and the imine is hydrolyzed.

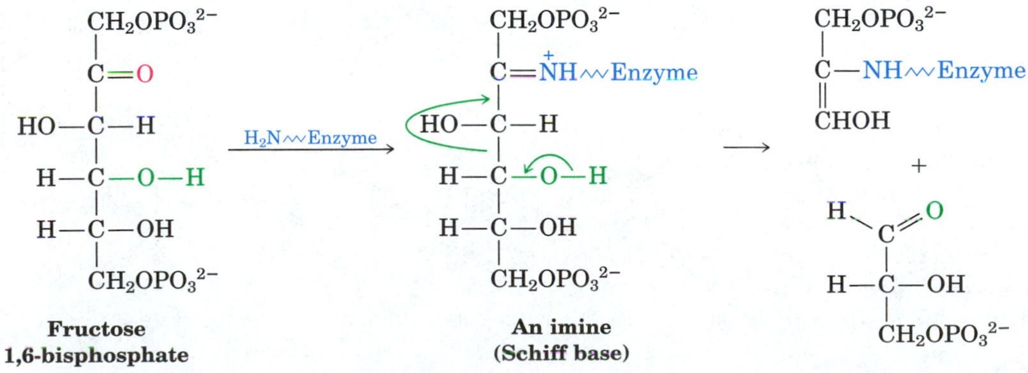

Step 1
Glucose is phosphorylated by reaction with ATP to yield glucose 6-phosphate.

Step 2
Glucose 6-phosphate is isomerized to fructose 6-phosphate.

Step 3
Fructose 6-phosphate is phosphorylated by reaction with ATP to yield fructose 1,6-bisphosphate.

Step 4
Fructose 1,6-bisphosphate is cleaved into two three-carbon pieces by the enzyme aldolase.

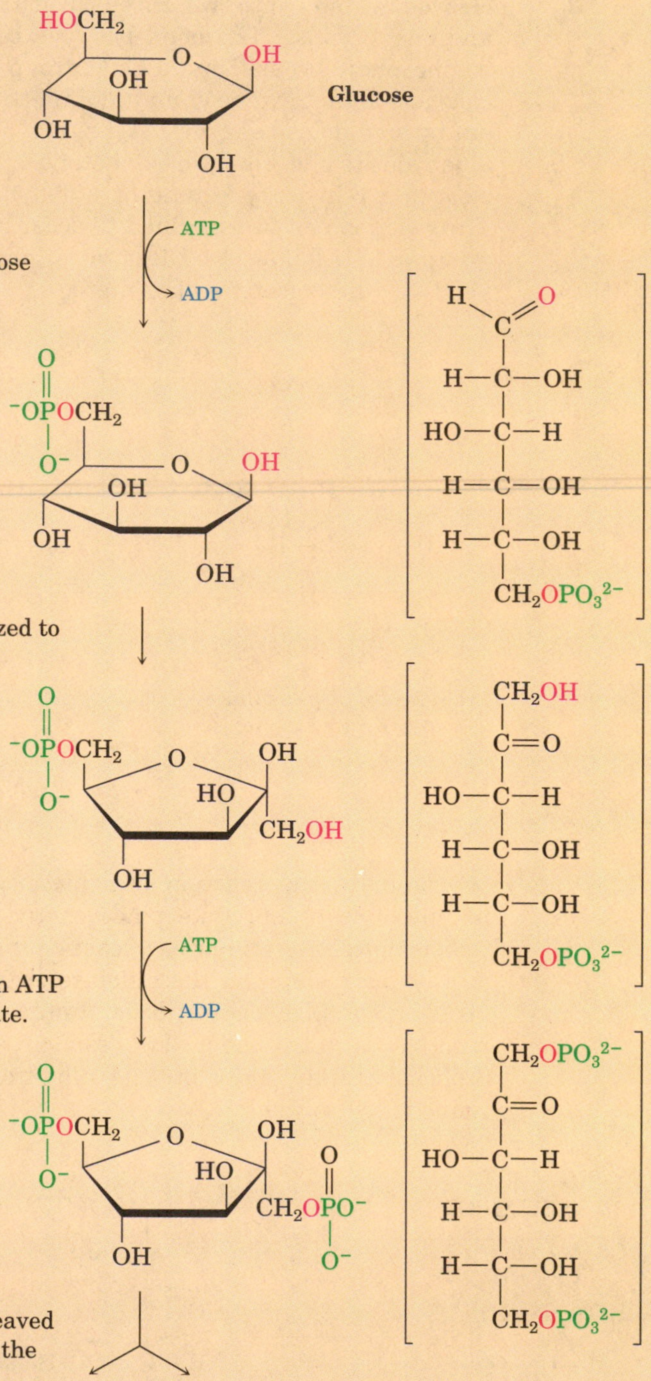

Figure 30.4 The ten-step glycolysis pathway for catabolizing glucose to pyruvate.

Step 5
Dihydroxyacetone phosphate, one of the products of step 4, is isomerized to glyceraldehyde 3-phosphate, the other product of step 4.

Step 6
Glyceraldehyde 3-phosphate is oxidized and phosphorylated to yield 3-phosphoglyceroyl phosphate.

Step 7
A phosphate is transferred from the carboxyl group to ADP, resulting in synthesis of an ATP and yielding 3-phosphoglycerate.

Step 8
A phosphate group is transferred from the C3 hydroxyl to the C2 hydroxyl, giving 2-phosphoglycerate.

Step 9
Dehydration occurs to yield phosphoenolpyruvate (PEP).

Step 10
A phosphate is transferred from PEP to ADP, yielding pyruvate and ATP.

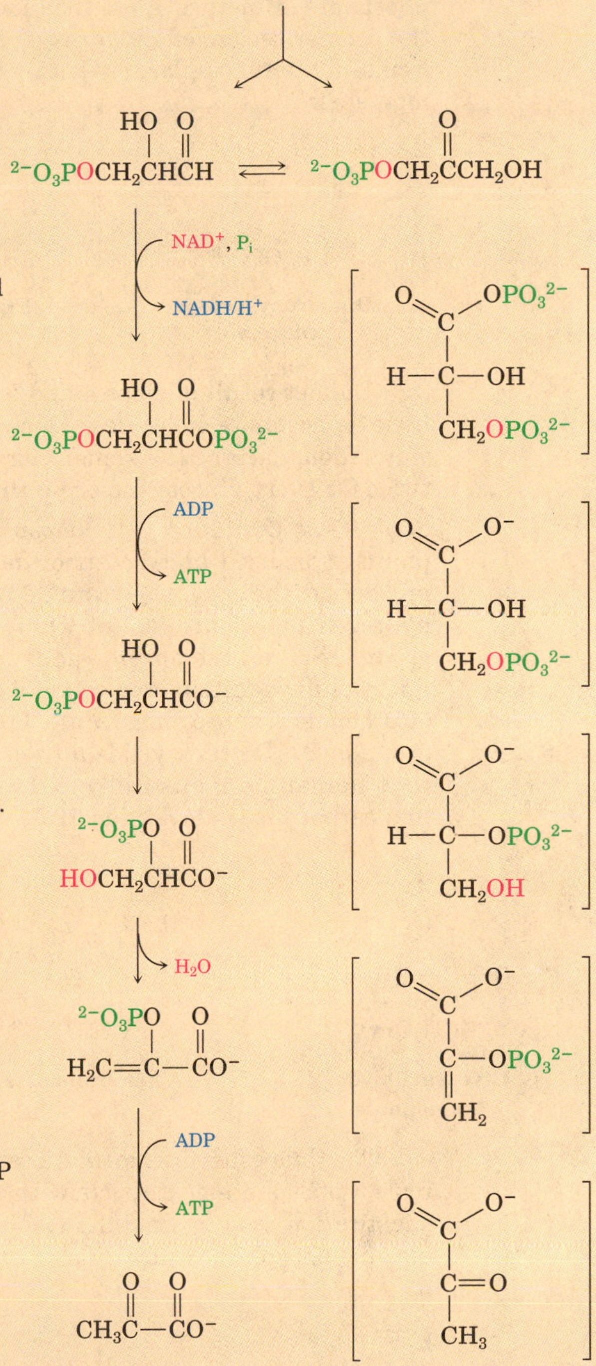

Figure 30.4 *(continued)*

Glyceraldehyde 3-phosphate continues on in the glycolysis pathway, but dihydroxyacetone phosphate is first isomerized by the enzyme triose phosphate isomerase. As in the glucose-to-fructose conversion of step 2, the isomerization of dihydroxyacetone phosphate to glyceraldehyde 3-phosphate takes place by keto–enol tautomerization through a common enol:

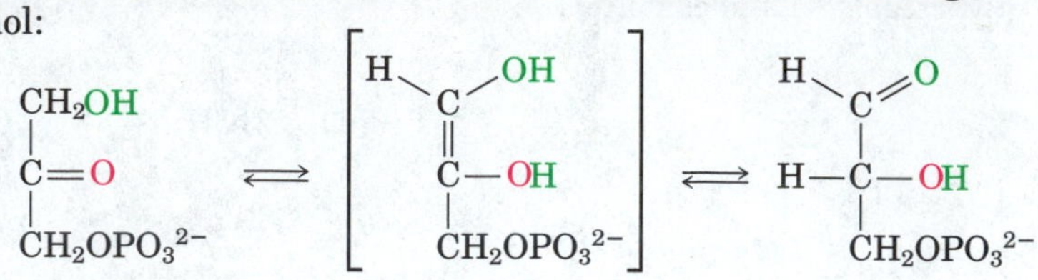

Dihydroxyacetone phosphate **Enol** **Glyceraldehyde 3-phosphate**

The net result of steps 4 and 5 is the production of two glyceraldehyde 3-phosphate molecules, which both pass down the rest of the pathway. Thus, each of the remaining five steps of glycolysis takes place twice for every glucose molecule that enters at step 1.

Steps 6–8: Oxidation and phosphorylation Glyceraldehyde 3-phosphate is oxidized and phosphorylated by the coenzyme NAD^+ in the presence of the enzyme glyceraldehyde 3-phosphate dehydrogenase and hydrogen phosphate ion, HPO_4^{2-}. The reaction occurs when a thiol group (–SH) on the enzyme adds to the aldehyde carbonyl group in a nucleophilic addition step to yield a *hemithioacetal*, the sulfur analog of a hemiacetal (Section 19.14). Oxidation of the hemithioacetal –OH group by NAD^+ then yields a thiol ester intermediate. The reaction is thus similar mechanistically to the laboratory oxidation of an aldehyde to a carboxylic acid (Section 19.5).

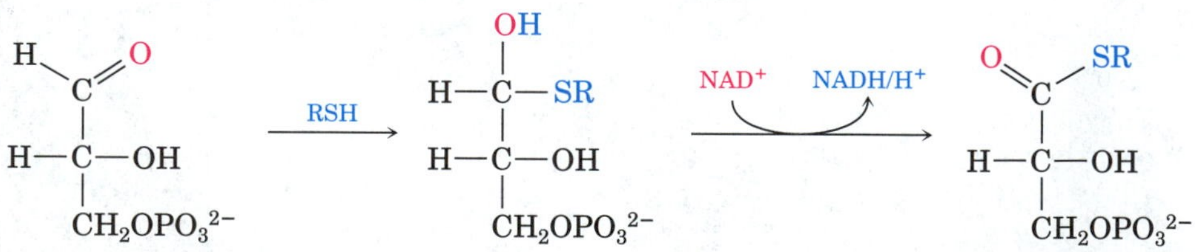

Glyceraldehyde 3-phosphate **A hemithioacetal** **A thiol ester**

The thiol ester intermediate resulting from oxidation of glyceraldehyde 3-phosphate next reacts with phosphate ion in a nucleophilic acyl substitution step to yield 3-phosphoglyceroyl phosphate, a mixed carboxylic–phosphoric anhydride:

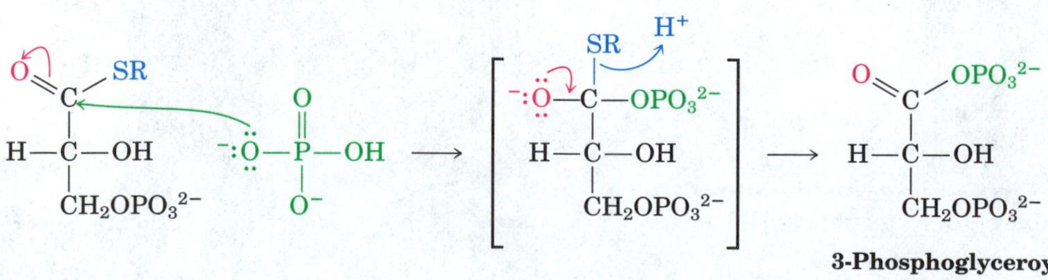

3-Phosphoglyceroyl phosphate

Like all anhydrides, the mixed carboxylic–phosphoric anhydride is a reactive substrate in nucleophilic acyl substitution reactions (Section 21.6). Reaction of 3-phosphoglyceroyl phosphate with ADP results in nucleophilic attack on phosphorus and transfer of a phosphate group to yield ATP and 3-phosphoglycerate. The process is catalyzed by the enzyme phosphoglycerate kinase.

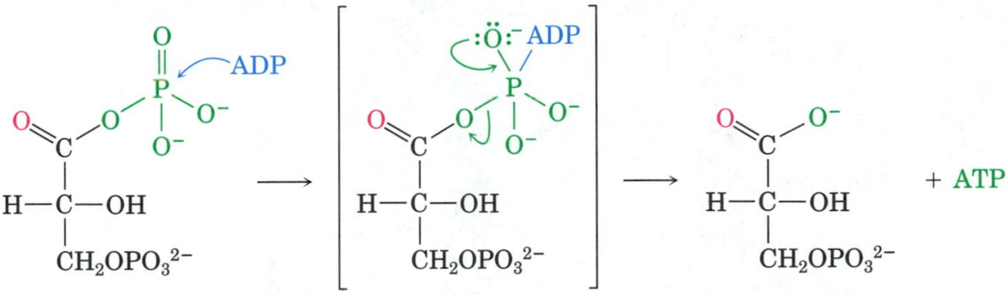

3-Phosphoglyceroyl phosphate **3-Phosphoglycerate**

Isomerization of 3-phosphoglycerate then gives 2-phosphoglycerate in a step catalyzed by the enzyme phosphoglyceromutase:

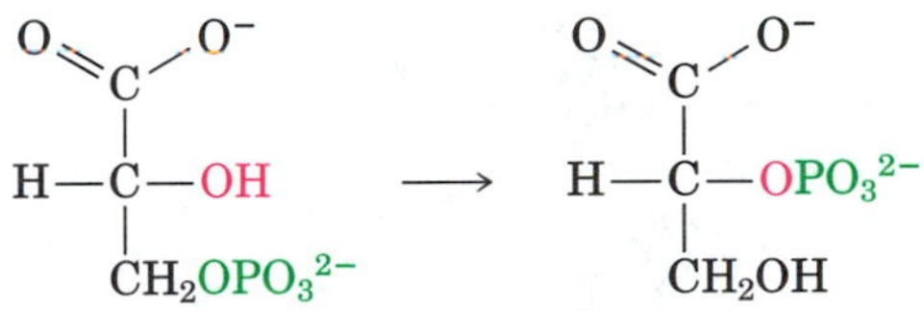

3-Phosphoglycerate **2-Phosphoglycerate**

Steps 9–10: Dehydration and dephosphorylation Like the β-hydroxy carbonyl compounds produced in aldol reactions (Section 23.4), 2-phosphoglycerate undergoes a ready dehydration in an E2-like reaction. The process is catalyzed by enolase, and the product is phosphoenolpyruvate, abbreviated PEP:

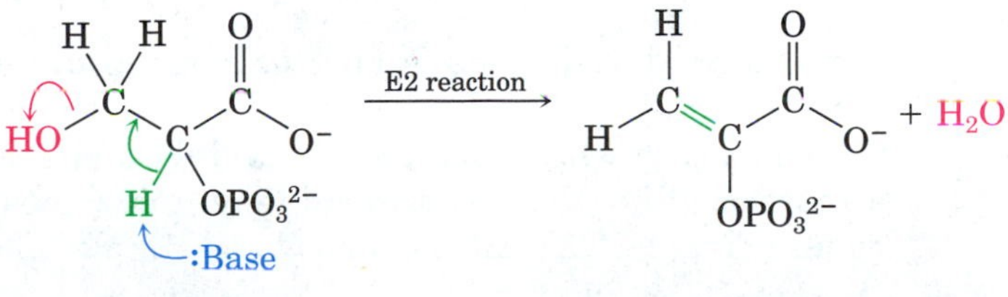

2-Phosphoglycerate (a β-hydroxy carbonyl compound) **Phosphoenolpyruvate (PEP)**

Transfer of the phosphate group to ADP then generates ATP and gives pyruvate, a reaction catalyzed by pyruvate kinase:

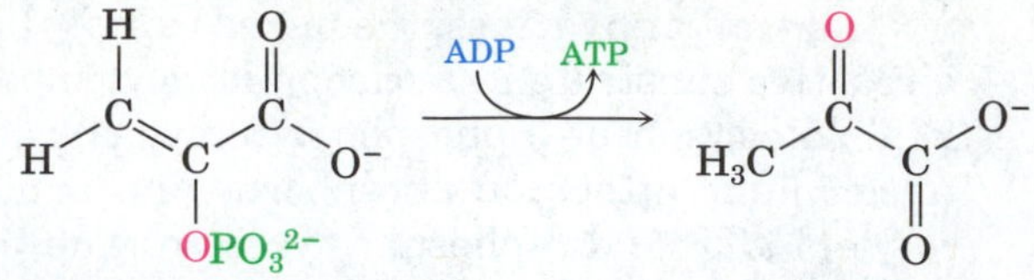

Phosphoenolpyruvate **Pyruvate**

The net result of glycolysis can be summarized by the following equation:

$$\underset{\textbf{Glucose}}{C_6H_{12}O_6} + 2\ NAD^+ + 2\ P_i + 2\ ADP$$

$$\longrightarrow 2\ \underset{\textbf{Pyruvate}}{CH_3\overset{O}{\overset{\|}{C}}-\overset{O}{\overset{\|}{C}}O^-} + 2\ NADH + 2\ ATP + 2\ H_2O + 2\ H^+$$

PROBLEM.

30.5 Identify the two steps in glycolysis in which ATP is produced.

PROBLEM.

30.6 Propose a mechanism for the isomerization of dihydroxyacetone phosphate to glyceraldehyde 3-phosphate in step 5 of glycolysis.

PROBLEM.

30.7 Look at the entire glycolysis pathway and make a list of the kinds of organic reactions that take place—nucleophilic acyl substitutions, aldol reactions, and so forth.

30.4 The Conversion of Pyruvate to Acetyl CoA

Pyruvate produced in the catabolism of glucose can undergo several further transformations depending on the conditions and on the organism. Most commonly, pyruvate is converted to acetyl CoA through a multistep sequence of reactions that requires three different enzymes and four different coenzymes (Figure 30.5). All the steps have simple laboratory analogies.

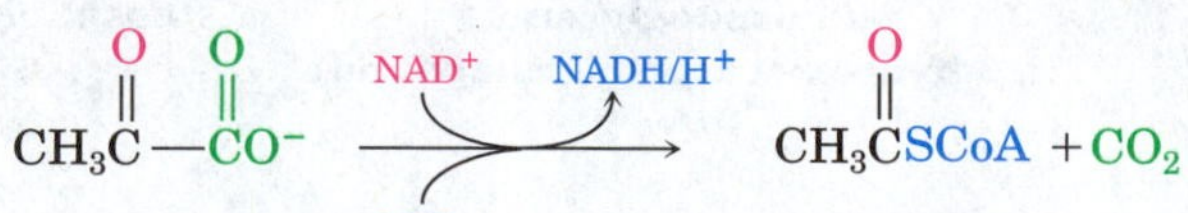

Pyruvate **Acetyl CoA**

Step 1
Nucleophilic addition of thiamine pyrophosphate to the ketone carbonyl group of pyruvate yields an intermediate addition product.

Step 2
Decarboxylation occurs, analogous to the loss of CO_2 from a β-keto acid, yielding an enamine intermediate.

Step 3
The nucleophilic enamine double bond attacks a sulfur atom of lipoamide and does an S_N2-like displacement of the second sulfur atom.

Step 4
Elimination of thiamine pyrophosphate from the tetrahedral intermediate then yields acetyl dihydrolipoamide.

Step 5
Reaction with coenzyme A exchanges one thiol ester for another, giving acetyl CoA and dihydrolipoamide.

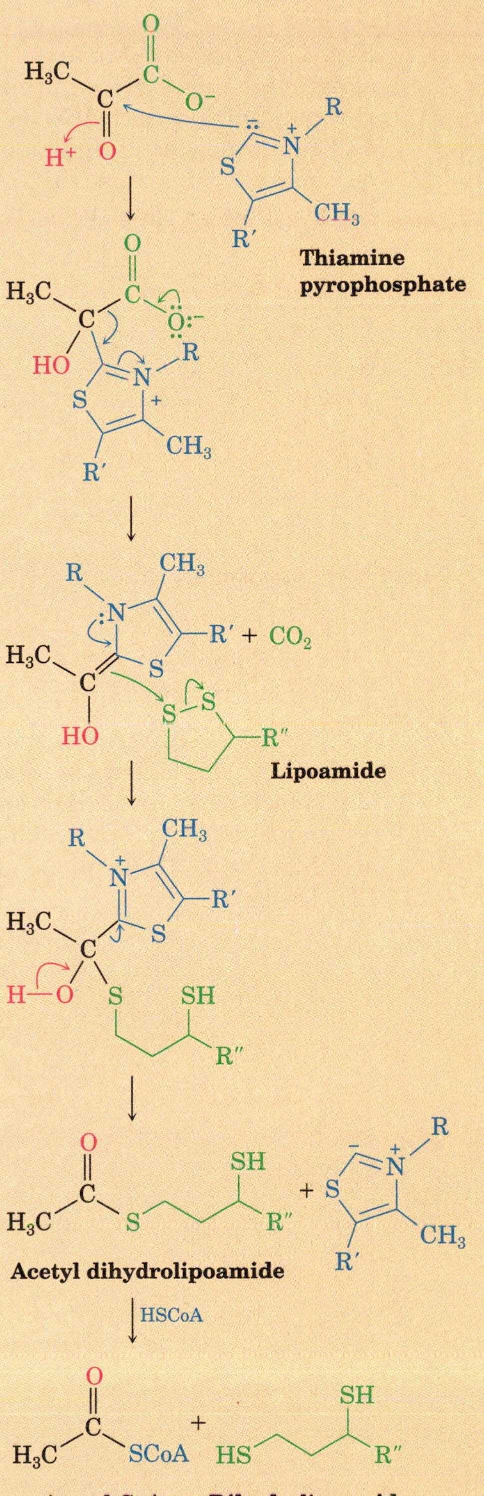

Figure 30.5 Mechanism of the conversion of pyruvate to acetyl CoA. The multistep sequence of reactions requires three different enzymes and four different coenzymes.

Step 1: Addition of thiamine As shown in Figure 30.5, the conversion of pyruvate to acetyl CoA begins by reaction of pyruvate with thiamine pyrophosphate, a derivative of vitamin B_1. The hydrogen on the heterocyclic thiazole ring of thiamine pyrophosphate is weakly acidic and can be removed by reaction with base to yield a nucleophilic ylide much like the phosphorus ylides used in Wittig reactions (Section 19.15):

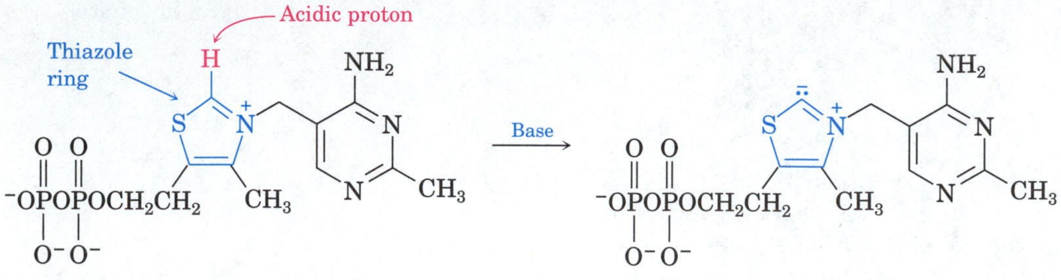

Thiamine pyrophosphate

This nucleophilic ylide adds to the ketone carbonyl of pyruvate to yield a tetrahedral intermediate.

Step 2: Decarboxylation Decarboxylation of the pyruvate/thiamine addition product occurs in much the same way that decarboxylation of a β-keto acid intermediate occurs in the acetoacetic ester synthesis (Section 22.8). The $C=N^+$ double bond of the pyruvate addition product acts like the C=O double bond of a β-keto acid to accept electrons as CO_2 leaves.

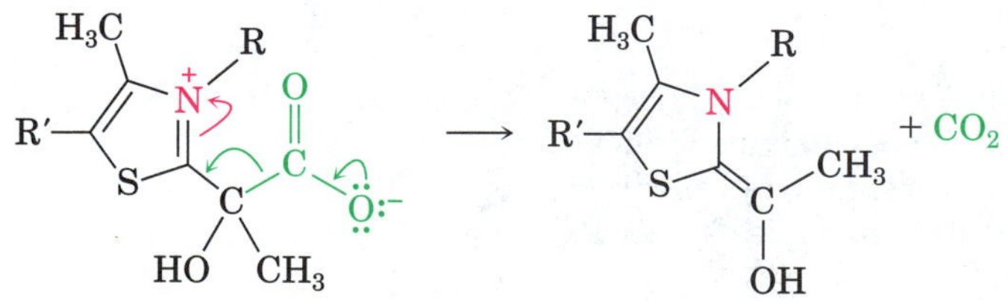

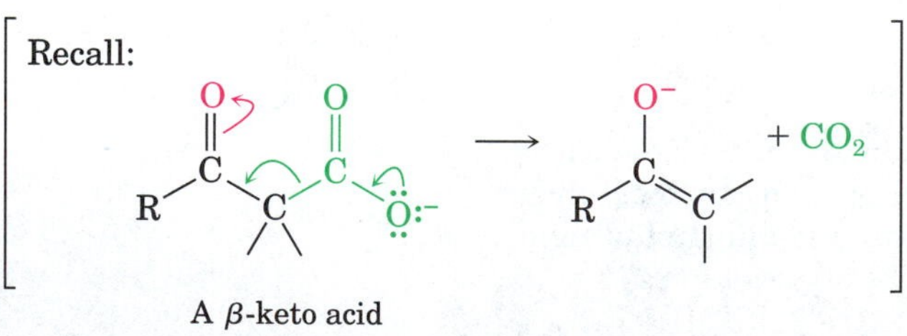

A β-keto acid

Step 3: Reaction with lipoamide The decarboxylation product is an enamine ($R_2N{-}C{=}C$), which, like all enamines, is strongly nucleophilic (Section 23.12). The enamine undergoes reaction with the cyclic disulfide lipoamide by nucleophilic attack on a sulfur atom, displacing the second sulfur in an S_N2-like process.

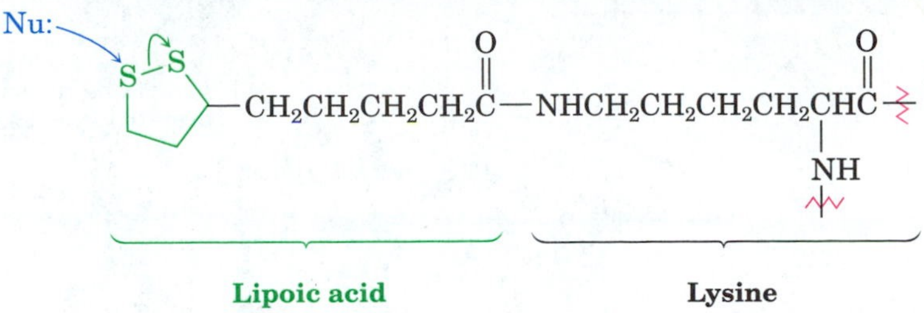

Lipoamide: Lipoic acid is linked through an amide bond to the side-chain NH_2 group of a lysine residue in dihydrolipoyl transacetylase

Step 4: Elimination of thiamine The product of the enamine reaction with lipoamide is itself a tetrahedral carbonyl addition product, which can eliminate thiamine pyrophosphate. This elimination, the exact reverse of step 1, generates the carbonyl compound acetyl dihydrolipoamide.

Step 5: Acyl transfer Acetyl dihydrolipoamide, a thiol ester, undergoes a nucleophilic acyl substitution reaction with coenzyme A in an ester exchange reaction to yield acetyl CoA plus dihydrolipoamide. The dihydrolipoamide is then oxidized back to lipoamide by FAD, and the $FADH_2$ that results is in turn oxidized back to FAD by NAD^+.

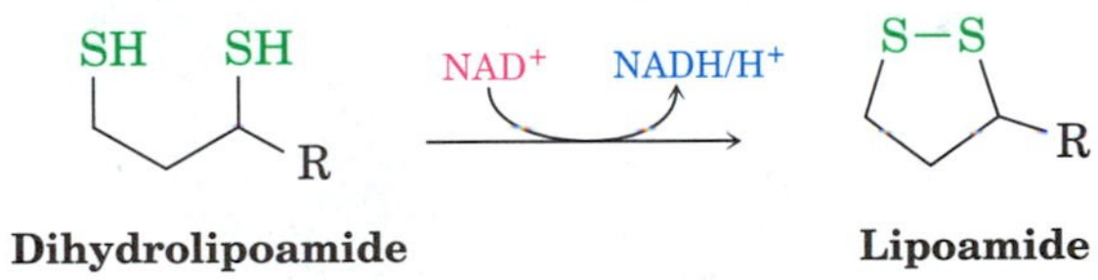

PROBLEM.

30.8 Show the mechanism of the thiol ester exchange reaction of acetyl dihydrolipoamide with coenzyme A to yield acetyl CoA in step 5 of Figure 30.5.

30.5 The Citric Acid Cycle

The first two stages of catabolism result in the conversion of fats and carbohydrates into acetyl groups that are bonded through a thiol ester link to coenzyme A. These acetyl groups then enter the third stage of catabolism—the **citric acid cycle**, also called the *tricarboxylic acid (TCA) cycle* or *Krebs cycle* (after Hans Krebs,[5] who unraveled its complexities in 1937). The steps of the citric acid cycle and a brief description of each are given in Figure 30.6.

[5]Hans Krebs (1900–1981); b. Hildesheim, Germany; M.D. Hamburg (1925); professor, Berlin, Freiburg, Sheffield (1935–1954), Oxford (1954–1967); Nobel Prize (1953).

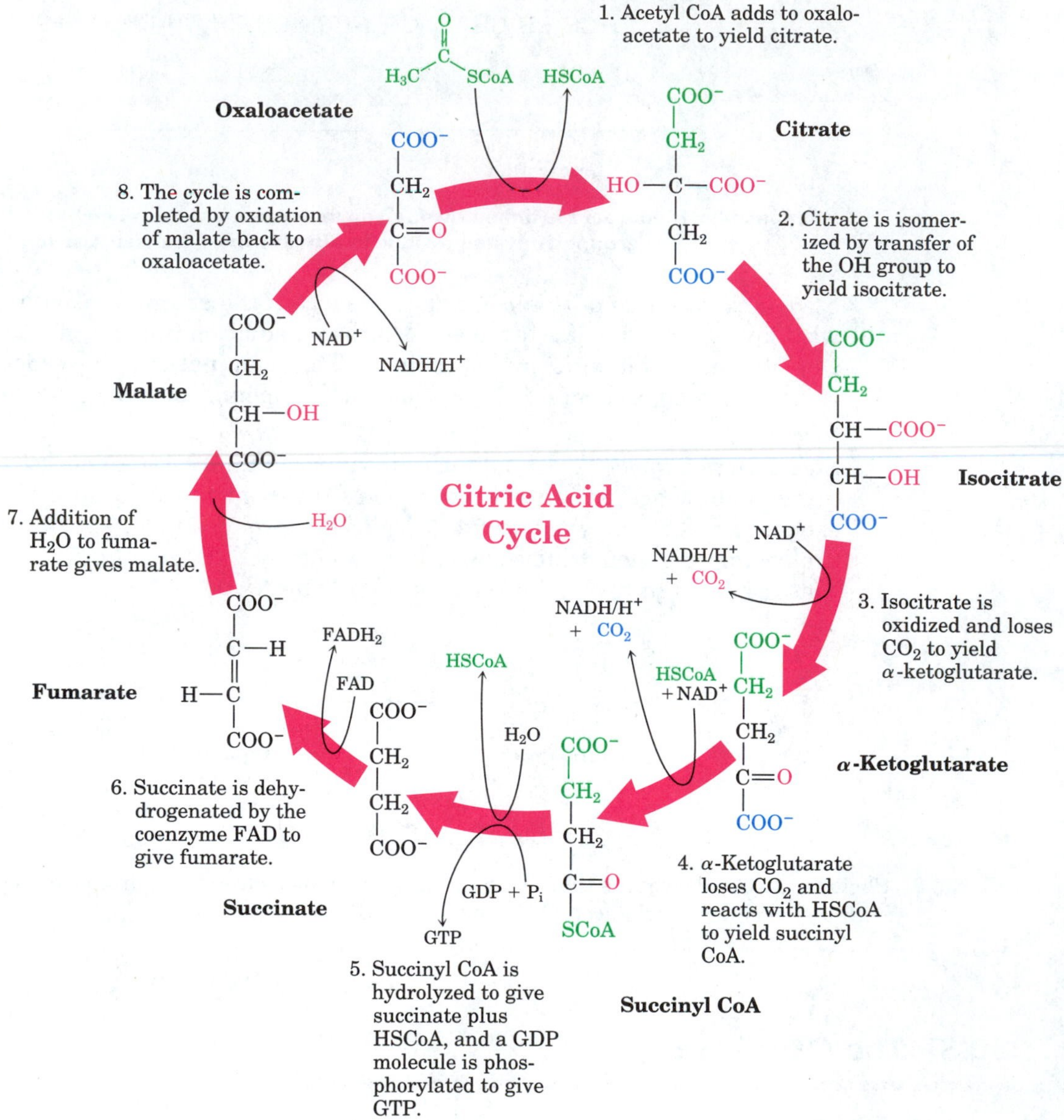

Figure 30.6 The citric acid cycle, an eight-step series of reactions that results in the conversion of an acetyl group into two molecules of CO_2 plus reduced coenzymes.

As its name implies, the citric acid *cycle* is a closed loop of reactions in which the product of the final step is a reactant in the first step. The intermediates are constantly regenerated and flow continuously through the cycle, which operates as long as the oxidizing coenzymes NAD^+ and FAD are available. To meet this condition, the reduced coenzymes NADH and $FADH_2$ must be reoxidized via the respiratory chain, which in turn relies on oxygen as the ultimate electron acceptor. Thus, the cycle is dependent on the availability of oxygen and on the operation of the respiratory chain.

Steps 1–2: Addition to oxaloacetate Acetyl groups from acetyl CoA enter the citric acid cycle in step 1 by nucleophilic addition to the ketone carbonyl group of oxaloacetate to give citryl CoA. The addition is an aldol reaction (Section 23.2) of an enolate ion from acetyl CoA, and is catalyzed by the enzyme citrate synthetase. Citryl CoA is then hydrolyzed to citrate.

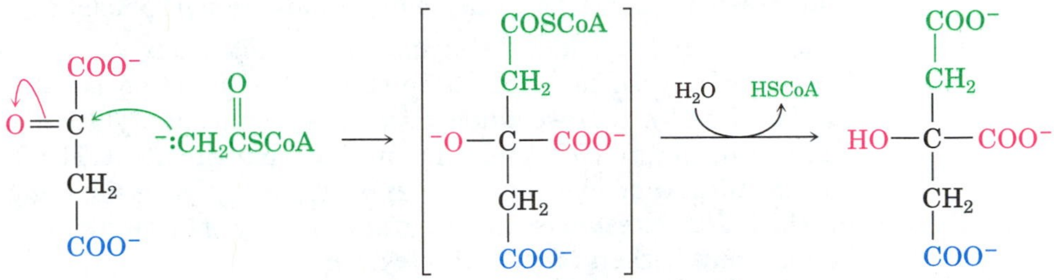

Oxaloacetate

Citrate

Citrate, a tertiary alcohol, is next converted into its isomer, isocitrate, a secondary alcohol that can be oxidized. The isomerization occurs in two steps, both of which are catalyzed by the same aconitase enzyme. The initial step is an E2 dehydration of a β-hydroxy acid, the same sort of reaction that occurs in step 9 of glycolysis (Figure 30.4). The second step is a conjugate nucleophilic addition of water, the same sort of reaction that occurs in step 2 of the β-oxidation pathway (Figure 30.2).

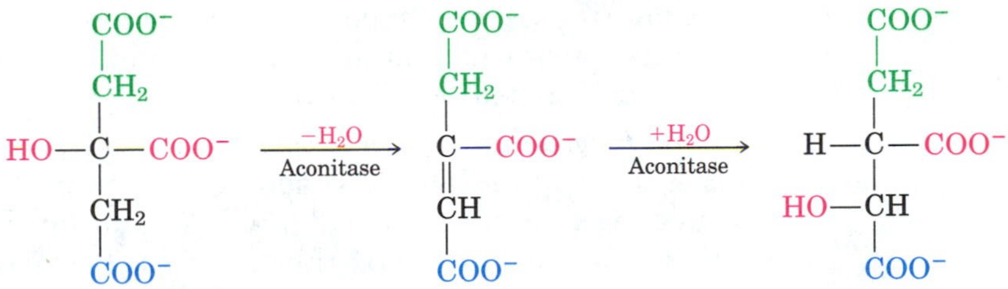

Citrate **Aconitate** **Isocitrate**

Note that the dehydration of citrate takes place specifically *away* from the carbon atoms of the acetyl group that added to oxaloacetate in step 1.

Steps 3–4: Oxidative decarboxylations Isocitrate, a secondary alcohol, is oxidized by NAD^+ in step 3 to give a ketone intermediate, which loses CO_2 to give α-ketoglutarate. Catalyzed by the enzyme isocitrate dehydrogenase, the decarboxylation is a typical reaction of a carboxylic acid that has a second carbonyl group two atoms away (a β-keto acid). A similar decarboxylation reaction occurs in the acetoacetic ester synthesis (Section 22.8).

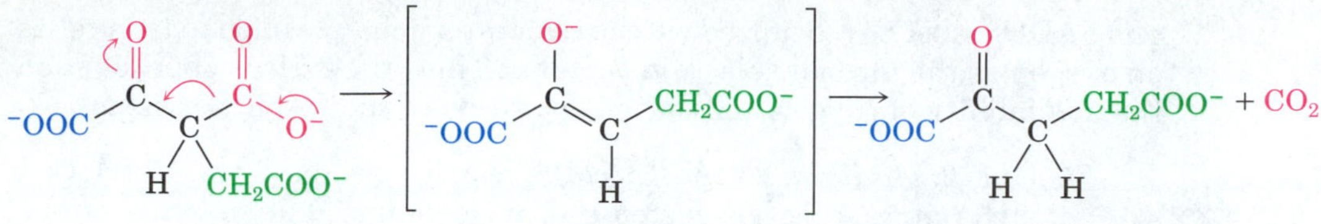

α-Ketoglutarate

The transformation of α-ketoglutarate to succinyl CoA in step 4 is a multistep process, analogous to the transformation of pyruvate to acetyl CoA that we saw in the previous section. In both cases, an α-keto acid loses CO_2 in a step catalyzed by thiamine pyrophosphate.

Steps 5–6: Hydrolysis and dehydrogenation of succinyl CoA Succinyl CoA is hydrolyzed to succinate in step 5. The reaction is catalyzed by succinyl CoA synthetase and is coupled with phosphorylation of guanosine diphosphate (GDP) to give guanosine triphosphate (GTP). The overall transformation is similar to that of step 8 in glycolysis (Figure 30.4), in which a thiol ester is converted into an acyl phosphate and a phosphate group is then transferred to ADP.

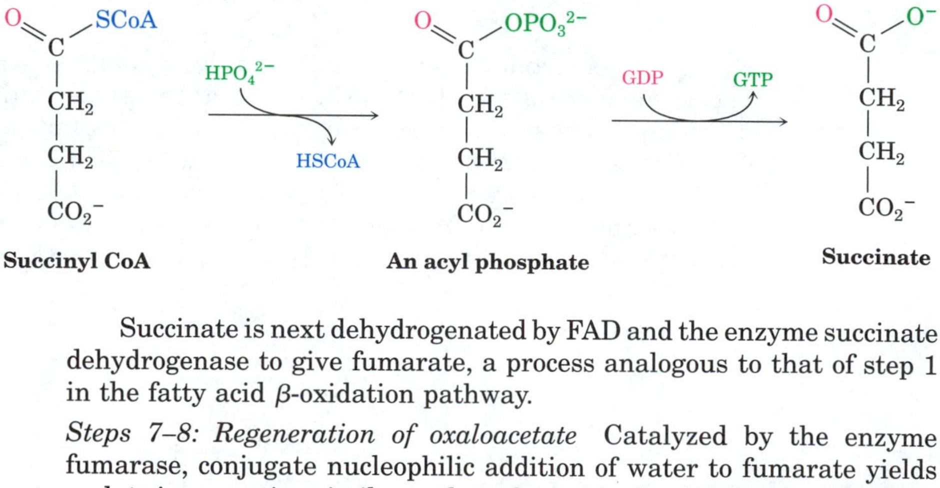

Succinyl CoA **An acyl phosphate** **Succinate**

Succinate is next dehydrogenated by FAD and the enzyme succinate dehydrogenase to give fumarate, a process analogous to that of step 1 in the fatty acid β-oxidation pathway.

Steps 7–8: Regeneration of oxaloacetate Catalyzed by the enzyme fumarase, conjugate nucleophilic addition of water to fumarate yields malate in a reaction similar to that of step 2 in the fatty acid β-oxidation pathway. Oxidation with NAD^+ then gives oxaloacetate in a step catalyzed by malate dehydrogenase, and the citric acid cycle has returned to its starting point, ready to revolve again. The net result of the cycle can be summarized as:

$$\text{Acetyl CoA} + 3\ NAD^+ + FAD + ADP + P_i + 2\ H_2O \longrightarrow HSCoA + 3\ NADH + 3\ H^+ + FADH_2 + ATP + 2\ CO_2$$

In looking at the metabolic pathways discussed so far, it's interesting to note that two are linear (fatty acid β-oxidation and glycolysis) but one is cyclic (citric acid cycle). Why does nature use different strategies for the different pathways? Although there's probably no one "right" answer, part of the reason may well be the fact that chemistry in living organisms and chemistry in the laboratory must follow the same rules of reactivity.

When a relatively large, multifunctional molecule such as a fatty acid or glucose is being degraded, the reaction choices are numerous, and an efficient linear pathway is possible. When a smaller, monofunctional molecule such as acetyl CoA is being degraded, however, the mechanistic options are limited. There are relatively few reaction choices available for degrading acetyl CoA, and a linear pathway may not be energetically feasible. By employing a cyclic pathway with multifunctional intermediates, the range of reaction choices available becomes much larger.

PROBLEM

30.9 Which of the substances in the citric acid cycle are tricarboxylic acids (thus giving the cycle its alternate name)?

PROBLEM

30.10 Write mechanisms for step 2 of the citric acid cycle, the dehydration of citrate and the addition of water to aconitate.

PROBLEM

30.11 Write a mechanism for the conversion of succinyl CoA to succinate in step 5 of the citric acid cycle.

30.6 Catabolism of Proteins: Transamination

The catabolism of proteins is more complex than that of fats and carbohydrates because each of the 20 amino acids is degraded through its own unique pathway. The general idea, however, is that the amino nitrogen atoms are removed and the substances that remain are converted into compounds that enter the citric acid cycle.

Most amino acids lose their nitrogen atom by a **transamination** reaction in which the $-NH_2$ group of the amino acid changes places with the keto group of α-ketoglutarate. The products are a new α-keto acid and glutamate:

$$\underset{\text{An amino acid}}{\mathrm{RCH(NH_3^+)COO^-}} + \underset{\alpha\text{-Ketoglutarate}}{\mathrm{^-OOCCH_2CH_2C(=O)COO^-}} \rightleftharpoons \underset{\text{An }\alpha\text{-keto acid}}{\mathrm{RC(=O)COO^-}} + \underset{\text{Glutamate}}{\mathrm{^-OOCCH_2CH_2CH(NH_3^+)COO^-}}$$

Transaminations use pyridoxal phosphate, a derivative of vitamin B_6, as cofactor. As shown in Figure 30.7 for the reaction of alanine, the key step in transamination is nucleophilic addition of the amino acid $-NH_2$ group to the pyridoxal aldehyde group to yield an imine (Section 19.12). Loss of a proton from the α position then results in a bond rearrangement to give a new imine, which is hydrolyzed (the exact reverse of imine formation) to yield pyruvate and a nitrogen-containing derivative of pyridoxal phosphate. Pyruvate is converted into acetyl CoA (Section 30.4), which enters the citric acid cycle for further catabolism. The pyridoxal phosphate derivative transfers its nitrogen atom to α-ketoglutarate by the reverse of the steps in Figure 30.7, thereby forming glutamate and regenerating pyridoxal phosphate for further use.

Glutamate, which now contains the nitrogen atom of the former amino acid, next undergoes an oxidative deamination to yield ammonium ion and regenerated α-ketoglutarate. The oxidation of the amine to an imine is mechanistically similar to the oxidation of a secondary alcohol to a ketone and is carried out by NAD^+. The imine is then hydrolyzed in the usual way.

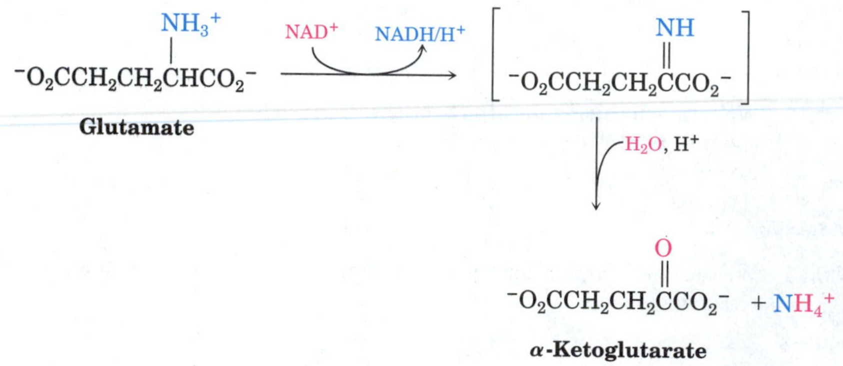

PROBLEM

30.12 Show the mechanism of the reaction of α-ketoglutarate with the nitrogen-containing pyridoxal by-product resulting from transamination of an amino acid. Glutamate and regenerated pyridoxal phosphate are produced.

PROBLEM

30.13 Show the product from transamination of leucine.

30.7 Anabolism of Fatty Acids

One of the most striking features of the common fatty acids is that all have an even number of carbon atoms (Table 28.1). This even number results because all fatty acids are derived biosynthetically from the simple two-carbon precursor, acetyl CoA. The anabolic pathway by which organisms synthesize fatty acids is shown in Figure 30.8 (p. 1198).

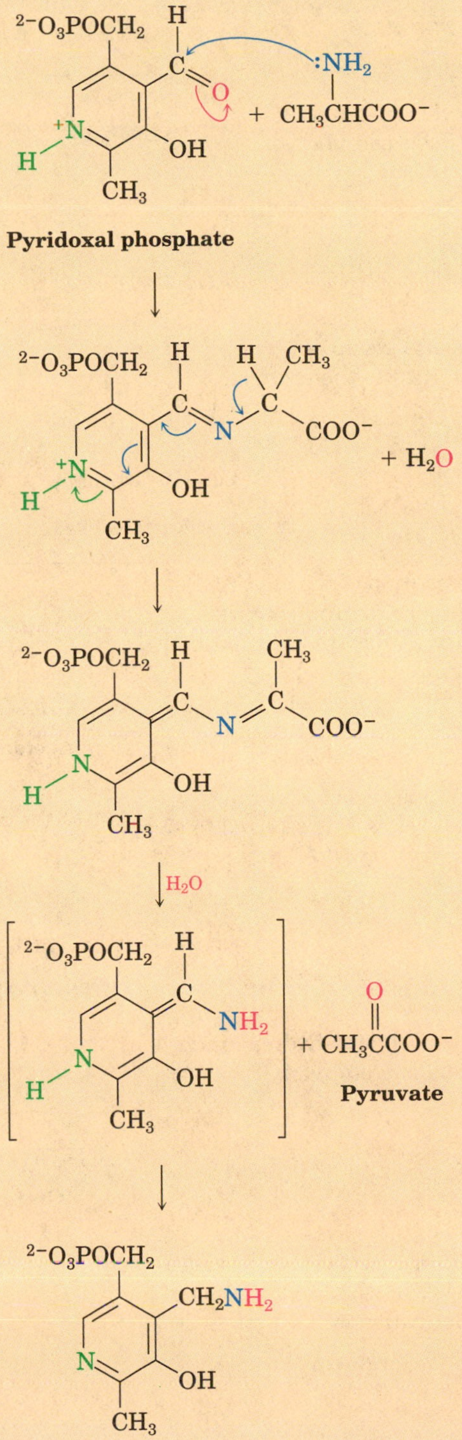

Nucleophilic attack of the amino acid on the pyridoxal phosphate carbonyl group gives an imine.

Loss of a proton moves the double bonds and gives a second imine intermediate.

Hydrolysis of the imine then yields an α-keto acid along with a nitrogen-containing pyridoxal phosphate derivative.

Bond tautomerization regenerates an aromatic pyridine ring.

Figure 30.7 Oxidative deamination of alanine requires the cofactor pyridoxal phosphate and yields pyruvate as product.

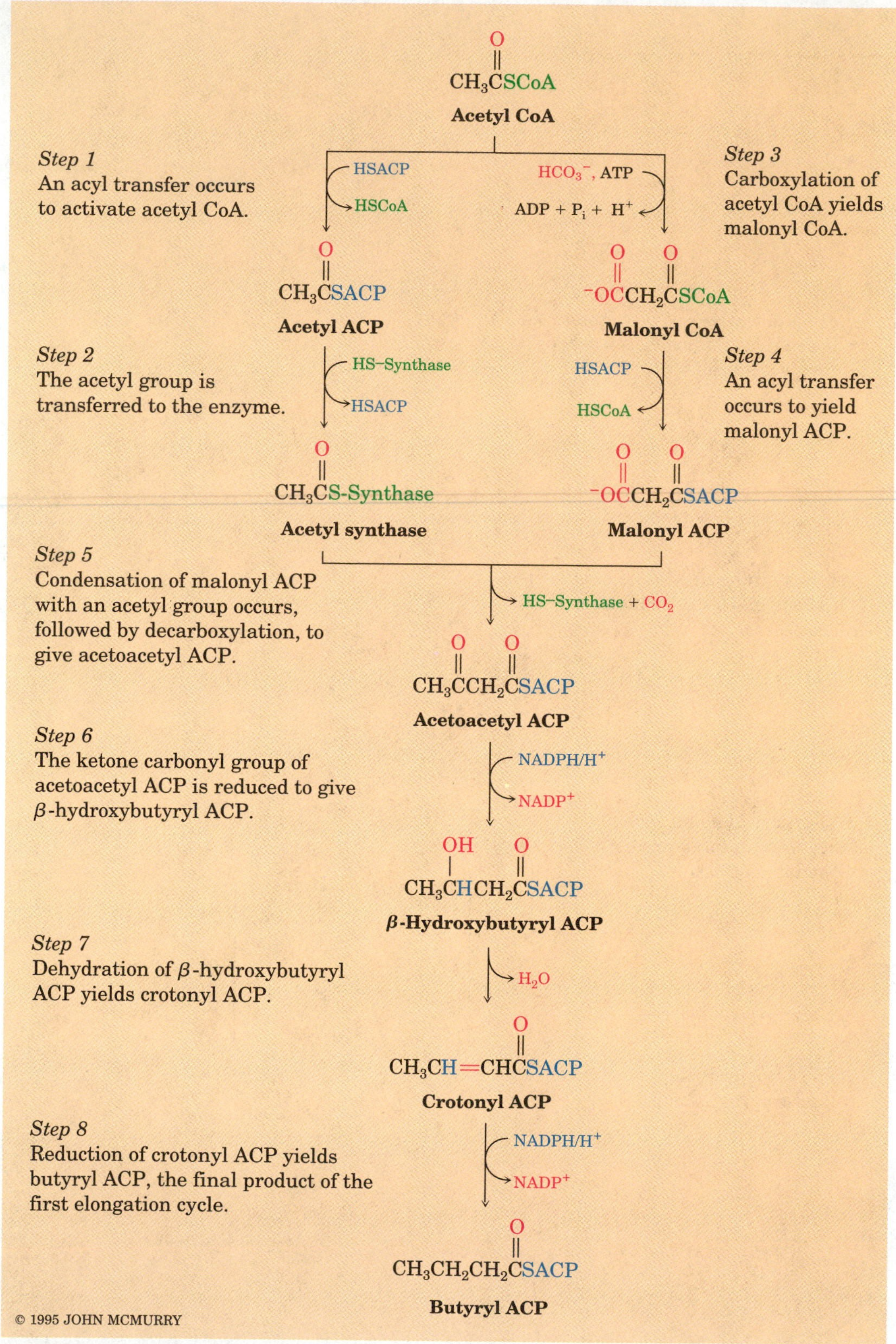

Figure 30.8 Biological pathway for fatty acid synthesis from the two-carbon precursor, acetyl CoA.

Steps 1–2: Acyl transfers The starting material for fatty acid synthesis is the thiol ester acetyl CoA, which is prepared in nature by decarboxylation of pyruvate. The fatty acid synthetic pathway begins with several *priming reactions*, which convert acetyl CoA into more reactive species. The first step in the priming sequence is a nucleophilic acyl substitution reaction that converts acetyl CoA into a different thiol ester acetyl ACP (acyl carrier protein). The reaction is catalyzed by the enzyme ACP transferase. Step 2 involves a further exchange of thiol ester linkages and results in covalent bonding of the acetyl group to a synthase enzyme that will catalyze the upcoming condensation step.

Steps 3–4: Carboxylation and acyl transfer The third step in the priming sequence again starts with acetyl CoA, which is carboxylated by reaction with HCO_3^- and ATP to yield malonyl CoA plus ADP. This step involves the coenzyme biotin, which is bonded to the lysine residue of the carboxylase enzyme and acts as a carrier of CO_2. The enolate ion of acetyl CoA reacts with carboxylated biotin and transfers the CO_2 group in a nucleophilic acyl substitution reaction.

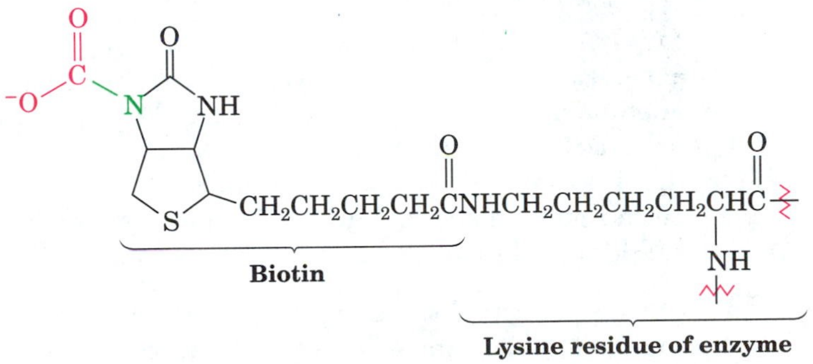

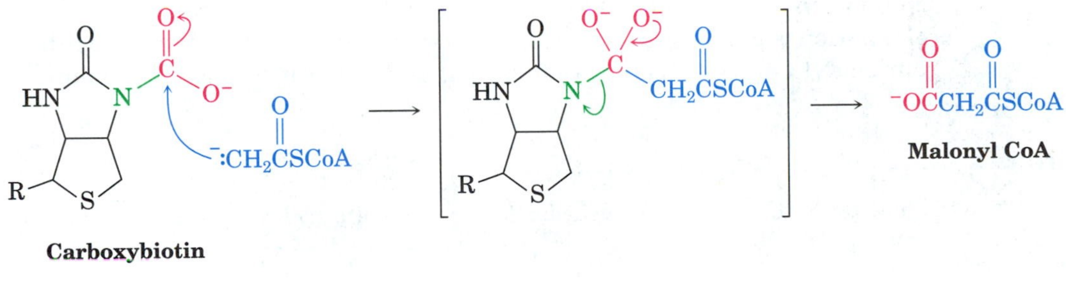

Step 4 is another thiol ester exchange reaction that converts malonyl CoA into the more reactive malonyl ACP.

Step 5: Condensation The key carbon–carbon bond-forming reaction that builds the fatty acid chain occurs in step 5. This step is simply a Claisen condensation (Section 23.8) between acetyl synthase as the electrophilic acceptor and malonyl ACP as the nucleophilic donor. An enolate ion derived from the doubly activated $-CH_2-$ group of malonyl ACP adds to the carbonyl group of acetyl synthase, yielding an intermediate β-keto acid that loses carbon dioxide to give the four-carbon product acetoacetyl ACP.

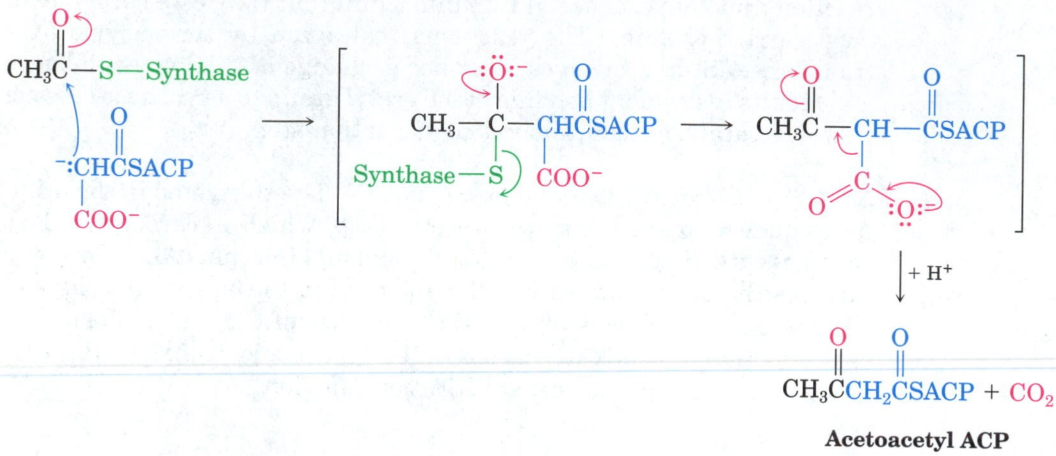

Steps 6–8: Reduction and dehydration The ketone carbonyl group in acetoacetyl ACP is next reduced to an alcohol by NADPH (nicotinamide adenine dinucleotide phosphate), a reducing coenzyme closely related to NADH. Subsequent dehydration of the resulting β-hydroxy thiol ester (E2 reaction) in step 7 yields crotonyl ACP, and the carbon–carbon double bond of crotonyl ACP is further reduced by NADPH in step 8 to yield butyryl ACP.

The net effect of these eight steps is to take two acetyl groups and combine them into a single four-carbon butyryl group. Further condensation of butyryl synthase with another malonyl ACP yields a six-carbon unit, and still further repetitions of the pathway add two more carbon atoms to the chain each time until the 16-carbon palmitic acid is reached. Further chain elongation of palmitic acid occurs by reactions similar to those just described, but acetyl CoA itself rather than malonyl ACP is the two-carbon donor.

PROBLEM

30.14 Show the mechanism of the dehydration reaction of β-hydroxybutyryl ACP to yield crotonyl ACP in step 7 of fatty acid synthesis.

PROBLEM

30.15 Evidence for the role of acetate in fatty acid biosynthesis comes from isotope-labeling experiments. If acetate labeled with ^{13}C in the methyl group ($^{13}CH_3COOH$) were incorporated into fatty acids, at what positions in the fatty acid chain would you expect the ^{13}C label to appear?

30.8 Anabolism of Carbohydrates: Gluconeogenesis

As a rule, the anabolic pathway by which an organism makes a substance is not the reverse of the catabolic pathway by which the organism degrades the same substance. For example, the β-oxidation pathway for fatty acid degradation (Figure 30.2) and the cycle for fatty acid synthesis (Figure 30.8) are clearly related, but one is not the exact reverse of the other. Fatty acid synthesis involves carboxylation and decarboxylation reactions, for example, but β-oxidation does not.

The differences between catabolic and anabolic pathways are due to differences in energy. As noted previously, the overall free-energy change ΔG for any reaction sequence must be negative for the sequence to proceed spontaneously. But if ΔG for a sequence is *negative* in one direction, it must be *positive* in the reverse direction, implying that the reverse sequence can't proceed spontaneously. To both catabolize and anabolize a substance, an organism must use different reaction sequences, both of which have favorable free-energy changes.

Just as fatty acids are catabolized and anabolized by different pathways, so too are carbohydrates. **Gluconeogenesis**, the anabolic pathway by which organisms make glucose from simple precursors, is related to glycolysis but is not its exact reverse. The gluconeogenesis pathway is shown in Figure 30.9 (pp. 1202–1203).

Step 1: Carboxylation Gluconeogenesis begins with the carboxylation of pyruvate to yield oxaloacetate. As in the third step of fatty acid synthesis (Figure 30.8), the reaction requires ATP and the coenzyme biotin, acting as a carrier of CO_2.

Step 2: Decarboxylation and phosphorylation Decarboxylation of oxaloacetate, a β-keto acid, occurs by a mechanism similar to that of step 4 in the citric acid cycle (Figure 30.6), but phosphorylation of the resultant pyruvate enolate ion occurs concurrently to give phosphoenolpyruvate.

Steps 3–4: Hydration and isomerization Conjugate addition of water to the double bond of phosphoenolpyruvate takes place in a process similar to that of step 2 in the β-oxidation pathway (Figure 30.2). Isomerization then occurs by transfer of a phosphate group from C2 to C3, yielding 3-phosphoglycerate.

Steps 5–7: Phosphorylation, oxidation, and tautomerization Reaction of 3-phosphoglycerate with ATP generates the corresponding acyl phosphate in which the carbonyl group is activated for reduction by NADH/H^+. Keto–enol tautomerization of the resultant aldehyde gives dihydroxyacetone phosphate, the same reaction as step 5 of glycolysis (Figure 30.4).

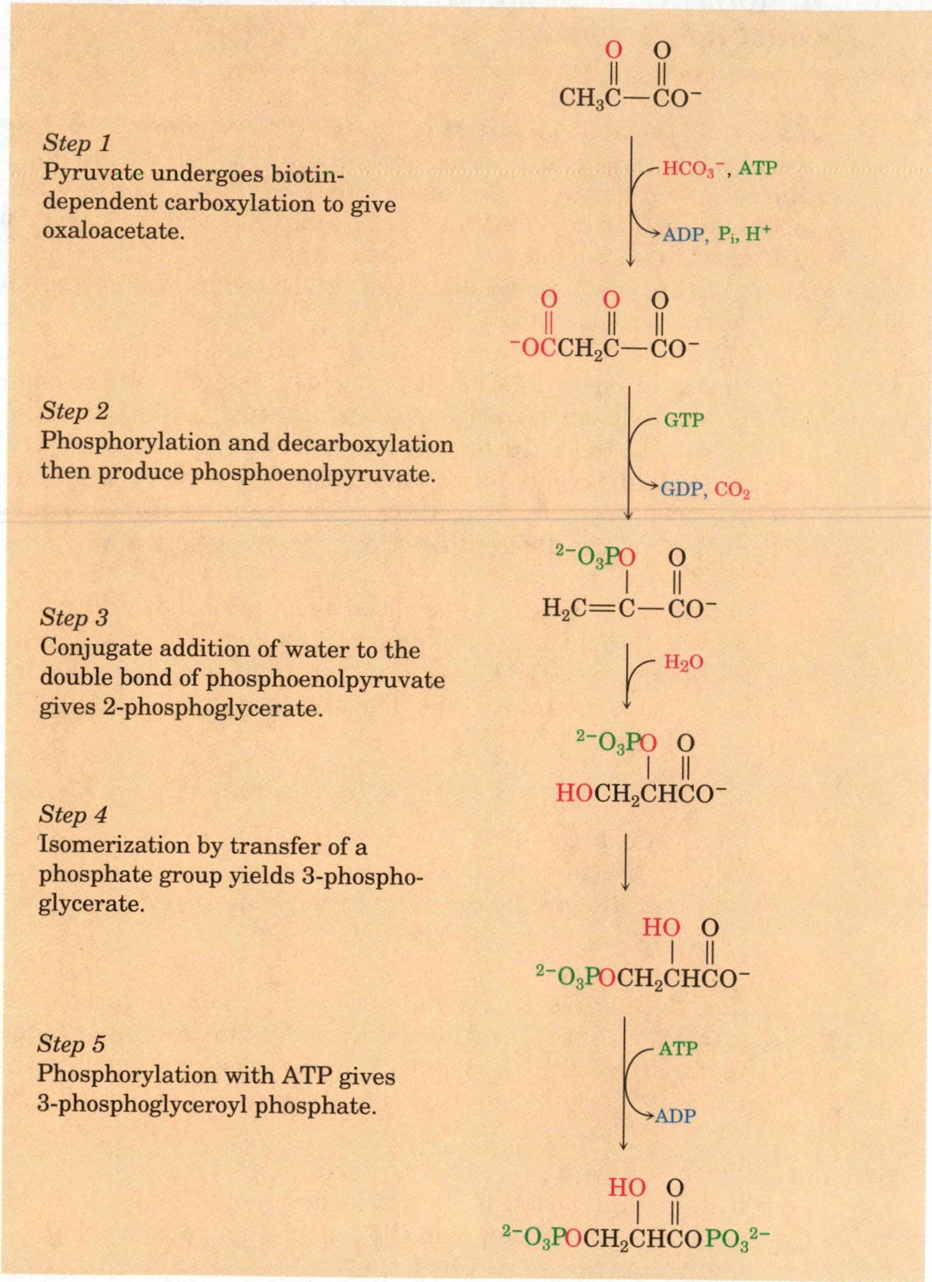

Figure 30.9 *The gluconeogenesis pathway for biosynthesis of glucose.*

Step 8: Aldol condensation Dihydroxyacetone phosphate and glyceraldehyde 3-phosphate, the two three-carbon units produced in step 7, join in step 8 to give fructose 1,6-bisphosphate. The reaction looks like an aldol condensation (Section 23.2) between the enolate ion from dihydroxyacetone phosphate and the carbonyl group of glyceraldehyde 3-phosphate:

Step 6–7
Reduction yields the aldehyde 3-phosphoglyceraldehyde, which undergoes keto–enol tautomerization to give dihydroxyacetone phosphate.

Step 8
Two three-carbon units join in an aldol reaction to yield fructose 1,6-bisphosphate.

Step 9
Hydrolysis of the phosphate group at C1 occurs, giving fructose 6-phosphate.

Step 10
Keto–enol tautomerization shifts the carbonyl group from C2 to C1, yielding glucose 6-phosphate.

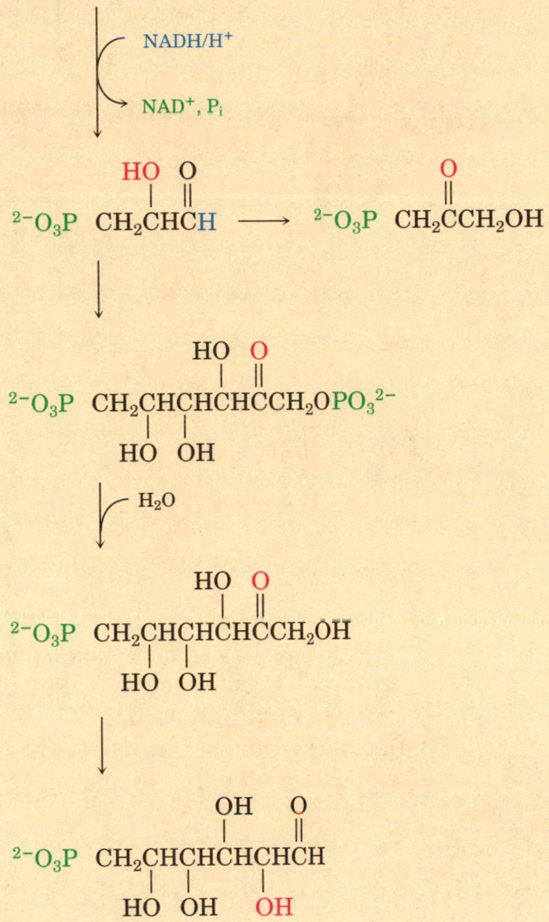

Figure 30.9 *(continued)*

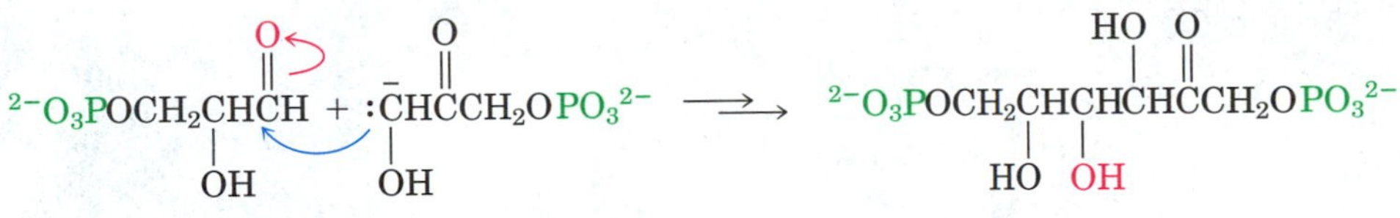

Glyceraldehyde 3-phosphate **Dihydroxyacetone phosphate** **Fructose 1,6-bisphosphate**

As was true in step 4 of glycolysis (Figure 30.4), this "aldol" reaction actually takes place not on the free ketone but on an imine (Schiff base) formed by reaction of dihydroxyacetone phosphate with a side-chain $-NH_2$ group on the enzyme. Loss of a proton from the neighboring carbon then generates an enamine (Section 19.12), an aldol-like reaction ensues, and the product is hydrolyzed to regenerate the ketone.

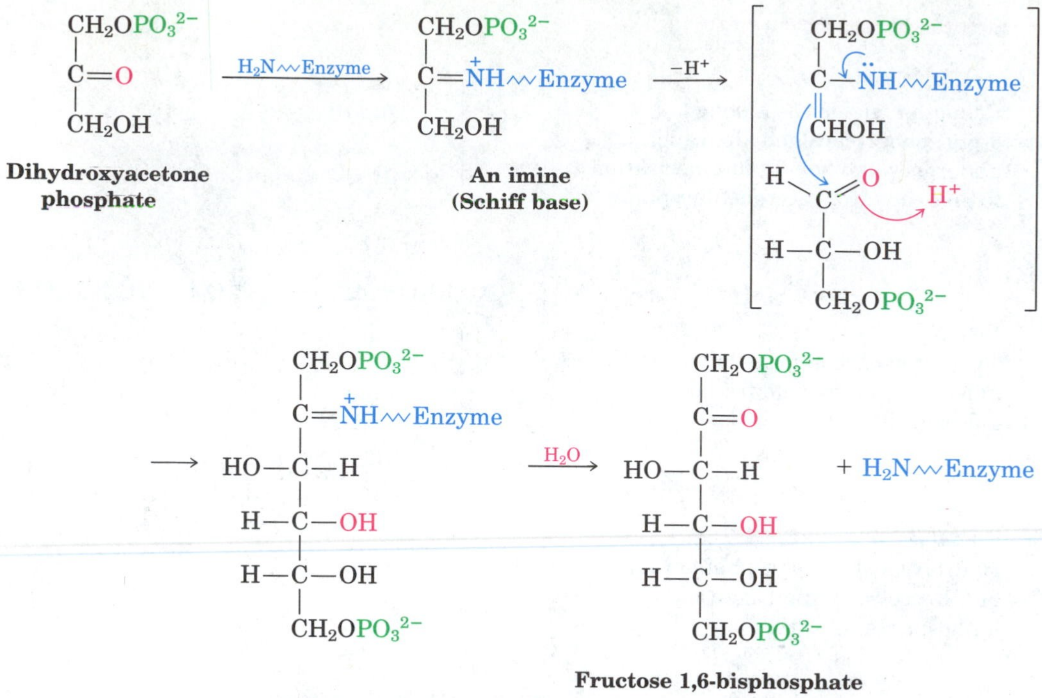

Steps 9–10: Hydrolysis and isomerization Hydrolysis of the phosphate group at C1 of fructose 1,6-bisphosphate, followed by isomerization of the carbonyl group from C2 to C1, then completes gluconeogenesis. The isomerization occurs through an enol intermediate and is the exact reverse of step 2 in glycolysis (Figure 30.4).

PROBLEM.

30.16 Write a mechanism for step 6 of gluconeogenesis, the reduction of 3-phosphoglyceroyl phosphate with NADH/H^+ to yield glyceraldehyde 3-phosphate.

INTERLUDE

Basal Metabolism

Ann Trason, perhaps the world's finest endurance runner, will use up to 10,000 kcal in completing the Western States 100 Mile Endurance Run.

(continued)➤

The minimum amount of energy per unit time an organism expends to stay alive is called the organism's **basal metabolic rate (BMR)**. This rate is measured by monitoring respiration and finding the rate of oxygen consumption, which is proportional to the amount of energy used. Assuming an average dietary mix of fats, carbohydrates, and proteins, approximately 4.82 kcal are required for each liter of oxygen consumed.

The average basal metabolic rate for humans is about 65 kcal/h, or 1600 kcal/day. Obviously, the rate varies for different people depending on sex, age, weight, and physical condition. As a rule, the BMR is lower for older people than for younger people, is lower for females than for males, and is lower for people in good physical condition than for those who are out of shape and overweight. A BMR substantially above the expected value indicates an unusually rapid metabolism, perhaps caused by a fever or some biochemical abnormality.

The total number of calories a person needs each day is the sum of the basal requirement plus the energy used for physical activities, as shown in Table 30.1. A relatively inactive person needs about 30% above basal requirements per day, a lightly active person needs about 50% above basal, and a very active person such as an athlete or construction worker may need 100% above basal requirements. Some endurance athletes in ultradistance events can use as many as 10,000 kcal/day above the basal level. Each day that your caloric intake is above what you use, fat is stored in your body and your weight rises. Each day that your caloric intake is below what you use, fat in your body is metabolized and your weight drops.

Table 30.1 Energy Cost of Various Activities[a]

Activity	*kcal / min*	*Activity*	*kcal / min*
Sleeping	1.2	Tennis	7–9
Sitting, reading	1.6	Basketball	9–10
Standing still	1.8	Walking up stairs	10–18
Walking	3–6	Running	9–22

[a]For a 70 kg man.

Summary and Key Words

Metabolism is the sum of all chemical reactions in the body. Reactions that break down large molecules into smaller fragments are called **catabolism**; reactions that build up large molecules from small pieces are called **anabolism**. Although the details of specific biochemical pathways are sometimes complex, all the reactions that occur follow the normal rules of organic chemical reactivity.

The catabolism of fats begins with **digestion**, in which ester bonds are hydrolyzed to give glycerol and fatty acids. The fatty acids are degraded in the four-step **β-oxidation pathway** by removal of two carbons at a time, yielding acetyl CoA. Catabolism of carbohydrates begins with the hydrolysis of glycoside bonds to give glucose, which is degraded in the ten-step **glycolysis** pathway. Pyruvate, the initial product of glycolysis, is then converted into acetyl CoA. The acetyl groups produced by degradation of fats and carbohydrates next enter the eight-step **citric acid cycle**, where they are further degraded into CO_2.

Catabolism of proteins is more complex than that of fats or carbohydrates because each of the 20 different amino acids is degraded by its own unique pathway. In general, though, the amino nitrogen atoms are removed and the substances that remain are converted into compounds that enter the citric acid cycle. Most amino acids lose their nitrogen atom by **transamination**, a reaction in which the $-NH_2$ group of the amino acid changes places with the keto group of an α-keto acid such as α-ketoglutarate. The products are a new α-keto acid and glutamate. Oxidation and hydrolysis of glutamate then yield ammonium ion.

The energy released in catabolic pathways is used in the respiratory chain to make molecules of adenosine triphosphate (ATP). ATP, the final result of food catabolism, couples to and drives many otherwise unfavorable reactions.

Biomolecules can be synthesized as well as degraded, although the pathways for anabolism and catabolism are not the exact reverse of one another. Fatty acids are biosynthesized from acetic acid by an eight-step pathway, and carbohydrates are made from pyruvate by the ten-step **gluconeogenesis** pathway.

ADDITIONAL PROBLEMS

30.17 What chemical events occur during the digestion of food?

30.18 What is the difference between digestion and metabolism?

30.19 What is the difference between anabolism and catabolism?

30.20 Draw the structure of adenosine *mono*phosphate (AMP), an intermediate in some biochemical pathways.

30.21 Cyclic adenosine monophosphate (cyclic AMP), a modulator of hormone action, is related to AMP (Problem 30.20) but has its phosphate group linked to *two* hydroxyl groups at C3′ and C5′ of the sugar. Draw the structure of cyclic AMP.

30.22 What general kind of reaction does ATP carry out?

30.23 What general kind of reaction does NAD^+ carry out?

30.24 What general kind of reaction does FAD carry out?

30.25 Why are the glycolysis and gluconeogenesis pathways not the exact reverse of one another?

30.26 Lactate, a product of glucose catabolism in oxygen-starved muscles, can be converted into pyruvate by oxidation. What coenzyme do you think is needed? Write the equation in the normal biochemical format using a curved arrow.

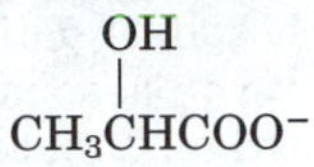
Lactate

30.27 How many moles of acetyl CoA are produced by catabolism of the following substances?
(a) 1.0 mol glucose (b) 1.0 mol palmitic acid ($C_{15}H_{31}CO_2H$)
(c) 1.0 mol maltose

30.28 How many grams of acetyl CoA (MW = 809.6 amu) are produced by catabolism of the following substances?
(a) 100.0 g glucose (b) 100.0 g palmitic acid (c) 100.0 g maltose

30.29 Which of the substances listed in Problem 30.28 is the most efficient precursor of acetyl CoA on a weight basis?

30.30 List the sequence of intermediates involved in the catabolism of glycerol from hydrolyzed fats to yield acetyl CoA.

30.31 Write the equation for the final step in the β-oxidation pathway of any fatty acid with an even number of carbon atoms.

30.32 Show the products of each of the following reactions:

(a) 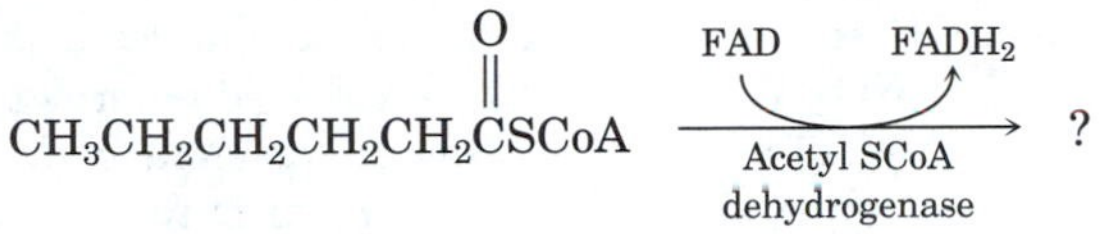

(b) Product of (a) + H_2O $\xrightarrow[\text{hydratase}]{\text{Enoyl SCoA}}$?

(c) Product of (b) 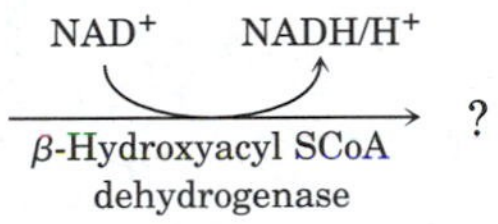

30.33 What is the structure of the α-keto acid formed by transamination of each of these amino acids?
(a) Valine (b) Phenylalanine (c) Methionine

30.34 What enzyme cofactor is associated with each of the following kinds of reactions?
(a) Transamination (b) Carboxylation of a ketone
(c) Decarboxylation of an α-keto acid

30.35 The glycolysis pathway shown in Figure 30.4 has a number of intermediates that contain phosphate groups. Why can 3-phosphoglyceroyl phosphate and phosphoenolpyruvate transfer a phosphate group to ADP while glucose 6-phosphate cannot?

30.36 In the *pentose phosphate* pathway for degrading sugars, ribulose 5-phosphate is converted to ribose 5-phosphate. Propose a mechanism for the isomerization.

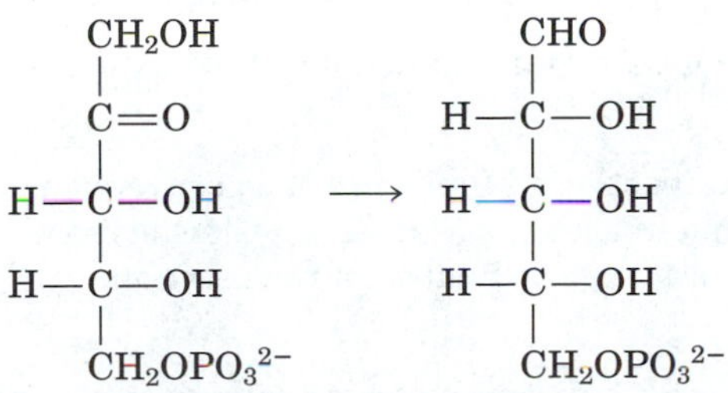

Ribulose 5-phosphate **Ribose 5-phosphate**

30.37 Another step in the pentose phosphate pathway for degrading sugars (see Problem 30.36) is the conversion of ribose 5-phosphate to glyceraldehyde 3-phosphate. What kind of organic process is occurring? Propose a mechanism for the conversion.

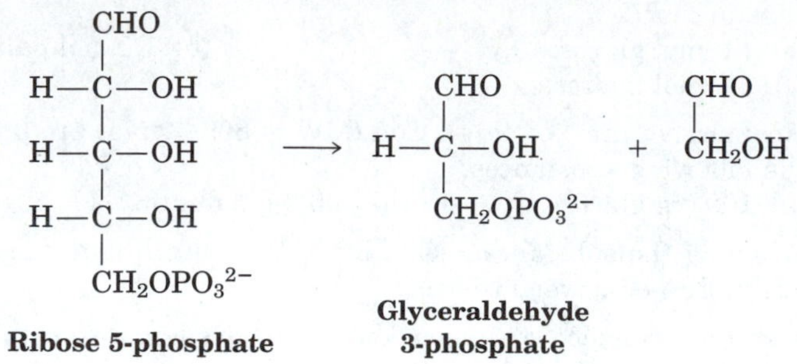

30.38 Write a mechanism for the conversion of α-ketoglutarate to succinyl CoA in step 4 of the citric acid cycle (Figure 30.6).

30.39 The primary fate of acetyl CoA under normal metabolic conditions is degradation in the citric acid cycle to yield CO_2. When the body is stressed by prolonged starvation, however, acetyl CoA is converted into compounds called *ketone bodies*, which can be used by the brain as a temporary fuel. The biochemical pathway for the synthesis of ketone bodies from acetyl CoA is shown:

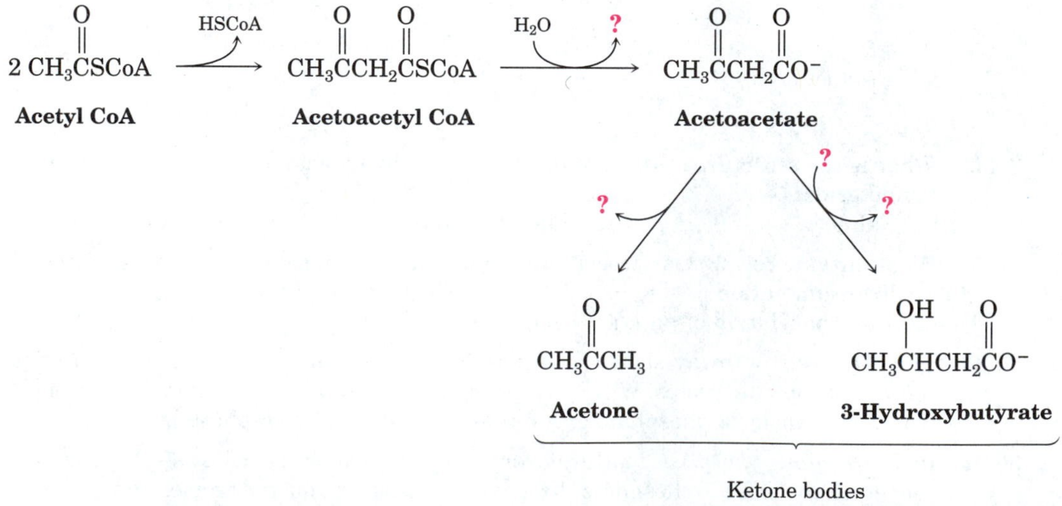

Fill in the missing information indicated by the four question marks.

30.40 The initial reaction in Problem 30.39, conversion of two molecules of acetyl CoA to one molecule of acetoacetyl CoA, is a Claisen reaction. Assuming that there is a base present, show the mechanism of the reaction.

30.41 Sedoheptulose 7-phosphate reacts with glyceraldehyde 3-phosphate in the presence of a transaldolase enzyme to yield erythrose 4-phosphate and fructose 6-phosphate. Propose a mechanism.

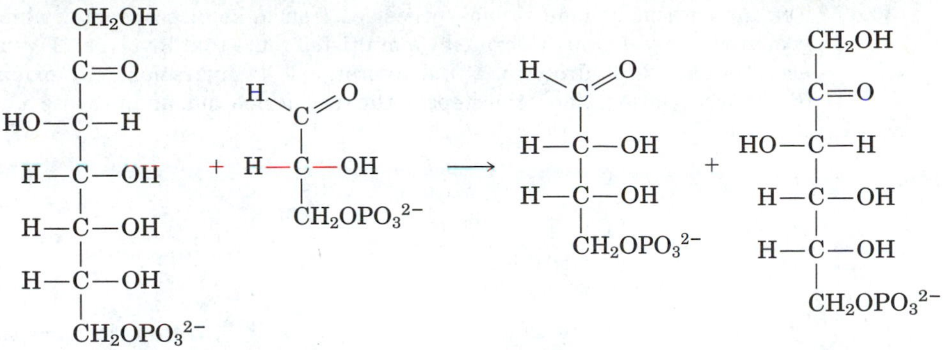

D-Sedoheptulose 7-phosphate | Glyceraldehyde 3-phosphate | D-Erythrose 4-phosphate | Fructose 6-phosphate

30.42 The amino acid tyrosine is biologically degraded by a series of steps that include the following transformations:

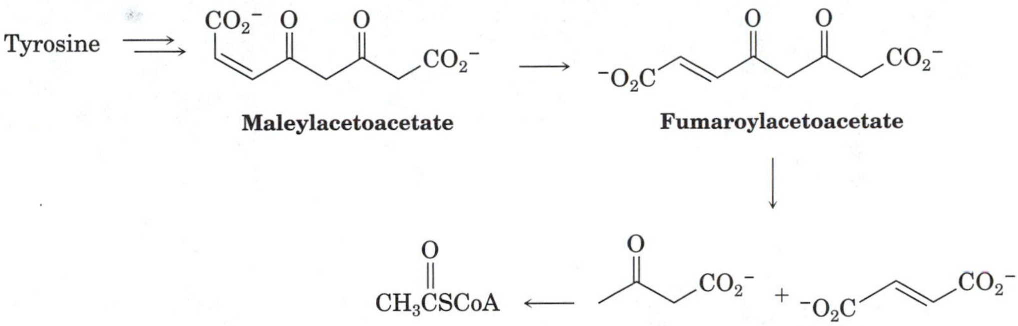

The double-bond isomerization of maleylacetoacetate to fumaroyl acetoacetate is catalyzed by practically any nucleophile, :Nu$^-$. Propose a mechanism.

30.43 Propose a chemically reasonable mechanism for the biological conversion of fumaroylacetoacetate to fumarate plus acetoacetate (Problem 30.42).

30.44 Propose a chemically reasonable mechanism for the biological conversion of acetoacetate to acetyl CoA (Problem 30.42).

30.45 Design your own degradative pathway. You know the rules (organic mechanisms) and you've seen the kinds of reactions that occur in the biological degradation of fats and carbohydrates into acetyl CoA. If you were Mother Nature, what series of steps would you use to degrade the amino acid serine into acetyl CoA?

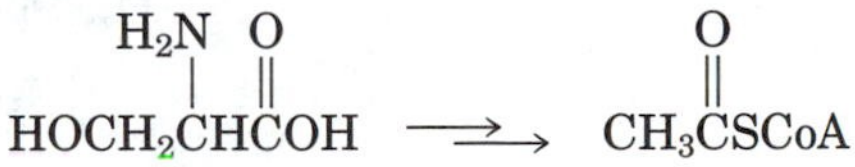

Serine | **Acetyl CoA**

30.46 The amino acid serine is biosynthesized by a route that involves reaction of 3-phosphohydroxypyruvate with glutamate. Propose a mechanism.

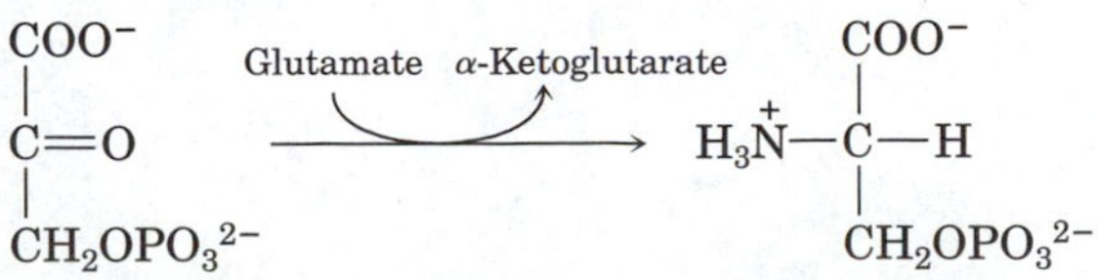

3-Phosphohydroxypyruvate | **3-Phosphoserine**

30.47 The amino acid leucine is biosynthesized from α-ketoisocaproate, which is itself prepared from α-ketoisovalerate by a multistep route that involves: (1) reaction with acetyl CoA, (2) hydrolysis, (3) dehydration, (4) hydration, (5) oxidation, and (6) decarboxylation. Show the steps in the transformation, and propose a mechanism for each.

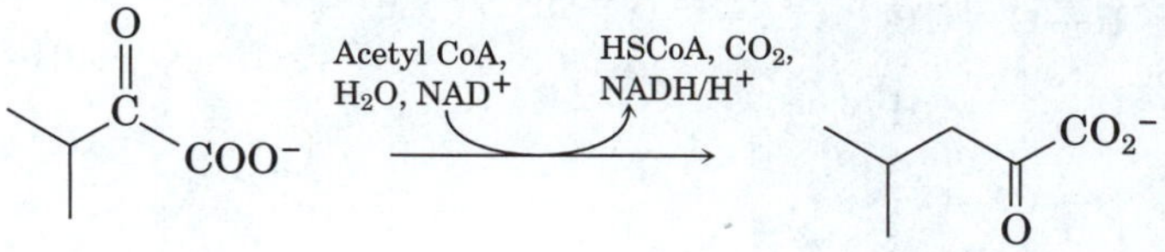

α-Ketoisovalerate **α-Ketoisocaproate**

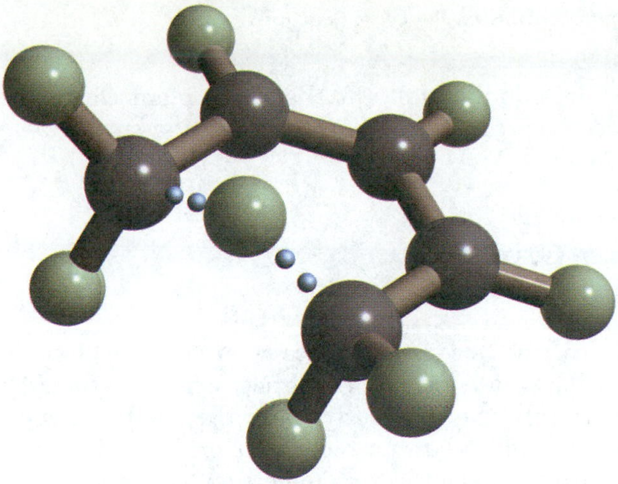

This [1,5] hydrogen shift is a common pericyclic reaction.

31

Orbitals and Organic Chemistry: Pericyclic Reactions

Most organic reactions take place by polar mechanisms, in which a nucleophile donates two electrons to an electrophile in forming a new bond. Other organic reactions take place by radical mechanisms, in which each of two reactants donates one electron in forming a new bond. Although much remains to be learned about them, both classes have been studied for many years and are relatively well understood.

By contrast, the fundamental principles of *pericyclic reactions*, the third major class of organic reaction mechanisms, have been worked out only recently. Many individuals have made major contributions, but it was the work of Robert Woodward[1] and Roald Hoffmann[2] in the mid-1960s that made most chemists aware of the principles of pericyclic reactions.

We previously defined a **pericyclic reaction** as one that occurs by a concerted process through a cyclic transition state. The word *concerted* means that all bonding changes occur at the same time and in a single step; no intermediates are involved. Rather than try to expand this definition now, let's begin by briefly reviewing some of the bonding concepts introduced

[1]Robert Burns Woodward (1917–1979); b. Boston; Ph.D. Massachusetts Institute of Technology (1937); professor, Harvard University (1941–1979); Nobel Prize (1965).

[2]Roald Hoffmann (1937–); b. Zloczow, Poland; Ph.D. Harvard University (1962); professor, Cornell University; Nobel Prize (1981).

in Section 1.7. We'll then look individually at the three main classes of pericyclic reactions: *electrocyclic reactions*, *cycloadditions*, and *sigmatropic rearrangements*.

31.1 Atomic and Molecular Orbitals

We've said that an *orbital* describes a region of space where a given electron is most likely to be found. When a bond is formed between two atoms by the interaction of atomic orbitals, the two electrons in the bond no longer occupy atomic orbitals but instead occupy a molecular orbital (MO). The interaction of two atomic orbitals leads to two molecular orbitals, one of which is lower in energy than either of the two atomic orbitals, and one of which is higher in energy.

As shown in Figure 31.1 for the interaction of two p orbitals, the lower-energy MO, denoted ψ_1 (Greek psi), is a **bonding molecular orbital** because it results from additive overlap of two lobes with the same algebraic sign. The higher-energy MO, denoted ψ_2*, is an **antibonding molecular orbital** because it results from subtractive overlap of two lobes with the opposite algebraic sign. Note that the antibonding orbital has a nodal plane between nuclei—a plane of zero electron density between lobes of opposite sign. Higher-energy orbitals always have more nodes between nuclei than lower-energy orbitals and thus have fewer favorable bonding interactions.

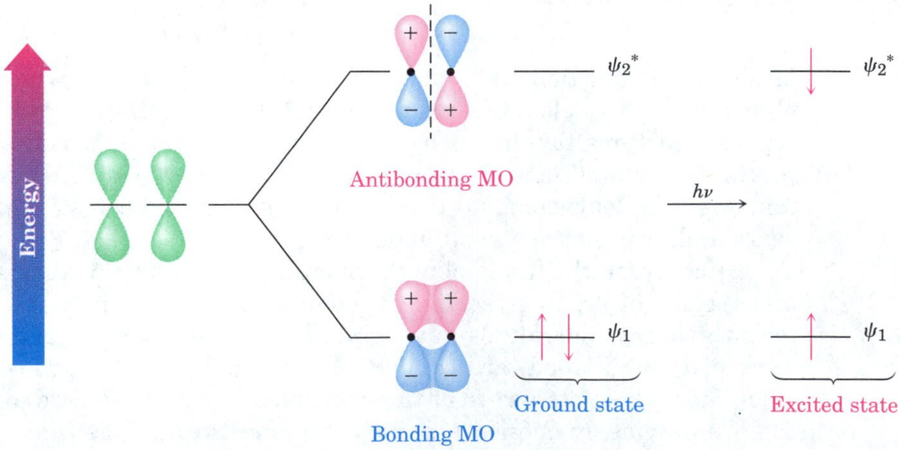

Figure 31.1 The π molecular orbitals resulting from the interaction of two p atomic orbitals. The bonding MO results from overlap of like lobes, whereas the antibonding MO results from overlap of unlike lobes. Both ground-state and excited-state electronic configurations of the ethylene π bond are indicated.

Like an atomic orbital, a molecular orbital can hold a maximum of two electrons. For an alkene π bond, the bonding MO ψ_1 is filled and the antibonding MO ψ_2* is vacant. Irradiation with ultraviolet light, however, excites an electron from ψ_1 to $\psi_2{}^*$, leaving each orbital with one electron (Section 14.12). The electronic configurations of both ground and excited states of ethylene are shown in Figure 31.1.

31.2 Molecular Orbitals of Conjugated π Systems

The molecular orbital description of a conjugated π system is more complex than that of a simple alkene because the π electrons are delocalized over more than two atoms. In 1,3-butadiene, for example, four 2p atomic orbitals combine into four diene molecular orbitals spanning the entire π system. Two of these MO's are bonding (lower-energy, ψ_1 and ψ_2), and two are antibonding (higher-energy, $\psi_3{}^*$ and $\psi_4{}^*$), as shown in Figure 31.2. In the ground state, the two bonding orbitals are occupied by four electrons, while the two antibonding orbitals are unoccupied. Note again that the higher the energy of a molecular orbital, the more nodes it has between nuclei.

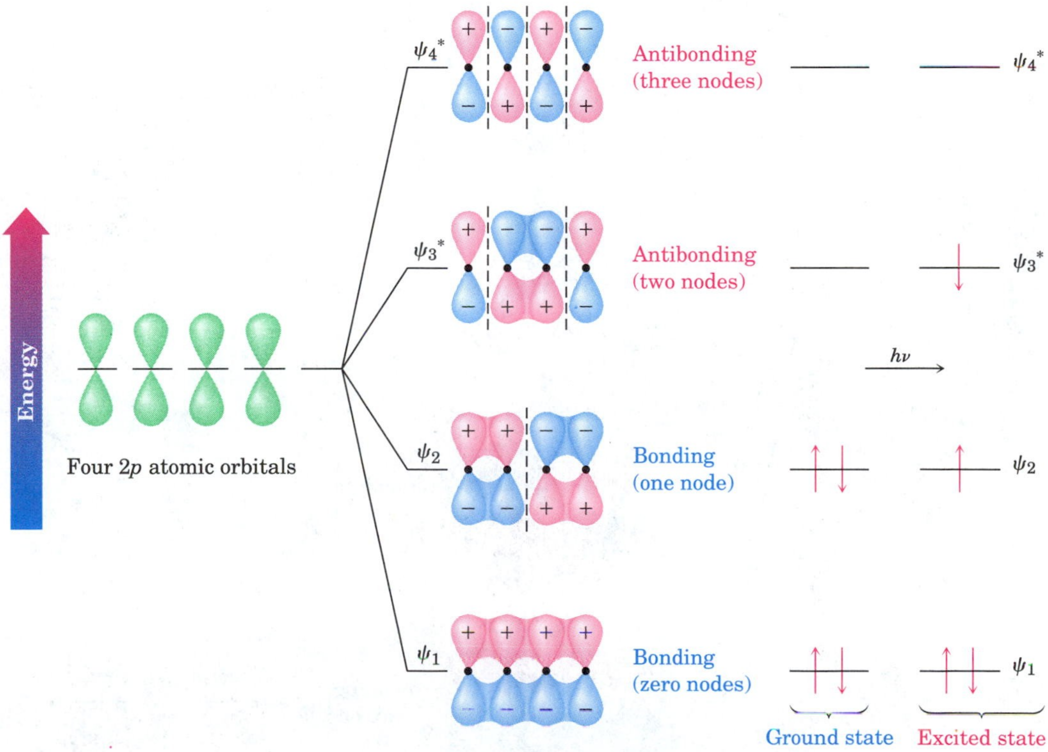

Figure 31.2 The π molecular orbitals of 1,3-butadiene. In the ground state, the two bonding orbitals are occupied and the two antibonding orbitals are vacant.

A similar sort of molecular orbital description can be derived for a conjugated triene, tetraene, or for *any* conjugated π electron system. The six MO's of 1,3,5-hexatriene are shown in Figure 31.3. Only the three bonding orbitals, ψ_1, ψ_2, and ψ_3, are occupied in the ground state, whereas ψ_3 and $\psi_4{}^*$ have one electron each in the excited state.

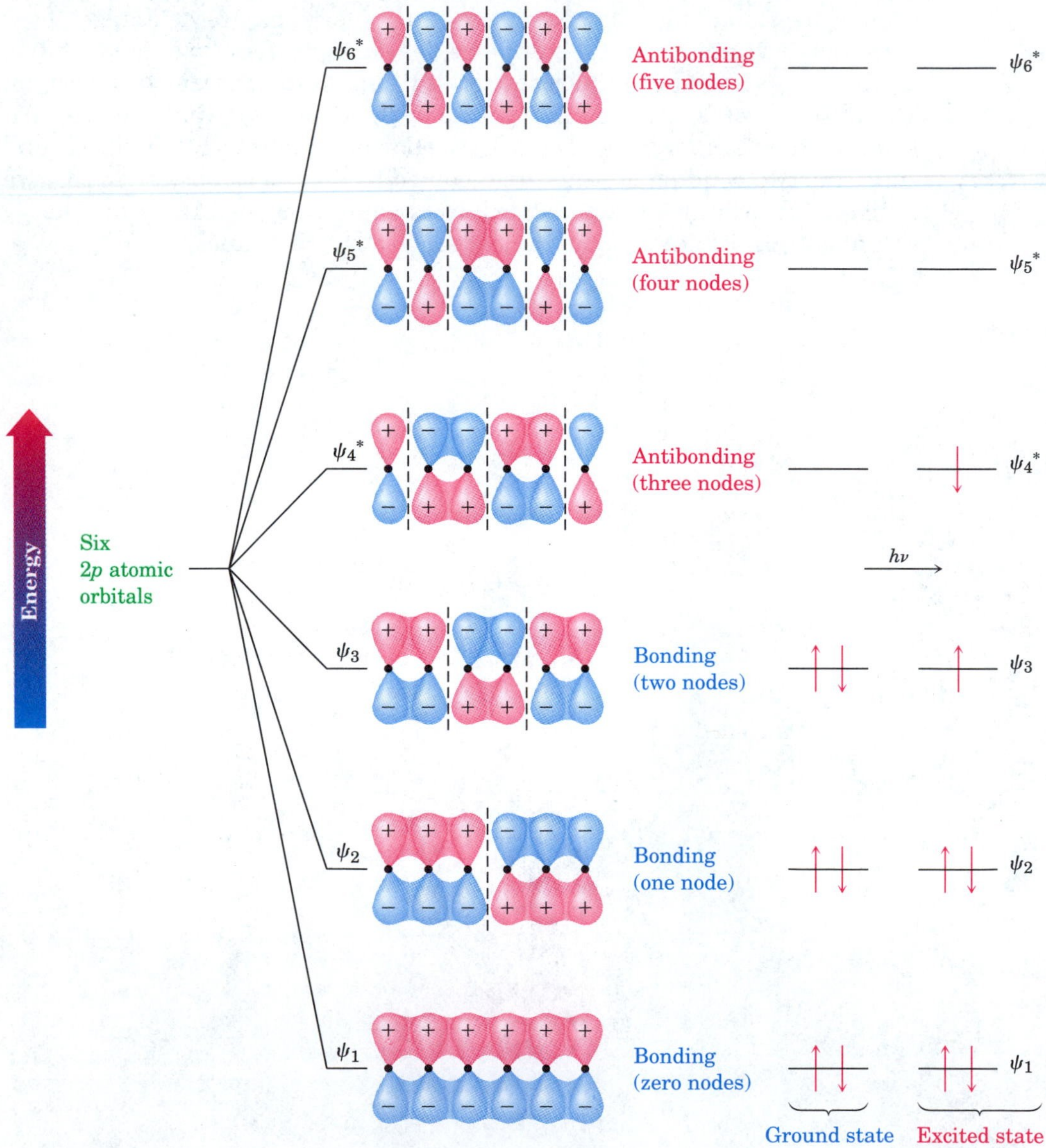

Figure 31.3 The π molecular orbitals of 1,3,5-hexatriene.

31.3 Molecular Orbitals and Pericyclic Reactions

What do molecular orbitals and the algebraic signs of their lobes have to do with pericyclic reactions? The answer is, *everything*. According to a series of rules formulated by Woodward and Hoffmann, a pericyclic reaction can take place only if the symmetry of the reactant MO's is the same as the symmetry of the product MO's. In other words, the lobes of reactant MO's must be of the correct algebraic sign for bonding overlap to occur in the transition state leading to product.

If the symmetries of both reactant and product orbitals match up, or "correlate," the reaction is said to be **symmetry-allowed**. If the symmetries of reactant and product orbitals don't correlate, the reaction is **symmetry-disallowed**. Symmetry-allowed reactions often occur under relatively mild conditions, but symmetry-disallowed reactions can't occur by concerted paths. Either they take place by nonconcerted, high-energy pathways, or they don't take place at all.

The Woodward–Hoffmann rules for pericyclic reactions require an analysis of all reactant and product molecular orbitals, but Kenichi Fukui[3] at Kyoto University in Japan introduced a simplified version. According to Fukui, we need consider only *two* molecular orbitals, called the **frontier orbitals**. These frontier orbitals are the **highest occupied molecular orbital (HOMO)** and the **lowest unoccupied molecular orbital (LUMO)**. In ground-state ethylene, for example, ψ_1 is the HOMO because it has two electrons, and $\psi_2{}^*$ is the LUMO because it is vacant (Figure 31.1). In ground-state 1,3-butadiene, ψ_2 is the HOMO and $\psi_3{}^*$ is the LUMO (Figure 31.2).

PROBLEM

31.1 Refer to Figure 31.3 to find the molecular orbitals of a conjugated triene, and tell which molecular orbital is the HOMO and which is the LUMO for both ground and excited states.

31.4 Electrocyclic Reactions

The best way to understand how orbital symmetry affects pericyclic reactions is to look at some examples. Let's look first at a group of polyene rearrangements called *electrocyclic reactions*. **Electrocyclic reactions** are pericyclic processes that involve the cyclization of conjugated polyenes. One π bond is broken, the other π bonds change position, a new σ bond is formed,

[3]Kenichi Fukui (1918–); b. Nara Prefecture, Japan; Ph.D. Kyoto University; professor, Kyoto University; Nobel Prize (1981).

and a cyclic compound results. For example, conjugated trienes can be converted into cyclohexadienes, and conjugated dienes can be converted into cyclobutenes.

Both these reactions are reversible, and the position of the equilibrium depends on the specific case. In general, the triene $\rightleftharpoons$ cyclohexadiene equilibrium favors the ring-closed product, whereas the diene $\rightleftharpoons$ cyclobutene equilibrium favors the ring-opened product.

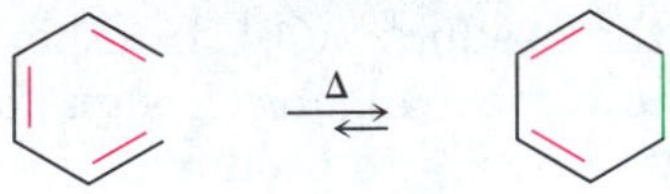

A conjugated triene A cyclohexadiene

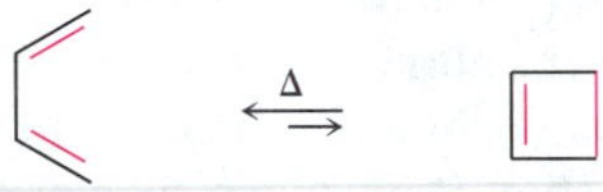

A conjugated diene A cyclobutene

The most important feature of electrocyclic reactions is their stereochemistry. For example, (2*E*,4*Z*,6*E*)-octatriene yields only *cis*-5,6-dimethyl-1,3-cyclohexadiene when heated, and (2*E*,4*Z*,6*Z*)-octatriene yields only *trans*-5,6-dimethyl-1,3-cyclohexadiene. Remarkably, however, the stereochemical results change completely when the reactions are carried out under *photochemical*, rather than thermal, conditions. Thus, irradiation of (2*E*,4*Z*,6*E*)-octatriene with ultraviolet light yields *trans*-5,6-dimethyl-1,3-cyclohexadiene (Figure 31.4).

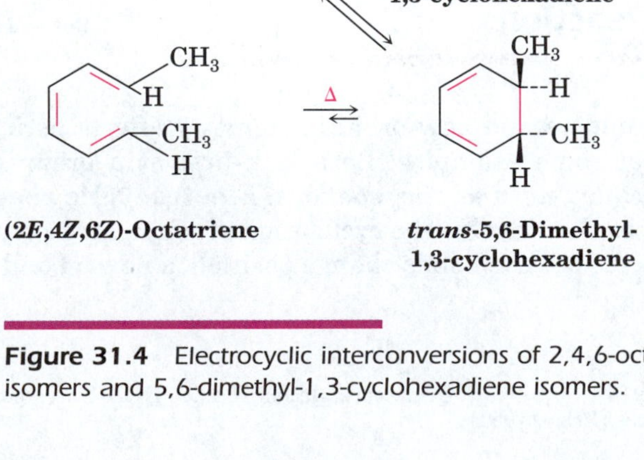

Figure 31.4 Electrocyclic interconversions of 2,4,6-octatriene isomers and 5,6-dimethyl-1,3-cyclohexadiene isomers.

A similar result is obtained for the thermal electrocyclic ring opening of 3,4-dimethylcyclobutene. The trans isomer yields only (2*E*,4*E*)-hexadiene when heated, and the cis isomer yields only (2*E*,4*Z*)-hexadiene. On UV irradiation, however, the results are opposite: Cyclization of the 2*E*,4*E* isomer under photochemical conditions yields cis product (Figure 31.5).

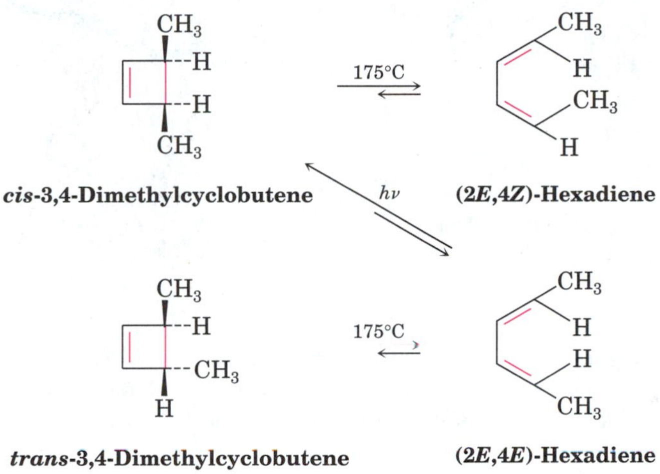

Figure 31.5 Electrocyclic interconversions of 2,4-hexadiene isomers and 3,4-dimethylcyclobutene isomers.

All these results can be accounted for by orbital-symmetry arguments. To do so, we need to look at the two outermost lobes of the polyene MO's. There are two possibilities: The lobes of like sign can be either on the same side or on opposite sides of the molecule.

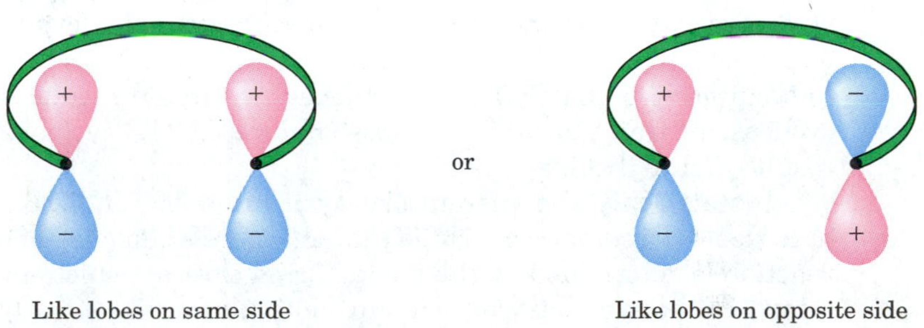

For a bond to form, the outermost π lobes must rotate so that favorable bonding overlap is achieved—a positive lobe overlapping a positive lobe or a negative lobe overlapping a negative lobe. If two lobes of like sign are on the *same* side of the molecule, the two orbitals must rotate in *opposite* directions—one clockwise and one counterclockwise. This kind of motion is referred to as **disrotatory**.

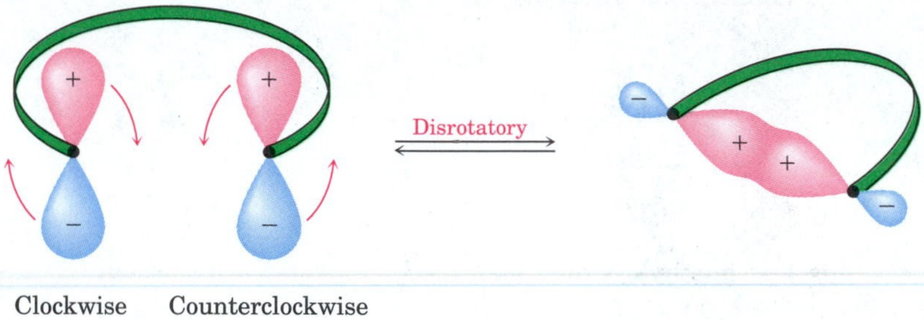

Conversely, if lobes of like sign are on *opposite* sides of the molecules, both orbitals must rotate in the *same* direction, either both clockwise or both counterclockwise. This kind of motion is called **conrotatory**.

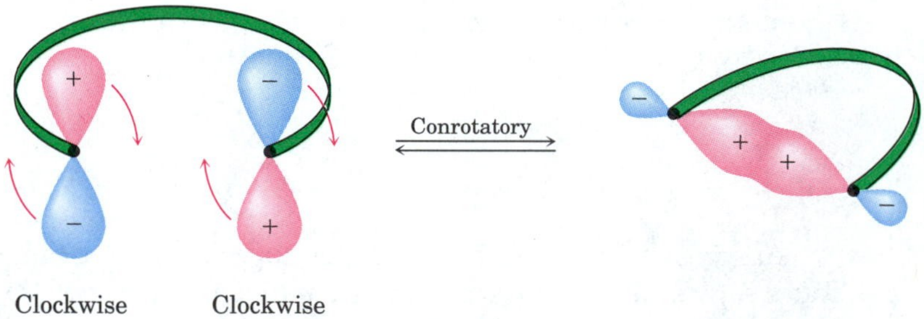

Now let's see what happens to substituents on the polyene carbon atoms when cyclization occurs. To choose a specific example, let's look at the thermal cyclization of (2*E*,4*Z*,6*E*)-octatriene. If a disrotatory cyclization were to occur, *cis*-5,6-dimethyl-1,3-cyclohexadiene would result. If a conrotatory cyclization were to occur, *trans*-5,6-dimethyl-1,3-cyclohexadiene would result (Figure 31.6).

In fact, only the disrotatory cyclization of (2*E*,4*Z*,6*E*)-octatriene is observed. We therefore conclude that the stereochemistry of an electrocyclic reaction is determined by the mode of ring closure, which, in turn, is determined by the symmetry of reactant molecular orbitals.

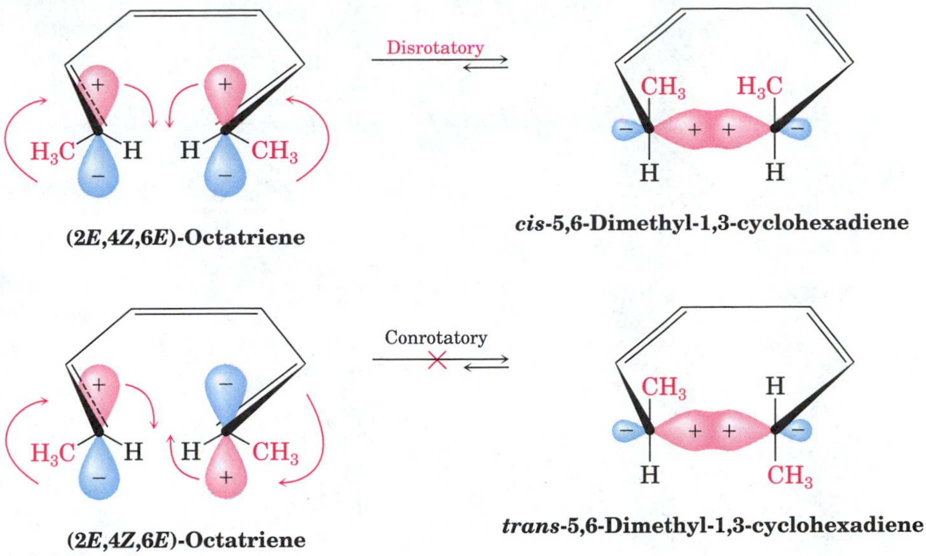

Figure 31.6 Stereochemistry of cyclization of (2*E*,4*Z*,6*E*)-octatriene. Disrotatory cyclization leads to the *cis* product, and conrotatory cyclization leads to the *trans* product.

31.5 Stereochemistry of Thermal Electrocyclic Reactions

How can we predict whether conrotatory or disrotatory motion will occur in a given case? How can we tell whether the terminal lobes of like sign will be on the same side or on opposite sides of the molecule?

According to frontier orbital theory, *the stereochemistry of an electrocyclic reaction is determined by the symmetry of the polyene's HOMO*. The electrons in the HOMO are the highest-energy, most loosely held electrons, and are therefore most easily moved during reaction. For thermal ring openings and closings, the ground-state electronic configuration is used to identify the HOMO; for photochemical ring openings and closings, the excited-state electronic configuration is used.

Let's look again at the thermal ring closure of conjugated trienes. According to Figure 31.3, the HOMO of a conjugated triene in its ground state has a symmetry that predicts a disrotatory ring closure. This disrotatory cyclization is exactly what is observed in the thermal cyclization of 2,4,6-octatriene. The 2*E*,4*Z*,6*E* isomer yields cis product; the 2*E*,4*Z*,6*Z* isomer yields trans product (Figure 31.7).

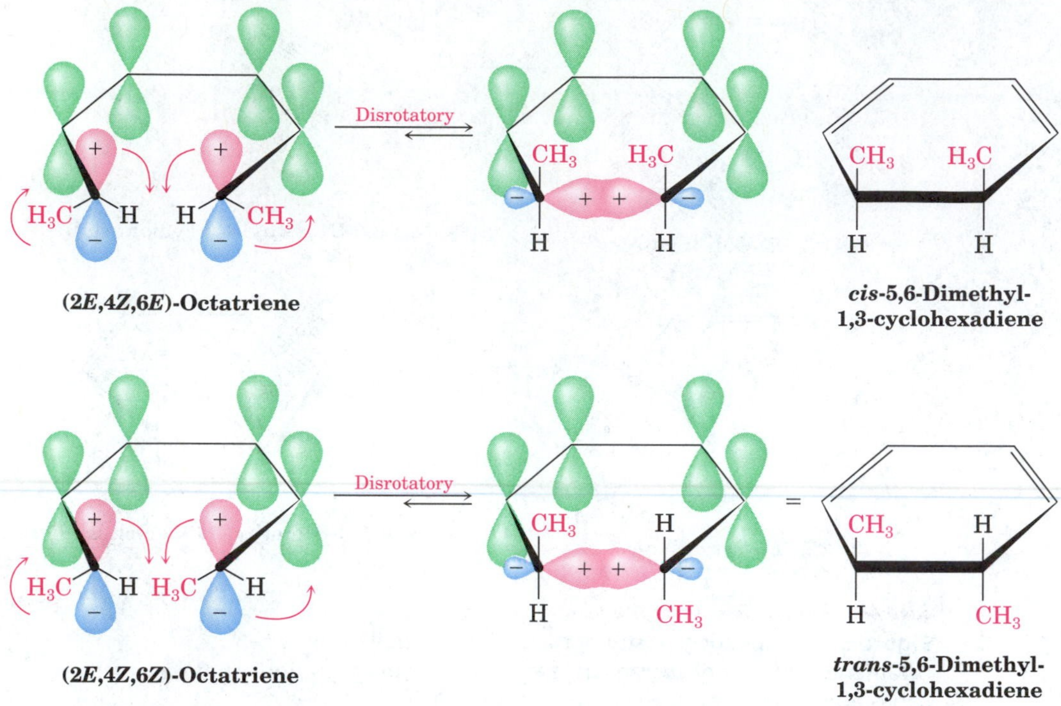

Figure 31.7 Thermal disrotatory ring closure of 2,4,6-octatrienes.

In a similar manner, the ground-state HOMO of a conjugated diene (Figure 31.2) has a symmetry that predicts conrotatory ring closure. In practice, however, the conjugated diene reaction can only be observed in the reverse direction (cyclobutene → butadiene) because of the position of the equilibrium. We therefore predict that the 3,4-dimethylcyclobutene ring will *open* in a conrotatory fashion, which is exactly what is observed. *cis*-3,4-Dimethylcyclobutene yields (2*E*,4*Z*)-hexadiene, and *trans*-3,4-dimethylcyclobutene yields (2*E*,4*E*)-hexadiene by conrotatory opening (Figure 31.8).

Note that a conjugated diene and a conjugated triene react in opposite stereochemical senses. The diene opens and closes by a conrotatory path, whereas the triene opens and closes by a disrotatory path. This difference, of course, is due to the different symmetries of the diene and triene HOMO's:

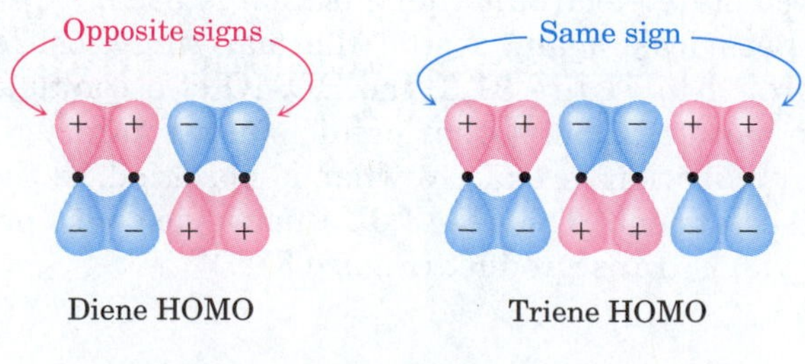

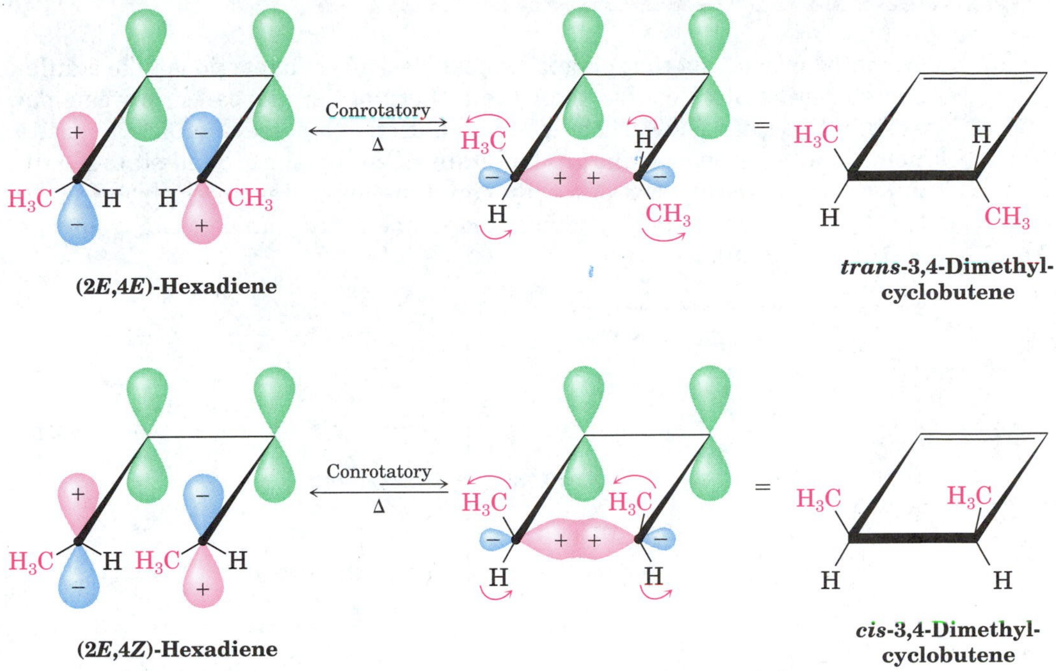

Figure 31.8 Thermal conrotatory ring opening of *cis*- and *trans*-dimethylcyclobutene.

There is an alternating relationship between the number of electron pairs (double bonds) undergoing bond reorganization and the mode of ring opening or closure. Polyenes with an even number of electron pairs undergo thermal electrocyclic reactions in a conrotatory sense, whereas polyenes with an odd number of electron pairs undergo the same reactions in a disrotatory sense.

PROBLEM

31.2 Draw the products you would expect from conrotatory and disrotatory cyclizations of (2*Z*,4*Z*,6*Z*)-octatriene. Which of the two paths would you expect the thermal reaction to follow?

PROBLEM

31.3 In principle, *trans*-3,4-dimethylcyclobutene can open by two conrotatory paths to give either (2*E*,4*E*)-hexadiene or (2*Z*,4*Z*)-hexadiene. Explain why both products are symmetry-allowed, and then account for the fact that only the 2*E*,4*E* isomer is obtained in practice.

31.6 Photochemical Electrocyclic Reactions

We noted previously that photochemical electrocyclic reactions take a different stereochemical course than their thermal counterparts. We can now explain this difference. Ultraviolet irradiation of a polyene causes an excitation of one electron from the ground-state HOMO to the ground-state LUMO. For example, irradiation of a conjugated diene excites an electron from ψ_2 to ψ_3^*, and irradiation of a conjugated triene excites an electron from ψ_3 to ψ_4^* (Figure 31.9).

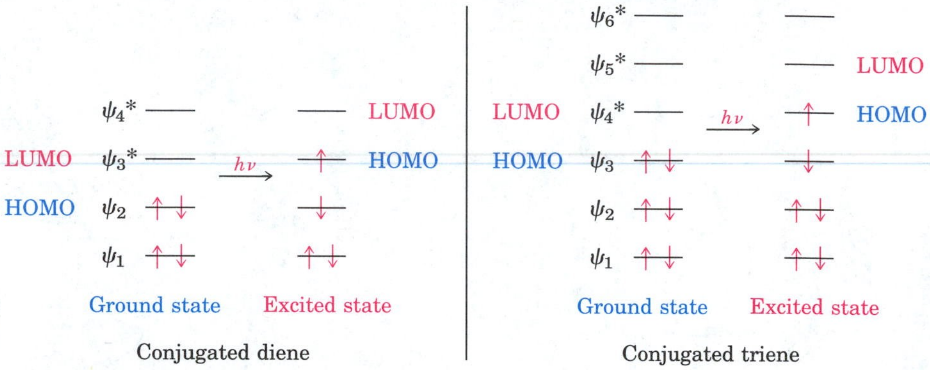

Figure 31.9 Ground-state and excited-state electron configurations of conjugated dienes and trienes.

Since electronic excitation changes the symmetries of HOMO and LUMO, it also changes the reaction stereochemistry. Thus, (2*E*,4*E*)-hexadiene undergoes *photochemical* cyclization by a disrotatory path, whereas the thermal reaction is conrotatory.

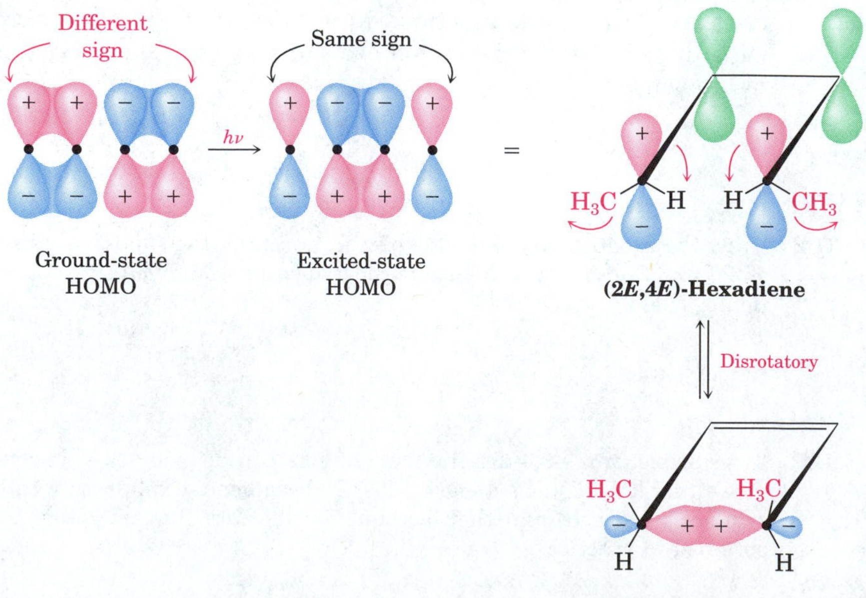

In the same way, (2*E*,4*Z*,6*E*)-octatriene undergoes photochemical cyclization by a conrotatory path, whereas the thermal reaction is disrotatory.

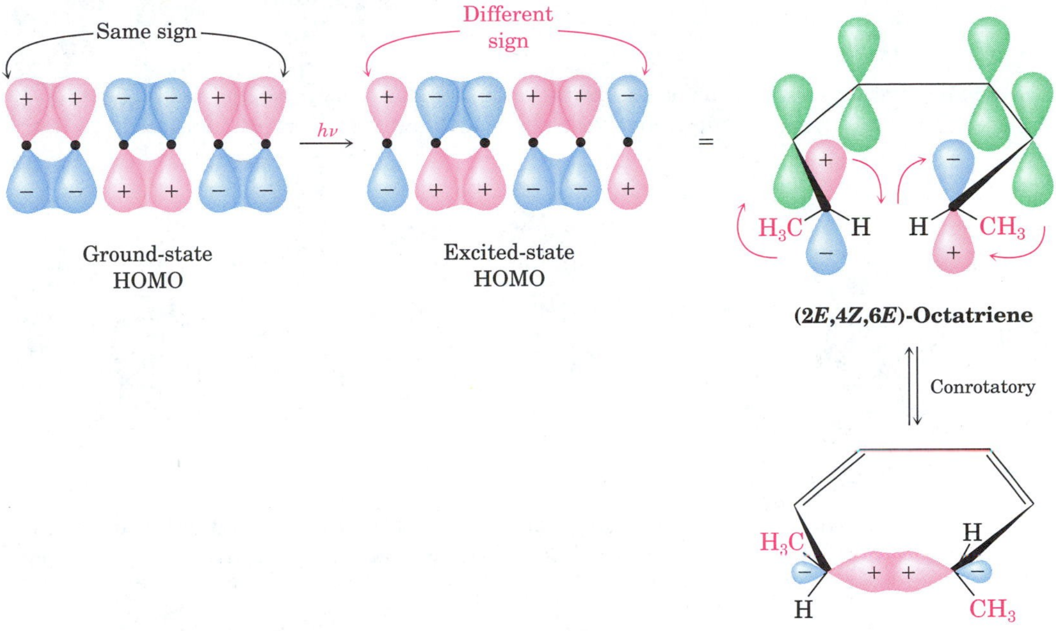

(2*E*,4*Z*,6*E*)-Octatriene

***trans*-5,6-Dimethyl-1,3-cyclohexadiene**

Thermal and photochemical electrocyclic reactions *always* take place with opposite stereochemistry because the symmetries of the frontier orbitals are always different. Table 31.1 gives some simple rules that make it possible to predict the stereochemistry of a large number of electrocyclic reactions.

Table 31.1 Stereochemical Rules for Electrocyclic Reactions

Electron pairs (double bonds)	*Thermal reaction*	*Photochemical reaction*
Even number	Conrotatory	Disrotatory
Odd number	Disrotatory	Conrotatory

PROBLEM.

31.4 What product would you expect to obtain from the photochemical cyclization of (2*E*,4*Z*,6*E*)-octatriene? Of (2*E*,4*Z*,6*Z*)-octatriene?

PROBLEM.

31.5 The following thermal isomerization occurs under relatively mild conditions. Identify the pericyclic reactions involved, and show how the rearrangement occurs.

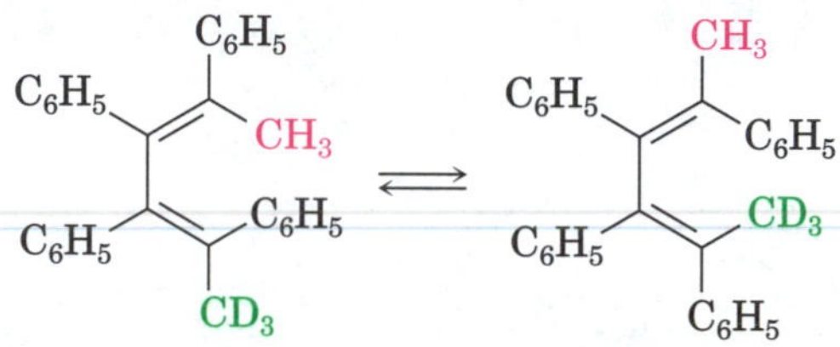

PROBLEM.

31.6 Would you expect the following reaction to proceed in a conrotatory or disrotatory manner? Show the stereochemistry of the cyclobutene product, and explain your answer.

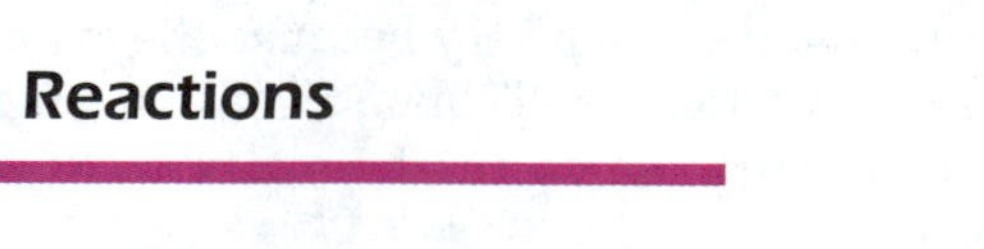

31.7 Cycloaddition Reactions

Cycloaddition reactions are intermolecular pericyclic processes in which two unsaturated molecules add to one another, yielding a cyclic product. As with electrocyclic reactions, cycloadditions are controlled by the orbital symmetry of the reactants. Symmetry-allowed processes often take place readily, but symmetry-disallowed processes take place with great difficulty, if at all, and then only by nonconcerted pathways. Let's look at two possible reactions to see how they differ.

The Diels–Alder cycloaddition reaction (Section 14.8) is a pericyclic process that takes place between a diene (4 π electrons) and a dienophile (2 π electrons) to yield a cyclohexene product. Thousands of examples of Diels–Alder reactions are known. They often take place under mild conditions (room temperature or slightly above), and they are stereospecific with respect to substituents. For example, room-temperature reaction between 1,3-butadiene and diethyl maleate (cis) yields exclusively the cis-disubstituted cyclohexene product. Similar reaction between 1,3-butadiene and diethyl fumarate (trans) yields exclusively the trans-disubstituted product (Figure 31.10).

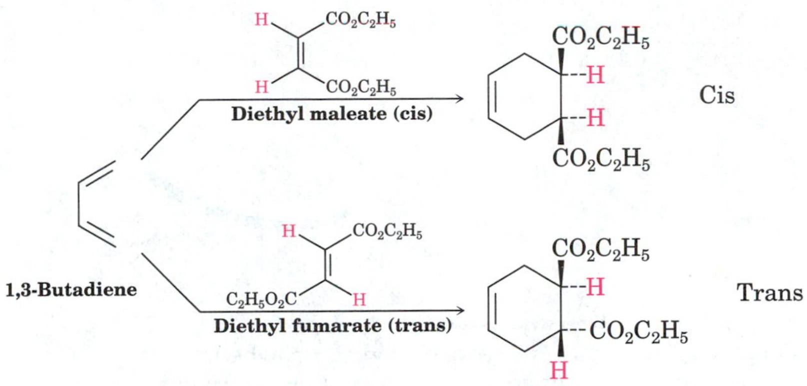

Figure 31.10 Diels–Alder cycloaddition reactions of diethyl maleate (cis) and diethyl fumarate (trans). The reactions are stereospecific.

In contrast to the [4 + 2]-π-electron Diels–Alder reaction, thermal cycloaddition between two alkenes (2 π + 2 π) does not occur. *Photochemical* [2 + 2] cycloadditions often take place readily, however, to yield cyclobutane products.

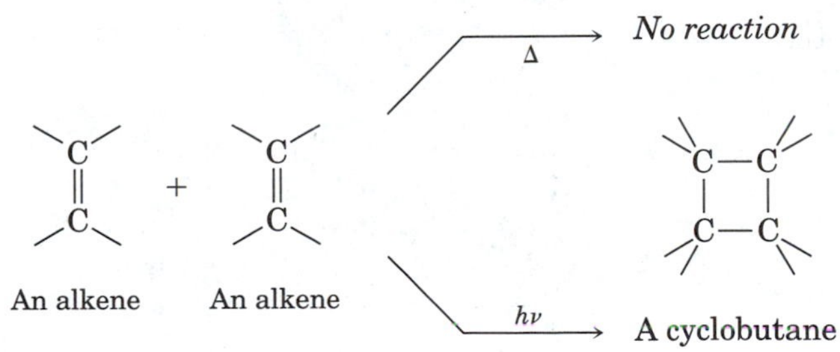

For a successful cycloaddition to take place, the terminal π lobes of the two unsaturated reactants must have the correct symmetry for bonding overlap to occur. This can happen in either of two ways, designated *suprafacial* and *antarafacial*. **Suprafacial** cycloadditions take place when a bonding interaction occurs between lobes on the *same* face of one reactant and lobes on the *same* face of the other reactant (Figure 31.11).

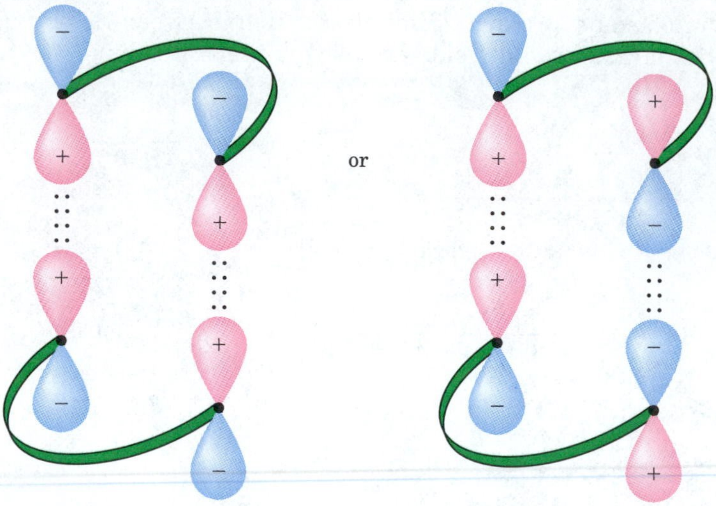

Figure 31.11 Suprafacial cycloaddition occurs when there is bonding between lobes on the same face of one reactant and lobes on the same face of the second reactant.

Antarafacial cycloadditions take place when a bonding interaction occurs between lobes on the *same* face of one reactant and lobes on *opposite* faces of the other reactant (Figure 31.12).

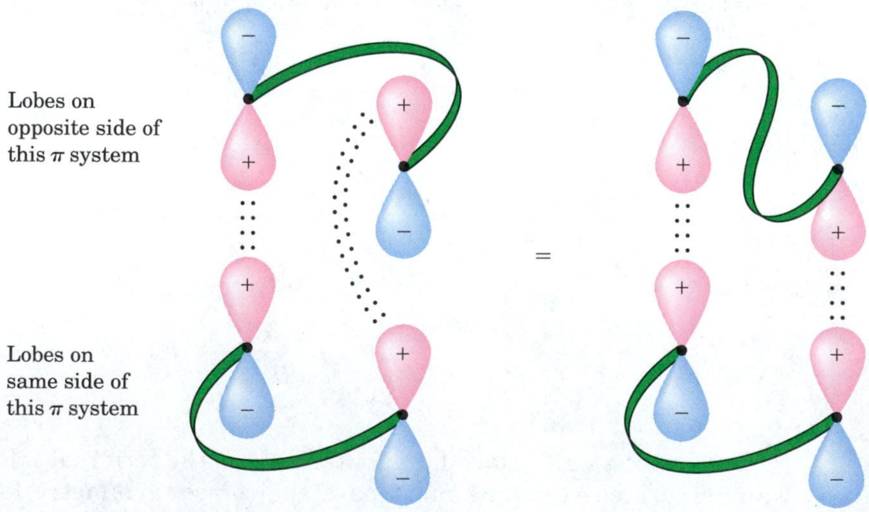

Figure 31.12 Antarafacial cycloaddition occurs when there is bonding between lobes on the same face of one reactant and lobes on opposite faces of the second reactant.

Note that both suprafacial and antarafacial cycloadditions are allowed on orbital-symmetry grounds. Geometric constraints often make antarafacial reactions difficult, however, because there must be twisting of the π orbital system. Thus, suprafacial cycloadditions are most common for small π systems.

31.8 Stereochemistry of Cycloadditions

How can we predict whether a given cycloaddition reaction will occur with suprafacial or with antarafacial geometry? According to frontier orbital theory, cycloaddition reactions take place when a bonding interaction occurs between the HOMO of one reactant and the LUMO of the other. A good intuitive explanation of this rule is to imagine that one reactant donates electrons to the other. As with electrocyclic reactions, it's the electrons in the HOMO of the first reactant that are least tightly held and most likely to be donated. But because only two electrons can occupy any one orbital, the donated electrons must go into a *vacant* orbital of the second reactant—the LUMO.

For the [4 + 2]-π-electron cycloaddition (Diels–Alder reaction), let's arbitrarily select the diene LUMO and the alkene HOMO. (We could equally well use the diene HOMO and the alkene LUMO.) The symmetries of the two ground-state orbitals are such that bonding overlap of the terminal lobes can occur with suprafacial geometry (Figure 31.13). The Diels–Alder

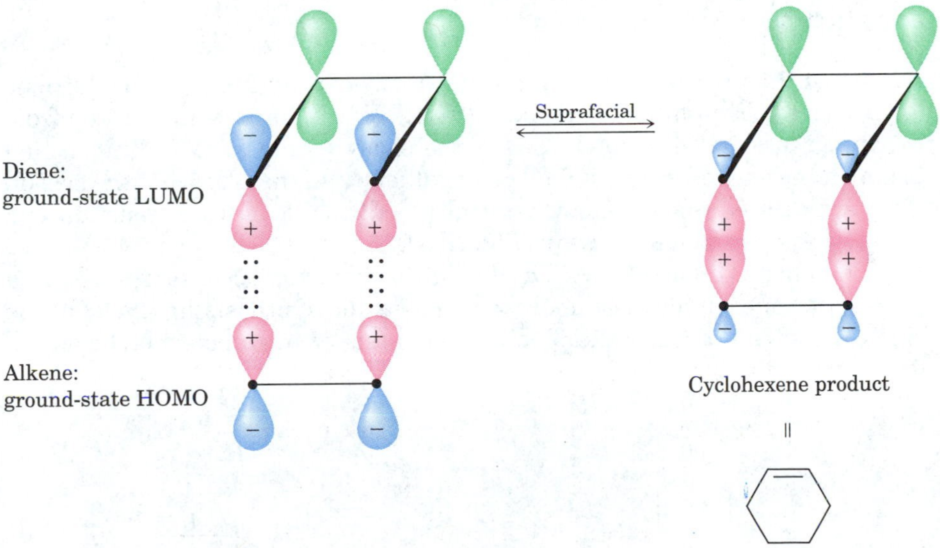

Figure 31.13 Interaction of diene LUMO and alkene HOMO in a suprafacial [4 + 2] cycloaddition reaction (Diels–Alder reaction).

reaction therefore takes place readily under thermal conditions. Note that, as with electrocyclic reactions, we need be concerned only with the *terminal* lobes. For purposes of prediction, it doesn't matter whether the interior lobes have a bonding or an antibonding interaction in the product.

In contrast to the [4 + 2] Diels–Alder reaction, the [2 + 2] cycloaddition of two alkenes to yield a cyclobutane does not occur thermally but can only be observed photochemically. The explanation follows from orbital-symmetry arguments. Looking at the ground-state HOMO of one alkene and the LUMO of the second alkene, it's apparent that a thermal [2 + 2] cycloaddition must take place by an antarafacial pathway (Figure 31.14). Geometric constraints make the antarafacial transition state difficult, however, and concerted thermal [2 + 2] cycloadditions are therefore rarely observed.

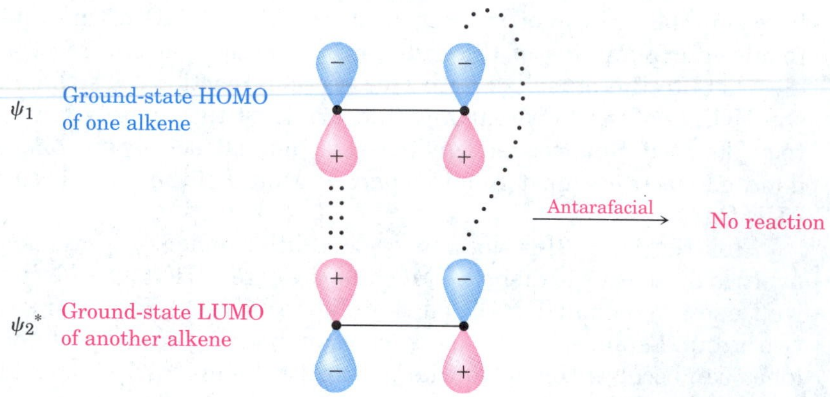

Figure 31.14 Interaction of HOMO and LUMO in a thermal [2 + 2] cycloaddition. The reaction rarely occurs because antarafacial geometry is too strained.

In contrast to the thermal process, *photochemical* [2 + 2] cycloadditions *are* observed. Irradiation of an alkene with UV light excites an electron from ψ_1, the ground-state HOMO, to ψ_2^*, the excited-state HOMO. Interaction between the excited-state HOMO of one alkene and the LUMO of the second alkene indicates that a photochemical [2 + 2] cycloaddition reaction can occur by a suprafacial pathway (Figure 31.15).

The photochemical [2 + 2] cycloaddition reaction occurs smoothly and represents one of the best methods known for synthesizing cyclobutane rings. The reaction can take place either inter- or intramolecularly, as the following examples show:

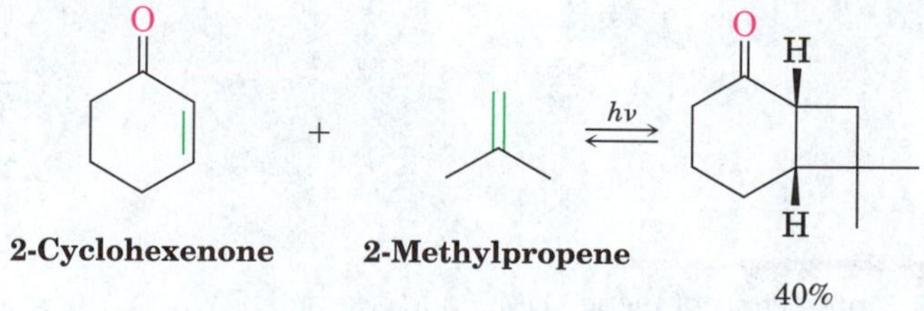

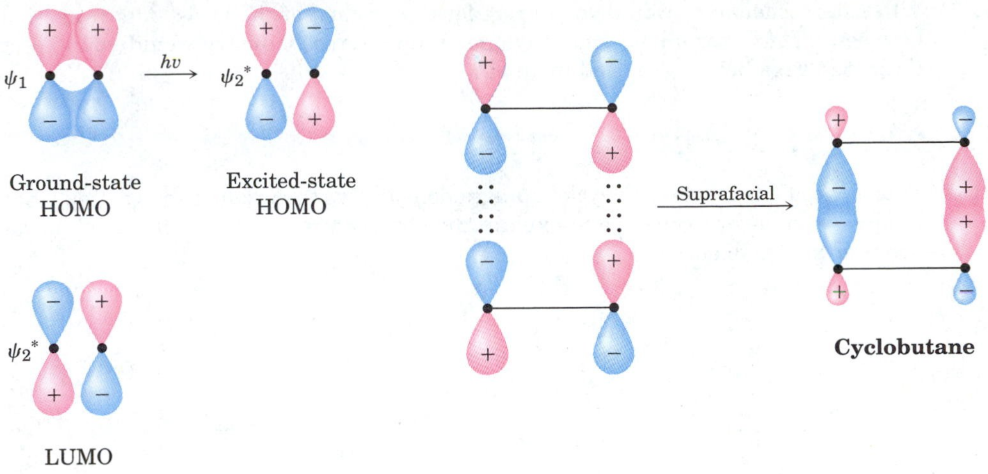

Figure 31.15 Interaction of excited-state HOMO and LUMO in photochemical [2 + 2] cycloaddition reactions. The reaction occurs with suprafacial geometry.

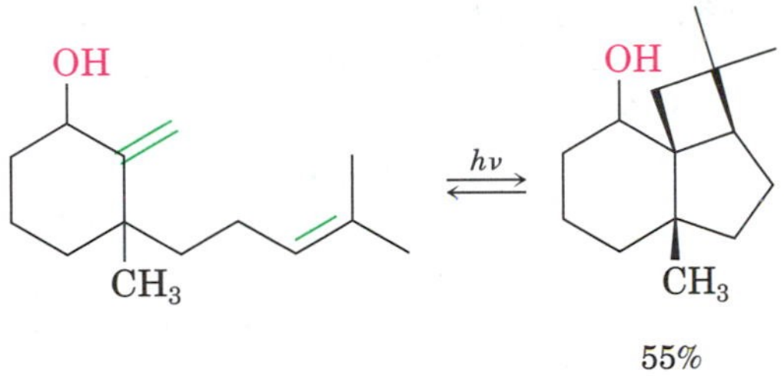

Thermal and photochemical cycloaddition reactions always take place by opposite stereochemical pathways. As with electrocyclic reactions, we can categorize cycloadditions according to the total number of electron pairs (double bonds) involved in the rearrangement. Thus, a Diels–Alder [4 + 2] reaction between a diene and a dienophile involves an odd number (three) of electron pairs and takes place by a ground-state suprafacial pathway. A [2 + 2] reaction between two alkenes involves an even number (two) of electron pairs and takes place by a ground-state antarafacial pathway. The general rules governing such processes are given in Table 31.2.

Table 31.2 Stereochemical Rules for Cycloaddition Reactions

Electron pairs (double bonds)	*Thermal reaction*	*Photochemical reaction*
Even number	Antarafacial	Suprafacial
Odd number	Suprafacial	Antarafacial

PROBLEM

31.7 What stereochemistry would you expect for the product of the Diels–Alder reaction between (2*E*,4*E*)-hexadiene and ethylene? What stereochemistry would you expect if (2*E*,4*Z*)-hexadiene were used instead?

PROBLEM

31.8 Cyclopentadiene reacts with cycloheptatrienone to give the product shown. Tell what kind of reaction is involved, and explain the observed result. Is the reaction suprafacial or antarafacial?

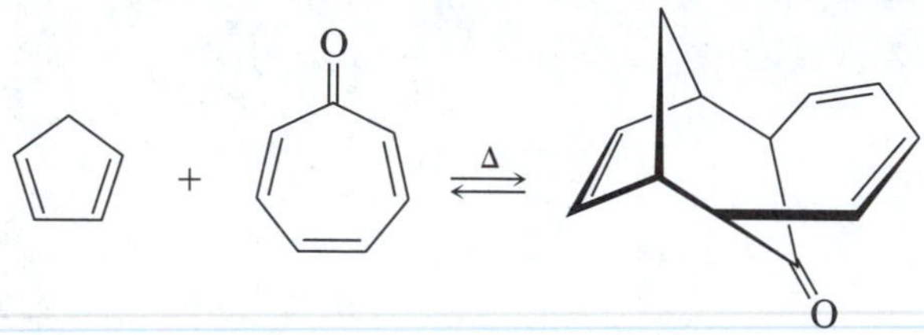

PROBLEM

31.9 The following reaction takes place in two steps, one of which is a cycloaddition and the other of which is a *reverse* cycloaddition. Identify the two pericyclic reactions, and show how they occur.

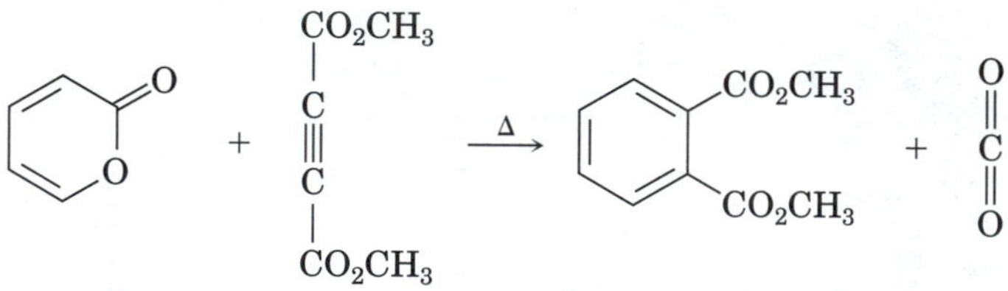

31.9 Sigmatropic Rearrangements

The third general kind of pericyclic reaction, **sigmatropic rearrangements**, are processes in which a σ-bonded substituent group (denoted below by a circled S) migrates across a π electron system. One σ bond is broken in the starting material, π bonds move, and a new σ bond is formed in the product. The σ-bonded group can be either at the end or in the middle of the π system, as the following [1,3] and [3,3] rearrangements illustrate:

A [1,3] rearrangement

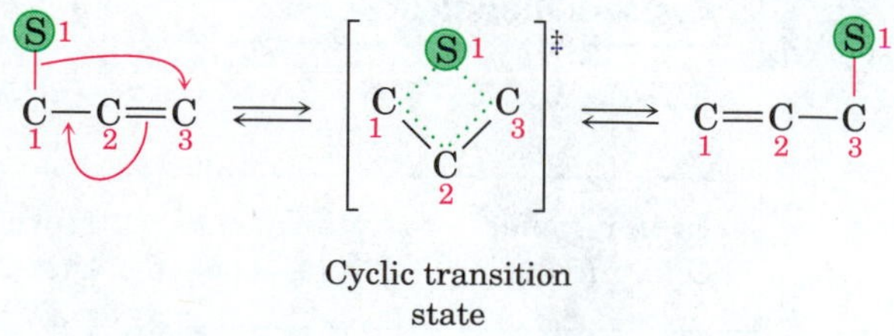

Cyclic transition state

A [3,3] rearrangement

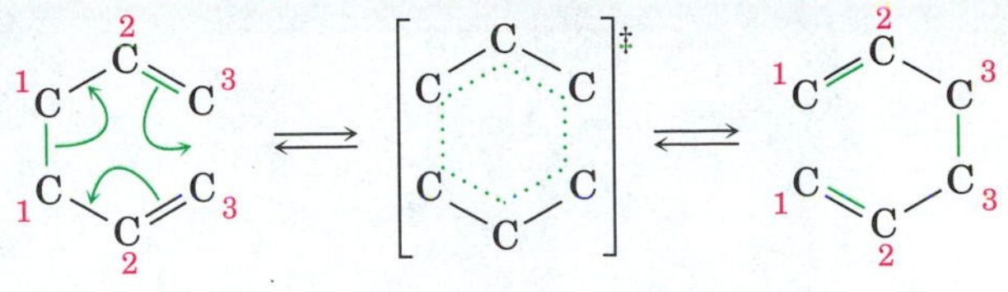

Cyclic transition state

The notations [1,3] and [3,3] describe the kind of rearrangement that has occurred. The two numbers in brackets refer to the two groups connected by the σ bond and designate the positions in those groups *to which migration occurs*. For example, in the [1,5] sigmatropic rearrangement of a diene, the two groups connected by the σ bond are a hydrogen atom and a pentadienyl fragment. Migration occurs to position 1 of the H group (the only possibility) and to position 5 of the pentadienyl group.

A [1,5] sigmatropic rearrangement

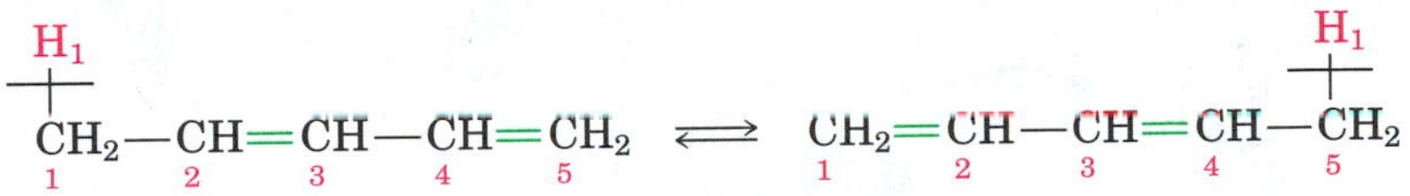

In the [3,3] Claisen rearrangement (Section 25.8), the two groups connected by the σ bond are an allyl group and a vinylic ether group. Migration occurs to position 3 of the allyl group and also to position 3 of the vinylic ether.

Claisen rearrangement, a [3,3] rearrangement

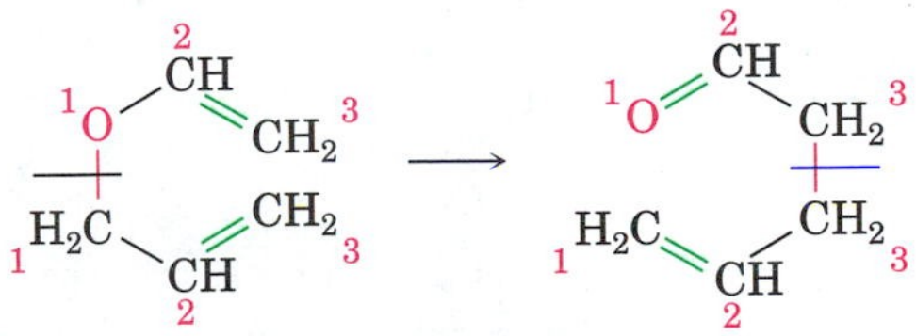

31.10 Stereochemistry of Sigmatropic Rearrangements

Sigmatropic rearrangements are more complex than either electrocyclic or cycloaddition reactions but are nonetheless controlled by orbital-symmetry considerations. There are two possible modes of reaction: Migration of a group across the same face of the π system is a *suprafacial* rearrangement, and migration of a group from one face of the π system to the other face is an *antarafacial* rearrangement (Figure 31.16).

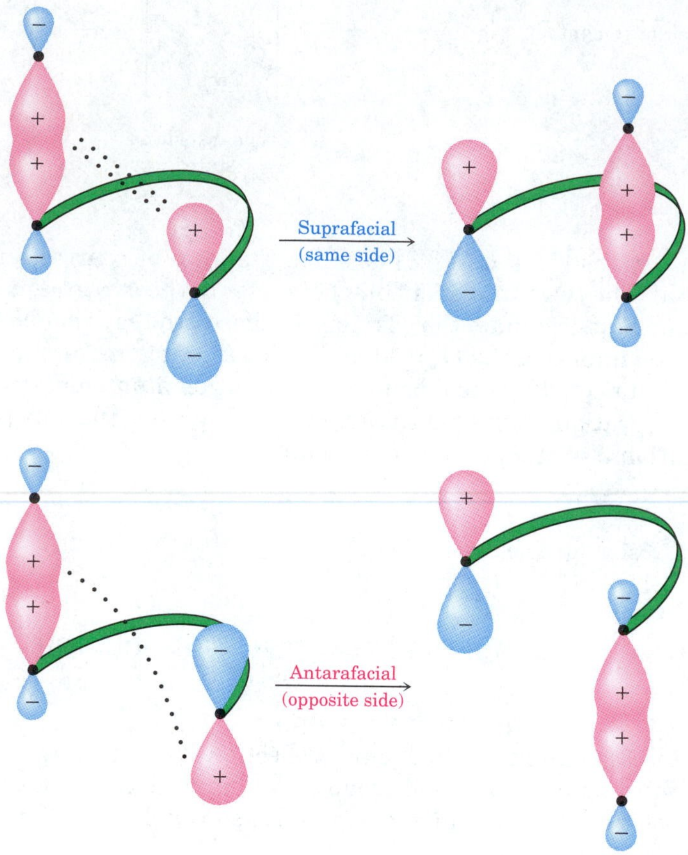

Figure 31.16 Suprafacial and antarafacial sigmatropic rearrangements.

The rules for sigmatropic rearrangements are identical to those for cycloaddition reactions (Table 31.3). Both suprafacial and antarafacial sigmatropic rearrangements are symmetry-allowed, but suprafacial rearrangements are often easier for geometric reasons.

Table 31.3 Stereochemical Rules for Sigmatropic Rearrangements

Electron pairs (double bonds)	*Thermal reaction*	*Photochemical reaction*
Even number	Antarafacial	Suprafacial
Odd number	Suprafacial	Antarafacial

31.11 Some Examples of Sigmatropic Rearrangements

Since a [1,5] sigmatropic rearrangement involves three electron pairs (two π bonds and one σ bond), the orbital-symmetry rules in Table 31.3 predict a suprafacial reaction. In fact, the [1,5] suprafacial shift of a hydrogen atom across two double bonds of a π system is one of the most commonly observed of all sigmatropic rearrangements. For example, 5-methylcyclopentadiene rapidly rearranges at room temperature to yield a mixture of 1-methyl-, 2-methyl-, and 5-methyl-substituted products.

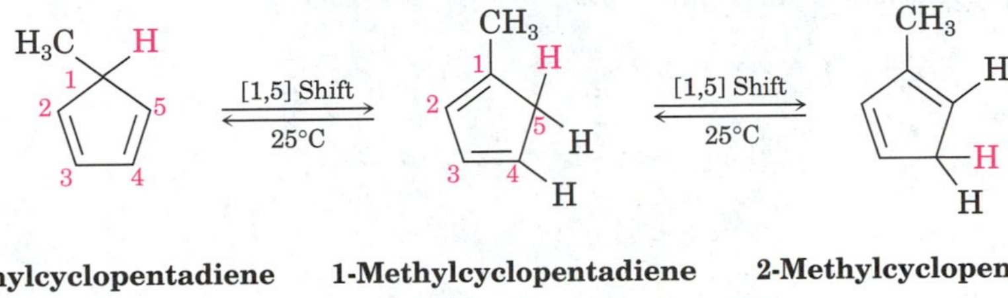

As another example, heating 5,5,5-trideuterio-(1,3*Z*)-pentadiene causes scrambling of deuterium between positions 1 and 5.

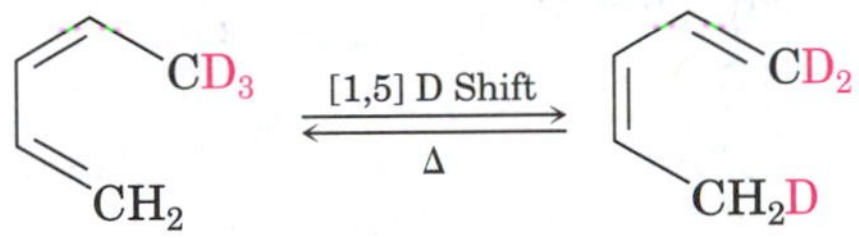

Both of these [1,5] hydrogen shifts occur by a symmetry-allowed suprafacial rearrangement, as illustrated in Figure 31.17.

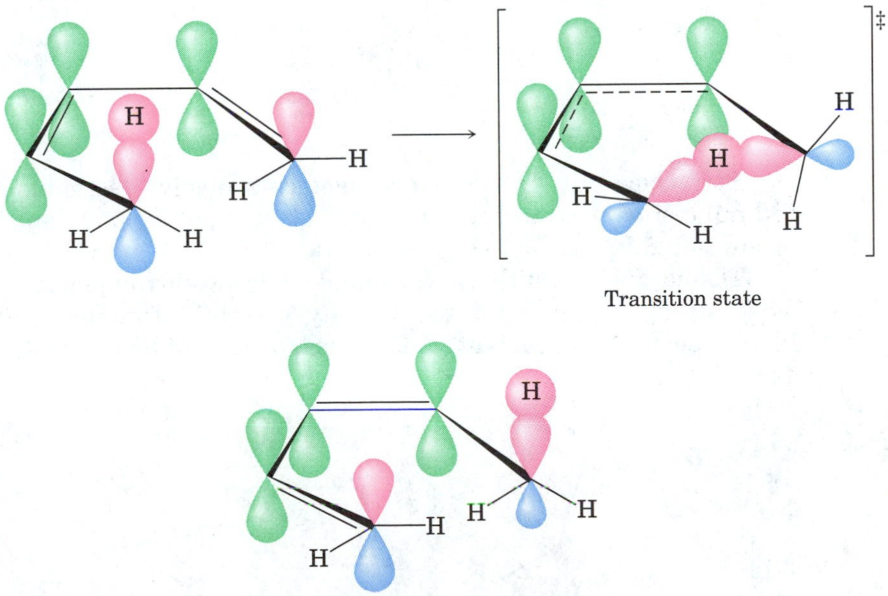

Figure 31.17 An orbital view of a suprafacial [1,5] hydrogen shift.

In contrast to the preceding two examples of [1,*5*] sigmatropic hydrogen shifts, thermal [1,*3*] hydrogen shifts are unknown. Were they to occur, they would have to proceed by a strained antarafacial reaction pathway.

Two other important sigmatropic reactions are the **Cope rearrangement** of a 1,5-hexadiene and the **Claisen rearrangement** of an allyl aryl ether (Section 25.8). These two, along with the Diels–Alder reaction, are the most useful pericyclic reactions for organic synthesis; many thousands of examples of all three are known. Note that the Claisen rearrangement works well with both allyl aryl ethers and with allyl vinylic ethers.

Claisen rearrangement

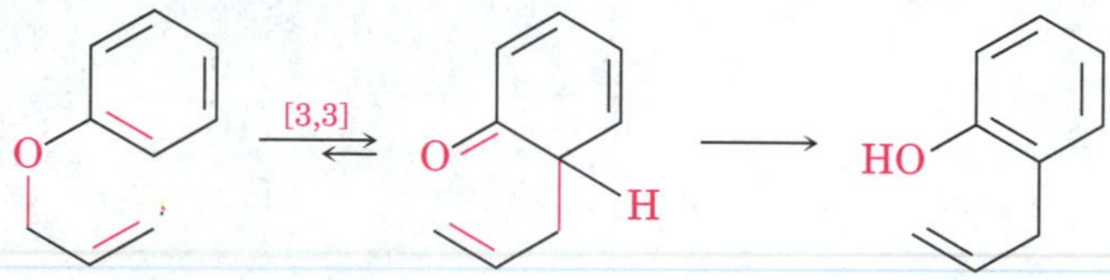

An allyl aryl ether   An *o*-allylphenol

Claisen rearrangement

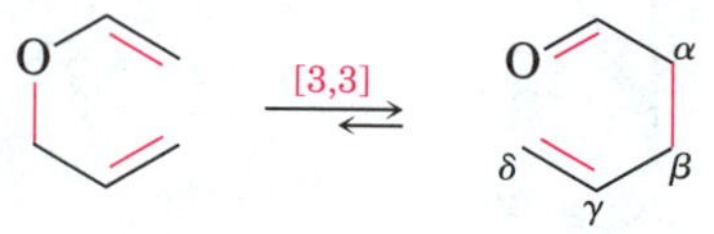

An allyl vinylic ether   A γ,δ-unsaturated carbonyl compound

Cope rearrangement

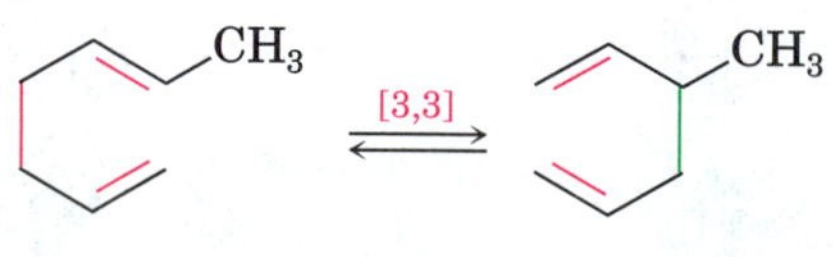

A 1,5 diene   A new 1,5 diene

Both Cope and Claisen rearrangements involve reorganization of an odd number of electron pairs (two π bonds and one σ bond), and both react by suprafacial pathways (Figure 31.18).

The ease with which the Cope and Claisen rearrangements occur provides strong evidence for the validity of orbital-symmetry predictions. Indeed, some Cope rearrangements, such as that of homotropilidene, occur as rapidly as *several hundred times each second* at room temperature.

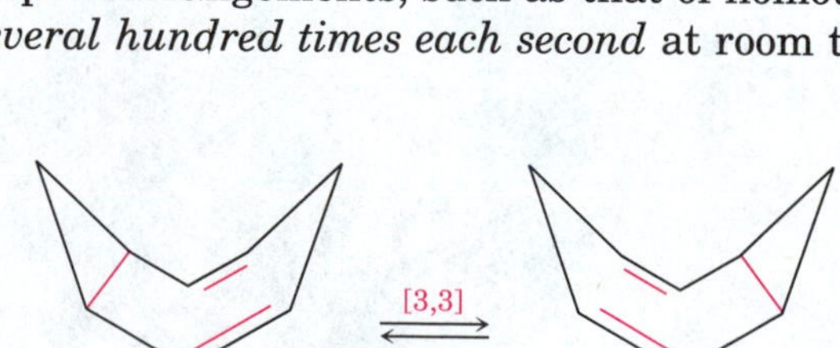

Homotropilidene

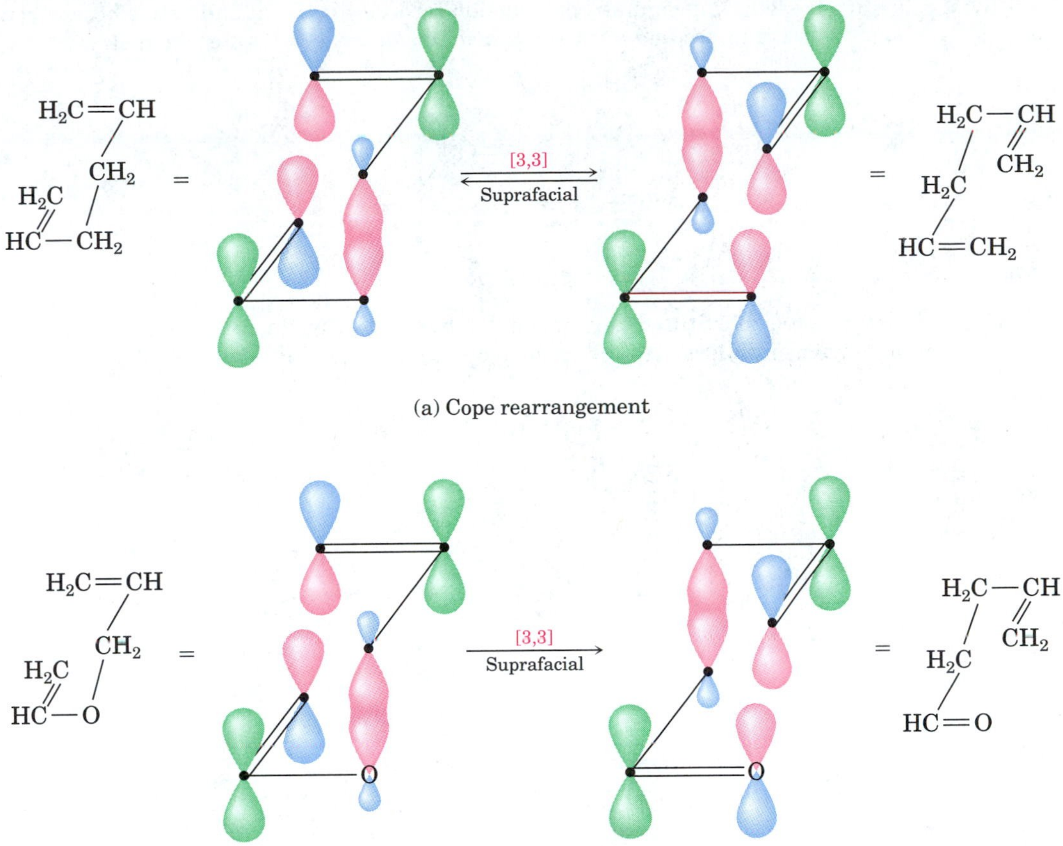

(a) Cope rearrangement

(b) Claisen rearrangement

Figure 31.18 Suprafacial [3,3] Cope and Claisen rearrangements.

PROBLEM...

31.10 The ^{13}C NMR spectrum of homotropilidene taken at room temperature shows only three peaks. Explain.

PROBLEM...

31.11 How can you account for the fact that heating 1-deuterioindene scrambles the isotope label to all three positions on the five-membered ring?

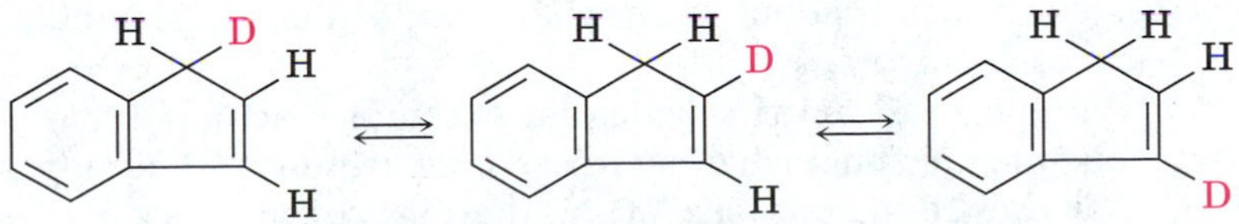

PROBLEM.

31.12 Classify the following sigmatropic reaction by order [x,y], and indicate whether you would expect it to proceed with suprafacial or antarafacial stereochemistry.

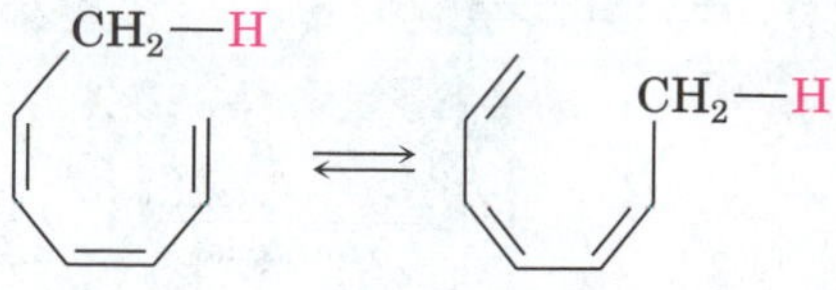

PROBLEM.

31.13 When a 2,6-disubstituted allyl phenyl ether is heated in an attempted Claisen rearrangement, migration occurs to give *p*-allyl product. Explain.

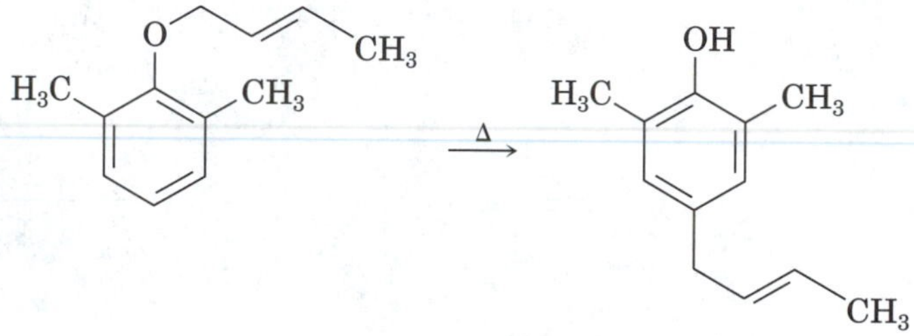

31.12 A Summary of Rules for Pericyclic Reactions

Pericyclic, electrocyclic, cycloaddition, sigmatropic, conrotatory, disrotatory, suprafacial, antarafacial . . . how can you keep it all straight?

The information provided in Tables 31.1, 31.2, and 31.3 to summarize the selection rules for electrocyclic, cycloaddition, and sigmatropic reactions leads to the conclusion that pericyclic processes can be grouped according to whether they involve the reorganization of an even number or an odd number of electron pairs (bonds). All this information can be distilled into one simple phrase that provides an easy and accurate way to predict the stereochemical outcome of any pericyclic reaction:

> For a ground-state (thermal) pericyclic reaction, the groupings are *odd-supra-dis* and *even-antara-con*.

Cycloaddition and sigmatropic reactions involving an odd number of electron pairs occur with suprafacial geometry; electrocyclic reactions involving an odd number of electron pairs occur with disrotatory stereochemistry. Conversely, pericyclic reactions involving an even number of electron pairs occur with either antarafacial geometry or conrotatory stereochemistry.

Once the selection rules for thermal reactions have been memorized, the rules for photochemical reactions are derived simply by remembering that they're the opposite of the thermal rules:

For an excited-state (photochemical) pericyclic reaction, the groupings are *odd-antara-con* and *even-supra-dis*.

Both rules are summarized in Table 31.4. Memorizing this table will give you the ability to predict the stereochemistry of literally thousands of pericyclic reactions.

Table 31.4 Stereochemical Rules for Pericyclic Reactions

Electron state	*Electron pairs*	*Stereochemistry*
Ground state (thermal)	Even number	Antara-con
	Odd number	Supra-dis
Excited state (photochemical)	Even number	Supra-dis
	Odd number	Antara-con

PROBLEM

31.14 Predict the stereochemistry of the following pericyclic reactions.
(a) The thermal cyclization of a conjugated tetraene
(b) The photochemical cyclization of a conjugated tetraene
(c) A photochemical [4 + 4] cycloaddition
(d) A thermal [2 + 6] cycloaddition
(e) A photochemical [3,5] sigmatropic rearrangement

INTERLUDE

Vitamin D, the Sunshine Vitamin

It may look easy, but these sunbathers are working hard to synthesize vitamin D.

Vitamin D, discovered in 1918, is a general name for two related compounds, *cholecalciferol* (vitamin D_3) and *ergocalciferol* (vitamin D_2). Both are steroids (Section 28.7) and differ only in the nature of the hydrocarbon side chain attached to the five-membered ring. Cholecalciferol comes from dairy products and fish; ergocalciferol comes from some vegetables. Their

(continued)➤

function in the body is to control the calcification of bones by increasing intestinal absorption of calcium. When sufficient vitamin D is present, approximately 30% of ingested calcium is absorbed, but in the absence of vitamin D, calcium absorption falls to about 10%. A deficiency of vitamin D thus leads to poor bone growth and to the childhood disease known as *rickets*.

Actually, the active forms of vitamins D_2 and D_3 are not present in foods. In the presence of sunlight, both are manufactured under the skin by photochemical reactions of the precursor molecules 7-dehydrocholesterol and ergosterol, hence the nickname "sunshine vitamin."

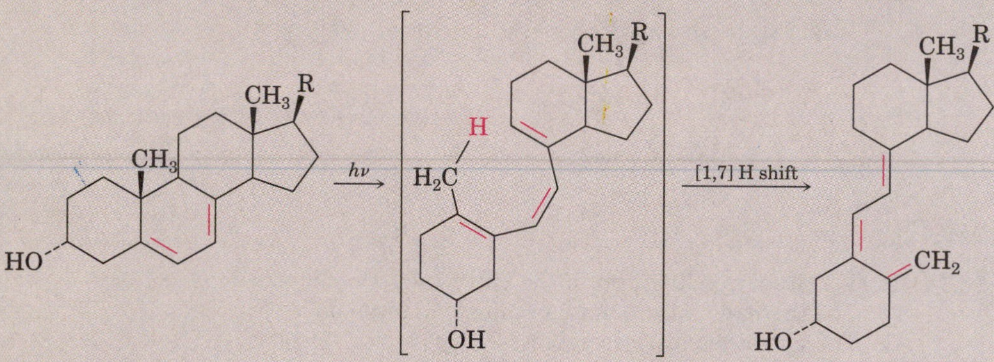

7-Dehydrocholesterol $R = CH(CH_3)CH_2CH_2CH_2CH(CH_3)_2$ **Cholecalciferol**
Ergosterol $R = CH(CH_3)CH{=}CHCH(CH_3)CH(CH_3)_2$ **Ergocalciferol**

Pericyclic reactions are unusual in living organisms, and the photochemical synthesis of vitamin D is one of the few well-studied examples. The reaction takes place in two steps: an electrocyclic ring opening of a cyclohexadiene to yield a hexatriene, followed by a sigmatropic [1,7] H shift to yield an isomeric hexatriene. Further metabolic processing in the liver and the kidney introduces several –OH groups to give the active form of the vitamin.

Summary and Key Words

A **pericyclic reaction** is one that takes place in a single step through a cyclic transition state; no intermediates are involved. There are three major classes of pericyclic processes: **electrocyclic reactions**, **cycloaddition reactions**, and **sigmatropic rearrangements**. The stereochemistry of these reactions is controlled by the symmetry of the orbitals involved in bond reorganization.

Electrocyclic reactions involve the cyclization of conjugated polyenes. For example, 1,3,5-hexatriene cyclizes to 1,3-cyclohexadiene on heating. Electrocyclic reactions can occur by either **conrotatory** or **disrotatory** paths, depending on the symmetry of the terminal lobes of the π system. Conrotatory cyclization requires that both lobes rotate in the same direction, whereas disrotatory cyclization requires that the lobes rotate in opposite directions. The reaction course in a specific case can be found by looking at the symmetry of the **highest occupied molecular orbital (HOMO)**.

Cycloaddition reactions are those in which two unsaturated molecules add together to yield a cyclic product. For example, Diels–Alder reaction between a diene (4 π electrons) and a dienophile (2 π electrons) yields a cyclohexene. Cycloadditions can take place either by **suprafacial** or **antarafacial** pathways. Suprafacial cycloaddition involves interaction between lobes on the same face of one component and on the same face of the second component. Antarafacial cycloaddition involves interaction between lobes on the same face of one component and on opposite faces of the other component. The reaction course in a specific case can be found by looking at the symmetry of the HOMO of one component and the **lowest unoccupied molecular orbital (LUMO)** of the other component.

Sigmatropic rearrangements involve the migration of a σ-bonded group across a π electron system. For example, **Claisen rearrangement** of an allyl vinylic ether yields an unsaturated carbonyl compound, and **Cope rearrangement** of a 1,5-hexadiene yields a new 1,5-hexadiene. Sigmatropic rearrangements can occur with either suprafacial or antarafacial stereochemistry; the selection rules for a given case are the same as those for cycloaddition reactions.

The stereochemistry of any pericyclic reaction can be predicted by counting the total number of electron pairs (bonds) involved in bond reorganization and then applying some simple rules. Thermal (ground-state) reactions involving an even number of electron pairs occur with either antarafacial or conrotatory stereochemistry (**even-antara-con**); thermal reactions involving an odd number of electron pairs occur with suprafacial or disrotatory stereochemistry (**odd-supra-dis**). Exactly the opposite rules apply to photochemical (excited-state) reactions.

ADDITIONAL PROBLEMS

31.15 Define the following terms:

(a) Electrocyclic reaction
(b) Conrotatory motion
(c) Suprafacial
(d) Antarafacial
(e) Disrotatory motion
(f) Sigmatropic rearrangement

31.16 Have the following reactions taken place in a conrotatory or disrotatory manner? Under what conditions, thermal or photochemical, would you carry out each reaction?

(a)

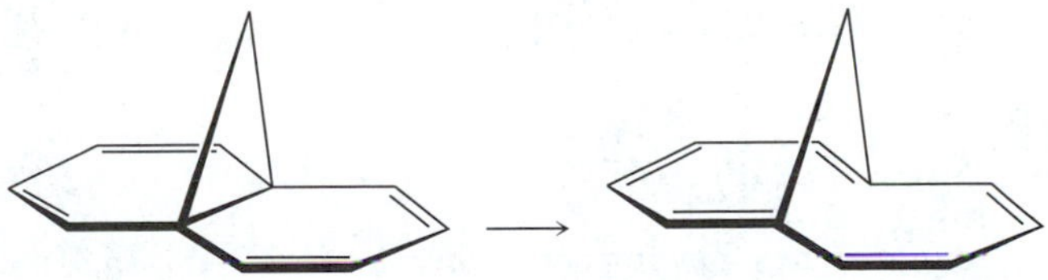

(b)

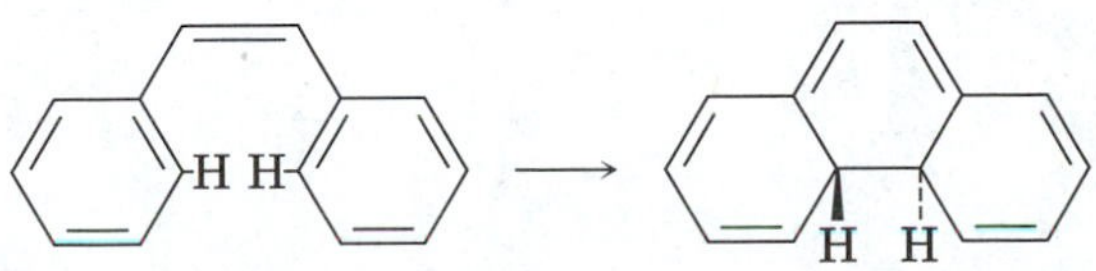

31.17 (2*E*,4*Z*,6*Z*,8*E*)-Decatetraene has been cyclized to give 7,8-dimethyl-1,3,5-cyclooctatriene. Predict the manner of ring closure—conrotatory or disrotatory—for both thermal and photochemical reactions, and predict the stereochemistry of the product in each case.

31.18 Answer Problem 31.17 for the thermal and photochemical cyclizations of (2*E*,4*Z*,6*Z*,8*Z*)-decatetraene.

31.19 What stereochemistry would you expect to observe in the following reactions?
(a) A photochemical [1,5] sigmatropic rearrangement
(b) A thermal [4 + 6] cycloaddition
(c) A thermal [1,7] sigmatropic rearrangement
(d) A photochemical [2 + 6] cycloaddition

31.20 The cyclohexadecaoctaene shown isomerizes to two different isomers, depending on reaction conditions. Explain the observed results, and indicate whether each reaction is conrotatory or disrotatory.

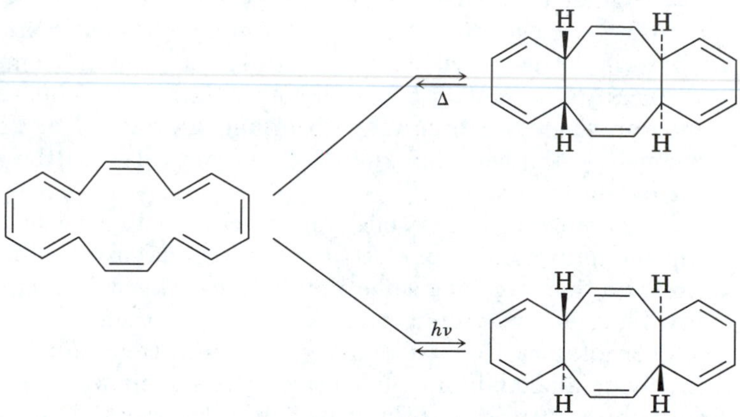

31.21 Which of the following reactions is more likely to occur? Explain.

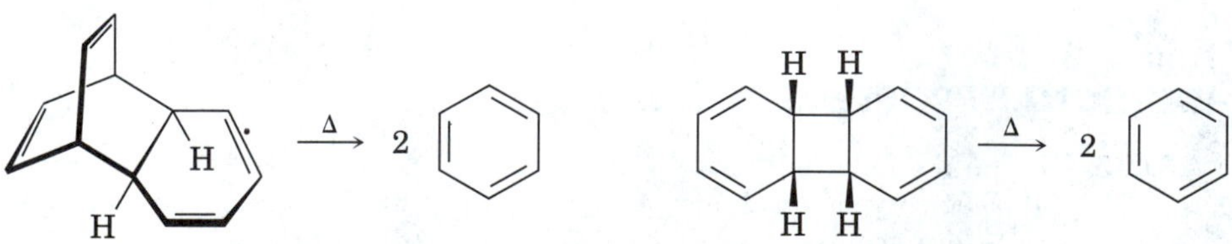

31.22 The following thermal rearrangement involves two pericyclic reactions in sequence. Identify them, and propose a mechanism to account for the observed result.

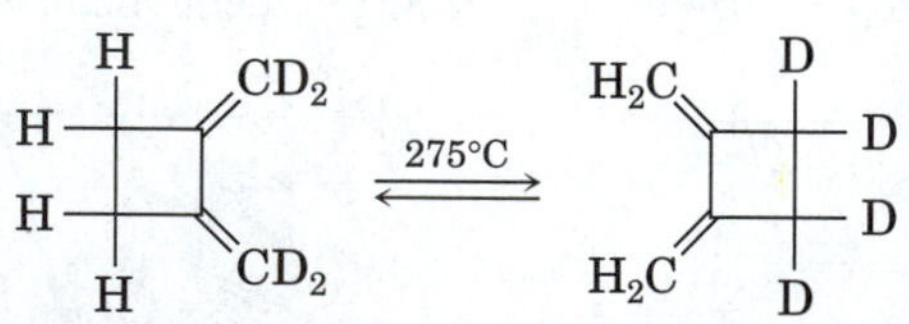

31.23 Predict the product of the following pericyclic reaction. Is this [5,5] shift a suprafacial or an antarafacial process?

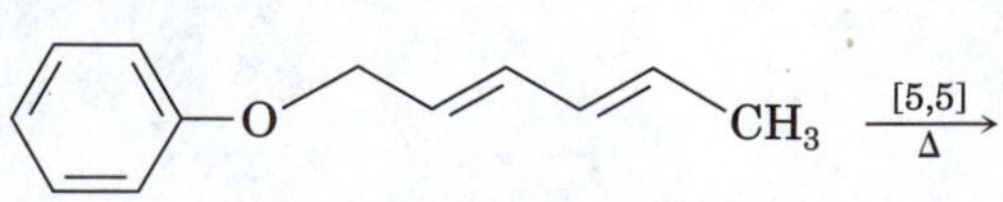

31.24 How can you account for the fact that ring opening of the *trans*-cyclobutene isomer shown takes place at much lower temperatures than a similar ring opening of the *cis*-cyclobutene isomer? Identify the stereochemistry of each reaction as either conrotatory or disrotatory.

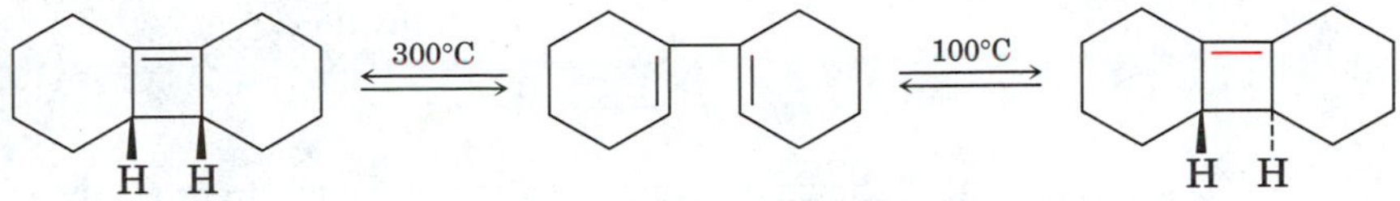

31.25 Photolysis of the *cis*-cyclobutene isomer in Problem 31.24 yields *cis*-cyclododecaen-7-yne, but photolysis of the trans isomer yields *trans*-cyclododecaen-7-yne. Explain these results, and identify the type and stereochemistry of the pericyclic reaction.

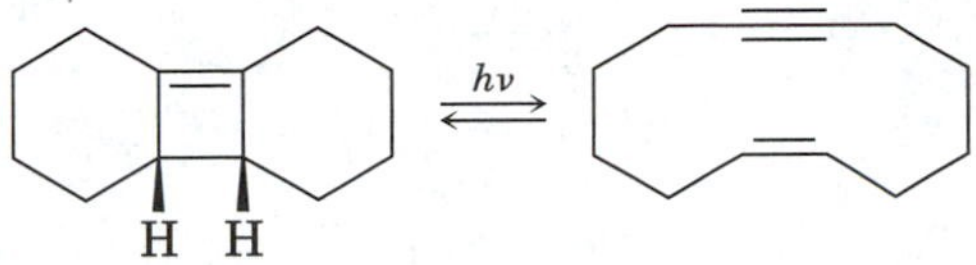

31.26 Two pericyclic reactions are involved in the furan synthesis shown. Identify them, and propose a mechanism for the transformation.

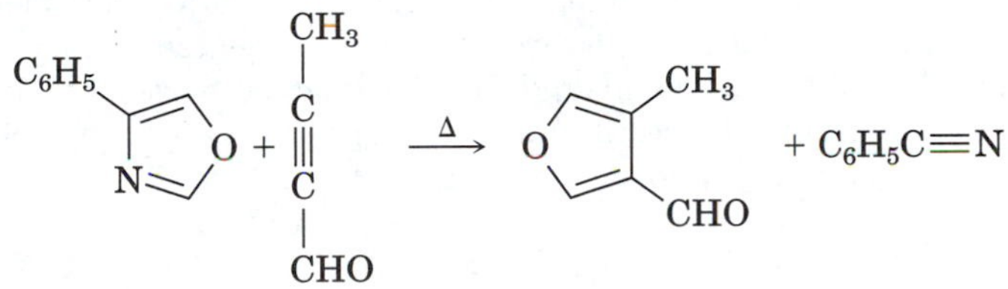

31.27 The following synthesis of dienones occurs readily. Propose a mechanism to account for the results, and identify the kind of pericyclic reaction involved.

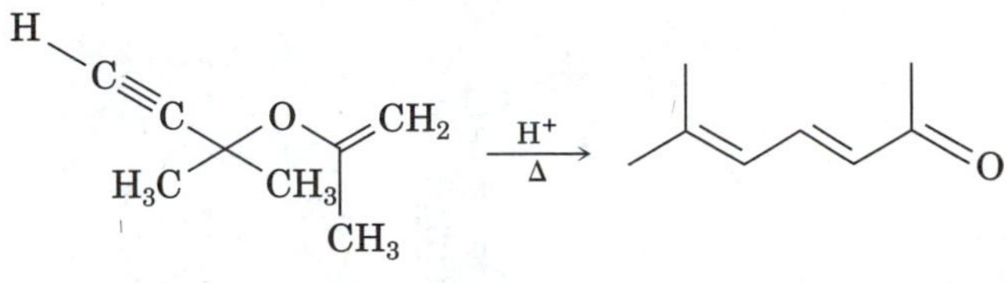

31.28 Karahanaenone, a terpene isolated from oil of hops, has been synthesized by the thermal reaction shown. Identify the kind of pericyclic reaction, and explain how karahanaenone is formed.

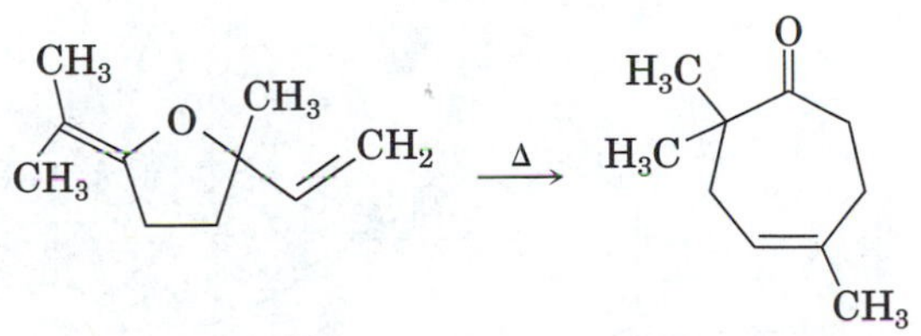

Karahanaenone

31.29 The ^{1}H NMR spectrum of bullvalene at 100°C consists only of a single peak at 4.22 δ. What conclusion can you draw from this? How can you account for this result?

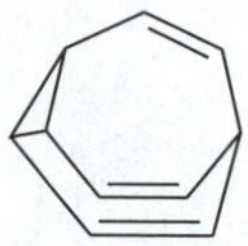

Bullvalene

31.30 The rearrangement shown was devised and carried out to prove the stereochemistry of [1,5] sigmatropic hydrogen shifts. Explain how the observed result confirms the predictions of orbital symmetry.

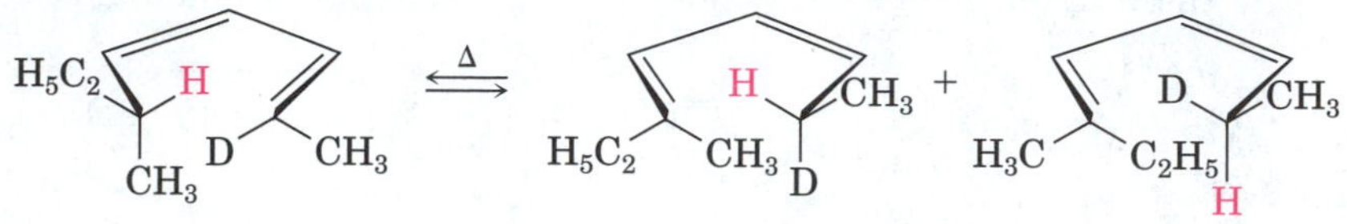

31.31 The following reaction is an example of a [2,3] sigmatropic rearrangement. Would you expect the reaction to be suprafacial or antarafacial? Explain.

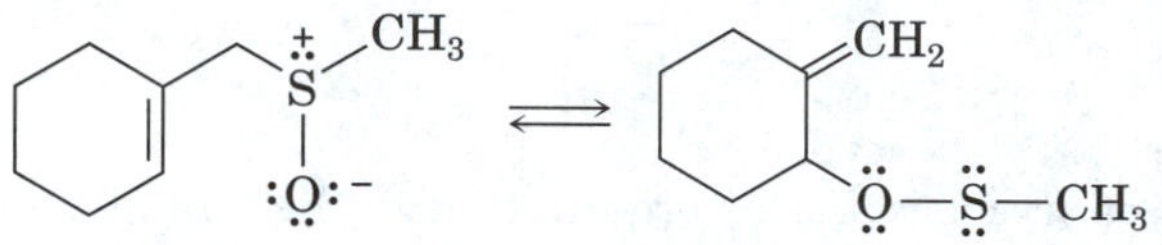

31.32 When the compound having a cyclobutene fused to a five-membered ring is heated, (1*Z*,3*Z*)-cycloheptadiene is formed. When the related compound having a cyclobutene fused to an eight-membered ring is heated, however, (1*E*,3*Z*)-cyclodecadiene is formed. Explain these results, and suggest a reason why the eight-membered-ring opening occurs at a lower temperature.

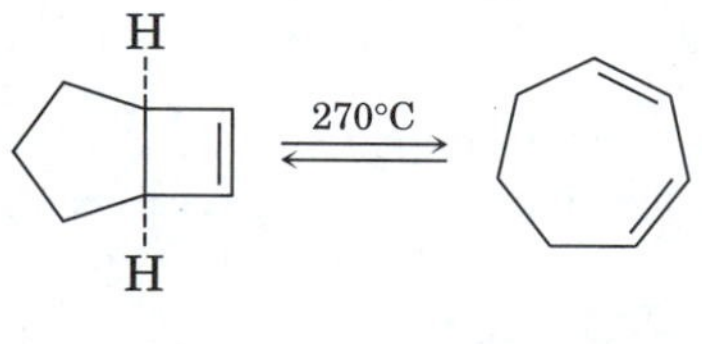

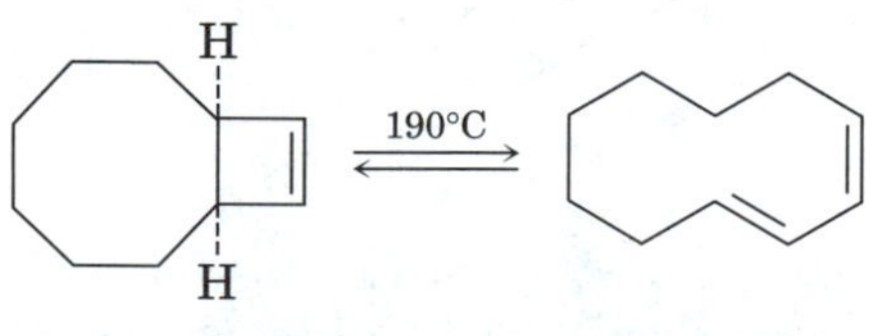

31.33 In light of your answer to Problem 31.32, explain why a mixture of products occurs in the following reaction:

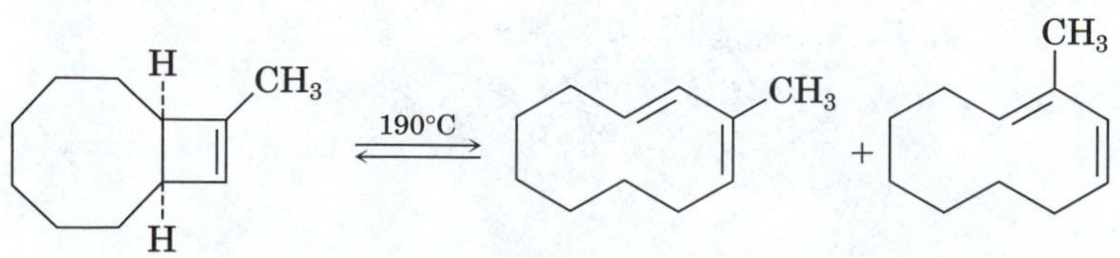

31.34 Estrone, an important female sex hormone, has been synthesized by a route that involves the following step. Identify the pericyclic reactions involved, and propose a mechanism.

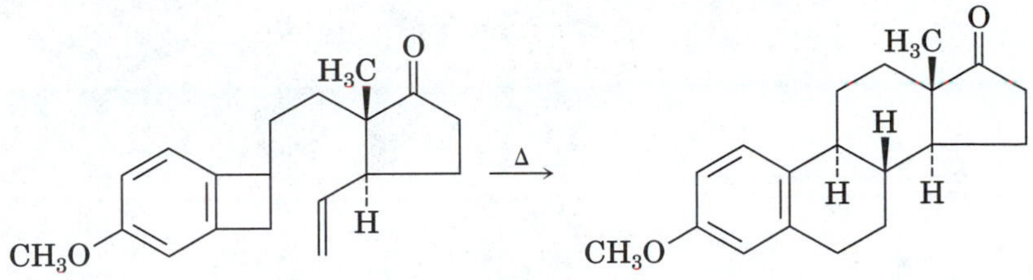

Estrone methyl ether

31.35 Coronafacic acid, a bacterial toxin, was synthesized using a key step that involves three sequential pericyclic reactions. Identify them, and propose a mechanism for the overall transformation. How would you complete the synthesis?

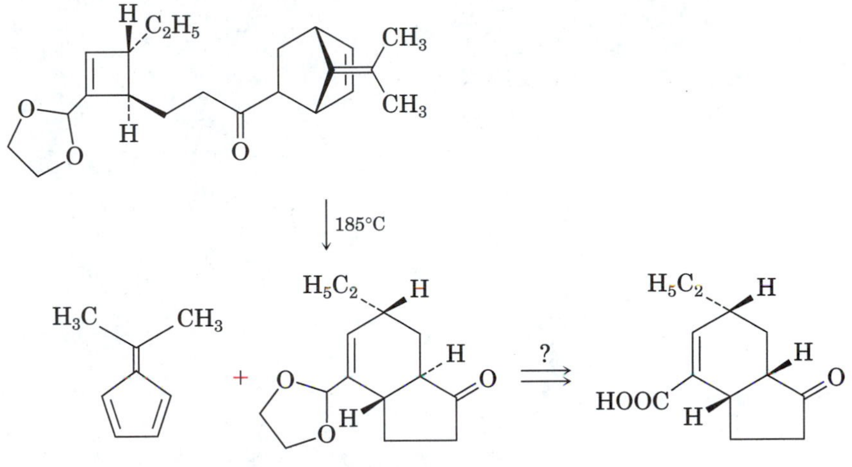

92%

Coronafacic acid

31.36 The following rearrangement of *N*-allyl-*N,N*-dimethylanilinium ion has been observed. Propose a mechanism to account for the reaction.

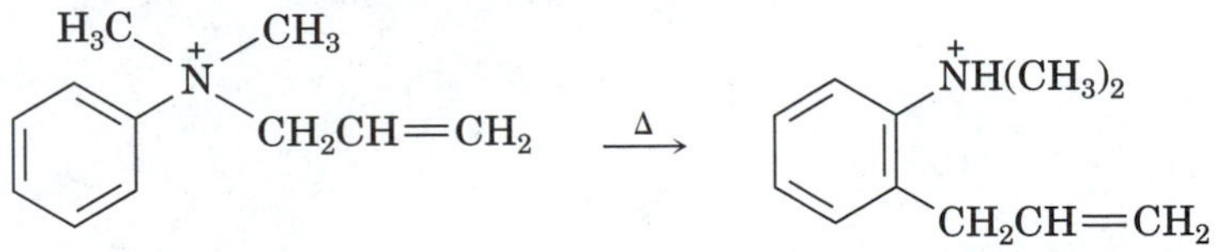

***N*-Allyl-*N,N*-dimethylanilinium ion** ***o*-Allyl-*N,N*-dimethylanilinium ion**

Appendix A Nomenclature of Polyfunctional Organic Compounds

Judging from the number of incorrect names that appear in the chemical literature, it's probably safe to say that relatively few practicing organic chemists are fully conversant with the rules of organic nomenclature. Simple hydrocarbons and monofunctional compounds present few difficulties because the basic rules for naming such compounds are logical and easy to understand. Problems, however, are often encountered with polyfunctional compounds. Whereas most chemists could correctly identify hydrocarbon **1** as 3-ethyl-2,5-dimethylheptane, rather few could correctly identify polyfunctional compound **2**. Should we consider **2** as an ether? As an ethyl ester? As a ketone? As an alkene? It is, of course, all four, but it has only one correct name: ethyl 3-(4-methoxy-2-oxo-3-cyclohexenyl)propanoate.

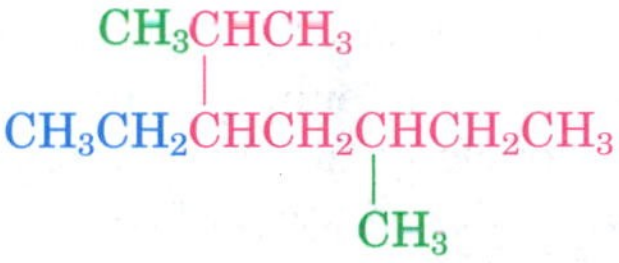

1. 3-Ethyl-2,5-dimethylheptane

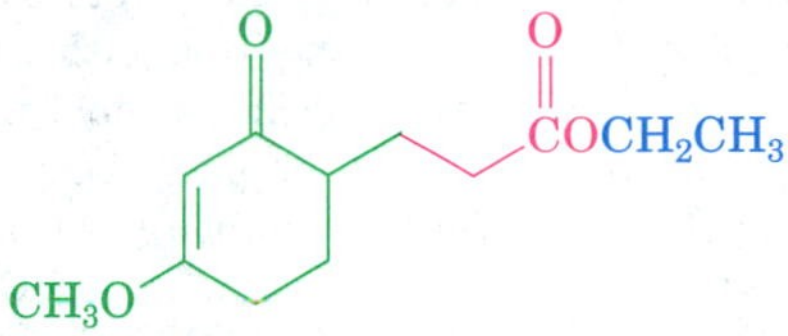

2. Ethyl 3-(4-methoxy-2-oxo-3-cyclohexenyl)propanoate

Naming polyfunctional organic compounds isn't really much harder than naming monofunctional ones. All that's required is a knowledge of nomenclature for monofunctional compounds and a set of additional rules. In the following discussion, it's assumed that you have a good command of the rules of nomenclature for monofunctional compounds that were given throughout the text as each new functional group was introduced. A list of where these rules can be found is shown in Table A.1.

Table A.1 Where to Find Nomenclature Rules for Simple Functional Groups

Functional group	*Text section*	*Functional group*	*Text section*
Acid anhydrides	21.1	Amines	24.1
Acid halides	21.1	Aromatic compounds	15.2
Alcohols	17.1	Carboxylic acids	20.1
Aldehydes	19.2	Cycloalkanes	3.7
Alkanes	3.4	Esters	21.1
Alkenes	6.3	Ethers	18.1
Alkyl halides	10.1	Ketones	19.2
Alkynes	8.2	Nitriles	21.1
Amides	21.1	Phenols	25.4

The name of a polyfunctional organic molecule has four parts:

1. **Suffix**—the part that identifies the principal functional-group class to which the molecule belongs.
2. **Parent**—the part that identifies the size of the main chain or ring.
3. **Substituent prefixes**—parts that identify what substituents are located on the main chain or ring.
4. **Locants**—numbers that tell where substituents are located on the main chain or ring.

To arrive at the correct name for a complex molecule, you must identify the four name parts and then express them in the proper order and format. Let's look at the four parts.

The Suffix—Functional-Group Precedence

A polyfunctional organic molecule can contain many different kinds of functional groups, but for nomenclature purposes, we must choose just one suffix. It's not correct to use two suffixes. Thus, keto ester **3** must be named either as a ketone with an *-one* suffix or as an ester with an *-oate* suffix but can't be named as an *-onoate*. Similarly, amino alcohol **4** must be named either as an alcohol (*-ol*) or as an amine (*-amine*) but can't properly be named as an *-olamine*. The only exception to this rule is in naming compounds that have double or triple bonds. For example, the unsaturated acid $H_2C{=}CHCH_2COOH$ is 3-butenoic acid, and the acetylenic alcohol $HC{\equiv}CCH_2CH_2CH_2CH_2OH$ is 5-hexyn-1-ol.

3. Named as an ester with a keto (oxo) substituent: **methyl 4-oxopentanoate**

4. Named as an alcohol with an amino substituent: **5-amino-2-pentanol**

How do we choose which suffix to use? Functional groups are divided into two classes, **principal groups** and **subordinate groups**, as shown in Table A.2. Principal groups are those that may be cited either as prefixes or as suffixes, whereas subordinate groups are those that may be cited only as prefixes. Within the principal groups, an order of precedence has been established. The proper suffix for a given compound is determined by identifying all of the functional groups present and then choosing the principal group of highest priority. For example, Table A.2 indicates that keto ester **3** must be named as an ester rather than as a ketone, since an ester functional group is higher in priority than a ketone is. Similarly, amino alcohol **4** must be named as an alcohol rather than as an amine. The correct name of **3** is methyl 4-oxopentanoate, and the correct name of **4** is 5-amino-2-pentanol. Further examples are shown below.

5. Named as a cyclohexanecarboxylic acid with an oxo substitutent: **4-oxocyclohexanecarboxylic acid**

Table A.2 Classification of Functional Groups for Purposes of Nomenclature[a]

Functional group	*Name as suffix*	*Name as prefix*
Principal groups		
Carboxylic acids	-oic acid -carboxylic acid	carboxy
Acid anhydrides	-oic anhydride -carboxylic anhydride	
Esters	-oate -carboxylate	alkoxycarbonyl
Acid halides	-oyl halide -carbonyl halide	halocarbonyl
Amides	-amide -carboxamide	amido
Nitriles	-nitrile -carbonitrile	cyano
Aldehydes	-al -carbaldehyde	oxo
Ketones	-one	oxo
Alcohols	-ol	hydroxy
Phenols	-ol	hydroxy
Thiols	-thiol	mercapto
Amines	-amine	amino
Imines	-imine	imino
Alkenes	-ene	alkenyl
Alkynes	-yne	alkynyl
Alkanes	-ane	alkyl
Subordinate groups		
Ethers		alkoxy
Sulfides		alkylthio
Halides		halo
Nitro		nitro
Azides		azido
Diazo		diazo

[a]Principal functional groups are listed in order of decreasing priority; subordinate functional groups have no established priority order.

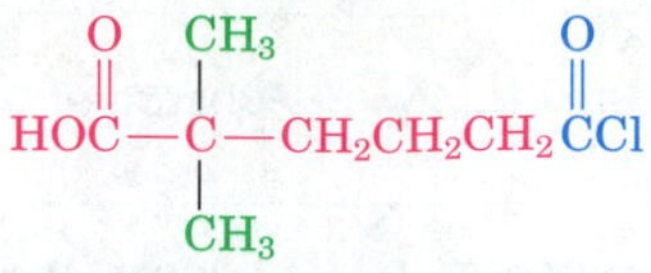

6. Named as a carboxylic acid with a chlorocarbonyl substituent:
5-chlorocarbonyl-2,2-dimethylpentanoic acid

7. Named as an ester with an oxo substituent:
methyl 5-methyl-6-oxohexanoate

The Parent—Selecting the Main Chain or Ring

The parent or base name of a polyfunctional organic compound is usually easy to identify. If the group of highest priority is *part of* an open chain, we simply select the longest chain that contains the largest number of principal functional groups. If the highest-priority group is *attached to* a ring, we use the name of that ring system as the parent. For example, compounds **8** and **9** are isomeric aldehydo acids, and both must be named as acids rather than as aldehydes according to Table A.2. The longest chain in compound **8** has seven carbons, and the substance is therefore named 6-methyl-7-oxoheptanoic acid. Compound **9** also has a chain of seven carbons, but the longest chain that contains both of the principal functional groups has only three carbons. The correct name of **9** is 3-oxo-2-pentylpropanoic acid.

8. Named as a substituted heptanoic acid:
6-methyl-7-oxoheptanoic acid

9. Named as a substituted propanoic acid:
3-oxo-2-pentylpropanoic acid

Similar rules apply for compounds **10–13**, which contain rings. Compounds **10** and **11** are isomeric keto nitriles, and both must be named as nitriles according to Table A.2. Substance **10** is named as a benzonitrile since the –CN functional group is a substituent on the aromatic ring, but substance **11** is named as an acetonitrile since the –CN functional group is part of an open chain. The correct names are 2-acetyl-4-methylbenzonitrile (**10**) and (2-acetyl-4-methylphenyl)acetonitrile (**11**). Compounds **12** and **13** are both keto acids and must be named as acids. The correct names are 3-(2-oxocyclohexyl)propanoic acid (**12**) and 2-(3-oxopropyl)cyclohexanecarboxylic acid (**13**).

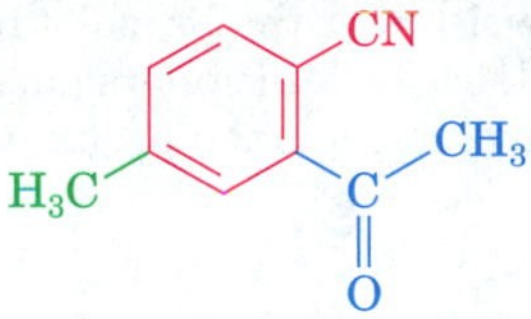

10. Named as a substituted benzonitrile:
2-acetyl-4-methylbenzonitrile

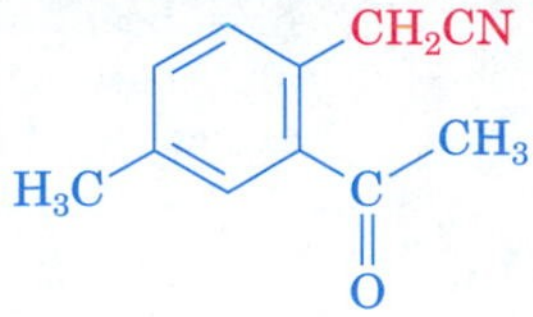

11. Named as a substituted acetonitrile:
(2-acetyl-4-methylphenyl)acetonitrile

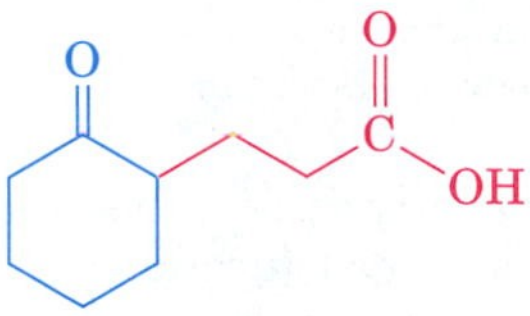

12. Named as a carboxylic acid:
3-(2-oxocyclohexyl)propanoic acid

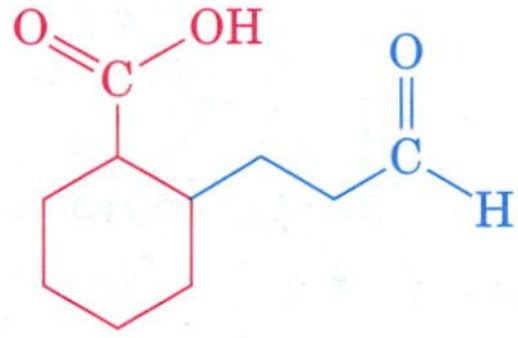

13. Named as a carboxylic acid:
2-(3-oxopropyl)cyclohexanecarboxylic acid

The Prefixes and Locants

With the suffix and parent name established, the next step is to identify and number all substituents on the parent chain or ring. These substituents include all alkyl groups and all functional groups other than the one cited in the suffix. For example, compound **14** contains three different functional groups (carboxyl, keto, and double bond). Because the carboxyl group is highest in priority, and because the longest chain containing the functional groups is seven carbons long, **14** is a heptenoic acid. In addition, the main chain has an oxo (keto) substituent and three methyl groups. Numbering from the end nearer the highest-priority functional group, we find that **14** is 2,5,5-trimethyl-4-oxo-2-heptenoic acid. Note that the final *-e* of heptene is deleted in the word *heptenoic*. This deletion occurs only when the name would have two adjacent vowels (thus, *heptenoic* has the final *e* deleted, but *heptenenitrile* retains the *-e*). Look back at some of the other compounds we've named to see other examples of how prefixes and locants are assigned.

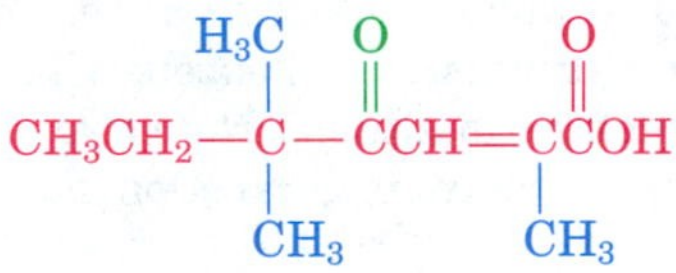

14. Named as a heptenoic acid:
2,5,5-trimethyl-4-oxo-2-heptenoic acid

Writing the Name

Once the name parts have been established, the entire name is written out. Several additional rules apply:

1. *Order of prefixes* When the substituents have been identified, the main chain has been numbered, and the proper multipliers such as *di-* and *tri-*

have been assigned, the name is written with the substituents listed in alphabetical, rather than numerical, order. Multipliers such as *di-* and *tri-* are not used for alphabetization purposes, but the prefix *iso-* is used.

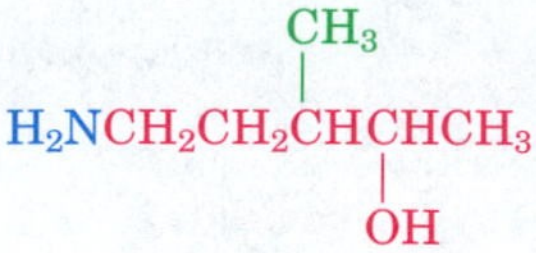

15. 5-Amino-3-methyl-2-pentanol
(not 3-methyl-5-amino-2-pentanol)

2. *Use of hyphens; single- and multiple-word names* The general rule in such cases is to determine whether the principal functional group is itself an element or compound. If it is, then the name is written as a single word; if it isn't, then the name is written as multiple words. For example, methylbenzene (one word) is correct because the parent—benzene—is itself a compound. Diethyl ether, however, is written as two words because the parent—ether—is a class name rather than a compound name. Some further examples are shown:

$H_3C—Mg—CH_3$

16. Dimethylmagnesium
(one word, since magnesium is an element)

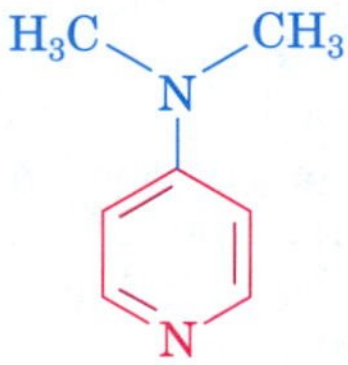

18. 4-(Dimethylamino)pyridine
(one word, since pyridine is a compound)

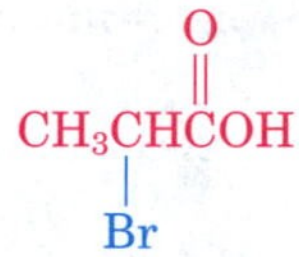

17. 2-Bromopropanoic acid
(two words, since "acid" is not a compound)

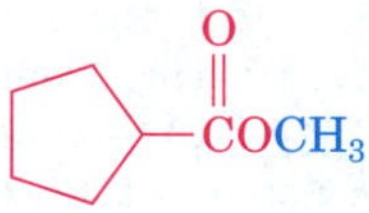

19. Methyl cyclopentanecarboxylate

3. *Parentheses* Parentheses are used to denote complex substituents when ambiguity would otherwise arise. For example, chloromethylbenzene has two substituents on a benzene ring, but (chloromethyl)benzene has only one complex substituent. Note that the expression in parentheses is not set off by hyphens from the rest of the name.

20. *p*-Chloromethylbenzene

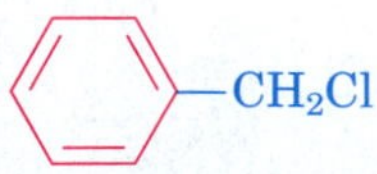

21. (Chloromethyl)benzene

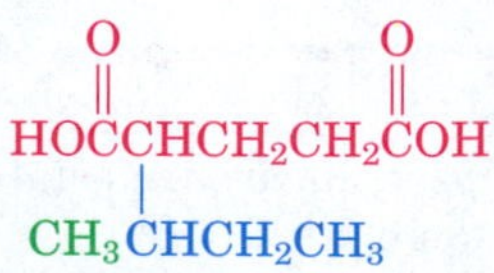

22. 2-(1-Methylpropyl)pentanedioic acid
(The 1-methylpropyl group is a complex substituent on C2 of the main chain.)

Additional Reading

Further explanations of the rules of organic nomenclature can be found in the following references:

1. "A Guide to IUPAC Nomenclature of Organic Compounds," CRC Press, Boca Raton, FL, 1993.
2. J. G. Traynham, "Organic Nomenclature: A Programmed Introduction," Prentice Hall, Englewood Cliffs, NJ, 1985.
3. "Nomenclature of Organic Chemistry, Sections A, B, C, D, E, F, and H," International Union of Pure and Applied Chemistry, Pergamon Press, Oxford, 1979.

Appendix B Acidity Constants for Some Organic Compounds

Compound	pK_a
CH_3SO_3H	−1.8
$CH(NO_2)_3$	0.1
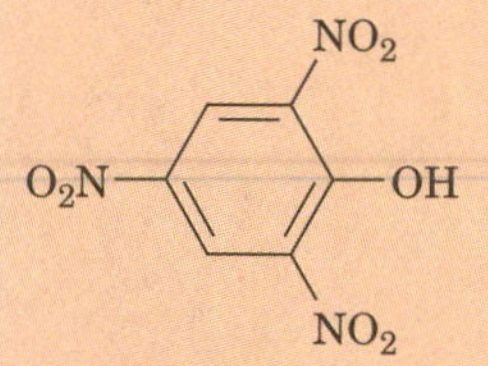	0.3
CCl_3COOH	0.5
CF_3COOH	0.5
CBr_3COOH	0.7
$HOOCC{\equiv}CCOOH$	1.2; 2.5
$HOOCCOOH$	1.2; 3.7
$CHCl_2COOH$	1.3
$CH_2(NO_2)COOH$	1.3
$HC{\equiv}CCOOH$	1.9
Z $HOOCCH{=}CHCOOH$	1.9; 6.3
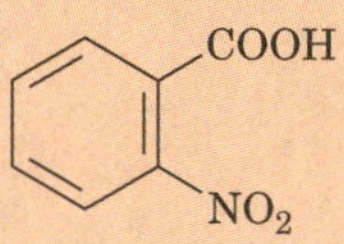	2.4
$CH_3COCOOH$	2.4
$NCCH_2COOH$	2.5
$CH_3C{\equiv}CCOOH$	2.6
CH_2FCOOH	2.7
$CH_2ClCOOH$	2.8
$HOOCCH_2COOH$	2.8; 5.6
$CH_2BrCOOH$	2.9
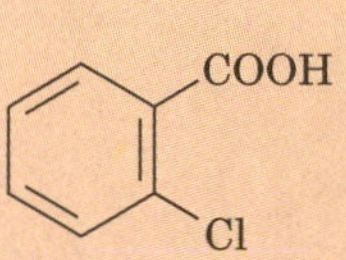	3.0
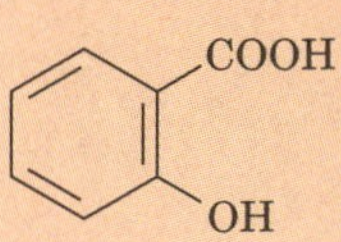	3.0

Compound	pK_a
CH_2ICOOH	3.2
$CHOCOOH$	3.2
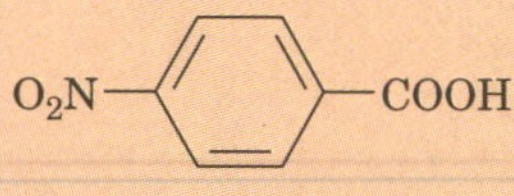	3.4
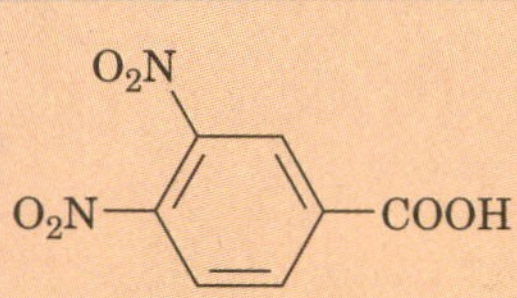	3.5
$HSCH_2COOH$	3.5; 10.2
$CH_2(NO_2)_2$	3.6
CH_3OCH_2COOH	3.6
CH_3COCH_2COOH	3.6; 7.8
$HOCH_2COOH$	3.7
$HCOOH$	3.7
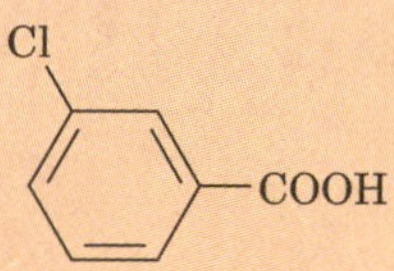	3.8
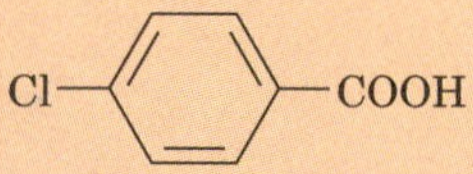	4.0
CH_2BrCH_2COOH	4.0
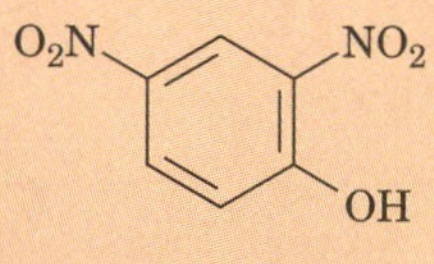	4.1
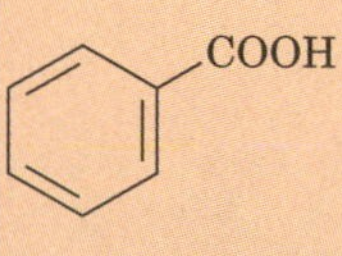	4.2
$H_2C{=}CHCOOH$	4.2
$HOOCCH_2CH_2COOH$	4.2; 5.7
$HOOCCH_2CH_2CH_2COOH$	4.3; 5.4

Compound	pK_a
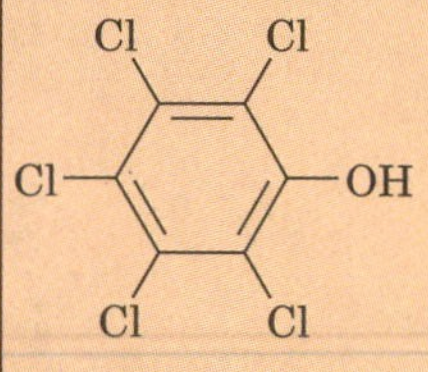	4.5
CH_3COOH	4.8
$H_2C{=}C(CH_3)COOH$	4.7
CH_3CH_2COOH	4.8
$(CH_3)_3CCOOH$	5.0
$CH_3COCH_2NO_2$	5.1
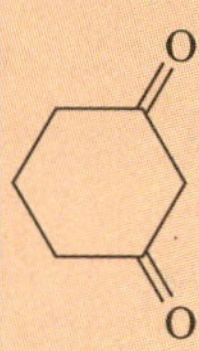	5.3
$O_2NCH_2COOCH_3$	5.8
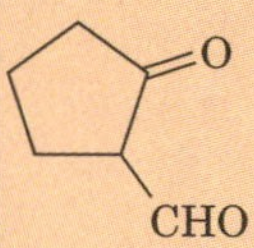	5.8
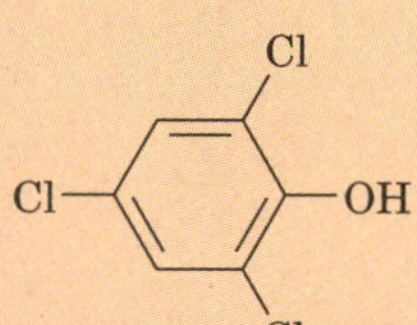	6.2
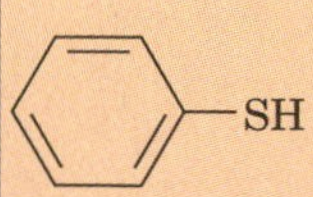	6.6
HCO_3H	7.1
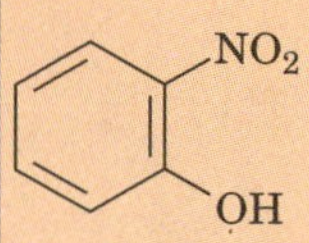	7.2

Compound	pK_a	Compound	pK_a	Compound	pK_a
$(CH_3)_2CHNO_2$	7.7	$CH_3COCH_2SOCH_3$	10.0	CH_3CH_2OH	16.0
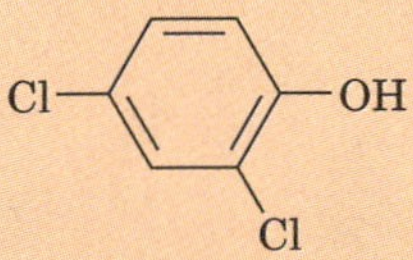	7.8	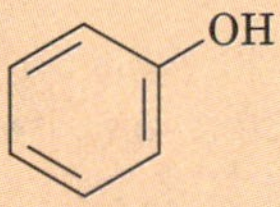	10.0	$CH_3CH_2CH_2OH$	16.1
CH_3CO_3H	8.2	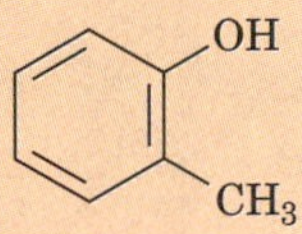	10.3	CH_3COCH_2Br	16.1
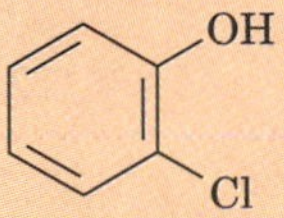	8.5	CH_3NO_2	10.3	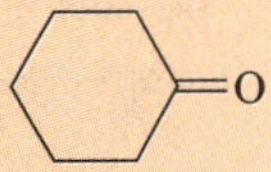	16.7
$CH_3CH_2NO_2$	8.5	CH_3SH	10.3	$(CH_3)_2CHOH$	17.1
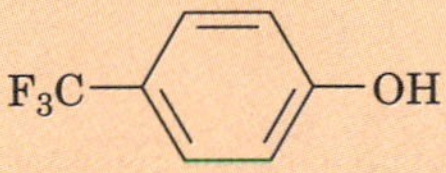	8.7	$CH_3COCH_2COOCH_3$	10.6	$(CH_3)_3COH$	18.0
$CH_3COCH_2COCH_3$	9.0	CH_3COCHO	11.0	CH_3COCH_3	19.3
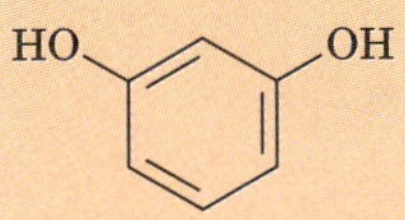	9.3; 11.1	$CH_2(CN)_2$	11.2	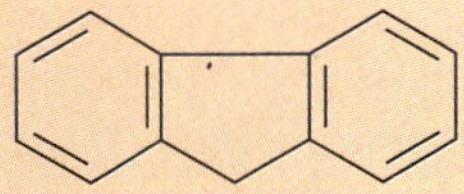	23
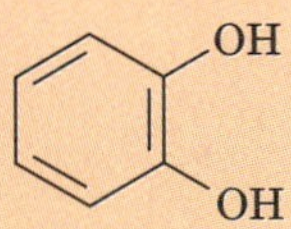	9.3; 12.6	CCl_3CH_2OH	12.2	$CH_3COOCH_2CH_3$	25
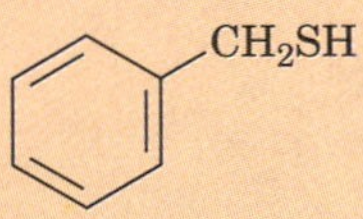	9.4	Glucose	12.3	$HC\equiv CH$	25
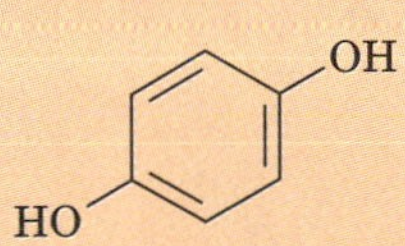	9.9; 11.5	$(CH_3)_2C=NOH$	12.4	CH_3CN	25
		$CH_2(COOCH_3)_2$	12.9	$CH_3SO_2CH_3$	28
		$CHCl_2CH_2OH$	12.9	$(C_6H_5)_3CH$	32
		$CH_2(OH)_2$	13.3	$(C_6H_5)_2CH_2$	34
		CH_3CHO	13.5	CH_3SOCH_3	35
		$(CH_3)_2CHCHO$	13.8	NH_3	35
		$HOCH_2CH(OH)CH_2OH$	14.1	$CH_3CH_2NH_2$	35
		CH_2ClCH_2OH	14.3	$(CH_3CH_2)_2NH$	36
		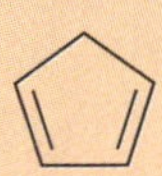	15.0	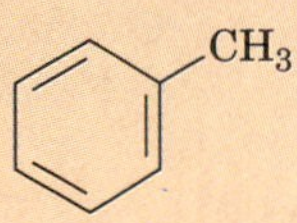	41
		CH_3OH	15.5	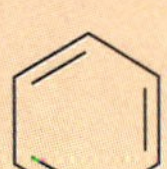	43
		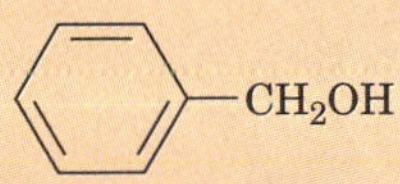	15.4	$H_2C=CH_2$	44
		$H_2C=CHCH_2OH$	15.5	CH_4	~60

An acidity list covering more than 5000 organic compounds has been published: E. P. Serjeant and B. Dempsey (eds.), "Ionization Constants of Organic Acids in Aqueous Solution," IUPAC Chemical Data Series No. 23, Pergamon Press, Oxford, 1979.

Appendix C
Reaction Mechanisms in Review

Appendix D
Glossary

Absolute configuration (Section 9.6): The exact three-dimensional structure of a chiral molecule. Absolute configurations are specified verbally by the Cahn–Ingold–Prelog *R*,*S* convention and are represented on paper by Fischer projections.

Absorption spectrum (Section 12.4): A plot of wavelength of incident light versus amount of light absorbed. Organic molecules show absorption spectra in both the infrared and the ultraviolet regions of the electromagnetic spectrum. By interpreting these spectra, useful structural information about the sample can be obtained.

Acetal (Section 19.14): A functional group consisting of two ether-like oxygen atoms bonded to the same carbon, $R_2C(OR')_2$. Acetals are often used as protecting groups for ketones and aldehydes because they are stable to basic and nucleophilic reagents but can be easily removed by acidic hydrolysis.

Acetyl group (Section 19.2): The CH_3CO- group.

Acetylide anion (Section 8.8): The anion formed by removal of a proton from a terminal alkyne.

Achiral (Section 9.2): Having a lack of handedness. A molecule is achiral if it has a plane of symmetry and is thus superimposable on its mirror image.

Acidity constant (K_a) (Section 2.6): A measure of acid strength. For any acid HA, the acidity constant is given by the expression

$$K_a = K_{eq}[H_2O] = \frac{[H_3O^+][A^-]}{[HA]}$$

Activating group (Section 16.5): An electron-donating group such as hydroxyl (–OH) or amino ($-NH_2$) that increases the reactivity of an aromatic ring toward electrophilic aromatic substitution. All activating groups are ortho- and para-directing.

Activation energy (Section 5.8): The difference in energy between ground state and transition state. The amount of activation energy required by a reaction determines the rate at which the reaction proceeds. Most organic reactions have activation energies of 40–100 kJ/mol.

Acyl group (Section 19.2): A –COR group.

Acylation (Section 16.4): The introduction of an acyl group, –COR, onto a molecule. For example, acylation of an alcohol yields an ester (R′OH → R′OCOR), acylation of an amine yields an amide ($R'NH_2$ → R′NHCOR), and acylation of an aromatic ring yields an alkyl aryl ketone (ArH → ArCOR).

Acylium ion (Section 16.4): A resonance-stabilized carbocation in which the positive charge is located at a carbonyl-group carbon, $R-\overset{+}{C}=O \leftrightarrow R-C\equiv O^+$. Acylium ions are strongly electrophilic and are involved as intermediates in Friedel–Crafts acylation reactions.

1,4 Addition (Sections 14.5, 19.17): Addition of a reagent to the ends of a conjugated π system. Conjugated dienes yield 1,4 adducts when treated with electrophiles such as HCl. Conjugated enones yield 1,4 adducts when treated with nucleophiles such as cyanide ion.

Addition reaction (Section 5.1): The reaction that occurs when two reactants add together to form a single new product with no atoms "left over."

Adrenocortical hormone (Section 28.7): A steroid hormone secreted by the adrenal glands. There are two types of adrenocortical hormones: mineralocorticoids and glucocorticoids.

Alcohol (Chapter 17 introduction): A compound with an –OH group bonded to a saturated, alkane-like carbon.

Aldaric acid (Section 26.8): The dicarboxylic acid resulting from oxidation of an aldose.

Aldehyde (Section 19.1): A compound containing the –CHO functional group.

Alditol (Section 26.8): The polyalcohol resulting from reduction of the carbonyl group of a sugar.

Aldonic acid (Section 26.8): The monocarboxylic acid resulting from mild oxidation of an aldose.

Aldose (Section 26.1): A carbohydrate with an aldehyde functional group.

Alicyclic (Section 3.6): An aliphatic cyclic hydrocarbon such as a cycloalkane or cycloalkene.

Aliphatic (Section 3.2): A nonaromatic hydrocarbon such as a simple alkane, alkene, or alkyne.

Alkaloid (Chapter 24 Interlude): A naturally occurring amine, such as morphine.

Alkane (Section 3.2): A compound of carbon and hydrogen that contains only single bonds.

Alkene (Chapter 6 introduction): A hydrocarbon that contains a carbon–carbon double bond.

Alkoxide ion (Section 17.4): The anion RO^- formed by deprotonation of an alcohol.

Alkyl group (Section 3.3): The partial structure that remains when a hydrogen atom is removed from an alkane.

Alkylation (Sections 16.3, 18.4, 22.8): Introduction of an alkyl group onto a molecule. For example, aromatic rings can be alkylated to yield arenes (ArH → ArR), alkoxide anions can be alkylated to yield ethers ($R'O^- \rightarrow R'OR$), and enolate anions can be alkylated to yield α-substituted carbonyl compounds ($R'_2C{=}C(R')O^- \rightarrow RR'_2C{-}COR'$).

Alkyne (Chapter 8 introduction): A hydrocarbon that contains a carbon–carbon triple bond.

Allyl group (Section 6.3): A $H_2C{=}CHCH_2{-}$ substituent.

Allylic (Section 10.5): The position next to a double bond. For example, $H_2C{=}CHCH_2Br$ is an allylic bromide, and an allylic radical is a conjugated, resonance-stabilized species in which the unpaired electron is in a *p* orbital next to a double bond (C=C–C· ↔ ·C–C=C).

Alpha (α) position (Chapter 22 introduction): The position next to a carbonyl group.

Amide (Chapter 21 introduction): A compound containing the $-CONR_2$ functional group.

Amine (Chapter 24 introduction): A compound containing one or more organic substituents bonded to a nitrogen atom, RNH_2, R_2NH, or R_3N.

α-Amino acid (Section 27.1): A difunctional compound with an $-NH_2$ group as a substituent on the carbon atom next to a $-COOH$ group, $RCH(NH_2)COOH$.

Amino sugar (Section 26.12): A sugar with one of its $-OH$ groups replaced by $-NH_2$.

Amplitude (Section 12.4): The height of a wave measured from the midpoint to the maximum. The intensity of radiant energy is proportional to the square of the amplitude of the wave.

Anabolism (Section 30.1): The group of metabolic pathways that build up larger molecules from smaller ones.

Androgen (Section 28.7): A male steroid sex hormone.

Angle strain (Section 4.4): The strain introduced into a molecule when a bond angle is deformed from its ideal value. Angle strain is particularly important in small-ring cycloalkanes, where it results from compression of bond angles to less than their ideal tetrahedral values.

Anomers (Section 26.6): Cyclic stereoisomers of sugars that differ only in their configuration at the hemiacetal (anomeric) carbon.

Antarafacial (Section 31.8): A word used to describe the geometry of pericyclic reactions. An antarafacial reaction is one that takes place on opposite faces of the two ends of a π electron system.

Anti conformation (Section 4.3): The geometric arrangement around a carbon–carbon single bond in which the two largest substituents are 180° apart as viewed in a Newman projection.

Anti stereochemistry (Section 7.2): Referring to opposite sides of a double bond or molecule. An anti addition reaction is one in which the two ends of the double bond are attacked from different sides. For example, addition of Br_2 to cyclohexene yields *trans*-1,2-dibromocyclohexane, the product of anti addition. An anti elimination reaction is one in which the two groups leave from opposite sides of the molecule.

Antiaromatic (Section 15.7): Referring to planar, conjugated molecules with $4n$ π electrons. Delocalization of the π electrons leads to an increase in energy.

Antibonding orbital (Section 1.7): A molecular orbital that is higher in energy than the atomic orbitals from which it is formed.

Anticodon (Section 29.14): A sequence of three bases on tRNA that read the codons on mRNA and bring the correct amino acids into position for protein synthesis.

Antisense strand (Section 29.13): The strand of double-helical DNA that does not contain the gene.

Apoenzyme (Section 27.15): The protein part of an enzyme that also contains a cofactor.

Arene (Section 15.2): An alkyl-substituted benzene.

Aromaticity (Chapter 15): The special characteristics of cyclic conjugated π electron systems that result from their electronic structures. These characteristics include unusual stability, the presence of a ring current in the ^1H NMR spectrum, and a tendency to undergo substitution reactions rather than addition reactions on treatment with electrophiles. Aromatic molecules are planar, cyclic, conjugated species that have $4n + 2$ π electrons.

Arylamine (Section 25.1): An amino-substituted aromatic compound, $ArNH_2$.

Atomic number (Z) (Section 1.1): The number of protons in the nucleus of an atom.

Atomic weight (Section 1.1): The average mass number of the atoms of an element.

Axial bond (Section 4.10): A bond to chair cyclohexane that lies along the ring axis perpendicular to the rough plane of the ring.

Axial bonds

Azo compound (Section 25.4): A compound with the general structure R–N=N–R′.

Basal metabolic rate (Chapter 30 Interlude): The minimum amount of energy per unit time an organism expends to stay alive.

Base peak (Section 12.1): The most intense peak in a mass spectrum.

Bent bonds (Section 4.7): The bonds in small rings such as cyclopropane that bend away from the internuclear line and overlap at a slight angle, rather than head-on. Bent bonds are highly strained and highly reactive.

Benzoyl group (Section 19.2): The C_6H_5CO- group.

Benzylic (Sections 11.9, 16.10): The position next to an aromatic ring. For example, a benzylic cation is a resonance-stabilized, conjugated carbocation having its positive charge located on a carbon atom next to the benzene ring in a π orbital that overlaps the aromatic π system.

Benzyne (Section 16.9): An unstable compound having a triple bond in a benzene ring. Benzynes are implicated as intermediates in certain nucleophilic aromatic substitution reactions of aryl halides with strong bases.

Betaine (Section 19.15): A neutral dipolar molecule with nonadjacent positive and negative charges. For example, the initial adduct of a Wittig reagent with a carbonyl compound is a betaine.

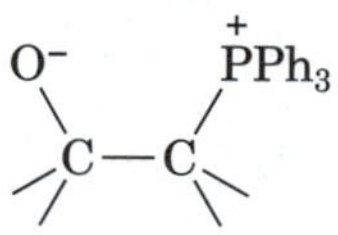

A Wittig betaine

Bimolecular reaction (Section 11.4): A reaction that occurs between two reagents.

Boat cyclohexane (Section 4.14): A conformation of cyclohexane that bears a slight resemblance to a boat. Boat cyclohexane has no angle strain, but has a large number of eclipsing interactions that make it less stable than chair cyclohexane.

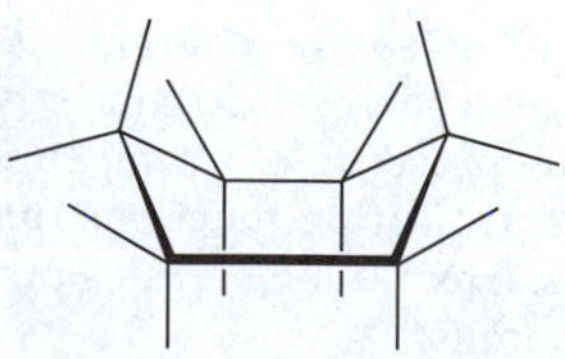

Boat cyclohexane

Bond angle (Section 1.8): The angle formed between two adjacent bonds.

Bond dissociation energy (Section 5.7): The amount of energy needed to break a bond homolytically and produce two radical fragments.

Bond length (Section 1.7): The equilibrium distance between the nuclei of two atoms that are bonded to each other.

Bond strength (Section 1.7): An alternative name for bond dissociation energy.

Bonding orbital (Section 1.7): A molecular orbital that is lower in energy than the atomic orbitals from which it is formed.

Branched-chain alkane (Section 3.2): An alkane that contains a branching connection of carbons as opposed to a straight-chain alkane.

Bromohydrin (Section 7.3): A 1,2-disubstituted bromoalcohol; obtained by addition of HOBr to an alkene.

Brønsted–Lowry acid (Section 2.6): A substance that donates a hydrogen ion (proton; H^+) to a base.

Brønsted–Lowry base (Section 2.6): A substance that accepts H^+ from an acid.

C-terminal amino acid (Section 27.6): The amino acid with a free –COOH group at the end of a protein chain.

Carbanion (Section 10.8): A carbon anion, or substance that contains a trivalent, negatively charged carbon atom ($R_3C{:}^-$). Carbanions are sp^3-hybridized and have eight electrons in the outer shell of the negatively charged carbon.

Carbene (Section 7.6): A neutral substance that contains a divalent carbon atom having only six electrons in its outer shell ($R_2C{:}$).

Carbinolamine (Section 19.12): A molecule that contains the $R_2C(OH)NH_2$ functional group. Carbinolamines are produced as intermediates during the nucleophilic addition of amines to carbonyl compounds.

Carbocation (Sections 5.5, 6.10): A carbon cation, or substance that contains a trivalent, positively charged carbon atom having six electrons in its outer shell (R_3C^+). Carbocations are planar and sp^2-hybridized.

Carbocycle (Section 15.9): A cyclic molecule that has only carbon atoms in its ring.

Carbohydrate (Section 26.1): A polyhydroxy aldehyde or ketone. Carbohydrates can be either simple sugars, such as glucose, or complex sugars, such as cellulose.

Carbonyl group (Section 2.1): The C=O functional group.

Carboxylation (Section 20.6): The addition of CO_2 to a molecule.

Carboxylic acid (Chapter 21 introduction): A compound containing the –COOH functional group.

Catabolism (Section 30.1): The group of metabolic pathways that break down larger molecules into smaller ones.

Chain-growth polymer (Section 21.11): A polymer whose bonds are produced by chain reactions. Polyethylene and other alkene polymers are examples.

Chain reaction (Section 5.3): A reaction that, once initiated, sustains itself in an endlessly repeating cycle of propagation steps. The radical chlorination of alkanes is an example of a chain reaction that is initiated by irradiation with light and then continues in a series of propagation steps.

Step 1.	Initiation:	$Cl_2 \rightarrow 2\ Cl\cdot$
Steps 2 and 3.	Propagation:	$Cl\cdot + CH_4 \rightarrow HCl + \cdot CH_3$
		$\cdot CH_3 + Cl_2 \rightarrow CH_3Cl + Cl\cdot$
Step 4.	Termination:	$R\cdot + R\cdot \rightarrow R–R$

Chair cyclohexane (Section 4.9): A three-dimensional conformation of cyclohexane that resembles the rough shape of a chair. The chair form of cyclohexane, which has neither angle strain nor eclipsing strain, represents the lowest-energy conformation of the molecule.

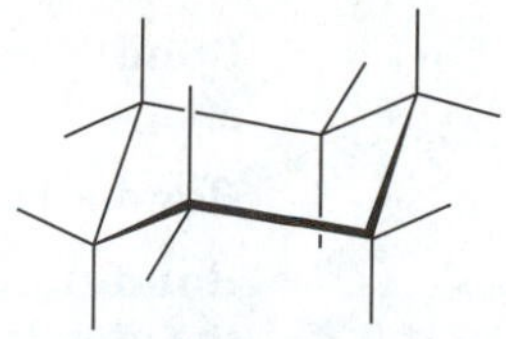

Chair cyclohexane

Chemical shift (Section 13.3): The position on the NMR chart where a nucleus absorbs. By convention, the chemical shift of tetramethylsilane (TMS) is arbitrarily set at zero, and all other absorptions usually occur downfield (to the left on the chart). Chemical shifts are expressed in delta units, δ, where 1 δ equals 1 ppm of the spectrometer operating frequency. The chemical shift of a given nucleus is related to the chemical environment of that nucleus in the molecule.

Chiral (Section 9.2): Having handedness. Chiral molecules are those that do not have a plane of symmetry and are therefore not superimposable on their mirror image. A chiral molecule thus exists in two forms, one right-handed and one left-handed. The most common (though not the only) cause of chirality in a molecule is the presence of a carbon atom that is bonded to four different substituents.

Chlorohydrin (Section 7.3): A 1,2-disubstituted chloroalcohol; obtained by addition of HOCl to an alkene.

Chromatography (Chapter 12 Interlude): A technique for separating a mixture of compounds into pure components. Chromatography operates on a principle of differential adsorption whereby different compounds adsorb to a stationary support phase and are then carried along at different rates by a mobile phase.

Cis–trans isomers (Sections 3.8, 6.5): Stereoisomers that differ in their stereochemistry about a double bond or ring. Cis–trans isomers are also called geometric isomers.

Citric acid cycle (Section 30.5): The metabolic pathway by which acetyl CoA is degraded to CO_2.

Coding strand (Section 29.13): The strand of double-helical DNA that contains the gene.

Codon (Section 29.14): A three-base sequence on a messenger RNA chain that encodes the genetic information necessary to cause a specific amino acid to be incorporated into a protein. Codons on mRNA are read by complementary anticodons on tRNA.

Coenzyme (Section 27.15): A small organic molecule that acts as a cofactor.

Cofactor (Section 27.15): A small nonprotein part of an enzyme that is necessary for biological activity.

Complex carbohydrate (Section 26.1): A carbohydrate that is made of two or more simple sugars linked together.

Concerted (Section 31.1): A reaction that takes place in a single step without intermediates. For example, the Diels–Alder cycloaddition reaction is a concerted process.

Condensed structure (Section 2.9): A shorthand way of writing structures in which carbon–hydrogen and carbon–carbon bonds are understood rather than shown explicitly. Propane, for example, has the condensed structure $CH_3CH_2CH_3$.

Configuration (Section 9.6): The three-dimensional arrangement of atoms bonded to a stereogenic center.

Conformation (Section 4.1): An exact three-dimensional shape of a molecule at any given instant, assuming that rotation around single bonds is frozen.

Conformational analysis (Section 4.13): A means of assessing the energy of a substituted cycloalkane by totaling the steric interactions present in the molecule. Conformational analysis is particularly useful in assessing the relative stabilities of different conformations of substituted cyclohexane rings.

Conjugate acid (Section 2.6): The product that results from protonation of a Brønsted base.

Conjugate addition (Section 19.17): Addition of a nucleophile to the β carbon atom of an α,β-unsaturated carbonyl compound.

Conjugate base (Section 2.6): The anion that results from deprotonation of a Brønsted acid.

Conjugated protein (Section 27.13): A protein that yields other compounds, such as carbohydrates, fats, or nucleic acids, in addition to amino acids on hydrolysis.

Conjugation (Section 14.1): A series of overlapping *p* orbitals, usually in alternating single and multiple bonds. For example, 1,3-butadiene is a conjugated diene, 3-buten-2-one is a conjugated enone, and benzene is a cyclic conjugated triene.

Conrotatory (Section 31.5): A term used to indicate the fact that *p* orbitals must rotate in the same direction during electrocyclic ring opening or ring closure.

Constitutional isomers (Sections 3.2, 9.12): Isomers that have their atoms connected in a different order. For example, butane and 2-methylpropane are constitutional isomers.

Coupling (Section 13.7): A magnetic interaction of the nuclear spins of nearby atoms in a molecule.

Coupling constant (*J*) (Section 13.7): The magnitude (expressed in hertz) of the interaction between nuclei whose spins are coupled.

Covalent bond (Section 1.6): A bond formed by sharing electrons between atoms.

Cracking (Chapter 3 Interlude): A process used in petroleum refining in which large alkanes are thermally cracked into smaller fragments.

Cycloaddition (Sections 14.8, 31.7): A pericyclic reaction in which two reactants add together in a single step to yield a cyclic product. The Diels–Alder reaction between a diene and a dienophile to give a cyclohexene is the best-known example of a cycloaddition.

Cycloalkane (Section 3.6): An alkane that contains a ring of carbons.

D sugar (Section 26.3): A sugar whose hydroxyl group at the stereogenic center farthest from the carbonyl group points to the right when drawn in Fischer projection.

***d,l* form** (Section 9.10): A shorthand way of indicating the racemic modification of a compound.

Deactivating group (Section 16.5): An electron-withdrawing substituent that decreases the reactivity of an aromatic ring toward electrophilic aromatic substitution. Most deactivating groups, such as nitro, cyano, and carbonyl, are meta-directors, but halogen substituents are ortho- and para-directors.

Decarboxylation (Section 22.8): A reaction that involves loss of carbon dioxide from the starting material. β-Keto acids decarboxylate particularly readily on heating.

Degenerate orbitals (Section 15.6): Two or more orbitals that have the same energy level.

Degree of unsaturation (Section 6.2): The number of rings and/or multiple bonds in a molecule.

Dehydration (Sections 7.1, 17.8): A reaction that involves loss of water from an alcohol. Most alcohols can be dehydrated to yield alkenes, but aldol condensation products (β-hydroxy ketones) dehydrate particularly readily.

Dehydrohalogenation (Sections 7.1, 11.10): A reaction that involves loss of HX from an alkyl halide. Alkyl halides undergo dehydrohalogenation to yield alkenes on treatment with strong base.

Delocalization (Section 10.6): A spreading out of electron density over a conjugated π electron system. For example, allylic cations and allylic anions are delocalized because their charges are spread out over the entire π electron system.

Delta scale (Section 13.3): An arbitrary scale used to calibrate NMR charts. One delta unit (δ) is equal to 1 part per million (ppm; one-millionth) of the spectrometer operating frequency.

Denaturation (Section 27.16): The physical changes that occur in a protein when secondary and tertiary structures are disrupted. Denaturation is usually brought about by heat treatment or by a change in pH and is accompanied by a loss of biological activity.

Deoxy sugar (Section 26.12): A sugar with one of its oxygen atoms "missing." In other words, an –OH group is replaced by an –H.

DEPT-NMR (Section 13.12): An NMR method for distinguishing among signals due to CH_3, CH_2, CH, and quaternary carbons. That is, the number of hydrogens attached to each carbon can be determined.

Deshielding (Section 13.2): An effect observed in NMR that causes a nucleus to absorb downfield (to the left) of tetramethylsilane (TMS) standard. Deshielding is caused by a withdrawal of electron density from the nucleus and is responsible for the observed chemical shifts of vinylic and aromatic protons.

Deuterium isotope effect (Section 11.13): A tool used in mechanistic investigations to establish whether a C–H bond is broken in the rate-limiting step of a reaction.

Dextrorotatory (Section 9.3): A word used to describe an optically active substance that rotates the plane of polarization of plane-polarized light in a right-handed (clockwise) direction. The direction of rotation is not related to the absolute configuration of the molecule.

Diastereomer (Section 9.7): A term that indicates the relationship between non-mirror-image stereoisomers. Diastereomers are stereoisomers that have the same configuration at one or more stereogenic centers but differ at other stereogenic centers.

1,3-Diaxial interaction (Section 4.12): The strain energy caused by a steric interaction between axial groups three carbon atoms apart in chair cyclohexane.

Diazonium salt (Section 25.4): A compound with the general structure $RN_2^+ X^-$.

Diazotization (Section 25.3): The conversion of a primary amine, RNH_2, into a diazonium ion, RN_2^+ by treatment with nitrous acid. Aryl diazonium salts are stable, but alkyl diazonium salts are extremely reactive and are rarely isolable.

Dielectric constant (Section 11.9): A measure of the ability of a solvent to act as an insulator of electric charge. Solvents that have high dielectric constants are highly polar and are particularly valuable in S_N1 reactions because of their ability to stabilize the developing positive charge of the intermediate carbocation.

Dienophile (Section 14.8): A compound containing a double bond that can take part in the Diels–Alder cycloaddition reaction. The most reactive dienophiles are those that have electron-withdrawing groups such as nitro, cyano, or carbonyl on the double bond.

Digestion (Section 30.1): The first stage of catabolism, in which food is broken down by hydrolysis of ester, glycoside (acetal), and peptide (amide) bonds to yield fatty acids, simple sugars, and amino acids.

Dipolar molecule (Section 2.3): A molecule that is neutral overall but has plus and minus charges on individual atoms.

Dipole moment, μ (Section 2.2): A measure of the net polarity of a molecule. A dipole moment arises when the centers of mass of positive and negative charges within a molecule do not coincide.

Disrotatory (Section 31.5): A term used to indicate the fact that *p* orbitals rotate in opposite directions during electrocyclic ring opening or ring closing.

Disulfide (Section 17.12): A compound of the general structure RSSR′.

DNA (Section 29.9): Deoxyribonucleic acid; the biopolymer consisting of deoxyribonucleotide units linked together through phosphate–sugar bonds. Found in the nucleus of cells, DNA contains the genetic information of an organism.

Doublet (Section 13.7): A two-line NMR absorption caused by spin–spin splitting when the spin of the nucleus under observation couples with the spin of a neighboring magnetic nucleus.

Downfield (Section 13.3): Referring to the left-hand portion of the NMR chart.

Eclipsed conformation (Section 4.1): The geometric arrangement around a carbon–carbon single bond in which the bonds to substituents on one carbon are parallel to the bonds to substituents on the neighboring carbon as viewed in a Newman projection.

Eclipsed conformation

Eclipsing strain (Section 4.1): The strain energy in a molecule caused by electron repulsions between eclipsed bonds. Eclipsing strain is also called torsional strain.

Electrocyclic reaction (Section 31.4): A unimolecular pericyclic reaction in which a ring is formed or broken by a concerted reorganization of electrons through a cyclic transition state. For example, the cyclization of 1,3,5-hexatriene to yield 1,3-cyclohexadiene is an electrocyclic reaction.

Electromagnetic spectrum (Section 12.4): The range of electromagnetic energy, including infrared, ultraviolet, and visible radiation.

Electron affinity (Section 1.5): The measure of the tendency of an atom to gain an electron and form an anion. Elements on the right side of the periodic table, such as the halogens, have higher electron affinities than do elements on the left side.

Electron-dot structure (Section 1.6): A representation of a molecule showing covalent bonds as a pair of electron dots between atoms.

Electronegativity (Section 2.1): The ability of an atom to attract electrons in a covalent bond. As a rule, electronegativity increases in going across the periodic table from right to left and in going from bottom to top.

Electrophile (Section 5.4): An "electron-lover," or substance that accepts an electron pair from a nucleophile in a polar bond-forming reaction.

Electrophilic addition reaction (Section 6.8): The addition of an electrophile to an alkene to yield a saturated product.

Electrophilic aromatic substitution (Chapter 16 introduction): A general reaction in which an electrophile (E^+) reacts with an aromatic ring and substitutes for one of the ring hydrogens.

Electrophoresis (Section 27.3): A technique used for separating charged organic molecules, particularly proteins and amino acids. The mixture to be separated is placed on a buffered gel or paper, and an electric potential is applied across the ends of the apparatus. Negatively charged molecules migrate toward the positive electrode, and positively charged molecules migrate toward the negative electrode.

Elimination reaction (Section 5.1): What occurs when a single reactant splits into two products.

Elution (Chapter 12 Interlude): The removal of a substance from a chromatography column.

Embden–Meyerhof pathway (Section 30.3): An alternative name for glycolysis.

Enantiomers (Sections 9.1, 9.5): Stereoisomers of a chiral substance that have a mirror-image relationship. Enantiomers must have opposite configurations at all stereogenic centers in the molecule.

Endothermic (Section 5.6): A term used to describe reactions that absorb heat and therefore have positive enthalpy changes. In reaction energy diagrams, the products of endothermic reactions have higher energy levels than the reactants.

Enol (Section 8.5): A vinylic alcohol that is in equilibrium with a carbonyl compound.

Entgegen (*E*) (Section 6.6): A term used to describe the stereochemistry of a carbon–carbon double bond. The two groups on each carbon are assigned priorities according to the Cahn–Ingold–Prelog sequence rules, and the two carbons are then compared. If the high-priority groups on each carbon are on opposite sides of the double bond, the bond has *E* geometry.

Enthalpy change, ΔH (Section 5.6): The heat of reaction. The enthalpy change that occurs during a reaction is a measure of the difference in total bond energy between reactants and products.

Entropy change, ΔS (Section 5.6): The change in amount of disorder. The entropy change that occurs during a reaction is a measure of the difference in disorder between reactants and products.

Enzyme (Section 27.15): A biological catalyst. Enzymes are large proteins that catalyze specific biochemical reactions.

Epoxide (Section 18.7): A three-membered-ring ether functional group.

Equatorial bond (Section 4.10): A bond to cyclohexane that lies along the rough equator of the ring.

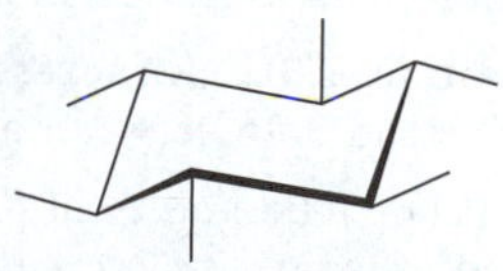

Equatorial bonds

Equilibrium constant, K_{eq} (Sections 2.6, 5.6): A measure of the equilibrium position for a reaction. The equilibrium constant for the reaction $aA + bB \rightarrow cC + dD$ is given by the expression

$$K_{eq} = \frac{[C]^c[D]^d}{[A]^a[B]^b}$$

where the numbers in brackets refer to the molar concentrations of the reactants and products.

Essential oil (Section 28.5): The volatile oil obtained by steam distillation of a plant extract.

Ester (Chapter 21 introduction): A compound containing the –COOR functional group.

Estrogen (Section 28.7): A female steroid sex hormone.

Ether (Chapter 18 introduction): A compound that has two organic substituents bonded to the same oxygen atom, ROR′.

Exon (Section 29.13): A section of DNA that contains genetic information.

Exothermic (Section 5.6): A term used to describe reactions that release heat and therefore have negative enthalpy changes. On reaction energy diagrams, the products of exothermic reactions have energy levels lower than those of reactants.

Fat (Section 28.1): A solid triacylglycerol derived from animal sources.

Fatty acid (Section 28.1): A long, straight-chain carboxylic acid found in fats and oils.

Fibrous protein (Section 27.13): A protein that consists of polypeptide chains arranged side by side in long threads. Such proteins are tough, insoluble in water, and are used in nature for structural materials such as hair, hooves, and fingernails.

Fingerprint region (Section 12.6): The complex region of the infrared spectrum from 1500 cm^{-1} to 400 cm^{-1}. If two substances have identical absorption patterns in the fingerprint region of the IR, they are almost certainly identical.

First-order reaction (Section 11.7): A reaction whose rate-limiting step is unimolecular and whose kinetics therefore depend on the concentration of only one reactant.

Fischer projection (Sections 9.13, 26.2): A means of depicting the absolute configuration of chiral molecules on a flat page. A Fischer projection uses a cross to represent the stereogenic center. The horizontal arms of the cross represent bonds coming out of the plane of the page, and the vertical arms of the cross represent bonds going back into the plane of the page.

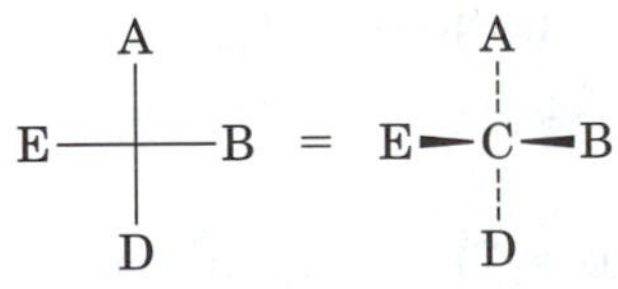

Fischer projection

Formal charge (Section 2.3): The difference in the number of electrons owned by an atom in a molecule and by the same atom in its elemental state. The formal charge on an atom is given by the following formula:

$$\text{Formal charge} = \begin{pmatrix}\text{Number of outer-shell} \\ \text{electrons in free atom}\end{pmatrix} - \begin{pmatrix}\text{Number of outer-shell} \\ \text{electrons in bonded atom}\end{pmatrix}$$

Frequency (Section 12.4): The number of electromagnetic wave cycles that travel past a fixed point in a given unit of time. Frequencies are expressed in units of cycles per second, or hertz.

Frontier orbitals (Section 31.1): The highest occupied (HOMO) and lowest unoccupied (LUMO) molecular orbitals.

Functional group (Section 3.1): An atom or group of atoms that is part of a larger molecule and that has a characteristic chemical reactivity. Functional groups display the same chemistry in all molecules of which they are a part.

Furanose (Section 26.5): The five-membered-ring form of a simple sugar.

Gauche conformation (Section 4.3): The conformation of butane in which the two methyl groups lie 60° apart as viewed in a Newman projection. This conformation has 3.8 kJ/mol steric strain.

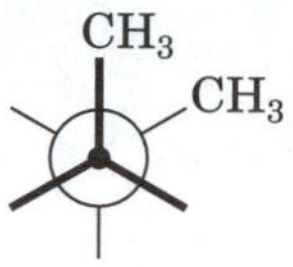

Gauche conformation

Geminal (Section 19.8): Referring to two groups attached to the same carbon atom. For example, 1,1-dibromopropane is a geminal dibromide.

Geometric isomers (Section 3.8, 6.5): An old term for cis–trans isomers.

Gibbs free-energy change, ΔG (Section 5.6): The free-energy change that occurs during a reaction. The Gibbs free-energy change for a reaction takes both enthalpy and entropy into account and is given by the formula $\Delta G = \Delta H - T\Delta S$.

Gilman reagent (Section 10.9): A diorganocopper reagent, R_2CuLi.

Globular protein (Section 27.13): A protein that is coiled into a compact, nearly spherical shape. These proteins, which are generally water-soluble and mobile within the cell, are the structural class to which enzymes belong.

Gluconeogenesis (Section 30.8): The anabolic pathway by which organisms make glucose from simple precursors.

Glycol (Section 7.8): A diol, such as ethylene glycol, $HOCH_2CH_2OH$.

Glycolysis (Section 30.3): A series of ten enzyme-catalyzed reactions that break down glucose into 2 equivalents of pyruvate, $CH_3COCO_2^-$.

Glycoside (Section 26.8): A cyclic acetal formed by reaction of a sugar with another alcohol.

Grignard reagent (Section 10.8): An organomagnesium halide, RMgX.

Ground state (Section 1.3): The most stable, lowest-energy electron configuration of a molecule.

Haloform (Section 22.8): A trihalomethane, such as $CHCl_3$, $CHBr_3$, or CHI_3.

Halohydrin (Section 7.3): A 1,2-disubstituted haloalcohol, such as is obtained on addition of HOBr to an alkene.

Halonium ion (Section 7.2): A species containing a positively charged, divalent halogen. Three-membered-ring bromonium ions are implicated as intermediates in the electrophilic addition of Br_2 to alkenes.

Hammond postulate (Section 6.11): A postulate stating that we can get a picture of what a given transition state looks like by looking at the structure of the nearest stable species. Exothermic reactions have transition states that resemble reactant; endothermic reactions have transition states that resemble product.

Haworth projection (Section 26.5): A means of viewing stereochemistry in cyclic hemiacetal forms of sugars. Haworth projections are drawn so that the ring is flat and is viewed from an oblique angle with the hemiacetal oxygen at the upper right.

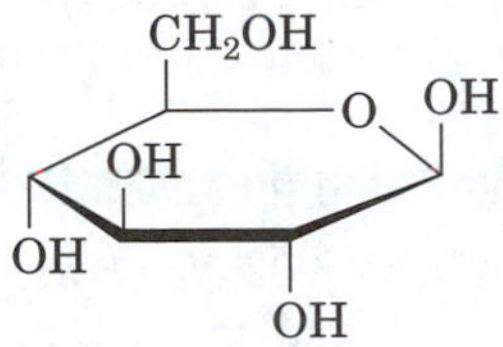

Haworth projection of glucose

Heat of combustion (Section 4.5): The amount of heat released when a compound is burned in a calorimeter, according to the equation

$$C_nH_m + O_2 \longrightarrow n\ CO_2 + \frac{m}{2}\ H_2O$$

Heat of hydrogenation (Section 6.7): The amount of heat released when a carbon–carbon double bond is hydrogenated. Comparison of heats of hydrogenation for different alkenes allows one to determine the stability of the different double bonds.

Heat of reaction (Section 5.6): An alternative name for the enthalpy change, ΔH, in a reaction.

Henderson–Hasselbalch equation (Section 27.3): An equation for determining the extent of deprotonation of a weak acid at various pH values.

Heterocycle (Sections 15.9, 29.1): A cyclic molecule whose ring contains more than one kind of atom. For example, pyridine is a heterocycle that contains five carbon atoms and one nitrogen atom in its ring.

Heterogenic bond formation (Section 5.2): What occurs when one reaction partner donates both electrons in forming a new bond. Polar reactions always involve heterogenic bond formation: $A^+ + B{:}^- \rightarrow A{:}B$

Heterolytic bond breakage (Section 5.2): The kind of bond breaking that occurs in polar reactions when one fragment leaves with both of the bonding electrons, as in the equation: $A{:}B \rightarrow A^+ + B{:}^-$

Holoenzyme (Section 27.15): The combination of apoenzyme plus cofactor.

HOMO (Sections 14.12, 31.3): An acronym for highest occupied molecular orbital. The symmetries of the HOMO and LUMO are important in pericyclic reactions.

Homogenic bond formation (Section 5.2): What occurs in radical reactions when each reactant donates one electron to the new bond: $A\cdot + B\cdot \rightarrow A{:}B$

Homolytic bond breakage (Section 5.2): The kind of bond breaking that occurs in radical reactions when each fragment leaves with one bonding electron, according to the equation: $A{:}B \rightarrow A\cdot + B\cdot$

Hückel's rule (Section 15.7): A rule stating that monocyclic conjugated molecules having $4n + 2$ π electrons (n = an integer) show the unusual stability associated with aromaticity.

Hund's rule (Section 1.3): If two or more empty orbitals of equal energy are available, one electron occupies each, with their spins parallel until all are half-full.

Hybrid orbital (Section 1.8): An orbital derived from a combination of ground-state (s, p, d) atomic orbitals. Hybrid orbitals, such as the sp^3, sp^2, and sp hybrids of carbon, are strongly directed and form stronger bonds than atomic orbitals do.

Hydration (Section 7.4): Addition of water to a molecule, such as occurs when alkenes are treated with aqueous sulfuric acid to give alcohols.

Hydride shift (Section 6.12): The shift of a hydrogen atom and its electron pair to a nearby cationic center.

Hydroboration (Section 7.5): Addition of borane (BH_3) or an alkylborane to an alkene. The resultant trialkylborane products are useful synthetic intermediates that can be oxidized to yield alcohols.

Hydrocarbon (Section 3.2): A compound that contains only carbon and hydrogen.

Hydrogen bond (Section 17.3): A weak (20 kJ/mol) attraction between a hydrogen atom bonded to an electronegative atom and an electron lone pair on another atom. Hydrogen bonding plays an important role in determining the secondary structure of proteins and in stabilizing the DNA double helix.

Hydrogenation (Section 7.7): Addition of hydrogen to a double or triple bond to yield a saturated product.

Hydroquinone (Section 25.8): A 1,4-dihydroxybenzene.

Hyperconjugation (Section 6.7): A weak interaction that results from overlap of a vacant *p* orbital on one atom with a neighboring σ bond. Hyperconjugation is important in stabilizing carbocations and in stabilizing substituted alkenes.

Inductive effect (Sections 2.1, 6.10, 16.5): The electron-attracting or electron-withdrawing effect that is transmitted through σ bonds. Electronegative elements have an electron-withdrawing inductive effect, whereas electropositive elements have an electron-donating inductive effect.

Infrared spectroscopy (Section 12.4): A kind of optical spectroscopy that uses infrared energy. IR spectroscopy is particularly useful in organic chemistry for determining the kinds of functional groups present in molecules.

Initiator (Section 5.3): A substance with an easily broken bond that is used to initiate a radical chain reaction. For example, radical chlorination of alkanes is initiated when light energy breaks the weak Cl–Cl bond to form Cl· radicals.

Integration (Section 13.6): A technique for measuring the area under an NMR peak to determine the relative number of each kind of proton in a molecule. Integrated peak areas are superimposed over the spectrum as a "stair-step" line, with the height of each step proportional to the area underneath the peak, and therefore proportional to the relative number of protons causing the peak.

Intermediate (Section 5.9): A species that is formed during the course of a multistep reaction but is not the final product. Intermediates are more stable than transition states, but may or may not be stable enough to isolate.

Intramolecular, intermolecular (Section 23.7): Reactions that occur within the same molecule are intramolecular; reactions that occur between two molecules are intermolecular.

Intron (Section 29.13): A nonsense section of DNA that does not contain genetic information.

Ion pair (Section 11.8): A loose complex between two ions in solution. Ion pairs are implicated as intermediates in S_N1 reactions to account for the partial retention of stereochemistry that is often observed.

Ionic bond (Section 1.5): A bond between two ions due to the electrical attraction of unlike charges. Ionic bonds are formed between strongly electronegative elements (such as the halogens) and strongly electropositive elements (such as the alkali metals).

Ionization energy (Section 1.5): The amount of energy required to remove an electron from an isolated atom in the gas phase. Elements on the far right of the periodic table have high ionization energies, and elements on the far left of the periodic table have low ionization energies.

Isoelectric point (Section 27.3): The pH at which the number of positive charges and the number of negative charges on a protein or an amino acid are equal.

Isomers (Section 3.2): Compounds that have the same molecular formula but have different structures.

Isoprene rule (Section 28.6): An observation to the effect that terpenoids appear to be made up of isoprene (2-methyl-1,3-butadiene) units connected in a head-to-tail fashion. Monoterpenes have two isoprene units, sesquiterpenes have three isoprene units, diterpenes have four isoprene units, and so on.

IUPAC nomenclature (Section 3.4): Rules for naming compounds, devised by the International Union of Pure and Applied Chemistry.

Kekulé structure (Section 1.6): A method of representing molecules in which a line between atoms indicates a bond.

Ketone (Section 19.1): A compound with two organic substituents bonded to a carbonyl group, $R_2C{=}O$.

Ketose (Section 26.1): A carbohydrate with a ketone functional group.

Kinetic control (Section 14.6): Reactions that follow the lowest activation energy pathway are said to be kinetically controlled. The product formed in a kinetically controlled reaction is the one that is formed most rapidly, but is not necessarily the most stable.

Kinetics (Section 11.3): Referring to reaction rates. Kinetic measurements can be extremely important in helping to determine reaction mechanisms.

Krebs cycle (Section 30.5): An alternative name for the citric acid cycle, by which acetyl CoA is degraded to CO_2.

L **sugar** (Section 26.3): A sugar whose hydroxyl group at the stereogenic center farthest from the carbonyl group points to the left when drawn in Fischer projection.

Lactam (Section 21.8): A cyclic amide.

Lactone (Section 21.7): A cyclic ester.

Leaving group (Section 11.5): The group that is replaced in a substitution reaction. The best leaving groups in nucleophilic substitution reactions are those that form the most stable, least basic, anions.

Levorotatory (Section 9.3): Used to describe an optically active substance that rotates the plane of polarization of plane-polarized light in a left-handed (counterclockwise) direction.

Lewis acid (Section 2.8): A substance having a vacant low-energy orbital that can accept an electron pair from a base. All electrophiles are Lewis acids.

Lewis base (Section 2.8): A substance that donates an electron lone pair to an acid. All nucleophiles are Lewis bases.

Lewis structure (Section 1.6): A representation of a molecule showing covalent bonds as a pair of electron dots between atoms.

Line-bond structure (Section 1.6): A representation of a molecule showing covalent bonds as lines between atoms.

Lipid (Section 28.1): A naturally occurring substance isolated from cells and tissues by extraction with nonpolar solvents. Lipids belong to many different structural classes, including fats, terpenes, prostaglandins, and steroids.

Lipid bilayer (Section 28.3): The ordered lipid molecules that form a cell membrane.

Lone-pair electrons (Section 1.6): Nonbonding electron pairs that occupy valence orbitals. It is the lone-pair electrons that are used by nucleophiles in their reactions with electrophiles.

LUMO (Sections 14.11, 31.3): An acronym for lowest unoccupied molecular orbital. The symmetries of the LUMO and HOMO are important in determining the stereochemistry of pericyclic reactions.

Markovnikov's rule (Section 6.9): A guide for determining the regiochemistry (orientation) of electrophilic addition reactions. In the addition of HX to an alkene, the hydrogen atom bonds to the alkene carbon that has fewer alkyl substituents. A modern statement of this rule is that electrophilic addition reactions proceed via the most stable carbocation intermediate.

Mass number (*A*) (Section 1.1): The total of protons plus neutrons in an atom.

Mass spectrometry (Section 12.1): A technique for measuring the mass, and therefore the molecular weight (MW), of a molecule.

Mechanism (Section 5.2): A complete description of how a reaction occurs. A mechanism must account for all starting materials and all products, and must describe the details of each individual step in the overall reaction process.

Meisenheimer complex (Section 16.8): An intermediate formed by addition of a nucleophile to a halo-substituted aromatic ring.

Mercapto group (Section 17.12): An alternative name for the thiol group, –SH.

Meso compound (Section 9.8): A compound that contains stereogenic centers but is nevertheless achiral by virtue of a symmetry plane. For example, (2*R*,3*S*)-butanediol has two stereogenic carbon atoms but is achiral because of a symmetry plane between carbons 2 and 3.

Metabolism (Section 30.1): A collective name for the many reactions that go on in the cells of living organisms.

Methylene group (Section 6.3): A $=CH_2$ substituent.

Micelle (Section 28.2): A spherical cluster of soap-like molecules that aggregate in aqueous solution. The ionic heads of the molecules lie on the outside where they are solvated by water, and the organic tails bunch together on the inside of the micelle.

Molar absorptivity (Section 14.7): A quantitative measure of the amount of UV light absorbed by a sample.

Molecular ion (Section 12.1): The cation produced in the mass spectrometer by loss of an electron from the parent molecule. The mass of the molecular ion corresponds to the molecular weight of the sample.

Molecular mechanics (Chapter 4 Interlude): A computer-based method for determining the minimum-energy conformation of a molecule.

Molecular orbital (MO) (Section 1.7): An orbital that is the property of the entire molecule rather than an individual atom. Molecular orbitals result from interaction of two or more atomic orbitals when bonds are formed. Bonding MO's are lower in energy than the starting atomic orbitals, nonbonding MO's are equal in energy to the starting orbitals, and antibonding orbitals are higher in energy.

Molecular orbital (MO) theory (Section 1.7): A description of covalent bond formation as resulting from a mathematical combination of atomic orbitals (wave functions) to form molecular orbitals.

Molecule (Section 1.6): A neutral collection of atoms held together by covalent bonds.

Molozonide (Section 7.8): The initial addition product of ozone with an alkene.

Monomer (Section 7.11): The simple starting unit from which a polymer is made.

Multiplet (Section 13.7): A pattern of peaks in an NMR spectrum that arises by spin–spin splitting of a single absorption because of coupling between neighboring magnetic nuclei.

Mutarotation (Section 26.6): The change in optical rotation observed when a pure anomer of a sugar is dissolved in water. Mutarotation is caused by the reversible opening and closing of the acetal linkage, which yields an equilibrium mixture of anomers.

***n* + 1 rule** (Section 13.7): A hydrogen with *n* other hydrogens on neighboring carbons shows *n* + 1 peaks in its ^{1}H NMR spectrum.

N-terminal amino acid (Section 27.6): The amino acid with a free $-NH_2$ group at the end of a protein chain.

Newman projection (Section 4.1): A means of indicating stereochemical relationships between substituent groups on neighboring carbons. The carbon–carbon bond is viewed end-on, and the carbons are indicated by a circle. Bonds radiating from the center of the circle are attached to the front carbon, and bonds radiating from the edge of the circle are attached to the rear carbon.

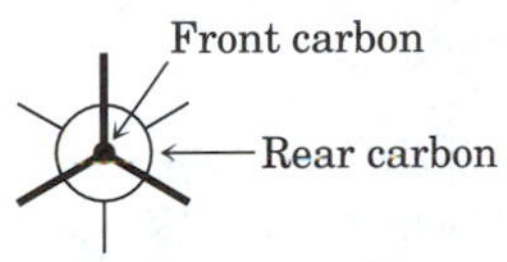

Nitrogen rule (Section 24.9): A compound with an odd number of nitrogen atoms has an odd-numbered molecular weight.

Node (Section 1.3): A surface of zero electron density within an orbital. For example, a *p* orbital has a nodal plane passing through the center of the nucleus, perpendicular to the axis of the orbital.

Nonbonding electrons (Section 1.6): Valence electrons that are not used in forming covalent bonds.

Normal alkane (Section 3.2): A straight-chain alkane, as opposed to a branched alkane. Normal alkanes are denoted by the suffix *n*, as in n-C_4H_{10} (*n*-butane).

NSAID (Chapter 16 Interlude): A nonsteroidal antiinflammatory drug, such as aspirin.

Nuclear magnetic resonance, NMR (Chapter 13): A spectroscopic technique that provides information about the carbon–hydrogen framework of a molecule. NMR works by detecting the energy absorption accompanying the transition between nuclear spin states that occurs when a molecule is placed in a strong magnetic field and irradiated with radiofrequency waves. Different nuclei within a molecule are in slightly different magnetic environments and therefore show absorptions at slightly different frequencies.

Nucleophile (Section 5.4): A "nucleus-lover," or species that donates an electron pair to an electrophile in a polar bond-forming reaction. Nucleophiles are also Lewis bases.

Nucleophilic acyl substitution reaction (Section 21.2): A reaction in which a nucleophile attacks a carbonyl compound and substitutes for a leaving group bound to the carbonyl carbon.

Nucleophilic addition reaction (Section 19.6): A reaction in which a nucleophile adds to the electrophilic carbonyl group of a ketone or aldehyde to give an alcohol.

Nucleophilic substitution reaction (Section 11.1): A reaction in which one nucleophile replaces another attached to a saturated carbon atom.

Nucleoside (Section 29.8): A nucleic acid constituent, consisting of a sugar residue bonded to a heterocyclic purine or pyrimidine base.

Nucleotide (Section 29.8): A nucleic acid constituent, consisting of a sugar residue bonded both to a heterocyclic purine or pyrimidine base and to a phosphoric acid. Nucleotides are the monomer units from which DNA and RNA are constructed.

Olefin (Chapter 6 introduction): An alternative name for an alkene.

Optical isomers (Section 9.5): An alternative name for enantiomers. Optical isomers are isomers that have a mirror-image relationship.

Optically active (Section 9.3): A substance that rotates the plane of polarization of plane-polarized light. Note that an optically active sample must contain chiral molecules, but that all samples with chiral molecules are not optically active. Thus, a racemic sample is optically inactive even though the individual molecules are chiral.

Orbital (Section 1.2): A wave function, which defines the volume of space around a nucleus in which an electron is most likely to be found.

Oxidation (Section 10.10): A reaction that causes a decrease in electron ownership by carbon, either by bond formation between carbon and a more electronegative atom (usually oxygen, nitrogen, or a halogen) or by bond breaking between carbon and a less electronegative atom (usually hydrogen).

β-Oxidation pathway (Section 30.2): The repetitive four-step sequence of enzyme-catalyzed reactions for catabolism of fatty acids.

Oxirane (Section 18.7): An alternative name for an epoxide.

Oxymercuration (Section 7.4): A method for double-bond hydration using aqueous mercuric acetate as the reagent.

Ozonide (Section 7.8): The product formed by addition of ozone to a carbon–carbon double bond. Ozonides are usually treated with a reducing agent, such as zinc in acetic acid, to produce carbonyl compounds.

Paraffin (Section 3.5): A common name for alkanes.

Parent peak (Section 12.1): The peak in a mass spectrum corresponding to the molecular ion. The mass of the parent peak therefore represents the molecular weight of the compound.

Pauli exclusion principle (Section 1.3): A statement of the fact that no more than two electrons can occupy the same orbital, and that those two must have spins of opposite sign.

Peptide (Section 27.6): An amino acid polymer in which the individual amino acid residues are linked by amide bonds.

Pericyclic reaction (Chapter 31): A reaction that occurs by a concerted reorganization of bonding electrons in a cyclic transition state.

Periplanar (Section 11.11): A conformation in which bonds to neighboring atoms have a parallel arrangement. In an eclipsed conformation, the neighboring bonds are syn periplanar; in a staggered conformation, the bonds are anti periplanar.

Anti periplanar   Syn periplanar

Peroxide (Section 18.2): A molecule containing an oxygen–oxygen bond functional group, ROOR′ or ROOH. The "peroxides" present as explosive impurities in ether solvents are usually of the latter type. Since the oxygen–oxygen bond is weak and easily broken, peroxides are often used to initiate radical chain reactions.

Phase-transfer catalysts (Section 25.8): Agents that cause the transfer of ionic reagents between water and organic phases, thus catalyzing reactions. Tetraalkylammonium salts, $R_4N^+ X^-$, are often used to transport inorganic anions from the aqueous phase to the organic phase where the desired reaction then occurs.

Phenol (Section 25.4): A compound with an –OH group directly bonded to an aromatic ring, ArOH.

Phenyl (Section 15.2): The name for the $-C_6H_5$ unit when the benzene ring is considered as a substituent. A phenyl group is abbreviated as –Ph or Φ (Greek phi).

Phospholipid (Section 28.3): A lipid that contains a phosphate residue. For example, phosphoglycerides contain a glycerol backbone linked to two fatty acids and a phosphoric acid.

Pi (π) bond (Section 1.10): The covalent bond formed by sideways overlap of atomic orbitals. For example, carbon–carbon double bonds contain a π bond formed by sideways overlap of two *p* orbitals.

Plane of symmetry (Section 9.2): A plane that bisects a molecule such that one half of the molecule is the mirror image of the other half. Molecules containing a plane of symmetry are achiral.

Plane-polarized light (Section 9.3): Ordinary light that has its electric vectors in a single plane rather than in random planes. The plane of polarization is rotated when the light is passed through a solution of a chiral substance.

Polar aprotic solvent (Section 11.5): A polar solvent that can't function as a hydrogen ion donor. Polar aprotic solvents such as dimethyl sulfoxide (DMSO), hexamethylphosphoramide (HMPA), and dimethylformamide (DMF) are particularly useful in S_N2 reactions because of their ability to solvate cations.

Polar covalent bond (Section 2.1): A covalent bond in which the electron distribution between atoms is unsymmetrical.

Polar reaction (Section 5.2): A reaction in which bonds are made when a nucleophile donates two electrons to an electrophile, and in which bonds are broken when one fragment leaves with both electrons from the bond.

Polarity (Section 2.1): The unsymmetrical distribution of electrons in a molecule that results when one atom attracts electrons more strongly than another.

Polarizability (Section 5.4): The measure of the change in a molecule's electron distribution in response to changing electric interactions with solvents or ionic reagents.

Polycyclic (Section 4.15): Referring to a compound that contains more than one ring.

Polycyclic aromatic compound (Section 15.11): A compound with two or more benzene-like aromatic rings fused together.

Polymer (Section 7.11): A large molecule made up of repeating smaller units. For example, polyethylene is a synthetic polymer made from repeating ethylene units, and DNA is a biopolymer made of repeating deoxyribonucleotide units.

Polysaccharide (Section 26.1): A carbohydrate that is made of many simple sugars linked together.

Polyunsaturated fatty acid (PUFA) (Section 28.1): A fatty acid containing two or more double bonds.

Primary, secondary, tertiary, quaternary (Section 3.4): Terms used to describe the substitution pattern at a specific site. A primary site has one organic substituent attached to it, a secondary site has two organic substituents, a tertiary site has three, and a quaternary site has four.

	Carbon	*Hydrogen*	*Alcohol*	*Amine*
Primary	RCH_3	RCH_3	RCH_2OH	RNH_2
Secondary	R_2CH_2	R_2CH_2	R_2CHOH	R_2NH
Tertiary	R_3CH	R_3CH	R_3COH	R_3N
Quaternary	R_4C			

Primary structure (Section 27.14): The amino acid sequence in a protein.

Propagation step (Section 5.3): The step or series of steps in a radical chain reaction that carry on the chain. The propagation steps must yield both product and a reactive intermediate.

Prostaglandin (Section 28.4): A lipid with the general carbon skeleton

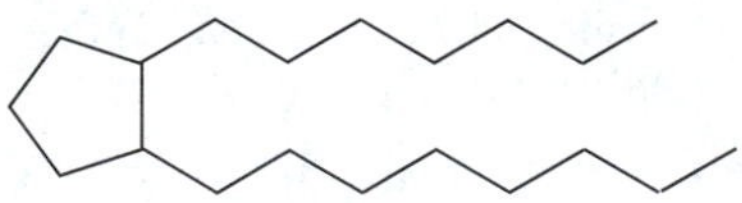

Prostaglandins are present in nearly all body tissues and fluids, where they serve a large number of important hormonal functions.

Prosthetic group (Section 27.14): A covalently bound organic group attached to a protein.

Protecting group (Section 17.10): A group that is introduced to protect a sensitive functional group toward reaction elsewhere in the molecule. After serving its protective function, the group is then removed. For example, ketones and aldehydes are often protected as acetals by reaction with ethylene glycol, and alcohols are often protected as trimethylsilyl ethers.

Protein (Section 27.13): A large peptide containing 50 or more amino acid residues. Proteins serve both as structural materials and as enzymes that control an organism's chemistry.

Protic solvent (Section 11.9): A solvent such as water or alcohol that can act as a proton donor. Protic solvents are particularly good at stabilizing anions by hydrogen bonding, thereby lowering their reactivity.

Pyranose (Section 26.5): The six-membered-ring form of a simple sugar.

Quantum mechanics (Section 1.2): A mathematical description of the electronic structure of atoms.

Quartet (Section 13.7): A set of four peaks in the NMR, caused by spin–spin splitting of a signal by three adjacent nuclear spins.

Quaternary (Section 3.4; *see* Primary)

Quaternary structure (Section 27.14): The highest level of protein structure, involving a specific aggregation of individual proteins into a larger cluster.

Quinone (Section 25.8): A 2,5-cyclohexadiene-1,4-dione.

R group (Section 3.4): A generalized abbreviation for an organic partial structure.

***R,S* convention** (Section 9.6): A method for defining the absolute configuration at stereogenic centers using the Cahn–Ingold–Prelog sequence rules.

Racemic mixture (Section 9.10): A mixture consisting of equal parts (+) and (−) enantiomers of a chiral substance. Even though the individual molecules are chiral, racemic mixtures are optically inactive.

Radical (Section 5.2): A species that has an odd number of electrons, such as the chlorine radical, Cl·.

Radical reaction (Section 5.2): A reaction in which bonds are made by donation of one electron from each of two reagents and in which bonds are broken when each fragment leaves with one electron.

Rate equation (Section 11.3): An equation that expresses the dependence of the rate of a reaction on the concentration of reactants.

Rate-limiting step (Section 11.7): The slowest step in a multistep reaction sequence. The rate-limiting step acts as a kind of bottleneck in multistep reactions and is observed by kinetics measurements.

Reaction energy diagram (Section 5.8): A pictorial representation of the course of a reaction, in which potential energy is plotted as a function of reaction progress. Starting materials, transition states, intermediates, and final products are represented, and their appropriate energy levels are indicated.

Rearrangement reaction (Section 5.1): What occurs when a single reactant undergoes a reorganization of bonds and atoms to yield an isomeric product.

Reducing sugar (Section 26.8): A sugar that reduces silver ion in the Tollens test or cupric ion in the Fehling or Benedict tests. All sugars that are aldehydes or can be readily converted into aldehydes are reducing. Glycosides, however, are not reducing sugars.

Reduction (Section 10.10): A reaction that causes an increase of electron ownership by carbon, either by bond breaking between carbon and a more electronegative atom or by bond formation between carbon and a less electronegative atom.

Refining (Chapter 3 Interlude): The process by which petroleum is converted into gasoline and other useful products.

Regiochemistry (Section 6.9): A term describing the orientation of a reaction that occurs on an unsymmetrical substrate. Markovnikov's rule, for example, predicts the regiochemistry of electrophilic addition reactions.

Regiospecific (Section 6.9): A term describing a reaction that occurs with a specific regiochemistry to give a single product rather than a mixture of products.

Replication (Section 29.12): The process by which double-stranded DNA uncoils and is replicated to produce two new copies.

Residue (Section 27.6): An amino acid in a protein chain.

Resolution (Section 9.10): The process by which a racemic mixture is separated into its two pure enantiomers. For example, a racemic carboxylic acid might be converted by reaction with a chiral amine base into a diastereomeric mixture of salts, which could be separated by fractional crystallization. Regeneration of the free acids would then yield the two pure enantiomeric acids.

Resonance effect (Section 16.5): The effect by which substituents donate or withdraw electrons through orbital overlap with neighboring π bonds. For example, an oxygen or nitrogen substituent donates electrons to an aromatic ring by overlap of the O or N orbital with the aromatic ring p orbitals. A carbonyl substituent, however, withdraws electron density from an aromatic ring by p orbital overlap.

Resonance hybrid (Section 2.4): A molecule, such as benzene, that can't be represented adequately by a single Kekulé structure but must instead be considered as an average of two or more resonance structures. The resonance structures themselves differ only in the positions of their electrons, not their nuclei.

Restriction endonuclease (Section 29.15): An enzyme that is able to cleave a DNA molecule at well-defined points in the chain where a specific base sequence occurs.

Retrosynthetic (Section 16.12): A technique for planning organic syntheses by working backward from the final product to the starting material.

Ring current (Section 15.12): The circulation of π electrons induced in aromatic rings by an external magnetic field. This effect accounts for the pronounced downfield shift of aromatic ring protons in the ^{1}H NMR spectrum.

Ring-flip (Section 4.11): The molecular motion that converts one chair conformation of cyclohexane into another chair conformation. The effect of a ring-flip is to convert an axial substituent into an equatorial substituent.

RNA (Section 29.8): Ribonucleic acid; the biopolymer found in cells that serves to transcribe the genetic information found in DNA and uses that information to direct the synthesis of proteins.

Saccharide (Section 26.1): A sugar.

Saponification (Section 21.7): An old term for the base-induced hydrolysis of an ester to yield a carboxylic acid salt.

Saturated (Section 3.2): A molecule that has only single bonds and thus can't undergo addition reactions. Alkanes are saturated, but alkenes are unsaturated.

Sawhorse structure (Section 4.1): A manner of representing stereochemistry that uses a stick drawing and gives a perspective view of the conformation around a single bond.

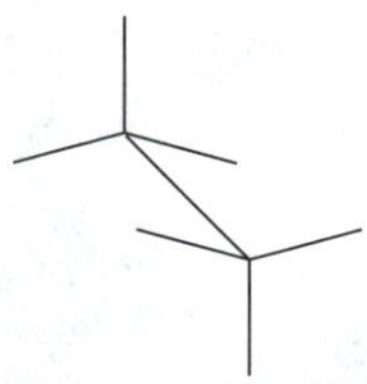

Sawhorse structure

Second-order reaction (Section 11.3): A reaction whose rate-limiting step is bimolecular and whose kinetics are therefore dependent on the concentration of two reagents.

Secondary (Section 3.4; *see* Primary)

Secondary structure (Section 27.14): The level of protein substructure that involves organization of chain sections into ordered arrangements such as β-pleated sheets or α-helices.

Sense strand (Section 29.13): The strand of double-helical DNA that contains the gene.

Sequence rules (Sections 6.6, 9.6): A series of rules for assigning relative priorities to substituent groups on a double-bond carbon atom or on a stereogenic center. Once priorities have been established, *E,Z* double-bond geometry and *R,S* configurational assignments can be made.

Shielding (Section 13.2): An effect observed in NMR that causes a nucleus to absorb toward the right (upfield) side of the chart. Shielding is caused by donation of electron density to the nucleus.

Sigma (σ) bond (Section 1.7): A covalent bond formed by head-on overlap of atomic orbitals.

Sigmatropic reaction (Section 31.9): A pericyclic reaction that involves the migration of a group from one end of a π electron system to the other, for example, the [1,5] sigmatropic rearrangement of a hydrogen atom in cyclopentadiene.

Simple protein (Section 27.13): A protein, such as blood serum albumin, that yields only amino acids on hydrolysis.

Skeletal structure (Section 2.9): A shorthand way of writing structures in which carbon atoms are assumed to be at each intersection of two lines (bonds) and at the end of each line.

Soap (Section 28.2): The mixture of long-chain fatty acid salts obtained on base hydrolysis of animal fat.

Solid-phase synthesis (Section 27.12): A technique of synthesis whereby the starting material is covalently bound to a solid polymer bead and reactions are carried out on the bound substrate. After the desired transformations have been effected, the product is cleaved from the polymer and isolated. This technique is particularly useful in peptide synthesis (Merrifield method) and in DNA synthesis.

Solvation (Section 11.5): The clustering of solvent molecules around a solute particle to stabilize it.

***sp* orbital** (Section 1.11): A hybrid orbital derived from the combination of an *s* and a *p* atomic orbital. The two *sp* orbitals that result from hybridization are oriented at an angle of 180° to each other.

sp^2 orbital (Section 1.10): A hybrid orbital derived by combination of an *s* atomic orbital with two *p* atomic orbitals. The three sp^2 hybrid orbitals that result lie in a plane at angles of 120° to each other.

sp^3 orbital (Section 1.8): A hybrid orbital derived by combination of an *s* atomic orbital with three *p* atomic orbitals. The four sp^3 hybrid orbitals that result are directed toward the corners of a regular tetrahedron at angles of 109° to each other.

Specific rotation, $[\alpha]_D$ (Section 9.4): The specific rotation of a chiral compound is a physical constant that is defined by the equation

$$[\alpha]_D = \frac{\text{Observed rotation}}{\text{Path length} \times \text{Concentration}} = \frac{\alpha}{l \times C}$$

where the path length of the sample solution is expressed in decimeters and the concentration of the sample solution is expressed in grams per milliliter.

Sphingolipids (Section 28.3): Phospholipids that have sphingosine or a related dihydroxyamine as their backbone and are constituents of plant and animal cell membranes.

Spin–spin splitting (Section 13.7): The splitting of an NMR signal into a multiplet because of an interaction between nearby magnetic nuclei whose spins are coupled. The magnitude of spin–spin splitting is given by the coupling constant, *J*.

Staggered conformation (Section 4.1): The three-dimensional arrangement of atoms around a carbon–carbon single bond in which the bonds on one carbon bisect the bond angles on the second carbon as viewed end-on.

Staggered conformation

Step-growth polymer (Section 21.11): A polymer in which each bond is formed independently of the others. Polyesters and polyamides (nylons) are examples.

Stereochemistry (Chapters 4, 9): The branch of chemistry concerned with the three-dimensional arrangement of atoms in molecules.

Stereogenic center (Section 9.2): An atom (usually carbon) that is bonded to four different groups.

Stereoisomers (Section 3.8): Isomers that have their atoms connected in the same order but have different three-dimensional arrangements. The term *stereoisomer* includes both enantiomers and diastereomers but does not include constitutional isomers.

Stereospecific (Section 7.6): A term indicating that only a single stereoisomer is produced in a given reaction rather than a mixture.

Steric strain (Sections 4.3, 4.12): The strain imposed on a molecule when two groups are too close together and try to occupy the same space. Steric strain is responsible both for the greater stability of trans versus cis alkenes and for the greater stability of equatorially substituted versus axially substituted cyclohexanes.

Steroid (Section 28.7): A lipid whose structure is based on the tetracyclic carbon skeleton:

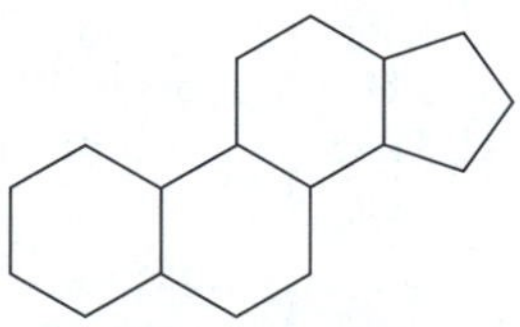

Steroids occur in both plants and animals and have a variety of important hormonal functions.

Straight-chain alkane (Section 3.2): An alkane whose carbon atoms are connected without branching.

Substitution reaction (Section 5.1): What occurs when two reactants exchange parts to give two new products. S_N1 and S_N2 reactions are examples.

Sulfide (Section 18.11): A compound that has two organic substituents bonded to the same sulfur atom, RSR′.

Sulfone (Section 18.11): A compound of the general structure RSO_2R'.

Sulfoxide (Section 18.11): A compound of the general structure RSOR′.

Suprafacial (Section 31.7): A word used to describe the geometry of pericyclic reactions. Suprafacial reactions take place on the same side of the two ends of a π electron system.

Symmetry-allowed, symmetry-disallowed (Section 31.3): A symmetry-allowed reaction is a pericyclic process that has a favorable orbital symmetry for reaction through a concerted pathway. A symmetry-disallowed reaction is one that does not have favorable orbital symmetry for reaction through a concerted pathway.

Symmetry plane (Section 9.2): A plane that bisects a molecule such that one half of the molecule is the mirror image of the other half. Molecules containing a plane of symmetry are achiral.

Syn stereochemistry (Section 7.2): A syn addition reaction is one in which the two ends of the double bond are attacked from the same side. For example, OsO_4 induced hydroxylation of cyclohexene yields *cis*-1,2-cyclohexanediol, the product of syn addition. A syn elimination is one in which the two groups leave from the same side of the molecule.

Tautomers (Sections 8.5, 22.1): Isomers that are rapidly interconverted. For example, enols and ketones are tautomers because they are rapidly interconverted by treatment with either acid or base.

Template strand (Section 29.13): The strand of double-helical DNA that does not contain the gene.

Terpene (Section 28.5): A lipid that is formally derived by head-to-tail polymerization of isoprene units.

Tertiary (Section 3.4; *see* Primary)

Tertiary structure (Section 27.14): The level of protein structure that involves the manner in which the entire protein chain is folded into a specific three-dimensional arrangement.

Thermodynamic control (Section 14.6): Equilibrium reactions that yield the lowest-energy, most stable product are said to be thermodynamically controlled. Even though it is the most stable, the product of a thermodynamically controlled reaction is not necessarily formed fastest.

Thiol (Section 17.12): A compound containing the –SH functional group.

Torsional strain (Section 4.1): The strain in a molecule caused by electron repulsion between eclipsed bonds. Torsional strain destabilizes boat cyclohexane relative to chair cyclohexane.

Transcription (Section 29.13): The process by which the genetic information encoded in DNA is read and used to synthesize RNA in the nucleus of the cell. A small portion of double-stranded DNA uncoils, and complementary ribonucleotides line up in the correct sequence for RNA synthesis.

Transition state (Section 5.8): An activated complex between reagents, representing the highest-energy point on a reaction curve. Transition states are unstable complexes that can't be isolated.

Translation (Section 29.14): The process by which the genetic information transcribed from DNA onto mRNA is read by tRNA and used to direct protein synthesis.

Tree diagram (Section 13.8): A diagram used in NMR to sort out the complicated splitting patterns that can arise from multiple couplings.

Triacylglycerol (Section 28.1): A lipid, such as found in animal fat and vegetable oil, that is a triester of glycerol with long-chain fatty acids.

Tricarboxylic acid cycle (Section 30.5): An alternative name for the citric acid cycle by which acetyl CoA is degraded to CO_2.

Triplet (Section 13.7): A symmetrical three-line splitting pattern observed in the ^1H NMR spectrum when a proton has two equivalent neighbor protons.

Twist-boat conformation (Section 4.14): A conformation of cyclohexane that is somewhat more stable than a pure boat conformation.

Ultraviolet (UV) spectroscopy (Section 14.11): An optical spectroscopy employing ultraviolet irradiation. UV spectroscopy provides structural information about the extent of π electron conjugation in organic molecules.

Unimolecular reaction (Section 11.7): A reaction that occurs by spontaneous transformation of the starting material without the intervention of other reactants. For example, the dissociation of a tertiary alkyl halide in the S_N1 reaction is a unimolecular process.

Unsaturated (Section 6.2): A molecule that has multiple bonds and can undergo addition reactions. Alkenes and alkynes, for example, are unsaturated.

Upfield (Section 13.3): Referring to the right-hand portion of the NMR chart.

Valence bond theory (Section 1.7): A bonding theory that describes a covalent bond as resulting from the overlap of two atomic orbitals.

Valence shell (Section 1.5): The outermost electron shell of an atom.

Van der Waals forces (Section 3.5): The attractive forces between molecules caused by dipole–dipole interactions. Van der Waals forces are one of the primary forces responsible for holding molecules together in the liquid and solid states.

Vicinal (Section 8.3): A term used to refer to a 1,2-disubstitution pattern. For example, 1,2-dibromoethane is a vicinal dibromide.

Vinyl group (Section 6.3): A $H_2C{=}CH{-}$ substituent.

Vinylic (Section 8.4): A term that refers to a substituent at a double-bond carbon atom. For example, chloroethylene is a vinylic chloride, and enols are vinylic alcohols.

Vitamin (Section 27.15): A small organic molecule that must be obtained in the diet and is required in trace amounts for proper growth.

Vulcanization (Section 14.7): A technique for heating a crude diene polymer with a few percent by weight of sulfur. Sulfur forms cross-links between polymer chains, thereby locking the chains together and hardening the polymer.

Wave equation (Section 1.2): A mathematical expression that defines the behavior of an electron in an atom.

Wave function (Section 1.2): A solution to the wave equation for defining the behavior of an electron in an atom. The square of the wave function defines the shape of an orbital.

Wavelength (Section 12.4): The length of a wave from peak to peak. The wavelength of electromagnetic radiation is inversely proportional to frequency and inversely proportional to energy.

Wavenumber (Section 12.5): The reciprocal of the wavelength, in centimeters.

Ylide (Section 19.15): A neutral dipolar molecule with adjacent positive and negative charges. The phosphoranes used in Wittig reactions are ylides.

Zaitsev's rule (Section 11.10): A rule stating that E2 elimination reactions normally yield the more highly substituted alkene as major product.

Zusammen (Z) (Section 6.6): A term used to describe the stereochemistry of a carbon–carbon double bond. The two groups on each carbon are assigned priorities according to the Cahn–Ingold–Prelog sequence rules, and the two carbons are compared. If the high-priority groups on each carbon are on the same side of the double bond, the bond has *Z* geometry.

Zwitterion (Section 27.2): A neutral dipolar molecule in which the positive and negative charges are not adjacent. For example, amino acids exist as zwitterions, $H_3N^+{-}CHR{-}COO^-$. Zwitterions are also called *betaines*.

Appendix E Answers to Selected In-Text Problems

The following answers are meant only as a quick check. Full answers and explanations for all problems are provided in the accompanying *Study Guide and Solutions Manual.*

Chapter 1

1.1 (a) 1 (b) 3 (c) 8

1.2 (a) Boron: $1s^2\ 2s^2\ 2p$ (b) Phosphorus: $1s^2\ 2s^2\ 2p^6\ 3s^2\ 3p^3$ (c) Oxygen: $1s^2\ 2s^2\ 2p^4$ (d) Chlorine: $1s^2\ 2s^2\ 2p^6\ 3s^2\ 3p^5$

1.3 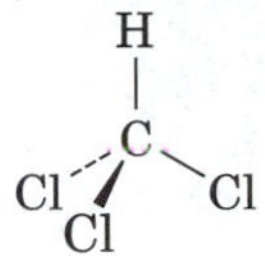

1.4 (a) CCl_4 (b) AlH_3 (c) CH_2Cl_2 (d) SiF_4 (e) CH_3NH_2

1.5 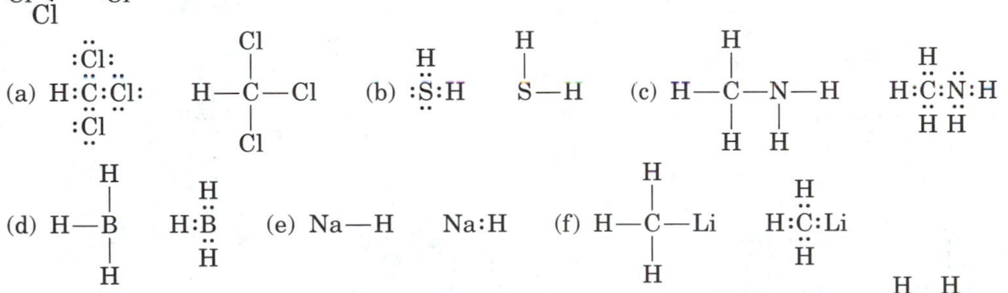

1.6 Ionic: LiI, KBr, $MgCl_2$ Covalent: CH_4, CH_2Cl_2, Cl_2

1.7 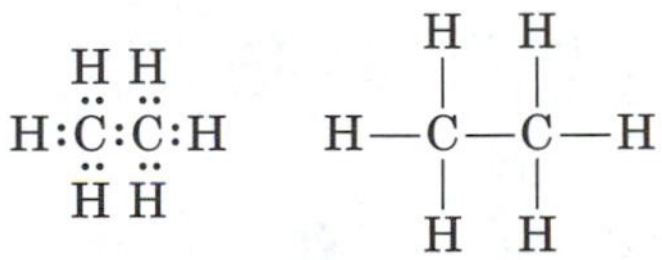

1.8 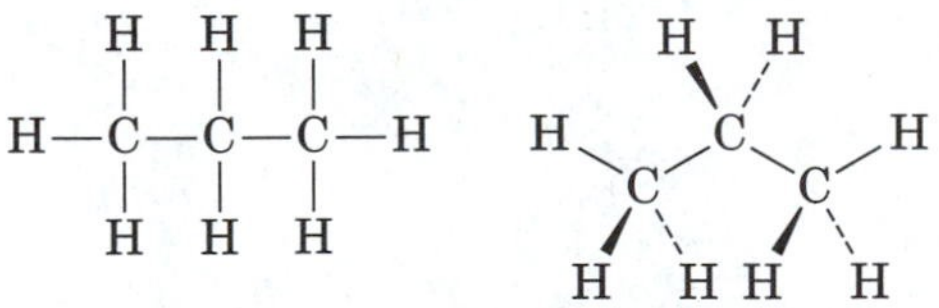 All bond angles are near 109°.

1.9 C_2H_7 has too many hydrogens for a compound with two carbons.

1.10 The CH_3 carbon is sp^3, the double-bond carbons are sp^2, and the C=C–C bond angle is approximately 120°.

1.11 All carbons are sp^2, and all bond angles are near 120°.

1.12 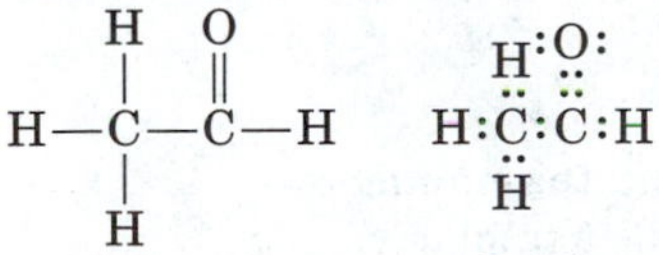

1.13 The CH_3 carbon is sp^3, the triple-bond carbons are sp, and the C≡C–C bond angle is approximately 180°.

1.14 Formaldimine has a C=N double bond, with four shared electrons.

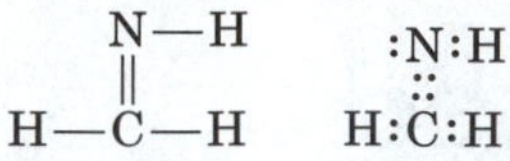

The nitrogen atom is sp^2.

1.15 All are sp^3 hybridized and have tetrahedral geometry.

Chapter 2

2.1 (a) H (b) Br (c) Cl (d) C

2.2 (a) C is $\delta+$, Br is $\delta-$ (b) C is $\delta+$, N is $\delta-$ (c) Li is $\delta+$, C is $\delta-$ (d) H is $\delta+$, N is $\delta-$ (e) C is $\delta+$, O is $\delta-$ (f) Mg is $\delta+$, C is $\delta-$ (g) C is $\delta+$, F is $\delta-$

2.3 $H_3C{-}OH < H_3C{-}MgBr < H_3C{-}Li < H_3C{-}F < H_3C{-}K$ **2.4**

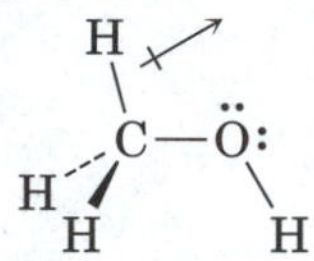

2.5 The two C=O dipoles cancel because of the 180° O=C=O bond angle.

2.6 (a)

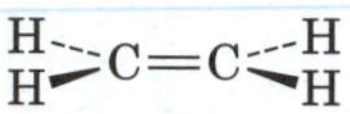

No dipole moment

(b)

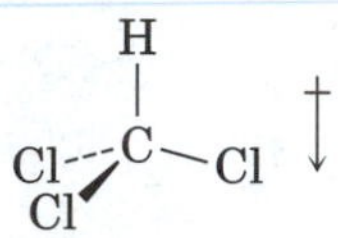

(c)

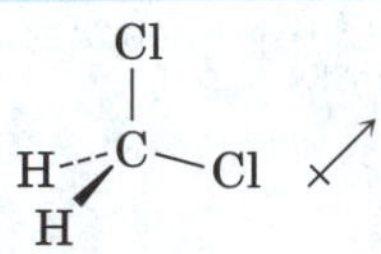

(d)

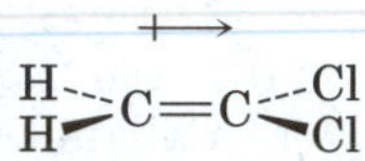

2.7 For the dimethylsulfoxide sulfur atom: $FC = 6 - \frac{6}{2} - 2 = +1$. For the dimethylsulfoxide oxygen atom: $FC = 6 - \frac{2}{2} - 6 = -1$

2.8 (a) For carbon: $FC = 4 - \frac{8}{2} - 0 = 0$. For the middle nitrogen: $FC = 5 - \frac{8}{2} - 0 = +1$.

For the end nitrogen: $FC = 5 - \frac{4}{2} - 4 = -1$.

(b) For nitrogen: $FC = 5 - \frac{8}{2} - 0 = +1$. For oxygen: $FC = 6 - \frac{2}{2} - 6 = -1$.

(c) For nitrogen: $FC = 5 - \frac{8}{2} - 0 = +1$. For the end carbon: $FC = 4 - \frac{6}{2} - 2 = -1$.

2.9 (a)

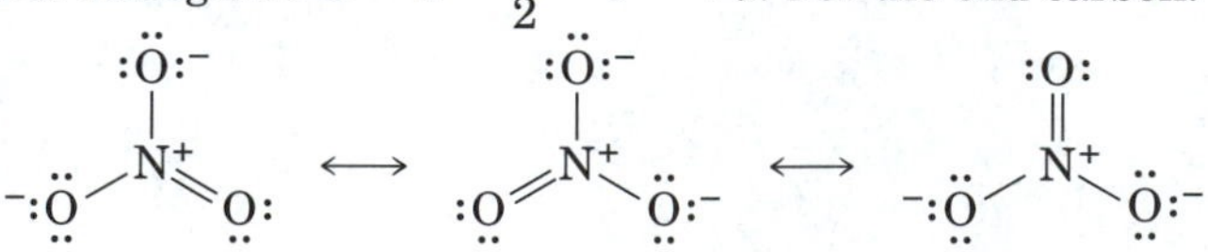

(b)

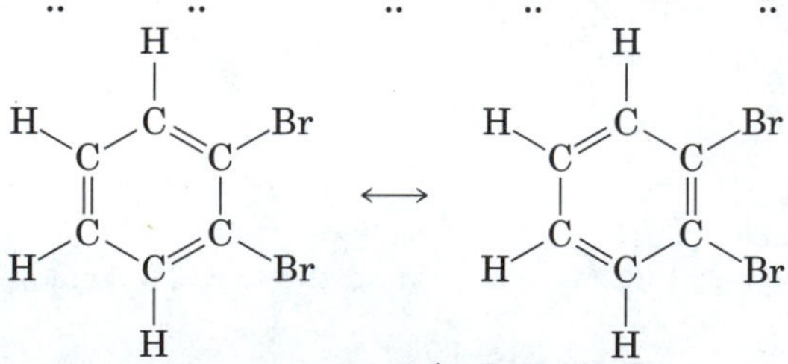

(c) $H_2C{=}CH{-}CH_2^+ \longleftrightarrow H_2\overset{+}{C}{-}CH{=}CH_2$

(d) $:N{\equiv}\overset{+}{N}{-}\ddot{\underset{\cdot\cdot}{N}}^{-}{-}H \longleftrightarrow {}^{-}:\ddot{N}{=}\overset{+}{N}{=}\ddot{N}{-}H$

2.10 Picric acid is stronger. **2.11** Water is a stronger acid than ammonia.

2.12 Neither reaction will take place. **2.13** Reaction will take place.

2.14 $K_a = 4.9 \times 10^{-10}$ **2.15** pH = 2.25

2.16 (a) $CH_3CH_2OH + HCl \rightarrow CH_3CH_2OH_2^+\ Cl^-$; $(CH_3)_2NH + HCl \rightarrow (CH_3)_2NH_2^+\ Cl^-$; $(CH_3)_3P + HCl \rightarrow (CH_3)_3PH^+\ Cl^-$

(b) $HO^- + CH_3^+ \rightarrow HO{-}CH_3$; $HO^- + B(CH_3)_3 \rightarrow HO{-}B(CH_3)_3^-$;
$HO^- + MgBr_2 \rightarrow HO{-}MgBr_2^-$

2.17 (a) For boron: $FC = 3 - \frac{8}{2} - 0 = -1$. For oxygen: $FC = 6 - \frac{6}{2} - 2 = +1$.

(b) For aluminum: $FC = 3 - \frac{8}{2} - 0 = -1$. For nitrogen: $FC = 5 - \frac{8}{2} - 0 = +1$.

2.18 BF_3 is the acid; formaldehyde is the base. **2.19** (a) C_5H_5N (b) $C_6H_{10}O$ (c) C_8H_7N

Chapter 3

3.2 (a) CH_3OH (b)  (c) 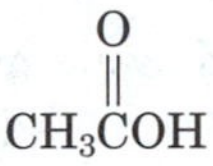 (d) CH_3NH_2 (e) 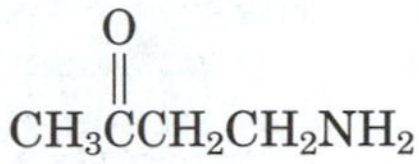

(f) $H_2C{=}CHCH{=}CH_2$

3.3 $CH_3CH_2CH_2CH_2CH_2CH_3$ 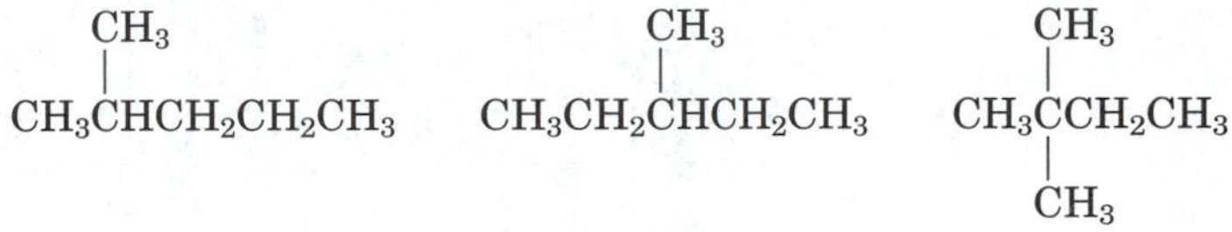

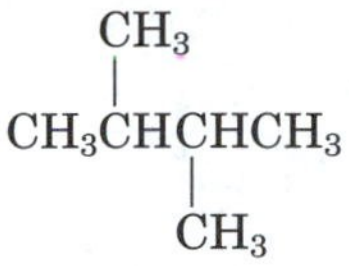

3.4 $CH_3CH_2CH_2CH_2OH$ 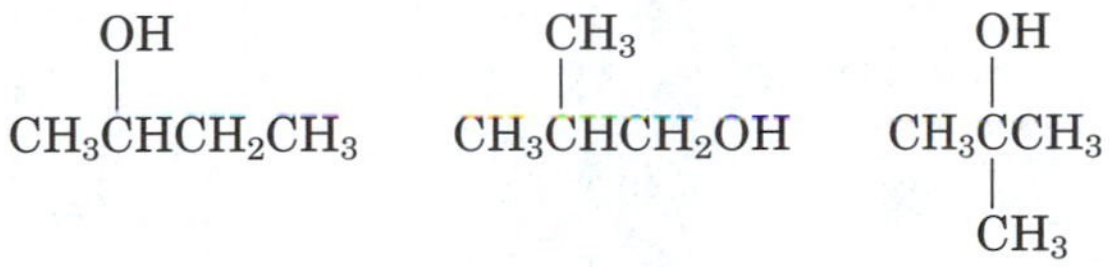

$CH_3OCH_2CH_2CH_3$ $CH_3CH_2OCH_2CH_3$ 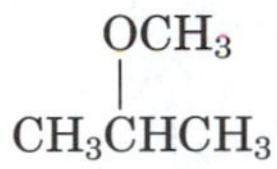

3.5 Part (a) has nine possible answers.

(a) 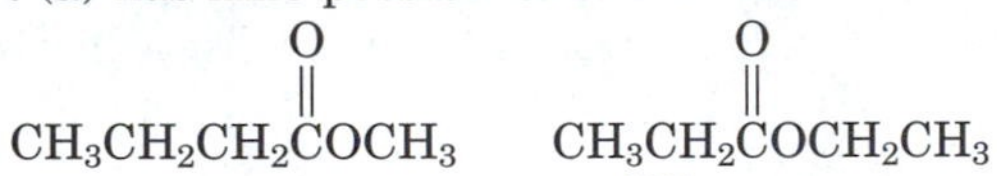

(b) $CH_3CH_2CH_2C{\equiv}N$ 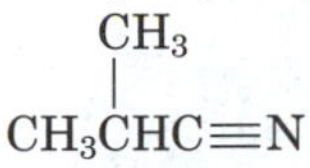

3.6 (a) Two (b) Four

3.7 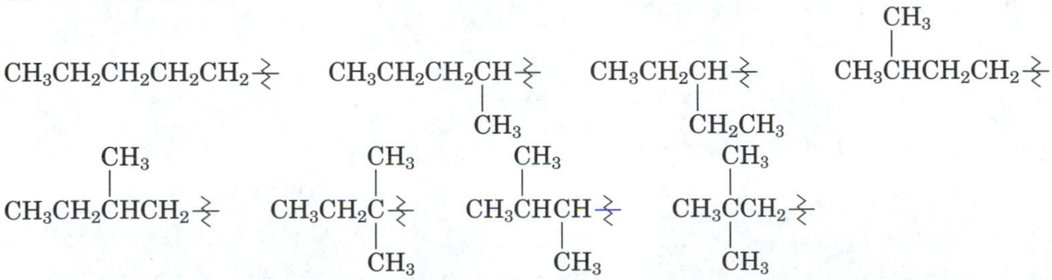

3.8 (a) 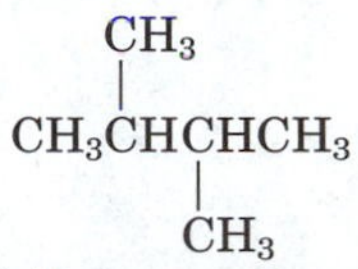 (b) 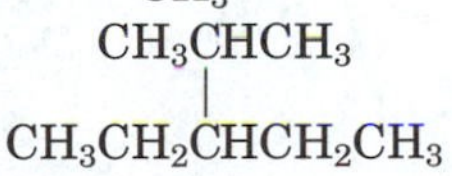 (c) 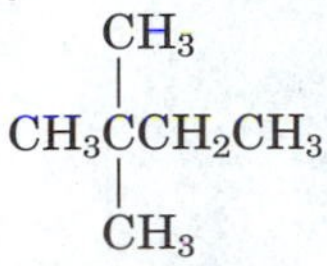

3.11 (a) Pentane, 2-methylbutane, 2,2-dimethylpropane (b) 3,4-Dimethylhexane
(c) 2,4-Dimethylpentane (d) 2,2,5-Trimethylheptane

3.12 (a) $CH_3CH_2CH(CH_3)CH(CH_3)CH_2CH_2CH_2CH_2CH_3$ (b) $CH_3CH_2CH_2C(CH_3)_2$—$CH(CH_2CH_3)CH_2CH_3$

(c) $CH_3C(CH_3)_2CH_2CH(CH_2CH_2CH_3)CH_2CH_2CH_2CH_3$ (d) $CH_3C(CH_3)_2CH_2CH(CH_3)CH_3$

3.13 (a) 2-Methylhexane (b) 4,5-Dimethyloctane (c) 3-Ethyl-2,2-dimethylpentane (d) 3,4-Dimethylheptane (e) 2,3-Dimethylhexane (f) 3,3-Dimethylpentane

3.14 Pentyl, 1-methylbutyl, 1-ethylpropyl, 2-methylbutyl, 3-methylbutyl, 1,1-dimethylpropyl, 1,2-dimethylpropyl, 2,2-dimethylpropyl

3.15 (a) 1,4-Dimethylcyclohexane (b) 1-Methyl-3-propylcyclopentane (c) 3-Cyclobutylpentane (d) 1-Bromo-4-ethylcyclodecane (e) 1-Isopropyl-2-methylcyclohexane (f) 4-Bromo-1-*tert*-butyl-2-methylcycloheptane

3.17 (a) (b) (c)

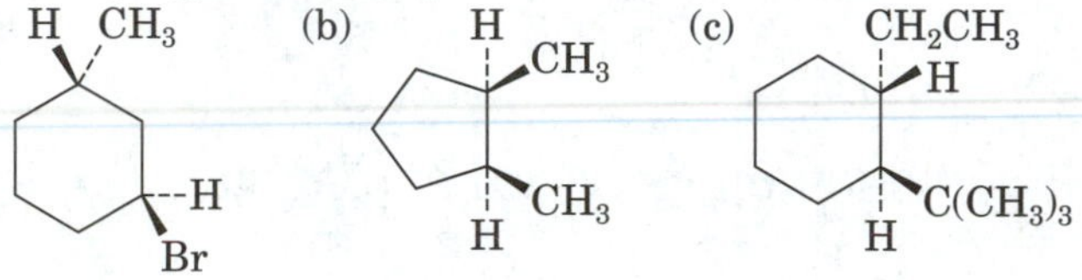

Chapter 4

4.3

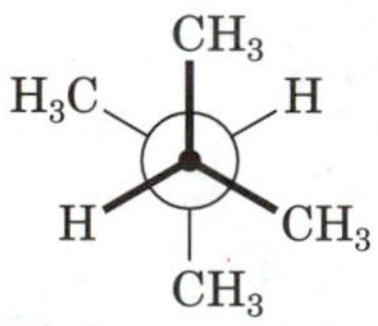

4.4 (a) (b)

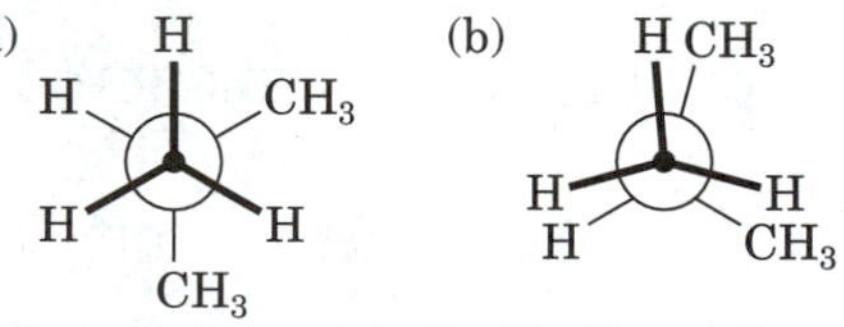

4.5 Cyclopropane is a more efficient fuel. **4.6** Six H↔H interactions; 21% of strain.

4.7 The trans isomer is more stable and has lower heat of combustion.

4.8 Methyl groups in *cis*-1,2-dimethylcyclobutane and in *trans*-1,3-dimethylcyclobutane have across-ring steric interactions.

4.9

4.10 Ten H↔H eclipsing interactions; 40 kJ/mol.

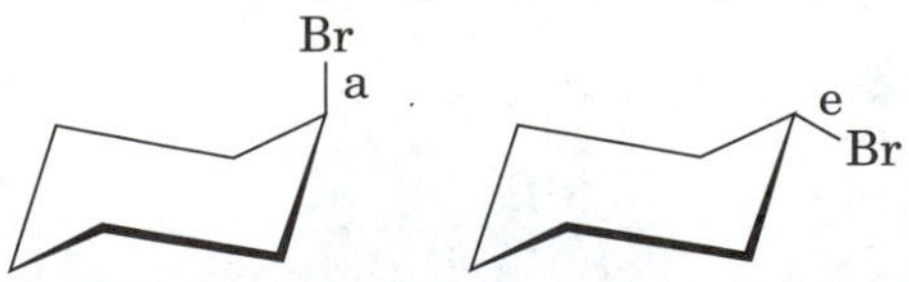

4.11

4.12–4.13 Axial and equatorial positions alternate around the ring on each side.

4.14

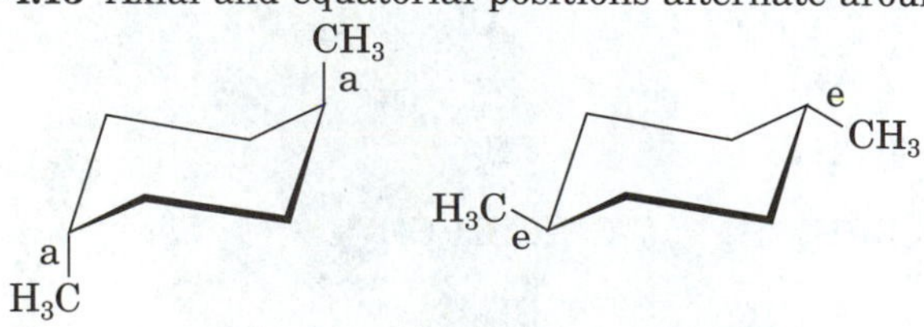

4.15 A hydrogen of an isopropyl group points in toward ring, as in methyl and ethyl. With a *tert*-butyl group, however, a methyl points in.

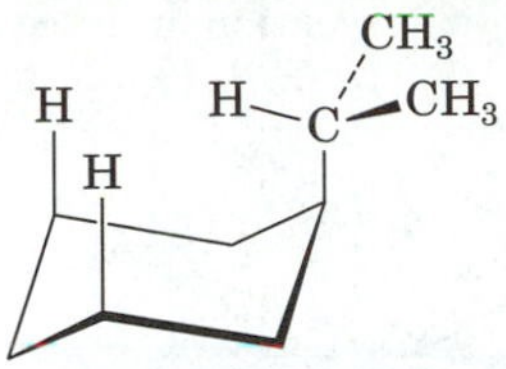

4.16 Cyano group points straight up. **4.17** Equatorial = 70%
4.18 (a) 2.0 kJ/mol (b) 11.4 kJ/mol (c) 2.0 kJ/mol (d) 8.0 kJ/mol
4.20 *trans*-Decalin is more stable because it has no 1,3-diaxial interactions.

Chapter 5

5.1 (a) Substitution (b) Elimination (c) Addition
5.2 1-Chloro-2-methylpentane, 2-chloro-2-methylpentane, 3-chloro-2-methylpentane, 2-chloro-4-methylpentane, 1-chloro-4-methylpentane
5.3 Pentane has three types of hydrogens; neopentane has only one.
5.5 Electrophiles: H^+, Mg^{2+}; nucleophiles: HO^-, NH_3 **5.6** Bromocyclohexane
5.7 Negative $\Delta G°$ more favored. **5.8** Larger K_{eq} more exothermic.
5.9 $\Delta G° = -17.1$ kJ/mol, 0 kJ/mol, +17.1 kJ/mol; $K_{eq} = 1.0 \times 10^7$, 1, 1.0×10^{-7}
5.10 (a) $\Delta H° = +243$ kJ/mol (b) $\Delta H° = +6$ kJ/mol (c) $\Delta H° = -108$ kJ/mol (d) $\Delta H° = -102$ kJ/mol
5.11 (a) $\Delta H° = -33$ kJ/mol (b) $\Delta H° = +33$ kJ/mol
5.12 Lower $\Delta G^\ddagger$ is faster. Can't predict K_{eq}.

Chapter 6

6.1 (a) 2 (b) 3 (c) 3 (d) 5 (e) 13 **6.2** (a) 1 (b) 2 (c) 2
6.3 (a) 5 (b) 5 (c) 3 (d) 1 (e) 6 (f) 5
6.4 (a) 3,4,4-Trimethyl-1-pentene (b) 3-Methyl-3-hexene (c) 4,7-Dimethyl-2,5-octadiene
6.5 (a) $H_2C{=}C(CH_3)CH_2CH_2CH{=}CH_2$ (b) $CH_3CH_2CH_2CH{=}C(CH_2CH_3)C(CH_3)_3$
(c) $CH_3CH{=}CHCH{=}CHC(CH_3)_2{-}C(CH_3){=}CH_2$ (d) $[(CH_3)_2CH]_2C{=}C[CH(CH_3)_2]_2$
(e) $CH_3CH(CH_3)CH_2CH(C(CH_3)_3)CH_2CH_2CH_3$
6.6 (a) 1,2-Dimethylcyclohexene (b) 4,4-Dimethylcycloheptene (c) 3-Isopropylcyclopentene
6.7 Compounds (c), (e), and (f) have cis–trans isomers.
6.8 *trans*-Cyclohexene is too strained.
6.9 (a) –Br (b) –Br (c) $-CH_2CH_3$ (d) –OH (e) $-CH_2OH$ (f) –CH=O
6.10 (a) –Cl, –OH, $-CH_3$, –H (b) $-CH_2OH$, $-CH{=}CH_2$, $-CH_2CH_3$, $-CH_3$
(c) –COOH, $-CH_2OH$, –CN, $-CH_2NH_2$ (d) $-CH_2OCH_3$, –CN, –C≡CH, $-CH_2CH_3$
6.11 (a) *Z* (b) *E* (c) *Z* (d) *E*
6.12 (a) 2-Methylpropene (b) *E* isomer (c) 1-Methylcyclohexene
6.14 (a) Chlorocyclohexane (b) 2-Bromo-2-methylpentane (c) 2-Iodopentane
(d) 1-Bromo-1-methylcyclohexane
6.15 (a) Cyclopentene (b) 3-Hexene (c) 1-Ethylcyclohexene or ethylidenecyclohexane
(d) Cyclohexylethylene
6.16 (a) $CH_3CH_2C^+(CH_3)CH_2CH(CH_3)CH_3$ (b) 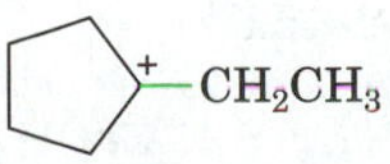

6.17 The second step is exothermic; transition state resembles the carbocation.

6.18 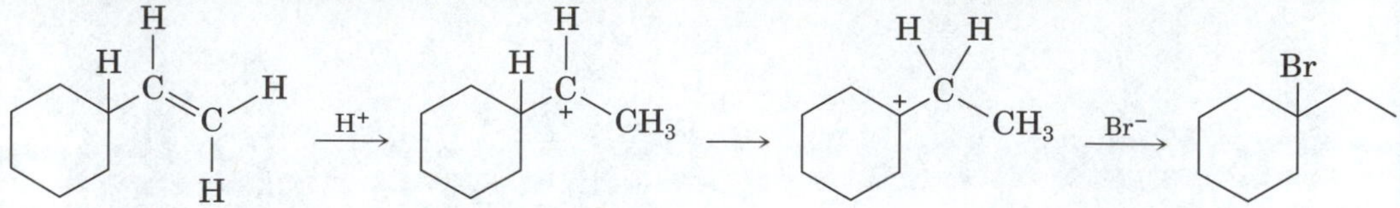

Chapter 7

7.1 2-Methyl-2-butene and 2-methyl-1-butene **7.2** Five

7.3 *trans*-1,2-Dichloro-1,2-dimethylcyclohexane

7.4 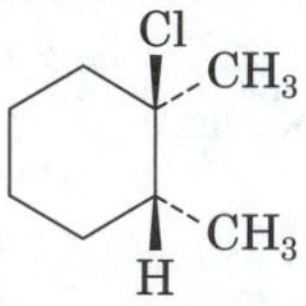 and 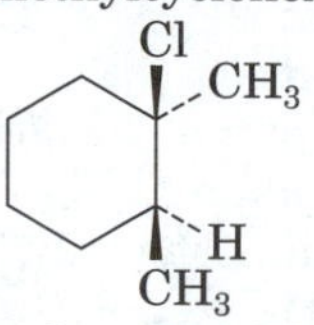

7.5 *trans*-2-Bromocyclopentanol **7.6** Markovnikov

7.7 (a) 2-Pentanol (b) 2-Methyl-2-pentanol

7.8 (a) 2-Methyl-1-hexene or 2-methyl-2-hexene (b) Cyclohexylethylene

7.9 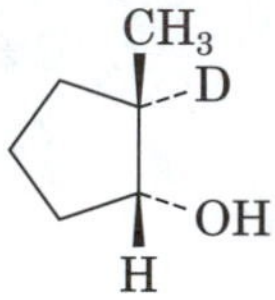

7.10 (a) 3-Methyl-1-butene (b) 2-Methyl-2-butene (c) Methylenecyclohexane

7.11 (a) 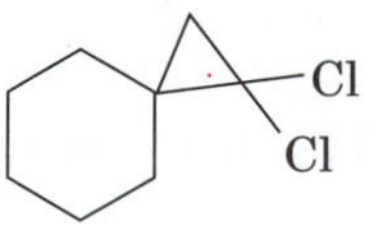 (b) 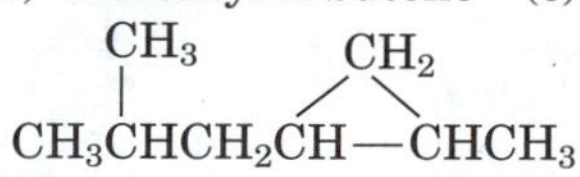

7.12 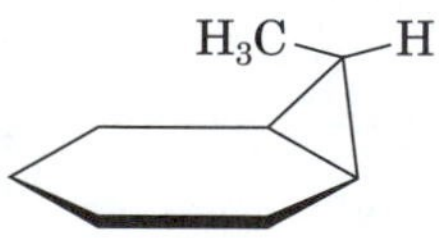 and 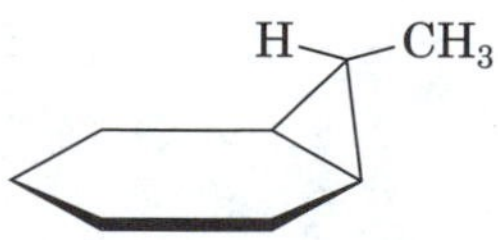

7.13 (a) 2-Methylpentane (b) 1,1-Dimethylcyclopentane

7.14 (a) 1-Methylcyclohexene + OsO_4 (b) 2-Methyl-2-pentene + OsO_4
(c) 1,3-Butadiene + OsO_4

7.15 (a) $CH_3COCH_2CH_2CH_2CH_2COOH$ (b) $CH_3COCH_2CH_2CH_2CH_2CHO$

7.16 (a) 2-Methylpropene (b) 3-Hexene

7.17 (a) 2-Methyl-1-butene + HBr (b) 1-Butene + HBr + Peroxides
(c) 3-Methyl-3-hexene + HBr + Peroxides

7.19 (a) $H_2C{=}CHOCH_3$ (b) $ClCH{=}CHCl$

Chapter 8

8.1 (a) 2,5-Dimethyl-3-hexyne (b) 3,3-Dimethyl-1-butyne (c) 2,4-Octadiene-6-yne
(d) 3,3-Dimethyl-4-octyne (e) 2,5,5-Trimethyl-3-heptyne (f) 6-Isopropylcyclodecyne

8.2 1-Hexyne, 2-hexyne, 3-hexyne, 3-methyl-1-pentyne, 4-methyl-1-pentyne,
4-methyl-2-pentyne, 3,3-dimethyl-1-butyne

8.3 (a) 1,1,2,2-Tetrachloropentane (b) 1-Bromo-1-cyclopentylethylene
(c) 2-Bromo-2-heptene and 3-bromo-2-heptene

8.4 4-Octanone, 2-methyl-4-octanone, and 7-methyl-4-octanone

8.5 (a) 1-Pentyne (b) 3-Hexyne **8.6** (a) $C_6H_5C{\equiv}CH$ (b) 2,5-Dimethyl-3-hexyne

8.7 (a) Reduce 2-octyne with Li/NH_3 (b) Reduce 3-heptyne with H_2/Lindlar catalyst
(c) Reduce 3-methyl-1-pentyne

8.8 (a) $C_6H_5C{\equiv}CH$ (b) $CH_3(CH_2)_7C{\equiv}C(CH_2)_7C{\equiv}C(CH_2)_7CH_3$

8.9 No: (a), (c), (d); Yes: (b)

8.10 (a) 1-Pentyne + CH_3I, or propyne + $CH_3CH_2CH_2I$
(b) 3-Methyl-1-butyne + CH_3CH_2I (c) Cyclohexylacetylene + CH_3I
(d) 4-Methyl-1-pentyne + CH_3I, or propyne + $(CH_3)_2CHCH_2I$
(e) 3-3-Dimethyl-1-butyne + CH_3CH_2I

8.11 $CH_3C{\equiv}CH \xrightarrow[2.\ CH_3I]{1.\ NaNH_2} CH_3C{\equiv}CCH_3 \xrightarrow[\text{Lindlar catalyst}]{H_2}$ *cis*-$CH_3CH{=}CHCH_3$

8.12 (a) $KMnO_4$, H_3O^+ (b) H_2/Lindlar catalyst (c) **1.** H_2/Lindlar catalyst; **2.** HBr
(d) **1.** H_2/Lindlar catalyst; **2.** BH_3; **3.** NaOH, H_2O_2 (e) **1.** H_2/Lindlar catalyst; **2.** Cl_2

8.13 (a) **1.** $HC{\equiv}CH + NaNH_2$; **2.** $CH_3(CH_2)_7I$; **3.** 2 H_2/Pd
(b) **1.** $HC{\equiv}CH + NaNH_2$; **2.** $(CH_3)_3CCH_2CH_2I$; **3.** 2 H_2/Pd
(c) **1.** $HC{\equiv}CH + NaNH_2$; **2.** $CH_3CH_2CH_2CH_2I$; **3.** BH_3; **4.** H_2O_2
(d) **1.** $HC{\equiv}CH + NaNH_2$; **2.** $CH_3CH_2CH_2CH_2CH_2I$; **3.** $HgSO_4$, H_3O^+

Chapter 9

9.1 Chiral: screw, beanstalk, shoe **9.2** Chiral: (b)

9.3 (a)

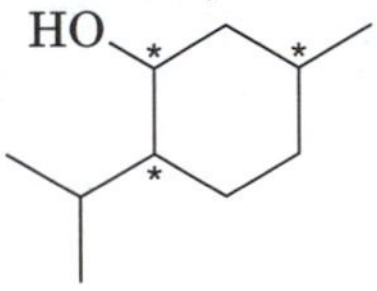

(b)

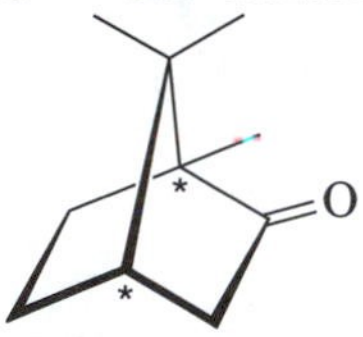

(c)

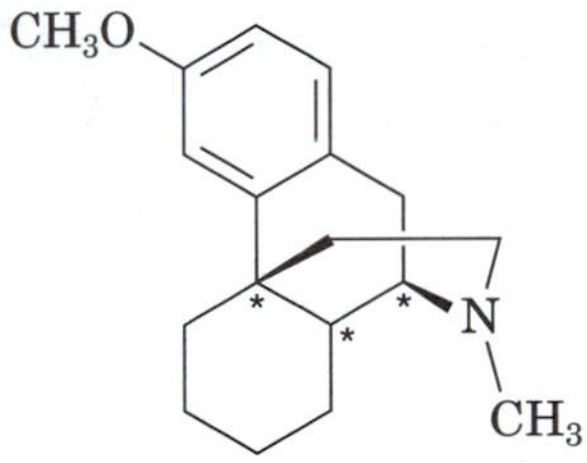

9.4

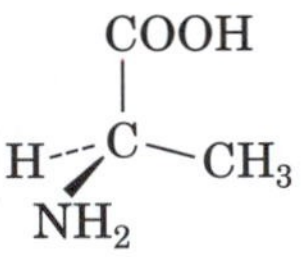

and

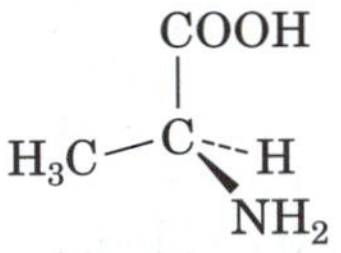

9.5 +16.1°

9.6 (a) $-Br$, $-CH_2CH_2OH$, $-CH_2CH_3$, $-H$ (b) $-OH$, $-CO_2CH_3$, $-CO_2H$, $-CH_2OH$
(c) $-NH_2$, $-CN$, $-CH_2NHCH_3$, $-CH_2NH_2$ (d) $-Br$, $-Cl$, $-CH_2Br$, $-CH_2Cl$

9.7 (a) *S* (b) *S* (c) *R* **9.8**

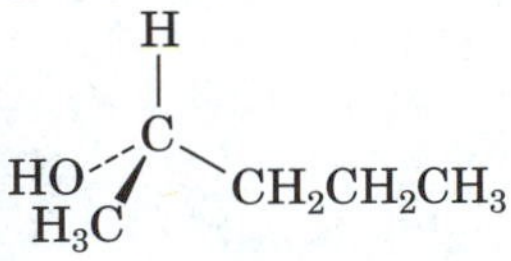

9.9 (a) *R,R* (b) *S,R* (c) *R,S*
(d) *S,S*; (a) and (d) are enantiomers and are diastereomeric to (b) and (c).

9.10 *R,R* **9.11** Meso: (a) and (d) **9.12** Meso: (a) and (c)

9.13 Five stereogenic centers; 32 stereoisomers

9.14 (a) Constitutional isomers (b) Diastereomers

9.15 A and B are identical; C and D are identical.

9.16 (a) Enantiomers (b) Enantiomers **9.17** (a) *S* (b) *S* (c) *R*

9.18 Racemic mixture **9.19** Non-50:50 mixture of two racemic pairs

9.20 Non-50:50 mixture of two racemic pairs

9.21 Non-50:50 mixture of two racemic pairs; optically inactive

9.22 50:50 mixture of two racemic pairs

9.23 Four stereoisomers:

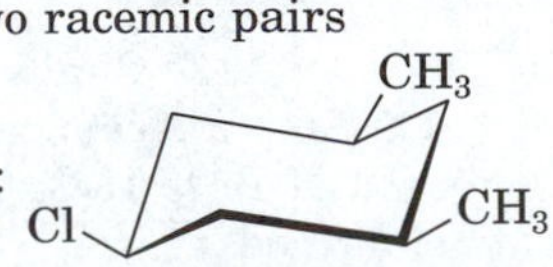

Most stable

9.24 Three stereoisomers: one pair of enantiomers and one meso compound

Chapter 10

10.1 (a) 1-Iodobutane (b) 1-Chloro-3-methylbutane (c) 1,5-Dibromo-2,2-dimethylpentane (d) 1,3-Dichloro-3-methylbutane (e) 1-Chloro-3-ethyl-4-iodopentane (f) 2-Bromo-5-chlorohexane

10.2 (a) $CH_3CH_2CH_2C(CH_3)_2CH(Cl)CH_3$ (b) $CH_3CH_2CH_2C(Cl)_2CH(CH_3)_2$ (c) $CH_3CH_2C(Br)(CH_2CH_3)_2$

(d) 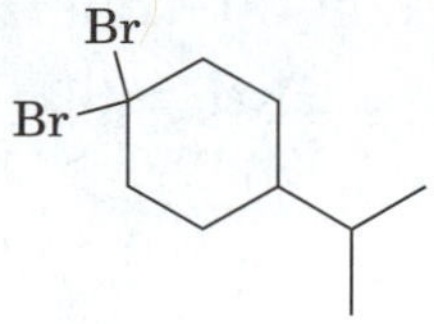

(e) $CH_3CH_2CH_2CH_2CH_2CH(CH(CH_3)CH_2CH_3)CH_2CH(Cl)CH_3$ 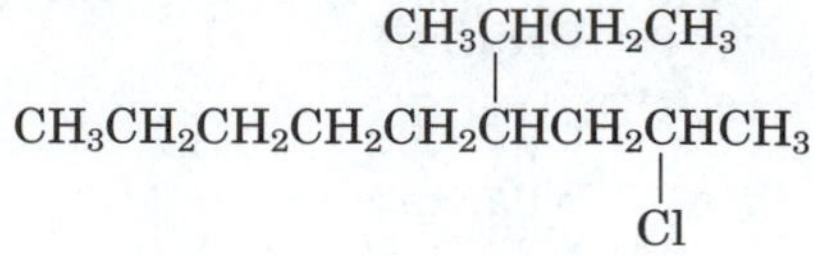

(f) 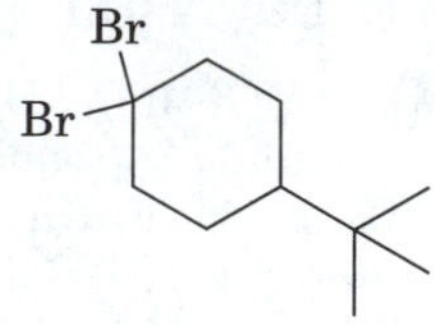

10.3 1-Chloro-2-methylpentane, 2-chloro-2-methylpentane, 3-chloro-2-methylpentane, 2-chloro-4-methylpentane, 1-chloro-4-methylpentane. The first, third, and fourth compounds are chiral.

10.4 1-Chloro-2-methylbutane (29%), 1-chloro-3-methylbutane (14%), 2-chloro-2-methylbutane (24%), 2-chloro-3-methylbutane (33%)

10.5 For Cl·, $\Delta H° = -31$ kJ/mol; for Br·, $\Delta H° = +35$ kJ/mol. Bromination is more selective.

10.6 The intermediate allylic radical reacts at the more accessible site and gives the more highly substituted double bond.

10.7 (a) 3-Bromo-5-methylcycloheptene and 3-bromo-6-methylcycloheptene (b) Four products

10.8 (a) 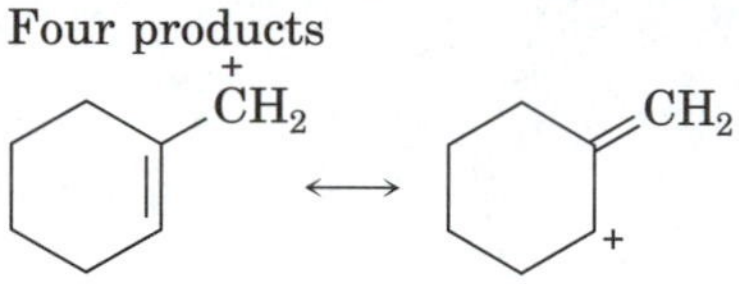

(b) 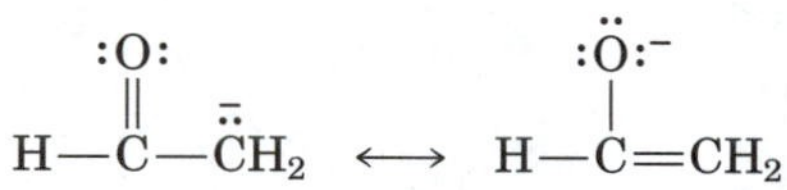

(c) 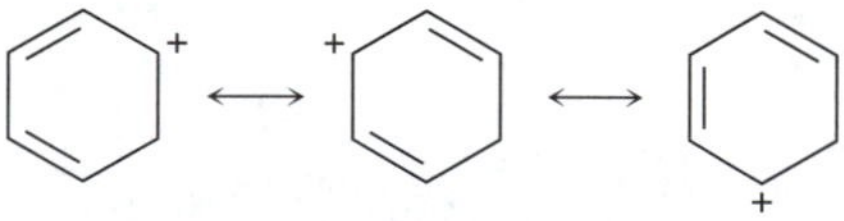

10.9 (a) 2-Methyl-2-propanol + HCl (b) 4-Methyl-2-pentanol + PBr_3 (c) 5-Methyl-1-pentanol + PBr_3 (d) 2,4-Dimethyl-2-hexanol + HCl

10.10 Both reactions occur. **10.11** React Grignard reagent with D_2O.

10.12 The –OH proton reacts with the Grignard reagent.

10.13 (a) **1.** NBS; **2.** $(CH_3)_2CuLi$ (b) **1.** Li; **2.** CuI; **3.** $CH_3CH_2CH_2CH_2Br$ (c) **1.** HBr, peroxides; **2.** Li; **3.** CuI; **4.** $CH_3(CH_2)_3CH_2Br$

10.14 (a) 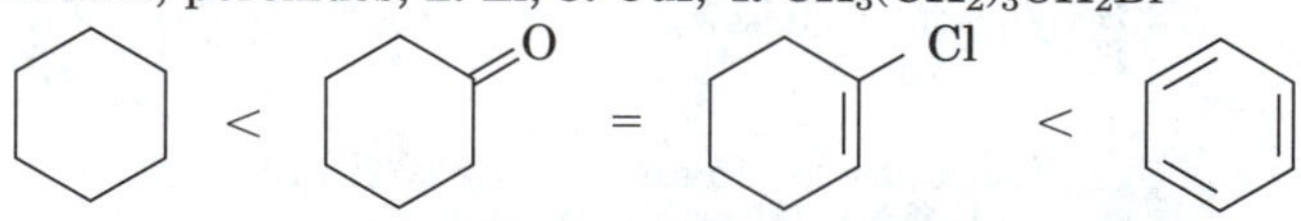

(b) $CH_3CH_2NH_2 < H_2NCH_2CH_2NH_2 < CH_3CN$

Chapter 11

11.1 (*R*)-1-Methylpentyl acetate, $CH_3CO_2CH(CH_3)CH_2CH_2CH_2CH_3$

11.2 (*S*)-2-Butanol **11.3** Back-side attack is too hindered.

11.4 (a) 1-Iodobutane (b) 1-Butanol (c) 1-Hexyne (d) Butylammonium bromide

11.5 (a) $(CH_3)_2N^-$ (b) $(CH_3)_3N$ (c) H_2S

11.6 $CH_3OTos > CH_3Br > (CH_3)_2CHCl > (CH_3)_3CCl$ **11.7** Similar to protic solvents.

11.8 Racemic 1-ethyl-1-methylhexyl acetate **11.9** 90.1% racemization, 9.9% inversion

11.10 $H_2C{=}CHCH(Br)CH_3 > CH_3CH(Br)CH_3 > CH_3CH_2Br > H_2C{=}CHBr$

11.11 The same allylic carbocation intermediate is formed.

11.12 The rate-limiting step does not involve the nucleophile.
11.13 The carbon atom cannot become planar.
11.14 (a) 2-Methyl-2-pentene (b) 2,3,5-Trimethyl-2-hexene (c) Ethylidenecyclohexane
11.15 (*Z*)-1-Bromo-1,2-diphenylethylene
11.16 Cis isomer reacts faster because the bromine is axial.
11.17 (a) S_N2 (b) E2 (c) S_N1

Chapter 12

12.1 (a) C_6H_{14}, $C_5H_{10}O$, $C_4H_6O_2$, $C_3H_2O_3$
(b) C_9H_{20}, $C_{10}H_8$, $C_8H_{16}O$, $C_7H_{12}O_2$, $C_6H_8O_3$, $C_5H_4O_4$
(c) $C_{11}H_{24}$, $C_{12}H_{12}$, $C_{11}H_8O$, $C_{10}H_{20}O$, $C_9H_{16}O_2$, $C_8H_{12}O_3$, $C_7H_8O_4$, $C_6H_4O_5$
12.2 $C_{15}H_{22}O$, $C_{14}H_{18}O_2$, $C_{13}H_{14}O_3$, $C_{12}H_{10}O_4$, $C_{11}H_6O_5$, $C_{16}H_{10}O$
12.3 100% and 6.75% **12.4** C_4H_9, *tert*-butyl cation
12.5 (a) 2-Methyl-2-pentene (b) 2-Hexene
12.6 X-ray energy is higher **12.7** $\lambda = 9.0 \times 10^{-4}$ cm is higher in energy.
12.8 (a) 2.4×10^6 kJ/mol (b) 4.0×10^4 kJ/mol (c) 2.4×10^3 kJ/mol
(d) 2.8×10^2 kJ/mol (e) 6.0 kJ/mol (f) 4.0×10^{-2} kJ/mol
12.9 (a) 3225 cm^{-1} (b) 1710 cm^{-1} (c) 4.44 μm (d) 10.3 μm
12.10 (a) Ketone or aldehyde (b) Nitro (c) Carboxylic acid
12.11 (a) CH_3CH_2OH has an –OH absorption
(b) 1-Hexene has a double-bond absorption
(c) CH_3CH_2COOH has a very broad –OH absorption
12.13 (a) 1715 cm^{-1} (b) 3300, 2100,1730 cm^{-1}

Chapter 13

13.1 2.2×10^{-5} kJ/mol for ^{19}F versus 2.4×10^{-5} kJ/mol for ^1H.
13.2 4.0×10^{-5} kJ/mol
13.3 (a) ^1H, one; ^{13}C, two (b) ^1H, four; ^{13}C, five (c) ^1H, one; ^{13}C, two
(d) ^1H, two; ^{13}C, four (e) ^1H, two; ^{13}C, three (f) ^1H, two; ^{13}C, three
13.4 The vinylic C–H protons are nonequivalent. **13.5** (a) 126 Hz (b) 2.1 δ (c) 210 Hz
13.6 (a) 7.27 δ (b) 3.05 δ (c) 3.47 δ (d) 5.30 δ
13.7 (a) Two (b) Four (c) Three (d) Four (e) Five (f) Three
13.8 (a) 1.43 δ (b) 2.17 δ (c) 7.37 δ (d) 9.70 δ (e) 5.30 δ (f) 2.12 δ
13.9 Seven kinds of protons **13.10** Two peaks; 3:2 ratio
13.11 (a) $-CHBr_2$, quartet; $-CH_3$, doublet
(b) CH_3O-, singlet; $-OCH_2-$, triplet; $-CH_2Br$, triplet
(c) $ClCH_2-$, triplet; $-CH_2-$, quintet
(d) CH_3-, triplet; $-CH_2-$, quartet; $-CH-$, septet; $(CH_3)_2$, doublet
13.12 (a) CH_3OCH_3 (b) $CH_3CHClCH_3$ (c) $ClCH_2CH_2OCH_2CH_2Cl$ (d) $CH_3CH_2COOCH_3$
13.13 $CH_3CH_2OCH_2CH_3$
13.15 1-Chloro-1-methylcyclohexane has a singlet methyl absorption.
13.16 $-CH_3$, 9.3 δ; $-CH_2-$, 27.6 δ; C=O, 174.6 δ; $-OCH_3$, 51.4 δ
13.17 (a) Four (b) Seven (c) Four (d) Five
13.18 (a) 1,3-Dimethylcyclopentene (b) 2-Methylpentane (c) 1-Chloro-2-methylpropane
13.20 $(CH_3)_2CHCH_2Br$ has three peaks in its ^{13}C NMR spectrum; $(CH_3)_3CBr$ has two.

Chapter 14

14.1 Conjugated: (b), (c), (d), (f)
14.2 Expected ΔH_{hydrog} for allene is -252 kJ/mol. Allene is less stable than either a conjugated or a nonconjugated diene.

14.3 1-Chloro-2-pentene, 3-chloro-1-pentene, 4-chloro-2-pentene
14.4 4-Chloro-2-pentene predominates in both.
14.5 ψ_1 is doubly occupied; ψ_2 is singly occupied.
14.6 Interconversion occurs by S_N1 dissociation to a common intermediate cation.
14.7 The double bond is more highly substituted. **14.8** $\text{+}CH_2C(C_6H_5){=}CHCH_2\text{+}$
14.10 Good dienophiles: (a), (d) **14.11** *s*-cis: (a); compound (c) can rotate to *s*-cis.
14.12 300–600 kJ/mol **14.13** UV energy is greater than IR or NMR energy.
14.14 1.47×10^{-5} M **14.15** UV absorptions: all except (a)

Chapter 15

15.1 (a) meta (b) para (c) ortho
15.2 (a) *m*-Bromochlorobenzene (b) (3-Methylbutyl)benzene
(c) *p*-Bromoaniline (d) 2,5-Dichlorotoluene
(e) 1-Ethyl-2,4-dinitrobenzene (f) 1,2,3,5-Tetramethylbenzene
15.4 (a) 1-Bromo-2-chlorobenzene, or *o*-bromochlorobenzene (b) 2,4-Dinitrotoluene
(c) 1-Bromo-4-methylbenzene or *p*-bromotoluene (d) 2–Chloro-1,4-dimethylbenzene
15.5 Four: 1,2,3-tribromobenzene, 1,3,5-tribromobenzene, and two 1,2,4-tribromobenzenes
15.6 One monobromo and three dibromo derivatives for Ladenburg benzene; two monobromo and six dibromo derivatives for Dewar benzene.
15.7 The 3:2:1 experimental ratio corresponds to the theoretical ratio if *o*-toluene is a resonance hybrid.
15.8 Pyridine has an aromatic sextet of electrons.
15.9 Cyclodecapentaene is not flat because of steric interactions.
15.10 Five resonance structures; all C–C bonds are equivalent; one resonance line in both ^{1}H and ^{13}C NMR spectra.
15.11 The cyclooctatetraenyl dianion is aromatic (8 π electrons) and flat.
15.12 The singly bonded nitrogen contributes two π electrons, and the doubly bonded nitrogen contributes one π electron to the aromatic sextet.
15.13 The oxygen contributes two π electrons to the aromatic sextet.
15.15

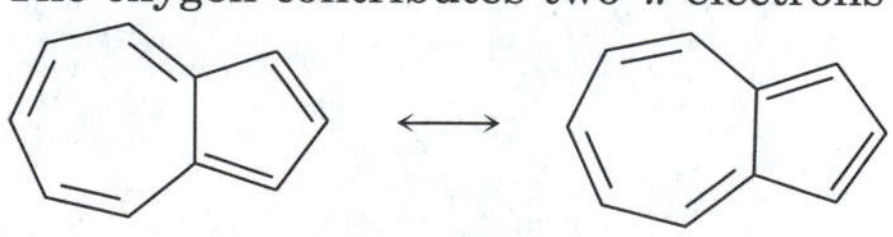

15.16 Five

Chapter 16

16.1 o-, *m*-, and *p*-bromotoluene
16.3 *o*-Xylene has two kinds of ring hydrogens that can be substituted, but *p*-xylene has only one.
16.4 Three **16.5** D^+ does electrophilic substitutions on the ring.
16.6 *tert*-Butylbenzene **16.7** No rearrangement: (a), (b), (e)
16.8 (a) *o*- and *p*-bromonitrobenzene (b) *m*-Bromonitrobenzene
(c) *o*- and *p*-chlorophenol (d) *o*- and *p*-bromoaniline
16.11 The intermediates from ortho and para attack are more stable.
16.12 (a) Phenol > Toluene > Benzene > Nitrobenzene
(b) Phenol > Benzene > Chlorobenzene > Benzoic acid
(c) Aniline > Benzene > Bromobenzene > Benzaldehyde
16.13 Alkylbenzenes are more reactive than benzene itself, but acylbenzenes are less reactive.
16.14 Toluene is more reactive; the trifluoromethyl group is electron-withdrawing.
16.15 The nitrogen electrons are donated to the nearby carbonyl group.
16.16 Phenoxide ion > Phenol > Phenyl acetate

16.17 The electronegativity of N and O make the nitroso group deactivating; the lone pair of electrons on N makes the group an ortho–para director.
16.18 Meta is most favored.
16.19 (a) Ortho and para to $-OCH_3$ (b) Ortho and para to $-NH_2$
(c) Ortho and para to –Cl
16.20 Addition of $^-OCH_3$ followed by elimination of Cl^-.
16.21 Only one benzyne intermediate can form from *p*-bromotoluene; two different benzyne intermediates can form from *m*-bromotoluene.
16.22 (a) 1,2-Benzenedicarboxylic acid (b) *m*-Nitrobenzoic acid
16.23 **1.** CH_3CH_2Cl, $AlCl_3$; **2.** NBS; **3.** KOH, ethanol
16.24 A benzyl radical is more stable than a primary alkyl radical by 52 kJ/mol and is similar in stability to an allyl radical.
16.25 **1.** PhCOCl, $AlCl_3$; **2.** H_2/Pd
16.26 (a) **1.** HNO_3, H_2/Pd; **2.** Cl_2, $FeCl_3$ (b) **1.** CH_3COCl, $AlCl_3$; **2.** Cl_2, $FeCl_3$; **3.** H_2/Pd
(c) **1.** Cl_2, $FeCl_3$; **2.** CH_3CH_2COCl, $AlCl_3$; **3.** H_2/Pd
16.27 (a) Friedel–Crafts acylation does not occur on a deactivated ring.
(b) Rearrangement occurs during Friedel–Crafts alkylation with primary halides; chlorination occurs ortho to the alkyl group.

Chapter 17

17.1 (a) 5-Methyl-2,4-hexanediol (b) 2-Methyl-4-phenyl-2-butanol
(c) 4,4-Dimethylcyclohexanol (d) *trans*-2-Bromocyclopentanol
17.3 Hydrogen bonding is more difficult in hindered alcohols.
17.4 HC≡CH < $(CH_3)_2CHOH$ < CH_3OH < $(CF_3)_2CHOH$
17.5 The electron-withdrawing nitro group stabilizes an alkoxide ion.
17.6 (a) 2-Methyl-4-phenyl-1-butanol (b) 2-Methyl-2-pentanol (c) *meso*-5,6-Decanediol
17.7 (a) $NaBH_4$ (b) $LiAlH_4$ (c) $LiAlH_4$ (d) H_2/Pd
17.8 (a) Benzaldehyde or benzoic acid (or ester) (b) Acetophenone (c) Cyclohexanone
(d) 2-Methylpropanal or 2-methylpropanoic acid (or ester)
17.9 (a) 1-Methylcyclopentanol (b) 1,1-Diphenylethanol (c) 3-Methyl-3-hexanol
17.10 (a) Acetone + CH_3MgBr, or ethyl acetate + 2 CH_3MgBr
(b) Cyclohexanone + CH_3MgBr
(c) 3-Pentanone + CH_3MgBr, or 2-butanone + CH_3CH_2MgBr, or ethyl acetate + 2 CH_3CH_2MgBr
(d) 2-Butanone + PhMgBr, or ethyl phenyl ketone + CH_3MgBr, or acetophenone + 2 CH_3CH_2MgBr
(e) Formaldehyde + PhMgBr
17.11 (a) 2-Methyl-2-pentene (b) 3-Methylcyclohexene (c) 1-Methylcyclohexene
17.13 (a) 1-Phenylethanol (b) 2-Methyl-1-propanol (c) Cyclopentanol
17.14 (a) Hexanoic acid, hexanal (b) 2-Hexanone (c) Hexanoic acid, no reaction
17.15 S_N2 reaction of F^- on silicon.
17.16 Disappearance of –OH absorption; appearance of C=O
17.17 (a) Singlet (b) Doublet (c) Triplet (d) Doublet (e) Doublet (f) Singlet
17.18 (a) 2-Butanethiol (b) 2,2,6-Trimethyl-4-heptanethiol (c) 2-Cyclopentene-1-thiol
17.19 (a) **1.** $LiAlH_4$; **2.** PBr_3; **3.** $(H_2N)_2C{=}S$; **4.** H_2O, NaOH
(b) **1.** HBr; **2.** $(H_2N)_2C{=}S$; **3.** H_2O, NaOH

Chapter 18

18.1 (a) Diisopropyl ether (b) Cyclopentyl propyl ether
(c) *p*-Bromoanisole or 4-bromo-1-methoxybenzene (d) 1-Methoxycyclohexene
(e) Ethyl isobutyl ether (f) Allyl vinyl ether

18.2 Mixtures of products result: diethyl ether, dipropyl ether, ethyl propyl ether.
18.4 (a) $CH_3CH_2CH_2O^- + CH_3Br$ (b) $PhO^- + CH_3Br$
(c) $(CH_3)_2CHO^- + PhCH_2Br$ (d) $(CH_3)_3CCH_2O^- + CH_3CH_2Br$
18.5 (a) Bromoethane > 2-Bromopropane > Bromobenzene
(b) Bromoethane > Chloroethane > 1-Iodopropene
18.7 (a) Either method (b) Williamson method
(c) Alkoxymercuration method (d) Williamson method
18.8 E1 reaction **18.9** Br^- and I^- are better nucleophiles than Cl^-.
18.10 *cis*-2,3-Epoxybutane **18.11** *trans*-2,3-Epoxybutane
18.12 (±)-5,6-Decanediol **18.13** *meso*-5,6-Decanediol
18.14 (a) $CH_3CH_2CH(OH)CH_2NH_2$ (b) $CH_3CH_2CH(OH)CH_2{}^{18}OH$
18.16 1,2-Epoxybutane
18.17 (a) Ethyl methyl sulfide (b) *tert*-Butyl ethyl sulfide
(c) *o*-(Dimethylthio)benzene (d) *p*-(Phenylthio)toluene
18.18 Dimethyl sulfoxide is highly polar.

Chapter 19

19.1 (a) 2-Methyl-3-pentanone (b) 3-Phenylpropanal (c) 2,6-Octanedione
(d) *trans*-2-Methylcyclohexanecarbaldehyde (e) Pentanedial
(f) *cis*-2,5-Dimethylcyclohexanone (g) 4-Methyl-3-propyl-2-hexanone (h) 4-Hexenal
19.3 (a) PCC (b) **1.** O_3; **2.** Zn (c) DIBAH
19.4 (a) $Hg(OAc)_2$, H_3O^+ (b) **1.** CH_3COCl, $AlCl_3$; **2.** Br_2, $FeBr_3$
(c) **1.** Mg; **2.** CH_3CHO; **3.** H_3O^+; **4.** PCC (d) **1.** BH_3; **2.** H_2O_2, NaOH; **3.** PCC
19.5 $(CH_3)_2C(OH)CN$ **19.6** The carbonyl carbon in benzaldehyde is less electrophilic.
19.7 *p*-Nitrobenzaldehyde is more reactive. **19.8** $CCl_3CH(OH)_2$
19.9 Labeled water adds reversibly to the carbonyl group.
19.10 The S_N2 substitution product is an aldehyde hydrate that loses water.
19.11 The equilibrium is unfavorable for sterically hindered ketones.
19.12 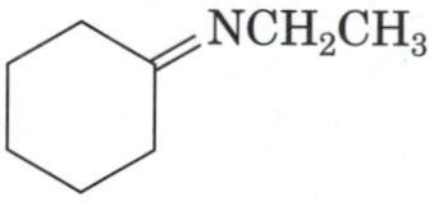

and 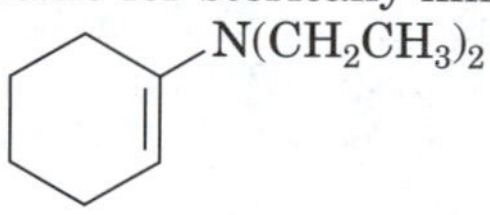

19.13 The steps are the exact reverse of the forward reaction.
19.15 (a) Cyclohexanone + $(Ph)_3P{=}CHCH_3$ (b) 2-Cyclohexenone + $(Ph)_3P{=}CH_2$
(c) Acetone + $(Ph)_3P{=}CHCH_2CH_2CH_3$ (d) Acetone + $(Ph)_3P{=}CHPh$
(e) Benzaldehyde + $(Ph)_3P{=}CHPh$
19.16 $(Ph)_3P{=}CHC(CH_3){=}CHCH{=}CHCH{=}C(CH_3)CH{=}P(Ph)_3$
19.17 Intramolecular Cannizzaro reaction.
19.18 (a) 3-Buten-2-one + $(CH_3CH_2CH_2)_2CuLi$ (b) 3-Methyl-2-cyclohexenone + $(CH_3)_2CuLi$
(c) 4-*tert*-Butyl-2-cyclohexenone + $(CH_3CH_2)_2CuLi$
(d) Unsaturated ketone + $(H_2C{=}CH)_2CuLi$
19.19 Look for appearance of either an alcohol or a saturated ketone in the product.
19.20 (a) 1715 cm^{-1} (b) 1685 cm^{-1} (c) 1750 cm^{-1} (d) 1705 cm^{-1}
(e) 1715 cm^{-1} (f) 1705 cm^{-1}
19.21 3-Phenyl-2-cyclohexenone

Chapter 20

20.1 (a) 3-Methylbutanoic acid (b) 4-Bromopentanoic acid (c) 2,4-Hexadienoic acid
(d) 2-Ethylpentanoic acid (e) *cis*-1,3-Cyclopentanedicarboxylic acid
(f) 2-Phenylpropanoic acid
20.3 +92 kJ/mol, +27.3 kJ/mol

20.4 Dissolve the mixture in ether, extract with aqueous NaOH, separate and acidify the aqueous layer, and extract with ether.
20.5 43%
20.6 (a) $CH_3CH_2COOH < BrCH_2COOH < FCH_2COOH$
(b) *p*-Methoxybenzoic acid < Benzoic acid < *p*-Nitrobenzoic acid
(c) $CH_3CH_2NH_2 < CH_3CH_2OH < CH_3CH_2COOH$
20.7 The dianion is destabilized by repulsion between charges. **20.8** More reactive.
20.9 (a) *p*-Methylbenzoic acid < Benzoic acid < *p*-Chlorobenzoic acid
(b) Acetic acid < Benzoic acid < *p*-Nitrobenzoic acid
20.10 The cyanide displacement method works with compounds that are sensitive to Grignard reagents but is limited to 1° halides. The Grignard method works with compounds that do not undergo S_N2 reactions.
20.11 (a) Grignard carboxylation (b) Grignard carboxylation (c) Either method

Chapter 21

21.1 (a) 4-Methylpentanoyl chloride (b) Cyclohexylacetamide (c) 2-Methylbutanenitrile
(d) Benzoic anhydride (e) Isopropyl cyclopentanecarboxylate
(f) Cyclopentyl 2-methylpropanoate (g) 4-Pentenamide (h) 2-Ethylbutanenitrile
(i) 2,3-Dimethyl-2-butenoyl chloride (j) Bis(trifluoroacetic) anhydride
21.2 (a) 2-Pentenenitrile (b) *N*-Methylbutanamide (c) 2,4-Dimethylpentanoyl chloride
(d) Methyl 1-methylcyclohexanecarboxylate
21.3 (a) Acetyl chloride > Methyl acetate > Acetamide
(b) Hexafluoroisopropyl acetate > 2,2,2-Trichloroethyl acetate > Methyl acetate
21.4 The electron-withdrawing trifluoromethyl group polarizes the carbonyl carbon.
21.5 (a) Acetic acid + 1-Butanol (b) Butanoic acid + Methanol **21.6**

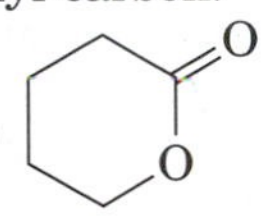

21.7 (a) Propanoyl chloride + Methanol (b) Acetyl chloride + Ethanol
(c) Benzoyl chloride + Ethanol
21.8 Benzoyl chloride + Cyclohexanol
21.10 The equilibrium favors reactants because there is no N–H hydrogen that can be removed.
21.11 (a) Propanoyl chloride + Methylamine (b) Benzoyl chloride + Dimethylamine
(c) Propanoyl chloride + Ammonia
21.12 (a) Benzoyl chloride + $[(CH_3)_2CH]_2CuLi$, or 2-methylpropanoyl chloride + Ph_2CuLi
(b) Propenoyl chloride + $(CH_3CH_2CH_2)_2CuLi$, or butanoyl chloride + $(H_2C{=}CH)_2CuLi$
21.13 Acid chlorides are more reactive than ketones toward Grignard reaction.
21.14 Monomethyl benzene-1,2-dicarboxylate
21.16 If no added base were present, half of the reactant amine would form a salt.
21.17 Reduce labeled acetic acid with $LiAlH_4$ to give $CH_3CH_2{}^{18}OH$, and then make ester.
21.18 Reaction of an acid with an alkoxide ion gives the unreactive carboxylate ion.
21.19 $HOCH_2CH_2CH_2CHO$
21.20 (a) $CH_3CH_2CH_2CH(CH_3)CH_2OH$ (b) $PhOH + PhCH_2OH$
21.21 (a) Ethyl benzoate + 2 CH_3MgBr (b) Ethyl acetate + 2 PhMgBr
(c) Ethyl butanoate + 2 CH_3CH_2MgBr
21.22 (a) H_2O, NaOH (b) Benzoic acid + BH_3 (c) $LiAlH_4$
21.23 **1.** Mg; **2.** CO_2, then H_3O^+; **3.** $SOCl_2$; **4.** $(CH_3)_2NH$; **5.** $LiAlH_4$
21.25 (a) $CH_3CH_2CN + CH_3CH_2MgBr$
(b) $CH_3CH_2CN + (CH_3)_2CHMgBr$ or $(CH_3)_2CHCN + CH_3CH_2MgBr$
(c) $(CH_3)_2CHCN$ + DIBAH (d) PhCN + CH_3MgBr or CH_3CN + PhMgBr
(e) Cyclohexanecarbonitrile + $C_6H_{11}MgBr$
21.26 **1.** NaCN; **2.** CH_3CH_2MgBr **21.29** The product has a large amount of cross-linking.

21.30 (a) Ester (b) Acid chloride (c) Carboxylic acid (d) Nitrile
(e) Aliphatic ketone or cyclohexanone

21.31 (a) CH_3CH_2CN (b) $CH_3CH_2CH_2COOCH_2CH_3$ and other possibilities
(c) $CH_3CON(CH_3)_2$ (d) $CH_3CH{=}CHCOCl$ or $H_2C{=}C(CH_3)COCl$

Chapter 22

22.2 (a) 4 (b) 3 (c) 3 (d) 2 (e) 4 (f) 5 (g) 3

22.3 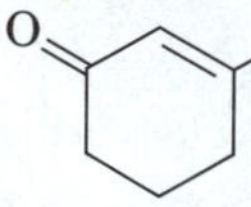

Most stable **22.4** **1.** Br_2; **2.** Pyridine, Δ

22.5 The enol intermediate is achiral. **22.6** Racemic

22.7 The intermediate α-bromo acid bromide undergoes a nucleophilic acyl substitution reaction.

22.8 (a) $CH_3C\underline{H}_2CHO$ (b) $(CH_3)_3CCOC\underline{H}_3$ (c) $CH_3COO\underline{H}$ (d) $PhCON\underline{H}_2$
(e) $CH_3CH_2C\underline{H}_2CN$ (f) $C\underline{H}_3CON(CH_3)_2$ (g) $-COC\underline{H}_2CO-$

22.9 $^-{:}CH_2C{\equiv}N{:} \longleftrightarrow H_2C{=}C{=}\ddot{N}{:}^-$ **22.10** The enolate ion intermediate is achiral.

22.11 No racemization occurs. **22.12** Enolate ion formation is rate-limiting.

22.13 Acid is regenerated, but base is used stoichiometrically.

22.14 Reaction of the trihalomethyl ketone with OH^- is a nucleophilic acyl substitution reaction.

22.15 (a) CH_3CH_2Br (b) $(CH_3)_2CHCH_2CH_2Br$

22.16 (a) Alkylate with $PhCH_2Br$.
(b) Alkylate first with $CH_3CH_2CH_2Br$ and then with CH_3Br.
(c) Alkylate with $(CH_3)_2CHCH_2Br$. (d) Alkylate with $BrCH_2CH_2CH_2Br$.

22.17 Malonic ester has only two acidic hydrogens.

22.18 (a) $(CH_3)_2CHCH_2Br$ (b) $PhCH_2CH_2Br$

22.19 Alkylate with $BrCH_2CH_2CH_2CH_2Br$. **22.20** (b), (c), and (d) can't be prepared.

22.21 (a) Alkylate phenylpropanone with CH_3I. (b) Alkylate pentanenitrile with CH_3CH_2I.
(c) Alkylate cyclohexanone with $H_2C{=}CHCH_2Br$.
(d) Alkylate cyclohexanone with excess CH_3I.

Chapter 23

23.1 (a) 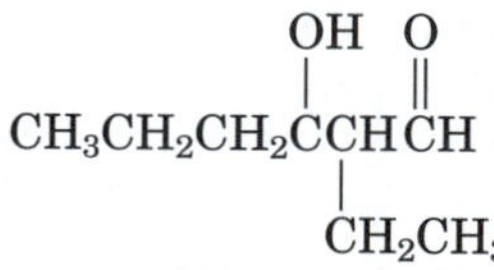

(b) 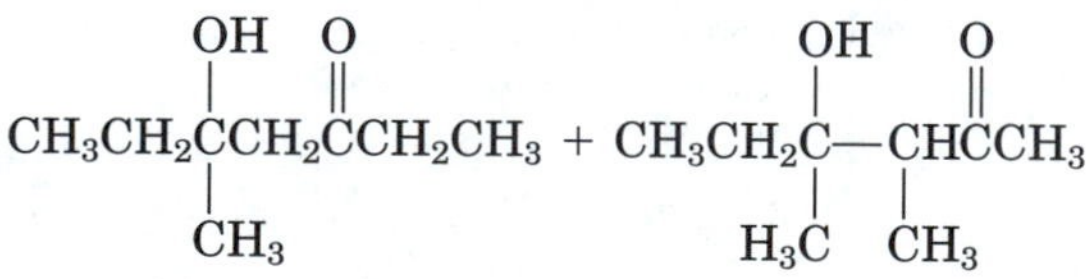

(c) 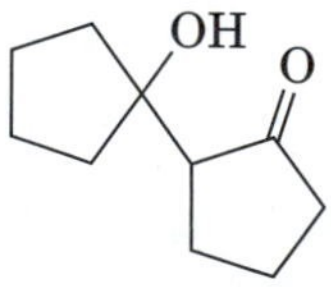

23.2 An enol **23.3** The reverse reaction is the exact opposite of the forward reaction.

23.4 (a) 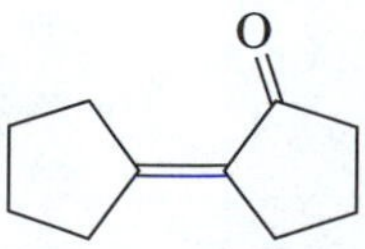

(b) 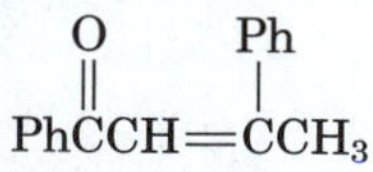

(c) 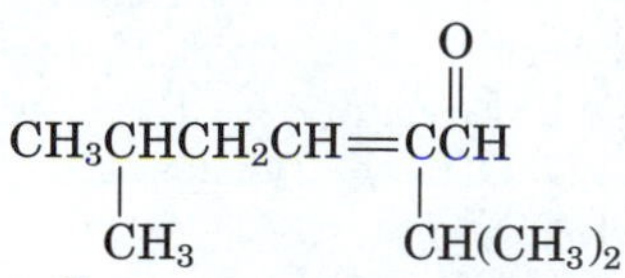

23.5 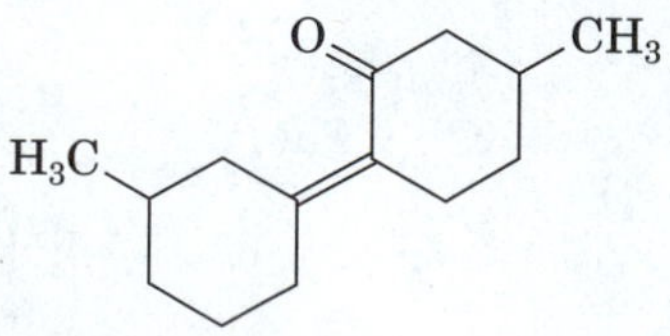

and 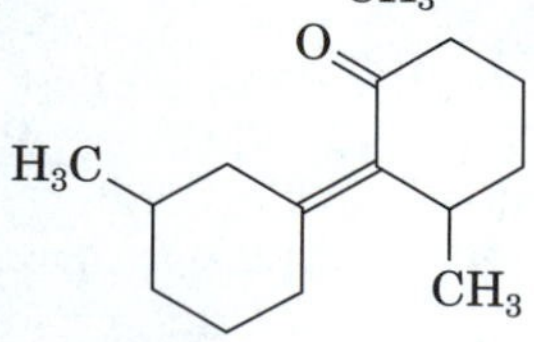

23.6 Self-condensation products: (c) **23.7** **1.** NaOH; **2.** $LiAlH_4$; **3.** H_2/Pd **23.8** (a), (d)

23.9 The cyclobutenone is unfavorable because of ring strain. **23.10** 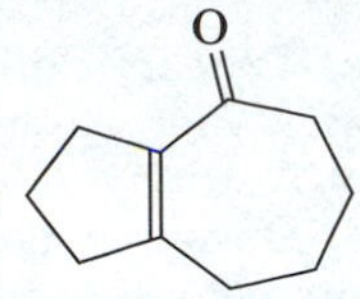

23.11 (a) 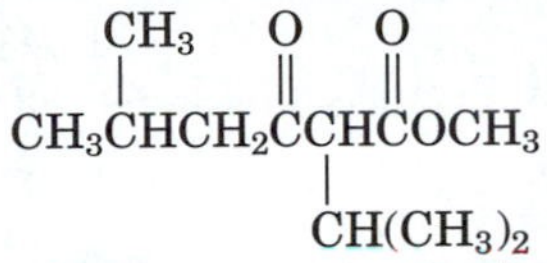 (b) 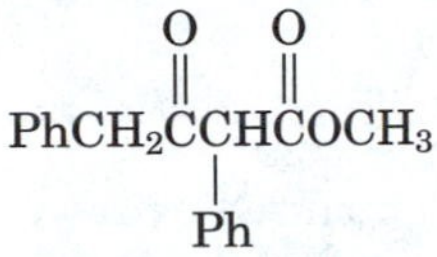 (c) 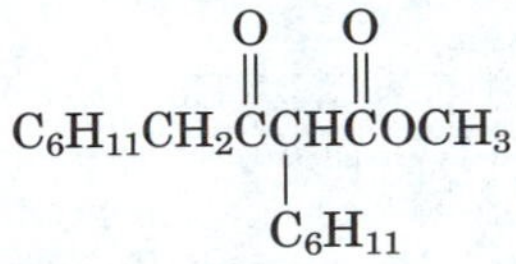

23.12 The cleavage reaction is the exact reverse of the forward reaction.

23.13 Yes; $EtO_2CCOCH_2CO_2Et$ **23.14** 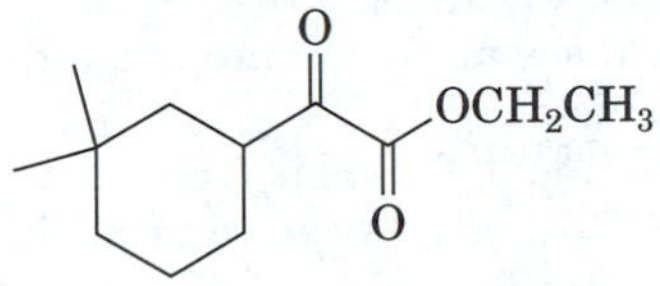

23.15 Ethyl 5-methyl-2-oxocyclohexanecarboxylate

23.16 Ethyl 4-methyl-2-oxocyclohexanecarboxylate and ethyl 6-methyl-2-oxocyclohexanecarboxylate

23.17 (a) 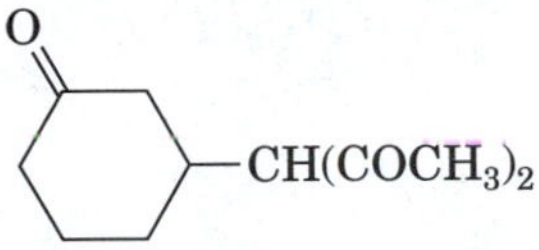 (b) $(CH_3CO)_2CHCH_2CH_2CN$

(c) 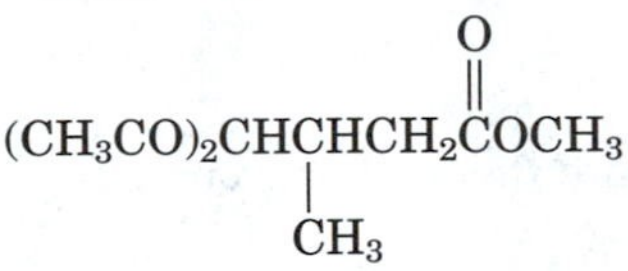

23.18 (a) 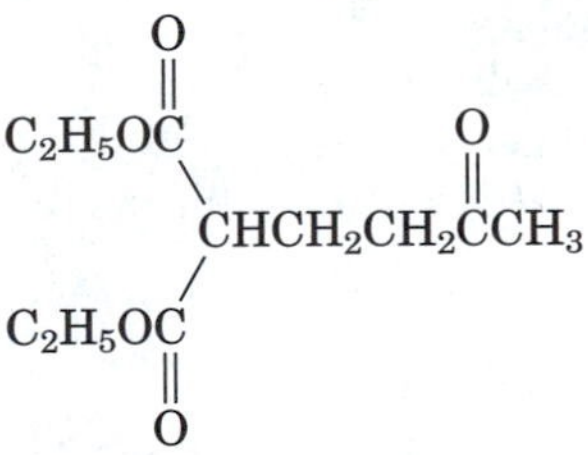 (b) 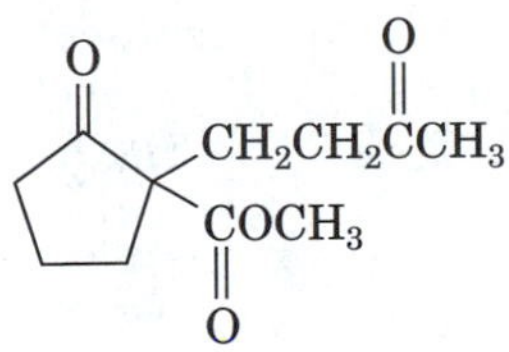 (c) 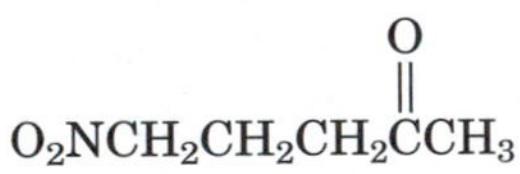

23.19 (a) 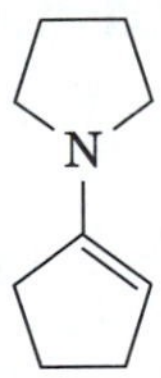 (b) 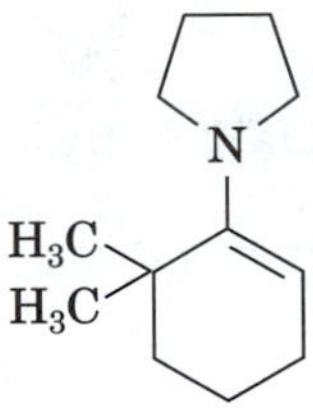

23.20 (a) 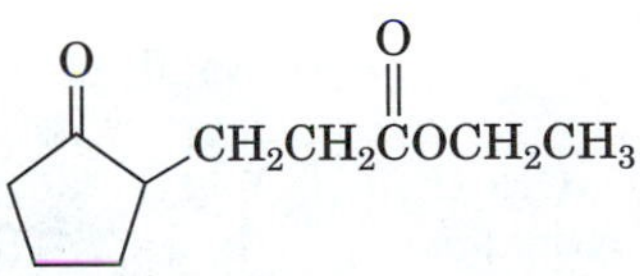 (b) 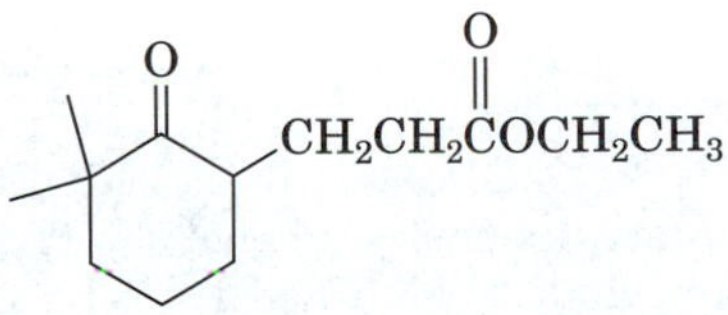

23.22 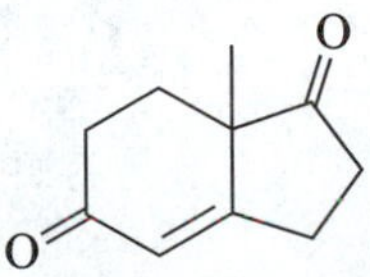

Chapter 24

24.1 (a) Primary (b) Tertiary (c) Quaternary (d) Secondary (e) Secondary
24.2 (a) *N*-Methylethylamine (b) Tricyclohexylamine
(c) *N*-Methyl-*N*-propylcyclohexylamine (d) *N*-Methylpyrrolidine
(e) Diisopropylamine (f) 1,3-Butanediamine
24.5 Interconverting enantiomers of a quaternary ammonium ion would require breaking bonds.
24.6 (a) $CH_3CH_2NH_2$ (b) NaOH (c) CH_3NHCH_3 **24.7** Propylamine is stronger.
24.8 Benzylamine, 4.67; propylamine, 3.29
24.9 The *O*-protonation product is stabilized by resonance.
24.12 (a) Propanenitrile or propanamide (b) *N*-Propylpropanamide
(c) Benzonitrile or benzamide (d) *N*-Phenylacetamide
24.13 (a) Ethylamine + acetone, or isopropylamine + acetaldehyde
(b) Aniline + acetaldehyde
(c) Cyclopentylamine + formaldehyde, or methylamine + cyclopentanone
24.14 An iminium ion, $R_2C{=}NR_2^+$ is an intermediate.
24.16 (a) 4,4-Dimethylpentanamide or 4,4-dimethylpentanoyl azide
(b) *p*-Methylbenzamide or *p*-methylbenzoyl azide
24.17 (a) 3-Octene and 4-octene (b) Cyclohexene (c) 3-Heptene
(d) Ethylene and cyclohexene
24.18 $H_2C{=}CHCH_2CH_2CH_2N(CH_3)_2$ **24.19** $(CH_3)_3CCOCH_3 \rightarrow (CH_3)_3CCH(NH_2)CH_3$

Chapter 25

25.1 The amine lone-pair electrons can be delocalized onto the *p*-nitro group but not the *m*-nitro group.
25.2 (a) *p*-Nitroaniline < *p*-Aminobenzaldehyde < *p*-Bromoaniline
(b) *p*-Aminoacetophenone < *p*-Chloroaniline < *p*-Methylaniline
(c) *p*-(Trifluoromethyl)aniline < *p*-(Fluoromethyl)aniline < *p*-Methylaniline
25.3 **1.** HNO_3, H_2SO_4; **2.** H_2/PtO_2; **3.** $(CH_3CO)_2O$; **4.** $HOSO_2Cl$; **5.** Aminothiazole; **6.** H_2O, NaOH
25.5 (a) **1.** HNO_3, H_2SO_4; **2.** H_2/PtO_2; **3.** 2 CH_3Br
(b) **1.** HNO_3, H_2SO_4; **2.** H_2/PtO_2; **3.** $(CH_3CO)_2O$; **4.** Cl_2; **5.** H_2O, NaOH
(c) **1.** HNO_3, H_2SO_4; **2.** Cl_2, $FeCl_3$; **3.** $SnCl_2$
(d) **1.** HNO_3, H_2SO_4; **2.** H_2/PtO_2; **3.** $(CH_3CO)_2O$; **4.** 2 CH_3Cl, $AlCl_3$; **5.** H_2O, NaOH
25.6 **1.** $NaNO_2$, H_2SO_4; **2.** CuCN; **3.** $LiAlH_4$, then H_2O
25.7 (a) **1.** CH_3Cl, $AlCl_3$; **2.** HNO_3, H_2SO_4; **3.** $SnCl_2$; **4.** $NaNO_2$, H_2SO_4; **5.** CuBr; **6.** $KMnO_4$, H_2O
(b) **1.** HNO_3, H_2SO_4; **2.** Br_2, $FeBr_3$; **3.** $SnCl_2$, H_3O^+; **4.** $NaNO_2$, H_2SO_4; **5.** CuCN; **6.** H_3O^+
(c) **1.** HNO_3, H_2SO_4; **2.** Cl_2, $FeCl_3$; **3.** $SnCl_2$; **4.** $NaNO_2$, H_2SO_4; **5.** CuBr
(d) **1.** CH_3Cl, $AlCl_3$; **2.** HNO_3, H_2SO_4; **3.** $SnCl_2$; **4.** $NaNO_2$, H_2SO_4; **5.** CuCN; **6.** H_3O^+
(e) **1.** HNO_3, H_2SO_4; **2.** H_2/PtO_2; **3.** $(CH_3CO)_2O$; **4.** 2 Br_2; **5.** H_2O, NaOH; **6.** $NaNO_2$, H_2SO_4; **7.** CuBr
25.10 **1.** 2 Cl_2; **2.** NaOH, then $BrCH_2COOEt$; **3.** H_2O, NaOH, then H_3O^+
25.11 (a) *p*-Methylphenol < Phenol < *p*-(Trifluoromethyl)phenol
(b) Benzyl alcohol < Phenol < *p*-Hydroxybenzoic acid
(c) *p*-Bromophenol < 2,4-Dibromophenol < 2,4,6-Tribromophenol
25.12 **1.** CH_3Cl, $AlCl_3$; **2.** SO_3, H_2SO_4; **3.** NaOH, 200°C
25.14 Alkylation occurs at the α carbon. **25.16** *o*-(1-Methylallyl)phenol

Chapter 26

26.1 (a) Aldotetrose (b) Ketopentose (c) Ketohexose (d) Aldopentose
26.2 A, B, and C are the same. **26.3** (a) *S* (b) *R* (c) S
26.4 (a) L-Erythrose; 2*S*,3*S* (b) D-Xylose; 2*R*,3*S*,4R (c) D-Xylulose; 3*S*,4*R*
26.5 L-Arabinose **26.7** There are 16 D and 16 L aldoheptoses.
26.9

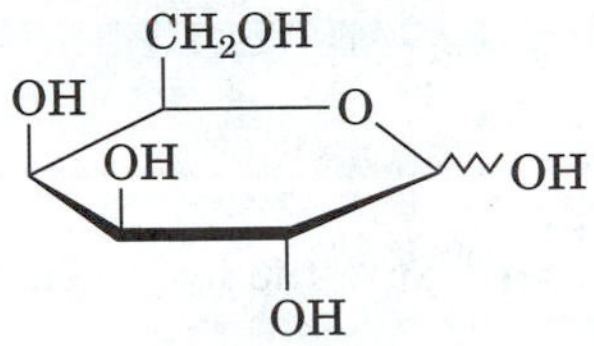

26.12 36% α anomer; 64% β anomer
26.13

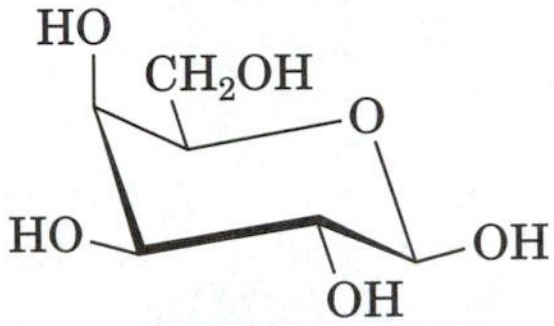

β-D-Galactopyranose

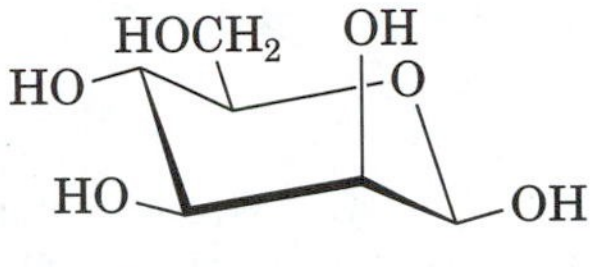

β-D-Mannopyranose

26.14

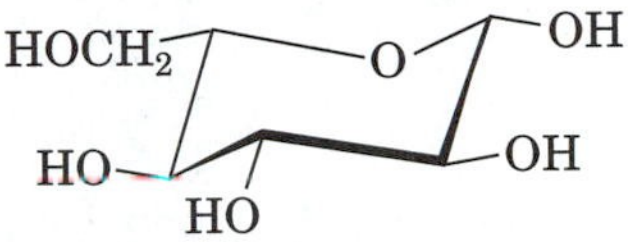

26.16 D-Galactitol has a plane of symmetry.
26.18 D-Allaric acid has a symmetry plane, but D-glucaric acid does not.
26.19 D-Allose and D-galactose yield meso aldaric acids; the other six D-hexoses yield optically active aldaric acids.
26.20 D-Allose + D-Altrose **26.21** L-Xylose **26.22** D-Xylose and D-lyxose

Chapter 27

27.1 Aromatic: Phe, Tyr, Trp, His; sulfur-containing: Cys, Met; alcohols: Ser, Trp; hydrocarbon side-chains: Ala, Ile, Leu, Val
27.2 The sulfur atom in the $-CH_2SH$ group of cysteine makes the side chain higher in priority than the $-COOH$ group.
27.5 (a) $PhCH_2CH(NH_3^+)C(=O)O^-$ (b) $HOCH_2CH(NH_3^+)C(=O)O^-$ (c)

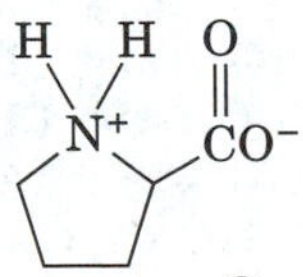

27.6 (a) $PhCH_2CH(NH_2)C(=O)O^-$ (b) $PhCH_2CH(NH_3^+)C(=O)O^-$ (c) $PhCH_2CH(NH_3^+)C(=O)OH$
27.7 pH 2.0; $H_3\overset{+}{N}CH_2COOH$; pH 6.0: $H_3\overset{+}{N}CH_2COO^-$; pH 10.0: $H_2NCH_2COO^-$
27.8 (a) Toward anode: Glu > Val; toward cathode: none
(b) Toward anode: Phe; toward cathode: Gly
(c) Toward anode: none; toward cathode: Gly > Ser
(d) Toward anode: Phe > Ser; toward cathode: none
27.9 At pH 1.5: 20% neutral and 80% protonated; at pH 10.0: 11% neutral and 89% deprotonated.

27.10 (a) Start with 3-phenylpropanoic acid: **1.** Br_2, PBr_3; **2.** NH_3
(b) Start with 3-methylbutanoic acid: **1.** Br_2, PBr_3; **2.** NH_3
27.12 **1.** NH_4Cl, KCN; **2.** H_3O^+ **27.13** (a) $(CH_3)_2CHCH_2Br$ (d) $CH_3SCH_2CH_2Br$
27.14 Use formaldehyde in an aldol-like reaction with diethyl acetamidomalonate.
27.15 Val-Tyr-Gly, Tyr-Gly-Val, Gly-Val-Tyr, Val-Gly-Tyr, Tyr-Val-Gly, Gly-Tyr-Val
27.18 An amine must be primary (RNH_2) to give the ninhydrin reaction.
27.19 Trypsin: Asp-Arg + Val-Tyr-Ile-His-Pro-Phe; Chymotrypsin: Asp-Arg-Val-Tyr + Ile-His-Pro-Phe
27.20 (a) Arg-Pro-Leu-Gly-Ile-Val (b) Val-Met-Trp-Asp-Val-Leu
27.21 Pro-Leu-Gly-Pro-Arg-Pro **27.22** The tripeptide is cyclic.
27.26 The amide nitrogen of proline has no hydrogen and thus can't form the hydrogen bond necessary for an α-helix.
27.27 (a) Lyase (b) Hydrolase (c) Oxidoreductase

Chapter 28

28.1 $CH_3(CH_2)_{18}COOCH_2(CH_2)_{30}CH_3$ **28.2** Glyceryl tripalmitate is higher melting.
28.3 $CH_3(CH_2)_7C{\equiv}C(CH_2)_7COOH$ **28.4** $[CH_3(CH_2)_7CH{=}CH(CH_2)_7COO^-]_2\ Mg^{2+}$
28.7 (a) Equatorial methyl (b) Axial methyl **28.8** Equatorial hydroxyl

Chapter 29

29.1 One lone pair is in an sp^2 orbital in the plane of the ring; the other lone pair is in a *p* orbital perpendicular to the plane of the ring.
29.2 Nitrogen is the positive end of the dipole.
29.3 Electrophilic aromatic substitution mechanism.
29.4 The pyridine-like, doubly bonded nitrogen is more basic.
29.5 The intermediate from attack at C3 does not have the positive charge on the electronegative nitrogen atom.
29.6 The intermediate from attack at C4 has the negative charge on nitrogen.
29.7 Reaction occurs by a benzyne mechanism.
29.8 The side-chain nitrogen atom is more basic.
29.9 The intermediate from attack at C3 is stabilized by resonance involving the nitrogen atom.
29.10 The lactam has more resonance forms than the lactim.
29.13 C-C-G-A-T-T-A-G-G-C-A **29.15** C-U-A-A-U-G-G-C-A-U
29.16 (a) GCU, GCC, GCA, GCG (b) UUU, UUC
(c) UUA, UUG, CUU, CUC, CUA, CUG (d) UAU, UAC
29.17 (a) CGA, CGG, CGU, CGC (b) AAA, AAG
(c) AAU, AAC, GAA, GAG, GAU, GAC (d) AUA, AUG
29.18 Leu-Met-Ala-Trp-Pro-Stop **29.19** GAA-UAC-CGA-ACC-GGG-AUU
29.20 GAA-TAC-CGA-ACC-GGG-ATT
29.21 A cleavage: ^{32}P-A, ^{32}P-A-A-C, ^{32}P-A-A-C-A-T-G-G-C-G-C-T-T, ^{32}P-A-A-C-A-T-G-G-C-G-C-T-T-A-T-G, ^{32}P-A-A-C-A-T-G-G-C-G-C-T-T-A-T-G-A-C-G
G cleavage: ^{32}P-A-A-C-A-T, ^{32}P-A-A-C-A-T-G, ^{32}P-A-A-C-A-T-G-G-C, ^{32}P-A-A-C-A-T-G-G-C-G-C-T-T-A-T, ^{32}P-A-A-C-A-T-G-G-C-G-C-T-T-A-T-G-A-C
C cleavage: ^{32}P-A-A, ^{32}P-A-A-C-A-T-G-G, ^{32}P-A-A-C-A-T-G-G-C-G, ^{32}P-A-A-C-A-T-G-G-C-G-C-T-T-A-T-G-A
C + T cleavage: ^{32}P-A-A, ^{32}P-A-A-C-A, ^{32}P-A-A-C-A-T-G-G, ^{32}P-A-A-C-A-T-G-G-C-G, ^{32}P-A-A-C-A-T-G-G-C-G-C, ^{32}P-A-A-C-A-T-G-G-C-G-C-T, ^{32}P-A-A-C-A-T-G-G-C-G-C-T-T-A, ^{32}P-A-A-C-A-T-G-G-C-G-C-T-T-A-T-G-A
29.23 T-C-G-G-T-A-C **29.24** The cleavage is an S_N1 reaction.
29.25 The cleavage is an E2 reaction.

Chapter 30

30.1 $HOCH_2CH(OH)CH_2OH + ATP \rightarrow HOCH_2CH(OH)CH_2OPO_3^{2-} + ADP$
30.2 The reaction is a nucleophilic acyl substitution reaction at the phosphorus atom.
30.4 (a) 8 acetyl CoA; 7 passages (b) 10 acetyl CoA; 9 passages
30.5 Steps 7 and 10 **30.8** Nucleophilic acyl substitution **30.9** Citrate and isocitrate
30.10 E2 elimination of water, followed by conjugate addition
30.13 $(CH_3)_2CHCH_2COCO_2^-$ **30.14** E2 reaction
30.15 At C2, C4, C6, C8, and so forth.
30.16 Nucleophilic acyl substitution of H^- for phosphate ion

Chapter 31

31.1 Ground state: ψ_3 is the HOMO; ψ_4 is the LUMO. Excited state: ψ_4 is the HOMO; ψ_5 is the LUMO.
31.2 Disrotatory: *cis*-5,6-dimethyl-1,3-cyclohexadiene; conrotatory: *trans*-5,6-dimethyl-1,3-cyclohexadiene. Disrotatory closure occurs.
31.3 The more stable of two allowed products is formed.
31.4 *trans*-5,6-Dimethyl-1,3-cyclohexadiene; *cis*-5,6-dimethyl-1,3-cyclohexadiene
31.5 Conrotatory cyclization to a cyclobutene, followed by conrotatory ring opening.
31.6 Disrotatory; the cyclobutene hydrogens are cis.
31.7 *cis*-3,6-Dimethylcyclohexene; *trans*-3,6-dimethylcyclohexene
31.8 A [6+4] suprafacial cycloaddition.
31.9 A Diels–Alder reaction followed by a retro Diels–Alder cleavage.
31.10 Cope rearrangement is so rapid that only a time-averaged NMR spectrum is seen.
31.11 A series of [1,5] hydrogen shifts occur.
31.12 An antarafacial [1,7] sigmatropic rearrangement
31.13 Claisen rearrangement is followed by a Cope rearrangement.
31.14 (a) Conrotatory (b) Disrotatory (c) Suprafacial (d) Antarafacial (e) Suprafacial

Index

The page references given in boldface type refer either to pages where terms are defined or to minibiographies.

Photo Credits

Page 31: Photo Disc, Inc.
Page 62: John Cullen
Page 63: F. N. Smith/TexStockPhoto Inc.
Page 99: B. J. Nixon/Tenneco, Inc.
Page 171: Wide World
Page 213: Lawrence Migdale/Stock, Boston
Page 252: Tom McCarthy/SKA
Page 333: Focus on Sports, Inc.
Page 363: Science Source/Photo Researchers, Inc.
Page 414: Charles Angelo/Photo Researchers, Inc.
Page 485: Mehav Kulyk/Photo Researchers, Inc.
Page 525: AT&T Bell Laboratories
Page 563: Ralf-Finn Hestoft/Third Coast Stock Source
Page 611: David Madison
Page 664: Rob Crandall/Stock, Boston
Page 695: Jim Corwin/Stock, Boston
Page 759: Science Source/Photo Researchers, Inc.
Page 792: Bob Daemmrich/Stock, Boston
Page 845: Larry Jensen/Visuals Unlimited
Page 923: ("Summer" by Giuseppe Arcimboldo) Erich Lessing/Art Resource, NY
Page 962: Drug Enforcement Agency
Page 1048: Tom McCarthy/SKA
Page 1091: Bob Daemmrich/Stock, Boston
Page 1121: SIU/Visuals Unlimited
Page 1165: David Parker/Photo Researchers, Inc.
Page 1204: David Madison
Page 1237: Jonathan Nourok/PhotoEdit

Credits

Sponsoring Editor: *Harvey C. Pantzis*
Marketing Team: *Connie Jirovsky, Margaret Parks, Tracey Armstrong*
Editorial Associate: *Beth Wilbur*
Production Service: *Phyllis Niklas*
Production Supervisor: *Jamie Sue Brooks*
Manuscript Editor: *Phyllis Niklas*
Art Coordination/Technical Illustration: *Jonathan Peck Typographers, Ltd.*
Text Design: *Nancy Benedict*
Cover Design: *Vernon T. Boes*
Cover Illustration and chapter opener illustrations: *Kenneth Eward, BioGrafx*
Photo Researcher: *Stuart Kenter*
Type Composition/Prepress: *Jonathan Peck Typographers, Ltd.*
Cover Printing: *Phoenix Color Corp.*
Printing and Binding: *R. R. Donnelley, Willard*

On the Cover: A molecular model of benzene showing electrostatic potential in the equatorial plane of the molecule (gold = carbon, white = hydrogen). The isobars, which denote electrostatic potential much as the contour lines on a geographic map denote altitude, were calculated using the AM1 semi-empirical method and later smoothed.

Periodic Table of the Elements

Key: Atomic number — 22; Name — Titanium; Symbol — **Ti**; Atomic weight — 47.90

[a]Mass number of most stable or best-known isotope

[b]Mass of the isotope of longest half-life

Atomic weights are based on carbon-12. Atomic weights in parentheses indicate the most stable or best-known isotope. Names and symbols for elements 104 through 111 are under consideration.

Period	1A	2A	3B	4B	5B	6B	7B	8B	8B	8B	1B	2B	3A	4A	5A	6A	7A	8A
1	1 Hydrogen **H** 1.0079																	2 Helium **He** 4.00260
2	3 Lithium **Li** 6.941	4 Beryllium **Be** 9.01218											5 Boron **B** 10.81	6 Carbon **C** 12.011	7 Nitrogen **N** 14.0067	8 Oxygen **O** 15.9994	9 Fluorine **F** 18.99840	10 Neon **Ne** 20.179
3	11 Sodium **Na** 22.98977	12 Magnesium **Mg** 24.305											13 Aluminum **Al** 26.98154	14 Silicon **Si** 28.086	15 Phosphorus **P** 30.97376	16 Sulfur **S** 32.06	17 Chlorine **Cl** 35.453	18 Argon **Ar** 39.948
4	19 Potassium **K** 39.098	20 Calcium **Ca** 40.08	21 Scandium **Sc** 44.9559	22 Titanium **Ti** 47.90	23 Vanadium **V** 50.9414	24 Chromium **Cr** 51.996	25 Manganese **Mn** 54.9380	26 Iron **Fe** 55.847	27 Cobalt **Co** 58.9332	28 Nickel **Ni** 58.69	29 Copper **Cu** 63.546	30 Zinc **Zn** 65.38	31 Gallium **Ga** 69.72	32 Germanium **Ge** 72.61	33 Arsenic **As** 74.9216	34 Selenium **Se** 78.96	35 Bromine **Br** 79.904	36 Krypton **Kr** 83.80
5	37 Rubidium **Rb** 85.4678	38 Strontium **Sr** 87.62	39 Yttrium **Y** 88.9059	40 Zirconium **Zr** 91.22	41 Niobium **Nb** 92.9064	42 Molybdenum **Mo** 95.94	43 Technetium **Tc** 98.9062[b]	44 Ruthenium **Ru** 101.07	45 Rhodium **Rh** 102.9055	46 Palladium **Pd** 106.4	47 Silver **Ag** 107.868	48 Cadmium **Cd** 112.40	49 Indium **In** 114.82	50 Tin **Sn** 118.71	51 Antimony **Sb** 121.75	52 Tellurium **Te** 127.60	53 Iodine **I** 126.9045	54 Xenon **Xe** 131.30
6	55 Cesium **Cs** 132.9054	56 Barium **Ba** 137.34	* 57 Lanthanum **La** 138.9055	72 Hafnium **Hf** 178.49	73 Tantalum **Ta** 180.9479	74 Wolfram (Tungsten) **W** 183.85	75 Rhenium **Re** 186.2	76 Osmium **Os** 190.2	77 Iridium **Ir** 192.22	78 Platinum **Pt** 195.09	79 Gold **Au** 196.9665	80 Mercury **Hg** 200.59	81 Thallium **Tl** 204.37	82 Lead **Pb** 207.2	83 Bismuth **Bi** 208.9804	84 Polonium **Po** (210)[a]	85 Astatine **At** (210)[a]	86 Radon **Rn** (222)[a]
7	87 Francium **Fr** (223)[a]	88 Radium **Ra** 226.0254[b]	** 89 Actinium **Ac** (227)[a]	104 (261)[a]	105 (262)[a]	106 (263)[a]	107 (262)	108 (265)	109 (268)	110 (269)	111 (272)							

Transition elements: groups 3B through 2B

Inner transition elements

* Lanthanide series 6	58 Cerium **Ce** 140.12	59 Praseodymium **Pr** 140.9077	60 Neodymium **Nd** 144.24	61 Promethium **Pm** (145)[a]	62 Samarium **Sm** 150.4	63 Europium **Eu** 151.96	64 Gadolinium **Gd** 157.25	65 Terbium **Tb** 158.9254	66 Dysprosium **Dy** 162.50	67 Holmium **Ho** 164.9304	68 Erbium **Er** 167.26	69 Thulium **Tm** 168.9342	70 Ytterbium **Yb** 173.04	71 Lutetium **Lu** 174.97
** Actinide series 7	90 Thorium **Th** 232.0381[b]	91 Protactinium **Pa** 231.0359[b]	92 Uranium **U** 238.029	93 Neptunium **Np** 237.0482	94 Plutonium **Pu** (242)[a]	95 Americium **Am** (243)[a]	96 Curium **Cm** (247)[a]	97 Berkelium **Bk** (249)[a]	98 Californium **Cf** (251)[a]	99 Einsteinium **Es** (254)[a]	100 Fermium **Fm** (253)[a]	101 Mendelevium **Md** (256)[a]	102 Nobelium **No** (254)[a]	103 Lawrencium **Lr** (257)[a]